HANDBOOK OF STATISTICS

VOLUME 13

Handbook of Statistics

VOLUME 13

General Editor

C.R. Rao

ELSEVIER
AMSTERDAM - LAUSANNE - NEW YORK - OXFORD - SHANNON - TOKYO

Design and Analysis of Experiments

Edited by

S. Ghosh

University of California
Riverside, CA, USA

and

C.R. Rao

The Pennsylvania State University
University Park, PA, USA

1996

ELSEVIER
AMSTERDAM - LAUSANNE - NEW YORK - OXFORD - SHANNON - TOKYO

ELSEVIER SCIENCE B.V.
Sara Burgerhartstraat 25
P.O. Box 211, 1000 AE Amsterdam, The Netherlands

ISBN: 0-444-82061-2

This book is printed on acid-free paper.

Printed in The Netherlands.

Preface

In scientific investigations, the collection of pertinent information and the efficient analysis of the collected information using statistical principles and methods are very important tasks. The purpose of this volume of the Handbook of Statistics is to provide the reader with the state-of-the-art of statistical design and methods of analysis that are available as well as the frontiers of research activities in developing new and better methods for performing such tasks. The volume is a tribute to all individuals who helped in developing this particular area of knowledge to be useful in our everyday life.

Scientific experiments in medicine, industry, agriculture, computer and many other disciplines are covered in this volume. Statistical methods like parametric, semiparametric, nonparametric, adaptive, univariate, multivariate, frequentist and Bayesian, are discussed. Block, row–column, nested, factorial, response surface, spatial, robust, optimum, search, singlestage, multistage, exact and approximate designs are presented. The chapters are written in expository style and should be of value to students, researchers, consultants and practitioners in universities as well as industries.

We would like to express our deep appreciation to the following colleagues for their cooperation and help in reviewing the chapters.

Deborah Best, Tadeusz Calinski, Kathryn Chaloner, Richard Cutler, Angela Dean, Benjamin Eichhorn, Richard Gunst, Sudhir Gupta, Linda Haines, Thomas Hettmansperger, Klaus Hinkelmann, Theodore Holford, Jason Hsu, Sanpei Kageyama, Andre Khuri, Dibyen Majumdar, John Matthews, Douglas Montgomery, Christine Muller, William Notz, Thomas Santner, Pranab Sen, Stanley Shapiro, Carl Spruill, Jagdish Srivastava, John Stufken, Ajit Tamhane, Isabella Verdinelli and Shelley Zacks.

We are very grateful to the authors for their contributions. Our sincere thanks go to Dr. Gerard Wanrooy and others at North-Holland Publishing Company, also to Dr. Rimas Maliukevičius and others of VTEX Ltd. Subir Ghosh would like to thank two very special individuals in his life, Mrs Benu Roy (the late) and Mr. Amalendu Roy as well as his wife Susnata and daughter Malancha for their patience, help and understanding.

S. Ghosh and C. R. Rao

Table of Contents

Preface v

Contributors xvii

Ch. 1. The Design and Analysis of Clinical Trials 1
P. Armitage

1. Introduction 1
2. The planning of a clinical trial 5
3. The conduct and monitoring of a clinical trial 13
4. The analysis of data 17
5. The reporting and interpretation of results 22
6. Conclusion 25

Acknowledgements 25
References 25

Ch. 2. Clinical Trials in Drug Development: Some Statistical Issues 31
H. I. Patel

1. Introduction 31
2. Pharmacokinetic studies 35
3. Dose-response studies 39
4. Mixed effects models 45
5. Markovian models 53
6. Discussion 56

Acknowledgements 57
References 57

Ch. 3. Optimal Crossover Designs 63
J. Stufken

1. Introduction 63
2. Terminology and notation 65
3. Optimal crossover designs when errors are uncorrelated 69
4. Optimal crossover designs when errors are uncorrelated: The special case of two treatments 77
5. Optimal crossover designs when errors are autocorrelated 81
6. Summary and discussion 88

Acknowledgement 89
References 89

Ch. 4. Design and Analysis of Experiments: Nonparametric Methods with Applications to Clinical Trials 91
P. K. Sen

1. Introduction 91
2. One-way layout nonparametrics 92
3. Two-way layouts nonparametrics 103
4. Nonparametric MANOVA 111
5. Nonparametrics for incomplete block designs 115
6. Nonparametrics in factorial designs 118
7. Paired comparisons designs: Nonparametrics 123
8. Nonparametrics for crossover designs 127
9. Clinical trials and survival analysis 131
10. Nonparametrics in incomplete multiresponse clinical designs 140
11. Concluding remarks 143
Acknowledgements 147
References 147

Ch. 5. Adaptive Designs for Parametric Models 151
S. Zacks

1. Introduction 151
2. Optimal designs with respect to the Fisher information 152
3. Adaptive designs for inverse regression problems 165
4. Sequential allocation of experiments 172
Acknowledgement 178
References 179

Ch. 6. Observational Studies and Nonrandomized Experiments 181
P. R. Rosenbaum

1. An informal history or observational studies in statistics 181
2. Inference in randomized experiments 182
3. Observational studies in the absence of hidden bias 183
4. Sensitivity to hidden bias in observational studies 188
5. Detection of hidden bias in observational studies 193
6. Coherence and focused hypotheses 194
7. Summary 194
References 195

Ch. 7. Robust Design: Experiments for Improving Quality 199
D. M. Steinberg

1. Introduction 199
2. Robust design and the quality revolution 202
3. Overview of robust design experiments 203
4. Examples 206
5. Experiments with design factors only 209
6. Noise factors and dispersion effects 217
7. Product array experiments 222
8. Further design issues 227
9. Experiments with signal factors 234
10. Conclusions 237
References 237

Ch. 8. Analysis of Location and Dispersion Effects from Factorial Experiments with a Directional Response 241
C. M. Anderson-Cook

1. Introduction 241
2. Location analysis 243
3. Dispersion analysis 251
4. Conclusion 258
References 259

Ch. 9. Computer Experiments 261
J. R. Koehler and A. B. Owen

1. Introduction 261
2. Goals in computer experiments 262
3. Approaches to computer experiments 264
4. Bayesian prediction and inference 265
5. Bayesian designs 281
6. Frequentist prediction and inference 290
7. Frequentist experimental designs 292
8. Selected applications 303
References 305

Ch. 10. A Critique of Some Aspects of Experimental Design 309
J. N. Srivastava

1. Introduction 309
2. Search linear model 312
3. Designs with one nuisance factor 315
4. Row–column designs 318
5. Factorial experiments 324
Acknowledgement 339
References 339

Ch. 11. Response Surface Designs 343
N. R. Draper and D. K. J. Lin

1. Response surfaces and models 343
2. Second order surfaces 345
3. Criteria for experimental designs 346
4. Sequential experimentation 347
5. "Value for money" in designs 348
6. Screening designs and projection properties 349
7. The central composite design 356
8. Small composite designs 358
9. Orthogonal blocking of second order designs 361
10. Rotatability 364
11. Variance, bias and lack of fit 365
12. Some other second order designs 370
References 373

Ch. 12. Multiresponse Surface Methodology 377
A. I. Khuri

1. Introduction 377
2. Plotting of multiresponse data 378
3. Estimation of parameters of a multiresponse model 380
4. Inference for multiresponse models 386
5. Designs for multiresponse models 389
6. Multiresponse optimization 398
7. Concluding remarks 402
References 403

Ch. 13. Sequential Assembly of Fractions in Factorial Experiments 407
S. Ghosh

1. Introduction 407
2. Normal probability plot in factor screening 408
3. Sequential assembly of fractions 411
4. Orthogonal and balanced plans 415
5. Search designs 417
6. Influential nonnegligible parameters 421
7. Parallel and intersecting flats fractions 423
8. Composite designs 427
9. Discussions and conclusions 432
Acknowledgements 433
References 433

Ch. 14. Designs for Nonlinear and Generalized Linear Models 437
A. C. Atkinson and L. M. Haines

1. Background 437
2. Nonlinear regression models 438
3. Generalized linear models 456
4. Model checking and compound design criteria 467
5. Designs for discriminating between models 469
6. Extensions 472
Acknowledgements 473
References 473

Ch. 15. Spatial Experimental Design 477
R. J. Martin

1. Introduction 477
2. Historical review 480
3. Spatial analysis 483
4. Designs with neighbour balance 489
5. Efficient designs for spatially dependent observations 499
Acknowledgement 511
References 511

Ch. 16. Design of Spatial Experiments: Model Fitting and Prediction 515
V. Fedorov

1. Introduction 515
2. Standard design problem 516
3. Optimal designs with bounded density 520

4. Correlated observational errors 523
5. Random coefficients regression models: Trend estimation 526
6. Random coefficient regression models: Prediction 531
7. Comparison with the methods based on the variance–covariance structure of observed random fields 534
8. Discrete case. Optimality criteria and the lower bounds 542
9. Unknown covariance function 545
10. Space and time 549
Acknowledgement 550
References 550

Ch. 17. Design of Experiments with Selection and Ranking Goals 555
S. S. Gupta and S. Panchapakesan

1. Introduction 555
2. Selecting the best treatment in single-factor experiments: Normal theory 558
3. Selection in experiments with blocking: Normal theory 565
4. Selection in factorial experiments: Normal theory 569
5. Selection with reference to a standard or a control: Normal theory 573
6. Selection in experiments involving other models 578
7. Concluding remarks 579
Acknowledgement 580
References 580

Ch. 18. Multiple Comparisons 587
A. C. Tamhane

1. Introduction 587
2. Basic concepts 588
3. Types of MCP's and methods of their construction 592
4. Modified Bonferroni procedures based on p-values 595
5. Normal theory procedures for inter-treatment comparisons 600
6. Multiple endpoints 615
7. Miscellaneous problems 623
8. Concluding remarks 624
Acknowledgement 625
References 625

Ch. 19. Nonparametric Methods in Design and Analysis of Experiments 631
E. Brunner and M. L. Puri

1. Introduction and notations 631
2. Fixed models 637
3. Random models 660
4. Mixed models 662
Acknowledgements 699
References 700

Ch. 20. Nonparametric Analysis of Experiments 705
A. M. Dean and D. A. Wolfe

1. Overview 705
2. One-way layout 713
3. Two-way layout 729
4. Higher way layouts 749
References 756

Ch. 21. Block and Other Designs Used in Agriculture 759
D. J. Street

1. Preliminaries 759
2. Incomplete block designs 761
3. Row–column designs 774
4. Repeated measurements 779
5. Neighbour designs for field trials 784
6. Factorial designs and fractional factorial designs 787
7. Response surface designs 795
8. Miscellaneous designs 802
Acknowledgement 803
References 803

Ch. 22. Block Designs: Their Combinatorial and Statistical Properties 809
T. Caliński and S. Kageyama

0. Introduction 809
1. Preliminaries 811
2. Concurrences and different notions of balance 823
3. Classification of block designs 830
4. Nested block designs and the concept of resolvability 840
5. Analysis of experiments in block designs under the randomization model 848
6. The concept of general balance 859
Acknowledgements 867
References 867

Ch. 23. Developments in Incomplete Block Designs for Parallel Line Bioassays 875
S. Gupta and R. Mukerjee

1. Introduction 875
2. Parallel line assays 876
3. Parallel line assays in block designs 882
4. Symmetric parallel line assays: L-designs 885
5. Symmetric parallel line assays: Further results 889
6. Symmetric parallel line assays: Q-designs 892
7. Asymmetric parallel line assays 894

8. On the role of non-equireplicate designs 896
9. Some open issues 899
Acknowledgement 900
References 900

Ch. 24. Row–Column Designs 903
K. R. Shah and B. K. Sinha

1. Introduction 903
2. Preliminaries: Model, estimability, connectedness and randomization 904
3. Analysis of data: Estimation of treatment contrasts 908
4. Further aspects of connectedness and analysis 911
5. Efficiency bounds for row–column designs 916
6. Optimality considerations 919
7. Nested row–column designs 921
8. Three-way balanced designs 921
9. Mixed effects model 923
10. Designs for factorial and quasi-factorial experiments 926
11. Miscellaneous topics 929
Acknowledgement 934
References 934

Ch. 25. Nested Designs 939
J. P. Morgan

1. Introduction 939
2. Nesting in block designs – Models and related considerations 941
3. Resolvable block designs 946
4. Nested BIBDs and related designs 952
5. Nesting of row and column designs – Models and related considerations 957
6. Binary block nested row and column designs 963
7. Nested row and column designs optimal for the bottom stratum analysis 968
8. Other nested designs 971
Acknowledgement 972
References 972

Ch. 26. Optimal Design: Exact Theory 977
C.-S. Cheng

1. Introduction 977
2. Information matrices and optimality criteria 978
3. Optimal symmetric designs 984
4. Optimality of asymmetrical designs 991
5. Using the approximate theory 999
6. Miscellaneous results and methods 1001
Acknowledgement 1004
References 1004

Ch. 27. Optimal and Efficient Treatment-Control Designs 1007
D. Majumdar

1. Introduction 1007
2. Early development 1007
3. Efficient block designs for estimation 1011
4. Efficient designs for confidence intervals 1023
5. Efficient row–column designs for estimation 1029
6. Bayes optimal designs 1033
7. On efficiency bounds of designs 1040
8. Optimal and efficient designs in other settings 1044
Acknowledgements 1049
References 1049

Ch. 28. Model Robust Designs 1055
Y-J. Chang and W. I. Notz

1. Introduction 1055
2. Notation and a useful model 1058
3. Designs for model fitting and parameter estimation 1061
4. Extrapolation 1071
5. Detection of model inadequacy 1079
6. Bayesian robust design 1084
7. Applications to computer experiments 1089
References 1096

Ch. 29. Review of Optimal Bayes Designs 1099
A. DasGupta

1. General introduction 1099
2. Outline 1105
3. Some history 1105
4. Alphabetic criteria and other formulations 1108
5. Mathematics of Bayes design 1112
6. Examples and other information of use to practitioners 1115
7. Critique of the optimality theory 1120
8. Nonconjugate priors and robust Bayes designs 1122
9. Miscellaneous other design problems 1125
10. Conditional formulations and sample size choice 1126
11. Nonlinear problems 1130
12. Future of Bayes design 1132
13. Appendix 1133
Acknowledgement 1141
References 1142

Ch. 30. Approximate Designs for Polynomial Regression: Invariance, Admissibility, and Optimality 1149
N. Gaffke and B. Heiligers

1. Introduction 1149
2. Invariance of designs 1156
3. Admissibility of designs 1165
4. Invariant and admissible multiple polynomial regression designs 1172
5. Optimality: Gradients and first order characterizations 1185
6. Reduction of dimensionality 1189
7. Conceptual algorithm 1195
References 1199

Subject Index 1201

Contents of Previous Volumes 1217

Contributors

C. M. Anderson-Cook, *Department of Statistical and Actuarial Sciences, University of Western Ontario, London, Ontario, N6A 5B7, Canada* (Ch. 8)

P. Armitage, *University of Oxford, 2 Reading Road, Wallingford, Oxon OX10 9DP, England* (Ch. 1)

A. C. Atkinson, *Department of Statistics, London School of Economics, Houghton Street, London WC2A 2AE, UK* (Ch. 14)

E. Brunner, *Abt. Med. Statistik, University of Goettingen, Humboldt Allee 32, 37073, Goettingen, Germany* (Ch. 19)

T. Caliński, *Department of Mathematics and Statistical Methods, Agricultural University of Poznan, Wojska Polskiego 28, PL 60-637 Poznan, Poland* (Ch. 22)

Y-J. Chang, *Department of Mathematical Sciences, Idaho State University, Pocatello, ID 83201, USA* (Ch. 28)

C.-S. Cheng, *Department of Statistics, University of California, Berkeley, CA 94720, USA* (Ch. 26)

A. DasGupta, *Department of Statistics, Purdue University, 1399 Mathematical Sciences Building, West Lafayette, IN 47907-1399, USA* (Ch. 29)

A. M. Dean, *Department of Statistics, Ohio State University, 141 Cockins Hall, 1958 Neil Avenue, Columbus, OH 43210, USA* (Ch. 20)

N. R. Draper, *Department of Statistics, University of Wisconsin, 1210 W. Dayton Street, Madison, WI 53706, USA* (Ch. 11)

V. Fedorov, *Oak Ridge National Laboratory, P.O. Box 2008, Oak Ridge, TN 37831-6367, USA* (Ch. 16)

N. Gaffke, *Fakultät für Mathematik, Institut für Mathematische Stochastik, Otto-von-Guericke-Universität Magdeburg, Postfach 4120, D-39016 Magdeburg, Germany* (Ch. 30)

S. Ghosh, *Department of Statistics, University of California, Riverside, CA 92521-0138, USA* (Ch. 13)

S. S. Gupta, *Department of Statistics, Purdue University, 1399 Mathematical Sciences Building, West Lafayette, IN 47907-1399, USA* (Ch. 17)

S. Gupta, *Division of Statistics, Northern Illinois University, DeKalb, IL 60115-2854, USA* (Ch. 23)

L. M. Haines, *Department of Statistics and Biometry, Faculty of Science, University of Natal, Private Bag X01 Scottsville, Pietermaritzburgh 3209, South Africa* (Ch. 14)

B. Heiligers, *Fakultät für Mathematik, Institut für Mathematische Stochastik, Otto-von-Guericke-Universität Magdeburg, Postfach 4120, D-39016 Magdeburg, Germany* (Ch. 30)

S. Kageyama, *Department of Mathematics, Faculty of School Education, Hiroshima University, 1-1-1 Kagamiyama, Higashi-Hiroshima 739, Japan* (Ch. 22)

A. I. Khuri, *Department of Statistics, The University of Florida, 103 Griffin-Floyd Hall, Gainesville, Florida 32611-8545, USA* (Ch. 12)

J. R. Koehler, *Department of Mathematics, University of Colorado, Denver, CO 80217-3364, USA* (Ch. 9)

D. K. J. Lin, *Department of Statistics, University of Statistics, University of Tennessee, Knoxville, TN 37996-0532, USA* (Ch. 11)

D. Majumdar, *Department of Mathematics, Statistics and Computer Science (M/C 249), University of Illinois, 851 South Morgan Street, Chicago, IL 60607-3041, USA* (Ch. 27)

R. J. Martin, *School of Mathematics and Statistics, University of Sheffield, P.O. Box 597, Sheffield S3 7RH, UK* (Ch. 15)

J. P. Morgan, *Department of Mathematics and Statistics, Old Dominion University, Norfolk, Virginia 23529-0077, USA* (Ch. 25)

R. Mukerjee, *Indian Institute of Management, P.O. Box 16757, Calcutta 700027, India* (Ch. 23)

W. I. Notz, *Department of Statistics, Ohio State University, 1958 Neil Avenue, Columbus, OH 43210, USA* (Ch. 28)

A. B. Owen, *Department of Statistics, Stanford University, Sequoia Hall, Stanford, CA 94305-4065, USA* (Ch. 9)

S. Panchapakesan, *Department of Mathematics, Southern Illinois University, Carbondale, IL 62901-4408, USA* (Ch. 17)

H. I. Patel, *Department of Epidemiology and Medical Affairs, Berlex Laboratories, Inc., 300 Fairfield Road, Wayne, NJ 074770-7358, USA* (Ch. 2)

M. L. Puri, *Department of Mathematics, Indiana University, Rawles Hall, Bloomington, IN 47405-5701, USA* (Ch. 19)

P. R. Rosenbaum, *Department of Statistics, The Wharton School, University of Pennsylvania, 3013 Steinberg-Dietrich Hall, Philadelphia, PA 19104-6302, USA* (Ch. 6)

P. K. Sen, *Department of Biostatistics, University of North Carolina at Chapel Hill, CB#7400, Chapel Hill, NC 27599-7400, USA* (Ch. 4)

K. R. Shah, *Statistics Department, University of Waterloo, Waterloo, N2L 3G1, Canada* (Ch. 24)

B. K. Sinha, *Stat.-Math. Division, Indian Statistical Institute, 203 B.T. Road, Calcutta 700035, India* (Ch. 24)

J. N. Srivastava, *Department of Statistics, Colorado State University, Fort Collins, CO 80523-1877, USA* (Ch. 10)

D. M. Steinberg, *Department of Statistics and Operations Research, Tel Aviv University, Ramat Aviv 69978, Tel Aviv, Israel* (Ch. 7)

D. J. Street, *University of Technology, Sydney, Broadway NSW 2007, Australia* (Ch. 21)

J. Stufken, *Department of Statistics, Iowa State University, 102-E Snedecor Hall, Ames, IA 50011-1210, USA* (Ch. 3)

A. C. Tamhane, *Department of Industrial Engineering, Northwestern University, Evanston, IL 60208-3119, USA* (Ch. 18)

D. A. Wolfe, *Department of Statistics, Ohio State University, 141 Cockins Hall, 1958 Neil Avenue, Columbus, OH 43210, USA* (Ch. 20)

S. Zacks, *Department of Mathematical Sciences, Binghamton University, P.O. Box 6000, Binghamton, NY 13902-6000, USA* (Ch. 5)

S. Ghosh and C. R. Rao, eds., *Handbook of Statistics, Vol. 13*

1

The Design and Analysis of Clinical Trials

Peter Armitage

1. Introduction

1.1. The background

Of all the current forms of statistical experimentation, the clinical trial is the most likely to be familiar to a layperson, and the most likely to engage him or her as a participant. If we date statistical experimentation back to the mid-1920s (Fisher, 1926), then clinical trials have been with us for some 50 of the subsequent 70 years, so they can reasonably claim to be one of the oldest as well as one of the most prolific and pervasive branches of statistical science.

That assessment makes the customary assumption that the first published clinical trial to meet modern standards of rigour is the trial of streptomycin in the treatment of pulmonary tuberculosis (Medical Research Council, 1948). This was indeed a landmark, but there were important forerunners, gradually preparing the ground for the new methods. An essential feature in a clinical trial is the *comparison* between the effects of different medical procedures. This principle had been recognized in the 18th and 19th centuries by many investigators, who had realised the need to provide a fair basis for the comparison; Bull (1959) presents an excellent survey of this early work. But the solution to the problem of ensuring fairness in the comparison had to await Fisher's exposition of the case for random assignment in the 1920s.

During the 1930s many statisticians and others who appreciated the value of randomized field studies in agricultural research may well have dismissed the possibility of applying these methods in medicine. For the experimental units were no longer inanimate objects or plots of ground, but were human beings – and sick ones at that. The early trials organized by the (British) Medical Research Council, and those run in the United States by the Veterans' Administration and, later, the National Institutes of Health, showed that such fears were unjustified. The developments were supported by the strong advocacy of statisticians such as A Bradford Hill (Hill, 1962) in the United Kingdom and Harold Dorn in the United States, and by many influential physicians. A glance at the current contents of the major medical journals shows how well established the method has now become. It seems at times to be threatened by financial or legal setbacks, but it is strongly supported by the requirements of statutory bodies for the regulation of new medical procedures, by the needs of the pharmaceutical industry, and by a growing realisation by the medical profession that there is no other reliable way of assessing the relative merits of competing medical treatments.

In this paper I shall outline some of the main features of clinical trials that need to be taken into account by a statistician involved in their planning, analysis and interpretation. Further details may be obtained from a number of excellent books, such as those by Schwartz et al. (1980), Shapiro and Louis (1983), Pocock (1983), Buyse et al. (1984), Friedman et al. (1985) and Meinert (1986).

1.2. The move towards randomization

The random assignment of treatments was advocated by Pierce and Jastrow for psychological experiments as early as 1883, but the nearest approximation to gain any currency during the first three decades of the 20th century was the idea of systematic assignment of two treatments to alternate cases in a series; (in most trials patients become available for treatment serially rather than simultaneously, so alternation is feasible). Alternation would usually be effective if there was no possibility that the order of presentation or assessment of eligibility were affected by a knowledge of the treatment to be applied. But with a systematic assignment schedule this type of *selection bias* is very difficult to avoid. For this reason, Hill was led to advocate strictly random assignment for the tuberculosis trial published in 1948, and for a trial of pertussis vaccines started earlier but published later (Medical Research Council, 1951).

These remarks might have been of purely historical interest, were it not that the principle of random assignment still comes under occasional attack. Gehan and Freireich (1974), for instance, sought to identify situations for which the use of historical controls was preferable to randomization. However, the impossibility of ensuring the absence of bias between the groups being compared, and of measuring the extent of that bias, has led most investigators to reject this approach, except in very special circumstances (see Section 1.3, p. 3). Similarly, Byar (1980) argued powerfully against the use of data banks which supposedly might provide substantial information about the effects of standard treatments, against which results with new treatments might be compared. Again, biases will almost certainly be present, and there are likely to be deficiencies in the recording of information about patient characteristics which is needed to validate the comparison. For an extended discussion, see Section 4.2 of Pocock (1983).

1.3. The medical context

The randomized trial has pervaded almost every branch of medical research. Indeed, it is unduly limiting to refer to 'clinical' trials (since clinical medicine is strictly concerned with patients). Many important randomized trials are conducted in preventive medicine, where the subjects have no specific illness and are receiving prophylactic agents such as vaccines with the hope of reducing the future occurrence of disease. Cochrane (1972) introduced the term *randomized controlled trial* (RCT), to include not only the traditional therapeutic and prophylactic trials, but also trials to compare the effectiveness of different measures of medical care, such as the choice between home or hospital treatment for patients of a particular type.

Trials occur in virtually every branch of clinical medicine, notable examples being in cardiovascular medicine, oncology, psychopharmacology, ophthalmology and rheumatic disease. Most trials involve comparisons of therapeutic drugs, but other forms of treatment or prophylaxis may be examined, such as surgical procedures, radiotherapy regimens, forms of psychotherapy, screening programs, medical devices or behavioural interventions. Within therapeutic medicine the aim may be to induce a rapid improvement in a patient's condition, as in the treatment of an infectious disease; to improve the long-term prognosis for patients with chronic disease, as in many trials in cancer medicine; or to prevent future exacerbations of disease, as in the treatment of patients who have had a myocardial infarction (heart attack) with the aim of reducing the risk of further attacks (so-called *secondary prevention*).

Trials fall broadly into three categories, according to their financial and scientific sponsorship. First, there are trials, usually relatively small, initiated by single research workers or small localized groups, supported by the resources of a local medical school or hospital. Many of the earliest examples of clinical trials were of this type. The second category, now more prominent in medical research, is that of larger trials, usually multicentre, supported financially by national or international research agencies or by fund-raising organizations concerned with specific branches of medicine.

The third category comprises trials supported, and often initiated, by the pharmaceutical industry. A convenient distinction may be drawn between trials conducted at different stages of drug development. *Phase I* studies, on healthy volunteers, are concerned with the establishment of a safe dosage, and need not be considered here. *Phase II* trials are small-scale studies to see whether *prima facie* evidence of efficacy exists. They are often regarded as the first stages in a screening process, the successful contenders proceeding to full-scale evaluation in the next phase. Since Phase II trials are not intended to provide conclusive evidence, they are often conducted without random assignment. *Phase III* trials are the main concern of the present review. They provide authoritative evidence of efficacy, and are an essential feature of submissions to regulatory bodies for marketing approval. They provide evidence also of safety, although adverse events occurring with low frequency, or after prolonged intervals of time, are likely to be missed. *Phase IV* studies, based on post-marketing experience, may reveal long-term safety and efficacy effects, but do not concern us here.

1.4. Objectives and constraints

Finally, in this introductory section, it is useful to review some of the broader aspects of clinical trials: the varying objectives of their investigators, and the constraints under which they must operate.

A simplistic view of the objective in a clinical trial (the *selection model*) is that the investigator seeks to choose the better, or best, of a set of possible treatments for some specific condition. This may sometimes be appropriate: the investigator may decide that treatment A is clearly better than treatment B, and immediately switch to using

A on the relevant patients. More often, though, the decision will not be so clear-cut. Even though a trial shows clear evidence of a difference in efficacy, the doctor may well need to balance this against the suspected or known influence of adverse effects, to take into account special problems presented by individual patients, and to consider whether the picture is affected by evidence emerging from other related studies. From this point of view the clinical trial is best regarded as a means of building up a reliable bank of information which enables individual physicians to make balanced and well-founded choices. A good deal of decision-theoretic work has been done on the selection model of clinical trials (see Armitage (1985) for a review); we shall not examine this in detail.

Another useful distinction is that drawn by Schwartz and Lellouch (1967) between *explanatory* and *pragmatic* attitudes to clinical trials. An explanatory trial is intended to be as closely analogous as possible to a laboratory experiment. Conditions in different groups of patients are rigidly controlled, so that a clear picture emerges of the relative efficacy of carefully defined treatments, all other circumstances being kept constant. For instance, one might attempt to compare the effects of two regimens of radiotherapy for patients with malignant disease, the two regimens being clearly defined and adhered to during the trial.

A pragmatic trial, by contrast, aims to simulate more closely the conditions of routine medical practice, where treatment regimens rarely receive 100% compliance. In the example cited above, the investigator might choose to compare two therapeutic strategies based on the two regimens, but recognizing that deviations from the prescribed regimens will occur, often for good reason and with the approval of the patient's physician. Most trials exhibit both explanatory and pragmatic features, but in practice the pragmatic attitude tends to dominate. The distinction has important consequences for the analysis of trial results, as will be seen later (Section 4.4).

In most trials the investigator will be primarily interested in differences in efficacy between different treatment regimens, but concern for improved efficacy – a higher proportion of rapid cures of an infectious disease, more rapid relief of pain, a reduced risk of reoccurrence of a myocardial infarction – must always be tempered by an awareness of adverse effects. Unwanted side-effects must always be recorded and their incidence compared between treatment groups. However, adverse effects may be exhibited only after the end of the trial, and the investigator may need to observe patients' progress during a subsequent follow-up period.

In some trials, particularly those carried out at an early stage of drug development, the aim may be not so much to show improved efficacy, but rather to show that one treatment has an efficacy very similar to a standard treatment. The new treatment may, for example, be a different formulation of a drug whose efficacy has already been established. In these *equivalence* trials, it will be necessary to define an acceptable range of equivalence, on some appropriate scale, and to show whether or not the responses to the two treatments are sufficiently similar to justify a claim of equivalence.

In the analysis of any clinical trial, a major aim will be to estimate the difference in efficacy between treatments, with an appropriate statement of uncertainty. In the typical efficacy trial, interest will often centre on a null hypothesis of no difference, but it should not be assumed that the establishment of a nonzero difference in a particular direction is necessarily of clinical importance. Adverse effects or other disadvantages may outweigh small benefits in efficacy, and it is useful to draw a distinction between *statistical significance* and *clinical significance* (see Section 2.3). In equivalence trials a test of significance of a null hypothesis of zero difference is pointless, and the essential question is whether, allowing for uncertainty, the true difference clearly lies within the range of equivalence. (A similar point arises in *bioequivalence studies*, where the equivalence of different formulations of the same active agent is assessed by comparison of blood levels measured at intervals after administration, rather than by their clinical effects. See Chapter 2.)

Clinical trials are feasible only insofar as their conduct is consistent with accepted standards of medical ethics. The crucial point is that, in general, physicians will not employ treatments which they believe to be inferior to others currently available. Normally, therefore, an investigator will be prepared to assign treatments randomly only if he or she is agnostic about their relative merits. In forming a judgement on such questions the investigator may strike a balance between prior evidence which may suggest superiority for one treatment in a specific therapeutic response, and uncertainty about long-term or other effects. In a multicentre trial each investigator will need to make an individual judgement on the ethical question, although taking into consideration the views of colleagues. Certainly, ethical issues must be considered in any trial, however small, and however mild the condition under study.

In recent decades it has become usual to require the informed consent of any patient entered into a trial, perhaps by a written statement. The justification for this procedure may seem self-evident, yet it remains anomalous that physicians are not required to obtain such consent for procedures (however weakly supported by research) employed in routine practice, whereas they need to do so for well-designed research studies. Fortunately, in most trials consent can be readily obtained.

2. The planning of a clinical trial

2.1. The protocol

It is important that the investigator(s) should draw up, in advance, a detailed plan of the study, to be documented in the *protocol.* This should include a summary of the case for undertaking the trial, and of the evidence already available on the relative merits of the treatments under study. There should be an unambiguous definition of the categories of patients to be admitted, and of the treatment schedules to be applied. The protocol should also include detailed instructions for the implementation of the trial, definitions of the outcome variables (or *endpoints*) to be used in the analysis, statistical details of the design, the number of patients and (where appropriate) the length of follow-up, and a brief outline of the proposed methods of analysis. The protocol should be accompanied by, or include, copies of the *case report form*(*s*) [CRF(s)], in the design of which great care must be exercised to avoid ambiguities and to enable information to be recorded as effectively as possible.

Most of these topics are discussed in later sections of this review, but we consider now the question of defining the patient categories and treatments.

In broad terms these features will have been agreed at an early stage. The investigators may, for instance, have agreed to compare a new and a standard drug for the relief of pain in patients with rheumatoid arthritis. The question may then arise as to whether the patients should have as uniform a pattern of disease as possible, say by restriction to a rather narrow age range, a specific combination of symptoms, and a minimum period since the disease was diagnosed. This would be a natural consequence of the 'explanatory' attitude referred to earlier, just as in laboratory experiments one tries to keep experimental units as uniform as possible. By contrast, the 'pragmatic' approach would be to widen the selection to all categories for whom the choice of these treatments is relevant, which might mean almost any patient with the disease in question. This has the merit of more nearly simulating the choice facing the physician in routine practice, and of enabling more patients to be recruited and thereby potentially improving the precision of the trial (although perhaps to a lesser extent than expected, since the variability of outcome variables might be increased). If the results for a narrowly defined type of patient are of particular interest, these can always be extracted from the larger set of data.

A similar contrast presents itself in the definition of treatments. The explanatory attitude would be to define precisely the treatment to be applied – for instance, in a drug trial, the dosage, frequency and time of administration – and to control strictly the use of concomitant treatments. Departures from this defined regimen would be regarded as lapses in experimental technique. The pragmatic attitude would be to recognise that in routine practice treatments are administered in a flexible way, schedules being changed from time to time by the physician's perception of the patient's progress or by the patient's own choice. A clear definition of each treatment regimen is required, otherwise the results of the trial are useless, but this definition may well be one of the strategy to be implemented for varying the schedule rather than of a single, unvarying regimen.

2.2. Some basic designs

The range of experimental designs used in clinical trials is much more restricted than in some other branches of technology, such as agricultural or industrial research. The reason is perhaps partly the need to avoid complexity in the context of medical care which often has to be applied urgently; and partly the fact that elaborate blocking systems are impracticable when the experimental units are patients who enter over a period of time and whose characteristics cannot, therefore, be listed at the outset.

All trials involve comparisons between different treatments, and an important distinction is between *parallel-group trials* in which each subject receives only one of the contrasted treatments, so that the comparison is *between* subjects, and *crossover trials* in which each subject receives different treatments on different occasions, so that the comparison is *within* subjects. We consider briefly some special cases.

Subtrials

In some multicentre studies, perhaps involving controversial treatments, it may happen that not all the investigators are willing to randomize between all the treatments.

For instance, if a trial is designed to compare three treatments, A, B and C, some investigators might be prepared to use any of these, others might object to A but be willing to compare B and C, while others might wish to compare A and B, or A and C. There could then be four *subtrials*, each investigator being free to join whichever subtrial is preferred. (In other circumstances, a similar range of choice might be offered to individual patients.) Note that information about the contrast of, say, A and B, is available (a) from the (A, B, C) subtrial; (b) from the (A, B) subtrial; and (c) from the contrast between A versus C in the (A, C) subtrial, and B versus C in the (B, C) subtrial. The design is not as efficient as an (A, B, C) trial alone, but it may enable more investigators and patients to enrol than would otherwise be possible. Note, however, that the effect of any treatment contrast may vary between subtrials, because of the different characteristics of patients choosing the various options, and it may not be possible to pool results across subtrials.

An example of this approach was the ISIS-3 trial of agents for the treatment of suspected myocardial infarction (ISIS-3 Collaborative Group, 1992). Here, patients for whom physicians believed that fibrinolytic therapy was required were randomized between three potentially active agents, whereas those for whom the indication for fibrinolytic therapy was regarded as uncertain were randomized between these three agents and a fourth treatment which was a placebo (Section 2.4). Jarrett and Solomon (1994) discuss a number of subtrial designs which might be used in a proposed trial to compare heroin, methadone, and their combination, for heroin-dependent users.

Factorial designs

The advantages of factorial designs are well-understood and these designs are increasingly widely used. There are many situations where different features of treatment can be varied and combined, and it makes sense to study them together in one trial rather than as separate projects.

An example is provided by the ISIS-3 trial referred to above (ISIS-3 Collaborative Group, 1992), which compared the effects of three levels of one factor (streptokinase, tPA and APSAC) in combination with two levels of a second factor (aspirin plus heparin and heparin alone). The recently completed ISIS-4 trial, again for the treatment of suspected myocardial infarction (ISIS-4 Collaborative Group, 1995), studied the eight combinations of three binary factors: oral mononitrate, oral captopril and intravenous magnesium.

The factors, the levels of which act in combination, may be different facets of treatment with the same agent, rather than essentially different agents. For example, Stephen et al. (1994) reported a trial to compare the anti-caries efficacy of different fluoride toothpastes in children initially aged 11–12, over a period of three years. There were three active compounds, each of which provided fluoride at one of two levels. The two factors were, therefore, the compound (three levels) and the fluoride dose (two levels).

Factorial designs are a powerful scientific tool for the simultaneous study of several questions. If attention focusses on one particular combination of factor levels, the data for that combination alone may be inadequate; the factorial design has overcome this inadequacy by amalgamating information at several combinations. This may be a problem in some drug trials, where a manufacturer may aim to submit a particular combination of levels for regulatory approval. In such cases it may be helpful to follow a factorial trial by a more clearly focussed trial using the combination in question, together with one control level.

Crossover trials

Crossover trials have the dual advantage that (a) treatment comparisons are affected by within-subject random variation, which is usually smaller than between-subject variation; and (b) they economize in the number of subjects required for a fixed number of observations. However, they are often impracticable – for instance, if the patient's condition is changing rapidly over time or may undergo irreversible deterioration, if the purpose of treatment is to produce a rapid cure, or if each treatment has to be used for a long period. The ideal situation is the comparison of methods for the palliative treatment of a chronic disease, where after administration of each treatment the patient's condition reverts to a fairly constant baseline. The temporary alleviation of chronic cough might be an example. Crossover designs are often used for Phase II trials of new drugs, where a rapid response to treatment may be available and the longer-term effects can be studied later if the drug is selected for a Phase III trial. They are useful also for bioequivalence studies, where the response may be derived from measurements of blood or urinary concentrations after ingestion of the drug.

Crossover trials have generated a formidable literature (see, for instance, Jones and Kenward, 1989; Senn, 1993; and an issue of *Statistical Methods in Medical Research*, 1994, devoted to crossover designs). In the simplest design, for two treatments (A and B) and two periods, subjects are assigned randomly to two sequences: AB and BA. For a standard linear model with Gaussian errors, the analysis proceeds straightforwardly, with treatment (T) and period (P) effects. Subjects are usually taken as a random effect. The T $\times$ P interaction plays a crucial role. It may be caused by a carryover of response, from the first to the second period, and investigators should try to minimize the chance of this happening by allowing an adequate wash-out period between the treatments. If the interaction exists, the estimation of treatment effect becomes difficult and ambiguous. Unfortunately, the interaction is aliased with the contrast between the two sequences, which is affected by between-subject error and may therefore be imprecisely measured. These difficulties have led many writers to deprecate the use of this simple design (Freeman, 1989; Senn, 1992), although it should perhaps retain its place in suitably understood situations. Armitage and Hills (1982) pointed out that 'a single crossover trial cannot provide the evidence for its own validity' and suggested that 'crossover trials should be regarded with some suspicion unless they are supported by evidence, from previous trials with similar patients, treatments and response variables, that the interaction is likely to be negligible'.

Some of the deficiencies of the simple two-treatment, two-period design can be reduced by extending the number of sequences, treatments and/or periods. The extensive literature on this broad class of repeated-measures designs has been reviewed by Matthews (1988, 1994b), Jones and Kenward (1989), Shah and Sinha (1989) and Afsarinejad (1990). Optimal design theory can be applied, but optimality depends on the assumed model for the response variable. Even with a Gaussian model, there are ambiguities about the modelling of carry-over effects (Fleiss, 1989; Senn, 1992), about possible serial correlation between successive observations, and in the method of estimation (for instance, ordinary or generalized least squares). Moreover, responses will often be clearly non-Gaussian, the extreme case being that of binary responses. It is, therefore, difficult to go beyond broad recommendations. For two treatments and periods, the simple design (AB and its dual) may be extended by adding AA and its dual (Balaam, 1968); these additional sequences provide no extra within-subject information on the treatment effect, but help to elucidate the interaction. Still with two treatments, the number of periods can be extended beyond two. In general a chosen sequence should be accompanied by equal assignment of subjects to the *dual sequence*: for example, the sequence ABB should be accompanied by BAA. Under a certain model the sequences ABB and its dual are optimal, but some robustness against different assumptions is provided by the addition of two other sequences: either ABA and its dual (Ebbutt, 1984) or AAB and its dual (Carrière, 1994). Exploration of larger designs, particularly for more than two treatments, has been somewhat tentative. One problem is uncertainty about the effects of subjects who drop out before their sequence of treatments is completed, particularly as drop-outs are likely to be nonrandom (see Section 4.4). Matthews (1994a) reports the successful completion of a trial with 12 periods and 12 subjects; however, the subjects were healthy volunteers, who may have been more likely than patients to cooperate fully.

2.3. The prevention of bias

Methods of randomization

The case for random assignment of treatments has already been discussed (Section 1.2), and need not be re-emphasized here. Randomization schedules are normally prepared in advance by computer routines, and (to avoid selection bias) individual assignments are revealed only after a patient has been formally entered into the trial.

Randomization will, of course, produce similarity between the characteristics of patients allotted to different treatments, within limits predictable by probability theory. In some trials it is thought advisable to ensure closer agreement between treatment groups, for instance by ensuring near-equality of numbers on different treatments within particular subgroups or *strata* defined by certain pre-treatment characteristics (or *baseline variables*). One method used in the past is that of *permuted blocks*, whereby numbers assigned to different treatments in a particular stratum are arranged to be equal at fixed intervals in the schedule. More recently methods of *minimization* are often used (Taves, 1974; Begg and Iglewitz, 1980). These assign treatments adaptively so as to minimize (in some sense) the current disparity between the groups, taking account simultaneously of a variety of baseline variables. In multicentre trials, such methods are often used to ensure approximate balance between treatment numbers within each centre. In any scheme of balanced assignment, the effect of the balanced factors in reducing residual variation should be taken into account, for example by an analysis of covariance. If this is done, the statistical advantage of balanced assignment schemes, over simple randomization with covariance adjustment, is very small (Forsythe and Stitt, 1977). Their main merit is probably psychological, in reassuring investigators in centres providing few patients that their contribution to a multicentre study is of value, and in ensuring that the final report of a trial can produce convincing evidence of similarity of groups.

In most parallel-group trials the numbers of patients assigned to different treatments are (apart from random variation) equal or in simple ratios which are retained throughout the trial. Many authors have argued that, on ethical grounds, the proportionate assignments should change during the trial, in such a way that more patients are gradually assigned to the apparently more successful treatments (see, for instance, Zelen, 1969; Chernoff and Petkau, 1981; Berry and Fristedt, 1985; Bather, 1985). Such *data-dependent allocation* schemes have rarely been used in practice, and can give rise to problems in implementation and analysis. See Armitage (1985) for a general discussion, and Ware (1989) and the published discussion following that paper for a case-study of a controversial trial of extracorporeal membrane oxygenation (ECMO) in which these methods were used.

The prevention of response bias

Randomization, properly performed, ensures the absence of systematic bias in the selection of subjects receiving different treatments. This precaution would, though, be useless if other forms of bias caused different standards to be applied in the assessment of response to different treatment regimens.

In a trial in which the use of a new treatment is compared with its absence, perhaps in addition to some standard therapy, bias may arise if the subjects are aware whether or not they receive the new treatment, for instance by taking or not taking additional numbers of tablets. The pharmacological effect of a new drug may then be confounded with the psychological effect of the knowledge that it is being used. For this reason, the patients in the control group may be given a *placebo*, an inert form of treatment formulated to be indistinguishable in taste and consistency from the active agents.

Placebos are commonly used in drug trials, although complete disguise may be difficult, particularly if the active agents have easily detectable side-effects. Their use is clearly much more difficult, perhaps impossible, in other fields such as surgery.

The principle of masking the identity of a treatment may be extended to trials in which different active agents are compared, for instance in a comparison of two drugs or of different doses of the same drug. Such steps would not be necessary if the endpoints used in the comparison of treatments were wholly objective. Unfortunately the only such endpoint is death. Any other measure of the progress of an illness, such as the reporting of symptoms by the patient, the eliciting of signs by a physician, or major recurrences of disease, may be influenced by knowledge of the treatment received. This includes knowledge by the patient, which may affect the reporting of symptoms or the general course of the disease, and also by medical and other staff whose recording of events may be affected and who may (perhaps subconsciously) transmit to the patient their own enthusiasm or scepticism about the treatment.

For these reasons, it is desirable where possible to arrange that treatments are administered not only *single-blind* (masked from the patient), but *double-blind* (masked also from the physician and any other staff concerned with the medical care or assessment of response). When different drugs are administered in essentially different ways, for instance by tablet or capsule, it may be necessary to use the *double-dummy* method. For instance, to compare drug A by tablet with drug B by capsule, the two groups would receive

Active A tablets, plus Placebo B capsules,

or

Placebo A tablets, plus Active B capsules.

In a drug trial, medicaments should be made available in packages labelled only with the patient's serial number, rather than by code letters such as A and B, since the latter system identifies groups of patients receiving the *same* treatment and can easily lead to a breakdown of the masking device.

2.4. *Trial size*

The precision of treatment comparisons is clearly affected by the size of the trial, the primary consideration being the number of subjects assigned to each treatment. Where repeated measurements are made on each subject, for instance with regular blood pressure determinations, the precision may be improved somewhat by increasing the number of repeated measurements, but the overriding factor will still be the number of subjects. In many chronic disease studies the response of interest is the incidence over time of some critical event, such as the relapse of a patient with malignant disease, the reoccurrence of a cardiovascular event, or a patient's death. Here the crucial parameter is the expected number of events in a treatment group, and this can be increased either by enrolling more patients or by lengthening the follow-up time for each patient. The latter choice has the advantage of extending the study over a longer portion of the natural course of the disease and the possible advantage of avoiding additional recruitment from less reliable centres, but the disadvantage of delaying the end of the trial.

The traditional methods for sample-size determination are described in detail in many books on statistics (for example, Armitage and Berry, 1994, Section 6.6), and tables for many common situations are provided by Machin and Campbell (1987). The details depend on the nature of the comparison to be made – for instance, the comparison of means, proportions or counts of events. The calculations usually refer to the comparison of two groups, and are essentially along the following lines:

(a) A parameter δ is chosen to represent the contrast between responses on the two treatments, taking the value 0 under the null hypothesis H_0 that there is no difference.

(b) A significance test (usually two-sided) of H_0 will be based on a statistic X.

(c) The sample size is chosen to ensure that the test has a specified power $1 - \beta$ for an alternative hypothesis H_1 that $\delta = \delta_A$, with a specified significance level α. Allowance may be made for a hypothetical rate at which patients might withdraw from prescribed treatment, thus diluting any possible effect.

The introduction of sample-size calculations into the planning stages of a trial, and their incorporation into the protocol, have the clear advantage of making investigators aware of some consequences of their choice of trial size, and perhaps of preventing the implementation of trials which are too small to be of any real value. There are, however, some problems about a rigid adherence to the formulation described above:

(i) The probabilities α and β are arbitrary. The power $1 - \beta$ is usually taken to be between 0.8 and 0.95, and the (usually two-sided) significance level α is usually 0.05 or 0.01. Different choices have substantial effects on the required group size.

(ii) The critical value δ_A is difficult to interpret. As pointed out by Spiegelhalter et al. (1994), it may represent (A) the smallest clinically worthwhile difference, (B) a difference considered 'worth detecting', or (C) a difference thought likely to occur. Whichever interpretation is preferred, the value chosen will be subjective, and agreement between investigators may not be easy to achieve.

(iii) For any choice of the parameters referred to in (a)–(c), the consequent group size will depend on other, unknown, parameters, such as the variability of continuous measurements, the mean level of success proportions or of incidence rates, and the withdrawal rate. Well-informed estimates of these quantities may be available from other studies, or from early data from the present study, but some uncertainty must remain.

(iv) The standard argument assumes a null hypothesis of zero difference, whereas (as implied in Section 1.4) a zero difference may have no clinical significance, and some nonzero value, trading off improved efficacy against minor inconvenience, may be more appropriate.

(v) More broadly, the standard approach is centred around the power of a significance test, although precision in estimation is sometimes more important.

(vi) The investigators may intend to combine the results of the trial with those of other studies (Section 5.3). The power of the current trial, considered in isolation, may then be less relevant than that of the complete data set.

These reservations suggest that sample-size calculations should be regarded in a flexible way, as providing guidance in the rational choice of group size, rather than as a rigid prescription. In any case, the outcome of formal calculations always needs to be balanced against practical constraints: any determination of trial size is likely to involve some degree of compromise.

Many authors have argued that uncertainties of the sort outlined above are best recognised and taken into account within a Bayesian framework (Spiegelhalter et al., 1986; Spiegelhalter and Freedman, 1986; Brown et al., 1987; Moussa, 1989; Spiegelhalter et al., 1994). The Bayesian approach will be discussed further in a later section (Section 5.2), but we note here some points relevant to the determination of trial size (Spiegelhalter et al., 1994). Some basic steps are:

(i) The choice of a prior distribution for the difference parameter δ. This may reasonably vary between investigators, and in a multicentre study some compromise may have to be reached. It may be useful to perform calculations based on alternative priors, representing different degrees of enthusiasm or scepticism about the possible difference.

(ii) The identification of an indifference interval (δ_I, δ_S) for the parameter. The implication is that for $\delta < \delta_I$ one treatment (the 'standard', say) is taken to be superior, and for $\delta > \delta_S$ the other treatment (the 'new') is superior. When the parameter falls within the indifference interval no clear preference is expressed.

(iii) For any given trial size, and with appropriate assumptions about nuisance parameters, the probability may be calculated at the outset that a central posterior interval for δ will exclude one of the critical values δ_I or δ_S. For instance, a high posterior probability that $\delta > \delta_I$ would provide assurance that the new treatment could not be ruled out of favour; a similar result for $\delta > \delta_S$ would imply that the new treatment was definitely superior.

With this approach, the trial size may be chosen with some assessment of the chance of obtaining a result favouring one treatment. It should be understood, though, that an outcome clearly in favour of one treatment is not necessary for the success of a trial. Reliable information is always useful, even when it suggests near-equivalence of treatment effects.

3. The conduct and monitoring of a clinical trial

3.1. The implementation of the protocol

The investigators should have been fully involved in the drawing up of the protocol, and will normally be anxious to ensure a high degree of compliance with the agreed procedure. Nevertheless, departures from the protocol are likely to occur from time to time. The impact of protocol departures by individual patients on the analysis of results is discussed in Section 4.4. The investigators will need to set up a procedure for checking that high standards are maintained in the administration of the trial, and in a multicentre study a separate committee is likely to be needed for this purpose. This *administrative monitoring* will check whether the intended recruitment rate is being achieved, detect violations in entry criteria, and monitor the accuracy of the information recorded.

The data processing (coding of information recorded on the CRFs, checking of data for internal consistency, computer entry and verification, and subsequent analysis) is often done by specialist teams with *standard operating procedures* (SOPs) which should be overseen as part of the administrative monitoring. A number of software packages for trial management are commercially available.

As the trial proceeds it may be clear that certain provisions in the protocol should be changed. The rate of accrual or of critical events may be lower than expected, leading to a decision to extend the trial. Experience with the clinical or other technical procedures may point to the desirability of changes. It may, for various reasons, become desirable to augment or reduce the entry criteria. Although investigators will clearly wish to avoid protocol changes without very strong reason, it is better to complete the study with a relevant framework rather than persist with an original protocol which has major deficiencies.

In deciding whether to enter patients into a trial an investigator will normally screen a larger number of patients for potential eligibility. There may be (a) some patients in broadly the right disease category, but found on examination to be ineligible for the trial, for instance on grounds of age or particular form of disease; and (b) some patients who would have been fully eligible according to the protocol, but were not admitted to the trial, either by error, for administrative convenience, because of individual characteristics not covered in the protocol but thought by the physician to be important, or through the patient's refusal. Patients in category (b) should ideally be very few, and their numbers and the reasons for exclusion should be recorded. Category (a) is often regarded as interesting, and many investigators insist that such patients be entered into a *patient log* which may indicate the extent to which the trial population is selected from the wider population of patients entering the relevant clinics. The importance of the patient log is perhaps exaggerated, since the populations screened are likely to differ considerably in number and characteristics in different clinics, and the numerical proportion of screened patients who are entered into the trial may be hard to interpret.

The compliance of patients with the treatment regimen required by the protocol is clearly important, and steps should be taken to make this as nearly complete as possible. Some aspects of compliance, such as attendance at follow-up clinics and acceptance of trial medication, are measurable by the doctor. The actual taking of trial medication is another matter, at any rate for non-hospitalized patients. Statements by patients of their consumption of medication are unreliable, but these may sometimes be augmented by biochemical tests of concentrations of relevant substances in blood or urine.

Occasionally, patients will cease to follow the regimen prescribed by the protocol. The reasons may be various: death or a severe deterioration in medical condition, change of address, an unspecific refusal to continue, the occurrence of severe adverse effects, the occurrence of a medical complication which leads the physician to advise a change in management, and so on. All such occurrences must be carefully documented. The physician will no doubt try to advise the patient to continue the prescribed treatment as long as possible, but in the last resort the patient has the right to opt out, and the physician has the right to advise whether or not this would be a sensible step to take. Any patient withdrawing from the protocol regimen should be regarded as still in the trial, and should be included in follow-up examinations as far as possible.

3.2. Data and safety monitoring and interim analyses

The administrative monitoring referred to in Section 3.1 is a form of quality assurance about the structure of the trial. A quite different form of monitoring is needed for the accumulating results of the trial. The issues here relate to the safety and efficacy of the treatments under study.

Many medical treatments produce minor side effects or affect the biochemistry of the body. Minor effects of this sort need not be a source of concern, particularly when their occurrence is expected from previous experience. *Serious adverse effects* (SAEs), of a potentially life-threatening nature, are a different matter, and must be monitored throughout the trial. An unduly high incidence of SAEs, not obviously balanced by advantages in survival, is likely to be a good reason for early stopping of the trial or at least for a change in one or more treatment regimens.

Differences in efficacy may emerge during the course of the trial and give rise to ethical concerns. The investigators will have been uncertain at the outset as to whether the differences, if any, were large enough to be of clinical importance, and this ignorance provided the ethical justification for the trial. If the emerging evidence suggests strongly that one treatment is inferior to another, there may be a strong incentive to stop the trial or at least to abandon the suspect treatment.

In a double-blind trial the investigators will be unaware of the treatment assignments and unable themselves to monitor the results. They could arrange to be given serial summaries of the data by the data-processing team, but it is usually better for the monitoring to be handled by an independent group, with a title such as *Data [and Safety] Monitoring Committee* (D[S]MC). A DMC would typically comprise one or more statisticians, some medical specialists in the field under investigation, and perhaps lay members such as ethicists or patient representatives. Commercial sponsors are not normally represented. The data summaries are usually presented by the trial data-processing team, although occasionally independent arrangements are made. The trial investigators are occasionally represented on the DMC, but otherwise are not made aware of the DMC's discussions. In a large multicentre trial designed to last several years, it would be usual for the DMC to meet several times a year. Its main function is to keep the accumulating data under continuous review, and to make recommendations to the investigators if at any time the DMC members think the trial should be terminated or the protocol revised. (For a different view, favouring data monitoring by the investigators, or a committee reporting direct to them, rather than by an independent DMC, see Harrington et al. (1994).)

Any analysis of accumulating data is a form of sequential analysis, and much of the relevant literature seeks to adapt standard methods of sequential analysis to the context presented in a clinical trial (Armitage, 1975; Whitehead, 1992). Two points commonly emphasized are (a) the need to have closed sequential designs, defining the maximum length of the trial; and (b) the merits of *group sequential* designs, in which the data are analysed at discrete points of time (as, for instance, at successive meetings of a DMC) rather than continuously after each new item of data.

In a sequential analysis of either safety or efficacy data, it may be useful to do *repeated significance tests* of some null hypothesis (often specifying no difference in effects of treatments, but perhaps allowing for a nonzero difference for reasons discussed in Section 2.4). The consequences of such repeated testing are well-known (Armitage, 1958, 1975; Pocock, 1977): the probability of rejecting a null hypothesis at some stage (the Type I error rate) is greater than the nominal significance level at any single stage.

A number of different schemes have been devised which provide for early stopping when a statistic measuring an efficacy difference exceeds a predetermined bound, and ensure an acceptable and known Type I error probability. Other properties such as the power function and sample size distribution are also of interest. See Whitehead (1992) and Jennison and Turnbull (1990) for general surveys.

Most of the statistical advantages of continuous sequential schemes are retained by group sequential schemes, in which the accumulating results are examined at a small number of time-points. Three systems have been particularly widely used: (a) Repeated significance tests at a constant nominal significance level (Pocock, 1977); (b) Tests at interim analyses at a very stringent level (say, differences of three standard errors), with a final test at a level close to the required Type I probability (Haybittle, 1971; Peto et al., 1976); (c) Even more stringent tests at earlier stages, with bounds for a cumulative difference determined by the test at the final stage (O'Brien and Fleming, 1979). The choice between these is perhaps best made by considering whether large effects are a priori likely. A trial showing a large difference would be more likely to stop early with scheme (a), and least likely with scheme (c). Sceptical investigators might therefore prefer (c), whereas those regarding a large difference as inherently plausible might prefer (a).

With these schemes the stopping rule is predetermined. Lan and DeMets (1983), Kim and DeMets (1992) and others have described methods by which a predetermined Type I error probability can be 'spent' in a flexible way, for instance to meet the schedule of a DMC. However, most workers prefer to regard the whole data-monitoring procedure as being too flexible to fit rigidly into any formal system of rules. A decision whether to recommend termination of a trial is likely to depend on analyses of many endpoints for both safety and efficacy, to be affected by levels of recruitment and compliance, and to take into account other ongoing research. According to this view (Armitage, 1993; Liberati, 1994), formal schemes for stopping rules provide guidelines, and usefully illustrate some of the consequences of data-dependent stopping, but should not be regarded as rigidly prescriptive.

A more radical attack on the formal stopping-rule schemes outlined above comes from a Bayesian approach (Berry, 1987; Spiegelhalter and Freedman, 1988; Spiegelhalter et al., 1994). In Bayesian inference, as distinct from frequency theory, probabilities over sample spaces, such as Type I error probabilities, play no part. Inferences at any stage in data collection are uninfluenced by stopping rules used earlier or by those which it is proposed to use in the future. With the sort of framework for Bayesian inference described in Section 2.4, a stopping rule might propose termination if, at any stage, the posterior probability of a definitive difference (i.e., one affecting clinical practice) was sufficiently high. As noted in the last paragraph, considerations of this sort would form merely one part of the DMC's deliberations.

Early stopping might occasionally be called for if the efficacy difference was remarkably small. This situation would present no ethical problems, but might suggest that research effort would be better concentrated on other projects. Such a decision might require a prediction of the likely upper limit to the size of the difference to be expected at the termination point originally intended. This prediction could be based on frequency theory, leading to methods of *stochastic curtailment* (Lan et al., 1982, 1984), or on Bayesian methods (Spiegelhalter and Freedman, 1988).

Early stopping because of the emergence of negligible differences is generally to be discouraged, unless there is urgent need to proceed to other projects. It is often important to show that only small differences exist, and to estimate these as precisely as possible, i.e., by continuing the study. Even 'negative' results of this sort may contribute usefully to an overview covering several studies (Section 5.3). Moreover, if a decision is to be taken to stop early because of small differences, it is preferable to base this on inferences from the data available, rather than introduce unnecessary random error by predicting the future.

More extensive discussion on all the topics covered in this section may be found in Armitage (1991) and in the two special issues of *Statistics in Medicine* edited by Ellenberg et al. (1993) and Souhami and Whitehead (1994).

4. The analysis of data

4.1. Basic concepts

Statistical analyses will be needed for any interim inspections required by the DMC, and, much more extensively, for the final report(s) on the trial results. Almost any standard statistical technique may come into play, and it is clearly impossible to provide a comprehensive review here. Instead, we shall outline some of the considerations which are especially important in the analysis of clinical trial data.

Descriptive statistics, mainly in the form of tables and diagrams, are needed to show the main *baseline* (i.e., before treatment) characteristics of the patients in the various groups. Statistical inference is needed to draw conclusions about treatment effects, for both efficacy and safety.

The principal contrasts under study, and the intended ways of measuring these, should have been defined in the protocol. As noted in Section 4.2, it is important to avoid the confusion created by the analysis of a very large number of response variables and/or a large number of different contrasts. For this reason it is usually advisable to define, in the protocol, a small number of *primary endpoints* and a small number of contrasts. The principal analysis will be confined to these. Any remarkable effects seen outside this set should be noted, reported with caution, and perhaps put forward for further study in another trial.

Statistical analyses will usually follow familiar lines: significance tests (perhaps of nonzero differences (Section 2.4)), point and interval estimation for measures of treatment effects. Alternatively, Bayesian methods may be used, whereby inferences are summarized by posterior distributions for the relevant treatment-effect parameters. Bayesian methods are discussed further in Section 5.2.

4.2. Multiplicity

Clinical trials often generate a vast amount of data for each patient: many measures of clinical response and biochemical test measurements, each perhaps recorded at several points of time. An uncritical analysis, placing all these variables on the same footing and emphasizing those which suggested the most striking treatment effects, would clearly be misleading. Attempts may be made to allow for multiplicity of response variables (Pocock et al., 1987), but methods of adjustment depend on the correlation between variables, and lead to conservative inferences which may obscure real effects on important variables.

The most satisfactory approach is that outlined in Section 4.1, to perform unadjusted analyses on a small number of primary endpoints, and to regard findings on other variables as hypothesis-generating rather than hypothesis-testing.

Multiplicity arises in other contexts. The problems of repeated inspections of data have been discussed in Section 3.2, and can be dealt with either informally, by conservative inferences, or by application of a formal sequential scheme. Jennison and Turnbull (1989) have described methods by which repeated confidence intervals for a treatment effect may be calculated at arbitrary stages during the accumulation of trial data, with a guaranteed confidence coefficient. However, these tend to be much wider than standard intervals and may be of limited value.

Multiple comparisons of the traditional type, arising through the multiplicity of treatment groups (Miller, 1981), are of less concern. Most clinical trials use only a few treatments, and contrasts between them are usually all of primary importance: their impact should not be reduced by conservative adjustments.

A more difficult question concerns interactions between treatments and baseline variables. It would clearly be important to know whether certain treatment effects, if they exist, apply only to certain subgroups of the patient population, or affect different subgroups differently. However, many baseline variables will have been recorded, and it will be possible to subdivide by baseline characteristics in many ways, each of which provides an opportunity to observe an apparent interaction. Such subgroup comparisons should be treated in the same light as the analysis of secondary endpoints: that is, to provide suggestions for further studies but to be interpreted with reserve at present. Some subgroup analyses may have been defined in the protocol as being of primary interest, and these, of course, may be reported without reserve.

In a multicentre study, an interaction may exist between treatments and centres. Like other interactions it may be regarded as being affected by multiplicity and therefore of secondary importance. It may, though, be too pronounced to be easily ignored. In that case the first step would be to seek rational explanations, such as known differences between centres in ancillary treatments or in the distribution of baseline characteristics. If no such explanation is found, there are two choices. The interaction may be ignored on the grounds that the relevant patient population is that provided by all centres combined (Fleiss, 1986). Alternatively, the variation in treatment effect between centres may be regarded as an additional component of random variation, to be taken into account in measuring the precision of the treatment effect (Grizzle, 1987). Essentially the same point arises in meta-analysis, and will be discussed further in Section 5.3.

4.3. Adjustment for covariates

Randomization should ensure that any baseline variables that are prognostic for outcome variables are similarly distributed in different treatment groups and therefore produce no systematic biases in treatment comparisons. Nevertheless, adjustment for prognostic baseline variables will tend to improve the precision of treatment comparisons, and will have the secondary effect of correcting for any major imbalance in baseline distributions that may have occurred by chance.

If the relation between an outcome variable and one or more baseline variables can be reasonably approximated by a standard linear model with normal errors, the correction can be done by the analysis of covariance, usefully implemented by a general linear model program. A similar approach may be made for binary outcomes using a logistic regression program, and for survival data using a proportional hazards or other standard program.

In some instances it may be hard to justify such models, and a more pragmatic approach is to estimate a treatment effect separately within each of a number of subgroups defined by combinations of baseline characteristics, and then combine these estimates by an appropriately weighted mean.

Methods of minimization referred to in Section 2.3 have the effect of making treatment groups more alike in baseline characteristics than would occur purely by chance. An analysis ignoring baseline characteristics therefore underestimates precision, although the effect may well be small. Ideally, therefore, adjustment for covariates should always be performed when minimization has been been used for treatment assignment.

4.4. Protocol deviations

However much effort is made to ensure adherence to the protocol, most trials will include some patients whose treatment does not conform precisely to the prescribed schedule. How should their results be handled in the analysis?

One approach is to omit the protocol deviants from the analysis. This is the *per protocol* or (misleadingly) *efficacy* analysis. It seeks to follow the explanatory approach (Section 1.4) by examining the consequences of precisely defined treatment regimens. The problem here is that protocol deviants are likely to be unrepresentative of the total trial population, and the deviants in different treatment groups may well differ both in frequency and in disease characteristics. A comparison of the residual groups of protocol compliers has therefore lost some of the benefit of randomization, and the extent of the consequent bias cannot be measured. A *per protocol* analysis may be useful as a secondary approach to the analysis of a Phase III trial, or to provide insight in an early-stage trial to be followed by a larger study. But it should never form the main body of evidence for a major trial.

The alternative is an *intention-to-treat* (ITT) analysis of all the patients randomized to the different groups. Clearly, if there is a high proportion of protocol deviations leading to cessation of the trial regimens, any real difference between active agents may be diluted and its importance underestimated. For this reason, every attempt should be made to minimize the frequency of deviations. An ITT analysis preserves the benefit of randomization, in that the treatment comparisons are made on groups sufficiently comparable for baseline characteristics. It follows the pragmatic approach to trial design (Section 1.4), in that the groups receive treatments based on ideal strategies laid down in the protocol, with the recognition that (as in routine medical practice) rigidly prescribed regimens will not always be followed.

In an ITT analysis, it may be reasonable to omit a very small proportion of patients who opted out of treatment before that treatment had started. But this should only be done when it is quite clear (as in a double-blind drug trial) that the same pressures to opt out apply to all treatment groups. In some cases, for instance in a trial to compare immediate with delayed radiotherapy, this would not necessarily be so.

A similar point arises in trials with a critical event as an endpoint. For some patients the critical event may occur before the trial regimen is expected to have had any effect. Unless it is quite clear that these expectations are the same for all treatments, and are defined as such in the protocol, such events should not be excluded from the analysis.

One form of protocol deviation is the loss of contact between the patient and the doctor, leading to gaps in the follow-up information. (Deaths can, however, often be traced through national death registration schemes.) In an ITT analysis of, say, changes in serum concentrations of some key substance 12 months after start of treatment, what should be done about the gaps? A common device is the *last observation carried forward* (LOCF) method, whereby each patient contributes the latest available record. This may involve an important bias if the groups are unbalanced in the timing of the drop-outs and if the response variable shows a time trend. Brown (1992) has suggested a nonparametric approach. An arbitrarily poor score is given to each missing response. This is chosen to be the median response in a placebo or other control group, and all patients with that response or worse are grouped into one broad response category. Treatment groups are then compared by a standard nonparametric method (the Mann-Whitney test), allowing for the broad grouping. Some information is lost by this approach, and estimation (as distinct from testing) is somewhat complicated. For another approach to the analysis of data with censoring related to changes in the outcome variable, see Wu and Bailey (1988, 1989) and Wu and Carroll (1988).

4.5. Some specific methods

It would be impracticable to discuss in detail the whole range of statistical methods which might be applied in the analysis of trial data. We merely comment here on a few types of data which commonly arise, and on some of the associated forms of analysis.

Survival data

In many trials the primary endpoint is the time to occurrence of some critical event. This may be death, but other events such as the onset of specific disease complications may also be of interest. The standard methods of survival analysis (Cox and Oakes, 1984) may be applied for comparison of treatment groups with any of these endpoints. For a simple comparison of two groups the logrank test and associated estimates of hazard ratio may be used, stratifying where appropriate by categories formed by baseline covariates. The Cox proportional-hazards model is a powerful way of testing for and estimating treatment effects in the presence of a number of covariates, which may be introduced in the linear predictor or by defining separate strata.

The survival curves for different subgroups are conveniently calculated, and displayed graphically, by the Kaplan–Meier method. Such plots will occasionally show that the proportional-hazards assumption underlying standard methods is inappropriate, and more research is needed to develop diagnostic methods and alternative models (Therneau et al., 1990; Chen and Wang, 1991).

Categorical data

In many analyses the endpoint will be binary: for instance the occurrence of a remission after treatment for a particular form of cancer. The analysis of binary data is now a well-researched area, based principally on the technique of logistic regression (Cox and Snell, 1989).

In many trials the endpoint is polytomous, i.e., with several response categories. This is often the case in Phase II and other small trials, where short-term changes in a patient's condition may be more appropriately measured by a subjective judgement by the patient or the doctor, using a small number of categories. There is often a natural ordering of categories in terms of the severity of the disease state or the patient's discomfort.

The analysis of ordered categorical data has been a subject of active research in recent years (Agresti, 1989, 1990). Models have now been developed which incorporate treatment and baseline effects, permitting significance tests and estimation of treatment effects.

Repeated measurements

In some trials repeated observations on the same physiological or biochemical variable are made, for each patient, over a period of time. The question of interest may be that of treatment effects on the trends in these variables during a prolonged period of treatment for a chronic disease. In bioequivalence studies attention may focus on the trend in serum concentrations during a fairly short period after ingestion of a drug.

Multivariate methods of handling repeated measures data concentrate on an efficient estimate of the time trend, but in trials the emphasis is on the estimation of treatment effects. A useful approach, described in detail by Matthews et al. (1990), is to summarize the essential features of an individual's responses by a small number of summary measures. Examples might be the mean response over time, and the linear regression on time. Each such summary measure is then analysed in a straightforward way.

The last example serves to remind us that multivariate methods do not play a major part in the analysis of trial results. The reason is partly the emphasis in multivariate analysis on significance tests, whereas it is important in trials to be able to estimate effects in a clinically meaningful way. Moreover, when a small number of primary variables have been defined at the outset, it is usually essential to provide clear and separate information about each, rather than to combine the variables in obscure ways. However, when there are many endpoints of interest, and perhaps many treatments, multivariate methods may help to solve some of the problems of multiplicity; see Geary et al. (1992) for an example of the use of multivariate analysis of variance in the analysis of four dental trials in which several measures of dental health were available.

5. The reporting and interpretation of results

5.1. The relevance of trial results

Whether the patients in a clinical trial were selected by narrow eligibility criteria, or, as advocated in Section 2.1, they were entered more liberally, the results of the trial will be of potential interest for the treatment of patients whose characteristics vary and do not necessarily replicate those of the trial population. How safe is it to generalize beyond the trial itself?

It will clearly be safer to do so if the entry criteria are broad rather than narrow, if the trial provides no suggestion of treatment interactions with relevant baseline variables, and if there is no external evidence, either empirical or theoretical, to suggest that important interactions exist. Cowan and Wittes (1994) suggest that extrapolation is safer for treatments having essentially biological effects than for those, such as methods of contraception, the success of which is likely to depend on social factors. Clearly, an element of faith is always needed. It would be impossible to provide strong evidence of treatment effects separately for every possible combination of baseline factors. For this reason, recent proposals that NIH-sponsored trials should aim to provide strong evidence for all minority groups are self-defeating (Piantadosi and Wittes, 1993).

There are other possible reasons for failures to convince other workers that the results of a clinical trial are widely applicable. Doubts may exist about some aspects of the treatments used, such as the details of drug administration. There is sometimes a tendency for groups of clinicians to want to verify, in their own environment, results reported elsewhere, even though no explicit doubts are raised.

One source of doubt about positive findings (i.e., those indicating real effect differences) should be taken seriously. It seems likely that most treatments compared in trials do not in fact differ greatly in their effects. Purely by chance a few of these will show significant differences. Of those trials reporting statistically significant differences, then, a certain proportion will be 'false positives'. The position is aggrevated by *publication bias* (Begg and Berlin, 1988), which ensures that, at least for relatively small trials, 'positive' results are more likely to be published than 'negative' ones. Such doubts may be difficult to overcome if reported effects are only marginally significant.

Questions may arise also about the clinical relevance of the endpoints reported in a trial. In many research investigations the primary outcome measure may involve long-term follow-up, perhaps to death, and in a trial of finite length the times to these events may be censored by termination of the trial observation period. It may be tempting to use a *surrogate endpoint*, namely a measure of response available more quickly (or in some cases more conveniently) than the main endpoint. Unfortunately, even when the two endpoints are clearly correlated, it cannot be assumed that treatment effects on one imply similar effects on the other. An example is provided by the report of the Concorde Coordinating Committee (1994) on the Concorde trial to compare immediate and deferred use of zidovudine in symptom-free individuals with HIV infection. The primary endpoints were survival, serious adverse events and progression to clinical disease (ARC or AIDS). Survival and progression are known to be related to levels of CD4 cell counts, and immediate use of zidovudine certainly caused CD4 counts to rise above the declining levels seen in the deferred group. Unfortunately, the benefit of immediate use was not reflected in the patterns of survival or progression, which remained similar for the two groups.

5.2. The Bayesian approach

Bayesian methods have already been referred to, particularly in relation to trial size (Section 2.4) and interim analyses (Section 3.2). It is, of course, common ground to take some account of prior evidence in assessing the plans for a new trial or the results from a recently completed trial. The discussion in Section 5.1, about the prior supposition that differences are likely to be small, is in principle Bayesian.

It has been argued strongly in recent years, by a number of statisticians experienced in clinical trials, that an explicitly and exclusively Bayesian approach to the design, analysis and interpretation of trials would overcome some of the ambiguities associated with traditional, frequency-theory, statistics (Berry, 1987, 1993; Spiegelhalter and Freedman, 1988; Spiegelhalter et al., 1993, 1994). The Bayesian approach enables alternative analyses to be explored, using a variety of prior distributions representing different degrees of scepticism or enthusiasm about the likely magnitude of treatment effects. Modern computing methods such as Gibbs sampling have greatly facilitated this process (Gilks et al., 1993; George et al., 1994). Spiegelhalter et al. (1994) suggest that the main analysis of a trial, usually presented in the Results section of a report, should be based on noninformative priors, so as to express as objective a summary as possible, essentially by means of likelihood functions. The effect of alternative priors could then be described in the Discussion section. The use of noninformative priors will usually lead to conclusions similar (apart from changes of nomenclature) to those from frequency analyses, and from this point of view the proposal is not as revolutionary as might at first appear.

Each of these two approaches has its own advantages. The frequency approach has the advantage of greater familiarity within the community of clinical trialists, of richness in the available choice of statistical methods, and of easy access to these through statistical packages. The Bayesian approach has the advantage of enabling the effect of different prior beliefs to be explored and presented for discussion. In my view both approaches will co-exist for some time, and many statisticians will make use of them in an eclectic, if not always consistent, manner. Spiegelhalter et al. (1994), for example, writing from a Bayesian standpoint, adduce pragmatic reasons for paying attention to the Type I error probability in a monitoring scheme, in order to avoid too many false positive results.

5.3. Meta-analysis and overviews

In most branches of science it is regarded as necessary for experiments to be repeated and confirmed by other workers before being accepted as well-established. During the first few decades of clinical trials, it was very rare to find any trial being replicated by other workers. A positive result, showing treatment differences, would usually inhibit random assignment of the weaker treatment in any future trial. A negative result would often discourage repetition of the same comparison, particularly if other potentially important questions remained to be studied. Unfortunately, early trials were often too small to permit detection of important differences, which may therefore have remained undisclosed (Pocock et al., 1978).

During the last 20 years or so, there have been many more instances of replicated trials. The protocols in any one such collection of trials will rarely, if ever, be exactly the same. The treatments may differ in some details, and the inclusion criteria for patients may vary. But it will often be reasonable to assume that the effects of these variations are small in comparison with sampling errors. It therefore seems reasonable to pool the information from a collection of similar trials, and to take advantage of the consequent increase in precision.

Such studies are called *meta-analyses* or *overviews*. Some examples are overviews of beta blockade in myocardial infarction (Yusuf et al., 1985); antiplatelet treatment in vascular disease (Antiplatelet Trialists' Collaboration, 1988); side-effects of non-steroidal anti-inflammatory drug treatment (Chalmers et al., 1988); and hormonal, cytotoxic or immune therapy for early breast cancer (Early Breast Cancer Trialists' Collaborative Group, 1992).

In a meta-analysis a separate estimate of some appropriate measure of treatment effect is obtained from each trial, and it is these estimates that are pooled rather than the original data. Suppose that for the ith of k trials the estimated treatment effect is y_i with variance $v_i = w_i^{-1}$. A standard method of combining the estimates is to take the weighted mean $\bar{y} = \sum w_i y_i / \sum w_i$. The homogeneity of the k estimates may be tested by the heterogeneity index $G = \sum w_i y_i^2 - (\sum w_i y_i)^2 / \sum w_i$, which is distributed approximately as χ^2 on $k - 1$ degrees of freedom. If homogeneity is assumed, $\mathrm{var}(\bar{y}) = (\sum w_i)^{-1}$ approximately.

In trials in which the primary endpoint is the rate of occurrence of critical events, it is often convenient to measure the treatment effect by the odds ratio of event rates. A commonly used approach (Yusuf et al., 1985) is to test the significance of the effect by a score statistic asymptotically equivalent to the weighted mean, and to use a related estimate and heterogeneity test. This method is somewhat biased when the effect is not small (Greenland and Salvan, 1990).

Meta-analysis is a powerful and important technique. Two aspects remain somewhat controversial. The first is the difficulty of deciding precisely which trials should be included. Some authors (Chalmers, 1991) emphasize that primary attention should be given to the quality of the research. Others (Peto, 1987) emphasize the importance of including unpublished as well as published results, in order to minimize publication bias and to ensure inclusion of very recent data.

The second question arises when a meta-analysis shows evidence of heterogeneity between treatment effects in different trials. Clearly, an effort should then be made to seek reasons, such as differences in protocols, which might plausibly explain the interaction. If no such reasons can be found, what should follow? One school of thought (Yusuf et al., 1985; Early Breast Cancer Trialists' Collaborative Group, 1990) leads to use of the estimates described above, as being appropriate for the particular mix of patients and study characteristics found in these trials. A different approach is to regard the variation in treatment effect between trials as a component of variation which is effectively random in the sense of being unexplained. According to this view, any generalization beyond the current trial population must take this variation into account, thereby leading to a less precise estimate of the main effect than would otherwise be obtained (DerSimonian and Laird, 1986). For further discussion of meta-analysis in clinical trials see the papers in the special issue of *Statistics in Medicine* edited by Yusuf et al. (1987); also Berlin et al. (1989), Pocock and Hughes (1990), Whitehead and Whitehead (1991) and Jones (1995).

6. Conclusion

This has been a deliberately wide-ranging account of the role of statistics in clinical trials. Design, execution, analysis and interpretation are inextricably interwoven, and each aspect gains by being considered in context rather than in isolation. Methodology moves rapidly and standards of performance become ever more rigorous. Provided that financial, legal and societal constraints are not unduly repressive we can look forward to further developments in trial methodology, a wider appreciation of the need for reliable comparisons of medical treatments, and a continued contribution of this important branch of statistical experimentation to the progress of medical science.

Acknowledgements

I am grateful to Ray Harris, John Matthews and a reviewer for helpful comments on a draft of this paper.

References

Afsarinejad, K. (1990). Repeated measurements designs – A review. *Comm. Statist. Theory Methods* **19**, 3985–4028.

Agresti, A. (1989). A survey of models for repeated ordered categorical response data. *Statist. Med.* **8**, 1209–1224.

Agresti, A. (1990). *Categorical Data Analysis*. Wiley, New York.

Antiplatelet Trialists' Collaboration (1988). Secondary prevention of vascular disease by prolonged antiplatelet treatment. *British Med. J.* **296**, 320–331.

Armitage, P. (1958). Sequential methods in clinical trials. *Amer. J. Public Health* **48**, 1395–1402.

Armitage, P. (1975). *Sequential Medical Trials*, 2nd edn. Blackwell, Oxford.

Armitage, P. (1985). The search for optimality in clinical trials. *Internat. Statist. Rev.* **53**, 15–24.

Armitage, P. (1991). Interim analysis in clinical trials. *Statist. Med.* **10**, 925–937.

Armitage, P. (1993). Interim analyses in clinical trials. In: F. M. Hoppe, ed., *Multiple Comparisons, Selection Procedures and Applications in Biometry*. Dekker, New York, 391–402.

Armitage, P. and G. Berry (1994). *Statistical Methods in Medical Research*, 3rd edn. Blackwell, Oxford.
Armitage, P. and M. Hills (1982). The two-period crossover trial. *Statistician* **31**, 119–131.
Balaam, L. N. (1968). A two-period design with t^2 experimental units. *Biometrics* **24**, 61–73.
Bather, J. A. (1985). On the allocation of treatments in sequential medical trials. *Internat. Statist. Rev.* **53**, 1–13.
Begg, C. B. and J. A. Berlin (1988). Publication bias: A problem in interpreting medical data (with discussion). *J. Roy. Statist. Soc. Ser. A* **151**, 419–453.
Begg, C. B. and B. Iglewicz (1980). A treatment allocation procedure for clinical trials. *Biometrics* **36**, 81–90.
Berlin, J. A., N. M. Laird, H. S. Sacks and T. C. Chalmers (1989). A comparison of statistical methods for combining event rates from clinical trials. *Statist. Med.* **8**, 141–151.
Berry, D. A. (1987). Interim analysis in clinical research. *Cancer Invest.* **5**, 469–477.
Berry, D. A. (1993). A case for Bayesianism in clinical trials. *Statist. Med.* **12**, 1377–1393.
Berry, D. A. and B. Fristedt (1985). *Bandit Problems: Sequential Allocation of Experiments.* Chapman and Hall, London.
Brown, B. M. (1992). A test for the difference between two treatments in a continuous measure of outcome when there are dropouts. *Cont. Clin. Trials* **13**, 213–225.
Brown, B. W., J. Herson, N. Atkinson and M. E. Rozell (1987). Projection from previous studies: A Bayesian and frequentist compromise. *Cont. Clin. Trials* **8**, 29–44.
Bull, J. P. (1959). The historical development of clinical therapeutic trials. *J. Chronic Dis.* **10**, 218–248.
Buyse, M. E., M. J. Staquet and R. J. Sylvester, eds. (1984). *Cancer Clinical Trials: Methods and Practice.* Oxford Univ. Press, Oxford.
Byar, D. P. (1980). Why data bases should not replace randomized clinical trials. *Biometrics* **36**, 337–342.
Carrière, K. C. (1994). Crossover designs for clinical trials. *Statist. Med.* **13**, 1063–1069.
Chalmers, T. C. (1991). Problems induced by meta-analysis. *Statist. Med.* **10**, 971–980.
Chalmers, T. C., J. Berrier, P. Hewitt, J. A. Berlin, D. Reitman, R. Nagalingam and H. S. Sacks (1988). Meta-analysis of randomized control trials as a method of estimating rare complications of non-steroidal anti-inflammatory drug therapy. *Alim. Pharm. Therapeut.* **2** (Suppl. 1), 9–26.
Chen, C.-H. and P. C. Wang (1991). Diagnostic plots in Cox's regression model. *Biometrics* **47**, 841–850.
Chernoff, H. and A. J. Petkau (1981). Sequential medical trials involving paired data. *Biometrika* **68**, 119–132.
Cochrane, A. L. (1972). *Effectiveness and Efficiency: Random Reflections on Health Services.* Nuffield Provincial Hospitals Trust, London.
Concorde Coordinating Committee (1994). Concorde: MRC/ANRS randomized double-blind controlled trial of immediate and deferred zidovudine in symptom-free HIV infection. *Lancet* **343**, 871–881.
Cowan, C. D. and J. Wittes (1994). Intercept studies, clinical trials, and cluster experiments: To whom can we extrapolate? *Cont. Clin. Trials* **15**, 24–29.
Cox, D. R. and D. Oakes (1984). *Analysis of Survival Data.* Chapman and Hall, London.
Cox, D. R. and E. J. Snell (1989). *Analysis of Binary Data.* Chapman and Hall, London.
DerSimonian, R. and N. Laird (1986). Meta-analysis in clinical trials. *Cont. Clin. Trials* **7**, 177–188.
Early Breast Cancer Trialists' Collaborative Group (1990). *Treatment of Early Breast Cancer, Vol. 1: Worldwide Evidence 1985–1990.* Oxford Univ. Press, Oxford.
Early Breast Cancer Trialists' Collaborative Group (1992). Systemic treatment of early breast cancer by hormonal, cytotoxic, or immune therapy. *Lancet* **339**, 1–15; 71–85.
Ebbutt, A. F. (1984). Three-period crossover designs for two treatments. *Biometrics* **40**, 219–224.
Ellenberg, S., N. Geller, R. Simon and S. Yusuf, eds. (1993). Proceedings of the Workshop on Practical Issues in Data Monitoring of Clinical Trials. *Statist. Med.* **12**, 415–616.
Fisher, R. A. (1926). The arrangement of field experiments. *J. Min. Agric. Great Britain* **33**, 503–513.
Fleiss, J. L. (1986). Analysis of data from multiclinic trials. *Cont. Clin. Trials* **7**, 267–275.
Fleiss, J. L. (1989). A critique of recent research on the two-treatment crossover design. *Cont. Clin. Trials* **10**, 237–243.
Forsythe, A. B. and F. W. Stitt (1977). Randomization or minimization in the treatment assignment of patient trials: Validity and power of tests. Tech. Report No. 28, BMDP Statistical Software. Health Sciences Computing Facility, University of California, Los Angeles.

Freeman, P. R. (1989). The performance of the two-stage analysis of two-treatment, two-period crossover trials. *Statist. Med.* **8**, 1421–1432.
Friedman, L. M., C. D. Furberg and D. L. DeMets (1985). *Fundamentals of Clinical Trials*, 2nd edn. Wright, Boston.
Geary, D. N., E. Huntington and R. J. Gilbert (1992). Analysis of multivariate data from four dental clinical trials. *J. Roy. Statist. Soc. Ser. A*, **155**, 77–89.
Gehan, E. A. and E. J. Freireich (1974). Non-randomized controls in cancer clinical trials. *New Eng. J. Med.* **290**, 198–203.
George, S. L., L. Chengchang, D. A. Berry and M. R. Green (1994). Stopping a clinical trial early: Frequentist and Bayesian approaches applied to a CALGB trial in non-small-cell lung cancer. *Statist. Med.* **13**, 1313–1327.
Gilks, W. R., D. G. Clayton, D. J. Spiegelhalter, N. G. Best, A. J. McNeil, L. D. Sharples and A. J. Kirby (1993). Modelling complexity: Applications of Gibbs sampling in medicine (with discussion). *J. Roy. Statist. Soc. Ser. B*, **55**, 39–102.
Greenland, S. and A. Salvan (1990). Bias in the one-step method for pooling study results. *Statist. Med.* **9**, 247–252.
Grizzle, J. E. (1987). Letter to the Editor. *Cont. Clin. Trials* **8**, 392–393.
Harrington, D., J. Crowley, S. L. George, T. Pajak, C. Redmond and S. Wieand (1994). The case against independent monitoring committees. *Statist. Med.* **13**, 1411–1414.
Haybittle, J. L. (1971). Repeated assessment of results in clinical trials of cancer treatment. *British J. Radiol.* **44**, 793–797.
Hill, A. Bradford (1962). *Statistical Methods in Clinical and Preventive Medicine*. Livingstone, Edinburgh.
ISIS-3 (Third International Study of Infarct Survival) Collaborative Group (1992). ISIS-3: A randomized trial of streptokinase vs tissue plasminogen activator vs anistreplase and of aspirin plus heparin vs aspirin alone among 41 299 cases of suspected acute myocardial infarction. *Lancet* **339**, 753–770.
ISIS-4 (Fourth International Study of Infarct Survival) Collaborative Group (1995). ISIS-4: A randomized factorial trial comparing oral captopril, oral mononitrate, and intravenous magnesium sulphate in 58 050 patients with suspected acute myocardial infarction. *Lancet* **345**, 669–685.
Jarrett, R. G. and P. J. Solomon (1994). An evaluation of possible designs for a heroin trial. In: *Issues for Designing and Evaluating a 'Heroin Trial'. Three Discussion Papers*. Working Paper Number 8. National Centre for Epidemiology and Population Health, Canberra, 11–30.
Jennison, C. and B. W. Turnbull (1989). Interim analyses: The repeated confidence interval approach (with discussion). *J. Roy. Statist. Soc. Ser. B.* **51**, 305–361.
Jennison, C. and B. W. Turnbull (1990). Interim monitoring of clinical trials. *Statist. Sci.* **5**, 299–317.
Jones, B. and M. G. Kenward (1989). *Design and Analysis of Cross-over Trials*. Chapman and Hall, London.
Jones, D. R. (1995). Meta-analysis: Weighing the evidence. *Statist. Med.* **14**, 137–149.
Kim, K. and D. L. DeMets (1992). Sample size determination for group sequential clinical trials with immediate response. *Statist. Med.* **11**, 1391–1399.
Lan, K. K. G. and D. L. DeMets (1983). Discrete sequential boundaries for clinical trials. *Biometrika* **70**, 659–663.
Lan, K. K. G., R. Simon and M. Halperin (1982). Stochastically curtailed tests in long-term clinical trials. *Comm. Statist. C* **1**, 207–219.
Lan, K. K. G., D. L. DeMets and M. Halperin (1984). More flexible sequential and non-sequential designs in long-term clinical trials. *Comm. Statist. Theory Methods* **13**, 2339–2353.
Liberati, A. (1994). Conclusions. 1: The relationship between clinical trials and clinical practice: The risks of underestimating its complexity. *Statist. Med.* **13**, 1485–1491.
Machin, D. and M. J. Campbell (1987). *Statistical Tables for the Design of Clinical Trials*. Blackwell, Oxford.
Matthews, J. N. S. (1988). Recent developments in crossover designs. *Internat. Statist. Rev.* **56**, 117–127.
Matthews, J. N. S. (1994a). Modelling and optimality in the design of crossover studies for medical applications. *J. Statist. Plann. Inference* **42**, 89–108.
Matthews, J. N. S. (1994b). Multi-period crossover trials. *Statist. Methods Med. Res.* **3**, 383–405.
Matthews, J. N. S., D. Altman, M. J. Campbell and P. Royston (1990). Analysis of serial measurements in medical research. *British Med. J.* **300**, 230–235.

Medical Research Council (1948). Streptomycin treatment of pulmonary tuberculosis: A report of the Streptomycin in Tuberculosis Trials Committee. *British Med. J.* **2**, 769–782.

Medical Research Council (1951). The prevention of whooping-cough by vaccination: A report of the Whooping-Cough Immunization Committee. *British Med. J.* **1**, 1463–1471.

Meinert, C. L. (1986). *Clinical Trials: Design, Conduct and Analysis*. Oxford Univ. Press, New York.

Miller, R. G., Jr. (1981). *Simultaneous Statistical Inference*. Springer, New York.

Moussa, M. A. A. (1989). Exact, conditional, and predictive power in planning clinical trials. *Cont. Clin. Trials* **10**, 378–385.

O'Brien, P. C. and T. R. Fleming (1979). A multiple testing procedure for clinical trials. *Biometrics* **35**, 549–556.

Peto, R. (1987). Why do we need systematic overviews of randomized trials? *Statist. Med.* **6**, 233–240.

Peto, R., M. C. Pike, P. Armitage, N. E. Breslow, D. R. Cox, S. V. Howard, N. Mantel, K. McPherson, J. Peto and P. G. Smith (1976). Design and analysis of randomized clinical trials requiring prolonged observation of each patient. I. Introduction and design. *British J. Cancer* **34**, 585–612.

Piantodosi, S. and J. Wittes (1993). Letter to the Editor: Politically correct clinical trials. *Cont. Clin. Trials* **14**, 562–567.

Pocock, S. J. (1977). Group sequential methods in the design and analysis of clinical trials. *Biometrika* **64**, 191–199.

Pocock, S. J. (1983). *Clinical Trials: A Practical Approach*. Wiley, Chichester.

Pocock, S. J. and M. D. Hughes (1990). Estimation issues in clinical trials and overviews. *Statist. Med.* **9**, 657–671.

Pocock, S. J., P. Armitage and D. A. G. Galton (1978). The size of cancer clinical trials: An international survey. *UICC Tech. Rep. Ser.* **36**, 5–34.

Pocock, S. J., N. L. Geller and A. A. Tsiatis (1987). The analysis of multiple endpoints in clinical trials. *Biometrics* **43**, 487–498.

Schwartz, D., R. Flamant and J. Lellouch (1980). *Clinical Trials* (transl. M. J. R. Healy). Academic Press, London.

Schwartz, D. and J. Lellouch (1967). Explanatory and pragmatic attitudes in therapeutic trials. *J. Chronic Dis.* **20**, 637–648.

Senn, S. J. (1992). Is the 'simple carry-over' model useful? *Statist. Med.* **11**, 715–726.

Senn, S. (1993). *Cross-over Trials in Clinical Research*. Wiley, Chichester.

Shah, K. R. and B. K. Sinha (1989). *Theory of Optimal Designs*. Springer, Berlin.

Shapiro, S. H. and T. A. Louis, eds. (1983). *Clinical Trials: Issues and Approaches*. Dekker, New York.

Souhami, R. L. and J. Whitehead (1994). Proceedings of the Workshop on Early Stopping Rules in Cancer Clinical Trials. *Statist. Med.* **13**, 1289–1500.

Spiegelhalter, D. J. and L. S. Freedman (1986). A predictive approach to selecting the size of a clinical trial, based on subjective clinical opinion. *Statist. Med.* **5**, 1–13.

Spiegelhalter, D. J. and L. S. Freedman (1988). Bayesian approaches to clinical trials. In: J. M. Bernado et al., eds., *Bayesian Statistics, 3*. Oxford Univ. Press, Oxford, 453–477.

Spiegelhalter, D. J., L. S. Freedman and P. R. Blackburn (1986). Monitoring clinical trials: Conditional or predictive power? *Cont. Clin. Trials* **7**, 8–17.

Spiegelhalter, D. J., L. S. Freedman and M. K. B. Parmar (1993). Applying Bayesian thinking in drug development and clinical trials. *Statist. Med.* **12**, 1501–1511.

Spiegelhalter, D. J., L. S. Freedman and M. K. B. Parmar (1994). Bayesian approaches to randomized trials (with discussion). *J. Roy. Statist. Soc. Ser. A* **157**, 357–416.

Statistical Methods in Medical Research (1994). Editorial and five review articles on Crossover Designs. **3**, 301–429.

Stephen, K. W., I. G. Chestnutt, A. P. M. Jacobson, D. R. McCall, R. K. Chesters, E. Huntington and F. Schäfer (1994). The effect of NaF and SMFP toothpastes on three-year caries increments in adolescents. *Internat. Dent. J.* **44**, 287–295.

Taves, D. R. (1974). Minimization: A new method of assigning patients to treatment and control groups. *Clin. Pharmacol. Ther.* **15**, 443–453.

Therneau, T. M., P. M. Grambsch and T. R. Fleming (1990). Martingale-based residuals for survival models. *Biometrika* **77**, 147–160.

Ware, J. H. (1989). Investigating therapies of potentially great benefit: ECMO (with discussion). *Statist. Sci.* **4**, 298–340.

Whitehead, J. (1992). *The Design and Analysis of Sequential Clinical Trials*, 2nd edn. Horwood, Chichester.

Whitehead, A. and J. Whitehead (1991). A general parametric approach to the meta-analysis of randomized clinical trials. *Statist. Med.* **10**, 1665–1677.

Wu, M. C. and K. Bailey (1988). Analyzing changes in the presence of informative right censoring caused by death and withdrawal. *Statist. Med.* **7**, 337–346.

Wu, M. C. and K. Bailey (1989). Estimation and comparison of changes in the presence of informative right censoring: Conditional linear model. *Biometrics* **45**, 939–955.

Wu, M. C. and R. J. Carroll (1988). Estimation and comparison of changes in the presence of informative right censoring by modelling the censoring process. *Biometrics* **44**, 175–188.

Yusuf, S., R. Peto, J. Lewis, R. Collins and P. Sleight (1985). Beta-blockade during and after myocardial infarction: An overview of the randomized clinical trials. *Prog. Cardiovasc. Dis.* **27**, 335–371.

Yusuf, S., R. Simon and S. Ellenberg, eds. (1987). Proceedings of the Workshop on Methodologic Issues in Overviews of Randomized Clinical Trials. *Statist. Med.* **6**, 217–409.

Zelen, M. (1969). Play the winner rule and the controlled clinical trial. *J. Amer. Statist. Assoc.* **64**, 131–146.

S. Ghosh and C. R. Rao, eds., *Handbook of Statistics, Vol. 13*

2

Clinical Trials in Drug Development: Some Statistical Issues

H. I. Patel

1. Introduction

1.1. Background

For improving the quality of health care, the search for better alternative or new therapies continues. To demonstrate efficacy and safety of a therapy the sponsor conducts a sequence of clinical trials. A clinical trial is generally understood as being a designed experiment in patients with the targeted disease to answer questions related to the efficacy and safety of a drug or an intervention therapy in comparison with a control group. In a broader sense a clinical trial is not limited to patients. For example, in the early stages of the drug development healthy volunteers are administered single or multiple doses of the drug to evaluate the drug tolerability. The field of clinical trials is very fascinating and challenging as it deals with complex scientific experiments conducted in human beings.

As in any scientific experiment, there exist many factors other than the treatments that influence the clinical responses which are commonly referred to as endpoints. Some of these factors cause confounding effects with the treatments and others behave as extraneous sources of variation. The factors of the first group cause a bias in the estimation of the treatment difference; often it is not easy to understand the nature of the bias. The factors of the second group can be divided into two subgroups: patients themselves and the environmental conditions during the course of a trial. The sources of between-patient variability include age, gender, race, disease severity, disease duration and genetic make-up among others. The environmental conditions include life style, diet, compliance, use of concomitant medications, etc. There are some ways (experimental designs) to reduce the impact of between-patient variability, but it is practically impossible to have a good control on environmental factors during the course of the trial.

There exists an extensive literature on general methodology for designing and analyzing clinical trials. An excellent paper on this topic, citing many books and review papers and giving historical perspective of clinical trials, is written by Professor Armitage in this book. His paper also nicely covers the implementation of the experimental design principles, randomization, local control and replication, in clinical trials. In the next subsection we describe the scope of this paper.

The randomized controlled trials have revolutionized the way we infer about the treatment benefit and risk. However, the evaluation of just efficacy and safety is not regarded enough. Attempts are made to integrate these aspects with patients' views on their quality of life and economic consequences of the treatment alternatives.

1.2. The scope of the paper

The purpose of this paper is to briefly describe the various steps of the drug development process in the industrial setting and highlight the commonly used statistical designs and their analyses. The clinical trials sponsored by the pharmaceutical companies are designed, conducted and analyzed under the umbrella of regulatory requirements. Therefore, the environment under which they are done is more restricted than that for the NIH or academic institution sponsored trials, even though they all use the same basic statistical principles. Since Professor Armitage's paper is leaning towards the latter trials, we emphasize on the industry side in this paper. All attempts are made to minimize the overlapping between this paper and the paper by Professor Armitage. Professor Sen's paper in this book emphasizes different aspects of clinical trials; it covers the use of nonparametric methods in clinical trials.

Despite having a common theme of developing new compounds, there exist differences in the general approach to selecting designs and their analysis models in the industry. In this context, it would not be unfair to say that the tone of this paper may have been influenced by the author's own experience.

Here we have adopted a broader definition of a clinical trial to include experiments done in healthy volunteers. Section 2 briefly summarizes the pharmacokinetic studies done in healthy volunteers. It emphasizes bioequivalence studies which are of primary interest for the approval of generic drugs. Dose-response studies are described in Section 3. Mixed effects linear and nonlinear models for applications in clinical trials are described in Chapter 4. Chapter 5 briefly covers Markovian models used for analyzing repeated measures designs, between-subject and within-subject designs. While attempting to tackle some practical problems faced by biostatisticians in the pharmaceutical industry, this chapter emphasizes on statistical modelling and analysis of some of the recent topics in clinical trials.

1.3. Clinical trial phases in industry

The experiments begin with pre-clinical research. The drug is tested *in vitro* (in an environment not involving living organisms) and *in vivo* (in an environment involving living organisms). Pharmacokinetic and pharmacologic profiles are studied in animals. Pharmacokinetic studies in human are described in Section 2. Pharmacology is a branch of science dealing with the (desirable or undesirable) effects of chemical substances on biological activities. Initial formulation is developed and its manufacturing process is documented. Physical and chemical properties of the formulation are studied. Such studies include drug stability studies which are conducted to project the shelf life of the drug. If the drug is perceived to be safe and effective in humans, the sponsor files an IND (Investigational New Drug) application, supported by the *in vitro* and *in vivo* pre-clinical data and clinical data if available from foreign countries along with a clinical program outline. If the sponsor does not hear from the US (United States) regulatory agency within a certain period of time, the clinical program may begin.

The drug testing in humans is generally divided into three phases. In Phase I small studies, generally using 8 to 10 healthy male volunteers per study, are conducted to investigate how well the subjects tolerate single and multiple doses of different strengths of the experimental drug. Pharmacokinetic studies are also conducted in this phase using single and multiple doses of different strengths. The purpose here is to project safety based on observed pharmacokinetic characteristics.

Phase II studies are conducted in patients with the targeted disease. The primary goal here is to see whether the therapy is efficacious and evaluate the dose-response relationship. Sometimes pharmacokinetic information is obtained along with efficacy measurements to understand their causal-effect relationship. Phase II studies are relatively small, but well controlled and monitored, evaluating patients from a tightly defined population. Typically fewer than 200 patients are studied in this phase. The management of the sponsor company uses the efficacy and safety results obtained from this phase to decide whether or not to proceed to Phase III. A good pilot study may play an important role for such decision making.

Phase III studies are large, involving several participating centers. A typical Phase III study enrolls several hundred to several thousand patients. These studies generally have longer treatment duration to evaluate safety and efficacy than Phase II studies. The entry criteria for the patients to be studied in the trial are generally much broader than those in Phase II studies. The clinical and statistical design issues depend on the therapeutic area. The knowledge gained from Phase II studies and historical studies conducted for other similar therapies is used in planning Phase III studies.

In the United States, after completion of all three phases, a New Drug Application (NDA) is submitted by the sponsor to the Food and Drug Administration (FDA) which is part of the US Government's Department of Health and Human Services. Similar submissions are made in European and other foreign countries by the sponsors of new drugs to their respective government agencies. After a thorough review the regulatory agency decides whether or not to approve the drug. Often the approval of the drug is subject to some restrictions. The FDA may require some additional studies to gain more information on long-term safety and overall benefits. In Phase III it may not have been possible to treat all segments of the patient population of the targeted disease. In this case the FDA may ask the sponsor to conduct additional studies. In general, the FDA may ask the sponsor to conduct additional studies to answer the questions that could not have been answered from the NDA submission. The studies conducted after NDA approval are referred to as Phase IV studies which are planned and conducted while the drug is available in the market for treating patients. Phase IV studies, conducted for long-term safety, are usually simple but involve thousands of patients. The sponsors usually maintain special safety databases to monitor the incidences of major adverse experiences.

Not too long ago, the FDA tried to convince the industry to collect pharmacokinetic information from Phase III trials and relate it to efficacy and safety measurements.

However, because of the cost and logistic problems involved in the data collection, this concept never became widely acceptable in the industry.

Recently, the cost-effectiveness of a marketed drug has become an issue. In the current competitive environment, showing that the drug is safe and efficacious is not considered enough. Patients should be able to feel that the drug improves their quality of life and also provides economic benefits.

1.4. The role of statistics in clinical trials

Typically we do randomized, double-blind, multicenter, comparative trials. Unlike experiments in physical sciences, it is practically impossible to control the sources of variation due to environmental factors during the course of the trial. Statistics plays a very important role in all stages of clinical trials: planning, data collection, analysis and interpretation of the results.

During the planning stage, a clinical researcher and a statistician jointly prepare a protocol detailing the objectives of the study, the patient population to be enrolled, clinical and statistical designs, sample size, measurement issues, rules of conduct, statistical analysis plan and many other aspects of the trial. During this stage, we face many hurdles constrained by science, resources, ethics, and regulatory requirements. In the presence of many sources of variation, estimating treatment effects as purely as possible and with as few units as possible becomes a special challenge to a statistician. Here the statistician tries to introduce both innovative designs and quality assurance measures. The sample size calculation is true only to the extent to which the premises under which it is calculated are true. One needs a good projection of the variance from historical data and a clinically meaningful difference to be detected between the two treatments. A critical thinking is required here to make sure that a study design would help answer the questions posed by clinical researchers. A bad design and poor planning cannot be rescued by the data.

A statistician also prepares an analysis plan as a part of the protocol. A few, usually not more than three, primary efficacy variables are identified. The remaining endpoints are regarded as tentative. Many other issues such as hypotheses formulation, modelling the endpoints, multiple comparisons, subgroup analysis, interim analysis, handling dropouts, etc. are outlined in the protocol. Unfortunately, some statisticians prefer to stick to the proposed analysis plan rigidly even though the data fail to satisfy important assumptions underlying the chosen models. This is probably for the fear of being criticized by FDA. One solution is not to give too much details while writing the methodology. What is written before examining the data should not let us undermine the importance of performing valid statistical analysis.

The statistician plays at least an equally important role in analyzing data and interpreting the results in the context of the data collected. We know the limitations of statistical methods; they cannot remove all biases in estimating the treatment differences. Sometimes the patient population actually studied is different from what is defined by the protocol. In this case the interpretation of the results is limited in scope. As professional statisticians we cannot afford to work in isolation; we must take an active role in the process of designing a trial, be prepared in presenting the truth, continue learning each others disciplines, and have better communication with our colleagues at FDA. This will certainly help earn respect for our profession.

2. Pharmacokinetic studies

2.1. Introduction

Gibaldi and Perrier (1982) describe pharmacokinetics as the study of the time course of drug absorption, distribution, metabolism and excretion. This branch of biological sciences describes the processes and rates of drug movement from the site of absorption into the blood stream, distribution to the tissues and elimination by metabolism or excretion. It deals with mathematical modelling of the kinetics, related inference problems and applications to pharmacology. There exists a vast literature on this topic, dealing with the kinetics under single and multicompartment models and under single and multiple dosing (see, e.g., Wagner, 1971; Metzler, 1994; Rowland and Tozer, 1980; Gibaldi and Perrier, 1982).

The usefulness of the pharmacokinetic models in patient monitoring in clinical practice is appreciated when the relationships between drug kinetics and the therapeutic and toxicologic effects are understood. Recently, some pharmacokinetists have advocated special studies called population kinetic studies. Here kinetic parameters are treated as random variables (varying from patient from patient) having some prior distributions rather than fixed quantities as treated in pharmacokinetic mathematical modelling. This topic is further discussed in Section 4.

Of special applications are bioavailability (the extent and rate at which the drug is available in blood over a period of time after administration of a given dose) and bioequivalence (assessment of equivalence of two formulations with respect to the drug bioavailability) studies in the development of new formulations. The effectiveness of the drug therapy depends on the percent of the administered dose that is available to the site of pharmacologic action. For a solid dosage form this amount depends on patient characteristics, the physicochemical properties of the drug and the manufacturing process. Factors such as particle size, salt form, the coating, and the conditions under which the dose is administered also play an important role. It has become practice to rely on the rate and extent of the absorption of the drug into the bloodstream to measure the drug availability. This assumes that the drug distribution to the site of pharmacologic action and its elimination from the body are proportional to the amount of the drug absorbed. For an IV (intravenous) dose the absorption is 100 percent because it directly enters into the blood stream and therefore it is regarded as yardstick to compute the (absolute) bioavailability of any other route of administration. Sometimes, the bioavailability of an oral formulation is computed relative to the solution given orally. Two formulations are called bioequivalent if their bioavailabilities are equivalent.

The rates and extents of the bioavailability of various doses are used in determining the appropriate dose and frequency of dosing that are expected to be optimal for efficacy and tolerability of the drug, depending on the type of disease. This requires a dose-response study in pharmacokinetics. For chronic and subchronic diseases when a patient receives multiple doses, the question of achieving a steady state situation arises. Bioequivalence and dose-response studies are further discussed in Sections 2.2 and 2.3, respectively.

2.2. *Bioavailability and bioequivalence studies*

Pharmacokinetic studies for bioavailability and bioequivalence are conducted, generally in normal subjects, to compare two or more formulations with respect to the extent and rate of the drug availability. The extent is measured by AUC (area under the plasma curve) and the rate by $C_{\max}$ (maximum concentration over the observation period), and $T_{\max}$ (time when $C_{\max}$ occurs). According to the current regulatory guidance (OGD, Office of Generic Drugs, FDA, 1993), two formulations are considered bioequivalent if the ratio of a location parameter of the test (new) formulation to that of the reference (existing and already tested) formulation for the responses, such as AUC and $C_{\max}$, falls between 0.8 and 1.25. The previous guidelines (FDA, Food and Drug Administration, 1985; WHO, World Health Organization, 1986) required that these limits be 0.8 and 1.20. If μ_{T} and μ_{R} are location parameters, generally the population means, for the test and reference formulations, the problem of bioequivalence essentially reduces to that of testing

$$H_0\colon \mu_{\mathrm{T}}/\mu_{\mathrm{R}} \leqslant 0.8 \quad \text{or} \quad \mu_{\mathrm{T}}/\mu_{\mathrm{R}} \geqslant 1.25$$

$$\text{vs.} \quad H_1\colon 0.8 < \mu_{\mathrm{T}}/\mu_{\mathrm{R}} < 1.25. \tag{2.1}$$

Obviously, H_0 is not a linear hypothesis; it becomes linear with known interval boundaries only after the log transformation.

We limit our attention to a 2×2 crossover design because of its wide use in bioequivalence studies. This design assigns each subject randomly to one of the two sequences TR and RT, where T and R represent the test and reference formulations, respectively. The subjects assigned to Sequence TR receive T during the first period and R during the second period and the subjects assigned to Sequence RT receive the formulations in the reverse order. A washout period of adequate length is generally considered in this design. The sample sizes are generally small; as few as 6 subjects per sequence. Let y_{ijk} be a derived measurement, e.g., AUC, from the plasma level profile of subject k of Sequence j, during Period i, $i = 1, 2$; $j = 1, 2$; $k = 1, \ldots, n_j$. We assume a bivariate normal distribution for the pairs (y_{1jk}, y_{2jk}), leading to a more general model than Grizzle's (1965) model.

There are two concepts of bioequivalence: average and individual. The average bioequivalence, commonly used in practice, is based on a statistic defined as a function of location parameter estimators. On the other hand, the individual bioequivalence emphasizes the likelihood of the ratio $y_{\mathrm{T}}/y_{\mathrm{R}}$ falling within specified interval boundaries, where y_{T} and y_{R} are the test and reference formulation responses from the same subject.

There exists a vast literature on the methods of evaluating the average bioequivalence. The original work of Westlake (1972) and Metzler (1974) in this field is noteworthy. A good reference book is by Chow and Liu (1993). The most commonly used methods in the pharmaceutical industry for assessing the bioequivalence are (1) interpretation of a 90% conventional CI (confidence interval), referred to as Westlake (1972) CI approach, and (2) an equivalent approach of performing two one-sided 5% level t-tests, referred to as Shuirmann (1987) method. These formulas, however, use incorrect variances.

Some statisticians have recently advocated a routine use of the log transformation of the data followed by normal-theory based methods. Of course, after the log transformation, one can linearize the hypotheses H_{01} and H_{02} of (2.1). But this is not a scientific approach because one then routinely assumes that the data are lognormal. For any continuous distribution, in the presence or absence of wild observations, one can perform two one-sided Wilcoxon rank-sum tests on log transformed data as proposed by Hauschke et al. (1990). However, the power of this procedure would not be satisfactory especially in small samples. Another problem is that the normal approximation for the rank-sum statistic may be poor for small sample sizes and the exact distribution of the test statistic introduces the discrete sample space which may not lead to an exact α-size test.

Recently Hsu et al. (1994) have introduced a $1 - \alpha$ level CI for the difference between the formulation means which is contained in Westlake's (1976) $1 - \alpha$ level symmetric CI and is therefore shorter than the latter. However, their CI cannot be converted into the CI for the ratio unless one assumes normality for the log transformed data. There exist many alternative procedures including the methods of Fieller (1954) and Anderson–Hauck (1983) and Bayesian methods for evaluating bioequivalence. The Anderson–Hauck method is approximate because it replaces a naturally resulting noncentral t-statistic with the central t-statistic. The empirical studies have shown that this approximation tends to inflate Type I error rate. Mandallaz and Mau (1981) showed, under the assumption of fixed subject effects, that the Fieller's method is highly unlikely to fail. They provided the exact symmetrical CI for μ_T/μ_R when specified interval boundaries for decision making are symmetric about 1 and compared it with Fieller's CI. Westlake (1976) symmetric CI is an approximation to their exact symmetric CI. While Bayesian CIs are appealing in the context (1.1), they have not become popular in the pharmaceutical industry primarily because of difficulties in justifying a chosen prior. On the other hand, it is difficult to interpret a nonBayesian CI in this context using fiducial argument.

Recently, Shen and Iglewicz (1994) have bootstrapped trimmed t-statistics, in the context of a two one-sided tests procedure, for the difference of location parameters rather than for the ratio. Their procedure is useful in the presence of outliers, but it is associated with loss of information.

Anderson and Hauck (1990) proposed a method for assessing individual bioequivalence after formulating the hypotheses on $\pi = \text{pr}\{|y_T/y_R - 1| < 0.2\}$, where y_T and y_R are responses from the same subject. This approach does measure, in some sense, a simultaneous shift in both location and scale parameters. However, π is not a good indicator of bioequivalence because even in the case of equal population means and variances it heavily depends on the size of the common variance. Furthermore the test (sign test) is approximate because of the discreteness of the sample space. Wellek

(1993) recently extended this concept to lognormal data and introduced a noncentral F-statistic for testing the null hypothesis. Although derived under different considerations, this is exactly the same test as was proposed by Patel and Gupta (1984) in the context equivalence in clinical trials where the sample sizes are usually large. Esinhart and Chinchilli (1994) extended the tolerance interval approach, originally proposed by Westlake (1988), to higher order designs for evaluating the individual bioequivalence. Schall and Luus (1993) provided a unified approach for jointly assessing average and individual bioequivalence through a less economical design. It should be noted that this area of research is still evolving from the regulatory view points.

2.3. Dose-linearity

Pharmacokinetic studies are done to examine whether the kinetic, measuring the bioavailability in terms of AUC or $C_{\max}$, of a drug is proportional to the dose within a certain range of doses. The dose proportionality implies that the dose response curve is linear and passes through the origin. However, this concept is commonly interpreted by the pharmacokineticists as dose linearity. The linearity in a statistical sense would mean that the dose-response curve is linear and the regression line has an arbitrary intercept.

Consider a design where the pharmacokinetic measurements AUC and $C_{\max}$ are assessed from plasma profiles over a certain period of time in a single dose or multiple dose setting. Because of a relatively large between-subject variability, a dose-proportionality study is done using a within-subject design, generally a Latin square or a balanced incomplete block (BIB) design which is replicated a certain number of times. Subjects are assigned randomly to the treatment sequences in such a way that each sequence is received by an equal number of subjects. Sufficiently long washout periods are planned between successive doses so that the carryover effects can be assumed zero. Suppose y_{ij} is the response corresponding to dose x for subject i $(i = 1, \ldots, n)$ and period j $(j = 1, \ldots, p)$, then the dose-response is linear if $\mathrm{E}(y_{ij} \mid x) = \alpha + \beta x$. We assume that y is normally distributed. There are situations where the plasma drug levels are endogenous. For example, hormones are naturally produced by the body. So in a hormone study even a placebo subject will show some levels of hormone in the plasma. In this case the intercept α is positive. Also, when the drug is exogenous, there exist situations where the dose-response is nonlinear at the first pass, i.e., within a small dose-interval near zero. This means even if the dose-response is linear over the observed dose-range, it cannot be assumed linear over the dose-range starting from zero, even though the plasma levels are zero at the zero dose.

Another problem in this type study is that for the variables AUC and $C_{\max}$ the variance of y is proportional to x^2. Hence, the conventional split-plot type analysis of a within-subject design is inappropriate. So the data analysts consider the transformation $z = y/x$ to make the variance of z independent of x. However, in this case the regression of z on x is $\mathrm{E}(z_{ij} \mid x) = \alpha/x + \beta$, which is nonlinear unless $\alpha = 0$. The data analysts assume that $\alpha = 0$ and test the hypothesis of equal z-means associated with the dose levels using a split-plot type analysis. If the null hypothesis is not rejected, they conclude the dose proportionality of the response. There are two problems, however, with this currently practiced method. First, α must be zero for $\mathrm{E}(z_{ij} \mid x)$ to be constant. The second problem is that the lack of significance of the hypothesis of equal z-means does not prove the null hypothesis. It may be that the design is not sensitive enough to pick the differences in the z-means. This argument is similar to what is made in the problem of showing equivalence of two populations.

Patel (1994a) has proposed a maximum likelihood procedure for stepwise fitting of polynomials over the observed dose-range. This procedure assumes multivariate normality of the y-responses of a subject with the dispersion matrix equal to $D_i \Omega D_i$, where $D_i = \operatorname{diag}(x_{i1}, \dots, x_{ip})$, x_{ij} is the dose received by the ith subject in the jth period and Ω has a compound symmetry structure, i.e., $\omega_{ii} = \delta + \eta > 0$ and $\omega_{ij} = \eta \geqslant 0$ for $i \neq j$. Patel has also shown by a simulation study that the successive tests in a stepwise polynomial fitting procedure can be regarded independent. The approach of polynomial regression fitting answers the question whether the dose-response is linear. If the dose-response is linear, one can express the predicted response as being proportional to the corrected dose, i.e., $y = \widehat{\beta}(x - \widehat{c})$, where $\widehat{c} = -\widehat{\beta}/\widehat{\alpha}$. However, this prediction is good only for a dose in the dose-range examined in the dose proportionality study.

3. Dose-response studies

3.1. Introduction

Before determining a narrow range of safe and efficacious doses that could be prescribed to patients with a given disease, a series of dose-response studies are done in animals and humans. Although the discussion in this section is primarily limited to dose-response studies in humans, we give some background of animal studies designed to help choose safe and efficacious doses for clinical development.

Following the drug screening program, special studies are done to assess the pharmacological effect at various doses of a chemical substance. Bioassay is a special field of research where several doses of test and standard preparations (in the form of drugs, vaccines, antibiotics, vitamins, etc.) are administered to living organisms (animals, animal tissues, or micro-organisms), relevant pharmacological responses are obtained and the potency of the test preparation relative to that of the standard preparation is estimated. The literature on bioassay designs and their analysis abounds. See Finney (1971), Carter and Hubert (1985), Srivastava (1986), Hubert et al. (1988), Meier et al. (1993), Yuan and Kshirsagar (1993), among others. A valid and efficient estimate of the relative potency from dose-response profiles of the two preparations is of primary concern in bioassay studies.

Potentially harmful effects of human exposure to hazardous chemicals in the environment are generally extrapolated from animal studies. A similar situation arises in a long-term patient treatment. From animal toxicology studies, we try to predict safety for patients who would be taking a pharmaceutical product especially for the treatment of a chronic disease. This is an extremely difficult problem because the extrapolation comes in two ways. The animal models could be quite different from human models and animal studies are generally conducted at very high doses relative to the doses patients are generally exposed to. There exists a vast literature on animal carcinogenicity studies. See, for example, Cornfield (1977), Armitage (1982), Krewski and Brown (1981), Van Ryzin and Rai (1987), Schoenfeld (1986) and Carr and Portier (1993).

Several short-term dose-response studies designed to evaluate safety in animals precede Phase I clinical studies. Long-term carcinogenicity studies in animals are conducted to compare the incidences of various types of tumors associated with different dose levels and a placebo. The regulatory agency has issued guidelines for the timing of these studies and design related issues. Selwyn (1988) has written an excellent review paper on design and analysis aspects of animal toxicology dose-response studies that are conducted by the pharmaceutical industry.

Phase I studies are primarily designed to address safety and tolerability whereas Phase II dose-response studies address both efficacy and safety issues. Except in cancer research, Phase I studies are done in healthy volunteers. Phase II studies, on the other hand, use patients of the targeted patient population. We describe these commonly used study designs in the following sub-sections.

3.2. *Phase I studies*

Phase I trials enroll a small number of subjects and are done in a relatively short time because of the time pressure for the drug development. The primary purpose of these studies is to find a range of doses that are tolerable or are associated with toxicity no higher than some acceptable level.

3.2.1. *Cancer chemotherapy studies*

Phase I cancer chemotherapy studies, where we expect severe toxic effects, are different in nature from the Phase I studies done for other therapeutic areas. Korn et al. (1994) have described what is called a "standard" Phase I design in cancer research as follows: First, there is a precise definition in the protocol of what is considered dose-limiting toxicity (DLT); it may differ in different settings. The dose levels are fixed in advance, with level 1 (the lowest dose) being the starting dose. Cohorts of three patients are treated at a time. Initially, 3 patients are treated with dose level 1. If none of them experiences DLT, one proceeds to the next higher level with a cohort of 3 patients. If 1 out of 3 patients experiences DLT, additional 3 patients are treated at the same dose level. If 1 out of 6 patients experiences DLT at the current level, the dose escalates for the next cohort of 3 patients. This process continues. If $\geqslant 2$ out of 6 patients experience DLT, or $\geqslant 2$ out of 3 patients experience DLT in the initial cohort treated at a dose level, then one has exceeded the MTD (maximum tolerable dose; note that the definition of MTD varies with investigators). Some investigations will, at this point, declare the previous dose level as the MTD, but a more common requirement is to have 6 patients treated at the MTD (if it is higher than level 0). To satisfy this requirement, one would treat another 3 patients at this previous dose level if there were only 3 already treated. The MTD is then defined as the highest dose level ($\geqslant 1$) in which 6 patients have been treated with $\leqslant 1$ instance of DLT, or dose level 0 if there were $\geqslant 2$ instances of DLT at dose level 1. The number of doses recommended in these studies is about 6. Once the MTD is estimated, Phase II trials will start to evaluate the efficacy with the doses lower or equal to MTD.

Storer (1989) defined MTD as the dose associated with some specified percentile of a tolerance distribution and derived its MLE (maximum likelihood estimate), assuming a logistic dose-toxicity curve. He also considered some designs which are variations on what is called an "up and down" method. O'Quigley et al. (1990) introduced the concept of CRM (continued reassessment method), where patients are studied one at a time. They used a Bayesian approach to compute the posterior probability at each step to choose a dose for the next patient. The objective in CRM is to estimate the dose associated with some acceptable targeted toxicity level. O'Quigley and Chevret (1991) have reviewed some dose finding designs and provided simulation results. Korn et al. (1994) compared the performances of the "standard" and CRM methods with simulations. They observed that with CRM, more patients will be treated at very high doses and the trial will take longer to complete then the "standard" design. They have also proposed some modifications in these two designs.

Because of small sample sizes, generally less than 30, the estimate of MTD can not be estimated with good precision. Because of ethical issues, it is desirable to be on the conservative side in estimating MTD.

3.2.2. Other therapeutic areas

The treatments for such diseases as hypertension or diabetes are not as toxic as in the cancer chemotherapy. Phase I studies are therefore done in healthy subjects without major ethical issues. The sample sizes are small, ranging generally from 8 to 20. Typically, these are single dose studies in young, healthy male volunteers. In a typical design, subjects are randomly assigned to $k+1$ sequences formed with a placebo and k (about 5) doses of drug, separated by washout periods, in such a way that the doses will appear in an increasing order and a placebo at the diagonal places. Each subject receives all doses and a placebo. Here the placebo helps maintain pseudo double-blinding and also serves as control. Various safety measurements, including vital signs and laboratory determinations (blood chemistry, hematology, etc.) are collected both before and after a treatment. This design meets ethical concerns in that the dose is not increased until the safety and tolerability of subject's current dose is assured. Sometimes a Latin square design with washout periods between successive doses is considered if the order of doses in a sequence does not pose an ethical concern.

An alternating panel design is sometimes used. In this design a group of subjects is randomly divided into two subgroups; one is assigned to the lowest dose and the other to a placebo. After safety and tolerability for these subjects is assured, another group of subjects is enrolled to the next higher dose and a placebo. After assuring the safety and tolerability of the second dose, the next higher dose and a placebo are given to another group of subjects. This process continues until all doses are tested or the study is terminated because of safety problems. The termination of the study depends on the investigator's judgement. These designs can be used for a multiple dose study, but the length of the study may be prohibitive.

It is becoming more and more common to take blood samples from these early phase studies for pharmacokinetic measurements and to associate them with safety measurements.

3.3. Dose-response studies for efficacy

After a range of safe and tolerable doses is determined from Phase I studies, Phase II efficacy studies in patients begin. Phase II studies are primarily done to evaluate efficacy of the drug in an exploratory manner and to further narrow down the range of doses for designing a Phase III study. Because of the time pressure, the management often wants to skip a good part of Phase II. There have been several cases where a wrong dose, generally higher than what would be necessary, has been on the market because of lack of or poor planning of a Phase II dose-response study. Note that finding a range of (reasonably) safe and efficacious doses is of primary interest throughout the development process.

3.3.1. Objectives

One should ideally consider fitting two dose-response curves, one for efficacy and the other for safety, to estimate such characteristics as a therapeutic window and CID (a clinically important dose). The lower limit of the therapeutic window is traditionally estimated as a minimum effective dose (MINED) and the upper limit as minimum of maximum effective dose (MAXED) and a maximum tolerable dose (MTD). A dose below the therapeutic window would be considered ineffective and a dose above it unsafe. The term MAXED is defined as the lowest dose associated with a maximum benefit that the drug can produce. As defined earlier, the highest dose for which no clinically important adverse experiences occur is chosen as MTD. The phrase "clinically important" is subjective and is often left to the discretion of individual investigators to interpret. The quantities MAXED and MTD are drug specific and therefore are estimated from the efficacy and safety dose-response profiles, respectively.

Although it is difficult to pinpoint a globally agreeable efficacy value that would correspond to CID, clinical researchers have better understanding of this concept than the efficacy corresponding MINED. Ekholm et al. (1990) have defined MINED as the lowest dose that produces a clinically meaningful response. This definition is confusing with the definition of CID. Logically it should be the dose that produces the efficacy that is similar to, but visibly distinct from, that of a placebo. This may be interpreted as an upper or lower bound of an interval consisting of efficacy values that are practically equivalent to the placebo effect. Like clinically important difference, one should *a priori* define for a given therapeutic area what minimal efficacy is regardless of the drug tested. As examples of MINED and CID, in a treatment for prevention of bone loss (osteoporosis) in postmenopausal women, a zero percent change in bone mineral density from baseline can be considered as MINED and a 2.5 to 3 percent change as CID. The bone mineral density decreases in the absence of any treatment.

The ultimate goal of a dose-response study is to recommend one or two doses of the drug that are safe and efficacious for further development in Phase III. Depending on the therapeutic area and availability of drugs in the market, one may judge the extent to which safety can be compromised over efficacy. The more potent the dose, the more unsafe it is likely to be.

3.3.2. Designs

Several designs, including parallel-group, Latin square, dose-escalation and dose-titration designs, are used for a dose-response study. Some are more commonly used than others. The treatment duration for each dose depends on the therapeutic area. When a good surrogate measurement for efficacy variable exists, it is used to reduce the length of the observation period for each dose. For example, the mean increase in gastric pH from baseline for a single dose is used as a surrogate measurement for the rate of ulcer healing. Sometimes repeated measurements (two or more observations at protocol scheduled time points) are obtained during the treatment period for each dose. Latin square and systematic dose-escalation designs would be acceptable for a chronic disease provided they have sufficiently long washout periods. However, they require a long time to finish. Also, the patient dropout rate increases with the length of a trial.

The purpose of a dose-response study is to estimate certain characteristics of the dose-response curve. Thus, how precisely the expected response is predicted at a given dose is more important in such studies than a test of significance of the difference between placebo and a dose. Tests of hypotheses have a place in Phase III trials as they are considered confirmatory.

In a dose-titration study, each patient starts the treatment with the lowest dose. After some minimum length of treatment, if the subject shows a response based on well defined criteria, the patient continues receiving the same dose as before. If a patient fails to show a response, the dose is increased to the next level. At any time during the study, if clinically important adverse events occur, the dose is lowered or patient is dropped from the study. Because of its resemblance to clinical practice, a dose-titration design is sometimes preferred to a parallel-group design where a patient is assigned randomly to one of the doses and continues receiving it throughout the study. A dose titration design has several problems. It does not clearly answer the safety question. Because of a limited study length, the successive doses are titrated faster than are normally done in clinical practice. As a result, the doses recommended for Phase III tend to be overestimated. The titration design is not suitable for fitting a dose-response curve for the following reasons:

(1) There is no clear causal-effect relationship between the dose and effect; the dose for the next period depends on the observed response of the preceding period.
(2) Carryover effects and confounding between time and doses cloud the inference.
(3) Only nonresponders continue receiving higher doses. As a result, the observed efficacy at the highest doses could, on average, be worse than that at the lower doses, which is contrary to what we would expect in a true dose-response profile.

Several attempts have been made to analyze a titration study using stochastic models (see, for example, Chuang, 1987; Shih et al., 1989). However, it is difficult to make sure that an assumed model adequately represents the true data-generating mechanism and to verify the underlying model assumptions.

Recently, Chi et al. (1994) have suggested a modified forced titration design as a compromise between a parallel group design and a systematic dose-escalation design. Their design is illustrated in Figure 1. In this design subjects are randomly assigned to

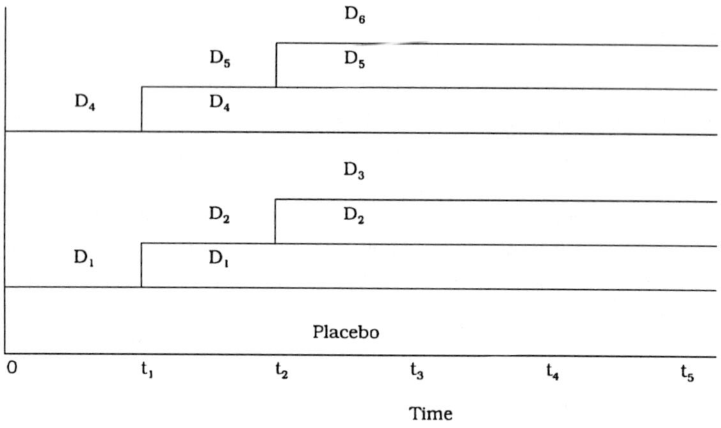

Fig. 1. Design proposed by Chi et al.

pre-defined dose-sequences (with placebo being Dose 0). Depending on the sequence assigned, a subject either continues receiving the same dose as received in the previous time interval or receives the next higher dose at a scheduled time point. The authors indicate that this design should help estimate the net incremental effect at each titration step when there is potential for carryover effect. Even for this design, the questions on dose-response cannot be clearly answered for the reasons stated earlier.

Although less ethical than a titration design, a parallel-group design should be preferred so that the questions on a therapeutic range can be answered with clarity. This, of course, requires the evaluations of both safety and efficacy. Given a well-established therapeutic range from a sufficiently large study, a physician would be comfortable to titrate the doses in actual practice.

3.3.3. Analysis

Some statisticians consider this an isotonic regression problem and apply Williams (1972) method for multiple comparisons (dose vs. placebo) assuming that the dose-response curve is monotonic. They declare the lowest dose showing a significantly greater (or lower) effect than a placebo as MINED. This approach focuses on statistical significance rather than clinical interpretation of efficacy. Another problem is that the dose-response curve may not be monotonic. Rom et al. (1994) mention some situations where the curve initially increases and then, after reaching to a peak, starts decreasing. They also proposed two closed test procedures for multiple comparisons which are based on p-values and are therefore not limited to any particular method of analysis.

Whatever multiple comparison method is used, it is not going to solve the basic problem of estimating certain characteristics of a dose-response profile. In practice, we cannot consider a continuum of doses. Only a finite number of doses are selected and the choice of dose intervals might be quite arbitrary. Therefore, the interpretation of the results based on tests of hypotheses are design dependent. Another problem is that clinical researchers take the p-values seriously and a p-value depends on the sample size. So two trials based on different sample sizes would lead to different conclusions on the therapeutic range.

If adequate number of doses are considered in the design, a suitable nonlinear or polynomial dose-response can be fitted. Some insight regarding the number of doses and computing a common sample size per dose for a binary response is given by Strijbosch et al. (1990). They considered a logit-linear model imposing a restriction that at least a prespecified number of doses would satisfy certain bounds on the response on a logit scale. Patel (1992) has provided a method to compute the sample size per dose so that a fitted logistic curve yields a desired level of precision for estimating CID. This method applies to both binary and normal data.

4. Mixed effects models

4.1. Introduction

Like in any experimental situation, linear models play an important role for making inference in clinical trials. A linear combination of fixed and random effects defines a linear model. Fixed effects represent the population mean effects associated with a finite number of levels of a factor. For example, in a dose-response study the effects of a placebo and low, medium, and high doses of a drug are considered fixed effects. The random effects for a given factor are regarded as realization of a random sample from the population of a large set of levels representing the factor. For example, in a simple, two-sequence, two-period, crossover trial, subjects (patients) are randomly assigned to one of the two sequences AB and BA, where A and B represent the treatments. Here subjects are assumed to represent the entire population of subjects described by the protocol. Another example of random effects is the centers in a multicenter trial. Some controversy still surrounds about labeling the centers as random effects. We discuss this point in more detail in Section 4.2.

The first example of a crossover design introduces nesting of some effects. The between-subject variation is partitioned as between sequences (fixed) and between subjects within sequences (random). The within-subject variation is partitioned as between periods (fixed), sequence-by-period interaction (fixed), and subjects within sequences-by-period interactions (random). The sequence-by-period interaction is also the between treatment difference. Thus this design is like a split-plot design, except that a unit (subject) cannot be split in order to control the time effect and this imposes some restriction on the analysis of a crossover design. In the second example of random effects, the treatments are crossed with the centers and the model includes the treatments (fixed), centers (random), treatment $\times$ center interactions (random), and the experimental error (random). Since the treatments are randomly assigned to subjects within each center, one may regard this as a nested design with treatments within centers as being the sum of the treatment effects and the treatment-by-center interactions.

When a subject is treated with one of randomly assigned treatments throughout the trial and observed repeatedly, the design is referred to as a parallel group study with repeated measurements or simply a longitudinal study. This design is also referred to as a between-subject design as opposed to a within-subject design where some or all subjects receive two or more treatments. In a within-subject design, a subject is assigned randomly to one of several sequences, formed after permuting the treatments with or without restrictions, and receives the treatments in the order of the assigned sequence. A Latin square design, balanced incomplete design, and Williams (1950) designs are a few examples of within-subject designs. The simple 2×2 (2 sequence, 2 period) design, a special case of a Latin square design, is widely used in early phases of clinical trials. It is not uncommon to have incomplete data in a long-term trial. A subject could be discontinued from the trial for one or more reasons. Occasionally, subjects miss a few intermediate visits and this leads to additional missing values. There are two types of early withdrawals: response related and response unrelated. However, this classification may not be clear-cut. Moreover, for the subjects who are just lost to follow-up, without informing the investigator, it is difficult to say whether the reason for discontinuation is response related. Rubin (1976) defined three types of missing data: missing completely at random (MCAR), missing at random (MAR) and informative. If the missing data (censoring) mechanism is independent of both observed and unobserved data, the missing values are MCAR. If the mechanism is independent of unobserved data, the missing values are MAR. If the mechanism depends on unobserved data, the missing values are informative. Rubin (1976) also showed that MCAR values are ignorable in the sense that the analysis based on existing data will give unbiased estimates of the location parameters. For MAR values he showed that likelihood based procedures give valid inference for the treatment comparisons. It is difficult to account for the influence of informative missing values in the analysis. Some attempts have been made to model the probability of informative censoring at successive time points either through a Markovian model or a random effects model (see, e.g., Wu and Carroll, 1988; Wu and Bailey, 1989; Diggle and Kenward, 1994). These methods are, however, model dependent and in practice it is extremely difficult to verify the goodness of fit of a given model.

Extensive literature, including books written by Jones and Kenward (1989), Jones (1993), Longford (1993), and Diggle et al. (1994), exist for analyzing repeated measures designs and cluster data. In this section, we limit our attention to mixed effects models, i.e., models having both fixed and random effects, and we emphasize on making inference on fixed effects rather than on estimating the variance components. The latter topic is covered extensively in the books by Searle et al. (1992) and Rao and Kleffe (1988). In Section 4.2, we consider a model for a multicenter trial when each subject has a single response. We introduce the Laird–Ware (Laird and Ware, 1982) model to represent a class of repeated measures designs in Section 4.3. Section 4.4 deals with non-normal distributions.

4.2. Multicenter trial

In a multicenter trial, several centers (clinics) participate in the trial following a common protocol to evaluate efficacy and safety of two or more treatments. The treatments are randomly assigned to subjects at each center. The purpose of designing a multicenter trial is to expedite the patient recruitment and to have a broad coverage of patient population, medical practice and environmental factors. Since the subjects within a center have something in common, they may be regarded as forming a cluster. The inference on treatment effects is of primary interest and it is made from the data supplied by all centers.

For a continuous response variable, suppose we write a linear model as

$$y_{ijk} = \tau_i + \delta_j + \gamma_{ij} + x'_{ijk}\beta + \varepsilon_{ijk}, \tag{4.1}$$

where y_{ijk} is the response from the kth ($k = 1, \ldots, n_{ij}$) subject on the ith ($i = 1, \ldots, t$) treatment in the jth ($j = 1, \ldots, m$) center, x_{ijk} is a $p \times 1$ vector of covariates, β is a $p \times 1$ vector of regression coefficients, τ's are the treatment effects, δ's are the center effects, γ's are the interaction effects, and ε's are residuals which are assumed to be $IN(0, \sigma^2)$, i.e., independent normal with means zero and variances σ^2. The treatment effects are clearly fixed effects. Regarding the center effects, however, some controversy exists. The interaction effects are obviously random if at least one of the factors involved is random. For the trials sponsored by the pharmaceutical industry the centers are traditionally considered as fixed effects. Statisticians first test the significance of the treatment-by-center interaction and then test the significance of the treatment difference if the interaction is nonsignificant. When the interaction is present, common practice is to compare the treatments in each center separately. Separate analyses obviously have low powers and do not help to reach an overall conclusion about the treatment effects. Thus current approach of analyzing a multicenter trial may defeat the purpose of designing a multicenter trial.

Khatri and Patel (1992) have argued in favor of treating the centers as random effects. We make the following arguments to justify the centers as random:

1. We are not interested in making inference for a particular set of centers.
2. While interpreting the treatment differences, we do not limit the inference to the subjects of a given set of centers. Instead, we extend the results to the general population.
3. Although the participating centers are not chosen randomly from the entire population of centers, this is not a good reason for not considering the centers as random. In biological experiments, for example, this type of situation is common. Whatever animals are at disposal of a researcher are used and they do not constitute a random sample from all animals.
4. For a given center, many factors, including type of patients (demography, genetics, socio-economic conditions, disease severity, etc.), clinical methodology, patient behavior (attitude, compliance, etc.), and environmental conditions, play a role in a complex manner. If a few prognostic factors can be identified as covariates, then what is left in the process after removing their influence is simply random noise generated by a center.
5. If we treated the treatment-by-center interaction as fixed, it would mean that in the presence of the interaction we cannot make valid inferences on treatments. This cannot be justified, however, because it would be unlikely to observe the same interaction profile if trials were repeatedly conducted under the same protocol but with different sets of centers. Even the same set of centers is unlikely to generate

the same interaction profile under repeated sampling because of many unknown or uncontrolled factors.

These reasons lead us to believe that the treatment-by-center interaction is a consequence of randomly occurring events. An excellent review on the topic of interaction is given by Cox (1984). He gave some criteria to judge whether a given interaction is fixed or random. With a mixed effects model the conventional interpretation of the treatment-by-center interaction no longer applies. If the estimate of the interaction variance component is large, we duly pay a penalty for comparing the treatments by increasing the variance. Chakravorti and Grizzle (1975) derived the likelihood ratio test for comparing the treatment effects for Model (4.1) with no covariates. They assumed that δ's $\sim IN(0, \sigma_\delta^2)$ and γ's $\sim IN(0, \sigma_\gamma^2)$. A unified theory of a more general problem is given by Harville (1977) which is further discussed in the context of a repeated measures design in Section 4.3. Exact analysis of Model (4.1) with no covariates is given by Gallo and Khuri (1990). As Scheffé (1959) pointed out, Model (4.1) is not general enough and therefore is limited in scope. For example, in a clinical trial designed to compare an active treatment with a placebo, the treatment groups may be associated with different variances.

Khatri and Patel (1992) considered a more general model for the $n_{.j}$ observations from the jth center as

$$y_j = A_j'\theta_j + X_j'\beta + \varepsilon_j, \quad \text{for } j = 1, \ldots, m, \tag{4.2}$$

where $y_j' = (y_{1j}', \ldots, y_{tj}')$, $y_{ij}' = (y_{ij1}, \ldots, y_{ijn_{ij}})$, $n_{ij} > 0$, for $i = 1, \ldots, t$, $A_j' = \text{diag}(1_{n_{1j}}, \ldots, 1_{n_{tj}})$ is an $n_{.j} \times t$ design matrix for the jth center, $X_j = (X_{1j}, \ldots, X_{tj})$ is a $p \times n_{.j}$ matrix of covariate values, $X_{ij} = (x_{ij1}, \ldots, x_{ijn_{ij}})$, θ_j and ε_j are independent random vectors such that $\theta_j \sim N_t(\mu, \Omega)$ and $\varepsilon_j \sim N_n(0, \sigma^2 I_{n_{.j}})$, and $\beta' = (\beta_1, \ldots, \beta_p)$ is a vector of regression slopes. Here N_q stands for a q-dimensional normal distribution. The covariates are centered at their respective means. The dispersion matrix of y_j is $\Phi_j = A_j'\Omega A_j + \sigma^2 I_{n_{.}}$. Khatri and Patel computed maximum likelihood estimates of the parameters under the conditions that $\sigma^2 > 0$ and $\Omega + \sigma^2 D_j^{-1}$ is non-negative definite for all j, where $D_j = A_j A_j'$ and derived the likelihood ratio test for comparing the treatment effects. When Ω has a compound symmetry structure (equal variances and equal covariances), Model (4.2) reduces to Model (4.1). If the assumption of compound symmetry is satisfied, Model (4.1) will be more powerful.

It is not clear as to what should be a minimum m, the number of centers, for a reasonably reliable estimate of Ω and thus for a satisfactory approximation for the likelihood ratio test for the treatment comparisons. The numerical examples in Chakravorti and Grizzle (1975), Gallo and Khuri (1990) and Khatri and Patel (1992) included 3, 9 and 6 centers, respectively. As a rule of thumb, we suggest that for $m \leqslant 5$, a fixed effects model be preferred to a mixed effects model. If $m \leqslant 5$, one approach is to choose *a priori* a fixed effects model without the interaction term. The alternative is to choose a fixed effects model with the interaction term and follow these steps:

(1) If the treatment-by-center interaction is not significant at 0.2 level of significance (preliminary test), drop it from the model to increase the efficiency of the test for treatments. The reduced model, however, includes the centers as a factor.

(2) If the interaction is significant at the 0.05 level of significance do an exploratory analysis to see if any covariate (in addition to those included in the model) can explain the interaction. Sometimes a few outlying observations play a major role in introducing inconsistency of treatment differences across centers. This is not to suggest that such exploratory analysis should be limited to a fixed effects model.

If nothing helps explain the interaction, one can follow the following strategies: (i) If the interaction is quantitative in nature, i.e., the treatment differences are in one direction, one can still use the original model to test the treatment effects after making adjustment for the interaction term. (ii) If the interaction is qualitative in nature, one may rely on the analysis in subgroups of the centers and interpret the results appropriately, making inference in relation to the center specific characteristics, if possible.

4.3. The Laird–Ware model

In this section we concentrate on longitudinal data. In seventies, the analysis of longitudinal data generally used a split-plot type model that required an assumption of equal variances and equal covariances for the repeated measurements, even though multivariate analysis of variance approach (Cole and Grizzle, 1966) and growth curve analysis methods (Rao, 1959; Potthoff and Roy, 1964; Rao, 1965; Khatri, 1966; and Grizzle and Allen, 1969) already existed. In a growth curve analysis, the expected response is modelled as a continuous function of time. In general, they are unsuitable for clinical trial data because of ill-behaved response profiles of subjects over the treatment period. A linear growth curve model has been used by Lee and DeMets (1991) in analyzing a group sequential trial, where the treatments are compared once or more than once during the course of the trial. Techniques for analyzing incomplete data from general (unstructured dispersion matrix) multivariate normal populations were developed (see, for example, Hartley and Hocking, 1971; Kleinbaum, 1973; Beale and Little, 1975; Dempster et al., 1977). However, these methods were not found suitable to clinical trial data because of over parameterization, especially when the number of measurements on a subject is large relative to the number of subjects.

The work having a major impact on clinical trials with repeated measures designs came from Laird and Ware (1982). Based on work of Harville (1977), they developed ML (maximum likelihood) and REML (restricted maximum likelihood) procedures for analyzing a general mixed effects model for repeated measurements. Suppose $y_i = (y_{i1}, \dots, y_{iT_i})'$ is a $T_i \times 1$ vector of responses from the ith ($i = 1, \dots, n$) subject at times $t_{i1}, \dots, t_{iT_i}$ with $T_i \leqslant T$. Then the Laird–Ware model is written as

$$y_i = X_i\alpha + Z_ib_i + \varepsilon_i, \tag{4.3}$$

where X_i is a $T_i \times p$ design matrix corresponding to the fixed effects vector α of order $p \times 1$, Z_i is a $T_i \times k$ design matrix associated with the random effects vector b_i of order $k \times 1$. The random effects b_i are assumed to be independent normal with mean vector 0 and dispersion matrix $\sigma^2 B$. We write this as $b_i \sim IN(0, \sigma^2 B)$. The residuals ε_i are assumed to be $IN(0, \sigma^2 W_i)$ and also independent of b's. The matrices B, representing between-subject variability, and W_i, representing within-subject variability, are positive definite. The design matrices X_i and Z_i are subject specific and may not be of full rank. The marginal distribution of y_i is $IN(X_i\alpha, W_i + Z_i B Z_i')$. This allows one to write the likelihood as the product of the marginal densities of the vectors $y_1, \ldots, y_n$. The dispersion matrices B and W_i are assumed to have some structures so that their elements can be written as functions of parameters on a lower dimensional space.

When one or more columns of X_i are functions of the time points t_{ij} $(1 \leqslant j \leqslant T_i)$ and $X_i = Z_i$, the model serves as a growth curve model. Other columns represent either the overall mean or changing covariates. For example, when

$$X_1' = Z_1' = \begin{pmatrix} 1 \cdots 1 \\ t_{i1} \cdots t_{iT_i} \end{pmatrix},$$

the model reduces to a linear growth curve model which is a special case of a random coefficient model described by Rao (1965). Suppose only one response is obtained from each subject and the centers are treated as clusters in a multicenter trial. Then writing the responses from the subjects of the jth cluster as an $n_{.j} \times 1$ vector y_j, the Laird–Ware model reduces to the Model (4.2) for a multicenter trial. For a within-subject design, suppose we write y_i as a vector of responses at successive periods from subject i and α as a vector whose elements are the sequence-by-period cell means. Then Model (4.3) reduces to $y_i = X_i\alpha + \varepsilon_i$ with $\text{var}(\varepsilon_i) = \Sigma_i$. For the conventional 2×2 (two-sequence, two-period) crossover design with no missing values, i.e., when Σ_i is a 2×2 matrix for all i, the assumption of compound symmetry is required for the split-plot type analysis proposed by Grizzle (1965). If this assumption does not hold, the method (Hills and Armitage, 1979) based on the sum and difference of the period responses can be used.

The Laird–Ware model can accommodate any missing data pattern. Their model is not limited to a design where the subjects must follow pre-specified visit schedules. For example, in a trial with seriously ill patients, it would be unethical if we did not allow patients to visit their clinics at times other than pre-specified visits.

Jennrich and Schluchter (1986) illustrated this model under different structures of

$$\Sigma_i = W_i + Z_i B Z_i'$$

and proposed computing algorithms for ML and REML estimates and the corresponding likelihood ratio tests under the multivariate normal distribution. They considered independence, compound symmetry, random-effects, factor-analytic, first order autoregressive [AR(1)], banded (general autoregressive) and unstructured models for Σ_i. They also compared Newton–Raphson, Fisher scoring and hybrid EM scoring algorithms for computing and made some recommendations. Jones (1987), Diggle (1988), and Chi and Reinsel (1989) among others analyzed longitudinal data after modelling b_i as random effects and $\varepsilon_{i1}, \ldots, \varepsilon_{iT}$ as a stochastic process representing an autoregressive model.

No systematic efforts have been made to suggest a practical structure for the covariance matrix Σ_i for a repeated measures design. Test of goodness of fit of a model with a particular covariance structure is not difficult, but its asymptotic likelihood ratio test is, in general, sensitive to the departure from multivariate normality. A SAS (1992) procedure PROC MIXED analyzes Model (4.3), allowing different structures for the matrices W_i and B. It does compute both ML and REML estimates.

4.4. Non-normal distributions

Recently, mixed effects models for a distribution from the exponential family have received considerable attention. We first discuss the analysis of a multicenter trial for binary data and then some development in repeated measures design for binary and Poisson data. Beitler and Landis (1985) considered a mixed effects model with no covariates directly for a binary response rather than for a link function of its expected value. They computed the variance components from the quadratic forms from the conventional ANOVA table as one would obtain for normal data. The validity and efficiency of these estimates are questionable. Dersimonian and Laird (1977) introduced a two-stage model for pooling empirical odds ratios in the context of meta analysis. Raghunathan and Yoichi (1993) examined this model closely and considered it inappropriate for analyzing a multicenter trial when the center sample sizes are small. They introduced a general linear model for the log odds for the treatment-by-center cells with center and center-by-treatment interaction effects as random effects. The model also included the center specific covariates. Assuming normality for the random effects they provided a likelihood based procedure for comparing the treatments using the EM algorithm and a procedure after conditioning on the center estimators. In another paper Raghunathan (1994) introduced a Bayesian approach after assuming a bivariate normal distribution for the logits associated with two treatments in each center for analyzing 2×2 tables. He modelled the expected values of logits as linear functions of covariates and the center dispersion matrix to allow the extra-binomial variation as well as a center specific random multiplier. He analyzed data using Gibbs and importance sampling methods, allowing different priors. This approach can also be used for case-control studies.

For analyzing longitudinal data satisfying a distribution from the exponential family, several methods have been developed. The empirical generalized least squares (EGLS) procedure developed by Koch et al. (1977) exploits full multinomial structure in computing the dispersion matrix for the estimates. However, it does not perform well when data are sparse. Also, it takes only discrete covariates. Although computationally more complicated, the generalized estimating equations (GEE) approach for marginal models, proposed by Liang and Zeger (1986) and Zeger and Liang (1986), has certain advantages. The population-averaged parameters are modelled as functions of covariates in marginal models. The main advantages of GEE method are that it accommodates continuous time-dependent or time-independent covariates and that the dispersion matrix can be modelled in terms of fewer parameters than the number of parameters in an unstructured dispersion matrix and thus the consequences of sparse data can be avoided. The GEE approach does not attempt to model the joint distribution of the repeated measurements. The marginal distribution at each time point is modelled as a function of covariates. Allowing a working correlation matrix among the subject responses the regression parameters and their dispersion matrix are estimated. These estimators are consistent as long as the population means are correctly specified and a robust estimate of the dispersion matrix is used. In case of incorrectly specified correlations, however, there is some loss in efficiency. This procedure is a multivariate extension of quasi-likelihood and is not a likelihood based procedure. Hence it gives valid inferences only when the missing data are MCAR (defined in Section 4.1). Prentice and Zhao (1991) extended this approach by simultaneously modelling and estimating the location and dispersion parameters to increase the efficiency of the estimators. Lipsitz et al. (1991) modelled odds ratios instead of the correlations and showed that their approach is slightly more efficient. Recently, Fitzmaurice et al. (1994) gave a likelihood-based procedure for analyzing incomplete longitudinal binary responses. This approach gives valid inferences when the missing data are MAR or MCAR, but is computationally more complicated than GEE approach. The readers are referred to the review papers by Kenward and Jones (1992) and Fitzmaurice et al. (1993). Recently, Lipsitz et al. (1994) have analyzed repeated ordered categorical data. There is a trade-off between GEE and its extensions: One has to choose between (i) estimation bias due to incorrectly specified models for the location and dispersion parameters and (ii) efficiency of the estimators. Among other models for repeated measures data are generalized linear models with random coefficients, leading to such approaches as logit-normal, probit-normal and log linear-normal. These models have different interpretation from marginal models, because, except for a linear link function, the expectation of a random effects model does not correspond to a marginal model. The analysis methods are Bayesian in nature and require extensive numerical computations. Several researchers including Stiratelli et al. (1984), Ochi and Prentice (1984), Breslow (1984), Crowder (1985), Zeger and Karim (1991), and Longford (1994) have studied such models and applied methods such as EM algorithm, Newton–Raphson and Gibbs sampling. In general, no closed form for the marginal mean can be obtained. However, some approximations can be used depending on the link function and the distribution of the random effects (see, for example, Gilmour et al., 1985; and Zeger et al., 1988). Schall (1991) provided the more general and yet simpler approach after linearizing a link function of the response variable. A good review paper on this topics is written by Rutter and Elashoff (1994).

Another approach for analyzing random effects models is to use the conditional likelihood given sufficient statistics for the subject effects. Diggle et al. (1994) describe this approach in the context of crossover designs and point out the disadvantage of losing some information as the method relies entirely on within-subject comparisons.

Nonlinear mixed effects models can also be analyzed using the methods developed in the context of population kinetics where pharmacokinetics models with random effects are fit. Attempts have been made to compute maximum likelihood estimates of the regression coefficients and the dispersion parameters following different methods of linearizing the expected values of the observations. The readers are referred to Sheiner and Beal (1983), Beal and Sheiner (1988), Lindstorm and Bates (1990), and Vonesh and Carter (1992) among others.

5. Markovian models

5.1. Introduction

Markovian models are of considerable practical importance in the analysis of a time series. A repeated measures design generates independent response profiles over time, one for each subject, whose length and observation times depend on a trial design. A markovian model for a Gaussian distribution is also referred to as an ante-dependence model. When the p ordered response variables are multivariate normal, the ante-dependence model of order g implies that the ith variable ($i > g$), given the preceding g variables, is uncorrelated with all further preceding variables (Gabriel, 1962).

If Σ is the dispersion matrix of a vector of successive measurements, then for an ante-dependence model of order g, $\Sigma^{-1} = (\sigma^{ij})$ with $\sigma^{ij} = 0$ for $|i-j| > g$ (Gabriel, 1962). For $g = p - 1$ the model corresponds to multivariate normal with unrestricted dispersion matrix and when $g = 0$ it leads to complete independence of responses. For $g = 1$, another related model for unequally spaced longitudinal data can be derived as a stationary continuous first-order autoregressive process, CAR(1), (Jones and Boadi-Boateng, 1991). Suppose the measurements $y_1, y_2, \ldots$ are obtained at time points $t_1, t_2, \ldots$, then CAR(1) has a covariance function $\text{cov}(y_j, y_k) = \sigma^2 \exp\{-\alpha|t_j - t_k|\}$, where $\alpha > 0$. For equally spaced measurements the process is referred to as AR(1) for which $\text{cov}(y_j, y_k) = \sigma^2 \rho^{|k-j|}$. This suggests that AR(1) is a special case of ante-dependence model of order 1 when the Markov process is stationary.

Because of natural chronological ordering of the variables in a repeated measures design ante-dependence models are more tenable than other commonly assumed simple structures such as compound symmetry and AR(1). An ante-dependence model provides a natural interpretation of the dependency among subject responses through successive regression models; the present observation is influenced by one or more immediate past observations and not by future observations. These models are therefore most suitable for analyzing data with a monotone pattern. A dataset is called monotone if a subject with a missing value at Visit i also has a missing value at Visit j for $j > i$. The computational methods for analyzing models of Section 4 cannot accommodate ante-dependence models. On the other hand, the models of Section 4 do not require the times of measurements to be common for all subjects. In Section 5.2, we review applications of ante-dependence models for analyzing between-subject designs (longitudinal data) with an emphasize on monotone patterned data. We also give a brief review of applications of Markovian models for analyzing non-normal data in this section. Markovian models for analyzing within-subject designs are reviewed in Section 5.3.

5.2. Between-subject designs

Anderson (1957) and Bhargava (1975) derived maximum likelihood estimators under a monotone pattern assuming multivariate normality for the repeated measurements. Patel (1979) extended this to a k-sample problem for analysis of covariance. All this work assumed the general dependence; the likelihood was written as the product of conditionally independent densities under an ante-dependence model of order $p - 1$. Patel and Khatri (1981) used the Markovian (ante-dependence of order one) normal density for analyzing one-way classification with fixed covariates and provided the likelihood ratio test statistics and their improved asymptotic distributions for testing various hypotheses. Byrne and Arnold (1983) studied this model for a one-sample problem and showed that it is more powerful than Hotelling's T^2 requiring an unstructured dispersion matrix. Ante-dependence models of order $g > 1$ for analyzing monotone data in a parallel group design have been applied by Kenward (1987) and Patel (1991). They also derived better approximations for the distributions of Likelihood ratio statistics. Murray and Findlay (1988) analyzed blood pressure data with a monotone pattern to compare two treatments using an ante-dependence model of order 1. Patel (1991) also provided a method of estimating the marginal treatment mean profiles after adjusting for time-independent covariates. Using the adjusted cell (time-by-treatment) means and their dispersion matrix, one can make inference on a set of between and within treatment contrasts. For example, a test for treatment-by-time interaction may be of interest. Software is written in PROC IML of SAS6.07. Likelihood ratio for testing the hypothesis that the order of the ante-dependence model is g versus that it is $h > g$ is derived by Kenward (1987). Although this test is sensitive to the departure from the assumption of normality, it can serve as a guideline in choosing an approximate value of g. If there exist a few intermediate missing values, they should be estimated first in order to analyze a monotone pattern. A method for this can be found in Patel (1991). Recently, Patel (1994b) applied an ante-dependence model for implementing a group sequential procedure for a longitudinal trial. This model is also used by Patel (1994c) in another application.

For longitudinal data representing a member of the exponential family, a generalized linear model can be written to represent the Markovian property. For example, for repeated binary responses we can write a logit regression model for each response using the past history as covariates. These models also allow time-dependent or time-independent covariates. Analogous to ante-dependence models, we have Markov chain of order g for a sequence of binary responses. Bristol and Patel (1990) used a Markov chain model for comparing k treatment profiles of incidences of AEs (adverse experiences), associated with successive visits, in a monotone dataset. They derived the likelihood ratio test allowing a non-stationary transition probabilities which does not require equally spaced observations. This method takes into account the fact that the AEs are recurrent events.

For a single sequence of binary responses y_t $(t = 1, \ldots, n)$, Zeger and Qaqish (1988) modelled a logit link function of the conditional mean μ_t at time t as

$$h(\mu_t) = x_t'\beta + \sum_{i=1}^{g} \theta_i y_{t-i},$$

where x_t is a vector of present and past covariates at time t and θ's are stationary regression coefficients associated with the past y-observations. This is the Markov chain of order g suggested by Cox (1970). Here $\exp(\theta_i)$ represents the odds of a positive response at time t given $y_{t-i} = 1$ relative to the odds given $y_{t-i} = 0$ after adjusting for the other covariates. Since θ's do not depend on t, this model represents a stationary Markov chain. This idea is extended to a repeated measures design in Diggle et al. (1994). For $g > 1$, the full likelihood cannot be written and therefore they relied on conditional likelihood and quasi-likelihood based inference. Since the conditional means rather than marginal means are modelled, the regression coefficients in such models are interpreted as being conditional on the past history and covariates. Zeger and Qaqish (1988) also applied the analysis of a time series to Poisson and gamma data. Recently, Azzalini (1994) considered a non-homogeneous Markov chain with time-varying transition probabilities for analyzing repeated measures binary data.

5.3. *Within-subject designs*

We introduced within-subject designs, commonly referred to as crossover designs, in Section 4. For a review of within-subject designs the readers are referred to Matthews (1988) and Jones and Kenward (1989). Primary advantages of allowing run-in and washout periods and obtaining baseline measurements at the start of successive treatment periods in a within-subject design are (1) to have a valid interpretation of carry-over effects; (2) to increase the power for testing the significance of their differences; and (3) to obtain valid estimates of the treatment differences. The issues related to baseline measurements in the context of the 2×2 design are discussed in Patel (1990). In an incomplete block design each subject receives a subset of all treatments and consequently the influence of between-subject variability may appear on a test for treatment comparisons. It would therefore be desirable to consider a few most important prognostic variables as between-subject (time-independent) covariates. For example, in the treatment of respiratory disease or hyperglycemia, smoking is considered a good prognostic variable.

Let us consider a design using a total of t treatments, p periods and m sequences where (a) the uth treatment sequence is administered on n_u subjects, allocations being random; (b) the uth sequence has k_u ($\leqslant p$) distinct treatments, some being possibly repeated; (c) baseline measurements are used as within-subject (time-dependent) covariates; and (d) one or more prognostic factors are used as between-subject covariates when every subject does not receive all t treatments.

We assume that the vector of period responses is multivariate normal. When the number of periods is large relative to the sample size, the power associated with the unrestricted dispersion matrix will not be satisfactory. We assume that the vector of period responses follows a multivariate normal distribution with an ante-dependence structure of order g. Matthews (1990) considered AR(1) and moving average process models to evaluate the efficiency of a within-subject design relative to the variance structure that assumes independence of the errors.

Let y_{ij} be the treatment response for Period i and Subject j and x_{ij} the corresponding baseline measurement, measured from the x-mean at Period i, $i = 1, \ldots, p$; $j = 1, \ldots, n$. We consider the model allowing unequal number of subjects per treatment sequence. Let u_{kj} be the value of the kth ($k = 1, \ldots, q$) between-subject covariate, measured from its mean, for the jth subject.

We define $Y = (y_{ij})$, $X = (x_{ij})$ and $U = (u_{kj})$ as known matrices of orders $p \times n$, $p \times n$, and $q \times n$, respectively. Now we consider the model as

$$Y = \mu A + \Gamma X + \beta U + E, \tag{5.1}$$

where μ is a $p \times m$ matrix of unknowns whose rows correspond to the periods and columns to the sequences, $A = (a_{ij})$ is a design matrix of order $m \times n$ so that $a_{ij} = 1$ if the jth subject is in the ith treatment sequence and 0 otherwise, $\Gamma = \text{diag}(\gamma_1, \ldots, \gamma_p)$ with γ_i being a within-subject regression parameter for the ith period, $\beta = (\beta_{ij})$ is a $p \times q$ matrix with β_{ij} being a regression parameter associated with the ith period response and the jth between-subject covariate, and E is a $p \times n$ error matrix. The model assumes that the columns of E are independently distributed with a common multivariate normal distribution with the $p \times 1$ mean vector of zeros and a $p \times p$ dispersion matrix Ω.

A model with several time-dependent and time-independent covariates for a repeated measures design with complete data was studied by Patel (1986). We did not allow missing values in Model (5.1). However, if missing values exist and if they are caused by premature withdrawals, i.e., when the data form a monotone pattern, this model can be modified and analyzed along the same lines as in Patel (1991). If a few intermediate values are missing, they can be imputed using the models at successive time points. This type of modelling is general for analyzing any within-subject design. Patel (1985) considered a special case for analyzing monotone data (missing data only at Period 2) without covariates from the 2×2 design assuming normality. Here it was assumed that the missing data at Period 2 are non-informative. We hope that the future research will tackle non-normal distributions from the exponential family to analyze a class of within-subject designs with incomplete data.

6. Discussion

A great deal of statistical expertise is required in both theory and computation for designing and analyzing today's clinical trials. Because of the space limitation we have briefly reviewed a few statistical topics. Such important topics as survival analysis, group sequential trials and multiple comparisons are widely applied in clinical trials, but are not included here. Several computer packages such as SAS, BMDP and S-PLUS are available for statistical analysis and are widely used by biostatisticians and data processing people. The industry, CROs (contract research organizations), academia and government employ a large number of biostatisticians whose primary responsibilities are to consult with clinical researchers on statistical designs, prepare analysis plans, analyze data and interpret results. Biostatisticians in the industry have to work in a somewhat restricted environment because of regulatory requirements.

There are some limitations of clinical trials. Because of a limited patient-time exposure to the treatment in clinical trials it is difficult to predict long term toxicity. There have been some instances where an unacceptable level of toxicity is observed after the drug is marketed. Sometimes a sub-optimal (for both efficacy and safety) dose enters the market. Another problem is confounding factors. For example, the drug-compliance and the use of concomitant medications are major factors that influence efficacy and safety measurements. Not only that the clinical trials cannot simulate the actual clinical practice, it is extremely complicated, if not impossible, to estimate the drug efficacy and safety after removing the influence of these factors.

The importance of a good dose-response study cannot be understated. More work is needed in both designing and analyzing such a study. Although some attempts have been made for comparing the treatment groups using incomplete longitudinal data, more research is needed in this area. When the patient recruitment becomes a major problem, innovative designs, including within-subject designs, should be considered for a chronic disease. Another approach would be to recruit a large number of centers without worrying about a minimum number of patients per center and treat the centers as random effects. A meta-analysis where several independent studies are combined to increase the power of the test can be considered in the same spirit.

The current environment of health policy demands an expanded role of clinical trials. Besides the drug efficacy and safety, the cost-effectiveness and quality of life are expected to play important roles in the future. There is an increasing interest in collecting data on medical care utilization and quality of life from clinical trials especially for disabling diseases. Two evolving fields, Pharmacoeconomics and Pharmacoepidemiology, will hopefully help in offering more clear choices of treatment strategies to patients under the managed health care system.

Acknowledgements

I am grateful to Professors Peter Armitage and Subir Ghosh and a referee for helpful suggestions.

References

Anderson, S. and W. W. Hauck (1983). A new procedure for testing equivalence in comparative bioavailability and other clinical trials. *Comm. Statist. Theory Methods* **12**, 2663–2692.

Anderson, S. and W. W. Hauck (1990). Consideration of individual bioequivalence. *J. Pharmacokinetics and Biopharmaceutics* **18**, 259–273.

Anderson, T. W. (1957). Maximum likelihood estimates for a multivariate distribution when some observations are missing. *J. Amer. Statist. Assoc.* **57**, 200–203.

Armitage, P. (1982). The assessment of low-dose carcinogenicity. In: *Proceedings of the Memorial Symposium in Honor of Jerome Cornfield. Biometrics* (Supplement: Current Topics in biostatistics and Epidemiology) **38**, 119–129.

Azzalini, A. (1994). Logistic regression for autocorrelated data with application to repeated measures. *Biometrika* **81**, 767–775.

Beal, S. L. and L. B. Sheiner (1988). Heteroscedastic nonlinear regression. *Technometrics* **30**, 327–338.

Beale, E. M. L. and R. J. A. Little (1975). Missing values in multivariate analysis. *J. Roy. Statist. Soc. Ser. B* **37**, 129–146.

Beitler, P. J. and J. R. Landis (1985). A mixed effects model for categorical data. *Biometrics* **41**, 991–1000.

Bhargava, R. P. (1975). Some one-sample testing problems when there is a monotone sample from a multivariate normal population. *Ann. Inst. Statist. Math.* **27**, 327–339.

Breslow, N. E. (1984). Extra-Poisson variation in log-linear models. *Appl. Statist.* **33**, 38–44.

Bristol, D. R. and H. I. Patel (1990). A Markovian model for comparing incidences of side effects. *Statist. Med.* **9**, 803–809.

Byrne, P. J. and S. F. Arnold (1983). Inference about multivariate means for a nonstationary Autoregressive model. *J. Amer. Statist. Assoc.* **78**, 850–856.

Carr, G. J. and C. J. Portier (1993). An evaluation of some methods for fitting dose-response models to quantal-response developmental toxicology data. *Biometrics* **49**, 779–792.

Carter, E. M. and J. J. Hubert (1985). Analysis of parallel-line assays with multivariate responses. *Biometrics* **41**, 703–710.

Chakravorti, S. R. and J. E. Grizzle (1975). Analysis of data from multiclinic experiments. *Biometrics* **31**, 325–338.

Chi, G. Y. H., H. M. J. Hung, S. D. Dubey and R. J. Lipicky (1994). Dose response studies and special populations. In: *Proc. Amer. Statist. Assoc., Biopharmaceutical Section.*

Chi, E. M. and G. C. Reinsel (1989). Models for longitudinal data with random effects and AR(1) errors. *J. Amer. Statist. Assoc.* **84**, 452–459.

Chow, S. C. and J. P. Liu (1992). *Design and Analysis of Bioavailability and Bioequivalence Studies*. Marcel Dekker, New York.

Chuang, C. (1987). The analysis of titration study. *Statist. Med.* **6**, 583–590.

Cole, J. W. L. and J. E. Grizzle (1966). Applications of multivariate analysis of variance to repeated measures experiments. *Biometrics* **22**, 810–828.

Cornfield, J. (1977). Carcinogenic risk assessment. *Science* **198**, 693–699.

Cox, D. R. (1970). *Analysis of Binary Data*. Chapman and Hall, London.

Cox, D. R. (1985). Interaction. *Internat. Statist. Rev.* **52**, 1–31.

Crowder, M. J. (1985). Gaussian estimation for correlated binary data. *J. Roy. Statist. Soc. Ser. B* **47**, 229–237.

Dempster, A. P., N. M. Laird and D. B. Rubin (1977). Maximum likelihood from incomplete data via the EM algorithm (with discussion). *J. Roy. Statist. Soc. Ser. B* **39**, 1–38.

Dersimonian, R. and N. M. Laird (1986). Meta-analysis in clinical trials. *Contr. Clin. Trials* **7**, 177–188.

Diggle, P. J. (1988). An approach to the analysis of repeated measurements. *Biometrics* **44**, 959–971.

Diggle, P. J. and M. G. Kenward (1994). Informative dropout in longitudinal data analysis (with discussion). *Appl. Statist.* **43**, 49–93.

Diggle, P. J., K.-Y. Liang and S. L. Zeger (1994). *Analysis of Longitudinal Data*. Oxford Univ. Press, Oxford.

Ekholm, B. R., T. L. Fox and J. A. Bolognese (1990). Dose-response: Relating doses and plasma levels to efficacy and adverse experiences. In: D. A. Berry, ed., *Statistical Methodology in the Pharmaceutical Sciences*. Marcel Dekker, New York.

Esinhart, J. D. and V. M. Chinchilli (1994). Extension to the use of tolerance intervals for the assessment of individual bioequivalence. *J. Biopharm. Statist.* **4**, 39–52.

Fieller, E. C. (1954). Some problems in interval estimation (with discussion). *J. Roy. Statist. Soc. Ser. B* **16**, 175–185.

Finney, D. J. (1971). *Statistical Methods in Biological Assay*, 2nd edn. Griffin, London.

Fitzmaurice, G. M., N. M. Laird and S. R. Lipsitz (1994). Analyzing incomplete longitudinal binary responses: A likelihood-based approach. *Biometrics* **50**, 601–612.

Fitzmaurice, G. M., N. M. Laird and N. M. Rotnitsky (1993). Regression models for discrete longitudinal data (with discussion). *Statist. Sci.* **8**, 248–309.

FDA (1985). *Code of Federal Regulations* 21 (Food and Drugs). Part 320.22.

Gabriel, K. R. (1962). Ante-dependence analysis of an ordered set of variables. *Ann. Math. Statist.* **33**, 201–212.

Gallo, J. and A. I. Khuri (1990). Exact tests for the random and fixed effects in an unbalanced mixed two-way cross-classification model. *Biometrics* **46**, 1087–1095.

Gibardi, M. and D. Perrier (1982). *Pharmacokinetics*, 2nd edn. Marcel Dekker, New York.

Gilmour, A. R., R. D. Anderson and A. L. Rae (1985). The analysis of binomial data by a generalized linear mixed model. *Biometrika* **72**, 593–599.

Grizzle, J. E. (1965). The two-period change-over design and its use in clinical trials. *Biometrics* **21**, 467–480.

Grizzle, J. E. and D. M. Allen (1969). Analysis of growth and dose-response curves. *Biometrics* **25**, 357–381.

Hartley, H. O. and R. R. Hocking (1971). The analysis of incomplete data. *Biometrics* **27**, 783–808.

Harville, D. A. (1977). Maximum likelihood approaches to variance component estimation and to related problems. *J. Amer. Statist. Assoc.* **72**, 320–340.

Hauschke, D., V. W. Steinijans and E. Diletti (1990). A distribution-free procedure for the statistical analysis of bioequivalence studies. *Internat. J. Clinical Pharmacology, Therapy and Toxicology* **28**, 72–78.

Hills, M. and P. Armitage (1979). The two period crossover clinical trial. *British J. Clinical Pharmacology* **8**, 7–20.

Hsu, J. C., J. T. Hwang, H. K. Liu and S. Ruberg (1994). Confidence intervals associated with tests for bioequivalence. *Biometrika* **81**, 103–114.

Hubert, J. J., N. R. Bohidar and K. E. Peace (1988). Assessment of Pharmacological activity. In: K. E. Peace, ed., *Biopharmaceutical Statistics for Drug Development*. Marcel Dekker, New York.

Jennrich, R. I. and M. D. Schluchter (1986). Unbalanced repeated measures models with structured covariance matrices. *Biometrics* **42**, 805–820.

Jones, B. and M. G. Kenward (1989). *Design and Analysis of Cross-over Trials*. Chapman and Hall, London.

Jones, R. H. (1987). Serial correlation in unbalanced mixed models. *Bull. Internat. Statist. Inst. Proc. 46th Session*, Tokyo, 8–16 Sept. 1987, Book 4, 105–122.

Jones, R. H. (1993). *Longitudinal Data with Serial Correlation: A State-Space Approach*. Chapman and Hall, London.

Jones, R. H. and F. Boardi-Boateng (1991). Unequally spaced longitudinal data with AR(1) serial correlation. *Biometrics* **47**, 161–175.

Kenward, M. G. (1987). A method for comparing profiles of repeated measurements. *Appl. Statist.* **36**, 296–308.

Kenward, M. G. and B. Jones (1992). Alternative approaches to the analysis of binary and categorical repeated measurements. *J. Biopharm. Statist.* **2**, 137–170.

Khatri, C. G. (1966). A note on a MANOVA model applied to problems in growth curves. *Ann. Inst. Statist. Math.* **18**, 75–86.

Khatri, C. G. and H. I. Patel (1992). Analysis of a multicenter trial using a multivariate approach to a mixed linear model. *Comm. Statist. Theory Methods* **21**, 21–39.

Kleinbaum, D. G. (1973). A generalization of the growth curve model which allows missing data. *J. Multivariate Anal.* **3**, 117–124.

Koch, G. G., J. R. Landis, J. L. Freeman, D. H. Freeman and R. G. Lehnen (1977). A general methodology for the analysis of experiments with repeated measurements of categorical data. *Biometrics* **33**, 133–158.

Korn, E. L., D. Midthune, T. T. Chen, L. V. Rubinstein, M. C. Christian and R. M. Simon (1994). A comparison of two Phase I trial designs. *Statist. Med.* **13**, 1799–1806.

Krewski, D. and C. Brown (1981). Carcinogenic risk assessment: A guide to the literature. *Biometrics* **37**, 353–366.

Laird, N. M. and J. H. Ware (1982). Random-effects models for longitudinal data. *Biometrics* **38**, 963–974.

Laska, E. M. and M. J. Meisner (1989). Testing whether the identified treatment is best. *Biometrics* **45**, 1139–1151.

Lee, J. W. and D. L. DeMets (1991). Sequential comparison of changes with repeated measurements data. *J. Amer. Statist. Assoc.* **86**, 757–762.

Liang, K.-Y. and S. L. Zeger (1986). Longitudinal data analysis using generalized linear models. *Biometrika* **73**, 13–22.

Lindstorm, M. J. and D. M. Bates (1990). Nonlinear mixed effects models for repeated measures data. *Biometrics* **46**, 673–687.

Lipsitz, S. R., K. Kim and L. Zhao (1994). Analysis of repeated categorical data using generalized estimating equations. *Statist. Med.* **13**, 1149–1163.

Lipsitz, S. R., N. M. Laird and D. P. Harrington (1991). Generalized estimating equations for correlated binary data: Using the odds ratio as a measure of association. *Biometrika* **78**, 153–160.

Longford, N. T. (1993). *Random Coefficient Models*. Oxford Univ. Press, Oxford.

Longford, N. T. (1994). Logistic regression with random coefficients. *Comput. Statist. Data Anal.* **17**, 1–15.
Mandallaz, D. and J. Mau (1981). Comparison of different methods for decision-making in bioequivalence assessment. *Biometrics* **37**, 213–222.
Matthews, J. N. S. (1988). Recent developments in crossover designs. *Internat. Statist. Rev.* **56**, 117–127.
Matthews, J. N. S. (1990). The analysis of data from crossover designs: The efficiency of ordinary least squares. *Biometrics* **46**, 689–696.
Meier, K. L., A. J. Bailer and C. J. Portier (1993). A measure of tumorigenic potency incorporating dose-response shape. *Biometrics* **49**, 917–926.
Metzler, C. M. (1974). Bioavailability: A problem in equivalence. *Biometrics* **30**, 309–317.
Murray, G. D. and J. G. Findlay (1988). Correcting for the bias caused by drop-outs in hypertension trials. *Statist. Med.* **7**, 941–946.
Ochi, Y. and R. L. Prentice (1984). Likelihood inference in a correlated probit regression model. *Biometrika* **71**, 531–543.
OGD (1993). In vivo bioequivalence guidences. *Pharmacopeial Forum* **19**, 6067–6077.
O'Quigley, J. and S. Chevret (1991). Methods for dose finding studies in cancer trials: A review and results of a Monte Carlo study. *Statist. Med.* **10**, 1647–1664.
O'Quigley, J., M. Pepe and L. Fisher (1990). Continual reassessment method: A practical design for Phase I clinical trials in cancer. *Biometrics* **46**, 33–48.
Patel, H. I. (1979). Analysis of covariance of incomplete data on experiments with repeated measurements in clinical trials. *Proc. Fourth Conference of the SAS Users Group, International* 84-92, SAS Institute, Inc., Cary, NC.
Patel, H. I. (1985). Analysis of incomplete data in a two-period crossover design with reference to clinical trials. *Biometrika* **72**, 411–418.
Patel, H. I. (1986). Analysis of repeated measures designs with changing covariates in clinical trials. *Biometrika* **73**, 707–715.
Patel, H. I. (1990). Baseline measurements in a 2×2 crossover trial. In: K. E. Peace, ed., *Statistical Issues in Pharmaceutical Development* Marcel Dekker, New York, 177–184.
Patel, H. I. (1991). Analysis of incomplete data from a clinical trial with repeated measurements. *Biometrika* **78**, 609–619.
Patel, H. I. (1992). Sample size for a dose-response study. *J. Biopharm. Statist.* **2**, 1–8.
Patel, H. I. (1994a). Dose-response in pharmacokinetics. *Comm. Statist. Theory Methods* **23**, 451–465.
Patel, H. I. (1994b). Group sequential analysis of a clinical trial with repeated measurements. *Comm. Statist. Theory Methods* **23**, 981–995.
Patel, H. I. (1994c). A repeated measures design with repeated randomization. *J. Statist. Plann. Inference* **42**, 257–270.
Patel, H. I. and G. D. Gupta (1984). A problem of equivalence in clinical trials. *Biom. J.* **5**, 471–474.
Patel, H. I. and C. G. Khatri (1981). Analysis of incomplete data in experiments with repeated measurements using a stochastic model. *Comm. Statist. Theory Methods* **22**, 2259–2277.
Potthoff, R. F. and S. N. Roy (1964). A generalized multivariate analysis of variance model useful especially for growth curve problems. *Biometrika* **51**, 313–326.
Prentice, R. L. and L. P. Zhao (1991). Estimating equations for parameters in means and covariances of multivariate discrete and continuous responses. *Biometrics* **47**, 825–839.
Raghunathan, T. E. (1994). Monte Carlo methods for exploring sensitivity to distributional assumptions in a Bayesian analysis of a series of 2×2 tables. *Statist. Med.* **13**, 1525–1538.
Raghunathan, T. E. and Yoichi I. (1993). Analysis of binary data from a multicenter clinical trial. *Biometrika* **80**, 127–139.
Rao, C. R. (1959). Some problems involving linear hypotheses in multivariate analysis. *Biometrika* **46**, 49–58.
Rao, C. R. (1965). A theory of least squares when the parameters are stochastic and its application to the analysis of growth curves. *Biometrika* **52**, 447–458.
Rao, C. R. and J. Kleffe (1988). *Estimation of Variance Components and Applications.* North-Holland, Amsterdam.
Rom, D. A., R. J. Costello and L. T. Connell (1994). On closed test procedures for dose-response analysis. *Statist. Med.* **13**, 1583–1596.

Rowland, M. and T. N. Tozer (1980). *Clinical Pharmacokinetics: Concepts and Applications.* Lea and Febiger, Philadelphia, PA.

Rubin, D. B. (1976). Inference and missing data. *Biometrika* **63**, 581–592.

Rutter, C. M. and R. M. Elashoff (1994). Analysis of longitudinal data: Random coefficient regression modelling. *Statist. Med.* **13**, 1211–1231.

Schall, R. (1991). Estimation in generalized linear models with random effects. *Biometrika* **78**, 719–727.

Schall, R. and H. G. Luus (1993). On population and individual bioequivalence. *Statist. Med.* **12**, 1109–1124.

Scheffé, H. (1959). *The Analysis of Variance.* Wiley, New York.

Schoenfeld, D. A. (1986). Confidence bounds for normal means under order restrictions, with application to dose-response curves, toxicological experiments, and low-dose extrapolation. *J. Amer. Statist. Assoc.* **81**, 186–195.

Searle, S. R., G. Casella and C. E. McCulloch (1992). *Variance Components.* Wiley, New York.

Selwyn, M. R. (1988). Preclinical safety development. In: K. E. Peace, ed., *Biopharmaceutical Statistics for Drug Development.* Marcel Dekker, New York.

Sheiner, L. B. and S. L. Beal (1983). Evaluation of methods for estimating population pharmacokinetic parameters. III. Nonexperimental model: Routine clinical pharmacokinetic data. *J. Pharmacokin. Biopharm.* **11**, 303–319.

Shen, C. F. and B. Iglewicz (1994). Robust and bootstrap testing procedures for bioequivalence. *J. Biopharmaceut. Statist.* **4**, 65–90.

Shih, W. J., A. L. Gould and I. K. Hwang (1989). The analysis of titration studies in Phase III clinical trials. *Statist. Med.* **8**, 583–591.

Shuirmann, D. J. (1987). A comparison of the two one-sided tests procedure and the power approach for assessing the equivalence of average bioavailability. *J. Pharmacokin. Biopharm.* **15**, 657–680.

Snapinn, S. M. (1987). Evaluating the efficacy of a combination therapy. *Statist. Med.* **6**, 657–665.

Srivastava, M. S. (1986). Multivariate bioassay, combination of bioassays, and Fieller's theorem. *Biometrics* **42**, 131–141.

SAS Institute Inc. (1992). *SAS Technical Report P-229. Software: Changes and Enhancements, Release 6.07.* Chapter 16: The MIXED procedure. Cary, NC.

SAS Institute Inc. (1990). *SAS/IML Software, Version 6.* Cary, NC.

Stiratelli, R., N. M. Laird and J. H. Ware (1984). Random-effects model for several observations with binary response. *Biometrics* **40**, 961–971.

Storer, B. E. (1989). Design and analysis of Phase I clinical trials. *Biometrics* **45**, 925–937.

Strijbosch, L. W. G., R. J. M. M. Does and W. Albers (1990). Design methods for some dose-response models. *Statist. Med.* **9**, 1353–1363.

Van Ryzin, J. and K. Rai (1987). A dose-response model incorporating nonlinear kinetics. *Biometrics* **43**, 95–105.

Vonesh, E. F. and R. L. Carter (1992). Mixed-effects nonlinear regression for unbalanced repeated measures. *Biometrics* **48**, 1–17.

Wagner, J. E. (1971). *Biopharmaceutics and Relevant Pharmacokinetics.* Drug Intelligence Publications, Hamilton, IL.

Wellek, S. (1993). Basing the analysis of comparative bioavailability trials on an individualized statistical definition of equivalence. *Biom. J.* **1**, 47–55.

Westlake, W. J. (1972). Use of confidence intervals in analysis of comparative bioavailability trials. *J. Pharmac. Sci.* **61**, 1340–1341.

Westlake, W. J. (1976). Symmetrical confidence intervals for bioequivalence trials. *Biometrics* **32**, 741–744.

Westlake, W. J. (1988). Bioavailability and bioequivalence of pharmaceutical formulations. In: K. E. Peace, ed., *Biopharmaceutical Statistics for Drug Development.* Marcel Dekker, New York, 329–352.

Williams, D. A. (1972). The comparison of several dose levels with a zero dose control. *Biometrics* **28**, 519–531.

Williams, E. J. (1950). Experimental designs balanced for pairs of residual effects. *Austral. J. Sci. Res.* **3**, 351–363.

WHO, Regional Office for Europe (1986). *Guidelines for the Investigation of Bioavailability.* Copenhagen.

Wu, M. C. and K. R. Bailey (1989). Estimation and comparison of changes in the presence of informative right censoring: Conditional linear model. *Biometrics* **45**, 939–955.

Wu, M. C. and R. J. Carroll (1988). Estimation and comparison of changes in the presence of right censoring by modeling the censoring process. *Biometrics* **44**, 175–188.

Yuan, W. and A. M. Kshirsagar (1993). Analysis of multivariate parallel-line bioassay with composite responses and composite doses, using canonical correlations. *J. Biopharm. Statist.* **3**, 57–72.

Zeger, S. L. and M. R. Karim (1991). Generalized linear models with random effects: A Gibbs sampling approach. *J. Amer. Statist. Assoc.* **86**, 79–86.

Zeger, S. L. and K.-Y. Liang (1986). Longitudinal data analysis for discrete and continuous outcomes. *Biometrics* **42**, 121–130.

Zeger, S. L., K.-Y. Liang and P. S. Albert (1988). Models for longitudinal data: A generalized estimating equation approach. *Biometrics* **44**, 1049–1060.

Zeger, S. L. and B. Qaqish (1988). Markov regression models for time series: A quasi-likelihood approach. *Biometrics* **44**, 1019–1032.

S. Ghosh and C. R. Rao, eds., *Handbook of Statistics, Vol. 13*

3

Optimal Crossover Designs

John Stufken

1. Introduction

As for many other designs, the use of crossover designs originated in the agricultural sciences. Crossover designs, also known as change-over or repeated measurements designs, were first used in animal feeding trials. Some early references and a small example, providing only part of the entire data set, are presented in Cochran and Cox (1957). Currently crossover designs have applications in many other sciences and research areas; examples are listed in Kershner and Federer (1981) and Afsarinejad (1990). The use of crossover designs in pharmaceutical studies and clinical trials receives now perhaps more attention than applications in any other area. For some examples and further discussion and references, the reader may want to consult the recent books by Jones and Kenward (1989), Ratkowsky et al. (1992) and Senn (1993).

The principal idea associated with crossover designs is to use a number of available units for several measurements at different occasions. We will refer to these units as subjects, and in many applications they are humans, animals or plots of land. The different occasions at which the subjects are used are known as periods. We will assume that the main purpose of the experiment consists of the comparison of t treatments. Each subject will receive a treatment at each of p periods, and a relevant measurement is obtained for each subject in each period. A subject may receive a different treatment in each period, but treatments may also be repeated on a subject.

If we denote the number of subjects by n, then we may think of a crossover design as a $p \times n$ matrix with entries from $\{1, \dots, t\}$, where the entry in position (i, j) denotes the treatment that subject j receives in the ith period. Corresponding to such a design we will also have an array of pn observable random variables y_{ij}, whose values will be determined by the measurements to be made. We will assume that these measurements are of a continuous nature.

One possible motive for using a crossover design, as opposed to using each of pn subjects for one measurement, is that a crossover design requires fewer subjects for the same number of observations. This can obviously be an important consideration when subjects are scarce and when including a large number of subjects in the experiment can be prohibitively expensive. Another possible motive for using crossover designs is that these designs provide within subject information about treatment differences. In many applications the different subjects would exhibit large natural differences, and inferences concerning treatment comparisons based on between subject information (available if subject effects are assumed to be random effects) would require a much larger replication of the treatments in order to achieve the same precision as inferences based on within subject information. Indeed, designs are at times chosen based on the within subject information that they provide, and the between subject information is conveniently ignored.

There are however also various potential problems with the use of crossover designs. Firstly, compared to using each subject only once, the duration of an experiment when using a crossover design may be considerably longer. Typically, it is therefore undesirable to have a large number of periods.

Secondly, we may have to deal with carry-over effects. Measurements may not only be affected by the treatment assigned most recently to a subject, but could also be affected by lingering effects of treatments that were assigned to the same subject in one of the previous periods. Such lingering effects are called carry-over effects. One way to avoid or reduce the problem of carry-over effects is to use wash-out periods between periods in which measurements are made. The idea is that the effect of a previously given treatment can wear out during this wash-out period. Use of wash-out periods will however further increase the duration of the experiment, and may in some cases meet with ethical objections. (It is hard to deny a pain killer to a suffering patient just because he or she happens to be in a wash-out period!)

A third potential problem with crossover designs is that an assumption of uncorrelated error terms may not always be reasonable. It may be more realistic to view the data as n short time series, one for each subject. Different error structures may affect recommendations concerning choice of design. We will return to this issue in Section 5.

Fourthly, the use of crossover designs is of course limited to situations where a treatment does not essentially alter a subject. Crossover designs may be fine if the treatments alleviate a symptom temporarily, but not if the treatments provide a cure for the condition that a subject suffers from.

This chapter will focus on the choice of design when a crossover design is to be used. Selected results concerning optimal design choices will be discussed. Of course, while the discussion will concentrate on statistical considerations for selecting a design, in any application there may be practical constraints that should be taken into account. The results concerning optimal designs should only be used as a guide to select good designs, or to avoid the selection of very poor designs.

The literature on crossover designs is quite extensive. Many different models have been considered, and different models may result in different recommendations concerning design selection. This chapter represents therefore, inevitably, a selection of available results that is biased by the author's personal interest. There are however a number of other recent review papers and book chapters on crossover designs; the interested reader is referred to these sources for further details, additional references, and, possibly, bias in a different direction. Good sources, in addition to the aforementioned recent books, are Afsarinejad (1990), Barker et al. (1982), Bishop and Jones (1984), Matthews (1988, 1994), Shah and Sinha (1989, Chapter 6) and Street (1989).

2. Terminology and notation

The approach throughout this chapter will be to assume a linear model for the observable random variables y_{ij}. While different options are possible for such a model, once we settle for a model we will want to address the question of selecting an optimal design for inferences concerning the treatment effects or the carry-over effects.

By $\underline{1}_a$ and $\underline{0}_a$ we will mean the $a \times 1$ vectors of 1's and 0's, respectively. By $\boldsymbol{I}_a$ and $\mathbf{0}_{a\times b}$ we will mean the $a \times a$ identity matrix and the $a \times b$ matrix with all entries equal to 0, respectively.

The basic linear model that we will use is the model

$$\begin{aligned} y_{ij} &= \mu + \alpha_i + \beta_j + \tau_{d(i,j)} + \gamma_{d(i-1,j)} + \varepsilon_{ij}, \\ &i \in \{1,2,\ldots,p\},\ j \in \{1,2,\ldots,n\}, \end{aligned}$$

where $d(i,j)$ stands for the treatment that is assigned to subject j in period i under design d. One may think of μ as a general mean, of α_i as an effect due to the ith period, of β_j as an effect due to the jth subject, of $\tau_{d(i,j)}$ as a treatment effect due to treatment $d(i,j)$, and of $\gamma_{d(i-1,j)}$ as a carry-over effect due to treatment $d(i-1,j)$. For the latter, we define $\gamma_{d(0,j)} = 0$. The ε_{ij} are the non-observable random error terms. In matrix notation we will write our model as

$$\underline{y} = \mu \underline{1}_{pn} + \boldsymbol{X}_1 \underline{\alpha} + \boldsymbol{X}_2 \underline{\beta} + \boldsymbol{X}_{d3} \underline{\tau} + \boldsymbol{X}_{d4} \underline{\gamma} + \underline{\varepsilon}, \tag{2.1}$$

where $\underline{y} = (y_{11}, y_{21}, \ldots, y_{pn})'$, $\underline{\alpha} = (\alpha_1, \ldots, \alpha_p)'$, $\underline{\beta} = (\beta_1, \ldots, \beta_n)'$, $\underline{\tau} = (\tau_1, \ldots, \tau_t)'$, $\underline{\gamma} = (\gamma_1, \ldots, \gamma_t)'$, $\underline{\varepsilon} = (\varepsilon_{11}, \varepsilon_{21}, \ldots, \varepsilon_{pn})'$, the $pn \times p$ and $pn \times n$ matrices $\boldsymbol{X}_1$ and $\boldsymbol{X}_2$ are

$$\boldsymbol{X}_1 = \begin{bmatrix} \boldsymbol{I}_p \\ \vdots \\ \boldsymbol{I}_p \end{bmatrix} = \underline{1}_n \otimes \boldsymbol{I}_p, \qquad \boldsymbol{X}_2 = \begin{bmatrix} \underline{1}_p & \underline{0}_p & \cdots & \underline{0}_p \\ \underline{0}_p & \underline{1}_p & \cdots & \underline{0}_p \\ \vdots & \vdots & \ddots & \vdots \\ \underline{0}_p & \underline{0}_p & \cdots & \underline{1}_p \end{bmatrix} = \boldsymbol{I}_n \otimes \underline{1}_p,$$

and the $pn \times t$ matrices $\boldsymbol{X}_{d3}$ and $\boldsymbol{X}_{d4}$, which are design dependent, are

$$\boldsymbol{X}_{d3} = \begin{bmatrix} \boldsymbol{X}_{d31} \\ \boldsymbol{X}_{d32} \\ \vdots \\ \boldsymbol{X}_{d3n} \end{bmatrix}, \qquad \boldsymbol{X}_{d4} = \begin{bmatrix} \boldsymbol{X}_{d41} \\ \boldsymbol{X}_{d42} \\ \vdots \\ \boldsymbol{X}_{d4n} \end{bmatrix},$$

where $\boldsymbol{X}_{d3j}$ stands for the $p \times t$ period-treatment incidence matrix for subject j under design d and where $\boldsymbol{X}_{d4j} = \boldsymbol{L}\boldsymbol{X}_{d3j}$ with the $p \times p$ matrix $\boldsymbol{L}$ defined as

$$\boldsymbol{L} = \begin{bmatrix} 0 & 0 & \cdots & 0 & 0 \\ 1 & 0 & \cdots & 0 & 0 \\ 0 & 1 & \cdots & 0 & 0 \\ \vdots & \vdots & \ddots & \vdots & \vdots \\ 0 & 0 & \cdots & 1 & 0 \end{bmatrix}.$$

For the model in (2.1) we will assume that $\underline{\varepsilon}$ follows a multivariate normal distribution with mean $\underline{0}_{pn}$ and variance-covariance matrix $\sigma^2\boldsymbol{V}$, for an unknown scalar σ^2 and a $pn \times pn$ positive definite matrix $\boldsymbol{V}$, to be specified later.

Some comments concerning the model in (2.1) are in order. Firstly, all of the effects, including subject effects, are assumed to be fixed effects. While for many applications it may be quite reasonable to take the subject effects as random effects, with a relatively large between subject variability it may also in those cases be quite reasonable to make a design choice based on within subject information only. As explained earlier, in most applications the latter would be by far the more precise source of information. For some references and results on optimal design choice when between subject information is also considered, see Mukhopadhyay and Saha (1983), Shah and Sinha (1989) and Carrière and Reinsel (1993).

Secondly, while the model includes carry-over effects, it only allows for the possibility of first-order carry-over effects; only the treatment that was used in the period immediately preceding the current period is considered to have a possible lingering effect on a measurement in the current period. The model also reflects the assumption that there is no carry-over effect for measurements in the first period. Some have called this a non-circular model (see Shah and Sinha, 1989), in contrast to a circular model (Magda, 1980). A circular model would be a model where there is also a carry-over effect for measurements in the first period, as a result of treatments given to the subjects in a preperiod. But the use of preperiods is rather uncommon and unintelligible in many applications. For some results on optimal choice of design in the presence of a preperiod see Afsarinejad (1990).

Thirdly, in some applications additional information on the subjects may be available through concomitant variables. Measurements could, for example, be taken on each subject at the beginning of the experiment or at the beginning of each period. Such so-called baseline measurements could be used in various ways when analyzing data from a crossover design. The model in (2.1) does not include use of such information. Use of baseline measurements can also be incorporated in the design selection problem; see, for example, Laska and Meisner (1985) and Carrière and Reinsel (1992) for the basic ideas.

Fourthly, the model in (2.1) does not include any interactions. It assumes, for example, that the period effects are the same for each of the subjects. It also assumes that the effect of a treatment is the same no matter which treatment contributes the carry-over

Table 2.1
Notation

Symbol	Description
n_{duj}	The number of times that treatment u is assigned to subject j
$\tilde{n}_{duj}$	The number of times that treatment u is assigned to subject j in the first $p-1$ periods
l_{dui}	The number of times that treatment u is assigned to a subject in period i
m_{duv}	The number of times that treatment u is immediately preceded by treatment v
r_{du}	The replication of treatment u throughout the design
$\tilde{r}_{du}$	The replication of treatment u restricted to the first $p-1$ periods

effect. If these assumptions seem unreasonable, recommendations concerning design choices in the next sections may also be quite unreasonable.

For some alternative models see, for example, Kershner and Federer (1981).

By Tr($\boldsymbol{A}$) we will mean the trace of a square matrix $\boldsymbol{A}$. We will say that an $a \times a$ matrix $\boldsymbol{A}$ is completely symmetric, abbreviated as c.s., if $\boldsymbol{A} = b_1\boldsymbol{I}_a + b_2\underline{1}_a\underline{1}_a'$ for some constants b_1, b_2. Following Kunert (1985), for any matrix $\boldsymbol{A}$ we will write $w(\boldsymbol{A})$ to denote the orthogonal projection matrix onto the column space of $\boldsymbol{A}$, that is, $w(\boldsymbol{A}) = \boldsymbol{A}(\boldsymbol{A}'\boldsymbol{A})^-\boldsymbol{A}'$. The following basic result on orthogonal projection matrices of partitioned matrices will be quite useful.

LEMMA 2.1. *For an $a \times b$ matrix $\boldsymbol{X} = [\boldsymbol{Y}\ \ \boldsymbol{Z}]$ we have that $w(\boldsymbol{X}) = w(\boldsymbol{Y}) + w((\boldsymbol{I}_a - w(\boldsymbol{Y}))\boldsymbol{Z})$.*

Our notation and terminology will to a large extent follow that of Cheng and Wu (1980). By $\Omega_{t,n,p}$ we will denote the class of all crossover designs for t treatments, n subjects and p periods. We will say that a design $d \in \Omega_{t,n,p}$ is *uniform on the periods* if d assigns each treatment to n/t subjects in each period. A design $d \in \Omega_{t,n,p}$ is *uniform on the subjects* if d assigns each treatment p/t times to each subject. A design is said to be *uniform* if it is uniform on the periods and uniform on the subjects. We will also make extensive use of the notation presented in Table 2.1.

We also define $l_{du0} = 0$.

Among the crossover designs with some desirable properties, as we will see, are those that are balanced or strongly balanced for carry-over effects. A crossover design is said to be *balanced for carry-over effects*, or *balanced* for brevity, if no treatment is immediately preceded by itself, and each treatment is immediately preceded by each of the other treatments equally often. A crossover design is called *strongly balanced for carry-over effects*, or just *strongly balanced*, if each treatment is immediately preceded by each of the treatments equally often.

Orthogonal arrays of Type I form a useful class of designs when searching for optimal crossover designs if $p \leqslant t$. These arrays were introduced by Rao (1961). Formally, we define an *orthogonal array of Type I and strength* 2 as a $p \times n$ array with entries from $\{1, 2, \ldots, t\}$ such that any $2 \times n$ subarray contains all $t(t-1)$ ordered 2-tuples without repetition equally often. We denote such an array by $\mathrm{OA_I}(n, p, t, 2)$. Clearly, such an array can only exist if n is a multiple of $t(t-1)$.

Table 2.2
Examples of orthogonal arrays of Type I

$$\begin{bmatrix} 1 & 2 \\ 2 & 1 \end{bmatrix} \quad \begin{bmatrix} 1 & 1 & 2 & 2 & 3 & 3 \\ 2 & 3 & 1 & 3 & 1 & 2 \\ 3 & 2 & 3 & 1 & 2 & 1 \end{bmatrix} \quad \begin{bmatrix} 1 & 1 & 1 & 2 & 2 & 2 & 3 & 3 & 3 & 4 & 4 & 4 \\ 2 & 3 & 4 & 1 & 3 & 4 & 1 & 2 & 4 & 1 & 2 & 3 \\ 4 & 2 & 3 & 3 & 4 & 1 & 4 & 1 & 2 & 2 & 3 & 1 \\ 3 & 4 & 2 & 4 & 1 & 3 & 2 & 4 & 1 & 3 & 1 & 2 \end{bmatrix}$$

Orthogonal arrays of Type I are closely related to orthogonal arrays. Orthogonal arrays were introduced by Rao (1947). An *orthogonal array of strength 2* is a $p \times n$ array with entries from $\{1, 2, \ldots, t\}$ such that any $2 \times n$ subarray contains all t^2 ordered 2-tuples equally often. We denote such an array by $\mathrm{OA}(n, p, t, 2)$; a necessary condition for its existence is that n is a multiple of t^2. A forthcoming book by Hedayat, Sloane and Stufken (1996) contains an overview of existence and construction results for orthogonal arrays. For the purpose of this chapter, the following two well known results are sufficient.

LEMMA 2.2. *A necessary condition for the existence of an* $\mathrm{OA}(t^2, p, t, 2)$ *is that* $p \leqslant t + 1$. *Moreover, if* t *is a prime power, such an orthogonal array exists for any such* p.

LEMMA 2.3. *An orthogonal array of Type I,* $\mathrm{OA_I}(t(t-1), p, t, 2)$, *can be constructed from an orthogonal array* $\mathrm{OA}(t^2, p + 1, t, 2)$.

Combining Lemmas 2.2 and 2.3 leads thus to the result that an $\mathrm{OA_I}(t(t-1), p, t, 2)$ exists for any $p \leqslant t$ if t is a prime power. (See also Rao, 1961.)

Examples of orthogonal arrays of Type I for $p = t = 2$, $p = t = 3$ and $p = t = 4$, each time with $n = t(t-1)$, are presented in Table 2.2. If smaller values of p are desired, one can simply delete one or more rows from the given arrays; if larger values of n are desired one can simply juxtapose multiple copies of the given arrays. The reader who is less familiar with these combinatorial structures can be assured that Table 2.2 is essentially all that is needed for the remainder of this chapter.

Choosing an optimal design is typically based on choosing a design with, in some sense, a large information matrix (see, for example, Shah and Sinha, 1989; or Silvey, 1980). The concept of universal optimality (Kiefer, 1975) has been considered extensively for crossover designs, and we will give a brief definition for completeness. Let $\mathcal{P}_a$ be the class of all nonnegative definite $a \times a$ matrices with entries summing to 0 in each row. Let Φ_a be the class of all functions $\phi: \mathcal{P}_a \to (-\infty, \infty]$ such that (i) $\phi(\boldsymbol{QCQ}') = \phi(\boldsymbol{C})$ for all $\boldsymbol{C} \in \mathcal{P}_a$ and all $a \times a$ permutation matrices $\boldsymbol{Q}$ (permutation invariance), (ii) $\phi(b\boldsymbol{C})$ is non-increasing in the scalar $b \geqslant 0$, for all $\boldsymbol{C} \in \mathcal{P}_a$, and (iii) $\phi(b\boldsymbol{C}_1 + (1-b)\boldsymbol{C}_2) \leqslant b\phi(\boldsymbol{C}_1) + (1-b)\phi(\boldsymbol{C}_2)$ for all $b \in (0, 1)$ and all $\boldsymbol{C}_1, \boldsymbol{C}_2 \in \mathcal{P}_a$ (convexity). If $\boldsymbol{C}_d$ and $\widetilde{\boldsymbol{C}}_d$ denote the information matrices for $\underline{\tau}$ and $\underline{\gamma}$ under design $d \in \Omega_{t,n,p}$, respectively, then $\boldsymbol{C}_d$, $\widetilde{\boldsymbol{C}}_d \in \mathcal{P}_t$. We say that $d^* \in \Omega$, where $\Omega \subseteq \Omega_{t,n,p}$, is *universally optimal* for $\underline{\tau}$ $(\underline{\gamma})$ in Ω if, for all $\phi \in \Phi_t$, $\phi(\boldsymbol{C}_d)$ $(\phi(\widetilde{\boldsymbol{C}}_d))$ is minimized over $d \in \Omega$ by d^*. It is well known that a design that is universally optimal in Ω is also optimal in Ω under the commonly used A-, D- and E-optimality criteria.

Except for a minor extension, there is only one major tool to establish whether a design is universally optimal. We formulate this result (due to Kiefer, 1975) with reference to the information matrix $\boldsymbol{C}_d$ for $\underline{\tau}$, but with the understanding that a similar result holds for $\widetilde{\boldsymbol{C}}_d$ and $\underline{\gamma}$.

THEOREM 2.1. *A design $d^* \in \Omega$ is universally optimal for $\underline{\tau}$ in Ω if it maximizes* $\mathrm{Tr}(\boldsymbol{C}_d)$ *over $d \in \Omega$ and if $\boldsymbol{C}_{d^*}$ is c.s.*

The result in Theorem 2.1 will be used extensively in the next sections. Preferably we would like to take $\Omega = \Omega_{t,n,p}$, but will not always be able to do this either because we are unable to show that a candidate design d^* maximizes $\mathrm{Tr}(\boldsymbol{C}_d)$ over that large a class of designs, or because we know that it only maximizes this trace over a subclass of $\Omega_{t,n,p}$ and not over the entire class.

3. Optimal crossover designs when errors are uncorrelated

3.1. Expressions for the information matrices

The additional assumption in Section 3 for model (2.1) is that the error variance-covariance matrix $\sigma^2\boldsymbol{V}$ is equal to $\sigma^2\boldsymbol{I}_{pn}$. If we define

$$\begin{aligned}
\boldsymbol{C}_{d11} &= \boldsymbol{X}'_{d3}(\boldsymbol{I}_{pn} - w([\boldsymbol{X}_1 \ \ \boldsymbol{X}_2]))\boldsymbol{X}_{d3}, \\
\boldsymbol{C}_{d22} &= \boldsymbol{X}'_{d4}(\boldsymbol{I}_{pn} - w([\boldsymbol{X}_1 \ \ \boldsymbol{X}_2]))\boldsymbol{X}_{d4}, \\
\boldsymbol{C}_{d12} &= \boldsymbol{X}'_{d3}(\boldsymbol{I}_{pn} - w([\boldsymbol{X}_1 \ \ \boldsymbol{X}_2]))\boldsymbol{X}_{d4}, \\
\boldsymbol{C}_{d21} &= \boldsymbol{C}'_{d12}
\end{aligned}$$

then we have, using Lemma 2.1, that the information matrix $\boldsymbol{C}_d$ for $\underline{\tau}$ is equal to

$$\begin{aligned}
\boldsymbol{C}_d &= \boldsymbol{X}'_{d3}(\boldsymbol{I}_{pn} - w([\boldsymbol{X}_1 \ \ \boldsymbol{X}_2 \ \ \boldsymbol{X}_{d4}]))\boldsymbol{X}_{d3} \\
&= \boldsymbol{X}'_{d3}(\boldsymbol{I}_{pn} - w([\boldsymbol{X}_1 \ \ \boldsymbol{X}_2]))\boldsymbol{X}_{d3} \\
&\quad -\boldsymbol{X}'_{d3}w((\boldsymbol{I}_{pn} - w([\boldsymbol{X}_1 \ \ \boldsymbol{X}_2]))\boldsymbol{X}_{d4})\boldsymbol{X}_{d3} \\
&= \boldsymbol{C}_{d11} - \boldsymbol{C}_{d12}\boldsymbol{C}^-_{d22}\boldsymbol{C}_{d21}.
\end{aligned}$$

It follows in a similar way that the information matrix $\widetilde{\boldsymbol{C}}_d$ for $\underline{\gamma}$ is equal to

$$\widetilde{\boldsymbol{C}}_d = \boldsymbol{C}_{d22} - \boldsymbol{C}_{d21}\boldsymbol{C}^-_{d11}\boldsymbol{C}_{d12}.$$

Many results on optimality of crossover designs are based on a careful study of the matrices $\boldsymbol{C}_{d11}$, $\boldsymbol{C}_{d22}$ and $\boldsymbol{C}_{d12}$, or on a more subtle application of Lemma 2.1.

The elements of the matrices $\boldsymbol{C}_{d11}$, $\boldsymbol{C}_{d22}$ and $\boldsymbol{C}_{d12}$ are readily expressed in terms of the quantities introduced in Section 2. Once more calling on Lemma 2.1, and writing

$$\begin{aligned}
\boldsymbol{I}_{pn} - w([\boldsymbol{X}_1 \ \ \boldsymbol{X}_2]) &= \boldsymbol{I}_{pn} - w(\boldsymbol{X}_2) - w((\boldsymbol{I}_{pn} - w(\boldsymbol{X}_2))\boldsymbol{X}_1) \\
&= \boldsymbol{I}_{pn} - w(\boldsymbol{X}_2) - w((\boldsymbol{I}_{pn} - w(\underline{1}_{pn}))\boldsymbol{X}_1),
\end{aligned}$$

it is easily seen that the diagonal elements for $\boldsymbol{C}_{d11}$, $\boldsymbol{C}_{d22}$ and $\boldsymbol{C}_{d12}$ in position (u, u) are equal to

$$r_{du} - \frac{1}{p}\sum_{j=1}^{n} n_{duj}^2 - \frac{1}{n}\sum_{i=1}^{p}(l_{dui} - r_{du}/p)^2, \tag{3.1}$$

$$\widetilde{r}_{du} - \frac{1}{p}\sum_{j=1}^{n} \widetilde{n}_{duj}^2 - \frac{1}{n}\sum_{i=1}^{p}(l_{du(i-1)} - \widetilde{r}_{du}/p)^2 \tag{3.2}$$

and

$$m_{duu} - \frac{1}{p}\sum_{j=1}^{n} n_{duj}\widetilde{n}_{duj} - \frac{1}{n}\sum_{i=1}^{p}(l_{dui} - r_{du}/p)(l_{du(i-1)} - \widetilde{r}_{du}/p), \tag{3.3}$$

respectively. The off-diagonal elements in position (u, v) for these matrices are

$$-\frac{1}{p}\sum_{j=1}^{n} n_{duj}n_{dvj} - \frac{1}{n}\sum_{i=1}^{p}(l_{dui} - r_{du}/p)(l_{dvi} - r_{dv}/p), \tag{3.4}$$

$$-\frac{1}{p}\sum_{j=1}^{n} \widetilde{n}_{duj}\widetilde{n}_{dvj} - \frac{1}{n}\sum_{i=1}^{p}(l_{du(i-1)} - \widetilde{r}_{du}/p)(l_{dv(i-1)} - \widetilde{r}_{dv}/p) \tag{3.5}$$

and

$$m_{duv} - \frac{1}{p}\sum_{j=1}^{n} n_{duj}\widetilde{n}_{dvj} - \frac{1}{n}\sum_{i=1}^{p}(l_{dui} - r_{du}/p)(l_{dv(i-1)} - \widetilde{r}_{dv}/p), \tag{3.6}$$

respectively.

3.2. Optimal strongly balanced crossover designs

We will first focus on finding optimal designs for $\underline{\tau}$. Since Theorem 2.1 will be our major tool, candidate designs d^* should have an information matrix $\boldsymbol{C}_{d^*}$ that is c.s. One way to achieve this is by selecting a design for which each of the matrices $\boldsymbol{C}_{d^*11}$, $\boldsymbol{C}_{d^*22}$ and $\boldsymbol{C}_{d^*12}$ is c.s. Since it is easily seen that all row sums in these matrices are 0, they will be c.s. if all off-diagonal elements are equal. Moreover, if d^* also maximizes $\mathrm{Tr}(\boldsymbol{C}_{d11})$ over all $d \in \Omega_{t,n,p}$ and if $\boldsymbol{C}_{d^*12} = 0$, then d^* is universally optimal for $\underline{\tau}$ in $\Omega_{t,n,p}$. The simple strategy of searching for such designs d^* works surprisingly well. The following results based on this strategy are due to Cheng and Wu (1980).

THEOREM 3.1. *A uniform strongly balanced crossover design is universally optimal for $\underline{\tau}$ in $\Omega_{t,n,p}$.*

THEOREM 3.2. *A strongly balanced crossover design that is uniform on the periods and uniform on the units when restricted to the first $p-1$ periods is universally optimal for $\underline{\tau}$ in $\Omega_{t,n,p}$.*

The proofs for the two theorems are based on the same ideas. If d^* is a design as in one of the previous theorems, then it is uniform on the periods. The expression in (3.6) reduces therefore to

$$m_{d^*uv} - \frac{1}{p}\sum_{j=1}^{n} n_{d^*uj}\widetilde{n}_{d^*vj}.$$

It is easily verified that this becomes 0 under the conditions in each of Theorems 3.1 and 3.2, so that $\boldsymbol{C}_{d^*12} = \boldsymbol{0}_{t\times t}$. Moreover, using (3.1) we have that

$$\mathrm{Tr}(\boldsymbol{C}_{d^*11}) = \sum_{u=1}^{t} r_{du} - \frac{1}{p}\sum_{u=1}^{t}\sum_{j=1}^{n} n_{duj}^2 - \frac{1}{n}\sum_{u=1}^{t}\sum_{i=1}^{p}(l_{dui} - r_{du}/p)^2.$$

The last term that is subtracted in this expression is 0 for a design that is uniform on the periods. The other term that is subtracted is, following arguments as in Cheng and Wu (1980), minimized over $\Omega_{t,n,p}$ by designs as described in the theorems. Thus, based on the strategy as outlined previously, the claims in the theorems follow.

A strategy analogous to that used for finding universally optimal designs for $\underline{\tau}$ can also be used for finding universally optimal designs for $\underline{\gamma}$. Thus, a design d^* that maximizes $\mathrm{Tr}(\boldsymbol{C}_{d22})$ over $\Omega_{t,n,p}$ and for which $\boldsymbol{C}_{d^*12} = \boldsymbol{0}_{t\times t}$ and $\boldsymbol{C}_{d^*22}$ is c.s. is universally optimal for $\underline{\gamma}$. This leads to the following results, also due to Cheng and Wu (1980).

THEOREM 3.3. *A uniform strongly balanced crossover design is universally optimal for $\underline{\gamma}$ in $\Omega_{t,n,p}$.*

THEOREM 3.4. *A strongly balanced crossover design that is uniform on the periods and uniform on the units when restricted to the first $p-1$ periods is universally optimal for $\underline{\gamma}$ in $\Omega_{t,n,p}$.*

For the proofs, the argument that $\boldsymbol{C}_{d^*12} = \boldsymbol{0}_{t\times t}$ for designs in Theorems 3.3 and 3.4 is similar as for Theorems 3.1 and 3.2. Further, using (3.2) we have that

$$\begin{aligned}\mathrm{Tr}(\boldsymbol{C}_{d^*22}) &= \sum_{u=1}^{t}\widetilde{r}_{du} - \frac{1}{p}\sum_{u=1}^{t}\sum_{j=1}^{n}\widetilde{n}_{duj}^2 - \frac{1}{n}\sum_{u=1}^{t}\sum_{i=1}^{p}(l_{du(i-1)} - \widetilde{r}_{du}/p)^2\\ &= n(p-1) - \frac{1}{p}\sum_{u=1}^{t}\sum_{j=1}^{n}\widetilde{n}_{duj}^2 - \frac{1}{n}\sum_{u=1}^{t}\widetilde{r}_{du}^{\,2}/(p(p-1))\\ &\quad - \frac{1}{n}\sum_{u=1}^{t}\sum_{i=2}^{p}(l_{du(i-1)} - \widetilde{r}_{du}/(p-1))^2.\end{aligned}$$

The last term that is subtracted in this expression is 0 for a design that is uniform on the first $p-1$ periods. The other two terms that are subtracted are, again based on arguments as in Cheng and Wu (1980), minimized over $\Omega_{t,n,p}$ by designs as described in the theorems.

Observe that the designs in Theorems 3.1 and 3.2 are the same as those in Theorems 3.3 and 3.4, respectively.

The previous theorems would be rather meaningless if crossover designs as described in the theorems would not exist. For the designs in Theorems 3.1 and 3.3 it is readily seen that necessary conditions for their existence are (i) $p \equiv 0 \pmod t$ and $p/t \geqslant 2$, (ii) $n \equiv 0 \pmod t$, and (iii) $n(p-1) \equiv 0 \pmod{t^2}$. Therefore, we need that $n = \lambda_1 t^2$ and $p = \lambda_2 t$ for integers $\lambda_1 \geqslant 1$ and $\lambda_2 \geqslant 2$. These conditions are actually also sufficient.

THEOREM 3.5. *A uniform strongly balanced crossover design in $\Omega_{t,n,p}$ exists whenever $n = \lambda_1 t^2$ and $p = \lambda_2 t$ for integers $\lambda_1 \geqslant 1$ and $\lambda_2 \geqslant 2$.*

PROOF. It suffices to show that the designs exist for $n = t^2$; for $n = \lambda_1 t^2$ one can then simply juxtapose λ_1 copies of the design for $n = t^2$. Let $\boldsymbol{A}_t$ be an $\mathrm{OA}(t^2, 3, t, 2)$ with entries from $\{1, 2, \ldots, t\}$. Such an orthogonal array exists for any $t \geqslant 2$, and can easily be obtained from a Latin square of order t. Let $\boldsymbol{B}_t$ be the $\mathrm{OA}(t^2, 2, t, 2)$ obtained from $\boldsymbol{A}_t$ by deleting the third row in A_t. For $i \in \{1, 2, \ldots, t-1\}$ let $\boldsymbol{A}_i = \boldsymbol{A}_t + i$ and $\boldsymbol{B}_i = \boldsymbol{B}_t + i$, where i is added to each element of $\boldsymbol{A}_t$ or $\boldsymbol{B}_t$ modulo t. Let $\boldsymbol{A}$ and $\boldsymbol{B}$ be defined as the $3t \times t^2$ and $2t \times t^2$ arrays

$$\boldsymbol{A} = \begin{bmatrix} \boldsymbol{A}_1 \\ \boldsymbol{A}_2 \\ \vdots \\ \boldsymbol{A}_t \end{bmatrix}, \qquad \boldsymbol{B} = \begin{bmatrix} \boldsymbol{B}_1 \\ \boldsymbol{B}_2 \\ \vdots \\ \boldsymbol{B}_t \end{bmatrix}.$$

With $p = \lambda_2 t$, $\lambda_2 \geqslant 2$, write $\lambda_2 = 3\delta_1 + 2\delta_2$ for nonnegative integers δ_1 and δ_2. The $p \times t^2$ array

$$\begin{bmatrix} \boldsymbol{A}' & \cdots & \boldsymbol{A}' & \boldsymbol{B}' & \cdots & \boldsymbol{B}' \end{bmatrix}'$$

consisting of δ_1 copies of $\boldsymbol{A}$ and δ_2 copies of $\boldsymbol{B}$ is a uniform strongly balanced crossover design in $\Omega_{t,t^2,p}$.

As a small example of the construction in the previous proof, let $t = 3$, $n = 9$ and $p = 9$. Since $\lambda_2 = 3$, we take one copy of $\boldsymbol{A}$ to form the desired crossover design. For $\boldsymbol{A}_3$ we can take the orthogonal array

$$\begin{bmatrix} 1 & 1 & 1 & 2 & 2 & 2 & 3 & 3 & 3 \\ 1 & 2 & 3 & 1 & 2 & 3 & 1 & 2 & 3 \\ 1 & 2 & 3 & 2 & 3 & 1 & 3 & 1 & 2 \end{bmatrix}.$$

The desired uniform strongly balanced crossover design looks now as follows:

$$\begin{bmatrix} 2&2&2&3&3&3&1&1&1\\ 2&3&1&2&3&1&2&3&1\\ 2&3&1&3&1&2&1&2&3\\ 3&3&3&1&1&1&2&2&2\\ 3&1&2&3&1&2&3&1&2\\ 3&1&2&1&2&3&2&3&1\\ 1&1&1&2&2&2&3&3&3\\ 1&2&3&1&2&3&1&2&3\\ 1&2&3&2&3&1&3&1&2 \end{bmatrix}.$$

Necessary conditions for the existence of crossover designs as described in Theorems 3.2 and 3.4 are (i) $p-1 \equiv 0 \pmod t$ and (ii) $n \equiv 0 \pmod t$. Cheng and Wu (1980) observed a simple way to construct some of these designs from uniform balanced designs. We will therefore return to the construction of these designs in Subsection 3.4, after a discussion on the existence of uniform balanced designs. □

3.3. *Optimal and efficient crossover designs when $p \leqslant t$*

Theorems 3.1 through 3.4 all require that the number of periods is larger than the number of treatments. This may not be desirable or possible in many practical problems. The larger the number of periods, the longer the experiment will have to last.

For $p \leqslant t$ the knowledge about universally optimal designs for $\underline{\tau}$ in $\Omega_{t,n,p}$ is rather limited. This is in part because the strategy that resulted in Theorems 3.1 and 3.2 does not work for this case. When $p \leqslant t$ and $n \equiv 0 \pmod t$ we would have to take $n_{duj} \in \{0,1\}$ for all u and j in order to maximize $\mathrm{Tr}(\boldsymbol{C}_{d11})$; but this would also imply that $m_{duu}=0$ for all u, and hence, using (3.3), that $\boldsymbol{C}_{d12}$ cannot be equal to $\mathbf{0}_{t\times t}$.

A result on universally optimal designs for $\underline{\tau}$ in $\Omega_{t,n,p}$ is however available for the case $p=2$. Hedayat and Zhao (1990) prove the following result.

THEOREM 3.6. *Any crossover design $d^* \in \Omega_{t,n,2}$ with* (i) $\widetilde{r}_{d^*u} \equiv 0 \pmod t$ *for all $u \in \{1,2,\ldots,t\}$ and* (ii) $m_{d^*uv} = \widetilde{r}_{d^*v}/t$ *for all $u,v \in \{1,2,\ldots,t\}$ is universally optimal for $\underline{\tau}$ in $\Omega_{t,n,2}$.*

Theorem 3.6 implies in particular that a design that uses only treatment 1 in the first period and that is uniform on the second period is a universally optimal design for $\underline{\tau}$. (Of course, such a design would provide absolutely no information about $\underline{\gamma}$.) As another consequence of Theorem 3.6, a design in the form of an $\mathrm{OA}(n,2,t,2)$ is also universally optimal for $\underline{\tau}$. Designs as in Theorem 3.6 exist if and only if $n \equiv 0 \pmod t$. For more discussion on the case $p=2$ see also Carrière and Reinsel (1993); they consider a model in which subject effects are assumed to be random effects, and find that some of the designs in Theorem 3.6 are not very efficient under that model. Designs in the form of an orthogonal array, however, are still universally optimal for $\underline{\tau}$ under their model. For $t = 2$ and $t = 3$, in both cases with $n = t^2$, universally optimal designs in the form of orthogonal arrays are:

$$\begin{bmatrix} 1 & 1 & 2 & 2 \\ 1 & 2 & 1 & 2 \end{bmatrix} \quad \text{and} \quad \begin{bmatrix} 1 & 1 & 1 & 2 & 2 & 2 & 3 & 3 & 3 \\ 1 & 2 & 3 & 1 & 2 & 3 & 1 & 2 & 3 \end{bmatrix}.$$

Stufken (1991) studies crossover designs for $p \leqslant t$ and $n = \lambda t(t-1)$. The designs studied there exist if an $\mathrm{OA_I}(t(t-1), p, t, 2)$ exists (see the discussion after Lemmas 2.2 and 2.3). While these designs are only shown to be universally optimal for $\underline{\tau}$ in a small subclass of $\Omega_{t,n,p}$, all indications are that they are at least very efficient and possibly universally optimal for $\underline{\tau}$ in the entire class $\Omega_{t,n,p}$. The designs are constructed as follows. Let $\boldsymbol{A}$ be an $\mathrm{OA_I}(t(t-1), p, t, 2)$ and let the $p \times t(t-1)$ array $\boldsymbol{B}$ be obtained from $\boldsymbol{A}$ by repeating period $p-1$ in $\boldsymbol{A}$ as period p, thereby deleting the original pth period in $\boldsymbol{A}$. With $n = \lambda t(t-1)$, compute $\lambda/(t(p-1))$. Let δ be the nearest integer to this ratio, or either one of the two nearest integers if there is a tie. Form a crossover design in $\Omega_{t,n,p}$ by juxtaposing $\lambda - \delta$ copies of $\boldsymbol{A}$ and δ copies of $\boldsymbol{B}$. For more discussion and precise results concerning the optimality for $\underline{\tau}$ of these designs see Stufken (1991).

Observe that the designs suggested by Stufken (1991) are, as a result of Theorem 3.6, universally optimal for $\underline{\tau}$ in $\Omega_{t,n,p}$ if $p = 2$ and if λ/t is an integer.

The following result (Kunert, 1984; Stufken, 1991) identifies universally optimal designs for $\underline{\gamma}$ when $p \leqslant t$.

THEOREM 3.7. *A crossover design for which the first $p-1$ periods are in the form of an $\mathrm{OA_I}(n, p-1, t, 2)$ and for which the pth period is identical to period $p-1$ is universally optimal for $\underline{\gamma}$ in $\Omega_{t,n,p}$.*

The proof for this result is analogous to those for Theorems 3.3 and 3.4. As previously explained, the strategy employed in these proofs fails for identifying universally optimal designs for $\underline{\tau}$ when $p \leqslant t$, but, as shown by Theorem 3.7, remains successful for $\underline{\gamma}$.

The existence of crossover designs as described in Theorem 3.7 has already been discussed briefly after Lemmas 2.2 and 2.3. A necessary condition for the existence of such crossover designs if $p \geqslant 3$, but certainly not a sufficient condition, is that $n \equiv 0 \pmod{t(t-1)}$. If $n = t(t-1)$ the design exists if and only if there are $p-1$ mutually orthogonal Latin squares of order t. If $p = 2$, all that is needed in Theorem 3.7 is that the first period is uniform on the treatments and that the second period is identical to the first. (See also Hedayat and Zhao, 1990.) While these designs for $p = 2$ are universally optimal for $\underline{\gamma}$, they provide no information at all for $\underline{\tau}$.

Referring back to the discussion before Theorem 3.7, note that $(\lambda - \delta)/\delta$ is approximately equal to $t(p-1) - 1$, so that the suggested efficient crossover designs for $\underline{\tau}$ use many more copies of $\boldsymbol{A}$ than of $\boldsymbol{B}$. Theorem 3.7 implies that designs that are universally optimal for $\underline{\gamma}$ only use copies of $\boldsymbol{B}$.

3.4. *Optimal and efficient balanced crossover designs*

Uniform balanced designs form a class of designs that have been in use for a long time. Existence of these designs was already studied by Williams (1949). Optimality properties of these designs were first studied by Hedayat and Afsarinejad (1978), and later by Cheng and Wu (1980) and Kunert (1984). While uniform balanced designs are universally optimal for $\underline{\tau}$ and $\underline{\gamma}$ in subclasses of $\Omega_{t,n,p}$, generally this is not true in the entire class. The following results are due to Cheng and Wu (1980).

THEOREM 3.8. *A uniform balanced crossover design with* $n = \lambda_1 t$ *and* $p = \lambda_2 t$ *is universally optimal for* $\underline{\tau}$ *in the subclass of* $\Omega_{t,n,p}$ *that consists of all designs d with* (i) $m_{duu} = 0$ *for all u,* (ii) *d is uniform on the units, and* (iii) *d is uniform on the last period.*

THEOREM 3.9. *A uniform balanced crossover design with* $n = \lambda_1 t$ *and* $p = \lambda_2 t$ *is universally optimal for* $\underline{\gamma}$ *in the subclass of* $\Omega_{t,n,p}$ *that consists of all designs d with* (i) $m_{duu} = 0$ *for all u, and* (ii) *if* $\lambda_2 \geqslant 2$, *then* $\tilde{r}_{du}$ *is the same for all u.*

Compared to Theorems 3.1 through 3.4 and 3.7, the additional difficulty in Theorems 3.8 and 3.9 is that for a uniform balanced crossover design the matrix $\boldsymbol{C}_{d12}$ is not $\mathbf{0}_{t\times t}$. In applying Theorem 2.1, this makes it tremendously difficult to show that $\mathrm{Tr}(\boldsymbol{C}_d)$ or $\mathrm{Tr}(\widetilde{\boldsymbol{C}}_d)$ are maximized by uniform balanced crossover designs. This leads to the restrictions for the subclasses as defined in Theorems 3.8 and 3.9. And, indeed, these restrictions are essential because the two traces are in general not maximized over the entire class $\Omega_{t,n,p}$ by uniform balanced designs. For example, if $p = t$ and $n \equiv 0 \pmod{t(t-1)}$, the universally optimal designs in Theorem 3.7 have a larger value for $\mathrm{Tr}(\widetilde{\boldsymbol{C}}_d)$ than uniform balanced designs. Under the same conditions, the efficient designs for $\underline{\tau}$ suggested in Stufken (1991), as discussed in Subsection 3.3, have a larger value for $\mathrm{Tr}(\boldsymbol{C}_d)$ than uniform balanced designs.

Uniform balanced designs for $p = t$ have, as an immediate consequence of the definition, all m_{duu} values equal to 0. The designs that improve on uniform balanced designs suggest that this is not ideal. They also suggest that good designs for $\underline{\gamma}$ tend to have larger values for m_{duu} than do good designs for $\underline{\tau}$ (see Subsection 3.3). As a result, the efficiency of uniform balanced designs for $p = t$ is smaller for $\underline{\gamma}$ than it is for $\underline{\tau}$. For some further discussion see Kunert (1984), who also shows the following result.

THEOREM 3.10. *A uniform balanced crossover design with* $n = \lambda_1 t$ *and* $p = t$ *is universally optimal for* $\underline{\tau}$ *in* $\Omega_{t,n,p}$ *if* (i) $\lambda_1 = 1$ *and* $t \geqslant 3$, *or* (ii) $\lambda_1 = 2$ *and* $t \geqslant 6$.

Thus for small values of n, some uniform balanced design are universally optimal for $\underline{\tau}$ in $\Omega_{t,n,p}$. This is also consistent with the recommendations in Stufken (1991).

Necessary conditions for the existence of a uniform balanced design in $\Omega_{t,n,p}$ are (i) $n = \lambda_1 t$ for a positive integer λ_1, (ii) $p = \lambda_2 t$ for a positive integer λ_2, and (iii) $\lambda_1(\lambda_2 - 1) \equiv 0 \pmod{t-1}$. The first two conditions are needed for the uniformity, while the last condition is a consequence of $n(p-1) \equiv 0 \pmod{t(t-1)}$ which is required for the balance. We will first address the existence question of uniform balanced designs for $\lambda_1 = \lambda_2 = 1$ and will then return to the more general case.

A uniform balanced crossover design in $\Omega_{t,t,t}$ exists for all even values of t, but only for some odd values. A construction for even values of t is due to Williams (1949). Take the first column, say $\underline{a}_t$, to be $(1, t, 2, t-1, \ldots, t/2-1, t/2+2, t/2, t/2+1)'$, and define $\underline{a}_l$, $l = 1, 2, \ldots, t-1$, by $\underline{a}_l = \underline{a}_t + l$, where l is added to every entry of $\underline{a}_t$ modulo t. The $t \times t$ array

$$\boldsymbol{A}_t = \left[\, \underline{a}_1 \ \cdots \ \underline{a}_t \,\right] \tag{3.7}$$

is then the desired uniform balanced crossover design. For odd values of t, it is, among others, known that a uniform balanced design in $\Omega_{t,t,t}$ does not exist for $t = 3, 5, 7$ and does exist for $t = 9$. (See Afsarinejad, 1990.) The existence problem is thus much more difficult for odd t than for even t. What is known for every odd value of t is that a uniform balanced design exists in $\Omega_{t,2t,t}$. To see this, let $\underline{b}_t = (1, t, 2, t-1, \ldots, (t+5)/2, (t-1)/2, (t+3)/2, (t+1)/2)'$. For $l = 1, 2, \ldots, t-1$, let $\underline{b}_l = \underline{b}_t + l$, and for $l = 1, \ldots, t$, let $\underline{c}_l = -\underline{b}_l$, where computations are again modulo t. The $t \times 2t$ array

$$\boldsymbol{B}_t = \left[\, \underline{b}_1 \ \cdots \ \underline{b}_t \ \underline{c}_1 \ \cdots \ \underline{c}_t \,\right] \tag{3.8}$$

is then the desired uniform balanced crossover design.

More general results concerning the existence of uniform balanced crossover designs are formulated in the following theorem.

THEOREM 3.11. *The necessary conditions for the existence of a uniform balanced crossover design in $\Omega_{t,n,p}$ are also sufficient if t is even. For odd t, a uniform balanced crossover design exists in $\Omega_{t,n,p}$ if, in addition to the necessary conditions, n/t is even.*

PROOF. Let $\lambda_1 = n/t$ and $\lambda_2 = p/t$ be integers with $\lambda_1(\lambda_2 - 1) \equiv 0 \pmod{t-1}$. If t is even, let $\boldsymbol{D} = (d_{ij})$ be a $\lambda_2 \times \lambda_1$ matrix with entries from $\{1, 2, \ldots, t\}$ such that, when computed modulo t, among the $\lambda_1(\lambda_2 - 1)$ differences $d_{ij} - d_{(i-1)j}$, $i = 2, \ldots, \lambda_2$, $j = 1, \ldots, \lambda_1$, the numbers $0, 1, \ldots, t/2-1, t/2+1, \ldots, t-1$ all appear equally often. (Clearly, a matrix $\boldsymbol{D}$ as required exists for all positive integers λ_1, λ_2 and even t as in this proof.) With $\boldsymbol{A}_t$ as in (3.7), define $\boldsymbol{A}_l = \boldsymbol{A}_t + l$ for $l = 1, 2, \ldots, t-1$, where the addition is modulo t. Define a $p \times n$ array $\boldsymbol{A}$ by

$$\boldsymbol{A} = \begin{bmatrix} \boldsymbol{A}_{d_{11}} & \cdots & \boldsymbol{A}_{d_{1\lambda_1}} \\ \vdots & \ddots & \vdots \\ \boldsymbol{A}_{d_{\lambda_2 1}} & \cdots & \boldsymbol{A}_{d_{\lambda_2\lambda_1}} \end{bmatrix}.$$

Using the definition of $\boldsymbol{A}_l$ and the properties of $\boldsymbol{D}$, it is easily verified that $\boldsymbol{A}$ is a uniform balanced crossover design.

If t is odd and λ_1 is even, construct a $\lambda_2 \times \lambda_1/2$ matrix $\boldsymbol{D} = (d_{ij})$ with entries from $\{1, 2, \ldots, t\}$ such that, when computed modulo t, among the $(\lambda_1/2)(\lambda_2 - 1)$

differences $d_{ij} - d_{(i-1)j}$, $i = 2, \ldots, \lambda_2$, $j = 1, \ldots, \lambda_1/2$, the $(t-1)/2$ numbers $(t+3)/2, (t+7)/2, \ldots, (t-3)/2$, reduced modulo t where needed, all appear equally often. It is again obvious that such a matrix $\boldsymbol{D}$ exists. With $\boldsymbol{B}_t$ as in (3.8), let $\boldsymbol{B}_l$, $l = 1, 2, \ldots, t-1$, be defined as $\boldsymbol{B}_t + l$, with addition again modulo t. The $p \times n$ array $\boldsymbol{B}$ defined by

$$\boldsymbol{B} = \begin{bmatrix} \boldsymbol{B}_{d_{11}} & \cdots & \boldsymbol{B}_{d_{1(\lambda_1/2)}} \\ \vdots & \ddots & \vdots \\ \boldsymbol{B}_{d_{\lambda_2 1}} & \cdots & \boldsymbol{B}_{d_{\lambda_2(\lambda_1/2)}} \end{bmatrix}$$

is then a uniform balanced crossover design. Verification of this result is straightforward, and therefore omitted.

As pointed out earlier, the condition that λ_1 is even when t is odd is not a necessary condition for the existence of uniform balanced crossover designs with $n = \lambda_1 t$.

We now return to the existence of strongly balanced crossover designs that are uniform on the periods and uniform on the units when restricted to the first $p-1$ periods. Cheng and Wu (1980) observed that such strongly balanced designs can be constructed if $p = t+1$ from a uniform balanced design with $p = t$ by repeating the tth period in the balanced design as period $t+1$ in the strongly balanced design. The following is a simple extension of their idea. □

THEOREM 3.12. *Let $n = \lambda_1 t$ and $p = \lambda_2 t + 1$, where λ_1 and λ_2 are positive integers with $\lambda_1(\lambda_2 - 1) \equiv 0 \ (mod\ t-1)$. Then there exists a strongly balanced crossover design in $\Omega_{t,n,p}$ that is uniform on the periods and uniform on the units when restricted to the first $p-1$ periods if* (i) *t is even, or* (ii) *t is odd and λ_1 is even.*

PROOF. For even t, let $\boldsymbol{A}_l$ be defined as in the proof of Theorem 3.11. Use $\boldsymbol{A}_t$ and $\boldsymbol{A}_{t/2}$ to form a $\lambda_2 t \times \lambda_1 t$ array as follows. For the first t rows juxtapose λ_1 copies of $\boldsymbol{A}_t$. For the next t rows juxtapose λ_1 copies of $\boldsymbol{A}_{t/2}$. Continue like this, alternating between $\boldsymbol{A}_t$ and $\boldsymbol{A}_{t/2}$, until $\lambda_2 t$ rows are obtained. Add one more row to this array, identical to row $\lambda_2 t$. This gives the desired strongly balanced crossover design.

If t is odd and λ_1 is even, let $\boldsymbol{B}_t$ be defined as in (3.8). We start by making a $\lambda_2 t \times \lambda_1 t$ array as follows. For the first t rows juxtapose $\lambda_1/2$ copies of $\boldsymbol{B}_t$. For the next t rows do the same thing, but permute the columns within each copy of $\boldsymbol{B}_t$ such that the first row of this juxtaposition is identical to the last row in the previous juxtaposition. Continue like this until $\lambda_2 t$ rows are obtained, each time, through permutations of columns within copies of $\boldsymbol{B}_t$, making sure that periods lt and $lt+1$, $l = 1, \ldots, \lambda_2 - 1$ are identical. Add one more period to this array, identical to period $\lambda_2 t$. This gives the desired strongly balanced crossover design. □

4. Optimal crossover designs when errors are uncorrelated: The special case of two treatments

While the previous section presents results on optimal crossover designs for uncorrelated errors and for general values of the number of treatments t, the special case

$t = 2$ deserves some extra consideration. Firstly, it is quite common that the number of treatments to be compared in a crossover design is small, including $t = 2$. Secondly, while the results in the previous section for general t apply also for $t = 2$, the problem of finding universally optimal designs for $t = 2$ can be simplified considerably. The considerations in this section will also reveal that we have, typically, a choice among several optimal designs. This choice can be important when practical constraints may make one optimal design preferable over another.

The basic ideas in this section appear in Matthews (1990); for our notation and presentation we will however heavily rely on the ideas and development in the previous section.

The model assumptions in this section are the same as those in Section 3. The two treatments will be denoted by 1 and 2, and the 2^p possible $p \times 1$ vectors with entries from $\{1, 2\}$ represent the sequences of treatments that can be assigned to the subjects. A universally optimal design for $\underline{\tau}$ in $\Omega_{2,n,p}$ is now simply one that minimizes the variance of the best linear unbiased estimator of $\tau_2 - \tau_1$ over all designs in $\Omega_{2,n,p}$. An analogous interpretation holds for universally optimal designs for $\underline{\gamma}$.

For a $p \times 1$ treatment sequence $\underline{T}$ we will say that $3\underline{1}_p - \underline{T}$ is its *dual sequence*; the dual sequence of a treatment sequence $\underline{T}$ is thus obtained by permuting 1's and 2's in $\underline{T}$. We will call a design dual balanced if every sequence is used equally often as its dual. If a design is considered to be a probability measure on all possible sequences, then it is easily seen that among the optimal designs, whether for $\underline{\tau}$ or for $\underline{\gamma}$, there is one that is dual balanced (Matthews, 1987). Thus, there is no loss of generality by restricting attention to dual balanced designs.

The strategy that we will employ to find universally optimal designs is analogous to that in Section 3. But because $t = 2$, the information matrices $\boldsymbol{C}_d$ and $\widetilde{\boldsymbol{C}}_d$ are 2×2 matrices, and, since their column and row sums are 0, will automatically be c.s. Hence, to find a universally optimal design for $\underline{\tau}$ we will search among the dual balanced designs for a design d^* that maximizes $\mathrm{Tr}(\boldsymbol{C}_{d11})$ over all designs, and for which $\boldsymbol{C}_{d^*12} = \mathbf{0}_{2\times 2}$.

Since we can restrict our attention to dual balanced designs, the expression in (3.1) reduces to

$$\frac{np}{2} - \frac{1}{p}\sum_{j=1}^{n} n_{duj}^2,$$

which implies that $\mathrm{Tr}(\boldsymbol{C}_{d11})$ is maximized if and only if for each sequence in the design the difference in replication for the two treatments in the sequence is at most 1. Thus, for example, if $p = 4$ only the sequences that replicate treatment 1 twice should be included; if $p = 5$ only the sequences in which treatment 1 appears twice or thrice should be included.

How often a sequence should be used is determined by the requirement that $\boldsymbol{C}_{d^*12} = \mathbf{0}_{2\times 2}$. The expression in (3.3) reduces for a dual balanced design to

$$m_{duu} - \frac{1}{p}\sum_{j=1}^{n} n_{duj}\widetilde{n}_{duj}. \tag{4.1}$$

Table 4.1
Pairs of treatment sequences for $p = 4$:
Searching for optimal designs for $\underline{\tau}$

	Pair 1		Pair 2		Pair 3	
	1	2	1	2	1	2
	1	2	2	1	2	1
	2	1	1	2	2	1
	2	1	2	1	1	2
v_l	.5		−1.5		−.5	

We only need to evaluate this expression for $u = 1$, and require that it reduces to 0. Since the design will be dual balanced, we will evaluate (4.1) for $n = 2$ for each of the pairs of dual sequences, only using those pairs with sequences that replicate treatment 1 as close to $p/2$ as possible. If there are s such pairs of dual sequences, and if $v_1, \ldots, v_s$ denote the values in (4.1) for these s pairs, then a universally optimal design for $\underline{\tau}$ is obtained by using the lth pair $nf_l/2$ times, where the f_l's are nonnegative numbers that add to 1 and for which $\sum_l f_l v_l = 0$. The latter condition is required in order that $\boldsymbol{C}_{d12} = \boldsymbol{0}_{2\times 2}$.

Of course, $nf_l/2$ will only give integer values for all l for certain values of n; but that is inevitable with the approach in this section.

As an example, consider the case of $p = 4$. There are only three pairs of dual sequences for which treatment 1 is replicated twice in each sequence. These pairs, with the corresponding values for the v_l's, are presented in Table 4.1. Since computation of the v_l's is trivial, so is finding optimal designs.

With f_l, $l = 1, 2, 3$, denoting the proportion of time that pair l is used, any design with $\sum_l f_l v_l = 0$, or equivalently $f_1 - 3f_2 - f_3 = 0$, is optimal for $\underline{\tau}$ in $\Omega_{2,n,4}$. Thus, we can take $f_2 \in [0, 1/4]$ and $f_1 = 1/2 + f_2$, $f_3 = 1/2 - 2f_2$. A popular solution is the one with $f_2 = 0$, and $f_1 = f_3 = 1/2$, using pairs 1 and 3 equally often; but, for example, $f_1 = 3/4$, $f_2 = 1/4$ and $f_3 = 0$ is another solution that gives an optimal design for $\underline{\tau}$. The optimal designs for $\underline{\tau}$ in $\Omega_{2,8,4}$ corresponding to these two solutions are as follows:

$$\begin{bmatrix} 1 & 1 & 2 & 2 & 1 & 1 & 2 & 2 \\ 1 & 1 & 2 & 2 & 2 & 2 & 1 & 1 \\ 2 & 2 & 1 & 1 & 2 & 2 & 1 & 1 \\ 2 & 2 & 1 & 1 & 1 & 1 & 2 & 2 \end{bmatrix} \quad \begin{bmatrix} 1 & 1 & 1 & 2 & 2 & 2 & 1 & 2 \\ 1 & 1 & 1 & 2 & 2 & 2 & 2 & 1 \\ 2 & 2 & 2 & 1 & 1 & 1 & 1 & 2 \\ 2 & 2 & 2 & 1 & 1 & 1 & 2 & 1 \end{bmatrix}.$$

Finding optimal designs for $\underline{\gamma}$ can proceed in a similar way. The only difference is that the set of pairs of dual sequences that can possibly be included in an optimal design will be different. The pairs that need to be considered now are those in which each sequence replicates treatment 1 in the first $p - 1$ periods as close to $(p - 1)/2$ times as possible. Using only such pairs will lead to a design that maximizes $\mathrm{Tr}(\boldsymbol{C}_{d22})$.

Table 4.2
Pairs of treatment sequences for $p = 4$: Searching for optimal designs for $\underline{\gamma}$

	Pair 1	Pair 2	Pair 3	Pair 4	Pair 5	Pair 6
	1 2	1 2	1 2	1 2	1 2	1 2
	1 2	2 1	2 1	1 2	2 1	2 1
	2 1	1 2	2 1	2 1	1 2	2 1
	2 1	2 1	1 2	1 2	1 2	2 1
v_l	.5	−1.5	−.5	−.75	−.75	.25

Table 4.3
Pairs of sequences for optimal two-treatment designs, $4 \leqslant p \leqslant 6$

p	Pairs	v_l	$\underline{\tau}$ or $\underline{\gamma}$	p	Pairs	v_l	$\underline{\tau}$ or $\underline{\gamma}$
4	1 1 2 2	1/2	both	6	1 1 1 2 2 2	3/2	both
	1 2 1 2	−3/2	both		1 1 2 1 2 2	−1/2	both
	1 2 2 1	−1/2	both		1 1 2 2 1 2	−1/2	both
	1 1 2 1	−3/4	$\underline{\gamma}$ only		1 1 2 2 2 1	1/2	both
	1 2 1 1	−3/4	$\underline{\gamma}$ only		1 2 1 1 2 2	−1/2	both
	1 2 2 2	1/4	$\underline{\gamma}$ only		1 2 1 2 1 2	−5/2	both
					1 2 1 2 2 1	−3/2	both
5	1 1 2 2 1	0	both		1 2 2 1 1 2	−1/2	both
	1 1 2 2 2	1	both		1 2 2 1 2 1	−3/2	both
	1 2 1 2 1	−2	both		1 2 2 2 1 1	1/2	both
	1 2 1 2 2	−1	both		1 1 1 2 2 1	1/3	$\underline{\gamma}$ only
	1 2 2 1 1	0	both		1 1 2 1 2 1	−5/3	$\underline{\gamma}$ only
	1 2 2 1 2	−1	both		1 1 2 2 1 1	1/3	$\underline{\gamma}$ only
	1 1 1 2 2	4/5	$\underline{\tau}$ only		1 2 1 1 2 1	−5/3	$\underline{\gamma}$ only
	1 1 2 1 2	−6/5	$\underline{\tau}$ only		1 2 1 2 1 1	−5/3	$\underline{\gamma}$ only
	1 2 1 1 2	−6/5	$\underline{\tau}$ only		1 2 2 1 1 1	1/3	$\underline{\gamma}$ only
	1 2 2 2 1	−1/5	$\underline{\tau}$ only		1 1 2 2 2 2	4/3	$\underline{\gamma}$ only
					1 2 1 2 2 2	−2/3	$\underline{\gamma}$ only
					1 2 2 1 2 2	−2/3	$\underline{\gamma}$ only
					1 2 2 2 1 2	−2/3	$\underline{\gamma}$ only

Other steps are the same as in the case of finding optimal designs for $\underline{\tau}$. The values of v_l are again computed for each pair of sequences from (4.1).

As an example, Table 4.2 presents the pairs of dual sequences that can be used when searching for optimal designs for $\underline{\gamma}$ for $p = 4$.

With the nonnegative numbers $f_1, \ldots, f_6$ associated with the six pairs as before, and with $\sum_l f_l = 1$, any design with $2f_1 - 6f_2 - 2f_3 - 3f_4 - 3f_5 + f_6 = 0$ is optimal for $\underline{\gamma}$.

If p is even, every pair of sequences that can be used to construct optimal dual balanced designs for $\underline{\tau}$ can also be used in the construction of optimal designs for $\underline{\gamma}$. Thus, any optimal design for $\underline{\tau}$ constructed by the method in this section is also optimal for $\underline{\gamma}$ if p is even. If p is odd, the roles for $\underline{\tau}$ and $\underline{\gamma}$ in the above statement need to be reversed: any optimal design for $\underline{\gamma}$ constructed by the method in this section is also optimal for $\underline{\tau}$ if p is odd.

As already alluded to in Section 3, the strategy used in this section does not work for $\underline{\tau}$ when $p = 2$. For $p = 3$ the unique optimal design for $\underline{\tau}$ and $\underline{\gamma}$ uses only one pair of dual sequences, namely the one consisting of $(1, 2, 2)'$ and $(2, 1, 1)'$. Table 4.3 presents the sequences that can be used to construct optimal designs with $t = 2$ for $4 \leqslant p \leqslant 6$. To save space, each pair is represented by only one of the sequences in the pair. The values for v_l are given for each pair. Information whether a pair can be used in an optimal design for $\underline{\tau}$ or for $\underline{\gamma}$ is also provided in the table. Optimal designs are obtained by finding nonnegative f_l's that add to 1 for which $\sum_l f_l v_l = 0$, only using eligible pairs of sequences.

As an example, if we are interested in an optimal design for $\underline{\tau}$ when $p = 6$, we can only use the 10 pairs for $p = 6$ that are labeled as 'both' in Table 4.3. With

$$(f_1, \ldots, f_{10}) = (1/4, 1/4, 1/4, 0, 1/4, 0, 0, 0, 0, 0)$$

or

$$(f_1, \ldots, f_{10}) = (1/2, 0, 0, 0, 0, 0, 1/2, 0, 0, 0),$$

for example, we have two vectors of nonnegative constants that add to 1 and for which $\sum_l f_l v_l = 0$. Hence, designs corresponding to these vectors of constants are optimal for $\underline{\tau}$ (and also for $\underline{\gamma}$). For $n = 8$ these designs are

$$\begin{bmatrix} 1 & 2 & 1 & 2 & 1 & 2 & 1 & 2 \\ 1 & 2 & 1 & 2 & 1 & 2 & 2 & 1 \\ 1 & 2 & 2 & 1 & 2 & 1 & 1 & 2 \\ 2 & 1 & 1 & 2 & 2 & 1 & 1 & 2 \\ 2 & 1 & 2 & 1 & 1 & 2 & 2 & 1 \\ 2 & 1 & 2 & 1 & 2 & 1 & 2 & 1 \end{bmatrix} \quad \text{and} \quad \begin{bmatrix} 1 & 1 & 2 & 2 & 1 & 1 & 2 & 2 \\ 1 & 1 & 2 & 2 & 2 & 2 & 1 & 1 \\ 1 & 1 & 2 & 2 & 1 & 1 & 2 & 2 \\ 2 & 2 & 1 & 1 & 2 & 2 & 1 & 1 \\ 2 & 2 & 1 & 1 & 2 & 2 & 1 & 1 \\ 2 & 2 & 1 & 1 & 1 & 1 & 2 & 2 \end{bmatrix},$$

respectively.

5. Optimal crossover designs when errors are autocorrelated

While there are many ways in which the assumption of uncorrelated errors can be violated, in the context of crossover designs it is often reasonable to assume that error terms are only correlated if they correspond to measurements on the same subject. One way to model this is by assuming that the errors follow for each subject a first-order autoregressive process.

The assumption of a first-order autoregressive error process for each subject has been made by various authors. It is not only intuitively appealing, but also just shy of being too complicated to address the question of optimal design choice. Among those that have considered such a structure for the error variance-covariance matrix are Azzalini and Giovagnoli (1987), Berenblut and Webb (1974), Bora (1984, 1985), Kunert (1985, 1991), Laska and Meisner (1985) and Matthews (1987). The discussion in this section is based on model (2.1). The results in this section are due to Matthews (1987), but we will again rely heavily on notation and concepts introduced in the previous sections. For some additional results under the same model, resulting in efficient crossover designs for $t = 2$ and larger values of p, see Kunert (1991).

The first-order autoregressive error process for each of the subjects leads to a variance-covariance matrix $\sigma^2 \boldsymbol{V}$ for the error vector $\underline{\varepsilon}$ that is most conveniently expressed by using a Kronecker product:

$$\sigma^2 \boldsymbol{V} = (\sigma^2 \boldsymbol{I}_n) \otimes \boldsymbol{W},$$

where $\boldsymbol{W} = (w_{ij})$ is the $p \times p$ matrix defined by $w_{ij} = \rho^{|i-j|}/(1-\rho^2)$ and $\rho \in (-1, 1)$ is a constant.

With $\boldsymbol{U}$ as a $p \times p$ matrix such that $\boldsymbol{UWU}' = \boldsymbol{I}_p$, consider the non-singular data transformation $\underline{z} = (\boldsymbol{I}_n \otimes \boldsymbol{U})\underline{y}$. The model for $\underline{z}$ induced by (2.1) is then the Gauss–Markov model

$$\begin{aligned}\underline{z} = \mu(\underline{1}_n \otimes \boldsymbol{U}\underline{1}_p) + (\underline{1}_n \otimes \boldsymbol{U})\underline{\alpha} + (\boldsymbol{I}_n \otimes \boldsymbol{U}\underline{1}_p)\underline{\beta} + (\boldsymbol{I}_n \otimes \boldsymbol{U})\boldsymbol{X}_{d3}\underline{\tau} \\ + (\boldsymbol{I}_n \otimes \boldsymbol{U})\boldsymbol{X}_{d4}\underline{\gamma} + (\boldsymbol{I}_n \otimes \boldsymbol{U})\underline{\varepsilon},\end{aligned} \tag{5.1}$$

with $\mathrm{Var}(\underline{z}) = \sigma^2 \boldsymbol{I}_{pn}$ as a consequence of the choice of $\boldsymbol{U}$. The matrix $\boldsymbol{U}$ depends of course on ρ; while ρ will be an unknown in applications, to study its effect on optimal design choice we will for the time being assume that it is known.

Let $\boldsymbol{C}_d(\rho)$ and $\widetilde{\boldsymbol{C}}_d(\rho)$ denote the information matrices under model (5.1) for $\underline{\tau}$ and $\underline{\gamma}$, respectively. With

$$\boldsymbol{A} = (\boldsymbol{I}_n \otimes \boldsymbol{U}')(\boldsymbol{I}_{pn} - w(\,[\underline{1}_n \otimes \boldsymbol{U} \quad \boldsymbol{I}_n \otimes \boldsymbol{U}\underline{1}_p]\,))(\boldsymbol{I}_n \otimes \boldsymbol{U})$$

it is then easily seen that

$$\boldsymbol{C}_d(\rho) = \boldsymbol{X}'_{d3}\boldsymbol{A}\boldsymbol{X}_{d3} - \boldsymbol{X}'_{d3}\boldsymbol{A}\boldsymbol{X}_{d4}(\boldsymbol{X}'_{d4}\boldsymbol{A}\boldsymbol{X}_{d4})^{-}\boldsymbol{X}'_{d4}\boldsymbol{A}\boldsymbol{X}_{d3}, \tag{5.2}$$

and

$$\widetilde{\boldsymbol{C}}_d(\rho) = \boldsymbol{X}'_{d4}\boldsymbol{A}\boldsymbol{X}_{d4} - \boldsymbol{X}'_{d4}\boldsymbol{A}\boldsymbol{X}_{d3}(\boldsymbol{X}'_{d3}\boldsymbol{A}\boldsymbol{X}_{d3})^{-}\boldsymbol{X}'_{d3}\boldsymbol{A}\boldsymbol{X}_{d4}. \tag{5.3}$$

By observing that

$$\begin{aligned}&w(\,[\underline{1}_n \otimes \boldsymbol{U} \quad \boldsymbol{I}_n \otimes \boldsymbol{U}\underline{1}_p]\,) \\ &= [\underline{1}_n \otimes \boldsymbol{U} \quad \boldsymbol{I}_n \otimes \boldsymbol{U}\underline{1}_p] \begin{bmatrix} n\boldsymbol{U}'\boldsymbol{U} & \boldsymbol{U}'\boldsymbol{U}\underline{1}_p\underline{1}'_n \\ \underline{1}_n\underline{1}'_p\boldsymbol{U}'\boldsymbol{U} & (\underline{1}'_p\boldsymbol{U}'\boldsymbol{U}\underline{1}_p)\boldsymbol{I}_n \end{bmatrix}^{-} \begin{bmatrix} \underline{1}'_n \otimes \boldsymbol{U}' \\ \boldsymbol{I}_n \otimes \underline{1}'_p\boldsymbol{U}' \end{bmatrix}\end{aligned}$$

$$= \left[\underline{1}_n \otimes \boldsymbol{U} \quad \boldsymbol{I}_n \otimes \boldsymbol{U}\underline{1}_p\right] \begin{bmatrix} \frac{1}{n}(\boldsymbol{W} - \underline{1}_p\underline{1}_p'/(\underline{1}_p'\boldsymbol{U}'\boldsymbol{U}\underline{1}_p) & \mathbf{0}_{p\times n} \\ \mathbf{0}_{n\times p} & \boldsymbol{I}_n/(\underline{1}_p'\boldsymbol{U}'\boldsymbol{U}\underline{1}_p) \end{bmatrix}$$

$$\times \begin{bmatrix} \underline{1}_n' \otimes \boldsymbol{U}' \\ \boldsymbol{I}_n \otimes \underline{1}_p'\boldsymbol{U}' \end{bmatrix}$$

$$= \boldsymbol{I}_n \otimes \left(\frac{\boldsymbol{U}\underline{1}_p\underline{1}_p'\boldsymbol{U}'}{\underline{1}_p'\boldsymbol{U}'\boldsymbol{U}\underline{1}_p}\right) + \left(\frac{1}{n}\underline{1}_n\underline{1}_n'\right) \otimes \left(\boldsymbol{I}_p - \frac{\boldsymbol{U}\underline{1}_p\underline{1}_p'\boldsymbol{U}'}{\underline{1}_p'\boldsymbol{U}'\boldsymbol{U}\underline{1}_p}\right),$$

it is easily seen that

$$\boldsymbol{A} = \left(\boldsymbol{I}_n - \frac{1}{n}\underline{1}_n\underline{1}_n'\right) \otimes \left(\boldsymbol{U}'(\boldsymbol{I}_p - w(\boldsymbol{U}\underline{1}_p))\boldsymbol{U}\right). \tag{5.4}$$

If $t = 2$, and it is to this case that we will restrict our attention henceforth, we can, as in the previous section, assume without loss of generality that the competing designs are dual balanced. In that case $\sum_{j=1}^{n} \boldsymbol{X}_{d3j} = \frac{n}{2}\underline{1}_p\underline{1}_2'$. Therefore,

$$\left[\left(\frac{1}{n}\underline{1}_n\underline{1}_n'\right) \otimes \left(\boldsymbol{U}'(\boldsymbol{I}_p - w(\boldsymbol{U}\underline{1}_p))\boldsymbol{U}\right)\right] \boldsymbol{X}_{d3}$$

$$= \frac{1}{n}\underline{1}_n \otimes \left[(\boldsymbol{U}'(\boldsymbol{I}_p - w(\boldsymbol{U}\underline{1}_p))\boldsymbol{U})\left(\sum_{j=1}^{n} \boldsymbol{X}_{d3j}\right)\right]$$

$$= \frac{1}{n}\underline{1}_n \otimes \left[(\boldsymbol{U}'(\boldsymbol{I}_p - w(\boldsymbol{U}\underline{1}_p))\boldsymbol{U})\left(\frac{n}{2}\underline{1}_p\underline{1}_2'\right)\right] = \mathbf{0}_{pn\times 2}. \tag{5.5}$$

Observe that $\boldsymbol{U}'\boldsymbol{U} = \boldsymbol{W}^{-1}$, and that

$$\boldsymbol{W}^{-1} = \begin{bmatrix} 1 & -\rho & 0 & \cdots & 0 & 0 & 0 \\ -\rho & 1+\rho^2 & -\rho & \cdots & 0 & 0 & 0 \\ 0 & -\rho & 1+\rho^2 & \cdots & 0 & 0 & 0 \\ \vdots & \vdots & \vdots & \ddots & \vdots & \vdots & \vdots \\ 0 & 0 & 0 & \cdots & 1+\rho^2 & -\rho & 0 \\ 0 & 0 & 0 & \cdots & -\rho & 1+\rho^2 & -\rho \\ 0 & 0 & 0 & \cdots & 0 & -\rho & 1 \end{bmatrix}.$$

From (5.4) and (5.5) it follows now easily that

$$\boldsymbol{X}_{d3}'\boldsymbol{A}\boldsymbol{X}_{d3} = \boldsymbol{X}_{d3}'\left(\boldsymbol{I}_n \otimes (\boldsymbol{U}'(\boldsymbol{I}_p - w(\boldsymbol{U}\underline{1}_p))\boldsymbol{U})\right)\boldsymbol{X}_{d3}$$

$$= \sum_{j=1}^{n} \boldsymbol{X}_{d3j}'\left(\boldsymbol{W}^{-1} - \frac{\boldsymbol{W}^{-1}\underline{1}_p\underline{1}_p'\boldsymbol{W}^{-1}}{\underline{1}_p'\boldsymbol{W}^{-1}\underline{1}_p}\right)\boldsymbol{X}_{d3j}. \tag{5.6}$$

It follows in a similar way that

$$X'_{d3}AX_{d4} = \sum_{j=1}^{n} X'_{d3j}\left(W^{-1} - \frac{W^{-1}\underline{1}_p\underline{1}'_pW^{-1}}{\underline{1}'_pW^{-1}\underline{1}_p}\right)X_{d4j}. \tag{5.7}$$

Furthermore, for a dual balanced design we have that $\sum_{j=1}^{n} X_{d4j} = \frac{n}{2}L\underline{1}_p\underline{1}'_2$, where L is as defined in Section 2. Using this, we obtain that

$$\begin{aligned}
&X'_{d4}\left[\left(\frac{1}{n}\underline{1}_n\underline{1}'_n\right)\otimes\left(U'(I_p - w(U\underline{1}_p))U\right)\right]X_{d4}\\
&= \left(\sum_{j=1}^{n} X'_{d4j}\right)\left(U'(I_p - w(U\underline{1}_p))U\right)\left(\sum_{j=1}^{n} X_{d4j}\right)\\
&= \frac{n}{4}\left(\underline{1}'_pL'W^{-1}L\underline{1}_p - \frac{(\underline{1}'_pW^{-1}L\underline{1}_p)^2}{\underline{1}'_pW^{-1}\underline{1}_p}\right)\underline{1}_2\underline{1}'_2.
\end{aligned}$$

With (5.4) this results in

$$\begin{aligned}
X'_{d4}AX_{d4} = &\sum_{j=1}^{n} X'_{d4j}\left(W^{-1} - \frac{W^{-1}\underline{1}_p\underline{1}'_pW^{-1}}{\underline{1}'_pW^{-1}\underline{1}_p}\right)X_{d4j}\\
&- \frac{n}{4}\left(\underline{1}'_pL'W^{-1}L\underline{1}_p - \frac{(\underline{1}'_pW^{-1}L\underline{1}_p)^2}{\underline{1}'_pW^{-1}\underline{1}_p}\right)\underline{1}_2\underline{1}'_2.
\end{aligned}$$

The matrices $X'_{d3}AX_{d3}$, $X'_{d3}AX_{d4}$ and $X'_{d4}AX_{d4}$ are all 2×2 matrices with zero row and column sums. Hence it suffices to compute just one entry of each of these matrices to know the entire matrices. To this purpose, let $\underline{x}_j$ denote the first column of X_{d3j}. For some j', a subject who receives the dual of the sequence that subject j receives, we have that $\underline{x}_{j'} = \underline{1}_p - \underline{x}_j$. Moreover, we have that the first column of X_{d4j} is $L\underline{x}_j$. The contribution that these two subjects make to the entry in position (1,1) of the matrix $X'_{d3}AX_{d3}$ is thus equal to

$$\begin{aligned}
&\underline{x}'_jW^{-1}\underline{x}_j - \frac{(\underline{x}'_jW^{-1}\underline{1}_p)^2}{\underline{1}'_pW^{-1}\underline{1}_p} + (\underline{1}_p - \underline{x}_j)'W^{-1}(\underline{1}_p - \underline{x}_j)\\
&\quad - \frac{((\underline{1}_p - \underline{x}_j)'W^{-1}\underline{1}_p)^2}{\underline{1}'_pW^{-1}\underline{1}_p}\\
&= 2\left(\underline{x}'_jW^{-1}\underline{x}_j - \frac{(\underline{x}'_jW^{-1}\underline{1}_p)^2}{\underline{1}'_pW^{-1}\underline{1}_p}\right).
\end{aligned} \tag{5.8}$$

If we use f_l, as in Section 4, to denote the proportion of time that the lth pair of dual sequences is used, where f_l's are nonnegative numbers that add to 1, and if we use s to denote the number of such pairs ($s = 2^{p-1}$), then we obtain from (5.8) that the entry in position $(1,1)$ of the matrix $\boldsymbol{X}'_{d3}\boldsymbol{A}\boldsymbol{X}_{d3}$ is given by

$$n\sum_{l=1}^{s} f_l\left(\underline{x}_l'\boldsymbol{W}^{-1}\underline{x}_l - \frac{(\underline{x}_l'\boldsymbol{W}^{-1}\underline{1}_p)^2}{\underline{1}_p'\boldsymbol{W}^{-1}\underline{1}_p}\right), \tag{5.9}$$

where we have made a slight change of notation by using $\underline{x}_l$ to denote the column that would appear as first column in a matrix $\boldsymbol{X}_{d3j}$ if subject j would receive the sequence in pair l that starts with treatment 1; we will refer to this sequence as the first sequence in the pair.

In a similar way we obtain the entries in position $(1,1)$ for the matrices $\boldsymbol{X}'_{d3}\boldsymbol{A}\boldsymbol{X}_{d4}$ and $\boldsymbol{X}'_{d4}\boldsymbol{A}\boldsymbol{X}_{d4}$. These are

$$\begin{aligned}\frac{n}{2}\sum_{l=1}^{s} f_l\Bigg[2\left(\underline{x}_l'\boldsymbol{W}^{-1}\boldsymbol{L}\underline{x}_l - \frac{(\underline{x}_l'\boldsymbol{W}^{-1}\underline{1}_p)(\underline{1}_p'\boldsymbol{W}^{-1}\boldsymbol{L}\underline{x}_l)}{\underline{1}_p'\boldsymbol{W}^{-1}\underline{1}_p}\right)\\ - \underline{x}_l'\boldsymbol{W}^{-1}\boldsymbol{L}\underline{1}_p + \frac{(\underline{x}_l'\boldsymbol{W}^{-1}\underline{1}_p)(\underline{1}_p'\boldsymbol{W}^{-1}\boldsymbol{L}\underline{1}_p)}{\underline{1}_p'\boldsymbol{W}^{-1}\underline{1}_p}\Bigg]\end{aligned} \tag{5.10}$$

and

$$\begin{aligned}\frac{n}{4}\left(\underline{1}_p'\boldsymbol{L}'\boldsymbol{W}^{-1}\boldsymbol{L}\underline{1}_p - \frac{(\underline{1}_p'\boldsymbol{W}^{-1}\boldsymbol{L}\underline{1}_p)^2}{\underline{1}_p'\boldsymbol{W}^{-1}\underline{1}_p}\right)\\ + n\sum_{l=1}^{s} f_l\Bigg[-\underline{x}_l'\boldsymbol{L}'\boldsymbol{W}^{-1}\boldsymbol{L}(\underline{1}_p - \underline{x}_l)\\ + \frac{(\underline{1}_p'\boldsymbol{W}^{-1}\boldsymbol{L}\underline{x}_l)(\underline{1}_p'\boldsymbol{W}^{-1}\boldsymbol{L}(\underline{1}_p - \underline{x}_l))}{\underline{1}_p'\boldsymbol{W}^{-1}\underline{1}_p}\Bigg],\end{aligned} \tag{5.11}$$

respectively.

Equations (5.9), (5.10) and (5.11) all depend on scalars of the form $\underline{z}_1'\boldsymbol{W}^{-1}\underline{z}_2$, for some vectors $\underline{z}_1$ and $\underline{z}_2$. These expressions are perhaps easiest evaluated by defining $\boldsymbol{N}$ as the $p \times p$ matrix with a 1 in positions (i,i) for $i \in \{2,\ldots,p-1\}$, and a 0 elsewhere, and by observing that

$$\boldsymbol{W}^{-1} = \boldsymbol{I}_p + \rho^2\boldsymbol{N} - \rho\boldsymbol{L} - \rho\boldsymbol{L}'.$$

Using this expression, all the relevant scalars of the form $\underline{z}_1'\boldsymbol{W}^{-1}\underline{z}_2$ can be further evaluated and the results are presented in Table 5.1. The notation used in this table is as follows: By r_l we denote the replication of treatment 1 for the first sequence in pair l; by $r_{(i)l}$ and $r_{(i,i')l}$ we denote this same replication after deleting period i and periods i and i', respectively; by m_l we denote the number of times that treatment 1 is immediately preceded by itself for the first sequence in pair l; by $m_{(p)l}$ we denote

Table 5.1
Searching for optimal designs for $t = 2$ when errors are autocorrelated

No.	Expression	Evaluation
1	$\underline{1}_p' \boldsymbol{W}^{-1} \underline{1}_p$	$(1-\rho)(p-\rho(p-2))$
2	$\underline{1}_p' \boldsymbol{W}^{-1} \boldsymbol{L} \underline{1}_p$	$(1-\rho)(p-1-\rho(p-2))$
3	$\underline{1}_p' \boldsymbol{L}' \boldsymbol{W}^{-1} \boldsymbol{L} \underline{1}_p$	$(1-\rho)^2(p-2)+1$
4	$\underline{1}_p' \boldsymbol{W}^{-1} \underline{x}_l$	$r_l + \rho^2 r_{(1,p)l} - \rho(r_{(1)l} + r_{(p)l})$
5	$\underline{1}_p' \boldsymbol{W}^{-1} \boldsymbol{L} \underline{x}_l$	$(1-\rho)(r_{(p)l} - \rho r_{(p-1,p)l})$
6	$\underline{x}_l' \boldsymbol{W}^{-1} \boldsymbol{L} \underline{1}_p$	$r_{(1)l} + \rho^2 r_{(1,p)l} - \rho(r_{(1,2)l} + r_{(p)l})$
7	$\underline{x}_l' \boldsymbol{L}' \boldsymbol{W}^{-1} \boldsymbol{L} \underline{1}_p$	$r_{(p)l} + \rho^2 r_{(p-1,p)l} - \rho(r_{(1,p)l} + r_{(p-1,p)l})$
8	$\underline{x}_l' \boldsymbol{W}^{-1} \underline{x}_l$	$r_l + \rho^2 r_{(1,p)l} - 2\rho m_l$
9	$\underline{x}_l' \boldsymbol{W}^{-1} \boldsymbol{L} \underline{x}_l$	$m_l + \rho^2 m_{(p)l} - \rho(m_l^{(2)} + r_{(p)l})$
10	$\underline{x}_l' \boldsymbol{L}' \boldsymbol{W}^{-1} \boldsymbol{L} \underline{x}_l$	$r_{(p)l} + \rho^2 r_{(p-1,p)l} - 2\rho m_{(p)l}$

a similar quantity, but now counted after deleting period p; finally, by $m_l^{(2)}$ we denote the number of periods i, $i \in \{1, \ldots, p-2\}$, such that treatment 1 appears in period i and period $i+2$ in the first sequence of pair l. As an example of the computation,

$$\begin{aligned}\underline{1}_p' \boldsymbol{W}^{-1} \boldsymbol{L} \underline{x}_l &= \underline{1}_p' \boldsymbol{L} \underline{x}_l + \rho^2 \underline{1}_p' \boldsymbol{N} \boldsymbol{L} \underline{x}_l - \rho \underline{1}_p' \boldsymbol{L}^2 \underline{x}_l - \rho \underline{1}_p' \boldsymbol{L}' \boldsymbol{L} \underline{x}_l \\ &= r_{(p)l} + \rho^2 r_{(p-1,p)l} - \rho r_{(p-1,p)l} - \rho r_{(p)l} \\ &= (1-\rho)(r_{(p)l} - \rho r_{(p-1,p)l}).\end{aligned}$$

Other expressions in Table 5.1 are evaluated in a similar way.

For $p = 3$ we will demonstrate how the preceding computations can be used to obtain optimal designs. First evaluate, as a function of ρ, the expressions numbered as 1 through 3 in Table 5.1 for $p = 3$. These expressions do not depend on the choice of design. Then, for each possible pair of treatment sequences, or equivalently for each 3×1 vector $\underline{x}_l$ with the first entry 1 and all other entries 0 or 1, evaluate the expressions numbered as 4 through 10 in Table 5.1. The results of this for $p = 3$ are given in Table 5.2, with each possible pair represented by the first sequence in the pair. The numbering in Table 5.2 corresponds to that in Table 5.1.

With the values in Table 5.2, the expressions in (5.9), (5.10) and (5.11) can be evaluated, all as a function of ρ and $f_1, \ldots, f_4$. If we call these three expressions v_{33}, v_{34} and v_{44}, respectively, then we should maximize

$$v_{33} - \frac{v_{34}^2}{v_{44}} \tag{5.12}$$

Table 5.2
Searching for optimal designs with $t = 2$ and $p = 3$

	Pairs			
No.	111	112	121	122
1	$(1-\rho)(3-\rho)$	$(1-\rho)(3-\rho)$	$(1-\rho)(3-\rho)$	$(1-\rho)(3-\rho)$
2	$(1-\rho)(2-\rho)$	$(1-\rho)(2-\rho)$	$(1-\rho)(2-\rho)$	$(1-\rho)(2-\rho)$
3	$(1-\rho)^2+1$	$(1-\rho)^2+1$	$(1-\rho)^2+1$	$(1-\rho)^2+1$
4	$(1-\rho)(3-\rho)$	$(1-\rho)(2-\rho)$	$2(1-\rho)$	$1-\rho$
5	$(1-\rho)(2-\rho)$	$(1-\rho)(2-\rho)$	$(1-\rho)^2$	$(1-\rho)^2$
6	$(1-\rho)(2-\rho)$	$(1-\rho)^2$	$1-2\rho$	$-\rho$
7	$(1-\rho)^2+1$	$(1-\rho)^2+1$	$1+\rho^2-\rho$	$1+\rho^2-\rho$
8	$(1-\rho)(3-\rho)$	$(1-\rho)^2+1$	2	1
9	$(1-\rho)(2-\rho)$	$(1-\rho)^2$	-2ρ	$-\rho$
10	$(1-\rho)^2+1$	$(1-\rho)^2+1$	$1+\rho^2$	$1+\rho^2$

over $f_1, \ldots, f_4$ to obtain proportions for an optimal design for $\underline{\tau}$, and

$$v_{44} - \frac{v_{34}^2}{v_{33}} \tag{5.13}$$

to obtain proportions for an optimal design for $\underline{\gamma}$.

The results in Tables 5.1 and 5.2 and equations (5.9), (5.10) and (5.11) give that

$$v_{33} = \frac{2n}{3-\rho}\left[f_2 + f_4 + (1+\rho)f_3\right],$$

$$v_{34} = \frac{n}{2(3-\rho)}\left[-(1-\rho)f_2 - 3(1+\rho)f_3 - 2\rho f_4\right],$$

and

$$v_{44} = \frac{n}{2(3-\rho)} + \frac{n}{3-\rho}\left[(1+\rho)(f_3+f_4)\right].$$

The expressions in (5.12) and (5.13) are then equal to

$$\frac{n}{2(3-\rho)}\left[4f_2 + 4f_4 + 4(1+\rho)f_3 - \frac{((1-\rho)f_2 + 3(1+\rho)f_3 + 2\rho f_4)^2}{1+2(1+\rho)(f_3+f_4)}\right]$$

and

$$\frac{n}{2(3-\rho)}\left[1 + 2(1+\rho)(f_3+f_4) - \frac{((1-\rho)f_2 + 3(1+\rho)f_3 + 2\rho f_4)^2}{4f_2 + 4f_4 + 4(1+\rho)f_3}\right],$$

respectively. To obtain optimal designs, for any given value of ρ it is rather simple to maximize the above quantities over f_1, f_2, f_3 and f_4, subject to the f_i's being nonnegative and adding to 1; in fact, explicit formulas for optimal f_l's as a function of ρ can be found for this simple case of $p = 3$ (see Matthews, 1987).

While the procedure described in this section works nicely for small values of p, it becomes quite cumbersome for larger values of p. See Kunert (1991) for efficient designs in that case.

Matthews gives optimal values for the f_l's as a function of ρ for the cases $p = 3$ and $p = 4$, both for $\underline{\tau}$ and for $\underline{\gamma}$.

One important final consideration is that we will typically not know what the correct value for ρ is. In such case, instead of choosing a design that is optimal for a particular value of ρ we may want to choose a design that is at least efficient over an interval that most likely contains the true value for ρ. The efficiency of a dual balanced design may be computed, for various values of ρ in such an interval, as the ratio of quantities as in (5.12) and (5.13), using the value for the design under consideration in the numerator and the optimal value in the denominator. Matthews (1987) shows that some of the commonly used designs for $p = 3$ and $p = 4$ have excellent efficiencies for $\underline{\tau}$ or for $\underline{\gamma}$ over large intervals for ρ, but that some other designs can have extremely poor efficiencies.

6. Summary and discussion

One of the early incentives for using uniform balanced and strongly balanced crossover designs was the relative simplicity of analysis of data obtained by using these designs, based on a model as in Section 3. While this is no longer a strong incentive in our era, Theorems 3.1 through 3.4 show that the classes of balanced and strongly balanced crossover designs remain appealing. Section 3 also identifies optimal and efficient designs for the case $p \leqslant t$ based on orthogonal arrays of Type I.

The case $t = 2$ has received special attention in the literature, for good reason, and results on optimal designs for this case under different assumptions for the error variance-covariance matrix are presented in Sections 4 and 5.

The many recent review articles, books, and book chapters on crossover designs are a reflection of the increased attention that these designs have received since the late seventies. Research continues to flourish, and several interesting and challenging problems require additional attention. Following are a few thoughts on directions for further research.

The results in the previous sections provide universally optimal designs for $\underline{\tau}$ or $\underline{\gamma}$ for selected values of t, n and p. The severity of the constraints on t, n and p is in part a problem that is inextricably tied to exact design theory, but in part also due to the use of universal optimality as the optimality criterion. For many values of t, n and p a universally optimal design will simply not exist; less stringent optimality criteria could be considered for such cases. When theoretical results are harder to obtain due to the increased mathematical difficulties under such an approach, development of efficient algorithms or general guidelines for obtaining good designs under specified optimality criteria for arbitrary values of t, n and p would be quite useful.

As alluded to previously, results from optimal design theory should be used as a guide towards obtaining good designs or to avoid use of extremely poor designs. It would be foolish to insist on using optimal designs only; the specified model under which an optimal design for $\underline{\tau}$ or $\underline{\gamma}$ is optimal represents at best an approximation of the true relationship between the response variable and the independent variables. In the worst case, the model may be entirely misspecified, and the optimal design under this misspecified model may be hopelessly inefficient under a better model. It is therefore important to identify designs and classes of designs that are highly efficient under various models. Alternative models to the one used in Section 3 could include models with some interactions, with random subject effects, or with different error variance-covariance matrices. While there are some results pertaining to identification of designs that possess a certain robustness to model misspecification (see, for example, the discussions concerning $p = 2$ and $t = 2$ in Sections 3 and 5, respectively), this direction of research is still in its infancy.

Extension of results for $t = 2$ as discussed in Sections 4 and 5 to other values of t is also desirable. Further, attention to designs for small values of p, possibly through additional considerations for $p \leqslant t$, is required. Some directions of research not addressed in this chapter include designs for special treatment structures (for example when one of the treatments is a control treatment or when the treatments possess a factorial structure) and optimal crossover designs for binary or discrete response variables. These are also areas that require further attention.

Finally, a question of a more philosophical nature that has recently been raised is whether the models that were originally developed for use in areas of agricultural sciences are still adequate in the many different fields where they are used nowadays. Are the traditional model assumptions reasonable in all these fields, or are they merely used because nobody has seriously thought about more appropriate alternatives?

Acknowledgement

Research for this chapter was partially supported by NSF grant DMS-9504882.

References

Afsarinejad, K. (1990). Repeated measurements designs – A review. *Comm. Statist. Theory Methods* **19**, 3985–4028.

Azzalini, A. and A. Giovagnoli (1987). Some optimal designs for repeated measurements with autoregressive errors. *Biometrika* **74**, 725–734.

Barker, N., R. J. Hews, A. Huitson and J. Poloniecki (1982). The two period cross over trial. *Bull. Appl. Statist.* **9**, 67–116.

Berenblut, I. I. and G. I. Webb (1974). Experimental design in the presence of autocorrelated errors. *Biometrika* **61**, 427–437.

Bishop, S. H. and B. Jones (1984). A review of higher order crossover designs. *J. Appl. Statist.* **11**, 29–50.

Bora, A. C. (1984). Change over designs with errors following a first order autoregressive process. *Austral. J. Statist.* **26**, 179–188.

Bora, A. C. (1985). Change over design with first order residual effects and errors following a first order autoregressive process. *The Statistician* **34**, 161–173.

Carrière, K. C. and G. C. Reinsel (1992). Investigation of dual-balanced crossover designs for two treatments. *Biometrics* **48**, 1157–1164.

Carrière, K. C. and G. C. Reinsel (1993). Optimal two-period repeated measurement designs with two or more treatments. *Biometrika* **80**, 924–929.

Cheng, C.-S. and C.-F. Wu (1980). Balanced repeated measurements designs. *Ann. Statist.* **8**, 1272–1283; correction **11** (1983), 349.

Cochran, W. G. and G. M. Cox (1957). *Experimental Designs*, 2nd edn. Wiley, New York.

Hedayat, A. and K. Afsarinejad (1978). Repeated measurements designs, II. *Ann. Statist.* **6**, 619–628.

Hedayat, A. S., N. J. A. Sloane and J. Stufken (1996). *Orthogonal Arrays: Theory and Applications*, to appear.

Hedayat, A. S. and W. Zhao (1990). Optimal two-period repeated measurements designs. *Ann. Statist.* **18**, 1805–1816.

Jones, B. and M. G. Kenward (1989). *Design and Analysis of Cross-over Trials*. Chapman and Hall, London.

Kershner, R. P. and W. T. Federer (1981). Two-treatment crossover designs for estimating a variety of effects. *J. Amer. Statist. Assoc.* **76**, 612–619.

Kiefer, J. (1975). Construction and optimality of generalized Youden designs. In: J. N. Srivastava, ed., *A Survey of Statistical Design and Linear Models*. North-Holland, Amsterdam, 333–353.

Kunert, J. (1984). Optimality of balanced uniform repeated measurements designs. *Ann. Statist.* **12**, 1006–1017.

Kunert, J. (1985). Optimal repeated measurements designs for correlated observations and analysis by weighted least squares. *Biometrika* **72**, 375–389.

Kunert, J. (1991). Cross-over designs for two treatments and correlated errors. *Biometrika* **78**, 315–324.

Laska, E. M. and M. Meisner (1985). A variational approach to optimal two-treatment crossover designs: Application to carryover-effect models. *J. Amer. Statist. Assoc.* **80**, 704–710.

Magda, G. C. (1980). Circular balanced repeated measurements designs. *Comm. Statist. Theory Methods* **A9**, 1901–1918.

Matthews, J. N. S. (1987). Optimal crossover designs for the comparison of two treatments in the presence of carryover effects and autocorrelated errors. *Biometrika* **74**, 311–320; correction **75** (1988), 396.

Matthews, J. N. S. (1988). Recent developments in crossover designs. *Internat. Statist. Rev.* **56**, 117–127.

Matthews, J. N. S. (1990). Optimal dual-balanced two treatment crossover designs. *Sankhya Ser. B* **52**, 332–337.

Matthews, J. N. S. (1994). Modelling and optimality in the design of crossover studies for medical applications. *J. Statist. Plann. Inference* **42**, 89–108.

Mukhopadhyay, A. C. and R. Saha (1983). Repeated measurement designs. *Calcutta Statist. Assoc. Bull.* **32**, 153–168.

Rao, C. R. (1947). Factorial experiments derivable from combinatorial arrangements of arrays. *J. Roy. Statist. Soc.* (Supplement) **9**, 128–139.

Rao, C. R. (1961). Combinatorial arrangements analogous to orthogonal arrays. *Sankhya Ser. A* **23**, 283–286.

Ratkowsky, D. A., M. A. Evans and J. R. Alldredge (1992). *Cross-over Experiments: Design, Analysis, and Application*. Marcel Dekker, New York.

Senn, S. (1993). *Cross-over Trials in Clinical Research*. Wiley, New York.

Shah, K. R. and B. K. Sinha (1989). *Theory of Optimal Designs*. Lecture Notes in Statistics. Springer, New York.

Silvey, S. D. (1980). *Optimal Design*. Chapman and Hall, London.

Street, D. J. (1989). Combinatorial problems in repeated measurements designs. *Discrete Math.* **77**, 323–343.

Stufken, J. (1991). Some families of optimal and efficient repeated measurements designs. *J. Statist. Plann. Inference* **27**, 75–83.

Williams, E. J. (1949). Experimental designs balanced for the estimation of residual effects of treatments. *Austral. J. Sci. Res.* **A2**, 149–168.

S. Ghosh and C. R. Rao, eds., *Handbook of Statistics, Vol. 13*

4

Design and Analysis of Experiments: Nonparametric Methods with Applications to Clinical Trials

Pranab K. Sen

1. Introduction

In a variety of experimental studies ranging from conventional agricultural, biological and industrial experiments to modern biomedical investigations and (multicenter) clinical trials, a sound statistical planning (design) of the experiment constitutes the first and foremost task.

In a conventional setup it is generally assumed that the response distribution is continuous, and further that it can be taken as normal whose mean (location parameter) depends *linearly* on the model based parameters, while the variance is a constant under varied experimental setups. The *linearity of the model* (i.e., the additivity of the effects), *homoscedasticity* and *normality* of the errors are the three basic regularity assumptions underlying the traditional parametric statistical analysis of experimental outcomes. In *biological assays*, biomedical and clinical experiments, the response variable is typically nonnegative and has a positively skewed distribution. For this reason, often, a logarithmic or square (or cubic) root transformation (belonging to the family of *Box–Cox* type transformations) is advocated to induce greater degree of symmetry of the distribution of the transformed variate (for which the normality assumption may be more likely). However, such a *nonlinear transformation* may generally affect the linearity of the regression model as well as the homoscedasticity condition. Thus, in general, there may not be any guarantee that a Box–Cox type transformation may simultaneously improve the normality of the transformed errors, their homoscedasticity and thc additivity of the model. To the contrary, generally, one of the two postulations: additivity of the model effects and normality of the error distribution, has to be compromised at the price for the other. Therefore, there is a genuine need to looking into the composite picture of linearity of regression, homoscedasticity and normality of errors from a broader perspective so as to allow plausible departures from each of these postulations. This, in turn, may dictate the need for more appropriate designs, and consequently, broader statistical analysis schemes. *Nonparametric* and *robust* statistical procedures have their genesis in this robustness and validity complex. In passing, we may remark that in biological assays, socio-economic and psychometric studies, often, a response may be *quantal* (i.e., all or nothing), *qualitative* (i.e., categorical) or *discrete* (viz., relating to count data), while in a majority of practical applications, one may also encounter *grouped data* typically arising when observations are recorded in suitable *class intervals*. Design and statistical analysis of such nonstandard experiments may call for a somewhat different statistical approach, and we may briefly mention some of the basic differences in the setups of such nonstandard statistical analyses contrasted with standard parametric formulations.

Pedagogically, developments in nonparametrics started in the hypotheses testing sector, where under suitable *hypotheses of invariance*, the sample observations have a joint distribution which remains invariant under appropriate *groups of transformations* which map the sample space onto itself. This characterization of the null hypothesis generates *exactly distribution-free* (EDF) tests for some simple hypotheses testing problems. However, in a more complex situation, such as arising in multivariate problems or in composite hypotheses testing problems, this EDF property may not be generally tenable. In such a case, often, it may be possible to have some conditionally distribution-free tests, although this prescrtiption may not always work out well. Moreover, even in simple hypotheses testing problems, the distribution theory of EDF tests may become prohibitively complicated for any algebraic manipulations as the sample size increases. This unpleasant feature is shared by nonparametric (point as well as interval) estimators which are based on such EDF statistics. In the later phase of developments on nonparametrics (viz., Sen (1991) for a review), *asymptotic methods* have invaded all walks of this field. Not only such asymptotics provide simpler distribution theory for large sample sizes, but also clearer motivations for the use of nonparametrics instead of their parametric counterparts. Indeed, at the current state of developments on nonparametrics, such asymptotics play a vital role in all aspects.

A significant amount of research work on design and analysis of experiments relates to the so called bio-assays. Statistical methodology for bio-assays, even in a parametric setup, has a somewhat different perspective, and nonparametrics as developed in the past three decades (viz., Sen, 1963, 1971b, 1972) are very relevant too. Bio-assays are the precursors of modern *clinical trials*, and during the past twenty five years, statistical perspectives in clinical trials and medical investigations have received due attention from clinical as well as statistical professionals (and *Regulatory Agencies* too). This has paved the way for development of some *clinical designs* with greater emphasis on *hazard models* and *relative risk* formulation; nonparametrics play a vital role in this context too. Perhaps, our treatise of this field would remain a bit incomplete without some discussion of such designs and their analysis schemes in the light of such recently developed nonparametrics. Statistical planning (design) and inference (analysis) remain pertinent to two other related areas, namely, (i) *epidemiological investigations* and (ii) *environmental studies*, and more work remains to be accomplished before they can be systematically reviewed.

2. One-way layout nonparametrics

Let us consider the following linear model related to completely randomized designs:

$$\boldsymbol{Y} = \boldsymbol{X}\boldsymbol{\beta} + \boldsymbol{e}; \quad \boldsymbol{e} = (e_1, \ldots, e_n)', \tag{2.1}$$

where $\boldsymbol{Y}$ is the n-vector of independent observations (response variables), $\boldsymbol{X}$ is an $n \times p$ matrix of known constants (design variables), $\boldsymbol{\beta} = (\beta_1, \ldots, \beta_p)'$ is a p-vector of unknown (fixed-effects) parameters, and the error variables e_i are assumed to be independent and identically distributed random variables (i.i.d.r.v.) with a continuous univariate distribution function (d.f.) F. In a parametric setup, for testing plausible hypotheses on $\boldsymbol{\beta}$ (or to construct suitable confidence sets), F is assumed to be normal with 0 mean and a finite, positive variance σ^2 (unknown), but in nonparametrics this normality or even the finiteness of the error variance is not that crucial. Two basic statistical issues arising in this context are the following: (i) To choose the design matrix $\boldsymbol{X}$ in such a way that for a given n, the number of observations, the *information* on $\boldsymbol{\beta}$ acquirable from the experiment is maximized (the *optimal design* problem), and (ii) for a chosen design, to draw statistical conclusions (on $\boldsymbol{\beta}$) in an optimal manner (the *optimal statistical inference* problem). For the linear model in (2.1), when F is normal, the information on $\boldsymbol{\beta}$ contained in the experiment is given by $\sigma^{-2}(\boldsymbol{X}'\boldsymbol{X})$, where $\boldsymbol{X}'\boldsymbol{X}$ is a positive semi-definite (p.s.d.) matrix. Therefore maximization of information on $\boldsymbol{\beta}$ relates to an optimal choice of the design matrix $\boldsymbol{X}$, such that $\boldsymbol{X}'\boldsymbol{X}$ is maximized in a meaningful manner. Kiefer's *universal optimality of designs* is a landmark in this context. Two recent monographs by Shah and Sinha (1989) and Pukelsheim (1993) are excellent sources for detailed coverage of such optimal designs. When the normality of the error components is not taken for granted, the information criterion is taken as $I(f)\boldsymbol{X}'\boldsymbol{X}$ where $I(f)$ is the *Fisher information* for the location parameter for the density f, when the latter belongs to the location-scale family. Thus, if $I(f)$ is substituted for σ^{-2}, the situation remains comparable, so that for a given density having a finite Fisher information $I(f)$, maximization of the information (on $\boldsymbol{\beta}$) through a choice of the design matrix remains isomorphic to the normal theory case.

The second problem has a somewhat different perspective. Optimal statistical inference may depend very much on the underlying density. The main point in this context is that for a possibly nonnormal d.f., having a finite variance σ^2, the dispersion matrix of the maximum likelihood estimator of $\boldsymbol{\beta}$ is given by $I^{-1}(f)(\boldsymbol{X}'\boldsymbol{X})^{-1}$, and the normal theory estimator has the dispersion matrix $\sigma^2(\boldsymbol{X}'\boldsymbol{X})^{-1}$, where by the classical *Cramér–Rao information-inequality*,

$$\sigma^2 \geqslant \{I^{-1}(f)\}, \tag{2.2}$$

so that the normal theory or linear inference procedures generally entail some loss of efficiency. This loss is measured by the difference of $I(f)$ and σ^{-2}, and for *heavy-tailed distributions*, this loss can be quite large. In the extreme case of infinite variance (viz., the Cauchy case), linear/normal theory inference procedures are inconsistent and totally inefficient too. On the top of that if in (2.1) the very additivity of the effects (i.e., linearity of the model) does not hold, the entire picture may be drastically different. Robust and nonparametric methods have mainly been considered to retain general validity as well as good efficiency properties under such plausible departures from the model based assumptions. Although such nonparametric procedures were proposed originally for hypotheses testing problems, the dual estimation problems have also received adequate attention, and we summarize them as follows.

In a completely randomized design, through a transformation on the design matrix $\boldsymbol{X} \to \boldsymbol{X}^* = \boldsymbol{X}\boldsymbol{\Lambda}$, for a suitable p.d. $\boldsymbol{A}$ and a conjugate transformation on $\boldsymbol{\beta} \to \boldsymbol{\beta}^* = \boldsymbol{A}^{-1}\boldsymbol{\beta}$, (2.1) can be *canonically* reduced to a multi-sample location model. In this canonical representation, the *homogeneity* of the location parameters constitute the basic null hypothesis, while estimable parameters relate to such location parameters as well as *contrasts* among them. Motivated by this canonical reduction, we partition $\boldsymbol{Y}$ into c ($\geqslant 2$) subvectors $\boldsymbol{Y}_k$, $k = 1, \ldots, c$, where $\boldsymbol{Y}_k$ is an n_k-vector $(Y_{k1}, \ldots, Y_{kn_k})'$ such that the Y_{ki} are i.i.d.r.v.'s with a d.f. $F_k(x) = F(x - \theta_k)$, for $k = 1, \ldots, c$. In this formulation, the d.f. F is assumed to be continuous, and it need not have a finite variance or first moment, and also the symmetry assumption is not crucial for testing the null hypothesis of equality of the θ_k. If, on the other hand, we want to estimate or test for the individual θ, then, the symmetry of the d.f. F is needed, and this will be discussed later on. Keeping this estimability and testability in mind, we may rewrite (2.1) in the following form:

$$Y_i = \theta + \boldsymbol{\beta}'\boldsymbol{t}_i + e_i, \quad i = 1, \ldots, n, \tag{2.3}$$

where the $\boldsymbol{t}_i$ are vectors of (known) regression constants, not all equal, θ is the *intercept* parameter, and $\boldsymbol{\beta} = (\beta_1, \ldots, \beta_q)'$ is a vector of unknown *regression* parameters where q is a positive integer and $n \geqslant q + 1$; the e_i are i.i.d.r.v.'s having the d.f. F. In this setup, the null and alternative hypotheses of interest are

$$H_0\colon \boldsymbol{\beta} = \boldsymbol{0} \quad \text{vs.} \quad H_1\colon \boldsymbol{\beta} \neq \boldsymbol{0}. \tag{2.4}$$

Note that for the canonical multisample model, (2.4) relates to the *homogeneity* of the location parameters, and we may also want to estimate the *paired differences* $\theta_k - \theta_q$, $k \neq q = 1, \ldots, c$, in a robust manner. Note that under the null hypothesis in (2.4), the Y_i are i.i.d.r.v.'s with a location parameter θ, so that their joint distribution remains invariant under any permutation of the Y_i among themselves. For this reason, the null hypothesis in (2.4) is termed the *hypothesis of randomness or permutation-invariance*. In the literature, H_1 in (2.4) is termed the *regression alternative*. For such alternatives, tests for (2.4) do not require the symmetry of the d.f. F, and the same regularity assumptions pertain to the estimation of $\boldsymbol{\beta}$. Further, it is possible to draw statistical inference on θ under an additional assumption that F is a symmetric d.f.

Let R_{ni} be the rank of Y_i among $Y_1, \ldots, Y_n$, for $i = 1, \ldots, n$ and $n \geqslant 1$. Also, let $a_n(1), \ldots, a_n(n)$ be a set of *scores* which depend on the sample size n and some chosen *score generating* function. Of particular interest are the following special scores: (i) *Wilcoxon scores*: $a_n(i) = i/(n+1)$, $i = 1, \ldots, n$; (ii) *Normal scores*: $a_n(i) = E(Z_{n:i})$, $i = 1, \ldots, n$, where $Z_{n:1} \leqslant \cdots \leqslant Z_{n:n}$ are the order statistics of a sample of size n drawn from a standard normal distribution; and (iii) *Log-rank scores*: $a_n(i) = \sum_{j=0}^{i-1}(n-j)^{-1} - 1$, $i = 1, \ldots, n$. With these notations, we define (a vector of) *linear rank statistics* by letting

$$\boldsymbol{L}_n = \sum_{i=1}^{n}(\boldsymbol{t}_i - \bar{\boldsymbol{t}}_n)a_n(R_{ni}); \tag{2.5}$$

where the $\boldsymbol{t}_i$ are defined as in (2.3) and $\bar{\boldsymbol{t}}_n = n^{-1}\sum_{i=1}^n \boldsymbol{t}_i$. Also, we define

$$\bar{a}_n = n^{-1}\sum_{i=1}^n a_n(i), \qquad A_n^2 = (n-1)^{-1}\sum_{i=1}^n [a_n(i) - \bar{a}_n]^2, \tag{2.6}$$

and

$$\boldsymbol{Q}_n = \sum_{i=1}^n (\boldsymbol{t}_i - \bar{\boldsymbol{t}}_n)(\boldsymbol{t}_i - \bar{\boldsymbol{t}}_n)'. \tag{2.7}$$

As has been noted earlier, under H_0 in (2.4), the Y_i are i.i.d., and hence, the vector $\boldsymbol{R}_n = (R_{n1}, \ldots, R_{nn})'$ takes on each permutation of $(1, \ldots, n)$ with the common probability $(n!)^{-1}$. If we denote this discrete uniform probability measure by $\mathcal{P}_n$, we obtain immediately that

$$E_{\mathcal{P}_n}(\boldsymbol{L}_n) = \boldsymbol{0} \quad \text{and} \quad E_{\mathcal{P}_n}(\boldsymbol{L}_n \boldsymbol{L}_n') = A_n^2 \cdot \boldsymbol{Q}_n. \tag{2.8}$$

A typical rank order test statistic is quadratic form in $\boldsymbol{L}_n$, and is given by

$$\mathcal{L}_n = A_n^{-2}[\boldsymbol{L}_n' \boldsymbol{Q}_n^- \boldsymbol{L}_n], \tag{2.9}$$

where $\boldsymbol{Q}_n^-$, a *generalized inverse* of $\boldsymbol{Q}_n$, is defined by $\boldsymbol{Q}_n^- \boldsymbol{Q}_n \boldsymbol{Q}_n^- = \boldsymbol{Q}_n^-$. For small values of n, the exact null hypothesis distribution of $\mathcal{L}_n$ can be obtained by direct enumeration of all possible $(n!)$ equally likely realizations of $\boldsymbol{R}_n$. Although this enumeration process becomes prohibitively laborious as the sample size becomes large, there are various *permutational central limit theorems* by which one may easily approximate this permutation distribution by a (central) *chi squared* distribution with *q degrees of freedom* (DF) when $\boldsymbol{Q}_n$ is of full rank (q). The right hand tail of the null distribution of $\mathcal{L}_n$ can therefore be used to find out a *critical level* $l_{n\alpha}$ corresponding to a given *level of significance* α: $0 \leqslant \alpha \leqslant 1$, such that if $\chi^2_{q,\alpha}$ stands for the upper *α-quantile* of the central chi squared d.f. with q DF, then

$$l_{n\alpha} \to \chi^2_{q,\alpha}, \quad \text{as} \quad n \to \infty. \tag{2.10}$$

Under quite general regularity conditions, when $\boldsymbol{\beta}$ is $\neq \boldsymbol{0}$, $\boldsymbol{Q}_n^{-1}\boldsymbol{L}_n$ converges to a nonnull vector, so that $n^{-1}\mathcal{L}_n$ converges to a positive number, while, by (2.10), $n^{-1}l_{n\alpha} \to 0$, as $n \to \infty$. Therefore, the test based on $\mathcal{L}_n$ is consistent against the entire class of alternatives for which the centering element of $\boldsymbol{Q}_n^{-1}\boldsymbol{L}_n$ is different from **o**. We may refer to Puri and Sen (1985) for some details. For this reason, *local alternatives* are chosen in such a way that under such alternatives, $\mathcal{L}_n$ has a nondegenerate asymptotic distribution which may then be incorporated in the study of various *asymptotic properties* of such rank order tests.

For a local (Pitman-type) shift alternative of the type

$$H_n\colon\ \boldsymbol{\beta} = \boldsymbol{\beta}_{(n)} = n^{-1/2}\lambda, \quad \text{for some fixed } \lambda \in R^q, \tag{2.11}$$

for large n, $\mathcal{L}_n$ follows closely a *noncentral chi squared* d.f. with q DF and noncentrality parameter $\Delta = \gamma^2\ \lambda' \boldsymbol{Q}^{-1}\lambda/A^2$, where γ is a functional of ϕ, F and is formally defined by (2.19). It is assumed that A_n^2 converges to a finite positive limit A^2 as n increases, and similarly, $n^{-1}\boldsymbol{Q}_n$ converges to a p.d. $\boldsymbol{Q}$. Also, it is assumed that the *generalized Noether-condition* holds for the $\boldsymbol{t}_i$. For the normal scores rank test, we have asymptotic optimality when the underlying F is itself a normal d.f.. Moreover, this particular test is asymptotically at least as efficient as the classical *variance ratio* (ANOVA) test for all F belonging to a general class $\mathcal{F}$; the equality of their *asymptotic efficacies* holds only for a normal F which belongs to this class. For a *logistic* d.f., the test based on the Wilcoxon scores is asymptotically optimal, while for an *exponential* d.f., the log-rank test is asymptotically optimal. In general, if the d.f. F admits an absolutely continuous density f having a finite *Fisher information* $I(f)$, and if we let $a_n^*(k) = E\phi_f(U_{n:k})$, $k = 1, \ldots, n$, where $U_{n:1} \leqslant \cdots \leqslant U_{n:n}$ are the order statistics of a sample of size n from a *uniform* $[0, 1]$ d.f., and the score generating function $\phi_f(u)$ is given by

$$\phi_f(u) = -f'(F^{-1}(u))/f(F^{-1}(u)), \quad u \in (0, 1), \tag{2.12}$$

then the rank test based on the scores $a_n^*(k)$ is asymptotically optimal when the underlying d.f. is F; Hájek and Šidák (1967) is an excellent source for the theoretical motivation for such local optimality properties of linear rank statistics. For the specific multisample model, as has been treated earlier, if we denote the average rank scores of the observations in the kth sample (with respect to the pooled sample observations) by $\bar{a}_{n,k}$, for $k = 1, \ldots, c$, then the rank order test statistic in (2.9) simplifies to

$$\mathcal{L}_n = \left[\sum_{k=1}^{c} n_k(\bar{a}_{n,k} - \bar{a}_n)^2\right] / A_n^2, \tag{2.13}$$

where $\bar{a}_n$ is defined by (2.6). For this model, $q = c-1$. For the Wilcoxon scores, (2.13) reduces to the well known *Kruskal–Wallis* test statistic, while the *Brown and Mood median* test statistic corresponds to the scores $a_n(k) = 0$ or 1, according as k is $\leqslant$ or $>$ $(n+1)/2$. In this multisample setup, $\mathcal{L}_n$ actually tests for the homogeneity of the d.f.'s $F_1, \ldots, F_c$ against alternatives which are more general than the simple location or shift ones treated earlier. For example, if we let $\pi_{rs} = P\{Y_{r1} \geqslant Y_{s1}\}$, for $r \neq s = 1, \ldots, c$, then the Kruskal–Wallis test is consistent against the broader class of alternatives that the π_{rs} are not all equal to $1/2$, which relate to the so called *stochastically larger (smaller) alternatives*. Thus, the linearity of the model is not that crucial in this context. In the multisample model, when the scores are monotone, $\mathcal{L}_n$ remains consistent against stochastically larger (smaller) class of alternatives, containing shift alternatives as a particular subclass of such a broader class of alternatives. This explains the robustness of such nonparametric tests. In this context, we may note that the ranks R_{ni} are *invariant* under any strictly monotone transformation $g(\cdot)$ on the Y_i, so that if the $g(Y_i) = Y_i^*$ follow a linear model for some $g(\cdot)$, a rank statistic based on the Y_i and Y_i^* being the same will pertain to such a *generalized linear model* setup. This invariance eliminates the need for *Box–Cox* type transformations on the Y_i and thereby adds further to the robustness of such rank tests against plausible departures from the model based assumptions. The situation is a little less satisfactory for testing *subhypotheses* or for *multiple comparisons*, and we shall discuss them later on.

We present *R-estimators* which are based on such rank tests. Although for one and two-sample location problems, such estimates were considered by Hodges and Lehmann (1963) and Sen (1963), for general linear models, the developments took place a few years later. Adichie (1967), Sen (1968d) and Jurečková (1971), among others, considered the simple regression model. For the linear model in (2.3), let $Y_i(\boldsymbol{b}) = Y_i - \boldsymbol{b}'\boldsymbol{t}_i$, $i = 1, \ldots, n$, where $\boldsymbol{b} \in \mathcal{R}^q$, and let $R_{ni}(\boldsymbol{b})$ be the rank of $Y_i(\boldsymbol{b})$ among the $Y_r(\boldsymbol{b})$, $r = 1, \ldots, n$, $\boldsymbol{b} \in \mathcal{R}^q$. In (2.5), replacing the R_{ni} by $R_{ni}(\boldsymbol{b})$, we define the linear rank statistics $\boldsymbol{L}_n(\boldsymbol{b})$, $\boldsymbol{b} \in \mathcal{R}^q$. As in Jaeckel (1972), we introduce a *measure of rank dispersion*:

$$D_n(\boldsymbol{b}) = \sum_{i=1}^{N} \{a_n(R_{ni}(\boldsymbol{b})) - \bar{a}_n\} Y_i(\boldsymbol{b}), \quad \boldsymbol{b} \in \mathcal{R}^q, \tag{2.14}$$

where we confine ourselves to monotone scores, so that $a_n(1) \leqslant \cdots \leqslant a_n(n)$, for every $n \geqslant 1$. An R-estimator of $\boldsymbol{\beta}$ is a solution to the minimization of $D_n(\boldsymbol{b})$ with respect to $\boldsymbol{b} \in \mathcal{R}^q$, so that we write

$$\widehat{\boldsymbol{\beta}}_n = \arg\min\{D_n(\boldsymbol{b})\colon\ \boldsymbol{b} \in \mathcal{R}^q\}. \tag{2.15}$$

It can be shown (viz., Jurečková and Sen, 1995, Chapter 6) that $D_n(\boldsymbol{b})$ is a *nonnegative, continuous, piecewise linear* and *convex function* of $\boldsymbol{b} \in \mathcal{R}^q$. Note that $D_n(\boldsymbol{b})$ is differentiable in $\boldsymbol{b}$ almost everywhere and

$$(\partial/\partial \boldsymbol{b}) D_n(\boldsymbol{b})\big|_{\boldsymbol{b}^o} = -\boldsymbol{L}_n(\boldsymbol{b}^o), \tag{2.16}$$

whenever $\boldsymbol{b}^o$ is a point of differentiability of $D_n(\cdot)$. At any other point, one may work with the subgradient $\nabla D_n(\boldsymbol{b}^o)$. Thus, essentially, the task reduces to solving for the following *estimating equations* with respect to $\boldsymbol{b} \in \mathcal{R}^q$:

$$\boldsymbol{L}_n(\boldsymbol{b}) = \boldsymbol{0}, \tag{2.17}$$

where to eliminate multiple solutions, adopt some convention. These R-estimators are generally obtained by iterative procedures (as in the case of maximum likelihood estimators for a density not belonging to the exponential family), and, often, a one or two-step procedure starting with a *consistent* and *asymptotically normal* (CAN) initial estimator serves the purpose very well; for some theoretical developments along with an extended bibliography, we refer to Jurečková and Sen (1995). It follows from their general methodology that under essentially the same regularity conditions as pertaining to the hypothesis testing problem (treated before), the following *first-order asymptotic distributional representation* (FOADR) result holds:

$$(\widehat{\boldsymbol{\beta}}_n - \boldsymbol{\beta}) = \gamma^{-1} \sum_{i=1}^{n} \boldsymbol{d}_{ni} \phi(F(e_i)) + \mathbf{o}_p(n^{-1/2}), \tag{2.18}$$

where $\boldsymbol{d}_{ni} = \boldsymbol{Q}_n^{-1}(\boldsymbol{t}_i - \bar{\boldsymbol{t}}_n)$, for $i = 1, \ldots, n$, $\phi(\cdot)$ is the score generating function for the rank scores,

$$\gamma = \int_0^1 \phi(u)\phi_f(u)\,du, \tag{2.19}$$

and the Fisher information score generating function $\phi_f(\cdot)$ is defined by (2.12). Note that under a generalized *Noether condition* on the $\boldsymbol{d}_{ni}$, the classical (multivariate) central limit theorem holds for the principal term on the right hand side of (2.18), so that on defining $\boldsymbol{Q}_n^*$ by $\boldsymbol{Q}_n^* \boldsymbol{Q}_n^{-1} \boldsymbol{Q}_n^* = \boldsymbol{I}$, we obtain from (2.18) and (2.19) that for large sample sizes,

$$\boldsymbol{Q}_n^*(\widehat{\boldsymbol{\beta}}_n - \boldsymbol{\beta}) \xrightarrow{\mathcal{D}} N(\boldsymbol{0}, \gamma^{-2}A^2\boldsymbol{I}), \tag{2.20}$$

where A^2 is the variance of the score function ϕ. For the normal theory model, the classical *maximum likelihood estimator* (MLE) agrees with the usual *least squares estimator* (LSE), and for this (2.20) hold with $\gamma^{-2}A^2$ being replaced by σ^2, the error variance. As such, the *asymptotic relative efficiency* (ARE) of the R-estimator, based on the score function ϕ, with respect to the classical LSE is given by

$$e(R; LS) = \gamma^2\sigma^2/A^2, \tag{2.21}$$

which does not depend on the design matrix $\boldsymbol{Q}_n$. In particular, if we use the normal scores for the derived R-estimators, then (2.21) is bounded from below by 1, where the lower bound is attained only when the underlying distribution is normal. This explains the robustness as well as asymptotic efficiency of the normal scores R-estimators in such completely randomized designs. From robustness considerations, often, it may be better to use the Wilcoxon scores estimators. Although for this particular choice of the score generating function, (2.21) is not bounded from below by 1, it is quite close to 1 for near normal distributions and may be high for heavy tailed ones. If the error density $f(\cdot)$ is of known functional form, one may use the MLE for that pdf, and in that case, in (2.20), we need to replace $\gamma^{-2}A^2$ by $\{I(f)\}^{-1}$, where $I(f)$ is the Fisher information for location of the density f. Thus, in this case, the ARE is given by

$$e(R; ML) = \gamma^2/\{I(f)A^2\}, \tag{2.22}$$

which by the classical Cramér–Rao inequality is always bounded from above by 1. Nevertheless, it follows from the general results in Hušková and Sen (1985) that if the score generating function is chosen adaptively, then the corresponding *adaptive* R-estimator is asymptotically efficient in the sense that in (2.22) the ARE is equal to 1. The same conclusion holds for adaptive rank tests for $\boldsymbol{\beta}$ as well.

As has been mentioned earlier, the ranks $R_{ni}(\boldsymbol{b})$ are *translation-invariant* so that they provide no information on the intercept parameter θ. Thus, for testing any plausible null hypothesis on θ or to estimate the same parameter, linear rank statistics are not of much use. This problem has been eliminated to a greater extent by the use of *signed rank statistics*, which is typically defined as

$$S_n = \sum_{i=1}^{n} \text{sign}(Y_i) a_n(R_{ni}^+), \tag{2.23}$$

where the rank scores $a_n(k)$ are defined as in before and R_{ni}^+ is the rank of $|Y_i|$ among the $|Y_r|$, $r = 1, \dots, n$. Under the null hypothesis of symmetry of the d.f. F about 0, the vector of the $|R_{ni}|$ and the vector of the sign(Y_i) are stochastically independent, so that the set of 2^n equally likely *sign-inversions* generates the exact null distribution of S_n. This may also be used to derive the related R-estimator of θ. Such a test and estimator share all the properties of the corresponding test and estimator for the regression parameter. But, in the current context, there is a basic problem. The test for $\boldsymbol{\beta}$ based on $\mathcal{L}_n$, being translation-invariant, does not depend on the intercept parameter (which is taken as a nuisance one). On the other hand, for testing a null hypothesis on θ or estimating the same, the parameter $\boldsymbol{\beta}$ is treated as a nuisance one, and the signed ranks are not *regression-invariant*. Thus, the *exact distribution-freeness* (EDF) property may have to be sacrifice in favor of *asymptotically distribution-free* (ADF) ones. An exception is the case when one wants to test simultaneously for $\theta = 0$ and $\boldsymbol{\beta} = \mathbf{0}$.

We denote a suitable R-estimator of $\boldsymbol{\beta}$ by $\widehat{\boldsymbol{\beta}}_n$, and incorporate the same to obtain the residuals:

$$\widehat{Y}_{ni} = Y_i - \widehat{\boldsymbol{\beta}}_n' \boldsymbol{t}_i, \quad i = 1, \dots, n. \tag{2.24}$$

For every real d, let $\widehat{R}_{ni}^+(d)$ be the rank of $|\widehat{Y}_{ni} - d|$ among the $|\widehat{Y}_{nr} - d|$, $r = 1, \dots, n$, for $i = 1, \dots, n$. Also let

$$\widehat{S}_n(d) = \sum_{i=1}^{n} \text{sign}(\widehat{Y}_{ni} - d) a_n(\widehat{R}_{ni}^+(d)), \quad d \in \mathcal{R}. \tag{2.25}$$

If the scores $a_n(k)$ are monotone (in k, for each n), then, it is easy to show that $\widehat{S}_n(d)$ is monotone in $d \in \mathcal{R}$, and hence, we may equate $\widehat{S}_n(d)$ to 0 (with respect to $d \in \mathcal{R}$) and the solution, say, $\widehat{\theta}_n$, is then taken as a *translation-equivariant* estimator of θ. In the particular case of the sign statistic, $\widehat{\theta}_n$ can bc expressed as the median of the residuals $\widehat{Y}_{ni}$, and for the case of the Wilcoxon signed-rank statistic, it is given by the median of the midranges of these residuals. In general, for other score functions, an iterative procedure is needed to solve for $\widehat{\theta}_n$, and in such a case, one may as well start with the Wilcoxon scores estimator as the preliminary one, and in a few steps converge to the desired one. There is a basic difference between this model and the simple location model where $\boldsymbol{\beta}$ is null: In the latter case, the signed rank statistics based on the true value $d = \theta$ are EDF, while in this case they are only ADF. To verify that they are ADF, one convenient way is to appeal to some *asymptotic uniform linearity* results on general signed rank statistics (in the location and regression parameters), and such results have been presented in a unified manner in Chapter 6 of Jurečková and Sen (1996), where pertinent references are also cited in detail.

Let us discuss briefly the subhypothesis testing problem for this simple design. A particular subhypothesis testing problem relates to the null hypothesis that $\theta = 0$ against $\theta \neq 0$, treating $\boldsymbol{\beta}$ as a nuisance parameter (vector). We have already observed that the basic hypothesis of sign- (or permutation) invariance does not hold when the above null hypothesis holds, and hence, EDF tests may not generally exist. However, ADF test can be considered by incorporating the residuals $\widehat{Y}_{ni}$ instead of the Y_i in the formulation of suitable signed rank statistics. Such tests were termed *aligned rank tests* by Hodges and Lehmann (1962) who considered the simplest ANOVA model. Here alignment is made by substituting the estimates of the nuisance parameters as is also done in the classical normal theory linear models. A very similar picture holds for a plausible subhypothesis testing problem on the regression parameter vector. To pose such a problem in a simple manner, we partition the parameter vector $\boldsymbol{\beta}$ as

$$\boldsymbol{\beta}' = (\boldsymbol{\beta}_1', \boldsymbol{\beta}_2'), \tag{2.26}$$

where $\boldsymbol{\beta}_j$ is a p_j-vector, $p_j \geqslant 1$, for $j = 1, 2$, and $p = p_1 + p_2$. Suppose now that we want to test for

$$H_0\colon\ \boldsymbol{\beta}_1 = \mathbf{0} \quad \text{vs.} \quad H_1\colon\ \boldsymbol{\beta}_1 \neq \mathbf{0}, \quad \text{treating } \boldsymbol{\beta}_2 \text{ as a nuisance parameter.} \tag{2.27}$$

Here also, under H_0 in (2.27), the hypothesis of permutational invariance may not be generally true, and hence, an EDF rank test may not generally exist. But, ADF rank tests based on aligned rank statistics can be constructed as follows. Note that if in (2.3), we partition the $\boldsymbol{t}_i$ as $(\boldsymbol{t}_{i1}', \boldsymbol{t}_{i2}')'$, involving p_1 and p_2 coordinates, then under H_0 in (2.27), we obtain that

$$Y_i = \theta + \boldsymbol{\beta}_2' \boldsymbol{t}_{i2} + e_i, \quad i = 1, \ldots, n. \tag{2.28}$$

Based on the model in (2.28), we denote the R-estimator of $\boldsymbol{\beta}_2$ by $\tilde{\boldsymbol{\beta}}_{n2}$, and we form the residuals

$$\tilde{Y}_{ni} = Y_i - \tilde{\boldsymbol{\beta}}_{n2}' \boldsymbol{t}_{i2}, \quad \text{for} \quad i = 1, \ldots, n. \tag{2.29}$$

As in after (2.4), we define the aligned ranks $\tilde{R}_{ni}$ wherein we replace the Y_i by the residuals $\tilde{Y}_{ni}$. The vector of linear rank statistics $\tilde{\boldsymbol{L}}_n$ is then defined as in (2.5) with the ranks R_{ni} being replaced by $\tilde{R}_{ni}$. Also, we partition this p-vector as $(\tilde{\boldsymbol{L}}_{n1}', \tilde{\boldsymbol{L}}_{n2}')'$, and our test is then based on the first component of this aligned rank statistics vector. This is given by

$$\tilde{\mathcal{L}}_{n1} = A_n^{-2} \{ \tilde{\boldsymbol{L}}_{n1}' \boldsymbol{Q}_{n11.2}^{-1} \tilde{\boldsymbol{L}}_{n1} \}, \tag{2.30}$$

where A_n^2 is defined as in (2.6), and defining $\boldsymbol{Q}_n$ as in (2.7) and partitioning it into four submatrices, we have

$$\boldsymbol{Q}_{n11.2} = \boldsymbol{Q}_{n11} - \boldsymbol{Q}_{n12}\boldsymbol{Q}_{n22}^{-1}\boldsymbol{Q}_{n21}. \tag{2.31}$$

It follows from the general results in Sen and Puri (1977), further streamlined and discussed in detail in Section 7.3 of Puri and Sen (1985) that under the null hypothesis in (2.27), $\tilde{\mathcal{L}}_{n1}$ has asymptotically chi-squared distribution with p_1 DF, so that an ADF test for H_0 in (2.27) can be based on the critical level given by $\chi^2_{p_1,\alpha}$, the upper α-percentile of this distribution. For local alternatives, the noncentral distribution theory runs parallel to the case of the null hypothesis of $\boldsymbol{\beta} = \mathbf{0}$, with the DF p being replaced by p_1 and an appropriate change in the noncentrality parameter as well. The regularity conditions governing these asymptotic distributional results have been unified and relaxed to a certain extent in Chapter 6 of Jurečková and Sen (1996).

Another important area where nonparametrics have played a vital role in such completely randomized designs is the so called *mixed effects* models. In this setup, we extend (2.3) as follows. Let $Y_1, \dots, Y_n$ be independent random variables, such that associated with the Y_i there are (i) given design (nonstochastic) (q-)vectors $\boldsymbol{t}_i$ and (ii) observable stochastic *concomitant* (p-)vectors $\boldsymbol{Z}_i$, $i = 1, \dots, n$. Then conditionally on $\boldsymbol{Z}_i = \boldsymbol{z}$, we have

$$\begin{aligned} F_i(y \mid \boldsymbol{z}) &= P\{Y_i \leqslant y \mid \boldsymbol{Z}_i = \boldsymbol{z}\} \\ &= F(y - \alpha - \boldsymbol{\beta}'\boldsymbol{t}_i - \boldsymbol{\gamma}'\boldsymbol{z}), \quad i = 1, \dots, n, \end{aligned} \tag{2.32}$$

where $\boldsymbol{\beta}$ and $\boldsymbol{\gamma}$ are respectively the regression parameter vector of Y on the design and concomitant variates, and α is the intercept parameter. In this linear model setup, in a nonparametric formulation, the d.f. F is allowed to be arbitrary (but, continuous), so that the finiteness of its second moment is not that crucial. In a parametric as well as nonparametric formulation a basic assumption on the concomitant variates is that they are not affected by the design variates, so that the $\boldsymbol{Z}_i$ are i.i.d.r.v.'s. Here also, in a nonparametric formulation, the joint distribution of $\boldsymbol{Z}_i$ is taken to be an arbitrary continuous one (defined on $\mathcal{R}^p$). The Chatterjee–Sen (1964) *multivariate rank permutation principle* plays a basic role in this nonparametric *analysis of covariance* (ANOCOVA) problem. Basically, for the $(p+1)$-variate observable stochastic vectors $(Y_i, \boldsymbol{Z}_i')'$, with respect to the q-variate design vectors $\boldsymbol{t}_i$, one can construct a $q \times (p+1)$ linear rank statistics vector with the elements

$$L_{njk}, \quad \text{for} \quad j = 0, 1, \dots, p, \ k = 1, \dots, q, \tag{2.33}$$

where $\boldsymbol{L}_{n0} = (L_{n01}, \dots, L_{n0q})'$ stands for the linear rank statistics vector for the primary variate (Y) and is defined as in (2.5) (with the R_{ni} being relabeled as R_{ni0}), while for the jth coordinate of the concomitant vectors, adopting the same ranking method as in before (2.5) and denoting these ranks by R_{nij}, $i = 1, \dots, n$, we define the linear rank statistics vector $\boldsymbol{L}_{nj} = (L_{nj1}, \dots, L_{njq})'$ as in (2.5), for $j = 1, \dots, p$. Note that ranking is done separately for each coordinate of the concomitant vector and the primary variate, so that we have a $(p+1) \times n$ *rank collection* matrix $\boldsymbol{R}_n$. The Chatterjee–Sen rank permutation principle applies to the $n!$ *column permutations* of $\boldsymbol{R}_n$ (which are conditionally equally likely), and this generates *conditionally distribution-free* (CDF) tests based on the linear rank statistics in (2.33). We may allow the scores (defined before (2.5)) to be possibly different for the primary and concomitant variates, so for the jth coordinate, these scores are taken as $a_{nj}(k)$, $k = 1, \ldots, n$; $j = 0, 1, \ldots, p$, and further, without any loss of generality, we may standardize these scores in such a way that adopting the definitions in (2.6), the $\bar{a}_{nj}$ are all equal to 0 and the A_{nj}^2 are all equal to one, $j = 0, 1, \ldots, p$. Consider then a $(p+1) \times (p+1)$ matrix $\boldsymbol{V}_n$ whose diagonal elements are all equal to one, and whose elements are given by

$$v_{njl} = \sum_{i=1}^{n} a_{nj}(R_{nij}) a_{nl}(R_{nil}), \quad j, l = 0, 1, \ldots, p. \tag{2.34}$$

We denote the cofactor of v_{noo} in $\boldsymbol{V}_n$ by $\boldsymbol{V}_{noo}$, $\boldsymbol{v}_{no} = (v_{no1}, \ldots, v_{nop})'$, and denote by

$$\boldsymbol{w}_n = \boldsymbol{V}_{noo}^{-1} \boldsymbol{v}_{no}, \tag{2.35}$$

$$v_{noo}^* = v_{noo} - \boldsymbol{v}_{no}' \boldsymbol{V}_{noo}^{-1} \boldsymbol{v}_{no}, \tag{2.36}$$

and

$$\boldsymbol{L}_{no}^* = \boldsymbol{L}_{no} - (\boldsymbol{L}_{n1}, \ldots, \boldsymbol{L}_{np}) \boldsymbol{w}_n. \tag{2.37}$$

Let us define $\boldsymbol{Q}_n$ as in (2.7) and consider the quadratic form

$$\mathcal{L}_{no}^* = \{(\boldsymbol{L}_{no}^*)' \boldsymbol{Q}_n^{-1} (\boldsymbol{L}_{no}^*)\} / v_{noo}^*, \tag{2.38}$$

which may be used as a test statistic for testing the null hypothesis H_0: $\boldsymbol{\beta} = \boldsymbol{0}$ against alternatives that $\boldsymbol{\beta} \neq \boldsymbol{0}$, treating θ and γ as nuisance parameters. Asymptotic nonnull distribution theory, power properties and optimality of such aligned rank order tests (for local alternatives), studied first by Sen and Puri (1977), can most conveniently be unified by an appeal to the *uniform asymptotic linearity* of aligned rank statistics, and the results presented in Section 7.3 of Puri and Sen (1985) pertain to this scheme; again, the linearity results in their most general form have been presented in a unified manner in Chapter 6 of Jurečková and Sen (1996). This latter reference also contains a good account of the recent developments on *regression rank scores* procedures which may have some advantages (in terms of computational simplicity) over the aligned rank tests.

In the above formulation of a mixed-effect model, the linearity of the regression of the primary response variate on the design and concomitant variates has been taken for granted, while the normality of the errors has been waived to a certain extent by less stringent assumptions. While this can, often, be done with appropriate transformations on primary and concomitant variates, there are certain cases where it may be more reasonable to allow the regression on the concomitant variate part to be rather of some arbitrary (unknown) functional form. That is, the regression on the design variates is taken to be of a parametric (viz., linear) form, while the regression on the covariates is taken as of a *nonparametric* form. In this formulation, for the conditional d.f.'s in (2.32), we take

$$F_i(y \mid \boldsymbol{z}) = F(y - \boldsymbol{\beta}'\boldsymbol{t}_i - \theta(\boldsymbol{z})), \quad i = 1, \ldots, n, \tag{2.39}$$

where the d.f. $F(\cdot)$, $\boldsymbol{\beta}$, etc. are all defined as in before, while $\theta(\boldsymbol{z})$ is a *translation-equivariant* (location-regression) functional, depicting the regression of the errors $Y_i - \boldsymbol{\beta}'\boldsymbol{t}_i$ on the concomitant vector $\boldsymbol{Z}_i$. The basic difference between (2.32) and (2.39) is that in the former case, the linear regression function $\boldsymbol{\gamma}'\boldsymbol{z}$ involves a finite dimensional parameter $\boldsymbol{\gamma}$, while in the latter case, the nonparametric regression function $\theta(\boldsymbol{z})$ may not be finite-dimensional, nor to speak of a linear one. Thus, here we need to treat $\theta(\boldsymbol{z})$ as a functional defined on the domain Z of the concomitant variate $\boldsymbol{Z}$. This formulation may generally entail extra regularity (smoothness) conditions on this nonparametric functional, and because of that, the estimation of $\theta(\boldsymbol{z})$, $\boldsymbol{z} \in Z$, may entail a comparatively slower rate of convergence. Nevertheless, as regards the estimation of the fixed-effects parameters (i.e., $\boldsymbol{\beta}$), the conventional $\sqrt{n}$-rate of convergence still holds, although these conventional estimators may not be fully efficient, even asymptotically. A complete coverage of nonparametric methods in this type of mixed-effects models is beyond the scope of this treatise; we may refer to Sen (1995a, c) where a detailed treatment is included.

3. Two-way layouts nonparametrics

The simplest kind of designs for two-way layouts are the so called *randomized block* or *complete block* designs. 5 equal number of times in each block, and the treatment combinations may Consider a randomized block design comprising n ($\geqslant 2$) blocks of p ($\geqslant 2$) plots each, such that p different treatments are applied to the p plots in each block. The allocation of the treatments into the plots in each block is made through randomization. Let Y_{ij} be the response of the plot in the ith block receiving the jth treatment, for $i = 1, \ldots, n$, $j = 1, \ldots, p$. In the normal theory model, it is assumed that

$$Y_{ij} = \mu + \beta_i + \tau_j + e_{ij}, \quad i = 1, \ldots, n;\ j = 1, \ldots, p, \tag{3.1}$$

where μ is the *mean effect*, β_i is the ith *block effect*, τ_j is the jth *treatment effect*, and the e_{ij} are the error components which are assumed to be independent and identically distributed according to a normal distribution with zero mean and a finite, positive variance σ^2. The block and treatment effects may either be fixed or random, resulting in the hierarchy of fixed-, mixed- and random-effects models. As in the case of one-way layouts, a departure from such model assumptions can take place along the routes of nonlinearity of the model, possible heteroscedasticity, dependence or nonnormality of the errors. It is quite interesting to note that the method of m-ranking, one of the earliest nonparametric procedures, has a basic feature that it does not need many of these regularity assumptions, and yet works out in a very simple manner. Suppose that we desire to test the null hypothesis of no treatment effect, treating the block effects as nuisance parameters. Under this hypothesis, in (3.1), the τ_j drop out, so that the observations within a block are i.i.d.r.v. We may even allow the errors to be exchangeable (instead of i.i.d.), and this implies that under the above hypothesis, the observations within a block are *exchangeable* or *interchangeable* r.v.'s. Therefore, if we denote by r_{ij} the rank of Y_{ij} among $Y_{i1}, \ldots, Y_{ip}$, for $j = 1, \ldots, p$, then, for each i $(= 1, \ldots, n)$, under the hypothesis of no treatment effect, the ranks $r_{i1}, \ldots, r_{ip}$ are interchangeable r.v.'s. Moreover, for different blocks, such *intra-block* rank-vectors are stochastically independent of each other. Therefore, the problem of testing the null hypothesis of no treatment effect in a randomized block design can be reduced to that of testing the interchangeability of the within block rankings. On the other hand, this hypothesis can also be stated in terms of the exchangeability of the within block response variables, and in that setup, the linearity of the block and treatment effects are not that crucial. This scenario leaves us to adopting either of the two routes for nonparametrics in two-way layouts: (i) Incorporate such intra-block rankings with the major emphasis on robustness against possible nonnormality of the errors as well as nonlinearity of the effects, and (ii) Deemphasize the normality of errors, but with due respect to the linearity of the model, incorporate inter-block comparisons in a more visible manner to develop appropriate rank procedures which are robust to possible nonnormality of errors. Aligned rank procedures are quite appropriate in this context, and we shall discuss them later on. For intra-block ranking procedures, we consider a set of scores $\{a(1), \ldots, a(p)\}$ which may depend on p and some underlying score generating function (but not on the number of blocks). In general these are different from the ones introduced in Section 2 (for one-way layouts). For optimal scores for specific types of local alternatives, we may refer to Sen (1968a). Then, we may define

$$T_{nj} = \sum_{i=1}^{n} a(r_{ij}), \quad j = 1, \ldots, p. \tag{3.2}$$

Moreover let

$$\bar{a} = p^{-1} \sum_{j=1}^{p} a(j), \qquad A^2 = (p-1)^{-1} \sum_{j=1}^{p} [a(j) - \bar{a}]^2. \tag{3.3}$$

Then, a suitable test statistic for testing the hypothesis of no treatment effect is the following:

$$\mathcal{L}_n = (nA^2)^{-1} \sum_{j=1}^{p} (T_{nj} - n\bar{a})^2. \tag{3.4}$$

In particular, if we let $a(j) = j$, $j = 1, \ldots, p$, the T_{nj} reduce to the rank sums, $\bar{a} = (p+1)/2$ and $A^2 = p(p+1)/12$, so that (3.4) reduces to the classical Friedman (1937) χ_r^2 test statistic:

$$\chi_r^2 = \frac{12}{np(p+1)} \sum_{j=1}^{p} \left(\sum_{i=1}^{n} r_{ij} - n(p+1)/2 \right)^2 . \tag{3.5}$$

Similarly, letting $a(j) = 0$ or 1 according as j is $\leqslant (p+1)/2$ or not, we obtain the well known Brown and Mood (1951) *median* test statistic. In either case, and in general, for (3.4), the exact distribution (under the null hypothesis) can be obtained by complete enumeration of all possible equally likely $(p!)^n$ permutations of the intra-block rank vectors, each over $(1, \ldots, p)$. This process may become quite cumbersome as p and/or n increase. Fortunately, the central limit theorems are adoptable for the intra-block rank vectors which are independent of each other, and hence, it follows that under the null hypothesis, $\mathcal{L}_n$ has closely the central chi squared distribution with $p-1$ DF when n is large. The main advantage of using an intra-block rank test, such as $\mathcal{L}^n$ in (3.4), is that it eliminates the need for assuming additive block effects, and also, the treatment effects may not be additive too. As in the case of the Kruskal–Wallis test, introduced for one-way layouts in the last section, stochastic ordering of the treatment responses (within each block) suffices for the consistency of the test based on such intra-block ranks. Thus, such tests are very robust. The Brown–Mood median test is asymptotically optimal for local shift alternatives when the underlying d.f. F is Laplace, while for a logistic F, the Friedman χ_r^2 is locally optimal. We may refer to Sen (1968a) for a detailed discussion of the choice of locally optimal intra-block rank tests in some specific models. The main drawback of such intra-block rank tests is that they may not adequately incorporate the *inter-block information* as is generally provided by comparisons of observations from different blocks. For example, if the block effects are additive then a contrast in the ith block observations has the same distribution as in any other block, and hence, some comparisons of such contrasts may provide additional information and may lead to more efficient tests.There are various ways of inducing such inter-block comparisons in rank tests, and among them the two popular ones are the following: (i) *Ranking after alignment*, and (ii) *weighted ranking*.

In a weighted ranking method, instead of having the sum statistics $\sum_{i=1}^{n} a(r_{ij})$, the intra-block rankings or rank scores are weighed to reflect possible inter-block variation, and such weights are typically inversely proportional to some measure of the within block dispersion of the observations (such as the range or standard deviation or even some rank measures of dispersion). Thus, we may use the statistics $\sum_{i=1}^{n} w_{ni} a(r_{ij}), j = 1, \ldots, p$, where the w_{ni} are nonnegative weights, and are typically random elements. The analysis can then be carried out in the same manner as in before. Note that such a measure of intra-block dispersion is typically independent of the ranks r_{ij}, so that given these weights, a very similar test statistic can be worked out by reference to the $(p!)^n$ permutations of the intra-block rankings. However, such a law is conditional on the given set of weights, so that we end up with conditionally distribution-free tests instead of EDF tests based on $\mathcal{L}_n$. One way of achieving the

EDF property of such weighted ranking procedures is to replace the w_{ni} by their ranks and allowing these ranks to have all possible $(n!)$ realizations. Since these have been pursued in some other chapters of this volume (and also presented in detail in Chapter 10 of *Handbook of Statistics*, Volume 4), we shall not go into further details. The main drawback of such weighted ranking procedures is that the choice of the weights (typically stochastic) retains some arbitrariness and thereby introduces some extra variability, which in turn may generally lead to some loss of efficiency when in particular the block effects are additive. This feature is shared by the other type of weighing where the ranks of the w_{ni} are used instead of their ordinary values. However, if the block effects are not additive and the intra-block error components have the same distribution with possibly different scale parameters, weighing by some measure of dispersion alone may not be fully rational, and hence, from that perspective, such weighing procedures are also subjected to criticism.

Ranking after alignment has a natural appeal for the conventional linear model even when the errors are not normally distributed. The basic idea is due to Hodges and Lehmann (1962) who considered a very simple setup, and it has been shown by Mehra and Sarangi (1967) and in a more general setup by Sen (1968b) that such procedures are quite *robust* under plausible departures from model based assumptions (including homoscedasticity, normality and independence of the errors). As such, we may like to provide more practical aspects of this methodology. To motivate the alignment procedure, we go back to the conventional linear model in (3.1) (sans the normality of the error components). Suppose further that the block-effects are either random variables (which may be taken as i.i.d.) or they are fixed, and the errors in the same block are interchangeable or exchangeable random variables. In this way, we are able to include both fixed- and mixed-effects models in our formulations. Let $\tilde{Y}_i$ be a *translation-equivariant* function of $(Y_{i1}, \ldots, Y_{ip})$, such that it is symmetric in its p arguments. Typically, we choose a robust estimator of the ith block mean response, and in order to preserve robustness, instead of the block average, median, trimmed mean or other measures of central tendency can be adopted. Define then the aligned observations as

$$\tilde{Y}_{ij} = Y_{ij} - \tilde{Y}_i, \quad j = 1, \ldots, p; \; i = 1, \ldots, n. \tag{3.6}$$

By (3.1) and (3.6), we may write

$$\tilde{Y}_{ij} = \tau_j - \tilde{\tau} + \tilde{e}_{ij}; \quad \tilde{e}_{ij} = e_{ij} - \tilde{e}_i, \tag{3.7}$$

for $j = 1, \ldots, p$; $i = 1, \ldots, n$, where $\tilde{\tau}$ and $\tilde{e}_i$ are defined by the same functional form as the $\tilde{Y}_i$. Note that for each i $(= 1, \ldots, n)$, the joint distribution of $(\tilde{e}_{i1}, \ldots, \tilde{e}_{ip})$ is symmetric in its p arguments, and moreover these vectors have the same joint distribution for all blocks. Therefore, it seems very logical to adopt an overall ranking of all the $N = np$ aligned observations $(Y_{11}, \ldots, Y_{np})$ and base a rank test statistic on such aligned ranks. The only negative feature is that the overall ranking procedure distorts the independence of the rank vectors from block to block; nevertheless they retain their permutability, and this provides the access to developing conditionally distribution-free tests for testing the null hypothesis of no treatment effect.

Let $\tilde{Y}_{i:1}, \ldots, \tilde{Y}_{i:p}$ be the order statistics corresponding to the aligned observations $\tilde{Y}_{i1}, \ldots, \tilde{Y}_{ip}$ in the ith block, for $i = 1, \ldots, n$. Then under the null hypothesis of interchangeability of the Y_{ij}, $j = 1, \ldots, p$, for each i $(= 1, \ldots, n)$, the $\tilde{Y}_{ij}$ has the (discrete) uniform distribution over the $p!$ possible permutations of the coordinates of $(\tilde{Y}_{i:1}, \ldots, \tilde{Y}_{i:p})$, and this permutation law is independent for different blocks. Thus, we obtain a group of $(p!)^n$ of permutations generated by the within block permutations of the aligned order statistics, and by reference to this (conditional) law, we can construct conditionally distribution-free tests for the hypothesis of interchangeability of the treatments. Under block-additivity, the vector of intra-block (aligned) order statistics are interchangeable, and hence, ranking after alignment (ignoring the blocks) remains rational. For the aligned observations, we define the ranks R_{ij} as in the preceding section, so that these R_{ij} take on the values $1, \ldots, N$, when ties among them are neglected, a case that may be done under very mild continuity assumptions on the error distributions. For the pooled sample size N, we introduce a set of scores $a_N(k)$, $k = 1, \ldots, N$, as in Section 2, and consider the *aligned rank statistics*:

$$T_{Nj} = n^{-1} \sum_{i=1}^{n} a_N(R_{ij}), \quad j = 1, \ldots, p. \tag{3.8}$$

Also, define

$$\bar{a}_{Ni} = p^{-1} \sum_{j=1}^{p} a_N(R_{ij}), \quad i = 1, \ldots, n; \tag{3.9}$$

$$V_N = \{n(p-1)\}^{-1} \sum_{i=1}^{n} \sum_{j=1}^{p} \{a_N(R_{ij}) - \bar{a}_{Ni}\}^2. \tag{3.10}$$

Then an aligned rank test statistic for testing the hypothesis of no treatment effect can be formulated as

$$\mathcal{L}_N^o = n\left\{ \sum_{j=1}^{p} [T_{Nj} - \bar{a}_N]^2 \right\} / V_N. \tag{3.11}$$

For small values of n (and p), the permutational (conditional) distribution of $\mathcal{L}_N^o$ can be incorporated to construct a conditionally distribution-free test for the above hypothesis, while, it follows from Sen (1968b) that for large sample sizes, under the null hypothesis, $\mathcal{L}_N^o$ has closely chi squared distribution with $p-1$ DF. Various robustness properties of such aligned rank tests have been studied in detail by Sen (1968c). It has been observed there that it may not be necessary that the aligned errors $\tilde{e}_{ij}$ have the common distribution for all i (i.e., blocks). In particular, for the heteroscedastic model, allowing the scale parameters to vary from block to block, it was observed that an aligned rank test may have greater ARE with respect to the classical ANOVA test than in the homoscedastic case. Some of these details are also reported in Chapter 7 of Puri and Sen (1971). Also, the alignment procedure remains applicable in the mixed-effects model too, where the block effects being stochastic or not drop out due to alignment, and hence, better robustness properties percolate. We shall discuss this aspect later on. More important is the fact that the ARE of aligned rank tests relative to the intra-block rank tests based on conjugate scores is generally greater than 1, particularly when p is not so large. For example, for the Wilcoxon score rank statistics, the ARE of the aligned rank test with respect to the Friedman χ_r^2 test is $\geqslant (p+1)/p$, so that for small values of p, there may be considerable gain in using an aligned rank test, albeit in terms of model robustness, the intra-block rank tests fare better.

In the above development, it has been assumed that each treatment is applied to one plot in each block. We may consider a more general case where the jth treatment is applied to m_j ($\geqslant 1$) plots in each block, for $j = 1, \ldots, p$. We let $M = \sum_{j \leqslant p} m_j$ and $N' = nM$. Thus, the aligned ranks span over the set $\{1, \ldots, N'\}$, and the average of the nm_j rank scores for the jth treatment is denoted by $T_{N',j}$, $j = 1, \ldots, p$. The definition of V_N is modified accordingly. Then, as a direct extension of (3.11), we may consider the following aligned rank test statistic:

$$\mathcal{L}_{N'}^{o} = [n/V_{N'}] \sum_{j=1}^{p} m_j [T_{N',j} - \bar{a}_{N'}]^2. \tag{3.12}$$

This test is also conditionally distribution-free and under the null hypothesis, it has asymptotically central chi squared distribution with $p-1$ DF. It enjoys all the robustness and asymptotic efficiency properties as in the particular case of all the m_j being equal to 1.

Next, as in the case of one-way layouts treated in Section 2, we consider the problem of simultaneous testing for all paired treatment differences. We may note here that there are some variations in the formulation of such tests. Treatments vs. control tests compare simultaneously all the treatments with a control, so that there is some asymmetry in this setup. A simultaneous test for all possible $\binom{p}{2}$ treatment differences preserves the symmetry to a greater extent. It is also possible to formulate such simultaneous tests based on intra-block rank statistics, but they may not incorporate inter-block information, and hence, may be a little less efficient than the ones based on aligned rank statistics. The basic results are due to Sen (1970a). We define the T_N, j and V_N as in before and let

$$W_N = \max\left\{n^{1/2}|T_{N,j} - T_{N,l}|/V_N^{1/2}\colon\ 1 \leqslant j \neq l \leqslant p\right\} \tag{3.13}$$

and let $R_{p,\alpha}$ be the upper $100\alpha\%$ point of the distribution of the sample range of a sample of size p drawn from a standard normal distribution. Then the simultaneous (aligned) rank test is based on the decision rule that rejects all pairs (j, l) of treatments as significantly different for which

$$n^{1/2}|T_{N,j} - T_{N,l}| \geqslant R_{p,\alpha} V_N^{1/2}. \tag{3.14}$$

This simultaneous test is also conditionally distribution-free, and asymptotically it has the level of significance α.

In many situations the treatments represent an increasing sequence of *doses*, and it may therefore be of natural interest to test for the null hypothesis of equality of treatment effects against an *ordered alternative* H_1: $\tau_1 \leqslant \cdots \leqslant \tau_p$ (with at least one strict inequality). Tests for such ordered alternatives may also be based on intra-block rank statistics, but they may be comparatively less efficient for the same reason (of not incorporating inter-block information that effectively). Hence, aligned rank tests are often advocated. We define the aligned rank statistics $T_{N,j}$, $j = 1, \ldots, p$, as in before, and let $\boldsymbol{a} = (a_1, \ldots, a_p)'$ be a vector of real coefficients satisfying the conditions: (i) $\sum_{j=1}^{p} a_j = 0$, (ii) $\sum_{j=1}^{p} a_j^2 = 1$ and (iii) $a_1 \leqslant a_2 \leqslant \cdots \leqslant a_p$ with at least one strict inequality. Then a test statistic may be defined conveniently as

$$Q_n = n\left[\sum_{j=1}^{p} a_j\{T_{N,j} - \bar{a}_N\}\right] / V_N^{1/2}. \tag{3.15}$$

The choice of $\boldsymbol{a}$ may be made on some heuristic considerations, and in this context, Bayesian solutions have also been incorporated. Among such possibilities, the following one (viz., Sen, 1968b) provides a simple interpretation from linear trend point of view. We let

$$a_j = \sqrt{12}\{j - (p+1)/2\}/\{p(p^2-1)\}^{1/2}, \quad j = 1, \ldots, p. \tag{3.16}$$

The resulting test statistic

$$Q_n^o = \sqrt{12n}\left\{\sum_{j=1}^{p}(j - (p+1)/2)T_{N,j}\right\}/\{p(p^2-1)V_N\}^{1/2} \tag{3.17}$$

is conditionally distribution-free (under H_0) and asymptotically normal. It provides a robust and efficient test for ordered alternatives. Some other nonparametric tests for ordered alternatives in randomized blocks are discussed in detail in Chapter 7 of Puri and Sen (1971).

As in Section 2, we will consider here R-estimators of the treatment effects $\tau_1, \ldots, \tau_p$, treating the block effects as nuisance parameters (or possibly random variables, in a mixed-effects setup). In this context, the aligned rank statistics introduced earlier play a vital role. For each pair (j, l), consider the paired differences $Y_{i,jl} = Y_{ij} - Y_{il}$, for $i = 1, \ldots, n$. We write $\Delta_{jl} = \tau_j - \tau_l$ and $e_{i,jl} = e_{ij} - e_{il}$, for $i = 1, \ldots, n; 1 \leqslant j < l \leqslant p$. Then the $e_{i,jl}$ have a d.f. symmetric about 0, so that we may use an aligned signed rank statistic $S_{n,jl}(d)$ as in (2.25) and equating this to 0 (with respect to *d*), we obtain an R-estimator $(\widehat{\Delta}_{n,jl})$ of Δ_{jl}. In this process, we obtain the set of $\binom{p}{2}$ estimators $\widehat{\Delta}_{n,jl}$, $1 \leqslant j < l \leqslant p$. We let $\Delta_{jj} = 0$ and note that by construction, $\widehat{\Delta}_{n,lj} = -\widehat{\Delta}_{n,jl}$, for all $j, l = 1, \ldots, p$. In the case of least squares estimation theory, the estimators are linear, so that an estimator of a contrast in the τ_j can be expressed in terms of such paired difference estimators in an arbitrary manner. But, the R-estimators may not be strictly linear, and hence, for small number of blocks at least, there remains some arbitrariness in combining such paired difference estimators to yield an estimator of an arbitrary contrast. Lehmann (1963a, b, 1964) suggested a simple modification to yield *compatible R-estimators* of contrasts in randomized block designs. For this define

$$\widehat{\Delta}_{n,j\cdot} = p^{-1} \sum_{l=1}^{p} \widehat{\Delta}_{n,jl}, \quad j = 1, \ldots, p; \tag{3.18}$$

$$\widehat{\Delta}^{o}_{n,jL} = \widehat{\Delta}_{n,j\cdot} - \widehat{\Delta}_{n,l\cdot}, \quad j, l = 1, \ldots, p. \tag{3.19}$$

Like R-estimators in Section 2, the compatible estimators are translation-equivariant, and they are robust, consistent and asymptotically normally distributed. However, the way (3.19) has been formulated, expressions for the asymptotic dispersion matrix of the compatible estimators are slightly different from the ones in Section 2 (cf. (2.20)). Towards this we introduce the following parameters. Let $G(\cdot)$ be the common marginal d.f. of the paired difference $e_{i,jl}$, and let $G^*(\cdot)$ be the bivariate d.f. of a pair $(e_{i,jl}, e_{i,jl'})$, where $l \neq l'$. Note that the process of alignment distorts the independence of the errors even when the original e_{ij} were stochastically independent. Also, let $\phi(\cdot)$ be the score generating function for the rank statistics. Then we define $A^2 = \int_0^1 \phi^2(u)\,du$ and

$$\gamma(\phi, G) = \int_{\mathcal{R}} (d/dx)\phi(G(x))\,\mathrm{d}G(x), \tag{3.20}$$

and

$$\lambda_\phi(G) = \int_{-\infty}^{\infty}\int_{-\infty}^{\infty} \phi(G(x))\phi(G(y))\,\mathrm{d}G^*(x, y). \tag{3.21}$$

The score generating function is taken to be skew-symmetric about $u = 1/2$, so that $\overline{\phi} = 0$. Moreover, by definition in (3.19), the $\widehat{\Delta}^o_{n,jl}$ are expressible as paired differences of the p statistics $\widehat{\Delta}_{n,j\cdot}$, $j = 1, \ldots, p$. Hence, to study the joint distribution of all these $\binom{p}{2}$ estimators, it suffices to consider only the $p-1$ vector $n^{1/2}(\widehat{\Delta}^o_{n,1p} - \Delta_{1p}, \ldots, \widehat{\Delta}^o_{n,p-1p} - \Delta_{p-1p})$. This vector has asymptotically a $p-1$ variate normal distribution with null mean vector and dispersion matrix $\sigma_o^2[\boldsymbol{I}_{p-1} + \mathbf{1}_{p-1}\mathbf{1}'_{p-1}]$, where $\mathbf{1}_{p-1} = (1, \ldots, 1)'$, and

$$\sigma_o^2 = [(p-1)/p]\{A^2 + (p-2)\lambda_\phi(G)\}/\gamma^2(\phi, G). \tag{3.22}$$

If we compare the dispersion matrix of the compatible R-estimators with that of the raw R-estimators derived earlier, we obtain that the ARE of the compatible estimators relative to the raw ones is given by

$$e(\widehat{\Delta}^{o}; \widehat{\Delta}) = pA^2/\{2[A^2 + (p-2)\lambda_\phi(G)]\}, \tag{3.23}$$

and using the easily verifiable inequality that $\lambda_\phi(G) \leqslant (1/2)A^2$, it readily follows that (3.23) is bounded from below by 1 (albeit, this is usually quite close to 1). This shows that such compatible R-estimators are also preferable on the ground of their asymptotic efficiency properties. For details, we may refer to Puri and Sen (1971, Chapter 7).

4. Nonparametric MANOVA

In a general multivariate setup, the response $\boldsymbol{Y}$ is a p-vector, for some $p \geqslant 1$, and as in Sections 2 and 3, these responses may be related to various factors (fixed-, random- or mixed-effects models). In the conventional case, one assumes that a linear model as in (2.1) or (3.1) holds where the e_i or e_{ij} are distributed according to a multinormal distribution with null mean vector and a positive definite (unknown) dispersion matrix $\boldsymbol{\Sigma}$. The scope for departures from this assumed model is even more in the multivariate case, as normality is even less likely to hold in the multivariate than univariate cases. In design aspects, of course, the situation is quite similar to that in the univariate case, but in statistical analysis, the situation is more complex in multivariate models. In nonparametrics, there are additional impasses in the multivariate case. In the conventional linear model, *affine transformations* are often used to simplify the distribution theory of appropriate test statistics which are invariant under affine transformations or of estimators which are *affine-equivariant*. In ranking procedures, usually ranking is made for each of the p coordinates separately. Thus, in one hand, such procedures are invariant for a larger group of (not necessarily linear) strictly monotone transformation for each coordinate, while, on the other hand, they are not affine-invariant. Therefore, in nonparametrics for multivariate analysis of variance (MANOVA), affine-invariance is not generally true. This does not, of course, pose a serious problem, as in many cases, the coordinate responses may not be quite conformable in a sense that an arbitrary linear compound will have a meaningful interpretation. In such a case, coordinatewise ranking with due emphasis on their dependence may serve the purpose much better. The other serious problem with nonparametrics in MANOVA is the lack of EDF property in a general setup. For examplc, in Section 2 we have posed the permutational invariance property which yields $n!$ equally likely permutations of $\{1,\ldots,n\}$ for the rank vector. In the multivariate case, we have the rank matrix of order $p \times n$, so that the total number of possible realizations of these matrices is equal to $(n!)^p$, and the distribution of the rank matrix over this set depends, in general, on the underlying (multivariate) distribution. Thus, a test based on the coordinatewise ranking, in general, may not be genuinely distribution-free. A very similar situation arises in two-way layouts when one uses the method of ranking separately for each coordinate. This drawback of multivariate nonparametrics has been eliminated largely by the *rank permutation principle* due to Chatterjee and Sen (1964). This rank permutation principle has been exploited in various directions, and up-to-date accounts of these developments are given by Puri and Sen (1971, 1985). We present here only a brief synopsis of the main highlights of these developments with due emphasis on the design aspects.

First, consider the multivariate analogue of (2.3) where the $\boldsymbol{Y}_i$ are p-vectors, so that $\boldsymbol{\theta}$ and the $\boldsymbol{e}_i$ are p-vectors too, while $\boldsymbol{\beta}$ is a $p \times q$ matrix. Then $\boldsymbol{e}_i$ is assumed to have a p-variate (continuous) d.f. F, which need not have independence structure. The hypotheses in (2.4) remain the same. Let us denote the rank of Y_{ij} among the Y_{rj}, $r = 1, \ldots, n$, by R_{nij}, for $i = 1, \ldots, n$; $j = 1, \ldots, p$. As in (2.5), for the jth coordinate, we denote the vector of linear rank statistics by $\boldsymbol{L}_{nj}$, for $j = 1, \ldots, p$, where, we may even take the scores $a_n(\cdot)$ possibly different for different j $(= 1, \ldots, p)$; we may add an additional subscript j to $a_n(\cdot)$ to do so, but for notational simplicity, this refinement is suppressed. We also define a $p \times p$ matrix $\boldsymbol{V}_n$ as in (2.34), and the rank collection matrix $\boldsymbol{R}_n$ is defined as in after (2.33). The permutational (conditional) probability law is then generated by the $n!$ (conditionally) equally likely permutations of the columns of the rank collection matrix. We define $\boldsymbol{Q}_n$ as in (2.7), and if we denote the $p \times q$ matrix of the L_{njl} by $\boldsymbol{L}_n$, as in (2.8), we will have here $E_{\mathcal{P}_n}(\boldsymbol{L}_n) = \boldsymbol{0}$ and

$$E_{\mathcal{P}_n}[\text{vec } \boldsymbol{L}_n \text{vec } \boldsymbol{L}_n'] = \boldsymbol{V}_n \otimes \boldsymbol{Q}_n, \tag{4.1}$$

where vec $\boldsymbol{L}_n$ is the pq-vector obtained from $\boldsymbol{L}_n$ by stacking the columns over each other and $\otimes$ stands for the Kronecker product of the two matrices. Then, as an extension of (2.9) to the multivariate case, we have the test statistic

$$\mathcal{L}_n = [\text{vec } \boldsymbol{L}_n'(\boldsymbol{V}_n \otimes \boldsymbol{Q}_n)^- \text{vec } \boldsymbol{L}_n]. \tag{4.2}$$

For small values of n, the exact (conditional) permutational distribution of $\mathcal{L}_n$ can be obtained by direct enumeration of all possible $n!$ column permutations of the rank collection matrix. This process becomes cumbersome as n increases, but, for large n, this permutation distribution as well as the unconditional null hypothesis distribution of $\mathcal{L}_n$ can be approximated by the central chi squared distribution with pq DF. Thus, we may proceed as in (2.10) with q replaced by pq, and also, we may consider, in the same vein, local Pitman-type alternatives as in (2.11) and consider asymptotic power properties of such multivariate rank tests. There are some difficulties concerning the characterization of asymptotically optimal rank tests (compare with (2.12)), and further regularity conditions are needed to establish such properties in a general multivariate setup. We may refer to Puri and Sen (1985, Chapters 5–7). For the specific multisample multivariate case, treated in the univariate case in Section 2, the expression for $\mathcal{L}_n$ in (4.2) simplifies to the following:

$$\mathcal{L}_n = \sum_{k=1}^{c} n_k(\overline{\boldsymbol{a}}_{n,k} - \overline{\boldsymbol{a}}_n)' \boldsymbol{V}_n^-(\overline{\boldsymbol{a}}_{n,k} - \overline{\boldsymbol{a}}_n), \tag{4.3}$$

where the $\overline{\boldsymbol{a}}_{n,k}$ and $\overline{\boldsymbol{a}}_n$ are defined as in (2.13), but for the vector case. The discussion following (2.13) pertaining to the specific choice of scores also applies here.

Let us consider the R-estimation problem for the parameter matrix $\boldsymbol{\beta}$. As a natural generalization of (2.14), we may consider here a measure of rank dispersion $D_n(\boldsymbol{B})$ defined for $\boldsymbol{B} \in \boldsymbol{R}^{pq}$, wherein we replace the $a_N(R_{ni}(\boldsymbol{b}))$ by $\boldsymbol{a}_n(\boldsymbol{R}_{ni}(\boldsymbol{B}))$, $\bar{a}_n$ by $\overline{\boldsymbol{a}}_n$ and $Y_i(\boldsymbol{b})$ by $\boldsymbol{Y}_i(\boldsymbol{B}) = \boldsymbol{Y}_i - \boldsymbol{B}(\boldsymbol{t}_i - \bar{\boldsymbol{t}}_n)$; also we need to introduce a matrix to depict the scale factors and possible dependence of the coordinates of the vector $\boldsymbol{Y}$. Because the rankings are made separately for each coordinate and affine invariance may not generally hold for such R-estimators, we find it convenient to adopt a coordinatewise R-estimation procedure. Based on the aligned scores and aligned observations on the jth coordinate, we define a measure of rank dispersion as in (2.14), for $j = 1, \ldots, p$. Then proceeding virtually as in (2.14) through (2.17), we arrive at the following:

$$\boldsymbol{L}_{nj}(\boldsymbol{b}_j) = \boldsymbol{0}, \quad j = 1, \ldots, p, \tag{4.4}$$

where each $\boldsymbol{b}_j$ is a q-vector. If we express $\boldsymbol{\beta}' = (\boldsymbol{\beta}'_1, \ldots, \boldsymbol{\beta}'_p)$, then the jth estimating equation in (4.4) yields the R-estimator $\widehat{\boldsymbol{\beta}}_{nj}$ of $\boldsymbol{\beta}_j$, for $j = 1, \ldots, p$. Each of these estimators (vectors) satisfies a FOADR result given in (2.18) where we need to attach a subscript j to each $\widehat{\boldsymbol{\beta}}_n$, $\boldsymbol{\beta}$, γ, ϕ, F and e_i to indicate their dependence on the jth coordinate, for $j = 1, \ldots, p$. We denote the dispersion matrix of the p-vector $(\gamma_1^{-1}\phi_1(F_{[1]}(e_{i1})), \ldots, \gamma_p^{-1}\phi(F_{[p]}(e_{ip})))$ by $\boldsymbol{\Psi}$, and then as a direct extension of (2.20) in the multivariate case, we arrive at the following: As $n \to \infty$,

$$(\widehat{\boldsymbol{\beta}}_n - \boldsymbol{\beta})\boldsymbol{Q}_n^* \xrightarrow{\mathcal{D}} \mathcal{N}(\boldsymbol{0}, \boldsymbol{\Psi} \otimes \boldsymbol{I}), \tag{4.5}$$

where $\boldsymbol{Q}_n^*$ is defined as in (2.20). In this matrix case, the definition of the ARE in (2.21) needs some modification, and the usual *A-*, *D-* and *E-optimality* (*efficiency*) criteria can be incorporated to suit the purpose. Again, we may refer to Puri and Sen (1985, Chapter 6) for some of these details.

Let us consider next the MANOVA nonparametrics for the two-way layouts. As in Section 3, here also, we may consider the intra-block ranking and ranking after alignment cases separately, and compare their merits and demerits. The intra-block ranking method was considered by Gerig (1969) who developed a multivariate extension of the Friedman χ_r^2 test statistic and used the same intra-block permutation groups to develop a permutationally (conditionally) distribution-free test for the null hypothesis of no treatment differences across the q variates. Let us go back to (3.5) and denote by $r_{ij}^{(k)}$ the rank on the kth variate for the jth treatment in the ith block, when ranking is made separately for each variate and within each block. Thus, we will have n rank matrices $\boldsymbol{R}_i$, $i = 1, \ldots, n$, where each $\boldsymbol{R}_i$ is a $q \times p$ matrix with the elements $r_{ij}^{(k)}$. We define a $q \times p$ matrix of rank statistics

$$\bar{r}_{nj}^{(k)} = n^{-1} \sum_{i=1}^{n} r_{ij}^{(k)}, \quad j = 1, \ldots, p; \ k = 1, \ldots, q,$$

express this into a pq-vector (as in earlier this section), and also define a $q \times q$ matrix $\boldsymbol{V}_n = ((v_{nkl}))$ by letting

$$v_{n,kl} = [n(p-1)]^{-1} \sum_{i=1}^{n} \sum_{j=1}^{p} \left\{ r_{ij}^{(k)} - (p+1)/2 \right\} \left\{ r_{ij}^{(l)} - (p+1)/2 \right\},$$
$$k, l = 1, \ldots, q. \tag{4.6}$$

Then, the multivariate analogue of (3.5) is given by

$$\mathcal{L}_n = n \left[\sum_{j=1}^{p} \sum_{k=1}^{q} \sum_{l=1}^{q} v_n^{kl} \left\{ \bar{r}_{nj}^{(k)} - (p+1)/2 \right\} \left\{ \bar{r}_{nj}^{(l)} - (p+1)/2 \right\} \right], \tag{4.7}$$

where the v_n^{kl} stand for the elements of the matrix $\boldsymbol{V}_n^{-1}$. The permutational (conditional) distribution of $\mathcal{L}_n$ over the set of $(p!)^n$ intra-block column permutations of the rank matrices can be approximated well by the chi squared distribution with $q(p-1)$ DF when n is large, and asymptotic power properties have been studied by Gerig (1969) and others. Gerig (1975) has also extended this test for the multivariate analysis of covariance (MANOCOVA) problem, and has studied its robustness properties too. The modifications follow along the lines of (2.35) through (2.39), and hence, we omit the details. As in the univariate case, such intra-block rank tests are generally not fully informative (as they may not recover the inter-block information to a satisfactory extent), and for this reason, aligned rank procedures in MANOVA and MANOCOVA are often preferred. The prospect for weighted rankings is somewhat less apparent here as the weight would depend on the intra-block dispersion matrices, and hence, more delicate considerations are needed in a rational formulation. Most of the developments on aligned rank procedures in MANOVA/MANOCOVA are due to Sen (1969b, 1984a), and presented succinctly here.

We start with the conventional linear model in (3.1), as extended to the multivariate case, and we drop the assumption of normality of errors. Thus, the response vector for the jth treatment in the ith block is a q-vector $\boldsymbol{Y}_{ij}$, and as such, in (3.1), we change the elements on the right hand side by appropriate q-vectors. For each coordinate $(k = 1 \ldots, q)$, we consider a suitable translation-equivariant intra-block measure of central tendency, and denote these by $\tilde{\boldsymbol{Y}}_i, i = 1, \ldots, n$. Then, the aligned response vectors are defined as in (3.7) as $\tilde{\boldsymbol{Y}}_{ij} = \boldsymbol{Y}_{ij} - \tilde{\boldsymbol{Y}}_i$, $j = 1, \ldots, p$; $i = 1, \ldots, n$. For each k $(= 1, \ldots, q)$, we introduce the aligned ranks $R_{ij}^{(k)}$, $j = 1, \ldots, p$; $i = 1, \ldots, n$, as in before (3.8), and we define the set of aligned rank statistics $T_{nj}^{(k)}$, $j = 1, \ldots, p$; $k = 1, \ldots, q$, as in (3.8) (where we may even allow the score function $a_n(\cdot)$ to vary from one coordinate to another, but for notational simplicity this is dropped). As a natural extension of (3.10) to the multivariate case, we define then $\boldsymbol{V}_N = ((v_{N,kl}))$, by letting for each $k, l = 1, \ldots, q$,

$$v_{N,kl} = [n(p-1)]^{-1}$$
$$\times \sum_{i=1}^{n} \sum_{j=1}^{p} \left\{ a_N(R_{ij}^{(k)}) - \bar{a}_{N,i}^{(k)} \right\} \left\{ a_N(R_{ij}^{(l)}) - \bar{a}_{N,i}^{(l)} \right\}, \tag{4.8}$$

where the intrablock average rank score vectors have the elements denoted by $\bar{a}_{N,i}^{(k)}$, for $k = 1, \ldots, q$; $i = 1, \ldots, n$. Then, a multivariate version of the general aligned rank statistic in (3.11) is given by

$$\mathcal{L}_N^o = n \sum_{j=1}^{p} \sum_{k=1}^{q} \sum_{l=1}^{q} v_N^{kl} \left\{ T_{N,j}^{(k)} - \bar{a}_N^{(k)} \right\} \left\{ T_{N,j}^{(l)} - \bar{a}_N^{(l)} \right\}, \tag{4.9}$$

where the v_N^{kl} are the elements of $\boldsymbol{V}_N^{-1}$. Here also, a conditionally (permutationally) distribution-free test based on $\mathcal{L}_N^o$ can be obtained by reference to the set of $(p!)^N$ intrablock column permutations of the aligned observation matrices, and for large n, this conditional distribution as well as the unconditional null hypothesis distribution can be well approximated by the central chi squared distribution with $q(p-1)$ DF. Various asymptotic (power and efficiency) properties of this nonparametric MANOVA procedure are studied by Sen (1969b). MANOCOVA nonparametrics also follow the same line of attack as in (2.35) through (2.38). Multivariate extensions of simultaneous (aligned) rank tests for all possible pairs of treatments and all possible coordinates follow by using the Roy (1953) *largest root* criterion.

5. Nonparametrics for incomplete block designs

In this section, we deal with a general subclass of two-way layouts where possibly due to a large number of treatments, blocks of smaller size are used, so that not all treatments are applied to all blocks. In the literature, these are referred to as *Incomplete Block Designs* (IBD). Consider n replications of an IBD consisting of b blocks of constant size k ($\geqslant 2$) to which r ($\geqslant k$) treatments are applied in such a way that (i) no treatment occurs more than once in any block, (ii) the jth treatment occurs in r_j ($\leqslant b$) blocks, and (iii) the (j, j')th treatments occur together in $r_{jj'}$ (> 0) blocks, for $j \neq j' = 1, \ldots, v$. Let then $\mathcal{S}_j$ stand for the set of treatments occurring in the ith block, $i = 1, \ldots, b$. In the sth replicate ($s = 1, \ldots, n$), the response of the plot in the ith block and receiving the jth treatment is a stochastic p-vector $\boldsymbol{X}_{sij}$, for $j \in \mathcal{S}_i$; $i = 1, \ldots, b$. In the univariate case (i.e., $p = 1$), intra-block rank tests for IBD's are due to Durbin (1951), Benard and Elteren (1953) and Bhapkar (1961a), among others. For some special IBD's, the studies made by Elteren and Noether (1959) and Bhapkar (1961a) reveal the low (Pitman-) efficiency of such tests, particularly when k is small. For this reason and motivated by the results in the preceding two sections, we shall mainly consider here suitable aligned rank tests for IBD's. We shall only summarize the results and for details, we may refer to Sen (1971a).

We consider the model

$$\boldsymbol{X}_{sij} = \boldsymbol{\mu}_s + \boldsymbol{\beta}_{si} + \boldsymbol{\tau}_j + \boldsymbol{\varepsilon}_{sij}, \quad j \in \mathcal{S}_i, \tag{5.1}$$

for $i = 1, \ldots, b$; $s = 1, \ldots, n$, where the $\boldsymbol{\mu}_s$ stand for the *replicate effects*, $\boldsymbol{\beta}_{si}$ for the block effects (nuisance parameters in fixed-effects models or spurious random vectors in mixed-effects models), $\boldsymbol{\tau}_1, \ldots, \boldsymbol{\tau}_v$ are the treatment effects (parameters of interest) and the $\boldsymbol{\varepsilon}_{sij}$ are the error vectors. We may set without any loss of generality $\sum_{j=1}^{v} \boldsymbol{\tau}_j = \mathbf{0}$. Instead of the specific multinormality assumption on the errors, it is assumed that for each (s,i), $\{\boldsymbol{\varepsilon}_{sij},\ j \in \mathcal{S}_i\}$ have jointly a continuous cumulative d.f. $G(\boldsymbol{x}_1, \ldots, \boldsymbol{x}_k)$ which is symmetric in its k argument vectors. This includes the conventional assumption of independence and identity of distributions of all the N $(= nbk)$ error vectors as a special case. As in Sections 3 or 4, we define the aligned observations by

$$\boldsymbol{Y}_{sij} = \boldsymbol{X}_{sij} - k^{-1} \sum_{l \in \mathcal{S}_i} \boldsymbol{X}_{sil}, \quad j \in \mathcal{S}_i,\ i = 1, \ldots, b,\ s = 1, \ldots, n, \tag{5.2}$$

and let $R_{sij}^{(k)}$ be the rank of $Y_{sij}^{(k)}$ among the N aligned observations on the kth response variate, for $j \in \mathcal{S}_i$, $i = 1, \ldots, b$, $s = 1, \ldots, n$; $k = 1, \ldots, q$. Also, for each i $(= 1, \ldots, b)$, let

$$\boldsymbol{\tau}_{j,i} = \boldsymbol{\tau}_j - k^{-1} \sum_{j \in \mathcal{S}_i} \boldsymbol{\tau}_l, \qquad \boldsymbol{e}_{i,j} = \boldsymbol{\varepsilon}_{ij} - k^{-1} \sum_{l \in \mathcal{S}_i} \boldsymbol{\varepsilon}_{il}, \tag{5.3}$$

for $j \in \mathcal{S}_i$, $i = 1, \ldots, b$; $s = 1, \ldots, n$. Then, we have

$$\boldsymbol{Y}_{sij} = \boldsymbol{\tau}_{j,i} + \boldsymbol{e}_{i,j}, \quad j \in \mathcal{S}_i,\ 1 \leqslant i \leqslant b;\ 1 \leqslant s \leqslant n. \tag{5.4}$$

We want to test the null hypothesis of no treatment effect, i.e.,

$$H_o\colon\ \boldsymbol{\tau}_1 = \cdots = \boldsymbol{\tau}_v = \mathbf{0}, \tag{5.5}$$

against the set of alternatives that at least one of the $\boldsymbol{\tau}_j$ is different from $\mathbf{0}$. Note that for each (s,i), the $\boldsymbol{e}_{sij}$, $j \in \mathcal{S}_i$ are exchangeable random vectors, so that under the null hypothesis, by (5.4), the aligned vectors within each block (in each replicate) are also exchangeable. This provides the same permutational invariance structure as in the case of complete blocks, and hence, similar conditionally (permutationally) distribution-free aligned rank tests can be constructed. Let us denote by

$$\mathcal{P}_j = \{i(\in [1,b])\colon\ j \in \mathcal{S}_i\}, \quad j = 1, \ldots, v. \tag{5.6}$$

Then, for each k $(= 1, \ldots, q)$, we introduce scores $a_{Nk}(\alpha)$, $\alpha = 1, \ldots, N$, as in Section 3 (or 4), and denote the block averages of these scores by $\bar{a}_{Nsi.}^{(k)}$, replicate averages by $\bar{a}_{Ns..}^{(k)}$ and the grand average by $\bar{a}_N^{(k)}$. Let then

$$T_{N,j}^{(k)} = n^{-1} \sum_{s=1}^{n} \sum_{i \in \mathcal{P}_j} a_{Nk}(R_{ij}^{(k)}), \quad j = 1, \ldots, v;\ k = 1, \ldots, q. \tag{5.7}$$

Recall that the r_j may not be all equal, and so may not be the $r_{jj'}$. This calls for some adjustments for the permutational covariance matrix. We define two matrices $\boldsymbol{V}_N^{(l)} = ((v_{N,kk'}^{(l)}))$, $l = 1, 2$, where for $k, k' = 1, \ldots, q$

$$v_{N,kk'}^{(1)} = (nbk)^{-1} \times \sum_{s=1}^{n} \sum_{i=1}^{b} \sum_{j \in \mathcal{S}_i} \{a_{Nk}(R_{ij}^{(k)} - \bar{a}_{Nsi.}^{(k)}\}\{a_{Nk}(R_{ij}^{(k')} - \bar{a}_{Ns.}^{(k')}\}; \quad (5.8)$$

and

$$v_{N,kk'}^{(2)} = (nb)^{-1} \sum_{s=1}^{n} \sum_{i=1}^{b} [\bar{a}_{Nsi.}^{(k)} - \bar{a}_{Ns..}^{(k)}][\bar{a}_{Nsi.}^{(k')} - \bar{a}_{Ns..}^{(k')}]. \quad (5.9)$$

Also, define two (design) matrices $\boldsymbol{A}^{(l)} = ((a_{jj'}^{(l)}))$, $l = 1, 2$, by letting

$$a_{jj'}^{(1)} = [kr_j\delta_{jj'} - r_{jj'}]/(k-1), a_{jj'}^{(2)} = [br_{jj'} - r_jR_{j'}]/(b-1), \quad (5.10)$$

for $j, j' = 1, \ldots, v$, where $\delta_{jj'}$ is the usual Kronecker delta, and $r_{jj} = r_j$. Further, let

$$\boldsymbol{W}_N = \boldsymbol{A}^{(1)} \otimes \boldsymbol{V}_N^{(1)} + \boldsymbol{A}^{(2)} \otimes \boldsymbol{V}_N^{(2)}. \quad (5.11)$$

Finally, let

$$\boldsymbol{T}_N = (T_{N,1}^{(1)}, \ldots, T_{N,v}^{(1)}, \ldots, T_{N,1}^{(q)}, \ldots, T_{N,v}^{(q)}), \quad (5.12)$$

and

$$\boldsymbol{\eta} = (\bar{a}_N^{(1)}(r_1, \ldots, r_v), \ldots, \bar{a}_N^{(q)}(r_1, \ldots, r_v)). \quad (5.13)$$

Then, as in Sen (1971a), we may consider the following aligned rank order test statistic:

$$\mathcal{L}_N^* = (\boldsymbol{T}_N - \boldsymbol{\eta})\boldsymbol{W}_N^-(\boldsymbol{T}_N - \boldsymbol{\eta})', \quad (5.14)$$

where $\boldsymbol{W}_N^-$ stands for a generalized inverse of $\boldsymbol{W}_N$. Keeping in mind the class of balanced, partially balanced and group divisible IBD's, as in Sen (1971), we may assume that

$$\text{Rank of } \boldsymbol{A}^{(1)} = v - 1, \quad (5.15)$$

and

$$\boldsymbol{A}^{(2)} \text{ and } b\boldsymbol{A}^{(1)} - (b-1)\boldsymbol{A}^{(2)} \text{ are positive semi-definite.} \quad (5.16)$$

For small n, the exact permutation distribution of $\mathcal{L}_N^*$ can be obtained by direct enumeration, but the task becomes prohibitively laborious as n increases. However, as in Sen (1971a), we claim that under the null hypothesis, the permutational (conditional) distribution of $\mathcal{L}_N^*$ as well as the unconditional distribution can be approximated by the central chi squared distribution with $q(v-1)$ DF. For studies of asymptotic power and relative efficiency properties of such aligned rank tests in IBD's, we may refer to Sen (1971a). In the univariate case (i.e., for $q = 1$), for balanced incomplete block designs, the ARE results for intra-block rank tests were studied by Elteren and Noether (1959). In general, aligned rank tests fare better for IBD's, particularly when k is small. It has been observed in this context that aligned rank tests are also robust to possible heteroscedasticity of the joint error distributions from replicate to replicate, a case that may often arise in practice when the replicates are not so homogeneous in a statistical sense. In the nonparametric case, so long as the linearity of the model can be assumed (but the errors need not be normally distributed), the same block totals can be put into an alignment scheme for generating aligned rank tests which provides additional information on the hypothesis testing problem. These details can be sketched as in Sen (1971a), although it would be more advantageous for us to report on this nonparametric recovery of interblock information in a comparatively more general setup in clinical trials in a later section. As such, we omit the details here. We conclude this section with a note that robust R-estimation of treatment effects (contrasts) in IBD's can be formulated very much in the same way as in Section 3 (viz., (3.18) through (3.23)). In an univariate setup, this was done by Greenberg (1966) and Puri and Sen (1967), while by virtue of the comments on R-estimation in the multivariate case made in Section 4, these findings extend readily to the general model treated in this section. Therefore, we omit these details.

6. Nonparametrics in factorial designs

Nonparametric ANOVA, MANOVA and (M)ANOCOVA models presented in the preceding sections relate mostly to the case where treatmentwise there is a one-way layout, although incorporating blockwise variations, it may be a two-factor model. There are many situations where the treatments represent the combinations (at two or more levels) of two or more factors, so that we may not only be interested in their *main effects* but also in their possible *interactions*. In a normal theory model, such interaction effects and main effects all can be handled by suitable (linear) transformations on the original response variables, and similar test statistics can be used to test for plausible null hypothesis of no interaction or no main effects. The situation is more complex in the nonparametric case. The primary impasse stems out of the fact that whereas the least squares methodology addresses well the invariance under affine transformations on the response vectors, their (coordinatewise) ranks are not affine-invariant. Moreover, for testing the null hypothesis of no interaction, it may be more reasonable to assume that the main effects of the various factors may not have insignificant differences, so that they should be treated as nuisance parameters. Although some people have tried the *rank transformation* approach, referred to in Section 3, to mimic the usual ANOVA tests based on such transformed vectors, there are some serious theoretical deficiencies of such procedures in a general multi-factor experiment. The formulation of null and alternative hypotheses requires a much more restricted setup for such rank transformed data sets (hinging on strictly monotone but arbitrary nonlinear transformations), and, often, there is a big compromise on the underlying robustness aspects (which were the original motivations for favoring a nonparametric approach). For these reasons, we shall not emphasize on such rank transformations in factorial designs, and we continue exploring aligned rank procedures in such designs. We shall mainly follow the approach of Mehra and Sen (1969).

We consider the case of replicated two-factor experiments with one observations per cell. Let Y_{ijk} be the response variate for the cell (j,k) in the ith replicate, and assume that the following fixed-effects factorial model holds:

$$Y_{ijk} = \mu_i + \nu_j + \tau_k + \gamma_{jk} + \omega_{ijk}, \tag{6.1}$$

for $i = 1,\ldots,n$; $j = 1,\ldots,p$; $k = 1,\ldots,q$, with $n \geqslant 2$, $p \geqslant 2$, $q \geqslant 2$. Here the μ_i relate to the replicate effects, ν_j and τ_k to the main effects for the two factors, γ_{jk} for the interaction effects of the two factors, and ω_{ijk} are the residual error components. We may set without any loss of generality

$$\sum_{i=1}^{n} \mu_i = 0, \qquad \sum_{j=1}^{p} \nu_j = 0, \qquad \sum_{k=1}^{q} \tau_k = 0, \tag{6.2}$$

and

$$\gamma_{j\cdot} = q^{-1} \sum_{k=1}^{q} \gamma_{jk} = 0, \quad j = 1,\ldots,p; \tag{6.3}$$

$$\gamma_{\cdot k} = p{-}1 \sum_{j=1}^{p} \gamma_{jk} = 0, \quad k = 1,\ldots,q. \tag{6.4}$$

It is further assumed that for each i, $(\omega_{i11},\ldots,\omega_{ipq})$ have a joint d.f. G which is a symmetric function of its pq arguments, and these n (pq-) vectors are independent. This includes the conventional assumption of i.i.d. structure of the ω_{ijk} as a particular case, and more generally, it allows each replicate error vector to have interchangeable components which may still be dependent, a case that may arise if we allow the replicate effects to be possibly stochastic, so that we would have then a mixed effects factorial model. The null hypothesis of interest is

$$H_0\colon\ \boldsymbol{\Gamma} = ((\gamma_{jk})) = \mathbf{0}, \tag{6.5}$$

against alternatives that $\boldsymbol{\Gamma}$ is non-null. We would like to formulate suitable aligned rank tests for this hypothesis testing problem.

For an m ($\geqslant 1$), let $\mathbf{1}_m = (1,\dots,1)'$, and consider the following intra-block transformations which eliminates the replicate and main effects. Let $\boldsymbol{Y}_i = ((Y_{ijk}))_{p\times q}$, $\boldsymbol{\Omega}_i$ be the corresponding matrix of the error components, and let

$$\boldsymbol{Z}_i = (\boldsymbol{I}_p - p^{-1}\mathbf{1}_p\mathbf{1}_p')\boldsymbol{Y}_i(\boldsymbol{I}_q - q^{-1}\mathbf{1}_q\mathbf{1}_q'), \quad i = 1,\dots,n; \tag{6.6}$$

$$\boldsymbol{E}_i = (\boldsymbol{I}_p - p^{-1}\mathbf{1}_p\mathbf{1}_p')\boldsymbol{\Omega}_i(\boldsymbol{I}_q - q^{-1}\mathbf{1}_q\mathbf{1}_q'), \quad i = 1,\dots,n. \tag{6.7}$$

Then from (6.4), (6.6) and (6.7), we have

$$\boldsymbol{Z}_i = \boldsymbol{\Gamma} + \boldsymbol{E}_i, \quad i = 1,\dots,n. \tag{6.8}$$

So that on this transformed model, the nuisance parameters are all eliminated. Note that the assumed interchangeability condition on the intra-block error components implies that for each i ($= 1,\dots,n$), the components of $\boldsymbol{E}_i$ remain interchangeable too. This provides the access to using permutationally distribution-free procedures based on the stochastic matrices $\boldsymbol{Z}_i$, $i = 1,\dots,n$. On the other hand, the $\boldsymbol{E}_i$ satisfy the same restrains as in (6.3) and (6.4), so that there are effectively only $(p-1)(q-1)$ linearly independent components among the pq ones (for each i).

It follows from (6.7) and the assumed interchangeability of the elements of $\boldsymbol{\Omega}_i$ that the joint distribution of $\boldsymbol{E}_i$ remains invariant under any of the possible $p!$ permutations of its columns, and also under any of the possible $q!$ permutations of its rows. Thus, there is a finite group G of $(p!q!)$ permutations which maps the sample space of $\boldsymbol{E}_i$ onto itself and leaves the joint distribution invariant, so that working with the n independent aligned error matrices, we arrive at a group $\mathcal{G}_n$ of transformations having $(p!q!)^n$ elements, and this provides the access to the exact permutation distribution of suitable test statistics based on these aligned observations. We may proceed as in Section 3 with intra-block rankings of these aligned observations and get a robust test, although it may not generally compare favorably in terms of power with aligned rank tests based on overall rankings, justifiable on the ground that the $\boldsymbol{Z}_i$ do not contain any block effect.

Let R_{ijk} be the rank of Z_{ijk} among the $N = npq$ aligned observations Z_{suv}, $s = 1,\dots,n$; $u = 1,\dots,p$; $v = 1,\dots,q$, and define the scores $a_N(r)$, $r = 1,\dots,N$ as in Section 3. For notational simplicity, we let $\eta_{ijk} = a_N(R_{ijk})$, $i = 1,\dots,n$; $j = 1,\dots,p$; $k = 1,\dots,q$, and let

$$\eta_{ij.} = q^{-1}\sum_{k=1}^{q}\eta_{ijk}, \quad j = 1,\dots,p; \tag{6.9}$$

$$\eta_{i.k} = p^{-1}\sum_{j=1}^{p}\eta_{ijk}, \quad k = 1,\dots,q; \tag{6.10}$$

$$\eta_{i..} = (pq)^{-1}\sum_{j=1}^{p}\sum_{k=1}^{q}\eta_{ijk}, \tag{6.11}$$

for $i = 1, \dots, n$, and let $\eta_{...} = n^{-1}\sum_{i=1}^{n} \eta_{i..}$. Define the aligned rank statistics as

$$\boldsymbol{L}_{N,jk} = n^{-1}\sum_{i=1}^{n} \eta_{ijk}; \qquad \boldsymbol{L}_N = ((\boldsymbol{L}_{N,jk})). \tag{6.12}$$

Then the rank-adjusted statistics are defined by

$$\boldsymbol{L}_N^* = (\boldsymbol{I}_p - p^{-1}\mathbf{1}_p\mathbf{1}_p')\boldsymbol{L}_N(\boldsymbol{I}_q - q^{-1}\mathbf{1}_q\mathbf{1}_q') = ((L_{Njk}^*)), \text{ say.} \tag{6.13}$$

Let us also define the rank measure of dispersion:

$$V_n = [n(p-1)(q-1)]^{-1}\sum_{i=1}^{n}\sum_{j=1}^{p}\sum_{k=1}^{q}(\eta_{ijk} - \eta_{ij.} - \eta_{i.k} + \eta_{i..})^2. \tag{6.14}$$

Then, as in Mehra and Sen (1969), we consider the following test statistic:

$$\mathcal{L}_N^* = [n/V_N]\sum_{j=1}^{p}\sum_{k=1}^{q}\{L_{Njk}^*\}^2, \tag{6.15}$$

which is analogous to the classical parametric test statistic based on the variance ratio criterion. It may be appropriate here to mention that as in the case of two-way layouts, if we have a mixed-effects model, where the treatment effects and their interactions are fixed effects, while the block effects are stochastic, the alignment process eliminates the block-effects (fixed or not), and hence, aligned rank tests are usable for such mixed-effects models too. At this stage, it may be appropriate to point out the basic difference between the current alignment procedure and an alternative one, the rank transformation procedure. In the latter case, one simply replaces the original Y_{ijk} by their ranks (within the overall set) and performs the usual ANOVA test for interactions based on such rank matrices. Basically, rank transformations relate to the sample counterpart of the classical *probability integral* transformation. For a continuous d.f. F, the latter is continuous, but still the former is a step function. Moreover, the latter is a bounded and typically nonlinear (monotone) function, so that the original linear model fitted to the Y_{ijk} may not fit to their transformed counterparts $F(Y_{ijk})$, when F is highly nonlinear. Thus, even if the block effects are eliminated by intra-block transformations, such nonlinearity effects are present in the foundation of rank transformations, and this makes them generally much less adoptable in factorial designs. In particular, if the main effects are not null, their latent effects in the rank transformation procedure may cause serious problems with respect to the validity and efficiency criteria. The aligned ranking procedure sketched here is free from this drawback as long as the basic linearity of the model in (6.1) is tenable.

For small values of n, p and q, the exact (conditional) permutational distribution of $\mathcal{L}_N^*$ can be obtained by considering the $(p!q!)^n$ (conditionally) equally likely row and column permutations of the matrices $\boldsymbol{H}_1 = ((\eta_{ijk}))$, $i = 1, \dots, n$, and as this process becomes unpracticable for large n, we appeal to the following large sample result: As n increases, the permutational (conditional) as well as the unconditional null distribution of $\mathcal{L}_N^*$ can be approximated by the central chi squared distribution with $(p-1)(q-1)$ DF. For various asymptotic properties we may refer to Mehra and Sen (1969). The procedure extends readily to more than two-factor designs, and all we have to do is to define the aligned observations first to eliminate the nuisance parameters, and on such aligned observations we need to incorporate appropriate groups of transformations preserving invariance of their joint distributions, and with respect to such a group, we can obtain the permutational (rank) measure of dispersion. This provides the access to constructing variance-ratio type statistics based on such aligned rank statistics. This prescription is of sufficient general form so as to include the general class of IBD's treated in Section 5. Moreover, the results discussed here for univariate response variates percolate through general *incomplete multiresponse designs* (IMD) pertaining to clinical trials and medical studies (viz., Sen, 1994a). At this stage we may refer to rank transformations as have been advocated by a host of researchers. However, one has to keep in mind that the scope of such procedures for blocked designs may be considerably less than the aligned ones presented here.

In practice, replicated m ($\geqslant 2$)-factor experimental designs crop up in a variety of ways, and in this setup, often, each of these factors is adapted at two levels, say 1 and 2. This way, we are led to a class of n *replicated* 2^m *factorial experiments*. For such designs, a similar ranking after alignment procedure, due to Sen (1970b), works out well. Let $\boldsymbol{j} = (j_1, \ldots, j_m)$ represent the combination of the *levels* $j_1, \ldots, j_m$ of m factors $(A_1, \ldots, A_m)$, where $j_k = 1, 2$, for $k = 1, \ldots, m$. We denote by $\boldsymbol{J}$ the set of all (2^m) realizations of $\boldsymbol{j}$. For the ith replicate, the response of the plot receiving the treatment combination $\boldsymbol{j}$ is denoted by $X_{i\boldsymbol{j}}$, and we consider the usual linear model (sans the normality of the errors):

$$X_{i\boldsymbol{j}} = \beta_i + \left[\sum_{\boldsymbol{r} \in \boldsymbol{R}} (-1)^{\langle \boldsymbol{j}, \boldsymbol{r} \rangle} \tau_{\boldsymbol{r}} \right] / 2 + e_{i\boldsymbol{j}}, \quad \boldsymbol{j} \in \boldsymbol{J},\ i = 1, \ldots, n, \tag{6.16}$$

where $\langle \boldsymbol{a}, \boldsymbol{b} \rangle = \boldsymbol{a}'\boldsymbol{b}$, the β_i represent the block effects, the $e_{i\boldsymbol{j}}$ are the error components, $\boldsymbol{r} = (r_1, \ldots, r_m)'$ with each r_j either 0 or 1, $\boldsymbol{R}$ is the set of all possible (2^m) realizations of $\boldsymbol{r}$, and the treatment effects $\tau_{\boldsymbol{r}}$ are defined as follows.

$$\tau_{\boldsymbol{r}} = \tau_{A_1^{r_1} \cdots A_m^{r_m}}, \quad \boldsymbol{r} \neq \boldsymbol{0};\ \tau_{\boldsymbol{0}} = 0, \tag{6.17}$$

where for each j $(= 1, \ldots, m)$, $A_j^0 = 0$. Thus, $\tau_{A_1} = \tau_{1,0,\ldots,0}, \ldots, \tau_{A_m} = \tau_{0,\ldots,0,1}$ represent the main effects, $\tau_{A_1,A_2} = \tau_{1,1,\mathbf{0}}$ etc. represent a two-factor interaction, and so on; $\tau_{\boldsymbol{r}}$ is a k-factor interaction effect if $\langle \boldsymbol{r}, \boldsymbol{1} \rangle = k$, for $k = 1, \ldots, m$. As in earlier sections, we assume here that for each i $(= 1, \ldots, n)$, the set $\{e_{i\boldsymbol{j}},\ \boldsymbol{j} \in \boldsymbol{J}\}$ consists of interchangeable r.v.'s, and the block-effects need not be fixed; they may as well be stochastic. Let now $\boldsymbol{P}$ be a subset of $\boldsymbol{R}$, and suppose that we want to test the null hypothesis

$$H_{0,\boldsymbol{P}}\colon \{\tau_{\boldsymbol{r}},\ \boldsymbol{r} \in \boldsymbol{P}\} = \boldsymbol{0}, \tag{6.18}$$

against the set of alternatives that these effects are not all equal to 0.

Since (6.16) involves the block effects as nuisance parameters (or spurious r.v.'s), by means of the following intra-block transformations, we obtain the aligned observations. These aligned observations provide both the least squares and R-estimators of the $\tau_{\boldsymbol{r}}$. Let

$$t_{i,\boldsymbol{r}} = 2^{-(m-1)} \sum_{j \in \boldsymbol{J}} (-1)^{\langle j, \boldsymbol{r} \rangle} X_{ij}, \quad \boldsymbol{r} \in \boldsymbol{R},\ i = 1, \ldots, n. \tag{6.19}$$

Then we may write $t_{i,\boldsymbol{r}} = \tau_{\boldsymbol{r}} + g_{i,\boldsymbol{r}}$, for every $\boldsymbol{r} \in \boldsymbol{R}$, where the $g_{i,\boldsymbol{r}}$ are the corresponding aligned error components. It is easy to verify that these $g_{i,\boldsymbol{r}}$ remain exchangeable r.v.'s too, within each block. Moreover, it has been shown by Sen (1970b) by simple arguments that univariate d.f.'s for these aligned errors are all symmetric about zero, and all their bivariate d.f.'s are diagonally symmetric about $\mathbf{o}$. Actually, the joint distribution of these aligned errors (within each block) is also diagonally symmetric about $\mathbf{0}$. Thus, for the R-estimation of the $\tau_{\boldsymbol{r}}$, we may use the (marginal) set $t_{i,\boldsymbol{r}}$, $i = 1, \ldots, n$, and as in Section 2 (see (2.25)), incorporate a general signed rank statistic to yield the desired estimator. These are based on i.i.d.r.v.'s, and hence, no residuals are needed to reconstruct the estimators. As regards rank tests for the null hypothesis in (6.18), we may consider the n i.i.d.r. vectors

$$(t_{i,\boldsymbol{r}},\ \boldsymbol{r} \in \boldsymbol{P}), \quad i = 1, \ldots, n, \tag{6.20}$$

and use multivariate signed-rank test statistics, displayed in detail in Chapter 4 of Puri and Sen (1971). Asymptotic properties of such tests, studied in detail there, remain in tact for such aligned rank tests in 2^m factorial experiments. Extensions to *confounded* or *partially confounded* designs have also been covered in Sen (1970b).

7. Paired comparisons designs: Nonparametrics

In order to compare a number (say, t ($\geqslant 2$)) of objects which are presented in pairs to a set of (say, n ($\geqslant 2$)) judges who verdict (independently) a relative preference of one over the other within each pair, the method of *paired comparisons* (PC), developed mostly by the psychologists, allows one to draw statistical conclusions on the relative positions of all the objects. *Paired comparisons designs* (PCD) are thus incomplete block designs with blocks of size two and a dichotomous response on the ordering of the intra-block plot yields. There are several detours from this simple description of PCD. For example, it may be possible to have observable responses (continuous variates) for each pair of objects: This will relate to the classical IMD with two plots in each block, so that, the results developed in earlier sections would be applicable here. Hence, we skip these details. Another route relates to *paired characteristics* so that ordering of the two objects within each pair may have four possible outcomes (instead of the two in the case of a single characteristic). Nonparametrics for such paired comparisons for paired characteristics were developed by Sen and David (1968) and Davidson and Bradley (1969, 1970), among others. A general account of such PCD methodology is given in David (1988) where other references are also cited. A general characteristics of such paired comparisons procedures is that *circular triads* may arise in a natural way, and this may lead to *intransitiveness* of statistical inference tools when viewed from a *decision theoretics* point; the problem becomes even more complex in a multivariate setup. However, following David (1988) we may say that it is a valuable feature of the method of paired comparisons that it allows such contradictions to show themselves ..., and hence, the methodology developed addresses this issue in a sound statistical manner. As in earlier sections, it is also possible to work out the (M)ANOVA and (M)ANOCOVA models side by side, and following Sen (1995b), we summarize the main results along the same vein.

Paired comparisons procedures in a multivariate setup rest on suitable representations of probability laws for multiple dichotomous attributes. Let us consider p ($\geqslant 1$) dichotomous attributes, and let $\boldsymbol{i} = (i_1, \ldots, i_p)'$, where each i_j can take only two values 0 and 1, for $j = 1, \ldots, p$. The totality of all such 2^p realizations of $\boldsymbol{i}$ is denoted by the set I, and consider a stochastic p-vector $\boldsymbol{X} = (X_1, \ldots, X_p)'$, such that

$$P\{\boldsymbol{X} = \boldsymbol{i}\} = \pi(\boldsymbol{i}), \quad \boldsymbol{i} \in \mathcal{I}. \tag{7.1}$$

This probability law is defined on a 2^p-simplex

$$\Pi = \left\{ \pi(\boldsymbol{i}) \geqslant 0, \ \forall \boldsymbol{i} \in \mathcal{I}; \ \sum_{\boldsymbol{i} \in \mathcal{I}} \pi(\boldsymbol{i}) = 1 \right\},$$

so that there are $2^p - 1$ linearly independent elements in Π. Since there ate t objects (forming $\binom{t}{2}$ pairs), the total number of linearly independent parameters is equal to $\{2^p - 1\}\binom{t}{2}$, and this is generally large when t and/or p is not small. We consider the following modification of the Bahadur (1961) representation for multiple dichotomous attributes. Let

$$\pi_{i*}^{(j)} = P\{X_j = i\}, \quad i = 0, 1; \ 1 \leqslant j \leqslant p. \tag{7.2}$$

We denote by $\theta_j = \pi_{0*}^{(j)}$, $j = 1, \ldots, p$. Also, for every l: $2 \leqslant l \leqslant p; 1 \leqslant i_1 < \cdots < i_l \leqslant p$, define an lth order *association parameter*

$$\theta_{i_1 \cdots i_l} = \theta_{i_1} \cdots i_l(\Pi), \tag{7.3}$$

where there are $\binom{p}{l}$ such parameters, for $l = 2, \ldots, p$. Taking into account the set of θ's, marginal and association parameters, we exhaust the totality of $2^p - 1$ linearly independent parameters. We denote this set by $\Theta = \{\theta_{i_1 \cdots i_l}, 1 \leqslant i_1 < \cdots < i_l \leqslant p; 1 \leqslant l \leqslant p\}$, and arrive at the following.

$$\pi(\boldsymbol{i}) = \prod_{j=1}^{p} \pi_{i_j *}^{(j)} + \sum_{1 \leqslant j_1 < j_2 \leqslant p} (-1)^{i_{j_1} + i_{j_2}} \theta_{j_1 j_2} \prod_{r=1}^{2} \theta_{j_r} \prod_{s=1, \neq j_1, j_2}^{p} \pi_{i_s *}^{s}$$

$$+ \sum_{1 \leqslant j_1 < j_2 < j_3 \leqslant p} (-1)^{i_{j_1}+i_{j_2}+i_{j_3}} \theta_{j_1 j_2 j_3} \prod_{r=1}^{3} \theta_{j_r} \prod_{s=1, \neq j_1, j_2, j_3}^{p} \pi_{i_s *}^{s}$$

$$+ \cdots + (-1)^{i_1+\cdots+i_p} \theta_{1\cdots p} \prod_{r=1}^{p} \theta_r. \tag{7.4}$$

The PCD models are atuned to such representations. By reference to the PCD model under consideration, for the pair (i, j), we denote the response vector by $\boldsymbol{X}'_{ij} = (X_{ij}^{(1)}, \ldots, X_{ij}^{(p)})$ (each coordinate being a dichotomous variable), and the probability law of $\boldsymbol{X}_{ij}$ over the 2^p-simplex is denoted by Π_{ij}, and its transformation to Θ is denoted by Θ_{ij}, for $1 \leqslant i < j \leqslant t$. Basically, PC methodology relates to comparing these $\binom{t}{2}$ probability laws inducing a reduction of the parameter space to a subset of the Θ_{ij} by an appeal to the Bahadur representation in (7.4). Thus, the basic null hypothesis of interest is

$$H_o\colon\ \Theta_{ij} = \Theta^o \text{ (unknown)}, \quad \forall 1 \leqslant i < j \leqslant t, \tag{7.5}$$

against the set of alternatives that the Θ_{ij} are not all the same. To motivate suitable nonparametric testing procedures, we make an appeal to the classical Bradley–Terry (1952) model in the univariate case, which has also been extended to the multivariate case by Davidson and Bradley (1969, 1970) and others. In the univariate case, dropping the superscript (j) in (7.2), we denote the corresponding probabilities for the $\binom{t}{2}$ pairs by π_{ij}, $1 \leqslant i < j \leqslant t$. Then, conceive of a set $\{\alpha_1, \ldots, \alpha_t\}$ of positive numbers, such that $\sum_{j=1}^{t} \alpha_j = 1$, and write

$$\pi_{ij} = \frac{\alpha_i}{\alpha_i + \alpha_j}, \quad 1 \leqslant i < j \leqslant t. \tag{7.6}$$

Thus, the set of $\binom{t}{2}$ unknown (probabilities) parameters is expressed in terms of $t-1$ unknown α's, and the null hypothesis can equivalently be expressed as $H^0\colon\ \alpha_1 = \cdots = \alpha_t = t^{-1}$. With this formulation, the (nonparametric) MLE of the α can be obtained from a given data set and can then be incorporated in the usual (likelihood ratio-) type tests for the null hypothesis of homogeneity of the α. In the multivariate case, for each of the p coordinates, we would have such α-parameters (in all $p(t-1)$ in number), and homogeneity of these (p-) vectors α_i, $i = 1, \ldots, t$, constitutes the null hypothesis of interest. In this formulation of the hypotheses, the association parameters (i.e., the $\boldsymbol{\Theta}_{ij}$) are to be treated as nuisance parameters, and usually they are assumed to be homogeneous. Davidson and Bradley (1969) in their formulation assumed that any third or higher order association parameter is null, and incorporated the classical likelihood ratio principle to formulate some large sample PC tests in a multivariate setup. Sen (1995b) has shown that part of this assumption may not be that crucial, and also the results extend directly to the MANOCOVA model. As in the case of multivariate nonparametric procedures, treated earlier, such tests may no longer be genuinely distribution-free, and hence, suitable permutational invariance structures are to be exploited to render them as permutationally (conditionally) distribution-free; such procedures have been considered by Sen and David (1968) and Sen (1995b), among others. We therefore summarize here these results in a general setup, and indicate the simplifications for simpler models.

In a MANOCOVA setup, we partition the p responses into two subsets: *primary* responses, p_1 ($\geqslant 1$) in number, and *concomitant* responses, p_2 ($\geqslant 1$) in number; $p = p_1 + p_2$. Consider then the component null hypotheses: H_{or}: $\alpha_{r1} = \cdots = \alpha_{rp} = t^{-1}$, for $r = 1, \ldots, p$. The intersection of the p_1 null hypotheses H_{or}, $r = 1, \ldots, p_1$, is denoted by H^*_{o1}, and similarly, H^*_{o2} denotes the intersection of the p_2 component null hypotheses for the concomitant responses. Then the MANOCOVAPC null hypothesis can be stated as

$$H^*_0 = H^*_{o1} \mid H^*_{o2}. \tag{7.7}$$

For every pair (j, l): $1 \leqslant j < l \leqslant t$, we denote the observed cell frequencies by $n_{jl}(\boldsymbol{i})$, where $\boldsymbol{i}$ ranges over the set I, and the total number of observations for this pair is n_{jl}. The marginal frequencies are denoted by $n_{jl,r}(i_r)$, for $i_r = 0, 1$; $r = 1, \ldots, p$, and the bivariate marginals by $n_{jl,rs}(i_r, i_s)$, for $i_r = 0, 1$, $i_s = 0, 1$; $r \neq s = 1, \ldots, p$. Then as in Sen and David (1968), we may obtain some partial maximum likelihood estimator (PMLE) of the association parameters as follows.

$$\widehat{\theta}_{n,rs} = n^{-1} \sum_{1 \leqslant j < l \leqslant t} \bigl(n_{jl,rs}(00) + n_{jl,rs}(11) - n_{jl,rs}(01) - n_{jl,rs}(10)\bigr), \tag{7.8}$$

for $r \neq s = 1, \ldots, p$. Conventionally, we let $\widehat{\theta}_{rr} = 1$, for $r = 1, \ldots, p$, and consider the following matrix, partitioned appropriately:

$$\widehat{\boldsymbol{\Theta}}_n = ((\widehat{\theta}_{n,rs})) = \bigl((\widehat{\boldsymbol{\Theta}}_{n,jl})\bigr)_{j,l=1,2}, \tag{7.9}$$

where the partitioned matrices are of the order $p_j \times p_l$, for $j, l = 1, 2$. Let us also introduce the statistics

$$T_{n,jr} = \sum_{l=1,\ l \neq j}^{t} n_{jl}^{-1/2} \bigl[n_{jl,r}(0) - n_{jl,r}(1)\bigr],$$
$$j = 1, \ldots, t;\ r = 1, \ldots, p, \tag{7.10}$$

and write

$$\boldsymbol{T}_{n,j} = (T_{n,j1}, \ldots, T_{n,jp})', \quad j = 1, \ldots, t. \tag{7.11}$$

For the MANOVA test, as in Sen and David (1968) and Sen (1995), we consider the following test statistic:

$$\mathcal{L}_n = t^{-1} \sum_{j=1}^{t} \boldsymbol{T}'_{n,j} \widehat{\boldsymbol{\Theta}}_n^{-1} \boldsymbol{T}_{n,j}. \tag{7.12}$$

In the univariate case (i.e., for $p = 1$), $\mathcal{L}_n$ is exactly distribution-free under the null hypothesis; for $p > 1$, this EDF property may not generally hold, but permutational (conditional) distribution-freeness holds. Thus, for small sample sizes, a finite group of 2^n possible sign-inversions can be incorporated to generate the permutational distribution of $\mathcal{L}_n$. This procedure has been elaborated in Sen and David (1968) in the bivariate case, and a very similar picture holds for general multivariate paired comparisons models. For large sample sizes, the null hypothesis distribution of $\mathcal{L}_n$ can be well approximated by the central chi squared distribution with $t-1$ DF. For local alternatives, noncentral chi-square approximations also hold.

Let us consider the MANOCOVAPC model in the same setup as explained before. We confine ourselves only to the set of (p_2) concomitant traits, and based on these responses, we construct a MANOVAPC test statistic in the same manner as in (7.12). Let us denote this test statistic by $\mathcal{L}_{n2}$. Under H_{02}, $\mathcal{L}_{n2}$ is permutationally (conditionally) distribution-free, and asymptotically, it has the chi squared distribution with $(t-1)p_2$ DF. Then for testing the null hypothesis H_{01} assuming that H_{02} holds, we consider the test statistic

$$\mathcal{L}_{n1}^* = \mathcal{L}_n - \mathcal{L}_{n2}. \tag{7.13}$$

It may be noted that $\mathcal{L}_{n1}^*$ may also be expressed in terms of concomitant variate adjusted $\boldsymbol{T}_{nj}$ and a similarly adjusted covariance matrix; for some details, we may refer to Sen (1995). It follows from the above discussion that under the MANOCOVA model null hypothesis, $\mathcal{L}_{n1}^*$ is also permutationally (conditionally) distribution-free. Moreover, under the same null hypothesis, it has asymptotically the central chi squared distribution with $(t-1)p_1$ DF. Asymptotic nonnull distribution theory (for local alternatives) for such MANOCOVAPC tests has been studied in detail by Sen (1995b), and it has been shown that the ARE of this MANOCOVA test with respect to the corresponding MANOVA test (ignoring the p_2 concomitant traits), in a general multivariate setup, is bounded from below by 1. Thus, at least asymptotically, the MANOCOVAPC tests are better alternatives than the corresponding MANOVA PC tests. A similar picture holds for the ARE of the MANOCOVAPC test with respect to the MANOVAPC test based on all the p traits (when, in fact, there are p_2 concomitant traits). The intuitive reason for this better ARE picture is that the concomitant traits do not contribute to the growth of the noncentrality of the MANOVAPC test (based on the entire set), so that a larger DF with a common noncentrality parameter leads to a decrease in the (asymptotic) power function.

8. Nonparametrics for crossover designs

In the context of clinical trials and/or biomedical studies, each experimental unit (or subject) receives several treatments at different time-periods, so that there is a *repeated measurement design* (RMD) flavor in the statistical modeling and analysis schemes pertaining to such experimental plans. The simplest situation relates to a two-period design where for some subjects two treatments are administered exclusively in these two periods in a specific order, while for others, it is done in the reverse order. For this reason, it is also called a *changeover* or *crossover design*. A basic feature of such RMD's is that *residual effects* or *carryover effects* are likely to be a vital part of the response pattern, and hence, in the modeling, such effects are to be incorporated in an appropriate manner. The more generality one may want to achieve in this formulation, the more complex may be the actual statistical modeling and analysis schemes, and hence, often, it is assumed that such residual effects are additive and have some structural form. While most of such technicalities are discussed in some other chapters of this volume, we may like to introduce only some specific crossover designs, and examine in that context how far nonparamterics can be accepted as an alternative to standard normal theory parametrics.

Let us consider a p ($\geqslant 2$) period model wherein n experimental units are used, in such a way that for the jth unit, in the ith period, treatment $d(i,j)$ is used, where the $d(i,j)$ belong to an index set relating to the treatments administered in the experiment. We denote the response of the jth unit at the ith period by Y_{ij}, and consider the conventional linear model incorporating the *first order carryover* effects $\rho(d(i-1,j))$:

$$Y_{ij} = \mu + \alpha_i + \beta_j + \tau_{d(i,j)} + \rho_{d(i-1,j)} + \varepsilon_{ij}, \tag{8.1}$$

for $i = 1,\ldots,p; j = 1,\ldots,n$, where μ is the mean effect, α_i are the period effects, β_j are the unit effects, $\tau_{d(i,j)}$ are the treatment effects, $\rho(d(0,j)) = 0$, and the errors ε_{ij} are assumed to be i.i.d.r.v.'s with zero mean and a finite positive variance σ^2. In the normal theory model, again their error distribution is assumed to be normal. This is the so called *fixed effects model with first-order residual effects* and it corresponds to the familiar completely randomized design in the conventional case. Motivated by the two-way layouts discussed earlier, we may also extend the model to randomized block designs where the experimental units may be blocked into relatively homogeneous groups, and for each group, we have a model as in above. Thus, there is a need to introduce block effects as well as interaction parameters with respect to block vs. carryover parameters. In this context, it may be remarked that in clinical trials or biological assays and other biomedical experiments, the very mechanism by which the blocks are formulated, these block effects may be random, and, in turn, the *carryover* $\times$ *block* interactions may also be random. Therefore, one may encounter a so called *mixed effects* model, which may be presented as:

$$Y_{ijk} = \mu + \alpha_i + \beta_j + \xi_{jk} + \tau_{d(i,j,k)} + \rho_{d(i-1,j,k)} + \varepsilon_{ijk}, \tag{8.2}$$

where i stand for the experimental unit, j for the period, and k for the block, with the parameters defined accordingly. A simplified version of this mixed-effects model was considered by Grizzle (1965). In the above formulation when we treat the ξ_{jk} as random, it is quite likely that the residual or carryover effects $\rho_{d(i-1,j,k)}$ are also to be treated as stochastic. In the normal theory model, all these stochastic elements are assumed to be independent and normally distributed with zero means and appropriate (unknown, positive) variances, so that the model can be interpreted in the light of conventional *variance components models*. Although such an independence assumption may not be that unrealistic in a practical application, the assumption of normality may, however, not be tenable in a variety of situations. Sans this normality assumption, the classical parametric procedures may lose their appeal on theoretical (viz., *optimality*) as well as practical *robustness*) grounds, and hence, there is a general feeling that nonparametric and robust statistical procedures are to be advocated in this context. The basic idea is to incorporate the *alignment* principle as far as possible, so as to reduce the number of estimable parameters and error components, so that classical multivariate nonparametric and robust methods discussed in earlier sections can be implemented successfully. This alignment principle is isomorphic to the one in the classical normal theory models. From more elementary practical considerations, Koch (1972) initiated the use of some nonparametric methods in the statistical analysis of two-period-change-over design with emphasis on applications, and Tudor and Koch (1994) have a recent review of applied works in this field, which cast additional light on related applicational developments. As such, instead of providing a general but abstract formulation, we shall try to motivate the basic ideas with specific simplifications.

Let there be a complete block design with p periods (indexed as $i = 1, \ldots, p$), b blocks (indexed as $j = 1, \ldots, b$), and m units in each block (indexed as $k = 1, \ldots, m$). Consider a contrast in the p measurements on the kth unit in the jth block:

$$Y^*_{jk} = \sum_{i=1}^{p} l_i Y_{ijk}, \quad \text{for} \quad j = 1, \ldots, b, \; k = 1, \ldots, m, \tag{8.3}$$

where $\sum_{i=1}^{p} l_i = 0$. Then by (8.2) and (8.3), we obtain that

$$\begin{aligned} Y^*_{jk} &= \sum_{i=1}^{p} l_i \{\alpha_i + \tau_{d(i,j,k)} + \rho_{d(i-1,j,k)}\} \\ &\quad + \sum_{i=1}^{p} l_i \varepsilon_{ijk} = \theta_{jk} + \varepsilon^*_{jk}, \quad \text{say.} \end{aligned} \tag{8.4}$$

In this formulation when the ε_{ijk} are not normally distributed, the distribution of the aligned ε^*_{jk} may depend on the chosen l_i, and may not even be symmetrical. To eliminate this drawback, we proceed as in the case of aligned rank procedures treated earlier and define the set of aligned observations as

$$Y^o_{ijk} = Y_{ijk} - p^{-1} \sum_{i=1}^{p} Y_{ijk}, \quad i = 1, \ldots, p, \tag{8.5}$$

for every $j = 1, \ldots, b$; $k = 1, \ldots, m$. In the case of complete *balanced* (B)RMD's, $\sum_{i=1}^{p} \{\alpha_i + \tau_{d(i,j,k)} + \rho_{d(i-1,j,k)}\} = 0$, for every $j = 1, \ldots, b$; $k = 1, \ldots, m$, so that if we consider the set of intra-block intra-unit aligned observations:

$$Y^o_{ijk} = Y_{ijk} - p^{-1} \sum_{i=1}^{p} Y_{ijk}, \quad k = 1, \ldots, m; \; j = 1, \ldots, b; \; i = 1, \ldots, p, \tag{8.6}$$

then for the jth block, we have a set of m independent p-vectors $\boldsymbol{Y}_{jk}^{o} = (Y_{1jk}^{o}, \ldots, Y_{pjk}^{o})'$, for $k = 1, \ldots, m$, and for different blocks too, these aligned vectors are stochastically independent. We denote the collection of m vectors $\boldsymbol{Y}_{jk}^{o}$, $k = 1, \ldots, m$, in the jth block by $\boldsymbol{Y}_{j\cdot}^{o}$ and the combined collection by $\boldsymbol{Y}_{\cdot\cdot}^{o}$. In a similar manner, we define the aligned error vectors $\boldsymbol{e}_{jk}^{o}$, for $k = 1, \ldots, m$; $j = 1, \ldots, b$. Further, for each j, k, we let $\boldsymbol{\Theta}_{jk} = (\theta_{1jk}, \ldots, \theta_{pjk})'$, where $\theta_{ijk} = \alpha_i + \tau_{d(i,j,k)} + \rho_{d(i-1,j,k)}$, for $i = 1, \ldots, p$; $j = 1, \ldots, b$; $k = 1, \ldots, m$. Let $n = km$ and $N = kpm$. Then we consider the aligned observation matrix $\boldsymbol{Y}^{o}$ of order $p \times n$, expressed as $(\boldsymbol{Y}_{11}^{o}, \ldots, \boldsymbol{Y}_{bm}^{o})$, and a similar representation is made for $\boldsymbol{e}^{o}$ and $\boldsymbol{\Theta}^{o}$. As such, we write

$$\boldsymbol{Y}^{o} = \boldsymbol{\Theta}^{o} + \boldsymbol{e}^{o}. \tag{8.7}$$

This representation enables us to incorporate the general theory and methodology of multivariate nonparametrics, discussed in Section 4. In this context, we may remark that for each j, k, whenever the ε_{ijk}, $1 \leqslant i \leqslant p$, are interchangeable r.v.'s (a condition implied by the usual assumption that they are i.i.d.), the e_{ijk}, $1 \leqslant i \leqslant p$, are also interchangeable, and for different j, k, these stochastic vectors are independent. Thus, for these aligned error vectors, we have the same exchangeability assumption as in the classical nonparametric (M-)ANOVA model. This intuitively suggests that aligned rank procedures for the MANOVA model, considered in Section 4, can be incorporated in the current context too. There is an additional simplification in this setup. The marginal d.f. of each e_{ijk}^{o} is the same, so that while ranking these aligned observations, we do not have to rank the elements separately for each of the p rows in the matrix $\boldsymbol{Y}^{o}$; rather, we consider the overall ranking of all the N aligned observations. In order to achieve this simplification, we need, however, to check a basic condition that estimable parameters among the sets of period-effects, treatment-effects and carryover-effects can be expressed in terms of $\boldsymbol{\Theta}^{o}$-contrasts. This can easily be verified in the case of balanced RMD's, while for general unbalanced RMD's, we may set appropriate design restraints which would ensure the same.

The aligned observation matrix $\boldsymbol{Y}^{o}$ plays also a vital role in robust estimation and testing procedures for RMD's. In the normal theory model, if we confine ourselves to the fixed-effects case in (8.1), then the classical weighted least squares estimation (WLSE) methodology can be incorporated to characterize the optimality of estimators and tests based on this aligned observation matrix. For the mixed-effects model in (8.2), the situation is somewhat more complex due to the random block and interaction effects, and the assumption of independence as well as normality of these components seems to be even more vulnerable in actual practical applications. On the other hand, a characterization of the optimality of the WLSE is, of course, limited to the basic assumption that all the stochastic elements in model (8.2) are normally distributed and their independence-homoscedasticity condition holds. Any departure from such model-assumptions can not only take away the optimality properties but also may signal lack of validity. As such, *robust methods* for drawing statistical conclusions are quite appealing in this context. While the aligned rank based estimates and tests are generally globally robust, they may not be (even asymptotically) fully efficient when the form of the underlying error distribution is not known. In a local robustness perspective, if only small departures from the assumed model are contemplated, it may be more appropriate to use *M-estimators* and related *M-tests* based on appropriate score functions. Generally, such score functions are smooth but non-linear, and based on consideration of robustness, they are usually bounded and monotone. For some detail discussion of such robust procedures for general linear models, we may refer to Hampel et al. (1986) and Jurečková and Sen (1996), among others.

9. Clinical trials and survival analysis

In clinical trials and life testing problems, although the setups are related to classical statistical designs, general objectives and operational constraints, often, call for different types of designs and appropriate statistical analysis schemes. Semi-parametrics and nonpametrics are more appealing than standard parametrics in such designs. We illustrate this with a very simple life-testing model, which may as well be adapted to a clinical trial setup. Suppose that we want to study the impact of *smoking* on longivity of human beings. Usually, before a clinical trial is initiated on a specific human sector, it is planned to have an *animal study* to study safe dosage, side effects and other causal effects, so that the actual clinical trial can be administered with less restraints from medical ethics and other humanitarian points. Suppose that two groups of monkeys are chosen, one for the *placebo* (no smoking) and the other for the *treatment* (smoking) group. After a period of study, say eighteen months, these monkeys are sacrificed, and their arterial cross-section at some specified location is examined for the constriction of the arterial channel. Thus, for this *arteriosclerosis* problem, the response variable is the ratio of the open space to the entire cross-section. It is hypothesized that smoking tends to make this ratio stochastically smaller, so that essentially we have a two sample model for testing homogeneity against a one-sided alternative. But, there are some basic differences between this life testing and the classical two-sample model. First, some animals may die before the study period is over, resulting in *censoring*. Such censoring may also arise due to dropouts or withdrawals of the subjects due to causes other than early failures. Secondly, the duration of the study period has to be decided on the basis of extraneous information on the associated inhalation toxicity problem, and the outcome response variable may depend in a rather complex manner on this study period, in the sense that a linear or log-linear regression may not be reasonable. Thirdly, in an agricultural experiment, the treatment can be applied in a reasonably controlled manner, whereas, in such life testing models, such a controlled experimental setup is not that expected. Finally, it may not be very reasonable to assume that the response variable is exponentially or normally or lognormally distributed, so that standard parametric procedures may not be very appealing in such a context. We may refer to Sen (1984a, b) for some discussion of these aspects of multivariate nonparametrics relating to medical studies. If the study has to cover human subjects, occupational factors, sex, diet, physical exercise, age and many other concomitant variates may make the model far more complex for adoption of parametric statistical analysis schemes. *Robustness* and *validity* are therefore of important considerations in this respect.

In designing such clinical trials and/or life testing models, it may therefore be necessary to incorporate various auxiliary and concomitant variates in the model, plan judiciously on the duration of the study, identify the follow-up nature of the study, and check the appropriateness of *repeated significance tests* and/or *interim analysis* for such experimental data sets. These factors have a great bearing on the formulation of appropriate statistical analysis schemes.

Consider a simple (say, placebo vs. treatment) clinical trial or life testing model which we put in a slightly more general regression model setup. Note that in survival analysis or life testing problems, the failure times are nonnegative random variables typically with skewed distribution, and hence, log-transformations are used to induce more symmetry (if not normality) on the distribution of the transformed response variable (termed), the *response metameter.* Similarly, on the dosage, often such a transformation is used to induce more linearity of the response-dosage regression, and such a transformed dose is referred to as a *dose metameter.* With such transformations, we may have greater confidence on the linearity of regression as well as symmetry of the response distribution, although it may not be entirely satisfactory to assume normality or logistic form for such a distribution, as is typically done in parametric analysis. Thus, from robustness and validity points of view, we shall allow the response distribution to be arbitrary to a greater extent. Let there be n subjects under study, and let their (transformed) responses be denoted by $X_1, \dots, X_n$ respectively. We assume that the X_i are independent with continuous distribution functions $F_1(x), \dots, F_n(x)$, $x \in \boldsymbol{R}$, respectively. These d.f.'s may depend on the dose levels to which the subjects are subjected, and we conceive of (nonstochastic) constants (regressors) $t_1, \dots, t_n$ (not all equal) which can be used to formulate a semi-parametric model:

$$F_i(x) = F_o(x - \beta t_i), \quad x \in \boldsymbol{R},\ i = 1, \dots, n, \tag{9.1}$$

where β stands for the regression parameter (unknown) and the d.f. F_o is assumed to be continuous but otherwise of arbitrary form. As in Section 2, we may also consider a more general linear model (viz., (2.3)) where the $\boldsymbol{t}_i$ are known vectors of regression constants and $\boldsymbol{\beta}$ is a vector of regression parameters. The classical placebo vs. treatment setup is thus a special case of (9.1) where the t_i are binary. While this model is isomorphic to the classical linear regression model treated in Section 2, there is a basic difference between the two setups. There, all the X_i were assumed to be observable at the same time, while here we have follow-up scheme. Let us denote the order statistics corresponding to $X_1, \dots, X_n$ by $X_{n:1} \leqslant \cdots \leqslant X_{n:n}$, where by virtue of the assumed continuity of F_o, ties among these observations may be neglected with probability 1. Let us also denote the *anti-ranks* by $S_1, \dots, S_n$, so that

$$X_{S_i} = X_{n:i}, \quad \text{for} \quad i = 1, \dots, n, \tag{9.2}$$

and $(S_1, \dots, S_n)$ is a (random) permutation of $(1, \dots, n)$. At a time point t within the study period, the observable random elements are the failures occurring before that time and the corresponding c_{S_i} values; although the entire set of $c_1, \dots, c_n$ is known at the beginning of the study. Thus, as we move along the study, we gather an accumulating data set

$$\{X_{n:j},\ c_{S_j};\ j \leqslant i\}, \quad \text{for} \quad i = 1, \ldots, n, \tag{9.3}$$

where generally we have a subset of these elements due to censoring and withdrawals/dropouts. For example, during the study period a (random) subset of failures may occur, while the other subjects have failure times larger than the set endpoint of the study; this is referred to as *right truncation*. Alternatively, the study may be so planned that it would be conducted until a prespecified number ($m \leqslant n$) of failures take place, resulting in a stochastic duration of the study; this is referred to as *right censoring*. In either case, there is incompleteness in the observed data set due to possible immature termination of the study. Actually, in biomedical studies, including clinical trials and epidemiological or environmental investigations, censoring may be more complex than the ones referred to above; we may refer to Sen (1995d) for some details of the interface of statistical censoring in practice and the controversies arising in such a context. Here we stick to some simple censoring patterns as have been described before. In the above setup, from medical ethics point of view, it seems reasonable to set an underlying condition that if there is a significant difference in the placebo vs. treatment group responses (in our setup, $\beta \neq 0$), then we should terminate the study as early as possible and switch all the subjects to the better group for better health prospects. For this reason, instead of waiting until the study period is over, it is often desirable to look into the accumulating data set at regular time intervals, resulting in the so called *interim analysis schemes*; at each such prechosen time-point, a test of significance (on β) is to be made for possible stopping of the trial. This results in the so called *repeated significance testing* (RST) schemes. Such statistical analysis schemes in a broader setup are also referred to as *time-sequential procedures* (Sen 1981b). Monitoring on a continual basis (i.e., at every failure point) is referred to as *progressively censored schemes* (PCS) (Chatterjee and Sen, 1973).

We start with the setup of (2.3)–(2.5). Let us examine the picture at the kth failure point $X_{n:k}$. We have the knowledge of the previous failure points as well as the corresponding anti-ranks $S_1, \ldots, S_k$; also, we know that the remaining observations are right-censored. Thus, it seems very natural to project the linear rank statistic $\boldsymbol{L}_n$, defined in (2.5), onto the subspace generated by the anti-ranks $S_1, \ldots, S_k$. For this let us define

$$a_{nk}^* = (n-k)^{-1} \sum_{j=k+1}^{n} a_n(j), \quad \text{for} \quad k = 0, \ldots, n-1, \tag{9.4}$$

and conventionally, we let $a_{nn}^* = 0$. Note that the ranks for the k observed failures are $1, \ldots, k$ respectively, while each of the remaining $(n-k)$ censored observation is given the average rank score a_{nk}^*, so that at the kth failure point, the censored linear rank statistic is given by

$$\begin{aligned} \boldsymbol{L}_{nk} &= \sum_{i \leqslant k} (\boldsymbol{t}_{S_i} - \bar{\boldsymbol{t}}_n) a_n(i) + a_{nk}^* \sum_{j > k} (\boldsymbol{t}_{S_j} - \bar{\boldsymbol{t}}_n) \\ &= \sum_{i \leqslant k} (\boldsymbol{t}_{S_i} - \bar{\boldsymbol{t}}_n)[a_n(i) - a_{nk}^*], \quad \text{for} \quad k = 1, \ldots, n, \end{aligned} \tag{9.5}$$

and conventionally, we let $\boldsymbol{L}_{n0} = \boldsymbol{0}$. We define $\boldsymbol{Q}_n$ as in (2.7), and in the case of scalar t_i, we denote this by Q_n^2. Let us also define $\bar{a}_n$ and A_n^2 as in (2.6), and for every k: $0 \leqslant k \leqslant n$, we let

$$A_{nk}^2 = A_n^2 - (n-1)^{-1}(n-k)[a_{nk}^* - \bar{a}_n]^2, \tag{9.6}$$

so that it follows by some simple arguments that

$$0 = A_{n0}^2 \leqslant A_{n1}^2 \leqslant \cdots \leqslant A_{nn}^2 = A_n^2. \tag{9.7}$$

Consider now an experimental scheme wherein the study is planned to be curtailed at the rth failure point, for some prefixed positive integer r ($\leqslant n$) (so that we have a Type II censoring scheme). In the case of scalar t_i, we consider the test statistic

$$Z_{nr} = \{L_{nr}\}/\{Q_n A_{nr}\}. \tag{9.8}$$

If the null hypothesis (H_0) relates to the homogeneity of the F_i (or, equivalently, $\beta = 0$), following the line of attack of Chatterjee and Sen (1973), we can claim that under H_0 Z_{nr} is EDF, and for large n, Z_{nr} has normal distribution with 0 mean and unit variance. Thus, an appropriate (one or two-sided) test can be based Z_{nr}. In the case of vector $\boldsymbol{t}_i$, parallel to (2.9), we define

$$\mathcal{L}_{nr} = A_{nr}^{-2}\{\boldsymbol{L}'_{nr}\boldsymbol{Q}_n^- \boldsymbol{L}_{nr}\}. \tag{9.9}$$

Here also, under the null hypothesis of homogeneity of the d.f.'s F_i, $\mathcal{L}_{nr}$ is EDF, and proceeding as in after (2.9), we may argue that under the null hypothesis, $\mathcal{L}_{nr}$ has asymptotically central chi squared distribution with q DF when $\boldsymbol{Q}$ is of rank q. Therefore an appropriate test for the null hypothesis H_0: $\boldsymbol{\beta} = \boldsymbol{0}$ can be based on the test statistic $\mathcal{L}_{nr}$. Let us now consider the Type I censoring or truncation case. Here, for some prefixed time-point T_o, the experiment is planned for the prefixed duration $(0, T_o]$. Let $r(T_o)$ be the number of failures occurring in the study period $(0, T_o]$. In this setup, $r(T_0)$ is a nonnegative integer valued random variable. It may be tempting to use the statistics $Z_{nr(T_o)}$ or $\mathcal{L}_{nr(T_o)}$ as appropriate test statistics. This is indeed possible. But, it may be kept in mind that such statistics are not generally EDF, even under the hypothesis of homogeneity of the underlying d.f.'s. However, as in Chatterjee and Sen (1973), we may argue that conditionally on $r(T_o) = r$, $\mathcal{L}_{nr(T_o)}$ or $Z_{nr(T_o)}$ is distribution-free under H_0, and they enjoy the same properties as in the case of Type II censoring. In clinical trials, often, an interim analysis scheme is adopted wherein one plans to review the accumulating dataset either at regular time intervals or after regular failure intervals. In that way, one has an extension of Type I or II censoring schemes. For such schemes, RST procedures are generally adopted to guard against an inflation of Type I error for the overall significance testing procedures. In an extreme case, one may also like to monitor the study more or less on a continual basis, and in that setup, a progressive censoring scheme (PCS) is more appropriate. With this motivation, we first consider some PCS schemes, mainly adapted from Chatterjee and Sen (1973) and Sen (1981b).

Consider the (double-)sequence of PCS linear rank statistics $\{\boldsymbol{L}_{nk};\ 0 \leqslant k \leqslant n\}$, defined by (9.9), and let $\mathcal{B}_{nk}$ be the sigma-field generated by $\{S_j;\ j \leqslant k\}$, for $k = 0, \ldots, n$. Then, it follows from Chatterjee and Sen (1973) and Majumdar and Sen (1978) that under the null hypothesis of homogeneity of the d.f.'s $F_1, \ldots, F_n$, for every n, $\{\boldsymbol{L}_{nk}, \mathcal{B}_{nk};\ 0 \leqslant k \leqslant n\}$ is a null mean (vector) martingale (array). Moreover, under the null hypothesis, these $\boldsymbol{L}_{nk}$ are (jointly) distribution-free, so that a test based on this collection is also distribution-free under the null hypothesis. With the possibility of early termination in mind, we may consider the following Kolmogorov–Smirnov type test statistics. First, consider the case of scalar t_i, so that the L_{nk} are scalar too. Let then

$$K_n^+ = \max\{Q_n^{-1} A_n^{-1} L_{nk}\colon\ 0 \leqslant k \leqslant n\}; \tag{9.10}$$

$$K_n = \max\{Q_n^{-1} A_n^{-1} |L_{nk}|\colon\ 0 \leqslant k \leqslant n\}. \tag{9.11}$$

It may be remarked that (9.10) is designed for testing against one-sided alternatives $\beta > 0$, while (9.11) is for the two-sided ones $\beta \neq 0$. The exact null hypothesis distribution of either of these statistics can be obtained by enumeration of all possible $n!$ realizations of $S_1, \ldots, S_n$ over the permutations of $1, \ldots, n$. This task becomes prohibitively laborious as n increases, and hence, for large values of n, suitable distributional approximations are generally used. Towards these limit laws, we may construct suitable stochastic processes $W_n = \{W_n(t),\ 0 \leqslant t \leqslant 1\}$, $n \geqslant 1$, by letting

$$W_n(t) = Q_n^{-1} A_n^{-1} L_{nk_n(t)}, \quad t \in [0, 1], \tag{9.12}$$

where

$$k_n(t) = \max\{k\colon\ A_{nk}^2 \leqslant t A_n^2\}, \quad t \in [0, 1], \tag{9.13}$$

and the A_{nk}^2 are all defined as in (9.6). Note that by definition,

$$K_n^+ = \sup\{W_n(t)\colon\ t \in [0, 1]\}; \tag{9.14}$$

$$K_n = \sup\{|W_n(t)|\colon\ t \in [0, 1]\}. \tag{9.15}$$

Incorporating the martingale property described earlier, it was shown by Chatterjee and Sen (1973) that under the null hypothesis,

$$W_n \text{ converges in law to } W, \text{ as } n \to \infty, \tag{9.16}$$

where $W = \{W(t),\ t \in [0,1]\}$ is a standard Brownian motion on the unit interval $[0,1]$. This weak convergence result in turn leads us to the following: Under the null hypothesis of homogeneity of the d.f.'s $F_1, \ldots, F_n$, for every $\lambda \geqslant 0$, as n increases,

$$P\{K_n^+ \geqslant \lambda \mid H_0\} \to 2\big[1 - \Phi(\lambda)\big], \tag{9.17}$$

$$P\{K_n \geqslant \lambda \mid H_0\} \to 2[1 - \Phi(\lambda)] \\ -2\sum_{k\geqslant 1}(-1)^k\big[\Phi((2k+1)\lambda) - \Phi((2k-1)\lambda)\big], \tag{9.18}$$

where $\Phi(\cdot)$ is the standard normal d.f. Therefore, the asymptotic critical levels for K_n^+ and K_n can be obtained from the above two expressions. Let us denote the actual α-level critical values for K_n^+ and K_n by $K_{n,\alpha}^+$ and $K_{n,\alpha}$ respectively. Then, we have the following *time-sequential* testing procedures: (i) To test the null hypothesis H_0 against the one-sided alternatives H_1, at the kth failure point $X_{n:k}$, compute the statistic $W_{nk} = Q_n^{-1}A_n^{-1}L_{nk}$, for $k \geqslant 0$. As long as these W_{nk} lie below the level $K_{n,\alpha}$, continue in having more accumulating data; if for the first time, at some $k = M$, say, W_{nM} is $\geqslant K_{n,\alpha}^+$, stop at this Mth failure point and reject the null hypothesis. If no such M exists, stop at the last failure point and accept the null hypothesis. (ii) For the two-sided alternatives, work with the $|W_{nk}|$ and have a similar procedure where the critical level is taken as $K_{n,\alpha}$.

Several modifications of this procedure are quite easy to workout. First, consider the case where the experiment is preplanned to a maximum of r out of n failures, so that a time-sequential procedure has a maximal duration $X_{n:r}$. To this end, one may simply define the statistics

$$W_{nk;r} = Q_n^{-1}A_{nr}^{-1}L_{nk}, \quad k = 0, 1, \ldots, r, \tag{9.19}$$

and define the sup-norm statistics as in before (with $k \leqslant r$). Thus, effectively, we shrink the range of k and rescale by the truncated variance function. In that way, the EDF character of the tests (under the null hypothesis) remains in tact, and the same limiting distributions hold. Secondly, often, this maximum duration of a time-sequential procedure is set in terms of a given time point. Borrowing the analogy with the Type I censoring scheme, we may again derive some conditionally distribution-free time sequential tests, for which in the above setup, we are to restrict ourselves to values of $k \leqslant R$, the number of failures within that set time period. Thirdly, instead of the unweighted sup-norm statistics, we could have used some weighted version wherein the $W_n(t)$ are to multiplied by a nonnegative scalar factor $q(t)$, such that $\int_0^1 q^2(t)\,\mathrm{d}t < \infty$. The choice of this weight function $q(\cdot) = \{q(t);\ t \in (0,1)\}$ may be made on the basis of the importance of early stopping (from clinical point of view). This will relate to the weak convergence to a general Gaussian process and would therefore rely on the boundary crossing probabilities for such a process. In some simple cases, such probabilities are known (viz., Chapter 2 of Sen, 1981b), and these can then be incorporated in the simplification of asymptotic critical levels for the test statistics. But, in general, for an arbitrary square integrable $q(\cdot)$, such algebraic expressions may not be available, and hence, numerical or simulation methodology may have to be used. We may refer to Sinha and Sen (1982) for some related results. Fourthly, instead of a more or less continuously monitoring, as is the case with PCS, it may be desirable to have a prefixed number of looks into the accumulating data set either on a calendar time basis or on the response outcome one; indeed this is usual for conventional interim analyses schemes. In this finite dimensional version, even for the asymptotic case, one needs to look into multivariate normal distributional probability (multiple) integrals, and exact evaluation seems to be rather impracticable. There are some simplifications when the number of looks is as small as 2 or 3, although they are to be obtained by numerical quadrature formulae, and we may refer to Flemming and Harrington (1991) for some related studies. Lan and DeMets (1983) introduced a novel concept of *spending function* in this context, and that can be used with some advantages, particularly when the interim time points are based on the response outcome. In passing, we may remark that if the number of looks exceeds 10 and these points are not too concentrated in any particular patch of the unit interval, then a solution obtained from the continuous process provides a close (upper) bound to the one from the finite discrete version. In long range, multi-center clinical trials, generally such an interim analysis involves a moderately large number of looks into the accumulating data set on a fairly regular basis, and hence, such weak convergence based approximations work out fairly well.

We consider next the general case of vector $\boldsymbol{t}_i$, and as in (9.9) define the statistics $\mathcal{L}_{nr}$, $r = 0, 1, \ldots, n$. Let then

$$\mathcal{L}^*_{nr} = \{A^2_{nr}/A^2_n\}\mathcal{L}_{nr}, \quad r \leqslant n. \tag{9.20}$$

In this case, as a test statistic, we consider the following:

$$K^*_n = \max\{\mathcal{L}^*_{nk}\colon\ k \leqslant n\}. \tag{9.21}$$

In order to express this statistic in terms of a suitable stochastic process, we adopt the same definition of $k_n(t)$, $t \in [0, 1]$, as in (9.13). Also, let $B^2_n = \{B^2_n(t); t \in [0, 1]\}$ be defined by

$$B^2_n(t) = \mathcal{L}^*_{nk_n(t)}, \quad t \in [0, 1]. \tag{9.22}$$

Then, we may write equivalently

$$K^*_n = \sup\{B^2_n(t)\colon\ t \in [0, 1]\}. \tag{9.23}$$

Let $W_1, \ldots, W_q$ be independent copies of a standardized Brownian motion W. Let us define then $B^2 = \{B^2(t),\ t \in [0, 1]\}$ by letting

$$B^2(t) = \sum_{j=1}^{q} W^2_j(t), \quad t \in [0, 1]. \tag{9.24}$$

In the literature these are known as the *Bessel (squared) processes*, and boundary crossing probabilities for them have been extensively studied by DeLong (1981) and others. It follows from Majumdar and Sen (1978) that under the null hypothesis of homogeneity of the d.f.'s $F_1, \ldots, F_n$, as n increases, B_n^2 converges weakly to B^2, so that K_n^* converges in law to a Bessel squared process functional $K^* = \sup\{B^2(t)\colon t \in [0, 1]\}$. This enables us to use the extensive tables provided by DeLong (1981) for asymptotic approximations for the exact null hypothesis distribution of K_n^*. In this respect we may note that this test is essentially against a multi-sided alternative, and hence, one-sided versions are not that suitable in this setup. Secondly, as in the case of K_n^+ or K_n, here also, we may work with suitably weighted versions of K_n^*. Thirdly, modifications for a Type II censoring scheme in connection with PCS can be made by replacing A_n^2 by the corresponding value at the target number r^*, and a conditionally distribution-free version may similarly be considered for the truncation PCS scheme.

In the discussion made above, we have mainly confined ourselves to identification of EDF structures and simplifications of the null hypothesis distributions of suitable rank based procedures. The study of their non-null distributions entails even more mathematical complexities where often the regularity assumptions may not match the reality of practical applications. However, under the usual local alternatives, such asymptotic distributions, albeit being more complex than in the conventional situations, have been studied in a unified manner, and a general account of these asymptotics is given in Sen (1981b, Chapter 11). In clinical trials or survival analysis, it is not uncommon to encounter *noncompliance* due to dropouts or withdrawals. One common approach to accommodating noncompliance in an objective perspective is to introduce the concept of *random censoring* wherein it is assumed that a *censoring variable* C_i is associated with the primary variate (Y_i), such that the observable r.v.'s are

$$T_i = \min\{Y_i, C_i\} \quad \text{and} \quad I_i = I\ (T_i = Y_i), \quad i \geqslant 1, \tag{9.25}$$

and there may be, in general, other concomitant variates too. In this setup, it is assumed that the C_i are stochastically independent of the Y_i (a condition that may not generally hold in practice), and further the distribution of C_i is not affected by the treatments to which the subjects are subjected in the study; this latter condition is known as *noninformative censoring*, and again, this may not meet the light of reality in all applications. In a nonparametric formulation, such a random censoring scheme may introduce complications beyond a simple amendment range, and may therefore require more structural assumptions on the survival functions of the associated variables. In this respect, a fundamental contribution is due to Cox (1972, 1975) who incorporated the novel idea of *partial likelihood functions* along with the basic assumption of *proportional hazards* (PH) for the survival functions for various treatment groups. In the simplest case of a placebo vs. treatment study, if F_0 and F_1 stand for the respective d.f.'s, we denote the corresponding survival functions by $\bar{F}_0$ and $\bar{F}_1$, and let f_j and h_j be the density and hazard functions corresponding to F_j, for $j = 0, 1$. Thus, $h_j(x) = f_j(x)/\bar{F}_j(x)$, for $x \in (0, \infty)$ and $j = 0, 1$. It is then assumed that

$$h_1(x) = c \cdot h_0(x), \quad \text{for all } x \in (0, \infty), \tag{9.26}$$

where c is a positive constant. In terms of the survival functions, this PH assumption leads to the following formulation:

$$\bar{F}_1(x) = [\bar{F}_0(x)]^c, \quad \text{for all } x \in (0, \infty). \tag{9.27}$$

In this formulation, one can allow the d.f. F_0 to be arbitrary (but absolutely continuous), so that the hazard function $h_0(x)$ is treated as an arbitrary function on $(0, \infty)$. If both F_0 and F_1 are exponential d.f.'s, then this PH condition is automatically true, and the above formulation extends this characterization to a wider class. Motivated by this simple formulation, let us consider a set of n d.f.'s F_i, $i = 1, \ldots, n$, denote the corresponding survival and hazard functions by $\bar{F}_i$, $i = 1, \ldots, n$, and $h_i(x)$, $i = 1, \ldots, n$, respectively, and consider a PH formulation incorporating a regression function $\boldsymbol{\beta}'\boldsymbol{t}_i$ as follows:

$$\begin{aligned} h_i(x) = h_i(x \mid \boldsymbol{t}_i) &= h_o(x) \cdot \exp\{\boldsymbol{\beta}'\boldsymbol{t}_i\}, \\ &x \in (0, \infty),\ i = 1, \ldots, n, \end{aligned} \tag{9.28}$$

where the $\boldsymbol{t}_i$ are given design variates, and $h_o(x)$, the base line hazard function, is treated as arbitrary. In this formulation, we may even include (stochastic) concomitant variates in the $\boldsymbol{t}_i$, so that we may term the h_i as conditional hazard function, and in terms of the log-hazard functions, we have then a linear regression model. Since the regression part of the model is of parametric form while the base line hazard function is nonparametric in character, this model is also referred to as a *semi-parametric model.* The ingenuity of the Cox formulation lies in incorporation of the partial likelihood formulation for drawing valid and efficient statistical conclusions on the *regression* parameter (vector) $\boldsymbol{\beta}$, treating the baseline hazard as a nuisance parameter (functional). A complete treatment of this novel methodology is beyond the scope of this article, although some other chapters in this volume are likely to deal with this in some details. In passing, we may refer to the recent monograph of Andersen et al. (1993) where this semi-parametric approach has been dealt with in a much more general and sound theoretical basis. Use of rank based procedures for this PH model is naturally appealing on the ground of *invariance* under strictly monotone transformations on the primary variate, and it is not surprising to see that the classical log-rank procedure described in Section 2 has a close affinity to this model as well. PCS modeling for such PH models has also been worked out along the same line as in before, and we may refer to Sen (1981b) for some details. There are a few points to ponder in this respect. First, the PH assumption itself may not hold in general in all applications; we may refer to Sen (1994b) for some exposition of the nonrobustness aspects of the classical PH models in survival analysis and clinical trials. Secondly, *staggered entry* or batch arrival models are usually encountered in practice, where the entry pattern may be quite arbitrary and stochastic in nature. In such a case, the partial likelihood approach may not lead to the usual log-rank procedures, and may call for more complex statistical analysis schemes; we refer to Sen (1985) for some details. Thirdly, if the study involves some concomitant variates (and usually they abound in practical applications), often, we have some of them as *time-dependent.* In such a case, the Cox formulation of the PH model may require more sophisticated statistical solutions, and much of the simplicities may be lost in this quest; we refer to Murphy and Sen (1991) for some accounts of such developments. Finally, we may remark that censoring in statistical theory and practice may not be very complementary to each other or coherent in a natural sense. Part of this difficulty (viz., Sen, 1995d) stems from the fact that the basic regularity assumptions, as are generally needed in biomedical studies (including clinical trials, and epidemiological and environmental investigations), may not provide an easy access to incorporating the simple censoring schemes referred to before; the more complex is the nature of such censoring the greater is the price one has to pay for implementation of suitable statistical designs and for developing suitable statistical parametric or nonparametric analysis schemes to accommodate such complexities. We complete this section with some discussion on the *staggered entry* plans which are common in medical studies and clinical trials. Recruitment of subjects for such investigations often require extensive search for valid subjects, and as these are generally people having some medical problems requiring some treatments, they may arrive at the clinics either in batches or sometimes sporadically over time. This may result in a differential exposure time if the study is planned for a fixed duration of time, and hence, the basic formulation of Type I, II or random censoring schemes may not be very appealing in such a case. To handle such a relatively more complex censoring pattern, one may need to look into the composite picture from a multiple time-parameter point of view. Sinha and Sen (1982) considered such a scheme based on the usual empirical distributions adjusted for staggered entries and formulated suitable nonparametric testing procedures. A more general treatment with rank statistics in a staggering entry plan is due to Sen (1985). Weak convergence to multi-dimensional Gaussian processes provides the desired statistical tools to implement such methodology in practical applications.

10. Nonparametrics in incomplete multiresponse clinical designs

In clinical trials, from epidemiological perspectives, often, information is gathered on more than one response variable, and, in addition, on relevant concomitant variables too. Nevertheless, from clinical or medical perspectives, it is not uncommon to single out one of the response variates as the primary endpoint and the others as auxiliary ones. In some cases, characterizing a primary endpoint in this conventional manner may entail stringent cost constraints primarily related to its precise measurement, and some *surrogate* endpoints are advocated to cast vital information at relatively lower levels of cost or time consumptions. The multitude of such response variates through their mutual statistical dependence can cast light on the primary endpoint and hence their simultaneous measurements generally lead to comparatively more precise statistical conclusions. On the other hand, based on other practical considerations, such as relative cost, ease and/or precision of measurement, it may not be very convenient to include the entire battery of simultaneous measurements on all the response as well as concomitant variates for all the subjects (units). For this reason, often, an *Incomplete Multiresponse Design* (IMD) or a *Hierarchical Design* (HD) is adopted.

In the latter design, there is a hierarchy on the response variates in the sense that there is a (partial) ordering with respect to the number of experimental units on which their measurements are recorded; for this reason, they are also termed *Nested Designs* (ND). To illustrate this point, suppose that there are p (> 1) response variates, denoted by $Y_1, \ldots, Y_p$ respectively. On a smallest set, sat $\mathcal{S}_1$, of experimental units, all these p responses are measured simultaneously; for a larger set $\mathcal{S}_2$, containing $\mathcal{S}_1$ as a subset, $Y_2, \ldots, Y_p$ (but not Y_1) are recorded on the subset $\mathcal{S}_2 \setminus \mathcal{S}_1$, and so on. For the largest set $\mathcal{S}_p$, containing $\mathcal{S}_{p-1}$ as a subset, Y_p alone is recorded on the subset $\mathcal{S}_p \setminus \mathcal{S}_{p-1}$. Such a multiresponse design, determined by the inherent nesting $\mathcal{S}_1 \subset \mathcal{S}_2 \subset \cdots \subset \mathcal{S}_p$ is termed a hierarchical design (viz., Roy et al., 1971, Chapter 8). It may not always be desirable or even practicable to impose this basic hierarchy condition. For example, the (random pattern) *missing observations* in multiresponse designs may distort this hierarchy condition to a certain extent. Nevertheless, it may be feasible to incorporate some IMD's wherein the set $\boldsymbol{Y} = \{Y_1, \ldots, Y_p\}$ can bc partitioned into various subsets $\{Y_{i_1}, \ldots, Y_{i_r}\}$, $1 \leqslant r \leqslant p$, $1 \leqslant i_1 < \cdots < i_r \leqslant p$, such that these subsets are not necessarily nested and they are adoptable for possibly different number of experimental units. For example, for $p = 2$, we have three possible subsets $\{Y_1\}$, $\{Y_2\}$ and $\{Y_1, Y_2\}$, and possibly different designs (say, $\mathcal{D}_1$, $\mathcal{D}_2$ and $\mathcal{D}_{12}$) may be chosen for these subsets. In this context it may be recalled that in clinical trials, often, the primary emphasis is on a comparative study of a placebo and one or more treatments, so that these designs are to be chosen in a conventional sense with due emphasis on these treatments.

In clinical trials, a primary endpoint, in spite of being the most relevant one, may encounter some basic problems regarding its precise measurement (due to possibly excessive cost or some other practical limitations); therefore, it is not uncommon to make use of a very closely related but presumably, relatively less expensive variate, termed a surrogate endpoint. Generally, surrogate endpoints may not contain as much information as contained in the primary endpoint, and such a substitution may have serious effects on valid and efficient statistical modeling and analysis, unless the surrogate variate has some statistical concordance with the primary one. The situation may particularly be very bleak when this statistical interface of surrogate and primary endpoints is not that clearly known, and this case arises typically when no specific data are available on simultaneous measurement of both these variables. Nevertheless, the use of such surrogate endpoints in clinical trials and medical investigations has generally been accepted by the allied medical community and has caught the attention of statisticians as well. A nice statistical account of such uses, and abuses too, is given in a set of articles published in *Statistics in Medicine*, Vol. 8, No. 2 (1989). More technical exposition of this field are due to Pepe (1992) and Sen (1994a), among others. Not all auxiliary variables qualify for surrogates, and for qualified ones, it seems very reasonable (if not essential) to design a study in such a way that for a majority of the experimental units, termed the *surrogate sample*, valid surrogate endpoints and concomitant variates are recorded, while for a smaller subset of experimental units, termed the *validation sample*, simultaneous recording of the primary and surrogate endpoints throws light on their statistical relation which enables us to combine the evidence from both the subsets of data and draw better statistical conclusions. If statistical conclusions are to be drawn only from the surrogate sample observations, some stronger regularity assumptions are generally needed to justify the conclusions, while the use of a validation sample may enhance the scope of the study considerably. We may refer to Prentice (1989) and Pepe (1992) for some useful accounts of these pros and cons of surrogate endpoints in clinical trials. The Prentice–Pepe setups can be characterized as both a hierarchical and incomplete multiresponse model with $p = 2$. Many clinical trials encounter a more complex setup involving multiple endpoints resulting in multiresponse primary variates. We may refer to Wei et al. (1989) and Prentice and Cai (1992) for some statistical treatments for such designs. A more comprehensive IMD/HD approach in a nonparametric setup is due to Sen (1994a), and we summarize these results here.

There may be in general more than one primary endpoints, and we denote this by $\boldsymbol{Y} = (Y_1, \ldots, Y_p)'$, where p ($\geqslant 1$) and there may be a partial ordering of the importance of these primary endpoints, which may also be taken into account in the design and statistical analysis of the study. Similarly, the surrogate endpoint may also be represented by a q-vector $\boldsymbol{Y}_o$, where q is a positive integer. Thus, in general, we have a set of $p + q$ responses, some of which may be costly to record. In order to extract information on the statistical relations between $\boldsymbol{Y}$ and $\boldsymbol{Y}_o$, and to incorporate the same in drawing statistical conclusions, it may be desirable to use IMD's or HD's. In this respect in a conventional approach, one adopts *Multivariate General Linear Models* (MGLM) for statistical modeling and analysis; however, the basic regularity assumptions are even more unlikely to be tenable in this multivariate situation. Thus, the appropriateness of MGLM's in clinical trials is questionable. Use of *Generalized Linear Models* (GLM) is also subject to similar limitations, and on model robustness grounds they are even more vulnerable to plausible departures from the assumed regularity assumptions. The Cox (1972) PHM based *partial likelihood* approach is also subject to serious nonrobustness constraints, and hence, sans sufficient confidence on such a parametric or semi-parametric model, such procedures should not be advocated in real applications. Some aspects of nonrobustness of the Cox PHM approach are discussed in Sen (1994a), and these remarks pertain to general IMD's as well. Nonparametrics, on the other hand, possess good robustness properties, and are better competitors to these alternative ones. This has been the main motivation of Sen (1994a) in pursuing general nonparametrics for such IMD's with adequate emphasis on the related asymptotics.

The basic rank procedures described in the earlier sections, particularly, for randomized blocks, incomplete block designs, factorial experiments and multivariate models, provide the necessary access for this development, and we shall unify these in a convenient mold. Aligned rank procedures are particularly useful in this context. It may be recalled that a surrogate endpoint is a qualified substitute for the primary endpoint only if it reflects a picture with reference to the treatment difference concordant with the primary endpoint; in the literature this condition is also referred to as the *validity* criterion for a surrogate. Such a condition can be tested if one has a validation sample where both the primary and surrogate endpoints are recorded. But, generally, such a validation sample has a smaller size compared to the surrogate sample. Hence, the general nonparametric approach is based on the following scheme: (i) For testing a plausible hypothesis relating to treatment differences construct a suitable nonparametric test statistic based on the surrogate sample observations and adjusted for covariates, if any. (ii) For the validation sample, construct a similar nonparametric statistic for both the primary and surrogate endpoints (using the multivariate approach treated in Section 4), also adjusted for concomitant variates, if any. (iii) Test for the concordance of the primary and surrogate endpoints with respect to treatment differences based on the statistics in Step (ii). Again, nonparametric tests can be used here. (iv) Regress the primary endpoint statistics on the surrogate endpoint ones (in the validation sample in Step (ii)), and obtain the aligned statistics for the primary endpoint as residuals from this fitted regression. (v) Combine the statistics in Step (i) and (iv) by the usual weighted least squares principle and use the same in the formulation of the actual test statistic to be used for testing a hypothesis on treatment differences.

In this context the joint (asymptotic) normality of multivariate rank statistics provides the theoretical justifications for the various steps sketched above, and also provides the foundation of general asymptotics relevant to this topics. These details are provided in Sen (1994a).

For linear models in the parametric case, IMD's entail a secondary task: *Recovery of interblock information* from the block totals. In the nonparametric case, although the basic linearity of the model may not be fully appreciated, such recovery of interblock information is possible. The basic motivation is the same, although a somewhat different alignment process is needed to incorporate this recovery in nonparametric analysis. This alignment procedure is very similar to the classical parametric case: Block averages or some other measure of central tendency are used for construction of interblock aligned rank statistics, while the residuals within each block are pooled together for all blocks and replicates to construct aligned rank statistics for intra-block analysis. These two sets of rank statistics are then combined (as in Step (v) above) in a convenient way to construct suitable test statistics which have greater power than the one based solely on the intra-block residuals. For clinical trials involving (a treatment-wise) IMD and a surrogate endpoint, recovery of interblock information in nonparametric analysis of covariance models has recently been treated in a unified manner by El-Moalem and Sen (1997). The idea is quite simple. In addition to using the aligned rank statistics (for the primary and surrogate endpoints as well as the concomitant variates), in a replicated IMD, it is also possible to use the within replicate block totals, align them with a view to eminiating the replicate effects, and then to use aligned rank statistics on such aligned block totals to extract further information on the treatment effects. Use of such aligned rank statistics eliminates the cruciality of the linearity of the model to a certain extent and makes it possible to use the usual weighted least squares methodology to construct suitable pooled aligned rank statistics which may be incorporated in the construction of a plausible test statistic having, at least asymptotically, better power properties. In this context it is not necessary to assume that the treatments are replicated equal number of times within each replicate or any pair of them are done so, and the treatise covers a general class of IMD's.

11. Concluding remarks

The current state of art with the developments on nonparametrics in design and analysis of various types of experiments really calls for a far more thorough treatise of the subject matter than presented in this writeup. For lack of space, it has not been possible to include the entire battery of topics in design and analysis of experiments where nonparametrics are relevant.

As regards the basic nonparametrics presented in the first eight sections of this writeup, the treatment here is fairly thorough. However, the last two sections are presented with more motivations from applications point of view but from methodological point of view somewhat less technical details are provided than they deserve. The nonparametric task remains as much more challenging, although some work has already been in progress in this direction. On the top of this there is another important consideration underlying practical adoption of statistical designs and analysis packages in clinical trials and medical studies in general. *Missing observations* may be a part of such experimental data, and statistical analysis should address this issue adequately with due considerations to practical adoptions. In clinical trials, such a missing pattern may be due to censoring of various types discussed in earlier sections, while in a general setup, it may be due to other factors as well. In epidemiological studies, it is not uncommon to encounter multiple causes of failures, and hence, a *competing risk* setup is often judged as appropriate. Again design and statistical analysis (parametrics as well as nonparametrics) for such studies follow somewhat different tracks, and it may be desirable to pay due attention to the developments of nonparametrics for competing risks models in as much generality as possible. *Random missing patterns* are often introduced as a part of the basic assumptions to deal with messy data sets arising in such studies. In a nonparametric MANOVA setup, for some developments on such random missing patterns, we may refer to Servy and Sen (1987), where other pertinent references are also cited. There remains much more to be accomplished in this direction. Competing risks models in a general multiresponse (or multiple end-point) clinical trial poses even more complex statistical designing and analysis tasks. Only in some simplest situations, some relevant nonparametrics have been developed; we may refer to DeMasi (1994) where other references are cited in detail. Since more complex censoring patterns may arise in this context, statistical modeling (underlying either parametric or robust procedures) needs to address the infrastructure in an adequate manner; this not only increases the number of parameters associated with the model, but also may raise some *identifiability* issues which call for more delicate treatments. These need to be addressed in a more general and integrated manner than done here.

Throughout this presentation the major emphasis has been on nonparametrics based on rank statistics and allied estimators. Although such rank procedures can mostly be justified from a *global robustness* point of view (with very little emphasis on the form of the underlying error distributions), there are some other situations where it may be wiser to take recourse to *local robustness* properties wherein only small departures from an assumed model are contemplated, so that high efficiency mingled with low sensitiveness to such local departures dominate the scenario. In this setup, as viable competitors to such nonparametrics, robust procedures based on suitable L- and M-statistics are often advocated. *Regression quantiles* have their genesis in this complex, but in the recent past, they have paved the way for the related *regression rank scores* estimators and test statistics which compare very favorably with nonparametric procedures based on R-estimators. Recently, Jurečková and Sen (1993) have established certain asymptotic equivalence results on the classical R-estimators and regression rank scores estimators in a linear model based on a common score generating function, and as such, taking into account the relative computational complexities of these two approaches, in some cases, we may advocate the use of such regression rank scores procedures as well. For some general accounts of these findings we may refer to Jurečková and Sen (1996, Chapter 6), where mostly fixed-effects models are considered, and to Sen (1995a) where some mixed-effects models have also been treated.

In the clinical and epidemiological sectors, due to medical ethics standards and current policies of some of the regulatory agencies in USA or other industrized nations, a *multi-phase design* approach for human usage is generally adopted. In Phase I, primary emphasis is on exploration of biochemical/biomedical effects, toxicity etc, while in Phase II, some *therapeutic* factors are taken into account. In this setup, it is quite common to have first some *animal studies*, and the conclusions as may be gathered from such studies are then to be incorporated in the design and general formulation of the main study: Phase III clinical trials. The emerging sub-discipline: *clinical epidemiology* has been geared to address more complex issues arising in this interdisciplinary field. Because of apparently conflicting attitudes of statisticians and epidemiologists to some of these clinical problems, in clinical epidemiology, there is, often, a blending of *ecology* and *etiology* for which the design as well as analysis aspects may differ drastically even in some simple parametric setups (see, e.g., Sen 1994c). Nonparametrics play a fundamental role in this setup too. For example, extrapolating the statistical findings from experiments conducted on subhuman primates to human beings raises the question of their validity and scope. In the statistical literature, such methodologies are categorized under the topic: *Accelerated Life Testing* (ATL) procedures. In parametric setups, the basic regularity assumptions appear to be quite stringent, and hence, nonparametrics are generally advocated for greater scope and reliability. However, in this context too, validity and reliability of statistical regularity assumptions need to be assessed properly. *Biological assays* are the main statistical assessment tools in this venture. Designs for such bio-assays may often be somewhat different, and we may refer to the classical text of Finney (1964) for a detailed account of such developments. His treatise has mainly been on a conventional parametric walk, wherein due emphasis has been laid down on *transformations* on the response and dose variables (termed the *response metameter* and *dosage* or *dose metameter* respectively) under which suitable parametric models can be justified. Nevertheless, in practice, such transformations may not simultaneously achieve the basic linearity of the model and normality (or logistic or some other simple form) of the tolerance distribution. As such, there is good scope for nonparametrics. Some developments in this sector took place during the sixties (viz., Sen, 1963) and early seventies, and a systematic review of this work is reported in Sen (1984a), where pertinent references are also cited.

A new area of statistical awareness relates to our endangered environment and the statistical endeavors to cope with such problems; these have led to the developments of another frontier of statistical sciences: the *Environmetrics*. The tasks are truly challenging and statistical considerations are overwhelming in this venture. Unlike the case of conventional agricultural experiments, animal studies or even the clinical trials, environmental problems are generally characterized by the lack of control in the conduct of a scientific study to a much greater extent. Also, a large number of factors contributes to unaccountable variations in the response patterns. Moreover, the response variables are often imprecisely defined and may also encounter serious measurement problems. For example, to assess the air pollution standard of various urban, suburban and rural areas in USA, the basic task may be to define precisely the response variables, identify their probable causes or factor variables (viz., auto-exhaustion, environmental smoking, industrial emissions, etc.), variation with the whether conditions, day-to-night variation, and many other factors which may not be properly defined and may hardly be controllable to a satisfactory extent. Some of these variables may even be binary or polychotomous in nature. On the top of that even when a variable is quantitative, it may usually be recorded in class intervals leading to the so called *interval censoring* schemes. Thus, *measurement errors* and to a certain extent *misclassifications* are usually encountered as a vital part of such response as well as dose variables. Even in the simplest case of two or several sample problems, for such *grouped data* ties among the observations may not be negligible, and there may not be a unique way to handle such ties; we may refer to Hájek and Šidak (1967, Chapter 3) for some treatment of ties for the exact null hypothesis distributions of rank statistics, and to Sen (1967) for asymptotic optimality of rank tests for grouped data in a simple regression model. These results extend directly to general linear models. However, linear models are hardly appropriate in this complex setup when the response-dose regression, subject to possible measurement errors/misclassifications, may be quite nonlinear in nature, and suitable transformations on the factor as well as response variables are generally used to induce more linearity in the models; their impacts on the distributional assumptions are needed to be assessed carefully. Assumptions of independence, homoscedasticity and even the symmetry of the error components are to be examined critically in the particular contexts, and for these reasons, statistical designs and analysis schemes are to be developed in more practically adoptable settings. Such environmental problems are not totally out of the reach of clinical trials and biomedical studies. The emerging field of *environmental health sciences* deals with the impact of the environment on human health and prospects for long-range healthy living. Environmental health effects have been identified to be far more outreaching than in a simple chemical or biochemical setting, and *Genotoxicity* has also been identified as an important ingredient in this phenomenon. In this quest, biological assays involving *biological markers* are vital tools for assessments on subhuman primates, and suitable design of (mutagenetic) experiments are generally advocated for extrapolation of the findings from animals to human beings. Because of the fundamental roles of *Molecular Biology* and *Human Genetics* in these complex experimental schemes, such designs are generally quite different from the conventional ones considered in this volume. The appropriateness of an interdisciplinary approach is crucial in this context. *Inhalation toxicology, water contamination, air pollution* and scores of other serious environmental threats are affecting the *Quality of Life* (QOL) and in many ways, endangering our lives too, and, in this respect, an *interdisciplinary approach* is very much needed to provide scientifically sound and operationally manageable solutions. We may refer to Sen and Margolin (1995) for some of the basic statistical issues in some environmental studies with major emphasis on inhalation toxicology, and conclude that statistical planning and analysis schemes are most vital in this venture. Parametrics or semi-parametrics are less likely to be appropriate in this emerging research field, and nonparametrics are indispensible in this context to a far greater extent.

12. Acknowledgements

I am grateful to Professor Subir Ghosh for critical reading of the manuscript which has eliminated numerous typos and some obscurities as well.

References

Adichie, J. N. (1967). Estimates of regression parameters based on rank tests. *Ann. Math. Statist.* **38**, 894–904.

Andersen, P. K., O. Borgan, R. D. Gill and N. Keiding (1993). *Statistical Models Based on Counting Processes.* Springer, New York.

Armitage, P. (1975). *Sequential Medical Trials*, 2nd edn. Blackwell, Oxford.

Bahadur, R. R. (1961). A representation of the joint distribution of responses to n dichotomous items. In: H. Solomon, ed., *Studies in Item Analysis and Prediction.* Stanford Univ. Press, CA.

Benard, A. and P. van Elteren (1953). A generalization of the method of m-rankings. *Indag. Math.* **15**, 358–369.

Bhapkar, V. P. (1961a). Some nonparametric median procedures. *Ann. Math. Statist.* **32**, 846–863.

Bhapkar, V. P. (1961b). A nonparametric test for the problem of several samples. *Ann. Math. Statist.* **32**, 1108–1117.

Bradley, R. A. and M. E. Terry (1952). Rank analysis of incomplete block designs, I. The method of paired comparison. *Biometrika* **39**, 324–345.

Brown, G. W. and A. M. Mood (1951). On median tests for linear hypotheses. *Proc. 2nd Berkeley Symp. Math. Statist. Probab.* 159–166.

Chatterjee, S. K. and P. K. Sen (1964). Nonparametric tests for the bivariate two-sample location problem. *Calcutta Statist. Assoc. Bull.* **13**, 18–58.

Chatterjee, S. K. and P. K. Sen (1973). Nonparametric testing under progressive censoring. *Calcutta Statist. Assoc. Bull.* **22**, 13–50.

Cox, D. R. (1972). Regression models and life tables (with discussion). *J. Roy. Statist. Soc. Ser. B* **34**, 187–220.

Cox, D. R. (1975). Partial likelihood. *Biometrika* **62**, 369–375.

David, H. A. (1988). *The Method of Paired Comparisons*, 2nd edn. Oxford Univ. Press, New York.

Davidson, R. R. and R. A. Bradley (1969). Multivariate paired comparisons: The extension of a univariate model and associated estimation and test procedures. *Biometrika* **56**, 81–94.

Davidson, R. R. and R. A. Bradley (1970). Multivariate paired comparisons: Some large sample results on estimation and test of equality of preference. In: M. L. Puri, ed., *Nonparametric Techniques in Statistical Inference.* Cambridge Univ. Press, New York, 111–125.

DeLong, D. M. (1981). Crossing probabilities for a square root boundary by a Bessel process. *Comm. Statist. Theory Methods* **A10**, 2197–2213.

DeMasi, R. A. (1994). Proportional Hazards models for multivariate failure time data with generalized competing risks. Unpublished Doctoral Dissertation, Univ. North Carolina, Chapel Hill.

Durbin, J. (1951). Incomplete blocks in ranking experiments. *Brit. J. Statist. Psychol.* **4**, 85–90.

El-Moalem, H. and P. K. Sen (1997). Nonparametric recovery of interblock information in clinical trials with a surrogate endpoint. *J. Statist. Plann. Inference* (to appear).

van Elteren, P. and G. E. Noether (1959). The asymptotic efficiency of the χ_r^2-test for a balanced incomplete block design. *Biometrika* **46**, 475–477.

Finney, D. J. (1964). *Statistical Methods in Biological Assay*, 2nd edn. Griffin, London.

Fleming, T. R. and D. P. Harrington (1991). *Counting Processes and Survival Analysis*. Wiley, New York.

Friedman, M. (1937). The use of ranks to avoid the assumption of normality implicit in the analysis of variance. *J. Amer. Statist. Assoc.* **32**, 675–701.

Gerig, T. M. (1969). A multivariate extension of Friedman's χ^2-test. *J. Amer. Statist. Assoc.* **64**, 1595–1608.

Gerig, T. M. (1975). A multivariate extension of Friedman's χ^2-test with random covariates. *J. Amer. Statist. Assoc.* **70**, 443–447.

Greenberg, V. L. (1966). Robust estimation in incomplete block designs. *Ann. Math. Statist.* **37**, 1331–1337.

Grizzle, J. E. (1965). The two-period change-over design and its use in clinical trials. *Biometrics* **21**, 467–480.

Hájek, J. and Z. Šidák (1967). *Theory of Rank Tests*. Academic Press, New York.

Hampel, F. R., E. M. Ronchetti, P. J. Rousseeuw and W. A. Stahel (1986). *Robust Statistics: The Approach Based on Influence Function*. Wiley, New York.

Hodges, J. L., Jr. and E. L. Lehmann (1962). Rank methods for combination of independent experiments in analysis of variance. *Ann. Math. Statist.* **33**, 487–497.

Hodges, J. L., Jr. and E. L. Lehmann (1963). Estimates of location based on rank tests. *Ann. Math. Statist.* **34**, 598–611.

Huber, P. J. (1981). *Robust Statistics*. Wiley, New York.

Jaeckel, L. A. (1972). Estimating regression coefficients by minimizing dispersion of the residuals. *Ann. Math. Statist.* **43**, 1449–1458.

Jurečková, J. (1971). Nonparametric estimate of regression coefficients. *Ann. Math. Statist.* **42**, 1328–1338.

Jurečková, J. (1977). Asymptotic relations of M-estimates and R-estimates in linear models. *Ann. Statist.* **5**, 464–472.

Jurečková, J. and P. K. Sen (1993). Asymptotic equivalence of regression rank scores estimators and R-estimators in linear models. In: J. K. Ghosh et al., eds., *Statistics and Probability: A Raghu Raj Bahadur Festschrift*, Wiley Eastern, New Delhi, 279–292.

Jurečková, J. and P. K. Sen (1996). *Robust Statistical Procedures: Asymptotics and Interrelations*. Wiley, New York.

Koch, G. G. (1972). The use of nonparametric methods in the statistical analysis of two-period change-over design. *Biometrics* **28**, 577–584.

Krishnaiah, P. R. (ed.) (1981). *Handbook of Statistics, Vol. 1: Analysis of Variance*. North-Holland, Amsterdam.

Krishnaiah, P. R. and P. K. Sen (eds.) (1984). *Handbook of Statistics, Vol. 4: Nonparametric Methods*. North-Holland, Amsterdam.

Lan, K. K. B. and D. L. DeMets (1983). Discrete sequential boundaries for clinical trials. *Biometrika* **70**, 659–663.

Lehmann, E. L. (1963a). Robust estimation in analysis of variance. *Ann. Math. Statist.* **34**, 957–966.

Lehmann, E. L. (1963b). Asymptotically nonparametric inference: An alternative approach to linear models. *Ann. Math. Statist.* **34**, 1494–1506.

Lehmann, E. L. (1964). Asymptotically nonparametric inference in some linear models with one observations per cell. *Ann. Math. Statist.* **35**, 726–734.

Majumdar, H. and P. K. Sen (1978). Nonparametric tests for multiple regression under progressive censoring. *J. Multivariate Anal.* **8**, 73–95.

Mantel, N. and W. Haenszel (1959). Statistical aspects of analysis of data from retrospective studies of disease. *J. Nat. Cancer Inst.* **22**, 719–748.

Mehra, K. L. and J. Sarangi (1967). Asymptotic efficiency of some rank tests for comparative experiments. *Ann. Math. Statist.* **38**, 90–107.

Mehra, K. L. and P. K. Sen (1969). On a class of conditionally distribution-free tests for interactions in factorial experiments. *Ann. Math. Statist.* **40**, 658–666.

Murphy, S. A. and P. K. Sen (1991). Time-dependent coefficients in a Cox-type regression model. *Stochast. Proc. Appl.* **39**, 153–180.

Pepe, M. S. (1992). Inference using surrogate outcome data and a validation sample. *Biometrika* **79**, 495–512.

Prentice, R. L. (1989). Surrogate endpoints in clinical trials: Definition and operational criteria. *Statist. Med.* **8**, 431–440.

Prentice, R. L. and J. Cai (1992). Covariance and survival function estimation using censored multivariate failure time data. *Biometrika* **79**, 495–512.

Pukelsheim, F. (1993). *Optimal Experimental Design*. Wiley, New York.

Puri, M. L. and P. K. Sen (1967). On robust estimation in incomplete block designs. *Ann. Math. Statist.* **38**, 1587–1591.

Puri, M. L. and P. K. Sen (1971). *Nonparametric Methods in Multivariate Analysis*. Wiley, New York.

Puri, M. L. and P. K. Sen (1985). *Nonparametric Methods in General Linear Models*. Wiley, New York.

Quade, D. (1984). Nonparametric methods in two-way layouts. In: P. R. Krisnaiah and P. K. Sen, eds., *Handbook of Statistics, Vol. 4: Nonparametric Methods*. North-Holland, Amsterdam, 185–228.

Roy, S. N. (1953). On a heuristic method of test construction and its use in multivariate analysis. *Ann. Math. Statist.* **24**, 220–238.

Roy, S. N., R. Gnanadesikan and J. N. Srivastava (1970). *Analysis and Design of Certain Quantitative Multiresponse Experiments*. Pergamon Press, New York.

Sen, P. K. (1963). On the estimation of relative potency in dilution (-direct) assays by distribution-free methods. *Biometrics* **19**, 532–552.

Sen, P. K. (1967a). Asymptotically mostpowerful rank order tests for grouped data. *Ann. Math. Statist.* **38**, 1229–1239.

Sen, P. K. (1967b). A note on the asymptotic efficiency of Friedman's χ_r^2-test. *Biometrika* **54**, 677–679.

Sen, P. K. (1968a). Asymptotically efficient test by the method of n-ranking. *J. Roy. Statist. Soc. Ser. B* **30**, 312–317.

Sen, P. K. (1968b). On a class of aligned rank order tests in two-way layouts. *Ann. Math. Statist.* **39**, 1115–1124.

Sen, P. K. (1968c). Robustness of some nonparametric procedures in linear models. *Ann. Math. Statist.* **39**, 1913–1922.

Sen, P. K. (1968d). Estimates of the regression coefficient based on Kendall's tau. *J. Amer. Statist. Assoc.* **63**, 1379–1389.

Sen, P. K. (1969a). On nonparametric T-method of multiple comparisons in randomized blocks. *Ann. Inst. Statist. Math.* **21**, 329–333.

Sen, P. K. (1969b). Nonparametric tests for multivariate interchangeability. Part II: The problem of MANOVA in two-way layouts. *Sankhyā Ser. A* **31**, 145–156.

Sen, P. K. (1970a). On the robust efficiency of the combination of independent nonparametric tests. *Ann. Inst. Statist. Math.* **22**, 277–280.

Sen, P. K. (1970b). Nonparametric inference in n replicated 2^m factorial experiments. *Ann. Inst. Statist. Math.* **22**, 281–294.

Sen, P. K. (1971a). Asymptotic efficiency of a class of aligned rank order tests for multiresponse experiments in some incomplete block designs. *Ann. Math. Statist.* **42**, 1104–1112.

Sen, P. K. (1971b). Robust statistical procedures in problems of linear regression with special reference to quantitative bio-assays, I. *Internat. Statist. Rev.* **39**, 21–38.

Sen, P. K. (1972). Robust statistical procedures in problems of linear regression with special reference to quantitative bio-assays, II. *Internat. Statist. Rev.* **40**, 161–172.

Sen, P. K. (1979). Rank analysis of covariance under progressive censoring. *Sankhyā Ser. A* **41**, 147–169.

Sen, P. K. (1980). Nonparametric simultaneous inference for some MANOVA models. In: P. R. Krishnaiah, ed., *Handbook of Statistics, Vol. 1: Analysis of Variance*. North-Holland, Amsterdam, 673–702.

Sen, P. K. (1981a). The Cox regression model, invariance principles for some induced quantile processes and some repeated significance tests. *Ann. Statist.* **9**, 109–121.

Sen, P. K. (1981b). *Sequential Nonparametrics: Invariance Principles and Statistical Inference*. Wiley, New York.

Sen, P. K. (1984a). Some miscellaneous problems in nonparametric inference. In: P. R. Krishnaiah and P. K. Sen, eds., *Handbook of Statistics, Vol. 4: Nonparametric Methods*. North-Holland, Amsterdam, 699–739.

Sen, P. K. (1984b). Multivariate nonparametric procedures for certain arteriosclerosis problems. In: P. R. Krishnaiah, ed., *Multivariate Analysis VI*. North-Holland, Amsterdam, 563–581.

Sen, P. K. (1985). *Theory and Applications of Sequential Nonparametrics*. SIAM, Philadelphia, PA.

Sen, P. K. (1988). Combination of statistical tests for multivariate hypotheses against restricted alternatives. In: S. Dasgupta and J. K. Ghosh, eds., *Statistics: Applications and New Directions*. Ind. Statist. Inst., Calcutta, 377–402.

Sen, P. K. (1991a). Nonparametrics: Retrospectives and perspectives (with discussion). *J. Nonparamert. Statist.* **1**, 3–53.

Sen, P. K. (1991b). Repeated significance tests in frequency and time domains. In: B. K. Ghosh and P. K. Sen, eds., *Handbook of Sequential Analysis*, Marcel Dekker, New York, 169–198.

Sen, P. K. (1993a). Statistical perspectives in clinical and health sciences: The broadway to modern applied statistics. *J. Appl. Statist. Sci.* **1**, 1–50.

Sen, P. K. (1993b). Perspectives in multivariate nonparametrics: Conditional functionals and ANOCOVA models. *Sankhyā Ser. A* **55**, 516–532.

Sen, P. K. (1994a). Incomplete multiresponse designs and surrogate endpoints in clinical trials. *J. Statist. Plann. Inference* **42**, 161–186.

Sen, P. K. (1994b). Some change-point problems in survival analysis: Relevance of nonparametrics in applications. *J. Appl. Statist. Sci.* **1**, 425–444.

Sen, P. K. (1994c). Bridging the biostatistics-epidemiology gap: The Bangladesh task. *J. Statist. Res.* **28**, 21–39.

Sen, P. K. (1995a). Nonparametric and robust methods in linear models with mixed effects. *Tetra Mount. Math. J.* **7**, 1–12.

Sen, P. K. (1995b). Paired comparisons for multiple characteristics: An ANOCOVA approach. In: H. N. Nagaraja, D. F. Morrison and P. K. Sen, eds., *H. A. David Festschrift*. Springer, New York, 237–264.

Sen, P. K. (1995c). Regression rank scores estimation in ANOCOVA. *Ann. Statist.* (to appear).

Sen, P. K. (1995d). Censoring in Theory and Practice: Statistical perspectives and controversies. *IMS Lecture Notes Monograph Ser.* 27, 175–192.

Sen, P. K. and H. A. David (1968). Paired comparisons for paired characteristics. *Ann. Math. Statist.* **39**, 200–208.

Sen, P. K. and B. H. Margolin (1995). Inhalation toxicology: Awareness, identifiability and statistical perspectives. *Sankhyā Ser. B* **57**, 253–276.

Sen, P. K. and M. L. Puri (1977). Asymptotically distribution-free aligned rank order tests for composite hypotheses for general linear models. *Zeit. Wahrsch. verw. Geb.* **39**, 175–186.

Senn, S. (1993). *Cross-over Trials in Clinical Research.* Wiley, New York.

Servy, E. C. and P. K. Sen (1987). Missing values in multisample rank permutation tests for MANOVA and MANOCOVA. *Sankhyā Ser. A* **49**, 78–95.

Shah, K. R. and B. K. Sinha (1989). *Theory of Optimal Designs*. Lect. Notes Statist. No. 54, Springer, New York.

Sinha, A. N. and P. K. Sen (1982). Tests based on empirical processes for progressive censoring schemes with staggering entry and random withdrawal. *Sankhyā Ser. B* **44**, 1–18.

Tudor, G. and G. G. Koch (1994). Review of nonparametric methods for the analysis of cross over studies. *Statist. Meth. Med. Res.* **3**, 345–381.

Wei, L. J., D. Y. Lin and L. Weissfeld (1989). Regression analysis of multivariate incomplete failure time data by modeling marginal distributions. *J. Amer. Statist. Assoc.* **84**, 1065–1073.

S. Ghosh and C. R. Rao, eds., *Handbook of Statistics, Vol. 13*

5

Adaptive Designs for Parametric Models

S. Zacks

1. Introduction

Adaptive designs are those performed in stages (sequentially) in order to correct in each stage the design level and approach the optimal level(s) as the number of stages grow. Adaptive designs are needed when the optimal design level(s) depend on some unknown parameter(s) or distributions. In the present chapter we discuss adaptive designs for parametric models. These models specify the families of distributions of the observed random variables at the various design levels. An important class of such design problems is that of quantal response analysis of bioassays (see Finney, 1964). A dosage d of some toxic material is administered to K biological subjects. In dilutive assays the probability of response (death, convulsion, etc.) is related often to the dose in a parametric model specifying that the distribution of the number of subjects responding, out of K, at dose d, Y_d, is the binomial

$$Y_d \sim \mathcal{B}(K, F(\alpha + \beta x)), \tag{1.1}$$

where $x = \log(d)$, α and β are some parameters, and $F(u)$ is a c.d.f., called the *tolerance distribution*. The functional form of F is specified. When the biological subjects are not human, and there are no ethical restrictions, adaptive designs may not be necessary. In such cases the optimal design levels (doses) for estimating the parameters of the model are quite different from the ones presented here. In clinical trials with human subjects, in particular in phase I studies (see Geller, 1984), the objective is to administer the largest possible dose for which the probability of life threatening toxicity is not greater than a prespecified level, γ, $0 < \gamma < 1$. In this case the optimal log-dose is $x_\gamma = (F^{-1}(\gamma) - \alpha)/\beta$. This optimal level depends on the unknown parameters α and β. Adaptive designs are performed in such a case, and the information on α and β, which is gathered sequentially, is used to improve the approximation to the optimal level x_γ. In clinical trials, when it is important not to exceed x_γ, a constraint is sometimes imposed on the designs, such that $P\{X_n \leqslant x_\gamma\} \geqslant 1 - \alpha$, for each $n \geqslant 1$ and all (α, β) (see Section 3).

In Section 2 we discuss optimal designs, when the objective is to maximize the Fisher information on the unknown parameters. In Sections 2.1 and 2.2 we develop locally optimal designs and Bayes designs for a single sample situation. Section 2.3

deals with adaptive designs based on the MLE and on Bayesian estimates of the parameters. Section 3 is devoted to adaptive designs for inverse regression problems, which contain the above mentioned phase I clinical trials as a special case. We present results on adaptive designs when the observed random variables have normal distributions, with means and variances which depend on the design levels. Section 4 discusses the important problem of optimal allocation of experiments. These problems are known under the name of "bandit problems". Allocation problems are one type of adaptive designs in which sequential stopping rules play an important role. We provide an example of a problem in which the optimal allocation is reduced to a stopping time problem. An important non-parametric methodology for attacking some of the problems discussed in the present chapter is that of *stochastic approximation*, introduced by Robbins and Monro (1951). This important methodology deserves a special exposition and is therefore not discussed here. We refer the reader to the papers of Anbar (1977, 1984), in which stochastic approximation methods have been used for problems similar to the ones discussed here.

2. Optimal designs with respect to the Fisher information

2.1. Locally optimal designs

Let $\mathcal{F} = \{F(\cdot; x, \boldsymbol{\theta}); \boldsymbol{\theta} \in \Theta, x \in \mathcal{X}\}$ be a regular family of distribution functions, of random variables Y_x. x is a known control or design variable. The *design space* $\mathcal{X}$ is a compact set in $\mathbb{R}^m$. $\boldsymbol{\theta}$ is an *unknown* parameter of the distribution. The parameter space Θ is an open set in $\mathbb{R}^k$. The regularity of $\mathcal{F}$ means that all its elements satisfy the well known Cramér–Rao regularity conditions (Wijsman, 1973). Let $f(y; x, \boldsymbol{\theta})$ be a p.d.f. of $F(\cdot; x, \boldsymbol{\theta})$ with respect to some σ-finite measure μ. The *Fisher information* function (matrix) is

$$I(\boldsymbol{\theta}; x) = \mathrm{Var}_{\boldsymbol{\theta},x}\left\{\frac{\partial}{\partial\theta}\log f(Y; x, \boldsymbol{\theta})\right\} \tag{2.1}$$

where, in the k-parameter case ($k \geqslant 2$), $I(\theta; x)$ denotes a $k \times k$ covariance matrix, and $\frac{\partial}{\partial\boldsymbol{\theta}}\log f(y; x, \boldsymbol{\theta})$ is a gradient vector (score vector). A design level $x^0(\boldsymbol{\theta})$ is called optimal with respect to the Fisher information, if it *maximizes* some functional of the information matrix.

EXAMPLE 2.1. Consider the simple normal linear regression, i.e., $Y_i \sim N(\beta_0 + \beta_1 x_i, \sigma^2)$, $i = 1, \ldots, n$. $Y_1, \ldots, Y_n$ are mutually independent. $x_1, \ldots, x_n$ are chosen from a closed interval $[x^*, x^{**}]$. The parameters $(\beta_0, \beta_1, \sigma^2)$ are unknown, $(\beta_0, \beta_1) \in \mathbb{R}^2$ and $\sigma^2 \in \mathbb{R}^+$. The Fisher information matrix is

$$\boldsymbol{I}(\beta_0, \beta_1, \sigma^2; x) = \begin{bmatrix} \frac{n}{\sigma^2} & \frac{n\bar{X}}{\sigma^2} & 0 \\ \frac{n\bar{X}}{\sigma^2} & \frac{\sum_{i=1}^n X_i^2}{\sigma^2} & 0 \\ 0 & 0 & \frac{n}{2\sigma^6} \end{bmatrix}, \tag{2.2}$$

where $\bar{X} = \frac{1}{n}\sum_{i=1}^{n} X_i$. It is interesting to realize that the information matrix is independent of (β_0, β_1). The determinant of the Fisher information matrix is $D_n = \frac{n^2}{2\sigma^{10}} Q_x$, where $Q_x = \sum_{i=1}^{n}(X_i - \bar{X})^2$. A design which maximizes D_n is called *D-optimal*. In the present case a D-optimal design, for $n = 2m$, exists independently of $(\beta_0, \beta_1, \sigma^2)$, and is given by the design which uses m values of x at x^* and m values of x at x^{**}. This is a special simple case of Elfing's optimal designs for the linear regression (see Chernoff, 1972, p. 13). This example can be generalized to the multiple linear regression in which $\boldsymbol{Y} \sim N(\boldsymbol{X\beta}, \sigma^2 \boldsymbol{I})$, and $\boldsymbol{X}$ is a design matrix.

Zacks (1977) considered the more general case in which the distribution of Y_x, is a one-parameter exponential type, with density functions

$$f(y; \alpha, \beta, x) = g(y; x) \exp\{yw(\alpha + \beta x) + \psi(\alpha + \beta x)\}, \tag{2.3}$$

where $g(y; x) > 0$ on a set $\mathcal{S}$ (support of f) independent of (α, β), and $w(u)$ and $\psi(u)$ are some analytic functions, such that $\int g(y; u) e^{yw(u)}\, dy < \infty$ for all u and $\psi(u) = -\log \int g(y; u) e^{yw(u)}\, dy$ for all u. Of special interest is the class of binomial models, where $Y_x \sim \mathcal{B}(K, \theta(\alpha + \beta x))$ where

$$\theta(\alpha + \beta x) = \frac{\exp\{w(\alpha + \beta x)\}}{1 + \exp\{w(\alpha + \beta x)\}}. \tag{2.4}$$

This class of models is applied in quantal response analysis of bioassays. In these models, $\theta(u)$ is a c.d.f. (tolerance distribution). The normal, logistic, Weibull, extreme-values, and other distributions have been applied in the literature. This type of analysis appears also in reliability life testing. The following example is a simple illustration.

EXAMPLE 2.2. K systems are subjected to life testing. The time till failure of each system is exponentially distributed with mean β, $0 < \beta < \infty$. β is an unknown parameter. The times till failure of the K system $T_1, T_2, \ldots, T_k$ are i.i.d. Y_x is the number of systems that fail in the time interval $(0, x]$, i.e., $Y_x \sim \mathcal{B}(K, 1 - \exp\{-x/\beta\})$. Ehrenfeld (1962) considered the problem of finding the optimal x for minimizing a certain expected loss. See also Zacks and Fenske (1973). Cochran (1973) discussed this example too, under different parameterization. This example can be traced back to Fisher (1922). The Fisher information function on β, based on the statistic Y_x, for $x^* \leqslant x \leqslant x^{**}$, is

$$I(\beta; x) = \frac{Kx^2}{\beta^4} \frac{e^{-x/\beta}}{1 - e^{-x/\beta}}, \quad 0 < \beta < \infty.$$

Let $u = x(\beta)/\beta$. The value of u which maximizes $I(\beta; x)$ is given by the root of equation

$$u = 2(1 - e^{-u}).$$

This equation has a unique root given approximately by $u^0 \cong 1.6$. Thus, the value of x which maximizes the Fisher information is approximately 1.6β. In *adaptive designs* discussed later, we consider the problem of determining a random sequence of x-values, $\{X_n\}_{n=1}^{\infty}$ such that $X_n \to x^0(\beta)$ in probability (or a.s.) for each β, and which satisfies certain additional optimality condition.

Returning to the exponential type p.d.f. given by equation (2.3), it is straightforward to verify that the Fisher information matrix corresponding to n independent observations at $(x_1, Y_1), (x_2, Y_2), \ldots, (x_n, Y_n)$ is

$$\boldsymbol{I}(\alpha, \beta; \boldsymbol{x}) = \begin{bmatrix} \sum_{i=1}^n W(\alpha + \beta x_i) & \sum_{i=1}^n x_i W(\alpha + \beta x_i) \\ \sum_{i=1}^n x_i W(\alpha + \beta x_i) & \sum_{i=1}^n x_i^2 W(\alpha + \beta x_i) \end{bmatrix} \tag{2.5}$$

where

$$W(\alpha + \beta x) = (w'(\alpha + \beta x))^2 V\{Y \mid \alpha + \beta x\}, \tag{2.6}$$

and

$$V\{Y \mid \alpha + \beta x\} = \frac{1}{w^3(\alpha + \beta x)} \{\psi'(\alpha + \beta x) w''(\alpha + \beta x) - \psi''(\alpha + \beta x) w'(\alpha + \beta x)\}. \tag{2.7}$$

$w'(u)$, $w''(u)$, $\psi'(u)$ and $\psi''(u)$ are the 1st and 2nd order derivatives of $w(u)$ and $\psi(u)$, respectively. Let $T(\alpha, \beta; \boldsymbol{x}) = \sum_{i=1}^n W(\alpha + \beta x_i)$, and

$$\mathrm{SSD}(\alpha, \beta; \boldsymbol{x}) = \sum_{i=1}^n W(\alpha + \beta x_i) x_i^2 - \left(\sum_{i=1}^n W(\alpha + \beta x_i) x_i \right)^2 / T(\alpha, \beta; \boldsymbol{x}).$$

The Fisher information matrix has a determinant

$$|I(\alpha, \beta; \boldsymbol{x})| = T(\alpha, \beta; \boldsymbol{x}) \mathrm{SSD}(\alpha, \beta; \boldsymbol{x}). \tag{2.8}$$

D-optimal design points $\boldsymbol{x}^0(\alpha, \beta)$ are those maximizing (2.8). These optimal points depend on (α, β).

Chernoff (1972, p. 29) shows how to use Elfving's approach to determine design points for minimizing the variance of the maximum likelihood estimator of a linear functional $\lambda_1 \alpha + \lambda_2 \beta$, where $\theta(\alpha + \beta x)$ is given by the normal model $\Phi(\alpha + \beta x)$. As usual, $\Phi(z)$ designates the standard normal integral.

2.2. *Bayesian designs for the Fisher information*

Whenever the design level $x^0(\boldsymbol{\theta})$, which is maximizing the Fisher information, depends on the unknown parameter(s), $\boldsymbol{\theta}$, one cannot determine the optimal design without knowledge of $\boldsymbol{\theta}$. The Bayesian approach seeks to optimize the expected Fisher information, according to some prior distribution $H(\boldsymbol{\theta})$. Thus, the Bayesian design, x_H^0, is the design maximizing the prior expectation $\mathrm{E}_H\{I(\boldsymbol{\theta}; x)\}$. In general, a Bayesian design minimizes some prior risk, for suitably chosen loss function. An excellent review of Bayesian experimental design has recently been written by Chaloner and Verdinelli (1995), who provide there an extensive list of references. In the following example we show this Bayesian optimization.

EXAMPLE 2.3. We consider the problem of Example 2.2. Let $\theta = 1/\beta$. The Fisher information function of θ is

$$I(\theta; x) = Kx^2 \frac{\mathrm{e}^{-\theta x}}{1 - \mathrm{e}^{-\theta x}}. \tag{2.9}$$

Let θ have a gamma prior distribution with scale parameter 1 and shape parameter ν. The prior expectation of $I(\theta; x)$ is

$$\begin{aligned}
\mathrm{E}_H\{I(\theta; x)\} &= \frac{Kx^2}{\Gamma(\nu)} \int_0^\infty \theta^{\nu-1} \frac{\mathrm{e}^{-\theta(x+1)}}{(1 - \mathrm{e}^{-\theta x})}\, \mathrm{d}\theta \\
&= Kx^2 \sum_{j=0}^{\infty} \frac{1}{\Gamma(\nu)} \int_0^\infty \theta^{\nu-1} \mathrm{e}^{-\theta(1+(j+1)x)}\, \mathrm{d}\theta \\
&= Kx^2 \sum_{j=1}^{\infty} (1 + xj)^{-\nu}.
\end{aligned} \tag{2.10}$$

Notice that for $\nu = 1$ the infinite sum in (2.10) diverges to infinity. We have to require that $\nu \geqslant 1 + \delta$. The derivative of $\mathrm{E}_H\{I(x;\theta)\}$ with respect to x is

$$\frac{\partial}{\partial x} \mathrm{E}_H\{I(x; \theta)\} = x \sum_{j=1}^{\infty} \frac{2 - (\nu - 2)xj}{(1 + jx)^{\nu+1}}. \tag{2.11}$$

Thus, if $1 < \nu \leqslant 2$ the function $\mathrm{E}_H\{I(x;\theta)\}$ is strictly increasing in x, and the optimal design is at x^{**}. If $\nu > 2$ there is a unique solution, which can be determined numerically. In the following table we give some of the values of $x^0(\nu)$.

Khan (1988) investigated the properties of the Bayesian solution for problems similar to those of Examples 2.2 and 2.3. Thus, let $J \mid x \sim \mathcal{B}(K, F(x - \theta))$, where $F(u)$ is a c.d.f., $-\infty < x < \infty$, $-\infty < \theta < \infty$. The unknown parameter, θ, is a location (translation) parameter. The Fisher information is

$$I(\theta; x) = K \frac{f^2(x - \theta)}{F(x - \theta)\bar{F}(x - \theta)}, \tag{2.12}$$

Table 2.1
Optimal Bayesian levels

ν	3	4	5	6
$x^0(\nu)$	1.516	0.775	0.519	0.390

where $f(u) = F'(u)$ is the p.d.f. of $F(u)$, and $\bar{F}(x-\theta) = 1-F(x-\theta)$. Let $u = x-\theta$, then we can write the Fisher information as a function of u, i.e.,

$$I(u) = K\frac{f^2(u)}{F(u)\bar{F}(u)}. \tag{2.13}$$

Notice that in the Logistics case, $F(u) = (1+\mathrm{e}^{-u})^{-1}$, $-\infty < u < \infty$, and $I(u) = Kf(u)$. Indeed,

$$f(u) = F(u)\bar{F}(u) = \frac{\mathrm{e}^{-u}}{(1+\mathrm{e}^{-u})^2}.$$

The density $f(u)$ is symmetric, unimodal, with maximum at $u = 0$. Thus, the x value maximizing $I(x-\theta)$ is $x^0(\theta) = \theta$. The same is true for the normal c.d.f. $F(u) = \Phi(u)$, i.e., $I(u)$ is symmetric around zero with maximum at $u = 0$. The question is, what prior density $h(\theta)$ can guarantee that the equation $\frac{\partial}{\partial x}\mathrm{E}_H\{I(x-\theta)\} \equiv 0$ has a unique solution, which is the point of maximum? Khan (1988) proved the following result:

If the Fisher information function $I(u)$ is continuously differentiable unimodal, and the prior density $h(\theta)$ is log-concave, then the maximum x_H^0 of $\mathrm{E}_H\{I(x-\theta)\}$ is the unique solution of the above equation.

Khan generalized this result also to stagewise designs. After k stages, the optimal x-level for the $(k+1)$st stage is obtained by the root of (2.13), where the expectation is with respect to the posterior distribution of θ, given $(x_1, J_1), \ldots, (x_k, J_k)$.

2.3. *Adaptive designs for the Fisher information*

We consider again the general class of problems, where the observed r.v. Y_x at level x has a binomial distribution with parameters K (known) and $F(\frac{x-\xi}{\sigma})$, where $F(u)$ is an absolutely continuous c.d.f. (tolerance distribution); ξ and σ are location and scale parameters; $x^* \leqslant x \leqslant x^{**}$, $-\infty < \xi < \infty$, $0 < \sigma < \infty$. We discuss in this section the simpler problem where ξ is known, and without loss of generality we set $\xi = 0$. The scale parameter σ is unknown. We assume that the experiment can be performed in stages. In each stage the number of trials K is fixed. All the trials in the nth stage are performed at the same level, x_n. The level x_1 is some constant. The levels x_n, $i \geqslant 2$, can be determined as functions of the results of the previous stages. Let $\mathcal{B}_n$ denote the σ-field generated by $(x_1, Y_{x_1}, \ldots, Y_{x_n})$. Accordingly, we assume that, for each $n = 1, 2, \ldots$ the level for the $(n+1)$st stage is a random variable X_{n+1}, measurable with respect to $\mathcal{B}_n$. In the following we discuss two different approaches for the determination of X_n $(n = 1, 2, \ldots)$.

2.3.1. Adaptive designs based on the MLE

We assume here that the Fisher information function $I(\sigma; x)$ has a maximum at $x^0(\sigma)$, where $x^* < x^0(\sigma) < x^{**}$. Moreover, we assume that the tolerance distribution $F(u)$ is sufficiently smooth, so that $x^0(\sigma)$ is continuous function of σ. Conditions on $F(u)$ will be given below. After n stages, let $\widehat{\sigma}_n$ denote the *maximum likelihood estimator* (MLE) of σ. The likelihood function of σ after n stages, is

$$L_n(\sigma) = \prod_{i=1}^{n} F\left(\frac{x_i}{\sigma}\right)^{Y_i} \left(1 - F\left(\frac{x_i}{\sigma}\right)\right)^{K-Y_i}, \tag{2.14}$$

where $Y_i = Y_{x_i}$ $(i = 1, \ldots, n)$. $\widehat{\sigma}_n$ is a positive real value which maximizes $L_n(\sigma)$. Notice that if $Y_i = 0$ for all $i = 1, \ldots, n$, or $Y_i = 1$ for all $i = 1, \ldots, n$, then $\widehat{\sigma}_n$ does not exist. The probabilities of these events are small, even for small values of n, if the value of K is not too small and $F(\frac{x_i}{\sigma})$ is not too close to 0 or 1.

Generally, the MLE is the root of the equation

$$\sum_{i=1}^{n} \left(\widehat{p}_i - F\left(\frac{x_i}{\sigma}\right)\right) \frac{x_i f\left(\frac{x_i}{\sigma}\right)}{F\left(\frac{x_i}{\sigma}\right) \bar{F}\left(\frac{x_i}{\sigma}\right)} = 0, \tag{2.15}$$

where $\widehat{p}_i = Y_i/K$, $i = 1, \ldots, n$, and $f(u) = F'(u)$ is the p.d.f. of $F(u)$, $\bar{F}(u) = 1 - F(u)$.

The adaptive procedure under consideration determines the level for the $(n+1)$st stage, as

$$X_{n+1} = \left[x^0(\widehat{\sigma}_n)\right]_{x^*}^{x^{**}}, \quad n \geqslant 1, \tag{2.16}$$

where

$$\left[a\right]_{x^*}^{x^{**}} = x^* I\{a \leqslant x^*\} + a I\{x^* < a < x^{**}\} + x^{**} I\{a \geqslant x^{**}\}.$$

If $\widehat{\sigma}_n$ does not exist, we set $X_{n+1} = X_n$. As seen previously, the Fisher information function is

$$I(\sigma; x) = K x^2 \frac{f^2\left(\frac{x}{\sigma}\right)}{F\left(\frac{x}{\sigma}\right) \bar{F}\left(\frac{x}{\sigma}\right)}. \tag{2.17}$$

If $f(u)$ is symmetric around $u = 0$, $u^2 f^2(u)/(F(u)\bar{F}(u))$ is a bimodal function symmetric about $u = 0$, with two points of maximum, u^0 and $-u^0$. Accordingly, define $x^0(\sigma) = u^0 \sigma$, where

$$u^0 = \text{first positive } \arg\max \left\{ \frac{u^2 f^2(u)}{F(u)\bar{F}(u)} \right\}. \tag{2.18}$$

Also,

$$I(\sigma; x^0(\sigma)) = \sigma^2 K I(1; u^0). \tag{2.19}$$

The above results imply that $X_{n+1} = u^0 \widehat{\sigma}_n$ is the MLE of $x^0(\sigma)$ after n stages. If $\widehat{\sigma}_n$ is a consistent estimator of σ, then $p\lim_{n\to\infty} X_{n+1} = x^0(\sigma)$ for each σ. The following theorem provides the conditions and proof of the consistency of the adaptive procedure based on the MLE. For this purpose we assume that $0 < \sigma_L \leqslant \sigma \leqslant \sigma_U < \infty$. Let $\boldsymbol{\theta} = (\sigma, x)$ and $\Theta = (\sigma_L, \sigma_U) \times (x^*, x^{**})$.

THEOREM. *Assume that*

(i) $0 < \delta \leqslant F\left(\dfrac{x^*}{\sigma_U}\right) < F\left(\dfrac{x^{**}}{\sigma_L}\right) < 1;$

(ii) $\inf_{\boldsymbol{\theta}\in\Theta} f\left(\dfrac{x}{\sigma}\right) \geqslant \eta > 0;$

(iii) $\sup_{\boldsymbol{\theta}\in\Theta} f\left(\dfrac{x}{\sigma}\right) \leqslant f^* < \infty;$

(iv) $f(u)$ *is continuously differentiable and*

$$-\infty < \inf_{\boldsymbol{\theta}\in\Theta} f'\left(\frac{x}{\sigma}\right) \leqslant \sup_{\boldsymbol{\theta}\in\Theta} f'\left(\frac{x}{\sigma}\right) < \infty$$

then $\widehat{\sigma}_n \to \sigma$ *in probability. Furthermore,*

$$\sqrt{n}(\widehat{\sigma}_n - \sigma) \underset{n\to\infty}{\overset{d}{\longrightarrow}} N(0, I^{-1}(\sigma; u^0\sigma)). \tag{2.20}$$

PROOF. For a given $\boldsymbol{\theta} \in \Theta$, define

$$D(x;\sigma) = -Kx\left[\widehat{p} - F\left(\frac{x}{\sigma}\right)\right] \frac{f\left(\frac{x}{\sigma}\right)}{F\left(\frac{x}{\sigma}\right)\bar{F}\left(\frac{x}{\sigma}\right)}$$

and let $S_n(\sigma) = \sum_{i=1}^n D(X_i;\sigma)$. $\{S_n(\sigma), \mathcal{B}_n; n \geqslant 1\}$ is a zero-mean martingale, under $\boldsymbol{\theta}$. Indeed, since

$$\widehat{p} \mid X = x \sim \frac{1}{K}\mathcal{B}\left(K, F\left(\frac{x}{\sigma}\right)\right),$$

$$\mathrm{E}_\sigma\{D(X_n;\sigma) \mid \mathcal{B}_{n-1}\} = 0 \quad \text{a.s.} \tag{2.21}$$

for all $n \geqslant 1$, where $S_0 = 0$. Moreover, according to assumptions (i)–(iv)

$$\mathrm{E}_\sigma\{D^2(X_n;\sigma) \mid \mathcal{B}_{n-1}\} = I(\sigma; X_n) \leqslant \frac{K(x^{**})^2(f^*)^2}{\delta\left(1 - F\left(\frac{x^{**}}{\sigma_U}\right)\right)} < \infty.$$

This implies that

$$\lim_{n\to\infty} \frac{1}{n^2} \sum_{i=1}^{n} \mathrm{E}_\sigma\{D^2(X_i;\sigma)\} = 0 \tag{2.22}$$

and for each σ in $[\sigma_L, \sigma_U]$,

$$\lim_{n\to\infty} \frac{1}{n} S_n(\sigma) = 0 \quad \text{a.s.} \tag{2.23}$$

(see Prakasa Rao, 1987, p. 47). Since, $S_n(\widehat{\sigma}_n) = 0$, for every n for which $\widehat{\sigma}_n$ exists, expansion of $S_n(\widehat{\sigma}_n)$ around σ yields

$$\frac{1}{n} S_n(\sigma) = -\frac{1}{n}(\widehat{\sigma}_n - \sigma) S_n'(\sigma) + \mathrm{o}_p\left(\frac{1}{n}\right), \tag{2.24}$$

as $n \to \infty$. One can immediately verify that

$$\mathrm{E}_\sigma\left\{ -\frac{\partial}{\partial\sigma} D(x;\sigma) \right\} = I(\sigma; x), \tag{2.25}$$

hence, from equation (2.24),

$$-\frac{1}{n} S_n'(\sigma) = \frac{1}{n} \sum_{i=1}^{n} I(\sigma; X_i) + \frac{1}{n} \sum_{i=1}^{n} \left\{ -\frac{\partial}{\partial\sigma} D(X_i;\sigma) - I(\sigma; X_i) \right\}. \tag{2.26}$$

Let $\frac{1}{n} R_n$ denote the second term on the L.H.S. of (2.26). The sequence $\{R_n, \mathcal{B}_n; n \geqslant 1\}$ is a zero-mean martingale. Hence $R_n/n \to 0$ a.s. as $n \to \infty$. Moreover

$$I(\sigma; X_n) \geqslant 4K(x^*)^2 \eta^2 > 0.$$

Thus,

$$\frac{1}{n} \varliminf_{n\to\infty} \sum_{i=1}^{n} I(\sigma; X_i) \geqslant 4K(x^*)^2 \eta^2 > 0. \tag{2.27}$$

Equations (2.23), (2.24) and (2.27) imply that $\widehat{\sigma}_n \to \sigma$ in probability, and

$$p \lim_{n\to\infty} \frac{1}{n} \sum_{i=1}^{n} I(\sigma; X_i) = I(\sigma; u^0 \sigma). \tag{2.28}$$

From (2.24) and (2.28) we obtain that, as $n \to \infty$,

$$\sqrt{n}(\widehat{\sigma}_n - \sigma) = \frac{\frac{1}{\sqrt{n}} S_n(\sigma)}{I(\sigma; u^0(\sigma))} + o_p\left(\frac{1}{\sqrt{n}}\right). \tag{2.29}$$

Let

$$W_{in} = \frac{1}{\sqrt{n}} D(X_i; \sigma), \quad i = 1, \ldots, n, \ n = 1, 2, \ldots. \tag{2.30}$$

For any given $\varepsilon > 0$,

$$\{|W_{in}| > \varepsilon\} \Leftrightarrow \left|\widehat{p}_i - F\left(\frac{X_i}{\sigma}\right)\right| > \varepsilon \frac{\sqrt{n} F\left(\frac{X}{\sigma}\right) \bar{F}\left(\frac{X}{\sigma}\right)}{K X_i^2 f_i^2 \left(\frac{X}{\sigma}\right)}.$$

But

$$\left|\widehat{p}_i - F\left(\frac{X_i}{\sigma}\right)\right| \leqslant 1$$

for all (X_i, σ). Hence, if $n \geqslant N_1(\varepsilon)$, where

$$N_1(\varepsilon) = \frac{K(x^{**})^2 (f^*)^2}{\varepsilon F\left(\frac{x^*}{\sigma_U}\right) \bar{F}\left(\frac{x^{**}}{\sigma_L}\right)} \tag{2.31}$$

then for all $i = 1, \ldots, n$, $n \geqslant 1$,

$$P_\sigma\{|W_{in}| > \varepsilon \mid \mathcal{B}_{i-1}\} = 0 \quad \text{a.s.} \tag{2.32}$$

Moreover, for all $n \geqslant N_2$ sufficiently large, $|W_{in}| \leqslant 1$ for all $i = 1, \ldots, n$. Hence,

$$\mathrm{E}_\sigma\big\{W_{in} I\{|W_{in}| > \varepsilon\} \mid \mathcal{B}_{i-1}\big\} = 0 \quad \text{a.s.}, \tag{2.33}$$

for all $n \geqslant \max\{N_1(\varepsilon), N_2\}$. Finally,

$$V_\sigma\big\{W_{in} I\{|W_{in}| \leqslant \varepsilon\} \mid \mathcal{B}_{i-1}\big\} = \frac{1}{n} I(\sigma; X_i) \quad \text{a.s., for all } n \geqslant 1. \tag{2.34}$$

Hence, by the Central Limit Theorem for martingales (see Shiryayev, 1984, p. 509),

$$\frac{1}{\sqrt{n}} S_n(\sigma) \xrightarrow[n \to \infty]{d} N(0, I(\sigma; u^0 \sigma)). \tag{2.35}$$

Equations (2.29) and (2.35) imply (2.20). □

EXAMPLE 2.4. Consider the logistic tolerance distribution, namely

$$F(u) = (1 + e^{-u})^{-1}, \quad -\infty < u < \infty.$$

The corresponding p.d.f. is

$$f(u) = \frac{e^{-u}}{(1 + e^{-u})^2}, \quad -\infty < u < \infty.$$

The MLE of σ, after the nth stage is the root of the equation

$$\sum_{i=1}^{n} x_i \big(\widehat{p}_i - (1 + e^{-x_i/\sigma})^{-1}\big) = 0.$$

This root exists and is unique, for all $n \geqslant n_0$, where n_0 is the first k s.t.

$$0 < \sum_{i=1}^{k} \widehat{p}_i < k.$$

u^0 is the positive argument maximizing the function $u^2 \exp\{-u\}/(1 + e^{-u})^2$. This is approximately $u^0 = 2.4$. In Table 2.1 we present the results of simulation runs of the adaptive sequence $\{X_n; n \geqslant 1\}$, where $X_{n+1} = [2.4\widehat{\sigma}_n]_{x^*}^{x^{**}}$, with $\sigma = 1$, $x^* = 1$, $x^{**} = 5$ and $X_1 = x^*$, for $K = 25, 50, 75, 100$. We see that the adaptive process corrects itself quite fast, and yields results within 10% of u^0 after a small number of stages. The convergence to u^0 is faster for large values of K.

Khan (1984) studied adaptive designs on a discrete set $\{d_i\colon a = 1, \ldots, k\}$, where $d_0 = x^*$ and $d_i = d_0 + i(x^{**} - x^*)/k$, $i = 1, 2, \ldots, k$. Let $M_{i,n} = \sum_{j=1}^{n} I\{X_j = d_i\}$. The estimator of σ is

$$\widehat{\sigma}_n = \frac{1}{n} \sum_{i=1}^{n} \frac{M_{i,n} d_i}{F^{-1}(\widehat{p}_{i,n})}, \tag{2.36}$$

where

$$\widehat{p}_{i,n} = \frac{1}{M_{i,n}} \sum_{j=1}^{M_{i,n}} Y_{d_i,j}. \tag{2.37}$$

If $M_{i,n} = 0$ for all $i = 1, \ldots, k$ then σ_n does not exist. The estimator exists, with probability 1, for all n sufficiently large. If $M_{i,n} = 0$ for some i, set above $\widehat{p}_{i,n} = 0$ and $F^{-1}(\widehat{p}_{i,n}) = x^*$. Furthermore, as long as $\widehat{\sigma}_n$ does not exist set $X_{n+1} = x^*$. As soon as $\widehat{\sigma}_n$ exists, define

$$\xi_n = u^0 \widehat{\sigma}_n,$$

Table 2.2
Simulations of X_n in an MLE adaptive procedure

$n\backslash K$	25	50	75	100	$n\backslash K$	25	50	75	100
1	1.00	1.00	1.00	1.00	26	2.42	2.55	2.44	2.42
2	2.08	1.44	3.18	1.73	27	2.40	2.52	2.44	2.41
3	2.69	2.49	2.79	2.00	28	2.38	2.53	2.44	2.38
4	2.93	2.47	2.59	2.54	29	2.45	2.50	2.43	2.37
5	2.92	2.54	2.58	2.40	30	2.51	2.51	2.42	2.39
6	2.76	2.40	2.52	2.48	31	2.53	2.53	2.41	2.41
7	2.86	2.44	2.45	2.61	32	2.54	2.55	2.42	2.42
8	2.65	.243	2.38	2.65	33	2.52	2.54	2.42	2.43
9	2.65	2.47	2.38	2.62	34	2.52	2.53	2.40	2.43
10	2.50	2.53	2.50	2.54	35	2.52	2.56	2.39	2.43
11	2.56	2.64	2.49	2.57	36	2.50	2.53	2.39	2.43
12	2.55	2.58	2.49	2.54	37	2.53	2.52	2.40	2.41
13	2.50	2.55	2.47	2.52	38	2.53	2.52	2.41	2.43
14	2.49	2.55	2.44	2.48	39	2.53	2.51	2.40	2.44
15	2.45	2.63	2.42	2.44	40	2.54	2.49	2.40	2.44
16	2.52	2.61	2.39	2.46	41	2.54	2.47	2.42	2.43
17	2.52	2.53	2.42	2.47	42	2.53	2.47	2.41	2.42
18	2.62	2.53	2.44	2.46	43	2.53	2.47	2.42	2.41
19	2.55	2.49	2.45	2.46	44	2.51	2.47	2.41	2.40
20	2.48	2.56	2.47	2.43	45	2.50	2.46	2.39	2.40
21	2.48	2.53	2.45	2.42	46	2.49	2.46	2.41	2.41
22	2.45	2.57	2.42	2.43	47	2.47	2.46	2.41	2.41
23	2.43	2.57	2.42	2.42	48	2.48	2.46	2.40	2.41
24	2.45	2.57	2.41	2.41	49	2.49	2.44	2.41	2.41
25	2.45	2.55	2.41	2.41	50	2.52	2.44	2.41	2.40

and

$$X_{n+1} = \sum_{i=1}^{k} d_{i-1} I\{d_{i-1} \leqslant \xi_n < d_i\}. \tag{2.38}$$

For a given $\varepsilon > 0$ let $k(\varepsilon) = \frac{(x^{**} - x^*)}{\varepsilon}$. Then for all $k \geqslant k(\varepsilon)$ Khan proved that

$$P_\sigma\Big\{ \limsup_{n\to\infty} |X_n - x^0(\sigma)| < \varepsilon \Big\} = 1. \tag{2.39}$$

In addition he proved that

$$\sqrt{n}(\widehat{\sigma}_n - \sigma) \xrightarrow{d} N(0, I^{-1}(\sigma; d_{i0})), \tag{2.40}$$

where $x^0(\sigma) \in [d_{i0}, d_{i0+1})$. This is the discrete analogue of equation (2.20).

2.3.2. *Adaptive Bayesian designs*

Adaptive Bayesian designs in N stages are designs in which, after each stage the posterior distribution of the unknown parameter is determined, on the basis of the observed results, and the remaining experimental levels are determined so as to maximize the expected payoff (utility), or minimize the total expected loss of the remaining stages. As in the previous section, we assume here that K Bernoulli trials are performed at each stage, with probability of success $F(\frac{x}{\sigma})$, where $x \in [x^*, x^{**}]$ and σ is an unknown parameter, $0 < \sigma < \infty$. Let $H(\sigma)$ denote a prior c.d.f. of σ, having a density function $h(\sigma)$ on $(0, \infty)$.

The likelihood function of σ, after n stages is given by equation (2.14). The posterior p.d.f. of σ after n stages is

$$h_n(\sigma) = \frac{h(\sigma)L_n(\sigma)}{\int_0^\infty h(\sigma)L_n(\sigma)\,\mathrm{d}\sigma}. \tag{2.41}$$

The experimental level for the n-th stage X_n is a function of the observations in the previous stages, i.e., $X_n \in \mathcal{B}_{n-1}$, $n \geqslant 1$, where X_1 depends only on the prior distribution of σ. Accordingly, the objective is to maximize $\mathrm{E}_H\{\sum_{n=1}^N I(\sigma; X_n)\}$, where $I(\sigma; x)$ is given by (2.17). As in the previous section, let $X_n = u^0\,\widehat{\sigma}_n$, $n \geqslant 1$, where $\widehat{\sigma}_n \in \mathcal{B}_{n-1}$ is a posterior estimator of σ. For a finite number of stages, N, define

$$M_N(x) = x^2 \mathrm{E}_H\left\{\frac{f^2\left(\frac{x}{\sigma}\right)}{F\left(\frac{x}{\sigma}\right)\bar{F}\left(\frac{x}{\sigma}\right)}\,\Big|\,\mathcal{B}_{N-1}\right\}. \tag{2.42}$$

The optimal design level, X_N^0, is the argument, in $[x^*, x^{**}]$, maximizing $M_N(x)$. Recursively we define, for each $j = 1, 2, \ldots, N-1$,

$$M_{N-j}(x) = x^2 \mathrm{E}_H\left\{\frac{f^2\left(\frac{x}{\sigma}\right)}{F\left(\frac{x}{\sigma}\right)\bar{F}\left(\frac{x}{\sigma}\right)} + M_{N-j+1}(X_{N-j+1}^0)\,\Big|\,\mathcal{B}_{N-j-1}\right\} \tag{2.43}$$

and X_{N-j}^0 is the argument in $[x^*, x^{**}]$ maximizing $M_{N-j}(x)$.

Generally it is impossible to obtain explicit expressions for X_n^0. Zacks (1977) discussed a two stage ($N = 2$) optimal Bayesian design. This will be shown in the following example.

EXAMPLE 2.5. Consider the problem of performing n^* trials in two stages ($N = 2$). In stage 1 we perform n trials at level x_1. The distribution of Y_1 is $\mathcal{B}(n, 1 - \exp\{-\theta x_1\})$. In stage 2 we perform $(n^* - n)$ trials at level x_2. θ is an unknown parameter. The problem is to determine (n, x_1, x_2) to maximize the prior expected total information

$$\begin{aligned} J_H(n, x_1, x_2) = \mathrm{E}_H\Big\{& n x_1^2 \mathrm{e}^{-\theta x_1}(1 - \mathrm{e}^{-\theta x_1})^{-1} \\ &+ \mathrm{E}_H\{(n^* - n)x_2^2 \mathrm{e}^{-\theta x_2}(1 - \mathrm{e}^{-\theta x_2})^{-1} \mid n, x_1, Y_1\}\Big\}. \end{aligned}$$

We determine first x_2^0 to maximize the posterior expectation $x_2^2 \mathrm{E}_H\{\mathrm{e}^{-\theta x_2}(1 - \mathrm{e}^{-\theta x_2})^{-1} \mid n, x_1, Y_1\}$, n and x_1 are determined so as to maximize

$$\mathrm{E}_H\Big\{n x_1^2 \mathrm{e}^{-\theta x_1}(1-\mathrm{e}^{-\theta x_1})^{-1} + (n^* - n)\mathrm{E}_H\big\{(x_2^0)^2 \mathrm{e}^{-\theta x_2^0}(1-\mathrm{e}^{-\theta x_2^0})^{-1} \,\big|\, n, x_1, Y_1\big\}\Big\}.$$

For a prior gamma distribution for θ, with scale parameter 1 and shape parameter ν, the posterior expectation of $I(\theta; x_2)$ is

$$\begin{aligned}&(n^* - n)\mathrm{E}\{x_2^2 \mathrm{e}^{-\theta x_2}(1-\mathrm{e}^{-\theta x_2})^{-1} \mid n, x_1, Y_1\}\\ &\quad = (n^* - n) x_2^2 \frac{\sum_{j=0}^{n-Y_1}(-1)^j\binom{n-Y_1}{j}\sum_{k=1}^{\infty}[1 + x_1(Y_1 + j) + x_2 k]^{-\nu}}{\sum_{j=0}^{n-Y_1}(-1)^j\binom{n-Y_1}{j}(1 + x_1(Y_1+j))^{-\nu}},\end{aligned}$$

$x_2^0(n, x_1, Y_1)$ is determined numerically. The predictive distribution of Y_1, given (n, x_1) has the p.d.f.

$$p(j; n, x_1) = \binom{n}{j}\sum_{i=0}^{n-j}(-1)^i\binom{n-j}{i}(1 + x_1(j+i))^{-\nu}.$$

Accordingly,

$$\begin{aligned}J_H(n, x_1, x_2^0) = n x_1^2 \sum_{k=1}^{\infty}(1 + x_1 k)^{-\nu} + (N - n)\sum_{j=0}^{n}\binom{n}{j}(x_2^0(n,j))^2\\ \times \sum_{i=0}^{n-j}(-1)^i\binom{n-j}{i}(1 + x_1(i+j) + x_2^0(n,j)k)^{-\nu}\end{aligned}$$

n and x_1 are determined by numerically maximizing the last expression. Thus, for $\nu = 3$ and $N = 50$, $n^0 = 10$, $x_1^0 = 1.518$.

In some situations, like clinical trials, two stage designs may be preferable to multi-stage designs, due to feasibility constraints. There are many papers in the literature in which the Bayesian adaptive sequence is not optimal but is given by $[X_n = u^0\widehat{\sigma}_n]_{x^*}^{x^{**}}$ (see, for example, Chevret, 1993) where $\widehat{\sigma}_n$ is the Bayesian estimator of σ for the squared-error loss, i.e., the posterior mean

$$\widehat{\sigma}_n = \frac{\int_0^\infty \sigma h(\sigma)L_n(\sigma)\,\mathrm{d}\sigma}{\int_0^\infty h(\sigma)L_n(\sigma)\,\mathrm{d}\sigma}. \tag{2.44}$$

We should comment in this connection that due to the complexity of the likelihood function $L_n(\sigma)$, when the x-levels vary, the evaluation of $\widehat{\sigma}_n$ by equation (2.44) generally requires delicate numerical integration.

A general proof of the consistency of $\widehat{\sigma}_n$, as $n \to \infty$, can be obtained by showing that the Bayesian estimator and the MLE are asymptotically equivalent.

3. Adaptive designs for inverse regression problems

Consider the following tolerance problem. Let Y_x be a normal random variable having mean $\beta_0 + \beta_1 x$ and variance $\sigma_x^2 = \sigma^2 x^2$. It is desired that $Y_x \leqslant \eta$ with probability γ. η is a specified threshold. The largest value of x satisfying this requirement is

$$x_\gamma(\boldsymbol{\theta}) = \frac{\eta - \beta_0}{\beta_1 + \sigma Z_\gamma}, \tag{3.1}$$

where $\boldsymbol{\theta} = (\beta_0, \beta_1, \sigma)$ and $Z_\gamma = \Phi^{-1}(\gamma)$ is the γth quantile of the standard normal distribution. If the parameters $(\beta_0, \beta_1, \sigma)$ are known, the optimal x-level is x_γ. When the parameters are unknown we consider adaptive designs, in which x levels are applied sequentially on groups or on individual trials. We consider here sequential designs in which one trial is performed at each stage. Let $\{X_n; n \geqslant 1\}$ designate the design levels of a given procedure. It is known a priori that $-\infty < x^* < x_\gamma(\boldsymbol{\theta}) < x^{**} < \infty$. Thus, all design level X_n are restricted to the closed interval $[x^*, x^{**}]$. We further require, that the random variables X_n will satisfy

$$P_{\boldsymbol{\theta}}\{X_n \leqslant x_\gamma(\boldsymbol{\theta})\} \geqslant 1 - \alpha, \tag{3.2}$$

for all $\boldsymbol{\theta} = (\beta_0, \beta_1, \sigma)$, and all $n \geqslant 1$. This requirement is called the *feasibility requirement.* We wish also that the sequence $\{X_n\}$ will be consistent, i.e.,

$$\lim_{n\to\infty} P_{\boldsymbol{\theta}}\{|X_n - x_\gamma(\boldsymbol{\theta})| > \varepsilon\} = 0, \tag{3.3}$$

for any $\varepsilon > 0$ and $\boldsymbol{\theta}$. In some cases we can also prove that a feasible sequence $\{X_n^0\}$ is optimal, in the sense that,

$$\sum_{n=1}^{N} \mathrm{E}_{\boldsymbol{\theta}}\{(X_n^0 - x_\gamma(\boldsymbol{\theta}))^-\} \leqslant \sum_{n=\gamma\theta}^{N} \{(X_n - x_\gamma(\boldsymbol{\theta}))^-\} \tag{3.4}$$

for $N \geqslant 1$, where $\{X_n\}$ is any other feasible sequence, and where $a^- = -\min(a, 0)$.

In the present section we present some of the results of Eichhorn (1973, 1974), Eichhorn and Zacks (1973, 1981). See also the review article of Zacks and Eichhorn (1975).

3.1. Non-Bayesian designs, σ known

To start, let us consider the special case where β_0 and σ are known. In this case the conditional distribution of $(Y_n - \beta_0)/X_n$, given X_n, is that of a $N(\beta_1, \sigma^2)$, almost surely. Hence, the random variables $U_n = (Y_n - \beta_0)/X_n$ are i.i.d. $N(\beta_1, \sigma^2)$ and

$$\widehat{\beta}_1^{(n)} = \frac{1}{n}\sum_{i=1}^{n}\frac{Y_i - \beta_0}{X_i}, \quad n \geqslant 1, \tag{3.5}$$

is the best unbiased estimator of β_1. A UMA $(1-\alpha)$-confidence interval for β_1 is $\widehat{\beta}_1^{(n)} + Z_{1-\alpha}\sigma/\sqrt{n}$. It follows that the sequence defined by

$$X_{n+1} = \left[\xi_n\right]_{x^*}^{x^{**}}, \tag{3.6}$$

where

$$\xi_n = \frac{\eta - \beta_0}{\widehat{\beta}_1^{(n)} + \left(\frac{Z_{1-\alpha}}{\sqrt{n}} + Z_\gamma\right)\sigma}, \quad n \geqslant, 1 \tag{3.7}$$

is feasible, strongly consistent and optimal. The feasibility is due to the fact that

$$P_{\boldsymbol{\theta}}\left\{\beta_1 \leqslant \beta_1^{(n)} + \frac{\sigma Z_{1-\alpha}}{\sqrt{n}}\right\} = 1 - \alpha$$

for all $\boldsymbol{\theta}$. Strong consistency is due to the fact that $\widehat{\beta}_1^{(n)} \xrightarrow{\text{a.s.}} \beta_1$. The optimality is proven using the UMA property of $\widehat{\beta}_1^{(n)} + (Z_{1-\alpha}\sigma/\sqrt{n})$, which means (see Lehmann, 1959, p. 177) that for any other $(1-\alpha)$ upper confidence limit for β_1, say $\widetilde{\beta}_{1,\alpha}^{(n)}$, and $\varepsilon > 0$

$$P_{\boldsymbol{\theta}}\left\{\widehat{\beta}_1^{(n)} + \frac{\sigma Z_{1-\alpha}}{\sqrt{n}} \geqslant \beta_1 + \varepsilon\right\} \leqslant P_{\boldsymbol{\theta}}\{\widetilde{\beta}_{1,\alpha}^{(n)} \geqslant \beta_1 + \varepsilon\}, \tag{3.8}$$

for all $\boldsymbol{\theta}$. Let

$$\xi' = \frac{\eta - \beta_0}{\beta_1 + \varepsilon + \sigma Z_\gamma} \quad \text{and} \quad y = x_\gamma(\boldsymbol{\theta}) - \xi'.$$

Let

$$\widetilde{X}_n = \frac{n - \beta_0}{\widetilde{\beta}_{1,\alpha}^{(n)} + \sigma Z_\gamma}.$$

Then, from (3.8) one obtains that, for all $y > 0$,

$$P_{\boldsymbol{\theta}}\{x_\gamma(\boldsymbol{\theta}) - X_n \geqslant y\} \leqslant P_{\boldsymbol{\theta}}\{x_\gamma(\boldsymbol{\theta}) - \widetilde{X}_n \geqslant y\}.$$

Finally,

$$\begin{aligned}\mathrm{E}_{\boldsymbol{\theta}}\{(x_\gamma(\boldsymbol{\theta}) - X_n)^+\} &= \int_0^\infty P_{\boldsymbol{\theta}}\{x_\gamma(\boldsymbol{\theta}) - X_n \geqslant y\}\,\mathrm{d}y \\ &\leqslant \int_0^\infty P_{\boldsymbol{\theta}}\{x_\gamma(\boldsymbol{\theta}) - \widetilde{X}_n \geqslant y\}\,\mathrm{d}y \\ &= E_{\boldsymbol{\theta}}\big\{(x_\gamma(\theta) - \widetilde{X}_n)^+\big\},\end{aligned}$$

for all $n \geqslant 1$. This proves the optimality of $\{X_n\}$ given by equations (3.6)–(3.7).

When the slope β_1 (and σ) are known, we can obtain a feasible, strongly consistent and optimal design, by similar methods (see Zacks and Eichhorn, 1975). On the other hand, when both β_0 and β_1 are unknown we can obtain a feasible, consistent design but not optimal in the previous sense. The reason for it will be explained below. Roughly speaking, one cannot estimate β_0 and β_1 if all observations are at one fixed value of x. In the adaptive procedure, to assure consistency, one needs to assume that

$$\sum_{i=1}^n \left(\frac{1}{X_i} - \frac{1}{n}\sum_{i=1}^n \frac{1}{X_i}\right)^2 \longrightarrow \infty \quad \text{as } n \to \infty.$$

Let $R_i = Y_i/X_i$ and $V_i = 1/X_i$, $i = 1, 2, \ldots$. After n trials, let $\boldsymbol{X}^{(n)} = (X_1, X_2, \ldots, X_n)$,

$$\bar{R}_n = \frac{1}{n}\sum_{i=1}^n R_i, \qquad \widehat{V}_n = \frac{1}{n}\sum_{i=1}^n V_i,$$

$$\mathrm{SSD}_V = \sum_{i=1}^n V_i^2 - n\bar{V}_n^2, \qquad Q_n = \sum_{i=1}^n V_i^2$$

and

$$\mathrm{SPD}_{VR} = \sum_{i=1}^n V_i R_i - n\bar{V}_n \bar{R}_n.$$

Notice that $R_i \mid X_i \sim N(\beta_1 + \beta_0 V_i, \sigma^2)$, $i = 1, 2, \ldots$. Let $\widehat{\beta}_{0,n}$ and $\widehat{\beta}_{1,n}$ be the least squares estimators of β_0 and β_1, for the linear regression

$$R_i = \beta_1 + \beta_0 V_i + e_i, \quad i = 1, \ldots, n \tag{3.9}$$

where $e_1, \ldots, e_n$ are i.i.d. $N(0, \sigma^2)$ random variables, independent of $\boldsymbol{X}^{(n)}$. Accordingly,

$$\begin{aligned}\widehat{\beta}_{0,n} &= \frac{\mathrm{SPD}_{VR}}{\mathrm{SSD}_V}, \\ \widehat{\beta}_{1,n} &= \bar{R}_n - \widehat{\beta}_{0,n}\bar{V}_n.\end{aligned} \tag{3.10}$$

Given $\boldsymbol{X}^{(n)}$, the conditional distribution of $(\widehat{\beta}_{0,n}, \widehat{\beta}_{1,n})$ is bivariate normal with mean (β_0, β_1) and covariance matrix

$$\mathbf{\Sigma}(\boldsymbol{X}^{(n)}) = \sigma^2 \begin{bmatrix} \frac{1}{\mathrm{SSD}_V} & -\frac{\bar{V}_n}{\mathrm{SSD}_V} \\ -\frac{\bar{V}_n}{\mathrm{SSD}_V} & \frac{1}{n} + \frac{\bar{V}_n^2}{\mathrm{SSD}_V} \end{bmatrix}. \tag{3.11}$$

To apply this approach, we have to require that $\mathrm{SSD}_V > 0$ for each $n \geqslant 2$ and $\mathrm{SSD}_V \xrightarrow{\text{a.s.}} \infty$ as $n \to \infty$. This will assure that $\lim_{n\to\infty} \mathbf{\Sigma}(\boldsymbol{X}^{(n)}) = 0$ in a proper metric (the largest eigenvalue).

Using Fieller's Theorem (Fieller, 1940) we obtain confidence intervals for $x_\gamma(\boldsymbol{\theta})$, in the following manner. Let

$$W_n = (\eta - \widehat{\beta}_{0,n}) - x_\gamma(\boldsymbol{\theta})(\widehat{\beta}_{1,n} + \sigma Z_\gamma). \tag{3.12}$$

The *conditional* distribution of W_n, *given* $\boldsymbol{X}^{(n)}$, is normal with mean zero and variance

$$V\{W_n \mid \boldsymbol{X}^{(n)}\} = \frac{\sigma^2}{\mathrm{SSD}_V}\left[1 + x_\gamma^2(\boldsymbol{\theta})\frac{Q_n}{n} - 2x_\gamma(\boldsymbol{\theta})\bar{V}_n\right]. \tag{3.13}$$

It follows that the conditional distribution of W_n^2, given $\boldsymbol{X}^{(n)}$, is like that of $V\{W_n \mid \boldsymbol{X}^{(n)}\}\chi^2[1]$. Hence, the two real roots of the quadratic equation

$$A_n\xi^2 - 2B_n\xi + C_n = 0 \tag{3.14}$$

are $(1-\alpha)$ level confidence limits for $x_\gamma(\boldsymbol{\theta})$. Here

$$A_n = (\widehat{\beta}_{1,n} + \sigma Z_\gamma)^2 - \frac{\sigma^2}{\mathrm{SSD}_V}\cdot\frac{Q_n}{n}\chi_{1-2\alpha}^2,$$

$$B_n = (\widehat{\beta}_{1,n} + \sigma Z_\gamma)(\eta - \widehat{\beta}_{0,n}) + \frac{\sigma^2}{\mathrm{SSD}_V}\bar{V}_n\chi_{1-2\alpha}^2,$$

and

$$C_n = (\eta - \widehat{\beta}_{0,n})^2 - \frac{\sigma^2}{\mathrm{SSD}_V}\chi_{1-2\alpha}^2, \tag{3.15}$$

and where $\chi_{1-2\alpha}^2$ is the $(1-2\alpha)$th quantile of the chi-squared distribution with 1 degree of freedom. Two real roots for equation (3.14) exist with probability $1-\alpha$.

Let $\boldsymbol{\xi}_{n,\alpha}$ and $\bar{\xi}_{n,\alpha}$ denote the two roots of equation (3.14). The sequence of dosages is defined as, $X_1 = x^*$ and

$$X_{n+1} = \left[\boldsymbol{\xi}_{n,\alpha}\right]_{x^*}^{x^{**}}, \quad n \geqslant 1. \tag{3.16}$$

If for some n_0, real roots of equation (3.14) do not exist, we set $X_{n_0+1} = X_{n_0}$. By construction, $P_{\boldsymbol{\theta}}\{X_n \leqslant x_\gamma(\boldsymbol{\theta})\} \geqslant 1-\alpha$, for all $\boldsymbol{\theta}$. That is, *the sequence* (3.16) *is feasible*. The sequence (3.16) might not be consistent. To secure the consistency of $\boldsymbol{\xi}_{n,\alpha}$, i.e., $\lim_{n\to\infty}\boldsymbol{\xi}_{n,\alpha} = x_\gamma(\boldsymbol{\theta})$ in probability, it is sufficient to require that $\mathrm{SSD}_V \to \infty$ a.s. To achieve this, we modify the sequence (3.16) slightly. Let $I^* = \{n_j = \mathrm{e}^j, j = 1, 2, \ldots\}$ and fix Δ, $0 < \Delta$ but small. Then define

$$\widehat{X}_n = \begin{cases} \max\{X_n - \Delta, x^*\}, & \text{if } n \in I^*, \\ X_n, & \text{if } n \notin I^*, \end{cases} \tag{3.17}$$

where $\{X_n\}$ is given by (3.16). The sequence $\{\widehat{X}_n\}$ is feasible. Moreover, the lower confidence limits $\widehat{\boldsymbol{\xi}}_{n,\alpha}$, based on $\{\widehat{X}_n\}$, converge in probability to $x_\gamma(\boldsymbol{\theta})$. The partial means $\widehat{\bar{X}}_N = \frac{1}{N}\sum_{n=1}^N \widehat{X}_n$ converge in probability to $x_\gamma(\boldsymbol{\theta})$.

Finally we mention that the requirement of σ^2 known is not essential since, after each stage one can estimate σ^2 by the usual regression estimator $\widehat{\sigma}_n^2$, which has a distribution like $\sigma^2\chi^2[n-2]/(n-2)$ independently of $\widehat{\beta}_{0,n}$ and $\widehat{\beta}_{1,n}$. Fieller's theorem can be modified by replacing $\sigma^2\chi_{1-\alpha}^2$ with $\widehat{\sigma}_n^2 \cdot F[1, n-2]$.

3.2. Bayesian adaptive designs, σ known

In a Bayesian framework the feasibility requirement is formulated in the following manner. Let $h(\beta_0, \beta_1)$ be a prior p.d.f. of (β_0, β_1). Let $\{X_n\}$ be a sequence of design levels. We say that $\{X_n\}$ is *Bayesian feasible* at level $1-\alpha$ if

$$P_H\{X_{n+1} \leqslant x_\gamma(\boldsymbol{\theta}) \mid \mathcal{B}_n\} \geqslant 1-\alpha, \quad n = 0, 1, 2, \ldots, \tag{3.18}$$

where $P_H\{\cdot \mid \mathcal{B}_n\}$ denotes the posterior probability under H, given $\mathcal{B}_n$. Eichhorn and Zacks (1981) investigated the properties of Bayesian designs, for prior bivariate normal distributions of (β_0, β_1).

Let the prior distribution of (β_0, β_1) be a bivariate normal, with mean $(\bar{\beta}_0, \bar{\beta}_1)$ and covariance matrix

$$\Sigma_0 = \begin{bmatrix} v_0 & c_0 \\ c_0 & w_0 \end{bmatrix}.$$

The posterior distribution of (β_0, β_1) after n trials, given $(x_1, Y_1), \ldots, (x_n, Y_n)$ is normal with mean vector and covariance matrix having recursive formulae:

$$\begin{pmatrix} \beta_0^{(n)} \\ \beta_1^{(n)} \end{pmatrix} = \begin{pmatrix} \beta_0^{(n-1)} \\ \beta_1^{(n-1)} \end{pmatrix} + \frac{Y_n - \beta_0^{(n-1)} - \beta_1^{(n-1)} x_n}{D_n} \times \begin{pmatrix} v_{n-1} + c_{n-1}x_n \\ c_{n-1} + w_{n-1}x_n \end{pmatrix} \tag{3.19}$$

with

$$D_n = \sigma^2 x_n^2 + v_{n-1} + 2x_n c_{n-1} + x_n^2 w_{n-1}, \tag{3.20}$$

and

$$\Sigma_n = \Sigma_{n-1} - \frac{1}{D_n}\Sigma_{n-1}\begin{bmatrix} 1 & x_n \\ x_n & x_n^2 \end{bmatrix}\Sigma_{n-1}. \tag{3.21}$$

Let v_n, c_n and w_n denote the posterior variance of β_0, covariance of β_0, β_1 and variance of β_1. Let $\boldsymbol{\xi}_{\alpha,n}$ denote a $(1-\alpha)$ lower credibility limit of $x_\gamma(\boldsymbol{\theta})$ given $\mathcal{B}_n$, i.e.,

$$P_H\left\{\frac{\eta - \beta_0}{\beta_1 + \sigma Z_\gamma} \geqslant \boldsymbol{\xi}_{\alpha,n} \,\middle|\, \mathcal{B}_n\right\} = 1 - \alpha. \tag{3.22}$$

From the posterior distribution of $\beta_0 + \boldsymbol{\xi}_{\alpha,n} \cdot (\beta_1 + \sigma Z_\gamma)$, given $\mathcal{B}_n$, we find that $\boldsymbol{\xi}_{\alpha,n}$ is the smallest root of the equation

$$\left((\eta - \beta_0^{(n)}) - \xi(\beta_1^{(n)} + \sigma Z_\gamma)\right)^2 = \chi_{1-2\alpha}^2(v_n + 2\xi c_n + \xi^2 w_n). \tag{3.23}$$

Equivalently, $\boldsymbol{\xi}_{\alpha,n}$ is the smallest real root of the quadratic equation

$$\xi^2 A_n^* - 2\xi B_n^* + C_n^* = 0, \tag{3.24}$$

where

$$\begin{aligned} A_n^* &= (\beta_1^{(n)} + \sigma Z_\gamma)^2 - \chi_{1-2\alpha}^2 w_n, \\ B_n^* &= (\eta - \beta_0^{(n)})(\beta_1^{(n)} + \sigma Z_\gamma) + \chi_{1-2\alpha}^2 c_n, \\ C_n^* &= (\eta - \beta_0^{(n)})^2 - \chi_{1-2\alpha}^2 v_n. \end{aligned} \tag{3.25}$$

The adaptive design given by the sequence $\{X_n\}$, where

$$X_{n+1} = [\boldsymbol{\xi}_{\alpha,n}]_{x^*}^{x^{**}}, \tag{3.26}$$

is Bayesian feasible.

Let $\boldsymbol{X}^{(n)} = (1, x_n)'$, and $\widehat{\boldsymbol{\beta}}^{(n)} = (\beta_0^{(n)}, \beta_1^{(n)})$. Define,

$$Z_n = (Y_n - \widehat{\boldsymbol{\beta}}^{(n-1)'}\boldsymbol{x}^{(n)})/D_n^{1/2}, \quad n = 1, 2, \ldots.$$

Let $\mathrm{E}_H\{Z_n \mid \mathcal{B}_{n-1}\}$ denote the predictive distribution of Z_n, under the prior H, given $\mathcal{B}_{n-1}$. Notice that $D_n = V_H\{Y_n \mid \mathcal{B}_{n-1}\}$. Since $\mathrm{E}_H\{Z_n \mid \mathcal{B}_{n-1}\} = 0$ a.s., we obtain

Table 3.1
Simulated Bayesian design

n	X_n	$\beta_0^{(n)}$	$\beta_1^{(n)}$	v_n	c_n	w_n
1	1.380	−8.850	2.921	1.098	−0.220	0.244
2	1.464	−8.848	2.921	0.775	−0.267	0.237
3	1.530	−8.955	2.899	0.677	−0.287	0.233
4	1.596	−9.111	2.854	0.635	−0.300	0.299
5	1.588	−9.028	2.889	0.677	−0.308	0.226
10	1.736	−9.317	2.728	0.589	−0.324	0.214
15	1.788	−9.353	2.656	0.587	−0.328	0.204
20	1.816	−9.352	2.609	0.587	−0.328	0.200
25	1.802	−9.369	2.685	0.587	−0.327	0.196
30	1.830	−9.359	2.617	0.587	−0.326	0.192
35	1.824	−9.375	2.657	0.586	−0.325	0.190
40	1.820	−9.384	2.683	0.586	−0.325	0.188
45	1.844	−9.352	2.607	0.586	−0.324	0.186
50	1.847	−9.357	2.610	0.585	−0.323	0.185

from (3.19) that $\mathrm{E}_H\{\widehat{\boldsymbol{\beta}}^{(n)} \mid \mathcal{B}_{n-1}\} = \widehat{\boldsymbol{\beta}}^{(n-1)}$ a.s. Moreover, $\mathrm{E}_H\{|Z_n| \mid \mathcal{B}_{n-1}\} = \sqrt{2/\pi}$. Thus, since all X_n are restricted to the closed interval $[x^*, x^{**}]$, (3.19)–(3.20) imply that

$$\sup_n E\{|Z_n|\boldsymbol{x}^{(n)\prime}\,\Sigma_n\boldsymbol{x}^{(n)}\} < \infty. \tag{3.27}$$

Thus, the Martingale Convergence Theorem (see Shiryayev, 1984, p. 476) implies that

$$\lim_{n\to\infty} \widehat{\boldsymbol{\beta}}^{(n)} = \boldsymbol{\beta}^{(\infty)} \quad \text{a.s. } [H] \quad \text{and} \quad \xi_{n,\alpha} \to \xi^{(\infty)} \quad \text{a.s. } [H].$$

$\widehat{\boldsymbol{\beta}}^{(\infty)}$ is a random vector having a singular bivariate normal distribution, concentrated on the line

$$\beta_0 + x_\gamma\beta_1 = \eta - x_\gamma\sigma Z_\gamma.$$

Eichhorn and Zacks (1981) showed that a discrete version of (3.26) converges a.s. to within a δ-neighborhood of $x_\gamma(\boldsymbol{\theta})$.

EXAMPLE 3.1. In the present example we illustrate the Bayesian adaptive procedure. Without loss of generality, assume that $\eta = 0$. The data is generated by simulation from a normal distribution $N(\beta_0+\beta_1 x, \sigma^2x^2)$ with $\beta_0 = -10$, $\beta_1 = 3$ and $\sigma = 1$. The design levels are restricted to the interval with $x^* = 1$ and $x^{**} = 10$. For $\alpha = .05$ and $\gamma = .99$ we have $x_\gamma(\boldsymbol{\theta}) = 1.878$. We start with the prior bivariate normal distribution with mean $\beta_0^{(0)} = -15$ and $\beta_1^{(0)} = 2.75$ and covariance matrix with $v_0 = 9$, $c_0 = 0$ and $w_0 = 0.25$. In the following table we provide the values of X_n, $\beta_0^{(n)}$, $\beta_1^{(n)}$, v_n, c_n, w_n.

Notice that after 50 stages, the posterior correlation $c_n/(v_n \cdot w_n)^{1/2} = -0.982$. The posterior correlation approaches -1 as n grows.

4. Sequential allocation of experiments

In the present section we discuss the problem of choosing for each trial one of a finite set of alternative experiments, in order to optimize some objective function. Design problems falling under this class of problems are known in the literature as *bandit problems*. A discrete version of the design level problem discussed in the previous sections, in which only K points $x^* \leqslant x_1 < x_2 < \cdots < x_k \leqslant x^{**}$ are available for choice in each trial, can be considered also as a bandit problem. Generally, suppose that K alternative experiments are available, $\mathcal{E}_1, \ldots, \mathcal{E}_K$. If $\mathcal{E}_i$ is performed, the yield is a random variable Y_i, having a distribution function F_i $(i = 1, \ldots, K)$. We assume that $\mathrm{E}\{|Y_i|\} < \infty$ for all F_i $(i = 1, 2, \ldots, K)$. The objective is to maximize the expected discounted yield namely, $\sum_{i=1}^{\infty} \alpha_i \mathrm{E}\{Y_{j_i} \mid F_{j_i}\}$, where $\alpha_i \geqslant 0$, $\sum_{i=1}^{\infty} \alpha_i < \infty$, and where Y_{j_i} is the random yield from $\mathcal{E}_{j_i}$, $j_i \in \{1, 2, \ldots, K\}$, $i = 1, 2, \ldots$. The sequence $\boldsymbol{\alpha} = (\alpha_1, \alpha_2, \ldots)$ is called a discounting sequence. If the distribution functions $F_1, \ldots, F_K$ are known, there is no problem in attaining the objective. One should perform in each trial an experiment $\mathcal{E}_{i0}$ for which $\mathrm{E}\{Y \mid F_{i0}\} = \max_{1 \leqslant i \leqslant K} \mathrm{E}\{Y \mid F_i\}$. The problem is what should be the optimal allocation of the trials, among the K experiments, when $F_1, \ldots, F_K$ are unknown. Berry and Fristedt (1985) published an excellent monograph on this subject, which presents, the theories available on the subject for Bayesian optimal allocation. It contains also an annotated bibliography of papers written on the subject up to 1985. In the following section we present a general formulation of the problem for exponential type distributions.

4.1. Bayesian sequential allocations

In the present section we focus attention on the case where each F_i $(i = 1, \ldots, K)$ is a one-parameter exponential type distribution, having a p.d.f. with respect to a σ-finite measure μ, of the form

$$f_i(y; \theta) = g_i(y) \exp\{\theta T_i(y) + C_i(\theta)\}, \quad i = 1, \ldots, k,$$

where θ is an unknown natural parameter, having a parameter space $\Theta_i \subset \mathbb{R}$, and $C_i(\theta)$ is an analytic function of θ. Furthermore, $g_i(y) > 0$ on a (support) set $\mathcal{S}_i$, which is independent of θ. Let $\{Y_{jn}\}$, $j = 1, \ldots, K$, $n = 1, 2, \ldots$, be independent sequences of i.i.d. random variables. Y_{jn} is associated with experiment $\mathcal{E}_j$ $(j = 1, \ldots, K)$, and has the p.d.f. $f_j(y; \theta)$.

Let $\boldsymbol{N}^{(n)} = (N_1^{(n)}, \ldots, N_K^{(n)})$ be a (random) vector counting the frequencies of applying the various experiments under a given strategy, among the last n trials. Given that $N_i^{(n)} = n_i$ $(n_i \geqslant 0, \sum_{i=1}^{K} n_i = n)$, define $T_i^{(n)} = 0$ if $n_i = 0$ and $T_i^{(n)} = \sum_{j=1}^{n_i} T_i(Y_{ij})$ if $n_i > 0$. $T_i^{(n)}$ is a minimal sufficient statistic for the parametric family $\mathcal{F}_i$ associated with $\mathcal{E}_i$. Let $\boldsymbol{T}^{(n)} = (T_1^{(n)}, \ldots, T_K^{(n)})$, and let $\mathcal{B}_n$ be the σ-field generated by the statistic $(\boldsymbol{N}^{(n)}, \boldsymbol{T}^{(n)})$. The sequence $\{(\boldsymbol{N}^{(n)}, \boldsymbol{T}^{(n)}); n \geqslant 1\}$ is sequentially sufficient and transitive for the model (see Zacks, 1971, p. 83).

A strategy is a sequence $\boldsymbol{\tau} = (\tau_1, \tau_2, \ldots)$ such that $\tau_n \in \mathcal{B}_{n-1}$ for each $n = 1, 2, \ldots$ and where, for each $n \geqslant 1$, the range of τ_n is $\{1, 2, \ldots, K\}$. In other words, τ_n is a discrete random variable such that $\{\tau_n = i\} \in \mathcal{B}_{n-1}$, $i = 1, \ldots, K$, $n \geqslant 1$. According to $\boldsymbol{\tau}$, if $\tau_n = i$ then experiment $\mathcal{E}_i$ is performed at the nth trial. The expected yield of a strategy $\boldsymbol{\tau}$ is

$$W(\boldsymbol{\tau}, \boldsymbol{\alpha}, \boldsymbol{\theta}) = \sum_{n=1}^{\infty} \alpha_n \sum_{j=1}^{k} \mathrm{E}_{\boldsymbol{\theta}}\{Y_{jn} I\{\tau_n = j\}\}, \tag{4.1}$$

where $\boldsymbol{\theta} = (\theta_1, \ldots, \theta_K)$ is the vector of parameters specifying $F_1, \ldots, F_K$.

Let H be a prior distribution on $\Omega = \Theta_1 \times \cdots \times \Theta_K$. The prior expected yield of a strategy with respect to H is called the *worth* of τ and is given by

$$W_H(\boldsymbol{\tau}, \boldsymbol{\alpha}) = \sum_{n=1}^{\infty} \alpha_n \sum_{j=1}^{K} \int_{\Omega} \mathrm{E}_{\boldsymbol{\theta}}\{Y_{jn} I\{\tau_n = j\}\} \, \mathrm{d}H(\boldsymbol{\theta}). \tag{4.2}$$

A strategy $\boldsymbol{\tau}_H^0$ is Bayesian (optimal) against H if

$$W_H(\boldsymbol{\tau}_H^0, \boldsymbol{\alpha}) = \sup_{\tau} \; W_H(\boldsymbol{\tau}, \boldsymbol{\alpha}). \tag{4.3}$$

Let $V(H, \boldsymbol{\alpha}) = W_H(\boldsymbol{\tau}_H^0, \boldsymbol{\alpha})$. Since $\tau_n \in \mathcal{B}_{n-1}$ and $\mathrm{E}\{Y_\tau \mid \tau_n = j\} = \mu_j(\theta_j)$ where

$$\begin{aligned} \mu_j(\theta_j) &= \int_{S_j} y g_j(y) \exp\{\theta_j T_j(y) + C_j(\theta_j)\} \, \mathrm{d}\mu(y) \\ &= -C_j'(\theta)\big|_{\theta=\theta_j}, \quad j = 1, \ldots, K, \end{aligned} \tag{4.4}$$

we obtain from (4.1) that

$$W(\boldsymbol{\tau}, \boldsymbol{\alpha}, \boldsymbol{\theta}) = \sum_{n=1}^{\infty} \alpha_n \sum_{j=1}^{K} \mu_j(\theta_j) P_{\boldsymbol{\theta}}\{\tau_n = j\}. \tag{4.5}$$

Furthermore,

$$W_H(\boldsymbol{\tau}, \boldsymbol{\alpha}) = \sum_{n=1}^{\infty} \alpha_n \sum_{j=1}^{K} \mathrm{E}_H\{\mu_j(\theta_j) P_{\boldsymbol{\theta}}\{\tau_n = j\}\}. \tag{4.6}$$

Notice that

$$\boldsymbol{N}^{(n)} = \boldsymbol{N}^{(n-1)} + \sum_{j=1}^{K} \boldsymbol{e}_j^{(K)} I\{\tau_n = j\}, \tag{4.7}$$

where $\boldsymbol{e}_j^{(K)}$ is a K-dimensional row vector whose jth component is 1 and all the other components are 0. Similarly,

$$\boldsymbol{T}^{(n)} = \boldsymbol{T}^{(n-1)} + \sum_{j=1}^{K} Y_{jn}\boldsymbol{e}_j^{(K)} I\{\tau_n = j\}. \tag{4.8}$$

We see that the amount of information on $(\theta_1, \dots, \theta_K)$ after n trials depends on the strategy $\boldsymbol{\tau}$.

The likelihood function of $(\theta_1, \dots, \theta_K)$, given $(\boldsymbol{N}^{(n)}, \boldsymbol{T}^{(n)})$ is equivalent to

$$L(\theta_1, \dots, \theta_k; \boldsymbol{N}^{(n)}, \boldsymbol{T}^{(n)}) = \exp\left\{\sum_{i=1}^{K} \theta_i T_i^{(n)} + \sum_{i=1}^{k} N_i^{(n)} C_i(\theta_i)\right\}, \tag{4.9}$$

$(\theta_1, \dots, \theta_k) \in \Omega$. Notice that the likelihood function $L(\boldsymbol{\theta}; \boldsymbol{N}^{(n)}, \boldsymbol{T}^{(n)})$ depends on the strategy $\boldsymbol{\tau}$ only through the statistic $(\boldsymbol{N}^{(n)}, \boldsymbol{T}^{(n)})$. Let $H(\boldsymbol{\theta})$ be a prior c.d.f. on Ω. By Bayes Theorem, the posterior distribution of $\boldsymbol{\theta}$, given $(\boldsymbol{N}^{(n)}, \boldsymbol{T}^{(n)})$ is given by

$$\begin{aligned} &\mathrm{d}H(\theta_1, \dots, \theta_k \mid \boldsymbol{N}^{(n)}, \boldsymbol{T}^{(n)}) \\ &\quad = \frac{L(\theta_1, \dots, \theta_k; \boldsymbol{N}^{(n)}, \boldsymbol{T}^{(n)})\,\mathrm{d}H(\theta_1, \dots, \theta_k)}{\int_\Omega L(\theta_1, \dots, \theta_k; \boldsymbol{N}^{(n)}, \boldsymbol{T}^{(n)})\,\mathrm{d}H(\theta_1, \dots, \theta_k)}. \end{aligned} \tag{4.10}$$

If for some j, $N_j^{(n)} = 0$, then the posterior marginal distribution of θ_j is its prior marginal distribution. The predictive p.d.f. of $Y_{j,n+1}$ given $\mathcal{B}_n$, is

$$f_j^*(y; \boldsymbol{N}^{(n)}, \boldsymbol{T}^{(n)}) = \int_{\Theta_j} f_j(y; \theta)\,\mathrm{d}H_j(\theta \mid \boldsymbol{N}^{(n)}, \boldsymbol{T}^{(n)}), \tag{4.11}$$

where $H_j(\theta \mid \boldsymbol{N}^{(n)}, \boldsymbol{T}^{(n)})$ is the marginal posterior c.d.f. of θ_j, given $\mathcal{B}_n$. Thus, the predictive expectation of $Y_{j,n+1}$, given $\mathcal{B}_n$, is for $j = 1, \dots, k$, $n \geqslant 1$,

$$M_j^{(n+1)}(\boldsymbol{N}^{(n)}, \boldsymbol{T}^{(n)}) = \int_{\Theta_j} \mu_j(\theta)\,\mathrm{d}H_j(\theta \mid \boldsymbol{N}^{(n)}, \boldsymbol{T}^{(n)}). \tag{4.12}$$

Without loss of generality, let $\alpha_1 = 1$. If $\alpha_2 > 0$ then, the function $V(H, \boldsymbol{\alpha})$ satisfies the functional equation

$$\begin{aligned} V(H, \boldsymbol{\alpha}) = \max_{1 \leqslant j \leqslant k} \Bigg\{ &M_j^{(1)}(\boldsymbol{N}^{(0)}, \boldsymbol{T}^{(0)}) \\ &+ \alpha_2 \int V\Big(H(\cdot \mid \boldsymbol{e}_j^{(K)}, y\boldsymbol{e}_j^{(K)}), \boldsymbol{\alpha}^{(2)}\Big) \\ &\times f_j^*(y \mid \boldsymbol{N}^{(0)}, \boldsymbol{T}^{(0)})\,\mathrm{d}\mu(y) \Bigg\}, \end{aligned} \tag{4.13}$$

where

$$\boldsymbol{\alpha}^{(2)} = \left(1, \frac{\alpha_3}{\alpha_2}, \frac{\alpha_4}{\alpha_2}, \ldots\right).$$

Under general conditions, if a policy $\boldsymbol{\tau}$ has a worth function $W(\boldsymbol{\tau}, H, \boldsymbol{\alpha})$ satisfying equation (4.13) then it is optimal. If one has to allocate only a finite number, n^*, of trials we say that the problem has a *finite horizon*. In this case the discounting sequence has zero values for all indices greater than n^*. In a finite horizon problem the optimal policy can be determined in principle by the backward induction method of *dynamic programming* (see Berry and Fristedt, 1985, p. 24). Generally it is very difficult to obtain, even numerically, the optimal solution for $\boldsymbol{\tau}_H^{(n^*)}$. For methods of solution and approximation see Ross (1983).

Let $\boldsymbol{\tau}_H^{(n^*)}$ be the optimal policy for a finite horizon problem of length n^*. For any H and n^*, the optimal allocation of the last trial is

$$\tau_{H,n^*}^{(n^*)} = \operatorname*{argmax}_{1 \leqslant j \leqslant K} \{M_j^{(n^*)}(\boldsymbol{N}^{(n^*-1)}, \boldsymbol{T}^{(n^*-1)})\}. \tag{4.14}$$

A strategy which allocates each trial to the experiment having the largest predicted expectation, $M_j^{(n)}(\boldsymbol{N}^{(n-1)}, \boldsymbol{T}^{(n-1)})$, is called *myopic*. Myopic strategies are generally non-optimal. In some special cases, like $K = 2$, $Y_{j,n}$, are Bernoulli with success probabilities (θ_1, θ_2), and prior distribution, H, such that

$$P_H\{(\theta_1, \theta_2) = (a, b)\} = \pi$$

and

$$P_H\{(\theta_1, \theta_2) = (b, a)\} = 1 - \pi,$$

with $0 \leqslant a < b \leqslant 1$, $0 \leqslant \pi \leqslant 1$ and n-horizon uniform discounting, the myopic policy is optimal. This result was first proven by Feldman (1962) and generalized by Kelly (1974), Rochman (1978) and others. Gittins and Jones (1974) provided a method of finding the optimal strategy in the special case of geometric discounting, i.e., $\boldsymbol{\alpha} = (1, \alpha, \alpha^2, \ldots)$, for some $0 < \alpha < 1$, and priorly *independent* $\theta_1, \ldots, \theta_K$. In this special case, one should compute, for each $j = 1, \ldots, K$, after the nth trial, a *dynamic allocation index* $A_j(\boldsymbol{N}^{(n)}, \boldsymbol{T}^{(n)})$. The optimal allocation for the $(n+1)$st trial is

$$\tau_{n+1}^0 = \operatorname*{argmax}_{1 \leqslant j \leqslant k} A_j(\boldsymbol{N}^{(n)}, \boldsymbol{T}^{(n)}). \tag{4.15}$$

Formulae for determining these dynamic allocation indices are generally very complicated (see Bather, 1983; Berry and Fristedt, 1985; Gittins, 1979).

A subclass of problems are those called the "one-armed bandit problems" (OAB). In this class of OAB, $K = 2$ and one experiment, $\mathcal{E}_2$ say, has a *known* expected yield μ. The expected yield of $\mathcal{E}_1$, $\mu_1(\theta_1)$, is unknown. Obviously, if we had known that $\mu_1(\theta_1) \geqslant \mu$ then all trials would be allocated to $\mathcal{E}_1$. Let H be a prior distribution of μ_1. A discounting sequence $\boldsymbol{\alpha}$ is called *regular* if

$$\frac{\alpha_{m+2}}{\alpha_{m+1}} \leqslant \frac{\alpha_{m+1}}{\alpha_m}, \quad \text{for all} \quad m = 1, 2, \ldots,$$

or $\alpha_{m-j} = 0$ for all $j = 0, \ldots, m-1$, or $\alpha_{m+j} = 0$ for all $j = 1, 2, \ldots$ (see Berry Fristedt, 1985). Berry and Fristedt (1985) prove that if $\boldsymbol{\alpha}$ is a regular discounting sequence then if $\boldsymbol{\tau}_H^0$ is the optimal strategy and $\tau_{H,n^*}^0 = 2$ for some n^* then $\tau_{H,n^*+j}^0 = 2$ for all $j \geqslant 0$. That is, once it is optimal to perform $\mathcal{E}_2$, one should stay with this experiment for the rest of the trials. Thus, the problem is reduced to that of finding an optimal stopping variable M, $M = 0, 1, 2, \ldots$, $M \in \mathcal{B}_{n-1}$, such that if $M \geqslant n$ one performs the nth trial on $\mathcal{E}_1$, $n = 1, 2, \ldots$. Let H_n denote the posterior distribution of θ_1 after n trials on $\mathcal{E}_1$. One has then to compute the value of a function $\mathcal{M}(H_n, \boldsymbol{\alpha}^{(n)})$. If $\mathcal{M}(H_n, \boldsymbol{\alpha}^{(n)}) \geqslant \mu$ then $\mathcal{E}_1$ is performed at the $(n+1)$st trial. Thus, the optimal stopping variable is

$$M^0 = \text{least } n = 0, 1, \ldots \quad \text{such that} \quad \mathcal{M}(H_n, \boldsymbol{\alpha}^{(n)}) < \mu, \tag{4.16}$$

where $H_0 = H$. Formulae for $\mathcal{M}(H_n, \boldsymbol{\alpha}^{(n)})$ are given by Berry and Fristedt (1985) for some special case of Bernoulli trials. Chernoff (1967) considered an n^*-horizon OAB Bernoulli trials, and applied continuous approximations which were previously used by Chernoff and Ray (1965). The approximation is based on transforming the problem to one of optimal stopping for a Wiener process. For details see Chernoff (1972, p. 92). Petkau (1978) used Chernoff's approximations to derive optimal stopping time for sequential medical trials, with finite horizon (large) and uniform discounting. See also Bather (1983).

Several papers on sequential medical trials (see Colton, 1963; Lai et al., 1983 and Bather and Simons, 1985) consider the problem as a two-armed bandit (TAB) Bernoulli trials. The approach in these papers is to use at the initial phase $r(n^*)$ observations on each experiment, and then use the experiment showing a higher mean for the rest of the trials. Bather and Simons (1985) showed that $r(n^*) \cong 0.3\sqrt{n}$ yields *minimax* risk. See also Chapter 4 of Siegmund (1985). Additional material on adaptive designs in clinical trials can be found in Flournoy and Rosenberger (1995).

4.2. Sequential estimation of the common mean of two normal distributions: One variance known

In the present section we present an example of a sequential estimation problem, in which two alternative experiments are available and the objective is to minimize the expected total number of trials. This problem can be considered a OAB problem. We follow the exposition and results of Zacks (1973).

Two experiments are available. $\mathcal{E}_1$ yields a sequence of i.i.d. $N(\mu, \sigma)$ random variables, while $\mathcal{E}_2$ yields a sequence of $N(\mu, 1)$ random variables. $-\infty < \mu < \infty$ and

$0 < \sigma < \infty$ are unknown parameters. The objective is to construct a fixed width confidence interval for the common mean μ, with the minimal expected total sample size. If one knows that $\sigma > 1$ then the solution is to perform

$$n_1(\delta) = \text{least integer } n \geqslant \frac{\chi^2_{1-\alpha}[1]}{\delta^2} \tag{4.17}$$

observations on $\mathcal{E}_2$ and construct the interval $(\bar{Y}^{(2)}_{n_1(\delta)} - \delta, \bar{Y}^{(2)}_{n_1(\delta)} + \delta)$, where $\bar{Y}^{(2)}_{n_1(\delta)} = \frac{1}{n_1(\delta)}\sum_{j=1}^{n_1(\delta)} Y_{2i}$. This interval has coverage probability of at least $(1-\alpha)$, $0 < \alpha < 1$, and $n_1(\delta)$ is the smallest integer for which the required coverage is attained. If $\sigma < 1$ and known, one can attain the objective by performing $n' = n_1(\delta)\sigma$ trials on $\mathcal{E}_1$ and using the interval $(\bar{Y}^{(1)}_{n'} - \delta, \bar{Y}^{(1)}_{n'} + \delta)$. The following is a procedure for unknown σ.

We start the trials on $\mathcal{E}_1$. After n observations we compute $\bar{Y}^{(1)}_n$ and

$$Q_n = \sum_{i=1}^{n} (Y_{1i} - \bar{Y}^{(1)}_n)^2.$$

Let N be a stopping variable, which depends on $\mathbf{Y}_1$ only through Q_n. Furthermore, if $\sigma < 1$, the number of observations required from $\mathcal{E}_1$ is not greater than $n_1(\delta)$. We therefore truncate N to be smaller or equal to $n_1(\delta)$. Moreover, when $\sigma^2 < 1$, N should not exceed much the value of $\sigma^2 n_1(\delta)$. Thus, we use for N a stopping variable of the form

$$N(\delta) = \min\{N_1(\delta), N_2(\delta), n_1(\delta)\}, \tag{4.18}$$

where $N_1(\delta)$ is a stopping variable for immediate estimation and $N_2(\delta)$ is a stopping variable for testing the hypotheses H_0: $\sigma \leqslant 1 - \delta$ against H_1: $\sigma \geqslant 1 + \delta$. If $N(\delta) = N_1(\delta)$ we stop and construct the interval $(\bar{Y}^{(1)}_{N_1(\delta)} - \delta, \bar{Y}^{(1)}_{N_1(\delta)} + \delta)$. If $N(\delta) = N_2(\delta)$ we stop the trials on $\mathcal{E}_1$ and add $M(\delta)$ observations on $\mathcal{E}_2$, where

$$\begin{aligned} M(\delta) = \text{least integer } & m \geqslant 0 \text{ such that} \\ & m \geqslant n_1(\delta) - N_2^2(\delta)/Q_{N_2(\delta)}. \end{aligned} \tag{4.19}$$

The interval estimator is $(\widehat{\mu} - \delta, \widehat{\mu} + \delta)$ where

$$\widehat{\mu} = \frac{M(\delta)\frac{Q_{N_2(\delta)}}{N_2(\delta)}\bar{Y}^{(2)}_{M(\delta)} + N_2(\delta)\bar{Y}^{(1)}_{N_2(\delta)}}{\frac{M(\delta)}{N_2(\delta)}Q_{N_2}(\delta) + N_2(\delta)}. \tag{4.20}$$

For $N_1(\delta)$ we use the Chow–Robbins stopping variable (see Chow and Robbins, 1965)

$$N_1(\delta) = \text{least integer } n \geqslant 1, \text{ such that } Q_n \leqslant \frac{\delta^2}{\chi^2_{1-\alpha}} n^2 \tag{4.21}$$

Table 4.1
Simulation estimates of $EN_2^{(\cdot)}(\delta)$, $E\{M(\delta)\}$ and coverage probability $n_1(\delta) = 50$, $\gamma = .95$

	$E\{N_2(\delta)\}$			$E\{M(\delta)\}$			$P\{\lvert\widehat{\mu} - \mu\rvert < \delta\}$		
σ	P	W	C	P	W	C	P	W	C
0.25	3.72	2.32	2.32	0.00	0.00	0.00	.9329	.8976	.8976
0.50	9.11	5.82	5.82	0.00	0.00	0.00	.8223	.7153	.7153
0.75	17.75	15.07	14.48	2.30	0.00	0.93	.8115	.6995	.6993
1.00	16.16	23.22	19.10	22.57	11.08	15.65	.8424	.7612	.7629
1.25	8.41	12.90	12.67	36.47	29.19	29.39	.8492	.7826	.7825
1.50	5.24	7.96	8.81	42.23	35.98	35.54	.8725	.7978	.7972
1.75	4.24	6.11	7.50	44.15	39.68	39.26	.8839	.8235	.8236
2.00	3.71	5.04	6.28	45.83	41.00	40.73	.8989	.8279	.8281

whose properties are well known. The question is what variable one should use for $N_2(\delta)$.

Zacks (1973) considered three types of stopping variables for $N_2(\delta)$, called parabolic (P), Wald (W) and Chernoff (C). For the rational leading to each one of these variables see Zacks (1973). The formulae of these stopping variables are:

$$N_2^{(P)}(\delta) = \text{least } n,\ n \geqslant 2, \quad \text{such that}$$
$$Q_n \geqslant n e^{1/n} \Phi\left(\sqrt{\frac{n}{2}} \ln\left(\frac{n_1(\delta) Q_n}{n^2}\right) - \sqrt{\frac{2}{n}}\right), \tag{4.22}$$

$$N_2^{(W)}(\delta) = \text{least } n,\ n \geqslant 2, \quad \text{such that}$$
$$Q_n \geqslant n \exp\{(\log(n_1, (\delta)))/n\}, \tag{4.23}$$

and

$$N_2^{(C)}(\delta) = \text{least } n,\ n \geqslant 2, \quad \text{such that}$$
$$Q_n \geqslant \begin{cases} n \exp\left\{\sqrt{\frac{2}{n}}\left[\log\left(\frac{n_1^3(\delta)}{n^3 8\pi}\right)\right]^{1/2}\right\}, & \text{if } n \leqslant \frac{n_1(\delta)}{(8\pi)^{1/3}}, \\ n, & \text{otherwise.} \end{cases}$$

The performance of these procedures with respect to coverage probability and expected sample size was evaluated by simulations. The results are given in the following table.

Acknowledgement

The author gratefully acknowledges the helpful discussions with Professor Anton Schick.

References

Anbar, D. (1977). The application of stochastic approximation procedures to the bioassay problem. *J. Statist. Plann. Inference* **1**, 191–206.

Anbar, D. (1984). Stochastic approximation methods and their use in bioassay and Phase I clinical trials. *Comm. Statist.* **13**, 2451–2467.

Bather, J. (1983). Optimal stopping of Brownian motion: A comparison technique, In: M. H. Rizvi, J. S. Rustagi and D. Siegmund, eds., *Recent Advances in Statistics*. Academic Press, New York, 19–49 .

Bather, J. and G. Simons (1985). The minimax risk for clinical trials. *J. Roy. Statist. Soc. Ser. B* **47**, 466–475.

Berry, D. A. and B. Fristedt (1985). *Bandit Problems: Sequential Allocation of Experiments*. Chapman and Hall, London.

Chaloner, K. and I. Verdinelli (1995). Bayesian experimental design: A review. *Statist. Sci.* **10**, 273–304.

Chernoff, H. (1967). Sequential models for clinical trials. In: *Proc. Fifth Berkeley Symp. of Math. Statist. and Probab.* Vol. 4, 805–812.

Chernoff, H. (1972). *Sequential Analysis and Optimal Design*. Regional Conferences Series in Applied Mathematics, Vol. 8. SIAM, Philadelphia, PA.

Chernoff, H. and S. N. Ray (1965). A Bayes sequential sampling inspection plan. *Ann. Math. Statist.* **36**, 1387–1407.

Chevret, S. (1993). The continual reassessment method in cancer Phase I clinical trials: A simulation study. *Statist. Med.* **12**, 903–1108.

Chow, Y. S. and H. Robbins (1965). On the asymptotic theory of fixed-width sequential confidence intervals for the mean. *Ann. Math. Statist.* **36**, 457–462.

Cochran, W. G. (1973). Experiments for non-linear functions, R. A. Fisher memorial lecture. *J. Amer. Statist. Assoc.* **68**, 771–781.

Colton, T. (1963). A model for selecting one of two medical treatments. *J. Amer. Statist. Assoc.* **58**, 388–400.

Ehrenfeld, S. (1962). Some experimental design problems in attribute life testing. *J. Amer. Statist. Assoc.* **57**, 668–679.

Eichhorn, B. H. (1973). Sequential search of an optimal dosage. *Naval Res. Logistics Quaterly* **20**, 729–736.

Eichhorn, B. H. (1974). Sequential search of an optimal dosage for cases of linear dosage-toxicity regression. *Comm. Statist. Theory Methods* **3**, 263–271.

Eichhorn, B. H. and S. Zacks (1973). Sequential search of an optimal dosage, I. *J. Amer. Statist. Assoc.* **68**, 594–598.

Feldman, D. (1962). Contributions to the two-armed bandit problem. *Ann. Math. Statist.* **33**, 847–856.

Fieller, E. C. (1940). The biological standardization of insulin, *Roy. Statist. Soc.* Supplement 7, 1–64.

Finney, D. J. (1964). *Statistical Methods in Biological Assay*, 2nd edn. Griffin, London.

Fisher, R. A. (1922). On the mathematical foundations of theoretical statistics. *Philos. Trans. Roy. Soc., London A* **222**, 309–368.

Flournoy, N. and W. F. Rosenberger (1995). *Adaptive Designs*. Selected Proceedings of a 1992 Joint AMS–IMS–SIAM Summer Conference, Lecture Notes – Monograph Series, Vol. 25. IMS.

Geller, N. (1984). Design of Phase I and Phase II clinical trials: A statistician's view. *Cancer Invest.* **2**, 483–491.

Gittins, J. C. (1979). Bandit processes and dynamic allocation indices (with discussion). *J. Roy. Statist. Soc. Ser. B* **41**, 148–177.

Gittins, J. C. and D. M. Jones (1974). A dynamic allocation index for the discounted multiarmed bandit problem. *Biometrika* **66**, 561–565.

Kelley, T. A. (1974). A note on the Bernoulli two-armed bandit problem. *Ann. Statist.* **2**, 1056–1062.

Khan, M. K. (1984). Discrete adaptive design in attribute life testing. *Comm. Statist. Theory Methods* **13**, 1423–1433.

Khan, M. K. (1988). Optimal Bayesian estimation of the median effective dose. *J. Statist. Plann. Inference* **18**, 69–81.

Lai, T. L., H. Robbins and D. Siegmund (1983). Sequential design of comparative clinical trials. In: M. H. Rizvi, J. S. Rustagi and D. Siegmund, eds., *Recent Advances in Statistics*. Academic Press, New York, 51–68.

Lehmann, E. L. (1959). *Testing Statistical Hypotheses*. Wiley, New York.

Petkau, A. J. (1978). Sequential medical trials for comparing an experiment with a standard treatment. *J. Amer. Statist. Assoc.* **73**, 328–338.

Prakasa Rao, B. L. S. (1987). *Asymptotic Theory of Statistical Inference*. Wiley, New York.

Robbins, H. and S. Monro (1951). A stochastic approximation method. *Ann. Math. Statist.* **22**, 400–407.

Rodman, L. (1978). On the many armed bandit problem. *Ann. Probab.* **6**, 491–498.

Ross, S. M. (1983). *Introduction to Stochastic Dynamic Programming*. Academic Press, New York.

Shiryayev, A. N. (1984). *Probability*. Springer, New York.

Siegmund, D. (1985). *Sequential Analysis: Tests and Confidence Intervals*. Springer, New York.

Wijsman, R. A. (1973). On the attainment of the Cramer–Rao lower bound. *Ann. Statist.* **1**, 538–542.

Zacks, S. (1971). *The Theory of Statistical Inference*. Wiley, New York.

Zacks, S. (1973). Sequential estimation of the common mean of two normal distributions. I: The case of one variance known. *J. Amer. Statist. Assoc.* **68**, 422–427.

Zacks, S. (1977). Problems and approaches in design of experiments for estimation and testing in non-linear models. *Multivariate Anal.* **IV**, 209–223.

Zacks, S. and B. H. Eichhorn (1975). Sequential search of optimal dosages: The linear regression case. In: J. N. Srivastava, ed., *Survey of Statistical Designs and Linear Models*. North-Holland, New York.

Zacks, S. and W. J. Fenske (1973). Sequential determination of inspection epochs for reliability systems with general life time distributions. *Naval Res. Logistics Quaterly* **3**, 377–385.

S. Ghosh and C. R. Rao, eds., *Handbook of Statistics, Vol. 13*

6

Observational Studies and Nonrandomized Experiments

Paul R. Rosenbaum

1. An informal history or observational studies in statistics

In an observational study or nonrandomized experiment, treated and control groups formed without random assignment are compared in an effort to estimate the effects of a treatment. The term "observational study" was first used in this way by William G. Cochran; see, for instance, Cochran (1965). In an observational study, the treated and control groups may have differed prior to treatment in ways that are relevant for the outcomes under study, and these pretreatment differences may be mistaken for an effect of the treatment. If the groups differed in terms of relevant pretreatment variables or covariates that were observed and recorded, then there is an overt bias. If the groups differed in terms of covariates that were not recorded, then there is a hidden bias. Controlling overt biases and addressing hidden biases are central concerns in an observational study.

In a sense, studies that would today be called observational studies have been conducted since the beginning of empirical science. In the field of epidemiology, a field that conducts many observational studies, attention is often called to the investigations in the mid 1850's of John Snow concerning the causes of cholera. Snow's observational studies compared cholera rates for individuals in London served by different water companies having different and changing sources of water. Snow's careful studies have a distinctly modern character and are still used to teach epidemiology; see MacMahon and Pugh (1970, pp. 6–11) for detailed discussion of Snow's work.

In another sense, observational studies were born at the same time as randomized experiments, with the work in the 1920's and 1930's of Sir Ronald Fisher and Jerzy Neyman on the formal properties of randomization. Prior to that time, there was no formal distinction between experiments and observational studies. However, the systematic development of statistical methods for observational studies came much later.

Many principles that structure the statistical theory of observational studies were created by individuals actively involved in the 1950s in the controversy about smoking as a cause of lung cancer. Sir Austin Bradford Hill and Sir Ronald Fisher, though taking opposing sides in the controversy, agreed that a key part of an observational study is the effort to detect hidden biases using an "elaborate theory" describing how the treatment produces its effects. See Hill (1965) for a concise statement of his general methodological principles and see Cochran (1965, Section 5) for discussion of

Fisher's general views. The first formal method for appraising sensitivity of conclusions to hidden bias came in a paper by Cornfield, Haenszel, Hammond, Lilienfeld, Shimkin and Wynder (1959) which surveyed the evidence linking smoking with lung cancer; see also Greenhouse (1982). Cochran had been an author of the U.S. Surgeon General's report on *Smoking and Health* (Bayne-Jones, et al., 1964) which reviewed the results of many observational studies. Cochran's survey articles on observational studies identified the common structure of most observational studies: (i) the control of visible biases using matching, stratification, or model based adjustments, and (ii) the devices used to distinguish actual treatment effects from hidden biases. Mantel and Haenszel (1959) introduced a test for treatment effect which, in effect, views a single observational study as a series of independent randomized experiments within strata defined observed covariates, where the randomization probabilities vary between strata but are constant within strata.

In addition to epidemiology, the statistical theory of observational studies also draws from the large literature concerning educational and public program evaluation. Examples of observational studies in this area are the studies concerning the relative effectiveness of public and Catholic private high schools – e.g., Coleman, Hoffer and Kilgore (1982) and Goldberger and Cain (1982) – and studies of the effects of various approaches to bilingual education – e.g., Meyer and Fienberg (1992). General discussions of methodology in this area include Campbell and Stanley (1963), Campbell (1969), Cook and Campbell (1979), Kish (1987) and Rossi and Freeman (1985). General discussions of statistics in observational studies are found in: Breslow and Day (1980, 1987), Cochran (1965, 1972), Cox (1992), Gastwirth (1988), Holland (1986), Holland and Rubin (1988), Rosenbaum (1995), Rubin (1974, 1977, 1978), Wold (1956). This chapter is a concise summary of a point of view that is developed in Rosenbaum (1995), though the chapter also discusses a few additional topics and a new example.

2. Inference in randomized experiments

In a randomized experiment, N units are divided into S strata with n_s units in stratum s, $N = n_1 + \cdots + n_S$. The strata are defined using observed covariates, that is, variables describing the units prior to treatment. In stratum s, a fixed number m_s of units are randomly selected to receive the treatment, where $0 \leqslant m_s \leqslant n_s$, and the remaining $n_s - m_s$ receive the control. If $S = 1$, this is a completely randomized experiment. If $n_s = 2$, $m_s = 1$ for $s = 1, \ldots, S$, then this is a paired randomized experiment. Write $\boldsymbol{m} = (m_1, \ldots, m_S)^\top$.

Write $Z_{si} = 1$ if unit i in stratum s receives the treatment and $Z_{si} = 0$ if this unit receives the control, so $m_s = \Sigma_i Z_{si}$. Write $\boldsymbol{Z}$ for the N-dimensional column vector containing the Z_{si}. Write $|A|$ for the number of elements of a set A. Let Ω be the set containing the $K = |\Omega| = \prod \binom{n_s}{m_s}$ possible values of $\boldsymbol{Z}$. In a conventional randomized experiment, $\boldsymbol{Z}$ is selected using a random device or random numbers that ensure that $\text{prob}(\boldsymbol{Z} = \boldsymbol{z}) = 1/K$ for each $\boldsymbol{z} \in \Omega$.

Unit i in stratum s exhibits a (vector) response $\boldsymbol{r}_{si}$. Write $\boldsymbol{R}$ for the matrix whose N rows are the $\boldsymbol{r}_{si}$'s. The null hypothesis of no treatment effect asserts that the responses that units exhibit do not change as the treatment assignment changes; that is, the null hypothesis asserts that $\boldsymbol{R}$ is fixed, so that this same $\boldsymbol{R}$ would have been observed no matter which treatment assignment $\boldsymbol{Z}$ had been randomly selected.

Let $T = t(\boldsymbol{Z}, \boldsymbol{R})$ be a test statistic, and consider the one-sided tail area of this test statistic under the null hypothesis of no effect. Under the null hypothesis, the chance that $T \geqslant k$ is

$$\text{prob}(T \geqslant k) = \frac{|\{\boldsymbol{z} \in \Omega\colon\ t(\boldsymbol{z}, \boldsymbol{R}) \geqslant k\}|}{|\Omega|}, \tag{1}$$

that is, the proportion of the equally probable treatment assignments $\boldsymbol{z} \in \Omega$ giving rise to values of the test statistic greater than or equal to k. This calculation makes use only of the probability distribution implied by the random assignment of treatments, so it involves no assumptions about distributions, and for this reason Fisher (1935) referred to randomization as the "reasoned basis for inference" in experiments. Many commonly used tests are of either of this form or are large sample approximations to (1), including nonparametric tests such as the rank sum and signed rank tests, tests for binary responses such as the Fisher's exact test, McNemar's test for paired binary responses, and the test of Mantel and Haenszel (1959), Birch (1964) and Cox (1966) for stratified binary responses, Mantel's (1963) test for discrete scores, as well as various rank tests for censored data.

Given the test (1) of the null hypothesis of no effect, a confidence interval for an additive or multiplicative effect is obtained in the usual way by inverting the test. Point estimates are often obtained by the method of Hodges and Lehmann (1963).

3. Observational studies in the absence of hidden bias

3.1. Overt and hidden bias

In an observational study, treatments are not randomly assigned, so there may be no reason to believe that all units have the same chance of receiving the treatment, and so no reason to trust the significance level (1). This section discusses inference when there are overt biases but no hidden biases, that is, the chance of receiving the treatment varies with the observed covariates used to define the strata but does not vary with unobserved covariates. Later sections discuss the more fundamental concern that hidden biases may be present.

Consider the following model for the distribution of treatment assignments $\boldsymbol{Z}$ in an observational study. Let Ω^* be the set containing the $|\Omega^*| = 2^N$ vectors of dimension N with coordinates equal to 1 or 0. The model asserts that the Z_{si} are independent with $\text{prob}(Z_{si} = 1) = \pi_{si}$, where the π_{si} are unknown with $0 < \pi_{si} < 1$; that is,

$$\text{prob}(\boldsymbol{Z} = \boldsymbol{z}) = \prod_{s=1}^{S} \prod_{i=1}^{n_s} \pi_{si}^{z_{si}} (1 - \pi_{si})^{1 - z_{si}} \quad \text{for all } \boldsymbol{z} \in \Omega^*. \tag{2}$$

By itself, (2) cannot be used as the basis for inference because the π_{si} are not known, so (2) cannot be calculated. Using (2), the terms "overt bias" and "hidden bias" will be *defined*. There is overt bias but no hidden bias if π_{si} is constant within each stratum; that is, there is *no hidden bias* if there exist probabilities $\bar{\pi}_s$ such that $\pi_{si} = \bar{\pi}_s$ for $i = 1, \ldots, n_s$. Under model (2), there is *hidden bias* if π_{si} varies within strata, that is, if $\pi_{si} \neq \pi_{sj}$ for some s, i, j. When (2) holds with no hidden bias, Rubin (1977) says there is "randomization on the basis of a covariate".

3.2. Permutation inference in the absence of hidden bias

If there is no hidden bias, then overt bias is not, in principle, a difficult problem. The following argument appears explicitly in Rosenbaum (1984a), but it appears to be implicit in Mantel and Haenszel (1959)'s choice of test statistic. If there is no hidden bias then the $\bar{\pi}_s$ are unknown, but $\boldsymbol{m}$ is a sufficient statistic for the $\bar{\pi}_s$, and

$$\text{prob}(\boldsymbol{Z} = \boldsymbol{z} \mid \boldsymbol{m}) = \frac{1}{|\Omega|} \quad \text{for all } \boldsymbol{z} \in \Omega. \tag{3}$$

In other words, if there is no hidden bias, the conditional distribution of treatment assignments given the numbers $\boldsymbol{m}$ treated in each stratum is the usual randomization distribution. Hence, if there is no hidden bias, conventional randomization inference may be applied to the stratified data. This leads, for instance, to the Mantel and Haenszel (1959) statistic when the responses are binary.

In some instances, there is a sufficient statistic for the $\bar{\pi}_s$ which is of lower dimension than $\boldsymbol{m}$, and this is useful when the data are thinly spread across many strata. Methods of inference in this case are discussed in Rosenbaum (1984a, 1988a). For instance, this approach may be applied to matched pairs when the matching has failed to control imbalances in certain observed covariates.

In short, if there is no hidden bias, the analysis of an observational study is not fundamentally different from the analysis of a stratified or matched randomized experiment. If there is no hidden bias, care is needed to appropriately adjust for observed covariates, but careful adjustments will remove overt biases. Addressing hidden bias is, therefore, the central concern in an observational study. The remainder of Section 3 discusses practical aspects of controlling overt biases, and later sections discuss hidden bias.

3.3. The propensity score

A practical difficulty in matching and stratification arises when there is a need to control for many observed covariates simultaneously. Suppose that there are P observed covariates. Even if each of the P covariates take on only two values, there will be 2^P possible values of the covariate, or more than a million possible values for $P = 20$ covariates. As a result, with the sample sizes typically available in observational studies, it may be impossible to find for each treated subject a control subject with exactly the same value of all covariates. Exact matching is rarely feasible when there are many covariates.

Fortunately, exact matching is not needed to control overt biases. Examination of the argument in Subsections 3.1 and 3.2 reveals that the argument is valid in the absence of hidden bias providing all subjects in the same stratum or matched pair have the same probability $\bar{\pi}_s$ of receiving the treatment; however, it is not necessary that subjects in the same stratum have identical values of observed covariates.

The propensity score is the conditional probability of receiving the treatment given the observed covariates. It has several useful properties. First, in the absence of hidden bias, strata or matched sets that are homogeneous in the one-dimensional propensity score control bias due to all observed covariates (Rosenbaum and Rubin, 1983, Theorem 4; Rosenbaum, 1984a, Theorem 1). Second, even in the presence of hidden bias, strata or matched sets that are homogeneous in the propensity score tend to *balance* observed covariates, that is, to produce treated and control groups with the same distribution of observed covariates (Rosenbaum and Rubin, 1983, Theorem 1). In other words, strata or matched sets that are homogeneous in the propensity score control overt biases even if hidden biases are present.

In practical work, the propensity score is unknown and must be estimated using a model, perhaps a logit model, predicting the binary treatment assignment from the P-dimensional vector of observed covariates. Estimated propensity scores appear to perform well. In empirical studies, Rosenbaum and Rubin (1984, 1985) found that estimated propensity scores produced greater balance in observed covariates than theory anticipates from true propensity scores. It appears that estimated propensity scores remove some chance imbalances in observed covariates. In a simulation study, Gu and Rosenbaum (1993) found that matching on an estimated propensity score produces greater covariate balance when $P = 20$ than multivariate distance based matching methods. Rosenbaum (1984a) discusses exact conditional inference given a sufficient statistic for the unknown parameter of the propensity score.

3.4. The structure of an optimal stratification

Stratifications and matchings in observational studies have varied structures. In pair matching, $n_s = 2$ and $m_s = 1$ for $s = 1, \dots, S = N/2$. In matching with a fixed number k of controls, $n_s = k + 1$ and $m_s = 1$ for $s = 1, \dots, S = N/(k + 1)$; see, for instance, Ury (1975). In matching with a variable number of controls, $n_s \geqslant 2$ and $m_s = 1$ for $s = 1, \dots, S$; see, for instance, the example presented by Cox (1966). Haphazard choices of n_s and m_s are common in practice. Does any particular structure have a claim to priority?

Consider M treated subjects and $N - M$ controls and a distance $\delta_{ij} \geqslant 0$ between treated subject i and control j. The distance δ_{ij} measures the difference between the values of the observed covariate vectors for i and j. A stratification is a division of the N subjects into S strata so that each stratum contains at least one treated subject and one control, that is, $m_s \geqslant 1$ and $n_s - m_s \geqslant 1$. For a given stratum s, let Δ_s be the average of the $m_s(n_s - m_s)$ distances δ_{ij} between the m_s treated and $n_s - m_s$ control subjects in this stratum. Let

$$\Delta = \sum_{s=1}^{S} w(m_s, n_s - m_s)\Delta_s,$$

where $w(a, b) \geqslant 0$ is a weight function defined for positive integers a and b. The weight function $w(a, b)$ is *neutral* if for all $a \geqslant 2$ and $b \geqslant 2$, $w(a, b) = w(a - 1, b - 1) + w(1, 1)$; that is, the weight function is neutral if the total weight neither increases nor decreases when a pair is separated from a stratum. For instance, three neutral weight functions are $w(m_s, n_s - m_s) = n_s/N$, $w(m_s, n_s - m_s) = m_s/M$, and $w(m_s, n_s - m_s) = (n_s - m_s)/(N - M)$.

A *full matching* is a stratification in which $\min(m_s, n_s - m_s) = 1$ for $s = 1, \ldots, S$. It may be shown that for any neutral weight function and any distance δ_{ij}, there is a full matching that minimizes Δ over the set of all stratifications. This is not true of pair matching, matching with k controls or matching with a variable number of controls. Indeed, as measured by Δ, pair matching may be arbitrarily poor compared to the best full matching. More than this, for a neutral weight function, if the covariates are multivariate Normal and δ_{ij} is the Mahalanobis distance then, with probability 1, a stratification that is not a full matching is not optimal. For proof of these claims and extensions, see Rosenbaum (1991a).

In this specific sense, the optimal form for stratification is a full matching. Optimality refers to one particular criterion, in this case distance or overt bias. Under simple constant variance models, full matching is not optimal in terms of minimizing the variance of the mean of the matched treated-minus-control differences; rather, $1 - k$ matching is optimal. As shown by elementary calculations, a bias that does not diminish as $N \to \infty$ quickly becomes more important to the mean squared error than a variance that is decreasing as $1/N$ as $N \to \infty$. Still, variance is a consideration. In addition, the optimal full matching may be highly unbalanced, which may be unpleasant for aesthetic reasons. In practice, it is probably best to limit the degree to which matched sets can be unbalanced while not requiring perfect balance. An example of doing this is discussed by Rosenbaum (1989a, Section 3.3).

In a simulation, Gu and Rosenbaum (1993, Section 3.3) compared: (i) optimal full matching with $M/N = 1/2$ to optimal pair matching and (ii) optimal full matching with $M/N = 1/4$ to optimal matching with $k = 3$ controls per treated subject. In other words, in both (i) and (ii), the number of controls was the same in full matching and $1 - k$ matching ($k = 1$ or $k = 3$), the difference being the flexible structure of full matching. Two distances were used, the difference in propensity score and the Mahalanobis distance. In terms of both distances and in terms of covariate imbalance, full matching was much better than $1 - k$ matching for both $k = 1$ and $k = 3$.

3.5. *Constructing optimal matched samples*

Matched pairs or sets that minimize the total distance within sets may be constructed using minimum cost network flow algorithms. This process is illustrated in Rosenbaum

(1989). Unlike many combinatorial optimization algorithms, the relevant network flow algorithms are relatively fast, and in particular, some are polynomial bounded. An excellent, modern general discussion of network flow algorithms is given by Bertsekas (1991) who also supplies FORTRAN code and performance comparisons on a Macintosh Plus.

Gu and Rosenbaum (1993, Section 3.1) use simulation to compare optimal pair matching and greedy pair matching when 50 treated subjects and 50 or 100 or 150 or 300 controls are available to form the 50 matched pairs. Greedy matching consists of forming pairs one at a time, the pair with the smallest distance being selected first, the remaining pair with the second smallest distance being selected second, and so on. Traditionally, some form of greedy matching has been used to create matched samples in statistics. Greedy matching does not generally minimize the total distance within pairs, and in theory it can be quite poor compared to optimal matching. The simulation suggests that optimal matching is sometimes noticeably better than greedy matching in terms of distance within pairs, sometimes only slightly better, but optimal matching is no better at producing balanced matched samples. Optimal and greedy matching seem to select roughly the same 50 controls, but optimal matching does a better job with the pairing, so distance within pairs is affected but the balance of the entire matched groups is not.

3.6. Matching and stratification followed by analytical adjustments

When matching or stratification provide imperfect control for observed covariates, analytical adjustments may by added. Rubin (1973b, 1979) uses simulation to study covariance adjustment for covariates that have been imperfectly controlled by matching. He finds that covariance adjustment of matched pair differences is more robust to misspecification of the form of the covariance model than is covariance adjustment of unmatched samples.

Holford (1978) and Holford, White and Kelsey (1978) perform model-based adjustments for matched pairs with a binary outcome in case-control studies. Rosenbaum and Rubin (1984, Section 3.3) use a log-linear model to adjust for covariates in conjunction with strata formed from an estimated propensity score. Rosenbaum (1989a) discusses permutation tests that stratify imperfectly matched pairs. This leads to generalizations of Wilcoxon's signed rank test and McNemar's test that control residual imbalances in observed covariates.

3.7. The bias due to incomplete matching

Incomplete matching occurs when some treated subjects are discarded because controls cannot be found. It is possible to show that incomplete matching can introduce a bias into a study that is free of hidden bias even if matched pairs are exactly matched for observed covariates; see Rosenbaum and Rubin (1985b). There, an example is presented in which unmatched treated subjects differ more from matched treated subjects than the treated subjects initially differed from the controls. Incomplete matching can substantially and unintentionally alter the nature of the treated group.

3.8. *Consequences of adjustment for affected concomitant variables*

Section 3 has discussed adjustments for covariates, that is, for variables measured prior to treatment and hence unaffected by the treatment. Adjustments are sometimes made for variables measured after treatment that may have been affected by the treatment. For instance, Coleman et al. (1982) compare senior year test scores in public and Catholic high schools after adjusting for sophomore year test scores, reasoning that most of the difference in sophomore test scores existed prior to high school and was primarily not an effect of the difference between public and Catholic high schools. They used the observed sophomore year test score in place of an important unobserved covariate, namely test scores immediately before high school.

When there is no hidden bias, adjustment for a variable affected by the treatment may remove part of the treatment effect, that is, it may introduce a bias that would not have been present had adjustments been confined to covariates. When hidden biases are present, adjustments for affected concomitant variables may at times introduce an avoidable bias, or it may reduce a hidden bias, or it may do both to some degree. See Rosenbaum (1984b) for general discussion including several analytical options.

4. Sensitivity to hidden bias in observational studies

4.1. *The first sensitivity analysis*

A sensitivity analysis asks how hidden biases of various magnitudes might alter the conclusions of an observational study. The first formal sensitivity analysis was conducted by Cornfield et al. (1959); see also Greenhouse (1982). They concluded that if one were to dismiss the observed association between heavy cigarette smoking and long cancer, if one were to assert that it does not reflect an effect of smoking but rather some hidden bias, then one would need to postulate a hidden bias of enormous proportions. Specifically, they concluded that an unobserved binary covariate would need to be nine times more common among heavy smokers than among nonsmokers and would need to be a near perfect predictor of lung cancer if it were to explain the observed association between smoking and lung cancer. There are many differences between smokers and nonsmokers, differences in diet, personality, occupation, social and economic status; however, it is difficult to imagine that these traits are nine times more common among smokers and are near perfect predictors of lung cancer.

A sensitivity analysis restricts the scope of debate about hidden biases by indicating the magnitude of bias needed to alter conclusions. However, a sensitivity analysis does not indicate whether bias is present, nor what magnitudes of bias are plausible. There are examples of observational studies which were contradicted by subsequent randomized experiments even though the observational studies were insensitive to moderately large biases. This suggests that some observational studies have been affected by hidden biases that are quite large. See Rosenbaum (1988b) for discussion of an example, including a sensitivity analysis and a discussion of the subsequent randomized experiment.

4.2. A general method of sensitivity analysis

The sensitivity analysis discussed here is similar in spirit to the analysis discussed by Cornfield et al. (1959); however, it is more general in that it is not confined to binary outcomes and it makes appropriate allowance for sampling error. The method is discussed in Rosenbaum (1987a, 1988b, 1991b) and Rosenbaum and Krieger (1990). Other methods of sensitivity analysis are discussed by Bross (1966), Gastwirth and Greenhouse (1987), Gastwirth (1992), and Schlesselmann (1978).

If there is hidden bias, then $\pi_{si} \neq \pi_{sj}$ for some s, i, j. To quantify the magnitude of the hidden bias, introduce a parameter $\Gamma \geqslant 1$ such that:

$$\frac{1}{\Gamma} \leqslant \frac{\pi_{si}(1-\pi_{sj})}{\pi_{sj}(1-\pi_{si})} \leqslant \Gamma \quad \text{for all } s,\ i,\ j. \tag{4}$$

Expression (4) says that two units, say i and j, in the same stratum, say s, may have different probabilities of receiving the treatment, $\pi_{si} \neq \pi_{sj}$; however, the odds that i receives the treatment are at most Γ times greater than the odds that j receives the treatment. In a randomized experiment, Γ is known to equal 1 because randomization has ensured that all units in each stratum had the same chance of receiving the treatment. A sensitivity analysis asks: How large must Γ be to alter the conclusions of the study?

If there is hidden bias of magnitude Γ then the distribution of treatment assignments $\boldsymbol{Z}$ given the numbers $\boldsymbol{m}$ of treated subjects in each stratum is not constant as in (3). It is not difficult to check that (2) and (4) imply that there is an N-dimensional vector $\boldsymbol{u}$ with coordinates u_{si} bounded by zero and one, $0 \leqslant u_{si} \leqslant 1$, such that:

$$\operatorname{prob}(\boldsymbol{Z} = \boldsymbol{z} \mid \boldsymbol{m}) = \frac{\exp(\gamma \boldsymbol{u}^\top \boldsymbol{z})}{\sum_{\boldsymbol{b}\in\Omega} \exp(\gamma \boldsymbol{u}^\top \boldsymbol{b})} \quad \text{for all } \boldsymbol{z} \in \Omega, \tag{5}$$

where $\gamma = \ln(\Gamma)$. Notice that (5) reduces to (3) when $\gamma = 0$ or $\Gamma = 1$, that is, when there is no hidden bias. In (5), u_{si} may be interpreted as an unobserved covariate describing the ith unit in stratum s. The restriction $0 \leqslant u_{si} \leqslant 1$ may be viewed in several ways: (i) as a formal reexpression of (4), (ii) as a scale restriction required so that the magnitude of the parameter γ has meaning, or (iii) in practical effect, as saying that the unobserved variable is a binary trait taking values 0 or 1. Point (iii) is true in effect because, while the model allows u_{si} to take values between 0 and 1, the sensitivity analysis works with bounds and extreme cases in which u_{si} turns out to be 0 or 1, so there would be no change if this were assumed at the outset. Let U be the set of all N-dimensional vectors with coordinates u_{si} between zero and one, $0 \leqslant u_{si} \leqslant 1$, so $\boldsymbol{u} \in U$.

Under the null hypothesis of no treatment effect, under model (4) or (5), the tail area of the test statistic T is:

$$\alpha_\gamma(\boldsymbol{u}) = \sum_{\boldsymbol{z}\in\Omega} \left[t(\boldsymbol{z}, \boldsymbol{R}) \geqslant k\right] \frac{\exp(\gamma \boldsymbol{u}^\top \boldsymbol{z})}{\sum_{\boldsymbol{b}\in\Omega} \exp(\gamma \boldsymbol{u}^\top \boldsymbol{b})}, \tag{6}$$

where [event] = 1 if the event occurs and [event] = 0 otherwise. The significance level $\alpha_\gamma(\boldsymbol{u})$ depends on the scalar γ and the N-dimensional vector $\boldsymbol{u}$, both of which are unknown, so $\alpha_\gamma(\boldsymbol{u})$ is unknown.

The sensitivity analysis considers several values of γ, and for each γ calculates the range of possible significance levels, thereby indicating the degree to which conclusions might change in the presence of hidden biases of various magnitudes.

4.3. An example of a sensitivity analysis

Thun et al. (1989) examined the possible effect of occupational exposure to cadmium on kidney function. They compared male workers exposed to cadmium at a metal recovery plant in Colorado to unexposed male workers at a local hospital. Clearly, a study of occupational exposures to cadmium cannot be conducted as a randomized experiment, in part because cadmium is potentially harmful and in part because occupational choices cannot be randomly assigned by the investigator.

Thun et al. (1989) "frequency matched" for age; however, in the discussion that follows, frequency matching has been replaced by pair matching for age. Table 1 is based on the report by Thun (1993). In Table 1, the 23 cadmium workers with

Table 1
Paired data on cadmium exposure and kidney dysfunction

Pair	Exposed B2PGC	Control B2PGC	Differences of base 2 logs	Absolute ranks	Signs
1	107,143	311	8.43	23	1
2	33,679	338	6.64	20	1
3	18,836	159	6.89	21	1
4	173	110	0.65	4	1
5	389	226	0.79	5	1
6	1,144	305	1.91	11	1
7	513	222	1.21	7	1
8	211	242	−0.20	1	0
9	24,288	250	6.60	19	1
10	67,632	256	8.05	22	1
11	488	135	1.85	10	1
12	700	96	2.87	15	1
13	328	142	1.21	6	1
14	98	120	−0.30	2	0
15	122	376	−1.62	9	0
16	2302	173	3.73	17	1
17	10,208	178	5.85	18	1
18	892	213	2.06	13	1
19	2803	257	3.45	16	1
20	201	81	1.30	8	1
21	148	199	−0.42	3	0
22	522	114	2.20	14	1
23	941	247	1.93	12	1

complete data and exposures to cadmium above 500 mg/m^3-day are paired for age with 23 unexposed hospital workers. The outcome is an indicator of a type of kidney dysfunction, namely beta-2-microglobulin expressed in $\mu g/g$ of creatinine or B2PGC. Table 1 gives values of B2PGC for exposed and control subjects together with the difference of their base 2 logarithms. For instance, in the first pair, the exposed subject had a $\beta - 2$ value of 107,143 $\mu g/g$ creatinine, which is $344.5 = 2^{8.43}$ times greater than value 311 for the control subject, so the control subject's value must be doubled 8.43 times to equal the exposed subject's value.

The last two columns of Table 1 give calculations for Wilcoxon's signed rank test. The absolute values of the differences in logarithms are ranked, the "signs" of the differences are determined, and the ranks for positive differences are summed to yield a signed rank statistic of 261. In a randomized experiment, in the absence of a treatment effect, the signed rank statistic would have expectation 138 and variance 1081, yielding a standardized deviate of $3.74 = (261 - 138)/\sqrt{1081}$, with approximate one-sided significance level of .0001. If this were a randomized experiment, there would be little doubt that cadmium exposure causes kidney dysfunctions apparent in elevated $\beta - 2$ levels. This is not, however, a randomized experiment.

The significance level just obtained for a randomized experiment is a large sample approximation to (1). This is equivalent to assuming there is no hidden bias beyond the overt bias in age controlled by matching, that is, to assuming $\Gamma = 1$ in (4), in which case the significance level (6) equals randomization significance level (1). The sensitivity analysis considers the range of possible significance levels for several values of Γ. For instance, for $\Gamma = 2$ two subjects matched together may differ in their odds of receiving the treatment by a factor of 2 because they differ in terms of the unobserved covariate u.

Table 2 presents the sensitivity analysis for the signed rank test for the data in Table 1. Specifically, Table 2 contains the range of one sided significance level (6) for $\boldsymbol{u} \in U$. For $\Gamma = 1$, there is a single significance level, namely the randomization significance level (1), because the chance π_{si} of receiving the treatment does not vary with u_{si}. For $\Gamma = 2$, the range of significance levels is from $< .0001$ to .006, so there is some uncertainly about the precise significance level because the value of $\boldsymbol{u} \in U$ is unknown, but it is clear that a highly significant effect is present. For hidden bias to explain the higher levels of kidney dysfunction among cadmium workers, the magnitude of the bias would need to be greater than $\Gamma = 3.5$ – that is, matched

Table 2
Sensitivity analysis for cadmium data

Γ	Range of significance levels
1	$< .0001$
2	$< .0001$ to .006
3	$< .0001$ to .029
3.5	$< .0001$ to .046
4	$< .0001$ to .063

subjects would have to differ in their odds of treatment by a factor of at least 3.5. The study is insensitive to small and moderate hidden biases, but it is more sensitive to bias than the studies linking smoking with lung cancer. In Rosenbaum (1991b), five observational studies are compared in terms of their degree of sensitivity to hidden bias. These and other examples are discussed in Rosenbaum (1995, Section 4).

If a study is sensitive to hidden biases of moderate size, it is especially important to detect biases that may exist; see Section 5.

4.4. Calculating sensitivity bounds

The calculations leading to Table 2 have a simple form. For the signed rank statistic, the distribution (6) is stochastically largest for one particular $\boldsymbol{u} \in U$ and stochastically smallest for another $\boldsymbol{u} \in U$. These distributions have expectation and variance

$$\mathrm{E}(T) = \frac{pS(S+1)}{2},$$

$$\mathrm{Var}(T) = \frac{p(1-p)S(S+1)(2S+1)}{6}$$

where $p = \Gamma/(1+\Gamma)$ for the largest distribution and $p = 1/(1+\Gamma)$ for the smallest distribution. When there is no hidden bias so $\Gamma = 1$, these are the usual expectation and variance of the signed rank statistic based on its randomization distribution (1). Approximate bounds on the significance level are found by referring $\{T - \mathrm{E}(T)\}/\sqrt{\mathrm{var}(T)}$ to the normal distribution for $p = \Gamma/(1+\Gamma)$ and $p = 1/(1+\Gamma)$.

Sensitivity bounds exist for many commonly used statistics; see Rosenbaum (1995, Section 4) for a survey. For matched pairs and matching with multiple controls, including the signed rank statistic, McNemar's statistic, the Mantel–Haenszel statistic, and certain statistics for censored data, see Rosenbaum (1988b). For the comparison of two unmatched groups using the rank sum statistic, Fisher's exact test for a 2×2 table, and certain rank statistics for censored data, see Rosenbaum and Krieger (1990). For matched case-referent studies, see Rosenbaum (1991b).

4.5. Sensitivity analysis for other hypotheses, for point estimates and for confidence intervals

Section 4.3 illustrated the sensitivity analysis for a significance level testing the null hypothesis of no treatment effect. The model of an additive effect τ asserts that each subject would have a response that is τ higher under treatment than under control. This model appears in analysis of covariance and in many nonparametric discussions. Under the model of an additive treatment effect, the hypothesis H_0: $\tau = \tau_0$ is tested by calculating adjusted responses in which τ_0 is subtracted from the observed response of each treated subject, and then testing the null hypothesis of no treatment effect for these adjusted responses.

A $(1-\alpha)$-confidence interval for τ is found as the set of all values of τ not rejected in this way by an α-level test. A Hodges–Lehmann point estimate τ is (essentially) that value $\hat{\tau}$ which, when subtracted from the responses of treated subjects, equates the test statistic T to its expectation under the null hypothesis $\mathrm{E}(T)$. In other words, the distribution of T in the absence of a treatment effect forms the basis for confidence intervals and point estimates of an additive effect. Similar considerations apply to other effects, such as a multiplicative effect. See Lehmann (1975) for general discussion of these standard devices.

Sensitivity analyses for confidence intervals and Hodges–Lehmann point estimates use the approach just described. The null distribution of T in the absence of a treatment effect is given by (6). Under the model of an additive effect, subtracting the true value τ from the responses of treated subjects before calculating T yields this null distribution. For details of confidence interval calculations, see Rosenbaum (1987, 1995) and for Hodges–Lehmann estimates see Rosenbaum (1993, 1995).

5. Detection of hidden bias in observational studies

The sensitivity analysis in Section 4 asked how hidden biases of various magnitudes might alter a study's conclusions, but it provided no indication of the presence or absence of hidden bias. Efforts to detect the presence of hidden bias and to appraise its magnitude when present are the topic of the current section. Generally, efforts of this sort require some additional information beyond the treated and control groups, their outcomes and covariates. Aspects of this subject are discussed by Campbell (1969a, b), Campbell and Stanley (1963), Cochran (1965, Section 5), Cook and Campbell (1979), Hill (1965), Lilienfeld and Lilienfeld (1980), MacMahon and Pugh (1970), Rosenbaum (1984c, 1987c, 1989b, 1995), Sartwell (1960), Yerushalmy and Palmer (1959).

Efforts to detect hidden biases are characterized by the type of additional information that is available and by the types of biases that this sort of information can hope to detect. Ideally, a study is designed so that actual treatment effects and hidden biases are expected to have different appearances in observable data. The power of a test for hidden bias is an important consideration in interpreting the results of the test. Two examples are sketched below.

- Two control groups are selected in such a way that they are known to differ with respect to a particular unobserved covariate. The treated group is compared to each control group and the control groups are compared to each other. If the control groups differ dramatically, then this is taken as evidence of hidden bias. Properties of this device for detecting hidden bias, including its limitations, are discussed in Campbell (1969a) and Rosenbaum (1987b, 1995).
- In addition to the outcomes of primary interest, the study includes outcomes known to be unaffected by the treatment but to be correlated with a particular unobserved covariate. Treated and control groups are compared with respect to all outcomes, those of primary interest and those known to be unaffected by the treatment. If treated and control groups differ with respect to outcomes the treatment should

not affect, then this is taken as evidence of hidden bias. Formal properties and limitations of this method are discussed in Rosenbaum (1989b, 1995).

An example using both multiple control groups and known effects is discussed in Rosenbaum (1984c). The example concerns the possible effects on childhood cancers of tests of nuclear weapons performed above ground. Efforts to detect hidden biases can inform sensitivity analyses; see Rosenbaum (1992).

6. Coherence and focused hypotheses

Consider a theory claiming that a particular treatment causes certain effects, for instance, that cigarette smoking causes lung cancer. Quite often, such a theory will yield many predictions for observable data. For instance, one may anticipate not only that smokers will have more lung cancer than nonsmokers, but also that heavy smokers will have more lung cancer than light smokers, that quitters will have less lung cancer than continuing smokers, and so on. There is a strong intuitive feeling that if many different predictions are made and each is confirmed in data, then the evidence in favor of causality is stronger. See the discussion of Fisher's view of "elaborate theories" in Cochran (1965, Section 5). Also, see the discussion of coherence of associations in Hill (1965), Susser (1977), and Lilienfeld and Lilienfeld (1980, p. 315).

In common practice, multiple predictions of a single causal theory are tested one at a time, and so it is difficult to formalize the intuition that repeated confirmations of multiple predictions strengthen the evidence of causality. An alternative approach is to develop a single, highly focused test of all predictions simultaneously. The hope would be that, when many predictions are supported by observable data, a focused, simultaneous test would capture this by exhibiting less sensitivity to hidden bias in the sense of Section 4. This hope is realized in part; see Rosenbaum (1994) for detailed development with an example in the comparison of two groups, and see Rosenbaum (1991c) for development of a desirable property of simultaneous tests.

7. Summary

Two tasks are central to the analysis of observational or nonrandomized comparisons of treated and control groups. First, if the groups are seen to differ prior to treatment – that is, if there is overt bias – then adjustments are made to correct for these initial differences. Adjustments may use matching, stratification, model based procedures such as covariance adjustment, or a combination of two or more techniques. Second, groups may differ prior to treatment in ways that were not observed – that is, there may be a hidden bias – and this possibility needs to be addressed. Typically, hidden biases are addressed though sensitivity analyses, through efforts to detect bias, and through concepts such as coherent associations.

References

Bayne-Jones, S., W. Burdette, W. Cochran, E. Farber, L. Fieser, J. Furth, J. Hickman, C. LeMaistre, L. Schuman and M. Seevers (1964). *Smoking and Health.* U.S. Department of Health, Education and Welfare, Washington, DC.

Bertsekas, D. (1991). *Linear Network Optimization: Algorithms and Codes.* MIT Press, Cambridge, MA.

Birth, M. W. (1964). The detection of partial association, I: The 2×2 case. *J. Roy. Statist. Soc. Ser. B* **26**, 313–324.

Breslow, N. and N. Day (1980). *Statistical Methods in Cancer Research, Vol. 1: The Analysis of Case-Control Studies.* International Agency for Research on Cancer of the World Health Organization, Lyon, France.

Breslow, N. and N. Day (1987). *The Design and Analysis of Cohort Studies.* International Agency for Research on Cancer, Lyon, France.

Bross, I. D. J. (1966). Spurious effects from an extraneous variable. *J. Chronic Dis.* **19**, 637–647.

Campbell, D. (1969a). Prospective: Artifact and control. In: R. Rosenthal and R. Rosnow, eds., *Artifact in Behavioral Research.* Academic Press, New York, 351–382.

Campbell, D. (1969b). Reforms as experiments. *Amer. Psychol.* **24**, 409–429.

Campbell, D. and J. Stanley (1963). *Experimental and Quasi-Experimental Designs for Research.* Rand McNally, Chicago, IL.

Carpenter, R. (1977). Matching when covariables are normally distributed. *Biometrika* **64**, 299–307.

Cochran, W. G. (1965). The planning of observational studies of human populations (with discussion). *J. Roy. Statist. Soc. Ser. A* **128**, 134–155.

Cochran, W. G. (1972). Observational studies. In: T. A. Bancroft, ed., *Statistical Papers in Honor of George Snedecor.* Iowa State Univ. Press, Ames, IA, 77–90.

Cochran, W. G. and D. B. Rubin (1973). Controling bias in observational studies: A review. *Sankhyā Ser. A* **35**, 417–446.

Coleman, J., T. Hoffer and S. Kilgore (1982). Cognitive outcomes in public and private schools. *Sociol. Educ.* **55**, 65–76.

Cook, T. and D. Campbell (1979). *Quasi-Experimentation.* Rand McNally, Chicago.

Cornfield, J., W. Haenszel, E. Hammond, A. Lilienfeld, M. Shimkin and E. Wynder (1959). Smoking and lung cancer: Recent evidence and a discussion of some questions. *J. Nat. Cancer Inst.* **22**, 173–203.

Cox, D. R. (1966). A simple example of a comparison involving quantal data. *Biometrika* **53**, 562–565.

Cox, D. R. (1992). Causality: Some statistical aspects. *J. Roy. Statist. Soc. Ser. A* **155**, 291–301.

Fisher, R. A. (1935). *The Design of Experiments.* Oliver and Boyd, Edinburgh.

Gastwirth, J. (1988). *Statistical Reasoning in Law and Public Policy.* Academic Press, New York.

Gastwirth, J. (1992). Method for assessing the sensitivity of statistical comparisons used in Title VII cases to omitted variables. *Jurimetrics* **33**, 19–34.

Gastwirth, J. and S. Greenhouse (1987). Estimating a common relative risk: Application in equal employment. *J. Amer. Statist. Assoc.* **82**, 38–45.

Goldberger, A. and G. Cain (1982). The causal analysis of cognitive outcomes in the Coleman, Hoffer, and Kilgore report. *Sociol. Educ.* **55**, 103–122.

Greenhouse, S. (1982). Jerome Cornfield's contributions to epidemiology. *Biometrics* **28** (Supplement), 33–46.

Gu, X. S. and P. R. Rosenbaum (1993). Comparison of multivariate matching methods: Structures, distances and algorithms. *J. Comput. Graphical Statist.* **2**, 405–420.

Hill, A. B. (1965). The environment and disease: Association or causation? *Proc. Roy. Soc. Medicine* **58**, 295–300.

Hodges, J. and E. Lehmann (1963). Estimates of location based on rank tests. *Ann. Math. Statist.* **34**, 598–611.

Holford, T. (1978). The analysis of pair-matched case-control studies: A multivariate approach. *Biometrics* **34**, 665–672.

Holford, T., C. White and J. Kelsey (1978). Multivariate analysis for matched case-control studies. *Amer. J. Epidemiol.* **107**, 245–256.

Holland, P. (1986). Statisctics and causal inference (with discussion). *J. Amer. Statist. Assoc.* **81**, 945–970.

Holland, P. and D. Rubin (1983). On Lord's Paradox. In: H. Wainer and S. Messick, eds., *Principle of Modern Psychological Measurement.* Lawrence-Erlbaum, Hillsdale, NJ, 3–26.

Holland, P. and D. Rubin (1988). Causal inference in retrospective studies. *Evaluation Rev.* **12**, 203–231.

Lehmann, E. (1975). *Nonparametrics: Statistical Methods Based on Ranks.* Holden-Day, San Francisco, CA.

Lilienfeld, A. and D. Lilienfeld (1980). *Foundations of Epidemiology*, 2nd edn. Oxford Univ. Press, New York.

Kish, L. (1987). *Statistical Design for Research.* Wiley, New York.

MacMahon, B. and T. Pugh (1970). *Epidemiology: Principles and Methods.* Little, Brown, Boston.

Mantel, N. (1963). Chi-square tests with one degree of freedom: Extensions of the Mantel–Haenszel procedure. *J. Amer. Statist. Assoc.* **58**, 690–700.

Mantel, N. (1976). A personal perspective on statistical techniques for quasi-experiments. In: D. B. Owen, ed., *On the History of Statistics and Probability.* Marcel Dekker, New York, 103–130.

Mantel, N. and W. Haenszel (1959). Statistical aspects of retrospective studies of disease. *J. Nat. Cancer Inst.* **22**, 719–748.

Meyer, M. and S. Fienberg, eds. (1992). *Assessing Evaluation Studies: The Case of Bilingual Education Strategies.* National Academy Press, New York.

Rossi, P. and H. Freeman (1985). *Evaluation.* Sage, Beverly Hills, CA.

Rosenbaum, P. R. (1984a). Conditional permutation tests and the propensity score in observational studies. *J. Amer. Statist. Assoc.* **79**, 565–574.

Rosenbaum, P. R. (1984b). The consequences of adjustment for a concomitant variable that has been affected by the treatment. *J. Roy. Statist. Soc. Ser. A* **147**, 656–666.

Rosenbaum, P. R. (1984c). From association to causation in observational studies: The role of tests of strongly ignorable treatment assignment. *J. Amer. Statist. Assoc.* **79**, 41–48.

Rosenbaum, P. R. (1987a). Sensitivity analysis for certain permutation inferences in matched observational studies. *Biometrika* **74**, 13–26.

Rosenbaum, P. R. (1987b). The role of a second control group in an observational study (with discussion). *Statist. Sci.* **2**, 292–316.

Rosenbaum, P. R. (1988a). Permutation tests for matched pairs with adjustments for covariates. *Appl. Statist.* **37**, 401–411.

Rosenbaum, P. R. (1988b). Sensitivity analysis for matching with multiple controls. *Biometrika* **75**, 577–581.

Rosenbaum, P. R. (1989a). Optimal matching for observational studies. *J. Amer. Statist. Assoc.* **84**, 1024–1032.

Rosenbaum, P. R. (1989b). The role of known effects in observational studies. *Biometrics* **45**, 557–569.

Rosenbaum, P. R. (1991a). A characterization of optimal designs for observational studies. *J. Roy. Statist. Soc. Ser. B* **53**, 597–610.

Rosenbaum, P. R. (1991b). Sensitivity analysis for matched case-control studies. *Biometrics* **47**, 87–100.

Rosenbaum, P. R. (1991c). Some poset statistics. *Ann. Statist.* **19**, 1091–1097.

Rosenbaum, P. R. (1992). Detecting bias with confidence in observational studies. *Biometrika* **79**, 367–374.

Rosenbaum, P. R. (1993). Hodges–Lehmann point estimates of treatment effect in observational studies. *J. Amer. Statist. Assoc.* **88**, 1250–1253.

Rosenbaum, P. R. (1994). Coherence in observational studies. *Biometrics* **50**, 368–374.

Rosenbaum, P. R. (1995). *Observational Studies.* Springer-Verlag, New York.

Rosenbaum, P. and A. Krieger (1990). Sensitivity analysis for two-sample permutation inferences in observational studies. *J. Amer. Statist. Assoc.* **85**, 493–498.

Rosenbaum, P. and D. Rubin (1983). The central role of the propensity score in observational studies for causal effects. *Biometrika* **70**, 41–55.

Rosenbaum, P. and D. Rubin (1984). Reducing bias in observational studies using subclassification on the propensity score. *J. Amer. Statist. Assoc.* **79**, 516–524.

Rosenbaum, P. and D. Rubin (1985a). Constructing a control group using multivariate matched sampling methods that incorporate the propensity score. *Amer. Statist.* **39**, 33–38.

Rosenbaum, P. and D. Rubin (1985b). The bias due to incomplete matching. *Biometrics* **41**, 106–116.

Rubin, D. B. (1973a). Matching to remove bias in observational studies. *Biometrics* **29**, 159–183. Correction: **30** (1974), 728.

Rubin, D. B. (1973b). The use of matched sampling and regression adjustment to remove bias in observational studies. *Biometrics* **29**, 185–203.

Rubin, D. B. (1974). Estimating the causal effects of treatments in randomized and nonramdomized studies. *J. Educ. Psychol.* **66**, 688–701.

Rubin, D. B. (1977). Assignment to treatment group on the basis of a covariate. *J. Educ. Statist.* **2**, 1–26.

Rubin, D. B. (1978). Bayesian inference for causal effects: The role of randomization. *Ann. Statist.* **6**, 34–58.

Rubin, D. B. (1979). Using multivariate matched sampling and regression adjustment to control bias in observational studies. *J. Amer. Statist. Assoc.* **74**, 318–328.

Sartwell, P. (1960). On the methodology of investigations of etiologic factors in chronic diseases: Further comments. *J. Chronic Dis.* **11**, 61–63.

Schlesselmann, J. J. (1978). Assessing the effects of confounding variables. *Amer. J. Epidemiol.* **108**, 3–8.

Smith, A., J. Kark, J. Cassel and G. Spears (1977). Analysis of prospective epidemiologic studies by minimum distance case-control matching. *Amer. J. Epidemiol.* **105**, 567–574.

Susser, M. (1973). *Causal Thinking in the Health Sciences: Concepts and Strategies in Epidemiology.* Oxford Univ. Press, New York.

Thun, M. (1993). Kidney dysfunction in cadmium workers. In: K. Steenland, ed., *Case Studies in Occupational Epidemiology.* Oxford Univ. Press, New York, 105–126.

Thun, M., A. Osorio, S. Schober, W. Hannon, B. Lewis and W. Halperin (1989). Nephropathy in cadmium workers: Assessment of risk from airborne occupational exposure to cadmium. *Br. J. Ind. Med.* **46**, 689–697.

Ury, H. K. (1975). Efficiency of case-control studies with multiple controls per case: Continuous or dichotomous data. *Biometrics* **31**, 643–649.

Wold, H. (1956). Causal inference from observational data. *J. Roy. Statist. Soc.* 28–50.

Yerushalmy, J. and C. Palmer (1959). On the methodology of investigations of etiologic factors in chronic diseases. *J. Chronic. Dis.* **10**, 27–40.

S. Ghosh and C. R. Rao, eds., *Handbook of Statistics, Vol. 13*

7

Robust Design: Experiments for Improving Quality

David M. Steinberg

1. Introduction

Robust design refers to quality engineering activities whose goal is the development of low-cost, yet high quality products and processes. Special emphasis is placed on reducing variation and the name "robust design" derives from the idea of making products insensitive, or "robust", to the effects of natural variations in their production and use environments. A key tool in robust design is the use of statistically planned experiments to identify factors that affect product quality and to optimize their nominal levels. This chapter will review the planning, analysis, and interpretation of robust design experiments and the role they play in quality improvement, with special emphasis on issues relating to experimental design.

Robust design was pioneered by Genichi Taguchi in Japan and is the central part of an innovative quality engineering strategy he developed to design quality in to products and processes. Taguchi began his work in the 1950's at the Electrical Communications Laboratories of Nippon Telephone and Telegraph Company, where he trained engineers in the use of effective research and development techniques. In 1962 he received the Deming Award for his contributions to quality in Japan. His ideas won many converts there, but remained virtually unknown in the United States and in Europe until 1980, when he visited AT&T. Several projects initiated during that visit led to major quality improvements (see, for example, Phadke, Kackar, Speeney and Grieco, 1983) and generated further interest at AT&T and at other companies, including Xerox, IT&T and Ford. Within just a few years, Taguchi's ideas were being taught to large numbers of engineers and applied in major companies throughout the world.

Taguchi's quality engineering scheme is based on a three stage model for development:

- *System Design*, in which a working prototype is developed;
- *Parameter Design*, in which the prototype is optimized by finding nominal levels of system parameters (factors) that maximize quality without excessively increasing cost;
- *Tolerance Design*, in which further quality improvement is achieved by scientifically tightening some tolerances, despite possible cost increases.

As Byrne and Taguchi (1989) observed, "the key element for achieving high quality and low cost is the stage called parameter design" (p. 75). "In the United States, most engineers ... jump from system design to tolerance design, omitting the step where they are likely to have the most to gain in terms of cost and quality. It is this step which the Japanese do so well – parameter design" (pp. 63–64). (Note that Taguchi uses the term *parameter design experiments*, since their purpose is to find settings for design parameters, rather than the more evocative term robust design experiments that I use here.) Robust design experiments are intended for use when products and processes are first being designed. However, many of the techniques can also be used to trouble shoot and improve existing processes.

Statistical methods, in particular designed experiments, play an important role in Taguchi's quality strategy. However, the reader is advised to bear in mind that Taguchi's approach is guided by *engineering* ideas, not statistical ones. The statistical methods involved are tools used to achieve engineering goals. For example, it seems evident that a key feature in guiding the design and analysis of robust design experiments has been the desire for simple application and interpretation, even when more informative analyses might be used. Simplicity makes the approach more accessible to engineers and thus facilitates widespread use of the methods.

Many engineers have enthusiastically embraced robust design and numerous successful applications have been reported. Yet robust design experiments have also generated controversy, confusion, and criticism, especially within the statistics community. Adherents of the Taguchi school have often downplayed the relevance of traditional statistical methods for industrial quality improvement. The fact is that many of the successful applications (for example, Quinlan, 1985) have used experimental plans long advocated by statisticians. The Taguchi school has been dogmatic in rejecting recent research that has shown that some of Taguchi's statistical methods are inefficient. Unfortunately, those arguments have led some statisticians to ignore Taguchi's important quality engineering ideas. See the recent essay by Greenfield (1995) in the February issue of News and Notes of the Royal Statistical Society, and the letters in the April, 1995, issue, for some interesting debate and especially sharp criticism.

My own view is that both statisticians and Taguchi devotees share a common goal in promoting the use of factorial experiments for improving quality in industry. Engineers who run factorial experiments will learn faster than those who do not. It is to the credit of Professor Taguchi that his approach, and his success in communicating it to engineers, has greatly increased the use of factorial experiments in industry. That said, there are certainly areas where methods can be improved and where continued statistical research will be valuable. It is unfortunate that bickering has at times overshadowed the shared interests and inhibited cooperative advance.

The quality engineering context in which Taguchi developed his approach led him to frame some new and important questions on how to effectively use experiments to promote quality improvement. This chapter will focus on several unique features of robust design experiments that have contributed to the great interest in them and served as the stimulus for much recent statistical research:

1. Variance is an important quality characteristic and specialized experimental plans are used for studying factor effects on variance. Since variance reduction can often be equated to quality improvement, this goal was ideally suited to the growing emphasis on quality in Western industries in the 1980s.

2. Experimental factors are classified into different types: design, signal, and noise, depending on the role they will play in actual production and product use. The experimental strategy is closely tied to the factor classification.

3. Orthogonal arrays are used to generate experimental plans. Often three levels are used for experimental factors.

4. Interactions are treated in a non-standard way, which has been a source of confusion and misunderstanding.

5. Taguchi has advocated some original methods of data analysis. Statistical research has shown that alternative methods of analysis are often superior.

6. Factorial experiments have been used to study purely deterministic systems in what have come to be known as "computer experiments". (For more detail, see the chapter on computer experiments in this volume.)

7. The methods have achieved widespread use in industry. In particular, they have made inroads in many engineering groups where use of statistical techniques had not previously enjoyed popularity.

The chapter is organized as follows. Section 2 provides perspective, showing how robust design experiments fit into the overall scheme of quality improvement. Section 3 presents an overview of robust design experimentation and Section 4 presents several examples that illustrate the ideas. Section 5 discusses the design and analysis of experiments that involve only design factors. Section 6 shows how the innovative idea of using *noise factors* can help to identify design factors that affect process variance and Section 7 discusses the use of *product array* experiments when noise factors are included. Section 8 considers some additional issues related to the design of experiments. Section 9 describes experiments that include *signal factors*. Some concluding remarks are made in Section 10.

This review chapter stands out from the several books on robust design that have appeared in recent years in two main respects. First, since the chapter is directed toward statisticians who are familiar with standard methods for the design and analysis of factorial experiments, it is limited to robust design experiments and emphasizes their special features. Second, the chapter presents both Taguchi's ideas and alternative suggestions, including the most recent proposals. I have tried to keep the presentation even-handed, pointing out the advantages and pitfalls of the various approaches. Most of the books have been written primarily for engineers and devote more space to the basics of factorial design than to robust design experiments. For example, the comprehensive two volume work on design by Taguchi (1987) includes many additional topics in experimental design. Only a small fraction of the book is devoted specifically to robust design experiments. Other books with broad coverage include Grove and Davis (1992), Lochner and Matar (1990), Logothetis and Wynn (1989) and Ross (1988). The book by Phadke (1989) also assumes limited background in experimental design but is much more focused on robust design experiments than are the books listed above. Phadke's book may be especially useful to statisticians because of its excellent discussion of the engineering context that motivates the use of robust design experiments. The treatment of robust design experiments in all of these books largely follows the methods proposed by Taguchi. Discussion of the ideas and philosophy of robust design with rather limited technical detail can be found in

Ealey (1988) and Peace (1993). Several collections of articles have appeared in book form (Bendell et al., 1989; Dehnad, 1989; Ghosh, 1993) and many interesting case studies can be found in the annual symposia proceedings published by the American Supplier Institute. The panel discussion edited by Nair (1992) is highly recommended for interested readers. Some background in robust design experiments (such as that provided in this chapter) is necessary to appreciate the discussion, which includes the views of a number of experts, some supporters and others critics of Taguchi's ideas. As might be expected, the discussion focuses more on points of controversy than on those of consensus, so some important topics are not presented at length. On those topics that are included, the panel discussion provides excellent insight into many of the ideas and methods surrounding robust design experiments.

2. Robust design and the quality revolution

Robust design experiments entered the statistical lexicon in the 1980's, following Genichi Taguchi's visit to the US. That visit came during a period of immense interest in quality control and improvement among Western industries and the excitement generated by robust design experiments is due in no small part to its role in the broader context of the "quality revolution". High quality of products and services has emerged in the last 15 years as one of the vital ingredients for success in competitive, global markets. In particular, superior quality has been recognized as one of the major reasons for the success of many Japanese companies (Hayes et al., 1988, Chapter 1) in sectors such as home electronics, automobiles, and microprocessors. In the late 1970's and early 1980's, as many Western companies began to lose market shares to Japanese competitors, the quality gap received much publicity and led to an explosion of interest in quality among Western industries.

Efforts to quality improvement have involved a broad spectrum of ideas. The genius of leaders like Deming and Juran, who helped catalyze the quality revolution in Japan, was their deep understanding of the symbiosis of management practices and technical tools that is needed to achieve high quality. Neither, by itself, is likely to succeed. The technical tools are needed to provide an objective analysis of systems and processes rooted in application of the scientific method. But they will be effective only if management is committed to quality and has created an environment of trust among employees that encourages open and constructive analysis and innovation. Deming (1982, 1986) has emphasized the crucial responsibility of management for achieving quality improvement. In Deming's view, product quality is primarily determined by the characteristics of the production system. Thus most quality improvement requires systemic changes which only management can direct and decide to implement. When and how to change the system require knowledge that can best be gained by careful data collection and analysis, so that statistical methods are essential tools in the improvement process. See Box and Bisgaard (1987) for a lucid account of the role of statistics in quality improvement.

Taguchi's quality engineering ideas mesh very well with Deming's quality improvement philosophy. Underlying Taguchi's approach is an assumption that poor quality can be measured by a loss function that increases monotonically with the distance of quality characteristics from their target values. He typically assumes that loss is quadratic, at least as a reasonable approximation, which immediately implies that high quality is achieved by minimizing variance around target values. By contrast, many engineers look only at the percent of product that meets specification limits, which is tantamount to adopting a 0–1 loss function. Taguchi's approach shows that continual improvement is possible, even if all production meets specification limits. Management decisions that define products and processes should design quality into the product, minimizing the need for subsequent inspection, rework, and screening of defects. Taguchi and Clausing (1990) summarized the single most important message of Taguchi's principles: "Quality is a virtue of design" (p. 65). Robust design experiments can provide the knowledge necessary to make effective design choices by discovering how a large number of input factors affect product quality. The rapid feedback possible from designed experiments enables engineers to accelerate the learning curve, reducing development time and bringing high quality products to the market ahead of the competition.

3. Overview of robust design experiments

Figure 1, adapted from Phadke (1989), shows a schematic diagram for robust design experiments. The focus of the experiment is on a quality characteristic Y which is affected by a variety of factors. The factors are classified into three types: Some of these, which I will call *design factors*, can be set by engineering specifications. These may represent the raw materials and equipment used to make the product, assembly instructions, settings of process parameters, etc. Other factors, known as *noise factors*, cannot be controlled. Noise factors may describe the use environment of the product or variable elements of the production process. For example, the performance of an automobile engine may depend on ambient temperature (use environment) and on the deviation of a machined part about its nominal value (production variation). In principle, of course, better control of the production process could eliminate the latter factors. But such control will often be expensive. Making the process "robust" to their effects may be a much more cost effective strategy. It is exactly in this way that robust design experiments achieve high quality without increasing costs. A *signal factor* is a factor whose value is set by the user of the product to obtain a desired response. Typical examples are the shutter speed on a camera and the pressure exerted on a car's brake pedal.

I will denote the design factors by X, the noise factors by Z, a signal factor by M and the random error by ε. Then a general mathematical model for the response is

$$Y = h(X, Z, M, \varepsilon). \tag{1}$$

All robust design experiments will include design factors. Noise factors and a signal factor may also be included. Physical experiments will have a random error component, but it will be absent in computer experiments.

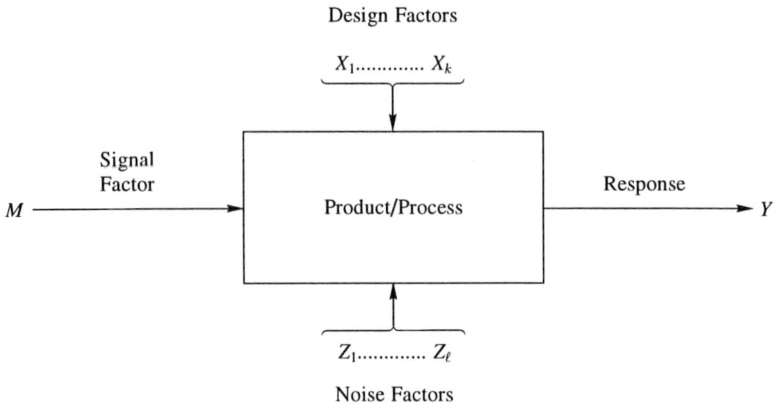

Fig. 1. Block diagram of product/process for a robust design experiment.

The classification of factors into these different categories plays an important role in robust design experimentation. Especially important is the realization that much of the random variation observed in product performance may be attributable to specific factors. Standard statistical models for experimental data include only design factors and a random error. In robust design, the variation is decomposed into two parts, one corresponding to noise factors that vary during production or use, and the other corresponding to further residual variation. One of the unique features of robust design is to systematically vary noise factors that can be controlled for the purposes of the experiment.

3.1. Experimental goals

The goal of a robust design experiment is to find settings of the design factors that achieve a particular response with high consistency. The most common objectives are:

- maximize the response,
- minimize the response,
- keep the response on target.

In all cases, especially the last, low variance is also desired. Including noise factors in an experiment helps achieve that goal. Since noise factors vary naturally in production or in product use, they transmit variation to the response Y. Thus the variation of Y can be reduced by finding settings of the design factors that reduce the effects of the noise factors and so minimize the variation transmitted by the noise factors. This search is facilitated by building the noise factor variation directly into the experiment. Statistical process control, by contrast, would likely adopt the much costlier strategy of directly reducing the variation of the noise factor.

To reduce variance, it will be important to identify factors that affect variance. Such a factor is said to have a *dispersion effect*. (Some authors use the term *control factors* to refer to factors with dispersion effects. Other authors use control factors as a synonym for design factors.) When the goal of the experiment is to keep a response on target with minimal variance, the hope is to set the design factors with dispersion effects so as to minimize variance, then to set the mean level on target with an *adjustment factor* that has a dominant effect on mean level. Sometimes engineering considerations suggest an obvious adjustment parameter. For example, to adjust the mean level of a layer of silicon deposited on a semi-conductor, the obvious factor to vary is the amount of silicon used by changing either the deposition rate or the deposition time. Otherwise, one may attempt to identify an adjustment factor from the experimental results.

When there is also a signal factor, the goal is typically to obtain a particular response curve, or *ideal function*, that expresses the desired dependence of Y on the signal. Taguchi uses the term *dynamic problems* for such engineering tasks and I will discuss them in Section 9. For example, the braking system of a car might be designed to achieve an ideal engineering relationship between force on the brake pedal and actual braking force. Consistency is again an important issue – the brakes should respond the same way regardless of the road surface, the amount of air in the tires, and any production tolerances.

3.2. Selecting quality characteristics and factors

Good engineering input is essential to the success of any industrial experiment. The selection of what response variables to record, what factors to vary, and over what ranges to vary them rely much more on engineering judgment than on statistical considerations. An engineer who selects the important factors is likely to find quick answers to her questions. A poor choice of factors is unlikely to provide a successful engineering solution, no matter what the statistical efficiency of the experimental plan or the ingenuity of the data analysis. Process analysis tools like fishbone diagrams and flow charts can be very helpful in guiding these choices. Team efforts, with open group discussion, are highly advisable. For an excellent discussion of these general issues in experimental planning, see Box (1992) and Coleman and Montgomery (1993). Nair (1992, Sections 1, 2, 3 and 5) contains much good advice on planning robust design experiments.

I will mention briefly some of the more statistical issues that should be considered. Response variables should be chosen to measure the actual process output. The most effective variables are typically continuous measures of physical properties. Measures of a go/no go variety are common but are much less informative. Consider, for example, an experiment to improve a machining process for a part with specifications of 49.9–50.1 centimeters. It may be easy to obtain measurements on the percent of out of spec parts. But it is important to know if the parts are out of spec because the average is too high or low, or because the variance is too large. That information can be obtained if actual length measurements are recorded.

Phadke (1989, Chapter 6) presented a number of interesting examples that provide insight into the choice of a response characteristic. In designing the automatic feed mechanism for a photocopier, for example, an important quality characteristic may be the mean time between feed failures. However, since failures are a rare occurrence, mean time between failures is not an informative response characteristic: large volumes of copying must be done to discern differences in mean time between failures. Examination of the engineering context suggests that most feed failures occur when the force applied to the next sheet is either excessive or insufficient. One can learn about and improve the feed mechanism much faster by measuring the force and reducing its variation about a target value.

On what scale should the response be measured? Ideally one would like to have a scale whose implications for quality are direct and easy to understand and on which the the relation between the response and the experimental factors is simple and parsimonious. Engineering considerations are important in both issues. The second can also be explored empirically. For example, it may be possible to achieve substantial simplification (and with it increased understanding) by making a transformation of Y. See Box and Cox (1964), Box (1988) and Logothetis (1990).

Which factors should be chosen? The major question to consider is "what drives the system under study?" An accurate answer to that question should suggest how to specify factors that will provide the most direct explanation of the responses. In the machining study, for example, two potential factors might be the pressure setting on the lathe and the machining time. However, it seems likely that the size of the finished part will be more closely related to the total pressure exerted, i.e., to the product of pressure and time. So it might be better to use the product as a single factor, or to use it in tandem with either time or pressure.

4. Examples

This section briefly reviews some applications of robust design experimentation.

4.1. INA Tile

Probably the best known example of a robust design experiment was performed at the INA Tile company in Japan in the 1950's. Many tiles failed to meet Japan's strict specification limits and had to be scrapped. An initial study suggested that the source of the problem was uneven heat distribution in the kiln where the tiles were baked: the size distribution of tiles at the periphery of each batch differed substantially from that of tiles inside the batch. As Kackar (1985) noted, a conventional process quality approach might call for improving control of the heat distribution in the kiln, a costly solution to the problem. Instead, INA Tile sought to identify, experimentally, process conditions for which the tile size distribution would be "robust" to the variations in kiln heat. The design factors they studied are listed in Table 1 and the experimental data are presented in Table 2. The response variable was the percentage of tiles outside specification limits. (It is curious that percent out of spec was used as the

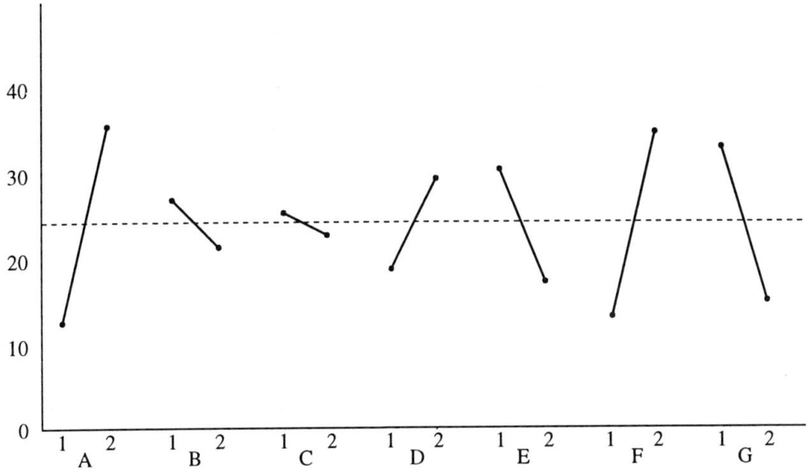

Fig. 2. Factor effects for the INA Tile experiment. The plot shows the average percent of defective tiles at each level of each factor. The dashed line running across the lot is the overall average.

Table 1
Design factors and levels of INA Tile experiment. Initial factor levels are marked*

Factors	Level 1	Level 2
A: Contents of certain lime	5%	1%*
B: Fineness of the additive	Coarser*	Finer
C: Content of agalmalotite	43%	53%*
D: Kind of agalmalotite	Current*	New
E: Raw material charging quantity	1300 kg	1200 *
F: Content of waste return	0%	4%*
G: Content of feldspar	0%	5%*

response variable here; the actual dimensions of the tiles would have been a much better choice.) A graphical summary of the factor effects is given in Figure 2. The major process change was to increase the lime content of the tiles. When each factor was set to its best experimental level, the scrap rate dropped to near 0. The quality improvement was achieved at almost no cost at all!

4.2. *Speedometer cable*

Quinlan (1985) described an experiment to reduce the shrinkage in plastic casing for speedometer cables produced for the automotive industry. Excessive shrinkage had been a problem for some time and, despite the concerted efforts of quality teams, no effective solution had been found. A team of engineers identified 15 design factors and planned a 16 run fractional factorial experiment. At each experimental setting,

Table 2
Data from the INA Tile experiment. The response is the number of defective tiles out of 100

A	B	C	D	E	F	G	No. of defectives
1	1	1	1	1	1	1	16
1	1	1	2	2	2	2	17
1	2	2	1	1	2	2	12
1	2	2	2	2	1	1	6
2	1	2	1	2	1	2	6
2	1	2	2	1	2	1	68
2	2	1	1	2	2	1	42
2	2	1	2	1	1	2	26

3000 feet of plastic casing were produced and four sample specimens were cut out. Then each specimen was heat treated and its percent shrinkage measured. Quinlan (1985) identified eight factors as having significant effects on shrinkage, noting that the type of wire braid (factor E) and the wire diameter (factor G) had the most important effects. Quinlan (1985) reported that setting all the significant factors to their optimal levels resulted in a large decrease in percent shrinkage. Subsequent analysis by Box and Bisgaard (1987) (also Box, 1988) suggested that wire braid and diameter were the only significant factors. So in this example, although there are some questions about the statistical conclusions from the experiment, a very effective engineering solution was found.

4.3. Diesel engine performance

Baranescu and Ginder (1988) described an investigation of a diesel engine. Their study was carried out by computer simulation rather than by physical experimentation. To physically vary engine parameters would require expensive retooling of production equipment. The computer simulation offers great economy. A number of different performance characteristics were of interest in this study, including power output, fuel economy, and emission of pollutants. Brainstorming among the engineers led to the selection of five engine parameters for inclusion as design factors in the experiment: turbocharger build, intake flow area, exhaust flow area, intake valve, and exhaust valve. Four key noise factors were included in the experiment: ambient temperature and fuel input (use conditions) and compression ratio and injection timing (which were subject to manufacturing tolerances). Separate experimental plans were laid out for the design factors (an 18 run orthogonal array) and the noise factors (a 9 run orthogonal array). The full experiment tested each of the 18 engine configurations at each of the 9 noise configurations, for a total of 162 runs. The results indicated that substantial improvements could be made by modifying the exhaust and intake valves.

4.4. *Differential operational amplifier circuit*

Phadke (1986; also 1989, Chapter 8) described a study conducted at AT&T Bell Laboratories to improve the quality of a differential op-amp circuit. The goal was to identify factor settings at which the offset voltage would be consistently close to 0.0 mV. The offset voltage is a function of current sources, resistors, transistors, and other electrical components. Two current sources and three of the resistors could be chosen independently by the circuit designer and served as design factors. The remaining parameters were then specified by simple linear relations dictated by the nature of the circuit. Given the values of all the parameters, the offset voltage could be computed exactly from laws of current flow. It might appear that there is no problem here: just specify the values so that the solved offset voltage is zero. In fact, all the circuit parameters could be specified only to within some tolerance of their nominal values. Variations about the nominal values created fluctuations in the offset voltage and it was important to minimize the extent of these fluctuations, i.e., to make the offset voltage robust to the input tolerances. The tolerances were represented in the experiment by noise factors.

Circuit performance was studied by crossing a 36 run array in the design factors with a 36 run array in the noise factors, for a total of 1296 runs. For each of these design points, the settings in the design array and the circuit constraints were used to determine the nominal value of each parameter. Then the noise array was used to convert the nominal values to actual values, in accord with the tolerance distributions. For example, suppose a resistor had two nominal values in the design array, 1,000 Ohms and 15,000 Ohms, and a tolerance of $\pm 1\%$. Since most resistors will not be at the extreme ends of the tolerance interval, one might represent actual conditions by, say, a two-level noise factor whose levels are $\pm 0.5\%$. When the design array called for a 1000 Ohm resistor, the values actually used for the resistor in the simulation would be 995 Ohms (99.5% of nominal) and 1005 Ohms (100.5% of nominal). When the design array called for a 1000 Ohm resistor, the corresponding values were 14,925 Ohms and 15,075 Ohms. Varying the input parameters across their tolerance ranges thus "built in" the variability in offset voltage that would be expected once the circuits went into full production.

The experiment identified changes in nominal values that led to a 63% reduction in the mean square offset voltage.

5. Experiments with design factors only

Many experiments run within the robust design framework do not use either noise or signal factors. That is true in particular when the goal is to minimize or maximize a particular response and the study of variability is of secondary importance. For example, the experiment at INA Tile (see Section 4) used a standard two-level fractional factorial design and the analysis was based on computation of main effects on the percentage of defective tiles. Moreover, the standard methods of analysis suggested by Taguchi take no account of noise factors even when they are included in the experiment and many of the alternative analyses that have been proposed are directed toward experiments without noise factors. The analysis of these experiments has been the subject of considerable debate so, before going on, it will be useful to focus on the analysis questions. The following two sections will then discuss in detail the use of noise factors. Section 9 will extend that discussion to experiments that also include signal factors.

5.1. *Experimental plans for design factors*

Taguchi (1987) advocated the use of orthogonal array (OA) plans of strength 2 for design factors and tabled many OA's to facilitate their use. The defining characteristic of these plans is pairwise balance: for each pair of factors, the possible combinations of their levels appear the same number of times. Thus, when projected down onto any pair of factors, an OA of strength 2 is a full factorial with equal replication in each cell. OA's were first described by Rao (1947) and include two-level fractional factorials (Finney, 1945), Plackett and Burman (1946) designs, and Graeco–Latin squares. For more detail, see the chapter on OA designs in this volume.

Taguchi denotes these plans by L_n, where n stands for the number of rows (i.e., the number of experimental runs) in the array. Table 3, adapted from Phadke (1989),

Table 3
List of standard orthogonal array designs. For each design, the table lists the number of rows (experimental points), the maximum number of factors that can be tested, and the maximum number of factors at each of 2, 3, 4 and 5 levels

Orthogonal array	Number of rows	Maximum number of factors	Maximum number of factors by number of levels			
			2	3	4	5
L_4	4	3	3	–	–	–
L_8	8	7	7	–	–	–
L_9	9	4	–	4	–	–
L_{12}	12	11	11	–	–	–
L_{16}	16	15	15	–	–	–
L'_{16}	16	5	–	–	5	–
L_{18}	18	8	1	7	–	–
L_{25}	25	6	–	–	–	6
L_{27}	27	13	–	13	–	–
L_{32}	32	31	31	–	–	–
L'_{32}	32	10	1	–	9	–
L_{36}	36	23	11	12	–	–
L'_{36}	36	16	3	13	–	–
L_{27}	27	13	–	13	–	–
L_{50}	50	12	1	–	–	11
L_{54}	54	26	1	25	–	–
L_{64}	64	63	63	–	–	–
L'_{64}	64	21	–	–	21	–
L_{81}	81	40	–	40	–	–

lists some common OA designs and the maximum number of factors that each can accommodate with a given number of levels. The L_{18} array, for example, can be used to plan an experiment with one two-level factor and as many as 7 three-level factors. In addition to the designs mentioned above, Taguchi's tables include designs that were derived by Fisher (1945) (L_{27}), Burman (1946) (L_{18}) and Seiden (1954) (L_{36}).

There seem to be two major arguments in support of using OA designs of strength 2. First, these designs enable one to study a large number of factors, with two or more levels, in a relatively small number of runs. When some factors have 3 levels, in particular, OA plans are often smaller than are fractions of the 3^k series. For most industrial experiments, many potential design factors exist and hard decisions must be made about which to include in the experiment, which to hold constant, and which to ignore. There is a pronounced tendency in robust design experiments to maximize the factors to runs ratio. Obviously, such a strategy increases the chance that the most important factor(s) will be among those included in the experiment. The second advantage of OA designs is that the orthogonality of main effects facilitates very simple analysis. The major drawback to OA designs of strength 2 is their inability to estimate interactions. I will discuss these issues further and also explore alternative experimental plans in Section 8.

5.2. Replicates and dispersion effects

Since an important goal of many robust design experiments is to identify dispersion effects and use them to reduce variance, it is common to include replicate observations at each design factor combination in the experiment. The replicates can be used to compute sample variances at each design point, which can then be related to the design factors. In Sections 7 and 8 I will discuss designs in which replicates are generated by deliberately varying noise factors. For now, though, I will assume that "simple replicates" have been obtained by repeat trials at the design factor combinations. Box and Meyer (1986) presented a method for estimating dispersion effects from an unreplicated design when the number of location effects is small.

5.3. Analysis and optimization by signal to noise ratio

Taguchi (1987) has proposed the following paradigm for analyzing robust design experiments and improving product/process quality:

1. Summarize the data for each design factor combination by calculating one or two summary statistics, usually called *Signal to Noise* (SN) *Ratios*.
2. Treat the summary statistics as response variables and analyze their relation to the design factors.
3. Set factors that affect the SN ratio to optimal levels based on their main effects.
4. If the goal is to achieve a target value, use one of the design factors to *adjust* the mean value to the target.
5. Make a *confirmation run* at the proposed optimal settings to verify the improvement.

Once the response variables have been defined, their analysis proceeds in a straightforward way. With OA designs, main effects can be assessed by simply comparing the averages at each level of a factor. It is common to graph the level averages for each factor in a single plot, like Figure 2, which provides quick insight into which factors have large effects. Taguchi generally recommends ignoring interactions between pairs of design factors unless they were anticipated in advance and the experimental plan has been specifically set up to accommodate them. (See Section 8.2 for discussion of interactions.) Analysis of variance (ANOVA) is used to summarize the results, to assess the percent contribution of each factor to the overall variance of Y, and to test the statistical significance of the factor effects. In many of the examples presented by Taguchi, small sums of squares were pooled to form an overall error sum of squares for the F-tests. Box (1988) and others have pointed out that this pooling procedure induces a "selection bias" and can result in an excessive number of "significant" effects.

Taguchi has devised a large number of SN ratios for use with differing experimental objectives. There has been considerable debate about the usefulness of analyzing SN ratios; see, for example, Box (1988), Box et al. (1988), León et al. (1987), Nair (1992), Shoemaker et al. (1991), Steinberg and Bursztyn (1993, 1994), Welch and Sacks (1991) and Welch et al. (1990). These articles have pointed out the following problems:

1. The blanket use of a performance criterion is unwise. Suitable criteria should be closely adapted both to the engineering context and to the statistical model for the experiment.
2. The SN ratios often do not efficiently summarize the information in the data.
3. The use of data transformations to simplify the analysis and improve understanding is not exploited.
4. The SN ratios take no account of the special structure of robust design experiments that include noise factors. When noise factors are used to generate replicates, the analysis should explicitly model their effects, rather than computing summary measures across the noise array.
5. Taguchi's approach encourages "one-shot" experimentation, in which a single experiment should lead to the optimum solution. Instead, many problems can be solved more successfully by sequential learning, in which a sequence of several experiments is used to identify the most important factors and converge toward their optimal settings, adapting the complexity of the designs and fitted models to match the degree of local approximation that is needed.

The following subsections will discuss alternative approaches to data analysis that take account of these criticisms. Experiments with noise factors will be the main topic of Sections 6 and 7.

Although I agree with the critics, I will present and discuss the most common SN ratios for completeness. Let y_j ($j = 1, \ldots, m$) denote the observations for a particular combination of the design factors and let $\bar{y}$ and s denote their average and standard deviation.

Goal	SN ratio
Minimize response	$\text{SN} = -10\log(\sum y_j^2/m)$
Maximize response	$\text{SN} = -10\log(\sum(1/y_j^2)/m)$
Response on target	$\text{SN} = 10\log(\bar{y}^2/s^2)$

All of the SN ratios should be maximized. Some articles use a slightly modified version of the SN ratio for "response on target" problems,

$$\text{SN} = 10\log\left[(\bar{y}^2 - s^2/m)/s^2\right]. \tag{2}$$

The modification makes the numerator (inside the log) an unbiased estimator of μ^2.

When the goal is to minimize Y, the motivation for the SN ratio is that it measures expected loss on a log scale. Since $\sum y_j^2/m = \bar{y}^2 + (m-1)s^2/m$, the SN ratio will favor design factor combinations that have low mean and low variance. The SN ratio for maximizing Y is justified by noting that maximizing Y is equivalent to minimizing $1/Y$, then using the previous SN ratio. In many examples, these SN ratios are highly correlated with $\bar{y}$, so that analyzing the SN ratios is essentially equivalent to analyzing the averages.

The response on target SN ratio is designed to assess dispersion effects. Analysis of this SN ratio is accompanied with an analysis of $\bar{y}$ to determine which design factors have location effects. The SN ratio is predicated on the assumption that, as a rule, dispersion will be an increasing function of mean level. If this is the case, then using s, say, as a dispersion summary would lead to the conclusion that the design factors should be set to levels that minimize the *mean*. But then most variance reduction will be due to the change in mean and will be lost when the mean is adjusted back to its target value. To prevent falling into this trap, simple dependence of the standard deviation on the mean level is removed by taking their ratio. The SN ratio is thus constructed to detect dispersion effects that are free of the relation to mean level and will preserve a reduction in variance even after the mean value is adjusted to target. See Section 5.6 for a more formal discussion of why it might be useful to use this SN ratio to summarize dispersion.

5.4. Optimizing design factor settings

Taguchi's strategy then calls for using the results of the analysis to select improved settings for the design factors. The most important contribution here is his strategy for "on target" problems, in which the goal is to obtain a process whose mean value attains a fixed target and which has minimal variance about the target. The clever insight in Taguchi's strategy is that variance reduction is usually much more difficult than mean level adjustment. Thus the first step is to find design factors with dispersion effects and set them to levels that will minimize the variance. Common engineering wisdom would reverse these steps, first adjusting the mean and then trying to reduce variance. Taguchi's prescription for minimizing variance is to set design factors that have large effects on the above SN ratio to that experimental level that had the largest average SN. Once these factors have been set, the mean level should be adjusted to the target using an *adjustment factor*, a design factor that affects the mean but has little or no dispersion effect. (The term "signal factor" has sometimes been used here, but "adjustment factor" is much more evocative and avoids confusion with the term signal factor in dynamic problems; see Section 9.) As noted earlier, a number of recent articles have cast doubts on the usefulness of analyzing the SN ratio for reducing variance; other analyses are preferable. But there is no doubt that Taguchi has made fundamental contributions in focusing attention on the importance of using experiments to reduce variation and through his two-stage optimization scheme.

When the goal is to minimize or maximize Y, the suggestion is to use whatever level had the largest average SN. Often the same recommendations would result by just analyzing the average responses.

5.5. *Confirmation experiment*

The final step in Taguchi's strategy is to perform a *confirmation experiment* in which all of the design factors are set to their recommended levels. This experiment enables the engineer to test whether the new levels actually improve the process/product. It is also intended as a test for interations among design factors. Since the initial design and analysis focus on main effects of the design factors, strong interactions may lead to sub-optimal settings for the design factors. The confirmation experiment may shed some light on this question by comparing the observed results to predicted values on the basis of the initial experiment.

5.6. *PerMIA's*

In "on target" problems the objective of a robust design experiment is to minimize mean squared error. But Taguchi's analysis, using the log of the squared coefficient of variation, makes no direct appeal to mean squared error. To help explain the connection, León et al. (1987) developed the concept of a *Performance Measure Independent of Adjustment* (PerMIA). Like Taguchi, they broke the minimization up into two steps, such that optimum values of some of the factors can be determined at the first step *independently* of the values of the remaining parameters. Factors in the second group are then used to adjust the process to minimum expected loss. The function that is optimized at the first step is called a PerMIA and its precise form will depend on the engineering context, the loss function and the statistical model assumed. Consider, for example, the "on target" problem with squared error loss and suppose that a reasonable model is

$$Y = \mu(X_1, X_2)\varepsilon(X_1), \tag{3}$$

where X_1 and X_2 are vectors containing the two sets of design factors and ε is a random error term whose distribution depends on X_1 but not on X_2. Then

$$\mathrm{Var}\{Y\} = [\mu(X_1, X_2)]^2 \mathrm{Var}\{\varepsilon(X_1)\}. \tag{4}$$

It is then easy to show that the expected loss can be minimized by first setting the factors in X_1 to minimize Taguchi's SN ratio and then setting the factors in X_2 to adjust the mean. Taguchi's SN is a PerMIA in this setting and his two-step optimization procedure will be effective. If, though, the error term were additive, not multiplicative, then $\text{Var}\{Y\}$ would be a PerMIA rather than SN. León et al. applied their ideas to a number of different problems, appealing to the engineering context to suggest appropriate loss functions and models and then deriving PerMIA's. They concluded that unquestioning use of Taguchi's SN ratio is not desirable.

Box (1988) generalized the above argument for "on target" problems. He assumed that mean level and variance are linked in such a way that

$$P(X_1) = \sigma^2(X)/\{f[\mu(X)]\}^2 \tag{5}$$

is a function of only a subset (X_1) of the design factors. Then $P(X_1)$ is a PerMIA and optimal values could be chosen first for X_1 and then for the remaining factors. A simple, but useful, special case is $f(\mu) = \mu^\alpha$. For $\alpha = 0$, $P(X_1) = \sigma^2(X_1)$ is the appropriate performance measure to optimize. For $\alpha = 1$, one obtains the squared coefficient of variation or, equivalently, Taguchi's SN ratio. These choices correspond to cases examined by León et al. that were mentioned above. Box (1988) also showed that, under these assumptions, the expected loss is

$$\{f[\mu(X)]\}^2 P(X_1) + [\mu(X) - T]^2, \tag{6}$$

where T is the target value. The expected loss is minimized by setting X_1 to minimize $P(X_1)$ and setting X_2 to adjust the mean to a level μ_0 that satisfies the equation

$$\mu_0 = T - f(\mu_0)f'(\mu_0)P(X_1). \tag{7}$$

He called the quantity $f(\mu_0)f'(\mu_0)P(X_1)$ the *aim-off factor* and noted that it will usually be small. The fact that adjusting the mean level to T does not necessarily minimize the expected loss was apparently first noted by Easterling (1987).

León and Wu (1992) developed a general theory for performance measures in robust design experiments. They derived some general conditions under which expected loss can be minimized in two steps, with the first step an unconstrained optimization of a PerMIA. They applied their theory to linear and quadratic loss functions in which the loss is not symmetric about the target value. Moorhead and Wu (1994) further extended this approach, showing how to apply it when the loss is linear on one side of the target but quadratic on the other side.

5.7. *Further issues in analysis*

The criticism of Taguchi's analysis paradigm has been accompanied by a variety of suggestions for alternative analyses. These have included: exploiting transformations, using different summary statistics, fitting response surface models, and using generalized linear models. We discuss some of these here.

One of the main points made by Box (1988) is that making a suitable transformation of the data can often lead to much simpler analyses and interpretations of robust design experiments. For example, in the setting described in Section 5.6, when $f(\mu) = \mu$, there may be great advantages to making a logarithmic transformation of the data. On the log scale, there will no longer be a functional relationship between mean and dispersion. Thus analyses of design factor combination averages and standard deviations can be used to identify which factors have location and dispersion effects, respectively. In general, it is useful to log the standard deviations before analyzing them (Bartlett and Kendall, 1946). The standard deviations of the logged data are in fact nearly a linear function of the coefficients of variation of the original data, so this procedure is essentially equivalent to an analysis of Taguchi's SN. Box concluded that indiscriminate use of SN is tantamount to a blanket assumption that a log transformation will be desirable. Box argued that other transformations must also be considered and showed how plots can effectively be used to identify a good choice from within the class of power transformations (Box and Cox, 1964). Logothetis (1990) presented an interesting example of the usefulness of these transformations in a robust design experiment.

When the goal is to minimize or maximize a response, Taguchi's SN ratio combines information on both the mean value and variance in a single performance statistic. However, as noted earlier, the mean value is typically quite dominant in determining the value of these summaries. Box (1988) argued that it would be more informative to examine mean value and variance separately.

Several authors have suggested alternative summary statistics for assessing dispersion effects and the use of more sophisticated techniques to model the dispersion summaries. Ghosh and Duh (1992) defined dispersion effects using ratios of variances at the factorial points and estimated them with corresponding ratios of sample variances. They also considered estimating interaction effects on dispersion and proposed a rule for using the effects to select design factor levels that minimize variance. Ghosh and Lagergren (1993) proposed several different dispersion summaries using different types of residuals. Ghosh and Lagergren (1990) and Rosenbaum (1994) investigated properties of the design factor arrays that permit unbiased estimation of dispersion effects that are defined as contrasts in the variances at the factor combinations.

Vining and Myers (1990) proposed the use of joint response surface models for the mean and the variance. The complexity of these models will, of course, depend on the design factor plan that has been used. Vining and Myers emphasized situations in which the design factor plan is large enough to enable estimation of a full second-degree model. They summarized the data at each design factor combination by their average and standard deviation, then modeled each as functions of the design factors, using least squares to fit the models. In fitting the coefficients for the mean value model, they advocated using weighted least squares, with weights that reflect differences in variance that were found from the model for the standard deviations. They then suggested selecting process settings via the dual response surface approach of Myers and Carter (1973), in which one response is optimized subject to the second response satisfying a constraint. Thus, for example, one could use this idea to minimize the estimated standard deviation subject to keeping the estimated mean value at a given target level. Chan and Mak (1995) proposed a similar approach, but included the estimated mean response and its interactions with the design factors as possible explanatory variables in modeling the logged residuals. The advantage of this model is its sensitivity to any gross dependence of dispersion on mean level. Estimated dispersion effects for the design factors will be adjusted accordingly. Lin and Tu (1995) presented a method for direct minimization of mean squared deviation from the target with a dual response surface model.

Nair and Pregibon (1988) advocated the use of generalized linear models to model the variance. They assumed that the data at each design factor combination have normal distributions, but may differ in both mean and variance. Then the sum of squared deviations at each design point, $\sum(y_j - \bar{y})^2$, has a scaled χ^2 distribution, with the scale proportional to the variance. Nair and Pregibon then assumed that, on a log scale, the variances are linearly related to the design factors and, possibly, their interactions. This setting corresponds exactly to a generalized linear model with a log link function. Nelder and Lee (1991) further developed the idea of using generalized linear models to assess dispersion effects. Rather than simply using sample averages, they computed the dispersion about the fitted mean value at each design point. The fitted mean values, in turn, are estimated by weighted least squares and so depend on the fitted variances. Thus it is necessary to invoke an iterative scheme in which each model is fit, conditional on the current fit for the other model, until the procedure converges. Engel (1992) added the idea that the variance at a design point might have two components: first, it might be proportional to some power of the mean level at that design point, second, it might have further dependence on the design factors above and beyond the simple relation to the mean. He showed how generalized linear models for the sample variances could be used to estimate both the power dependence on the mean value and the subsequent dependence on the design factors.

6. Noise factors and dispersion effects

In this section I explore the relationship between noise factors and dispersion effects and show why including noise factors in an experiment is an effective way to identify dispersion effects. A number of models linking the response Y to the design and noise factors will be considered. Of course, one never knows in advance what model will be appropriate for a system that is being studied and good models evolve from an interplay of theory, experiment, and data analysis. So my intent here is to describe some models that are potentially useful, not to argue for their unquestioning use. For convenience, I will write these models as equalities; in fact, they are really meant just as good approximations to the true relationship.

6.1. Interactions and dispersion effects

Consider first a simple setting with just one design factor X and one noise factor Z. The discussion here follows ideas in Shoemaker et al. (1991). Suppose a reasonable model for Y, perhaps after suitable transformation, includes main effects of both X and Z and their interaction. Then for fixed values of X and Z,

$$Y(X, Z) = \mu + \beta X + \alpha Z + \gamma XZ + \varepsilon. \tag{8}$$

The level of X can be set by engineering specifications, but Z will vary randomly in actual production or use; assume that it will have mean value 0 and standard deviation σ_Z. Then the distribution of Y for a given value of X will have

$$\mathrm{E}\{Y(X)\} = \mu + \beta X, \tag{9}$$

$$\mathrm{Var}\{Y(X)\} = (\alpha + \gamma X)^2 \sigma_Z^2 + \sigma_\varepsilon^2. \tag{10}$$

Thus X will have a dispersion effect if it *interacts* with Z (i.e., if $\gamma \neq 0$). Selecting a value of X for which $\alpha + \gamma X \approx 0$ effectively neutralizes the variation transmitted from Z to Y and makes the product/process robust to variations in Z.

The above argument is illustrated in Figure 3, which graphs $\mathrm{E}\{Y\}$ against Z for two values of X, as in the interaction plot from a two-level factorial design. The transmitted variation will be proportional to the slope of the line and so will be much smaller when $X = -1$ than when $X = 1$ in the figure. If X is a continuous factor, then further variance reduction may be possible by choosing a still smaller value for X (see Lorenzen and Villalobos, 1990).

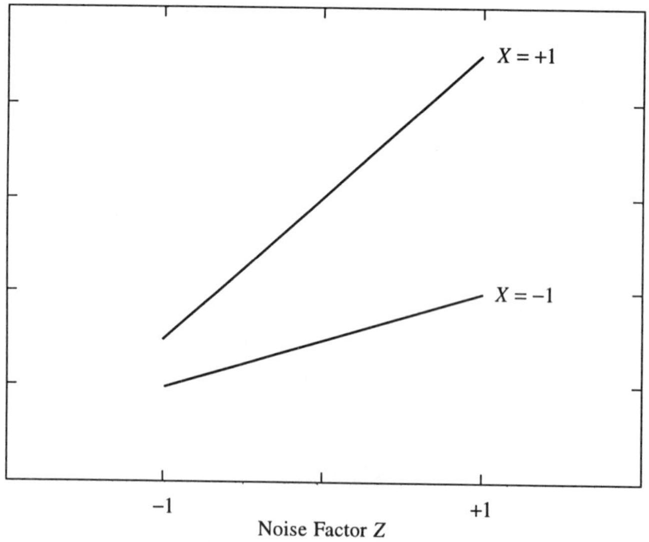

Fig. 3. Interaction plot for the design factor X and the noise factor Z, as might be generated from a two-level design. The variance transmitted to Y by the noise factor is proportional to the square of the slope. Thus the transmitted variation can be reduced by setting X to -1, where the slope is "flatter".

The use of noise factors in an experiment is an example of Taguchi's clever engineering insight. Without the noise factor, one would observe only the projection of Figure 3 on the Y axis, with somewhat different levels of variability at each level of the design factor. The crucial engineering question is to ask what is responsible for the variation in Y. If an appropriate noise factor Z can be found, we can then explain the variation in Y in terms of its regression on Z. In terms of Figure 3, this means obtaining the full two-dimensional plot rather than just its projection on the Y axis. The information in the full plot provides a much better basis for reducing variance.

Some of the assumptions made above deserve discussion.

1. The response function is a first order model with an interaction between the two factors. If the interaction were not present, then Var $\{Y\}$ would be independent of X. Were such a model appropriate, there would be no hope for reducing the variance by adjusting X. One might argue that the model should be expanded to a full quadratic in both factors. In that case, the process variance of Y has two additional terms. One of those terms is independent of X and the other is proportional to $\alpha + \gamma X$. So the key to variance reduction is still the presence of an $X \times Z$ interaction. Although the full second order model will give a better approximation to the response function, the simpler model is often just as effective for achieving variance reduction.

2. The distribution of Z is independent of the nominal level chosen for X. Whether or not this assumption is reasonable will depend on the particular factors in question. Of particular concern is the case when one of the noise factors is the tolerance associated with one of the design factors, which is discussed in more detail below.

3. The error standard deviation σ_ε does not depend on X. Certainly this assumption might be false. For example, there might be another noise factor, unknown to the experimenter, which affects Y and interacts with X. Such a possibility should be considered. However, Gunter (1988) and Carroll and Ruppert (1988) pointed out that, in a small experiment, dispersion effects associated with residual error variance are extremely difficult to detect unless they are quite large. Steinberg and Bursztyn (1993) showed that dispersion effects associated with design by noise factor interactions are much easier to detect. The intuitive explanation of this result is that it is easier to estimate means than variances. Including noise factors in the experiment converts an effect on variances (at the process level) to a regression interaction coefficient (at the experimental level). So the assumption that σ_ε does not depend on X is not meant to rule out dependence, but rather to focus attention on dispersion effects that are related to interactions.

6.2. Nonlinearity and dispersion effects

Dispersion effects also arise when the response is a nonlinear function of design factors that are subject to manufacturing tolerances. Consider a single input X whose nominal level is X_0 and suppose that

$$Y(X) = g(X) + \varepsilon. \tag{11}$$

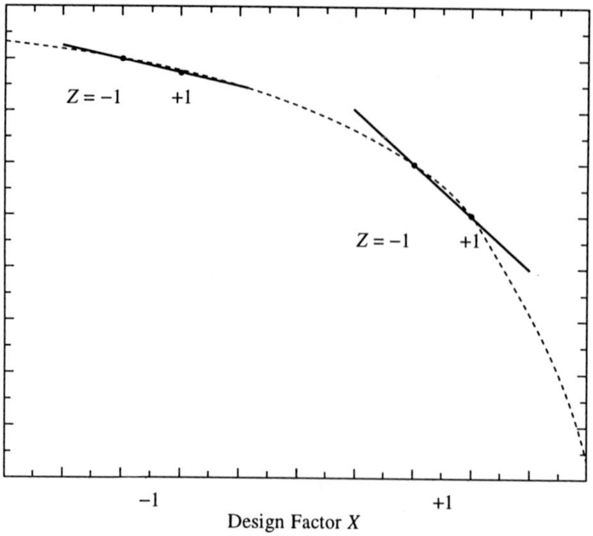

Fig. 4. Nonlinearity and dispersion effects. The response Y depends in a nonlinear way on the input factor X. Variation will be transmitted to Y if X varies about its nominal value. The transmitted variation can be reduced by setting X to -1, where the slope is "flatter". This situation can be thought of as an "interaction" between a design factor that represents the nominal level of X and a noise factor Z that represents the variation about the nominal.

Due to manufacturing imprecisions, X actually follows a distribution with mean value X_0 and standard deviation $\sigma(X_0)$. Then standard results for first-order transmitted variation imply that

$$\text{Var}\,\{Y(X_0)\} \approx [g'(X_0)]^2\sigma^2(X_0) + \sigma_\varepsilon^2. \tag{12}$$

If $g(X)$ is nonlinear, then it may be possible to reduce the variance of Y by choosing X_0 in a region where $g(\cdot)$ is "flat", relative to the variation in X. This situation is illustrated in Figure 4. The transmitted variance will be lower at $X = -1$ than at $X = 1$.

What is the relationship between dispersion effects due to nonlinearity and those due to design factor by noise factor interactions? Suppose X_0 is the nominal level of the design factor but, due to manufacturing tolerances, the value for a particular item is actually $X_0 + Z$. We can think of Z, which represents the deviation of the design factor from nominal, as a noise factor. The observed response for the item will be

$$Y(X_0, Z) = g(X_0 + Z) + \varepsilon \tag{13}$$

$$\approx g(X_0) + g'(X_0)Z + \varepsilon. \tag{14}$$

Now suppose that the response function is roughly linear, so that

$$g(X_0) \approx \mu + \beta X_0, \tag{15}$$

and that the slope is also approximately linear, so that

$$g'(X_0) \approx \alpha + \gamma X_0. \tag{16}$$

Then

$$Y(X_0, Z) \approx \mu + \beta X_0 + (\alpha + \gamma X_0)Z + \varepsilon, \tag{17}$$

exactly the model that was considered in Section 6.1. The nonlinearity leads to an interaction between the nominal level of the design factor and the noise factor that represents its deviation from nominal.

Relations (15) and (16) above cannot, of course, hold exactly. They are meant only as a useful approximation that focuses attention on the critical issues for robust design:

- How does the overall response level change as a function of the nominal value?
- What is the typical slope, and hence the degree of transmitted variation due to the tolerance about the nominal?
- How will changes in the nominal value alter the slope and the transmitted variation?

Rough answers to these questions are obtained by estimating the design factor main effect, the noise factor main effect, and the interaction, respectively.

The use of variance transmission formulas to study variability in engineering design also has early forerunners in the statistical literature. Morrison (1957) is a noteworthy example. He correctly pointed out that valid conclusions depend critically on knowing how the standard deviation due to manufacturing tolerances depends on the nominal level. Box and Fung (1993) presented an example in which completely different conclusions are reached about the optimal setting of X as the form of $\sigma(X)$ is modified.

6.3. General setting

The simple model above can be easily generalized to include a large number of factors. Denote the design factors by the vector X and the noise factors by the vector Z. The noise factors might represent either environmental factors or deviations of design factors from their nominal levels. Assume for now that there is no signal factor. As above, the simplest model of interest must include main effects and design factor by noise factor interactions, so that

$$Y(X, Z) = \mu + \beta' X + \alpha' Z + X' \Gamma Z + \varepsilon. \tag{18}$$

Here β and α represent the main effects of the design and noise factors, respectively, and Γ is a matrix of design by noise factor interaction coefficients. This type of model was studied by Box and Jones (1992) and Myers et al. (1992).

At the process level, Z will vary at random. We may assume that Z has been coded so that $\mathrm{E}\{Z\} = 0$. Let V denote the covariance matrix of Z. Then the process characteristics will be

$$\mathrm{E}\{Y(X)\} = \mu + \beta X, \tag{19}$$

$$\mathrm{Var}\{Y(X)\} = (\alpha' + X'\Gamma)V(\alpha + \Gamma'X) + \sigma_\varepsilon^2. \tag{20}$$

Note that the vector $\alpha' + X'\Gamma$ gives the slopes of the expected response with respect to each of the noise factors when the design factors are set to X. As before, the effects of the noise factors will be effectively neutralized if $\alpha' + X'\Gamma \approx 0$. Myers and Kim (1994) considered this problem and showed how effective design factor settings can be found by considering confidence intervals for the estimated variance.

When some noise factors represent deviations from nominal values of design factors, it might be that the extent of the deviations would also depend on the nominal value. This would occur, for example, if the deviation was naturally thought of as a fixed percent of the nominal. In that case, $V(X)$ should be used rather than V. As noted earlier, the nature of this dependence can have important implications for the selection of design factor levels and so should be given careful consideration.

More complicated models can also be considered. For example, one could include all quadratic effects, not just the design factor by noise factor interactions. Box and Jones (1993) showed that, to find settings that minimize the expected squared deviation from a target value, it is not necessary to estimate all the coefficients in the full second degree model. They derived some specialized designs to capitalize on this situation. Lucas (1994) and Myers et al. (1992) also examined the consequences of using more elaborate models.

7. Product array experiments

This section discusses the design and analysis of product array experiments, which have been popularized by Taguchi (1987) for finding dispersion effects. Alternative experimental layouts may offer advantages and I will discuss a variety of issues relating to the selection of an experimental plan in Section 8. For now I will limit the presentation to designs that include only design and noise factors. Section 9 will discuss the use of product array experiments when signal factors are also present.

7.1. Construction of product array designs

Product array experiments are constructed by setting out separate design matrices for the design factors and the noise factors and then "crossing" the two. For example, if there are 6 design factors arrayed in a 2^{6-3} design and 3 noise factors arrayed in a 2^{3-1} plan, the total experiment will have 32 runs: each of the 8 design factor combinations will be paired with each of the 4 noise factor combinations. I will refer to the individual design matrices as the *design array* and the *noise array*; Taguchi uses

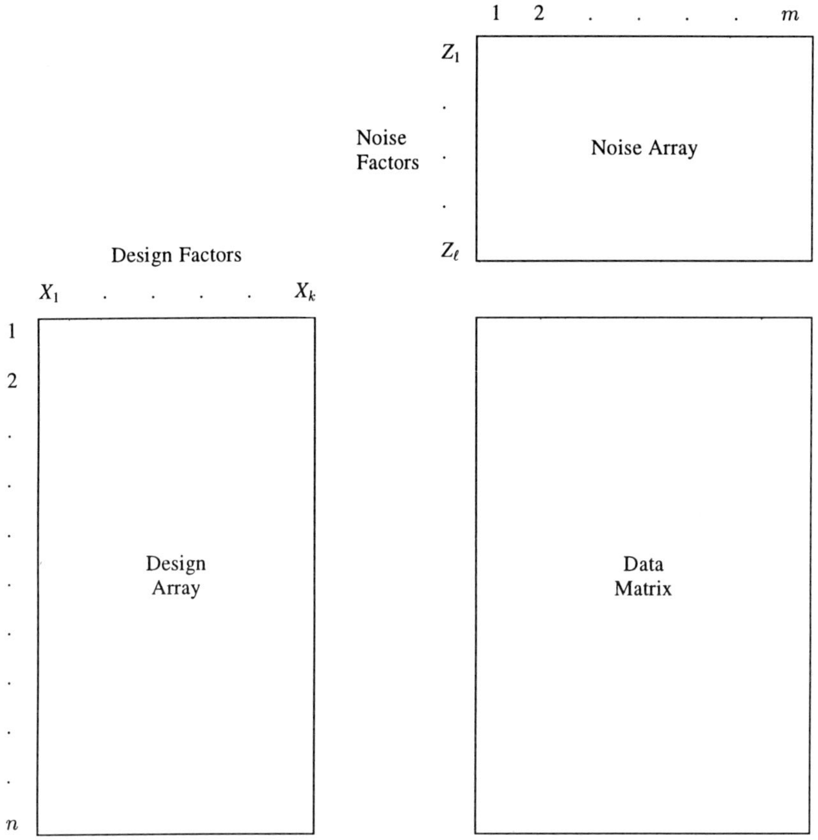

Fig. 5. Experimental layout for a product array design. The design array has k factors and n experimental points. The noise array has ℓ factors and m points. The full layout includes all $n * m$ combinations of the design and noise factors.

the terms *inner array* and *outer array*, respectively. Product arrays are naturally written in matrix form, as in Figure 5. Each row corresponds to a design factor combination and each column to a noise factor combination. Taguchi recommends using OA plans of strength 2 for both the design array and the noise array.

The clever and innovative feature of product arrays is the use of noise factors to force variability into the replicates. The advantages of this approach are most evident when compared with the alternative of obtaining "simple" replicates at each design factor combination, with no attempt made to control the noise factors. From Figure 1, the conceptual picture underlying a robust design study attributes much observed variability in product performance to the natural variation of noise factors. In obtaining simple replicates, noise factors will vary randomly. The extent to which they vary will differ from one design combination to the next and so may bias attempts to compare sample variances. These biases can be mitigated to some extent by randomizing the run order (in particular, not making all the runs on the same design combination consecutively), but cannot be avoided. Generating replicates with a noise array guarantees "fair" comparisons because the same noise conditions were used for each design factor combination. A more important advantage is that the noise factors can be made to vary substantially. Were they allowed to vary naturally, some noise factors would attain values that cover only a fraction of their true ranges. For example, the performance of a wave soldering machine may depend on the type of printed circuit boards that are processed. Unless circuit board type is consciously varied, there is no assurance that the full range of boards will be included in the experiment and important sources of variability may be completely overlooked. When processes are studied by computer simulation, repeating the same design factor settings produces identical outputs. The only way to examine variability is to force it into the results by varying noise factors.

7.2. Analysis by signal to noise ratio

Taguchi (1987) proposed that product array experiments be analyzed in exactly the same way as experiments with simple replicates. Although the noise factors are included in the design, they are ignored in the analysis.

7.3. Response model analysis

A number of articles have argued that the analysis of product array experiments should explicitly model the effects of the noise factors. The resulting model can then be used to suggest optimal settings for the design factors. This approach is favored by most of the discussants in the panel discussion edited by Nair (1992). Perhaps the first full application of this approach was made by Welch et al. (1990), but the idea is implicit in Easterling (1985). This approach has become known as *response model* analysis, by contrast with Taguchi's strategy of analyzing summary measures such as SN ratios. Ghosh and Derderian (1993) and Steinberg and Bursztyn (1994) compared the analysis methods on actual data sets.

Following the discussion in the previous section, the simplest model of interest for a robust design application will include main effects for all the design and noise factors as well as design factor by noise factor interactions:

$$Y(X, Z) = \mu + \beta' X + \alpha' Z + X' \Gamma Z + \varepsilon, \tag{21}$$

where β and α represent the main effects of the design and noise factors, respectively, and Γ is a matrix of design by noise factor interaction coefficients.

Shoemaker et al. (1991) showed that all of these effects are estimable from a product array experiment. The basic idea is easy to understand from the matrix form of the experiment. Each row of data corresponds to a particular design factor combination, varied over all the conditions in the noise factor array. Since the noise factor array has strength 2, row contrasts can be computed for the main effects of each noise factor.

For example, if a noise factor has 3 levels, one can compute its linear and quadratic effects. Once an effect has been computed for each row, it can be treated as a response variable and examined for possible dependence on the design factors. Since the design array also has strength 2, one can compute the main effect contrasts of each design factor on the noise contrasts. These, of course, are precisely interaction effects of the design and noise factors. The idea of treating noise factor effects as response variables was proposed by Easterling (1985).

The response model analysis makes no explicit use of the matrix layout of a product array design. Instead, the design is simply treated as a large factorial in all the factors and all relevant effects are estimated from it. Standard tools like normal (or half-normal) plots of effects will be useful for identifying the most important effects. If some factors have 3 levels, it will be possible to compute pure quadratic contrasts for them. When there are both design and noise factors with 3 levels, one can also estimate their linear by quadratic and quadratic by quadratic interactions. (This follows from the result of Shoemaker et al. (1991) described above.)

Obviously, the above "paradigm" for analyzing the experiment and selecting settings for the design factors must be complemented by common sense and process knowledge. For example, a design factor might have a large effect on the mean level and also interact with a noise factor. Choosing the level of that factor to minimize the effect of the noise factor may move the mean level far from its target value. So the final choice of this factor's setting should involve careful consideration of the tradeoffs involved between process mean and process variance. An advantage of the response model approach is that these tradeoffs become explicit and are easy to compute and analyze.

Other standard data analysis procedures are also important, It will often be desirable to transform the data, as noted in the last section. Logothetis and Wynn (1989) made a number of good practical suggestions for analyzing data from robust design experiments.

Once a response model has been fit to the experimental data, it can be used to select optimal levels for the design factors. For example, the model can be used together with a loss function to compute expected loss with respect to the distribution of the noise factors. It is then possible to find settings of the design factors that minimize the expected loss. In particular, when the model includes both design and noise factors, the mean level and the variance can be estimated as functions of the design factors using equations (19) and (20). Shoemaker and Tsui (1993) presented a useful paradigm for using an estimated response model to find design factor settings that minimize variance and keep the response on target. Welch and Sacks (1991) provided a concise description of the response model strategy in the context of computer experiments. They made a convincing case that it is much more efficient to first model the response and then to compute expected loss from the model than to compute a summary of expected loss and then model the summary, as advocated by Taguchi.

7.4. Example: Injection molding experiment

Engel (1992) described an experiment to improve an injection molding process. The key quality characteristic was the percent shrinkage of parts produced by the process.

Table 4
Factors in the injection molding experiment

Design factors	Noise factors
A: Cycle time	M: Percentage regrind
B: Mold temperature	N: Moisture content
C: Cavity thickness	O: Ambient temperature
D: Holding pressure	
E: Injection speed	
F: Holding time	
G: Gate size	

Table 5
Observed percent shrinkages from the injection molding experiment, with averages, standard deviations and "response on target" SN ratios for each design factor combination

Design factors							Noise factors (M, N, O)						
A	B	C	D	E	F	G	(1, 1, 1)	(1, 2, 2)	(2, 1, 2)	(2, 2, 1)	Average	SD	SN
1	1	1	1	1	1	1	2.2	2.1	2.3	2.3	2.225	0.10	26.95
1	1	1	2	2	2	2	0.3	2.5	2.7	0.3	1.450	1.33	0.75
1	2	2	1	1	2	2	0.5	3.1	0.4	2.8	1.700	1.45	1.38
1	2	2	2	2	1	1	2.0	1.9	1.8	2.0	1.925	0.10	25.69
2	1	2	1	2	1	2	3.0	3.1	3.0	3.0	3.025	0.05	29.61
2	1	2	2	1	2	1	2.1	4.2	1.0	3.1	2.600	1.37	5.78
2	2	1	1	2	2	1	4.0	1.9	4.6	2.2	3.175	1.33	7.76
2	2	1	2	1	1	2	2.0	1.9	1.9	1.8	1.900	0.08	27.51

Excessive variation in shrinkage resulted in poor quality and the goal was to find process settings that would result in more uniform shrinkage. The experiment included 7 design factors and 3 noise factors, which are listed in Table 4.

The experimental design was a product array experiment that crossed a 2^{7-4} plan for the design factors with a 2^{3-1} plan for the noise factors. The 32 observed percent shrinkages are listed in Table 5, along with the average, standard deviation, and "response on target" SN ratio for each design factor combination.

Analysis of the SN ratio indicates a major effect for factor F, holding time. Analysis of the standard deviations leads to the same conclusion. In fact, simple inspection of Table 5 shows that four design points had standard deviations of at most 0.1, the other four had standard deviations of about 1.4, and this division corresponds exactly to the levels of holding time. These analyses both indicate that holding time should be set to its low level to minimize variance in the percent shrinkage.

Three factors had large effects on the mean level: cycle time, holding pressure, and gate size. Setting cycle time to its low level and the other two factors to their high levels will minimize the expected shrinkage.

Steinberg and Bursztyn (1994) reanalyzed these data using the response model approach. Estimates of the main effects of all ten factors and the 21 design factor by noise factor interactions exhaust all the orthogonal contrasts in the data. This analysis finds the same three design factors affecting the mean level. None of the noise factors have large main effects. However, there are two large interactions: cavity thickness by moisture content (CN = 0.90) and injection speed by moisture content (EN = −0.84). This finding implies that moisture content is the only one of the three noise factors that transmits variation to the percent shrinkage and that the extent of the variance transmission depends on the settings of both the cavity thickness and the injection speed. Closer inspection shows that, when these two design factors are at the same level, the interactions essentially cancel one another, and the estimated effect of moisture content is near its main effect, 0.28. However, when the two design factors are at opposite levels, the interactions amplify one another, producing a large positive effect in one case and a large negative effect in the other.

The results from the response model analysis also provide a simple explanation for the seeming dispersion effect of holding time that was identified by the SN analysis. Large (small) standard deviations should be found in those rows in which cavity thickness and injection speed have opposite (same) levels. Thus we should expect to find a cavity thickness by injection speed (CE) interaction when analyzing the standard deviations (or the SN ratios). However, the design factor plan is saturated, so all interactions are aliased with main effects. In particular, the CE interaction is aliased with factor F, holding time. Thus the evident dispersion effect of holding time is an illusion, generated by the effects of cavity thickness and injection speed and the saturated design. If holding time really were responsible for the differences in standard deviations, we would have found a sizeable holding time by moisture content interaction in the response model analysis. The SN analysis is unable to distinguish between these two possible explanations of the data; the response model analysis is able to do so.

8. Further design issues

In this section I will discuss a number of additional issues related to the arrangement of robust design experiments. Many arise from the need to limit the size and expense of these experiments. Decisions about which effects to estimate and which to ignore will have important ramifications for the choice of an experimental plan.

8.1. Interactions, linear graphs and resolution

Taguchi (1987) recommends selecting experimental plans that are geared primarily for the estimation of main effects. Interactions tend to be ignored. Similarly, no attention is paid to design resolution. A number of authors have pointed out examples in which resolution III designs were used but resolution IV designs were available.

Actually, Taguchi's attitude toward interactions has a somewhat schizophrenic character to it. It is assumed that there are design factor by noise factor interactions, since they are crucial to variance reduction, but these interactions are not explicitly estimated. All other interactions are ignored. To justify this approach, Taguchi (1987, 1991) argued that engineers are best served by linear relationships which, it is hoped, will remain accurate under extrapolation to wider factor levels and to related applications. Moreover, he claimed that judicious choice of a performance statistic and definition of the design factors should remove interactions.

The idea of ignoring interactions on the grounds that relationships *should* be linear has been greeted with skepticism. See, for example, Box (1988) and comments by Box and by Lorenzen in Nair (1992). Moreover, as noted by Lorenzen, the tendency to discount interactions among design factors stands in sharp contrast to the exploitation of design factor by noise factor interactions to achieve product robustness.

My own reading of the situation is that practical considerations underlie the decision to downplay interactions. First, focusing on main effects alone makes it possible to use a much smaller array than would be needed to estimate both main effects and interactions. The dictum is that design factor contrasts should be invested in extra factors, *not* in estimating interactions. Screening experiments have always followed this same strategy. However, such designs have typically been regarded by statisticians as first stages in an experimental program, to be followed up by designs that study the important factors in greater detail. In Taguchi's scheme, these designs are expected to provide a "one-shot" solution to the engineering problem. Second, and perhaps more important, the assumption of an additive model makes it much easier to analyze the experimental data and to select improved factor settings. Here the desire to "keep things simple" may be the overriding concern and also serves to explain why Taguchi recommends the analysis of summary measures like SN ratios rather than the response model approach.

Taguchi (1987) did include two elements in his experimental program to take account of interactions. One element, as I noted in Section 5, is to make a confirmation run at the suggested optimal factor settings. If main effects of the design factors are dominant, the confirmation run should not differ much from a prediction based on the original experiment. However, Bisgaard and Diamond (1991) showed that confirmation trials are not an efficient way to reveal interactions.

The second element is a set of tools that can facilitate the estimation of *particular* interactions that are specified in advance by the experimental team. The basic idea is to assign the factors to columns in a tabled OA plan in such a way that no factor is assigned to a column that is aliased with a suspected interaction. Suppose, for example, that a 2^{5-2} experiment is to be run on a chemical process and it is thought that temperature may interact with run time and with pressure. Then the factors should be assigned to columns in such a way that each of the two *unassigned* columns is associated with one of the anticipated interactions. To help with this problem, Taguchi provided two tools:

1. A table that shows, for each pair of columns, the column(s) associated with the interaction of the factors assigned to them.
2. *Linear graphs*, which illustrate the aliasing structure. In these graphs, some of the columns in an OA are represented by vertices. Certain pairs of vertices are connected by edges, which represent the column associated with the interaction of the vertex columns.

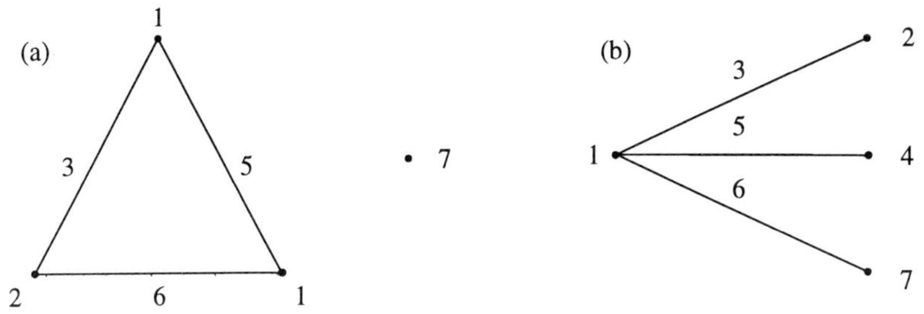

Fig. 6. Two standard linear graphs for L_8.

Wu (see Nair, 1992) remarked that "dot and line graphs" is a more accurate translation of Taguchi's Japanese terminology. But the term "linear graphs" is, by now, so commonly used that it is unlikely to be amended.

Figure 6 is a typical example and shows two linear graphs for the L_8 array used in the INA tile experiment of Section 4. In panel (a), we see that column 3 is associated with the interaction of columns 1 and 2, column 5 with that of columns 1 and 4, etc. To solve the design problem posed earlier, we can construct a similar graph in which all the factors are vertices and edges are added for each anticipated interaction. Our graph will have edges between temperature and time and between temperature and pressure. Matching that graph to panel (a), we see that a successful design can be achieved by assigning temperature to column 1, time and pressure to columns 2 and 4, respectively, and the two remaining factors to columns 6 and 7. Columns 3 and 5 are left "open" to accommodate the interactions.

Are linear graphs useful? I have doubts. Ultimately, practitioners will decide through their use of or disregard for the graphs. A number of reservations have been put forth. First, Box (1988) (see also Box's comments in Nair, 1992) has questioned whether experimenters are really able to specify, at the design stage, which factors are likely to interact and which are not. Second, Tsui (1988), Li et al. (1990) and Wu and Chen (1992) have noted that some resolution III designs derived from Taguchi's graphs could be replaced by resolution IV designs that permit estimation of the same interactions. Tsui (1988) developed an improved set of interaction tables. Wu and Chen (1992) derived new sets of graphs that assure, if possible, a minimum aberration (Hunter and Fries, 1982), and hence maximum resolution, design given the specified set of interactions. They also developed software that helps to find such designs. Finally, there is a practical concern. When a small number of interactions is specified, the allocation problem is trivial and the graphs are unnecessary. When the design must accommodate more than five or six interactions, the graphs can be so complex that they are almost unreadable. For these problems, a software solution would surely be preferable.

8.2. OA's or response surface methods

Many of the applications described by Taguchi (1987) use OA plans like L_{18}, L_{27} and L_{36}, in which most or all of the factors have three levels. By contrast, statistical researchers have emphasized the use of 2-level factors, especially at early stages of investigation, followed by the use of response surface designs for full second-order models when the response is too complex for a first-order model.

There seem to be four reasons that support the use of OA plans with 3-level factors:

1. They permit the investigation of non-linearity with respect to continuous design factors.
2. When a current setting is included as an experimental level, settings both above and below it can be examined.
3. They do not require substantially more runs than comparable 2-level designs.
4. Since Taguchi generally recommends that optimal levels be chosen from among those that appeared in the experiment, using three levels rather than two considerably expands the optimization space.

A number of articles have commented critically on the first reason, noting that the same rationale should then require the study of other second order effects, i.e., interactions. I believe that reasons (2)–(4) have more practical importance. However, better solutions to all these issues can often be achieved by a sequential approach, in which extra levels are added for factors only when initial data indicate that they are likely to be necessary. Taguchi seems not to have appreciated the great advantages offered by this sort of sequential experimentation.

Robust design will benefit by incorporating elements from response surface methodology (RSM) (Box and Draper, 1989; Khuri and Cornell, 1987), which is built around such a sequential approach. The strategy common in RSM is to begin with a screening design whose goal is to identify the most important factors. Typically all factors will be at two levels to keep run size to a minimum. If the factors are quantitative, a center point can be included as a global check of the need for pure quadratics. The results of that experiment may lead to changing the ranges of some of the factors, changing the factors themselves, or extending the design to permit estimation of some second-order effects. See Box (1992) for an excellent discussion of the varied routes that such a follow-up experiment might take.

The idea of linking RSM to robust design experiments with noise factors presents some interesting questions. In many applications, for example, error transmission from important noise factors will be dominated by their linear effects and it will be wasteful to use three levels for noise factors. Useful research might be done on designs that explore second order effects of design factors but only linear effects of noise factors. The works by Box and Jones (1992), Lucas (1994), Myers et al. (1992), Myers and Kim (1994) and Tuck et al. (1993) have begun to pursue this connection. More research in this area will be useful.

8.3. *Sliding levels and compound factors*

Sometimes there are natural links between input factors. Taguchi has proposed adapting designs to this feature through the use of *sliding levels*. Suppose x_1 and x_2 must be changed jointly and let $f(x_1)$ denote the nominal level of x_2 that seems best suited to x_1. The idea of sliding levels is to center the levels of x_2 about $f(x_1)$. For example, the op-amp circuit example (Section 4) includes two pairs of transistors as noise factors. Within each pair, the resistors may vary quite widely about their nominal values, but it is anticipated that they will do so in a highly correlated fashion, so that either both will exceed, or be below, their nominal values. This type of variation is reflected in the design by assuming a wide range for the levels of the first transistor. The nominal value for the second transistor is taken to be the value of the first transistor, about which there is a small amount of additional variation.

The use of sliding levels is closely related to the discussion in Section 3 of how to define factors that are most directly related to the system's outputs. In effect, the experimental factor x_1 really represents a "compound factor" in which x_1 and x_2 move in unison along the curve $x_2 = f(x_1)$.

Phadke (1989) (see also Nair, 1992) stated that sliding levels can be used to ameliorate interactions. However, Hamada and Wu (1995) showed that this cannot occur if the new design is a linear transformation of the two factors, so that $f(x_1)$ is a linear function and the further deviations in x_2 are additive; in that case, it is easy to show that an interaction in x_1 and x_2 will also be present in the experimental factors. Some interactions can be absorbed by sliding factor levels when a nonlinear transformation is used to define the nominal levels of x_2 or when the additional variation in x_2 is not additive. However, Hamada and Wu (1995) found that the interactions will be eliminated only for particular choices of the slid levels – other choices will not succeed. Moreover, they showed that some important consequences for robustness may be missed by ignoring the new scaling that has been forced on the slid factor. They concluded that the sliding level technique can be effective, but only to provide better coverage of a skewed factor space, not to eliminate interactions.

An additional, and more important, benefit of sliding levels is to eliminate a *main effect*, which will happen if the "compound factor" summarizes the effect of both factors on the response. In that case, the "slid" factor, which relates to further variation about the compound factor, may have little further relation to the response. The experiment might then proceed using *only* the compound factor, with potential savings in the number of experimental runs needed.

8.4. Collapsing levels and idle columns

It may occur that no established OA design exactly matches the number of factors and levels desired for an experiment. Taguchi has recommended using the method of *collapsing levels* to adapt an OA to the setting at hand without recourse to developing a specialized plan. The idea is to use an s-level column in the array for a t-level factor (when $t < s$) by assigning some of the factor levels to more than one symbol in the OA column. For example, a 2-level factor could be assigned to a 3-level column, using the factor at level 2 whenever the column specified either 2 or 3.

Addelman (1962, 1963) first developed this method. He showed that main effects remain orthogonal. Since the levels of a collapsed factor will not be equi-replicated, there is some loss of statistical efficiency in assessing its effect.

A somewhat more sophisticated application of the idea of collapsing levels is found in Taguchi's *idle column* technique, which is used to assign a small number of three-level factors to one of the orthogonal arrays that are constructed strictly from two-level factors. First one column in the OA is chosen as the "idle column". That column will not be assigned to any factor. Then take any two columns whose interaction is given by the idle column and use them to define a new, three level column, by collapsing the four combinations onto three levels. If a single three-level column is added in this way, a design like those proposed by Addelman is obtained, in which all main effects are orthogonal. However, the technique can be used to add further three-level columns to the design, generating each three-level column from two new columns whose interaction is given by the idle column. When several factors are added in this way, the main effects will no longer be orthogonal. Moreover, Grove and Davis (1991) showed that these designs can also be quite sensitive to the particular scheme used to collapse the pairs of OA columns onto the levels of the new factor. They concluded that the idle column technique could be useful, but must be used with care.

8.5. Hard-to-change factors

A strong sense of practicality pervades much of Taguchi's work on experimental design. A common problem in industrial experimentation is that some factors may be difficult or expensive to change. Taguchi's OA tables for 2-level designs order the columns from those with the fewest level changes to those with the most level changes. The organization makes it easy to match hard-to-change factors with appropriate columns.

Cheng (1985) provided a thorough description of how to set up 2-level designs in which certain factors have few level changes. His method cannot match the simplicity of Taguchi's, but it is more complete and can be easily adapted to generate run orders that are also free of polynomial time trends.

8.6. Combined array experiments

Including noise factors in an experiment creates some interesting problems for experimental design. Product arrays (Section 7) have been criticized for requiring too many runs (Shoemaker et al., 1991; Welch et al., 1991). These designs actually generate large factorials in the full set of factors. So it is natural to inquire whether a different factorial arrangement (involving all the factors) might not be a better plan. Experimental plans that include both design and noise factors, but without restriction to the product array form, have come to be known as *combined array experiments.*

The use of a combined array is closely linked to use of the response model analysis (Section 7). The analysis of signal to noise ratios effectively requires a product array, so that each design factor combination is varied across the same noise factor settings. With the response model analysis, though, the emphasis is on estimating main effects and design factor by noise factor interactions. A variety of designs may be able to accomplish this.

As a simple example, suppose there are four design factors and one noise factor. A product array experiment might cross a 2^{4-1} plan in the design factors with the two levels of the noise factor. This plan uses eight different design factor combinations, each at both levels of the noise factor. Design factor by noise factor interactions are estimable, but interactions among pairs of design factors are aliased with one another. Alternatively, one could run a 2^{5-1} plan, which would permit estimation of all the 2-factor interactions. This plan would use all 16 design factor combinations. Half would be paired with the low level of the noise factor and half with the high level.

The primary motivation for the combined array approach was to reduce the run size required by product array experiments. Whether or not that is possible depends on the effects that are to be estimated. If interest focuses on main effects and on all design factor by noise factor interactions, a product array with minimal Plackett and Burman (1946) designs used for the design and noise arrays offers near minimal run size. However, combined arrays can sometimes provide a more favorable aliasing structure, as in the example presented earlier.

Combined arrays can reduce run size if the experimenters are prepared to confound some of the design factor by noise factor interactions with one another. Suppose, for example, that there are four design factors $(X_1, \ldots, X_4)$ and two noise factors (Z_1, Z_2). The smallest two-level product array experiment will cross a 2^{4-1} plan in the design factors with a 2^2 design in the noise factors, for a total of 32 runs. One alternative would be to use a 16 run combined array experiment with defining relation $I = X_1X_2X_3 = X_1X_2X_4Z_1 = X_3X_4Z_1$. This design permits estimation of all the main effects and all the design by noise interactions except X_3Z_1, which is aliased with X_4, and X_4Z_1, which is aliased with X_3. Another alternative would be to use the defining relation $I = X_1X_2X_3Z_1 = X_1X_2X_4Z_2 = X_3X_4Z_1Z_2$. This plan has resolution IV at the expense of confounding two pairs of design by noise factor interactions with one another ($X_3Z_1 = X_4Z_2$ and $X_3Z_2 = X_4Z_1$).

Lucas (1994) presented a variety of response surface designs for experiments with both design and noise factors. He showed that the number of runs can often be reduced by using "mixed resolution" designs, in which the extent to which main effects are free of interactions may differ among the two sets of factors.

8.7. *Split plot designs*

Box and Jones (1992) pointed out that product array experiments will often involve split plot structure. This will certainly be the case when the noise factors define test conditions for prototypes that have been constructed according to the design array. Box and Jones (1992) showed how effective use of split plotting could greatly reduce the overall experimental effort and thus mitigate the criticism that product array designs are too costly. They also cautioned that the use of standard analyses for split plot experiments could lead to invalid conclusions, in which too many significant "whole plot" effects are found because of improper comparison with the "sub-plot" variance. Bisgaard and Steinberg (1993) showed how one can analyze these experiments using standard tools like normal probability plots in a way that takes account of the split plotting.

8.8. *Observable noise factors*

All the discussion thus far has been predicated on the assumption that important noise factors can be controlled for the purpose of an experiment. What should one do if some noise factors are impossible to control? An interesting strategy was proposed by Freeny and Nair (1992) for situations in which the noise factor can be observed and recorded. They applied this approach to an experiment for thermal design of cabinets for telecommunications equipment. They discovered that the primary response variable, the surface temperature, was quite sensitive to the ambient temperature, which could not be fully controlled. They fit response models, as in the last section, with the observed value of ambient temperature as a covariate. Of course, the observed noise factor was not orthogonal to the design factors, as would typically occur if it could be controlled, but the same models are still relevant. Freeny and Nair (1992) found that the response models effectively explained the data and identified important dispersion effects via interactions between design factors and the ambient temperature. Moreover, the ambient temperature had behaved quite differently at the various design factor settings used in the experiment, so that analyses that did not model its effect identified nonexistent dispersion effects. The approach of Freeny and Nair (1992) marks an important extension of robust design methodology. It applies the concepts behind robust design studies to settings in which an actual robust design experiment, with control of the noise factor, cannot be carried out. Their work also provides further evidence in favor of the response model approach, since there is no way to account for observable noise factors like the ambient temperature using the SN analysis.

9. Experiments with signal factors

In many engineering systems, a *signal factor* is used to direct the output or response. Simple examples are the light-dark switch on a photocopier, the amount of material deposited on a silicon wafer, or the size of a mold in an injection molding process. Interest in these systems focuses on the signal-output relationship. Taguchi (1987) has called these *dynamic problems*, by contrast with *static problems,* in which a set design configuration is established and a fixed output desired. I provide a brief introduction here to robust design experiments in the dynamic setting.

An important first step in studying dynamic systems is to derive an *ideal function* that describes the desired engineering relationship between the output characteristic(s) and the signal. Taguchi (1991) has argued that studying the energy flow of the system is the key to obtaining meaningful response variables and their relationships to signal factors, and that such relationships should typically be linear, which corresponds to energy transmission and also provides a basis for extrapolation to levels of the signal factor not studied in the experiment. However, each application must be considered on its own merits in deriving an ideal function.

A measurement system provides a simple example of a dynamic system. Here the signal factor (M) is the known value of a test quantity, Y is the measured response from the system, and the ideal function is $Y = M$. A good measurement system should stay close to the ideal function even in sub-optimal conditions. Research on measurement systems at the US National Bureau of Standards included many elements analogous to those in robust design. The squared ratio of slope to standard deviation, whose logarithm gives Taguchi's SN ratio for dynamic problems, was introduced by

Mandel and Stiehler (1954) as a criterion for assessing measurement systems. The idea of a *rugged* measurement system (Wernimont, 1977) exactly parallels the use of robustness here.

The goal of robust design experiments for dynamic systems, as with static systems, is to determine optimal settings of the design factors that define the product or process. The signal factor is assumed to be known in advance from the engineering context. There are two engineering goals:

1. Make the expected response equal to the ideal function.
2. Minimize variance about the expected response.

Thus experiments for dynamic systems must provide information about how both the signal-response relationship and the variation about that response curve depend on the design factors. One way to do this is to use a product array design in which the signal factor is added to the noise array and this is the approach that has been advocated by Taguchi (1987, 1991). As with static experiments, the study of variability is facilitated by including relevant noise factors.

How many levels should be used for the signal factor? Many published studies of dynamic system experiments have used three levels for the signal factor. This choice appears to be linked to the assumption that the signal-response relationship should be linear, since three levels is the minimal number to fit and criticize a straight line. Including more levels permits more detailed study of the relationship but with a corresponding increase in experimental cost and burden.

With respect to both design and analysis, many of the same issues that were discussed for static systems are relevant for dynamic systems. These include, for example, how many levels to include for the design and noise factors, whether to allow for interactions among design factors, and whether to analyze the data via a response model or to use summary statistics. The remainder of this section will focus on the question of the appropriate analysis.

The analysis proposed by Taguchi for dynamic problems is analogous to that for static "on target" problems. It is assumed in this standard analysis that the ideal signal-response function is linear; modifications will be necessary if that is not the case.

1. Summarize the data at each design factor combination. The summary is based on a linear regression of the response on the signal factor. The strength of the signal-response relationship for the ith combination is summarized by the least squares slope $\widehat{\beta}_i$; the consistency of the relationship is summarized by the SN ratio $\mathrm{SN}_i = 10\log(\widehat{\beta}_i^2/s_i^2)$, where s_i is the estimated standard deviation about the regression line.
2. The SN ratio is analyzed for dependencc on the design factors. Factors that have large effects are set to levels that maximize the SN ratio.
3. The slope β is adjusted to the desired value by changing an appropriate design factor. This adjustment factor may be suggested *a priori* by engineering considerations (for example a gear ratio) or may be chosen on the basis of the experiment, using a factor that has a large effect on the estimated slopes but little effect on the SN ratio.
4. Conduct a confirmation experiment at the suggested optimal settings of the design factors.

The selection of optimal process settings is again achieved in two steps, following the idea that reducing variance is more difficult than adjusting the slope and thus should be tackled first. The form of the SN ratio reflects a tacit assumption that dispersion, in general, will be an increasing function of the slope. Since the slope will be adjusted to a target valuc, the dispersion summary is adjusted for dependence on slope before relating it to the design factors.

Taguchi's analysis paradigm for dynamic problems suffers from the same drawbacks cited in Section 5 regarding the analysis of static problems. For example, there is no guarantee that dispersion will be related to the slope, nor is this assumption ever examined in the analysis. No use is made of transformations, which might simplify the analysis if dispersion is related to slope. Noise factors that were used in the design are ignored in the analysis. In addition, the analysis relies heavily on the assumed linear signal-response relationship. Variation about the fitted straight lines might be due either to curvature in the response (model inadequacy), to the effects of noise factors, or to residual variation. No attempt is made to separate and exploit these separate sources of variation. Instead, they are all combined in the residual standard deviation s_i.

The response model approach can be applied to dynamic problems in much the same way as to static problems. The major extension is that the model will now involve terms for the signal-response relationship and its dependence on the design factors. There are many possible ways to model the signal-response relationship. Whenever possible, theoretical considerations should be used to suggest a mechanistic model. Otherwise, a variety of different empirical models might be useful. For purpose of illustration, consider the use of orthogonal polynomials as in Bisgaard and Steinberg (1993). Let $f(M) = (f_1(M), \ldots, f_p(M))'$ be the vector of polynomials. Then a useful model will be

$$\begin{aligned} Y(X,Z,M) = \mu + \beta' X + \alpha' Z + \delta' f(M) + X' \Gamma Z \\ + X' \Theta' f(M) + \varepsilon. \end{aligned} \tag{22}$$

As before, β and α represent the main effects of the design and noise factors, respectively, and Γ is a matrix of design by noise factor interaction coefficients. The matrix Θ contains interaction coefficients for each design factor by each component of the response curve. If the design factors are set at X, the expected response curve is

$$\mathrm{E}\{Y(X,M)\} = \mu + (\delta' + X'\Theta') f(M) \tag{23}$$

and the variance about the curve is

$$\mathrm{Var}\{Y(X,M)\} = (\alpha' + X'\Gamma) V (\alpha + \Gamma' X) + \sigma_\varepsilon^2, \tag{24}$$

where V denotes the covariance matrix of the noise factors. It is now possible to select design factor settings by examining jointly how the response curve and the process variance depend on X.

The response model approach for dynamic problems has been studied by Miller and Wu (1991) in the context of improving a calibration system. They showed that maximizing Taguchi's SN ratio in this setting minimizes the expected length of the Fieller confidence interval for the true value as a function of the measured value. Lumani et al. (1995) considered a variety of useful data analytic procedures for dynamic problems, in particular with regard to estimating and exploiting a power relationship between dispersion and the slope. They showed that Taguchi's analysis could be misleading if the standard deviation is not proportional to the slope. Grove and Davis (1992) made a number of useful suggestions for analyzing data from dynamic experiments and illustrated them on an interesting application from the automotive industry.

10. Conclusions

Efforts to achieve high quality have moved steadily upstream: from end-of-the-line inspection to on-line process control to off-line study aimed at building quality in to products and processes from the beginnings of the design stage. As Box and Bisgaard (1987) aptly observed, the key to quality by design is to learn about processes and designed experiments are a great tool to speed up the learning curve. This is the arena in which robust design experiments can make a great contribution to quality engineering.

Genichi Taguchi and his innovative quality engineering plan have had a major impact in industry. He deserves great credit for focusing attention on variance reduction, for helping to define the engineering context of this problem, for realizing the potential of factorial experiments in reducing variance, for keen insights like the use of noise factors, and for invigorating the use of planned experiments in industry. Many of his engineering goals are achieved with the aid of data collection and analysis. As such, it is natural that statistical research should complement his ideas by developing better experimental strategies and more revealing analyses for robust design studies. Some of the initial research efforts have already moved in that direction: certainly response model analyses should supplant automatic reliance on less informative SN ratios and the importance of data transformations should be recognized. The use of split plot designs, combined arrays and sequential experimentation will help reduce the size of robust design experiments. Recent initiatives to use response surface methodology in robust design are also a step in the right direction. The end result will be more informative robust design experiments and greater gains in quality.

References

Addelman, S. (1962). Orthogonal main effect plans for asymmetrical factorial experiments. *Technometrics* **4**, 21–46.

Addelman, S. (1963). Techniques for constructing fractional replicate plans. *J. Amer. Statist. Assoc.* **58**, 45–71.

Baranescu, R. A. and D. A. Ginder (1988). Engine performance optimization – A computer simulation enhanced by Taguchi methods. In: *Sixth Symposium on Taguchi Methods*. American Supplier Institute, Dearborn, MI, 394–405.

Bartlett, M. S. and D. G. Kendall (1946). The statistical analysis of variance-heterogeneity and the logarithmic transformation. *J. Roy. Statist. Soc. Ser. B* **8**, 128–138.

Bendell, A., J. Disney and W. A. Pridmore, eds. (1989). *Taguchi Methods: Applications in World Industry*. IFS Publications, Bedford, UK.

Bisgaard, S. and N. Diamond (1991). An analysis of Taguchi's method of confirmatory trials. Tech. Report No. 60, Center for Quality and Productivity Improvement, University of Wisconsin, Madison, WI.

Bisgaard, S. and D. M. Steinberg (1993). The design and analysis of $2^{k-p} \times S$ prototype experiments. Tech. Report No. 96, Center for Quality and Productivity Improvement, University of Wisconsin, Madison, WI.

Box, G. E. P. (1988). Signal-to-noise ratios, performance criteria, and transformations (with discussion). *Technometrics* **30**, 1–40.

Box, G. E. P. (1992). Sequential experimentation and sequential assembly of designs. *Quality Engineering* **5**, 321–330.

Box and Bisgaard (1987). The scientific context of quality improvement. *Quality Progr.* **20**, 54–61.

Box, G. E. P., S. Bisgaard and C. Fung (1988). An explanation and critique of Taguchi's contributions to quality engineering. *Quality Reliability Internat.* **4**, 123–131.

Box, G. E. P. and D. R. Cox (1964). An analysis of transformations (with discussion). *J. Roy. Statist. Soc. Ser. B* **26**, 211–252.

Box, G. E. P. and C. A. Fung (1993). Is your robust design procedure robust? Tech. Report No. 101, Center for Quality and Productivity Improvement, University of Wisconsin, Madison, WI.

Box, G. E. P. and S. Jones (1992). Split-plot designs for robust product experimentation. *J. Appl. Statist.* **19**, 3–26.

Box, G. E. P. and S. Jones (1993). Designing products that are robust to the environment. *Total Quality Management* **3**.

Box, G. E. P. and R. D. Meyer (1986). Dispersion effects from fractional designs. *Technometrics* **28**, 19–27.

Byrne, D. M. and S. Taguchi (1989). The Taguchi approach to parameter design. In: Bendell, Disney and Pridmore, eds., *Taguchi Methods: Applications in World Industry*, 57–76.

Burman, J. P. (1946). Note added in proof to the paper by Plackett and Burman (1946) (see below).

Carroll, R. J. and D. Ruppert (1988). Discussion of the paper by Box (1988). *Technometrics* **30**, 30–31.

Chan, L. K. and T. K. Mak (1995). A regression approach for discovering small variation around a target. *Appl. Statist.* **44**, 369–377.

Cheng, C. S. (1985). Run orders of factorial designs. In: L. M. Le Cam and R. A. Olshen, eds., *Proc. Berkeley Conf. in Honor of Jerzy Neyman and Jack Kiefer*, Vol. 1. Wadsworth, Belmont, CA, 33–47.

Coleman, D. E. and D. C. Montgomery (1993). A systematic approach to planning for a designed experiment (with discussion). *Technometrics* **35**, 1–28.

Dehnad, K., ed. (1989). *Quality Control, Robust Design and the Taguchi Method*. Wadsworth & Brooks/Cole, Pacific Grove, CA.

Deming, W. E. (1982). *Quality, Productivity and Competitive Position*. Cambridge, MIT-Center for Advanced Engineering Study, MA.

Deming, W. E. (1986). *Out of the Crisis*. Cambridge, MIT-Center for Advanced Engineering Study, MA.

Ealey, L. A. (1988). *Quality By Design: Taguchi Methods and US Industry*. ASI Press, Dearborn, MI.

Easterling, R. G. (1985). Discussion of the paper by Kackar (1985). *J. Quality Technol.* **17**, 191–192.

Easterling, R. G. (1987). Discussion of the paper by León, Shoemaker and Kacker (1987). *Technometrics* **29**, 267–269.

Engel, J. (1992). Modeling variation in industrial experiments. *Appl. Statist.* **41**, 579–593.

Finney, D. J. (1945). Fractional replication of factorial arrangements. *Ann. Eugenics* **12**, 291–301.

Fisher, R. A. (1945). A system of confounding for factors with more than two alternatives giving completely orthogonal cubes and higher powers. *Ann. Eugenics* **12**, 283–290.

Freeny, A. E. and V. N. Nair (1992). Robust parameter design with uncontrolled noise variables. *Statist. Sinica* **2**, 313–334.

Ghosh, S., ed. (1993). *Statistical Design and Analysis of Industrial Experiments*. Marcel Dekker, New York.

Ghosh, S. and E. Derderian (1993). Robust experimental plan and its role in determining robust design against noise factors. *The Statistician* **42**, 19–28.

Ghosh, S. and Y.-J. Duh (1992). Determination of optimum experimental conditions using dispersion main effects and interactions of factors in replicated factorial experiments. *J. Appl. Statist.* **19**, 367–378.

Ghosh, S. and E. S. Lagergren (1990). Dispersion models and estimation of dispersion effects in replicated factorial experiments. *J. Statist. Plann. Inference* **26**, 253–262.

Ghosh, S. and E. S. Lagergren (1993). Measuring dispersion effects of factors in factorial experiments. In: S. Ghosh, ed., *Statistical Design and Analysis of Industrial Experiments*, Marcel Dekker, New York, 459–478.

Greenfield, T. (1995). Taguchi policy: A ridiculous and dangerous fashion. *Roy. Statist. Soc. News Notes* **22**(6), 6–7.

Grove, D. M. and T. P. Davis (1991). Taguchi's idle column method. *Technometrics* **33**, 349–354.

Grove, D. M. and T. P. Davis (1992). *Engineering Quality and Experimental Design*. Longman Scientific and Technical, Essex, England.

Gunter, B. A. (1988). Discussion of the paper by Box (1988). *Technometrics* **30**, 32–35.

Hamada, M. and C. F. J. Wu (1995). The treatment of related experimental factors by sliding levels. *J. Quality Technol.* **27**, 45–55.

Hayes, R. H., S. C. Wheelwright and K. B. Clark (1988). *Dynamic Manufacturing*. The Free Press, New York.

Kackar, R. N. (1985). Off-line quality control, parameter design, and the Taguchi method (with discussion). *J. Quality Technol.* **17**, 176–209.

Khuri, A. I. and J. A. Cornell (1987). *Response Surfaces*. Marcel Dekker, New York.

León, R. V., A. C. Shoemaker and R. N. Kacker (1987). Performance measures independent of adjustment (with discussion). *Technometrics* **29**, 253–285.

León, R. V. and C. F. J. Wu (1992). A theory of performance measures in parameter design. *Statist. Sinica* **2**, 335–358.

Li, C. C., Y. Washio, T. Iida and S. Tanimoto (1990). New linear graphs for orthogonal array $L_{16}(2^{15})$. *J. Chinese Inst. Industr. Eng.* **7**, 17–23 (in Chinese).

Lin, D. K. J. and W. Tu (1995). Dual response surface optimization. *J. Quality Technol.* **27**, 34–39.

Lochner, R. H. and J. E. Matar (1990). *An Introduction to the Best of Taguchi and Western Methods of Statistical Experimental Design*. Quality Resources, New York.

Logothetis, N. (1990). Box–Cox transformations and the Taguchi method. *Appl. Statist.* **39**, 31–48.

Logothetis, N. and H. P. Wynn (1989). *Quality Through Design: Experimental Design, Off-line Quality Control, and Taguchi's Contributions*. Clarendon Press, Oxford.

Lorenzen, T. J. and M. A. Villalobos (1990). Understanding robust design, loss functions, and signal to noise ratios. General Motors Report GMR-7118, Mathematics Department.

Lucas, J. M. (1994). How to achieve a robust process using response surface methodology. *J. Quality Technol.* **26** 248–260.

Lunani, M., V. Nair and G. S. Wasserman (1995). Robust design with dynamic characteristics: A graphical approach to identifying suitable measures of dispersion. Tech. Report No. 253, University of Michigan.

Mandel, J. and R. D. Stiehler (1954). Sensitivity – A criterion for the comparison of methods of test. *J. Res. Nat. Bureau Standards* **53**, 155–159.

Miller, A. E. and C. F. J. Wu (1991). Improving a calibration system through designed experiments. Research Report 91-06, Institute for Improvement in Quality and Productivity, University of Waterloo.

Moorhead, P. R. and C. F. J. Wu (1994). Cost-driven parameter design. Tech. Report No. 222, Department of Statistics, University of Michigan.

Morrison, S. J. (1957). The study of variability in engineering design. *Appl. Statist.* **6**, 133–138.

Myers, R. H. and W. H. Carter, Jr. (1973). Response surface techniques for dual response systems. *Technometrics* **15**, 301–317.

Myers, R. H., A. I. Khuri and G. Vining (1992). Response surface alternatives to the Taguchi robust parameter design approach. *Amer. Statist.* **46**, 131–139.

Myers, R. H. and Y. Kim (1994). Response surface methods and the use of noise variables. Draft manuscript.

Nair, V. N., ed. (1992). Taguchi's parameter design: A panel discussion. *Technometrics* **34**, 127–161.

Nair, V. N. and D. Pregibon (1988). Analyzing dispersion effects from replicated factorial experiments. *Technometrics* **30**, 247–257.

Nelder, J. A. and Y. Lee (1991). Generalized linear models for the analysis of Taguchi-type experiments. *Appl. Stochast. Models Data Anal.* **7**, 107–120.

Peace, G. S. (1993). *Taguchi Methods: A Hands-On Approach.* Addison-Wesley, Reading, MA.

Phadke, M. S. (1986). Design optimization case studies. *AT&T Tech. J.* **65**, 51–68.

Phadke, M. S. (1989). *Quality Engineering Using Robust Design.* Prentice-Hall, Englewood Cliffs, NJ.

Phadke, M. S. and K. Dehnad (1988). Optimization of product and process design for quality and cost. *Quality Reliability Internat.* **4**, 105–112.

Phadke, M. S., R. N. Kackar, D. V. Speeney and M. J. Grieco (1983). Off-line quality control in integrated circuit fabrication using experimental design. *Bell Syst. Tech. J.* **62**, 1273–1309.

Plackett, R. P. and J. P. Burman (1946). The design of optimum multifactorial experiments. *Biometrika* **33**, 305–325.

Quinlan, J. (1985). Product improvement by application of the Taguchi methods. *3rd Suppl. Symp. on Taguchi Methods.* American Supplier Institute, Dearborn MI.

Rao, C. R. (1947). Factorial experiments derivable from combinatorial arrangements of arrays. *J. Roy. Statist. Soc.* (Supplement) **9**, 128–139.

Rosenbaum, P. (1994). Dispersion effects from fractional factorials in Taguchi's method of quality design. *J. Roy. Statist. Soc. Ser. B* **56**, 641–652.

Ross, P. J. (1988). *Taguchi Techniques for Quality Engineering.* McGraw-Hill, New York.

Seiden, E. (1954). On the problem of construction of orthogonal arrays. *Ann. Math. Statist.* **25**, 151–156.

Shoemaker, A. C. and K.-L. Tsui (1993). Response model analysis for robust design experiments. *Comm. Statist. Ser. A* **22**, 1037–1064.

Shoemaker, A. C., K.-L. Tsui and C. F. J. Wu (1991). Economical experimentation methods for robust design. *Technometrics* **33**, 415–427.

Steinberg, D. M. and D. Bursztyn (1993). Noise factors, dispersion effects, and robust design. Tech. Report No. 107, Center for Quality and Productivity Improvement, University of Wisconsin, Madison, WI..

Steinberg, D. M. and D. Bursztyn (1994). Dispersion effects in robust design experiments with noise factors. *J. Quality Technol.* **26**, 12–20.

Taguchi, G. (1987). *System of Experimental Design*, Vols. 1 and 2. UNIPUB, White Plains, NY.

Taguchi, G. (1991). *Taguchi Methods, Vol. 1: Research and Development*; *Vol. 3: Signal-to-Noise Ratio for Quality Evaluation.* ASI Press, Dearborn, MI.

Taguchi, G. and D. Clausing (1990). Robust quality. *Harvard Business Rev.* January–February, 65–75.

Tsui, K.-L. (1988). Strategies for planning experiments using orthogonal arrays and confounding tables. *Quality Reliability Internat.* **4**, 113–122.

Tuck, M. G., S. M. Lewis and J. I. L. Cottrell (1993). Response surface methodology and Taguchi: A quality improvement study from the milling industry. *Appl. Statist.* **42**, 671–681.

Vining, G. G. and R. H. Myers (1990). Combining Taguchi and response surface philosophies: A dual response approach. *J. Quality Technol.* **22**, 38–45.

Welch, W. J. and J. Sacks (1991). A system for quality improvement via computer experiments. *Comm. Statist. Ser. A* **20**, 477–495.

Welch, W. J., T. K. Yu, S. M. Kang and J. Sacks (1990). Computer experiments for quality control by parameter design. *J. Quality Technol.* **22**, 15–22.

Wernimont, G. (1977). Ruggedness evaluation of test procedures. *Standardization News* **5**, 13–16.

Wu, C. F. J. and Y. Chen (1992). A graph-aided method for planning two-level experiments when certain interactions are important. *Technometrics* **34**, 162–175.

S. Ghosh and C. R. Rao, eds., *Handbook of Statistics, Vol. 13*

8

Analysis of Location and Dispersion Effects from Factorial Experiments with a Directional Response

C. M. Anderson-Cook

1. Introduction

Directional data has long been studied in a wide variety of applications including biology (wind direction, bird and plant migration patterns) (see Batschelet, 1981), geology (magnetic polarity and orientation of rock samples) and astronomy (location of celestial bodies) (see Mardia, 1972). The common attribute of this type of data is that it can be expressed as an angle or direction, rather than measured on the usual linear scale. In this chapter we consider only *circular data*, a common case of directional data. Here the data is restricted to lie on a unit circle and can be summarized by either a vector of unit length or an angle. Because of the data sources, experiments to determine the causal effect of experimenter-controlled input factors have been quite rare. Rather the form of data collection has generally been *observational*, meaning that the input factors thought to influence the directional response were not controlled in the study, but rather collected in parallel with the angular measure.

As a result, factorial experiments were considered unusual, and techniques for studying the effect of several factors simultaneously on the response were largely unavailable. Motivating the new interest in multi-way designs with a directional response are a number of problems arising in industrial settings. For example during automotive production, large numbers of rotation parts are produced, including tires, brake rotors, engine flywheels, and crankshafts. All of these parts must be precisely balance to avoid undue vibration or wear, since the parts are designed to spin at high speeds during operation of the vehicle. Mass production of the parts does not yield exactly balanced parts, and hence it is necessary to quantify and correct the direction of the imbalance. Measuring the imbalance yields a bivariate response: angular data (the direction in which a correction should be made) and linear data (the magnitude of the imbalance). A sizeable portion of the cost of repairing the imbalance is attributed to positioning the part to make the corrective adjustment. Therefore a designed experiment to determine which production factors significantly alter the location of the imbalance can provide strategies to reduce production costs.

Other applications of designed experiments yielding a directional response also exist. For example, for some processes, it may be more helpful to consider the connection between a target location and the actual placement of a mark or hole, in directional terms. It can be quantified as a direction of the miss and the distance of imperfection, as an alternative to the more traditional approach of measuring the difference on the usual x- and y-axes. In addition, the study of the location patterns of surface defects on circular or cylindrical parts require the analysis of a directional response given several input factors.

In this chapter, techniques for measuring and comparing the effect of several input factors on a directional response are presented. We focus our attention on studying a circular response resulting from a factorial experiment with replicates at each combination of factors. One of the purposes of robust parameter design as pioneered by Taguchi (1986) was to highlight the importance of categorizing the nature of factor effects as those affecting the mean and those influencing variability. Two separate analyses are presented. The first enables us to study how factors influence the location (or mean) of the directions, while the second considers how the spread (or dispersion) of the angles is altered by different factor combinations. The following subsection presents some of the required background for the remainder of the chapter. Section 2 gives details of an exploratory data analysis strategy for location, while Section 3 outlines the method for studying dispersion.

1.1. Background and notation

Before considering the particulars of either method, we provide some underlying framework for the basic measures and techniques used in directional data. For further details about any of this background information, the book by Fisher (1993) is an excellent source.

It will be useful to first discuss the one sample case, where data is thought to come from a single population (say n observations characterized by the angles $\theta_1, \ldots, \theta_n$, or by the unit vectors from the origin $u_1, \ldots, u_n$). Summarizing directional data is frequently done by calculating the *resultant vector*, the sum of the unit vectors, $\sum_i u_i$. This is a vector with length between 0 and n, and pointing in the *average direction* of the sample, $\theta_.$, where this direction is defined by

$$\theta_. = \begin{cases} \arctan(s/c), & \text{if } c > 0, \\ \frac{\pi}{2}, & \text{if } c = 0 \text{ and } s > 0, \\ \frac{-\pi}{2}, & \text{if } c = 0 \text{ and } s < 0, \\ \pi + \arctan(s/c), & \text{otherwise}, \end{cases} \tag{1}$$

where $c = \sum \cos\theta_i$ and $s = \sum \sin\theta_i$. We use the notation "." to denote summarizing over that index of the angles or resultant lengths. We standardize the length of this vector to lie in the interval [0, 1] by dividing by the number of observations to obtain

$$\bar{R} = \frac{1}{n}(c^2 + s^2)^{1/2}. \tag{2}$$

This standardized length will be close to one, if the data are closely clustered around the mean. However, if the data are spread evenly around the circle, then $\bar{R}$ will be near zero. Hence, this quantity gives us a natural measure of spread.

We define the unit vector associated with the average direction to be

$$\underline{u}_{\cdot} = \frac{u_{\cdot}}{||u_{\cdot}||}, \tag{3}$$

where $||\cdot||$ measures the length of the vector.

By examining the resultant vector as an average direction, an obvious measure of location, and its length, a dispersion measure, we obtain a non-parametric decompositions of the summary statistic into two components that quantify attributes of interest for our analysis. The components of this decomposition will be used subsequently as the starting points of both location and dispersion analyses.

For the balanced one-way situation with p groups, each with m sample values, $\theta_{i1}, \ldots, \theta_{im}$. The standardized resultant vector for each group is obtained as above, giving the following summary statistics, $\theta_{i\cdot}$ and $\bar{R}_{i\cdot}$ for each group with the "$\cdot$" again indicating averaging over of all the observations under that index. For the two-way situation, the resultant vectors for all of the rows, columns and cells are obtained. Hence, $\theta_{i\cdot\cdot}$ and $\theta_{\cdot j\cdot}$ are the resultant angles for the ith row and the jth column, respectively, and $\bar{R}_{ij\cdot}$ is the standardized length of $u_{ij\cdot}$ for the ijth cell.

A common choice of distribution for circular data is the von Mises distribution with mean direction μ and concentration parameter k, denoted by $VM(\mu, k)$, which has density of the form:

$$f(\theta) = \frac{1}{2\pi I_0(k)} \exp\{k \cos(\theta - \mu)\}, \tag{4}$$

where $I_0(k)$ is the modified Bessel function with $k > 0$ and $\theta \in (-\pi, \pi)$. It is roughly similar to the normal distribution for traditional data measured on a linear scale, in that it is a natural choice of distribution for a wide range of applications. For more details about the desirable properties associated with the von Mises, see Mardia (1972, pp. 55–58).

2. Location analysis

In this section, we describe a method for determining the size of a location effect on a directional response and then for obtaining its relative ranking with approximate significance levels for all of the main and interaction factor effects.

Before discussing the measurement of effects, it is helpful to consider what null hypothesis being tested. If a single factor has no significant effect, we expect that the average direction of the resultant vector for each level of the factor will be approximately the same. This parallels closely what is meant by no main effect with data measured on a linear scale. The Watson–Williams method (1956) is a useful directional decomposition of variation for one-way designs and nested two-way designs (Stephens, 1982). It also uses this definition of no main effect. However, it does not extend to the multi-way factorial setting.

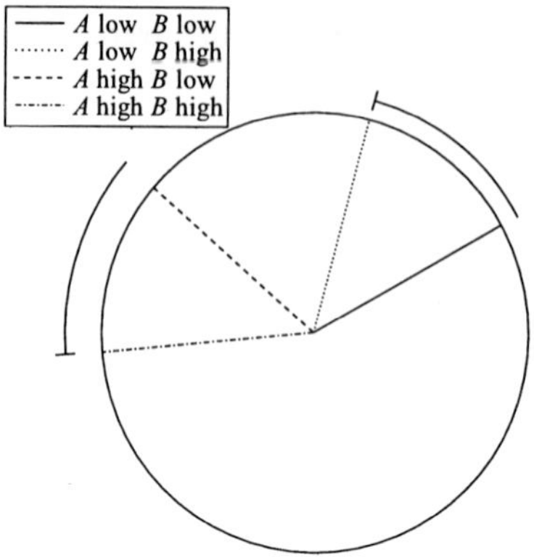

Fig. 1. Example of no interaction.

The matching definition for no interaction (of two or more factors) follows less directly from the linear case, because there is, at present, no underlying model for multi-way designs with a circular data response comparable to the linear regression model,

$$Y_{ijk} = \mu + A_i + B_j + AB_{ij} + \varepsilon_{ijk}. \tag{5}$$

For more details about the modelling difficulties see Anderson (1993) or Anderson and Wu (1995). As a result, several possible definitions of interaction are possible in different contexts. One that is relatively appealing for a wide variety of applications gives no interaction if there is a fixed angular rotation between levels of one factor for all levels of the other factors when they are held fixed. Figure 1 gives an example of no interaction for a two-way interaction, using a circular interaction plot presented in Anderson (1994). The four radial lines give the means of the four factor combinations, while the arcs show the angular rotations between levels of one factor at a fixed level of the other factor. An interaction occurs if the size or direction of the angular rotation differs across levels of one factor. The roles of the two factors are interchangeable.

2.1. *Test statistics*

The Anderson–Wu (1995) analysis method was developed starting from the one-way design and uses the likelihood ratio test statistic assuming data for a von Mises distribution with unknown means for each group. The likelihood function for an observation from this distribution has the form,

$$l(\mu) = 2k\cos(\theta - \mu), \tag{6}$$

where θ is the observed value. Hence, the test statistic for the null hypothesis of no difference between groups in the one-way design with m observations per group is

$$\begin{aligned}\text{LRTS} &= k\sum_i m\bar{R}_{i\cdot}\,\|\underline{u}_{i\cdot} - \underline{u}_{\cdot\cdot}\|^2 \\ &= 2k\sum_i m\bar{R}_{i\cdot}\,\big(1-\cos(\theta_{i\cdot}-\theta_{\cdot\cdot})\big)\end{aligned} \tag{7}$$

where $\bar{R}_{i\cdot}$ is the length of the standardized resultant vector for the ith group, $\theta_{i\cdot}$ and $\theta_{\cdot\cdot}$ are the average angle of the ith group and the overall average of all observation, and $\underline{u}_{i\cdot}$ and $\underline{u}_{\cdot\cdot}$ are the unit vectors pointed in the direction of their average angles. It is easier to picture what is being measured with the first form of the test statistic, while the second form is computationally more advantageous.

Interpreting the pieces shows that the test statistic is comparing the square of the linear distance between the points on the unit circle that correspond to the individual group means and the overall mean. Intuitively, if there is no difference between groups, then all of the average directions will be close to each other and the linear distances between them will be zero. If the distance is 'large', then the average directions are not close to each other and a significant difference between groups exists. The weighting factor, $m\bar{R}_{i\cdot}$, is an adjustment for the different levels of confidence that we have in the correct identification of the individual group mean. Since $\bar{R}_{i\cdot}$ is a measure of dispersion for the group, the more closely concentrated the data in group i, the larger the value of the weight. If the observations in the group are located in a wide range of the circle, then the weight of that term of the sum is reduced to near zero, since we have relatively little confidence in the average direction identified.

Usually k, the concentration parameter, is not known, and hence we use the maximum likelihood estimator, $\hat{k} = A^{-1}(\frac{1}{p}\sum_i \bar{R}_i)$, where $A(k) = \frac{I_1(k)}{I_0(k)}$ is the ratio of modified Bessel functions of order 1 and 0. When this is substituted into equation (7) for k it gives LRTS*. Hence, this is the form of the test statistic used for measuring the main effects of the multi-way design,

$$\text{LRTS}^\star = 2\hat{k}\sum_i m\bar{R}_{i\cdot}(1-\cos(\theta_{i\cdot}-\theta_{\cdot\cdot})). \tag{8}$$

To adapt this test statistic to deal with the multi-way case and study the influence of two or more factors on the directional response, we note that no interaction for the linear case can be equivalently described by three equations. For example for the case with two factors each at two levels, we obtain no interaction effect when

$$\begin{aligned}\bar{Y}_{12} - \bar{Y}_{11} &= \bar{Y}_{22} - \bar{Y}_{21}, \\ \bar{Y}_{21} - \bar{Y}_{11} &= \bar{Y}_{22} - \bar{Y}_{12}, \\ \bar{Y}_{11} + \bar{Y}_{22} &= \bar{Y}_{12} + \bar{Y}_{21}.\end{aligned} \tag{9}$$

However, for the circular data case, the analogous first two equations when we substitute in the circular averages may not be equivalent to the third, since the distributive property ($a(b+c) = ab+ac$) does not hold universally when we do arithmetic modulo 2π or work with the directional averages of the groups (see Anderson, 1993). As a result, by starting with the third equation of (9), a test statistic for quantifying interaction effects can be formulated. We construct a surrogate main effect by grouping together the factor combinations into new factor levels. For example, for the 2^2 factorial case with main effects A and B, the new surrogate factor will have a high value for $(A, B) \in \{(L, L), (H, H)\}$ and low for $(A, B) \in \{(L, H), (H, L)\}$. If the main effect using the LRTS* is zero, than this corresponds with a zero interaction effect as illustrated by Figure 1. However, it is possible to have a non-zero main interaction for the surrogate and still have a zero interaction as defined by the fixed angular rotation criteria. This may occur if the main effects A or B are themselves large and the resultant vectors for the surrogate factors point in exactly opposite directions, instead of in the same direction. This is purely a function of the "wrap-around" nature of directional data which does not have a natural maximum or minimum value. With a slight modification of the LRTS* we can accommodate this situation and obtain a measure of the interaction effect using this definition, which considers both the possibility of the group averages pointing in the same direction or 180° opposite.

$$\begin{aligned}\mathrm{LRTS}^{\star}_{\mathrm{int}} &= \hat{k}\sum_{l} m\bar{R}_l \min\left\{1-\cos(\theta_l-\theta_{..}),\ 1-\cos(\theta_l-\theta_{..}+\pi)\right\}\\ &= \hat{k}\sum_{l} m\bar{R}_l\left(1-|\cos(\theta_l-\theta_{..})|\right) \qquad (10)\end{aligned}$$

where the subscript l indicates the different levels of the constructed surrogate variable.

For a general 2^r factorial design, the interaction terms are constructed by grouping observations into two mutually exclusive groups, combining all observations with an odd number of high levels into one group and the remaining observations into the second group. For example, for a 3-way interaction (of factors A, B and C), the high level of the surrogate factor D would include observations with factor combinations HHH, HLL, LHL, and LLH, while the low level would consist of HHL, HLH, LHH and LLL, where HLH means that factor A is at the high level, factor B low and factor C high. Once the observations are divided into groups, the form of the test statistic is the same as the two-way interaction. A zero three-way interaction means the same AB interaction is observed at both levels of factor C (i.e., the same difference in angular rotation between levels of factor A, when B is fixed, would hold for both levels of C). The roles of the three factors can be arbitrarily interchanged. The use of a surrogate variable with the interaction test statistic gives a zero effect for these circumstances.

By choosing this form of the test statistic for quantifying interactions, we obtain a simple exploratory means of calculating and ranking the factor effects for a factorial design. Analysis of location-only directional interactions was previously not possible as the extension to the Watson–Williams method (Underwood and Chapman, 1985) does not measure interaction terms in an easily interpreted manner. The Anderson–Wu method parallels our understanding of interactions gained from linear models, and hence is a good exploratory tool. Because of the orthogonal structure of a factorial experiment, we might hope that the test statistics for all the factors are minimally correlated. This is in fact true if all of the effects are near zero. However, if one or more factor effects are large, then some negative correlation does exist. While uncorrelated test statistics would be optimal, the negative correlations do not prove to be problematic, as the ranking of effects is well preserved. The only influence of the correlation is to influence any predicted significance levels. See Anderson and Wu (1995) for further details.

2.2. *Procedure*

We now turn out attention to the complete analysis of a factorial design with a directional response. For the one-way analysis with the assumption that the data comes from a von Mises distribution, the likelihood ratio test statistic is asymptotically chi-squared with $p-1$ degrees of freedom, where p is the number of groups being considered. If the distributional assumptions are relaxed, this approximation remains reasonable for many data sets. Therefore, for the 2^r factorial case with replicates, the individual test statistics for any main or interaction effect are approximately chi-squared with one degree of freedom and a half-normal plot of square root of the LRTS* (either the form for the main effect or the interaction, as appropriate) will provide a framework for comparing the size of the effects relative to each other. This graphical assessment of the contributions of the factors to the location used in conjunction with the likelihood ratio test statistic values and their estimated significance levels directly can give insight into the influence of the factors. Once influential main or interaction effects of the factors have been identified, then particular combinations of factors can be considered to provide future observations from the process which are close to a desired target.

2.3. *Industrial example*

To illustrate the details of the method for a multi-way experiment, the analysis of an industrial data set is provided. The procedure for analyzing a 2^3 factorial experiment with replicates is outlined below.

The data set provided in Table 1 comes from an automotive production plant, and considers the possible effect of three explanatory variables on the direction of imbalance for brake rotors. In this case the machine used to determine the location of the imbalance was restricted in precision to 15° increments on the circle, hence only 24 different location values were possible. The three factors thought to influence directional location are as follows:

A – *Gang Core type* either type 4 (0) or type 6 (1).

B – *Core position* either normal (0) or offset (1).

C – *Core thickness* either normal (0) or reduced (1).

Table 1
Automotive brake rotor data

Group	Factors A	B	C	Response in degrees							
1	0	0	0	240	270	270	270	255	255	75	255
2	0	0	1	180	345	330	180	195	120	60	345
3	0	1	0	30	180	90	135	165	165	45	120
4	0	1	1	180	240	180	135	180	180	105	150
5	1	0	0	225	180	0	195	285	195	120	180
6	1	0	1	255	240	75	195	165	180	285	150
7	1	1	0	225	150	135	165	150	180	300	180
8	1	1	1	180	345	30	345	150	195	30	195

If these parts are not precisely balanced, then under normal operating conditions they may cause vibration to be transmitted throughout the system and result in poor performance and premature wear of other parts.

After the production of the parts, each one was tested to determine the magnitude of the imbalance and the direction of the imbalance. The size of the imbalance is a scalar measure, while the particular location on the edge of the part where a corrective action must be taken (either weight being added or subtracted, depending on the application) is directional in nature. Hence a typical measurement from this type of analysis might be a 1.25 foot-pound imbalance in the direction 135° from the "home" location on the part. In this analysis, we consider only the directional component of this bivariate response, since a significant portion of the cost of correcting a defective part is associated with the location of the imbalance and some portions of the part may be more cost effective to correct than others.

To begin the analysis, we calculate the resultant vectors for the 8 combinations of factors. Table 2 shows the angle, θ_i, and unstandardized length, R_i, of the resultant vector for each combination of factors. In addition, the fourth and fifth columns of Table 2 show the x and y components of the unstandardized resultant vectors (obtained using the usual Polar to Cartesian coordinate conversion equations, $x_i = R_i \cos\theta_i$ and $y_i = R_i \sin\theta_i$) for each group of 8 observations. These components will be helpful later for the calculation of other resultant vectors on interest.

In the experiment, all 7 factor effects (A, B, C, D, AB, AC, BC, and ABC) were considered of potential importance, and hence in the next step of the analysis 15 resultant vectors (2 for each of the main and interaction effects and 1 for the overall data) were calculated by summing up the relevant x and y entries for groups included in a particular combination of factors. For each of the factor effects, there were 32 observations at each level of the effect, while for the overall resultant vector all 64 were used. The calculation of the standardized resultant vector lengths, $\bar{R}$, and the average angles, $\theta.$, can be easily implemented into most computer packages for greater efficiency.

Once, these quantities have been calculated the test statistics, either equation (8) for main effects or equation (10) for interactions are used to obtain a value for the

Table 2
Group summaries of brake rotor data

Group	θ_i	R_i	x_i	y_i
1	5.887	−1.745	−1.018	−5.798
2	0.486	1.924	−0.168	0.456
3	5.006	2.109	−2.566	4.298
4	6.465	2.938	−6.332	1.307
5	4.100	−2.813	−3.880	−1.325
6	4.253	−2.823	−4.039	−1.332
7	5.626	3.072	−5.612	0.393
8	0.484	1.851	−0.134	0.465

Table 3
Main and interaction effects for brake rotor data

Effect	LRTS*	S.L.	Rank
A	0.002	0.96	(7)
B	0.159	0.69	(1)
C	0.008	0.93	(6)
AB	0.056	0.81	(4)
AC	0.012	0.91	(5)
BC	0.063	0.80	(3)
ABC	0.068	0.80	(2)

likelihood ratio test statistic for the null hypothesis of any particular effect being zero. The value of $\hat{k}$ can be estimated from the average of the individual group resultant vector lengths using Appendix 2.3 of Mardia (1972). In this case the concentration parameter value is estimated to be $\hat{k} = 1.174$. Table 3 shows the test statistic and approximate significance level for effects in the brake rotor data set. While the significance levels of any of these test statistics are only rough approximations, simulation results have confirmed that the relative size of the effects is well preserved if we rank the likelihood ratio test statistics (or their approximate significance levels) (Anderson and Wu, 1995). Therefore, from the ranking and the half-normal plot, it is possible to determine which effects are 'large' relative to the others and should be considered genuine effects. Figure 2 shows a half-normal plot of the square root of the LRTS* values.

Here the information in the half-normal plot and the summary table of values of factor effects seem to provide conflicting results. If we examine only the half-normal plot, we would likely conclude that factor B is influential in affecting the location of the response. On the other hand, if we look at the significance levels of the test statistics, our conclusion would likely be that no factors are influential (in fact, the results imply that even at the 10% level, nothing is close). While we know a priori that the asymptotic results for the distribution of the test statistic will not hold exactly

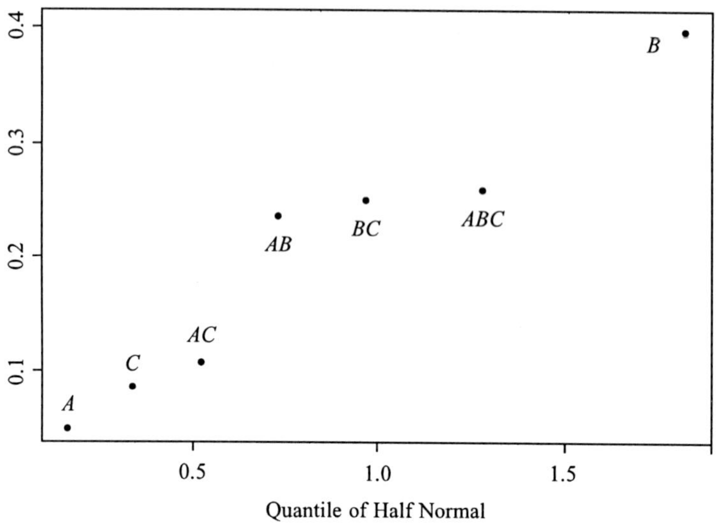

Fig. 2. Half-normal plot of location effects.

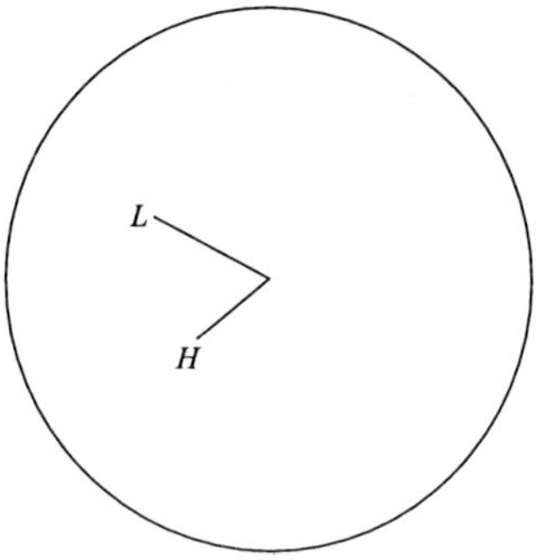

Fig. 3. Resultant vectors of factor B.

for small samples, it is not clear how much weight to give to the interpretation of the half-normal plot versus the approximate significance levels. To help resolve the conflict of whether Factor B should be considered important, plots of the factor can be considered. Figure 3 shows the resultant vectors for Factor B for both the high and low levels. The angular distance between resultant vectors appears to be relatively large, but without scaling for how disperse the original data are, it is difficult to draw meaningful conclusions.

To incorporate the spread of the underlying distribution of the brake data, we plot two densities in Figure 4. We preserve the difference in average angle, and use the estimated concentration parameter value of $\hat{k} = 1.174$. From this plot we see that the difference between the high and low levels of the factor are indeed small relative to the total variability of the data, and hence a conclusion of no significant factor influencing the location of the directional response is most appropriate for this data

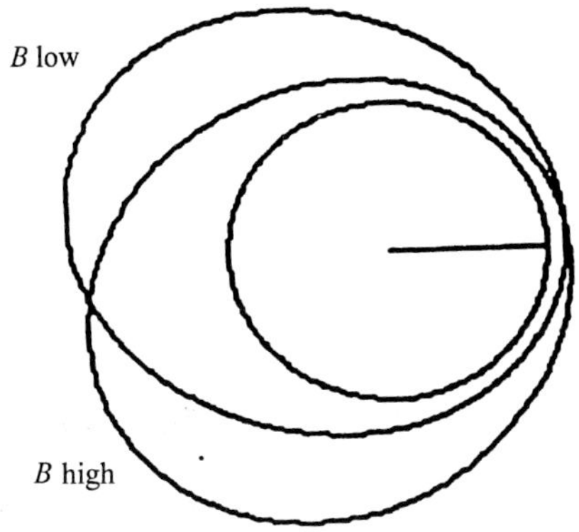

Fig. 4. Densities of factor B.

set. The key to note with this data set is that the original data is quite disperse, and hence for a factor to have a significant effect on the response the angular difference between resultant vectors would have to be very large.

For an analysis of factor effects on the response where significant effects are observed, see the automotive flywheel data in Anderson and Wu (1995).

3. Dispersion analysis

In the previous section, we considered how factors might affect the mean of a directional response. Next we study how the spread of the data is influenced by changes in the factor levels of a factorial experiment with replicates.

To motivate the need for such an analysis, we return to the industrial example of rotating parts. For this class of problem, corrections may uniformly expensive around the entire circumference of the part, but savings could be realized if all of the corrections are localized to a small segment of the part. Alternately, we may wish to target a specific region of the part (location analysis) and then subsequently reduce the variability around that target (dispersion analysis) by adjusting the levels of some factors that do not influence the mean. Methods for incorporating the study of noise factors are described in Anderson and Wu (1994).

For linear data, analysis strategies exist for looking for an optimal combination of factors to minimize the spread of the data (usually measured by the variance). The strategy presented in this section is modelled after a popular variation reduction method (Nair and Pregibon, 1988; Box, 1988), but has been adapted to deal with the special needs of directional data. One aspect of directional data that makes it different from traditional linear data is that it is confined to lie on a limited range (namely the circle) which does not have a natural maximum or minimum. Hence, there is an automatic upper bound on how disperse data can be, which occurs when the data is spread uniformly around the entire circumference of the circle. The new method adapts to the peculiarities of directional data and gives a method for determining which factor are most influential on the spread of the data.

3.1. Circular variance and standard deviation

Before we can develop a method for studying dispersion, a summary statistic or statistics similar in properties to the variance for linear data must be found. It is desirable that it is simple, intuitively pleasing and computationally convenient for data from any distribution. In addition, it would be beneficial if the measure has known distributional properties under more restrictive assumptions.

As mentioned in Section 1, the decomposition of the resultant vector into two non-parametric components, the average direction and the resultant length, provides a starting point for the analysis. The resultant vector is a simple measure of dispersion, and hence the natural choice for quantifying the spread of the data. Fisher and Lee (1992) propose a method for studying dispersion using the von Mises concentration parameter, $\hat{k}$, but since this does not have a sensible interpretation for non von Mises data and is computationally more difficult to work with, it is not pursued here.

The circular variance is a common dispersion statistic defined in terms of the length of the standardized resultant vector,

$$S_0 = 1 - \bar{R}. \tag{11}$$

Since $\bar{R}$ is restricted to the range $[0, 1]$, so is the circular variance. Minimum variation occurs when $S_0 = 0$ ($\bar{R} = 1$) and corresponds to all of the observations in a given sample occurring at precisely the same location. As noted above, for circular data there is a natural upper limit to the possible variation. This occurs for data uniformly distributed around the circle, and corresponds to the circular variance being equal to one.

Calculation of $\bar{R}$, and hence the variance is straightforward, and interpretation of the results does not depend on distributional assumptions about the original data. Various authors, including Watson and Williams (1956) and Mardia (1972, p. 113) describe the asymptotic properties of S_0 under the more restrictive assumption that the original data comes from a concentrated von Mises distribution. Anderson and Wu (1996) show that even from small samples and moderate concentration parameter values for the von Mises distribution, the chi-squared approximation of the circular variance is good. Hence, S_0 satisfies our requirements for an ideal dispersion summary statistic.

In addition, another candidate which possesses some of the same properties as the circular variance is the circular standard deviation. Unlike the linear case, the standard deviation is not defined as the square root of the variance, but rather as a different function of the standardized length of the resultant vector,

$$s_0 = (-2 \log \bar{R})^{1/2}. \tag{12}$$

This summary statistic is non-negative, ranging from zero when there is no variation in the sample, to infinity when the data are uniformly distributed around the circle. Less is known about its distributional properties, but for concentrated samples the following relationship between the circular standard deviation and the variance is known to exist

$$s_0 \approx \sqrt{2S_0}\left\{1+\frac{1}{4}S_0+\frac{13}{96}(S_0)^2+\frac{43}{384}(S_0)^3\right\}, \tag{13}$$

where the right hand bracket of (13) will approach one as the data becomes more concentrated and S_0 shrinks. Hence the circular standard deviation is another possible candidate for summary statistic which is easily calculated and can be interpreted for all data sets. We will subsequently exploit the approximate relationship between these two test statistics.

3.2. Analysis procedure

This section describes in detail the analysis method for determining which factor effects influence the spread of the directional response from a factorial experiment.

A good first test to determine if there are significant differences between the dispersions for the factorial combinations, is to use Bartlett's test of homogeneity of concentration parameters for von Mises data (Stephens, 1982). For each of the 2^r groups with m replicates in a 2^r factorial experiment, define $Q_l = mS_l$ and $q_l = m-1$ where l ranges from 1 to 2^r for the different groups and S_l is the circular variance as defined in (11) for group l. We also define $T=\sum_l Q_l$ and $t=\sum_l q_l$. Then the test statistic for testing if a difference between groups exists is Z/C, where

$$Z = t\log T - \sum_l (q_l \log Q_l) - t\log t + \sum_l (q_l \log q_l) \tag{14}$$

and

$$C = 1 + \frac{1}{3(s-1)}\left(\sum_l \frac{1}{q_l} - \frac{1}{t}\right), \tag{15}$$

where $s = 2^r$ is the number of groups. Z/C is approximately chi-squared with $(s-1)$ degrees of freedom under the null hypothesis of no difference between groups. Therefore, for a test of size $1-\alpha$, we would reject that hypothesis if

$$P\left(\chi^2_{(s-1)} \geqslant \frac{Z}{C}\right) \leqslant \alpha. \tag{16}$$

Bartlett's test is generally sensitive to assumptions of normality, which in this case corresponds to the data originating from a von Mises distribution, and hence this test should be viewed primarily as a diagnostic tool to gain a feel for the data and used only for identifying large variation differences. If there is no evidence against the hypothesis that variance estimates are constant, then the remaining analysis will likely not bear fruitful results.

If Bartlett's test does identify differences, we proceed with further analysis. Since the circular variance has all of the desirable properties that we seek in a dispersion summary statistic, we use it as our starting point. The Box–Cox (1964) data transformation modelling approach strives to balance three separate goals: simplicity of structure, variance homogeneity, and normality. A suitable transformation of the data is sought using the one parameter Box–Cox power transformation family of the form:

$$S_0^\lambda = X\Upsilon + \varepsilon, \tag{17}$$

where S_0 is the vector of circular variances for each combination of factors, and λ is the transformation power. X is the design matrix (comprised of -1's and 1's for a two-level factorial design), Υ is the vector of parameters, and $\varepsilon \sim MVN(0, \sigma^2 I_n)$ is the vector of error terms.

It should be noted that to use this method for a full factorial with replicates, it is necessary to simplify the model used to compare all of the different possible transformation values. A useful first attempt is to consider only the main effects and the two-way interactions for different λ values, and use this as a basis for choosing the optimal value or values. Once the transformation has been selected, then reverting to the full model with all higher order interaction terms is possible.

Since we are using a dispersion summary measured on a linear scale, the model has the same basic form as dispersion analyses for traditional linear data and normality can be assumed for the error term. This transformation to a linear scale is essential, because we have little intuitive feel for directional dispersion measures, and we are able to use the existing methods for a linear response. While the transformation selected for $\lambda = 0$ is the natural logarithm as in the linear case, we make a special choice of transformation if $\lambda = \frac{1}{2}$. In this case, it is more desirable to select the circular standard deviation as the left-hand side of equation (17) rather than the square root of the circular variance, because it gives a more easily interpreted model in many cases, because the influence of the factor effects is being expressed in terms of a natural measure of dispersion. Recall the connection between the two choices (as summarized in equation (13)) for concentrated von Mises data.

Because the circular variances are frequently close in distribution to chi-squared, the optimal choice of the power transformation parameter will commonly lie in $\lambda \in (-1, 1)$. In addition, the special cases of $\lambda = 0$ and $\lambda = \frac{1}{2}$ are often contained in the 95% confidence interval. We now consider these models in greater detail, since their interpretation in terms of directional quantities is straight-forward.

We obtain a model with multiplicative effects and multiplicative errors on the circular variance for situations where $\lambda = 0$, and the model takes the form $\log S_0 = X\Upsilon + \varepsilon$. For example for a two-way design we obtain

$$\log S_{ijk} = \sigma_0 + A_i + B_j + AB_{ij} + \varepsilon_{ijk}. \tag{18}$$

This can be interpreted as the circular variance being influenced by the factors in the following way

$$S_{ijk} = \mathrm{e}^{\sigma_0}\, \mathrm{e}^{A_i}\, \mathrm{e}^{B_j}\, \mathrm{e}^{AB_{ij}}\, \mathrm{e}^{\varepsilon_{ijk}} = \sigma_0^*\, A_i^*\, B_j^*\, AB_{ij}^*\, \varepsilon_{ijk}^*, \tag{19}$$

where σ_0^* is the baseline measure of variability of the data, and A_i^*, B_j^*, and AB_{ij}^* are the main and interaction effects of the factors, respectively. If $A_i < 0$, and hence $A_i^* < 1$, then level i of the factor reduces the variance, and should be selected to satisfy our goal of minimizing variation.

The second frequently selected model occurs if $\lambda = \frac{1}{2}$. It yields the additive model with an additive error,

$$s_0 = X\Upsilon + \varepsilon. \tag{20}$$

Again illustrating with the same two-way design, we obtain the following model:

$$s_{ijk} = \sigma_0 + A_i + B_j + AB_{ij} + \varepsilon_{ijk}, \tag{21}$$

where σ_0 is the baseline estimate of the circular standard deviation of the data, and A_i, B_j, and AB_{ij} are the main and interaction effects of the factors, respectively. Again, $A_i < 0$ corresponds to level i of factor A reducing the spread of the data.

In different industrial applications, one of the models described above may concur more closely with the physical understanding of the process, and hence be preferable. After a suitable transformation has been found, the size of the factor effects for the full model can be calculated and "large" effects identified by using a half-normal plot. Based on the results of this analysis, an optimal set of factor levels can be identified for minimizing the variation of the directional response. In addition, the procedure can be refined to consider parameter choices robust to noise effects in a control-by-noise array design. For further details of this type of analysis see Anderson and Wu (1996).

3.3. *Industrial example*

An automotive example involving the balancing engine flywheels is given here to illustrate the process of selecting an optimal combination of factors to minimize the spread of imbalance angles. The four factors considered potentially important for describing the variation of the directional response are as follows:

Buttweld location either fixed (F) or random (R).

Radius tolerance either high (H) or low (L) quality.

Thickness tolerance either high (H) or low (L) quality.

Counter-weight size either high (H) or low (L).

The 160 observations are designed as a 2^4 factorial with 10 observations at each combination of factors. In reality the 10 observations have some additional structure, but this will be ignored here to simplify the analysis presented in Anderson and Wu (1996). Table 4 shows the data with the factor levels.

When Bartlett's test for homogeneity of concentration parameters was carried out, a test statistic near 44 was obtained with an approximate significance level of 0.0001. Hence dramatic difference between groups exist.

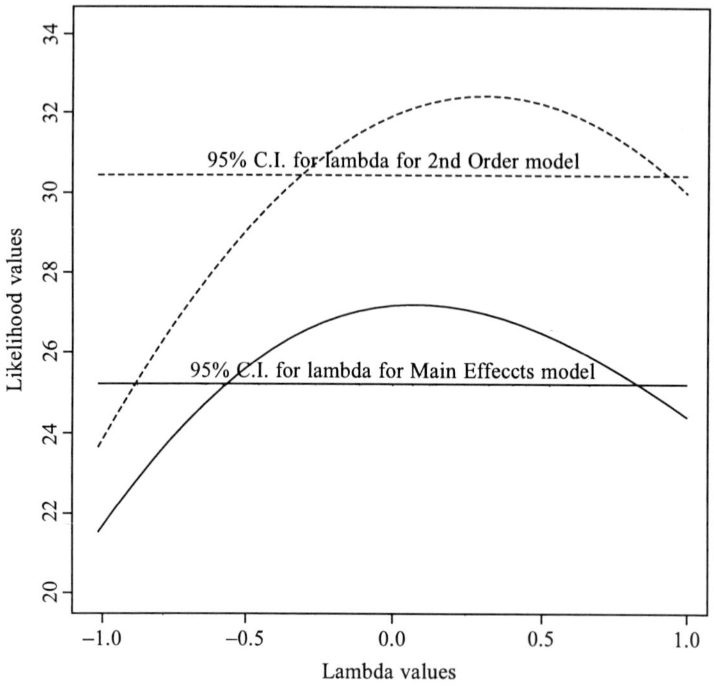

Fig. 5. Box–Cox maximum likelihood estimates of lambda.

Figure 5 shows the result of the Box–Cox procedure for the main effects only model and the 2nd order model. Clearly, both $\lambda = 0$ and $\lambda = \frac{1}{2}$ could be considered acceptable models. The project engineers felt that the multiplicative effects with a multiplicative error model better matched their understanding of the process, and hence this model was pursued. It should be noted that the analysis for $\lambda = \frac{1}{2}$ was also considered and gave many of the same conclusions as the natural logarithm transformation.

Table 5 shows the sums of squares for each of the factor main and interaction effects with their relative ranking. Figure 6 gives the half-normal plot of the square root of these sums of squares.

Factor "B" is the most influential effect, with the two-way interaction "CD" also appearing to be significant. Assuming that a hierarchical model is suitable, we would include the main effects "B", "C" and "D", with the two-way interactions "CD" and "BC". These are the five largest effects and give a final model of

$$\log S_0 = -1.34 - 0.94B_i - 0.52C_j + 0.48D_k + 0.78(CD)_{jk} + 0.56(BC)_{ij} + \varepsilon_{ijkl} \tag{22}$$

where the indicator variables have value 0 at the low level and 1 at the high level of each factor. The optimal choice is high-high-high for "BCD", which yields a

Table 4
Flywheel data

Group	A	B	C	D	Data									
1	R	L	L	L	133	175	178	178	153	190	221	177	281	190
2	R	L	L	H	139	61	109	187	74	351	309	236	69	320
3	R	L	H	L	111	122	105	49	189	188	177	151	62	329
4	R	L	H	H	170	162	19	337	171	114	341	10	266	201
5	R	H	L	L	127	215	125	188	187	175	162	172	169	82
6	R	H	L	H	150	84	113	318	84	353	301	12	82	351
7	R	H	H	L	152	164	180	187	159	149	127	148	175	201
8	R	H	H	H	184	128	177	186	163	178	196	155	150	120
9	F	L	L	L	154	200	147	133	171	318	100	108	86	73
10	F	L	L	H	198	165	31	51	314	84	267	135	318	14
11	F	L	H	L	345	43	4	295	75	138	149	141	198	175
12	F	L	H	H	153	194	207	136	144	206	151	202	104	188
13	F	H	L	L	140	134	170	62	109	127	132	116	94	183
14	F	H	L	H	340	111	128	327	81	301	3	335	215	334
15	F	H	H	L	160	152	187	158	143	91	200	143	84	191
16	F	H	H	H	171	156	171	195	159	153	188	125	107	98

Table 5
Flywheel analysis

	$\text{Log}(S_0)$	
Source	Sum of squares	Rank
A	0.069	(12)
B	3.563	(1)
C	1.071	(4)
D	0.924	(5)
AB	0.263	(10)
AC	0.015	(15)
AD	0.632	(7)
BC	1.123	(3)
BD	0.091	(11)
CD	2.004	(2)
ABC	0.872	(6)
ABD	0.418	(8)
ACD	0.371	(9)
BCD	0.016	(14)
$ABCD$	0.021	(13)

predicted value of -2.325. If these levels of the factors are selected, we would expect the circular variance to be 0.098. Alternately, if the present production levels ("BCD" at the low-low-low levels) are used the resultant length is 0.812. By changing from

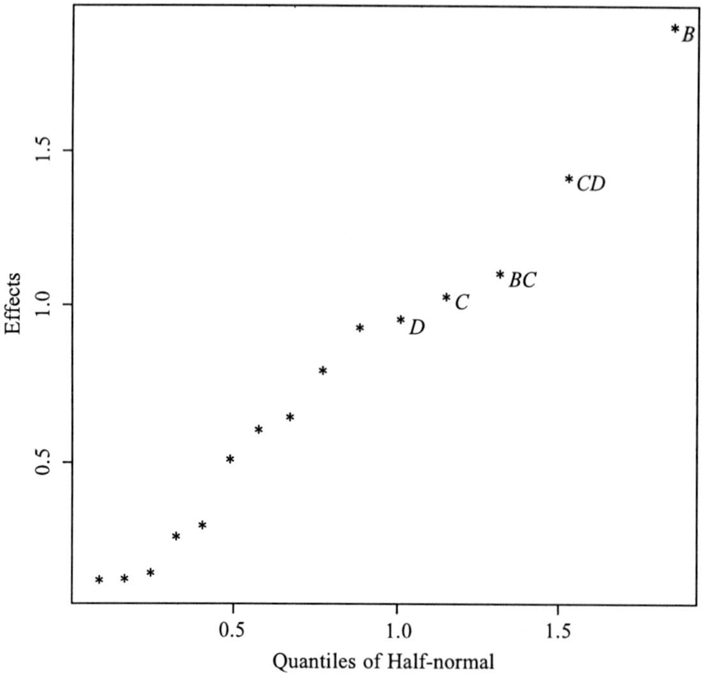

Fig. 6. Half-normal plot of dispersion effects.

the current production levels to the new set of factor combinations, we would be able to reduce variation from 0.188 to 0.098, a dramatic 48% reduction. By examining the two combinations of factors with the HHH for "BCD" we find that their average circular variance is 0.109, still a 42% reduction from current levels. Therefore, a substantial savings can be realized as a result of this study.

4. Conclusion

In the previous sections we have described how to study both the influence of factors on location and the dispersion of a circular response. While the goals of the two analyses are very different, it should be noted, that quite often the optimization of a process involves simultaneous goal for both of these aspects of the data. For example, it may be desirable to minimize the cost of production by examining both the location and the dispersion influences. In these cases, the procedures outlined in Sections 2 and 3 can be followed and the results combined to try to simultaneously optimize the location goal and the minimization of the dispersion.

While in some situations it may not be possible to make optimal factor combination choices to satisfy both goals, a compromise which balances the parallel objectives of the two analyses may be possible.

Regardless of the conclusions resulting from each study, a better understanding of both location and dispersion effects is possible and can be used to influence the direction of future investigations.

References

Anderson, C. M. (1993). Location and dispersion analyses for factorial experiments with directional data. Unpublished Thesis, Department of Statistics and Actuarial Science, University of Waterloo.

Anderson, C. M. (1994). Graphical methods for circular and cylindrical data. Tech. Report TR-94-05, Department of Statistical and Actuarial Sciences, University of Western Ontario.

Anderson, C. M. and C. F. J. Wu (1994). Studying noise factors and dispersion of directional data. *ASA Proceedings, Section on Quality and Productivity*, 265–268.

Anderson, C. M. and C. F. J. Wu (1995). Measuring location effects from factorial experiments with a directional response, *Internat. Statist. Rev.* **63**, 345–363.

Anderson, C. M. and C. F. J. Wu (1996). Dispersion measures and analysis of factorial directional data with replicates. *Appl. Statist.* **45**, 47–61.

Batschelet, E. (1981). *Circular Statistics in Biology.* Academic Press, London.

Best, D. J. and N. I. Fisher (1979). Efficient simulation of the von Mises distribution. *Appl. Statist.* **28**, 152–157.

Box, G. E. P. (1988). Signal-to-noise ratios, performance criteria, and transformations (with discussion). *Technometrics* **30**, 1–40.

Fisher, N. I. (1993). *Statistical Analysis of Circular Data.* Cambridge Univ. Press, Cambridge.

Fisher, N. I. and A. J. Lee (1992). Regression models for an angular response. *Biometrics* **48**, 665–677.

Mardia, K. V. (1972). *Statistics of Directional Data.* Academic Press, London.

Nair, V. N. and D. Pregibon (1988). Analyzing dispersion effects from replicated factorial experiments. *Technometrics* **30**, 247–257.

Stephens, M. A. (1969). Tests for the von Mises distribution. *Biometrika* **56**, 149–160.

Stephens, M. A. (1982). Use of the von Mises distribution to analyse continuous proportions. *Biometrika* **69**, 197–203.

Taguchi, G. (1986). *Introduction to Quality Engineering: Designing Quality Into Products and Processes.* Asian Productivity Organization, Tokyo, Japan.

Underwood, A. J. and M. G. Chapman (1985). Multifactorial analyses of directions of movements of animals, *J. Exp. Marine Biol. Ecol.* **91**, 17–43.

Watson, G. S. and E. J. Williams (1956). On the construction of significance tests on the circle and the sphere. *Biometrika* **43**, 344–352.

S. Ghosh and C. R. Rao, eds., *Handbook of Statistics, Vol. 13*

9

Computer Experiments

J. R. Koehler and A. B. Owen

1. Introduction

Deterministic computer simulations of physical phenomena are becoming widely used in science and engineering. Computers are used to describe the flow of air over an airplane wing, combustion of gases in a flame, behavior of a metal structure under stress, safety of a nuclear reactor, and so on.

Some of the most widely used computer models, and the ones that lead us to work in this area, arise in the design of the semiconductors used in the computers themselves. A process simulator starts with a data structure representing an unprocessed piece of silicon and simulates the steps such as oxidation, etching and ion injection that produce a semiconductor device such as a transistor. A device simulator takes a description of such a device and simulates the flow of current through it under varying conditions to determine properties of the device such as its switching speed and the critical voltage at which it switches. A circuit simulator takes a list of devices and the way that they are arranged together and computes properties of the circuit as a whole.

In each of these computer simulations, the user must specify the values of some governing variables. For example in process simulation the user might have to specify the duration and temperature of the oxidation step, and doses and energies for each of the ion implantation steps. These are continuously valued variables. There may also be discrete variables, such as whether to use wet or dry oxidation. Most of this chapter treats the case of continuous variables, but some of it is easily adaptable to discrete variables, especially those taking only two values.

Let $X \in R^p$ denote the vector of input values chosen for the computer program. We will write X as the row vector $(X^1, \ldots, X^p)$ using superscripts to denote components of X. We assume that each component X^j is continuously adjustable between a lower and an upper limit, which after a linear transformation can be taken to be 0 and 1 respectively. (For some results where every input is dichotomous see Mitchel et al. (1990).) The computer program is denoted by f and it computes q output quantities, denoted by $Y \in R^q$.

$$Y = f(X), \quad X \in [0, 1]^p. \tag{1}$$

Some important quantities describing a computer model are the number of inputs p, the number of outputs q and the speed with which f can be computed. These vary enormously in applications. In the semiconductor problems we have considered p is usually between 4 and 10. Other computer experiments use scores or even hundreds of input variables. In our motivating applications q is usually larger than 1. For example interest might center on the switching speed of a device and also on its stability as measured by a breakdown voltage. For some problems f takes hours to evaluate on a supercomputer and for others f runs in milliseconds on a personal computer.

Equation (1) differs from the usual $X - Y$ relationship studied by statisticians in that there is no random error term. If the program is run twice with the same X, the same Y is obtained both times. Therefore it is worth discussing why a statistical approach is called for.

These computer programs are written to calculate Y from a known value of X. The way they are often used however, is to search for good values of X according to some goals for Y. Suppose that $X_1 = (X_1^1, \ldots, X_1^p)$ is the initial choice for X. Often X_1 does not give a desirable $Y_1 = f(X_1)$. The engineer or scientist can often deduce why this is, from the program output, and select a new value, X_2 for which $Y_2 = f(X_2)$ is likely to be an improvement. This improvement process can be repeated until a satisfactory design is found. The disadvantage of this procedure is that it may easily miss some good designs X, because it does not fully explore the design space. It can also be slow, especially when p is large, or when improvements of Y^1 say, tend to appear with worsenings of Y^2 and vice versa.

A commonly used way of exploring the design space around X_1 is to vary each of the X_1^j one at a time. As is well known to statisticians, this approach can be misleading if there are strong interactions among the components of X. Increasing X^1 may be an improvement and increasing X^2 may be an improvement, but increasing them both together might make things worse. This would usually be determined from a confirmation run in which both X^1 and X^2 have been increased. The greater difficulty with interactions stems from missed opportunities: the best combination might be to increase X^1 while decreasing X^2, but one at a time experimentation might never lead the user to try this. Thus techniques from experimental design may be expected to help in exploring the input space.

This chapter presents and compares two statistical approaches to computer experiments. Randomness is required in order to generate probability or confidence intervals. The first approach introduces randomness by modeling the function f as a realization of a Gaussian process. The second approach does so by taking random input points (with some balance properties).

2. Goals in computer experiments

There are many different but related goals that arise in computer experiments. The problem described in the previous section is that of finding a good value for X according to some criterion on Y. Here are some other goals in computer experimentation: finding a simple approximation $\hat{f}$ that is accurate enough over a region A of X values, estimating the size of the error $\hat{f}(X_0) - f(X_0)$ for some $X_0 \in A$, estimating $\int_A f \, \mathrm{d}X$, sensitivity analysis of Y with respect to changes in X, finding which X^j are most important for each response Y^k, finding which competing goals for Y conflict the most, visualizing the function f and uncovering bugs in the implementation of f.

2.1. *Optimization*

Many engineering design problems take the form of optimizing Y^1 over allowable values of X. The problem may be to find the fastest chip, or the least expensive soda can. There is often, perhaps usually, some additional constraint on another response Y^2. The chip should be stable enough, and the can should be able to withstand a specified internal pressure.

Standard optimization methods, such as quasi-Newton or conjugate gradients (see for example Gill et al., 1981) can be unsatisfactory for computer experiments. These methods usually require first and possibly second derivatives of f, and these may be difficult to obtain or expensive to run. The standard methods also depend strongly on having good starting values. Computer experimentation as described below is useful in the early stages of optimization where one is searching for a suitable starting value. It is also useful when searching for several widely separated regions of the predictor space that might all have good Y values. Given a good starting value, the standard methods will be superior if one needs to locate the optimum precisely.

2.2. *Visualization*

As Diaconis (1988) points out, being able to compute a function f at any given value X does not necessarily imply that one "understands" the function. One might not know whether the function is continuous or bounded or unimodal, where its optimum is or whether it has asymptotes.

Computer experimentation can serve as a primitive way to visualize functions. One evaluates f at a well chosen set of points $x_1, \ldots, x_n$ obtaining responses $y_1, \ldots, y_n$. Then data visualization methods may be applied to the $p+q$ dimensional points (x_i, y_i), $i = 1, \ldots, n$. Plotting the responses versus the input variables (there are pq such plots) identifies strong dependencies, and plotting residuals from a fit can show weaker dependencies. Selecting the points with desirable values of Y and then producing histograms and plots of the corresponding X values can be used to identify the most promising subregion of X values. Sharifzadeh et al. (1989) took this approach to find that increasing a certain implant dose helped to make two different threshold voltages near their common targets *and* nearly equal (as they should have been). Similar exploration can identify which input combinations are likely to crash the simulator.

Roosen (1995) has used computer experiment designs for the purpose of visualizing functions fit to data.

2.3. *Approximation*

The original program f may be exceedingly expensive to evaluate. It may however be possible to approximate f by some very simple function $\hat{f}$, the approximation holding adequately in a region of interest, though not necessarily over the whole domain of f. If the function $\hat{f}$ is fast to evaluate, as for instance a polynomial, neural network or a MARS model (see Friedman, 1991), then it may be feasible to make millions of $\hat{f}$

evaluations. This makes possible brute force approximations for the other problems. For example, optimization could be approached by finding the best value of $\hat{f}(x)$ over a million random runs x.

Approximation by computer experiments involves choosing where to gather $(x_i, f(x_i))$ pairs, how to construct an approximation based on them and how to assess the accuracy of this approximation.

2.4. *Integration*

Suppose that X^* is the target value of the input vector, but in the system being modeled the actual value of X will be random with a distribution $\mathrm{d}F$ that hopefully is concentrated near X^*. Then one is naturally interested in $\int f(X)\,\mathrm{d}F$, the average value of Y over this distribution. Similarly the variance of Y and the probability that Y exceeds some threshold can be expressed in terms of integrals. This sort of calculation is of interest to researchers studying nuclear safety. McKay (1995) surveys this literature.

Integration and optimization goals can appear together in the same problem. In robust design problems (Phadke, 1988), one might seek the value X_0 that minimizes the variance of Y as X varies randomly in a neighborhood of X_0.

3. Approaches to computer experiments

There are two main statistical approaches to computer experiments, one based on Bayesian statistics and a frequentist one based on sampling techniques. It seems to be essential to introduce randomness in one or other of these ways, especially for the problem of gauging how much an estimate $\hat{f}(X_0)$ might differ from the true value $f(X_0)$.

In the Bayesian framework, surveyed below in Sections 4 and 5, f is a realization of a random process. One sets a prior distribution on the space of all functions from $[0, 1]^p$ to R^q. Given the values $y_i = f(x_i)$, $i = 1, \ldots, n$, one forms a posterior distribution on f or at least on certain aspects of it such as $f(x_0)$. This approach is extremely elegant. The prior distribution is usually taken to be Gaussian so that any finite list of function values has a multivariate normal distribution. Then the posterior distribution, given observed function values is also multivariate normal. The posterior mean interpolates the observed values and the posterior variance may be used to give 95% posterior probability intervals. The method extends naturally to incorporate measurement and prediction of derivatives, partial derivatives and definite integrals of f.

The Bayesian framework is well developed as evidenced by all the work cited below in Sections 4 and 5. But, as is common with Bayesian methods there may be difficulty in finding an appropriate prior distribution. The simulator output might not have as many derivatives as the underlying physical reality, and assuming too much smoothness for the function can lead to Gibbs-effect overshoots. A numerical difficulty also arises: the Bayesian approach requires solving n linear equations in

n unknowns when there are n data points. The effort involved grows as n^3 while the effort in computing $f(X_1), \ldots, f(X_n)$ grows proportionally to n. Inevitably this limits the size of problems that can be addressed. For example, suppose that one spends an hour computing $f(x_1), \ldots, f(x_n)$ and then one minute solving the linear equations. If one then finds it necessary to run 24 times as many function evaluations, the time to compute the $f(x_i)$ grows from an hour to a day, while the time to solve the linear equations grows from one minute to over nine and a half days.

These difficulties with the Bayesian approach motivate a search for an alternative. The frequentist approach, surveyed in Sections 6 and 7, introduces randomness by taking function values $x_1, \ldots, x_n$ that are partially determined by pseudo-random number generators. Then this randomness in the x_i is propagated through to randomness in $\hat{f}(x_0)$. This approach allows one to consider f to be deterministic, and in particular to avoid having to specify a distribution for f. The material given there expands on a proposal of Owen (1992a). There is still much more to be done.

4. Bayesian prediction and inference

A Bayesian approach to modeling simulator output (Sacks et al., 1989a, b; Welch et al., 1990) can be based on a spatial model adapted from the geo-statistical Kriging model (Matheron, 1963; Journel and Huibregts, 1978; Cressie, 1986, 1993; Ripley, 1981). This approach treats the bias, or systematic departure of the response surface from a linear model, as the realization of a stationary random function. This model has exact predictions at the observed responses and predicts with increasing error variance as the prediction point moves away from all the design points.

This section introduces the Kriging (or Bayesian) approach to modeling the response surfaces of computer experiments. Several correlation families are discussed as well as their effect on prediction and error analysis. Additionally, extensions to this model are presented that allow the use and the modeling of gradient information.

4.1. The Kriging model

The Kriging approach uses a two component model. The first component consists of a general linear model while the second (or lack of fit) component is treated as the realization of a stationary Gaussian random function. Define $S = [0,1]^p$ to be the design space and let $x \in S$ be a scaled p-dimensional vector of input values. The Kriging approach models the associated response as

$$Y(x) = \sum_{j=1}^{k} \beta_j h_j(x) + Z(x) \tag{2}$$

where the h_j's are known fixed functions, the β_j's are unknown coefficients to be estimated and $Z(x)$ is a stationary Gaussian random function with $\mathrm{E}[Z(x)] = 0$ and covariance

$$\mathrm{Cov}[Z(x_i), Z(x_j)] = \sigma^2 R(x_j - x_i). \tag{3}$$

For any point $x \in S$, the simulator output $Y(x)$ at that point has the Gaussian distribution with mean $\sum \beta_j h_j(x)$ and variance σ^2. The linear component models the drift in the response, while the systematic lack-of-fit (or bias) is modeled by the second component. The smoothness and other properties of $Z(\cdot)$ are controlled by $R(\cdot)$.

Let design $D = \{x_i,\ i = 1, \ldots, n\} \subset S$ yield responses $y'_D = \{y(x_1), \ldots, y(x_n)\}$ and consider a linear predictor

$$\widehat{y}(x_0) = \lambda'(x_0) y_D$$

of an unobserved point x_0. The Kriging approach of Matheron (1963) treats $\widehat{y}(x_0)$ as a random variable by substituting Y_D for y_D where

$$Y'_D = (Y(x_1), \ldots, Y(x_n)).$$

The best linear unbiased predictor (BLUP) finds the $\lambda(x_0)$ that minimizes

$$\mathrm{MSE}[\widehat{Y}(x_0)] = \mathrm{E}[\lambda' Y_D - Y(x_0)]^2$$

subject to the unbiasedness condition

$$\mathrm{E}[\lambda' Y_D] = \mathrm{E}[Y(x_0)].$$

The BLUP of $Y(x_0)$ is given by

$$\widehat{Y}(x_0) = h'(x_0)\widehat{\beta} + v'_{x_0} V_D^{-1}(Y_D - H_D \widehat{\beta}) \tag{4}$$

where

$$\begin{aligned} h'(x_0) &= (h_1(x_0), \ldots, h_k(x_0)), \\ (H_D)_{ij} &= h_j(x_i), \\ (V_D)_{ij} &= \mathrm{Cov}[Z(x_i), Z(x_j)], \\ v'_{x_0} &= \big(\mathrm{Cov}[Z(x_0), Z(x_1)], \ldots, \mathrm{Cov}[Z(x_0), Z(x_n)]\big) \end{aligned}$$

and

$$\widehat{\beta} = [H'V^{-1}H]^{-1} H'V^{-1} Y_D$$

is the generalized least squares estimate of β. The mean square error of $\widehat{Y}(x_0)$ is

$$\mathrm{MSE}[\widehat{Y}(x_0)] = \sigma^2 - (h'(x_0), v'_{x_0}) \begin{pmatrix} 0 & H'_D \\ H_D & V_D \end{pmatrix}^{-1} \begin{pmatrix} h(x_0) \\ v_{x_0} \end{pmatrix}.$$

The first component of equation (4) is the generalized least squares prediction at point x_0 given the design covariance matrix V_D, while the second component

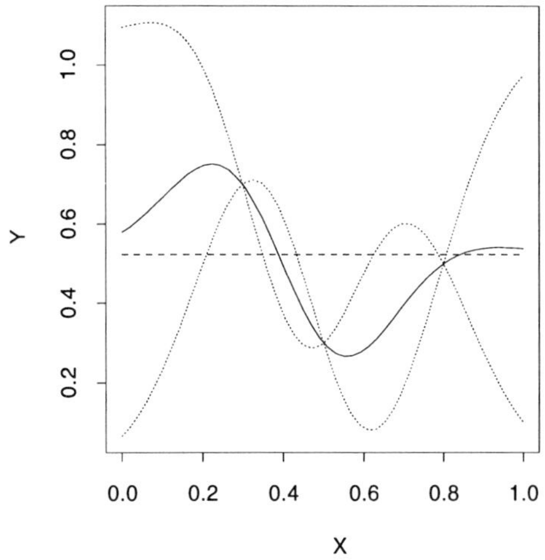

Fig. 1. A prediction example with $n = 3$.

"pulls" the generalized least squares response surface through the observed data points. The elasticity of the response surface "pull" is solely determined by the correlation function $R(\cdot)$. The predictions at the design points are exactly the corresponding observations, and the mean square error equals zero. As a prediction point x_0 moves away from all of the design points, the second component of equation (4) goes to zero, yielding the generalized least squares prediction, while the mean square error at that point goes to $\sigma^2 + h'(x_0)\left[H'V_D^{-1}H\right]^{-1} h(x_0)$. In fact, these results are true in the wide sense if the Gaussian assumption is removed.

As an example, consider an experiment where $n = 3$, $p = 1$, $\sigma^2 = .05$, $R(d) = \exp(-20d^2)$ and $D = \{.3,\ .5,\ .8\}$. The response of the unknown function at the design is $y_D' = (.7,\ .3,\ .5)$. The dashed line of Figure 1 is the generalized least squares prediction surface for $h(\cdot) \equiv 1$ where $\widehat{\beta} = .524$. The effect of the second component of equation (4) is to pull the dashed line through the observed design points as shown by the solid line. The shape of the surface or the amount of elasticity of the "pull" is determined by the vector $v_x' V_D^{-1}$ as a function of x and therefore is completely determined by $R(\cdot)$. The dotted lines are $\pm 2\sqrt{\mathrm{MSE}[\widehat{Y}(x)]}$ or 95% pointwise confidence envelopes around the prediction surface. The interpretation of these pointwise confidence envelopes is that for any point x_0, if the unknown function is truly generated by a random function with constant mean and correlation function $R(d) = \exp(-20d^2)$, then approximately 95% of the sample paths that go through the observed design points would be between these dotted lines at x_0. The predictions and confidence intervals can be very different for different σ^2 and $R(\cdot)$. The effect of different correlation functions is discussed in Section 4.3. Clearly, the true function is not "generated" stochastically. The above model is used for prediction and to quantify the uncertainty of the prediction. This naturally leads to a Bayesian interpretation of this methodology.

4.2. A fully Bayesian interpretation

An alternative to the above interpretation of equation (2) is the fully Bayesian interpretation which uses the model as a way of quantifying the uncertainty of the unknown function. The Bayesian approach (Currin et al., 1991; O'Hagan, 1989) uses the same model but has a different interpretation of the β_j's. Here the β_j's are random variables with prior distribution π_j. The effect of these prior distributions is to quantify the prior belief of the unknown function or to put a prior distribution on a large class of functions $\mathcal{G}$. Hence hopefully the true function $y(\cdot) \in \mathcal{G}$. The mixed convolution of the π_j's and $\pi(Z)$ yield the prior distribution $\Pi(G)$ for subsets of functions $G \subset \mathcal{G}$.

Once the data $Y_D = y_D$ has been observed, the posterior distribution $\Pi(G \mid Y_D)$ is calculated. The mean

$$\widehat{Y}(x_0) = \int g(x_0)\Pi(g \mid Y_D = y_D)\,\mathrm{d}g$$

and variance

$$\mathrm{Var}(\widehat{Y}(x_0) \mid Y_D = y_D) = \int (g(x_0) - \widehat{Y}(x_0))^2 \Pi(g \mid Y_D = y_D)\,\mathrm{d}g$$

of the posterior distribution at each input point are then used as the predictor and measure of error, respectively, at that point. In general, the Kriging and Bayesian approaches will lead to different estimators. However, if the prior distribution of $Z(\cdot)$ is Gaussian and if the prior distribution of the β_j's is diffuse, then the two approaches yield identical estimators.

As an example, consider the case where the prior distribution of the vector of β's is

$$\beta \sim N_k(b, \tau^2\Sigma)$$

and the prior distribution of $Z(\cdot)$ is a stationary Gaussian distribution with expected value zero and covariance function given by equation (3). After the simulator function has been evaluated at the experimental design, the posterior distribution of β is

$$\beta \mid Y_D \sim N_k(\widetilde{\beta}, \widetilde{\Sigma})$$

where

$$\widetilde{\beta} = \widetilde{\Sigma}\left[H'V_D^{-1}Y_D + \tau^{-2}\Sigma^{-1}b\right]$$

and

$$\widetilde{\Sigma} = \left[H'V_D^{-1}H + \tau^{-2}\Sigma^{-1}\right]^{-1}$$

and the posterior distribution of $Y(x_0)$ is

$$Y(x_0) \mid Y_D \sim N\left(v'_{x_0}V_D^{-1}Y_D + c'_{x_0}\widetilde{\beta}, \sigma^2 - v'_{x_0}V_D^{-1}v_{x_0} + c'_{x_0}\widetilde{\Sigma}c_{x_0}\right)$$

where

$$c'_{x0} = h' - v'_{x0} V_D^{-1} H.$$

Hence the posterior distribution is still Gaussian but it is no longer stationary. Now if $\tau^2 \to \infty$ then

$$\widetilde{\beta} \to \widehat{\beta},$$
$$\widetilde{\Sigma} \to \left[H' V_D^{-1} H\right]^{-1}$$

and hence the posterior variance of $Y(x_0)$ is

$$\begin{aligned}
\mathrm{Var}(Y(x_0) \mid Y_D) &= \sigma^2 - v'_{x0} V_D^{-1} v_{x0} + c'_{x0} \left[H' V_D^{-1} H\right]^{-1} c_{x0} \\
&= \sigma^2 - v'_{x0} V_D^{-1} v_{x0} + h' \left[H' V_D^{-1} H\right]^{-1} h \\
&\quad - 2h' \left[H' V_D^{-1} H\right]^{-1} H' V_D^{-1} v_{x0} \\
&\quad + v'_{x0} V_D^{-1} H \left[H' V_D^{-1} H\right]^{-1} H' V_D^{-1} v_{x0} \\
&= \sigma^2 - \Big[- h' \left[H' V_D^{-1} H\right]^{-1} h \\
&\quad + 2h' \left[H' V_D^{-1} H\right]^{-1} H' V_D^{-1} v_{x0} \Big] \\
&\quad - \left[v'_{x0} (V_D^{-1} - V_D^{-1} H \left[H' V_D^{-1} H\right]^{-1} H' V_D^{-1}) v_{x0} \right] \\
&= \sigma^2 - (h'(x_0), v'_{x0}) \begin{pmatrix} 0 & H'_D \\ H_D & V_D \end{pmatrix}^{-1} \begin{pmatrix} h(x_0) \\ v_{x0} \end{pmatrix}
\end{aligned}$$

which is the same variance as the BLUP in the Kriging approach. Therefore, if $Z(\cdot)$ has a Gaussian prior distribution and if the β's have a diffuse prior, the Bayesian and the Kriging approaches yield identical estimators.

Currin et al. (1991) provide a more in depth discussion of the Bayesian approach for the model with a fixed mean ($h \equiv 1$). O'Hagan (1989) discusses Bayes Linear Estimators (BLE) and their connection to equations (2) and (4). The Bayesian approach, which uses random functions as a method of quantifying the uncertainty of the unknown simulator function $Y(\cdot)$, is more subjective than the Kriging or frequentist approach. While both approaches require prior knowledge or an objective method of estimating the covariance function, the Bayesian approach additionally requires knowledge of parameters of the prior distribution of β (b and Σ). For this reason, the Kriging results and Bayesian approach with diffuse prior distributions and the Gaussian assumption are widely used in computer experiments.

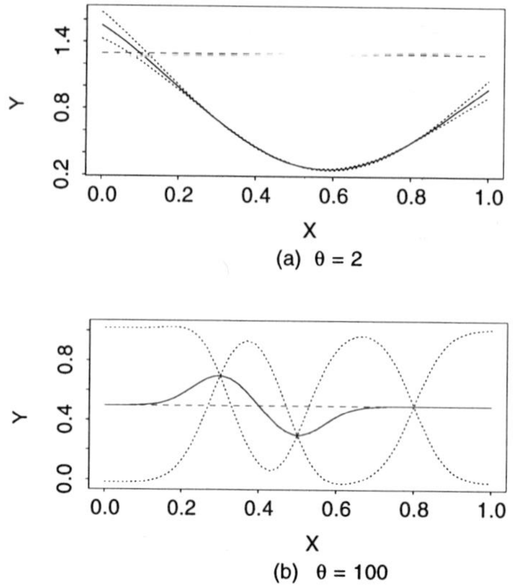

Fig. 2. The effects of θ on prediction.

4.3. Correlation functions

As discussed above, the selection of $R(\cdot)$ plays a crucial role in constructing designs and in the predictive process. Consider the example of Section 4.1 where $n = 3$, $p = 1$, $D = \{.3,\ .5,\ .8)$, $y_d' = \{.7,\ .3,\ .5\}$, $R(d) = \exp\{-\theta d^2\}$ and $\theta = 20$. Figure 2(a) shows the effect on prediction for $\theta = 2$. Now $\widehat{\beta} = 1.3$ and the surface elasticity is very low. The predictions outside of the design are actually higher than the observed surface since the convex nature of the observed response indicate that the design range contains a local minimum for the total process. Eventually, the extrapolations would return to the value of $\widehat{\beta}$. Additionally, the 95% pointwise confidence intervals are much narrower within the range of the design than in Figure 1. Figure 2(b) displays the prediction when $\theta = 100$. Here $\widehat{\beta} = .5$ and the surface elasticity is very high. The prediction line is typically .5 with smooth curves pulling the surface through the design points. The 95% pointwise confidence intervals are wider than before.

This section presents some simplifying restrictions on $R(\cdot)$ and four families of univariate correlation functions used in generating the simplified correlation functions. Examples of realization of these families will be shown to explain the effect on prediction by varying the parameter of these families. Furthermore, the maximum likelihood method for estimating the parameters of a correlation family along with a technique for implementation will be discussed in Section 4.4.

4.3.1. Restrictions on $R(\cdot)$

Any positive definite function R with $R(x,x) = 1$ could be used as a correlation function, but for simplicity, it is common to restrict $R(\cdot)$ such that for any $x_1, x_2 \in S$

$$R(x_1, x_2) = R(x_1 - x_2)$$

so that the process $Z(\cdot)$ is stationary. Some types of nonstationary behavior in the mean function of $Y(\cdot)$ can be modeled by the linear term in equation (2). A further restriction makes the correlation function depend only on the magnitude of the distance.

$$R(x_1, x_2) = R(|x_1 - x_2|).$$

In higher dimensions ($p \geqslant 2$) a product correlation function,

$$R(x_1, x_2) = \prod_{j=1}^{p} R_j(|x_{1j} - x_{2j}|)$$

is often used for mathematical convenience. That is, $R(\cdot)$ is a product of univariate correlation functions and, hence, only univariate correlation functions are of interest. The product correlation function has been used for prediction in spatial settings (Ylvisaher, 1975; Curin et al., 1991; Sacks et al., 1989a, b; Welch et al., 1990, 1992).

Several choices for the factors in the product correlation function are outlined below.

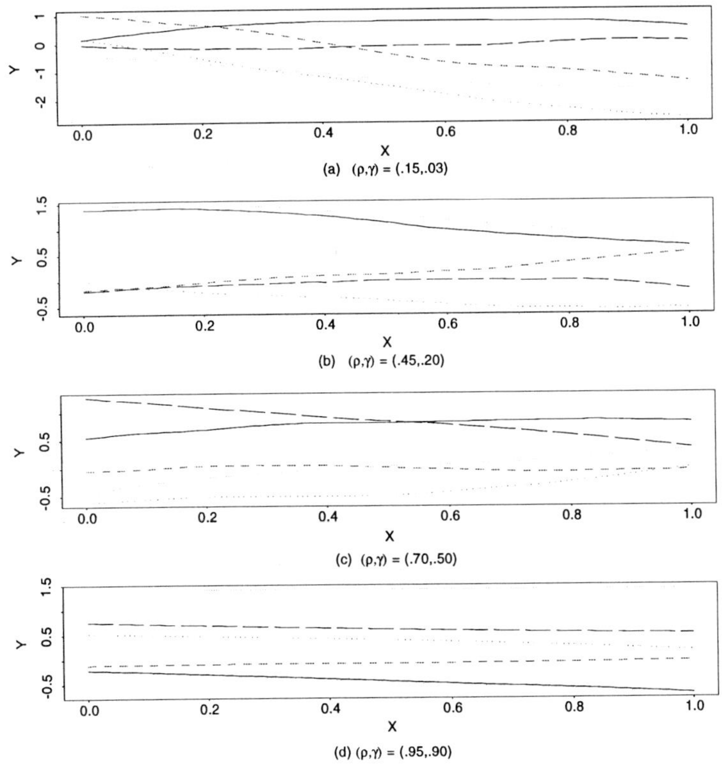

Fig. 3. Realizations for the cubic correlation function $(\rho, \gamma) =$ (a) (.15, .03), (b) (.45, .20), (c) (.70, .50), and (d) (.95, .90).

4.3.2. Cubic

The (univariate) cubic correlation family is parameterized by $\rho \in [0, 1]$ and $\gamma \in [0, 1]$ and is given for $d \in [0, 1]$ by

$$R(d) = 1 - \frac{3(1-\rho)}{2+\gamma} d^2 + \frac{(1-\rho)(1-\gamma)}{2+\gamma} |d|^3$$

where ρ and γ are restricted by

$$\rho \geqslant \frac{5\gamma^2 + 8\gamma - 1}{\gamma^2 + 4\gamma + 7}$$

to ensure that the function is positive definite (see Mitchell et al., 1990). Here $\rho = \text{corr}(Y(0), Y(1))$ is the correlation between endpoint observations and $\gamma = \text{corr}(Y'(0), Y'(1))$ is the correlation between endpoints of the derivative process. The cubic correlation function implies that the derivative process has a linear correlation process with parameter γ.

A prediction model in one dimension for this family is a cubic spline interpolator. In two dimensions, when the correlation is a product of univariate cubic correlation functions the predictions are piece-wise cubic in each variable.

Processes generated with the cubic correlation function are once mean square differentiable. Figure 3 shows several realizations of processes with the cubic correlation function and parameter pairs (.15, .03), (.45, .20), (.70, .50), (.95, .9). Notice that the realizations are quite smooth and almost linear for parameter pair (.95, .90).

4.3.3. Exponential

The (univariate) exponential correlation family is parameterized by $\theta \in (0, \infty)$ and is given by

$$R(d) = \exp(-\theta|d|)$$

for $d \in [0, 1]$. Processes with the exponential correlation function are Ornstein–Uhlenbeck processes (Parzen, 1962). The exponential correlation function is not mean square differentiable.

Figure 4 presents several realizations of one dimensional processes with the exponential correlation function and $\theta = 0.5,\ 2.0,\ 5.0,\ 20$. Figure 4(a) is for $\theta = 0.5$ and these realizations have very small global trends but much local variation. Figure 4(d) is for $\theta = 20$, and is very jumpy. Mitchell et al. (1990) also found necessary and sufficient conditions on the correlation function so that the derivative process has an exponential correlation function. These are called smoothed exponential correlation functions.

4.3.4. Gaussian

Sacks et al. (1989b) generalized the exponential correlation function by using

$$R(d) = \exp(-\theta|d|^q)$$

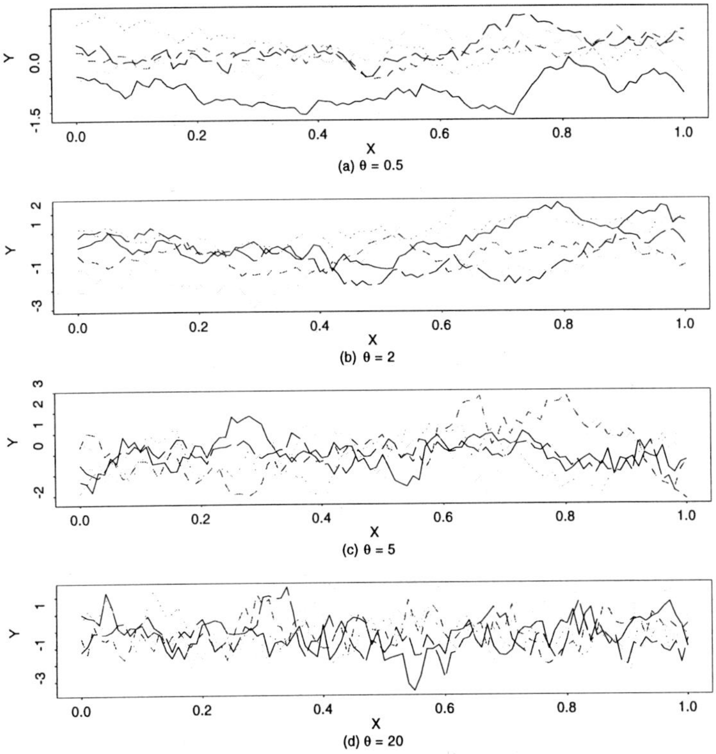

Fig. 4. Realizations for the exponential correlation function with $\theta =$ (a) 0.5, (b) 2.0, (c) 5.0, and (d) 20.0.

where $0 < q \leqslant 2$ and $\theta \in (0, \infty)$. Taking $q = 1$ recovers the exponential correlation function. As q increases, this correlation function produces smoother realizations. However, as long as $q < 2$, these processes are not mean square differentiable.

The Gaussian correlation function is the case $q = 2$ and the associated processes are infinitely mean square differentiable. In the Bayesian interpretation, this correlation function puts all of the prior mass on analytic functions (Currin et al., 1991). This correlation function is appropriate when the simulator output is known to be analytic. Figure 5 displays several realizations for various θ for the Gaussian correlation function. These realizations are very smooth, even when $\theta = 50$.

4.3.5. Matérn

All of the univariate correlation functions described above are either zero, once or infinitely many times mean square differentiable. Stein (1989) recommends a more flexible family of correlation function (Matérn, 1947; Yaglom, 1987). The Matérn correlation function is parameterized by $\theta \in (0, \infty)$ and $\nu \in (-1, \infty)$ and is given by

$$R(d) = \frac{(\theta|d|)^{\nu}}{\Gamma(\nu)2^{\nu-1}} K_{\nu}(\theta|d|)$$

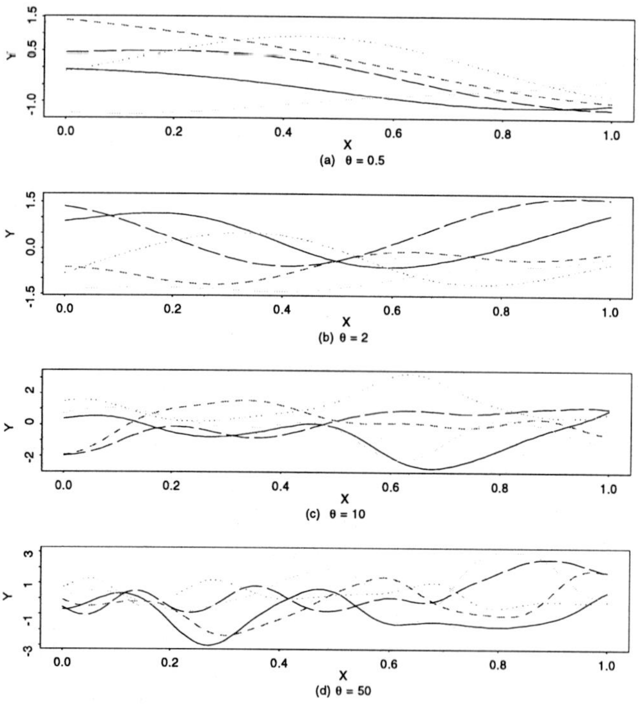

Fig. 5. Realizations for the Gaussian correlation function with $\theta =$ (a) 0.5, (b) 2.0, (c) 10.0, and (d) 50.0.

where $K_\nu(\cdot)$ is a modified Bessel function of order ν. The associated process will be m times differentiable if and only if $\nu > m$. Hence, the amount of differentiability can be controlled by ν while θ controls the range of the correlations. This correlation family is more flexible than the other correlation families described above due to the control of the differentiability of the predictive surface.

Figure 6 displays several realizations of processes with the Matérn correlation function with $\nu = 2.5$ and various values of θ. For small values of θ, the realizations are very smooth and flat while the realizations are erratic for large values of θ.

4.3.6. *Summary*

The correlation functions described above have been applied in computer experiments. Software for predicting with them is described in Koehler (1990). The cubic correlation function yields predictions that are cubic splines. The exponential predictions are non-differentiable while the Gaussian predictions are infinitely differentiable. The Matérn correlation function is the most flexible since the degree of differentiability and the smoothness of the predictions can be controlled. In general, enough prior information to fix the parameters of a particular correlation family and σ^2 will not be available. A pure Bayesian approach would place a prior distribution on the parameters of a family and use the posterior distribution of the parameter in the estimation

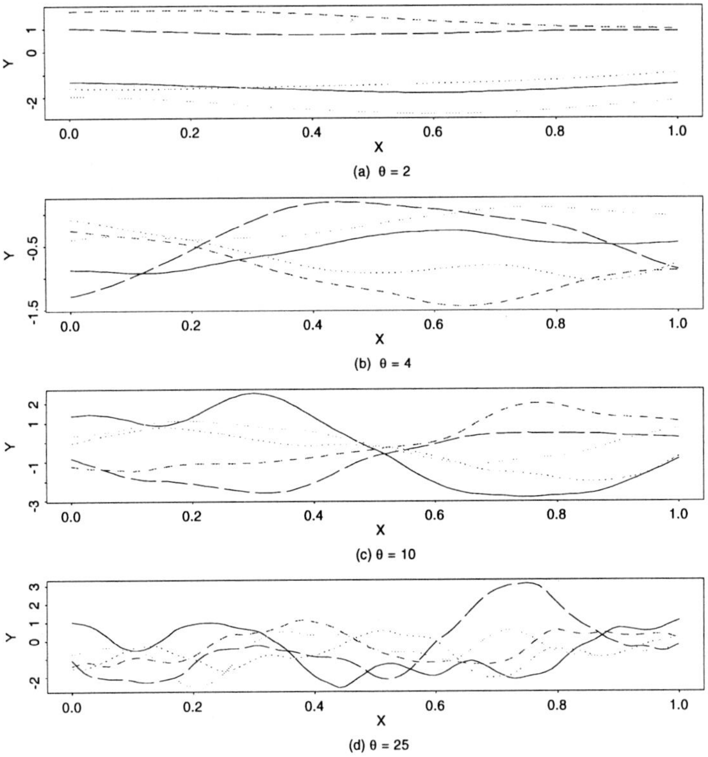

Fig. 6. Realizations for the Matérn correlation function with $\nu = 2.5$ and $\theta =$ (a) 2.0, (b) 4.0, (c) 10.0, and (d) 25.0.

process. Alternatively, an empirical Bayes approach which uses the data to estimate the parameters of a correlation family and σ^2 is often used. The maximum likelihood estimation procedure will be presented and discussed in the next section.

4.4. Correlation function estimation – maximum likelihood

The previous subsections of this section presented the Kriging model, and families of correlation functions. The families of correlations are all parameterized by one or two parameters which control the range of correlation and the smoothness of the corresponding processes. This model assumes that σ^2, the family and parameters of $R(\cdot)$ are known. In general, these values are not completely known a priori. The appropriate correlation family might be known from the simulator's designers experience regarding the smoothness of the function. Also, ranges for σ^2 and the parameters of $R(\cdot)$ might be known if a similar computer experiment has been performed. A pure Bayesian approach is to quantify this knowledge into a prior distribution on σ^2 and $R(\cdot)$. How to distribute a non-informative prior across the different correlation families and within each family is unclear. Furthermore, the calculation of the posterior distribution is generally intractable.

An alternative and more objective method of estimating these parameters is an empirical Bayes approach which finds the parameters which are most consistent with the observed data. This section presents the maximum likelihood method for estimating β, σ^2 and the parameters of a fixed correlation family when the underlying distribution of $Z(\cdot)$ is Gaussian. The best parameter set from each correlation family can be evaluated to find the overall "best" σ^2 and $R(\cdot)$.

Consider the case where the distribution of $Z(\cdot)$ is Gaussian. Then the distribution for the response at the n design points Y_D is multinormal and the likelihood is given by

$$\begin{aligned}\text{lik}(\beta, \sigma^2, R \mid Y_D) &= (2\pi)^{-n/2} \sigma^{-n} |R_D|^{-1/2} \\ &\quad \times \exp\left\{-\frac{1}{2\sigma^2}(Y_D - H\beta)' R_D^{-1} (Y_D - H\beta)\right\}\end{aligned}$$

where R_D is the design correlation matrix. The log likelihood is

$$\begin{aligned}l_{ml}(\beta, \sigma^2, R_D \mid Y_D) &= -\frac{n}{2}\ln(2\pi) - \frac{n}{2}\ln(\sigma^2) - \frac{1}{2}\ln(|R_D|) \\ &\quad - \frac{1}{2\sigma^2}(Y_D - H\beta)' R_D^{-1} (Y_D - H\beta).\end{aligned} \tag{5}$$

Hence

$$\frac{\partial l_{ml}(\beta, \sigma^2, R \mid Y_D)}{\partial \beta} = -\frac{1}{\sigma^2}\left(H' R_D^{-1} Y_D - H' R_D^{-1} H\beta\right)$$

which when set to zero yields the maximum likelihood estimate of β that is the same as the generalized least squares estimate,

$$\widehat{\beta}_{ml} = \left[H' R_D^{-1} H\right]^{-1} H' R_D^{-1} Y_D. \tag{6}$$

Similarly,

$$\frac{\partial l_{ml}(\beta, \sigma^2, R_D \mid Y_D)}{\partial \sigma^2} = -\frac{n}{2\sigma^2} + \frac{1}{2\sigma^4}(Y_D - H\beta)' R_D^{-1} (Y_D - H\beta)$$

which when set to zero yields the maximum likelihood estimate of σ^2

$$\widehat{\sigma}^2_{ml} = \frac{1}{n}(Y_D - H\widehat{\beta})' R_D^{-1} (Y_D - H\widehat{\beta}). \tag{7}$$

Therefore, if R_D is known, the maximum likelihood estimates of β and σ^2 are easily calculated. However, if $R(\cdot)$ is parameterized by $\theta = (\theta_1, \ldots, \theta_s)$,

$$\begin{aligned}\frac{\partial l_{ml}(\beta, \sigma^2, R_D \mid Y_D)}{\partial \theta_i} &= -\frac{1}{2}\frac{\partial |R_D|}{\partial \theta_i} - \frac{1}{2\sigma^2}(Y_D - H\beta)'\frac{\partial R_D^{-1}}{\partial \theta_i}(Y_D - H\beta) \\ &= -\frac{1}{2}\operatorname{tr}\left\{R_D^{-1}\frac{\partial R_D}{\partial \theta_i}\right\} \\ &\quad + \frac{1}{2\sigma^2}(Y_D - H\beta)'R_D^{-1}\frac{\partial R_D}{\partial \theta_i}R_D^{-1}(Y_D - H\beta) \qquad (8)\end{aligned}$$

does not generally yield an analytic solution for θ when set to zero for $i = 1, \ldots, s$. (Commonly $s = p$ or $2p$, but this need not be assumed.)

An alternative method to estimate θ is to use a nonlinear optimization routine using equation (5) as the function to be optimized. For a given value of θ, estimates of β and σ^2 are calculated using equations (6) and (7), respectively. Next, equation (8) is used in calculating the partial derivatives of the objective function. See Mardia and Marshall (1984) for an overview of the maximum likelihood procedure.

4.5. Estimating and using derivatives

In the manufacturing sciences, deterministic simulators help describe the relationships between product design, and the manufacturing process to the product's final characteristics. This allows the product to be designed and manufactured efficiently. Equally important are the effects of uncontrollable variation in the manufacturing parameters to the end product. If the product's characteristics are sensitive to slight variations in the manufacturing process, the yield, or percentage of marketable units produced, may decrease. Furthermore, understanding the sensitivities of the product's characteristics can help design more reliable products and increase the overall quality of the product.

Many simulators need to solve differential equations and can provide the gradient of the response at a design point with little or no additional computational cost. However, some simulators require that the gradient be approximated by a difference equation. Then the cost of finding a directional derivative at a point is equal to evaluating an additional point while approximating the total gradient requires p additional runs.

Consider Figure 7 for an example in $p = 1$ showing the effects of including gradient information on prediction. The solid lines, Y in Figure 7(a) and Y' in Figure 7(b), are the true function and it's derivative, respectively, while the long dashed lines are Kriging predictors $\widehat{Y}_3$ and $\widehat{Y}_3'$ based on $n = 3$ observations. As expected $\widehat{Y}_3$ goes through the design points, $D = \{.2,\ .5,\ .8\}$, but $\widehat{Y}_3'$ is a poor predictor of Y'. The short dashed lines are the $n = 3$ predictors with derivative information $\widehat{Y}_{3'}$ and $\widehat{Y}_{3'}'$. Notice that this predictor now matches Y' and Y at D and the interpolations are over all much better. The addition of gradient information substantially improves the fits of both Y and Y'. The dotted lines are the $n = 6$ predictors $\widehat{Y}_6$ and $\widehat{Y}_6'$ and is a fairer comparison if the derivative costs are equal to the response cost. The predictor $\widehat{Y}_6$ is a little better on the interior of S but $\widehat{Y}_6'$ is worse at $x = 0$ than $\widehat{Y}_{3'}'$.

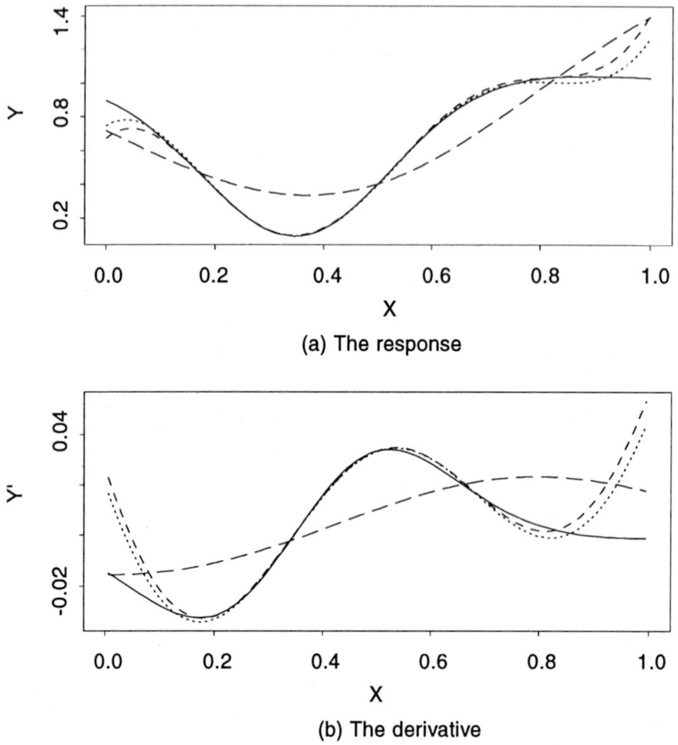

Fig. 7. (a) An example of a response (Y) and three predictors ($\widehat{Y}_3, \widehat{Y}_{3'}, \widehat{Y}_6$). (b) An example of a derivative (Y') and three predictors ($\widehat{Y}'_3, \widehat{Y}'_{3'}, \widehat{Y}'_6$).

The Kriging methodology easily extends to model gradients. To see this for $p = 1$, let $\mathrm{E}[Y(\cdot)] = \mu$ and $d = t_2 - t_1$, then

$$\mathrm{Cov}\left[Y(t_1), Y'(t_2)\right] = \mathrm{E}\left[Y(t_1)Y'(t_2)\right] - \mathrm{E}\left[Y(t_1)\right]\mathrm{E}\left[Y'(t_2)\right].$$

Now due to the stationarity of $Y(\cdot)$, $\mathrm{E}[Y'(\cdot)] = 0$ and

$$\begin{aligned}
\mathrm{Cov}\left[Y(t_1), Y'(t_2)\right] &= \mathrm{E}\left[Y(t_1)Y'(t_2)\right] \\
&= \mathrm{E}\left[Y(t_1) \lim_{\delta \to 0} \frac{Y(t_2+\delta) - Y(t_2)}{\delta}\right] \\
&= \mathrm{E}\left[\lim_{\delta \to 0} \frac{Y(t_1)Y(t_2+\delta) - Y(t_1)Y(t_2)}{\delta}\right] \\
&= \sigma^2 \lim_{\delta \to 0} \frac{R(d+\delta) - R(d)}{\delta} \\
&= \sigma^2 R'(d)
\end{aligned}$$

for differentiable $R(\cdot)$. Similarly,

$$\mathrm{Cov}\left[Y'(t_1), Y(t_2)\right] = -\sigma^2 R'(d)$$

and

$$\mathrm{Cov}\left[Y'(t_1), Y'(t_2)\right] = -\sigma^2 R''(d)$$

For more general p and for higher derivatives, following Morris et al. (1993) let

$$Y^{(a_1,\ldots,a_p)}(t) = \frac{\partial^a}{\partial t_1^{(a_1)} \cdots \partial t_p^{(a_p)}} Y(t)$$

where $a = \sum_{j=1}^p a_j$ and t_j is the jth component of t. Then $\mathrm{E}[Y^{(a_1,\ldots,a_p)}] = 0$ and

$$\mathrm{Cov}\left[Y^{(a_1,\ldots,a_p)}(t_1), Y^{(b_1,\ldots,b_p)}(t_2)\right] = (-1)^a \sigma^2 \prod_{j=1}^p R_j^{(a_j+b_j)}(t_{2j} - t_{1j})$$

for $R(d) = \prod_{j=1}^p R_j(d_j)$.

Furthermore, for directional derivatives, let $Y_\nu'(t)$ be the directional derivative of $Y(t)$ in the direction $\nu = (\nu_1, \ldots, \nu_p)'$, $\sum_{j=1}^p \nu_j^2 = 1$,

$$Y_\nu'(t) = \sum_{j=1}^p \frac{\partial Y(t)}{\partial t_j} \nu_j = \langle \nabla Y(t), \nu \rangle .$$

Then $\mathrm{E}[Y_\nu'(t)] = 0$ and for $d = t - s$,

$$\begin{aligned}
\mathrm{Cov}\left[Y(s), Y_\nu'(t)\right] &= \mathrm{E}\left[Y(s) Y_\nu'(t)\right] \\
&= \sum_{j=1}^p \mathrm{E}\left[Y(s) \frac{\partial Y(t)}{\partial t_j} \nu_j\right] \\
&= \sum_{j=1}^p \mathrm{Cov}\left[Y(s), \frac{\partial Y(t)}{\partial t_j}\right] \nu_j \\
&= \sigma^2 \sum_{j=1}^p \frac{\partial \dot{R}(d)}{\partial d_j} \nu_j \\
&= \sigma^2 \langle \dot{R}(d), \nu \rangle
\end{aligned} \tag{9}$$

where $\dot{R}(d) = [\partial R(d)/\partial d_1, \ldots, \partial R(d)/\partial d_p]'$. Similarly,

$$\mathrm{Cov}\left[Y_\nu'(s), Y(t)\right] = -\sigma^2 \langle \dot{R}(d), \nu \rangle \tag{10}$$

and

$$\text{Cov}\left[Y'_{\nu_s}(s), Y'_{\nu_t}(t)\right] = -\sigma^2 \nu'_s \ddot{R}(d) \nu_t \tag{11}$$

where

$$\left(\ddot{R}(d)\right)_{kl} = \frac{\partial^2 R(d)}{\partial d_k \partial d_l}$$

is the matrix of 2nd partial derivatives evaluated at d.

The Kriging methodology is modified to model gradient information by letting

$$y_D^* = \left[y(x_1), \ldots, y(x_n), y'_{\nu_{11}}(x_1), y'_{\nu_{12}}(x_1), \ldots, y'_{\nu_{nm}}(x_n)\right]'$$

where ν_{il} is the direction of the lth directional derivative at x_i. Also let

$$\mu^* = \left(\mu, \mu, \ldots, \mu, 0, 0, \ldots, 0\right)'$$

with n μs and mn 0s and let V^* be the combined covariance matrix for the design responses and derivatives with the entries as prescribed above (equations (9), (10), and (11)). Then

$$\widehat{Y}(x_0) = \mu + v'^*_{x_0} V^{*^{-1}} \left(y_D^* - \mu^*\right)$$

and

$$\widehat{Y}'_\nu(x_0) = v'^*_{x_0,\nu} V^{*^{-1}} \left(y_D^* - \mu^*\right)$$

where $v'^*_{x_0} = \text{Cov}[Y(x_0), Y_D^*]$, and $v'^*_{x_0,\nu} = \text{Cov}[Y'_\nu(x_0), Y_D^*]$.

Notice that once differentiable random functions need twice differentiable correlation functions. One problem with using the total gradient information is the rapid increase in the covariance matrix. For each additional design point, V^* increases by $p+1$ rows and columns. Fortunately, these new rows and columns generally have lower correlations than the corresponding rows and columns for an equal number of response. The inversion of V^* is more computationally stable than for an equally sized V_D. More research is needed to provide general guidelines for using gradient information efficiently.

4.6. *Complexity of computer experiments*

Recent progress in complexity theory, a branch of theoretical computer science, has shed some light on computer experiments. The dissertation of Ritter (1995) contains an excellent summary of this area. Consider the case where $Y(x) = Z(x)$, that is where there is no regression function. If for $r \geqslant 1$ all of the r'th order partial derivatives of $Z(x)$ exist in the mean square sense and obey a Holder condition of order β, then it is possible (see Ritter et al., 1993) to approximate $Z(x)$ with an L^2 error that decays as $\mathrm{O}(n^{-(r+\beta)/p})$. This error is a root mean square average over randomly generated functions Z.

When the covariance has a tensor product form, like those considered here, one can do even better. Ritter et al. (1995) show that the error rate for approximation in this case is $n^{-r-1/2}(\log n)^{(p-1)(r+1)}$ for products of covariances satisfying Sacks–Ylvisaker conditions of order $r \geqslant 0$. When Z is a p dimensional Wiener sheet process, for which $r = 0$, the result is $n^{-1/2}(\log n)^{(p-1)}$ which was first established by Wozniakowski (1991).

In the general case, the rate for integration is $n^{-1/2}$ times the rate for approximation. A theorem of Wasilkowski (1994) shows that a rate n^{-d} for approximation can usually be turned into a rate $n^{-d-1/2}$ for integration by the simple device of fitting an approximation with $n/2$ function evaluations, integrating the approximation, and then adjusting the result by the average approximation error on $n/2$ more Monte Carlo function evaluations. For tensor product kernels the rate for integration is $n^{-r-1}(\log n)^{(p-1)/2}$ (see Paskov, 1993), which has a more favorable power of $\log n$ than would arise via Wasilkowski's theorem.

The fact that much better rates are possible under tensor product models than for general covariances suggests that the tensor product assumption may be a very strong one. The tensor product assumption is at least strong enough that under it, there is no average case curse of dimensionality for approximation.

5. Bayesian designs

Selecting an experimental design, D, is a key issue in building an efficient and informative Kriging model. Since there is no random error in this model, we wish to find designs that minimize squared-bias. While some experimental design theories (Box and Draper, 1959; Steinberg, 1985) do investigate the case where bias rather than solely variance plays a crucial role in the error of the fitted model, how good these designs are for the pure bias problem of computer experiments is unclear. Box and Draper (1959) studied the effect of scaling factorial designs by using a first order polynomial model when the true function is a quadratic polynomial. Box and Draper (1983) extended the results to using a quadratic polynomial model when the true response surface is a cubic polynomial. They found that mean squared-error optimal designs are close to bias optimal designs. Steinberg (1985) extended these ideas further by using a prior model proposed by Young (1977) that puts prior distributions on the coefficients of a sufficiently large polynomial. However, model (2) is more flexible than high ordered polynomials and therefore better designs are needed.

This section introduces four design optimality criteria for use with computer experiments: entropy, mean squared-error, maximin and minimax designs. Entropy designs maximize the amount of information expected for the design while mean squared-error designs minimize the expected mean squared-error. Both these designs require a priori knowledge of the correlation function $R(\cdot)$. The design criteria described below are for the case of fixed design size n. Simple sequential designs, where the location of

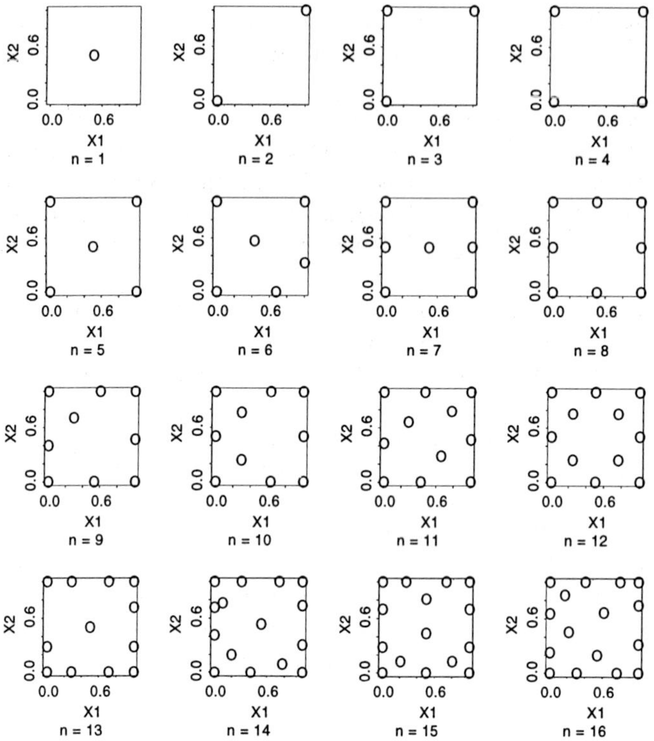

Fig. 8(a). Maximum entropy designs for $p = 2$, $n = 1$–16, and the Gaussian correlation function with $\theta = (0.5, 0.5)$.

the nth design point is determined after the first $n-1$ points have been evaluated, will not be presented due to their tendencies to replicate (Sacks et al., 1989b). However, sequential block strategies could be used where the above designs could be used as starting blocks. Depending upon the ultimate goal of the computer experiment, the first design block might be utilized to refine the design and reduce the design space.

5.1. Entropy designs

Lindley (1956) introduced a measure, based upon Shannon's entropy (Shannon, 1948), of the amount of information provided by an experiment. This Bayesian measure uses the expected reduction in entropy as a design criterion. This criterion has been used in Box and Hill (1967) and Borth (1975) for model discrimination. Shewry and Wynn (1987) showed that, if the design space is discrete (i.e., a lattice in $[0, 1]^p$), then minimizing the expected posterior entropy is equivalent to maximizing the prior entropy.

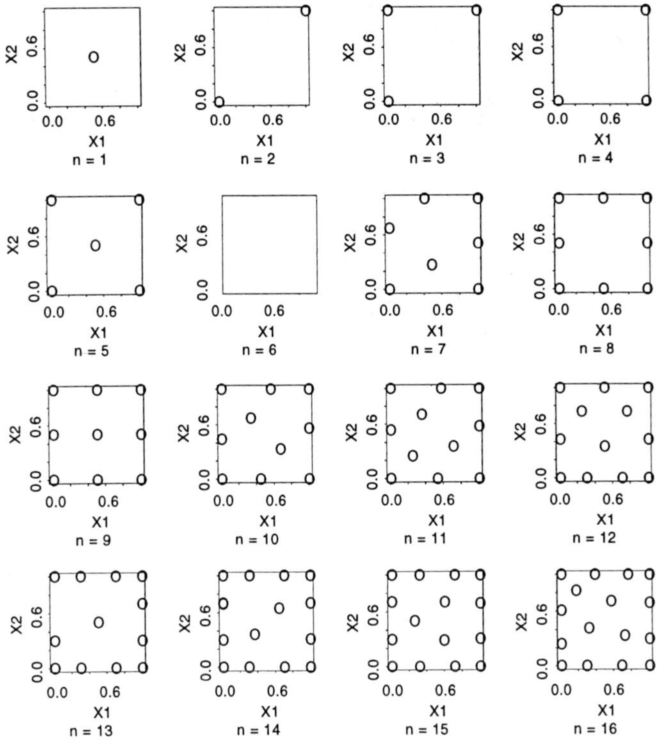

Fig. 8(b). Maximum entropy designs for $p = 2$, $n = 1$–16, and the Gaussian correlation function with $\theta = (2, 2)$.

DEFINITION 1. A design D_E is a *Maximum Entropy Design* if

$$\mathrm{E}_Y\left[-\ln P(Y_{D_E})\right] = \min_D \mathrm{E}_Y\left[-\ln P(Y_D)\right]$$

where $P(Y_D)$ is the density of Y_D.

In the Gaussian case, this is equivalent to finding a design that maximizes the determinant of the variance of Y_D. In the Gaussian prior case, where $\beta \sim N_k(b, \tau^2\Sigma)$, the determinant of the unconditioned covariance matrix is

$$\left|V_D + \tau^2 H\Sigma H'\right| = \begin{vmatrix} V_D + \tau^2 H\Sigma H' & H \\ 0 & I \end{vmatrix}$$

$$= \left|\begin{pmatrix} V_D & H \\ -\tau^2\Sigma H' & I \end{pmatrix}\begin{pmatrix} I & 0 \\ \tau^2\Sigma H' & I \end{pmatrix}\right|$$

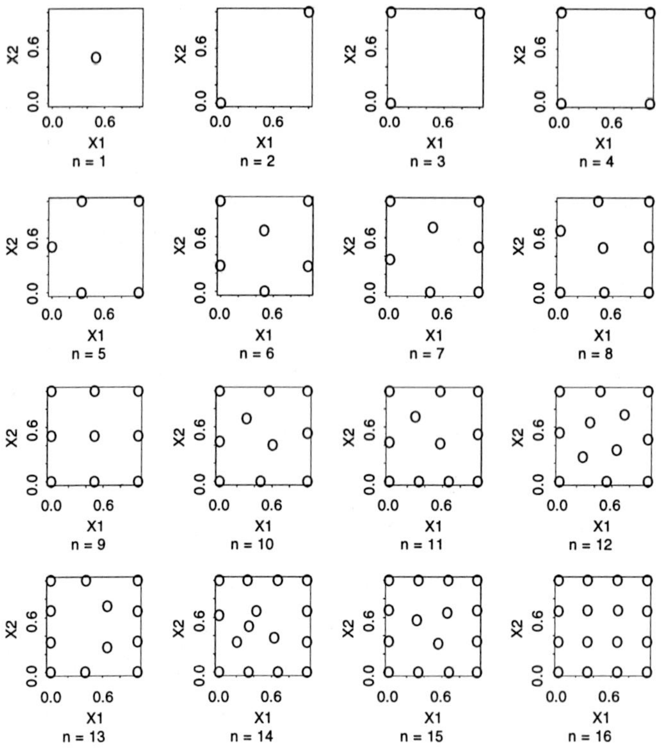

Fig. 8(c). Maximum entropy designs for $p = 2$, $n = 1$–16, and the Gaussian correlation function with $\theta = (10, 10)$.

$$
\begin{aligned}
&= \begin{vmatrix} V_D & H \\ -\tau^2\Sigma H' & I \end{vmatrix} \\
&= \left| \begin{pmatrix} I & 0 \\ \tau^2\Sigma H' V_D^{-1} & I \end{pmatrix} \begin{pmatrix} V_D & H \\ -\tau^2\Sigma H' & I \end{pmatrix} \right| \\
&= \begin{vmatrix} V_D & H \\ 0 & \tau^2\Sigma H' V_D^{-1} H + I \end{vmatrix} \\
&= |V_D| \left| \tau^2\Sigma H' V_D^{-1} H + I \right| \\
&= |V_D| \left| H' V_D^{-1} H + \tau^{-2}\Sigma^{-1} \right| \left| \tau^2 \Sigma \right| .
\end{aligned}
$$

Since $\tau^2\Sigma$ is fixed, the maximum entropy criterion is equivalent to finding the design D_E that maximizes

$$|V_D| \left| H' V_D^{-1} H + \tau^{-2}\Sigma^{-1} \right| .$$

If the prior distribution is diffuse, $\tau^2 \to \infty$, the maximum entropy criterion is equivalent to

$$|V_D|\,\left|H'V_D^{-1}H\right|$$

and if β is treated as fixed, then the maximum entropy criterion is equivalent to $|V_D|$.

Shewry and Wynn (1987, 1988) applied this measure in designs for spatial models. Currin et al. (1991) and Mitchell and Scott (1987) have applied the entropy measure to finding designs for computer experiments. By this measure, the amount of information in experimental design is dependent on the prior knowledge of $Z(\cdot)$ through $R(\cdot)$. In general, $R(\cdot)$ will not be known a priori. Additionally, these optimal designs are difficult to construct due to the required $n \times p$ dimensional optimization of the n design point locations. Currin et al. (1991) describe an algorithm adopted from DETMAX (Mitchell, 1974) which successively removes and adds points to improve the design.

Figure 8(a) shows the optimal entropy designs for $p = 2$, $n = 1, \ldots, 16$, $R(d) = \exp\{-\theta \sum d_j^2\}$ where $\theta = 0.5,\ 2,\ 10$. The entropy designs tend to spread the points out in the plane and favor the edge of the design space over the interior. For example, the $n = 16$ designs displayed in Figure 8(a) have 12 points on the edge and only 4 points in the interior. Furthermore, most of the designs are similar across the different correlation functions although there are some differences. Generally, the ratio of the edge to interior points are constant. The entropy criterion appears to be insensitive to changes in the location of the interior points. Johnson et al. (1990) indicate that entropy designs for extremely "weak" correlation functions are in a limiting sense maximin designs (see Section 5.3).

5.2. *Mean squared-error designs*

Box and Draper (1959) proposed minimizing the normalized integrated mean squared-error (IMSE) of $\widehat{Y}(x)$ over $[0,1]^p$. Welch (1983) extended this measure to the case when the bias is more complicated. Sacks and Schiller (1988) and Sacks et al. (1989a) discuss in more detail IMSE designs for computer experiments.

DEFINITION 2. A design D_I is an *Integrated Mean Squared-Error (IMSE) design* if

$$J(D_I) = \min_D J(D)$$

where

$$J(D) = \frac{1}{\sigma^2}\int_{[0,1]^p} \mathrm{E}\big[Y(x) - \widehat{Y}(x)\big]^2\,\mathrm{d}x.$$

$J(D)$ is dependent on $R(\cdot)$ through $Y(x)$. For any design, $J(D)$ can be expressed as

$$J(D) = \sigma^2 - \operatorname{trace}\left\{\begin{bmatrix} 0 & H' \\ H & V_D \end{bmatrix}^{-1} \int \begin{bmatrix} h(x)h'(x) & h(x)v_x' \\ v_x h'(x) & v_x v_x' \end{bmatrix} \mathrm{d}x\right\}$$

and, as pointed out by Sacks et al. (1989a), if the elements of $h(x)$ and V_x are products of functions of a single input variable, the multidimensional integral simplifies to products of one-dimensional integrals. As in the entropy design criterion, the minimization of $J(D)$ is a optimization in $n \times p$ dimensions and is also dependent on $R(\cdot)$.

Sacks and Schiller (1988) describe the use of a simulated annealing method for constructing IMSE designs for bounded and discrete design spaces. Sacks et al. (1989b) use a quasi-Newton optimizer on a Cray X-MP48. They found that optimizing a $n = 16$, $p = 6$ design with $\theta_1 = \cdots = \theta_6 = 2$ took 11 minutes. The PACE program (Koehler, 1990) uses the optimization program NPSOL (Gill et al., 1986) to solve the IMSE optimization for a continuous design space. For $n = 16$, $p = 6$, this optimization requires 13 minutes on a DEC3100, a much less powerful machine than the Cray. Generally, these algorithms can find only local minima and therefore many random

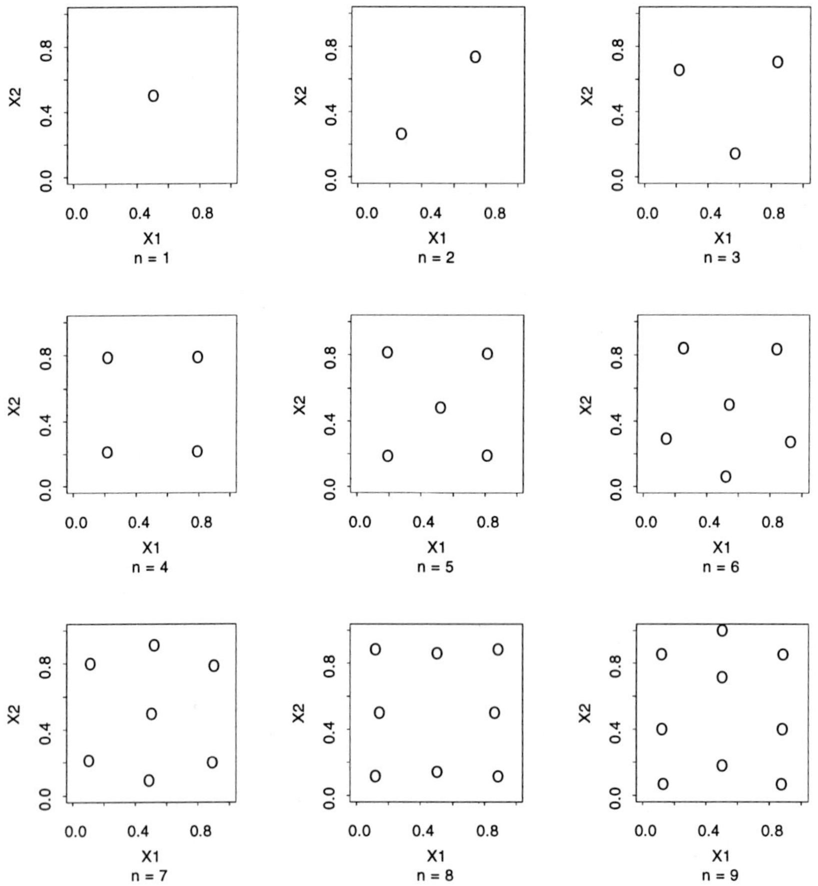

Fig. 9(a). Minimum integrated mean square error designs for $p = 2$, $n = 1$–9, and the Gaussian correlation function with $\theta = (.5,\ .5)$.

starts are required.

Since $J(D)$ is dependent on $R(\cdot)$, robust designs need to be found for general $R(\cdot)$. Sacks et al. (1989a) found that for $n = 9,\ p = 2$ and $R(d) = \exp\{-\theta \sum_{j=1}^{2} d_j^2\}$ (see Section 4.3.4 for details on the Gaussian correlation function) the IMSE design for $\theta = 1$ is robust in terms of relative efficiency. However, this analysis used a quadratic polynomial model and the results may not extend to higher dimensions nor different linear model components. Sacks et al. (1989b) used the optimal design for the Gaussian correlation function with $\theta = 2$ for design efficiency-robustness.

Figure 9(a) displays IMSE designs for $p = 2$ and $n = 1, \ldots, 9$ for $\theta = .5,\ 2,\ 10$. The designs, in general lie in the interior of S. For fixed design size n, the designs usually are similar geometrically for different θ values with the scale decreasing as θ increases. They have much symmetry for some values of n, particularly $n = 12$. Notice that for the case when $n = 5$ that the design only takes on three unique values for each of the input variables. These designs tend to have clumped projections onto

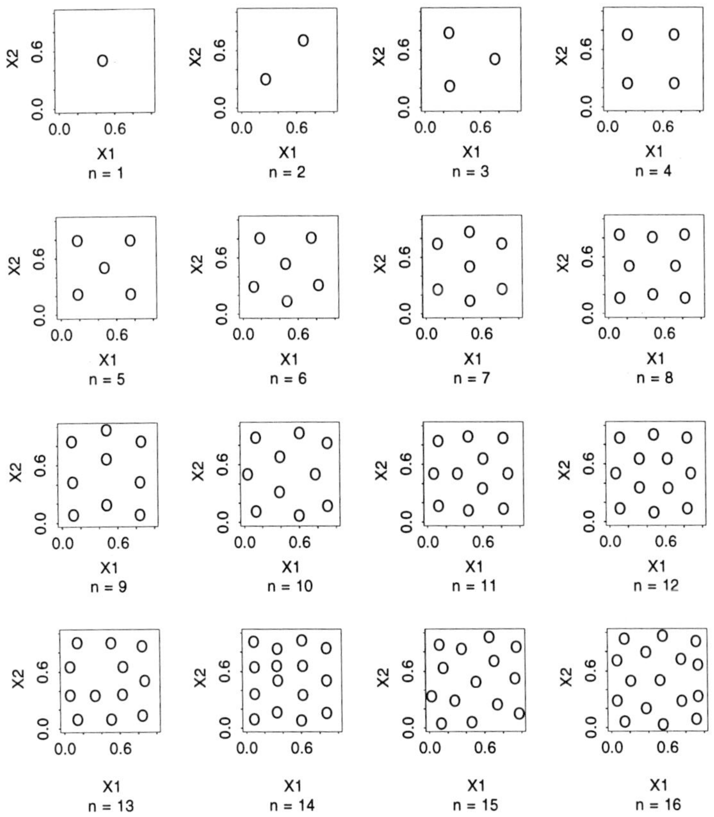

Fig. 9(b). Minimum integrated mean square error designs for $p = 2$, $n = 1$–16, and the Gaussian correlation function with $\theta = (2, 2)$.

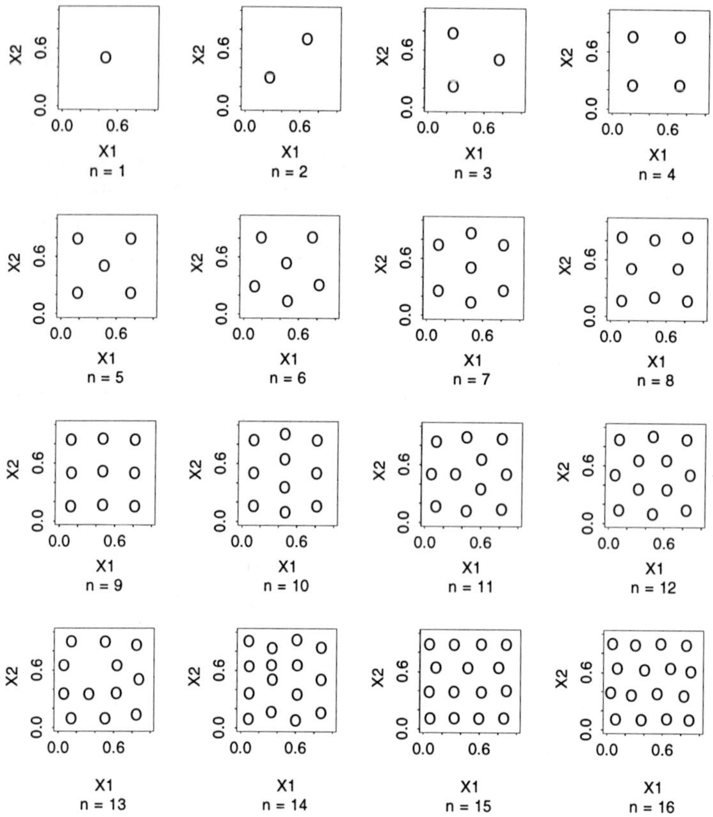

Fig. 9(c). Minimum integrated mean square error designs for $p = 2$, $n = 1$–16, and the Gaussian correlation function with $\theta = (10, 10)$.

lower dimension marginals of the input space. Better projection properties are needed when the true function is only dependent on a subset of the input variables.

5.3. Maximin and minimax designs

Johnson et al. (1990) developed the idea of minimax and maximin designs. These designs are dependent on a distance measure or metric. Let $d(\cdot, \cdot)$ be a metric on $[0, 1]^p$. Hence $\forall x_1, x_2, x_3 \in [0, 1]^p$,

$$d(x_1, x_2) = d(x_2, x_1),$$
$$d(x_1, x_2) \geqslant 0,$$
$$d(x_1, x_2) = 0 \Leftrightarrow x_1 = x_2,$$
$$d(x_1, x_2) \leqslant d(x_1, x_3) + d(x_3, x_2).$$

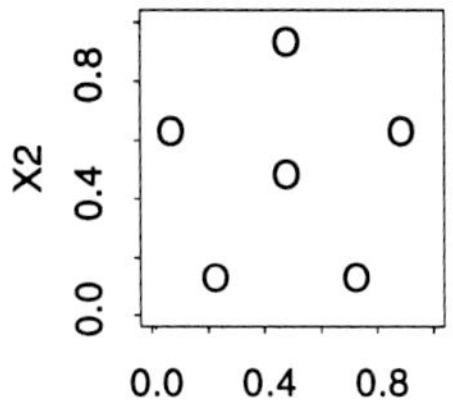

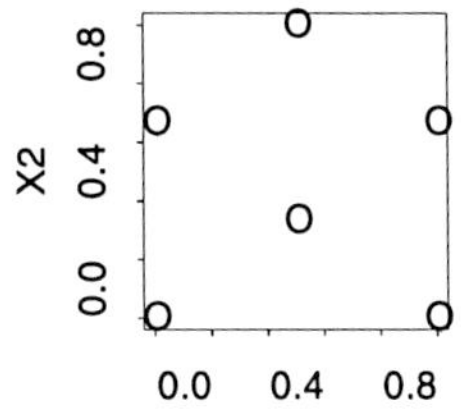

Fig. 10. (a) Minimax and (b) Maximin designs for $n = 6$ and $p = 2$ with Euclidean distance.

DEFINITION 3. Design D_{MI} is a *Minimax Distance Design* if

$$\min_D \max_x d(x, D) = \max_x d(x, D_{MI})$$

where

$$d(x, D) = \min_{x_0 \in D} d(x, x_0).$$

Minimax distance designs ensure that all points in $[0, 1]^p$ are not too far from a design point. Let $d(\cdot, \cdot)$ be Euclidean distance and consider placing a p-dimensional sphere with radius r around each design point. The idea of a minimax design is to place the n points so that the design space is covered by the spheres with minimal r. As an illustration, consider the owner of a petroleum corporation who wants to open some franchise gas stations. The gas company would like to locate the stations in the most convenient sites for the customers. A minimax strategy of placing gas stations would ensure that no customer is too far from one of the company's stations.

Figure 10(a) shows a minimax design for $p = 2$ and $n = 6$ with $d(\cdot, \cdot)$ being Euclidean distance. The maximum distance to a design point is .318. For small n, minimax designs will generally lie in the interior of the design space.

DEFINITION 4. A design D_{MA} is a *Maximin Distance Design* if

$$\max_D \min_{x_1, x_2 \in D} d(x_1, x_2) = \min_{x_1, x_2 \in D_{MA}} d(x_1, x_2).$$

Again, let $d(\cdot,\cdot)$ be Euclidean distance. Maximin designs pack the n design points, with their associated spheres, into the design space, S, with maximum radius. Parts of the sphere may be out of S but the design points must be in S. Analogous to the minimax illustration above is the position of the owners the gas station franchises. They wish to minimize the competition from each other by locating the stations as far apart as possible. A maximin strategy for placing the franchises would ensure that no two stations are too close to each other.

Figure 10(b) shows a maximin design for $p = 2$, $n = 6$ and $d(\cdot,\cdot)$ Euclidean distance. For small n, maximin designs will generally lie on the exterior of S and fill in the interior as n becomes large.

5.4. Hyperbolic cross points

Under the tensor product covariance models, it is possible to approximate and integrate functions with greater accuracy than in the general case. One gets the same rates of convergence as in univariate problems, apart from a multiplicative penalty that is some power of $\log n$. Hyperbolic cross point designs, also known as sparse grids have been shown to achieve optimal rates in these cases. See Ritter (1995). These point sets were first developed by Smolyak (1963). They were used in interpolation by Wahba (1978) and Gordon (1971) and by Paskov (1993) for integration. Chapter 4 of Ritter (1995) gives a good description of the construction of these points and lists other references.

6. Frequentist prediction and inference

The frequentist approach to prediction and inference in computer experiments is based on numerical integration. For a scalar function $Y = f(X)$, consider a regression model of the form

$$Y = f(X) \doteq Z(X)\beta \tag{12}$$

where $Z(X)$ is a row vector of predictor functions and β is a vector of parameters. Suitable functions Z might include low order polynomials, trigonometric polynomials wavelets, or some functions specifically geared to the application. Ordinarily $Z(X)$ includes a component that is always equal to 1 in order to introduce an intercept term into equation (12).

It is unrealistic to expect that the function f will be exactly representable as the finite linear combination given by (12), and it is also unrealistic to expect that the residual will be a random variable with mean zero at every fixed X_0. This is why we only write $f \doteq Z\beta$. There are many ways to define the best value of β, but an especially natural approach is to choose β to minimize the mean squared error of the approximation, with respect to some distribution F on $[0,1]^p$. Then the optimal value for β is

$$\beta_{LS} = \left(\int Z(X)'Z(X)\,\mathrm{d}F\right)^{-1}\int Z(X)'f(X)\,\mathrm{d}F.$$

So if one can integrate over the domain of X then one can fit regression approximations there.

The quality of the approximation may be assessed globally by the integrated mean squared error

$$\int (Y - Z(X)\beta)^2 \, dF.$$

For simplicity we take the distribution F to be uniform on $[0, 1]^p$. Also for simplicity the integration schemes to be considered usually estimate $\int g(X)\, dF$ by

$$\frac{1}{n}\sum_{i=1}^{n} g(x_i)$$

for well chosen points $x_1, \ldots, x_n$. Then β_{LS} may be estimated by linear regression

$$\widehat{\beta} = \left(\frac{1}{n}\sum_{i=1}^{n} Z(x_i)'Z(x_i)\right)^{-1} \frac{1}{n}\sum_{i=1}^{n} Z(x_i)'f(x_i),$$

or when the integrals of squares and cross products of Z's are known by

$$\widetilde{\beta} = \left(\int Z(X)'Z(X)\, dF\right)^{-1} \frac{1}{n}\sum_{i=1}^{n} Z(x_i)'f(x_i). \tag{13}$$

Choosing the components of Z to be an orthogonal basis, such as tensor products of orthogonal polynomials, multivariate Fourier series or wavelets, equation (13) simplifies to

$$\widetilde{\beta} = \frac{1}{n}\sum_{i=1}^{n} Z(x_i)'f(x_i) \tag{14}$$

and one can avoid the cost of matrix inversion. The computation required by equation (14) grows proportionally to nr not n^3, where $r = r(n)$ is the number of regression variables in Z. If $r = O(n)$ then the computations grow as n^2. Then, in the example from Section 3, an hour of function evaluation followed by a minute of algebra would scale into a day of function evaluation followed by 9.6 hours of algebra, instead of the 9.6 days that an n^3 algorithm would require. If the $Z(x_i)$ exhibit some sparsity then it may be possible to reduce the algebra to order n or order $n \log n$.

Thus the idea of turning the function into data and making exploratory plots can be extended to turning the function into data and applying regression techniques. The theoretically simplest technique is to take X_i iid $U[0, 1]^p$. Then (X_i, Y_i) are iid pairs with the complication that Y has zero variance given X. The variance matrix of $\widetilde{\beta}$ is then

$$\frac{1}{n}\left(\int Z'Z\,\mathrm{d}F\right)^{-1} \mathrm{Var}\left(Z(X)'Y(X)\right)\left(\int Z'Z\,\mathrm{d}F\right)^{-1} \tag{15}$$

and for orthogonal predictors this simplifies further to

$$\frac{1}{n}\mathrm{Var}\left(Z(X)'Y(X)\right). \tag{16}$$

Thus any integration scheme that allows one to estimate variances and covariances of averages of Y times components of Z allows one to estimate the sampling variance matrix of the regression coefficients $\widetilde{\beta}$. For iid sampling one can estimate this variance matrix by

$$\frac{1}{n-r-1}\sum_{i=1}^{n}\left(Z(x_i)Y(x_i)-\widetilde{\beta}\right)'\left(Z(x_i)Y(x_i)-\widetilde{\beta}\right)$$

when the row vector Z comprises an intercept and r additional regression coefficients.

This approach to computer experimentation should improve if more accurate integration techniques are substituted for the iid sampling. Owen (1992a) investigates the case of Latin hypercube sampling for which a central limit theorem also holds.

Clearly more work is needed to make this method practical. For instance a scheme for deciding how many predictors should be in Z, or otherwise for regularizing $\widetilde{\beta}$ is required.

7. Frequentist experimental designs

The frequentist approach proposed in the previous section requires a set of points $x_1, \ldots, x_n$ that are good for numerical integration and also allow one to estimate the sampling variance of the corresponding integrals. These two goals are somewhat at odds. Using an iid sample makes variance estimation easier while more complicated schemes described below improve accuracy but make variance estimation harder.

The more basic goal of getting points x_i into "interesting corners" of the input space, so that important features are likely to be found is usually well served by point sets that are good for numerical integration.

We assume that the region of interest is the unit cube $[0,1]^p$, and that the integrals of interest are with respect to the uniform distribution over this cube. Other regions of interest can usually be reduced to the unit cube and other distributions can be changed to the uniform by a change of variable that can be subsumed into f.

Throughout this section we consider an example with $p = 5$, and plot the design points x_i.

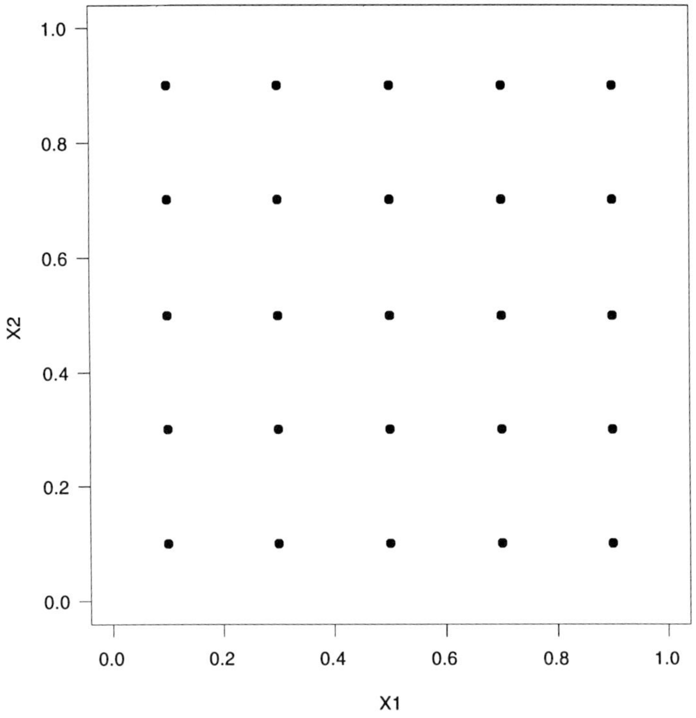

Fig. 11. 25 distinct points among 625 points in a 5^5 grid.

7.1. *Grids*

Since varying one coordinate at a time can cause one to miss important aspects of f, it is natural to consider instead sampling f on a regular grid. One chooses k different values for each of X^1 through X^p and then runs all k^p combinations. This works well for small values of p, perhaps 2 or 3, but for larger p it becomes completely impractical because the number of runs required grows explosively.

Figure 11 shows a projection of $5^5 = 625$ points from a uniform grid in $[0,1]^5$ onto two of the input variables. Notice that with 625 runs, only 25 distinct values appear in the plane, each representing 25 input settings in the other three variables. Only 5 distinct values appear for each of input variable taken singly. In situations where one of the responses Y^k depends very strongly on only one or two of the inputs X^j the grid design leads to much wasteful duplication.

The grid design does not lend itself to variance estimation since averages over the grid are not random. The accuracy of a grid based integral is typically that of a univariate integral based on $k = n^{1/p}$ evaluations. (See Davis and Rabinowitz, 1984.) For large p this is a severe disadvantage.

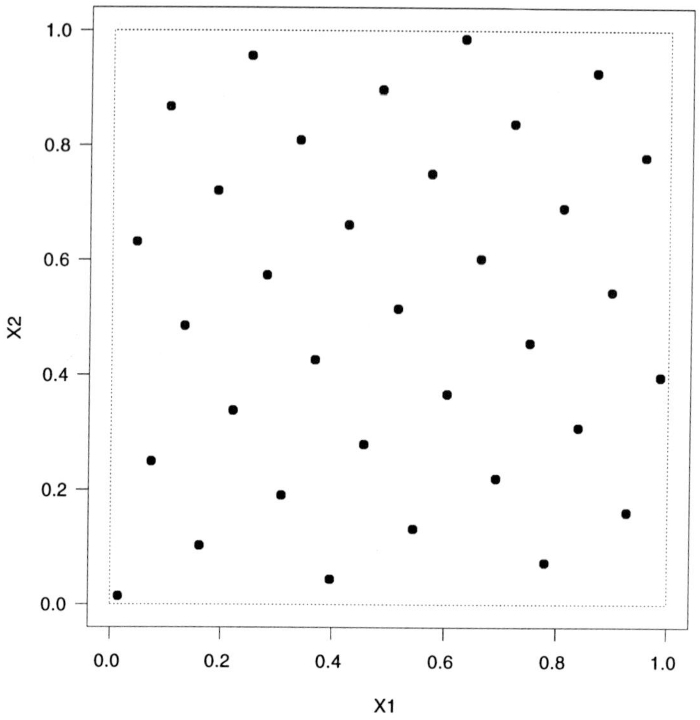

Fig. 12. A 34 point Fibonacci lattice in $[0, 1]^2$.

7.2. *Good lattice points*

A significant improvement on grids may be obtained in integration by the method of good lattice points. (See Sloan and Joe (1994) and Niederreiter (1992) for background and Fang and Wang (1994) for applications to statistics.)

For good lattice points

$$x_i^j = \left\{ \frac{h_j(i-1) + 0.5}{n} \right\}$$

where $\{z\}$ is z modulo 1, that is, z minus the greatest integer less than or equal to z and h_j are integers with $h_1 = 1$. The points v_i with $v_i^j = ih_j/n$ for integer i form a lattice in R^p. The points x_i are versions of these lattice points confined to the unit cube, and the term "good" refers to a careful choice of n and h_j usually based on number theory.

Figure 12 shows the Fibonacci lattice for $p = 2$ and $n = 34$. For more details see Sloan and Joe (1994). Here $h_1 = 1$ and $h_2 = 21$. The Fibonacci lattice is only available in 2 dimensions. Appendix A of Fang and Wang (1994) lists several other choices for good lattice points, but the smallest value of n there for $p = 5$ is 1069.

Hickernell (1996) discusses greedy algorithms for finding good lattice points with smaller n.

The recent text (Sloan and Joe, 1994) discusses lattice rules for integration, which generalize the method of good lattice points. Cranley and Patterson (1976) consider randomly perturbing the good lattice points by adding, modulo 1, a random vector uniform over $[0,1]^p$ to all the x_i. Taking r such random offsets for each of the n data points gives nr observations with $r-1$ degrees of freedom for estimating variance.

Lattice integration rules can be extraordinarily accurate on smooth periodic integrands and thus an approach to computer experiments based on Cranley and Patterson's method might be expected to work well when both $f(x)$ and $Z(x)$ are smooth and periodic. Bates et al. (1996) have explored the use of lattice rules as designs for computer experiments.

7.3. *Latin hypercubes*

While good lattice points start by improving the low dimensional projections of grids, Latin hypercube sampling starts with iid samples. A Latin hypercube sample has

$$X_i^j = \frac{\pi^j(i) - U_j^i}{n} \tag{17}$$

where the π^j are independent uniform random permutations of the integers 1 through n, and the U_i^j are independent $U[0,1]$ random variables independent of the π_j.

Latin hypercube sampling was introduced by McKay et al. (1979) in what is widely considered to be the first paper on computer experiments. The sample points are stratified on each of p input axes. A common variant of Latin hypercube sampling has centered points

$$X_i^j = \frac{\pi^j(i) - 0.5}{n}. \tag{18}$$

Point sets of this type were studied by Patterson (1954) who called them lattice samples.

Figure 13 shows a projection of 25 points from a (centered) Latin hypercube sample over 5 variables onto two of the coordinate axes. Each input variable gets explored in each of 25 equally spaced bins.

The stratification in Latin hypercube sampling usually reduces the variance of estimated integrals. Stein (1987) finds an expression for the variance of a sample mean under Latin hypercube sampling. Assuming that $\int f(X)^2\,\mathrm{d}F < \infty$ write

$$f(X) = \mu + \sum_{j=1}^{p} \alpha_j(X^j) + e(X) \tag{19}$$

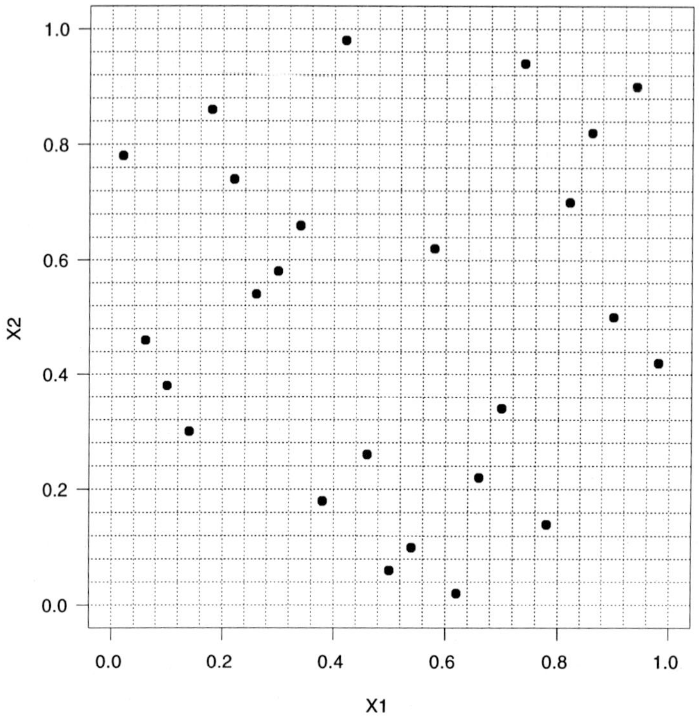

Fig. 13. 25 points of a Latin hypercube sample. The range of each input variable may be partitioned into 25 bins of equal width, drawn here with horizontal and vertical dotted lines, and each such bin contains one of the points.

where $\mu = \int f(X)\,\mathrm{d}F$ and $\alpha_j(x) = \int_{X:X^j=x}(f(X)-\mu)\,\mathrm{d}F_{-j}$ in which $\mathrm{d}F_{-j} = \prod_{k\neq j} dX^k$ is the uniform distribution over all input variables except the j'th. Equation (19) expresses f as the sum of a grand mean μ, univariate main effects α_j and a residual from additivity $e(X)$.

Stein shows that under Latin hypercube sampling

$$\mathrm{Var}\left(\frac{1}{n}\sum_{i=1}^{n} f(x_i)\right) = \frac{1}{n}\int e(X)^2\,\mathrm{d}F + \mathrm{o}\left(\frac{1}{n}\right) \tag{20}$$

whereas under iid sampling

$$\mathrm{Var}\left(\frac{1}{n}\sum_{i=1}^{n} f(x_i)\right) = \frac{1}{n}\left(\int e(X)^2\,\mathrm{d}F + \sum_{j=1}^{p}\int \alpha_j(X^j)^2\,\mathrm{d}F\right). \tag{21}$$

By balancing the univariate margins, Latin hypercube sampling has removed the main effects of the function f from the error variance.

Owen (1992a) proves a central limit theorem for Latin hypercube sampling of bounded functions and Loh (1993) proves a central limit theorem under weaker conditions. For variance estimation in Latin hypercube sampling see (Stein, 1987; Owen, 1992a).

7.4. Better Latin hypercubes

Latin hypercube samples look like random scatter in any bivariate plot, though they are quite regular in each univariate plot. Some effort has been made to find especially good Latin hypercube samples.

One approach has been to find Latin hypercube samples in which the input variables have small correlations. Iman and Conover (1982) perturbed Latin hypercube samples in a way that reduces off diagonal correlation. Owen (1994b) showed that the technique in Iman and Conover (1982) typically reduces off diagonal correlations by a factor of 3, and presented a method that empirically seemed to reduce the off diagonal correlations by a factor of order n from $O(n^{-1/2})$ to $O(n^{-3/2})$. This removes certain bilinear terms from the lead term in the error. Dandekar (1993) found that iterating the method in Iman and Conover (1982) can lead to large improvements.

Small correlations are desirable but not sufficient, because one can construct centered Latin hypercube samples with zero correlation (unless n is equal to 2 modulo 4) which are nonetheless highly structured. For example the points could be arranged in a diamond shape in the plane, thus missing the center and corners of the input space.

Some researchers have looked for Latin hypercube samples having good properties when considered as designs for Bayesian prediction. Park (1994) studies the IMSE criterion and Morris and Mitchell (1995) consider entropy.

7.5. Randomized orthogonal arrays

An orthogonal array A is an n by p matrix of integers $0 \leqslant A_i^j \leqslant b-1$. The array has strength $t \leqslant p$ if in every n by t submatrix of A all of the b^t possible rows appear the same number λ of times. Of course $n = \lambda b^t$.

Independently Owen (1992b, 1994a) and Tang (1992, 1993) considered using orthogonal arrays to improve upon Latin hypercube samples.

A randomized orthogonal array (Owen, 1992b) has two versions,

$$X_i^j = \frac{\pi_j(A_i^j) + U_i^j}{b} \tag{22}$$

and

$$X_i^j = \frac{\pi_j(A_i^j) + 0.5}{b} \tag{23}$$

just as Latin hypercube sampling has two versions. Indeed Latin hypercube sampling corresponds to strength $t = 1$, with $\lambda = 1$. Here the π_j are independent uniform

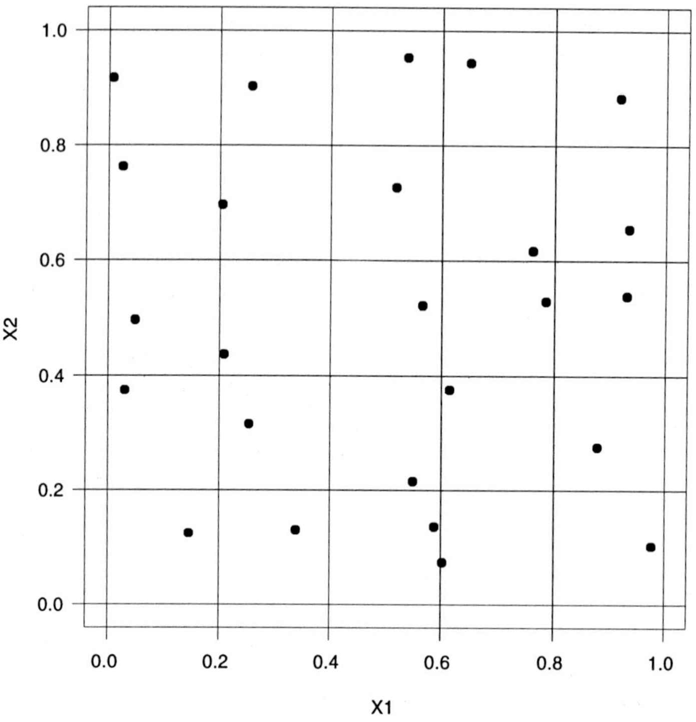

Fig. 14. 25 points of a randomly centered randomized orthogonal array. For whichever two (of five) variables that are plotted, there is one point in each reference square.

permutations of $0, \ldots, b-1$. Patterson (1954) considered some schemes like the centered version.

If one were to plot the points of a randomized orthogonal array in t or fewer of the coordinates, the result would be a regular grid. The points of a randomized orthogonal array of strength 2 appear to be randomly scattered in 3 dimensions.

Figure 14 shows a projection of 25 points from a randomly centered randomized orthogonal array over 5 variables onto two of the coordinate points. Each pair of variables gets explored in each of 25 square bins. The plot for the centered version of a randomized orthogonal array is identical to that for a grid as shown in Figure 11.

The analysis of variance decomposition used above for Latin hypercube sampling can be extended to include interactions among 2 or more factors. See Efron and Stein (1981), Owen (1992b) and Wahba (1990) for details. Gu and Wahba (1993) describe how to estimate and form confidence intervals for these main effects in noisy data.

Owen (1992b) shows that main effects and interactions of t or fewer variables do not contribute to the asymptotic variance of a mean over a randomized orthogonal array, and Owen (1994a) shows that the variance is approximately n^{-1} times the sum of integrals of squares of interactions among more than t inputs.

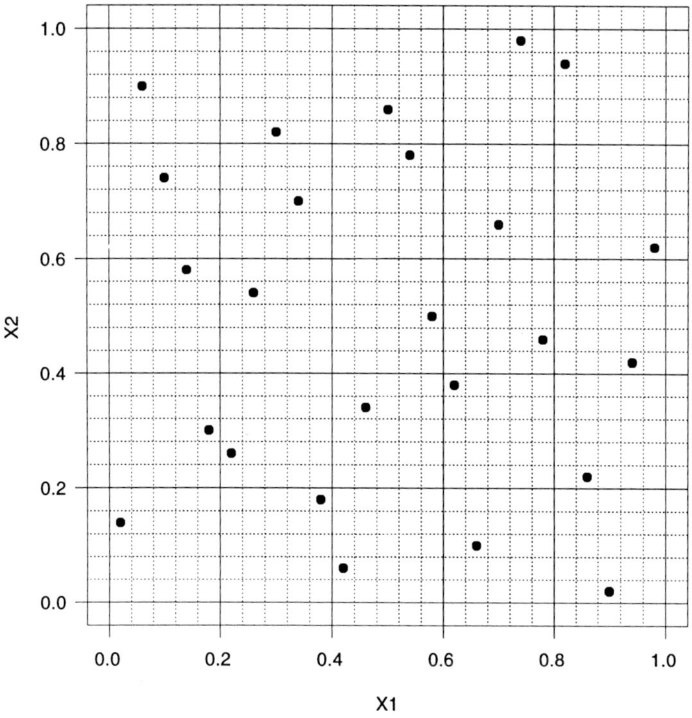

Fig. 15. 25 points of an orthogonal array based Latin hypercube sample. For whichever two (of five) variables that are plotted, there is one point in each reference square bounded by solid lines. Each variable is sampled once within each of 25 horizontal or vertical bins.

Tang (1993) introduced orthogonal array based Latin hypercube samples. The points of these designs are Latin hypercube samples X_i^j, such that $\lfloor bX_i^j \rfloor$ is an orthogonal array. Here b is an integer and $\lfloor z \rfloor$ is the smallest integer less than or equal to z. Tang (1993) shows that for a strength 2 array the main effects and two variable interactions do not contribute to the integration variance.

Figure 15 shows a projection of 25 points from an orthogonal array based Latin hypercube sample over 5 variables onto two of the coordinate points. Each variable individually gets explored in each of 25 equal bins and each pair of variables gets explored in each of 25 squares.

7.6. Scrambled nets

Orthogonal arrays were developed to balance discrete experimental factors. As seen above they can be embedded into the unit cube and randomized with the result that sampling variance is reduced. But numerical analysts and algebraists have developed some integration techniques directly adapted to balancing in a continuous space. Here we describe (t, m, s)-nets and their randomizations. A full account of (t, m, s)-nets

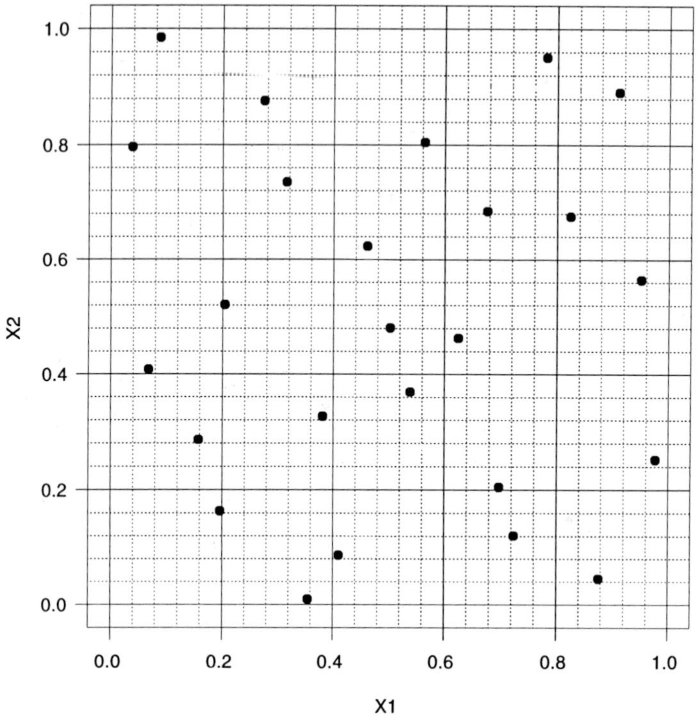

Fig. 16. 25 points of a scrambled $(0, 2, 5)$-net in base 5. For whichever two (of five) variables that are plotted, there is one point in each reference square. Each variable is sampled once within each of 25 equal bins.

is given by Niederreiter (1992). Their randomization is described by Owen (1995, 1996a).

Let $p = s \geqslant 1$ and $b \geqslant 2$ be integers. An elementary subcube in base b is of the form

$$E = \prod_{j=1}^{s} \left[\frac{c_j}{b^{k_j}}, \frac{c_j + 1}{b^{k_j}}\right)$$

for integers k_j, c_j with $k_j \geqslant 0$ and $0 \leqslant c_j < b^{k_j}$.

Let $m \geqslant 0$ be an integer. A set of points X_i, $i = 1, \ldots, b^m$, of from $[0, 1)^s$ is a $(0, m, s)$-net in base b if every elementary subcube E in base b of volume b^{-m} has exactly 1 of the points. That is, every cell that "should" have one point of the sequence does have one point of the sequence.

This is a very strong form of equidistribution and by weakening it somewhat, constructions for more values of s and b become available. Let $t \leqslant m$ be a nonnegative integer. A finite set of b^m points from $[0, 1)^s$ is a (t, m, s)-net in base b if every elementary subcube in base b of volume b^{t-m} contains exactly b^t points of the sequence.

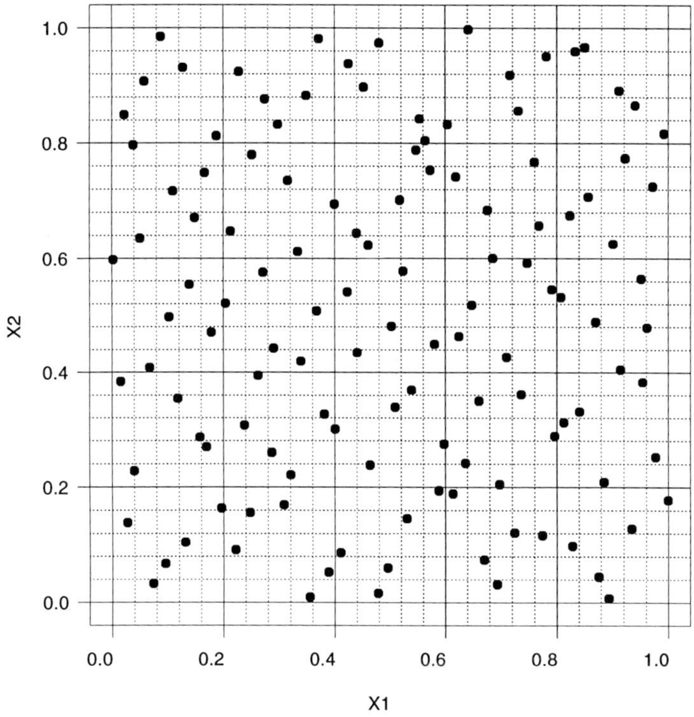

Fig. 17. The 125 points of a scrambled $(0, 3, 5)$-net in base 5. For whichever two (of five) variables that are plotted, the result is a 5 by 5 grid of 5 point Latin hypercube samples. Each variable is sampled once within each of 125 equal bins. Each triple of variables can be partitioned into 125 congruent cubes, each of which has one point.

Cells that "should" have b^t points do have b^t points, though cells that "should" have 1 point might not.

By common usage the name (t, m, s)-net assumes that the letter s is used to denote the dimension of the input space, though one could speak of (t, m, p)-nets. Another convention to note is that the subcubes are half-open. This makes it convenient to partition the input space into congruent subcubes.

The balance properties of a (t, m, s)-net are greater than those of an orthogonal array. If X_i^j is a (t, m, s)-net in base b then $\lfloor bX_i^j \rfloor$ is an orthogonal array of strength $\min\{s, m - t\}$. But the net also has balance properties when rounded to different powers of b on all axes, so long as the powers sum to no more than $m - t$. Thus the net combines aspects of orthogonal arrays and multi-level orthogonal arrays all in one point set.

In the case of a $(0, 4, 5)$-net in base 5, one has 625 points in $[0, 1)^5$ and one can count that there are 43750 elementary subcubes of volume $1/625$ of varying aspect ratios each of which has one of the 625 points.

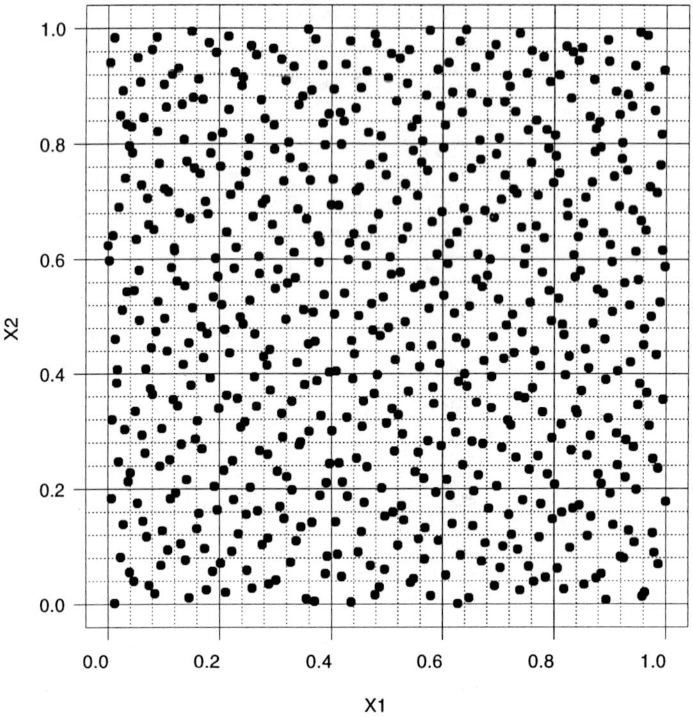

Fig. 18. The 625 points of a scrambled $(0, 4, 5)$-net in base 5. For whichever two (of five) variables that are plotted, the square can be divided into 625 squares of side $1/25$ or into 625 rectangles of side $1/5$ by $1/125$ or into 625 rectangles of side $1/125$ by $1/5$ and each such rectangle has one of the points. Each variable is sampled once within each of 625 equal bins. Each triple of variables can be partitioned into 625 hyperrectangles in three different ways and each such hyperrectangle has one of the points. Each quadruple of variables can be partitioned into 625 congruent hypercubes of side $1/5$, each of which has one point.

For $t \geqslant 0$, an infinite sequence $(X_i)_{i \geqslant 1}$ of points from $[0, 1)^s$ is a (t, s)-sequence in base b if for all $k \geqslant 0$ and $m \geqslant t$ the finite sequence $(X_i)_{i=kb^m+1}^{(k+1)b^m}$ is a (t, m, s)-net in base b.

The advantage of a (t, s)-sequence is that if one finds that the first b^m points are not sufficient for an integration problem, one can find another b^m points that also form a (t, m, s)-net and tend to fill in places not occupied by the first set. If one continues to the point of having b such (t, m, s)-nets, then the complete set of points comprises a $(t, m + 1, s)$-net.

The theory of (t, m, s)-nets and (t, s)-sequences is given in Niederreiter (1992). A famous result of the theory is that integration over a (t, m, s)-net can attain an accuracy of order $O(\log(n)^{s-1}/n)$ while restricting to (t, s)-sequences raises this slightly to $O(\log(n)^s/n)$. These results require that the integrand be of bounded variation in the sense of Hardy and Krause. For large s, it takes unrealistically large n for these rates to be clearly better than $n^{-1/2}$ but in examples they seem to outperform simple Monte Carlo.

The construction of (t, m, s)-nets and (t, s)-sequences is also described in Niederreiter (1992). Here we remark that for prime numbers s a construction by Faure (1982) gives $(0, s)$-nets in base s and Niederreiter extended the method to prime powers s. (See Niederreiter, 1992.) Thus one can choose b to be the smallest prime power greater than or equal to s and use the first s variables of the corresponding $(0, b)$-sequence in base b.

Owen (1995) describes a scheme to randomize (t, m, s)-nets and (t, s)-sequences. The points are written in a base b expansion and certain random permutations are applied to the coefficients in the expansion. The result is to make each permuted X_i uniformly distributed over $[0, 1)^s$ while preserving the (t, m, s)-net or (t, s)-sequence structure of the ensemble of X_i. Thus the sample estimate $n^{-1}\sum_{i=1}^{n} f(X_i)$ is unbiased for $\int f(X)\,\mathrm{d}F$ and the variance of it may be estimated by replication. On some test integrands in (Owen, 1995) the randomized nets outperformed their unrandomized counterparts. It appears that the unscrambled nets have considerable structure, stemming from the algebra underlying them, and that this structure is a liability in integration.

Figure 16 shows the 25 points of a scrambled $(0, 2, 5)$-net in base 5 projected onto two of the five input coordinates. These points are the initial 25 points of a $(0, 5)$-sequence in base 5. This design has the equidistribution properties of an orthogonal array based Latin hypercube sample. Moreover every consecutive 25 points in the sequence $X_{25a+1}, X_{25a+2}, \ldots, X_{25(a+1)}$ has these equidistribution properties. The first 125 points, shown in Figure 17 have still more equidistribution properties: any triple of the input variables can be split into 125 subcubes each with one of the X_i, in any pair of variables the points appear as a 5 by 5 grid of 5 point Latin hypercube samples and each individual input variable can be split into 125 cells each having one point. The first 625 points, are shown in Figure 18.

Owen (1996a) finds a variance formula for means over randomized (t, m, s)-nets and (t, s)-sequences. The formula involves a wavelet-like anova combining nested terms on each coordinate, all crossed against each other. It turns out that for any square integrable integrand, the resulting variance is $o(n^{-1})$ and it therefore beats any of the usual variance reduction techniques, which typically only reduce the asymptotic coefficient of n^{-1}.

For smooth integrands with $s = 1$, the variance is in fact $O(n^{-3})$ and in the general case Owen (1996b) shows that the variance is $O(n^{-3}(\log n)^{s-1})$.

8. Selected applications

One of the largest fields using and developing deterministic simulators is in the designing and manufacturing of VLSI circuits. Alvarez et al. (1988) describe the use of SUPREM-III (Ho et al., 1984) and SEDAN-II (Yu et al., 1982) in designing BIMOS devices for manufacturability. Aoki et al. (1987), use CADDETH a two dimensional device simulator, for optimizing devices and for accurate prediction of device sensitivities. Sharifzadeh et al. (1989) use SUPREME-III and PISCES-II (Pinto et al., 1984) to compute CMOS device characteristics as a function of the designable technology parameters. Nasif et al. (1984) describe the use of FABRICS-II to estimate circuit delay times in integrated circuits.

The input variables for the above work are generally device sizes, metal concentrations, implant doses and gate oxide temperatures. The multiple responses are threshold voltages, subthreshold slopes, saturation currents and linear transconductance although the output variables of concern depend on the technology under investigation. The engineers use the physical/numerical simulators to assist them in optimizing process, device, and circuit design before the costly step of building prototype devices. They are also concerned with minimizing transmitted variability as this can significantly reduce the performance of the devices and hence reduce yield. For example, Welch et al. (1990), Currin et al. (1991) and Sacks et al. (1989b) discuss the use of simulators to investigate the effect of transistor dimensions on the asynchronization of two clocks. They want to find the combination of transistor widths that produce zero clock skews with very small transmitted variability due to uncontrollable manufacturing variability in the transistors.

TIMS, a simulator developed by T. Osswald and C. L. Tucker III, helps in optimizing a compression mold filling process for manufacturing automobiles (Church et al., 1988). In this process a sheet of molding compound is cut and placed in a heated mold. The mold is slowly closed and a constant force is applied during the curing reaction. The controlling variables of the process are the geometry and thickness of the part, the compound viscosity, shape and location within the charge, and the mold closing speed. The simulator then predicts the position of the flow front as a function of time.

Miller and Frenklach (1983) discuss the use of computers to solve systems of differential equations describing chemical kinetic models. In their work, the inputs to the simulator are vectors of possibly unknown combustion rate constants and the outputs are induction-delay times and concentrations of chemical species at specified reaction times. The objectives of their investigations are to find values of the rate constants that agree with experimental data and to find the most important rate constant to the process. Sacks et al. (1989a) explore some of the design issues and applications to this field.

TWOLAYER, a thermal energy storage model developed by Alan Solomon and his colleagues at the Oak Ridge National Laboratory, simulates heat transfer through a wall containing two layers of different phase change material. Currin et al. (1991) utilize TWOLAYER in a computer experiment. The inputs into TWOLAYER are the layers dimensions, the thermal properties of the materials and the characteristics of the heat source. The object of interest was finding the configuration of the input variables that produce the highest value of a heat storage utility index.

FOAM (Bartell et al., 1981) models the transport of polycyclic aromatic hydrocarbon spills in streams using structure activity relationships. Bartell et al. (1983) modified this model to predict the fate of anthracene when introduced into ponds. This model tracks the "evaporation and dissolution of anthracene from a surface slick of synthetic oil, volatilization and photolytic degradation of dissolved anthracene, sorption to suspended particulate matter and sediments and accumulation by pond biota" (Bartell, 1983). They used Monte Carlo error analyses to assess the effect of the uncertainty in model parameters on their results.

References

Alvarez, A. R., B. L. Abdi, D. L. Young, H. D. Weed, J. Teplik and E. Herald (1988). Application of statistical design and response surface methods to computer-aided VLSI device design. *IEEE Trans. Comput. Aided Design* **7**(2), 271–288.

Aoki, Y., H. Masuda, S. Shimada and S. Sato (1987). A new design-centering methodology for VLSI device development. *IEEE Trans. Comput. Aided Design* **6**(3), 452–461.

Bartell, S. M., R. H. Gardner, R. V. O'Neill and J. M. Giddings (1983). Error analysis of predicted fate of anthracene in a simulated pond. *Environ. Toxicol. Chem.* **2**, 19–28.

Bartell, S. M., J. P. Landrum, J. P. Giesy and G. J. Leversee (1981). Simulated transport of polycyclic aromatic hydrocarbons in artificial streams. In: W. J. Mitch, R. W. Bosserman and J. M. Klopatek, eds., *Energy and Ecological Modelling*. Elsevier, New York, 133–143.

Bates, R. A., R. J. Buck, E. Riccomagno and H. P. Wynn (1996). Experimental design and observation for large systems (with discussion). *J. Roy. Statist. Soc. Ser. B* **58**(1), 77–94.

Borth, D. M. (1975). A total entropy criterion for the dual problem of model discrimination and parameter estimation. *J. Roy. Statist. Soc. Ser. B* **37**, 77–87.

Box, G. E. P. and N. R. Draper (1959). A basis for the selection of a response surface design. *J. Amer. Statist. Assoc.* **54**, 622–654.

Box, G. E. P. and N. R. Draper (1963). The choice of a second order rotatable design. *Biometrika* **50**, 335–352.

Box, G. E. P. and W. J. Hill (1967). Discrimination among mechanistic models. *Technometrics* **9**, 57–70.

Church, A., T. Mitchell and D. Fleming (1988). Computer experiments to optimize a compression mold filling process. Talk given at the Workshop on Design for Computer Experiments in Oak Ridge, TN, November.

Cranley, R. and T. N. L. Patterson (1976). Randomization of number theoretic methods for multiple integration. *SIAM J. Numer. Anal.* **23**, 904–914.

Cressie, N. A. C. (1986). Kriging nonstationary data. *J. Amer. Statist. Assoc.* **81**, 625–634.

Cressie, N. A. C. (1993). *Statistics for Spatial Data (Revised edition)*. Wiley, New York.

Currin, C., M. Mitchell, M. Morris and D. Ylvisaker (1991). Bayesian prediction of deterministic functions, with applications to the design and analysis of computer experiments. *J. Amer. Statist. Assoc.* **86**, 953–963.

Dandekar, R. (1993). Performance improvement of restricted pairing algorithm for Latin hypercube sampling Draft Report, Energy Information Administration, U.S.D.O.E.

Davis, P. J. and P. Rabinowitz (1984). *Methods of Numerical Integration*, 2nd. edn. Academic Press, San Diego.

Diaconis, P. (1988). Bayesian numerical analysis In: S. S. Gupta and J. O. Berger, eds., *Statistical Decision Theory and Related Topics IV*, Vol. 1. Springer, New York, 163–176.

Efron, B. and C. Stein (1981). The jackknife estimate of variance. *Ann. Statist.* **9**, 586–596.

Fang, K. T. and Y. Wang (1994). *Number-theoretic Methods in Statistics*. Chapman and Hall, London.

Faure, H. (1982). Discrépances des suites associées à un système de numération (en dimension s). *Acta Arithmetica* **41**, 337–351.

Friedman, J. H. (1991). Multivariate adaptive regression splines (with Discussion). *Ann. Statist.* **19**, 1–67.

Gill, P. E., W. Murray, M. A. Saunders and M. H. Wright (1986). User's guide for npsol (version 4.0): A Fortran package for nonlinear programming. SOL 86-2, Stanford Optimization Laboratory, Dept. of Operations Research, Stanford University, California, 94305, January.

Gill, P. E., W. Murray and M. H. Wright (1981). *Practical Optimization*. Academic Press, London.

Gordon, W. J. (1971). Blending function methods of bivariate and multivariate interpolation and approximation. *SIAM J. Numer. Anal.* **8**, 158–177.

Gu, C. and G. Wahba (1993). Smoothing spline ANOVA with component-wise Bayesian "confidence intervals". *J. Comp. Graph. Statist.* **2**, 97–117.

Hickernell, F. J. (1996). Quadrature error bounds with applications to lattice rules. *SIAM J. Numer. Anal.* **33** (in press).

Ho, S. P., S. E. Hansen and P. M. Fahey (1984). Suprem III – a program for integrated circuit process modeling and simulation. TR-SEL84 1, Stanford Electronics Laboratories.

Iman, R. L. and W. J. Conover (1982). A distributon-free approach to inducing rank correlation among input variables. *Comm. Statist.* **B11**(3), 311–334.

Johnson, M. E., L. M. Moore and D. Ylvisaker (1990). Minimax and maximin distance designs. *J. Statist. Plann. Inference* **26**, 131–148.

Journel, A. G. and C. J. Huijbregts (1978). *Mining Geostatistics*. Academic Press, London.

Koehler, J. R. (1990). Design and estimation issues in computer experiments. Dissertation, Dept. of Statistics, Stanford University.

Lindley, D. V. (1956). On a measure of the information provided by an experiment. *Ann. Math. Statist.* **27**, 986–1005.

Loh, W.-L. (1993). On Latin hypercube sampling. Tech. Report No. 93-52, Dept. of Statistics, Purdue University.

Loh, W.-L. (1994). A combinatorial central limit theorem for randomized orthogonal array sampling designs. Tech. Report No. 94-4, Dept. of Statistics, Purdue University.

Mardia, K. V. and R. J. Marshall (1984). Maximum likelihood estimation of models for residual covariance in spatial regression. *Biometrika* **71**(1), 135–146.

Matérn, B. (1947). Method of estimating the accuracy of line and sample plot surveys. *Medd. Skogsforskn Inst.* **36**(1).

Matheron, G. (1963). Principles of geostatistics. *Econom. Geol.* **58**, 1246–1266.

McKay, M. (1995). Evaluating prediction uncertainty. Report NUREG/CR-6311, Los Alamos National Laboratory.

McKay, M., R. Beckman and W. Conover (1979). A comparison of three methods for selecting values of input variables in the analysis of output from a computer code. *Technometrics* **21**(2), 239–245.

Miller, D. and M. Frenklach (1983). Sensitivity analysis and parameter estimation in dynamic modeling of chemical kinetics. *Internat. J. Chem. Kinetics* **15**, 677–696.

Mitchell, T. J. (1974). An algorithm for the construction of 'D-optimal' experimental designs. *Technometrics* **16**, 203–210.

Mitchell, T., M. Morris and D. Ylvisaker (1990). Existence of smoothed stationary processes on an interval. *Stochastic Process. Appl.* **35**, 109–119.

Mitchell, T., M. Morris and D. Ylvisaker (1995). Two-level fractional factorials and Bayesian prediction. *Statist. Sinica* **5**, 559–573.

Mitchell, T. J. and D. S. Scott (1987). A computer program for the design of group testing experiments. *Comm. Statist. Theory Methods* **16**, 2943–2955.

Morris, M. D. and T. J. Mitchell (1995). Exploratory designs for computational experiments. *J. Statist. Plann. Inference* **43**, 381–402.

Morris, M. D., T. J. Mitchell and D. Ylvisaker (1993). Bayesian design and analysis of computer experiments: Use of derivative in surface prediction. *Technometrics* **35**(3), 243–255.

Nassif, S. R., A. J. Strojwas and S. W. Director (1984). FABRICS II: A statistically based IC fabrication process simulator. *IEEE Trans. Comput. Aided Design* **3**, 40–46.

Niederreiter, H. (1992). *Random Number Generation and Quasi-Monte Carlo Methods*. SIAM, Philadelphia, PA.

O'Hagan, A. (1989). Comment: Design and analysis of computer experiments. *Statist. Sci.* **4**(4), 430–432.

Owen, A. B. (1992a). A central limit theorem for Latin hypercube sampling. *J. Roy. Statist. Soc. Ser. B* **54**, 541–551.

Owen, A. B. (1992b). Orthogonal arrays for computer experiments, integration and visualization. *Statist. Sinica* **2**, 439–452.

Owen, A. B. (1994a). Lattice sampling revisited: Monte Carlo variance of means over randomized orthogonal arrays. *Ann. Statist.* **22**, 930–945.

Owen, A. B. (1994b). Controlling correlations in latin hypercube samples. *J. Amer. Statist. Assoc.* **89**, 1517–1522.

Owen, A. B. (1995). Randomly permuted (t, m, s)-nets and (t, s)-sequences. In: H. Niederreiter and P. J.-S. Shiue, eds., *Monte Carlo and Quasi-Monte Carlo Methods in Scientific Computing*. Springer, New York, 299–317.

Owen, A. B. (1996a). Monte Carlo variance of scrambled net quadrature. *SIAM J. Numer. Anal.*, to appear.

Owen, A. B. (1996b). Scrambled net variance for integrals of smooth functions. Tech. Report Number 493, Department of Statistics, Stanford University.

Paskov, S. H. (1993). Average case complexity of multivariate integration for smooth functions. *J. Complexity* **9**, 291–312.

Park, J.-S. (1994) Optimal Latin-hypercube designs for computer experiments. *J. Statist. Plann. Inference* **39**, 95–111.

Parzen, A. B. (1962). *Stochastic Processes*. Holden-Day, San Francisco, CA.

Patterson, H. D. (1954). The errors of lattice sampling. *J. Roy. Statist. Soc. Ser. B* **16**, 140–149.

Phadke, M. (1988). *Quality Engineering Using Robust Design*. Prentice-Hall, Englewood Cliffs, NJ.

Pinto, M. R., C. S. Rafferty and R. W. Dutton (1984). PISCES-II–posson and continuity equation solver. DAGG-29-83-k 0125, Stanford Electron. Lab.

Ripley, B. (1981). *Spatial Statistics*. Wiley, New York.

Ritter, K. (1995). Average case analysis of numerical problems. Dissertation, University of Erlangen.

Ritter, K., G. Wasilkowski and H. Wozniakowski (1993). On multivariate integration for stochastic processes. In: H. Brass and G. Hammerlin, eds., *Numerical Integration*, Birkhauser, Basel, 331–347.

Ritter, K., G. Wasilkowski and H. Wozniakowski (1995). Multivariate integration and approximation for random fields satisfying Sacks–Ylvisaker conditions. *Ann. Appl. Prob.* **5**, 518–540.

Roosen, C. B. (1995). Visualization and exploration of high-dimensional functions using the functional ANOVA decomposition. Dissertation, Dept. of Statistics, Stanford University.

Sacks, J. and S. Schiller (1988). Spatial designs. In: S. S. Gupta and J. O. Berger, eds., *Statistical Decision Theory and Related Topics IV*, Vol. 2. Springer, New York, 385–399.

Sacks, J., S. B. Schiller and W. J. Welch (1989). Designs for computer experiments. *Technometrics* **31**(1), 41–47.

Sacks, J., W. J. Welch, T. J. Mitchell and H. P. Wynn (1989). Design and analysis of computer experiments. *Statist. Sci.* **4**(4), 409–423.

Shannon, C. E. (1948). A mathematical theory of communication. *Bell Syst. Tech. J.* **27**, 379–423, 623–656.

Sharifzadeh, S., J. R. Koehler, A. B. Owen and J. D. Shott (1989). Using simulators to model transmitted variability in IC manufacturing. *IEEE Trans. Semicond. Manufact.* **2**(3), 82–93.

Shewry, M. C. and H. P. Wynn (1987). Maximum entropy sampling. *J. Appl. Statist.* **14**, 165–170.

Shewry, M. C. and H. P. Wynn (1988). Maximum entropy sampling and simulation codes. In: *Proc. 12th World Congress on Scientific Computation*, Vol. 2, IMAC88, 517–519.

Sloan, I. H. and S. Joe (1994). *Lattice Methods for Multiple Integration*. Oxford Science Publications, Oxford.

Smolyak, S. A. (1963). Quadrature and interpolation formulas for tensor products of certain classes of functions. *Soviet Math. Dokl.* **4**, 240–243.

Stein, M. L. (1987). Large sample properties of simulations using Latin hypercube sampling. *Technometrics* **29**(2), 143–151.

Stein, M. L. (1989). Comment: Design and analysis of computer experiments. *Statist. Sci.* **4**(4), 432–433.

Steinberg, D. M. (1985). Model robust response surface designs: Scaling two-level factorials. *Biometrika* **72**, 513–26.

Tang, B. (1992). Latin hypercubes and supersaturated designs. Dissertation, Dept. of Statistics and Actuarial Science, University of Waterloo.

Tang, B. (1993). Orthogonal array-based Latin hypercubes. *J. Amer. Statist. Assoc.* **88**, 1392–1397.

Wahba, G. (1978). Interpolating surfaces: High order convergence rates and their associated designs, with applications to X-ray image reconstruction. Tech. report 523, Statistics Department, University of Wisconsin, Madison.

Wahba, G. (1990). *Spline Models for Observational Data*. CBMS-NSF Regional Conference Series in Applied Mathematics, Vol. 59. SIAM, Philadelphia, PA.

Wasilkowski, G. (1993). Integration and approximation of multivariate functions: Average case complexity with Wiener measure. *Bull. Amer. Math. Soc. (N. S.)* **28**, 308–314. Full version *J. Approx. Theory* **77**, 212–227.

Wozniakowski H. (1991). Average case complexity of multivariate integration. *Bull. Amer. Math. Soc. (N. S.)* **24**, 185–194.

Welch, W. J. (1983). A mean squared error criterion for the design of experiments. *Biometrika* **70**(1), 201–213.

Welch, W. Yu, T. Kang and J. Sacks (1990). Computer experiments for quality control by parameter design. *J. Quality Technol.* **22**, 15–22.

Welch, W. J., J. R. Buck, J. Sacks, H. P. Wynn, T. J. Mitchell and M. D. Morris. Screening, prediction, and computer experiments. *Technometrics* **34**(1), 15–25.

Yaglom, A. M. (1987). *Correlation Theory of Stationary and Related Random Functions*, Vol. 1. Springer, New York.

Ylvisaker, D. (1975). Designs on random fields. In: J. N. Srivastava, ed., *A Survey of Statistical Design and Linear Models*. North-Holland, Amsterdam, 593–607.

Young, A. S. (1977). A Bayesian approach to prediction using polynomials. *Biometrika* **64**, 309–317.

Yu, Z., G. G. Y. Chang and R. W. Dutton (1982). Supplementary report on sedan II. TR-G201 12, Stanford Electronics Laboratories.

S. Ghosh and C. R. Rao, eds., *Handbook of Statistics, Vol. 13*

10

A Critique of Some Aspects of Experimental Design

J. N. Srivastava

1. Introduction

Comparative experiments form a very large part of the subject of statistical design of scientific experiments. There are v "treatments", and we need to compare their "yields", the terminology being borrowed from agricultural experiments, in the context of which the subject of experimental design initially grew. This essentially subsumes a larger class of experiments where the objective is to study the "response surface", i.e., the general nature of dependence of the yields on the treatments. Also, the set of treatments (i.e., the "factor space") could be (discretely or continuously) of infinite cardinality, although in most cases, only a "sample" of such points can be included in the experiment. It may be stressed that in the above, the "treatments" are well defined entities, being for example, varieties of wheat, levels of a fertilizer, amounts of a chemical, a set of drugs, various settings of a particular machine, the presence or absence of a particular action, and so on.

The experiment is done using "experimental material", this being divided into "units". Usually, on each unit, one of the treatments is applied, and a variable y is measured, which in some sense denotes the "effect" of the treatment applied to it "plus" (i.e., "confounded with") the effect of its own "innate characteristics" (i.e., its "individual effect"). To help "average out" over such individual effects, the principle of "replication" (i.e., try each treatment on several units) was founded. Notice that "plus" does not necessarily mean that the two effects are additive. If t is a treatment, and ϕ_t is its effect, then the effect of a unit u may change ϕ_t to $g_u(\phi_t)$, where g_u is some function, and where $g_u(\phi_t)$ is not necessarily of the additive form $(\alpha_u + \phi_t)$, where α_u denotes the effect of u. Usually, additivity is assumed, but not because it is established to be valid. The simple fact is that, except for agriculture, it has not been much studied in the various subject matter fields. The author feels that the "nature" of the experimental material should be studied in its own right in any given science before major experiments are laid out. Also, we do not have much information on the kind of studies needed on a given experimental material in a given field so as to throw light on the intrinsic usefulness of the material in the study of the phenomena that are of interest.

Again, to avoid biases arising out of a systematic choice of units to which a treatment is applied, the concept of "randomization" was introduced. This, indeed, is valuable and necessary. But, this should not be confused with "randomization analysis" of designs. This analysis is based on an averaging over the universe (of placements of treatments onto the units) generated by the act of randomization. It is shown that, approximately, it gives the same answers concerning (analysis of variance) test statistics as the ones based on linear models and normality. Some people regard this as an added justification of, or even a more desirable basis of, the usual analysis. However, the latter is controversial. Consider two *different* acts of randomization A_i $(i = 1, 2)$ with two different systems of probabilities P_i (of placements of treatments into units). Since P_1 and P_2 are different, A, and A_2 will generate two different universes (say, U_1, and U_2). However, A_1 and A_2 could both result in the *same* placement Π of treatments on the units. Now, the experiment depends only on Π, not on A_1 and A_2 (or P_1 and P_2). However, the analysis of the experiment using A_i $(i = 1, 2)$ would be done by averaging out U_i, which would generally lead to two *different* conclusions. This is a bit discomforting, since there is no reason to accept one analysis over the other (Srivastava, 1975).

A third principle put forth by Fisher is "local control". This refers to the division of units into groups (called "blocks"), such that each block is relatively more homogeneous than the set of units as a whole. Such blocks may correspond to "levels" of a "nuisance" factor (i.e., a factor not of interest in itself, but which enters into the units nevertheless). Note that the levels are either known (like litters of animals), or assumed known (strips of land in an agricultural experiment). In the usual analysis, the differences between the blocks are eliminated. Since the differences between the treatments are compared with the "random fluctuations" on the units (measured by the "error variance" σ^2) the elimination of blocks leads to the use of a σ^2, which is smaller than the one that would result if the blocks were ignored. Thus, the treatments are compared with greater precision by eliminating the effect of the nuisance factor.

Three criteria of goodness of a design have been put forth (Srivastava, 1984). These are (i) variance-optimality, (ii) sensitivity, and (iii) revealing power. For simplicity, assume that y's are independent with variance σ^2. Variance-optimality relates to the situation where σ^2 is fixed (though unknown) and the expected values of the y's are known except for unknown parameters, which need to be determined. Variance-optimality seeks to determine a design which maximizes some functions of the information matrix of the estimates of the parameters.

However, a little thought would reveal that such a situation (where the model is known except for parameters) would, relatively, arise only toward the end of an investigation. In other words, before seeking an optimal design, one would need to determine, as well as possible the following: (A_1) the phenomenon to be studied, the purpose of such study, the mode of study that is relevant (leading perhaps to a set of well-defined treatments to be compared), and the experimental material to be used, and the response variable(s) to be measured, (A_2) the nature of this material with respect to heterogeneity and nuisance factors etc., the units to be used, and finally (and very importantly) (A_3) to explore and determine what model does arise in this situation.

Assuming that all this has been satisfactorily done, the optimal design problem is to decide how to (a) sample the factor space to determine which points (i.e., treatments) are to be chosen for the experiment, including the number of units on which each such point is to be tried, and (b) the actual assignment of the treatments to the units. Fisher's principle of replication is thus subsumed under variance-optimality.

Sensitivity is a generalization of the concept of local control. It deals with the determination of grouping of units, so that the effects of known (and, possibly, even of unknown) nuisance factors are eliminated as much as possible. Designs with incomplete blocks, or those where heterogeneity is sought to be eliminated two-way or multi-way, have arisen out of this context. Thus, in a sensitive experiment, σ^2 is as small as it is possible to have.

Revealing power refers to the ability of the design to find out what the true model might be. The danger of being satisfied with a given model, without knowing how misleading it is (i.e., in which directions, and to what extent), is that the very purpose of conducting the experiment may be defeated. It may indeed be worse than doing no experiment. For example, consider the study of dependence of y on x, in the x-interval $[0, 1]$. Suppose that it is assumed that a first degree polynomial is adequate, and that four observations will be made. The optimal designs will put two observations at each of the points $x = 0$ and 1. Suppose the experiment is done, and y values are found which are nearly zero for each of the four points. It will be concluded that x has no effect on y. But, this may happen also if $\mathrm{E}(y(x)) = ax^2 - ax$. A design with at least one point in the interior of $(0, 1)$ may help to reveal this fact, and thus may be said to possess more revealing power.

Above, we mentioned procedural steps $(A_1)-(A_3)$ for determining a suitable experimental design. Of these, A_1 is to be done primarily by the scientist (with enlightened interference by the statistician). Also, A_2 relates to sensitivity, and A_3 to revealing power. Seeking variance-optimality may be referred to as step A_4. If, on the basis of past knowledge or intuition, the correct model is known, then A_3 is already done, and one may proceed to A_4. However, in other situations, the correct model may not be knowable a priori. In such cases, we will have to combine A_3 and A_4 in the light of whatever knowledge is available, and pursue A_3 and A_4 simultaneously. This last category of cases is expected to be the more common one, and here we shall seek a design with the two fold property that it helps us to determine the true model as well as allow accurate estimation of parameters under the model that is the true one.

The four steps $A_1, \ldots, A_4$ are necessary for a good experiment. Of these, only A_2 and A_4 have generally been pursued. Steps A_2, A_3, A_4 are usually considered to be the main body of the design problem. However, A_1 is even more important; we illustrate this fact by some brief remarks on clinical trials, a field which we can not go into in detail in this paper because of lack of space.

Since we wish to be brief, we shall take up an issue out of A_1, namely the question of "treatments" and "responses", the experimental unit being a human being. We shall illustrate by contrasting the situation with an agronomical experiment, where a plot of land is typically the experimental unit. One basic difference between the two units is that while a person can speak out his reaction at anytime regarding how he "feels" after taking a particular treatment, the plot cannot do so. In the latter case, the scientist must decide on a set of responses which are to be observed, plus a plan of observing them. This can be done with respect to a man also, but a man is able to feel and communicate responses that the scientist may not have even thought of, and which may indeed provide crucial information. For, example, a pneumonia patient, after being treated by a particular broncho-dilator, may suddenly inform: "I feel that the inner muscles of my neck around my throat area are rapidly getting paralysed, and I will choke to death in 15 minutes". Now, the scientist-in-charge, may not have imagined such a response, because it may not be "common". However, this response is still there, and is a consequence of the total body-condition of the experimental unit (i.e, the patient), of which pneumonia is only one symptom. The fact that this symptom is considered "major" by the scientist, does not mean that it is really so all the time.

The above example shows that in certain clinical trials, we may run into a problem, namely, how to cope with the multitude of responses possible, which may initially not be even well defined!

What about the concept of "comparing treatments", or "applying a treatment to an experimental unit"? Consider the latter first. In ordinary statistical design, it simply means apply the treatment to the unit; i.e., there is no question of desirability of "continued application". But, in clinical trials, "treatments" (i.e., drugs) are to be taken by the subjects (i.e., the patients) for a relatively long time. If, after taking the drug a few times, the subject feels that it is hurting in some way (i.e., if a response is found whose value is quite negative), then common sense would dictate that the "treatment" be stopped or modified in some way. In other words, the concept of a "treatment", as we know in agronomic experiments, is not quite valid here. Because of this, "comparing treatments" in the medical field, has to be, should be, a different game than what agronomy-born statistical design is used to play. In actual practice, the above troubles are further confounded by the current medical philosophy, where the goal appears to be more the suppression of symptoms rather than understanding and curing the disease.

A discussion of these issues will be made elsewhere by the author. Here, the above example is included just to show the nature and importance of A_1.

The next section deals with "Search Linear Models", a topic useful in studies on how to reveal the model. This is needed in the latter sections, the first two of which deal with designs with nuisance factors, and the third and last one with factorial experiments.

2. Search linear model

In this section, we briefly recall certain elements of the theory of search linear models since it appears to arise in an intrinsic way when we attempt to identify the correct model.

Let $\underline{y}(N \times 1)$ be a vector of observations such that

$$\mathrm{E}(\underline{y}) = A_1\underline{\xi}_1 + A_2\underline{\xi}_2, \tag{2.1}$$

where E denotes expected value, $A_1(N \times \nu_1)$ and $A_2(N \times \nu_2)$ are known matrices, $\underline{\xi}_1(\nu_1 \times 1)$ is a vector of unknown parameters, and $\underline{\xi}_2(\nu_2 \times 1)$ is a vector of parameters whose values are unknown but about which partial information is available as follows. It is known that there is a number k, such that at most k elements of $\underline{\xi}_2$ are nonnegligible; but it is not known which elements of $\underline{\xi}_2$, are nonnegligible, and what are the values of the nonnegligible elements. The problem is to identify the nonnegligible elements of ξ_2, and to estimate them along with estimating the elements of $\underline{\xi}_1$.

The following result is fundamental in this theory (Srivastava, 1975).

THEOREM 2.1. (a) *A necessary condition that the above problem (i.e., the identification of the nonnegligible element of $\underline{\xi}_2$, and the estimation of these and of the nonzero elements of $\underline{\xi}_1$) can be solved (for all values of $\underline{\xi}_1$ and $\underline{\xi}_2$, subject to the above conditions) is that the following rank condition is satisfied:*

$$\text{Rank}(A_1 : A_{20}) = \nu_1 + 2k, \tag{2.2}$$

for all submatrices $A_{20}(N \times 2k)$ of A_2.

(b) *If $\sigma^2 = 0$, then the rank condition (2.2) is also sufficient. In this case, the identification can be done with probability one, and the estimation with variance zero. If $\sigma^2 > 0$, this probability is, of course, less then one and the variances are positive.*

How are the identification and estimation actually done? Though many methods have been proposed, the following one appears to be most successful. Let $S_e^2(\underline{\xi}_1, \underline{\xi}_{20})$ denote the sum of squares due to error under the model

$$\text{E}(\underline{y}) = A_1\underline{\xi}_1 + A_{20}\underline{\xi}_{20}, \quad V(\underline{y}) = \sigma^2 I_N, \tag{2.3}$$

where $\underline{\xi}_{20}(k \times 1)$ is a subvector of $\underline{\xi}_2$, and A_{20} is the submatrix of A_2 whose columns correspond to $\underline{\xi}_{20}$. Let $\underline{\xi}_{20}^*$ be the value of $\underline{\xi}_{20}$ such that $S_e^2(\underline{\xi}_1, \underline{\xi}_{20}^*)$ is minimum among all values of $S_e^2(\underline{\xi}_1, \underline{\xi}_{20})$ obtained by varying $\underline{\xi}_{20}$ over the set of $\binom{\nu_2}{k}$ possible choices (as a subvector of $\underline{\xi}_2$). Then, $\underline{\xi}_{20}^*$ is taken to be the (possibly) nonzero set of parameters in $\underline{\xi}_2$, and $\underline{\xi}_1$ and $\underline{\xi}_{20}^*$ are estimated as usual, ignoring the other parameters in $\underline{\xi}_2$. If $\sigma^2 = 0$, and (2.2) is satisfied, this procedure will lead to correct identification with probability one, and the parameters will be estimated with variance zero; indeed, in this case, the procedure is equivalent to projecting $\underline{y}$ on the columns of $(A_1 : A_{20})$, for various A_{20} until we find an A_{20} (say A_{20}^*) for which the projection is perfect. The rank condition (2.2) ensures that this will be possible (for a *unique* A_{20}^*).

It is important to study what happens when (2.2) is not satisfied. In this case, there exists a submatrix $A_{21}(N \times 2k)$ of A_2, and vectors $\underline{\theta}_1(\nu_1 \times 1)$ and $\underline{\theta}_{21}(2k \times 1)$ such that

$$A_1\underline{\theta}_1 + A_{21}\underline{\theta}_{21} = \underline{O}_{N1}, \tag{2.4}$$

where, throughout this paper, O_{pq}, J_{pq} and I_p will respectively denote the $(p \times q)$ zero matrix, the $(p \times q)$ matrix with 1 everywhere, and the $(p \times p)$ identity matrix.

Now, let the columns of $A_{21}(N \times 2k)$ correspond to the elements of $\underline{\xi}_{21}(2k \times 1)$, where $\underline{\xi}_{21}$ is a subvector of $\underline{\xi}_2$. Furthermore, let $\underline{\xi}_{22}(k \times 1)$ be a subvector of $\underline{\xi}_{21}$, and let $\underline{\xi}_{23}(k \times 1)$ be the vector consisting of the elements of $\underline{\xi}_{21}$ which are not in $\underline{\xi}_{22}$. Also, let $A_{22}(N \times k)$ and $A_{23}(N \times k)$ be the submatrices of A_{21} corresponding to the subvectors $\underline{\xi}_{22}$ and $\underline{\xi}_{23}$. Similarly, let $\underline{\theta}_{22}(k \times 1)$ and $\underline{\theta}_{23}(k \times 1)$ be the subvectors of $\underline{\theta}_{21}$ corresponding to the A_{22} and A_{23}. Then, for *all* $\underline{\theta}_1^*$, (2.4) gives

$$A_1\underline{\theta}_1^* + A_{22}\underline{\theta}_{22} = A_1(\underline{\theta}_1^* - \underline{\theta}_1) + A_{23}(-\underline{\theta}_{23}). \tag{2.5}$$

This shows that $\underline{y}$ would equal each side of (2.5), under two different situations, where each situation satisfies the condition of the model; these two situations are given by (i) $\underline{\xi}_1 = \underline{\theta}_1^*$, and $\underline{\xi}_{22}$ is the nonzero set of parameters, with value $\underline{\theta}_{22}$, and (ii) $\underline{\xi}_1 = (\underline{\theta}_1^* - \underline{\theta}_1)$, and $\underline{\xi}_{23}$ is the nonzero set of parameters, with value $(-\underline{\theta}_{23})$. Thus, in this case, there is a confounding between the parameter sets $\underline{\xi}_{22}$ and $\underline{\xi}_{23}$.

However, the important point to note is that this confounding is of a limited kind. Indeed, it is *not* true that for *every* $\underline{\theta}_{22}^*(k \times 1)$ there exists a $\underline{\theta}_{23}^*(k \times 1)$, and two $(\nu_1 \times 1)$ vectors $\underline{\theta}_{11}^*$ and $\underline{\theta}_{12}^*(\nu_1 \times 1)$ such that

$$A_1\underline{\theta}_{11}^* + A_{22}\underline{\theta}_{22}^* = A_1\underline{\theta}_{12}^* + A_{23}\underline{\theta}_{23}^*. \tag{2.6}$$

This fact leads to the following important class of observations, whose implication seems to have been missed thus far in much of the literature on Search Linear Models (and Search Designs). Basically, the observation is that even if (2.2) is not satisfied, very often, $\mathrm{E}(\underline{y})$ will have a *unique* projection on the columns of $(A_1 : A_{20})$, where $A_{20}(N \times k)$ is some submatrix of A_2. To elaborate, consider (2.5). Suppose, for some vectors $\underline{\theta}_{11}^0(\nu_1 \times 1)$, and $\underline{\theta}_{22}^0(k \times 1)$, we have

$$\mathrm{E}(\underline{y}) = A_1\underline{\theta}_{11}^0 + A_{22}\underline{\theta}_{22}^0. \tag{2.7}$$

Then, it may indeed be true that there does not exist a submatrix $A_{24}(N \times k)$ in A_2, such that the columns of A_{24} correspond to a set of parameters $\underline{\xi}_{24}(k \times 1)$ (this being a subvector of $\underline{\xi}_2$) such that $\underline{\xi}_{24} \neq \underline{\xi}_{22}$, and furthermore, such that $\mathrm{E}(\underline{y})$ can be expressed as a linear combination of the columns of $(A_1 : A_{24})$. If this happens, the projection of $\mathrm{E}(\underline{y})$ in (2.7) is unique. This means that even though (2.2) is not satisfied, in case $\sigma^2 = 0$, the (identification and estimation) problem is still totally solvable (for the value of $\underline{y}$ under consideration), and for $\sigma^2 > 0$, $S_e^2(\underline{\xi}_1, \underline{\xi}_{22})$ would be expected to attain the minimum value (which is the best one could hope for).

The above discussion gives a valuable hint concerning the *analysis* of the observation vector $\underline{y}$. For example, in the application of the Search Linear Models to scientific investigations, the value of k would usually be unknown. Suppose k_0 is the maximum value of k for which (2.2) holds. Then, given $\underline{y}$, one should try projecting it on the columns of $[A_1 : A_{20}]$, where $A_{20}(N \times k)$ are the various submatrices of A_2 with $k \geqslant k_0$. In this way, there is a possibility that one may discover the entire set of nonnegligible parameters. We shall elaborate this remark later on in the paper.

Before closing this section, it is important to note the effect of reparameterisation on the model (2.1). The problem of identification of the nonnegligible elements of $\underline{\xi}_2$ is unique to the model (2.1), and is not invariant if in place of $\underline{\xi}_2$ some other parameter vector is used. To, elaborate, note that as an ordinary linear model, (2.1) is equivalent to

$$\mathrm{E}(\underline{y}) = A_1\underline{\xi}_1^* + A_3\underline{\xi}_3, \tag{2.8}$$

where

$$A_3\underline{\xi}_3 = A_1(\underline{\xi}_1 - \underline{\xi}_1^*) + A_2\underline{\xi}_2. \tag{2.9}$$

However, in (2.8), the identification is of nonzero elements of $\underline{\xi}_3$, whereas in (2.1) it relates to the elements of $\underline{\xi}_2$, so that as search linear models, (2.1) and (2.8) are not equivalent. Examples of this are given by the two models I and II to be discussed in Section 4.

3. Designs with one nuisance factor

In this paper we shall limit our discussion to one or two nuisance factors, and the usual block designs and their analyses. First, consider one nuisance factor. In the general case, there are v treatments, and b blocks, with sizes say $k_1, \ldots, k_b$ with $k_1 \leqslant k_2 \leqslant \cdots \leqslant k_1$ Let N be the incidence matrix with elements n_{ij}, which denotes the number of units in block i to which jth treatment is applied.

In choosing our design, we have a few liberties. Firstly, we don't have to use all the b blocks; if we so wish, we can use only a subset of b' $(< b)$ blocks, and ignore the others. Also, if a block has k units in it, we do not have to use all of them; if we so wish, we may use a subset of the available units and discard the rest. Also, if we so wish, then instead of using a binary design (i.e., one in which each treatment occurs in any given block only zero times or once), we may relax this and allow the n_{ij} to take values larger than 1.

In the analysis of designs, what restrictions do we have? The assumptions are that each block is homogeneous, that the variance σ^2 (of the observations on a unit) depends on block size (but, somehow, is reasonably constant for some range say (k_{01}, k_{02})) of block sizes (which we are going to use), that σ^2 is constant from one treatment to another and from one block to another, that the block effects and treatment effects are mutually additive, and that with respect to the (unknown location) parameters the model is linear.

Now, two situations may arise. One is where the experimental material has been studied thoroughly previous to the experiment being planned, and it has been affirmed that the above assumptions hold to a reasonable degree. The other situation is where the first one is negated.

Consider the first situation. The block sizes range from k_{01} to k_{02}. If $k_{01} = k_{02}$, then, we select a design hopefully optimal *w.r.t.* the criteria set forth. Such topics have been studied extensively in the subject.

Similar studies have been done for the case when $k_{01} < k_{02}$. However, here, a new feature arises. Suppose that, approximately, n_0 units are needed for the experiment. The following may happen: (i) There are enough blocks available with size k_{02}, so as to accommodate the whole experiment (with n_0 units); (ii) condition (i) does not hold, but there exists k'_{01} ($\geqslant k_{01}$, and $< k_{02}$) such that the experiment can be accommodated in blocks of sizes ranging in the interval (k'_{01}, k_{02}), but not in the interval $(k'_{01}+1, k_{02})$.

Now, generally, it is true that if σ^2 is constant for two block sizes k'_1 and k'_2 with $k'_1 < k'_2$, then the information matrix is "large" if blocks of size k'_2 only are used, rather than using some or all blocks of size k'_1. This last statement is subject to many obvious constraints, such as the availability of the same 'kind' of design (for example, same association scheme (in case a PBIBD is used)) for the different block sizes, and various other combinatorial restrictions (in case different efficiencies are desired on the various treatment contrasts).

In view of the above, under condition (i), we would generally use a design with fixed block size k'_2. On the other hand, if condition (ii) holds, then we have the following situation: we are given a fixed number b_k of blocks for each size k in the range $k = k'_{01}, \ldots, k_{02}$. We have v treatments. We may have variance conditions to satisfy, such as having different variances for various (normalized) treatment contrasts. Subject to all such conditions and constraints, the problem is to obtain an optimal or near-optimal design.

Now, the number of combinations of such variance conditions, block size ranges, numbers of the block sizes, and the values of v are potentially very large. Even if a large number of such designs are catalogued, still nothing may be available when the need arises in a particular instance. It is therefore suggested that instead of developing individual designs of this kind, we produce classes of algorithms which generate them. It is likely that most designs so produced may not be optimal and some may be far from optimal. This will be due to the fact that the algorithm addresses a relatively large group of design situations. To get closer to optimality, each such group may be divided into suitable subgroups, and a series of (second step) algorithms may be developed for each subgroup. Even this may not do the job for all subgroups. Such subgroups may be divided into subsubgroups, each one of which having a new third-step algorithm. This process may be continued, if necessary.

It may seem that the above may necessitate the development of too many algorithms. However, the author feels that it should still be manageable. It seems necessary for use in scientific situations where condition (ii) arises frequently. As such algorithms are developed, they should be well publicized through design-oriented journals, so that the users are kept well informed.

Next, we consider situations where the experimental material has not been (and, cannot be) studied adequately prior to the intended experiment. Here, depending on the nature of the inadequacy, an appropriate procedure will have to be determined. We discuss a few issues of this kind.

Take the case where, in order to accommodate the experiment, we do need to use blocks of unequal sizes (with a given number of blocks of each size), and it is not known (or, not expected) that σ^2 will stay constant with change in block size. We need to decide how to select the design, and when the experiment is done, how to analyze the results.

Let U_k $(k \in K)$ denote the set of units in blocks of size k, M_k the corresponding information matrix for $\bar{\tau}$ (the $(v \times 1)$ vector of treatment effects), and σ_k^2 the value of σ^2, where K is the set of distinct block sizes that need to be used. To select the design, a Bayesian approach is highly justified. Depending upon the knowledge we have of the experimental material, we select a model for the dependence of C_k on k, where we assume $\sigma_k^2 = C_k\sigma^2$, σ^2 being unknown. (The selected model should completely specify C_k.) This, in effect reduces the situation to one where the data from the whole experiment can be expressed as a linear model in which the observations have a variance known up to a constant (unknown) multiplier (σ^2). The information matrix M^* of such an experiment can be written down.

Now, assuming that a good optimal design algorithm is available, an appropriate design may be selected, and the experiment conducted.

From the data, an estimate $\widehat{\sigma}_k^2$ of σ_k^2 will be available for each $k \in K$. The problem is to combine the M_k $(k \in K)$, and obtain a combined estimate of $\bar{\tau}$ for the whole experiment. Some ideas of this kind will be found in Srivastava and Beaver (1986).

For further discussion, it will be instructive to note a result on the comparison of two designs D_1 and D_2, where D_1 is a BIBD with parameters (v, b, r, k, λ) with $k = k_o m$, where m and k_o are positive integers. Also, D_2 is a design with v treatments, arranged in b blocks each of size k, obtained from a BIBD D_{20} with parameters $(v, b, r_o, k_o, \lambda_o)$ as follows. If a treatment x occurs in a block of D_{20}, then x occurs on exactly m units in the corresponding block of D_2; if x does not occur in a block of D_{20}, then it also does not occur in the corresponding block of D_2. (In this paper, designs like D_2 will be referred to as m-BIBD's.) We assume that v, b, k_o and m are such that both D_1 and D_{20} exist. Let M_1^* and M_2^* denote the information matrices respectively for D_1 and D_2. Then, it can be checked that

$$M_2^* = \theta M_1^*, \tag{3.1}$$

where

$$\theta = m(k_o - 1)/(k_o m - 1). \tag{3.2}$$

Thus, the loss of information which equals $(1 - \theta)$ is given by

$$1 - \theta = (m - 1)/(k_o m - 1) = (m - 1)/(k - 1). \tag{3.3}$$

This means, for example, that if block size k equals 12, and we allow 0 or 3 repetitions of each treatment in a block, then we lose only $(2/11)$ ($\approx$ 18.2%) of the information.

We now consider the case where we are not sure whether the block and treatment effects are additive, or whether the variance of an observation depends upon the treatment or block to which it corresponds. Here, the comparison between D_1 and D_2 in the last paragraph is pertinent. Suppose $k = 12$, $m = 3$, $k_o = 4$. Assume v, b are such that D_1 and D_2 exist. Then, two avenues are open. We may use D_1 or D_2. If the assumptions are true, then in using D_2 over D_1, we lose only about 18.2% of the information. But, if the assumptions are false, and we use D_1, our data are subject to unknown biases in the location parameters, and we are not sure which observations are more reliable (i.e., have less variance) and which are not, because D_1 does not help in exploring these questions, its revealing power with respect to such issues being zero. Furthermore, its sensitivity will also be negatively affected. On the other hand, each block of D_2 provides k_o sets of $(m-1)$ orthogonal contrast belonging to the error space. Thus, D_2 provides a wealth of information which will throw light on the basic assumptions.

The author believes that designs such as D_2 are preferable over those like D_1 for the above reasons. However, it will be useful to provide more methodology for the full analysis of data from D_2 than is currently available.

4. Row–column designs

Consider two nuisance factors, at p and q levels, respectively. We shall assume a $(p \times q)$ 'Latin Rectangle Design' defined as follows. In this design, each treatment occurs (p/v) times in each column. Also, with respect to the rows considered as blocks, the design is an m-BIBD with v treatments, p blocks and block size q. Clearly, (p/v), and (q/vm) are integers. Many well known designs such as Latin Square, Youden Square etc. are special cases of this design.

In this section, we shall briefly discuss the problem of nonadditivity (of the two nuisance factors) in relation to the designs and their analysis, and its misleading effect on the inferences drawn concerning treatment effects. We shall largely confine ourselves to the Latin Square design. Detailed studies of this type will be found in Srivastava (1993), Wang (1995), and Srivastava and Wang (1996), where other designs such as Lattice Squares are also covered. Although occasionally the discussion will be in general terms, we shall illustrate mostly through the 4×4 Latin Square design.

Consider a $(p \times p)$ Latin Square design (so that $v = q = p, m = 1$). Let y_{ij} $(i, j = 1, \ldots, p)$ denote the observation in the (i, j) cell. The model is

$$y_{ij} = \gamma_{ij} + \tau'_k + e_{ij}, \tag{4.1}$$

where γ_{ij} denotes the effect of the (i, j) cell, τ'_k the effect of the kth treatment (assumed to be assigned to the (i, j) cell), and where e_{ij} denotes the random fluctuation on the observations in the (i, j) cell, these being independent from cell to cell, with mean zero, and variance σ^2. Let τ' be the mean of the τ'_k and let $\gamma_{\cdot\cdot}$, $\gamma_{i\cdot}$, and $\gamma_{\cdot j}$ denote respectively the overall mean (of the γ_{ij}), and the mean for the $i$the row and jth column. Let

$$\mu = \gamma_{\cdot\cdot} + \tau'_{\cdot}, \quad \delta_{ij} = \gamma_{ij} - \gamma_{i\cdot} - \gamma_{\cdot j} + \gamma_{\cdot\cdot},$$

$$\alpha_i = \gamma_{i\cdot} - \gamma_{\cdot\cdot}, \quad \beta_j = \gamma_{\cdot j} - \gamma_{\cdot\cdot}, \quad \tau_k = \tau'_k - \tau'_{\cdot}. \tag{4.2}$$

Clearly, (4.1) and (4.2) give

$$y_{ij} = \mu + \alpha_i + \beta_j + \delta_{ij} + \tau_k + e_{ij}, \tag{4.3a}$$

$$\sum_i \delta_{ij} = \sum_j \delta_{ij} = 0. \tag{4.3b}$$

The above will be referred to as the "general model"; this model is to be said to be additive if and only if

$$\delta_{ij} = 0, \quad \text{for all } (i,j); \tag{4.4}$$

otherwise, it is called "nonadditive". The additive model thus has the form

$$y_{ij} = \mu + \alpha_i + \beta_j + \tau_k + e_{ij}. \tag{4.5}$$

Clearly, there is no reason to believe why arbitrary cross classification nuisance factors arising in nature should be additive.

Let $\sum^u$ $(u = 1, \ldots, v)$ denote the sum over all cells (i,j) to which treatment u is assigned. Then, it is easy to see that (4.3a, b) implies that there is an intrinsic confounding between the δ_{ij} and the τ_k; this confounding arises in nature, and can not be 'undone' (for a Latin rectangle design with $v > 1$) even when $\sigma^2 = 0$. Because of this, the usual estimate of an elementary contrast $(\tau_h - \tau_k)$ is biased by an amount $(\delta_h - \delta_k)$, where

$$\delta_h = \sum\nolimits^h \delta_{ij}, \quad \text{for } h = 1, \ldots, v. \tag{4.6}$$

In the design class under consideration, this bias can not be prevented. Indeed, even if σ^2 equals zero so that $e_{ij} = 0$), and the y_{ij} are representable by the additive model (4.5), the condition (4.4) may still not be satisfied, and (4.3a, b) may hold with some δ_{ij} being nonzero. This happens when for all h, δ_{ij} equals a constant δ_h^o for all cells (i,j) to which treatment h is assigned. It should be added, however that if we take a Bayesian point of view, and assume that the δ_{ij} have a joint a priori distribution which is absolutely continuous, then the probability of the event (in the last sentence) occurring is zero. In future discussion, we shall ignore this event.

If the nuisance factors (i.e., their levels, and the nature and amount of nonadditivity) remain relatively constant, a uniformity trial (before the main experiment) would be valuable. Here, we shall have $v = 1$, the (single) treatment being chosen so as to minimize any interaction it may have with the levels of the nuisance factors. The author believes that, where ever possible, such trials should be considered as a must.

We shall now briefly summarize some investigations on nonadditivity started in Srivastava (1993), and further developed in Wang (1995), and Srivastava and Wang (1996). We shall then discuss the situation in the light of this.

It is clear that the δ's and e's are confounded, and that small values of the δ's will be indistinguishable from the e's. To get rid of the δ's completely, a design ought to be used in which at least two observations are taken (usually on two different treatments) within each cell (i,j) of nuisance factors, which is included in the experiment. Then the difference between the two observations will provide an unbiased estimate of the difference between the corresponding two treatment effects. However, such a design may not be possible in many situations. We shall therefore, discuss the latin rectangle design in detail, because of its current popularity.

In view of the remarks above, it is important to identify cells with "large" δ's. So, we pretend that most of the δ's are negligible, except possibly for a few which we wish to identify. (This is somewhat like our attitude in testing of hypothesis in general. For example, when we test the null hypotheses H_0: $\tau_h - \tau_k = 0$, we know that it is almost certain that the null hypothesis is false. So, when we test H_0, we are really looking for a difference $(\tau_h - \tau_k)$ that is "large" in some sense.)

So, we assume that the δ's are negligible, except possibly a few. Let m denote the number of nonadditive (NA) cells, i.e., m is the smallest value of m', such that there exist $(pq - m')$ cells (i, j) for which $\mathrm{E}(y_{ij})$ satisfies the model (4.3). Of course, we hope that m is not too large. Let M be a set of m cells ignoring which, the remaining $(pq - m)$ cells satisfy (4.3). A fundamental question is this: Can there be two distinct sets M_1 and M_2, each with m cells satisfying the last conditions. The answer is yes, as will be seen later.

It is easy to see that, because of the above situation, the model (4.3) is a special case of the Search Linear Model (2.1). Here, we have

$$N = pq, \qquad \underline{\xi}_1' = \{\mu, \alpha_1, \ldots, \alpha_{p-1}, \beta_1, \ldots, \beta_{q-1}, \tau_1, \ldots, \tau_{v-1}\},$$

$$\nu_1 = p + q + v - 2. \tag{4.7}$$

Also, $\underline{y}$ is the vector of observations $(y_{11}, y_{12}, \ldots, y_{1q}, y_{21}, \ldots, y_{pq})$, and $A_1(N \times \nu_1)$ has 1 everywhere in the first column. The second column of A_1 corresponds to α_1, and consists of "0" everywhere except for (i) a "1" against the q observations $y_{11}, \ldots, y_{1q}$ which are from the first row of the latin rectangle and (ii) a "-1" against the q observations $y_{p1}, \ldots, y_{pq}$. (The reason for the "minus one" is that the y_{pj} $(j = 1, \ldots, q)$ involve α_p which equals $(-\alpha_1 - \alpha_2 - \cdots - \alpha_{p-1})$.) The other columns of A_1 can be similarly defined.

There are two models (called I and II) which arise here with respect to the δ's. Model II corresponds to (4.3); here, $\nu_2 = (p-1)(q-1)$, $\underline{\xi}_2$ consist of the δ_{ij} $(i < p, j < q)$, and the row for y_{ij} $(i < p, j < q)$ has "1" in the columns for δ_{ij} and zero elsewhere, the row for y_{iq} $(i < p)$ has "-1" in the columns for δ_{ij} $(j < q)$ and zero elsewhere, the row for y_{pj} has "-1" in the columns for δ_{ij} $(i < p)$ and zero elsewhere, and y_{pq} has 1 in the columns for all δ_{ij} $(i < p, j < q)$. (In defining A_2 and $\underline{\xi}_2$, we have implicitly assumed that "all the nonzero δ's have indices (i, j) with $i < p$, $j < q$. This can be done when m is reasonably small.)

In model I, we take $\nu_2 = pq$, $A_2 = I_{pq}$, and $\underline{\xi}_2$ has all the γ_{ij} as the elements (i.e., without the γ's) being reduced to the δ's by (4.2). Models I and II are different as Search Linear Models (as remarked in Section 2 around (2.8), (2.9)), but are the same considered as Linear Models. Some studies on Model II have been reported in Wang (1995). The discussion here relates to Model I, but also occasionally includes

Model II (as, for example, in the remarks on the probability of correct identification.) Notice that Model I can be written as

$$\mathrm{E}(\underline{y}) = A_1\underline{\xi}_1 + \underline{\xi}_2. \tag{4.8}$$

Consider a model of the form (4.8) in general; here we have a Search Model with $\underline{\xi}_2(N \times 1)$ having at most m nonzero elements. The idea is that we wish to explain $\mathrm{E}(\underline{y})$ as much as possible using $\underline{\xi}_1(\nu_1 \times 1)$, and we wish to use $\underline{\xi}_2(\nu_2 \times 1)$ as little as possible. Now, if A_1 does not have full (column) rank ν_1, even $\underline{\xi}_1$ can not be uniquely estimated. So, let us assume $\mathrm{Rank}(A_1) = \nu_1$. We now consider confounding of parameters in $\underline{\xi}_2$, i.e., non-unique solution for $\underline{\xi}_2$.

Let there be a set of u ($\leqslant 2m$) rows in A_1, such that $\mathrm{Rank}(A_1^*) < \nu_1$, where A_1^* is the matrix obtained from A_1 by deleting these u rows. Then there exists a non-null value $\underline{\xi}_1^*$ for $\underline{\xi}_1$ such that $A_1^*\underline{\xi}_1^* = O_{N-u,1}$. Hence, if $\underline{\theta}_1$ and $\underline{\theta}_2$ are such that $\underline{\theta}_2 = \underline{\theta}_1 + \underline{\xi}_1^*$, then $A_1^*\underline{\theta}_1 = A_1^*\underline{\theta}_2$. Now, partition A_1 as $A_1' = [A_1^{*\prime}, A_{11}', A_{12}']$, where A_{1j} ($j = 1, 2$) is $(N_j \times \nu_1)$ and $N_1 + N_2 = u$ with $N_1, N_2 \leqslant (u+1)/2$. Also, corresponding to the partition of A_1, let

$$\underline{y}' = \left[\underline{y}^{*\prime}((N-u) \times 1),\ \underline{y}_1'(N_1 \times 1),\ \underline{y}_2'(N_2 \times 1)\right],$$

and similarly, let

$$\underline{\xi}_2' = (\underline{\xi}_2^{*\prime}, \underline{\xi}_{21}', \underline{\xi}_{22}').$$

Suppose

$$\mathrm{E}(\underline{y}) = \mathrm{E}\begin{bmatrix} \underline{y}^* \\ \underline{y}_1 \\ \underline{y}_2 \end{bmatrix} = \begin{bmatrix} A_1^*\underline{\theta}_1 \\ A_{11}\underline{\theta}_1 \\ A_{12}\underline{\theta}_2 \end{bmatrix}. \tag{4.9}$$

Then, (4.9) can arise under (4.8) in two ways. We may have either one of the following two competing values of the parameters:

$$\underline{\xi}_1 = \underline{\theta}_1, \qquad \underline{\xi}_{21} = \underline{O}, \qquad \underline{\xi}_{22} = A_{12}(\underline{\theta}_2 - \underline{\theta}_1), \tag{4.10a}$$

$$\underline{\xi}_1 = \underline{\theta}_2, \qquad \underline{\xi}_{21} = A_{11}(\underline{\theta}_1 - \underline{\theta}_2), \qquad \underline{\xi}_{22} = \underline{O}. \tag{4.10b}$$

This explains why the question asked earlier (concerning two different sets of cells M_1 and M_2) has an affirmative answer (under certain conditions) for the case of the LR Design. The existence of two different sets of cells and two different solutions would indeed imply a wasted experiment, and is therefore to be avoided. So, we must hope that m is not so large that (4.10) may occur.

Generally, whether the answer is affirmative or not depends upon how large the value of m is. For the LS design (under Model I) with $p = 3$, no positive value of m will work! (Srivastava, 1993). For $p = 4$, the maximum value of m is 1. For $p = 5$, if the LS is cyclic, then the maximum value of m is 2. Determination of such maximum values for different designs is an important and nontrivial question, but it can be approached through the theory of search linear models. Wang (1995) has studied this question for many designs under Model II. Some studies for Model I are included in Srivastava and Wang (1996).

We now consider the probability of correct identification of the nonnegligible cells when $\sigma^2 > 0$, which is the realistic case. Since both models I and II are special cases of the search linear model, the methods of Section 2 based on the computation of $S_e^2 (= S_e^2(\underline{\xi}_1, \underline{\xi}_{20})$ under (2.1)) are applicable. For $p \times p$ Latin Squares, with $p = 4, 5, 6$ Wang (1995) studied this method and also various other methods available in the literature for the identification of nonnegligible parameters, and found this method to be the best. For this comparison, simulation was used. For $p = 4, m = 1$, Wang also computed the probability theoretically. This shows that the theoretical comparison of the probability is possible. But, it is extremely messy at present. The region of integration is very complicated, being the intersection of $\binom{pq}{m}$ quadratic regions. Using these ideas, however, it should be possible to develop systematic theoretical procedures using the algebra of Bose and Mesner (and its generalization to the multidimensional partially balanced association scheme, i.e., the algebra of Bose and Srivastava), and the theory of the complex multinormal distribution. (The complex case arises because of the need to diagonalize the covariance matrix using a Hermitian matrix.) Though this field is very complicated, it has a certain elegance, and many researchers would enjoy developing it. It may be added that the theory so developed would be useful in studies not only of the nonadditivity of nuisance factors, but in other cases as well, where the actual model deviates from the model usually assumed. Using simulation, the probability was computed (for $p = 4, m = 1$, and various values of δ). Such computations were done using increasing sample sizes. Using this, the sample size needed so that the simulation results match the theoretical results to within 0.1 percent, was determined. Next, using such (or, larger) sample sizes, the probability of correct identification was determined for a few other values p, m, and δ's.

The results (Wang, 1995) obtained are quite discouraging. For $m = 1$, and $\delta/\sigma = 3$, the probability ranged between 0.36 for $p = 4$ to 0.43 for $p = 6$; for $\delta/\sigma = 4$, this range was 0.54 to 0.67. (Recall that, for the normal distribution, 2σ limits correspond to 95.46% of the area, and 3σ limits to 99.73% of the area. Thus, the identification probabilities encountered are too small.)

For $m = 2$, and $\delta/\sigma = 4$ and 6, the probability was 0.19, 0.56 and 0.64 respectively for $p = 4, 5$, and 6. For $m = 3$, and $\delta/\sigma = 4, 6$, and 8, the values were 0.34, and 0.60, for $p = 5$ and 6.

The above should be an eye opener to all concerned. The probabilities are simply too low to allow the nonnegligible cells to be identified for reasonable values of δ/σ. How much δ/σ should be so that the probability is about 99%? Again, the answers are disappointing. For $m = 1$, $p = 4$, δ/σ should be about 9 and for $p = 5$, about 8. For $m = 2$, $p = 5$, one pair of values (of δ/σ, so that the probability is about 99%) is $(8, 12)$, for $p = 6$, it is $(7, 10.5)$.

What about uniformity trials (i.e., $v = 1$)? For $m = 1$, $p = 4$, the value of δ/σ (so that the probability is about 99%) is about 7, and the answer is about 6 for $p = 5$ and 6.

The above shows that it is difficult to pinpoint nonadditive cell values unless δ/σ is extremely large. Now, if a nonadditive cell is correctly identified, then it is clear that the observation from this cell must simply be ignored in the analysis, because the most we can get out of the observation is an estimate of the value of the δ corresponding to that cell. On the other hand, if there is a nonadditive cell which is not ignored in the analysis (irrespective of whether or not it has been identified), then it contributes towards the bias in the estimates of treatment contrasts, the larger the δ the more (potentially) being the bias.

Hence, it is obvious that at least the cells which have large δ's be correctly identified, and ignored in the analysis. However, the above discussion shows that the probability that they will get correctly identified is too small to be practical. Thus, except for cells with extremely large δ's, others will probably be not ignored, and will contribute to a tremendous bias.

Thus, the row–column designs (like the LR designs) have a great draw back associated with them. Notice that (from the results for $v = 1$) even a uniformity trial may not help much.

The question arises as to whether the above discussion is merely a theoretical thunderbolt, and in practice nonadditivity might seldom arise. To answer this, Srivastava and Wang (1996) studied examples of real data (for 4×4 LS) from various books. Nine cases were located, and in a majority of them, there were cells with *extremely* large values of δ's. (For example, in an experiment (4 LS of size 4 each) reported in Bliss (1967), we found 3 cells with δ-values whose probability is of the order 3.7×10^{-7}.) From the above theory, it is clear that cells with even moderately large δ's are unlikely to be identified. Thus, if in a situation, too many cells were not identified, it does *not* mean that too many cells with moderately large δ's do not exist. This point is also supported from another angle. As stated above, data from published books showed the existence of cells with extremely large δ's. Hence, from the branch of probability theory known as "Extreme Value Theory", it is clear that if in 50% of the situations there exist cells with extremely high δ's, then cells with moderately high δ's must be relatively very common.

It should be useful to explain here the method used in Srivastava and Wang (1996) to analyse the real data such as those in the book of Bliss. The method appears to be general though we used it only for simple designs. Let D be any row–column design and let Ω be the set of cells used in D. Let Ω_0 be a class of subsets of cells in Ω, such that each subset contains m cells in it. Suppose we wish to investigate whether there exists a subset of m cells in Ω_0 which is nonadditive. Then, a procedure (using simulation) is as follows.

Consider the model M under which D is being studied. Under the model M, generate a random sample of data (for all cells Ω under the design D). For this data, compute r, where

$$r = S_e^2 / S_e^2(\Omega_0), \tag{4.12a}$$

S_e^2 = sum of squares due to error when data from D (from all cells Ω) is analyzed under M, (4.12b)

$S_e^2(w)$ = sum of squares due to error when the data is analyzed under model M and design D, and the set of m cells w is ignored, where $w \in \Omega_0$,

$$S_e^2(\Omega_0) = \min_{w \in \Omega_0} S_e^2(w).$$

We draw n independent samples, and the value of r is computed for each case. The number n is taken to be large, say $10,000$ so as to give a fairly good estimate of the distribution of r. Of course, the distribution of r could possibly be theoretically determined, but we have not attempted it yet.

The above distribution is used as follows. To analyze a real set of data which is from an experiment under a design D and model M, and in which we suspect a set of m cells (belonging to some class Ω_0) to be nonadditive, we compute r for this data. The value of r is then matched against the distribution of r obtained above, and the probability of obtaining a value this large or larger is estimated. (In the example from Bliss, the value of r obtained was so large that it was far outside the range of values found in the simulation distribution. So an estimate of the probability was obtained by using Thebychev's inequality.)

The question now arises: What should be done in view of the above facts, if it is not possible to take more than one observation per cell. It is not clear what is the best solution. However, one could possibly divide the rectangle into smaller ones and use a Nested Multidimensional Block Design (Srivastava and Beaver, 1986) with blocks of small size. For example, instead of using one 8×8 Latin Square, we could use 16 squares each of size 2×2. This should considerably reduce the danger of bias. Clearly, much further research is warranted.

5. Factorial experiments

For simplicity, we shall restrict to the 2^m case. Let $\underline{t} = (t_1, \ldots, t_m)'$ denote a treatment (where $t_i = 0$ or 1, for $i = 1, \ldots, m$), and $\phi(\underline{t})$ its true effect. Let $\underline{\phi}(2^m \times 1)$ be the vector of the true effects of all the 2^m distinct treatments. Let $H_m(\overline{2}^m \times 2^m)$ denote the symmetric Hadamard matrix of order 2^m, so that

$$H_m = H_1 \otimes H_1 \otimes \cdots \otimes H_1 \quad (m \text{ times}), \tag{5.1}$$

where $\otimes$ denotes the (left Kronecker product), and

$$H_1 = \begin{bmatrix} 1 & 1 \\ 1 & -1 \end{bmatrix}. \tag{5.2}$$

Let $A_1^{j_1} \cdots A_m^{j_m}$ ($j_u = 0$ or $1, u = 1, \ldots, m$) denote an interaction; it is a k-factor interaction if k of the j's are nonzero, and equals μ (the general 'mean') if

$k = 0$. Let $\underline{\alpha}(k)$ $(k = 0, 1, \ldots, m)$ denote the $\binom{m}{k}$-vector of all k-factor interactions. Let $\underline{\alpha}(2^m \times 1)$ be the vector of all the 2^m distinct interactions, with the elements $A_1^{j_1} \cdots A_m^{j_m}$, so that the $\underline{\alpha}(k)$ are subvectors of $\underline{\alpha}$. If $\underline{\alpha}$ and $\underline{\phi}$ are both arranged in Yates order, then it is well known that

$$\underline{\alpha} = 2^{-m/2} H_m \underline{\phi} \tag{5.3}$$

where the constant in (5.3) is chosen so that $(2^{-m/2} H_m)$ be an orthogonal matrix, in which case

$$\underline{\phi} = 2^{-m/2} H_m \underline{\alpha}. \tag{5.4}$$

It is known from empirical observations in the various sciences that very often, a large number of elements of $\underline{\alpha}$, particularly interactions of high order, are negligible. (Of course, in many situations, both for symmetrical and asymmetrical experiments, for various reasons, the full experiment may be called for. Indeed, several replications may be attempted. Nuisance factor(s) may be present. This leads to confounded designs, which has a considerable literature. Some of the pertinent problems here are covered under block designs.) Let $\underline{L}(\nu \times 1)$ be the vector of nonnegligible effects. Then $\underline{L}_1$ is a subvector of $\underline{\alpha}$. We shall assume, for simplicity, that $\mu \in \underline{L}$. Now, the set of elements of $\underline{L}$ (i.e., which members of $\underline{\alpha}$ are also in $\underline{L}$) may or may not be known.

Several cases arise: (Q1) The elements of $\underline{L}$ are known without any reasonable doubt, i.e., which elements of $\underline{\alpha}$ are in $\underline{L}$ is known. Note that the *value* of the elements of $\underline{L}$ is still unknown. (Q2) It is known beyond any reasonable doubt that $\underline{L}$ is a subvector of $\underline{L}^*(\nu^* \times 1)$ (which is a subvector of $\underline{\alpha}$), and the elements of $\underline{L}^*$ are known. Here, ν^* is not large, but is a reasonably small integer. (Q3) In Q2, it is very plausible to believe that most elements of $\underline{L}$ are contained in $\underline{L}^*$, but we can not be certain that $\underline{L}$ is wholly contained in $\underline{L}^*$; in other words, a few elements of $\underline{L}$ are probably not in $\underline{L}^*$, but are in $\underline{L}^{**}((2^m - \nu^*) \times 1)$, where L^{**} is the vector of elements of $\underline{\alpha}$ which are not in $\underline{L}^*$. (Q4) It is very plausible to assume that though a few elements of $\underline{L}$ may be contained in $\underline{L}^*$ (which is known), most elements of $\underline{L}$ are not known and may not be in $\underline{L}^*$. (Q5) Elements of $\underline{L}$ are unknown, except for μ. In Q4 and Q5, two subcases arise according as a reasonable estimate ν_0 of ν (the number of elements in $\underline{L}$) is (a) available, and (b) not available. (By 'reasonable', we mean that we have $(3\nu_0/2) \geqslant \nu \geqslant (\nu_0/2)$.) Consider Q1 and Q2. The author believes Q1 would arise rarely. The case Q2 is important, however, since at some stage we will have to answer the question as to which design is appropriate for a given $\underline{L}^*$. Obviously, variance-optimality is called for. It is quite nontrivial to obtain an optimal design for arbitrary $\underline{L}^*$ in a given number of assemblies (treatments). This problem, in general, is far from solved. Let T be a design, i.e., a set of assemblies, and let M $(\equiv M_T(\underline{L}^*))$ be the information matrix for $\underline{L}^*$ if T is used. If M is diagonal, T is an "orthogonal design"; such designs enjoy optimality from several angles. Hence, from the beginning, orthogonal designs have attracted considerable attention.

If $\underline{L}^*$ consists only of $\underline{\alpha}(k)$, for $k = 0, 1, \ldots, \ell$, so that $\nu = \sum_{k=0}^{\ell} \binom{m}{k}$, and T allows estimation of $\underline{L}^*$, then T is said to be of resolution $(2\ell + 1)$; in this same situation, if we are interested in estimating all elements of $\underline{L}^*$ except those in $\underline{\alpha}(\ell)$, and T allows this, then T is of resolution 2ℓ. A design of resolution q is orthogonal if and only if it is an orthogonal array (OA) of strength $(q-1)$("OA$(q-1)$").

The simplest case is $\ell = 1$, and corresponds to "main effect plans" (MEP). Here, OA's of strength two are needed. Orthogonal saturated (and non-saturated) designs (i.e., those where N, the number of assemblies, equals ν, the number of parameters) with $4g$ (where g is an integer) assemblies (and $m \leqslant 4g-1$) can be readily made by taking Hadamard matrix H of size $(4g \times 4g)$ with one row consisting of ones only, and then by deleting this row (of ones). It should be stressed that before using MEP's, one should be certain of being under Q2.

It should be recorded that "Addelman's Orthogonal main effect plans" (see Srivastava and Ghosh, 1996) are not orthogonal with respect to the regular parameters. Several authors have been misled by these plans. They are orthogonal under a new set of parameters, which are to be obtained by a transformation of the regular parameters, the transformation being dependent on the design. However, if we allow transformation of parameters, allowing the transformation to be dependent upon the design, then *every* design should be considered orthogonal, because for every design such transformation can be done. Indeed, such transformation corresponds to the diagonalization of the information matrix. There are other such loose "concepts" (for example, a "screening design") appearing in less rigorous journals, which need correction (or sharpening).

The case $\ell = 2$ is more reasonable, and arises far more than $\ell = 1$. But then an orthogonal design needs to be an OA(4). Unfortunately, OA(4)'s require an unreasonably large value of N (the number of assemblies in T). In the mid-fifties, this led researchers such as Connor to study "irregular designs" (i.e., those which are not 2^{m-q} type of fractions). One class introduced by him was that of PF (parallel flats) designs, where T consists of a few parallel flats in the finite Euclidean geometery EG(m, s), of m dimensions based on GP(s), the finite field with s elements. Addelman and Patel obtained 2-level, 3-level, and mixed designs of this type. P.W.M. John's three-quater replicate of 2^5 design belongs to this class. The PF designs were theoretically studied by Srivastava, and then by Anderson and Mardekian. These authors (Srivastava et al., 1983), for the s^m experiment (s, a prime number) connected the theory with cyclotomic fields, and obtained an efficient reduced representation of the information matrix over such fields which is easy to compute for a given T which is of the PF-type. The theory has been further extended by Srivastava (1987), Buhamra and Anderson (1996); a theory of orthogonal designs of this type (for general $\underline{L}^*$) has been developed by Li (1990), Srivastava and Li (1993), and extensions have been made by Liao (1994), Liao and Iyer (1994), Liao et al. (1996). Much further useful work (for general $\underline{L}^*$) can be done in this area which is quite promising.

Balanced designs is another class of irregular designs which has been extensively studied. Defined initially by Chakravarti (1956), they were studied in detail principally by Srivastava (and co-authors Bose, Chopra, Anderson), and by Yamamoto (and co-authors Shirakura, Kuwada, and others). For balanced designs, the information matrix is invariant under a permutation of the name of the factors. Let the design T be written as an $(N \times m)$ matrix, so that the rows of T represent the N treatments, and the columns of T represent the m factors. Let T^* be obtained from T by permuting the columns. Then T is a balanced design for estimating $\underline{L}^*$, if $M_T(\underline{L}^*) = M_{T^*}(\underline{L}^*)$, for all T^* obtainable from T in the above way. It is well known that a balanced design of resolution q is also a balanced array (BA) of strength $(q-1)$("BA$(q-1)$"). For all q, an OA(q) is also a BA(q), but not visa versa.

Optimal balanced designs of resolution V were obtained by Srivastava and/or Chopra in a series of papers; some of these designs were shown to be optimal in the class of all designs by Cheng (1980). However, some designs in this class have a rather low efficiency, and could be improved. Nguyen and Miller (1993) have obtained (unbalanced) designs which improve upon these. However, optimal designs are not known in general. Indeed, in general, we do not have designs which have been shown to be near-optimal (i.e., whose efficiency is, say, within 95% of that of the (unknown but) optimal design. For the case of the mixed factorials, the situation is even worse than that for the 2^n case.

In this connection, mention should be made of the work of Kuwada (1988) on "partially balanced arrays". These arrays should not be confused with those of Chakravarti (1956), which were re-named balanced arrays by the author because they correspond to balanced designs. The information matrix they give rise to belongs to the (non commutative) algebras of Bose and Srivastava, which arises from the multidimensional partially balanced association scheme (Srivastava, 1961; Bose and Srivastava, 1964a, b). But this is not because the design is less than balanced. Rather, it arises because many sets of interactions (i.e., $\underline{\alpha}(u)$, for several values of u) are being considered. For a single value of u, the information matrix for $\underline{\alpha}(u)$ belongs to the Bose Mesner algebra; but the "partial balance" arising here arises because of the structure of $\underline{\alpha}(u)$. Indeed, if $\theta, \phi_1, \phi_2 \in \underline{\alpha}(u)$, then the number of factors in common between θ and ϕ_1 may be different from that for θ and ϕ_2. For example, if $u = 2$, $\theta = A_1A_2$, $\phi_1 = A_1A_3$, $\phi_2 = A_3A_4$, then the number of common factors between θ and ϕ_1 is 1, but for θ and ϕ_2, it is 0. This feature does lead to a partially balanced association scheme, but this scheme arises out of a balanced design.

On the other hand, Kuwada's designs are not balanced, to begin with. They are a generalization of B-arrays, in a sense similar to the PBIB designs being a generalization of the BIB designs. The information matrix of Kuwada's PB-arrays is much more complex than the one for B-arrays, since here there is "partial balance" arising both in the design and in the set of parameters $\underline{\alpha}(u)$. Because of this reason, unfortunately, a lot of symbolism is needed to describe and comprehend their structure. Still, however, the situation is much better than those of unbalanced designs, since they have no structure. If their efficiency is close to that of a competing but best known unbalanced design, then they may be preferable. We will return to this point later.

Very little work has been done on balanced designs for the asymmetical case, or for the symmetrical case with general $\underline{L}^*$. Some cases worthy of attention include those where $\underline{L}^*$ contains μ, main effects, and two factor interactions of the form A_iA_j, where $i \in \theta_1$, $j \in \theta_2$ and θ_1 and θ_2 are two (not necessarily mutual exclusive) groups of factors. Another practical case is where, in the above, both (i, j) belong to θ_1, or both belong to θ_2, but one (of (i, j)) from θ_1 and the other from θ_2 is not allowed. Many other similar important cases could be produced.

Balanced designs have a certain case in the analysis and interpretation of results, and so are attractive. It would be useful to first construct an optimal balanced design (say, for a given $\underline{L}^*$ of the type discussed in the last paragraph), and check its efficiency. If it is optimum or near-optimum, then general unbalanced irregular designs may not be attractive. It may be remarked that for general $\underline{L}^*$, orthogonal PF designs may be investigated even before balanced designs, since the authors' experience shows that for many types of cases of "intermediate" resolutions, such designs have a high chance of being existent.

A great deal of interesting work has been done lately on OA(2)'s, both for symmetrical and mixed cases. This should be quite interesting from the general combinatorial angle. For statistical applications, before these arrays are used it will be important to ensure that the situation is under Q2, and there too, under the situation where the interactions are all negligible.

It will be useful now to define a property of $\underline{\alpha}$, the vector of factorial effects, known as "tree-structure"; this requires that if any k-factor interaction ($k \geqslant 2$) $A_{i_1} A_{i_2} \cdots A_{i_k}$ belongs to $\underline{L}$, then a $(k-1)$-factor effect $A_{j_1} A_{j_2} \cdots A_{j_{k-1}}$ also belongs to $\underline{L}$, where $(j_1, j_2, \ldots, j_{k-1})$ is a subset of $(i_1, i_2, \ldots, i_k)$. Thus, for example, if $A_2 A_3 \in \underline{L}$, then one of the main effects A_2 or A_3 also belongs to $\underline{L}$. This property was noticed by the author (Srivastava and Srivastava, 1976), in a review of papers published in some journals in social sciences for a period of ten years (ending aroung 1975). In many of these papers, the set of all significant factorial effects were reported, and in almost all cases, the author noticed that the tree structure was present.

Among designs of (even) resolutions 2ℓ, studies have been made mostly on the case $\ell = 2$, where we wish to estimate $\{\mu, \underline{\alpha}(1)\}$, when elements of $\underline{\alpha}(2)$ may not be negligible. The foldover method produces such designs, but it is only a sufficient condition. The necessary and sufficient condition is "zero-one symmetry with respect to triplets" (Bose and Srivastava, 1964a); i.e., we need a BA(3) in which in every 3-rowed submatrix, every (3×1) column vector occured the same number of times as its $(0, 1)$-complement. Further research on such BA's would enable us to obtain improved designs.

However, the basic concept behind such designs requires justification; here we are basically in Q2, with $\underline{L}^* = \{\mu, \underline{\alpha}(1), \underline{\alpha}(2)\}$, but we wish to estimate only $\{\mu, \underline{\alpha}(1)\}$, even though some (unknown) elements of $\underline{\alpha}(2)$ are nonnegligible. Now, it is clear that unless the elements of $\underline{L}$ (and their values) are known, we do not know $\underline{\alpha}$ and hence we can not estimate $\underline{\phi}$, which is the "response surface" in this discrete case. Unless we know $\underline{\phi}$, it is difficult to find $\underline{t}$ for which $\phi(\underline{t})$ may be maximal or minimal, a problem which is occasionally of great interest. Thus, the above $\underline{L}^*$ needs to be estimated in any case!

The usual justification of a resolution IV design is that if we do not have money to do a larger (resolution V) experiment, we may first do a resolution IV experiment, to find out which main effects are significant. We can later do a smaller experiment involving principally the factors that turn out to be important. Here, we must notice that this logic is not quite correct, because an interaction may be larger even when a main effect involved in it is small. However, such logic does have some support from the tree-structure concept, since such structure may be present, and if it is present then if *both* main effects A_i and A_j are insignificant, the interacation A_iA_j is also likely to be so.

We now consider Q3. Two cases arise according as the elements of $\underline{L}$ not in $\underline{L}^*$ are (i) ignored, (ii) not ignored. In case (i), after the experiment, we would have the situation where (a) some elements of $\underline{L}$ are unknown, (b) the estimate of $\underline{L}^*$ is biased, the sources and the amount of bias in the estimates of various elements of $\underline{L}^*$ being unknown, and (i) the estimate of $\underline{\phi}$ is biased similarly, which may in turn lead to a wrong conclusion about the $\underline{t}$ for which $\phi(\underline{t})$ is extremal (in some sense).

The question arises: If case (ii) is chosen, don't we have to do the whole experiment. The answer is no, because then the situation falls under the theory of search linear models. Let $\underline{L}_o^{**}$ be a subvector of $\underline{L}^{**}$ such that we are quite certain that the elements of $\underline{L}$ not in $\underline{L}^*$ are in $\underline{L}_o^{**}$. Then, if $T(N \times m)$ is the design chosen, and $\underline{y}(T)$, the corresponding vector of observations, we can express $\mathrm{E}[\underline{y}(T)]$ by a model of the form (2.1), where $\underline{\xi}_1$ and $\underline{\xi}_2$ are respectively $\underline{L}^*$ and $\underline{L}_o^{**}$, and where the corresponding matrices A_1 and A_2 can be easily written down. The number k (of model (2.1)) for this case will not be known if (as will often be the case) we do not know how many elements of $\underline{L}$ are in $\underline{L}_o^{**}$.

Designs T through which one may tackle the situation in the last paragraph are called Search Designs, on which considerable literature exists. Some important cases considered include search designs of resolution 3.1, 3.2, 5.1 and 5.2, where "resolution $(2\ell + 1)$. k" means that the design satisfies the rank condition (2.2), when $(\underline{\xi}_1 =)\underline{L}^*$ includes $\underline{\alpha}(k')$, for $k' = 0, 1, \ldots, \ell$, and $(\underline{\xi}_2 =)\underline{L}_o^{**}$ equals $\underline{L}^{**}$, and up to k parameters in $\underline{\xi}_2$ could be nonnegligible. Such designs were developed in the work (singly or jointly) of, among others, Srivastava, Gupta, and Ghosh. Ohnishi and Shirakura (1985) obtained AD-optimal designs in this class, the criterion of AD-optimality being developed earlier by Srivastava (1977).

Another case studied which is combinatorially significant is where, in the above, $\ell = 0, k \geqslant 2$, (Katona and Srivastava, 1983); the case $k = 3, 4$ has been studied but is unpublished yet. In this connection, the concept of a q-covering of $\mathrm{EG}(m, s)$ was introduced by Srivastava (1978), and many studies on this line were made in Jain (1980) (under the direction of R.C. Bose) using quadrics and tangent spaces. (A set of N points T in $\mathrm{EG}(m, s)$ is said to constitute a q-covering of $\mathrm{EG}(m, s)$ if and only if every $(m - q)$-dimensional hyperplane of $\mathrm{EG}(m, s)$ has a non-empty intersection T.)

Anderson and Thomas (1980), and Mukerjee and Chaterjee (1986) have developed search designs for the situations where the number of levels of a factor is not necessarily equal to 2. Studies on minimal designs of resolution 3.1 and 3.2 (for the 2^m case) have been made; these include particularly the work of Gupta (1988), and of Srivastava and Arora (1987, 1988, 1991).

Unfortunately, most of the studies on search designs have concentrated on producing designs which satisfies (2.2), thereby guaranteeing that irrespective of which parameters are nonnegligible (and what their values are), the correct identification will be possible (with probability 1 if $\sigma^2 = 0$). But, if we recall the remarks made around the equations (2.6) and (2.7), instead of worrying too much about (2.2), we should try to see what sets of parameters the design can help estimate. This is particularly important, since the value of k would, in most cases, be known only approximately, and may actually be larger than the value which was guessed initially (and corresponding to which the design is to be made). There is some discussion in Srivastava (1992) in this spirit, but nothing substantial has yet been done. It should be remarked here that there is another reason why we should emphasize (2.2) less, and the remarks around (2.6) and (2.7) more. It is that in order to satisfy (2.2), even for small k, it turns out that very often, relatively too many treatments are needed. On the other hand, in most situations, the value of k will usually be unknown, and indeed may be more than what we believe it to be. Thus, the matter boils down more to what should be done in order to identify all the nonnegligible parameters, rather than to have a design which will be uniformly good irrespective of which set of k parameters (out of $\underline{L}_o^{**}$ is nonnegligible), where k is some small integer.

This brings us to the concept of "revealing power of a design", introduced in Srivastava (1984), and used in a few contexts since then. Broadly speaking, it refers to the ability of a design to reveal or identify the correct model. To discuss this further, we explain some terminology first. A design is called single-stage, if a single experiment is planned, and the data from this experiment is all that we shall have, in order to throw light on the model and the phenomenon under study. A multistage design, on the other hand, envisions the possibility of a series of experiments, the second experiment being planned in the light of the results of the first experiment, the third being planned in light of the results of the second, and so on. Such a design may end up having several (but, a relatively small number of) experiments, each one being substantial. The term sequential will be used, where a large sequence of experiments is envisaged, each experiment being an observation on a small number of treatments, possibly, even of a single treatment only. It is intuitively clear that if (in a discreate factorial experiment) the total number of observations is fixed, then generally, a multistage design will have more revealing power than a single stage design.

In conjunction with Q3, we shall now consider Q4 and Q5 as well. All of these are characterised by the fact that there are nonnegligible unknown parameters (say k, in number) which need to be identified and estimated. Under Q3, ν^* (the number of elements in $\underline{L}^*$ (which we *do* wish to estimate)) is given, and k (whether known or unknown) is relatively small. In Q4, ν^* is given, but k may be quite large, and in Q5, ν^* is rather small, and k is nearly equal to ν.

We shall occasionally discuss response surface experiments, which are factorial experiments with continuous factors. From the view point of variance-optimality, recent work of Ghosh and Al-Sabah (1994), has shown that the classical designs in this field could be *vastly* improved; new designs are hundreds of times more efficient than the old ones! Obviously, this line of work has great potential and promise, and should be vigorously pursued; there is no reason to leave the response surface field underdeveloped.

With respect to the nature of the models arising in statistical design, three situations need to be considered: (M1) The model is (functionally) unknown and not known to be representable as a linear model (with unknown paramters, but known coefficients), (M2) the model is representable as a linear model, but the coefficients of some of the unknown parameters are unknown, and (M3) the model is representable as a linear model with unknown paramters, where the coefficients of the parameters are known. (In general, the coefficient will depend upon the points in the factor space.)

Now, usually, the purpose of the experiment is either to study the response surface itself, or to find points in the factor space for which the expected response is extremal (i.e, maximum or minimum), or both. Clearly, in the first case, we need to find the model itself, and determine it completely, i.e., obtain the values of all parameters involved. In case we wish to find the extremal points, a knowledge of the complete model may still be needed except in certain special situations. Even in these special cases, a knowledge of certain crucial features of the model will be necessary.

On the other hand, the purpose of the subject of statistical design of scientific experiments must be to do the best (from the statistical view point) to help achieve the purpose of the experiment itself. Hence, from the above discussion, it is clear that the purpose of statistical design must be to completely determine the model (or, in some special cases, at least certain crucial features of the model).

This brings us to the cases M1, M2, and M3. Of these, design theory has dealt largely with M3 only, which we consider first. This entails the situations (Q1)–(Q5), of which we have been considering the last three. In these cases, there are a number of parameters that have to be identified. Hence, from the above discussions, it follows that under M3, the purpose of statistical design must be to identify these parameters, and to estimate the values of these and other possible nonnegligible parameters. Thus, under M3, (Q3)–(Q5), one purpose of statistical design must be to help identify the nonnegligible parameters as accurately as possible, the ability of a design to help us do such identification is refered to as the "revealing power" of the design.

It is clear that "revealing power" is a general concept, which will have to be sharply defined in each special situation, if we are interested in a quantitative assessment of the same. Several measures of "revealing power" for the case of a sequantial design, are introduced in Srivastava and Hveberg (1992).

Consider (M3, Q3), where search design theory is pertinent. In a design of resolution $\{2\ell+1, k\}$, the estimation of $\underline{L}^{*\prime}(=(\mu, \underline{\alpha}'(1), \ldots, \underline{\alpha}'(\ell)))$ plus any set of k interactions in $\underline{L}^{**}$, is guaranteed. Thus, with respect to this $\underline{L}^{*}$, and this class of situations in $\underline{L}^{**}$, the design's revealing power may be said to be 100%.

As a contrast, we may say that the revealing power of a 2^{4t-1} orthogonal main effect plan (OMEP) in $4t$ runs is zero, where $\underline{L}^{*} = \{\mu, \underline{\alpha}(1)\}$, $\underline{L}^{**}$ is the set of all interactions, and the class of situations is where a single element of $\underline{L}^{**}$ is nonzero!

As another instructive class of examples, consider Hadamard matrices of size $(N \times N)$, with N not equal to a power of 2. Thus, let $m = 5$, $N = 12$, $\underline{L}^{*'} = \{\mu, \underline{\alpha}'(1)\}$. Suppose $\underline{L}_o^{**} = \{\underline{\alpha}'(2), \underline{\alpha}'(3)\}$, and the class of situations of interest is where at most three elements of $\underline{L}_o^{**}$ are nonnegligible, and that these elements obey the tree structure among themselves. (For example, the nonnegligible elements could be the sets $\{A_1A_2,\ A_1A_2A_4,\ A_1A_2A_5\}$ or $\{A_1A_2,\ A_1A_3,\ A_1A_2A_4\}$ but not the set $\{A_1A_2,\ A_1A_3,\ A_2A_3A_5\}$, since the last one does not have tree structure.) An interesting question is: What is the "revealing power" of the (5×12) OA under consideration? The author believes that such OA's (which are obtained from Hadamard matrices but are not obtainable as Kronecker Products of smaller Hadamard matrices, and are obtained using quadratic residues, or other techniques) should have good "revealing power" with respect to certain classes of situations where $\underline{L}^* = \{\mu, \underline{\alpha}'(1)\}$; however the corresponding $\underline{L}_o^{**}$ and the class of situations need to be identified. Minimum aberration designs may also be investigated from this angle.

The question still remains: how shall we assess the revealing power of a (single stage) design. We illustrate the answer by the example in the last paragraph. In this case, since $m = 5$, both $\underline{\alpha}(2)$ and $\underline{\alpha}(3)$ have 10 elements each. Now, we can have (for the nonnegligible elements), one of three situations; (i) all belong to $\underline{\alpha}(2)$, there are $\binom{10}{3} = 120$ cases here; (ii) two are from $\underline{\alpha}(2)$ and one is from $\underline{\alpha}(3)$, there are $[\binom{5}{3}\binom{3}{2}\binom{3}{3} + \binom{5}{4} \times 3 \times 6] = 120$ cases here as well, and (iii) one element from $\underline{\alpha}(2)$ and two from $\underline{\alpha}(3)$, there are $\binom{5}{2}\binom{3}{2} = 30$ cases here. Let $\underline{\xi}_{20}(k \times 1)$ denote the vector of nonnegligible parameters. In the last example, we have $k = 3$. Then, $\underline{\xi}_{20}$ is a subvector of $\underline{\xi}_2$ in (2.2). Let $\underline{\xi}_{21}(k \times 1)$ be any other subvector of $\underline{\xi}_2$. We will say that $\underline{\xi}_{20}$ is "clear" if and only if $[A_1 : A_{20} : A_{21}]$ is of rank $(\nu_1 + k_o)$, for all $\underline{\xi}_{21}$ in $\underline{\xi}_2$, where k_o depends upon $\underline{\xi}_{21}$ (with $k \leqslant k_o \leqslant 2k$) and equals the number of distinct elements of $\underline{\xi}_2$ contained in the two vectors $(\underline{\xi}_{20}, \underline{\xi}_{21})$. With this definition, for the above example we can consider (ρ_1, ρ_2, ρ_3) to be a measure of the "zero-order revealing power"of the design, where ρ_j $(j = 1, 2, 3)$ is the number of cases under situation j which are clear.

To understand this definition, first notice that if $\underline{\xi}_{20}$ is not clear, then there will be a $\underline{\xi}_{21}$ such that $\mathrm{E}(\underline{y})$ will be a linear function of $(\underline{\xi}_1, \underline{\xi}_{20})$, and also of $(\underline{\xi}_1, \underline{\xi}_{21})$, simultaneously. In this case, even if $\sigma^2 = 0$, there is no hope of distinguishing between $\underline{\xi}_{20}$ and $\underline{\xi}_{21}$ on the basis of our experiment; in other words, this design can not reveal $\underline{L}$. On the other hand, if $\underline{\xi}_{20}$ is clear, then at least for small σ^2, the vector $\underline{\xi}_{20}$ could be determined (i.e., the design is able to reveal $\underline{\xi}_{20}$) with a relatively high probability. Thus, in the last example, if $\rho_1 = 15$, we know that 15 cases under situation (i) can be resolved in the above sense. The larger the value of ρ_1 (and, similarly, ρ_2 and ρ_3), the happier we may be. On the other hand, for example, if we feel that situations (i) and (iii) have little to chance to arise, then we may go after a design for which ρ_2 is large.

What is the meaning of "zero order" in "zero order revealing power"? This phrase refers to the fact that we are counting the number of vectors $\underline{\xi}_{20}(k \times 1)$ which are clear. We will now define "jth order revealing power" $(j = 0, 1, 2, \ldots, k)$ as the vector $(\rho_{1j}, \rho_{2j}, \rho_{3j})$ such that ρ_{uj} $(u = 1, 2, 3)$ is the number of values of $\underline{\xi}_{20}$ under situation u, for which $[A_1 : A_{20} : A_{21}]$ has rank $(\nu_1 + k_o - j)$ or more, for all $\underline{\xi}_{21}$ (and rank $(\nu_1 + k_o - j)$ for at least one value of $\underline{\xi}_{21}$) where k_o is defined as before. Thus, ρ_{31} counts the number of sets $\underline{\xi}_{20}$ which are confounded with one extra parameter in $\underline{\xi}_2$. It should be remarked that there may be complex patterns of confounding present, and for $j \geqslant 1$, this definition of resolving power may be found too simplistic in many cases. Also, we must caution ourselves for relatively large values of k and j, in the sense that even if a subvector $\underline{\xi}_{20}$ is clear, the probability that $\underline{\xi}_{20}$ will be correctly identified may be quite small, even for moderately small values of σ^2. Thus, it appears that in general, resolving power of zero and first orders only may be of interest.

This brings us to an important point; it is that the revealing power of a single stage design is limited. To elaborate, if we want the design to be reasonably small, then the identification of the unknown nonnegligible paramters is likely only if k is quite small; the value of k for which correct identification may be possible with relatively high degree of certainty being a decreasing function of σ^2. In view of this, it is clear that the ordinary search designs (which are single stage designs) can help only when k is actually very small.

However, k will usually be unknown. So, after the experiment is done, and the analysis of its data is completed, there will arise the problem of checking somehow whether the major nonnegligible parameters have all been identified or some are still lurking. Thus, the following problem (or, rather, class of problems) is significant. We are given the data Δ_1 from an experiment E_1, and a vector of parameters $\tilde{\underline{L}}$, and we are asked to determine whether $\tilde{\underline{L}}$ contains as a subvector the (unknown) vector $\underline{L}$. In particular, as a first step, we need to decide whether this question can be answered adequately using Δ_1, or some further experiment E_2 is acutally needed. (Note that E_1 itself could be a series of experiments, rather than a single one.)

Again, it is clear that the larger σ^2 is, the harder it would be to resolve this problem in the light of any given set of data. Two cases are important: (a) σ^2 is very small, (b) σ^2 is not very small. Consider (a); it is instructive from the view point of theoretical insight to first look into this case. This case may arise in practice in situations where, once a treatment is included in the experiment, a large number of independent observations can be produced on this treatment without much cost; the mean of all such observations will then have a negligible variance. One such situation is that of a factory, where the experimental factors represent the settings of different machines. To try a new treatment, the factory may need to be closed for a while in order to set the different machines at the settings represented by the treatment; such closure would, of course, be very expensive, since production will have to be shut down. However, once the settings have been fixed, the product will be produced, resulting in a very large number of observations for this treatment. Thus, this is a situation where we would like to try a very small number of treatments, but where σ^2 may be assumed to be negligible.

Clearly, if in a particular situation, a method of identification would fail when $\sigma^2 = 0$, it would also fail if $\sigma^2 > 0$. Thus, it is important to examine some identification techinques for the case $\sigma^2 = 0$ in more detail. In Srivastava (1992), a case study was made for a 2^6 experiment, with $\nu = 12$. Two designs were compared, an OA(5) with $N = 32$ and a BA(6) with $N = 22$. The elements of $\underline{L}$ obeyed a tree structure. It was seen there that the OA(5) failed to reveal $\underline{L}$, because some elements of $\underline{L}$ were mutually confounded. It is easy to see that this indeed would be a difficulty in general, with the classical designs obtained using the Bose–Kishen–Fisher theory; since the 2^{m-q} type of fractions necessarily give rise to 2^q interactions which are mutually confounded. On the other hand, an irregular design does not necessarily have such confounding, and thus may have much more revealing power, as is demonstrated in the above case study.

Even when $\sigma^2 = 0$, and the revealing power of a design is 100% with respect to a set of parameters, it may be quite a difficult task to retrieve the parameters. In this connection, two techniques (called 'temporary elimination principle' (TEP) and 'intersection sieve' (IS)) were introduced in Srivastava (1992) and have been found useful not only in the factorial design area but also in determining non-additive cells in row–column designs. Let θ_i $(i = 1, 2, 3)$ be linear functions of parameters, such that the parameters occurring in each θ_i are all distinct, and such that for $i \neq j$ $(i, j = 1, 2, 3)$, θ_i and θ_j do not involve any common parameters. Then, TEP says that if $\theta_1 = 0$, then "temporarily", we can assume all parameters occurring in θ_1 to be individually zero. Similarly, if $\theta_1 + \theta_2 = \theta_1 + \theta_3$, then IS says that we should "temporarily" regard *each* parameter in θ_2 and in θ_3 to be zero. Here, 'temporarily' refers to the fact that TEP and IS are "tentative decision rules", i.e., the decision made by using them is made only for a while during the analysis of the data, and that at a later stage in this analysis, such decision could possibly be reversed. The purpose of this analysis is to somehow identify the elements of $\underline{L}$, and the tentative rules often help simplify the complexity arising out of the presence of a large number of parameters.

Many other useful identification techniques were introduced in Srivastava (1987b), of which we discuss one, namely a 'balanced view-field'. A view field is just a fancy name for a bunch of linear functions of parameters, which we wish to examine. A 'balanced' view field is the collection of linear dunctions $\{p(\theta) \mid p \in P\}$ where P is the set of all $(m!)$ permutations (for a 2^m experiment) of the factor symbols, and θ is some linear function of the parameters. For example, for $m = 4$, and $\theta = A_1 - A_{12}$, the view field is the set of 12 linear functions $\{A_1 - A_{12}, A_1 - A_{13}, A_1 - A_{14}, A_2 - A_{12}, A_2 - A_{23}, A_2 - A_{24}, A_3 - A_{13}, A_3 - A_{23}, A_3 - A_{34}, A_4 - A_{41}, A_4 - A_{42}, A_4 - A_{43}\}$; where, for example, $A_2 - A_{24}$ is obtained from θ_1 by using the permutation (2413) or (2431). Notice that if the view field is balanced, then any linear function in it could generate the rest.

If the design is a B-array of full strength, and a linear function θ is estimable, then the whole (balanced) view field generated by θ is estimable, and is available for inspection. Now, if ν is relatively small, and θ is appropriately selected, then for small σ^2, patterns would appear in the view field, in the sense that the (estimated) value of $p(\theta)$ would be roughly constant for p ranging over a certain subset P_1 of P. In other words, P may break up into disjoint set $P_1, \ldots, P_q$, such that for $p \in P_j$ $(j = 1, \ldots, q)$, the value of $p(\theta)$ is some constant C_j. By comparing the $p(\theta)$ within the class $p \in P_j$, and using the IS and TEP often leads to identification of the large nonnegligible parameters. The use of IS and TEP is more convincing and effective when the view field is balanced. Also, because of the presence of symmetry, the view field can be grasped and inspected more easily. Indeed, for many cases, this appears to be the most powerful technique known. No doubt, the method based on the minimum error sum of squares (discussed in Section 2 for identifying parameters under the SLM) is still there, and has a more decisive role to play. However, it suffers from the draw back that such a sum of squares has to be computed for all possible competing $\underline{L}$ which would be a bit tiresome even for the computers. One needs faster methods, and the balanced view-fields do appear to be useful in at least some situations. Investigations are continuing.

It would be useful to describe a few B-arrays which may serve as good first stage designs. Let Ω_{mj} denote the set of $\binom{m}{j}$ treatments in a 2^m experiment, in which exactly j factors are at level zero. Let T_{mo} be the design with $(N = 1 + \binom{m}{1} + \binom{m}{2})$

treatments consisting of Ω_{m0}, Ω_{m1}, and $\Omega_{m,m-2}$. As a resolution V design, it was studied in Srivastava (1961), and shown to be asymptotically orthogonal. We may use T_{m0} as a first stage design, or preferably T_m (which is T_{m0} plus the single treatment Ω_{mm}). For $5 \leqslant m \leqslant 8$, the author believes that the design will be good if ν lies in the following ranges ($m = 5$, $N = 17$, $7 \leqslant \nu \leqslant 10$; $m = 6$, $N = 23$, $9 \leqslant \nu \leqslant 14$; $m = 7$, $N = 30$, $12 \leqslant \nu \leqslant 18$; $m = 8$, $N = 38$, $15 \leqslant \nu \leqslant 23$). If ν is below the range shown, the design will still be good, but may be a bit too large in some"cases". Here, 'case' refers to the structure, i.e., the nature of elements of $\underline{L}$. If ν is above the range, the design may work in some situations, but as ν increases the number of patterns seen in view fields would decrease, and further supplementation by a second stage design will be needed. As we shall discuss later, in all cases, a second stage design will be needed at least for checking the model.

What about other values of m, and for ν outside the range in the above cases? According to author's intuition, the value of N for the first stage design should be roughly double the value of ν. It is easy to check that we do not have B-arrays of full strength for m in the practical range, other than what was presented above. (An exception is the design for $m = 6$, $N = 22$, $8 \leqslant \nu \leqslant 14$, consisting of Ω_{60}, Ω_{66}, and Ω_{63}. This could work in some cases.) If the first stage design is unbalanced, then we do not yet have interesting view-fields to examine. Here, it should be worth while to investigate Kuwada's PB-arrays, since these are not totally unbalanced. Indeed, they have a large amount of balance, and are available in a variety of forms. Some interesting cases (particularly, where $\underline{L}$ is thought to be close to intermediate resolution) include those where the PB-array corresponds to intra-and inter-group balanced designs, and group divisible partially balanced designs. Parallel flats designs also need to be investigated; they should be useful for certain types of vectors $\underline{L}$. Of course $\underline{L}$ is not known. However, if there are some guesses, the same could be utilized. It should be remarked that view fields other than balanced ones have not been investigated. Thus, basic work is needed for the development of techniques which may be used in conjuction with unbalanced designs. Certain algorithms and other techniques are being developed by the author for cases under Q4 and Q5, in some of these, foldover designs (and their generalization) are being employed.

We now discuss the question of checking the model. Thus, suppose, after the (analysis of the data from the) first stage design (say D_1), we somehow believe that a (known) vector $\tilde{\underline{L}}$ contains $\underline{L}$. The question is; how to verify this. Two cases arise: (i) $\tilde{\underline{L}}$ is estimable from the data of D_1, and (ii) the negation of (i) is true. Similarly, two cases arise according is σ^2 is known or unknown. We will assume σ^2 known unless otherwise stated. We can obtain the MSE (mean square due to error) under D_1 (when parameter $\tilde{\underline{L}}$ are fitted) and compare it with σ^2, using a χ^2-test. If the MSE is of a reasonable magnitude, we may consider $\underline{L}$ to be contained in $\tilde{\underline{L}}$. If, on the other hand, the χ^2-test is not too decisive, we may need to use a second stage design D_2 to increase the size of the data. Also, even if the χ^2-test is "decisive" in the usual sense, all it implies is that $\tilde{\underline{L}}$ fits well the model so far as the treatments in D_1 are concerned. It does not necessarily imply that treatments not in D_1 will also be fitted equally well by $\tilde{\underline{L}}$. Thus, again, there arises the desirability of having supplementary information from a second stage design D_2.

How to choose D_2? We will assume that there is no need to repeat the treatments from D_1 in D_2, so that the treatments in D_2 will all be distinct from those in D_1. Now, the treatments in D_2 can be selected (i) systematically, (ii) randomly, (iii) by selecting a few systematically, and a few others randomly. The idea of a systematic selection will be attractive if the overall purpose of experimenation is to find a set of treatments which have specific properties. (For example, we may be looking for treatment $\underline{t}$ for which $\phi(\underline{t})$ is maximum, or is minimum, or lies in a specified interval.) In such a case, using $\underline{\tilde{L}}$ from D_1, we can estimate ϕ, and determine the set of treatments which do have the specific properties; out of this set, those treatments that are not in D_1 are then included in D_2. The reason why systematic selection is attractive is that it allows the value of $\phi(\underline{t})$ for treatments $\underline{t}$ with the specific properties to be estimated directly (i.e., by observing the corresponding 'yield' $y(t)$), rather than be estimated indirectly from other treatments.

If D_2 is chosen systematically, then D_1 and D_2 together form a design to which the previous criticism still applies, namely, that $\underline{\tilde{L}}$ may not properly fit some treatments. Such criticism could be quite valid in many situations with continuous factors, where simplistic designs like Box–Behnken design with $(1 + 2m + 2^m)$ points are used. For $m = 2$, this design reduces to the four corners of a square plus the center point, plus the midpoints of the four edges. This design can easily miss the characteristics of a lot of functions. Indeed, any function which is roughly constant on the edges of the square and the center, but different elsewhere will be missed.

Since a design has only a finite set of points for observations, each design will suffer from the above criticism. Perhaps the only thing which could be done is to choose the points randomly in some sense; in that case, we can make some probabilistic statement that we "captured" the "main" features of the response surface with a sufficiently high degree of certainty. This shows that in the theory of identification of the model, an element of "Sampling" is inherently present. If $\underline{\tilde{L}}$ contains $\underline{L}$, $\underline{t}$ is a new treatment, and $\widehat{\phi}(\underline{t})$ is the estimate of $\phi(\underline{t})$ based on D_1, then $\mathrm{E}[y(\underline{t}) - \widehat{\phi}(\underline{t})]^2$ equals $\{\sigma^2 + \mathrm{Var}(\widehat{\phi}(\underline{t}))]$, which is known except for the multiplier σ^2. When $\underline{\tilde{L}}$ does not contain $\underline{L}$, the same quantity equals $\{\sigma^2 + \mathrm{Var}(\tilde{\phi}(\underline{t})) + (\mathrm{E}[\tilde{\phi}(\underline{t}) - \phi(\underline{t})]^2)\}$. Thus, using an F-test, the ratio of the two quantities gives an insight into whether $\underline{\tilde{L}}$ does or does not contain $\underline{L}$. Clearly, the larger the σ^2 is, the smaller the insight, since the bias part does not involve σ^2. If D_2 is to contain more than one treatment, the above idea can be extended and a test developed either by pooling or by considering each $\underline{t} \in D_2$ separately, using Roy's union-interaction principle; the author believes the latter test may be preferable. Separate study of each $\underline{t}$ should be done anyway so as to obtain insight into which elements of $\underline{L}$ are not in $\underline{\tilde{L}}$. Clearly, if D_2 is obtained, say, by simple random sampling, a statement concerning the "probability" that "$\underline{\tilde{L}}$ contains $\underline{L}$" (i.e., what we observe is not a result of the choice of D_1 and D_2), will be possible. Investigations on these lines as well are being made by the author.

It is instructive to discuss briefly the information contained in a single observation (say) on a treatment from a 2^m factorial experiment. From the purely information theoretic viewpoint, each such observation gives an equal amount of information. What is the nature of such information? Suppose D is a set of distinct treatments, and we take an observation on each of them. Then, clearly, all of the information supplied is concerning the model, since the observations give us an unbiased estimate of $\phi(\underline{t})$, for $\underline{t} \in D$. (Note that the model will be fully known if we know $\phi(\underline{t})$, for all $\underline{t}$.) The information supplied is made up of two parts. The first concerns the nature of the model. In the continuous case, this relates to deciding a model, within a group of competing models (suggested by prior knowledge, etc.). In the discrete case, this has to do with deciding which parameters in $\underline{\alpha}$ might be nonnegligible, i.e., deciding on an $\underline{L}^*$ (which would contain $\underline{L}$). The second part of the information concerns the actual value of the parameters which occur inside the model that is decided upon.

In the light of the heuristic discussion of the last section, it appears that under Q5, it may be a good idea to obtain the design randomly at least for the first stage experiment. On the other hand, under Q4, if it is felt that many main effects and two-factor interactions are significant, it may be better to use a balanced design (as suggested earlier for $5 \leqslant m \leqslant 8$) for the first stage experiment, and use a random design for the second stage. Studies on both Q4 and Q5 are currently being made by the author. The criterion of "revealing power" used here, is "the number of treatments needed to find $\underline{L}$".

The last criterion for revealing power, and some other criteria, were also used by Srivastava and Hveberg (1992), in studies on sequential designs for the 2^m case, with $m = 3$ and 4. (For some farther work, see Hveberg (1989) and Li (1990).)

What about the information contained in repitition of the same treatment. Of course, this gives information on $\sigma^2(\underline{t})$, where $\underline{t}$ is the treatment which is repeated. One of the more important discoveries of Taguchi in experimental design in industry is that $\sigma^2(\underline{t})$ quite often changes with $\underline{t}$. Significant studies, such as that by Ghosh (1987) have been made on the dispersion effects. The author would like to add that if in an experiment we encounter treatments $\underline{t}$ for which $\sigma^2(\underline{t})$ is relatively large, then there exist factors which have significant effect on the response under consideration, but which have been missed by the experimenter, and have not been included in the expeiment. Attempt should then be made by the experimenter to pin point such factors, and conduct an expanded experiment.

What about the so-called "screening designs"? This is a phrase which some people have used to refer to designs (such as Plackett–Burman designs), which allow μ and the main effects to be estimated (orthogonally). It is not clear to the author what the logic (behind the phrase) is. The sentiment that is sometimes expressed with this phrase is somewhat like this: By using a screening design, we can screen out unimportant factors (presumably, those with small estimates of main effects), and then study the remaining ones in more detail. However, estimating main effects unbiasedly is not possible with out strong assumptions on $\underline{\phi}$; even a foldover resolution IV design can not allow it if, for example, $(2k+1)$-factor effects $(k \geqslant 1)$ are present. Furthermore, it is not necessary that if a factor has small main effect, then it would also not be involved in any significant interactions. Thus, the above logic is untenable. It is unfortunate that phrases are introduced which are not justifiable and are misleading to unsuspecting workers in other disciplines.

What about assuming that the chosen $\underline{L}^*$ contains $\underline{L}$, and staying happy with it. The answer is that one should try to be happy under all circumstances about which one is helpless; however, making assumptions that may be unjustified carries the associated risks. It is said that it has been declared that, for example, in industry, interactions occur seldom, and are always known in advance (except for their value). Again, in any discipline or subdiscipline, unless one has established that his $\underline{L}^*$ contains $\underline{L}$, the above kind of declarations have the status only of sentimental assertions. The author believes, however, that in many cases, such statements will have some truth in them. Still, however, it is imperative that they be actually checked somehow. Once the appropriate studies have been made, sentimental declarations will not be needed.

Contrary to the above assertions, the author feels that in some fields (such as nutrition), interactions (even high order ones) would be significant quiet often. The spirit in which Q3, Q4 and Q5 have been discussed here is that ν is reasonably small. If ν is large, methods discussed so far may generally fail.

Models which involve too many unknown parameters are cumbersome. In such situations, smaller experiments (in which some factors are kept at constant levels) would be useful. In some areas (including certain fields of nutrition), a different kind of situation may hold. In this paper, we considered the situation where $\underline{\alpha}$ (the vector of factorial effects) has a relatively small number ν of significant elements. However, the other kind of situation would be where $\underline{\phi}$ (the vector of treatment effects) may have a relatively small number (say, λ) of elements such that these λ elements have values significantly different from the remaining $(2^m - \lambda)$ elements whose values are roughly the same. Let ψ be the set of the λ significant treatments. The author believes that the set ψ may have interesting structures, whose exploration may be quite fruitful scientifically. No theory is available yet for the exploration of ψ.

For Q3 and Q4, another sentiment is this: What about "balancing" the bias, or "minimising" it? The answer is that "balancing" may be illusory; we can't balance things that are unknown. Also "minimising" could decrease optimality of the design with respect to parameters being estimated. Furthermore, ignoring bias (like in Resolutin IV desigsns), even if possible, is of limited value, since, as argued earlier, unless $\underline{L}$ is known, $\underline{\phi}$ can not be estimated. The same holds true for balancing (even if that was possible), since $\underline{L}$ would still remain undetermined.

Another question is this: If the factors are continuous, should we go for a discrete model, or use a continuous response surface model (for example, a polynomial). Here, the main point is that the equivalence between the two models can be made on the discrete set of points (i.e., the factor space of the discrete model) by choosing an appropriate continuous function. The parameters in the discrete model will then be the same as in the continuous model. However, still the two models are not equivalent. In the discrete case, the assumption is that the model holds *only* on the discrete set of points; in the continuous case, the model applies to an entire (continuous) factor space which contains the discrete set of points. Furthermore, in the discrete case, there is basically *no* assumption, since we are merely making an orthogonal transformation from $\underline{\phi}$ to $\underline{\alpha}$. In the continuous case, there is an assumption being made about the response function. Thus, the two are nonequivalent.

If we are working in a fairly small region of the factor space, where (say, because of the Taylor's theorem) the response function is representable (near the "origin") by a low degree polynomial (of known degree), we can try the continuous model. However, when the factor space is fairly large and the response function unknown, such assumptions could be invalid and misleading. "Optimal designs" under such assumed model will be illusory. In such a situation, one could go for the discrete case first. After we develop a good idea of the values of $\phi(\underline{t})$, for $\underline{t}$ belonging to the discrete factor space, we can formulate more cogent continuous models.

We now consider sequential procedures under Q5. We will assume σ^2 is very small; to do this, sometimes, it may be a good idea to include as many factors as possible. Three cases arise, according as ν is quite small, moderately large, or quite large. If ν is very small, many treatments $\underline{t}$ would tend to have the same value $\phi(\underline{t})$. For somewhat larger ν, there may be simple linear combinations which may take a constant value. For example, we may find pairs of treatments $\underline{t}_1, \underline{t}_2$, such that $\{\phi(\underline{t}_1) - \phi(\underline{t}_2)\}$ is constant for each pair. So, it is advisable to try some treatments (selected randomly), and look for much constancy. This would help to reveal $\underline{L}$. After some insight into $\underline{L}$ is obtained, treatment may be tried using a systematic selection procedure.

For ν moderately large, it may be a good idea to start with a balanced design first. This would help gain insight into $\underline{L}$ since balanced view-fields will be available. After that, random or systematic selection of treatments may be tried, depending upon whether a smaller or larger amount of insight has been gained. Based on author's experience, it will be good to have the basic (balanced, or partially balanced) design such that its size is roughly twice the value of ν.

In cases, where ν is relatively large, it may be a good idea to start with a treatment $\underline{t}_1$ $(= (t_1, \ldots, t_m)$, say) which is most important in some sense. Then, the next treatment $\underline{t}_2$ should be different from $\underline{t}_1$ in only one factor. The factor whose level is to be changed should be selected by Bayesian considerations. We wish to select treatments, so that sucessive treatment differ from each other in as few places as possible, and $\{\phi(\underline{t}_{i+1}) - \phi(\underline{t}_i)\}$ should be as numerically large as possible. The rule about the choice of successive treatments may be broken as more insight in the nature of $\underline{L}$ is obtained, whch may indicate that certain specific treatments may be interesting and should be tried next.

The above discussions are based on author's current experience. Further insights will be offered elsewhere.

Acknowledgement

The author is thankful to Jeanie Weitzel for the excellent typing of the paper, and to Subir Ghosh, Usha Srivastava, and Jui-Yuan Chu who helped in proof-reading during the author's sickness.

References

Addleman, S. (1962). Orthogonal main-effect plans for asymmetrical factorial experiments. *Technometrics* **4**, 21–46.

Anderson, D. A. and A. M. Thomas (1980). Weakly resolvable IV. 3 search designs for the p^n factorial experiment. *J. Statist. Plann. Inference* **4**, 299–312.

Bliss, C. I. (1967). *Statistics in Biology*, Vols. 1 and 2. McGraw-Hill, New York.

Bose, R. C. and J. N. Srivastava (1964a). Analysis of irregular factorial fractions. *Sankhyā Ser. A* **26**, 117–144.

Bose, R. C. and J. N. Srivastava (1964b). Multidimensional partially balanced designs and their analysis with applications to partially balanced factorial fractions. *Sankhyā Ser. A* **26**, 145–168.

Buhamra, S. (1987). Theory of parallel flat fractions for s^n factorial experiments. Ph.D. Thesis, Department of Statistics, University of Wyoming, Laramie.

Chakravarti, I. M. (1956). Fractional replication in asymmetrical factorial designs and partially balanced arrays. *Sankhyā* **17**, 143–164.

Chatterjee, K. and R. Mukherjee (1986). Some search designs for symmetric and asymmetric factorials. *J. Statist. Plann. Inference* **13**, 357–363.

Cheng, C.-S. (1980). Optimality of some weighing and 2^m fractional-factorial designs. *Ann. Statist.* **8**, 436–446.

Ghosh, S. (1987). Non-orthogonal designs for measuring dispersion effects in sequential factor screening experiments using search linear model. *Comm. Statist. Theory Methods* **16**(10), 2839–2850.

Ghosh, S. (1990). *Statistical Design and Analysis of Industrial Experiments*. Marcel Dekker, New York.

Ghosh, S. and W. S. Al-Sabah (1994). Efficient composite designs with small number of runs. *J. Statist. Plann. Inference*, to appear.

Gupta, B. C. (1988). A bound connected with factorial search designs of resolution III, I. *Comm. Statist.*

Gupta, S. C. and R. Mukherjee (1989). A calculus for factorial arrangements.

Hveberg, R. (1989). Sequential factorial probing designs for the 2^3 case, when the three factor interaction may not be negligible. *Bull. Intern. Statist. Inst.*, Paris, 1989.

Jain, N. C. (1980). Investigations on 2-covering of a finite Euclidean space based on GF(2). Ph.D. Dissertation, under R. C. Bose, Colorado State University, Fort Collins.

Katona, G. and J. N. Srivastava (1983). Minimal 2 coverings of a finite affine space based on GF(2). *J. Statist. Plann. Inference* **8**, 375–388.

Kuwada, M. (1988). A-optimal partially balanced $2^{m_1+m_2}$ factorial designs of resolution V, with $4 \leqslant m_1 + m_2 \leqslant 6$. *J. Statist. Plann. Inference* **18**, 177–193.

Li, J. F. (1991). Sequential and optimal single stage factorial designs, with industrial applications. Ph.D. Dissertation, Colorado State University, Fort Collins.

Liao, C. T. (1994). Fractional factorial designs for estimating location effects and screening dispersion effects. Ph.D. Thesis, Department of Statistics, Colorado State University, Fort Collins.

Liao, C. T. and H. K. Iyer (1995). Orthogonal parallel flats designs for s^n and $s_1^{n_1} \times s_2^{n_2}$ factorial experiments. *J. Statist. Plann. Inference*, under revision.

Liao, C. T., H. K. Iyer and D. F. Vecchia (1996). Construction of orthogonal two-level designs of user-specified resolution where $N \neq 2^k$. *Technometrics*, accepted for publication.

Nguyen, N. and A. J. Miller (1993). 2^m fractional factorial designs of resolution V and high A-efficiency, $7 \leqslant m \leqslant 10$, to appear in *J. Statist. Plann. Inference.*

Ohnishi, T. and T. Shirakura (1985). Search designs for 2^m factorial experiments. *J. Statist. Plann. Inference* **11**, 241–245.

Rao, C. R. (1946). Hypercubes of strength d leading to confounded designs in factorial experiments. *Bull. Calcutta Math. Soc.* **38**, 67–73.

Srivastava, J. N. (1961). Contributions to the construction and analysis of designs. Institute of Statistics, Mimeo Ser. No. 301, University of North Carolina, Chapel Hill.

Srivastava, J. N. (1975). Designs for searching nonnegligible effects. In: J. N. Srivastava, ed., *Survey of Statistical Designs and Linear Models*. North-Holland, Amsterdam, 507–519.

Srivastava, J. N. (1977). Optimal search designs or designs optimal under bias-free optimality criteria. In: S. S. Gupta and D. S. Moore, eds., *Statistical Decision Theory and Related Topics*, Vol. 2, 375–409.

Srivastava, J. N. (1978). On the linear independence of sets of 2^q columns of certain $(1, -1)$ matrices with a group structure, and its connection with finite geometries. In: D. A. Holton and J. Seberry, eds., *Combinatorial Mathematics*. Springer, Berlin.

Srivastava, J. N. (1984). Sensitivity and revealing power: Two fundamental statistical criteria other than optimality arising in discrete experimentation. In: K. Hinkelmann, ed., *Experimental Designs, Statistical Models, and Genetic Statistics Models, and Genetic Statistics*. Marcel Dekker, New York, 95–117.

Srivastava, J. N. (1987a). Advances in the general theory of factorial designs based on partial pencils in Euclidean n-space. *Utilitas Math.* **32**, 75–94.

Srivastava, J. N. (1987b). On the inadequacy of the customary orthogonal arrays in quality control and scientific experimention, and the need of probing designs. *Comm. Statist.* **16**, 2901–2941.

Srivastava, J. N. (1990). Modern factorial design theory for experimenters. In: S. Ghosh, ed., *Design and Analysis of Industrial Experiments*. Marcel Dekker, New York, 311–406.

Srivastava, J. N. (1992). A 2^8 factorial search design with good revealing power. *Sankhyā* **54**, 461–474.

Srivastava, J. N. (1993). Nonadditivity in row–column designs. *J. Combin. Inform. System Sci.* **18**, 85–96.

Srivastava, J. N., D. A. Anderson and J. Mardekian (1984). Theory of factorial designs of the parallel flats type 1: The coefficient matrix. *J. Statist. Plann. Inference* **9**, 229–252.

Srivastava, J. N. and S. Arora (1987). On a minimal resolution 3.2 design for the 2^4 factorial experiment. *Indian J. Math.* **29**, 309–320.

Srivastava, J. N. and S. Arora (1988). Minimal search designs of resolution 3.1 and 3.2 for the 2^4 experiment. In: D. Raychaudhuri, ed., *Coding Theory and Design Theory*, Vol. 2. Springer, Berlin, 336–361.

Srivastava, J. N. and S. Arora (1991). An infinite class of 2^m factorial designs of resolution 3.2 for general m. *Discrete Math.* **98**, 35–56.

Srivastava, J. N. and R. J. Beaver (1986). On the superiority of the nested multidimensional block design, relative to the classical incomplete block designs. *J. Statist. Plann. Inference* **13**, 133–150.

Srivastava, J. N. and S. Ghosh (1996). On nonorthogonality and nonoptimality of Addelman's main-effect plans satisfying the condition of proportional frequencies, *Statist. Probab. Lett.*, to appear.

Srivastava, J. N. and R. Hveberg (1992). Sequential factorial probing designs for identifying and estimating nonnegligible parameters. *J. Statist. Plann. Inference* **30**, 141–162.

Srivastava, J. N. and Junfang Li (1994). Orthogonal designs of parallel flats type. *J. Statist. Plann. Inference*, to appear.

Srivastava, J. N. and P. Srivastava (1977). Examination of results of experiments in social sciences using factorial designs (unpublished report).

Srivastava, J. N. and Y. C. Wang (1996). Row–column designs: how and why non-additivity may make them unfit for use. Submitted.

S. Ghosh and C. R. Rao, eds., *Handbook of Statistics, Vol. 13*

11

Response Surface Designs

Norman R. Draper and Dennis K. J. Lin

1. Response surfaces and models

Suppose we have a set of data containing observations on a response variable y and k predictor variables $\xi_1, \xi_2, \ldots, \xi_k$. A response surface model is a mathematical model fitted to y as a function of the ξ's in order to provide a summary representation of the behaviour of the response, as the predictor variables are changed. This might be done in order to (a) optimize the response (minimize a cost, maximize a percentage yield, minimize an impurity, for example), (b) find what regions of the ξ-space lead to a desirable product (viscosity within stated bounds, transparency not worse than a standard, appropriate color maintained, for example), or (c) gain knowledge of the general form of the underlying relationship with a view to describing options such as (a) and (b) to customers.

When the mechanism that produced the data is either unknown or poorly understood, so that the mathematical form of the true response surface is unknown, an *empirical model* is often fitted to the data. An empirical model is usually linear in the parameters and often of polynomial form, either in the basic predictor variables or in transformed entities constructed from these basic predictors. The purpose of fitting empirical models is to provide a mathematical French curve that will summarize the data. This chapter will discuss only design of experiments for such empirical models.

There is another useful type of model, however, the mechanistic model. If knowledge of the underlying mechanism that produced the data *is* available, it is sometimes possible to construct a model that represents the mechanism reasonably well. An empirical model usually contains fewer parameters, fits the data better, and extrapolates more sensibly. (Polynomial models often extrapolate poorly.) However, mechanistic models are often nonlinear in the parameters, and more difficult to formulate, to fit, and to evaluate. For information on this topic, see Bates and Watts (1988) and Seber and Wilde (1989).

When little is known of the nature of the true underlying relationship, the model fitted will usually be a polynomial in the ξ's. The philosophy applied here is that we are approximating the true but unknown surface by low-order (equivalently: low degree) terms in its Taylor's series expansion. Most used in practice are polynomials of first and second order. The first-order model is

$$y_u = \beta_0' + \beta_1'\xi_{1u} + \beta_2'\xi_{2u} + \cdots + \beta_k'\xi_{ku} + \varepsilon_u, \tag{1}$$

where $(y_u, \xi_{1u}, \xi_{2u}, \ldots, \xi_{ku}), u = 1, 2, \ldots, n$, are the available data and where it is usually tentatively assumed that the errors $\varepsilon_u \sim N(0, \sigma^2)$ and are independent. Such assumptions are always carefully checked by examining the residuals (the differences between observed and predicted values of y) for possible contradictory patterns. The second-order model contains additional terms

$$\beta'_{11}\xi^2_{1u} + \beta'_{22}\xi^2_{2u} + \cdots + \beta'_{kk}\xi^2_{ku} + \beta'_{12}\xi_{1u}\xi_{2u} + \cdots + \beta'_{k-1,k}\xi_{k-1,u}\xi_{ku}. \tag{2}$$

Polynomial models of order higher than 2 are rarely fitted, in practice. This is partially because of the difficulty of interpreting the form of the fitted surface, which, in any case, produces predictions whose standard errors are greater than those from the lower-order fit, and partly because the region of interest is usually chosen small enough for a first- or second-order model to be a reasonable choice. Exceptions occur only when two or three ξ's are used. When a second-order polynomial is not adequate, and often even when it is, the possibility of making a simplifying transformation in y or in one or more of the ξ's would usually be explored before reluctantly proceeding to higher order. A more parsimonious representation involving fewer terms is generally more desirable.

Coding

In actual applications, it is common practice to code the ξ's via $x_{iu} = (\xi_{iu} - \xi_{i0})/S_i$, $i = 1, 2, \ldots, k$, where ξ_{i0} is some selected central value of the ξ_i range to be explored, and S_i is a selected scale factor. For example, if a temperature (T) range of 150–170°C is to be covered using three levels 150, 160, 170°C, the coding $x = (T - 160)/10$ will code these levels to $x = -1, 0, 1$, respectively. The second-order model would then be recast as

$$\begin{aligned} y_u = {} & \beta_0 + \beta_1 x_{1u} + \cdots + \beta_k x_{ku} \\ & + \beta_{11} x^2_{1u} + \cdots + \beta_{kk} x^2_{ku} \\ & + \beta_{12} x_{1u} x_{2u} + \cdots + \beta_{k-1,k} x_{k-1,u} x_{ku} + \varepsilon_u \end{aligned} \tag{3}$$

or

$$y = X\beta + \varepsilon$$

in matrix form, and would usually be fitted by least squares in that form. Substitution of the coding formulas into (3) enables the β''s to be expressed in terms of the β's, if desired.

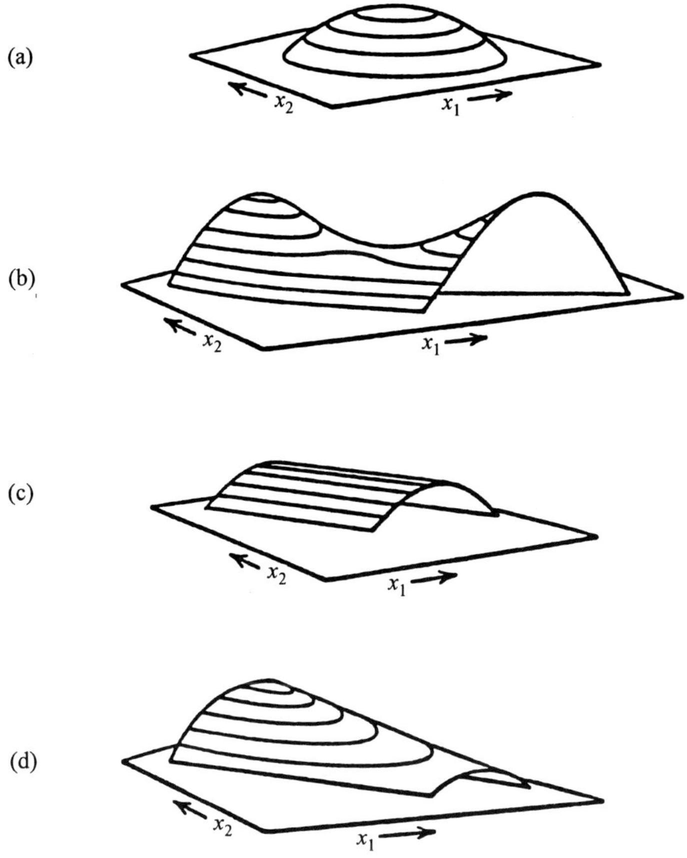

Fig. 1. Examples of surfaces representable by a second-degree equation: (a) simple maximum, (b) saddle (or col or minimax), (c) stationary ridge, and (d) rising ridge.

2. Second order surfaces

A model of the form (3) can represent a variety of surfaces, one of which will best fit a given set of data. Figure 1 shows examples of the four basic types that occur when $k = 2$. "Upside down" versions can also occur. For example, linked with the simple maximum of Figure 1(a) (a hill) is a simple minimum (a hollow). The "upside down" version of a rising ridge (Figure 1(d)) is a falling valley, and so on. In higher dimensions the drawings become more complicated, but the two dimensional sections of higher dimensional surfaces are always one of the basic types illustrated in Figure 1, together with their upside down versions. For additional details on surface types in various dimensions, including the important reduction of such surfaces to canonical form, see Davies (1978) or Box and Draper (1987).

3. Criteria for experimental designs

First and second order models have proved valuable in a variety of subject areas. Sometimes they are fitted to data that have been obtained by observing a running process. More typically, the data will result from a carefully planned series of experimental runs (individual experiments) which, taken as a whole, are called the *experimental design*, and denoted, in the coded x-space by the n sets of values $(x_{1u}, x_{2u}, \ldots, x_{ku})$, $u = 1, 2, \ldots, n$. These coordinates define a pattern of n points in a k-dimensional space. A response surface design is simply an experimental arrangement of points in x-space that permits the fitting of a response surface to the corresponding observations y_u. We thus speak of first-order designs (if a first-order surface can be fitted), second-order designs, and so on. Obviously, a design of a particular order is also necessarily a design of lower order.

The choice of a response surface design is thus one of selecting a set of suitable points in k-dimensional x-space according to some preselected criterion or criteria of goodness. The technical literature of experimental design contains many discussions of so-called "optimal designs". However, skepticism is called for in reading many books and papers, because their authors often concentrate on one criterion only (and sometimes one that by practical experimental standards is inappropriate) and then derive the best designs under that single criterion. While this often provides interesting mathematics, it does not necessarily constitute useful practical advice. There are many possible desirable characteristics for a "good" response surface design. Box and Draper (1987) gave 14 such characteristics. The design should:

1. Generate a satisfactory distribution of information about the behaviour of the response variable throughout a region of interest, R;
2. Ensure that the fitted value at $\boldsymbol{x}, \widehat{y}(\boldsymbol{x})$, be as close as possible to the true value at $\boldsymbol{x}, \eta(\boldsymbol{x})$;
3. Give good detectability of lack of fit;
4. Allow transformations to be estimated;
5. Allow experiments to be performed in blocks;
6. Allow designs of increasing order to be built up sequentially;
7. Provide an internal estimate of error;
8. Be insensitive to wild observations and to violation of the usual normal theory assumptions;
9. Require a minimum number of experimental points;
10. Provide simple data patterns that allow ready visual appreciation;
11. Ensure simplicity of calculation;
12. Behave well when errors occur in the settings of the predictor variables, the x's;
13. Not require an impractically large number of predictor variable levels;
14. Provide a check on the "constancy of variance" assumption.

It is impossible for a design to satisfy all the characteristics simultaneously. Indeed some characteristics work against others, for example, (9) conflicts with the need to add extra points to attain (3). In any given experimental situation, certain characteristics will loom larger than others, depending on what the desired objectives are. If we wish to examine a large number of variables and pick the most effective few, for example, criteria (1)–(4) may not be of much interest temporarily. A good statistician will be able to size up the current situation and give emphasis to the various criteria accordingly. Moreover, certain types of designs satisfy many, if not all of the criteria. Such designs are especially valuable.

4. Sequential experimentation

Although each experimental design is an important step in itself, experimentation is rarely a one-step process. Typically one proceeds through a series of steps:

1. Identify all the variables currently of interest. If there are many, because current knowledge is sparse, consider a screening experiment that will enable us to eliminate unimportant predictor (x) variables and retain influential ones. (The WHICH? stage.) We would usually initially consider the possibility that a first-order model might be satisfactory, and perform a first-order design. A simple but good choice (see Box, 1952) would be a regular simplex design with one or more center points. The general regular simplex in k dimensions has $n = k+1$ points (runs) and can be oriented to have its coordinates given as in Table 1, where $a_i = \{cn/[i(i+1)]\}^{1/2}$, and c is a scaling constant to be selected. Alternatively, a two-level factorial or fractional factorial, or a Plackett and Burman design with added center point(s) would be excellent. In all cases, the center point(s) average response can be compared to the average response at the noncentral points to give a measure of nonplanarity. For additional details, see Box et al. (1978, p. 516) or Box and Draper (1987), as well as Section 6 below.

2. If only a few (of many) x-variables were effective, the results could be projected into those fewer x-dimensions, and a first order surface could be refitted. Then, if

Table 1
The rows are the coordinates of the $(k+1)$ points of a simplex design in k dimensions

x_1	x_2	x_3	$\cdots$	x_i	$\cdots$	x_k
$-a_1$	$-a_2$	$-a_3$	$\cdots$	$-a_i$	$\cdots$	$-a_k$
a_1	$-a_2$	$-a_3$	$\cdots$	$-a_i$	$\cdots$	$-a_k$
0	$2a_2$	$-a_3$	$\cdots$	$-a_i$	$\cdots$	$-a_k$
0	0	$3a_3$	$\cdots$	$-a_i$	$\cdots$	$-a_k$
·	·	0		·		·
·	·	·		·		·
·	·	·		·		·
				ia_i		
				0		
				·		
				·		
				·		
0	0	0		0		ka_k

the reduced first-order surface fitted well, one could either interpret its nature if the local relationship were being sought, or else move out along a path of steepest ascent (or descent) if improved conditions were sought; see Box and Draper (1987). If the first-order surface were an inadequate representation of the local data, either initially or after one or more steepest ascent(s) (or descent(s)), it would be sensible to consider transformations of the response and/or predictor variables that *would* allow a first-order representation. When the possibilities of using first-order surfaces had been exhausted, one would then consider the need for a second-order surface. Second order designs will be discussed in Sections 7, 8 and 12.

3. If a second order surface were deemed inadequate, it would again be sensible to seek suitable variable transformations. Proceeding to models of order higher than second would be a last resort in most applications.

5. "Value for money" in designs

When we have a limited budget for experimentation (typically the case in practice), we wish to choose our experimental design to get full value for what we spend. We can assess value in a design by considering what the degrees of freedom (the "money") buy for us, that is, what benefits they provide. Consider the first order simplex design for k factors with an additional n_0 center points, a total of $n = k + 1 + n_0$ degrees of freedom (df) available. Of these, $(k+1)$ are used to estimate the coefficients of the first order model, $(n_0 - 1)$ provide a pure error estimate of σ^2, and 1 df provides a test for non-planarity. So we have "good value for money" here, in the sense that the design performs well on criteria 1, 3, 6, 7, 9 and 11. Of course, the design can be criticised, in various degrees, with respect to other criteria. It makes excellent sense, in considering any specific design to evaluate exactly what the available degrees of freedom will provide in terms of model estimation, pure error, and lack of fit, particularly when a choice between competing designs needs to be made.

When considering the next experimental design we most often choose the location of its center as the point representing the current "best" (whatever that is defined to mean) conditions. Three common general objectives of response surface methodology are:

1. To find the local nature of the relationship between the response and the predictors and so "explain" the response's behavior. It may, for example, be desired to keep the response within specifications requested by a customer, and/or to check whether the predictor variable settings are critical and sensitive.
2. To move from the current "best" conditions to better conditions (lower cost, higher yield, improved tear resistance, and so on).
3. To use the fitted surface as an intermediate step to mechanistic understanding of the underlying process.

For other possible objectives, see Herzberg (1982).

6. Screening designs and projection properties

Practitioners projection properties are constantly faced with distinguishing between the factors that have an actual effect and those factors whose effects are due to random error. Typically, many possible factors are suggested for investigation, but it is often anticipated that only a "small" subset of these (k, say) will be effective, the so called "effect sparsity" situation. Thus, it is believed that perhaps even a smaller subset of the specific terms in (1) and (2) are actually needed to describe the behaviour of the response. A design suitable for screening out the k relevant factors from the q total factors is called a screening design. See Box at al. (1978, pp. 545–546).

Screening designs are typically used in the initial stages of an experimental investigation. (Sometimes, several responses are measured in each experiment.) Because of their relative simplicity of use, two-level screening designs are very popular in practice. For example, the 2^{q-p} fractional factorial designs (that is, a 2^{-p} fraction of a 2^q two-level factorial design; see Box et al., 1978) and the Plackett and Burman (1946) designs are widely used. When such an n-run screening design is employed, it is not expected that every factor will show up as important, merely a subset. This permits the use of fractionated designs with complicated alias structures. After the initial analysis, the whole design is then projected into a lower dimensional space which contains only the k apparently important factors.

We employ throughout the standard notation introduced by Box and Hunter (1961) in which I represents an n-run column of plus signs and, for example, 123 represents a column of signs determined by taking the product of the signs $\pm$ in columns 1, 2, and 3 of the two-level factorial design, where $-$ and $+$ denote the two levels of the factor allocated to any column.

Consider, for a simple example, the eight run, four factor 2_{IV}^{4-1} design $I = 1234$ consisting of the runs $(x_1, x_2, x_3, x_4) = (- - - -), (+ - - +), (- + - +), (+ + - -), (- - + +), (+ - + -), (- + + -), (+ + + +)$. Suppose one of the four factors is inactive; we do not know which factor it might be. No matter *which* factor it is, if we drop that variable from the design, the remaining three variables are represented by a full 2^3 factorial. For example, let us drop variable x_1 by removing the first sign from the parentheses above. We are left with $(- - -), (- - +), (+ - +), (+ - -), (- + +), (- + -), (+ + -), (+ + +)$, a 2^3 factorial design in (x_2, x_3, x_4). Such a design will remain no matter which variable is dropped from the design.

Details of the projection properties of two-level designs will be discussed next. This knowledge also allows us to see what *additional* runs can be of value, after the results of the initial screening are available. In addition, it provides insight into how to allocate the design variables to the factors that are thought, a priori, to be important.

Projections of 2^{q-p} designs

When a 2^{q-p} screening design is used, all projections are either standard two-level full factorials or fractional factorials. For $k = 2$ and $n \geqslant 4$, the projection is always a 2^2 design, with multiplicity $n/4$. For $k = 3$ and $n \geqslant 8$, there are two types of projections: a 2^3 design, with multiplicity $n/8$, or a 2_{III}^{3-1} design ($I = \pm 123$), with multiplicity $n/4$. For $k = 4$ and $n \geqslant 16$, there are three possibilities: a 2^4 design, with multiplicity $n/16$, or a 2_{IV}^{4-1} design ($I = \pm 1234$), with multiplicity $n/8$, or a 2_{III}^{4-1} design ($I = \pm 123$), with multiplicity $n/8$.

For $k = 5$, the possibilities for the projected designs are given below:

Type	Generators	Multiplicity
2^5	–	$n/32$
2_{V}^{5-1}	$I = \pm 12345$	$n/16$
2_{IV}^{5-1}	$I = \pm 1234$	$n/16$
2_{III}^{5-1}	$I = \pm 123$	$n/16$
2_{III}^{5-2}	$I = \pm 124 = \pm 1235$	$n/8$

The extension to $k \geqslant 6$ is similar and straightforward. However, if the projected design remains resolution III, estimated main effects are confounded with two-factor interactions. The usual advice given in such circumstances to eliminate such blurring is to "fold over" the design (Box et al., 1978, pp. 340, 399), that is, repeat the projected design with all signs reversed. Foldover always converts a resolution III design into a resolution IV design (see Box et al., 1978, p. 398). It also doubles the size of the experiment, however, which can be disadvantageous.

Plackett and Burman screening designs

Table 2 shows a 12-run Plackett and Burman design, obtained as follows.

(a) Write down the set of signs $++-+++---+-$, provided by Plackett and Burman (1946).

Table 2
A 12-run Plackett and Burman design

Run No.	Factors										
	1	2	3	4	5	6	7	8	9	10	11
1	+	+	−	+	+	+	−	−	−	+	−
2	−	+	+	−	+	+	+	−	−	−	+
3	+	−	+	+	−	+	+	+	−	−	−
4	−	+	−	+	+	−	+	+	+	−	−
5	−	−	+	−	+	+	−	+	+	+	−
6	−	−	−	+	−	+	+	−	+	+	+
7	+	−	−	−	+	−	+	+	−	+	+
8	+	+	−	−	−	+	−	+	+	−	+
9	+	+	+	−	−	−	+	−	+	+	−
10	−	+	+	+	−	−	−	+	−	+	+
11	+	−	+	+	+	−	−	−	+	−	+
12	−	−	−	−	−	−	−	−	−	−	−

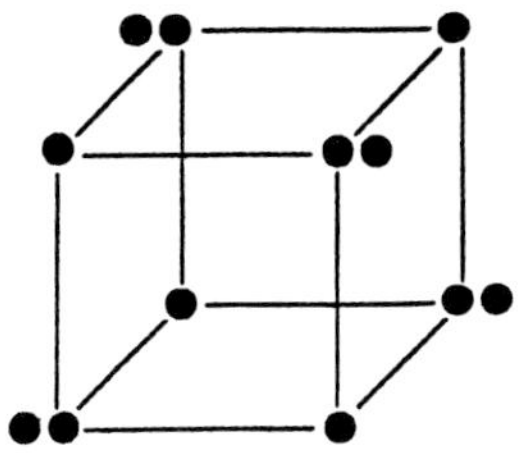

Fig. 2. A 12-run, 11 factor Plackett and Burman design projected into any three dimensions.

(b) Permute the signs in 11 rows total, by taking the sign from the right hand side and moving it to the left hand side.

(c) Add a 12th row of all minus signs.

For $n \leqslant 24$, all of the Plackett and Burman designs can be obtained by such a cyclic permutation. The signs for the first rows are:

$n = 8$: $+ + + - + - -$,

$n = 12$: $+ + - + + + - - - + -$,

$n = 16$: $+ + + + - + - + + - - + - - -$,

$n = 20$: $+ + - - + + + + - + - + - - - - + + -$,

$n = 24$: $+ + + + + - + - + + - - + + - - + - + - - - -$.

For $n = 8$ and 16, we obtain a standard 2^{q-p} design, so these cases are covered by the previous section. For the projection of the 12-run design in any k of the 11 factor dimensions, we select k columns and examine the design that results by ignoring the other $11 - k$ columns. For example, suppose $k = 3$ and we select the 1, 2, 3 columns. The reduced 12-point design consists of a 2^3 design plus a 2^{3-1} design with $I = 123$, shown in Figure 2. This very desirable arrangement provides complete coverage of all the factorial effects plus additional pure error information obtained at four different locations well spread out over the experimental region. Moreover, **no matter which three factors are designated as the survivor columns**, a similar design is always obtained, that is, a 2^3 plus a 2^{3-1} with $I = \pm ABC$ where A, B, C represent any three of the eleven factors. See, also, Lin and Draper (1992) and Box and Bisgaard (1993).

For $k = 3$, and $n = 20$, two types of projections can occur:

1. A 2 : 3 type. (This means two full 2^3 factorials and an additional 2^{3-1}. At the corners of the cube there are either two or three points.)

2. A 1 : 4 type. (This means a 2^3 factorial and three identical 2^{3-1} designs. At each corner of the cube there is either one point or there are four points.)

The notation "$(r : s)$" means r points lie at four of the "2^{3-1} locations" and s points lie at the other four. 2^{3-1} locations are always defined by $I = \pm$ the relevant three factor interaction. Another point to note is that, in some cases, we can proceed from a three column n runs projection to a three columns $(n + 4)$ runs projection by simply adding a 2^{3-1} design. For $n = 20$, a $(2 : 3)$ can be converted into either a

(3 : 3) or a (2 : 4) with $n + 4 = 24$, depending on which 2^{3-1} is added. Similarly, a (1 : 4) can become a (2 : 4) or a (1 : 5); however, the latter is not a three-column projection of a 24-run Plackett and Burman design. Other possibilities for $k = 3$ are given in Lin and Draper (1991).

Projections into $k = 4$ dimensions

For $n = 12$, one could complete a 2_{IV}^{4-1} design by adding one run; there are five additional runs as well. For example, suppose we use columns 1–4 of Table 3. The eight runs 1, 3, 5, 6, 7, 9, 10, and 11 all have a negative product of signs, but runs 3 and 11 are identical, $(+ - + +)$. The new run $(- + - -)$ completes the 2_{IV}^{4-1} with $I = -1234$. This additional point is always uniquely determined as the foldover complement of the duplicate point. An alternative to this one run addition would be to fit a "main effects plus two factor interactions" model to all the initially available data.

For $n = 20$, only three types of projections exist (apart from sign changes in the columns, permutations of the columns, and rearrangements of the rows). For $n = 24$, there are four types of projections. Two types of projections provide a full 2^4 design plus a 2_R^{4-1}; for one R = IV, and for the other R = III. Another two types of projections require two additional runs to complete a full 2^4 design. Which projection is actually attained depends on the specific four factors retained after analysis.

Projections into $k = 5$ dimensions

For $n = 12$, two types of designs are possible, one with a repeat run pair ("type 5.1", say) and one with a mirror image pair ("type 5.2"); see Draper (1985, Table 2). A number of possibilities exist for supplementing these designs. To determine which of the two design types has been obtained via projection, one must check to see if the specific design has a repeat run pair, or a mirror image run pair, an easy thing to do.

For example, if we choose columns 1–5 of Table 2, we see that runs 7, $(+ - - - +)$, and 10, $(- + + + -)$, are mirror image runs. Thus we have a design of type 5.2 which we could convert to a standard form, in which the mirror image runs are $(- - - - -)$ and $(+ + + + +)$, by changing the signs in either columns 1 and 5 or in 2, 3, and 4 and perhaps rearranging the columns appropriately. Even without making those changes, it is clear that the product of signs in the columns 1, 2, 3, 4, and 5 is "$-$" for runs 1, 5, 7, 8, 11, and 12, and "$+$" for the remaining six runs. A 2_{V}^{5-1} can thus be produced in two alternative ways, by adding 10 runs with the same signed products in each case.

In examining the possibilities for Design 5.2, we discover that the 32 runs of a 2^5 design can be divided as follows:

(a) Into a 12-run portion and a 20-run portion so that the two portions are the projections into five dimensions of (respectively) 12-run and 20-run Plackett and Burman designs.

(b) Into 8-run and 24-run portions which are projections into five dimensions of 8 and 24 run Plackett and Burman designs.

In all such cases, the model appropriate to the completed 2_R^{q-p} design could be fitted by least squares, and the runs already made in addition to the 2_R^{q-p} runs will provide some residual degrees of freedom in an analysis of variance table.

All the nine possible different projected designs for $n = 20$ as well as the nine projections that occur for $n = 24$ are given in detail in Lin and Draper (1991, 1994).

Non-equivalent Hadamard matrices for $n = 16, 20$

By adding a column of 1's to a Plackett and Burman design, we obtain a Hadamard matrix H which satisfies $H'H = nI$. For $n = 12$, H is unique, but for higher n this is not true. Non-equivalent Hadamard matrices have different projection properties. We illustrate using the cases $n = 16$ and $n = 20$.

There are five non-equivalent Hadamard matrices for $n = 16$; see Hall (1961). Only one of these corresponds to a Plackett and Burman design, that is, only one (called H16-1 here, and I by Hall) provides a 2^{q-p} 16-run design of the type whose projections were studied in the previous section. We now briefly discuss the projection patterns of the other four types, which we designate as H16-2, H16-3, H16-4, and H16-5. These designs are, respectively designs II, III, IV, and V in Hall (1961, pp. 23–24).

For $k = 3$, there are three different possible projections. Two of these arise from *all* the five Hadamard matrices. In addition, H16-5 produces projected designs of type 1 : 3.

For $k = 4$, five projections occur, as shown in Figure 3. Two of these, (a) and (c), arise from *all* of the Hadamard matrices. In all parts of Figure 3, the cube represents the space of three of the four factors and the fourth is represented by open dots for the lower level, and solid dots for the upper level. Note the "unbalanced" structure of designs (d) and (e). Whereas the other three designs (a), (b), and (c) are replicated 2_{III}^{4-1}, replicated 2_{IV}^{4-1} and a full 2^4 respectively, designs (d) and (e) are not of this form.

For $k = 5$, there are eight different projections; see Figure 4. Two of these, (a) and (b) arise from *all* of the Hadamard matrices. In all parts of Figure 4, the cube represents the space of three of the five factors. Each circle represents a run of the projected design and in each circle, the left portion is for the fourth factor and the right portion is for the fifth factor. An open half-circle represents the lower level of a factor and a solid (black) half-circle represents the upper level of a factor. Note the "unbalanced" structure of projections (e), (f), (g), and (h). Designs (a), (b), (c), and (d) are, respectively, a replicated 2_{III}^{5-2}, a 2_{III}^{5-1}, a 2_{IV}^{5-1}, and a 2_{V}^{5-1}. There are three non-equivalent Hadamard matrices for $n = 20$ (Hall, 1965), only one of which is equivalent to a Plackett and Burman design. The other two give exactly similar projections for $k = 3$ and 4. For $k = 5$, however, there is one projection additional to the nine listed previously. For additional details, see Lin and Draper (1991, 1994) and Wang and Wu (1995).

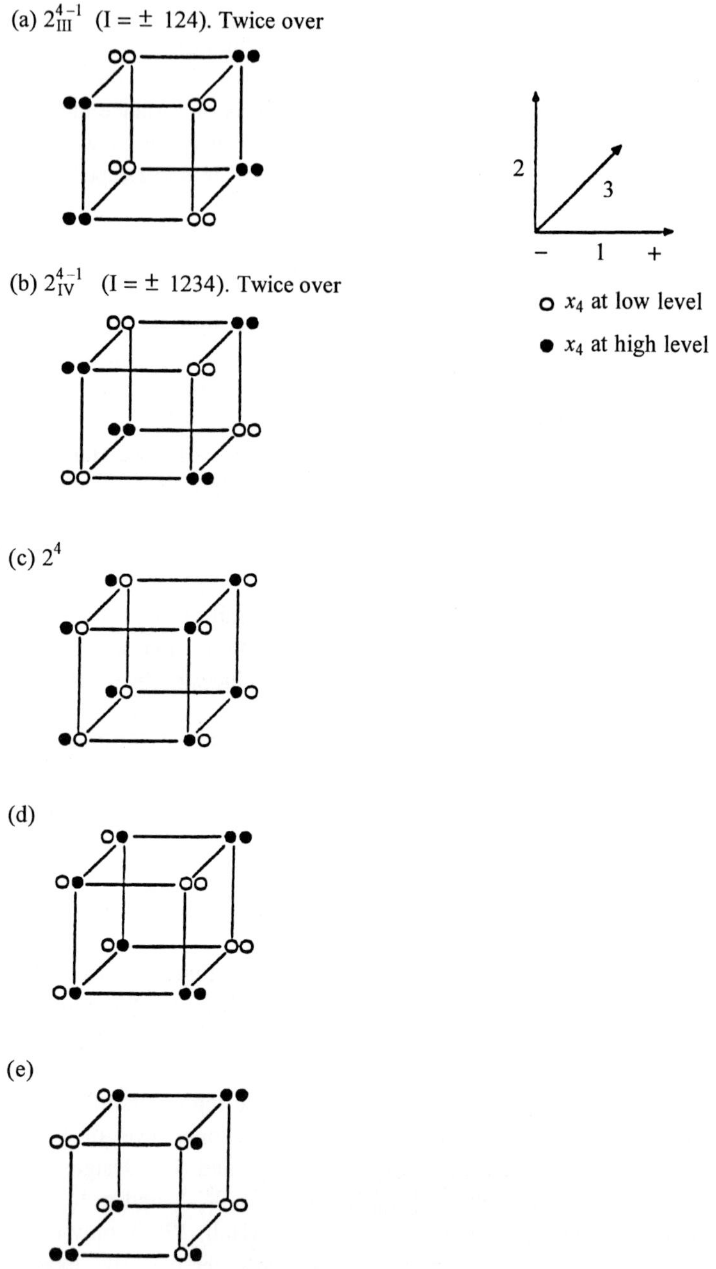

Fig. 3. Projections of 16-Run Hadamard matrix type designs into four dimensions.

(a) 2_{III}^{5-2} $(I = \pm 124 = \pm 135)$ (e)

(b) 2_{III}^{5-1} $(I = \pm 125)$ (f)

(c) 2_{IV}^{5-1} $(I = \pm 1235)$ (g)

(d) 2_V^{5-1} $(I = \pm 12345)$ (h)

x_4 x_5	:	●	◐	◑	◎
(x_4, x_5)	=	$(+,+)$	$(+,-)$	$(-,+)$	$(-,-)$

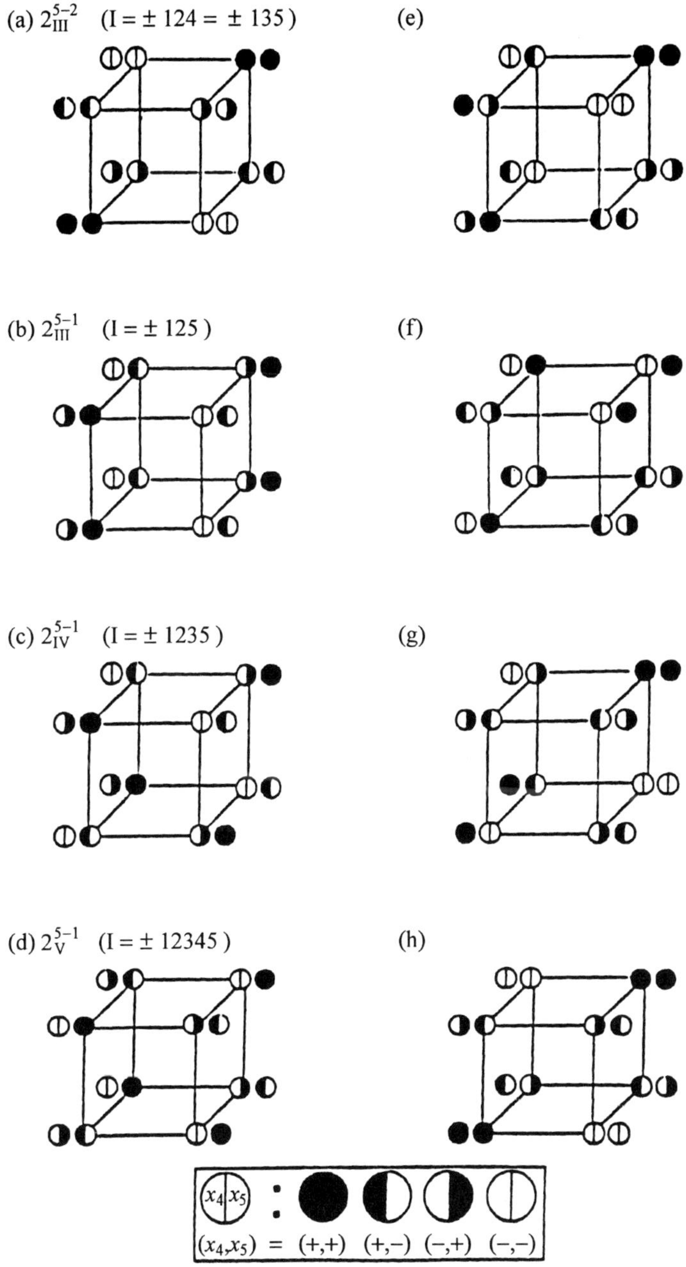

Fig. 4. Projections of 16-run Hadamard matrix type designs into five dimensions.

7. The central composite design

The *central composite design* was one of the early design suggestions (see Box and Hunter, 1957) for obtaining data for fitting a second order surface. It turned out that this design (often called just the *composite design*) satisfied many of the desirable criteria previous listed, and so it has become a cornerstone of response surface methodology. It is constructed from three sets of points which can be described in the coded x-space as follows.

(a) the 2^k vertices $(\pm 1, \pm 1, \ldots, \pm 1)$ of a k-dimensional "cube" ($k \leqslant 4$), or a fraction of it ($k \geqslant 5$);
(b) the $2k$ vertices $(\pm\alpha, 0, \ldots, 0), (0, \pm\alpha, \ldots, 0), \ldots, (0, 0, \ldots, \pm\alpha)$ of a k-dimensional cross-polytope or "star";
(c) a number, n_0, of "center points", $(0, 0, \ldots, 0)$.

Set (a) is simply a full 2^k factorial design or a 2^{k-p} fractional factorial if $k \geqslant 5$. The notation $(\pm 1, \pm 1, \ldots, \pm 1)$ means that 2^k points obtained by taking all possible combinations of signs are used for full factorial cases. (In response surface applications, these points are often referred to as a "cube", whatever the number of factors.)

Set (b) consists of pairs of points on the coordinate axes all at a distance α from the origin. (The quantity α has yet to be specified; according to its value the points may lie inside or outside the cube.) In three dimensions the points are the vertices of an octahedron and this word is sometimes used for other values of $k \neq 3$. However, a more convenient name for such a set of points in k dimensions is "star" or, more formally, cross-polytope.

These sets and the complete design (the n_0 center points represented by a single center point) are shown diagrammatically in Figures 5 and 6 for the cases $k = 2$ and 3.

Fractionation of the cube is possible whenever the resulting design will permit individual estimation of all the coefficients in equation (3). This is guaranteed for fractions of resolution $\leqslant 5$. The smallest usable fraction is then a 2^{k-1} design (a half-fraction) for $k = 5, 6, 7$, a 2^{k-2} design (a quarter-fraction) for $k = 8, 9$, a 2^{k-3} for $k = 10$, and so on. (See Box et al., 1978, p. 408.) Table 3, adapted from Box and Hunter (1957, p. 227) shows the number of parameters in equation (3) and the number of noncentral design points in the corresponding composite design for $k = 2, \ldots, 9$. The values to be substituted for p are $p = 0$ for $k = 2, 3$, and 4; $p = 1$ for $k = 5, 6$, and 7; and $p = 2$ for $k = 8$ and 9; they correspond to the fraction, $1/2^p$, of the cube used for the design.

Points 1, 4, 5, 6, 7, 9, 10, 11 and 12 in Section 3 can all be satisfied by the composite design. Satisfaction of some requires suitable choices of α, n_0, and shrinking or expanding all the design points relative to the region R (see Box and Draper, 1959, 1963; see also, Welch, 1984). Overall the composite design is an excellent choice for many investigations.

Table 3
Features of certain composite designs

No. of variables	k	2	3	4	5	6	7	8	9
No. of parameters	$(k+1)(k+2)/2$	6	10	15	21	28	36	45	55
Cube + star	$2^k + 2k$	8	14	24	–	–	–	–	–
$\frac{1}{2}$(cube) + star	$2^{k-1} + 2k$	–	–	–	26	44	78	–	–
$\frac{1}{4}$(cube) + star	$2^{k-2} + 2k$	–	–	–	–	–	–	80	130
α (rotatable)	$2^{(k-p)/4}$	1.414	1.682	2	2	2.378	2.828	2.828	3.364
Suggested n_0		2–4	2–4	2–4	0–4	0–4	2–4	2–4	2–4

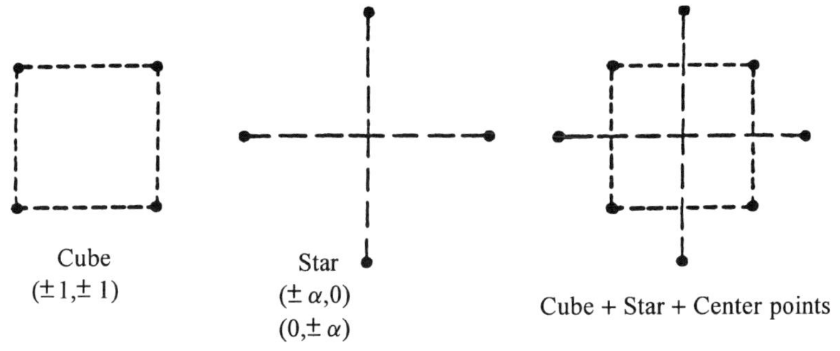

Fig. 5. Composite design for $k = 2$ variables.

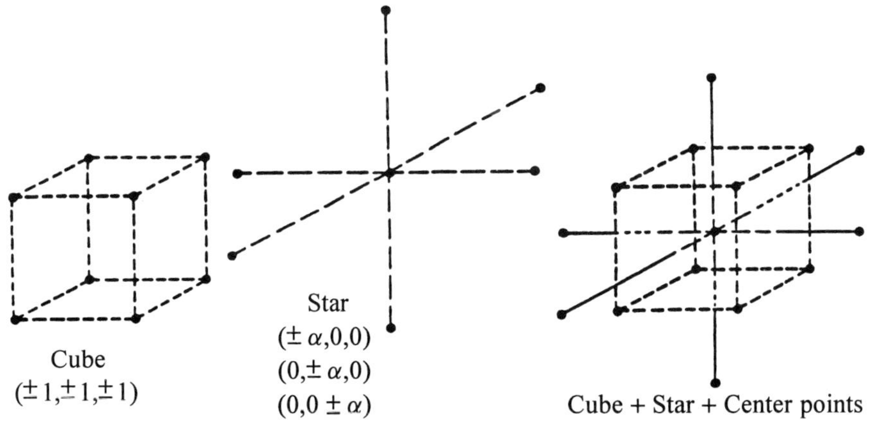

Fig. 6. Composite design for $k = 3$ variables.

Choices of α and n_0

What values should be chosen for α and n_0? The value of α determines if the star points fall inside the cube ($\alpha < 1$), outside the cube ($\alpha > 1$), or on the faces of the cube ($\alpha = 1$). Note that when $\alpha = 1$ only three experimental levels $(-1, 0, 1)$ are required, which may be an advantage or necessity in some experimental situations. For additional comments and specific designs see De Baun (1956) and Box and Behnken (1960), also discussed in Section 12.

If three levels are not essential, what value of α should be selected? One criterion that can be applied to decide this is that of *rotatability*. A design (of any order) is rotatable when the contours of the variance function $V\{\widehat{y}(\boldsymbol{x})\}$ are spheres about the origin in the k-dimensional factor space defined by variables $x_1, x_2, \ldots, x_k$. Box and Hunter (1957) showed that the required values (given in Table 3) are $\alpha = 2^{(k-p)/4}$, where $p = 0, 1$, or 2 according to the fraction of the cube used in the design.

Note that the rotatability property is specifically related to the codings chosen for the x's. It is usually assumed that these codings have been chosen in a manner that anticipates (roughly speaking) that one unit of change in any x will have about the same effect on the response variable. In such a case, obtaining equal information at the same radial distance in any direction (which is what rotatability implies) is clearly sensible. Codings are rarely perfect; the codings are adjusted in future designs as a result of information gained in current and past experiments. Exact rotatability is not a primary consideration. However, knowledge of the tabulated values provides a target to aim at, while attempting to satisfy other desirable design features.

How large a value should be selected for n_0? There are many possible criteria to apply; these are summarized by Draper (1982, 1984). The suggested values in the table appear to be sensible with respect to many criteria, the overall message being that only a few center points are usually needed. (Whenever α is chosen so that all the design points lie on a sphere, at least one center point is needed so that all of the coefficients can be individually estimated.) A few additional center points will do no harm. Nevertheless, additional runs are probably better used to duplicate selected noncentral design points, unless special considerations apply, as below. Repeated points spread over the design provide a check of the usual "homogeneous variance" assumption; see Box (1959) and Dykstra (1959, 1960).

For a wide variety of examples, see Box and Draper (1987).

8. Small composite designs

When experimentation is expensive, difficult or time-consuming, small designs might be appropriate, especially when an independent estimate of experimental error is available. Hartley (1959) pointed out that, for estimation of the quadratic surface, the cube portion of the composite design need not be of resolution V. It could be of resolution as low as III, provided that two-factor interactions were not aliased with other two-factor interactions. Hartley employed a smaller fraction of the 2^k factorial than is used in the original Box–Wilson designs and so reduced the total number of design points. Hartley's cubes may be designated *resolution* III*, meaning a design of resolution III but with no words of length four in the defining relation; see Draper and Lin (1990b). Hartley thus obtained minimal- or near-minimal-point second-order designs for $k = 2, 3, 4$ and 6. For $k = 5, 7, 9$, and higher numbers, there was then the possibility that a worthwhile improvement could be made.

Table 4
Numbers of cube points in some small composite designs

	Factors, k							
	3	4	5	6	7	8	9	10
Coefficients $p = (k+1)(k+2)/2$	10	15	21	28	36	45	55	66
Star points $2k$	6	8	10	12	14	16	18	20
Minimal points in cube	4	7	11	16	22	29	37	46
Box and Hunter (1957)	8	16	16	32	64	64	128	128
	(2^3)	(2^4)	(2_{V}^{5-1})	(2_{V}^{6-1})	(2_{V}^{7-1})	(2_{V}^{8-2})	(2_{V}^{9-2})	(2_{V}^{10-3})
Hartley (1959)	4	8	–	16	32	–	64	–
	(2_{III*}^{3-1})	(2_{III*}^{4-1})	–	(2_{III*}^{8-2})	(2_{III*}^{7-2})	–	(2_{III*}^{9-3})	–
Westlake (1965)	–	–	12	–	26	–	44	–
			$(3/8 \times 2^5)$	–	$(13/64 \times 2^7)$	–	$(11/128 \times 2^9)$	–
Draper (1985)	–	–	12	–	28	–	44	–
Minimal runs via Plackett and Burman	4	8	12	16	24	36	40	48
After elimination of repeat	4	8	11	16	22	30	38	46

Westlake (1965) provided a method for generating composite designs based on irregular fractions of the 2^k factorial system rather than using the complete factorials or regular fractions of factorials employed by Box and Wilson (1951) and Hartley (1959). Westlake gave designs for the following:

1. $k = 5$, based on a 3/8 fraction of the 2^5 factorial;
2. $k = 7$, based on a 13/64 fraction of the 2^7 factorial;
3. $k = 9$, based on a 11/128 fraction of the 2^9 factorial.

An alternative approach to obtaining small composite designs was used by Draper (1985), who employed columns of the Plackett and Burman designs rather than regular or irregular fractions. An advantage of this Plackett and Burman type of approach is that the designs are easy to construct. Specifically, (a) we can use, for the cube portion of the design, k columns of a Plackett and Burman (1946) design, and (b) where repeat runs exist, we can remove one of each set of duplicates if we wish to reduce the number of runs required.

Applying this method, Draper (1985) used 12-run, 28-run, and 44-run Plackett and Burman designs and obtained second-order response-surface designs with 22, 42, and 62 total runs (i.e., cube plus star points) for $k = 5, 7$, and 9, respectively. Deleting one of each duplicate pair gave 21 runs for $k = 5$ (a minimal-point design, beating Westlake's design by one run), 39 runs for $k = 7$ (again, one run fewer than Westlake's), and 60 runs for $k = 9$ (two runs fewer than Westlake's designs). A subsequent paper, Draper and Lin (1990a), provided new designs for $k = 7, 8, 9$, and 10, with improvements for cases $k = 7$ and $k = 9$.

Table 4 summarizes the major results related to fitting a second-order model, listing the cube points needed for the composite designs discussed previously. In all cases, center point and star points have been omitted from the table.

Comments on Table 4

Case $k = 3$. As discussed previously, the four-run Plackett and Burman design is a minimal-point design. It is equivalent to Hartley's design and is a 2^{3-1}_{III} design.

Case $k = 4$. The minimum possible number of cube points required is 7, so the eight-run Plackett and Burman design is considered. Columns (1, 2, 3, 6) give the highest D value, where $D = |X'X|^{1/p}/n$, known as the "information per point", is a popular measure of goodness for experimental designs. This 2^{4-1}_{III} design is equivalent to Hartley's design. There is one run more than the minimum number required in the cube.

Case $k = 5$. Five columns of the 12-run Plackett and Burman design are used, because 11 is the minimum possible number of cube points required. As Draper (1985, p. 174) showed, there are two basic types of designs, one with a repeat pair and one with a mirror-image pair. All other choices are equivalent to one of these. The columns (1, 2, 3, 4, 5) produce a mirror-image pair and the higher D value. The columns (1, 2, 3, 9, 11) produce a repeat pair, leading to a minimal-point design with 11 runs in the cube portion after removal of a duplicate run.

Case $k = 6$. Again, a minimal-point design is automatically obtained when six appropriate columns are chosen from a 16-run Plackett and Burman design. Based on the D criterion, the choice of columns (1, 2, 3, 4, 5, 14) is recommended. This is equivalent to Hartley's 2^{6-2}_{III} design.

Case $k = 7$. There are 36 coefficients to estimate and 14 star points. Thus a minimum of 22 cube points is required. The smallest Plackett and Burman design that can be used is thus the one with 24 runs. We wish to pick seven columns. There are 12 possible projection patterns, 5 of which produce nonsingular second-order $X'X$ matrices (see Draper and Lin, 1990a). The choice of columns (1, 2, 3, 5, 6, 7, 9) will give the highest D value. The choice of columns (1, 2, 5, 6, 7, 9, 10), however, will produce two repeat pairs, permitting the elimination of two runs, one from each pair. This minimal-point 22-run design is not only smaller than Hartley's 32-run design, but it is also smaller than Westlake's 26-run design.

Case $k = 8$. There are 45 coefficients to estimate, and 16 star points, so a minimum of 29 cube points is required. The 32-run Plackett and Burman design thus suggests itself. The choice of eight columns from this design constitutes a 2^{8-3} design. There is no 2^{8-3} design of resolution III*, however. (The table by Westlake (1965, p. 325) incorrectly suggested that there is.) Thus we use the 36-run Plackett and Burman design instead. Columns (1, 3, 4, 5, 6, 7, 8, 9) will give the highest D value. Columns (1, 3, 4, 6, 8, 10, 16, 17) will produce six repeat pairs, of which one run each can be eliminated to obtain only 30 runs in the cube portion, one run more than the minimum number required.

Case $k = 9$. There are 55 coefficients to estimate and 18 star points. Thus a minimum of 37 cube points is required. This suggests use of nine columns of the 40-run Plackett and Burman design. There are at least 50 different projection patterns (see Draper and Lin, 1990a). The highest D value found is obtained by choosing columns

Table 5
Columns that provide the highest relative D values found

k	p	n_{pb}	Columns chosen	Total points N
3	10	4	(1, 2, 3)	10
4	15	8	(1, 2, 3, 6)	16
5	21	12	(1, 2, 3, 4, 5)	22
6	28	16	(1, 2, 3, 4, 5, 14)	28
7	36	24	(1, 2, 3, 5, 6, 7, 9)	38
8	45	36	(1, 3, 4, 5, 6, 7, 8, 9)	52
9	55	40	(1, 2, 5, 6, 8, 21, 22, 23, 26)	58
10	66	48	(1, 2, 3, 4, 5, 6, 7, 11, 12, 25)	68

(1, 2, 5, 6, 8, 21, 22, 23, 26). Columns (1, 2, 3, 33, 34, 35, 36, 37, 38) provide two repeat pairs, however, in each of which one run could be eliminated to give a two-level design of 38 points. This compares with 128 runs for Box and Hunter (1957, p. 233), 64 runs for Hartley (1959), 44 runs for Westlake (1965, p. 331) and 44 runs for Draper (1985, p. 179).

Case $k = 10$. For $k = 10$ factors, the smallest 2_{V}^{k-p} design requires 128 runs. There are 66 coefficients to estimate and 20 star points, so a minimum of 46 cube points is required. The obvious choice is to try 10 columns of the 48-run Plackett and Burman design. At least 32 types of projected designs exist, and the highest D value among them is obtained by choosing the columns (1, 2, 3, 4, 5, 6, 8, 9, 17, 18). Choice of the columns (1, 4, 5, 7, 10, 11, 14, 16, 17, 20), however, produces two repeat pairs, permitting elimination of one run from each pair to obtain a minimal-point design.

Table 5 summarizes, for $3 \leqslant k \leqslant 10$, those column choices already described that provide the highest relative D values.

Repeat runs provide information on pure error. Some repeat runs can be eliminated, however, if reduction in the total number of runs is critical. For more details, see, Draper and Lin (1990a). Note that when repeats runs are eliminated, the orthogonality is lost, causing correlations among the estimates.

9. Orthogonal blocking of second order designs

When experiments are spread out over space or time or material or equipment, it is possible for extraneous changes to occur which affect the response values, over and above the effects induced by changes in the predictor variables. For example, test bread ovens may hold only a few loaves, so that several baking sessions are needed. Or, a run might take nearly two hours, so that only four runs are possible on an eight hour shift. Or, the raw material for the experiment may come from two manufacturers. Or the response values at given conditions might drift over time periods, or be affected by changing weather conditions. In all these cases, it is usually desirable to divide the whole experiment up into blocks of runs in such a way that the responses within a block are "consistent", apart, of course, from the effects of the predictor variables. In some circumstances, second order designs can be *orthogonally blocked*, that is, divided into two or more sections or blocks in such a manner that this split does not affect the estimates (of the second order model parameters) that are obtained from an ordinary least squares regression analysis. The necessary and sufficient conditions were given by Box and Hunter (1957); see also De Baun (1956) and Box (1959). They are:

1. Each block must, by itself, be a first-order orthogonal design. Thus for $i \neq j = 1, 2, \ldots, k$, $\sum_u x_{iu} x_{ju} = 0$, for each block.

2. The fraction of total sum of squares of each variable x_i contributed by every block must be equal to the fraction of the total observations allotted to the block. Thus, for each block,

$$\frac{\sum_u x_{iu}^2}{\sum_{u=1}^n x_{iu}^2} = \frac{n_b}{n}, \tag{4}$$

where n_b denotes the number of runs in the block under consideration, $\sum_u$ denotes summation *only in that block*, and the denominators of (4) refer to the entire design.

The simplest orthogonal block division of the composite design is into the orthogonal design pieces:

Block 1. Cube portion (2^{k-p} points) plus c_0 center points.
Block 2. Star portion ($2k$ points) plus s_0 center points.

Application of (4) then implies that

$$\alpha = \left\{2^{k-p-1}(2k + s_0)/(2^{k-p} + c_0)\right\}^{1/2}. \tag{5}$$

For example, if $k = 4$ and $p = 0$, so that the first block is a 2^4 factorial plus c_0 center points and the second block is an eight-point octahedron plus s_0 center points, then

$$\alpha = \left\{8(8 + s_0)/(16 + c_0)\right\}^{1/2} \tag{6}$$

achieves orthogonal blocking. Rotatability is achieved when $\alpha = 2^{(k-p)/4} = 2$. Substituting this in (6) shows that this design is both rotatable *and* orthogonally blocked whenever $c_0 = 2s_o$. The satisfaction of *both* criteria is not possible in general. Consider the case $k = 3$ and $p = 0$, so that the first block is a 2^3 factorial plus c_0 center points and the second block is a six-point octahedron plus s_0 center points. Then for orthogonal blocking, we need

$$\alpha = \left\{4(6 + s_0)/(8 + c_0)\right\}^{1/2}. \tag{7}$$

If $c_0 = 4$ center points are added to the cube and no center points are added to the star ($s_0 = 0$), then $\alpha = 2^{1/2} = 1.414$. This design is orthogonally blocked but is *not* rotatable. However, values of α closer to the rotatable value 1.682 are possible. For example, if $c_0 = 0$, $s_0 = 0$, and $\alpha = (24/8)^{1/2} = 1.732$ or if $c_0 = 4$, $s_0 = 2$, and $\alpha = (32/12)^{1/2} = 1.633$. The choices are, of course, limited by the fact that c_0 and s_0 must be integers. Generally, orthogonal blocking (α from (5)) takes precedence over rotatability, for which $\alpha = 2^{(k-p)/4}$ is needed. The general condition for both to be achieved simultaneously is

$$2^{k-p} + c_0 = 2^{0.5(k-p)-1}(2k + s_0) \tag{8}$$

for integer (k, p, c_0, s_0). Some possibilities are $(2, 0, s_0, s_0), s_0 \geqslant 1; (4, 0, 2s_0, s_0)$, already discussed below (6); $(5, 1, (4 + 2s_0), s_0); (7, 1, 4(s_0 - 2), s_0), s_0 \geqslant 2$; and $(8, 2, 4s_0, s_0)$, where $s_0 = 0, 1, 2, \ldots$, unless otherwise specified. (Note that some of these arrangements call for more center points than recommended in the table, an example of how applications of different criteria can produce conflicting conclusions.)

Further division of the star will not lead to an orthogonally blocked design. However, it is possible to divide the cube portion into smaller blocks and still maintain orthogonal blocking if $k > 2$. As long as the pieces that result are fractional factorials of resolution III or more (see Box et al., 1978, p. 385), each piece will be an orthogonal design. All fractional factorial pieces *must* contain the same number of center points or else (4) cannot be satisfied. Thus c_0 must be divisible by the number of blocks.

Replication of point sets

In a composite design, replication of either the cube portion or the star portion, or both can be chosen if desired. An attractive example of such possibilities is given by Box and Draper (1987, p. 362). This is a 24-run second-order design for three factors that is both rotatable *and* orthogonally blocked into four blocks of equal size. It consists of a cube (fractionated via $x_1x_2x_3 = \pm 1$) plus replicated (doubled) star plus four center points, two in each 2^{3-1} block. This particular design also provides an interesting example of estimating σ^2 in the situation where center points in *different* blocks of the design are no longer directly comparable due to possible block effects.

Obtaining the block sum of squares

When a second-order design is orthogonally blocked, one can

1. Estimate the β coefficients of the second-order model in the usual way, ignoring blocking.

2. Calculate pure error from repeated points *within* the same block only, and then combine these contributions in the usual way. Runs in different blocks cannot be considered as repeats.

3. Place an extra term

$$SS\text{ (blocks)} = \sum_{w=1}^{m} \frac{B_w^2}{n_w} - \frac{G^2}{n},$$

with $(m-1)$ degrees of freedom in the analysis of variance table, where B_w is the total of the n_w observations in the wth block and G is the grand total of all the observations in all the m blocks.

If a design is not orthogonally blocked, the sum of squares for blocks is conditional on terms taken out before it. An "extra" sum of squares calculation is needed; see Draper and Smith (1981).

10. Rotatability

Rotatability is a useful property of an experimental design. Any given design produces an X matrix whose columns are generated by the x-terms in the model to be fitted (e.g., (3)) and whose rows correspond to values from the n given design points. If z' is a vector of the form of a row of X but generated by a selected point at which a predicted response is required after estimation of the model's coefficients, then it can be shown that the variance of that prediction is $\mathrm{V}(\widehat{y}(x)) = z'(X'X)^{-1}z\sigma^2$ where σ^2 is the variance of an observed response value, assumed to be constant. For any given design, contours of $\mathrm{V}(\widehat{y}(x)) =$ constant can be plotted in the k-dimensional x-space. If those contours are *spherical*, the design is said to be *rotatable*. In practice, *exact* rotatability is not important, but it is a plus if the design is at least "close to being rotatable" in the sense that $\mathrm{V}\{\widehat{y}(x)\}$ changes little for points that are a constant distance from the origin in the region covered by the design points. For more on rotatability, see Box and Draper (1987).

To assess how close a design is to being rotatable, we can use a criterion of Draper and Pukelsheim (1990). We describe this in the context of second order designs, although the concept is completely general for any order. For easy generalization a special expanded notation is needed. Let $\boldsymbol{x} = (x_1, x_2, \ldots, x_k)'$. We shall denote the terms in the second-order model by a vector with elements

$$1;\ \boldsymbol{x}';\ \boldsymbol{x}' \otimes \boldsymbol{x}',$$

where the symbol $\otimes$ denotes the Kronecker product. Thus there are $(1+k+k^2)$ terms,

$$1;\ x_1, x_2, \ldots, x_k;\ x_1^2, x_1x_2, \ldots, x_1x_k;$$

$$x_2x_1, x_2^2, \ldots, x_2x_k; \ldots; x_kx_1, x_kx_2, \ldots, x_k^2.$$

(An obvious disadvantage of this notation is that all cross-product terms occur twice, so the corresponding $\boldsymbol{X}'\boldsymbol{X}$ matrix is singular. A suitable generalized inverse is obvious, however, and this notation is very easily extended to higher orders. For example, third order is added via $\boldsymbol{x}' \otimes \boldsymbol{x}' \otimes \boldsymbol{x}'$, and so on.)

Consider any second-order rotatable design with second-order moments $\lambda_2 = N^{-1}\sum_u x_{iu}^2$ and $\lambda_4 = N^{-1}\sum_u x_{iu}^2 x_{ju}^2$, for $i, j = 1, 2, \ldots, k$ and $i \neq j$. We can write its moment matrix $\boldsymbol{V}$ of order $(1+k+k^2) \times (1+k+k^2)$, in the form

$$\boldsymbol{V} = \boldsymbol{V}_0 + \lambda_2(3k)^{1/2}\boldsymbol{V}_2 + \lambda_4[3k(k+2)]^{1/2}\boldsymbol{V}_4, \tag{9}$$

where V_0 consists of a one in the $(1, 1)$ position and zeros elsewhere, where V_2 consists of $(3k)^{-1/2}$ in each of the $3k$ positions corresponding to pure second-order moments in $\boldsymbol{V}$ and zeros elsewhere, and V_4 consists of $3[3k(k+2)]^{-1/2}$ in the k positions corresponding to pure fourth-order moments, $[3k(k+2)]^{-1/2}$ in the $3k(k-1)$ positions corresponding to mixed even fourth-order moments in $\boldsymbol{V}$, and zeros elsewhere. Note that V_0, V_2, and V_4 are symmetric and orthogonal so that $V_iV_j = \mathbf{0}$, and also the V_i have norms $\| V_i \| = [\mathrm{tr}(V_iV_i)]^{1/2} = 1$.

Suppose we now take an arbitrary design with moment matrix $\boldsymbol{A}$, say. Draper et al. (1991) showed that, by averaging $\boldsymbol{A}$ over all possible rotations in the x space, we obtain

$$\overline{\boldsymbol{A}} = V_0 + V_2\mathrm{tr}(\boldsymbol{A}V_2) + V_4\mathrm{tr}(\boldsymbol{A}V_4). \tag{10}$$

We call $\overline{\boldsymbol{A}}$ the *rotatable component of* $\boldsymbol{A}$. The measure of rotatability is

$$\begin{aligned} Q^* &= \| \overline{\boldsymbol{A}} - V_0 \|^2 / \| \boldsymbol{A} - V_0 \|^2 \\ &= \{\mathrm{tr}(\overline{\boldsymbol{A}} - V_0)^2\}/\{\mathrm{tr}(\boldsymbol{A} - V_0)^2\}. \end{aligned} \tag{11}$$

The rotatability measure Q^* is essentially an R^2 statistic for the regression of the design moments of second and fourth order in $\boldsymbol{A}$ onto the "ideal" design moments represented by $\boldsymbol{V}$. Such a criterion is easy to compute and is invariant under design rotation. It enables us to say how rotatable a design is, and to improve the design's Q^* value by adding new design points. For examples, see Draper and Pukelsheim (1990).

11. Variance, bias and lack of fit

Suppose that $\mathrm{E}(\boldsymbol{y}) = \boldsymbol{\eta}(\boldsymbol{\xi})$ where $\boldsymbol{\xi}$ is a vector of predictor variables and let $\boldsymbol{f}(\boldsymbol{\xi})$ be the vector with polynomial elements used to approximate $\boldsymbol{y}$. We choose the form of f in the hope that it will provide a good approximation to η over some limited region of interest, R say. Two type of errors then need to be considered:

1. Systematic, or bias, errors $\boldsymbol{\delta}(\boldsymbol{\xi}) = \boldsymbol{\eta}(\boldsymbol{\xi}) - \boldsymbol{f}(\boldsymbol{\xi})$, the difference between the expected value of the response, $\mathrm{E}(\boldsymbol{y}) = \boldsymbol{\eta}(\boldsymbol{\xi})$ and the approximating function $\boldsymbol{f}(\boldsymbol{\xi})$.
2. Random errors $\boldsymbol{\varepsilon}$.

Although the above implies that systematic errors $\boldsymbol{\delta}(\boldsymbol{\xi})$ are always to be expected, they are often wrongly ignored. Yet it is only rarely true that bias can be totally ignored. Suppose that $\widehat{y}(\boldsymbol{\xi})$ is the fitted value obtained at a general point $\boldsymbol{\xi}$ in the experimental space, when the function $f(\boldsymbol{\xi})$ is fitted to available data on y and $\boldsymbol{\xi}$, then the associated mean square error, standardized for N, the number of observations and σ^2, the error variance is

$$\begin{aligned} &(N/\sigma^2)\mathrm{E}\{\widehat{y}(\boldsymbol{\xi}) - \eta(\boldsymbol{\xi})\}'\{\widehat{y}(\boldsymbol{\xi}) - \eta(\boldsymbol{\xi})\} \\ &\quad = (N/\sigma^2)\mathrm{E}\{\widehat{y}(\boldsymbol{\xi}) - \mathrm{E}\widehat{y}(\boldsymbol{\xi}) + \mathrm{E}\widehat{y}(\boldsymbol{\xi}) - \eta(\boldsymbol{\xi})\}^2 \\ &\quad = (N/\sigma^2)\mathrm{V}\{\widehat{y}(\boldsymbol{\xi})\} + (N/\sigma^2)\{\mathrm{E}\widehat{y}(\boldsymbol{\xi}) - \eta(\boldsymbol{\xi})\}^2 \end{aligned}$$

after some reduction. We can write this as

$$\mathrm{M}(\boldsymbol{\xi}) = \mathrm{V}(\boldsymbol{\xi}) + \mathrm{B}(\boldsymbol{\xi})$$

and describe it as "the standardized mean square error at a point $\boldsymbol{\xi}$ is equal to the variance $\mathrm{V}(\boldsymbol{\xi})$ of prediction plus the squared bias $\mathrm{B}(\boldsymbol{\xi})$". We can also make an assessment of variance and bias over any given region of interest R by averaging (and normalizing) $\mathrm{V}(\boldsymbol{\xi})$ and $\mathrm{B}(\boldsymbol{\xi})$ over R. More generally, if $\omega(\boldsymbol{\xi})$ is a weight function, we can write

$$V = \int \omega(\boldsymbol{\xi})\mathrm{V}(\boldsymbol{\xi})\,\mathrm{d}\boldsymbol{\xi} \Big/ \int \omega(\boldsymbol{\xi})\,\mathrm{d}\boldsymbol{\xi} \quad \text{and} \quad B = \int \mathrm{B}(\boldsymbol{\xi})\,\mathrm{d}\boldsymbol{\xi} \Big/ \int \omega(\boldsymbol{\xi})\,\mathrm{d}\boldsymbol{\xi}$$

and integrate it over the entire $\boldsymbol{\xi}$-space. Most often in practice we would have

$$\omega(\boldsymbol{\xi}) = \begin{cases} 1 & \text{within } R, \\ 0 & \text{outside } R, \end{cases}$$

whereupon V and B would represent integrals taken over R. If we denote the integrated mean squared error by M, we can write

$$M = V + B.$$

In practice, of course, the true relationship $\eta(\boldsymbol{\xi})$ would be unknown. To make further progress, we can proceed as follows:

1. Given that we are going to fit a polynomial of degree d_1 (say) to represent the function over some interval R, we can suppose that the true function $\eta(\boldsymbol{\xi})$ is a polynomial of degree d_2, greater than d_1.

2. We need also to say something about the *relative* magnitudes of systematic (bias) and random (variance) errors that we could expect to meet in practical cases. An investigator might typically employ a fitted approximating function such as a straight line, if he believed that the average departure from the truth induced by the approximating function were no worse than that induced by the process of fitting. We shall suppose this to be so, and will assume, therefore, that the experimenter will tend to choose the weight function $\omega(\boldsymbol{\xi})$, the size of his region R, and the degree of his approximating function in such a way that the integrated random error and the integrated systematic error are about equal. Thus we shall suppose that the situation of typical interest is that where B is roughly equal to V.

All-bias designs

If the problem of choosing a suitable experiment design is considered in the context described above, a major result can be deduced. An appropriate experimental design for an "average situation" when V and B are roughly equal has size roughly 10%

greater than the *all bias design*, appropriate when $V = 0$. This result is important because the moments of the all-bias design are easily determined. Suppose that we now work in terms of variables $\boldsymbol{x}$, where the x's are coded forms of the ξ's, and centered around the origin, a conventional step. Suppose further that a polynomial model of degree d_1

$$\widehat{y}(\boldsymbol{x}) = \boldsymbol{x}_1'\boldsymbol{b}_1$$

is fitted to the data, while the true model is a polynomial of degree d_2,

$$\eta(\boldsymbol{x}) = \boldsymbol{x}_1'\boldsymbol{\beta}_1 + \boldsymbol{x}_2'\boldsymbol{\beta}_2.$$

Thus, for the complete set of N data points

$$\widehat{y}(\boldsymbol{x}) = \boldsymbol{X}_1\boldsymbol{b}_1,$$

$$\eta(\boldsymbol{x}) = \boldsymbol{X}_1\boldsymbol{\beta}_1 + \boldsymbol{X}_2\boldsymbol{\beta}_2.$$

Quite often, it would reasonable to choose $d_2 = d_1 + 1$. Let us now write

$$\boldsymbol{M}_{11} = N^{-1}\boldsymbol{X}_1'\boldsymbol{X}_1, \qquad \boldsymbol{M}_{12} = N^{-1}\boldsymbol{X}_1'\boldsymbol{X}_2,$$

$$\boldsymbol{\mu}_{11} = \int_O w(\boldsymbol{x})\boldsymbol{x}_1\boldsymbol{x}_1'\,\mathrm{d}\boldsymbol{x}, \qquad \boldsymbol{\mu}_{12} = \int_O w(\boldsymbol{x})\boldsymbol{x}_1\boldsymbol{x}_2'\,\mathrm{d}\boldsymbol{x}.$$

It can now be shown that, whatever the values of $\boldsymbol{\beta}_1$ and $\boldsymbol{\beta}_2$, a necessary and sufficient condition for the squared bias B to be minimized is that

$$\boldsymbol{M}_{11}^{-1}\boldsymbol{M}_{12} = \boldsymbol{\mu}_{11}^{-1}\boldsymbol{\mu}_{12}.$$

A sufficient (but not necessary) condition for B to be minimised is that

$$\boldsymbol{M}_{11} = \boldsymbol{\mu}_{11} \quad \text{and} \quad \boldsymbol{M}_{12} = \boldsymbol{\mu}_{12}.$$

Now the elements of $\boldsymbol{\mu}_{11}$ and $\boldsymbol{\mu}_{12}$ are of the form

$$\int_O w(\boldsymbol{x})x_1^{\alpha_1}x_2^{\alpha_2}\cdots x_k^{\alpha_k}\,\mathrm{d}\boldsymbol{x}$$

and the elements of $\boldsymbol{M}_{11}$ and $\boldsymbol{M}_{12}$ are of the form

$$N^{-1}\sum_{u=1}^{N} x_{1u}^{\alpha_1}x_{2u}^{\alpha_2}\cdots x_{ku}^{\alpha_k}.$$

These typical elements are, respectively, moments of the weight function and moments of the design points of order

$$\alpha = \alpha_1 + \alpha_2 + \cdots + \alpha_k.$$

Thus, the sufficient condition above states that, up to and including order $d_1 + d_2$, all the moments of the design are equal to all the moments of the weight function.

EXAMPLE 1. Suppose we wish to fit a straight line $y = \beta_0 + \beta_1 x + \varepsilon$ to data to be taken over the region R, $-1 \leqslant x \leqslant 1$, where the weight function $\omega(x)$ is uniform within R and zero outside R. Suppose quadratic bias is slightly feared. Then the all-bias design is obtained when the design moments m_1, m_2, m_3, where

$$m_i = N^{-1} \sum_{u=1}^{N} x_u^i$$

are chosen to be $m_1 = m_3 = 0$, because $\mu_1 = \mu_3 = 0$, and

$$m_2 = \mu_2 = \int_{-1}^{1} x^2 \, dx \bigg/ \int_{-1}^{1} dx = \frac{1}{3}.$$

It follows that, if we use a three-site, three-point design at positions $x = -a, 0, a$, we must choose $2a^2/3 = 1/3$ or $a = 2^{-1/2} = 0.707$. For a typical case where $V = B$ roughly, we could increase a slightly to (say) 0.75 or 0.80, about 10% or so.

EXAMPLE 2. In k dimensions, fitting a plane and fearing a quadratic, with R the unit sphere, an all bias design is a 2_{III}^{k-p} design with points $(\pm a, \pm a, \ldots, \pm a)$ such that

$$2^{k-p} a^2 / n = k/(k+2)$$

which implies, if n_0 center points are used that

$$a = \left\{ \frac{k(2^{k-p} + n_0)}{(k+2)2^{k-p}} \right\}^{1/2}.$$

Note that the special case $k = 1, p = 0, n_0 = 1$ is Example 1. For $k = 4, p = 0, n_0 = 2$, we have $a = 0.866$ for the all bias case. Note that this places the factorial points at distances $r = (4a^2)^{1/2} = 3^{1/2}$ from the origin, that is, outside R.

Detecting lack of fit

Consider the mechanics of making a test of goodness of fit using the analysis of variance. Suppose we are estimating p parameters, observations are made at $p + f$ distinct points, and repeated observations are made at certain of these points to provide e pure error degrees of freedom, so that the total number of observations is $N = p + f + e$.

The expectation of the unbiased pure error mean square is σ^2, the experimental error variance, and the expected value of the lack of fit mean square equals $\sigma^2 + \Lambda^2/f$ where Λ^2 is a noncentrality parameter. The test for goodness of fit is now made by comparing the mean square for lack of fit against the mean square for pure error, via an $F(f, e)$ test.

In general, the noncentrality parameter takes the form

$$\Lambda^2 = \sum_{u=1}^{N} \{\mathrm{E}(\widehat{y}_u) - \eta_u\}^2 = \mathrm{E}(S_L) - f\sigma^2, \tag{12}$$

where S_L is the lack of fit sum of squares. Thus, good detectability of general lack of fit can be obtained by choosing a design that makes Λ^2 large. It turns out that this requirement of good detection of model inadequacy can, like the earlier requirement of good estimation, be achieved by certain conditions on the design moments. Thus, under certain sensible assumptions, it can be shown that a $(d_1 =)$ dth order design would provide high detectability for terms of order $(d_2 =)(d+1)$ if (1) all odd design moments of order $(2d+1)$ or less are zero, and (2) the ratio

$$\frac{N^d \sum_{u=1}^{N} r_u^{2(d+1)}}{\left\{\sum_{u=1}^{N} r_u^2\right\}^{d+1}} \tag{13}$$

is large, where

$$r_u^2 = x_{1u}^2 + x_{2u}^2 + \cdots + x_{ku}^2.$$

In particular, this would require that, for a first-order design $(d = 1)$ the ratio $N\sum r_u^4/\{\sum r_u^2\}^2$ should be large to provide high detectability of quadratic lack of fit; for a second-order design $(d = 2)$, the ratio $N^2 \sum r_{iu}^6/\{\sum r_{iu}^2\}^3$ should be large to provide high detectability of cubic lack of fit.

Note that, for the 2^{k-p} design of Example 2 above, the detectability criterion is independent of the size of a, which cancels out. Increasing the number of center points slightly increases detectability, however, since this is determined by contrasting the factorial point average response minus the center point average response.

Further reading

For additional commentary, see Chapter 13 of Box and Draper (1987) and Chapter 6 of Khuri and Cornell (1987). Related work includes Draper and Sanders (1988), DuMouchel and Jones (1994), and Wiens (1993).

12. Some other second order designs

The central composite design is an excellent design for fitting a second-order response surface, but there are also other useful designs available. We now mention some of these briefly.

The 3^k factorial designs

The 3^k factorial design series consists of all possible combinations of three levels of k input variables. These levels are usually coded to $-1, 0$, and 1. For the $k = 2$ case, the design matrix is

$$D = \begin{array}{c} \begin{matrix} x_1 & x_2 \end{matrix} \\ \begin{bmatrix} -1 & -1 \\ 0 & -1 \\ 1 & -1 \\ -1 & 0 \\ 0 & 0 \\ 1 & 0 \\ -1 & 1 \\ 0 & 1 \\ 1 & 1 \end{bmatrix} \end{array}$$

Such a design can actually be used to fit a model of form $E(y) = \beta_0 + \beta_1 X_1 + \beta_2 X_2 + \beta_{11} X_1^2 + \beta_{22} X_2^2 + \beta_{12} X_1 X_2 + \beta_{122} X_1 X_2^2 + \beta_{112} X_1^2 X_2 + \beta_{1122} X_1^2 X_2^2 + \varepsilon$, although the cubic and quartic terms would usually be associated with "error degrees of freedom".

To reduce the total number of experimental design points when k is large, fractional replications 3^{k-p} would often be employed if the number of runs were enough to fit the full second-order model. An extended table of fractional 3^{k-p} designs is given by Connor and Zelen (1959).

The Box–Behnken designs

The Box–Behnken (1960) designs were constructed for situations in which it was desired to fit a second-order model (3), but only three levels of each predictor variable $x_1, x_2, \ldots, x_k$, coded to $-1, 0$, and 1, could be used. The design points are carefully chosen subsets of the points of 3^k factorial designs, and are generated through balanced incomplete block designs (BIBD) or partially balanced incomplete block designs (PBIBD). They are available for k = 3–7, 9–12, and 16. (See Table 6 for k = 3–7.) They are either rotatable (for k = 4 and 7) or close to rotatable. Except for the designs for which $k = 3$ and 11, all can be blocked orthogonally. The designs

Table 6
The Box–Behnken (1960) designs, $3 \leqslant k \leqslant 7$

Number of factors k	Design matrix	No. of points	Blocking and association schemes
3	±1 ±1 0		No orthogonal blocking
	±1 0 ±1	12	BIB (one associate class)
	0 ±1 ±1		
	0 0 0	3	
		$N = 15$	
4	±1 ±1 0 0	8	3 blocks of 9
	0 0 ±1 ±1		BIB (one associate class)
	0 0 0 0	1	
	±1 0 0 ±1	8	
	0 ±1 ±1 0		
	0 0 0 0	1	
	±1 0 ±1 0	8	
	0 ±1 0 ±1		
	0 0 0 0	1	
		$N = 27$	
5	±1 ±1 0 0 0		2 blocks of 23
	0 0 ±1 ±1 0		BIB (one associate class)
	0 ±1 0 0 ±1	20	
	±1 0 ±1 0 0		
	0 0 0 ±1 ±1		
	0 0 0 0 0	3	
	0 ±1 ±1 0 0		
	±1 0 0 ±1 0		
	0 0 ±1 0 ±1	20	
	±1 0 0 0 ±1		
	0 ±1 0 ±1 0		
	0 0 0 0 0	3	
		$N = 46$	
6	±1 ±1 0 ±1 0 0		2 blocks of 27
	0 ±1 ±1 0 ±1 0		First associates:
	0 0 ±1 ±1 0 ±1	48	$(1,4); (2,5); (3,6)$
	±1 0 0 ±1 ±1 0		
	0 ±1 0 0 ±1 ±1		
	±1 0 ±1 0 0 ±1		
	0 0 0 0 0 0	6	
		$N = 54$	
7	0 0 0 ±1 ±1 ±1 0		2 blocks of 31
	±1 0 0 0 0 ±1 ±1		BIB (one associate class)
	0 ±1 0 0 ±1 0 ±1		
	±1 ±1 0 ±1 0 0 0	56	
	0 0 ±1 ±1 0 0 ±1		
	±1 0 ±1 0 ±1 0 0		
	0 ±1 ±1 0 0 ±1 0		
	0 0 0 0 0 0 0	6	
		$N = 62$	

Table 7
Rechtschaffner's (1967) point sets

Number	Points	Design generator (point set)	Typical point
I	1	$(+1, +1, \ldots, +1)$ or $(-1, -1, \ldots, -1)$	$(+1, +1, \ldots, +1)$
II	k	One $+1$ and all other -1	$(+1, -1, \ldots, -1)$
III	$k(k-1)/2$	Two $+1$ and all other -1	$(+1, +1, -1, \ldots, -1)$
IV	k	One $+1$ and all other 0	$(+1, 0, \ldots, 0)$

Table 8
Point sets of Box and Draper (1972, 1974)

Number	Points	Design generator (point set)	Typical point
I	1	$(+1, +1, \ldots, +1)$ or $(-1, -1, \ldots, -1)$	$(-1, -1, \ldots, -1)$
II	k	One $+1$ and all other -1	$(+1, -1, \ldots, -1, \ldots, -1)$
III	$k(k-1)/2$	Two λ and all other -1	$(\lambda, \lambda, -1, \ldots, -1)$
IV	k	One μ and all other 1	$(\mu, 1, \ldots, 1)$

have a relatively modest number of runs compared to the number of parameters in the corresponding second-order models. For additional appreciation of the usefulness of these designs, see Draper, Davis, Pozueta and Grove (1994).

Some minimal-point second-order designs

Lucas (1974) gave minimal-point designs not of composite type that he called "smallest symmetric composite designs," which consist of one center point, $2k$ star points, and $\binom{k}{2}$ "edge points." An edge point is a $k \times 1$ vector having ones in the ith and jth location and zeros elsewhere. Note that the edge point designs do not contain any two-level factorial points.

Rechtschaffner (1967) used four different so-called *design generators* (actually point sets) to construct minimal-point designs for estimating a second-order surface (see Table 7). The signs of design generators I, II, and III can be varied (e.g., we may have one -1 and all other $+1$ in design generator II, say). Rechtschaffner's designs are available for $k = 2, 3, 4, \ldots$, but, as pointed out out by Notz (1982), they have an asymptotical D efficiency of 0 as $k \to \infty$ with respect to the class of saturated designs.

Box and Draper (1971, 1974) provided other minimal-point designs for $k = 2, 3, 4$, and 5, made up from the design generators (point sets) shown in Table 8. Values for λ and μ were tabulated in the 1974 article. Kiefer, in unpublished correspondence, established, via an existence result, that this type of design cannot be optimal for $k \geqslant 7$, however. Box and Draper's designs were given for $k \leqslant 5$, though they can be generated for any k.

Mitchell and Bayne (1976) used a computer algorithm called DETMAX that Mitchell (1974) developed earlier to find an n-run design that maximizes $|\boldsymbol{X}'\boldsymbol{X}|$, given n, a specified model, and a set of "candidate" design points. For each value of $k = 2, 3, 4$, and 5, they ran the algorithm 10 times, each time starting with a different randomly selected initial n-run design. The algorithm then improved the starting design by adding or removing points according to a so-called "excursion" scheme until no further improvement was possible.

Notz (1982) studied designs for which $p = n$. He partitioned $\boldsymbol{X}$ so that

$$\boldsymbol{X} = \begin{bmatrix} \boldsymbol{Z}_1 \\ \boldsymbol{Z}_2 \end{bmatrix} = \begin{bmatrix} \boldsymbol{Y}_{11} & \boldsymbol{Y}_{12} \\ \boldsymbol{Y}_{21} & \boldsymbol{Y}_{22} \end{bmatrix},$$

where $\boldsymbol{Z}_1$ is $(p-k) \times p$ and $\boldsymbol{Z}_2$ is $(p-k) \times k$. Note that $\boldsymbol{Y}_{11}$ is $(p-k) \times (p-k)$, $\boldsymbol{Y}_{12}$ is $(p-k) \times k$, $\boldsymbol{Y}_{21}$ is $k \times (p-k)$, and $\boldsymbol{Y}_{22}$ is $k \times k$, and we can think of $\boldsymbol{Z}_1$ as representing the cube points and $\boldsymbol{Z}_2$ the star points; $\boldsymbol{Y}_{12}$ over $\boldsymbol{Y}_{22}$ consists of the columns $(x_1^2, x_2^2, \ldots, x_k^2)$. Thus (a) all elements in $\boldsymbol{Y}_{11}$ are either $+1$ or -1, (b) all elements in $\boldsymbol{Y}_{22}$ are either 1 or 0, and, more important, (c) all elements in $\boldsymbol{Y}_{12}$ are $+1$. It follows that $|\boldsymbol{X}| = |\boldsymbol{X}'\boldsymbol{X}|^{1/2} = |\boldsymbol{Y}_{11}| \cdot |\boldsymbol{Y}_{22} - \boldsymbol{J}_{k,k}|$, where $\boldsymbol{J}_{k,k}$ is a $k \times k$ matrix with all of its elements equal to 1. Maximization of $|\boldsymbol{X}'\boldsymbol{X}|$ is now equivalent to maximization of $|\boldsymbol{Y}_{11}|$ and $|\boldsymbol{Y}_{22} - \boldsymbol{J}_{k,k}|$ separately. Notz found new saturated designs for $k \leqslant 5$ and extended his result to the $k = 6$ case.

Most of the minimal-point designs available for $k \geqslant 7$ comprise the extensions of Lucas's (1974) or Rechtschaffner's (1967) or Box and Draper's (1971, 1974) designs. Minimal-point designs can also be obtained by using the methods given in Draper (1985) and Draper and Lin (1990a), employing projections of Plackett and Burman designs for $k = 3, 5, 6, 7$, and 10. See Section 8. Their main virtues are that they are easy to construct and of composite form, providing orthogonal or near orthogonal designs and including other previously known small composite designs as special cases. For other designs and related considerations, see Khuri and Cornell (1987). A comparison of all the designs we have discussed in this section is made in Draper and Lin (1990a).

References

Bates, D. M. and D. G. Watts (1988). *Nonlinear Regression Analysis and Its Applications.* Wiley, New York.

Box, G. E. P. (1952). Multifactor designs of first order. *Biometrika* **39**, 40–57. (See also p. 189, note by Tocher.)

Box, G. E. P. (1959). Answer to query: Replication of non-center points in the rotatable and near-rotatable central composite design. *Biometrics* **15**, 133–135.

Box, G. E. P. and D. W. Behnken (1960). Some new three level designs for the study of quantitative variables. *Technometrics* **2**, 455–475.

Box, G. E. P. and S. Bisgaard (1993). What can you find out from 12 experimental runs. *Quality Eng.* **4**.

Box, G. E. P. and N. R. Draper (1959). A basis for the selection of a response surface design. *J. Amer. Statist. Assoc.* **54**, 622–654.

Box, G. E. P. and N. R. Draper (1963). The choice of a second order rotatable design. *Biometrika* **50**, 335–352.

Box, G. E. P. and N. R. Draper (1987). *Empirical Model-Building and Response Surfaces*. Wiley, New York.

Box, G. E. P. and J. S. Hunter (1957). Multifactor experimental designs for exploring response surfaces. *Ann. Math. Statist.* **28**, 195–241.

Box, G. E. P. and J. S. Hunter (1961). The 2^{k-p} fractional factorial designs, Parts I and II. *Technometrics* **3**, 311–351 and 449–458.

Box, G. E. P., W. G. Hunter and J. S. Hunter (1978). *Statistics for Experimenters*. Wiley, New York.

Box, G. E. P. and K. B. Wilson (1951). On the experimental attainment of optimum conditions. *J. Roy. Statist. Soc. Ser. B* **13**, 1–38, discussion 38–45.

Box, M. J. and N. R. Draper (1971). Factorial designs, the $|\boldsymbol{X}'\boldsymbol{X}|$ criterion and some related matters. *Technometrics* **13**, 731–742. Corrections **14** (1972), 511 and **15** (1973), 430.

Box, M. J. and N. R. Draper (1974). Some minimum point designs for second order response surfaces. *Technometrics* **16**, 613–616.

Connor, W. S. and M. Zelen (1959). Fractional factorial experimental designs for factors at three levels. U.S. Dept. of Commerce, National Bureau of Standards, Applied Math. Series No. 54.

Davies, O. L., ed. (1978). *Design and Analysis of Industrial Experiments*. Longman Group, New York.

De Baun, R. M. (1956). Block effects in the determination of optimum conditions. *Biometrics* **12**, 20–22.

De Baun, R. M. (1959). Response surface designs for three factors at three levels. *Technometrics* **1**, 1–8.

Draper, N. R. (1982). Center points in response surface designs. *Technometrics* **24**, 127–133.

Draper, N. R. (1984). Schläflian rotatability. *J. Roy. Statist. Soc. Ser. B* **46**, 406–411.

Draper, N. R. (1985). Small composite designs. *Technometrics* **27**, 173–180.

Draper, N. R., T. P. Davis, L. Pozueta and D. M. Grove (1994). Isolation of degrees of freedom for Box–Behnken designs. *Technometrics* **36**, 283–291.

Draper, N. R. and H. Smith (1981). *Applied Regression Analysis*. Wiley, New York.

Draper, N. R. and D. K. J. Lin (1990a). Small response surface designs. *Technometrics* **32**, 187–194.

Draper, N. R. and D. K. J. Lin (1990b). Connections between two-level designs of resolution III^* and V. *Technometrics* **32**, 283–288.

Draper, N. R., N. Gaffke and F. Pukelsheim (1991). First and second-order rotatability of experimental designs, moment matrices, and information functions. *Metrika* **38**, 129–161.

Draper, N. R. and F. Pukelsheim (1990). Another look at rotatability. *Technometrics* **32**, 195–202.

Draper, N. R. and E. R. Sanders (1988). Designs for minimum bias estimation. *Technometrics* **30**, 319–325.

DuMouchel, W. and B. Jones (1994). A simple Bayesian modification of D-optimal designs to reduce dependence on an assumed model. *Technometrics* **36**, 37–47.

Dykstra, O. (1959). Partial duplication of factorial experiments. *Technometrics* **1**, 63–75.

Dykstra, O. (1960). Partial duplication of response surface designs. *Technometrics* **2**, 185–195.

Hall, M. J. (1961). Hadamard matrix of Order 16. Jet Propulsion Laboratory Summary **1**, 21–26.

Hall, M. J. (1965). Hadamard matrix of Order 20. Jet Propulsion Laboratory Technical Report **1**, 32–76.

Hartley, H. O. (1959). Small composite designs for quadratic response surfaces. *Biometrics* **15**, 611–624.

Herzberg, A. M. (1982). The robust design of experiments: A review. *Serdica Bulgaricae Math. Publ.* **8**, 223–228.

Khuri, A. I. and J. A. Cornell (1987). *Response Surfaces, Designs and Analyses*. Marcel Dekker/ASQC Quality Press, New York/Milwaukee.

Lin, D. K. J. and N. R. Draper (1991). Projection properties of Plackett and Burman designs. Tech. Report 885, Department of Statistics, University of Wisconsin.

Lin, D. K. J. and N. R. Draper (1992). Projection properties of Plackett and Burman designs. *Technometrics* **34**, 423–428.

Lin, D. K. J. and N. R. Draper (1995). Screening properties of certain two level designs. *Metrika* **42**, 99–118.

Lucas, J. M. (1974). Optimum composite designs. *Techometrics* **16**, 561–567.

Mitchell, T. J. (1974). An algorithm for the construction of D-optimal experimental designs. *Technometrics* **16**, 203–210.

Mitchell, T. J. and C. K. Bayne (1978). D-optimal fractions of three-level fractional designs. *Technometrics* **20**, 369–380, discussion 381–383.

Notz, W. (1982). Minimal point second order designs. *J. Statist. Plann. Inference* **6**, 47–58.

Plackett, R. L. and J. P. Burman (1946). The design of optimum multifactorial experiments. *Biometrika* **33**, 305–325.

Rechtschaffner, R. L. (1967). Saturated fractions of 2^n and 3^n fractional designs. *Technometrics* **9**, 569–575.

Seber, G. A. F. and C. J. Wild (1989). *Nonlinear Regression*. Wiley, New York.

Wang, J. C. and C. F. J. Wu (1995). A hidden projection property of Plackett–Burman and related designs. *Statist. Sinica* **5**, 235–250.

Welch, W. J. (1984). Computer-aided design of experiments for response estimation. *Technometrics* **26**, 217–224.

Westlake, W. J. (1965). Composite designs based on irregular fractions of factorials. *Biometrics* **21**, 324–336.

Wiens, D. P. (1993). Designs for approximately linear regression: Maximizing the minimum coverage probability of confidence ellipsoids. *Canad. J. Statist.* **21**, 59–70.

S. Ghosh and C. R. Rao, eds., *Handbook of Statistics, Vol. 13*

12

Multiresponse Surface Methodology

André I. Khuri

1. Introduction

The formal development of response surface methodology (RSM) was initiated by the work of Box and Wilson (1951), which introduced the sequential approach in an experimental investigation. This particular approach became the cornerstone and one of the characteristic trademarks of RSM. It was effectively utilized in many applications, particularly in the chemical industry. The article by Myers et al. (1989) provides a broad review of RSM. Earlier, Hill and Hunter (1966) emphasized practical applications of RSM in the chemical industry. This was followed by another review article by Mead and Pike (1975) where the emphasis was on biological applications of RSM. In addition to these review articles, the four books by Box and Draper (1987), Khuri and Cornell (1996), Myers (1976), and Myers and Montgomery (1995) give a comprehensive coverage of the various techniques used in RSM.

In the early development of RSM, only single-response variables were considered. Quite often, however, several response variables may be of interest in an experimental situation. By definition, an experiment in which a number of response variables are measured for each setting of a group of input (control) variables is called a *multiresponse experiment*. The analysis of such experiments did not receive much attention until the publication of Zellner (1962) and Box and Draper (1965). Both articles addressed the problem of estimation of parameters for several response models. This, however, does not mean that no multiresponse experiments were performed prior to 1962. Lind et al. (1960), for example, considered a two-response experiment in an attempt to determine conditions that were favorable to both responses.

Data obtained in a multiresponse experiment (multiresponse data) are multivariate in character. It is therefore necessary that multivariate techniques be deployed for the analysis of such data. This way, information from several response variables can be combined. As we shall see later in this chapter, the multivariate approach has several advantages over the univariate approach. The latter is based on one-response-at-a-time treatment of a multiresponse system. This amounts to a total disregard of the multivariate nature of the experiment. On the other hand, the former enables us to gain a better understanding of the underlying mechanism as well as acquire information about any relationships that may exist among the response variables. This added information can be instrumental in providing more precise estimates of optimal process conditions, for example, and more accurate analyses of the multiresponse data.

The purpose of this chapter is to provide a thorough coverage of the basic methods used in RSM for the design and analysis of multiresponse experiments. An earlier review of this subject was given in Khuri (1990a). The present chapter is more up-to-date and more extensive in scope and coverage. It should therefore be useful to those who have an interest in RSM, and would like to explore extending its applicability to multiresponse situations. It is hoped that the information given in this chapter will provide a stimulus to the reader to pursue further research in the multiresponse area.

The main topics covered in this chapter include the following:

- Plotting of multiresponse data.
- Estimation of parameters of a multiresponse model.
- Inference for multiresponse models.
- Designs for multiresponse models.
- Multiresponse optimization.

2. Plotting of multiresponse data

A graphical display of a data set provides a visual perception of its structure. It can convey a variety of information and may reveal some salient features of the data. Plotting is therefore an effective tool in data analysis. This is evidently true in two-dimensional and, to a lesser extent, three-dimensional plots. For example, in regression analysis, a plot of residuals is a useful diagnostic tool for checking the validity of assumptions made about the data and the fitted model.

While two-dimensional and three-dimensional plots are easy to visualize and interpret, the same is not true in higher dimensions. For example, it is difficult to form a mental image of a four-dimensional scatter plot. Yet, if we were to acquire a better insight and extract more information from a multiresponse data set, an efficient scheme for producing a multidimensional plot would be essential (see Tukey, 1962, Section 4).

There are several techniques for graphing multiresponse data. Some techniques plot projections of the data on subspaces of dimensions three or fewer. For example, if variables are considered two at a time, pairs of values can be obtained and used to produce two-dimensional scatter plots. This generates an array of plots that can be arranged in a *scatter plot matrix* (see Cleveland, 1993, Chapter 5), or in a so-called *generalized draftsman's display* (see Chambers et al., 1983, Section 5.4). Other techniques use symbolic coding schemes. For example, Anderson (1960) proposed using small circles of fixed radius, instead of points in a two dimensional scatter plot, which he called *glyphs*. Several rays of various lengths emanate from each glyph representing values of different variables, which may be quantitative or qualitative (e.g., low, medium, high). Anderson extended the use of glyphs to situations other than scatter diagrams and subsequently used the more general term *metroglyphs*. Friedman et al. (1972) introduced a similar coding scheme called *stars* where the rays of a glyph are joined up to form polygons. Several other variants of the glyph technique are reported in Fienberg (1979). Chernoff (1973) used cartoon *faces* to represent data values by means of different facial features such as the length of a nose, shape of a smile, and size of eyes. This technique is not easy to use or interpret and has several major problems that were pointed out by Fienberg (1979). Kleiner and Hartigan (1981) introduced a method in which each point is represented by a *tree*. Variables are assigned in a particular manner to the branches of a tree whose lengths determine the values of the variables. Additional examples of symbolic plotting can be found in Everitt (1978), Tukey and Tukey (1981), and Chambers et al. (1983).

Other techniques for viewing multidimensional data are more mathematical in nature. They are based on projecting the data orthogonally onto low-dimensional subspaces, for example, of dimension two. Prominent among such techniques are *projection pursuit* by Friedman and Tukey (1974) and Huber (1985) (see also Jones and Sibson, 1987), and the *grand tour* by Asimov (1985). In the latter, a sequence of projections, chosen to be dense in the set of all projections, are selected for viewing. These techniques can lead to important insights concerning data. It is difficult, however, to interpret and draw conclusions regarding the structure of data from knowledge of projections only. A great deal of experience is therefore needed in order to extract meaningful information from the projections.

Another scheme for multidimensional plotting is based on representing each point by means of a curve or a broken line in a two- dimensional space. Andrews (1972) proposed a method for representing a point $\boldsymbol{x} = (x_1, x_2, \ldots, x_k)'$ in a k-dimensional space by a finite series of the form

$$f_{\boldsymbol{x}}(t) = \frac{1}{\sqrt{2}} x_1 + x_2 \sin t + x_3 \cos t + x_4 \sin 2t + x_5 \cos 2t + \cdots. \tag{2.1}$$

Thus $\boldsymbol{x}$ is mapped into the function $f_{\boldsymbol{x}}(t)$, which is then plotted over the interval $-\pi < t < \pi$. We note that the k terms in the sequence of functions,

$$\{g_1(t), g_2(t), \ldots, g_k(t)\} = \left\{\frac{1}{\sqrt{2}}, \sin t, \cos t, \sin 2t, \cos 2t, \ldots\right\}, \tag{2.2}$$

are orthonormal on $(-\pi, \pi)$ in the sense that

$$\int_{-\pi}^{\pi} g_i(t) g_j(t)\, \mathrm{d}t = 0, \quad i \neq j,$$

$$\frac{1}{\pi}\int_{-\pi}^{\pi} g_i^2(t)\, \mathrm{d}t = 1, \quad i = 1, 2, \ldots, k.$$

We also note that $f_{\boldsymbol{x}}(t)$ is the dot product $\boldsymbol{x}'\boldsymbol{g}(t)$, where $\boldsymbol{g}(t)$ is a vector consisting of the terms of the sequence in (2.2). Thus for a specified value of t, say t_0,

$$f_{\boldsymbol{x}}(t_0) = ||\boldsymbol{g}(t_0)||\, ||\boldsymbol{x}|| \cos \phi_{\boldsymbol{x}, t_0},$$

where $|| \cdot ||$ denotes the Euclidean norm of a vector and $\phi_{\boldsymbol{x}, t_0}$ is the angle between the vectors $\boldsymbol{x}$ and $\boldsymbol{g}(t_0)$. Hence, the absolute value of $f_{\boldsymbol{x}}(t_0)$ is proportional to the length of the orthogonal projection of $\boldsymbol{x}$ on the vector $\boldsymbol{g}(t_0)$. It follows that if we have a data set consisting of n points, $\boldsymbol{x}_1, \boldsymbol{x}_2, \ldots, \boldsymbol{x}_n$, in a k-dimensional space, then the values of $|f_{\boldsymbol{x}_i}(t_0)|$ are proportional to the lengths of the orthogonal projections of the data points on a one-dimensional subspace determined by $\boldsymbol{g}(t_0)$. Several such one-dimensional views of the data can be obtained by varying the value of t_0 over the interval $(-\pi, \pi)$. Andrews' plots can therefore be classified along with the *grand tour* and the *projection pursuit* techniques.

Andrews (1972) outlined several useful statistical properties of the function $f_{\boldsymbol{x}}(t)$. These properties are also described in Gnanadesikan (1977, Section 6.2), which gives a detailed discussion and several illustrations of this class of plots. It should be noted that Andrews' plots depend on the order of the variables in formula (2.1). They also depend on the scale of the variables since large values of x_i can mask the effects of other smaller values. Embrechts and Herzberg (1991) investigated the effects of reordering and rescaling of the variables on the plots.

Another related plotting technique is the *method of parallel coordinates*, which was suggested by Wegman (1990). He proposed to draw the coordinate axes in a k-dimensional Euclidean space as parallel. Using this scheme, a point such as $\boldsymbol{x} = (x_1, x_2, \ldots, x_k)'$ is represented by plotting x_1 on the first axis, x_2 on the second axis, $\ldots, x_k$ on the kth axis. The resulting points on the k axes are then connected by a broken line. Thus each point in a data set is mapped into a broken line in a two-dimensional space. Gennings et al. (1990) used this method to plot the dose- response surface and its contours of constant response for a combination of drugs.

It can be seen that the available techniques for plotting multidimensional data are, in general, not easy to implement and interpret, even by an expert user. Advances in computer technology should make this task easier. In this respect, it can be stated that Andrews' plots are considered to be easier to use and interpret than most existing techniques.

Plotting of multiresponse data serves as an exploratory tool in a response surface investigation. It is not, however, a substitute for a formal analysis of the data. The next four sections provide a broad coverage of methods available for the design and analysis of multiresponse experiments. More specifically, Section 3 addresses the fitting of a multiresponse model. Section 4 discusses inference making procedures, mainly for linear multiresponse models. Section 5 describes several criteria that can be used for the choice of a multiresponse design. Finally, Section 6 is concerned with the problem of multiresponse optimization. Some concluding remarks are given in Section 7.

3. Estimation of parameters of a multiresponse model

Suppose that a number of response variables are measured for each setting of a group of k input variables denoted by $x_1, x_2, \ldots, x_k$. Let $\{y_{ui}\}$ $(i = 1, 2, \ldots, r;$ $u = 1, 2, \ldots, n)$ represent n sets of observations on each of r response variables denoted by $y_1, y_2, \ldots, y_r$. The setting of x_i at the uth experimental run is denoted by x_{ui} $(i = 1, 2, \ldots, k;\ u = 1, 2, \ldots, n)$. The relationship between the response variables and the input variables is given by the model

$$y_{ui} = f_i(\boldsymbol{x}_u, \boldsymbol{\theta}) + \varepsilon_{ui}, \quad i = 1, 2, \ldots, r;\ u = 1, 2, \ldots, n, \tag{3.1}$$

where $\boldsymbol{x}_u = (x_{u1}, x_{u2}, \ldots, x_{uk})'$, $\boldsymbol{\theta} = (\theta_1, \theta_2, \ldots, \theta_p)'$ is a vector of p unknown parameters, ε_{ui} is a random error, and f_i is a response function of known form. Model (3.1) can be represented in matrix form as

$$\boldsymbol{Y} = \boldsymbol{F}(\mathcal{D}, \boldsymbol{\theta}) + \boldsymbol{\varepsilon}, \tag{3.2}$$

where $\boldsymbol{Y} = [\boldsymbol{y}_1 : \boldsymbol{y}_2 : \cdots : \boldsymbol{y}_r]$ is an $n \times r$ matrix whose ith column is $\boldsymbol{y}_i$, the vector of observations from the ith response $(i = 1, 2, \ldots, r)$, $\boldsymbol{\varepsilon} = [\boldsymbol{\varepsilon}_1 : \boldsymbol{\varepsilon}_2 : \cdots : \boldsymbol{\varepsilon}_r]$ is the corresponding matrix of random errors, $\mathcal{D}$ denotes the design matrix, which is of order $n \times k$ with rows equal to $\boldsymbol{x}'_1, \boldsymbol{x}'_2, \ldots, \boldsymbol{x}'_n$, and $\boldsymbol{F}(\mathcal{D}, \boldsymbol{\theta})$ is a matrix of order $n \times r$ whose ith column consists of the values of $f_i(\boldsymbol{x}_u, \boldsymbol{\theta})$ for $u = 1, 2, \ldots, n$ $(i = 1, 2, \ldots, r)$. It is assumed that the rows of $\boldsymbol{\varepsilon}$ are independently and identically distributed as $N(\boldsymbol{0}, \boldsymbol{\Sigma})$, where $\boldsymbol{\Sigma}$ is an unknown variance-covariance matrix for the r response variables.

3.1. The Box–Draper determinant criterion

Estimates of the elements of $\boldsymbol{\theta}$ can be obtained by using the *method of maximum likelihood*. This, however, requires knowledge of the value of $\boldsymbol{\Sigma}$, which, in general, is unknown. Box and Draper (1965) proposed a Bayesian approach for estimating $\boldsymbol{\theta}$ without knowing $\boldsymbol{\Sigma}$. A summary of their approach follows:

The likelihood function for the $n \times r$ data matrix is of the form

$$L(\boldsymbol{Y} \mid \boldsymbol{\theta}, \boldsymbol{\Sigma}) = \frac{1}{(2\pi)^{nr/2} |\boldsymbol{\Sigma}|^{n/2}} \exp\left\{ -\frac{1}{2} \mathrm{tr}[\boldsymbol{\Sigma}^{-1} \boldsymbol{V}(\boldsymbol{\theta})] \right\}, \tag{3.3}$$

where tr(·) denotes the trace of a square matrix and $\boldsymbol{V}(\boldsymbol{\theta})$ is the $r \times r$ matrix

$$\boldsymbol{V}(\boldsymbol{\theta}) = [\boldsymbol{Y} - \boldsymbol{F}(\mathcal{D}, \boldsymbol{\theta})]'[\boldsymbol{Y} - \boldsymbol{F}(\mathcal{D}, \boldsymbol{\theta})]. \tag{3.4}$$

Box and Draper (1965) assumed noninformative prior distributions for $\boldsymbol{\theta}$ and $\boldsymbol{\Sigma}$. By combining these distributions with the likelihood function in formula (3.3) they obtained the posterior density function $\pi(\boldsymbol{\theta}, \boldsymbol{\Sigma} \mid \boldsymbol{Y})$. The marginal posterior density function for $\boldsymbol{\theta}$ can be obtained from $\pi(\boldsymbol{\theta}, \boldsymbol{\Sigma} \mid \boldsymbol{Y})$ by integrating the elements of $\boldsymbol{\Sigma}$ out. An estimate of $\boldsymbol{\theta}$ can then be obtained by maximizing this marginal posterior density. Box and Draper (1965) showed that this process is equivalent to finding the value of $\boldsymbol{\theta}$ that minimizes the determinant

$$h(\boldsymbol{\theta}) = |\boldsymbol{V}(\boldsymbol{\theta})|. \tag{3.5}$$

This estimation rule is called the *Box–Draper determinant criterion*. Bates and Watts (1984, 1987) presented a method for the minimization of $h(\boldsymbol{\theta})$ along with a computer algorithm for its implementation.

Several variants of the *Box–Draper determinant criterion* were considered. Box et al. (1970) discussed the implementation of this criterion when the multiresponse data are incomplete due to the missing of only few values. The missing values were treated as additional parameters to be estimated. An alternative procedure was proposed by Box (1971) when there are a large number of missing values. Stewart and Sorensen (1981) also addressed the missing data problem. Their approach is based on using the posterior density function $\pi(\boldsymbol{\theta}, \boldsymbol{\Sigma} \mid \boldsymbol{Y})$ in which only genuine multiresponse values appear. Box (1970) considered a situation in which the variance-covariance matrix $\boldsymbol{\Sigma}$ is not constant for all the experimental runs.

3.2. *The problem of linear dependencies among the responses*

Box et al. (1973) cautioned that the use of the *Box–Draper determinant criterion* can lead to meaningless results when exact linear relationships exist among the responses. The reason for this is the following: Suppose that there are m linearly independent relationships among the responses of the form

$$\boldsymbol{A}\boldsymbol{Y}' = \boldsymbol{C}, \tag{3.6}$$

where $\boldsymbol{A}$ is a constant $m \times r$ matrix of rank m $(< r)$ and $\boldsymbol{C}$ is another constant matrix of order $m \times n$. In this case, since $E(\boldsymbol{Y}) = \boldsymbol{F}(\mathcal{D}, \boldsymbol{\theta})$,

$$\boldsymbol{A}\boldsymbol{F}'(\mathcal{D}, \boldsymbol{\theta}) = \boldsymbol{C}. \tag{3.7}$$

From (3.6) and (3.7) it follows that

$$\boldsymbol{A}[\boldsymbol{Y}' - \boldsymbol{F}'(\mathcal{D}, \boldsymbol{\theta})] = \boldsymbol{0}.$$

This implies the existence of m linearly independent relationships among the rows of the matrix $\boldsymbol{Y}' - \boldsymbol{F}'(\mathcal{D}, \boldsymbol{\theta})$. Consequently, the matrix $\boldsymbol{V}(\boldsymbol{\theta})$ in formula (3.4) is singular, that is, its determinant is equal to zero for any value of $\boldsymbol{\theta}$. Since this determinant is nonnegative by the fact that $\boldsymbol{V}(\boldsymbol{\theta})$ is positive semidefinite, minimizing $|\boldsymbol{V}(\boldsymbol{\theta})|$ would be meaningless in this case.

Box et al. (1973) developed a technique for identifying linear dependencies among the responses. It is based on an examination of the eigenvalues of the matrix $\boldsymbol{D}\boldsymbol{D}'$, where

$$\boldsymbol{D} = \boldsymbol{Y}'\left(\boldsymbol{I}_n - \frac{1}{n}\boldsymbol{J}_n\right) \tag{3.8}$$

and $\boldsymbol{I}_n$, $\boldsymbol{J}_n$ denote the identity matrix and the matrix of ones of order $n \times n$. If there are linear relationships of the type given in formula (3.6), and if the columns of $\boldsymbol{C}$ are identical, then

$$\boldsymbol{A}\boldsymbol{D} = \boldsymbol{A}\boldsymbol{Y}'\left(\boldsymbol{I}_n - \frac{1}{n}\boldsymbol{J}_n\right) = \boldsymbol{C}\left(\boldsymbol{I}_n - \frac{1}{n}\boldsymbol{J}_n\right) = \boldsymbol{0}. \tag{3.9}$$

This indicates that there are m linearly independent relationships among the rows of $\boldsymbol{D}$. Hence, the rank of $\boldsymbol{D}$ is $r - m$ and the matrix $\boldsymbol{DD}'$ has a zero eigenvalue of multiplicity m. Vice versa, if $\boldsymbol{DD}'$ has a zero eigenvalue of multiplicity m, then formula (3.9) is true for some constant matrix $\boldsymbol{A}$ of order $m \times r$ and rank m. Thus,

$$\boldsymbol{AY}' = \frac{1}{n}\boldsymbol{AY}'\boldsymbol{J}_n, \tag{3.10}$$

which looks like formula (3.6). Note that the matrix $\frac{1}{n}\boldsymbol{AY}'\boldsymbol{J}_n$ has identical columns. We conclude that m linearly independent relationships, of the type given in (3.6) with $\boldsymbol{C}$ having identical columns, exist among the responses if and only if $\boldsymbol{DD}'$ has a zero eigenvalue of multiplicity m. It is easy to see that such linear relationships can be defined in terms of m orthonormal eigenvectors of $\boldsymbol{DD}'$ corresponding to its zero eigenvalue.

McLean et al. (1979) noted that if the columns of $\boldsymbol{C}$ in (3.6) are not identical, then $\boldsymbol{AD} \neq \boldsymbol{0}$; consequently, no linear relationships exist among the rows of $\boldsymbol{D}$. In this case, $\boldsymbol{DD}'$ does not have zero eigenvalues, even though the matrix $\boldsymbol{V}(\boldsymbol{\theta})$ in (3.4) is singular with m zero eigenvalues.

In practice, the observed values of the responses are rounded off to a certain number of decimal places. In this case, it is possible that none of the eigenvalues of $\boldsymbol{DD}'$ is exactly zero, even if the responses are linearly related. Let us therefore suppose that λ is a "small" eigenvalue of $\boldsymbol{DD}'$, which, if it were not for the rounding errors, would be equal to zero. If the rounding errors are distributed independently and uniformly over the interval $(-\delta, \delta)$, where δ is small enough, then the expected value of λ is approximately equal to

$$\mathrm{E}(\lambda) = (n-1)\sigma_{re}^2, \tag{3.11}$$

where $\sigma_{re}^2 = \delta^2/3$ is the rounding error variance. Here, δ is equal to one half of the last digit reported when the multiresponse values are rounded off to the same number of decimal places. Formula (3.11) was given by Box et al. (1973). Khuri and Conlon (1981) showed that an upper bound on the variance of λ is approximately given by

$$\mathrm{Var}(\lambda) \leqslant \left[\frac{9nr}{5} + nr(nr-1) - (n-1)^2\right]\sigma_{re}^4. \tag{3.12}$$

Formulas (3.11) and (3.12) can be used to determine if a "small" eigenvalue of $\boldsymbol{DD}'$ is actually zero. If, for example, such an eigenvalue falls within two or three standard deviations from $\mathrm{E}(\lambda)$, then it can be regarded as equal to zero. A conservative value for the standard deviation is the square root of the upper bound in inequality (3.12). If m "small" eigenvalues of $\boldsymbol{DD}'$ are labeled as zero according to this procedure, then $\boldsymbol{DD}'$ is considered to have m zero eigenvalues.

For example, consider the chemical experiment described in Box et al. (1973, Section 6), where $r = 5$, $n = 8$, and the multiresponse data are rounded off to one decimal place. Then, $\delta = 0.05$, $\sigma_{re}^2 = \delta^2/3 = 0.0008$, and from (3.11), $\mathrm{E}(\lambda) = 0.0056$.

The upper bound in (3.12) has the value 0.001. The eigenvalues of $\boldsymbol{DD}'$ are $\lambda_1 = 0.0013, \lambda_2 = 0.0168, \lambda_3 = 1.21, \lambda_4 = 25.0$, and $\lambda_5 = 9660$. Using the value $\sqrt{0.001} = 0.0316$ as a conservative estimate of the standard deviation of λ, we find that λ_1 and λ_2 fall within one standard deviation from $\mathrm{E}(\lambda)$. However, the next smallest eigenvalue of $\boldsymbol{DD}'$, namely λ_3, is more than 38 standard deviations away. We conclude that λ_1 and λ_2 correspond to a zero eigenvalue of $\boldsymbol{DD}'$ of multiplicity $m = 2$.

In general, the response variables may be measured in different units. As a result, the round-off errors for the various responses can vary appreciably. This makes it difficult to interpret the results of the eigenvalue analysis described earlier. Furthermore, large differences in the responses' scales of measurements can cause problems in identifying those responses that are multicollinear, if any. To avoid such problems, Khuri (1990b) proposed that the responses be scaled first so that each row of $\boldsymbol{D}$ has a unit length. Using this scaling convention, he developed new expressions concerning the expected value and variance of a "small" eigenvalue of $\boldsymbol{DD}'$ in lieu of (3.11) and (3.12).

3.3. *Confidence region on* θ

Let $\widehat{\boldsymbol{\theta}}$ denote an estimator of $\boldsymbol{\theta}$ that minimizes the function $h(\boldsymbol{\theta})$ in formula (3.5). Assuming that a second-order Taylor series approximation of $h(\boldsymbol{\theta})$ is adequate in a neighborhood of $\widehat{\boldsymbol{\theta}}$, Bates and Watts (1985) obtained an approximate variance-covariance matrix of $\widehat{\boldsymbol{\theta}}$ given by

$$\mathrm{Var}(\widehat{\boldsymbol{\theta}}) = 2s^2\boldsymbol{\Gamma}^{-1}, \tag{3.13}$$

where $s^2 = h(\widehat{\boldsymbol{\theta}})/(n-p)$, p is the number of elements of $\boldsymbol{\theta}$, and $\boldsymbol{\Gamma}$ is an approximate Hessian matrix of $h(\boldsymbol{\theta})$ evaluated at $\widehat{\boldsymbol{\theta}}$. Hence, an approximate $(1-\alpha)$ 100% confidence region on $\boldsymbol{\theta}$ can be expressed as

$$(\boldsymbol{\theta} - \widehat{\boldsymbol{\theta}})\frac{1}{2}\boldsymbol{\Gamma}(\boldsymbol{\theta} - \widehat{\boldsymbol{\theta}}) \leqslant ps^2F_{\alpha,p,n-p}, \tag{3.14}$$

where $F_{\alpha,p,n-p}$ is the upper $100\alpha\%$ point of the F distribution with p and $n-p$ degrees of freedom.

One of the important advantages of multiresponse analysis is increased precision of parameter estimates by comparison to the estimates obtained when the responses are considered individually. Bates and Watts (1985) remarked that "the increased precision is due to the combination of different types of information from the responses". Ziegel and Gorman (1980) demonstrated the utility of multiresponse estimation for studying kinetic reaction models using data from an oil shale pyrolysis experiment.

3.4. *The special case of linear multiresponse models*

Suppose that model (3.1) is linear in the parameters, then it can be expressed in vector form as

$$\boldsymbol{y}_i = \boldsymbol{X}_i\boldsymbol{\phi}_i + \boldsymbol{\varepsilon}_i, \quad i = 1, 2, \ldots, r, \tag{3.15}$$

where, if we recall, $\boldsymbol{y}_i$ is the vector of observations from the ith response and $\boldsymbol{\varepsilon}_i$ is the corresponding error vector, $\boldsymbol{X}_i$ is a matrix of order $n \times p_i$ and rank p_i, and $\boldsymbol{\phi}_i$ is a vector of p_i unknown parameters ($i = 1, 2, \ldots, r$). The models in (3.15) can be combined into a single linear multiresponse model of the form

$$\boldsymbol{y} = \boldsymbol{X}\boldsymbol{\Phi} + \widetilde{\boldsymbol{\varepsilon}}, \tag{3.16}$$

where

$$\boldsymbol{y} = [\boldsymbol{y}_1' : \boldsymbol{y}_2' : \cdots : \boldsymbol{y}_r']', \qquad \widetilde{\boldsymbol{\varepsilon}} = [\boldsymbol{\varepsilon}_1' : \boldsymbol{\varepsilon}_2' : \cdots : \boldsymbol{\varepsilon}_r']',$$

$$\boldsymbol{\Phi} = [\boldsymbol{\phi}_1' : \boldsymbol{\phi}_2' : \cdots : \boldsymbol{\phi}_r']',$$

and $\boldsymbol{X}$ is the block-diagonal matrix $\boldsymbol{X} = \operatorname{diag}(\boldsymbol{X}_1, \boldsymbol{X}_2, \ldots, \boldsymbol{X}_r)$. Model (3.16) is called a *multiple design multivariate model* because it can accommodate different design matrices for the different responses (see McDonald, 1975).

If the assumptions made earlier concerning the distribution of the rows of the error matrix $\boldsymbol{\varepsilon}$ in model (3.2) are valid, then $\widetilde{\boldsymbol{\varepsilon}}$ has the multivariate normal distribution $N(\boldsymbol{0}, \boldsymbol{\Sigma} \otimes \boldsymbol{I}_n)$, where $\boldsymbol{\Sigma}$ is the variance-covariance matrix for the r responses, and $\boldsymbol{\Sigma} \otimes \boldsymbol{I}_n$ is its direct product with $\boldsymbol{I}_n$. The *best linear unbiased estimator* (BLUE) of $\boldsymbol{\Phi}$ is the generalized least-squares estimator

$$\widehat{\boldsymbol{\Phi}} = [\boldsymbol{X}'(\boldsymbol{\Sigma}^{-1} \otimes \boldsymbol{I}_n)\boldsymbol{X}]^{-1}\boldsymbol{X}'(\boldsymbol{\Sigma}^{-1} \otimes \boldsymbol{I}_n)\boldsymbol{y}. \tag{3.17}$$

Since $\widehat{\boldsymbol{\Phi}}$ depends on $\boldsymbol{\Sigma}$, which, in general, is unknown, a two-stage estimation procedure proposed by Zellner (1962) can be used. In the first stage, $\boldsymbol{\Sigma}$ is estimated by $\widehat{\boldsymbol{\Sigma}} = (\widehat{\sigma}_{ij})$, where

$$\widehat{\sigma}_{ij} = \frac{1}{n}\boldsymbol{y}_i'[\boldsymbol{I}_n - \boldsymbol{X}_i(\boldsymbol{X}_i'\boldsymbol{X}_i)^{-1}\boldsymbol{X}_i'][\boldsymbol{I}_n - \boldsymbol{X}_j(\boldsymbol{X}_j'\boldsymbol{X}_j)^{-1}\boldsymbol{X}_j']\boldsymbol{y}_j,$$
$$i, j = 1, 2, \ldots, r.$$

This estimate is obtained by using the residuals that result from fitting each individual response by the method of ordinary least-squares. In the second stage, an estimate of $\boldsymbol{\Phi}$ is obtained by replacing $\boldsymbol{\Sigma}$ with $\widehat{\boldsymbol{\Sigma}}$ in formula (3.17). The resulting estimate, denoted by $\widehat{\boldsymbol{\Phi}}_e$, does not have the optimal properties of $\widehat{\boldsymbol{\Phi}}$. It does, however, have certain asymptotic properties as was shown in Zellner (1962).

In particular, if in formula (3.15), $\boldsymbol{X}_i = \boldsymbol{X}_0$ for $i = 1, 2, \ldots, r$, then $\boldsymbol{X} = \operatorname{diag}(\boldsymbol{X}_1, \boldsymbol{X}_2, \ldots, \boldsymbol{X}_r)$ reduces to $\boldsymbol{X} = \boldsymbol{I}_r \otimes \boldsymbol{X}_0$. In this case, it is easy to show that

$$\widehat{\boldsymbol{\Phi}} = [\boldsymbol{I}_r \otimes (\boldsymbol{X}_0'\boldsymbol{X}_0)^{-1}\boldsymbol{X}_0']\boldsymbol{y}.$$

Thus the *BLUE* of $\boldsymbol{\Phi}$ does not depend on $\boldsymbol{\Sigma}$ and is therefore the same as the ordinary least-squares estimator obtained from fitting the r response models individually.

4. Inference for multiresponse models

The approximate confidence region defined by inequality (3.14) can be used to test a hypothesis concerning the parameter vector $\boldsymbol{\theta} = (\theta_1, \theta_2, \ldots, \theta_p)'$ for model (3.2). By projecting this region on the p coordinate axes, approximate simultaneous confidence intervals on the elements of $\boldsymbol{\theta}$ can be obtained. These intervals are conservative in the sense that their joint confidence coefficient is at least as large as the confidence coefficient for the confidence region.

4.1. Tests concerning the linear multiresponse model

We recall that a linear multiresponse model is expressible as in formula (3.16). Alternatively, the r response models in (3.15) can be combined into a single model of the form

$$\boldsymbol{Y} = \widetilde{\boldsymbol{X}}\widetilde{\boldsymbol{\Phi}} + \boldsymbol{\varepsilon}, \tag{4.1}$$

where $\boldsymbol{Y}$ and $\boldsymbol{\varepsilon}$ are the same as in model (3.2), $\widetilde{\boldsymbol{X}} = [\boldsymbol{X}_1 : \boldsymbol{X}_2 : \cdots : \boldsymbol{X}_r]$, and $\widetilde{\boldsymbol{\Phi}} = \operatorname{diag}(\boldsymbol{\phi}_1, \boldsymbol{\phi}_2, \ldots, \boldsymbol{\phi}_r)$.

Let us now consider testing the hypothesis

$$H_0\colon\ \boldsymbol{G}\widetilde{\boldsymbol{\Phi}} = \boldsymbol{C}_0,$$

where $\boldsymbol{G}$ and $\boldsymbol{C}_0$ are known matrices with $\boldsymbol{G}$ being of full row rank q. If $\boldsymbol{G}\widetilde{\boldsymbol{\Phi}}$ is estimable, then H_0 is said to be *testable*. This occurs if and only if the rows of $\boldsymbol{G}$ are spanned by the rows of $\widetilde{\boldsymbol{X}}$. Following Roy et al. (1971), a test statistic for testing H_0 is given by the largest eigenvalue, $e_{\max}(\boldsymbol{S}_h\boldsymbol{S}_e^{-1})$, of the matrix $\boldsymbol{S}_h\boldsymbol{S}_e^{-1}$, where

$$\boldsymbol{S}_h = \left[\boldsymbol{G}(\widetilde{\boldsymbol{X}}'\widetilde{\boldsymbol{X}})^{-}\widetilde{\boldsymbol{X}}'\boldsymbol{Y} - \boldsymbol{C}_0\right]'\left[\boldsymbol{G}(\widetilde{\boldsymbol{X}}'\widetilde{\boldsymbol{X}})^{-}\boldsymbol{G}'\right]^{-1}\left[\boldsymbol{G}(\widetilde{\boldsymbol{X}}'\widetilde{\boldsymbol{X}})^{-}\widetilde{\boldsymbol{X}}'\boldsymbol{Y} - \boldsymbol{C}_0\right],$$
$$\boldsymbol{S}_e = \boldsymbol{Y}'\left[\boldsymbol{I}_n - \widetilde{\boldsymbol{X}}(\widetilde{\boldsymbol{X}}'\widetilde{\boldsymbol{X}})^{-}\widetilde{\boldsymbol{X}}'\right]\boldsymbol{Y},$$

and $(\widetilde{\boldsymbol{X}}'\widetilde{\boldsymbol{X}})^{-}$ denotes a g-inverse of $\widetilde{\boldsymbol{X}}'\widetilde{\boldsymbol{X}}$. The matrices $\boldsymbol{S}_h$ and $\boldsymbol{S}_e$ are called the *matrix due to the hypothesis* and the *matrix due to the error*, respectively. The matrix $\boldsymbol{S}_e$ is positive definite with probability 1 if $n - \rho \geqslant r$, where ρ is the rank of $\widetilde{\boldsymbol{X}}$ (see Roy et al., 1971, p. 35), and has the central Wishart distribution with $n - \rho$ degrees of freedom. Furthermore, $\boldsymbol{S}_h$ is independent of $\boldsymbol{S}_e$ and has the noncentral Wishart distribution with q degrees of freedom and a noncentrality parameter matrix $\boldsymbol{\Omega}$ given by (see, for example, Seber, 1984, p. 414)

$$\boldsymbol{\Omega} = \boldsymbol{\Sigma}^{-1/2}(\boldsymbol{G}\widetilde{\boldsymbol{\Phi}} - \boldsymbol{C}_0)'\left[\boldsymbol{G}(\widetilde{\boldsymbol{X}}'\widetilde{\boldsymbol{X}})^{-}\boldsymbol{G}'\right]^{-1}(\boldsymbol{G}\widetilde{\boldsymbol{\Phi}} - \boldsymbol{C}_0)\boldsymbol{\Sigma}^{-1/2}. \tag{4.2}$$

The statistic $e_{\max}(\boldsymbol{S}_h\boldsymbol{S}_e^{-1})$ is called *Roy's largest root*. The hypothesis H_0 is rejected for large values of $e_{\max}(\boldsymbol{S}_h\boldsymbol{S}_e^{-1})$. Critical values of this test statistic can be obtained, for example, from Roy et al. (1971, Appendix B), Seber (1984, Appendix D14), and Morrison (1976).

Other test statistics for testing H_0 include *Wilks' likelihood ratio* $|\boldsymbol{S}_e|/|\boldsymbol{S}_e + \boldsymbol{S}_h|$, *Hotelling–Lawley's trace*, $\mathrm{tr}(\boldsymbol{S}_h\boldsymbol{S}_e^{-1})$, and *Pillai's trace*, $\mathrm{tr}[\boldsymbol{S}_h(\boldsymbol{S}_e + \boldsymbol{S}_h)^{-1}]$. Small values of *Wilks' likelihood ratio* are significant, whereas large values of the other two test statistics are significant.

More detailed discussions concerning the aforementioned multivariate tests can be found in Muirhead (1982, Chapter 10) and Seber (1984, Section 8.6.2.). Khuri (1986) used these tests to compare the parameter vectors for several correlated linear response models. The models are of the first order and depend on the same set of input variables, but can have different design matrices. Khuri investigated the effects of multicollinearity in the linearly independent columns of the $\widetilde{\boldsymbol{X}}$ matrix (see model 4.1), as well as the structure of the variance- covariance matrix $\boldsymbol{\Sigma}$, on the power of the multivariate tests. He concluded that the higher the degree of multicollinearity and the more correlated the responses are, the smaller the power. Khuri also presented an example to illustrate the application of the multivariate tests using investment data from three American corporations in the electronic industry. The purpose of the application was to compare the investment models for these corporations.

4.2. *Testing lack of fit of a linear multiresponse model*

Estimates of the parameters of the linear multiresponse model (4.1) are obtained under the assumption that the component models in (3.15) adequately represent the true means of the r responses. If, however, at least one model is inadequate, then its lack of fit can influence the fit of the remaining models since the responses are correlated in general. The adequacy of model (4.1) should therefore be checked before any formal analysis of the multiresponse data is attempted. Box and Draper (1965) advocated such a practice and stressed the need for a formal multivariate lack of fit test.

A multivariate lack of fit test for model (4.1) was presented by Khuri (1985). The test requires the availability of replicated observations on all the responses at some points in the experimental region. Furthermore, the rows of the error matrix $\boldsymbol{\varepsilon}$ are assumed to be independently distributed as $N(\boldsymbol{0}, \boldsymbol{\Sigma})$. Without any loss of generality, we consider that the replicated runs are taken at each of the first n_0 points of the design ($n_0 \leqslant n$) with ν_i observations at the ith run ($i = 1, 2, \ldots, n_0$). The test is based on the notion that model (4.1) is correct if and only if all linear combinations of the models in (3.15) are correct. These combinations are modeled as

$$\boldsymbol{y}_c = \widetilde{\boldsymbol{X}}\widetilde{\boldsymbol{\Phi}}_c + \boldsymbol{\varepsilon}_c, \tag{4.3}$$

where $\boldsymbol{y}_c = \boldsymbol{Y}\boldsymbol{c}$, $\widetilde{\boldsymbol{\Phi}}_c = \widetilde{\boldsymbol{\Phi}}\boldsymbol{c}$, $\boldsymbol{\varepsilon}_c = \boldsymbol{\varepsilon}\boldsymbol{c}$, and $\boldsymbol{c}$ is an arbitrary nonzero vector of order $r \times 1$. We note that $\boldsymbol{\varepsilon}_c \sim N(\boldsymbol{0}, \sigma_c^2\boldsymbol{I}_n)$, where $\sigma_c^2 = \boldsymbol{c}'\boldsymbol{\Sigma}\boldsymbol{c}$. By using the usual lack of fit test statistic for the single-response model in (4.3) along with *Roy's union intersection principle*, Khuri (1985) showed that *Roy's largest root*, $e_{\max}(G_1G_2^{-1})$, can be used as a multivariate lack of fit test statistic for model (4.1), where

$$G_1 = Y'\big[I_n - \widetilde{X}(\widetilde{X}'\widetilde{X})^{-}\widetilde{X}' - K\big]Y, \tag{4.4}$$

$$G_2 = Y'KY, \tag{4.5}$$

and

$$K = \mathrm{diag}(K_1, K_2, \ldots, K_{n_0}, 0),$$
$$K_i = I_{\nu_i} - \frac{1}{\nu_i}J_{\nu_i}, \quad i = 1, 2, \ldots, n_0. \tag{4.6}$$

In formula (4.6), $\mathbf{0}$ is a zero matrix of order $(n-\sum_{i=1}^{n_0}\nu_i)\times(n-\sum_{i=1}^{n_0}\nu_i)$. Lack of fit is significant at the α-level if $e_{\max}(G_1G_2^{-1}) > \lambda_\alpha^*$, where λ_α^* is the upper $100\alpha\%$ point of *Roy's largest root* test statistic. As was stated earlier in Section 4.1, other multivariate test statistics can also be used, which include *Wilks' likelihood ratio*, $|G_2|/|G_1 + G_2|$, *Hotelling–Lawley's trace*, $\mathrm{tr}(G_1G_2^{-1})$, and *Pillai's trace*, $\mathrm{tr}[G_1(G_1 + G_2)^{-1}]$.

Having a significant multivariate lack of fit test means that for some c, model (4.3) is inadequate. In other words, model (4.1) is inadequate if and only if there exists a linear combination of the responses that is not adequately represented by model (4.3). The responses involved in such a linear combination are considered to contribute significantly to lack of fit of model (4.1). Khuri (1985) proposed the following procedure for detecting such responses: let $c^* = (c_1^*, c_2^*, \ldots, c_r^*)'$ be an eigenvector of $G_1G_2^{-1}$ corresponding to $e_{\max}(G_1G_2^{-1})$. Consider the linear combination, $\sum_{i=1}^r d_i^* z_i$, where $d_i^* = c_i^*||y_i||$, $z_i = \frac{y_i}{||y_i||}$, and $||y_i||$ denotes the Euclidean norm $(y_i'y_i)^{1/2}$ of the vector of observations on the ith response ($i = 1, 2, \ldots, r$). Large absolute values of d_i^* indicate which responses are influential contributors to lack of fit. Alternatively, all possible nonempty subsets of the r responses can be checked to identify those subsets that individually produce a significant lack of fit. This is determined by computing $e_{\max}(G_1G_2^{-1})$ using only the responses that belong to a given subset. If the resulting value exceeds λ_α^*, the critical value mentioned earlier, then the responses in the subset under consideration are considered to contribute significantly to lack of fit.

In the example given in Khuri (1985), measurements were made on three responses of interest to the food industry. A second-order model in five input variables was fitted to the data from each response using a central composite design with five replications at the center. The application of Roy's largest root test produced the value, $e_{\max}(G_1G_2^{-1}) = 245.518$, which is significant at the 10% level ($\lambda_{0.10}^* = 85.21$). The corresponding eigenvector of $G_1G_2^{-1}$ is $c^* = (3.2659, 0.0385, -0.0904)'$. Furthermore, the Euclidean norms of the three response vectors are $||y_1|| = 27.60$, $||y_2|| = 5929.27$, and $||y_3|| = 517.49$. In this case,

$$\sum_{i=1}^{3} d_i^* z_i = 90.139z_1 + 228.277z_2 - 46.781z_3 \sim 0.395z_1 + z_2 - 0.205z_3,$$

where $\sim$ indicates that the first linear combination is proportional to the second. From the size of the absolute values of the coefficients of z_1, z_2, and z_3 in the latter combination, we conclude that the responses y_1 and y_2 are the main contributors to a significant lack of fit test. This conclusion was subsequently corroborated by an examination of the values of $e_{\max}(\boldsymbol{G}_1\boldsymbol{G}_2^{-1})$ using all possible nonempty subsets of the three responses (see Table 4 in Khuri, 1985).

More recently, Levy and Neill (1990) considered additional multivariate lack of fit tests based on the determinant and trace of $\boldsymbol{G}_1\boldsymbol{G}_2^{-1}$. They compared the power functions of these tests by using simulation, and extended the testing for lack of fit to situations in which no exact replications are available on all the responses.

5. Designs for multiresponse models

A design for fitting a multiresponse model in k input variables is a collection of points in a k-dimensional Euclidean space that specify experimental (design) settings for the input variables. These settings are chosen on the basis of certain criteria that pertain to all the responses. Some typical design criteria include (a) obtaining good estimates, in some well-defined sense, of the parameters in a model such as (3.2) or (4.1) (b) increasing the sensitivity (power) of the lack of fit test (c) improving robustness of tests with respect to outliers or against particular departures from model assumptions (d) achieving reliable predictions from the fitted models.

One of the earliest articles on multiresponse design is the one by Draper and Hunter (1966). They developed a criterion for selecting additional experimental runs after a certain number of runs have already been chosen. They used a Bayesian approach, which can be summarized as follows: A locally uniform prior distribution is assumed for the parameter vector $\boldsymbol{\theta}$ (see formula 3.1). The posterior density of $\boldsymbol{\theta}$, which can be obtained after the additional runs have been taken, is then maximized with respect to $\boldsymbol{\theta}$ and the additional experimental runs. The variance-covariance matrix $\boldsymbol{\Sigma}$ for the responses is assumed known. Draper and Hunter (1967) extended this Bayesian approach to situations in which the prior distribution for $\boldsymbol{\theta}$ is multinormal. Box and Draper (1972) introduced another extension of Draper and Hunter's (1966) criterion: The design is divided into blocks, each having a different variance-covariance structure, which may be unknown. In particular, if the variance-covariance matrix is the same for all runs, then their criterion reduces to the one given by Draper and Hunter (1967).

5.1. Locally D-optimal designs

Consider the general multiresponse model (3.1). Suppose that $f_i(\boldsymbol{x}_u, \boldsymbol{\theta})$ is approximately linear in $\boldsymbol{\theta}$ in a neighborhood of $\boldsymbol{\theta} = \boldsymbol{\theta}_0$. In this case, a first-order Taylor's series expansion of $f_i(\boldsymbol{x}_u, \boldsymbol{\theta})$ in a neighborhood of $\boldsymbol{\theta}_0$ yields the following approximation:

$$f_i(\boldsymbol{x}_u, \boldsymbol{\theta}) \approx f_i(\boldsymbol{x}_u, \boldsymbol{\theta}_0) + \sum_{j=1}^{p}(\theta_j - \theta_{j0})\frac{\partial f_i(\boldsymbol{x}_u, \boldsymbol{\theta})}{\partial \theta_j}\bigg|_{\boldsymbol{\theta}=\boldsymbol{\theta}_0},$$

$$i = 1, 2, \ldots, r; \; u = 1, 2, \ldots, n, \tag{5.1}$$

where θ_j and θ_{j0} are the jth elements of $\boldsymbol{\theta}$ and $\boldsymbol{\theta}_0$, respectively $(j = 1, 2, \ldots, p)$. From (3.1) and (5.1) we obtain the approximate linear model

$$\widetilde{\boldsymbol{y}}_i = \boldsymbol{F}_i(\mathcal{D}, \boldsymbol{\theta}_0)\boldsymbol{\theta} + \boldsymbol{\varepsilon}_i, \quad i = 1, 2, \ldots, r, \tag{5.2}$$

where $\widetilde{\boldsymbol{y}}_i = (\widetilde{y}_{1i}, \widetilde{y}_{2i}, \ldots, \widetilde{y}_{ni})'$ with

$$\widetilde{y}_{ui} = y_{ui} - f_i(\boldsymbol{x}_u, \boldsymbol{\theta}_0) + \sum_{j=1}^{p} \theta_{j0} \frac{\partial f_i(\boldsymbol{x}_u, \boldsymbol{\theta})}{\partial \theta_j}\Big|_{\boldsymbol{\theta}=\boldsymbol{\theta}_0},$$
$$i = 1, 2, \ldots, r; \; u = 1, 2, \ldots, n,$$

and $\boldsymbol{F}_i(\mathcal{D}, \boldsymbol{\theta}_0)$ is a matrix of order $n \times p$ whose (u, j)th element is

$$\frac{\partial f_i(\boldsymbol{x}_u, \boldsymbol{\theta})}{\partial \theta_j}\Big|_{\boldsymbol{\theta}=\boldsymbol{\theta}_0}.$$

From (5.2) we derive an approximate linear multiresponse model of the form

$$\widetilde{\boldsymbol{y}} = \widetilde{\boldsymbol{F}}\boldsymbol{\theta} + \widetilde{\boldsymbol{\varepsilon}}, \tag{5.3}$$

where $\widetilde{\boldsymbol{y}} = [\widetilde{\boldsymbol{y}}_1' : \widetilde{\boldsymbol{y}}_2' : \cdots : \widetilde{\boldsymbol{y}}_r']'$,

$$\widetilde{\boldsymbol{F}} = [\boldsymbol{F}_1'(\mathcal{D}, \boldsymbol{\theta}_0) : \boldsymbol{F}_2'(\mathcal{D}, \boldsymbol{\theta}_0) : \cdots : \boldsymbol{F}_r'(\mathcal{D}, \boldsymbol{\theta}_0)]', \tag{5.4}$$

and $\widetilde{\boldsymbol{\varepsilon}} = [\boldsymbol{\varepsilon}_1' : \boldsymbol{\varepsilon}_2' : \cdots : \boldsymbol{\varepsilon}_r']'$. Note that

$$\mathrm{Var}(\widetilde{\boldsymbol{\varepsilon}}) = \boldsymbol{\Sigma} \otimes \boldsymbol{I}_n,$$

where $\boldsymbol{\Sigma}$ is the variance-covariance matrix for the responses.

On the basis of model (5.3), the variance-covariance matrix of the *BLUE* of $\boldsymbol{\theta}$ is approximately equal to $[\widetilde{\boldsymbol{F}}'(\boldsymbol{\Sigma}^{-1} \otimes \boldsymbol{I}_n)\widetilde{\boldsymbol{F}}]^{-1}$. Following the D-optimality approach, traditionally used in selecting designs for nonlinear models (see, for example, Box and Lucas, 1959), a design $\mathcal{D}$ for model (3.2) can be obtained by maximizing the determinant of the matrix $\widetilde{\boldsymbol{F}}'(\boldsymbol{\Sigma}^{-1} \otimes \boldsymbol{I}_n)\widetilde{\boldsymbol{F}}$, where $\widetilde{\boldsymbol{F}}$ is given by formula (5.4). Note that the resulting design depends on $\boldsymbol{\theta}_0$, as is usually the case with nonlinear models. It also depends on having an adequate first-order approximation of the nonlinear function $f_i(\boldsymbol{x}_u, \boldsymbol{\theta})$ in a neighborhood of $\boldsymbol{\theta}_0$. Such a design is appropriately described as *locally D-optimal* (see Chernoff, 1953). Hatzis and Larntz (1992) constructed locally D-optimal designs for a nonlinear multiresponse model describing the behavior of a biological system.

5.2. Designs for linear multiresponse models

5.2.1. D-optimal designs

Consider the linear multiresponse model (3.16). We recall that the *BLUE* of the parameter vector $\boldsymbol{\Phi}$ is $\widehat{\boldsymbol{\Phi}}$ given by formula (3.17). The predicted ith response value at the point $\boldsymbol{x} = (x_1, x_2, \ldots, x_k)'$ is expressed as

$$\widehat{y}_i(\boldsymbol{x}) = \boldsymbol{w}_i'(\boldsymbol{x})\widehat{\boldsymbol{\phi}}_i, \quad i = 1, 2, \ldots, r, \tag{5.5}$$

where $\boldsymbol{w}_i'(\boldsymbol{x})$ is a vector of order $1 \times p_i$ that has the same form as a row of $\boldsymbol{X}_i$, but is evaluated at the point $\boldsymbol{x}$, and $\widehat{\boldsymbol{\phi}}_i$ is the portion of $\widehat{\boldsymbol{\Phi}}$ corresponding to the ith response $(i = 1, 2, \ldots, r)$. Let $\widehat{\boldsymbol{y}}(\boldsymbol{x}) = [\widehat{y}_1(\boldsymbol{x}), \widehat{y}_2(\boldsymbol{x}), \ldots, \widehat{y}_r(\boldsymbol{x})]'$. Then,

$$\widehat{\boldsymbol{y}}(\boldsymbol{x}) = \boldsymbol{W}'(\boldsymbol{x})\widehat{\boldsymbol{\Phi}}, \tag{5.6}$$

where

$$\boldsymbol{W}'(\boldsymbol{x}) = \operatorname{diag}[\boldsymbol{w}_1'(\boldsymbol{x}), \boldsymbol{w}_2'(\boldsymbol{x}), \ldots, \boldsymbol{w}_r'(\boldsymbol{x})].$$

Thus, the variance-covariance matrix of $\widehat{\boldsymbol{y}}(\boldsymbol{x})$ is given by

$$\operatorname{Var}[\widehat{\boldsymbol{y}}(\boldsymbol{x})] = \boldsymbol{W}'(\boldsymbol{x})[\boldsymbol{X}'(\boldsymbol{\Sigma}^{-1} \otimes \boldsymbol{I}_n)\boldsymbol{X}]^{-1}\boldsymbol{W}(\boldsymbol{x}), \tag{5.7}$$

which follows from the fact that

$$\operatorname{Var}[\widehat{\boldsymbol{\Phi}}] = [\boldsymbol{X}'(\boldsymbol{\Sigma}^{-1} \otimes \boldsymbol{I}_n)\boldsymbol{X}]^{-1}.$$

If the error vector $\widetilde{\boldsymbol{\varepsilon}}$ in model (3.16) is distributed as $N(\boldsymbol{0}, \boldsymbol{\Sigma} \otimes \boldsymbol{I}_n)$, then a $(1-\alpha)100\%$ confidence region on $\boldsymbol{\Phi}$ is given by the ellipsoid

$$\{\boldsymbol{\Phi}\colon (\widehat{\boldsymbol{\Phi}} - \boldsymbol{\Phi})'[\boldsymbol{X}'(\boldsymbol{\Sigma}^{-1} \otimes \boldsymbol{I}_n)\boldsymbol{X}](\widehat{\boldsymbol{\Phi}} - \boldsymbol{\Phi}) \leqslant \chi^2_{\alpha,p}\},$$

where $p = \sum_{i=1}^{r} p_i$ with p_i being the number of elements of $\boldsymbol{\phi}_i$, and $\chi^2_{\alpha,p}$ is the upper $100\alpha\%$ point of the chi-squared distribution with p degrees of freedom.

Let $\mathcal{D}_n$ denote an n-point design matrix for the r responses. As was seen earlier in Section 5.1, this design is *D-optimal* if it maximizes the determinant of $\boldsymbol{X}'(\boldsymbol{\Sigma}^{-1} \otimes \boldsymbol{I}_n)\boldsymbol{X}$. This is equivalent to minimizing the size of the aforementioned confidence region since its volume is proportional to $|\boldsymbol{X}'(\boldsymbol{\Sigma}^{-1} \otimes \boldsymbol{I}_n)\boldsymbol{X}|^{-1/2}$. Such a design is more accurately called a *discrete D-optimal design* to distinguish it from the D-optimal design based on continuous design measure theory. By definition, a continuous design is a probability measure, $\kappa(\boldsymbol{x})$, defined on a region χ and satisfies the conditions,

$$\int_{\chi} \mathrm{d}\kappa(\boldsymbol{x}) = 1, \quad \kappa(\boldsymbol{x}) \geqslant 0 \quad \text{for all } \boldsymbol{x} \in \chi$$

(see, for example, Kiefer, 1959; Wynn, 1970). Let $\mathcal{H}$ denote the class of all such design measures. For each $\kappa \in \mathcal{H}$ and a given variance-covariance matrix $\boldsymbol{\Sigma}$, the moment matrix $\mathcal{M}(\kappa, \boldsymbol{\Sigma})$ is defined as

$$\mathcal{M}(\kappa, \boldsymbol{\Sigma}) = \int_{\chi} \boldsymbol{W}(\boldsymbol{x})\boldsymbol{\Sigma}^{-1}\boldsymbol{W}'(\boldsymbol{x})\,\mathrm{d}\kappa(\boldsymbol{x}).$$

In particular, a design measure is discrete if the design consists of points selected in χ. A rational number is assigned to each point representing the fraction of times the point is replicated. If n is the total number of replications, then $\mathcal{D}_n$ is used to denote such a design. In this case, the moment matrix takes the form

$$\mathcal{M}(\mathcal{D}_n, \boldsymbol{\Sigma}) = \frac{1}{n}\boldsymbol{X}'(\boldsymbol{\Sigma}^{-1} \otimes \boldsymbol{I}_n)\boldsymbol{X}.$$

In general, a design measure κ^* is *D-optimal* if it maximizes the determinant of $\mathcal{M}(\kappa, \boldsymbol{\Sigma})$ with respect to $\kappa \in \mathcal{H}$, that is,

$$|\mathcal{M}(\kappa^*, \boldsymbol{\Sigma})| = \sup_{\kappa \in \mathcal{H}} |\mathcal{M}(\kappa, \boldsymbol{\Sigma})|.$$

Fedorov (1972, p. 212) presented two conditions, each of which is necessary and sufficient for D-optimality:

(i) κ^* is D-optimal if and only if it is G-optimal, that is, it minimizes

$$\max_{\boldsymbol{x} \in \chi} \operatorname{tr}\big[\boldsymbol{\Sigma}^{-1}\boldsymbol{\Psi}(\boldsymbol{x}, \kappa, \boldsymbol{\Sigma})\big] \quad \text{with respect to } \kappa \in \mathcal{H},$$

where

$$\boldsymbol{\Psi}(\boldsymbol{x}, \kappa, \boldsymbol{\Sigma}) = \boldsymbol{W}'(\boldsymbol{x})\mathcal{M}^{-1}(\kappa, \boldsymbol{\Sigma})\boldsymbol{W}(\boldsymbol{x}), \quad \boldsymbol{x} \in \chi,\ \kappa \in \mathcal{H}.$$

(ii) κ^* is D-optimal if and only if

$$\max_{\boldsymbol{x} \in \chi} \operatorname{tr}[\boldsymbol{\Sigma}^{-1}\boldsymbol{\Psi}(\boldsymbol{x}, \kappa^*, \boldsymbol{\Sigma})] = p.$$

Conditions (i) and (ii) represent an extension of the Kiefer–Wolfowitz (1960) *Equivalence Theorem* in the single-response case.

Fedorov (1972, Chapter 5) used condition (ii) to generate a D-optimal design in a sequential manner. His procedure, however, requires knowledge of $\boldsymbol{\Sigma}$, which, in general, is unknown. A modification of this procedure was proposed by Wijesinha and Khuri (1987a). They used an estimate of $\boldsymbol{\Sigma}$ at each step of the sequential process, thus eliminating the need to know $\boldsymbol{\Sigma}$. A summary of their procedure follows:

(a) An initial discrete design, $\mathcal{D}_0$, consisting of m_0 points, is chosen. The only condition on $\mathcal{D}_0$ is that $\mathcal{M}(\mathcal{D}_0, \boldsymbol{I}_r)$ be nonsingular. Note that $\boldsymbol{I}_r$ is used here in place of $\boldsymbol{\Sigma}$.

(b) A series of experimental runs are conducted at the m_0 points of $\mathcal{D}_0$, and the resulting data are used to obtain an estimate of $\boldsymbol{\Sigma}$, namely $\widehat{\boldsymbol{\Sigma}}_0 = (\widehat{\sigma}_{0ij})$, where

$$\widehat{\sigma}_{0ij} = \frac{1}{m_0} \boldsymbol{y}_i' [\boldsymbol{I}_{m_0} - \boldsymbol{X}_i(\boldsymbol{X}_i'\boldsymbol{X}_i)^{-1}\boldsymbol{X}_i'][\boldsymbol{I}_{m_0} - \boldsymbol{X}_j(\boldsymbol{X}_j'\boldsymbol{X}_j)^{-1}\boldsymbol{X}_j']\boldsymbol{y}_j.$$

This estimate is the same as the one described in Section 3.4,

(c) A point $\boldsymbol{x}_1$ in the region χ is selected by maximizing the so-called *trace function*, $\operatorname{tr}[\widehat{\boldsymbol{\Sigma}}_0^{-1} \boldsymbol{\Psi}(\boldsymbol{x}, \mathcal{D}_0, \widehat{\boldsymbol{\Sigma}}_0)]$, with respect to $\boldsymbol{x}$ over χ. A new discrete design, $\mathcal{D}_1$, is then constructed by augmenting $\mathcal{D}_0$ with $\boldsymbol{x}_1$. Measurements are subsequently obtained on the r responses using the settings of $\boldsymbol{x}_1$, and the resulting data, along with the data generated by $\mathcal{D}_0$, are used to obtain an updated estimate, $\widehat{\boldsymbol{\Sigma}}_1$, of $\boldsymbol{\Sigma}$.

(d) Step (c) is repeated with $\widehat{\boldsymbol{\Sigma}}_0$ and $\mathcal{D}_0$ replaced by $\widehat{\boldsymbol{\Sigma}}_1$ and $\mathcal{D}_1$, respectively, a new point $\boldsymbol{x}_2$ is found in χ, and then added to $\mathcal{D}_1$. This sequential process continues until at a certain stage, the maximum of the aforementioned trace function becomes close enough to p. This stopping rule is based on Fedorov's condition (ii), which was mentioned earlier.

Several examples were presented by Wijesinha and Khuri (1987a) to illustrate the implementation of their sequential procedure.

More recently, Bischoff (1993) showed that if $\boldsymbol{\Sigma}$ belongs to a special class of positive definite matrices, then the D-optimal design does not depend on the value of $\boldsymbol{\Sigma}$. In this case, the design will be the same as when the responses are uncorrelated and have a common variance.

5.2.2. Designs to increase the power of tests

The D-optimality criterion discussed in Section 5.2.1 is variance related since it improves the precision of estimating the parameter vector for a multiresponse model. In this section, we present another criterion that provides a better detection of model inadequacy.

We recall that in Section 4.2, a multivariate lack of fit test statistic, namely *Roy's largest root*, was presented to check the adequacy of model (4.1). Suppose that in reality the true mean of $\boldsymbol{Y}$ is represented by the model

$$\mathrm{E}(\boldsymbol{Y}) = \widetilde{\boldsymbol{X}}\widetilde{\boldsymbol{\Phi}} + \widetilde{\boldsymbol{Z}}\widetilde{\boldsymbol{\Delta}}, \tag{5.8}$$

where $\widetilde{\boldsymbol{X}}$ and $\widetilde{\boldsymbol{\Phi}}$ are the same as in (4.1),

$$\widetilde{\boldsymbol{Z}} = [\boldsymbol{Z}_1 : \boldsymbol{Z}_2 : \cdots : \boldsymbol{Z}_r], \qquad \widetilde{\boldsymbol{\Delta}} = \operatorname{diag}(\boldsymbol{\Delta}_1, \boldsymbol{\Delta}_2, \ldots, \boldsymbol{\Delta}_r);$$

and $\boldsymbol{Z}_i, \boldsymbol{\Delta}_i$ are such that

$$\mathrm{E}(\boldsymbol{y}_i) = \boldsymbol{X}_i\boldsymbol{\phi}_i + \boldsymbol{Z}_i\boldsymbol{\Delta}_i, \quad i = 1, 2, \ldots, r.$$

The matrix $\boldsymbol{Z}_i$, which is of order $n \times q_i$, along with the unknown vector $\boldsymbol{\Delta}_i$ represent the portion of the "true" model for the ith response that was left out from the fitted model ($i = 1, 2, \ldots, r$). In this case, the matrix $\boldsymbol{G}_1$ in formula (4.4) has the noncentral Wishart distribution with $\nu_{LF} = n - \rho - \nu_{PE}$ degrees of freedom, where, if we recall, ρ is the rank of $\widetilde{\boldsymbol{X}}$ and $\nu_{PE} = \sum_{i=1}^{n_0} (\nu_i - 1)$ with ν_i being the number of replications at the ith of n_0 runs. The noncentrality parameter matrix associated with $\boldsymbol{G}_1$ is of the form

$$\widetilde{\boldsymbol{\Omega}} = \boldsymbol{\Sigma}^{-1/2} \widetilde{\boldsymbol{\Delta}}' \widetilde{\boldsymbol{Z}}' \left[\boldsymbol{I}_n - \widetilde{\boldsymbol{X}} (\widetilde{\boldsymbol{X}}' \widetilde{\boldsymbol{X}})^- \widetilde{\boldsymbol{X}}' \right] \widetilde{\boldsymbol{Z}} \widetilde{\boldsymbol{\Delta}} \boldsymbol{\Sigma}^{-1/2}. \tag{5.9}$$

The power of the lack of fit test is a monotone increasing function of the eigenvalues of $\widetilde{\boldsymbol{\Omega}}$ (see Roy et al., 1971, p. 68). Note that

$$\operatorname{tr}(\widetilde{\boldsymbol{\Omega}}) \geqslant e_{\min}(\boldsymbol{\Sigma}^{-1}) \operatorname{tr}\left\{ \widetilde{\boldsymbol{\Delta}}' \widetilde{\boldsymbol{Z}}' \left[\boldsymbol{I}_n - \widetilde{\boldsymbol{X}} (\widetilde{\boldsymbol{X}}' \widetilde{\boldsymbol{X}})^- \widetilde{\boldsymbol{X}}' \right] \widetilde{\boldsymbol{Z}} \widetilde{\boldsymbol{\Delta}} \right\},$$

where $e_{\min}(\cdot)$ denotes the smallest eigenvalue of a symmetric matrix.

Since the power increases with $\operatorname{tr}(\widetilde{\boldsymbol{\Omega}})$, a reasonable design criterion is the maximization of the trace of the matrix

$$\boldsymbol{M}(\widetilde{\boldsymbol{\Delta}}) = \widetilde{\boldsymbol{\Delta}}' \widetilde{\boldsymbol{Z}}' \left[\boldsymbol{I}_n - \widetilde{\boldsymbol{X}} (\widetilde{\boldsymbol{X}}' \widetilde{\boldsymbol{X}})^- \widetilde{\boldsymbol{X}}' \right] \widetilde{\boldsymbol{Z}} \widetilde{\boldsymbol{\Delta}}. \tag{5.10}$$

This matrix, however, depends on $\widetilde{\boldsymbol{\Delta}}$, which is unknown. Wijensinha and Khuri (1987b) developed a procedure to resolve this dependency problem. Their procedure was motivated by the *maximin method* proposed by Atkinson and Fedorov (1975) and used by Jones and Mitchell (1978) to obtain a design for a single response model.

Wijesinha and Khuri's (1987b) procedure can be summarized as follows: It is shown that

$$\operatorname{tr}\left[\boldsymbol{M}(\widetilde{\boldsymbol{\Delta}}) \right] = \boldsymbol{\gamma}' \boldsymbol{H} (\boldsymbol{I}_r \otimes \boldsymbol{S}) \boldsymbol{H}' \boldsymbol{\gamma},$$

where

$$\begin{aligned}
\boldsymbol{\gamma}' &= [\boldsymbol{\Delta}_1' : \boldsymbol{\Delta}_2' : \cdots : \boldsymbol{\Delta}_r'], \\
\boldsymbol{H} &= \operatorname{diag}(\boldsymbol{H}_1', \boldsymbol{H}_2', \ldots, \boldsymbol{H}_r'), \\
\boldsymbol{S} &= \widetilde{\boldsymbol{Z}}_0' \left[\boldsymbol{I}_n - \widetilde{\boldsymbol{X}} (\widetilde{\boldsymbol{X}}' \widetilde{\boldsymbol{X}})^- \widetilde{\boldsymbol{X}}' \right] \widetilde{\boldsymbol{Z}}_0.
\end{aligned}$$

Here, $\widetilde{\boldsymbol{Z}}_0$ is a matrix whose columns form a basis for the column space of $\widetilde{\boldsymbol{Z}}$ (see model 5.8), and the matrix $\boldsymbol{H}_i$ is such that

$$\boldsymbol{Z}_i = \widetilde{\boldsymbol{Z}}_0 \boldsymbol{H}_i, \quad i = 1, 2, \ldots, r.$$

Consider now two alternative design strategies:

Strategy 1. Let Π denote the region

$$\Pi = \{\gamma\colon\ \gamma' \boldsymbol{T} \gamma \geqslant \tau\},$$

where $\boldsymbol{T}$ is a known positive definite matrix chosen appropriately, and τ is some positive constant. One possible choice for $\boldsymbol{T}$ is based on using region moment matrices corresponding to a certain region of interest (see Wijesinha and Khuri, 1987b, p. 182). The quantity $\gamma' \boldsymbol{T} \gamma$ provides a measure of inadequacy of the fitted model. A positive value of this quantity is an indication that $\gamma \neq \mathbf{0}$ and, therefore, the model is inadequate.

Under Strategy 1, a design is chosen so as to maximize the quantity Λ_1 given by

$$\Lambda_1 = \inf_{\gamma \in \Pi} \{\gamma' \boldsymbol{H}(\boldsymbol{I}_r \otimes \boldsymbol{S})\boldsymbol{H}' \gamma\}.$$

This is a multiresponse extension of the Λ_1-optimality criterion used by Jones and Mitchell (1978). It can be shown that

$$\Lambda_1 = \tau e_{\min}\{\boldsymbol{T}^{-1}\boldsymbol{H}(\boldsymbol{I}_r \otimes \boldsymbol{S})\boldsymbol{H}'\}.$$

A design that maximizes $e_{\min}\{\boldsymbol{T}^{-1}\boldsymbol{H}(\boldsymbol{I}_r \otimes \boldsymbol{S})\boldsymbol{H}'\}$ is called a *Λ_1-optimal design.* It should be noted that this criterion is not meaningful if Λ_1 is zero for any design. This occurs, for example, when $r(n-\rho) < \sum_{i=1}^r q_i$, where ρ is the rank of $\widetilde{\boldsymbol{X}}$. In this case, the matrix $\boldsymbol{H}(\boldsymbol{I}_r \otimes \boldsymbol{S})\boldsymbol{H}'$, which is of order $\sum_{i=1}^r q_i \times \sum_{i=1}^r q_i$ and rank not exceeding $r(n-\rho)$, will be singular causing Λ_1 to have the value zero no matter what design is used.

Strategy 2. An alternative design strategy is to maximize the quantity Λ_2, the average of $\gamma' \boldsymbol{H}(\boldsymbol{I}_r \otimes \boldsymbol{S})\boldsymbol{H}' \gamma$ over the boundary of the region Π, that is,

$$\Lambda_2 = \frac{1}{\int_{\Pi_0} \mathrm{d}\sigma} \int_{\Pi_0} \gamma' \boldsymbol{H}(\boldsymbol{I}_r \otimes \boldsymbol{S})\boldsymbol{H}' \gamma \,\mathrm{d}\sigma,$$

where $\mathrm{d}\sigma$ is the differential of the surface area of the ellipsoid $\Pi_0 = \{\gamma\colon\ \gamma' \boldsymbol{T} \gamma = \tau\}$. Using an identity given in Jones and Mitchell (1978, p. 544), we can write Λ_2 as

$$\Lambda_2 = \frac{\tau}{\sum_{i=1}^r q_i} \mathrm{tr}\big[\boldsymbol{T}^{-1}\boldsymbol{H}(\boldsymbol{I}_r \otimes \boldsymbol{S})\boldsymbol{H}'\big].$$

A design that maximizes Λ_2 is called a *Λ_2-optimal design.* Equivalently, such a design maximizes Λ_2', where

$$\Lambda_2' = \mathrm{tr}\big[\boldsymbol{T}^{-1}\boldsymbol{H}(\boldsymbol{I}_r \otimes \boldsymbol{S})\boldsymbol{H}'\big].$$

This criterion can be applied even when the matrix $\boldsymbol{H}(\boldsymbol{I}_r \otimes \boldsymbol{S})\boldsymbol{H}'$ is singular.

Wijesinha and Khuri (1987b) presented a sequential procedure for the generation of Λ_2-optimal designs. As in Section 5.2.1, an initial design $\mathcal{D}_0$ is chosen. Henceforth, design points are selected one at a time and added to the previous design. The number of points of $\mathcal{D}_0$ must at least be equal to the rank of the matrix $\widetilde{\boldsymbol{X}}$ (see model 4.1). The stopping rule for this sequential procedure is based on making the so-called Fréchet derivative of Λ_2' arbitrarily close to zero (for a definition of this derivative, see Wijesinha and Khuri, 1987b, p. 184, and Theorem 2, p. 186).

EXAMPLE (Wijesinha and Khuri, 1987b, pp. 188–191). Consider a multiresponse experiment with three responses, y_1, y_2, y_3, and three input variables, x_1, x_2, x_3, coded so that $-1 \leqslant x_i \leqslant 1$ $(i = 1, 2, 3)$. The fitted models are

$$y_1 = \beta_{10} + \beta_{11}x_1 + \beta_{13}x_3 + \beta_{113}x_1x_3 + \varepsilon_1,$$

$$y_2 = \beta_{20} + \sum_{i=1}^{3} \beta_{2i}x_i + \beta_{213}x_1x_3 + \beta_{211}x_1^2 + \beta_{233}x_3^2 + \varepsilon_2,$$

$$y_3 = \beta_{30} + \sum_{i=1}^{3} \beta_{3i}x_i + \varepsilon_3.$$

The true means of the three responses, however, are represented by the models

$$\mathrm{E}(y_1) = \beta_{10} + \beta_{11}x_1 + \beta_{13}x_3 + \beta_{113}x_1x_3 + \beta_{111}x_1^2 + \beta_{133}x_3^2,$$

$$\mathrm{E}(y_2) = \beta_{20} + \sum_{i=1}^{3} \beta_{2i}x_i + \beta_{213}x_1x_3 + \beta_{211}x_1^2 + \beta_{233}x_3^2 + \beta_{212}x_1x_2 + \beta_{223}x_2x_3 + \beta_{222}x_2^2,$$

$$\mathrm{E}(y_3) = \beta_{30} + \sum_{i=1}^{3} \beta_{3i}x_i + \beta_{312}x_1x_2 + \beta_{313}x_1x_3 + \beta_{323}x_2x_3 + \beta_{311}x_1^2 + \beta_{322}x_2^2 + \beta_{333}x_3^2.$$

The initial design, $\mathcal{D}_0$, chosen for this example is

x_1	x_2	x_3
0.5	−0.6	0.3
1.0	1.0	−1.0
0.8	0.7	−1.0
1.0	0.3	−1.0
0.4	0.6	0.7
0.5	−0.8	0.9
1.0	−1.0	0.4

For this design, the rank of $\widetilde{\boldsymbol{X}}$ is equal to 7, which is no larger than the number of points in $\mathcal{D}_0$, as it should be.

Table 5.1

n	x_1	x_2	x_3	Λ_2'
8	−1.000	−1.000	−1.000	0.0000
9	−1.000	0.880	0.901	0.6116
10	−1.000	1.000	−1.000	0.8424
11	0.620	−1.000	−1.000	4.0733
12	−1.000	−1.000	1.000	7.4027
13	1.000	0.995	1.000	8.5210
14	−1.000	−1.000	−0.120	8.3795
15	1.000	−1.000	−1.000	10.4716
16	−1.000	0.880	1.000	11.6581
17	−1.000	−1.000	0.995	12.0597
18	−1.000	0.880	−1.000	12.1594
19	−1.000	−1.000	−1.000	12.1818
20	1.000	−1.000	0.894	12.1063
21	1.000	1.000	1.000	12.7122
22	1.000	−1.000	−1.000	13.1270
23	−1.000	0.880	1.000	13.2994
24	−0.940	−1.000	0.899	13.4066
25	1.000	1.000	−1.000	13.3929
26	−1.000	0.898	−1.000	13.8408
27	−1.000	−1.000	−1.000	13.8770
28	1.000	−1.000	−1.000	13.8495
29	1.000	−1.000	1.000	14.0416
30	1.000	1.000	1.000	14.2208
31	−1.000	0.894	1.000	14.3827
32	−1.000	−1.000	0.880	14.4395
33	−1.000	−1.000	1.000	14.4925
34	1.000	1.000	−1.000	14.3978
35	−1.000	1.000	−1.000	14.6391
36	−1.000	−1.000	−1.000	14.7352
37	0.910	−1.000	−1.000	14.7178
38	1.000	−1.000	1.000	14.7342
39	1.000	1.000	1.000	14.8585
40	1.000	1.000	1.000	14.9712
41	−1.000	−1.000	−1.000	14.9461
42	1.000	−1.000	−1.000	14.9103

Several points, selected one at a time, were added to $\mathcal{D}_0$. The process was terminated when the Fréchet derivative of Λ_2' reached a value close enough to zero. The augmented points are shown in Table 5.1. Note that the maximum value attained by Λ_2' is 14.9712.

More recently, Wijesinha (1993) applied the maximin technique used by Jones and Mitchell (1978) to improve the power of a test concerning a general linear hypothesis for model (4.1) (see Section 4.1). In another related work, Wijesinha and Khuri (1991) considered the effect of nonnormality on *Pillai's trace*, which, if we recall, is one of four multivariate test statistics for testing a general linear hypothesis. They used robustness of the test to nonnormality of the error distribution as a criterion for the choice of a design. A design obtained under this criterion is called *robust*. It is used to reduce the sensitivity of the test to a possible failure of the normality assumption.

5.2.3. The integrated mean squared error criterion

Box and Draper (1959) considered the minimization of the *integrated mean squared error*, denoted by J, as a criterion for the choice of a design in the single-response case. Recently, Kim and Draper (1994) presented an extension of this criterion to the case of two responses. Their multiresponse version of J is a matrix $\boldsymbol{J}$ of order 2×2, which, among other things, depends on $\boldsymbol{\Sigma}$, the variance-covariance matrix for the two responses. The fitted models were assumed to be of the first order in only one input variable x. The choice of design is based on the minimization of the trace of $\boldsymbol{J}$. This, however, requires knowledge of $\boldsymbol{\Sigma}$ as well as the values of the "true" models' parameters.

5.2.4. Rotatable multiresponse designs

Box and Hunter (1957) introduced the design criterion of *rotatability* for a single-response model. By definition, a design is said to be *rotatable* if the prediction variance is constant at all points that are equidistant from the design center. An extension of this criterion to linear multiresponse models was presented by Khuri (1990c). The following is a brief account of the so-called *multiresponse rotatability*: Consider the linear multiresponse model (3.16). Suppose that all the models in (3.15) depend on the same set of input variables, $x_1, x_2, \ldots, x_k$, with the ith model being a complete e_ith-order model ($i = 1, 2, \ldots, r$). The vector $\widehat{\boldsymbol{y}}(\boldsymbol{x})$ of predicted responses is given by formula (5.6), and the corresponding variance-covariance matrix $\mathrm{Var}[\widehat{\boldsymbol{y}}(\boldsymbol{x})]$ is shown in formula (5.7). As in the single-response case, if $\mathrm{Var}[\widehat{\boldsymbol{y}}(\boldsymbol{x})]$ is constant at all points that are equidistant from the design center, then the design is said to be *multiresponse rotatable*. In this case, the variances and covariances of the predicted responses are constant on hyperspheres (in a k-dimensional Euclidean space) centered at the design center. Khuri (1990c) presented necessary and sufficient conditions for *multiresponse rotatability* similar to those developed by Box and Hunter (1957). He concluded that *multiresponse rotatability* can be achieved if and only if the design is rotatable for a single-response model of degree $e_0 = \max_{1 \leqslant i \leqslant r}(e_i)$. This is a model having the highest degree of all the response models under consideration.

6. Multiresponse optimization

One primary objective of a response surface investigation is the determination of operating conditions that yield an optimum response. For example, in an industrial process, it may be of interest to determine the levels of temperature, pressure, and flow rate that maximize the product yield. This aspect of response surface methodology received a great deal of attention in the fifties and sixties (see, for example, Box and Wilson, 1951; Box and Hunter, 1954; Davies, 1956; Draper, 1963; Hunter, 1958).

Optimization in a multiresponse setting was addressed somewhat belatedly. By multiresponse optimization we mean determining operating conditions on a set of input variables that yield optimal (or near optimal) values for several response variables considered simultaneously. For example, in a chemical process, it may be of interest to determine operating conditions on a group of process variables that result in a product with superior (optimal) characteristics. Graphical techniques were initially used in the early development of response surface methodology. By superimposing response contours and visually searching for a common region where the responses achieve near optimal values, it was possible to arrive at a location (or locations) of a so-called *compromise optimum*. This technique was described in Hill and Hunter (1966) in reference to an article by Lind et al. (1960). Obviously, such a technique is not feasible in situations involving several responses and more than three input variables. Furthermore, no account is given of possible correlations among the responses.

6.1. The desirability function approach

In an industrial setting, a product may be required to have several characteristics in order to be marketable. Each characteristic is represented by a certain response function whose value goes up as the desirability of the characteristic increases. Thus the problem of finding a set of conditions that result in a product with a desirable combination of marketable characteristics is equivalent to a simultaneous maximization of the corresponding response functions. Harrington (1965), and later Derringer and Suich (1980), proposed a solution to this problem by using the so-called *desirability function approach*. In this approach, multiresponse data are used to obtain adequate model fits for all the responses under consideration. Each predicted response is then transformed into a desirability index having values inside the closed interval $[0, 1]$. Thus the ith predicted response $\widehat{y}_i(\boldsymbol{x})$ is transformed into $\eta_i(\boldsymbol{x})$, where $0 \leqslant \eta_i(\boldsymbol{x}) \leqslant 1$ $(i = 1, 2, \ldots, r)$. The value of $\eta_i(\boldsymbol{x})$ increases with the desirability of the corresponding response (product characteristic). The choice of a transformation is subjective and depends on how the user assesses the desirability of a given product characteristic. Harrington (1965) used exponential-type transformations of the form

$$\eta_i(\boldsymbol{x}) = \exp[-|\widehat{y}_i(\boldsymbol{x})|^a],$$

where a is a positive number specified by the user. Derringer and Suich (1980) used power transformations that allow the user to specify minimum and maximum acceptable values for each response.

The individual desirability indices are combined using the geometric mean, namely

$$\eta(\boldsymbol{x}) = \left[\prod_{i=1}^{r} \eta_i(\boldsymbol{x})\right]^{1/r},$$

where r is the number of responses, or product characteristics, considered. The rationale for choosing the geometric mean is the following: If the product does not meet a specified characteristic, that is, $\eta_i = 0$ for some $i = 1, 2, \ldots, r$, then the product is considered unacceptable. Thus $\eta(\boldsymbol{x})$ measures the overall desirability of the product.

Note that $0 \leqslant \eta(\boldsymbol{x}) \leqslant 1$ and large values of $\eta(\boldsymbol{x})$ correspond to a highly desirable product. Consequently, maximizing $\eta(\boldsymbol{x})$ over a region of interest can produce desirable conditions for a marketable product. We note that the multiresponse optimization problem in this case has been reduced to the maximization of the single function $\eta(\boldsymbol{x})$.

This approach is simple, but is subjective since it depends on how the user interprets the desirability of a given product characteristic. It is therefore possible that different users may arrive at different sets of optimum conditions. Furthermore, this approach does not account for any correlations that may exist among the responses, or the possibility of having heteroscedastic errors associated with the responses. We recall that the variance-covariance matrix $\boldsymbol{\Sigma}$ for the responses affect the fit of the models. It should therefore be taken into consideration in any simultaneous optimization of the responses.

6.2. *The generalized distance approach*

Khuri and Conlon (1981) considered the simultaneous optimization of several response functions represented by models of the form (3.15) having equal $\boldsymbol{X}_i$ matrices. From Section 3.4 it may be recalled that when $\boldsymbol{X}_1 = \boldsymbol{X}_2 = \cdots = \boldsymbol{X}_r = \boldsymbol{X}_0$, the *BLUE* of $\boldsymbol{\Phi}$ in model (3.16) is expressible as

$$\widehat{\boldsymbol{\Phi}} = \left[\boldsymbol{I}_r \otimes (\boldsymbol{X}_0'\boldsymbol{X}_0)^{-1}\boldsymbol{X}_0'\right]\boldsymbol{y}.$$

The vector of predicted responses at a point $\boldsymbol{x}$ in a region of interest R is given by formula (5.6), which, in this case, can be written as

$$\widehat{\boldsymbol{y}}(\boldsymbol{x}) = \left[\boldsymbol{I}_r \otimes \boldsymbol{w}_0'(\boldsymbol{x})(\boldsymbol{X}_0'\boldsymbol{X}_0)^{-1}\boldsymbol{X}_0'\right]\boldsymbol{y},$$

since $\boldsymbol{W}'(\boldsymbol{x}) = \boldsymbol{I}_r \otimes \boldsymbol{w}_0'(\boldsymbol{x})$, where $\boldsymbol{w}_0'(\boldsymbol{x})$ is the common value of $\boldsymbol{w}_i'(\boldsymbol{x})$ $(i = 1, 2, \ldots, r)$. Thus

$$\operatorname{Var}[\widehat{\boldsymbol{y}}(\boldsymbol{x})] = \boldsymbol{w}_0'(\boldsymbol{x})(\boldsymbol{X}_0'\boldsymbol{X}_0)^{-1}\boldsymbol{w}_0(\boldsymbol{x})\boldsymbol{\Sigma}.$$

An unbiased estimate of $\operatorname{Var}[\widehat{\boldsymbol{y}}(\boldsymbol{x})]$ is then given by

$$\widehat{\operatorname{Var}}[\widehat{\boldsymbol{y}}(\boldsymbol{x})] = \boldsymbol{w}_0'(\boldsymbol{x})(\boldsymbol{X}_0'\boldsymbol{X}_0)^{-1}\boldsymbol{w}_0(\boldsymbol{x})\widehat{\boldsymbol{\Sigma}}_u,$$

where

$$\widehat{\boldsymbol{\Sigma}}_u = \frac{1}{n - p_0}\boldsymbol{Y}'\left[\boldsymbol{I}_n - \boldsymbol{X}_0(\boldsymbol{X}_0'\boldsymbol{X}_0)^{-1}\boldsymbol{X}_0'\right]\boldsymbol{Y},$$

$\boldsymbol{Y}$ is the multiresponse data matrix, $[\boldsymbol{y}_1 : \boldsymbol{y}_2 : \cdots : \boldsymbol{y}_r]$, and p_0 is the number of columns of $\boldsymbol{X}_0$.

Let $\widehat{\mu}_i$ denote the optimal value of $\widehat{y}_i(\boldsymbol{x})$ over R, obtained individually and independently of the remaining responses. We refer to this value as the *individual optimum* of the ith predicted response $(i = 1, 2, \ldots, r)$. Let $\widehat{\boldsymbol{\mu}} = (\widehat{\mu}_1, \widehat{\mu}_2, \ldots, \widehat{\mu}_r)'$. In general, these individual optima are not necessarily attained at the same set of conditions. The term *ideal optimum* is used to refer to a situation in which all such optima are achieved at the same value of $\boldsymbol{x}$ in R. Khuri and Conlon (1981) introduced a *metric* (distance function), denoted by $\rho[\widehat{\boldsymbol{y}}(\boldsymbol{x}), \widehat{\boldsymbol{\mu}}]$, which measures the deviation of $\widehat{\boldsymbol{y}}(\boldsymbol{x})$ from the ideal optimum. They proposed minimizing this metric with respect to $\boldsymbol{x}$ over R in order to arrive at conditions under which the predicted responses deviate as little as possible from the ideal optimum. Optimal conditions found in this manner lead to a so-called *compromise optimum.*

Several metrics were presented in Khuri and Conlon (1981), namely

$$\rho_1[\widehat{\boldsymbol{y}}(\boldsymbol{x}), \widehat{\boldsymbol{\mu}}] = \left\{(\widehat{\boldsymbol{y}}(\boldsymbol{x}) - \widehat{\boldsymbol{\mu}})' \left(\widehat{\mathrm{Var}}[\widehat{\boldsymbol{y}}(\boldsymbol{x})]\right)^{-1} (\widehat{\boldsymbol{y}}(\boldsymbol{x}) - \widehat{\boldsymbol{\mu}})\right\}^{1/2}$$

$$= \left[\frac{(\widehat{\boldsymbol{y}}(\boldsymbol{x}) - \widehat{\boldsymbol{\mu}})' \widehat{\boldsymbol{\Sigma}}_u^{-1} (\widehat{\boldsymbol{y}}(\boldsymbol{x}) - \widehat{\boldsymbol{\mu}})}{\boldsymbol{w}_0'(\boldsymbol{x})(\boldsymbol{X}_0'\boldsymbol{X}_0)^{-1}\boldsymbol{w}_0(\boldsymbol{x})}\right]^{1/2},$$

$$\rho_2[\widehat{\boldsymbol{y}}(\boldsymbol{x}), \widehat{\boldsymbol{\mu}}] = \left\{\sum_{i=1}^{r} \frac{(\widehat{y}_i(\boldsymbol{x}) - \widehat{\mu}_i)^2}{[\widehat{\sigma}_{uii}\boldsymbol{w}_0'(\boldsymbol{x})(\boldsymbol{X}_0'\boldsymbol{X}_0)^{-1}\boldsymbol{w}_0(\boldsymbol{x})]}\right\}^{1/2},$$

where $\widehat{\sigma}_{uii}$ is the ith diagonal element of $\widehat{\boldsymbol{\Sigma}}_u$ $(i = 1, 2, \ldots, r)$,

$$\rho_3[\widehat{\boldsymbol{y}}(\boldsymbol{x}), \widehat{\boldsymbol{\mu}}] = \left[\sum_{i=1}^{r} \frac{(\widehat{y}_i(\boldsymbol{x}) - \widehat{\mu}_i)^2}{\widehat{\mu}_i^2}\right]^{1/2}.$$

The metric ρ_2 is appropriate whenever the responses are believed to be statistically independent. The third metric measures the total relative deviation from the individual optima and can be used when $\widehat{\boldsymbol{\Sigma}}_u$ is ill conditioned.

It should be noted that in the process of minimizing any of the aforementioned metrics, $\widehat{y}_i(\boldsymbol{x})$ is treated as fixed (not random) for $i = 1, 2, \ldots, r$. To account for the variation caused by the randomness of $\widehat{y}_i(\boldsymbol{x})$ $(i = 1, 2, \ldots, r)$, Khuri and Conlon (1981) obtained a confidence region C_μ on $\boldsymbol{\mu} = (\mu_1, \mu_2, \ldots, \mu_r)'$, where μ_i denotes the individual optimum of the true mean of the ith response $(i = 1, 2, \ldots, r)$ over the region R. For a fixed $\boldsymbol{x} \in R$, the maximum of $\rho[\widehat{\boldsymbol{y}}(\boldsymbol{x}), \boldsymbol{\zeta}]$ with respect to $\boldsymbol{\zeta} \in C_\mu$ is obtained. This maximum, which provides a conservative estimate of $\rho[\widehat{\boldsymbol{y}}(\boldsymbol{x}), \boldsymbol{\mu}]$, is then minimized with respect to $\boldsymbol{x}$ over R. A computer algorithm for the implementation of this procedure was written by Conlon (1988). An electronic copy of this algorithm, complete with examples and code, can be downloaded from the Internet at: ftp://ftp.stat.ufl.edu/pub/mr.tar.Z.

6.3. *Other related procedures*

There are other optimization procedures that involve a group of response variables. However, these procedures are not concerned with simultaneous optimization as was the case in Sections 6.1 and 6.2. Instead, one response is identified as a *primary response* to be optimized subject to constraints placed on the remaining responses, which are labeled as *secondary responses.* Myers and Carter (1973) considered only one secondary response. Their optimization procedure was developed in a manner similar to that of *ridge analysis* (see Draper, 1963). The more general situation involving several secondary responses was treated by Biles (1975). His procedure employed a modification of the *method of steepest ascent* initially described in Box and Wilson (1951). Roth and Stewart (1969) presented a practical example concerning a chemical reaction which involved two secondary responses.

More recently, Böckenholt (1989) proposed a method for testing a hypothesis concerning the equality of the individual optima of several response functions represented by second-order models. He also showed how to obtain a confidence region on the location of the common optimum if the hypothesis is not rejected.

7. Concluding remarks

The use of multiresponse techniques in response surface methodology is considered to be a relatively novel endeavor. Both the design and analysis of multiresponse experiments have received precious little attention even though they are sorely needed. It is a common knowledge that experiments with two or more responses are quite prevalent. I believe that there are several reasons for this seemingly unsettling phenomenon. Multiresponse surface methodology is still in its early stages of development. The few techniques that have been introduced in this subject area are not well known, and consequently, not frequently used by research workers. This may be partially attributed to a feeling of "insecurity" which keeps some potential users from venturing beyond the familiar realm of the first dimension. Such a statement, however, is not meant to place the blame entirely on the users. Developers of multiresponse surface methodology bear some of the blame too. They have the responsibility of making their procedures easily accessible to the users. Unfortunately, this has not been the case, in general, due primarily to a severe software shortage in the multiresponse area. Data analysts and potential users are attracted mostly to statistical techniques for which software is available. This problem should therefore be rectified in order to facilitate and promote the applicability of the multiresponse approach.

There are several areas in multiresponse surface methodology that need further development and expansion. Here are some examples:

1. In the design area, the work by Kim and Draper (1994) should be extended so that it applies to more general models (of order higher than one with several input variables). Furthermore, since Kim and Draper's generated designs depend on knowing the value of the variance-covariance matrix $\boldsymbol{\Sigma}$, a procedure that does not require such a knowledge is needed. This can be achieved by generating designs in a sequential manner that allows the estimation of $\boldsymbol{\Sigma}$.

2. The optimization procedure by Khuri and Conlon (1981), which was described in Section 6.2, assumes that all response models are of the same form and use the same

design. This restricts the applicability of the procedure to more general experimental situations.

3. There is a need to develop multiresponse techniques that apply when some of the responses have discrete distributions, rather than the usual normal distribution. Such a multiresponse situation would more appropriately be handled by using *generalized linear models* (see McCullagh and Nelder, 1989, Chapter 6). The construction of designs and determination of optimum operating conditions under this more general setting can be a challenging problem.

4. There is a need to consider multiresponse models that contain random effects, such as block effects, in addition to the usual fixed polynomial effects. The presence of random effects results in *mixed-effects models*. Khuri (1996) considered such models in the single-response case.

5. The determination of optimum conditions should take into account the size of the prediction variance. This is true in the single-response case and even more so in the multiresponse case. Existing multiresponse optimization techniques should be modified in order to attain adequate precisions for the estimated optima.

6. The recent push for quality in industry has brought response surface methodology to the attention of many users. This is certainly true in the single-response case, but, sadly enough, however, not in the multiresponse case. The review article by Myers, Khuri and Vining (1992) made reference to response surface alternatives to the *Taguchi robust parameter design* approach. The need to extend this methodology to the multiresponse case is not only interesting, but absolutely imperative. It is hoped that a concerted effort in this direction will make multiresponse surface methodology an indispensable tool in the field of industrial research.

References

Anderson, E. (1960). A semigraphical method for the analysis of complex problems. *Technometrics* **2**, 387–391.

Andrews, D. F. (1972). Plots of high-dimensional data. *Biometrics* **28**, 125–136.

Asimov, D. (1985). The grand tour: A tool for viewing multidimensional data. *SIAM J. Sci. Statist. Comput.* **6**, 128–143.

Atkinson, A. C. and V. V. Fedorov (1975). Optimal design: Experiments for discriminating between several models. *Biometrika* **62**, 289–303.

Bates, D. M. and D. G. Watts (1984). A multi-response Gauss–Newton algorithm. *Comm. Statist. Simulation Comput.* **13**, 705–715.

Bates, D. M. and D. G. Watts (1985). Multiresponse estimation with special application to linear systems of differential equations. *Technometrics* **27**, 329–339.

Bates, D. M. and D. G. Watts (1987). A generalized Gauss–Newton procedure for multi-response parameter estimation. *SIAM J. Sci. Statist. Comput.* **8**, 49–57.

Biles, W. E. (1975). A response surface method for experimental optimization of multi-response processes. *Ind. Eng. Chem. Process Des. Dev.* **14**, 152–158.

Bischoff, W. (1993). On *D*-optimal designs for linear models under correlated observations with an application to a linear model with multiple response. *J. Statist. Plann. Inference* **37**, 69–80.

Böckenholt, U. (1989). Analyzing optima in the exploration of multiple response surfaces. *Biometrics* **45**, 1001–1008.

Box, G. E. P. and N. R. Draper (1959). A basis for the selection of a response surface design. *J. Amer. Statist. Assoc.* **54**, 622–654.

Box, G. E. P. and N. R. Draper (1965). The Bayesian estimation of common parameters from several responses. *Biometrika* **52**, 355–365.

Box, G. E. P. and N. R. Draper (1987). *Empirical Model-Building and Response Surfaces*. Wiley, New York.

Box, G. E. P. and J. S. Hunter (1954). A confidence region for the solutoin of a set of simultaneous equations with an application to experimental design. *Biometrika* **41**, 190–199.

Box, G. E. P. and J. S. Hunter (1957). Multi-factor experimental designs for exploring response surfaces. *Ann. Math. Statist.* **28**, 195–241.

Box, G. E. P. and H. L. Lucas (1959). Design of experiments in nonlinear situations. *Biometrika* **46**, 77–90.

Box, G. E. P. and K. B. Wilson (1951). On the experimental attainment of optimum conditions. *J. Roy. Statist. Soc. Ser. B* **13**, 1–45.

Box, G. E. P., W. G. Hunter, J. F. MacGregor and J. Erjavec (1973). Some problems associated with the analysis of multiresponse data. *Technometrics* **15**, 33–51.

Box, M. J. (1970). Improved parameter estimation. *Technometrics* **12**, 219–229.

Box, M. J. (1971). A parameter estimation criterion for multiresponse models applicable when some observations are missing. *J. Roy. Statist. Soc. Ser. C* **20**, 1–7.

Box, M. J. and N. R. Draper (1972). Estimation and design criteria for multiresponse non-linear models with non-homogeneous variance. *J. Roy. Statist. Soc. Ser. C* **21**, 13–24.

Box, M. J., N. R. Draper and W. G. Hunter (1970). Missing values in multiresponse nonlinear model fitting. *Technometrics* **12**, 613–620.

Chambers, J. M., W. S. Cleveland, B. Kleiner and P. A. Tukey (1983). *Graphical Methods for Data Analysis*. Wadsworth, Belmont, CA.

Chernoff, H. (1953). Locally optimal designs for estimating parameters. *Ann. Math. Statist.* **24**, 586–602.

Chernoff, H. (1973). The use of faces to represent points in k-dimensional space graphically. *J. Amer. Statist. Assoc.* **68**, 361–368.

Cleveland, W. S. (1993). *Visualizing Data*. AT&T Bell Laboratories, Murray Hill, NJ.

Conlon, M. (1988). MR: Multiple response optimization. Report No. 322, Department of Statistics, University of Florida, Gainesville, FL.

Davies, O. L. (1956). *The Design and Analysis of Industrial Experiments*. Hafner, New York.

Derringer, G. and R. Suich (1980). Simultaneous optimization of several response variables. *J. Quality Tech.* **12**, 214–219.

Draper, N. R. (1963). Ridge analysis of response surfaces. *Technometrics* **5**, 469–479.

Draper, N. R. and W. G. Hunter (1966). Design of experiments for parameter estimation in multiresponse situations. *Biometrika* **53**, 525–533.

Draper, N. R. and W. G. Hunter (1967). The use of prior distributions in the design of experiments for parameter estimation in non-linear situations: Multiresponse case. *Biometrika* **54**, 662–665.

Embrechts, P. and A. M. Herzberg (1991). Variations of Andrews' plots. *Internat. Statist. Rev.* **59**, 175–194.

Everitt, B. (1978). *Graphical Techniques for Multivariate Data*. North-Holland, New York.

Fedorov, V. V. (1972). *Theory of Optimal Experiments*. Academic Press, New York.

Fienberg, S. E. (1979). Graphical methods in statistics. *Amer. Statist.* **33**, 165–178.

Friedman, H. P., E. S. Farrell, R. M. Goldwyn, M. Miller and J. H. Siegel (1972). A graphic way of describing changing multivariate patterns. In: *Proc. Sixth Int. Symposium on Computer Science and Statistics*. University of California, Berkeley, CA, 56–59.

Friedman, J. H. and J. W. Tukey (1974). A projection pursuit algorithm for exploratory data analysis. *IEEE Trans. Comput.* **C-23**, 881–890.

Gennings, C., K. S. Dawson, W. H. Carter, Jr. and R. H. Myers (1990). Interpreting plots of a multidimensional dose-response surface in a parallel coordinate system. *Biometrics* **46**, 719–735.

Gnanadesikan, R. (1977). *Methods for Statistical Data Analysis of Multivariate Observations*. Wiley, New York.

Harrington, E. C., Jr. (1965). The desirability function. *Indust. Quality Contr.* **21**, 494–498.

Hatzis, C. and K. Larntz (1992). Optimal design in nonlinear multiresponse estimation: Poisson model for filter feeding. *Biometrics* **48**, 1235–1248.

Hill, W. J. and W. G. Hunter (1966). A review of response surface methodology: A literature review. *Technometrics* **8**, 571–590.

Huber, P. J. (1985). Projection pursuit (with discussion). *Ann. Statist.* **13**, 435–525.
Hunter, J. S. (1958). Determination of optimum operating conditions by experimental methods: Part II-1. Models and methods. *Indust. Quality Contr.* **15**, 16–24.
Jones, E. R. and T. J. Mitchell (1978). Design criteria for detecting model inadquacy. *Biometrika* **65**, 541–551.
Jones, M. C. and R. Sibson (1987). What is projection pursuit? *J. Roy. Statist. Soc. Ser. A* **150**, 1–36.
Khuri, A. I. (1985). A test for lack of fit of a linear multiresponse model. *Technometrics* **27**, 213–218.
Khuri, A. I. (1986). Exact tests for the comparison of correlated response models with an unknown dispersion matrix. *Technometrics* **28**, 347–357.
Khuri, A. I. (1990a). Analysis of multiresponse experiments: A review. In: S. Ghosh, ed., *Statistical Design and Analysis of Industrial Experiments*. Dekker, New York, 231–246.
Khuri, A. I. (1990b). The effect of response scaling on the detection of linear dependencies among multiresponse data. *Metrika* **37**, 217–231.
Khuri, A. I. (1990c). Multiresponse rotatability. *J. Statist. Plann. Inference* **25**, 1–6.
Khuri, A. I. (1996). Response surface models with mixed effects. *J. Quality Tech.* **28**, 177–186.
Khuri, A. I. and M. Conlon (1981). Simultaneous optimization of multiple responses represented by polynomial regression functions. *Technometrics* **23**, 363–375.
Khuri, A. I. and J. A. Cornell (1996). *Response Surfaces*, 2nd edn. Dekker, New York.
Kiefer, J. (1959). Optimum experimental designs (with discussion). *J. Roy. Statist. Soc. Ser. B* **21**, 272–319.
Kiefer, J. and J. Wolfowitz (1960). The equivalence of two extremum problems. *Canad. J. Math.* **12**, 363–366.
Kim, W. B. and N. R. Draper (1994). Choosing a design for straight line fits to two correlated responses. *Statist. Sinica* **4**, 275–280.
Kleiner, B. and J. A. Hartigan (1981). Representing points in many dimensions by trees and castles (with comments). *J. Amer. Statist. Assoc.* **76**, 260–276.
Levy, M. S. and J. W. Neill (1990). Testing for lack of fit in linear multiresponse models based on exact or near replicates. *Comm. Statist. Theory Methods* **19**, 1987–2002.
Lind, E. E., J. Goldin and J. B. Hickman (1960). Fitting yield and cost response surfaces. *Chem. Eng. Progr.* **56**, 62–68.
McCullagh, P. and J. A. Nelder (1989). *Generalized Linear Models*, 2nd edn. Chapman and Hall, London.
McDonald, L. (1975). Tests for the general linear hypothesis under the multiple design multivariate linear model. *Ann. Statist.* **3**, 461–466.
McLean, D. D., D. J. Pritchard, D. W. Bacon and J. Downie (1979). Singularities in multiresponse modeling. *Technometrics* **21**, 291–298.
Mead, R. and D. J. Pike (1975). A review of response surface methodology from a biometric viewpoint. *Biometrics* **31**, 803–851.
Morrison, D. F. (1976). *Multivariate Statistical Methods*, 2nd edn. McGraw-Hill, New York.
Muirhead, R. J. (1982). *Aspects of Multiviarate Statistical Theory*. Wiley, New York.
Myers, R. H. (1976). *Response Surface Methodology*. Author, Blacksburg, VA.
Myers, R. H. and W. H. Carter, Jr. (1973). Response surface techniques for dual response systems. *Technometrics* **15**, 301–317.
Myers, R. H., A. I. Khuri and W. H. Carter, Jr. (1989). Response surface methodology: 1966–1988. *Technometrics* **31**, 137–157.
Myers, R. H., A. I. Khuri and G. G. Vining (1992). Response surface alternatives to the Taguchi robust parameter design approach. *Amer. Statist.* **46**, 131–139.
Myers, R. H. and D. C. Montgomery (1995). *Response Surface Methodology*. Wiley, New York.
Roth, P. M. and R. A. Stewart (1969). Experimental studies with multiple responses. *J. Roy. Statist. Soc. Ser. C* **18**, 221–228.
Roy, S. N., G. Gnanadesikan and J. N. Srivastava (1971). *Analysis and Design of Certain Quantitative Multiresponse Experiments*. Pergamon Press, Oxford.
Seber, G. A. F. (1984). *Multivariate Observations*. Wiley, New York.
Stewart, W. E. and J. P. Sorensen (1981). Bayesian estimation of common parameters from multiresponse data with missing observations. *Technometrics* **23**, 131–141.
Tukey, J. W. (1962). The future of data analysis. *Ann. Math. Statist.* **33**, 1–67.

Tukey, P. A. and J. W. Tukey (1981). Graphical display of data sets in 3 or more dimensions. In: V. Barnett, ed., *Interpreting Multivariate Data*. Wiley, New York, Chapters 10, 11, and 12.

Wegman, E. J. (1990). Hyperdimensional data analysis using parallel coordinates. *J. Amer. Statist. Assoc.* **85**, 664–675.

Wijesinha, M. C. (1993). Optimal designs to improve the power of multiresponse hypothesis tests. *Comm. Statist. Theory Methods* **22**, 587–602.

Wijesinha, M. C. and A. I. Khuri (1987a). The sequential generation of multiresponse D-optimal designs when the variance-covariance matrix is not known. *Comm. Statist. Simulation Comput.* **16**, 239–259.

Wijesinha, M. C. and A. I. Khuri (1987b). Construction of optimal designs to increase the power of the multiresponse lack of fit test. *J. Statist. Plann. Inference* **16**, 179–192.

Wijesinha, M. C. and A. I. Khuri (1991). Robust designs for first-order multiple design multivariate models. *Comm. Statist. Theory Methods* **20**, 2987–2999.

Wynn, H. P. (1970). The sequential generation of D-optimum experimental designs. *Ann. Math. Statist.* **41**, 1655–1664.

Zellner, A. (1962). An efficient method of estimating seemingly unrelated regressions and tests for aggregation bias. *J. Amer. Statist. Assoc.* **57**, 348–368.

Ziegel, E. R. and J. W. Gorman (1980). Kinetic modelling with multiresponse data. *Technometrics* **22**, 139–151.

S. Ghosh and C. R. Rao, eds., *Handbook of Statistics, Vol. 13*

13

Sequential Assembly of Fractions in Factorial Experiments

Subir Ghosh

1. Introduction

A complete factorial experiment with all possible runs is not cost-and-time effective when the number of factors is not small. Fractional factorial plans are developed with smaller number of runs for performing tasks depending on the available resources. For example, a task may be drawing inferences on the main effects assuming two-factor and higher order interactions to be zero. Using an efficient fractional factorial plan for performing this task, the data are collected and the inferences are drawn on the main effects. Some main effects are found to be large compared to the rest. The experimenter may suspect that a weak main effect is due to the presence of an interaction between the corresponding factor and another factor with a strong main effect. A second fraction may be assembled to the first fraction to investigate this point. This technique of assembling fractions is called the sequential assembly of fractions. Some authors prefer to call this technique as the multistage assembly of fractions. Notice that the technique here is different from the commonly used methods in sequential analysis with well defined stopping rules. Another task may be representing the response surface locally in a small region. It is decided that a first order model is to be fitted first. Using an efficient fractional factorial plan for performing this task, the data are collected and the inferences are drawn on fitting of the first order model. A lack of fit is detected due to the presence of curvature. A second fraction may be assembled to the first fraction to investigate the nature and significance of the detected curvature. This is again the sequential assembly of fractions. Researchers and practitioners have worked on such exploratory problems over the years. The purpose of this contribution is to present the developments in this area of research.

Sequential assembly of fractions appeared in Box and Wilson (1951), Box and Hunter (1957), Hartley (1959), Westlake (1965), Draper (1985), Draper and Lin (1990), Ghosh and Al-Sabah (1996), and numerous other contributions in response surface experiments. Sequential assembly of fractions also appeared in Connor (1960), Connor and Young (1961), Dykstra (1959), Patel (1963), Daniel (1962), John (1962), Addleman (1961, 1963, 1969), Srivastava (1975), Srivastava and Ghosh (1976), Srivastava and Hueberg (1992), Srivastava and Li (1994), Shirakura (1991), Anderson and Thomas (1979, 1980), Box, Hunter and Hunter (1978), and numerous other contributions in factorial experiments. Section 2 presents a critical examination of the normal probability plot for factor screening. Normal probability plot is a useful tool but should be used with discretion in factor screening as demonstrated in Section 2. Section 3 introduces the sequential assembly of fractions with an illustrative example and gives the options that are available in the literature. Section 4 presents the classifications of plans in terms of orthogonality and balance. Section 5 describes the search linear models and presents search designs that are available in the literature. Search designs are obtained by the sequential assembly of runs. Section 6 gives the inference procedures under the search linear model. Deficiencies identified in Section 2 in the normal probability plot are possible to overcome by the use of search procedure and the designs described in Sections 5 and 6. In Section 7, parallel and intersecting flats fractions are given as a very useful technique in the sequential assembly of fractions. In Section 8, the sequential assembly of fractions in response surface experiments is discussed. Composite designs are given with some striking results in their efficiency comparisons. In composite designs, there are factorial points, axial points, and center points. The axial points are needed for drawing inferences on the pure quadratic terms in the model. Factorial points and center points permit us to fit the first order model with the detection of lack of fit due to the second order terms. When the axial points are assembled to the factorial and center points, the resulting composite design permits us to fit the second order model. Section 9 summarizes this contribution with some concluding remarks. Throughout the paper, '$+$' represents '1' and '$-$' represents '-1'.

2. Normal probability plot in factor screening

In factorial experiments, the factor screening is routinely done using the normal probability plot. This technique is claimed to be particularly valuable when a single replicate of the experiment is available. The purpose of this section is to examine critically the method of normal probability plot. An illustrative example is presented for this purpose in Montogomery (1991, p. 291). A complete 2^4 factorial experiment is considered with the four factors. A: Temperature, B: Pressure, C: Concentration of formal dehyde, and D: Stirring rate. The response variable is the filtration rate of a chemical product produced in a pressure vessel. The goal is to maximize the filtration rate. The data for 16 runs are given in Table 9–7 on p. 291, the estimates of factorial effects are given on p. 292 and the normal probability plot is displayed on p. 293. The important effects found from the analysis are the main effects A, C, and D and the interaction effects AC and AD. The factor B is then completely screened out for the subsequent analyses given on pp. 294–296. It is demonstrated in the analysis given below that screening out of the factor B using the normal probability plot may not be the best thing to do from statistical considerations.

Consider the following three models with the vectors of parameters $\underline{\beta}$'s as

Model 1.1. $\underline{\beta}' = (\mu, A, C, D, AC, AD, ACD, CD)$,
Model 1.2. $\underline{\beta}' = (\mu, A, C, D, AC, AD, ACD)$,
Model 1.3. $\underline{\beta}' = (\mu, A, C, D, AC, AD)$.

Table 1
The $\underline{b}$'s, p-values, R^2, R^2_{adj}, SSE, DF, and MSE for Models 1.1–1.3

	Model 1.1			Model 1.2			Model 1.3		
	$\underline{\beta}$	$\underline{b}$	p-value	$\underline{\beta}$	$\underline{b}$	p-value	$\underline{\beta}$	$\underline{b}$	p-value
	μ	70.1	0.000	μ	70.1	0.000	μ	70.1	0.000
	A	10.8	0.000	A	10.8	0.000	A	10.8	0.000
	C	4.94	0.003	C	4.94	0.002	C	4.94	0.001
	D	7.31	0.000	D	7.31	0.000	D	7.31	0.000
	AC	−9.06	0.000	AC	−9.06	0.000	AC	−9.06	0.000
	AD	8.31	0.000	AD	8.31	0.000	AD	8.31	0.000
	ACD	−0.81	0.521	ACD	−0.81	0.491			
	CD	−0.56	0.642						
R^2		96.9%			96.8%			96.6%	
R^2_{adj}		94.1%			94.6%			94.9%	
SSE		179.50			184.56			195.10	
DF		8			9			10	
MSE		22.44			20.51			19.51	

Notice that Model 1.1 has all the parameters of a complete 2^3 factorial experiment with 3 factors A, C and D. In Table 1, Models 1.1–1.3 can be compared with respect to the values of R^2, R^2_{adj}, Sum of Squares due to Error (SSE), Mean Square Error (MSE), Degrees of Freedom (DF) due to error. The vector of estimates $\underline{b}$ of parameters and the p-values for each model are also given. Clearly, Model 1.1 does not provide much of an improvement over Model 1.3 in terms of R^2, R^2_{adj}, and MSE.

It is now important to compare Models 1.1–1.3 with the following models that do not screen out the factor B.

Model 2.1. $\underline{\beta} = (\mu, A, B, C, D, AC, AD, ABC, BCD, BC, ABC, ACD, ABCD)$,

Model 2.2. $\underline{\beta}' = (\mu, A, B, C, D, AC, AD, ABD, BCD, BC, ABC, ACD)$,

Model 2.3. $\underline{\beta}' = (\mu, A, B, C, D, AC, AD, ABD, BCD, BC, ABC)$,

Model 2.4. $\underline{\beta}' = (\mu, A, B, C, D, AC, AD, ABD, BCD, BC)$,

Model 2.5. $\underline{\beta}' = (\mu, A, B, C, D, AC, AD, ABD, BCD)$,

Model 2.6. $\underline{\beta}' = (\mu, A, B, C, D, AC, AD, ABD)$,

Model 2.7. $\underline{\beta}' = (\mu, A, B, C, D, AC, AD)$.

In Table 2, Models 2.1–2.7 can be compared with respect to R^2, R^2_{adj}, SSE, DF, and MSE. The p-values of B and ABD in Models 2.1–2.6 suggest that they are important and should not be ignored. Models 2.1, 2.2 and 2.3 are comparable to each other and they are considerably better than Models 1.1, 1.2 and 1.3 with respect to R^2, R^2_{adj}, SSE and MSE.

Normal probability plot must be cautiously used for screening out factors. As it is explained with the illustrative example, the factor B becomes significant when the interactions ABD, BCD, BC, ABC, and ACD are included in the model. The factor

Table 2
The $\underline{b}$, p-values, R^2, R^2_{adj}, SSE, DF, and MSE for Models 2.1–2.3

	Model 2.1			Model 2.2		Model 2.3	
	$\underline{\beta}$	$\underline{b}$	p-value	$\underline{\beta}$	p-value	$\underline{\beta}$	p-value
	μ	70.1	0.000	μ	0.000	μ	0.000
	A	10.8	0.000	A	0.000	A	0.000
	B	1.56	0.020	B	0.026	B	0.035
	C	4.94	0.001	C	0.000	C	0.000
	D	7.31	0.000	D	0.000	D	0.000
	AC	−9.06	0.000	AC	0.000	AC	0.000
	AD	8.31	0.000	AD	0.000	AD	0.000
	ABD	2.06	0.009	ABD	0.011	ABD	0.013
	BCD	−1.31	0.032	BCD	0.045	BCD	0.061
	BC	1.19	0.041	BC	0.059	BC	0.081
	ABC	0.937	0.072	ABC	0.108	ABC	0.146
	ACD	−0.812	0.099	ACD	0.149		
	$ABCD$	0.688	0.140				
R^2		99.9%			99.8%		99.6%
R^2_{adj}		99.5%			99.1%		98.8%
SSE		5.69			13.25		23.81
DF		3			4		5
MSE		1.90			3.31		4.76

The $\underline{b}$, p-values, R^2, R^2_{adj}, SSE, DF, and MSE for Models 2.4–2.7

	Model 2.4		Model 2.5		Model 2.6		Model 2.7	
	$\underline{\beta}$	p-value	$\underline{\beta}$	p-value	$\underline{\beta}$	p-value	$\underline{\beta}$	p-value
	μ	0.000	μ	0.000	μ	0.000	μ	0.000
	A	0.000	A	0.000	A	0.000	A	0.000
	B	0.047	B	0.071	B	0.096	B	0.168
	C	0.000	C	0.000	C	0.000	C	0.000
	D	0.000	D	0.000	D	0.000	D	0.000
	AC	0.000	AC	0.000	AC	0.000	AC	0.000
	AD	0.000	AD	0.000	AD	0.000	AD	0.000
	ABD	0.017	ABD	0.026	ABD	0.038		
	BCD	0.082	BCD	0.117				
	BC	0.108						
R^2	99.3%		98.9%		98.5%		97.3%	
R^2_{adj}	98.3%		97.7%		97.1%		95.5%	
SSE	37.88		60.44		88.0		156.06	
DF	6		7		8		9	
MSE	6.31		8.63		11.00		17.34	

B may not be as important as A, C, and D, but it should not be ignored particularly because of its main effect and interactions with factors A, C, and D.

Practical considerations like cost, time, convenience etc. may dominate over statistical benefits in the decision making to ignore the main effect of B and the interaction effects ABC, BCD, BC, ABC and ACD. Even then strong recommendations should be made in keeping those main effect and interactions in the model.

Normal probability plot requires the least squares estimation of all parameters in the model. In the example discussed above, a single replicate of a complete factorial experiment is performed and as a result the estimation of the general mean, all main effects and all interactions are possible. When the number of factors is more than four, fractional factorial plans of various resolutions are normally performed. For example, in a Resolution V plan, three factor and higher order interactions are assumed to be zero. The plan allows us to estimate the general mean, main effects and two factor interactions. Normal probability plot is used to screen out main effects and two factor interactions. There is a possibility of having a few important factorial effects from three factor and higher order interactions and moreover, in their presence, some of the main effects and two factor interactions screened out by the normal probability plot may turn out to be significant. The practice of factor screening using the normal probability plot is even more dangerous in an experiment with a Resolution III plan assuming two factor and higher order interactions to be zero.

3. Sequential assembly of fractions

Main effects or Resolution III plans are very common in scientific experiments. Most scientific experiments are done sequentially. Sequential assembly of fractions is a technique of augmenting two or more fractions together in resolving ambiguities (see Box et al., 1978, p. 396) or in getting designs with higher revealing power (Srivastava, 1975, 1984). The fold-over technique for 2^m factorial experiments is a sequential augmentation of a Resolution III plan with its complement and the resulting plan is a Resolution IV plan which allows the estimation of the general mean and the main effects in the presence of two factor interactions assuming three factor and higher order interactions to be zero. If there is a significant three factor interaction present, then of course the use of Resolution IV plan is not recommended. The purpose of this section is to present the options that are available in the literature on sequential assembly of fractions in the context of an example.

Consider the Bicycle Example given in Box et al. (1978, p. 390). There are seven factors each at 2 levels, namely seat (up/down), dynamo (off/on), handlebars (up/down), gear (low/medium), raincoat (on/off), breakfast (yes/no), tires (hard/soft). The response variable is the time in seconds for a particular person to run up a hill between fixed marks. The design given below can be seen in Table 12.5 of Box et al. (1978, p. 391).

$$T_1 = \begin{bmatrix} - & - & - & + & + & + & - \\ + & - & - & - & - & + & + \\ - & + & - & - & + & - & + \\ + & + & - & + & - & - & - \\ - & - & + & + & - & - & + \\ + & - & + & - & + & - & - \\ - & + & + & - & - & + & - \\ + & + & + & + & + & + & + \end{bmatrix}.$$

In T_1, there are 8 runs (rows) and 7 factors (columns). The first 7 runs in T_1 represent the incidence matrix of a symmetric BIB design with parameters $v = b = 7$, $r = k = 3$ and $\lambda = 1$. In another representation, '$-$' corresponds to 0 and '$+$' corresponds to 1. The design T_1 is in fact a Resolution III plan. The following options are available in the literature on sequential assembly of fractions.

Option 1

The following set of 8 runs are given in Table 12.7 of Box et al. (1978, p. 394).

$$T_2^{(1)} = \begin{bmatrix} - & - & - & - & + & + & - \\ + & - & - & + & - & + & + \\ - & + & - & + & + & - & + \\ + & + & - & - & - & - & - \\ - & - & + & - & - & - & + \\ + & - & + & + & + & - & - \\ - & + & + & + & - & + & - \\ + & + & + & - & + & + & + \end{bmatrix}.$$

Notice that $T_2^{(1)}$ is obtained from T_1 by switching the signs of the column 4. The design $T_2^{(1)}$ is also a Resolution III plan.

The fraction resulting from the sequential assembly of T_1 and $T_2^{(1)}$, consists of 16 runs. The 16 runs permit us to estimate the main effect and all two factor interactions involving the factor 4 under the assumption that three factor and higher order interactions are zero.

Option 2 (Fold-Over)

The following set of 8 runs are obtained from T_1 by switching the signs of the columns 1–7.

$$T_2^{(2)} = \begin{bmatrix} + & + & + & - & - & - & + \\ - & + & + & + & + & - & - \\ + & - & + & + & - & + & - \\ - & - & + & - & + & + & + \\ + & + & - & - & + & + & - \\ - & + & - & + & - & + & + \\ + & - & - & + & + & - & + \\ - & - & - & - & - & - & - \end{bmatrix}.$$

The design $T_2^{(2)}$ is also a Resolution III plan.

The 16 runs obtained from the sequential assembly of T_1 and $T_2^{(2)}$, permit us to estimate all main effects assuming three factor and higher order interactions to be zero. The resulting design is therefore a Resolution IV plan.

The example, bottleneck of the filtration stage of an industrial plant, given in Box et al. (1978, pp. 424–429), presents an experiment with Resolution IV plan and the data in Tables 13.4 and 13.6.

Option 3 (Ghosh, 1993b)

The following set of 8 runs are given in Ghosh (1993b).

$$T_2^{(3)} = \begin{bmatrix} + & + & - & + & + & - & - \\ + & - & + & - & + & + & - \\ + & - & + & + & - & - & + \\ - & - & - & + & + & + & + \\ + & + & - & - & - & + & + \\ - & + & + & - & + & - & + \\ - & + & + & + & - & + & - \\ - & - & - & - & - & - & - \end{bmatrix}.$$

Note that in Ghosh (1993b), the above matrix $T_2^{(3)}$ is presented as T_2 with the columns $(1, 2, 3)$ as $(4, 5, 6)$, $(4, 5, 6)$ as $(1, 2, 3)$ and the runs (rows) are also in different order. The same is true for T_1. The first 7 runs in $T_2^{(3)}$ represent the incidence matrix of a BIB design with parameters $v = b = 7$, $r = k = 4$, $\lambda = 2$. The 16 runs obtained from the sequential assembly of T_1 and $T_2^{(3)}$, permit us to estimate the general mean, all main effects and furthermore, search and estimate one nonzero interaction from

2-factor and 3-factor interactions. These designs are known as Search Designs (see Srivastava, 1975) and will be discussed in more detail in the subsequent sections.

Option 3.1 (Ghosh, 1993b)
The sets of 15 runs are obtained from the sequential assembly of T_1 and $T_2^{(3)}$ in Option 3, by deleting the ith run. Each set of 15 runs permits us to estimate the general mean, all main effects and furthermore, search and estimate one nonzero interaction from 2-factor and 3-factor interactions. This is known as the Robustness Property of 16 runs in Option 3 against the deletion of any run and will be discussed in more detail in subsequent sections.

Option 3.2 (Shirakura, 1991)
In Option 3.1, the set of 15 runs obtained by deleting the last run in $T_2^{(3)}$ from the 16 runs in Option 3, is due to Shirakura (1991).

Option 4 (Ghosh, 1993a)

Denote $T_3 = 2I_7 - J_7$ and $T_4 = J_7 - 2I_7$, where I_7 is the 7×7 identity matrix and J_7 is the 7×7 matrix with all elements unity. The 30 runs obtained from the sequential assembly of the runs in T_1, $T_2^{(3)}$, T_3 and T_4 permit us to estimate the general mean, main effects and furthermore, search and estimate two nonzero interactions from 2-factor and 3-factor interactions.

Option 4.1
The 14 sets of 29 runs obtained by deleting the ith run from T_3 or T_4 permit us to do the same in terms of search and estimation.

Option 4.2
The 2 sets of 27 runs obtained by deleting the run 23 from T_3 and the runs 24 and 25 from T_4, and deleting the run 18 from T_3 and the runs 24 and 25 from T_4, also permit us to do the same in terms of search and estimation.

Option 5 (Shirakura, 1991)

The 21 runs obtained from the sequential assembly of the runs in $T_1, T_2^{(3)}$ and the 6 runs from T_3 by ignoring the run 1, permit us to estimate the general mean, main effects and furthermore, search and estimate two nonzero interactions from only 2-factor interactions.

Option 6 (Ghosh, 1993a)

The 23 runs obtained from the sequential assembly of the runs in T_1, $T_2^{(3)}$ and T_3, permit us to estimate the general mean, main effects and, furthermore, search and estimate one nonzero interaction from 2-factor and higher order interactions.

Option 6.1
The 8 sets of 22 runs obtained by deleting the ith run in T_2 permit us to do the same in terms of search and estimation.

Option 6.2
The 7 sets of 21 runs obtained by deleting the last run in T_2 and any other run in T_2, also permit us to do the same.

4. Orthogonal and balanced plans

The plans presented in Section 3 for sequential assembly of fractions are not all identical statistically. The purpose of this section is first to highlight distinguishing features of these plans and then to demonstrate how they can be used in constructing plans for other experiments.

Consider the sets of 8 runs in T_1, $T_2^{(1)}$, and $T_2^{(2)}$ given in Section 3. These are Orthogonal Resolution III plans for a 2^7 factorial experiment and also Orthogonal Arrays of strength 2 and with 2 symbols, 8 rows (runs) and 7 columns (factors) (see Rao, 1947). Note that some authors prefer to express Orthogonal Arrays in the transposed form of the above matrices, i.e., their rows are factors and columns are runs.

Let T_3^* and T_4^* be 8×7 matrices obtained from T_3 and T_4 by adding one more run $(+++++++)$ and $(-------)$, respectively. The sets of 8 runs in T_3^* and T_4^* form Balanced Resolution III plans for a 2^7 factorial experiment and also Balanced Arrays of strength 7 and with 2 symbols, 8 rows (runs) and 7 columns (factors) (see Srivastava and Chopra, 1973; Srivastava, 1990, p. 339).

Let $\underline{y}(8 \times 1)$ be the vector of observations for a set of 8 runs, $\underline{\beta}$ be the vector of the general mean and 7 main effects. The standard model is then

$$\mathrm{E}(\underline{y}) = X\underline{\beta} \quad \text{and} \quad \mathrm{V}(\underline{y}) = \sigma^2 I,$$

where $X(8 \times 8)$ depends on the runs and σ^2 is an unknown constant. For T_1, the matrix X takes the form $X = [\underline{j}, T_1]$, where $\underline{j}$ is a column vector of $+$'s. Thus $X'X = 8I$. The same is true for $T_2^{(1)}$, $T_2^{(2)}$, and $T_2^{(3)}$. For T_3^*, the matrix X takes the form $X = [\underline{j}, T_3^*]$. Thus

$$X'X = \begin{bmatrix} 8 & -4\underline{j}' \\ -4\underline{j} & 4I + 4J \end{bmatrix}.$$

For T_4^*, the matrix X takes the form $X = [\underline{j}, T_4^*]$. Hence

$$X'X = \begin{bmatrix} 8 & 4\underline{j}' \\ 4\underline{j} & 4I + 4J \end{bmatrix}.$$

Consider now a 2^6 factorial experiment in 7 runs of Resolution III. There is no Orthogonal Resolution III plan for such a situation. However there are Balanced Resolution III plans available and four such plans are given below.

$$T_5 = \begin{bmatrix} - & - & - & + & + & + \\ + & - & - & - & - & + \\ - & + & - & - & + & - \\ + & + & - & + & - & - \\ - & - & + & + & - & - \\ + & - & + & - & + & - \\ - & + & + & - & - & + \end{bmatrix}, \quad T_6 = \begin{bmatrix} + & + & - & + & + & - \\ + & - & + & - & + & + \\ + & - & + & + & - & - \\ - & - & - & + & + & + \\ + & + & - & - & - & + \\ - & + & + & - & + & - \\ - & + & + & + & - & + \end{bmatrix},$$

$$T_7 = \begin{bmatrix} + & - & - & - & - & - \\ - & + & - & - & - & - \\ - & - & + & - & - & - \\ - & - & - & + & - & - \\ - & - & - & - & + & - \\ - & - & - & - & - & + \\ + & + & + & + & + & + \end{bmatrix}, \quad T_8 = \begin{bmatrix} - & + & + & + & + & + \\ + & - & + & + & + & + \\ + & + & - & + & + & + \\ + & + & + & - & + & + \\ + & + & + & + & - & + \\ + & + & + & + & + & - \\ - & - & - & - & - & - \end{bmatrix}.$$

It is to be noted that T_5, T_6, T_7, and T_8 are obtained from T_1, $T_2^{(3)}$, T_4^*, and T_5^* by deleting the last column for all of them and the last row for T_1 and $T_2^{(3)}$ and the second row from the bottom for T_4^* and T_5^*. The T_5 and T_6 are Balanced Arrays of strength 6, and with 2 symbols, 7 rows (runs) and 6 columns (factors). The following major options are available on sequential assembly of fractions T_5, T_6, T_7, and T_8 for 2^6 factorial experiments.

Option 1

The 14 runs obtained from the sequential assembly of T_5 and T_6, permit us to estimate the general mean, all main effects and furthermore, search and estimate one nonzero interaction from 2-factor and 3-factor interactions.

Option 2

The 28 runs obtained from the sequential assembly of the runs in T_5, T_6, T_7, and T_8, permit us to estimate the general mean, main effects, and moreover, search and estimate two non-zero interactions from 2-factor and 3-factor interactions.

Option 3

The 21 runs obtained from the sequential assembly of the runs in T_5, T_6, and T_7, permit us to estimate the general mean, main effects and furthermore, search and estimate one nonzero interaction from 2-factor and higher order interactions.

There are of course some other possible options using the ideas described in Section 3.

5. Search designs

In various Resolution plans, the higher order interactions are all assumed to be zero. For example, in Resolution III plans, the 2-factor and higher order interactions are all assumed to be negligible. In reality such an assumption may or may not be valid. There may be a few significant higher order interaction present resulting in a distorted inference. Search designs are introduced in Srivastava (1975) to search and identify those significant interactions with small number of runs. The purpose of this section is to describe the search designs given in various Options of Sections 3 and 4 and also present some other designs that are available in the literature.

Let $\underline{y}(N \times 1)$ be the vector of observations for N runs in the experiment. The runs may or may not be distinct. The search linear model is given below.

$$\mathrm{E}(\underline{y}) = X_1\underline{\beta}_1 + X_2\underline{\beta}_2, \qquad \mathrm{V}(\underline{y}) = \sigma^2 I,$$

where $\underline{\beta}_1(p_1 \times 1)$ is a vector of fixed unknown parameters, $\underline{\beta}_2(p_2 \times 1)$ is a vector of fixed parameters with partial information on them, and σ^2 is an unknown constant. It is known that there are k (small relative to p_2) elements of $\underline{\beta}_2$ are nonzero but it is not known which are these nonzero elements. Search designs permit us to search and identify k nonzero elements of $\underline{\beta}_2$ and then to draw inference on them in addition to the elements in $\underline{\beta}_1$.

For a search design, the matrices X_1 and X_2 satisfy the following rank condition,

$$\mathrm{Rank}[X_1 : X_{2j}] = p_1 + 2k, \quad j = 1, \ldots, \binom{p_2}{2k},$$

where X_{2j} is an $(N \times 2k)$ submatrix of X_2. The rank condition allows us to find the least squares estimators of the parameters in every pair of models given below and also to discriminate between models in every pair.

$$\mathrm{E}(\underline{y}) = X_1\underline{\beta}_1 + X_{2j_1}, \qquad \mathrm{V}(\underline{y}) = \sigma^2 I,$$

$$\mathrm{E}(\underline{y}) = X_1\underline{\beta}_1 + X_{2j_2}, \qquad \mathrm{V}(\underline{y}) = \sigma^2 I,$$

where $\underline{\beta}_{2j_1}$ and $\underline{\beta}_{2j_2}$ are two sets of k elements in $\underline{\beta}_2$, $j_1 \neq j_2$, $j_1, j_2 \in \{1, \ldots, \binom{p_2}{k}\}$. Note that $X_{2j} = [X_{2j_1} : X_{2j_2}]$ for some j_1 and j_2. The condition in fact allows to discriminate between models and to identify the true model(s) with probability one in the less noisy cases. Checking the condition is an arduous task. Various techniques are developed in Srivastava and Gupta (1979), Srivastava and Ghosh (1977), and Srivastava (1977). All these techniques involve enormous computations.

In Option 3 of Section 3, $\underline{\beta}_1$ consists of the general mean and main effects, and $\underline{\beta}_2$ consists of 2-factor and 3-factor interactions. The 16 runs from the sequential assembly of T_1 and $T_2^{(3)}$ is a search design with $k = 1$. Note that $N = 16$, $p_1 = 8$, and $p_2 = 56$. In Option 4 of Section 3, the 30 runs from the sequential assembly of T_1, $T_2^{(3)}$, T_3 and T_4 is a search design with $k = 2$. Note that $\binom{p_k}{k} = \binom{56}{2} = 1{,}540$, and $\binom{p_2}{2k} = \binom{56}{4} = 367{,}290$. This gives a sense on the degree of complexity. In Option 6 of Section 3, $\underline{\beta}_1$ consists of the general mean and main effects, and $\underline{\beta}_2$ consists of 2-factor and higher order interactions. The 23 runs from the sequential assembly of the runs in T_1, $T_2^{(3)}$, and T_3 is a search design with $k = 1$. Note that $N = 23, p_1 = 8$, and $p_2 = 120$.

In Sections 3 and 4, search designs are given for 2^7 and 2^6 factorial experiments. Search designs are now considered for experiments with m ($\geqslant 2$) factors and the number of levels not necessarily two.

I. Consider the search linear model where $\underline{\beta}_1$ consists of the general mean and main effects and $\underline{\beta}_2$ consists of 2-factor interactions.

I.1. *Shirakura plan* ($k = 1$)

For a 2^m factorial experiment with $m = 2^h - 1$ and h ($\geqslant 3$) being an integer, the $(m+1)$ runs in an $(m+1) \times m$ matrix T_{1m} are obtained by filling in the entries of the first h columns and 2^h rows with elements '+' and '–', and then the remaining $(2^h - h - 1)$ columns as the interaction columns for a 2^h factorial. Note that T_{1m} for $h = 3$ and $m = 7$ is in fact T_1 in Section 3.

Consider a BIB design (v, b, r, k, λ) with $v = m = 2^h - 1, b = \frac{1}{3}(2^{h-1} - 1)(2^h - 1), r = 2^{h-1} - 1, k = 3$, and $\lambda = 1$. Let $T_{2m}(b \times m)$ be a matrix such that the elements in the jth row and ith column is '+' if the ith treatment does not occur in the jth block and is '–' if the ith treatment does occur in the jth block. The rows of T_{2m} represent b runs. Note that T_{2m} is the complement of the transpose of the incidence matrix for the BIB design.

In Shirakura (1991), it is shown that the sequential assembly of T_{1m} and T_{2m} results in a Search Design with $(m + 1 + b)$ runs and $k = 1$.

I.2. *Ghosh–Zhang plan* ($k = 1$)

For a 3^m factorial experiment, with $m \geqslant 3$, let $T_{3m}((1+2m) \times m)$ be a matrix with the first row (run) having all elements '+' and the remaining rows (runs) having the ith element ($i = 1, \ldots, m$) and '–' and '0' and other elements as '+'. The set of $(1 + 2m)$ runs in T_{3m} is a Resolution III plan.

Let $T_{4m}(m \times m)$ be a matrix where the ith row (run) has the ith element '–' and the remaining elements are 0.

It can be seen in Ghosh and Zhang (1987) that the sequential assembly of T_{3m} and T_{4m} is a search design with $(1 + 3m)$ runs and $k = 1$.

I.3. *Anderson–Thomas plans*

Anderson and Thomas (1980) presented search designs for general symmetrical factorial experiments with the number of levels is a prime or a prime power. These designs are able to search any combination of three or fewer pairs of factors that interact.

I. *Chatterjee–Mukerjee and Chatterjee plans*

Chatterjee and Mukerjee (1986) and Chatterjee (1989) presented search designs with $k = 1$ in general symmetrical and asymmetrical factorial experiments. These designs are able to search one set of nonzero interactions between two factors.

II. Consider the search linear model where $\underline{\beta}_1$ consists of the general mean and main effects and $\underline{\beta}_2$ consists of 2-factor and 3-factor interactions.

II.1. *Ohnishi–Shirakura plans* $(k = 1)$

Ohnishi and Shirakura (1985) presented search designs for $3 \leqslant m \leqslant 8$ and with $k = 1$.

II.2. *Chatterjee plans*

Chatterjee (1990) presented search designs for general symmetric and asymmetric factorial experiments. These designs are able to search one non-zero effect from 2-factor and 3-factor interactions.

II.3. *Ghosh–Talebi plan* $(k = 1)$

Let $T_{5m}((m+1) \times m)$ be a set of $(m+1)$ distinct runs with the run having all $+$'s and m runs having one $+$ and $(m-1)$ $-$'s in each of them. Note that T_{5m} is in fact T_3^* in Section 4 for $m = 7$. Consider a symmetric BIB design with parameters $v = b = 4\lambda - 1$, $r = k = 2\lambda$, and λ, where λ $(\geqslant 1)$ is a positive integer.

Let $T_{6m}((m+1) \times m)$ with $m = v = b$, be a set of $(m+1)$ runs. The first row of T_{6m} is having all elements '$-$'. The element of the jth $(\geqslant 2)$ row and the ith column is '$+$' if the ith treatment occurs in the jth block and is '$-$' if the ith treatment does not occur in the jth block.

In Ghosh and Talebi (1993), it is shown that the sequential assembly of the runs in T_{5m} and T_{6m} results in a search design with $(2m+2)$ runs and $k = 1$.

III. Consider the search linear model with $\underline{\beta}_1$ consists of the general mean and main effects and $\underline{\beta}_2$ of 2-factor and higher order interactions.

III.1. *Srivastava–Arora plan* $(k = 2)$

Let $T_{7m}((1 + \binom{m}{2}) \times m)$ be a set of distinct runs with the run having all $-$'s and $\binom{m}{2}$ runs having two $+$'s and $(m-2)$ $-$'s in each of them.

It can be seen in Srivastava and Arora (1991) that the sequential assembly of T_{5m} and T_{7m} is a search design with $(2 + m + \binom{m}{2})$ runs and $k = 2$.

III.2. *Gupta and Gupta–Carvazal plans* ($k = 1$ *and* 2)

Gupta (1981) showed that Srivastava–Arora plan in III.1 with deletion of the run having all $-$'s is a search design with $k = 1$. Gupta and Carvazal (1984, 1987), Gupta (1992) presented tables of search designs with small number of runs for $m = 4, 5, 6, 7$ and with $k = 1$ and 2.

IV. Consider the search linear model with $\underline{\beta}_1$ consists of the general mean, main effects, and 2-factor interactions, and $\underline{\beta}_2$ consists of 3-factor and higher order interactions.

IV.1. *Srivastava–Ghosh plans* ($k = 1$)

Let $T_{8m}(m \times m)$ be a set of m runs obtained from T_{5m} by deleting the run having all $+$'s and $T_{9m}(m \times m)$ be the complement of T_{8m}.

The sequential assembly of T_{7m}, T_{8m}, and T_{9m} is a search design with $(1 + 2m + \binom{m}{2})$ runs and $k = 1$. The proof is given in Srivastava and Ghosh (1976) for the complement. Many other search designs for $4 \leqslant m \leqslant 8$ and with $k = 1$ are given in Srivastava and Ghosh (1977).

V. Consider the search linear model where $\underline{\beta}_1$ consists of the general mean, main effects, and 2-factor interactions, and $\underline{\beta}_2$ consists of only 3-factor interactions.

V.1. *Shirakura–Tazawa plans* ($k = 1$ *and* 2)

Let $T_{10m}(\binom{m}{2} \times m)$ be a set of $\binom{m}{2}$ runs obtained from T_{7m} by deleting the run having all $-$'s and $T_{11m}((1 + \binom{m}{2}) \times m)$ be the complement of T_{7m}. The following plans are given in Shirakura and Tazawa (1992).

The sequential assembly of T_{8m}, T_{9m}, and T_{10m} is a search design with $(2m + \binom{m}{2})$ runs $m \geqslant 7$ and $k = 1$. Notice that the difference between this plan and Srivastava–Ghosh plan in IV.1 is in a single run and Srivastava–Ghosh plan has higher revealing power in terms of search. Shirakura and Tazawa presented their plan as a complement of the above plan.

The sequential assembly of T_{10m} and T_{11m} is a search design with $(1 + 2\binom{m}{2})$ runs $m \geqslant 7$ and $k = 2$. The sequential assembly of T_{7m} and T_{11m} for $m = 5$ is a search design with 22 runs and $k = 2$. The sequential assembly of T_{10m}, the complement of T_{10m}, and T_{8m} for $m = 6$ is a search design with 36 runs and $k = 2$.

V.2. *Ghosh–Zhang plan* ($k = 1$)

For a 3^m factorial experiment with $m \geqslant 4$, let $T_{12m}(4\binom{m}{2} \times m)$ be a matrix having the rows (runs) with the (i, j)th elements ($i, j = 1, \ldots, m,\ i < j$) as $(-, -), (-, 0), (0, -)$, and $(0, 0)$ and the other elements as $+$.

The sequential assembly of T_{3m}, T_{4m} given in I.2, and T_{12m} is a search design with $(1 + 3m + 4\binom{m}{2})$ runs and $k = 1$. The plan is given in Ghosh and Zhang (1987).

VI. *Optimal search designs*

The concept of optimal search designs and the optimality criteria are introduced in Srivastava (1977).

VI.1. *Shirakura–Ohnishi plans* $(k = 1)$

In the setting of IV, the AD-optimal search designs are given in Shirakura and Ohnishi (1985) for 2^m factorial experiments, $6 \leqslant m \leqslant 8$, and for various number of runs.

VII. *Robust search designs*

The robustness property of designs against the unavailability of data introduced in Ghosh (1979a) can be stated for search designs as follows.

Definition: A search design with certain search property is said to be robust against deletion of any set of s (a positive integer) runs if the resulting plan after deletion of s runs remains a search design with the same search property.

VII.1. *Ghosh–Talebi and Ghosh plans* $(k = 1)$

Ghosh and Talebi (1993), Ghosh (1993a and b) presented several search designs in the setting of II for 2^m factorial experiments.

VIII. *Factor screening*

Consider the search linear model where $\underline{\beta}_1$ consists of the general mean and $\underline{\beta}_2$ consists of the main effects. The directions of factors are known to be nonnegative.

VIII.1. *Srivastava plans*

The plans are given in Srivastava (1976).

VIII.2. *Ghosh and Ghosh–Avila plans*

Ghosh (1979b), Ghosh and Avila (1985) presented plans for $k = 2$ and 3.

VIII.3. *Müller plans*

Müller (1993) presented plans for $k = 1$ and 2.

IX. *Sequential factorial probing designs*

In Srivastava and Hveberg (1992), a set of "tree structure" factorial effects is introduced. Instead of considering the search linear model, sequential factorial probing designs are presented for identifying a true set of nonzero tree structured factorial effects from a class of possible such effects.

6. Influential nonnegligible parameters

In the search linear model discussed in Section 5, it is assumed that there are k nonzero elements in $\underline{\beta}_2$ but it is not known which are these nonzero elements. The purpose of this Section is to present statistical methods for estimating the value of k and also searching the influential nonnegligible elements of $\underline{\beta}_2$.

Consider a general factorial experiment with t runs, n_u ($\geqslant 1$) observations for the uth run and

$$\sum_{u=1}^{t} n_u = N.$$

Let y_{uv} be the observation for the vth replication of the uth run and $\bar{y}_u$ be the mean of all observations for the uth run. Suppose that k_1 is an initial guess on k. There are three possibilities $k_1 > k$, $k_1 = k$, and $k_1 < k$. Consider $\binom{p_2}{k_1}$ models,

$$\mathrm{E}(\underline{y}) = X_1\underline{\beta}_1 + X_{2j}\underline{\beta}_{2j}, \qquad \mathrm{V}(\underline{y}) = \sigma^2 I,$$

$$j = 1, \ldots, \binom{p_2}{k_1}, \qquad \mathrm{Rank}\,[X_1, X_{2j}] = p_1 + k_1.$$

Let $\widehat{\underline{\beta}}_{1j}$ and $\widehat{\underline{\beta}}_{2j}$ be the least squares estimators of $\underline{\beta}_1$ and $\underline{\beta}_{2j}$ for the jth model,

$$\widehat{\underline{y}}_j = X_1\widehat{\underline{\beta}}_{1j} + X_{2j}\widehat{\underline{\beta}}_{2j}, \qquad \underline{R}_j = \underline{y} - \widehat{\underline{y}}_j,$$

F_j be the F-statistic for testing H_0: $\underline{\beta}_{2j} = \underline{0}$ under the normality assumption, F_j^{LOF} be the F-statistic for testing H_0: No Lack of Fit, SSE_j be the sum of squares due to error, SSLOF_j be the sum of squares due to lack of fit, SSPE be the sum of squares due to pure error, $\widehat{\underline{y}}_0$ be the fitted values when $\underline{\beta}_{2j} = \underline{0}$, $\underline{R}_0$ be the residuals when $\underline{\beta}_{2j} = \underline{0}$, SSE_{ij} be the sum of squares due to error when the ith elements of $\underline{\beta}_{2j}$ denoted by $\beta_{2j}^{(i)}$ is zero, t_{ij} be the t-statistic for testing H_0: $\beta_{2j}^{(i)} = 0$. The following results are given in Ghosh (1987).

I. For $\ell \in \{1, \ldots, \binom{p_2}{k_1}\}$, the following statements are equivalent.

(a) SSE_ℓ is a minimum,
(b) F_ℓ is a maximum,
(c) SSLOF_ℓ is a minimum,
(d) F_ℓ^{LOF} is a minimum,
(e) The Euclidean distance between $\widehat{\underline{y}}_\ell$ and $\widehat{\underline{y}}_0$ is a maximum,
(f) The square of the simple correlation coefficient between the elements of $\underline{R}_\ell$ and $\underline{R}_0$ is a minimum.

II. For $\ell \in \{1, \ldots, \binom{p_2}{k_1}\}$, $q \in \{1, \ldots, k_1\}$, the following statements are equivalent.

(a) $\mathrm{SSE}_{\ell q}$ is a minimum,
(b) $t_{\ell q}$ is a maximum.

A set of nonnegligible parameters $\underline{\beta}_{2\ell}$ are said to be influential if the sum of squares due to error (SSE_ℓ) is minimum for the model with those parameters. A nonnegligible parameter $\beta_{2\ell q}$ is said to be significant if the value $t_{\ell q}$ is large. Under the assumption of normality and the null hypothesis H_0: $\underline{\beta}_{2j} = \underline{0}$, F_j has the central F distribution with $(k_1, N - p_1 - k_1)$ d.f. and under the null hypothesis H_0: $\beta_{2j}^{(i)} = 0, t_{ij}$ has the central t distribution with $(N - p_1 - k_1)$ d.f. The following method presented in Ghosh (1987) is due to J. N. Srivastava.

Srivastava method

Case I. If $\max_j F_j \leqslant F_{\alpha;k_1,N-p_1-k_1}$, then there is no significant nonnegligible parameter. The $F_{\alpha;k_1,N-p_1-k_1}$ is the upper α percent point of the central F distribution with $(k_1, N - p_1 - k_1)$ d.f.

Case II. Suppose for $j = j_1, \ldots, j_s$, $F_j > F_{\alpha;k_1,N-p_1-k_1}$. For $w = 1, \ldots, p_2$,

$$\delta_w = \text{ the number of } j \text{ in } \{j_1, \ldots, j_s\} \text{ for which } |t_{wj}| > t_{\alpha/2,N-p_1-k_1}.$$

Clearly, $0 \leqslant \delta_w \leqslant s$. The δ_w's are now arranged in decreasing order of magnitude and write $\delta_{(1)} \geqslant \delta_{(2)} \geqslant \cdots \geqslant \delta_{(p_2)}$. If there are k $(\geqslant k_1)$ nonzero $\delta_{(w)}$'s, the influential significant parameters are $\delta_{(1)}, \ldots, \delta_{(k)}$, otherwise the influential significant $\beta_{(w)}$'s correspond to nonzero $\delta_{(w)}$'s and the number of influential parameters is then less than k_1. In case of a tie, the values of $|t_{wj}|$ may be used as guidelines. The parameter $\beta_{(1)}$ is the most influential significant nonnegligible parameter. An estimator of the unknown k is

$$\widehat{k} = \text{ the number of nonzero } \delta_w\text{'s}, \quad w = 1, \ldots, p_2.$$

The probability of correct search by the above method is very high and in fact close to one as it is found in Monte Carlo studies done by Professor Srivastava and his students. In a special situation of pure search (i.e., $\underline{\beta}_1 = \underline{0}$), Srivastava and Mallenby (1985) presented a method of search with the minimum amount of computation and a method for computing the probability that the correct parameter $(k = 1)$ is identified. Professor T. Shirakura and his coworkers have recently made some important contribution in this area.

7. Parallel and intersecting flats fractions

Sequential assembly of fractions occurs naturally in parallel and intersecting flats fractions. The purpose of this section is to present developments for these types of fractions. Fractional factorial plans of parallel and intersecting flats types appeared in Connor (1960), Connor and Young (1961), Patel (1963), Daniel (1962), John (1962), Addelman (1961, 1963, 1969). In recent years, a lot of work is done in this area by J. N. Srivastava, D. A. Anderson and their coauthors. Some plans and discussions are given below to illustrate the ideas.

7.1. John plans

Consider the following parallel flat fraction of Resolution IV with 12 runs for a 2^4 factorial experiment. Three sets of 4 runs (x_1, x_2, x_3, x_4) with $x_i = +$ or $-, i = 1, 2, 3, 4$, satisfying

$$\begin{aligned} \text{a:} &\quad + = -x_1x_2 = -x_1x_3 = x_2x_3, \\ \text{b:} &\quad + = -x_1x_2 = x_1x_3 = -x_2x_3, \\ \text{c:} &\quad + = x_1x_2 = -x_1x_3 = -x_2x_3. \end{aligned}$$

Each set represents a flat and there is no common run between any two flats. From the above plan, a parallel flat fraction of Resolution IV with 12 runs is constructed for a 2^6 factorial experiment by taking $x_5 = x_1x_2x_4$ and $x_6 = x_1x_3x_4$. The 4 runs in sets a, b, and c are then

$$\text{Set a:} \begin{bmatrix} + & - & - & + & - & - \\ + & - & - & - & + & + \\ - & + & + & + & - & - \\ - & + & + & - & + & + \end{bmatrix}, \qquad \text{Set b:} \begin{bmatrix} + & - & + & + & - & + \\ + & - & + & - & + & - \\ - & + & - & + & - & + \\ - & + & - & - & + & - \end{bmatrix},$$

$$\text{Set c:} \begin{bmatrix} + & + & - & + & + & - \\ + & + & - & - & - & + \\ - & - & + & + & + & - \\ - & - & + & - & - & + \end{bmatrix}.$$

Several such plans of Resolutions III, IV and V are given in John (1962).

7.2. Anderson–Thomas plans

Consider the following intersecting flat fraction of Resolution IV with 30 runs for a 3^5 factorial experiment. Five sets of 9 runs $(x_1, x_2, x_3, x_4, x_5)$ with $x_i = +, 0, -, i = 1, \ldots, 5$, are as follows.

$$\text{Set a:} \begin{bmatrix} - & - & - & - & - \\ 0 & 0 & 0 & 0 & 0 \\ + & + & + & + & + \\ - & + & + & + & + \\ 0 & - & - & - & - \\ + & 0 & 0 & 0 & 0 \\ - & 0 & 0 & 0 & 0 \\ 0 & + & + & + & + \\ + & - & - & - & - \end{bmatrix}, \qquad \text{Set b:} \begin{bmatrix} - & 0 & 0 & 0 & 0 \\ 0 & + & + & + & + \\ + & - & - & - & - \\ - & + & 0 & 0 & 0 \\ 0 & - & + & + & + \\ + & 0 & - & - & - \\ - & - & 0 & 0 & 0 \\ 0 & 0 & + & + & + \\ + & + & - & - & - \end{bmatrix},$$

$$
\text{Set c:}\ \begin{bmatrix} - & - & 0 & 0 & 0 \\ 0 & 0 & + & + & + \\ + & + & - & - & - \\ - & - & + & 0 & 0 \\ 0 & 0 & - & + & + \\ + & + & 0 & - & - \\ - & - & - & 0 & 0 \\ 0 & 0 & 0 & + & + \\ + & + & + & - & - \end{bmatrix}, \qquad \text{Set d:}\ \begin{bmatrix} - & - & - & 0 & 0 \\ 0 & 0 & 0 & + & + \\ + & + & + & - & - \\ - & - & - & + & 0 \\ 0 & 0 & 0 & - & + \\ + & + & + & 0 & - \\ - & - & - & - & 0 \\ 0 & 0 & 0 & 0 & + \\ + & + & + & + & - \end{bmatrix},
$$

$$
\text{Set e:}\ \begin{bmatrix} - & - & - & - & 0 \\ 0 & 0 & 0 & 0 & + \\ + & + & + & + & - \\ - & - & - & - & + \\ 0 & 0 & 0 & 0 & - \\ + & + & + & + & 0 \\ - & - & - & - & - \\ 0 & 0 & 0 & 0 & 0 \\ + & + & + & + & + \end{bmatrix}.
$$

Note that the sets a–e represent 5 flats given below.

$$
\begin{aligned}
&\text{Set a:} \quad x_2 = x_3 = x_4 = x_5, \\
&\text{Set b:} \quad 1 + x_1 = x_3 = x_4 = x_5, \\
&\text{Set c:} \quad 1 + x_1 = 1 + x_2 = x_4 = x_5, \\
&\text{Set d:} \quad 1 + x_1 = 1 + x_2 = 1 + x_3 = x_5, \\
&\text{Set e:} \quad x_1 = x_2 = x_3 = x_4.
\end{aligned}
$$

The above 5 flats are not parallel. The last three runs of the sets a, b, c, d, and e are identical to the first three runs of the sets b, c, d, e, and a, respectively. There are 30 distinct runs in the sets a–c and they form a minimal Resolution IV plan. Anderson and Thomas (1979) gave a series of Resolution IV plans for the s^n factorial where s is a prime power, in $s(s-1)n$ runs and also a series of generalized foldover designs with $s \geqslant 3$ and $n \geqslant 3$, in $s(s-1)n + s$ runs.

7.3. *Srivastava–Li plans*

Consider the following parallel flat fraction with 16 runs for a 2^6 factorial experiment. There are 4 sets (flats) with 4 runs each as given below.

$$\begin{aligned}
&\text{Set a:} \quad x_1 = x_2 = x_3 = +,\ x_4x_5x_6 = -,\\
&\text{Set b:} \quad x_1 = x_3 = -,\ x_2 = +,\ x_4x_5x_6 = +,\\
&\text{Set c:} \quad x_1 = x_2 = x_3 = -,\ x_4x_5x_6 = +,\\
&\text{Set d:} \quad x_1 = x_2 = -,\ x_3 = +,\ x_4x_5x_6 = +.
\end{aligned}$$

The 16 runs allow us to orthogonally estimate the general mean, all the main effects and the two-factor interactions between the subgroup of factors $(1, 2, 3)$ and the subgroup $(4, 5)$. Varieties of orthogonal plans of parallel flats type are presented in Srivastava and Li (1994) for an s^n factorial experiment, s is a prime or prime power.

7.4. *Krehbiel–Anderson plans*

Krehbiel and Anderson (1991) presented a series of determinant optimal designs within the class of parallel flats fractions, for the 3^m factorial, $m = 6, \dots, 14$. These designs permit us to estimate all main effects and the interactions of one factor with all others.

7.5. *Patel plans*

Patel (1963) gave fractional factorial plans for 2^m factorial experiments, $m = 5, 6, 8, 9,$ and 10, using the parallel flats and then replicating just one flat. The corresponding runs are called Partially Duplicated. Each flat represented an Orthogonal Resolution III plan and the sequential assembly of plans resulted in a Resolution V plan which even permits the estimation of some three factors and higher order interactions.

7.6. *Addelman plans*

Addelman (1961, 1963) presented parallel flat fractions for 2^m factorial experiments, $m = 3, \dots, 9$. These nonorthogonal plans allow us to estimate the general mean, main effects, some two and three factor interactions assuming the remaining interactions to be zero. Addelman (1969) presented sequential assemblies of fractional factorial plans for the 2^m factorial, $m = 3, \dots, 11$. For example, the following parallel flat fraction of a 2^5 factorial experiment permits us to estimate the general mean, main effects, and the 2-factor interactions between the first factor and the remaining factors. Two sets of 8 runs $(x_1, \dots, x_5)$ with $x_i = +$ or $-$, $i = 1, \dots, 5$, satisfying

$$\begin{aligned}
&\text{a:} \quad + = -x_1x_2x_3 = -x_1x_4x_5 = x_2x_3x_4x_5,\\
&\text{b:} \quad + = x_1x_2x_3 = x_1x_4x_5 = x_2x_3x_4x_5.
\end{aligned}$$

7.7. *Daniel plans*

Daniel (1962) gave the sequential assemblies of fractional factorial plans for 2^m factorial experiments, $m = 4, 7, 8, 16$. The construction of fractions for the 2^{m+1} factorial from the 2^m factorial is used in this context.

7.8. *Dykstra plans*

Dykstra (1959) presented the sequential assemblies of fractional factorial plans with partial duplication of flats for 2^m factorial experiments, $m = 6, \ldots, 11$. For example, the following parallel flat fraction of a 2^6 factorial experiment with 48 runs. Two sets of 16 runs $(x_1, \ldots, x_6)$ with $x_i = +$ or $-$, $i = 1, \ldots, 6$, satisfying

$$\begin{aligned}
\text{a:}\quad & + = -x_1x_2x_3 = -x_4x_5x_6 = x_1x_2x_3x_4x_5x_6,\\
\text{b:}\quad & + = x_1x_2x_3 = x_4x_5x_6 = x_1x_2x_3x_4x_5x_6.
\end{aligned}$$

The 16 runs of the set a is replicated twice.

7.9. *Connor plans and Connor–Young plans*

Connor (1960), Connor and Young (1961) gave the sequential assemblies of parallel flat fractions for $2^m \times 3^n$ factorial experiments taking m and n within practical ranges.

7.10. *Ghosh plans and Ghosh–Lagergren plans*

Ghosh (1987), Ghosh and Lagergren (1991) gave the sequential assemblies of fractional factorial plans for 2^m factorial experiments in estimating dispersion effects.

Detailed theoretical developments in the area of parallel flat fractions are available in Srivastava and Chopra (1973), Srivastava et al. (1984), Srivastava (1987, 1990), and the unpublished work of Anderson with his students Mardekian (1979), Bu Hamra (1987), and Hussain (1986).

8. Composite designs

Sequential assembly of fractions is very common in response surface designs. In the study of dependence of a response variable y on k explanatory variables, coded as $x_1, \ldots, x_k$, the unknown response surface is approximated by a first or a second order polynomial in a small region with the center being the point of maximum interest. The first order model is tried first with a small number of runs. If the first order model gives a significant lack of fit indicating the presence of a surface curvature, a second order response surface model is then fitted by augmenting the first set of runs with another set of runs (see Box and Draper, 1987; Khuri and Cornell, 1987). One of the most useful second order designs is called a composite design (CD), consists of F

factorial points (FP's) which are a fraction of the 2^k points $(\pm, \ldots, \pm), 2k$ axial points (AP's) $(\pm\alpha, \ldots, 0), \ldots, (0, \ldots, \pm\alpha)$ where α is a given constant, and n_0 $(\geqslant 0)$ center points (CP's) $(0, \ldots, 0)$. The total number of points is $N = F + 2k + n_0$. Note that the points are in fact the runs. Box and Wilson (1951) introduced such designs, also known as central composite designs. A lot of research has been done on the choice of F FP's in composite designs. The purpose of this section is to present developments in this area of research.

For N points in the design $(x_{u1}, \ldots, x_{uk})$, the observations are

$$y(x_{u1}, \ldots, x_{uk}) = y_u, \quad u = 1, \ldots, N.$$

The expectation of y_u under the second order model is

$$\mathrm{E}(y_u) = \beta_0 + \sum_{i=1}^{k} \beta_i x_{ui} + \sum_{i=1}^{k} \beta_{ii} x_{ui}^2 + \sum_{\substack{i=1 \\ i<j}}^{k} \sum_{j=1}^{k} \beta_{ij} x_{ui} x_{uj},$$

where the intercept β_0, the linear coefficients β_i, the pure quadratic coefficients β_{ii}, and the interaction coefficients β_{ij} are unknown constants. The y_u's are assumed to be uncorrelated with variance σ^2, an unknown constant. The number of β's in (1) is $1 + 2k + \binom{k}{2}$. For a second order design with N points $(x_{u1}, \ldots, x_{uk}), u = 1, \ldots, N$, all β's are unbiasedly estimable. The expectation of y_u under the first order model is

$$\mathrm{E}(y_u) = \beta_0 + \sum_{i=1}^{k} \beta_i x_{ui}.$$

Different composite second order designs are given below.

8.1. Box–Hunter plans

Box and Hunter (1957) suggested FP's as the complete set of 2^k points or an orthogonal Resolution V plan which permits the unbiased estimation of $\beta_0 + \sum_{i=1}^{k} \beta_{ii}, \beta_i$'s and β_{ij}'s under the second order model and, moreover, the estimators are uncorrelated. Composite designs with such FP's and $\alpha = F^{1/4}$ in AP's give the variance of the predicted response dependent on the point only through its distance from the origin. The variance structure is achieved at the cost of a large number of FP's. Efforts are then made to reduce the number of FP's.

Box and Hunter (1957), Hartley (1959) plans for FP's which are of Resolution V can be expressed in general form as given below for $k = 3t + 2, t = 1, 2, \ldots,$

$$x_1x_2x_3x_4x_5 = + \text{ or } -,$$

$$x_4x_5x_6x_7x_8 = + \text{ or } -,$$

$$x_{k-4}x_{k-3}x_{k-2}x_{k-1}x_k = + \text{ or } - .$$

The number of FP's is then $F = 2^{2t+2}$. Note that the same plan can be used for $k = 3t + 3$ and $3t + 4$ with $F = 2^{2t+3}$ and $F = 2^{2t+4}$.

8.2. *Hartley plans*

Hartley (1959) pointed out that FP's need not be of Resolution V. It could be as low as of Resolution III which permits the unbiased estimation of $\beta_0 + \sum_{i=1}^{k} \beta_{ii}$ and β_i's assuming β_{ij}'s are known, with an additional condition that the unbiased estimation of the β_{ij}'s is possible assuming the other β's are known. Draper and Lin (1990) named such FP's as Resolution III*. Resolution III* plans in Hartley (1959) can be expressed in general form as given below for $k = 3t,\ t = 1, 2, \ldots,$

$$x_1x_2x_3 = + \text{ or } -,$$

$$x_4x_5x_6 = + \text{ or } -,$$

$$\ldots$$

$$x_{k-2}x_{k-1}x_k = + \text{ or } - .$$

The number of FP's is then $F = 2^{2t}$. Again, the same plan can be used for $k = 3t + 1$ and $3t + 2$ with $F = 2^{2t+1}$ and $F = 2^{2t+2}$. Resolution III* plans of Hartley (1959) are in fact orthogonal Resolution III plan with the additional property discussed earlier.

8.3. *Westlake plans*

Westlake (1965) presented FP's as parallel flat fractions discussed in Section 7 for $k = 5, 7$, and 9. For example, the 12 FP's for $k = 5$ are as follows.

Set a: $x_1x_2x_3x_4x_5 = -1,\ x_2 = 1,\ x_5 = -1,$

Set b: $x_1x_2x_3x_4x_5 = -1,\ x_2 = -1,\ x_5 = 1,$

Set c: $x_1x_2x_3x_4x_5 = -1,\ x_2 = 1,\ x_5 = 1.$

Sets a–c consisting of 4 points are representing 3 parallel flats. Note that Box and Hunter (1957) and Hartley (1959) plans both of Resolution V and Resolution III* have 16 FP's. The above Westlake plan for $k = 5$ is therefore an improvement over Box–Hunter plan and Hartley plan in reducing the number of FP's.

8.4. Draper plans and Draper–Lin plans

Draper (1985), Draper and Lin (1990) presented FP's for $k = 5, 7, 8, 9$ and 10 with small number of points. These FP's are of Resolution III* and are obtained from Plackett and Burman designs. For example, Draper (1985) presented two nonisomorphic choices of FP's for $k = 5$ in Table 2 on p. 175 obtained from 12 run Plackett and Burman design. Both Plans a and b of Draper (1985) consist of 12 FP's and Plan a has in fact 11 distinct FP's

8.5. Ghosh–Al-Sabah plans

Ghosh and Al-Sabah (1996) observed that the following submodel is needed for the choice of FP's. The expectation of y_u under the submodel is

$$\mathrm{E}(y_u) = \beta_0 + \sum_{i=1}^{k} \sum_{\substack{j=1 \\ i<j}}^{k} \beta_{ij} x_{ui} x_{uj}, \quad u = 1, \ldots, F,$$

and y_u's are assumed to be uncorrelated with variance σ^2. For composite designs, the FP's must be chosen so that β_0 and β_{ij}'s are unbiasedly estimable under the above submodel. It is first observed that the nonisomorphic Plans a and b of Draper (1985) for $k = 5$ are in fact isomorphic under the above model. It is also observed that the FP's do not have to be of Resolution III* and not even of Resolution III. Ghosh and Al-Sabah (1996) started with a set of FP's permitting the orthogonal estimation of β_0 and β_{ij}'s in the above submodel. The following robustness property against the deletion of points is used in reducing the number of FP's.

DEFINITION. A set of F FP's is said to be robust against deletion of t (a positive integer) points if the parameters β_0 and β_{ij}'s in the submodel remain unbiasedly estimable with the remaining $(F - t)$ FP's.

Ghosh and Al-Sabah (1996) presented techniques for efficient deletion and gave many plans for $k = 4, \ldots, 10$, with small number of runs. Efficiency comparisons of these plans with their competitors Westlake plans, Draper plans and Draper–Lin plans revealed striking results.

Let b_0, b_i, b_{ii}, and b_{ij} be the least squares estimators of $\beta_0, \beta_i, \beta_{ii}$, and β_{ij} in the second order model. Let $\underline{b} = (b_0; b_1, \ldots, b_k; b_{11}, \ldots, b_{kk}; b_{12}, \ldots, b_{\overline{k-1\ k}})'$ and $\mathrm{V}(\underline{b}) =$ the variance-covariance matrix of $\underline{b}$. Denote

$$\begin{aligned} \mathrm{MCR} &= \text{Maximum Characteristic Root of } \sigma^{-2}\mathrm{V}(\underline{b}), \\ \mathrm{T} &= \text{Trace of } \sigma^{-2}\mathrm{V}(\underline{b}), \\ \mathrm{D} &= \text{Determinant of } \sigma^{-2}\mathrm{V}(\underline{b}). \end{aligned}$$

Table 3
Efficiency comparison for $k = 5$

				V(r)	
Plan	MCR	T	D $\times 10^{15}$	$r = 1$	$r = \sqrt{2}$
Ghosh–Al-Sabah.1	1.5	8.5	1.3	12.8	28.2
Westlake	8.7	17.8	3.0	16.0	41.2
Draper.1	2.2	12.3	1.3	14.1	33.6
Draper.2	1.7	10.3	1.3	13.4	30.8

Table 4
Efficiency comparison for $k = 7$

				V(r)	
Plan	MCR	T	D $\times 10^{31}$	$r = 1$	$r = \sqrt{2}$
Ghosh–Al-Sabah.1	5.5	16.2	4.3×10^{-2}	22.8	54.2
Westlake	12.8	29.3	3.0	27.7	73.8
Draper–Lin.1	58.8	77.4	2.5	45.7	145.6
Draper–Lin.2	24.4	40.9	1.8	32.0	91.0

Table 5
Efficiency comparison for $k = 8$

				V(r)	
Plan	MCR	T	D $\times 10^{43}$	$r = 1$	$r = \sqrt{2}$
Ghosh–Al-Sabah.1	4.5	20.6	1.4×10^{-1}	28.9	69.9
Draper–Lin.1	9.0	29.8	4.2	32.4	83.8

Table 6
Efficiency comparison for $k = 9$

				V(r)	
Plan	MCR	T	D $\times 10^{56}$	$r = 1$	$r = \sqrt{2}$
Ghosh–Al-Sabah.1	1.9	18.0	2.3×10^{-2}	32.8	75.5
Draper–Lin.1	60.2	97.2	6.5	63.8	120.0

The predicted response $\widehat{y}(x_1, \ldots, x_k)$ at the point $(x_1, \ldots, x_k)$ is

$$\widehat{y}(x_1, \ldots, x_k) = b_0 + \sum_{i=1}^{k} b_i x_i + \sum_{i=1}^{k} b_{ii} x_i^2 + \sum_{i=1}^{k} \sum_{j=1}^{k} b_{ij} x_i x_j,$$

Table 7
Efficiency comparison for $k = 10$

Plan	MCR	T	$D \times 10^{70}$	V(r) $r = 1$	$r = \sqrt{2}$
Ghosh–Al-Sabah.1	7.1	31.2	3.3×10^{-2}	43.4	107.6
Draper–Lin.1	6791.4	6850.0	1.6	2762.8	10985.0

and its variance is $V(\widehat{y}(x_1, \ldots, x_k))$. The average predicted variance over a region $R = \{(x_1, \ldots, x_k) \mid x_1^2 + \ldots + x_k^2 \leqslant r^2\}$ is given by

$$\frac{\sigma^2}{N} V(r) = \frac{\int_R V(\widehat{y}(x_1, \ldots, x_k))\, dx_1 \cdots dx_k}{\int_R dx_1 \cdots dx_k}.$$

Efficiency comparisons of the plans are made with respect to MCR, T, D and $V(r)$. Tables 3, 4, 5, and 6, present excerpts from such comparisons for $n_0 = 1$ and $\alpha = 1$. For $k = 5$, Draper plans perform closely to Ghosh–Al-Sabah plan but as k increases Draper–Lin plans perform very poorly compared to Ghosh–Al-Sabah plans. Westlake plan for $k = 5$ is in fact worst in performance. Westlake plan for $k = 7$ performs better than Draper–Lin plans but worse than Ghosh–Al-Sabah plan.

9. Discussions and conclusions

There are numerous other designs that are obtained by sequential assembly of fractions. For example, Box–Behnken plans (see Box and Behnken, 1960), Koshal plans, and others in response surface experiments (see Box and Draper, 1987; Khuri and Cornell, 1987), Scheffé Simplex–Lattice and Simplex–Centroid plans (see Scheffé, 1958, 1963) and their variations in mixture experiments (see Cornell, 1990), Webb's saturated sequential factorial designs (see Webb, 1968) and others. It is impossible to discuss everything in detail and therefore this article is focused on the research areas that the author has made contributions.

Sequential assembly of fractions are discussed in both discrete and continuous factorial experiments. The goals are in searching and identifying the significant non-negligible parameters, fitting models under different resolution plans, and fitting the first and second order response surface models. The objective functions in designs are to reduce the number of runs, to achieve the optimality in terms of estimation and prediction, and to reduce the overall effect of uncontrollable noise variables. The technique of Normal Probability Plot that is commonly used in practice for identifying the influential parameters is informative but requires a large number of runs and does not provide complete information in factor screening situations as demonstrated with an illustrative example in Section 2. Sections 3–7 provide the methods with small number of runs and of higher revealing power. Not all composite designs with small number of runs are equally efficient and the performances of some are in fact dismal relative to the others as can be seen in Section 8.

In conclusion, it is to be remarked that the sequential assembly of fractions in factorial experiments is an effective tool in gathering pertinent information economically if it is done efficiently.

Acknowledgements

This work is sponsored by the Air Force Office of Scientific Research under grant F49620-95-1-0094. The author would like to thank Professors D. Best, T. Calinski, R. F. Gunst, K. Hinkelmann, S. Kageyama, A. Khuri, D. C. Montgomery, J. N. Srivastava, and S. Zacks for their valuable comments on the first draft of this paper.

References

Addleman, S. (1961). Irregular fractions of the 2^n factorial experiments. *Technometrics* **3**, 479–496.

Addleman, S. (1963). Techniques for constructing fractional replicate plans. *J. Amer. Statist. Assoc.* **58**, 45–71.

Addleman, S. (1969). Sequences of two-level fractional factorial plans. *Technometrics* **11**, 477–509.

Anderson, D. A. and A. M. Thomas (1979). Near minimal resolution IV designs for the S^n factorial experiment. *Technometrics* **21**, 331–336.

Anderson, D. A. and A. M. Thomas (1980). Weakly resolvable IV. 3 search designs for the p^n factorial experiment. *J. Statist. Plann. Inference* **4**, 299–312.

Bose, R. C. (1947). Mathematical theory of the symmetrical factorial designs. *Sankhyā* **8**, 107–166.

Box, G. E. P. and K. B. Wilson (1951). On the experimental attainment of optimum conditions. *J. Roy. Statist. Soc. Ser. B* **13**, 1–45 (with discussions).

Box, G. E. P. and J. S. Hunter (1957). Multi-factor experimental designs for exploring response surfaces. *Ann. Math. Statist.* **28**, 195–241.

Box, G. E. P. and D. W. Behnken (1960). Some new three level designs for the study of quantitative variables. *Technometrics* **2**, 455–475.

Box, G. E. P., W. G. Hunter and J. S. Hunter (1978). *Statistics for Experiments: An Introduction to Design, Data Analysis, and Model Building*. Wiley, New York.

Box, G. E. P. and N. R. Draper (1987). *Empirical Model-Buidling and Response Surfaces.* John Wiley, New York.

Bu Hamra, S. (1987). Theory of Parallel Flat Fractions for S^n Factorial Experiments. Ph.D. Thesis. Department of Statistics, University of Wyoming.

Chatterjee, K. and R. Mukerjee (1986). Some search designs for symmetric and asymmetric factorials. *J. Statist. Plann. and Inference* **13**, 357–363.

Chatterjee, K. (1989). Search designs for general asymmetric factorials. *Comm. Statist. Theory Methods* **18**, 2189–2223.

Chatterjee, K. (1990). Search designs for searching for one among the two- and three-factor interaction effects in the general symmetric and asymmetric factorials. *Ann. Inst. Statist. Math.* **42**, 783–803.

Connor, W. S. (1960). Construction of the fractional factorial designs of the mixed $2^m \times 3^n$ series. *Contributions to Probability and Statistics – Essays in Honor of Harold Hotelling*. Stanford Univ. Press, CA, 168–181.

Connor, W. S. and S. Young (1961). Fractional factorial designs for experiments with factors at two and three levels. *Appl. Math. Series*, National Bureau of Standards, No. 58.

Cornell, J. A. (1990). *Experiments with Mixtures: Designs, Models, and the Analysis of Mixture Data.* 2nd edn. Wiley, New York.

Daniel, C. (1962). Sequences of fractional replicates in the 2^{p-q} series. *J. Amer. Statist. Assoc.* **27**, 403–429.
Draper, N. R. (1985). Small composite designs. *Technometrics* **27**, 173–180.
Draper, N. R. and D. K. J. Lin (1990). Small response-surface designs. *Technometrics* **32**, 187–194.
Dykstra, O. (Jr.) (1959). Partial duplication of factorial experiments. *Technometrics* **1**, 63–75.
Ghosh, S. (1979a). On robustness of designs against incomplete data. *Sankhyā Ser. B* **40**, Parts 3 and 4, 204–208.
Ghosh, S. (1979b). On single and multistage factor screening procedures. *J. Combin. Inform. System Sci.* **4**, 275–284.
Ghosh, S. and D. Avila (1985). Some new factor screening designs using the search linear model. *J. Statist. Plann. Inference* **11**, 259–266.
Ghosh, S. (1987). Influential nonnegligible parameters under the search linear model. *Comm. Statist. Theory Methods* **16**, 1013–1025.
Ghosh, S. and X. D. Zhang (1987). Two new series of search designs for 3^m factorial experiments. *Utilitas Mathematica* **32**, 245–254.
Ghosh, S. (1987). Non-orthogonal designs for measuring dispersion effects in sequential factor screening experiments using search linear models. *Comm. Statist. Theory Methods* **16**, 2839–2850.
Ghosh, S. and E. Lagergren (1991). Dispersion models and estimation of dispersion effects in replicataed factorial experiments. *J. Statist. Plann. Inference* **26**, 253–262.
Ghosh, S. and H. Talebi (1993). Main effect plans with additional search property for 2^m factorial experiments. *J. Statist. Plann. Inference* **36**, 367–384.
Ghosh, S. (1993a). New main effect plus one plans for 2^7 factorial experiments and their robustness property against deletion of runs. In: K. Matusita et al., eds., *Statistical Sciences and Data Analysis*, VSP, The Netherlands, 529–542.
Ghosh, S. (1993b). Sequential construction of new main effect plans with higher revealing power for 2^7 factorials. *J. Combin. Inform. System Sci.* **18**, 217–231.
Ghosh, S. and W. S. Al-Sabah (1996). Efficient composite designs with small number of runs. *J. Statist. Plann. Inference*, to appear.
Gupta, B. C. (1981). Main effect plus one plan for 2^m factorial designs. Memorias de Matematica 12J, IM-UFRJ, Rio de Janeiro, Brazil.
Gupta, B. C. (1984). A lower bound for number of treatments in a main effect plus one plan for 2^4 factorials. *Utilitas Math.* **26**, 259–267.
Gupta, B. C. and S. S. R. Carvazal (1987). A lower bound for the number of treatments in a main effect plus one plan for 2^5 factorials. *Utilitas Math.* **31**, 33–43.
Gupta, B. C. (1988). A bound connected with 2^6 factorial search designs of Resolution 3.1. *Comm. Statist. Theory Methods* **17**, 3137–3144.
Gupta, B. C. (1992). On the existence of main effect plus k plans for 2^m factorials and tables for main effect plus 1 and 2 plans for 2^7 factorials. *Comm. Statist. Theory Methods* **21**, 1137–1143.
Hartley, H. O. (1959). Smallest composite designs for quadratic response surface. *Biometrics* **15**, 611–624.
Hussain, A. H. M. (1986). Sequential Designs for the 3^n Factorial. Ph.D. Thesis. Department of Statistics, University of Wyoming.
John, P. W. M. (1962). Three-quarter replicates of 2^n designs. *Biometrics* **18**, 172–184.
Khuri, A. I. and J. A. Cornell (1987). *Response Surfaces: Designs and Analyses*. Wiley, New York.
Krehbiel, T. C. and D. A. Anderson (1991). Optimal fractional factorial designs for estimating interactions of one factor with all others: 3^m series. *Comm. Statist. Theory Methods* **20**, 1055–1072.
Margolin, B. H. (1969). Resolution IV fractional factorial designs. *J. Roy. Statist. Soc. Ser. B* **31**, 514–523.
Margolin, B. H. (1969). Results on factorial designs of resolution IV for the 2^n and 2^n3^m series. *Technometrics* **11**, 431–444.
Montogomery, D. C. (1991). *Design and Analysis of Experiments*, 3rd edn. Wiley, New York.
Müller, M. (1993). Supersaturated designs for one or two effective factors. *J. Statist. Plann. Inference* **37**, 237–244.
Ohnishi, T. and T. Shirakura (1985). Search designs for 2^m factorial experiments. *J. Statist. Plann. Inference* **11**, 241–245.
Patel, M. S. (1963). Partially duplicated fractional factorial designs. *Technometrics* **5**, 71–83.

Plackett, R. L. and J. P. Burman (1946). The design of optimum multifactorial experiments. *Biometrika* **33**, 305–325.

Rao, C. R. (1947). Factorial arrangements derivable from combinatorial arrangements of arrays. Supplement to *J. Roy. Statist. Soc.* **9**, 128–139.

Scheffé, H. (1958). Experiments with mixtures. *J. Roy. Statist. Soc. Ser. B* **20**, 344–360.

Scheffé, H. (1963). The simplex-controid design for experiments with mixtures. *J. Roy. Statist. Soc. Ser. B* **25**, 235–263.

Shirakura, T. (1991). Main effect plus one or two plans for 2^m factorials. *J. Statist. Plann. Inference* **27**, 65–74.

Shirakura, T. and S. Tazawa (1992). A series of search designs for 2^m factorial designs fo Resolution V which permit search of one or two unknown extra three-factor interactions. *Ann. Inst. Statist. Math.* **44**, 185–196.

Shirakura, T. and T. Ohnishi (1985). Search designs for 2^m factorials derived from balanced arrays of strength $2(\ell + 1)$ and AD-optimal search designs. *J. Statist. Plann. Inference* **11**, 247–258.

Srivastava, J. N. and D. V. Chopra (1973). Balanced arrays and orthogonal arrays. In: J. N. Srivastava et al., eds., *A Survey of Combinatorial Theory*. North-Holland, Amsterdam, 411–428.

Srivastava, J. N. (1975). Designs for searching nonnegligible effects. In: J. N. Srivastava, ed., *A Survey of Statistical Designs and Linear Models*. North-Holland, Amsterdam, 507–519.

Srivastava, J. N. and S. Ghosh (1976). A series of 2^m factorial designs of Resolution V which allow search and estimation of one extra unknown effect. *Sankhyā Ser. B* **38**, 280–289.

Srivastava, J. N. and S. Ghosh (1977). Balanced 2^m factorial designs of Resolution V which allow search and estimation of one extra unknown effect, $4 \leqslant m \leqslant 8$. *Comm. Statist. Theory Methods* **A6**(2), 141–166.

Srivastava, J. N. (1977). Optimal search designs or designs optimal under bias free optimality criteria. In: S. S. Gupta and D. S. Moore, eds., *Statistical Decision Theory and Related Topics*, Vol. 2, 375–409.

Srivastava, J. N. and B. C. Gupta (1979). Main effect plan for 2^m factorials which allow search and estimation of one non-negligible effect. *J. Statist. Plann. Inference* **3**, 259–265.

Srivastava, J. N., D. A. Anderson and J. Mardekian (1984). Theory of factorial designs of the parallel flats type 1: The coefficient matrix. *J. Statist. Plann. Inference* **9**, 229–252.

Srivastava, J. N. (1984). Sensitivity and revealing power: Two fundamental statistical criteria other than optimality arising in discrete experimentation. In: K. Hinkelmann, ed., *Experimental Designs, Statistical Models and Genetic Statistics Models, and Genetic Statistics*. Marcel Dekker, New York, 95–117.

Srivastava, J. N. and D. M. Mallenby (1985). On a decision rule using dichotomies for identifying non-negligible parameters in certain linear models. *J. Multivariate Anal.* **16**, 318–334.

Srivastava, J. N. (1987). On the inadequacy of the customary orthogonal arrays in quality control and scientific experimentation, and the need of probing designs. *Comm. Statist.* **16**, 2901–2941.

Srivastava, J. N. (1987). Advances in the general theory of factorial designs based on partial pencils in Euclidean n-space. *Utilitas Math.* **32**, 75–94.

Srivastava, J. N. (1990). Modern factorial design theory for experimenters and statisticians. In: S. Ghosh, ed., *Statistical Design and Analysis of Industrial Experiments*. Marcel Dekker, New York, 311–406.

Srivastava, J. N. and S. Arora (1991). An infinite series of Resolution III. 2 designs for the 2^m factorial experiment. *Discrete Math.* **98**, 35–56.

Srivastava, J. N. and R. Hveberg (1992). Sequential factorial probing designs for identifying and estimating non-negligible factorial effects for the 2^m experiment under the tree structure. *J. Statist. Plann. Inference* **30**, 141–162.

Srivastava, J. N. and J. Li (1994). Orthogonal designs of parallel flats type. *J. Statist. Plann. Inference*, to appear.

Webb, S. R. (1968). Saturated sequential factorial designs. *Technometrics* **10**, 535–550.

Westlake, W. J. (1965). Composite designs based on irregular fractions of factorials. *Biometrics* **21**, 324–336.

Zacks, S. (1968). Bayes sequential design of fractional factorial experiments for the estimation of a subgroup of pre-assigned parameters. *Ann. Math. Statist.* **38**(3), 973–982.

S. Ghosh and C. R. Rao, eds., *Handbook of Statistics, Vol. 13*

14

Designs for Nonlinear and Generalized Linear Models

A. C. Atkinson and L. M. Haines

1. Background

The theory and practice of optimum experimental design has received several book length treatments including those of Fedorov (1972), Silvey (1980), Pazman (1986), Shah and Sinha (1989), Atkinson and Donev (1992) and Pukelsheim (1993). With the exception of Atkinson and Donev (1992), the emphasis in these texts is on linear models: for these models the optimum designs do not depend, at least in general, on the values of the parameters of the model. A survey of the use and usefulness of optimum experimental designs is given by Atkinson (1996), and contains examples of models of importance in medicine, technology, biology and agriculture, which are either nonlinear regression models or which belong to the class of generalized linear models described by McCullagh and Nelder (1989). Optimum experimental designs for these nonlinear models depend on the values of the unknown parameters and the problem of their construction is therefore necessarily more complicated than that for linear models. This problem is the focus of the present chapter.

One approach to the problem of design for nonlinear models is to adopt a best guess for the parameters. The simplicity of optimum design for linear models is then recovered and, in particular, locally D-optimum designs for the precise estimation of all the parameters in the nonlinear model, and locally c-optimum designs for the precise estimation of specified linear combinations of these parameters, are readily constructed and their optimality confirmed. An alternative and, in a sense, a more realistic approach to this design problem is to introduce a prior distribution on the parameters and to incorporate this prior into appropriate design criteria, usually by integrating D- or c-optimality criteria over the prior distribution. Designs optimizing these criteria are termed Bayesian optimum designs and are usually constructed numerically.

Optimum designs for nonlinear regression models are discussed in Section 2. In particular the theory pertaining to both locally and Bayesian optimum designs for these models is developed and illustrative examples, including examples involving compartmental models and a fertilizer experiment, are presented. For the locally optimum designs emphasis is placed on geometrical insights and for the Bayesian optimum designs on numerical methods for computing the required integrals. The theory of optimum design for generalized linear models is analogous to that for nonlinear regression models and is outlined in Section 3. In addition, examples of designs for logistic regression models with one and two explanatory variables are described in some detail in that section.

Optimum designs for checking the adequacy of a nonlinear model are discussed in Section 4 and the use of compound design criteria to accommodate both efficient parameter estimation and efficient model checking is highlighted. A simple, illustrative example, that of first-order decay, is also included in that section. Locally and Bayesian optimum designs for discriminating between two nonlinear regression models are described in Section 5. The ideas discussed there are also extended to include designs for discriminating between generalized linear models. The final section of the chapter contains some general remarks.

2. Nonlinear regression models

2.1. Linear models

The response, Y, for a linear regression model has mean, $\mathrm{E}(Y) = f^T(x)\theta$, and constant variance, σ^2, where x is a vector of m explanatory variables defined on a design region, $\mathcal{X}$, $f(x)$ represents a k-valued vector function of x, and θ is a vector of k unknown parameters. Most commonly the mean, $f^T(x)\theta$, is taken to be a low-order polynomial in the elements of x. For responses which are normally distributed the information matrix for θ from a design comprising n observations at the points, $x_i, i = 1, \ldots, n$, is given by $\sum_{i=1}^n f(x_i)f^T(x_i)$, where σ^2 is taken, without loss of generality, to be 1. This matrix can be expressed more succinctly as F^TF, where F is an $n \times k$ design matrix with ith row given by $f^T(x_i)$ for $i = 1, \ldots, n$. In the approximate or continuous design theory of Kiefer (1985) such an n-trial design is replaced by a probability measure, ξ, over the design region, $\mathcal{X}$, termed a design measure, and the information matrix for θ is then given by

$$M(\xi) = \int_{\mathcal{X}} f(x)f^T(x)\xi(\mathrm{d}x).$$

Exact n-trial D-optimum designs are designs which maximize the determinant of the information matrix, $|F^TF|$, or, equivalently, which minimize the generalized variance of the parameter estimates. In the same way, continuous D-optimum designs, with which this chapter is concerned, maximize the determinant, $|M(\xi)|$, or equivalently the function, $\log|M(\xi)|$. There are two results which are central to the optimum experimental design theory of Kiefer (1985) , and which are introduced here in the context of D-optimality for linear models. Firstly, it follows from Carathéodory's Theorem (Silvey, 1980, p. 72) that a continuous D-optimum design is based on a finite number of support points and, more specifically, on at most $k(k+1)/2$ such points. As an aside, it is interesting to observe that D-optimum designs are in fact very commonly based on exactly k points of support and that in such cases the weights associated with the support points are necessarily equal and thus equal to $1/k$ (Silvey, 1980, p. 42). Secondly, the celebrated General Equivalence Theorem of Kiefer and Wolfowitz (1960) makes it possible to check the global optimality or otherwise of a candidate continuous D-optimum design. In particular the standardized variance of the prediction at x for a linear regression model and a design, ξ, is given by

$$d(x, \xi) = f^T(x)M^{-1}(\xi)f(x),$$

and the relevant part of the Equivalence Theorem states that, for the D-optimum design, ξ^*, the maximum value of $d(x, \xi^*)$ over the design region, $\mathcal{X}$, is k, the number of parameters in the model, and further that this maximum value is attained at the points of support of ξ^*. The theorem also provides a basis for algorithmic construction of D-optimum designs.

Geometrical insight into the structure of D-optimum designs is found by considering the set, $\mathcal{F}$, generated by the vector, $f(x)$, as x varies over the design space, $\mathcal{X}$. This set is termed the induced design space or the design locus and was introduced by Box and Lucas (1959). The support points of the D-optimum design correspond to the points of contact between the boundary of $\mathcal{F}$ and the ellipsoid, $E_{\mathcal{F}}$, with minimum content centered at the origin and containing $\mathcal{F}$ (Sivley, 1980, p. 41). In fact this ellipsoid, which is termed the minimal ellipsoid, is defined by the equation,

$$f^T(x)[M(\xi^*)]^{-1}f(x) = k,$$

where ξ^* is the D-optimum design and the result just cited constitutes an alternative representation of the Equivalence Theorem to that based on the standardized variance function (Haines, 1993). In addition, for exact designs based on k points of support, the determinant of the information matrix, $\sum_{i=1}^{k} f(x_i)f^T(x_i)$, is the square of the volume of the simplex with vertices the points, $f(x) \in \mathcal{F}$, corresponding to the k design points together with the origin and the exact k- point D-optimum design is therefore the design for which this simplex has maximum volume.

Example 1. Quadratic regression
As a first example consider the linear model with

$$E(Y) = \sum_{j=1}^{3} \theta_j f_j(x) = \theta_1 + \theta_2 x + \theta_3 x^2$$

and with design region, $\mathcal{X} = [-1, 1]$. The continuous D-optimum design has equally weighted support points at the extremities and centre of the one dimensional design space and can be written succinctly as

$$\xi^* = \begin{pmatrix} -1 & 0 & 1 \\ 1/3 & 1/3 & 1/3 \end{pmatrix}.$$

Figure 1(a) shows the induced design space for this quadratic model, together with the 3 points in $\mathcal{F}$ corresponding to the support points of the optimum design, ξ^*. It is clear firstly that these points are the points of contact of the minimal ellipsoid, $E_{\mathcal{F}}$,

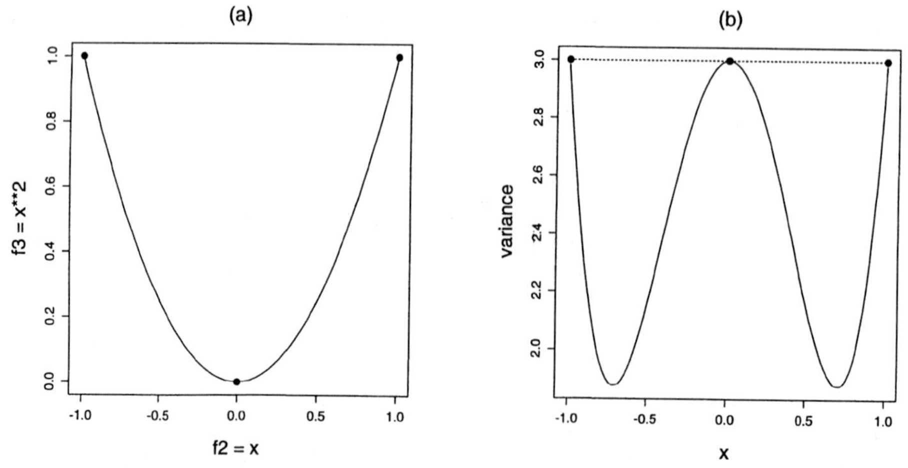

Fig. 1. Example 1. Quadratic regression. (a) induced design space in the plane $f_1(x) = 1$ showing that the points of the optimum design, •, form a triangle of maximum area; (b) the variance $d(x, \xi^*)$ showing that the three-point design of (a) is optimum.

with the induced design space, $\mathcal{F}$, and secondly that they form a triangle of maximum area in the plane, $f_1(x) = 1$, and hence, together with the origin, form a simplex of maximum volume over all such sets of 3 points in $\mathcal{F}$. These features serve to confirm that ξ^* is indeed globally D-optimum. Another way of confirming the optimality of the design is to look at the behaviour of the standardized variance over the design region and, in particular, to plot $d(x, \xi^*)$ against $x \in [-1, 1]$. Figure 1(b) is such a plot – it shows that the maximum of the variance is three and that the maxima occur at the points of the design, ξ^*.

2.2. *Locally optimum designs*

We now turn to the main subject of this paper, designs for nonlinear models. In particular the response, Y, for a nonlinear regression model has mean, $\mathrm{E}(Y) = \eta(x, \theta)$, where η is function nonlinear in at least one of the k parameters, and constant variance, which is again taken without loss of generality to be 1. The information matrix for θ obtained from an n-trial design is given by

$$I(x, \theta) = \sum_{i=1}^{n} f(x_i, \theta) f^T(x_i, \theta) = F^T F$$

where the vector, $f(x_i, \theta)$, has jth element

$$f_j(x_i, \theta) = \frac{\partial \eta(x_i, \theta)}{\partial \theta_j}, \quad \text{for } j = 1, \dots, k,$$

and where

$$F = \begin{pmatrix} f^T(x_1, \theta) \\ \vdots \\ f^T(x_n, \theta) \end{pmatrix},$$

and from a continuous design, ξ, by

$$M(\xi, \theta) = \int_{\mathcal{X}} f(x, \theta) f^T(x, \theta) \xi(\mathrm{d}x).$$

The information matrix thus depends on the unknown parameter, θ, and a natural way of accommodating the obvious problems which accrue from this dependence is to adopt a best guess for the parameters, say θ^o, and to consider designs which maximize an appropriate function of $M(\xi, \theta)$ evaluated at $\theta = \theta^o$ (Chernoff, 1953). Such designs are termed locally optimum and the theory pertaining to optimum designs for linear models outlined in Section 2.1 clearly applies to them.

2.2.1. Locally D-optimum designs

For locally D-optimum designs, the powerful results provided by Carathéodory's Theorem and by the General Equivalence Theorem, with the standardized variance of the prediction at x now defined by

$$d(x, \xi, \theta) = f^T(x, \theta) M^{-1}(\xi, \theta) f(x, \theta), \tag{1}$$

hold. Furthermore the geometry relating to the induced design space can be invoked. These and other features are illustrated by means of the following example.

Example 2. A simple compartmental model (a)

Compartmental models are frequently used to model the passage of a drug through a subject and Atkinson et al. (1993) consider in particular the nonlinear regression model with expected response,

$$\eta(x, \theta) = \theta_3(\mathrm{e}^{-\theta_1 x} - \mathrm{e}^{-\theta_2 x}), \quad x \geqslant 0,\ \theta_2 > \theta_1, \tag{2}$$

which arises from such a model. For θ_3 known and equal to one, the model simplifies to the difference of two exponentials,

$$\eta(x, \theta) = \mathrm{e}^{-\theta_1 x} - \mathrm{e}^{-\theta_2 x}, \tag{3}$$

and this latter two-parameter form is investigated in the present example. The three sets of values for θ_1 and θ_2 given in Table 1 are introduced in order to illustrate the properties and scope of the model and its associated locally optimum designs and Figure 2 shows plots of the response $\eta(x, \theta)$ against time, x, for each of these sets.

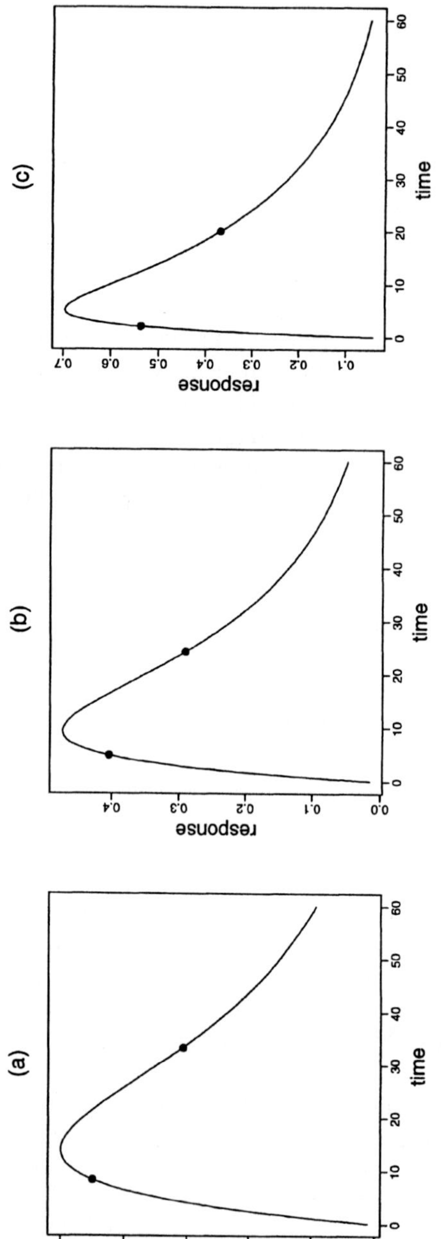

Fig. 2. Example 2. A simple compartmental model. The response $E(Y)$ for the parameter values (a) $\theta_1 = 0.05, \theta_2 = 0.1$; (b) $\theta_1 = 0.05, \theta_2 = 0.2$; (c) $\theta_1 = 0.05, \theta_2 = 0.5$.

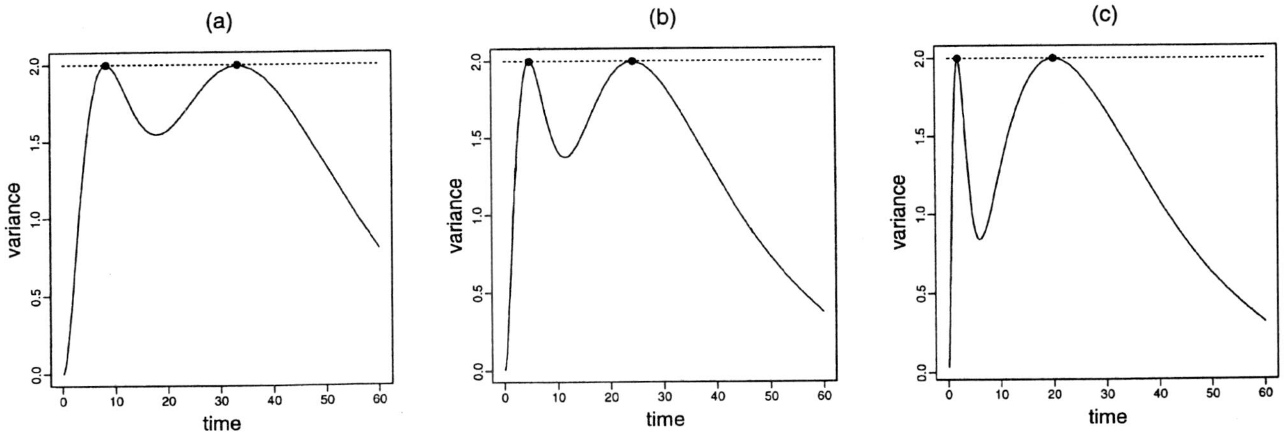

Fig. 3. Example 2. A simple compartmental model. The variance $d(x, \xi^*, \theta)$ demonstrating the optimality of the locally D-optimum designs for the parameter values (a) $\theta_1 = 0.05, \theta_2 = 0.1$; (b) $\theta_1 = 0.05, \theta_2 = 0.2$; (c) $\theta_1 = 0.05, \theta_2 = 0.5$.

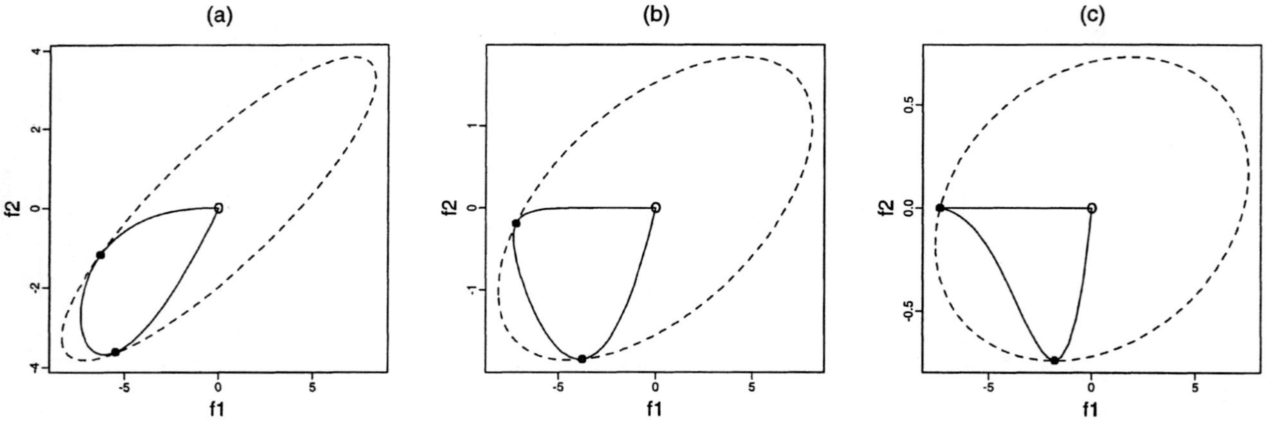

Fig. 4. Example 2. A simple compartmental model. The induced design space and locally D-optimum design for the parameter values (a) $\theta_1 = 0.05$, $\theta_2 = 0.1$; (b) $\theta_1 = 0.05$, $\theta_2 = 0.2$; (c) $\theta_1 = 0.05$, $\theta_2 = 0.5$.

Table 1
Example 2. The simple compartmental model: Locally D-optimum designs illustrated in Figures 3 and 4 (for condition (c) $x_j^* = 1/\theta_j$)

Conditions	θ_1^o	θ_2^o	x_2^*	x_1^*
(a)	0.05	0.1	8.34	33.4
(b)	0.05	0.2	4.79	24.2
(c)	0.05	0.5	2.00	20.0

Note that for θ_1 fixed at 0.05, the peak of the response curve becomes sharper as θ_2 increases, reflecting the fact that the second exponential term in (3) decays more rapidly with time as θ_2 increases.

D-optimum designs for this model and the chosen parameter settings comprise two equally weighted support points, x_1^* and x_2^*, and are given in Table 1. Figure 3 shows that these designs are indeed globally optimal, with the variance function (1) achieving its maximum value of 2 at the two design points. The figure also indicates, through the flatness of the variance function around the larger of the design points, x_1^*, that the D-optimum design is relatively insensitive to the exact value of that point. Explicit formulae for x_1^* and x_2^* are not available. However it is interesting to observe that for values of θ_1 and θ_2 which are widely separated, such as 0.05 and 0.5, the two exponential terms in the expected response (3) interact very little over the design region and, as a consequence, the optimal design points, x_1^* and x_2^*, are very close in value to $1/\theta_1$ and $1/\theta_2$, the support points of the D-optimum designs for the individual one-parameter models, $\exp(-\theta_1 x)$ and $\exp(-\theta_2 x)$, respectively.

The induced design space, $\mathcal{F}$, for model (3) is the parametric curve described by the elements, $f_1(x,\theta) = -x\exp(-\theta_1 x)$ and $f_2(x,\theta) = -x\exp(-\theta_2 x)$, of $f(x,\theta)$ as x varies between 0 and infinity and is shown, together with the minimal ellipse, $E_{\mathcal{F}}$, for the three parameter settings of Table 1 in Figure 4. The values of $f(x,\theta)$ corresponding to the D-optimum design points, x_1^* and x_2^*, coincide with the points of contact between the curve, $\mathcal{F}$, and the ellipse, $E_{\mathcal{F}}$, and in addition form a triangle with the origin which has maximum area over all such triangles. These features are illustrated in Figure 4 and serve as further confirmation of the global D-optimality of the exact two-point design with support at x_1^* and x_2^*. It is also interesting to note that the shape of the curve, $\mathcal{F}$, changes as the parameters, θ_1 and θ_2, change and that, for $\theta_1 = 0.05$ and $\theta_2 = 0.5$, the points, $f(x_1^*, \theta)$ and $f(x_2^*, \theta)$, occur close to the minima of $f_1(x,\theta)$ and $f_2(x,\theta)$ in accord with the nature of the optimum design.

2.2.2. *Locally c-optimum designs*

For the linear model a c-optimum design is a design which minimizes the variance of the least squares estimator of a linear combination of the parameters, $c^T\widehat{\theta}$, and thus, equivalently, which minimizes the criterion, $c^T M^{-1}(\xi)\, c$. This notion can be extended to a nonlinear function of the parameters by considering the variance of the linearized form of its least squares estimator obtained by expanding that estimator in a Taylor series about θ, and it can be further extended to include nonlinear regression models by considering the asymptotic variance of the appropriate least squares estimator or its linearized form. Specifically, a locally c-optimum design for a nonlinear regression model is a design which minimizes the criterion,

$$c^T(\theta)M^{-1}(\xi,\theta)c(\theta),$$

where $c(\theta)$ represents the vector of derivatives, $\partial b(\theta)/\partial\theta$, for a function, $b(\theta)$, of interest and where all terms are evaluated at a best guess of the parameters, $\theta = \theta^o$.

Carathéodory's Theorem provides an upper bound of $k(k+1)/2$ on the number of support points of a locally c-optimum design but this bound can in fact be sharpened to k, the number of parameters in the model (Chernoff, 1979, p. 28; Pukelsheim, 1993, p. 188). In many instances this latter bound is not attained and the c-optimum design has an information matrix which is singular and is itself said to be singular. For nonsingular locally c-optimum designs there exists a powerful Equivalence Theorem which can be used to confirm the global optimality or otherwise of a candidate design. Specifically, this theorem states that a design, ξ^*, is locally c-optimum if and only if the condition,

$$\{f^T(x,\theta)M^{-1}(\xi^*,\theta)c(\theta)\}^2 - c^T(\theta)M^{-1}(\xi^*,\theta)c(\theta) \leqslant 0, \qquad (4)$$

holds for all points, $x \in \mathcal{X}$, with equality being attained at the support points of ξ^*. For singular locally c-optimum designs the situation is less straightforward, firstly in that the Equivalence Theorem requires modification to accommodate singular information matrices and secondly in that problems relating to estimability can arise.

Geometric insight into locally c-optimum designs for nonlinear models with a *single* explanatory variable is obtained by considering the Elfving set, $\mathcal{F}^*$, which is the convex hull of the set, $\mathcal{F}$, and its inversion through the origin (Elfving, 1952; Chernoff, 1979, p. 30). In particular, the support points and associated weights for a c-optimum design involving the vector, $c(\theta)$, are readily derived from the point of intersection of a ray through the origin in the direction of $c(\theta)$ and the boundary of the Elfving set, $\mathcal{F}^*$ (see Kitsos et al., 1988; Ford et al., 1992).

We now consider examples of nonlinear models in which both c- and D-optimum designs are of interest.

Example 2. A simple compartmental model (b)
Locally c-optimum designs for model (3) are readily found. In particular, for the best guess of the parameters, $\theta_1^o = 0.05$ and $\theta_2^o = 0.1$, the c-optimum design for the function, $2\theta_1 + \theta_2$, comprises a single point at 13.863, while in contrast that for the function, $\theta_1 + \theta_2$, puts weights of 0.7964 and 0.2036 on the two support points, 6.154 and 39.912 respectively. Insight into these designs is obtained from the Elfving set, $\mathcal{F}^*$, for the model shown in Figure 5. For the function, $2\theta_1 + \theta_2$, the associated vector, $c(\theta) = (2,1)^T$, intersects the boundary of $\mathcal{F}^*$ at a point in the set, $\mathcal{F}$, itself, thus demonstrating that the c-optimum design is based on a single point, while for the function, $\theta_1 + \theta_2$, the corresponding ray, $c(\theta) = (1,1)^T$, intersects this boundary in a line segment whose end-points correspond to the support points of the design.

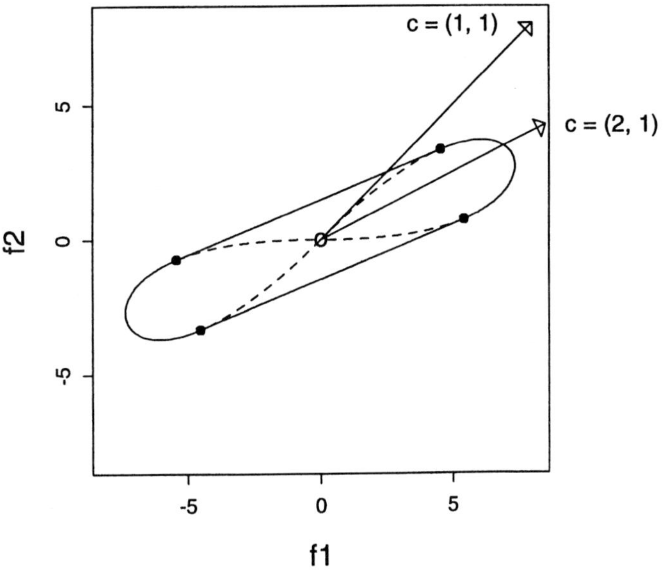

Fig. 5. Example 2. A simple compartmental model. The Elfving set, $\mathcal{F}^*$, for the parameter values $\theta_1 = 0.05, \theta_2 = 0.1$, together with the rays $c^T(\theta) = (2,1)$, $c^T(\theta) = (1,1)$ associated with the locally c-optimum designs for the functions, $2\theta_1 + \theta_2$ and $\theta_1 + \theta_2$ respectively.

Example 3. Optimum return from fertilizer dressing

The expected yield of wheat resulting from the application of an amount of fertilizer, x, is well described, over a series of trials, by the nonlinear function, $\theta_1 + \theta_2 e^{\theta_3 x}$, with θ_1 positive and θ_2 and θ_3 negative. Clearly this expected yield is given by $\theta_1 + \theta_2$ in the absence of fertilizer and tends towards a limiting value of θ_1 as x increases. The economic return, Y, which is the response of interest, is found by subtracting the price of fertilizer from the yield, leading to a nonlinear regression model with deterministic component,

$$\mathrm{E}(Y \mid x) = \eta(x, \theta) = \theta_1 + \theta_2 e^{\theta_3 x} + \theta_4 x, \tag{5}$$

where the parameter, θ_4, represents cost and is negative. Since there is usually a limit on the amount of fertilizer, x, which can be applied, it is convenient to scale x so that it lies between 0 and 1, and thus to take the design region, $\mathcal{X}$, to be the interval $[0, 1]$. A typical plot of the expected yield, $\eta(x, \theta)$, against x is shown in Figure 6(a) and is approximately quadratic in form.

The vector, $f(x, \theta)$, is given by

$$\frac{\partial \eta(x, \theta)}{\partial \theta} = (1,\ e^{\theta_3 x},\ \theta_2 x e^{\theta_3 x},\ x)^T$$

and it is interesting to note that although the model (5) contains only one nonlinear parameter, θ_3, two of the elements of $f(x, \theta)$ depend on that parameter. The determi-

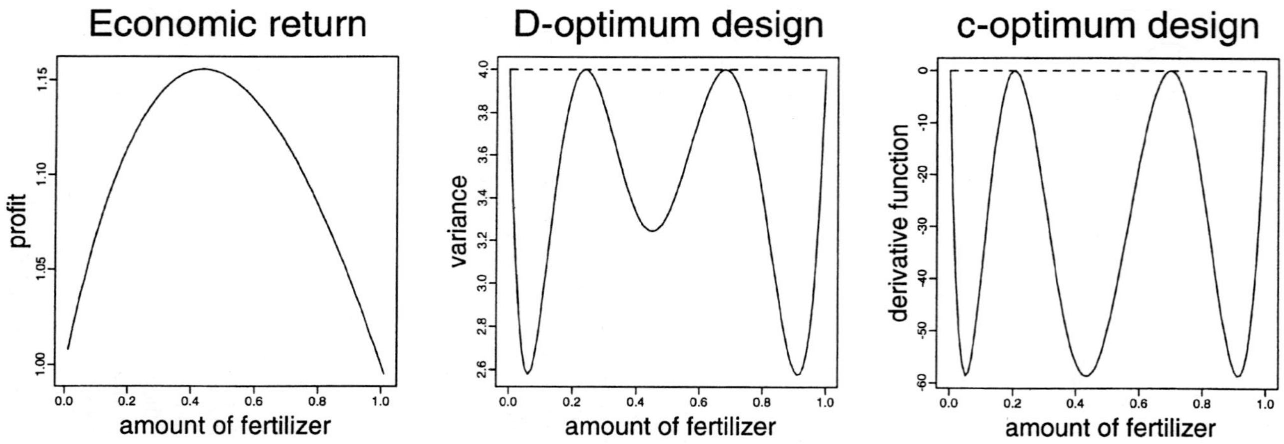

Fig. 6. Example 3. Optimum return from fertilizer dressing. (a) economic return; (b) equivalence theorem for the locally D-optimum design; (c) equivalence theorem for the locally c-optimum design for the economic optimum.

Table 2
Example 3. Optimum return from fertilizer dressing. Locally c-optimum designs for $\widehat{x}$ with $\log(\theta_3) = 0.3$

θ_4/θ_2	$\widehat{x}$			Experimental design		
1.2	0.0027	x	0	0.2147	0.71025	1
		$\xi(x)$	0.3546	0.4459	0.1454	0.0541
0.8	0.3395	x	0.142	0.680	1	
		$\xi(x)$	0.4	0.5	0.1	
0.7	0.4504	x	0	0.2147	0.71025	1
		$\xi(x)$	0.0558	0.4447	0.4442	0.0553
0.6	0.5785	x	0	0.244	0.820	
		$\xi(x)$	0.1	0.5	0.4	
0.2	1.4910	x	0	0.2147	0.71025	1
		$\xi(x)$	0.1053	0.2512	0.3947	0.2488

nant of the information matrix, $M(\xi, \theta)$, depends on θ_2 solely through a multiplicative constant and thus it follows that locally D-optimum designs for model (5) depend only on θ_3. For all values of θ_3 examined here this optimum design is an exact 4-point design, and in all cases the support points include 0 and 1. The two other points of support on the interval, $[0, 1]$, vary slightly over the range of θ_3 from $\log(0.1)$ to $\log(0.9)$, increasing from 0.208 and 0.641 for $\theta_3 = \log(0.1)$ to 0.272 and 0.720 for $\theta_3 = \log(0.9)$, but are very different for large negative values of θ_3 outside this range. The optimality of these designs can be checked using the Equivalence Theorem and, in particular, the plot of standardized variance against x for $\theta_3 = \log(0.8)$ shown in Figure 6(b) confirms the global D-optimality of the design with equally weighted support points at 0, 0.269, 0.716 and 1.

The purpose of the experimental programme leading to the formulation of model (5) was to determine the economically optimum amount of fertilizer dressing and this is given by the value of x satisfying $\partial\eta(x, \theta)/\partial x = 0$, and thus by

$$\widehat{x} = \left\{ \log(-\theta_4) - \log(\theta_2\theta_3) \right\} / \theta_3.$$

The locally D-optimum design for model (5) provides an efficient design for the precise estimation of all of the parameters together but not necessarily for the precise estimation of $\widehat{x}$. Attention therefore focuses here on finding locally c-optimum designs for the function, $\widehat{x}$, and the relevant coefficient, $c(\theta)$, for such designs is found by differentiation to be

$$c(\theta) = \frac{\partial \widehat{x}}{\partial \theta} = \frac{1}{\theta_3}\left(0,\ -\frac{1}{\theta_2},\ -\frac{\{1 + \log(-\theta_4/\theta_2\theta_3)\}}{\theta_3},\ \frac{1}{\theta_4}\right)^T.$$

Locally c-optimum designs for $\widehat{x}$ thus depend on the values of both the linear and the nonlinear parameters of the model, and in fact can be shown to depend on θ_3 and on the ratio, θ_4/θ_2, only (Haines, 1996). Table 2 summarizes these optimum designs for $\log(\theta_3) = 0.3$ and for a range of values of θ_4/θ_2 over which the corresponding values of $\widehat{x}$ vary from near zero to a value which is greater than 1 and thus outside the experimental region.

It is clear from this table that certain of these locally c-optimum designs are based on 4 points of support and that in such cases these points are identical but their associated weights are very different, while other optimum designs are based on 3 points of support and are therefore singular. The nature of these designs appears to be related to the associated values of $\widehat{x}$. Thus for cases in which the value of $\widehat{x}$ falls within the design region, $\mathcal{X}$, the experimental effort is concentrated on points close to that value but, for the case in which $\widehat{x}$ falls well outside this region, the design weights are spread evenly over the three highest design points, thus indicating that precise estimation of all parameters is required for efficient extrapolation. The global optimality of the locally c-optimum design for $\widehat{x}$ and for the best guess of the parameter values, $\theta_2 = -1, \theta_4 = 0.2$ and $\theta_3 = \log(0.3)$, is demonstrated by the plot of the function specified by the left hand side of the inequality (4) against x shown in Figure 6(c).

2.2.3. *Other locally optimum designs*

The notion of locally c-optimum designs can be extended in a natural way to incorporate more than one function of the parameters. In particular an omnibus criterion, that of average variance, $\mathrm{tr}[C^T(\theta)M^{-1}(\xi,\theta)C(\theta)]$, where $C(\theta)$ is an assembly of linearly independent coefficients which may or may not depend on the best guess of the parameters, is usually adopted (Atkinson et al., 1993). For designs minimizing this criterion, an upper bound on the number of support points is given by $r(2k - r + 1)/2$, where r is the rank of the matrix, $C(\theta)$, and an Equivalence Theorem analogous to that for locally c-optimum designs can be formulated and used to confirm the optimality of a design. Another criterion closely related to c-optimality is that of E-optimality, for which the maximum variance of all possible normalized linear combinations of the parameter estimates is minimized or, equivalently, the maximum eigenvalue of the information matrix, $M(\xi,\theta)$, is maximized. However E-optimality is not as mathematically tractable as its c-optimality counterpart and, while there are a number of studies of E- optimum designs for linear models, there is only one reported thus far for nonlinear models, that of Dette and Haines (1994). In particular, the support points and attendant weights for locally E-optimum designs can be deduced from the geometry of the Elfving set, $\mathcal{F}^*$, but the procedure is not entirely straightforward.

2.3. *Bayesian optimum designs*

In general locally optimum designs exhibit an appreciable dependence on the choice of best guess for the parameters. This is clearly evidenced here by the locally D-optimum designs for the compartmental model (3) shown in Table 1 and by the locally c-optimum designs for the economic return model (5) displayed in Table 2. This sensitivity of locally optimum designs to the choice of the parameter value, θ^o, presents problems since on the one hand a poor guess for the true parameter value will lead to an inefficient design and on the other hand a guess close to the true value is overprecise and may result in a non-informative c-optimum design.

An obvious way of accommodating this dependence of optimum design on the chosen parameter values is to introduce a prior distribution on the parameters and to incorporate this distribution into appropriate design criteria. The criterion most widely used in this regard is that of Bayesian D-optimality, for which the expectation of the logarithm of the determinant of the information matrix,

$$\mathrm{E}_\theta \log|M(\xi,\theta)| = \int \log|M(\xi,\theta)|p(\theta)\,\mathrm{d}\theta, \tag{6}$$

where $p(\theta)$ is the prior distribution on θ, is maximized. A formal justification for this criterion within the Bayesian paradigm is provided by Chaloner and Verdinelli (1995) and by DasGupta in another chapter of this volume and is based on an approximation to the expected Shannon information. An alternative, Bayesian approach to justifying (6) is to accept the implied dichotomy of taking a specific prior distribution on the parameters in the design stage and a vague prior for inference. However a more direct, classical interpretation of this criterion is simply that the prior represents the experimenter's prior belief in the adequacy of the model over a specified range of parameter values.

For discrete priors based on p points, Carathéodory's Theorem provides an upper bound of $pk(k+1)/2$ on the number of support points for a Bayesian D-optimum design (Läuter, 1974b), but for a continuous prior distribution no such bound exists. The incorporation of $\log|M(\xi,\theta)|$, rather than $|M(\xi,\theta)|$ itself, into the Bayesian D-optimum design criterion (6) ensures that this criterion is concave on the set of all possible information matrices and hence that a General Equivalence Theorem can be formulated and used to confirm the global optimality or otherwise of candidate designs and to form a basis for algorithmic design construction. This Equivalence Theorem is based on the expectation of the standardized variance (1), which is given by $\bar{d}(x,\xi) = \mathrm{E}_\theta d(x,\xi,\theta)$, and states that for a design, ξ^*, to be Bayesian D-optimum the condition, $\bar{d}(x,\xi^*) \leqslant k$, must hold for all x in the design region, $\mathcal{X}$.

The extension of the criterion of c-optimality in order to incorporate prior information follows in an analogous way to that of D-optimality. In particular, a Bayesian c-optimum design is a design which minimizes the expectation of the asymptotic variance of the least squares estimate of a function of the parameters, written as

$$\mathrm{E}_\theta c^T(\theta)M^{-1}(\xi,\theta)c(\theta) = \int c^T(\theta)M^{-1}(\xi,\theta)c(\theta)p(\theta)\,\mathrm{d}\theta$$

and the relevant Equivalence Theorem states that a design, ξ^*, is Bayesian c-optimal provided the condition,

$$\mathrm{E}_\theta\{f^T(x,\theta)M^{-1}(\xi^*,\theta)c(\theta)\}^2 - \mathrm{E}_\theta c^T(\theta)M^{-1}(\xi^*,\theta)c(\theta) \leqslant 0, \tag{7}$$

is satisfied for all x in the design region, $\mathcal{X}$.

Bayesian optimum designs are found, in almost all cases, by using numerical methods. Exceptions include the optimum designs for one-parameter nonlinear models when the prior has two points of support derived analytically by Haines (1995) and those for a family of exponential models presented by Dette and Sperlich (1994) and Mukhopadhyay and Haines (1995). The overriding difficulty in constructing Bayesian optimum designs for continuous prior distributions numerically is that of evaluating the integrals embedded in the criteria. One approach to this problem is to discretize the prior and thus to maximize a weighted sum of terms of the form, $\log |M(\xi, \theta)|$, while an alternative and more obvious procedure is to use established methods of numerical integration, such as Gaussian quadrature or the Romberg method. It is important to note however that quadrature methods do in fact involve specific discretizations of the prior. An interesting Monte Carlo approach to evaluating the required integrals is provided by Atkinson et al. (1995), and involves drawing a sample of parameter values from the continuous prior distribution and maximizing the appropriate criterion evaluated over that set of sample values. In this context it should be observed that it is a straightforward matter to sample from prior distributions which are multivariate normal and also that it is possible to sample from other multivariate priors either by transforming the distributions to near-normality using Box–Cox transformations (Andrews et al., 1971) or, more importantly, by invoking Markov chain Monte Carlo techniques.

Example 2. A simple compartmental model (c)
As an example of the use of the Monte-Carlo sampling method for the calculation of Bayesian D-optimum designs consider again the simple compartmental model (3). For a best guess of the parameters values given by $\theta_1^o = 0.05$ and $\theta_2^o = 0.5$, the locally D-optimum design is the exact 2-point design with support at $x_1^* = 20$ and $x_2^* = 2$, as noted earlier. Suppose now that there is uncertainty concerning this best guess and suppose further that this uncertainty can be quantified by introducing a prior distribution on the parameters which is lognormal and which has the form,

$$\begin{bmatrix} \log \theta_1 \\ \log \theta_2 \end{bmatrix} \sim N\left(\begin{bmatrix} \log \theta_1^o \\ \log \theta_2^o \end{bmatrix}, \tau^2 (\log 2)^2 \begin{bmatrix} 1 & 0.5 \\ 0.5 & 1 \end{bmatrix} \right), \tag{8}$$

where τ^2 is a constant. Then Bayesian D-optimum designs for this prior can be constructed numerically and compared with the locally optimum design which corresponds to the point prior, $\theta = \theta^o$, with $\tau^2 = 0$. In the present example two priors of the form (8) with $\tau^2 = 0.5$ and $\tau^2 = 1$ were selected for study and the Monte Carlo method of Atkinson et al. (1995), implemented by drawing samples of 100 parameter pairs from each distribution, was used to obtain Bayesian D-optimum designs. These designs are summarized in Table 3 and their global optimality confirmed from the plots of the expected standardized variance function, $\bar{d}(x, \xi^*)$, against x shown in Figures 7(a) and (b).

It is interesting to observe that the number of support points for the Bayesian D-optimum design for the prior with $\tau^2 = 0.5$ is three, whereas that for the more diffuse prior with $\tau^2 = 1$ is four. It is also interesting to note that the variance function, $\bar{d}(x, \xi^*)$, becomes flatter as prior uncertainty increases. This is made particularly clear from a comparison of the plot of the variance function for the point prior, $\theta = \theta^o$, given

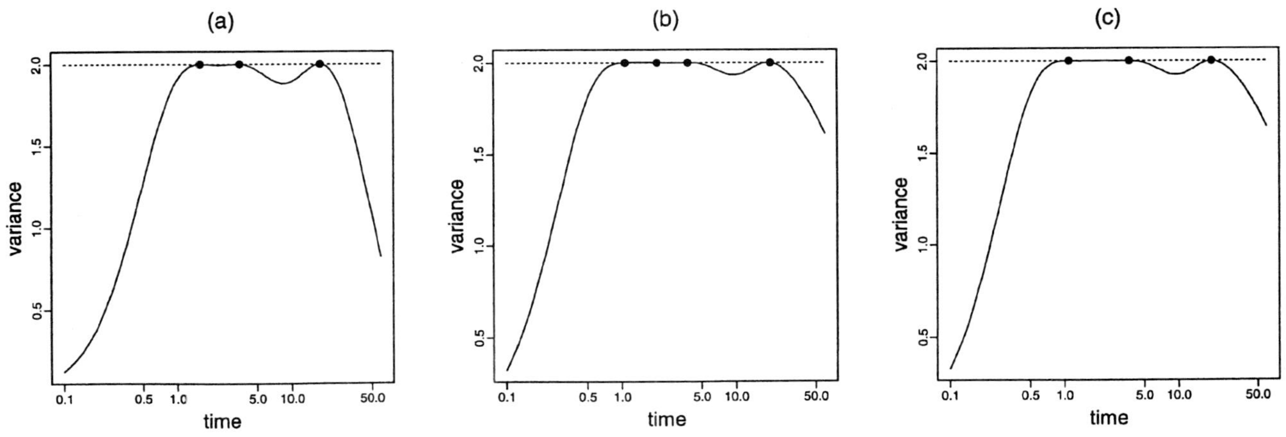

Fig. 7. Example 2. A simple compartmental model. Bayesian D-optimum designs of Table 3. (a) $\tau^2 = 0.5$; (b) $\tau^2 = 1$; (c) three-point approximation to the four-point optimum design in (b).

Table 3
Example 2. Simple compartmental model. Bayesian D-optimum designs: when $\tau^2 = 1$, 68.3% of the prior distribution lies within the interval $(\theta/2, 2\theta)$. Design (c) is a three-point approximation to the four-point design (b)

Design	τ^2		Experimental design			
(a)	0.5	x	17.9	3.55	1.60	
		$\xi(x)$	0.481	0.077	0.442	
(b)	1.0	x	19.9	3.78	2.04	1.08
		$\xi(x)$	0.427	0.249	0.022	0.302
(c)	1.0	x	19.8	3.72	1.09	
		$\xi(x)$	0.428	0.258	0.314	

in Figure 3(c) with the corresponding plots for the more diffuse priors with $\tau^2 = 0.5$ and $\tau^2 = 1$ shown in Figures 7(a) and (b) respectively. It is tempting to conclude from this observation that Bayesian D-optimum designs are less sensitive to changes in design than their locally D-optimum counterparts. To reinforce this suggestion, the three-point design maximizing the Bayesian D-optimum criterion for the prior (8) with $\tau^2 = 1$ was found, again using the Monte Carlo method, and compared with the globally optimum 4-point design.

As shown in Table 3, these two designs differ in that the point at 2.04 with an associated weight of 0.022 in the 4-point design is absent from the 3-point design. It is clear however from the fact that the criterion values for the two designs agree to within four figures and from the fact that the variance functions for the designs, displayed in Figures 7(b) and (c), are virtually indistinguishable, that the best 3-point design is near-optimal.

This example of the simple compartmental model shows the dependence of the D-optimum design on the prior distribution of the parameters. We now use the full three-parameter compartmental model to illustrate further properties of Bayesian designs, particularly those that are c-optimum.

Example 4. A full compartmental model

Atkinson et al. (1993) fitted the compartmental model (2) to data obtained by monitoring the passage of the drug, theophylline, through a horse. Maximum likelihood estimates of

$$\theta_1 = 0.05884, \qquad \theta_2 = 4.298, \qquad \theta_3 = 21.80, \quad \text{and} \quad \sigma = 1.126$$

were obtained, and the values for θ adopted as the best guess, θ^o, for the purposes of design construction. It is interesting to note that for the parameter value, θ^o, the expected response rises rapidly to a peak at $x = 1.01$ and then decays approximately exponentially towards a limiting value of zero as x increases.

The locally D-optimum design for θ^o is the exact 3-point design shown in Table 4 and provides an efficient design for the precise estimation of the parameters, θ, overall.

Table 4
Example 4. Full compartmental model. Locally D- and c-optimum designs for $\widehat{x}$, the time to maximum concentration

Criterion		Experimental design		
D-optimum	x	0.2288	1.3886	18.417
	$\xi(x)$	$\frac{1}{3}$	$\frac{1}{3}$	$\frac{1}{3}$
c-optimum	x	0.1793	3.5671	
	$\xi(x)$	0.6062	0.3938	

However for the precise estimation of specific properties of the model which are of interest, such as those relating to the efficacy of the drug, locally c-optimum designs are to be preferred. As an example, consider the time to maximum concentration, which is found by setting the derivative of (2) with respect to x equal to zero and which is given by

$$\widehat{x} = \frac{\log\theta_2 - \log\theta_1}{\theta_2 - \theta_1}.$$

The coefficient, $c(\theta)$, embedded in the criterion of c-optimality for $\widehat{x}$ is obtained as

$$c(\theta) = \frac{\partial\widehat{x}}{\partial\theta}$$
$$= \frac{1}{(\theta_2-\theta_1)^2}\left(-\frac{(\theta_2-\theta_1)}{\theta_1} + \log\frac{\theta_2}{\theta_1},\ \frac{(\theta_2-\theta_1)}{\theta_2} - \log\frac{\theta_2}{\theta_1},\ 0\right)^T,$$

and the locally c-optimum design, which is shown in Table 4, puts observations at only two values of time and is therefore singular.

The effect on design construction of introducing uncertainty into the parameter values is now investigated. Since θ_3 enters the model linearly, locally D-optimum designs and locally c-optimum designs for $\widehat{x}$ depend on θ_3 only through a multiplicative constant (Atkinson et al., 1993). As a consequence no assumptions about the value of θ_3 need to be made in constructing the corresponding Bayesian optimum designs. For the parameters, θ_1 and θ_2, two prior distributions are considered. The first, identified as Prior 1, takes θ_1 to be uniform on 0.05884 ± 0.01 and, independently, θ_2 to be uniform on 4.298 ± 1.0, and the second, more diffuse prior, labelled Prior 2, takes θ_1 to be uniform on 0.05884 ± 0.04 and, independently, θ_2 to be uniform on 4.298 ± 4.0. These intervals correspond to the maximum likelihood estimates of the parameters plus or minus approximately two standard errors for Prior 1 and plus or minus approximately eight standard errors for Prior 2. The number of support points for the Bayesian D-optimum and the Bayesian c-optimum designs for $\widehat{x}$ for these two priors are summarized in Table 5 and it is again evident from this table that the number of points of support for Bayesian optimum designs increases as the uncertainty encapsulated in the prior increases.

Table 5
Example 4. Full compartmental model. Increase in number of support points for optimum designs as prior information becomes less precise: c-optimum designs are for $\widehat{x}$, the time to maximum concentration

Criterion	Local	Prior 1	Prior 2
D-optimum	3	3	5
c-optimum	2	3	5

Full details of these and other related Bayesian optimum designs are given in Atkinson et al. (1993).

3. Generalized linear models

3.1. Basic ideas

For a generalized linear model, the response, Y, follows a distribution from the exponential family, such as normal, binomial, Poisson or gamma (McCullagh and Nelder, 1989). The mean of this response, $\mathrm{E}(Y) = \mu$, depends on the vector of unknown parameters, θ, and the vector of explanatory variables, x, through a usually nonlinear function of the linear predictor, $\eta = f^T(x)\theta$, termed the link function, and the variance of this response, written $\mathrm{var}(Y \mid x)$, similarly depends only on this linear predictor. The information matrix for θ at the design point, x, can be expressed as

$$I(x,\theta) = w(x,\theta)f(x)f^T(x) \quad \text{with} \quad w(x,\theta) = \frac{1}{\mathrm{var}(Y \mid x)}\left(\frac{\partial\mu}{\partial\eta}\right)^2,$$

where the term, $w(x,\theta)$, is referred to as a weight and depends only on the linear predictor, η. It thus follows that the information matrix of θ from a design comprising the points, x_i, $i = 1, \ldots, n$, is given by

$$\sum_{i=1}^{n} w(x_i,\theta)f(x_i)f^T(x_i) = F^T W F,$$

where the matrix, W, is a diagonal matrix with ith diagonal element equal to $w(x_i, \theta)$ for $i = 1, \ldots, n$, and that, more generally, the information matrix from the continuous design, ξ, can be expressed succinctly as

$$M(\xi,\theta) = \int_{\mathcal{X}} w(x,\theta)f(x)f^T(x)\xi(\mathrm{d}x). \tag{9}$$

The information matrix for a generalized linear model thus depends on the unknown parameters, θ, and the same considerations regarding optimum design as those described for nonlinear regression models in the previous section apply. In particular, both locally and Bayesian optimum designs can be constructed and the powerful results relating to Carathéodory's Theorem and to the General Equivalence Theorems can be invoked. Furthermore geometrical insight into the locally optimum designs can again be obtained by considering the induced design space, $\mathcal{F}$, which is now defined as the space generated by the vector, $f_w(x,\theta) = w^{1/2}(x,\theta)f(x)$, as x varies over the design region, $\mathcal{X}$. It is interesting to observe, as an aside, that the information matrix (**??**) is identical to that for a weighted linear regression model with precision, $w(x,\theta)$, attached to an observation at x, to that for linear regression with $f(x)$ coinciding with $f_w(x,\theta)$ for fixed θ, and also to that for a nonlinear regression model with the derivative, $\partial\eta(x,\theta)/\partial\theta$, equal to $f_w(x,\theta)$.

A particularly appealing feature of optimum design theory for generalized linear models is that there exists a powerful mechanism for transforming certain locally optimum design problems into concise, canonical forms, which are often solved more readily than the original problem, and which in any case provide some valuable insights (Ford et al., 1992). To be specific, suppose that the vector, $f(x)$, is transformed into a vector, $z = Bf(x)$, where B is a nonsingular matrix depending only on the parameter, θ, and having as its last row the vector, θ^T, so that the element, z_k, is equal to the linear predictor, $\eta = f^T(x)\theta$. Then as x varies over the design region, $\mathcal{X}$, so it generates a transformed or canonical design region of z values, denoted $\mathcal{Z}$. Furthermore a design, ξ, defined on $\mathcal{X}$ can be transformed to a design, ξ_z, on $\mathcal{Z}$, and *vice versa*, by transforming the support points and retaining their associated weights. It now follows that the information matrix for θ from the design, ξ, on $\mathcal{X}$ can be expressed as

$$M(\xi,\theta) = B^{-1}\, M_z(\xi_z)\, (B^T)^{-1},$$

where the matrix,

$$M_z(\xi_z) = \int_{\mathcal{Z}} w(z_k) z z^T \xi_z(\mathrm{d}z),$$

is the information matrix for a weighted linear regression model with explanatory variables, z, and associated weights, $w(z_k)$. Thus the problem of constructing a locally D-optimum design by maximizing the determinant, $|M(\xi,\theta)|$, over all designs on the region, $\mathcal{X}$, is equivalent to the canonical problem of maximizing the determinant, $|M_z(\xi_z)|$, over all designs on $\mathcal{Z}$. Similarly the problem of finding a locally c-optimum design by minimizing the criterion, $c^T(\theta)M^{-1}(\xi,\theta)c(\theta)$, is equivalent to the canonical problem of minimizing the criterion, $c_z^T M^{-1}(\xi_z)c_z$, where the vector, c_z, is given by $Bc(\theta)$. It is particularly interesting to observe that the canonical problem for D-optimality depends on the parameters of the original model only through the transformed design region, $\mathcal{Z}$.

The construction of locally and Bayesian D- and c-optimum designs for generalized linear models is now illustrated by means of examples. Particular emphasis is placed on the use of canonical forms to obtain locally optimum designs and on the use of the induced design space to gain geometrical insight into the nature of these designs.

3.2. Optimum designs for models with one explanatory variable

The first example presented here involves the usual logistic regression model with two parameters and the second example provides a particularly interesting extension of that model to include a qualitative factor.

Example 5. Logistic model for binary data
Consider an experiment for which the data are binary with R successes out of n trials and with a probability of success that depends on a single explanatory variable, x. Such data can arise, for example, from counting the number of insects which die when a batch is exposed to a particular dose level of insecticide and the probability of an insect dying depends on that dose level. For the logistic model with probability of success, $\mathrm{E}(R/n) = \mu$, the linear predictor is given by

$$\eta = \log\{\mu/(1-\mu)\} = \theta_1 + \theta_2 x, \tag{10}$$

for x defined on the real line and corresponding to log-dose, and the weight, $w(x, \theta)$, is given by $\mu(1-\mu)$.

Locally D-optimum designs for this model are most easily obtained by solving the canonical form of the design problem. In particular, the transformation, $z = \theta_1 + \theta_2 x$, can be invoked and the information matrix, $M(\xi_z)$, is then given by $\sum_{i=1}^{n} \xi_z(z_i) I_z(z_i)$, where

$$I_z(z) = \frac{\mathrm{e}^z}{(1+\mathrm{e}^z)^2} \begin{bmatrix} 1 & z \\ z & z^2 \end{bmatrix}.$$

The design maximizing $|M(\xi_z)|$ puts equal weights on the points, $z = \pm 1.5434$, and the locally D-optimum design can be recovered by transformation as the design which is symmetric about the LD50 and which puts half the trials at the point corresponding to $\mu = 0.1760$ and half at the point where $\mu = 0.8240$. Locally c-optimum designs for the dose required for a specified probability of success, p, expressed as

$$x_p = \frac{\log(p/(1-p)) - \theta_1}{\theta_2},$$

are also most readily obtained by solving the canonical problem of minimizing the criterion, $c_z^T M^{-1}(\xi_z) c_z$, where the coefficient, c_z is given by

$$-\frac{1}{\theta_2}\left(1, \log\frac{p}{(1-p)}\right)^T.$$

In particular, it follows from the solutions to this canonical problem that locally c-optimum designs for the LD50 comprise a single point corresponding to $\mu = 0.5$, while in contrast c-optimum designs for the LD95 are those designs which put weights of 0.0926 and 0.9074 on the support points corresponding to μ values of 0.0832 and

0.9168 respectively. Further insight into these and related designs is provided by the induced design space and the Elfving set of the canonical form and details are to be found in Chernoff (1979, p. 30) and Ford et al. (1992).

Bayesian D- and c-optimum designs for the logistic model can also be constructed using the methodology discussed in Section 2.3 and are fully described in the seminal paper of Chaloner and Larntz (1989).

Example 6. Logistic model when males and females differ in response
Atkinson et al. (1995) describe an extension to the logistic model of Example 5 which arose in an experiment on flies. The novel feature of the design problem was that it was not practically feasible to sex live flies. The specified experimental treatments had therefore to be applied to a mixture of males and females, although it was known that the two sexes responded differently. In the experiment approximately 100 live flies from a breeding chamber were caught in a net. These flies were exposed to a chosen dose of insecticide and the numbers of dead males and dead females after a specified time, in this case 14 hours, were counted. The surviving flies were given a lethal dose of alcohol and sexed, so giving the total numbers of males and females that were at risk.

The linear logistic model specified by (10) was extended to accommodate the different responses of male and female flies by expressing the linear predictor as

$$\eta = \log\{\mu/(1-\mu)\} = \theta_1 + \theta_2 x + \theta_3 u,$$

where u is an indicator variable which is 0 for males and 1 for females. Specifically, for a female fly the linear predictor can be expressed as

$$\eta_F = \log\{\mu_F/(1-\mu_F)\} = \theta_1 + \theta_2 x + \theta_3$$

with associated weight, $w_F = \mu_F(1-\mu_F)$, and for a male fly as

$$\eta_M = \log\{\mu_M/(1-\mu_M)\} = \theta_1 + \theta_2 x$$

with associated weight, $w_M = \mu_M(1-\mu_M)$. As already stated, the numbers of male and female flies is unknown at the design stage of the experiment. However it is not unreasonable to assume that these numbers are binomially distributed with the probability of a male or a female fly being one half. Then the information matrix per observation can be expressed as an expectation over this binomial distribution and is given by

$$I(x,\theta) = \frac{1}{2} w_M \begin{pmatrix} 1 & x & 0 \\ x & x^2 & 0 \\ 0 & 0 & 0 \end{pmatrix} + \frac{1}{2} w_F \begin{pmatrix} 1 & x & 1 \\ x & x^2 & x \\ 1 & x & 1 \end{pmatrix}, \tag{11}$$

where dependence on the parameter, θ, is solely through the weights, w_M and w_F. The information for the experiment is thus obtained by integrating the expression (11) over a design measure, ξ, on $\mathcal{X}$ as

$$M(\xi,\theta) = \int_{\mathcal{X}} I(x,\theta)\xi(\mathrm{d}x).$$

Locally D-optimum designs for the experiment are found by maximizing the determinant of this information matrix, $|M(\xi,\theta)|$, in the usual way. Furthermore, a modification of the Equivalence Theorem, which involves taking the expectation of the standardized variance over the binomial distribution on the sexes, can be invoked to confirm the global D-optimality or otherwise of a candidate design. Specifically, the design, ξ^*, is locally D-optimum provided the condition,

$$\begin{aligned} d(x,\xi^*,\theta) &= \frac{1}{2} w_M f_M^T(x) M^{-1}(\xi^*,\theta) f_M(x) \\ &\quad + \frac{1}{2} w_F f_F^T(x) M^{-1}(\xi^*,\theta) f_F(x) \leqslant 3, \end{aligned}$$

where $f_F(x) = (1,x,1)^T$ and $f_M(x) = (1,x,0)^T$, holds for all x in the design region, $\mathcal{X}$. Related results for multivariate normal models are given in Chapter 5 of Fedorov (1972).

The construction of locally c-optimum designs for a specified probability of death for either male or female flies is very similar to that described in the previous example, but finding such designs for the population as a *whole* is a little more involved. Specifically, the dose required for a probability of death, p, for the entire population, written x_p, satisfies the equation, $\frac{1}{2}(\mu_F + \mu_M) = p$, and is thus defined implicitly by

$$g(x_p,\theta) = \mu_F + \mu_M - 2p = 0. \tag{12}$$

It then follows by partial differentiation of $g = g(x_p,\theta)$ with respect to x_p and to θ that the coefficient, $c(\theta)$, has elements which are given by

$$c_1(\theta) = \frac{\partial x_p}{\partial \theta_1} = -\frac{\partial g/\partial \theta_1}{\partial g/\partial x_p} = -\frac{1}{\theta_2},$$

$$c_2(\theta) = \frac{\partial x_p}{\partial \theta_2} = -\frac{\partial g/\partial \theta_2}{\partial g/\partial x_p} = -\frac{x_p}{\theta_2},$$

and

$$c_3(\theta) = \frac{\partial x_p}{\partial \theta_3} = -\frac{\partial g/\partial \theta_3}{\partial g/\partial x_p} = -\frac{w_F}{\theta_2(w_M + w_F)}.$$

Thus the locally c-optimum design for x_p can be found by solving equation (12) using, for example, the Newton–Raphson method, by computing $c(\theta)$ and by minimizing the criterion, $c^T(\theta)M^{-1}(\xi,\theta)c(\theta)$, in the usual way. In addition the global optimality of a candidate design can be confirmed by an extension of the Equivalence Theorem, with the condition for optimality of a design, ξ^*, being that

$$\sum_S \frac{1}{2} w_S(x,\theta)\{f_S^T(x)M^{-1}(\xi^*,\theta)c(\theta)\}^2 \leqslant c^T(\theta)M^{-1}(\xi^*,\theta)c(\theta),$$

where the summation is over males and females, holds for all x in $\mathcal{X}$.

Bayesian D- and c-optimum designs for this experiment can be found by taking expectations over a prior distribution on θ. In the example given by Atkinson et al. (1995) sufficient sample information was available for the prior to be assumed to be normal. The criteria, $\mathrm{E}_\theta \log|M(\xi,\theta)|$ and $\mathrm{E}_\theta c^T(\theta)M^{-1}(\xi^*,\theta)c(\theta)$, were evaluated by taking a Monte Carlo sample of size 100 from this prior distribution, evaluating the criterion of interest at each sample point, and averaging. Full details of the resultant Bayesian optimum designs, including comparisons with the locally optimum designs, are given in Atkinson et al. (1995).

3.3. Locally optimum designs with more than one explanatory variable

The construction of locally optimum designs for generalized linear models with more than one explanatory variable is not entirely straightforward. In particular, the algebraic and numerical calculations involved in design construction are necessarily cumbersome and furthermore the geometric insights associated with the design problem for one variable are not easily extended to higher dimensions. These and other issues are examined in the following example.

Example 7. Logistic regression with two explanatory variables
Consider the logistic regression model with linear predictor,

$$\eta = \log\{\mu/(1-\mu)\} = \theta_1 + \theta_2 x_1 + \theta_3 x_2, \tag{13}$$

involving the two explanatory variables, x_1 and x_2, and with the weight, w, equal to $\mu(1-\mu)$. The induced design space, $\mathcal{F}$, is generated by the vector,

$$f_w(x,\theta) = \left(\sqrt{w},\ \sqrt{w}\,x_1,\ \sqrt{w}\,x_2\right)^T,$$

as x_1 and x_2 vary over the design region, $\mathcal{X}$, and for all x_1 and x_2 such that $\theta_1 + \theta_2 x_1 + \theta_3 x_2 = 0$ this vector is equal to $\frac{1}{2}(1, x_1, x_2)^T$. It thus follows that if the design region, $\mathcal{X}$, is assumed to be unbounded, then the induced design space, $\mathcal{F}$, is also unbounded, and as a consequence the locally D-optimum design does not exist. To obviate this difficulty, the design region, $\mathcal{X}$, is therefore constrained to be bounded, and in the present example, to be the square, $[-1,1] \times [-1,1]$. In contrast, it is not necessary to place such constraints on the design region, $\mathcal{X}$, for the logistic model of Example 5 with a single explanatory variable since the induced design space, $\mathcal{F}$, generated as x varies over the real line, is bounded.

Table 6
Example 7. Logistic regression for two variables. Locally D-optimum designs for various parameter settings

θ^T		Locally D-optimum design					
$(0,1,1)$	x	$(-1,-1)$	$(-1,1)$	$(1,-1)$	$(1,1)$		
	$\xi(x)$	0.2041	0.2959	0.2959	0.2041		
	μ	0.1192	0.5	0.5	0.8808		
$(0,2,2)$	x	$(-1,1)$	$(-1,0.118)$	$(0.118,-1)$	$(-0.118,1)$	$(1,-1)$	$(1,-0.118)$
	$\xi(x)$	0.2674	0.1063	0.1341	0.1063	0.2518	0.1341
	μ	0.5	0.1462	0.1462	0.8538	0.5	0.8538
$(2,2,2)$	x	$(-1,-0.737)$	$(-1,0.737)$	$(-0.737,-1)$	$(0.737,-1)$		
	$\xi(x)$	0.1686	0.3314	0.1686	0.3314		
	μ	0.1863	0.8137	0.1863	0.8137		
$(0,\frac{3}{2},\frac{1}{2})$	x	$(-1,1)$	$(-0.575,-1)$	$(0.575,1)$	$(1,-1)$		
	$\xi(x)$	0.2619	0.2381	0.2381	0.2619		
	μ	0.2689	0.2038	0.7962	0.7311		
$(0,3,1)$	x	$(-0.074,-1)$	$(0.074,1)$	$(0.741,-1)$	$(-0.741,1)$		
	$\xi(x)$	$\frac{1}{4}$	$\frac{1}{4}$	$\frac{1}{4}$	$\frac{1}{4}$		
	μ	0.2274	0.7726	0.7726	0.2274		
$(3,3,1)$	x	$(-1,-1)$	$(-1,1)$	$(-0.068,-1)$			
	$\xi(x)$	$\frac{1}{3}$	$\frac{1}{3}$	$\frac{1}{3}$			
	μ	0.2689	0.7311	0.8577			

Locally D-optimum designs for the logistic model (13) with the three parameter settings, $\theta^T = (0,1,1)$, $\theta^T = (0,2,2)$, and $\theta^T = (2,2,2)$, were constructed using constrained numerical optimization over the design region, $\mathcal{X} = [-1,1] \times [-1,1]$, and are summarized in Table 6 and illustrated in Figure 8. It is interesting to observe that for $\theta^T = (0,1,1)$ the D-optimum design is a 2^2 factorial with points not quite equally weighted, while for $\theta^T = (0,2,2)$ and $\theta^T = (2,2,2)$ the optimum designs are based on 6 and 4 points of support respectively. The dependence of these designs on the parameter values is clarified by examining the corresponding induced design spaces. In particular, Figure 9 shows the spaces, $\mathcal{F}$, for the three chosen parameter settings, viewed down the $f_{w1}(x,\theta)$ axis, and with the D-optimum design points superimposed. It is clear from these figures that the design points correspond to the points of contact of the minimal ellipsoid, $E_{\mathcal{F}}$, and the induced design space, $\mathcal{F}$, and thus that the designs constructed numerically are indeed globally D-optimum. Furthermore, for $\theta^T = (0,1,1)$, the space, $\mathcal{F}$, is not much distorted from the square design region, $\mathcal{X}$, and as a consequence the optimum design points, which necessarily correspond to extreme points of $\mathcal{F}$, also coincide with the extreme points of $\mathcal{X}$. In contrast the induced design spaces for $\theta^T = (0,2,2)$ and $\theta^T = (2,2,2)$ appear to be "folded over", and thus the design points corresponding to the extreme points of $\mathcal{F}$ do not all coincide with the extreme points of $\mathcal{X}$, in accord with the optimum designs presented in Table 6.

For the parameter settings investigated thus far the values of θ_2 and θ_3 are equal and as a consequence the locally D-optimum designs are symmetric with respect to

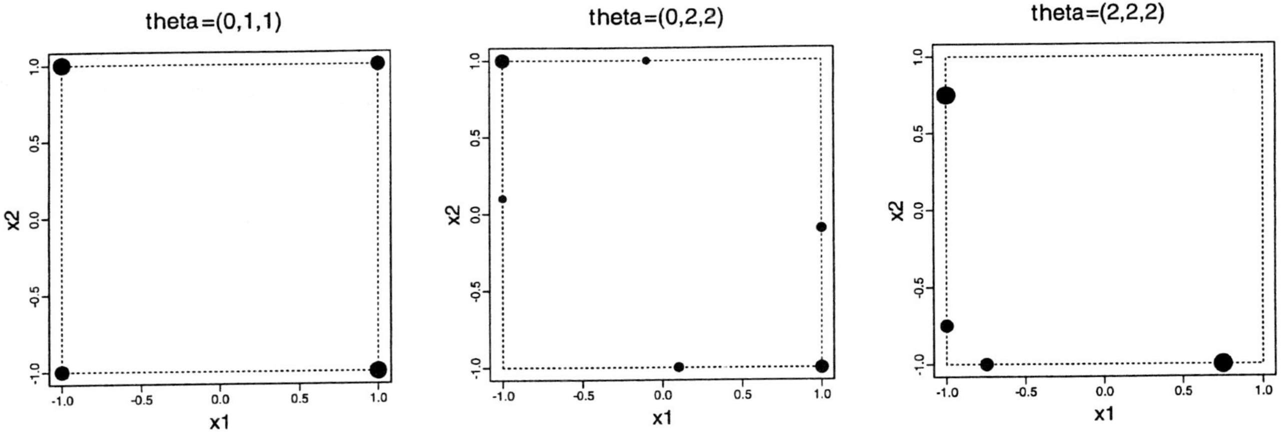

Fig. 8. Example 7. Logistic regression with two explanatory variables. D-optimum designs for parameter values (a) $\theta_1 = 0, \theta_2 = 1, \theta_3 = 1$; (b) $\theta_1 = 0, \theta_2 = 2, \theta_3 = 2$; (c) $\theta_1 = 2, \theta_2 = 2, \theta_3 = 2$. (The design weights are proportional to the areas of the plotting symbols.)

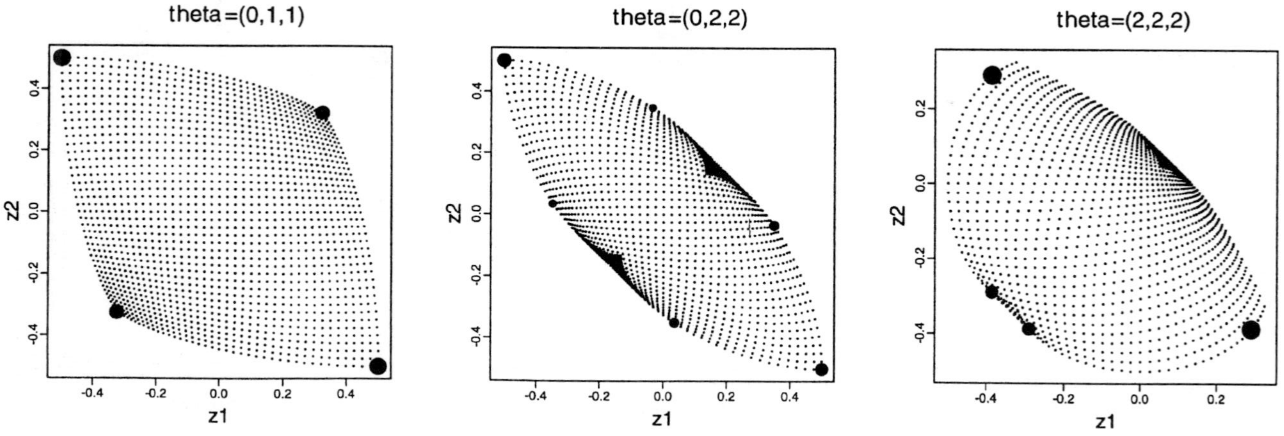

Fig. 9. Example 7. Logistic regression with two explanatory variables. D-optimum designs in the induced design space, $\mathcal{F}$, for parameter values (a) $\theta_1 = 0, \theta_2 = 1, \theta_3 = 1$; (b) $\theta_1 = 0, \theta_2 = 2, \theta_3 = 2$; (c) $\theta_1 = 2, \theta_2 = 2, \theta_3 = 2$. (The design weights are proportional to the areas of the plotting symbols.)

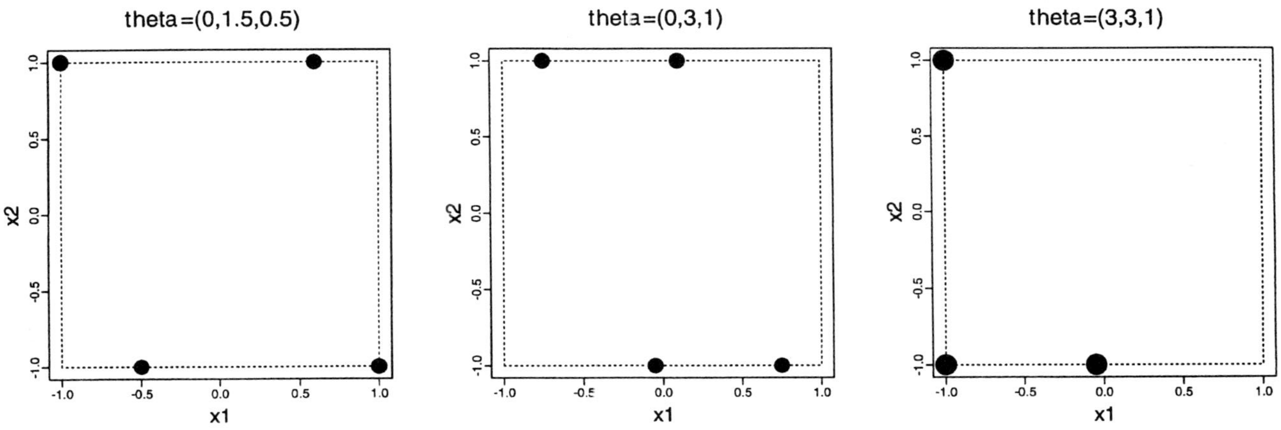

Fig. 10. Example 7. Logistic regression with two explanatory variables. D-optimum designs for parameter values (a) $\theta_1 = 0, \theta_2 = 1.5, \theta_3 = 0.5$; (b) $\theta_1 = 0, \theta_2 = 3, \theta_3 = 1$; (c) $\theta_1 = 3, \theta_2 = 3, \theta_3 = 1$. (The design weights are proportional to the areas of the plotting symbols.)

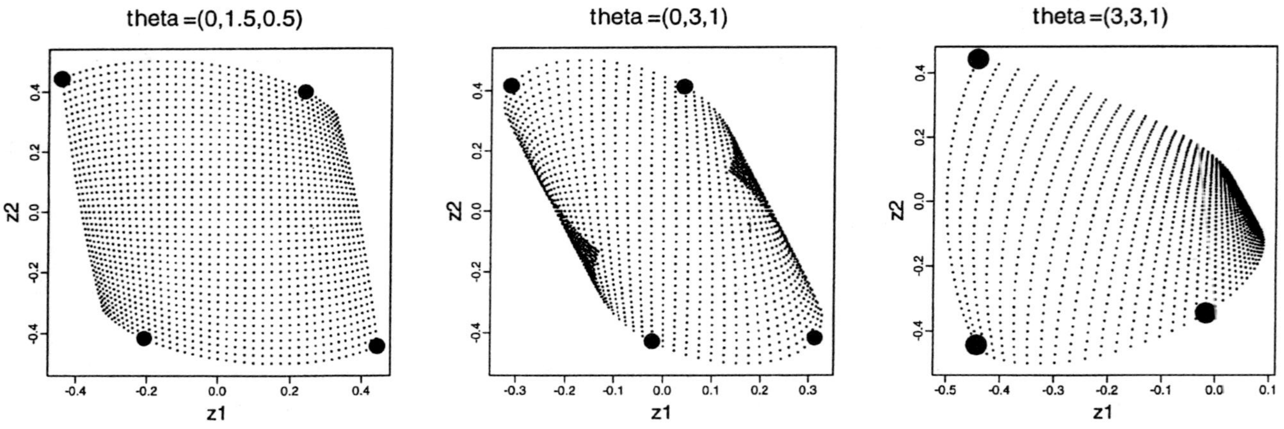

Fig. 11. Example 7. Logistic regression with two explanatory variables. D-optimum designs in the induced design space, $\mathcal{F}$, for parameter values (a) $\theta_1 = 0, \theta_2 = 1.5, \theta_3 = 0.5$; (b) $\theta_1 = 0, \theta_2 = 3, \theta_3 = 1$; (c) $\theta_1 = 3, \theta_2 = 3, \theta_3 = 1$. (The design weights are proportional to the areas of the plotting symbols.)

the explanatory variables, x_1 and x_2. It is therefore of interest to examine D-optimum designs for other parameter settings and to this end values of $\theta^T = (0, \frac{3}{2}, \frac{1}{2})$, $\theta^T = (0, 3, 1)$, and $\theta^T = (3, 3, 1)$ were selected for study. The resultant D-optimum designs are asymmetric and are presented in Table 6 and represented pictorially in Figure 10. The design for $\theta^T = (3, 3, 1)$ is of particular interest in that it comprises exactly three support points which are necessarily equally weighted. The induced design spaces, $\mathcal{F}$, for these parameter settings are shown in Figure 11, together with the D-optimum design points, and it is again evident from these figures that the spaces are "folded over" and that the optimum support therefore includes very few factorial points.

Overall it is clear that both the number and the location of the design points for the locally D-optimum designs of model (13) are highly dependent upon the best guess adopted for the parameters values. In addition it is interesting to note that, for the parameter settings examined here, the D-optimum design points correspond to values of the mean response, μ, spanning a wide range from low to high, as the values did for the logistic model with one explanatory variable. In all cases except one, if a design includes a point giving a value of μ^* for μ, equal weight will be given to a point giving a value of $1 - \mu^*$. The exception is the three-point design, for which this relationship holds for two of the design points. The symmetry to success and failure of the one-factor design is thus maintained.

Broad, analytical results for the construction of locally D-optimum designs for generalized linear models with two or more explanatory variables would seem to be particularly difficult to develop. Both Sitter and Torsney (1995) and Burridge and Sebastiani (1994) report conditions for certain simple designs, such as factorials or designs in which one factor at a time is changed, to be D-optimum but these studies would seem to be isolated ones.

4. Model checking and compound design criteria

Many locally D-optimum designs for nonlinear models with k parameters comprise only k points of support and do not therefore allow any checks to be made on the appropriateness or otherwise of the model. Furthermore their Bayesian counterparts, which are very often based on more than k points, are efficient for parameter estimation but not necessarily for model checking. It is therefore of interest to develop criteria specifically for checking the adequacy of a nonlinear model.

The usual approach to this problem involves, as a first stage, embedding the nonlinear model of interest in a more general model structure from which it can be recovered by setting certain of the parameters equal to specified values, frequently zero. Suitable choices of this general model form for nonlinear regression are suggested by Atkinson (1972) and include the following.

1. Add low order polynomial terms in the m explanatory variables to the nonlinear model.
2. Add squared and interaction terms in the first order derivatives, $\partial\eta(x, \theta)/\partial\theta$, to the nonlinear model.

3. Select a suitable parametric model into which the nonlinear model can be nested.

For generalized linear models the last of these options is the most viable. The second stage in this approach involves estimating as precisely as possible those parameters of the general model which are associated with departures of interest from the nonlinear model. This can be achieved by using D_s-optimum designs which minimize the generalized variance of the appropriate subset of parameters for the general model and also by constructing Bayesian optimum designs of the type developed by DuMouchel and Jones (1994). These latter designs are particularly attractive in that they provide a parsimonious way of checking model adequacy but the methodology is at present only established for linear models.

In practice the experimenter is very commonly concerned with both model checking and parameter estimation, and compromise criteria which embrace these two, possibly conflicting, goals can be constructed in the following way. The criteria for constructing locally optimum designs which are described in this chapter are concave, or can be replaced by criteria which are concave, on the set of all possible information matrices. It thus follows that any non-negatively weighted combination of these criteria will also be concave and that the powerful results pertaining to such criteria, including the formulation of Equivalence Theorems, will apply. These weighted criteria are termed composite or compound design criteria and are widely used. In particular it is possible, by a suitable combination of individual criteria, to construct designs which are efficient both for model checking and for the estimation of parameters. Several examples involving nonlinear models are given in Chapter 21 of Atkinson and Donev (1992) and one of these is presented here.

Example 8. First-order decay

For a first-order chemical reaction, represented as $A \to B$, the amount of compound A remaining at time x is described by a nonlinear regression model with

$$\mathrm{E}(Y) = \eta(x, \theta) = \mathrm{e}^{-\theta x}, \tag{14}$$

where θ represents the rate constant and is nonnegative. For this model, the information matrix of θ for a design, ξ, is scalar and is given by

$$M_1(\xi, \theta) = \sum_{i=1}^{n} \xi_i f^2(x_i, \theta)$$

where $f(x, \theta) = \partial\eta(x, \theta)/\partial\theta = -x\mathrm{e}^{-\theta x}$. The locally D-optimum design maximizes the information, $M_1(\xi, \theta)$, and comprises a single point at $x = 1/\theta$ which clearly provides no information for model checking.

A criterion for checking the adequacy of model (14) can be developed within the framework of the linear model with deterministic component,

$$\beta_1 f(x, \theta) + \beta_2 f^2(x, \theta), \tag{15}$$

where θ is taken to be fixed at a best guess. If the parameter, β_2, in this linear model is ignored the information on β_1 is equal to $M_1(\xi, \theta)$. It thus follows that β_2 is associated with departures from model (14) and hence that a suitable optimum design for model checking is the D_s-optimum design for this parameter. Specifically, this design minimizes the variance of the least squares estimate of β_2, and hence, equivalently, maximizes the ratio, $|M(\xi, \theta)|/M_1(\xi, \theta)$, where $M(\xi, \theta)$ is the information matrix for the full linear model (15).

Compromise designs for both parameter estimation and model checking for model (14) are obtained by maximizing the compound design criterion,

$$\Psi(\xi, \theta) = \alpha \log M_1(\xi, \theta) + (1 - \alpha) \log \{|M(\xi, \theta)|/M_1(\xi, \theta)\}, \tag{16}$$

where $0 \leqslant \alpha \leqslant 1$ and logarithms are introduced to ensure the concavity of the criterion. The coefficient, α, can be interpreted in a qualitative way as representing the relative importance of parameter estimation and model checking, but its precise meaning is difficult to quantify. A sensible strategy therefore is to find optimum designs for the compound criterion (16) for a range of α values, to calculate the efficiencies of the resultant designs with respect to the component criteria for parameter estimation and model checking, and to select a value for α on the basis of these efficiencies.

Finally it should be emphasized that compound design criteria are used in contexts other than model checking and include, *inter alia*, the weighting of D-optimality criteria for families of models and the weighting of different criteria, such as D- and c-optimality, for a specified model of interest. A comprehensive study of compound design criteria for linear models is given by Cook and Wong (1994) and a review of the results established thus far for nonlinear models is presented by Atkinson and Donev (1992). A point of particular import is that the Bayesian optimum design criteria introduced in Section 2 are themselves composite criteria, a fact emphasized in the seminal work of Läuter (1974a).

5. Designs for discriminating between models

It has so far been assumed that the model generating the data is known and that the purpose of the experimental programme is to estimate the parameters of the model, or certain functions of them, as precisely as possible. In this section a brief survey of designs for discriminating between two competing models is given. A fuller discussion of the problem, including examples and the extension to three or more models, is given in Chapter 20 of Atkinson and Donev (1992).

We begin by considering two competing nonlinear regression models which are conveniently identified as model 1 and model 2 and which have expected responses denoted by $\eta_1(x, \theta_1)$ and $\eta_2(x, \theta_2)$ respectively. Optimum designs for discriminating between these two models depend on which model is assumed to be the true one and, possibly, on the parameters of that true model. The simplest situation is that in which the true model is assumed to be known and is taken, without loss of generality, to be model 1 with $\eta_1(x, \theta_1)$ written as $\eta_t(x, \theta_1)$ to emphasize this assumption. Then the deviation of model 2 from the true model for a design, ξ, can be quantified as the sum of squares,

$$\int_{\mathcal{X}} \{\eta_t(x, \theta_1) - \eta_2(x, \theta_2)\}^2 \xi(\mathrm{d}x), \tag{17}$$

and for θ_2 unknown this can be estimated as the residual sum of squares,

$$\Delta_2(\xi, \theta_1) = \int_{\mathcal{X}} \{\eta_t(x, \theta_1) - \eta_2(x, \widetilde{\theta}_2)\}^2 \xi(\mathrm{d}x), \tag{18}$$

where $\widetilde{\theta}_2$ is that value of θ_2 for which (17) is a minimum. Clearly designs maximizing the residual sum of squares for lack of fit, $\Delta_2(\xi, \theta_1)$, are optimum for discriminating between models 1 and 2 when model 1 is known and such designs are termed locally T-optimum designs.

The function, $\Delta_2(\xi, \theta_1)$, is concave over the set of all possible designs, ξ, and it thus follows that a General Equivalence Theorem for locally T-optimum designs can be formulated and used to confirm global optimality and to provide a basis for algorithmic design construction. Specifically, the appropriate directional derivative is given by the squared difference between the true and the predicted responses as

$$\psi_2(x, \xi, \theta_1) = \{\eta_t(x, \theta_1) - \eta_2(x, \widetilde{\theta}_2)\}^2,$$

and the necessary and sufficient condition for a design, ξ^*, to be locally T-optimum is that $\psi_2(x, \xi^*, \theta_1) \leqslant \Delta_2(\xi^*, \theta_1)$ holds for all x in the design region, $\mathcal{X}$, with equality being attained at the support points of ξ^*. Full details of locally T-optimum designs for a range of examples are provided in the seminal paper of Atkinson and Fedorov (1975).

Bayesian versions of this T-optimum criterion follow naturally from the assumptions regarding the true model and are described by Ponce de Leon and Atkinson (1991). In particular, uncertainty in the value of the parameter, θ_1, for the true model, $\eta_t(x, \theta_1)$, can be quantified by introducing a prior distribution on that parameter. A Bayesian T-optimality criterion can then be formulated by taking the expectation of the residual sum of squares (18) over the prior distribution as

$$\Delta_2(\xi) = \mathrm{E}_{\theta_1}\{\Delta_2(\xi, \theta_1)\}$$

and furthermore an Equivalence Theorem for this criterion, based on the directional derivative, $\psi_2(x, \xi) = \mathrm{E}_{\theta_1}\{\psi_2(x, \xi, \theta_1)\}$, can be developed. A more realistic scenario however is that in which the true model is not known but prior probabilities, π_1 and π_2, are available for models 1 and 2 respectively, with $\pi_1 + \pi_2 = 1$. The appropriate Bayesian T-optimum criterion can then be formulated as the expected value of the residual sum of squares criterion taken over models and over parameters within models, and is given explicitly by

$$\Delta(\xi) = \pi_1 \mathrm{E}_{\theta_1}\{\Delta_2(\xi, \theta_1)\} + \pi_2 \mathrm{E}_{\theta_2}\{\Delta_1(\xi, \theta_2)\}.$$

The Equivalence Theorem for this criterion is based on the derivative function,

$$\psi(x,\xi) = \pi_1 \mathrm{E}_{\theta_1}\{\psi_2(x,\xi,\theta_1)\} + \pi_2 \mathrm{E}_{\theta_2}\{\psi_1(x,\xi,\theta_2)\},$$

and states that a design, ξ^*, is T-optimum if and only if the condition, $\psi(x,\xi^*) \leqslant \Delta(\xi^*)$, holds for all x in the design region, $\mathcal{X}$.

There are a number of interesting features relating to these Bayesian T-optimum designs. In particular, in the examples provided by Ponce de Leon and Atkinson (1991) the number of design points was found to increase with increasing prior uncertainty, as is the case for the Bayesian optimum designs of Section 2.3. Also, the choice of a joint prior distribution for the parameters, θ_1 and θ_2, which is required in the formulation of the criterion, $\Delta(\xi)$, is a particularly subtle one. Specifically, parameters for models 1 and 2 leading to similar response curves should carry similar weights and the priors on these parameters should not therefore be assumed to be independent.

Locally T-optimum designs for discriminating between generalized linear models are based on criteria involving the deviance in much the same way as those for nonlinear regression models are based on criteria involving the residual sum of squares. To fix ideas, suppose that there are two competing models for fitting binary data, identified as model 1 and model 2, and suppose further that the expected responses, $\mu_1(x,\theta_1)$ and $\mu_2(x,\theta_2)$ respectively, differ either in the linear predictor or in the link function or both. One example, given by Ponce de Leon and Atkinson (1992), involves models with a probit and a complementary log-log link function and a linear predictor of the form, $\eta = \theta_1 + \theta_2 x$. If model 1 is known to be the true model and $\mu_1(x,\theta_1)$ is written as $\mu_t(x,\theta_1)$, then it follows from likelihood theory that the difference in the contribution to the deviance of the two models for a design, ξ, is given by

$$\Delta_2^{(b)}(\xi,\theta_1) = \int_{\mathcal{X}} \left\{ \mu_t(x,\theta_1) \log \frac{\mu_t(x,\theta_1)}{\mu_2(x,\widehat{\theta}_2)} \right.$$
$$\left. + [1-\mu_t(x,\theta_1)] \log \frac{1-\mu_t(x,\theta_1)}{1-\mu_2(x,\widehat{\theta}_2)} \right\} \xi(\mathrm{d}x)$$

where $\widehat{\theta}_2$ is the maximum likelihood estimator of θ_2. Designs maximizing this criterion are thus locally T-optimum for discriminating between models 1 and 2 when model 1 is known to be the true one and an Equivalence Theorem for confirming their optimality can be readily formulated and used. Furthermore, the T-optimality criterion, $\Delta_2^{(b)}(\xi,\theta_1)$, can be extended to accommodate prior information on the parameters and on the models in a manner analogous to that described above for nonlinear regression models. Similar considerations apply to other generalized linear models and full details are given in Ponce de Leon and Atkinson (1992).

The notion of T-optimality for discriminating between competing models is applicable to both nested and separate nonlinear models. An alternative approach for discriminating between separate models is to identify a general parametric model which contains the separate ones as special cases. Designs are then constructed to estimate as precisely as possible those parameters of the general model specifying model form. For example, generalized linear models for binary data with logistic and complementary log-log link functions are special cases of the model with link function

$$\eta(\lambda) = \log\left\{\frac{1/(1-\mu)^{\lambda} - 1}{\lambda}\right\}, \tag{19}$$

obtained by setting λ to 1 and by letting λ tend to 0 respectively (McCullagh and Nelder, 1989, p. 378). Thus an optimum design for discriminating between the logistic and the complementary log–log models is the design which minimizes the variance of the parameter, λ, in the general form (19), i.e., the D_s-optimal design for λ (Ponce de Leon and Atkinson, 1993).

6. Extensions

This section contains brief comments on related design problems for nonlinear models which require extensions of the methods described here.

One general point is that the emphasis in this chapter has been on continuous designs, design measures and the equivalence theorem: little attention has been given to exact designs which are, of course, required in practice. The properties and construction of exact designs are covered in another chapter of this book. A second general point is that, if the design is sensitive to prior assumptions, it will make sense to use sequential procedures, so that accumulating experimental evidence can be used to improve the design, perhaps through the revision of prior distributions. From the point of view of minimizing the number of observations for a given precision, it is most efficient if the information can be incorporated after each new observation. However, if there is a delay between experimentation and the results becoming available, batches of a few experiments may have to be designed at once in a sequential framework. Agriculture is an obvious example. Industrial examples include failure time experiments and experiments on photographic emulsions which have to be processed into film and used before they can be assessed. A discussion of the properties of sequential, and batch sequential, designs for nonlinear models is given by Ford et al. (1989).

Sequential experimentation also simplifies the construction of the T-optimum designs of Section 5. The criterion (18) is defined for model 1 being true. In the sequential version θ_1 and θ_2 are estimated from the results of n trials, giving estimated responses $\eta_1(x, \widehat{\theta}_{1n})$ and $\eta_2(x, \widehat{\theta}_{2n})$. The $(n+1)$st trial is performed at the value of x maximising the estimated directional derivative

$$\widehat{\psi}_n(x) = \{\eta_1(x, \widehat{\theta}_{1n}) - \eta_2(x, \widehat{\theta}_{2n})\}^2.$$

If one of the models is true, for instance the first, then the estimated response $\eta_1(x, \widehat{\theta}_{1n})$ will converge to $\eta_t(x, \theta_1)$ and the sequential design procedure converges to the sequential construction of the T-optimum design for known θ_1. Examples are given by Atkinson and Fedorov (1975).

In this form of sequential experiment it is assumed that a series of single independent observations is taken at a specified set of time points. But sometimes, for example in kinetic experiments, the data arise as a series from observations at a specified number of distinct time points. One example is given by Morris and Solomon (1995), who find the optimum design by numerical maximization for a given correlation function and for numbers of time points from one to ten. Bohachevsky et al. (1986) find another method for distributing points in time by assuming that there is a specified minimum time interval between successive observations. They again find the optimum design by numerical optimization. In neither example is any mention made of equivalence theory and it is clear that much remains to be done in this area.

The designs in this chapter can be thought of as being generated by choosing experimental units with replacement. A consequence is that replication is possible. An example of an experiment in economics in which replication is not possible is described by Müller and Ponce de Leon (1996). The purpose of the experiment is to determine personal utility functions for decisions under risk. Subjects are asked to choose between pairs of gambles defined by probability distributions, the results being intended to discriminate between two rival theories for the utilities of such gambles. The experimental design problem is to choose the probabilities associated with the pairs of gambles so as to give the best discrimination between the theories. Because of the short time span of the particular experiment, it is clearly not advisable to repeat questions, since experimentees will recognize a replication. Designs are therefore constructed in which units are removed from the set of candidate units once they have been employed. Although the construction of the design requires modification of the algorithms for exact designs, rather than continuous designs, the problem does indicate another way in which the theory of this chapter needs extension for a practical problem.

Acknowledgements

L. M. Haines would like to thank the University of Natal and the Foundation for Research Development, South Africa, for financial support. We are grateful to Dr. Barbara Bogacka for helpful comments.

References

Andrews, D. F., R. Gnanadesikan and J. L. Warner (1971). Transformations of multivariate data. *Biometrics* **27**, 825–840.

Atkinson, A. C. (1972). Planning experiments to detect inadequate regression models. *Biometrika* **59**, 275–293.

Atkinson, A. C. (1996). The usefulness of optimum experimental designs (with discussion). *J. Roy. Statist. Soc. Ser. B* **58**, 59–76.

Atkinson, A. C., K. Chaloner, A. M. Herzberg and J. Juritz (1993). Optimum experimental designs for properties of a compartmental model. *Biometrics* **49**, 325–337.

Atkinson, A. C., C. G. B. Demetrio and S. Zocchi (1995). Optimum dose levels when males and females differ in response. *Appl. Statist.* **44**, 213–226.

Atkinson, A. C. and A. N. Donev (1992). *Optimum Experimental Designs*. Oxford Univ. Press, Oxford.

Atkinson, A. C. and V. V. Fedorov (1975). The design of experiments for discriminating between two rival models. *Biometrika* **62**, 57–70.

Bohachevsky, I. O., M. E. Johnson and M. L. Stein (1986). Generalized simulated annealing for function optimization. *Technometrics* **28**, 209–217.

Box, G. E. P. and H. L. Lucas (1959). Design of experiments in nonlinear situations. *Biometrika* **46**, 77–90.

Burridge, J. and P. Sebastiani (1994). D-optimal designs for generalised linear models with variance proportional to the square of the mean. *Biometrika* **81**, 295–304.

Chaloner, K. and K. Larntz (1989). Optimal Bayesian design applied to logistic regression experiments. *J. Statist. Plann. Inference* **21**, 191–208.

Chaloner, K. and I. Verdinelli (1995). Bayesian experimental design: A review, submitted.

Chernoff, H. (1953). Locally optimal designs for estimating parameters. *Ann. Math. Statist.* **24**, 586–602.

Chernoff, H. (1979). *Sequential Analysis and Optimal Design*. SIAM, Philadelphia, PA.

Cook, R. D. and W. K. Wong (1994). On the equivalence between constrained and compound optimal designs. *J. Amer. Statist. Assoc.* **89**, 687–692.

Dette, H. and L. M. Haines (1994). E-optimal designs for linear and nonlinear models with two parameters. *Biometrika* **81**, 739–754.

Dette, H. and S. Sperlich (1994). A note on Bayesian D-optimal designs for generalization of the simple exponential growth model. *South African Statist. J.* **28**, 103–117.

DuMouchel, W. and B. Jones (1994). A simple Bayesian modification of D-optimal designs to reduce dependence on an assumed model. *Technometrics* **36**, 37–47.

Elfving, G. (1952). Optimum allocation in linear regression theory. *Ann. Math. Statist.* **23**, 255–262.

Fedorov, V. V. (1972). *Theory of Optimal Experiments*. Academic Press, New York.

Ford, I., D. M. Titterington and C. P. Kitsos (1989). Recent advances in nonlinear experimental design. *Technometrics* **31**, 49–60.

Ford, I., B. Torsney and C. F. J. Wu (1992). The use of a canonical form in the construction of locally optimal designs for non-linear problems. *J. Roy. Statist. Soc. Ser. B* **54**, 569–583.

Haines, L. M. (1993). Optimal design for nonlinear regression models. *Comm. Statist. Theory Methods* **22**, 1613–1627.

Haines, L. M. (1995). A geometrical approach to optimal Bayesian designs for one-parameter models. *J. Roy. Statist. Soc. Ser. B* **57**, 575–598.

Haines, L. M. (1996). Discussion of 'The usefulness of optimum experimental designs' by A. C. Atkinson. *J. Roy. Statist. Soc. Ser. B* **58**, in press.

Kiefer, J. (1985). *Jack Carl Kiefer Collected Papers III*. Springer, New York.

Kiefer, J. and J. Wolfowitz (1960). The equivalence of two extremum problems. *Canad. J. Math.* **12**, 363–366.

Kitsos, C. P., D. M. Titterington and B. Torsney (1988). An optimal design problem in rhythmometry. *Biometrics* **44**, 657–671.

Läuter, E. (1974a). Experimental design in a class of models. *Math. Operat. Statist. Ser. Statistik* **5**, 379–398.

Läuter, E. (1974b). Method of experimental planning for the case of nonlinear models (in Russian). *Math. Operat. Statist. Ser. Statistik* **5**, 625–636.

McCullagh, P. and J. A. Nelder (1989). *Generalized Linear Models*, 2nd edn. Chapman and Hall, London.

Morris, M. D. and A. D. Solomon (1995). Design and analysis for an inverse problem arising from an advection-dispersion process. *Technometrics* **37**, 293–302.

Mukhopadhyay, S. and L. M. Haines (1995). Bayesian D-optimal designs for the exponential growth model. *J. Statist. Plann. Inference* **44**, 385–397.

Müller, W. G. and A. C. M. Ponce de Leon (1996). Optimal design of an experiment in economics. *J. Econ.*, to appear.

Pazman, A. (1986). *Foundations of Optimum Experimental Design*. Reidel, Dordrecht.

Ponce de Leon, A. M. and A. C. Atkinson (1991). Optimum experimental design for discriminating between two rival models in the presence of prior information. *Biometrika* **78**, 601–608.

Ponce de Leon, A. M. and A. C. Atkinson (1992). The design of experiments to discriminate between two rival generalized linear models. In: L. Fahrmeir, B. Francis, R. Gilchrist and G. Tutz, eds., *Advances in GLIM and Statistical Modelling: Proc. GLIM92 Conf., Munich*. Springer, New York, 159–164.

Ponce de Leon, A. M. and A. C. Atkinson (1993). Designing optimal experiments for the choice of link function for a binary data model. In: W. G. Müller, H. P. Wynn and A. A. Zhigljavsky, eds., *Model-Oriented Data Analysis*. Physica-Verlag, Heidelberg, 25–36.

Pukelsheim, F. (1993). *Optimal Design of Experiments*. Wiley, New York.

Shah, K. R. and B. K. Sinha (1989). *Theory of Optimal Design*. Lecture Notes in Statistics 54. Springer, Berlin.

Silvey, S. D. (1980). *Optimum Design*. Chapman and Hall, London.

Sitter, R. S. and B. Torsney (1995). D-optimal designs for generalized linear models. In: C. P. Kitsos and W. G. Müller, eds., *MODA 4 – Advances in Model-Oriented Data Analysis*. Physica-Verlag, Heidelberg, 87–102.

S. Ghosh and C. R. Rao, eds., *Handbook of Statistics, Vol. 13*

15

Spatial Experimental Design

R. J. Martin

1. Introduction

The statistical theory of experimental design is concerned with obtaining as much information as possible from the resources available. In some applications, of which agriculture is the most obvious, the experimental material consists of (possibly disjoint) sets of spatial (i.e., two-dimensional) material. Other spatial applications include sheet metal production and paper making.

In this chapter, the design of comparative experiments is considered for the case that the units (or subsets of units) are spatially arranged, and the spatial or directional layout is itself important. Research in this area has greatly expanded recently, and is having an increasing impact on practical experimentation. The bibliography of Gill (1991) lists 83 references on neighbour methods in design and analysis. There are currently around that number on analysis, and around 130 on design. The aims of this chapter are to give a broad up-to-date discussion of the area of spatial experimental design, to provide links to related introductory articles, and to indicate some very recent results.

This chapter is organized as follows. The rest of this section outlines the areas to be discussed, and contains some preliminary definitions. The history of spatial experimental design is reviewed in Section 2. Section 3 contains an outline of the spatial analysis of an experiment. Section 4 is concerned with designs which have some sort of neighbour balance. Lastly, in Section 5, designs that are efficient under spatial dependence are discussed, concentrating on the case that all contrasts are of equal interest (treatments of equal status). Two other situations of interest are briefly discussed: test-treatment designs under dependence, and factorial experiments under dependence.

In this chapter, spatial means that the layout of the experimental units is in one or two spatial dimensions, and that the spatial locations of the units, and the spatial relationships between different units, are relevant. A layout in one dimension is included provided the dimension is spatial. Indeed, at present this is the most important case in practice. There are experiments in which the units are arranged in three spatial dimensions – see Section 2.6 of Pearce (1983), but if the spatial locations of the units are taken into account, the principles of two-dimensional design would hold. Only rectangular units arranged in a rectangular array are considered here; for some results and references on triangular or hexagonal units, and the beehive and honeycomb designs, see Street and Street (1987, §§ 15.5–15.7).

Although the positional effects of the units can be relevant, the most important consideration with spatial layouts is usually the effects of the *neighbours* – those units which are adjacent to a unit. For spatial layouts, it is often reasonable from symmetry considerations to assume that there are no directional effects, so that, for example, a left neighbour is equivalent to a right neighbour. However, there are cases when it may be reasonable to postulate directional effects in one or both directions. This differs from measurements in time, when it is often reasonable to assume a unidirectional effect from past to future, resulting, for example, in a carry-over effect from one time point to the next. These fixed effects are usually included in models associated with *repeated-measurements designs* (or change-over, or cross-over designs) – see Shah and Sinha (1989, Chapter 6); Street (1996, § 4); and other chapters in this Handbook. Models in this case may also include dependent errors, which may be modelled so that the dependence decreases with the difference in time. If dependent errors are included, and the carry-over effects are not included, the situation is essentially the same as the one-dimensional case of spatial dependence.

The subject of choosing the size and shape of the plots will not be discussed here. There is a large statistical literature on this in agriculture, usually using data from uniformity trials. Recent investigations of this are discussed in Section 1.5 of Pearce (1983), and by Reddy and Chetty (1982). Other topics that will not be discussed here are spatial designs for continuous variables – see Fedorov (1996), computer experiments – see Morris and Mitchell (1995), and designs for spacing experiments – see Pearce (1983, § 9.1), and Street and Street (1987, § 15.8).

As noted earlier, there is now a large set of references in this area. Only a selection of these are given here. Details of other work, or original sources, can be found in one of Dagnelie (1987), Gill (1991), Martin (1986), Pearce (1983), Street and Street (1987), or in the recent references given here.

The following notation and terminology will be used. There will be n *units* or *plots* – the basic (smallest) division of the experimental material. There will be t *treatments* – what is applied to the unit. The term treatments here includes actual treatments, varieties, factorial combinations, etc. The number of times treatment i occurs is r_i, with $\sum r_i = n$, and $\boldsymbol{r} = (r_i)$ is the *replicate* vector. If $r_i = n/t\ \forall i$, the design is *equireplicate*, and r denotes n/t.

Unless otherwise stated, a simple block experiment (i.e., there is one blocking factor) with b blocks is assumed. The size of block i is k_i, with $\sum k_i = n$, and $\boldsymbol{k} = (k_i)$ the block-size vector. If $k_i = n/b\ \forall i$, then k denotes n/b. A block design is *binary* if all blocks contain distinct treatments – no treatment occurs more than once in any block. The units in each block may be arranged in a one- or two-dimensional layout. The same notation (with $k = pq$) can be used for the blocks in a nested $p \times q$ row-column design (see § 3 of Street, 1996). The extension to the rows and columns of a (nested) row-column design can easily be made.

Unless otherwise stated, it will be assumed that designs are equireplicate, with equal block sizes, so that $n = rt = bk$. Extensions to general $\boldsymbol{r}$ and $\boldsymbol{k}$ are possible, but may not always be straightforward. If $k = t$, the block is *complete* if it contains each treatment (once). If $k < t$, the block is *incomplete*. It is convenient here to use the term Balanced Incomplete Block Design (BIBD) (see § 2.1 of Street, 1996) for $k \leqslant t$ to include the Complete Block Design (the binary design with $k = t$). If $k > t$, each block of an *extended complete block* design contains $\mathrm{int}(k/t)$ complete blocks and a binary incomplete block of size $k - t \times \mathrm{int}(k/t)$. This is balanced if the incomplete part is a BIBD.

Treatments are represented by numerals $1, 2, 3, \ldots$, plus 0 and t. For a simple block experiment, y_{ij} (regarded as a random variable here) denotes the measured response on unit j of block i. The units are assumed to be ordered lexicographically in two dimensions. Let $\boldsymbol{y} = (y_{11}, \ldots y_{1k}, y_{21}, \ldots, y_{bk})'$ be the data vector.

In the following $\mathbf{1}_t$ denotes a t-vector of ones, I_t the $t \times t$ identity matrix, $J_t = \mathbf{1}_t \mathbf{1}_t'$ a $t \times t$ matrix of ones, and E_t denotes $I_t - t^{-1} J_t$.

Then a linear model for the mean of $\boldsymbol{y}$, $\mathrm{E}(\boldsymbol{y})$, is

$$\mathrm{E}(\boldsymbol{y}) = A\boldsymbol{\theta} + B\boldsymbol{\beta} + T\boldsymbol{\phi}, \tag{1}$$

where A, B, T are design matrices for treatment effects, block effects, and other fixed effects (such as covariates and interference effects, see Section 3.1) respectively, and $\boldsymbol{\theta}$, $\boldsymbol{\beta}$, $\boldsymbol{\phi}$ are the corresponding parameters. The usual *block-treatment model*, which has the term $T\boldsymbol{\phi}$ omitted, is

$$\mathrm{E}(\boldsymbol{y}) = A\boldsymbol{\theta} + B\boldsymbol{\beta}. \tag{2}$$

For a simple block experiment, B has the form $I_b \otimes \mathbf{1}_k$, where $\otimes$ is the Kronecker product. The $\{\theta_i\}$ are uncorrected treatment effects. Let τ_i denote the mean corrected effect of treatment i, $\tau_i = \theta_i - t^{-1} \sum \theta_j$, or $\boldsymbol{\tau} = E_t \boldsymbol{\theta}$.

The dispersion matrix of $\boldsymbol{y}$ is given by

$$\mathrm{var}(\boldsymbol{y}) = V\sigma^2,$$

where in general V is not diagonal, and is a function of a parameter vector $\boldsymbol{\lambda}$ which relates to the dependence structure (see Section 3.1). Henceforth, assume there is no between-block dependence, and that the within-block dependence is the same in each block. These assumptions are usually associated with blocks of the same size, shape and orientation that are in the same area but reasonably well separated, and with the same plot structure in each block. Thus, V has the form $I_b \otimes \Lambda$, where Λ is the within-block dispersion matrix which depends on $\boldsymbol{\lambda}$.

The usual ordinary least-squares (ols) analysis estimates the vector $\boldsymbol{\eta} = (\boldsymbol{\theta}', \boldsymbol{\beta}', \boldsymbol{\phi}')'$ by

$$\widehat{\boldsymbol{\eta}} = (X'X)^{-} X'\boldsymbol{y}, \tag{3}$$

where C^{-} denotes a generalized inverse of C, and $X = [A\ B\ T]$. Generalised least-squares (gls) using a non-singular V estimates the vector $\boldsymbol{\eta}$ by

$$\widehat{\boldsymbol{\eta}} = (X'V^{-1}X)^{-} X'V^{-1}\boldsymbol{y}. \tag{4}$$

A general extension for a singular V is possible, but the extension is easy in the cases considered in Section 3. In practice, $\boldsymbol{\lambda}$ is usually unknown, and estimated from the data. If $\boldsymbol{y}$ is normal, maximum likelihood estimation leads to (4) with V replaced by its estimate. It may still be reasonable to use this estimate of $\boldsymbol{\eta}$ when $\boldsymbol{y}$ is non-normal.

2. Historical review

The fact that neighbouring plots tend to be more alike than ones further apart, in other words that the (mean-corrected) observations are spatially dependent, was well known in the early days of agricultural experimentation (see also § 1.4 of Pearce, 1983). A simple way of partially overcoming this was to ensure that neighbouring plots should contain different treatments, and that plots having the same treatment should be well separated. This goes back at least to J. F. W. Johnson in 1849 (Cochran, 1976), who suggested that repetitions of a treatment should be a knight's move apart, an idea that was popularised in the1920s by Knut Vik. These layouts are *systematic*, with each subsequent row being formed by cyclically shifting the treatments in the previous row by two plots. Note that systematic is used in general to mean that the design was selected purposively, and was not the outcome of a valid randomization scheme.

These systematic designs were very popular, but had the defect that there was no correct statistical analysis, that is no correct way to compare treatments. Treatment effects were estimated by means, and variation was estimated using the (corrected) total sum of squares.

R. A. Fisher introduced several important ideas into experimental design and analysis in the 1920s and 1930s, including replication, error control by blocking (grouping homogeneous units together), crossed (row and column) blocking, factorial treatments, randomization and randomization tests. For simple designs, such as complete blocks, Latin squares, etc., he showed how to partition the total sum of squares to obtain the residual and treatment sums of squares. Then, valid randomization, within a given block structure, allows a correct analysis in which the errors are treated as independent. A rigorous treatment which clearly shows that the effect of randomization is to induce constant correlations for different observations within the same stratum was begun by J. A. Nelder, and continued by others – see, for example, Bailey and Rowley (1987).

The arguments in favour of randomization are long term ones. It had always been recognized that randomization might produce an obviously undesirable design, but that discarding the outcome would bias the analysis. A useful extension is to restricted or constrained randomization (Yates, 1948). This is a valid randomization using a subset of all the possible designs. The subset can be chosen by the experimenter to suit the particular application. The method has been considerably extended – see Bailey (1986, 1987). Fisher also introduced randomization tests as an added justification for the usual analysis. These usually produce a null distribution for the usual F statistic that is close to the null F distribution.

A drawback with randomized designs is that, when positive dependence exists, they are on average less efficient than good systematic designs. Because the sum of treatment and residual sums of squares is fixed, any efficient design will have the standard errors for treatment contrasts higher than they should be. This leads to conservative tests and confidence intervals. There had been considerable controversy in Britain in the 1930s – W. S. Gosset ("Student") was a leading proponent of the use of the more efficient systematic designs, and was still recommending them in 1937 (Pearson, 1938). Eventually Fisher's practices spread widely in the English speaking world. It was clear to Pearson (1938) that Fisher's views would predominate for some time: 'The ultimate decision can hardly be expected as yet; it will come in time, perhaps after 10 or 20 years, when, ..., freed from the weight of authority, ...'

A 1931 empirical study by O. Tedin, cited in Martin (1986), compared both forms of two systematic 5 by 5 Latin squares, the Diagonal (1-shift) and the Knight's Move (2-shift), with seven randomly selected ones, and one other purposively selected. It confirmed that the Knight's Move square was more efficient than the randomization average, and showed that the Diagonal square was less efficient. It is interesting that the four squares formed from these two systematic squares constitute a valid restricted randomization set (they are mutually orthogonal), and hence the set of 5 by 5 Latin squares without these also does.

Other aspects of Tedin's work, which show considerable foresight, were neglected. He recognized that the systematic Knight's Move and Diagonal designs might not have the most extreme efficiencies (which is why the last square was included), and tried to predict the efficiency using 'coefficients of spreading'. He was also interested in finding 5 by 5 Latin squares which would be valid without randomization.

The principles of Fisher were extended, in particular, by F. Yates in the 1930s and 1940s. The (balanced) incomplete block designs allow blocking to be much more effective. Smaller block sizes will usually ensure more homogeneous experimental material within blocks. The ensuing reduction in error variance will usually outweigh any loss of efficiency caused by the lack of complete replicates. However, the simple estimates based on treatment means cannot be used.

In other parts of the world, particularly Continental Europe, Fisher's ideas were not so widely accepted, and systematic designs continued to be used. As late as 1956, designs ('gerechte' designs) were proposed by W. U. Behrens (cited in Martin, 1986). These designs have the treatments distributed 'fairly' over a row-column design by using extra non-orthogonal blocking factors in the design (but not necessarily in the analysis). The extra blocks are chosen to be spatially compact. They could be rectangular, and could be congruent, but neither were necessary. The Magic Latin squares of G. M. Cox (see Section XV-3 of Federer, 1955) are a special case, using congruent rectangular blocks. Although randomization was recommended, there was no valid randomization theory given. In some cases, the suggested analysis is clearly not correct (Bailey et al., 1990, 1991). These designs were until recently still being used and recommended in text books in Germany.

Even in Britain, some aspects of randomization were beginning to be questioned by its strongest proponents. Yates (1967) said on the validity of the randomization that 'The condition...can scarcely be regarded as sufficient to ensure a good design', 'Fisher ... tended to lay undue emphasis on the importance of formal tests of significance ... What is ... required is an efficient estimate of these effects and a valid and reasonably accurate estimate of the error', and 'As soon as... an experimental layout is determined, the layout is known, and can be treated as supplementary information in the subsequent analysis if this appears relevant'.

Methods to overcome the drawbacks of the biased analysis of systematic designs were also being proposed. The most notable was the 1937 method proposed by J. S. Papadakis in Greece, which attempted to take account of the spatial dependence. Although examined theoretically then by M. S. Bartlett, the method was largely neglected in the English-speaking world. Interest was kept alive there by some, in particular, D. R. Cox and S. C. Pearce. The current revival in the spatial analysis of experiments began with the papers of Bartlett (1978) and Wilkinson et al. (1982). Most modern methods of spatial analysis can be regarded as using generalised least-squares for a given or estimated dependence structure – this is discussed in Section 3.1.

In the classical randomization approach, spatial information had been used for many years in the choice of plot size and shape, the choice of design type (blocking structure), and the choice of blocks. However, it was accepted that in some situations, such as when time was an important effect, little randomization was possible. The use of fixed one-dimensional polynomial trends to model fertility variation, etc., had been suggested in § 48 of later editions of Statistical Methods for Research Workers by R. A. Fisher (e.g., Fisher, 1950). Further early work on trend modelling by J. Neyman, R. A. Fisher, A. Hald and others, including the use of moving-averages to remove the variation, and the two-dimensional case, is mentioned in the Discussion of Cox (1950) – see also Cox (1951, 1952). An early published example of using a two-dimensional surface as a covariate is Federer and Schlottfeldt (1954). The study of designs that are efficient under polynomial trends began with Cox (1951, 1952), although it is interesting to note that in Section 48 of early editions of Statistical Methods for Research Workers (e.g., Fisher, 1932), Fisher gave an example of a non-systematic linear trend-free design, but concluded that the analysis was biased. There have been many developments since then (see Section 5.2).

Designs with some kind of neighbour balance were first considered by E. J. Williams and, independently, B. R. Bugelski – see Street and Street (1987, § 14.7), as balanced designs for models which included a carry-over effect. Since then, other types of designs with some neighbour balance have arisen, sometimes independently of other work, in many areas where full randomization may not be desirable. Some types have been studied extensively in Combinatorial Design. Areas of application include serology, interference and competition. Neighbour designs are discussed in Section 4.

The study of designs that are efficient when it is assumed that the analysis will take account of the dependence structure was begun by Williams (1952) (see Section 5.1 for further details), but in this case, apart from some optimality results by Kiefer (1961), there were few developments in the ensuing 25 years. Recent results are discussed in Section 5.

Finally, check plots, usually standard varieties forming a regular array, were often used in early agricultural experimentation. Although their use has declined in replicated experiments, they are still used in large unreplicated variety trials, and interest is developing in the spatial design and analysis of such experiments – see Besag and Kempton (1986), Cullis et al. (1989).

Further related historical discussion can be found in Cochran (1976), Pearce (1983, § 1.1), Kempthorne (1986), Freeman (1988).

3. Spatial analysis

Many types of designs have been proposed to take account of the spatial nature of the experimental material. Some spatial designs have arisen for particular circumstances, such as spacing trials, and genotype experiments. In these cases, the design is usually chosen on intuitive grounds, rather than for its efficiency under an assumed model. In other cases, efficiency is sought in terms of a prior assumed model for the unobserved data. Clearly this assumed model may turn out to be inappropriate in many different ways once the data are collected, and a different model may be used for the analysis. However, at the design stage, if an efficient design is required, it is necessary to make assumptions about the model and the analysis that will be used. In this section, models and spatial analysis, and measures of design efficiency are discussed.

3.1. Models and analysis

Consider first the usual block-treatment model (2), and the usual ordinary least-squares (ols) analysis. Using the usual residual sum of squares is liable to bias with systematic designs because of the underlying spatial dependence. Randomization is often regarded as a way of removing the dependence. Thus the data $\boldsymbol{y}$ are analysed as though $\boldsymbol{y}$ is normal with $\mathrm{var}(\boldsymbol{y}) = I\sigma^2$. It is more revealing to regard randomization as a method for replacing a dependence that depends on distance with a dependence that is constant within blocks. Thus the V that results from randomization is highly structured, as for the random blocks model (but negative correlations are possible), and generalised least-squares (gls) essentially gives the usual ols analysis.

Papadakis' original 1937 proposal to take account of the dependence was for a covariance analysis which adjusted for the effects of neighbours, estimated by the plot residual. This is now known as the *Papadakis method.* Since then (up to 1994), he has proposed modifications to his method. For the original method, the additional T matrix in (1) thus contains a covariate, but one that depends on the data $\boldsymbol{y}$. For further details and references, see Martin (1985) and Dagnelie (1987).

Fixed effects $T\boldsymbol{\phi}$ are also included in the model (1) to allow for fixed trends, for example polynomials in one- and two-dimensions, and Fourier terms.

Other important cases when model (1) with $T\boldsymbol{\phi}$ included is used are interference (and carry-over), and competition. Interference is when the treatment applied to one unit may affect the response on neighbouring units. Examples are experiments to compare weed or insect control. In the simplest case, the extra design matrix T is equal to NA, where N is a neighbour matrix – usually a $0/1$ matrix showing which sites are neighbours. Sometimes, N contains neighbour weights. A simplified model assumes that $\boldsymbol{\phi}$ is proportional to $\boldsymbol{\theta}$, $\boldsymbol{\phi} = \alpha\boldsymbol{\theta}$, where α is a scalar parameter. Several sets of neighbours could be included with corresponding matrices N_i, so that the model has the extra terms $\{N_i T\boldsymbol{\phi}_i\}$. For example, in cross-over trials the usual $N = N_1$

will have entries for the immediately preceding time, but sometimes N_2 for the time two periods before may also be included. In one-dimensional space, one N matrix may be used for both left and right neighbours, or separate matrices N_1, N_2 could be used. In two-dimensions, adjacent neighbours in both directions can be included, plus, perhaps, diagonal neighbours, etc.

Competition is when the response on one unit may affect the response on neighbouring units. An example is an experiment to compare varieties which may compete with each other through height or root growth. The usual competition model is similar to models used in econometrics, in that it expresses $\boldsymbol{y}$ in terms of (for one set of neighbours) $N\boldsymbol{y}$. It is written

$$\boldsymbol{y} = B\boldsymbol{\beta} + A\boldsymbol{\theta} + \alpha N\boldsymbol{y} + \boldsymbol{\varepsilon},$$

where $\boldsymbol{\varepsilon}$ is assumed to have mean $\mathbf{0}$ and variance $I_n\sigma^2$. A more revealing way of writing the model is as

$$E(\boldsymbol{y}) = GB\boldsymbol{\beta} + GA\boldsymbol{\theta}, \quad \text{with } \operatorname{var}(\boldsymbol{y}) = GG'\sigma^2,$$
$$\text{where } G = (I_n - \alpha N)^{-1}.$$

Since for small $|\alpha|$, $G \approx I + \alpha N$, the model is, to first order, assuming a similar mean to that for the proportional interference model, but it is postulating a dependence structure. However, the competition effect implies a negative α, which then implies negative dependence between neighbours. Negative dependence between neighbours is usually not an appropriate assumption. Further neighbour matrices can be included in the model.

Interference and competition models with positive dependence have been suggested (Correll and Anderson, 1983; Pithuncharurnlap et al., 1993), but their use is still at an early stage. For further details of the interference and competition models, see Besag and Kempton (1986).

Following Bartlett (1978) and Wilkinson et al. (1983), a large number of methods of analysis have been proposed that use spatial information. Although they can be derived and presented in different ways, most can be represented in the form of a model for $\Delta\boldsymbol{y}$, where Δ is a difference operator (a matrix with zero row sums). Given the parameters of the spatial variation, the treatment effects $\boldsymbol{\theta}$ are then estimated by generalized least-squares (gls). In practice, the spatial parameters are usually also estimated, either by exact Gaussian maximum likelihood, or, increasingly, by restricted maximum likelihood (REML) – see Gleeson and Cullis (1987).

With the assumptions of Section 1, when the variances are finite $V = I_b \otimes \Lambda$, for Λ positive definite. The generalization that within-block differencing in one or both directions is required for finite variances or simple modelling, requires that

$$\operatorname{var}\{(I_b \otimes \nabla)\boldsymbol{y}\} = I_b \otimes \nabla\Lambda\nabla'\sigma^2 \tag{5}$$

where ∇ is a $k-d$ by k difference matrix, $0 < d < k$, and $\nabla\Lambda\nabla'$ is a positive definite covariance matrix. Sometimes $\nabla\Lambda\nabla'$, rather than Λ, is specified directly.

The block-treatment model (2) now becomes, for $d > 0$,

$$\mathrm{E}\{(I_b \otimes \nabla)\boldsymbol{y}\} = (I_b \otimes \nabla)A\boldsymbol{\theta}. \tag{6}$$

If $d = 0$ and block effects are included, the estimator of $\boldsymbol{\tau}$ is a function of within-block differences. Thus, with simple block effects, the model with mean and variance given by (5) and (6) when $d = 1$, is equivalent to model (2) with $V = I_b \otimes \Lambda$. The model with (5) and (6) is essentially that which results from using the usual sweep operator for blocks. There is therefore no loss in generality in taking $d > 0$ henceforth and using (5) and (6). If model (1) is used, and $T\boldsymbol{\phi}$ corresponds to a fixed one-dimensional polynomial trend of degree $d - 1$, then taking the dth difference removes this term (and a random polynomial), leaving (5) and (6) again. In particular, in one dimension, $d = 2$ removes a fixed or random linear trend. There is thus a link between the theory for design under dependence and the theory for trend-free designs.

In two dimensions, having fixed row and column effects and no differencing is equivalent to taking first differences in both the rows and columns. There are similar extensions of further row and column differencing to include model (1) with $T\boldsymbol{\phi}$ representing a two-dimensional polynomial surface.

The reduced least-squares equations for estimating $\boldsymbol{\tau}$ are taken as

$$C\widehat{\boldsymbol{\tau}} = \boldsymbol{q}.$$

Under ordinary least-squares (ols)

$$C = A'(I_b \otimes E_k)A \quad \text{and} \quad \boldsymbol{q} = A'(I_b \otimes E_k)\boldsymbol{y}. \tag{7}$$

For a one-dimensional layout, provided Λ does not require more than $d = 1$ for finite variances, then, in the usual sense,

$$\mathrm{var}(\widehat{\boldsymbol{\tau}}) = C^+ A'(I_b \otimes \Lambda_0)AC^+\sigma^2, \tag{8}$$

where C^+ denotes the Moore–Penrose generalized inverse of C, and $\Lambda_0 = E_k \Lambda E_k$. For a nested row-column design with up to single differencing in each direction, equation (8) holds with $\Lambda_0 = (E_p \otimes E_q)\Lambda(E_p \otimes E_q)$.

Under generalized least-squares (gls) with $V = I_b \otimes \Lambda$ used and assumed correct:

$$C = A'(I_b \otimes \Lambda^*)A, \qquad \boldsymbol{q} = A'(I_b \otimes \Lambda^*)\boldsymbol{y} \quad \text{and} \quad \mathrm{var}(\widehat{\boldsymbol{\tau}}) = C^+\sigma^2, \tag{9}$$

where $\Lambda^* = \nabla'(\nabla\Lambda\nabla')^{-1}\nabla$ is the *block-corrected dependence matrix*. In this case, C is the *information matrix* for $\boldsymbol{\tau}$. If Λ is non-singular, Λ^* can also be written $\Lambda^{-1} - (\mathbf{1}_k'\Lambda^{-1}\mathbf{1}_k)^{-1}\Lambda^{-1}J_k\Lambda^{-1}$.

In one dimension, the general case of an ARIMA(p, d, q) with independent white noise (errors-in-variables) has $\Lambda^* = \nabla'(\psi H + \nabla\nabla')^{-1}\nabla$, where H is the covariance matrix of $n - d$ observations from a stationary ARMA(p, q) process, and $\psi \geqslant 0$.

Steinberg (1988) gives further equivalent models and representations. An important special case is the *Cullis–Gleeson* CG(d) model, which assumes that

$$\text{var}\{(I_b \otimes \nabla)\boldsymbol{y}\} = I_b \otimes (\psi I_{k-d} + \nabla\nabla')\sigma^2,$$

for $\psi \geqslant 0$, so that $H = I_{k-d}$. It is a special case of an ARIMA$(0, d, d)$. The extreme $\psi = 0$ corresponds to the undifferenced data being independent, and so is appropriate if there is a fixed polynomial trend of degree $d - 1$. The extreme $\psi = \infty$ corresponds to the differenced data being independent, the ARIMA$(0, d, 0)$. The CG(1) model is often known as the *linear variance* (LV) model.

Simple special cases that have been used for prior models in choosing designs are the MA(q) and the AR(p) processes. For the MA(q), V is banded (Toeplitz), and is 0 except for the central $2q + 1$ diagonals (upper-left to lower-right). For the AR(p), V^{-1} is approximately banded, and is 0 except for the central $2p + 1$ diagonals.

Stationary two-dimensional spatial models for dependence can be specified in several ways. Simple models that have been used for design purposes can essentially be specified by a sparse form for V (a *finite-lag process*, with only a finite set of non-zero correlations), or a sparse form for V^{-1} (a *conditional autoregression*, CAR, with only a finite set of non-zero inverse correlations). For most CAR processes, the precise form of the elements of V^{-1} for boundary sites is intractable, so that a boundary-corrected (non-stationary) version of the process is usually used – see, for example, Cressie (1991, § 6.6). A model that is being used for analysis is the separable AR(1) $*$ AR(1) with independent white noise (Cullis and Gleeson, 1991).

For further details on proposed methods of analysis, see the reviews of Azaïs et al. (1990), and Cressie (1991, § 5.7); the bibliography of Gill (1991); and the recent publications of Martin (1990b), Cullis and Gleeson (1991), Zimmerman and Harville (1991), Besag and Higdon (1993), Taplin and Raftery (1994), Besag et al. (1995).

3.2. *Efficiency measures and optimality*

A good design estimates as well as possible those contrasts $\{\boldsymbol{c}_j'\boldsymbol{\tau}\}$ of interest, where $\boldsymbol{c}_j'\mathbf{1}_k = 0$. Thus, the efficiencies of designs are usually compared through their values for some combination of the variances of the estimated contrasts, $\text{var}(\boldsymbol{c}_j'\widehat{\boldsymbol{\tau}})/\sigma^2 = \boldsymbol{c}_j'D^+\boldsymbol{c}_j$, where D^+ denotes $\text{var}(\widehat{\boldsymbol{\tau}})/\sigma^2$, which is given by equations (8) or (9). Often, functions of the matrix $D = (D^+)^+$ can be used, where $D = C$ for gls estimation.

When all contrasts are of equal interest, there are three commonly used measures, the A-, D- and E-values, which can be defined in various ways. They can be regarded as special cases ($p = 1, 0, \infty$ respectively) of the Φ_p-value. Assume that $\text{rank}(D) = t-1$, so that all contrasts are estimable. Let $\xi_1, \ldots, \xi_{t-1}$ be the non-zero eigenvalues of D. Then, for $0 < p < \infty$, the Φ_p-value is defined as $\{(t-1)^{-1}\sum \xi_i^{-p}\}^{1/p}$. The values for $p = 0, \infty$ are taken as the appropriate limits, $\{\prod \xi_i^{-1}\}^{1/(t-1)}$ and $\max \xi_i^{-1}$, respectively. Unless the continuity of p is important, the Φ_p-value can be taken as $\sum \xi_i^{-p}$ for $0 < p < \infty$, $\max \xi_i^{-1}$ for $p = \infty$, and $\prod \xi_i^{-1}$ or $-\sum \ln \xi_i$ for $p = 0$. A design is then *optimal* amongst all competing designs, under the Φ_p-criterion for a given $p \geqslant 0$, if it has the smallest Φ_p-value. Note that D-optimality is often defined by maximising $\prod \xi_i$ or $\{\prod \xi_i\}^{1/(t-1)}$ or $\sum \ln \xi_i$, and E-optimality can be defined by maximising $\min \xi_i$. The D-value is related to the volume of the confidence ellipsoid for $\widehat{\boldsymbol{\tau}}$ under normality. The A-value $\sum \xi_i^{-1}$ is $\mathrm{tr}(D^+)$, and is proportional to the average variance of an estimated pairwise contrast,

$$\frac{2}{t(t-1)} \sum\sum_{i<j} \mathrm{var}(\widehat{\tau}_i - \widehat{\tau}_j).$$

The E-value is the maximum variance of an estimated standardised contrast, $\max\{\mathrm{var}(\boldsymbol{c}'\widehat{\boldsymbol{\tau}})/(\boldsymbol{c}'\boldsymbol{c}\sigma^2)\} = \max(\boldsymbol{c}'D^+\boldsymbol{c}/\boldsymbol{c}'\boldsymbol{c})$.

Kiefer (1975) unified various optimality criteria under the term *universal optimality* (u.o.) This criterion requires the minimization of $\Phi(C)$ for all non-increasing, convex and orthogonally invariant Φ. It includes Φ_p-optimality for all $p \geqslant 0$. Kiefer essentially showed that, under gls estimation, the design is u.o. if the C-matrix in (9) has maximal trace and is *completely-symmetric*, that is all the diagonal elements are equal, and all the off-diagonal elements are equal (and, for C, all $t-1$ non-zero eigenvalues are equal). For cases when $D \neq C$, as when using ordinary least-squares estimation, Kiefer and Wynn (1981) introduced *weak universal optimality* (w.u.o.). This includes Φ_p-optimality for all $p \geqslant 1$, but does not include $0 \leqslant p < 1$, and hence excludes D-optimality. They showed that if $\mathrm{var}(\widehat{\boldsymbol{\tau}})$ is completely-symmetric with minimal trace then the design is w.u.o.

Extensions to Kiefer's result on universal optimality are possible – for example, Cheng (1987) considered the case of two distinct non-zero eigenvalues. Specific techniques may also be available in some cases, in particular with E-optimality. For further details on optimality criteria, see Chapter 1 of Shah and Sinha (1989).

Bounds and approximations to these Φ_p-values can be obtained. For a given $\mathrm{tr}(D)$, a simple lower bound for the Φ_p-value in the general formula is $\bar{\xi}^{-1}$, where $\bar{\xi} = (t-1)^{-1}\sum \xi_i = (t-1)^{-1}\mathrm{tr}(D)$. The bound is attained if D is completely symmetric. A global bound is obtained if the maximum, over competing designs, of $\mathrm{tr}(D)$ is used. In either case, the Φ_p-efficiency of a design can be defined by comparing the Φ_p-value to this, usually unattainable, bound.

Provided $\xi_i < 2\bar{\xi}\ \forall i$ (so that D is not too 'far' from complete-symmetry), the series expansion about $\bar{\xi}$ for the Φ_p-value can be used (Martin, 1990a). Thus, reasonable approximations are often given by

$$\sum \xi_i^{-p} \approx \bar{\xi}^{-p}\{(t-1) + p(p+1)S_{\xi\xi}/(2\bar{\xi}^2)\} \quad \text{for } p > 0$$

and not too large, and

$$\prod \xi_i^{-1} \approx \bar{\xi}^{-(t-1)} \exp\{S_{\xi\xi}/(2\bar{\xi}^2)\} \quad (p = 0,\ \text{the } D\text{-value}),$$

where $S_{\xi\xi} = \sum(\xi_i - \bar{\xi})^2 = \mathrm{tr}(D^2) - \{\mathrm{tr}(D)\}^2/(t-1)$. The approximation to the A-value is just $\mathrm{tr}(D^2)/\bar{\xi}^3 = (t-1)^3\mathrm{tr}(D^2)/\{\mathrm{tr}(D)\}^3$. A small $S_{\xi\xi}$ is associated with D being close to complete-symmetry. These approximations show that when $\mathrm{tr}(D)$ can vary over competing designs, and universally optimal designs do not exist, efficiency is a compromise between D having a high trace and being close to complete-symmetry. The weight given to the two components depends on p, with most weight going to a high trace if p is small. The conclusion that efficiency depends on a compromise between a high $\mathrm{tr}(D)$ and D close to complete-symmetry still holds when the expansion is not valid. Example 5.2 in Section 5.3 illustrates this compromise.

Note that

$$A'(I_b \otimes \Gamma)A = \sum_{i=1}^{b} A_i' \Gamma A_i,$$

where $A' = (A_1', \ldots, A_b')$ and A_i is the k by t treatment design matrix for block i. Under gls, using (9) gives $C = \sum A_i' \Lambda^* A_i$, which shows how the elements of Λ^* affect the efficiency of designs. It shows that $\mathrm{tr}(C)$ is constant for all binary designs, and is maximal over all designs if $(\Lambda^*)_{i,j} \leqslant 0 \; \forall i \neq j$. If $(\Lambda^*)_{i,j} > 0$ for some $i \neq j$, then, for example, $\mathrm{tr}(C)$ is increased for a non-binary design with a treatment occurring twice in a block on units i and j.

Under ols, if the design is variance-balanced, C in (7) is proportional to E_t, and $\mathrm{var}(\widehat{\boldsymbol{\tau}})$ in (8) is proportional to $A'(I_b \otimes E_k \Lambda E_k)A = \sum A_i' E_k \Lambda E_k A_i$. Thus the corresponding matrix to consider for comparing design efficiencies is $E_k \Lambda E_k$.

Two other cases, with different contrasts of interest, will be considered. In the *test-control* case, for the usual A_{tc}-criterion a design is optimal if it minimises the average variance of estimated pairwise contrasts with the control (0), $(t-1)^{-1} \sum_j \mathrm{var}(\widehat{\tau}_0 - \widehat{\tau}_j)$. If $\boldsymbol{\tau} = (\tau_0, \tau_1, \ldots)'$, and C is partitioned in the form

$$\begin{pmatrix} c_{ss} & \mathbf{c}_{ns}' \\ \mathbf{c}_{ns} & C_{nn} \end{pmatrix},$$

where c_{ss} is a scalar, $\mathbf{c}_{ns}$ is a $(t-1)$-vector, and C_{nn} is $(t-1)\times(t-1)$, the A_{tc}-criterion can be taken as $\mathrm{tr}(C_{nn}^{-1})$. Optimality is much harder to show in this situation. In some cases the optimal design has *supplemented balance*, that is $\mathbf{c}_{ns} = -(t-1)^{-1} c_{ss} \mathbf{1}_{t-1}$ is a constant vector, and $C_{nn} = a_0 I_{t-1} - a_1 J_{t-1}$ is completely symmetric, with $a_0 = (t-1)a_1 + (t-1)^{-1} c_{ss}$. Under ordinary least-squares and independence, this form results if an augmented design is used in which the original design is balanced. It also occurs for the more general balanced treatment incomplete block (BTIB) designs. In these cases, the test treatments are equally-replicated, but the control occurs more often. See § 7.4 of Shah and Sinha (1989) for further details of the usual case.

For *factorial* experiments, contrasts of interest are usually grouped into main effects, two-factor interactions, etc. In the case of 2-level factorial experiments, all contrasts can be represented by a vector with $n/2$ elements of $+1$, and $n/2$ elements of -1. An interaction contrast is the componentwise product of the contrasts for its constituent main effects. If the chosen t contrasts are all of equal interest, the Φ_p-value can be used as above. The mean can be included, but it is more usual for the C matrix to be only for the t contrasts. Then C will be of rank t, and the formulae for the Φ_p-value are modified accordingly.

4. Designs with neighbour balance

4.1. Definitions

There are several situations when it is important that there is some sort of neighbour balance. In the simplest case, neighbour balance means that each treatment occurs the same number of times next to every other treatment. Like pairs of treatments can also be included. All such designs will be called *neighbour* designs here. They can be regarded as an extension of a block design, using overlapping blocks of size 2.

Neighbour designs arise in (at least) three different cases. The first is in plant breeding trials, where it is desirable that each genotype can be pollinated by every available genotype, with each pair of genotypes having an equal chance of crossing. Airborne pollen may travel predominantly in one direction (or more), or all directions may be equally likely. The type of design is chosen for prior balance, rather than for efficiency under a model.

The second case is where the balance is needed to avoid possible bias arising because of interference or competition, where the treatment applied to one unit (interference) or the response on that unit (competition) may affect the response on neighbouring units. For interference, there may be a prior model including fixed neighbour effects, but the design is usually chosen on intuitive grounds (except in the case of repeated measurements designs, where design optimality has been considered). Although there has been some work on including spatial dependence in the analysis of experiments with competition, these models have not yet been used to suggest optimal designs. Again, designs are usually chosen on intuitive grounds.

The third case is when neighbour designs are used for efficiency under spatial dependence, which is usually modelled by a covariance structure. This case will be covered in more detail in Section 5, but it will be seen then that designs similar or identical to neighbour designs may arise.

Some of the neighbour designs with exact balance have received considerable attention in the Combinatorial Design literature, and the the term neighbour design is sometimes used for a particular design (the circular block design of Section 4.2.5). Two earlier reviews of neighbour designs are Street and Street (1987, Chapter 14) and Afsarinejad and Seeger (1988). See also § 5 of Street (1996). Clearly exact neighbour balance restricts the possible values of t, b and k, and approximate balance may be adequate in practice.

There are several distinctions to be made in discussing neighbour designs and neighbour balance:

(i) *Directional* or *non-directional* neighbour balance. Directional balance is needed when left to right, say, differs from right to left. It requires equal occurrences of ordered pairs of treatments. Non-directional balance is needed when left to right, say, is equivalent to right to left. It requires equal occurrences of unordered pairs of treatments. A design with directional neighbour balance automatically has non-directional balance.

(ii) *Distinct pairs* only or *Like pairs* included. The $t(t-1)$ ordered $\{$or $t(t-1)/2$ unordered$\}$ distinct pairs may only be considered. However, the t like pairs AA, etc., may also be included (self adjacencies). It may then be required that each of the t like pairs occur equally often, and as often as the unlike pairs. Note that AA is counted as 2 adjacencies between A and itself, one from left to right, and one from right to left.

(iii) *Blocked* or *single-block* design. For a blocked design, the *end design* is the design for the first and last plots of each block.

(iiia) If blocked, the blocks may be *contiguous* or *distinct*. When there is no blocking structure, the design can be regarded as having one distinct block. Blocks may be *contiguous*, when the last unit in block i is adjacent to the first unit in the block $i+1$, $i = 1, \ldots, b-1$. Otherwise, they are *disjoint*. In the designs below, contiguous blocks are denoted ( ); while disjoint blocks are denoted [ ].

(iiib) If blocked and $k \leqslant t$, the design may be *binary* or *non-binary*. If blocked and binary with $k|t$, the design may be *resolvable*, when the b blocks can be grouped into non-overlapping sets of t/k blocks each of which forms a complete replicate.

(iv) A one-dimensional design or block may be *linear* or *circular*. Units may be arranged in a linear sequence, or may form (an annulus of) a circle, where the units are sectors, with the last adjoining the first.

(iva) If linear, there may or may not be *border plots* (at both ends). If there are border plots, they are denoted { } in the designs below, and are always contiguous to the next unit.

Border plots are often included in neighbour designs to ensure balance, but may be necessary in practice to avoid edge effects, such as unshielded exposure of the edge plots to the weather. The responses on border plots are not regarded as part of the experimental results (except, where used, for adjusting responses on neighbouring plots), and a treatment occurrence on a border plot is not counted as an additional replicate. Neighbour counts involving border plots are from an interior plot to a border plot.

A neighbour-balanced circular design can be used to obtain a neighbour-balanced linearly-arranged design with border plots by cutting the circular design between two units, and adding sufficient border plots (but not all neighbour-balanced linearly-arranged designs can be obtained this way). For adjacent neighbour balance, the first border plot contains the treatment on the last unit, and the second border plot has the treatment on the first unit. For example, Figure 1 shows the first circular block of a neighbour balanced design (see Section 4.2.5).

The corresponding linear block with border plots (cut between the units with treatments 4 and 1, and read anticlockwise) is:

$$[\{4\}\ (1\ 2\ 4)\ \{1\}].$$

The first border plot gives a $(1, 4)$ adjacency, and the second gives a $(4, 1)$ adjacency.

A second method for obtaining a linear neighbour-balanced design from a circular one can also be used. In this case, extra units are added at one end of the linear sequence to keep the balance. The treatments on the extra units are those at the other end of the sequence. The responses on these units are regarded as part of the experimental results, so that an equireplicate circular design may lead to an unequally

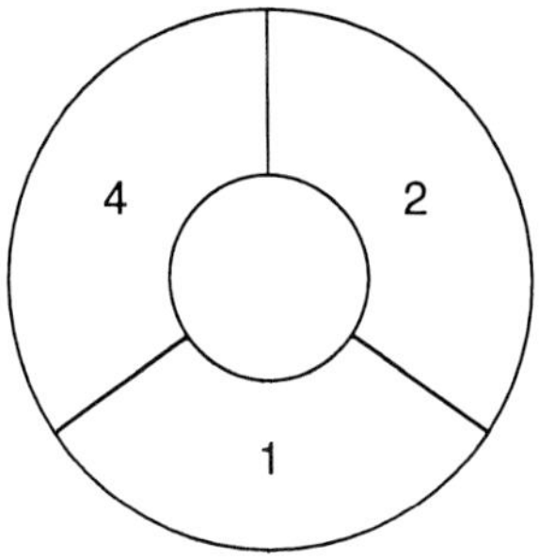

Fig. 1. A circular block.

replicated linear design. However, the end units may be downweighted, so that statistically the design is essentially equireplicate – see Sections 4.2.7, 5.1. In the example, the corresponding linear block would be:

[1 2 4 1] or [4 1 2 4].

In two dimensions, the different types of neighbour balance and design can hold or not hold in each of the two directions. If the two directions are regarded as equivalent, neighbour balance may only be required over both directions (either directed or undirected). Designs may be planar, regarded as on a cylinder if the ends of one dimension are joined, or regarded as on a torus if the ends of both dimensions are joined. Cylindrical designs can be used to obtain planar designs with border plots along 2 parallel sides (*side-bordered designs*), while toroidal designs can lead to planar designs with border plots along all sides (*fully-bordered designs*). Row and/or column blocking can be used.

More general definitions of neighbour can also be used. For example, in one-dimension, it may be convenient to also consider second-neighbours (units with one unit between them), or to pool first- and second-neighbours. In general, units are *neighbours at level* g, or *lag* g*-neighbours*, if they are g units apart, or equivalently, if there are $g - 1$ units between them. Consecutive triples of treatments may also be of interest. In two-dimensions, neighbours at lags g (horizontal) and h (vertical) can be considered. Along with adjacent neighbours, $(g, h) = (1, 0)$ or $(0, 1)$, it is often desirable to include diagonal neighbours, $(g, h) = (1, 1)$ or $(1, -1)$. Note that equivalence of the two diagonal directions is a separate assumption from equivalence of the two axial directions – either can hold without the other holding. If both these equivalences hold, the neighbours can be grouped into orders. For example, the order could increase with $g^2 + h^2$. Then, the first-order neighbours are the adjacent units, the second-order neighbours are the diagonally adjacent units, etc. The diagram below shows this order of neighbours up to 5 from the central point $\times$.

5	4	3	4	5
4	2	1	2	4
3	1	$\times$	1	3
4	2	1	2	4
5	4	3	4	5

4.2. *Neighbour designs*

4.2.1. *One-dimensional linearly-arranged distinct-block distinct-directional-pair designs*

E. J. Williams and, separately, B. R. Bugelski (see § 14.7 of Street and Street, 1987), considered *row-complete Latin squares*. A Latin square is *row-complete* if each ordered pair of distinct treatments occurs equally often (once) in adjacent positions in a row. The rows of these can be regarded as forming one-dimensional linearly-arranged distinct-block distinct-directional-pair designs for complete blocks of size t, and b a multiple of t (an even multiple if t is odd). These designs are *cross-over designs balanced for residual effects*. Extensions to other b and k require $t(t-1)|b(k-1)$, and are discussed in § 4.2 of Street (1996) and other chapters in this Handbook. An example for $t = 4$, $b = 4$ is:

[1 2 3 4]; [2 4 1 3]; [3 1 4 2]; [4 3 2 1].

4.2.2. *One-dimensional linearly-arranged distinct-block distinct-non-directional-pair (binary) designs*

These designs are the non-directional versions of the designs in Section 4.2.1, and include those in Section 4.2.1. The extension to quasi-complete Latin squares (see Section 4.3) which exist for all t, allows non-directional designs corresponding to the complete block designs in Section 4.2.1. An example for $t = 5$, $b = 5$ is:

[1 2 3 4 5]; [2 4 5 3 1]; [3 5 2 1 4]; [5 1 4 2 3]; [4 3 1 5 2].

These designs can be extended to other b and k, where it is necessary that $t(t-1)|2b(k-1)$. In particular, the designs are complete-block examples of *equi-neighboured (balanced incomplete) block designs* (EBIBD), which were introduced by Kiefer and Wynn (1981). These were considered in Combinatorial Design as *hand-cuffed designs* (introduced by Hell and Rosa, 1972). For a discussion of the existence and construction of these designs, see Street and Street (1987, § 14.6). Further constructions, and an extension to 2-class PBIBDs, are in Morgan (1988b). An example of an EBIBD with $t = 6$, $b = 15$, $k = 4$ is:

[5 2 4 3]; [6 2 4 3]; [6 1 4 3]; [6 1 5 3]; [6 1 5 2]; [6 3 5 4];
[1 3 5 4]; [1 2 5 4]; [1 2 6 4]; [1 2 6 3]; [4 1 3 2]; [5 1 3 2];
[5 6 3 2]; [5 6 4 2]; [5 6 4 1].

4.2.3. Designs of the type in Section 4.2.2 with higher-order neighbour balance
Morgan and Chakravarti (1988) gave constructions for complete block designs with first- and second-order neighbour balance, and also considered BIBDs with this balance. Ipinyomi (1986) generalized EBIBDs to *equineighboured designs*. These are binary incomplete or complete block designs, which are neighbour balanced at each level g, $g = 1, \dots, k-1$. These must have b a multiple of $t(t-1)/2$. Designs which satisfy these conditions are *semi-balanced arrays of strength* 2 (henceforth semi-balanced arrays), also known as *orthogonal arrays of type II of strength* 2 (introduced by Rao, 1961). These designs actually have a much stronger neighbour balance in that each treatment occurs equally often in each position in a block, and each unordered pair of treatments occurs equally often within the same block in each unordered pair of plot positions. They were used by Morgan and Chakravarti for the balance they needed, and implicitly by Ipinyomi. Further types of designs derived from the semi-balanced array are presented in Section 5.2. An example of an equineighboured design with $t = 5$, $b = 20$, $k = 4$, which is not a semi-balanced array, is:

[1 2 4 3]; [1 2 5 4]; [1 2 4 5]; [1 3 2 4]; [1 3 2 5]; [1 4 5 2];
[1 5 2 3]; [1 5 3 2]; [2 1 3 5]; [2 1 4 3]; [2 1 5 4]; [2 3 5 4];
[2 4 3 5]; [2 4 1 3]; [3 1 4 5]; [3 4 2 5]; [3 5 1 4]; [3 5 1 4];
[4 3 1 5]; [4 3 2 5].

An example of a semi-balanced array with t=5, b=10, k=4 is:

[1 2 3 4]; [2 3 4 5]; [3 4 5 1]; [4 5 1 2]; [5 1 2 3];
[1 3 5 2]; [2 4 1 3]; [3 5 2 4]; [4 1 3 5]; [5 2 4 1].

4.2.4. Partially neighbour balanced designs
Because the conditions for exact neighbour balance are very restrictive, Wilkinson et al. (1983) suggested *partially neighbour balanced designs* for use with a large t and small r. For a given level of neighbours, there should be no self adjacencies up to that level, and each unordered pair should occur at most once as neighbours. The latter can be generalized to the requirement that the number of distinct pairwise occurrences up to the given level should be m or $m + 1$ for some m. An example with $t = 10$, $b = 5$, $k = 4$ and level-1 partial balance is:

[1 2 3 4]; [5 6 7 8]; [7 1 0 9]; [0 4 5 2]; [6 9 8 3].

An example with $t = 10$, $b = 5$, $k = 4$ and level-3 partial balance is:

$$[1\ 2\ 3\ 4];\quad [2\ 5\ 6\ 7];\quad [3\ 7\ 8\ 9];\quad [5\ 0\ 4\ 8];\quad [6\ 1\ 9\ 0].$$

The last example still holds under any within-block permutations, since the design is a $(0, 1)$-design (all treatment concurrences are 0 or 1). Clearly, any $(0, 1)$-design has partial neighbour balance to level $k - 1$. Methods of construction, and a valid randomization, are discussed in Azaïs et al. (1993).

4.2.5. One-dimensional circular (and linear-bordered) distinct-block distinct-non-directional-pair designs

D. H. Rees considered the circular form of the designs in Section 4.2.2, called *circular block designs* (see § 14.7 of Street and Street, 1987). In Combinatorial Design, these are called *balanced cycle designs*. The blocks need not be binary. It is necessary that $(t - 1)|2r$. For a discussion of the existence and construction of these designs, and an example of a non-binary design, see Street and Street (1987, Table 14.1). An example of a binary circular block design with $t = 7$, $b = 7$, $k = 3$ is:

$$[1\ 2\ 4];\quad [2\ 3\ 5];\quad [3\ 4\ 6];\quad [4\ 5\ 7];\quad [5\ 6\ 1];\quad [6\ 7\ 2];\quad [7\ 1\ 3].$$

These designs can be adapted to become linearly-arranged neighbour balanced designs with border plots at each end of each block, by using on the left (right) border plot the treatment on the last (first) unit in the block (see Section 4.1).

4.2.6. One-dimensional circular (and linear-bordered) distinct-block distinct-directional-pair designs

The directional form of the circular block design in Section 4.2.5 (or the circular form of the designs in Section 4.2.1) is called a *balanced circuit* or *Mendelsohn* design in Combinatorial Design. It is necessary that $(t - 1)|r$. An example of a binary balanced circuit design with $t = 7$, $b = 14$, $k = 3$ is obtained by using the seven blocks of the first circular block design example ($t = 7$, $b = 7$, $k = 3$) above twice, once as they are and once with the within-block order reversed (so the eighth block is $[4\ 2\ 1]$, etc.).

Linear bordered designs which result from circular BIBD designs, have a valid randomization, and are efficient under competition models, are given in the catalogue of Azaïs et al. (1993). Some of their designs have (linear) level-2 neighbour balance also. An example for $t = 5$, $b = 5$, $k = 4$ with level-1 and level-2 balance is:

$$[\{3\}\ (1\ 2\ 4\ 3)\ \{1\}];\quad [\{4\}\ (2\ 3\ 5\ 4)\ \{2\}];\quad [\{5\}\ (3\ 4\ 1\ 5)\ \{3\}];$$
$$[\{1\}\ (4\ 5\ 2\ 1)\ \{4\}];\quad [\{2\}\ (5\ 1\ 3\ 2)\ \{5\}].$$

4.2.7. One-dimensional linearly-arranged contiguous-block non-directional designs

R. M. Williams (1952) introduced *Williams designs* and gave a method of construction using circular designs. Note that 'Williams designs' (from E. J. Williams) has also been used for row-complete designs with an extra property (a balanced end design). The designs are essentially neighbour-balanced circular designs, from which linear designs are obtained by adding an extra unit at one end which has the treatment that is on the other end (see Section 4.1). Thus, these designs are not equally replicated: $t-1$ treatments occur r times and one occurs $r+1$ times (and $n = rt+1$). However, in the gls analysis for which these designs were proposed (see Section 5.1), the extra unit has a reduced weight, so that the two end units in the linear sequence have a combined weight equal to that of any interior unit. If the responses are uncorrelated, the extra unit has zero weight. Thus, in the gls analysis the design is effectively equally-replicated.

Williams considered both the distinct-pair (his II(a) design), and the like-pair (his II(b) design) designs for r complete blocks of size t (with one extra unit at an end). For the like-pair designs, the like pairs occur as often as the distinct ones, and straddle adjacent blocks. It is necessary that $(t-1)|2r$ (distinct pairs) or $t|2r$ (like pairs). An example of a distinct-pair Williams design with $t = 4$, $b = 3$ is:

$$(1\ 2\ 3\ 4)\quad (2\ 3\ 1\ 4)\quad (3\ 1\ 4\ 2)\quad (1).$$

An example of a like-pair Williams design with $t = 3$, $b = 3$ is:

$$(1\ 2\ 3)\quad (3\ 1\ 2)\quad (2\ 3\ 1)\quad (1).$$

Generalized Williams designs, which are as above but not blocked, were considered by Kunert and Martin (1987a). Examples of generalized Williams designs are:

$$(1\ 2\ 1\ 3\ 2\ 3)\quad (1)\qquad \text{for } t = 3,\ r = 2 \text{ (distinct-pair);}$$

and

$$(1\ 1\ 2\ 2\ 1\ 3\ 3\ 2\ 3)\quad (1)\qquad \text{for } t = 3,\ r = 3 \text{ (like-pair).}$$

Williams additionally suggested designs (type III) which extend the distinct-pair Williams designs to have level-2 balance for distinct pairs also. These designs require two extra units at one end (and $n = rt+2$), although again in Williams' gls analysis these extra units combine in weight with the other two end units to give an effectively equally-replicated design. An example of a type III design for $t = 4$, $b = 3$ is:

$$(1\ 2\ 3\ 4)\quad (2\ 1\ 3\ 4)\quad (1\ 3\ 2\ 4)\quad (1\ 2).$$

Butcher (1956) considered an extension of the Williams designs to level-p balance, but the designs are unappealing, with blocks of size pt containing strings of p repeated treatments. For example, the design with $t = 3$, $p = 2$, $b = 2$ is:

$$(1\ 1\ 2\ 2\ 3\ 3)\quad (2\ 2\ 1\ 1\ 3\ 3)\quad (1\ 1).$$

4.2.8. One-dimensional linearly-arranged contiguous-block directional designs

The directional forms of the Williams designs in Section 4.2.7 are the *serially balanced sequences* due to D. J. Finney and A. D. Outhwaite. They are again essentially circular designs adapted to be linear, as in Section 4.2.7, and have $n = rt + 1$. It is necessary that $(t - 1)|r$ (distinct pairs) or t|r (like pairs). Clearly a serially balanced sequence is a Williams design of the same type. The designs were intended for use with repeated measurements when there are residual effects. For a discussion of the existence and construction of these designs, see Street and Street (1987, § 14.5). An example of a distinct-pair serially balanced sequence with $t = 5$, $b = 4$ is:

(1 2 3 4 5) (2 4 1 3 5) (4 3 2 1 5) (3 1 4 2 5) (1).

An example of a like-pair serially balanced sequence with $t = 3$, $b = 6$ is:

(1 2 3) (3 1 2) (2 3 1) (1 3 2) (2 1 3) (3 2 1) (1).

The serially balanced sequences were extended by Nair (1967) to *Nair designs* for which all $t(t - 1)(t - 2)$ distinct ordered triples occur equally often. They are again essentially circular designs. They require $(t - 1)(t - 2)|r$, and have two extra plots at one end (so that $n = rt + 2$). The designs need not be blocked. Nair gave some constructions. Examples of Nair designs for $t = 4$, $n = 24$ are:

(2 4 3 1) (4 3 2 1) (3 4 2 1) (4 2 3 1) (2 3 4 1) (3 2 4 1) (2 4)

(complete blocks);

(1 2 3 4 1 2 4 3 2 4 1 3 2 1 3 4 2 1 4 2 3 1 4 3) (1 2) (single block).

If only the triples are relevant, one extra plot can be at each end. The complete-block example becomes:

(1) (2 4 3 1) (4 3 2 1) (3 4 2 1) (4 2 3 1) (2 3 4 1)

(3 2 4 1) (2).

Dyke and Shelley (1976) sought extensions of the complete-block Nair designs in which the $t(t - 1)$ triples ABA, etc., occur equally often, and as often as the distinct triples. They put one extra unit at each end, and regard these extra units as border plots. An example with $t = 4$, $r = 9$, found by computer search, is:

{1} (2 1 3 4) (1 2 3 4) (2 4 1 3) (2 3 1 4) (1 4 3 2) (2 4 3 1)

(3 1 2 4) (2 1 4 3) (4 3 1 2) {2}.

4.3. Two-dimensional designs

All the one-dimensional designs in Section 4.2 can be extended in an obvious way to two dimensions, but few have been explicitly examined. Other types of designs can arise if the two directions are regarded as equivalent. Some of the designs that have been proposed are discussed here.

The two-dimensional extension of the row-complete Latin square is the *complete Latin square*. This is both row- and column-complete, and thus has distinct-pair directional neighbour balance in both directions. Directional row- and column-neighbour-balanced designs are examples of 'directional' rectangular lattice *polycross designs* (note that in the context of polycross designs, 'directional' has a different meaning from that used here – only some directions are relevant). See Street and Street (1987, § 14.4) for some 'directional' polycross designs, in which only some directions are important. An example of a complete Latin square with $t = 4$ is

1	2	3	4
3	1	4	2
2	4	1	3
4	3	2	1

.

The non-directional form of the complete Latin square is the *quasi-complete Latin square* (Freeman, 1981), in which each of the unordered pairs of distinct treatments are adjacent equally often (twice) in the rows and in the columns. Clearly, a complete Latin square is also quasi-complete. For a discussion of the existence and construction of complete and quasi-complete designs see Street and Street (1987, § 14.3). Bailey (1984) has shown that a valid randomization exists for Abelian group-based quasi-complete Latin squares. An example of a quasi-complete (but not complete) Latin square is any $t = 3$ Latin square, e.g.

1	2	3
3	1	2
2	3	1

.

If the two directions are equivalent, it is not necessary to have neighbour balance in each direction. A Latin square is *(nearest-) neighbour balanced* (Freeman, 1979) if each of the unordered pairs of distinct treatments are adjacent equally often (four times) over the rows and columns. Clearly, a quasi-complete Latin square is also neighbour balanced. These designs are examples of non-directional polycross designs, which can treat the first- and second-order neighbours jointly (row, column, and diagonal). Some constructions for polycross designs with like-pair diagonal neighbours are given by

Morgan (1988a, 1988b). An example of a neighbour balanced (but not quasi-complete) Latin square with $t = 4$ is

1	2	3	4
3	4	1	2
2	1	4	3
4	3	2	1

.

Neighbour-balanced nested row-column designs which are unbordered, side-bordered or fully bordered, and with or without balanced diagonal neighbours have been considered. For some constructions, see Morgan and Uddin (1991) and the references therein. A non-binary example of $t = 9$ in 9 blocks of 3 by 3 on a torus (giving a fully-bordered planar design), which has combined row and column (distinct pairs) neighbour balance and combined like-pairs diagonal neighbour balance is:

1	2	3		3	8	4		4	5	1		6	3	7		7	4	9
4	5	1		1	2	3		3	8	4		9	1	6		6	3	7
6	3	7		7	4	9		9	1	6		5	7	2		2	9	8

9	1	6		5	7	2		2	9	8		8	6	5
7	4	9		8	6	5		5	7	2		2	9	8
8	6	5		1	2	3		3	8	4		4	5	1

.

Street and Street (1987, Table 14.6) give a fully-bordered design for $t = 7$ in a (interior) 3 by 7 array with combined row and column (distinct pairs) neighbour balance, which does not come from a torus design.

Single-block torus designs that are neighbour balanced for all neighbours up to a given spatial order, were considered by Martin (1982). Fully-bordered planar designs can be obtained from these. An example which is neighbour-balanced at all orders if the two axial directions and the two diagonal directions are both equivalent is the torus $t = 5$ Knight's Move Latin square:

1	2	3	4	5
4	5	1	2	3
2	3	4	5	1
5	1	2	3	4
3	4	5	1	2

.

5. Efficient designs for spatially dependent observations

5.1. Introduction

As discussed in Section 2, spatial information was used in early experiments in the separation principle, and subsequently in choosing a blocking structure and forming blocks. More formal methods for choosing designs that are efficient under spatial dependence began in 1952, but for many years only simple situations and simple models were considered. It is only quite recently that computing power has allowed realistic models to be fitted, and that the need for efficient designs under these models has arisen.

It is often assumed that the neighbour-balanced designs of Section 4 must be efficient for spatial dependence, but the situation is usually more complicated. It will be shown in this section that it is rarely easy to obtain designs that are highly efficient under a spatial dependence model, but that the factors affecting efficiency are the positional balance of the treatments, the (non-directional) neighbour balance, and edge effects. Of these, low-order neighbour balance is often the most important, so there is indeed a link with the neighbour designs of Section 4.

In the following, the model given by equations (5) and (6) is assumed to hold after sufficient differencing.

The first investigation of design for a particular intended generalized least-squares analysis appears to be that of R. M. Williams (1952). It is instructive to discuss Williams' investigation, and subsequent work on it, in more detail.

EXAMPLE 5.1 (Designs for a linearly-arranged sequence of units under an AR(1) or AR(2)). Williams was interested in complete-block designs for linearly-arranged units in contiguous blocks when (a variant of) an AR(1) or AR(2) is assumed, and gls is used. The grouping into complete blocks, was not used in the analysis. He began by considering two systematic designs and the AR(1). His I(a) design regularly (with the same treatment order) repeated the complete block $(1\ 2\ \cdots\ t)$; whilst the I(b) (r even) regularly repeated the two complete blocks $(1\ 2\ \cdots\ t)\ (t\ \cdots\ 2\ 1)$, where the second block is the reflection of the first. He essentially noted that I(a) is poor under a linear trend, but I(b) is linear trend-free. However, his major objection to both designs was that the variances of estimated treatment differences were unequal.

The next designs he proposed, the II(a) and II(b) (see Section 4.2.7), were chosen to have equal variances of estimated treatment differences under an AR(1). For an AR(1), V^{-1} is tridiagonal, with constant elements on the diagonals, apart from the first and last diagonal elements. Williams in fact used a variant on the stationary AR(1) which gave less weight (λ^2, where $\lambda = \rho_1$) to the extra unit, so that the combined weight of the two end units $(1 + \lambda^2)$ is the same as the weight of an interior unit. Some of the possible AR(1) variants are discussed in Kunert and Martin (1987a). It is the facts that the leading off-diagonal of V^{-1} is constant, and all other off-diagonal terms are 0, that lead to the need for the first-order (and only first-order) neighbour balance. Note that equal variances of estimated treatment differences implies that the C matrix (9) is completely-symmetric (in this case $B = \mathbf{1}_n$).

The type III design (see Section 4.2.7) was chosen for the same reason for a similar variant of an AR(2). The form of V^{-1} leads to the need for the first- and second-order neighbour balance. Although he noted that the II(a) design was efficient for an AR(1) with positive λ (positive decaying correlations), and the II(b) design was efficient for a negative λ (alternating decaying correlations), Williams made no claims about optimality of the designs. Cox (1952) did conjecture optimality of the II(a) designs.

Kiefer (1961) investigated asymptotically optimal designs, which need not be in complete blocks, for a sequence under an AR(1) or AR(2). He proved that the II(a) (and the generalized II(a)) designs are asymptotically universally optimal (u.o.) for an AR(1) with positive dependence. Exact optimality under the stationary AR(1) is difficult to show, but D- and A-optimality was proved by Kunert and Martin (1987b).

Kiefer also noted that the II(b) designs are not optimal for an AR(1) with a negative λ. In this case, the asymptotically u.o. designs consist of t adjoining sets of r repeated treatments

$$(1\ 1\ \cdots\ 1)\quad (2\ 2\ \cdots\ 2)\ \ldots\ (t\ t\ \cdots\ t)$$

– a design that would never be used in practice. He showed that the Williams type II(b) (and the generalized II(b)) designs have asymptotic universal minimax optimality if the sign of λ is unknown. A design is minimax optimal under the Φ_p-criterion, say, if its largest Φ_p-value as λ varies is the smallest over all competing designs. A point not noted by Kiefer is that the II(b) designs are asymptotically u.o. complete block designs for the AR(1) with $\lambda < 0$, and so are then optimal among the class of designs considered by Williams.

Kiefer then gave a thorough (asymptotic) investigation of the case of AR(2) dependence, and showed that the Williams type III (and the generalized type III) designs are asymptotically u.o. under certain conditions (corresponding to positive, but non-monotonically decreasing, correlations). Apart from boundaries in the (λ_1, λ_2) region, there are four different types of (asymptotic) optimal design. Which is optimal does not simply depend on the signs of the AR(2) parameters λ_1, λ_2 nor on the signs of the correlations ρ_1, ρ_2 (note that Kiefer's ρ_1, ρ_2 are not correlations, but AR(2) model parameters). The above case is the only one for which a sequence of complete blocks can be optimal (Type III designs). The asymptotic u.o. designs for decreasing positive correlations have first-order neighbour balance for distinct pairs, but also have a maximal number of lag-2 like pairs. Kiefer also found the (asymptotic) universally minimax designs, which have first- and second-order like-pair neighbour balance.

Designs that have asymptotic minimax optimality for higher order autoregressions were investigated by Kiefer and Wynn (1984). As an aside on sequences of complete blocks, Martin et al. (1996) have shown that with $n = rt$, ols estimation and the LV process, any sequence of complete blocks is A-optimal.

Example 5.1 illustrates some of the difficulties in finding efficient designs under dependence, and some of the methods that can be used. In particular:

I. Designs that are intuitively appealing can be constructed, and their efficiency evaluated.

II. Designs can be sought for which the C-matrix is (close to being) completely-symmetric, and their efficiency evaluated.

III. Designs which are (weakly) universally optimal can be sought.

IV. Optimality may be very difficult to prove.

V. Optimality will usually depend on the model that is assumed for the dependence, and which estimator is used (usually ols, or gls with the assumed Λ).

VI. Optimality may depend not just on the model assumed, and whether the dependence is positive or not, but on actual values of the parameter λ.

VII. The structure of Λ_0 of (8) for ols estimation or Λ^* of (9) for gls estimation can be used to suggest what design features lead to efficiency.

VIII. If $k > t$, extended complete block designs may not be optimal; and if $k \leqslant t$, binary designs may not be optimal.

Two other points that did not arise in Williams' problem are:

IX. If there are few competing designs, a complete enumeration and evaluation may be possible.

X. Well-structured searches, or other algorithmic methods, can be used.

Some of these are related to problems with design optimality under independence, where for many $(t, b, \boldsymbol{k}, \boldsymbol{r})$ the optimal design is not known even when all contrasts are of equal interest.

In the rest of this section, some results on efficient spatial design are discussed. Section 5.2 considers the case that results in the theory of optimal design can be used to show that optimal designs exist for some (t, b, k), or that there are designs with criterion values very close to the best possible. In Section 5.3, the case that theory may be useful in showing what sort of approximate properties a design should have, and in furnishing efficiency bounds, is discussed. Sections 5.4 and 5.5 briefly consider test-treatment designs and factorial designs, respectively. For some other discussions on optimal design under dependence, see Shah and Sinha (1989, § 7.2) and Cressie (1991, § 5.6.2); and references in the bibliography of Gill (1991).

5.2. *Optimal designs*

In the following discussion, 'optimal' on its own means universally optimal for gls estimation (with the postulated Λ assumed correct) or weakly universally optimal for ols estimation (see Section 3.2). In most cases, optimal designs have only been sought for positive dependence. This is because negative dependence is unlikely in practice, and leads to (near-) optimal designs which would not be used in practice. These containing groups (runs in one dimension) of like treatments, or treatments only occurring in like pairs, e.g., 1 1 2 2 $\cdots$. Similarly, minimax optimality is only useful if there is no prior information on the parameters of the dependence structure. If n is large, asymptotic optimality can be sought (as for the Williams II(a) and III designs, and the EBIBDs – see below).

In determining the optimal design, the elements of Λ_0 of (8) are relevant for ols estimation, and those of Λ^* of (9) for gls (see Section 3.1). These matrices are symmetric, but in general are not banded (even for a stationary process). If with gls estimation an off-diagonal element of Λ^* is positive and $k \leqslant t$, $\mathrm{tr}(C)$ can be increased by using a non-binary design. In some cases of strong positive dependence, lag two like neighbours are desirable, and the optimal designs are non-binary. If $k > t$, the design must be non-binary, but the optimal design may not be an extended complete block design. With a one dimensional layout of units, the optimal extended block design may not use a sequence of complete blocks. When $k > t$, $\mathrm{tr}(C)$ is maximised if within-block replicates are positioned to take advantage of the largest positive or least negative sums of off-diagonal elements of Λ^*. If enough off-diagonal elements of Λ^* are positive, $\mathrm{tr}(C)$ can be increased by having additional within-block replication. Similar results can occur with ols estimation. In practice, it is likely that only binary incomplete designs or extended complete block designs would be used.

Apart from those arising in repeated-measurements, cases where optimal designs are known to exist include the nine cases given below. The first eight cases use the Kiefer, and Kiefer and Wynn results on optimality – see Section 3.2. Then the number of blocks must be such that it is possible for the appropriate positional and/or neighbour balance to ensure matrix D of Section 3.2 is completely-symmetric.

(1) *One-dimensional linearly-arranged contiguous-block designs under an* AR(1) *or* AR(2), *and gls estimation.* The Williams type II(a) design (Section 4.2.7) is asymptotically optimal under an AR(1) with $\lambda > 0$, and the Williams type III design (Section 4.2.7) is asymptotically optimal under an AR(2) with $\lambda_1, \lambda_2 > 0$ – see Section 5.1. See Kiefer (1961) for other cases, and minimax optimality. The optimality is exact for the circular designs with the circular stationary process. It is also exact for the variant used by Williams. In this case, only low-order neighbour balance is important. The extension to minimax optimal designs under an AR(p) process was given by Kiefer and Wynn (1984).

(2) *One-dimensional trend-free designs.* There is a relatively large literature on trend-free and nearly-trend-free (trend-resistant) designs under independence in various situations – see, for example, Bradley and Yeh (1988), and Lin and Dean (1991), and the references therein. A design is *trend-free* if the trend is orthogonal (given blocks) to the treatments, so that, with $V = I$, the C-matrix from model (1) is the same as that from model (2), i.e., $T'A = T'B(B'B)^{-1}B'A$. If a trend-free design is optimal for model (2) under independence, as for a BIBD, it is optimal under model (1) with the trend. Simple necessary conditions for a design to be trend-free can be found. For example, a block design in which each treatment occupies each plot position equally often must be trend-free under a straight line trend. Recall that having a (within block) polynomial trend of order d in model (1) in one dimension is equivalent to taking (within block) differences of order d in model (2) with $V = I$, which is equivalent to using a particular Λ^* in (5) and (6). Thus, trend-free designs are a limiting case of optimal designs under dependence when differencing is used. In particular, model (2) with the CG(d) process and $\psi = 0$ is equivalent to model (1) and independence with a polynomial trend of order d. Note that in this case, positional balance is important, and neighbour balance plays no part.

(3) *One-dimensional linearly-arranged distinct-block binary designs* ($k \leqslant t$) *with ols estimation (BIBD designs), or gls estimation.* A necessary condition for optimality among BIBD designs for ols estimation under an MA(1) was given by Kiefer and Wynn (1981). It is satisfied by designs which have non-directional neighbour balance, and a balanced end design, which requires $t(t-1)|2b$. The EBIBDs (Section 4.2.2) do not necessarily have C completely-symmetric, and so are not necessarily w.u.o. for the MA(1). However, they have efficiencies close to the theoretical optimal, and are asymptotically optimal. The extension to optimality under the MA(2) was considered by Morgan and Chakravarti (1988). They gave necessary conditions, and showed that the semi-balanced arrays (see Section 4.2.3) satisfied these.

Kunert (1987a) showed that for gls under an AR(1) with $\lambda > 0$, a neighbour-balanced design with a balanced end-design (see Section 4.1) is optimal. A neighbour-balanced design in which each treatment occurs equally often on an end unit (1 or k) will be optimal under gls estimation for the first-difference ARIMA$(0, 1, 0)$ model.

Ipinyomi (1986) proposed his equineighboured designs (see Section 4.2.3) believing they have a completely-symmetric C-matrix for ols estimation under a stationary process (which would imply they were optimal binary designs). However, this is false in general for equineighboured designs since $E_k \Lambda E_k$ is not, in general, banded. It has been shown that, provided they exist, the semi-balanced arrays (see Section 4.2.3) are optimal for all stationary processes among BIBDs for ols estimation (Kunert, 1985), and among all binary designs for gls estimation (Cheng, 1988). This follows from the pairwise positional balance of the treatments in the semi-balanced array.

Martin and Eccleston (1991) noted that most one-dimensional dependence structures that are proposed for use in design are *time-reversible* (left to right is equivalent to right to left), so that Λ and Λ^* are *centro-symmetric* (symmetric under a half-turn). They also noted that the arguments for optimality of the semi-balanced array are not affected by the dependence process being non-stationary (either because the variances differ, or because differencing is required). They generalized the equineighboured designs to *strongly equineighboured (SEN) designs*. The SEN designs include the semi-balanced arrays. They have the properties that (i) each treatment occurs equally often in each position in a block, except that position i is equivalent to position $k-i+1$, and (ii) each unordered pair of treatments occurs equally often ($m = 2b/\{t(t-1)\}$ or $2m$ times) within the same block in each unordered pair of plot positions j, j', $j \neq j'$, except that a $(k+1-j,\ k+1-j')$ plot pair is equivalent to a (j, j') plot pair. The number of times is m if $j + j' = k+1$, and $2m$ otherwise. In this case, m can be odd if k is even. The optimality properties of the semi-balanced array then hold for the SEN design if Λ and Λ^* are centro-symmetric. Some constructions of SEN designs are given by Street (1992). An example of a SEN design ($m = 1$) with $t = 4$, $b = 6$, $k = 4$, which is half of a semi-balanced array is:

$$[1\ 2\ 3\ 4];\quad [2\ 1\ 4\ 3];\quad [1\ 3\ 4\ 2];\quad [3\ 1\ 2\ 4];\quad [1\ 4\ 2\ 3];\quad [4\ 1\ 3\ 2].$$

In this example, each treatment occurs three times on an end plot (1 or 4), and three times in an interior plot (2 or 3). Each pair of treatments occurs twice as an end pair

$((1,2)$ or $(3,4))$, once as an interior pair $(2,3)$, twice two apart $((1,3)$ or $(2,4))$, and once three apart $(1,4)$.

Note that Kunert (1987b, § 3.1) showed that there may be non-BIBD binary unequally-replicated designs that are better than the semi-balanced arrays for some criteria with ols estimation.

(4) *Two-dimensional planar distinct-block binary designs* ($k \leqslant t$) *with ols estimation (BIBD designs), or gls estimation.* Extensions of the SEN designs to include blocks in which the units are arranged in a regular two-dimensional array were proposed by Martin and Eccleston (1993). Essentially, when it exists, a semi-balanced array can be used row by row (say), and the required balance properties for optimality still hold. Just as when the units are linearly arranged, other designs may be possible which use symmetries in the spatial dependence structure. The additional designs will not be precisely defined here – see Martin and Eccleston (1993) for the details. A *spatial strongly equineighboured* (SSEN) design is obtained from a SEN design by filling the block row by row (say), and is relevant when Λ^* is centro-symmetric. However, there are further symmetries in the plane. A *spatial strongly equineighboured under reflection-symmetry* (SSENR) design has the SSEN property except that a pair of units is assumed to be equivalent to the pairs which result by reflecting the block along either diagonal. It is relevant when the process has diagonal (or reflection) symmetry. A *spatial strongly equineighboured under axial-symmetry* (SSENA) design has the SSEN property except that a pair of units is assumed to be equivalent to the pairs which result by rotating the block by quarter turns. It is relevant when the process has axial symmetry. A *spatial strongly equineighboured under complete-symmetry* (SSENC) design satisfies both the SSENR and SSENA properties. This is relevant when the process has complete symmetry (both diagonal and axial symmetries). In each case, the optimality properties of the semi-balanced array carry over to the designs provided the spatial dependence structure has the corresponding symmetry.

As an example of a SSENR design, consider $t = 8$ with blocks of size 8 and plots arranged in a 4 by 2 rectangle. Then a SSENR$(8, 14, 8)$ design with 14 blocks is obtained by taking two columns from each of seven orthogonal Latin squares, for example 8 by 8 squares I to VII of Table XVI in Fisher and Yates (1963):

1 2	3 4	1 5	2 6	1 7	5 3	1 8
3 4	1 2	2 6	1 5	5 3	1 7	7 2
5 6	7 8	7 3	8 4	8 2	4 6	4 5
7 8	5 6	8 4	7 3	4 6	8 2	6 3

7 2	1 4	8 5	1 6	4 7	1 3	6 8
1 8	8 5	1 4	4 7	1 6	6 8	1 3
6 3	6 7	3 2	3 8	2 5	2 4	5 7
4 5	3 2	6 7	2 5	3 8	5 7	2 4

If the plots are ordered in the same way as the treatments in the first block, the reflection-symmetry means that the corner plots 1, 2, 7, 8 are equivalent, and the interior plots 3, 4, 5, 6 are equivalent. Also, for instance, the diagonal pairs $(1,4)$, $(2,3)$, $(5,8)$, $(6,7)$ are equivalent, but differ from the equivalent pairs $(3,6)$ and $(4,5)$. It is possible to split each block in half horizontally, and then from each consecutive pair to take one top and one bottom half to obtain a 14 block SSENR$(8,14,4)$ with blocks of size 4 arranged in a 2 by 2 square.

As an example of a SSENC design, consider $t = 5$ with blocks of size 4 and plots arranged in a 2 by 2 square. Then a SSENC$(5,5,4)$ design with 5 blocks is obtained from the first four columns of a Diagonal or Knight's Move Latin square, for example 5 by 5 squares I, II, III, and IV of Table XVI in Fisher and Yates (1963):

$$\begin{array}{|cc|}\hline 1 & 5 \\ 4 & 3 \\ \hline\end{array}\quad \begin{array}{|cc|}\hline 2 & 1 \\ 5 & 4 \\ \hline\end{array}\quad \begin{array}{|cc|}\hline 3 & 2 \\ 1 & 5 \\ \hline\end{array}\quad \begin{array}{|cc|}\hline 4 & 3 \\ 2 & 1 \\ \hline\end{array}\quad \begin{array}{|cc|}\hline 5 & 4 \\ 3 & 2 \\ \hline\end{array}\ .$$

The additional symmetries over the SSENR mean that horizontal and vertical adjacencies are equivalent. Although the necessary conditions for a SSENR are satisfied, it is not possible to obtain a SSENR from this design by rotating units in blocks.

(5) *One- or two-dimensional distinct-block non-binary designs with ols or gls estimation.* Cheng (1988) showed for the MA(1) that with either ols or gls there may be non-binary designs when $k \leqslant t$ that are universally better than a semi-balanced array. For example, consider gls estimation, $k = 3$ and $(\Lambda^*)_{13} > 0$, which holds for a stationary process if $1-2\rho_1+\rho_2 < 0$. Then, a design that has C completely-symmetric and each block of the form $[a\ b\ a]$ will be optimal. Cheng noted that such designs can be constructed by using two rows from a semi-balanced array, and repeating the first row.

Some extensions to general one- and two-dimensional dependence structures with small k were given by Martin and Eccleston (1991, 1993). For example, for a completely-symmetric stationary process, the design in which each unordered pair (a,b) occurs in a block as

$$\begin{array}{|cc|}\hline a & b \\ b & a \\ \hline\end{array}$$

is universally better than the SSEN design ($t \geqslant 4$) if $3 - 14\rho_{1,0} + 11\rho_{1,1} < 0$. Such non-binary designs can be optimal in the usual case of uncorrelated errors, fixed row and column effects, and an ols analysis (Bagchi et al., 1990).

If $k > t$, the design must be non-binary. In this case, a regular layout of the treatments in each block is not usually optimal. Two recent studies consider non-binary designs based on semi-balanced arrays. Martin (1996b) has listed the optimal one-dimensional designs for block sizes up to 6, all $t \geqslant 2$, using either ols or gls, and a range of dependence structures. Uddin and Morgan (1996) have thoroughly investigated the case of spatial blocks with two columns under gls with the AR(1) $*$ AR(1) process for models that may or may not include row and column effects.

(6) *Latin square designs with ols estimation.* Kiefer and Wynn (1981) showed that Latin squares with combined row- and column-neighbour balance (see Section 4.3) are optimal under the nearest-neighbour correlation structure (all correlations zero except for immediate neighbours, when it is ρ). Note that Kiefer and Wynn (1981) do consider the extension to $d > 2$ dimensions.

(7) *Two-dimensional single-block designs on a torus with gls estimation.* In two dimensions, the torus designs of Martin (1982), which extend the Williams II(a) and III designs (see Section 4.2.7) to two dimensions, are optimal for a conditional autoregressive process (CAR) with positive coefficients. If the process has positive non-zero coefficients as far as spatial lags g and h, then the design must be neighbour balanced to all lags as far as g and h. If any coefficients are equal, only sums of corresponding neighbour adjacencies need to be balanced. An example with $t = 3$ in a 6 by 6 array which is optimal for a two-parameter first-order CAR with positive coefficients is:

1	2	3	2	1	3
2	3	2	1	3	1
3	2	1	3	1	2
2	1	3	1	2	3
1	3	1	2	3	2
3	1	2	3	2	1

.

(8) *Two-dimensional block designs on a torus with gls estimation.* Block designs that are optimal on the torus for the completely-symmetric second-order CAR with positive parameters have been constructed by Morgan and Uddin (1991), and the efficiencies of planar versions calculated. These designs have combined row- and column-neighbour balance and combined diagonal neighbour balance (see Section 4.3). Uddin and Morgan (1991) give constructions for optimal and near-optimal planar block designs under a boundary-corrected version of the same process. These designs have extra balance requirements for edge and corner plots.

(9) *Small-order $t \times t$ designs.* For small designs, complete enumeration and comparison may be possible. Optimality results usually depend on the criterion used, and the actual parameter values. Martin (1986) considered $t \times t$ designs with $r = t$ for $t = 3, 4, 5$ using the D- and A-criteria. There are 42 spatially-distinct 3×3 designs for completely-symmetric processes. Usually the Latin square is optimal, but for some processes with gls the following design is optimal:

1	2	3
3	1	2
1	2	3

.

There are too many designs for $t > 3$ to consider all. Of the 12 spatially-distinct 4×4 Latin squares for completely-symmetric processes, the two complete ones are either optimal or very efficient under gls. Of the 192 spatially-distinct 5×5 Latin squares for completely-symmetric processes, the Knight's Move square is optimal under ols, and the seven quasi-complete ones are usually either optimal or very efficient under gls. Becher (1988) considered these designs for $t = 3, 4$ with different strengths of correlations in the two directions.

Latin squares that are resistant to a quadratic trend surface (i.e., to the linear-by-linear interaction) have been considered by Edmondson (1993). For $t \leqslant 6$, comparison with all Latin squares shows that the designs found are optimal (Martin, 1996a).

5.3. *Efficient designs*

The designs in Section 5.2 usually require a large number of replicates unless t is small. In practice, designs are usually required with much smaller numbers of replicates, and exact optimality cannot be shown. Various approaches for finding designs are possible. For a fully specified dependence structure (including all parameter values), a given optimality criterion, and a given intended analysis, an efficient design can be sought. This may be by using theoretical ideas on neighbour balance and, if the design is non-binary, a high $\mathrm{tr}(C)$ (see Section 3.2), or by an algorithmic search. A lower bound for the criterion can be used to see if acceptable efficiency has been achieved, but good lower bounds are often difficult to obtain. Examples of algorithms are Wild and Williams (1987), the use of exchange and interchange algorithms (Russell and Eccleston, 1987a, 1987b), and the use of simulated annealing algorithms (Martin and Eccleston, 1996).

Several of these approaches are used in Example 5.2.

EXAMPLE 5.2 (Four treatments in a 4×4 array under an $\mathrm{AR}(1) * \mathrm{AR}(1)$). Consider $t = 4$ in a 4 by 4 array, and suppose gls estimation is used for a separable equal-parameter $\mathrm{AR}(1) * \mathrm{AR}(1)$ process with positive correlations. This example was considered by Martin (1986). One of the 12 spatially-distinct Latin squares might be preferred because of their optimality under independence. They were compared with three other designs, D4.17 to D4.19, which have perfect positional balance. The first two were chosen to have a maximal (12) like-diagonal-neighbour count (since the corresponding elements of Λ^* are positive). The third had fewer diagonal self-adjacencies (8), but better neighbour balance. The best designs under the A-criterion for small $\lambda = \rho_{1,0}$ were the Latin squares D4.2 and D4.7 (equally). As λ increases, all three of D4.17 to D4.19 become better than the best Latin square, but D4.19 is always the best of them. This shows that good neighbour balance can compensate for a sub-optimal $\mathrm{tr}(C)$ (see the discussion in Section 3.2).

Kunert (1987, § 2.3; 1988) re-examined this case, looking at designs with perfect positional balance and first-order neighbour balance. He then sought designs with as many diagonal self-adjacencies as possible. He conjectured that one of the designs he found, D4.21 (with 7 diagonal self-adjacencies), is A-optimal for $\lambda > 0$. However, a simulated annealing algorithm (Martin and Eccleston, 1996) has found many designs (with little obvious symmetry) that are more efficient than D4.21. The best found so far is:

1	2	3	4
4	1	2	3
2	4	3	1
4	2	1	3

It has complete blocks in the rows, and 9 diagonal self-adjacencies. It does not have positional balance (e.g., 4 is in two corners), nor does it have first-order neighbour balance (the adjacencies are 4, 5, 3, 4, 5, 3). The design appears to be more efficient than D4.21 for all $\lambda > 0$. For example, it has, when $\lambda = 0.5$, a relative efficiency with respect to D4.21 of 1.039, and an efficiency of 0.934 compared with an unattainable lower bound. Another design found which is almost as good has positional balance and 10 diagonal self-adjacencies, but more unequal neighbour adjacencies (5, 5, 2, 2, 5, 5).

The above approaches can be unsatisfactory because, in many cases, the design obtained may be very specific to the assumptions made. Even a seemingly small change in parameter values may mean the chosen design is less efficient than another. This suggests another approach, which is to seek designs that are not necessarily optimal under any assumptions, but perform well under a range of likely conditions.

This approach was followed by Martin et al. (1993) for agricultural variety trials which use block designs in which the plots are arranged linearly, or, if arranged spatially, the predominant dependence is in one direction. There is empirical evidence from Australia that the LV (or CG(1) – see Section 3.1) model is acceptable for a large proportion of field experiments, with the parameter ψ usually between 0 and 5.

The approach was therefore to assume that the LV model will usually be reasonable, with ψ in the range $[0, 5]$, and to seek designs that are efficient in this range under gls estimation, but also efficient under some reasonable alternative models. These included the CG(2) and low-order stationary ARMA models. Looking at the form of the Λ^* matrix for the LV model and other models suggested some general principles:

(a) The design should be binary. Both the end design (plots 1 and k) and the interior design (plots 2 to $k - 1$) should approximately contain each treatment equally often.

(b) If possible, the design should have the low order (usually to at most lag 3) neighbour adjacency matrices N_g close to complete symmetry, and the end to next-to-end neighbours should also be as balanced as possible. If r is too small for this to be possible, try to get a weighted combination of the N_g, where the weights depend on the likely size of ψ, close to complete symmetry.

Examples of designs that are efficient and robust are, for $t = 6$, $k = 6$, $b = 3$:

[1 2 3 4 5 6]; [3 6 2 5 1 4]; [2 4 6 1 3 5];

and for $t = 10$, $k = 4$, $b = 5$:

[1 2 3 4]; [2 5 6 7]; [3 7 9 8]; [0 4 8 5]; [6 0 1 9].

5.4. *Treatment-control (test-control) designs*

The optimal design theory for this case is considerably more complicated than when the treatments are of equal status, even with independent observations. It has received little attention so far for dependent observations. Martin and Eccleston (1993) briefly compared some designs using their SEN designs. A more thorough investigation was carried out by Cutler (1993).

Cutler looked mainly at one-dimensional circular designs, and an assumed gls analysis under an AR(1). Exact A_{tc}-optimality (see Section 3.2) results are hard to obtain, but Cutler gave conditions under which a (circular) neighbour-balanced BTIB design (see Section 3.2) is optimal for some λ. Neighbour balance here means that the test treatments are neighbour balanced, and that, separately, the control treatment is neighbour balanced with the test treatments. Note that the optimal design may depend on the value of λ. Cutler briefly considered linearly-arranged designs with the AR(1). He proposed augmented neighbour balanced incomplete block designs as building blocks which may lead to efficient designs. These will have the C-matrix in the desired supplemented balance form (see Section 3.2), and can be constructed from semi-balanced arrays in the test treatments by inserting the control between fixed units in each block. If the control occurs more than once in each block, the number of adjacent occurrences should be minimized. Designs constructed this way will have the much stronger neighbour balance needed for the C-matrix to have the supplemented balance form for any dependence structure. However, the optimal design of this type still needs to be found.

Cutler's Example 5.1 compared two designs of this type with $t = 4$, $k = 3$, $b = 6$ (0 is the control):

D1: [1 0 2]; [2 0 3]; [3 0 1]; [0 1 2]; [0 2 3]; [0 3 1];
D2: [1 0 2]; [2 0 3]; [3 0 1]; [1 0 2]; [2 0 3]; [3 0 1].

D1 and D2 are equivalent and A_{tc}-optimal under independence. Cutler claims that D1 is A_{tc}-optimal for small $\lambda = \rho_1$ ($\leqslant 0.23$), while D2 is optimal for large λ ($\geqslant 0.72$). In fact, these are the wrong way round (and Cutler's paragraph above his Example 5.1 is in error). For stationary processes, D1 is better than D2 if $1 - 4\rho_1 + 3\rho_2 < 0$, which for the AR(1) gives $\lambda > 1/3$. Of all designs formed from D2 by within-block permutation, it appears that D2 is A_{tc}-optimal for $\lambda \leqslant 0.3185$, while D1 is optimal for $\lambda \geqslant 0.3652$. Between these values, the A_{tc}-optimal design appears to be the design obtained from D1 by replacing the fourth block by [1 0 2] (Martin and Eccleston, 1996).

Martin and Eccleston (1993) essentially looked at the spatial version of Cutler's linear designs which use semi-balanced arrays (they use a SSEN design, or a reduced form of Section 5.2, for the test treatments). As an example, consider a 3 by 3 square with $t = 6$, $b = 5$, and a completely-symmetric dependence structure. There are two ways of separating the controls in each block. Using a cross for the test treatments gives:

$$\begin{array}{|ccc|}\hline 0&1&0\\ 2&3&4\\ 0&5&0\\ \hline\end{array}\quad \begin{array}{|ccc|}\hline 0&2&0\\ 3&4&5\\ 0&1&0\\ \hline\end{array}\quad \begin{array}{|ccc|}\hline 0&3&0\\ 4&5&1\\ 0&2&0\\ \hline\end{array}\quad \begin{array}{|ccc|}\hline 0&4&0\\ 5&1&2\\ 0&3&0\\ \hline\end{array}\quad \begin{array}{|ccc|}\hline 0&5&0\\ 1&2&3\\ 0&4&0\\ \hline\end{array}\,.$$

Taking a diagonal cross for the test treatments gives:

$$\begin{array}{|ccc|}\hline 1&0&2\\ 0&3&0\\ 4&0&5\\ \hline\end{array}\quad \begin{array}{|ccc|}\hline 2&0&3\\ 0&4&0\\ 5&0&1\\ \hline\end{array}\quad \begin{array}{|ccc|}\hline 3&0&4\\ 0&5&0\\ 1&0&2\\ \hline\end{array}\quad \begin{array}{|ccc|}\hline 4&0&5\\ 0&1&0\\ 2&0&3\\ \hline\end{array}\quad \begin{array}{|ccc|}\hline 5&0&1\\ 0&2&0\\ 3&0&4\\ \hline\end{array}\,.$$

If the dependence between horizontal and vertical neighbours is largest and positive, then usually the first design is better for comparing the test treatments. It appears that the second design is A_{tc}-better if the dependence is not strong, but that the first is better for stronger dependence (this corrects the remark in Martin and Eccleston, 1993). The change-over occurs at $\rho_{1,0} \approx 0.253$ for the $AR(1)*AR(1)$. In this example, the control could be replaced by four different controls, one in each position. Each comparison between a control and the new treatments would be equally accurate, and comparisons between test treatments would be as accurate as before. The comparisons between the controls would not be equally accurate, but would probably be of little interest.

5.5. *Factorial designs*

Run orders for factorial designs have previously been considered to minimise the cost of changing levels, and, more recently, for the design to be trend-free. Only very recently have run orders been considered which are efficient under dependence. Cheng and Steinberg (1991) considered the run order of two-level main-effects single-block factorial experiments under one-dimensional dependence. Their reverse-foldover algorithm leads to run orders with a maximal number of level changes. These designs are usually very efficient under positive dependence, but the individual numbers of sign changes can vary markedly. For example, the minimum aberration resolution IV 2^{7-2} reverse foldover design has 183 level changes, distributed as 31, 30, 29, 28, 27, 23, 15. In some cases, designs with the same total number of level changes, but with them spread more equally, can be found. These will be only marginally more efficient under the D- and A-criteria, but may be much more desirable in practice.

As examples, consider a 2^4 design. The maximal number of sign changes is 53. The reverse-foldover algorithm gives

$$1,\ abcd,\ a,\ bcd,\ ab,\ cd,\ b,\ acd,\ bc,\ ad,\ abc,\ d,\ ac,\ bd,\ c,\ abd;$$

with individual sign changes of 15, 14, 13, 11 whilst the following design has individual sign changes of 15, 13, 13, 12 and is slightly more efficient – see Table 1 of Cheng and Steinberg (1991):

$$1,\ abcd,\ a,\ bcd,\ ab,\ cd,\ abc,\ d,\ ac,\ bd,\ c,\ acd,\ bc,\ ad,\ b,\ acd.$$

Saunders et al. (1995) show that there is no design with individual sign changes of 14, 13, 13, 13, but that there are two designs with individual sign changes of 14, 14, 13, 12. These are:

$$1,\ abcd,\ a,\ bcd,\ ab,\ cd,\ abc,\ d,\ ac,\ bd,\ acd,\ b,\ ad,\ bc,\ acd,\ c;$$

$$1,\ abcd,\ a,\ bcd,\ ad,\ bc,\ acd,\ b,\ cd,\ ab,\ d,\ abc,\ bd,\ ac,\ abd,\ c.$$

In the multi-level case with qualitative factors, and interest in main effects (all contrasts of equal interest), the principles of Section 5.3 hold for each factor. Efficient orthogonal run orders which are robust under dependence are those in which each factor is in complete blocks, has good neighbour balance, and has as few self-adjacencies as possible. An example of such a run order for a 5^2 design (Martin et al., 1995) is:

$$1,\ ab,\ a^2b^2,\ a^3b^3,\ a^4b^4,\ a^2b^3,\ b^4,\ a^4b,\ ab^2,\ a^3,\ ab^4,\ a^3b^2,\ b^3,\ a^4,\ a^2b,$$
$$a^4b^2,\ a^2,\ a^3b^4,\ b,\ ab^3,\ a^3b,\ a^4b^3,\ a,\ a^2b^4,\ b^2.$$

Acknowledgement

I would like to thank all those who have helped and stimulated me over the years on experimental design and analysis. I would particularly like to thank Dr. B. J. N. Blight, who introduced me to the problem of design with dependent errors, and Professors R. A. Bailey, J. A. Eccleston, J. Kunert for their help and support.

References

Afsarinejad, K. and P. Seeger (1988). Nearest neighbour designs. In: Y. Dodge, V. V. Fedorov and H. P. Wynn, eds., *Optimal Design and Analysis of Experiments*. North-Holland, Amsterdam, 99–113.

Azaïs, J. M., R. A. Bailey and H. Monod (1993). A catalogue of efficient neighbour-designs with border plots. *Biometrics* **49**, 1252–1261.

Azaïs, J. M., J.-B. Denis, T. D. Horne and A. Kobilinsky (1990). Neighbour analysis of plot experiments: A review of the different approaches. *Biom. Prax.* **30**, 15–39.

Bagchi, S., A. C. Mukhopadhay and B. K. Sinha (1990). A search for optimal nested row–column designs. *Sankhyā Ser. B* **52**, 93–104.

Bailey, R. A. (1984). Quasi-complete Latin squares: Construction and randomization. *J. Roy. Statist. Soc. Ser. B* **46**, 323–334.

Bailey, R. A. (1986). Randomization, constrained. In: S. Kotz and N. L. Johnson, eds., *Encyclopedia of Statistical Sciences*, Vol. 7. Wiley, New York, 524–530.

Bailey, R. A. (1987). Restricted randomization: A practical example. *J. Amer. Statist. Assoc.* **82**, 712–719.

Bailey, R. A., J. Kunert and R. J. Martin (1990). Some comments on gerechte designs. I. Analysis for uncorrelated errors. *J. Agron. Crop Sci.* **165**, 121–130.

Bailey, R. A, J. Kunert and R. J. Martin (1991). Some comments on gerechte designs. II. Randomization analysis, and other methods that allow for inter-plot dependence. *J. Agron. Crop Sci.* **166**, 101–111.

Bailey, R. A. and C. A. Rowley (1987). Valid randomization. *Proc. Roy. Soc. Lond. Ser. A* **410**, 105–124.

Bartlett, M. S. (1978). Nearest neighbour models in the analysis of field experiments (with discussion). *J. Roy. Statist. Soc. Ser. B* **40**, 147–174.

Becher, H. (1988). On optimal experimental design under spatial correlation structures for square and nonsquare plot designs. *Comm. Statist. Simulation Comput.* **17**, 771–780.

Besag, J., P. Green, D. Higdon, C. Kooperberg and K. Mengersen (1995). Spatial statistics, image analysis and Bayesian inference (with discussion). *Statist. Sci.* **10**, 3–66.

Besag, J. and D. Higdon (1993). Bayesian inference for agricultural field experiments. *Bull. Int. Statist. Inst.* **55**(1), 121–137.

Besag, J. and R. Kempton (1986). Statistical analysis of field experiments using neighbouring plots. *Biometrics* **42**, 231–251.

Bradley, R. A. and C.-M. Yeh (1988). Trend-free block designs. In: S. Kotz and N. L. Johnson, eds., *Encyclopedia of Statistical Sciences*, Vol. 9, Wiley, New York, 324–328.

Butcher, J. C. (1956). Treatment variances for experimental designs with serially correlated observations. *Biometrika* **43**, 208–212.

Cheng, C.-S. (1987). An optimization problem with applications to optimal design theory. *Ann. Statist.* **15**, 712–723.

Cheng, C.-S. (1988). A note on the optimality of semibalanced arrays. In: Y. Dodge, V. V. Fedorov and H. P. Wynn, eds., *Optimal Design and Analysis of Experiments*. North-Holland, Amsterdam, 115–122.

Cheng, C.-S. and D. M. Steinberg (1991). Trend robust two-level factorial designs. *Biometrika* **78**, 325–336.

Cochran, W. G. (1976). Early development of techniques in comparative experimentation. In: D. B. Owen, ed., *On the History of Statistics and Probability*. Dekker, New York, 1–26.

Correll, R. L. and R. B. Anderson (1983). Removal of intervarietal competition effects in forestry varietal trials. *Silvae Genetica* **32**, 162–165.

Cox, D. R. (1951). Some systematic experimental designs. *Biometrika* **38**, 312–323.

Cox, D. R. (1952). Some recent work on systematic experimental designs. *J. Roy. Statist. Soc. Ser. B* **14**, 211–219.

Cox, G. M. (1950). A survey of types of experimental designs. *Biometrics* **6**, 305–306. Discussion 317–318.

Cressie, N. A. C. (1991). *Statistics for Spatial Data*. Wiley, New York.

Cullis, B. R. and A. C. Gleeson (1991). Spatial analysis of field experiments – an extension to two dimensions. *Biometrics* **47**, 1449–1460.

Cullis, B. R., W. J. Lill, J. A. Fisher, B. J. Read and A. C. Gleeson (1989). A new procedure for the analysis of early generation variety trials. *Appl. Statist.* **38**, 361–375.

Cutler, D. R. (1993). Efficient block designs for comparing test treatments to a control when the errors are correlated. *J. Statist. Plann. Inference* **36**, 107–125 (and **37** (1993), 393–412).

Dagnelie, P. (1987). La méthode de Papadakis en expérimentation agronomique: Considérations historiques et bibliographiques. *Biom. Prax.* **27**, 49–64.

Dyke, G. V. and C. F. Shelley (1976). Serial designs balanced for effects of neighbours on both sides. *J. Agric. Sci.* **87**, 303–305.

Edmondson, R. N. (1993). Systematic row-and-column designs balanced for low order polynomial interactions between rows and columns. *J. Roy. Statist. Soc. Ser. B* **55**, 707–723.

Federer, W. T. (1955). *Experimental Design – Theory and Applications*. Macmillan, New York.

Federer, W. T. and C. S. Schlottfeldt (1954). The use of covariance to control gradients in experiments. *Biometrics* **10**, 282–290.

Fedorov, V. V. (1996). Design of spatial experiments. In: S. Ghosh and C. R. Rao, eds., *Handbook of Statistics*, Vol. 13, Design and Analysis of Experiments. North-Holland, Amsterdam, 515–553.

Fisher, R. A. (1932). *Statistical Methods for Research Workers*, 4th edn. Oliver and Boyd, Edinburgh.

Fisher, R. A. (1950). *Statistical Methods for Research Workers*, 11th edn. Oliver and Boyd, Edinburgh.

Fisher, R. A. and F. Yates (1963). *Statistical Tables for Biological, Agricultural and Medical Research*, 6th edn. Oliver and Boyd, Edinburgh.

Freeman, G. H. (1979). Complete Latin squares and related experimental designs. *J. Roy. Statist. Soc. Ser. B* **41**, 253–262.
Freeman, G. H. (1981). Further results on quasi-complete Latin squares. *J. Roy. Statist. Soc. Ser. B* **43**, 314–320.
Freeman, G. H. (1988). Systematic designs. In: S. Kotz and N. L. Johnson, eds., *Encyclopedia of Statistical Sciences*, Vol. 8. Wiley, New York, 143–147.
Gill, P. S. (1991). A bibliography of nearest neighbour methods in design and analysis. *Biom. J.* **4**, 455–459.
Gleeson, A. C. and B. R. Cullis (1987). Residual maximum likelihood (REML) estimation of a neighbour model for field experiments. *Biometrics* **43**, 277–287.
Hell, P. and A. Rosa (1972). Graph decomposition, handcuffed prisoners and balanced P-designs. *Discrete Math.* **2**, 229–252.
Ipinyomi, R. A. (1986). Equineighboured experimental designs. *Austral. J. Statist.* **28**, 79–88.
Kempthorne, O. (1986). Randomization-II. In: S. Kotz and N. L. Johnson, eds., *Encyclopedia of Statistical Sciences*, Vol. 7. Wiley, New York, 519–524.
Kiefer, J. (1961). Optimum experimental designs, V, with applications to systematic and rotatable designs. In: J. Neyman, ed., *Proc. 4th Berkeley Symposium*, Vol. 1. Univ. of California Press, Berkeley, CA, 381–405.
Kiefer, J. (1975). Construction and optimality of generalized Youden designs. In: J. N. Srivastava, ed., *A Survey of Statistical Design and Linear Models*. North-Holland, Amsterdam, 333–353.
Kiefer, J. and H. P. Wynn (1981). Optimum balanced block and Latin square designs for correlated observations. *Ann. Statist.* **9**, 737–757.
Kiefer, J. and H. P. Wynn (1984). Optimum and minimax exact treatment designs for one-dimensional autoregressive processes. *Ann. Statist.* **12**, 431–450.
Kunert, J. (1985). Optimal repeated measurements designs for correlated observations and analysis by weighted least squares. *Biometrika* **72**, 375–389.
Kunert, J. (1987a). Neighbour balanced block designs for correlated errors. *Biometrika* **74**, 717–724.
Kunert, J. (1987b). Recent results on optimal designs for correlated observations. Arbeitsberichte, Universität Trier.
Kunert, J. (1988). Considerations on optimal design for correlations in the plane. In: Y. Dodge, V. V. Fedorov and H. P. Wynn, eds., *Optimal Design and Analysis of Experiments*. North-Holland, Amsterdam, 123–131.
Kunert, J. and R. J. Martin (1987a). Some results on optimal design under a first-order autoregression and on finite Williams' type II designs. *Comm. Statist. Theory Methods* **16**, 1901–1922.
Kunert, J. and R. J. Martin (1987b). On the optimality of finite Williams II(a) designs. *Ann. Statist.* **15**, 1604–1628.
Lin, M. and A. M. Dean (1991). Trend-free block designs for varietal and factorial experiments. *Ann. Statist.* **19**, 1582–1596.
Martin, R. J. (1982). Some aspects of experimental design and analysis when errors are correlated. *Biometrika* **69**, 597–612.
Martin, R. J. (1985). Papadakis method. In: S. Kotz and N. L. Johnson, eds., *Encyclopedia of Statistical Sciences*, Vol. 6. Wiley, New York, 564–568.
Martin, R. J. (1986). On the design of experiments under spatial correlation. *Biometrika* **73**, 247–277. Correction (1988) **75**, 396.
Martin, R. J. (1990a). Some results on the Φ_p-value when errors are correlated. *Comput. Statist. Data Anal.* **9**, 113–121.
Martin, R. J. (1990b). The use of time series models and methods in the analysis of agricultural field trials. *Comm. Statist. Theory Methods* **19**, 55–81.
Martin, R. J. (1996a). Low-order spatially distinct Latin squares. *Comm. Statist. Theory Methods*, to appear.
Martin, R. J. (1996b). Optimal small-sized block designs under dependence. Preprint.
Martin, R. J. and J. A. Eccleston (1991). Optimal incomplete block designs for general dependence structures. *J. Statist. Plann. Inference* **28**, 67–81.
Martin, R. J. and J. A. Eccleston (1993). Incomplete block designs with spatial layouts when observations are dependent. *J. Statist. Plann. Inference* **35**, 77–91.
Martin, R. J. and J. A. Eccleston (1996). Construction of optimal and near-optimal designs for dependent observations using simulated annealing. Preprint.

Martin, R. J., J. A. Eccleston and A. C. Gleeson (1993). Robust designs when observations within a block are correlated. *J. Statist. Plann. Inference* **34**, 433–450.

Martin, R. J., J. A. Eccleston and G. Jones (1995). Some results on multi-level factorial designs with dependent observations. Preprint.

Morgan, J. P. (1988a). Terrace constructions for bordered, two-dimensional neighbor designs. *Ars Combinatoria* **26**, 123–140.

Morgan, J. P. (1988b). Balanced polycross designs. *J. Roy. Statist. Soc. Ser. B* **50**, 93–104.

Morgan, J. P. and I. M. Chakravarti (1988). Block designs for first and second order neighbor correlations. *Ann. Statist.* **16**, 1206–1224.

Morgan, J. P. and N. Uddin (1991). Two-dimensional design for correlated errors. *Ann. Statist.* **19**, 2160–2182.

Morris, M. D. and T. J. Mitchell (1995). Exploratory designs for computational experiments. *J. Statist. Plann. Inference* **43**, 381–402.

Nair, C. R. (1967). Sequences balanced for pairs of residual effects. *J. Amer. Statist. Assoc.* **62**, 205–225.

Pearce, S. C. (1983). *The Agricultural Field Experiment.* Wiley, Chichester.

Pearson, E. S. (1938). “Student” as statistician. *Biometrika* **30**, 210–250.

Pithuncharurnlap, M., K. E. Basford and W. T. Federer (1993). Neighbour analysis with adjustment for interplot competition. *Austral. J. Statist.* **35**, 263–270.

Rao, C. R. (1961). Combinatorial arrangements analogous to orthogonal arrays. *Sankhyā A* **23**, 283–286.

Reddy, M. N. and C. K. R. Chetty (1982). Effect of plot shape on variability in Smith's variance law. *Experimental Agriculture* **18**, 333–338.

Russell, K. G. and J. Eccleston (1987a). The construction of optimal balanced incomplete block designs when adjacent observations are correlated. *Austral. J. Statist.* **29**, 84–90.

Russell, K. G. and J. Eccleston (1987b). The construction of optimal incomplete block designs when observations within a block are correlated. *Austral. J. Statist.* **29**, 293–302.

Saunders, I. W., J. A. Eccleston and R. J. Martin (1995). An algorithm for the design of 2^p factorial experiments on continuous processes. *Austral. J. Statist.* **37**, 353–365.

Shah, K. R. and B. K. Sinha (1989). *Theory of Optimal Designs.* Springer, New York.

Steinberg, D. M. (1988). Factorial experiments with time trends. *Technometrics* **30**, 259–269.

Street, A. P. and D. J. Street (1987). *Combinatorics of Experimental Design.* Clarendon Press, Oxford.

Street, D. J. (1992). A note on strongly equineighboured designs. *J. Statist. Plan. Inference* **30**, 99–105.

Street, D. J. (1996). Block and other designs used in agriculture. In: S. Ghosh and C. R. Rao, eds., *Handbook of Statistics*, Vol. 13. Design and Analysis of Experiments. North-Holland, Amsterdam, 759–808.

Taplin, R. and A. E. Raftery (1994). Analysis of agricultural field trials in the presence of outliers and fertility jumps. *Biometrics* **50**, 764–781.

Uddin, N. and J. P. Morgan (1991). Optimal and near optimal sets of Latin squares for correlated errors. *J. Statist. Plann. Inference* **29**, 279–290.

Uddin, N. and J. P. Morgan (1996). Universally optimal two-dimensional block designs for correlated observations. Preprint.

Wild, P. R. and E. R. Williams (1987). The construction of neighbour designs. *Biometrika* **74**, 871–876.

Wilkinson, G. N., S. R. Eckert, T. W. Hancock and O. Mayo (1983). Nearest neighbour (NN) analysis of field experiments (with discussion). *J. Roy. Statist. Soc. Ser. B* **45**, 151–211.

Williams, R. M. (1952). Experimental designs for serially correlated observations. *Biometrika* **39**, 151–167.

Yates, F. (1948). Contribution to the discussion of a paper by F. J. Anscombe. *J. Roy. Statist. Soc. Ser. A* **111**, 204–205.

Yates, F. (1967). A fresh look at the basic principles of the design and analysis of experiments. In: L. M. Le Cam and J. Neyman, eds., *Proc. 5th Berkeley Symposium*, Vol. 4. Univ. of California Press, Berkeley, CA, 777–790.

Zimmerman, D. L. and D. A. Harville (1991). A random field approach to the analysis of field-plot experiments and other spatial experiments. *Biometrics* **47**, 223–239.

S. Ghosh and C. R. Rao, eds., *Handbook of Statistics, Vol. 13*
1996 Elsevier Science B.V.

16

Design of Spatial Experiments: Model Fitting and Prediction*

Valerii Fedorov[1]

1. Introduction

Since the earliest days of the experimental design theory, a number of concepts like split plots, strips, blocks, Latin squares, etc. (see Fisher, 1947), were strongly related to experiments with spatially distributed or allocated treatments and observations. In this survey we confine ourselves to what can be considered as an intersection of ideas developed in the areas of response surface design of experiments and spatial statistics.

The results which we are going to consider are also related to the results developed by Cambanis (1985), Cambanis and Su (1993), Matern (1986), Micchelli and Wahba (1981), Sacks and Ylvisaker (1966, 1968, 1970) and Ylvisaker (1975, 1987). What differs in the approach of this paper from those cited? We intend to use the techniques which are based on the concept of regression models while the cited studies are based on the ideas developed in the theory of stochastic processes and the theory of integral approximation.

If this survey were to be written for a very applied audience, the title "optimal allocations of sensors" or "optimal allocation of observing stations" could be more appropriate. Environmental monitoring, meteorology, surveillance, some industrial experiments and seismology are the most typical areas in which the considered results may be applied. What are the most common features of the experiments to be discussed?

1. There are variables $x \in X \subset R^k$, which can be controlled. Usually $k = 2$, and in the observing station problem, x_1 and x_2 are coordinates of stations and X is a region where those stations may be allocated.

*Research sponsored by the Applied Mathematical Sciences Research Program Office of Energy Research, U.S. Department of Energy under contract number DE-AC05-96OR22464 with Lockheed Martin Energy Systems Corporation.

[1]This submitted manuscript has been authored by a contractor of the U.S. Government under contract No. DE-AC05-96OR22464. Accordingly, the U.S. Government retains a nonexclusive, royalty-free license to publish or reproduce the published form of this contribution, or allow others to do so, for U.S. Government purposes.

2. There exists a model describing the observed response(s) or dependent variable(s) y. More specifically y and x are linked together by a model, which may contain some stochastic components.

3. An experimenter or a practitioner can formulate the quantitative objective function.

4. Once a station or a sensor is allocated a response y can be observed either continuously or according to any given time schedule without any additional significant expense.

5. Observations made at different sites may be correlated.

Assumptions 1–5 are very loosely formulated and they will be justified when needed. In the subsequent sections the term "sensor" stands for what could be an observing station, meteorological station, radiosonde or well in the particular applied problem.

2. Standard design problem

In what follows we will mostly refer to experiments which are typical in environmental monitoring setting as a background for the exposition of the main results. We hope that the reader will be able to apply the ideas and techniques to other types of experiments.

When Assumptions 4 and 5 are not considered we have what will be be called, the "standard design problem". The problem was extensively discussed (see for instance, Atkinson and Donev, 1992; Fedorov, 1972; Pazman, 1986; Pukelsheim, 1994; Silvey, 1980), and it is difficult to add anything new in this area of experimental design theory. Theorem 1 which follows, is a generalized version of the Kiefer–Wolfowitz equivalence theorem (see Kiefer, 1959) and stated here for the reader's convenience. It also serves as an opportunity to introduce the notation, which is sometimes different from that used in other articles of this volume.

Let

$$y_{ij} = \eta(x_i, \theta) + \varepsilon_{ij}, \quad i = 1, \ldots, n, \; j = 1, \ldots, r_i, \; \sum r_i = N, \qquad (2.1)$$
$$\eta(x, \theta) = \theta^T f(x),$$

where $\theta \in R^m$ are unknown parameters, $f^T(x) = (f_1(x), \ldots, f_m(x))$ are given functions, supporting points x_i are chosen from some set X, and the ε_{ij} are uncorrelated random errors with zero means and variances equal to one. We do not make distinctions in notation for random variables and their realizations when it is not confusing.

For the best linear unbiased estimator of unknown parameters the accumulated "precision" is described by the information matrix:

$$M(\xi) = \sum p_i f(x_i) f^T(x_i), \quad p_i = r_i/N, \qquad (2.2)$$

which is completely defined by the design $\xi = \{x_i, p_i\}_i^n$. In the context of the standard design theory

$$M(\xi) = \int f(x) f^T(x) \xi(\mathrm{d}x) = \int m(x) \xi(\mathrm{d}x), \tag{2.3}$$

where $\xi(\mathrm{d}x)$ is a probability measure with the supporting set belonging to X: $\operatorname{supp} \xi \subset X$, and

$$m(x) = f(x) f^T(x)$$

is the information matrix of an observation made at point x.

Regression model (2.1) and the subsequent comments do satisfy Assumptions 1 and 2 from the previous section. To be consistent with Assumption 3 let us introduce a function $\Psi(M)$, which is called the "criterion of optimality" in experimental design literature. A design

$$\xi^* = \arg\min_{\xi} \Psi\left[M(\xi)\right], \qquad \int \xi(\mathrm{d}x) = 1, \tag{2.4}$$

is called (Ψ-) optimal.

Minimization must be over the set of all possible probability measures Ξ with supporting sets belonging to X. Now let us assume that:

(a) X is compact;
(b) $f(x)$ are continuous functions in X, $f \in R^m$;
(c) $\Psi(M)$ is a convex function and

$$\Psi(M) \leqslant \Psi(M + \Delta), \quad M \geqslant 0, \ \Delta \geqslant 0,$$

i.e., matrices M and Δ are nonnegative definite.
(d) there exists a real number q such that

$$\{\xi\colon \Psi[M(\xi)] \leqslant q < \infty\} = \Xi(q) \neq \emptyset;$$

(e) for any $\xi \in \Xi(q)$ and $\bar{\xi} \in \Xi(q)$:

$$\Psi\left[(1-\alpha)M(\xi) + \alpha M(\bar{\xi})\right] = \Psi\left[M(\xi)\right] + \alpha \int \psi(x,\xi)\bar{\xi}(\mathrm{d}x) + \tau(\alpha,\xi,\bar{\xi}),$$

where $\tau(\alpha,\xi,\bar{\xi}) = \mathrm{o}(\alpha)$.

Here and in what follows we use $\Psi(\xi)$ for $\Psi[M(\xi)], \Psi^*$ for $\Psi(\xi^*)$ and $\min_x, \min_\xi, \int$, and so on, instead of $\min_{x \in X}$, $\min_{\xi \in \Xi}$, $\int_X$, respectively, if it does not lead to ambiguity.

THEOREM 1. *If* (a)–(e) *hold, then*

(1) *For any optimal design there exists a design with the same information matrix which contains no more than $n = m(m+1)/2$ supporting points.*

(2) *A necessary and sufficient condition for a design* ξ^* *to be optimal is fulfillment of the inequality:*

$$\min_x \psi(x, \xi^*) \geqslant 0.$$

(3) *The set of optimal designs is convex.*
(4) $\psi(x, \xi^*)$ *achieves zero almost everywhere in* $\operatorname{supp} \xi^*$,
where $\operatorname{supp} \xi$, *stands for supporting set of the design (measure)* ξ.

Functions $\psi(x, \xi)$ for the most popular criteria of optimality may be found, for instance, in Atkinson and Fedorov (1984). Theorem 1 provides a starting point for analytical exercises with various relatively simple regression problems and makes possible the development of a number of simple numerical procedures for the optimal design construction in more complicated and more realistic situations. Most of these procedures are based on the following iterative scheme:

- (a) There is a design $\xi_s \in \Xi(q)$. Find

$$x_s = \arg\min \left\{ \psi(x^+, \xi_s), \psi(x^-, \xi_s) \right\}, \tag{2.5}$$

$$x^+ = \arg\min_{x \in X} \psi(x, \xi_s),$$

$$x^- = \arg\max_{x \in X_s} \psi(x, \xi_s),$$

where $X_s = \operatorname{supp} \xi_s$.

- (b) Choose $0 < \beta_s < 1$ and construct

$$\xi_{s+1} = (1 - \beta_s)\xi_s + \beta_s \xi(x_s), \tag{2.6}$$

where $\xi(x_s)$ is a measure atomized at x_s.

The choice of a sequence β_s defines a variety of the algorithms; specific examples are given by Atkinson and Donev (1992), Cook and Nachtsheim (1989), Fedorov (1972, 1975) and Silvey (1980). The following sequences are most popular:

- $\beta_s = \alpha_s$, if $x_s = x^+$, and

$$\beta_s = \min \left\{ \alpha_s, \xi_s(x^-)/(1 - \xi_s(x^-)) \right\}, \quad \text{if } x_s = x^-,$$

where $\lim \alpha_s = 0$, $\sum_{s=0}^{\infty} \alpha_s = \infty$;

- $x = \arg\min_x \psi\left[M(\xi_{s+1}(x))\right]$,
  where $\xi_{s+1}(x) = (1 - \beta)\xi_s + \beta\xi(x_s)$;

- $\alpha_s = \begin{cases} \beta_{s-1}, & \text{if } \psi\left[M(\xi_s(\beta_{s-1}))\right] < \psi[M(\xi_{s-1})]; \\ \beta_{s-1}/\gamma, \gamma > 1, & \text{otherwise.} \end{cases}$

Theorem 1 together with iterative procedure (2.5), (2.6) provides quite powerful tools for constructing optimal design. The existing software products, see, for instance, Mitchell (1974), Nguen and Miller (1992), Nachtsheim (1987), SAS/QC Software (1995), Wheeler (1994), confirm this statement. Unfortunately, there are a few hurdles, which do not allow the direct use of the results reported above. The first one is that optimal designs defined by (2.4) may have unequal weights. What does this mean in the context of observing stations allocation? If we have N available stations or sensors, then $r_i^* = [p_i^* N]$ stations must be allocated at x_i^*, where $[p_i^* N]$ is some reasonable integer approximation of $p_i^* N$. It is obvious that in many cases (but not always) two or more stations sited in the immediate vicinity of each other will not give essentially more information than a single station. There are some arguments in favor of this statement, which can be expressed economically in colloquial statistical terminology: observations from these stations are strongly correlated. However, frequently weights p_i^* may be considered as the desirable precision of measurements taken at the ith station. The corresponding precision can be achieved through the proper technical steps or through controlling the longevity of the observational process.

Probably, Gribik et al. (1976), were the first to use the optimal experimental design methods for environmental monitoring. They analyzed the problem of allocating measuring resources to aid in accurately estimating ground level pollution concentrations throughout a region X. The regression model was the linearized version of the diffusion model for four pollution sources and unknown background source. Since the diffusion model used in the study was a large scale model, measurements separated by distances smaller than a threshold value distance appeared to be correlated in the corresponding parameter estimation problem. At the same time the design method was a particular case of the method discussed in this section, where the independence of observational errors is essential. To avoid a contradiction the authors imposed the additional constraint: the distance between any two observing sites must be greater than the characteristic distance:

$$(x_i - x_j)^T (x_i - x_j) > d^2.$$

Imposing constraints of that type is one of the simplest way to handle possible correlation between the observed values at neighboring stations. Obviously the approach does not work for long-range correlation, when the widely separated observations are correlated.

It was assumed that the ground level pollution is of the prime interest. The authors proposed to use the weighted average variance of the best linear unbiased estimator of the ground level pollution:

$$\Psi\left[M(\xi)\right] = \int_X w(x) f^T(x) M^{-1}(\xi) f(x)\, \mathrm{d}x$$

as the criterion of optimality. The weight function $w(x)$ was selected proportional to the population density in the considered region.

A rather detailed discussion of applicability of the standard design technique for spatial experiments may be found in Fedorov et al. (1988).

3. Optimal designs with bounded density

Gribik et al. (1976), used a very simple and transparent idea to avoid clustering of sensors at particular points. This idea can be exploited in a more general and formal setting. Let the number of sensors N be sufficiently large and the density of stations per square unit be introduced into consideration:

$$\xi(\mathrm{d}x) = \lim_{\Delta X \to 0} \frac{N(\Delta X)}{N}. \tag{3.1}$$

Introduction of (3.1) is very reasonable when the sensor allocation is considered in technological experiments. In the network allocation problem it is probably less realistic. Nevertheless, the results considered in this section help to explain why some intuitive approaches, similar to what was done by Gribik et al. (1976), do work well in most cases.

If X is not uniform (as might be appropriate say, with different topography for different parts of X), then it is natural to assume that the sensor density has to be constrained:

$$\Phi_1(\mathrm{d}x) \leqslant \xi(\mathrm{d}x) \leqslant \Phi_2(\mathrm{d}x).$$

With obvious redefining of the design measure $\xi(\mathrm{d}x)$ and the upper bound $\Phi_2(\mathrm{d}x)$ the latter may be reduced to a simpler statement:

$$0 \leqslant \xi(\mathrm{d}x) \leqslant \Phi(\mathrm{d}x).$$

Thus, the following optimization problem must be considered (we skip the evident left hand side constraint):

$$\xi^* = \arg\min_{\xi} \Psi\left[M(\xi)\right], \tag{3.2}$$

$$\int \xi(\mathrm{d}x) = 1, \quad \xi(\mathrm{d}x) \leqslant \Phi(\mathrm{d}x). \tag{3.3}$$

This optimization problem was discussed by Wynn (1982), and Fedorov (1989). To avoid unnecessary technical complications let us assume additionally to (a)–(e) from Section 2 that

(f) $\Phi(dx)$ is atomless, i.e.,

$$\lim_{\Delta X \to 0} \int_{\Delta X} \Phi(dx) = 0.$$

The following theorem summarizes the most important properties of designs with bounded density.

THEOREM 2. *Let Ξ_0 be a set of design ξ such that $\xi(dx) = \Phi(dx)$, when $\xi(dx) > 0$, and $\xi(dx) = 0$ otherwise, and let Assumptions* (a)–(f) *hold. Then:*

- *There exists an optimal designs $\xi^* \in \Xi_0$.*
- *A necessary and sufficient condition for this design to be optimal is that $\psi(x, \xi^*)$ separates the two sets $X^* = \operatorname{supp} \xi^*$ and its complement.*

In the above formulation "separate" means that there is a constant C such that $\psi(x, \xi^*) \leqslant C$ on X^* and $\psi(x, \xi*) > C$ on its complement.

Theorem 1 tells us that supporting sets of optimal designs must coincide with the points where $\psi(x, \xi^*)$ achieves its minimum. Therefore, in most cases the supporting set for the standard optimal design consists of a finite number of supporting points.

Theorem 2 forces $\operatorname{supp} \xi^*$ to occupy the subsets of X. How is $\xi^*(dx)$ to be realized by a practitioner? One of the possibilities is to replace $\xi^*(\Delta X)$ for relatively small areas ΔX by $N^*(\Delta X) = [\xi^*(\Delta X)N]$. When $N^*(\Delta X)$ is defined then the corresponding number of sensors have to be allocated in ΔX. For instance, they can be sited at the nodes of some uniform grid. Generally, that allocation has to guarantee a reasonable approximation of the integral

$$\int_{\Delta X} \psi(x, \xi^*) \xi^*(dx)$$

by the sum

$$\sum_{x_i \in \Delta X} \psi(x_i, \xi^*) \Delta X_i, \quad x_i \in \Delta X_i,$$

$$\bigcup_i \Delta X_i = \Delta X, \quad \Delta X_i \bigcap_{i \neq j} \Delta X_j = \emptyset.$$

The properties described by Theorem 2 allow us to formulate a simple numerical algorithm to construct optimal designs (see Fedorov, 1989). Let $\Phi(dx) = \phi(x)\, dx$ and

$$\lim_{s \to \infty} \alpha_s = 0, \qquad \lim_{s \to \infty} \sum_{s'=1}^{s} \alpha_{s'} = \infty \quad \text{and} \quad \lim_{s \to \infty} \sum_{s'=1}^{s} \alpha_{s'}^2 < \infty.$$

(a) There is a design $\xi_s \in \Xi_0$. Let $X_{1s} = \operatorname{supp} \xi_s$ and $X_{2s} = X \setminus X_{1s}$. Two sets $D_s \subset X_{1s}$ and $E_s \subset X_{2s}$ with equal measure,

$$\int_{D_s} \phi(x)\,\mathrm{d}x = \int_{E_s} \phi(x)\,\mathrm{d}x = \alpha_s,$$

and, correspondingly, including the points

$$x_{1s} = \arg\max_{x \in X_{1s}} \psi(x, \xi_s) \quad \text{and} \quad x_{2s} = \arg\min_{x \in X_{2s}} \psi(x, \xi_s).$$

(b) The design ξ_{s+1} with the supporting set

$$\operatorname{supp} \xi_{s+1} = X_{1(s+1)} = (X_{1s} \setminus D_s) \cup E_s$$

is constructed.

Usually $\phi(x)$ is assumed to be constant. All other cases may be converted to this one with the proper coordinate transformation. In the computerized version of the algorithm integrals in (a) are replaced with sums over some grid elements. If these elements and subsequently α_s are fixed and elements of both D_s and E_s coincide with the grid elements, then (a), (b) becomes an exchange type algorithm (see, for instance, Mitchell, 1974) with the simple constraint: every grid element cannot contain more than one supporting point and the weights of all supporting points are the same, i.e., N^{-1}. In practice it is sometimes convenient to consider grids of varying density, which has to be proportional to $\phi(x)$. While it can be shown that the exchange algorithm (a), (b) converges to an optimal design for properly diminishing α_s, it is not generally true for finite α_s and, in particular, when $\alpha_s \equiv N^{-1}$. The accuracy of the limit designs is defined by the accuracy of the approximation (see Assumption (e) from Section 2):

$$\Psi\left[(1-\alpha_s)M(\xi_s) + \alpha_s M(\xi)\right] \cong \Psi\left[M(\xi_s)\right] + \alpha_s \int \psi(x, \xi_s)\xi(\mathrm{d}x),$$

$$\int_{\Delta X_i} \psi(x, \xi_s)\xi(\mathrm{d}x) \simeq \psi(x_i, \xi_s)\Delta X_i.$$

When these approximations are reliable enough then we can hope that the limit designs do not deviate too much from the optimal ones. The term "limit design" must be used with some reservation when $\alpha_s \equiv N^{-1}$: instead of convergence some minor oscillations of $\Psi[M(\xi_s)]$ may be observed. Practical aspects of the iterative procedure (a), (b) were discussed by Fedorov and Müller (1989b) in the air pollution network design setting.

4. Correlated observational errors

Let us assume now that the random errors in model (2.1) are correlated and that the covariance structure is known, i.e., either the covariance matrix V or the covariance function $V(x, x')$ is given. There is no need to use the second subscript indicating the repeated observations and we consider

$$y_i = \eta(x_i, \theta) + \varepsilon_i, \tag{4.1}$$

where $i = 1, \dots, N$, $\mathrm{E}(\varepsilon_i) \equiv 0$ and

$$V(\xi_N) = \{\mathrm{E}(\varepsilon_i \varepsilon_l)\}_{1,1}^{N} = \{V(x_i, x_l)\}_{1,1}^{N}.$$

For the obvious reason, in this section we will use the simplified notation:

$$\xi_N = \{x_1, \dots, x_N\}.$$

In the case of correlated observations the best linear unbiased estimator is defined as (see, for instance, Rao, 1973):

$$\widehat{\theta} - \underline{M}^{-1}(\xi_N) F(\xi_N) V^{-1}(\xi_N) Y, \tag{4.2}$$

and its dispersion matrix equals

$$\mathrm{Var}(\widehat{\theta}) = \underline{M}^{-1}(\xi_N), \tag{4.3}$$

where

$$Y^T = (y_1, \dots, y_N), \quad F(\xi_N) = (f(x_1), \dots, f(x_N)),$$

$$\underline{M}(\xi_N) = F(\xi_N) V^{-1}(\xi_N) F^T(\xi_N) = N M(\xi_N). \tag{4.4}$$

The best linear unbiased predictor at a point x is

$$\widehat{y}(x) = f^T(x)\widehat{\theta} + V^T(x, \xi_N) V^{-1}(\xi_N) \left(Y - F^T(\xi_N)\widehat{\theta}\right), \tag{4.5}$$

$$\mathrm{E}\left[(y(x) - \widehat{y}(x))^2\right] = S^2(x, \xi_N) + \phi^T(x, \xi_N) \underline{M}^{-1}(\xi_N) \phi(x, \xi_N), \tag{4.6}$$

where

$$S^2(x, \xi_N) = V(x, x) - V^T(x, \xi_N) V^{-1}(\xi_N) V(x, \xi_N),$$
$$\phi(x, \xi_N) = f(x) - F(\xi_N) V^{-1}(\xi_N) V(x, \xi_N).$$

Unlike (2.2), the information matrix (4.4) is not a sum of information matrices of single observations. Therefore we cannot use directly the results of the convex design theory, which is essentially based on the additivity of information matrices. Actually, we have to consider the optimization problem

$$\xi_N^* = \arg\min_{\xi_N} \Psi\left[M(\xi)\right], \tag{4.7}$$

which does not have too much in common with (5) besides notation. For instance, the convexity of Ψ is not very helpful anymore.

In most studies authors try to imitate the iterative methods of optimal design construction considered in two previous sections. For instance, computations become similar to the standard (uncorrelated) case, if the following recursion formula is used (Brimkulov et al., 1986):

$$\underline{M}(\xi_{N+1}) = \underline{M}(\xi_N) + w(x, \xi_N)\phi(x, \xi_N)\phi^T(x, \xi_N), \tag{4.8}$$

where

$$\xi_{N+1} = \{\xi_N, x\} \quad \text{and} \quad w^{-1}(x, \xi_N) = S^2(x, \xi_N).$$

We can easily derive, for instance, that

$$|\underline{M}(\xi_{N+1})| = |\underline{M}(\xi_N)| \left[1 + w(x, \xi_N)\phi^T(x, \xi_N)\underline{M}^{-1}(\xi_N)\phi(x, \xi_N)\right]. \tag{4.9}$$

Subsequently, for the D-criterion the point

$$x_{N+1} = \arg\max_{x \in X} w(x, \xi_N)\phi^T(x, \xi_N)\underline{M}^{-1}(\xi_N)\phi(x, \xi_N), \tag{4.10}$$

must be added to the design ξ_N. That is an imitation of step (a) from the iterative procedures considered in the two previous sections.

There exists a simple intuitive explanation why iterative procedures based on (4.10) provide "good" supporting points in the sense of the D-criterion. First, let us recollect that in the no-correlation case accordingly to stage (a) of the iterative procedure from Section 2 the additional observation(s) must be allocated at point(s), where the ratio

$$\frac{\text{variance of prediction with the estimated } \theta}{\text{variance of prediction with the given } \theta}$$

$$= \frac{\sigma^2(x) + f^T(x)\underline{M}^{-1}(\xi)f(x)}{\sigma^2(x)} \tag{4.11}$$

is maximal. This follows, for instance, from (2.5) when

$$\Psi(M) = -\ln|M|$$

and

$$\psi(x, \xi) = m - f^T(x)M^{-1}(\xi)f(x) \quad \text{for } \sigma^2(x) \equiv 1$$

or

$$\psi(x, \xi) = m - \sigma^{-2}(x)f^T(x)M^{-1}(\xi)f(x),$$

in the more general case (see Fedorov, 1972) for details. In the case of correlated observations we are looking for a maximum of the same ratio

$$\frac{\text{variance of prediction with the estimated } \theta}{\text{variance of prediction with the given } \theta}$$

$$= \frac{S^2(x, \xi_N) + \psi^T(x, \xi_n)\underline{M}^{-1}(\xi_N)\psi(x, \xi_N)}{S^2(x, \xi_N)}. \tag{4.12}$$

When $x \to x_j \in \operatorname{supp} \xi_N$, then

$$w(x, \xi_N)\psi^T(x, \xi_N)\underline{M}^{-1}(\xi_N)\psi(x, \xi_N)) \to 0 \tag{4.13}$$

for $f(x)$ and $V(x, \xi_N)$ continuous in the vicinity of x. In other words the iterative procedure defined by (4.10) does not admit coinciding supporting points. The result follows from the definitions of $\psi(x, \xi_N)$ and $S^2(x, \xi_N)$, and the fact that

$$\left\{V^{-1}(\xi_N)V(x_i, \xi_N)\right\}_j = \delta_{ij},$$

where δ_{ij} is the Kronecker symbol. Formula (4.10) can be easily rewritten for the deleting procedure. From (4.9) it follows that in the case of the D-criterion candidates for deleting are defined by the equation:

$$x_{N+1}^{-} = \arg\min_{\ell} w(x_\ell, \xi_{N+1})\phi^T(x_\ell, \xi_{N+1})\underline{M}^{-1}(\xi_{N+1})\phi(x_\ell, \xi_{N+1}). \tag{4.14}$$

We are not aware of any results on the properties of the iterative procedures based on (4.10) and (4.14) for the D-criterion or similar procedures for other criteria. There are empirical confirmations that the exchange-type algorithms lead to a significant improvement of the starting design. For instance, Rabinowitz and Steinberg (1990) applied that type of algorithm to the problem of selecting sites for a seismographic network. They have shown that the computed designs are relatively efficient and are better than the standard D-optimal designs constructed for models with uncorrelated observations. It is reasonable to note that computationally (4.9) and (4.14) are much more demanding than their counterparts in the standard design theory. There exist a number of studies where the optimization problem (4.1) is considered for some special and relatively simple covariance functions, for instance, generated by autoregressive models. Various details and further references may be found in Bickel and Herzberg (1979), Bishoff (1992), Kunert (1988), Martin (1986), Müller and Pázman (1995) and Näther (1985).

In conclusion of this section let us emphasize again the significant difference between the case with correlated observations and the standard case. For uncorrelated observations the additiveness of the normalized information matrix ($\sigma(x) = 1$):

$$M(\xi) = N^{-1} \sum_{i=1}^{N} f(x_i) f^T(x_i) = N^{-1} \underline{M}(\xi_N)$$

leads to many simple and elegant theoretical results initiated by Kiefer's pioneering findings. Very frequently normalized information matrices may be treated as a limit, i.e.:

$$M(\xi) = \lim_{N \to \infty} N^{-1} \underline{M}(\xi_N). \tag{4.15}$$

In many cases for correlated observations the corresponding limit does not exist and the matrix $M(\xi)$ cannot be introduced. One of the most successful attempts to replace (4.15) was due to by Sacks and Ylvisaker (1966, 1968); see more in Section 7.

5. Random coefficients regression models: Trend estimation

In what follows we intend to consider some simple models for the random component in (2.1). It is convenient to partition "intrinsic" or "process", and "observational" sources of randomness:

$$y_{ij} = \eta(x_i, \theta) + u_{ij} + \varepsilon_{ij}. \tag{5.1}$$

Values u_{ij} describes deviations of the observed response from $\eta(x_i, \theta)$ due to some causes which are independent of an observer. For instance, an average wind velocity may be disturbed by various local micro-eddies. The term ε_{ij} describes "observational" errors. Sometimes these errors are defined by the selected observational technique and, at least partly, they are controlled by an observer. The proposed partitioning is very conditional, and the reader may use a different one, which is more compatible with the corresponding experimental situation.

Let us assume that

$$y_{ij} = \theta_1^T f_1(x_i) + \theta_{2j}^T f_2(x_i) + \varepsilon_{ij}, \quad i = 1, \ldots, N, \ j = 1, \ldots, k, \tag{5.2}$$

or

$$Y_j = F_1(\xi_N)^T \theta_1 + F_2(\xi_N)^T \theta_{2j} + \varepsilon_j, \tag{5.3}$$

where

$$\theta_1 \in R^t, \quad \theta_{2j} \in R^l, \quad t + l = m,$$

$$F_\alpha(\xi_N) = \left(f_\alpha(x_1), \dots, f_\alpha(x_N)\right), \quad \alpha = 1, 2,$$

$$\varepsilon_j^T = (\varepsilon_{1j}, \dots, \varepsilon_{Nj}).$$

Vector θ_2 is random with

$$\mathrm{E}(\theta_2) = 0, \qquad \mathrm{E}(\theta_2\theta_2^T) = \mathrm{Var}(\theta_2) = \Lambda, \tag{5.4}$$

vector ε describes the observational errors, which are random and

$$\mathrm{E}(\varepsilon) = 0, \qquad \mathrm{E}(\varepsilon\varepsilon^T) = \sigma^2 I.$$

We assume that θ_2 and ε are uncorrelated. In terms of (5.1) we have

$$\mathrm{E}(u) = 0, \qquad \mathrm{E}(uu^T) = F_2^T(\xi_N)\Lambda F_2(\xi_N).$$

If $e = u + \varepsilon$, then

$$\mathrm{E}(e) = 0, \qquad \mathrm{E}(ee^T) = V(\xi_N) = \sigma^2 I + F_2^T(\xi_N)\Lambda F_2(\xi_N). \tag{5.5}$$

Thus we are going to consider a very special case of (4.1) with $V(\xi_N)$ defined by (5.5). It may be illuminating to associate index "j" with time (hour, day, ...) and "i" with location (x_i is a vector of coordinates of a particular site).

Model (5.5) gives an opportunity to introduce criteria of optimality which provide a very reasonable description of various experimental situations. Those criteria may be divided in two main groups. The first group is related to the "average over time" behavior of the observed response. The corresponding criteria depend upon the precision of estimators of θ_1. This means that we consider some functions of $\mathrm{Var}(\widetilde{\theta}_1)$, where $\widetilde{\theta}_1$ is an estimator of θ_1.

The second group deals with "instant" responses and the corresponding criteria are based on $\mathrm{Var}(\widetilde{\theta})$, $\theta^T = (\theta_1^T, \theta_2^T)$.

Let us start with the first group, i.e., with estimating the subvector θ_1. The best linear unbiased estimator is (compare with (4.3)):

$$\widehat{\theta}_1 = \underline{M}_{11}^{-1}(\xi_N)F_1(\xi_N)V^{-1}(\xi_N)\sum_{j=1}^{k} k^{-1}Y_j. \tag{5.6}$$

The dispersion matrix of $\widehat{\theta}$ is

$$k \cdot \mathrm{Var}(\widehat{\theta}_1) = \underline{M}_{11}^{-1}(\xi_N), \tag{5.7}$$

where

$$\underline{M}_{11}(\xi_N) = F_1(\xi_N)V^{-1}(\xi_N)F_1^T(\xi_N). \tag{5.8}$$

An optimal design (or optimal observational network) is defined as

$$\xi_N^* = \arg\min_{\xi_N} \Psi\left[\underline{M}_{11}(\xi_N)\right], \tag{5.9}$$

which differs from (4.7) only by the more detailed information about $V(\xi_N)$. It may be expedient to note that unlike the situation described in comments accompanying (4.13) the covariance is not anymore a continuous function at the diagonal:

$$\lim_{x \to x'} \mathrm{E}(e(x)e(x')) \neq \sigma^2 + f_2^T(x)\Lambda f_2(x'). \tag{5.10}$$

Therefore (5.9) may admit designs with repeated observations, i.e., it could be that $x \in \operatorname{supp} \xi_N$ and $M(\xi_N + x)$ is better than $M(\xi_N)$.

Using the identity

$$(A + BDB^T)^{-1} = A^{-1} - A^{-1}B(B^TA^{-1}B + D^{-1})^{-1}B^TA^{-1} \tag{5.11}$$

and assuming the existence of the Λ^{-1}, one can find that

$$\underline{M}_{11}(\xi_N) = \overline{M}_{11}(\xi_N) - \overline{M}_{12}(\xi_N)\left[\overline{M}_{22}(\xi_N) + \Lambda^{-1}\right]^{-1}\overline{M}_{21}(\xi_N), \tag{5.12}$$

where $\overline{M}_{\alpha\beta}(\xi_N) = \sigma^{-2}F_\alpha(\xi_N)F_\beta^T(\xi_N)$.

Now let us consider the matrix

$$D(\xi_N) = \begin{pmatrix} D_{11}(\xi_N) & D_{12}(\xi_N) \\ D_{21}(\xi_N) & D_{22}(\xi_N) \end{pmatrix} = \begin{pmatrix} \overline{M}_{11}(\xi_N) & \overline{M}_{12}(\xi_N) \\ \overline{M}_{21}(\xi_N) & \overline{M}_{22}(\xi_N) + \Lambda^{-1} \end{pmatrix}^{-1}.$$

From the Frobenius formula it follows that

$$D_{11}(\xi_N) = \underline{M}_{11}^{-1}(\xi_N).$$

The matrix $D(\xi_N)$ can be also considered as the dispersion matrix of the best linear unbiased estimator of parameter $\theta^T = (\theta_1^T, \theta_2^T)$ for the regression model.

$$y = \theta^T F(\xi_N) + \varepsilon,$$

where $F^T(\xi_N) = (F_1^T(\xi_N), F_2^T(\xi_N))$, $\mathrm{E}(\varepsilon) = 0$, $\mathrm{E}(\varepsilon\varepsilon^T) = \sigma^2 I$, with the prior information about parameters θ_2 described by a prior distribution $P(\theta_2)$ such that

$$\mathrm{E}(\theta_2) = \int \theta_2 \, \mathrm{d}P(\theta_2) = 0$$

and

$$\Lambda = \mathrm{E}(\theta_2 \theta_2^T) = \int \theta_2 \theta_2^T \, \mathrm{d}P(\theta_2).$$

See, for instance, Fedorov (1972), Pilz (1991) and Seber (1977). Thus, the optimization problem (4.1) may be embedded in the framework of convex design theory. For instance, for the D-criterion, when $|\underline{M}_{11}(\xi_N)|^{-1} = |D_{11}(\xi_N)|$ must be minimized, one can use any algorithm developed for the construction of "exact" or "discrete" optimal designs; see, for instance, Cook and Nachtsheim (1980), Fedorov (1972), Ermakov (1983), and Pukelsheim (1993) when only the subvector θ_1 has to be estimated. More generally we can now describe experimental design as the following optimization problem

$$\xi^* = \arg\min_{\xi} \Psi \left(M(\xi) + M_0 \right), \tag{5.13}$$

where

$$M_0 = \sigma^2 N^{-1} \begin{pmatrix} 0 & 0 \\ 0 & \Lambda^{-1} \end{pmatrix},$$

N is now the total number of possible observations, and

$$M(\xi) = \int f(x) f^T(x) \xi(\mathrm{d}x),$$

$$f^T(x) = \left(f_1^T(x), f_2^T(x) \right).$$

Let us note that occasionally the total number of observations and the number of supporting points may coincide (like in (5.2)). Then N stands for both. The results of Sections 2 and 3 may routinely be applied to (5.13) when Ψ is properly defined and σ^2 and Λ are known.

Subset D-optimality. According to (4.10) we have to minimize some function of the matrix $\underline{M}_{11}^{-1}$, when the parameters θ_1, are of prime interest. In terms of (5.13) it means that the objective function Ψ must depend upon elements of the matrix $\Delta_{11}(\xi)$, which may be defined as follows:

$$\Delta(\xi) = \begin{pmatrix} \Delta_{11}(\xi) & \Delta_{12}(\xi) \\ \Delta_{21}(\xi) & \Delta_{22}(\xi) \end{pmatrix} = \begin{pmatrix} M_{11}(\xi) & M_{12}(\xi) \\ M_{21}(\xi) & M_{22}(\xi) + M_{022} \end{pmatrix}^{-1} \tag{5.14}$$

where

$$M_{022} = \sigma^2 N^{-1} \Lambda^{-1}.$$

One of the possibilities is to select

$$\Psi\left(M(\xi)+M_0\right)=\ln|\Delta_{11}(\xi)|\,.$$

From Theorem 1 it immediately follows that a necessary and sufficient condition for ξ^* to be optimal is that (compare with Fedorov, 1972):

$$\begin{aligned}&f^T(x)\Delta(\xi^*)f(x)-f_2^T(x)\left(M_{22}(\xi^*)+M_{022}\right)^{-1}f_2(x)\\&\quad\leqslant \operatorname{tr}\Delta(\xi^*)M(\xi^*)-\operatorname{tr}\left(M_{22}(\xi^*)+M_{022}\right)^{-1}M_{22}(\xi^*).\end{aligned}\tag{5.15}$$

Notice that

$$v_1(x,\xi)=f^T(x)\Delta(\xi)f(x)$$

is the normalized variance of $\widehat{\theta}^T f(x)$, where $\widehat{\theta}$ is the best linear unbiased estimator of θ from model (5.2), and

$$v_2(x,\xi)=f_2^T(x)\left(M_{22}(\xi^*)+M_{022}\right)^{-1}f_2(x)$$

may be considered as a normalized variance of the best linear estimator for the regression model with the same observational errors but with the response $\theta_2^T f_2(x)$. To get non-normalized values we have to multiply the normalized values by σ^2N^{-1}. Simple, but rather long matrix calculations show that

$$\begin{aligned}v_1(x,\xi)=&\,d(x,\xi)+R(x,x)-R^T(x,\xi)R(\xi)R(x,\xi)\\&+\beta^T(x,\xi)\left(R^{-1}(\xi)+\rho^{-1}(\xi)\right)^{-1}\beta(x,\xi),\end{aligned}\tag{5.16}$$

where

$$\begin{aligned}&d(x,\xi)=f_1^T(x)M_{11}^{-1}(\xi)f_1(x), \qquad R(x,x')=f_2^T(x)M_{022}^{-1}f_2(x'),\\&R(x,\xi)=F_2^TM_{022}^{-1}f_2(x), \qquad R(\xi)=F_2^TM_{022}^{-1}F_2,\\&\rho^{-1}(\xi)=W-Wd(\xi)W, \qquad \beta(x,\xi)=W\underline{d}(x,\xi)-R^{-1}(\xi)R(x,\xi),\\&d(\xi)=F_1^TM_{11}^{-1}F_1, \qquad \underline{d}(x,\xi)=F_1^TM_{11}^{-1}f_1(x),\\&F_1=\left(f_1(x_1),\dots,f_1(x_N)\right), \qquad F_2=\left(f_2(x_1),\dots,f_2(x_N)\right),\\&W_{ik}=\delta_{ik}p_i, \quad p_i=\xi(x_i), \ \{x_1,\dots,x_n\}=\operatorname{supp}\xi.\end{aligned}\tag{5.17}$$

This collection of formulae looks much more complicated than the similar terms in (5.15). However, that presentation has one remarkable feature: it does not depend upon functions $f_2(x)$ and matrix Λ explicitly. All elements in (5.16) are completely defined by the covariance function

$$\sigma^{-2}NR(x,x')=\mathrm{E}\left(f_2(x)\theta_2^T\theta_2f_2^T(x')\right)=V(x,x'), \quad x\neq x'.$$

Similar to (5.16)

$$v_2(x,\xi) = R(x,x) - R^T(x,\xi)\left(R(\xi) + W^{-1}\right)^{-1} R(x,\xi). \tag{5.18}$$

Thus, when the covariance function is known directly, i.e., we do not use (5.5) to get it, one can use (5.16) and (5.18) to construct optimal design. Moreover, the cases, in which $l = \dim\theta_2 \to \infty$ may be considered. The first attempt in this direction was done by Müller–Gronbach (1993).

6. Random coefficient regression models: Prediction

The presentation of the design problem for model (5.2)–(5.4) in the form (5.13) allows us to develop a rather simple technique for experimental design when the objective is the prediction of observed values. For the sake of simplicity, let $\eta(x,\theta) \equiv 0$ in (5.2) and

$$u_j(x_i) = u_{ij} = \theta_j^T f(x_i). \tag{6.1}$$

Then the corresponding optimal designs are defined as

$$\xi^* = \arg\min_{\xi} \Psi\left(M(\xi) + M_0\right), \tag{6.2}$$

where

$$M(\xi) = \int f(x) f^T(x)\xi(\mathrm{d}x) \quad \text{and} \quad M_0 = \sigma^2 N^{-1} \Lambda^{-1}.$$

It is expedient to note that

$$\min \mathrm{E}\left[(\widetilde{\theta} - \theta)(\widetilde{\theta} - \theta)^T\right] = D(\xi) = \sigma^2 N^{-1}\left(M(\xi) + M_0\right)^{-1},$$

where the expectation operator E takes into account randomness of both observational errors and regression parameters. Minimization is taken with respect to all linear estimators, see Gladitz and Pilz (1982), Fedorov and Müller (1989), Pilz (1991). The best linear estimator is

$$\widehat{\theta}_j = N^{-1}\left(M(\xi) + M_0\right)^{-1} \sum_{i=1}^{N} y_{ij} f(x_i).$$

Similar to arguments in Section 5 we can apply the equivalence theorem to (6.2) to find, for instance, necessary and sufficient conditions for a design ξ^* to be optimal. Leaving to the reader the possibility to formulate them for the general case we focus only on three simple and very popular criteria.

Minimax and D-criterion. For the D-criterion, when

$$\Psi\left(M(\xi)+M_0\right) = -\ln\left|M(\xi)+M_0\right|, \tag{6.3}$$

one can easily derive from Theorem 1 that a necessary and sufficient condition for ξ^* to be optimal is that

$$f^T(x)\left(M(\xi^*)+M_0\right)^{-1} f(x) \leqslant \operatorname{tr}\left(M(\xi^*)+M_0\right)^{-1} M(\xi^*) \tag{6.4}$$

for all $x \in X$.

This inequality appears, especially when the dimension of f is large, more attractive and meaningful in notation described in comments accompanying (5.11):

$$\begin{aligned} &R(x,x) - R^T(x,\xi)\left(W^{-1}+R(\xi)\right)^{-1} R(x,\xi) \\ &\quad\leqslant \operatorname{tr}\left(R(\xi) - R(\xi)\left(W^{-1}+R(\xi)\right)^{-1} R(\xi)\right). \end{aligned} \tag{6.5}$$

The variance of the best linear unbiased predictor for $u(x)$ equals

$$\begin{aligned} &\operatorname{Var}\left(u(x)-\widehat{u}(x)\mid\xi\right) \\ &\quad= \sigma^2 + \sigma^2 N^{-1}\left[R(x,x) - R^T(x,\xi)\left(W^{-1}+R(\xi)\right)^{-1} R(x,\xi)\right], \end{aligned} \tag{6.6}$$

where

$$\widehat{u}(x) = \widehat{\theta}^T f(x) = R^t(x,\xi)\left(W^{-1}+R(\xi)\right)^{-1}\overline{Y} \tag{6.7}$$

$$= V^T(x,\xi)\left(\sigma^2 W^{-1} N_{-1} + V(\xi)\right)^{-1}\overline{Y}, \tag{6.8}$$

and the components of the vector $\overline{Y}$ are averages of observations at the corresponding points. For the continuously changing weights p_i the ith component of $\overline{Y}$ may be considered as the observation made with a precision $\sigma/p_i N = {\sigma_i}^2$. If one introduce the covariance function

$$\widetilde{V}(x,x') = \begin{cases} V(x,x'), & x \neq x', \\ {\sigma_i}^2 + V(x,x), & \text{otherwise}, \end{cases} \tag{6.9}$$

then (compare with (4.6) or coming later (7.1)) predictor $\widehat{u}(x)$ coincides with the best linear unbiased predictor for $y(x) = u(x)+\varepsilon(x)$ everywhere except $x \in \operatorname{supp}\xi$. At the design points x_i the realization(s) of $y(x)$ are measured directly and are not needed to be predicted, i.e., one may select $y(x) = \widehat{y}(x)$ and $\operatorname{Var}(y(x) - \widehat{y}(x) \mid \xi) = 0$. Obviously

$$\operatorname{Var}\left(y(x)-\widehat{y}(x)\mid\xi\right) = \sigma^2 + \operatorname{Var}\left(u(x)-\widehat{u}(x)\mid\xi\right) \tag{6.10}$$

otherwise.

Using Theorem 1 together with (6.5) and (6.10) we can formulate the analogue of the Kiefer–Wolfowitz equivalence theorem:

THEOREM 3. *The following two design problems are equivalent:*

- $\min_\xi |D(\xi)|$,
- $\min_\xi \max_{x\in X} \mathrm{Var}(y(x) - \widehat{y}(x) \mid \xi)$.

There is one significant difference between this result and the original equivalence theorem: an optimal design generally depends upon the number of observations N to be used.

Theorem 3 and formula (6.5) give another insight into numerical procedures from Sections 2 and 3: at every stage one has to relocate the design measure from the point(s) where $y(x)$ may be predicted easily (small $\mathrm{Var}(y(x) - \widehat{y}(x) \mid \xi_s)$) to the point(s), where the prediction is poor (large $\mathrm{Var}(y(x) - \widehat{y}(x) \mid \xi_s)$) .

Two linear criteria. Two objective functions which are very popular in spatial statistics are the weighted average variance of prediction:

$$Q_1(\xi) = \mathrm{E}\left[\int_Z w^2(x)\left(y(x) - \widehat{y}(x)\right)^2 \mathrm{d}x\right],$$

and the variance of the weighted average of prediction:

$$Q_2(\xi) = \mathrm{E}\left[\int_Z w(x)y(x)\,\mathrm{d}x - \int_Z w(x)\widehat{y}(x)\,\mathrm{d}x\right]^2,$$

where Z is the "prediction" set or the area of interest. Using (6.6) one can find that minimization of $Q_1(\xi)$ and $Q_2(\xi)$ is equivalent to minimization of

$$\Psi\left[M(\xi)\right] = \mathrm{tr}\, A\left(M(\xi) + M_0\right), \tag{6.11}$$

where in the first case

$$A = \int_Z w^2(x) f(x) f^T(x)\,\mathrm{d}x,$$

and in the second one

$$A = aa^T, \quad a = \int_Z w(x) f^T(x)\,\mathrm{d}x.$$

From part 2 of Theorem 1 it is easy to conclude that

THEOREM 4. *The design ξ^* is linear optimal if and only if*

$$\phi(x, \xi^*) \leqslant \int_X \phi(x, \xi^*)\xi^*(\mathrm{d}x), \quad \textit{for all } x \in X, \tag{6.12}$$

where

$$\phi(x,\xi) = f^T(x)\left(M(\xi^*) + M_0\right)^{-1} A \left(M(\xi^*) + M_0\right)^{-1} f(x).$$

Similar to the D-criterion we can show that for the average variance of prediction

$$\phi(x,\xi) = \phi_1(x,\xi) = \int_Z \mathrm{Cov}^2(x,x' \mid \xi) w^2(x')\,\mathrm{d}x', \tag{6.13}$$

while for the variance of the weighted average

$$\phi(x,\xi) = \phi_2(x,\xi) = w^2(x)\left(\int_Z \mathrm{Cov}(x,x' \mid \xi) w(x')\,\mathrm{d}x'\right)^2, \tag{6.14}$$

where

$$\mathrm{Cov}(x,x' \mid \xi) = R(x,x') - R^T(x,\xi)\left(W^{-1} + R(\xi)\right)^{-1} R(x',\xi).$$

The counterparts of Theorems 3 and 4 may be formulated for optimal designs with bounded density. To do this function $\psi(x,\xi^*)$ in Theorem 2 should be replaced either with $\mathrm{Var}(y(x) - \widehat{y}(x) \mid \xi)$, or with $\phi_1(x,\xi)$, or with $\phi_2(x,\xi)$.

Remarks on applicability of the results. Let us note that the introduction of model (5.2)–(5.4) to generate correlated observations allows us to use the convex design theory for regression problems with correlated observations. Moreover, all results may be presented in a form which does not demand any direct knowledge of the functions $f(x)$. We can formulate results for a particular criterion using only information about the covariance function.

In this and in the previous section we have discussed only the properties of optimal designs. We hope, that having the sensitivity function $\psi(x,\xi)$ represented for various criteria in terms of the normalized covariance function $\mathrm{Cov}(x,x' \mid \xi)$, the reader can easily construct numerical procedures similar to those discussed in Sections 2 and 3.

7. Comparison with the methods based on the variance–covariance structure of observed random fields

Sacks–Ylvisaker approach. Let us suppose that in model (4.1) there is no trend, i.e., $\eta(x,\theta) \equiv 0$, and the covariance function $V(x,x')$ is defined and known for all $x, x' \in X$. The objective of an experiment is to predict $y(x)$ at a given set of points Z, which can be either discrete or continuous.

The best linear unbiased predictor for $y(x)$ may be presented as follows (compare with (4.5 and (4.6)):

$$\widehat{y}(x) = V^T(x,\xi_N) V^{-1}(\xi_N) Y_N, \tag{7.1}$$

$$\mathrm{Var}\big[\big(\widehat{y}(x) - y(x)\big)\big] = V(x,x) - V^T(x,\xi_N)V^{-1}(\xi_N)V(x,\xi_N). \tag{7.2}$$

We again use notation $\xi_N = (x_1, \ldots, x_N)$ to emphasize that there is only one observation at every point x_i. Criteria $Q_1(\xi_N)$ and $Q_2(\xi_N)$ introduced in the previous section have been most intensively analyzed in the studies related to the design problem with correlated errors. Usually it has been assumed that $Z = X$.

A very good summary of the main results for the criterion $Q_1(\xi_N)$ may be found in Micchelli and Wahba (1981). The criterion $Q_2(\xi_N)$ was analyzed by Sacks and Ylvisaker (1970) and Ylvisaker (1987). Further references and comments may be found in Cambanis and Benhenni (1992), Cambanis (1985), Cambanis and Su (1993).

Noting (see (7.1)) that

$$\int_X \widehat{y}(x)\,\mathrm{d}x = q^T(\xi_N)Y_N,$$

where

$$q^T(\xi_N) = \int_X V(x,\xi_N)V^{-1}(\xi_N)\,\mathrm{d}x,$$

we can consider minimization of either $Q_1(\xi_N)$ or $Q_2(\xi_N)$ as a problem of finding an optimal basis for a quadrature formula in approximation theory, with a rather specific objective function; see Karlin and Studden (1966), Sacks and Ylvisaker (1970), Stroud (1975).

When $N \to \infty$, both $Q_1(\xi_N)$ and $Q_2(\xi_N)$ converge to zero for "smooth" covariance functions $V(x,x')$ and for any atomless sequence ξ_N. As in Section 3 we may introduce the limit design measure that defines ξ_N. How a sequence ξ_N may be generated with a particular $\xi(\mathrm{d}x)$ is discussed in details by Cambanis (1985) for $X \subset R^1$. For instance, the so called regular median sequence or design ξ_N is defined as

$$x_{Ni} = \arg\left(\int_a^{x_{Ni}} \xi(\mathrm{d}x) = \frac{2i-1}{2N}\right), \quad i = 1,2,\ldots,n,\ X = [a,b]. \tag{7.3}$$

When X is hypercube and $V(x,x')$ is separable with respect to all components of x, then design ξ_N may be defined as a direct product of univariate designs (see Ylvisaker (1975) for details). Thus the design problem is reduced to the search of the limiting measures providing the best convergence rate for the selected optimality criterion. The rather elaborate technique, a close sibling of the classical approximation theory, leads to a very special minimization problem. Introducing the design density $\xi(\mathrm{d}x) = h(x)\,\mathrm{d}x$ we may state this problem as follows

$$h^* = \arg\min_h Q\left[B(h)\right], \tag{7.4}$$

where

$$B_{\alpha\beta}(h) = \int_X \varphi_\alpha(x)\varphi_\beta(x)h^{-2k-2}(x)\,\mathrm{d}x, \quad \alpha,\beta = 1,\ldots,m, \ \xi(\mathrm{d}x) = h(x)\,\mathrm{d}x,$$

m is the number of estimated integrals (for instance, integrals of $Q_2(\xi_N)$-type with various weight functions), functions $\varphi_\alpha(x)$ and integer k are defined by a covariance function and by an optimality criterion.

Similar problems were considered in studies concerned with simultaneous calculation of m integrals by Monte Carlo method; see Mikhailov and Zhigljavsky (1989), Zhigljavsky (1988), for details and further references. Actually, (7.4) being an optimization problem in a space of probability measures has many features in common with the standard design problem. Some interesting results including the analogue of the iterative numerical procedure from Section 2 are summarized and discussed by Zhigljavsky (1988).

Analytical solutions of (7.4) for the one-dimension case were proposed in the pioneering papers by Sacks and Ylvisaker (1966, 1968, 1970). Various generalizations may be found in Hajek and Kimeldorf (1976) and Wahba (1971, 1974). Following Cambanis (1985), the essence of those findings can be formulated as follows:

If there exist exactly k quadratic mean derivatives of the random process $y(x)$, then under certain regularity conditions (see details in the cited publications)

$$h^*(x) \sim \left[\alpha_k(x)w^2(x)\right]^{1/(2k+3)}, \tag{7.5}$$

where

$$\alpha_k(x) = V^{(k,k+1)}(x, x'-0) - V^{(k,k+1)}(x, x'+0)$$

and superscripts indicate the order of partial derivatives. For any design with density separated from zero the integral $Q(\xi_N)$ diminishes as $\mathrm{O}(N^{-2k-2})$ and $h^*(x)$ minimizes

$$\lim_{N\to\infty} N^{2k+2}Q_2(\xi_N). \tag{7.6}$$

In fact, the immediate objective of Sacks and Ylvisaker (1966, 1968, 1970) was minimization of some function of the dispersion matrix of estimators of parameters θ describing a linear trend $\theta^T f(x)$ in model (4.1). They reduced the corresponding minimization problem to minimization of objective functions similar to $Q_2(\xi_N)$. For instance, when $\theta \in R^1$, then one has to minimize $Q_2(\xi_N)$ with a weight function which is a solution of the following integral equation:

$$f(x) = \int_X V(x, x')w(x')\,\mathrm{d}x'.$$

Evidently, for stationary covariance functions $\alpha_k(x) \equiv$ constant. Subsequently the optimal limiting density $h^*(x)$ is completely defined by the weight function $w(x)$. In other words, only the behavior of $V(x, x')$ at its diagonal influences the solution! The Sacks–Ylvisaker approach (at least in its current form) cannot be used in two cases of practical importance. First, it does not work for "infinitely" smooth covariance functions, when $\alpha_k(x) \equiv 0$ for any k. The covariance function

$$V(x, x') \sim \exp\left[-(x - x')^2/2\right] \tag{7.7}$$

is a popular example (see Sacks and Ylvisaker (1966). The presence of the "white" noise (see model (5.1)) in observed variables gives another example, when the approach does not work. The latter case is of interest for many applications being a very reasonable model when a random process is observed with some instrumental error. In conclusion of this subsection let us note that the concept asymptotically optimal design ξ_N^* based on the existence of a continuous limit density $h^*(x)$ and assumption that

$$\lim_{N\to\infty} \max_i (x_{iN} - x_{(i-1)N}) = 0;$$

see Sacks and Ylvisaker (1966). The definition (7.3) of design ξ_N is one-dimension by its nature and that makes the approach difficult for spatial applications; see Ylvisaker (1975) for further details.

Random parameters approach. To understand the advantages and disadvantages of approach proposed in Sections 5 and 6 relative to to the Sacks–Ylvisaker approach let us introduce the following model:

$$y_{n\sigma}(x) = \sum_{\alpha=1}^{n} \theta_\alpha f_\alpha(x) + \sigma\varepsilon(x), \tag{7.8}$$

where $f_\alpha(x)$ are eigenfunctions of the covariance kernel $V(x, x')$

$$\lambda_\alpha f_\alpha(x) = \int_X V(x, x') f_\alpha(x')\, \mathrm{d}x', \tag{7.9}$$

θ_α are random with zero means and diagonal covariance matrix, such that $\mathrm{Var}(\theta_\alpha) = \lambda_\alpha$, $\lambda_1 \geqslant \lambda_2 \geqslant \cdots \geqslant \lambda_\alpha \geqslant \cdots$, $\varepsilon(x)$ is white noise with the variance 1, and σ^2 is a normalizing constant.

It is well known (Mercer's theorem; see for instance, Kanwal, 1971) that under very mild conditions the series

$$V_n(x, x') = \sum_{\alpha=1}^{n} \lambda_\alpha f_\alpha(x) f_\alpha(x'), \quad x, x' \in X,$$

is uniformly and absolutely convergent and subsequently under very mild assumptions $\{\lambda_\alpha\}$ must diminish not slower then $\mathrm{O}(1/\alpha)$. For many widely used stochastic processes or fields the rate is significantly faster; see Micchelli and Wahba (1981, Theorem 3).

Therefore, for sufficiently large n the kernel $V_n(x, x')$ may be a very reasonable approximation of $V(x, x')$. Allowing $\sigma \to 0$ we may hope that the process $y_{n,\sigma}(x)$ is "close" to $y(x)$ in the sense of their second moments. Subsequently, we might expect the closeness of the corresponding optimal designs. This probably holds for designs ξ_N with relatively small N. However, for $N \to \infty$ the diminishing σ does not guarantee closeness of optimal designs with $\sigma = 0$ and $\sigma > 0$. First, formally the Sacks–Ylvisaker approach does not work for any model with additive "white" noise, because it causes discontinuity of a covariance kernel at its diagonal:

$$\mathrm{E}\left[y(x)y(x')\right] = \begin{cases} V(x, x'), & x \neq x', \\ V(x, x) + \sigma^2, & x = x'. \end{cases}$$

Secondly, for any $\sigma > 0$ and any ξ_N the rate of convergence for either $Q_1(\xi_N)$ or for $Q_2(\xi_N)$ will not be generally better than $\mathrm{O}(N^{-1})$. This is slower than for any continuous covariance kernel.

Thus, for large N the Sacks–Ylvisaker approach and results from Sections 5 and 6 may lead to the different asymptotically optimal designs. If one believes that there is no instrument or any other observation error, then the Sacks–Ylvisaker approach leads to the better limit designs.

When the contribution of observation errors is significant then approximation (7.8) becomes very realistic and allows the use of methods from Sections 5 and 6, which usually produce optimal designs with very moderate numbers of supporting points. Usually these designs have about n supporting points. The existence of well developed numerical procedures and software allows the construction of optimal designs for any reasonable covariance function $V(x, x')$ and various design regions X, including two and three dimension cases. Let us notice that the function $\mathrm{Cov}(x, x' \mid \xi)$ used in Theorem 4 may be presented in the following form:

$$\begin{aligned} &\sigma^2 N^{-1} \mathrm{Cov}(x, x' \mid \xi) \\ &= V_n(x, x') - V_n^T(x, \xi)\left(\sigma^2 P + V_n(\xi)\right)^{-1} V_n(x, \xi), \qquad (7.10) \\ &P = \delta_{ij} r_i, \; r_i = p_i N. \end{aligned}$$

This formula is convenient for some theoretical exercises. For more applied objectives and for development of numerical algorithms based on the iterative procedures from Sections 2 and 3, the direct use of eigenfunctions $f_\alpha(x)$ is more convenient.

Popular kernels. There are several show-case processes and design regions for which analytic expressions for the covariance kernel exist, and the corresponding eigenvalues and eigenfunctions are known:

For the Brownian motion the kernel is

$$V(x, x') = \min(x, x'), \quad 0 \leqslant x, \; x' \leqslant 1,$$

and its eigenvalues and eigenfunctions are

$$\lambda_\alpha = (\alpha - 1/2)^{-2}\pi^{-2}, \qquad f_\alpha(x) = \sqrt{2}\sin(\alpha - 1/2)\pi x, \quad \alpha = 1, 2, \ldots .$$

For the Brownian bridge,

$$V(x, x') = \min(x, x')\left[1 - \max(x, x')\right], \quad 0 \leqslant x,\ x' \leqslant 1,$$

and

$$\lambda_\alpha = \alpha^{-2}\pi^{-2}, \qquad f_\alpha(x) = \sqrt{2}\sin\alpha\pi x, \quad \alpha = 1, 2, \ldots .$$

Both kernels are not differentiable on the diagonal (see comments to (7.5)) and their "jump" functions $\alpha_k(x)$ are easy to calculate. These two kernels or some simple functionals of them (compare with Wahba, 1971) are convenient candidates for the Sacks–Ylvisaker approach. For the Poisson kernel

$$V(x, x') = \frac{1 - \beta^2}{1 - 2\beta\cos 2\pi(x - x') + \beta^2}, \qquad 0 \leqslant x,\ x' \leqslant 1,\ 0 < \beta < 1,$$

and

$$\lambda_0 = 1, \qquad \lambda_{2\alpha-1} = \lambda_{2\alpha} = \beta^\alpha,$$

$$f_0(x) = 1, \qquad f_{2\alpha-1} = \sqrt{2}\cos 2\alpha\pi x, \qquad f_{2\alpha} = \sqrt{2}\sin 2\alpha\pi x,$$

$$\alpha = 1, 2, \ldots .$$

The shape of the Poisson kernel may be controlled by the parameter β. It is "smooth" at the diagonal and the Sacks–Ylvisaker approach cannot be used.

For most real-world problems it is impossible to represent covariance kernels in a simple closed form. However, a representation in the form of an infinite series is standard. For instance, in many experiments related to either diffusion or heat conduction the covariance kernel may be expressed in the the two dimensional finite domain case (see Butkovskiy, 1982) as

$$V(x, x') = 4\sum_{\alpha,\beta=1}^{\infty} \lambda_{\alpha\beta}\sin\alpha\pi x_1 \sin\beta\pi x_2 \sin\alpha\pi x_1' \sin\beta\pi x_2', \tag{7.11}$$

where $X = \{0 \leqslant x_1, x_2 \leqslant 1\}$. Evidently,

$$f_{\alpha\beta}(x) = 2\sin\alpha\pi x_1 \sin\beta\pi x_2, \qquad \lambda_{\alpha\beta} = \exp\left[-a^2\pi^2(\alpha^2 + \beta^2)\right],$$

$$\alpha, \beta = 1, 2, \ldots ,$$

and a is some constant. Representation (7.11) is very natural and convenient for the techniques considered in Sections 5 and 6. Note that the physical problems mentioned above lead to the Gaussian type kernels (compare with (7.7)) when X becomes infinite with respect to any or both coordinates.

The curious reader will find more covariance kernels in any serious book on integral equations containing a chapter on definite kernels or describing Green's functions (see, for instance, Kanwal, 1971; Butkovskiy, 1982).

Selecting the number of terms in (7.8) sufficiently large assures the closeness of $V_n(x, x')$ and $V(x, x')$ may be assured. In many cases it is convenient to assume that coefficients $\{\theta_\alpha\}_1^\infty$ are normally distributed. This assumption does not help in the present problem and in fact can cause some theoretical difficulties. To avoid that we suppose that for any α distribution $P(\theta_\alpha)$ has a finite support set $[a_\alpha, b_\alpha]$ in R^1. For instance, we may select some simple symmetric distribution $P(\theta)$ defined on $[-a, a]$ with $\mathrm{E}(\theta^2) = 1$ and use $P(\theta_\alpha\sqrt{\lambda_\alpha})$ as a distribution for θ_α. Selection of distributions with finite supporting sets assures not only closeness of an exact kernel and its approximation but proximity of $\sum_{\alpha=1}^{\infty} \theta_\alpha f_\alpha(x)$ and $\sum_{\alpha=1}^{n} \theta_\alpha f_\alpha(x)$ if the sequence $\{f_\alpha(x)\}_1^\infty$ satisfies some routine assumptions from the approximation theory. The basic idea of using model (7.8) is in deriving optimal designs for the approximate model $\sum_{\alpha=1}^{n} \theta_\alpha f_\alpha(x)$ and verifying the fact that these designs are optimal or close to optimal for $V_n(x, x')$ or $V(x, x')$. Furthermore, if the objective function is uniquely defined by a dispersion matrix of estimated parameters, then the constructed design is optimal for any model identical to the used one in terms of the first and second moments.

Using (7.8) with $\sigma = 0$ we may immediately conclude that minimization of $Q_2(\xi_N)$ is a rather standard problem from the approximate integration theory, see Davis and Rabinowitz (1985), Stroud (1975).

For instance, it is known (the Gauss–Jacobi Theorem) that for any polynomial $p(x)$ of degree $k \leqslant 2N - 1$ the exact equality

$$\int_a^b w(x)p(x)\,\mathrm{d}x = \sum_{i=1}^{N} q_i p(x_i), \tag{7.12}$$

can be achieved the properly selected weights and supporting points. If $\{p_\alpha(x)\}_0^N$ are orthogonal polynomials with the weight function $w(x) > 0$, then $\xi_N = \{x_i\}_1^N$ are zeros of $p_N(x)$, and

$$q_i^{-1} = q_i^{-1}(\xi_N) = \sum_{\alpha=0}^{N-1} p_\alpha^2(x_i). \tag{7.13}$$

For example, let us consider the Brownian bridge kernel. Note that

$$-f_\alpha(x)\,\mathrm{d}x = \frac{\sqrt{2}\sin\alpha\pi x}{\pi\sin\pi x}\,\mathrm{d}\cos\pi x = \frac{\sqrt{2}}{\pi}U_{\alpha-1}(t)\,\mathrm{d}t, \tag{7.14}$$

where $t = \cos \pi x$ and $U_{\alpha-1}(t)$ is the second kind Tschebysheff polynomial. Now, with $x(t) = \pi^{-1} \arccos t$, we have for a proper $\xi_N = \{x_i\}_1^N$

$$\int_0^1 w(x)y_n(x)\,\mathrm{d}x = \int_{-1}^1 w\left(x(t)\right) \frac{y_n\left(x(t)\right)}{\pi\sqrt{1-t^2}}\,\mathrm{d}t = \sum_{i=1}^N q_i(\xi_N) \frac{y_n\left(x(t_i)\right)}{\pi\sqrt{1-t_i^2}},$$

where $y_n(x) = \sum_{\alpha=1}^n \theta_\alpha f_\alpha(x)$, and it follows from (7.14) that $y_n(x(t))/\sqrt{1-t^2}$ is a polynomial of degree not higher than $n-1$. If $w(x(t)) = \sqrt{1-t^2}$, then ξ_N^* must coincide with zeros of $U_N(t)$, which are

$$x_i = x(t_i) = \frac{i}{N+1}, \quad i = 1, \ldots, N,\ 2N \geqslant n;$$

compare with Müller–Gronbach (1993). Accordingly to (7.13) weights are $q_i(\xi_N) = \pi \sin^2 \pi x_i/(N+1)$, and finally

$$\int_0^1 y_n(x) \sin \pi x\,\mathrm{d}x - \frac{1}{N+1} \sum_{i=1}^N y_n(x_i) \sin \pi x_i.$$

The solution is extremely simple, but it could be more complicated for other weight functions, see Davis and Rabinowitz (1986). The value of $Q_2(\xi_N^*)$ is of order $\mathrm{O}(\lambda_{2N})$. Various results about remainders in approximate integration theory may lead to the better estimates, but the corresponding technique is beyond the scope of this paper. Further details and related results may be found for instance, for the one dimension case in Davis and Rabinowitz (1986), Szegö (1959) and for the multi-dimension case in Stroud (1975). Similar exercises may be done for the criterion $Q_1(\xi_N)$ with $\sigma = 0$. The minimization of $Q_1(\xi_N)$ now becomes a problem from the function approximation theory. When $Z = X$ and $w(x) \equiv 1$, then it follows from (7.1) that

$$Q_1(\xi_N) = \mathrm{E} \int_X \left(y(x) - \widehat{y}(x)\right)^2 \mathrm{d}x = \mathrm{E} \int_X \left(y(x) - \sum_{i=1}^N y(x_i) v_i(x) \right)^2 \mathrm{d}x, \tag{7.15}$$

where $v^T(x) = V(x, \xi_N) V^{-1}(\xi_N)$.

It is known (see, e.g., Micchelli and Wahba, 1981), that

$$Q(\xi_N) \geqslant \mathrm{E}\left[\min_\gamma \int_X \left(y(x) - \sum_{\alpha=1}^N \gamma_\alpha f_\alpha(x) \right)^2 \mathrm{d}x \right] = \sum_{\alpha=N+1}^\infty \lambda_\alpha. \tag{7.16}$$

This lower bound may be used to evaluate the efficiency of ξ_N and can be achieved for any singular kernel, i.e., when

$$V(x,x') = \sum_{\alpha=1}^{n} \lambda_\alpha f_\alpha(x) f_\alpha(x').$$

To verify the latter conjecture one has to select the design ξ_N coinciding with all zeros of $f_n(x)$; see some additional details in Fedorov and Hackl (1994). In cases when eigenfunctions cannot be found analytically the use of the remainder theory is probably one of the most reliable ways to construct satisfactory designs; see, e.g., Davis and Rabinowitz (1984) or Achieser (1956). The ideas discussed in this sub-section help to generate effective designs with very moderate number of observations N, obviously much less than we need using the Sacks–Ylvisaker approach based on the local approximation of $y(x)$. The author is not familiar with any studies where the connection between the classical approximation theory and the Sacks–Ylvisaker approach were analyzed systematically for models of type (7.8) with $n \to \infty$. Perhaps Micchelli and Wahba (1981) and Müller–Gronbach (1993) considered the closest ideas and models.

Again, we would like to note that in most cases measurement errors may contribute substantially to the randomness of observations. The rule of thumb in selection of the number of terms in (7.8) is that the least eigenvalue λ_N should be significantly less than σ.

8. Discrete case. Optimality criteria and the lower bounds

When the design region X and the set of interest Z are discrete and contain N_X and N_Z points correspondingly, then the covariance matrix of the vector $\{y(x_i) - \widehat{y}(x_i)\}_1^{N_Z}$, $x \in Z$, completely describes any objective function based on the second moments. We use the notation $D_Z(\xi_N)$, when the latter matrix consists of elements

$$D(\xi_N)_{ij} = V(x_i,x_j) - V^T(x_i,\xi_N)V^{-1}(\xi_N)V(x_j,\xi_N), \quad x_i, x_j, \in Z.$$

The discrete versions of $Q_1(\xi_N)$ and $Q_2(\xi_N)$ are correspondingly

$$\operatorname{tr} D_Z(\xi_N) \quad \text{and} \quad L^T D_Z(\xi_N) L, \tag{8.1}$$

where $L^T = (1,\ldots,1)$. We will introduce any weights as we did in the continuous case, to keep notations simple. In the discrete case we may introduce a very special version of D-optimality

$$Q_3(\xi_N) = \ln |D_Z(\xi_N)|. \tag{8.2}$$

It is assumed that there are no points in common for Z and $\operatorname{supp}\xi_N$. Otherwise the determinant equals zero, because

$$\mathrm{E}\big[\big(\widehat{y}(x_i) - y(x_i)\big)^2\big] = 0, \quad \text{when } x_i \in \operatorname{supp}\xi_N.$$

Criterion (8.2) is very popular in the statistical literature related to the optimization of monitoring networks; see, for instance, Guttorp et al. (1993), Carelton et al. (1992), Schumacher and Zidek (1993), Shewry and Wynn (1987). In the cited papers the authors talk about either entropy or information. After the assumption of multivariate normality of the corresponding distributions is made, all approaches lead to various modifications of D-optimality; compare with Lindley (1956).

In addition to (8.1) and (8.2) a number of other criteria were introduced for application in monitoring network improvement. A good collection of them can be found in Megreditchan (1979, 1989); see also Fedorov and Hackl (1994).

As soon as the criterion of optimality $Q(\xi_N)$ (we use this notation to emphasize that only the criteria of optimality related to the problem of interpolation or extrapolation are considered in this section) and the kernel $V(x, x')$ are defined, we have to find

$$\xi_N^* = \arg\min_{\xi_N} Q(\xi_N). \tag{8.3}$$

It is interestingly to note that optimization problem (8.3) was considered in a very different setting by Currin et al. (1991) and by Morris et al. (1993); see also Sacks et al. (1989) for older references. They considered the Bayesian approach to design of computer experiments and introduced $Q(\xi_N)$ as a measure of discrepancy between a computer model and its approximation based on some prior knowledge expressed through the smoothness of the exact response. The latter was defined by a covariance function.

When N_Z is relatively "small" and N_X is not very "large" then exhaustive search may be a proper numerical procedure for a modern computer. With increase of N_Z and N_X one can use the exchange type algorithms discussed in Shewry and Wynn (1987) and in Fedorov and Hackl (1994), which are similar to those discussed in Sections 2–4.

An alternative approach may be based on the introduction of a model similar to (7.8). For the sake of simplicity of notations, let $Z \subset X$, and let

$$\lambda_\alpha f_\alpha = V f_\alpha, \tag{8.4}$$

where $\lambda_1 \geqslant \lambda_2 \geqslant \cdots \geqslant \lambda_{N_X}, V_{ij} = V(x_i, x_j)$ and $x_i, x_j \in X$. Then one may consider the following approximate model

$$Y \simeq Y_N = \sum_{\alpha=1}^{N} \theta_\alpha f_\alpha \tag{8.5}$$

where all vectors Y, Y_N and f_α have N_X components and

$$Y_i = y(x_i), \quad x_i \in X,$$

$$\mathrm{E}(\theta_\alpha) = 0, \quad \mathrm{E}(\theta_\alpha^2) = \lambda_\alpha, \; \alpha = 1, \ldots, N.$$

Similar to (7.16)

$$Q_1(\xi_N) \geqslant \left[\min_{\gamma}(Y - F\gamma^T)^T(Y - F\gamma^T)\right]$$

$$= [(Y - Y_N)^T(Y - Y_N)] = \sum_{\alpha=N+1}^{N_X} \lambda_\alpha, \tag{8.6}$$

where $F = (f_1, \ldots, f_N)$. There exists another result that can help to evaluate the closeness of Y_N and Y. Let

$$V_N = \mathrm{E}(Y_n Y_n^T) = \sum_{\alpha=1}^{N} \lambda_\alpha f_\alpha f_\alpha^T.$$

Then (compare with Rao, 1973, Chapter 8g)

$$V_N = \arg\min_A \|V - A\|,$$

where rank $A = N$ and symbol $\|B\|$ denotes the Frobenius norm of B defined by $(\operatorname{tr} B^2)^{1/2} = (\sum_{ij} B_{ij}^2)^{1/2}$. Moreover

$$\|V - V_N\|^2 = \sum_{\alpha=N+1}^{N_X} \lambda_\alpha^2. \tag{8.7}$$

Thus, the vector Y_N is the best (maybe not unique) approximation of Y in the sense of two criteria (8.6) and (8.7). In fact, it is the best one for any strictly increasing function of $D = \mathrm{E}[(Y - \widetilde{Y}_N)(Y - \widetilde{Y}_N)^T]$ which is invariant under orthogonal transformations, where $\widetilde{Y}_N = BY$, rank $B \leqslant N$; see Seber (1984, Chapter 5.2). The vector

$$\widehat{Y}_N = V(X, \xi_N)V^{-1}(\xi_N)Y,$$

where $V^T(X, \xi_N) = (V(x_1, \xi_N), \ldots, V(x_N, \xi_N))$, is one of the above linear estimates. Therefore (8.6) and (8.7) help to find the lower bounds for criteria depending upon

$$D(\xi_N) = V - V(X, \xi_N)V^{-1}(\xi_N)V(X, \xi_N) = \mathrm{E}\big[(Y - \widehat{Y}_N)(Y - \widehat{Y}_N)^T\big].$$

Model (8.5) helps to understand some features of optimal designs and lead to some interesting numerical procedures (see next section). Adding the "white" noise, i.e., introducing the following model

$$Y_n = \sum_{\alpha=1}^{N} \theta_\alpha f_\alpha + \sigma\varepsilon, \tag{8.8}$$

where $\mathrm{E}(\varepsilon) = 0$ and $\mathrm{E}(\varepsilon\varepsilon^T) = I$, allows us to use all the tools discussed in Sections 2, 3, 5 to generate optimal designs.

9. Unknown covariance function

All the results discussed in the previous sections have essentially used the fact that either a covariance function $V(x, x')$ or a matrix Λ is known. That is possible but unfortunately uncommon in practice. In this section we explore two approaches to estimate the covariance structure.

Direct estimation of a covariance matrix. Let us start with a discrete design region X and assume there exist repeated observations at every point of X. Meteorological and environmental networks provide the most typical examples; see, e.g., Megreditchan (1979, 1989) and Oehlert (1995a, b).

Let us define (compare with the previous section) the dispersion matrix of residuals $Y - \widehat{Y}$ as

$$D(\xi_N) = \min_B \mathrm{E}\big[(Y - BY(\xi_N))\,(Y - BY(\xi_N))^T\big], \tag{9.1}$$

where $\xi_N = (x_1, \ldots, x_N)$, $Y^T(\xi_N) = (y(x_1), \ldots, y(x_N))$, and B is an $N_X \times N$ matrix. For the sake of simplicity we assume that $\mathrm{E}(Y) = 0$ and that this fact is *a priori* known. Minimization is understood in the matrix ordering sense. A solution of (9.1) is

$$B^* = V(X, \xi_N)V^{-1}(\xi_N)$$

and

$$D(\xi_N) = V(X) - V(X, \xi_N)V^{-1}(\xi_N)V^T(X, \xi_N).$$

When sufficiently many observations are accumulated at every point of X the strong law of large numbers assures us that

$$\begin{aligned}\frac{1}{k}\sum_{j=1}^{k} &(Y_j - BY_j(\xi_N))\,(Y_j - BY_j(\xi_N))^T \\ &\simeq \mathrm{E}\Big[(Y - BY(\xi_N))\,(Y - BY(\xi_N))^T\Big]\end{aligned} \tag{9.2}$$

and subsequently $\widehat{B}^*$ and B^*, which minimize correspondingly the left and right hand-sides, are close to each other. Straightforward minimization gives

$$\widehat{B}^* = \widehat{V}(X, \xi_N)\widehat{V}^{-1}(\xi_N), \tag{9.3}$$

where both matrices with caps are evident partitions of

$$\widehat{V}(X) = k^{-1} \sum_{j=1}^{k} Y_j Y_j^T. \tag{9.4}$$

When there are missing observations, then it is better to use instead of (9.4) pairwise estimates

$$\widehat{V}_{il}(X) = k_{il}^{-1} \sum_{j=1}^{k_{il}} Y_{ji} Y_{jl},$$

where k_{il} is the number of cases when the response variable was measured at x_i and x_l simultaneously.

Thus, (9.2)–(9.4) lead us to a very simple and widely used recipe: replace unknown parameters by their estimates and use methods developed for cases in which all parameters are known. Together with this simple recommendation (9.2) helps to generate other versions of numerical algorithms considered earlier. Let us introduce matrix $I(\xi_N)$ with the following elements:

$$I_{il}(\xi_N) = \begin{cases} \delta_{il}, & \text{when } x_i \in \operatorname{supp} \xi_N, \\ 0, & \text{otherwise.} \end{cases}$$

The left-hand side of (9.2) may be represented now as

$$\widehat{D}(\xi_N, B) = k^{-1} \sum_{j=1}^{k} \left(Y_j - BI(\xi_N)Y_j\right) \left(Y_j - BI(\xi_N)Y_j\right)^T,$$

and the design problem may be viewed now as

$$\xi_N^* = \arg \min_{\xi_N} \min_{B} Q\left[\widehat{D}(\xi_N, B)\right]. \tag{9.5}$$

From the numerical point of view (9.5) may be considered as a multi-dimension version of the best regression selection problem. Stepwise regression and best subset selection are the popular algorithms and can be easily adopted to solve (9.5). In fact the same methods may be used when the matrix $\widehat{D}(\xi_N, B)$ is replaced by its true value; see comment in the conclusion of Section 8. Let

$$Q\left[\widehat{D}(\xi_N, B)\right] = \widehat{D}_{11}(\xi_N, B),$$

i.e., we want to minimize the variance of prediction at point x_1. In this case

$$\xi_N^* = \arg \min_{\xi_N} \min_{b} \sum_{j=1}^{k} \left(Y_{j1} - \sum_{i=2}^{N_X} b_i I_{1i}(\xi_N) Y_{ji} \right)^2, \tag{9.6}$$

and it is a very standard problem of selection of N predictors from $N_X - 1$ candidates and there exist a numerous number of the statistical packages which can be used to do that. The author is not familiar with multi-dimension versions of the corresponding software products, which are needed for more complicated criteria. The idea to use the least squares technique for selection of the most informative subset of sensors was probably initiated by Megreditchan (1979).

The search for an optimal design ξ_N may be viewed in this setting as a partitioning of X into two sets of given size $N_X - N$ and N. The latter must contain the most information about the whole set X; see, for instance, Shewry and Wynn (1987), who proposed using

$$\xi_N^* = \arg\min_{\xi_N} \ln |D_p(\xi_N)| , \tag{9.7}$$

where the subscript "p" indicates that the matrix contains only elements corresponding to the points (sites) with no observations. When (9.7) is replaced by its empirical version

$$\xi_N^* = \arg\min_{\xi_N} \min_{B} \ln \left|\widehat{D}(\xi_N, B)\right|, \tag{9.8}$$

then the following simple and intuitively attractive exchange-type procedure may be used to construct ξ_N^*; see Fedorov and Hackl (1994):

(a) Given $\xi_{Ns} = \{x_{i_s}\}_1^N$ find

$$i^+ = \arg\max_{i} \min_{b} \sum_{j=1}^{k} \left(Y_{ji} - \sum_{i_s=1}^{N} b_{i_s} Y_{ji_s} \right)^2 .$$

Add the point x_{i^+} to the design: $\xi_{(N+1)s} = \xi_{Ns} + x_{i^+}$.

(b) Find

$$i^- = \arg\min_{i_s} \min_{b} \sum_{j=1}^{k} \left(Y_{ji_s} - \sum_{l \neq i_s} b_l Y_{jl} \right)^2 ,$$

where $x_l \in \xi_{(N+1)s}$ and delete the point x_{i^-} from the design, i.e., construct $\xi_{N(s+1)} = \xi_{(N+1)s} - x_{i^-}$. Return to (a).

Briefly, the exchange procedure (a), (b) may be spelled out in the following way: add to the design the worst explained sites and delete from it the best explained sites. Apparently, the approach may be called "model free": only existence of first two moments of observed Y is assumed. That may attract many practitioners. However in the search for an optimal network we are confined to sites where the measurements have been previously made. In other words, the selection of the most informative subsets of sites (sensors, observing stations) may be discussed, but we cannot consider the problem of optimal extension.

Estimation of a parameterized covariance. In many practical cases the design region X is a continuous set and the covariance function has to be known everywhere at X. The most popular approach is based on the assumption that this function is homogeneous and isotropic, i.e.,

$$V(x, x') = V(r), \quad r^2 = (x - x')^T \, (x - x')$$

with the subsequent parsimonious approximation of function $V(r)$; see, e.g., Cressie (1991), Marshall and Mardia (1985), Matérn (1986), Ying (1995). The approach is frequently used in geostatistics, where a single realization of a random field is available, and in particular in the "kriging method" paradigm.

Methods from in Sections 5–8 are essentially based on approximation of the observed random fields by regression models with random coefficients. When prior to design of a network there exist some historical observations, then one may use the technique, which was developed for these models.It is expedient to note that accurate knowledge of Λ or Λ_n is useful but it is not as crucial as the knowledge of a covariance function in the Sacks–Ylvisaker approach. In fact, in basic optimization problems (5.13) and (6.2) the objective functions depend upon the sum $M(\xi) + M_0$, where M_0 is defined by Λ. For instance, in the case of (6.2)

$$M_0 = \sigma^2 N^{-1} \Lambda$$

and therefore the role of Λ diminishes when either $\sigma^2 \to 0$ or $N \to \infty$. Moreover, the simple dependence upon Λ allows to construct numerically optimal designs for different matrices Λ to learn about their sensitivity with respect to Λ.

In the simplest case, when the observational errors are negligible, the following estimators may be used:

$$\widehat{\Lambda}_n = k^{-1} \sum_{j=1}^{k} \widehat{\theta}_j \widehat{\theta}_j^T, \tag{9.9}$$

$$\widehat{\theta}_j = \arg\min_{\theta} \sum_{x_i \in X_j} \left(y_j(x_i) - \theta^T f(x_i)\right)^2,$$

where $\theta \in R^n$, X_j is the set of points with observations $y_j(x_i)$, and

$$\operatorname{rank}\left(\sum_{x_i \in X_j} f(x_i) f^T(x_i) \right) = n$$

for all $j = 1, \ldots, k$. It is assumed that functions $f(x)$ are known and

$$V(x, x') \simeq V_n(x, x') = f^T(x) \Lambda_n f(x').$$

Subsequently,

$$\widehat{V}(x, x') = f^T(x)\widehat{\Lambda}_n f(x').$$

Actually, it is more convenient to use the matrix Λ_n directly than the function $\widehat{V}(x, x')$ in all numerical procedures discussed in Sections 5, 6.

When the observational errors are comparable with the variations of θ, then (9.9) must be replaced with more sophisticated estimators, which are computationally much more demanding and complicate. Details and references may be found in Spjotvill (1977) and Fedorov et al. (1993).

10. Space and time

In most spatial experiments, after the sites are selected measurements are usually taken on some regular schedule, for instance, several times a day, or they are continuously recorded. Generally, the response function may depend upon time. Random errors can be correlated both in time and space. We consider only the simplest case, where there is no spatial correlation, following the ideas from Section 2. The generalization for more general models considered in Sections 3–6 is straightforward.

To adopt (2.1) for the time dependent response we assume that

$$\eta(x, t, \theta) = \theta^T f(x, t), \quad 0 \leqslant t \leqslant T,$$

and

$$\mathrm{E}\left[\varepsilon(x_i, t_j)\varepsilon(x_{i'}, t_{j'})\right] = \delta_{ii'}\rho_{jj'}(x).$$

When $\rho_{jj'} = \delta_{jj'}$, then the information matrix of observations made accordingly to the time schedule $\zeta(\mathrm{d}t/x)$ may be presented in the following form

$$m(x) = \int_0^T f(x, t) f^T(x, t)\zeta(\mathrm{d}t/x).$$

For measurements which are correlated in time,

$$m(x) = \sum_{jj'} f(x, t_j) R_{jj'}^{-1}(x) f^T(x, t_{j'}),$$

where, for the sake of simplicity, we assume that $\operatorname{supp}\zeta(\mathrm{d}t/x)$ is a discrete set $t_1, t_2, \ldots, t_r$, and $R(x) = \rho_{jj'}(x)_1^r$.

When the measure $\zeta(\mathrm{d}t/x)$ is fixed for each given x, then all the results from Section 2 may be used, with obvious replacement the function $\psi(x, \xi)$, which in the standard case has the form

$$\Phi(\xi) - f^T(x) A(\xi) f(x)$$

for all criteria satisfying Assumptions (a)–(e), by the function

$$\psi(x,\xi) = \Phi(\xi) - \text{tr } m(x)A(\xi).$$

For instance, for the D-criterion the sensitivity function $m - f^T(x)M^{-1}(\xi)f(x)$ must be replaced by $m - \text{tr } m(x)M^{-1}(\xi)$. More details may be found in Atkinson and Fedorov (1988), Fedorov and Nachtsheim (1995), and Spruil and Studden (1979). Formally the time dependent observation may be treated as a vector-observation case (see, for instance, Fedorov, 1972, Chapter 5).

Evidently, introducing the time variable does not change the basic theory, but makes all techniques, including computing of optimal designs, more time and effort consuming. However there exist models and optimality criteria for which optimal designs are the same both for the static and for the time dependent cases. For instance, the latter is true for models with uncorrelated observations and with separable variables, when

$$f(x,t) = f_1(x) \otimes f_2(t)$$

or where

$$\eta(x,\theta) = \theta_0 + \theta_1^T f_1(x) + \theta_2^T f_2(t), \quad \theta^T = (\theta_0, \theta_1^T, \theta_2^T),$$

and the selected criterion satisfies Assumptions (a)–(e) from Section 2; see Cook and Thibodeau (1980), Hoel (1965), Huang and Hsu (1993), Schwabe (1994, 1995).

When time is included explicitly in model, then the concept of sensor allocation can be extended and "mobile" sensors may be introduced. In this case design consists of trajectories $x_i(t) \in X$, $0 \leqslant t \leqslant T$. The topic is beyond the scope of this survey. A reader can find the results and references in Chang (1979), Fedorov and Nachtsheim (1995), Titterington (1980) and Zarrop (1979).

Acknowledgement

I am most grateful to my immediate colleagues D. Downing and M. Morris for their very constructive and effective help in preparing this paper. I thank B. Wheeler for his numerous and very useful comments and suggestions.

References

Achieser, N. I. (1956). *Theory of Approximation*. Frederick Ungar, New York.

Atkinson, A. C. and V. V. Fedorov (1988). Optimum design of experiments. In: S. Kotz and N. I. Johnson, eds., *Encyclopedia of Statistics, Supplemental Volume*. Wiley, New York, 107–114.

Atkinson, A. C. and A. N. Donev (1992). *Optimum Experimental Design*. Clarendon Press, Oxford.

Benhenni, K. and S. Cambanis (1992). Sampling designs for estimating integrals of stochastic process. *Ann. Statist.* **20**, 161–196.

Bickel, P. J. and A. M. Herzberg (1979). Robustness of design against autocorrelation in time, I. *Ann. Statist.* **7**, 77–95.

Bishoff, W. (1992). On exact D-optimal designs for models with correlated observations, *Ann. Inst. Statist. Math.* **44**, 229–238.

Bishoff, W. (1993). On D-optimal designs for linear models under correlated observations with applications to a linear model with multiple response. *J. Statist. Plann. Inference.* **37**, 69–80.

Brimkulov, U., G. Krug and V. Savanov (1986). *Design of Experiments for Random Fields and Processes.* Nauka, Moscow.

Butkovskiy, A. G. (1982). *Green's Functions and Transfer Functions Handbook*, Wiley, New York.

Cambanis, S. (1985). Sampling designs for time series. In: *Handbook of Statistics*, Vol. 5. North-Holland, New York, 337–362.

Cambanis, S. and Y. Su (1993). Sampling design for estimation of a random process. *Stochastic Process. Appl.* **46**, 47–89.

Chang, D. (1979). Design of optimal control for a regression problem. *Ann. Statist.* **7**, 1078–1085.

Cook, R. D. and L. A. Thibodau (1980). Marginally restricted D-optimal designs. *J. Amer. Statist. Assoc.* **75**, 366–371.

Cook, R. D. and C. J. Nachtsheim (1980). A comparison of algorithms for constructing exact D-optimal designs. *Technometrics* **24**, 315–324.

Cook, R. D. and V. V. Fedorov (1995). Constrained optimization of experimental design. *Statistics* **26**, 129–178.

Cressie, N. A. (1991). *Statistics for Spatial Data.* Wiley, New York.

Currin, C., T. Mitchell, M. Morris and D. Ylvisaker (1991). Bayesian prediction of deterministic functions, with applications to the design and analysis of computer experiments. *J. Amer. Statist. Assoc.* **31**, 953–963.

Davis, P. J. and Ph. Rabinowitz (1984). *Methods of Numerical Integration*, 2nd edn. Academic Press, New York.

Eaton, J. L., A. Giovagnoli and P. Sebastiani (1994). Tech. Report 598. School of Statistics, University of Minnesota.

Ermakov, S. M., ed. (1983). *Mathematical Theory of Experimental Design.* Nauka, Moscow.

Ermakov, S. M. (1989). Random interpolation in the theory of experimental design. *Comput. Statist. Data Anal.* **8**, 75–80.

Fedorov, V. V. (1972). *Theory of Optimal Experiments.* Academic Press, New York.

Fedorov, V. V. (1989). Optimum design of experiments with bounded density: Optimization algorithms of the exchange type. *J. Statist. Plann. Inference* **24**, 1–13.

Fedorov, V. V. and A. B. Uspensky (1975). *Numerical Aspects of Design and Analysis of Experiments.* Moscow State University Press, Moscow.

Fedorov, V. V., S. L. Leonov and S. A. Pitovranov (1988). Experimental design technique in the optimization of a monitoring network. In: V. V. Fedorov and M. Lauter, eds., *Model-Oriented Data Analysis.* Springer, New York, 165–175

Fedorov, V. V. and W. Müller (1989). Comparison of two approaches. In the optimal design of an observation network. *Statistics* **19**, 339–351.

Fedorov, V. V. and W. Müller (1989). Design of an air-pollution monitoring network. An application of experimental design theory. *Österreich. Z. Statist. Informatik* **19**, 5–17.

Fedorov, V. V., P. Hackl and W. Müller (1993). Estimation and experimental design for second kind regression models. *Informatik Biometr. Epidemiol.* **24**, 134–151.

Fedorov, V. V. and P. Hackl (1994). Optimal experimental design: Spatial sampling. *Calcutta Statist. Assoc. Bull.*

Fedorov, V. V. and C. J. Nachtsheim (1995). Optimal designs for time – dependent responses. In: L. P. Kitsos and W. G. Mueller, eds., *Advanced in Model-Oriented Data Analysis.* Springer, New York, 3–12. **44**, 57–81.

Fisher, R. A. (1947). *The Design of Experiments*, 4th edn. Hafner, New York.

Gribik, P. R., K. O. Kortanek and J. R. Sweigart (1976). Design a regional air pollution monitor network: An appraisal of a regression experimental approach. In: *Proc. Conf. Environmental Modeling and Simulation.* EPA 600/9-76-016. Research Triangle Park, NC, 86–91.

Guttorp, P., N. D. Le, P. D. Sampson and J. V. Zidek (1993). Using entropy in the redesign of environmental monitoring network. In: G. P. Patil and C. R. Rao, eds., *Multivariate Environmental Statistics.* Elsevier, New York, 175–202.

Hoel, P. G. (1965). Minimax designs in two dimensional regression. *Ann. Math. Statist.* **29**, 1134–1145.
Huang, M. L. and M. C. Hsu (1993). Marginally restricted linear-optimal designs. *J. Amer. Statist. Assoc.* **35**, 251–266.
Kanwal R. P. (1971). *Linear Integral Equations, Theory and Technique*. Academic Press, New York.
Karlin, S. and W. Studden (1966). *Tchebysheff Systems: With Applications in Analysis and Statistics*. Wiley, New York.
Kiefer, J. (1959). Optimal experimental design. *J. Roy. Statist. Soc. Ser. B* **21**, 272–319.
Kiefer, J. (1974). General equivalence theory for optimum designs (Approximate Theory). *Ann. Statist.* **2**, 849–879.
Kunert, J. (1988). Optimal designs for correlation in the plane. In: Dodge, Fedorov and Wynn, eds., *Optimal Design and Analysis of Experiments*. North-Holland, New York, 123–131.
Lindley, D. (1956). On a measure of information provided by an experiment. *Ann. Math. Statist.* **27**, 986–996.
Marshall, R. J. and K. V. Mardia (1985). Minimum norm quadratic estimation of components of spatial covariance. *Math. Geol.* **17**, 517–525.
Martin, R. J. (1986). On the design of experiments under spatial correlation. *Biometrika* **73**, 247–277.
Matèrn, B. (1986). *Spatial Variation*, Lecture Notes in Statistics. Springer, Berlin.
Megreditchan, G. (1979). L'optimization des reseaux d'observation des champs meterologiques. *La Meterologie* **6**, 51–66.
Megreditchan, G. (1989). Statistical redundancy as a criterion for meterological network optimization. *Österreich. Z. Statist. Informatik* **19**, 18–29.
Micchelli, C. A. and G. Wahba (1981). Design problems for optimal surface interpolation. In: Z. Ziegler, ed., *Approximation Theory and Applications*. Academic Press, New York, 329–348.
Mikhailov, G. A. and A. A. Zhigljavskii (1989). Uniform optimization of weighted estimates of the Monte Carlo method. *Soviet Math. Dokl.* **38**, 523–526.
Mitchell, T. J. (1974). (a) An algorithm for construction of D-optimal experimental design. *Technometrics* **16**, 203–210; (b) Computer construction of D-optimal "first-order" designs. *Technometrics* **16**, 211–220.
Morris, M. D., T. B. Mitchell and D. Ylvisaker (1993). Bayesian design and analysis of computer experiments: Use of derivatives in surface prediction. *Technometrics* **35**, 243–255.
Müller, W. and A. Pázman (1995). Design measures and extended information matrices for optimal designs when the observations are correlated, TR No. 47, Dept. Statistics, Univ. of Economics, Vienna.
Müller-Gronbach, Th. (1993). Optimal designs for approximating the path of stochastic process. Preprint Nr. A-93-14, Freie Universität, Berlin.
Nachtsheim, C. J. (1987). Tools for computer-aided design of experiments. *J. Quality Contr.* **19**, 132–160.
Näther, W. (1985). *Effective Observation of Random Fields*, Teubner Texte Zur Mathematik-Band 72. Teubner, Leipzig.
Nguen, N. K. and A. J. Miller (1992). A review of some exchange algorithms for constructing discrete D-optimal designs. *Comp. Statist. Data Anal.* **14**, 489–498.
Oehlert, G. W. (1995). The ability of wet deposition network to detect temporal trends. *Environmetrics* **6**, 327–339.
Oehlert, G. W. (1995). Shrinking a wet deposition network. *Atmospheric Environm.*, to appear.
Pázman, A. (1986). *Foundations of Optimum Experimental Design*. Reidel, Dordrecht.
Pilz, T. (1991). *Bayesian Estimation and Experimental Design in Linear Regression Models*. Wiley, New York.
Puckelsheim, F. (1993). *Optimal Design of Experiments*. Wiley, New York.
Rabinowitz, N. and D. M. Steinberg (1990). Optimal configuration of a seismographic network: A statistical approach. *Bull. Seismol. Soc. Amer.* **80**, 187–196.
Rao, C. R. (1973). *Linear Statistical Inference and its Applications*, 2nd edn. Wiley, New York.
Ripley, B. D. (1981). *Spatial Statistics*. Wiley, New York.
Sacks, J., W. J. Welch, T. J. Mitchell and H. P. Wynn (1989). Design and analysis for computer experiments. *Statist. Sci.* **4**, 409–423.
Sacks, J. and D. Ylvisaker (1966). Designs for regression problems with correlated errors. *Ann. Math. Statist.* **37**, 66–89.

Sacks, J. and D. Ylvisaker (1968). Designs for regression problems with correlated errors, many parameters. *Ann. Math. Statist.* **39**, 49–69.

Sacks, J. and D. Ylvisaker (1970). Designs for regression problems with correlated errors, III. *Ann. Math. Statist.* **41**, 2057–2074.

Sacks, J. and D. Ylvisaker (1970). Statistical designs and integral approximation. In: *Proc. 12th Bieu. Sem. Canad. Math. Cong.* Canad. Math. Cong., Montreal, 115–136.

SAS/QC Software: Design of Experiments Tools (1995). SAS Institute.

Seber, G. A. F. (1977). *Linear Regression Analysis.* Wiley, New York.

Seber, G. A. F. (1984) *Multivariate Observations.* Wiley, New York.

Shewry, M. C. and H. P. Wynn (1987). Maximum entropy sampling. *J. Appl. Statist.* **14**, 165–170.

Schumacher, P. and J. V. Zidek (1993). Using prior information in designing intervention detection experience. *Ann. Statist.* **21**, 447–463.

Schwabe, R. (1994). Optimal designs for additive linear models. Preprint No. 1-5-94, Freie Universität Berlin, Fachbereich Mathematik.

Schwabe, R. (1995). Designing experiments for additive nonlinear models. In: C. P. Kitsos and W. G. Mueller, *Advances in Model-Oriented Data Analysis.* Springer, New York, 77–85.

Silvey, S. D. (1980). *Optimal Design.* Chapman and Hall, London.

Spjotvill, E. (1977). Random coefficient regression models – A review. *Statistics* **8**, 69–93.

Spruill, M. C. and W. J. Studden (1979). A Kiefer–Wolfowitz theorem in stochastic process setting. *Ann. Statist.* **7**, 1329–1332.

Stroud, A. H. (1975). *Approximate Calculation of Multiple Integrals.* Prentice-Hall, Englewood Cliffs, NJ.

Szegö, G. (1959). *Orthogonal Polynomials.* Amer. Math. Soc., New York.

Titterington, D. M. (1980). Aspects of optimal design in dynamic systems. *Technometrics* **22**, 287–300.

Wahba, G. (1971). On the regression design problem of Sacks and Ylvisaker. *Ann. Math. Statist.* **42**, 1035–1053.

Wahba, G. (1974). Regression design for some equivalence classes of kernels. *Ann. Statist.* **2**, 925–934.

Wheeler, B. (1994). *ECHIP: Version 6.0 for Windows.* ECHIP, Hockessin.

Wynn, H. (1982). Optimum submeasures with applications to finite population sampling. In: *Statistical Decision Theory and Related Topics III*, Vol. 2. Academic Press, New York, 485–495.

Ying, Z. (1993). Maximum likelihood estimation of parameters under a spatial sampling scheme. *Ann. Statist.* **21**, 1567–1590.

Ylvisaker, D. (1975). Design on random fields. In: *A Survey of Statistical Design and Linear Models.* North-Holland, New York.

Ylvisaker, D. (1987). Prediction and design. *Ann. Statist.* **15**, 1–18.

Ylvisaker, D. (1988). Bayesian interpolation schemes. In: V. Dodge, V. V. Fedorov and M. Wynn, eds., *Optimal Design and Analysis of Experiments.* North-Holland, New York, 169–278.

Zarrop, M. B. (1979). *Optimal Experimental Design for Dynamic System Identification.* Springer, New York.

Zhigljavsky, A. A. (1988). Optimal designs for estimating several integrals. In: Y. Dodge, V. V. Fedorov and H. P. Wynn, eds., *Optimal Design and Analysis of Experiments.* North-Holland, New York.

Zimmerman, D. L. and M. B. Zimmerman (1991). A comparison of spatial semivariagram estimators and corresponding ordinary kriging predictors. *Technometrics* **33**, 77–91.

S. Ghosh and C. R. Rao, eds., *Handbook of Statistics, Vol. 13*

Design of Experiments with Selection and Ranking Goals

Shanti S. Gupta and S. Panchapakesan

1. Introduction

In many practical situations in everyday life, the experimenter is faced with the problem of comparing k ($\geqslant 2$) alternatives with a view to select the "best" among them. These may, for example, be different varieties of wheat in an agricultural experiment, or different coherent systems in engineering models, or different drugs prescribed for treatment of a certain disease. In all these problems, each alternative is characterized by the value of a parameter θ. In the above-mentioned situations, this parameter may be the average yield of a variety of wheat, or the reliability function of a system, or a measure of the effectiveness of a drug.

Consider the well-known balanced one-way layout given by the model:

$$Y_{ij} = \mu_i + \varepsilon_{ij}, \quad i = 1, \ldots, k; \; j = 1, \ldots, n, \tag{1.1}$$

where Y_{ij} is the jth response on the ith treatment, the μ_i are *unknown* treatment means, and the ε_{ij} are the measurement errors assumed to be independent and normally distributed with mean zero and variance σ^2. The classical approach in this case is generally to test the so-called homogeneity hypothesis H_0: $\mu_1 = \mu_2 = \cdots = \mu_k$ using the analysis of variance (ANOVA) approach. However, this does not serve the experimenter's real purpose which is not just to accept or reject the homogeneity hypothesis of no treatment differences. In practice, the experimenter will be considering treatments that are indeed different, and with sufficiently large sample, will be rejecting H_0 at any specified level. The real goal of the experimenter to be addressed then is to identify the best alternative (the variety with the largest average yield, the most reliable system, the most effective drug, and so on). Thus the inadequacy of the ANOVA lies in the types of decisions that are made on the basis of the data and not in the design aspects of the procedure. The method of estimating the sizes of the differences between the treatments was often used as an indirect way of reaching a decision regarding the best treatment(s). The attempts to formulate the decision problem to achieve this realistic goal set the stage for the development of the selection and ranking theory.

Selection and ranking problems have generally been formulated adopting one of two main approaches now familiarly known as the *indifference-zone formulation* and the *subset selection formulation*. Consider an experiment with k treatments in which we have n responses for each treatment. We assume the model (1.1).

Let the ordered μ_i be denoted by $\mu_{[1]} \leqslant \cdots \leqslant \mu_{[k]}$. It is assumed that there is no information regarding the correct pairing between the ordered and the unordered μ_i.

In the indifference-zone approach due to Bechhofer (1954), the goal is to select the treatment associated with the largest mean $\mu_{[k]}$ (called the best treatment). A selection satisfying the goal is defined to be a *correct selection* (CS). It is required that a correct selection be guaranteed with a probability P^* $(1/k < P^* < 1)$ whenever $\mu_{[k]} - \mu_{[k-1]} \geqslant \delta^*$, where δ^* is a positive constant. Here δ^* and P^* are specified *in advance* by the experimenter, and P^* is chosen greater than $1/k$ because otherwise we can make a no-data decision by selecting one of the treatments randomly as the best. When $\mu_{[k]} - \mu_{[k-1]} < \delta^*$, two or more treatments including the best are *sufficiently close* and the experimenter is assumed to be indifferent as to setting probability requirement in this case. The region $\Omega_{\delta^*} = \{(\underset{\sim}{\mu}, \sigma^2) \mid \underset{\sim}{\mu} = (\mu_1, \ldots, \mu_k), \sigma^2 > 0,\ \mu_{[k]} - \mu_{[k-1]} \geqslant \delta^*\}$ is called the *preference-zone* and its complement w.r.t. the entire parameter space $\Omega = \{(\underset{\sim}{\mu}, \sigma^2) \mid -\infty < \mu_i < \infty,\ i = 1, \ldots, k,\ \sigma^2 > 0\}$ is the *indifference-zone*. Denoting the probability of a correct selection (PCS) using the rule R by $P(\text{CS} \mid R)$, it is required that any valid rule R satisfy the condition:

$$P(\text{CS} \mid R) \geqslant P^* \quad \text{whenever } (\underset{\sim}{\mu}, \sigma^2) \in \Omega_{\delta^*}. \tag{1.2}$$

The design aspect of this basic setup is the determination of the minimum (common) sample size n so that the probability requirement (1.2) is satisfied.

In the subset selection approach developed by Gupta (1956), the goal is to select a nonempty subset of the k treatments so that the best treatment will be included in the selected subset with a guaranteed minimum probability P^*. The subset size is not specified in advance; it is random and determined by the data. Formally, any valid rule should satisfy the condition:

$$P(\text{CS} \mid R) \geqslant P^* \quad \text{for all } (\underset{\sim}{\mu}, \sigma^2) \in \Omega. \tag{1.3}$$

It is obvious that the requirement (1.3) can always be met by including all the treatments in the selected subset. So the performance of a rule is studied usually in terms of the expected size B of the selected subset. It is expected that a reasonable procedure will tend to select only one treatment when $\mu_{[k]} - \mu_{[k-1]}$ gets large.

Besides being a goal in itself, selecting a subset containing the best can also be considered as the first-stage screening in a two-stage procedure designed to select one treatment as the best; see, for example, Tamhane and Bechhofer (1977, 1979).

The probability requirements (1.2) and (1.3) are also known as the P^*-conditions. An important step in obtaining the constant(s) associated with a proposed rule R so that the P^*-condition is satisfied is to evaluate the infimum of the PCS over Ω or Ω_{δ^*} depending on the approach. Any configuration of $(\underset{\sim}{\mu}, \sigma^2)$ for which the infimum is attained is called a *least favorable configuration* (LFC).

Although we have discussed the selection problem in terms the normal means, the problem in general is to select from k populations $\Pi_1, \ldots, \Pi_k$ characterized by the distribution functions $F_{\theta_i}, i = 1, \ldots, k$, respectively, where the θ_i are unknown parameters taking values in the set Θ. The populations are ranked in terms of the θ_i (there may be other nuisance parameters). The ordered θ_i are denoted by $\theta_{[1]} \leqslant \cdots \leqslant \theta_{[k]}$ and the population associated with $\theta_{[k]}$ is defined to be the best. To define the preference-zone, one has to define a suitable nonnegative measure $\delta(\theta_i, \theta_j)$ of the separation between the populations Π_i and Π_j. Then $\Omega_{\delta^*} = \{\underset{\sim}{\theta} \mid \underset{\sim}{\theta} = (\theta_1, \ldots, \theta_k), \delta(\theta_{[k]}, \theta_{[k-1]}) \geqslant \delta^* > 0\}$.

There are several variations and generalizations of the basic goal in both indifference-zone and subset selection formulations. One can generalize the goal to select at least s of the t best populations with $1 \leqslant s \leqslant t \leqslant k-1$. In the subset selection approach, the size of the selected subset can be random subject to a specified maximum m ($1 \leqslant m \leqslant k$). This approach of *restricted subset selection* studied by Gupta and Santner (1973) and Santner (1975) combines the features of the indifference-zone and subset selection formulations. There have been other attempts in this direction of integrated formulations; for example, see Sobel (1969), Chen and Sobel (1987a, b). An important modification of the goal of selecting the best population is selecting a good population or a subset containing only good populations or containing all the good populations. A good population is defined as one which is close enough to the best within a specified threshold value.

There is now a vast literature on selection and ranking procedures. Several aspects of the theory and associated methodology of these and related procedures have been dealt with in the books by Bechhofer et al. (1968), Büringer et al. (1980), Gibbons et al. (1977), Gupta and Huang (1981), Gupta and Panchapakesan (1979) and Mukhopadhyay and Solanky (1994). A very recent book is by Bechhofer et al. (1995). A categorical bibliography is provided by Dudewicz and Koo (1982). Besides these books, there have been published review articles dealing with several specific aspects of selection and ranking. The reader is specially referred to Gupta and Panchapakesan (1985, 1988, 1991, 1993), Panchapakesan (1992, 1995a, b), and Van der Laan and Verdooran (1989).

In spite of the vast published literature, there have been only a few papers until the recent years devoted to design models beyond single-factor experiments. In this paper, besides the most common single-factor experiments involving mainly normal distributions, we review significant results involving blocking and factorial designs. The emphasis is not on a total coverage but enough to provide a focus on these problems to help assess the current status and potential for applications and further investigations.

Sections 2 through 5 discuss selection procedures and simultaneous confidence intervals under the assumption that the observed responses for treatments are normally distributed. Selection procedures under both the indifference-zone and the subset formulations are discussed in each section. Section 2 deals with selecting the best treatment and simultaneous confidence statements for comparisons with best in single-factor experiments. Section 3 considers these procedures in experiments with blocking while Section 4 deals with these procedures in factorial experiments. Selection with respect to a standard or control is discussed in Section 5. Selection in experiments involving other models is briefly discussed in Section 6. The models discussed are: Bernoulli, multinomial and restricted families such as IFR and IFRA.

2. Selecting the best treatment in single-factor experiments: Normal theory

Consider $k \geqslant 2$ treatments $\Pi_1, \ldots, \Pi_k$, where Π_i represents a normal population with mean μ_i and variance σ_i^2. The means μ_i are *unknown*. Different assumptions can be made about the σ_i^2 depending on the context of the experiment. As before, the ordered μ_i are denoted by $\mu_{[1]} \leqslant \cdots \leqslant \mu_{[k]}$ and no prior information is available regarding the true pairing of the ordered and unordered μ_i.

2.1. *Indifference-zone approach*

As described in Section 1, the goal is to identify one of the k treatments as the best (the one associated with $\mu_{[k]}$) with a guaranteed minimum probability P^* of a correct selection whenever $\mu_{[k]} - \mu_{[k-1]} \geqslant \delta^*$, where $\delta^* > 0$ and $1/k < P^* < 1$ are specified in advance.

Under the assumption that $\sigma_1^2 = \cdots = \sigma_k^2 = \sigma^2$ (*known*), Bechhofer (1954) proposed the following single-stage procedure based on samples of common size n. Let $\overline{Y}_i$ denote the mean of the sample responses Y_{ij}, $j = 1, \ldots, n$, from $\Pi_i, i = 1, \ldots, k$. His rule is

$$R_1: \quad \text{Select the treatment } \Pi_i \text{ that yields the largest } \overline{Y}_i. \tag{2.1}$$

The LFC for this rule is given by $\mu_{[1]} = \cdots = \mu_{[k-1]} = \mu_{[k]} - \delta^*$. For given $(k, \delta^*/\sigma, P^*)$, the minimum sample size n required to meet the P^*-condition is given by

$$n = \left\langle 2\left(\frac{\sigma H}{\delta^*}\right)^2 \right\rangle, \tag{2.2}$$

where $\langle x \rangle$ denotes the smallest integer $\geqslant x$, H satisfies

$$\Pr\{Z_1 \leqslant H, \ldots, Z_{k-1} \leqslant H\} = P^*, \tag{2.3}$$

and the Z_i are standard normal variates with equal correlation $\rho = 1/2$. Values of H can be obtained for several selected values of k and P^* from Bechhofer (1954), Gibbons et al. (1977), Gupta (1963), Gupta et al. (1973) and Milton (1963).

Hall (1959) and Eaton (1967) have shown that the rule R_1 in (2.1) is the most economical in the sense of requiring fewest observations per treatment among all single-stage location invariant procedures satisfying the P^*-condition.

Two stage procedures for the problem of selecting the normal treatment with the largest mean assuming a common known variance σ^2 have been studied by Cohen (1959), Alam (1970) and Tamhane and Bechhofer (1977, 1979). These procedures use the subset selection procedure of Gupta (1956, 1965) to eliminate inferior treatments at the first stage and select the best from among the remaining ones at the second stage. We describe the Tamhane–Bechhofer procedure R_2 below.

R_2: Take a random sample of n_1 observations from each Π_i, $i = 1, \ldots, k$. Eliminate from further consideration all treatments Π_i for which $\overline{Y}_i < \overline{Y}_{[k]} - h\sigma/\sqrt{n_1}$, where $\overline{Y}_{[1]} \leqslant \cdots \leqslant \overline{Y}_{[k]}$ are the ordered sample means $\overline{Y}_i$, and h is a constant to be determined. If only one treatment remains, then it is selected as the best. If more than one treatment remain, then proceed to the second stage by taking an additional random sample of size n_2 from each of these remaining treatments. Select the treatment that yields the largest sample mean based on the combined sample of $n_1 + n_2$ observations.

The above procedure R_2 of Tamhane and Bechhofer (1977, 1979) involves constants (n_1, n_2, h) to be determined in order to satisfy the P^*-condition. These constants are determined by using a minimax criterion (in addition to the P^*-condition) which minimizes the maximum over the entire parameter space Ω of the expected total sample size required by the procedure. The LFC for this procedure was first established only for $k = 2$. The constants (n_1, n_2, h) tabulated by Tamhane and Bechhofer (1979) for selected values of k, P^*, and δ^*/σ are conservative since they are based on the LFC for a lower bound of the PCS. The fact that the LFC for the PCS is $\mu_{[1]} = \cdots = \mu_{[k-1]} = \mu_{[k]} - \delta^*$ was proved by Sehr (1988) and Bhandari and Chaudhuri (1990).

A truncated sequential procedure for this problem has been investigated by Bechhofer and Goldsman (1987, 1989). It is designed to have improved performance over an earlier procedure of Bechhofer et al. (1968) which is an *open non-eliminating* sequential procedure as opposed to the Bechhofer–Goldsman procedure which is a *closed* but also a non-eliminating procedure. A multi-stage or sequential procedure is called *open* if, in advance of the experiment, no fixed upper bound is set on the number of observations to be taken from each treatment; otherwise, it is called *closed.* An *eliminating* procedure is one which excludes treatments from further sampling if they are removed from further consideration at any stage prior to taking the terminal decision. A non-eliminating procedure, on the other hand, samples from each treatment at each stage whether or not any treatment is removed from the final consideration of selection.

Another well-known procedure for the problem under discussion is that of Paulson (1964) which is a closed procedure with elimination. This procedure was successively improved (by changing the choices for certain constants) by Fabian (1974) and Hartman (1988). For further details regarding various procedures and their performance, see Gupta and Panchapakesan (1991).

We now consider the case of *unknown* common variance σ^2. This is the classical problem of the one-way ANOVA model. If one chooses to define the preference-zone as $\Omega_{\delta^*} = \{(\underset{\sim}{\mu}, \sigma) \mid \mu_{[k]} - \mu_{[k-1]} \geqslant \delta^*\sigma\}$, then the single-stage procedure R_1 in (2.1) can still be used with the minimum required sample size n given by (2.2) with $\sigma = 1$. If we continue with the preference-zone $\Omega_{\delta^*} = \{(\underset{\sim}{\mu}, \sigma) \mid \mu_{[k]} - \mu_{[k-1]} \geqslant \delta^*\}$ as before, it is not possible to devise a single-stage procedure that satisfies the P^*-condition. This is intuitively clear from the fact that the determination of the minimum sample size required depends on the knowledge of σ. In this case of unknown σ^2, Bechhofer et al. (1954) proposed an *open two-stage non-eliminating* procedure for selecting the best treatment. The first stage is used to estimate σ^2 and determine the total sample size needed to guarantee the probability requirement. The second stage, if necessary, is used to make the terminal decision. Using in addition the idea of screening, Tamhane

(1976) and Hochberg and Marcus (1981) have studied three-stage procedures where the first stage is utilized to determine the additional sample sizes necessary in the subsequent stages, the second stage is used to eliminate inferior populations by a subset rule, and the third stage (if necessary) to make the final decision. Tamhane (1976) also considered a two-stage eliminating procedure which was found to be inferior to the non-eliminating procedure of Bechhofer et al. (1954). Later, Gupta and Kim (1984) proposed a *two-stage eliminating* procedure with a new design criterion and obtained a sharp lower bound on the PCS. Gupta and Miescke (1984) studied two-stage eliminating procedures using a Bayes approach. Here we will describe the procedure of Gupta and Kim (1984).

R_3: Take a random sample of size n_1 ($\geqslant 2$) from each treatment. Let $\overline{X}_i$ be the sample mean associated with treatment Π_i, $i = 1, \ldots, k$, and S_ν^2 denote the usual pooled sample variance based on $\nu = k(n_1 - 1)$ degrees of freedom. Determine the subset I of $\{\Pi_1, \ldots, \Pi_k\}$ given by

$$I = \{\Pi_i \mid \overline{X}_i \geqslant \overline{X}_{[k]} - (dS/\sqrt{n_1} - \delta^*)^+\},$$

where $a^+ = \max(a, 0)$ and d is a constant to be chosen to satisfy the P^*-condition. If I consists of only one treatment, then select it as the best; otherwise, take an additional sample of size $N - n_1$ from each treatment in I, where

$$N = \max\{n_1, \langle (hS_\nu/\delta^*)^2 \rangle\},$$

$\langle y \rangle$ denotes the smallest integer $\geqslant y$, and h is a positive constant to be suitably chosen to satisfy the P^*-condition. Now, select as the best the population in I which yields the largest sample mean based on the combined sample of size N.

There are several possible choices for (n_1, d, h) to satisfy the P^*-condition. Gupta and Kim (1984) used the requirement that

$$\Pr\{\text{the best population is included the subset } I\} \geqslant P_1^*, \tag{2.4}$$

where $P_1^* (P^* < P_1^* < 1)$ is pre-assigned. Evaluation of these constants is based on a lower bound for the PCS. The Monte Carlo study of Gupta and Kim (1984) shows that their procedure R_3 performs much better than that of Bechhofer et al. (1954) in terms of the expected total sample size.

Recently, there have been a series of papers regarding the conjecture of the LFC for the Tamhane–Bechhofer procedure R_2 and some other related procedures for selecting the best normal treatment when the common variance σ^2 is known. As mentioned earlier, the conjecture that the LFC is $\mu_{[1]} = \cdots = \mu_{[k-1]} = \mu_{[k]} - \delta^*$ has been proved by Sehr (1988) and Bhandari and Chaudhuri (1990). The LFC's for two-stage procedures for more generalized goals have been established by Santner and Hayter (1993) and Hayter (1994). It will be interesting to reexamine the performance of the concerned procedures by using the exact infimum of the PCS.

2.2. Subset selection approach

As in Section 2.1, we are still interested in selecting the best treatment. However, we do not set in advance the number of treatments to be included in the selected subset. It is expected that a good rule will tend to select only one population as $\mu_{[k]} - \mu_{[k-1]}$ gets sufficiently large. Gupta (1956) considered the case of known as well as unknown σ^2. Based on samples of size n from each population, his rule, in the case of known σ^2, is R_4: Select the treatment Π_i if and only if

$$\overline{X}_i \geqslant \max_{1 \leqslant j \leqslant k} \overline{X}_j - \frac{d\sigma}{\sqrt{n}}, \tag{2.5}$$

where $\overline{X}_i$ is the sample mean from Π_i and d is the smallest positive constant for which the PCS $\geqslant P^*$ for all $(\underset{\sim}{\mu}, \sigma) \in \Omega$. (Any larger d would obviously satisfy the P^*-condition but would, if anything, only increase the size of the selected subset.) This smallest d is given by $\sqrt{2H}$ where H is the solution of (2.3). Thus d can be obtained from the tables mentioned previously.

When σ^2 is unknown, Gupta (1956) proposed the rule R_5 which is R_4 with σ replaced by S_ν, where S_ν^2 is the usual pooled estimator of σ^2 with $\nu = k(n-1)$ degrees of freedom. To keep the distinction between the two cases, we use d' in the place of d. The smallest d' needed to satisfy the P^*-condition is the one-sided upper $(1 - P^*)$ equicoordinate point of the equicorrelated $(k-1)$-variate central t-distribution with the equal correlation $\rho = 0.5$ and the associated degrees of freedom $\nu = k(n-1)$. The values of d' have been tabulated by Gupta and Sobel (1957) for selected values of k, n, and P^*. They are also available from the tables of Gupta et al. (1985) corresponding to correlation $\rho = 0.5$.

It should be noted that, unlike in the case of the indifference-zone approach, we do have a single-stage procedure for any specified n when σ^2 is unknown.

We may not have a common variance σ^2 (the heteroscedasticity case) or a common sample size (unbalanced design). These cases have been studied by Gupta and Huang (1976), Chen et al. (1976) and Gupta and Wong (1982). In all these cases, the authors have used lower bounds for the infimum of the PCS to meet the P^*-condition.

When the variances are unknown and unequal, and the sample sizes are unequal, Dudewicz and Dalal (1975) proposed a two-stage procedure using both the indifference-zone and subset selection approaches. Sequential subset selection procedures have also been studied which are applicable to the normal model. For a review of these, the reader is referred to Gupta and Panchapakesan (1991).

As we have pointed out previously, several modifications of the basic goal have been investigated. In particular, we mention here the restricted subset selection approach which includes a specified upper bound for the expected subset size which is otherwise random. Procedures of this type have been proposed by Gupta and Santner (1973) and Santner (1975).

Several authors have also studied the modified goal of selecting good populations. Reference can be made to Gupta and Panchapakesan (1985) and Gupta and Panchapakesan (1991).

2.3. Simultaneous confidence intervals for comparisons with the best

Related to the selection and ranking objectives is the multiple comparison approach in which one seeks simultaneous confidence sets for meaningful contrasts among a set of given treatments. A comprehensive treatment of this topic can be found in the text by Hochberg and Tamhane (1987). Our main interest here is simultaneous comparisons of all treatments with the best among them. In other words, we are interested in simultaneous confidence intervals for $\mu_i - \max_{j\neq i}\mu_j$, taking a larger treatment effect to imply a better treatment. If $\mu_i - \max_{j\neq i}\mu_j < 0$, then treatment i is not the best and the difference represents the amount by which treatment i is inferior to the best. On the other hand, if the difference is positive, then treatment i is the best and the difference is the amount by which it is better than the second best.

Assume that all (normal) treatments Π_i have a common unknown variance σ^2. Let $\overline{Y}_i$ denote the mean of n independent responses on treatment $\Pi_i, i = 1,\ldots,k$. Let S_ν^2 denote the usual pooled (unbiased) estimator of σ^2 based on $\nu = k(n-1)$ degrees of freedom. Hsu (1984) showed that the intervals

$$\Big[-\Big(\overline{Y}_i - \max_{j\neq i}\overline{Y}_j - C\Big)^-, \Big(\overline{Y}_i - \max_{j\neq i}\overline{Y}_j + C\Big)^+\Big], \quad i = 1,\ldots,k, \tag{2.6}$$

form $100(1-\alpha)\%$ simultaneous confidence intervals for $\mu_i - \max_{j\neq i}\mu_j$, $i = 1,\ldots,k$. Here $-x^- = \min(x,0)$, $x^+ = \max(x,0)$, and $C = cS_\nu/\sqrt{n}$, where c satisfies

$$P\Big[\overline{Y}_k - \max_{j\neq k}\overline{Y}_j > -cS_\nu/\sqrt{n}\Big] = 1-\alpha \tag{2.7}$$

under the assumption that $\mu_1 = \cdots = \mu_k = 0$.

The intervals in (2.6) are closely related to the selection and ranking methods discussed previously. It was shown by HSU (1984) that the upper bounds of these intervals imply the subset selection inference of Gupta (1956) and the lower bounds imply the indifference-zone selection inference of Bechhofer (1954). Any treatment i for which the upper bound of $\mu_i - \max_{j\neq i}\mu_j$ is zero can be inferred to be not the best. Similarly, any treatment i for which the lower bound of $\mu_i - \max_{j\neq i}\mu_j$ is zero can be inferred to be the best.

The intervals in (2.6) are "constrained" in the sense that the lower bounds are nonpositive and the upper bounds are nonnegative. Removing these constraints would require an increase in the critical value. The nonnegativity constraint on the upper bounds does not present a great disadvantage as one will not normally be interested in knowing how bad is a treatment that is rejected as not the best. On the other hand, it will be of interest to assess how much better than others is a treatment inferred to be the best. Motivated by these considerations Hsu (1985) provided a method of unconstrained multiple comparisons with the best which removes the nonpositivity constraint on the lower bounds in (2.6) by increasing the critical value slightly. We describe these simultaneous intervals below.

Let $D = dS_\nu/\sqrt{n}$ where d is the solution of

$$\Pr\{Z_i \leqslant Z_k + dS_\nu/\sqrt{n}, i = 1, \ldots, k-2, |Z_{k-1} - Z_k| \leqslant dS_\nu/\sqrt{n}\}$$
$$= 1 - \alpha, \tag{2.8}$$

where the Z_i are independent $N(0,1)$ variables. The $100(1-\alpha)\%$ simultaneous confidence intervals $[D_i^*, D_i^{**}]$ of Hsu (1985) for the differences $\mu_i - \max_{j \neq i} \mu_j$ are defined as follows:

$$D_i^* = \overline{Y}_i - \max_{j \neq i} \overline{Y}_j - D \quad \text{for } i = 1, \ldots, k,$$

$$D_i^{**} = \begin{cases} \left(\overline{Y}_i - \max_{j \neq i} \overline{Y}_j + D\right)^+ \wedge \left(-D_{[k]}^*\right), & \text{if } \overline{Y}_i \neq \overline{Y}_{[k]}, \\ \overline{Y}_i - \max_{j \neq i} \overline{Y}_j + D, & \text{if } \overline{Y}_i = \overline{Y}_{[k]}. \end{cases} \tag{2.9}$$

Here $\overline{Y}_{[k]}$ denotes the largest $\overline{Y}_i$ and $a \wedge b = \min(a, b)$. The constrained simultaneous confidence intervals can be implemented by using a computer SAS package; see Gupta and Hsu (1984, 1985), and Aubuchon et al. (1986). Hsu (1985) has tabulated the values of $d/\sqrt{2}$ for $k = 2(1)5$, $\nu = 5(1)20, 24, 30, 40, 60, 120, \infty$, and $\alpha = .01, .05$.

2.4. *Estimation after selection*

Consider the selection rule R_1 of Bechhofer (1954) defined in (2.1) for selecting the best treatment, namely, the one associated with the largest μ_i. This rule selects the population that yields the largest sample mean $\overline{Y}_i$. Let μ_S denote the treatment mean of the selected population. Then μ_S is a random variable and

$$\Pr\{\mu_S = \mu_i\} = \Pr\{\overline{Y}_i \geqslant \overline{Y}_j, j \neq i\}, \quad i = 1, \ldots, k.$$

The experimenter not only wishes to select the treatment with the highest mean but also wants an estimate of the mean for the treatment selected. Of course, the "natural" estimator of μ_S is $\overline{Y}_{[k]}$. For $k = 2$, $\overline{Y}_{[k]}$ is admissible and minimax under the squared error loss. For $k = 2$ and especially for $k > 2$, $\overline{Y}_{[k]}$ is highly unsatisfactory. It is highly positively biased when the μ_i are equal or close. The bias becomes more severe as k increases and, in fact, it tends to infinity as $k \to \infty$. Sarkadi (1967) and Dahiya (1974) have studied this problem for $k = 2$ and known common variance σ^2. Hsieh (1981) also discussed the $k = 2$ case but with unknown σ^2. Cohen and Sackrowitz (1982) considered the case $k \geqslant 3$ with known σ^2. They have given an estimator which is a convex weighted combination of the ordered sample means $\overline{Y}_{[i]}$ where the weights depend on the adjacent differences in the ordered means.

Jeyaratnam and Panchapakesan (1984) discussed estimation after selection associated with the subset selection rule R_4 defined in (2.5) for selecting a subset containing the best treatment. They considered estimating the average worth of the selected subset defined by $M = \sum_{i \in S} \mu_i / |S|$, where S denotes the selected subset and $|S|$ denotes the size of S. For the case of $k = 2$ and known σ^2, Jeyaratnam and Panchapakesan (1984) considered the natural estimator which is positively biased and some modified estimators with reduced bias.

Cohen and Sackrowitz (1988) have presented a decision-theoretic framework for the combined decision problem of selecting the best treatment and estimating the mean of the selected treatment and derived results for the case of $k = 2$ and known σ^2 with common sample size. Gupta and Miescke (1990) extended this study in several directions. They have considered $k > 2$ treatments, different loss components, and both equal and unequal sample sizes. As pointed out by Cohen and Sackrowitz (1988), the decision-theoretic treatment of the combined selection-estimation problem leads to "selecting after estimation" rather than "estimating after selection".

Estimation after selection is a meaningful problem which needs further study. The earlier papers of Dahiya (1974), Hsieh (1981), and Jeyaratnam and Panchapakesan (1984) dealt with several modified estimators in the case of $k = 2$ treatments. These have not been studied in detail as regards their desirable properties. The decision-theoretic results require too many details to provide a comprehensive view. For a list of references, see Gupta and Miescke (1990).

2.5. *Estimation of PCS*

Consider the selection rule R_1 of Bechhofer (1954) defined in (2.1) for selecting the treatment with the largest mean μ_i. This rule is designed to guarantee that PCS $\geqslant P^*$ whenever $\mu_{[k]} - \mu_{[k-1]} \geqslant \delta^*$. However, the true parametric configuration is unknown. If $\mu_{[k]} - \mu_{[k-1]} < \delta^*$, the minimum PCS cannot be guaranteed. Thus a retrospective analysis regarding the PCS is of importance.

For any configuration of $\underset{\sim}{\mu}$,

$$\text{PCS} = \int_{-\infty}^{\infty} \prod_{i=1}^{k-1} \Phi\left(t + \frac{\sqrt{n}(\mu_{[k]} - \mu_{[i]})}{\sigma}\right) \phi(t)\, \mathrm{d}t, \tag{2.10}$$

where Φ and ϕ are the standard normal cdf and density function, respectively. Olkin et al. (1976, 1982) considered the estimator $\widehat{P}$ obtained by replacing the $\mu_{[i]}$ by $\overline{Y}_{[i]}$ in (2.10). This estimator $\widehat{P}$ is consistent, but its evaluation is not easy. Olkin, Sobel and Tong (1976) have given upper and lower bounds for the PCS that hold for any true configuration, with no regard to any least favorable configuration. They have also obtained the asymptotic distribution of $\widehat{P}$ which is a function of $\widehat{\delta}_i = \overline{Y}_{[k]} - \overline{Y}_{[i]}, i = 1, \ldots, k$; however, the expression for the variance of the asymptotic distribution is complicated.

Faltin and McCulloch (1983) have studied the small-sample performance of the Olkin–Sobel–Tong estimator $\widehat{P}$ of the PCS in (2.10), analytically for $k = 2$ populations and via Monte Carlo simulation for $k \geqslant 2$. They have found that the estimator tends to overestimate PCS (getting worse when $k > 2$) when the means are close together and tends to underestimate when $\sqrt{n}\delta/\sigma$ is large.

Anderson et al. (1977) first gave a lower confidence bound for PCS in the case of the selection rule R_1 of Bechhofer (1954) defined in (2.1). Faltin (1980) provided, in the case of $k = 2$ treatments, a quantile unbiased estimator of PCS which can be regarded as a lower confidence bound for PCS. Later Kim (1986) obtained a lower confidence bound on PCS which is sharper than that of Anderson et al. (1977) and reduces to that of Faltin (1980) in the special case of $k = 2$ treatments. Recently, Gupta et al. (1994), using a new approach, derived a confidence region for the differences $\mu_{[k-i+1]} - \mu_{[k-i]}$, $i = 1, \ldots, k-1$, and then obtained a lower bound for PCS which is sharper than that of Kim (1986). They also derived some practical lower bounds by reducing the dimensionality of $\underset{\sim}{\delta} = (\delta_1, \ldots, \delta_{k-1})$, where $\delta_i = \mu_{[k-i+1]} - \mu_{[k-i]}$, $i = 1, \ldots, k-1$. The lower bound improves as this free-to-choose dimensionally q $(1 \leqslant q \leqslant k-1)$ increases and the result for $q = 1$ coincides with that of Kim (1986).

Gupta and Liang (1991) obtained a lower bound for PCS by deriving simultaneous lower confidence bounds on $\mu_{[k]} - \mu_{[i]}$, $i = 1, \ldots, k-1$, where a range statistic was used. Of the two methods of Gupta and Liang (1991) and Gupta et al. (1994), one does not dominate the other in the sense of providing larger PCS values. Generally speaking, for moderate k (say, $k \geqslant 5$), the Gupta–Liang method tends to underestimate PCS.

Finally, Gupta and Liang (1991) obtained a lower bound for PCS also in the case of the two-stage procedure of Bechhofer et al. (1954) for selecting the treatment with the largest mean when the common variance σ^2 is unknown.

2.6. Notes and comments

For the rule R_1 of Bechhofer (1954) defined in (2.1), Fabian (1962) has shown that a stronger assertion can be made without decreasing the infimum of the PCS. We define a treatment Π_i to be *good* if $\mu_i \geqslant \mu_{[k]} - \delta^*$ and modify the goal to be selection of a good treatment. Now a CS occurs if the selected treatment is a good treatment. Then the procedure R_1 guarantees with a minimum probability P^* that the selected treatment is good no matter what the configuration of the μ_i is.

We have not discussed sequential procedures for selecting the best treatment. There is a vast literature available in this regard. Reference can be made to Gupta and Panchapakesan (1991) besides the books mentioned in Section 1.

For the problem of estimating the PCS, Olkin et al. (1976), Kim (1986), Gupta and Liang (1991), and Gupta et al. (1994) have obtained their general results for location parameters with special discussion of the normal means case. Gupta et al. (1990) have discussed the case of truncated location parameter models.

Finally, robustness of selection procedures is an important aspect. This has been examined in the past by a few authors and there is a renewed interest in recent years. A survey of these studies is provided by Panchapakesan (1995a).

3. Selection in experiments with blocking: Normal theory

There may not always be sufficient quantities of homogeneous experimental material available for an experiment using a completely randomized design. However, it may be possible to group experimental units into blocks of homogeneous material. Then one can employ a traditional blocking design which minimizes possible bias and reduces the error variance.

3.1. Indifference-zone approach

Assume that there are sufficient experimental units so that each treatment can be used at least once in each block. Consider the *randomized complete block design* with fixed treatment effects, namely,

$$Y_{ij\ell} = \mu + \tau_i + \beta_j + \varepsilon_{ij\ell}, \tag{3.1}$$

where $Y_{ij\ell}$ is the ℓth observation $(1 \leqslant \ell \leqslant n)$ on treatment i $(1 \leqslant i \leqslant k)$ in block j $(1 \leqslant j \leqslant b)$. Here μ is the over-all mean, the τ_i are the treatment effects, the β_j are the block effects, and the errors ε_{ijk} are assumed to be iid $N(0, \sigma^2)$. It is assumed without loss of generality that

$$\sum_{i=1}^{k} \tau_i = \sum_{j=1}^{b} \beta_j = 0.$$

There is no interaction between blocks and treatments.

Let $\tau_{[1]} \leqslant \cdots \leqslant \tau_{[k]}$ denote the ordered τ_i. The goal is to select the best treatment, namely, the one associated B with $\tau_{[k]}$. Let us assume that σ^2 is *known*. Then the procedure R_1 of Bechhofer defined in (2.1) for the completely randomized design can easily be adapted here. We take n independent observations $Y_{ij\ell}$ $(1 \leqslant \ell \leqslant n)$ on treatment Π_i $(1 \leqslant i \leqslant k)$ in block j $(1 \leqslant j \leqslant b)$. The procedure is based on the estimates $\widehat{\tau}_i = \overline{Y}_{i..} - \overline{Y}_{...}$, where $\overline{Y}_{i..}$ and $\overline{Y}_{...}$ are the averages of $Y_{ij\ell}$ over the suffixes replaced by dots. The $\widehat{\tau}_i$ are the best linear unbiased estimates (BLUE's) of the treatment effects τ_i. Then the adapted procedure is

$$R_6\text{:} \quad \text{Select the treatment } \Pi_i \text{ that yields the largest } \widehat{\tau}_i. \tag{3.2}$$

In order to guarantee that the PCS is at least P^* whenever $\tau_{[k]} - \tau_{[k-1]} \geqslant \delta^*$, the minimum sample size n required is given by

$$n = \left\langle \frac{2}{b} \left(\frac{\sigma H}{\delta^*} \right)^2 \right\rangle, \tag{3.3}$$

where $\langle x \rangle$ denotes the smallest integer $\geqslant x$, and H is given by (2.3).

One can modify the basic procedure R_1 defined in (2.1) to select the treatment associated with the largest effect $\tau_{[k]}$ in other cases such as the balanced incomplete block design (BIBD) and Latin Square design. However, when the error variance σ^2 is *unknown*, there is no single stage procedure which will accomplish our goal with a guaranteed PCS. Also, Driessen (1992) and Dourleijn (1993) have discussed in detail selection of the best treatment in experiments using more general types of block designs called *connected designs*.

3.2. Subset selection approach

Consider model (3.1) with one observation per cell, which we rewrite as

$$Y_{ij} = \mu + \tau_i + \beta_j + \varepsilon_{ij} \tag{3.4}$$

where Y_{ij} is the observation on treatment Π_i $(1 \leqslant i \leqslant k)$ in block j $(1 \leqslant j \leqslant b)$ and the ε_{ij} are iid $N(0, \sigma^2)$. Our goal is to select a non-empty subset containing the treatment associated with $\tau_{[k]}$.

When σ^2 is *known*, we can use the procedure R_4 defined in (2.5), which will be in this model, R_7: Select treatment Π_i if and only if

$$\widehat{\tau}_i \geqslant \widehat{\tau}_{[k]} - \frac{d\sigma}{\sqrt{b}}, \tag{3.5}$$

where $\widehat{\tau}_i = \overline{Y}_{i.} - \overline{Y}_{..}$ $(1 \leqslant i \leqslant k)$ and the constant d is the smallest positive constant for which the PCS $\geqslant P^*$. This constant $d = \sqrt{2}\,H$, where H is the solution of (2.3).

When σ^2 is *unknown*, we use the procedure R_8 which is R_7 with σ replaced by S_ν, where S_ν^2 is given by

$$S_\nu^2 = \sum_{i=1}^{k}\sum_{j=1}^{b}(Y_{ij} - \overline{Y}_{i.} - \overline{Y}_{.j} + \overline{Y}_{..})^2/\nu \tag{3.6}$$

based on $\nu = (k-1)(b-1)$ degrees of freedom. To keep the distinction between R_7 and R_8, we denote the constant needed by d' instead of d. The values of d' (as mentioned in the case of R_5 of Section 2.2) have been tabulated for selected values of k, b, and P^* by Gupta and Sobel (1957). They can also be obtained from the tables of Gupta et al. (1985) corresponding to correlation coefficient $\rho = 0.5$. Gupta and Hsu (1980) have applied the procedure R_8 and its usual analogue for selecting the treatment associated with $\tau_{[1]}$ to a data set relating to motor vehicle traffic fatality rates (MFR) for the forty-eight contiguous states and the District of Columbia for the years 1960 to 1976. Their goal is to select a subset of best (worst) states in terms of MFR.

As in the case of indifference-zone approach, the basic procedures R_4 and R_5 of Gupta (1956) discussed in Section 2.2 can be adapted for other designs such as the BIBD and Latin Square. Driessen (1992) and Dourleijn (1993) have discussed in detail subset selection in experiments involving connected designs.

3.3. Simultaneous inference with respect to the best

In the model (3.4), let us assume that the error variance σ^2 is unknown. Hsu (1982) gave a procedure for selecting a subset C of the k treatments that includes the treatment associated with $\tau_{[k]}$ and at the same time providing simultaneous upper confidence bounds $D_1, \ldots, D_k$ for $\tau_{[k]} - \tau_1, \ldots, \tau_{[k]} - \tau_k$. His procedure is based on the sample treatment means $\overline{Y}_{i.}$ and S_ν^2 given in (3.6). The procedure R_9 of Hsu (1982) defines

$$C = \left\{ \Pi_i\colon \overline{Y}_{i.} \geqslant \max_{j \neq i} \overline{Y}_{j.} - (dS_\nu/\sqrt{b}) \right\} \tag{3.7}$$

and

$$D_i = \max \left\{ \max_{j \neq i} \overline{Y}_{j.} - \overline{Y}_{i.} + (dS_\nu/\sqrt{b}) \right\}, \quad i = 1, \ldots, k, \tag{3.8}$$

where the constant $d = (k, b, P^*)$ is to be chosen so that

$$\Pr\left\{ \Pi_{(k)} \epsilon C \text{ and } \theta_{[k]} - \theta_{[i]} \leqslant D_i \text{ for } i = 1, \ldots, k \right\} = P^*$$

and $\Pi_{(k)}$ is the treatment associated with $\tau_{[k]}$. The constant $d = d(k, b, P^*)$ turns out to be the constant d' of the procedure R_5 and it can be obtained for selected values of k, b, and P^* from the tables of Gupta and Sobel (1957) and Gupta et al. (1985).

For further detailed treatment of multiple comparisons with and selection of the best treatment, the reader is referred to Driessen (1991, 1992).

3.4. Notes and comments

Rasch (1978) has discussed selection problems in balanced designs. Wu and Cheung (1994) have considered subset selection for normal means in a two-way design. Given b groups, each containing the same k treatments, their goal is to select a non-empty subset from each group so that the probability of simultaneous correct selection is at least P^*. Dourleijn and Driessen (1993) have discussed four different subset selection procedures for randomized designs with illustrated applications to a plant breeding variety trial.

Gupta and Leu (1987) have investigated an asymptotic distribution-free subset selection procedure for the two-way model (3.4) with the assumption that the $\underset{\sim}{\varepsilon}_i = (\varepsilon_{i1}, \ldots, \varepsilon_{i\ell})$ are iid with cdf $F(\underset{\sim}{\varepsilon})$ symmetric in its arguments. Their procedure is based on the Hodges–Lehmann estimators of location parameters. For a Bayesian treatment of ranking and selection problems in two-way models, reference should be made to Fong (1990) and Fong and Berger (1993).

Hsu (1982) has discussed simultaneous inference with respect to the best treatment in block designs in more generality than what was described previously. He assumes that the ε_{ij} are iid with an absolutely continuous cdf F with some regularity conditions. Besides the procedure R_9 based on sample means discussed previously, he considered two procedures based on signed ranks. Finally, Hsu and Nelson (1993) have surveyed, unified and extended multiple comparisons for the General Linear Model.

4. Selection in factorial experiments: Normal theory

Factorial experimentation when employed in ranking and selection problems can produce considerable savings in total sample size relative to independent single-factor experimentation when the probability requirements are comparable in both cases. This was in fact pointed out by Bechhofer (1954) who proposed a single-stage procedure for ranking normal means when no interaction is present between factor-level effects and common known variance is assumed. In this section, we will be mainly concerned with the two-factor model.

Consider a two-factor experiment involving factors A and B at a and b levels, respectively. The treatment means are μ_{ij} $(1 \leqslant i \leqslant a, 1 \leqslant j \leqslant b)$ are defined by

$$\mu_{ij} = \mu + \alpha_i + \beta_j + (\alpha\beta)_{ij}, \tag{4.1}$$

where μ is the over-all mean, the α_i $(1 \leqslant i \leqslant a)$ are the so-called row factor main effects, the β_j $(1 \leqslant j \leqslant b)$ are the column factor main effects, the $(\alpha\beta)_{ij}$ are the two-way interactions subject to the conditions

$$\sum_{i=1}^{a} \alpha_i = \sum_{j=1}^{b} \beta_j = 0, \qquad \sum_{j=1}^{b} (\alpha\beta)_{ij} = 0 \quad \text{for all } i,$$

and

$$\sum_{i=1}^{a} (\alpha\beta)_{ij} = 0 \quad \text{for all } j.$$

The factors A and B are said to be *additive* if $(\alpha\beta)_{ij} = 0$ for *all* i and j, and to *interact* otherwise. In this section, we discuss selection problems under the indifference-zone as well as subset selection approach both when the factors A and B are additive and interacting. Deciding whether or not the additive model holds is an important problem to be handled with caution. We will be content with just referring to Fabian (1991).

4.1. Indifference-zone approach

We will first assume an *additive* model. Independent random samples Y_{ijm}, $m = 1, 2, \ldots$, are taken from normal treatments Π_{ij} $(1 \leqslant i \leqslant a, 1 \leqslant j \leqslant b)$ with associated unknown means $\mu_{ij} = \mu + \alpha_i + \beta_j$ and common variance σ^2. Here $a \geqslant 2$ and $b \geqslant 2$. The goal is to select the treatment combination associated with $\alpha_{[a]}$ and $\beta_{[b]}$; in other words, we seek simultaneously the best levels of both factors. The probability requirement is:

$$P\{\mathrm{CS} \mid \underset{\sim}{\mu}\} \geqslant P^* \quad \text{whenever } \underset{\sim}{\mu} \in \Omega_{\alpha,\beta}, \tag{4.2}$$

where $\Omega_{\alpha,\beta} = \{\underset{\sim}{\mu} \mid \alpha_{[a]} - \alpha_{[a-1]} \geqslant \delta^*_\alpha, \beta_{[b]} - \beta_{[b-1]} \geqslant \delta^*_\beta\}$.

For the common *known* σ^2 case, Bechhofer (1954) proposed a single-stage procedure based on n independent observations from each Π_{ij}. Let $\overline{Y}_{i..}$ and $\overline{Y}_{.j.}$ denote the means of the observations corresponding to the levels i and j of the factors A and B, respectively. Then the procedure of Bechhofer (1954) is

$$R_{10}\text{: Select the treatment combination of levels associated with the largest } \overline{Y}_{i..} \text{ and the largest } \overline{Y}_{.j.} \tag{4.3}$$

The LFC for this procedure is given by

$$\alpha_{[1]} = \cdots = \alpha_{[a-1]} = \alpha_{[a]} - \delta^*_\alpha;$$

$$\beta_{[1]} = \cdots = \beta_{[b-1]} = \beta_{[b]} - \delta^*_\beta. \tag{4.4}$$

The PCS for the rule R_{10} at the LFC can be written as a product of the PCS's at the LFC's when the rule R_1 in (2.1) is applied marginally to each factor. This fact enables one to determine the smallest n to guarantee the minimum PCS.

Bechhofer et al. (1993) have studied the performances of the single-stage procedure R_{10} of Bechhoffer (1954) described previously and two other sequential procedures (not discussed here). One of these is a truncated sequential procedure of Bechhofer and Goldsman (1988b) and the other is a closed sequential procedure with elimination by Hartmann (1993).

The procedure R_{10} can be easily generalized to the case of r factors with levels $k_1, \ldots, k_r$. One is naturally interested in examining the efficiency of an r-factor experiment relative to that of r independent single-factor experiments in the absence of interaction when both guarantee the same minimum PCS P^*. Let n_f and n_r denote the total numbers of observations required for the r-factor experiment and r single-factor experiments, respectively. Then Bawa (1972) showed that the asymptotic relative efficiency (ARE) of the r-factor experiment, defined by ARE $= \lim_{p^* \to 1}(n_f/n_r)$, is given by

$$\text{ARE} = \frac{\max_{1 \leqslant j \leqslant r}\{k_j/\delta_j^{*2}\}}{\sum_{j=1}^k (k_j/\delta_j^{*2})}, \tag{4.5}$$

where δ_j^* is the threshold for the preference-zone for the levels of factor j.

Under the additive model, assuming common known σ^2, procedures can be developed for RCBD and split-plot design experiments. The latter has been studied by Pan and Santner (1993).

When the common variance σ^2 is *unknown*, as in the case of a single-factor experiment, we cannot design a single-stage r-factor experiment that will guarantee the probability requirement. Assuming still an *additive* model, Bechhofer and Dunnett (1986) proposed a two-stage procedure which is a straightforward generalization of the open two-stage non-eliminating procedure of Bechhofer et al. (1954). It is also assumed by Bechhofer and Dunnett (1986) that the indifference-zone widths are the same for the factors A and B, i.e., $\delta^*_\alpha = \delta^*_\beta$.

When σ^2 is unknown, Pan and Santner (1993) have discussed a procedure for the two-factor additive model using a split-plot design.

Factorial experiments with interaction

We now consider the model (4.1) in which $(\alpha\beta)_{ij}$ is not zero for some or all pairs (i, j), and assume that σ^2 is *known*. Dudewicz and Taneja (1982) have discussed selecting the combination of the factor levels with the largest mean in an r-factor experiment. Later, Taneja and Dudewicz (1987) discussed in detail the two-factor case. When the interaction is *known* to exist, their procedure is based on the means $\overline{Y}_{ij.}$ of n independent observations taken from each Π_{ij}. Their procedure is

$$R_{11}\text{: Select the } \Pi_{ij} \text{ that yields the largest } \overline{Y}_{ij.}. \tag{4.6}$$

We note that the rule R_{11} can be used even when there is no interaction in which case, however, R_{10} is the best to use (see Dudewicz and Taneja, 1982). This motivates the procedure with a preliminary test, referred to as R_{12} here, proposed by Taneja and Dudewicz (1987) when there is no prior information regarding the presence of interaction. The preliminary test is the likelihood ratio test for H_0: $(\alpha\beta)_{ij} = 0$ for all pairs (i, j) versus H_1: Not H_0 at level α. If H_0 is rejected, we proceed by applying the selection rule R_{11} in (4.6); otherwise, we apply the rule R_{10} in (4.3).

Borowiak and De los Reyes (1992) considered the class C of rules R_{12} obtained by varying α in $[0, 1]$. In the special case of $a = b = 2$, they showed that the rule R_{11} (which corresponds to $\alpha = 1$) maximizes the minimum PCS in the class C.

When the common variance σ^2 is unknown, as in the case of single-factor experiments, no single-stage procedure can guarantee the probability requirement. In the case of additive model, Bechhofer (1977) proposed a two-stage procedure analogous to that of Bechhofer et al. (1954). The decision rule of this procedure parallels R_{10}. In the case of interaction being present, Taneja (1986) proposed a similar two-stage procedure whose decision rule parallels R_{11}.

For the remainder of this subsection, we consider the model in (4.1) with interaction and let, for convenience, $\gamma_{ij} = (\alpha\beta)_{ij}$. Assume that the common variance σ^2 is known. We consider a new goal, namely, selecting the treatment combination associated with the largest γ_{ij}. Let $\gamma_{[1]} \leqslant \cdots \leqslant \gamma_{[ab]}$ denote the ordered interaction effects.

It is required that

$$P(\text{CS} \mid \gamma) \geqslant P^* \quad \text{whenever } \gamma_{[ab]} \geqslant \Delta^* \text{ and } \gamma_{[ab]} - \gamma_{[ab-1]} \geqslant \delta^*, \tag{4.7}$$

where the event [CS] occurs if and only if the treatment combination corresponding to $\gamma_{[ab]}$ is selected, and the constants δ^*, Δ^* and P^* satisfy

$$\delta^* > 0, \qquad \frac{(a-1)(b-1)}{(a-1)(b-1)-1}\delta^* < \Delta^* \quad \text{and} \quad \frac{1}{ab} < P^* < 1.$$

This problem was first considered by Bechhofer et al. (1977) in the case of $a = 2$ with an arbitrary b. Their study was expanded by Santner (1981) to arbitrary levels a and b. Take n observations on each treatment combination and let $\widehat{\gamma}_{ij} = \overline{Y}_{ij.} - \overline{Y}_{i..} - \overline{Y}_{.j.} + \overline{Y}_{...}$ $(1 \leqslant i \leqslant a, 1 \leqslant j \leqslant b)$, where a dot replacing a subscript indicates that an average has been computed over the elements for that subscript. The single-stage procedure of Santner (1981) is

$$R_{13}: \text{ Select the treatment combination that yields the largest } \widehat{\gamma}_{ij}. \tag{4.8}$$

The LFC for this procedure has been shown by Santner (1981) to be a solution to a non-linear programming problem. However, the computation of the PCS is difficult. The LFC can be determined more easily in some special cases.

4.2. Subset selection approach

Consider the model in (4.1) with no interaction, i.e., $(\alpha\beta)_{ij} = 0$ for all pairs (i, j). We assume that the common variance σ^2 is *unknown*. Our goal is to select a subset of the treatment combinations which contains the combination of levels associated with $\alpha_{[a]}$ and $\beta_{[b]}$. A correct selection occurs if a subset is selected consistent with this goal. Generalizing the single-factor procedure of Gupta (1956), the following procedure was proposed by Bechhofer and Dunnett (1987).

R_{14}: Include in the selected subset all treatment combinations for which

$$\overline{Y}_{i..} \geqslant \max_{1 \leqslant \ell \leqslant a} \overline{Y}_{\ell..} - C_A S_\nu / \sqrt{bn}$$

and

$$\overline{Y}_{.j.} \geqslant \max_{1 \leqslant \ell \leqslant b} \overline{Y}_{.\ell.} - C_B S_\nu / \sqrt{an}. \tag{4.9}$$

Here n is the number of observations taken on each treatment combination, $\overline{Y}_{i..}$ and $\overline{Y}_{.j.}$ are the appropriate means (as defined for the rule R_{13}), and S_ν^2 is the usual unbiased estimator of σ^2 on $\nu = abn - a - b + 1$ degrees of freedom. The LFC for the rule R_{14} is:

$$\alpha_1 = \cdots = \alpha_a \quad \text{and} \quad \beta_1 = \cdots = \beta_b. \tag{4.10}$$

For given n, a, b and P^*, equating the PCS at the LFC in (4.10) to P^*, one can solve for C_A and C_B. The solution is not unique. Tables C.1 through C.3 of Bechhofer and Dunnett (1987) give a particular solution $g_1 = C_A\sqrt{2} = C_B\sqrt{2}$ for $P^* = 0.80, 0.90, 0.95$ respectively, for $a = 2(1)5, b = a(1)7, 10$ and $\nu = 5(1)30(5)50, 60(20)120, 200, \infty$.

Pan and Santner (1993) have discussed subset selection procedures under the additive model when a split-plot design is used.

One can consider subset selection procedures when interaction is present. A natural goal to consider is selecting a subset of the ab treatment combinations which contains the one associated with the largest γ_{ij}. If the experimenter is unsure whether or not interaction is present, then the selection can be done by using an appropriate procedure depending on the conclusion of a preliminary test for interaction. These problems have not been studied in detail.

4.3. Notes and comments

Pan and Santner (1993) have also discussed selection procedures under the indifference-zone approach as well as the subset selection approach for blocked strip-plot experiments.

When interaction is present in (4.1), one may be interested in selecting the treatment combination associated with the largest absolute interaction. Procedures for this goal have been considered by Bechhofer et al. (1977) and Santner (1981) for completely randomized two-factor experiments.

Bechhofer and Goldsman (1988a) have studied sequential selection procedures for multi-factor experiments involving Koopman–Darmois population under the additivity assumption. Federer and McCulloch (1984, 1993) obtained simultaneous confidence intervals for the sets of differences $\alpha_i - \alpha_{[a]}$, $i = 1, \ldots, a$, and $\beta_j - \beta_{[b]}$, $j = 1, \ldots, b$, when the experiment is conducted using split-plot, split-split-plot and split-block designs. Taneja (1987) considered nonparametric selection procedures in complete factorial experiments. Finally, reference should be made to Driessen (1992).

5. Selection with reference to a standard or a control: Normal theory

In the preceding sections we discussed decision procedures for choosing the best from among a given set of k ($\geqslant 2$) treatments. Although the experimenter is generally interested in selecting the best one of the competing treatments, in certain situations even the best may not be good enough to warrant its selection. This typically happens in experiments involving comparison of a set of experimental treatments with a standard or a control treatment.

Let $\Pi_1, \ldots, \Pi_k$ be k experimental (normal) treatments with unknown means $\mu_1, \ldots, \mu_k$, respectively. These mean responses are compared with μ_0 which is either a *specified* (known) standard or the *unknown* mean of a control treatment Π_0. When the comparison is with a standard, the decision is based on data from the k experimental treatments. When the comparison involves a control population, we take samples from all the $k+1$ treatments. For convenience, we will refer to the standard also as Π_0.

5.1. Indifference-zone approach

Let $\mu_{[1]} \leqslant \cdots \leqslant \mu_{[k]}$ denote the ordered means of the k experimental treatments. Our goal here is to select the treatment associated with $\mu_{[k]}$ if $\mu_{[k]} > \mu_0$, or to select Π_0

if $\mu_{[k]} \leqslant \mu_0$. Any selection consistent with the goal is a correct selection (CS). The probability requirement for any valid procedure is:

$$\Pr\{\Pi_0 \text{ is selected}\} \geqslant P_0^* \quad \text{whenever } \mu_0 \geqslant \mu_{[k]} + \delta_0^* B \tag{5.1}$$

and

$$\Pr\{\Pi_{(k)} \text{ is selected}\} \geqslant P_1^* \quad \text{whenever } \mu_{[k]} \geqslant \mu_0 + \delta_1^* \text{ and} \\ \mu_{[k]} \geqslant \mu_{[k-1]} + \delta_2^*, \tag{5.2}$$

where $\delta_0^*, \delta_1^*, \delta_2^*, P_0^*$, and P_1^* are constants with $0 < \{\delta_1^*, \delta_2^*\} < \infty, -\delta_1^* < \delta_0^* < \infty$, $2^{-k} < P_0^* < 1$, $(1 - 2^{-k})/k < P_1^* < 1$, and $\Pi_{(k)}$ denotes the treatment associated with $\mu_{[k]}$.

We assume that the treatments have a common *known* variance σ^2. Let μ_0 be the specified standard. In this case, Bechhofer and Turnbull (1978) proposed a single-stage procedure based on $\overline{Y}_i$, $i = 1, \ldots, k$, the means of sample of size n from each treatment. Their procedure is

$$R_{15}\text{: Choose } \Pi_0 \text{ if } \overline{Y}_{[k]} < \mu_0 + c\text{; otherwise, choose the population} \\ \text{that yields } \overline{Y}_{[k]} \text{ as the one associated with } \Pi_{(k)}. \tag{5.3}$$

Bechhofer and Turnbull (1978) have given simultaneous equations for obtaining (n_e, c) such as the requirements (5.1) and (5.2) are satisfied for the common sample size $n = [n_e] + 1$, where $[x]$ denotes the largest integer $\leqslant x$.

In many applications, we take $\delta_0^* = 0$ and $\delta_1^* = \delta_2^* = \delta^*$ (say). This is the formulation considered earlier by Paulson (1952). If we let $h = \sqrt{n_e}\, c/\sigma$ and $g = \sqrt{n_e}\, \delta^*/\sigma$, then h and g satisfy:

$$\Phi^k(h) = P_0^* \tag{5.4}$$

and

$$\int_{h-g}^{\infty} \Phi^{k-1}(y+g)\phi(y)\,\mathrm{d}y = P_1^* \tag{5.5}$$

where $\Phi(y)$ and $\phi(y)$ denote the standard normal c.d.f. and density function, respectively. The values of h and c satisfying (5.4) and (5.5) are tabulated by Bechhofer and Turnbull (1978) for $k = 2(1)15$ and selected values of P_0^* and P_1^*.

When σ^2 is unknown, Bechhofer and Turnbull (1978) proposed a two-stage sampling procedure which is an analogue of the two-stage sampling procedure of Bechhofer et al. (1954). This procedure of Bechhofer and Turnbull (1978) and other early procedures for comparisons with a control are discussed in Gupta and Panchapakesan (1979, Chapter 20).

5.2. *Subset selection approach*

Here our goal is to select a subset of the treatments that includes *all* those treatments which are better than the standard or the control treatment, i.e., all those experimental treatments for which $\mu_i > \mu_0$.

Let us first consider a specified standard μ_0 and assume that the common variance σ^2 is *known*. Any valid rule R is required to satisfy

$$P(\mathrm{CS} \mid R) \geqslant P^* \quad \text{for all } \underset{\sim}{\mu} = (\mu_1, \ldots, \mu_k). \tag{5.6}$$

Let Y_{ij}, $j = 1, \ldots, n_i$, be independent sample responses from treatment Π_i ($i = 1, \ldots, k$). Based on the sample means $\overline{Y}_i$, Gupta and Sobel (1958) proposed the rule

R_{16}: Include Π_i in the selected subset if and only if

$$\overline{Y}_i > \mu_0 - d\sigma/\sqrt{n_i} \tag{5.7}$$

where $d > 0$ is the smallest number so that the requirement on the PCS can be met, and it is given by $\Phi(d) = (P^*)^{1/k}$.

When σ^2 is *unknown*, we replace σ in (5.7) by S_ν where S_ν^2 is the usual pooled estimator of σ^2 with $\nu = \sum_{i=1}^{k}(n_i - 1)$ degrees of freedom and $n_i \geqslant 2$ for $1 \leqslant i \leqslant k$. In this case, the constant d is given by

$$\int_0^\infty \Phi^k(yd) q_\nu(y)\, dy = P^* \tag{5.8}$$

where $q_\nu(y)$ is the density of $Y = S_\nu/\sigma$. The d-values satisfying (5.8) can be obtained from the tables of Dunnett (1955) for selected values of k, ν, and P^*. It should be noted that d is the one-sided upper-$(1-P^*)$ equicoordinate point of the equicorrelated $(k-1)$-variate central t-distribution with the equal correlation $\rho = 0$ and ν degrees of freedom.

Now, let μ_0 be the *known* mean of the control treatment. We assume that all treatments have a common variance σ^2 . Let $\overline{Y}_i$, $i = 0, 1, \ldots, n$, be the means of random samples of size n_0 from the control treatment and of size n from each of the experimental treatments. When σ^2 is *unknown*, the Gupta–Sobel procedure is

R_{17}: Include Π_i in the selected subset if and only if

$$\overline{Y}_i > \overline{Y}_0 - d\sigma\sqrt{\frac{1}{n} + \frac{1}{n_0}}, \tag{5.9}$$

where the smallest $d > 0$ for which the minimum PCS is guaranteed to be P^* is the solution of

$$\Pr\{Z_1 \leqslant d, \ldots, Z_k \leqslant d\} = P^*, \tag{5.10}$$

and the Z_i are equicorrelated standard normal variables with equal correlation $\rho = n/(n + n_0)$. The d-values are tabulated by Gupta et al. (1973) for selected values of

k, P^* and ρ. When $n = n_0$, then $d = H$, given by (2.3) with $k-1$ replaced by k, and thus can be obtained from the tables mentioned in that case.

When σ^2 is *unknown*, we use rule R_{17} with σ replaced by s_ν, where s_ν^2 is the usual pooled estimator of σ^2 based on $\nu = k(n-1) + (n_0 - 1)$ degrees of freedom. In this case, d is the one-sided upper-$(1-P^*)$ equicoordinate point of the equicorrelated (k)-variate central t-distribution with the equal correlation $\rho = n/(n+n_0)$ and ν degrees of freedom. Values of d are tabulated by Gupta et al. (1985) and Bechhofer and Dunnett (1988) for selected values of k, P^*, ν and ρ.

5.3. Simultaneous confidence intervals

In some applications, the experimenter may be interested in the differences between the experimental treatments and μ_0, which is either a specified standard or the unknown mean of a control treatment. Let us first consider the case of comparisons with a standard μ_0. Assume that the treatments have a common *known* variance σ^2. Let $\overline{Y}_i$ $(i = 1, \ldots, k)$ denote the mean of a random sample of size n_i from Π_i. Define

$$I_i = \left(\overline{Y}_i - \mu_0 - d\sigma/\sqrt{n_i}, \overline{Y}_i - \mu_0 + d\sigma/\sqrt{n_i}\right) \tag{5.11}$$

for $i = 1, \ldots, k$, where d is the α-quantile of the standard normal distribution with $\alpha = [1 + (P^*)^{1/k}]/2$. Then

$$\Pr\{\mu_i - \mu_0 \in I_i,\ i = 1, \ldots, k\} \geqslant P^*. \tag{5.12}$$

When σ^2 is *unknown*, let I_i' be the interval obtained by replacing σ in (5.11) with S_ν, where S_ν^2 is the pooled estimator of σ^2 with $\nu = \sum_{i=1}^k (n_i - 1)$ degrees of freedom. In this case, the joint confidence statement (5.12) holds by taking d as the two-sided upper-$(1-P^*)$ equicoordinate point of the equicorrelated k-variate central t-distribution with the equal correlation $\rho = 0$ and ν degrees of freedom.

When μ_0 is the *unknown* mean of a control treatment Π_0, let $\overline{Y}_0$ be the mean of a random sample from Π_0. When the common variance σ^2 is *unknown*, Dunnett (1955) obtained one-sided and two-sided confidence intervals for $\mu_i - \mu_0, i = 1, \ldots, k$, with joint confidence coefficient P^*. The lower joint confidence limits are given by

$$\overline{Y}_i - \overline{Y}_0 - d_i S_\nu \sqrt{\frac{1}{n_i} + \frac{1}{n_0}}, \quad i = 1, \ldots, k, \tag{5.13}$$

where s_ν^2 is the pooled estimator σ^2 based on $\nu = \sum_{i=0}^k (n_i - 1)$ degrees of freedom and the constants d_i are chosen such that

$$\Pr\{t_1 < d_1, \ldots, t_k < d_k\} = P^* \tag{5.14}$$

where the joint distribution of the t_i is the multivariate t. If $n_1 = \cdots = n_k = n$, then the t_i are equicorrelated with correlation $\rho = n/(n+n_0)$. In this case, $d_1 = \cdots =$

$d_k = d$. For selected values of k, P^*, ν and ρ, the value of d can be obtained from the tables of Gupta et al. (1985). Dunnett (1955) has tabulated d-values in the case of $n_0 = n_1 = \cdots = n_k$ (i.e., $\rho = 0.5$).

Similar to (5.14), we can write upper confidence limits and two-sided limits. In the equal sample sizes case, Dunnett (1964) has tabulated the constant needed.

When the n_i are unequal, there arises a problem of optimal allocation of observations between the control and the experimental treatments. The optimality is in the sense of maximizing the confidence coefficient for fixed $N = \sum_{i=0}^{k} n_i$. This problem has been studied by several authors. For a detailed discussion, see Gupta and Panchapakesan (1979, Chapter 20, Section 10).

5.4. Notes and comments

Chen and Hsu (1992) proposed a two-stage procedure which involves selecting in the first stage the best treatment provided it is better than a control and testing a hypothesis in the second stage between the best treatment selected (if any) at the first stage and the control.

There are studies in which several treatments and a control are administered to the same individuals (experimental units) at different times. The observations collected from the same unit under these treatments are no longer independent. This type of design is called repeated measurements design. Chen (1984) has considered selecting treatments better than a control under such a design.

Bechhofer et al. (1989) have studied two-stage procedures for comparing treatment with a control. In the first stage, they employ the subset selection procedure of Gupta and Sobel (1958) to eliminate "inferior" treatments. In the second stage, joint confidence statement is made for the treatment versus control differences (for those treatments retained after the first stage) using Dunnett's (1955) procedure.

Bechhofer and Tamhane (1981) developed a theory of optimal incomplete block designs for comparing several treatments with a control. They proposed a general class of designs that are balanced with respect to test treatments (BTIB).

Bofinger and Lewis (1992) considered simultaneous confidence intervals for normal treatments versus control differences allowing unknown and unequal treatment variances. Gupta and Kim (1980) and Gupta and Hsiao (1983) have studied subset selection with respect to a standard or control using decision-theoretic and Bayesian formulations. Hoover (1991) generalized the procedure of Dunnett (1955) to comparisons with respect to two controls.

In our discussion of selecting treatments that are better than a standard or control, we assumed that there was no information about the ordering of the treatment means μ_i. In some situations, we may have partial prior information in the form of a simple or partial order relationship among the unknown means μ_i of the experimental treatments. For example, in experiments involving different dose levels of a drug, the treatment effects will have a known ordering. In other words, we know that $\mu_1 \leqslant \mu_2 \leqslant \cdots \leqslant \mu_k$ even though the μ_i are unknown. For the goal of selecting all populations for which $\mu_i \geqslant \mu_0$, we would expect any reasonable procedure R to have the property: If R selects Π_i, then it selects all treatments Π_j with $j > i$. This is the *isotonic* behavior of R. Naturally, such a procedure will be based on the isotonic estimators of the μ_i. Such procedures have been investigated by Gupta and Yang (1984) in the case of normal treatment means allowing the common variance σ^2 to be known or unknown.

6. Selection in experiments involving other models

Thus far we discussed selection procedures and simultaneous confidence intervals under the assumption that the treatment responses are normally distributed. In this section, we briefly mention some other models for which these problems have been investigated.

6.1. Single-factor Bernoulli models

The Bernoulli distribution serves as an appropriate model in experiments involving manufacturing processes and clinical trials. In these experiments, response variables are qualitative giving rise to dichotomous data such as defective-nondefective or success-failure. Thus we are interested in comparing Bernoulli populations in terms of their success probabilities. The initial and basic contributions to this problem were made by Sobel and Huyett (1957) under the indifference-zone formulation and Gupta and Sobel (1960). There are many interesting aspects of the Bernoulli selection problems. For specification of the preference-zone one can use different measures for the separation between the best and the next best population, namely, $p_{[k]} - p_{[k-1]}, p_{[k]}/p_{[k-1]}$ and $[p_{[k]}(1 - p_{[k-1]})]/[(1 - p_{[k]})p_{[k-1]}]$. The last measure is the *odds ratio* used in biomedical studies. Besides the usual fixed sample size procedures and purely sequential procedures, the literature includes inverse sampling procedures and so-called Play-the-Winner sampling rules. For a detailed review of these procedures, reference may be made to Gupta and Panchapakesan (1979, 1985).

6.2. Multinomial models

The multinomial distribution, as a prototype for many practical problems, is a very useful model. When observations from a population are classified into a certain number of categories, it is natural to look for categories that occur very often or rarely. Consider a multinomial distribution on m cells with probabilities $p_1, \ldots, p_m$. Selecting the most and the least probable cells are two common goals. The early investigations of Bechhofer et al. (1959), Gupta and Nagel (1967), and Cacoullos and Sobel (1966) set the pace for a considerable number of papers that followed. The investigations of multinomial selection problems reveal an interesting picture regarding the structure of the LFC which, it turns out, is not similar for the two common goals mentioned previously and also depends on whether a ratio or a difference is used to define the preference-zone. For further discussion and additional references, see Gupta and Panchapakesan (1993).

Although selecting the best cell from a single multinomial population has been investigated over a period of close to forty years, selecting the best of several multinomial populations has not received enough attention until recently except for the paper by Gupta and Wong (1977). For ranking multinomial populations, we need a measure of diversity within a population. Selection procedures have been studied in terms of diversity measures such as Shannon's entropy and the Gini–Simpson index. An account of these procedures is given in Gupta and Panchapakesan (1993).

6.3. Reliability models

In experiments involving life-length distributions, many specific distributions such as the exponential, Weibull and gamma have been used to characterize the life-length. Panchapakesan (1995b) provides a review of selection procedures for the one- and two-parameter exponential distributions. How the life-length distribution is described as a member of a family characterized in terms of failure rate properties. The IFR (increasing failure rate) and IFRA (increasing failure rate on the average) families are well-known examples of such families. Selection procedures for distributions belonging to such families have been investigated substantially by several authors. A review of these investigations is provided by Gupta and Panchapakesan (1988).

7. Concluding remarks

As we have pointed out in Section 1, our review of design of experiments with selection and ranking goals covers mainly basic normal theory for single-factor experiments with and without blocking and 2-factorial experiments with and without interaction. We have referred to a few authors who have studied the problem using a Bayesian approach. There have also been a number of investigations under an empirical Bayes approach. Some useful additional references in this connection are: Berger and Deely (1988), Fong (1992), Gupta and Liang (1987) and Gupta et al. (1994).

In Section 2.4, we referred to a computer SAS package of Aubuchon et al. (1986) for implementing simultaneous confidence intervals for the difference between each treatment mean and the best of the other treatment means. This package can also be used for selecting the best treatment using the indifference-zone and the subset selection approaches. There are a few other statistical packages such as CADEMO and MINITAB which contain modules for selection procedures. A commercially distributed package exclusively devoted to selection procedures is RANKSEL; see Edwards (1985, 1986) for details. There are also programs developed by several researchers in the course of their investigations. Rasch (1995) has given a summary of available software for selection procedures with specific description of each. Several FORTRAN programs needed for investigation and implementation of selection procedures are given in the recent book by Bechhofer et al. (1995).

In the foregoing sections, we have discussed, as alternatives to tests of hypotheses among treatment means, three types of formulations: indifference-zone approach, subset selection approach, and multiple comparisons approach. We have mainly considered single-stage fixed sample size procedures. In some cases we have described two-stage procedures. Sequential procedures have only been referred to. In all these cases, we have not described every available procedure for a given goal. As such we have not gone into efficiency comparisons of competing procedures. However, brief comments have been made in certain cases where a procedure was improved upon or bested by another at a later date. It should however be emphasized that the procedures we have described are viable as yet.

Acknowledgement

This research was supported in part by US Army Research Office Grant DAAH04-95-1-0165.

References

Alam, K. (1970). A two-sample procedure for selecting the populations with the largest mean from k normal populations. *Ann. Inst. Statist. Math.* **22**, 127–136.

Anderson, P. O., T. A. Bishop and E. J. Dudewicz (1977). Indifference-zone ranking and selection: Confidence intervals for true achieved $P(CD)$. *Comm. Statist. Theory Methods* **6**, 1121–1132.

Aubuchon, J. C., S. S. Gupta and J. C. Hsu (1986). PROC RSMCB: A procedure for ranking, selection, and multiple comparisons with the best. *Proc. SAS Users Group Internat. Conf.* **11**, 761–765.

Bawa, V. S. (1972). Asymptotic efficiency of one R-factor experiment relative to R one-factor experiments for selecting the best normal population. *J. Amer. Statist. Assoc.* **67**, 660–661.

Bechhofer, R. E. (1954). A single-sample multiple decision procedure for ranking means of normal populations with known variances. *Ann. Math. Statist.* **25**, 16–39.

Bechhofer, R. E. (1977). Selection in factorial experiments. In: H. J. Highland, R. G. Sargent and J. W. Schmidt, eds., *Proc. 1977 Winter Simulation Conf.* National Bureau of Standards, Gaitherburg, MD, 65–77.

Bechhofer, R. E. and C. W. Dunnett (1986). Two-stage selection of the best factor-level combination in multi-factor experiments: Common known variance. In: C. E. McCulloch, S. J. Schwager, G. Casella and S. R. Searle, eds., *Statistical Design: Theory and Practice.* Biometrics Unit, Cornell University, Ithaca, NY, 3–16.

Bechhofer, R. E. and C. W. Dunnett (1987). Subset selection for normal means in multifactor experiments. *Comm. Statist. Theory Methods* **16**, 2277–2286.

Bechhofer, R. E. and C. W. Dunnett (1988). Percentage points of multivariate Student t distributions. In: *Selected Tables in Mathematical Statistics*, Vol. 11. Amer. Mathematical Soc., Providence, RI.

Bechhofer, R. E., C. W. Dunnett and M. Sobel (1954). A two-sample multiple-decision procedure for ranking means of normal populations with a common known variance. *Biometrika* **41**, 170–176.

Bechhofer, R. E., C. W. Dunnett and A. C. Tamhane (1989). Two-stage procedures for comparing treatments with a control: Elimination at the first stage and estimation at the second stage. *Biometr. J.* **5**, 545–561.

Bechhofer, R. E., S. Elmaghraby and N. Morse (1959). A single-sample multiple decision procedure for selecting the multinomial event which has the highest probability. *Ann. Math. Statist.* **30**, 102–119.

Bechhofer, R. E. and D. M. Goldsman (1987). Truncation of the Bechhofer–Kiefer–Sobel sequential procedure for selecting the normal population which has the largest mean. *Comm. Statist. Simulation Comput.* **16**, 1067–1092.

Bechhofer, R. E. and D. M. Goldsman (1988a). Sequential selection procedures for multi-factor experiments involving Koopman–Dormois populations with additivity. In: S. S. Gupta and J. O. Berger, eds., *Statistical Decision Theory and Related Topics IV*, Vol. 2. Springer, New York, 3–21.

Bechhofer, R. E. and D. M. Goldsman (1988b). Truncation of the Bechhofer–Kiefer–Sobel sequential procedure for selecting the normal population which has the largest mean (II): 2-factor experiments with no interaction. *Comm. Statist. Simulation Comput.* **17**, 103–128.

Bechhofer, R. E. and D. M. Goldsman (1989). A comparison of the performances of procedures for selecting the normal population having the largest mean when the variances are known and equal. In: L. J. Gleser et al., eds., *Contributions to Probability and Statistics: Essays in Honor of Ingram Olkin*. Springer, New York, 303–317.

Bechhofer, R. E., D. M. Goldsman and M. Hartmann (1993). Performances of selection procedures for 2-factor additive normal populations with common known variance. In: F. M. Hoppe, ed., *Multiple Comparisons, Selection, and Applications in Biometry*. Dekker, New York, 209–224.

Bechhofer, R. E., J. Keifer and M. Sobel (1986). *Sequential Identification and Ranking Procedures (with Special Reference to Koopman–Dormois Populations)*. Univ. of Chicago Press, Chicago, IL.

Bechhofer, R. E., T. J. Santner and D. M. Goldsman (1995). *Design and Analysis of Experiments for Statistical Selection, Screening and Multiple Comparisons*. Wiley, New York.

Bechhofer, R. E., T. J. Santner and B. W. Turnbull (1977). Selecting the largest interaction in a two-factor experiment. In: S. S. Gupta and D. S. Moore, eds., *Statistical Decision Theory and Related Topics II*. Academic Press, New York, 1–18.

Bechhofer, R. E. and A. C. Tamhane (1981). Incomplete block designs for comparing treatments with a control: General theory. *Technometrics* **23**, 45–57.

Bechhofer, R. E. and B. W. Turnbull (1978). Two $(k+1)$-decision selection procedure for comparing k normal means with a specified standard. *J. Amer. Statist. Assoc.* **73**, 385–392.

Berger, J. and J. J. Deely (1988). A Bayesian approach to ranking and selection of related means with alternatives to analysis-of-variance methodology. *J. Amer. Statist. Assoc.* **83**, 364–373.

Bhandari, S. K. and A. R. Chaudhuri (1990). On two conjectures about two-stage selection problems. *Sankhyā Ser. A.* **52**, 131–141.

Bofinger, E. and G. J. Lewis (1992). Simultaneous comparisons with a control and with the best: Two stage procedures (with discussion). In: E. Bofinger et al., eds., *The Frontiers of Modern Statistical Inference Procedures*. American Sciences Press, Columbus, OH, 25–45.

Borowiak, D. S. and J. P. De Los Reyes (1992). Selection of the best in 2×2 factorial designs. *Comm. Statist. Theory Methods* **21**, 2493–2500.

Büringer, H., H. Martin and K.-H. Schriever (1980). *Nonparametric Sequential Selection Procedures*. Birkhauser, Boston, MA.

Cacoullos, T. and M. Sobel (1966). An inverse-sampling procedure for selecting the most probable event in multinomial distribution. In: P. R. Krishnaiah, ed., *Multivariate Analysis*. Academic Press, New York, 423–455.

Chen, H. J. (1984). Selecting a group of treatments better than a standard in repeated measurements design. *Statist. Decisions* **2**, 63–74.

Chen, H. J., E. J. Dudewicz and Y. J. Lee (1976). Subset selection procedures for normal means under unequal sample sizes. *Sankhyā Ser. B* **38**, 249–255.

Chen, P. and L. Hsu (1992). A two-stage design for comparing clinical trials. *Biometr. J.* **34**, 29–35.

Chen, P. and M. Sobel (1987a). An integrated formulation for selecting the t best of k normal populations. *Comm. Statist. Theory Methods* **16**, 121–146.

Chen, P. and M. Sobel (1987b). A new formulation for the multinomial selection problem. *Comm. Statist. Theory Methods* **16**, 147–180.

Cohen, A. and H. B. Sackrowitz (1982). Estimating the mean of the selected population. In: S. S. Gupta and J. O. Berger, eds., *Statistical Decision Theory and Related Topics III*, Vol. 1. Academic Press, New York, 243–270.

Cohen, A. and H. B. Sackrowitz (1988). A decision theory formulation for population selection followed by estimating the mean of the selected population. In: S. S. Gupta and J. O. Berger, eds., *Statistical Decision Theory and Related Topics IV*, Vol. 2. Springer, New York, 33–36.

Cohen, D. D. (1959). A two-sample decision procedure for ranking means of normal populations with a common known variance. M.S. Thesis. Dept. of Operations Research, Cornell University, Ithaca, NY.

Dahiya, R. C. (1974). Estimation of the mean of the selected population. *J. Amer. Statist. Assoc.* **69**, 226–230.

Dourleijn, C. J. (1993). On statistical selection in plant breeding. Ph.D. Dissertation. Agricultural University, Wageningen, The Netherlands.

Dourleijn, C. J. and S. G. A. J. Driessen (1993). Subset selection procedures for randomized designs. *Biometr. J.* **35**, 267–282.

Driessen, S. G. A. J. (1991). Multiple comparisons with and selection of the best treatment in (incomplete) block designs. *Comm. Statist. Theory Methods* **20**, 179–217.

Driessen, S. G. A. J. (1992). Statistical selection: Multiple comparison approach. Ph.D. Dissertation. Eindhovan University of Technology, Eindhoven, The Netherlands.

Dudewicz, E. J. and S. R. Dalal (1975). Allocation of observations in ranking and selection with unequal variances. *Sankhyā Ser. B* **37**, 28–78.

Dudewicz, E. J. and J. O. Koo (1982). *The Complete Categorized Guide to Statistical Selection and Ranking Procedures*. Series in Mathematical and Management Sciences, Vol. 6. American Sciences Press, Columbus, OH.

Dudewicz, E. J. and B. K. Taneja (1982). Ranking and selection in designed experiments: Complete factorial experiments. *J. Japan Statist. Soc.* **12**, 51–62.

Dunnett, C. W. (1955). A multiple comparison procedure for comparing several treatments with a control. *J. Amer. Statist. Assoc.* **50**, 1096–1121.

Dunnett, C. W. (1964). New tables for multiple comparisons with a control. *Biometrics* **20**, 482–491.

Eaton, M. L. (1967). Some optimum properties of ranking procedures. *Ann. Math. Statist.* **38**, 124–137.

Edwards, H. P. (1985). RANKSEL – An interactive computer package of ranking and selection procedures (with discussion). In: E. J. Dudewicz, ed., *The Frontiers of Modern Statistical Inference Procedures*. American Sciences Press, Columbus, OH, 169–184.

Edwards, H. P. (1986). The ranking and selection computer package RANKSEL. *Amer. J. Math. Management Sci.* **6**, 143–167.

Fabian, V. (1962). On multiple decision methods for ranking population means. *Ann. Math. Statist.* **33**, 248–254.

Fabian, V. (1974). Note on Anderson's sequential procedures with triangular boundary. *Ann. Statist.* **2**, 170–176.

Fabian, V. (1991). On the problem of interactions in the analysis of variance (with discussion). *J. Amer. Statist. Assoc.* **86**, 362–375.

Faltin, F. W. (1980). A quantile unbiased estimator of the probability of correct selection achieved by Bechhofer's single-stage procedure for the two population normal means problem. Abstract. *IMS Bull.* **9**, 180–181.

Faltin, F. W. and C. E. McCulloch (1983). On the small-sample properties of the Olkin–Sobel–Tong estimator of the probability of correct selection. *J. Amer. Statist. Assoc.* **78**, 464–467.

Federer, W. T. and C. E. McCulloch (1984). Multiple comparison procedures for some split plot and split block designs. In: T. J. Santner and A. C. Tamhane, eds., *Design of Experiments: Ranking and Selection*. Dekker, New York, 7–22.

Federer, W. T. and C. E. McCulloch (1993). Multiple comparisons in split block and split-split plot designs. In: F. M. Hoppe, ed., *Multiple Comparisons, Selection, and Applications in Biometry*. Dekker, New York, 47–62.

Fong, D. K. H. (1990). Ranking and estimation of related means in two-way models – A Bayesian approach. *J. Statist. Comput. Simulation* **34**, 107–117.

Fong, D. K. H. (1992). Ranking and estimation of related means in the presence of a covariance – A Bayesian approach. *J. Amer. Statist. Assoc.* **87**, 1128–1135.

Fong, D. K. H. and J. O. Berger (1993). Ranking, estimation and hypothesis testing in unbalanced two-way additive models – A Bayesian approach. *Statist. Decisions* **11**, 1–24.

Gibbons, J. D., I. Olkin and M. Sobel (1977). *Selecting and Ordering Populations: A New Statistical Methodology*. Wiley, New York.

Gupta, S. S. (1956). On a decision rule for a problem in ranking means. Mimeo, Ser. 150, Institute of Statistics, University of North Carolina, Chapel Hill, NC.

Gupta, S. S. (1963). Probability integrals of the multivariate normal and multivariate t. *Ann. Math. Statist.* **34**, 792–828.

Gupta, S. S. (1965). On some multiple decision (selection and ranking) rules. *Technometrics* **7**, 225–245.

Gupta, S. S. and P. Hsiao (1981). On gamma-minimax, minimax, and Bayes procedures for selecting populations close to a control. *Sankhyā Ser. B* **43**, 291–318.

Gupta, S. S. and J. C. Hsu (1980). Subset selection procedures with application to motor vehicle fatality data in a two-way layout. *Technometrics* **22**, 543–546.

Gupta, S. S. and J. C. Hsu (1984). A computer package for ranking, selection, and multiple comparisons with the best. In: S. Sheppard, U. Pooch and C. D. Pegden, eds., *Proc. 1984 Winter Simulation Conf.* Institute of Electrical and Electronics Engineers, Piscataway, NJ, 251–257.

Gupta, S. S. and J. C. Hsu (1985). RS-MCB: Ranking, selection, and multiple comparisons with the best. *Amer. Statist.* **39**, 313–314.

Gupta, S. S. and D.-Y. Huang (1976). On subset selection procedures for the means and variances of normal populations: Unequal sample sizes case. *Sankhyā Ser. A* **38**, 153–173.

Gupta, S. S. and D.-Y. Huang (1981). *Multiple Decision Theory: Recent Developments*, Lecture Notes in Statistics, Vol. 6. Springer, New York.

Gupta, S. S. and W.-C. Kim (1980). Gamma-minimax and minimax decision rules for comparison of treatments with a control. In: K. Matusita, ed., *Recent Developments in Statistical Inference and Data Analysis*. North-Holland, Amsterdam, 55–71.

Gupta, S. S. and W.-C. Kim (1984). A two-stage elimination type procedure for selecting the largest of several normal means with a common unknown variance. In: T. J. Santner and A. C. Tamhane, eds., *Design of Experiments: Ranking and Selection*. Dekker, New York, 77–93.

Gupta, S. S. and L.-Y. Leu (1987). An asymptotic distribution-free selection procedure for a two-way layout problem. *Comm. Statist. Theory Methods* **16**, 2313–2325.

Gupta, S. S., L.-Y. Leu and T. Liang (1990). On lower confidence bounds for PCS in truncated location parameter models. *Comm. Statist. Theory Methods* **19**, 527–546.

Gupta, S. S. and T. Liang (1987). On some Bayes and empirical Bayes selection procedures. In: R. Viertl, ed., *Probability and Bayesian Statistics*. Plenum Press, New York, 233–246.

Gupta, S. S. and T. Liang (1991). On a lower confidence bound for the probability of a correct selection: Analytical and simulation studies. In: Aydin Öztürk and E. C. van der Meulen, eds., *The Frontiers of Statistical Scientific Theory and Industrial Applications*. American Sciences Press, Columbus, OH, 77–95.

Gupta, S. S., T. Liang and R.-B. Rau (1994). Bayes and empirical Bayes rules for selecting the best normal population compared with a control. *Statist. Decisions* **12**, 125–147.

Gupta, S. S., Y. Liao, C. Qiu and J. Wang (1994). A new technique for improved confidence bounds for the probability of correct selection. *Statist. Sinica* **4**, 715–727.

Gupta, S. S. and K. J. Miescke (1984). On two-stage Bayes selection procedures. *Sankhyā Ser. B* **46**, 123–134.

Gupta, S. S. and K. J. Miescke (1990). On finding the largest mean and estimating the selected mean. *Sankhyā Ser. B* **52**, 144–157.

Gupta, S. S. and K. Nagel (1967). On selection and ranking procedures and order statistics from the multinomial distribution. *Sankhyā Ser. B* **29**, 1–34.

Gupta, S. S., K. Nagel and S. Panchapakesan (1973). On the order statistics from equally correlated normal random variables. *Biometrika* **60**, 403–413.

Gupta, S. S. and S. Panchapakesan (1979). *Multiple Decision Procedures: Theory and Methodology of Selecting and Ranking Populations*. Wiley, New York.

Gupta, S. S. and S. Panchapakesan (1985). Subset selection procedures: Review and assessment. *Amer. J. Math. Management Sci.* **5**, 235–311.

Gupta, S. S. and S. Panchapakesan (1988). Selection and ranking procedures in reliability models. In: P. R. Krishnaiah and C. R. Rao, eds., *Handbook of Statistics*, Vol. 7. Quality Control and Reliability. North-Holland, Amsterdam, 131–156.

Gupta, S. S. and S. Panchapakesan (1991). On sequential ranking and selection procedures. In: B. K. Ghosh and P. K. Sen, eds., *Handbook of Sequential Analysis*. Dekker, New York, 363–380.

Gupta, S. S. and S. Panchapakesan (1993). Selection and screening procedures in multivariate analysis. In: C. R. Rao, ed., *Multivariate Analysis: Future Directions*, North-Holland Series in Statistics and Probability, Vol. 5. Elsevier, Amsterdam, 223–262.

Gupta, S. S., S. Panchapakesan and J. K. Sohn (1985). On the distribution of the studentized maximum of equally correlated normal random variables. *Comm. Statist. Simulation Comput.* **14**, 103–135.

Gupta, S. S. and T. J. Santner (1973). On selection and ranking procedures – A restricted subset selection rule. In: *Proc. 39th Session of the Internat. Statist. Institute* **45**, Book I, 478–486.

Gupta, S. S. and M. Sobel (1957). On a statistic which arises in selection and ranking problems. *Ann. Math. Statist.* **28**, 957–967.

Gupta, S. S. and M. Sobel (1958). On selecting a subset which contains all populations better than a standard. *Ann. Math. Statist.* **29**, 235–244.

Gupta, S. S. and M. Sobel (1960). Selecting a subset containing the best of several binomial populations. In: I. Olkin et al., eds., *Contributions to Probability and Statistics*. Stanford Univ. Press, Stanford, CA, 224–248.

Gupta, S. S. and W.-Y. Wong (1977). Subset selection for finite schemes in information theory. In: I. Csisár and P. Elias, eds., *Colloquia Mathematica Societatis János Bolyai*, 16. Topics in Information Theory, 279–291.

Gupta, S. S. and W.-Y. Wong (1982). Subset selection procedures for the means of normal populations with unequal variances: Unequal sample sizes case. *Sel. Statist. Canad.* **6**, 109–149.

Gupta, S. S. and H.-M. Yang (1984). Isotonic procedures for selecting populations better than a control under ordering prior. In: J. K. Ghosh and J. Roy, eds., *Statistics: Applications and New Directions*. Indian Statistical Soc., Calcutta, 279–312.

Hall, W. J. (1959). The most economical character of Bechhofer and Sobel decision rules. *Ann. Math. Statist.* **30**, 964–969.

Hartmann, M. (1988). An improvement on Paulson's sequential ranking procedure. *Sequential Anal.* **7**, 363–372.

Hartmann, M. (1993). Multi-factor extensions of Paulson's procedures for selecting the best normal population. In: F. M. Hoppe, ed., *Multiple Comparisons, Selection, and Applications in Biometry.* Dekker, New York, 225–245.

Hayter, A. J. (1994). On the selection probabilities of two-stage decision procedures. *J. Statist. Plann. Inference* **38**, 223–236.

Hochberg, Y. and R. Marcus (1981). Three stage elimination type procedures for selecting the best normal population when variances are unknown. *Comm. Statist. Theory Methods* **10**, 597–612.

Hochberg, Y. and A. C. Tamhane (1987). *Multiple Comparison Procedures*. Wiley, New York.

Hoover, D. R. (1991). Simultaneous comparisons of multiple treatments to two (or more) controls. *Biometr. J.* **33**, 913–921.

Hsieh, H.-K. (1981). On estimating the mean of the selected population with unknown variance. *Comm. Statist. Theory Methods* **10**, 1869–1878.

Hsu, J. C. (1982). Simultaneous inference with respect to the best treatment in block designs. *J. Amer. Statist. Assoc.* **77**, 461–467.

Hsu, J. C. (1984). Constrained simultaneous confidence intervals for multiple comparisons with the best. *Ann. Statist.* **12**, 1136–1144.

Hsu, J. C. (1985). A method of unconstrained multiple comparisons with the best. *Comm. Statist. Theory Methods* **14**, 2009–2028.

Hsu, J. C. and B. Nelson (1993). Multiple comparisons in the general linear models. Unpublished Report.

Jeyarantnem, S. and S. Panchapakesan (1984). An estimation problem relating to subset selection for normal populations. In: T. J. Santner and A. C. Tamhane, eds., *Design of Experiments: Ranking and Selection*. Dekker, New York, 287–302.

Kim, W.-C. (1986). A lower confidence bound on the probability of a correct selection. *J. Amer. Statist. Assoc.* **81**, 1012–1017.

Milton, R. C. (1963). Tables of equally correlated multivariate normal probability integral. Tech. Report No. 27, Department of Statistics, University of Minnesota, Minneapolis, MN.

Mukhopadhyay, N. and T. K. S. Solanky (1994). *Multistage Selection and Ranking Procedures: Second-Order Asymptotics*. Dekker, New York.

Olkin, I., M. Sobel and Y. L. Tong (1976). Estimating the true probability of correct selection for location and scale parameter families. Tech. Report No. 174, Department of Operation Research and Department of Statistics, Stanford University, Stanford, CA.

Olkin, I., M. Sobel and Y. L. Tong (1982). Bounds for a k-fold integral for location and scale parameter models with applications to statistical ranking and selection problems. In: S. S. Gupta and J. O. Berger, eds., *Statistical Decision Theory and Related Topics III*, Vol. 2. Academic Press, New York, 193–212.

Pan, G. and T. J. Santner (1993). Selection and screening in additive two-factor experiments using randomization restricted designs. Tech. Report 523, Department of Statistics, The Ohio State University, Columbus, OH.

Panchapakesan, S. (1992). Ranking and selection procedures. In: N. Balakrishnan, ed., *Handbook of the Logistic Distribution*. Dekker, New York, 145–167.

Panchapakesan, S. (1995a). Robustness of selection procedures. *J. Statist. Plann. Inference*, to appear.

Panchapakesan, S. (1995b). Selection and ranking procedures. In: N. Balakrishnan and A. P. Basu, eds., *The Exponential Distribution: Method, Theory and Applications*. Gordon and Breach, New York, to appear.

Paulson, E. (1952). On the comparison of several experimental categories with a control. *Ann. Math. Statist.* **23**, 239–246.

Rasch, D. (1978). Selection problems in balanced designs. *Biometr. J.* **20**, 275–278.

Rasch, D. (1995). Software for selection procedures. *J. Statist. Plann. Inference*, to appear.

Santner, T. J. (1975). A restricted subset selection approach to ranking and selection problem. *Ann. Statist.* **3**, 334–349.

Santner, T. J. (1981). Designing two factor experiments for selecting interactions. *J. Statist. Plann. Inference* **5**, 45–55.

Santner, T. J. and A. J. Hayter (1993). The least favorable configuration of a two-stage procedure for selecting the largest normal mean. In: F. M. Hoppe, ed., *Multiple Comparisons, Selection and Applications in Biometry*. Dekker, New York, 247–265.

Sarkadi, K. (1967). Estimation after selection. *Studia Sci. Math. Hungr.* **2**, 341–350.

Sehr, J. (1988). On a conjecture concerning the least favorable configuration of a two-stage selection procedure. *Comm. Statist. Theory Methods* **17**, 3221–3233.

Sobel, M. (1969). Selecting a subset containing at least one of the t best populations. In: P. R. Krishnaiah, ed., *Multivariate Analysis II*. Academic Press, New York, 515–540.

Sobel, M. and M. J. Huyett (1957). Selecting the best one of several binomial populations. *Bell Syst. Tech. J.* **36**, 537–576.

Tamhane, A. C. (1976). A three-stage elimination type procedure for selecting the largest normal mean (common unknown variance case). *Sankhyā Ser. B* **38**, 339–349.

Tamhane, A. C. and R. E. Bechhofer (1977). A two-stage minimax procedure with screening for selecting the largest normal mean. *Comm. Statist. Theory Methods* **6**, 1003–1033.

Tamhane, A. C. and R. E. Bechhofer (1979). A two-stage minimax procedure with screening for selecting the largest normal mean (II): An improved PCS lower bound and associated tables. *Comm. Statist. Theory Methods* **8**, 337–358.

Taneja, B. K. (1986). Selection of the best normal mean in complete factorial experiments with interaction and with common unknown variance. *J. Japanese Statist. Soc.* **16**, 55–65.

Taneja, B. K. (1987). Nonparametric selection procedures in complete factorial experiments. In: W. Sendler, ed., *Contributions to Statistics*. Physica-Verlag, Heidelberg, 214–235.

Taneja, B. K. and E. J. Dudewicz (1987). Selection in factorial experiments with interaction, especially the 2×2 case. *Acta Math. Sinica, New Series* **3**, 191–203.

Van der Laan, P. and L. B. Verdooren (1989). Selection of populations: An overview and some recent results. *Biometr. J.* **31**, 383–420.

Wu, K. H. and S. H. Cheung (1994). Subset selection for normal means in a two-way design. *Biometr. J.* **36**, 165–175.

S. Ghosh and C. R. Rao, eds., *Handbook of Statistics, Vol. 13*

18

Multiple Comparisons

Ajit C. Tamhane

1. Introduction

The pitfalls inherent in making multiple inferences based on the same data are well-known, and have concerned statisticians for a long time. This problem was noted by Fisher (1935) in the context of making pairwise comparisons using multiple α-level two-sample t tests in a one-way ANOVA setting. To alleviate the resulting inflated type I error probability of falsely declaring at least one pairwise difference significant, he suggested the Bonferroni and the protected least significant difference (LSD) methods as simple ad-hoc solutions.

Multiple comparisons as a separate field started in the late 40's and early 50's with the fundamental works of Duncan (1951, 1955), Dunnett (1955), Roy and Bose (1953), Scheffé (1953) and Tukey (1949). Since then the field has made tremendous progress and is still extremely active with research motivated by many real life problems stemming mainly from applications in medical, psychological and educational research. A number of books and monographs have been written on the subject beginning with the mimeographed notes of Tukey (1953) (recently published by Braun, 1994) followed by Miller (1966, 1981), Hochberg and Tamhane (1987), Toothaker (1991, 1993) and Westfall and Young (1989). Recent review articles are by Bauer (1991) and Shaffer (1994); see Hochberg and Tamhane (1987) (hereafter referred to as HT) for references to earlier review articles. The aim of the present article is to give an overview of the subject, focusing more on important developments since the publication of HT. All mathematical proofs are omitted, references being given wherever needed.

The outline of the paper is as follows: Section 2 discusses the basic concepts of multiple comparisons. Section 3 gives some methods for constructing multiple testing procedures. Section 4 gives p-value based procedures, which are modifications of the simple Bonferroni procedure. Section 5 covers classical normal theory procedures for inter-treatment comparisons with emphasis on two families: comparisons with a control and pairwise comparisons. Section 6 is devoted to the problem of multiple endpoints; continuous and discrete endpoints are covered in two separate subsections. Section 7 reviews several miscellaneous problems. The paper ends with some concluding remarks in Section 8.

2. Basic concepts

2.1. Family

A *family* is a set of contextually related inferences (comparisons) from which some common conclusions are drawn or decisions are made. Often we refer to the collection of parameters on which hypotheses are tested or confidence statements are made as a family. Some examples of families are pairwise comparisons between a set of treatments, comparisons of test treatments with a control, comparison of two treatments based on multiple endpoints, and significance tests on pairwise correlations among a set of variables. These are all *finite* families. An example of an *infinite* family is the collection of all contrasts among a set of treatment means.

Which inferences to include in a family can be a difficult question. The following guidelines are useful to resolve this question:

- Contextual relatedness (not statistical dependence) is a primary consideration for grouping a set of inferences into a single family. Generally, not all inferences made in a single experiment may constitute a family and a single experiment may involve more than one family. For an example from pharmaceutical industry, see Dunnett and Tamhane (1992b).
- On the other hand, an individual experiment should be considered as a unit for forming a family. In other words, a family should not extend over inferences made in several different experiments.
- A family should include not just the inferences actually made, but also all other similar inferences that potentially could have been made had the data turned out differently. This is especially necessary if an MCP is used for data-snooping.
  For example, suppose that a particular pairwise contrast among the treatment means is tested not because it was of *a priori* interest, but rather because it turned out to be the largest. Since this difference is selected by data-snooping from the set of all pairwise comparisons, this latter set constitutes the appropriate family. Note that although only one test is conducted *explicitly*, all pairwise tests are conducted *implicitly*, the most significant one (or possibly more than one) being reported. Thus pre-hoc *multiple* inferences and post-hoc *selective* inferences are two sides of the same coin. In the latter case, in order to specify a family, one must be able to delineate the set of potential comparisons from which the ones actually reported are selected.
- To achieve reasonable power, it is advisable that the search for interesting contrasts should be narrowed down to some finite family based on substantive questions of interest. The larger the family, the less power does the MCP have. An infinite family such as the family of all contrasts or all linear combinations of the treatment means should generally be avoided.

2.2. Error rates and their control

An *error rate* is a probabilistic measure of erroneous inferences (restricted here only to type I errors) in a family. A *multiple comparison procedure* (MCP) is a statistical procedure for making all or selected inferences from a given family while controlling or adjusting for the increased incidence of type I errors due to multiplicity of inferences. For this purpose, usually the *familywise error rate* (FWE) (also called the *experimentwise error rate*) is used, where

$$\text{FWE} = P\{\text{At least one wrong inference}\}.$$

Control of the FWE for an MCP used for multiple null hypotheses testing (called a *multiple test procedure* (MTP)) means that

$$P\{\text{At least one true null hypothesis is rejected}\} \leqslant \alpha \qquad (2.1)$$

for a stated α. For an MCP used for confidence estimation (called a *simultaneous confidence procedure* (SCP)), the corresponding requirement is

$$P\{\text{All parameters are included in their confidence intervals}\} \geqslant 1 - \alpha. \qquad (2.2)$$

Both MTP's and SCP's will be referred to as MCP's unless a distinction must be made between the two.

Two main reasons for controlling the FWE are:

- Control of the FWE means control of the probability of an error in any subset of inferences, however selected (pre-hoc or post-hoc) from the specified family. This is useful in *exploratory studies* since it permits data-snooping.
- Control of the FWE guarantees that *all* inferences in the family are correct with probability $\geqslant 1 - \alpha$. This is useful in *confirmatory studies* where the correctness of an overall decision depends on the simultaneous correctness of all individual inferences, e.g., in selection type problems.

The FWE can be controlled strongly or weakly. To explain this, consider the problem of testing a family of null hypotheses H_i for $i \in K$, where K is a finite or an infinite index set. By $H = \bigcap_{i\in K} H_i$ we denote the *overall null hypothesis*. If any hypothesis in this family implies any other hypothesis then the family is said to be *hierarchical*; otherwise it is said to be *non-hierarchical*. If the family includes the overall null hypothesis H then it is clearly hierarchical since H implies all the H_i's.

Strong control of the FWE means that (2.1) is satisfied under all *partial null hypotheses* of the type $H_I = \bigcap_{i\in I} H_i$ where $I \subseteq K$. *Weak control* of the FWE means that (2.1) is satisfied only under some H_I, typically only under $H = H_K$. Note that since the simultaneous confidence statement (2.2) is satisfied under all configurations, an MTP derived by inverting an SCP controls the FWE strongly. A majority of applications require strong control of the FWE. Henceforth, by control of the FWE we will always mean strong control unless specified otherwise.

An alternative error rate used is the *per-family error rate* (PFE) (also called the *per-experiment error rate*):

$$\text{PFE} = \text{E}\{\text{No. of wrong inferences}\},$$

which is defined only for finite families. If E_i is the event that the ith inference is wrong, $i \in K$, then using the Bonferroni inequality we obtain

$$\text{FWE} = P\left(\bigcup_{i \in K} E_i\right) \leqslant \sum_{i \in K} P(E_i) = \text{ PFE}. \tag{2.3}$$

Therefore control of the PFE implies conservative control of the FWE. This is the basis of the *Bonferroni method* of multiple comparisons. If the events E_i are positively dependent (Tong, 1980) then a sharper upper bound on the FWE is provided by the following multiplicative inequality:

$$\text{FWE} = P\left(\bigcup_{i \in K} E_i\right) \leqslant 1 - \prod_{i \in K} \{1 - P(E_i)\}\,. \tag{2.4}$$

Many statisticians are philosophically opposed to the idea of lumping inferences together into a family and controlling the FWE; see, e.g., Carmer and Walker (1982), O'Brien (1983), the discussion following O'Neill and Wetherill (1971), Perry (1986), Rothman (1990) and Saville (1990). The thrust of their argument is that if a separate experiment is conducted for each inference then no adjustment for multiplicity is required, therefore the same should apply even when multiple inferences are made in a single experiment. Thus the error rate should be controlled separately for each inference. This is equivalent to controlling the *per-comparison error rate* (PCE) (also called the *comparisonwise error rate*) where

$$\text{PCE} = \frac{\text{E}\{\text{No. of wrong inferences}\}}{\text{Total no. of inferences}}.$$

Note that PCE $\leqslant$ FWE. Controlling PCE $= \alpha$ often results in the FWE being far in excess of α. For example, if the inferences are statistically independent and if H is true then

$$\text{FWE } = 1 - (1 - \alpha)^k,$$

where $k = \text{card}(K)$ is the (finite) number of inferences in the family. Other statisticians argue that such excessively high and uncontrolled incidence of type I errors is unacceptable; therefore an MCP that controls the FWE should be used.

Benjamini and Hochberg (1995) have recently proposed another intermediate error rate, called the *false discovery rate* (FDR), as a possible candidate for control; see also Sorić (1989). Let T and F be the (random) number of true and false null hypotheses rejected. Then the FDR is defined as

$$\text{FDR } = \text{E}\left(\frac{T}{T + F}\right), \tag{2.5}$$

where $0/0$ is defined as 0. FDR is the expected value of the ratio of the number of false rejections (discoveries) to the number of total rejections. When all null hypotheses are true, the FDR equals FWE. Therefore control of the FDR implies weak control of the FWE. Since control of the FDR is less stringent than control of the FWE, it generally results in more rejections. Benjamini and Hochberg's approach involves determining the α-level for individual tests so as to maximize $\mathrm{E}(T+F)$ subject to controlling the FDR below a specified upper bound.

2.3. *Directional decisions*

Let $\theta_1, \theta_2, \ldots, \theta_k$ be $k \geqslant 2$ parameters on which two-sided hypotheses H_i: $\theta_i = 0$ vs. A_i: $\theta_i \neq 0$ $(1 \leqslant i \leqslant k)$ are to be tested. The directional decision problem arises when one needs to decide the signs of the θ_i's ($\theta_i < 0$ or $\theta_i > 0$) for the rejected H_i's. The directional decisions are made depending on the signs of the test statistics. The new feature of this problem is that one must consider type III errors, which are the errors of misclassifying the signs of the nonnull θ_i's.

To avoid the type III error probability from approaching $1/2$ for any H_i as θ_i approaches zero, it is necessary to allow a third decision for each hypothesis, namely that the data are "inconclusive". Subject to control of a suitable error rate, we need to maximize the number of correct directional decisions (referred to as *confident directions* by Tukey (1991)).

Two approaches to error rate control are generally adopted. One approach is to control the type III FWE, which is the probability of making any type III error. This approach ignores type I errors altogether, which is justified if there is very little loss associated with them or because the null values $\theta_i = 0$ are *a priori* unlikely. As an example of this approach, see Bofinger (1985). The second approach is to control the type I and III FWE, which is the probability of making any type I or type III error. An MTP obtained by inverting simultaneous confidence intervals (SCI's) controls this latter FWE. As we shall see in the sequel, for the common types of SCI's, such MTP's are of the single-step type. It is not known whether the MTP's that do tests in a stepwise fashion (e.g., Fisher's protected LSD) have this property in general.

2.4. *Adjusted p-values*

As in the case of a single hypothesis test, it is more informative to report a p-value for each hypothesis instead of simply stating whether the hypothesis can be rejected or not at some preassigned α. This also has the advantage that table look-up is avoided. A complicating factor for multiple hypotheses is that each p-value must be adjusted for multiplicity of tests. The *adjusted p-value* of a hypothesis H_i (denoted by p_{ai}) is defined as the smallest value of the familywise level α at which H_i can be rejected by a given MCP. Once the p_{ai}'s are computed, tests can be conducted at any desired familywise level α by simply rejecting any H_i if $p_{ai} < \alpha$ for $i \in K$.

The idea of adjusted p-values has been around for a long time, but the terminology and their use has been made more popular recently by Westfall and Young (1992) and

Wright (1992). Formulas for calculating the adjusted p-values for different MCP's are given when those MCP's are discussed in the sequel; for example, see (4.1). A resampling approach to obtain simulation estimates of the adjusted p-values is the main theme of Westfall and Young (1993).

2.5. Coherence and consonance

In a hierarchical family the following logical requirement must be satisfied: If any hypothesis is accepted then all hypotheses implied by it must also be accepted. This requirement is called *coherence* (Gabriel, 1969). Another requirement that is sometimes desirable, but not absolutely required is that if any hypothesis is rejected then at least one hypothesis implied by it must also be rejected. This requirement is called *consonance* (Gabriel, 1969). For example, if in a one-way layout the family consists of the overall null hypothesis H that all treatment means are equal and hypotheses of pairwise equality of treatment means, and if H is rejected then the consonance requires that at least one pair of treatment means must be declared significantly different. Not all MCP's satisfy this requirement. For example, Fisher's LSD may find the F test of H significant, but none of the pairwise t tests to be significant.

3. Types of MCP's and methods of their construction

3.1. Types of MCP's

MCP's can be classified as shown in Figure 1.

A *single-step MCP* tests the hypotheses H_i's simultaneously without reference to one another, typically using a common critical constant. An example of a single-step MCP is the Bonferroni procedure, which tests each H_i at level α/k. A single-step MCP can be readily inverted to obtain an SCP. Such is generally not the case with stepwise MCP's.

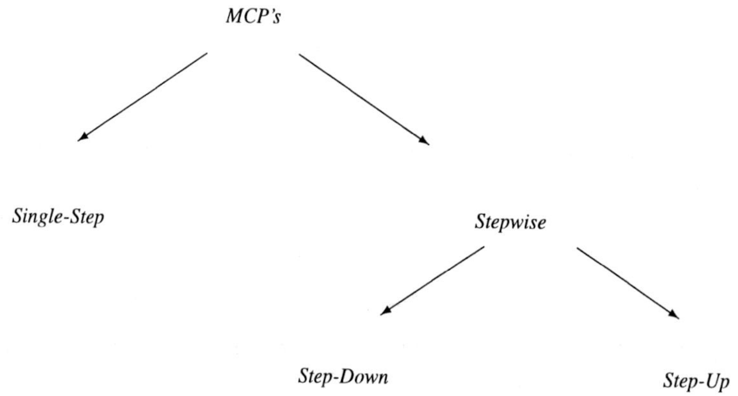

Fig. 1. Types of MCP's.

A *step-down MCP* tests the H_i's in the decreasing order of their significance or in the hierarchical order of implication with the most significant or the most implying hypotheses (e.g., the one that implies all other hypotheses in the family) being tested first, using a monotone set of critical constants. Testing continues until a hypothesis is accepted; the remaining hypotheses are accepted by implication without further tests. The protected LSD provides an example of a step-down MCP.

A *step-up MCP* tests the H_i's in the increasing order of their significance or in the hierarchical order of implication with the least significant or the least implying hypotheses (e.g., the set of hypotheses that do not imply each other, but are implied by one or more hypotheses in the family that are outside this set) being tested first, using a monotone set of critical constants. Testing continues until a hypothesis is rejected; the remaining hypotheses are rejected by implication without further tests. Hochberg's (1988) procedure discussed in Section 4.2.2 and Welsch's (1977) procedure discussed in Section 5.2.3 are examples of step-up MCP's.

3.2. *Union–intersection method*

The union–intersection (UI) method was proposed by Roy (1953) as a heuristic method to test a single hypothesis H, which can be expressed as an intersection of some component hypotheses H_i, $i \in K$. Suppose that tests are available for testing H_i versus A_i for $i \in K$. Then the UI test of

$$H = \bigcap_{i \in K} H_i \quad \text{versus} \quad A = \bigcup_{i \in K} A_i \tag{3.1}$$

rejects if *at least one* of the component hypotheses H_i is rejected.

Roy and Bose (1953) showed how a single-step MTP can be derived from a UI test. To make the ideas concrete, let t_i be a test statistic for H_i with the corresponding rejection region being $t_i > c$, where c is a critical constant. Then the UI test rejects if $t_{\max} = \max_{i \in K} t_i > c$. The critical constant c is determined so that the UI test has a designated level α, i.e., c is the $(1 - \alpha)$-quantile (the upper α point) of the distribution of $t_{\max}$ under H. The corresponding MTP uses the same constant c to test the individual H_i by rejecting H_i if $t_i > c$. This MTP controls the FWE strongly if the FWE is maximum under H. The well-known procedures of Dunnett (1955), Scheffé (1953) and Tukey (1949) which control the FWE strongly are based on the UI method. Krishnaiah (1979) gave a general UI method for deriving MTP's for finite families which he called *finite intersection tests*.

3.3. *Intersection–union method*

In some problems the overall null hypothesis can be expressed as a union rather than as an intersection of hypotheses, i.e., the testing problem can be formulated as

$$H = \bigcup_{i \in K} H_i \quad \text{versus} \quad A = \bigcap_{i \in K} A_i. \tag{3.2}$$

This formulation is applicable when all component hypotheses, H_i, must be rejected in order to reject H and take an action appropriate to the alternative hypothesis A. For example, a combination drug must be shown to be superior to all of its subcombinations for it to be acceptable.

It can be shown (Berger, 1982) that an α-level test of H rejects iff *all* tests of component hypotheses reject at level α. If the test statistics t_i have the same marginal distribution under H_i and c is the $(1-\alpha)$-quantile of that distribution, then H is rejected at level α if $t_{\min} = \min_{i \in K} t_i > c$. This is the MIN test of Laska and Meisner (1989). For example, if each t_i has marginally Student's t-distribution with ν degrees of freedom (d.f.) then $c = t_\nu^{(\alpha)}$, the $(1-\alpha)$-quantile of that distribution.

It is interesting to note that although this problem appears to be a multiple testing problem, no multiplicity adjustment is needed for doing individual tests, which are all done at PCE $= \alpha$. The IU nature of the problem can be easily overlooked and one of the standard MCP's might be recommended when all that is needed is separate α-level tests; see, e.g., the problem considered in D'Agostino and Heeren (1991) for which a correct solution was given in Dunnett and Tamhane (1992b).

3.4. Closure method

Marcus et al. (1976) gave a general formulation of the *closure method* originally proposed by Peritz (1970). Given a finite family of k null hypotheses $\{H_i \ (1 \leqslant i \leqslant k)\}$, first form a *closure family* by including in it all intersections $H_I = \bigcap_{i \in I} H_i$ for $I \subseteq K = \{1, 2, \ldots, k\}$. Reject any null hypothesis H_I (including any elementary hypothesis H_i) at level α iff all null hypotheses $H_{I'}$ implying it, i.e., all $H_{I'}$ for $I' \supseteq I$, are rejected by their individual α-level tests. This method controls the FWE strongly (see Theorem 4.1 in Chapter 2 of HT) and is also coherent since if any null hypothesis H_J is accepted then all $H_{J'}$ for $J' \subseteq J$ are accepted without further tests.

A practical implementation of the closure method generally requires a step-down algorithm. It begins by testing the overall null hypothesis $H = H_K$; if H is accepted by its α-level test then all null hypotheses are accepted without further tests. Otherwise one tests H_I for the next lower (i.e., $k-1$) dimensional subsets $I \subseteq K$ by their α-level tests, and so on. Sometimes shortcut methods can be used to reduce the number of tests as we shall see in the sequel.

3.5. Projection method

Let $\boldsymbol{\theta} = (\theta_1, \theta_2, \ldots, \theta_k)'$ be a vector of parameters and let C be a $(1-\alpha)$-level confidence region for $\boldsymbol{\theta}$. SCI's with joint confidence level $1-\alpha$ for all linear combinations $\boldsymbol{a}'\boldsymbol{\theta}$ where $\boldsymbol{a} = (a_1, a_2, \ldots, a_k)' \in \mathcal{R}^k$ with $\|\boldsymbol{a}\| = 1$ can be obtained by projecting C onto the line spanned by $\boldsymbol{a}$. If C is convex then the CI for each $\boldsymbol{a}'\boldsymbol{\theta}$ corresponds to the line segment intercepted by the two supporting hyperplanes orthogonal to the line spanned by $\boldsymbol{a}$. The SCI's can be used to perform hypotheses tests on $\boldsymbol{a}'\boldsymbol{\theta}$ in the usual manner with FWE controlled at level α. If SCI's are desired for linear combinations $\boldsymbol{\ell}'\boldsymbol{\theta}$ for $\boldsymbol{\ell} = (\ell_1, \ell_2, \ldots, \ell_k)'$ belonging to some subspace $\mathcal{L}$ of $\mathcal{R}^k$ of dimension

$m < k$ (e.g., the contrast space $\mathcal{C}^k = \{\boldsymbol{c} = (c_1, c_2, \ldots, c_k)' \colon \sum_{i=1}^k c_i = 0\}$ of dimension $k-1$), then the above procedure can be applied by starting with a $(1-\alpha)$-level confidence region for $\boldsymbol{\gamma} = \boldsymbol{L}\boldsymbol{\theta}$ where $\boldsymbol{L}$ is an $m \times k$ matrix whose rows form a basis for $\boldsymbol{L}$. Then SCI's for all linear combinations $\boldsymbol{b}'\boldsymbol{\gamma}$ (where $\boldsymbol{b} = (b_1, b_2, \ldots, b_m)' \in \mathcal{R}^m$) are equivalent to those for all $\boldsymbol{\ell}'\boldsymbol{\theta}$, $\boldsymbol{\ell} \in \mathcal{L}$. The Scheffé procedure discussed in Section 5.2.1 is a classic example of the projection method. A related method is the confidence region tests of Aitchinson (1964). Refer to HT, Chapter 2, Section 2.2 and Section 3.3.2 for additional details.

4. Modified Bonferroni procedures based on p-values

Reporting p-values for individual hypotheses is a standard practice. These p-values may be obtained by using different tests for different hypotheses, e.g., t, χ^2, Wilcoxon, logrank, etc. Therefore MCP's based on p-values are especially useful. They also are very simple (involving minimal computation other than what is required by a package to produce the p-values).

Throughout this section we assume a non-hierarchical finite family, $\{H_i \ (1 \leqslant i \leqslant k)\}$, and denote the unadjusted (raw) p-values of the hypotheses by $p_i \ (1 \leqslant i \leqslant k)$.

4.1. Single-step procedures

4.1.1. Bonferroni procedure

The Bonferroni procedure is the simplest MCP. It rejects any hypothesis H_i if $p_i < \alpha/k$. By the UI method it rejects the overall null hypothesis $H = \bigcap_{i=1}^k H_i$ if $p_{\min} < \alpha/k$. It controls the FWE strongly since for any number $k_0 \leqslant k$ of true hypotheses from (2.3) we have

$$\mathrm{FWE} \leqslant \mathrm{PFE} = k_0(\alpha/k) \leqslant \alpha.$$

The adjusted p-values are given by

$$p_{ai} = \min(kp_i, 1) \quad (1 \leqslant i \leqslant k). \tag{4.1}$$

Being a single-step MTP, it can be inverted to obtain an SCP; this SCP makes a $(1-\alpha/k)$-level confidence statement on each of the k parameters, thus conservatively guaranteeing a simultaneous confidence level of $1-\alpha$.

The Bonferroni procedure is a completely general procedure requiring no assumptions concerning the joint distribution or the correlations among the test statistics. If it is assumed that the test statistics have nonnegative correlations (more generally if they are positive orthant dependent, see Tong (1980), then a slight improvement is offered by using the multiplicative inequality (2.4) which yields

$$p_{ai} = 1 - (1 - p_i)^k \quad (1 \leqslant i \leqslant k). \tag{4.2}$$

In practice, the values of the p_{ai} obtained using (4.1) and (4.2) are quite close (especially when the p_i are small and k is large), but the latter involves an additional assumption. Therefore the basic Bonferroni procedure is generally preferred.

The omnibus applicability of the Bonferroni procedure is at the expense of the following drawbacks:

1. It is too conservative if k is large and/or if the test statistics are highly correlated (Pocock et al., 1987).

2. It is also conservative if the data are discrete because then the p_i-values are discrete, and $p_i < \alpha/k$ may not even be achievable for some i (Tarone, 1990; see Section 6.2.1.2 below).

3. The rejection decision on H is based only on $p_{\min}$. Thus it ignores information in other p_i's (O'Brien, 1984).

To overcome these drawbacks, many modifications of the basic Bonferroni method have been proposed some of which are discussed in the remainder of this section.

4.1.2. Simes procedure

Simes (1986) gave a procedure for testing a *single* hypothesis $H = \bigcap_{i=1}^{k} H_i$ which uses all the p-values, thus overcoming the last mentioned drawback of the Bonferroni procedure. Let $P_1, P_2, \ldots, P_k$ be the random variables (r.v.'s) corresponding to the observed p-values $p_1, p_2, \ldots, p_k$. The Simes procedure assumes that the P_i are independent and form a random sample from a uniform distribution over $(0,1)$ (the $\mathcal{U}(0,1)$ distribution) under the overall null hypothesis H. Let $P_{(1)} \geqslant P_{(2)} \geqslant \cdots \geqslant P_{(k)}$ be the ordered values of the P_i. Simes showed that

$$P\left[\text{At least one } P_{(i)} < \frac{k-i+1}{k}\alpha \quad (1 \leqslant i \leqslant k) \,\middle|\, H\right] = \alpha.$$

From this it follows that the Simes procedure, which rejects H if

$$p_{(i)} < \frac{k-i+1}{k}\alpha \quad \text{for some } i = 1, 2, \ldots, k,$$

is an α-level test.

EXAMPLE 4.1. Suppose $k = 5$ and $\alpha = .05$. Then the Simes procedure rejects H if

$$p_{(1)} < .05 \text{ or } p_{(2)} < .04 \text{ or } p_{(3)} < .03 \text{ or } p_{(4)} < .02 \text{ or } p_{(5)} < .01,$$

while the Bonferroni procedure rejects H if $p_{(5)} < .01$.

It is clear that the Simes procedure offers more opportunities for rejection of H, and is thus more powerful than the Bonferroni procedure. However, it has two limitations:

1. The independence assumption is severely restrictive. For correlated test statistics, the error probability can be as high as $\alpha \sum_{i=1}^{k}(1/i)$ (Hommel, 1986). Simulation results show that the α-level is controlled for positively correlated multivariate normal and gamma distributed test statistics (Hochberg and Rom, 1994).

2. The procedure tests only a single hypothesis H unlike the Bonferroni procedure, which also provides decisions on the individual H_i. However, it can be used in conjunction with the closure method to develop step-up MCP's, which are discussed below in Section 4.2.2.

4.2. *Stepwise procedures*

4.2.1. *Step-down procedures*

If the Bonferroni procedure is used to test each intersection hypothesis

$$H_I = \bigcap_{i \in I} H_i \quad \text{for } I \subseteq K = \{1, 2, \ldots, k\}$$

at level α then we obtain the step-down MTP proposed by Holm (1979a) which operates as follows: Order the p-values: $p_{(1)} \geqslant p_{(2)} \geqslant \cdots \geqslant p_{(k)}$, and denote the corresponding hypotheses by $H_{(1)}, H_{(2)}, \ldots, H_{(k)}$. Begin by testing the most significant hypothesis $H_{(k)}$ by comparing $p_{(k)}$ with α/k. If $p_{(k)} < \alpha/k$ then reject $H_{(k)}$ and test $H_{(k-1)}$ by comparing $p_{(k-1)}$ with $\alpha/(k-1)$; otherwise accept all null hypotheses and stop testing. In general, reject $H_{(i)}$ iff $p_{(j)} < \alpha/j$ for $j = k, k-1, \ldots, i$.

It is clear that the Holm procedure rejects the same null hypotheses rejected by the Bonferroni procedure, and possibly more. Thus it is at least as powerful as the Bonferroni procedure. The adjusted p-values for the Holm procedure are given by

$$p_{a(i)} = \min\left[1, \max\left\{kp_{(k)}, (k-1)p_{(k-1)}, \ldots, ip_{(i)}\right\}\right] \quad (1 \leqslant i \leqslant k).$$

These adjusted p-values are never larger than the adjusted p-values for the Bonferroni procedure. Holland and Copenhaver (1987) have given a slightly improved Holm procedure based on the sharper multiplicative Bonferroni inequality (2.4).

4.2.2. *Step-up procedures*

The use of the Simes procedure instead of the Bonferroni procedure to provide α-level tests of intersection hypotheses in the closure method results in a somewhat complicated procedure derived by Hommel (1988). Hommel's procedure finds the largest value of m $(1 \leqslant m \leqslant k)$ such that

$$p_{(j)} > (m - j + 1)\alpha/m, \quad j = 1, 2, \ldots, m. \tag{4.3}$$

If there is no such m then it rejects all hypotheses; else it rejects those H_i with $p_i < \alpha/m$. A simpler closed procedure was derived by Hochberg (1988) which is slightly conservative (Dunnett and Tamhane, 1993; Hommel, 1989). It operates in a step-up manner as follows: Order the p-values and the hypotheses as in the Holm procedure. Begin by testing the least significant hypothesis $H_{(1)}$ by comparing $p_{(1)}$ with $\alpha/1$. If $p_{(1)} \geqslant \alpha/1$ then accept $H_{(1)}$ and test $H_{(2)}$ by comparing $p_{(2)}$ with $\alpha/2$; otherwise reject all hypotheses and stop testing. In general, accept $H_{(i)}$ iff $p_{(j)} \geqslant \alpha/j$ for $j = 1, 2, \ldots, i$.

Both Holm's and Hochberg's procedures use the same critical constants, but the former is step-down while the latter is step-up. It is easy to see that Hochberg's procedure offers more opportunities for rejecting the hypotheses than does Holm's. Thus Hochberg's procedure is at least as powerful as Holm's, which in turn dominates the Bonferroni procedure. This is also clear from the adjusted p-values for the Hochberg procedure given by

$$p_{a(i)} = \min[1, \min\{p_{(1)}, 2p_{(2)}, \ldots, ip_{(i)}\}] \quad (1 \leqslant i \leqslant k).$$

Recently, Liu (1996) has given a general closure method of multiple testing of which the Hommel, Hochberg and Holm are special cases. In this general framework it is easy to see by comparing the critical constants used by these procedures that Hommel's procedure is more powerful than Hochberg's, which in turn is more powerful than Holm's. However, it must be remembered that both the Hommel and Hochberg procedures rest on the same assumption as does the Simes procedure which is that the P_i are independent. On the other hand, Holm's procedure needs no such assumption.

Rom (1990) provided a method for computing sharper critical constants, c_i, in place of the α/i used in Hochberg's procedure. Let the P_i be as defined in the Simes procedure. Let $P_{1,m} \geqslant P_{2,m} \geqslant \cdots \geqslant P_{m,m}$ be the ordered values of $P_1, P_2, \ldots, P_m$ $(1 \leqslant m \leqslant k)$. Then the c_i are obtained by solving the following system of recursive equations:

$$P\{P_{1,m} \geqslant c_1, \ldots, P_{m,m} \geqslant c_m\} = 1 - \alpha \quad (1 \leqslant m \leqslant k). \tag{4.4}$$

This equation simplifies to

$$\sum_{i=1}^{m-1} c_1^i = \sum_{i=1}^{m-1} c_{i+1}^{m-i} \quad (2 \leqslant m \leqslant k),$$

where $c_1 = \alpha$. Rom has given a table of the c_i for $k = 1(1)10$ and $\alpha = .01, .05$. These critical constants are fractionally larger than α/i as can be seen from Table 1. Thus both the Hommel and Rom procedures offer only slightly more powerful improvements of the simpler Hochberg procedure (Dunnett and Tamhane, 1993).

4.2.3. Example

This example is adapted from Hommel (1988). Suppose there are $k = 10$ hypotheses with the associated raw p-values given in the second column of Table 1.

From the adjusted p-values given in the last three columns of the table we see that at $\alpha = .05$, Bonferroni rejects only H_{10}, Holm rejects H_9 and H_{10}, while Hochberg rejects H_7 through H_{10} (the $p_{ai} < .05$ are marked with asterisks). Use of Rom's critical constants c_i in Hochberg's procedure does not result in additional rejections in the present example.

Table 1
Illustration of the Bonferroni, Holm and Hochberg procedures

				Rom's	p_{ai}		
H_i	p_i	ip_i	α/i	c_i	Bonferroni	Holm	Hochberg
H_1	.0605	.0605	.0500	.0500	.6050	.1090	.0605
H_2	.0466	.0932	.0250	.0250	.4660	.1090	.0605
H_3	.0352	.1056	.0167	.0169	.3520	.1090	.0605
H_4	.0238	.0952	.0125	.0127	.2380	.1090	.0605
H_5	.0218	.1090	.0100	.0102	.2180	.1090	.0605
H_6	.0093	.0558	.0083	.0085	.0930	.0558	.0558
H_7	.0069	.0483	.0071	.0073	.0690	.0520	.0483*
H_8	.0065	.0520	.0062	.0064	.0650	.0520	.0483*
H_9	.0053	.0477	.0055	.0057	.0530	.0477*	.0477*
H_{10}	.0021	.0210	.0050	.0050	.0210*	.0210*	.0210*

For the Hommel procedure the largest m that satisfies (4.3) is found from the following sequence of inequalities:

$m = 1:$	$.0605 \geqslant .05$				
$m = 2:$	$.0605 \geqslant .05$	$.0466 \geqslant .025$			
$m = 3:$	$.0605 \geqslant .05$	$.0466 \geqslant .033$	$.0352 \geqslant .017$		
$m = 4:$	$.0605 \geqslant .05$	$.0466 \geqslant .0375$	$.0352 \geqslant .025$	$.0238 \geqslant .0125$	
$m = 5:$	$.0605 \geqslant .05$	$.0466 \geqslant .04$	$.0352 \geqslant .03$	$.0238 \geqslant .02$	$.0218 \geqslant .01$
$m = 6:$	$.0605 \geqslant .05$	$.0466 \geqslant .0417$	$.0352 \geqslant .033$	$.0238 < .025$	...

Therefore the desired $m = 5$, and Hommel's procedure rejects all H_i with $p_i < .05/5 = .01$, i.e., $H_6, \ldots, H_{10}$. Note that one additional hypothesis (H_6) is rejected compared to the Hochberg procedure.

4.2.4. *Other modifications*

The stepwise procedures discussed above assume that the truth or falsity of any hypothesis does not imply that of any other hypothesis. Stated differently, this means that all combinations $\{H_i, i \in I$ are true and $H_i, i \notin I$ are false$\}$ are possible for $I \subseteq K$. This is called the *free combinations condition* by Holm (1979a). This condition is not satisfied by the family of pairwise comparisons. As an example, consider three null hypotheses, H_{ij}: $\theta_i = \theta_j$ for $1 \leqslant i < j \leqslant 3$. If one of them is false, say H_{12}, then at least one of H_{13} and H_{23} must also be false. Such hypotheses are known as *logically related hypotheses*.

Shaffer (1986) proposed two modifications of the Holm MTP which utilize this information. These modifications use the critical value α/k_i at the ith step of testing, where k_i is the maximum possible number of true hypotheses at that step. In Holm's unmodified procedure, $k_i = k - i + 1$. Shaffer gave tables of $k_i \leqslant k - i + 1$ for the pairwise comparisons family when (i) it is only known that $i - 1$ hypotheses have been rejected prior to the ith step, and (ii) it is also known which *specific* $i - 1$ hypotheses have been rejected prior to the ith step. These modifications can also be applied to

Hochberg's step-up procedure among others (Hochberg and Rom, 1994; Rom and Holland, 1994).

In some situations it may bc desirable to weigh the evidence against different hypotheses differently. This can be done by choosing weights w_i (subject to $w_i > 0$ and $\sum_{i=1}^{k} w_i = k$) and using weighted p-values, $p_{wi} = p_i/w_i$ in place of the p_i in the above procedures (Hochberg and Liberman, 1994).

5. Normal theory procedures for inter-treatment comparisons

5.1. Comparisons with a control and orthogonal contrasts

We consider the following distributional setting. For $i = 1, 2, \ldots, k$, let the θ_i be contrasts of interest in a general linear model and let the $\widehat{\theta}_i$ be their least squares (LS) estimators. We assume that the $\widehat{\theta}_i$ are jointly normally distributed with means θ_i, variances $\sigma^2\tau_i^2$ and $\mathrm{corr}(\widehat{\theta}_i, \widehat{\theta}_j) = \rho_{ij}$. The τ_i^2 and ρ_{ij} are known design-dependent constants, while σ^2 is an unknown scalar. We assume that S^2 is an unbiased mean square error estimator of σ^2 with ν d.f. and $\nu S^2/\sigma^2 \sim \chi_\nu^2$ independent of the $\widehat{\theta}_i$. Simultaneous inferences (confidence intervals and/or hypothesis tests) are desired on the θ_i.

Two examples of this setting are:

EXAMPLE 5.1 (Comparisons with a control in a one-way layout). Let the treatments be labelled $0, 1, 2, \ldots, k$ where 0 is a control treatment and $1, 2, \ldots, k$ are $k \geqslant 2$ test treatments. Denote by $\bar{Y}_i$ the sample mean based on n_i observations from the ith treatment $(0 \leqslant i \leqslant k)$. The $\bar{Y}_i$ are assumed to be independent $N(\mu_i, \sigma^2/n_i)$ random values and $\nu S^2/\sigma^2 \sim \chi_\nu^2$ independently of the $\bar{Y}_i$ where $\nu = \sum_{i=0}^{k} n_i - (k+1)$. The contrasts of interest are $\theta_i = \mu_i - \mu_0$, their LS estimators being $\bar{Y}_i - \bar{Y}_0$ $(1 \leqslant i \leqslant k)$. In this case we have

$$\tau_i^2 = \frac{1}{n_i} + \frac{1}{n_0} \quad \text{and} \quad \rho_{ij} = \lambda_i \lambda_j$$

$$\text{where } \lambda_i = \sqrt{\frac{n_i}{n_i + n_0}} \quad (1 \leqslant i \leqslant k).$$

EXAMPLE 5.2 (Orthogonal contrasts in balanced designs). Consider a balanced factorial experiment with m factorial combinations (cells) and n normal independent and identically distributed (i.i.d.) replicates per cell. Let μ_j and $\bar{Y}_j \sim N(\mu_j, \sigma^2/n)$ denote the true and the sample means, respectively, for the jth cell. The factorial effects (main effects and interactions) can be expressed as $\theta_i = \sum_{j=1}^{m} c_{ij}\mu_j$, which are estimated by $\widehat{\theta}_i = \sum_{j=1}^{m} c_{ij}\bar{Y}_j$ where $(c_{i1}, c_{i2}, \ldots, c_{im})$ are mutually orthogonal contrast vectors, $1 \leqslant i \leqslant k$. In this case we have

$$\tau_i^2 = \frac{\sigma^2}{n}\sum_{j=1}^{m} c_{ij}^2 \quad \text{and} \quad \rho_{ij} \propto \sum_{\ell=1}^{m} c_{i\ell}c_{j\ell} = 0.$$

5.1.1. Single-step procedures

Single-step MCP's for the contrasts θ_i are based on the upper α critical points of $\max_{1\leqslant i\leqslant k} T_i$ and $\max_{1\leqslant i\leqslant k} |T_i|$, where $(T_1, T_2, \ldots, T_k)$ have a joint k-variate t-distribution with ν d.f. and correlation matrix $\{\rho_{ij}\}$. Denote these critical points by $t^{(\alpha)}_{k,\nu,\{\rho_{ij}\}}$ and $|t|^{(\alpha)}_{k,\nu,\{\rho_{ij}\}}$, respectively. The $100(1-\alpha)\%$ one-sided SCI's are given by

$$\theta_i \geqslant \widehat{\theta}_i - t^{(\alpha)}_{k,\nu,\{\rho_{ij}\}} S_{T_i} \quad (1 \leqslant i \leqslant k). \tag{5.1}$$

The $100(1-\alpha)\%$ two-sided SCI's are given by

$$\theta_i \in \left[\widehat{\theta}_i \pm |t|^{(\alpha)}_{k,\nu,\{\rho_{ij}\}} S_{T_i}\right] \quad (1 \leqslant i \leqslant k). \tag{5.2}$$

Dunnett (1955) derived these SCI's for the comparisons with a control problem. When $n_1 = \cdots = n_k = n$, we have $\rho_{ij} = \rho = n/(n + n_0)$. For this equicorrelated case, the tables of the critical points $t^{(\alpha)}_{k,\nu,\rho}$ and $|t|^{(\alpha)}_{k,\nu,\rho}$ are given in Bechhofer and Dunnett (1988). When the n_i's are unequal, a good approximation to the exact critical points can be found by using the arithmetic average $\bar{\rho}$ of the correlations ρ_{ij} (HT, p. 146) and interpolating in the tables. Alternatively, the exact critical points can be computed without too much difficulty since the multivariate t integral can be reduced to a bivariate iterated integral because of the product correlation structure, $\rho_{ij} = \lambda_i \lambda_j$. For example, $t^{(\alpha)}_{k,\nu,\{\rho_{ij}\}}$ is the solution in t of the equation:

$$\int_0^{\infty} \int_{-\infty}^{\infty} \prod_{i=1}^{k} \Phi\left[\frac{\lambda_i z + tu}{\sqrt{1-\lambda_i^2}}\right] \mathrm{d}\Phi(z)\, \mathrm{d}F_\nu(u) = 1 - \alpha,$$

where $\Phi(\cdot)$ is the standard normal cdf and $F_\nu(\cdot)$ is the cdf of a $\sqrt{\chi^2_\nu/\nu}$ r.v. Dunnett's (1989) computer program can be used to evaluate this integral.

For orthogonal contrasts, the common correlation $\rho = 0$. In that case the critical points $t^{(\alpha)}_{k,\nu,0} = m^{(\alpha)}_{k,\nu}$ and $|t|^{(\alpha)}_{k,\nu,0} = |m|^{(\alpha)}_{k,\nu}$ are known as Studentized maximum and Studentized maximum modulus critical points, respectively.

Single-step MTP's can be based on the SCI's (5.1) and (5.2) in the usual manner. For example, for testing H_i: $\theta_i = 0$ vs. A_i: $\theta_i > 0$, the test statistics are

$$t_i = \frac{\widehat{\theta}_i}{S_{T_i}} \quad (1 \leqslant i \leqslant k).$$

The MTP based on (5.1)

$$\text{rejects } H_i \text{ if } t_i > t^{(\alpha)}_{k,\nu,\{\rho_{ij}\}} \quad (1 \leqslant i \leqslant k). \tag{5.3}$$

The adjusted p-values for this MTP can be found from the equation

$$p_{ai} = P\left\{\max_{1\leqslant j\leqslant k} T_j \geqslant t_i\right\} \quad (1 \leqslant i \leqslant k).$$

Two-sided tests can be done in a similar manner. Being based on the SCI's (5.2), this two-sided MTP permits directional decisions following rejections of the H_i: $\theta_i = 0$ (i.e., it controls the type I and III FWE).

5.1.2. Step-down procedures

Marcus et al. (1976) (see also Naik, 1975) applied the closure method to test the hypotheses H_i: $\theta_i = 0$ vs. A_i: $\theta_i > 0$ $(1 \leqslant i \leqslant k)$ in the context of the comparisons with a control problem. They applied Dunnett's test given by (5.3) to test intersection hypotheses $H_I = \bigcap_{i \in I} H_i$. The resulting test procedure is a step-down procedure, which takes a particularly simple form when the correlations are equal, $\rho_{ij} \equiv \rho$ (e.g., when $n_1 = n_2 = \cdots = n_k$ in the comparisons with a control problem). The steps in this step-down MTP are as follows:

Step 0. Order the test statistics t_i: $t_{(1)} \leqslant t_{(2)} \leqslant \cdots \leqslant t_{(k)}$. Let $H_{(1)}, H_{(2)}, \ldots, H_{(k)}$ be the corresponding hypotheses.

Step 1. Reject $H_{(k)}$ if $t_{(k)} > t^{(\alpha)}_{k,\nu,\rho}$ and go to Step 2. Otherwise accept all hypotheses and stop testing.

Step 2. Reject $H_{(k-1)}$ if $t_{(k-1)} > t^{(\alpha)}_{k-1,\nu,\rho}$ and go to Step 3. Otherwise accept $H_{(k-1)}, \ldots, H_{(1)}$ and stop testing, etc.

In general, reject $H_{(i)}$ iff $t_{(j)} > t^{(\alpha)}_{j,\nu,\rho}$ for $j = k, k-1, \ldots, i$.

This MTP is clearly more powerful than the single-step MTP. It requires 10%–25% less observations to control the same type I and type II error rates (Hayter and Tamhane, 1990). The adjusted p-values for this step-down MTP are given by

$$p_{a(i)} = \max\left(p'_{(k)}, p'_{(k-1)}, \ldots, p'_{(i)}\right),$$

where

$$p'_{(i)} = P\left\{\max(T_1, \ldots, T_i) \geqslant t_{(i)}\right\}.$$

Holm's procedure can be viewed as a Bonferroni approximation to this step-down MTP.

Bofinger (1987) and Stefánsson et al. (1988) derived SCI's by inverting this one-sided step-down MTP. This is the first known example where SCI's associated with a step-down MTP have been obtained, thus destroying the myth that stepwise MTP's are for "testing only." If $p \leqslant k$ is the number of accepted hypotheses then the SCI's are given by

$$\mu_i - \mu_0 \geqslant \begin{cases} \bar{Y}_i - \bar{Y}_0 - t^{(\alpha)}_{p,\nu,\rho} S\sqrt{\frac{1}{n} + \frac{1}{n_0}} & \text{for accepted hypotheses,} \\ 0 & \text{for rejected hypotheses.} \end{cases} \tag{5.4}$$

Note that for the accepted hypotheses, the above SCI's are at least as sharp as the single-step SCI's (5.1) since $t^{(\alpha)}_{p,\nu,\rho} \leqslant t^{(\alpha)}_{k,\nu,\rho}$ for $p \leqslant k$. But, for any rejected null hypothesis H_i: $\mu_i = \mu_0$, we are only able to make a weak claim that $\mu_i \geqslant \mu_0$ using (5.4), while with the single-step SCI's (5.1) we are able to give a sharper positive lower bound on $\mu_i - \mu_0$. Generally, one wants sharper confidence bounds on the parameters corresponding to rejected hypotheses. Therefore the practical usefulness of (5.4) is somewhat questionable. See Hayter and Hsu (1994) for a further discussion of stepwise MTP's and SCI's associated with them. These authors discuss tradeoffs between SCI's and multiple testing demonstrated with special examples for $k = 2$.

The two-sided step-down MTP is analogous, with the t_i replaced by the $|t_i|$ and the one-sided critical points $t^{(\alpha)}_{i,\nu,\rho}$ replaced by the two-sided critical points $|t|^{(\alpha)}_{i,\nu,\rho}$. It is not known whether this MTP permits directional decisions except for the cases $k > 2$, $\rho = 0$ (Holm, 1979b; Shaffer, 1980) and $k = 2$, ρ arbitrary (Finner, 1990).

Dunnett and Tamhane (1991) have given an extension of this step-down MTP for the case of unequal sample sizes when comparing treatments with a control in a one-way layout setting.

5.1.3. Step-up procedures

Dunnett and Tamhane (1992a) proposed a step-up MTP analogous to Hochberg's step-up MTP for the case of equal correlation. The steps in the procedure are as follows:

Step 0. Order the test statistics t_i: $t_{(1)} \leqslant t_{(2)} \leqslant \cdots \leqslant t_{(k)}$. Let $H_{(1)}, H_{(2)}, \ldots, H_{(k)}$ be the corresponding hypotheses. Choose critical constants $c_1 \leqslant c_2 \leqslant \cdots \leqslant c_k$ as indicated below.

Step 1. Accept $H_{(1)}$ if $t_{(1)} \leqslant c_1$ and go to Step 2. Otherwise reject all H_i and stop testing.

Step 2. Accept $H_{(2)}$ if $t_{(2)} \leqslant c_2$ and go to Step 3. Otherwise reject $H_{(2)}, \ldots, H_{(k)}$ and stop testing, etc.

In general, accept $H_{(i)}$ iff $t_{(j)} \leqslant c_j$ for $j = 1, 2, \ldots, i$.

The critical constants $c_1 \leqslant c_2 \leqslant \cdots \leqslant c_k$ are the solutions to the following recursive equations: Let $(T_1, T_2, \ldots, T_k)$ have a multivariate t-distribution given before. Let $T_{1,m} \leqslant T_{2,m} \leqslant \cdots \leqslant T_{m,m}$ be the order statistics of $T_1, T_2, \ldots, T_m$ $(1 \leqslant m \leqslant k)$. Then

$$P\{T_{1,m} \leqslant c_1, \ldots, T_{m,m} \leqslant c_m\} = 1 - \alpha \quad (1 \leqslant m \leqslant k).$$

Note the similarity of this equation with Rom's equation (4.4). The c_i can be calculated recursively starting with $c_1 = t^{(\alpha)}_{\nu}$. An algorithm for this purpose is given in Dunnett and Tamhane (1992a) where the tables of the constants c_i are also provided for selected values of k, ν, ρ and α. It is easy to show that $c_i > t^{(\alpha)}_{i,\nu,\rho}$ (which are the critical constants used by the step-down MTP) for $i > 1$. Therefore the step-up MTP is not uniformly more powerful than the step-down MTP; however, if most hypotheses are false then the step-up MTP is moderately more powerful than the step-down MTP

(Dunnett and Tamhane, 1993). The adjusted p-values are calculated as follows: Let $c_i = t_{(i)}$ and find $c_1, \ldots, c_{i-1}$ such that

$$P\{T_{1,j} \leqslant c_1, \ldots, T_{j,j} \leqslant c_j\} = 1 - p'_{(i)} \quad \text{for } j = 1, \ldots, i.$$

Then

$$p_{a(i)} = \min\left(p'_{(1)}, p'_{(2)}, \ldots, p'_{(i)}\right).$$

Hochberg's procedure can be viewed as a Bonferroni approximation to this step-up MTP.

Dunnett and Tamhane (1995) have given an extension of this procedure for the case of unequal sample sizes when comparing treatments with a control in a one-way layout setting. The difficulty with extending step-down and step-up MTP's to unbalanced designs is not the unequal variances of the $\widehat{\theta}_i$, but the unequal correlations ρ_{ij}. The correlation matrix at each step of testing is a submatrix of the correlation matrix at the previous step. When using the average correlation approximation, $\bar{\rho}$ changes at each step which increases the burden of the computational effort.

5.2. *Pairwise and more general comparisons*

Throughout this section we assume the same one-way layout setting discussed in Example 5.1 except that there is no control treatment labelled 0, all k treatments being test treatments. The d.f. for estimating the common error variance σ^2 is $\nu = \sum_{i=1}^{k} n_i - k$. The most common type of comparisons of interest in this setting are pairwise mean differences $\mu_i - \mu_j$ $(1 \leqslant i < j \leqslant k)$. This is known as the family of *pairwise comparisons*. A more general family of comparisons is the family of *all contrasts* where a contrast is a linear parametric function $\sum_{i=1}^{k} c_i \mu_i$ with $\sum_{i=1}^{k} c_i = 0$. We first discuss two classical single-step MCP's for these two families. Then, in the following subsection, we discuss stepwise MCP's.

5.2.1. *Single-step procedures*

The two classical MCP's for the families of pairwise comparisons and contrasts are the Tukey–Kramer (TK) and Scheffé (S) procedures, respectively. The TK procedure gives the following $100(1-\alpha)\%$ SCI's for all pairwise contrasts:

$$\mu_i - \mu_j \in \left[\bar{Y}_i - \bar{Y}_j \pm q_{k,\nu}^{(\alpha)} S \sqrt{\frac{1}{2}\left(\frac{1}{n_i} + \frac{1}{n_j}\right)}\right] \quad (1 \leqslant i < j \leqslant k), \qquad (5.5)$$

where $q_{k,\nu}^{(\alpha)}$ is the upper α point of the Studentized range distribution of dimension k and d.f. $\nu = \sum_{i=1}^{k} n_i - k$. When all the n_i's are equal, these SCI's are exact and the corresponding MCP is known as the Tukey (T) procedure. The above modification (5.5) for unequal sample sizes is known to be slightly conservative (shown using simulations by Dunnett (1980a) and analytically by Hayter (1984)). For pairwise comparisons the TK procedure generally provides the shortest SCI's. (For equal sample sizes, the T intervals are the shortest in the class of all $(1-\alpha)$-level SCI's of equal width; see Gabriel (1970).)

Next consider the family of all contrasts, the family of pairwise comparisons being a subfamily of this. Generally, the only higher order contrasts (i.e., contrasts involving three or more treatment means) of any interest in practice are the differences of average means of two subsets of treatments of the type

$$\frac{1}{|I|}\sum_{i\in I}\mu_i - \frac{1}{|J|}\sum_{j\in J}\mu_j$$

for two disjoint subsets I and J of $K = \{1,2,\ldots,k\}$. Most other contrasts are too complex and uninteresting. The TK intervals in (5.5) can be extended to the family of all contrasts (see equation (3.3) of Chapter 3 in HT). It gives rather wide intervals for these uninteresting higher order contrasts, but short intervals for interesting pairwise contrasts. This is a profitable trade-off in most applications.

The $100(1-\alpha)\%$ SCI's according to the S procedure are given by

$$\sum_{i=1}^{k} c_i\mu_i \in \left[\sum_{i=1}^{k} c_i\bar{Y}_i \pm \sqrt{(k-1)F_{k-1,\nu}^{(\alpha)}}\sqrt{S^2\sum_{i=1}^{k}\frac{c_i^2}{n_i}}\right] \quad \text{for all contrasts,} \qquad (5.6)$$

where $F_{k-1,\nu}^{(\alpha)}$ is the upper α point of the F-distribution with $k-1$ and ν d.f. The S procedure is related to the F test of the ANOVA as follows: The α-level F test rejects the overall null hypothesis H: $\mu_1 = \mu_2 = \cdots = \mu_k$ if at least one contrast, $\sum_{i=1}^{k} c_i\bar{Y}_i$, is significantly different from zero, i.e., the interval (5.6) excludes zero for at least one contrast. This relationship comes from the fact that the S-intervals can be obtained by the projection method from the confidence ellipsoid for any $k-1$ linearly independent contrasts among the μ_i.

For higher order contrasts, the S intervals are shorter than the TK intervals, but at the expense of substantially wider intervals for pairwise contrasts. Therefore the S procedure should be used only when data-snooping for *any* contrast that might seem interesting. If a finite family of, say, m potentially interesting contrasts can be *a priori* specified, then usually a better alternative is to use the Bonferroni method that tests each contrast at level α/m, unless m is very large. For the family of pairwise comparisons, of course, the TK procedure should be used. On the other hand, the TK procedure charges a well-deserved penalty of wide intervals to a researcher who goes on a fishing expedition to data-snoop for any significant contrast (Tukey, 1995).

Both the TK and S procedures extend to other designs following the general linear model. Suppose that $\widehat{\boldsymbol{\mu}}$ is the LS estimator of a k-dimensional parameter vector $\boldsymbol{\mu}$ (e.g., treatment effects in an ANCOVA design), $\widehat{\boldsymbol{\mu}} \sim N(\boldsymbol{\mu}, \sigma^2\boldsymbol{V})$ where $\boldsymbol{V} = \{v_{ij}\}$

is a known design-dependent matrix. Then, analogous to (5.5), the TK intervals are given by

$$\mu_i - \mu_j \in \left[\widehat{\mu}_i - \widehat{\mu}_j \pm q_{k,\nu}^{(\alpha)} S \sqrt{\frac{d_{ij}}{2}}\right] \quad (1 \leqslant i < j \leqslant k), \tag{5.7}$$

where $d_{ij} = v_{ii} + v_{jj} - 2v_{ij}$. As yet, no general proof is available to show that these intervals are conservative (although they are conjectured to be so) except for the cases $d_{ij} = a_i + a_j$ for some positive real numbers $a_1, a_2, \ldots, a_k$ (Hayter, 1989) and $d_{ij} \equiv d$ for all i, j; in this latter case the TK procedure is the same as the exact T procedure (Hochberg 1974a).

Analogous to (5.6), the S intervals are given by

$$\sum_{i=1}^{k} c_i \mu_i \in \left[\sum_{i=1}^{k} c_i \widehat{\mu}_i \pm \sqrt{(k-1) F_{k-1,\nu}^{(\alpha)}} \sqrt{S^2 \boldsymbol{c}' \boldsymbol{V} \boldsymbol{c}}\right] \quad \text{for all contrasts.} \tag{5.8}$$

These SCI's are too conservative for the family of pairwise comparisons. In this case one could use the TK intervals (5.7) without a guarantee of the FWE control (except for the cases noted). If guaranteed control of the FWE is desired then a more powerful method than the S procedure is Hochberg's (1974b) GT2. This latter procedure uses the critical constant $|m|_{k^*,\nu}^{(\alpha)}$ (the upper α point of the Studentized maximum modulus distribution of dimension $k^* = \binom{k}{2}$ and d.f. ν) in place of $q_{k,\nu}^{(\alpha)}/\sqrt{2}$ used in (5.7). The so-called Dunn–Šidák method (Dunn, 1958, 1961; Šidák, 1967), which is a sharpened version of the Bonferroni method and which uses the critical point $t_\nu^{(\alpha^*/2)}$ where $\alpha^* = 1 - (1-\alpha)^{1/k^*}$ is less powerful than GT2 and should not be used.

The T and S procedures can be extended to the family of *subset hypotheses*:

$$H_P\colon\ \mu_i = \mu_j \quad \forall\, i, j \in P \subseteq K = \{1, 2, \ldots, k\}. \tag{5.9}$$

This is a hierarchical family. For a balanced one-way layout, the T procedure rejects H_P if

$$Q_P = \frac{(\max_{i\in P} \bar{Y}_i - \min_{i \in P} \bar{Y}_i)\sqrt{n}}{S} > q_{k,\nu}^{(\alpha)}.$$

For a general one-way layout, the S procedure rejects H_P if

$$(p-1) F_P = \frac{\sum_{i\in P} n_i (\bar{Y}_i - \bar{Y}_P)^2}{S^2} > (k-1) F_{k-1,\nu}^{(\alpha)},$$

where $\bar{Y}_P = \sum_{i \in P} n_i \bar{Y}_i / \sum_{i\in P} n_i$. Gabriel (1969) referred to these as *simultaneous test procedures* (STP's) because they test the hypotheses H_P simultaneously without reference to one another (referred to as single-step MTP's here). Both the S and T procedures are coherent, but only the T procedure is consonant.

5.2.2. Step-down procedures

5.2.2.1. Fisher's protected LSD. Fisher's protected LSD provides the simplest example of a step-down MTP consisting of only two steps. The family consists of pairwise null hypotheses H_{ij}: $\mu_i = \mu_j$ $(1 \leqslant i < j \leqslant k)$ and their intersection H: $\mu_1 = \mu_2 = \cdots = \mu_k$. At the first step, LSD does an α-level F test of H, which implies all pairwise null hypotheses. If H is accepted then all H_{ij} are accepted by implication, and testing stops. If H is rejected then at the second step all H_{ij} are tested individually at level α using the usual two-sided t tests.

It is easy to see that the maximum FWE of this MTP exceeds α by considering the configuration $\mu_1 = \cdots = \mu_{k-1}, \mu_k \to \infty$. In this case, H is rejected with probability $\to 1$, effectively resulting in an *unprotected* LSD to make pairwise comparisons among $k-1$ equal means at the second step. Thus the protected LSD provides only weak control of the FWE for $k > 3$ and strong control of the FWE for $k = 3$ (which also follows from the fact that it is a closed procedure for $k = 3$). Hayter (1986) showed that if the pairwise two-sided t tests at the second step are done using the critical constant $q^{(\alpha)}_{k-1,\nu}/\sqrt{2}$ instead of the usual critical constant $t^{(\alpha/2)}_{\nu}$ then the FWE is controlled at level α.

5.2.2.2. Duncan and Newman–Keuls type procedures. Consider a one-way layout setting and the family of subset hypotheses (5.9). Let Z_P be a test statistic and c_p, a critical constant ($p = \mathrm{card}(P)$) for testing H_P.

The classical step-down procedures of Duncan (1955) and Newman–Keuls (Newman, 1939; Keuls, 1952) are special cases of the following general step-down testing scheme.

Step 1. Test H_K: $\mu_1 = \cdots = \mu_k$. Reject H_K if $Z_K > c_k$ and go to Step 2; else accept all hypotheses H_P and stop.

Step 2. Test H_P for all subsets $P \subset K$ with $\mathrm{card}(P) = k-1$. Reject H_P if $Z_P > c_{k-1}$. If any hypothesis is not rejected, accept it and also accept all hypotheses implied by it without actually testing them. Stop if there are no more hypotheses to be tested; otherwise go to the next step to test H_P with $\mathrm{card}(P) = k-2$.

In general, test and reject H_P iff it is not accepted at an earlier step (i.e., if $Z_{P'} > c_{p'}$ $\forall P' \supset P$ where $p' = \mathrm{card}(P')$) and $Z_P > c_p$. Stop if there are no more hypotheses to be tested; otherwise go to the next step to test non-accepted hypotheses H_P for P of the next lower cardinality.
The critical constants c_p are determined from the equation

$$\alpha_p = P\{Z_P > c_p \mid H_P\},$$

where α_p is a specified *nominal significance level* for testing H_P. The reason α_p is called "nominal" is that it does not represent the actual type I error probability; that probability is the supremum over all configurations satisfying H_P of the probability of the event $Z_{P'} > c_{p'}$ $\forall P' \supseteq P$, and is less than α_p except when $P = K$. How the

FWE depends on the α_p will be seen in the sequel. Duncan's choice for the nominal significance level was

$$\alpha_p = 1 - (1-\alpha)^{p-1} \quad (2 \leqslant p \leqslant k), \tag{5.10}$$

while Newman–Keuls' choice was

$$\alpha_p \equiv \alpha \quad (2 \leqslant p \leqslant k). \tag{5.11}$$

As will be seen, neither of these choices control the FWE.

A multiple F procedure uses

$$Z_P = F_P = \frac{\sum_{i \in P} n_i (\bar{Y}_i - \bar{Y}_P)^2}{(p-1)S^2},$$

and $c_p = F_{p,\nu}^{(\alpha_p)}$. A multiple range procedure uses (for a balanced one-way layout),

$$Z_P = Q_P = \frac{(\max_{i \in P} \bar{Y}_i - \min_{i \in P} \bar{Y}_i)\sqrt{n}}{S},$$

and $c_p = q_{p,\nu}^{(\alpha_p)}$.

The general step-down testing scheme involves a large number of tests. However, it is possible to develop a shortcut for a multiple range procedure in the case of a balanced one-way layout. This shortcut procedure is based on the following result of Lehmann and Shaffer (1977): Rejection of H_P implies rejection of $H_{P'}$ for all $P' \subseteq P$ containing the two treatments that yielded $\max_{i \in P} \bar{Y}_i$ and $\min_{i \in P} \bar{Y}_i$ iff $c_2 \leqslant c_3 \leqslant \cdots \leqslant c_k$. Thus, if the critical constants are monotone then whenever H_P is rejected, the two sample means corresponding to $\max_{i \in P} \bar{Y}_i$ and $\min_{i \in P} \bar{Y}_i$ can be immediately declared to be significantly different, and any subsets containing these two treatments need not be separately tested. This results in a tremendous reduction in the number of tests. The monotonicity condition on the $c_p = F_{p,\nu}^{(\alpha_p)}$ or $c_p = q_{p,\nu}^{(\alpha_p)}$ is satisfied by the Newman–Keuls choice $\alpha_p \equiv \alpha$, but not by the Duncan choice (although the constants in Duncan's tables are adjusted upward so as to satisfy the condition). The resulting shortcut procedure is as follows:

Step 0. Order the $\bar{Y}_i$: $\bar{Y}_{(1)} \leqslant \bar{Y}_{(2)} \leqslant \cdots \leqslant \bar{Y}_{(k)}$, and let $\mu_{(1)}, \mu_{(2)}, \ldots, \mu_{(k)}$ be the true means of the corresponding treatments.

Step 1. Test $H_K = H$: $\mu_{(1)} = \cdots = \mu_{(k)}$. Reject H_K if

$$Q_K = \frac{(\bar{Y}_{(k)} - \bar{Y}_{(1)})\sqrt{n}}{S} > q_{k,\nu}^{(\alpha_k)},$$

conclude $\mu_{(1)} \neq \mu_{(k)}$ and go to Step 2. Else accept H_K and all of its subsets and stop.

Step 2 (a). Test H_{P_1}: $\mu_{(1)} = \cdots = \mu_{(k-1)}$. Reject H_{P_1} if

$$Q_{P_1} = \frac{(\bar{Y}_{(k-1)} - \bar{Y}_{(1)})\sqrt{n}}{S} > q_{k-1,\nu}^{(\alpha_{k-1})},$$

and conclude $\mu_{(1)} \neq \mu_{(k-1)}$. Else accept H_{P_1} and all of its subsets.

Step 2 (b). Test H_{P_2}: $\mu_{(2)} = \cdots = \mu_{(k)}$. Reject H_{P_2} if

$$Q_{P_2} = \frac{(\bar{Y}_{(k)} - \bar{Y}_{(2)})\sqrt{n}}{S} > q_{k-1,\nu}^{(\alpha_{k-1})},$$

and conclude $\mu_{(2)} \neq \mu_{(k)}$. Else accept H_{P_2} and all of its subsets. Go to Step 3 if either test rejects, else stop, etc.

Note that for unbalanced designs, the multiple range procedure cannot be used in a shortcut manner. For this case, Kramer (1956) and Duncan (1957) proposed two different procedures. Duncan's range statistic is more appropriate, and is given by

$$Q_P = \max_{i,j \in P} \left\{ \frac{|\bar{Y}_i - \bar{Y}_j|}{S\sqrt{\frac{1}{2}(1/n_i + 1/n_j)}} \right\}.$$

For further details see HT, Chapter 4, Section 2.2.

We next discuss the problem of control of the FWE using these step-down MTP's. For this purpose we need to consider null hypotheses of the type

$$H_{\boldsymbol{P}} = H_{P_1} \cap H_{P_2} \cap \cdots \cap H_{P_r}, \tag{5.12}$$

where $\boldsymbol{P} = (P_1, P_2, \ldots, P_r)$ is some partition of $K = \{1, 2, \ldots, k\}$ (some of the P_j may be singletons in which case the corresponding H_{P_j} are vacuous). We refer to $H_{\boldsymbol{P}}$ as a *multiple subset hypothesis*. Denoting the FWE under $H_{\boldsymbol{P}}$ by $\alpha^*(\boldsymbol{P})$, for strong FWE control we must have

$$\alpha^* \equiv \max_{H_{\boldsymbol{P}}} \alpha^*(\boldsymbol{P}) \leqslant \alpha.$$

Tukey (1953) showed that, under a so-called separability condition,

$$\alpha^*(\boldsymbol{P}) \leqslant 1 - \prod_{j=1}^{r} (1 - \alpha_{p_j}),$$

where $p_j = \text{card}(P_j) \geqslant 1, \sum_{j=1}^{r} p_j = k$ and $\alpha_1 = 0$. Hence

$$\alpha^* \leqslant \max_{(p_1, p_2, \ldots, p_r)} \left[1 - \prod_{j=1}^{r} (1 - \alpha_{p_j}) \right]. \tag{5.13}$$

The following results can be shown using this formula (see HT, Chapter 2, Section 4.3.3.3).

- For Duncan's choice (5.10) of the α_p, max FWE $= 1 - (1-\alpha)^{k-1}$ which exceeds α for $k > 2$. Therefore Duncan's procedure does not control the FWE. However, it can be modified to do so by using $\alpha_p = 1 - (1-\alpha)^{(p-1)/(k-1)}$ (Einot and Gabriel, 1975).
- For Newman–Keuls' choice (5.11) of the α_p, max FWE = $1 - (1-\alpha)^{[k/2]}$ where $[k/2]$ denotes the integer part of $k/2$. Therefore max FWE exceeds α for $k > 3$. Thus the Newman–Keuls procedure does not control the FWE except for $k = 3$ (in which case it can be shown to be a closed procedure). However, it can be modified to do so by using $\alpha_p \equiv \alpha' = 1 - (1-\alpha)^{1/[k/2]}$ (Einot and Gabriel, 1975).
- Using (5.13), Einot and Gabriel (1975) suggested the choice $\alpha_p = 1 - (1-\alpha)^{p/k}$ $(2 \leqslant p \leqslant k)$ to control the FWE. Ryan's (1960) choice $\alpha_p = p\alpha/k$ is slightly more conservative. It was improved (to yield higher power while still maintaining the FWE control) by Welsch (1972) who suggested to use $\alpha_{k-1} = \alpha$; this modification was suggested earlier by Tukey (1953). From (5.13) we can see that this modification can be used also in conjunction with the Einot and Gabriel choice (since $\alpha_{k-1} = \alpha, \alpha_1 = 0$ still satisfies $\alpha^* \leqslant \alpha$) resulting finally in the choice

$$\alpha_p = 1 - (1-\alpha)^{p/k} \quad (2 \leqslant p \leqslant k-2),$$

$$\alpha_{k-1} = \alpha_k = \alpha. \tag{5.14}$$

This is often referred to as the *Ryan–Einot–Gabriel–Welsch* (REGW) choice (referred to as the Tukey–Welsch choice in HT). In SAS statistical software, the corresponding procedures based on the Q and F statistics are available under REGWQ and REGWF options in PROC MEANS. Lehmann and Shaffer (1979) showed that the REGW choice approximately maximizes the power subject to the FWE control requirement (2.1).

Shaffer (1979) proposed an improvement that can be applied to any step-down MTP in the above class. This improvement entails performing a preliminary F test as in Fisher's protected LSD. If the F test is nonsignificant, all subset null hypotheses are accepted and testing stops. If the F test is significant then the implication is that at least one mean is different from the others. Therefore at the first step of whichever step-down MTP is employed, H_K is tested using the critical constant appropriate for the second step, i.e., for testing the homogeneity of a subset of size $k-1$. For example, in Holm's step-down MTP, instead of using the critical constant α/k at the first step, the critical constant $\alpha/(k-1)$ is used. The testing proceeds in the usual manner from this point onward.

5.2.2.3. Peritz's closed procedure. Peritz (1970) was the first to propose the closure method, and suggest its application to the family of subset hypotheses, $\{H_P\}$. The closure of this family is the family of *multiple subset hypotheses*, $\{H_{\mathbf{P}}\}$, defined in (5.12). The procedure is to test each $H_{\mathbf{P}}$ at level α subject to the coherence requirement: if any $H_{\boldsymbol{P}}$ is accepted then accept all $H_{\boldsymbol{P}'}$ implied by it. For example, for $k = 6$, if $H_{\boldsymbol{P}}$ is accepted for $\boldsymbol{P} = \{P_1 = (1,2,3), P_2 = (4), P_3 = (5,6)\}$ then $H_{\boldsymbol{P}'}$ is accepted by implication (i.e., without a test) for $\boldsymbol{P}' = \{P_1' = (1,2), P_2' = (3), P_3' = (4), P_4' = (5,6)\}$.

The two main questions are: (i) how to test each $H_{\boldsymbol{P}}$ at level α, and (ii) how to insure coherence? Begun and Gabriel (1981) showed that the first question is answered by using an UI test of $H_{\boldsymbol{P}}$ whose component tests are obtained as follows: Let $r' \leqslant r$ be the number of non-vacuous component hypotheses H_{P_i} (i.e., with $p_i = \text{card}(P_i) \geqslant 2$) in $H_{\boldsymbol{P}}$. Test each such H_{P_i} at level

$$\alpha_{p_i} = 1 - (1 - \alpha)^{p_i/k} \simeq \frac{p_i \alpha}{k} \quad \text{if } r' > 1,$$

and at level α if $r' = 1$. Thus Peritz's procedure can be implemented by testing $H_{\boldsymbol{P}}$ iff it is not accepted earlier by implication and rejecting it iff at least one of the non-vacuous H_{P_i} is significant at level α_{p_i} given by the above formula.

Begun and Gabriel (1981) answered the second question by giving an algorithm for its implementation, thus making Peritz's procedure practicable. This algorithm uses the general step-down testing scheme of Section 5.2.2.1 for testing subset hypotheses H_P with two different choices of α_p in tandem, the Newman–Keuls (NK) choice (5.11) and the REGW choice (5.14). The algorithm tests the hypotheses H_P using the NK and REGW step-down MTP's. It accepts all H_P's that are accepted by the NK procedure and rejects all H_P's that are rejected by the REGW procedure. It makes decisions on the remaining (contentious) hypotheses as follows: a hypothesis H_P is accepted if (i) a hypothesis $H_{P'}$, which implies H_P (i.e., $P \subseteq P'$) is accepted, or (ii) H_P is nonsignificant at level $\alpha_p = 1-(1-\alpha)^{p/k}$ and for some P' in the complement of P with $\text{card}(P') = p' \geqslant 2$, $H_{P'}$ is nonsignificant at level $\alpha_{p'} = 1 - (1 - \alpha)^{p'/k}$; otherwise H_P is rejected. Once the decisions on the individual subset hypotheses are available, decisions on multiple subset hypotheses can be made using the UI method as indicated above. See HT, Chapter 4, Section 2.2 for further details. Either Q or F statistics can be used in the above procedure. A Fortran program for the Peritz procedure using the range statistics is given in Appendix B of Toothaker (1991). Competitors to Peritz's procedure have been proposed by Ramsey (1981), and Braun and Tukey (1983).

5.2.3. A step-up procedure

Welsch (1977) proposed a step-up procedure based on range statistics in the case of a balanced one-way layout with n observations per treatment. The procedure uses critical constants $c_2 \leqslant c_3 \leqslant \cdots \leqslant c_k$, which are determined (Welsch used Monte Carlo simulation for this purpose) to control the FWE. First, the treatment means are ordered: $\bar{Y}_{(1)} \leqslant \bar{Y}_{(2)} \leqslant \cdots \leqslant \bar{Y}_{(k)}$. The procedure begins by testing "gaps" or "2-ranges", $(\bar{Y}_{(i+1)} - \bar{Y}_{(i)})$ for $1 \leqslant i \leqslant k-1$. If any gap exceeds the critical value $c_2 S/\sqrt{n}$ then the corresponding treatment means are declared significantly different. Furthermore, all subsets P containing that pair of treatments are declared heterogeneous (i.e., the hypotheses H_P are rejected) and the corresponding p-ranges significant by implication.

In general, a p-range, $(\bar{Y}_{(p+i-1)} - \bar{Y}_{(i)})$ is tested iff it is not already declared significant by implication. If that p-range is tested and exceeds the critical value $c_p S/\sqrt{n}$ then it is declared significant.

5.3. Comparisons of MCP's and recommendations for use

In this subsection, we provide guidelines for selecting from the wide array of MCP's discussed above when the assumptions of normality and homogeneous variances are satisfied approximately in practice. See the next subsection for the performance of these MCP's and modifications needed when these assumptions are not satisfied. Comparisons are confined to MCP's that strongly control the FWE.

The two major considerations when selecting an MCP are:

1. What is the family of comparisons?

2. Are only multiple tests needed or are simultaneous confidence intervals also needed?

If the family consists of comparisons with a control or orthogonal contrasts then an MCP from Section 5.1 should be used; if the family consists of pairwise comparisons or subset hypotheses then an MCP from Section 5.2 should be used. MCP's for other special families can be constructed using the UI method; see, e.g., Hochberg and Rodriguez (1977). The Bonferroni method is always available as a general omnibus method.

If only multiple tests are needed then one of the stepwise MTP's should be used since they are more powerful than the corresponding single-step MTP's. If SCI's are needed (based on which tests can always be made) then single-step SCP's must be used. Two other advantages of the MTP's corresponding to these SCP's are that (i) they permit directional decisions in cases of rejections of null hypotheses in favor of two-sided alternatives, and (ii) they are simpler to apply than stepwise MTP's.

We now focus on MCP's of Section 5.2. If SCI's for pairwise comparisons are needed then the clear choice is the T procedure for pairwise balanced designs (e.g., balanced one-way layouts) and the TK procedure for unbalanced designs. It should be noted, however, that there is no analytical proof yet to show that the latter controls the FWE in all cases. The S procedure is too conservative in most applications, and should be used only when it is desired to select *any* contrast that seems interesting in light of the data, the price paid being the lack of power for simple comparisons. In other cases, the Bonferroni method may be used as long as the family of comparisons is prespecified and the number of comparisons is not unduly large.

For multiple testing, the criterion for the choice of an MTP is usually power. At least three different types of power have been studied in the literature for pairwise comparisons using Monte Carlo simulation; see Einot and Gabriel (1975) and Ramsey (1978, 1981). They are: all pairs power, any pair power and per pair power. The *all pairs power* is the probability of detecting all true pairwise differences, $\mu_i \neq \mu_j$. The *any pair power* is the probability of detecting at least one true pairwise difference. The *per pair power* is the average probability of detecting each true pairwise difference, averaged over all differences. The all pairs power has the drawback that it is vitiated by low power for even one pair even if all the pairs have high power (Tukey, 1995).

The any pair and per pair powers are more stable measures of power and hence may be preferred.

Another difficulty in comparing the powers of the MTP's is that, for a given procedure, any type of power depends on the true means configuration, and one procedure may be more powerful than another for certain configurations, but less powerful for others. In other words, except for some obvious cases (e.g., the step-down REGWQ procedure versus the single-step T procedure), rarely does one procedure uniformly dominate another. The following recommendations must be taken with these provisos.

Using the criterion of the all pairs power, Ramsey (1978) found that Peritz's procedure based on F statistics is the best choice. Peritz's procedure based on Q statistics and the REGWQ procedure are also good choices. Welsch's step-up procedure also has good power, but its applicability is limited to balanced one-way layouts; also, its critical constants are available only for a few cases thus further limiting its use. Another procedure whose power has not been studied in detail, but which is likely to have good power and which is very easy to apply even by hand is Hayter's (1986) modification of Fisher's LSD. For implementing other stepwise procedures, a computer program is almost always a must.

5.4. Violations of homoscedasticity and normality assumptions

Homoscedasticity refers to the assumption that the treatment variances, σ_i^2, are equal. When this assumption is violated, the performances (the FWE and power) of the MCP's for pairwise comparisons which all assume a common σ^2 is not severely affected if the sample sizes, n_i, are equal for all treatments. (However, the performances of the MCP's for comparisons with a control could be seriously affected if the control variance is different from the treatment variance.) The lack of robustness of the MCP's becomes more evident when the n_i's are unequal. Generally, the MCP's are conservative (FWE $< \alpha$) when the n_i's are directly paired with the σ_i^2 (e.g., when the n_i and σ_i^2 are highly positively correlated), and are liberal (FWE $> \alpha$) when they are inversely paired (e.g., when the n_i and σ_i^2 are highly negatively correlated).

A common solution to deal with unequal σ_i^2 is to use separate variance estimates, S_i^2, in conjunction with the Welch–Satterthwaite formula for the d.f. separately for each pairwise comparison:

$$\widehat{\nu}_{ij} = \frac{(S_i^2/n_i + S_j^2/n_j)^2}{(S_i^2/n_i)^2/(n_i - 1) + (S_j^2/n_j)^2/(n_j - 1)} \qquad (1 \leqslant i < j \leqslant k).$$

Games and Howell (1976) proposed a direct extension of the Tukey–Kramer procedure using this modification resulting in the following $100(1-\alpha)\%$ SCI's for all pairwise comparisons:

$$\mu_i - \mu_j \in \left[\bar{Y}_i - \bar{Y}_j \pm q_{k,\widehat{\nu}_{ij}}^{(\alpha)} \sqrt{\frac{1}{2}\left(\frac{S_i^2}{n_i} + \frac{S_j^2}{n_j}\right)}\,\right] \qquad (1 \leqslant i < j \leqslant k).$$

We refer to this as the GH procedure. Tamhane (1977) proposed an alternative procedure (called the T2 procedure) based on the Šidák (1967) inequality that uses the critical point $t_{\hat{\nu}_{ij}}^{(\alpha^*/2)}$ in place of $q_{k,\hat{\nu}_{ij}}^{(\alpha)}/\sqrt{2}$ used in the GH procedure. Simulations performed by Tamhane (1979) showed that the T2 procedure is always conservative, while the GH procedure can be slightly liberal under inverse pairings of the n_i with the σ_i^2. To remove the conservatism of T2, Dunnett (1980b) proposed the use of the Kimball (1951) inequality instead of the Šidák inequality. This leads to the T3 procedure, which uses the critical point $|m|_{k^*,\hat{\nu}_{ij}}^{(\alpha)}$ where $k^* = \binom{k}{2}$. The T3 procedure can be viewed as an extension of the GT2 procedure to the unequal variances case. Dunnett's simulations showed that T3 is always conservative. In conclusion, T3 is recommended if guaranteed control of the FWE is desired; GH may be used with a slight gain in power if some excess FWE can be tolerated in extreme cases.

Nonnormality of the data is a less serious concern than heteroscedasticity because many MCP's are based on sample means, which are approximately normally distributed thanks to the central limit theorem. However, one difficulty is that MCP's such as the Bonferroni procedures use the critical points from the extreme tail portions of the t distribution, and this is the portion that is most sensitive to nonnormality. Therefore one still needs to be concerned about the normality assumption. Distribution-free and robust procedures to deal with nonnormality are given in Chapter 9 of HT. Westfall and Young's (1993) bootstrap (resampling) approach discussed in Section 6.1 is also distribution-free, but more computer intensive.

5.5. *Multiple comparisons with the best*

Often the benchmark for comparison of treatments is the "best" treatment rather than an external control or standard. Hsu (1981, 1984) was the first to consider this multiple comparisons problem, and relate it to an extensive body of literature on ranking and selection (Gupta and Panchapakesan, 1979).

For simplicity, consider a balanced one-way layout with k treatments having n observations each. The notation is the same as that given in Example 5.1 except that there is no control treatment. Hsu considered the problem of making simultaneous confidence statements on $\mu_i - \mu_{\max}$ or equivalently on $\mu_i - \max_{j \neq i} \mu_j$ $(1 \leqslant i \leqslant k)$. He derived the following $100(1-\alpha)\%$ SCI's:

$$\mu_i - \max_{j \neq i} \mu_j \in \left[\left(\bar{Y}_i - \max_{j \neq i} \bar{Y}_j - t_{k-1,\nu,1/2}^{(\alpha)} S\sqrt{2/n} \right)^-, \right.$$
$$\left. \left(\bar{Y}_i - \max_{j \neq i} \bar{Y}_j + t_{k-1,\nu,1/2}^{(\alpha)} S\sqrt{2/n} \right)^+ \right] \quad (1 \leqslant i \leqslant k),$$

where x^- denotes the negative part of x, x^+ denotes the positive part of x and $t_{k-1,\nu,1/2}^{(\alpha)}$ is the upper α equicoordinate critical point of the $(k-1)$-variate t distribution with ν d.f. and common correlation $1/2$. For further extensions of these SCI's to unbalanced one-way layouts and other cases, see Edwards and Hsu (1983) and Hsu (1982, 1984).

6. Multiple endpoints

Thus far we have mostly focused on the problem of inter-treatment comparisons. Another common problem arising in biomedical studies is that of comparing two groups of patients (a treatment group and a control group) based on multiple responses (called *endpoints*).

Suppose there are $k \geqslant 2$ endpoints, $Y_1, Y_2, \ldots, Y_k$. Denote by $\boldsymbol{Y}_0 = (Y_{01}, Y_{02}, \ldots, Y_{0k})$ and $\boldsymbol{Y}_1 = (Y_{11}, Y_{12}, \ldots, Y_{1k})$ the vectors of observations on a typical patient from the control group and the treatment group, respectively. Let $\boldsymbol{\mu}_0 = (\mu_{01}, \mu_{02}, \ldots, \mu_{0k})$ and $\boldsymbol{\mu}_1 = (\mu_{11}, \mu_{12}, \ldots, \mu_{1k})$ be the mean vectors of the two groups, and let $\boldsymbol{\theta} = \boldsymbol{\mu}_1 - \boldsymbol{\mu}_0$ be the difference vector.

Two different types of questions are often posed:

1. Is there at least one endpoint for which the treatment is more effective than the control? Identify all such endpoints.

2. Do different endpoints point in the same direction with regard to the superiority of the treatment over the control? If so, does the combined evidence support the treatment's superiority?

Assuming that larger values represent higher efficacy for each endpoint, the first question is often formulated as a multiple hypotheses testing problem:

$$H_i\colon\ \theta_i = 0 \quad \text{vs.} \quad A_i\colon\ \theta_i > 0 \quad (1 \leqslant i \leqslant k). \tag{6.1}$$

The second question is often formulated as a single hypothesis testing problem:

$$H\colon\ \boldsymbol{\theta} = \mathbf{0} \quad \text{vs.} \quad A\colon\ \boldsymbol{\theta} \geqslant \mathbf{0}, \tag{6.2}$$

where $\mathbf{0}$ denotes the null vector and $\boldsymbol{\theta} \geqslant \mathbf{0}$ means that $\theta_i \geqslant 0$ for all i with a strict inequality for at least some i; equivalently, $\boldsymbol{\theta}$ is in the positive orthant.

To address the testing problem (6.1), the usual approach is to evaluate the adjusted (for multiplicity) p-values of the tests comparing the two groups on the k endpoints. The p-value based MCP's of Section 3 (e.g., Bonferroni, Holm, Hochberg) are applicable here. However, those MCP's do not capitalize on any information on the correlations between the Y_i's. Also, if the Y_i's are discrete then those MCP's can be unduly conservative. Section 6.1 discusses different methods that take into account arbitrary correlations between continuous endpoints; some comments on the tests for the testing problem (6.2) are also given. Section 6.2 discusses improved Bonferroni and exact procedures for discrete (binary) endpoints.

6.1. *Continuous endpoints*

Let $\boldsymbol{Y}_{0m}$, $m = 1, 2, \ldots, n_0$, be n_0 i.i.d. observations from the control group and $\boldsymbol{Y}_{1m}$, $m = 1, 2, \ldots, n_1$, be n_1 i.i.d. observations from the treatment group. A common assumption is that the covariance matrices of the $\boldsymbol{Y}_{\ell m}$ in each group $\ell = 0, 1$ are equal. Let ρ_{ij} be the correlation coefficient between Y_i and Y_j (the ith and jth endpoint) for $1 \leqslant i < j \leqslant k$. The ρ_{ij} are generally unknown and must be estimated from data, although this will not be made explicit notationally. Let p_i be the raw p-value obtained using some test statistic for the ith endpoint $(1 \leqslant i \leqslant k)$. First we mention some ad-hoc methods for adjusting the p_i. These methods are based only on the raw p-values, and do not make any assumption on the joint distribution of the test statistics.

Tukey, Ciminera and Heyse (1985) suggested the following formula as an *ad-hoc* approximation when the ρ_{ij} are completely unknown:

$$p_{ai} = 1 - (1 - p_i)^{\sqrt{k}} \quad (1 \leqslant i \leqslant k).$$

We can generalize this formula to depend on the ρ_{ij} (which will need to be estimated if unknown) as follows:

$$p_{ai} = 1 - (1 - p_i)^{k^{(1-\bar{\rho})}} \quad (1 \leqslant i \leqslant k),$$

where $\bar{\rho}$ is the average of all the ρ_{ij}. Note that if $\bar{\rho} = 0$ then this reduces to the formula (4.2), while if $\bar{\rho} = 1$ then $p_{ai} = p_i$, i.e., no adjustment is made. Dubey (1985) suggested a similar formula, but in which $\bar{\rho}$ is replaced by $\bar{\rho}_i$, which is the average of the ρ_{ij} for $j \neq i$. However, this modification is inappropriate, as can be seen from the fact that the p_{ai} depend on the joint distribution of *all* the P_j since

$$p_{ai} = P\{\text{At least one } P_j \leqslant p_i\} = 1 - P\left\{\bigcap_{j=1}^{k}(P_j > p_i)\right\};$$

therefore any adjustment should involve all ρ_{ij}, not just the ρ_{ij} for $j \neq i$.

In the equal correlation case James (1991) gave an analytic approximation for p_{ai} under the multivariate normality assumption; in the unequal correlation case she suggested using the arithmetic average of the absolute values of the ρ_{ij}. Armitage and Parmar (1986) have suggested a more complicated empirical formula.

Westfall and Young (1993) proposed a general bootstrap (resampling) approach, which can be used to estimate the p_{ai} without making any parametric distributional or correlational assumptions. Let $\boldsymbol{y}_{01}, \boldsymbol{y}_{02}, \ldots, \boldsymbol{y}_{0n_0}$ and $\boldsymbol{y}_{11}, \boldsymbol{y}_{12}, \ldots, \boldsymbol{y}_{1n_1}$ be the observed data vectors from the control and the treatment groups, respectively. Let $p_1, p_2, \ldots, p_k$ be the observed raw p-values obtained using appropriate two-sample tests for each endpoint. The bootstrap procedure operates as follows:

1. Pool the two samples together.
2. Draw bootstrap samples $\boldsymbol{y}^*_{01}, \boldsymbol{y}^*_{02}, \ldots, \boldsymbol{y}^*_{0n_0}$ and $\boldsymbol{y}^*_{11}, \boldsymbol{y}^*_{12}, \ldots, \boldsymbol{y}^*_{1n_1}$ *with replacement* from the pooled sample. (See Section 6.2.2.3 for a *without replacement* version of this procedure.)
3. Apply the appropriate two-sample tests to each of the k endpoints using the bootstrap samples and calculate bootstrap p-values $p^*_1, p^*_2, \ldots, p^*_k$.
4. Repeat steps 2 and 3 some large number (N) of times.
5. The bootstrap estimates of the adjusted p-values are then

$$\widehat{p}_{ai} = \frac{\sharp(\min p^*_j \leqslant p_i)}{N} \quad (1 \leqslant i \leqslant k) \tag{6.3}$$

where $\sharp(\min p_j^* \leqslant p_i)$ is the number of simulations resulting in $\min p_j^* \leqslant p_i$.

Step-down versions of this procedure are given by Westfall and Young (1993) and Troendle (1995a). Step-up versions are given by Troendle (1995b).

The advantages of the bootstrap approach are that (i) it is distribution-free, (ii) it accounts for the dependence structure automatically from the observed data, and (iii) it is very flexible in accommodating different tests for different endpoints. However, the justification for the FWE control is only in the asymptotic sense.

Now suppose that the $\boldsymbol{Y}_{0m}$ and $\boldsymbol{Y}_{1m}$ are multivariate normal. For testing H_{0i}: $\theta_i = 0$, consider the usual test statistic

$$Z_i = \frac{\bar{Y}_{1i} - \bar{Y}_{0i}}{\sigma_i \sqrt{1/n_1 + 1/n_0}} \tag{6.4}$$

where $\bar{Y}_{1i}$ and $\bar{Y}_{0i}$ are the corresponding sample means and σ_i is the standard deviation of Y_i (usually estimated from data) $(1 \leqslant i \leqslant k)$. Note that $\text{corr}(Z_i, Z_j) = \rho_{ij}$ $(1 \leqslant i < j \leqslant k)$. The raw p-values are given by

$$p_i = P\{Z_i \geqslant z_i \mid \theta_i = 0\},$$

where z_i is the observed value of Z_i $(1 \leqslant i \leqslant k)$. To obtain the p_{ai} we need to compute multivariate normal probabilities for which Schervish's (1984) algorithm can be used. Unfortunately, this algorithm is not practically feasible for large k (e.g., $k > 5$). It is possible to use the average ρ approximation and the faster algorithm of Dunnett (1989).

Recently Mosier et al. (1995) have developed the following ad-hoc method, which is a hybrid of the multivariate normal and the p-value based methods. The focus of this method is to find the effective number of comparisons, k', to be used in the formula (4.2). Let $z^{(\alpha)}$ be the upper α critical point of the univariate standard normal distribution. Then k' is found from

$$1 - (1 - \alpha)^{k'} = P\Big\{\max_{1 \leqslant i \leqslant k} Z_i \geqslant z^{(\alpha)}\Big\}.$$

The average ρ approximation is suggested for evaluating the right hand side of this equation if the correlation structure of the Z_i is not simple. Having found k', the adjusted p-values are calculated using

$$p_{ai} = 1 - (1 - p_i)^{k'} \quad (1 \leqslant i \leqslant k).$$

Next we discuss the hypothesis testing problem (6.2). Although this is a multivariate testing problem, Hotelling's T^2 test is inappropriate since here the alternative is one-sided, while the T^2 test is for a two-sided omnibus alternative. One-sided global tests have been proposed by O'Brien (1984), Tang et al. (1989a, b) and Tang et al. (1993). We skip the discussion of these tests because of lack of space; also, these are single tests and as such they do not address not a multiple comparisons problem. However, once H: $\boldsymbol{\theta} = \mathbf{0}$ is rejected, follow-up tests may be required to determine which endpoints or sets of endpoints caused rejection. Lehmacher et al. (1991) have given a closure method for this purpose that tests subset hypotheses (e.g., H_I: $\theta_i = 0$ for $i \in I$ vs. A_I: $\theta_i \geqslant 0$ with at least one strict inequality for $i \in I \subseteq K = \{1, 2, \ldots, k\}$) in a step-down manner. Subsets of endpoints found to yield significance may have interpretation as representing certain syndromes (e.g., pain) for which the treatment is more effective than the control. Tang and Lin (1994) report further refinements of this approach. Stepanavage et al. (1994) have given a review of the literature in this area.

6.2. *Discrete endpoints*

Binary outcomes are common in medical studies. One example is an animal carcinogenic study in which about 100 rats/mice are randomly divided into a control and a treatment group. The animals in the control group are fed a normal diet, while those in the treatment group are fed a diet containing the compound under test for carcinogenicity. For each animal, k different sites (e.g., liver, thyroid, skin) are examined for the occurrence of tumors. The k 0–1 outcomes for each animal can be regarded as multiple endpoints. Based on these data, it is of interest to determine if there is an increased incidence of tumors in the treatment group at certain sites. If π_{0i} and π_{1i} denote the true tumor incidence rates at site i for the control and treatment groups, respectively, then this can be formulated as a multiple hypotheses testing problem:

$$H_i\colon \pi_{0i} = \pi_{1i} \quad \text{vs.} \quad A_i\colon \pi_{0i} < \pi_{1i} \quad (1 \leqslant i \leqslant k).$$

Let $H = \bigcap_{i=1}^{k} H_i$ and $A = \bigcup_{i=1}^{k} A_i$.

Suppose there are n_0 animals in the control group and n_1 in the treatment group. Let Y_{0i} and Y_{1i} be the numbers of animals in each group with tumors at site i ($1 \leqslant i \leqslant k$). Then $\mathbf{Y}_0 = (Y_{01}, Y_{02}, \ldots, Y_{0k})$ and $\mathbf{Y}_1 = (Y_{11}, Y_{12}, \ldots, Y_{1k})$ are independent multivariate binomial vectors with correlated components. Let $\boldsymbol{y}_0 = (y_{01}, y_{02}, \ldots, y_{0k})$ and $\boldsymbol{y}_1 = (y_{11}, y_{12}, \ldots, y_{1k})$ be the corresponding observed data vectors. For each site i we have a 2×2 table:

	Tumor	No Tumor	
Control	y_{0i}	$n_0 - y_{0i}$	n_0
Treatment	y_{1i}	$n_1 - y_{1i}$	n_1
	m_i	$n - m_i$	n

where $n = n_0 + n_1$ is the total number of animals in the study. The raw p_i can be obtained by conditioning on m_i and using Fisher's exact test:

$$p_i = \sum_{y \leqslant y_{0i}} \frac{\binom{n_0}{y}\binom{n_1}{m_i - y}}{\binom{n}{m_i}} = \sum_{y \geqslant y_{1i}} \frac{\binom{n_0}{m_i - y}\binom{n_1}{y}}{\binom{n}{m_i}}, \qquad i = 1, \ldots, k.$$

One may consider using the p_i to test the H_i and (by the UI method) $p_{\min}$ to test H. However, to account for the multiplicity of the tests, the adjusted p-values,

p_{ai} and p_{amin}, must be used. For this purpose, the Bonferroni methods for continuous data are generally too conservative. In the following two subsections we discuss modified Bonferroni and exact permutational procedures which take into account the discreteness of the test statistics.

6.2.1. *Modified Bonferroni procedures*

6.2.1.1. Tukey–Mantel procedure. The usual Bonferroni adjusted p-value, $p_{amin} = \min(kp_{min}, 1)$, is often too large because the r.v.'s P_i have discrete (not continuous $\mathcal{U}(0,1)$) distributions under H. Therefore,

$$P_H\{P_i \leqslant p_{min} \mid m_i\} \equiv p_i^* \leqslant p_{min} \quad (1 \leqslant i \leqslant k). \tag{6.5}$$

In fact, some of the p_i^* could be zero. Based on this observation, Mantel (1980) gave (based on an earlier suggestion by Tukey) the following formula:

$$p_{amin} = \min\left(\sum_{i=1}^{k} p_i^*, 1\right). \tag{6.6}$$

A slightly sharper bound is provided by assuming that the P_i are independent which leads to

$$p_{amin} = 1 - \prod_{i=1}^{k}(1 - p_i^*). \tag{6.7}$$

These formulas are easily generalized to calculate the p_{ai}.

6.2.1.2. Tarone's procedure. Discreteness of P_i means that its minimum achievable value is bounded below by a positive constant, $p_{i,min}$. If $p_{i,min}$ is greater than the level used to test H_i then H_i is not rejectable at that level. Therefore the number of hypotheses could be reduced by eliminating those which are not rejectable at a given level. Tarone (1990) used this idea to sharpen the Bonferroni procedure as follows: Calculate the minimum value of p_i for each i. If $m_i \leqslant n_1$ then

$$p_{i,min} = \frac{\binom{n_1}{m_i}}{\binom{n}{m_i}} \quad (1 \leqslant i \leqslant k).$$

1. First check whether the Bonferroni procedure can be used with level α for each hypothesis. Since the FWE must be controlled at level α, this is possible only if there is at most one rejectable hypothesis, i.e., if

$$k_1 = \sharp\{i\colon\ p_{i,min} < \alpha\} \leqslant 1.$$

If there are no rejectable hypotheses ($k_1 = 0$) then accept all H_i's. If $k_1 = 1$ then test that rejectable hypothesis at level α.

2. If $k_1 > 1$ then check whether the Bonferroni procedure can be used with level $\alpha/2$ for each hypothesis. Since the FWE must be controlled at level α, this is possible only if there are at most two rejectable hypotheses, i.e., if

$$k_2 = \sharp\{i\colon\ p_{i,\min} < \alpha/2\} \leqslant 2.$$

If $k_2 = 0$ then accept all H_i's. If $k_2 = 1$ or 2 then test those rejectable hypotheses each at level $\alpha/2$. If $k_2 > 2$ go to the next step.

3. In general, let

$$k_j = \sharp\{i\colon\ p_{i,\min} < \alpha/j\}, \quad j = 1, 2, \ldots, k.$$

Note $k_1 \geqslant k_2 \geqslant \cdots \geqslant k_k$. Find the smallest $j = j^*$ such that $k_j \leqslant j$. Then test the rejectable H_i at level α/j^*.

Roth (1996) has given additional improvements of Tarone's method.

6.2.2. Permutational procedures

6.2.2.1. Brown and Fears procedure, Brown and Fears (1981) gave a method for evaluating p_{amin} by using the exact joint distribution of Fisher's test statistics for the k 2×2 tables. To explain this method, introduce the notation $Y_0(S)$ and $Y_1(S)$ where $Y_0(S)$ (resp., $Y_1(S)$) is the number of animals in the control group (resp., treatment group) with at least one tumor at each site $i \in S \subseteq K = \{1, 2, \ldots, k\}$; if S is an empty set then the notation stands for animals with no tumors at any of the sites. Note

$$Y_{0i} = \sum_{S\colon\ i \in S} Y_0(S) \quad \text{and} \quad Y_{1i} = \sum_{S\colon\ i \in S} Y_1(S).$$

Let $Y_0(S) + Y_1(S) = m(S)$ be the total number of animals with at least one tumor at each site $i \in S$. The Brown and Fears method is based on the permutational (randomization) joint distribution of $\boldsymbol{Y}_1 = (Y_{11}, Y_{12}, \ldots, Y_{1k})$ (or equivalently $\boldsymbol{Y}_0 = (Y_{01}, Y_{02}, \ldots, Y_{0k})$) conditioned on *all* $m(S), S \subseteq \{1, 2, \ldots, k\}$ (not just the marginal totals m_i). Under

$$H\colon\ \pi_{0i} = \pi_{1i}\ (1 \leqslant i \leqslant k),$$

this distribution is multivariate hypergeometric:

$$P\{\boldsymbol{Y}_1 = \boldsymbol{y}_1\} = \sum \prod_{S \subseteq K} \binom{m(S)}{y_1(S)} \Big/ \binom{n}{n_1}, \tag{6.8}$$

where $\boldsymbol{y}_1 = (y_{11}, y_{12}, \ldots, y_{1k})$ and the sum is over all $y_1(S),\ S \subseteq K$ such that

$$y_{1i} = \sum_{S\colon\ i \in S} y_1(S) \quad (1 \leqslant i \leqslant k).$$

Using this distribution, p_{amin} is obtained from

$$p_{\text{amin}} = P_H\left\{\bigcup_{i=1}^{k}(Y_{1i} \geqslant c_i)\,\Big|\, m(S) \quad \forall S \subseteq K\right\}, \tag{6.9}$$

where c_i is the largest integer such that

$$P_H\{Y_{1i} \geqslant c_i \mid m_i\} = p_i^* \leqslant p_{\min},$$

where p_i^* is defined in (6.5). If no c_i exists satisfying the above inequality (in which case p_i^* is defined to be zero in (6.5)), the corresponding event is vacuous in (6.9).

Brown and Fears (1981) also considered extensions of this permutational approach to the case of more than one dose of the compound. In this case, two kinds of multiple comparison problems arise. One problem is that of comparing various dose levels to the control (zero dose) group, each comparison involving k sites. The other problem is that of dose response involving a test of trend at each site. They suggested approximate and simulation methods to address these problems. The same permutational approach was proposed by Heyse and Rom (1988) and Farrar and Crump (1988). Heyse and Rom gave an exact procedure for the trend test problem.

6.2.2.2. Rom procedure. Rom (1992) proposed to test the overall null hypothesis H based on the adjusted p-value (denoted by p_a) that takes into account all the p-values instead of only the $p_{\min}$. Let $p_{(1)} \geqslant p_{(2)} \geqslant \cdots \geqslant p_{(k)}$ be the ordered p-values and let $P_{(i)}$ be the r.v. corresponding to $p_{(i)}$. Then p_a is the probability of the event that

$$\{P_{(k)} < p_{(k)}\} \quad \text{or} \quad \{P_{(k)} = p_{(k)}\} \cap \{P_{(k-1)} < p_{(k-1)}\} \quad \text{or} \ \cdots \ \text{or}$$

$$\{P_{(k)} = p_{(k)}\} \cap \cdots \cap \{P_{(2)} = p_{(2)}\} \cap \{P_{(1)} < p_{(1)}\}.$$

Clearly, this probability is never larger (and often much smaller) than $p_{\text{amin}} = P\{P_{\min} \leqslant p_{\min}\}$. Therefore the test of H based on p_a is more powerful than the test based on p_{amin}. However, this procedure is computationally more complicated.

Chang and Rom (1994) have given a test of H based on the exact permutational distribution of Fisher's combination statistic, $-2\sum_{i=1}^{k}\log(p_i)$. This distribution takes into account the binary nature of the endpoints and the dependence between them.

6.2.2.3. Westfall and Young procedure. Let $\boldsymbol{x}_{hi} = (x_{hi1}, \ldots, x_{hik})$ be the data vector for the ith animal from the hth group ($h = 0, 1; i = 1, 2, \ldots, n_h$). Then the $\boldsymbol{x}_{hi}$ are multivariate Bernoulli vectors and $\boldsymbol{y}_h = \sum_{i=1}^{n_h} \boldsymbol{x}_{hi}$. Westfall and Young (1989) proposed to apply the bootstrap procedure of Section 6.1 by sampling *with replacement* from the $\boldsymbol{x}_{hi}$ and calculating the p-values, $p_1^*, p_2^*, \ldots, p_k^*$ using Fisher's exact tests on k marginal 2×2 tables for the bootstrapped data. By repeating this procedure some large number (N) of times, the p_{ai} can be estimated using the formula (6.3). This method does not fix the marginal totals m_i. Therefore the p_{ai} estimated are *unconditional.* If the bootstrap sampling is done *without replacement* then the bootstrap procedure corresponds to the Brown and Fears *conditional* permutation procedure.

6.2.3. Example

The following example is from Rom (1992). In a hypothetical study 100 animals are randomly assigned with 50 each to the control and the treatment group. Only $k = 2$ tumor sites, A and B, are examined with the following results:

Site	Control	Treatment	Total
A only	0	3	3
B only	1	6	7
A and B	1	2	3
No Tumor	48	39	87
Total	50	50	100

The marginal 2×2 tables are as follows:

Site A		Tumor	No Tumor	Total
	Control	1	49	50
	Treatment	5	45	50
	Total	6	94	100

Site B		Tumor	No Tumor	Total
	Control	2	48	50
	Treatment	8	42	50
	Total	10	90	100

The marginal p-values using Fisher's exact test are: $p_1 = P\{Y_{11} \geqslant 5 \mid m_1 = 6\} = .1022$ and $p_2 = P\{Y_{12} \geqslant 8 \mid m_2 = 10\} = .0457$ (see Table 2 which gives the joint as well as the marginal distributions of the Fisher exact test statistics, Y_{11}, Y_{12}); hence $p_{\min} = .0457$. We shall now calculate p_{amin} using the methods discussed above.

First, for the Bonferroni Procedure we have $p_{\text{amin}} = 2 \times .0457 = .0914$. Next, to apply the Tukey–Mantel procedure we need to calculate p_1^* and p_2^*. We have $P\{Y_{11} \geqslant 6 \mid m_1 = 6\} = .0133 < p_{\min}$ and $P\{Y_{11} \geqslant 5 \mid m_1 = 6\} = .1022 > p_{\min}$; therefore $p_1^* = .0133$. Next, $p_2^* = .0457$. Substituting in (6.6) gives $p_{\text{amin}} = .0133 + .0457 = .0590$. If (6.7) is used then we get $p_{\text{amin}} = 1 - (1 - .0133)(1 - .0457) = .0584$.

To apply the Tarone procedure, first calculate $p_{1,\min} = .0133$ and $p_{2,\min} = .0005$. Therefore $k_1 = 2, k_2 = 2$ and $j^* = 2$; thus no reduction in the number of rejectable hypotheses is achieved. Comparing the observed p_1 and p_2 with $\alpha/j^* = .025$, we find that neither site has a significant result at $\alpha = .05$

To apply the Brown and Fears procedure we need the joint distribution of $\mathbf{Y}_1 = (Y_{11}, Y_{12})$ which is given in Table 2 (calculated using (6.8)). From the marginal distributions of Y_{11} and Y_{12} we see that the largest values c_i such that $P\{Y_{1i} \geqslant c_i \mid m_i\} \leqslant p_{\min}$ are $c_1 = 6$ and $c_2 = 8$. Therefore

$$
\begin{aligned}
p_{\text{amin}} &= P\{(Y_{11} \geqslant 6) \cup (Y_{12} \geqslant 8)\} \\
&= P\{Y_{11} \geqslant 6\} + P\{Y_{12} \geqslant 8\} - P\{(Y_{11} \geqslant 6) \cap (Y_{12} \geqslant 8)\} \\
&= .0133 + .0457 - .022 = .0568.
\end{aligned}
$$

Table 2
Joint distribution of Y_{11} and Y_{12}

		Y_{11} 0	1	2	3	4	5	6	Marginal Prob.
	0	.0000	.0002	.0002	.0001	.0000	.0000	.0000	.0005
	1	.0005	.0019	.0028	.0017	.0003	.0000	.0000	.0072
	2	.0017	.0080	.0136	.0107	.0036	.0004	.0000	.0380
	3	.0035	.0182	.0366	.0353	.0164	.0031	.0001	.1132
	4	.0040	.0250	.0600	.0698	.0409	.0109	.0009	.2115
Y_{12}	5	.0026	.0212	.0622	.0872	.0622	.0212	.0026	.2592
	6	.0009	.0109	.0409	.0698	.0600	.0250	.0040	.2115
	7	.0001	.0031	.0164	.0353	.0366	.0182	.0035	.1132
	8	.0000	.0004	.0036	.0107	.0136	.0080	.0017	.0380
	9	.0000	.0000	.0003	.0017	.0028	.0019	.0005	.0072
	10	.0000	.0000	.0000	.0001	.0002	.0002	.0000	.0005
Marginal Probab.		.0133	.0889	.2366	.3224	.2366	.0889	.0133	1.000

Notice that the Mantel–Tukey approximations, namely .0590 and .0584, are quite close to the exact p_{amin}. However, they are all greater than $\alpha = .05$ and so H cannot be rejected.

Finally we apply the Rom procedure to these data. The boundary of the region whose probability is Rom's overall adjusted p-value is shown by a solid line in Table 2. Adding up the probabilities we find that $p_a = .0285$. Thus, in this example, only the Rom procedure yields a significant result.

7. Miscellaneous problems

Multiple comparisons is a vast subject. Because of lack of space we have only presented here some selected topics based on personal taste. There are large areas that have been left untouched by the present overview. In this section we briefly comment on a few of these.

Duncan (1965) formulated a Bayesian approach to the problem of pairwise comparisons in a balanced one-way layout setting. A prior distribution on the means and the type I and type II error losses are assumed to be given. The goal is to minimize the Bayes risk. This approach is further extended in Waller and Duncan (1969). The resulting multiple tests are called k-ratio t-tests whose critical values depend on the F-statistic for the overall null hypothesis in addition to the ratio of the type I and type II error losses. Duncan and Dixon (1983) have given a summary of work in this area.

Another area of interest is comparisons among ordered treatments, e.g., increasing dose levels of a drug. In such dose response studies it is common to include a zero dose treatment, which serves as a control. Suppose that the responses are normally distributed with a common variance σ^2 and means $\mu_0, \mu_1, \ldots, \mu_k$ corresponding to the increasing dose levels $d_0 = 0, d_1, \ldots, d_k$. The μ_i may be monotonically ordered (say,

$\mu_0 \leqslant \mu_1 \leqslant \cdots \leqslant \mu_k$) or may have a point of reversal (e.g., due to increasing toxicity). Several papers have addressed the problem of identifying the minimum effective dose (MED) defined as the lowest dose level for which $\mu_i > \mu_0$. This problem could be formulated as an estimation problem or as a multiple testing problem. In the latter case the identified MED is not really an estimate of the true MED; rather it is the lowest dose that is statistically significantly different from the zero dose.

For the multiple testing formulation, Williams (1971, 1972) proposed a step-down procedure which compares isotonic estimates of the μ_i $(1 \leqslant i \leqslant k)$ with the zero dose sample mean. Ruberg (1989) considered step-down and single-step procedures based on selected contrasts among sample means. Rom et al. (1994) gave closed procedures also based on selected contrasts among sample means that allow for additional comparisons among sets of successive dose levels. Tamhane et al. (1996) have compared these and other procedures using Monte Carlo simulation.

For other multiple comparison problems involving ordered means, see HT, Chapter 10, Section 4. Hayter (1990) has given a one-sided analog of Tukey's Studentized range test for comparing ordered means, μ_j versus μ_i $(1 \leqslant i < j \leqslant k)$.

A third area of interest is the problem of designing experiments for multiple comparisons. The problem of sample size determination is addressed in various papers; a summary of the work can be found in HT, Chapter 6. Some recent papers are Bristol (1989), Hsu (1988), and Hayter and Liu (1992). There also has been a large body of work on block designs for comparing treatments with a control beginning with a paper by Bechhofer and Tamhane (1981). This work is summarized in the article by Hedayat et al. (1988) and a review article by Majumdar (1996) in the present volume.

8. Concluding remarks

As stated at the beginning of this paper, research in multiple comparisons is still progressing at full steam even though the field has certainly matured since its inception more than forty years ago. This is a good point to take a pause and ask the question "Where should multiple comparisons go next?" This is precisely the title of a recent paper by John Tukey (Tukey, 1993), one of the founding fathers of the field. Many interesting avenues of research and problems are outlined in this paper. A theme that Tukey emphasizes is the use of graphical methods (see also Tukey, 1991). These methods are discussed in HT, Chapter 10, Section 3, but more work is needed. In this connection, see Hochberg and Benjamini (1990), who use the graphical method of Schweder and Spjotvoll (1982) to improve the power of some p-value based MCP's.

Another promising line of work is the choice of alternative error rates to control with the goal of developing more powerful MCP's. False discovery rate (FDR) proposed by Benjamini and Hochberg (1995) has been mentioned in Section 2.2. Halperin et al. (1988) made another proposal that has not been followed up.

Finally, it is important to remember that the field of multiple comparisons owes its origin to real problems arising in statistical practice. The various MCP's will remain unused unless they are made available in a user-friendly software. Although many statistical packages do have the options to carry out some of the MCP's discussed here, until recently there was not a single package that did a majority of the most useful ones. This gap has been filled by a SAS application package called *MultComp* by Rom (1994). In conclusion, we have attempted to present a comprehensive overview of selected developments in multiple comparisons since the publication of HT and also of much background material (discussed in more detail in HT). Our hope is that the researchers will find this paper a useful springboard for pursuing new and relevant practical problems, thus maintaining the vitality of the field and helping it to grow into a healthy and essential subdiscipline of statistics.

Acknowledgement

For comments on the first draft I am indebted to Charlie Dunnett, Tony Hayter, Ludwig Hothorn, a referee, and especially to John Tukey whose comments led me to revise several statements and opinions expressed in that draft.

Over the last about seven years, it has been my privilege and good fortune to have collaborated closely with Charlie Dunnett from whom I have learned a great deal. As a token of my gratitude, I dedicate this paper to him.

References

Aitchinson, J. (1964). Confidence-region tests. *J. Roy. Statist. Soc. Ser. B* **26**, 462–476.

Armitage, P. and M. Parmar (1986). Some approaches to the problem of multiplicity in clinical trials. In: *Proc. 13th Internat. Biometric Conf.* Biometric Society, Seattle.

Bauer, P. (1991). Multiple testing in clinical trials. *Statist. Medicine* **10**, 871–890.

Bechhofer, R. E. and C. W. Dunnett (1988). Tables of percentage points of multivariate Student t distributions. In: *Selected Tables in Mathematical Statistics*, Vol. 11, 1–371.

Bechhofer, R. E. and A. C. Tamhane (1981). Incomplete block designs for comparing treatments with a control. *Technometrics* **23**, 45–57.

Begun, J. and K. R. Gabriel (1981). Closure of the Newman–Keuls multiple comparison procedure. *J. Amer. Statist. Assoc.* **76**, 241–245.

Benjamini, Y. and Y. Hochberg (1995). Controlling the false discovery rate: A practical and powerful approach to multiple testing. *J. Roy. Statist. Soc. Ser. B* **57**, 289–300.

Berger, R. L. (1982). Multiparameter hypothesis testing and acceptance sampling. *Technometrics* **24**, 295–300.

Bofinger, E. (1985). Multiple comparisons and type III errors. *J. Amer. Statist. Assoc.* **80**, 433–437.

Bofinger, E. (1987). Step-down procedures for comparison with a control. *Austral. J. Statist.* **29**, 348–364.

Braun, H. I. (1994). *The Collected Works of John W. Tukey*, Vol. VIII. Multiple Comparisons: 1948–1983. Chapman and Hall, New York.

Braun, H. I. and J. W. Tukey (1983). Multiple comparisons through orderly partitions: The maximum subrange procedure. In: H. Wainer and S. Messick, eds., *Principles of Modern Psychological Measurement: A Festschrift in Honor of Frederick M. Lord.* Lawrence Erlbaum Associates, Hillsdale, NJ, 55–65.

Bristol, D. R. (1989). Designing clinical trials for two-sided multiple comparisons with a control. *Controlled Clinical Trials* **10**, 142–152.

Brown, C. C. and T. R. Fears (1981). Exact significance levels for multiple binomial testing with application to carcinogenicity screens. *Biometrics* **37**, 763–774.

Carmer, S. G. and W. M. Walker (1982). Baby bear's dilemma: A statistical tale. *Agronomy J.* **74**, 122–124.

Chang, C.-K. and D. M. Rom (1994). On the analysis of multiple correlated binary endpoints in medical studies. Unpublished manuscript.

D'Agostino, R. B. and T. C. Heeren (1991). Multiple comparisons in over-the-counter drug clinical trials with both positive and placebo controls (with comments and rejoinder). *Statist. Medicine* **10**, 1–31.

Dubey, S. D. (1985). On the adjustment of P-values for the multiplicities of the intercorrelated symptoms. Presented at the Sixth Annual Meeting of the International Society for Clinical Biostatistics, Dusseldorf, W. Germany.

Duncan, D. B. (1951). A significance test for differences between ranked treatments in an analysis of variance. *Virginia J. Sci.* **2**, 171–189.

Duncan, D. B. (1955). Multiple range and multiple F tests. *Biometrics* **11**, 1–42.

Duncan, D. B. (1957). Multiple range tests for correlated and heteroscedastic means. *Biometrics* **13**, 164–176.

Duncan, D. B. (1965). A Bayesian approach to multiple comparisons. *Technometrics* **7**, 171–222.

Duncan, D. B. and D. O. Dixon (1983). k-ratio t-tests, t intervals and point estimates for multiple comparisons, In: S. Kotz and N. L. Johnson, eds., *Encyclopedia of Statistical Sciences*, Vol. 4. Wiley, New York, 403–410.

Dunn, O. J. (1958). Estimation of means of dependent variables. *Ann. Math. Statist.* **29**, 1095–1111.

Dunn, O. J. (1961). Multiple comparisons among means. *J. Amer. Statist. Assoc.* **56**, 52–64.

Dunnett, C. W. (1955). A multiple comparison procedure for comparing several treatments with a control. *J. Amer. Statist. Assoc.* **50**, 1096–1121.

Dunnett, C. W. (1980a). Pairwise multiple comparisons in the homogeneous variances, unequal sample size case. *J. Amer. Statist. Assoc.* **75**, 789–795.

Dunnett, C. W. (1980b). Pairwise multiple comparisons in the unequal variance case. *J. Amer. Statist. Assoc.* **75**, 796–800.

Dunnett, C. W. (1989). Multivariate normal probability integrals with product correlation structure, Algorithm AS251. *Appl. Statist.* **38**, 564–579; Correction note **42**, 709.

Dunnett, C. W. and A. C. Tamhane (1991). Step-down multiple tests for comparing treatments with a control in unbalanced one-way layouts. *Statist. Medicine* **11**, 1057–1063.

Dunnett, C. W. and A. C. Tamhane (1992a). A step-up multiple test procedure. *J. Amer. Statist. Assoc.* **87**, 162–170.

Dunnett, C. W. and A. C. Tamhane (1992b). Comparisons between a new drug and placebo controls in an efficacy trial. *Statist. Medicine* **11**, 1057–1063.

Dunnett, C. W. and A. C. Tamhane (1993). Power comparisons of some step-up multiple test procedures. *Statist. Probab. Lett.* **16**, 55–58.

Dunnett, C. W. and A. C. Tamhane (1995). Step-up multiple testing of parameters with unequally correlated estimates. *Biometrics* **51**, 217–227.

Edwards, D. G. and J. C. Hsu (1983). Multiple comparisons with the best treatment. *J. Amer. Statist. Assoc.* **78**, 965–971.

Einot, I. and K. R. Gabriel (1975). A study of the powers of several methods in multiple comparisons. *J. Amer. Statist. Assoc.* **70**, 574–583.

Farrar, D. B. and K. S. Crump (1988). Exact statistical tests for any carcinogenic effect in animal bioassays. *Fund. Appl. Toxicol.* **11**, 652–663.

Finner, H. (1990). On the modified S-method and directional errors. *Comm. Statist. Ser. A* **19**, 41–53.

Fisher, R. A. (1935). *The Design of Experiments*. Oliver and Boyd, Edinburgh, UK.

Gabriel, K. R. (1969). Simultaneous test procedures – Some theory of multiple comparisons. *Ann. Math. Statist.* **40**, 224–250.

Gabriel, K. R. (1970). On the relationship between union-intersection tests. In: R. C. Bose et al., eds., *Essays in Probability and Statistics*. Univ. of North Carolina Press, Chapel Hill, NC, 251–266.

Games, P. A. and J. F. Howell (1976). Pairwise multiple comparison procedures with unequal N's and/or variances: A Monte Carlo study. *J. Educ. Statist.* **1**, 113–125.

Gupta, S. S. and S. Panchapakesan (1979). *Multiple Decision Procedures*. Wiley, New York.

Halperin, M., K. K. G. Lan and M. I. Hamdy (1988). Some implications of an alternative definition of the multiple comparisons problem. *Biometrika* **75**, 773–778.

Hayter, A. J. (1984). A proof of the conjecture that the Tukey–Kramer multiple comparisons procedure is conservative. *Ann. Statist.* **12**, 61–75.

Hayter, A. J. (1986). The maximum familywise error rate of Fisher's least significant difference test. *J. Amer. Statist. Assoc.* **81**, 1000–1004.

Hayter, A. J. (1989). Pairwise comparisons of generally correlated means. *J. Amer. Statist. Assoc.* **84**, 208–213.

Hayter, A. J. (1990). A one-sided Studentized range test for testing against a simple ordered alternative. *J. Amer. Statist. Assoc.* **85**, 778–785.

Hayter, A. J. and J. C. Hsu (1994). On the relationship between stepwise decision procedures and confidence sets. *J. Amer. Statist. Assoc.* **89**, 128–137.

Hayter, A. J. and W. Liu (1992). A method of power assessment for comparing several treatments with a control. *Comm. Statist. Theory Methods* **21**, 1871–1889.

Hayter, A. J. and A. C. Tamhane (1990). Sample size determination for step-down multiple comparison procedures: Orthogonal contrasts and comparisons with a control. *J. Statist. Plann. Inference* **27**, 271–290.

Hedayat, A. S., M. Jacroux and D. Majumdar (1988). Optimal designs for comparing test treatments with a control (with discussion). *Statist. Sci.* **4**, 462–491.

Heyse, J. and D. Rom (1988). Adjusting for multiplicity of statistical tests in the analysis of carcinogenicity studies. *Biometr. J.* **8**, 883–896.

Hochberg, Y. (1974a). The distribution of the range in general unbalanced models. *Amer. Statist.* **28**, 137–138.

Hochberg, Y. (1974b). Some generalizations of the T-method in simultaneous inference. *J. Multivariate Anal.* **4**, 224–234.

Hochberg, Y. (1988). A sharper Bonferroni procedure for multiple tests of significance. *Biometrika* **75**, 800–802.

Hochberg, Y. and Y. Benjamini (1990). More powerful procedures for multiple significance testing. *Statist. Medicine* **9**, 811–818.

Hochberg, Y. and U. Liberman (1994). An extended Simes test. *Statist. Probab. Lett.* **21**, 101–105.

Hochberg, Y. and G. Rodriguez (1977). Intermediate simultaneous inference procedures. *J. Amer. Statist. Assoc.* **72**, 220–225.

Hochberg, Y. and D. Rom (1996). Extensions of multiple testing procedures based on Simes' test. *J. Statist. Plann. Inference* **48**, 141–152.

Hochberg, Y. and A. C. Tamhane (1987). *Multiple Comparison Procedures*. Wiley, New York.

Holland, B. S. and M. D. Copenhaver (1987). An improved sequentially rejective Bonferroni test procedure. *Biometrics* **43**, 417–424.

Holm, S. (1979a). A simple sequentially rejective multiple test procedure. *Scand. J. Statist.* **6**, 65–70.

Holm, S. (1979b). A stagewise directional test based on t statistic. Unpublished manuscript.

Hommel, G. (1986). Multiple test procedures for arbitrary dependence structures. *Metrika* **33**, 321–336.

Hommel, G. (1988). A stagewise rejective multiple test procedure based on a modified Bonferroni test. *Biometrika* **75**, 383–386.

Hommel, G. (1989). A comparison of two modified Bonferroni procedures. *Biometrika* **76**, 624–625.

Hsu, J. C. (1981). Simultaneous confidence intervals for all distances from the 'best'. *Ann. Statist.* **9**, 1026–1034.

Hsu, J. C. (1982). Simultaneous inference with respect to the best treatment in block designs. *J. Amer. Statist. Assoc.* **77**, 461–467.

Hsu, J. C. (1984). Ranking and selection and multiple comparisons with the best. In: T. J. Santner and A. C. Tamhane, eds., *Design of Experiments: Ranking and Selection (Essays in Honor of Robert E. Bechhofer)*. Marcel Dekker, New York, 23–33.

Hsu, J. C. (1988). Sample size computation for designing multiple comparison experiments, *Comput. Statist. Data Anal.* **7**, 79–91.

James, S. (1991). Approximate multinormal probabilities applied to correlated multiple endpoints. *Statist. Medicine* **10**, 1123–1135.

Krishnaiah, P. R. (1979). Some developments on simultaneous test procedures. In: P. R. Krishnaiah, ed., *Developments in Statistics*, Vol. 2. North-Holland, Amsterdam, 157–201.

Keuls, M. (1952). The use of the 'Studentized range' in connection with an analysis of variance. *Euphytica* **1**, 112–122.

Kimball, A. W. (1951). On dependent tests of significance in the analysis of variance. *Ann. Math. Statist.* **22**, 600–602.

Kramer, C. Y. (1956). Extension of multiple range tests to group means with unequal number of replications. *Biometrics* **12**, 307–310.

Laska, E. M. and M. J. Meisner (1989). Testing whether an identified treatment is best. *Biometrics* **45**, 1139–1151.

Lehmacher, W., G. Wassmer and P. Reitmeir (1991). Procedures for two-sample comparisons with multiple endpoints controlling the experimentwise error rate. *Biometrics* **47**, 511–521.

Lehmann, E. L. and J. P. Shaffer (1977). On a fundamental theorem in multiple comparisons. *J. Amer. Statist. Assoc.* **72**, 576–578.

Lehmann, E. L. and J. P. Shaffer (1979). Optimum significance levels for multistage comparison procedures. *Ann. Statist.* **7**, 27–45.

Liu, W. (1996). Multiple tests of a nonhierarchical finite family of hypotheses. *J. Roy. Statist. Soc. Ser. B* **58**, 455–461.

Majumdar, D. (1996). Treatment-control designs. Chapter 27 in this Handbook.

Mantel, N. (1980). Assessing laboratory evidence for neoplastic activity. *Biometrics* **36**, 381–399; Corrig., *Biometrics* **37**, 875.

Marcus, R., K. R. Gabriel and E. Peritz (1976). On closed testing procedures with special reference to ordered analysis of variance. *Biometrika* **63**, 655–660.

Miller, R. G., Jr. (1966). *Simultaneous Statistical Inference*, McGraw-Hill, New York.

Miller, R. G., Jr. (1981). *Simultaneous Statistical Inference*. 2nd edn. Springer, New York.

Mosier, M., Y. Hochberg and S. Ruberg (1995). Simple multiplicity adjustments to correlated tests, in preparation.

Naik, U. D. (1975). Some selection rules for comparing p processes with a standard. *Comm. Statist. Ser. A* **4**, 519–535.

Newman, D. (1939). The distribution of the range in samples from a normal population, expressed in terms of an independent estimate of the standard deviation. *Biometrika* **31**, 20–30.

O'Brien, P. C. (1983). The appropriateness of analysis of variance and multiple comparison procedures. *Biometrics* **39**, 787–788.

O'Brien, P. C. (1984). Procedures for comparing samples with multiple endpoints. *Biometrics* **40**, 1079–1087.

O'Neill, R. T. and G. B. Wetherill (1971). The present state of multiple comparisons methods (with discussion). *J. Roy. Statist. Soc. Ser. B* **33**, 218–241.

Peritz, E. (1970). A note on multiple comparisons. Unpublished manuscript, Hebrew University.

Perry, J. N. (1986). Multiple comparison procedures: A dissenting view. *J. Econ. Entom.* **79**, 1149–1155.

Pocock, S. J., N. L. Geller and A. A. Tsiatis (1987). The analysis of multiple endpoints in clinical trials. *Biometrics* **43**, 487–498.

Ramsey, P. H. (1978). Power differences between pairwise multiple comparison procedures. *J. Amer. Statist. Assoc.* **73**, 479–485.

Ramsey, P. H. (1981). Power of univariate pairwise multiple comparison procedures. *Psychol. Bull.* **90**, 352–366.

Rom, D. (1990). A sequentially rejective test procedure based on a modified Bonferroni inequality. *Biometrika* **77**, 663–665.

Rom, D. (1992). Strengthening some common multiple test procedures for discrete data. *Statist. Medicine* **11**, 511–514.

Rom, D. (1994). *MultComp2.0 for PC*, User's manual. ProSoft, Philadelphia, PA.

Rom, D. and B. W. Holland (1994). A new closed multiple testing procedure for hierarchical family of hypotheses. *J. Statist. Plann. Inference*, to appear.

Rom, D., R. J. Costello and L. T. Connell (1994). On closed test procedures for dose-response analysis. *Statist. Medicine* **13**, 1583–1596.

Roth, A. J. (1996). Multiple comparison procedures for discrete test statistics. Talk presented at the International Conference on Multiple Comparisons. Tel Aviv, Israel.

Rothman, K. J. (1990). No adjustments are needed for multiple comparisons. *Epidemiology* **1**, 43–46.

Roy, S. N. (1953). On a heuristic method of test construction and its use in multivariate analysis. *Ann. Math. Statist.* **24**, 220–238.

Roy, S. N. and R. C. Bose (1953). Simultaneous confidence interval estimation. *Ann. Math. Statist.* **24**, 513–536.

Ruberg, S. J. (1989). Contrasts for identifying the minimum effective dose. *J. Amer. Statist. Assoc.* **84**, 816–822.

Ryan, T. A. (1960). Significance tests for multiple comparison of proportions, variances and other statistics. *Psychol. Bull.* **57**, 318–328.

Saville, D. J. (1990). Multiple comparison procedures: The practical solution. *Amer. Statist.* **44**, 174–180.

Scheffé, H. (1953). A method for judging all contrasts in the analysis of variance. *Biometrika* **40**, 87–104.

Schervish, M. (1984). Multivariate normal probabilities with error bound, Algorithm AS 195. *Appl. Statist.* **33**, 89–94; Corrig., *Appl. Statist.* **34**, 103–104.

Schweder, T. and E. Spjotvoll (1982). Plots of p-values to evaluate many tests simultaneously. *Biometrika* **69**, 493–502.

Shaffer, J. P. (1979). Comparison of means: An F test followed by a modified multiple range procedure. *J. Educ. Statist.* **4**, 14–23.

Shaffer, J. P. (1980). Control of directional errors with stagewise multiple test procedures. *Ann. Statist.* **8**, 1342–1348.

Shaffer, J. P. (1986). Modified sequentially rejective multiple test procedures. *J. Amer. Statist. Assoc.* **81**, 826–831.

Shaffer, J. P. (1994). Multiple hypothesis testing: A review. Tech. Report 23, National Inst. Statist. Sci., Research Triangle Park, NC.

Šidák, Z. (1967). Rectangular confidence regions for the means of multivariate normal distributions. *J. Amer. Statist. Assoc.* **62**, 626–633.

Simes, R. J. (1986). An improved Bonferroni procedure for multiple tests of significance. *Biometrika* **73**, 751–754.

Sorić, B. (1989). Statistical 'discoveries' and effect size estimation. *J. Amer. Statist. Assoc.* **84**, 608–610.

Stefánsson, G., W. C. Kim and J. C. Hsu (1988). On confidence sets in multiple comparisons. In: S. S. Gupta and J. Berger, eds., *Statistical Decision Theory and Related Topics IV*, Vol. 2. Springer, New York, 89–104.

Stepanavage, M., H. Quan, J. Ng and J. Zhang (1994). A review of statistical methods for multiple endpoints in clinical trials. Unpublished manuscript.

Tamhane, A. C. (1977). Multiple comparisons in model I one-way ANOVA with unequal variances. *Comm. Statist. Ser. A* **6**, 15–32.

Tamhane, A. C. (1979). A comparison of procedures for multiple comparisons of means with unequal variances. *J. Amer. Statist. Assoc.* **74**, 471–480.

Tamhane, A. C., Y. Hochberg and C. W. Dunnett (1996). Multiple test procedures for dose finding. *Biometrics* **52**, 21–37.

Tang, D. I. and S. Lin (1994). On improving some methods for multiple endpoints. Unpublished manuscript.

Tang, D. I., N. L. Geller and S. J. Pocock (1993). On the design and analysis of randomized clinical trials with multiple endpoints. *Biometrics* **49**, 23–30.

Tang, D. I., C. Gnecco and N. L. Geller (1989a). Design of group sequential clinical trials with multiple endpoints. *J. Amer. Statist. Assoc.* **84**, 776–779.

Tang, D. I., C. Gnecco and N. L. Geller (1989b). An approximate likelihood ratio test for a normal mean vector with nonnegative components with application to clinical trials. *Biometrika* **76**, 577–583.

Tarone, R. E. (1990). A modified Bonferroni method for discrete data. *Biometrics* **46**, 515–522.

Tong, Y. L. (1980). *Probability Inequalities in Multivariate Distributions*. Academic Press, New York.

Toothaker, S. E. (1991). *Multiple Comparisons for Researchers*. Sage, Newberry Park, CA.

Toothaker, S. E. (1993). *Multiple Comparison Procedures*. Sage, Newberry Park, CA.

Troendle, J. F. (1995a). A stepwise resampling method of multiple hypothesis testing. *J. Amer. Statist. Assoc.* **90**, 370–378.

Troendle, J. F. (1995b). A permutational step-up method of testing multiple outcomes. Preprint.

Tukey, J. W. (1949). Comparing individual means in the analysis of variance. *Biometrics* **5**, 99–114.

Tukey, J. W. (1953). *The Problem of Multiple Comparisons*. Mimeographed Notes, Princeton University, Princeton, NJ.

Tukey, J. W. (1991). The philosophy of multiple comparisons. *Statist. Sci.* **6**, 100–116.

Tukey, J. W. (1993). Where should multiple comparisons go next? In: F. M. Hoppe, ed., *Multiple Comparisons, Selection and Biometry*. Marcel-Dekker, New York, 187–207.

Tukey, J. W. (1995). Personal communication.

Tukey, J. W., J. L. Ciminera and J. F. Heyse (1985). Testing the statistical certainty of a response to increasing doses of a drug. *Biometrics* **41**, 295–301.

Waller, R. A. and D. B. Duncan (1969). A Bayes rule for symmetric multiple comparisons. *J. Amer. Statist. Assoc.* **64**, 1484–1503; Corrig., *J. Amer. Statist. Assoc.* **67**, 253–255.

Welsch, R. E. (1972). A modification of the Newman–Keuls procedure for multiple comparisons. Working Paper 612-72, Sloan School of Management, M.I.T., Boston, MA.

Welsch, R. E. (1977). Stepwise multiple comparison procedures. *J. Amer. Statist. Assoc.* **72**, 566–575.

Westfall, P. H. and S. S. Young (1989). p value adjustments for multiple tests in multivariate binomial models. *J. Amer. Statist. Assoc.* **84**, 780–786.

Westfall, P. H. and S. S. Young (1993). *Resampling Based Multiple Testing*. Wiley, New York.

Williams, D. A. (1971). A test for differences between treatment means when several dose levels are compared with a zero dose control. *Biometrics* **27**, 103–117.

Williams, D. A. (1972). The comparison of several dose levels with a zero dose control. *Biometrics* **28**, 519–531.

Wright, S. P. (1992). Adjusted P-values for simultaneous inference. *Biometrics* **48**, 1005–1014.

S. Ghosh and C. R. Rao, eds., *Handbook of Statistics, Vol. 13*

19

Nonparametric Methods in Design and Analysis of Experiments

Edgar Brunner and Madan L. Puri

1. Introduction and notations

1.1. Historical background

The two main assumptions underlying the classical analysis of variance (ANOVA) models are the linear model and the normal distribution of the error term. One of the first attempts to relax these assumptions was made by Friedman (1937) where in a two-way layout with one observation per cell the observations are 'replaced' by their 'place numbers', called 'ranks' within the block factor. The next step was taken by Kruskal and Wallis (1952) in the one-way layout where the observed random variables are replaced by their ranks.

Designs in practical data analysis however, are more complex than these two simple designs and the demand for the analysis of such designs without the restrictive assumptions of the ANOVA was one of the most burning problems in applied statistics. The similarity of both the Friedman statistic and the Kruskal–Wallis statistic to their parametric counterparts gave raise to the hope that also nonparametric statistics for more complex designs would look similar to their parametric counterparts. One difficulty however is that the variance of a rank statistic depends on the alternative and the covariance matrix of the vector of rank means in the one-way layout has a rather difficult form (see, e.g., Puri, 1964) and it seems prohibitive to estimate the matrix for two- or higher-way layouts.

Thus, the way for the development of nonparametric procedures in the two-way layout was determined by the need to circumvent this problem. Lemmer and Stoker (1967) assumed that all distribution functions in the design were identical and proposed statistics for main effects and interactions which had a similar form like the statistics for the two-way ANOVA with fixed effects. However it seems to be rather unrealistic to develop a statistic for the interaction, e.g., under the assumption that all distribution functions are identical. So it was natural to remove first the nuisance parameters from the model by subtracting consistent estimates of them from the data and then replacing the residuals by their ranks. This concept of 'ranking after alignment' (RAA) was introduced by Hodges and Lehmann (1962) and further developed by Sen (1968), Puri and Sen (1969, 1973), Sen and Puri (1977) and Adichie (1978), among others.

For a comprehensive treatment of aligned rank tests in the context of time series analysis, the reader is referred to Hallin and Puri (1994) and the references cited therein.

The main effect of one factor may also be removed by ranking within the levels of this factor, like in the Friedman test. The remaining nuisance parameters, the interactions, either must be excluded from the model or a 'joint hypothesis' should be tested, i.e., the interaction and the main effect be tested together. Both concepts were used by Koch (1969, 1970) in a split-plot and in a complex split-plot design and Mehra and Sen (1969) used the RAA-technique to develop a test for the interaction in a two-factor block design. Parallel to this development, first steps were made by Sen (1967) and by Koch and Sen (1968) to formulate hypotheses for treatment effects and to derive statistics in a nonparametric mixed model and the ideas of these papers were later applied by Koch (1969, 1970) in partially nested designs.

All procedures using the RAA-techniques are clearly restricted to linear models and are not applicable to pure ordinal data where it is not reasonable to compute sums or differences of the observations. Moreover, only one of the main assumptions of ANOVA, the normality of the error distribution is relaxed with this approach. Although some hypotheses in the papers of Koch and Sen (1968) and Koch (1969, 1970) are formulated in a nonparametric setup, no unified theory for nonparametric procedures in two- or higher-way fixed models or in the mixed model was developed.

Patel and Hoel (1973) seem to be the first to leave the floor of the linear model in a two-way-layout and defined a nonparametric interaction in this design by the difference of the probabilities underlying the two Wilcoxon–Mann–Whitney statistics in a 2×2 design. It seems to be the first time that in a two-way layout a nonparametric effect was defined and a consistent estimator of this effect was used as a basis for the test statistic rather than replacing observations by ranks. The ranks of the observations came out as a convenient tool to estimate the nonparametric effect. This concept was further developed for main effects and interactions in some fixed and mixed 2×2 designs by Hilgers (1979), Brunner and Neumann (1984, 1986a) for fixed and mixed 2×2 designs and by Boos and Brownie (1992) and by Brunner, Puri and Sun (1995) for fixed and mixed $2 \times b$ designs.

The next step in the development of the nonparametric mixed model was taken by Hollander, Pledger and Lin (1974) who considered the robustness of the Wilcoxon–Mann–Whitney statistic in the paired two-sample design. Govindarajulu (1975b) derived the asymptotic distribution of the Wilcoxon rank sum in this design and indicated how to estimate the asymptotic variance of the statistic under the hypothesis. A more general result in the mixed model with one fixed factor, one random factor and an equal number of replications per cell was given by Brunner and Neumann (1982) where also random interactions were included into the model. The general form of the covariance matrix of the rank means and explicit estimators for the unknown variance terms under the hypothesis were given in this paper and the results were applied to different mixed models. In all the three aforementioned papers, only the ranks over all observations were used.

During the same time, McKean and Hettmansperger (1976) and Hettmansperger and McKean (1977) developed asymptotically distribution free statistics in general fixed models based on minimizing Jaeckel's dispersion function (Jaeckel, 1972). However the proposed statistics are not pure rank statistics. For a description of this method, we refer to Hettmansperger and McKean (1983).

Parallel to this development, the idea of the 'rank transform' (RT) was born (Conover and Iman, 1976, 1981) and some empirical results with this method were given (Lemmer, 1980). The simple technique to replace the observations in the respective parametric statistic by their ranks and then assume that the asymptotic distributions of both statistics are same, has been criticized by Fligner (1981), Brunner and Neumann (1986a), Blair, Sawilowski and Higgens (1987), Akritas (1990), Thompson and Ammann (1989, 1990), Akritas (1990, 1991, 1993) and Thompson (1991a). Analytic counter examples showing that RT-statistics may become degenerate under the hypothesis, have been given by Brunner and Neumann (1986a). Blair, Sawilowski and Higgens (1987) showed by a simulation study that RT-statistics may have undesirable properties.

Hora and Conover (1984), Iman, Hora and Conover (1984) and Hora and Iman (1988) derived RT-statistics in the two-way layout without interactions and with equal number of replications where the assumption of no interactions was made for the formulation of the hypotheses as well as for the derivation of the asymptotic distributions of the statistics.

Akritas (1990) showed that the RT was not 'valid' to test a main effect or the interaction in a linear model since it is a nonlinear transformation of the data. He also showed that the homoscedasticity of the error terms was not transferred to the ranks in general unless all distribution functions in the design are identical. Thus, the RT is valid in the two-way linear model with crossed fixed factors to test the joint hypothesis of no main effect and no interaction together. It is also valid in the two-way hierarchical design to test the hypothesis of no nested effect. Akritas (1991, 1993) derived similar results for repeated measurements models. Kepner and Robinson (1988) derived a rank test for the mixed model with one observation per cell.

Thompson (1990) generalized the results of Brunner and Neumann (1982) and derived the asymptotic distribution of a linear rank statistic for independent vectors of equal fixed length and she applied the result to various balanced mixed models. The assumption of equal fixed length of the vectors was used in the proofs of the asymptotic normality of the statistics in both papers and no general theoretical result for the asymptotic normality of a linear rank statistic in an unbalanced mixed model was available. Akritas (1993) derived a rank test in a special unbalanced mixed model. A general result was derived by Brunner and Denker (1994) for vectors of unequal length and was applied to various unbalanced mixed models.

It is rather astonishing that most of the theoretical results as well as the derivations of rank statistics were based on the assumption of the continuity of the underlying distribution functions and ties were excluded although some theoretical results (e.g., Conover, 1973; Lehmann, 1975; Behnen, 1976) regarding ties were available in literature for linear rank statistics of independent observations. Koch and Sen (1968) and Koch (1970) recommended to brake ties by using the 'mid-rank-method'. This method was used by Boos and Brownie (1992) to formulate hypotheses and to derive estimators of the asymptotic variances of rank statistics using the U-statistic representation. Munzel (1994) generalized the results of Brunner and Denker (1994) to the case of ties and the same method was used by Brunner, Puri and Sun (1995) to derive the asymptotic normality of linear rank statistics in mixed models and consistent estimators of the asymptotic variances for fixed and mixed $2 \times b$ designs.

The problem to derive (pure) rank tests for interactions and main effects separately in two- or higher-way layouts remained open until Akritas and Arnold (1994) had the simple idea to formulate the hypotheses in a two-way repeated measurements model in terms of the distribution functions simply by replacing the vector of the expectations in the linear model by the vector of the distribution functions. They showed that the covariance matrix has a simple form under these hypotheses. It is easy to see that the hypotheses in the linear model are implied by theses hypotheses. The acceptance and interpretation of such hypotheses by applied researchers has to be seen in future.

The idea to formulate the hypotheses in two- and higher-way layouts in terms of the distribution functions is investigated in the general model with fixed factors (Akritas et al., 1994) and in the general mixed model (Akritas and Brunner, 1995). The purpose of this paper is to provide a unified theory of rank tests in some aspects of the design and analysis of experiments, exploiting the ideas of Akritas and Arnold (1994) and Akritas and Brunner (1995). For lack of space, we mainly restrict ourselves to discussing the test procedures and the rationale underlying the hypotheses.

1.2. Aim of the paper

In this paper, we consider only pure rank statistics in factorial designs. On the one hand, such pure rank statistics are invariant under any strict monotone transformation of the data, on the other hand they are robust against outliers. Moreover, they are applicable to ordinal data such as scores in psychological tests, grading scales in order to describe the degree of the damage of plants or trees in ecological or environmental studies. With such data, ties must also be taken into account.

Thus we shall not consider procedures which need sums or differences of the original data and which are therefore restricted to linear models. Our aim is to generalize the classical models of ANOVA in such a way that not only the assumption of normality of the error terms is relaxed but also the structure of the designs is introduced in a broader framework. In addition, the concept of treatment effects is redefined within this framework. We formulate hypotheses in a nonparametric setup for generalized treatment effects in various designs and derive nonparametric tests for these hypotheses in such a way that the common rank tests existing in literature will come out as special cases. Moreover, new procedures are presented within this unified approach. In order to identify the testing problem underlying the different rank procedures, the relations between the hypotheses in the general model and in the standard linear model are investigated. We do not assume the continuity of the underlying distribution functions so that data with ties can be evaluated with the procedures given in this paper.

In the next section, we give some general notations which are used throughout the paper in order to avoid unnecessary repetitions. In Section 2, models with fixed factors are considered and the concept of nonparametric hypotheses introduced by Akritas and Arnold (1994) is explained. Also in this section, following Fligner and Policello (1981), Brunner and Neumann (1982, 1986a, b) and Brunner et al. (1995), rank procedures for heteroscedastic models (nonparametric Behrens–Fisher problem) for two-sample designs and for the stratified two-sample design (fixed-factor model) are discussed.

In Section 3, the random-factor model is briefly discussed, and in Section 4, the mixed model is considered. All procedures are derived from the general Theorems given in Section 4.2. Rank procedures for heteroscedastic mixed models are considered separately for the two-sample design and the stratified two-sample designs, following the ideas in Brunner et al. (1995).

Procedures for ordered alternatives, many-one and multiple comparisons follow from the general approach presented in the paper. However, they are not considered in this paper. For reasons of brevity and the editor's suggestions, the proofs of the results are either given very briefly or are omitted. They will appear somewhere else.

1.3. Notations used throughout the paper

Distribution functions. The distribution function of a random variable X_i is denoted by

$$F_i(x) = \frac{1}{2}\left[F_i^+(x) + F_i^-(x)\right] \tag{1.1}$$

where $F_i^+(x) = P(X_i \leqslant x)$ is the right continuous version and $F_i^-(x) = P(X_i < x)$ is the left continuous version of the distribution function. This definition of the distribution function includes the case of ties and, moreover discrete ordinal data are included in this setup. We exclude only the trivial case when $F_i(x)$ is a one-point distribution function. The vector of the distribution functions is denoted by $\boldsymbol{F} = (F_1, \ldots, F_d)'$. If a random variable X_i is distributed according to a distribution function $F_i(x)$, this will be denoted by $X_i \sim F_i(x)$ and if the random variables X_N and Y_N are asymptotically equivalent, this will be written as $X_N \doteq Y_N$.

Let $X_{ij} \sim F_i(x)$, $i = 1, \ldots, d$ and $j = 1, \ldots, n_i$. Then the empirical distribution function of F_i is denoted by

$$\widehat{F}_i(x) = \frac{1}{2}\left[\widehat{F}_i^+(x) + \widehat{F}_i^-(x)\right] = \frac{1}{n_i}\sum_{j=1}^{n_i} c(x - X_{ij}) \tag{1.2}$$

where $c(u) = \frac{1}{2}[c^+(u) + c^-(u)]$ denotes the counting function and $c^+(u) = 0$ or 1 according as $u <$ or $\geqslant 0$ and $c^-(u) = 0$ or 1 according as $u \leqslant$ or > 0. The vector of the empirical distribution functions is written as $\widehat{\boldsymbol{F}}(x) = (\widehat{F}_1(x), \ldots, \widehat{F}_d(x))'$ or shortly $\widehat{\boldsymbol{F}} = (\widehat{F}_1, \ldots, \widehat{F}_d)'$.

Ranks. Let $X_1, \ldots, X_N$ be N random variables and let

$$\widehat{H}(x) = \frac{1}{N}\sum_{i=1}^{N} c(x - X_i)$$

be the empirical distribution function of $X_1, \ldots, X_N$. Then the rank of X_i among the N random variables $X_1, \ldots, X_N$ is defined as

$$R_i = \frac{1}{2} + N\widehat{H}(X_i) = \frac{1}{2} + \sum_{s=1}^{N} c(X_i - X_s). \tag{1.3}$$

Note that $1/2$ has to be added to $N\widehat{H}(X_i)$ in order to get the position numbers of the ordered observations in case of no ties since $c(0) = 1/2$. Note also that R_i is the midrank in case of ties.

Matrices. Let $\boldsymbol{\mu} = (\mu_1, \ldots, \mu_d)'$ be a d-dimensional vector of constants. To formulate hypotheses concerning the components of $\boldsymbol{\mu}$, we will use the contrast matrices

$$\boldsymbol{C}_1 = \left(\boldsymbol{1}_{d-1} \vdots - \boldsymbol{I}_{d-1}\right) \quad \text{and} \quad \boldsymbol{C}_2 = \boldsymbol{P}_d = \boldsymbol{I}_d - d^{-1}\boldsymbol{J}_d \tag{1.4}$$

where $\boldsymbol{1}_d = (1, \ldots, 1)'$ denotes the d-dimensional vector of 1's, $\boldsymbol{I}_d$ denotes the d-dimensional unit matrix and $\boldsymbol{J}_d = \boldsymbol{1}_d\boldsymbol{1}_d'$ denotes the d-dimensional square matrix of 1's. (Any matrix $\boldsymbol{C}_{r\times d}$ is called a *contrast matrix* if $\boldsymbol{C}_{r\times d}\boldsymbol{1}_d = \boldsymbol{0}_{r\times 1}$.) Note that $\text{rank}(\boldsymbol{C}_1) = d - 1$ and so, $\boldsymbol{C}_1\boldsymbol{C}_1'$ is nonsingular. The matrix $\boldsymbol{C}_2$ is a d-dimensional projector of rank $d - 1$. We will also use the contrast matrix

$$\boldsymbol{W} = \boldsymbol{V}^{-1}\left(\boldsymbol{I}_d - \boldsymbol{J}_d\boldsymbol{V}^{-1}/(\boldsymbol{1}_d'\boldsymbol{V}^{-1}\boldsymbol{1}_d)\right) \tag{1.5}$$

where $\boldsymbol{V}$ is a nonsingular covariance matrix. Note that $\boldsymbol{WVW} = \boldsymbol{W}$.

Factors (in the sense of experimental design) will be denoted with capital letters $A, B, C, \ldots$, and the levels of A are numbered by $i = 1, \ldots, a$, the levels of B are numbered by $j = 1, \ldots, b$, etc. If factor B is nested under factor A, this is denoted by $B(A)$.

For a convenient and technically simple description of hypotheses in two- and higher-way layouts, we will use the Kronecker-product and the Kronecker-sum of matrices. The Kronecker-product of two matrices

$$\boldsymbol{A}_{p\times q} = \begin{pmatrix} a_{11} & \cdots & a_{1q} \\ \vdots & & \vdots \\ a_{p1} & \cdots & a_{pq} \end{pmatrix} \quad \text{and} \quad \boldsymbol{B}_{r\times s} = \begin{pmatrix} b_{11} & \cdots & b_{1s} \\ \vdots & & \vdots \\ b_{r1} & \cdots & b_{rs} \end{pmatrix}$$

is defined as

$$\boldsymbol{A} \otimes \boldsymbol{B} = \begin{pmatrix} a_{11}\boldsymbol{B} & \cdots & a_{1q}\boldsymbol{B} \\ \vdots & & \vdots \\ a_{p1}\boldsymbol{B} & \cdots & a_{pq}\boldsymbol{B} \end{pmatrix}_{pr\times qs}.$$

The Kronecker-product of the matrices $\boldsymbol{A}_i$, $i = 1, \ldots, a$, is written as $\bigotimes_{i=1}^{a} \boldsymbol{A}_i$.

The Kronecker-sum of the two matrices $\boldsymbol{A}$ and $\boldsymbol{B}$ is defined as

$$\boldsymbol{A} \oplus \boldsymbol{B} = \left(\begin{array}{c|c} \boldsymbol{A} & \boldsymbol{0} \\ \hline \boldsymbol{0} & \boldsymbol{B} \end{array}\right)_{(p+r)\times(q+s)}$$

and the Kronecker-sum of the matrices $\boldsymbol{A}_i$, $i = 1, \ldots, a$, is written as $\bigoplus_{i=1}^{a} \boldsymbol{A}_i$. Note that

$$\left(\bigotimes_{i=1}^{a} \boldsymbol{A}_i\right)\left(\bigotimes_{i=1}^{a} \boldsymbol{B}_i\right) = \bigotimes_{i=1}^{a} \boldsymbol{A}_i \boldsymbol{B}_i$$

and

$$\left(\bigoplus_{i=1}^{a} \boldsymbol{A}_i\right)\left(\bigoplus_{i=1}^{a} \boldsymbol{B}_i\right) = \bigoplus_{i=1}^{a} \boldsymbol{A}_i \boldsymbol{B}_i$$

if the matrices $\boldsymbol{A}_i$ and $\boldsymbol{B}_i$ are conformable with respect to multiplication. We also will use the two rules

$$\left(\bigotimes_{i=1}^{a} \boldsymbol{A}_i\right)^{-1} = \bigotimes_{i=1}^{a} \boldsymbol{A}_i^{-1} \quad \text{and} \quad \left(\bigoplus_{i=1}^{a} \boldsymbol{A}_i\right)^{-1} = \bigoplus_{i=1}^{a} \boldsymbol{A}_i^{-1}$$

if the inverses exist.

2. Fixed models

2.1. One-factor designs

2.1.1. Models and hypotheses

In this section, we consider independent random variables X_{ij}, $i = 1, \ldots, a$, $j = 1, \ldots, n_i$. In the classical normal theory model it is assumed that $X_{ij} \sim F_i = N(\mu_i, \sigma^2)$, $j = 1, \ldots, n_i$. If we relax the assumption of normality then we will loose the 'natural' parameters μ_i and σ^2 and therefore, we have to introduce 'artificial' parameters in order to describe a treatment effect. Such models are called 'semiparametric' models and will be described first in the sequel in order to get some motivation for a reasonable generalization of these models to the nonparametric case. In general, semiparametric models are constructed by eliminating some of the restrictions which specify a parametric model, e.g., by eliminating the requirement that the errors are normally distributed in the classical normal theory model. Theses models (i.e., semiparametric models) arise by parametrically restricting some aspects of a nonparametric model. Here, e.g., we assume that the distribution functions $F_i(x)$ are generated by some distribution function $F(x)$ and some parameters λ_i. A suitable representation of a semiparametric model may be stated by considering the generalized mean $\boldsymbol{\nu}(H, \lambda_1, \ldots, \lambda_a) = \int H(x)\,\mathrm{d}\boldsymbol{F}(x)$ where $\boldsymbol{F} = (F_1, \ldots, F_a)'$ and $F_i(x)$ is a function of $F(x)$ and parameters λ_i, $i = 1, \ldots, a$, and $H(x)$ is a suitable function of x.

Shift model. In the linear model,

$$F_i(x) = F(x - \mu_i), \tag{2.1}$$

the treatment effects are defined by the differences $\mu_i - \mu$ where $\mu = \int x\,\mathrm{d}F(x)$ is the overall mean. Here, $H(x) = x$ and $\nu_i(H, \mu_i) = \int x\,\mathrm{d}F(x - \mu_i) = \mu + \mu_i$. The hypotheses considered in this model are commonly written as $H_0^{\mu}(\boldsymbol{C})$: $\boldsymbol{C}\boldsymbol{\mu} = \boldsymbol{0}$ where $\boldsymbol{C}$ is a contrast matrix and $\boldsymbol{\mu} = (\mu_1, \ldots, \mu_a)' = \int x\,\mathrm{d}\boldsymbol{F}$. For the overall hypothesis 'no treatment effect at all', $\boldsymbol{C}$ is chosen as $\boldsymbol{C} = \boldsymbol{P}_a = \boldsymbol{I}_a - \frac{1}{a}\boldsymbol{J}_a$.

Lehmann model. Here, $F_i(x) = [F(x)]^{\alpha_i}$ where $[F(x)]^{\alpha_i}$ is defined as

$$[F(x)]^{\alpha_i} = \frac{1}{2}\left([F^+(x)]^{\alpha_i} + [F^-(x)]^{\alpha_i}\right) \quad \text{and} \quad H(x) = F(x). \tag{2.2}$$

Then $\nu_i(F, \alpha_i) = \int F(x)\,\mathrm{d}[F(x)]^{\alpha_i} = \Psi(\alpha_i)$. If $F(\cdot)$ is continuous, then $\Psi(\alpha_i) = \alpha_i/(1 + \alpha_i)$. The mean $\Psi(\alpha_i)$ is strictly increasing in α_i and a reasonable treatment effect is defined as $\delta_i = \Psi(\alpha_i) - 1/2$.

In this paper, we will use only the shift model in order to interpret the treatment effects and hypotheses in the general models.

General (nonparametric) model. In the general model, we assume that the random variables $X_{ij} \sim F_i(x)$, $i = 1, \ldots, a$, $j = 1, \ldots, n_i$, are all independent and $F_i(x)$ is an arbitrary distribution function (excluding the trivial case of a one-point distribution). Since here the distribution functions $F_i(x)$ are not assumed to be generated by any particular distribution function $F(x)$, a reasonable function $H(x)$ for a generalized mean $\boldsymbol{\nu}(H)$ can be taken to be the mean of all $N = \sum_{i=1}^{a} n_i$ distribution functions in the experiment, namely, $H(x) = \frac{1}{N}\sum_{i=1}^{a} n_i F_i(x)$. The vector of the generalized means is then $\boldsymbol{\nu}(H) = \boldsymbol{p} = (p_1, \ldots, p_a)' = \int H(x)\,\mathrm{d}\boldsymbol{F}(x)$. The component $p_i = \int H\,\mathrm{d}F_i$ can be regarded in some sense a relative effect of treatment i with respect to all other treatments. If $F_1 = \cdots = F_a = H$, then there is no treatment effect at all and $\boldsymbol{p} = \frac{1}{2}\boldsymbol{1}_a$. Hypotheses in this general model may be formulated either in terms of the distribution functions F_i or by the generalized mean vector $\boldsymbol{p}$. The simplest (and strongest) hypotheses are written in form of the distribution functions, viz. H_0^F: $F_1 = \cdots = F_a$ or $H_0^F(\boldsymbol{P}_a)$: $\boldsymbol{P}_a\boldsymbol{F} = \boldsymbol{0}$. If the linear model (2.1) is assumed, then $H_0^F(\boldsymbol{P}_a)$ implies $H_0^{\mu}(\boldsymbol{P}_a)$ since it follows from $\boldsymbol{P}_a\boldsymbol{F} = \boldsymbol{0}$ that $\boldsymbol{P}_a\boldsymbol{\mu} = \int x\,\mathrm{d}(\boldsymbol{P}_a\boldsymbol{F}) = \boldsymbol{0}$.

The formulation of the hypotheses by the generalized means p_i needs some clarification. Note that in general, $\boldsymbol{p}$ depends on the sample sizes n_i through $H(x)$. Therefore, $p_i - 1/2$ is not a reasonable treatment effect in the general nonparametric model. To avoid the dependence on sample sizes, the function $H(x)$ has to be defined in a different way. This approach will lead to unweighted generalized means the estimation of which requires other techniques than ranking observations. This problem will not be discussed in this paper.

For the case of $a = 2$, a treatment effect in the general model can be defined as $p = \int F_1 \, dF_2$ which evidently does not depend on sample sizes. If $F_1 = F_2$, then $p = 1/2$ and the hypothesis is formulated as H_0^p: $p = \int F_1 \, dF_2 = 1/2$. Note that the hypothesis H_0^p includes also the hypothesis of 'no treatment effect' in a heteroscedastic model and for the nonparametric Behrens–Fisher problem.

We now state the relationships between the hypotheses in the linear model.

PROPOSITION 2.1. *In the linear model* (2.1), $H_0^\mu(\boldsymbol{P}_a)$: $\boldsymbol{P}_a \boldsymbol{\mu} = \boldsymbol{0} \Leftrightarrow H_0^F(\boldsymbol{P}_a)$: $\boldsymbol{P}_a \boldsymbol{F} = \boldsymbol{0}$ *and if* $a = 2$, *then* H_0^μ: $\mu_1 = \mu_2 \Leftrightarrow H_0^F$: $F_1 = F_2 \Leftrightarrow H_0^p$: $p = 1/2$.

PROOF. Let $F_i(x) = F(x - \mu_i)$, then it is easy to see that $H_0^\mu(\boldsymbol{P}_a) \Rightarrow H_0^F(\boldsymbol{P}_a)$. On the other hand, it follows from $H_0^F(\boldsymbol{P}_a)$: $\boldsymbol{P}_a \boldsymbol{F} = \boldsymbol{0}$ that $\boldsymbol{P}_a \boldsymbol{\mu} = \boldsymbol{P}_a \int x \, d\boldsymbol{F} = \int x \, d(\boldsymbol{P}_a \boldsymbol{F}) = \boldsymbol{0}$. For $a = 2$, we note that H_0^p: $p = 1/2$ follows from H_0^F: $F_1 = F_2$. To show that $H_0^p \Rightarrow H_0^\mu$, define $G(\delta) = \int F(x + \delta) \, dF(x)$ and note that $G(\delta)$ is strictly increasing in the neighborhood of 0. □

To derive tests for $H_0^F(\boldsymbol{P}_a)$: $\boldsymbol{P}_a \boldsymbol{F} = \boldsymbol{0}$, we will consider the asymptotic distribution of a consistent estimator for the generalized mean vector $\boldsymbol{p}$ under $H_0^F(\boldsymbol{P}_a)$. In the next section, we will give some asymptotic results for the one-factor fixed model. The considerations regarding the hypothesis H_0^p for heteroscedastic models will be restricted to the case of $a = 2$ samples and asymptotic results for this case will be given separately.

2.1.2. General asymptotic results

Here, the asymptotic results for the one-factor fixed model including the case of ties are given. They are derived using the ideas in Brunner et al. (1995). We consider the vector $\boldsymbol{p} = (p_1, \ldots, p_a)' = \int H \, d\boldsymbol{F}$ of the generalized means $p_i = \int H \, dF_i$, $i = 1, \ldots, a$.

PROPOSITION 2.2. *Let* $X_{ij} \sim F_i$, $i = 1, \ldots, a$, $j = 1, \ldots, n_i$, *be independent random variables and let* $H = \frac{1}{N} \sum_{i=1}^{a} n_i F_i$ *and* $\overline{R}_{i\cdot} = n_i^{-1} \sum_{j=1}^{n_i} R_{ij}$ *where* R_{ij} *is the rank of* X_{ij} *among all the* N *observations* $X_{11}, \ldots, X_{an_a}$. *Then* $\widehat{\boldsymbol{p}} = (\widehat{p}_1, \ldots, \widehat{p}_a)' = \int \widehat{H} \, d\widehat{\boldsymbol{F}} = N^{-1}(\overline{R}_{1\cdot} - 1/2, \ldots, \overline{R}_{a\cdot} - 1/2)'$ *is unbiased and consistent for* $\boldsymbol{p} = \int H \, d\boldsymbol{F}$.

It follows from Proposition 2.2 that the ranks R_{ij} arise in a natural way and they should be considered as a convenient tool for estimating the generalized means p_i rather than replacing observations by 'place-numbers'. In this paper, we will also derive other ranking methods which arise by estimating generalized treatment effects and unknown variances of test statistics.

The asymptotic distribution of $\sqrt{N} \boldsymbol{C} \widehat{\boldsymbol{p}}$ is derived in Theorems 2.3 to 2.6 given below where $\boldsymbol{C} = \boldsymbol{I}_a - \frac{1}{a} \boldsymbol{J}_a$. (It is shown in Akritas et al. (1996) how the following results will fit into a general approach for fixed models with two or more crossed or nested factors.)

THEOREM 2.3. *Under the assumptions of Proposition 2.2, the statistics*

$$\sqrt{N}\int \widehat{H}\,\mathrm{d}(\widehat{\boldsymbol{F}}-\boldsymbol{F}) \quad \text{and} \quad \sqrt{N}\int H\,\mathrm{d}(\widehat{\boldsymbol{F}}-\boldsymbol{F})$$

are asymptotically equivalent, i.e., both statistics have asymptotically the same distribution.

This result is a special case of Theorem 3.1 in Brunner and Denker (1994) and is derived by letting $J(u) \equiv u$ and $m_i \equiv 1$. For the case of ties, the statement follows from Munzel (1994). Note that the statistic $\sqrt{N}\,\overline{\boldsymbol{Y}} = \sqrt{N}\int H\,\mathrm{d}\widehat{\boldsymbol{F}}$ is a vector, the components of which can be written as

$$\sqrt{N}\,\overline{Y}_{i\cdot} = \sqrt{N}\int H\,\mathrm{d}\widehat{F}_i = \frac{\sqrt{N}}{n_i}\sum_{j=1}^{n_i} Y_{ij} \tag{2.3}$$

where $Y_{ij} = H(X_{ij})$ are independent by assumption and the asymptotic multinormality of $\sqrt{N}\,\overline{\boldsymbol{Y}}$ will follow from the Lindeberg–Feller Central Limit Theorem.

THEOREM 2.4. *Let* $\sigma^2 = \mathrm{Var}(H(X_{11})) > 0$. *If* $\min_{1\leqslant i\leqslant a} n_i \to \infty$ *and if* $0 < \lambda_0 \leqslant n_i/N \leqslant 1-\lambda_0 < 1$, $\forall i = 1,\ldots,a$, *then under* $H_0^F\colon F_1 = \cdots = F_a$, *the statistic* $\sqrt{N}\int H\,\mathrm{d}(\widehat{\boldsymbol{F}}-\boldsymbol{F})$ *has asymptotically a multivariate normal distribution with mean* $\mathbf{0}$ *and covariance matrix* $\boldsymbol{V} = N\sigma^2\boldsymbol{\Lambda}^{-1}$ *where* $\boldsymbol{\Lambda} = \mathrm{diag}\{n_1,\ldots,n_a\}$.

The unknown variance σ^2 can be estimated consistently by the ranks R_{ij}.

THEOREM 2.5. *Under* $H_0^F\colon F_1 = \cdots = F_a$ *and under the assumptions of Theorem 2.4,*

(1) $\widehat{\sigma}_N^2 = \dfrac{1}{N^2(N-a)}\displaystyle\sum_{i=1}^{a}\sum_{j=1}^{n_i}\left(R_{ij}-\overline{R}_{i\cdot}\right)^2 \xrightarrow{p} \sigma^2,$

(2) $\widehat{\sigma}_0^2 = \dfrac{1}{N^2(N-1)}\displaystyle\sum_{i=1}^{a}\sum_{j=1}^{n_i}\left(R_{ij}-\frac{N+1}{2}\right)^2 \xrightarrow{p} \sigma^2.$

Next, the asymptotic distribution of the a-sample rank statistic $\sqrt{N}\boldsymbol{P}_a(\widehat{\boldsymbol{p}}-\boldsymbol{p})$ is given under the hypothesis $H_0^F\colon F_1 = \cdots = F_a$.

THEOREM 2.6. *Let* σ^2 *and* $\boldsymbol{\Lambda}$ *be given as in Theorem 2.4,* $\widehat{\sigma}_N^2$ *as in Theorem 2.5 and assume that* $\widehat{\sigma}_N^2 > 0$. *Then, under the assumptions of Theorem 2.4 and under* $H_0^F\colon \boldsymbol{P}_a\boldsymbol{F} = \mathbf{0}$,

(1) *The statistic* $\sqrt{N}\boldsymbol{P}_a\widehat{\boldsymbol{p}} = \sqrt{N}\boldsymbol{P}_a\int \widehat{H}\,\mathrm{d}\widehat{\boldsymbol{F}}$ *has asymptotically a multivariate normal distribution with mean* $\mathbf{0}$ *and covariance matrix* $N\sigma^2\boldsymbol{P}_a\boldsymbol{\Lambda}^{-1}\boldsymbol{P}_a$,

(2) *The quadratic form* $\widehat{Q}(\boldsymbol{P}_a) = \widehat{\boldsymbol{p}}'(\boldsymbol{\Lambda} - \frac{1}{N}\boldsymbol{\Lambda}\boldsymbol{J}_a\boldsymbol{\Lambda})\widehat{\boldsymbol{p}}/\widehat{\sigma}_N^2$ *has asymptotically a central* χ_f^2*-distribution with* $f = \mathrm{rank}(\boldsymbol{P}_a) = a-1$.

Next, we give the asymptotic results for the case of heteroscedastic models. For simplicity, we restrict ourselves to the case of $a = 2$ samples. The quantity $p = \int F_1 \, \mathrm{d}F_2$ is estimated by

$$\widehat{p} = \frac{1}{n_1}\left(\overline{R}_{2\cdot} - \frac{n_2+1}{2}\right) \tag{2.4}$$

where $\overline{R}_{2\cdot} = n_2^{-1}\sum_{j=1}^{n_2} R_{2j}$ is the mean of the ranks R_{ij} in sample $i = 2$. Note that $\widehat{p}$ is unbiased and consistent for p if $\min(n_1, n_2) \to \infty$. The asymptotic normality of $\widehat{p}$ is stated in the next theorem.

THEOREM 2.7. *Let* $L_N = \widehat{p} - p = \int \widehat{F}_1 \, \mathrm{d}\widehat{F}_2 - \int F_1 \, \mathrm{d}F_2$ *and* $s_N^2 = \mathrm{Var}(\frac{n_1 n_2}{N} L_N) > 0$ *where* $N = n_1 + n_2$. *If* $\min(n_1, n_2) \to \infty$ *and if* $F_1(X_{21})$ *and* $F_2(X_{11})$ *are not both one-point distributions, then*

$$\frac{n_1 n_2}{N s_N} L_N \xrightarrow{\mathcal{L}} U \sim N(0,1) \quad \text{as } N \to \infty.$$

The asymptotic variance s_N^2 can be estimated consistently using the (overall) ranks R_{ij} and the (within) ranks $R_{ij}^{(i)}$ which are defined in the next theorem.

THEOREM 2.8. *Let* s_N^2 *be given as in Theorem* 2.7, *let* R_{ij} *be the rank of* X_{ij} *among all the* $N = n_1 + n_2$ *observations* $X_{11}, \ldots, X_{2n_2}$ *and let* $R_{ij}^{(i)}$, $i = 1, 2$, *be the rank of* X_{ij} *among the* n_i *observations within sample* i. *If* $\min(n_1, n_2) \to \infty$ *and if* $F_1(X_{21})$ *and* $F_2(X_{11})$ *are not both one-point distributions, then*

$$\widehat{s}_N^2 = \frac{1}{N^2}\sum_{i=1}^{2}\frac{n_i}{n_i - 1}\sum_{j=1}^{n_i}\left(R_{ij} - R_{ij}^{(i)} - \overline{R}_{i\cdot} + \frac{n_i+1}{2}\right)^2$$

is a consistent estimator for s_N^2. *Moreover, under* H_0^p: $p = 1/2$,

$$\frac{1}{N\widehat{s}_N}\left(R_{2\cdot} - n_2\frac{N+1}{2}\right) \xrightarrow{\mathcal{L}} U \sim N(0,1), \quad \text{as } N \to \infty.$$

For the proof of this theorem, see Brunner et al. (1995). The asymptotic results given here are applied in the next Section to derive statistics for the one-factor designs.

2.1.3. Derivation of the statistics

A consistent estimator $\widehat{\boldsymbol{p}} = (\widehat{p}_1, \ldots, \widehat{p}_a)'$ for the vector of the generalized means $\boldsymbol{p} = (p_1, \ldots, p_a)'$ is taken as a convenient statistic and the distribution of $\widehat{\boldsymbol{p}}$ is considered under $H_0^F(\boldsymbol{P}_a)$: $\boldsymbol{P}_a \boldsymbol{F} = \boldsymbol{0}$.

Hypothesis H_0^F; small sample sizes. To test H_0^F in the case of small sample sizes, it suffices to work with the quadratic form $Q = \sum_{i=1}^a R_{i\cdot}^2$ where $R_{i\cdot} = \sum_{j=1}^{n_i} R_{ij} = n_i(N\widehat{p}_i + \frac{1}{2})$. Under H_0^F, the N random variables X_{ij} are independent and identically distributed, i.e., $X_{ij} \sim F_i(x) = H(x)$. It is well known that the vector of ranks $\boldsymbol{R}_N = (R_{11}, \ldots, R_{an_a})'$ has a discrete uniform distribution on the orbit of $\{1, \ldots, N\}$ if $F(\cdot)$ is continuous (see, e.g., Randles and Wolfe, 1979, p. 37). Therefore, under H_0^F, the exact distribution of Q can be quickly computed by a multivariate shift algorithm (see, e.g., Streitberg and Roehmel, 1986). In case of ties, the integers $1, \ldots, N$ are replaced by the observed midranks and again, the same algorithm can be used to compute the exact distribution of $4Q$ conditioned on the observed (mid)-ranks. Note that the statistic Q has to be multiplied by 4 in this case since the algorithm works only with integers. The special case of $a = 2$ will be considered separately in Section 2.1.3.3. For large sample sizes however, the asymptotic results given in Section 2.1.2 are needed.

Hypothesis H_0^F; large sample sizes. Let $\overline{\boldsymbol{Y}} = (\overline{Y}_{1\cdot}, \ldots, \overline{Y}_{a\cdot})' = \int H \, d\widehat{\boldsymbol{F}}$ where $\overline{Y}_{i\cdot}$ is given in (2.3). Then the covariance matrix of $\sqrt{N}\,\overline{\boldsymbol{Y}}$ is $\boldsymbol{V}$ given in Theorem 2.4 and a consistent estimator for σ^2 is $\widehat{\sigma}_0^2$ given in Theorem 2.5.

In the asymptotic case, for testing H_0^F, we take $\sqrt{N}\boldsymbol{P}_a\widehat{\boldsymbol{p}}$ as a statistic which under H_0^F: $\boldsymbol{P}_a\boldsymbol{F} = \boldsymbol{0}$ has asymptotically the same distribution as $\sqrt{N}\boldsymbol{P}_a\overline{\boldsymbol{Y}}$. Then under H_0^F, it follows from Theorem 2.6, 2. that the quadratic form

$$
\begin{aligned}
Q_H &= \frac{1}{\widehat{\sigma}_0^2}\widehat{\boldsymbol{p}}'\left(\boldsymbol{\Lambda} - \frac{1}{N}\boldsymbol{\Lambda}\boldsymbol{J}_a\boldsymbol{\Lambda}\right)\widehat{\boldsymbol{p}} \\
&= \frac{N-1}{\sum_{i=1}^a \sum_{j=1}^{n_i} \left(R_{ij} - (N+1)/2\right)^2} \sum_{i=1}^a n_i \left(\overline{R}_{i\cdot} - \frac{N+1}{2}\right)^2 \qquad (2.5)
\end{aligned}
$$

has asymptotically a central χ_f^2-distribution with $f = a - 1$. If $F_i \equiv H$ is continuous, then $\sum_{i=1}^a \sum_{j=1}^{n_i} (R_{ij} - \frac{N+1}{2})^2 = N(N^2 - 1)/12$ and Q_H becomes the Kruskal–Wallis statistic

$$
Q_W = \frac{12}{N(N+1)} \sum_{i=1}^a n_i(\overline{R}_{i\cdot} - \overline{R}_{\cdot\cdot})^2 = \frac{12}{N(N+1)} \sum_{i=1}^a \frac{R_{i\cdot}^2}{n_i} - 3(N+1)
$$

(see Kruskal, 1952; Kruskal and Wallis, 1952, 1953). We note that another consistent estimator for σ^2 is given in Theorem 2.5, 1., namely $\widehat{\sigma}_N^2 = [N^2(N - a)]^{-1} \sum_{i=1}^a \sum_{j=1}^{n_i} (R_{ij} - \overline{R}_{i\cdot})^2$ and under H_0^F, the statistic

$$
\widehat{Q}_W = \frac{N-a}{\sum_{i=1}^a \sum_{j=1}^{n_i} \left(R_{ij} - \overline{R}_{i\cdot}\right)^2} \sum_{i=1}^a n_i \left(\overline{R}_{i\cdot} - \frac{N+1}{2}\right)^2 \qquad (2.6)
$$

has also asymptotically a central χ_f^2-distribution with $f = a - 1$. For small sample sizes, the distribution of $\widehat{Q}_W/(a-1)$ may be approximated by the central F_{f_1,f_2}-distribution with $f_1 = a - 1$ and $f_2 = (N - a) - 1$. For another approximation, we refer to Wallace (1959) and Iman and Davenport (1976). Boos and Brownie (1994) recommend to compare $\widehat{Q}_W/(a-1)$ with the central F_{f_1,f_2}-distribution with $f_1 = c(a-1)$ and $f_2 = c(N-a)$ where $c = 1 - [6(N+1)/(5(N-1)(N+6/5))]$.

2.1.3.1. The 'Rank Transform' (RT) property. The statistic $\widehat{Q}_W$ in (2.6) can formally be derived from the statistic of the one-way ANOVA by replacing the original observations X_{ij} by their ranks R_{ij} which has been called 'rank transform' (RT) statistic (Conover and Iman, 1976, 1981). However, one should be cautious to use this heuristic technique to derive rank tests for more complex experimental designs where this technique may lead to incorrect procedures. The RT may work in special cases; the reason for this is clear from Theorems 2.3, 2.4 and 2.6 in Section 2.1.2.

It follows from Theorems 2.3 and 2.6 that the rank statistic $\sqrt{N}\boldsymbol{P}_a \int \widehat{H}\, d\widehat{\boldsymbol{F}}$ and the statistic $\sqrt{N}\boldsymbol{P}_a \int H\, d\widehat{\boldsymbol{F}} = \sqrt{N}\boldsymbol{P}_a\overline{\boldsymbol{Y}}$ given in (2.3) are asymptotically equivalent under the hypothesis $H_0^F(\boldsymbol{P}_a)$: $\boldsymbol{P}_a\boldsymbol{F} = \boldsymbol{0}$ formulated in terms of the distribution functions. The statistic $\overline{\boldsymbol{Y}}$ is called 'asymptotic rank transform' (see Akritas, 1990). If the hypothesis is formulated in a linear model as $H_0^\mu(\boldsymbol{P}_a)$: $\boldsymbol{P}_a\boldsymbol{\mu} = \boldsymbol{0}$, then it follows from Proposition 2.1 that the hypotheses $H_0^F(\cdot)$ and $H_0^\mu(\cdot)$ are equivalent in the one-way layout. The second reason why the RT does not work in general has been pointed out by Akritas (1990). The homoscedasticity of the error term is not transferred to the asymptotic RT in general. If all distribution functions are equal, i.e., $H = F_1 = \cdots = F_a$, then $\boldsymbol{V} = N\sigma^2\boldsymbol{\Lambda}^{-1}$. This shows why the RT 'works' in the one-factor fixed design under the hypothesis in the linear model.

If a rank statistic has the property of being a 'RT-statistic', then this is of importance for computational purposes. The parametric counterpart of a RT-statistic which may be available in a statistical software package can be applied to the ranked data. Only the quality of approximation to the asymptotic distribution or some finite approximation has to be taken care of (e.g., by performing simulation studies). In any case, it is necessary to identify the statistical model of the asymptotic RT under H_0^F. The statistical model underlying the parametric counterpart for which a procedure should be available in the software package should correspond to the statistical model of the asymptotic RT under H_0^F. The RT should not be regarded as a technique to derive statistics; it should rather be considered a property of a statistic which can be useful for computational purposes. We prefer to derive rank tests in general models from the estimation of fixed quantities by which treatment effects and hypotheses are described.

2.1.3.2. Large number of treatments. If the number a of treatments is large while the number of replications per treatment is small, then the approximation of the distribution of the quadratic forms given in (2.5) and (2.6) may be very poor. In such a case, the asymptotics should be based on the large number of treatments rather than on the small number of replications. This situation has been considered by Boos and Brownie (1994, 1995). For details, we refer to these papers. Let $\widehat{Q}_W$ be as given in (2.6). Then under H_0^F, the statistic $T_a = \sqrt{a(n-1)/(2n)}(\widehat{Q}_W/(a-1) - 1)$ has asymptotically (as $a \to \infty$) a standard normal distribution where equal sample sizes are assumed for simplicity.

2.1.3.3. Tests for $a = 2$ samples. In this section, we consider tests for the hypotheses H_0^F: $F_1 = F_2$ and H_0^p: $p = 1/2$ in the two-sample case.

Hypothesis H_0^F. First we consider the hypothesis H_0^F: $F_1 = F_2 = H$. In the case of continuous distribution functions, the exact distribution of $R_{2\cdot} = \sum_{j=1}^{n_2} R_{2j}$, the Wilcoxon rank sum statistic, can easily be computed by means of a well known recursion formula (see, e.g., Randles and Wolfe, 1979, p. 46). For quick computation, one may use a shift algorithm, a program for which is given in literature by Zimmermann (1985b). In case of ties, the integers $1, \ldots, N$ are replaced by the observed (mid)-ranks R_{ij} and the same shift algorithm can be used for the computation of the exact (conditional) permutation distribution of the statistic $2R_{2\cdot}$. Here, $R_{2\cdot}$ is multiplied by 2 in order to have integer values for the shift algorithm. For larger sample sizes, the asymptotic distribution of the standardized Wilcoxon rank sum statistic may be used to test the hypothesis H_0^F (see Theorem 2.7). If $F(\cdot)$ is continuous, then it is well known that under H_0^F, the Wilcoxon–Mann–Whitney statistic

$$W_N = \sqrt{\frac{12}{n_1 n_2 (N+1)}} \left(R_{2\cdot} - n_2 \frac{N+1}{2} \right) \tag{2.7}$$

has asymptotically a standard normal distribution if $\min(n_1, n_2) \to \infty$. In case of ties, let s_N^2 denote the variance of $\frac{1}{N} R_{2\cdot}$ under H_0^F: $F_1 = F_2$. Then, it follows from Theorem 2.5 that

$$\widetilde{s}_{N,0}^2 = \frac{n_1 n_2}{N^3 (N-1)} \sum_{i=1}^{2} \sum_{j=1}^{n_i} \left(R_{ij} - \frac{N+1}{2} \right)^2 \quad \text{and}$$

$$\widehat{s}_{N,0}^2 = \frac{n_1 n_2}{N^3 (N-2)} \sum_{i=1}^{2} \sum_{j=1}^{n_i} (R_{ij} - \overline{R}_{i\cdot})^2$$

are consistent estimators for s_N^2 where 'consistent' means that $\widetilde{s}_{N,0}^2 / s_N^2 \xrightarrow{p} 1$ and $\widehat{s}_{N,0}^2 / s_N^2 \xrightarrow{p} 1$ as $N \to \infty$. Moreover, it follows that the statistics

$$W_N^* = \frac{1}{N \widetilde{s}_{N,0}} \left(R_{2\cdot} - n_2 \frac{N+1}{2} \right) \tag{2.8}$$

and

$$\begin{aligned} W_N^{\mathrm{RT}} &= \frac{1}{N \widehat{s}_{N,0}} \left(R_{2\cdot} - n_2 \frac{N+1}{2} \right) \\ &= \frac{\overline{R}_{2\cdot} - \overline{R}_{1\cdot}}{\sqrt{\sum_{i=1}^{2} \sum_{j=1}^{n_i} (R_{ij} - \overline{R}_{i\cdot})^2 / (N-2)}} \cdot \sqrt{\frac{n_1 n_2}{N}} \end{aligned} \tag{2.9}$$

have asymptotically standard normal distributions if $\min(n_1, n_2) \to \infty$. Note that W_N^{RT} has the RT-property with respect to t-test statistic for homoscedastic errors. Approximations of the distributions of W_N^* and W_N^{RT} have been studied by Iman (1976) and references cited therein.

Hypothesis H_0^p. Under the hypothesis H_0^p: $p = 1/2$, the distribution functions F_1 and F_2 are not necessarily identical and no permutation argument is valid. Note that the hypothesis H_0^p includes all heteroscedastic models, especially all symmetric distribution functions with the same center of symmetry. Therefore, this problem has been called 'generalized Behrens–Fisher problem' (see Potthoff, 1963; Fligner and Policello, 1981; Brunner and Neumann, 1986b).

From Theorem 2.8, under H_0^p: $p = 1/2$, the statistic

$$W_N^{\mathrm{BF}} = (N\widehat{s}_N)^{-1}\left(R_{2\cdot} - n_2(N+1)/2\right)$$

has asymptotically a standard normal distribution if $\min(n_1, n_2) \to \infty$ and $F_1(X_{21})$ and $F_2(X_{11})$ are not both one-point distributions.

REMARK. For estimating the variance s_N^2, the ranks R_{ij} among all the N observations (overall ranks) as well as the ranks $R_{ij}^{(j)}$ within treatment j (within ranks) are needed. The reason for this is that the variances of the unobservable random variables $F_1(X_{21})$ and $F_2(X_{11})$ have to be estimated consistently. They are replaced by the observable random variables $\widehat{F}_1(X_{21})$ and $\widehat{F}_2(X_{11})$. Note, e.g., that $N_1\widehat{F}_1(X_{21}) = N\widehat{H}(X_{21}) - N_2\widehat{F}_2(X_{22}) = R_{21} - R_{21}^{(2)}$. Therefore, $\widehat{s}_N^2$ given in Theorem 2.8 has not the RT-property so that the RT-method fails to 'work' in this simple case since the hypothesis $\int F_1\, \mathrm{d}F_2 = 1/2$ is considered and not the hypothesis $F_1 = F_2$ (see the remark on the RT-property in Section 2.1.3.1). For details, we refer to Brunner and Neumann (1986a) and to Brunner et al. (1995).

2.2. *Two-factor cross-classified designs*

Here we consider models with two crossed factors, factor A, the row-factor with $i = 1, \ldots, a$ levels and factor B, the column-factor with $j = 1, \ldots, b$ levels. In this setup, the simplest models are the models without interaction. Many rank procedures for such models have been considered in literature where two types of ranking the observations are used. Procedures which use the *ranks over all observations* have been described by Hora and Conover (1984), Iman et al. (1984), Hora and Iman (1988) and Akritas (1990) and Thompson (1991a). Statistics based on *ranks within each level of one factor* have been described by Friedman (1937), Durbin (1951), Benard and van Elteren (1953), Mack and Skillings (1980), Rinaman (1983), Haux et al. (1984), Groggle and Skillings (1986), Hettmansperger and Norton (1987), and Wittkowski (1988) among others (see the references cited in these papers). In order to explain the meaning of the two types of ranking the observations and to identify the underlying testing problems, the hypotheses connected with these procedures shall be worked out in the next subsection.

For models with interactions (which are more realistic in practice), some rank procedures in literature have been proposed by intuition ('rank transform statistics') rather than by theoretical considerations. In many cases, these statistics do not have the 'desired' asymptotic χ^2-distribution under the hypotheses stated in those papers. It has been shown by counterexamples (see Brunner and Neumann, 1984, 1986a) that the distribution of rank transform statistics may become degenerate under the hypotheses of 'no main effect' or of 'no interaction' in the linear model. Akritas (1990) showed that the rank transform fails to work in general for the hypotheses of 'no main effect' or of 'no interaction' in the linear model and derived a rank test for a joint hypothesis (see Section 2.2.2). Another statistic for the same hypothesis has been given by Thompson (1991a). Nonparametric interactions and main effects in 2×2 and $2 \times b$ designs have been defined and analyzed by Patel and Hoel (1973), Hilgers (1979), Brunner and Neumann (1984, 1986a), Boos and Brownie (1992) and Brunner et al. (1995).

2.2.1. Models and hypotheses

2.2.1.1. Models.

Linear model. In the linear model it is assumed that the random variables $X_{ijk} \sim F_{ij}$, $i = 1, \ldots, a$, $j = 1, \ldots, b$ and $k = 1, \ldots, n_{ij}$, are independent and $F_{ij}(x) = F(x - \mu_{ij})$ where for models without interaction, μ_{ij} is decomposed as

$$\mu_{ij} = \mu + \zeta_i + \tau_j, \quad i = 1, \ldots, a, \ j = 1, \ldots, b, \tag{2.10}$$

with the constraints $\sum_{i=1}^{a} \zeta_i = \sum_{j=1}^{b} \tau_j = 0$ and for models with interaction, μ_{ij} is decomposed as

$$\mu_{ij} = \mu + \zeta_i + \tau_j + \theta_{ij}, \quad i = 1, \ldots, a, \ j = 1, \ldots, b, \tag{2.11}$$

with the constraints $\sum_{i=1}^{a} \zeta_i = \sum_{j=1}^{b} \tau_j = \sum_{i=1}^{a} \theta_{ij} = \sum_{j=1}^{b} \theta_{ij} = 0$.

In the sequel, we denote by $\boldsymbol{F} = (F_{11}, \ldots, F_{1b}, \ldots, F_{a1}, \ldots, F_{ab})'$ the vector of the distribution functions in the two-way layout and by $\boldsymbol{\mu} = (\mu_{11}, \ldots, \mu_{1b}, \ldots, \mu_{a1}, \ldots, \mu_{ab})' = \int x \, \mathrm{d}\boldsymbol{F}$ the vector of the expectations.

General model. In the general model, we assume that

$$X_{ijk} \sim F_{ij}(x), \quad i = 1, \ldots, a, \ j = 1, \ldots, b, \ k = 1, \ldots, n_{ij}, \tag{2.12}$$

are independent random variables. As in the case of the one-factor designs, we take as a reasonable function $H(x)$ for a generalized mean $\boldsymbol{\nu}(H)$ the mean of all distribution functions in the experiment, namely $H(x) = \frac{1}{N} \sum_{i=1}^{a} \sum_{j=1}^{b} n_{ij} F_{ij}(x)$. Let $\boldsymbol{p} = (p_{11}, \ldots, p_{ab})' = \int H \, \mathrm{d}\boldsymbol{F}$ where $p_{ij} = \int H \, \mathrm{d}F_{ij}$ and let $\overline{F}_{i\cdot} = b^{-1} \sum_{j=1}^{b} F_{ij}$, $\overline{F}_{\cdot j} = a^{-1} \sum_{i=1}^{a} F_{ij}$ and $\overline{F}_{\cdot\cdot} = (ab)^{-1} \sum_{i=1}^{a} \sum_{j=1}^{b} F_{ij}$. Recall that $\boldsymbol{p}$ depends on the sample sizes n_{ij} through $H(x)$ and therefore $\boldsymbol{p}$ will not be used to formulate hypotheses.

The effects underlying the procedures using ranks within the levels of one factor can be described by generalized 'conditional' means, i.e., by generalized means within the levels of one factor.

As in the case of the one-factor models, we take as a reasonable function $H_j^{A|B}(x)$ for a generalized conditional mean $\nu(H_j^{A|B})$ within level j of factor B the mean of all distribution functions within this level, namely $H_j^{A|B}(x) = N_j^{-1} \sum_{i=1}^a n_{ij} F_{ij}(x)$ where $N_j = \sum_{i=1}^a n_{ij}$ and similarly $H_i^{B|A}(x) = N_i^{-1} \sum_{i=1}^b n_{ij} F_{ij}(x)$ for $\nu(H_i^{B|A})$ within level i of factor A where $N_i = \sum_{j=1}^b n_{ij}$. Let $\boldsymbol{w}^{A|B} = (w_{11}^{A|B}, \ldots, w_{ab}^{A|B})'$ where $w_{ij}^{A|B} = \int H_j^{A|B} \, \mathrm{d}F_{ij}$, $i = 1, \ldots, a$ and $j = 1, \ldots, b$. These generalized means shall be referred to as 'conditional means'.

Since the construction of the design is symmetric in the factors A and B, we consider only $H_j^{A|B}(x)$ in order to formulate hypotheses on factor A and on the interaction between A and B. However we note that the definition of an interaction in this setup is not symmetric in A and B (see also the remark on the rank interaction in de Kroon and van der Laan (1981).

Note also that $\boldsymbol{w}^{A|B}$ depends on the sample sizes n_{ij} through $H_j^{A|B}(x)$. For the case of $a = 2$ samples, the generalized mean within level j of factor B can be expressed as $w_j = \int F_{1j} \, \mathrm{d}F_{2j}$ which obviously does not depend on the sample sizes. Since many procedures considered in literature are based on the conditional means, we will define the hypotheses needed for these procedures and investigate the relations between them in order to explain the meaning of these tests.

2.2.1.2. Hypotheses. In the theory of linear models, the formulations of the hypotheses for fixed factors are identical for the fixed model and for the mixed model. In a nonparametric setup, Akritas and Arnold (1994) introduced the formulation of the hypotheses in terms of the distribution functions for one and two fixed factors in the context of repeated measurements designs with one and two crossed fixed factors. They replaced the vector of expectations which is used to formulate the hypotheses in the theory of linear models by the vector of the marginal distribution functions to formulate nonparametric hypotheses for main and interaction effects separately. In the linear model (2.11), the expectations μ_{ij} of the random variables X_{ijk} are linearly decomposed in order to define main effects and interactions of the factors. If the floor of the linear model is left and if only the distribution functions F_{ij} are used to describe the outcome of the experiment, then the distribution functions should be used to define treatment effects, main effects and interactions. So it seems to be natural to express hypotheses in a general model by a linear decomposition of the distribution functions F_{ij}.

Analogous to the theory of linear models, the dependence or independence of the random variables in a statistical model is not involved in the formulation of the hypotheses. Therefore, since the idea introduced by Akritas and Arnold (1994) is valid for all fixed and mixed models with crossed, nested or partially nested effects, we discuss it in detail only once in the paper. The considerations made in the sequel are also valid for the mixed models discussed in Sections 4.4.1.1 and 4.4.2.1. In these Sections, we will refer to the formulations and results given below.

We shall not only consider the common main and interaction effects, but also the 'simple factor effects' where a main effect and an interaction is tested together. These effects have been called 'joint effects' (Koch, 1969, 1970) and 'simple main effects' (Kirk, 1982). The reasons to consider also simple factor effects are that on the one hand the interpretation of a (pure) main effect is difficult in the presence of interactions. On the other hand, the information that there is a simple factor effect at least on one level of the other fixed factor may be a useful information for the researcher. Rank tests in fixed and mixed models for simple factor effects have also been considered in literature by Thompson and Ammann (1990), Thompson (1991a), Akritas (1990, 1991, 1993) and Akritas and Arnold (1994). For the derivation of rank tests in balanced and unbalanced two- and higher-way layouts based on the nonparametric hypotheses discussed below, we refer to Akritas et al. (1996).

The cross-classified two-way layout is symmetric in A and B and therefore we shall consider only the hypotheses $H_0(A), H_0(AB)$ and $H_0(A \mid B)$.

Linear model

$$
\begin{aligned}
H_0^\mu(A)\colon &\quad \text{main effect } A\colon \ \left(\boldsymbol{P}_a \otimes \frac{1}{b}\mathbf{1}_b'\right)\boldsymbol{\mu} = \mathbf{0} \quad \text{or} \\
&\quad \overline{\mu}_{i\cdot} - \mu = \zeta_i = 0, \ i = 1,\dots,a, \\
H_0^\mu(AB)\colon &\quad \text{interaction between } A \text{ and } B\colon \ (\boldsymbol{P}_a \otimes \boldsymbol{P}_b)\,\boldsymbol{\mu} = \mathbf{0} \quad \text{or} \\
&\quad \mu_{ij} - \overline{\mu}_{i\cdot} - \overline{\mu}_{\cdot j} + \mu = \theta_{ij} = 0, \ i = 1,\dots,a, \quad j = 1,\dots,b, \\
H_0^\mu(A \mid B)\colon &\quad \text{simple factor effect } A \text{ within } B\colon \ (\boldsymbol{P}_a \otimes \boldsymbol{I}_b)\,\boldsymbol{\mu} = \mathbf{0} \quad \text{or} \\
&\quad \mu_{ij} - \overline{\mu}_{\cdot j} = 0, \quad i = 1,\dots,a, \ j = 1,\dots,b.
\end{aligned}
$$

General model. In the general model, we formulate the hypotheses H_0^F based on the distribution functions and H_0^w based on the generalized conditional means.

The hypotheses based on the distribution functions are

$$
\begin{aligned}
H_0^F(A)\colon &\ \left(\boldsymbol{P}_a \otimes \frac{1}{b}\mathbf{1}_b'\right)\boldsymbol{F} = \mathbf{0} \quad \text{or} \quad \overline{F}_{i\cdot} - \overline{F}_{\cdot\cdot} = 0, \ i = 1,\dots,a, \\
H_0^F(AB)\colon &\ (\boldsymbol{P}_a \otimes \boldsymbol{P}_b)\boldsymbol{F} = \mathbf{0} \quad \text{or} \quad F_{ij} - \overline{F}_{i\cdot} - \overline{F}_{\cdot j} + \overline{F}_{\cdot\cdot} = 0, \\
&\qquad i = 1,\dots,a, \ j = 1,\dots,b, \\
H_0^F(A \mid B)\colon &\ (\boldsymbol{P}_a \otimes \boldsymbol{I}_b)\boldsymbol{F} = \mathbf{0} \quad \text{or} \quad F_{ij} - \overline{F}_{\cdot j} = 0, \ i = 1,\dots,a, \ j = 1,\dots,b.
\end{aligned}
$$

The hypotheses based on the generalized conditional means are

$$
\begin{aligned}
H_0^w(A)\colon &\ \left(\boldsymbol{P}_a \otimes \frac{1}{b}\mathbf{1}_b'\right)\boldsymbol{w}^{A|B} = \mathbf{0} \quad \text{or} \quad \overline{w}_{i\cdot}^{A|B} - \overline{w}_{\cdot\cdot}^{A|B} = 0, \ i = 1,\dots,a, \\
H_0^w(AB)\colon &\ (\boldsymbol{P}_a \otimes \boldsymbol{P}_b)\boldsymbol{w}^{A|B} = \mathbf{0} \quad \text{or} \quad w_{ij}^{A|B} - \overline{w}_{i\cdot}^{A|B} - \overline{w}_{\cdot j}^{A|B} + \overline{w}_{\cdot\cdot}^{A|B} = 0, \\
&\qquad i = 1,\dots,a, \ j = 1,\dots,b,
\end{aligned}
$$

$$H_0^w(A \mid B)\colon \big(\boldsymbol{P}_a \otimes \boldsymbol{I}_b\big)\boldsymbol{w}^{A|B} = \boldsymbol{0} \quad \text{or} \quad w_{ij}^{A|B} - \overline{w}_{\cdot j}^{A|B} = 0,$$
$$i = 1, \ldots, a, \; j = 1, \ldots, b.$$

Note that the hypotheses $H_0^F(\cdot)$ and $H_0^w(\cdot)$ in the general model are formally derived from the hypotheses $H_0^\mu(\cdot)$ in the linear model by replacing the vector $\boldsymbol{\mu}$ of the expectations by the vector $\boldsymbol{F}$ of the distribution functions F_{ij} or by the vector $\boldsymbol{w}^{A|B}$ of the generalized conditional means $w_{ij}^{A|B}$, respectively. The relations of the nonparametric hypotheses $H_0^F(\cdot)$ and $H_0^w(\cdot)$ to the hypotheses $H_0^\mu(\cdot)$ in the linear model are stated below for models with interactions (Proposition 2.9) and for models without interactions (Proposition 2.10).

PROPOSITION 2.9.

(1) *In the general model* (2.12), *the following holds:*

$$H_0^F(A \mid B) \Rightarrow H_0^F(A), \quad H_0^F(A \mid B) \Rightarrow H_0^w(A \mid B) \Rightarrow H_0^w(A).$$

(2) *In the linear model with interactions* (2.11),

$$H_0^F(A) \Rightarrow H_0^\mu(A), \quad H_0^F(AB) \Rightarrow H_0^\mu(AB), \quad H_0^F(A \mid B) \Longleftrightarrow H_0^\mu(A \mid B).$$
$$\textit{If } n_{ij} \equiv n, \quad \textit{then } H_0^w(A \mid B) \Longleftrightarrow H_0^\mu(A \mid B).$$

(3) *Moreover, if* $a = b = 2$ *and if* $F(\cdot)$ *is strictly increasing, then in the linear model with interactions* (2.11),

$$H_0^\mu(A) \Longleftrightarrow H_0^w(A)\colon \int F_{11}\, \mathrm{d}F_{21} + \int F_{12}\, \mathrm{d}F_{22} = 1,$$

$$H_0^\mu(AB) \Longleftrightarrow H_0^w(AB)\colon \int F_{11}\, \mathrm{d}F_{21} - \int F_{12}\, \mathrm{d}F_{22} = 0.$$

PROOF. (1) is evident. To prove (2), we note that $\boldsymbol{CF} = \boldsymbol{0} \Rightarrow \boldsymbol{C\mu} = \boldsymbol{C}\int x\, \mathrm{d}\boldsymbol{F}(x) = \int x\, \mathrm{d}(\boldsymbol{CF}(x)) = \boldsymbol{0}$ and the hypotheses are written by a suitable choice of the contrast matrix $\boldsymbol{C}$. Simple counter examples show that the converse will not be true. The conditional effect for $A \mid B$ however, can be regarded as a one-factor design within each level of the factor B. Thus, the equivalence of the hypotheses $H_0^F(A \mid B)$ and $H_0^\mu(A \mid B)$ follows from Proposition 2.1. It is easy to see that $H_0^\mu(A \mid B) \Rightarrow H_0^w(A \mid B)$ and $\overline{w}_{\cdot j}^{A|B} = 1/2$. The converse is proved by contradiction. Define $G(\delta) = \int F(x+\delta)\, \mathrm{d}F(x)$ and note that $G(\delta)$ is strictly increasing in the neighborhood of 0. Let $\alpha_i + (\alpha\beta)_{ij} = \delta_{ij}$ and assume that $\delta_{1j} = \cdots = \delta_{aj}$ is not true under $H_0^w(A \mid B)\colon w_{ij}^{A|B} - \overline{w}_{\cdot j}^{A|B} = 0$. Then there exists at least one index i, say, such that $\delta_{ij} \geqslant \delta_{sj}$ $\forall s = 1, \ldots, a$ and $\delta_{ij} > \delta_{sj}$ for at least one $s = 1, \ldots, a$ and $G(\delta_{ij} - \delta_{sj}) > 1/2$ since $G(\cdot)$ is strictly increasing in the neighborhood of 0 and $G(0) = 1/2$. Thus, $w_{ij}^{A|B} = \frac{1}{a}\sum_{s=1}^{a} G(\delta_{ij} - \delta_{sj}) > 1/2$ which contradicts $H_0^w(A \mid B)$. For the proof of (3), note that $G(\delta) = \int F(x+\delta)\,\mathrm{d}F(x)$ is symmetric about 0 and strictly increasing $\forall y \in \mathbb{R}$ if $F(\cdot)$ is strictly increasing and the results follow using the constraints for the linear model. □

REMARK. For the case of $a = b = 2$, note that

(1) the hypotheses $H_0^w(A)$ and $H_0^w(A \mid B)$ are formulated separately so that no sample sizes are involved in the hypotheses,

(2) the assumption that $F(\cdot)$ is strictly increasing can be relaxed if the statement is restricted to those values $\delta = \mu_{ij} - \mu_{i'j'}$ of the distribution for which $G(\delta)$ is strictly increasing.

Next, the relations between the hypotheses for models without interactions are stated.

PROPOSITION 2.10.

(1) *In the general model* (2.12), *the following holds:*

$$\begin{aligned}&\text{(a) If } (\boldsymbol{P}_a \otimes \boldsymbol{P}_b)\boldsymbol{F} = \boldsymbol{0} \quad \text{then } H_0^F(A \mid B) \iff H_0^F(A),\\ &\text{(b) if } (\boldsymbol{P}_a \otimes \boldsymbol{P}_b)\boldsymbol{w}^{A|B} = \boldsymbol{0} \text{ then } H_0^w(A \mid B) \iff H_0^w(A).\end{aligned}$$

(2) *In the linear model without interactions* (2.10), *the following holds:*

$$\begin{aligned}H_0^F(A \mid B) \iff H_0^F(A) \iff H_0^w(A \mid B) \iff H_0^w(A) \iff H_0^\mu(A \mid B)\\ \iff H_0^\mu(A).\end{aligned}$$

The proofs follows by the definitions of 'no interaction' in different models and by the results of Proposition 2.9. □

Equivalence of the hypotheses. The hypotheses $H_0^F(\cdot)$ and $H_0^\mu(\cdot)$ are equivalent in the linear model in the following cases:

(1) in the one factor design,

(2) in the two-factor design $(a \times b)$ without interaction,

(3) in the two-factor design $(a \times b)$ with interaction for the simple factor effects,

(4) in the two-factor design (2×2) with interaction, $H_0^\mu(\cdot)$ and $H_0^w(\cdot)$ are equivalent for all effects if $H_0^w(\cdot)$ is formulated as in Proposition 2.9 (3).

REMARK. In case of interactions in a linear model, a nonparametric main effect may be different from the main effect defined in a linear model. Note however, that the interpretation of a main effect in a linear model is difficult when interactions are present. In this case, it is easy to see that a main effect in a linear model may be changed or may even disappear by a suitable monotone transformation of the data. The generalized means $\boldsymbol{p}$ and $\boldsymbol{w}^{A|B}$, by which nonparametric effects can be defined, are invariant under any strictly monotone transformation. This is not true for the means in the linear model. If there are no interactions, then the parametric and nonparametric hypotheses of 'no main effect' are identical.

Comments on existing procedures. Many special procedures have been developed in literature for models without interaction. See, e.g., Mack and Skillings (1980), Groggle and Skillings (1986), Wittkowski (1988), Hora and Conover (1984), Haux et al. (1984), Rinaman (1983), Akritas (1990) and Thompson (1991a). The number of different procedures as well as the number of different hypotheses underlying these test procedures is rather confusing. Many procedures use the equivalence of $H_0^F(B \mid A)$ and $H_0^F(B)$ and $H_0^\mu(B)$, respectively, in a model without interaction. This assumption is crucial for the derivation of the asymptotic distribution of the proposed statistics. Therefore, these statistics should be applied with extreme caution since interactions can be excluded very seldom in practical problems. (See also the comment on the Hora–Conover statistic in Akritas (1990).) For this reason, we will restrict our considerations regarding models without interaction to the case of one observation per cell (i, j) where interactions are not included in a general model.

Tests for models with interactions have been developed for various hypotheses under different assumptions. See, e.g., Lemmer and Stoker (1967), Patel and Hoel (1973), Conover and Iman (1976), Lemmer (1980), de Kroon and van der Laan (1981), Brunner and Neumann (1986a), Akritas (1990), Thompson (1991a, b, 1993), Boos and Brownie (1992) and Brunner et al. (1995).

The crucial problem in designs with fixed interactions is the structure of the covariance matrix under the hypothesis. Several approaches to overcome this problem have been made:

1. **'Joint hypotheses'**. The set of parameters (or distribution functions) in the hypothesis is restricted such that all distribution functions are equal. In this case, the covariance matrix of the statistic has a simple form. (See, e.g., Akritas, 1990; Thompson, 1991a.) Although a 'joint hypothesis' may be useful in practice, this approach can only be considered a partial solution of the problem. Note that in Thompson (1991a), the contrast matrix used for the statistic is different from the contrast matrix used to formulate the hypothesis.

2. **'Additional restriction'**. For the derivation of the covariance matrix of the statistic, the set of parameters (or distribution functions) in the hypothesis is restricted such that all distribution functions are equal and the structure of the covariance matrix becomes simple. The distribution of the proposed statistic under the 'non-restricted hypothesis' remains open with this approach (see, e.g., Lemmer and Stoker, 1967; de Kroon and van der Laan, 1981; Thompson, 1991b, 1993).

3. **'$a = 2$'**. In the $2 \times b$ designs, the treatment effects and interactions are defined by the 'conditional means' $w_j = \int F_{1j}\, dF_{2j}$ so that only the variances of two-sample statistics have to be estimated (see Patel and Hoel, 1973; Brunner and Neumann, 1986a; Boos and Brownie, 1992; Brunner et al., 1995). Although these procedures are useful for many practical cases, this approach also can only be considered a partial solution of the problem.

4. The **'rank transform'** technique does not take care of computing the covariance matrix under H_0. Therefore, it will work in some cases but it will fail in other cases. (See also the remark on the RT-property in Section 2.1.3.1.)

5. The idea of Akritas and Arnold (1994), introduced in the context of multivariate repeated measures designs, seems to be a breakthrough in solving this crucial problem.

Under the hypothesis, the covariance matrix of the rank vector has a rather simple form if the hypothesis is formulated in terms of the distribution functions. The application of this idea to balanced and unbalanced fixed and mixed models is currently under research.

In the next section, we give a test for the hypothesis $H_0^F(A \mid B)$ in the $a \times b$-design ($a > 2$, $b > 2$) derived by Akritas (1990) and Thompson (1991a). For the 2×2-design and the $2 \times b$-design, we derive statistics based on ranks within the levels of factor B for the hypotheses $H_0^w(A \mid B)$, $H_0^w(AB)$ and $H_0^w(A)$. In all cases, unequal sample sizes are allowed. The implications for the linear model (2.11) can easily be seen from Proposition 2.9.

We also consider models without interaction where the number of replications per factor level combination (i,j) is $n = 1$ (complete designs) and either the number of levels of factor A or of factor B is large, i.e., $a \to \infty$ and b fixed or vice versa. These are the so-called 'block designs' where the block effect is assumed to be fixed. In addition, incomplete balanced designs are briefly considered.

2.2.2. Models with interactions

2.2.2.1. A test for $H_0^F(A \mid B)$ in the $a \times b$ design. The two-factor design with crossed fixed factors A and B is symmetric in A and B. For the analysis of the simple factor effects, it suffices therefore to consider only the hypothesis $H_0^F(A \mid B)$. The random variables $X_{ijk} \sim F_{ij}(x)$, $i = 1, \ldots, a$, $j = 1, \ldots, b$ and $k = 1, \ldots, n_{ij}$, in this design are assumed to be independent. We will use the notation of Section 2.1.1 and estimate $\boldsymbol{p} = \int H \, \mathrm{d}\boldsymbol{F}$ by $\widehat{\boldsymbol{p}} = \int \widehat{H} \, \mathrm{d}\widehat{\boldsymbol{F}}$ where $H = N^{-1} \sum_{i=1}^a \sum_{j=1}^b n_{ij} F_{ij}$, $\widehat{H} = N^{-1} \sum_{i=1}^a \sum_{j=1}^b n_{ij} \widehat{F}_{ij}$, $\widehat{\boldsymbol{F}} = (\widehat{F}_{11}, \ldots, \widehat{F}_{ab})'$ and $\widehat{F}_{ij}$ is the empirical distribution function of the observations $X_{ij1}, \ldots, X_{ijn_{ij}}$. Let R_{ijk} be the rank of X_{ijk} among all the $N = \sum_{i=1}^a \sum_{j=1}^b n_{ij}$ observations and let $\overline{R}_{ij\cdot} = n_{ij}^{-1} \sum_{k=1}^{n_{ij}} R_{ijk}$, and $\overline{R}_{\cdot j \cdot} = N_j^{-1} \sum_{i=1}^a \sum_{k=1}^{n_{ij}} R_{ijk}$ where $N_j = \sum_{i=1}^a n_{ij}$.

To derive a rank test for $H_0^F(A \mid B)$: $(\boldsymbol{P}_a \otimes \boldsymbol{I}_b)\boldsymbol{F} = \boldsymbol{0}$, we consider the asymptotic distribution of the estimator $(\boldsymbol{P}_a \otimes \boldsymbol{I}_b)\widehat{\boldsymbol{p}} = (\boldsymbol{P}_a \otimes \boldsymbol{I}_b) \int \widehat{H} \, \mathrm{d}\widehat{\boldsymbol{F}}$. It follows from Proposition 2.2 that the estimator

$$
\begin{aligned}
(\boldsymbol{P}_a \otimes \boldsymbol{I}_b) \int \widehat{H} \, \mathrm{d}\widehat{\boldsymbol{F}} &= (\widehat{p}_{11} - \overline{\widehat{p}}_{\cdot 1}, \ldots, \widehat{p}_{ab} - \overline{\widehat{p}}_{\cdot b})' \\
&= \frac{1}{N} \left(\overline{R}_{11\cdot} - \overline{R}_{\cdot 1 \cdot}, \ldots, \overline{R}_{ab\cdot} - \overline{R}_{\cdot b \cdot} \right)'
\end{aligned}
$$

is unbiased and consistent for $(\boldsymbol{P}_a \otimes \boldsymbol{I}_b)\boldsymbol{p}$. Under $H_0^F(A \mid B)$, the statistic is the same as b times the statistic for $H_0^F(A)$ in the one-factor design based on the ranks over all observations R_{ijk} since $H_0^F(A \mid B)$ in a two-factor design is equivalent to $H_0^F(A)$ in a one-factor design which is replicated b times. The variances of the statistics within each level of factor B are estimated separately by $\widehat{\tau}_j^2 = [N^2(N_j - a)]^{-1} S_j^2$ where

$S_j^2 = \sum_{i=1}^{a} \sum_{k=1}^{n_{ij}} (R_{ijk} - \overline{R}_{ij\cdot})^2$ (see Theorem 2.5). Then under $H_0^F(A \mid B)$, the quadratic form

$$Q^{A|B} = \sum_{j=1}^{b} \frac{N_j - a}{S_j^2} \sum_{i=1}^{a} n_{ij} \left(\overline{R}_{ij\cdot} - \overline{R}_{\cdot j\cdot} \right)^2 \tag{2.13}$$

has asymptotically a central χ_f^2-distribution with $f = (a-1)b$. (For a derivation in the context of a linear model with equal sample sizes, see Akritas (1990).)

2.2.2.2. *Tests for H_0^w in $2 \times b$ designs.*

$H_0^w(A)$ and $H_0^w(AB)$ in the 2×2 design. The 2×2-design is – in a certain sense – the natural extension of the one-factor two sample design to the case of two-factor designs. It shall be considered separately since in the 2×2-design the hypotheses for the main effects $H_0^w(A)$ and $H_0^w(B)$ as well as for the interaction $H_0^w(AB)$ in the general model (2.12) are identical to the parametric hypotheses and in such cases, heteroscedastic distributions may be assumed. At the present state, this seems to be the only way to provide rank tests for heteroscedastic models.

We consider the general model $X_{ijk} \sim F_{ij}(x)$, $i = 1, 2$, $j = 1, 2$, $k = 1, \ldots, n_{ij}$, where the X_{ijk}'s are independent random variables. The linear model simplifies to $\mu_{11} = \zeta + \tau + \theta$, $\mu_{12} = \zeta - \tau - \theta$, $\mu_{21} = -\zeta + \tau - \theta$ and $\mu_{22} = -\zeta - \tau + \theta$.

Let $w_j = \int F_{1j}\, \mathrm{d}F_{2j}$, $j = 1, 2$. Then the hypothesis for the main effect A is formulated as $H_0^w(A)$: $w^A = (w_1 + w_2)/2 = 1/2$ and the hypothesis for the interaction is formulated as $H_0^w(AB)$: $w^{AB} = w_1 - w_2 = 0$. The hypothesis $H_0^w(B)$ follows by the symmetry of the design. The generalized conditional mean $w^A = (w_1 + w_2)/2$ is estimated by $\widehat{w}^A = (\widehat{w}_1 + \widehat{w}_2)/2$ where

$$\widehat{w}_j = \left(\overline{R}_{2j\cdot}^B - (n_{2j} + 1)/2 \right) / n_{1j} \tag{2.14}$$

and $\overline{R}_{2j\cdot}^B = n_{2j}^{-1} \sum_{k=1}^{n_{2j}} R_{ijk}^B$ as in the one-factor design. Here, R_{ijk}^B is the rank of X_{ijk} among all the observations within level j of factor B. In the same way, w^{AB} is estimated by $\widehat{w}^{AB} = \widehat{w}_1 - \widehat{w}_2$. Let $T^A = \widehat{w}^A - 1/2$, $T^{AB} = \widehat{w}^{AB}$ and $N_j = n_{1j} + n_{2j}$, $j = 1, 2$, and let $R_{ijk}^{(ij)}$ be the rank of X_{ijk} within level i of factor A and within level j of factor B. The variance estimators $\widehat{\sigma}_j^2$ for the variances of the statistics T^A under $H_0^w(A)$ and for T^{AB} under $H_0^w(AB)$ are easily derived from Theorem 2.8 for $\widehat{\sigma}^2 = \sum_{j=1}^{2} \widehat{\sigma}_j^2$ where

$$\widehat{\sigma}_j^2 = \frac{1}{n_{1j}^2 n_{2j}^2} \sum_{i=1}^{2} \frac{n_{ij}}{n_{ij} - 1} \sum_{k=1}^{n_{ij}} \left[R_{ijk}^B - R_{ijk}^{(ij)} - \overline{R}_{ij\cdot}^B + (n_{ij} + 1)/2 \right]^2. \tag{2.15}$$

Then under $H_0^w(A)$: $w^A = 1/2$, the statistic

$$\frac{T^A}{\widehat{\sigma}} = \frac{1}{\widehat{\sigma}} \cdot \sum_{j=1}^{2} \frac{1}{n_{1j}} \left(\overline{R}_{2j\cdot}^B - \frac{N_j + 1}{2} \right) \tag{2.16}$$

has asymptotically ($\min_{i,j}(n_{ij}) \to \infty$) a standard normal distribution. Under $H_0^w(AB)$: $w^{AB} = 0$, the statistic

$$\frac{T^{AB}}{\widehat{\sigma}} = \frac{1}{\widehat{\sigma}} \cdot \left(\frac{1}{n_{11}} \overline{R}^B_{21\cdot} - \frac{1}{n_{12}} \overline{R}^B_{22\cdot} - \frac{n_{11}(n_{22}+1) - n_{12}(n_{21}+1)}{2 n_{11} n_{12}} \right) \tag{2.17}$$

has asymptotically ($\min_{i,j}(n_{ij}) \to \infty$) a standard normal distribution. For small sample sizes, both statistics may be approximated by a central t_f-distribution with $f = \min_{i,j}(n_{11}, n_{21}) + \min_{i,j}(n_{12}, n_{22}) - 2$ d.f. For details, we refer to Patel and Hoel (1973), Brunner and Neumann (1986a) and Brunner et al. (1995).

$H_0^w(A \mid B)$ in the $2 \times b$ design. For the $2 \times b$ design with independent random variables $X_{ijk} \sim F_{ij}(x)$, $i = 1, 2$, $j = 1, \ldots, b$ and $k = 1, \ldots, n_{ij}$, we derive a rank test based on the conditional means $w_j = \int F_{1j}\, \mathrm{d}F_{2j}$. The hypothesis is then stated as $H_0^w(A \mid B)$: $w = \sum_{j=1}^b (w_j - 1/2)^2 = 0$. The quantity w is estimated by $\widehat{w} = \sum_{j=1}^b (\widehat{w}_j - 1/2)^2$ where $\widehat{w}_j$ is given in (2.14). Under $H_0^w(A \mid B)$, it follows that the statistic

$$Q_b^w = \sum_{j=1}^b \frac{1}{n_{1j}^2 \widehat{\sigma}_j^2} \left(\overline{R}^B_{2j\cdot} - (N_j + 1)/2 \right)^2 \tag{2.18}$$

has asymptotically (as $\min_{i,j} n_{ij} \to \infty$) a central χ_b^2-distribution where

$$\widehat{\sigma}_j^2 = \frac{1}{n_{1j} n_{2j}} \left(\sum_{i=1}^2 \frac{N_j - n_{ij} - 1}{n_{ij}(N_j - n_{ij})^2} \times \sum_{k=1}^{n_{ij}} \left[R^B_{ijk} - R^{(ij)}_{ijk} - \overline{R}^B_{ij\cdot} + \frac{n_{ij}+1}{2} \right]^2 + \frac{1}{4} \right) \tag{2.19}$$

which follows from the variance estimator given in (2.15).

$H_0^w(AB)$ in the $2 \times b$ design. To derive a test for the interaction, let $\boldsymbol{w} = (w_1, \ldots, w_b)'$ be the vector of the effects within the levels of factor B. The hypothesis of no interaction is stated as $H_0^w(AB)$: $\boldsymbol{P}_b \boldsymbol{w} = \boldsymbol{0}$. The vector $\boldsymbol{w}$ is estimated by $\widehat{\boldsymbol{w}} = (\widehat{w}_1, \ldots, \widehat{w}_b)'$ where $\widehat{w}_j$ is given in (2.14). Then $\widehat{\boldsymbol{w}}$ has asymptotically (as $\min_{i,j} n_{ij} \to \infty$) a multinormal distribution with mean $\boldsymbol{w}$ and covariance matrix $\boldsymbol{V}_b = \operatorname{diag}\{\sigma_1^2, \ldots, \sigma_b^2\}$, and σ_j^2 is the variance of the statistic $\widehat{w}_j$ given in (2.14). Let $\boldsymbol{A}_b = \boldsymbol{V}_b^{-1}[\boldsymbol{I}_b - \boldsymbol{J}_b \boldsymbol{V}_b^{-1}/\mathrm{trace}(\boldsymbol{V}_b^{-1})]$. Then $\boldsymbol{A}_b \boldsymbol{V}_b \boldsymbol{A}_b = \boldsymbol{A}_b$ and $\boldsymbol{P}_b \boldsymbol{w} = \boldsymbol{0} \Leftrightarrow \boldsymbol{A}_b \boldsymbol{w} = \boldsymbol{0}$. Consequently, the quadratic form

$$Q^{AB} = \widehat{\boldsymbol{w}}' \boldsymbol{A}_b \widehat{\boldsymbol{w}} = \sum_{j=1}^b \frac{1}{\widehat{s}_j^2} \left(\widehat{w}_j - \frac{1}{\sum_{r=1}^b (1/s_r^2)} \sum_{r=1}^b \frac{\widehat{w}_r}{\widehat{s}_r^2} \right)^2 \tag{2.20}$$

has asymptotically a noncentral χ^2_f-distribution with $f = \text{rank}(\boldsymbol{A}_b\boldsymbol{V}_b) = b-1$ d.f. and noncentrality parameter $\lambda = \boldsymbol{w}'\boldsymbol{A}_b\boldsymbol{w}$ and under $H_0^w(AB)$, it has central χ^2_{b-1}-distribution. The variance estimators $\widehat{s}_j^2$ are given in (2.15).

$H_0^w(A)$ in the $2 \times b$ design. Finally we consider a statistic for the mean treatment effect $m^A = b^{-1}\sum_{j=1}^b w_j$ if no interactions are assumed. In this case, $p_1 = \cdots = p_b$ and the hypothesis reduces to $H_0^w(A)$: $p_1 = \cdots = p_b = 1/2$. The treatment effect m^A is estimated by an unweighted sum of the $\widehat{w}_j$'s, which leads to the statistic

$$\widehat{m}^A = \frac{1}{S_b}\sum_{j=1}^{b}\frac{1}{n_{1j}}\left(\overline{R}^B_{2j\cdot} - (N_j+1)/2\right) \tag{2.21}$$

where $S_b^2 = \sum_{j=1}^b \widehat{\sigma}_j^2$ and $\widehat{\sigma}_j^2$ is given in (2.19). Under $H_0^w(A)$: $m^A = 1/2$, the statistic $\widehat{m}^A$ has asymptotically a standard normal distribution if $\min_{i,j} n_{ij} \to \infty$ and $\sigma_j^2 > 0$, $i = 1,\ldots,2$, $j = 1,\ldots,b$. We like to mention that the statistic $\widehat{m}^A$ can also be used to test the hypothesis $H_0^w(A)$ in the presence of interactions if the variances σ_j^2 are estimated by $\widehat{\sigma}_j^2$ given in (2.15). We refer to Proposition 2.9 for the implication of the hypotheses in different models. It may be noted that the statistic $\widehat{m}^A$ is also appropriate to test the same hypothesis if the number of centers is large and the number of replications per center is small. In this case, $\widehat{m}^A$ has asymptotically the standard normal distribution under $H_0^w(A)$ provided $\sum_{j=1}^b (N_j+1)/(n_{1j}n_{2j}) \to \infty$ as $b \to \infty$.

Note that the statistics given in this subsection do not have the RT-property which is clear from the variance estimators given in (2.15) and (2.19).

EXAMPLE 1. To show the applicability of the procedures derived in this section, we give an example where the data consist of the grading scales of an improvement. It is not appropriate to assume a shift model for such data and ties have to be taken into account. We are grateful to Dr. E. Römer (Institut für biologische Forschung, Köln) for making available the data set.

Two inhalable test substances, drug 1 and drug 2 (factor A), are to be compared with regard to their irritative activity in the respiratory tract of the rat after subchronic inhalative exposure. Reserve cell hyperplasia in the respiratory epithelium of the nose after exposure to 2, 5, and 10[*ppm*] of the test substances served as a criterion for irritation. The result was histopathologically evaluated by the quality scales: 0 = 'no changes', 1 = 'slight changes', 2 = 'distinct changes', 3 = 'severe changes'. The results for the two drugs and three concentration groups with 20 rats each, are given in Table 2.1.

Two questions shall be answered:

(1) Is there an interaction between the test substances and the concentration?

(2) Is one test substance more effective than the other one?

To answer question (1), we apply the test for $H_0^w(AB)$ given in (2.20) and for question (2), we apply the test for $H_0^w(A)$ given in (2.21). The data are ranked within

Table 2.1
Data, rank means and results for the reserve cell hyperplasia trial

	Testsubstance								Results		
	Drug 1				Drug 2				$\widehat{m}^A = 2.135$		$Q^{AB} = 1.057$
Concentration	No. of rats with scale				No. of rats with scale				Rank means $\overline{R}^B_{ij\cdot}$		Effect w_j
	0	1	2	3	0	1	2	3	Drug 1	Drug 2	
2	18	2	0	0	16	3	1	0	19.45	21.55	0.553
5	12	6	2	0	8	8	3	1	18.25	22.75	0.613
10	3	7	6	4	1	5	8	6	18.20	22.80	0.615

each concentration and the rank means for each concentration and drug as well as the estimators for w_j, $j = 1, 2, 3$, are given in Table 2.1. The result for $H_0^w(AB)$ is $Q^{AB} = 1.057$ ($p = 0.590$) which indicates that the results are quite homogeneous within the three concentrations (no interaction). The result for $H_0^w(A)$ is $\widehat{m}^A = 2.135$ ($p = 0.0328$) and a significant treatment effect is proved at the 5% level.

2.2.3. Models without interactions

Models without (fixed) interaction do not share the difficulties of models with interaction since it follows from Proposition 2.9 that the hypotheses for the main effects are equivalent to the hypotheses for the simple factor effects. Therefore, the statistics for testing a simple factor effect in a model with interaction may also be used to test the main effect in a model without interaction. This fact has been used in many papers where the hypothesis of 'no treatment effect' is formulated as $H_0^F(A \mid B)$: $F_{1j} = \cdots = F_{aj}$, $j = 1, \ldots, b$. It has already been pointed out that the assumption of no interaction is crucial for the derivation of the null distribution of these statistics.

For this reason, we restrict our considerations regarding models without interaction to the case of one observation per cell (i, j) where interactions are not included in a general model. For asymptotic considerations, we have to assume that either $b \to \infty$ and a is fixed or that $a \to \infty$ and b is fixed. In the sequel, we consider test statistics for the main effect A for both types of ranking the observations, i.e., the treatment effect is either based on the generalized mean vector $\boldsymbol{p}$ (ranking over all observations) or on the conditional generalized mean vector $\boldsymbol{w}^{A|B}$ (ranking within the levels of factor B, sometimes called 'n-rankings').

2.2.3.1. Statistics based on $\boldsymbol{w}^{A|B}$. The statistics are based on the conditional generalized means $\boldsymbol{w}^{A|B}$. We consider the hypothesis $H_0^w(A)$: $(\boldsymbol{P}_a \otimes \frac{1}{b}\boldsymbol{1}_b')\boldsymbol{w}^{A|B} = \boldsymbol{0}$, where $(\boldsymbol{P}_a \otimes \frac{1}{b}\boldsymbol{1}_b')\boldsymbol{w}^{A|B} = (\overline{w}_{1\cdot}^{A|B} - \overline{w}_{\cdot\cdot}^{A|B}, \ldots, \overline{w}_{a\cdot}^{A|B} - \overline{w}_{\cdot\cdot}^{A|B})'$ and $\overline{w}_{\cdot\cdot}^{A|B} = a^{-1}\sum_{i=1}^{a} \overline{w}_{i\cdot}^{A|B}$. The quantities $w_{ij}^{A|B}$ are estimated by $\widehat{w}_{ij}^{A|B} = \int \widehat{H}_j^{A|B}\, d\widehat{F}_{ij}$ and thus, $\widehat{w}_{ij}^{A|B} - \widehat{\overline{w}}_{\cdot j}^{A|B} = a^{-1}(R_{ij}^B - (a+1)/2)$ where R_{ij}^B is the rank of X_{ij} within level j of factor B and a is fixed. Let $S_w^2 = \sum_{i=1}^a \sum_{j=1}^b (R_{ij}^B - (a+1)/2)^2$ and $\overline{R}_{i\cdot}^B = b^{-1}\sum_{j=1}^b R_{ij}^B$. Then the statistic

$$F_b^A = \frac{b^2(a-1)}{S_w^2} \sum_{i=1}^a \left(\overline{R}_{i\cdot}^B - (a+1)/2\right)^2 \tag{2.22}$$

has asymptotically under $H_0^w(A)$ a central χ_f^2-distribution with $f = a-1$ d.f. For the approximation in the case of small samples, see Iman and Davenport (1980).

In case of no ties,

$$F_b^A = \frac{12}{ba(a+1)} \sum_{i=1}^a (R_{i\cdot}^B)^2 - 3b(a+1) \tag{2.23}$$

is the well known Friedman (1937) statistic which is the simplest (and perhaps oldest) rank statistic for testing a treatment effect in this design.

Next, we consider the case where the number a of levels of factor A is large ($a \to \infty$) and the number of levels b of factor B is assumed to be small. Therefore, asymptotic considerations should be based on the large number of the levels of factor A. (For details, we refer to Boos and Brownie, 1994, 1995.)

Let $\widetilde{F}_b^A$ denote the statistic of the Friedman test (2.22) where S_w^2 is replaced by $bS_R^2/(b-1)$ and

$$S_R^2 = \sum_{i=1}^a \sum_{j=1}^b \left(R_{ij}^B - \overline{R}_{i\cdot}^B - \overline{R}_{\cdot j}^B + \frac{a+1}{2}\right)^2.$$

Here, R_{ij}^B is the rank of X_{ij} as in the Friedman test. Then it can easily be shown that $bS_R^2/[(b-1)S_w^2] \xrightarrow{p} 1$ and the statistics F_b^A and $\widetilde{F}_b^A = (b-1)S_w^2 F_b^A/(bS_R^2)$ are asymptotically equivalent. Under $H_0^F(A \mid B)$: $F_{1j} = \cdots = F_{aj}$, $j = 1, \ldots, b$, the statistic $T_b^A = \sqrt{a(b-1)/(2b)}(\widetilde{F}_b^A/(a-1) - 1)$ has asymptotically ($a \to \infty$) a standard normal distribution.

We mention briefly the incomplete balanced designs which have been considered by Durbin (1951) for the case of no ties. In these designs, it is assumed that within each level j of factor B, there are $k < a$ observed random variables X_{ij}, $j = 1, \ldots, b$. Let $r < b$ denote the number of observed random variables X_{ij} within each level i of factor A, $i = 1, \ldots, a$, and let $\lambda_{i,i'}$ denote the number of levels of factor B where the pair $(X_{ij}, X_{i'j})$, $i, i' = 1, \ldots, a$, $j = 1, \ldots, b$, has been observed. Within level j of factor B, let I_j be the set of i's where $n_{ij} = 1$, i.e., where a treatment has been applied to the level j of factor B. Within level i of factor A, let similarly J_i be the set of j's where a treatment has been applied. Let R_{ij}^B be the rank of X_{ij} among all the k observations within level j of factor B, i.e., $1 \leqslant R_{ij}^B \leqslant k$. Finally let

$$R_{i\cdot}^B = \sum_{j \in J_i} R_{ij}^B \quad \text{and} \quad \widehat{\sigma}^2 = [r(k-1)]^{-1} \sum_{j \in J_i} \sum_{i \in I_j} \left(R_{ij}^B - (k+1)/2\right)^2.$$

Then under $H_0^F(A)$: $F_{1j} = \cdots = F_{aj}$, $j = 1, \ldots, b$, the statistic

$$D_r^A = \frac{k}{a\lambda\hat{\sigma}^2} \sum_{i=1}^{a} \left(R_{i\cdot}^B\right)^2 - \frac{r^2(k+1)^2 k}{4\lambda\hat{\sigma}^2} \tag{2.24}$$

has asymptotically (i.e., for large values of r) a central χ_f^2-distribution with $f = a - 1$ d.f.

In case of no ties, D_r^A simplifies to the statistic given by Durbin (1951), namely

$$D_r^A = \frac{12}{a\lambda(k+1)} \sum_{i=1}^{a} \left(R_{i\cdot}^B\right)^2 - \frac{3r^2(k+1)}{\lambda}. \tag{2.25}$$

For the approximation of Durbin's statistic in case of small designs, see van der Laan and Prakken (1972).

2.2.3.2. Statistics based on $\boldsymbol{p}$. We develop a test for the hypothesis $H_0^F(A)$: $(\boldsymbol{P}_a \otimes \frac{1}{b}\boldsymbol{1}_b')\boldsymbol{F} = \boldsymbol{0}$ and we base the statistic on the generalized mean vector $\boldsymbol{p} = \int H \,\mathrm{d}\boldsymbol{F}$ where $H = N^{-1}\sum_{i=1}^{a}\sum_{j=1}^{b} F_{ij}$ and $N = ab$. Thus, we consider the asymptotic distribution of $(\boldsymbol{P}_a \otimes \frac{1}{b}\boldsymbol{1}_b')\widehat{\boldsymbol{p}} = (\boldsymbol{P}_a \otimes \frac{1}{b}\boldsymbol{1}_b')\int \widehat{H}\,\mathrm{d}\widehat{\boldsymbol{F}}$ under $H_0^F(A)$ where $\widehat{\boldsymbol{F}} = (\widehat{F}_{11}, \ldots, \widehat{F}_{ab})'$, $\widehat{F}_{ij}(x) = c(x - X_{ij})$ and $\widehat{H}(x) = N^{-1}\sum_{i=1}^{a}\sum_{j=1}^{b} c(x - X_{ij})$. Let

$$\overline{R}_{i\cdot} = b^{-1}\sum_{j=1}^{b} R_{ij}$$

where R_{ij} is the rank of X_{ij} among all the N observations. Then it follows that the statistic

$$\sqrt{b}\left(\boldsymbol{P}_a \otimes \frac{1}{b}\boldsymbol{1}_b'\right)\widehat{\boldsymbol{p}} = \frac{\sqrt{b}}{N}\left(\overline{R}_{1\cdot} - \frac{N+1}{2}, \ldots, \overline{R}_{a\cdot} - \frac{N+1}{2}\right)'$$

has asymptotically ($b \to \infty$) under $H_0^F(A)$ a multivariate normal distribution with mean $\boldsymbol{0}$ and covariance matrix $\boldsymbol{V} = \tau^2\boldsymbol{P}_a$ where $\tau^2 = b^{-1}\sum_{j=1}^{b}\sigma_j^2$ and $\sigma_j^2 = \sigma_{1j}^2 = \cdots = \sigma_{aj}^2 = \mathrm{Var}(H(X_{1j}))$, $j = 1, \ldots, b$. The last step follows from the equivalence of the hypotheses $H_0^F(A)$: $(\boldsymbol{P}_a \otimes \frac{1}{b}\boldsymbol{1}_b')\boldsymbol{F} = \boldsymbol{0}$ and $H_0^F(A \mid B)$: $(\boldsymbol{P}_a \otimes \boldsymbol{I}_b)\boldsymbol{F} = \boldsymbol{0}$ in the two-factor model without interaction. The estimator

$$\widehat{\tau}^2 = \frac{1}{N^2 b(a-1)}\sum_{j=1}^{b}\sum_{i=1}^{a}\left(R_{ij} - \overline{R}_{\cdot j}\right)^2 = \frac{1}{N^2 b(a-1)} S_N^2 \tag{2.26}$$

is consistent for τ^2 in the sense that $\widehat{\tau}^2/\tau^2 \xrightarrow{p} 1$ as $b \to \infty$. Combining these results, it follows under $H_0^F(A)$ that the statistic

$$K^A = \frac{b^2(a-1)}{S_N^2} \sum_{i=1}^{a} \left(\overline{R}_{i\cdot} - \frac{N+1}{2} \right)^2 \tag{2.27}$$

has asymptotically a central χ_f^2-distribution with $f = a - 1$ d.f. We note that the same statistic has been given by Brunner and Neumann (1982) and by Kepner and Robinson (1988) for the case where the factor B is random.

Conover and Iman (1976) proposed a RT-version of the ANOVA F-statistic for this design, namely

$$F_R = \frac{b(b-1)\sum_{i=1}^{a} \left(\overline{R}_{i\cdot} - (N+1)/2 \right)^2}{\sum_{i=1}^{a} \sum_{j=1}^{b} \left(R_{ij} - \overline{R}_{i\cdot} - \overline{R}_{\cdot j} + (N+1)/2 \right)^2}. \tag{2.28}$$

Iman et al. (1984) showed for the case of no ties that $(a-1)F_R$ has a central χ_{a-1}^2-distribution under $H_0^F(A \mid B)$. They showed also by a simulation study that the power of the F_R-statistic is higher than the power of the Friedman statistic given in (2.23) which is based on ranks within the levels of factor B. There is only a slight loss in power when a normal distribution is assumed. If the underlying distribution function is log-normal or Cauchy, this simulation study showed a considerable gain in power compared with the results of the ANOVA. Hora and Iman (1988) derived asymptotic relative efficiencies of the RT-procedure for this design. For details, we refer to these articles.

2.3. Special procedures for linear models

In this paper, we are considering only pure rank statistics which have the property of being invariant under monotone transformations of the data and which may also be applied to pure ordinal data. We do not like to restrict our considerations to linear models since in many experiments, shift effects are not realistic. The nonparametric analysis of multi-factor experiments in literature is mainly related to linear models. Based on the idea of Hodges and Lehmann (1962), first to estimate the nuisance parameters and then rank the residuals, Sen (1968) developed a class of aligned rank order tests in two-way layouts without interactions and Puri and Sen (1969, 1973), Sen and Puri (1977) and Adichie (1978) developed aligned rank tests for general linear hypotheses. For a unified description of this method, see Puri and Sen (1985). Aligned rank tests for linear models with autocorrelated errors are considered in Hallin and Puri (1994).

Another approach is based on minimizing Jaeckel's dispersion function (see Jaeckel, 1972) and has been developed by McKean and Hettmansperger (1976), and Hettmansperger and McKean (1977). Note that the statistics given there are not pure rank statistics. For a further approach and an excellent description and comparison of these methods, see Hettmansperger and McKean (1983).

3. Random models

3.1. One-factor random models

3.1.1. Models and hypotheses

Linear model. The one-way layout linear model for random effects is

$$X_{ij} = \mu + A_i + \varepsilon_{ij}, \quad i = 1, \ldots, a; \; j = 1, \ldots, n_i, \tag{3.1}$$

where μ is the overall mean, A_i are i.i.d. $N(0, \sigma_A^2)$ random variables and ε_{ij} are i.i.d. $N(0, \sigma_\varepsilon^2)$ random variables independent of A_i, $i = 1, \ldots, a$ and $j = 1, \ldots, n_i$. By assumption, $\mathrm{Var}(X_{11}) = \sigma_A^2 + \sigma_\varepsilon^2$, $\mathrm{Cov}(X_{11}, X_{21}) = 0$ and $\mathrm{Cov}(X_{11}, X_{12}) = \sigma_A^2$.

General model. Let $\boldsymbol{X} = (\boldsymbol{X}_1', \ldots, \boldsymbol{X}_a')'$ where $\boldsymbol{X}_i = (X_{i1}, \ldots, X_{in_i})'$ are independent random vectors with common distribution functions $G_i(\boldsymbol{x})$ and identical marginal distribution functions $F(x)$, $i = 1, \ldots, a$, $j = 1, \ldots, n_i$, and $N = \sum_{i=1}^a n_i$. It is reasonable to assume that the random variables within each vector $\boldsymbol{X}_i$ are interchangeable. Thus, (X_{i1}, X_{i2}) have identical bivariate marginal distribution functions $F_i^{**}(x, y)$ and it is assumed that $F_i^{**}(x, y)$ does not depend on i, i.e., $F_i^{**}(x, y) = F^{**}(x, y)$, $i = 1, \ldots, a$. The bivariate marginal distribution function of $(X_{i1}, X_{i'1})$, $i \neq i' = 1, \ldots, a$, is denoted by $F^*(x, y)$ and by independence, $F^*(x, y) = F(x)F(y)$. Thus, the covariance matrices of $\boldsymbol{X}_i$ and $\boldsymbol{X}$ are $\boldsymbol{S}_i = \mathrm{Cov}(\boldsymbol{X}_i) = (\sigma^2 - c)\boldsymbol{I}_{n_i} + c\boldsymbol{J}_{n_i}$ and

$$\boldsymbol{S}_N = \mathrm{Cov}(\boldsymbol{X}) = \bigoplus_{i=1}^{a} \boldsymbol{S}_i$$

respectively, where $\sigma^2 = \mathrm{Var}(X_{11})$ and $c = \mathrm{Cov}(X_{11}, X_{12})$. The common distribution function $G_i(\cdot)$ is called 'compound symmetric' and the covariance matrix of a random vector with a 'compound symmetric' common distribution function has the structure of the matrix $\boldsymbol{S}_i$.

Hypothesis. In the linear model, the hypothesis of no effect of the random factor A is usually stated as H_0: $\sigma_A^2 = 0$. In the general model, by assumption, the random variables X_{ij} are interchangeable within each level i of factor A. Intuitively, the hypothesis of no effect of the random factor A is equivalent to the condition that all random variables X_{ij}, $i = 1, \ldots, a$, $j = 1, \ldots, n_i$, are interchangeable, and thus in turn, they have the same bivariate marginal distribution functions and the hypothesis of no random effect in the general model is formulated as H_0^*: $F^{**}(x, y) = F^*(x, y) = F(x)F(y)$. It is easy to see that $H_0^* \Rightarrow H_0$: $\sigma_A^2 = 0$ if the linear model (3.1) is assumed. Note that, by assumption, $\boldsymbol{P}_a \boldsymbol{F} = \boldsymbol{0}$ since $\boldsymbol{F} = F\boldsymbol{1}_a$, and that $\boldsymbol{P}_a \boldsymbol{p} = \boldsymbol{0}$ where $\boldsymbol{p} = \int F \, \mathrm{d}\boldsymbol{F} = \frac{1}{2}\boldsymbol{1}_a$.

3.2. Test for H_0^*

We proceed as in the linear model and consider the asymptotic distribution of $\sqrt{N}\,\widehat{\boldsymbol{p}} = \sqrt{N}\int \widehat{H}\,\mathrm{d}\widehat{\boldsymbol{F}}$ under the assumption that factor A is random. Let $\boldsymbol{Y} = (Y_{11}, \dots, Y_{an_a})'$ where $Y_{ij} = H(X_{ij})$ and let

$$\overline{\boldsymbol{Y}}. = \left(\bigoplus_{i=1}^{a} \frac{1}{n_i} \boldsymbol{1}'_{n_i} \right) \boldsymbol{Y},$$

$$\sigma_H^2 = \mathrm{Var}(Y_{11}) = \int H^2\,\mathrm{d}H - 1/4 \quad (= 1/12 \text{ if } F \text{ is continuous}).$$

Then by Theorem 2.4, the statistic $\sqrt{N}\,\overline{\boldsymbol{Y}}. = \sqrt{N}\int H\,\mathrm{d}\widehat{\boldsymbol{F}}$ has asymptotically a multivariate normal distribution with mean $\boldsymbol{0}$ and covariance matrix $\boldsymbol{V}_a = N\sigma_H^2 \mathrm{diag}\{n_1^{-1}, \dots, n_a^{-1}\}$. Let $\boldsymbol{W}_a = \boldsymbol{V}_a^{-1}[\boldsymbol{I}_a - \boldsymbol{J}_a \boldsymbol{V}_a^{-1} \sigma_H^2]$ be the contrast matrix defined in (1.5) and let $\widehat{\boldsymbol{W}}_a = \widehat{\boldsymbol{V}}_a^{-1}[\boldsymbol{I}_a - \boldsymbol{J}_a \widehat{\boldsymbol{V}}_a^{-1} \widehat{\sigma}_H^2]$ where $\widehat{\boldsymbol{V}}_a = N\widehat{\sigma}_H^2 \mathrm{diag}\{n_1^{-1}, \dots, n_a^{-1}\}$ and $\widehat{\sigma}_H^2 = (N-a)^{-1} \sum_{i=1}^{a} \sum_{j=1}^{n_i} \left(R_{ij} - \overline{R}_{i\cdot}\right)^2$. In the same way as for the one-factor fixed design in Section 2.1.3, it follows that the quadratic form

$$Q_H = \frac{N-a}{\sum_{i=1}^{a} \sum_{j=1}^{n_i} \left(R_{ij} - \overline{R}_{i\cdot}\right)^2} \sum_{i=1}^{a} n_i \left(\overline{R}_{i\cdot} - \frac{N+1}{2} \right)^2 \tag{3.2}$$

has asymptotically a central χ_f^2-distribution with $f = a - 1$ under H_0^*.

The above derivation of the asymptotic distribution of Q_H under H_0^* in the general model is simple. Under the alternative H_1^*: $F^*(x, y) \neq F(x)F(y)$ however, additional assumptions or restrictions of the model are necessary to derive the asymptotic distribution of Q_H if a is fixed and $n_i \to \infty$.

In literature, nonparametric procedures for the one-way random effects model have been given for the linear model $X_{ij} = \mu + \theta Y_i + \varepsilon_{ij}$, $i = 1, \dots, a$, $j = 1, \dots, n_i$, where μ is the overall mean, Y_i and ε_{ij} are mutually independent random variables and the constant $\theta \geqslant 0$ represents the degree of the random treatment effect. The hypothesis of 'no random effect' is expressed in this model as H_0^r: $\theta = 0$. Greenberg (1964) considered the case where Y_i is normally distributed and ε_{ij} has an arbitrary continuous distribution. Govindarajulu and Deshpande (1972) and Govindarajulu (1975a) relaxed the assumption that Y_i is normally distributed and derived locally most powerful tests for H_0^r. Shetty and Govindarajulu (1988) studied the asymptotic distribution of these statistics under local alternatives and the power properties were later studied by Clemmens and Govindarajulu (1990). Shirahata (1985) derived the asymptotic distribution of the Kruskal–Wallis statistic in the random effects model when a is fixed and $n_i \to \infty$. The case of n_i fixed and $a \to \infty$ was earlier considered by Shirahata (1982) where the statistic is based on some measures of intraclass correlation.

A two-way layout with random effects without interaction $X_{ij} = \mu + Y_i + \theta Z_j + \varepsilon_{ij}$, $i = 1, \dots, n \to \infty$, $j = 1, \dots, b$, is considered by Shirahata (1985) where the asymptotic distribution of the Friedman statistic is considered.

The nonparametric treatment of the random model needs further research to develop a unified theory. On the one hand, procedures for two- and higher-way layouts including interaction should be investigated for linear models. On the other hand, it is necessary to examine to what extent the assumption of the additivity of the random effects can be relaxed.

4. Mixed models

4.1. Background and examples

In a mixed model, randomly chosen subjects are observed repeatedly under the same or under different treatments. Such designs occur in many biological experiments and medical or psychological studies. The subjects are the levels of the random factor(s) and the subject effects are regarded as unobservable random variables. Here, we shall consider four different designs.

(1) **Two-factor mixed models:**

- (a) random factor and fixed factor crossed,
- (b) random factor nested under the fixed factor.

(2) **Three-factor mixed models with two factors fixed:**

- (a) repeated measurements on one fixed factor, the random factor is crossed with the fixed factor B and is nested under the fixed factor A,
- (b) repeated measurements on both crossed fixed factors.

The hypotheses in these designs are formulated in the same way as for the fixed models. The statistics are based on consistent estimators for the generalized means. These estimators are vectors the components of which are linear rank statistics.

We distinguish two models:

(I) The repeated measurements model where in the case of 'no treatment effect' the common distribution function of the observed random variables on subject i is invariant under the numbering of the treatment levels. This means that the random variables within one subject are interchangeable and the common distribution function is compound symmetric. Note that in the linear mixed model with independent random effects, the compound symmetry of the common distribution functions of the subjects is preserved under the alternative. In general models however (to be stated in the following sections), it seems to be unrealistic to assume that compound symmetry is preserved if treatment effects are present.

(II) The multivariate model allows arbitrary dependencies between the observed random variables within one subject. This is typically the case with longitudinal data or inhomogeneous materials. A multivariate model is also assumed if a treatment effect is present in a repeated measurements model. We do not consider special patterns of dependencies such as an autocorrelation structure, for example.

Example for $(1, a)$/compound symmetry. A cell culture, some tissue or a blood sample is split into $j = 1, \dots, b$ homogeneous parts and each part receives one of the b levels of the treatment. This experiment is repeated for $i = 1, \dots, n$ independent randomly chosen subjects (cultures, tissues, blood samples etc.). For this design, we will also consider the case where $m_{ij} \geqslant 1$ repeated observations are taken for each subject i and for each treatment j.

Example for $(1, a)$/multivariate model. Growth curves of subjects or any observations taken at different (closely distant) timepoints such that observations taken at close timepoints are 'more dependent' than observations from timepoints which are far spaced.

Example for $(1, b)$. The level i of the fixed treatment is applied to n_i subjects which are observed repeatedly under the same treatment.

Example for $(2, a)$/compound symmetry. In the experiment described in example $(1, a)$, there exist $i = 1, \dots, a$ different groups of subjects where all the homogeneous parts of the subjects in group i are treated with level i of the fixed factor A and the homogeneous part j of a subject within group i is treated with treatment j. The subjects are nested under factor A while they are crossed with factor B. Therefore, this design is sometimes called 'partially nested'.

Example for $(2, a)$/multivariate model. Two groups of subjects are given different treatments $(i = 1, 2)$ and the outcome X_{ijk} is observed at b fixed timepoints $j = 1, \dots, b$ for the subjects $k = 1, \dots, n_i$ where the observations for one subject may be arbitrary dependent.

Example for $(2, b)$/multivariate model. Each subject receives all $i = 1, \dots, a$ levels of the fixed treatment A and the outcome is observed at $j = 1, \dots, b$ fixed timepoints for each subject.

Some historical remarks. Nonparametric hypotheses and tests for the mixed model have already been considered by Sen (1967), Koch and Sen (1968) and by Koch (1969, 1970). In the latter article, a complex split-plot design is considered and different types of ranks are given to aligned and original observations and the asymptotic distributions of univariate and multivariate rank statistics are given. Mainly joint hypotheses in the linear model are considered, i.e., main effects and certain interactions are tested together. However, no unified theory for the derivation of rank tests in mixed models is presented. Moreover, some of the statistics are not pure rank statistics rather than aligned rank statistics and therefore they are restricted to linear models.

For the simple mixed model with two treatments for paired observations, rank tests using overall ranks on the original observations have been considered by Hollander et al. (1974) and Govindarajulu (1975b). In the former paper, the robustness of the Wilcoxon–Mann–Whitney statistic with respect to deviation from independence is studied. In the latter paper, a rank statistic is derived and it is indicated how the unknown variance may be estimated. Lam and Longnecker (1983) derived an estimator for the unknown variance based on Spearman's rank correlation. Brunner and Neumann (1982, 1984, 1986a, b) derived the asymptotic distribution of rank statistics in two-factor mixed models with an equal number of replications and applied the results to different mixed models. The asymptotic variances and covariance matrices of the statistics were estimated using ranks over all the observations and ranks within the treatments. Rank tests for the mixed model with $m = 1$ replication were considered by Kepner and Robinson (1988). Thompson (1990) considered the asymptotic distribution of linear rank statistics in mixed models with an equal number of replications and applied the results to different balanced repeated measurements designs (Thompson and Ammann, 1989, 1990; Thompson, 1991a) for joint hypotheses where main effects and interactions are tested together. The asymptotic distribution of linear rank statistics for vectors of different lengths was derived by Brunner and Denker (1994) and the results were applied to rank tests for nonparametric hypotheses in unbalanced mixed models. Akritas (1991, 1993) considered rank tests for joint hypotheses in the linear model for balanced repeated measurements designs and in an unbalanced design.

Nonparametric hypotheses in mixed models based on the generalized mean vectors have been considered by Brunner and Neumann (1986a, b) for paired observations and in 2×2 designs and are further developed for $2\times b$ designs by Boos and Brownie (1992) and Brunner et al. (1995). We note that in the last two papers no continuity of the distribution functions has been assumed. The results of Brunner and Denker (1994) which were derived under the assumption of continuous distribution functions are generalized to the case of ties by Munzel (1994) where the score function is assumed to have a bounded second derivative. For simplicity, we consider here only the Wilcoxon scores. In mixed models with two factors fixed, nonparametric hypotheses based on the distribution functions have been introduced by Akritas and Arnold (1994) and were used to provide a unified approach to rank test for mixed models by Akritas and Brunner (1995). In the next sections we discuss rank procedures for these hypotheses.

4.2. General asymptotic results

Here we give the general asymptotic results for mixed models and we will show how to apply these results to the different designs.

The general mixed model can be formulated by independent random vectors

$$\boldsymbol{X}_{ik} = \left(\boldsymbol{X}'_{i1k}, \ldots, \boldsymbol{X}'_{ick}\right)', \quad i = 1, \ldots, r \text{ and } k = 1, \ldots, n_i, \tag{4.1}$$

where $\boldsymbol{X}_{ijk} = (X_{ijk1}, \ldots, X_{ijkm_{ijk}})'$, $j = 1, \ldots, c$, and $X_{ijks} \sim F_{ij}$, $k = 1, \ldots, n_i$ and $s = 1, \ldots, m_{ijk}$.

The row-factor with r levels is applied to all c parts of one subject and the subjects are nested under this factor. For each level i, there are n_i independent subjects (replications). If more than one factor is applied to the subjects, then the r levels may be regarded as a lexicographic ordering of all factor level combinations of the factors.

The column-factor with c levels is applied to all subjects. However, the level j of this factor is applied only to the jth part of the subject (which is split into c homogeneous parts). This factor is crossed with the subjects. If more than one factor is applied to each subject, then the c levels may be regarded as a lexicographic ordering of all factor level combinations of the factors.

Examples

In the matched pairs design, we have n independent random vectors $\boldsymbol{X}_i = (X_{i1}, X_{i2})'$ where $X_{ij} \sim F_j$, $i = 1, \ldots, n$ and $j = 1, 2$. This design is derived from the general mixed model (4.1) by letting $r = 1$, $n_1 = n$, $c = 2$ and $m_{ijk} = 1$. The one-factor hierarchical design is a special case of (4.1) if $r = a$, $c = 1$, $m_{ijk} = m_{ik}$ and $X_{ijks} = X_{iks} \sim F_i$, $i = 1, \ldots, a$, $k = 1, \ldots, n_i$ and $s = 1, \ldots, m_{ik}$. For the one-factor block design with b treatment levels and with n blocks, we choose $r = 1$, $c = b$, $m_{ijk} = 1$, $X_{ijks} = X_{jk} \sim F_j$, $j = 1, \ldots, b$ and $k = 1, \ldots, n$. The $a \times b$ split-plot design is derived from (4.1) by letting $r = a$, $c = b$, $m_{ijk} = 1$, $X_{ijks} = X_{ijk} \sim F_{ij}$, $i = 1, \ldots, a$, $j = 1, \ldots, b$ and $k = 1, \ldots, n_i$. For the two-factor block design with n blocks and where factor A has a levels and factor B has b levels, we choose $r = 1$, $n_1 = n$, $c = ab$ and $m_{ijk} = 1$. The index j is split into $u = 1, \ldots, a$ and $v = 1, \ldots, b$. Then $X_{uvk} \sim F_{uv}$, $k = 1, \ldots, n$.

To state the asymptotic results, we introduce some notations. The vector of the distribution functions is denoted by $\boldsymbol{F} = (F_{11}, \ldots, F_{1c}, \ldots, F_{r1}, \ldots, F_{rc})'$ and we define $\widetilde{\boldsymbol{F}} = (\widetilde{F}_{11}, \ldots, \widetilde{F}_{rc})'$ where $\widetilde{F}_{ij} = n_i^{-1} \sum_{k=1}^{n_i} \widehat{F}_{ijk}$ and $\widehat{F}_{ijk}(x) = m_{ijk}^{-1} \sum_{s=1}^{m_{ijk}} c(x - X_{ijks})$ is the empirical distribution function within the cell (i, j, k). Let

$$H = N^{-1} \sum_{i=1}^{r} \sum_{j=1}^{c} \sum_{k=1}^{n_i} m_{ijk} F_{ij}$$

and

$$\widehat{H}(x) = N^{-1} \sum_{i=1}^{r} \sum_{j=1}^{c} \sum_{k=1}^{n_i} m_{ijk} \widehat{F}_{ijk}(x)$$

where $N = \sum_i \sum_j \sum_k m_{ijk}$. The vector of the generalized means $\boldsymbol{p} = \int H \, \mathrm{d}\boldsymbol{F}$ is estimated by $\widetilde{\boldsymbol{p}} = \int \widehat{H} \, \mathrm{d}\widetilde{\boldsymbol{F}}$.

PROPOSITION 4.1. *Let $\boldsymbol{X}_{ik} = (\boldsymbol{X}'_{i1k}, \ldots, \boldsymbol{X}'_{ick})'$ be independent random vectors as defined in* (4.1) *and assume that the number of replications m_{ijk} for a subject k is uniformly bounded, i.e., $m_{ijk} \leqslant M < \infty$. Then $\widetilde{\mathbf{p}} = \int \widehat{H} \, \mathrm{d}\widetilde{\boldsymbol{F}}$ is asymptotically unbiased and consistent for $\mathbf{p} = \int H \, \mathrm{d}\boldsymbol{F}$ in the sense that $\widetilde{\mathbf{p}} - \mathbf{p} \xrightarrow{p} \mathbf{0}$ as* $\min n_i \to \infty$.

The next result is analogous to Theorem 2.3 in Section 2.1.2.

THEOREM 4.2. *Let $\boldsymbol{X}_{ik}$ be as in Proposition* 4.1 *and let $Y_{ijks} = H(X_{ijks})$, $\overline{Y}_{ijk\cdot} = m_{ijk}^{-1} \sum_{s=1}^{m_{ijk}} Y_{ijks}$ and assume that $m_{ijk} \leqslant M < \infty$. If* $\min n_i \to \infty$, *then $\sqrt{N} \int \widehat{H} \, \mathrm{d}(\widetilde{\boldsymbol{F}} - \boldsymbol{F}) \doteq \sqrt{N} \int H \, \mathrm{d}(\widetilde{\boldsymbol{F}} - \boldsymbol{F}) = \sqrt{N}(\widetilde{\mathbf{Y}}_{\cdot\cdot} - \mathbf{p})$ where $\widetilde{\mathbf{Y}}_{\cdot\cdot} = (\widetilde{\mathbf{Y}}'_{1\cdot\cdot}, \ldots, \widetilde{\mathbf{Y}}'_{r\cdot\cdot})'$, $\widetilde{\mathbf{Y}}_{i\cdot\cdot} = (\widetilde{Y}_{i1\cdot\cdot}, \ldots, \widetilde{Y}_{ic\cdot\cdot})'$ and $\widetilde{Y}_{ij\cdot\cdot} = n_i^{-1} \sum_{k=1}^{n_i} \overline{Y}_{ijk\cdot}$.*

(Note that the vectors $\widetilde{\boldsymbol{Y}}_{i\cdot\cdot}$ are independent.) Denote $\boldsymbol{V}_i = \mathrm{Cov}(\sqrt{N}\ \widetilde{\boldsymbol{Y}}_{i\cdot\cdot})$. Then

$$\boldsymbol{V} = \mathrm{Cov}\big(\sqrt{N}\ \widetilde{\boldsymbol{Y}}_{\cdot\cdot}\big) = \bigoplus_{i=1}^{r} \boldsymbol{V}_i. \tag{4.2}$$

The covariance matrix $\boldsymbol{V}$ can be estimated by the ranks R_{ijks}.

THEOREM 4.3. *Let* $\overline{R}_{ijk\cdot} = m_{ijk}^{-1} \sum_{s=1}^{m_{ijk}} R_{ijks}$, $\overline{\boldsymbol{R}}_{ik\cdot} = (\overline{R}_{i1k\cdot}, \ldots, \overline{R}_{ick\cdot})'$ *and* $\widetilde{\boldsymbol{R}}_{\cdot\cdot} = (\widetilde{R}_{i1\cdot\cdot}, \ldots, \widetilde{R}_{ic\cdot\cdot})'$ *where* $\widetilde{R}_{ij\cdot\cdot} = n_i^{-1} \sum_{k=1}^{n_i} \overline{R}_{ijk\cdot}$ *and* R_{ijks} *is the rank of* X_{ijks} *among all the* N *observations. Let*

$$\widehat{\boldsymbol{V}} = \bigoplus_{i=1}^{r} \widehat{\boldsymbol{V}}_i = \bigoplus_{i=1}^{r} \frac{1}{Nn_i(n_i-1)} \sum_{k=1}^{n_i} \big(\overline{\boldsymbol{R}}_{ik\cdot} - \widetilde{\boldsymbol{R}}_{i\cdot\cdot}\big)\big(\overline{\boldsymbol{R}}_{ik\cdot} - \widetilde{\boldsymbol{R}}_{i\cdot\cdot}\big)'. \tag{4.3}$$

Then, under the assumptions of Theorem 4.2, $\|\widehat{\boldsymbol{V}} - \boldsymbol{V}\| \xrightarrow{p} 0$ *as* $\min n_i \to \infty$.

To state the results for the compound symmetry model where the covariance matrix has a simpler form, we need some further notations. Let $\boldsymbol{Y}_{ik} = (Y_{i1k1}, \ldots, Y_{i1km_{i1k}}, \ldots, Y_{ick1}, \ldots, Y_{ickm_{ick}})'$ and $\overline{\boldsymbol{Y}}_{ik\cdot} = (\overline{Y}_{i1k\cdot}, \ldots, \overline{Y}_{ick\cdot})'$ where $\overline{Y}_{ijk\cdot} = m_{ijk}^{-1} \sum_{s=1}^{m_{ijk}} Y_{ijks}$ and let $M_{ik} = \sum_{j=1}^{c} m_{ijk}$. In the compound symmetry model it follows under $H_0^F\colon F_{i1} = \cdots = F_{ic}$ that the variances $\sigma_i^2 = \mathrm{Var}(Y_{ijks})$, $i = 1, \ldots, r$, do not depend on j, k or s. It is assumed that the covariances $c_i^* = \mathrm{Cov}(Y_{ijks}, Y_{ij'ks'})$, $j \neq j' = 1, \ldots, c$, do not depend on j, j', k, s or s' and $c_i^{**} = \mathrm{Cov}(Y_{ijks}, Y_{ijks'})$, $i = 1, \ldots, r$ do not depend on j, k, s or s'. Thus under $H_0^F\colon F_{i1} = \cdots = F_{ic}$,

$$\mathrm{Cov}\,(\boldsymbol{Y}_{ik}) = \boldsymbol{\Sigma}_{ik} = \bigoplus_{j=1}^{c} \big[(\sigma_i^2 - c_i^{**})\boldsymbol{I}_{m_{ijk}} + (c_i^{**} - c_i^*)\boldsymbol{J}_{m_{ijk}}\big] + c_i^* \boldsymbol{J}_{M_{ik}},$$

$$\boldsymbol{V}_i = \frac{N}{n_i}\big(\mathrm{diag}\{\tau_{i1}, \ldots, \tau_{ic}\} + c_i^* \boldsymbol{J}_c\big) = \frac{N}{n_i}\big(\boldsymbol{D}_i + c_i^* \boldsymbol{J}_c\big)$$

where

$$\tau_{ijk} = \frac{1}{m_{ijk}}\big(\sigma_i^2 + (m_{ijk}-1)c_i^{**}\big) - c_i^* \quad \text{and} \quad \tau_{ij} = \frac{1}{n_i}\sum_{k=1}^{n_i} \tau_{ijk}. \tag{4.4}$$

Compound symmetry will only be assumed for hypotheses regarding the sub-plot factor, i.e., for hypotheses that can be written as $H_0^F\colon (\boldsymbol{I}_r \otimes \boldsymbol{C}_c)\,\boldsymbol{F} = \boldsymbol{0}$ where $\boldsymbol{C}_c$ is a suitable contrast matrix for the sub-plot factor. Thus we need only to estimate $\boldsymbol{D}_i$ since

$$\big(\boldsymbol{I}_r \otimes \boldsymbol{C}_c\big)\ \boldsymbol{V}\ \big(\boldsymbol{I}_r \otimes \boldsymbol{C}_c'\big) = \bigoplus_{i=1}^{r} \boldsymbol{C}_c \boldsymbol{D}_i \boldsymbol{C}_c'.$$

THEOREM 4.4. *Let* R_{ijks} *be the rank of* X_{ijks} *as in Theorem* 4.3, *let* $\overline{R}_{i\cdot k\cdot} = M_{ik}^{-1}\sum_{j=1}^{c}\sum_{s=1}^{m_{ijk}} R_{ijks}$ *and let* $\boldsymbol{D}_i = \operatorname{diag}\{\tau_{i1},\ldots,\tau_{ic}\}$ *where* τ_{ij} *is given in* (4.4) *and assume that* $|\boldsymbol{D}_i| \neq 0$. *Let* $\widehat{\boldsymbol{D}}_i$ *denote the matrix corresponding to* $\boldsymbol{D}_i$ *where* τ_{ij} *is replaced by* $\widehat{\tau}_{ij} = n_i^{-1}\sum_{k=1}^{n_i}\widehat{\tau}_{ijk}$ *and* $\widehat{\tau}_{ijk}$ *is defined below. If* $m_{ijk} \leqslant M < \infty$ *and if* $0 < \lambda_0 \leqslant n_i/N \leqslant 1-\lambda_0 < 1$, $i = 1,\ldots,r$, *then in the compound symmetry model under* H_0^F: $F_{i1} = \cdots = F_{ic}$, $\|\widehat{\boldsymbol{D}}_i - \boldsymbol{D}_i\| \xrightarrow{p} 0$ *as* $\min n_i \to \infty$. *If* m_{ijk} *is not equivalent to a constant, then*

$$\widehat{\tau}_{ijk} = \left[\frac{1}{m_{ijk}} - \frac{M_{ik}(c-1)}{M_{ik}^2 - \sum_{l=1}^{c} m_{ilk}^2}\right]\frac{S_{i1k}^2}{N^2\,(M_{ik}-c)} + \frac{M_{ik}S_{i2k}^2}{N^2\left(M_{ik}^2 - \sum_{l=1}^{c} m_{ilk}^2\right)}$$

where

$$S_{i1k}^2 = \sum_{j=1}^{c}\sum_{s=1}^{m_{ijk}}\left(R_{ijks} - \overline{R}_{ijk\cdot}\right)^2 \quad \text{and} \quad S_{i2k}^2 = \sum_{j=1}^{c} m_{ijk}\left(\overline{R}_{ijk\cdot} - \overline{R}_{i\cdot k\cdot}\right)^2.$$

If $m_{ijk} \equiv m$, *then*

$$\widehat{\tau}_{ijk} \equiv \widehat{\tau}_{ik} = \frac{1}{N^2(c-1)}\sum_{l=1}^{c}\left(\overline{R}_{ilk\cdot} - \overline{R}_{i\cdot k\cdot}\right)^2.$$

Finally, we give the asymptotic distribution of $\sqrt{N}\boldsymbol{C}\widetilde{\boldsymbol{p}}$ under the hypothesis $H_0^F(\boldsymbol{C})$: $\boldsymbol{C}\boldsymbol{F} = \boldsymbol{0}$.

THEOREM 4.5. *Let* $\boldsymbol{X}_{ik}$ *be as in Proposition* 4.1, $\widetilde{\boldsymbol{Y}}_{i\cdot\cdot}$ *as in Theorem* 4.2, $\boldsymbol{V}_i$ *as defined below Theorem* 4.2. *Let* $\boldsymbol{C}$ *be a contrast matrix and assume that* $|\boldsymbol{V}_i| \geqslant k_0 > 0$ *and* $|\widehat{\boldsymbol{V}}_i| \geqslant k_0 > 0$, $i = 1,\ldots,r$.

If

(1) $m_{ijk} \leqslant M < \infty$,

(2) $0 < \lambda_0 \leqslant n_i/N \leqslant 1-\lambda_0 < 1$ *and*

(3) $\min n_i \to \infty$, $i = 1,\ldots,r$, *then under the hypothesis* $H_0^F(\boldsymbol{C})$: $\boldsymbol{C}\boldsymbol{F} = \boldsymbol{0}$,

  (i) *the statistic* $\sqrt{N}\boldsymbol{C}\widetilde{\boldsymbol{p}} = \sqrt{N}\boldsymbol{C}\int \widehat{H}\,\mathrm{d}\widetilde{\boldsymbol{F}}$ *has asymptotically a multivariate normal distribution with mean* $\boldsymbol{0}$ *and covariance matrix* $\boldsymbol{C}\boldsymbol{V}\boldsymbol{C}'$,

  (ii) *the quadratic form* $Q(\boldsymbol{C}) = N\widetilde{\boldsymbol{p}}'\boldsymbol{C}'[\boldsymbol{C}\boldsymbol{V}\boldsymbol{C}']^{-}\boldsymbol{C}\widetilde{\boldsymbol{p}}$ *has asymptotically a central* χ_f^2*-distribution with* $f = \operatorname{rank}(\boldsymbol{C})$ *and where* $[\boldsymbol{C}\boldsymbol{V}\boldsymbol{C}']^{-}$ *denotes a generalized inverse of* $[\boldsymbol{C}\boldsymbol{V}\boldsymbol{C}']$.

  (iii) *If* $\boldsymbol{C}$ *is of full row rank, then* $\widehat{Q}(\boldsymbol{C}) = N\widetilde{\boldsymbol{p}}'\boldsymbol{C}'[\boldsymbol{C}\widehat{\boldsymbol{V}}\boldsymbol{C}']^{-1}\boldsymbol{C}\widetilde{\boldsymbol{p}}$ *has asymptotically a central* χ_f^2*-distribution with* $f = \operatorname{rank}(\boldsymbol{C})$ *where* $\widehat{\boldsymbol{V}}$ *is given in Theorem* 4.3.

(iv) *Let* $\boldsymbol{W} = \boldsymbol{V}^{-1}[\boldsymbol{I} - \boldsymbol{J}\boldsymbol{V}^{-1}/\boldsymbol{1}'\boldsymbol{V}^{-1}\boldsymbol{1}]$ *and let* $\widehat{\boldsymbol{W}} = \widehat{\boldsymbol{V}}^{-1}[\boldsymbol{I} - \boldsymbol{J}\widehat{\boldsymbol{V}}^{-1}/\boldsymbol{1}'\widehat{\boldsymbol{V}}^{-1}\boldsymbol{1}]$ *where* $\widehat{\boldsymbol{V}}$ *is given in Theorem* 4.3. *Then* $\widehat{Q}(\boldsymbol{W}) = N\widetilde{\boldsymbol{p}}'\widehat{\boldsymbol{W}}\widetilde{\boldsymbol{p}}$ *has asymptotically a central* χ_f^2*-distribution with* $f = \operatorname{rank}(\boldsymbol{W})$.

The results for the heteroscedastic models for $c = 2$ treatments are given separately for the different models and hypotheses. In the next section, the general results given here will be applied to the two-factor mixed model with one factor fixed.

4.3. *Two-factor mixed models*

4.3.1. *Cross-classified designs*

4.3.1.1. *Models and hypotheses*

Models. In a cross-classified mixed model, the random variables X_{ijk} are observed on the ith randomly chosen subject (or block), $i = 1, \ldots, n$ which is repeatedly observed (or measured) under treatment $j = 1, \ldots, b$ and $k = 1, \ldots, m_{ij}$ repeated observations are made on the same subject i under treatment j. In the classical linear model theory, this is described as

$$X_{ijk} = \mu_j + A_i + W_{ij} + \varepsilon_{ijk} \tag{4.5}$$

where $\mu_j = \mathrm{E}(X_{ijk})$, the A_i's are i.i.d. random variables with $\mathrm{E}(A_1) = 0$ and $\operatorname{Var}(A_1) = \sigma_A^2$. W_{ij} are i.i.d. random variables independent of A_i with $\mathrm{E}(W_{11}) = 0$ and $\operatorname{Var}(W_{11}) = \sigma_{AB}^2$; ε_{ijk} are i.i.d. $N(0, \sigma^2)$ random variables independent of A_i and W_{ij}, $i = 1, \ldots, n$, $j = 1, \ldots, b$ and $k = 1, \ldots, m_{ij}$. (For a discussion of different assumptions on A_i and W_{ij}, see Hocking (1973).)

Let $\boldsymbol{X}_i = (X_{i11}, \ldots, X_{ibm_{ib}})'$ be the vector of observations for block i. Then

$$\operatorname{Cov}(\boldsymbol{X}_i) = \boldsymbol{\Sigma}_i = \bigoplus_{j=1}^{b} \left[(\sigma_x^2 - c_x^{**})\boldsymbol{I}_{m_{ij}} + (c_x^{**} - c_x^*)\boldsymbol{J}_{m_{ij}}\right] + c_x^*\boldsymbol{J}_{M_i} \tag{4.6}$$

where $\sigma_x^2 = \operatorname{Var}(X_{111})$, $c_x^* = \operatorname{Cov}(X_{111}, X_{121})$, $c_x^{**} = \operatorname{Cov}(X_{111}, X_{112})$ and $M_i = \sum_{j=1}^{b} m_{ij}$.

This is the usual linear block (or repeated measurements) model which is appropriate if a subject is split into b homogeneous parts and each part is observed repeatedly m_{ij} times. In a multivariate model where, e.g., the observations are taken at different time points (not necessarily equidistant), $\boldsymbol{\Sigma}_i$ is an arbitrary positive definite covariance matrix. We note that the terminology is not unique in literature for these two models. We will use the terms repeated measurements design if $\boldsymbol{\Sigma}_i$ has the compound symmetry form given in (4.6) and multivariate design if this is not the case.

In this setup, the random variables X_{ijk} and $X_{i'jk'}$ are identically distributed according to a distribution function $F_j(x)$, $j = 1, \ldots, b$, and they are assumed to be independent for $i \neq i'$ but they may be dependent for $i = i'$ since they are observed on the same (random) subject and the random variables $X_{ijk} \sim F_j(x)$ and

$X_{ij'k'} \sim F_{j'}(x)$ may also be dependent. In a general model, we need also the bivariate common distribution functions of two random variables within treatment j (denoted by $F_j^{**}(x,y)$) and of two random variables between two treatments j and j' (denoted by $F_{jj'}^{*}(x,y)$). Thus, the general two-factor mixed model can be described by independent random vectors

$$\begin{aligned} \boldsymbol{X}_i &= (\boldsymbol{X}_{i1}', \ldots, \boldsymbol{X}_{ib}')', \ i = 1, \ldots, n, \ \text{with common distribution} \\ &\qquad\qquad\qquad\qquad\qquad \text{functions } G_i(\boldsymbol{x}), \qquad (4.7) \\ \boldsymbol{X}_{ij} &= (X_{ij1}, \ldots, X_{ijm_{ij}})' \quad X_{ijk} \sim F_j(x), \quad i = 1, \ldots, n; \ k = 1, \ldots, m_{ij}, \\ &\quad (X_{ijk}, X_{ijk'})' \sim F_j^{**}(x,y), \quad i = 1, \ldots, n; \ k \neq k' = 1, \ldots, m_{ij}, \\ &\quad (X_{ij1}, X_{ij'1})' \sim F_{jj'}^{*}(x,y), \quad i = 1, \ldots, n; \ j \neq j' = 1, \ldots, b. \end{aligned}$$

It is reasonable to assume in this general model the compound symmetry if a subject is split into homogeneous parts and if there is no treatment effect. This means that the parts of each subject are 'interchangeable' under the hypothesis. However under a treatment effect, the compound symmetry structure may not be preserved in the general model. The property of interchangeable parts of the subjects is reflected by the interchangeability of the random variables X_{ijk} and $X_{ij'k'}$ for $j \neq j' = 1, \ldots, b$ and $\forall k, k'$ under the hypothesis of no treatment effect. The random variables X_{ijk} and $X_{ijk'}$, $k, k' = 1, \ldots, m_{ij}$, are always interchangeable since they describe replications of the same experiment under the same treatment. Thus it follows for this model that $\mathrm{Var}(X_{1j1}) = \sigma_{x,j}^2 = \sigma_x^2$, $F_j^{**} = F^{**}$, and $F_{jj'}^{*} = F^{*}$, $j, j' = 1, \ldots, b$. For the multivariate model, no special assumptions on the bivariate distribution functions are made. Treatment effects in the general model (4.7) are described by the generalized means $p_j = \int H \, \mathrm{d}F_j$, $j = 1, \ldots, b$, where $H = N^{-1} \sum_{j=1}^{b} N_j F_j$ and $N = \sum_{j=1}^{b} N_j$, $N_j = \sum_{i=1}^{n} m_{ij}$.

Hypotheses. We are mainly interested in analysing the fixed treatment effect which is defined for the linear model as in the one-factor fixed model. The hypothesis of no treatment effect is written as H_0^{μ}: $\boldsymbol{P}_b \boldsymbol{\mu} = \boldsymbol{0}$ where $\boldsymbol{\mu} = (\mu_1, \ldots, \mu_b)'$. In the general model (4.7), we consider two hypotheses. For the case of equal cell frequencies ($m_{ij} \equiv m$), the common distribution function $G_i(\boldsymbol{x}_i) = G(\boldsymbol{x}_i)$ is assumed to be independent of i and the hypothesis is formulated as H_0^{π}: $G(\boldsymbol{x}_i) = G(\pi(\boldsymbol{x}_i))$ where $\pi(\boldsymbol{x}_i) = (\boldsymbol{x}_{i\pi_1}', \ldots, \boldsymbol{x}_{i\pi_b}')'$ and $\pi_1, \ldots, \pi_b$ is any permutation of the first b positive integers. 'No treatment effect' means that the outcome of the experiment is independent of the numbering of the treatments. This hypothesis is used for small samples where the permutation distribution of the test statistic is computed. The other hypothesis H_0^F: $\boldsymbol{P}_b \boldsymbol{F} = \boldsymbol{0}$ is the same as in the one-factor fixed model and is only related to the one-dimensional marginal distributions. Therefore, this hypothesis is appropriate for the compound symmetry model as well as for the multivariate model. The relations between the hypotheses are stated in the next Proposition.

PROPOSITION 4.6.

(1) *In the general two-factor mixed model* (4.7), $H_0^{\pi} \Rightarrow H_0^F$.

(2) *In the linear two-factor mixed model* (4.5), $H_0^{\pi} \Leftrightarrow H_0^F \Leftrightarrow H_0^{\mu}$.

PROOF. (1) is evident and (2) follows by the additivity, the independence and the identical distribution functions of the random variables in (4.5). □

For $b \geqslant 2$ treatments, we derive tests for H_0^F in the compound symmetry model as well as in the multivariate model. For the case of $b = 2$ treatments, which is the most important one for applications, tests for the nonparametric hypotheses H_0^π, H_0^F and H_0^p: $\int F_1 \, \mathrm{d}F_2 = 1/2$ are given where an unequal number m_{ij} of replications is allowed for H_0^F and H_0^p. We need only the assumption of the independence of the vectors $\boldsymbol{X}_i = (\boldsymbol{X}_{i1}', \boldsymbol{X}_{i2}')'$. The so-called 'matched-pairs-design' $(m = 1)$ which is a special case of this model, is considered separately.

4.3.1.2. Tests for $b \geqslant 2$ samples

Notations. The statistics are based on a consistent estimator of the vector of the generalized means $\boldsymbol{p} = \int H \, \mathrm{d}\boldsymbol{F}$ where $\boldsymbol{F} = (F_1, \ldots, F_b)'$, $H = N^{-1} \sum_{j=1}^b N_j F_j$ and $N = \sum_{j=1}^b N_j = \sum_{j=1}^b \sum_{i=1}^n m_{ij}$. The vector $\boldsymbol{p}$ is estimated by an unweighted mean of the cell means. Let $\widetilde{\boldsymbol{F}} = (\widetilde{F}_1, \ldots, \widetilde{F}_b)'$ and

$$\widetilde{F}_j(x) = \frac{1}{n} \sum_{i=1}^{n} \frac{1}{m_{ij}} \sum_{k=1}^{m_{ij}} c(x - X_{ijk}),$$

$$\widehat{H}(x) = \frac{1}{N} \sum_{j=1}^{b} N_j \widetilde{F}_j(x) = \frac{1}{N} \sum_{i=1}^{n} \sum_{j=1}^{b} \sum_{k=1}^{m_{ij}} c(x - X_{ijk}).$$

Then $\widetilde{\boldsymbol{p}} = (\widetilde{p}_1, \ldots, \widetilde{p}_b)' = \int \widehat{H} \, \mathrm{d}\widetilde{\boldsymbol{F}}$ is consistent for $\boldsymbol{p}$ (see Proposition 4.1). The estimators for the components $\widetilde{p}_j$ are computed from the ranks R_{ijk} of X_{ijk} among all the N random variables

$$\begin{aligned} \widetilde{p}_j = \int \widehat{H} \, \mathrm{d}\widetilde{F}_j &= \frac{1}{n} \sum_{i=1}^{n} \frac{1}{m_{ij}} \sum_{k=1}^{m_{ij}} \frac{1}{N} \left(R_{ijk} - \frac{1}{2} \right) \\ &= \frac{1}{N} \left(\widetilde{R}_{\cdot j \cdot} - \frac{1}{2} \right). \end{aligned} \tag{4.8}$$

Denote by $\boldsymbol{R}_i = (\overline{R}_{i1\cdot}, \ldots, \overline{R}_{ib\cdot})'$, $i = 1, \ldots, n$, the vectors of the rank means for subject i where $\overline{R}_{ij\cdot} = m_{ij}^{-1} \sum_{k=1}^{m_{ij}} R_{ijk}$. Let further $\widetilde{\boldsymbol{R}}_{\cdot} = n^{-1} \sum_{i=1}^n \boldsymbol{R}_i$ denote the unweighted mean of the vectors $\boldsymbol{R}_i$.

It follows from Theorem 4.2 that the statistics $\sqrt{n} \int \widehat{H} \, \mathrm{d}(\widetilde{\boldsymbol{F}} - \boldsymbol{F})$ and $\sqrt{n} \int H \, \mathrm{d}(\widetilde{\boldsymbol{F}} - \boldsymbol{F}) = \sqrt{n}(\widetilde{\boldsymbol{Y}}_{\cdot} - \boldsymbol{p})$ are asymptotically equivalent. Moreover, under H_0^F: $\boldsymbol{C}\boldsymbol{F} = \boldsymbol{0}$, it follows that $\sqrt{n}\boldsymbol{C}\widetilde{\boldsymbol{p}}$ and $\sqrt{n}\boldsymbol{C}\widetilde{\boldsymbol{Y}}_{\cdot}$ are asymptotically equivalent. Here $\widetilde{\boldsymbol{Y}}_{\cdot} = n^{-1} \sum_{i=1}^n \boldsymbol{Y}_i$ is the mean of $\boldsymbol{Y}_i = (\overline{Y}_{i1\cdot}, \ldots, \overline{Y}_{ib\cdot})'$ where $\overline{Y}_{ij\cdot} = m_{ij}^{-1} \sum_{k=1}^{m_{ij}} Y_{ijk}$ and $Y_{ijk} = H(X_{ijk})$. Note that the result of Theorem 4.2 remains true if the statistic is multiplied by $\sqrt{n}$ instead of $\sqrt{N}$. Let $\boldsymbol{S}_i = \operatorname{Cov}(\boldsymbol{Y}_i)$. Then $\boldsymbol{V} = \operatorname{Cov}(\sqrt{n}\,\widetilde{\boldsymbol{Y}}_\cdot) = n^{-1}\sum_{i=1}^{n} \boldsymbol{S}_i$ since the $\boldsymbol{Y}_i$ are independent random vectors. A consistent estimate

$$\widehat{\boldsymbol{V}}_n = \frac{1}{N^2(n-1)} \sum_{i=1}^{n} \left(\boldsymbol{R}_i - \widetilde{\boldsymbol{R}}_\cdot\right)\left(\boldsymbol{R}_i - \widetilde{\boldsymbol{R}}_\cdot\right)' \tag{4.9}$$

for $\boldsymbol{V}$ follows directly from Theorem 4.3.

Statistic for H_0^F: $\boldsymbol{P}_b\boldsymbol{F} = \boldsymbol{0}$; multivariate model. Tests for this model have been considered by Thompson (1991a) and Akritas and Arnold (1994). In both these papers, ties are excluded. Thompson derived a statistic using a generalized inverse while Akritas and Arnold used a contrast matrix of full row rank and the quadratic form for the statistic is written in terms of an inverse containing the contrast matrix. In what follows, these results are generalized to an unequal number of replications m_{ij} per treatment j and block i. Moreover, ties are allowed.

We choose the contrast matrix $\boldsymbol{W} = \boldsymbol{V}^{-1}[\boldsymbol{I}_b - \boldsymbol{J}_b\boldsymbol{V}^{-1}/\boldsymbol{1}_b'\boldsymbol{V}^{-1}\boldsymbol{1}_b]$ and we note that $\boldsymbol{W}\boldsymbol{F} = \boldsymbol{0}$ iff $\boldsymbol{P}_b\boldsymbol{F} = \boldsymbol{0}$ and that $\boldsymbol{W}\boldsymbol{V}\boldsymbol{W} = \boldsymbol{W}$. The statistic $\sqrt{n}\,\boldsymbol{W}(\widetilde{\boldsymbol{p}} - \boldsymbol{p})$ is asymptotically equivalent to $\sqrt{n}\,\boldsymbol{W}\,\widetilde{\boldsymbol{Y}}_\cdot$ under H_0^F: $\boldsymbol{P}_b\boldsymbol{F} = \boldsymbol{0}$. Let $\widehat{\boldsymbol{W}} = \widehat{\boldsymbol{V}}_n^{-1}[\boldsymbol{I}_b - \boldsymbol{J}_b\widehat{\boldsymbol{V}}_n^{-1}/\boldsymbol{1}_b'\widehat{\boldsymbol{V}}_n^{-1}\boldsymbol{1}_b]$ where $\widehat{\boldsymbol{V}}_n$ is given in (4.9). Then $Q(\boldsymbol{W}) = n(\boldsymbol{W}\widetilde{\boldsymbol{p}})'(\boldsymbol{W}\boldsymbol{V}\boldsymbol{W})^{-}(\boldsymbol{W}\widetilde{\boldsymbol{p}}) = n\,\widetilde{\boldsymbol{p}}'\,\boldsymbol{W}\,\boldsymbol{p}$. Denote the (i,j)-element of $\widehat{\boldsymbol{V}}_n^{-1}$ by $\widehat{s}_{ij}$, $i,j = 1,\ldots,b$, and let $\widehat{s}_{\cdot j} = \sum_{i=1}^{b} \widehat{s}_{ij}$ and $\widehat{s}_{\cdot\cdot} = \sum_{j=1}^{b} \widehat{s}_{\cdot j}$. Then it follows from Theorem 4.5 that the statistic

$$Q_n^M(B) = n\,\widetilde{\boldsymbol{p}}'\,\widehat{\boldsymbol{W}}\,\widetilde{\boldsymbol{p}} = \frac{n}{N^2}\left[\sum_{i=1}^{b}\sum_{j=1}^{b} \widetilde{R}_{\cdot i\cdot}\widehat{s}_{ij}\widetilde{R}_{\cdot j\cdot} - \frac{1}{\widehat{s}_{\cdot\cdot}}\left(\sum_{j=1}^{b} \widehat{s}_{\cdot j}\widetilde{R}_{\cdot j\cdot}\right)^2\right] \tag{4.10}$$

has asymptotically ($n \to \infty$) a central χ_f^2-distribution with $f = \operatorname{rank}(\boldsymbol{W}) = b - 1$ under H_0^F: $\boldsymbol{W}\boldsymbol{F} = \boldsymbol{0}$ which is equivalent to $\boldsymbol{P}_b\boldsymbol{F} = \boldsymbol{0}$. (Note that $Q_n^M(B)$ has the RT-property with respect to a parametric statistic for a repeated measurements model with an unspecified structure of the covariance matrix $\boldsymbol{V}$.)

Statistic for H_0^F: $\boldsymbol{P}_b\boldsymbol{F} = \boldsymbol{0}$; compound symmetry model. We have only to apply Theorem 4.4 to the special design considered in this subsection by letting $r = 1, n_i = n, m_{ijk} = m_{ij}$ and $c = b$. Under H_0^F, it follows from (4.4) and Theorem 4.4 that $\boldsymbol{V} = \operatorname{Cov}(\sqrt{n}\,\widetilde{\boldsymbol{Y}}_\cdot) = \boldsymbol{D} + c^*\boldsymbol{J}_b$ where $\boldsymbol{D} = \operatorname{diag}\{\tau_1,\ldots,\tau_b\}$, $\tau_j = n^{-1}\sum_{i=1}^{n}\tau_{ij}$ and $\tau_{ij} = m_{ij}^{-1}\left(\sigma^2 + (m_{ij} - 1)c^{**}\right) - c^*$. The estimators for τ_j given in Theorem 4.4 simplify to $\widehat{\tau}_j = n^{-1}\sum_{i=1}^{n}\widehat{\tau}_{ij}$ where

$$\widehat{\tau}_{ij} = \left[\frac{1}{m_{ij}} - \frac{M_i(b-1)}{M_i^2 - \sum_{t=1}^{b} m_{it}^2}\right]\frac{S_{1i}^2}{N^2(M_i - b)} + \frac{M_i S_{2i}^2}{N^2\left(M_i^2 - \sum_{t=1}^{b} m_{it}^2\right)},$$

$$S_{1i}^2 = \sum_{j=1}^{b}\sum_{k=1}^{m_{ij}}\left(R_{ijk} - \overline{R}_{ij\cdot}\right)^2, \quad S_{2i}^2 = \sum_{j=1}^{b} m_{ij}\left(\overline{R}_{ij\cdot} - \overline{R}_{i\cdot\cdot}\right)^2, \quad M_i = \sum_{j=1}^{b} m_{ij}.$$

Let

$$\widetilde{R}_{\cdot j\cdot} = \frac{1}{n}\sum_{i=1}^{n} \overline{R}_{ij\cdot} \quad \text{and} \quad \widetilde{R} = \frac{1}{\sum_{j=1}^{b}(1/\widehat{\tau}_j)} \sum_{j=1}^{b} \frac{\widetilde{R}_{\cdot j\cdot}}{\widehat{\tau}_j}.$$

Then it follows from Theorem 4.5 that the quadratic form

$$Q_n^{CS}(B) = n \sum_{j=1}^{b} \frac{1}{\widehat{\tau}_j} \left(\widetilde{R}_{\cdot j\cdot} - \widetilde{R}\right)^2 \tag{4.11}$$

has asymptotically a central χ_f^2-distribution with $f = b - 1$ under H_0^F. In case of equal cell frequencies $m_{ij} \equiv m$,

$$\widehat{\tau}_j \equiv \widehat{\tau} = \frac{1}{n(b-1)} \sum_{i=1}^{n}\sum_{t=1}^{b} (\overline{R}_{it\cdot} - \overline{R}_{i\cdot\cdot})^2,$$

$$\widetilde{R}_{\cdot j\cdot} = \overline{R}_{\cdot j\cdot}, \qquad \widetilde{R} = \frac{nmb+1}{2} \tag{4.12}$$

and $Q_n^{CS}(B)$ given in (4.11) simplifies to

$$Q_n^{CS}(B) = \frac{n^2(b-1)}{\sum_{i=1}^{n}\sum_{j=1}^{b}(\overline{R}_{ij\cdot} - \overline{R}_{i\cdot\cdot})^2} \sum_{j=1}^{b} \left(\overline{R}_{\cdot j\cdot} - \frac{nmb+1}{2}\right)^2 \tag{4.13}$$

which has been given by Brunner and Neumann (1982). A special case of this, a model without random interaction and $m = 1$ has been considered by Kepner and Robinson (1988) for the hypothesis H_0^{π} which implies H_0^F. Two consistent estimators for the unknown variances τ_j ($\equiv \tau$ under H_0^F) are given there, namely

$$\widehat{\tau}_{n,1}^2 = \frac{1}{n(b-1)} \sum_{i=1}^{n}\sum_{j=1}^{b} \left(R_{ij} - \overline{R}_{i\cdot}\right)^2$$

and

$$\widehat{\tau}_{n,2}^2 = \frac{1}{(n-1)(b-1)} \sum_{i=1}^{n}\sum_{j=1}^{b} \left(R_{ij} - \overline{R}_{i\cdot} - \overline{R}_{\cdot j} + (nb+1)/2\right)^2.$$

The estimator $\widehat{\tau}_{n,1}^2$ is identical to the estimator used in (4.12) when specialized to $m = 1$. The second estimator $\widehat{\tau}_{n,2}^2$ has the RT-property with regard to the variance estimator of the ANOVA F-test.

Table 4.1
Ranks and rank means of the observations 'ratio'

Pair	Type of cell			
	N	MIT	MIC	S
1	17	4	30	14.5
2	25.5	29	27.5	19
3	6	8.5	10	8.5
4	12	18	13	16
5	11	32	27	23
6	31	7	5	14.5
7	20	21	24	25.5
8	22	3	1	2
Means	18.06	15.31	17.25	15.38

EXAMPLE 2. In this example, we re-analyse the data given by Koch (1969), Example 1. For the description of the experiment and the data set, we refer to this article. The two factors 'Pair' (random, with $n = 8$ levels) and 'Cell' (fixed, with $b = 4$ levels: N, MIT, MIC, and S) are crossed and $m_{ij} = 1$ replication is observed for the variable 'ratio'. The ranks R_{ij} of the observations and the means are given in Table 4.1.

For the multivariate model, the estimated covariance matrix is

$$\widehat{\boldsymbol{V}}_n = \frac{1}{1024}\begin{pmatrix} 67.89 & -11.88 & -15.13 & 1.26 \\ . & 128.64 & 77.91 & 66.40 \\ . & . & 129.00 & 63.96 \\ . & . & . & 57.41 \end{pmatrix}$$

and the statistic $Q_n^M(B)$ given in (4.10) is $Q_n^M(B) = 1.33$ $(p = 0.722)$, when compared with the χ_3^2-distribution. The inspection of the estimated covariance matrix $\widehat{\boldsymbol{V}}_n$ recommends the multivariate model rather than the compound symmetry model for this experiment. Because the sample size is small, $(n-b+1)Q_n^M(B)/[(b-1)(n-1)]$ is compared with the F-distribution with $f_1 = b - 1 = 3$ and $f_2 = n - b + 1 = 5$ resulting a p-value of $p = 0.8138$. The small sample distribution is motivated by the distribution of Hotelling's T^2 under the assumption of multivariate normality where the hypothesis H_0^μ: $\mu_1 = \cdots = \mu_b$ is tested. Since the statistic $Q_n^M(B)$ has the RT-property, the results given here can be computed by an appropriate statistical software package.

Tests for $b = 2$ samples

Statistics for H_0^F and H_0^p. Here we consider the special case of $b = 2$ treatments. In this case, explicit statistics can be given for the hypotheses H_0^F and H_0^p. For $b = 2$ samples, the hypotheses H_0^F and H_0^p are formulated as H_0^F: $F_1 = F_2$ and H_0^p: $p = \int F_1 \, dF_2 = 1/2$. Therefore, a consistent estimator $\widehat{p}$ for p can be written as a linear rank statistic which does not require the inverse of a covariance matrix. Let $\boldsymbol{X}_i = (\boldsymbol{X}_{i1}', \boldsymbol{X}_{i2}')'$, $i = 1, \ldots, n$, be independent random vectors, $\boldsymbol{X}_{ij} = (X_{ij1}, \ldots, X_{ijm_{ij}})'$, $j = 1, 2$, and assume that $X_{ijk} \sim F_j(x)$, $i = 1, \ldots, n$, $k = 1, \ldots, m_{ij}$, $j = 1, 2$. Let $\widehat{F}_j(x) = N_j^{-1} \sum_{i=1}^{n} \sum_{k=1}^{m_{ij}} c(x - X_{ijk})$, $j = 1, 2$, where $N_j = \sum_{i=1}^{n} m_{ij}$ and let $N = N_1 + N_2$. Then an asymptotically unbiased and consistent estimator of p is

$$\widehat{p} = \int \widehat{F}_1 \, \mathrm{d}\widehat{F}_2 = \frac{1}{N_1 N_2} \left(R_{\cdot 2 \cdot} - N_2 \frac{N_2 + 1}{2} \right)$$

where $R_{\cdot 2 \cdot} = \sum_{i=1}^{n} \sum_{k=1}^{m_{i2}} R_{i2k}$, and R_{ijk} is the rank of X_{ijk} among all N observations. Let $s_{N,0}^2$ denote the variance of $N_1 N_2 \widehat{p}/N$ under H_0^F: $F_1 = F_2$ and let

$$\widehat{s}_{N,0}^2 = \frac{n}{N^4(n-1)} \sum_{i=1}^{n} \left[N_1 (R_{i2\cdot} - m_{i2} \overline{R}_{\cdot 2 \cdot}) - N_2 (R_{i1\cdot} - m_{i1} \overline{R}_{\cdot 1 \cdot}) \right]^2$$

where $\overline{R}_{\cdot j \cdot} = N_j^{-1} R_{\cdot 1 \cdot}$, $j = 1, 2$. Then under H_0^F, $\widehat{s}_{N,0}^2$ is a consistent estimator of $s_{N,0}^2$ in the sense that $E[\widehat{s}_{N,0}^2 / s_{N,0}^2 - 1]^2 \to 0$ and the statistic

$$T_n^F = \frac{N_1 N_2}{N \widehat{s}_{N,0}} (\widehat{p} - 1/2) = \frac{R_{\cdot 2 \cdot} - N_2(N+1)/2}{N \widehat{s}_{N,0}} \tag{4.14}$$

has asymptotically ($n \to \infty$) a standard normal distribution. For small samples, the distribution of T_n^F may be approximated by the central t_f-distribution with $f = n - 1$.

For testing the hypothesis H_0^p: $p = 1/2$ in the heteroscedastic case, the estimator $\widehat{s}_{N,0}^2$ is replaced by

$$\widehat{s}_N^2 = \frac{n}{N^2(n-1)} \sum_{i=1}^{n} \Bigg[(R_{i2\cdot} - R_{i2\cdot}^{(2)}) - (R_{i1\cdot} - R_{i1\cdot}^{(1)}) - m_{i2} \left(\overline{R}_{\cdot 2 \cdot} - \frac{N_2 + 1}{2} \right) + m_{i1} \left(\overline{R}_{\cdot 1 \cdot} - \frac{N_1 + 1}{2} \right) \Bigg]^2$$

where $R_{ij\cdot}^{(j)} = \sum_{k=1}^{m_{ij}} R_{ijk}^{(j)}$ and $R_{ijk}^{(j)}$ is the rank of X_{ijk} among all the N_j observations under treatment j. Under H_0^p, the statistic

$$T_n^p = \frac{N_1 N_2}{N \widehat{s}_N} (\widehat{p} - 1/2) = \frac{R_{\cdot 2 \cdot} - N_2(N+1)/2}{N \widehat{s}_N} \tag{4.15}$$

has asymptotically ($n \to \infty$) a standard normal distribution. For small samples, the distribution of T_n^p may be approximated by the central t_f-distribution with $f = n - 1$. For details, see Brunner et al. (1995).

Table 4.2
Ranks and means for the sunburn data

Lotion	Ranks of the sunburn degree for subject 1	2	3	4	5	6	7	8	9	10	Mean
Old	16	17	20	12.5	12.5	18	4	6	7	2	11.5
New	11	14.5	9	14.5	10	19	3	8	5	1	9.5

To derive the exact (conditional) distribution for small samples, we have to restrict the considerations to an equal number of replications $m_{ij} \equiv m$ in order to apply a permutation argument. For testing the null hypothesis of no treatment effect, either the linear model (4.5) is assumed and the hypothesis H_0^{μ}: $\boldsymbol{P}_b\boldsymbol{\mu} = \mathbf{0}$ is considered, or in the general model (4.7) the hypothesis H_0^{π} is tested. In both cases under the hypothesis, the common distribution function of $\boldsymbol{X} = (\boldsymbol{X}_1', \dots, \boldsymbol{X}_n')'$ is invariant under all 2^n equally likely permutations of the vectors $\boldsymbol{X}_{i1} = (X_{i11}, \dots, X_{i1m})'$ and $\boldsymbol{X}_{i2} = (X_{i21}, \dots, X_{i2m})'$, $i = 1, \dots, n$. Therefore, the null distribution of the difference of the rank sums $R_{\cdot 2\cdot} - R_{\cdot 1\cdot}$ can easily be computed by a shift algorithm based on a recurrence relation identical to that one for the Wilcoxon signed rank statistic if the integers $1, \dots, n$ in the recursion formula for the latter one are replaced by $A_i = |R_{i2\cdot} - R_{i1\cdot}|$, $i = 1, \dots, n$. In case of ties, the statistic $2(R_{\cdot 2\cdot} - R_{\cdot 1\cdot})$ is used in order to have integers $2A_i$ for the shift algorithm. For details, see Brunner and Compagnone (1988) and Zimmermann (1985a).

EXAMPLE 3. In this example, we analyse the data given by Gibbons and Chakraborti (1992), Problem 6.18 where the degree of sunburn after the application of two suntan lotions is measured for 10 randomly chosen subjects. For the description of the data set, we refer to Gibbons and Chakraborti (loc. cit.). The $N = 20$ ranks of the 10 paired observations are listed in Table 4.2.

The results for testing H_0^F and H_0^p are $T_n^F = -1.642$ ($p = 0.135$), and $T_n^p = -1.43$, ($p = 0.186$). Both p-values are obtained from the approximation by the t_f-distribution with $f = n - 1 = 9$.

The matched pairs design with missing observations. The so-called 'matched-pairs-design' ($m = 1$) is a special case of the model considered in the previous paragraph. In this case we have independent vectors $\boldsymbol{X}_i = (X_{i1}, X_{i2})'$, $i = 1, \dots, n$, and the statistics for testing H_0^F or H_0^p are easily derived from (4.14) and (4.15). However, we shall consider separately the case of missing observations which is of some importance in practice. We denote by $\boldsymbol{X}_{1i} = (X_{1i1}, X_{1i2})'$ the n_c complete observed vectors $\boldsymbol{X}_{ki}$, $k = 1$; $i = 1, \dots, n_c$. Let X_{2ij}, $i = 1, \dots, u_j$; $j = 1, 2$, denote the u_j incomplete observations where the matched pair has only been observed under treatment j, $j = 1, 2$, and the paired observation is missing. In order to test the hypothesis H_0^F or H_0^p, also these incomplete observations can be used. In total, there are $N = N_1 + N_2$ observations where $N_j = n_c + u_j$, $j = 1, 2$, is the total number of observations under treatment j. Let R_{kij} be the rank of X_{kij} among all the N observations, let $R_{kij}^{(j)}$ be the rank of X_{kij} among all the N_j observations under treatment j and let $R_{\cdot\cdot 2} = \sum_{i=1}^{n_c} R_{1i2} + \sum_{i=1}^{u_2} R_{2i2}$ and let

$$\widetilde{s}_{N,0}^2 = \frac{1}{N^4}\left[\sum_{i=1}^{n_c}\left([N_1 R_{1i2} - N_2 R_{1i1}] - [N_1 \overline{R}_{1\cdot 2} - N_2 \overline{R}_{1\cdot 1}]\right)^2 + \sum_{j=1}^{2}(N - N_j)^2 \sum_{i=1}^{u_j}\left(R_{2ij} - \overline{R}_{2\cdot j}\right)^2\right]$$

where

$$\overline{R}_{1\cdot j} = \frac{1}{n_c}\sum_{i=1}^{n_c} R_{1ij} \quad \text{and} \quad \overline{R}_{2\cdot j} = \frac{1}{u_j}\sum_{i=1}^{u_j} R_{2ij}, \quad j = 1, 2.$$

Then, under H_0^F, the statistic

$$\widetilde{T}_N^F = \frac{R_{\cdot\cdot 2} - N_2(N+1)/2}{N\widetilde{s}_{N,0}} \tag{4.16}$$

has asymptotically $(\min(n_c + u_1, n_c + u_2) \to \infty)$ a standard normal distribution.

In the heteroscedastic case, let

$$\widetilde{s}_N^2 = \frac{1}{N^2}\left[\sum_{i=1}^{n_c}\left(\left[R_{1i2} - R_{1i2}^{(2)}\right] - \left[\overline{R}_{1\cdot 2} - \overline{R}_{1\cdot 2}^{(2)}\right] - \left[R_{1i1} - R_{1i1}^{(1)}\right] + \left[\overline{R}_{1\cdot 1} - \overline{R}_{1\cdot 1}^{(1)}\right]\right)^2 + \sum_{j=1}^{2}\sum_{i=1}^{u_j}\left(\left[R_{2ij} - R_{2ij}^{(j)}\right] - \left[\overline{R}_{2\cdot j} - \overline{R}_{2\cdot j}^{(j)}\right]\right)^2\right]$$

where

$$\overline{R}_{1\cdot j} = \frac{1}{n_c}\sum_{i=1}^{n_c} R_{1ij}, \qquad \overline{R}_{1\cdot j}^{(j)} = \frac{1}{n_c}\sum_{i=1}^{n_c} R_{1ij}^{(j)},$$

$$\overline{R}_{2\cdot j} = \frac{1}{u_j}\sum_{i=1}^{u_j} R_{2ij}, \qquad \overline{R}_{2\cdot j}^{(j)} = \frac{1}{u_j}\sum_{i=1}^{u_j} R_{2ij}^{(j)}, \quad j = 1, 2.$$

Then, under H_0^p, the statistic

$$\widetilde{T}_N^p = \frac{R_{\cdot\cdot 2} - N_2(N+1)/2}{N\widetilde{s}_N} \tag{4.17}$$

has asymptotically (as $\min(n_c + u_1, n_c + u_2) \to \infty$) a standard normal distribution.

For small samples, the null distribution of $R_{1\cdot 2}$ (within the complete observations) is computed for the ranks of the complete observations $R_{111}, \ldots, R_{1n_c2}$ as described in case (1). The null distribution of $R_{\cdot\cdot 2}$ (within the incomplete observations) is computed as for the Wilcoxon–Mann–Whitney statistic under H_0^F where the integers $1, \ldots, N$ are replaced by the ranks of the incomplete observations R_{2ij}, $i = 1, \ldots, n_j$, $j = 1, 2$. Since the incomplete observations are independent from the complete observations, the desired null distribution of $R_{\cdot\cdot 2}$ is the convolution of the two null distributions.

4.3.2. Nested designs

4.3.2.1. Models and hypotheses

Linear model. The random variables X_{ijk}, $i = 1, \ldots, a$; $j = 1, \ldots, n_i$; $k = 1, \ldots, m_{ij}$, are written as

$$X_{ijk} = \mu_i + B_{j(i)} + \varepsilon_{ijk} \tag{4.18}$$

where $\mu_i = \mathrm{E}(X_{ijk})$ are unknown parameters, $B_{j(i)} \sim N(0, \sigma_B^2)$ are i.i.d. random variables and $\varepsilon_{ijk} \sim N(0, \sigma^2)$ are i.i.d. random variables independent from $B_{j(i)}$. The notation $j(i)$ means that the random subjects $B_{1(i)}, \ldots, B_{n_i(i)}$ are nested under treatment i. Such designs are typically used when $N = \sum_{i=1}^{a} n_i$ randomly chosen subjects are repeatedly ($k = 1, \ldots, m_{ij}$) observed or measured under the same treatment.

General model. In the linear model (4.18), the random variables X_{ijk} and $X_{i'j'k'}$ are independent if $i \neq i'$ or if $j \neq j'$ where X_{ijk} and $X_{ij'k'}$ are identically distributed according to $F_i(x)$. Note that the random variables X_{ijk} and $X_{ijk'}$ may be dependent. Model (4.18) is generalized in the following way:

Let

$$\boldsymbol{X}_{ij} = (X_{ij1}, \ldots, X_{ijm})', \quad i = 1, \ldots, a; \quad j = 1, \ldots, n_i, \tag{4.19}$$

be independent random vectors of the observations for block j under treatment i where $X_{ijk} \sim F_i(x)$, $j = 1, \ldots, n_i$; $k = 1, \ldots, m_{ij}$. In the linear model (4.18), $\boldsymbol{S}_{ij} = \mathrm{Cov}(\boldsymbol{X}_{ij}) = (\sigma_x^2 - c_x)\boldsymbol{I}_{m_{ij}} + c_x \boldsymbol{J}_{m_{ij}}$, where $\sigma_x^2 = \mathrm{Var}(X_{111})$ and $c_x = \mathrm{Cov}(X_{111}, X_{112})$. In the general model (4.19), the assumption of a special structure of the covariance matrix of $\boldsymbol{X}_{ij}$ in case of no treatment effect is not necessary. This will be explained in Section 4.3.2.2 when the test statistic is derived. The shift model $F_i(x) = F(x - \mu_i)$ is easily derived from the general model (4.19).

Hypotheses. The definition of the treatment effects and the hypotheses are the same as for the model with independent observations considered in Section 2.1.1. The hypothesis for the linear model is H_0^μ: $\boldsymbol{P}_a \boldsymbol{\mu} = \boldsymbol{0}$ where $\boldsymbol{\mu} = (\mu_1, \ldots, \mu_a)'$. In the general model, we consider the hypothesis H_0^F: $\boldsymbol{P}_a \boldsymbol{F} = \boldsymbol{0}$ where $\boldsymbol{F} = (F_1, \ldots, F_a)'$. The statistics are based on a consistent estimator of $\boldsymbol{p} = (p_1, \ldots, p_a)' = \int H \, \mathrm{d}\boldsymbol{F}$ where $H(x) = N^{-1} \sum_{i=1}^{a} \sum_{j=1}^{n_i} m_{ij} F_i(x)$ and $N = \sum_{i=1}^{a} \sum_{j=1}^{n_i} m_{ij}$. We restrict the consideration of the hypothesis H_0^p to the case of $a = 2$ samples where the hypothesis is formulated as H_0^p: $p = \int F_1 \, \mathrm{d}F_2 = 1/2$.

It is obvious that the same relations between the hypotheses in the semiparametric models and the general model as given in Proposition 2.1 are also valid in the hierarchical model.

4.3.2.2. Tests for $a \geqslant 2$ samples. Here, as in the case of the cross-classification, we admit unequal numbers m_{ij} of replications for block j under treatment i. Therefore, the components p_i of the generalized mean vector $\boldsymbol{p} = (p_1, \ldots, p_a)'$ are estimated by an unweighted sum of cell means. We will use the notation of Subsection 4.3.1.2. Let $\widetilde{\boldsymbol{F}} = (\widetilde{F}_1, \ldots, \widetilde{F}_a)'$, then $p_i = \int H \, \mathrm{d}F_i$ is estimated by

$$\widetilde{p}_i = \int \widehat{H} \, \mathrm{d}\widetilde{F}_i = \frac{1}{N}(\widetilde{R}_{i\cdot\cdot} - 1/2)$$

where $\widetilde{R}_{i\cdot\cdot} = n_i^{-1} \sum_{j=1}^{n_i} \overline{R}_{ij\cdot}$ and $\overline{R}_{ij\cdot} = m_{ij}^{-1} \sum_{k=1}^{m_{ij}} R_{ijk}$. Here, R_{ijk} is the rank of X_{ijk} among all the $N = \sum_{i=1}^{a} \sum_{j=1}^{n_i} m_{ij}$ observations. Let $\widetilde{\boldsymbol{Y}}_{\cdot} = (\widetilde{Y}_{1\cdot\cdot}, \ldots, \widetilde{Y}_{a\cdot\cdot})'$ where $\widetilde{Y}_{i\cdot\cdot} = n_i^{-1} \sum_{j=1}^{n_i} \overline{Y}_{ij\cdot}$, then

$$\boldsymbol{V}_a = \operatorname{Cov}\big(\sqrt{N}\, \widetilde{\boldsymbol{Y}}_{\cdot}\big) = \bigoplus_{i=1}^{a} \frac{N}{n_i} \sigma_i^2$$

where $\sigma_i^2 = n_i^{-1} \sum_{j=1}^{n_i} \sigma_{ij}^2$ and $\sigma_{ij}^2 = \operatorname{Var}(\overline{Y}_{ij\cdot})$. It is not necessary to make special assumptions on the bivariate distribution functions of $\boldsymbol{X}_{ij}$ under the hypothesis since the random variables $\overline{Y}_{ij\cdot}$ are independent and only the variances σ_i^2 have to be estimated. The equality of certain two-dimensional marginal distribution functions (which generates a convenient form of the covariance matrix) is only needed if different treatments are applied to the same subject as in the cross-classification.

Let $S_i^2 = \sum_{j=1}^{n_i} (\overline{R}_{ij\cdot} - \widetilde{R}_{i\cdot\cdot})^2$ and $\widehat{\sigma}_i^2 = [N^2(n_i - 1)]^{-1} S_i^2$ where $\widetilde{R}_{i\cdot\cdot} = n_i^{-1} \sum_{j=1}^{n_i} \overline{R}_{ij\cdot}$. Then $\widehat{\sigma}_i^2$ is consistent for σ_i^2 in the sense that $\widehat{\sigma}_i^2/\sigma_i^2 \xrightarrow{p} 1$.

Let $\boldsymbol{W} = \boldsymbol{V}_a^{-1}(\boldsymbol{I}_a - \boldsymbol{J}_a \boldsymbol{V}_a^{-1}/\operatorname{trace}(\boldsymbol{V}_a^{-1}))$ be the contrast matrix defined in (1.5) and let

$$\widehat{\boldsymbol{V}}_a = \bigoplus_{i=1}^{a} \frac{N}{n_i} \widehat{\sigma}_i^2$$

and $\widehat{\boldsymbol{W}} = \widehat{\boldsymbol{V}}_a^{-1}(\boldsymbol{I}_a - \boldsymbol{J}_a \widehat{\boldsymbol{V}}_a^{-1}/\operatorname{trace}(\widehat{\boldsymbol{V}}_a^{-1}))$ then $\widehat{\boldsymbol{W}}\, \widehat{\boldsymbol{V}}_a \widehat{\boldsymbol{W}} = \widehat{\boldsymbol{W}}$ and $\boldsymbol{W}\boldsymbol{F} = \boldsymbol{0}$ iff $\boldsymbol{P}_a \boldsymbol{F} = \boldsymbol{0}$. It follows from Theorem 4.5 that the quadratic form

$$\begin{aligned} Q_N^H &= N \widetilde{\boldsymbol{p}}' \, \widehat{\boldsymbol{W}} \, \widetilde{\boldsymbol{p}} \\ &= \sum_{i=1}^{a} \frac{n_i(n_i - 1)}{S_i^2} \left(\widetilde{R}_{i\cdot\cdot} - \frac{1}{\sum_{r=1}^{a} (n_r/\widehat{\sigma}_r^2)} \sum_{r=1}^{a} \frac{n_r \widetilde{R}_{r\cdot\cdot}}{\widehat{\sigma}_r^2} \right)^2 \end{aligned} \tag{4.20}$$

has asymptotically ($n_i \to \infty$) a χ_f^2-distribution with $f = a - 1$ under H_0^F: $\boldsymbol{P}_a \boldsymbol{F} = \boldsymbol{0}$.

In case of an equal number of replications $m_{ij} \equiv m$, $\overline{R}_{ij\cdot} = m^{-1}\sum_{k=1}^{m} R_{ijk}$, $\widetilde{R}_{i\cdot\cdot} = (mn_i)^{-1}R_{i\cdot\cdot} = \overline{R}_{i\cdot\cdot}$, $\widetilde{R} = \overline{R}_{i\cdot\cdot} = (N+1)/2$ and

$$\widehat{\sigma}^2 = (n. - a)^{-1}\sum_{i=1}^{a}\sum_{j=1}^{n_i}\left(\overline{R}_{ij\cdot} - \overline{R}_{i\cdot\cdot}\right)^2$$

where $n. = \sum_{i=1}^{a} n_i$. The quadratic form Q_N^H given in (4.20) reduces to

$$Q_N^H = (n. - a)\cdot\frac{\sum_{i=1}^{a} n_i\left(\overline{R}_{i\cdot\cdot} - (N+1)/2\right)^2}{\sum_{i=1}^{a}\sum_{j=1}^{n_i}\left(\overline{R}_{ij\cdot} - \overline{R}_{i\cdot\cdot}\right)^2}.$$

This statistic has been given by Brunner and Neumann (1982) for the case of no ties.

For small sample sizes (and an equal number of replications), the exact permutation distribution of $\sum_{i=1}^{a} R_{i\cdot\cdot}^2$ can be computed. Under H_0^F: $\boldsymbol{P}_a\boldsymbol{F} = \boldsymbol{0}$, the marginal distribution functions of all vectors $\boldsymbol{X}_{ij}$ are identical and, since the vectors are independent, the common distribution function of $\boldsymbol{X} = (\boldsymbol{X}_{11}', \ldots, \boldsymbol{X}_{an_a}')'$ remains invariant under any permutation of the vectors $\boldsymbol{X}_{ij}$, $i = 1, \ldots, a$, $j = 1, \ldots, n_i$. Therefore, the exact permutation distribution of $\sum_{i=1}^{a} R_{i\cdot\cdot}^2$ can be quickly computed by a multivariate shift algorithm (Streitberg and Roehmel, 1986) as in the case of the statistic Q given in Section 2.1.3. However, the integers $1, \ldots, N$ are replaced by the rank sums $R_{11}, \ldots, R_{an_a}$. In case of ties, the same remark applies as in Section 2.1.3.

4.3.2.3. Tests for $a = 2$ samples. In the case of two samples, we give test statistics for both hypotheses H_0^F: $F_1 = F_2$ and H_0^p: $p = 1/2$. Here we have independent random vectors $\boldsymbol{X}_{ij} = (X_{ij1}, \ldots, X_{ijm_{ij}})'$, $i = 1, 2$, $j = 1, \ldots, n_i$, with uniformly bounded length $m_{ij} \leqslant M < \infty$. For the case that the number of replications m_{ij} also tends to ∞ (however in a certain rate depending on $n. = n_1 + n_2$), we refer to Brunner and Denker (1994). The other assumptions on X_{ij} and the marginal distribution functions $F_1(x)$ and $F_2(x)$ are the same as in the previous paragraph.

As in Section 2.1.3.3, the treatment effect for $a = 2$ samples is $p = \int F_1\,\mathrm{d}F_2$ and an asymptotically unbiased and consistent estimator for p is given by $\widehat{p} = (N_1N_2)^{-1}(R_{2\cdot\cdot} - N_2(N_2+1)/2)$ where $N_i = \sum_{j=1}^{n_i} m_{ij}$ and $N = N_1 + N_2$. Under H_0^F: $F_1 = F_2$, the statistic

$$H_N^F = \frac{R_{2\cdot\cdot} - N_2\frac{N+1}{2}}{\sqrt{\sum_{i=1}^{2}\left(1 - \frac{N_i}{N}\right)^2\sum_{j=1}^{n_i} m_{ij}^2\left(\overline{R}_{ij\cdot} - \overline{R}_{i\cdot\cdot}\right)^2}} \tag{4.21}$$

has asymptotically a standard normal distribution if $\min(n_1, n_2) \to \infty$. Here R_{ijk} is the rank of X_{ijk} among all the N observations, $R_{i\cdot\cdot} = \sum_{j=1}^{n_i} R_{ij\cdot} = \sum_{j=1}^{n_i}\sum_{k=1}^{m_{ij}} R_{ijk}$, $\overline{R}_{ij\cdot} = m_{ij}^{-1}R_{ij\cdot}$ and $\overline{R}_{i\cdot\cdot} = N_i^{-1}R_{i\cdot\cdot}$, $i = 1, 2$.

Under H_0^p: $\int F_1 \, dF_2 = 1/2$, the statistic

$$H_N^p = \frac{R_{2\cdot\cdot} - N_2 \frac{N+1}{2}}{\sqrt{\sum_{i=1}^{2} \sum_{j=1}^{n_i} m_{ij}^2 \left(\overline{R}_{ij\cdot} - \overline{R}_{ij\cdot}^{(i)} - \overline{R}_{i\cdot\cdot} + \frac{N_i+1}{2} \right)^2}} \tag{4.22}$$

has asymptotically a standard normal distribution if $\min(n_1, n_2) \to \infty$. Here $R_{ijk}^{(i)}$ is the rank of X_{ijk} among all the N_i observations under treatment i and $\overline{R}_{ij\cdot}^{(i)} = m_{ij}^{-1} \sum_{k=1}^{m_{ij}} R_{ijk}^{(i)}$. For details, we refer to Brunner and Denker (1994).

4.4. Three-factor mixed models

4.4.1. Partially nested designs/repeated measurements on one fixed factor

4.4.1.1. Models and hypotheses. In this section, we consider two-factor repeated measurements designs where the repeated measurements are only taken on one factor, factor B say, and the subjects are nested within the levels of factor A. Therefore, this design is called 'partially nested' design. In medical and psychological studies it appears either when different groups of subjects are observed under the same treatments for each subject or when subjects are divided randomly into several treatment groups and the outcomes are observed consecutively at several time points.

Nonparametric procedures for this design have been considered by Brunner and Neumann (1984, 1986a) for the 2×2 design with heteroscedastic distributions. Designs with $a, b \geqslant 2$ levels have been considered by Thompson and Ammann (1990), Thompson (1991a) and Akritas (1993). Rank tests for the hypotheses considered in these papers as well as rank tests for other hypotheses are derived below. First we state the models.

Models. The linear model is commonly written as

$$X_{ijk} = \mu + \alpha_i + \beta_j + (\alpha\beta)_{ij} + S_{k(i)} + (BS)_{kj(i)} + \varepsilon_{ijk} \tag{4.23}$$

where μ is the overall mean, α_i is the treatment effect of level i of the fixed factor A, β_j is the treatment effect of level j of the fixed factor B on which the repeated measurements are taken, $(\alpha\beta)_{ij}$ is the effect of combination (i, j) of the fixed factors (interaction); $S_{k(i)}$ are i.i.d. $N(0, \sigma_S^2)$ random variables representing the random subject effect which is nested under the (group)-factor A; $(BS)_{jk(i)}$ are i.i.d. $N(0, \sigma_{BS}^2)$ random variables representing the random interaction between subjects and factor B, and the error terms ε_{ijk} are i.i.d. $N(0, \sigma_\varepsilon^2)$. All random effects are assumed to be independent from all other random variables. The compound symmetry of the covariance matrix of $\boldsymbol{X}_{ik} = (X_{i1k}, \ldots, X_{ibk})'$ has already been discussed in Section 4.1.

The general model is derived from the general mixed model (4.1) by letting $r = a$ and $c = b$. For simplicity, we consider only designs with $m_{ij} = 1$ replication for subject k and treatment j. However, the formulas given in this section can be generalized easily to the case of $m_{ij} \geqslant 1$ replications using the results from Section 4.2.

Thus, we consider independent vectors

$$X_{ik} = (X_{i1k}, \ldots, X_{ibk})', \quad i = 1, \ldots, a, \; k = 1, \ldots, n_i, \tag{4.24}$$

where $X_{ijk} \sim F_{ij}(x)$, $i = 1, \ldots, a$, $j = 1, \ldots, b$. In a general model it is not appropriate to assume that the covariance matrix $\boldsymbol{S}_{ik} = \mathrm{Cov}(\boldsymbol{X}_{ik})$ is not changed under the treatment. Therefore, it is only reasonable to assume a compound symmetry of $\boldsymbol{S}_{ik}$ if there is no treatment effect within group i, i.e., under $H_0^F(B \mid A)$: $F_{i1} = \cdots = F_{ib}$, $i = 1, \ldots, a$.

Hypotheses. The treatment effect is formulated in terms of the marginal distribution functions and no special structure of $\boldsymbol{S}_{ik}$ is assumed. The hypotheses are therefore essentially the same as in the two-way layout with independent observations (see Section 2.2.1). Let $\boldsymbol{F} = (F_{11}, \ldots, F_{ab})'$. Then the hypotheses are formulated as

$$\begin{aligned}
&H_0^F(A)\colon \left(\boldsymbol{P}_a \otimes \frac{1}{b}\boldsymbol{1}_b'\right)\boldsymbol{F} = \boldsymbol{0} \quad \text{(no main group effect)},\\
&H_0^F(B)\colon \left(\frac{1}{a}\boldsymbol{1}_a' \otimes \boldsymbol{P}_b\right)\boldsymbol{F} = \boldsymbol{0} \quad \text{(no main treatment effect)},\\
&H_0^F(AB)\colon (\boldsymbol{P}_a \otimes \boldsymbol{P}_b)\boldsymbol{F} = \boldsymbol{0} \quad \text{(no interaction)},\\
&H_0^F(A \mid B)\colon (\boldsymbol{P}_a \otimes \boldsymbol{I}_b)\boldsymbol{F} = \boldsymbol{0} \quad \text{(no simple group effect within the treatments)},\\
&H_0^F(B \mid A)\colon (\boldsymbol{I}_a \otimes \boldsymbol{P}_b)\boldsymbol{F} = \boldsymbol{0} \quad \text{(no simple treatment effect within the groups)}.
\end{aligned}$$

Note that the model is not symmetric in the factors A and B since the subjects are nested under the groups. Therefore, two hypotheses for the simple factor effects are stated. We like to point out that for the general model, the covariances between the observations X_{ijk} and $X_{ij'k}$ within one subject k may be arbitrary. It is only assumed that they do not depend on k (independent replications) and that the covariance matrix is non-singular. Thus, $\boldsymbol{S}_{ik} = \boldsymbol{S}_i$, $k = 1, \ldots, n_i$, and $|\boldsymbol{S}_i| \neq 0$, $i = 1, \ldots, a$. The implications of the hypotheses stated above for the general model (4.24) will follow from Proposition 2.9.

Rank tests for the simple factor B effect $H_0^F(B \mid A)$ for the compound symmetry model have been considered by Thompson and Ammann (1990) under the assumption of absolutely continuous distribution functions. For the general multivariate model, Thompson (1991a) derived a rank test for the same hypothesis, and Akritas (1993) derived rank tests for both simple factor effects, i.e., for $H_0^F(A \mid B)$ and $H_0^F(B \mid A)$. In both papers, ties were excluded.

Here we shall derive nonparametric tests for the partially nested design from the unified approach given in Section 4.2 for all the five hypotheses stated above and we do not assume that the distribution functions are continuous.

For the $a \times 2$ design, we also consider the hypotheses $H_0^w(B)$, $H_0^w(B \mid A)$ and $H_0^w(AB)$ including heteroscedastic models. In the 2×2 design, we especially concentrate on the two-period cross-over design, a design which is frequently used in psychological and medical studies.

4.4.1.2. Derivations of the statistics, $a, b \geqslant 2$

Estimators and notations. The statistics for the repeated measurements design are based on a consistent estimator of the vector of the generalized means $\boldsymbol{p} = (p_{11}, \ldots, p_{ab})' = \int H \, \mathrm{d}\boldsymbol{F}$ where $H = N^{-1} \sum_{i=1}^{a} \sum_{j=1}^{b} n_i F_{ij}$ and $\boldsymbol{F} = (F_{11}, \ldots, F_{ab})'$ is the vector of all the distribution functions. Let $\widehat{\boldsymbol{F}} = (\widehat{F}_{11}, \ldots, \widehat{F}_{ab})'$ where $\widehat{F}_{ij}(x) = n_i^{-1} \sum_{k=1}^{n_i} c(x - X_{ijk})$ is the empirical distribution function of F_{ij}.

Let R_{ijk} be the rank of X_{ijk} among all the $N = b \sum_{i=1}^{a} n_i$ observations, let $\boldsymbol{R}_{ik} = (R_{i1k}, \ldots, R_{ibk})'$ be the vector of the ranks for subject k within group i. Let $\overline{R}_{ij\cdot} = n_i^{-1} \sum_{k=1}^{n_i} R_{ijk}$ and let $\overline{\boldsymbol{R}}_{i\cdot} = (\overline{R}_{i1\cdot}, \ldots, \overline{R}_{ib\cdot})'$ be the vector of the rank means within group $i = 1, \ldots, a$ and $\overline{\boldsymbol{R}}_{\cdot} = (\overline{\boldsymbol{R}}_{1\cdot}', \ldots, \overline{\boldsymbol{R}}_{a\cdot}')'$. Furthermore, let $\overline{\boldsymbol{R}}_{\cdot\cdot} = (\overline{R}_{\cdot 1 \cdot}, \ldots, \overline{R}_{\cdot b \cdot})'$ where $\overline{R}_{\cdot j \cdot} = n_{\cdot}^{-1} \sum_{i=1}^{a} \sum_{k=1}^{n_i} R_{ijk}$ and $n_{\cdot} = \sum_{i=1}^{a} n_i = N/b$.

Let $\boldsymbol{C}$ be any suitable contrast matrix. Then, under H_0^F: $\boldsymbol{CF} = \boldsymbol{0}$, the statistics $\sqrt{N}\, \boldsymbol{C} \int \widehat{H} \, \mathrm{d}\widehat{\boldsymbol{F}} = N^{-1/2} \boldsymbol{C} \overline{\boldsymbol{R}}_{\cdot}$ and $\sqrt{N}\, \boldsymbol{C} \int H \, \mathrm{d}\widehat{\boldsymbol{F}} = \sqrt{N}\, \boldsymbol{C} \overline{\boldsymbol{Y}}_{\cdot}$ are asymptotically equivalent. (This follows from Theorem 4.2.) Here, $\overline{\boldsymbol{Y}}_{\cdot} = (\overline{\boldsymbol{Y}}_{1\cdot}', \ldots, \overline{\boldsymbol{Y}}_{a\cdot}')'$ where $\overline{\boldsymbol{Y}}_{i\cdot} = (\overline{Y}_{i1\cdot}, \ldots, \overline{Y}_{ib\cdot})'$, $i = 1, \ldots, a$, and $\overline{Y}_{ij\cdot} = n_i^{-1} \sum_{k=1}^{n_i} Y_{ijk}$ where $Y_{ijk} = H(X_{ijk})$. Note that the vectors $\boldsymbol{Y}_{ik} = (Y_{i1k}, \ldots, Y_{ibk})'$ are independent by assumption.

We need to estimate the covariance matrix $\boldsymbol{V}_i = \mathrm{Cov}(\sqrt{N}\, \overline{\boldsymbol{Y}}_{i\cdot})$. A consistent estimator

$$\widehat{\boldsymbol{V}}_i = \frac{1}{N n_i (n_i - 1)} \sum_{k=1}^{n_i} \left(\boldsymbol{R}_{ik} - \overline{\boldsymbol{R}}_{i\cdot}\right)\left(\boldsymbol{R}_{ik} - \overline{\boldsymbol{R}}_{i\cdot}\right)' \tag{4.25}$$

of $\boldsymbol{V}_i$ follows from Theorem 4.3 by letting $m_{ij} = 1$. We denote by

$$\boldsymbol{W}_i = \boldsymbol{V}_i^{-1} \left(\boldsymbol{I}_b - \boldsymbol{J}_b \boldsymbol{V}_i^{-1} / \boldsymbol{1}_b' \boldsymbol{V}_i^{-1} \boldsymbol{1}_b'\right) \tag{4.26}$$

the contrast matrix defined in 1.5. Note that $\boldsymbol{W}_i \boldsymbol{V}_i \boldsymbol{W}_i = \boldsymbol{W}_i$ and $\boldsymbol{W}_i \boldsymbol{F}_i = \boldsymbol{0}$ iff $\boldsymbol{P}_b \boldsymbol{F}_i = \boldsymbol{0}$, $i = 1, \ldots, a$, where $\boldsymbol{F}_i = (F_{i1}, \ldots, F_{ib})'$. An estimator of $\boldsymbol{W}_i$ obtained by replacing $\boldsymbol{V}_i^{-1}$ by $\widehat{\boldsymbol{V}}_i^{-1}$ is given by

$$\widehat{\boldsymbol{W}}_i = \widehat{\boldsymbol{V}}_i^{-1} \left(\boldsymbol{I}_b - \boldsymbol{J}_b \widehat{\boldsymbol{V}}_i^{-1} / \widehat{s}_{\cdot\cdot}^{(i)}\right) \tag{4.27}$$

where $\widehat{s}_{\cdot\cdot}^{(i)} = \sum_{j'=1}^{b} \widehat{s}_{\cdot j'}^{(i)} = \sum_{j=1}^{b} \sum_{j'=1}^{b} \widehat{s}_{jj'}^{(i)}$ and $\widehat{s}_{jj'}^{(i)}$ is the (j, j')-element of $\widehat{\boldsymbol{V}}_i^{-1}$. Further notation will be explained when it appears to be necessary for the derivation of the statistics.

Statistics

Test for the treatment effect B. A test statistic for $H_0^F(B)$: $(\frac{1}{a} \boldsymbol{1}_a' \otimes \boldsymbol{P}_b)\boldsymbol{F} = \boldsymbol{0}$ is derived here. It follows from Theorem 4.2 that the statistics $\sqrt{N} \frac{1}{a} \boldsymbol{1}_a' \otimes \boldsymbol{I}_b)\widehat{\boldsymbol{p}}$ and

$\sqrt{N}(\frac{1}{a}\mathbf{1}_a' \otimes \boldsymbol{I}_b) \int H \, d\widehat{\boldsymbol{F}} = \sqrt{N}(\frac{1}{a}\mathbf{1}_a' \otimes \boldsymbol{I}_b)\overline{\boldsymbol{Y}}_{\cdot}$ are asymptotically equivalent under H_0^F. The covariance matrix of $\sqrt{N}(\frac{1}{a}\mathbf{1}_a' \otimes \boldsymbol{I}_b)\overline{\boldsymbol{Y}}_{\cdot}$ is

$$\boldsymbol{V} = \operatorname{Cov}(\sqrt{N}\,\overline{\boldsymbol{Y}}_{\cdot\cdot}) = \frac{1}{a^2}\sum_{i=1}^{a} \operatorname{Cov}(\sqrt{N}\,\overline{\boldsymbol{Y}}_{i\cdot}) = \frac{1}{a^2}\sum_{i=1}^{a} \boldsymbol{V}_i.$$

Let $\boldsymbol{W} = \boldsymbol{V}^{-1}(\boldsymbol{I}_b - \boldsymbol{J}_b\boldsymbol{V}^{-1}/\mathbf{1}_b'\boldsymbol{V}^{-1}\mathbf{1}_b)$. Then $\boldsymbol{W}\boldsymbol{V}\boldsymbol{W} = \boldsymbol{W}$ and $(\frac{1}{a}\mathbf{1}_a' \otimes \boldsymbol{P}_b)\boldsymbol{F} = \mathbf{0}$ iff $(\frac{1}{a}\mathbf{1}_a' \otimes \boldsymbol{W})\boldsymbol{F} = \mathbf{0}$. Note that $\operatorname{rank}(\boldsymbol{W}) = b - 1$. A consistent estimator of $\boldsymbol{V}$ follows from (4.25), namely

$$\widehat{\boldsymbol{V}} = \frac{1}{a^2}\sum_{i=1}^{a} \frac{1}{Nn_i(n_i - 1)} \sum_{k=1}^{n_i} (\boldsymbol{R}_{ik} - \overline{\boldsymbol{R}}_{i\cdot})(\boldsymbol{R}_{ik} - \overline{\boldsymbol{R}}_{i\cdot})'$$

since the vectors $\boldsymbol{Y}_{ik}$ are independent. Let $\widehat{\boldsymbol{W}} = \widehat{\boldsymbol{V}}^{-1}(\boldsymbol{I}_b - \boldsymbol{J}_b\widehat{\boldsymbol{V}}^{-1}/\widehat{s}_{\cdot\cdot})$ be an estimator of $\boldsymbol{W}$ where $\boldsymbol{V}^{-1}$ is replaced by $\widehat{\boldsymbol{V}}^{-1}$ and $\widehat{s}_{\cdot\cdot} = \sum_{j=1}^{b}\sum_{j'=1}^{b} \widehat{s}_{jj'}$ where $\widehat{s}_{jj'}$ is the (j,j')-element of $\widehat{\boldsymbol{V}}^{-1}$.

The statistic for testing $H_0^F(B)$ is based on the vector $N^{-1/2}(\frac{1}{a}\mathbf{1}_a' \otimes \widehat{\boldsymbol{W}})\overline{\boldsymbol{R}}_{\cdot} = N^{-1/2}\,\widehat{\boldsymbol{W}}\,\overline{\boldsymbol{R}}_{\cdot\cdot}$. Note that under $H_0^F(B)$, $N^{-1/2}\,\widehat{\boldsymbol{W}}\,\overline{\boldsymbol{R}}_{\cdot\cdot}$ is asymptotically equivalent to $N^{1/2}\,\widehat{\boldsymbol{W}}\,\overline{\boldsymbol{Y}}_{\cdot\cdot}$. Thus, under $H_0^F(B)$, the quadratic form

$$\begin{aligned} Q_N(B) &= \frac{1}{N}\overline{\boldsymbol{R}}_{\cdot\cdot}'\,\widehat{\boldsymbol{W}}\,\overline{\boldsymbol{R}}_{\cdot\cdot} \\ &= \frac{1}{N}\left[\sum_{j=1}^{b}\sum_{j'=1}^{b} \overline{R}_{\cdot j\cdot}\,\widehat{s}_{jj'}\,\overline{R}_{\cdot j'\cdot} - \frac{1}{\widehat{s}_{\cdot\cdot}}\left(\sum_{j=1}^{b}\widehat{s}_{j\cdot}\overline{R}_{\cdot j\cdot}\right)^2\right] \end{aligned} \tag{4.28}$$

has asymptotically a χ_f^2-distribution with $f = b - 1$ d.f. and $Q_N(B)$ has the RT-property with respect to a multivariate parametric statistic with covariance matrix $\boldsymbol{V}$.

Test for the group effect A. The hypothesis of no main effect A is formulated as $H_0^F(A)$: $(\boldsymbol{P}_a \otimes \frac{1}{b}\mathbf{1}_b')\boldsymbol{F} = \mathbf{0}$ and the asymptotic distribution of $\sqrt{N}(\boldsymbol{I}_a \otimes \frac{1}{b}\mathbf{1}_b') \int \widehat{H}\, d\widehat{\boldsymbol{F}}$ is considered under $H_0^F(A)$. Note that the statistics $\sqrt{N}(\boldsymbol{I}_a \otimes \frac{1}{b}\mathbf{1}_b')\widehat{\boldsymbol{p}}$ and $\sqrt{N}(\boldsymbol{I}_a \otimes \frac{1}{b}\mathbf{1}_b') \int H\, d\widehat{\boldsymbol{F}} = \sqrt{N}(\boldsymbol{I}_a \otimes \frac{1}{b}\mathbf{1}_b')\overline{\boldsymbol{Y}}_{\cdot}$ are asymptotically equivalent under H_0^F. Let $\sigma_i^2 = \operatorname{Var}(\overline{Y}_{i\cdot\cdot})$ where $\overline{Y}_{i\cdot\cdot} = n_i^{-1}\sum_{k=1}^{n_i} \overline{Y}_{i\cdot k}$ and note that the random variables $\overline{Y}_{i\cdot k} = b^{-1}\sum_{j=1}^{b} Y_{ijk}$ are independent. Then, $\boldsymbol{V} = \operatorname{Cov}(\sqrt{N}(\boldsymbol{I}_a \otimes \frac{1}{b}\mathbf{1}_b')\overline{\boldsymbol{Y}}_{\cdot}) = \operatorname{diag}\{\tau_1^2, \ldots, \tau_a^2\}$ where $\tau_i^2 = N\sigma_i^2$ and σ_i^2 is estimated consistently by $\widehat{\sigma}_i^2 = [N^2 n_i(n_i - 1)]^{-1} S_i^2$ where $S_i^2 = \sum_{k=1}^{n_i}(\overline{R}_{i\cdot k} - \overline{R}_{i\cdot\cdot})^2$. Thus, $\widehat{\tau}_i^2 = [Nn_i(n_i - 1)]^{-1} S_i^2$ is consistent for τ_i^2. Let $\boldsymbol{W} = \boldsymbol{V}^{-1}(\boldsymbol{I}_a - \boldsymbol{J}_a\boldsymbol{V}^{-1}/\sum_{j=1}^{a}[1/\tau_j^2])$ and let $\widehat{\boldsymbol{W}}$ be the matrix corresponding to $\boldsymbol{W}$ where τ_i^2 is replaced by $\widehat{\tau}_i^2$. Let

$\overline{R}_{i\cdot\cdot} = (bn_i)^{-1} \sum_{k=1}^{n_i} \sum_{j=1}^{b} R_{ijk}$. By the same arguments as in the previous paragraphs, it follows under H_0^F that the quadratic form

$$Q_N(A) = \frac{1}{N}\left(\overline{R}_{1\cdot\cdot}, \ldots, \overline{R}_{a\cdot\cdot}\right) \widehat{\boldsymbol{W}} \left(\overline{R}_{1\cdot\cdot}, \ldots, \overline{R}_{a\cdot\cdot}\right)'$$
$$= \sum_{i=1}^{a} \frac{n_i(n_i-1)}{S_i^2} \left(\overline{R}_{i\cdot\cdot} - \frac{1}{\sum_{r=1}^{a} [n_r(n_r-1)/S_r^2]} \times \sum_{r=1}^{a} \frac{n_r(n_r-1)\overline{R}_{r\cdot\cdot}}{S_r^2} \right)^2 \tag{4.29}$$

has asymptotically a χ_f^2-distribution with $f = a - 1$ d.f. and $Q_N(A)$ has the RT-property with respect to a parametric statistic with heteroscedastic errors.

Test for the interaction AB. For the hypothesis of no interaction, $H_0^F(AB)$: $F_{ij} = \overline{F}_{i\cdot} + \overline{F}_{\cdot j} - \overline{F}_{\cdot\cdot}$, we use the contrast matrix $\boldsymbol{C} = \boldsymbol{C}_A \otimes \boldsymbol{C}_B$, where

$$\boldsymbol{C}_A = \left(\mathbf{1}_{a-1} \vdots - \boldsymbol{I}_{a-1}\right) \quad \text{and} \quad \boldsymbol{C}_B = \left(\mathbf{1}_{b-1} \vdots - \boldsymbol{I}_{b-1}\right).$$

Both contrast matrices are of full row rank. Thus, the hypothesis is written as $H_0^F(AB)$: $(\boldsymbol{C}_A \otimes \boldsymbol{C}_B)\boldsymbol{F} = \mathbf{0}$ and we consider the asymptotic distribution of $\sqrt{N}(\boldsymbol{C}_A \otimes \boldsymbol{C}_B)\widehat{\boldsymbol{p}}$ which is asymptotically equivalent (see Theorem 4.2) to the statistic $\sqrt{N}(\boldsymbol{C}_A \otimes \boldsymbol{C}_B) \int H \, d\widehat{\boldsymbol{F}} = \sqrt{N}(\boldsymbol{C}_A \otimes \boldsymbol{C}_B)\overline{\boldsymbol{Y}}_{\cdot}$ under $H_0^F(AB)$. The covariance matrix

$$\boldsymbol{V} = \mathrm{Cov}\left(\sqrt{N}\, \overline{\boldsymbol{Y}}_{\cdot}\right) = \bigoplus_{i=1}^{a} \boldsymbol{V}_i$$

is estimated consistently by

$$\widehat{\boldsymbol{V}} = \bigoplus_{i=1}^{a} \widehat{\boldsymbol{V}}_i,$$

where $\widehat{\boldsymbol{V}}_i$ is given in (4.25). Let $\overline{\boldsymbol{R}}_{\cdot} = (\overline{R}_{11\cdot}, \ldots, \overline{R}_{ab\cdot})'$. Then the quadratic form

$$Q_N(AB) = \frac{1}{N} \overline{\boldsymbol{R}}_{\cdot}' \left(\boldsymbol{C}_A' \otimes \boldsymbol{C}_B'\right) \times \left[(\boldsymbol{C}_A \otimes \boldsymbol{C}_B) \bigoplus_{i=1}^{a} \widehat{\boldsymbol{V}}_i \left(\boldsymbol{C}_A' \otimes \boldsymbol{C}_B'\right) \right]^{-1} (\boldsymbol{C}_A \otimes \boldsymbol{C}_B) \overline{\boldsymbol{R}}_{\cdot} \tag{4.30}$$

has a central χ_f^2-distribution with $f = (a-1)(b-1)$ d.f. under $H_0^F(AB)$ and $Q_N(AB)$ has the RT-property with respect to a parametric statistic with heteroscedastic errors.

Test for $H_0^F(B \mid A)$; compound symmetry. Here we derive a test for the simple factor effect B within the levels of A in the compound symmetry model. We consider the hypothesis $H_0^F(B \mid A)$: $(\boldsymbol{I}_a \otimes \boldsymbol{P}_b)\boldsymbol{F} = \boldsymbol{0}$. It follows from Theorem 4.2 that the statistics $\sqrt{N}(\boldsymbol{I}_a \otimes \boldsymbol{P}_b) \int \widehat{H}\, \mathrm{d}\widehat{\boldsymbol{F}}$ and $\sqrt{N}(\boldsymbol{I}_a \otimes \boldsymbol{P}_b) \int H\, \mathrm{d}\widehat{\boldsymbol{F}} = \sqrt{N}(\boldsymbol{I}_a \otimes \boldsymbol{P}_b)\overline{\boldsymbol{Y}}_{\cdot}$ are asymptotically equivalent under $H_0^F(B \mid A)$. In the compound symmetry model, it follows from (4.4) that the covariance matrix of $\boldsymbol{Y}_{ik} = (Y_{i1k}, \ldots, Y_{ibk})'$ is $\boldsymbol{\Sigma}_i = \mathrm{Cov}(\boldsymbol{Y}_{ik}) = \tau_i \boldsymbol{I}_b + c_i^* \boldsymbol{J}_b$ where $\tau_i = \sigma_i^2 - c_i^*$ and thus, $\mathrm{Cov}(\sqrt{N}\boldsymbol{P}_b \overline{\boldsymbol{Y}}_{i\cdot}) = N n_i^{-1} \tau_i \boldsymbol{P}_b$ and

$$\boldsymbol{V} = \mathrm{Cov}\big(\sqrt{N}(\boldsymbol{I}_a \otimes \boldsymbol{P}_b)\overline{\boldsymbol{Y}}_{\cdot}\big) = \bigoplus_{i=1}^{a} \frac{N}{n_i} \tau_i \boldsymbol{P}_b.$$

Let

$$\boldsymbol{V}^- = \bigoplus_{i=1}^{a} \frac{n_i}{N\tau_i} \boldsymbol{P}_b$$

and note that $\mathrm{rank}(\boldsymbol{I}_a \otimes \boldsymbol{P}_b) = a(b-1)$. Let $S_i^2 = \sum_{k=1}^{n_i} \sum_{j=1}^{b} (R_{ijk} - \overline{R}_{i\cdot k})^2$ where $\overline{R}_{i\cdot k} = b^{-1} \sum_{j=1}^{b} R_{ijk}$. Then it follows from Theorem 4.4 that $\widehat{\tau}_i = [N^2 n_i (b-1)]^{-1} S_i^2$ is consistent for τ_i and thus, under $H_0^F(B \mid A)$, the quadratic form

$$\begin{aligned} Q_N^{CS}(B \mid A) &= \frac{1}{N} \left[(\boldsymbol{I}_a \otimes \boldsymbol{P}_b)\overline{\boldsymbol{R}}_{\cdot}\right]' \boldsymbol{V}^- \left[(\boldsymbol{I}_a \otimes \boldsymbol{P}_b)\overline{\boldsymbol{R}}_{\cdot}\right] \\ &= \sum_{i=1}^{a} \frac{n_i(b-1)}{S_i^2} \sum_{j=1}^{b} (\overline{R}_{ij\cdot} - \overline{R}_{i\cdot\cdot})^2 \end{aligned} \tag{4.31}$$

has asymptotically a χ_f^2-distribution with $f = a(b-1)$ d.f.

Note that the statistic given by Thompson and Ammann (1990) is based on $(\frac{1}{a}\boldsymbol{1}_a' \otimes \boldsymbol{P}_b)\widehat{\boldsymbol{p}}$ rather than on $(\boldsymbol{I}_a \otimes \boldsymbol{P}_b)\widehat{\boldsymbol{p}}$ although the hypothesis is formulated as $(\boldsymbol{I}_a \otimes \boldsymbol{P}_b)\boldsymbol{F} = \boldsymbol{0}$. For this reason, the statistic given by Thompson and Ammann (1990) may be rather insensitive to interactions and more sensitive to alternatives of the form $(\frac{1}{a}\boldsymbol{1}_a' \otimes \boldsymbol{P}_b)\boldsymbol{p} \neq \boldsymbol{0}$ which means that there is a main treatment effect.

Test for $H_0^F(B \mid A)$; multivariate model. Next we derive a test for the simple factor effect B, i.e., for the hypothesis $H_0^F(B \mid A)$: $F_{i1} = \cdots = F_{ib} = \overline{F}_{i\cdot}$, $i = 1, \ldots, a$, in the multivariate model. As in the case of the interaction, we consider the asymptotic distribution of $\sqrt{N} \int H\, \mathrm{d}\widehat{\boldsymbol{F}} = \sqrt{N}(\overline{Y}_{11\cdot}, \ldots, \overline{Y}_{ab\cdot})' = \sqrt{N}\, \overline{\boldsymbol{Y}}_{\cdot}$ with covariance matrix

$$\boldsymbol{V} = \bigoplus_{i=1}^{a} \boldsymbol{V}_i.$$

We estimate $\boldsymbol{V}_i$ by $\widehat{\boldsymbol{V}}_i$ given in (4.25). Let $\boldsymbol{W}_i$ be as given in (4.26) and let

$$\boldsymbol{W} = \bigoplus_{i=1}^{a} \boldsymbol{W}_i.$$

Then

$$\boldsymbol{W}\boldsymbol{V}\boldsymbol{W} = \boldsymbol{W} \quad \text{and} \quad \boldsymbol{W}\boldsymbol{F} = \bigoplus_{i=1}^{a} \boldsymbol{W}_i\boldsymbol{F} = \boldsymbol{0}$$

iff $(\boldsymbol{I}_a \otimes \boldsymbol{P}_b)\boldsymbol{F} = \boldsymbol{0}$. The statistic for testing $H_0^F(B \mid A)$ is based on $N^{-1/2}\,\widehat{\boldsymbol{W}}\,\overline{\boldsymbol{R}}_{.}$ where

$$\widehat{\boldsymbol{W}} = \bigoplus_{i=1}^{a} \widehat{\boldsymbol{W}}_i$$

and $\widehat{\boldsymbol{W}}_i$ is given in (4.27). The quadratic form

$$\begin{aligned} Q_N(B \mid A) &= \frac{1}{N}\overline{\boldsymbol{R}}_{.}'\,\widehat{\boldsymbol{W}}\,\overline{\boldsymbol{R}}_{.} \\ &= \frac{1}{N}\sum_{i=1}^{a}\left[\sum_{j=1}^{b}\sum_{j'=1}^{b}\overline{R}_{ij.}\,\widehat{s}_{jj'}^{(i)}\,\overline{R}_{ij'.} - \frac{1}{\widehat{s}_{..}^{(i)}}\left(\sum_{j=1}^{b}\overline{R}_{ij.}\,\widehat{s}_{j.}^{(i)}\right)^2\right] \end{aligned} \tag{4.32}$$

has asymptotically a χ_f^2-distribution with $f = a(b-1)$ under $H_0^F(B \mid A)$. The quantities $\widehat{s}_{jj'}^{(i)}$ are given in (4.27). Note that $Q_N(B \mid A)$ has the RT-property with respect to a multivariate parametric statistic with covariance matrix $\boldsymbol{V}$.

Test for $H_0^F(A \mid B)$. Finally, we derive a test for the simple factor effect A within the levels of B, i.e., for $H_0^F(A \mid B)$: $F_{1j} = \cdots = F_{aj} = \overline{F}_{.j}$, $j = 1, \ldots, b$, and we write the hypothesis as $H_0^F(A \mid B)$: $(\boldsymbol{P}_a \otimes \boldsymbol{I}_b)\boldsymbol{F} = \boldsymbol{0}$. As in the previous paragraphs, the asymptotic distribution of $\sqrt{N}\int H\,d\widehat{\boldsymbol{F}} = \sqrt{N}\,\overline{\boldsymbol{Y}}_{.}$ is considered. The covariance matrix of $\sqrt{N}\,\overline{\boldsymbol{Y}}_{.}$ under $H_0^F(A \mid B)$ is

$$\operatorname{Cov}\big(\sqrt{N}\,\overline{\boldsymbol{Y}}_{.}\big) = \bigoplus_{i=1}^{a} \boldsymbol{V}_i = \boldsymbol{I}_a \otimes \boldsymbol{V}$$

where $\boldsymbol{V}_i \equiv \boldsymbol{V}$, $i = 1, \ldots, a$, since there is no main effect A and no interaction AB. The covariance matrix $\boldsymbol{V}$ is consistently estimated by $\widehat{\boldsymbol{V}} = a^{-1}\sum_{i=1}^{a}\widehat{\boldsymbol{V}}_i$ where $\widehat{\boldsymbol{V}}_i$ is given in (4.25). Then, it follows from Theorem 4.5 that the quadratic form

$$\begin{aligned} Q_N(A \mid B) &= \frac{1}{N}\overline{\boldsymbol{R}}_{.}'\left[\boldsymbol{P}_a \otimes \widehat{\boldsymbol{V}}^{-1}\right]\overline{\boldsymbol{R}}_{.} \\ &= \frac{1}{N}\sum_{i=1}^{a}\left(\overline{\boldsymbol{R}}_{i.} - \overline{\boldsymbol{R}}_{..}\right)'\,\widehat{\boldsymbol{V}}^{-1}\,\left(\overline{\boldsymbol{R}}_{i.} - \overline{\boldsymbol{R}}_{..}\right) \end{aligned} \tag{4.33}$$

Table 4.3
Rank means for the PIP data

Group	Time points							
	1	2	3	4	5	6	7	8
Control	185.3	99.6	47.7	42.6	79.0	109.6	122.7	170.7
Obese	212.0	180.0	150.5	123.5	95.1	115.3	144.0	172.3

has asymptotically a χ_f^2-distribution with $f = b(a-1)$ under $H_0^F(A \mid B)$ and the statistic $Q_N(A \mid B)$ has the RT-property with respect to a multivariate parametric statistic having covariance matrix $\boldsymbol{I}_a \otimes \boldsymbol{V}$.

EXAMPLE 4. We apply the procedures derived in this section to the data given by Zerbe (1979) where for 13 control and 20 obese patients plasma inorganic phosphate (PIP) was measured $0, \frac{1}{2}, 1, 1\frac{1}{2}, 2, 3, 4$ and 5 hours after a standard dose oral glucose challenge. For the description of the trial and the data set, we refer to Zerbe (1979). An adequate model for this trial is the 3-factor mixed model with two crossed fixed factors, namely the factor 'group' (control, obese) and the factor 'time' (8 fixed time points). The 33 subjects are the levels of the random factor and are nested under the factor 'group'. Since PIP is 8 times repeatedly measured during 5 hours, a multivariate model is appropriate. Two standard questions are commonly to be answered for such a trail: (1) Do the two time curves have the same shape? – (2) Is there any influence of the factor time?

The first question can be answered by testing the interaction (4.30) between the two fixed factors. The time effect is analysed by testing the treatment effect B (4.28). Since both quadratic forms have the RT-property, they can be easily computed from the overall ranks. The rank means and the results are given below.

If the results $Q_N(AB) = 60.29$ and $Q_N(B) = 329.44$ are compared with the asymptotic χ^2-distribution, the p-values may be liberal. An approximation of the p-values for small sample sizes is motivated by the small sample distribution of Hotelling's T^2. The null distribution of $(n_1+n_2-b)Q_N(AB)/[(b-1)(n_1+n_2-2)]$ is approximated by an F-distribution with $f_1 = b-1 = 7$ and $f_2 = n_1+n_2-b = 25$ which yields $p = 0.00012$ and a highly significant different pattern for the two time curves is detected. Similarly, $25 \cdot Q_N(B)/217 = 37.95$ is compared with the same F-distribution which yields $p < 10^{-5}$ and a highly significant influence of the time on PIP is detected.

4.4.1.3. Tests for heteroscedastic models, $a \times 2$-designs

Models and hypotheses. For heteroscedastic models, we restrict our considerations to $a \times 2$ designs since the estimation and inversion of the covariance matrix in the case of $b > 2$ requires tedious computations. We consider the general model

$$\begin{aligned} \boldsymbol{X}_{ik} &= (X_{i1k}, X_{i2k})', \quad X_{ijk} \sim F_{ij}(x), \\ & i = 1, \ldots, a, \ j = 1, 2, \ k = 1, \ldots, n_i, \end{aligned} \tag{4.34}$$

where the random vectors $\boldsymbol{X}_{ik}$ are assumed to be independent. This is typically the design in a clinical trial where a new drug (n_1 patients) is compared to a standard (n_2 patients) and the patients are observed prior to treatment (baseline values, level 1 of factor B) and after treatment (endpoint values, level 2 of factor B). Moreover, paired comparisons for $a \geqslant 2$ subgroups of experimental units are covered by this model. The well known two-period cross-over design (which is frequently used in pharmaceutical studies with volunteers or in psychological learning experiments) is studied separately in the last subsection.

To define treatment effects and to formulate hypotheses, we will use the conditional generalized means $w_i = \int F_{i1}\,\mathrm{d}F_{i2}$, which are generalized treatment effects within group $i = 1, \ldots, a$. For the case of $a = b = 2$, we consider also the group effect (main effect of factor A) and we define the generalized group effect as $g = g_1 + g_2$ where $g_j = \int F_{1j}\,\mathrm{d}F_{2j}$, $j = 1, 2$. In what follows, we consider the hypotheses

1. $H_0^w(B)$: $\overline{w} = \frac{1}{a}\sum_{i=1}^{a} w_i = 1/2$, no treatment effect (main effect B),
2. $H_0^w(AB)$: $\rho = \sum_{i=1}^{a}(w_i - \overline{w})^2 = 0$, no AB interaction,
3. $H_0^w(B \mid A)$: $\eta = \sum_{i=1}^{a}(w_i - 1/2)^2 = 0$, no simple factor B effect,
4. $H_0^w(A)$: $g = 1$, no group effect (main effect A), if $a = 2$.

The relations to the hypotheses in the linear model follow in the same way as in the 2×2 design (Proposition 2.9) in Section 2.2.1.

Estimators and notations

The quantities w_i are estimated consistently by

$$\widehat{w}_i = \int \widehat{F}_{i1}\,\mathrm{d}\widehat{F}_{i2} = \frac{1}{n_i}\left(\overline{R}_{i2\cdot}^{A} - \frac{n_i + 1}{2}\right), \quad i = 1, \ldots, a, \tag{4.35}$$

where $\overline{R}_{i2\cdot}^{A} = n_i^{-1}\sum_{k=1}^{n_i} R_{i2k}^{A}$ and R_{i2k}^{A} is the rank of X_{i2k} among all the $2n_i$ observations within group i (factor A), $i = 1, \ldots, a$. The quantities g_j (in the 2×2 design) are estimated consistently by

$$\widehat{g}_j = \int \widehat{F}_{1j}\,\mathrm{d}\widehat{F}_{2j} = \frac{1}{n_1}\left(\overline{R}_{2j\cdot}^{B} - \frac{n_2 + 1}{2}\right), \quad j = 1, 2, \tag{4.36}$$

where $\overline{R}_{2j\cdot}^{B} = n_2^{-1}\sum_{k=1}^{n_2} R_{2jk}^{B}$ and R_{2jk}^{B} is the rank of X_{2jk} among all the $N = n_1 + n_2$ observations within level j of factor B (treatments), $j = 1, 2$. Note that the estimators $\widehat{w}_i$, $i = 1, \ldots, a$, are independent whereas the estimators $\widehat{g}_1$ and $\widehat{g}_2$ may be dependent.

The statistic $\sqrt{n_i}(\widehat{w}_i - w_i)$ is asymptotically equivalent to

$$\frac{1}{\sqrt{n_i}}\sum_{k=1}^{n_i}\left[F_{i1}(X_{i2k}) - F_{i2}(X_{i1k})\right] - \sqrt{n_i}(2w_i - 1)$$

and the statistic $\sqrt{N}(\widehat{g}-g)=\sqrt{N}(\widehat{g}_1+\widehat{g}_2-g_1-g_2)$ is asymptotically equivalent to

$$\frac{\sqrt{N}}{n_1n_2}\left(\sum_{k=1}^{n_2} n_1\left[F_{11}(X_{21k})+F_{12}(X_{22k})\right]-\sum_{k=1}^{n_1} n_2\left[F_{21}(X_{11k})+F_{22}(X_{12k})\right]\right.$$
$$\left. - 2n_1n_2[g_1+g_2-1]\right).$$

For details, we refer to, e.g., Brunner et al. (1995) and Brunner and Neumann (1986a). Note that $2n_i\widehat{H}_i(X_{ijk})=n_i\widehat{F}_{i1}(X_{ijk})+n_i\widehat{F}_{i2}(X_{ijk})=R^A_{ijk}-1/2$ where R^A_{ijk} is the rank of X_{ijk} among all the $2n_i$ observations within group i (level i of factor A), $i=1,\dots,a$. Let $R^{(ij)}_{ijk}=n_i\widehat{F}_{ij}(X_{ijk})+1/2$ be the rank of X_{ijk} within group i and treatment j and let $R^{A*}_{ijk}=R^A_{ijk}-R^{(ij)}_{ijk}$. Then the estimator

$$\widehat{\tau}^2_i=\frac{1}{n_i^2(n_i-1)}\sum_{k=1}^{n_i}\left[\left(R^{A*}_{i2k}-R^{A*}_{i1k}\right)-\left(\overline{R}^A_{i2\cdot}-\overline{R}^A_{i1\cdot}\right)\right]^2 \tag{4.37}$$

is consistent for $\tau^2_i=\mathrm{Var}(\sqrt{n_i}\widehat{w}_i)$. In the same way, the variance $v^2=\mathrm{Var}(\widehat{g})$ is estimated consistently by

$$\widehat{v}^2=\sum_{i=1}^{2}\frac{1}{(N-n_i)^2n_i(n_i-1)}\sum_{k=1}^{n_i}\left[\sum_{j=1}^{2}\left(R^{B*}_{ijk}-\overline{R}^{B*}_{ij\cdot}\right)\right]^2 \tag{4.38}$$

where $N=n_1+n_2$, $R^{B*}_{ijk}=R^B_{ijk}-R^{(ij)}_{ijk}$, $R^{(ij)}_{ijk}$ is the rank of X_{ijk} within group i and treatment j and R^B_{ijk} is the rank of X_{jk} within treatment j.

Derivation of the statistics

Test for $H^w_0(B)$. Since the observations X_{ijk} are ranked separately for each group $i=1,\dots,a$, the estimators $\widehat{w}_i$ given in (4.35) are independent and the variance $\sigma^2=\mathrm{Var}(\sum_{i=1}^a\widehat{w}_i)$ is estimated by $\widehat{\sigma}^2=\sum_{i=1}^a n_i^{-1}\widehat{\tau}^2_i$ where $\widehat{\tau}^2_i$ is given in (4.37). Thus, the statistic

$$T(B)=\frac{1}{\widehat{\sigma}}\sum_{i=1}^{a}\frac{1}{n_i}\left(\overline{R}^A_{i2\cdot}-\frac{2n_i+1}{2}\right) \tag{4.39}$$

has asymptotically a standard normal distribution under $H^w_0(B)$.

Test for $H^w_0(AB)$. The hypothesis for the interaction is $H^w_0(AB)$: $\boldsymbol{P}_a\boldsymbol{w}=\boldsymbol{0}$ where $\boldsymbol{w}=(w_1,\dots,w_a)'$ is the vector of the generalized means w_i. The vector $\boldsymbol{w}$ is estimated by $\widehat{\boldsymbol{w}}=(\widehat{w}_1,\dots,\widehat{w}_a)'$ and $\widehat{w}_i$ is given in (4.35). Let $\boldsymbol{S}=\mathrm{Cov}(\sqrt{N}\,\widehat{\boldsymbol{w}})$ where $N=\sum_{i=1}^a n_i$. Then, $\boldsymbol{S}=N\mathrm{diag}\{\sigma^2_1,\dots,\sigma^2_a\}$ where $\sigma^2_i=\mathrm{Var}(\widehat{w}_i)=n_i^{-1}\tau^2_i$ is consistently estimated by $\widehat{\boldsymbol{S}}=N\mathrm{diag}\{\widehat{\tau}^2_1/n_1,\dots,\widehat{\tau}^2_a/n_a\}'$ and $\widehat{\tau}^2_i$ is given in formula (4.37). Let $\boldsymbol{A} = \boldsymbol{S}^{-1}(\boldsymbol{I}_a - \boldsymbol{J}_a\boldsymbol{S}^{-1}/\sum_{i=1}^{a}(1/\sigma_i^2))$. Then $\boldsymbol{ASA} = \boldsymbol{A}$ and $\boldsymbol{Aw} = \mathbf{0}$ iff $\boldsymbol{P}_a\boldsymbol{w} = \mathbf{0}$. Thus the quadratic form

$$Q_N(AB) = \sum_{i=1}^{a} \frac{n_i}{\widehat{\tau}_i^2}\left(\widehat{w}_i - \frac{1}{\sum_{s=1}^{a}(n_s/\widehat{\tau}_s^2)}\sum_{s=1}^{a}\frac{n_s\widehat{w}_s}{\widehat{\tau}_s^2}\right)^2 \tag{4.40}$$

has asymptotically a χ_f^2-distribution with $f = a - 1$ under $H_0^w(AB)$.

Test for $H_0^w(B \mid A)$. The hypothesis for no simple factor B effect is formulated as $H_0^w(B \mid A)$: $\sum_{i=1}^{a}(w_i - 1/2)^2 = 0$ and under this hypothesis, the statistic $(\widehat{w}_i - 1/2)^2/\widehat{\sigma}_i^2$ has asymptotically a χ_1^2-distribution, $i = 1, \ldots, a$. Therefore, the statistic

$$Q_N(B \mid A) = \sum_{i=1}^{a}(\widehat{w}_i - 1/2)^2/\widehat{\sigma}_i^2 = \sum_{i=1}^{a}\frac{n_i}{\widehat{\tau}_i^2}\left(\overline{R}_{i2\cdot}^A - \frac{2n_i + 1}{2}\right)^2 \tag{4.41}$$

has asymptotically a χ_a^2-distribution under $H_0^w(B \mid A)$. (Note that separate rankings within each group i are used.)

Test for the group effect A in the 2×2 design. The hypothesis of no group effect is $H_0^w(A)$: $g = g_1 + g_2 = 1$ and g is estimated consistently by $\widehat{g} = \widehat{g}_1 + \widehat{g}_2 = n_1^{-1}(\overline{R}_{21\cdot}^B + \overline{R}_{22\cdot}^B - (n_2 + 1))$ where $\overline{R}_{2j\cdot}^B = n_2^{-1}\sum_{k=1}^{n_2} R_{2jk}^B$ and R_{2jk}^B is the rank of X_{2jk} among all the $N = n_1 + n_2$ observations within level j of factor B, $j = 1, 2$. A consistent estimator $\widehat{v}^2$ for $v^2 = \mathrm{Var}(\widehat{g})$ is given in (4.38). Under $H_0^w(A)$, the statistic

$$T_N(A) = \frac{1}{n_1\widehat{v}}(\overline{R}_{21\cdot}^B + \overline{R}_{22\cdot}^B - (N + 1)) \tag{4.42}$$

has asymptotically a standard normal distribution.

The statistics given in this subsection have been considered by Brunner and Neumann (1986a) for the 2×2 design and $n_1 = n_2$ under the assumption of no ties. The results given here are also valid in the case of ties.

Note that the statistics given in this section are also applicable for heteroscedastic models. Moreover, they are invariant under monotone transformations of the data and they are robust to outliers since the original observations are ranked within each group.

4.4.1.4. The two-period cross-over design. The two-period cross-over design is rather often used in medical trials, especially with chronic diseases or in drug experiments with volunteers. With this design, the error variance is reduced by applying both treatments to the same subject. To exclude time (period) effects, one group of subjects receives the treatments in the sequence AB and the other group receives the treatments in the sequence BA. The disadvantage of this design is that in the second period, there may be a residual effect of the treatment given in the first period.

A linear model for the analysis of this design has been considered by Grizzle (1965, 1974). Koch (1972) considered rank tests for the linear model where the sums and differences of the observations were ranked. In this setup, the 'rank statistics' are not invariant under monotone transformations of the data. Moreover, these tests cannot be applied to pure ordinal data (like, e.g., scores in psychological tests or grading scales in clinical trials or Ecology) since sums or differences of such data are not reasonable.

A nonparametric model for the cross-over design was first considered by Brunner and Neumann (1987) where nonparametric hypotheses were stated and their properties in the linear model were analysed. Rank tests under the assumption of no ties have been derived in this paper. In the sequel, we use the generalized effects and the statistics of the previous section to analyse a general model in the two-period cross-over design (TPCOD) including the case of ties.

First we write the linear model for the TPCOD.

$$\begin{aligned} X_{11k} &= \mu + \Phi_1 + \pi_1 + S_{1k} + \epsilon_{11k}, \\ X_{12k} &= \mu + \Phi_2 + \pi_2 + \lambda_1 + S_{1k} + \epsilon_{12k}, \\ X_{21k} &= \mu + \Phi_2 + \pi_1 + S_{2k} + \epsilon_{21k}, \\ X_{22k} &= \mu + \Phi_1 + \pi_2 + \lambda_2 + S_{2k} + \epsilon_{22k}, \end{aligned} \tag{4.43}$$

where Φ_i is the effect of treatment $i = 1, 2$, λ_i is the residual effect of treatment i in period 2, π_i is the effect of period i, S_{ik} are i.i.d. $N(0, \sigma_S^2)$ random variables, representing the random subjects, and ϵ_{ijk} are i.i.d. $N(0, \sigma^2)$ random variables, representing the error terms. We may reparametrize this model by letting $\Phi = \Phi_1 = -\Phi_2$, (treatment effect), $\lambda = \lambda_1 = -\lambda_2$, (residual effect), $\pi = \pi_1 = -\pi_2$, (period effect) and $\delta = 2\Phi - \lambda$, (cross-over effect).

The hypotheses of interest in this model are:

1. $H_0(\delta)$: $\delta = 0$, no cross-over effect,
2. $H_0(\lambda)$: $\lambda = 0$, no residual effect,
3. $H_0(\pi)$: $\pi = 0$, no period effect

and the hypothesis of no treatment effect $H_0(\Phi)$: $\Phi = 0$ is 'tested' by the conclusion $\Phi = 0$ if $\delta = 0$ and $\lambda = 0$.

The nonparametric (or general) model is given in (4.34) and nonparametric effects in the TPCOD are related to the hypotheses $H_0^w(A)$, $H_0^w(B)$ and $H_0^w(AB)$ (see Section 4.4.1.3) in the following way:

DEFINITION 4.7 (Nonparametric effects in the TPCOD).
(1) $w^\delta = w_1 - w_2$, *where* $w_i = \int F_{i1}\,\mathrm{d}F_{i2}$, $i = 1, 2$, *is called 'nonparametric cross-over effect'*,
(2) $w^\pi = w_1 + w_2$ *is called 'nonparametric period effect'*,
(3) $g^\lambda = g_1 + g_2$, *where* $g_j = \int F_{1j}\,\mathrm{d}F_{2j}$, $j = 1, 2$, *is called 'nonparametric residual effect'*.

The properties and interpretation of these effects in the linear model (4.43) are given in the next Proposition.

PROPOSITION 4.8. *If $F(\cdot)$ is strictly increasing, then in the linear model* (4.43),
(1) w^δ *is strictly decreasing in* δ *and* $w^\delta = 0 \Longleftrightarrow \delta = 0$,
(2) w^π *is strictly decreasing in* π *and* $w^\pi = 1 \Longleftrightarrow \pi = 0$,
(3) g^λ *is strictly decreasing in* λ *and* $g^\lambda = 1 \Longleftrightarrow \lambda = 0$.

PROOF. (1) In the linear model, $\int F_{11}\,\mathrm{d}F_{12} = \int F(x-\Phi_1-\pi_1+\Phi_2+\pi_2+\lambda_1)\,\mathrm{d}F(x) = G(\Phi_2 - \Phi_1 + \pi_2 - \pi_1 + \lambda_1)$ and $\int F_{21}\,\mathrm{d}F_{22} = G(\Phi_1 - \Phi_2 + \pi_2 - \pi_1 + \lambda_2)$ where $G(u) = \int F(x+u)\,\mathrm{d}F(x)$ is a distribution function which is symmetric about 0, i.e., $G(u) = 1 - G(-u)$. If $F(\cdot)$ is strictly increasing, then $G(\cdot)$ is strictly increasing and $w_1 - w_2 = G(-2\Phi - 2\pi + \lambda) - 1 + G(-2\Phi + 2\pi + \lambda)$ is strictly decreasing in $\delta = 2\Phi - \lambda$. Let $\delta = 2\Phi - \lambda = 0$. Then $w_1 - w_2 = G(-2\pi) - 1 + G(2\pi) = 0$ and the solution $\delta = 0$ is unique since $G(\cdot)$ is strictly decreasing in δ. The proofs for (2) and (3) are similar. □

Tests for the hypotheses in the TPCOD

The test for $H_0(\delta)$: $w^\delta = 0$ (no cross-over effect) is a special case of the test for no interaction, $H_0^w(AB)$. Let

$$\widehat{\sigma}^2 = \sum_{i=2}^{2} \frac{1}{n_i^3(n_i-1)} \sum_{k=1}^{n_i} \left[(R_{i2k}^{A*} - R_{i1k}^{A*}) - (\overline{R}_{i2\cdot}^{A} - \overline{R}_{i1\cdot}^{A})\right]^2 \tag{4.44}$$

where R_{ijk}^{A*} is defined in (4.37) and R_{ijk}^{A} is the rank of X_{ijk} among all the $N_i = 2n_i$ observations within group $i = 1,2$. Then, the statistic $Q_N(AB)$ given in (4.40) simplifies to

$$\widehat{w}^\delta = \frac{1}{n_1 n_2 \widehat{\sigma}}\left(n_2 \overline{R}_{12\cdot}^{A} - n_1 \overline{R}_{22\cdot}^{A} - (n_2 - n_1)/2\right) \tag{4.45}$$

where $\overline{R}_{i2\cdot}^{A} = n_i^{-1}\sum_{k=1}^{n_i} R_{i2k}^{A}$, $i = 1,2$. Under $H_0(\delta)$: $w^\delta = 0$, the statistic $\widehat{w}^\delta$ has asymptotically a standard normal distribution.

The test for $H_0(\pi)$: $w^\pi = 1$ (no period effect) is a special case of the test for no main effect B, $H_0^w(B)$. Here, the statistic $T(B)$ given in (4.39) simplifies to

$$T^\pi = \frac{1}{n_1 n_2 \widehat{\sigma}}\left(n_2 \overline{R}_{12\cdot}^{A} + n_1 \overline{R}_{22\cdot}^{A} - (4n_1 n_2 + n_1 + n_2)/2\right) \tag{4.46}$$

where $\widehat{\sigma}^2$ is given in (4.44). Under $H_0(\pi)$: $w^\pi = 1$, the statistic T^π has asymptotically a standard normal distribution.

The test for $H_0(\lambda)$: $g^\lambda = 1$ (no residual effect) is the test for no group effect A, i.e., for $H_0^w(A)$. The statistic for this test is

$$T^\lambda = \frac{1}{n_1 \widehat{v}}\left(\overline{R}_{21\cdot}^{B} + \overline{R}_{22\cdot}^{B} - (N+1)\right) \tag{4.47}$$

where $\overline{R}_{2j\cdot}^{B}$, $j = 1,2$, and $\widehat{v}$ are given in (4.36) and (4.38) respectively. Under $H_0(\lambda)$, the statistic T^λ has asymptotically a standard normal distribution.

An application for the above procedures is given by the following example.

Table 4.4
Ranks and rank means for the residual effect g^λ, for the period effect w^π and for the cross-over effect w^δ

Group	Patient number	Period 1 R^A_{ijk}	R^B_{ijk}	$R^{(ij)}_{ijk}$	Period 2 R^A_{ijk}	R^B_{ijk}	$R^{(ij)}_{ijk}$
	1	8	11	6	11	11	5
$i = 1$	3	1.5	3.5	1.5	12	12	6
	5	4	6	4	9	9	3
A/B	6	5	7	5	7	8	2
	10	1.5	3.5	1.5	10	10	4
	12	3	5	3	6	6	1
Means		3.83	6		9.17	9.33	
	2	12	12	6	8	7	6
$i = 2$	4	1.5	1	1	3.5	2.5	2.5
	7	11	10	5	5	4	4
B/A	8	6	2	2	3.5	2.5	2.5
	9	10	9	4	7	5	5
	11	9	8	3	1.5	1	1
Means		8.25	7		4.75	3.67	

EXAMPLE 5. We analyse the data of a TPCOD given by Senn (1993) in Table 4.2 where visual analogue scale scores for the patients' judgement of the efficacy of two drugs are displayed. Note that a linear model is not appropriate for such score data. For the description of the trial as well as for the original data set, we refer to the above quoted book.

The residual effect g^λ is analysed by ranking the data within each period ($j = 1, 2$) separately over the two groups (R^B_{ijk}), $i = 1, 2$, and within each treatment group and period ($R^{(ij)}_{ijk}$). The period effect w^π and the conditional treatment effect (cross-over effect) w^δ are analysed by ranking the observations within each group ($i = 1, 2$) separately over both periods (R^A_{ijk}), $j = 1, 2$ and within each treatment group and period ($R^{(ij)}_{ijk}$). The ranks R^A_{ijk}, R^B_{ijk} and $R^{(ij)}_{ijk}$ as well as the rank means $\overline{R}^A_{ij\cdot}$ and $\overline{R}^B_{ij\cdot}$ are given in Table 4.4.

The results for $\widehat{v}^2$ given in (4.38) and T^λ given in (4.47) are $\widehat{v}^2 = 52.67/1080 = 0.049$ and $T^\lambda = (7 + 3.67 - 13)/1.33 = -1.76$ and $p = 0.139$ (two-sided) where the t_f-distribution with $f = \min(n_1, n_2) - 1 = 5$ is used as a small sample approximation for the distribution of T^λ under H^P_0. At the 5%-level, no significant residual effect is detected.

The results for $\widehat{\sigma}^2$ given in (4.44) and T^π given in (4.46) are $\widehat{\sigma}^2 = 26.33/1080 = 0.024$ and $T^\pi = 6(9.17 + 4.75 - 13)/5.62 = 0.98$ and $p = 0.350$ (two-sided) where the t_f-distribution with $f = 2(n - 1) = 10$ is used as a small sample approximation for the distribution of T^π under H^P_0. At the 5%-level, no significant period effect is detected.

The result for the cross-over effect is $\widehat{w}^\delta = 6(9.17 - 4.75)/5.62 = 4.72$ and $p = 0.00082$ (two-sided) where the t_f-distribution with $f = 2(n-1) = 10$ is used as a small sample approximation for the distribution of $\widehat{w}^\delta$ under H_0^p. At the 5%-level, a significant treatment effect is detected.

4.4.2. Cross-classified designs/repeated measurements on both fixed factors
4.4.2.1. Models and hypotheses
Linear model. Here we consider models where each randomly chosen subject k receives all the ab combinations (i,j), $i = 1,\ldots,a$, $j = 1,\ldots,b$, of two treatments A and B. The linear model of this design is

$$X_{ijk} = \mu + \alpha_i + \beta_j + (\alpha\beta)_{ij} + S_k + (AS)_{ik} + (BS)_{jk} + \varepsilon_{ijk}, \qquad (4.48)$$
$$i = 1,\ldots,a,\ j = 1,\ldots,b,\ k = 1,\ldots,n,$$

where μ is the overall mean, α_i is the treatment effect of level i of the fixed factor A, β_j is the treatment effect of level j of the fixed factor B, $(\alpha\beta)_{ij}$ is the effect of combination (i,j) of the fixed factors (interaction), S_k are i.i.d. $N(0,\sigma_S^2)$ random variables representing the random subject effect, $(AS)_{ik}$ are i.i.d. $N(0,\sigma_{AS}^2)$ random variables representing the random interaction between subjects and factor A, and $(BS)_{jk}$ are i.i.d. $N(0,\sigma_{BS}^2)$ random variables representing the random interaction between subjects and factor B. Finally, ε_{ijk} are i.i.d. $N(0,\sigma_\varepsilon^2)$. All random effects are assumed to be independent from all other random variables. Note that the three-way interaction $(ABS)_{ijk}$ is mixed up with the error term ε_{ijk} since there is only $m = 1$ replication for each combination (i,j). The vector of all ab observations on subject k is denoted by $\boldsymbol{X}_k = (X_{11k},\ldots,X_{abk})'$, $k = 1,\ldots,n$, and the vectors $\boldsymbol{X}_k$ are independent. The covariance matrix $\boldsymbol{S} = \mathrm{Cov}(\boldsymbol{X}_k)$ follows from the restrictive assumptions for the random effects. In the linear model (4.48), the structure of $\boldsymbol{S}$, namely the two-way compound symmetry, is not changed under the treatment.

Two-way layouts where the covariance matrix has the two-way compound symmetry are called 'two-factor repeated measurements designs'. If the compound symmetry of the covariance matrix $\boldsymbol{S}$ is not assumed, the model is called 'two-way multivariate model'. For a discussion of these models, see Arnold (1981).

General model. We relax the assumption of normality, of additive effects and of the compound symmetry of the common distribution function and write the general model as

$$\boldsymbol{X}_k = (X_{11k},\ldots,X_{abk})', \quad X_{ijk} \sim F_{ij}(x),$$
$$i = 1,\ldots,a,\ j = 1,\ldots,b,\ k = 1,\ldots,n, \qquad (4.49)$$

where the vectors $\boldsymbol{X}_k$ are independent and the covariance matrix $\boldsymbol{S} = \mathrm{Cov}(\boldsymbol{X}_k)$ is only assumed to be non-singular and not to depend on k. The assumption that the covariance matrix $\boldsymbol{S}$ in the general model is not changed under the treatment seems to be unrealistic. Therefore, the multivariate model with an arbitrary dependence structure of the observations within one subject seems to be the only reasonable assumption for the covariance matrix in the general model (4.49).

Hypotheses. Rank tests for the two-way multivariate model have been considered by Thompson (1991a) and Akritas and Arnold (1994). The first named author considered a rank test for the hypothesis H_0: $F_{ij} = F_i$, $j = 1, \dots, b$, $i = 1, \dots, a$, which is in fact the hypothesis $H_0(B \mid A)$: $(\boldsymbol{I}_a \otimes \boldsymbol{P}_b)\boldsymbol{F} = \boldsymbol{0}$ where $\boldsymbol{F} = (F_{11}, \dots, F_{ab})'$. Note that this hypothesis is a 'joint hypothesis', i.e., main and interaction effects are tested together. Such joint hypotheses are useful in practice and have been considered already by Koch (1970) in the context of a complex split-plot design. Akritas and Arnold (1994) stated the hypotheses of the linear two-way layout model in terms of distribution functions $H_0^F(A)$: $(\boldsymbol{P}_a \otimes \frac{1}{b}\boldsymbol{1}_b')\boldsymbol{F} = \boldsymbol{0}$, $H_0^F(AB)$: $(\boldsymbol{P}_a \otimes \boldsymbol{P}_b)\boldsymbol{F} = \boldsymbol{0}$ and $H_0^F(A \mid B)$: $(\boldsymbol{P}_a \otimes \boldsymbol{I}_b)\boldsymbol{F} = \boldsymbol{0}$, and based the statistics on a consistent estimator of $\boldsymbol{p} = \int H \, \mathrm{d}\boldsymbol{F}$ where $H = (ab)^{-1} \sum_{i=1}^a \sum_{j=1}^b F_{ij}$ and $\boldsymbol{F} = (F_{11}, \dots, F_{ab})'$. They showed that under the hypotheses, the covariance matrices of these statistics can easily be computed and estimated by ranks. They also pointed out that the hypotheses of the linear model are implied by the nonparametric hypotheses based on the distribution functions. By means of these hypotheses it is possible to test nonparametric main effects, interactions and simple factor effects separately in the general model (4.49).

The hypotheses for the three-way mixed model with two factors fixed as well as the relations between the hypotheses in different models are essentially the same as for the two-way layout with fixed factors and independent observations (see Section 2.2.1) since the formulation of the hypotheses does not imply the structure of the covariance matrix $\boldsymbol{S}$. The results given in Proposition 2.9 complete the discussion of the hypotheses for the two-way layout in Thompson (1991a, Section 7.2) and Akritas and Arnold (1994, Section 4.2).

Convenient rank statistics for general heteroscedastic models can only be derived for special cases, i.e., for the case of the 2×2-design. A Rank procedure for this design has been developed by Brunner and Neumann (1986a). In the next section, we derive rank statistics for $H_0^F(A)$, $H_0^F(AB)$ and $H_0^F(A \mid B)$ for $a, b \geqslant 2$ and for $H_0^w(A)$, $H_0^w(AB)$ and $H_0^w(A \mid B)$ in the 2×2 design.

4.4.2.2. Derivation of the statistics, $a, b \geqslant 2$

Estimators and notations

Let $\boldsymbol{X} = (\boldsymbol{X}_1', \dots, \boldsymbol{X}_n')'$ where $\boldsymbol{X}_k = (X_{11k}, \dots, X_{abk})'$, be the vector of the observed random variables where $X_{ijk} \sim F_{ij}(x)$, $k = 1, \dots, n$, and assume that the vectors $\boldsymbol{X}_k$ of the observations for subject k are independent. The observations X_{ijk} and $X_{i'j'k}$ within subject k however, may be arbitrarily dependent such that the covariance matrix of $\boldsymbol{X}_k$ is of full rank. As in the previous section, we will base the statistics on a consistent estimator of the vector of the generalized means $\boldsymbol{p} = \int H \, \mathrm{d}\boldsymbol{F}$ where $H = (ab)^{-1} \sum_{i=1}^a \sum_{j=1}^b F_{ij}$. The vector $\boldsymbol{p}$ is estimated consistently by $\widehat{\boldsymbol{p}} = \int \widehat{H} \, \mathrm{d}\widehat{\boldsymbol{F}}$ where $\widehat{\boldsymbol{F}} = (\widehat{F}_{11}, \dots, \widehat{F}_{ab})'$ and $\widehat{F}_{ij}(x) = n^{-1} \sum_{k=1}^n c(x - X_{ijk})$.

Let $\boldsymbol{Y}_k = (Y_{11k}, \dots, Y_{abk})'$ where $Y_{ijk} = H(X_{ijk})$ and let $\overline{\boldsymbol{Y}}_{\cdot} = (\overline{Y}_{11\cdot}, \dots, \overline{Y}_{ab\cdot})'$ where $\overline{Y}_{ij\cdot} = n^{-1} \sum_{k=1}^n Y_{ijk}$. Then it follows from Theorem 4.2 that the statistics $\sqrt{n} \int \widehat{H} \, \mathrm{d}(\widehat{\boldsymbol{F}} - \boldsymbol{F})$ and $\sqrt{n} \int H \, \mathrm{d}(\widehat{\boldsymbol{F}} - \boldsymbol{F}) = \sqrt{n}(\overline{\boldsymbol{Y}}_{\cdot} - \boldsymbol{p})$ are asymptotically equivalent. We denote by $\boldsymbol{V}$ the covariance matrix of $\sqrt{n}\, \overline{\boldsymbol{Y}}_{\cdot}$ and assume that $\boldsymbol{V}$ is nonsingular. Let R_{ijk} be the rank of X_{ijk} among all the $N = abn$ observations, and let $\boldsymbol{R}_k = (R_{11k}, \ldots, R_{abk})'$ be the vector of the ranks for subject k, and denote by $\overline{\boldsymbol{R}}_{\cdot} = (\overline{R}_{11\cdot}, \ldots, \overline{R}_{ab\cdot})'$ the vector of the rank means over all subjects. It then, follows from Theorem 4.3 that

$$\widehat{\boldsymbol{V}} = \frac{1}{N^2(n-1)} \sum_{k=1}^{n} (\boldsymbol{R}_k - \overline{\boldsymbol{R}}_{\cdot}) (\boldsymbol{R}_k - \overline{\boldsymbol{R}}_{\cdot})' \tag{4.50}$$

is a consistent estimator of $\boldsymbol{V}$ and thus, $\boldsymbol{C}\widehat{\boldsymbol{V}}\boldsymbol{C}'$ is a consistent estimator of $\mathrm{Cov}\left(\sqrt{n}\, \boldsymbol{C}\, \overline{\boldsymbol{Y}}_{\cdot}\right)$ where $\boldsymbol{C}$ is a suitable contrast matrix needed for the formulation of the hypothesis.

Statistics

Test for $H_0^F(A)$. We note that the design is symmetric in A and B and therefore, only a test for the main effect A is derived. The test for B will follow by interchanging the indices i and j. The hypothesis is formulated as $H_0^F(A)$. $(\boldsymbol{P}_a \otimes \frac{1}{b}\boldsymbol{1}_b')\boldsymbol{F} = \boldsymbol{0}$. As in the previous section, we consider first the asymptotic distribution of $\sqrt{n}(\boldsymbol{I}_a \otimes \frac{1}{b}\boldsymbol{1}_b')\widehat{\boldsymbol{p}}$ which is asymptotically equivalent to the statistic $\sqrt{n}(\boldsymbol{I}_a \otimes \frac{1}{b}\boldsymbol{1}_b')\overline{\boldsymbol{Y}}_{\cdot}$ under $H_0^F(A)$. The covariance matrix of $\sqrt{n}(\boldsymbol{I}_a \otimes \frac{1}{b}\boldsymbol{1}_b')\overline{\boldsymbol{Y}}_{\cdot}$ is $\boldsymbol{V}_a = (\boldsymbol{I}_a \otimes \frac{1}{b}\boldsymbol{1}_b')\boldsymbol{V}(\boldsymbol{I}_a \otimes \frac{1}{b}\boldsymbol{1}_b)$.

Let $\boldsymbol{W} = \boldsymbol{V}_a^{-1}(\boldsymbol{I}_a - \boldsymbol{J}_a\boldsymbol{V}_a^{-1}/\boldsymbol{1}_a'\boldsymbol{V}_a^{-1}\boldsymbol{1}_a)$. Then $\boldsymbol{W}\ \boldsymbol{V}_a\ \boldsymbol{W} = \boldsymbol{W}$ and $(\boldsymbol{W} \otimes \frac{1}{b}\boldsymbol{1}_b')\boldsymbol{F} = \boldsymbol{0}$ iff $(\boldsymbol{P}_a \otimes \frac{1}{b}\boldsymbol{1}_b')\boldsymbol{F} = \boldsymbol{0}$. The covariance matrix $\boldsymbol{V}_a$ is estimated by

$$\begin{aligned}\widehat{\boldsymbol{V}}_a &= \left(\boldsymbol{I}_a \otimes \frac{1}{b}\boldsymbol{1}_b'\right) \widehat{\boldsymbol{V}} \left(\boldsymbol{I}_a \otimes \frac{1}{b}\boldsymbol{1}_b\right) \\ &= \frac{1}{N^2(n-1)} \sum_{k=1}^{n} (\overline{\boldsymbol{R}}_k^A - \overline{\boldsymbol{R}}_{\cdot}^A)(\overline{\boldsymbol{R}}_k^A - \overline{\boldsymbol{R}}_{\cdot}^A)'\end{aligned}$$

where $\overline{\boldsymbol{R}}_k^A = (\boldsymbol{I}_a \otimes \frac{1}{b}\boldsymbol{1}_b')\boldsymbol{R}_k = (\overline{R}_{1\cdot k}, \ldots, \overline{R}_{a\cdot k})'$ and $\overline{\boldsymbol{R}}_{\cdot}^A = n^{-1}\sum_{k=1}^{n} \overline{\boldsymbol{R}}_k^A$. Let $\widehat{\boldsymbol{W}}$ denote an estimator of $\boldsymbol{W}$ where $\boldsymbol{V}_a$ is replaced by $\widehat{\boldsymbol{V}}_a$ and denote the elements of $\widehat{\boldsymbol{V}}_a^{-1}$ by $s_{ii'}$, $i, i' = 1, \ldots, a$, let $s_{i\cdot} = \sum_{i'=1}^{a} s_{ii'}$ and $s_{\cdot\cdot} = \sum_{i=1}^{a} s_{i\cdot}$. Then, under $H_0^F(A)$, the statistic

$$\begin{aligned}Q_n^A &= \sqrt{n}\left[\left(\boldsymbol{I}_a \otimes \frac{1}{b}\boldsymbol{1}_b'\right)\widehat{\boldsymbol{p}}\right]' \widehat{\boldsymbol{W}} \left[\left(\boldsymbol{I}_a \otimes \frac{1}{b}\boldsymbol{1}_b'\right)\widehat{\boldsymbol{p}}\right]\sqrt{n} \\ &= \frac{n}{N^2}\left[\sum_{i=1}^{a}\sum_{i'=1}^{a} \overline{R}_{i\cdot\cdot}s_{ii'}\overline{R}_{i'\cdot\cdot} - \frac{1}{s_{\cdot\cdot}}\left(\sum_{i=1}^{a} s_{i\cdot}\overline{R}_{i\cdot\cdot}\right)^2\right]\end{aligned} \tag{4.51}$$

has asymptotically (as $n \to \infty$) a central χ_f^2-distribution with $f = a - 1$ and Q_n^A has the RT-property with respect to a multivariate statistic with an unspecified covariance matrix $\boldsymbol{V}$.

Test for $H_0^F(A \mid B)$. The hypothesis $H_0^F(A \mid B)$: $(\boldsymbol{P}_a \otimes \boldsymbol{I}_b)\boldsymbol{F} = \mathbf{0}$ is equivalently formulated as

$$H_0^F(A \mid B)\colon\ (\boldsymbol{C}_A \otimes \boldsymbol{I}_b)\boldsymbol{F} = \mathbf{0} \quad \text{where } \boldsymbol{C}_A = \left(\mathbf{1}_{a-1} \vdots - \boldsymbol{I}_{a-1}\right)$$

is a $(a-1) \times a$ contrast matrix to describe the hypothesis H_0: $F_{1j} = \cdots = F_{aj} = F_j$, $j = 1, \ldots, b$. Note that $(\boldsymbol{C}_A \otimes \boldsymbol{I}_b)\boldsymbol{F} = \mathbf{0}$ iff $(\boldsymbol{P}_a \otimes \boldsymbol{I}_b)\boldsymbol{F} = \mathbf{0}$. Unlike as before, we do not use the $a \times a$ contrast matrix $\boldsymbol{P}_a$ which is not of full row rank. The reason for this is that we like to avoid the computation of a generalized inverse, since there is no simple structured symmetric matrix $\boldsymbol{W}$ such that $\boldsymbol{W}\boldsymbol{V}\boldsymbol{W} = \boldsymbol{W}$ and $\boldsymbol{W}\boldsymbol{F} = \mathbf{0}$ iff $(\boldsymbol{P}_a \otimes \boldsymbol{I}_b)\boldsymbol{F} = \mathbf{0}$.

We consider the asymptotic distribution of $\sqrt{n}(\boldsymbol{C}_A \otimes \boldsymbol{I}_b)\widehat{\boldsymbol{p}}$ and we use the notation of the previous section. Under $H_0^F(A \mid B)$, the statistics $\sqrt{n}(\boldsymbol{C}_A \otimes \boldsymbol{I}_b)\widehat{\boldsymbol{p}} = \sqrt{n}(\boldsymbol{C}_A \otimes \boldsymbol{I}_b) \int \widehat{H}\, \mathrm{d}\widehat{\boldsymbol{F}}$ and $\sqrt{n}(\boldsymbol{C}_A \otimes \boldsymbol{I}_b) \int H\, \mathrm{d}\widehat{\boldsymbol{F}} = \sqrt{n}(\boldsymbol{C}_A \otimes \boldsymbol{I}_b)\overline{\boldsymbol{Y}}_{.}$ are asymptotically equivalent. Assume that $\boldsymbol{V} = \mathrm{Cov}(\sqrt{n}\,\overline{\boldsymbol{Y}}_{.})$ is a non-singular $(ab \times ab)$ covariance matrix. A consistent estimator of $\boldsymbol{V}$ is given in (4.50). Then, under $H_0^F(A \mid B)$, the quadratic form

$$\begin{aligned} Q_n^{A|B} = {} & \frac{n}{N^2}\overline{\boldsymbol{R}}_{.}'(\boldsymbol{C}_A' \otimes \boldsymbol{I}_b)\left[(\boldsymbol{C}_A \otimes \boldsymbol{I}_b)\ \widehat{\boldsymbol{V}}\ (\boldsymbol{C}_A' \otimes \boldsymbol{I}_b)\right]^{-1} \\ & \times (\boldsymbol{C}_A \otimes \boldsymbol{I}_b)\overline{\boldsymbol{R}}_{.} \end{aligned} \tag{4.52}$$

has a central χ_f^2-distribution with $f = (a-1)b$ and $Q_n^{A|B}$ has the RT-property with respect to a multivariate statistic with an unspecified covariance matrix $\boldsymbol{V}$.

Test for $H_0^F(AB)$. The hypothesis $H_0^F(AB)$: $(\boldsymbol{P}_a \otimes \boldsymbol{P}_b)\boldsymbol{F} = \mathbf{0}$ is equivalently formulated as $H_0^F(AB)$: $(\boldsymbol{C}_A \otimes \boldsymbol{C}_B)\boldsymbol{F} = \mathbf{0}$ where $\boldsymbol{C}_A$ and $\boldsymbol{C}_B$ are both contrast matrices of full row rank. The statistic for the interaction is derived in the same way as for the conditional effect in the previous paragraph. Under $H_0^F(AB)$, the quadratic form

$$\begin{aligned} Q_n^{AB} = {} & \frac{n}{N^2}\overline{\boldsymbol{R}}_{.}'\ (\boldsymbol{C}_A' \otimes \boldsymbol{C}_B')\left[(\boldsymbol{C}_A \otimes \boldsymbol{C}_B)\ \widehat{\boldsymbol{V}}\ (\boldsymbol{C}_A' \otimes \boldsymbol{C}_B')\right]^{-1} \\ & \times (\boldsymbol{C}_A \otimes \boldsymbol{C}_B)\overline{\boldsymbol{R}}_{.} \end{aligned} \tag{4.53}$$

has a central χ_f^2-distribution with $f = (a-1)(b-1)$ and Q_n^{AB} has the RT-property with respect to a multivariate statistic with an unspecified covariance matrix $\boldsymbol{V}$.

For simplicity, we considered only the case of one observation per subject and treatment combination (i, j). It is straightforward to generalize the results to the case of several observations m_{ij} for each treatment combination using the general results in Section 4.2.

EXAMPLE 6. We apply the procedures derived in this section to the data given by Koch (1969) in Example 2, where repeated measurements on two crossed fixed factors for 8 pairs of animals are observed. Both fixed factors have two levels, namely

Table 4.5
Ranks, rank means and results for the diet data

	Diet (E) Gas		Diet (C) Gas		Results		
Pair	O	N	O	N	Effect	Statistic	p-value
1	30	7	26.5	12			
2	26.5	5	19	1	Diet effect	$Q_n^A = 2.19$	0.182
3	21	22	28	3	Gas effect	$Q_n^B = 46.62$	0.00025
4	29	31	32	15	Interaction	$Q_n^{AB} = 6.47$	0.0385
5	23	13.5	18	4			
6	11	13.5	25	2			
7	9	10	24	8			
8	17	16	20	6			
Mean	20.81	14.75	24.06	6.38			

'ethionine' (E) and 'control' (C) for factor A (diet) and 'oxygen' (O) and 'nitrogen' (N) for factor B (gas). For the description of the trial and for the data set, we refer to Koch (1969), Example 2. Note that in this paper only the hypothesis of equality of all 4 treatment combinations (EO, EN, CO, CN) is considered. The statistics given in this section in (4.51) and (4.53) however allow us to consider separately the non-parametric hypotheses of no diet effect, of no gas effect and of no interaction. The ranks R_{ijk} of all 32 observations, the rank means for the four treatment combinations and the results are given in Table 4.5. Since the statistics Q_n^A, Q_n^{AB} and Q_n^B have the RT-property, they can be computed by an appropriate statistical software package. The p-values are obtained from the F-distribution with $f_1 = 1$ and $f_2 = n - 1 = 7$ which is used as a small sample approximation of the null distributions of the statistics.

4.4.2.3. Tests for heteroscedastic models, $a = b = 2$. As in the two-factor design with independent observations, we consider separately, the case of two levels for each factor. The implications of the hypotheses are covered by Proposition 2.9. We extend the considerations of the previous subsection to heteroscedastic models and we derive a test for the hypotheses $H_0^w(A)$. The test for $H_0^w(B)$ will follow by the symmetry of the design. The AB interaction is tested by interchanging the cells $(2,1)$ and $(2,2)$ and then applying the test for $H_0^w(A)$. This follows also from the simple structure of the design.

Test for $H_0^w(A)$. Let $\mathbf{X}_k = (X_{11k}, X_{12k}, X_{21k}, X_{22k})'$ be $k = 1, \ldots, n$, independent random vectors where $X_{ijk} \sim F_{ij}(x)$, $i, j = 1, 2$, and $k = 1, \ldots, n$. Similar to the considerations in Section 4.4.1.3, we consider the generalized means $w_j = \int F_{1j}\, dF_{2j}$ within level j of factor B, $j = 1, 2$, and we base the statistic on $2w^A = w_1 + w_2$. A consistent estimate of $2w^A$ is apparently given by

$$2\widehat{w}^A = \widehat{w}_1 + \widehat{w}_2 = n^{-1}\left[\overline{R}_{21\cdot}^B + \overline{R}_{22\cdot}^B - (n+1)\right],$$

where $\overline{R}^B_{2j\cdot} = n^{-1}\sum_{k=1}^n R^B_{2jk}$ and R^B_{2jk} is the rank of X_{2jk} among all the $N = 2n$ observations within level j of factor B.

Under the hypothesis $H_0^w(A)$: $2w^A = 1$, the statistic $\sqrt{n}(\widehat{w}_1+\widehat{w}_2)$ is asymptotically equivalent to $(1/\sqrt{n})\sum_{k=1}^n Y_k$ where $Y_k = Y_{1k} + Y_{2k}$ and $Y_{jk} = F_{1j}(X_{2jk}) - F_{2j}(X_{1jk})$, $j = 1, 2$. Let $n\widehat{Y}_{jk} = (R^B_{2jk} - R^{(2j)}_{2jk}) - (R^B_{1jk} - R^{(1j)}_{1jk})$, $j = 1, 2$ where $R^{(ij)}_{ijk}$ is the rank of X_{ijk} among all the n observations within level i of factor A and level j of factor B over all n blocks. Then $\widehat{\sigma}^2_n = (n-1)^{-1}\sum_{k=1}^n[\widehat{Y}_{1k} + \widehat{Y}_{2k} - (\overline{\widehat{Y}}_{1\cdot} + \overline{\widehat{Y}}_{2\cdot})]^2$ is a consistent estimator for $\sigma^2 = \mathrm{Var}(n^{-1/2}\sum_{k=1}^n Y_k) = n^{-1}\sum_{k=1}^n \mathrm{Var}(Y_k)$ where $\overline{\widehat{Y}}_{j\cdot} = n^{-1}\sum_{k=1}^n \widehat{Y}_{jk}$. Thus, under $H_0^w(A)$: $2w^A = 1$, the statistic

$$T_n^A = \sqrt{n(n-1)} \cdot \frac{\overline{R}^B_{21\cdot} + \overline{R}^B_{22\cdot} - (2n+1)}{\sqrt{\sum_{k=1}^n n^2\left[\widehat{Y}_{1k} + \widehat{Y}_{2k} - (\overline{\widehat{Y}}_{1\cdot} + \overline{\widehat{Y}}_{2\cdot})\right]^2}} \tag{4.54}$$

has asymptotically (as $n \to \infty$) a standard normal distribution. For details, we refer to Brunner and Neumann (1986a).

4.4.3. Special procedures for linear models and other approaches

Nonparametric procedures for different mixed models have been considered in literature. Koch and Sen (1968) studied ranking methods in the mixed model under different assumptions. The results of Koch and Sen, (loc. cit.) were later extended by Koch (1969) to the partially nested design (4.19) for a multivariate model as well as for a model where the compound symmetry is assumed. These authors considered a multivariate rank test for $H_0(A \mid B)$ as well as ranks tests for some other hypotheses. Not all their procedures are applicable to pure ordinal data and most of the procedures considered by them are only restricted to the linear models.

Later, Koch (1970) studied a more complex partially nested design with three fixed factors. He formulated the hypotheses in a linear model setup and provided several rank tests using different ranking methods for joint hypotheses. In the cross-classified repeated measurements model (4.48), Mehra and Sen (1969) considered a ranking after alignment test for the interaction in the linear model. A rank test (using ranks within blocks) for an unbalanced mixed model with random interaction has been considered by Brunner and Dette (1992). For details, we refer to these articles.

Acknowledgements

The authors like to express their sincere gratitude to Prof. Manfred Denker and Prof. Michael G. Akritas for many helpful discussions and valuable comments during the preparation of the paper. The authors are also grateful to Ullrich Munzel for his very thorough examination of the manuscript and to Lars Pralle for providing the computations of the examples. The research of Madan L. Puri was supported by the Office of Naval Research Contract N00014-91-J-1020.

References

Adichie, J. N. (1978). Rank tests of sub-hypotheses in the general linear regression. *Ann. Statist.* **6**, 1012 1026.

Akritas, M. G. (1990). The rank transform method in some two-factor designs. *J. Amer. Statist. Assoc.* **85**, 73–78.

Akritas, M. G. (1991). Limitations on the rank transform procedure: A study of repeated measures designs, Part I. *J. Amer. Statist. Assoc.* **86**, 457–460.

Akritas, M. G. (1993). Limitations of the rank transform procedure: A study of repeated measures designs, Part II. *Statist. Probab. Lett.* **17**, 149–156.

Akritas, M. G. and S. F. Arnold (1994). Fully nonparametric hypotheses for factorial designs, I: Multivariate repeated measures designs. *J. Amer. Statist. Assoc.* **89**, 336–343.

Akritas, M. G., S. F. Arnold and E. Brunner (1996). Nonparametric hypotheses and rank statistics for unbalanced factorial designs. *J. Amer. Statist. Assoc.*, to appear.

Akritas, M. G. and E. Brunner (1995). A unified approach to ranks tests in mixed models. Preprint.

Arnold, S. F. (1981). *The Theory of Linear Models and Multivariate Analysis*. Wiley, New York.

Behnen, K. (1976). Asymptotic comparison of rank tests for the regression problem when ties are present. *Ann. Statist.* **4**, 157–174.

Benard, A. and Ph. van Elteren (1953). A generalisation of the methods of m rankings. *Proc. Kon. Nederl. Akad. Wetensch. Math.* **15**, 358–369.

Blair, R. C., S. S. Sawilowski and J. J. Higgens (1987). Limitations of the rank transform statistic in tests for interactions. *Comm. Statist. Simulation Comput.* **16**, 1133–1145.

Boos, D. D. and C. Brownie (1992). Rank based mixed model approach to multisite clinical trials. *Biometrics* **48**, 61–72.

Boos, D. D. and C. Brownie (1994). Type I error robustness of ANOVA and ANOVA on ranks when the number of treatments is large. *Biometrics* **50**, 542–549.

Boos, D. D. and C. Brownie (1995). ANOVA and rank tests when the number of treatments is large. *Statist. Probab. Lett.* **23**, 183–192.

Brunner, E. and D. Compagnone (1988). Two sample rank tests for repeated observations: The distribution for small sample sizes. *Statist. Softw. Newsletter* **14**, 36–42.

Brunner, E. and M. Denker (1994). Rank statistics under dependent observations and applications to factorial designs. *J. Statist. Plann. Inference* **42**, 353–378.

Brunner, E. and H. Dette (1992). Rank procedures for the two factor mixed model. *J. Amer. Statist. Assoc.* **87**, 884–888.

Brunner, E. and N. Neumann (1982). Rank tests for correlated random variables. *Biometr. J.* **24**, 373–389.

Brunner, E. and N. Neumann (1984). Rank tests for the 2×2 split plot design. *Metrika* **31**, 233–243.

Brunner, E. and N. Neumann (1986a). Rank tests in 2×2 designs. *Statist. Neerlandica* **40**, 251–271.

Brunner, E. and N. Neumann (1986b). Two-sample rank tests in general models. *Biometr. J.* **28**, 395–402.

Brunner, E. and N. Neumann (1987). Non-parametric methods for the 2-period-cross-over design under weak model assumptions. *Biometr. J.* **29**, 907–920.

Brunner, E., M. L. Puri and S. Sun (1995). Nonparametric methods for stratified two-sample designs with application to multi clinic trials. *J. Amer. Statist. Assoc.* **90**, 1004–1014.

Clemmens, A. E. and Z. Govindarajulu (1990). A certain locally most powerful test for one-way random effects model: Null distribution and power considerations. *Comm. Statist. Theory Methods* **19**, 4139–4151.

Conover, W. J. (1973). Rank tests one sample, two samples, and k samples without the assumtion of continuos function. *Ann. Statist.* **1**, 1105–1125.

Conover, W. J. and R. L. Iman (1976). On some alternative procedures using ranks for the analysis of experimental designs. *Comm. Statist. Theory Methods* **14**, 1349–1368.

Conover, W. J. and R. L. Iman (1981). Rank transformations as a bridge between parametric and nonparametric statistics (with discussion). *Amer. Statist.* **35**, 124–133.

de Kroon, J. P. M. and P. van der Laan (1981). Distribution-free test procedures in two-way layouts: A concept of rank interaction. *Statist. Neerlandica* **35**, 189–213.

Durbin, J. (1951). Incomplete blocks in ranking experiments. *British J. Psychol. (Statist. Section)* **4**, 85–90.

Fligner, M. A. (1981). Comment on "Rank transformations as a bridge between parametric and nonparametric statistics" (by W. J. Conover and R. L. Iman). *Amer. Statist.* **35**, 131–132.

Fligner, M. A. and G. E. Policello II (1981). Robust rank procedures for the Behrens–Fisher problem. *J. Amer. Statist. Assoc.* **76**, 162–168.

Friedmann, M. (1937). The use of ranks to avoid the assumption of normality implicit in the analysis of variance. *J. Amer. Statist. Assoc.* **32**, 675–699.

Gibbons, J. D. and S. Chakraborti (1992). *Nonparametric Statistical Inference*, 3rd edn. Marcel Dekker, New York.

Govindarajulu, Z. (1975a). Locally most powerful rank-order test for one-way random effects model. *Studia Sci. Math. Hungar.* **10**, 47–60.

Govindarajulu, Z. (1975b). Robustness of Mann–Whitney–Wilcoxon test to dependence in the variables. *Studia Sci. Math. Hungar.* **10**, 39–45.

Govindarajulu, Z. and J. V. Deshpande (1972). Random effects model: Nonparametric case. *Ann. Inst. Statist. Math.* **24**, 165–170.

Greenberg, V. L. (1964). Robust inference on some experimental designs. Ph.D. Thesis, University of California at Berkeley, CA.

Grizzle, J. (1965). The two-period change-over design and its use in clinical trials. *Biometrics* **21**, 462–480.

Grizzle, J. (1974). Corrections. *Biometrics* **30**, 727.

Groggle, D. J. and, J. H. Skilling (1986). Distribution-free tests for main effects in multifactor designs. *Amer. Statist.* **40**, 99–102.

Hallin, M. and M.L. Puri (1994). Aligned rank tests for linear models with autocorrelated error terms. *J. Multivariate Anal.* **50**, 175–237.

Haux, R. M., M. Schumacher and G. Weckesser (1984). Rank tests for complete block designs. *Biometr. J.* **26**, 567–582.

Hettmansperger, T. P. and J. W. McKean (1977). A robust alternative based on ranks to least squares in analyzing linear models. *Technometrics* **19**, 275–284.

Hettmansperger, T. P. and W. McKean (1983). A geometric interpretation of inferences based on ranks in the linear model. *J. Amer. Statist. Assoc.* **78**, 885–893.

Hettmansperger, T. P. and R. M. Norton (1987). Tests for patterned alternatives in k-sample problems. *J. Amer. Statist. Assoc.* **82**, 292–299.

Hilgers, R. (1979). Ein asymptotisch verteilungsfreier Wechselwirkungstest in zweifaktoriellen vollständigen Zufallsplänen. Ph.D. Thesis, Universität Dortmund.

Hocking, R. R. (1973). A discussion of the two-way mixed model. *Amer. Statist.* **27**, 148–152.

Hodges, J. L., Jr. and E. L. Lehmann (1962). Rank methods for combination of independent experiments in analysis of variance. *Ann. Math. Statist.* **33**, 482–497.

Hollander, M., G. Pledger and P. E. Lin (1974). Robustness of the Wilcoxon test to a certain dependency between samples. *Ann. Statist.* **2**, 177–181.

Hora, S. C. and W. J. Conover (1984). The F-statistic in the two-way layout with rank score transformed data. *J. Amer. Statist. Assoc.* **79**, 668–673.

Hora, S. C. and R. L. Iman (1988). Asymptotic relative efficiencies of the rank-transformation procedure in randomized complete block designs. *J. Amer. Statist. Assoc.* **83**, 462–470.

Iman, R. L. (1976). An approximation to the exact distribution of the Wilcoxon–Mann–Whitney rank sum test statistic. *Comm. Statist. Theory Methods* **5**(7), 537–598.

Iman, R. L. and J. M. Davenport (1976). New approximations to the exact distribution of the Kruskal–Wallis test statistic. *Comm. Statist. Theory Methods* **5**, 1335–1348.

Iman, R. L. and J. M. Davenport (1980). Approximations of the critical region of the Friedman statistic. *Comm. Statist. Theory Methods* **9**, 571–595.

Iman, R. L., S. C. Hora and W. J. Conover (1984). A comparison of asymptotically distribution-free procedures for the analysis of complete blocks. *J. Amer. Statist. Assoc.* **79**, 674–685.

Jaeckel, L. A. (1972). Estimating regression coefficients by minimizing the dispersion of the residuals. *Ann. Math. Statist.* **43**, 1449–1458.

Kepner, J. L. and D. H. Robinson (1988). Nonparametric methods for detecting treatment effects in repeated measures designs. *J. Amer. Statist. Assoc.* **83**, 456–461.

Kirk, R. (1982). *Experimental Design*, 2nd edn. Brooks/Cole, Monterey.

Koch, G. (1969). Some aspects of the statistical analysis of 'split-plot' experiments in completely randomized layouts. *J. Amer. Statist. Assoc.* **64**, 485–506.

Koch, G. (1970). The use of nonparametric methods in the statistical analysis of a complex split plot experiment. *Biometrics* **26**, 105–128.

Koch, G. (1972). The use of nonparametric methods in the statistical analysis of the two-period change-over design. *Biometrics* **28**, 577–584.

Koch, G. G. and P. K. Sen (1968). Some aspects of the statistical analysis of the 'mixed model'. *Biometrics* **24**, 27–48.

Kruskal, W. H. (1952). A nonparametric test for the several sample problem. *Ann. Math. Statist.* **23**, 525–540.

Kruskal, W. H. and W. A. Wallis (1952). The use of ranks in one-criterion variance analysis. *J. Amer. Statist. Assoc.* **47**, 583–621.

Kruskal, W. H. and W. A. Wallis (1953). Errata in: The use of ranks in one-criterion variance analysis. *J. Amer. Statist. Assoc.* **48**, 907–911.

Lam, F. C. and M. T. Longnecker (1983). A modified Wilcoxon rank sum test for paired data. *Biometrika* **70**, 510–513.

Lehmann, E. L. (1975). *Nonparametrics: Statistical Methods Based on Ranks.* Holden-Day, San Francisco, CA.

Lemmer, H. H. (1980). Some empirical results on the two-way analysis of variance by ranks. *Comm. Statist. Theory Methods* **14**, 1427–1438.

Lemmer, H. H. and D. J. Stoker (1967). A distribution-free analysis of variance for the two-way classification. *South African Statist. J.* **1**, 67–74.

Mack, G. A. and J. H. Skillings (1980). A Friedman-type rank test for main effects in a two-factor ANOVA. *J. Amer. Statist. Assoc.* **75**, 947–951.

McKean, J. W. and T. P. Hettmansperger (1976). Tests of hypotheses based on ranks in the general linear model. *Comm. Statist. Theory Methods* **5**(8), 693–709.

Mehra, K. L. and P. K. Sen (1969). On a class of conditionally distribution-free tests for interactions in factorial experiments. *Ann. Math. Statist.* **38**, 658–664.

Munzel, U. (1994). Asymptotische Normalität linearer Rangstatistiken bei Bindungen unter Abhängigkeit. Diplomarbeit Inst. Math. Stochastik der Univ. Göttingen.

Patel, K. M. and D. G. Hoel (1973). A nonparametric test for interaction in factorial experiments. *J. Amer. Statist. Assoc.* **68**, 615–620.

Potthoff, R. F. (1963). Use of the Wilcoxon statistic for a generalized Behrens–Fisher problem. *Ann. Math. Statist.* **34**, 1596–1599.

Puri, M. L. (1964). Asymptotic efficiency of a class of c-sample tests. *Ann. Math. Statist.* **35**, 102–121.

Puri, M. L. and P. K. Sen (1969). A class of rank order tests for a general linear hypothesis. *Ann. Math. Statist.* **40**, 1325–1343.

Puri, M. L. and P. K. Sen (1973). A note on ADF-test for subhypotheses in multiple linear regression. *Ann. Statist.* **1**, 553–556.

Puri, M. L. and P. K. Sen (1985). *Nonparametric Methods in General Linear Models.* Wiley, New York.

Randles, R. H. and D. A. Wolfe (1979). *Introduction to the Theory of Nonparametric Statistics.* Wiley, New York.

Rinaman, W. C., Jr. (1983). On distribution-free rank tests for two-way layouts. *J. Amer. Statist. Assoc.* **78**, 655–659.

Sen, P. K. (1967). On some non-parametric generalizations of Wilk's test for H_M, H_{VC} and H_{MVC}. *Ann. Inst. Statist. Math.* **19**, 541–571.

Sen, P. K. (1968). On a class of aligned rank order tests in two-way layouts. *Ann. Math. Statist.* **39**, 1115–1124.

Sen, P. K. and M. L. Puri (1977). Asymptotically distribution-free aligned rank order tests for composite hypotheses for general multivariate linear models. *Z. Wahr. verwandte Gebiete* **39**, 175–186.

Senn, S. (1993). *Cross-Over Trials in Clinical Research.* Wiley, New York.

Shetty, I. and Z. Govindarajulu (1988). An LMP rank test for random effects: Its asymptotic distribution under local alternatives and asymptotic relative efficiency. In: K. Matusita, ed., *Statistical Theory and Data Analysis*, Vol. 2. North-Holland, New York, 171–190.

Shirahata, S. (1982). Nonparametric measures of interclass correlation. *Comm. Statist. Theory Methods* **11**, 1707–1721.

Shirahata, S. (1985). Asymptotic properties of Kruskal–Wallis test and Friedman test in the analysis of variance models with random effects. *Comm. Statist. Theory Methods* **14**, 1685–1692.

Streitberg, B. and J. Roehmel (1986). Exact distribution for permutation and rank tests: An introduction to some recently published algorithms. *Statist. Software Newsletter* **12**, 10–17.

Thompson, G. L. (1990). Asymptotic distribution of rank statistics under dependencies with multivariate applications. *J. Multivariate Anal.* **33**, 183–211.

Thompson, G. L. (1991a). A unified approach to rank tests for multivariate and repeated measures designs. *J. Amer. Statist. Assoc.* **86**, 410–419.

Thompson, G. L. (1991b). A note on the rank transform for interactions. *Biometrika* **78**, 697–701.

Thompson, G. L. (1993). A note on the rank transform for interactions, amendments and corrections. *Biometrika* **80**, 711–713.

Thompson, G. L. and L. P. Ammann (1989). Efficacies of rank-transform statistics in two-way models with no interaction. *J. Amer. Statist. Assoc.* **84**, 325–330.

Thompson, G. L. and L. P. Ammann (1990). Efficiencies of interblock rank statistics for repeated measures designs. *J. Amer. Statist. Assoc.* **85**, 519–528.

van der Laan, P. and J. Prakken (1972). Exact distribution of Durbin's distribution-free test statistic for balanced incomplete block designs, and comparison with the chi-square and F approximation. *Statist. Neerlandica* **26**, 155–164.

Wallace, D. L. (1959). Simplified beta-approximations to the Kruskal–Wallis H-test. *J. Amer. Statist. Assoc.* **54**, 225–230.

Wittkowski, K. M. (1988). Friedman-type statistics and consistent multiple comparisons for unbalanced designs with missing data. *J. Amer. Statist. Assoc.* **83**, 1163–1170.

Zerbe, G. O. (1979). Randomization analysis of completely randomized design extended to growth and response curves. *J. Amer. Statist. Assoc.* **74**, 215–221.

Zimmermann, H. (1985a). Exact calculation of permutational distributions for two dependant samples I. *Biometr. J.* **27**, 349–352.

Zimmermann, H. (1985b). Exact calculation of permutational distributions for two independent samples. *Biometr. J.* **27**, 431–434.

S. Ghosh and C. R. Rao, eds., *Handbook of Statistics, Vol. 13*

Nonparametric Analysis of Experiments

A. M. Dean and D. A. Wolfe

1. Overview

1.1. Introduction

Traditional analysis of a factorial experiment is based on the assumption of an underlying normal distribution for the error variables. Data transformation procedures may help to correct for deviations from normality, but possibly at the expense of equality of variances of the error variables. Similarly, data transformation procedures commonly used for equalizing the error variances may result in the violation of the normality assumption. A nonparametric analysis avoids the necessity of specifying any particular error distribution, although most such procedures do require equality of error variances. Nonparametric procedures are generally easy to apply and, even when the normality assumption is valid, they are reasonably powerful.

In this article, we illustrate some simple nonparametric procedures that can be used to analyse factorial experiments. Some of these techniques are available in texts such as that by Hollander and Wolfe (1973), while others have appeared more recently in the literature. Most of the procedures that we shall discuss are hypothesis testing or decision procedures. Where confidence intervals are available, we shall discuss those as well. For simplicity, we have chosen to concentrate mostly on procedures applied to *ranks* of the data values. For procedures applied to other types of scores (functions of ranks), the reader is referred to Randles and Wolfe (1979) or Hettmansperger (1984), and for other rank-based methods, see Draper (1988).

For all of the situations we shall consider, the model is of the form

Model: $Y_i = \sum_{j=1}^{p} x_{ij}\gamma_j + E_i$, $i = 1, \ldots, N$, where Y_i is the ith observation, $\gamma_1, \ldots, \gamma_p$ are unknown parameters, x_{ij} is the known design variable, E_i is the random error associated with the ith observation, and where, in general,

Assumptions:
1. $E_1, E_2, \ldots, E_N$ are mutually independent.
2. The distributions of $E_1, E_2, \ldots, E_N$ are the same (in particular, the variances are equal).
3. E_i has a continuous distribution, $i = 1, \ldots, N$.

In particular, the assumption of continuous error distributions ensures that the relevant permutations of ranks are equally likely. However, this uniform distribution on the permutations is also implied, to a reasonable approximation, by most discrete distributions with adequate support (that is, sufficiently large number of possible values for the errors). Therefore, the procedures we discuss are also approximately valid for most discrete data.

The material in this article is organized around a series of experiments which are described in the following subsections. The reader is invited to look at these descriptions first, and then to follow the references to the sections containing the details of the various nonparametric procedures.

Section 2 is devoted to the analysis of the one-way layout; that is, experiments with responses which can be modelled by the effect of a single factor, called the treatment factor. In the case of factorial experiments, the one-way layout analysis can be used to compare the effects of different treatment combinations. We shall use the data from Experiments 1.2 and 1.3, described in the following subsections, to illustrate various nonparametric procedures applicable to the one-way layout. Experiment 1.2 is used to illustrate a factorial experiment with three factors and a control treatment combination. Experiment 1.3 involves a single quantitative treatment factor with unequally spaced levels.

Experiments 1.4, 1.5 and 1.6 have two treatment factors each and are analysed as two-way layouts in Section 3. The first two of these experiments have two observations per treatment combination (i.e., per cell), while the third has only one. The procedures illustrated for two treatment factors are equally relevant when one factor is a block factor, that is, for complete block designs. Experiment 1.7 is used to illustrate the balanced incomplete block design setting.

Experiment 1.2 is used again in Section 4 to illustrate analysis procedures for three or more treatment factors. Fewer nonparametric procedures are currently available for three-or-more-factor settings than for the one- or two-factor settings. However, we will discuss a rank-based general linear model procedure which is quite flexible.

1.2. Experiment – one-way layout; general alternatives

The following experiment was described by Raghu Kackar and Anne Shoemaker in the book edited by Dehnad (1989). The experiment was set up to investigate the robustness of the settings of eight "design" factors (A–H) to the thickness of deposition of an epitaxial layer on a polished silicon wafer. Thickness measurements were taken on wafers in 14 different positions in the bell jar in which the deposition took place and at 5 places on each wafer (see the original reference for details). The position variables were treated as noise factors and the mean thickness and the log sample variance of the thickness were measured over the 70 position combinations (noise levels) for each combination of the eight design factors. An analysis of the thickness variability was done to investigate which settings of the design factors were more sensitive to the noise factors, and an analysis of the mean thickness was done to investigate which factors could be adjusted to produce the target thickness.

Table 1
Thickness of epitaxial layer, Dehnad (1989)

Mean thickness	Log variance of thickness	A	B	C	D	E	F	G	H
14.821	−0.443	−1	−1	−1	−1	−1	−1	−1	−1
14.888	−1.199	−1	−1	−1	−1	1	1	1	1
14.037	−1.431	−1	−1	1	1	−1	−1	1	1
13.880	−0.651	−1	−1	1	1	1	1	−1	−1
14.165	−1.423	−1	1	−1	1	−1	1	−1	1
13.860	−0.497	−1	1	−1	1	1	−1	1	−1
14.757	−0.327	−1	1	1	−1	−1	1	1	−1
14.921	−0.627	−1	1	1	−1	1	−1	−1	1
13.972	−0.347	1	−1	−1	1	−1	1	1	−1
14.032	−0.856	1	−1	−1	1	1	−1	−1	1
14.843	−0.437	1	−1	1	−1	−1	1	−1	1
14.415	−0.313	1	−1	1	−1	1	−1	1	−1
14.878	−0.615	1	1	−1	−1	−1	−1	1	1
14.932	−0.229	1	1	−1	−1	1	1	−1	−1
13.907	−0.119	1	1	1	1	−1	−1	−1	−1
13.914	−0.863	1	1	1	1	1	1	1	1

The design was a fractional factorial experiment with 16 observations in total. Contrast estimates for main effects were calculated in the original article, but no formal testing was done. The design and response variables are reproduced in Table 1. The codes -1 and 1 mean that the factor is at its first and second levels, respectively.

To illustrate techniques for the one-way layout, in Section 2.2.6 we consider just factors A and H – a total of 4 treatment combinations, each observed four times, and we use log variance thickness as the response variable. Factor A was rotation method (continuous, oscillating) and factor H was nozzle position (2, 6).

If we represent the effect on the response of the jth treatment combination as τ_j $(j = 1, 2, 3, 4)$ then a starting point in the analysis of this set of data might be to test the null hypothesis that the four treatment combinations give rise to the same log variance thickness against a general alternative hypothesis; that is,

$$H_0\colon\ [\tau_1 = \tau_2 = \tau_3 = \tau_4] \quad \text{versus} \quad H_1\colon\ [\tau_1, \tau_2, \tau_3, \tau_4 \text{ are not all equal}]\,.$$

The well-known Kruskall–Wallis procedure can be used to test H_0 against H_1, and this is described in Section 2.2.1. The Kruskall–Wallis procedure requires equality of variances of the error variables. The Rust–Fligner procedure for the Behrens–Fisher problem of testing H_0 against H_1 when the error variances differ is presented in Section 2.2.2.

A follow-up (or alternative) to the Kruskall–Wallis test is the simultaneous investigation of the equality or non-equality of each pair of treatment effects τ_v, τ_u. A multiple comparison decision procedure which controls the experimentwise error rate (the probability of making at least one incorrect decision) is given in Section 2.2.3. Simultaneous confidence intervals for the pairwise comparisons $\tau_v - \tau_u$ are discussed

Table 2
Number of revertant colonies for Acid Red 114, TA98, hamster liver activation, Simpson and Margolin (1986)

Dose (μg/ml)					
0	100	333	1000	3333	10000
22	60	98	60	22	23
23	59	78	82	44	21
35	54	50	59	33	25

in Section 2.2.5. General contrasts $\sum_{j=1}^{m} c_j\tau_j$, $\sum_{j=1}^{m} c_j = 0$, are often of interest, and these include the pairwise comparisons as a subset. Point estimators for these contrasts are available and, for these, the reader is referred to Spjøtvoll (1968) or Hollander and Wolfe (1973, Section 6.4).

Prior to this particular experiment, factors A and H had been set at the levels "oscillating and 4", respectively. Let us suppose that "oscillating, 6", which is the treatment combination $(1, 1)$, is the combination which requires least change to the original manufacturing process. We may then regard this combination as a control, or standard, treatment. In this case, comparisons of the other three treatments with the control treatment would be of particular interest. This is called the *treatment versus control* multiple comparison problem and is discussed in Section 2.2.4. All of these procedures are illustrated for the wafer experiment in Section 2.2.6.

1.3. Experiment – one-way layout, ordered alternatives

Simpson and Margolin (1986) present data from three replicated Ames tests in which plates containing Salmonella bacteria of strain TA98 were exposed to $m = 6$ different doses of Acid Red 114. The response variable is the number of visible revertant colonies observed on each plate. The data from the first replicate are reproduced in Table 2.

If we represent the effect on the response of the jth dose of Acid Red 114 as τ_j, then a starting point in the analysis of these data might be to test the null hypothesis

$$H_0\colon\ [\tau_1 = \cdots = \tau_6]$$

against the general alternative hypothesis that the different doses do not give the same average number of visible revertant colonies. An appropriate test for this hypothesis is discussed in Section 2.2.1.

On the other hand, one may expect *a priori* that the number of colonies will increase as the dose increases, at least over the lower dose range. Thus, if we just consider the part of the experiment up to and including 1000 μg/ml, we may prefer to test the null hypothesis against the specific *ordered alternative* hypothesis

$$H_2\colon\ [\tau_1 \leqslant \tau_2 \leqslant \tau_3 \leqslant \tau_4,\ \text{with at least one strict inequality}]\,.$$

The Terpstra–Jonckheere test of H_0 against H_2 is presented in Section 2.3.1. A follow-up to such a test is to investigate simultaneously the possible equality versus preconceived ordering of each pair of treatment effects τ_j, τ_q. A multiple comparison decision procedure which controls the experimentwise error rate in this ordered setting is the Hayter–Stone procedure discussed in Section 2.3.2. The corresponding procedure for obtaining simultaneous confidence bounds for pairwise comparisons $\tau_v - \tau_u (1 \leqslant u < v \leqslant 4)$ is given in Section 2.3.3. All of these procedures are illustrated for the Acid Red experiment in Section 2.3.5.

With further knowledge about drug testing, it may be reasonable to expect that, eventually, a drug dose will become toxic and the response will begin to decline. In this case, an *umbrella alternative* hypothesis may be of interest; that is

$$H_3\colon\ [\tau_1 \leqslant \cdots \leqslant \tau_{p-1} \leqslant \tau_p \geqslant \cdots \geqslant \tau_6, \text{with at least one strict inequality}],$$

where the location of the "peak", p, of the umbrella (turning point) may or may not be known. Tests of H_0 against an umbrella alternative hypothesis H_3 will be discussed in Section 2.3.4, and illustrated for the Acid Red experiment in Section 2.3.5.

1.4. Experiment – two-way layout, with interaction, multiple observations per cell

The data shown in Table 3 were given by Anderson and McLean (1974) and show the strength of a weld in a steel bar. The two factors of interest were time of welding (total time of the automatic weld cycle) and gage bar setting (the distance the weld die travels during the automatic weld cycle).

We let η_{ij} denote the expected response at the ith gage bar setting and jth welding time. It is often convenient to express this as $\eta_{ij} = \beta_i + \tau_j + (\beta\tau)_{ij}$, where β_i represents the effect of the ith gage bar setting, τ_j represents the effect of the jth time of welding, and $(\beta\tau)_{ij}$ represents the effect of the ith gage bar setting and jth welding time that is not accounted for by the "additive model" $\eta_{ij} = \beta_i + \tau_j$. The term $(\beta\tau)_{ij}$ is often called an interaction term.

In Section 3.2.1, we discuss procedures that have been proposed for testing the hypothesis of additivity; that is, H_0^{add}: $[\eta_{ij} = \beta_i + \tau_j$, for all i and $j]$ against an alternative hypothesis of non-additivity. We also present there an alternative procedure that tests for *rank interaction* in the two-way layout. Rank interaction is present if the ranking of the levels of one of the factors is not the same for each of the levels of

Table 3
Strength of weld, Anderson and McLean (1974)

		Time of welding (j)									
	i	1		2		3		4		5	
Gage	1	10	12	13	17	21	30	18	16	17	21
Bar	2	15	19	14	12	30	38	15	11	14	12
Setting	3	10	8	12	9	10	5	14	15	19	11

the other factor. In terms of the general model for a two-way layout, a test for rank interaction is a test of the pair of null hypotheses

$$H_0^{\tau*(\beta)}\colon \big[\tau_1 + (\beta\tau)_{i1} = \tau_2 + (\beta\tau)_{i2} = \cdots = \tau_m + (\beta\tau)_{im},\ \text{for } i = 1,\ldots,b\big]$$

and

$$H_0^{\beta*(\tau)}\colon \big[\beta_1 + (\beta\tau)_{1j} = \beta_2 + (\beta\tau)_{2j} = \cdots = \beta_b + (\beta\tau)_{bj},\ \text{for } j = 1,\ldots,m\big].$$

The general alternative hypotheses are that the corresponding parameters are not all equal. It is possible for rank interaction to show in one direction but not the other; that is, $H_0^{\tau*(\beta)}$ might be rejected but not $H_0^{\beta*(\tau)}$, or vice versa.

1.5. Experiment – two-way layout, no interaction, multiple observations per cell

Wood and Hartvigsen (1964) present the design and analysis of a qualification test program for a small rocket engine. In Table 4, we reproduce their data for the thrust duration of a rocket engine under absence or presence of altitude cycling (factor C at levels 1 or 2) and four different temperature levels (factor D at levels 1, 2, 3 or 4). We have ignored two other factors (which their analysis showed to be nonsignificant) for the purposes of illustration of the nonparametric methods.

We let β_i, τ_j and $(\beta\tau)_{ij}$ denote, respectively, the effect on thrust duration of the ith level of altitude cycling, the jth level of temperature and their interaction. The RGLM test described in Section 4.3, can be used to show that the interaction between the two factors is negligible. Therefore it is of interest to test the main effect hypotheses

$$H_0^{\beta}\colon [\beta_1 = \beta_2] \quad \text{and} \quad H_0^{\tau}\colon [\tau_1 = \tau_2 = \tau_3 = \tau_4]$$

against their respective general alternative hypotheses. Main effect tests against a general alternative hypothesis in the two-way layout are discussed in Section 3.2.3, and corresponding multiple comparison procedures are presented in Section 3.2.4.

Table 4
Thrust duration of a rocket engine under varying conditions of altitude cycling and temperature

		Temperature							
		1		2		3		4	
	1	21.60	21.09	11.54	11.14	19.09	21.31	13.11	11.26
Altitude		21.60	19.57	11.75	11.69	19.50	20.11	13.72	12.09
Cycling	2	21.60	22.17	11.50	11.32	21.08	20.44	11.72	12.82
		21.86	21.86	9.82	11.18	21.66	20.44	13.03	12.29

1.6. Experiment – two-way layout, one observation per cell

The data given in Table 5 form part of an experiment, described by Wilkie (1962), to examine the position of maximum velocity of air blown down the space between a roughened rod and a smooth pipe surrounding it. The two factors of interest were the height of ribs on the roughened rod at equally spaced levels 0.010, 0.015 and 0.020 inches (coded 1, 2, 3) and Reynolds number at six levels (coded 1–6) equally spaced logarithmically over the range 4.8 to 5.3. The responses (positions of maximum air velocity) were measured as $y = (d - 1.4) \times 10^3$, where d is the distance in inches from the center of the rod.

Let η_{ij} denote the expected response at the ith rib height and jth Reynolds number. Then, as in Experiment 1.4, a null hypothesis of additivity can be expressed as

$$H_0^{\text{add}}\colon\ [\eta_{ij} = \beta_i + \tau_j,\ \text{for all } 1 \leqslant i \leqslant 3 \text{ and } 1 \leqslant j \leqslant 6].$$

The general alternative hypothesis of non-additivity is

$$H_1^{\text{add}}\colon\ [\eta_{ij} \neq \beta_i + \tau_j,\ \text{for at least one } (i,j) \text{ combination},\\ 1 \leqslant i \leqslant 3 \text{ and } 1 \leqslant j \leqslant 6],$$

where β_i represents the effect of the ith rib height and τ_j the jth Reynolds number.

A test for non-additivity in the two-way layout with one observation per cell is discussed in Section 3.3.1. When non-additivity is present, the multiple comparisons of most interest are usually comparisons between the treatment combinations, and the one-way layout procedures described in Section 2 can be used.

If an additive model is indicated, the response can be expressed as the sum of the effects of the two treatment factors. In the air velocity experiment, the amount of non-additivity is small, and a starting point in the analysis of main effects might be to test

$$H_0^{\beta}\colon\ [\beta_1 = \beta_2 = \beta_3] \quad \text{and} \quad H_0^{\tau}\colon\ [\tau_1 = \cdots = \tau_6]$$

against their respective general alternative hypotheses H_1^{β}: [β_i not all equal] and H_1^{τ}: [τ_j not all equal]. The null hypothesis H_0^{β} is the hypothesis that the effects of the rib heights are the same (taken as an average over Reynolds numbers) and the null

Table 5
Position of maximum air velocity, Wilke (1962)

		Reynolds number (j)					
	i	1	2	3	4	5	6
Rib	1	−24	−23	1	8	29	23
Height	2	33	28	45	57	74	80
	3	37	79	79	95	101	111

hypothesis $H_0^{\mathcal{T}}$ is the hypothesis that the effects of the different Reynolds numbers are the same (taken as an average over rib heights). The Friedman test for these hypotheses is described in Section 3.3.2, and a multiple comparisons procedure for comparing the effects of pairs of rib heights or pairs of Reynolds numbers is given in Section 3.3.3. The procedures are illustrated in Section 3.3.4.

Since the levels of Reynolds number are ordered, an ordered alternative main effect hypothesis might be of interest rather than a general alternative hypothesis; that is, one may wish to see if the location of maximum velocity is further from the center of the rod as the Reynolds number increases. The alternative hypothesis would then be

$$H_1^{\mathcal{T}}\colon\ [\tau_1 \leqslant \cdots \leqslant \tau_6, \text{with at least one strict inequality}].$$

A test of $H_0^{\mathcal{T}}$ against $H_1^{\mathcal{T}}$ is discussed in Section 3.3.5, and a multiple comparisons procedure in Section 3.3.6. The procedures are illustrated in Section 3.3.7.

All of the procedures discussed in Section 3 are also applicable to the randomized block design setting, where an additive model is often reasonable.

1.7. *Experiment – incomplete layout*

The nonparametric procedures used for experiments with two treatment factors or for randomized block designs are not suitable when one or more of the cells are empty. We shall use the following experiment, described by Kuehl (1994) and run by J. Berry and A. Deutschman at the University of Arizona, to illustrate the test of the hypothesis of no treatment effects against a general alternative hypothesis in the special case of a balanced incomplete block design.

The experiment was run to obtain specific information about the effect of pressure on percent conversion of methyl glucoside to monovinyl isomers. The conversion is achieved by addition of acetylene to methyl glucoside in the presence of a base under high pressure. Five pressures were examined in the experiment, but only three could be examined at any one time under identical experimental conditions, resulting in a balanced incomplete block design. The data and design are shown in Table 6.

Under the assumption of additivity (no interaction between the treatment and the block factors), we discuss in Section 3.4.1 a procedure that can be used to test the

Table 6
Percent conversion of methyl glucoside to monovinyl isomers, Keuhl (1994)

	Block									
Pressure (psi)	I	II	III	IV	V	VI	VII	VIII	IX	X
250	16	19				20	13	21	24	
325	18		26		19		13		10	24
400			39	21		33	34	30		31
475	32	46		35	47	31				37
550		45	61	55	48			52	50	

null hypothesis of no difference in percent conversion of the different pressures, that is

$$H_0^{\tau}: [\tau_1 = \cdots = \tau_5],$$

against a general alternative hypothesis that at least two of the pressures differ. In Section 3.4.3, we give a multiple comparison decision procedure for detecting differences between treatments in the balanced incomplete block design setting. These procedures are illustrated for the above experiment in Section 3.4.4.

A test for H_0^{τ} against a general alternative in an arbitrary two-way layout (that is, a two-way layout that does not necessarily satisfy the conditions of a balanced incomplete block design) is described in Section 3.4.2.

1.8. Experiment – higher way layouts

For an example of a three-way layout, we revisit Experiment 1.2. The currently available nonparametric procedures for analyzing factorial experiments with three or more factors all require at least one observation per cell. Consequently, for illustration purposes, we will analyze factors C, D and E only, allowing two observations in each of the eight cells.

All factorial experiments can be analysed using the techniques of the one-way layout, as illustrated in Section 2.2.6 with factors A and H from Experiment 1.2 and the log variance response. In Section 4.2, we show a method of using two-way layout techniques to analyse factors C, D and E of this experiment using the average thickness response. The RGLM method, discussed in Section 4.3, is completely general, and can be used for any factorial experiment with multiple observations per treatment combination.

2. One-way layout

2.1. Introduction

In the setting of the one-way layout, we are generally interested in detecting whether there are any differences in the probability distributions of m populations based on data contained in independent random samples from each of them. These populations may correspond to m levels of a single treatment factor or m treatment combinations in a factorial experiment.

Let $Y_{1j}, \ldots, Y_{n_j j}, j = 1, \ldots, m$, represent independent random samples from the m populations, where the jth population has cumulative distribution function (c.d.f.) F_j. Then, the null hypothesis that the distributions of the treatment populations are identical corresponds to $F_1 \equiv \cdots \equiv F_m$. In the one-way layout setting, the variety of possible alternative hypotheses almost always correspond to differences in locations for the m population distributions. As a result of this, we specialize our model assumptions and assume that the distribution functions $F_1, \ldots, F_m$ are connected by the relationship $F_j(t) = F(t - \tau_j), -\infty < t < \infty$, for $j = 1, \ldots, m$, where F is a distribution function for a continuous probability distribution with unknown median θ and where τ_j represents the unknown "treatment effect" for the jth population. These assumptions are equivalent to the model formulation

$$Y_{ij} = \theta + \tau_j + E_{ij}, \quad i = 1, \ldots, n_j \text{ and } j = 1, \ldots, m, \tag{1}$$

where θ is the common median, τ_j is the effect of the jth treatment and the error variables E_{ij} form a random sample from **any** continuous distribution with median 0.

In this section and using model (1), we are interested in testing the "location" null hypothesis

$$H_0\colon [\tau_1 = \cdots = \tau_m] \tag{2}$$

versus alternative hypotheses that correspond to a variety of different (non-null) relationships between the τ_j's. As mentioned in Section 1.1, the procedures discussed in this section are also approximately valid for discrete data, provided that the support of the error distribution is sufficiently large.

Distribution-free procedures in the one-way layout setting are based on the fact that, under the hypothesis that $F_1 \equiv \cdots \equiv F_m \equiv F$ (which is implied by (1) and (2)), every ordered arrangement of the sample observations $Y_{11}, \ldots, Y_{n_m m}$ is equally likely.

Let $N = n_1 + \cdots + n_m$ denote the total number of observations and let R_{ij} denote the rank (from least to greatest) of Y_{ij} among all N observations. Let $\boldsymbol{R} = (R_{11}, \ldots, R_{n_1 1}, R_{12}, \ldots, R_{n_2 2}, \ldots, R_{1m}, \ldots, R_{n_m m})$ be the vector of these "joint ranks." Thus, in the absence of ties, $\boldsymbol{R}$ is a random permutation of the available joint ranks $(1, 2, \ldots, N)$. Under $F_1 \equiv \cdots \equiv F_m \equiv F$, any test statistic that is a function of the response variables Y_{ij} only through their joint ranks $\boldsymbol{R}$ will have the same probability distribution under H_0 (2), regardless of the form of the common underlying continuous F. Such procedures based on the joint ranks $\boldsymbol{R}$ of the Y_{ij}'s are, therefore, *distribution-free* tests of the null hypothesis H_0 (2) and they control the type I error rate to be equal to the specified rate α over the **entire class** of continuous distributions F.

2.2. *One-way layout – general alternatives*

2.2.1. *Test designed for general alternatives when the variances are equal*

Kruskal and Wallis (1952) proposed a test procedure that is designed to detect the broad class of alternatives to H_0 (2) corresponding to

$$H_1\colon [\text{not all } \tau_j\text{'s are equal}]. \tag{3}$$

The associated test statistic is a direct analogue of the usual classical one-way layout F-test with the original observations replaced by their joint ranks.

We use the notation $R_{.j} = (R_{1j}+\cdots+R_{n_j j})$ for the sum of the joint ranks assigned to the n_j observations on the jth treatment ($j = 1,\ldots,m$) and $\bar{R}_{.j} = R_{.j}/n_j$ for their average. The Kruskal–Wallis statistic H for testing H_0 (2) versus the general alternative H_1 (3) is based on the sum of the squared distances of the average ranks from their common null expected value $(N+1)/2$. The specific form of the statistic is

$$H = [12/N(N+1)] \sum_{j=1}^{m} n_j (\bar{R}_{.j} - (N+1)/2)^2$$

$$= \left[(12/N(N+1)) \sum_{j=1}^{m} R_{.j}^2/n_j \right] - 3(N+1). \tag{4}$$

The associated test of H_0 versus H_1 at significance level α is

$$\text{reject } H_0 \text{ if and only if } H \geqslant h_\alpha(m, n_1, \ldots, n_m), \tag{5}$$

where $h_\alpha(m, n_1, \ldots, n_m)$ is the upper αth percentile for the distribution of the statistic H under the null hypothesis H_0 (2). The complete null distribution of H is available in Hollander and Wolfe (1973) for $m = 3$ and $n_i = 1,\ldots,5$, $i = 1,2,3$. Additional tables of $h_\alpha(m, n_1, \ldots, n_m)$ values for $\alpha \leqslant 0.10$ are provided in Iman et al. (1975) for $m = 3$, $n_i \leqslant 6$, $i = 1,2,3$; $m = 3$, $n_1 = n_2 = n_3 = 7,8$; $m = 4$, $n_i \leqslant 4$, $i = 1,2,3,4$; $m = 5$, $n_i \leqslant 3$, $i = 1,\ldots,5$. Finally, values of $h_\alpha(m, n_1, \ldots, n_m)$ for $\alpha = 0.01,\ 0.05$, and 0.10 are presented in Mosteller and Rourke (1973) for $m = 3$, $n_1 + n_2 + n_3 \leqslant 13$ and $m = 4$, $n_1 + n_2 + n_3 + n_4 \leqslant 9$.

When either the number of treatments or the sample size configurations are beyond the limits of these existing tables, we use the asymptotic distribution of H to provide approximate critical values for this test procedure. When the null hypothesis H_0 (2) is true, the statistic H has an asymptotic distribution [as $\min(n_1,\ldots,n_m) \to \infty$] which is chi-square with $m-1$ degrees of freedom. Thus the large sample approximation to procedure (5) is

$$\text{reject } H_0 \text{ if and only if } H \geqslant \chi^2_\alpha(m-1), \tag{6}$$

where $\chi^2_\alpha(m-1)$ is the upper αth percentile of the chi-square distribution with $m-1$ degrees of freedom.

Calculation of the Kruskal–Wallis statistic H is available in a variety of software packages. These include, but are not limited to, `Minitab`'s `KRUSKAL–WALLIS` command, `StatXact`'s `KW/MO` command, `SAS`'s `NPAR1WAY` or `PROC RANK–PROC ANOVA` commands, and `IMSL`'s `KRSKL` command. An example of the Kruskal–Wallis test using Experiment 1.2 is given in Section 2.2.6.

2.2.2. Test designed for general alternatives when the variances may not be equal; The k-sample Behrens–Fisher setting

Two of the basic assumptions associated with the test procedure discussed in Section 2.2.1 are that the underlying distributions belong to the same common family (F) and that they differ within this family at most in their medians. The less restrictive setting where these assumptions are relaxed to permit the possibility of differences in scale parameters as well as medians (but still requiring the same common family F) is known as the k-sample Behrens–Fisher problem. Rust and Fligner (1984) proposed a modification of the Kruskal–Wallis statistic to deal with this broader Behrens–Fisher setting when general alternatives (3) are of interest. Defining

$$d_{jj'} = P(Y_{1j} > Y_{1j'}), \quad \text{for all } j \neq j' = 1, \ldots, m, \tag{7}$$

the Rust–Fligner procedure is designed as a test of the less restrictive null and alternative hypotheses

$$H_0^*\colon \left[d_{jj'} = \tfrac{1}{2} \quad \text{for all } j \neq j' = 1, \ldots, m\right] \tag{8}$$

versus

$$H_1^*\colon \left[d_{jj'} \neq \tfrac{1}{2} \quad \text{for at least one } j \neq j' = 1, \ldots, m\right]. \tag{9}$$

Define

$$U_j = \left[n_j/N(N+1)\right]\left[\bar{R}_{.j} - (N+1)/2\right], \quad \text{for } j = 1, \ldots, m,$$

where $\bar{R}_{.j}$ is the average rank (from the joint ranking of all N observations) for the jth treatment. Set $\boldsymbol{U} = [U_1, \ldots, U_m]'$ and let Condition C denote the sample outcome that the minimum value in each of the m samples is less than the maximum value in each of the other samples (that is, every pair of samples overlaps). Then, the Rust–Fligner test statistic is defined to be

$$\begin{aligned} Q^* &= \left[\prod_{j=1}^{m} (n_j - 1)/n_j\right] N\boldsymbol{U}'\widehat{\boldsymbol{A}}^{-}\boldsymbol{U}, \text{ provided Condition C occurs,} \\ &= \infty, \text{ if Condition C does not occur,} \end{aligned} \tag{10}$$

where $\widehat{\boldsymbol{A}}$ is a sample estimator for the asymptotic covariance matrix for the vector $N^{1/2}\boldsymbol{U}$ and $\widehat{\boldsymbol{A}}^{-}$ is the Moore–Penrose generalized inverse of $\widehat{\boldsymbol{A}}$ (see Rust and Fligner, 1984, for computational details for $\widehat{\boldsymbol{A}}^{-}$). The statistic Q^* is exactly distribution-free under the more restrictive null hypothesis H_0 (2), but exact null distribution critical values for small sample sizes are not available in the literature. The large sample approximate level α Rust–Fligner test of the relaxed hypothesis H_0^* (8) versus H_1^* (9) is given by

$$\text{reject } H_0^* \text{ if and only if } Q^* \geqslant \chi^2_{\alpha}(m-1), \tag{11}$$

where $\chi^2_\alpha(m-1)$ is the upper αth percentile for the chi-square distribution with $m-1$ degrees of freedom.

As far as we know, the Rust–Fligner statistic Q^* is not an option on any available software package. An example illustrating the calculation of Q^* is not included in this paper because of space limitations.

2.2.3. Two-sided all treatments multiple comparisons based on joint rankings

In this subsection, we present a multiple comparison procedure that is designed to make two-sided decisions about all $m(m-1)/2$ pairs of treatment effects. The procedure was suggested by Nemenyi (1963) and critical values for equal sample sizes were supplied by Nemenyi (1963) and McDonald and Thompson (1967). Additional critical values for more general sample sizes were later given by Damico and Wolfe (1987). In the context of this paper, the procedure is to be viewed as an appropriate follow-up procedure to the general alternatives Kruskal–Wallis test of Section 2.2.1.

Since multiple comparison procedures are often used following the rejection of H_0, it is not uncommon to see the use of experimentwise error rates larger (e.g., .10 or .15) than would be used as significance levels for hypothesis tests. It often requires an experimentwise error rate higher than the p-value associated with the previous test in order to find such differences among the individual pairs of treatments.

At an experimentwise error rate of α, the Nemenyi–McDonald–Thompson–Damico–Wolfe two-sided all-treatments multiple comparison procedure based on joint rankings reaches its $m(m-1)/2$ decisions through the criterion

$$\text{decide } \tau_u \neq \tau_v \text{ if and only if } N^*|\bar{R}_{.u} - \bar{R}_{.v}| \geqslant g_\alpha(m, n_1, \ldots, n_m), \quad (12)$$

where $\bar{R}_{.j}$ is the average rank for the jth treatment in the joint ranking of all N observations, for $j = 1, \ldots, m$, and N^* is the least common multiple of $n_1, \ldots, n_m$, and the critical value $g_\alpha(m, n_1, \ldots, n_m)$ satisfies the probability restriction

$$\begin{aligned} &P_0(N^*|\bar{R}_{.u} - \bar{R}_{.v}| < g_\alpha(m, n_1, \ldots, n_m),\\ &\quad u = 1, \ldots, m-1; v = u+1, \ldots, m) = 1 - \alpha, \end{aligned} \quad (13)$$

with the probability $P_0(\cdot)$ being computed under H_0 (2). Equation (13) corresponds to the guarantee that the probability is $1 - \alpha$ that we will make all of the correct decisions under the strict null hypothesis that **all** of the treatment effects are equal.

Values of $g_\alpha(m, n_1, \ldots, n_m)$ can be found in Damico and Wolfe (1987) for available experimentwise error rates α closest to but not exceeding .001, .005, .01(.005).05(.01).15 and useful combinations of either $m = 3$ and $1 \leqslant n_1 \leqslant n_2 \leqslant n_3 \leqslant 6$ or $m = 4$ and $1 \leqslant n_1 \leqslant n_2 \leqslant n_3 \leqslant n_4 \leqslant 6$. When either the number of treatments or the sample size configurations are beyond these existing tables for $g_\alpha(m, n_1, \ldots, n_m)$, we must consider a large sample approximation to (12). In the case of equal numbers n of observations on each of the treatments, Miller (1966) suggested the large sample approximation to procedure (12) given by

$$\text{decide } \tau_u \neq \tau_v \text{ if and only if } |\bar{R}_{.u} - \bar{R}_{.v}| \geqslant q_\alpha[m(mn+1)/12]^{1/2}, \quad (14)$$

where q_α is the upper αth percentile for the distribution of the range of m independent $N(0,1)$ variables. Values of q_α for $\alpha = .0001, .0005, .001, .005, .01, .025, .05, .10$, and $.20$ and $m = 3(1)20(2)40(10)100$ can be found in Hollander and Wolfe (1973, Table A.10).

If the sample sizes are not equal, approximation (14) is no longer appropriate and we turn to Dunn's (1964) suggestion of using Bonferroni's Inequality to approximate the value of $g_\alpha(m, n_1, \dots, n_m)$. The Dunn approximation to procedure (12) is

$$\text{decide } \tau_u \neq \tau_v \text{ if and only if}$$
$$|\bar{R}_{.u} - \bar{R}_{.v}| \geqslant z_{\alpha/m(m-1)}\big[N(N+1)/12\big]^{1/2}\big[(n_u + n_v)/n_u n_v\big]^{1/2}, \tag{15}$$

where z_a is the upper ath percentile of the standard normal distribution. The Dunn approximation (15) could also be applied to the case of equal sample sizes. However, it can be quite conservative in such settings and the Miller approximation (14) is preferred.

The Nemenyi–McDonald–Thompson–Damico–Wolfe two-sided multiple comparison procedure based on joint rankings is illustrated in Section 2.2.6 for Experiment 1.2.

2.2.4. *One-sided treatment versus control multiple comparisons based on joint rankings*

In this subsection, we present a multiple comparison procedure that is designed to make one-sided decisions about which treatment populations have higher median effects than that of a single control population. In the context of this paper, it is to be viewed as an appropriate follow-up procedure to rejection of H_0 (2) against a general alternative hypothesis when one of the populations corresponds to a control treatment. The procedure was suggested by Nemenyi (1963) and additional critical values were obtained by Damico and Wolfe (1989).

For simplicity of notation, we let treatment 1 correspond to the single control population. Let N^* be the least common multiple of the sample sizes $n_1, n_2, \dots, n_m$, and let $\bar{R}_{.j}$ be the average rank for the jth treatment in the joint ranking of all N observations, for $j = 1, \dots, m$. At an experimentwise error rate of α, the Nemenyi–Damico–Wolfe one-sided treatment versus control multiple comparison procedure based on joint rankings reaches its $m-1$ decisions through the criterion

$$\text{decide } \tau_u > \tau_1 \text{ if and only if } N^*(\bar{R}_{.u} - \bar{R}_{.1}) \geqslant g^*_\alpha(m, n_1, \dots, n_m), \tag{16}$$

where the critical value $g^*_\alpha(m, n_1, \dots, n_m)$ satisfies the probability restriction

$$P_0\big(N^*(\bar{R}_{.u} - \bar{R}_{.1}) < g^*_\alpha(m, n_1, \dots, n_m),\ u = 2, \dots, m\big) = 1 - \alpha, \tag{17}$$

with the probability $P_0(\cdot)$ being computed under H_0 (2). Values of $g^*_\alpha(m, n_1, \dots, n_m)$ are available in Damico and Wolfe (1989) for achievable experimentwise error rates α closest to but not exceeding .001, .005, .01(.005).05(.01).15 and useful combinations of either $m = 3$ and $n_1 = 1(1)6$, $1 \leqslant n_2 \leqslant n_3 \leqslant 6$ or $m = 4$ and $n_1 = 1(1)6$, $1 \leqslant n_2 \leqslant n_3 \leqslant n_4 \leqslant 6$.

Note that we can make decisions regarding which treatments lead to a *smaller* response than does the control treatment by interchanging the subscripts u and 1 in (16).

When either the number of experimental treatments or the sample size configurations are beyond the existing tables for $g^*_\alpha(m, n_1, \ldots, n_m)$, we consider two possible large sample approximations. First, if $n_2 = n_3 = \cdots = n_m = n$, and both n_1 and n are large, Miller (1966) suggested the large sample approximation to procedure (16) given by

$$\begin{aligned}&\text{decide } \tau_u > \tau_1 \text{ if and only if}\\ &(\bar{R}_{.u} - \bar{R}_{.1}) \geqslant q^*_\alpha[N(N+1)(n_1+n)/12n_1n]^{1/2},\end{aligned} \tag{18}$$

where q^*_α is the upper αth percentile for the distribution of the maximum of $(m-1)$ standard normal random variables with common correlation $\rho = [n/(n_1+n)]$. Selected values of q^*_α for $m = 3(1)13$ and $\rho = .1, .125, .2, .25, .3, .333, .375, .4, .5, .6, .625, .667, .7, .75, .8, .875$, and $.9$ can be found in Hollander and Wolfe (1973, Table A.13).

When the sample sizes of the experimental treatments are not equal, Dunn (1964) suggested using the approximation to procedure (16) given by

$$\begin{aligned}&\text{decide } \tau_u > \tau_1 \text{ if and only if}\\ &(\bar{R}_{.u} - \bar{R}_{.1}) \geqslant z_{\alpha/(m-1)}[N(N+1)(n_1+n)/12n_1n]^{1/2},\end{aligned} \tag{19}$$

where z_a is the upper ath percentile of the standard normal distribution. The Dunn approximation (19) could be applied in the case of equal sample sizes, but it can be quite conservative in such settings and the Miller approximation (18) is preferred.

The Nemenyi–Damico–Wolfe one-sided treatment versus control multiple comparison procedure is illustrated in Section 2.2.6 in the context of Experiment 1.2.

2.2.5. *Simultaneous confidence intervals or bounds for simple contrasts*

Nonparametric point estimates for contrasts $\sum_{j=1}^m c_j\tau_j(\sum_{j=1}^m c_j = 0)$ are available and, for these, the reader is referred to Spjøtvoll (1968) or Hollander and Wolfe (1973, Section 6.4). Here, we discuss simultaneous confidence intervals and bounds for the collection of simple contrasts (pairwise comparisons) $\tau_v - \tau_u$, $1 \leqslant u < v \leqslant m$. The method was proposed by Critchlow and Fligner (1991).

For each pair of treatments (u, v), rank the $n_u + n_v$ observations in the combined uth and vth samples, and let $R_{u1}, R_{u2}, \ldots, R_{un_v}$ be those ranks assigned to the observations $Y_{1v}, \ldots, Y_{n_v v}$ in the vth sample. Let

$$W_{uv} = \sum_{j=1}^{n_v} R_{uj}, \quad \text{for } 1 \leqslant u < v \leqslant m,$$

and standardize W_{uv} as follows,

$$\begin{aligned}W^*_{uv} &= (2)^{1/2}[W_{uv} - \mathrm{E}_0(W_{uv})]/[\mathrm{var}_0(W_{uv})]^{1/2}\\ &= [W_{uv} - n_v(n_u+n_v+1)/2]/\{n_un_v(n_u+n_v+1)/24\}^{1/2}.\end{aligned} \tag{20}$$

Let the upper αth percentile for the distribution of maximum$\{|W^*_{uv}|,\ u \neq v = 1,\ldots,m\}$ under H_0 (2) be denoted by $d_\alpha(m, n_1, \ldots, n_m)$. Values of $d_\alpha(m, n_1, \ldots, n_m)$ are given in Critchlow and Fligner (1991) for $m = 3$ and useful combinations of sample sizes in the range $2 \leqslant n_1 \leqslant n_2 \leqslant n_3 \leqslant 7$, and all achievable values of $\alpha \leqslant 0.2$. Set

$$a_{uv} = (n_u n_v/2) - d_\alpha(m, n_1, \ldots, n_m)[n_u n_v(n_u + n_v + 1)/24]^{1/2} + 1 \qquad (21)$$

and

$$b_{uv} = a_{uv} - 1. \qquad (22)$$

For each pair of treatments (u, v), $u \neq v = 1, \ldots, m$, define the sample differences to be

$$D^{uv}_{ij} = Y_{jv} - Y_{iu}, \quad i = 1, \ldots, n_u;\ j = 1, \ldots, n_v, \qquad (23)$$

and let $D^{uv}_{(1)} \leqslant D^{uv}_{(2)} \leqslant \cdots \leqslant D^{uv}_{(n_u n_v)}$ denote the ordered values of the $n_u n_v$ sample differences, for $u \neq v = 1, \ldots, m$. Then the simultaneous $100(1-\alpha)\%$ confidence intervals for the collection of all simple contrasts $\{\tau_v - \tau_u\colon\ 1 \leqslant u < v \leqslant m\}$ are given by

$$\left[D^{uv}_{(\langle a_{uv}\rangle)},\ D^{uv}_{(n_u n_v - \langle b_{uv}\rangle)}\right), \quad 1 \leqslant u < v \leqslant m, \qquad (24)$$

where $\langle t\rangle$ denotes the greatest integer less than or equal to t.

When either the number of treatments or the sample size configurations are beyond the limits of the exact tables provided in Critchlow and Fligner (1991), the exact value $d_\alpha(m, n_1, \ldots, n_m)$ can be approximated by q_α, the upper αth percentile for the distribution of the range of m independent $N(0, 1)$ variables. Values of q_α for $\alpha = .0001, .0005, .001, .005, .01, .05, .10$, and $.20$ and $m = 3(1)20(2)40(10)100$ can be found in Hollander and Wolfe (1973, Table A.10). (When using this large sample approximation, the approximate value for a_{uv} can be less than 1. In such cases, take $\langle a_{uv}\rangle = 1$ and $\langle b_{uv}\rangle = 0$.)

The Critchlow–Fligner procedure for simultaneous confidence intervals is illustrated in the next section in the context of Experiment 1.2.

2.2.6. Example – one-way layout, general alternatives

We illustrate the various one-way layout procedures using the log variance wafer data from Experiment 1.2. We consider only factors A and H, as though these were the only factors in the experiment. Thus we wish to compare the effects on log variance response of the $m = 4$ treatment combinations listed in Table 7. There are $n = 4$ observations per treatment combination. The joint ranks of the $N = 16$ data values (from least to greatest) are shown in parentheses in Table 7.

Table 7
Log variance responses for factors A and H, Experiment 1.2 (joint ranks in parentheses)

A	H	Treatment j	$\ln S^2_{1j}$ (R_{1j})	$\ln S^2_{2j}$ (R_{2j})	$\ln S^2_{3j}$ (R_{3j})	$\ln S^2_{4j}$ (R_{4j})	$R_{.j}$
−1	−1	1	−0.443 (10)	−0.327 (13)	−0.651 (6)	−0.497 (9)	38
−1	1	2	−1.199 (3)	−0.627 (7)	−1.431 (1)	−1.423 (2)	13
1	−1	3	−0.313 (14)	−0.229 (15)	−0.347 (12)	−0.119 (16)	57
1	1	4	−0.437 (11)	−0.615 (8)	−0.856 (5)	−0.863 (4)	28

First we test the null hypothesis H_0 (2) against the general alternative hypothesis H_1 (3), using the Kruskall–Wallis test. The value of the Kruskall–Wallis statistic (4) is

$$\begin{aligned} H &= \left[(12/(16 \times 17))(38^2/4 + 13^2/4 + 57^2/4 + 28^2/4)\right] - 3(17) \\ &= 11.272. \end{aligned}$$

We compare the value of $H = 11.272$ with the percentile $h_{.01}(4,4,4,4,4)$ given by Iman et al. (1975). Now, $h_{.00999} = 9.287$. Since the observed value of H is larger than $h_{.00999}$, there is evidence of a difference among treatments in terms of log variance response.

If we apply the Nemenyi–McDonald–Thompson–Damico–Wolfe multiple comparisons procedure of Section 2.2.3, we

$$\text{decide } \tau_u \neq \tau_v \text{ if and only if } |R_{.u} - R_{.v}| \geqslant g_\alpha(4,4,4,4,4).$$

Selecting $\alpha = .099$, we obtain $g_\alpha(4,4,4,4,4) = 30$ from Damico and Wolfe (1987, Table II). Thus, using the values of $R_{.j}$ $(j = 1,2,3,4)$ from Table 7, our 6 decisions at an experimentwise error rate of 0.099 are $\tau_1 = \tau_2$, $\tau_1 = \tau_3$, $\tau_1 = \tau_4$, $\tau_2 \neq \tau_3, \tau_2 = \tau_4, \tau_3 = \tau_4$, and only treatments 2 and 3 are judged to have different effects on log variance response. We note that treatment 2 has both factors A and H at different levels from treatment 3. The lower variability is given by treatment 2. At an experimentwise error rate of 0.1239, treatment 3 would also be declared as having a different effect from treatment 4. These two treatments have factor H at different levels.

Treatments 3 and 4 are the treatments closest to the operating conditions that were in place prior to the experiment. If treatment 4 were to be regarded as the control, then it is of interest to know if any other treatment would reduce the log variance response. Using the Nemenyi–Damico–Wolfe procedure (16) for the treatment versus control problem, we decide

$$\tau_u < \tau_4 \text{ if and only if } (R_{.4} - R_{.u}) \geqslant g^*_\alpha(4,4,4,4,4).$$

The critical value $g^*_\alpha(4,4,4,4,4)$ is not among those listed by Damico and Wolfe (1989). For illustration, we use Miller's large sample approximation (18) even though the sample sizes are not particularly large. Selecting $\alpha = .1067$, the critical value q^*_α

for $m-1=3$ experimental treatments and common correlation $\rho=0.5$ is $q^*_\alpha=1.70$. We decide

$$\begin{aligned}&\tau_u<\tau_4 \text{ if and only if}\\&\bar{R}_{.4}-\bar{R}_{.u}\geqslant(1.70)\big[(16\times17\times8)/(12\times4\times4)\big]^{1/2}=5.723\end{aligned}$$

or, equivalently, if and only if $R_{.4}-R_{.u}\geqslant 22.89$. This, at significance level $\alpha=.1067$, we cannot conclude that any treatment gives a significantly lower log variance response than control treatment 4.

Finally, we illustrate the calculation of simultaneous 90% confidence intervals for simple contrasts $\tau_u-\tau_v$. For illustration, let $u=3$ and $v=4$, then the sample differences $D^{3,4}_{ij}$ (23) are

$$D^{3,4}_{11}=0.124,\quad D^{3,4}_{12}=0.302,\quad \ldots,\quad D^{3,4}_{43}=0.737,\quad D^{3,4}_{44}=0.744,$$

giving the ordered values

$$D^{3,4}_{(1)}=0.090,\quad D^{3,4}_{(2)}=0.124,\quad \ldots,\quad D^{3,4}_{(15)}=0.737,\quad D^{3,4}_{(16)}=0.744.$$

The value of $d_\alpha(4,4,4,4,4)$ required for (21) and (22) is not available in the literature, and we use the large sample approximation for $a_{3,4}$. Selecting $\alpha=0.1$, we have

$$\begin{aligned}a_{3,4}&=(4)(4)/2-q_{.1}\big[(4)(4)(4+4+1)/24\big]^{1/2}+1\\&=8-(3.24)[6]^{1/2}+1=1.06,\\b_{3,4}&=0.06.\end{aligned}$$

Thus, as part of a set of simultaneous approximate 90% confidence intervals, we have

$$\tau_3-\tau_4\in\big[D^{3,4}_{(1)},D^{3,4}_{(16)}\big)=[0.090,0.744).$$

2.3. *One-way layout – ordered alternatives*

2.3.1. *Test designed for monotonically ordered alternatives*

In this subsection, we discuss a test procedure proposed independently by Terpstra (1952) and Jonckheere (1954) that is especially effective at detecting restricted alternatives to H_0 (2) that are ordered with respect to the treatment labels, corresponding to

$$H_2\colon\ [\tau_1\leqslant\cdots\leqslant\tau_m,\ \text{with at least one strict inequality.}]\tag{25}$$

For each pair of treatments (u, v), let U_{uv} be the number of pairs of responses from treatments u and v for which treatment u has the smaller response (tied responses contribute a value 1/2 to the total); that is,

$$U_{uv} = \sum_{i=1}^{n_u} \sum_{j=1}^{n_v} \phi(X_{iu}, X_{jv}), \quad 1 \leqslant u < v \leqslant m, \tag{26}$$

where $\phi(a, b) = 1, 1/2, 0$ if $a <, =, > b$. Then, U_{uv} is the Mann–Whitney statistic between responses on the uth and vth treatments.

The Jonckheere–Terpstra statistic J for testing H_0 (2) versus the ordered alternative hypothesis H_2 (25) is the sum of these $k(k-1)/2$ Mann–Whitney counts, namely,

$$J = \sum_{u=1}^{v-1} \sum_{v=2}^{m} U_{uv}, \tag{27}$$

and the associated test of H_0 versus H_2 at significance level α is

$$\text{reject } H_0 \text{ if and only if } J \geqslant j_\alpha(m, n_1, \ldots, n_m), \tag{28}$$

where $j_\alpha(m, n_1, \ldots, n_m)$ is the upper αth percentile of the null sampling distribution of the statistic J (27). Tables of $j_\alpha(m, n_1, \ldots, n_m)$ for selected significance levels α and $m = 3, 2 \leqslant n_1 \leqslant n_2 \leqslant n_3 \leqslant 8$ and $m = 4, 5, 6$ and $n_1 = \cdots = n_m = 2(1)6$ are available in Odeh (1971) and Jonckheere (1954) or Hollander and Wolfe (1973, Table A.8).

When either the number of treatments or the sample size configurations are beyond the limits of these existing tables, the asymptotic behavior of a standardized version of the J statistic can be used to provide approximate critical values for the test procedure (28). When the null hypothesis H_0 (2) is true, the standardized statistic

$$J^* = [J - \mathrm{E}_0(J)]/[\mathrm{var}_0(J)]^{1/2} \tag{29}$$

has an asymptotic distribution [as $\min(n_1, \ldots, n_m) \to \infty$] that is standard normal, where

$$\mathrm{E}_0(J) = \left[N^2 - \sum_{j=1}^{m} n_j^2\right]/4 \quad \text{and}$$

$$\mathrm{var}_0(J) = \left[N^2(2N+3) - \sum_{j=1}^{m} n_j^2(2n_j+3)\right]/72 \tag{30}$$

are the expected value and variance, respectively, of J under the null hypothesis H_0. The large sample approximation to procedure (28) is, then,

$$\text{reject } H_0 \text{ if and only if } J^* \geqslant z_\alpha, \tag{31}$$

where z_α is the upper αth percentile of the standard normal distribution.

Calculation of the Jonckheere–Terpstra statistic J is also available in a variety of software packages. These include, but are not limited to, `StatXact`'s `JT/MO` command and `IMSL`'s `KTRND` command. The Jonckheere–Terpstra test is illustrated in the context of Experiment 1.3 in Section 2.3.5.

2.3.2. One-sided all treatments multiple comparisons using pairwise rankings

In this subsection, we present a multiple comparison procedure suggested by Hayter and Stone (1991) that is designed to make *one-sided* decisions about all $m(m-1)/2$ pairs of treatment effects. In the context of this paper, it is to be viewed as an appropriate follow-up procedure to the Jonckheere–Terpstra test for monotonically ordered alternatives discussed in Section 2.3.1.

Let W^*_{uv} be the standardized two-sample rank sum statistic for the uth and vth samples, as defined explicitly in (20). At an experimentwise error rate of α, the Hayter–Stone one-sided all-treatments multiple comparison procedure based on pairwise rankings reaches its $m(m-1)/2$ decisions through the criterion

$$\begin{aligned}&\text{decide } \tau_v > \tau_u \text{ if and only if } W^*_{uv} \geqslant c_\alpha(m, n_1, \ldots, n_m),\\ &1 \leqslant u < v \leqslant m,\end{aligned} \tag{32}$$

where the critical value $c_\alpha(m, n_1, \ldots, n_m)$ satisfies the probability restriction

$$\begin{aligned}&P_0\big(W^*_{uv} < c_\alpha(m, n_1, \ldots, n_m), u = 1, \ldots, m-1; v = u+1, \ldots, m\big)\\ &= 1 - \alpha,\end{aligned} \tag{33}$$

with the probability $P_0(\cdot)$ being computed under H_0 (2). Values of $c_\alpha(m, n_1, \ldots, n_m)$ are given in Hayter and Stone (1991) for $m = 3$, $3 \leqslant n_1, n_2, n_3 \leqslant 7$, and all achievable experimentwise error rates less than .20.

When either the number of treatments exceeds 3, or $m = 3$ and one of the sample sizes is larger than 7, Hayter and Stone (1991) proposed approximating the exact critical value $c_\alpha(m, n_1, \ldots, n_m)$ by $k_\alpha(m, n_1, \ldots, n_m)$, where $k_\alpha(m, n_1, \ldots, n_m)$ is the upper αth percentile point for the distribution of

$$K = \max_{1 \leqslant i < j \leqslant m} \big[(Z_j - Z_i)/\{(n_i + n_j)/2n_in_j\}^{1/2}\big], \tag{34}$$

where $Z_1, \ldots, Z_m$ are mutually independent and Z_i has a $N(0, n_i^{-1})$ distribution, for $i = 1, \ldots, m$. Thus, the large sample approximation to procedure (32) is

$$\begin{aligned}&\text{decide } \tau_v > \tau_u \text{ if and only if } W^*_{uv} \geqslant k_\alpha(m, n_1, \ldots, n_m),\\ &1 \leqslant u < v \leqslant m.\end{aligned} \tag{35}$$

Values of $k_\alpha(m, n_1, \ldots, n_m)$ for large equal sample sizes $n_1 = \cdots = n_m = n, m = 3(1)9$ and experimentwise error rate $\alpha = .01, .05$, and $.10$ can be found in Hayter and Stone (1991). Values of $k_\alpha(m, n_1, \ldots, n_m)$ for large, but unequal, sample sizes $n_1, \ldots, n_m$ are not presently available in the literature.

If one of the treatments is a control treatment, the Nemenyi–Damico–Wolfe multiple comparison procedure (Section 2.2.4) can be used instead of the Hayter–Stone procedure. The Hayter–Stone one-sided multiple comparison procedure based on pairwise rankings is illustrated for the Acid Red Experiment 1.3 data in Section 2.3.5, and the Nemenyi–Damico–Wolfe procedure is illustrated in Section 2.2.6 in the setting of an unordered alternative hypothesis for the wafer data of Experiment 1.2.

2.3.3. Simultaneous confidence bounds for simple contrasts in the ordered setting
In this subsection, we operate under the ordered restriction $\tau_1 \leqslant \tau_2 \leqslant \cdots \leqslant \tau_m$, so that we are assuming *a priori* that each of the simple contrasts $\tau_v - \tau_u$, $1 \leqslant u < v \leqslant m$, is non-negative. Therefore, in this setting, we will be interested only in simultaneous lower confidence bounds for this set of simple contrasts.

Let $D_{(1)}^{uv} \leqslant D_{(2)}^{uv} \leqslant \cdots \leqslant D_{(n_u n_v)}^{uv}$ denote the ordered values of the $n_u n_v$ sample differences defined in (23), for $u \neq v = 1, \ldots, m$. Let $c_\alpha(m, n_1, \ldots, n_m)$ be the critical value for the Hayter–Stone one-sided all-treatments multiple comparison procedure with experimentwise error rate α as given in (33). Set

$$h_{uv} = (n_u n_v/2) - c_\alpha(m, n_1, \ldots, n_m)[n_u n_v(n_u + n_v + 1)/24]^{1/2} + 1. \tag{36}$$

Then, the simultaneous $100(1-\alpha)\%$ lower confidence bounds for the collection of all simple contrasts $\{\tau_v - \tau_u\colon\ 1 \leqslant u < v \leqslant m\}$ are given by

$$\left[D_{(\langle h_{uv}\rangle)}^{uv}, \infty\right), \quad 1 \leqslant u < v \leqslant m, \tag{37}$$

where $\langle t \rangle$ denotes the greatest integer less than or equal to t.

When either the number of treatments or the sample sizes are beyond the extent of the exact critical value tables provided in Hayter and Stone (1991), the constant $c_\alpha(m, n_1, \ldots, n_m)$ can be approximated by $k_\alpha(m, n_1, \ldots, n_m)$, the upper αth percentile point for the distribution of K (34) in the previous subsection. (When using this large sample approximation, the approximate value for h_{uv} can be less than 1. In such cases, take $\langle h_{uv} \rangle = 1$.) An example illustrating the Hayter–Stone procedure for simultaneous confidence bounds is not included in this paper because of space limitations.

2.3.4. Tests designed for umbrella alternatives
In this subsection, we consider test procedures that address problems where we have *a priori* knowledge that enables us to label the treatments in such a way that the treatment effects will be in a monotonically increasing relationship with the treatment labels (as in Section 2.3.1) or will be ordered so that there is an initial period of increases in the treatment effects, but with an eventual downturn for the later treatments.

Such alternative hypotheses are generally referred to as "umbrella alternatives" in the literature and they can be written as

$$H_3\colon\ [\tau_1 \leqslant \tau_2 \leqslant \cdots \leqslant \tau_{p-1} \leqslant \tau_p \geqslant \tau_{p+1} \geqslant \cdots \geqslant \tau_m,$$
$$\text{with at least one strict inequality, for some } p \in \{1,2,\ldots,m\}]. \tag{38}$$

The treatment label "p" is referred to as the peak of the umbrella and may either be known or unknown in practical settings. Note that, if we are interested in alternative hypotheses corresponding to "inverted umbrellas" of the form $\tau_1 \geqslant \tau_2 \geqslant \cdots \geqslant \tau_{p-1} \geqslant \tau_p \leqslant \tau_{p+1} \leqslant \cdots \leqslant \tau_m$, with at least one strict inequality, we can apply the procedures described in this section to the negatives of the sample data. The negation of the sample values turns an inverted umbrella into an umbrella.

The test procedure proposed by Mack and Wolfe (1981), is a direct extension of the Jonckheere–Terpstra procedure for ordered alternatives. The umbrella alternatives (38) with known peak can be viewed as two sets of ordered alternatives, one increasing monotonically from the initial treatment level up to the known level, p, and the second set beginning at treatment level p and decreasing monotonically to the final level m. For each pair of treatments (u, v), let $U_{uv} (1 \leqslant u < v \leqslant m)$ be the Mann–Whitney statistic defined in (26). The Mack-Wolfe statistic A_p for testing H_0 (2) versus the umbrella alternatives H_3 (38), with peak known *a priori* to be at p, is given by

$$A_p = \sum_{u=1}^{v-1} \sum_{v=2}^{p} U_{uv} + \sum_{u=p}^{v-1} \sum_{v=p+1}^{m} U_{vu}, \tag{39}$$

and the associated test of H_0 versus H_3 at significance level α is

$$\text{reject } H_0 \text{ if and only if } A_p \geqslant a_\alpha(p, m, n_1, \ldots, n_m), \tag{40}$$

where $a_\alpha(p, m, n_1, \ldots, n_m)$ is the upper αth percentile of the null sampling distribution of the statistic A_p (39).

Values of $a_\alpha(p, m, n_1, \ldots, n_m)$ are generally not available in the published literature and, therefore, a large sample approximation to the exact null distribution of A_p is essential for its use in applications. Let $N_1 = n_1 + \cdots + n_p$ and $N_2 = n_p + \cdots + n_m$. When the null hypothesis H_0 (2) is true, the standardized statistic

$$A_p^* = [A_p - \mathrm{E}_0(A_p)] / [\mathrm{var}_0(A_p)]^{1/2} \tag{41}$$

has an asymptotic distribution [as $\min(n_1, \ldots, n_m) \to \infty$] that is standard normal, where

$$\mathrm{E}_0(A_p) = \left[N_1^2 + N_2^2 - \sum_{i=1}^{m} n_i^2 - n_p^2\right] / 4 \tag{42}$$

and

$$\mathrm{var}_0(A_p) = \left\{ 2\left(N_1^3 + N_2^3\right) + 3\left(N_1^2 + N_2^2\right) - \sum_{i=1}^{m} n_i^2(2n_i + 3) \right.$$

$$\left. - n_p^2(2n_p + 3) + 12 n_p N_1 N_2 - 12 n_p^2 N \right\} / 72 \qquad (43)$$

are the expected value and variance, respectively, of A_p under the null hypothesis H_0. The large sample approximation to procedure (40) is then

$$\text{reject } H_0 \text{ if and only if } A_p^* \geqslant z_\alpha, \qquad (44)$$

where z_α is the upper αth percentile of the standard normal distribution.

For the case of umbrella alternatives with unknown peak, we first estimate the peak of the umbrella configuration. To accomplish this, we calculate the m peak-picking statistics

$$U_{.v} = \sum_{u \neq v}^{m} U_{uv}, \quad \text{for } v = 1, \ldots m, \qquad (45)$$

where U_{uv} is the Mann–Whitney statistic (26) computed for the observations on the uth and vth treatments. Next, each $U_{.v}$ is standardized to obtain

$$\begin{aligned} U_{.v}^* &= \left[U_{.v} - \mathrm{E}_0(U_{.v})\right] / \left\{\mathrm{var}_0(U_{.v})\right\}^{1/2} \\ &= \left[U_{.v} - n_v(N - n_v)/2\right] / \left\{n_v(N - n_v)(N + 1)/12\right\}^{1/2}. \end{aligned} \qquad (46)$$

Let r be the number of treatments that are tied for the maximum $U_{.v}^*$ value and let M be the subset of $\{1, 2, \ldots, m\}$ that corresponds to these r treatments tied for the maximum. The Mack–Wolfe peak-unknown statistic $A_{\hat{p}}^*$ for testing H_0 (2) versus the umbrella alternatives H_3 (38) with unknown peak p is then given by

$$A_{\hat{p}}^* = (1/r) \sum_{j \in M} \left[A_j - \mathrm{E}_0(A_j)\right] / \left\{\mathrm{var}_0(A_j)\right\}^{1/2}, \qquad (47)$$

where A_j (39) is the peak-known statistic with peak at the jth treatment group and $\mathrm{E}_0(A_j)$ and $\mathrm{var}_0(A_j)$ are given in equations (42) and (43), respectively. The associated test of H_0 versus H_3 at significance level α is

$$\text{reject } H_0 \text{ if and only if } A_{\hat{p}}^* \geqslant a_\alpha^*(m, n_1, \ldots, n_m), \qquad (48)$$

where $a_\alpha^*(m, n_1, \ldots, n_m)$ is the upper αth percentile of the null sampling distribution of the statistic $A_{\hat{p}}^*$ (41). This sampling distribution properly takes into account the fact that we have first used the sample data to estimate the unknown peak. Tables of $a^*_\alpha(m, n_1, \ldots, n_m)$ for $\alpha \cong .01, .05$, and $.10, m = 3(1)10$, and equal sample sizes $n_1 = \cdots = n_m = 2(1)10$ are available in Mack and Wolfe (1981).

For unequal sample sizes $(n_1, \ldots, n_m)$, Mack and Wolfe (1981) suggest approximating $a^*_\alpha(m, n_1, \ldots, n_m)$ by the corresponding critical value for equal sample sizes $n_1 = \cdots = n_m = q$, namely, $a^*_\alpha(m, q, \ldots, q)$, where q is the integer closest to the average sample size $(n_1 + \cdots + n_m)/m$. If the value of q is greater than 10, they suggest using $a^*_\alpha(m, 10, \ldots, 10)$. Finally, if the number of treatments m is greater than 10, Mack and Wolfe suggest using the $m = 10$ critical value.

An example of the Mack–Wolfe test using the data from the Acid Red Experiment 1.3 is presented in the next section.

2.3.5. Example – one-way layout, ordered and umbrella alternatives

We illustrate various one-way layout analyses, when prior knowledge about ordering of the treatment effects is available. For the Acid Red data of Experiment 1.3, we first test the null hypothesis H_0 (2) of no differences in the treatment effects against the ordered alternative hypothesis H_2 (25), using the Jonckheere–Terpstra procedure of Section 2.3.1. For this experiment, we have $m = 6, n_j = 3, j = 1, \ldots, 6$, and $N = 18$. We compute the $m(m-1)/2 = 15$ individual Mann–Whitney statistics U_{uv} to be

$$U_{25} = U_{26} = U_{35} = U_{36} = U_{45} = U_{46} = 0, \quad U_{12} = U_{13} = U_{14} = 9,$$

$$U_{15} = 5.5, \quad U_{16} = 3.5, \quad U_{23} = 6, \quad U_{24} = 7, \quad U_{34} = 4, \quad U_{56} = 2.$$

Combining these counts in expression (27), we obtain $J = 55$. If we compare this value with the null distribution tables for J (Hollander and Wolfe, 1973, Table A.8), we find that the p-value for these data is greater than .50. Thus, there is no evidence of an increase in the number of revertant colonies as the dosage of Acid Red 114 increases. This might lead to the conclusion that we should not suspect the compound of being a potential mutagen. However, as Simpson and Margolin (1986) point out in their original discussion of these data, the Salmonella bacteria of strain TA 98 may actually succumb to the toxic effects of the higher doses of Acid Red 114, resulting in a reduction in the number of organisms at risk of mutation and leading to a downturn in the dose-response curve.

To address this issue, we now illustrate the application of the Mack–Wolfe peak-unknown procedure of Section 2.3.4 to the same data. We assume that, if there is a carcinogenic effect of the Acid Red 114, it will result in an umbrella pattern, but with unknown peak. First we need to estimate the peak of the umbrella through computation of the $m = 6$ combined-samples peak pickers $U_{.1}, \ldots, U_{.m}$, as given by equation (45). After calculating the necessary Mann–Whitney counts U_{uv}, $1 \leqslant v \neq u \leqslant 6$, we find that

$$U_{.1} = 9, \quad U_{.2} = 32, \quad U_{.3} = 38, \quad U_{.4} = 38, \quad U_{.5} = 12.5, \quad U_{.6} = 5.5.$$

Since we have equal sample sizes $n_j = 3$ ($j = 1, \ldots, 6$), each of the combined-samples Mann–Whitney statistics $U_{.1}, \ldots, U_{.6}$ has the same null mean and variance. As a result, we do not need to compute the standardized forms $U_{.q}^*$ in (46), since the treatment group with the largest $U_{.q}$ value will also be the one with the largest $U_{.q}^*$ value. Therefore, for these Acid Red 114 data, we have a tie between dosage levels 333 μg/ml and 1000 μg/ml (treatments 3 and 4) for the maximum $U_{.q}$ value, giving $r = 2$ and $M = \{3, 4\}$. The resulting form of the Mack–Wolfe peak-unknown statistic is then given by

$$A_{\hat{p}}^* = \{[A_3 - E_0(A_3)]/[\mathrm{var}_0(A_3)]^{1/2} + [A_4 - E_0(A_4)]/[\mathrm{var}_0(A_4)]^{1/2}\}/2$$
$$= \{[76 - 40.5]/[96.75]^{1/2} + [69 - 40.5]/[96.75]^{1/2}\}/2 = 3.25.$$

Comparing this value of $A_{\hat{p}}^*$ with the null distribution in Mack and Wolfe (1981) we find that the p-value for these data is much smaller than .01, since $\hat{a}^*(.01, 6, 3, 3, 3, 3, 3, 3) \cong 2.733$. Thus, there is substantial evidence in favor of an umbrella effect in the number of revertant colonies over the studied range of Acid Red 114 doses, with the peak effect estimated to be at either the dosage 333 μg/ml or the dosage 1000μg/ml.

To illustrate the Hayter–Stone one-sided multiple comparison procedure (32), we consider only the first three dosage levels (0, 100, and 333 μg/ml) so that we can make use of an exact critical value. With an experimentwise error rate of $\alpha = .1262$, we obtain the value $c_{.1262}(3, 3, 3, 3) = 2.7775$ from Table 1 in Hayter and Stone (1991). Since the sample size is 3 for each of the three dosages being considered, the standardizing constants in the calculation of W_{uv}^* (20) are the same for all of the pairwise rank sums, that is,

$$W_{uv}^* = (2)^{1/2}(W_{uv} - 10.5)/(5.25)^{1/2}, \quad 1 \leqslant u < v \leqslant 3.$$

As a result, the decision criterion for this setting can be rewritten to be

$$\text{decide } \tau_v > \tau_u \text{ if and only if } W_{uv} \geqslant 15.$$

We find the individual sums of pairwise ranks (using average ranks to break ties) to be $W_{12} = 15$, $W_{13} = 15$, and $W_{23} = 12$. Hence our three decisions at experimentwise error rate $\alpha = .1262$ are that $\tau_2 > \tau_1, \tau_3 > \tau_1$, and $\tau_3 = \tau_2$, indicating that, while both dosage levels 100 μg/ml and 333 μg/ml produce significantly greater numbers of revertant colonies than does the 0 dosage, there is no significant difference in the effects of the two larger doses. (We note that .1262 is the smallest possible exact experimentwise error rate for $m = 3$ and $n_1 = n_2 = n_3 = 3$.)

3. Two-way layout

3.1. Introduction

Let $Y_{ij1}, Y_{ij2}, \ldots, Y_{ijn_{ij}}$ ($i = 1, \ldots, b; j = 1, \ldots, m$) represent independent random samples from bm populations, where the (ij)th population has cumulative distribution function (cdf) F_{ij}. Then Y_{ijk} is the response variable corresponding to the kth observation on the ith level of a factor B and jth level of a factor U. Specifying the assumptions as for the one-way layout, we write our model as

$$Y_{ijk} = \theta + \beta_i + \tau_j + (\beta\tau)_{ij} + E_{ijk} \qquad (49)$$
$$i = 1, \ldots, b; j = 1, \ldots, m;\ k = 1, \ldots, n_{ij},$$

where θ is the common median, β_i is the effect of the ith level of factor B, τ_j is the effect of the jth level of factor U, $(\beta\tau)_{ij}$ is the effect of the interaction between the ith level of B and the jth level of U, and the $N(= n_{11} + \cdots + n_{bm})$ error variables E_{ijk} form a random sample from any continuous distribution with median 0. In the sections that follow, factor U is regarded as a treatment factor and factor B may be either a second treatment factor or a block factor.

In some experiments, the response can be modelled as the sum of the effects of the two factors; that is

$$Y_{ijk} = \theta + \tau_j + \beta_i + E_{ijk},$$
$$i = 1, \ldots, b;\ j = 1, \ldots, m;\ k = 1, \ldots, n_{ij}. \qquad (50)$$

This is known as an *additive model* and, in this setting, main effect comparisons of the two factors are of interest. Tests for additivity are discussed in Sections 3.2.1 and 3.3.1. Main effect tests and multiple comparison procedures are described in Sections 3.2.3, 3.2.4 and Sections 3.3.2–3.3.6.

Distribution-free procedures in this two-way layout additive model setting are most often based on the fact that under H_0^{τ}: $[\tau_j$ are all equal, $j = 1, \ldots, m]$, every ordered arrangement of the $n_{i.} = \sum_{j=1}^{m} n_{ij}$ sample observations within the ith level of factor B is equally likely, separately and independently for each level $i = 1, \ldots, b$ of factor B. This leads directly to the use of the joint ranks of the sample observations within levels of factor B. Any test procedure that is a function of the observations only through their joint ranks within levels of factor B, will have the same probability distribution under H_0^{τ} (61) regardless of the form of the common underlying continuous distribution for the error variables E_{ijk}.

An alternative design for the two-way layout is that of the *nested design* where the levels of one factor (say U) are observed at only one of the levels of the other factor (B). The model for the two-way nested layout is

$$Y_{ijk} = \theta + \beta_i + \tau(\beta)_{j(i)} + E_{ijk}$$
$$i = 1, \ldots, b;\ j = 1, \ldots, m;\ k = 1, \ldots, n_{ij}, \qquad (51)$$

where θ is the common median, β_i is the effect of the ith level of factor B, $\tau(\beta)_{j(i)}$ is the effect of the jth level of U observed at the ith level of B, and the N error variables E_{ijk} form a random sample from a continuous distribution with median 0. The null hypothesis

$$H_0^{\tau(\beta)}\colon\ [\tau(\beta)_{i(j)} \text{ are all equal},\ i = 1, \ldots, b,\ j = 1, \ldots, m] \qquad (52)$$

is equivalent to a global hypothesis of no main effect of U and no interaction between B and U for the two-way crossed model; that is,

$$H_0^T\colon\ [\tau_j + (\beta\tau)_{ij} \text{ are all equal, } i = 1,\ldots,b,\ j = 1,\ldots,m] \tag{53}$$

(see Section 3.5). Consequently, the same testing procedures can be used and we will not discuss the nested case further.

3.2. Two-way layout – multiple observations per cell

3.2.1. Testing for non-additivity – multiple observations per cell

Under the two-way model (49), a test of additivity is a test of the null hypothesis

$$H_0^{\text{add}}\colon\ [\eta_{ij} = \beta_i + \tau_j, \text{ for all } 1 \leqslant i \leqslant b \text{ and } 1 \leqslant j \leqslant m] \tag{54}$$

against the general alternative hypothesis

$$H_1^{\text{add}}\colon\ [\eta_{ij} \neq \beta_i + \tau_j, \text{ for at least one } 1 \leqslant i \leqslant b, 1 \leqslant j \leqslant m], \tag{55}$$

where η_{ij} is the expected response when factor B is at level i and factor U is at level j.

Conover and Iman (1976, 1981) suggested the use of the *rank transform* approach for testing for additivity in the two-way model with multiple observations per cell. Such tests have also been advocated in some user manuals for statistical computer packages and libraries. The rank transform procedure first ranks all of the observations from 1 to N without regard for the corresponding levels of the factors, and then uses the usual analysis of variance formulae on the ranks. The procedures are simple to use, but unfortunately, recent studies have shown that, in the presence of two large main effects, the significance levels of the rank transform test for testing H_0^{add} (54) can be severely inflated (for example, see Blair et al. (1987) for simulation studies, and see Akritas (1990) and Thompson (1991), for theoretical investigations). Consequently, the rank transform procedure is not recommended for testing H_0^{add} (54) (the only exception being when both factors have two levels). Akritas and Arnold (1994) show that the rank transform test of additivity does not, in fact, test H_0^{add} (54). Instead, it tests the null hypothesis that the common distribution, F_{ij}, of the response variables in the ijth cell is a mixture of two distributions, one depending on the level i of factor B and the other depending on the jth level of factor U; that is,

$$F_{ij}(y) = aD_i(y) + (1-a)H_j(y)\,,$$

where a is the same for every cell. Consequently, when such a hypothesis is of interest, the rank transform method would appear to give a valid test.

Sawilowsky (1990), in a review article, discusses a number of different tests for H_0^{add} (54), but many have drawbacks such as being conditional on the data collected or requiring the data to be collected replication by replication (which is generally only done when the replications are synonymous with blocks). Hettmansperger and McKean (1983), Aubuchon and Hettmansperger (1984) and Draper (1988) all give details of some of these tests. The only method that we shall discuss in this article is the rank-based robust general linear model (RGLM) method of McKean and Hettmansperger (1976).

The RGLM method possesses many of the features of the method of least squares and is very flexible. It can be used for testing a null hypothesis of additivity or for testing null hypotheses of no main effects (both against general alternatives). It can be used for any number of factors and for fractional factorial experiments provided that there are sufficient degrees of freedom to provide an estimate of the error variability. In addition, the method allows contrast estimates and confidence intervals to be calculated. It should be noted, however, that in the case of a fully parameterized model with two observations per cell, the contrast estimates are identical to those provided by the method of least squares.

Since the RGLM method can be applied in many different settings, we postpone discussion of the details to Section 4.3. We illustrate the procedure for the two-factor welding experiment (Experiment 1.4) in Section 4.3.2.

When factor B is a block factor, an alternative to testing H_0^{add} (54) is to test for a common ordering of treatment effects within the blocks. "Rank interaction" is said to occur when the orderings fail to be the same in all blocks simultaneously. Rank interaction can also occur when both factors are treatment factors, although it is perhaps less likely to be of interest in this setting. The null hypothesis $H_0^{\tau*(\beta)}$ of no rank interaction of treatments with blocks is

$$H_0^{\tau*(\beta)}\colon \big[\tau_1 + (\beta\tau)_{i1} = \tau_2 + (\beta\tau)_{i2} = \cdots = \tau_m + (\beta\tau)_{im}, \quad \text{for } i = 1, \ldots, b\big] \tag{56}$$

and the general alternative hypothesis is

$$H_1^{\tau*(\beta)}\colon \big[\tau_1 + (\beta\tau)_{i1}, \tau_2 + (\beta\tau)_{i2}, \ldots, \tau_m + (\beta\tau)_{im}, \quad \text{are not all equal}\big]. \tag{57}$$

A test for the null hypothesis $H_0^{\tau*(\beta)}$ (56) was developed by De Kroon and Van der Laan (1981) for equal numbers n of observations per cell as follows. Let R_{ijk} denote the rank of Y_{ijk} among the mn observations $Y_{i11}, Y_{i12}, \ldots, Y_{i1n}, \ldots, Y_{im1}, Y_{im2}, \ldots, Y_{imn}$ associated with the ith level of factor B ($i = 1, \ldots, b$) and let $\boldsymbol{R}_i = (R_{i11}, R_{i12}, \ldots, R_{i1n}, \ldots, R_{im1}, \ldots, R_{imn})$ be the vector of these joint ranks for level i of factor B. The test statistic developed by De Kroon and Van der Laan (1981) is

$$T_2 = \frac{12}{n^2 m(nm+1)} \left\{ \sum_{j=1}^{m} \sum_{i=1}^{b} R_{ij.}^2 - b^{-1} \sum_{j=1}^{m} R_{.j.}^2 \right\}, \tag{58}$$

where $R_{ij.} = \sum_{k=1}^{n} R_{ijk}$ and $R_{.j.} = \sum_{i=1}^{b} \sum_{k=1}^{n} R_{ijk}$. Their test of the null hypothesis $H_0^{\tau*(\beta)}$ (56) of no rank interaction against the alternative hypothesis $H_1^{\tau*(\beta)}$ (57) is

$$\text{reject } H_0^{\tau*(\beta)} \text{ if and only if } T_2 \geqslant t_\alpha^{(2)}(m, b, n, \ldots, n), \tag{59}$$

where $t_\alpha^{(2)}$ is the upper αth percentile of the null distribution of T_2. The percentiles are tabulated by De Kroon and Van der Laan (1981) for various values of m, b and n in the range $2 \leqslant m \leqslant 4$, $2 \leqslant b \leqslant 10$, and $2 \leqslant n \leqslant 4$. These authors state that the asymptotic distribution [as $n \to \infty$] of T_2 is chi-square with $(b-1)(m-1)$ degrees of freedom and, consequently, the large sample approximation to procedure (59) is

$$\text{reject } H_0^{\tau*(\beta)} \text{ if and only if } T_2 \geqslant \chi_\alpha^2((m-1)(b-1)), \tag{60}$$

where $\chi_\alpha^2(q)$ is the upper αth percentile of the chi-square distribution with q degrees of freedom.

When a null hypothesis of additivity or of no rank interaction is rejected, the individual cells can be compared using the techniques of the one-way layout. If the null hypothesis is not rejected, then analysis of the main effects is of interest (Sections 3.2.3 and 3.2.4).

3.2.2. Example – testing for non-additivity, multiple observations per cell.
We use the welding data in Table 3 of Experiment 1.4 in order to illustrate the analysis of a two-way layout with an equal number of observations per cell. The hypothesis of additivity H_0^{add} (54) can be tested using the RGLM method. Since this method can be used for higher-way layouts also, we have postponed the illustration until Section 4.3.2.

As an alternative, we may wish to test the hypothesis $H_0^{\tau*(\beta)}$ (56) of no rank interaction. Ranking the observations in Table 3 within the $i = 3$ levels of the gage bar settings (and using average ranks to break the ties), we obtain the rank vectors

$$\begin{array}{rcl} \boldsymbol{R}_1 &=& (1,\ 2,\ 3,\ 5.5,\ 8.5,\ 10,\ 7,\ 4,\ 5.5,\ 8.5), \\ \boldsymbol{R}_2 &=& (6.5,\ 8,\ 4.5,\ 2.5,\ 9,\ 10,\ 6.5,\ 1,\ 4.5,\ 2.5), \\ \boldsymbol{R}_3 &=& (4.5,\ 2,\ 7,\ 3,\ 4.5,\ 1,\ 8,\ 9,\ 10,\ 6\ ), \end{array}$$

giving

$$\begin{aligned} T_2 &= \frac{12}{(4)(5)(11)} \left\{ \left(3^2 + 14.5^2 + \cdots + 16^2\right) - \frac{1}{3}\left(24^2 + 25.5^2 + \cdots + 37^2\right) \right\} \\ &= 15.38. \end{aligned}$$

No tables are given by De Kroon and Van der Laan (1981) for $m = 5, b = 3$, and $n = 2$. Here, n is not large, and therefore we should not expect the chi-square distribution with $(b-1)(m-1) = 8$ degrees of freedom to be an accurate approximation to the true distribution of T_2. However, if we do compare the value $T_2 = 15.38$ with the percentiles of the $\chi^2(8)$ distribution, we see that 15.38 is around the 95th percentile. Table 5.3.6 of De Kroon and Van der Laan (1981) suggests that the percentile of the exact distribution is likely to be slightly lower, so there is some evidence at a significance level of just under $\alpha = 0.05$ to reject $H_0^{\tau*(\beta)}$ (56) and to conclude that there is rank interaction of U with B.

3.2.3. *Main effect tests designed for general alternatives – at least one observation per cell*

In this subsection, we consider two-way layout settings with at least one observation per cell (that is, $n_{ij} \geqslant 1$ for every $i = 1, \ldots, b$ and $j = 1, \ldots, m$). We assume that the additive model (50) holds and we wish to test the null hypothesis of no effect of factor U,

$$H_0^{\tau}\colon\ [\tau_j \text{ are all equal, } j = 1, \ldots, m] \tag{61}$$

against the general class of alternatives

$$H_1^{\tau}\colon\ [\text{not all } \tau_j\text{'s equal}]. \tag{62}$$

Let R_{ijk} be the rank of Y_{ijk} within the $n_{i.} = \sum_{j=1}^{m} n_{ij}$ observations on the ith level of B, for $j = 1, \ldots, m$ and $k = 1, \ldots, n_{ij}$. For each level of U ($j = 1, \ldots, m$), compute the sum of 'cell-wise weighted' average ranks, given by

$$S_j = \sum_{i=1}^{b} \left[\frac{R_{ij.}}{n_{i.}}\right], \tag{63}$$

where $R_{ij.} = \sum_{k=1}^{n_{ij}} R_{ijk}$. Define the vector $\boldsymbol{R}$ to be

$$\begin{aligned} \boldsymbol{R} &= (S_1 - \mathrm{E}_0[S_1], \ldots, S_{m-1} - \mathrm{E}_0[S_{m-1}]) \\ &= \Bigg(S_1 - \sum_{i=1}^{b}[n_{i1}(n_{i.}+1)/2n_{i.}], \ldots, \\ &\qquad S_{m-1} - \sum_{i=1}^{b}[n_{i,m-1}(n_{i.}+1)/2n_{i.}]\Bigg). \end{aligned} \tag{64}$$

Note that we have chosen to define $\boldsymbol{R}$ without a term for the mth level of U. The S_j's are linearly dependent, since a weighted linear combination of all m of them is a constant. We could omit any one of them in the definition of $\boldsymbol{R}$ and the test would be exactly the same no matter which one was chosen for omission.

The covariance matrix for $\boldsymbol{R}$ under H_0^{τ} (61) has the form $\boldsymbol{\Sigma}_0 = ((\sigma_{st}))$, where

$$\sigma_{st} = \sum_{i=1}^{b} \left[n_{is}(n_{i.} - n_{is})(n_{i.}+1)/12n_{i.}^2\right], \qquad \text{for } s = t = 1, \ldots, m-1$$

$$= -\sum_{i=1}^{b} \left[n_{is}n_{it}(n_{i.}+1)/12n_{i.}^2\right], \qquad \text{for } s \neq t = 1, \ldots, m-1. \tag{65}$$

Letting $\boldsymbol{\Sigma}_0^{-1}$ denote the inverse of $\boldsymbol{\Sigma}_0$, a general test statistic proposed by Mack and Skillings (1980) is

$$\mathrm{MS} = \boldsymbol{R}'\boldsymbol{\Sigma}_0^{-1}\boldsymbol{R}, \tag{66}$$

and the associated level α test of $H_0^{\mathcal{T}}$ (61) versus $H_1^{\mathcal{T}}$ (62) is

$$\text{reject } H_0^{\mathcal{T}} \text{ if and only if } \mathrm{MS} \geqslant w_\alpha(b, m, n_{11}, \ldots, n_{bm}), \tag{67}$$

where $w_\alpha(b, m, n_{11}, \ldots, n_{bm})$ is the upper αth percentile for the null sampling distribution of MS (66). Values of $w_\alpha(b, m, n_{11}, \ldots, n_{bm})$ are available in the literature for equal n_{ij} (see the subsection below) but not for arbitrary replications. However, the asymptotic distribution [as $N \to \infty$] of MS under $H_0^{\mathcal{T}}$ is chi-square with $m-1$ degrees of freedom where $N = \sum_{i=1}^{b}\sum_{j=1}^{m} n_{ij}$. Thus, when N is large, the approximation to procedure (67) is

$$\text{reject } H_0^{\mathcal{T}} \text{ if and only if } \mathrm{MS} \geqslant \chi_\alpha^2(m-1), \tag{68}$$

where $\chi_\alpha^2(m-1)$ is the upper αth percentile of the chi-square distribution with $m-1$ degrees of freedom. Mack and Skillings (1980) note that this chi-square approximation tends to be conservative, especially for small levels of α.

Equal cell replications

In the special case where we have an equal number of replications for each of the bm combinations of levels of factors U and B, the statistic MS (66) permits a closed form expression. Let $n_{11} = n_{12} = \cdots = n_{bm} = n$. Then $N = bmn$ and $n_{i.} = mn$ for $i = 1, \ldots, b$, and the Mack–Skillings statistic MS (66) can be written in closed form as

$$\mathrm{MS} = \left[12/m(N+b)\right] \sum_{j=1}^{m} \left(mS_j - (N+b)/2\right)^2. \tag{69}$$

The associated level α test of $H_0^{\mathcal{T}}$ versus $H_1^{\mathcal{T}}$ is, then,

$$\text{reject } H_0^{\mathcal{T}} \text{ if and only if } \mathrm{MS} \geqslant w_\alpha(b, m, n, \ldots, n). \tag{70}$$

The exact critical values $w_\alpha(b, m, n, \ldots, n)$, some obtained via simulation of the exact null distribution, can be found in Mack and Skillings (1980) for selected significance levels α and $b = 2(1)5$, $m = 2(1)5$, and common number of replications $n = 2(1)5$.

3.2.4. Two-sided all treatments multiple comparisons – equal numbers of observations per cell

In this subsection, we discuss a multiple comparison procedure that is designed to make two-sided decisions about all $m(m-1)/2$ pairs of treatment effects when we have an equal number n of replications from each combination of factors B and U in a two-way layout. It is appropriate as a follow-up procedure to the equal-replication Mack–Skillings test in the previous subsection and was proposed by Mack and Skillings (1980).

Let S_j be as given in (63) with $n_{i.} = mn$ for $j = 1, \ldots, m$. At an experimentwise error rate no greater than α, the Mack–Skillings two-sided all-treatments multiple comparison procedure reaches its $m(m-1)/2$ decisions through the criterion

$$\text{decide } \tau_u \neq \tau_v \text{ if and only if } |S_u - S_v| \geqslant \left[w_\alpha(N+b)/6\right]^{1/2}, \tag{71}$$

where $N = bmn$ is the total number of observations and $w_\alpha = w_\alpha(b, m, n, n, \ldots, n)$ is the upper αth percentile for the null sampling distribution of the Mack–Skillings statistic MS (69). The multiple comparison procedure in (71) guarantees that the probability is at least $1 - \alpha$ that we will make all of the correct decisions under the strict null hypothesis that **all** of the treatment effects are equal and controls the experimentwise error rate over the entire class of continuous distributions for the error terms. The procedure (71) often requires an experimentwise error rate higher than the p-value associated with any previous test in order to find the most important differences between the various treatments.

When the total number of observations is large, the critical value $[w_\alpha(N+b)/6]^{1/2}$ can be approximated by $[(N+b)/12]^{1/2}q_\alpha$, where q_α is the upper αth percentile for the distribution of the range of m independent $N(0,1)$ variables. Thus, the approximation to procedure (71) for N large is

$$\text{decide } \tau_u \neq \tau_v \text{ if and only if } |S_u - S_v| \geqslant q_\alpha[(N+b)/12]^{1/2}. \tag{72}$$

Values of q_α ($\alpha = .0001, .0005, .001, .005, .01, .025, .05, .10, .20$; $m = 3(1)20(2)$ $40(10)100$) can be found, for example, in Hollander and Wolfe (1973, Table A10).

Mack and Skillings (1980) note that the procedure (71) is rather conservative; that is, the true experimentwise error rate might be a good deal smaller than the bound α provided by (71). As a result, they recommend using the approximation (72) whenever the number of observations is reasonably large.

3.2.5. Example – main effects tests and multiple comparisons, two-way layout, multiple observations per cell

We use the rocket thrust duration data in Table 4 of Experiment 1.5 to illustrate the Mack–Skillings test of the hypothesis H_0 (61) against the general alternative hypothesis H_1 (62). For these data, the RGLM test (Section 4.3) would indicate no interaction between the factors. Similarly, the test (59) would indicate no rank interaction. Consequently, the additive model (50) is an adequate representation of the response.

Table 8
Ranks within levels of altitude cycling of the rocket thrust duration data

		Temperature							
		1		2		3		4	
Altitude Cycling	1	15.5	13	3	1	9	14	7	2
		15.5	11	5	4	10	12	8	6
	2	12	16	4	3	11	9.5	5	6
		14.5	14.5	1	2	13	9.5	8	7

We let factor B denote the $i = 2$ levels of altitude cycling and let the levels of factor U denote the $m = 4$ levels of temperature. The ranks of the data in Table 4 within levels of B (with average ranks used to break the ties) are shown in Table 8.

To test the hypothesis $H_0^{\mathcal{T}}$ (61) against the alternative hypothesis $H_1^{\mathcal{T}}$ (62), we first form the sum of cell-wise weighted average ranks (63), with $n_{i.} = mn = 16$, and obtain

$$S_1 = 7, \quad S_2 = 1.4375, \quad S_3 = 5.5, \quad S_4 = 3.0625.$$

Since we have an equal number $n = 4$ observations per cell, we use the closed form Mack–Skillings statistic (69) with $N = bmn = 32$, giving

$$\begin{aligned} \mathrm{MS} &= [12/4(32+2)]\left\{(4 \times 7 - 17)^2 + \cdots + (4 \times 3.0625 - 17)^2\right\} \\ &= 26.04\,. \end{aligned}$$

Comparing this value with the exact null distribution tables for MS in Mack and Skillings (1980) with $m = 4, b = 2$ and $n = 4$ we find the p-value for these data is considerably smaller than 0.01. Thus there is strong evidence to suggest that temperature has an effect on the rocket thrust duration.

In order to determine which temperatures differ in their effects on the thrust duration, we use the Mack–Skillings multiple comparison procedure (71) and, selecting a significance level of $\alpha = 0.0994$, we

$$\text{decide } \tau_u \neq \tau_v \text{ if and only if } |S_u - S_v| \geqslant [6.243(32+2)/6]^{1/2} = 5.948.$$

Thus, using the above values of S_j $(j = 1, 2, 3, 4)$, our 6 decisions, at an experimentwise error rate of 0.0994, are that no two treatments differ in their effect on the rocket thrust duration (although at a slightly larger experimentwise error rate, we would conclude a difference between the effects of the first two levels of temperature). This indicates the conservative nature of the decision procedure as pointed out by Mack and Skillings (1980).

3.3. Two-way layout – one observation per cell

3.3.1. Testing for non-additivity – one observation per cell

The test statistic MCRA was proposed by Hartlaub et al. (1993) for testing H_0^{add} (54) under the two-way model (49) with one observation per cell. The procedure is as follows. Arrange the data in a two-way table, with the levels of factor B defining the rows and those of factor U defining the columns. Subtract the jth column average from the data values in the jth column ($j = 1, \ldots, m$) to give $Y_{ij} - \bar{Y}_{.j}$ in the (ij)th cell ($i = 1, \ldots, b; j = 1, \ldots, m$). This is called *aligning in the columns* and removes the effect of factor U from the data in the table. (The column median can be subtracted instead, but this does not completely remove the effect of factor U from the test statistic.) Then, rank the aligned data values $Y_{i1} - \bar{Y}_{.1}, Y_{i2} - \bar{Y}_{.2}, \ldots, Y_{im} - \bar{Y}_{.m}$ in the ith row to give the rank vector $\boldsymbol{R}_i = (R_{i1}, R_{i2}, \ldots, R_{im})$, for each $i = 1, \ldots, b$. This removes the effect of factor B from the aligned data values in the cells. Any remaining variation is due to non-additivity. The $m(m-1)/2$ statistics $W_{jj'}$ are computed as

$$W_{jj'} = \sum_{1 \leqslant i < i' \leqslant b} \left(R_{ij} - R_{i'j} - R_{ij'} + R_{i'j'}\right)^2 \tag{73}$$

for each $1 \leqslant j < j' \leqslant m$. The test statistic is then

$$\text{MCRA} = \max_{1 \leqslant j < j' \leqslant m} W_{jj'} \,. \tag{74}$$

The statistic MCRA is more sensitive to some types of non-additivity than others. In particular, MCRA may detect non-additivity more easily when the rows and columns of the table are interchanged before the aligning and ranking takes place. If we let M_1 be the statistic MCRA (74) with the levels of factors B and U defining the rows and columns, respectively, and let M_2 be MCRA with the levels of factors U and B defining the rows and columns, respectively, the rejection rule suggested by Hartlaub et al. (1993) is

$$\text{reject } H_0^{\text{add}} \text{ if and only if } M_1 > h_{\alpha/2}(m, b) \text{ or } M_2 > h_{\alpha/2}(b, m) \tag{75}$$

where $h_{\alpha/2}(b, m)$ is the upper $(\alpha/2)$th percentile of the distribution of MCRA under a null hypothesis of no interaction and no main effects, and computed when the errors have a normal distribution. A FORTRAN computer program for calculating the critical values is available from the authors.

The test (75) is not strictly distribution free, and the significance level can rise a little above its nominal level for skewed distributions (such as the exponential). For heavy-tailed distributions (such as the Cauchy distribution), a more powerful test can be obtained by subtracting the column medians rather than the column averages before ranking in the rows. (However, as mentioned above, this does not completely remove the effect of the factor defining the columns).

3.3.2. Main effect test for general alternatives – one observation per cell

In this subsection, we consider two-way layout settings with a single observation for each combination of levels of two factors B and U and where an additive model (50) holds. We are interested in testing the hypothesis $H_0^{\mathcal{T}}$ (61) of no effect of factor U against the general class of alternatives given by $H_1^{\mathcal{T}}$ (62). The most commonly used test statistic for this setting, proposed independently by Friedman (1937) and Kendall and Babington Smith (1939), but commonly known as the Friedman statistic, is closely related to the usual classical two-way layout F-statistic with the original observations replaced by their within-block joint ranks.

For each $i = 1, \dots, b$ separately, let $R_{i1}, R_{i2}, \dots, R_{im}$ be the ranks of the observations $Y_{i1}, Y_{i2}, \dots, Y_{im}$ for the ith level of factor B. Let $R_{.j} = R_{1j} + \cdots + R_{bj}$ be the sum of these ranks that are assigned to the observations on the jth level of factor U. Let $\bar{R}_{.j} = R_{.j}/b$ be the corresponding average, for $j = 1, \dots, m$. The Friedman statistic for testing $H_0^{\mathcal{T}}$ (61) versus the general alternative $H_1^{\mathcal{T}}$ (62) is then given by

$$S = [12b/m(m+1)] \sum_{j=1}^{m} (\bar{R}_{.j} - (m+1)/2)^2, \tag{76}$$

and the associated level α test of $H_0^{\mathcal{T}}$ versus $H_1^{\mathcal{T}}$ is

$$\text{reject } H_0^{\mathcal{T}} \text{ if and only if } S \geqslant s_\alpha(m, b), \tag{77}$$

where $s_\alpha(m, b)$ is the upper αth percentile for the null sampling distribution of the statistic S (76). The complete null distribution of S is available in Hollander and Wolfe (1973) for $m = 3, b = 2(1)13$; $m = 4$, $b = 2(1)8$; and $m = 5$, $b = 3, 4, 5$. Additional tables for $m = 5, b = 6(1)8$ and $m = 6, b = 2(1)6$ can be found in Odeh (1977a).

When the number of levels of factor B is large, the appropriate asymptotic distribution of S can be used to provide approximate critical values for the test procedure. When the null hypothesis $H_0^{\mathcal{T}}$ (61) is true, the statistic S has an asymptotic distribution [as $b \to \infty$] that is chi-square with $m-1$ degrees of freedom. Thus, the approximation to procedure (77) for a large number of levels of B is

$$\text{reject } H_0^{\mathcal{T}} \text{ if and only if } S \geqslant \chi^2_\alpha(m-1), \tag{78}$$

where $\chi^2_\alpha(m-1)$ is the upper αth percentile of the chi-square distribution with $m-1$ degrees of freedom.

We note that the chi-square approximation (78) can, for smaller values of b, be rather conservative in the sense that the stated upper-tail probabilities are larger than the true ones. For this reason, Iman and Davenport (1980) proposed the alternative approximation that compares the statistic $Q = (b-1)S/[b(m-1) - S]$ with critical values from the F distribution having $m-1$ and $(b-1)(m-1)$ degrees of freedom in order to assess the approximate significance of the observed data.

Calculation of the Friedman statistic S is available on a variety of software packages. These include `Minitab`'s `FRIEDMAN` command, `SAS`'s `PROC RANK/PROC ANOVA` commands when applied to within-blocks ranks, and `IMSL`'s `FRDMN` command.

3.3.3. Two-sided all-treatments multiple comparisons – no replications
Nonparametric point estimates for contrasts $\sum_{j=1}^{m} c_j\tau_j$ ($\sum_{j=1}^{m} c_j = 0$) are available and, for these, the reader is referred to Doksum (1967) or Hollander and Wolfe (1973, Section 7.4). Here, we present a multiple comparison procedure that is designed to make two-sided decisions about all $m(m-1)/2$ pairs of effects of factor U in a two-way layout with one observation per cell and which is an appropriate follow-up procedure to the Friedman test of Section 3.3.2. The procedure was described by Nemenyi (1963) and by McDonald and Thompson (1967), who attributed it to Wilcoxon.

Let $R_{i1},\ldots,R_{im}$ be the ranks of the data within the ith level of factor $B, i = 1,\ldots,b$ and let $R_{.j}$ be the sum of these ranks assigned to the jth level of U, for $j = 1,\ldots,m$. At an experimentwise error rate of α, the Wilcoxon–Nemenyi–McDonald–Thompson two-sided all-treatments multiple comparison procedure reaches its $m(m-1)/2$ decisions through the criterion

$$\text{decide } \tau_u \neq \tau_v \text{ if and only if } |R_{.u} - R_{.v}| \geqslant r_\alpha(m,b), \tag{79}$$

where the critical value $r_\alpha(m,b)$ satisfies the probability restriction

$$\begin{aligned} &P_0\big(|R_{.u} - R_{.v}| \geqslant r_\alpha(m,b),\ u = 1,\ldots,m-1;\ v = u+1,\ldots,m\big) \\ &\quad = 1-\alpha, \end{aligned} \tag{80}$$

with the probability $P_0(.)$ being computed under H_0^τ (61). As with previously presented multiple comparison procedures, experimentwise error rates larger than those used as significance levels for hypothesis tests (e.g., .10 or .15) are frequently used in practice. Values of $r_\alpha(m,b)$ can be found in McDonald and Thompson (1967) or in Hollander and Wolfe (1973, Table A.17) for selected experimentwise error rates α and $m = 3$, $b = 3(1)15$ or $m = 4(1)15$, $b = 2(1)15$.

When the number of levels of factor B is large, the critical value $r_\alpha(m,b)$ can be approximated by the constant $[bm(m+1)/12]^{1/2}q_\alpha$, where q_α is the upper αth percentile for the distribution of the range of m independent $N(0,1)$ variables. Thus, the approximation to procedure (79) for b large is

$$\text{decide } \tau_u \neq \tau_v \text{ if and only if } |R_{.u} - R_{.v}| \geqslant q_\alpha[bm(m+1)/12]^{1/2}. \tag{81}$$

Values of q_α for $\alpha = .0001, .0005, .001, .005, .01, .025, .05, .10$, and $.20$ and $m = 3(1)20(2)40(10)100$ can be found, for example, in Hollander and Wolfe (1973, Table A.10).

3.3.4. Example – two-way layout, no replications, general alternatives
To illustrate the analysis of a two-way layout with one observation per cell, we use the air velocity data of Experiment 1.6 in Table 5. The response variable Y_{ij} is the position of maximum velocity of air blown down the space between a rib-roughened rod and a smooth pipe surrounding it. Let factor B denote the rib height, which has $b = 3$ levels, and let factor U denote the Reynolds number which has $m = 6$ levels.

Table 9
Aligned ranks for position of maximum air velocity. Column aligned data prior to ranking are shown in brackets

		Reynolds number (j)					
	i	1	2	3	4	5	6
Rib	1	5 (−39.3)	1 (−51.0)	4 (−40.7)	3 (−45.3)	6 (−39.0)	2 (−48.3)
Height	2	6 (17.7)	1 (0.0)	2 (3.3)	3 (3.7)	4 (6.0)	5 (8.7)
	3	1 (21.7)	6 (51.0)	3 (37.3)	5 (41.7)	2 (33.0)	4 (39.7)

We use the MCRA procedure to test the hypothesis of additivity (54) against the general alternative hypothesis (55) of no additivity. Aligning the data in the columns of Table 5 by subtracting column averages, and then ranking in the rows, gives the aligned ranks of Table 9.

If we calculate the statistics $W_{jj'}$ (73) and the test statistic MCRA (74), we obtain MCRA $= M_1 = 182$. When the table is transposed, and the new aligned ranks obtained, the test statistic (74) is MCRA $= M_2 = 89$. Critical values obtained from the program of Hartlaub, Dean and Wolfe are

$$m_{3,6,.013} = 200, \quad m_{3,6,.072} = 182, \quad m_{6,3,.0075} = 128, \quad m_{6,3,.034} = 125.$$

If we select an overall significance level of $\alpha \leqslant 0.026$, we have that $M_1 < m_{3,6,.013}$ and $M_2 < m_{6,3,.0075}$. Consequently, the hypothesis H_0^{add} (54) of no interaction would not be rejected at level $\alpha = .026$. However, at a higher overall significance level, say $\alpha \leqslant 0.144$, we have $M_1 \geqslant m_{3,6,.072}$ and the hypothesis H_0^{add} would be rejected.

A plot of the data indicates only a small amount of interaction between factors, and this is caused mostly by the small value recorded for the combination of rib height 1 and Reynold's number 3. Accordingly, the main effects are likely to be of interest here.

In order to use the Friedman test as described in Section 3.3.2, we let the $m = 3$ levels of factor U represent the rib heights and the $b = 6$ levels of factor B represent the Reynolds numbers. Ranking the three observations from least to greatest within each of the levels of B (Reynolds number), we find that $R_{i1} = 1$, $R_{i2} = 2$, $R_{i3} = 3$ for each level i of B. So the average "within-B rank" for each of the three rib heights, is $\bar{R}_{.1} = 1$, $\bar{R}_{.2} = 2$ and $\bar{R}_{.3} = 3$. Using these values in expression (76) for S, we obtain $S = 12$.

If we compare this value with the exact null distribution tables for S (Table A.15 in Hollander and Wolfe, 1973, for example) with $m = 3$ and $b = 6$, we find that the p-value for these data is less than .0005. This indicates strong evidence that there is a difference in the effects on position of maximum air velocity of the three rib heights.

In order to detect which of the three rib heights (levels of factor U) differ in their effects on the position of maximum air velocity, we apply the Wilcoxon–Nemenyi–McDonald–Thompson multiple comparison procedure (79). With an experimentwise error rate of $\alpha = .009$, we see from Table A.17 in Hollander and Wolfe (1973), with

$m = 3$ and $b = 6$, that $r_{.009}(3,6) = 10$. Using the above values of the within-B ranks we see that

$$|R_{.1} - R_{.2}| = |6 - 12| = 6, \qquad |R_{.1} - R_{.3}| = |6 - 18| = 12,$$

$$|R_{.2} - R_{.3}| = |12 - 18| = 6.$$

Comparing these observed differences in the treatment rank sums with the critical value 10, we see that our $3(2)/2 = 3$ two-sided decisions are $\tau_1 = \tau_2$, $\tau_2 = \tau_3$, and $\tau_1 \neq \tau_3$. Thus, there is a statistically significant difference in the locations of maximum air velocity for the first and third rib heights.

3.3.5. Test for monotonically ordered alternatives – one observation per cell

In this subsection, we consider the case of a two-way layout where there is a single observation for each combination of levels of the two factors B and U, and where we have *a priori* knowledge that enables us to label the levels of factor U so that their effects will be in a monotonically increasing relationship with the treatment labels. The test procedure that we now present was initially proposed by Page (1963) and is especially effective at detecting alternatives to H_0^τ (61) that are ordered with respect to the treatment labels corresponding to

$$H_2^\tau\colon\ [\tau_1 \leqslant \tau_2 \leqslant \cdots \leqslant \tau_m,\ \text{with at least one strict inequality}]. \tag{82}$$

Let $R_{i1}, R_{i2}, \ldots, R_{im}$ be the ranks of the data for the ith level of factor $B, i = 1, \ldots, b$, and let $R_{.j}$ be the sum of the ranks assigned to the jth level of U, $j = 1, \ldots, m$. The Page statistic for testing H_0^τ (61) versus the ordered alternative hypothesis H_2^τ (82) is defined by

$$L = \sum_{j=1}^{m} jR_{.j}, \tag{83}$$

and the associated level α test of H_0^τ versus H_2^τ is

$$\text{reject } H_0^\tau \text{ if and only if } L \geqslant l_\alpha(m,b), \tag{84}$$

where $l_\alpha(m,b)$ is the upper αth percentile for the null distribution of the statistic L (83). The critical values $l_\alpha(m,b)$ are given for $\alpha = .05, .01$ and .001 by Page (1963), and also by Hollander and Wolfe (1973, Table A16), for $m = 3, b = 2(1)20$ and $m = 4(1)8, b = 2(1)12$. Additional critical values are given by Odeh (1977b) for $\alpha =$ .2, .1, .025, .005 and $m = 3(1)8, b = 2(1)10$.

When the number of levels of B is large, the appropriate asymptotic distribution of a standardized form of L can be used to provide approximate critical values for the test procedure. When the null hypothesis H_0^τ (61) is true, the standardized statistic

$$L^* = [L - \mathrm{E}_0(L)]/[\mathrm{var}_0(L)]^{1/2} \tag{85}$$

has an asymptotic distribution [as $b \to \infty$] that is standard normal, where

$$\mathrm{E}_0(L) = bm(m+1)^2/4 \quad \text{and}$$

$$\mathrm{var}_0(L) = bm^2(m+1)^2(m-1)/144 \tag{86}$$

are the expected value and variance, respectively, of L under the null hypothesis $H_0^{\mathcal{T}}$. Thus the approximation to procedure (85) for b large is

$$\text{reject } H_0^{\mathcal{T}} \text{ if and only if } L^* \geqslant z_\alpha, \tag{87}$$

where z_α is the upper αth percentile of the standard normal distribution.

3.3.6. One-sided treatment versus control multiple comparisons – one observation per cell

In this subsection, we present a multiple comparison procedure that is designed for the two-way layout to make one-sided decisions about individual differences in the effects of levels $2, \ldots, m$ of factor U relative to the effect of a single control population corresponding to level 1 of U. In the context of this paper, it is to be viewed as an appropriate follow-up procedure to rejection of $H_0^{\mathcal{T}}$ (61) with either the Friedman procedure of Section 3.3.2 or the Page ordered alternatives procedure of Section 3.3.5 when one of the populations corresponds to a control population. Let $R_{i1}, \ldots, R_{im}$ be the ranks of the data within the ith level of factor B, $i = 1, \ldots, b$ and let $R_{.j}$ be the sum of these ranks assigned to the jth level of U, for $j = 1, \ldots, m$. At an experimentwise error rate of α, the procedure described by Nemenyi (1963), Wilcoxon–Wilcox (1964), and Miller (1966), reaches its $m - 1$ decisions through the criterion

$$\text{decide } \tau_u > \tau_1 \text{ if and only if } (R_{.u} - R_{.1}) \geqslant r_\alpha^*(m, b), \tag{88}$$

where the critical value $r_\alpha^*(m, b)$ satisfies the probability restriction

$$P_0\big([R_{.u} - R_{.1}] \geqslant r_\alpha^*(m, b), u = 2, \ldots, m\big) = 1 - \alpha, \tag{89}$$

with the probability $P_0(.)$ computed under $H_0^{\mathcal{T}}$ (61). Values of $r_\alpha^*(m, b)$ can be found in Hollander and Wolfe (1973, Table A.18) for $m = 3, b = 2(1)18$ and $m = 4, b = 2(1)5$, and additional tables for $m = 2(1)5, b = 2(1)8$ and $m = 6, b = 2(1)6$ in Odeh (1977c).

When the number of levels of factor B is large, the critical value $r_\alpha^*(m, b)$ can be approximated by the constant $q_{\alpha,1/2}^*[bm(m+1)/6]^{1/2}$, where $q_{\alpha,1/2}^*$ is the upper αth percentile for the distribution of the maximum of $m - 1$ $N(0, 1)$ variables with common correlation $\rho = 1/2$. Thus the approximation to procedure (88) for b large is

$$\text{decide } \tau_u > \tau_1 \text{ if and only if}$$
$$(R_{.u} - R_{.1}) \geqslant q_{\alpha,1/2}^*[bm(m+1)/6]^{1/2}. \tag{90}$$

Selected values of $q_{\alpha,1/2}^*$ for $m = 3(1)13$ can be found in Hollander and Wolfe (1973, Table A.13).

3.3.7. Example – two-way layout, no replications, ordered alternatives

To illustrate the test of $H_0^{\mathcal{T}}$ (61) against the ordered alternative $H_2^{\mathcal{T}}$ (82) we use the air velocity data from Experiment 1.6. We are interested in testing for possible differences in effects among the $m = 6$ Reynolds numbers which we regard as levels of factor U. In this setting, it is reasonable to expect that, if there is a significant treatment effect associated with the Reynolds number, this effect will be a monotonically increasing function of the number. Therefore, we are interested in testing $H_0^{\mathcal{T}}$: $[\tau_1 = \tau_2 = \cdots = \tau_6]$ versus $H_2^{\mathcal{T}}$: $[\tau_1 \leqslant \tau_2 \leqslant \cdots \leqslant \tau_6$, with at least one strict inequality$]$.

The $b = 3$ rib heights are the levels of factor B. We rank the $m = 6$ observations from least to greatest within each of the rib height levels (using average ranks to break the one tie). This gives the following rank sums for each of the six Reynolds numbers:

$$R_{.1} = 4, \quad R_{.2} = 5.5, \quad R_{.3} = 8.5, \quad R_{.4} = 12, \quad R_{.5} = 16, \quad \text{and}$$

$$R_{.6} = 17.$$

Page's test statistic (83) is then $L = 270.5$. If we compare this value with the exact tables for L (Table A.16 in Hollander and Wolfe (1973), for example) with $m = 6$ and $b = 3$, we see that $270.5 > l_{.001}(6, 3) = 260$, so that the p-value for these data is less than .001. Thus, there is strong evidence that the distance (from the center of the rod) of maximum air velocity is a monotonically increasing function of the Reynolds number.

In order to detect which of the Reynolds numbers yield maximum air velocities further from the center of the rod than for a baseline Reynolds number (designated as level 1 of U), we apply the Nemenyi–Wilcoxon–Wilcox–Miller one-sided treatment versus control procedure (88). Taking our experimentwise error rate to be $\alpha = .03475$, we obtain $r^*_{.03475}(6, 3) = 11$ from Table I in Odeh (1977c), with $m = 6$ and $b = 3$. Using the above values of the within-B rank sums, we see that

$$(R_{.2} - R_{.1}) = 1.5, \quad (R_{.3} - R_{.1}) = 4.5, \quad (R_{.4} - R_{.1}) = 8,$$

$$(R_{.5} - R_{.1}) = 12, \quad (R_{.6} - R_{.1}) = 13.$$

If we compare these observed differences in the rank sums with the critical value 11, we see that our one-sided treatment versus control decisions are $\tau_2 = \tau_1, \tau_3 = \tau_1, \tau_4 = \tau_1, \tau_5 > \tau_1$, and $\tau_6 > \tau_1$. Thus there is a statistically significant increase (over the effect of the baseline control Reynolds number corresponding to level 1 of U) in the position of maximum air velocity for the Reynolds numbers corresponding to levels 5 and 6 of U.

3.4. Two-way incomplete layout

3.4.1. Test for general alternatives in a balanced incomplete block design

In this subsection, we discuss a distribution-free procedure that can be used to analyze data that arise from a balanced incomplete block design. A balanced incomplete block design has b blocks with s $(< m)$ treatments observed per block and no treatment observed more than once per block. Every pair of treatments occurs together in λ blocks, and each of the m treatments is observed p times. The parameters of a balanced incomplete block design satisfy $p(s-1) = \lambda(m-1)$. We are interested in testing $H_0^{\mathcal{T}}$ (61) against the general alternatives $H_1^{\mathcal{T}}$ (62), under the assumption of no treatment-block interaction.

An appropriate nonparametric test statistic for a balanced incomplete block design was first proposed by Durbin (1951), with Skillings and Mack (1981) providing additional critical values. For this setting, we rank the observations within each block from 1 to s. Let $R_{.1}, \ldots, R_{.m}$ denote the sums of the within-blocks ranks for treatments $1, \ldots, m$, respectively. The Durbin statistic for testing $H_0^{\mathcal{T}}$ (61) versus the general alternative $H_1^{\mathcal{T}}$ (62) is defined to be

$$T = [12/\lambda m(s+1)] \sum_{j=1}^{m} (R_{.j} - p(s+1)/2)^2, \tag{91}$$

and the associated level α test of $H_0^{\mathcal{T}}$ versus $H_1^{\mathcal{T}}$ is

$$\text{reject } H_0^{\mathcal{T}} \text{ if and only if } T \geqslant t_\alpha(m, s, \lambda, p), \tag{92}$$

where $t_\alpha(m, s, \lambda, p)$ is the upper αth percentile for the null sampling distribution of the statistic T (91). Values (some of them obtained via simulation) of $t_\alpha(m, s, \lambda, p)$ for a variety of balanced incomplete block designs and significance levels closest to .10, .05, and .01 have been tabulated by Skillings and Mack (1981).

When the number of blocks is large, the statistic T has an asymptotic distribution [as $b \to \infty$] under $H_0^{\mathcal{T}}$ (61), that is chi-square with $m-1$ degrees of freedom. Thus the approximation to procedure (92) for large number of blocks is

$$\text{reject } H_0^{\mathcal{T}} \text{ if and only if } T \geqslant \chi_\alpha^2(m-1), \tag{93}$$

where $\chi_\alpha^2(m-1)$ is the upper αth percentile of the chi-square distribution with $m-1$ degrees of freedom.

Skillings and Mack (1981) have noted that the chi-square approximation (93) can be quite conservative when $\alpha = .01$ and either b or λ is small. In particular, they suggest that the approximation is not adequate when λ is either 1 or 2. In such cases, they strongly recommend the use of the exact tabulated values of $t_\alpha(m, s, \lambda, p)$ or the generation of 'exact' values via simulation in lieu of the chi-square approximation.

3.4.2. Main effect tests for general alternatives in arbitrary two-way incomplete layouts – no replications

Not all incomplete block designs satisfy the necessary constraints to be balanced incomplete block designs. In this subsection, we present a procedure for dealing with data from a general two-way layout with at most one observation per cell.

Let q_i denote the number of treatments observed in block i, for $i = 1, \ldots, b$. (If $q_i = 1$ for any block i, remove that block from the analysis and let b correspond to the number of blocks for which $q_i > 1$.) We are interested in testing $H_0^{\mathcal{T}}$ (61) against the general alternatives $H_1^{\mathcal{T}}$ (62), under the assumption of no treatment-block interaction.

Skillings and Mack (1981) proposed a test procedure that is appropriate for any incomplete two-way layout. We rank the data within block i from 1 to q_i. For $i = 1, \ldots, b$ and $j = 1, \ldots, m$, let

$$\begin{aligned} R_{ij} &= \text{ rank of } Y_{ij} \text{ among the observations present in block } i, \text{ if } n_{ij} = 1, \\ &= (q_i + 1)/2, \text{ if } n_{ij} = 0. \end{aligned}$$

Next, we compute the adjusted sum of ranks for each of the m treatments as follows

$$A_j = \sum_{i=1}^{b} \left[12/(q_i + 1)\right]^{1/2} \left[R_{ij} - (q_i + 1)/2\right], \quad j = 1, \ldots, m. \tag{94}$$

Set

$$\boldsymbol{A} = [A_1, \ldots, A_{m-1}]'. \tag{95}$$

Without loss of generality, we have chosen to omit A_m from the vector $\boldsymbol{A}$. The A_j's are linearly dependent, since a weighted linear combination of all m of them is a constant. We could omit any one of the A_j's and the approach we now discuss would lead to the same test statistic. Now, the covariance matrix for $\boldsymbol{A}$ under $H_0^{\mathcal{T}}$ (61) is given by

$$\boldsymbol{\Sigma}_0 = \begin{bmatrix} \sum_{t=2}^{m} \lambda_{1t} & -\lambda_{12} & -\lambda_{13} & \cdots & -\lambda_{1,m-1} \\ -\lambda_{12} & \sum_{t \neq 2}^{m} \lambda_{2t} & -\lambda_{23} & \cdots & -\lambda_{2,m-1} \\ \vdots & \vdots & \vdots & \vdots & \vdots \\ -\lambda_{1,m-1} & -\lambda_{2,m-1} & -\lambda_{3,m-1} & \cdots & \sum_{t \neq m-1}^{m} \lambda_{m-1,t} \end{bmatrix}, \tag{96}$$

where, for $t \neq j = 1, \ldots, m$,

$$\lambda_{jt} = [\text{number of blocks in which both treatments } j \text{ and } t \text{ are observed}]. \tag{97}$$

Let $\boldsymbol{\Sigma}_0^-$ be any generalized inverse for $\boldsymbol{\Sigma}_0$. The Skillings–Mack statistic SM for testing $H_0^{\mathcal{T}}$ (61) versus the general alternatives $H_1^{\mathcal{T}}$ (62) is given by

$$\mathrm{SM} = \boldsymbol{A}' \boldsymbol{\Sigma}_0^- \boldsymbol{A}. \tag{98}$$

If $\lambda_{jt} > 0$ for all $j \neq t$, then the covariance matrix $\boldsymbol{\Sigma}_0$ (96) has full rank and we can use the ordinary inverse $\boldsymbol{\Sigma}_0^{-1}$ in the definition of SM (98). The level α test of $H_0^{\mathcal{T}}$ (61) versus $H_1^{\mathcal{T}}$ (62) is

$$\text{reject } H_0^{\mathcal{T}} \text{ if and only if } \mathrm{SM} \geqslant s_\alpha^*(b, m, n_{11}, \ldots, n_{bm}), \tag{99}$$

where $s^*_\alpha(b, m, n_{11}, \ldots, n_{bm})$ is the upper αth percentile for the null sampling distribution of the statistic SM (98).

Values of $s^*_\alpha(b, m, n_{11}, \ldots, n_{bm})$ are not available in the literature for arbitrary incomplete block configurations. However, when every pair of treatments occurs in at least one block and the number of blocks is large, the statistic SM has an asymptotic distribution [as $b \to \infty$] under H_0^τ that is chi-square with $m - 1$ degrees of freedom. Thus, when $\lambda_{jt} > 0$ for all $j \neq t = 1, \ldots, m$, the approximation to procedure (99) for large b is

$$\text{reject } H_0^\tau \text{ if and only if SM} \geqslant \chi^2_\alpha(m-1), \tag{100}$$

where $\chi^2_\alpha(m-1)$ is the upper αth percentile of the chi-square distribution with $m-1$ degrees of freedom. Skillings and Mack (1981) have pointed out that the chi-square approximation (100) can be quite conservative when α is smaller than .01 and, in such cases, they recommend generation of 'exact' critical values $s^*_\alpha(b, m, n_{11}, \ldots, n_{bm})$ via simulation in lieu of the chi-square approximation.

Note that if $\lambda_{jt} = 0$ for a particular pair of treatments j and t, so that j and t never appear together in a block, then H_0^τ (61) could fail to be rejected even when τ_j and τ_t are quite different. Consequently, we recommend removing any such pairs of treatments from the analysis.

If the sample size configuration satisfies the constraints for a balanced incomplete block design, then the Skillings–Mack statistic SM (98) is identical to the Durbin statistic T (91). If the sample sizes are all 1, so that we have a randomized complete block design, the Skillings–Mack statistic SM (98) is identical to the Friedman statistic S (76).

3.4.3. Two-sided all treatments multiple comparisons for data from a balanced incomplete block design

In this subsection, we discuss a multiple comparison procedure that is designed to make two-sided decisions about all $m(m-1)/2$ pairs of treatment effects when we have data from a balanced incomplete block design. It is appropriate as a follow-up procedure to the general alternatives Durbin–Skillings–Mack test for equality of treatment effects in a balanced incomplete block design, as discussed in Section 3.4.1.

Let $R_{.1}, \ldots, R_{.m}$ be the sums of the within-blocks ranks for treatments $1, \ldots, m$, respectively. Let s denote the number of observations present in each of the blocks, let λ denote the number of blocks in which each pair of treatments occurs together, and let p denote the total number of observations on each of the m treatments. At an experimentwise error rate no greater than α, the Skillings–Mack two-sided all-treatments multiple comparison procedure reaches its $m(m-1)/2$ decisions through the criterion

$$\text{decide } \tau_u \neq \tau_v \text{ if and only if } |R_{.u} - R_{.v}| \geqslant [t_\alpha \lambda m(s+1)/6]^{1/2}, \tag{101}$$

where $t_\alpha = t_\alpha(m, s, \lambda, p)$ is the upper αth percentile for the null sampling distribution of the Durbin test statistic T (91), as discussed in Section 3.4.1. The associated multiple comparison procedure (101) controls the experimentwise error rate over the entire class of continuous distributions for the error terms. Values of $t_\alpha = t_\alpha(m, s, \lambda, p)$ for a variety of balanced incomplete block designs and experimentwise error rates closest to .10, .05, and .01 are available in Skillings and Mack (1981).

When the number of blocks is large, the critical value $[t_\alpha \lambda m(s+1)/6]^{1/2}$ can be approximated by $[(s+1)(ps-p+\lambda)/12]^{1/2} q_\alpha$, where q_α is the upper αth percentile for the distribution of the range of m independent $N(0,1)$ variables. Thus, the approximation to procedure (101) for b large is

$$\text{decide } \tau_u \neq \tau_v \text{ if and only if}$$
$$|R_{.u} - R_{.v}| \geqslant q_\alpha[(s+1)(ps-p+\lambda)/12]^{1/2}. \tag{102}$$

Values of q_α for $\alpha = .0001, .0005, .001, .005, .01, .025, .05, .10$, and .20 and $m = 3(1)20(2)40(10)100$ can be found, for example, in Hollander and Wolfe (1973, Table A.10).

Skillings and Mack (1981) note that the procedure (101) is rather conservative; that is, the true experimentwise error rate might be a good deal smaller than the bound α provided by (101). As a result, they recommend using the approximation (102) whenever the number of blocks is reasonably large.

3.4.4. Example – incomplete two-way layout

We apply the Durbin procedure (92) for testing H_0^τ (61) against a general alternative H_1^τ (62) to the monovinyl isomers data in the balanced incomplete block design of Experiment 1.7. The blocks (levels of B) are shown as columns in Table 6, and the levels of treatment factor U ("pressure" measured in pounds per square inch) indicate the rows. If we rank the data within each of the ten blocks (using average ranks to break the one tie), we obtain the following treatment rank sums:

$$R_{.1} = 7.5, \quad R_{.2} = 7.5, \quad R_{.3} = 13, \quad R_{.4} = 15, \quad R_{.5} = 17.$$

Using these treatment rank sums in expression (91), we obtain $T = 15.1$. If we compare this value with the null distribution for T (in Table 2 of Skillings and Mack, 1981) with $m = 5, b = 10, p = 6, s = 3$ and $\lambda = 3$, we find that the p-value for these data is less than .0105. This indicates strong evidence that there is a significant difference in the effects of the various pressure levels on the percent conversion of methyl glucoside to monovinyl isomers.

In order to detect which of the five pressure levels differ, we apply the Skillings–Mack multiple comparison procedure, as given in (101). Taking our experimentwise error rate to be $\alpha = .0499$, we see from Table 2 in Skillings and Mack (1981), with $m = 5, b = 10, \lambda = 3, p = 6$, and $s = 3$, that $t_{.0499} = 9.200$. Thus, our decision criterion is

$$\text{decide } \tau_u \neq \tau_v \text{ if and only if}$$
$$|R_{.u} - R_{.v}| \geqslant [(9.20)(3)(5)(4)/6]^{1/2} = 9.59.$$

Comparing the differences in the above treatment rank sums with the critical value 9.59, we see that, with an experimentwise error rate of $\alpha = .0499$, we cannot find any pairwise differences between the treatment effects, despite the fact that the p-value for the Durbin hypothesis test was less than .0105. This clearly illustrates the conservative nature of procedure (101), as noted by Skillings and Mack (1981).

If we choose instead to use approximation (102), taking $b = 10$ to be 'large', our critical value for approximate experimentwise error rate $\alpha = .05$ would be lower; that is,

$$q_{.05}[4\{6(3) - 6 + 3\}/12]^{1/2} = 3.858(2.236) = 8.627,$$

where $q_{.05} = 3.858$ is obtained from Hollander and Wolfe (1973, Table A.10) with $m = 5$. The decisions associated with this approximate procedure (102) at $\alpha = .05$ are that $\tau_1 = \tau_2, \tau_1 = \tau_3, \tau_1 = \tau_4, \tau_1 \neq \tau_5, \tau_2 = \tau_3, \tau_2 = \tau_4, \tau_2 \neq \tau_5, \tau_3 = \tau_4, \tau_3 = \tau_5$, and $\tau_4 = \tau_5$. Thus, with the approximate procedure (102), we are able to distinguish that pressures of 475 and 550 psi each lead to different percent conversions of methyl glucoside to monovinyl isomers than does the pressure 250 psi.

3.5. Other tests in the two-way layout

Mack (1981) describes a test of H_0^τ (61) against the general class of alternatives H_1^τ (62) for the incomplete two-way layout with some cells empty and other cells with more than one observation per cell. A test against the class of ordered alternatives H_2^τ (83) for this same setting is given by Skillings and Wolfe (1977, 1978).

For equal sample sizes, Mack's statistic is identical to a statistic proposed by De Kroon and Van der Laan (1981) for testing the global hypothesis H_0^T (53) of no main effect of factor U and no interaction between factors B and U.

We refer the reader to the original papers for details of each of these tests.

4. Higher way layouts

4.1. Introduction

Specialized procedures for the two-way layout were discussed in Section 3. In this section, we consider general procedures for experiments with two or more factors. Most of the techniques for the one-way layout, as described in Section 2, can be used for comparing the effects on the response of the various treatment combinations. However, there are available some specialized nonparametric techniques for investigating the effects of the factors separately. We discuss two such procedures in this section.

In Section 4.2, we discuss a technique for analyzing a factorial experiment with any number of factors when there is at least one observation per cell and all interactions among the factors are expected to be negligible. The method investigates the main effects, one factor at a time, grouping the levels of the other factors together and using the techniques of the two-way layout.

A robust version of the general linear model least squares analysis (RGLM) was developed by McKean and Hettmansperger (1976). This procedure, which is described in Section 4.3, is very general, but requires a computer algorithm for its implementation. Hettmansperger and McKean (1983), Aubuchon and Hettmansperger (1984), and Draper (1988) describe and discuss other methods available in the literature for testing parameters in the general linear model. The results of a small power study run by Hettmansperger and McKean (1983) suggest that the RGLM method, with the traditional type of test statistic based on the difference of error sums of squares for the reduced and full models, gives tests at least as powerful as the competitors. In addition, the RGLM procedure has most of the features of least squares analysis and allows estimation of parameters and multiple comparison procedures to be implemented. Consequently, we only discuss the RGLM method of handling the general linear model in this article, and refer the reader to the above mentioned papers for aligned rank and other related procedures.

4.2. Using two-way layout techniques

4.2.1. Main effect tests – multi-factor experiment

Groggel and Skillings (1986) suggested a method for testing individually the hypotheses of no main effects in a multi-factor experiment which utilizes the two-way analysis of Mack and Skillings (see Section 3.2.3). The method requires that there are no interactions between the factors and that all treatment combinations are observed at least once.

First of all, focus on just one of the factors and label it factor U with m levels, and regard the combinations of the levels of the other factors as the b levels of a combined factor B. If there are n_{ij} observations in a particular cell in the original experiment, then there are still n_{ij} observations in an analogous cell in the experiment involving factors B and U. We write the model as an additive two-way layout model (50) and we wish to test the hypothesis H_0^τ: $[\tau_j$ are all equal, $j = 1, \dots, m]$, where τ_j is the effect of the jth level of factor U, against a general alternative hypothesis that the effects of factor U are not all equal. The general formula for the Mack–Skillings statistic is given in (66). For factorial experiments with a large number of factors, it is usual that a small equal number n of observations are taken per treatment factor combination, in which case the test statistic reduces to (69). We take this equal number n of cell observations to be the case throughout the rest of this section.

Let R_{ijk} be the rank of Y_{ijk} within the mn observations on the ith level of the combined factor B. Let

$$S_j = (mn)^{-1} \sum_{i=1}^{b} \sum_{k=1}^{n} R_{ijk}.$$

The decision rule for testing H_0^τ against a general alternative hypothesis, as given by (70), is reject H_0^T if and only if

$$\frac{12}{m(N+b)}\sum_{j=1}^{m}\left[mS_j-\frac{(N+b)}{2}\right]^2 \geqslant w_\alpha(b,m,n,\ldots,n)\,, \tag{103}$$

where $N = bmn$ and $w_\alpha(b,m,n,\ldots,n)$ can be obtained from the table in Mack and Skillings (1980) for small values of b, m, and n, and can be approximated by $\chi^2_\alpha(m-1)$ if either n or b is large (see Groggel and Skillings, 1986). The test is applied to each factor in turn, the remaining factors forming the combined factor B.

4.2.2. Example – multi-factor experiment, main effect tests

We use factors C, D and E from Experiment 1.2 as an illustration of Groggel and Skilling's two-way procedure, and we ignore the other five factors as though they had not been part of the experiment. (Since this method is not appropriate for fractional factorial experiments, it cannot be used to test the significance of the eight factors in the entire experiment.) We use the mean thickness response (averaged over seventy levels of noise variables) for this illustration. The levels of the three factors C, D and E and the observed mean thicknesses are shown in Table 10. (Note that the treatment combinations have been reordered from Table 1.)

Table 10
Ranks of mean thickness of epitaxial layer – factors C, D and E

Mean thickness		C	D	E	R_{ijk}		$R_{ij.}$
14.821	14.878	−1	−1	−1	2	4	6
14.757	14.843	1	−1	−1	1	3	4
14.888	14.932	−1	−1	1	2	4	6
14.921	14.415	1	−1	1	3	1	4
14.165	13.972	−1	1	−1	4	2	6
14.037	13.907	1	1	−1	3	1	4
13.860	14.032	−1	1	1	1	4	5
13.880	13.914	1	1	1	2	3	5

First, let U be factor C with $m = 2$ levels, and take the $b = 4$ levels of factor B to be the four levels $(-1,-1), (-1,1), (1,-1)$ and $(1,1)$ of the combined factors D and E. Then the ranks assigned to the $m = 2$ levels of factor U (C) at each of the $b = 4$ levels of the combined factor B (D and E) are as shown in Table 10. The corresponding cell-wise weighted average ranks (63) for the two levels of U are $S_1^C = 23/4 = 5.75$ and $S_2^C = 17/4 = 4.25$. The decision rule (103) for testing H_0^T against a general alternative hypothesis is to reject H_0^T if and only if

$$\frac{12}{2(16+4)}\left[(11.5-10)^2+(8.5-10)^2\right] = 1.35 > w_\alpha(4,2,2,\ldots,2)\,.$$

From the table of Mack and Skillings (1980) for $m = 2, b = 4$, and $n = 4$, we obtain $w_{.0077}(4, 2, 2, \ldots, 2) = 7.35$. Thus at level $\alpha = .0077$ there is not sufficient evidence to reject the null hypothesis of no effect of the levels of C on the average thickness of the epitaxial layer.

On the other hand, if we let U be factor D and let B represent the combined factors C and E, similar calculations lead to the rank sums for the two levels of U being $S_1^D = 28/4 = 7$ and $S_2^D = 12/4 = 3$. Using the decision rule (103) for testing H_0^τ against a general alternative hypothesis for factor D, we reject H_0^τ since

$$\frac{12}{2(16+4)}\left[(14-10)^2 + (6-10)^2\right] = 9.6 > 7.35 = w_{.0077}(4, 2, 2, \ldots, 2)\,.$$

The hypothesis that the levels of E have no effect on the average thickness of the epitaxial layer would not be rejected since the value of the associated test statistic is 0.

4.3. *Robust general linear model*

4.3.1. *Tests and confidence intervals – multi-factor experiments*

Hettmansperger and McKean (1977, 1978) describe a robust alternative (RGLM) to the method of least squares which gives a general theory for estimation and testing of parameters in the linear model without requiring assumptions about the common distribution of the independent error variables. Recent discussion articles on the method are given by Draper (1988), Aubuchon and Hettmansperger (1984) and McKean and Vidmar (1994).

Let γ be the vector of parameters corresponding to a complete set of linearly independent contrasts in all treatment and block effects (but excluding the general mean) and let $\boldsymbol{X}_c$ be the corresponding design matrix, whose columns contain the contrast coefficients (so that the column sums of the $\boldsymbol{X}_c$ matrix are zero). We can then write the general linear model in matrix terms as

$$\boldsymbol{Y} = \boldsymbol{1}\theta + \boldsymbol{X}_c\gamma + \boldsymbol{E}$$

where $\boldsymbol{Y}' = [Y_1, \ldots, Y_N]$ and $\boldsymbol{E}' = [E_1, \ldots, E_N]$ are the vectors of response and error variables, respectively, $\boldsymbol{1}$ is a vector of 1's, and γ is the vector of location parameters. Temporarily, we assume that the common distribution of the errors is symmetric around zero.

We write the qth row of the matrix model as $Y_q = \theta + \boldsymbol{x}_q'\gamma + E_q$ with $\boldsymbol{x}_q'$ representing the qth row of the design (contrast) matrix $\boldsymbol{X}_c$. We let $\boldsymbol{y}$ and $\boldsymbol{e}$ represent the "observed" values of $\boldsymbol{Y}$ and $\boldsymbol{E}$, respectively. The traditional method of least squares selects an estimator $\widehat{\gamma}$ of γ to minimize the sum of squares of the errors; that is, to minimize

$$\boldsymbol{e}'\boldsymbol{e} = \sum_{q=1}^{N} e_q^2 = \sum_{q=1}^{N} (y_q - \theta - \boldsymbol{x}_q'\gamma)^2\,.$$

In a similar fashion, the RGLM method selects an estimator $\widetilde{\gamma}$ of γ to minimize the sum of the weighted errors,

$$D(\gamma) = \sum_{q=1}^{N} \left[a\left(R(e_q)\right)\right] e_q = \sum_{q=1}^{N} a\left(R_q\right) e_q, \tag{104}$$

where, for $q = 1, \ldots, N$, $R_q = R(e_q)$ is the rank of e_q among all Ne_t's and $a\left(R_q\right)$ is some function of R_q, called the *score*. The most commonly used scores are the *Wilcoxon scores*, defined as

$$a(R_q) = \sqrt{12}\left(\frac{R_q}{N+1} - \frac{1}{2}\right). \tag{105}$$

These scores ensure that negative and positive residuals of the same magnitude have equal weight in the minimization and that larger residuals have more weight than smaller ones (in the same way that a large squared residual has more weight than a small one in the method of least squares). It is possible to modify the scores so that residuals corresponding to outlying observations play little or no part in the minimization, as discussed by Hettmansperger and McKean (1977) and by Draper (1988), who also lists several other types of scores.

Using the Wilcoxon scores, the RGLM estimate of γ is then the value $\widetilde{\gamma}$ that minimizes

$$D(\gamma) = \sum_{q=1}^{N} \sqrt{12}\left(\frac{R_q}{N+1} - \frac{1}{2}\right)\left(y_q - \boldsymbol{x}_q'\gamma\right). \tag{106}$$

An experimental version of an algorithm which minimizes this function is available in the `rreg` routine of all versions of `MINITAB` since version 8. McKean and Vidmar (1994) state that a `SAS` algorithm is currently under development and that a `PC` algorithm can be obtained from those authors.

For a single replicate or fractional factorial experiment, the robust contrast estimates coincide with the estimates obtained from the method of least squares, although the test statistic, discussed below, differs. The same is true of any *fully parameterized* model with two observations per cell. The test was originally developed as an asymptotic test and, as such, is best with large numbers of observations per cell. In factorial experiments, large sample sizes rarely occur, and an approximation for small samples will be discussed below. In such cases, stated significance levels should be interpreted with caution.

Suppose that the null hypothesis H_0^{γ}: $[\boldsymbol{H}\gamma = 0]$ is to be tested against the general alternative hypothesis H_1^{γ}: $[\boldsymbol{H}\gamma \neq 0]$, where $\boldsymbol{H}\gamma$ represents a vector of h linearly independent estimable functions of the parameters. Then, as in the general linear model approach, a test statistic can be obtained by comparing the value $D(\widetilde{\gamma}_R)$ of

$D(\boldsymbol{\gamma})$ (106) for the reduced model under H_0^γ with $D(\widetilde{\boldsymbol{\gamma}}_F)$ for the full model. The test statistic is

$$F_R = \frac{[D(\widetilde{\boldsymbol{\gamma}}_R) - D(\widetilde{\boldsymbol{\gamma}}_F)]/h}{\tilde{\delta}/2} \tag{107}$$

where $\tilde{\delta}$ is an estimator of δ. For symmetric error distributions, $\tilde{\delta}$ is taken by Hettmansperger and McKean (1977) to be

$$\tilde{\delta} = \frac{\sqrt{N}}{2z_{\alpha/2}\sqrt{N-h-1}} \left[W_{(\frac{N(N+1)}{2}-t)} - W_{(t+1)} \right], \tag{108}$$

where $W_1, \ldots, W_{N(N+1)/2}$ are the pairwise *Walsh averages* $(\tilde{e}_q + \tilde{e}_p)/2$, with $\tilde{e}_i = (y_i - \boldsymbol{x}_i'\widetilde{\boldsymbol{\gamma}})$. Also, $W_{(i)}$ is the ith largest among $W_1, \ldots, W_{N(N+1)/2}$, and t is the lower critical point of a two-sided level α Wilcoxon signed-rank test, which is approximated by

$$t \approx -z_{\alpha/2}\sqrt{\frac{N(N+1)(2N+1)}{24}} + \frac{N(N+1)}{4} - \frac{1}{2}. \tag{109}$$

The estimator $\tilde{\delta}$ is called the *Hodges–Lehmann estimator* and is calculated in the `MINITAB rreg` routine using $\alpha = 0.10$ in (108).

When the error distribution is not symmetric, the Wilcoxon scores can be replaced by the sign scores and the estimator of δ by a kernel density estimator or a two-sample Hodges–Lehmann estimator (see Hettmansperger, 1984, 244–250 and Draper, 1988, 253).

For a large number of observations per cell, the test statistic F_R (107) has an approximate chi-square distribution divided by its degrees of freedom, h. However, for small numbers of observations per cell, the true distribution of F_R is better approximated by an $F(h, N-h-1)$ distribution. The test of the null hypothesis H_0^γ against the general alternative hypothesis H_1^γ at significance level α is then

$$\text{reject } H_0 \text{ if } F_R \geqslant F_\alpha(h, N-h-1), \tag{110}$$

where $F_\alpha(h, N-h-1)$ is the upper αth percentile of an F-distribution with h and $N-h-1$ degrees of freedom.

The sample estimates $\widetilde{\boldsymbol{\gamma}}$ of a set of linearly independent contrasts $\boldsymbol{\gamma}$ and their corresponding estimated standard errors (given by the diagonal elements of $\tilde{\delta}\sqrt{(\boldsymbol{X}_c'\boldsymbol{X}_c)^{-1}}$) can be obtained from the `MINITAB rreg` routine. A standard set of linearly independent contrasts that can be used in a model with two factors are the treatment-control contrasts $\beta_1 - \beta_i$, $\tau_1 - \tau_j$ and the interaction contrasts $(\beta\tau)_{11} - (\beta\tau)_{i1} - (\beta\tau)_{1j} + (\beta\tau)_{ij}$ $(i = 2, \ldots, b; j = 2, \ldots, m)$. If there are more than two factors, then the standard set of contrasts would be extended by including similar sets of contrasts involving the extra factors, together with higher order interaction contrasts. Any other contrast can be written as a linear combination, $\boldsymbol{l}'\boldsymbol{\gamma}$, of the standard contrasts. The contrast estimator

Table 11
Design matrix $\boldsymbol{X_c}$ for the welding experiment

1	1	1	1	1	1	2	−2	−2	2
1	1	−1	0	0	0	1	1	−1	−1
1	1	0	−1	0	0	0	2	0	−2
1	1	0	0	−1	0	−1	1	1	−1
1	1	0	0	0	−1	−2	−2	2	2
−1	0	1	1	1	1	0	0	4	−4
−1	0	−1	0	0	0	0	0	2	2
−1	0	0	−1	0	0	0	0	0	4
−1	0	0	0	−1	0	0	0	−2	2
−1	0	0	0	0	−1	0	0	−4	−4
0	−1	1	1	1	1	−2	2	−2	2
0	−1	−1	0	0	0	−1	−1	−1	−1
0	−1	0	−1	0	0	0	−2	0	−2
0	−1	0	0	−1	0	1	−1	1	−1
0	−1	0	0	0	−1	2	2	2	2

would then be $\boldsymbol{l}'\tilde{\gamma}$, with the estimated standard error $\tilde{\delta}\ \sqrt{\boldsymbol{l}'(\boldsymbol{X}_c'\boldsymbol{X}_c)^{-1}\boldsymbol{l}}$. Confidence intervals for a set of contrasts can be calculated using the usual Scheffé method of multiple comparisons; that is, a set of $100(1-\alpha)\%$ simultaneous confidence intervals for any number of contrasts of the form $\boldsymbol{l}'\gamma$ is given by

$$\boldsymbol{l}'\tilde{\gamma} \pm \sqrt{hF_\alpha(h, N-h-1)}\,\tilde{\delta}\sqrt{\boldsymbol{l}'(\boldsymbol{X}_c'\boldsymbol{X}_c)^{-1}\boldsymbol{l}} \tag{111}$$

(see Hettmansperger and McKean, 1978).

4.3.2. Example – RGLM method, multi-factor experiment

We use the welding data of Experiment 1.4 to illustrate the RGLM approach. There are two treatment factors with 3 and 5 levels, respectively, and an interaction is expected. A fully parameterized model with two observations per cell will lead to contrast estimates identical to the least squares estimates. For illustration purposes we include in the matrix model just the treatment-control contrasts $\beta_1 - \beta_i$ and $\tau_1 - \tau_j$, and the linearly independent trend contrasts representing the linear × linear, linear × quadratic, quadratic × linear and quadratic × quadratic components of the interaction (obtained from a table of orthogonal polynomials). The $\boldsymbol{X_c}$ matrix is then as shown in Table 11, but with each row repeated twice, corresponding to the two observations per cell.

If the columns of $\boldsymbol{X_c}$ are entered into the `Minitab rreg` command, the minimum value of $D(\gamma)$ (106) for the full model (with main effects and the four interaction trend contrasts) is obtained as $D(\tilde{\gamma}_F) = 118.06$. Dropping the last 4 columns of the $\boldsymbol{X_c}$ matrix (which correspond to the four interaction contrasts) gives the the minimum value of $D(\gamma)$ for the reduced model under the hypothesis of negligible interaction contrasts as $D(\tilde{\gamma}_R) = 136.44$. These values are obtained, together with the value of $\tilde{\delta} = 5.359$, from `MINITAB rreg` by specifying that the hypothesis of interest is that the last 4 parameters (contrasts) are all zero. The statistic F_R for testing the hypothesis

H_0^{add} (54) against an alternative hypothesis that the linear × linear, linear × quadratic, quadratic × linear and quadratic × quadratic components of the interaction are not all zero is then given by (107) as

$$F_R = \frac{(136.44 - 118.06)/4}{(5.359/2)} = 1.71.$$

Since $F_{0.05}(4,25) = 3.06$, the hypothesis that the low order interaction trend contrasts are negligible is not rejected at significance level $\alpha = 0.05$. In Section 3.2.2, the hypothesis of no rank interaction was rejected at significance level just under 0.05. The reason for the apparent contradiction is that the strongest interaction trend contrasts are the quartic contrasts, and these were assumed negligible in the above model (as is often done in practice when low order trends are fitted).

The main effect treatment-control contrasts are estimated by the `Minitab rreg` routine as

$$\widehat{\beta_1 - \beta_2} = -2.11 \quad \widehat{\beta_1 - \beta_3} = 3.82$$
$$\widehat{\tau_1 - \tau_2} = 2.73 \quad \widehat{\tau_1 - \tau_3} = -6.67 \quad \widehat{\tau_1 - \tau_4} = 0.65 \quad \widehat{\tau_1 - \tau_5} = 0.03.$$

Other contrast estimates can be obtained as linear combinations of these basic contrast estimates. For example, the estimate of the contrast which compares the third gage bar setting with the average of the other two is

$$\widehat{\beta_3 - \tfrac{1}{2}(\beta_1 + \beta_2)} = \tfrac{1}{2}\left[(\widehat{\beta_1 - \beta_2}) - 2(\widehat{\beta_1 - \beta_3})\right] = -4.88.$$

If a confidence interval is required for this contrast as part of a simultaneous set of 95% Scheffé intervals, we use (111) and obtain

$$\begin{aligned}\widehat{\beta_3 - \tfrac{1}{2}(\beta_1 + \beta_2)} &\in -4.88 \pm \sqrt{10F_{0.05}(4,25)}\,(5.359)\sqrt{(5/16)}\\ &= -4.88 \pm 8.17\,.\end{aligned}$$

References

Akritas, M. G. (1990). The rank transform method in some two-factor designs. *J. Amer. Statist. Assoc.* **85**, 73–78.

Akritas, M. G. and S. F. Arnold (1994). Fully nonparametric hypotheses for factorial designs I: Multivariate repeated measures designs. *J. Amer. Statist. Assoc.* **89**, 336–343.

Anderson, V. L. and R. A. McLean (1974). *Design of Experiments: A Realistic Approach.* Marcel Dekker, New York.

Aubuchon, J. C. and T. P. Hettmansperger (1984). On the use of rank tests and estimates in the linear model. In: P. R. Krishnaiah and P. K. Sen, eds., *Handbook of Statistics*, Vol. 4. Elsevier, Amsterdam, 259–274.

Blair, R. C., S. S. Sawilowsky and J. J. Higgins (1987). Limitations of the rank transform statistic in tests for interaction. *Comm. Statist. Simulation Comput.* **16**, 1133–1145.

Conover, W. J. and R. L. Iman (1976). On some alternative procedures using ranks for the analysis of experimental designs. *Comm. Statist. Theory Methods* **5**, 1349–1368.

Conover, W. J. and R. L. Iman (1981). Rank transformations as a bridge between parametric and nonparametric statistics (with discussion). *Amer. Statist.* **35**, 124–133.

Critchlow, D. E. and M. A. Fligner (1991). On distribution-free multiple comparisons in the one-way analysis of variance. *Comm. Statist. Theory Methods* **20**, 127–139.

Damico, J. A. and D. A. Wolfe (1987). Extended tables of the exact distribution of a rank statistic for all treatments multiple comparisons in one-way layout designs. *Comm. Statist. Theory Methods* **16**, 2343–2360.

Damico, J. A. and D. A. Wolfe (1989). Extended tables of the exact distribution of a rank statistic for treatments versus control multiple comparisons in one-way layout designs. *Comm. Statist. Theory Methods* **18**, 3327–3353.

De Kroon, J. and P. Van der Laan (1981). Distribution-free test procedures in two-way layouts; a concept of rank interaction. *Statist. Neerlandica* **35**, 189–213.

Dehnad, K. (1989). *Quality Control, Robust Design and the Taguchi Method.* Wadsworth and Brooks/Cole, Pacific Grove.

Doksum, K. (1967). Robust procedures for some linear models with one observation per cell. *Ann. Math. Statist.* **38**, 878–883.

Draper, D. (1988). Rank-based robust analysis of linear models I: Exposition and review. *Statist. Sci.* **3**, 239–271.

Durbin, J. (1951). Incomplete blocks in ranking experiments. *British J. Statist. Psychol.* **4**, 85–90.

Dunn, O. J. (1964). Multiple comparions using rank sums. *Technometrics* **6**, 241–252.

Friedman, M. (1937). The use of ranks to avoid the assumption of normality implicit in the analysis of variance. *J. Amer. Statist. Assoc.* **32**, 675–701.

Groggel, D. J. and J. H. Skillings (1986). Distribution-free tests for main effects in multifactor designs. *Amer. Statist.* **40**, 99–102.

Hartlaub, B. A., A. M. Dean and D. A. Wolfe (1993). Nonparametric aligned-rank test procedures for the presence of interaction in the two-way layout with one observation per cell. Tech. Report 525, Department of Statistics, The Ohio State University.

Hayter, A. J. and G. Stone (1991). Distribution free multiple comparisons for monotonically ordered treatment effects. *Austral. J. Statist.* **33**, 335–346.

Hettmansperger, T. P. (1984). *Statistical Inference Based on Ranks.* Wiley, New York.

Hettmansperger, T. P. and J. W. McKean (1977). A robust alternative based on ranks to least squares in analyzing linear methods. *Technometrics* **19**, 275–284.

Hettmansperger, T. P. and J. W. McKean (1978). Statistical inference based on ranks. *Psychometrika* **43**, 69–79.

Hettmansperger, T. P. and J. W. McKean (1983). A geometric interpretation of inferences based on ranks in the linear model. *J. Amer. Statist. Assoc.* **78**, 885–893.

Hollander, M. and D. A. Wolfe (1973). *Nonparametric Statistical Methods.* Wiley, New York.

Iman, R. L. and J. M. Davenport (1980). Approximations of the critical region of the Friedman statistic. *Comm. Statist. Theory Methods* **9**, 571–595.

Iman, R. L., D. Quade and D. A. Alexander (1975). Exact probability levels for the Kruskal–Wallis test. In: H. L. Harter and D. B. Owen, eds., *Selected Tables in Mathematical Statistics*, Vol. III. Markham Press, Chicago, 329–384.

Jonckheere, A. R. (1954). A distribution-free k-sample test against ordered alternatives. *Biometrika* **41**, 133–145.

Kendall, M. G. and B. Babington Smith (1939). The problem of m rankings. *Ann. Math. Statist.* **10**, 275–287.

Kruskal, W. H. and W. A. Wallis (1952). Use of ranks in one-criterion variance analysis. *J. Amer. Statist. Assoc.* **47**, 583–621.

Kuehl, R. O. (1994). *Statistical Principles of Research Design and Analysis.* Duxbury Press, Belmont, CA.

Mack, G. A. (1981). A quick and easy distribution-free test for main effects in a two-factor ANOVA. *Comm. Statist. Simulation Comput.* **10**, 571–591.

Mack, G. A. and J. H. Skillings (1980). A Friedman-type rank test for main effects in a two-factor ANOVA. *J. Amer. Statist. Assoc.* **75**, 947–951.

Mack, G. A. and D. A. Wolfe (1981). K-sample rank tests for umbrella alternatives. *J. Amer. Statist. Assoc.* **76**, 175–181.

McDonald, B. J. and W. A. Thompson, Jr. (1967). Rank sum multiple comparisons in one- and two-way classifications. *Biometrika* **54**, 487–497.

McKean, J. W. and T. P. Hettmansperger (1976). Tests of hypotheses of the general linear model based on ranks. *Comm. Statist. Theory Methods* **5**, 693–709.

McKean, J. W. and T. J. Vidmar (1994). A comparison of two rank-based methods for the analysis of linear models. *Amer. Statist.* **48**, 220–229.

Miller, R. G., Jr. (1966). *Simultaneous Statistical Inference.* McGraw-Hill, New York.

Mosteller, F. and R. E. K. Rourke (1973). *Sturdy Statistics.* Addison-Wesley, Reading, MA.

Nemenyi, P. (1963). Distribution-free multiple comparisons. Ph.D. Thesis, Princeton University, Princeton, NJ.

Odeh, R. E. (1971). On Jonckheere's k-sample test against ordered alternatives. *Technometrics* **13**, 912–918.

Odeh, R. E. (1977a). Extended tables of the distribution of Friedman's S-statistic in the two-way layout. *Comm. Statist. Simulation Comput.* **6**, 29–48.

Odeh, R. E. (1977b). The exact distribution of Page's L-statistic in the two-way layout. *Comm. Statist. Simulation Comput.* **6**, 49–61.

Odeh, R. E. (1977c). Extended tables of the distributions of rank statistics for treatment versus control in randomized block designs. *Comm. Statist. Simulation Comput.* **6**, 101–113.

Page, E. B. (1963). Ordered hypotheses for multiple treatments: A significance test for linear ranks. *J. Amer. Statist. Assoc.* **58**, 216–230.

Randles, R. H. and D. A. Wolfe (1979). *Introduction to the Theory of Nonparametric Statistics.* John Wiley, New York (reprinted in 1991 by Krieger Publishing, Malabar, FL).

Rust, S. W. and M. A. Fligner (1984). A modification of the Kruskal–Wallis statistic for the generalized Behrens–Fisher problem. *Comm. Statist. Theory Methods* **13**, 2013–2027.

Sawilowsky, S. S. (1990). Nonparametric tests of interaction in experimental design. *Rev. Ed. Res.* **60**, 91–126.

Simpson, D. G. and B. H. Margolin (1986). Recursive nonparametric testing for dose-response relationships subject to downturns at high doses. *Biometrika* **73**, 589–596.

Skillings, J. H. and G. A. Mack (1981). On the use of a Friedman-type statistic in balanced and unbalanced block designs. *Technometrics* **23**, 171–177.

Skillings, J. H. and D. A. Wolfe (1977). Testing for ordered alternatives by combining independent distribution-free block statistics. *Comm. Statist. Theory Methods* **6**, 1453–1463.

Skillings, J. H. and D. A. Wolfe (1978). Distribution-free tests for ordered alternatives in a randomized block design. *J. Amer. Statist. Assoc.* **73**, 427–431.

Spjøtvoll, E. (1968). A note on robust estimation in analysis of variance. *Ann. Math. Statist.* **39**, 1486–1492.

Terpstra, T. J. (1952). The asymptotic normality and consistency of Kendall's test against trend, when ties are present in one ranking. *Indag. Math.* **14**, 327–333.

Thompson, G. L. (1991). A unified approach to rank tests for multivariate and repeated measures designs. *J. Amer. Statist. Assoc.* **86**, 410–419.

Wilcoxon, F. and R. A. Wilcox (1964). *Some Rapid Approximate Statistical Procedures*, 2nd edn. American Cyanamid Co., Lederle Laboratories, Pearl River, NY.

Wilkie, D. (1962). A method of analysis of mixed level factorial experiments. *Appl. Statist.* **11**, 184–195.

Wood, S. R. and D. E. Hartvigsen (1964). Statistical design and analysis of qualification test program for a small rocket engine. *Industr. Quality Contr.* **20**, 14–18.

S. Ghosh and C. R. Rao, eds., *Handbook of Statistics, Vol. 13*

21

Block and Other Designs Used in Agriculture

Deborah J. Street

1. Preliminaries

1.1. Overview

This chapter brings together information on designs that are currently being used in agriculture. We survey block designs, including α designs; Latin square and other row–column designs; cross-over designs and split-plot designs; neighbour designs for field trials; factorial designs and response surface designs. For each design we give information on construction, published tabulations and computer programs available to generate the designs.

It appears that none of the standard reference books covers all of the designs discussed here. More details on design constructions can be found in John (1971), Montgomery (1991) and Street and Street (1987). More details on general design considerations, some specific to aspects of agricultural experimentation, can be found in Dyke (1988), Mead (1988) and Pearce (1983).

In the next subsection we review concepts important to all designs and then define the various designs that we discuss in detail in subsequent sections.

1.2. Definitions

We begin by reviewing various concepts related to the block structure, the treatment structure and the error structure of a design. We assume that the *response* of an experiment, be it the yield of crop or the gain in weight of animals, is a familiar concept but we do define what we mean by a factor.

A *factor* is any attribute of the experimental units which may affect the response observed in the experiment. Any factor may take one of several values which are called the *levels* of the factor. For example, in an experiment to compare three varieties of wheat, the factor is 'wheat' and it has three levels.

The *block structure* of the design concerns how the experimental units are related to each other by factors inherent to the units. These factors may be sources of variation, and thus must be allowed for in the design and analysis of the experiment, but are not of intrinsic interest to the experimenter. For example, if the experiment to compare the three varieties of wheat is conducted on five different farms then these farms are a potential source of variation and must be allowed for as such in the analysis. But the farms are not of intrinsic interest to the experimenter.

The *treatment structure* of the experiment is allocated by the experimenter and investigating this structure is the point of the experiment. The treatment structure may consist of one treatment factor and the experiment is then designed to compare the levels of this factor. But sometimes more than one treatment factor is included in an experiment. Then the experimenter is interested in the effects of these factors, both singly and in combination. Thus the design must be chosen so that all the interaction effects of interest can be independently estimated. This is the topic of Section 6, factorial designs.

The *error structure* of the design is determined by the randomisation that is used to obtain the design. If the treatment factors need to be allocated to the units in turn, as might happen in an experiment to investigate several methods of ploughing and several different varieties of wheat, then there are two, or more, randomisations and each of these gives rise to an error term in the model. This is discussed briefly in Section 4.

We assume that the experimenter knows that random allocation of treatments to experimental units is necessary; the layouts that we construct need to be randomised before use. A very practical account of this may be found in Dyke (1988).

We now define designs with zero, one and two blocking factors.

A *completely randomised design* is carried out on a set of homogeneous units to which are randomly allocated a set of treatments. Hence there is no blocking factor and any random method of allocation of treatments to plots may be used. GENSTAT and SAS have commands to produce such random allocations.

If we can not get a sufficiently large set of homogeneous plots but we can get several smaller sets of homogeneous plots then these sets of homogeneous plots are called *blocks* and the resulting design is called a *block design*. If each block has as many plots as treatments then we (usually) allocate each of the treatments to a plot chosen at random in each block and refer to the resulting design as a *randomised complete block design*. Again the construction problem is easy; there is one plot receiving each of the t treatments in each block and these must be randomly allocated within each block. GENSTAT and SAS have commands to produce such random allocations.

These two designs are used frequently in agriculture. For instance, slightly over half the papers in the 1994 issues of *Experimental Agriculture* and *Journal of Agricultural Science* used one of these two designs at some point in the work reported.

If the blocks have fewer plots than treatments then each of the blocks is allocated a subset of the treatments and the design is said to be an *incomplete block design*. Various incomplete block designs have been proposed in the literature. We talk about balanced incomplete block designs, lattice designs, α designs and partially balanced designs in Section 2.

Sometimes the blocks of incomplete block designs can be further grouped into sets of blocks so that each treatment appears in precisely one block of each set. These sets of blocks are called *resolution classes* and the incomplete block design is said to be *resolvable*. Resolvability is useful when managing the experiment provided that one, or more, resolution classes can be sown, harvested, or otherwise treated, at each session. Then complete replicates of treatments are represented on each occasion. We indicate the resolvability properties of the incomplete block designs we discuss.

Incomplete block designs allow for one extraneous source of variation, termed the blocks, in an experiment. Row–column designs and nested row–column designs can be used to allow for more than one blocking factor.

The most familiar row–column designs are the Latin squares. A *Latin square of order* k is a $k \times k$ array of the numbers $1, 2, \ldots, k$ arranged such that each symbol appears exactly once in each row of the square and exactly once in each column of the square. Hence Latin squares require that the number of treatments and the number of blocks in each of the two orthogonal systems of blocks be equal. This requirement is often too restrictive and so more general row–column designs have been developed.

In a *row–column design* there are v treatments laid out in b blocks with each block having p rows and q columns. We say that the row–column design is *nested* if $b > 1$. We discuss Latin squares, row–column α-designs, nested row–column designs and other related designs in Section 3.

A *repeated measurements study* is one in which each experimental unit is observed under at least two conditions; we consider split-plot designs and change-over designs. In a *split-plot design* there are at least two independent stages of randomisation. In the first stage of a typical field experiment fertilisers are randomly assigned to fields (whole plots) and in the second stage varieties are randomly assigned to sub-plots within each field. The sub-plots are the observational units. They are used when some treatment factors must, of necessity, be applied to rather large experimental units, whereas others can be applied to smaller ones. A *cross-over design* (or *change-over design*) (COD) is a design in which t treatments are compared using n experimental units and the experiment lasts for p periods. During the course of the experiment, each unit receives a sequence of p, not necessarily distinct, treatments and a response is recorded for each unit in each period. These designs are discussed in Section 4.

Designs may need to be balanced for neighbouring treatments in several different applications, notably for field experiments where the nearest-neighbour methodology is to be employed, when studying intergenotypic competition, when designing the collection of seed from seed orchards and in bioassay work when using Ouchterlony gel diffusion tests. Designs with the requisite balance properties are discussed in Section 5.

As mentioned, we discuss factorial designs in general in Section 6. In Section 7 we discuss a particular sort of factorial design, the response surface design. In these designs there are one or more treatment factors of interest, and these have quantitative levels. The aim of a response surface design is to predict the observed response as a polynomial in the levels of the factors. The choice of the factor levels to include and some of the commonly used designs are discussed in Section 7.

The final section briefly describes a number of other designs that have been proposed, and used, in an agricultural setting.

2. Incomplete block designs

Recall that incomplete block designs are used when there is one blocking factor and each block has fewer plots than treatments. In this section we consider balanced incomplete block designs, lattice designs, α designs and partially balanced incomplete block designs. In the final subsection we look at measures that have been proposed to compare and rank various incomplete block designs.

Randomisation of an incomplete block design is achieved by randomly allocating treatments to plots within each block. The blocks of the design are allocated to the blocks of experimental material at random. If the design is resolvable (see Subsection 1.2 for a definition and Subsections 2.1, 2.2, 2.3 and 2.4 for examples of resolvable designs) and the resolution classes correspond to attributes of the experiment then the randomisation of the blocks of the design to the blocks of experimental material is done at random within each resolution class. This would be appropriate if, say, one resolution class was to be harvested on each day.

2.1. Balanced incomplete block designs

The first of the incomplete block designs to be proposed was the balanced incomplete block design, introduced by Yates in 1935 as 'symmetrical incomplete randomised blocks'; the current name is due to Bose (1939). Subbarayan (1992) shows how designs with blocks of size three can be used as partial triallel mating designs.

A *balanced incomplete block design* (BIBD) is a set of v treatments arranged in b blocks of size k in such a way that every treatment appears r times in the design and any two treatments appear in the same block on precisely λ occasions. The design in Table 2.1.1 has $v = 9$, $b = 12$, $r = 4$, $k = 3$ and $\lambda = 1$. We usually write the parameters of the design as (v, b, r, k, λ). We refer to r as the *replication number* and to λ as the *index* of the design.

The parameters of a BIBD are not independent of each other. Counting the total number of plots in a design we see that there are b blocks each with k plots or, equivalently, that there are v treatments each replicated r times. Thus $bk = vr$, the total number of responses in the experiment. Counting pairs we see that each treatment appears with $(k - 1)$ other treatments in each of the r blocks in which it appears. Equivalently there are $(v - 1)$ other treatments and there are λ occurrences of each pair. Hence $r(k-1) = \lambda(v-1)$. These relationships can be used to determine whether a given set of parameters could be the parameters of a BIBD. For example, if $v = k^2$ and $\lambda = 1$ then $k^2 - 1 = r(k - 1)$. Hence $r = k + 1$ whence $b = k(k + 1)$. Because of these relationships, we often write (v, k, λ) BIBD instead of (v, b, r, k, λ) BIBD. Designs with $k = 3$ are often called *triple systems* and with $k = 4$ *quadruple systems*.

There are a number of methods to construct BIBDs described in the literature. Here we discuss one based on difference sets and one that uses sets of Latin squares. An

Table 2.1.1
A $(9, 12, 4, 3, 1)$ design (rows are blocks)

1 2 3	1 4 7	1 5 9	1 6 8
4 5 6	2 5 8	3 4 8	2 4 9
7 8 9	3 6 9	2 6 7	3 5 7

elementary account of these and other construction methods for BIBDs can be found in Street and Street (1987) and Anderson (1990). Colbourn and Dinitz (1996) provide an encyclopaedic coverage.

The *method of differences* is one of the most fruitful methods of constructing BIBDs. In outline, and in its simplest form, this method involves finding a set of numbers which can be used as the first block of the design and constructing every other block in the BIBD systematically from the first. The idea is quite old and goes back at least to Netto (1893); Bose (1939) studied cyclic difference sets and their extensions extensively.

For instance, suppose that we do all the arithmetic modulo 7 so that when we add two numbers together we record only the number between 0 and 6 to which the sum corresponds (so 7 and 0 are the same, as are 1 and 8 and 15 and so forth). We do the same thing for differences. Then consider the set $\{0, 1, 3\}$. The distinct differences from this set are $3 - 0 = 3$, $0 - 3 = 7 - 3 = 4$, $1 - 0 = 1$, $0 - 1 = 7 - 1 = 6$, $3 - 1 = 2$, and $1 - 3 = 8 - 3 = 5$ and we see that each non-zero difference modulo 7 occurs exactly once. We can then use $\{0, 1, 3\}$ as the *initial block* of a design. We get the other blocks by adding each of the numbers 1 to 6, in turn, to the original block and recording the results modulo 7. Thus the blocks are $(0, 1, 3)$, $(1, 2, 4)$, $(2, 3, 5)$, $(3, 4, 6)$, $(4, 5, 0)$, $(5, 6, 1)$, and $(6, 0, 2)$. We say that the design has been *developed* from the initial block. We see that the design then has 7 blocks and 7 treatments, each block is of size 3, each treatment appears in 3 blocks and any pair of treatments appear in 1 block.

Formally a *cyclic difference set* is a k-set B of the integers modulo v, $Z_v = \{0, 1, 2, \ldots, v - 1\}$, with the property that, given any non-zero integer d modulo v, there are precisely λ ordered pairs of elements of B whose difference is d. Again the differences are elements of Z_v so 0 and $-v$ are the same, as are -1 and $v - 1$ and so on. B is referred to as a (v, k, λ) *difference set*. Designs developed from a cyclic difference set always have $b = v$ and $r = k$ and are said to be *symmetric*.

It is possible for groups other than Z_v to be used. For instance consider 16 treatments represented by the ordered pairs (x, y), where $0 \leqslant x, y \leqslant 3$. These 16 pairs form the set $Z_4 \times Z_4$. Suppose we perform arithmetic on the ordered pairs componentwise so that $(x, y) + (w, z) = (x + w, y + z)$ and each sum is recorded as a number between 0 and 3. Consider the set $\{00, 02, 10, 30, 11, 33\}$, where we have written xy for (x, y). The set of differences is given in Table 2.1.2(a). We see that each non-zero difference appears exactly twice and so $\{00, 02, 10, 30, 11, 33\}$ is a $(16, 6, 2)$ difference set in $Z_4 \times Z_4$. We get the blocks from the difference set by adding each of the 16 elements of $Z_4 \times Z_4$ in turn to the difference set. The blocks are listed in Table 2.1.2(b).

Some small difference sets are given in Table 2.1.3. Extensive tables appear in Colbourn and Dinitz (1995).

The idea of a difference set can be extended to that of *supplementary difference sets* (SDS) over Z_v. This is a set of m k-sets, $B_1, B_2, \ldots, B_m$, such that, given any non-zero integer d modulo v, there are precisely λ ordered pairs of elements, with difference d, with each pair coming from one of the sets B_i, $1 \leqslant i \leqslant m$. For instance let $v = 13$ and consider the two sets $\{0, 1, 4\}$ and $\{0, 2, 7\}$. Then the set of distinct differences, modulo 13, from these two sets is $0 - 1 = 12$, $1 - 0 = 1$, $0 - 4 = 9$,

Table 2.1.2
The $(16, 6, 2)$ difference set and the corresponding design

(a)

	00	02	10	30	11	33
00	–	02	10	30	11	33
02	02	–	12	32	13	31
10	30	32	–	20	01	23
30	10	12	20	–	21	03
11	33	31	03	23	–	22
33	11	13	21	01	22	–

(b)

00, 02, 10, 30, 11, 33	20, 22, 30, 10, 31, 13
01, 03, 11, 31, 12, 30	21, 23, 31, 11, 32, 10
02, 00, 12, 32, 13, 31	22, 20, 32, 12, 33, 11
03, 01, 13, 33, 10, 32	23, 21, 33, 13, 30, 12
10, 12, 20, 00, 21, 03	30, 32, 00, 20, 01, 23
11, 13, 21, 01, 22, 00	31, 33, 01, 21, 02, 20
12, 10, 22, 02, 23, 01	32, 30, 02, 22, 03, 21
13, 11, 23, 03, 20, 02	33, 31, 03, 23, 00, 22

Table 2.1.3
Some small difference sets

(v, k, λ)	Group	Difference set
7, 3, 1	Z_7	0, 1, 3
7, 4, 2	Z_7	0, 2, 3, 4
13, 4, 1	Z_{13}	0, 1, 3, 9
21, 5, 1	Z_{21}	1, 2, 7, 9, 19
11, 5, 2	Z_{11}	1, 3, 4, 5, 9
16, 6, 2	$Z_4 \times Z_4$	00, 02, 10, 30, 11, 33
15, 7, 3	Z_{15}	0, 1, 2, 4, 5, 8, 10

$4 - 0 = 4$, $1 - 4 = 10$, $4 - 1 = 3$, $0 - 2 = 11$, $2 - 0 = 2$, $0 - 7 = 6$, $7 - 0 = 7$, $2 - 7 = 8$ and $7 - 2 = 5$. These are all the non-zero integers modulo 13 and so the 26 blocks developed from this pair of SDS, and given in Table 2.1.4, contain each ordered pair exactly once.

Again modifications of the basic idea are possible. One is to have an element, typically denoted by ∞, which remains invariant as the block is developed. For example consider the two sets $\{0, 2, 3, 4\}$ and $\{0, 1, 3, \infty\}$. Then each non-zero difference modulo 7 appears exactly three times and in the final design ∞ appears with each treatment three times also. Hence these blocks give an $(8, 4, 3)$ design.

Another useful idea is that of a *short block*. For instance, let $v = 15$ and consider the sets $\{0, 1, 4\}$, $\{0, 2, 8\}$ and $\{0, 5, 10\}$. Then we see that the third block has the differences 5 and 10 only. Adding 5 to this block gives $\{5, 10, 0\}$ and so instead of getting 15 blocks from this block we take only the 5 distinct blocks to get 35 blocks in the full $(15, 3, 1)$ design.

Table 2.1.4
Blocks of a $(13, 3, 1)$

0	1	4	0	2	7
1	2	5	1	3	8
2	3	6	2	4	9
3	4	7	3	5	10
4	5	8	4	6	11
5	6	9	5	7	12
6	7	10	6	8	0
7	8	11	7	9	1
8	9	12	8	10	2
9	10	0	9	11	3
10	11	1	10	12	4
11	12	2	11	0	5
12	0	3	12	1	6

Table 2.1.5
Some small supplementary difference sets

(v, b, r, k, λ)	Group	SDS		
15, 35, 7, 3, 1	Z_{15}	0, 1, 4	0, 2, 8	0, 5, 10 (short)
13, 26, 6, 3, 2	Z_{13}	0, 1, 4	0, 2, 7	
25, 50, 8, 4, 1	$Z_5 \times Z_5$	00, 01, 10, 11	00, 20, 02, 33	
8, 14, 7, 4, 3	$Z_7 \cup \{\infty\}$	0, 2, 3, 4	0, 1, 3, ∞	
9, 18, 8, 4, 3	$Z_3 \times Z_3$	11, 02, 22, 01	10, 21, 20, 12	
13, 26, 12, 6, 5	Z_{13}	1, 3, 4, 9, 10, 12	2, 5, 6, 7, 8, 11	

Some small supplementary difference sets appear in Table 2.1.5. Extensive tables appear in Colbourn and Dinitz (1995).

Resolvable BIBDs can be constructed easily from Latin squares and sets of mutually orthogonal Latin squares. Recall that a *Latin square of order* k is a $k \times k$ array of the numbers $1, 2, \ldots, k$ arranged such that each symbol appears exactly once in each row of the square and exactly once in each column of the square. A pair of Latin squares are said to be *orthogonal* if, when superimposed, each ordered pair of treatments appears exactly once. A set of Latin squares is said to be *mutually orthogonal* if any two squares in the set are orthogonal. We often write *MOLS* for a set of mutually orthogonal Latin squares. Theorem 3.4.1 gives a construction for sets of MOLS of prime and prime-power order. The following result has become part of the folklore.

CONSTRUCTION 2.1. Let $L_1, \ldots, L_{k-2}$ be a set of MOLS of order k. Then a resolvable BIBD with $v = k^2$, $b = k(k+1)$, $r = k+1$ and $\lambda = 1$ exists.

PROOF. Let A be a $k \times k$ array with entries $1, 2, \ldots, k^2$ in some order. The blocks of the first resolution class constitute the rows of A and those of the second resolution class constitute the columns of A. Third and subsequent resolution classes are obtained by superimposing each of the Latin squares in turn on A and using as block i those symbols in A which appear in the same position as does i in the Latin square. □

Table 2.1.6
MOLS of order 3

(a)		(b)
1 2 3	1 2 3	1 2 3
3 1 2	2 3 1	4 5 6
2 3 1	3 1 2	7 8 9

Table 2.1.7
A $(13, 13, 4, 4, 1)$ design (rows are blocks)

1 2 3 ∞_1	1 4 7 ∞_2	1 5 9 ∞_3	1 6 8 ∞_4	∞_1 ∞_2 ∞_3 ∞_4
4 5 6 ∞_1	2 5 8 ∞_2	3 4 8 ∞_3	2 4 9 ∞_4	
7 8 9 ∞_1	3 6 9 ∞_2	2 6 7 ∞_3	3 5 7 ∞_4	

The design in Table 2.1.1 was constructed from a pair of orthogonal Latin squares of order three in this way. The squares are given in Table 2.1.6(a) and the array A in Table 2.1.6(b).

The designs obtained from Construction 1 can be extended to give BIBDs with $v = k^2 + k + 1$ points by adjoining a new point, ∞_i, say, to each block in the ith resolution class, and adjoining a new block containing all the new points. The design in Table 2.1.1 becomes the $(13, 4, 1)$ of Table 2.1.7, for instance.

Ways to compare designs are discussed in Subsection 2.5. Here it is sufficient to note that for the purposes of treatment estimation and comparison, all BIBDs with the same parameters are equally 'good'. For other applications, for instance in survey work, this need not be the case (see Hedayat and Sinha, 1991).

Because of the number of experimental units required, it is not always possible for each treatment to appear an equal number of times with every other in blocks of constant size. Designs in which pairs of treatments need not appear equally often are frequently used in practice. Such designs are said to have *partial balance*. The three most common appear to be the lattice designs, the α designs and some of the so-called partially balanced designs. The first two types of designs are always resolvable. We discuss all three in turn below.

2.2. *Lattice designs*

A *lattice design* has $v = k^2$ treatments arranged in $b = rk$ blocks of size k such that the set of blocks can be subdivided into r resolution classes of k blocks and such that any two varieties appear together in at most one block. These designs were introduced by Yates (1940). Recent trials designed as lattice designs include those of Jackson and Hogarth (1992) and Taylor and Smith (1992).

A lattice design with $r = 2$ is called a *simple* lattice, with $r = 3$ is called a *triple* lattice and with $r = k + 1$ is called a *balanced* lattice. By definition any two varieties appear together in at most one block. If in fact any two varieties appear together in exactly one block then the lattice design is as large as it can be and is a BIBD with $\lambda = 1$. From the previous subsection we see that in this case $r = k + 1$. Hence if $r > k + 1$ then no lattice design can exist. The design in Table 2.1.1 is a balanced lattice with the blocks written down in the four resolution classes. Any two resolution classes of this design form a simple lattice and any three form a triple lattice. Any r resolution classes of one of the resolvable BIBDs constructed in Construction 2.1.1 form a lattice design.

A *rectangular lattice design* is an arrangement of $k(k-h)$ varieties into blocks of size $k-h$ such that the set of blocks can be partitioned into $r \leqslant (v-1)/(k-h-1)$ resolution classes of k blocks each and such that any two varieties appear together in at most one block. (Note that for a BIBD to exist we would require that $k(k-h)-1$ be a multiple of $(k-h-1)$.) Rectangular lattices with $h = 1$ were introduced by Harshbarger (1949); the extension to larger h appears in Kempthorne (1952).

Construction 2.1.1 can be modified to give rectangular lattices as follows. Let A be a $k \times k$ array with entries $1, 2, \ldots, k(k-h)$ in some order and such that each row and column of A has h empty cells. Then the rows and columns of A give the blocks of the first two resolution classes. We can construct a third resolution class by imposing on A a Latin square in which h entries have been deleted from each row, each column and from the set of occurrences of each symbol. For example, let $k = 5$ and $h = 2$. Consider the Latin square in Table 2.2.1(a). The array in Table 2.2.1(b) has been constructed from the array in Table 2.2.1(a) by omitting two cells from each row and two from in each column in such a way that the final array has only three occurrences of each symbol. This array can be superimposed on the array

Table 2.2.1
A rectangular lattice with $v = 15$, $k = r = 3$

(a)					(b)					(c)					(d)											
1	2	3	4	5	1	2	3	*	*	1	2	3	*	*	1	2	3	1	4	7	1	5	B
5	1	2	3	4	5	1	*	*	4	4	5	*	*	6	4	5	6	2	5	D	2	8	C
4	5	1	2	3	4	*	*	2	3	7	*	*	8	9	7	8	9	3	A	E	3	9	D
3	4	5	1	2	*	*	5	1	2	*	*	A	B	C	A	B	C	8	B	F	6	7	E
2	3	4	5	1	*	3	4	5	*	*	D	E	F	*	D	E	F	6	9	C	4	A	F

Table 2.2.2
A rectangular lattice with $v = 6$ and $r = 5$

1	2	*	1 2	1 3	1 4	1 5	1 6
3	*	4	3 4	2 5	2 6	2 4	2 3
*	5	6	5 6	4 6	3 5	3 6	4 5
	A			Resolution classes			

	1	2	*		1	2	*		1	2	*
$P_1 =$	3	*	1	$P_2 =$	3	*	2	$P_3 =$	2	*	3
	*	3	2		*	1	3		*	3	1

Partial Latin squares

of treatments given in Table 2.2.1(c) to get the rectangular lattice design with $v = 15$ and with blocks of size 3 listed in Table 2.2.1(d). (Here $*$ denotes an empty cell.) General constructions like this example have been presented by Afsarinejad (1990).

However it is unnecessarily restrictive to insist that the arrays that are superimposed on the array of treatments be Latin squares with some entries deleted. It is possible to use 'partial Latin squares', which can not be completed to Latin squares, to get third, and subsequent, resolution classes, as the following example from Street and Street (1987) shows. A rectangular lattice with $k = 3$, $h = 1$, $v = 3(3-1) = 6$ and $r = 5$ is given in Table 2.2.2. We give the array of treatments, A, say, the five resolution classes and the 'partial Latin squares' corresponding to the final three resolution classes. These squares are mutually orthogonal but none can be completed to a Latin square.

2.3. α designs

These designs were introduced by Patterson and Williams (1976) for the field trials required for statutory variety testing in the United Kingdom but they are now used in other areas of agriculture as the recent example in Ellis et al. (1993) shows.

An $\alpha(\lambda_1, \lambda_2, \ldots, \lambda_n)$-*design* is a design on $v = sk$ varieties, with r replicates of each variety, and with blocks of size k, arranged in r resolution classes of s blocks each. Any pair of varieties appears in either λ_1 or λ_2 or ... or λ_n blocks. The blocks in each resolution class are obtained from one initial block.

The design in Table 2.3.1 is an $\alpha(0,1,2)$-design with $v = 20$, $s = 4$, $k = 5$, $r = 3$. Varieties 0 and 1 never appear together in a block, varieties 0 and 4 appear together in one block and varieties 4 and 9 appear together in two blocks, for example. The design was constructed from the initial blocks $(0,4,8,12,16)$, $(0,5,10,15,19)$ and $(0,6,11,13,18)$, given in the tables in Patterson and Williams (1976), by a modification of the method used with difference sets. To describe this, let σ_i denote the permutation $(4(i-1),\ 4(i-1)+1,\ 4(i-1)+2,\ 4(i-1)+3)$, $1 \leqslant i \leqslant 5$. So $\sigma_1 = (0,1,2,3)$ and $\sigma_3 = (8,9,10,11)$, for instance. Then given the first block in a resolution class the element in position i of block j of that class is found by applying σ_i to the ith element of the first block $(j-1)$ times. Thus in the second resolution class the initial block is $(0,5,10,15,19)$ and the third block, say, is obtained by applying σ_1 to 0 twice, giving 2, σ_2 to 5 twice, giving 7, σ_3 to 10 twice, giving 8, and so on until all the elements in the block are obtained. The block is $(2,7,8,13,17)$.

Table 2.3.1
An $\alpha(0,1,2)$-design with $v = 20$, $s = 4$, $k = 5$, $r = 3$

0	4	8	12	16	0	5	10	15	19	0	6	11	13	18
1	5	9	13	17	1	6	11	12	16	1	7	8	14	19
2	6	10	14	18	2	7	8	13	17	2	4	9	15	16
3	7	11	15	19	3	4	9	14	18	3	5	10	12	17
Replicate 1					Replicate 2					Replicate 3				

Various results on the existence of α designs have been obtained; two are given below.

THEOREM 1 (Patterson and Williams, 1976). *If an* $\alpha(0,1)$*-design exists then* $k \leqslant s$, *provided that* $r \geqslant 2$.

THEOREM 2 (Patterson and Williams, 1976). *The following four series of* $\alpha(0,1)$*-designs are known:*

(i) $r = 2$, $k \leqslant s$;
(ii) $r = 3$, s *odd*, $k \leqslant s$;
(iii) $r = 3$, s *even*, $k \leqslant s - 1$;
(iv) $r = 4$, $s \equiv 1$ *or* $5 \pmod 6$, $k \leqslant s$.

Proofs of these theorems can be found in the original paper and in Street and Street (1987) and John (1987), where further results are also given.

One extension that we will mention briefly increases the range of available designs by allowing blocks of two different sizes. Suppose that $v = s_1k_1 + s_2k_2$, where s_1, k_1, s_2, k_2 are all integers and k_1 and k_2 differ by only 1. Then a resolvable α-design on v treatments can be obtained by constructing an α-design on $v + s_2$ treatments with $s_1 + s_2$ blocks of k_1 plots in each resolution class. Deleting a set of s_2 treatments, no two of which ever occur together in a block, gives the design.

2.4. Partially balanced incomplete block designs

These designs are used both in plant breeding work, for diallel crosses, and in areas like field trials.

A partially balanced incomplete block design (PBIBD) is a design defined on a set of treatments which have been divided into subsets. The subsets may be determined by an inherent attribute of the treatments, as happens in factorial experiments (see Section 6), but may be artificially imposed for convenience. PBIBDs are particularly easy to analyse and this was a major consideration in pre-computer days. Now that computers are available to perform the analysis, it can be argued that designs like α-designs are more appropriate for field trials if the association scheme of the PBIBD does not correspond in any natural way to the structure of the treatments actually used (see Bailey, 1985).

If a set of v treatments has an *association scheme with* m *associate classes* defined on it then:

(i) any two distinct treatments are ith associates for exactly one value of i, $1 \leqslant i \leqslant m$;

(ii) each treatment has exactly n_i ith associates, $1 \leqslant i \leqslant m$;

(iii) for any pair of ith associates, x and y, say, there are a fixed number of treatments which are both jth associates of x and hth associates of y and this number is independent of the particular pair of ith associates chosen.

The *parameters of the association scheme* are v and n_i $(1 \leqslant i \leqslant m)$.

Table 2.4.1
A group divisible PBIBD(2)

1 2	1 3	1 5	1 6
3 5	2 4	2 6	2 3
4 6	5 6	3 4	4 5

Table 2.4.2
A triangular PBIBD(2)

(a)	(b)
$*$ 1 2 3 4	0 1 2 7 8
1 $*$ 5 6 7	0 1 3 5 9
2 5 $*$ 8 9	0 2 4 5 6
3 6 8 $*$ 0	1 4 6 8 9
4 7 9 0 $*$	2 3 6 7 9
	3 4 5 7 8

A *partially balanced incomplete block design with m associate classes* (PBIBD(m)) is a design based on a set of v treatments with an m-associate class association scheme defined on them. There are b blocks of size k and each treatment is replicated r times. If treatments x and y are ith associates then there are λ_i blocks containing both x and y. The PBIBD determines the association scheme but not conversely.

Counting arguments show that $vr = bk$, as for BIBDs. Every treatment is an associate of every other and so $\sum_i n_i = v-1$. Counting pairs we have that $\sum_i n_i\lambda_i = r(k-1)$.

There are a number of different association schemes in common use. The two associate class schemes are usually subdivided into the following six types: group divisible; triangular; Latin-square-type; cyclic; partial geometry type; miscellaneous. We describe some of these six types below and indicate natural extensions to more than two associate classes when appropriate.

In the *group divisible scheme* the set of $v = mn$ treatments is subdivided into m subsets (or groups, in the non-mathematical sense) of n treatments each. Treatments in the same group are first associates; treatments in different groups are second associates. Hence we have that $n_1 = n - 1$ and $n_2 = n(m - 1)$.

The design in Table 2.4.1 is a PBIBD(2) based on a group divisible association scheme. There are $v = 6$ treatments, arranged in 3 groups of size 2. The groups are $\{1, 4\}$, $\{2, 5\}$ and $\{3, 6\}$. There are 12 blocks of size 2 and each treatment is replicated 4 times. We see that $\lambda_1 = 0$ and $\lambda_2 = 1$.

This scheme can be generalised to the rectangular association scheme (Raghavarao, 1971) and the extended-group-divisible association scheme (Chang and Hinkelmann, 1987). These are both examples of the factorial association scheme (Bailey, 1985).

To get a *triangular association scheme*, let $v = n(n-1)/2$, $n \geqslant 5$. Arrange the treatments in a symmetrical $n \times n$ array with the diagonal entries empty. First associates are treatments in the same row or column and all other treatments are second associates. Hence $n_1 = 2(n-2)$ and $n_2 = (n-2)(n-3)/2$.

Table 2.4.3
A PBIBD(2) of L_3 type

(a)					(b)														
1a	2b	3c	4d	5e	1	2	3	4	5	1	6	11	16	21	1	7	13	19	25
6e	7a	8b	9c	10d	6	7	8	9	10	2	7	12	17	22	2	8	14	20	21
11d	12e	13a	14b	15c	11	12	13	14	15	3	8	13	18	23	3	9	15	16	22
16c	17d	18e	19a	20b	16	17	18	19	20	4	9	14	19	24	4	10	11	17	23
21b	22c	23d	24e	25a	21	22	23	24	25	5	10	15	20	25	5	6	12	18	24

The design in Table 2.4.2 is a PBIBD(2) based on the triangular association scheme. The array of treatments is given in Table 2.4.2(a) and the blocks of the design are the rows of Table 2.4.2(b). Observe that $\lambda_1 = 1$ and $\lambda_2 = 2$.

A generalised triangular scheme is known as a Johnson scheme (Bailey, 1985).

The *Latin-square-type (or L_i-type) association scheme* is related to the lattice designs described in Subsection 2.3. There are $v = s^2$ treatments arranged in a square array. There are also $i - 2$ MOLS of order s. Then first associates appear in the same row, the same column or in a cell with the same symbol when any of the Latin squares are superimposed on the square array. Hence $n_1 = i(n-1)$ and $n_2 = (n-1)(n-i+1)$ for $i \leqslant n - 1$. If $i = n$ then the scheme is a group divisible one and if $i = n + 1$ then all varieties are first associates of each other and so the PBIBD will in fact be a BIBD.

Lattice designs are examples of L_2-type designs. An example of a design with the L_3-type scheme appears in Table 2.4.3. There are $v = 25$ treatments and $b = 15$ blocks. The association scheme is given in Table 2.4.3(a); treatments in the same row, or column, or in cells with the same letter are first associates. Thus the first associates of 1, for example, are 2, 3, 4, 5, 6, 11, 16, 21, 7, 13, 19 and 25. The blocks are given in Table 2.4.3(b). Observe that $\lambda_1 = 1$ and $\lambda_2 = 0$. This is design LS52 from Clatworthy's (1973) tables.

Hamming schemes are generalisations of the Latin-square-type association scheme (Bailey, 1985).

The *cyclic association scheme* is an extension of the difference sets defined in Subsection 2.1. Let $\{1, 2, \ldots, v-1\} = \{d_1, \ldots, d_{n_1}\} \cup \{e_1, \ldots, e_{n_2}\} = D \cup E$, where $n_1 + n_2 = v - 1$. Then a non-group divisible association scheme defined on Z_v is *cyclic* if $D = -D$ and if the non-zero differences of distinct elements of D contain each element of D an equal number of times and each element of E an equal number of times. The first associates of i are $i + D$.

For example, let $v = 17$ and let $D = \{3, 5, 6, 7, 10, 11, 12, 14\}$. Then the non-zero differences of distinct elements of D contain each element of D 3 times and each element of E 4 times. The sets $\{0, 1, 4, 5\}$ and $\{1, 8, 10, 16\}$ are the starter blocks of a PBIBD(2) with a cyclic association scheme with $v = 17$, $k = 4$, $\lambda_1 = 1$ and $\lambda_2 = 2$, as the elements of D arise once each as differences from the starter blocks and those of E twice.

2.5. Comparing block designs

Often several essentially different designs of the same size can be found. Is there any sensible way to decide which design is to be preferred? We have suggested already that resolvability is a desirable feature. Another useful criterion is how well treatment differences are estimated. One reasonable measure of good estimation is small variance. Ways to compare designs based on this idea include the pairwise efficiency factors of the design and the optimality of the design. We now develop the necessary notation to define these terms for designs in which every treatment difference can be estimated. Such a design is said to be *connected.*

In a completely randomised design with each treatment replicated r times the variance of any treatment difference is $2\sigma^2/r$. The *pairwise efficiency factor* is the ratio of $2\sigma^2/r$ to $\text{Var}(\widehat{\tau}_i - \widehat{\tau}_j)$. The ratio of $2\sigma^2/r$ to the average variance of the treatment differences is the *average efficiency factor.*

To calculate these estimates, and hence the corresponding estimated variances, we need to consider the appropriate linear model. Suppose we let Y_{ij} be the response observed in block j on the plot receiving treatment i (if treatment i is applied in block j, of course). We write $\text{E}(Y_{ij}) = \mu + \tau_i + \beta_j$, where we have assumed that the treatment and block effects do not interact. If, in fact, there is interaction present then the analysis may lead one to draw incorrect conclusions; interpretation is certainly more difficult. If interaction is suspected, it is sometimes possible to use a test for non-additivity to determine if interaction is present; see Preece (1983) and Read (1988). If non-additivity is found, Read suggests that a transformation to remove the non-additivity be applied. We then need to solve the normal equations associated with this model to obtain estimates of the treatment effects. Let N be a $v \times b$ treatment-block incidence matrix, with the (i, j) entry being the number of times that treatment i is in block j, A, the *information matrix*, be given by $A = rI - NN^T/k$, $\boldsymbol{q} = \boldsymbol{T} - N\boldsymbol{B}/k$, $\boldsymbol{T}$ be the vector of treatment totals and $\boldsymbol{B}$ be the vector of block totals. Then solving the normal equations gives (see John, 1987; or Johnson and Leone, 1977, for example) $A\widehat{\boldsymbol{\tau}} = \boldsymbol{q}$. These expressions are correct for any incomplete block design in which all treatments are replicated equally often and all blocks are the same size. Otherwise we modify them as follows. Let R be a diagonal matrix with the treatment replication numbers on the diagonal and let K be a matrix with the block sizes on the diagonal. Then in general $A = R - NK^{-1}N^T$ and $\boldsymbol{q} = \boldsymbol{T} - NK^{-1}\boldsymbol{B}$.

In either case A is not of full rank. Indeed it can be shown that a design is connected if and only if the rank of A is $v - 1$. To get an expression for $\widehat{\tau}$ we need to calculate a generalised inverse of A. A *generalised inverse* of A is a matrix G such that $AGA = A$. One way to get G is to calculate the eigenvalues and corresponding eigenvectors of A. Suppose that ν_i is an eigenvalue of A with corresponding eigenvector $\boldsymbol{x}_i$. These eigenvectors can be normalised so that $\boldsymbol{x}_i^T\boldsymbol{x}_j = 1$ unless $i = j$ when it equals 1. Then we can write A in *canonical form* as $A = \sum_i \nu_i \boldsymbol{x}_i \boldsymbol{x}_i^T$ and a generalised inverse of A is $G = \sum_i \nu_i^{-1} \boldsymbol{x}_i \boldsymbol{x}_i^T$, where this summation is over the non-zero eigenvalues only. Then $\widehat{\boldsymbol{\tau}} = G\boldsymbol{q}$ and $\text{Var}(\widehat{\boldsymbol{\tau}}) = \sigma^2 G$.

In the case of a BIBD it is possible to evaluate the entries in G in general. We know that the entries in NN^T are the number of times that treatments i and j appear together in the same block. Thus $NN^T = (r - -\lambda)I + \lambda J$. Substituting we have that $A = \lambda v/k(I - J/v)$. Now $(I - J/v)^2 = (I - J/v)$ and so $G = k/\lambda v(I - J/v)$. Hence all estimates of treatment differences have the same variance, which is $2\sigma^2 k/\lambda v$. Thus each pairwise efficiency factor is $\lambda v/rk$ and hence so is the average efficiency factor. Note that this expression depends only on the parameters of the BIBD and not on which particular BIBD with those parameters we have.

For PBIBD(2)s, a similar result can be obtained. In this case there are two pairwise efficiency factors, one for comparing first associates and one for comparing second associates. To give these factors, we first need to define a number of constants. For any pair of first associates, say x and y, let f be the number of first associates of x that are also second associates of y and for any pair of second associates, say x and y, let g be the number of treatments that are first associates of x and second associates of y. From the definition of an association scheme, these numbers are constant, independent of the particular pair of first or second associates chosen. Let $a = r(k-1)$ and define d by $k^2 d = (a + \lambda_1)(a + \lambda_2) + (\lambda_1 - \lambda_2)\{a(f - g) + f\lambda_2 - g\lambda_1\}$. Then define c_i by $kdc_i = \lambda_i(a + \lambda_{i+1}) + (\lambda_1 - \lambda_2)(f\lambda_2 - g\lambda_1)$, $i = 1, 2$ (subscript addition modulo 2). Then the pairwise efficiency factor for comparing two ith associates is $(k-1)/(k-c_i)$. For every design in Clatworthy (1973), these values are given.

For example, consider the design for Table 2.4.1. For any pair of first associates, there are 6 treatments that are first associates of one and second associates of the other. So $f = 6$. As it happens, for any pair of second associates there are also 6 treatments that are first associates of one and second associates of the other. So $g = 6$. Now $a = r(k-1) = 3(5-1) = 12$ and d is defined by $25d = (12+1)(12+0) + (1-0)\{12(6-6) + 6(0) - 6(1)\} = 150$. Then $5dc_1 = 1(12+0) + (1-0)(6(0) - 6(1)) = 6$. Thus $c_1 = 1/5$. Similarly $c_2 = -1/5$. The pairwise efficiency factors are then $0.83 = (5-1)/(5-0.2)$ and 0.77 respectively.

For a PBIBD(2) the average efficiency factor is $(v-1)/\sum_i (r/(r - \theta_i/k))\alpha_i$, where $\theta_i = r - \{(\lambda_1 - \lambda_2)[-(g-f) + (-1)^i\sqrt{((g-f)^2 + 2(f+g) + 1)}] + (\lambda_1 + \lambda_2)\}/2$ and $\alpha_i = (v-1)/2 + (-1)^i[(n_1 - n_2) + (g - f)(v-1)]/2\sqrt{((g-f)^2 + 2(f+g) + 1)}$, $i = 1, 2$.

For the PBIBD of Table 2.4.3, we have that $\theta_i = 3 - \{(1-0)[-(6-6) + (-1)^i\sqrt{((6-6)^2 + 2(6+6) + 1)}] + (1+0)\}/2$ and so $\theta_1 = 5$ and $\theta_2 = 0$. Also $\alpha_i = (25-1)/2 + (-1)^i[(12-12) + (6-6)(25-1)]/2\sqrt{((6-6^2 + 2(6+6) + 1)} = 12$, $i = 1, 2$. Thus the average efficiency factor is $24/\{12(3/(3-1)) + 12(3/(3-0))\} = 24/30 = 0.8$.

The average efficiency factor for a simple lattice is $(k+1)/(k+3)$, for a triple lattice is $(2k+2)/(2k+5)$ and for a balanced lattice is $k/(k+1)$ (see John, 1987).

The optimality of the design is based on the canonical efficiency factors of the design. Consider $\text{Var}(\boldsymbol{x}_i^T\hat{\boldsymbol{\tau}}) = \boldsymbol{x}_i^T G \boldsymbol{x}_i \sigma^2 = \sigma^2/\nu_i$. In a completely randomised design $\text{Var}(\boldsymbol{x}_i^T\hat{\boldsymbol{\tau}}) = (\boldsymbol{x}_i^T I \boldsymbol{x}_i/r)\sigma^2 = \sigma^2/r$. The *efficiency factor* of the contrast $\boldsymbol{x}_i^T\hat{\boldsymbol{\tau}}$ is ν_i/r. The values ν_i/r are called the *canonical efficiency factors* of the design with information matrix A.

For a BIBD the canonical efficiency factors are $\lambda v/kr$; for a PBIBD(2) they are $(1 - \theta_i/rk)$, with multiplicities α_i, $i = 1, 2$; for lattice designs they are $(r-1)/r$ and 1, with multiplicities $r(k-1)$ and $v - b + r - 1$ respectively; for rectangular lattices with $h = 1$, they are $(r(k-1)-k)/r(k-1)$, $k(r-1)/r(k-1)$ and 1 with multiplicities $(k-1)(r-1)$, $k-1$ and $v-b+r-1$, respectively (see John, 1987).

Various optimality criteria have been proposed. The most common are A-, D-, E- and (M,S)-optimality. Consider a set of possible designs for v treatments all with block size k. A design in the set is said to be *A-optimal* if it has minimum $\mathrm{tr}(G) = \sum_i (1/\nu_i)$; that is, if the sum of variance of the parameter estimates is minimised. It is said to be *D-optimal* if it has minimum $\det(G) = \prod_i (1/\nu_i)$; that is, if the generalised variance of the parameter estimates is minimised. It is said to be *E-optimal* if the maximum eigenvalue of G is a minimum; that is, choose the design with the minimum (largest value of $1/\nu_i$). An E-optimal design minimises the variance of the least well-estimated normalised contrast. *(M,S)-optimality* is a two-stage criterion based on the idea that the canonical efficiency factors should be as close to equal as possible. So at the first stage a class of designs which maximises $\sum_i \nu_i/r$ is formed. At the second stage the design, or designs, within this class which minimise $\sum_i (\nu_i/r)^2$ are obtained. These designs are the (M,S)-optimal designs. (Note that in all cases these expressions are in terms of the eigenvalues of the information matrix A and so do not depend on the particular generalised inverse chosen.)

When the canonical efficiency factors are all equal then the design is optimal under all four criteria. Hence BIBDs are A-, D-, E- and (M,S)-optimal.

Algorithms, such as GENDEX (1993) and that in Paterson et al. (1988) for constructing α-designs, look for designs which are close to the maximum possible average efficiency. See Jarrett (1989) for bounds on the average efficiency of pairwise comparisons of various designs. The tables by Clatworthy (1973) give the pairwise and average efficiency factors for the various PBIBDs that are tabulated.

2.6. Tables and algorithms

Programs to generate incomplete block designs are available. They include GENDEX (1993) for any incomplete block design and one given in Paterson et al. (1988) for α designs. Tjur (1993) gives an algorithm to find optimal block designs.

Tables of incomplete block designs exist. They include Colbourn and Dinitz (1996) for balanced incomplete block designs and for lattice designs (by selecting some resolution classes from appropriate balanced incomplete block designs as described in Subsection 2.2), Cochran and Cox (1957) for lattice designs as such, Patterson and Williams (1976) for α designs and Clatworthy (1973) for partially balanced incomplete block designs.

3. Row–column designs

Incomplete block designs allow for one extraneous source of variation, termed the blocks, in an experiment. Row–column designs and nested row–column designs can be used to allow for more than one extraneous source of variation, as was pointed out first by Fisher (1926) although earlier sporadic examples exist, most notably de

Palluel (1788) (see also Street and Street, 1988). One example of a recently conducted experiment using a row–column design is that of Azam-Ali et al. (1990).

See Subsection 4.2 for a discussion of row–column designs as cross-over designs. See Subsection 5.2 for row–column designs used as layouts with balanced adjacencies in both row and column directions.

The most familiar row–column designs are the Latin squares but these require that the number of treatments, the number of rows and the number of columns all be equal. This requirement is often unnecessarily restrictive and so more general row–column designs have been developed.

In a *row–column design* (RCD) there are v treatments laid out in b blocks with each block having p rows and q columns. We say that the row–column design is *nested* if $b > 1$. We discuss Latin squares, row–column α-designs, nested row–column designs and other related designs, and the available algorithms and tabulations below. We begin by discussing the linear model for a row–column design so that we can define the appropriate optimality values and efficiency factors.

3.1. Optimality and efficiency

The exposition below is based on John (1987) and the reader is referred there for more details.

Assume that treatment m is allocated to the plot in row j and column h of block i and suppose that the observed response is y_{ijhm}. Then the model is

$$\begin{aligned}\mathrm{E}(y_{ijhm}) &= \mu + \beta_i + \rho_{ij} + \kappa_{ih} + \tau_m,\\ &\quad i = 1,\ldots,b;\; j = 1,\ldots,p;\; h = 1,\ldots,q \text{ and } m = 1,\ldots,v,\end{aligned}$$

where β_i is the effect of block i, ρ_{ij} is the effect of the jth row in the ith block, κ_{ih} is the effect of the hth column in the ith block and τ_m is the effect of the mth treatment.

Notice that we are assuming that the block, row within block and column within block effects are additive and do not interact, and that none of these effects interact with the treatment effects. See Read (1988) for a description of tests to detect non-additivity and for some discussion of selecting transformations to remove it if found.

Let N_b be the treatment-block incidence matrix (ignoring rows and columns), let N_p be the treatment-row incidence matrix (ignoring blocks and columns) and let N_q be the treatment-column incidence matrix (ignoring blocks and rows). Let A_b, A_p and A_q be the corresponding information matrices. Then $A\hat{\boldsymbol{\tau}} = \boldsymbol{q}$ where it can be shown that $A = A_p + A_q - A_b$ and $\boldsymbol{q} = \boldsymbol{q}_b + \boldsymbol{q}_p + \boldsymbol{q}_q$.

Given A, the canonical efficiency factors, and the various optimality values, can be calculated for a row–column design, nested or not, exactly as described in Subsection 2.5 for incomplete block designs. The average efficiency factor can also be calculated; bounds for these are given in John and Street (1992).

On occasion the matrices A_b, A_p and A_q will have the same eigenvectors. Then the canonical efficiency factor, e_j, can be expressed in terms of the canonical efficiency

Table 3.1.1
A disconnected row–column design

(a)	(b)
$\begin{matrix} 0\,0\,2\,3 \\ 0\,0\,3\,2 \\ 2\,3\,1\,1 \\ 3\,2\,1\,1 \end{matrix}$	$N_p = N_q = \begin{matrix} 2\,0\,1\,1 \\ 2\,0\,1\,1 \\ 0\,2\,1\,1 \\ 0\,2\,1\,1 \end{matrix} \quad N_b = \begin{matrix} 4 \\ 4 \\ 4 \\ 4 \end{matrix}$

(c)	(d)
$A_q = A_p = \begin{matrix} 2.5 & -1.5 & -0.5 & -0.5 \\ -1.5 & 2.5 & -0.5 & -0.5 \\ -0.5 & -0.5 & 2.5 & -1.5 \\ -0.5 & -0.5 & -1.5 & 2.5 \end{matrix}$	$A_b = 4I - J \quad A = \begin{matrix} 2 & -2 & 0 & 0 \\ -2 & 2 & 0 & 0 \\ 0 & 0 & 2 & -2 \\ 0 & 0 & -2 & 2 \end{matrix}$

factors of the row, column and block designs as $e_j = e_{pj} + e_{qj} - e_{bj}$. This happens if the component designs are complete block designs or BIBDs or are derived from difference sets or sets of supplementary difference sets.

Some designs have estimates of treatment effects which are the same as the ones obtained if there are no block, row or column terms in the model. Such designs are said to be *orthogonal*. Let $\boldsymbol{r}$ be the vector of treatment replications. Indeed the requirement for orthogonality is that $N_b = \boldsymbol{r}\boldsymbol{j}^T/b$, $N_p = \boldsymbol{r}\boldsymbol{j}^T/pb$, and $N_q = \boldsymbol{r}\boldsymbol{j}^T/qb$, where $\boldsymbol{j}$ is a vector of the appropriate size with every entry 1.

Finally it is important to realise that the component designs may have nicer properties than the row–column design has. The design in Table 3.1.1(a) is given by Pearce (1975). There are $v = 4$ treatments arranged in $b = 1$ blocks with $p = 4$ rows and $q = 4$ columns. The matrices N_b, N_p and N_q are given in Table 3.1.1(b) and the matrices A_b, A_p and A_q in Table 3.1.1(c). These matrices are all of rank $v - 1 = 3$. The matrix $A = A_p + A_q - A_b$ is given in Table 3.1.1(d) and is of rank 2. Hence all three of the component designs are connected but the row–column design is not.

3.2. *Latin square designs*

Recall from Subsection 1.2 that a *Latin square of order* k is a $k \times k$ array of the numbers $1, 2, \ldots, k$ arranged such that each symbol appears exactly once in each row of the square and exactly once in each column of the square.

When used in field trials, a Latin square typically has the rows and columns as strips of land at right angles to each other and the plots are the intersections of the strips. The strips in the two directions need not be of the same size and so the whole design need not be square. As we have seen in the previous subsection, in this type of application the model of the design has parameters for the rows, the columns and the treatments and the corresponding effects are assumed to be additive.

Latin squares are orthogonal designs in the sense of the previous subsection and this property is still true of larger designs obtained by placing one, or more, Latin squares side-by-side, thereby increasing the number of rows, or columns or both. The canonical efficiency factors of a Latin square are all 1.

One useful generalisation of a Latin square is the concept of a *frequency square* or F *square*, due to MacMahon (1898, referenced in Denes and Keedwell, 1991). In these $k \times k$ squares there are $n < k$ symbols and the ith symbol appears μ_i times in each row and column of the square. Hence F squares are still orthogonal. These squares have been used to provide the layout for a second set of treatments added as an afterthought to a 10×10 orchard trial (Kirton and Seberry, 1978). The original treatments are then a blocking factor for the second set of treatments and this second set need not have the same number of treatments as the original set.

Further theoretical results on Latin squares and MOLS can be found in Denes and Keedwell (1991) and Street and Street (1987).

3.3. Row–column designs with $p < v$ and $q = v$

In these designs the rows will have one copy of each treatment and the columns will be incomplete blocks.

Here we see that $A = A_p + A_q - A_b = A_q$ since $A_p = A_b = pI - (p/v)J$. Hence an A-optimal row–column design is obtained from an A-optimal column component design, provided that the blocks of the column component design can be rearranged so that the rows of the design have one copy of each treatment. For example, the design in Table 3.3.1 is a 3×7 row–column design obtained from the $(7, 3, 1)$ BIBD of Subsection 2.1.

Youden (1937; referenced in Freeman, 1988b) showed how to rearrange any BIBD with the same number of blocks as treatments as a row–column design (now called *Youden squares*) thereby generalising a result in Yates (1936) who gave an example of a row–column design obtained by deleting one row from a Latin square design. Kiefer (1975) extended the idea of Youden squares to generalised Youden designs; a tabulation appears in Ash (1981). Other work on Youden designs may be found in Preece (1994) and references cited therein.

Some of the PBIBDs tabulated in Clatworthy (1973) can be used as row–column designs of this type. The designs in this tabulation are always written appropriately when this is possible.

On occasion α-designs can be rearranged suitably but each design needs to be considered individually as no tabulations are available.

Table 3.3.1
A 3×7 row–column design

0	1	2	3	4	5	6
1	2	3	4	5	6	0
3	4	5	6	0	1	2

If neither p nor q is divisible by v then no general results are available but row–column α-designs are examples of designs that may be useful. They were introduced by John and Eccleston (1986) who give some ideas for constructing these designs.

3.4. Related designs

A pair of Latin squares is said to be *orthogonal* if, when superimposed, each ordered pair of treatments appears exactly once. A set of Latin squares is said to be *mutually orthogonal* if any two squares in the set are orthogonal. We often write *MOLS* for a set of mutually orthogonal Latin squares. A pair of MOLS can be used to allow for a third blocking factor. In a trial on orchard trees the trees may be used in experiments for which they were not originally planned. Then the treatments of the original experiment become a blocking factor in the new experiment (see Kirton and Seberry, 1978; Pearce, 1983). So the blocks are the rows, the columns and the letters from the first square and the treatments are the letters of the second square.

Here we will give one well-known construction for a complete set of MOLS for prime power order.

THEOREM (Raghavarao, 1971, Section 1.3). *Let* $n = p^k$ *for some prime* p. *In the field* $GF[n]$, *let the elements be labelled* $f_0 = 0$, $f_1 = 1$, $f_2, \ldots, f_{n-1}$, *in some arbitrary, but fixed, order. Let the* $n \times n$ *arrays* $A_1, A_2, \ldots, A_{n-1}$ *be defined by* $(A_m)_{ij} = [f_m f_i + f_j]$, $i, j = 0, 1, \ldots, n-1$, $m = 1, \ldots, n-1$. *Then* $A_1, A_2, \ldots, A_{n-1}$ *are a set of* $n-1$ *MOLS of order* n.

For instance, to construct a set of four MOLS of order 5, let $f_i = i$, $i = 0, 1, \ldots, 4$, since the field of order 5 is just the integers mod 5. Then $A_1 = [f_i + f_j]$, $A_2 = [2f_i + f_j]$, $A_3 = [3f_i + f_j]$, and $A_4 = [4f_i + f_j]$. The resulting squares are in Table 3.4.1.

On occasion one, or more, cells of the array may be left empty. Then the rows and columns are said to have a *non-orthogonal structure* and the question arises as to the optimal structure for such designs. Shah and Sinha (1993) have developed some designs which balance the occurrence of each of the treatments as far as possible in both rows and columns; this is often referred to as combinatorial balance. Kunert (1993) has shown that these conditions are more restrictive than necessary and develops some conditions for optimal designs based on the A matrix.

Another area of development has been the investigation of the structure of row–column designs when the rows may be of different sizes, as may the columns. Some theoretical results and constructions can be found in Jacroux and Ray (1991).

Table 3.4.1
Four MOLS of order 5

	0 1 2 3 4		0 1 2 3 4		0 1 2 3 4		0 1 2 3 4
	1 2 3 4 0		2 3 4 0 1		3 4 0 1 2		4 0 1 2 3
$A_1 =$	2 3 4 0 1	$A_2 =$	4 0 1 2 3	$A_3 =$	1 2 3 4 0	$A_4 =$	3 4 0 1 2
	3 4 0 1 2		1 2 3 4 0		4 0 1 2 3		2 3 4 0 1
	4 0 1 2 3		3 4 0 1 2		2 3 4 0 1		1 2 3 4 0

3.5. *Tables and algorithms*

Algorithms to construct row–column designs include Venables and Eccleston (1993), Nguyen and Williams (1993) and GENDEX (1993). Tables of Latin squares of different orders can be found in Fisher and Yates (1963; orders 4, 5 and 6) and in Denes and Keedwell (1974; all non-isomorphic squares of orders 3, 4, 5 and 6). Tables giving sets of MOLS have been provided by Colbourn and Dinitz (1996).

The algorithms will provide a random layout. For designs obtained from tables, randomisation of row–column designs is similar to that for block designs. The order of the rows and columns in the actual layout is determined by randomly permuting the rows and the columns of the initial RCD. The initial RCD should be chosen at random from all RCDs of a particular size. The rows and columns are then randomly permuted before the actual layout to use is obtained. Sometimes smaller sets of RCDs provide a valid set from which to choose. Such sets are called valid randomisation sets; see Bailey (1986, 1987).

4. Repeated measurements

The term 'repeated measurements' has come to mean rather different things to different groups over the course of time. Koch et al. (1988) take the view that the essential feature of a repeated measurements study is that each experimental unit is observed under at least two conditions. They then define four general classes of repeated measurements design – the split-plot design, the cross-over (or change-over) design, longitudinal studies and studies to determine sources of variability. We discuss split-plot designs (briefly) and cross-over designs below. Row–column designs in general are discussed in Section 3. Designs with neighbour balance in both rows and columns are discussed in Section 5.

4.1. *Split-plot designs*

In a split-plot design there are at least two independent stages of randomisation. In the first stage of a typical field experiment fertilisers are randomly assigned to fields (whole plots) and in the second stage varieties are randomly assigned to sub-plots within each field. The sub-plots are the observational units. Examples of recently conducted split-plots experiments include Ishag and Ageeh (1991) and Fukai et al. (1991).

Split-plot designs are used when it is necessary to have some treatment factors applied to rather large units but other factors can be applied to smaller units. This means that the comparisons of the factor applied to whole plots is not made with the same precision as are comparisons of the factor applied to the split-plots. Spilt-plot designs have been the subject of much discussion because of the chance for confusion in the analysis of the split-plot design due to the double randomisation that is involved. It is important to realise that the layout of the whole-plots and of the split-plots are done independently. They have been well described by Federer (1975).

4.2. Cross-over designs

A *cross-over design* (or *change-over design*) (COD) is a design in which t treatments are compared using n experimental units and the experiment lasts for p periods. During the course of the experiment, each unit receives a sequence of p, not necessarily distinct, treatments and a response is recorded for each unit in each period. These designs are commonly written as a $p \times n$ array with entries from $\{1, 2, \ldots, t\}$. Thus any row–column design with $b = 1$ could be used as the layout of a COD (where now $q = n$). CODs were used by Khalili et al. (1992) and Khalili (1993).

Two linear models have been investigated extensively in connection with CODs. The first assumes that all responses are independent, but that there is a carry-over effect from one period to the next, the second that responses on the same unit are correlated. The second model can be transformed to the first but we will discuss optimal designs for both explicitly below.

A model with carryover effects is not appropriate for field trials as there are likely to be effects of any treatment on all neighbouring plots; see Section 5. Carryover effects are appropriate when the responses are measured in a time order, for instance. Sometimes one model may contain both terms; see Martin (1995).

Let y_{ij} be the response observed on unit i in period j. In the first model we have that

$$Y_{ij} = \alpha_i + \beta_j + \tau_{d[i,j]} + \rho_{d[i,j-1]} + E_{ij},$$

where α_i is the effect of unit i, $i = 1, \ldots, n$; β_j is the effect of period j, $j = 1, \ldots, p$; $\tau_{d[i,j]}$ is the effect of the treatment allocated to unit i in period j (and $1 \leqslant d[i,j] \leqslant 1$) and $\rho_{d[i,j-1]}$ is the residual effect of the treatment applied to unit i in period $j - 1$ to the response observed in period j (with $\rho_{d[i,0]} = 0$). The responses are assumed to be independent with the residual treatment term accounting for any effect of using units more than once.

In the second model we have that

$$Y_{ij} = \alpha_i + \beta_j + \tau_{d[i,j]} + E_{ij},$$

where the parameters in the model are the same as before but we assume now that the responses on the same unit are correlated with $\mathrm{Cov}(E_{ij}, E_{ih}) = \sigma^2 \rho^{|j-h|}/(1 - \rho^2)$ and responses on different units are uncorrelated. It is possible to transform the errors so that they are independent. If $\boldsymbol{Y} = (Y_{11}, \ldots, Y_{1p}, Y_{21}, \ldots, Y_{2p}, \ldots, Y_{np})^T$ then the model can be written as

$$\boldsymbol{Y} = X\theta + \boldsymbol{E}$$

and the variance matrix may be written as $\sigma^2 I_n \otimes W$, where $W = (w_{jh}) = (\rho^{|j-h|}/(1 - \rho^2))$. There is a matrix Z such that $ZWZ^T = I_p$. Then $(I_n \otimes Z)(I_n \otimes W)(I_n \otimes Z^T) = I_n \otimes I_p = I_{np}$. Hence we can write the model as

$$(I_n \otimes Z)\boldsymbol{Y} = (I_n \otimes Z)X\theta + (I_n \otimes Z)\boldsymbol{E},$$

Table 4.2.1
A uniform, balanced COD with $t = p = n = 6$

1	2	3	4	5	6
2	3	4	5	6	1
6	1	2	3	4	5
3	4	5	6	1	2
5	6	1	2	3	4
4	5	6	1	2	3

and the new error terms are independent. The estimates that result from this approach are weighted least squares estimates.

As with block designs, it is possible to evaluate the information matrices associated with a COD and hence determine the structure of optimal CODs. Note that for any model there are two information matrices involved here. One is the information matrix for the estimation of direct treatment effects and the other is the information matrix for the estimation of residual treatment effects. Expressions for these matrices may be found in Cheng and Wu (1980), Kunert (1983) or Street (1989), for instance. Here we merely summarise the results about the structure of optimal CODs. To do this we need some more definitions.

A design is said to be *uniform on the rows* (or periods) (*columns* (or units)) if each treatment appears equally often in each row (column) of the array. A design is said to be *uniform* if it is uniform on both rows and columns. Thus a design which is uniform on rows must have n a multiple of t (we write $t|n$) and one which is uniform on columns has $t|p$. A design is said to be *balanced* if each treatment precedes every other equally often (and hence $n(p-1)/t(t-1)$ times) and to be *strongly balanced* if every treatment precedes every treatment, including itself, equally often (and hence $n(p-1)/t^2$ times). The design in Table 4.2.1 is a uniform, balanced design and is an example of a Latin square of order 6. Adjoining a seventh row, equal to the sixth, would give a design that is strongly balanced and uniform on the rows but not on the columns.

Cheng and Wu (1980) have shown that the universally optimal CODs for estimating both direct, and residual, treatment effects under the first model above, are the strongly balanced, uniform CODs. The next result gives their construction for such designs. As before, we let the integers modulo t, $\{0, 1, 2, \ldots, t-1\}$, be represented by Z_t.

CONSTRUCTION 4.1 (Cheng and Wu, 1980). Let A be a $2 \times t^2$ array in which each of the t^2 ordered pairs appears as a column exactly once and such that the rows of A are uniform. Construct t further arrays by adding, in turn, each of the integers in Z_t to the entries in A. Then these t arrays, placed one above the other, give a strongly balanced, uniform COD with $p = 2t$ and $n = t^2$.

The design for 3 treatments tested on 9 units over 6 periods given in Table 4.2.2 was constructed in this way.

Table 4.2.2
A strongly balanced, uniform COD with $t = 3$, $p = 6$ and $n = 9$

0	0	0	1	1	1	2	2	2
0	1	2	0	1	2	0	1	2
1	1	1	2	2	2	0	0	0
1	2	0	1	2	0	1	2	0
2	2	2	0	0	0	1	1	1
2	0	1	2	0	1	2	0	1

Table 4.2.3
A strongly balanced, uniform COD with $t = 3$, $p = n = 9$ and its component arrays

$$H_3 = \begin{matrix} 1\,1\,1\,2\,2\,2\,3\,3\,3 \\ 3\,2\,1\,1\,3\,2\,2\,1\,3 \\ 2\,3\,1\,3\,1\,2\,1\,2\,3 \end{matrix} \qquad H_1 = \begin{matrix} 2\,2\,2\,3\,3\,3\,1\,1\,1 \\ 1\,3\,2\,2\,1\,3\,3\,2\,1 \\ 3\,1\,2\,1\,2\,3\,2\,3\,1 \end{matrix} \qquad H_2 = \begin{matrix} 3\,3\,3\,1\,1\,1\,2\,2\,2 \\ 2\,1\,3\,3\,2\,1\,1\,3\,2 \\ 1\,2\,3\,2\,3\,1\,3\,1\,2 \end{matrix}$$

1	3	2	2	1	3	3	2	1
1	2	3	2	3	1	3	1	2
1	1	1	2	2	2	3	3	3
2	1	3	3	2	1	1	3	2
2	3	1	3	1	2	1	2	3
2	2	2	3	3	3	1	1	1
3	2	1	1	3	2	2	1	3
3	1	2	1	2	3	2	3	1
3	3	3	1	1	1	2	2	2

The designs of Construction 1 can be extended both vertically and horizontally. Hence there is a strongly balanced, uniform COD with $p = 2\lambda t$ and n any multiple of t^2. Sen and Mukerjee (1987) have given a construction for strongly balanced, uniform designs when the number of periods is an odd multiple of t and $p \geqslant 3t$. (Recall that a construction for MOLS of prime and prime-power order is given in Subsection 3.4.)

CONSTRUCTION 4.2 (Sen and Mukerjee, 1987). Let L and N be two MOLS of order t. Let the ith column of L be L_i and the ith column of N be N_i and let $G_h = [L_h\ N_h\ hj]$, $1 \leqslant h \leqslant t$. Let $G = [G_1, G_2, \ldots, G_t]$ and $H_i = G + i \pmod t$ where this addition is performed on each entry of the matrix. Then the strongly balanced, uniform COD is $[H_1^T, H_2^T, \ldots, H_t^T]$.

The Latin squares of order 3 in Table 2.1.6(a) give the arrays in Table 4.2.3.

Balanced uniform designs can be smaller than strongly balanced uniform designs and these designs are universally optimal for the estimation of direct treatment effects when no treatment may be applied in successive periods and the designs are uniform on the units and the last period. The classic example is the balanced Latin square; it is usually called a *column-complete* Latin square in the combinatorial literature. If each treatment need only be adjacent to every other equally often (without regard to order), as might be the case in a field trial, for instance, then the squares are said to *column-quasi-complete* (or *partially balanced*). A similar nomenclature is used for row–column designs. (Extensions to non-square arrays are described in Subsection 5.2.)

Constructions of column-complete and column-quasi-complete Latin squares have been given by E. J. Williams (1949); extensions to row–column designs have been given by Freeman (1979) and Street (1986). E. J. Williams (1949) construction is particularly easy to describe.

CONSTRUCTION 4.3 (E. J. Williams, 1949).

(a) There is a column-complete Latin square of order $2m$ for every integer $m \geqslant 1$.

(b) There is a column-quasi-complete Latin square of order $2m+1$ for every integer $m \geqslant 1$ and a pair of Latin squares of order $2m+1$ which are together column-complete for every integer $m \geqslant 1$.

PROOF. (a) Let the Latin square be $L = (l_{ij})$, $1 \leqslant i, j \leqslant 2m$ and let the first row and column of L be

$$1\ 2\ 2m\ 3\ 2m-1\ 4\ \ldots\ m\ m+2\ m+1.$$

For $i > 1$, $j > 1$, let $l_{ij} = l_{i1} + l_{1j} - 1 \pmod{2m}$. Clearly L is a Latin square. The sequence of differences from adjacent positions is

$$\begin{aligned}&\{2-1,\ 3-2m,\ 4-(2m-1), \ldots, (m+1)-(m+2),\\ &\quad 2m-2,\ 2m-1-3, \ldots, m+2-m\}\\ &\quad = \{1, 3, 5, \ldots, 2m-1, 2m-2, 2m-4, \ldots, 2\}\end{aligned}$$

which contains each of the non-zero elements modulo $2m$ precisely once. Thus every ordered pair of distinct treatments appears in adjacent positions exactly once in the columns of L (and, as it happens, exactly once in the rows of L).

(b) If the order of the square is odd, then it is possible to construct a column-quasi-complete Latin square using the sequence $1\ 2\ 2m{+}1\ 3\ 2m\ 4\ \ldots\ m\ m{+}3\ m{+}1\ m{+}2$. To construct a pair of squares which are together column-complete use the sequence above and the sequence

$$1\ 2m+1\ 2\ 2m\ 3\ \ldots\ m+3\ m\ m+2\ m+1.$$

These sequences can be checked as in part (a). □

Table 4.2.4 contains a pair of squares of order 5 constructed in this way.

Variants of Williams designs are optimal designs for the second model above, the one with correlated errors. In particular let w_{ij} be the number of times that treatments i and j are adjacent in a uniform COD with $t = p$. If the w_{ij} $(i \neq j)$ are all equal then the COD is said to be a *Williams design*. Suppose that $d = (d_{ij})$ is a Williams design. Let B be a block design with blocks (d_{1j}, d_{pj}), $j = 1, 2, \ldots, n$. B is called the

Table 4.2.4
A pair of column-quasi-complete Latin squares of order 5

1	2	5	3	4	1	5	2	4	3
2	3	1	4	5	5	4	1	3	2
5	1	4	2	3	2	1	3	5	4
3	4	2	5	1	4	3	5	2	1
4	5	3	1	2	3	2	4	1	5

end-pair design. If $t = n$ and if the end-pair design is connected then d is a *Williams design with circular structure*. Either of the squares in Table 4.2.4 is an example of a Williams design with circular structure. If B is a BIBD then d is a *Williams design with balanced end-pairs*.

Kunert (1985) has established various results about the optimality of Williams designs. For example, a Williams design with balanced end-pairs is universally optimal for the estimation of treatment effects over the class of uniform CODs with $t = p$ and is universally optimal for over all CODs if $t \geqslant 4$ and $\rho \geqslant (t-2-\sqrt{(t^2-2)})/(2(t-3))$ and for all ρ if $t = 3$. Other results are summarised in Street (1989).

Gill (1992) has found the optimal designs when the observations are correlated and t divides both n and p if the model is $Y_{ij} = \alpha_i + \beta_j + \tau_{d[i,j]} + \theta Y_{i,j-1} + E_{ij}$, and θ and ρ are both known.

For some cautionary comments about change-over designs see Senn (1988, 1993). For change-over designs when the treatments have a factorial structure see Shing and Hinkelmann (1992) (for n factors all with a prime-power number of levels), Fletcher and John (1985) (who argue that a modification of the usual definition of balance is appropriate to allow for the fact that interaction effects are not as important as main effects) and Fletcher (1987) (who gives generalised cyclic designs for two and three factor experiments with each factor having up to four levels).

5. Neighbour designs for field trials

Designs may need to be balanced for neighbouring treatments in several different applications. We have seen that this is an important consideration in the design of change-over designs (see Subsection 4.2). It is also important in other areas such as in field trials, in designs for studying intergenotypic competition, in bioassay work when using Ouchterlony gel diffusion tests and when designing the collection of seed from seed orchards.

The nearest-neighbour method for the analysis of field trials was originally proposed by Papadakis (1937) as a way of making allowance for local variation in factors such as soil fertility. Because of this variation, responses on adjacent plots tend to be more alike than do responses on plots which are well-separated. Papadakis proposed that, for each response, the response of neighbouring plots be used as a covariate. Variations of this general idea have been proposed and investigated; a thorough survey of the area has been given in Martin (1985).

One-dimensional trials are ones in which the plots tend to be long and narrow and so each plot may be regarded as having only two neighbours, those adjacent on the long sides. Blocks in such designs are often called *linear blocks*. Gleeson and Eccleston (1992) have considered the design of field experiments for one-dimensional field trials and have found that partial neighbour balance (where any two treatments are neighbours zero or one times) is more important than having an incomplete block structure. *Two-dimensional trials*, in which plots are more square and so have four neighbouring plots, have not been investigated in this way but it is assumed that similar results will apply. Gleeson and Cullis (1991) describe over 1000 trials which have been conducted and analysed using the method of Papadakis or a variant and for which neighbour balanced designs are appropriate.

Issues relating to the randomisation of neighbour designs are discussed by Monod and Bailey (1993).

We consider both block designs, and row–column designs, with neighbour balance below.

5.1. Circular block designs

In gel diffusion tests the treatments appear in wells around a circle on an agar plate and so every treatment has two neighbours and the blocks may be written circularly. However circular blocks may be used as a way of representing linear block designs with edge effects where the edge effect can either be allowed for directly in the model or can be eliminated by having border plots; Martin (1985).

A *neighbour design* (or *circular block design*) for v treatments is a layout arranged in b blocks of size k such that each treatment appears r times and such that any two distinct treatments appear as neighbours (that is, in adjacent positions) λ' times. We do not require that the entries in a block be distinct, merely that a treatment never appears as its own neighbour. If a block in a neighbour design is written as $(a_1, a_2, \dots, a_k)$ then we say that the neighbours of a_i are a_{i-1} and a_{i+1}, where the subscripts are calculated modulo k. We will denote a neighbour design by $N[k, \lambda'; v]$. If the design is to be used in a field trial then, to preserve neighbour balance, the linear block used must be $a_k, a_1, a_2, \dots, a_k, a_1$.

An example of a neighbour design with $v = 11$, $k = 5$ and $\lambda' = 1$ is given in Table 5.1.1.

If the blocks are of size two then we say that each block has two pairs of neighbours. Thus by counting plots and pairs of treatments we get that $vr = bk$ and $\lambda'(v-1) = 2r$ for any neighbour design.

Table 5.1.1
An $N[5, 1; 11]$

$(1,2,0,3,7)$	$(2,3,1,4,8)$	$(3,4,2,5,9)$	$(4,5,3,6,10)$	$(5,6,4,7,0)$	$(6,7,5,8,1)$
$(7,8,6,9,2)$	$(8,9,7,10,3)$	$(9,10,8,0,4)$	$(10,0,9,1,5)$	$(0,1,10,2,6)$	

In a series of papers (see Hwang and Lin (1978) and references therein) the necessary conditions were shown to be sufficient for the existence of a neighbour design for all values of v and k by the actual construction of the designs concerned. These have been summarised in Street and Street (1987) and further constructions appear in Preece (1994). Azais et al. (1993) give a catalogue of neighbour designs with border plots.

For designs with relatively few replicates, Wilkinson et al. (1983) suggested that a good layout would have each pair of treatments on adjacent plots at most once and each pair of treatments on plots of distance two apart at most once also. They also have some desirable properties in terms of the intersections of the neighbour lists of the treatments. Design which satisfy these properties have been given by Street and Street (1985) and Azais et al. (1993).

5.2. *Row–column designs*

The properties of complete and quasi-complete Latin squares, defined in Subsection 4.2, have been extended to larger, not necessarily square, arrays by Freeman (1979). He considers $p \times q$ arrays of v treatments in which:

(i) each treatment appears equally often in the array, perhaps further restricted to appear equally often in rows, or columns or both;

(ii) no treatment appears adjacent to itself in either rows or columns;

(iii) the array has either directional balance (each ordered pair of distinct treatments appears adjacent equally often in rows and in columns) or non-directional balance (each pair of distinct treatments appears adjacent equally often in rows and columns considered together).

Thus a complete Latin square has $p = q = v$ and has directional balance while a quasi-complete Latin square has non-directional balance. Table 5.2.1 gives another example of an array with non-directional balance. For instance, the neighbours of treatment 1, in row 1, are 2, 2 and 4. Note that each treatment appears 9 times in the array but that the treatments can not appear equally often in each row (or column) of the array.

On occasion balance is obtained by including some border plots, where varieties are grown but the responses are only used as 'neighbour plots'. Constructions for (non-) directionally balanced designs, with and without borders, have been given by Freeman

Table 5.2.1
A non-directionally balanced design with $p = q = 6$ and $v = 4$

1	2	1	4	3	2
3	1	2	3	4	1
1	3	1	4	2	4
4	2	4	1	3	2
2	4	3	2	1	3
3	1	4	3	2	4

(1979) and Street (1986). Freeman (1988a) has studied nearest neighbour row–column designs for three and four treatments in detail. Morgan (1988, 1990) gives designs with directional balance and he also balances for diagonal neighbours as well. This is important in polycross experimentation for adequate mixing of the genotypes.

Other designs with neighbour balance exist, for instance the serially balanced sequences of R. M. Williams (1956), described in Street and Street (1987), and extended by Monod (1991).

Wild and E. R. Williams (1987) give an algorithm for making neighbour designs from α-arrays.

6. Factorial designs and fractional factorial designs

Factorial designs have been much used in agriculture since their introduction by Fisher in 1923. One recent example may be found in Iremiren, Osara and Okiy (1991).

We consider designs with k treatment factors where factor i has s_i levels, $1 \leqslant i \leqslant k$. We call this design a $s_1 \times s_2 \times \cdots \times s_k$ *factorial design*. If $s_1 = s_2$, say, then we write $s_1^2 \times \cdots \times s_k$. We call any combination of treatment factor levels a *treatment combination* and write it as a k-tuple with the ith entry being from a set of s_i symbols. If all the factors have the same number of levels we say that the factorial design is *symmetrical* (or sometimes *pure*); otherwise we say it is *asymmetrical* (or sometimes *mixed*). A *complete factorial design* has each treatment combination appearing an equal number of times; an *incomplete factorial design* (sometimes called a *fractional factorial design*) has only some of the treatment combinations appearing in it.

An effect is said to be *confounded* if it can not be distinguished from the effect of blocks and two treatment effects that can not be distinguished are said to be *aliased*. The *defining contrasts subgroup* is the set of all contrasts which are confounded with blocks in a factorial design. The choice of the fraction to use is non-trivial, as it is necessary to be sure that no effects of importance are confounded. It is possible to block both complete and fractional factorial designs. Again the allocation of treatment combinations to blocks needs to be done with a view to the confounding, and aliasing, that will result.

The construction of fractional factorial designs, and incomplete blocks for complete or fractional factorial designs, can be done either by using the theory that has been developed by various authors or by making use of published tables and algorithms (see Subsection 6.4).

6.1. Symmetrical prime-power factorials

There is an extensive theoretical development of these designs available. We begin by considering a small example.

Consider a 3^2 factorial design with factors A and B. Then the treatment levels for each factor are represented by the elements of $\{0, 1, 2\}$. Hence we get the 9 treatment combinations $\{(00), (01), (02), (10), (11), (12), (20), (21), (22)\}$. We would like to subdivide the treatment sum of squares into three parts, corresponding to the main effect of each of the factors and the interaction effect of the factors.

To calculate the sum of squares of the main effect of A we need to compare the responses to the three levels of A, independently of the level of B applied. Hence we want to compare the responses of the treatment combinations in the three sets $\{(00),(01),(02)\}, \{(10),(11),(12)\}$ and $\{(20),(21),(22)\}$. In each of these sets the level of A is constant. We write $P(1,0)$ or A to represent these three sets. To calculate the A sum of squares we choose any two orthogonal contrasts on these sets and calculate the corresponding sum of squares. The sum of these two contrast sum of squares is the A sum of squares and it is independent of the particular two orthogonal contrasts chosen. For example we might use the orthogonal contrasts $(-1,0,1)$ and $(-1,2,-1)$. Then, if y_{ij} is the response observed for treatment combination (i,j), the observed A sum of squares is

$$(y_{20}+y_{21}+y_{22}-(y_{00}+y_{01}+y_{02}))^2/2+(2(y_{10}+y_{11}+y_{12}) \\ -(y_{20}+y_{21}+y_{22}+y_{00}+y_{01}+y_{02}))^2/6.$$

We define the main effect of B similarly using the sets $\{(00),(10),(20)\}, \{(01),(11),(21)\}$ and $\{(02),(12),(22)\}$, the entries of $P(0,1)$ or B.

To calculate the interaction sum of squares we must first define the sets of treatment combinations which are the same. We would like to compare the responses in sets in which the pair of levels of A and B are in some sense constant. For instance, one set of three sets could be $\{(00\},(12),(21)\}, \{(01),(10),(22)\}, \{(02),(11),(20)\}$. In these sets the sum of the levels of A and B are always equal. We represent these sets as $P(1,1)$ or AB. Then the other set of sets has $\{(00),(11),(22)\}, \{(01),(12),(20)\}$ and $\{(02),(10),(21)\}$. In these sets the sum of the level of A and twice the level of B is a constant so we represent these by $P(1,2)$ or AB^2. Again we calculate the sum of squares of each partition using a pair of orthogonal contrasts; the interaction sum of squares is the sum of these two partition sum of squares. Observe also that any two sets in each of the four partitions, $P(1,0)$, $P(0,1)$, $P(1,1)$ and $P(1,2)$, of the treatment combinations have only one treatment combination in common.

We now formalise the ideas of the preceding example. To do this we let GF$[s]$ denote the Galois field of order s, where s must be a prime or a prime power. If s is a prime then GF$[s]$ is Z_s, the integers modulo s. If s is a prime power then a more complicated set must be used; see Street and Street (1987), for example, for an elementary account of finite fields.

Consider an s^k factorial design where s is a prime or a prime power. We define the partition $P(1,0,\ldots,0)=\{\{(x_1,\ldots,x_k) \mid x_1=\theta\}, \theta\in \text{GF}[s]\}$. Sometimes we write A for $P(1,0,\ldots,0)$. Each set in the partition has s^{k-1} elements in it and the sum of squares of the main effect of the first factor is calculated by using a set of $s-1$ orthogonal contrasts on the sets in the partition. For the interaction effect of the first two factors, say, we need to calculate $(s-1)$ partitions, $P(1,\alpha,0,\ldots,0)=AB^{\alpha}$, $\alpha\in \text{GF}[s]$, $\alpha\neq 0$, where $P(1,\alpha,0,\ldots,0)=\{\{(x_1,\ldots,x_k) \mid x_1+\alpha x_2=\theta\}, \theta\in \text{GF}[s]\}$. The sum of squares of the interaction effect is calculated by using a set of $s-1$ orthogonal contrasts on the sets in each of the partitions. Higher order interactions are defined similarly. For example, to estimate the interaction effect of the first three factors we need to calculate the $(s-1)^2$ partitions, $P(1,\alpha,\beta,0,\ldots,0) = AB^{\alpha}C^{\beta}$, $\alpha,\beta \in \mathrm{GF}[s]$, $\alpha \neq 0$, $\beta \neq 0$, where $P(1,\alpha,\beta,0,\ldots,0) = \{\{(x_1,\ldots,x_k) \mid x_1+\alpha x_2+\beta x_3 = \theta\}, \theta \in \mathrm{GF}[s]\}$. (Note that $P(\alpha,\beta,\gamma,0,\ldots,0) = P(1,\alpha^{-1}\beta,\alpha^{-1}\gamma,0,\ldots,0)$, as $\alpha x_1 + \beta x_2 + \gamma x_3 = \tau$ if and only if $x_1 + \alpha^{-1}\beta x_2 + \alpha^{-1}\gamma x_3 = \alpha^{-1}\tau$. Hence we may assume without loss of generality that any partition has a 1 as the first non-zero entry.)

When it is necessary to block the treatment combinations in an s^k factorial design, the sensible approach is to use as blocks the sets of one of the partitions corresponding to part of a high order interaction. This would give blocks of size s^{k-1} and the component of the effect which provided the blocks could not be estimated. Hence we say that the effect and the blocks are *confounded.* If the partition corresponding to a main effect is chosen to provide the blocks then it can not be estimated. If one of the partitions corresponding to an interaction effect is chosen to provide the blocks then an estimate of the interaction effect can be obtained, but only from the other partitions. In the 3^2 example, for instance, use the partition $P(1,1) = AB$ to give the three blocks $\{(00)\},(12),(21)\},\{(01),(10),(22)\},\{(02),(11),(20)\}$, each of size 3. Then the interaction effect can only be estimated from $P(1,2) = AB^2$ and so it is estimated with 50% efficiency. More generally if an interaction effect of h factors is chosen to provide the blocks then that interaction effect can be estimated with efficiency $1-(s-1)/(s-1)^h$ (Yates, 1937).

If smaller blocks are required then the usual approach is to use the intersections of sets from two, or more, partitions, as blocks. For example, in a 3^3 design suppose we want to use blocks of size 3. There are 13 partitions associated with this design; one for each of the three main effects, two for each of the three two-factor interactions and four for the three-factor interaction. Suppose we decide to use the intersections of the sets in $P(1,1,1) = ABC$ and $P(1,1,2) = ABC^2$ to be the blocks. Now $P(1,1,1) = \{\{(x_1,x_2,x_3) \mid x_1+x_2+x_3 = \theta\}, \theta = 0,1,2\}$ and $P(1,1,2) = \{\{(x_1,x_2,x_3) \mid x_1+x_2+2x_3 = \theta\}, \theta = 0,1,2\}$. Thus the sets in $P(1,1,1)$ are $\{000,012,021,102,201,120,210,111,222\}$, $\{001,010,022,100,202,121,211,112,220\}$ and $\{002,011,020,101,200,122,212,110,221\}$. The three sets in $P(1,1,2)$ are $\{000,120,210,101,202,112,221,011,022\}$, $\{(001,121,211,102,200,110,222,012,020\}$ and $\{002,122,212,100,201,111,220,010,021\}$. We get the blocks in the design by intersecting, in turn, each set from $P(1,1,1)$ with each set from $P(1,1,2)$. Thus the blocks obtained by intersecting the first set of $P(1,1,1)$ with each of the sets in $P(1,1,2)$ in turn are $\{000,120,210\}$, $\{102,222,012\}$ and $\{201,111,021\}$. The remaining 6 blocks are obtained in the same way from the other two sets of $P(1,1,1)$. The blocks are $\{202,112,022\}$, $\{001,121,211\}$, $\{100,220,010\}$, $\{101,221,011\}$, $\{200,110,020\}$ and $\{002,122,212\}$.

Observe that the intersection always gives a set of size 3. Also note that we have 9 blocks and so the block sum of squares has 8 degrees of freedom. There are 2 degrees of freedom associated with each of $P(1,1,1)$ and $P(1,1,2)$. To identify the other degrees of freedom suppose that (x_1,x_2,x_3) satisfies $x_1+x_2+x_3 = \alpha$ (corresponding to a set in $P(1,1,1)$) and $x_1+x_2+2x_3 = \beta$ (corresponding to a set in $P(1,1,2)$). Then we know that $x_3 = \beta-\alpha$ and $x_1+x_2 = \alpha-\beta+\alpha = 2\alpha-\beta$. Hence all the treatment combinations in a block have a given level of factor C and the sum of the levels of factors A and B is a constant in each block. Hence the main effect of the third factor and the two factor interaction of the first two factors have also been confounded with blocks. We call these effects, that have been confounded as a consequence of the two effects that we have chosen to confound, *generalised interactions.* Note that had we chosen to use the partitions $P(0,0,1)$ and $P(1,1,1)$, say, to obtain the blocks then the generalised interactions would be $x_1 + x_2 + 2x_3 = \alpha + \beta$ and $x_1 + x_2 = \alpha - \beta$, since we know that $x_1 + x_2 + x_3 = \alpha$ and $x_3 = \beta$, say.

Three effects are *independent* of each other if the third effect is not the generalised interaction of the first two. (We have seen that the ordering of the effects is not important.) When constructing an s^k factorial design in blocks of size s^m we need to choose m independent effects to confound. We then need to calculate the $s^{k-m} - m$ generalised interactions of these effects to be sure that we have not inadvertently confounded an effect of interest. Unfortunately there is no fast way of doing this in general, although for some small designs tables exist; see Subsection 6.4. The set of the m independent effects and the $s^{k-m} - m$ generalised interactions together form the defining contrasts subgroup mentioned earlier.

It is possible to get the blocks directly from the equations of the confounded effects. Consider the 3^3 example again. The block $B_0 = \{000, 120, 210\}$ has the treatment combinations which satisfy $x_1 + x_2 + x_3 = 0$ and $x_1 + x_2 + 2x_3 = 0$. Suppose that $\boldsymbol{x}$ and $\boldsymbol{y}$ are both in B_0. Then $\boldsymbol{x} + \boldsymbol{y} = (x_1 + y_1, x_2 + y_2, x_3 + y_3)$ is also in B_0 since $x_1 + y_1 + x_2 + y_2 + x_3 + y_3 = 0$ and $x_1 + y_1 + x_2 + y_2 + 2(x_3 + y_3) = 0$. We say that B_0 is closed under component-wise addition and forms a subgroup of the set of all treatment combinations. Now consider $B_0 + (001) = \{001, 121, 211\}$, say. This set consists of all the treatment combinations in which $x_1 + x_2 + x_3 = 1$ and $x_1 + x_2 + 2x_3 = 2$ and so is the intersection of the second set of $P(1,1,1)$ with the third set of $P(1,1,2)$. All the other blocks can be obtained from B_0 by adding another treatment combination that has not already appeared in the design. The other blocks are called the cosets of B_0.

In general the block in which the treatment combination $(00\cdots0)$ occurs is called the *principal block.* All other blocks are obtained from it by the addition of a treatment combination which has not yet appeared in the design.

On occasion it is not possible to find enough experimental units to allow every treatment combination to appear in the experiment. Hence a fractional factorial experiment, in which only a subset of the treatment combinations appear, is conducted. Most commonly the fraction that is used is obtained by using one block chosen at random from a confounded factorial with blocks of the right size. Now there will be some interaction effects which can not be distinguished from each other. Such effects are said to be *aliased.*

Suppose that for a 3^3 factorial design we could only find three experimental units. Using the principal block of the design we discussed above, the treatment combinations that appear are 000, 120 and 210. We know that the interactions confounded with blocks in the original design have partitions $P(1,1,1)$, $P(1,1,2)$, $P(1,1,0)$ and $P(0,0,1)$. Hence for these effects we have no information at all from this design since the three treatment combinations appear in the same set in each of these partitions. These effects are said to form the *defining contrast.* For every other partition we have one treatment combination from each set in the partition. For instance, for $P(1,0,0)$ we get the sets $\{000\}$, $\{120\}$ and $\{210\}$. This is the same as the partition for $P(0,1,0)$, for $P(1,1,0)$ and so on for each of the effects not in the defining contrast. Hence all the effects not in the defining contrast are aliased with each other.

Suppose that for a 3^3 factorial design we could find nine experimental units. If we use the partition $P(1,1,1)$ as the defining contrast and again choose the principal block, then we have the treatment combinations 000, 012, 021, 102, 201, 120, 210, 111 and 222 in the experiment. To determine the aliasing here we note that every treatment combination in the experiment satisfies $x_1 + x_2 + x_3 = 0$. Suppose that we want to determine the sets in $P(1,0,0)$. Then we also require that $x_1 = \theta$, $\theta = 0,1,2$. Hence we see that we also know that $x_2 + x_3 = -\theta = 2\theta$ and that $2x_1 + x_2 + x_3 = \theta$. But this second equation is just saying that $x_1 + 2x_2 + 2x_3 = 2\theta$. Thus the sets in $P(1,0,0)$, $P(0,1,1)$ and $P(1,2,2)$ are the same in this fraction and so these effects are aliased. Similarly the sets in $P(0,1,0)$, $P(1,2,1)$ and $P(1,0,1)$ are the same, as are the sets in $P(0,0,1)$, $P(1,1,0)$ and $P(1,1,2)$ and the sets in $P(1,2,0)$, $P(1,0,2)$ and $P(0,1,2)$. There are 9 treatment combinations in the experiment. There are therefore 8 degrees of freedom available and every partition corresponds to two degrees of freedom. Hence we expect that there will be 4 independent partitions. As there are $(3^3-1)/2 = 13$ partitions and one is used as the defining contrast, there will be four sets, each with three aliased partitions. Thus we have determined the aliasing scheme for this design.

In general one determines the aliasing scheme by first determining the confounded effects, including the generalised interactions, for the corresponding blocked complete factorial and then systematically calculating all the partitions which are the same in the fractional design. Note that the aliasing does not depend on the particular block chosen from the original, blocked design.

We close this subsection with an example of constructing a fractional factorial design in which there are four factors, each with 7 levels. Suppose that 49 units are available for the experiment. We decide that the partitions $P(a,b,c,d)$ and $P(w,x,y,z)$ will correspond to the defining contrast. The generalised interactions that are also confounded are given by $\alpha(a,b,c,d) + \beta(w,x,y,z)$, $\alpha, \beta \in \mathrm{GF}[7]$, α, β not both 0. Ideally we would like all of the defining contrasts and generalised interactions to involve four factors. However this is not possible since if w, say, is non-zero then as β takes each of the values in GF[7] so does βw. Hence for any choice of a, there will be a value of α such that $\alpha a + \beta w = 0$. So assume that $d = 0$ and that $a = b = c = 1$. Then the generalised interactions are of the form $(\alpha + \beta w, \alpha + \beta x, \alpha + \beta y, \beta z)$. We want z to be non-zero. Then we do not want all of w, x and y to be zero, otherwise we will be confounding the main effect of the fourth factor. Try $w = 1$. Then $\alpha + \beta w = \alpha + \beta$. If $\alpha + \beta = 0$ then we want $\alpha + \beta x$ and $\alpha + \beta y$ to be non-zero (so that the generalised interaction involves three factors). Hence we want that $\beta(x-1) \neq 0$ and $\beta(y-1) \neq 0$. Thus $x \neq 1$ and $y \neq 1$. Try $z = 1$, $x = y = 2$. Then the defining contrasts are $(1,1,1,0)$ and $(1,2,2,1)$. The generalised interactions are $(0,1,1,1) = 6(1,1,1,0) + (1,2,2,1)$ and $(1,0,0,6) = 2(1,1,1,0) + 6(1,2,2,1)$. This confounds part of a two-factor interaction. Can we do better? Try $x = 2$ and $y = 3$. Then the generalised interactions are of the form $(\alpha+\beta, \alpha+2\beta, \alpha+3\beta, \beta)$. If

$\alpha+\beta=0$ then this is $(0,\beta,2\beta,\beta)$ and if $\alpha+\beta=1$ then this is $(1,1+\beta,1+2\beta,\beta)$. These always involve at least three factors. Hence the treatment combinations to be used in the experiment are those in one block of the blocked design with principal block $\{(x_1,x_2,x_3,x_4) \mid x_1+x_2+x_3=0,\ x_1+2x_2+3x_3+x_4=0\}$.

6.2. *Single replicate generalised cyclic designs*

There are various families of asymmetrical factorial designs available in the literature. One of the largest and easiest to construct is the generalised cyclic designs.

Let Z_{s_i} be the integers modulo s_i. In a $s_1 \times s_2 \times \cdots \times s_k$ factorial design use the integers modulo s_i to represent the s_i levels of factor i, $1 \leqslant i \leqslant k$. Let G_t be the abelian group of treatment combinations with addition defined component-wise. A single replicate factorial design in incomplete blocks is said to be a *generalised cyclic design* if one block, B_0, say, is a subgroup of G_t and the other blocks are the cosets of B_0 in G_t. B_0 is called the *principal block* of the design. If B_0 can be generated by g treatment combinations then the generalised cyclic design is called a *g-generator design*. We see that the blocked factorial designs of the previous subsection are all examples of generalised cyclic designs.

For instance consider a 3^3 factorial design in which B_0 has the treatment combinations 000, 012, 021, 102, 201, 120, 210, 111 and 222. So B_0 is one set in $P(1,1,1)$. Then B_0 is also the principal block of a 2-generator design, as each treatment combination in B_0 is a linear combination of 012 and 102. We write $B_0=\langle 012,102\rangle$.

In a $2^3 \times 3^3$ design, where by convention the first three factors have 2 levels and the next three factors have 3 levels, with $B_0=\langle 011120,110120\rangle=\{000000,000120,$ $000210,011000,011120,011210,110000,110120,110210,101000,101120,101210\}$, another block of the design is $B_0+000001 = \{000001,000121,000211,011001,$ $011121,011211,110001,110121,110211,101001,101121,101211\}$.

We need to be able to calculate the defining contrasts subgroup of any generalised cyclic design. To do this we let a be the least common multiple of $s_1,s_2,\ldots,s_k$, written $a=lcm(s_1,s_2,\ldots,s_k)$. Let $G_c=\{(c_1,\ldots,c_k) \mid c_i \in Z_{s_i}\}$ be an abelian group with addition defined component-wise. For any treatment combination $\boldsymbol{x}$ and any $\boldsymbol{c} \in G_c$, define $[\boldsymbol{c},\boldsymbol{x}]=\sum_i(c_ix_ia/s_i)$ (modulo a). Then the *annihilator* of B_0, $(B_0)^0$, is $(B_0)^0=C=\{\boldsymbol{c} \mid \boldsymbol{c}\in G_c, [\boldsymbol{c},\boldsymbol{x}]=0\ \forall \boldsymbol{x}\in B_0\}$. The confounding scheme of the design constructed from B_0 is completely specified by the elements in C as each element $\boldsymbol{c}\in C$ represents a single degree of freedom from the interaction of the factors corresponding to the non-zero elements of $\boldsymbol{c}$. C is called the *defining contrasts subgroup*.

Consider the 3^3 design with $B_0=\langle 012,102\rangle$. Then $C=(B_0)^0=\{000,111,222\}$. This is what we expect since B_0, the principal block, is one of the sets in the partition $P(1,1,1)$.

Various results about the calculation of the defining contrasts subgroup for a generalised cyclic design have been obtained by Voss and Dean (1988). We summarise these below and refer the reader to Voss and Dean (1988) for the proofs.

The first result is for symmetrical one-generator generalised cyclic designs.

THEOREM 1 (Voss and Dean, 1988). *Let B_0 be the principal block of a one-generator s^k factorial design with generator $\boldsymbol{x} = (x_1, \ldots, x_k)$. Define the following three subsets of G_c by*

(i) $C_1 = \bigcup_{i:\, x_i=0}\{\mathbf{c} \mid c_i = 1, c_j = 0, i \neq j\}$,

(ii) $C_2 = \bigcup_{i:\, x_i\neq 0}\{\mathbf{c} \mid c_i = s/\gcd(x_i, s), c_j = 0, i \neq j\}$,

(iii) $C_3 = \bigcup_{i:\, x_i\neq 0,\ x_j\neq 0}\{\mathbf{c} \mid c_i = s - (x_j/g), c_j = x_i/g, c_m = 0$ *otherwise, where* $g = \gcd(x_i, x_j)\}$.

Then $C = (B_0)^0$ is generated by $C_1 \cup C_2 \cup C_3$.

Consider the principal block $B_0 = \langle 120\rangle$ of a 3^3 factorial design. Then $C_1 = \{001\}$, $C_2 = \varnothing$ and $C_3 = \{110\}$. Thus $C = \{000, 001, 002, 110, 220, 111, 112, 221, 222\}$. Hence we see that we have confounded two degrees of freedom of the main effect of the third factor, two degrees of freedom of the two factor interaction between factors 1 and 2, and four degrees of freedom of the three factor interaction. This is the same as we found in terms of the partitions of the previous subsection.

The next result extends Theorem 1 to symmetrical g-generator generalised cyclic designs.

THEOREM 2 (Voss and Dean, 1988). *Let B_0 be the principal block of a g-generator s^k factorial design with generators $\boldsymbol{x}_1, \ldots, \boldsymbol{x}_g$. Let $\langle \boldsymbol{x}_i\rangle$ be the block generated by $\boldsymbol{x}_i$, $i = 1, \ldots, g$. Let $D_i = \langle \boldsymbol{x}_i\rangle^0$. Then $C = (B_0)^0 = D_1 \cap \cdots \cap D_g$.*

Returning to the 3^3 design with $B_0 = \langle 012, 102\rangle$, let $\boldsymbol{x}_1 = 012$ and $\boldsymbol{x}_2 = 102$. Then $D_1 = \langle 100, 011\rangle = \{000, 100, 200, 011, 022, 111, 122, 211, 222\}$ and $D_2 = \langle 010, 101\rangle = \{000, 010, 020, 101, 202, 111, 212, 121, 222\}$. Thus $C = \{000, 111, 222\}$ and this is consistent with the previous discussion of this example.

The final result in this subsection extends the preceding theorem to g-generator generalised cyclic designs. Let B_0 be the principal block of a g-generator $s_1^{n_1} \times s_2^{n_2} \times \cdots \times s_k^{n_k}$ factorial design with generators $\boldsymbol{x}_1, \ldots, \boldsymbol{x}_g$. Let $\boldsymbol{x}_{i1}$ be the first n_1 positions of $\boldsymbol{x}_i$, let $\boldsymbol{x}_{i2}$ be the next n_2 positions of $\boldsymbol{x}_i$ and so on. Then B_0 is also generated by gk generators $(\boldsymbol{x}_{i1}, 0, \ldots, 0)$, $(0, \boldsymbol{x}_{2i}, 0, \ldots, 0), \ldots, (0, \ldots, 0, \boldsymbol{x}_{ik})$, $1 \leqslant i \leqslant g$, provided that $\gcd(s_i, s_j) = 1$ for $i \neq j$.

THEOREM 3 (Voss and Dean, 1988). *Let B_0 be the principal block of a g-generator $s_1^{n_1} \times s_2^{n_2} \times \cdots \times s_k^{n_k}$ factorial design with $\gcd(s_i, s_j) = 1$ for $i \neq j$ and with generators $\boldsymbol{x}_1, \ldots, \boldsymbol{x}_g$. Let $B_{0i} = \langle \boldsymbol{x}_{1i}, \boldsymbol{x}_{2i}, \ldots, \boldsymbol{x}_{gi}\rangle$ and let $D_i = (B_{0i})^0$, $i = 1, \ldots, k$. Then $C = (B_0)^0 = \{\mathbf{c} \mid \mathbf{c} = \mathbf{c}_1, \ldots, \mathbf{c}_g,\ \mathbf{c}_i \in D_i\}$.*

Recall the $2^3 \times 3^3$ design with $B_0 = \langle 011120, 110120\rangle$ discussed above. Since B_0 has 12 treatment combinations in it, there are 18 blocks in the design altogether. Hence 17 degrees of freedom are confounded in blocks. Now $B_{01} = \langle 011, 110\rangle$ and $B_{02} = \langle 120\rangle$. Hence $D_1 = \{(c_1, c_2, c_3) \mid c_2 + c_3 = 0, c_1 + c_2 = 0\}$ (where the additions are done modulo 2, of course). Thus $D_1 = \{000, 111\} = \langle 111\rangle$. We have seen above that $D_2 = \{000, 001, 002, 110, 220, 111, 112, 221, 222\} = \langle 001, 110\rangle$. Hence $(B_0)^0 = \langle 111001, 111110\rangle$.

It is possible to use the results above to construct designs with $\gcd(s_i, s_j) \neq 1$ by replacing factors with pseudo-factors with co-prime levels. Pseudo-factors are introduced for convenience in constructing the design and do not correspond to actual factors in the experiment.

For instance, consider a 3×6^2 design. Replace each of the two factors with 6 levels by two factors, one with 2 levels and one with 3 levels. We then have a $2^2 \times 3^3$ design and we look for a suitable design within this framework. We will order the pseudo-factors so that the first and third pseudo-factors correspond to the first factor, X, with 6 levels, the second and fourth pseudo-factors correspond to the second factor, Y, with 6 levels and the fifth pseudo-factor corresponds to the third factor, Z, with 3 levels. Suppose we want blocks of size 18. Then we need 6 blocks so C has 6 elements in it. We do not want to confound main effects so we do not want C to contain any elements of the form $(a, 0, b, 0, 0)$ (part of the main effect of X), $(0, a, 0, b, 0)$ (part of the main effect of Y) or $(0, 0, 0, 0, a)$ (part of the main effect of Z). Consider $C = \langle 11000, 00111 \rangle = \{00000, 11000, 00111, 00222, 11111, 11222\}$. Thus we have confounded, in order, 1 degree of freedom from the XY interaction and 4 degrees of freedom from the XYZ interaction. $B_0 = \{(x_1, x_2, x_3, x_4, x_5) \mid x_1 + x_2 = 0, x_3 + x_4 + x_5 = 0\} = \langle 11000, 00120, 00111 \rangle$.

Generators for some generalised cyclic designs may be found in John and Dean (1975) and Dean and John (1975).

6.3. *Comparing factorial designs*

We now define the concepts of resolution (introduced by Box and J. S. Hunter (1961)) and minimum aberration (introduced by Fries and W. G. Hunter (1980)), both useful when comparing two, or more, factorial designs of the same size.

A fractional factorial design is said to be of *resolution R* if no p factor interaction is aliased with another effect containing less than $R - p$ factors. We let $R_s(k, m)$ denote the maximum possible resolution of an s^{k-m} factorial design.

Thus we see that the higher the resolution the better the design, since it will be possible to estimate more of the interaction effects. For example, consider a 2^{3-1} design with treatment combinations $\{000, 011, 101, 110\}$. The defining contrasts subgroup for this design is $C = \{000, 111\}$ and so we have that $P(100)$ and $P(011)$ give equal partitions, as do $P(010)$ and $P(101)$, and $P(001)$ and $P(110)$. Hence we see that main effects and two-factor interactions are aliased and so the design is of resolution III. Now consider a 2^{3-1} design with treatment combinations $\{000, 001, 110, 111\}$. Here the defining contrasts subgroup is $C = \{000, 110\}$ and so the partitions $P(100)$ and $P(010)$ are equal. Hence the main effects of the first two factors are aliased and so the design is not of resolution III but of resolution II.

Consider the defining contrasts subgroup C. The *word length* of any effect in this subgroup is the number of non-zero entries that it has. If a_i is the number of vectors with word length i, $1 \leqslant i \leqslant k$, in the defining contrasts subgroup then the *word length pattern* of the design is given by $w = (a_1, \ldots, a_k)$. Hence we see that the resolution of the design is the smallest i such that a_i is positive. If two designs, D_1 and D_2, are both resolution R then we say that D_1 has *less aberration* than D_2 if the value of a_R

for D_1 is less than the value of a_R for D_2. An s^{k-m} design has *minimum aberration* if no other s^{k-m} design has less aberration.

For the 2^{3-1} designs discussed above we have that the word length patterns are $(0, 0, 1)$ and $(0, 1, 0)$ respectively, confirming the resolutions that we calculated above.

The concept of aberration was introduced by Fries and Hunter (1980) in an attempt to distinguish between designs of the same resolution. The general idea is that resolution tells you only that at least one interaction of the appropriate order has been confounded; aberration gives an indication of how many effects of the appropriate order are confounded. Fries and Hunter (1980) give some constructions for minimum aberration two-level designs and Franklin (1984) improves on them, and includes some small tables.

6.4. Tables and algorithms

Published tables are limited in extent. McLean and Anderson (1984) give tables of factorial designs in incomplete blocks, and the associated confounding schemes, for 2^k, 3^k and 2^k3^p designs for up to 10 factors. Similar tables appear in Colbourn and Dinitz (1996), although the minimum aberration designs are given where possible. Montgomery (1991) gives alias relationships, including some blocking options, for two-level designs with up to 11 factors and at most 64 units. Bisgaard (1994) gives blocked fractions for 2-level designs with up to 15 factors. Chen et al. (1993) give some 2 and 3 level fractional factorials with small numbers of runs. Some of the designs advocated by Taguchi are related to conventional two- and three-level factorial designs and hence some of the tables in Taguchi (1987) may be useful. See Box et al. (1988) for a discussion of the relationship between the 'conventional' factorial designs and the arrays, like L_8, advocated by Taguchi.

Generators for some generalised cyclic designs may be found in John and Dean (1975) and Dean and John (1975).

A number of algorithms are available. Turiel (1988) gives one to determine defining contrasts and treatment combinations for two-level fractional factorial designs of small resolution. Mount–Campbell and Neuhardt (1981) give one for enumerating the fractional two-level factorials of resolution III. Cook and Nachtsheim (1989) give one to assist in blocking factorial designs with pre-specified block sizes. Franklin (1984) gives an algorithm to construct minimum aberration designs. The DSIGN method, due to Patterson (1965), is a systematic approach to factorial design construction and is described in elementary terms in Street and Street (1987). For more details see Patterson and Bailey (1978).

7. Response surface designs

A *response surface model* is a model fitted to a response y as a function of predictors $x_1, x_2, \ldots, x_k$ in an attempt to get a 'mathematical French curve' (Box and Draper, 1987) that will summarise the data. Thus a *response surface design* is a selection of points in x-space that permit the fitting of a response surface to the corresponding observations y. The $\boldsymbol{x}$-points need to be chosen so that all the parameters in the model can be estimated and the significance, or otherwise, of each one established.

Response surface designs have been used by various workers in the agricultural area including Huett and Dettmann (1992), Thayer and Boyd (1991) and Pesti (1991).

Much of the current interest in response surface methodology (RSM) in the statistical literature has focussed on the industrial applications to process improvement, and comparing the benefits of the RSM approach to the approach advocated by Taguchi; Lucas (1994) is one such example.

In this section we will describe some of the designs commonly used in RSM and consider the attributes of 'good' response surface designs. For suggestions on analysis, graphical representation of the surface and so on see, e.g., Box and Draper (1987).

We begin by describing the standard response surface model.

7.1. Polynomial approximations to the response function

In this setting we assume that we can approximate the response, y, as a polynomial function of k predictors $x_1, \ldots, x_k$, where the x_i may be coded values of the original predictors. Typically we code the predictors so that the region of interest is centered at the origin and has a range ± 1 on each axis, but this is not mandatory. In any case the range of possible values for the x_i's, considered as a k-tuple, forms the *feasible region* for the experiment.

If there are k variables then a *first order polynomial model* is one where

$$\mathrm{E}(y) = \beta_0 + \beta_1 x_1 + \cdots + \beta_k x_k = \beta_0 + \sum_i \beta_i x_i.$$

The β_i are called the *linear coefficients* in the model.

A *second order model* is one where

$$\begin{aligned}\mathrm{E}(y) &= \beta_0 + \beta_1 x_1 + \cdots + \beta_k x_k + \beta_{11} x_1^2 + \cdots + \beta_{kk} x_k^2 \\ &\quad + \beta_{12} x_1 x_2 + \cdots + \beta_{k-1,k} x_{k-1} x_k \\ &= \beta_0 + \sum_i \beta_i x_i + \sum_i \beta_{ii} x_i^2 + \sum_{i<j} \beta_{ij} x_i x_j.\end{aligned}$$

The β_{ii} are called *pure quadratic coefficients* and the β_{ij} are called *mixed quadratic coefficients.*

Higher order polynomials can be fitted but we will not consider designs for these in the sequel.

We can write either of these models in matrix form as $\mathrm{E}(\boldsymbol{Y}) = X\boldsymbol{\beta}$, for suitable $\boldsymbol{\beta}$. The values $(x_{1u}, \ldots, x_{ku})$, $1 \leqslant u \leqslant n$, give the, perhaps coded, settings at which a response is observed. We call these n points the *design points* and the $n \times k$ matrix, D, say, with the design points as rows the *design matrix*. Clearly the matrices X and D are related. In the first order model $X = [\boldsymbol{j}\ D]$, a column of ones is adjoined to D. In the second order model $X = [\boldsymbol{j}\ D\ \boldsymbol{x}_1^2\ \boldsymbol{x}_2^2\ \ldots\ \boldsymbol{x}_k^2\ \boldsymbol{x}_1\boldsymbol{x}_2\ \boldsymbol{x}_1\boldsymbol{x}_3\ \ldots\ \boldsymbol{x}_{k-1}\boldsymbol{x}_k]$, where, for instance, $\boldsymbol{x}_1\boldsymbol{x}_2$ is a column vector of length n with $x_{1u}x_{2u}$ in the uth position.

7.2. *Good designs for first-order polynomial models*

The most commonly used first order designs are 2^k factorial and fractional factorial designs (discussed in Subsections 6.1 and 6.2). In this section we will use -1 and 1 as the two levels for each factor (or predictor variable) rather than 0 and 1. A centre point is a point with all x_i at level 0, the midpoint of the two original levels of the factor. The 2^k factorial and fractional factorial designs can be used to check for curvature by adding one, or more, centre points. If curvature seems to be present then this design can often be used as one of the blocks of a design for estimating the second order polynomial model; see the next subsection.

Other designs can be used to estimate parameters in first order polynomial models. These include the Plackett–Burman designs, the regular simplex designs and the Koshal-type first order designs.

Plackett–Burman designs (Plackett and Burman, 1946) are two-level fractional factorial designs in which the number of runs, n, is a multiple of 4 and the number of predictors, k, is $n-1$. If n is a power of 2 then these designs are just the fractional factorial designs discussed in Subsection 6.2 and can be constructed as described there. Otherwise the Plackett–Burman designs are most easily constructed from one initial row which is cyclically shifted to obtain the first $n-1$ runs; the final run is $(-1, \ldots, -1)$. An example for 12 runs, with initial row $1\ -1\ 1\ -1\ -1\ -1\ 1\ 1\ 1\ -1\ 1$ is given in Table 7.2.1; design points are the rows of the table.

These designs, like the fractional factorial designs of Subsection 6.2, involve a large number of factors and few runs. There is, therefore, much aliasing of interaction effects and information about aliased effects is usually not available in the tables of initial rows. Montgomery (1991) has a table of small Plackett–Burman designs and some warnings. Some idea of the aliasing issues can be gained by looking at Lin and Draper (1992). See Hamada and Wu (1992) for a defense of the Plackett–Burman designs.

The regular simplex designs were introduced by Box (1952). They are so-named since the design points form a regular simplex and hence the region of interest is

Table 7.2.1
A 12 run Plackett–Burman design (rows are runs)

1	−1	1	−1	−1	−1	1	1	1	−1	1
1	1	−1	1	−1	−1	−1	1	1	1	−1
−1	1	1	−1	1	−1	−1	−1	1	1	1
1	−1	1	1	−1	1	−1	−1	−1	1	1
1	1	−1	1	1	−1	1	−1	−1	−1	1
1	1	1	−1	1	1	−1	1	−1	−1	−1
−1	1	1	1	−1	1	1	−1	1	−1	−1
−1	−1	1	1	1	−1	1	1	−1	1	−1
−1	−1	−1	1	1	1	−1	1	1	−1	1
1	−1	−1	−1	1	1	1	−1	1	1	−1
−1	1	−1	−1	−1	1	1	1	−1	1	1
−1	−1	−1	−1	−1	−1	−1	−1	−1	−1	−1

Table 7.2.2
A regular simplex design with $n=6$ (rows are design points)

−1	−1	−1	−1	−1
1	−1	−1	−1	−1
0	2	−1	−1	−1
0	0	3	−1	−1
0	0	0	4	−1
0	0	0	0	5

Table 7.2.3
A first-order Koshal design with $k=3$ factors

−1	−1	−1
1	−1	−1
−1	1	−1
−1	−1	1

uniformly covered by the design points. One general way to construct these designs is to take any orthogonal matrix Q, of order n, with all elements in the first column equal to $1/\sqrt{n}$. Thus $QQ^T = Q^TQ = I$ and so the sum of the entries in any column of Q, other than the first, is 0. For a regular simplex design let $X = \sqrt{n}\,Q$. This is then the X matrix for a design involving $n-1$ factors in n runs. As usual the first column contains only 1s. The entries in the remaining columns give the coded levels for $(x_{1u}, \ldots, x_{ku})$, $1 \leqslant u \leqslant n$. A design with fewer than $n-1$ predictor variables can be obtained by deleting columns of X, except the first.

One easy way to get columns 2 to n of X (that is, the design points), described in Box and Draper (1987, p. 508), is to write down a row of -1's. In the second row make the first entry equal to 1 and all others equal to -1. In the third row make the first entry 0, the second entry 2 and all others -1. In the fourth row make the first two entries 0, the third 3 and all other entries -1. Continue in this way. An example for $n=6$ is given in Table 7.2.2. Here the rows give the coded levels of each of the $k=5$ predictor variables for each of the $n=6$ runs.

Koshal designs (Koshal, 1933) also have $k=n-1$ factors. If the two levels of each factor are represented by -1 and 1 then the Koshal designs have a row of -1's for the first treatment combination in the design. Every other treatment combination in the design has only one factor at the high level; all other factors are at the low level. An example appears in Table 7.2.3.

Orthogonality, the ability to estimate parameters independently of each other, has long been recognised as an important design principal. The notion of rotatability is an extension of this. We do not want the orientation of the response surface to affect our conclusions since we usually have no idea of the orientation of the response surface when we design the experiment. Hence we would like the variance of the predicted response to be constant at a given distance from the design origin. (The design origin is typically chosen to be the current best diet, say.) A design with this property is said to be *rotatable*.

For first order designs orthogonality and rotatability are the same We can see this by writing the first order model as $\mathrm{E}(\boldsymbol{Y}) = X\boldsymbol{\beta}$. Then $\widehat{\boldsymbol{\beta}} = (X^T X)^{-1} X^T \boldsymbol{Y}$ and $\mathrm{Cov}(\widehat{\boldsymbol{\beta}}) = \sigma^2 (X^T X)^{-1}$. Hence we see that the entries in $\boldsymbol{\beta}$ will be independently estimated if $(X^T X)$ is diagonal.

We can code the predictors in any way we like. Code them so that $(X^T X) = nI$. Let $y_{\boldsymbol{x}}$ be the predicted response at $\boldsymbol{x}$. Then $y_{\boldsymbol{x}} = \boldsymbol{x}^T \widehat{\boldsymbol{\beta}}$ and $\mathrm{Var}(y_{\boldsymbol{x}}) = \sigma^2 \boldsymbol{x}^T (X^T X)^{-1} \boldsymbol{x}$. For a rotatable design this will be constant for all $\boldsymbol{x}$ such that $\boldsymbol{x}^T \boldsymbol{x} = d$, say. With the recommended coding we see that for an orthogonal design, $\sigma^2 \boldsymbol{x}^T (X^T X)^{-1} \boldsymbol{x} = \sigma^2 d$. Hence a first order orthogonal design is rotatable.

Now consider a rotatable design. Since $\boldsymbol{x}^T (X^T X)^{-1} \boldsymbol{x}$ is constant for all $\boldsymbol{x}$ such that $\boldsymbol{x}^T \boldsymbol{x} = d$, consider $\boldsymbol{x} = (\sqrt{d}, 0, \ldots, 0)^T$. Then $\boldsymbol{x}^T (X^T X)^{-1} \boldsymbol{x} = d(X^T X)^{-1}_{11}$ and so all the diagonal entries of $(X^T X)^{-1}$ are equal. Considering $\boldsymbol{x} = (\sqrt{d}/2, \sqrt{d}/2, 0, \ldots, 0)^T$ we see that all the off-diagonal entries of $(X^T X)^{-1}$ are zero. Hence a rotatable first order design is orthogonal.

It is also possible to use first order designs to indicate the direction in which improved response may be expected to occur. After conducting the original design, a contour plot of the estimated response function can be drawn. This plot will indicate the direction in which better response can be expected to occur. Assuming that high responses are the aim of the experiment then the *method of steepest ascent* involves conducting experiments at points along the indicated direction until either the response plateaus, or the edge of the experimental region is reached.

If the response plateaus then another first order design is conducted, centered on the best point on the first path of steepest ascent. The second experiment is used to decide if another direction for steepest ascent is indicated or whether it is time to use a design capable of estimating second order terms.

If the boundary of the region is reached then a constrained direction of steepest ascent is followed, again looking for where the response plateaus. More details on the method of steepest ascent can be found in Box and Draper (1987, Chapter 6).

7.3. Good designs for second order polynomial models

To estimate the parameters in a second-order polynomial model, at least three levels are required for each factor. One obvious class of designs consists of the three level factorial and fractional factorial designs discussed in Subsection 6.2. Other classes of three-level designs are the Box–Behnken designs and the second-order Koshal designs (see Box and Draper, 1987).

A more commonly used design is the central composite design. These designs require that each factor have 5 levels and they are obtained by augmenting a two level factorial or fractional factorial design. This makes them particularly appropriate when using the method of steepest ascent, as in most cases useful blocking arrangements can also be obtained so that the first level design can be viewed as one block and later runs form second, and subsequent, blocks.

Table 7.2.4
A central composite design for three factors

x_1	x_2	x_3
-1	-1	-1
-1	-1	1
-1	1	-1
-1	1	1
1	-1	-1
1	-1	1
1	1	-1
1	1	1
α	0	0
$-\alpha$	0	0
0	α	0
0	$-\alpha$	0
0	0	α
0	0	$-\alpha$
0	0	0
0	0	0
0	0	0
0	0	0

A *central composite design* has three components. There is a two level factorial or fractional factorial design of resolution at least V (defined in Subsection 6.3), with levels represented by ± 1, a set of $2kr_a$ axial (or star) points, with coordinates $(\pm\alpha, 0, \ldots, 0)$, $(0, \pm\alpha, 0, \ldots, 0), \ldots, (0, 0, \ldots, 0, \pm\alpha)$ and n_0 centre points with coordinates $(0, 0, \ldots, 0)$. We will use r_f for the replication of the factorial points in the design and r_a for the replication of the axial points. An example of a central composite design with $k = 3$ factors, $r_f = r_a = 1$, is given in Table 7.2.4. Here $\alpha = (8)^{0.25}$.

There are three issues that we would like to consider in the context of composite designs. They are

(1) can the designs be made orthogonal?

(2) can the designs be made rotatable?

(3) can the designs be blocked orthogonally (so that the parameter estimates are independent of the block effects)?

We address them in turn below.

It is not possible to make central composite designs orthogonal. The best one can hope to do is to have the terms of different orders independently estimated and rotatable designs achieve this.

For any second order design we can decide whether or not the design is rotatable by evaluating the $\text{var}(y_{\boldsymbol{x}})$, just as we did for the first order design. An expression for $\text{var}(y_{\boldsymbol{x}})$ is derived in Box and Hunter (1957). From it we can establish when a second order design with k factors is rotatable. Assume that there are n treatment combinations in the experiment and let the actual levels of the ith factor be x_{iu},

$u = 1, 2, \ldots, n$. Assume that the levels have been scaled so that $\sum_u x_{iu}^2 = n$. Then, considering the entries in $X^T X$, we see that a second order design is rotatable if:

$$\sum_u x_{iu} = 0, \quad \text{for } i = 1, 2, \ldots, k;$$

$$\sum_u x_{iu} x_{ju} = 0, \quad \text{for } i, j = 1, 2, \ldots, k,\ i \neq j;$$

$$\sum_u x_{iu}^2 = n, \quad \text{for } i = 1, 2, \ldots, k;$$

$$\sum_u x_{iu}^4 = 3 \sum_u x_{iu}^2 x_{ju}^2 = 3n\lambda_4, \quad \text{for } i, j = 1, 2, \ldots, k,\ i \neq j;$$

and all sums of powers and products of order 3 are zero as are the other sums of products of order 4. The constant λ_4 is defined by these equations. For $(X^T X)$ to be non-singular we require that $\lambda_4 > k/(k+2)$. The minimum value of $k/(k+2)$ is attained if all the points lie on a sphere centered at the origin in which case the addition of centre points to the design will mean that the conditions for rotatability are still satisfied. (The first three equations are all that is required for a design to be a first order orthogonal, and hence rotatable, design since these are the conditions for $X^T X$ to be diagonal.)

For a central composite design we have, by construction, the sum of the factor levels for each factor is zero and that, for any two factors, $\sum_u x_{iu} x_{ju} = 0$. In general we have that $\sum_u x_{iu}^2 = r_f f + 2r_a \alpha^2$, that $\sum_u x_{iu}^4 = r_f f + 2r_a \alpha^4$, and that $3 \sum_u x_{iu}^2 x_{ju}^2 = 3r_f f$. The various other sums are zero by the properties of the (fractional) factorial design. Hence we require that $r_f f + 2r_a \alpha^2 = r_f f + 2r_a k + n_0$ and $3r_f f = r_f f + 2r_a \alpha^4 = 3(r_f f + 2r_a k + n_0)\lambda_4$. Thus we see that $2r_a \alpha^4 = 2(r_f f + 2r_a k + n_0)\lambda_4 = 2r_f f$ so $\alpha^4 = r_f f / r_a$. Given values for λ_4, suitable values of n_0 can then be determined. Box and Hunter (1957) express the Var($y_{\boldsymbol{x}}$) as a function of the distance of $\boldsymbol{x}$ from the origin. They give values of λ_4 so that the variance at the origin is the same as the variance at distance 1 from the origin in a rotatable design. Using these values of λ_4 we get the values in Table 7.2.5 for designs with $r_f = r_a = 1$.

Central composite designs can be orthogonally blocked on occasion. To do this requires that each block be a first order orthogonal design and the factors must, in some sense, appear in the blocks proportionally. Specifically we require that, if there are b_j runs in block j, and n runs in the experiment as a whole, $\sum_{u:\, u \in \text{block } j} x_{iu}^2 / \sum_u x_{iu}^2 = b_j / n$.

Consider a central composite design with $r_f = r_a = 1$. Then one natural allocation of treatment combinations to blocks is to allocate the factorial points to one block and the star points to another and to put 'some' centre points in each. Clearly each of these blocks is a first order orthogonal design. The proportionality requirement says that $f/(f + 2k\alpha^2) = (f + n_{0f})/(f + 2k + n_0)$ or, equivalently, $2k\alpha^2/(f + 2k\alpha^2) = (2k + n_{0a})/(f + 2k + n_0)$. Rearranging we see that $\alpha^2 = (2k + n_{0a})f/(f + n_0)2k$. The

Table 7.2.5
Second-order central composite designs

k	2	3	4	5	6
f	4	8	16	32	32
n_0	5	6	7	10	9
$\alpha\ (= f^{0.25})$	$\sqrt{2}$	1.68	2	2.38	2.38

Table 7.2.6
Blocked second-order central composite designs

k	2	2	2	3	3	4	5	6
f	4	4	4	8	8	16	32	32
n_{0f}	3	1	2	4	2	2	2	1
n_{0a}	3	6	8	2	16	5	8	8
α^2	1	2	2	4/3	3	1.4	1.5	1

number of centre points, and the corresponding value of α^2, for some small designs, are given in Table 7.2.6.

Sometimes smaller composite designs can be obtained. Examples have been given by Westlake (1965) and Draper (1985). There is a discussion in Box and Draper (1987).

7.4. Tables and algorithms

The algorithm of Cook and Nachtsheim (1989) can help to block response surface designs when the block sizes are pre-specified. Atkinson and Donev (1992, Chapter 14) have a brief discussion of blocking response surface designs. Tables of response surface designs, including blocked central composite designs, can be found in Box and Draper (1987). Montgomery (1991) has a table of small Plackett–Burman designs.

8. Miscellaneous designs

Inevitably there are designs that we have not been able to discuss above. Brief references for some of these will be given below.

We have not considered designs for intercropping work. These are reviewed by Mead in this volume. Federer (1994) has written a volume on intercropping designs for two crops and has promised a second volume on designs for more than two crops. Kahurananga (1991) uses an intercropping with a split-plot design.

There are designs to investigate hedgerow-alley plantings; see Nester (1994) and papers therein referred to. Atta-Krah (1990) discusses an actual trial.

There are designs appropriate for agroforestry situations discussed in Williams and Matheson (1994).

Planar grid designs, and variants, have been developed to investigate aspects of plant competition following a discussion by Cormack (1979). The constructions available have been summarised in Street and Street (1987).

The fan designs of Nelder (1962) are used to investigate the effect of spacing on one variety; an example may be found in Ellis et al. (1987). The plants are planted at the intersection of radial lines and arcs of concentric circles. They have been extended to two varieties by Wahau and Miller (1978). The designs are discussed in Street and Street (1987).

By construction, in a fan design the plants are planted more densely at the centre of the circle than at the perimeter. We say that the allocation of the planting density has been done *systematically*. The fan designs are just one example of a systematic experiment, however. Systematic experiments in general have been the subject of sometimes bitter controversy (much of it summarised in Wilkinson and Mayo (1982)) but they have adherents in some situations (Pearce, 1983). Edmonson (1993) gives some systematic row–column designs.

Another example of systematic designs are the trend-free factorial designs. In these designs the allocation of treatment combinations to plots is done systematically so that polynomial trend effects (if present) are orthogonal to parameter estimates. See Bailey et al. (1992) and references therein for constructions for these designs.

Designs optimal for general dependence structures (and so extending the ideas in the change-over design section) have been considered by Martin and Eccleston (1991) and Street (1992).

Acknowledgement

This work was commenced while the author was at the University of New South Wales and completed while a visitor at the University of Queensland. Many of the references in the statistical literature were found using the Current Index to Statistics. I thank an anonymous referee for a number of suggestions which improved the presentation.

References

Afsarenijad, K. (1990). Rectangular lattice designs. *Statist. Probab. Lett.* **10**, 279–281.

Anderson, I. (1990). *Combinatorial Designs: Construction Methods*. Halsted Press, New York.

Ash, A. (1981). Generalized Youden designs: Construction and tables. *J. Statist. Plann. Inference* **5**, 1–25.

Atkinson, A. C. and A. N. Donev (1992). *Optimum Experimental Designs*. Oxford Univ. Press, Oxford.

Atta-Krah, A. N. (1990). Alley farming with Leucaena: Effects of short grazed fallows in soil fertility and crop yields. *Experiment. Agriculture* **26**, 1–10.

Azais, J.-M., R. A. Bailey and H. Monod (1993). A catalogue of efficient neighbour designs with border plots. *Biometrics* **49**, 1252–1261.

Azam-Ali, S. N., R. B. Matthews, J. H. Williams and J. M. Peacock (1990). Light use, water uptake and performance of individual components of a sorghum/groundnut intercrop. *Experiment. Agriculture* **26**, 413–427.

Bailey, R. A. (1985). Partially balanced designs. In: S. Kotz and N. L. Johnson, eds., *Encyclopedia of Statistical Science*, Vol. 6. Wiley, New York, 593–610.

Bailey, R. A. (1986). Constrained randomization. In: S. Kotz and N. L. Johnson, eds., *Encyclopedia of Statistical Science*, Vol. 7. Wiley, New York, 524–530.

Bailey, R. A. (1987). Restricted randomization: A practical example. *J. Amer. Statist. Assoc.* **82**, 712–719.

Bailey, R. A., C. S. Cheng and P. Kipnis (1992). Construction of trend-resistant factorial designs. *Statist. Sinica* **2**, 393–411.

Bisgaard, S. (1994). Blocking generators for small 2^{k-p} designs. *J. Quality Technol.* **26**, 288–296.

Bose, R. C. (1939). On the construction of balanced incomplete block designs. *Ann. Eugenics* **9**, 353–399.

Box, G. E. P. (1952). Multifactor designs of first order. *Biometrika* **39**, 49–57.

Box, G. E. P., S. Bisgaard and C. A. Fung (1988). An explanation and critique of Taguchi's contribution to quality engineering. *Quality Reliability Eng. Internat.* **4**, 123–131.

Box, G. E. P. and N. R. Draper (1987). *Empirical Model-Building and Response Surfaces*. Wiley, New York.

Box, G. E. P. and J. S. Hunter (1957). Multifactor experimental designs for exploring response surfaces. *Ann. Math. Statist.* **28**, 195–241.

Box, G. E. P. and J. S. Hunter (1961). The 2^{k-p} fractional factorial designs, Part I. *Technometrics* **3**, 311–351.

Chang, C.-T. and K. Hinkelmann (1987). A new series of EGD-PBIB designs. *J. Statist. Plann. Inference* **16**, 1–13.

Chen, J., D. X. Sun and C. F. J. Wu (1993). A catalogue of two-level and three-level fractional factorial designs with small runs. *Internat. Statist. Rev.* **61**, 131–145.

Cheng, C. S. and C. F. Wu (1980). Balanced repeated measurements designs. *Ann. Statist.* **8**, 1272–1283.

Clatworthy, W. H. (1973). *Tables of Two-Associate-Class Partially Balanced Designs*. Applied Mathematics Series No. 63, National Bureau of Standards (US).

Cochran, W. G. and G. M. Cox (1992). *Experimental Designs*. Wiley, New York.

Colbourn, C. J. and J. H. Dinitz (1996). *Handbook of Combinatorial Design*. CRC Press, Boca Raton.

Cook, R. D. and C. J. Nachtsheim (1989). Computer-aided blocking of factorial and response-surface designs. *Technometrics* **31**, 339–346.

Cormack, R. M. (1979). Spatial aspects of competition between individuals. In: R. M. Cormack and J. K. Ord, eds., *Spatial and Temporal Analysis in Ecology*. International Publishing House, Fairfield, MD, 151–212.

De Palluel, C. (1788). Sur les advantages and l'economie que procurent les racines employees a l'engrais des moutons a l'etable. *Memoires d'Agriculture*, trimestre d'ete, 17–23. English translation (1790): *Annals of Agriculture* **14**, 133–139.

Dean, A. M. and J. A. John (1975). Single replicate factorial designs in generalised cyclic designs: II. Asymmetrical arrangements. *J. Roy. Statist. Soc. Ser. B* **37**, 72–76.

Denes, J. and A. D. Keedwell (1974). *Latin Squares and Their Applications*. English Universities Press, London.

Denes, J. and A. D. Keedwell (1991). *Latin Squares: New Developments in the Theory and Applications*. Elsevier, Amsterdam.

Draper, N. R. (1985). Small composite designs. *Technometrics* **27**, 173–180.

Dyke, G. V. (1988). *Comparative Experiments With Field Crops*. Griffin Bucks, London.

Edmonson, R. N. (1993). Systematic row–column designs balanced for low order polynomial interactions between rows and columns. *J. Roy. Statist. Soc. Ser. B* **55**, 707–723.

Ellis, P. R., G. H. Freeman, B. D. Dowker, J. A. Hardman and G. Kingswell (1987). The influence of plant density and position in field trials designed to evaluate the resistance of carrots to carrot fly (*Psila rosae*) attack. *Ann. Appl. Biol.* **111**, 21–31.

Ellis, P. R., J. A. Hardman, T. C. Crowther and P. L. Saw (1993). Exploitation of the resistance to carrot fly in the wild carrot species *Daucus capillifolius*. *Ann. Appl. Biol.* **122**, 79–91.

Federer, W. T. (1975). The misunderstood split-plot. In: *Applied Statistics, Proc. Conf. at Dalhousie University*. Halifax, 9–40.

Federer, W. T. (1993). *Statistical Design and Analysis for Intercropping Experiments*. Springer, New York.

Fisher, R. A. (1923). Studies on crop variation II: The manurial response of different potato varieties. *J. Agricult. Sci.* **13**, 311–320.

Fisher, R. A. (1926). The arrangement of field trials. *J. Ministry of Agriculture* **33**, 503–513.

Fisher, R. A. and F. Yates (1963). *Statistical Tables for Biological, Agricultural and Medical Research.* Oliver and Boyd, Edinburgh.

Fletcher, D. J. (1987). A new class of change-over designs for factorial experiments. *Biometrika* **74**, 649–654.

Fletcher, D. J. and J. A. John (1985). Change-over designs and factorial structure. *J. Roy. Statist. Soc. Ser. B* **47**, 117–124.

Franklin, M. F. (1984). Constructing tables of minimum aberration p^{n-m} designs. *Technometrics* **26**, 225–232.

Freeman, G. H. (1979). Some two-dimensional designs balanced for nearest neighbours. *J. Roy. Statist. Soc. Ser. B* **41**, 88–95.

Freeman, G. H. (1988a). Nearest neighbour designs for three or four treatments in rows and columns. *Utilitas Math.* **34**, 117–130.

Freeman, G. H. (1988b). Row and column designs. In: S. Kotz and N. L. Johnson, eds., *Encyclopedia of Statistical Science*, Vol. 8. Wiley, New York, 201–206.

Fries, A. and W. G. Hunter (1980). Minimum aberration 2^{k-p} designs. *Technometrics* **22**, 601–608.

Fukai, S., L. Li, P. T. Vizmonte and K. S. Fischer (1991). Control of grain yield by sink capacity and assimilate supply in various rice (*Oryza satuva*) cultivars. *Experiment. Agriculture* **27**, 127–135.

GENDEX (1993). GENDEX: An algorithmic toolkit for designers of experiments. Proceedings of Statistics, 73, University of Wollongong.

Gill, P. S. (1992). Balanced change-over designs for autocorrelated observations. *Austral. J. Statist.* **34**, 415–420.

Gleeson, A. C. and B. R. Cullis (1991). Spatial analysis of field experiments – An extension to two dimensions. *Biometrics* **47**, 1449–1460.

Gleeson, A. C. and J. A. Eccleston (1992). On the design of field experiments under a one-dimensional neighbour model. *Austral. J. Statist.* **34**, 91–97.

Hamada, M. and C. F. J. Wu (1992). Analysis of designed experiments using complex aliasing. *J. Quality Technol.* **24**, 130–137.

Harshbarger, B. (1949). Triple rectangular lattices. *Biometrics* **5**, 1–13.

Hedayat, A. S. and B. K. Sinha (1991). *Design and Inference in Finite Population Sampling.* Wiley, New York.

Huett, D. O. and E. B. Dettmann (1992). Nutrient uptake and partitioning by zucchini squash, head lettuce and potato in response to nitrogen. *Austral. J. Agricultural Res.* **43**, 1653–1665.

Hwang, F. K. and S. Lin (1978). Distributions of integers into k-tuples with prescribed conditions. *J. Combin. Theory Ser. A* **25**, 105–116.

Iremiren, G. O., A. W. Osara and D. A. Okiy (1991). Effects of age of harvesting after pod set on the growth, yield and quality of okra (*Abelmoschus esculentus*). *Experiment. Agriculture* **27**, 33–37.

Ishag, H. M. and O. A. A. Ageeh (1991). The physiology of grain yield in wheat in irrigated tropical environment. *Experiment. Agriculture* **27**, 71–77.

Jackson, P. A. and P. M. Hogarth (1992). Genotype $\times$ environment interactions in sugarcane. I. Patterns of response across sites and crop-years in North Queensland. *Austral. J. Agricultural Res.* **43**, 1447–1459.

Jacroux, M. and R. S. Ray (1991). On the determination and construction of optimal row–column designs having unequal row and column sizes. *Ann. Inst. Statist. Math.* **43**, 377–390.

Jarrett, R. (1989). A review of bounds for the efficiency factor of block designs. *Austral. J. Statist.* **31**, 118–129.

John, J. A. (1987). *Cyclic Designs*. Chapman and Hall, London.

John, J. A. and A. M. Dean (1975). Single replicate factorial designs in generalised cyclic designs: I. Symmetrical arrangements. *J. Roy. Statist. Soc. Ser. B* **37**, 63–71.

John, J. A. and J. A. Eccleston (1986). Row–column α-designs. *Biometrika* **73**, 301–306.

John, J. A. and D. J. Street (1992). Bounds for the efficiency factor of row–column designs. *Biometrika* **79**, 658–661.

John, P. W. M. (1971). *Statistical Design and Analysis of Experiments.* Macmillan, New York.

Johnson, N. L. and F. C. Leone (1977). *Statistics and Experimental Design in Engineering and the Physical Sciences.* Wiley, New York.

Kahurananga, J. (1991). Intercropping Ethiopian Trifolium species with wheat. *Experiment. Agriculture* **27**, 385–390.

Kempthorne, O. (1952). *The Design and Analysis of Experiments*. Wiley, New York.

Khalili, H. (1993). Supplementation of grass hay with molasses in crossbred (*Bos taurus* × *Bos indicus*) non-lactating cows: Effect of level of molasses on feed intake, digestion, rumen fermentation and rumen digesta pool size. *Animal Feed Sci. Technol.* **41**, 23–38.

Khalili, H., T. Varvikko and S. Crosse (1992). The effects of forage type and level of concentrate supplementation on food intake, diet digestibility and milk production of crossbred cows (*Bos taurus* × *Bos indicus*). *Animal Product.* **54**, 183–189.

Kiefer, J. (1975). Construction and optimality of generalized Youden designs. *Statist. Des. Linear Models*, 333–354.

Kirton, H. C. and J. Seberry (1978). Generation of a frequency square orthogonal to a 10 × 10 Latin square. In: *Combinatorial Mathematics*, Proc. Internat. Conf., Canberra. Lecture Notes in Mathematics 686. Springer, Berlin, 193–198.

Koch, G. G., J. D. Elashoff and I. A. Amara (1988). Repeated measurements: Design and analysis. In: S. Kotz and N. L. Johnson, eds., *Encyclopedia of Statistical Science*, Vol. 8. Wiley, New York, 46–73.

Koshal, R. S. (1933). Application of the method of maximum likelihood to the improvement of curves fitted by the method of moments. *J. Roy. Statist. Soc. Ser. A* **96**, 303–313.

Kunert, J. (1983). Optimal design and refinement of the linear model with applications to repeated measurements designs. *Ann. Statist.* **11**, 247–257.

Kunert, J. (1985). Optimal repeated measurements designs for correlated observations and analysis by weighted least squares. *Biometrika* **72**, 375–389.

Kunert, J. (1993). A note on optimal designs with a non-orthogonal row–column-structure. *J. Statist. Plann. Inference* **37**, 265–270.

Lin, D. K. J. and N. R. Draper (1992). Projection properties of Plackett and Burman designs. *Technometrics* **34**, 423–428.

Lucas, J. M. (1994). How to achieve a robust process using response surface methodology. *J. Quality Technol.* **26**, 248–260.

Martin, R. J. (1985). Papadakis method. In: S. Kotz and N. L. Johnson, eds., *Encyclopedia of Statistical Science*, Vol. 6. Wiley, New York, 564–568.

Martin, R. J. and J. A. Eccleston (1991). Optimal incomplete block designs for general dependence structures. *J. Statist. Plann. Inference* **28**, 67–81.

McLean, R. A. and V. L. Anderson (1984). *Applied Fractional Factorial Designs*. Marcel Dekker, New York.

Mead, R. (1988). *The Design of Experiments*. Cambridge Univ. Press, Cambridge, UK.

Monod, H. (1991). Serially balanced factorial sequences. *J. Statist. Plann. Inference* **27**, 325–333.

Monod, H. and R. A. Bailey (1993). Valid restricted randomization for unbalanced designs. *J. Roy. Statist. Soc. Ser. B* **55**, 237–251.

Montgomery, D. C. (1991). *Design and Analysis of Experiments*. Wiley, New York.

Morgan, J. P. (1988). Balanced polycross designs. *J. Roy. Statist. Soc. Ser. B* **50**, 93–104.

Morgan, J. P. (1990). Some series constructions for two-dimensional neighbor designs. *J. Statist. Plann. Inference* **24**, 37–54.

Mount-Campbell, C. A. and J. B. Neuhardt (1981). On the number of 2^{n-p} fractional factorials of resolution III. *Comm. Statist. Theory Methods* **10**, 2101–2111.

Nelder, J. A. (1962). New kinds of systematic designs for spacing experiments. *Biometrics* **18**, 283–307.

Nester, M. (1994). HAHA designs. *Austral. J. Combinat.* **9**, 261–274.

Netto, E. (1893). Zur Theorie der Tripelsysteme. *Math. Ann.* **42**, 143–152.

Nguyen, N. K. and E. R. Williams (1993). An algorithm for constructing optimal resolvable row–column designs. *Austral. J. Statist.* **35**, 363–370.

Papadakis, J. (1937). Methode statistique pour des experiences sur champ. *Bull. Inst. Amel. Plantes, Salonique (Greece)* **23**.

Paterson, L. J., P. Wild and E. R. Williams (1988). An algorithm to generate designs for variety trials. *J. Agricult. Sci., Cambridge* **111**, 133–136.

Patterson, H. D. (1965). The factorial combination of treatments in rotation experiments. *J. Agricult. Sci.* **65**, 171–182.

Patterson, H. D. and R. A. Bailey (1978). Design keys for factorial experiments. *Appl. Statist.* **27**, 335–343.

Patterson, H. D. and E. R. Williams (1976). A new class of resolvable incomplete block designs. *Biometrika* **63**, 83–92.

Pearce, S. C. (1975). Row-and-column designs. *Appl. Statist.* **24**, 60–74.

Pearce, S. C. (1983). *The Agricultural Field Experiment.* Wiley, New York.

Pesti, G. M. (1991). Response surface approach to studying the protein and energy requirements of laying hens. *Poultry Sci.* **70**, 103–114.

Plackett, R. L. and J. P. Burman (1946). The design of optimum multifactorial experiments. *Biometrika* **33**, 305–325 and 328–332.

Preece, D. A. (1983). Latin squares, Latin cubes, Latin rectangles, etc. In: S. Kotz and N. L. Johnson, eds., *Encyclopedia of Statistical Science*, Vol. 4. Wiley, New York, 504–510.

Preece, D. A. (1994). Balanced Ouchterlony neighbour designs and quasi-Rees designs. *J. Combinat. Math. Combinat. Computing* **15**, 197–219.

Preece, D. A. (1994). Triple Youden rectangles: A new class of fully balanced combinatorial arrangements. *Ars Combinatoria* **37**, 175–182.

Read, C. B. (1988). Tukey's test for nonadditivity. In: S. Kotz and N. L. Johnson, eds., *Encyclopedia of Statistical Science*, Vol. 9. Wiley, New York, 364–366.

Sen, M. and R. Mukerjee (1987). Optimal repeated measurements designs under interaction. *J. Statist. Plann. Inference* **17**, 81–91.

Senn, S. J. (1988). Cross-over trials, carry-over effects and the art of self-delusion. *Statist. Medicine* **7**, 1099–1101.

Senn, S. J. (1993). *Cross-over Trials in Clinical Research.* Wiley, New York.

Shah, K. R. and B. K. Sinha (1993). Optimality aspects of row–column designs with non-orthogonal structure. *J. Statist. Plann. Inference* **36**, 331–346.

Shing, C. C. and K. Hinkelmann (1992). Repeated measurement designs for factorial treatments when high order interactions are negligible. *J. Statist. Plann. Inference* **31**, 81–91.

Street, A. P. and D. J. Street (1987). *Combinatorics of Experimental Design.* Clarendon Press, Oxford.

Street, A. P. and D. J. Street (1988). Latin squares and agriculture: The other bicentennial. *Math. Scientist* **13**, 48–55.

Street, D. J. (1986). Unbordered two-dimensional nearest neighbour designs. *Ars Combinatoria* **22**, 51–57.

Street, D. J. (1989). Combinatorial problems in repeated measurements designs. *Discrete Math.* **77**, 323–343.

Street, D. J. (1992). A note on strongly equi-neighboured designs. *J. Statist. Plann. Inference* **30**, 99–105.

Street, D. J. and A. P. Street (1985). Designs with partial nearest neighbour balance. *J. Statist. Plann. Inference* **12**, 47–59.

Subbarayan, A. (1992). On the application of a pure cyclic triple system for plant breeding experiments. *J. Appl. Statist.* **19**, 489–500.

Taguchi, G. (1987). *System of Experimental Design: Engineering Methods to Optimize Quality and Minimize Costs.* Kraus International Publications, White Plains, NY and American Supplier Institute, Dearborn, MI.

Taylor, A. J. and C. J. Smith (1992). Effect of sowing date and seeding rate on yield and yield components of irrigated canola (Brassica napus L.) grown on a red-brown earth in south-eastern Australia. *Austral. J. Agricultural Res.* **43**, 1629–1641.

Thayer, D. W. and R. A. Boyd (1991). Survival of *Salmonella typhimurium* ATCC 14028 on the surface of chicken legs or in mechanically deboned chicken meat gamma irradiated in air or vacuum at temperatures of $-20°$ to $20°$C. *Poultry Sci.* **70**, 1026–1033.

Tjur, T. (1993). An algorithm for the construction of block designs. *J. Statist. Plann. Inference* **36**, 277–282.

Turiel, T. P. (1988). A computer program to determine the defining contrasts and factor combinations for the two-level fractional factorial designs of resolution III, IV and V. *J. Quality Technol.* **20**, 267–272.

Venables, W. N. and J. A. Eccleston (1993). Randomized search strategies for finding optimal or near optimal block and row–column designs. *Austral. J. Statist.* **35**, 371–382.

Voss, D. T. and A. M. Dean (1988). On confounding in single replicate factorial experiments. *Utilitas Math.* **33**, 59–64.

Wahua, T. A. T. and D. A. Miller (1978). Relative yield totals and yield components of intercropped sorghum and soybeans. *Agronomy J.* **70**, 287–291.

Westlake, W. J. (1965). Composite designs based on irregular fractions of factorials. *Biometrics* **21**, 324–336.

Wild, P. R. and E. R. Williams (1987). The construction of neighbour designs. *Biometrika* **74**, 871–876.

Wilkinson, G. N., S. R. Eckert, T. W. Hancock and O. Mayo (1983). Nearest neighbour (NN) analysis of field experiments (with discussion). *J. Roy. Statist. Soc. Ser. B* **45**, 151–211.

Wilkinson, G. N. and O. Mayo (1982). Control of variability in field trials: An essay on the controversy between 'Student' and Fisher, and a resolution of it. *Utilitas Math.* **21B**, 169–188.

Williams, E. J. (1949). Experimental designs balanced for the estimation of residual effects of treatments. *Austral. J. Sci. Res. Ser. A* **2**, 149–168.

Williams, E. R. and A. C. Matheson (1994). Experimental design and analysis for use in tree improvement. CSIRO, Melbourne.

Williams, R. M. (1952). Experimental designs for serially correlated observations. *Biometrika* **39**, 151–167.

Yates, F. (1935). Complex experiments (with discussion). *J. Roy. Statist. Soc., Suppl.* **2**, 181–247.

Yates, F. (1936). Incomplete Latin squares. *J. Agricul. Sci.* **26**, 301–315.

Yates, F. (1937). *The Design and Analysis of Factorial Experiments*, 35. Imperial Bureau of Soil Science, Technical Communication.

Yates, F. (1940). Lattice squares. *J. Agricul. Sci.* **30**, 672–687.

S. Ghosh and C. R. Rao, eds., *Handbook of Statistics, Vol. 13*

22

Block Designs: Their Combinatorial and Statistical Properties

Tadeusz Caliński and Sanpei Kageyama

0. Introduction

The basic principles of experimental design as we know them today, "replication", "randomization" and "local control", were formulated by Sir Ronald Fisher in his famous book, Fisher (1925), and in his manual for "The arrangement of field experiment", Fisher (1926). Replication of treatments was considered as a way of increasing accuracy and providing a basis for the estimation of error, its variance. The need for randomization (a most fruitful Fisher's original contribution) was recognized as a necessary condition for obtaining a valid estimate of error and, consequently, for a valid use of the test of significance provided by the analysis of variance (another of Fisher's brilliant contributions, according to Yates (1965)). Awarness of the undesirable effects of uncontrollable variations in the environment of the experiment led to the request that the experimental units are chosen as alike as possible for the treatment comparisons within each replicate. Clearly, in addition to replication, this local control of the units provided a means of increasing accuracy with a given amount of experimental material (cf. Yates, 1975). In an agricultural field trial this was achievable by recognizing the main "gradient" of the existing variation of soil fertility and stratifying the material accordingly, i.e., by grouping experimental units (contiguous field-plots) into blocks, each comprising possibly uniform units in a number equal to the number of treatments. Later incomplete block designs were introduced for field trials with large numbers of treatments. Frank Yates (1936a, b), trying to overcome the difficulty of including all experimental treatments in each of the blocks formed according to the principle of within-block homogeneity, initiated the use of designs with incomplete blocks to which the treatments are so allotted that every two treatments occur together in a block with equal frequency. These designs have been called balanced incomplete block (BIB) designs.

The idea of having fewer plots per block than the total number of treatments gave rise to an enormous development of the theory and practice of block designs. In fact, over the past half century, there has been a considerable amount of research activity devoted to the invention of experimental designs of a certain kind, such as the BIB designs, partially balanced incomplete block (PBIB) designs, square and rectangular lattice designs and cyclic designs. In particular, much work on problems of general designs itself stems from the sort of activity that arose in connection with BIB design theory.

Though interest in BIB designs was greatly enhanced in the thirtieths, their examples appeared much earlier in mathematical literature (Woolhouse, 1844; Kirkman, 1847, 1850a, b; Steiner, 1853). Relatively recent development in the theory of BIB designs with constant coincidence number being one (Steiner systems) and its generalization (t-designs), have been surveyed in books edited by Colbourn and Mathon (1987) and by Lindner and Rosa (1980). The family of block designs having constant coincidence number is one of the most important families of experimental designs. Their importance is due to their statistical optimalities, desirable symmetries for analyses and interpretations, and their usefulness for constructing other important designs and combinatorial structures. Research in the area of block designs has been steadily and rapidly growing, especially during the last four decades. Hence, it is a right time to review the basic knowledge on block designs, i.e., to give a sound and possibly thorough up-to-date exposition of the present state of affair in the theory of block designs.

Not all problems and solutions concerning block designs can be reported fully in the present chapter. It seems important, however, at least to draw attention to the main concepts underlying the development of the theory, to indicate the leading contributions, and to clarify the relevant terminology. For that, the most basic tools and terms are exposed in Section 1. Those concerning the commonly used notions of balance are presented and discussed in Section 2. The problem of classification of block designs is considered in Section 3, where also a new proposal is made towards a unified concept of classification based on efficiency factors of the design. Extension of the ordinary block designs to those with two strata of blocks used to control the variation in experimental material is taken into consideration in Section 4. Most of the concepts and problems considered in the first four sections of the chapter are related to constructional aspects of block designs. As combinatorial in their nature, they call for solutions within the theory of combinatorics, whose current development itself owes a great deal to the interest in block designs.

In general, the theory of discrete designs is intimately related to many other areas of combinatorial theory. Block designs have opened up many interesting and challenging problems in combinatorial mathematics. The fundamental problems related to block designs, as considered from a combinatorial point of view, are their *existence*, *uniqueness*, *extendability* and *characterization* (cf. Kageyama, 1993, Section 5).

Most of the problems on block designs, with the exception of group-theoretical problems, belong to one of the above categories. Here only problems related to the first and fourth are occasionally considered. Since the constructional problems have been discussed thoroughly in several books, for example, in Raghavarao (1971), Clatworthy (1973), John (1980), Lindner and Rosa (1980), Beth, Jungnickel and Lenz (1985), Hall (1986), there is no need to repeat the discussion here. Regarding the characterization of block designs in general, one can be referred to Nigam, Puri and Gupta (1988). More full treatments of the combinatorial design theory appear in texts by Beth, Jungnickel and Lenz (1985), by Hughes and Piper (1985), by Street and Street (1987), and by Wallis (1988). A wide survey on combinatorial applications of block designs can be found in Colbourn and Van Oorschot (1988).

On the side of the statistical analysis of results obtainable from experiments conducted in block designs, vast literature is available, starting from the early works of Yates (1936a, b). Most of the literature makes references to the so-called intra-block analysis only. This analysis, however, does not provide full information on treatment differences, unless the within-block homogeneity is completely achieved by a successful choice of the experimental units and their block structure. Such ideal situation is seldom met in practice and, therefore, a more realistic model is to be considered as a basis for the statistical analysis of experimental data. In deriving an appropriate model, the randomization procedures involved in laying out the experiment should definitely be taken into account, particularly as the main purpose of the randomization is to allow the experimental data to be considered as observations on random variables of certain homogeneous distributions. Several authors have tried to build models that fully recognize the randomization techniques employed. Among the various approaches of particular interest here are two main lines of development, different in attitude though not entirely disjoint in results, along which the randomization is incorporated into the model building. One, as indicated by Scheffé (1959, p. 291), originates from Neyman (1923, 1935) and was later extended by others, particularly by Kempthorne (1952, 1955). Another was initiated by Nelder (1954), under some influence of Anscombe (1948). Most of the references concerning the first development can be found in the papers by White (1975) and Kempthorne (1977), while the essential references concerning the second line has been given by Bailey (1981), and those related to further developments of this line, by Bailey and Rowley (1987).

With regard to block designs, the randomization model leads to an analytical procedure which takes into account also the information on treatment differences that are partially or totally confounded with block differences. After Yates (1939, 1940), this procedure is called the recovery on inter-block information. It is presented here for a general block design in Section 5, the earlier sections of the chapter being related mainly to the intra-block analysis, according to the most commonly adopted attitude towards the theory of block designs. Finally, since the property of simple combinability of the intra-block and inter-block information (as considered by Martin and Zyskind (1966)) appears to hold only under general balance of the design, some attention to this concept is given in Section 6. In its conclusion, the special role of basic contrasts of the design (introduced in Section 1.5) played in the evaluation of design properties is strongly emphasized.

1. Preliminaries

1.1. General terminological remarks

As already mentioned (in Section 0), block designs were originally used in agricultural experiments with the aim of allowing all treatments to be compared within similar conditions. A compact set of experimental units (plots) possibly uniform in their conditions and in a number equal to the number of treatments was called a block and, since the treatments were assigned to units within such blocks at random, the resulting experiment was said to be designed in "randomized blocks" (cf., e.g., Cochran and

Cox, 1957, Section 4.2). The design has also been called "the randomized-blocks design" (cf., e.g., Scheffé, 1959, Section 4.2) or "the randomized block design" (cf., e.g., Finney, 1960, Section 3.1) or "the randomized complete block design" (cf., e.g., Federer, 1955, Chapter 5; Margolin, 1982, p. 289). Although blocks of plots were first used in connection with this design, nowadays it is no more the only one design that utilizes blocks of experimental units.

In the history of developments of experimental designs, the system of blocking has been extended in various ways. One essential extension is that connected with the difficulty of comprising all the experimental treatments under study in each of the blocks formed according to the principle of within-block homogeneity. This difficulty has arisen originally in the context of crop variety trials, in which usually a large number of treatments, i.e., varieties, are to be compared. The solution found by Yates (1936a, b) led to the use of "balanced incomplete blocks" with fewer units (plots) per block than the total number of treatments or varieties. With this development many new ideas and terms have emerged. They have been introduced to describe the various properties of block designs, which have become much more flexible in their construction once the main condition of the randomized blocks, that all treatments are to be compared in each block, had been abandoned. Many of the new terms are connected merely with the constructional aspects, as already indicated in Section 0, but several terms concern certain important statistical properties of the designs. All these terms, and those of the latter type in particular, deserve unambiguous definitions and explanations. In this section and also throughout the subsequent sections (Sections 2–4) attempt will be made to give clear meaning and practical sense to some of the most basic terms and concepts that are used in connection with block designs in general. Some other terms will be explained and discussed later (in Sections 5 and 6), when considering a general theory of the statistical analysis for experiments in block designs based on the randomization model. It will, however, be assumed from the very beginning that the reader is well acquainted with such elementary terms as experimental unit (plot), treatment, treatment replication, etc., that is with terms which are usually used in basic texts on experimental design, one of the most elucidating being the book by Cox (1958).

At this point one should also be aware of the distinction between the term "randomized block design" (design in randomized blocks) and the term "block design". The former is reserved solely for the classical design composed of "complete" blocks, i.e., such which allows all the treatments to be allocated in each block. The latter term is of a broader sense. It covers all designs that utilize experimental units grouped into blocks. Although in these designs randomization is applied according to the same original principle, the adjective "randomized" is commonly used only for naming the classical design of randomized blocks (cf. Pearce, 1983, p. 96).

One has also to distinct between the description of a block design indicating the allocation of treatments in different blocks, and the actual assignment of treatments to the units within blocks, obtained after completing the randomizations. The first may be given either in form of a plan, as it is frequently shown in classical books on experimental designs, such as, e.g., that by Cochran and Cox (1957), or in form of the "incidence matrix", as it is applied, e.g., by Raghavarao (1971) and by Pearce (1983).

As to the second, its description is important for building a linear model on which the statistical analysis is to be based. In this, the appropriate design matrices, for blocks and for treatments, allow to indicate relations between the observed responses of units in different blocks to the treatments applied, on one side, and the corresponding unknown parameters or random variables for which statistical inferences are to be drawn, on the other. All the three matrices will be defined precisely in Section 1.2.

1.2. Basic descriptive tools and terms

Suppose that v treatments are to be compared in an experiment on n units (plots) which are arranged in b groups called blocks. A design which determines the allocation of the treatments to units grouped in such blocks is called a block design.

Any block design can be described by its $v \times b$ incidence matrix $\boldsymbol{N} = [n_{ij}]$, with a row for each treatment and a column for each block, where n_{ij} is the number of units in the jth block that receive the ith treatment ($i = 1, 2, \ldots, v$; $j = 1, 2, \ldots, b$). Therefore, a block design may be called by its incidence matrix, and thus the expression "a design $\boldsymbol{N}$" will be used sometimes for convenience. To match the design to variables observable on the experimental units, two other matrices are needed when modelling the analysis. Let $\boldsymbol{D}$ be a $b \times n$ matrix, with a row for each block and a column for each unit, and with elements 0 and 1, an element being 1 if the unit to which the column corresponds is in the block corresponding to the row, and 0 otherwise. The transposed matrix $\boldsymbol{D}'$ is called the "design matrix for blocks". Similarly, let Δ be a $v \times n$ matrix, with a row for each treatment and a column for each unit, and with elements 0 and 1, an element being 1 if the unit to which the column corresponds receives the treatment corresponding to the row, and 0 otherwise. The transposed matrix Δ' is then called the "design matrix for treatments". Furthermore, let $\boldsymbol{k} = [k_1, k_2, \ldots, k_b]'$ be the vector of block sizes and $\boldsymbol{r} = [r_1, r_2, \ldots, r_v]'$ the vector of treatment replications. Then it can easily be checked that the following relations hold:

$$\Delta \boldsymbol{D}' = \boldsymbol{N},$$

$$\boldsymbol{D}'\mathbf{1}_b = \mathbf{1}_n, \quad \boldsymbol{D}\mathbf{1}_n = \boldsymbol{k} = \boldsymbol{N}'\mathbf{1}_v, \quad \boldsymbol{D}\boldsymbol{D}' = \boldsymbol{k}^\delta = \operatorname{diag}[k_1, k_2, \ldots, k_b],$$

$$\Delta'\mathbf{1}_v = \mathbf{1}_n, \quad \Delta\mathbf{1}_n = \boldsymbol{r} = \boldsymbol{N}\mathbf{1}_b, \quad \Delta\Delta' = \boldsymbol{r}^\delta = \operatorname{diag}[r_1, r_2, \ldots, r_v] \tag{1.1}$$

and

$$n = \mathbf{1}_b'\boldsymbol{k} = \mathbf{1}_v'\boldsymbol{r} = \mathbf{1}_v'\boldsymbol{N}\mathbf{1}_b.$$

The numbers v, b, r_i, k_j, for $i = 1, 2, \ldots, v$ and $j = 1, 2, \ldots, b$, are called the parameters of the design. They all result from the incidence matrix $\boldsymbol{N} = [n_{ij}]$. If $n_{ij} = 0$ or 1, for any i and j, the design is called binary (in that case $r_i \leqslant b$ for all i and $k_j \leqslant v$ for all j). If all r_i are equal, the design is called equireplicate (or equireplicated). If all k_j are equal, the design is called proper (or equiblock-sized). If the set of v

treatments can be divided into two or more subsets such that for any two treatments not belonging to the same subset, say treatments ith and i'th, the inequality $n_{ij} > 0$ implies $n_{i'j} = 0$ for each j, then the design is called disconnected. Otherwise, it is called connected.

Note that the disconnectedness, as defined above, appears when different subsets of treatments are allotted to different blocks so that none of the treatments from one subset can occur in a block together with a treatment from another subset. Obviously, in such a case not all comparisons (contrasts) between treatments can be made intra-block.

In the analysis of data obtained from an experiment in a block design several matrices related to those already given above are helpful. In particular, use is often made of the $n \times n$ projection matrix

$$\boldsymbol{\phi} = \boldsymbol{I}_n - \boldsymbol{D}'\boldsymbol{k}^{-\delta}\boldsymbol{D} \tag{1.2}$$

(the symbol $\boldsymbol{\phi}$ being adopted from Pearce (1983, p. 59)). Note that

$$\boldsymbol{\phi}' = \boldsymbol{\phi}, \quad \boldsymbol{\phi}\boldsymbol{\phi} = \boldsymbol{\phi}, \quad \boldsymbol{\phi}\boldsymbol{1}_n = \boldsymbol{0}, \quad \boldsymbol{\phi}\boldsymbol{D}' = \boldsymbol{O} \tag{1.3}$$

and that $\boldsymbol{\phi}$ is of rank $n - b$. The matrix $\boldsymbol{\phi}$ appears often in conjunction with the matrix $\boldsymbol{\Delta}$, in particular in the product

$$\boldsymbol{\Delta}\boldsymbol{\phi} = \boldsymbol{\Delta} - \boldsymbol{\Delta}\boldsymbol{D}'\boldsymbol{k}^{-\delta}\boldsymbol{D} = \boldsymbol{\Delta} - \boldsymbol{N}\boldsymbol{k}^{-\delta}\boldsymbol{D}. \tag{1.4}$$

Its transpose, $\boldsymbol{\phi}\boldsymbol{\Delta}'$, is an $n \times v$ matrix which may be called the "adjusted design matrix for treatments", adjusted in the sense of eliminating block contributions. An analogous matrix,

$$\boldsymbol{\phi}_* = \boldsymbol{I}_n - \boldsymbol{\Delta}'\boldsymbol{r}^{-\delta}\boldsymbol{\Delta}, \tag{1.5}$$

with the properties

$$\boldsymbol{\phi}_*' = \boldsymbol{\phi}_*, \quad \boldsymbol{\phi}_*\boldsymbol{\phi}_* = \boldsymbol{\phi}_*, \quad \boldsymbol{\phi}_*\boldsymbol{1}_n = \boldsymbol{0}, \quad \boldsymbol{\phi}_*\boldsymbol{\Delta}' = \boldsymbol{O} \tag{1.6}$$

and of rank $n - v$, gives

$$\boldsymbol{D}\boldsymbol{\phi}_* = \boldsymbol{D} - \boldsymbol{D}\boldsymbol{\Delta}'\boldsymbol{r}^{-\delta} = \boldsymbol{D} - \boldsymbol{N}'\boldsymbol{r}^{-\delta}\boldsymbol{\Delta}. \tag{1.7}$$

Its transpose, $\boldsymbol{\phi}_*\boldsymbol{D}'$, is an $n \times b$ matrix which may be called the "adjusted design matrix for blocks", the adjustment being in the sense of eliminating treatment contributions.

The meaning of all these matrices becomes clear when referring to a linear model underlying the statistical analysis of the experimental data. The most standard model will be discussed in Section 1.4. Here it will be sufficient to notice that, after eliminating block effects, the expectation vector of the data is as in (1.22), with the dispersion (variance-covariance) matrix of the form (1.23). Thus, in describing properties of a block design, the matrices (1.2) and (1.4) play an essential role, and also the dual matrices, (1.5) and (1.7).

If the adjusted design matrices for blocks and for treatments, $\boldsymbol{\phi}\boldsymbol{\Delta}'$ and $\boldsymbol{\phi}_*\boldsymbol{D}'$, are mutually orthogonal, then the design is called orthogonal (see also Caliński, 1993a, Definition 2.7 and the discussion following it). From this definition, an evident statistical property of an orthogonal design is that, under it, the estimates of the intra-block contrasts of treatment effects are uncorrelated with the estimates of the dual contrasts of block effects.

Another matrix that is important, both in the analysis of data and in studying design properties, is the $v \times v$ matrix

$$\boldsymbol{\Delta}\boldsymbol{\phi}\boldsymbol{\Delta}' = \boldsymbol{r}^{\delta} - \boldsymbol{N}\boldsymbol{k}^{-\delta}\boldsymbol{N}' = \boldsymbol{C} \quad \text{(say)}, \tag{1.8}$$

called the "coefficient matrix" (Pearce, 1983, p. 59) or, simply, the "C-matrix" (Raghavarao, 1971, p. 49) or the "(reduced) intra-block matrix", due to its relation to the so-called "(reduced) intra-block equations", i.e., normal equations leading to the least squares intra-block estimates of the treatment parameters (Kempthorne, 1952, Section 6.3; Chakrabarti, 1962; John, 1980, Section 2.2). It is also called the "information matrix" (John, 1987, p. 8; Shah and Sinha, 1989, p. 2). Some authors use $\boldsymbol{A}$ rather than $\boldsymbol{C}$ to denote the coefficient matrix (e.g., John, 1987; Nigam et al., 1988, p. 11). In the present text the term C-matrix and the symbol $\boldsymbol{C}$, exchangeably with $\boldsymbol{\Delta}\boldsymbol{\phi}\boldsymbol{\Delta}'$, will be used throughout.

It may be mentioned now that the majority of literature on block designs concerns binary designs (see, e.g., Raghavarao, 1971), and among these designs the connected proper equireplicate designs are the most common (see also John, 1980). However, there are situations when certain contrasts (one or more) have deliberately to be totally confounded with blocks in order to permit the gaining of the advantages of small blocks for the remaining contrasts of interest. This idea of confounding gives rise to disconnected block designs. (See, e.g., Finney, 1960, p. 67; also Pearce, 1983, Section 3.7).

There are several equivalent definitions of connectedness (see Caliński, 1993a, Section 2). Originally the connectedness was defined as follows:

DEFINITION 1.1 (Bose, 1950; Raghavarao, 1971; Ogawa, 1974). A design is said to be connected if, given any two treatments i and i', it is possible to construct a chain of treatments $i = i_0, i_1, \ldots, i_n = i'$ such that every consecutive two treatments in the chain occur together in a block.

Certainly, this definition is equivalent to that already given above, and thus it has the same implications.

A portion of a block design is said to be connected if it is connected in the above sense with regard to the corresponding submatrix of the $v \times b$ incidence matrix of the original design. Any block design must break up into a number of connected portions such that any treatment belonging to one portion is disconnected from every treatment belonging to any of the other portions. That is, in terms of the incidence matrix, the original design $\boldsymbol{N}$ may be expressed (possibly after reordering its rows and columns) as $\boldsymbol{N} = \text{diag}[\boldsymbol{N}_1 : \boldsymbol{N}_2 : \cdots : \boldsymbol{N}_g]$, where $\boldsymbol{N}_i$'s correspond to connected portions for all $i = 1, 2, \ldots, g$ ($\geqslant 1$).

Now, another equivalent and convenient way of defining disconnectedness is the following (cf. Baksalary and Tabis, 1985).

DEFINITION 1.2. A design is said to be disconnected of degree $g - 1$ if, after an appropriate ordering of rows and columns, its incidence matrix $\boldsymbol{N}$ can be written as $\boldsymbol{N} = \text{diag}[\boldsymbol{N}_1 : \boldsymbol{N}_2 : \cdots : \boldsymbol{N}_g]$ for $g > 1$, where $\boldsymbol{N}_i$'s are the $v_i \times b_i$ incidence submatrices corresponding to connected portions for all $i = 1, 2, \ldots, g$, in accordance with partitions $v = v_1 + \cdots + v_g$ of treatments and $b = b_1 + \cdots + b_g$ of blocks.

Note that in Definition 1.2 $g = 1$ means that the design is connected, and if $g > 1$ then the design is strictly disconnected of degree $g - 1$. When this is the case, then to make the design connected one needs to add at least $g - 1$ experimental units, each with a suitable treatment applied. The problem of determining the minimum number of total experimental units for an experiment to be connected under given v and b has been studied by Kageyama (1985) and Baksalary and Tabis (1985). A review of the results is given in Kageyama (1993).

1.3. Some auxiliary results

In this section several simple results that will be often utilized subsequently are given. They mainly concern the matrices defined in Section 1.2, the matrices (1.4) and (1.8) in particular.

LEMMA 1.1. *The matrix $\boldsymbol{C} = \Delta\phi\Delta'$ is positive semidefinite and of rank equal to that of $\phi\Delta'$, (say) $h \leqslant v - 1$.*

For a proof see Caliński (1993a, Lemma 2.1).

LEMMA 1.2. *The rank of $\boldsymbol{C}$ is equal to $v - 1$ if the design is connected, and is equal to $v - g$ if it is disconnected of degree $g - 1$.*

For a proof see Caliński (1993a, Lemma 2.2).

LEMMA 1.3. *If for a block design the incidence matrix $\boldsymbol{N}$ is of rank $v - \rho$ and the coefficient matrix $\boldsymbol{C}$ is of rank h, then*

$$\boldsymbol{N}\boldsymbol{k}^{-\delta}\boldsymbol{N}' = \boldsymbol{r}^{\delta}\left(\sum_{i=\rho+1}^{h} \mu_i \boldsymbol{s}_i\boldsymbol{s}_i' + \sum_{i=h+1}^{v} \boldsymbol{s}_i\boldsymbol{s}_i'\right)\boldsymbol{r}^{\delta} \tag{1.9}$$

and

$$\boldsymbol{C} = \boldsymbol{r}^{\delta} - \boldsymbol{N}\boldsymbol{k}^{-\delta}\boldsymbol{N}' = \boldsymbol{r}^{\delta}\left(\sum_{i=1}^{\rho} \boldsymbol{s}_i\boldsymbol{s}_i' + \sum_{i=\rho+1}^{h} \varepsilon_i \boldsymbol{s}_i\boldsymbol{s}_i'\right)\boldsymbol{r}^{\delta}, \tag{1.10}$$

where

$$0 = \mu_1 = \cdots = \mu_\rho < \mu_{\rho+1} \leqslant \cdots \leqslant \mu_h < \mu_{h+1} = \cdots = \mu_v = 1$$

are the eigenvalues of $\boldsymbol{N}\boldsymbol{k}^{-\delta}\boldsymbol{N}'$ *with respect to* $\boldsymbol{r}^\delta$, *and the vectors* $\boldsymbol{s}_1, \ldots, \boldsymbol{s}_\rho, \boldsymbol{s}_{\rho+1}, \ldots, \boldsymbol{s}_h, \boldsymbol{s}_{h+1}, \ldots, \boldsymbol{s}_v$ *are the corresponding* $\boldsymbol{r}^\delta$*-orthonormal eigenvectors of* $\boldsymbol{N}\boldsymbol{k}^{-\delta}\boldsymbol{N}'$ *with respect to* $\boldsymbol{r}^\delta$, *i.e., such that* $\boldsymbol{N}\boldsymbol{k}^{-\delta}\boldsymbol{N}'\boldsymbol{s}_i = \mu_i \boldsymbol{r}^\delta \boldsymbol{s}_i$ *and* $\boldsymbol{s}_i'\boldsymbol{r}^\delta\boldsymbol{s}_{i'} = \delta_{ii'}$ *(the Kronecker delta) for* $i, i' = 1, 2, \ldots, v$, *with* $\boldsymbol{s}_v = n^{-1/2}\boldsymbol{1}_v$, *and where* $0 < \varepsilon_i = 1 - \mu_i$ *for* $i = \rho + 1, \rho + 2, \ldots, h$.

The results in Lemma 1.3 follow from the singular value decomposition

$$\boldsymbol{r}^{-\delta}\boldsymbol{N}\boldsymbol{k}^{-\delta} = \sum_{i=\rho+1}^{v} \mu_i^{1/2}\boldsymbol{s}_i\boldsymbol{t}_i'. \tag{1.11}$$

For a detailed proof see Caliński (1993a, Lemma 2.3).

REMARK 1.1. On account of (1.11), expressions analogous to (1.9) and (1.10) can be given for $\boldsymbol{N}'\boldsymbol{r}^{-\delta}\boldsymbol{N}$ and $\boldsymbol{k}^\delta - \boldsymbol{N}'\boldsymbol{r}^{-\delta}\boldsymbol{N}$, respectively, the nonzero eigenvalues of $\boldsymbol{N}'\boldsymbol{r}^{-\delta}\boldsymbol{N}$ with respect to $\boldsymbol{k}^\delta$ being exactly the same as those of $\boldsymbol{N}\boldsymbol{k}^{-\delta}\boldsymbol{N}'$ with respect to $\boldsymbol{r}^\delta$.

COROLLARY 1.1. *A block design is orthogonal if and only if the multiplicity of the unit eigenvalue of its matrix* $\boldsymbol{C}$ *with respect to* $\boldsymbol{r}^\delta$ *is equal to the rank of the matrix, i.e.,* $\rho = h$ *in* (1.10).

For a proof see Caliński (1993a, Corollary 2.1).

COROLLARY 1.2. *A block design satisfies Fisher's inequality* $v \leqslant b$ *if any of the following two equivalent conditions hold:*
(a) *the matrix* $\boldsymbol{N}\boldsymbol{k}^{-\delta}\boldsymbol{N}'$ *has no zero eigenvalue,*
(b) *the matrix* $\boldsymbol{C}$ *has no unit eigenvalue with respect to* $\boldsymbol{r}^\delta$.

PROOF. Since $\operatorname{rank}(\boldsymbol{N}) = v - \rho \leqslant b$, the conditions follow from Lemma 1.3. □

Since both (a) and (b) in Corollary 1.2 are equivalent to the condition that the rows of the incidence matrix $\boldsymbol{N}$ are linearly independent, the inequality $v \leqslant b$, originally established by Fisher (1940) for BIB designs, may be replaced by the row independence condition (see Baksalary, Dobek and Kala, 1980; Baksalary and Puri, 1988).

The following results are related to a generalized inverse (a g-inverse) of the matrix $\boldsymbol{C}$, i.e., to a matrix, denoted by $\boldsymbol{C}^-$, which satisfies the condition

$$\boldsymbol{C}\boldsymbol{C}^-\boldsymbol{C} = \boldsymbol{C} \tag{1.12}$$

(see Rao and Mitra, 1971, Section 2.2).

LEMMA 1.4. *Regardless of the choice of* $\boldsymbol{C}^-$,

$$\boldsymbol{C}\boldsymbol{C}^-\Delta\boldsymbol{\phi} = \Delta\boldsymbol{\phi}. \tag{1.13}$$

PROOF. The equality (1.13) follows directly from Lemma 2.2.6(b) of Rao and Mitra (1971) when taking $A = \phi\Delta'$. (See also Pcarcc, 1983, Lemma 3.2.D.) □

LEMMA 1.5. *The matrix* $\phi\Delta' C^- \Delta\phi$ *is invariant for any choice of* C^-, *is idempotent, symmetric and of rank* $h = \text{rank}(C)$.

PROOF. This result follows from Lemma 2.2.6(d) of Rao and Mitra (1971), on account of (1.3), and from Lemmas 1.1 and 1.4. (See also Pearce, 1983, Lemma 3.2.E.) □

From Lemma 1.3 the following useful result is evidently obtainable, by referring to (1.12).

COROLLARY 1.3. *One possible choice of* C^- *is the matrix*

$$\sum_{i=1}^{\rho} s_i s_i' + \sum_{i=\rho+1}^{h} \varepsilon_i^{-1} s_i s_i',$$

where the notation is as in Lemma 1.3.

A matrix which often will be applied is

$$\psi = \phi - \phi\Delta' C^- \Delta\phi = \phi(I_n - \Delta' C^- \Delta)\phi. \tag{1.14}$$

It can easily be shown, on account of (1.3), that

$$\psi\phi = \psi, \quad \psi 1_n = 0, \quad \psi D' = O$$

and, applying Lemmas 1.4 and 1.5, that

$$\psi' = \psi, \quad \psi\psi = \psi \quad \text{and} \quad \psi\Delta' = O. \tag{1.15}$$

Further result concerning the matrix (1.14) is the following.

LEMMA 1.6. *The matrix* ψ *is invariant for any choice of* C^-, *and is of rank (and trace) equal to* $n - b - h$.

PROOF. This follows immediately from Lemma 1.5, from (1.14) and from the readily seen result that $\text{tr}(\phi) = n - b$, on account of (1.2) and (1.3). □

Now, from (1.1), Lemma 1.5, the formulae (1.3), (1.6) and (1.15), and on account of Chapter 5 in Rao and Mitra (1971), the following can be said.

COROLLARY 1.4. *All the five matrices* $D' k^{-\delta} D$, $\phi\Delta' C^- \Delta\phi$, ϕ, ϕ_* *and* ψ *are orthogonal projectors.*

Thus, the following notation can be used:

$$D' k^{-\delta} D = P_{D'}, \tag{1.16}$$

$$\phi\Delta' C^- \Delta\phi = P_{\phi\Delta'}, \tag{1.17}$$

$$\phi = I_n - P_{D'} = P_{(D')^\perp}, \tag{1.18}$$

$$\phi_* = I_n - P_{\Delta'} = P_{(\Delta')^\perp} \tag{1.19}$$

and

$$\psi = P_{(D')^\perp} - P_{\phi\Delta'}. \tag{1.20}$$

Formula (1.20) shows that ψ is the orthogonal projector on the subspace $\mathcal{C}^\perp(D') \cap \mathcal{C}^\perp(\phi\Delta')$, since $P_{(D')^\perp} P_{\phi\Delta'} = P_{\phi\Delta'} P_{(D')^\perp} = P_{\phi\Delta'}$ (see Rao and Mitra, 1971, Theorem 5.1.3). On the other hand, since $\mathcal{C}^\perp(A) = \mathcal{N}(A')$, the null space of A (see Seber, 1980, Lemma 1.2.1), the subspace on which ψ projects can, equivalently, be written as

$$\mathcal{N}(\Delta\phi) \cap \mathcal{N}(D) = \mathcal{N}\begin{bmatrix} \Delta\phi \\ D \end{bmatrix}. \tag{1.21}$$

But, since $\Delta\phi = \Delta - \Delta D' k^{-\delta} D$, the subspace (1.21) is in fact equal to the null space of $(\Delta' : D')'$, and thus is the orthogonal complement of $\mathcal{C}(\Delta' : D')$, i.e., $\mathcal{C}^\perp(\Delta' : D')$. (By a similar argument it can be shown that $\psi_* = P_{(\Delta')^\perp} - P_{\phi_* D'}$ is the orthogonal projector on the subspace $\mathcal{C}^\perp(\Delta') \cap \mathcal{C}^\perp(\phi_* D')$, which in turn is equal to $\mathcal{C}^\perp(D' : \Delta')$, and thus $\psi_* = \psi$.) The above orthogonal projectors will be applied in the next section and also in Sections 5 and 6.

1.4. Intra-block analysis

The meaning of the various definitions and algebraical formulae introduced so far will become apparent when referring to a model on which the statistical analysis of the experimental data is to be based. Most of the statistical literature on block designs makes reference to a model in which the probabilistic assumptions are such that the $n \times 1$ vector y of the variables observable on the n units of the experiment has, after eliminating the block effects by the projection ϕy, the properties

$$\mathrm{E}(\phi y) = \phi\Delta'\tau = \Delta'\tau - D' k^{-\delta} D\Delta'\tau \tag{1.22}$$

and

$$\mathrm{Var}(\phi y) = \phi\sigma^2, \tag{1.23}$$

where $\tau = [\tau_1, \tau_2, \ldots, \tau_v]'$ is a vector of treatment parameters and σ^2 is a common variance. The model obtained from the projection ϕy, with the above properties, is usually called the "intra-block submodel". On it the so-called "intra-block" analysis is based (see, e.g., John, 1980, Chapter 2; John, 1987, Chapter 1; also Caliński and Kageyama, 1991, Section 3.1.)

The main result concerning estimation under the intra-block submodel can be stated as follows.

THEOREM 1.1. *Under the submodel* $\phi\boldsymbol{y}$, *with the properties* (1.22) *and* (1.23), *a function* $\boldsymbol{w}'\phi\boldsymbol{y}$ *is uniformly the best linear unbiased estimator* (BLUE) *of* $\boldsymbol{c}'\boldsymbol{\tau}$ *if and only if* $\phi\boldsymbol{w} = \phi\Delta'\boldsymbol{s}$, *where the vectors* $\mathbf{c}$ *and* $\boldsymbol{s}$ *are in the relation* $\mathbf{c} = \Delta\phi\Delta'\boldsymbol{s}$, *i.e.,* $\mathbf{c} = \boldsymbol{C}\boldsymbol{s}$, *where* $\boldsymbol{C}$ *is as in* (1.8).

For a proof, following from Theorem 3 of Zyskind (1967), see Caliński and Kageyama (1991, Theorem 3.1).

REMARK 1.2. Since $\mathbf{1}_v'\Delta\phi = \mathbf{0}'$, from (1.1) and (1.3), the only parametric functions for which the BLUEs may exist under $\phi\boldsymbol{y}$ are contrasts, i.e., functions of the type $\boldsymbol{c}'\boldsymbol{\tau}$ with $\boldsymbol{c}'\mathbf{1}_v = 0$.

If $\boldsymbol{c}'\boldsymbol{\tau}$ is a contrast, and the condition of Theorem 1.1 is satisfied, then its BLUE under $\phi\boldsymbol{y}$ is of the form

$$\widehat{\boldsymbol{c}'\boldsymbol{\tau}} = \boldsymbol{s}'\Delta\phi\boldsymbol{y} \tag{1.24}$$

and its variance is of the form

$$\mathrm{Var}(\widehat{\boldsymbol{c}'\boldsymbol{\tau}}) = \boldsymbol{s}'\Delta\phi\Delta'\boldsymbol{s}\sigma^2 = \boldsymbol{c}'(\Delta\phi\Delta')^-\boldsymbol{c}\sigma^2, \tag{1.25}$$

i.e., $\boldsymbol{c}'\boldsymbol{C}^-\boldsymbol{c}\sigma^2$ (cf. Pearce, 1983, p. 62).

REMARK 1.3. Since $\mathrm{Var}(\phi\boldsymbol{y})\phi\Delta' = \phi\Delta'\sigma^2$, which implies that both $\phi\Delta'\boldsymbol{s}$ and $\mathrm{Var}(\boldsymbol{y})\phi\Delta'\boldsymbol{s}$ belong to $\mathcal{C}(\phi\Delta')$ for any $\boldsymbol{s}$, it follows from Theorem 1.1, on account of Theorem 4 of Zyskind (1967), that the BLUEs under the intra-block submodel $\phi\boldsymbol{y}$, with the moments (1.22) and (1.23), can equivalently be obtained under a simple alternative model in which the variance-covariance matrix (1.23) is replaced by that of the form $\boldsymbol{I}_n\sigma^2$, i.e., that $\boldsymbol{s}'\Delta\phi\boldsymbol{y}$ is both the simple least square estimator (SLSE) and the BLUE of its expectation, for any $\boldsymbol{s}$. (See also Rao and Mitra, 1971, Section 8.2; Kala, 1991, p. 20.)

From Remark 1.3 it follows, in particular, that the BLUE of the expectation vector (1.22) can be obtained by a simple least squares procedure (see, e.g., Seber, 1980, Chapter 3), in the form

$$\mathrm{E}(\widehat{\phi\boldsymbol{y}}) = \boldsymbol{P}_{\phi\Delta'}\boldsymbol{y}.$$

Furthermore, it follows that the vector $\phi\boldsymbol{y}$ can be decomposed as (cf. Rao, 1974, Section 3)

$$\phi\boldsymbol{y} = \boldsymbol{P}_{\phi\Delta'}\boldsymbol{y} + (\phi - \boldsymbol{P}_{\phi\Delta'})\boldsymbol{y}. \tag{1.26}$$

Taking the squared norm on both sides of the decomposition (1.26), one can write

$$\|\phi y\|^2 = \|P_{\phi\Delta'} y\|^2 + \|(\phi - P_{\phi\Delta'})y\|^2.$$

This provides the intra-block analysis of variance, which in terms of the observed vector y can be expressed in a more customary way as

$$y'\phi y = y'\phi\Delta' C^- \Delta\phi y + y'(\phi - \phi\Delta' C^- \Delta\phi)y = Q' C^- Q + y'\psi y,$$

where $Q = \Delta\phi y$ (according to the notation of Pearce, 1983, Section 3.1) and ψ is as defined in (1.14) (the "residual matrix" used by Pearce, 1983, Section 3.3). The quadratic form $y'\phi y$ can be called the "intra-block total sum of squares", and its components, $Q'C^-Q$ and $y'\psi y$, can then be called the "intra-block treatment sum of squares" and the "intra-block residual sum of squares", respectively. The corresponding degrees of freedom (d.f.) are $n - b = \text{rank}(\phi)$ for the total, $h = \text{rank}(C)$ for the treatment component and $n - b - h = \text{rank}(\psi)$ for the residual component. It can easily be proved (e.g., by Theorem 9.4.1 of Rao and Mitra, 1971) that the two component sums of squares are distributed independently, if ϕy has a multivariate normal distribution. The expectations of these component sums of squares are $\text{E}(Q'C^-Q) = h\sigma^2 + \tau' C \tau$ and $\text{E}(y'\psi y) = (n - b - h)\sigma^2$. It follows from the second equation that the intra-block residual mean square $s^2 = y'\psi y/(n - b - h)$ is an unbiased estimator of σ^2. Moreover, s^2 is the minimum norm quadratic unbiased estimator (MINQUE) of σ^2 under the submodel ϕy, as it may be seen from Theorem 3.4 of Rao (1974).

Thus, s^2 can be used to obtain an unbiased estimator of the variance (1.25), in the form

$$\widehat{\text{Var}(c'\tau)} = s' C s s^2 = c' C^- c s^2.$$

Furthermore, since under the multivariate normal distribution of ϕy both $Q'C^-Q/\sigma^2$ and $y'\psi y/\sigma^2$ have χ^2 distributions, the first on h d.f. with the non-centrality parameter $\delta = \tau' C \tau/\sigma^2$, the second on $n - b - h$ d.f. with $\delta = 0$ (as can be proved applying, e.g., Theorem 9.2.1 of Rao and Mitra, 1971), the hypothesis $\tau' C \tau = 0$, equivalent to $\text{E}(\phi y) = \mathbf{0}$, can be tested by the variance ratio criterion $Q'C^-Q/(hs^2)$, which under the normality assumption has then the F distribution with h and $n - b - h$ d.f., central when the hypothesis is true.

1.5. Basic contrasts

Consider again the spectral decompositions (1.9) and (1.10). Note that they can also be written in the form

$$N k^{-\delta} N' = r^{\delta}\left(\sum_{\beta=1}^{m} \mu_\beta H_\beta + H_{m+1}\right) r^{\delta}$$

and

$$C = r^{\delta} \sum_{\beta=0}^{m-1} \varepsilon_\beta H_\beta r^{\delta}, \tag{1.27}$$

respectively, where the $v \times v$ matrices H_β are such that $H_\beta r^{\delta} H_\beta = H_\beta$, $H_\beta r^{\delta} H_{\beta'} = O$ for $\beta \neq \beta'$, $H_{m+1} = n^{-1}\mathbf{1}_v \mathbf{1}_v'$ and

$$\sum_{\beta=0}^{m+1} H_\beta = r^{-\delta},$$

and where $\varepsilon_\beta = 1 - \mu_\beta$, $\beta = 0, 1, \ldots, m$, are the distinct eigenvalues of the nonnegative definite matrix C with respect to r^{δ} such that $1 = \varepsilon_0 > \varepsilon_1 > \cdots > \varepsilon_m = 0$; among them ε_0 and ε_m may or may not exist, while $\varepsilon_{m+1} = 1 - \mu_{m+1} = 0$ always exists as the eigenvalue of C corresponding to $C\mathbf{1}_v = \mathbf{0}$. Denoting by ρ_β the multiplicity of ε_β, so that $\sum_{\beta=0}^{m} \rho_\beta = v - 1$, one can also write, for $\beta = 0, 1, \ldots, m, m+1$,

$$H_\beta = S_\beta S_\beta' = \sum_{j=1}^{\rho_\beta} s_{\beta j} s_{\beta j}', \qquad S_\beta = [s_{\beta 1} : s_{\beta 2} : \cdots : s_{\beta \rho_\beta}], \tag{1.28}$$

where the vectors $s_{\beta j}, j = 1, 2, \ldots, \rho_\beta$, are r^{δ}-orthonormal eigenvectors of the matrix C with respect to r^{δ}, corresponding to the common eigenvalue ε_β, i.e., are such that $C s_{\beta j} = \varepsilon_\beta r^{\delta} s_{\beta j}$, with $s_{\beta j}' r^{\delta} s_{\beta j} = 1$ and $s_{\beta j}' r^{\delta} s_{\beta' j'} = 0$ whenever $(\beta, j) \neq (\beta', j')$. Since there are always v such r^{δ}-orthonormal vectors $\{s_{\beta j}\}$, it will be justified and convenient to denote the single one which forms the matrix H_{m+1} by s_v, i.e., $s_v = n^{-1/2}\mathbf{1}_v$. It is evident that all the other satisfy the condition $s_{\beta j}' r = 0$ (cf. Lemma 1.3, where $\rho \equiv \rho_0$).

Contrasts of treatment parameters represented by the vectors $\{s_{\beta j}\}$, i.e., contrasts $c_{\beta j}'\tau = s_{\beta j}' r^{\delta}\tau$, for $\beta = 0, 1, \ldots, m$, have been termed basic contrasts of the design (cf. Pearce et al., 1974, Section 4; Pearce, 1983, Section 3.6). This concept applies to any block design. An essential feature of this concept is that the eigenvalue ε_β is the efficiency factor of the analysed design for each contrast $c_{\beta j}'\tau = s_{\beta j}' r^{\delta}\tau$, $j = 1, 2, \ldots, \rho_\beta$, when estimated in the intra-block analysis.

On account of (1.24) and (1.25), for any proper block design the BLUE of $c_{\beta j}'\tau = s_{\beta j}' r^{\delta}\tau$, under the submodel ϕy is

$$\widehat{c_{\beta j}'\tau} = \varepsilon_{\beta_j}^{-1} s_{\beta j}' Q, \quad \text{with} \quad \mathrm{Var}(\widehat{c_{\beta j}'\tau}) = \varepsilon_\beta^{-1}\sigma^2. \tag{1.29}$$

Since $0 \leqslant \varepsilon_\beta \leqslant 1$, for $\beta = 0, 1, \ldots, m$, it is justified to interpret ε_β as the efficiency factor of the design for $c_{\beta j}'\tau$ estimated in the intra-block analysis (see also Caliński, 1993a, Section 5).

2. Concurrences and different notions of balance

As already mentioned in Section 1.2, and shown with regard to the intra-block analysis in Section 1.4, the statistical analysis of any experiment in a block design involves the C-matrix of the design. Therefore, this matrix to a great extent determines the statistical properties of the design. In connection with this, the pattern of that matrix must be taken into account in any considerations concerning a concept of balance.

2.1. *Balance based on equality of weighted concurrences*

If a design is proper and equireplicate, then its C-matrix is clearly of the form

$$C = r\boldsymbol{I}_v - \frac{1}{k}\boldsymbol{NN}', \tag{2.1}$$

where r is the common treatment replication number and k is the common block size. Any further simplification of the matrix (2.1) depends on the pattern of the $v \times v$ matrix $\boldsymbol{NN}' = [\sum_{j=1}^{b} n_{ij}n_{i'j}]$, called the "concordance matrix" (by John, 1980, p. 30) or "concurrence matrix" (by Pearce, 1963, p. 353). This matrix has certain informative properties, particularly interesting for binary designs. For any binary design the ith diagonal element of $\boldsymbol{NN}'$ is equal to r_i, the number of blocks in which the ith treatment "occurs", and the off-diagonal element in the ith row and the i'th column is equal to the number of blocks in which the ith and the i'th treatments "concur", i.e., are both present (cf. Pearce, 1983, p. 4). Hence, the off-diagonal elements are usually called "concurrences" (cf. Pearce, 1976, p. 106), and are denoted sometimes by $\lambda_{ii'}$ (cf., e.g., John, 1980, p. 13; John, 1987, p. 17). Thus, the pattern of the C-matrix of a binary proper design depends on $\{r_i\}$ and $\{\lambda_{ii'}\}$, and that of a binary proper and equireplicate design depends on the latter only. Obviously, a disconnected design cannot have all $\lambda_{ii'}$ positive, while on the other hand the classical randomized complete block design has all r_i and all $\lambda_{ii'}$ equal to b, the number of blocks.

These considerations lead to the following concept of balance.

DEFINITION 2.1. A block design is said to be pairwise balanced if the off-diagonal elements of its concurrence matrix, $\boldsymbol{NN}'$, are all equal.

Originally, this definition was introduced by Bose and Shrikhande (1960) for binary block designs (cf. Raghavarao, 1971, p. 36). But Definition 2.1 need not be restricted to binary designs and is, in this form, due to Hedayat and Federer (1974). The pairwise balance is sometimes called combinatorial balance (cf., e.g., Puri and Nigam, 1977b, p. 1174). In fact, the pattern of the C-matrix depends directly on the pattern of the matrix $\boldsymbol{NN}'$ in the case of a proper design only. Moreover, if the design is proper and binary, then equal concurrences imply also equal occurrences, i.e., equal treatment replications, and hence equal diagonal elements of the matrix $\boldsymbol{C}$. This can be seen from the equality $k\boldsymbol{r} = \boldsymbol{r} + (v-1)\lambda\boldsymbol{1}_v$ obtainable when taking $\boldsymbol{NN}'\boldsymbol{1}_v$ for any proper block design of block size k (left-hand side) and for any such design that in addition is binary and pairwise balanced (right-hand side).

REMARK 2.1. Designs of this type, i.e., proper binary equireplicate pairwise balanced designs, were introduced by Yates (1936a) and listed by Fisher and Yates (1938), though their combinatorial idea was known much earlier (see Section 0). These designs are traditionally called balanced incomplete block (BIB) designs (cf. Raghavarao, 1971, Definition 4.3.2; Caliński, 1993a, Definition 3.2).

To extend the concept of balance based on the equality of concurrences from proper designs to designs of unequal block sizes, one may wish to consider the equality of the off-diagonal elements in $\boldsymbol{N}\boldsymbol{k}^{-\delta}\boldsymbol{N}'$ instead of that in $\boldsymbol{N}\boldsymbol{N}'$, since if the off-diagonal elements of $\boldsymbol{N}\boldsymbol{k}^{-\delta}\boldsymbol{N}'$ are equal, so too are the off-diagonal elements of the C-matrix, this being in general not true for the off-diagonal elements of $\boldsymbol{N}\boldsymbol{N}'$. Pointing to the difference between $\boldsymbol{N}\boldsymbol{k}^{-\delta}\boldsymbol{N}'$ and $\boldsymbol{N}\boldsymbol{N}'$, Pearce (1964, 1976) introduced the term "weighted concurrences" for the off-diagonal elements of the former, just as "concurrences" had been used for those elements of the latter. By this he extended the concept of pairwise balance from binary proper designs to all connected block designs.

DEFINITION 2.2. A connected block design is said to be totally balanced (or of Type T_0) if the off-diagonal elements of its matrix $\boldsymbol{N}\boldsymbol{k}^{-\delta}\boldsymbol{N}'$ (i.e., the weighted concurrences) are all equal.

The term total balance (Type T_0) used here is due to Pearce (1976, Section 4.A, and 1983, Section 5.2). It should not be confused with the notion of total balance (Type T) used by him earlier (Pearce, 1963, Section 3.2), and after him by some other authors, e.g., Graf-Jaccottet (1977) and Nigam et al. (1988, Section 2.6.2). Also one must note that Definition 2.2 is equivalent to Definition 4.3.1 of Raghavarao (1971), on account of Theorem 1 of Puri and Nigam (1977b). This definition of balance, originally used by Tocher (1952) and Rao (1958), is also equivalent to that of "balanced block (BB) designs" considered by Kageyama (1974), but it does not coincide with that used by Kiefer (1958) for his BB designs, except when the design is binary and proper.

The following result related to Definition 2.2 is due to Rao (1958).

THEOREM 2.1. *A connected block design is totally balanced (in the sense of Definition 2.2) if and only if the $v-1$ nonzero eigenvalues of its C-matrix are all equal.*

To prove this theorem note that the condition of Definition 2.2 can be written as

$$\boldsymbol{N}\boldsymbol{k}^{-\delta}\boldsymbol{N}' = \boldsymbol{r}^{\delta} - \theta\Big[\boldsymbol{I}_v - \frac{1}{v}\boldsymbol{1}_v\boldsymbol{1}_v'\Big] \tag{2.2}$$

or, equivalently, as

$$\boldsymbol{C} = \theta\Big(\boldsymbol{I}_v - \frac{1}{v}\boldsymbol{1}_v\boldsymbol{1}_v'\Big), \tag{2.3}$$

where θ is some positive scalar. For details of the proof see Caliński (1993a, Theorem 3.1).

REMARK 2.2. In general, $\theta = [n - \text{tr}(\boldsymbol{N}\boldsymbol{k}^{-\delta}\boldsymbol{N}')]/(v-1) \leqslant n/v$, whereas $\theta = (n - b)/(v-1)$ if the design is binary, while $\theta = r$ for an equireplicate connected orthogonal block design (see Remark 2.2 in Caliński, 1993a). Therefore, θ has been called by Pearce (1964, 1976, 1983, pp. 74–75) the "effective replication". That the property of the pairwise balance and its generalization apply to connected block designs only, can be seen directly from Definition 1.2.

Now the question arises whether the concept of balance expressed in Definition 2.2 can be generalized further, to cover disconnected block designs as well.

It follows from Definition 1.2 that the C-matrix of a disconnected design of degree $g-1$ can be written (after an appropriate ordering of its rows and columns) as

$$\boldsymbol{C} = \text{diag}[\boldsymbol{C}_1 : \boldsymbol{C}_2 : \cdots : \boldsymbol{C}_g] \tag{2.4}$$

with $\boldsymbol{C}_\ell = \boldsymbol{r}_\ell^\delta - \boldsymbol{N}_\ell \boldsymbol{k}_\ell^{-\delta}\boldsymbol{N}_\ell'$ ($\ell = 1, 2, \ldots, g$) as the C-matrix of the ℓth connected subdesign described by the incidence matrix $\boldsymbol{N}_\ell$. This presentation allows the following extension of Definition 2.2.

DEFINITION 2.3. A block design with disconnectedness of degree $g-1$ (connected, when $g = 1$) is said to be balanced (totally, when $g = 1$) if the off-diagonal elements of its matrices $\boldsymbol{N}_\ell \boldsymbol{k}_\ell^{-\delta}\boldsymbol{N}_\ell'$, $\ell = 1, 2, \ldots, g$, are all equal.

The concept of balance expressed in Definition 2.3 coincides with that originally used by Vartak (1963), who also proved the following.

THEOREM 2.2. *A block design is balanced* (*in the sense of Definition* 2.3) *if and only if the nonzero eigenvalues of its C-matrix are all equal. There are $v - g$ of them, if the degree of disconnectedness is $g - 1$.*

For a proof see Caliński (1993a, Theorem 3.2).

2.2. *Balance based on proportionality of weighted concurrences*

The concept of balance that has led to Definition 2.3 originates from the concept of pairwise balance (Definition 2.1) based on the equality of all concurrences, $\{\lambda_{ii'}\}$. An alternative concept of balance is related to the idea of making the concurrences not equal but proportional to the products of the relevant treatment replications, i.e., it is based on the condition $\lambda_{ii'} = \zeta r_i r_{i'}$, where ζ is some positive scalar, constant for all i and i'. For a proper binary design this condition can be written as $\boldsymbol{N}\boldsymbol{N}' = \boldsymbol{r}^\delta + \zeta(\boldsymbol{r}\boldsymbol{r}' - \boldsymbol{r}^{2\delta})$ (with $\boldsymbol{r}^{2\delta} = \boldsymbol{r}^\delta\boldsymbol{r}^\delta$), which, however, implies that the design is equireplicate. Thus, for proper binary block designs the two concepts of balance, that based on equal concurrences and that based on proportional concurrences, both lead to BIB designs. But in general these two concepts differ.

In a manner similar to that previously used in connection with the concept of balance based on the equality of concurrences, the alternative concept of balance, based on the proportionality of concurrences, can suitably be generalized by considering the weighted concurrences as those which are to be proportional to the products of the relevant treatment replications. For general connected block designs this concept is due to Jones (1959) and can be expressed as follows.

DEFINITION 2.4. A connected block design is totally balanced in the sense of Jones if the off-diagonal elements of its matrix $\boldsymbol{N}\boldsymbol{k}^{-\delta}\boldsymbol{N}'$ are proportional to the products of the relevant treatment replications, i.e., $\sum_{j=1}^{b} n_{ij}n_{i'j}/k_j = zr_ir_{i'}$, where z is some positive scalar, constant for any pair i, i' $(= 1, 2, \ldots, v)$.

This concept of balance was implicitly used by Nair and Rao (1948) and called "total balance" by Caliński (1971). Graf-Jaccottet (1977) introduced the term "balanced in the Jones sense" or "J-balanced".

The condition of Definition 2.4 can also be written as

$$\boldsymbol{N}\boldsymbol{k}^{-\delta}\boldsymbol{N}' = (1-\varepsilon)\boldsymbol{r}^{\delta} + \frac{\varepsilon}{n}\boldsymbol{r}\boldsymbol{r}', \quad \text{where } \varepsilon = zn. \tag{2.5}$$

Comparison of this with the condition (2.2) of Definition 2.2 shows that the two definitions coincide if and only if the design is equireplicate. This was originally noticed by Williams (1975).

It follows from (2.5) that the C-matrix of a connected block design totally balanced in the sense of Jones (Definition 2.4) is of the form

$$\boldsymbol{C} = \varepsilon\left(\boldsymbol{r}^{\delta} - \frac{1}{n}\boldsymbol{r}\boldsymbol{r}'\right), \tag{2.6}$$

where $\varepsilon = [n - \operatorname{tr}(\boldsymbol{N}\boldsymbol{k}^{-\delta}\boldsymbol{N}')]/(n - n^{-1}\boldsymbol{r}'\boldsymbol{r}) \leqslant 1$. This in fact is an equivalent condition to that of Definition 2.4 and may be traced back to Jones (1959) (cf. Caliński, 1971, p. 281). Now, the following result is easily obtainable (cf. Puri and Nigam, 1975a, b; also Caliński, 1977, Corollary 7).

THEOREM 2.3. *A connected block design is totally balanced in the sense of Jones (Definition 2.4) if and only if the $v-1$ nonzero eigenvalues of its C-matrix with respect to $\boldsymbol{r}^{\delta}$ are all equal.*

For a proof see Caliński (1993a, Theorem 3.3).

To generalize this concept of balance to disconnected block designs, the following definition can be adopted.

DEFINITION 2.5. A block design with disconnectedness of degree $g-1$ (connected, when $g = 1$) is balanced in the sense of Jones (totally, when $g = 1$) if the off-diagonal elements of its matrices $\boldsymbol{N}_\ell\boldsymbol{k}_\ell^{-\delta}\boldsymbol{N}_\ell'$, $\ell = 1, 2, \ldots, g$, are proportional to the products of the relevant treatment replications, with the same proportionality factor.

Now, Theorem 2.3 can be extended as follows.

THEOREM 2.4. *A block design is balanced in the sense of Jones (Definition 2.5) if and only if the nonzero eigenvalues of its C-matrix with respect to $\boldsymbol{r}^{\delta}$ are all equal. There are $v - g$ of them, if the degree of disconnectedness is $g - 1$.*

For a proof see Caliński (1993a, Theorem 3.4).

REMARK 2.3. It follows from Theorem 2.4 and Corollary 1.1 that an orthogonal block design is balanced in the sense of Jones (Definition 2.5), whether it is connected or not. The unique nonzero eigenvalue of its C-matrix with respect to $\boldsymbol{r}^{\delta}$ is 1, with multiplicity $v-g$ if the disconnectedness is of degree $g-1$. Furthermore, it follows that an equireplicate orthogonal block design is simultaneously balanced in the sense of Definition 2.3, with the unique nonzero eigenvalue of $\boldsymbol{C}$ equal to r, with multiplicity again $v - g$.

2.3. *Concepts of variance balance, efficiency balance and generalized efficiency balance*

The various definitions considered in Sections 2.1 and 2.2 have been formulated in terms of constructional features of the design, without mentioning any statistical property. Certainly, the latter depends on the assumed model. Therefore, the statistical meaning of any concept of balance can be discussed only when the model is specified.

In most of the statistical literature on block designs, the design properties are usually referred to the intra-block submodel considered in Section 1.4, i.e., to $\boldsymbol{\phi y}$ with the moments as in (1.22) and (1.23). In the remainder of the present section, the estimation will be considered under that submodel.

It follows from Theorem 1.1 that for any contrast $\boldsymbol{c}'\boldsymbol{\tau}$ the BLUE is of the form (1.24), provided that $\boldsymbol{c} = \boldsymbol{C}\boldsymbol{s}$, and its variance is then of the form (1.25).

Now, if the design is balanced in the sense of Definition 2.3, then on account of Theorem 2.2 the C-matrix can be represented as $\boldsymbol{C} = \boldsymbol{\Delta}\boldsymbol{\phi}\boldsymbol{\Delta}' = \theta(\sum_{i=1}^{v-g} \boldsymbol{c}_i\boldsymbol{c}_i')$, where the vectors $\{\boldsymbol{c}_i\}$ are orthonormal eigenvectors of $\boldsymbol{C}$ corresponding to the unique nonzero eigenvalue θ of multiplicity $v - g$. This also implies that $\boldsymbol{c}_i'\boldsymbol{1}_v = 0$ for all $i \leqslant v - g$, and so they represent contrasts, called by Pearce (1983, p. 73) the "natural contrasts" of the design. Hence, for any contrast $\boldsymbol{c}'\boldsymbol{\tau}$ such that $\boldsymbol{c} = \theta(\sum_{i=1}^{v-g} \ell_i\boldsymbol{c}_i)$, where $\ell_i = \boldsymbol{c}_i'\boldsymbol{s}$ for some $\boldsymbol{s}$, the BLUE is, from (1.24) and (1.25), of the form

$$\widehat{\boldsymbol{c}'\boldsymbol{\tau}} = \boldsymbol{s}'\boldsymbol{\Delta\phi y} = \frac{1}{\theta}\boldsymbol{c}'\boldsymbol{\Delta\phi y}, \quad \text{with} \quad \mathrm{Var}(\widehat{\boldsymbol{c}'\boldsymbol{\tau}}) = \frac{1}{\theta}\boldsymbol{c}'\boldsymbol{c}\sigma^2$$

(cf. Caliński, 1977; Pearce, 1983, Section 5.2).

Thus, a statistical consequence of the balance based on equal weighted concurrences is that any contrast $\boldsymbol{c}'\boldsymbol{\tau}$ for which the BLUE exists is estimated with precision $\theta/\boldsymbol{c}'\boldsymbol{c}$, this having the same value for all contrasts that have the same norm $\|\boldsymbol{c}\| = (\boldsymbol{c}'\boldsymbol{c})^{1/2}$. In particular, all the elementary contrasts, $\tau_i - \tau_{i'},\ i, i' = 1, 2, \ldots, v\ (i \neq i')$, for which the BLUEs exist are then estimated with the same precision $\theta/2$. This is why that concept of balance is called "variance balance" (cf. Raghavarao, 1971, p. 54; Hedayat and Federer, 1974, p. 333). The above statistical consequences of this variance balance are indeed so attractive that some authors use the term balance exclusively in the sense of variance balance (cf. Pearce, 1983, p. 121).

Also (see Theorem 2.1 and Remark 2.2) for any equireplicate connected orthogonal block design the unique nonzero eigenvalue of its C-matrix, with multiplicity $v - 1$, is $\theta = r$. Hence, the precision of estimating $\boldsymbol{c}'\boldsymbol{\tau}$ in such a design is $r/\boldsymbol{c}'\boldsymbol{c}$. Therefore, the efficiency factor, defined as the ratio of the precision of the BLUE obtainable in the design under consideration to that obtainable in an equireplicate connected orthogonal block design of the same n, i.e., the efficiency factor used by Kempthorne (1956), Kshirsagar (1958), Atiqullah (1961) and Raghavarao (1971, p. 58), is for the so-called variance-balanced design equal to θ/r, where $r = n/v$, for any contrast for which the BLUE exists. Thus, from this point of view a variance-balanced design is *simultaneously* balanced with regard to the efficiency.

On the other hand if the design is balanced in the sense of Jones (Definition 2.5), then, by Theorem 2.4, its C-matrix can be represented as $\boldsymbol{C} = \varepsilon \boldsymbol{r}^{\delta}(\sum_{i=1}^{v-g} \boldsymbol{s}_i\boldsymbol{s}_i')\boldsymbol{r}^{\delta}$, where the vectors $\{\boldsymbol{s}_i\}$ are defined as in Lemma 1.3 and correspond to the unique nonzero eigenvalue ε of multiplicity $v - g$. It follows that $\boldsymbol{s}_i'\boldsymbol{r} = 0$ for all $i \leqslant v - g$, and so the vectors $\boldsymbol{c}_i = \boldsymbol{r}^{\delta}\boldsymbol{s}_i$, $i = 1, 2, \ldots, v - g$, represent contrasts, called the "basic contrasts" (see Section 1.5). Hence, for any contrast $\boldsymbol{c}'\boldsymbol{\tau}$ such that $\boldsymbol{c} = \varepsilon \boldsymbol{r}^{\delta}(\sum_{i=1}^{v-g} \ell_i \boldsymbol{s}_i)$, where $\ell_i = \boldsymbol{s}_i'\boldsymbol{r}^{\delta}\boldsymbol{s}$ for some $\boldsymbol{s}$, the BLUE is, from (1.24) and (1.25),

$$\widehat{\boldsymbol{c}'\boldsymbol{\tau}} = \boldsymbol{s}'\boldsymbol{\Delta}\boldsymbol{\phi}\boldsymbol{y} = \frac{1}{\varepsilon}\boldsymbol{c}'\boldsymbol{r}^{-\delta}\boldsymbol{\Delta}\boldsymbol{\phi}\boldsymbol{y}, \quad \text{with} \quad \mathrm{Var}(\widehat{\boldsymbol{c}'\boldsymbol{\tau}}) = \frac{1}{\varepsilon}\boldsymbol{c}'\boldsymbol{r}^{-\delta}\boldsymbol{c}\sigma^2.$$

Thus, a statistical consequence of the balance based on the proportionality of the weighted concurrences to the products of the relevant treatment replications is that any contrast $\boldsymbol{c}'\boldsymbol{\tau}$ for which the BLUE exists is estimated with precision $\varepsilon/\boldsymbol{c}'\boldsymbol{r}^{-\delta}\boldsymbol{c}$, which has a common value for all contrasts of the same $\boldsymbol{r}^{-\delta}$-norm,

$$\|\boldsymbol{c}\|_{\boldsymbol{r}^{-\delta}} = (\boldsymbol{c}'\boldsymbol{r}^{-\delta}\boldsymbol{c})^{1/2}. \tag{2.7}$$

But (see Remark 2.3), for any connected orthogonal block design with a given replication vector $\boldsymbol{r}$, the unique nonzero eigenvalue of its C-matrix with respect to $\boldsymbol{r}^{\delta}$ is $\varepsilon = 1$. Hence, the precision of estimating $\boldsymbol{c}'\boldsymbol{\tau}$ in such a design is $1/\boldsymbol{c}'\boldsymbol{r}^{-\delta}\boldsymbol{c}$. Now, if the efficiency factor is defined as the ratio of the precision of the BLUE obtainable in the design under consideration to that obtainable in a connected orthogonal block design with the same treatment replications, i.e., according to the definition used by Pearce (1970), James and Wilkinson (1971), Pearce et al. (1974), Jarrett (1977), Ceranka and Mejza (1979), then ε is exactly the efficiency factor of a design balanced in the sense of Jones for any contrast $\boldsymbol{c}'\boldsymbol{\tau}$ for which the BLUE exists, in particular for all elementary contrasts for which the BLUEs exist. This is why this concept of balance has been called, by Williams (1975) and by Puri and Nigam (1975a, b), the "efficiency balance".

Thus, the two commonly used notions of balance, one referring to equal variances the other to equal efficiencies, can be formalized as follows, assuming that the estimation is confined to the intra-block submodel (Section 1.4).

DEFINITION 2.6. A block design is said to be variance-balanced (VB) if it provides for every contrast $\boldsymbol{c}'\boldsymbol{\tau}$ normalized with regard to the norm $\|\boldsymbol{c}\| = (\boldsymbol{c}'\boldsymbol{c})^{1/2}$, for which the BLUE in the design exists, the same variance of the BLUE.

DEFINITION 2.7. A block design is said to be efficiency-balanced (EB) if it provides for every contrast $\boldsymbol{c}'\boldsymbol{\tau}$, for which the BLUE in the design exists, the same efficiency factor, defined as the ratio of the precision of the BLUE obtainable in the design under consideration to that obtainable in a connected orthogonal block design with the same replication vector $\boldsymbol{r}$.

From the preceded discussion, and Theorems 2.1–2.4, it is evident that if the design is connected, then Definition 2.6 is equivalent to Definition 2.2, while in general it is equivalent to Definition 2.3. Also, it is evident that Definition 2.7 is equivalent to Definition 2.4 in the connected case, and is equivalent to Definition 2.5 in general.

However, there is some ambiguity with regard to Definitions 2.6 and 2.7, as they can equivalently be expressed in the following way.

DEFINITION 2.6*. A block design is said to be VB if it provides for every contrast $\boldsymbol{c}'\boldsymbol{\tau}$, for which the BLUE in the design exists, the same efficiency factor, defined as the ratio of the precision of the BLUE obtainable in the design under consideration to that obtainable in an equireplicate connected orthogonal block design with the common replication number $r = n/v$.

DEFINITION 2.7*. A block design is said to be EB if it provides for every contrast $\boldsymbol{c}'\boldsymbol{\tau}$ normalized with regard to the norm (2.10), for which the BLUE in the design exists, the same variance of the BLUE.

This shows that the introduction of the terms variance balance and efficiency balance has been rather arbitrary. Nevertheless, since these terms have been so widely adopted in the literature, it may be justified to use them also here, following the recommendations given by Preece (1982, p. 151). But, it may also be useful to indicate that both of these concepts of balance are in fact particular cases of a more general notion of balance, of the following form.

DEFINITION 2.8. Given a diagonal $v \times v$ matrix $\boldsymbol{X}$ with the diagonal elements all positive, a block design is said to be generalized efficiency-balanced (GEB), or $\boldsymbol{X}^{-1}$-balanced, if it provides for every contrast $\boldsymbol{c}'\boldsymbol{\tau}$ normalized with regard to the norm $\|\boldsymbol{c}\|_{\boldsymbol{X}^{-1}} = (\boldsymbol{c}'\boldsymbol{X}^{-1}\boldsymbol{c})^{1/2}$, for which the BLUE in the design exists, the same variance of the BLUE.

The term $\boldsymbol{X}^{-1}$-balance was introduced by Caliński (1977), while Das and Ghosh (1985) used the term GEB. The latter can be justified by the fact that an $\boldsymbol{X}^{-1}$-balanced design provides for every contrast $\boldsymbol{c}'\boldsymbol{\tau}$, for which the BLUE in the design exists, the same efficiency factor, defined as the ratio of the precision of the BLUE obtainable in the design under consideration to that obtainable in a connected orthogonal block design with the replication vector $\boldsymbol{r} = n(\boldsymbol{1}_v'\boldsymbol{X}\boldsymbol{1}_v)^{-1}\boldsymbol{X}\boldsymbol{1}_v$ (cf. Caliński et al., 1980).

3. Classification of block designs

Classification problems related to block designs will now be discussed and subsequently, a new terminology suggested. Before this, one of the most popular class of block designs appearing, besides the BIB designs, in the traditional classification of block designs, i.e., that of the partially balanced incomplete block (PBIB) designs, will be briefly explained. The PBIB designs were first defined by Bose and Nair (1939). Later on, Nair and Rao (1942a) generalized the original definition to include some confounded factorial designs, as well as many others, in the class of PBIB designs whose definition is based on the concept of association schemes. Hence, it is helpful to introduce this notion here (see Bose and Shimamoto, 1952).

DEFINITION 3.1. Given v treatments $1, 2, \ldots, v$, a relation satisfying the following three conditions is said to be an association scheme with m^* associate classes:

(a) Any two treatments are either 1st, 2nd, ..., or m^*th associates, the relation of association being symmetric, i.e., if treatment α is the ith associate of treatment β, then β is the ith associate of treatment α.

(b) Each treatment has n_i ith associates, the number n_i being independent of the treatment taken.

(c) If any two treatments α and β are ith associates, then the number of treatments which are jth associates of α and kth associates of β is p^i_{jk} and is independent of the pair of ith associates α and β.

REMARK 3.1. In an association scheme with two associate classes (i.e., $m^* = 2$), the condition (c) above can be much simplified (see Bose and Clatworthy, 1955; Raghavarao, 1971, Section 8.1).

The numbers $v, n_i, p^i_{jk},\ i, j, k = 1, 2, \ldots, m^*$, are called the parameters of the association scheme; all must be non-negative integers.

Now an m^*-associate PBIB design will be defined as follows (cf. Raghavarao, 1971, Chapter 8).

DEFINITION 3.2. Given an association scheme with m^* associate classes, a PBIB design with parameters v, b, r, k, λ_i based on the association scheme is a block design such that

(i) each treatment occurs at most once in a block (i.e., the design is binary);

(ii) each block contains k treatments;

(iii) each treatment occurs in exactly r blocks;

(iv) if two treatments α and β are ith associates, then they occur together (concur) in λ_i blocks (not all λ_i's are equal), the number λ_i being independent of the particular pair of ith associates α and β $(1 \leqslant i \leqslant m^*)$.

The numbers $v, b, r, k, \lambda_i,\ i = 1, 2, \ldots, m^*$, are called the parameters of the design. The λ_i's and p^i_{jk}'s are called the coincidence numbers and the second kind of parameters of the design, respectively.

Note that a BIB design is a particular case of the above PBIB designs with $m^* = 1$, i.e., $\lambda_1 = \lambda_2 = \cdots = \lambda_{m^*}$ (see Remark 2.1). In case when the parameters λ_i,

$i = 1, 2, \ldots, m^*$, of an m^*-associate PBIB design are not all different, the m^* associate classes of the design based on the association scheme may not be all distinct. The design may then become an m^{**}-associate PBIB design for some $m^{**} < m^*$ (see also Kageyama, 1974).

Among the association schemes with m^* associate classes, and the PBIB designs based on them, the case of $m^* = 2$ has received most attention. To this case mostly the group divisible, simple, triangular, Latin-square and cyclic schemes, and the corresponding PBIB designs, belong. Among them, a group divisible design is of particular interest here.

DEFINITION 3.3. A PBIB design with two associate classes is said to be group divisible (GD) if there are $v = st$ treatments, divided into s groups of t treatments each, such that any two treatments of the same group are first associates and two treatments from different groups are second associates (which is called the GD association scheme).

The parameters of the GD association scheme are given by $v = st, n_1 = t-1, n_2 = t(s-1)$ and the matrices

$$\boldsymbol{P}_1 = [p^1_{jk}] = \begin{bmatrix} t-2 & 0 \\ 0 & t(s-1) \end{bmatrix}, \qquad \boldsymbol{P}_2 = [p^2_{jk}] = \begin{bmatrix} 0 & t-1 \\ t-1 & t(s-2) \end{bmatrix}.$$

It is also known (cf. Raghavarao, 1971, p. 127) that in a GD design, having parameters $v = st, b, r, k, \lambda_1, \lambda_2$, with an incidence matrix $\boldsymbol{N}$ the eigenvalues of $\boldsymbol{N}\boldsymbol{N}'$ other than rk are $r - \lambda_1$ and $rk - v\lambda_2$, with respective multiplicities $s(t-1)$ and $s-1$. Depending upon the eigenvalues, the GD designs are further divided into three disjoint cases by Bose and Connor (1952) as follows: Singular if $r - \lambda_1 = 0$; Semi-regular if $r - \lambda_1 > 0$ and $rk - v\lambda_2 = 0$; Regular if $r - \lambda_1 > 0$ and $rk - v\lambda_2 > 0$.

For details see Raghavarao (1971), Clatworthy (1973) and John (1980).

3.1. Usual concepts of classification

In the literature, there are two different approaches to the problem of classifying block designs. The first is based on combinatorial properties of block designs and is aimed at classifying equireplicate and proper block designs only. In this most popular classification, block designs are divided into two groups: BIB designs including t-designs, due to Woolhouse (1844), Kirkman (1847), Steiner (1853) and Carmichael (1956), and m^*-associate PBIB designs. The 2-associate PBIB designs are further classified by Bose and Shimamoto (1952). A general classification of m^*-associate PBIB designs for $m^* > 2$ has not yet been made. Moreover not all block designs can be classified using this approach. The second approach is based on the efficiency of block designs for estimating contrasts of treatment parameters under the intra-block submodel (see Section 1.4). Pearce (1963) showed a classification of connected block designs which incorporates both of these two approaches. The basis for his classification is to consider various patterns of Tocher's (1952) matrix $\Omega^{-1} = \boldsymbol{r}^{\delta} - \boldsymbol{N}\boldsymbol{k}^{-\delta}\boldsymbol{N}' + n^{-1}\boldsymbol{r}\boldsymbol{r}'$. Again, not all available block designs are covered by this classification.

In the context of a general block design, various kinds of balancing, such as variance-balance, efficiency-balance, partially efficiency-balance, etc., are well known (cf. Section 2.3, also Puri et al., 1977; Caliński et al., 1980; Kageyama, 1980; Puri and Kageyama, 1985). Das and Ghosh (1985) considered GEB designs (Definition 2.8) which are the same as $\boldsymbol{X}^{-1}$-balanced designs introduced by Caliński (1977). Furthermore, Kageyama et al. (1988) introduced the concept of $\boldsymbol{X}^{-1}$-partially efficiency-balanced designs to unify these different notions of balancing. Let this be described now.

In the following, for any $v \times v$ diagonal matrix $\boldsymbol{X}$ with the diagonal elements $x_1, x_2, \ldots, x_v$ all positive, the notation $\boldsymbol{X}^{1/2} = \mathrm{diag}[x_1^{1/2}, x_2^{1/2}, \ldots, x_v^{1/2}]$ and $\boldsymbol{X}^{-1/2} = (\boldsymbol{X}^{1/2})^{-1}$ will be used. Note that one can always write $\boldsymbol{U} = \boldsymbol{X}^{-1/2}\boldsymbol{C}\boldsymbol{X}^{-1/2} = \sum_{i=1}^{m} \alpha_i \boldsymbol{A}_i$, where $\alpha_1, \alpha_2, \ldots, \alpha_m$ are the distinct positive eigenvalues of $\boldsymbol{U}$, and the corresponding $v \times v$ matrices $\boldsymbol{A}_i$ are symmetric, idempotent and such that $\sum_{i=1}^{m+1} \boldsymbol{A}_i = \boldsymbol{I}_v$ and $\boldsymbol{A}_i \boldsymbol{A}_{i'} = \boldsymbol{O}$ for all $i \neq i'$ $(= 1, 2, \ldots, m, m+1)$. Certainly, the matrix $\boldsymbol{A}_{m+1}$ corresponds to the zero eigenvalue of $\boldsymbol{U}$, of multiplicity 1 if the design is connected ($\boldsymbol{A}_{m+1}$ is then often denoted by $\boldsymbol{A}_0$). (Cf. Caliński, 1977, formula (2.5).)

DEFINITION 3.4. Given a diagonal matrix $\boldsymbol{X}$ with diagonal elements all positive, a connected block design is called an $\boldsymbol{X}^{-1}$-partially efficiency-balanced design with m efficiency classes, or simply a $\boldsymbol{X}^{-1}$-PEB(m) design, if the matrix $\boldsymbol{X}^{-1/2}\boldsymbol{C}\boldsymbol{X}^{-1/2}$ has exactly m distinct positive eigenvalues.

Note that the matrix $\boldsymbol{X}^{-1/2}\boldsymbol{C}\boldsymbol{X}^{-1/2}$ is a generalization of the matrix $\boldsymbol{F}$ used by Pearce et al. (1974). Clearly for any positive definite $\boldsymbol{X}$, every connected block design is a $\boldsymbol{X}^{-1}$-PEB(m) design with some $m \leqslant v-1$ (cf. Caliński and Ceranka, 1974, who used $\boldsymbol{X} = \boldsymbol{r}^{\delta}$). From a practical point of view, $\boldsymbol{X}^{-1}$-PEB(m) designs with $m \leqslant 2$ have nice statistical properties (cf. Pearce, 1970; Caliński, 1977; Gupta, 1987). A study of $\boldsymbol{X}^{-1}$-PEB(m) designs for larger values of m may also be interesting, at least from a purely mathematical perspective.

The class of $\boldsymbol{X}^{-1}$-PEB(m) designs with varying $\boldsymbol{X}$ is fairly large even if $m \leqslant 2$. In fact, this class includes the following classes of block designs (cf. Kageyama, 1993):

(1) Variance-balanced (VB) designs, when $\boldsymbol{X} = \boldsymbol{I}$ and $m = 1$ (Definition 2.6), due to Tocher (1952) and Rao (1958);

(2) BIB designs, when $\boldsymbol{X} = \boldsymbol{I}$ and $m = 1$, and the design is connected, binary and proper (Definition 2.1 and Remark 2.1);

(3) Two-associate PBIB designs, when $\boldsymbol{X} = \boldsymbol{I}$ and $m = 2$ (Definition 3.2);

(4) Efficiency-balanced (EB) designs, when $\boldsymbol{X} = \boldsymbol{r}^{\delta}$ and $m = 1$ (Definition 2.7), due to Jones (1959), Caliński (1971), and Puri and Nigam (1975a, b);

(5) Partially efficiency-balanced (PEB) designs with two efficiency classes, when $\boldsymbol{X} = \boldsymbol{r}^{\delta}$ and $m = 2$, due to Puri and Nigam (1977a);

(6) Simple PEB or C-designs, when $\boldsymbol{X} = \boldsymbol{r}^{\delta}$ and $m = 2$, with one of the eigenvalues equal to 1, due to Caliński (1971), Puri and Nigam (1977a), and Saha (1976);

(7) Generalized efficiency balanced (GEB) designs (or $\boldsymbol{X}^{-1}$-balanced designs), with any $\boldsymbol{X}$ and $m = 1$ (Definition 2.8), due to Caliński (1977) and Das and Ghosh (1985), i.e., precisely the $\boldsymbol{X}^{-1}$-PEB(1) designs.

Note that although any two-associate PBIB design is an $\boldsymbol{I}^{-1}$-PEB(2) design (cf. Puri and Nigam, 1977a; Nigam et al., 1988, Section 3.5), the reverse is not true. Also, some special two-associate PBIB designs belong to the class of C-designs (cf. Nigam et al., 1988, Section 3.5.2.1), but, on the other hand, that class includes many designs which are not PBIB designs (cf. Saha, 1976; Ceranka, 1983).

A characterization for connected $\boldsymbol{X}^{-1}$-PEB(m) designs with $m \leqslant 2$ is known. In particular, the following results are essential (given here without proofs, which can be found in Kageyama et al., 1988).

THEOREM 3.1. *For a matrix* $\boldsymbol{X} = \mathrm{diag}[x_1, x_2, \ldots, x_v]$ *with all* x_i *positive, a connected block design is a* $\boldsymbol{X}^{-1}$-PEB(m) *design with* $m \leqslant 2$ *if and only if there exist constants* γ, δ $(\delta > 0)$ *such that* $\boldsymbol{U}^2 - \gamma\boldsymbol{U} + \delta(\boldsymbol{I} - x_0^{-1}\boldsymbol{x}^{1/2}\boldsymbol{x}^{1/2'}) = \boldsymbol{O}$, *where* $\boldsymbol{U} = \boldsymbol{X}^{-1/2}\boldsymbol{C}\boldsymbol{X}^{-1/2}$, $x_0 = \mathrm{tr}(\boldsymbol{X})$ *and* $\boldsymbol{x}^{1/2} = [x_1^{1/2}, x_2^{1/2}, \ldots, x_v^{1/2}]'$.

COROLLARY 3.1. *For a matrix* $\boldsymbol{X} = \mathrm{diag}[x_1, x_2, \ldots, x_v]$ *with all* x_i *positive, a connected block design is a* $\boldsymbol{X}^{-1}$-PEB(1) *design if and only if there exists a positive constant* γ *such that* $\boldsymbol{U}^2 - \gamma\boldsymbol{U} = \boldsymbol{O}$, *where* $\boldsymbol{U} = \boldsymbol{X}^{-1/2}\boldsymbol{C}\boldsymbol{X}^{-1/2}$.

In a connected $\boldsymbol{X}^{-1}$-PEB(m) design with $m \leqslant 2$, the spectral decomposition of $\boldsymbol{U}$ yields $\boldsymbol{U} = \alpha_1\boldsymbol{A}_1 + \alpha_2\boldsymbol{A}_2$, where α_1 and α_2 are positive not necessarily distinct and where the matrices $\boldsymbol{A}_0 = x_0^{-1}\boldsymbol{x}^{1/2}\boldsymbol{x}^{1/2'}, \boldsymbol{A}_1, \boldsymbol{A}_2$ are symmetric, idempotent and such that $\boldsymbol{A}_0 + \boldsymbol{A}_1 + \boldsymbol{A}_2 = \boldsymbol{I}$, $\boldsymbol{A}_0\boldsymbol{A}_1 = \boldsymbol{A}_0\boldsymbol{A}_2 = \boldsymbol{A}_1\boldsymbol{A}_2 = \boldsymbol{O}$. The following gives a compact formula for a g-inverse of the C-matrix in a $\boldsymbol{X}^{-1}$-PEB(m) design with $m \leqslant 2$. The result can be checked directly.

THEOREM 3.2. *Consider a* $\boldsymbol{X}^{-1}$-PEB(m) *design with* $m \leqslant 2$. *Let* γ, δ $(\delta > 0)$ *be constants such that* $\boldsymbol{U}^2 - \gamma\boldsymbol{U} + \delta(\boldsymbol{I} - x_0^{-1}\boldsymbol{x}^{1/2}\boldsymbol{x}^{1/2'}) = \boldsymbol{O}$, *where* $\boldsymbol{U}$ *is as defined in Theorem* 3.1. *Then a* g*-inverse of* $\boldsymbol{C}$ *is given by*

$$\begin{aligned}\boldsymbol{C}^* &= \frac{\gamma}{\delta}\left(\boldsymbol{X}^{-1} - x_0^{-1}\boldsymbol{J}\right) - \frac{1}{\delta}\boldsymbol{X}^{-1}\boldsymbol{C}\boldsymbol{X}^{-1}\\ &= \boldsymbol{X}^{-1/2}\left(\alpha_1^{-1}\boldsymbol{A}_1 + \alpha_2^{-1}\boldsymbol{A}_2\right)\boldsymbol{X}^{-1/2}.\end{aligned}$$

Theorems 3.1 and 3.2 make the task of analyzing a $\boldsymbol{X}^{-1}$-PEB(m) design with $m \leqslant 2$ rather simple. Once $\boldsymbol{X}$ is chosen, it is not necessary to have a spectral decomposition of $\boldsymbol{U}$ and Theorem 3.2 is enough to yield an explicit expression for a g-inverse of the C-matrix of the design. The resulting computational simplicity is likely to be helpful in the statistical analysis.

Also the C-designs have the desirable property of admitting a simple analysis. Unfortunately, however, the class of C-designs is not very large. In particular, there are many two-associate PBIB designs which are not C-designs. In this regard, the concept of $\boldsymbol{X}^{-1}$-PEB(m) designs with $m \leqslant 2$ is helpful, as it extends the class of C-designs to a much wider class, while retaining simplicity in the analysis.

It is thus realized that the class of $\boldsymbol{X}^{-1}$-PEB(m) designs include most of the statistically balanced designs as special cases.

On the other hand, there are other combinatorially balanced designs that may have some statistical advantage, as the following:

(1) Pairwise balanced designs (Definition 2.1), due to Bose and Shrikhande (1960);
(2) (r, λ)-systems, due to Stanton and Mullin (1966);
(3) Balanced bipartite block designs, due to Nair and Rao (1942b), and Corsten (1962);
(4) Balanced treatment incomplete block designs, or type S designs, due to Pearce (1960), Bechhofer and Tamhane (1981), Hedayat et al. (1988);
(5) Symmetrical unequal-block arrangements with two unequal block sizes, due to Kishen (1940–1941) and Raghavarao (1962).

The construction problems for these designs are interesting in both the combinatorial and the statistical sense.

3.2. *Classification based on the concept of partially efficiency balance*

3.2.1. *Partially efficiency balance in connected designs*

As already mentioned in Section 3.1(5), Puri and Nigam (1977a) have introduced a broad class of connected block designs having simple analysis and allowing the important contrasts to be estimated with desired efficiency, termed as partially efficiency balanced (PEB) designs with m efficiency classes, which can be made available in varying replications and/or unequal block sizes. Such designs may be particularly useful for bio-assays, comparative varietal trials and factorial experiments. However, any block design is a PEB design with m efficiency classes for some $m \leqslant v - 1$. So the term "PEB" itself is not much informative in a statistical sense. More information may be added to this term. Here a new terminology will be suggested, aimed at giving more statistical meaning to the PEB designs, which may or may not be connected.

In the original definition of a PEB design with m efficiency classes (Puri and Nigam, 1977a, p. 755), the distinct eigenvalues μ_β of the matrix $\boldsymbol{M} = \boldsymbol{r}^{-\delta}\boldsymbol{N}\boldsymbol{k}^{-\delta}\boldsymbol{N}'$ have been ordered (at least implicitly) as

$$1 = \mu_0 > \mu_1, \ldots, \mu_{m-1} > \mu_m \geqslant 0,$$

assuming that $\mu_0 = 1$ has multiplicity 1. This might be convenient as long as only connected designs are considered. For a general block design it will be more convenient to have them ordered increasingly (as in Lemma 1.3), i.e., as

$$0 = \mu_0 < \mu_1 < \cdots < \mu_{m-1} < \mu_m = \mu_{m+1} = 1, \tag{3.1}$$

with multiplicities ρ_β for $\beta = 0, 1, \ldots, m, m + 1$. Here, in a connected design $\mu_m = 1$ does not appear, i.e., $\rho_m = 0$, but $\mu_{m+1} = 1$ always exists with $\rho_{m+1} = 1$. Thus, $m - 1$ is the number of distinct, less than 1, positive eigenvalues in (1.9). In this case the efficiency factors $\varepsilon_\beta = 1 - \mu_\beta$ are ordered as

$$1 = \varepsilon_0 > \varepsilon_1 > \cdots > \varepsilon_{m-1} > \varepsilon_m = \varepsilon_{m+1} = 0,$$

with multiplicities ρ_β for $\beta = 0, 1, \ldots, m, m+1$ (cf. formula (1.10), where $\rho_0 \equiv \rho$).

With this notation (adopted in Section 1, Section 1.5 in particular) the definition of Puri and Nigam (1977a) can be expressed for any connected block design as follows.

DEFINITION 3.5. An arrangement of v treatments into b blocks of sizes $k_1, k_2, \ldots, k_b$ is said to be a (connected) PEB design with m efficiency classes if

(i) the ith treatment is replicated r_i times, $i = 1, 2, \ldots, v$;

(ii) the efficiency factor associated with every contrast of the βth class is $1 - \mu_\beta$, where μ_β, $0 \leqslant \mu_\beta < 1$, $\beta = 0, 1, \ldots, m-1$, are the distinct eigenvalues of the matrix $\boldsymbol{M}_0$, with multiplicities ρ_β, such that $\sum_\beta \rho_\beta = v - 1$ (i.e., except the zero eigenvalue corresponding to $\boldsymbol{M}_0 \mathbf{1}_v = \mathbf{0}$);

(iii) the matrix $\boldsymbol{M}_0$, defined as $\boldsymbol{M}_0 = \boldsymbol{r}^{-\delta} \boldsymbol{N} \boldsymbol{k}^{-\delta} \boldsymbol{N}' - n^{-1} \mathbf{1}_v \boldsymbol{r}'$, has the spectral decomposition

$$\boldsymbol{M}_0 = \sum_{\beta=0}^{m-1} \mu_\beta \boldsymbol{L}_\beta, \tag{3.2}$$

where $\boldsymbol{L}_\beta, \beta = 0, 1, \ldots, m-1$, are mutually orthogonal idempotent matrices of ranks ρ_β such that $\sum_{\beta=0}^{m-1} \boldsymbol{L}_\beta = \boldsymbol{I}_v - n^{-1} \mathbf{1}_v \boldsymbol{r}'$, and $\mu_0 = 0$.

Note that in the case of $\rho_0 = 0$ the design will be called a PEB design with $m - 1$ efficiency classes. In this notation, corresponding to that in Section 1.5, a contrast belonging to the βth class is understood as $\boldsymbol{c}'_{\beta j} \boldsymbol{\tau} = \boldsymbol{s}'_{\beta j} \boldsymbol{r}^\delta \boldsymbol{\tau}$, where $\boldsymbol{s}_{\beta j}$ $(\mathbf{1}'_v \boldsymbol{s}_{\beta_j} = 0)$ is an eigenvector of $\boldsymbol{M}_0$ corresponding to the eigenvalue μ_β, and $\boldsymbol{\tau}$, as usual, denotes the vector of treatment parameters. Such contrasts, usually $\boldsymbol{r}^\delta$-orthonormalized, are called basic contrasts and are simply denoted by $\boldsymbol{s}_{\beta j}$, if convenient.

In other words, for a connected PEB design with m efficiency classes there exists a set of $v - 1$ linearly independent basic contrasts $\{\boldsymbol{s}_{\beta j}\}$ which can be partitioned into m $(< v)$ disjoint classes such that all the ρ_β members of the βth class, and their linear combinations, are estimated in the intra-block analysis with the same relative loss of information $\mu_\beta < 1$, i.e., are estimated under the intra-block submodel with the efficiency factor $\varepsilon_\beta = 1 - \mu_\beta$, respectively. If $\boldsymbol{s}_{\beta j}$ is $\boldsymbol{r}^\delta$-normalized (as in Section 1.5), then the BLUE in the intra-block analysis has its variance as in (1.29).

A special class of PEB designs is considered by Kageyama and Puri (1985a).

Alternatively to (3.2), one can use the spectral decomposition

$$\boldsymbol{M} = \boldsymbol{r}^{-\delta} \boldsymbol{N} \boldsymbol{k}^{-\delta} \boldsymbol{N}' = \sum_{\beta=0}^{m-1} \mu_\beta \boldsymbol{L}_\beta + \frac{1}{n} \mathbf{1}_v \boldsymbol{r}' \tag{3.3}$$

with

$$\sum_{\beta=0}^{m-1} \boldsymbol{L}_\beta + \frac{1}{n} \mathbf{1}_v \boldsymbol{r}' = \boldsymbol{I}_v.$$

Note that the non-symmetric matrix $\boldsymbol{M}$ comes from Jones (1959) and $\boldsymbol{M}_0$ from Caliński (1971).

More on the properties of connected PEB designs have been given by Kageyama and Puri (1985b). Since the spectral decomposition (3.3) can be extended to cover any block design, whether connected or disconnected, any block design is PEB in the sense that if it is connected, then all the ρ_β basic contrasts of the βth class are estimated intra-block with the efficiency $\varepsilon_\beta = 1 - \mu_\beta$ (> 0), but if the design is disconnected, then $\varepsilon_\beta = 0$ for one β. Therefore, for a classification of block designs in general, the specification of the multiplicities $\{\rho_\beta\}$ is important in the statistical sense. To give merely the number of efficiency classes is not enough to indicate the practical suitability of a PEB design. One wants to know how large the multiplicities are and to which efficiency factors they correspond, e.g., whether the design allows a number of contrasts to be estimated intra-block with full efficiency. In the next subsection more informative terms are suggested for a statistical characterization of block designs. Also the original definition of a PEB design is extended so to cover disconnected block designs as well. Some special illustrations are also given.

3.2.2. Generalization and a new terminology

Now let the new ordering (3.1) of the eigenvalues μ_β be used to cover connected as well as disconnected block designs.

Note first that for any block design one can write, with the vectors $\{\boldsymbol{s}_{\beta j}\}$ all $\boldsymbol{r}^\delta$-orthonormalized (as in Lemma 1.3), the decomposition

$$\boldsymbol{M} = \boldsymbol{r}^{-\delta}\boldsymbol{N}\boldsymbol{k}^{-\delta}\boldsymbol{N}' = \sum_{\beta=0}^{m+1}\sum_{j=1}^{\rho_\beta} \mu_\beta \boldsymbol{s}_{\beta j}\boldsymbol{s}'_{\beta j}\boldsymbol{r}^\delta = \sum_{\beta=0}^{m+1} \mu_\beta \boldsymbol{L}_\beta \tag{3.4}$$

and, using still the same definition of $\boldsymbol{M}_0$ as in (3.2), also

$$\boldsymbol{M}_0 = \sum_{\beta=0}^{m} \mu_\beta \boldsymbol{L}_\beta, \tag{3.5}$$

where

$$\boldsymbol{L}_\beta = \sum_{j=1}^{\rho_\beta} \boldsymbol{s}_{\beta j}\boldsymbol{s}'_{\beta j}\boldsymbol{r}^\delta, \quad \text{rank}(\boldsymbol{L}_\beta) = \rho_\beta \quad (\beta = 0, 1, \ldots, m), \tag{3.6}$$

with $\boldsymbol{L}_m$ appearing only in a disconnected block design and $\boldsymbol{L}_{m+1} = n^{-1}\boldsymbol{1}_v\boldsymbol{r}'$ corresponding to $\mu_{m+1} = 1$. (The definition of $\boldsymbol{M}_0$ used here is different from that used for the disconnected case by Pearce et al., 1974, p. 455.) Note that $\boldsymbol{L}_\beta^2 = \boldsymbol{L}_\beta$ and $\boldsymbol{L}_\beta\boldsymbol{L}_{\beta'} = \boldsymbol{O}$ if $\beta \neq \beta'$, for $\beta = 0, 1, \ldots, m+1$, and that $\sum_{\beta=0}^{m} \boldsymbol{L}_\beta = \boldsymbol{I}_v - n^{-1}\boldsymbol{1}_v\boldsymbol{r}'$. Also note that $\boldsymbol{L}_\beta = \boldsymbol{H}_\beta\boldsymbol{r}^\delta$ for any β, with $\boldsymbol{H}_\beta$ defined in (1.28). Thus, $\{\varepsilon_\beta\}$ are the eigenvalues and $\{\boldsymbol{s}_{\beta j}\}$ are the corresponding $\boldsymbol{r}^\delta$-orthonormal eigenvectors of $\boldsymbol{C}$ with respect to $\boldsymbol{r}^\delta$, representing the basic contrasts of the design (see Section 1.5).

The eigenvalues $\{\varepsilon_\beta\}$ are then the efficiency factors of the design for estimating the corresponding basic contrasts in the intra-block analysis. With (3.6), the spectral decomposition (1.27) of $\boldsymbol{C}$ can alternatively be written as

$$\boldsymbol{C} = \boldsymbol{r}^\delta \sum_{\beta=0}^{m-1} \varepsilon_\beta \boldsymbol{L}_\beta. \tag{3.7}$$

Evidently, the term for $\beta = 0$ can be deleted from (3.5) and (3.6), as $\mu_0 = 0$, but not from (3.7), where $\varepsilon_0 = 1$.

If the experimental problem has its reflection in distinguishing certain subsets of contrasts, ordered according to their importance or interest, the experiment should be designed in such a way that all members of a specified subset of basic contrasts receive a common efficiency factor, of the higher value the more important the contrasts of the subset are. If possible, the design should allow to estimate in the intra-block analysis the most important subset of contrasts with full efficiency, i.e., with $\varepsilon_0 = 1$. In this context, one is interested in knowing how many basic contrasts are estimated with the same efficiency. So the information about the multiplicity ρ_β is essential. It can form a basis for a classification of block designs. Thus, the characterization of a block design by the triples $(\mu_\beta, \rho_\beta, \boldsymbol{L}_\beta)$ for $\beta = 0, 1, \ldots, m$, i.e., by the idempotent matrices $\{\boldsymbol{L}_\beta\}$ defined in (3.6), their ranks $\{\rho_\beta\}$ and the corresponding eigenvalues $\{\mu_\beta\}$ appearing in (3.5), is very informative.

With the present notation the following modification of the original Definition 3.5 of partially efficiency balance for any connected block design can be given (modifying also the proposal of Ceranka and Mejza, 1980).

DEFINITION 3.6. A connected block design is said to be $(\rho_0; \rho_1, \ldots, \rho_{m-1})$-efficiency balanced (EB) if a complete set of its $v-1$ basic contrasts can be partitioned into at most m disjoint and non-empty subsets such that all the ρ_β basic contrasts of the βth subset correspond to a common efficiency factor $\varepsilon_\beta = 1 - \mu_\beta$, for $\beta = 0, 1, \ldots, m-1$, i.e., so that the matrix $\boldsymbol{M}_0$ $(= \boldsymbol{r}^{-\delta}\boldsymbol{N}\boldsymbol{k}^{-\delta}\boldsymbol{N}' - n^{-1}\boldsymbol{1}_v\boldsymbol{r}')$ has the spectral decomposition (3.2), where the eigenvalues $0 = \mu_0 < \mu_1 < \cdots < \mu_{m-1} < 1$ are distinct with multiplicities $\rho_0 \geqslant 0$, $\rho_1 \geqslant 1, \ldots, \rho_{m-1} \geqslant 1$, respectively, and where the matrices $\{\boldsymbol{L}_\beta\}$ are defined as in (3.6).

The parameters of a $(\rho_0; \rho_1, \ldots, \rho_{m-1})$-EB design can be written as v, b, $\boldsymbol{r}$, $\boldsymbol{k}$, $\varepsilon_\beta = 1 - \mu_\beta$, ρ_β, $\boldsymbol{L}_\beta$ for $\beta = 0, 1, \ldots, m-1$. Note that $\rho_0 = 0$ if no basic contrast is estimated in the intra-block analysis with full efficiency.

Definition 3.6 can be extended so to cover also disconnected block designs. For this recall Definition 1.2.

DEFINITION 3.7. A block design with disconnectedness of degree $g-1$ (connected when $g = 1$) is said to be $(\rho_0; \rho_1, \ldots, \rho_{m-1}; \rho_m)$-EB if a complete set of its $v-1$ basic contrasts can be partitioned into at most $m+1$ disjoint and non-empty subsets such that all ρ_β basic contrasts of the βth subset correspond to a common efficiency factor $\varepsilon_\beta = 1 - \mu_\beta$, for $\beta = 0, 1, \ldots, m-1, m$, i.e., so that the matrix $\boldsymbol{M}_0$ has the spectral decomposition (3.5), where the eigenvalues $0 = \mu_0 < \mu_1 < \cdots < \mu_{m-1} < \mu_m = 1$ are distinct with multiplicities $\rho_0 \geqslant 0$, $\rho_1 \geqslant 1, \ldots, \rho_{m-1} \geqslant 1$, $\rho_m = g - 1 \geqslant 0$, respectively, and where the matrices $\{\boldsymbol{L}_\beta\}$ are defined as in (3.6).

The parameters of a $(\rho_0; \rho_1, \ldots, \rho_{m-1}; \rho_m)$-EB design can be written as v, b, $\boldsymbol{r}$, $\boldsymbol{k}$, $\varepsilon_\beta = 1 - \mu_\beta$, ρ_β, $\boldsymbol{L}_\beta$ for $\beta = 0, 1, \ldots, m-1, m$. Hence (i) ρ_0 shows the number of basic contrasts estimated with full efficiency, i.e., not confounded with blocks, (ii) $\{\rho_1, \ldots, \rho_{m-1}\}$ shows the numbers of basic contrasts estimated with efficiencies $\{\varepsilon_\beta, \beta = 1, \ldots, m-1\}$ less than 1, i.e., partially confounded with blocks, (iii) ρ_m shows the number of basic contrasts with zero efficiency, i.e., totally confounded with blocks. In a connected design $\rho_m = 0$, i.e., no basic contrast is totally confounded with blocks.

It should be noted that since the decomposition (3.4) holds for any block design, whether connected, i.e., with $\rho_m = 0$, or disconnected, i.e., with $\rho_m \geqslant 1$, any block design satisfies the condition of Definition 3.7. Therefore, for a classification of block designs the specification of the multiplicities $\{\rho_\beta\}$ appearing in the definition is essential. To see it better, it may be interesting to indicate some special classes of block designs specified from the general Definition 3.7 point of view.

A block design belongs to the class of $(0; v-1; 0)$-EB designs if it is connected and totally balanced in the sense of Jones (1959) (Definition 2.4), or efficiency-balanced in the terminology of Williams (1975), and Puri and Nigam (1975a, b) (Definition 2.7), i.e., satisfies the conditions

$$\boldsymbol{M}_0 = \mu_1 \boldsymbol{L}_1 \quad \text{and} \quad \boldsymbol{C} = \varepsilon_1 \left(\boldsymbol{r}^\delta - \frac{1}{n} \boldsymbol{r}\boldsymbol{r}' \right).$$

In particular, any BIB design belongs to this class.

A block design belongs to the class of $(0; v-g; g-1)$-EB designs if it is disconnected of degree $g-1$ and is balanced in the sense of Jones (1959) (Definition 2.5), i.e., satisfies the conditions

$$\boldsymbol{M}_0 = \mu_1 \boldsymbol{L}_1 + \boldsymbol{L}_2 \quad \text{and} \quad \boldsymbol{C} = \varepsilon_1 \boldsymbol{r}^\delta \boldsymbol{L}_1.$$

A block design belongs to the class of $(v-1; 0)$-EB designs if it is connected and orthogonal, i.e., satisfies the conditions

$$\boldsymbol{M}_0 = \boldsymbol{O} \quad \text{and} \quad \boldsymbol{C} = \boldsymbol{r}^\delta - \frac{1}{n} \boldsymbol{r}\boldsymbol{r}'$$

(cf. Lemma 1.2 and Corollary 1.1).

A block design belongs to the class of $(\rho_0; v-1-\rho_0; 0)$-EB designs if it is a simple PEB design in the terminology of Puri and Nigam (1977a), or belongs to the class of C-designs according to Saha (1976), which covers the class D_1 of designs used by Shah (1964), i.e., satisfies the conditions

$$\boldsymbol{M}_0 = \mu_1 \boldsymbol{L}_1 \quad \text{and} \quad \boldsymbol{C} = \boldsymbol{r}^\delta (\boldsymbol{L}_0 + \varepsilon_1 \boldsymbol{L}_1).$$

In particular, any connected two-associate PBIB design (Definition 3.2), with its concurrence matrix NN' having a zero eigenvalue (of multiplicity ρ_0), belongs to this class. There are many such designs in the classes of group divisible (Definition 3.3), triangular, or Latin-square type designs (see Raghavarao, 1971, Section 8.4). For example, any singular group divisible and any semi-regular group divisible design is such. (See also Ceranka, 1983.)

On the other hand, any connected two-associate PBIB design, with NN' having only positive eigenvalues, belongs to the class of $(0;\ \rho_1, \rho_2; 0)$-EB designs. For example, any regular group divisible design is such.

Regarding designs with supplemented balance (see Pearce, 1960; Caliński, 1971; Caliński and Ceranka, 1974), i.e., designs obtained from a block design by adding to each block one or more supplementary treatments, one can say more. If the supplementary treatments on their own are arranged in an orthogonal design, Caliński (1971) calls the entire design an "orthogonally supplemented block design". An orthogonally supplemented BIB design belongs to the class of $(\rho_0; \rho_1; 0)$-EB designs. An orthogonally supplemented connected two-associate PBIB design with an incidence matrix N belongs to the class of $(\rho_0; \rho_1, \rho_2; 0)$-EB or $(\rho_0; \rho_1; 0)$-EB designs, according as both the eigenvalues of NN' are positive or one of them is zero.

Thus, the present terminology, using the multiplicities ρ_β, may be much more suitable for indicating the statistical advantage and utility of a block design, then the original reference to a PEB design with m efficiency classes. Note, particularly, the difference between a $(0; \rho_1, \rho_2; 0)$-EB design and a $(\rho_0; \rho_1; 0)$-EB design, both being PEB designs with two efficiency classes according to Definition 3.5. Also note that in the present terminology the use of the term "partial efficiency balance" is avoided, thus meeting the demand of Preece (1982) that "partial balance" should be used only in the sense of Nair and Rao (1942a).

3.3. Remarks on constructions

The construction procedures for and combinatorial properties of some X^{-1}-PEB(m) designs with $m \leqslant 2$ have been discussed in detail by Kageyama et al. (1988). See also Caliński (1977), Das and Ghosh (1985), and Kageyama and Mukerjee (1986), who have proposed a method of reinforcement of BIB designs. In particular, they have shown that if a single new treatment is added then the reinforcement of a BIB design always leads to a X^{-1}-PEB(1) design. Also, they have shown that it is impossible to construct a VB or an EB design in $v + 2$ or more treatments through reinforcement of a BIB design in v treatments. However, X^{-1}-PEB(1) designs in $v + 2$ or more treatments can be constructed in abundance. Thus, reinforcement is a powerful tool in the construction of X^{-1}-PEB(1) designs.

One of the useful methods of constructing C-designs is to utilize the treatment merging method (cf. Nigam et al., 1988) which, in fact, preserves the structure of units and leads to a new design to be a C-design with a less number of treatments as a starting design. Several methods of constructing C-designs are also given by Saha (1976), Nigam and Puri (1982), Ceranka (1983), Ceranka and Kozlowska (1984). The

EB designs are particular cases of C-designs. Hence, one can use them to construct C-designs.

For constructions and other properties of PEB designs see, for example, Ceranka (1984), Dey and Gupta (1986), Kageyama and Puri (1985a, b), Nigam and Puri (1982), Pal (1980), Puri and Kageyama (1985), Puri and Nigam (1977a, 1983), Puri et al. (1987).

4. Nested block designs and the concept of resolvability

Although a large number of block designs suitable for the local control of experimental material is available in the literature, there exist some situations where there are more sources of variation than can be controlled by ordinary blocking. Such sources of variation, though not of main interest to the experimenter, do contribute substantially to the variability in the experimental material, and should be controlled, whenever possible. If the sources are hierarchical in nature, designs with a block structure of two strata can be useful.

Situations where there is a need for such structure of blocks appear quite often in practice, particularly in agricultural and industrial experiments. Common examples are the lattice designs or, more generally, the so-called resolvable block designs (cf. John, 1987, Section 3.4).

4.1. Nested balanced incomplete block designs and their generalization

Preece (1967) introduced a class of nested balanced incomplete block (NBIB) designs for statistical situations where there are two sources of variability and one source is nested within the other. That is, an NBIB design has two systems of blocks, the second nested within the first (each block from the first system, called "superblock", containing some blocks, called "sub-blocks", from the second) such that ignoring either system leaves a BIB design whose blocks are those of the other system. This idea has been generalized to nested partially balanced incomplete block designs by Homel and Robinson (1975) and discussed by Banerjee and Kageyama (1990, 1993). The statistical justification of NBIB designs was shown by Yates (1940), Kleczkowski (1960), and Kassanis and Kleczkowski (1965). An interesting application of NBIB designs has recently been given by Gupta and Kageyama (1994) who showed that some NBIB designs provide universally optimal complete diallel crosses. Systematic methods of constructing NBIB designs were considered by Agrawal and Prasad (1983), Jimbo and Kuriki (1983), and Dey et al. (1986). It should be remarked that another type of nested designs has been also discussed in Longyear (1981), Colbourn and Colbourn (1983), Gupta (1984), and others.

Here a nested block design of a wider sense than the usual one will be introduced, and some statistical and combinatorial properties of such designs will be given, following Mejza and Kageyama (1994). The usual NBIB design is a special case of the designs considered here.

Consider a block design with two systems of blocks, the second nested within the first, such that ignoring either system leaves a block design whose blocks are those of the other system. Let the first system contain a blocks, further called as superblocks, and let the hth superblock contain b_h sub-blocks (second system of blocks) of sizes $k_{h1}, k_{h2}, \ldots, k_{hb_h}, h = 1, 2, \ldots, a$. This means that the hth superblock contains $k_{h\cdot} = \sum_{j=1}^{b_h} k_{hj}$ units, while $n = \sum_{h=1}^{a} k_{h\cdot}$ stands for the whole number of units. A design with such unit structure will be called a "nested block design" (NB design for short).

Let v denote, as usual, the number of treatments. Moreover, let D denote a block design with respect to superblocks (i.e., with respect to the first system of blocks), and let D^* denote the block design with respect to sub-blocks (i.e., with respect to the second system of blocks). The design D has a blocks of sizes $\boldsymbol{k}_D = (k_{1\cdot}, k_{2\cdot}, \ldots, k_{a\cdot})'$, while D^* has $b = \sum_{h=1}^{a} b_h$ blocks of sizes $\boldsymbol{k}_{D^*} = [\boldsymbol{k}_1', \boldsymbol{k}_2', \ldots, \boldsymbol{k}_a']'$, where $\boldsymbol{k}_h = [k_{h1}, k_{h2}, \ldots, k_{hb_h}]'$.

Let D_h denote a block design of the hth superblock of an NB design and let $\boldsymbol{N}_h$ be its incidence matrix for $h = 1, 2, \ldots, a$. Then $\boldsymbol{N} = [\boldsymbol{N}_1 : \boldsymbol{N}_2 : \cdots : \boldsymbol{N}_a]$ $(= [n_{is}]$, say) is the incidence matrix of D^*, while $\boldsymbol{R} = [\boldsymbol{N}_1\mathbf{1}_{b_1} : \boldsymbol{N}_2\mathbf{1}_{b_2} : \cdots : \boldsymbol{N}_a\mathbf{1}_{b_a}] = [\boldsymbol{r}_1 : \boldsymbol{r}_2 : \cdots : \boldsymbol{r}_a]$ $(= [r_{ih}]$, say) is the incidence matrix of D. The designs D and D^* will be called superblock and sub-block designs of an NB design (SN design and BN design for short), respectively.

Let $\boldsymbol{r} = \boldsymbol{R}\mathbf{1}_a = \boldsymbol{N}\mathbf{1}_b = [r_1, r_2, \ldots, r_v]'$ denote the vector of treatment replications. An NB design is said to be equireplicate if $r_i = r$, say, for all i, superblock proper if $k_{h\cdot} = k_D$, say, for all h, sub-block proper if $k_{hj} = k$, say, for all h and j, superblock binary if $r_{ih} = 1$ or 0 and sub-block binary if $n_{is} = 1$ or 0, for $i = 1, 2, \ldots, v$, $h = 1, 2, \ldots, a$, $j = 1, 2, \ldots, b_h$, $s = 1, 2, \ldots, b$.

The properties of a block design are related to the pattern of its C-matrix (cf. Sections 1 and 2). In the present case two such notions are available,

$$\boldsymbol{C}_D = \boldsymbol{r}^\delta - \boldsymbol{R}\boldsymbol{k}_D^{-\delta}\boldsymbol{R}' \quad \text{and} \quad \boldsymbol{C}_{D^*} = \boldsymbol{r}^\delta - \boldsymbol{N}\boldsymbol{k}_{D^*}^{-\delta}\boldsymbol{N}'.$$

There are many ways to introduce such properties as connectedness, balance, etc.

DEFINITION 4.1. An NB design is said to be superblock (sub-block) connected if its SN design (BN design) is connected (Definition 1.1).

DEFINITION 4.2. An NB design is said to be connected if both its SN design and its BN design are connected.

The following can be obtained.

THEOREM 4.1. *In an* NB *design, the sub-block connectedness implies the superblock connectedness.*

PROOF. This immediately follows from the meaning of the nested structure in an NB design and the definition of connectedness in the light of chains of treatments (Definition 1.1). □

DEFINITION 4.3. An NB design is said to be superblock (sub-block) variance-balanced (VB) (Definition 2.6) if its SN design (BN design) is VB, i.e., in case of connectedness, if its matrix $\boldsymbol{C}_D$ ($\boldsymbol{C}_{D^*}$) has the form

$$\boldsymbol{C}_D = \theta\left(\boldsymbol{I} - \frac{1}{v}\boldsymbol{1}\boldsymbol{1}'\right) \left[\boldsymbol{C}_{D^*} = \theta^*\left(\boldsymbol{I} - \frac{1}{v}\boldsymbol{1}\boldsymbol{1}'\right)\right],$$

where $\theta(\theta^*)$ is the unique nonzero eigenvalue of $\boldsymbol{C}_D$ ($\boldsymbol{C}_{D^*}$), respectively. If $\theta = r$ ($\theta^* = r$), then the design is called superblock orthogonal (sub-block orthogonal).

Referring to block designs D_h, $h = 1, 2, \ldots, a$, of a connected NB design, one can have the following.

THEOREM 4.2. *If the design D^* of an* NB *design is connected and all its designs D_h's are connected and* VB, *then the* NB *design is sub-block* VB.

PROOF. From the structure (2.3) of a connected VB block design, it follows that for each D_h $\boldsymbol{N}_h\boldsymbol{k}_h^{-\delta}\boldsymbol{N}_h' = \boldsymbol{r}_h^{\delta} - \theta_h(\boldsymbol{I} - v^{-1}\boldsymbol{1}\boldsymbol{1}')$. Hence, $\boldsymbol{C}_{D^*} = \boldsymbol{r}^{\delta} - \sum_{h=1}^{a}\boldsymbol{N}_h\boldsymbol{k}_h^{-\delta}\boldsymbol{N}_h' = \boldsymbol{r}^{\delta} - \boldsymbol{r}^{\delta} + \sum_{h=1}^{a}\theta_h(\boldsymbol{I} - v^{-1}\boldsymbol{1}\boldsymbol{1}') = \theta(\boldsymbol{I} - v^{-1}\boldsymbol{1}\boldsymbol{1}')$, i.e., the $\boldsymbol{C}_{D^*}$ is the C-matrix of a VB design. □

DEFINITION 4.4. A connected NB design is said to be VB if both its SN design and its BN design are VB.

DEFINITION 4.5. An NB design is said to be superblock (sub-block) efficiency-balanced (EB) (Definition 2.7) if its SN design (BN design) is EB, i.e., in case of connectedness if its matrix $\boldsymbol{C}_D$ ($\boldsymbol{C}_{D^*}$) has the form

$$\boldsymbol{C}_D = \varepsilon_D\left(\boldsymbol{r}^{\delta} - \frac{1}{n}\boldsymbol{r}\boldsymbol{r}'\right) \left[\boldsymbol{C}_{D^*} = \varepsilon_{D^*}\left(\boldsymbol{r}^{\delta} - \frac{1}{n}\boldsymbol{r}\boldsymbol{r}'\right)\right],$$

respectively, where $\varepsilon_D = [n - \text{tr}(\boldsymbol{R}\boldsymbol{k}_D^{-\delta}\boldsymbol{R}')]/c$ $\{\varepsilon_{D^*} = [n - \text{tr}(\boldsymbol{N}\boldsymbol{k}_{D^*}^{-\delta}\boldsymbol{N}')]/c\}$, with $c = n - n^{-1}\boldsymbol{r}'\boldsymbol{r}$. If $\varepsilon_D = 1$ ($\varepsilon_{D^*} = 1$), then the design is called superblock orthogonal (sub-block orthogonal).

DEFINITION 4.6. A connected NB design is said to be EB if both its SN design and its BN design are EB.

Consider a class of nested BIB designs, in which the incidence matrices $\boldsymbol{R}$ and $\boldsymbol{N}$ are incidence matrices of some BIB designs. Of course, these NB designs are VB and EB at the same time.

4.2. Resolvable block designs

The concept of resolvability introduced by Bose (1942) was generalized to μ-resolvability by Shrikhande and Raghavarao (1964) in a combinatorial sense. The concept of μ-resolvability can be further generalized to $(\mu_1, \mu_2, \ldots, \mu_a)$-resolvability as follows (Kageyama, 1976b).

DEFINITION 4.7. A block design with parameters v, b, r is said to be $(\mu_1, \mu_2, \ldots, \mu_a)$-resolvable if the blocks can be separated into a $(\geqslant 2)$ sets of $b_1, b_2, \ldots, b_a$ $(\geqslant 1)$ blocks such that the set consisting of b_h blocks contains every treatment $\mu_h (\geqslant 1)$ times for $h = 1, 2, \ldots, a$, i.e., the set of b_h blocks forms a μ_h-replication set of each treatment (called a resolution set). Here $r = \sum_{h=1}^{a} \mu_h$. Furthermore, when $\mu_1 = \mu_2 = \cdots = \mu_a$ $(= \mu$, say), it is simply called μ-resolvable for $\mu \geqslant 1$.

Note that Definition 4.7 corresponds to that of μ-resolvability introduced by Shrikhande and Raghavarao (1964). A 1-resolvable block design is simply called resolvable in the sense of Bose (1942). One of the earliest examples of a resolvable BIB design is related to the Kirkman (1950a, b) school girl problem, which attracted many mathematicians in the late 19th and early 20th centuries. A good bibliography can be found in Eckenstein (1912). However, no complete solution was known until Ray-Chaudhuri and Wilson (1971) completely solved the problem.

For practical applications, refer to John (1961) and Kageyama (1976b). These papers indicate the importance of $(\mu_1, \mu_2, \ldots, \mu_a)$-resolvable block designs, with possibly varying block sizes and having $\mu_1, \mu_2, \ldots, \mu_a$ not necessarily all equal.

It can easily be seen that any $(\mu_1, \mu_2, \ldots, \mu_a)$-resolvable block design is a particular NB design. This can be formalized as follows.

DEFINITION 4.8. An NB design is said to be $(\mu_1, \mu_2, \ldots, \mu_a)$-resolvable if its $\boldsymbol{r}_h = \mu_h \mathbf{1}_v$ for $h = 1, 2, \ldots, a$.

Since the class of NB designs includes all block designs satisfying Definition 4.7, a resolution set being a superblock, a $(\mu_1, \mu_2, \ldots, \mu_a)$-resolvable NB design can be called simply $(\mu_1, \mu_2, \ldots, \mu_a)$-resolvable block design.

Next definition concerns only those $(\mu_1, \mu_2, \ldots, \mu_a)$-resolvable block designs which have a constant block size within each resolution set. The constant block size within the hth set (superblock) is denoted by k_h^* for $h = 1, 2, \ldots, a$ (cf. Mukerjee and Kageyama, 1985).

DEFINITION 4.9. A $(\mu_1, \mu_2, \ldots, \mu_a)$-resolvable block design with a constant block size in each resolution set (superblock) is said to be affine $(\mu_1, \mu_2, \ldots, \mu_a)$-resolvable if:

(i) for $h = 1, 2, \ldots, a$, every two distinct blocks from the hth set intersect at the same number, say q_{hh}, of treatments;

(ii) for $h \neq h' = 1, 2, \ldots, a$, every block from the hth set intersect every block of the h'th set at the same number, say $q_{hh'}$, of treatments.

It is evident that for an affine $(\mu_1, \mu_2, \ldots, \mu_a)$-resolvable block design the equalities $q_{hh}(b_h - 1) = k_h^*(\mu_h - 1)$, $q_{hh'} b_{h'} = k_h^* \mu_{h'}$, $(h \neq h' = 1, 2, \ldots, a)$ hold.

Some properties of NB designs discussed earlier will now be specified for the resolvable block designs.

THEOREM 4.3. *A* $(\mu_1, \mu_2, \ldots, \mu_a)$*-resolvable block design is superblock orthogonal.*

PROOF. Note that $\boldsymbol{R} = (\mu_1\mathbf{1} : \mu_2\mathbf{1} : \cdots : \mu_a\mathbf{1})$, $\boldsymbol{k}_D = v(\mu_1, \mu_2, \ldots, \mu_a)', \boldsymbol{r} = r\mathbf{1}$ with $r = \sum_{h=1}^{a} \mu_h$. Then $\boldsymbol{C}_D = r(\boldsymbol{I} - v^{-1}\mathbf{1}\mathbf{1}')$ which, on account of Theorem 2.1 and Remark 2.2, completes the proof. □

Theorem 4.3 can yield, on account of Definition 4.4, the following.

COROLLARY 4.1. *If a* $(\mu_1, \mu_2, \ldots, \mu_a)$*-resolvable block design is sub-block* VB, *then it is* VB.

Since a $(\mu_1, \mu_2, \ldots, \mu_a)$-resolvable block design is equireplicate, the two notions of balance (VB and EB) are identical (cf. Williams, 1975). This leads to the following.

COROLLARY 4.2. *If a* $(\mu_1, \mu_2, \ldots, \mu_a)$*-resolvable block design is sub-block* VB, *then it is* EB.

An extension of Fisher's inequality (Corollary 1.2) for a $(\mu_1, \mu_2, \ldots, \mu_a)$-resolvable VB design can be given as follows.

THEOREM 4.4. *For a* $(\mu_1, \mu_2, \ldots, \mu_a)$*-resolvable* VB *block design, the inequality* $b \geqslant v + a - 1$ *holds.*

For a proof see Kageyama (1993, Theorem 3.1).

As special cases, Theorem 4.4 yields some results available in Hughes and Piper (1976), Kageyama (1973, 1976b, 1978, 1984), Raghavarao (1962, 1971) as follows.

COROLLARY 4.3. *The inequality* $b \geqslant v + a - 1$ *holds for each class of the following block designs:*

(i) μ*-resolvable* BIB *design* $(\mu_1 = \mu_2 = \cdots = \mu_a = \mu)$;
(ii) μ*-resolvable* VB *block design* $(\mu_1 = \mu_2 = \cdots = \mu_a = \mu)$;
(iii) $(\mu_1, \mu_2, \ldots, \mu_a)$*-resolvable BIB designs.*

In particular, a $(\mu_1, \mu_2, \ldots, \mu_a)$*-resolvable* BIB *design with* $b = v + a - 1$ *is affine* μ*-resolvable with* $\mu_1 = \mu_2 = \cdots = \mu_a = \mu$.

When $b = v + a - 1$, one can have the following.

THEOREM 4.5. *In a* $(\mu_1, \mu_2, \ldots, \mu_a)$*-resolvable* VB *block design with* $b = v + a - 1$, *except when* $\mu_1 = \mu_2 = \cdots = \mu_a = 1$, *block sizes of blocks belonging to the same resolution set (superblock) are equal.*

THEOREM 4.6. *A* μ*-resolvable* VB *block design satisfying* $b = v + a - 1$ *for* $\mu \geqslant 2$ *is an affine* μ*-resolvable* BIB *design.*

For proofs of these theorems see Kageyama (1993, Theorems 3.2 and 3.3).

Theorem 4.6 is an interesting result similar to the theorem due to Rao (1966) that an equireplicate binary VB block design with $b = v$ is a symmetric BIB design. It is remarkable that Theorem 4.6 shows that there does not exist a μ-resolvable VB block design with unequal block sizes satisfying $b = v + a - 1$ for a positive integer $\mu \geqslant 2$; the result gives a complete solution to the open problem proposed by Kageyama (1974, p. 610). Note that there exists a 1-resolvable VB block design with unequal block sizes satisfying $b = v + a - 1$.

By a direct calculation due to Shrikhande and Raghavarao (1964), the following can be shown.

THEOREM 4.7. *For a μ-resolvable incomplete block design involving b blocks in a resolution sets (superblocks) and v treatments with a constant block size, any two of the following imply the third:*

(a) *affine μ-resolvability,*
(b) VB,
(c) $b = v + a - 1$.

THEOREM 4.8. *A $(\mu_1, \mu_2, \ldots, \mu_a)$-resolvable* VB *block design with $b = v + a - 1$ and a constant block size within the hth resolution set (superblock), k_h^* $(h = 1, 2, \ldots, a)$, must be affine $(\mu_1, \mu_2, \ldots, \mu_a)$-resolvable with*

$$q_{hh} = (k_h^{*2}/v)[1 - (b - r)\{\mu_h(v - 1)\}]$$

provided that $b_h \geqslant 2$, and

$$q_{hh'} = k_h^* k_{h'}^*/v \quad (h \neq h' = 1, 2, \ldots, a).$$

THEOREM 4.9. *An incomplete block affine $(\mu_1, \mu_2, \ldots, \mu_a)$-resolvable* VB *block design must have $b = v + a - 1$.*

Theorems 4.8 and 4.9 extend respectively the two implications '(b), (c) imply (a)' and '(a), (b) imply (c)' contained in Theorem 4.7. Thus, Theorem 4.7 can be partially extended to $(\mu_1, \mu_2, \ldots, \mu_a)$-resolvable block designs. The result '(a), (c) imply (b)' of Theorem 4.7 cannot, however, be extended in general. That is, an incomplete block affine $(\mu_1, \mu_2, \ldots, \mu_a)$-resolvable block design with $b = v + a - 1$ is not necessarily VB. This point is illustrated by the following example.

EXAMPLE 4.1. Consider an affine $(2, 2, 1, 1)$-resolvable incomplete block design with parameters $v = 9$, $b = 12$, $r = 6$, $k_1^* = k_2^* = 6$, $k_3^* = k_4^* = 3$, $a = 6$, given by the following blocks $[(4,5,6,7,8,9)(1,2,3,7,8,9)(1,2,3,4,5,6)][(2,3,5,6,8,9)(1,3,4,6,7,9)(1,2,4,5,7,8)][(1,6,8)(2,4,9)(3,5,7)][(1,5,9)(2,6,7)(3,4,8)]$. Clearly, here $b = v + a - 1$, but the design is not VB, as can be checked easily.

In view of the above example, it would be interesting to determine necessary and sufficient conditions under which an incomplete block $(\mu_1, \mu_2, \ldots, \mu_a)$-resolvable block design with $b = v + a - 1$ becomes VB. One such condition is given by the following result.

THEOREM 4.10. *An incomplete block affine* $(\mu_1, \mu_2, \ldots, \mu_a)$*-resolvable design satisfying* $b = v + a - 1$ *is* VB *if and only if* $(\mu_h - 1)/(b_h - 1) = (r - a)/(v - 1)$, $h = 1, 2, \ldots, a$.

EXAMPLE 4.2. There exists an affine 1-resolvable VB design with unequal block sizes and $b = v + a - 1$, whose blocks are as given by $[(1,5)(2,6)(3,7)(4,8)][(1,2,3,4)(5,6,7,8)][(1,3,6,8)(2,4,5,7)][(1,2,7,8)(3,4,5,6)][(1,4,6,7)(2,3,5,8)]$, in which $C = 4(I_8 - \frac{1}{8}J_8)$.

The literature of block designs contains many articles exclusively related to VB designs with resolvability. As a characterization of the saturated case of Theorem 4.4, it can be shown that in a $(\mu_1, \mu_2, \ldots, \mu_a)$-resolvable VB design with $b = v + a - 1$, except for the case $\mu_1 = \mu_2 = \cdots = \mu_a = 1$, sizes of blocks belonging to the same resolution set (superblock) are always equal (Theorem 4.5). Whether the above holds for the case $\mu_1 = \mu_2 = \cdots = \mu_a = 1$ as well, is an open problem.

In the light of Corollaries 4.1 and 4.2, the μ-resolvable block designs play an important role. These designs may be treated as the starting designs for some constructions of designs with desirable property. In general, given a set of parameters, the construction of resolvable block designs with unequal block sizes, having some balancing property, is not so simple. There is not much in the literature devoted to designs with unequal block sizes. A paper on such designs by Ceranka et al. (1986) presented four different techniques for constructing μ-resolvable C-designs. The four construction techniques are based on dualization; merging of treatments and dualization; complementation; and juxtaposition. For a class of block designs with a constant block size, Shrikhande (1976) gave an excellent survey of known combinatorial results on affine resolvable BIB designs. For resolvable t-designs, the reader can be referred to, for example, Sprott (1955), Hedayat and Kageyama (1980), Kageyama (1976a), Kageyama and Hedayat (1983), Kimberley (1971), Lindner and Rosa (1980, Chapter 4), and Mavron (1972).

For the combinatorial discussions on $(\mu_1, \mu_2, \ldots, \mu_a)$-resolvability, the reader can be also referred to Bose (1942), Ceranka et al. (1986), Hughes and Piper (1976), Kageyama (1973, 1977, 1984), Kageyama and Sastry (1993), Mukerjee and Kageyama (1985), Raghavarao (1971), Ray-Chaudhuri and Wilson (1971), Shrikhande (1953, 1976), and Shrikhande and Raghavarao (1964).

4.3. *α-designs*

Resolvable incomplete block designs are extensively used in statutory trials of agricultural crop varieties in the United Kingdom (cf. Patterson and Silvey, 1980) and elsewhere. Numbers of varieties in these trials are fixed, i.e., not at the choice of the statistician, and large enough to require the use of incomplete block designs. Numbers of replications are usually fixed and must be the same for all treatments (varieties).

Thus, preference is given to resolvable block designs. For example, some important disease measurements are expensive and have, therefore, to be restricted to one or two replications. Again, large trials cannot always be completely drilled or harvested in a single session. Use of resolvable designs allows these operations to be done in stages, with one or more complete replications dealt with at each stage.

Yates (1939, 1940) has pointed to other advantages of resolvable designs. On page 325 of his 1940 paper he noted that "cases will arise in which the use of ordinary randomized blocks will be more efficient than the use of incomplete blocks, whereas lattice designs can never be less efficient than ordinary randomized blocks." This advantage of lattice designs is shared by all other resolvable incomplete block designs. Yates (1940) further stated that "incomplete block designs which cannot be arranged in complete replications are likely to be of less value in agriculture than ordinary lattice designs. Their greatest use is likely to be found in dealing with experimental material in which the block size is definitely determined by the nature of the material." In variety trials, of course, a wide choice of block size is open to the experimenter.

A general algorithm for constructing designs for this purpose has been described by Patterson and Williams (1976a). The algorithm is able to produce large numbers of designs but only the most efficient are adopted for practical application. In these designs the number of varieties is a multiple of the block size. These designs are called α-designs. They include as special cases some lattice and resolvable cyclic designs. Their method has been developed to provide a simple computer algorithm for automatic production of plans for variety trials. (See also Patterson et al., 1978).

When seed is in short supply or some other economy is enforced, the trials sometimes have to be conducted in only two replications. It is important in these circumstances to use the most efficient possible designs. Bose and Nair (1962) have described and tabulated a series of two-replicate resolvable designs. Another series of designs is given by the duals of the resolvable paired-comparison designs of Williams (1976). Both series involve the use of symmetric block designs. Patterson and Williams (1976b) have shown that every two-replicate resolvable design is uniquely associated with a symmetric incomplete block design. Williams et al. (1976) used this relationship to examine the efficiency of the designs given by Bose and Nair (1962) and Williams (1976). Furthermore, John and Mitchell (1977) gave a listing of optimal binary incomplete block designs obtained from an exhaustive computer search of possible designs. Using and extending their results on symmetric designs, Williams et al. (1977) obtained a series of optimal two-replicate resolvable designs and the identification of these designs was considered. Also, on the basis of a comparison of these designs with those of Bose and Nair (1962) and Williams (1976) recommendations were made on the choice of efficient designs.

Bose and Nair's (1962) designs very usefully augment the simple square and rectangular lattices but there appears to have been no parallel development for higher order lattices. David's (1967) construction for cyclic designs is capable of producing a large number of resolvable designs but again there are restrictions; this time block size k must equal either r, the number of replications, or a multiple of r.

REMARK 4.1. (i) Some other block designs are also resolvable. Clatworthy (1973) provided information on the resolvability, or otherwise, of most of the PBIB designs with two associate classes in his extensive tables. The resolvable designs are, however, usually either square lattices or less efficient alternatives. (ii) The α-designs, introduced by Patterson and Williams (1976a), are useful to construct NB designs. Moreover, only some of them are C-designs.

5. Analysis of experiments in block designs under the randomization model

One of the problems of interest related to experiments in block designs is the utilization of between block information for estimating treatment parametric contrasts. This problem, after Yates (1939, 1940) known in the literature as the recovery of inter-block information, is connected with the principle of randomizing the experimental material before subjecting it to experimental treatments. Therefore, methods dealing with the recovery of inter-block information have to be based on a properly derived randomization model. The purpose of the present section is to reconsider one of such methods and to re-examine the principles underlying the recovery of inter-block information in case of a general block design.

5.1. The randomization model and its statistical implications

According to one of the basic principles of experimental design (cf. Fisher, 1925, Section 48), the units are to be randomized before they enter the experiment. Suppose that the randomization is performed by randomly permuting labels of the available blocks, say N_B in number, and by randomly permuting labels of the available units in each given block, all the $1 + N_B$ permutations being carried out independently (cf. Nelder, 1954; White, 1975). Then, assuming the usual unit-treatment additivity (cf. Nelder, 1965b, p. 168; White, 1975, p. 560; Bailey, 1981, p. 215; Kala, 1991, p. 7), and also assuming, as usual, that the technical errors are uncorrelated, with zero expectation and a constant variance, and independent of the unit responses to treatments (cf. Neyman, 1935, pp. 110–114 and 145; Kempthorne, 1952, p. 132 and Section 8.4; Ogawa, 1963, p. 1559), the model of the variables observed on the n units actually used in the experiment can be written in matrix notation, with the matrices $\boldsymbol{\Delta}$ and $\boldsymbol{D}$ defined in Section 1.2, as

$$\boldsymbol{y} = \boldsymbol{\Delta}'\boldsymbol{\tau} + \boldsymbol{D}'\boldsymbol{\beta} + \boldsymbol{\eta} + \boldsymbol{e}, \tag{5.1}$$

where $\boldsymbol{y}$ is an $n \times 1$ vector of observed variables $\{y_{\ell(j)}(i)\}$, $\boldsymbol{\tau}$ is a $v \times 1$ vector of treatment parameters $\{\tau_i\}$, $\boldsymbol{\beta}$ is a $b \times 1$ vector of block random effects $\{\beta_j\}$, $\boldsymbol{\eta}$ is an $n \times 1$ vector of unit errors $\{\eta_{\ell(j)}\}$ and $\boldsymbol{e}$ is an $n \times 1$ vector of technical errors $\{e_{\ell(j)}\}$, $\ell(j)$ denoting the unit ℓ in block j ($i = 1, 2, \ldots, v$; $\ell = 1, 2, \ldots, k_j$; $j = 1, 2, \ldots, b$).

Properties of the model (5.1) have been studied for a general block design by Caliński and Kageyama (1988, 1991). These properties are different from those of the usually assumed model (see, e.g., John, 1987, Section 1.2). It has been found that the expectation vector and the dispersion matrix for $\boldsymbol{y}$ are

$$\mathrm{E}(\boldsymbol{y}) = \boldsymbol{\Delta}'\boldsymbol{\tau} \tag{5.2}$$

and

$$\mathrm{Var}(\boldsymbol{y}) = \left(\boldsymbol{D}'\boldsymbol{D} - \frac{1}{N_B}\mathbf{1}_n\mathbf{1}_n'\right)\sigma_B^2 + \left(\boldsymbol{I}_n - \frac{1}{K_H}\boldsymbol{D}'\boldsymbol{D}\right)\sigma_U^2 + \boldsymbol{I}_n\sigma_e^2, \tag{5.3}$$

respectively, where K_H is a weighted harmonic average of the available numbers of units within the N_B available blocks, from which units in numbers $\{k_j\}$ have been chosen for the experiment after the randomization, and where the variance components σ_B^2, σ_U^2, and σ_e^2 are related to the random vectors $\boldsymbol{\beta}, \boldsymbol{\eta}$ and $\boldsymbol{e}$, respectively.

To see this more precisely, suppose that the available N_B blocks are originally labelled $\xi = 1, 2, \ldots, N_B$, and that block ξ contains K_ξ units (plots), which are originally labelled $\pi = 1, 2, \ldots, K_\xi$. The label may also be written as $\pi(\xi)$ to denote that unit π is in block ξ. The randomization of blocks can then be understood as choosing at random a permutation of $\{1, 2, \ldots, N_B\}$, and then renumbering the blocks with $j = 1, 2, \ldots, N_B$ accordingly. Similarly, the randomization of experimental units within block ξ can be seen as selecting at random a permutation of $\{1, 2, \ldots, K_\xi\}$, and then renumbering the units of the block with $\ell = 1, 2, \ldots, K_\xi$ accordingly. It will be assumed here that any permutation of block labels can be selected with equal probability, as well as that any permutation of unit labels within a block can be selected with equal probability. Furthermore, it will be assumed that the randomizations of units within the blocks are among the blocks independent, and that they are also independent of the randomization of blocks.

The above randomization procedure reflects the practical instruction given by Nelder (1954): "choose a block at random and reorder its members at random ...;" then "repeat the procedure with one of the remaining blocks chosen at random ..., and so on". This can be accomplished whether the available blocks are of equal or unequal number of members, i.e., their units. The purpose of this randomization is not only to "homogenize" the within block variability, but also to "average out" the possibly heterogeneous variance among the experimental units from different blocks to a common value (cf. White, 1975, Sections 2 and 7).

Now, following Nelder (1965a), assume for a while that all the available units in the N_B available blocks receive the same treatment, no matter which. For this "null" experiment let the response of the unit $\pi(\xi)$ be denoted by $\mu_{\pi(\xi)}$, and let it be denoted by $m_{\ell(j)}$ if by the randomizations the block originally labelled ξ receives in the design label j and the unit originally labelled π in this block receives in the design label ℓ. With this, introducing the identity

$$\mu_{\pi(\xi)} = \mu_{.(.)} + (\mu_{.(\xi)} - \mu_{.(.)}) + (\mu_{\pi(\xi)} - \mu_{.(\xi)}),$$

where (according to the usual dot notation)

$$\mu_{.(\xi)} = \frac{1}{K_\xi}\sum_{\pi(\xi)=1}^{K_\xi} \mu_{\pi(\xi)} \quad \text{and} \quad \mu_{.(.)} = \frac{1}{N_B}\sum_{\xi=1}^{N_B} \mu_{.(\xi)},$$

one can write the linear model

$$m_{\ell(j)} = \mu + \beta_j + \eta_{\ell(j)}, \tag{5.4}$$

for any ℓ and j, where $\mu = \mu_{.(.)}$ is a constant parameter, the mean, while β_j and $\eta_{\ell(j)}$ are random variables, the first representing a block random effect, the second a unit error. The following moments of the random variables in (5.4) are easily obtainable:

$$\mathrm{E}(\beta_j) = 0, \qquad \mathrm{E}(\eta_{\ell(j)}) = 0,$$

$$\mathrm{Cov}(\beta_j, \eta_{\ell(j')}) = 0, \quad \text{whether } j = j' \text{ or } j \neq j',$$

$$\mathrm{Cov}(\beta_j, \beta_{j'}) = \begin{cases} N_B^{-1}(N_B - 1)\sigma_B^2, & \text{if } j = j', \\ -N_B^{-1}\sigma_B^2, & \text{if } j \neq j', \end{cases}$$

and

$$\mathrm{Cov}(\eta_{\ell(j)}, \eta_{\ell'(j')}) = \begin{cases} K_H^{-1}(K_H - 1)\sigma_U^2, & \text{if } j = j' \text{ and } \ell = \ell', \\ -K_H^{-1}\sigma_U^2, & \text{if } j = j' \text{ and } \ell \neq \ell', \\ 0, & \text{if } j \neq j', \end{cases}$$

where

$$\sigma_B^2 = (N_B - 1)^{-1} \sum_{\xi=1}^{N_B} (\mu_{.(\xi)} - \mu_{.(.)})^2$$

and

$$\sigma_U^2 = N_B^{-1} \sum_{\xi=1}^{N_B} \sigma_{U,\xi}^2, \quad \text{with} \quad \sigma_{U,\xi}^2 = (K_\xi - 1)^{-1} \sum_{\pi(\xi)=1}^{K_\xi} (\mu_{\pi(\xi)} - \mu_{.(\xi)})^2,$$

and where the weighted harmonic average K_H is defined as

$$K_H^{-1} = N_B^{-1} \sum_{\xi=1}^{N_B} K_\xi^{-1} \sigma_{U,\xi}^2 / \sigma_U^2.$$

Further, denoting the technical error affecting the observation of the response on the (randomized) unit $\ell(j)$ by $e_{\ell(j)}$, note that the model of the variable observed on that unit in the null experiment can be extended from (5.4) to

$$y_{\ell(j)} = m_{\ell(j)} + e_{\ell(j)} = \mu + \beta_j + \eta_{\ell(j)} + e_{\ell(j)}, \tag{5.5}$$

for any j and ℓ. It may usually be assumed that the technical errors $\{e_{\ell(j)}\}$ are uncorrelated, with zero expectation and a constant variance, σ_e^2 in (5.3), and that they are independent of the block effects $\{\beta_j\}$ and of the unit errors $\{\eta_{\ell(j)}\}$.

It follows from the model (5.5) and its properties that in the null experiment the moments of the random variables $\{y_{\ell(j)}\}$ do not depend on the labels received by the blocks and their units in result of the randomizations. This means that the randomized units within any of the randomized blocks can be considered as homogeneous in the sense that the variables observed on them have a common mean, variance and covariance. Moreover, this means that also the randomized blocks themselves can be considered as homogeneous, in the sense that the variables observed on equal subsets of units from different blocks have a common mean vector and a common dispersion matrix of relevant order. These homogeneities hold regardless of any heterogeneity of the variance $\sigma_{U,\xi}^2$ for different ξ ($\xi = 1, 2, \ldots, N_B$) (cf. White, 1975, Section 2).

Now, to adopt the results obtained for the null experiment to a real experiment designed according to a chosen incidence matrix $\boldsymbol{N}$, two questions are to be answered. How the randomized units within the randomized blocks are to be assigned to different treatments, and how the model (5.5) is to be adjusted accordingly.

As to the first question, the following rule can be used. Assign the block which due to the randomization is labelled j to the jth column of $\boldsymbol{N}$. Then assign the units of this block to treatments in numbers indicated by the elements of the jth column of $\boldsymbol{N}$ and in order determined by the labels the units have received due to the randomization. Repeat this procedure for $j = 1, 2, \ldots, b$. Note, however, that this rule requires not only that $b \leqslant N_B$, but also that none of the elements of the vector $\boldsymbol{N}'\mathbf{1}_v = \boldsymbol{k} = [k_1, k_2, \ldots, k_b]'$ exceeds the smallest K_ξ. Otherwise, an adjustment of $\boldsymbol{N}$ is to be made, as suggested by White (1975, pp. 558 and 561). For more discussion on this see Caliński and Kageyama (1996, Section 2.2).

With regard to the second question, note that from the assumption of the complete additivity mentioned earlier, which is equivalent to the assumption that the variances and covariances of the random variables $\{\beta_j\}, \{\eta_{\ell(j)}\}$ and $\{e_{\ell(j)}\}$ do not depend on the treatment applied, the adjustment of the model (5.5) to a real situation, of comparing v treatments on different units of the same experiment, can be made by changing the constant term μ only. Thus, the model gets the form

$$y_{\ell(j)}(i) = \mu(i) + \beta_j + \eta_{\ell(j)} + e_{\ell(j)},$$

which in matrix notation can be written as in (5.1), with $\tau_i = \mu(i)$, and the corresponding moments of $\{y_{\ell(j)}(i)\}$ as in (5.2) and (5.3).

The model (5.1), with properties (5.2) and (5.3), is exactly the same as that obtained by Kala (1991, Section 5) under more general considerations. It also coincides with the model (2) of Patterson and Thompson (1971), as will be shown in Lemma 5.1. Furthermore, if $k_1 = k_2 = \cdots = k_b = k$ (say), then (5.3) can be written as

$$\begin{aligned}\operatorname{Var}(\boldsymbol{y}) = {} & \left(\boldsymbol{I}_n - \frac{1}{k}\boldsymbol{D}'\boldsymbol{D}\right)\sigma_1^2 + \left(\frac{1}{k}\boldsymbol{D}'\boldsymbol{D} - \frac{1}{n}\mathbf{1}_n\mathbf{1}_n'\right)\sigma_2^2 \\ & + \left(\frac{1}{n}\mathbf{1}_n\mathbf{1}_n'\right)\sigma_3^2, \end{aligned} \tag{5.6}$$

where

$$\sigma_1^2 = \sigma_U^2 + \sigma_e^2, \qquad \sigma_2^2 = k\sigma_B^2 + \left(1 - \frac{k}{K_H}\right)\sigma_U^2 + \sigma_e^2$$

and

$$\sigma_3^2 = \left(1 - \frac{b}{N_B}\right)k\sigma_B^2 + \left(1 - \frac{k}{K_H}\right)\sigma_U^2 + \sigma_e^2,$$

which shows that the design has then the orthogonal block structure of Nelder (1965a, b) with three "strata", the "intra-block", the "inter-block" and the "total area" stratum (to be discussed in Section 6). The variances σ_1^2, σ_2^2 and σ_3^2 are called the "stratum variances" (cf. Houtman and Speed, 1983, Section 2.2). Evidently, they become smaller when $b = N_B$ and $k = K_H$. Then the present model becomes equivalent to that considered by Rao (1959) and recently by Shah (1992).

It should be emphasized here that the described randomizations of blocks and units are aimed not at the selection of experimental material, which is to be accomplished before, but at assigning the appropriately prepared experimental units, all or some of them, to the experimental treatments according to the chosen design. By these randomizations the responses of units to treatments become random variables of certain uniform dispersion properties, even if the desirable unification of units within blocks is not fully achieved (cf. White, 1975, p. 558).

It has been shown by Caliński and Kageyama (1991) that under the model (5.1), with its properties (5.2) and (5.3), a function $\boldsymbol{w}'\boldsymbol{y}$ is uniformly the BLUE of $\boldsymbol{c}'\boldsymbol{\tau}$ if and only if $\boldsymbol{w} = \boldsymbol{\Delta}'\boldsymbol{s}$, where $\boldsymbol{s} = \boldsymbol{r}^{-\delta}\boldsymbol{c}$ satisfies the condition

$$(\boldsymbol{k}^{\delta} - \boldsymbol{N}'\boldsymbol{r}^{-\delta}\boldsymbol{N})\boldsymbol{N}'\boldsymbol{s} = \boldsymbol{0}, \tag{5.7}$$

i.e., is related to the incidence matrix $\boldsymbol{N}$ of the design by either the condition (a) $\boldsymbol{N}'\boldsymbol{s} = \boldsymbol{0}$, the estimated function being then a contrast (i.e., $\boldsymbol{c}'\boldsymbol{1}_v = 0$), or (b) $\boldsymbol{N}'\boldsymbol{s} \neq \boldsymbol{0}$, with the elements of the vector $\boldsymbol{N}'\boldsymbol{s}$ all equal if the design is connected, and equal within any connected subdesign otherwise.

Condition (5.7) implies also that under the considered model any function $\boldsymbol{w}'\boldsymbol{y} = \boldsymbol{s}'\boldsymbol{\Delta}\boldsymbol{y}$, i.e., with any $\boldsymbol{s}$, is uniformly the BLUE of $\mathrm{E}(\boldsymbol{w}'\boldsymbol{y}) = \boldsymbol{s}'\boldsymbol{r}^{\delta}\boldsymbol{\tau}$, if and only if (i) the design is orthogonal and (ii) the block sizes of the design are constant within any of its connected subdesigns (recall Section 1.2 and Corollary 1.1).

5.2. Combined analysis utilizing informations from different strata

The results presented in Section 5.1 show that unless the function $\boldsymbol{c}'\boldsymbol{\tau}$ satisfies the condition (5.7), there does not exist the BLUE of it under the randomization model (5.1). However (as shown by Caliński and Kageyama, 1991, Section 3), for a contrast $\boldsymbol{c}'\boldsymbol{\tau}$ that does not satisfy (5.7) there exists the BLUE under the intra-block submodel, considered in Section 1.4, if $\boldsymbol{c} = \boldsymbol{\Delta}\boldsymbol{\phi}_1\boldsymbol{\Delta}'\boldsymbol{s}$ for some $\boldsymbol{s}$ and, simultaneously, also under the inter-block submodel, obtained by the projection $(\boldsymbol{D}'\boldsymbol{k}^{-\delta}\boldsymbol{D} - n^{-1}\mathbf{1}_n\mathbf{1}_n')\boldsymbol{y}$, if $\boldsymbol{c} = \Delta(\boldsymbol{D}'\boldsymbol{k}^{-\delta}\boldsymbol{D} - \mathbf{1}_n\mathbf{1}_n')\Delta'\boldsymbol{s}$ for the same (or proportional) $\boldsymbol{s}$, provided that the latter satisfies some additional condition involving block sizes. This means that in many, but not all, cases the estimation of a contrast can be based on information available in two strata of the experiment, the intra-block and the inter-block stratum. Unfortunately, each of them gives a separate estimate of the contrast, usually different in their actual values, sometimes even quite diverse. Therefore, a natural question arising in this context is whether and how it is possible to combine the information from both strata to obtain a single estimate in a somehow optimal way. Methods dealing with this problem are known in the literature as procedures of the recovery of inter-block information. An update of some of these methods for proper (i.e., equiblock-sized) block designs has been given by Shah (1992). Here a general theory underlying the recovery of inter-block information will be presented, in a way that is applicable to any block design, whether proper or not.

In an attempt to solve the problem of combining information from different strata on the basis of the randomization model (5.1), it is instructive to make provisionally an unrealistic assumption that the variance components appearing in the dispersion matrix (5.3) are known. Then the following results are essential.

LEMMA 5.1. *Let the model be as in* (5.1), *with the expectation vector* (5.2) *and the dispersion matrix* (5.3), *the latter written equivalently as*

$$\mathrm{Var}(\boldsymbol{y}) = \sigma_1^2(\boldsymbol{D}'\Gamma\boldsymbol{D} + \boldsymbol{I}_n), \tag{5.8}$$

where

$$\Gamma = \gamma\boldsymbol{I}_b - \frac{1}{N_B}\frac{\sigma_B^2}{\sigma_1^2}\mathbf{1}_b\mathbf{1}_b', \tag{5.9}$$

with $\sigma_1^2 = \sigma_U^2 + \sigma_e^2$ *and* $\gamma = (\sigma_B^2 - K_H^{-1}\sigma_U^2)/\sigma_1^2$. *Further, suppose that the true value of* γ *is known. Then,*

(a) *any function* $\boldsymbol{w}'\boldsymbol{y}$ *which is the BLUE of its expectation,*

(b) *a vector which is the BLUE of the expectation vector* $\Delta'\boldsymbol{\tau}$ *and, hence,*

(c) *a vector which gives the residuals,*

all remain unchanged when altering the present model by deleting $N_B^{-1}(\sigma_B^2/\sigma_1^2)\mathbf{1}_b\mathbf{1}_b'$ *in* (5.9), *i.e., by reducing* (5.8) *to*

$$\mathrm{Var}(\boldsymbol{y}) = \sigma_1^2(\gamma\boldsymbol{D}'\boldsymbol{D} + \boldsymbol{I}_n) = \sigma_1^2\boldsymbol{T}, \tag{5.10}$$

where $\boldsymbol{T} = \gamma\boldsymbol{D}'\boldsymbol{D} + \boldsymbol{I}_n$. *The matrix* $\boldsymbol{T}$ *in* (5.10) *is positive definite* (p.d.) *if* $\gamma > -1/k_{\max}$, *where* $k_{\max} = \max_j k_j$.

To prove this lemma note that from $\mathbf{1}_n'(\boldsymbol{I}_n - \boldsymbol{P}_{\Delta'}) = \mathbf{0}'$, where $\boldsymbol{P}_{\Delta'} = \Delta'\boldsymbol{r}^{-\delta}\Delta$, it follows that

$$(\boldsymbol{D}'\Gamma\boldsymbol{D} + \boldsymbol{I}_n)(\boldsymbol{I}_n - \boldsymbol{P}_{\Delta'}) = (\gamma\boldsymbol{D}'\boldsymbol{D} + \boldsymbol{I}_n)(\boldsymbol{I}_n - \boldsymbol{P}_{\Delta'}). \tag{5.11}$$

For details of the proof see Caliński and Kageyama (1996, Lemma 3.1). (See also Patterson and Thompson, 1971, p. 546, and Kala, 1981, Theorem 6.2.)

Now, on account of Lemma 5.1, a general theory due to Rao (1974) gives the following.

THEOREM 5.1. *Under the model and assumptions as in Lemma 5.1, including the assumption that* γ *is known and exceeds* $-1/k_{\max}$,

(a) *the* BLUE *of* $\boldsymbol{\tau}$ *is of the form*

$$\widehat{\boldsymbol{\tau}} = (\Delta \boldsymbol{T}^{-1}\Delta')^{-1}\Delta \boldsymbol{T}^{-1}\boldsymbol{y} \tag{5.12}$$

with

$$\boldsymbol{T}^{-1} = (\gamma \boldsymbol{D}'\boldsymbol{D} + \boldsymbol{I}_n)^{-1} = \boldsymbol{\phi} + \boldsymbol{D}'\boldsymbol{k}^{-\delta}(\boldsymbol{k}^{-\delta} + \gamma \boldsymbol{I}_b)^{-1}\boldsymbol{k}^{-\delta}\boldsymbol{D}; \tag{5.13}$$

(b) *the dispersion matrix of* $\widehat{\boldsymbol{\tau}}$ *is*

$$\mathrm{Var}(\widehat{\boldsymbol{\tau}}) = \sigma_1^2(\Delta \boldsymbol{T}^{-1}\Delta')^{-1} - \frac{1}{N_B}\sigma_B^2 \mathbf{1}_v\mathbf{1}_v'; \tag{5.14}$$

(c) *the* BLUE *of* $\mathbf{c}'\boldsymbol{\tau}$ *for any* $\mathbf{c}$ *is* $\mathbf{c}'\widehat{\boldsymbol{\tau}}$, *with the variance* $\mathbf{c}'\mathrm{Var}(\widehat{\boldsymbol{\tau}})\mathbf{c}$, *which reduces to*

$$\mathrm{Var}(\widehat{\mathbf{c}'\boldsymbol{\tau}}) = \sigma_1^2\mathbf{c}'(\Delta \boldsymbol{T}^{-1}\Delta')^{-1}\mathbf{c}, \tag{5.15}$$

if $\mathbf{c}'\boldsymbol{\tau}$ *is a contrast;*

(d) *the minimum norm quadratic unbiased estimator* (MINQUE) *of* σ_1^2 *is*

$$\widehat{\sigma_1^2} = \frac{1}{n-v}\|\boldsymbol{y} - \Delta'\widehat{\boldsymbol{\tau}}\|^2_{\boldsymbol{T}^{-1}} = \frac{1}{n-v}(\boldsymbol{y} - \Delta'\widehat{\boldsymbol{\tau}})'\boldsymbol{T}^{-1}(\boldsymbol{y} - \Delta'\widehat{\boldsymbol{\tau}}). \tag{5.16}$$

To prove this theorem note that from Theorem 3.2(c) of Rao (1974), and Lemma 5.1, the BLUE of $\Delta'\boldsymbol{\tau}$ is,

$$\widehat{\Delta'\boldsymbol{\tau}} = \boldsymbol{P}_{\Delta';\boldsymbol{T}^{-1}}\boldsymbol{y},$$

where

$$\boldsymbol{P}_{\Delta';\boldsymbol{T}^{-1}} = \Delta'(\Delta \boldsymbol{T}^{-1}\Delta')^{-1}\Delta \boldsymbol{T}^{-1}, \tag{5.17}$$

that (5.8) can be written as

$$\mathrm{Var}(\boldsymbol{y}) = \sigma_1^2\boldsymbol{T} - \frac{1}{N_B}\sigma_B^2\mathbf{1}_n\mathbf{1}_n', \tag{5.18}$$

that the residual sum of squares is of the form

$$\begin{aligned}&\|(\boldsymbol{I}_n - \boldsymbol{P}_{\Delta';\boldsymbol{T}^{-1}})\boldsymbol{y}\|^2_{\boldsymbol{T}^{-1}}\\&\quad = \boldsymbol{y}'(\boldsymbol{I}_n - \boldsymbol{P}_{\Delta';\boldsymbol{T}^{-1}})'\boldsymbol{T}^{-1}(\boldsymbol{I}_n - \boldsymbol{P}_{\Delta';\boldsymbol{T}^{-1}})\boldsymbol{y},\end{aligned} \tag{5.19}$$

and that the p.d. matrix $\boldsymbol{T}$ used in (5.18) satisfies the conditions required for the matrix $\boldsymbol{T}$ used in Rao's (1974) Theorem 3.2. For details of the proof see Caliński and Kageyama (1996, Theorem 3.1).

Since, for $\gamma > -1/k_{\max}$ (as assumed),

$$(\boldsymbol{k}^{-\delta} + \gamma\boldsymbol{I}_b)^{-1} = \begin{cases} \boldsymbol{k}^{\delta} - \boldsymbol{k}^{\delta}(\boldsymbol{k}^{\delta} + \gamma^{-1}\boldsymbol{I}_b)^{-1}\boldsymbol{k}^{\delta}, & \text{if } \gamma \neq 0,\\ \boldsymbol{k}^{\delta}, & \text{if } \gamma = 0,\end{cases}$$

it is permissible, on account of (5.13), to write

$$\boldsymbol{T}^{-1} = \boldsymbol{I}_n - \boldsymbol{D}'\boldsymbol{k}_*^{-\delta}\boldsymbol{D}, \tag{5.20}$$

where $\boldsymbol{k}_*^{-\delta}$ is a diagonal matrix with its jth diagonal element equal to

$$k_{*j}^{-1} = \frac{\gamma}{k_j\gamma + 1}. \tag{5.21}$$

With this notation, formulae (5.12), (5.13) and (5.14) can be written, respectively, as $\widehat{\boldsymbol{\tau}} = \boldsymbol{C}_c^{-1}\boldsymbol{Q}_c$, $\mathrm{Var}(\widehat{\boldsymbol{\tau}}) = \sigma_1^2\boldsymbol{C}_c^{-1} - N_B^{-1}\sigma_B^2\mathbf{1}_v\mathbf{1}_v'$ and, for $\boldsymbol{c}'\mathbf{1}_v = 0$, $\mathrm{Var}(\widehat{\boldsymbol{c}'\boldsymbol{\tau}}) = \sigma_1^2\boldsymbol{c}'\boldsymbol{C}_c^{-1}\boldsymbol{c}$, where, on account of (5.20),

$$\boldsymbol{C}_c = \boldsymbol{r}^{\delta} - \boldsymbol{N}\boldsymbol{k}_*^{-\delta}\boldsymbol{N}' \quad \text{and} \quad \boldsymbol{Q}_c = (\boldsymbol{\Delta} - \boldsymbol{N}\boldsymbol{k}_*^{-\delta}\boldsymbol{D})\boldsymbol{y}. \tag{5.22}$$

Note that $\boldsymbol{C}_c$ and $\boldsymbol{Q}_c$ in (5.22) are similar, respectively, to $\boldsymbol{C}$ and $\boldsymbol{Q}$ used in the intra-block analysis (Section 1.4), where σ^2 is to be replaced by σ_1^2 appearing in (5.8), but they are more general, in the sense of *combining* the intra-block and inter-block information (hence the subscript c). From (5.21) it is evident that a maximum recovery of the inter-block information is achieved in the case of $\gamma \leqslant 0$, i.e., when the blocking is completely unsuccessful, while there is no recovery of that information in the case of $\gamma \to \infty$, i.e., when the formation of blocks is fully successful, in the sense of eliminating the intra-block variation of experimental units, measured by σ_1^2. In the latter case $\boldsymbol{C}_c \to \boldsymbol{C}$ and $\boldsymbol{Q}_c \to \boldsymbol{Q}$. Also note that if the design is proper, i.e., if $k_j = k$ for all j, then (as in John, 1987, p. 193)

$$\boldsymbol{C}_c = \boldsymbol{r}^{\delta} - \frac{1}{k_*}\boldsymbol{N}\boldsymbol{N}' \quad \text{and} \quad \boldsymbol{Q}_c = \left(\boldsymbol{\Delta} - \frac{1}{k_*}\boldsymbol{N}\boldsymbol{D}\right)\boldsymbol{y}, \tag{5.23}$$

where $k_*^{-1} = k^{-1}(1 - \sigma_1^2/\sigma_2^2)$, with σ_2^2 defined as in (5.6). Moreover, in this case, any contrast $\boldsymbol{c}'\boldsymbol{\tau} = \boldsymbol{s}'\boldsymbol{r}^{\delta}\boldsymbol{\tau}$ satisfying the condition $\boldsymbol{\Delta}\boldsymbol{\phi}\boldsymbol{\Delta}'\boldsymbol{s} = \varepsilon\boldsymbol{r}^{\delta}\boldsymbol{s}$, with $0 < \varepsilon < 1$, has the "simple combinability" property of Martin and Zyskind (1966). Evidently this is secured if and only if the design is proper.

5.3. Estimation of stratum variances

All the results established in Section 5.2 are based on the assumption that the ratio γ is known (see Lemma 5.1). In practice, however, this is usually not the case. Therefore, to make the theory applicable, estimators of both, σ_1^2 and γ, are needed. There may be various approaches adopted for finding these estimators. To choose a practically suitable one, let first the residual sum of squares (5.19) be written as

$$\|(\boldsymbol{I}_n - \boldsymbol{P}_{\Delta';\boldsymbol{T}^{-1}})\boldsymbol{y}\|^2_{\boldsymbol{T}^{-1}} = \boldsymbol{y}'\boldsymbol{RTRy} = \gamma\boldsymbol{y}'\boldsymbol{RD}'\boldsymbol{DRy} + \boldsymbol{y}'\boldsymbol{RRy}, \quad (5.24)$$

where

$$\boldsymbol{R} = \boldsymbol{T}^{-1} - \boldsymbol{T}^{-1}\Delta'(\Delta\boldsymbol{T}^{-1}\Delta')^{-1}\Delta\boldsymbol{T}^{-1}. \quad (5.25)$$

Now, generalizing Nelder's (1968) approach, the simultaneous estimators of σ_1^2 and γ can be obtained by equating the partial sums of squares in (5.24) to their expectations. This leads to the equations

$$\begin{bmatrix} \mathrm{tr}(\boldsymbol{RD}'\boldsymbol{DRD}'\boldsymbol{D}) & \mathrm{tr}(\boldsymbol{RD}'\boldsymbol{DR}) \\ \mathrm{tr}(\boldsymbol{RD}'\boldsymbol{DR}) & \mathrm{tr}(\boldsymbol{RR}) \end{bmatrix} \begin{bmatrix} \sigma_1^2\gamma \\ \sigma_1^2 \end{bmatrix} = \begin{bmatrix} \boldsymbol{y}'\boldsymbol{RD}'\boldsymbol{DRy} \\ \boldsymbol{y}'\boldsymbol{RRy} \end{bmatrix}. \quad (5.26)$$

A solution of (5.26) gives estimators of $\sigma_1^2\gamma$ and σ_1^2, and hence of γ. However, the equations (5.26) clearly have no direct analytic solution, as the matrix $\boldsymbol{R}$ also contains the unknown parameter γ. Before considering their solution in detail, note that exactly the same equations as in (5.26) result from the MINQUE approach of Rao (1970, 1971, 1972). Also, note that the equations obtained from equating the sums of squares $\boldsymbol{y}'\boldsymbol{RD}'\boldsymbol{DRy}$ and $\boldsymbol{y}'\boldsymbol{RRy}$ to their expectations can be written equivalently as

$$\boldsymbol{y}'\boldsymbol{RD}'\boldsymbol{DRy} = \sigma_1^2\mathrm{tr}(\boldsymbol{RD}'\boldsymbol{D}) \quad \text{and} \quad \boldsymbol{y}'\boldsymbol{RRy} = \sigma_1^2\mathrm{tr}(\boldsymbol{R}), \quad (5.27)$$

since $\mathrm{tr}(\boldsymbol{RD}'\boldsymbol{DRT}) = \mathrm{tr}(\boldsymbol{RD}'\boldsymbol{D})$ and $\mathrm{tr}(\boldsymbol{RRT}) = \mathrm{tr}(\boldsymbol{R})$. Now, since $(\sigma_1^2)^{-1}\boldsymbol{R}$ can be shown to be the Moore-Penrose inverse of the matrix $\sigma_1^2\boldsymbol{\phi}_*\boldsymbol{T}\boldsymbol{\phi}_*$, where $\boldsymbol{\phi}_* = \boldsymbol{I}_n - \Delta'\boldsymbol{r}^{-\delta}\Delta$, the equations (5.27) coincide with the equations (5.1) of Patterson and Thompson (1975) on which the so-called modified (or marginal) maximum likelihood (MML) estimation method is based. The method is also known as the restricted maximum likelihood (REML) approach (see, e.g., Harville, 1977). Thus not only the original approach of Nelder (1968), but also its generalization presented here must, in principle, give the same results as those obtainable from the MML (REML) equations derived under the multivariate normality assumption (as indicated by Patterson and Thompson, 1971, p. 552, 1975, p. 206).

Note that for solving the equations (5.26), or their equivalences, an iterative procedure is to be used. According to the original MINQUE principle of Rao (1970, 1971), a solution of (5.26) would be obtained under some a priori value of γ, and thus should coincide with results obtained from the other two approaches when restricted to a single iteration, provided, however, that in all the three methods the same a priori or preliminary estimate of γ in $\boldsymbol{R}$ is used (cf. Patterson and Thompson, 1975, p. 204). But, as indicated by Rao (1972, p. 113), the MINQUE method can also be used iteratively, and then the three approaches will lead to the same results (cf. Rao, 1979, p. 151).

A suitable computational procedure for obtaining a practical solution of equations (5.26) is that given by Patterson and Thompson (1971, Section 6), who also illustrated its use by an example. The procedure starts with a preliminary estimate γ_0 of γ, usually with $\gamma_0 \geqslant 0$. Incorporating it into the coefficients of (5.26), the equations can be written as

$$\begin{bmatrix} \operatorname{tr}(\boldsymbol{R}_0\boldsymbol{D}'\boldsymbol{D}\boldsymbol{R}_0\boldsymbol{D}'\boldsymbol{D}) & \operatorname{tr}(\boldsymbol{R}_0\boldsymbol{D}'\boldsymbol{D}\boldsymbol{R}_0) \\ \operatorname{tr}(\boldsymbol{R}_0\boldsymbol{D}'\boldsymbol{D}\boldsymbol{R}_0) & \operatorname{tr}(\boldsymbol{R}_0\boldsymbol{R}_0) \end{bmatrix} \begin{bmatrix} \sigma_1^2\gamma \\ \sigma_1^2 \end{bmatrix} = \begin{bmatrix} \boldsymbol{y}'\boldsymbol{R}_0\boldsymbol{D}'\boldsymbol{D}\boldsymbol{R}_0\boldsymbol{y} \\ \boldsymbol{y}'\boldsymbol{R}_0\boldsymbol{R}_0\boldsymbol{y} \end{bmatrix}, \tag{5.28}$$

where $\boldsymbol{R}_0$ is defined as in (5.25), but with $\boldsymbol{T}^{-1}$ replaced by $\boldsymbol{T}_0^{-1}$ obtained after replacing γ in (5.13) by γ_0, i.e., with

$$\boldsymbol{T}_0^{-1} = \boldsymbol{\phi}_1 + \boldsymbol{D}'\boldsymbol{k}^{-\delta}(\boldsymbol{k}^{-\delta} + \gamma_0\boldsymbol{I}_b)^{-1}\boldsymbol{k}^{-\delta}\boldsymbol{D} \tag{5.29}$$

(with $\boldsymbol{T}_0^{-1} = \boldsymbol{\phi}_1$ if $\gamma_0 \to \infty$). However, instead of the equations (5.28), it is more convenient to consider their transformed form

$$\begin{bmatrix} \operatorname{tr}(\boldsymbol{R}_0\boldsymbol{D}'\boldsymbol{D}\boldsymbol{R}_0\boldsymbol{D}'\boldsymbol{D}) & \operatorname{tr}(\boldsymbol{R}_0\boldsymbol{D}'\boldsymbol{D}) \\ \operatorname{tr}(\boldsymbol{R}_0\boldsymbol{D}'\boldsymbol{D}) & n - v \end{bmatrix} \begin{bmatrix} \sigma_1^2\gamma - \sigma_1^2\gamma_0 \\ \sigma_1^2 \end{bmatrix} = \begin{bmatrix} \boldsymbol{y}'\boldsymbol{R}_0\boldsymbol{D}'\boldsymbol{D}\boldsymbol{R}_0\boldsymbol{y} \\ \boldsymbol{y}'\boldsymbol{R}_0\boldsymbol{y} \end{bmatrix}. \tag{5.30}$$

The solution of (5.30) gives a revised estimate of γ of the form

$$\widehat{\gamma} = \gamma_0 + \frac{(n-v)\boldsymbol{y}'\boldsymbol{R}_0\boldsymbol{D}'\boldsymbol{D}\boldsymbol{R}_0\boldsymbol{y} - \operatorname{tr}(\boldsymbol{D}\boldsymbol{R}_0\boldsymbol{D}')\boldsymbol{y}'\boldsymbol{R}_0\boldsymbol{y}}{\operatorname{tr}[(\boldsymbol{D}\boldsymbol{R}_0\boldsymbol{D}')^2]\boldsymbol{y}'\boldsymbol{R}_0\boldsymbol{y} - \operatorname{tr}(\boldsymbol{D}\boldsymbol{R}_0\boldsymbol{D}')\boldsymbol{y}'\boldsymbol{R}_0\boldsymbol{D}'\boldsymbol{D}\boldsymbol{R}_0\boldsymbol{y}}. \tag{5.31}$$

Thus, a single iteration of Fisher's iterative method of scoring, suggested by Patterson and Thompson (1971, p. 550), consists here of the following two steps:

(0) One starts with a preliminary estimate γ_0 ($> -1/k_{\max}$) of γ to obtain the equations (5.30).

(1) By solving (5.30), a revised estimate $\widehat{\gamma}$ of γ is obtained, of the form (5.31), and this is then used as a new preliminary estimate in step (0) of the next iteration.

One should, however, watch that γ_0 remains always above the lower bound $-1/k_{\max}$, as required. If $\widehat{\gamma} \leqslant -1/k_{\max}$, it cannot be used as a new γ_0 in step (0). In such a case formula (5.31) is to be replaced by

$$\widehat{\gamma} = \gamma_0 + \alpha\frac{(n-v)\boldsymbol{y}'\boldsymbol{R}_0\boldsymbol{D}'\boldsymbol{D}\boldsymbol{R}_0\boldsymbol{y} - \operatorname{tr}(\boldsymbol{D}\boldsymbol{R}_0\boldsymbol{D}')\boldsymbol{y}'\boldsymbol{R}_0\boldsymbol{y}}{\operatorname{tr}[(\boldsymbol{D}\boldsymbol{R}_0\boldsymbol{D}')^2]\boldsymbol{y}'\boldsymbol{R}_0\boldsymbol{y} - \operatorname{tr}(\boldsymbol{D}\boldsymbol{R}_0\boldsymbol{D}')\boldsymbol{y}'\boldsymbol{R}_0\boldsymbol{D}'\boldsymbol{D}\boldsymbol{R}_0\boldsymbol{y}},$$

with $\alpha \in (0, 1)$ chosen so that $\widehat{\gamma} > -1/k_{\max}$ (as suggested by Rao and Kleffe, 1988, p. 237).

The above iteration is to be repeated until convergence, i.e., until the equality

$$\frac{\boldsymbol{y}'\boldsymbol{R}_0\boldsymbol{D}'\boldsymbol{D}\boldsymbol{R}_0\boldsymbol{y}}{\mathrm{tr}(\boldsymbol{D}\boldsymbol{R}_0\boldsymbol{D}')} = \frac{\boldsymbol{y}'\boldsymbol{R}_0\boldsymbol{y}}{n-v} \tag{5.32}$$

is reached. The solution of (5.26) can then be written as $[\widehat{\sigma}_1^2\widehat{\gamma}, \widehat{\sigma}_1^2]'$, where $\widehat{\gamma}$ is the final estimate of γ obtained after the convergence, and

$$\widehat{\sigma}_1^2 = \frac{\boldsymbol{y}'\widehat{\boldsymbol{R}}\boldsymbol{y}}{n-v}, \tag{5.33}$$

where $\widehat{\boldsymbol{R}}$ is defined according to (5.25) in the same way as $\boldsymbol{R}_0$ has been, but now with $\boldsymbol{T}_0^{-1}$ in (5.29) becoming

$$\widehat{\boldsymbol{T}}^{-1} = \boldsymbol{\phi}_1 + \boldsymbol{D}'\boldsymbol{k}^{-\delta}(\boldsymbol{k}^{-\delta} + \widehat{\gamma}\boldsymbol{I}_b)^{-1}\boldsymbol{k}^{-\delta}\boldsymbol{D}$$

(provided that $\widehat{\gamma} \neq 0$, otherwise reducing to $\widehat{\boldsymbol{T}}^{-1} = \boldsymbol{I}_n$). Hence, the so obtained estimator of $\boldsymbol{\tau}$ can be written as

$$\tilde{\boldsymbol{\tau}} = (\boldsymbol{\Delta}\widehat{\boldsymbol{T}}^{-1}\boldsymbol{\Delta}')^{-1}\boldsymbol{\Delta}\widehat{\boldsymbol{T}}^{-1}\boldsymbol{y}, \tag{5.34}$$

which may be called (after Rao and Kleffe, 1988, p. 274) an "empirical" estimator of $\boldsymbol{\tau}$. Of course, it is not the same as the BLUE obtainable with the exact value of γ.

It can be shown (see, e.g., Kackar and Harville, 1981; Klaczynski et al., 1994) that if the unknown ratio γ appearing in the matrix $\boldsymbol{T}$ defined in (5.10) is replaced by its estimator $\widehat{\gamma}$ obtained under the assumption that y has a multivariate normal distribution, then the unbiasedness of the estimators of $\boldsymbol{\tau}$ and $\boldsymbol{c}'\boldsymbol{\tau}$ established in Theorem 5.1 is not violated. However, as to the variance of the estimator of $\boldsymbol{c}'\boldsymbol{\tau}$, the replacement of γ by $\widehat{\gamma}$ causes an increase of the variance (5.15). Unfortunately, the exact formula of that increased variance is in general intractable. But it can be approximated as suggested by Kackar and Harville (1984). For details and discussion see Caliński and Kageyama (1996).

5.4. Remarks concerning nested block designs

The randomization model for an NB design has been considered recently by Caliński (1994b). It differs from the model (5.1) by the addition of the component $\boldsymbol{G}'\boldsymbol{\alpha}$, where $\boldsymbol{G}'$ is the $n \times a$ "design matrix for superblocks" and $\boldsymbol{\alpha}$ is an $a \times 1$ vector of superblock random effects $\{\alpha_i\}$. The expectation vector (5.2) remains unchanged, while the dispersion matrix is of the form

$$\begin{aligned}\mathrm{Var}(\boldsymbol{y}) = {} & \left(\boldsymbol{G}'\boldsymbol{G} - \frac{1}{N_A}\mathbf{1}_n\mathbf{1}_n'\right)\sigma_A^2 + \left(\boldsymbol{D}'\boldsymbol{D} - \frac{1}{B_H}\boldsymbol{G}'\boldsymbol{G}\right)\sigma_B^2 \\ & + \left(\boldsymbol{I}_n - \frac{1}{K_H}\boldsymbol{D}'\boldsymbol{D}\right)\sigma_U^2 + \boldsymbol{I}_n\sigma_e^2, \end{aligned} \tag{5.35}$$

where N_A is the number of available superblocks, and B_H is a weighted harmonic average of the available numbers of blocks within the N_A available superblocks, similarly as K_H is such an average of the available numbers of units within the $B_1 + B_2 + \cdots + B_{N_A}$ available blocks, and where $\sigma_A^2, \sigma_B^2, \sigma_U^2$ and σ_e^2 are the variance components related to the random vectors $\boldsymbol{\alpha}, \boldsymbol{\beta}, \boldsymbol{\eta}$ and $\boldsymbol{e}$, respectively, all these quantities being defined similarly as in Section 5.1. Since (5.35) can also be written as

$$\mathrm{Var}(\boldsymbol{y}) = \sigma_1^2(\boldsymbol{G}'\Gamma_1\boldsymbol{G} + \boldsymbol{D}'\Gamma_2\boldsymbol{D} + \boldsymbol{I}_n),$$

where

$$\Gamma_1 = \boldsymbol{I}_a\gamma_1 - \frac{1}{N_A}\mathbf{1}_a\mathbf{1}_a'\sigma_A^2/\sigma_1^2,$$

$$\gamma_1 = \left(\sigma_A^2 - \frac{1}{B_H}\sigma_B^2\right)/\sigma_1^2, \qquad \sigma_1^2 = \sigma_U^2 + \sigma_e^2$$

and

$$\Gamma_2 = \boldsymbol{I}_b\gamma_2, \qquad \gamma_2 = \left(\sigma_B^2 - \frac{1}{K_H}\sigma_U^2\right)/\sigma_1^2,$$

the analysis of an experiment in an NB design can be seen as a straightforward extension of that described in Sections 5.2 and 5.3. For more details see Patterson and Thompson (1971, Section 10). If the NB design is resolvable, the analysis is more simple. It is described by Speed et al. (1985).

6. The concept of general balance

6.1. Submodels of the randomization model

As shown in Section 5.1, under the model (5.1) the BLUEs of linear treatment parametric functions exist in very restrictive circumstances only. Therefore, the usual procedure is to resolve the model (5.1) into three submodels (two for contrasts), in accordance with the stratification of the experimental units. This can be represented by the decomposition

$$\boldsymbol{y} = \boldsymbol{y}_1 + \boldsymbol{y}_2 + \boldsymbol{y}_3, \tag{6.1}$$

resulting from orthogonal projections of $\boldsymbol{y}$ on subspaces related to the three "strata":
1st – of units within blocks, the "intra-block" stratum,
2nd – of blocks within the total area, the "inter-block" stratum,
3rd – of the total area

(using the terminology of Pearce, 1983, p. 109). Explicitly (see Caliński and Kageyama, 1991, Section 3),

$$\boldsymbol{y}_1 = \boldsymbol{\phi}_1 \boldsymbol{y}, \qquad \boldsymbol{y}_2 = \boldsymbol{\phi}_2 \boldsymbol{y}, \qquad \boldsymbol{y}_3 = \boldsymbol{\phi}_3 \boldsymbol{y}, \tag{6.2}$$

where the projectors $\boldsymbol{\phi}_\alpha, \alpha = 1, 2, 3$, are defined as

$$\boldsymbol{\phi}_1 = \boldsymbol{I}_n - \boldsymbol{D}'\boldsymbol{k}^{-\delta}\boldsymbol{D}, \qquad \boldsymbol{\phi}_2 = \boldsymbol{D}'\boldsymbol{k}^{-\delta}\boldsymbol{D} - \frac{1}{n}\mathbf{1}_n\mathbf{1}_n' \quad \text{and} \quad \boldsymbol{\phi}_3 = \frac{1}{n}\mathbf{1}_n\mathbf{1}_n'.$$

They satisfy the conditions

$$\boldsymbol{\phi}_\alpha = \boldsymbol{\phi}_\alpha', \quad \boldsymbol{\phi}_\alpha\boldsymbol{\phi}_\alpha = \boldsymbol{\phi}_\alpha, \quad \boldsymbol{\phi}_\alpha\boldsymbol{\phi}_{\alpha'} = \boldsymbol{O} \quad \text{if } \alpha \neq \alpha', \quad \text{and}$$

$$\boldsymbol{\phi}_1 + \boldsymbol{\phi}_2 + \boldsymbol{\phi}_3 = \boldsymbol{I}_n, \tag{6.3}$$

also

$$\boldsymbol{\phi}_1\boldsymbol{D}' = \boldsymbol{O} \quad \text{and} \quad \boldsymbol{\phi}_\alpha\mathbf{1}_n = \mathbf{0} \quad \text{for } \alpha = 1, 2, \tag{6.4}$$

and are called the "stratum projectors" (cf. Houtman and Speed, 1983, p. 1070).

The submodels (6.2), called "intra-block", "inter-block" and "total-area", respectively, have the following properties:

$$\mathrm{E}(\boldsymbol{y}_1) = \boldsymbol{\phi}_1\boldsymbol{\Delta}'\boldsymbol{\tau}, \qquad \mathrm{Var}(\boldsymbol{y}_1) = \boldsymbol{\phi}_1\left(\sigma_U^2 + \sigma_e^2\right), \tag{6.5}$$

$$\mathrm{E}(\boldsymbol{y}_2) = \boldsymbol{\phi}_2\boldsymbol{\Delta}'\boldsymbol{\tau},$$

$$\mathrm{Var}(\boldsymbol{y}_2) = \boldsymbol{\phi}_2\boldsymbol{D}'\boldsymbol{D}\boldsymbol{\phi}_2\left(\sigma_B^2 - \frac{1}{K_H}\sigma_U^2\right) + \boldsymbol{\phi}_2\left(\sigma_U^2 + \sigma_e^2\right), \tag{6.6}$$

$$\mathrm{E}(\boldsymbol{y}_3) = \boldsymbol{\phi}_3\boldsymbol{\Delta}'\boldsymbol{\tau},$$

$$\mathrm{Var}(\boldsymbol{y}_3) = \boldsymbol{\phi}_3\left[\left(\frac{1}{n}\boldsymbol{k}'\boldsymbol{k} - \frac{n}{N_B}\right)\sigma_B^2 + \left(1 - \frac{1}{nK_H}\boldsymbol{k}'\boldsymbol{k}\right)\sigma_U^2 + \sigma_e^2\right]. \tag{6.7}$$

Evidently, if the block sizes are all equal, i.e., the design is proper, then the dispersion matrices of the inter-block and total-area submodels are simplified. However, the properties of the intra-block submodel remain the same, whether the design is proper or not. This means that the intra-block analysis (as that based on the intra-block submodel) can be considered generally, for any block design (see Section 1.4).

6.2. Proper block designs and the orthogonal block structure

As noticed in Sections 5.1 and 5.2, for proper block designs some advantageous simplifications occur. This is essential in particular with regard to the inter-block submodel, for which the structure of the covariance matrix, shown in (6.6), is not very satisfactory from the application point of view. Therefore, further development of the theory related to the model (5.1), with the properties (5.2) and (5.3), will be confined in the present section to proper designs, i.e., designs with

$$k_1 = k_2 = \cdots = k_b = k \quad \text{(say)}. \tag{6.8}$$

This will allow the theory to be presented in a unified form.

It can easily be shown that the dispersion matrices in (6.6) and (6.7) are simplified to

$$\operatorname{Var}(\boldsymbol{y}_2) = \boldsymbol{\phi}_2\left[k\sigma_B^2 + \left(1 - \frac{k}{K_H}\right)\sigma_U^2 + \sigma_e^2\right] \tag{6.9}$$

and

$$\operatorname{Var}(\boldsymbol{y}_3) = \boldsymbol{\phi}_3\left[\left(1 - \frac{b}{N_B}\right)k\sigma_B^2 + \left(1 - \frac{k}{K_H}\right)\sigma_U^2 + \sigma_e^2\right], \tag{6.10}$$

respectively, for any proper block design.

Thus, in case of a proper block design, the expectation vector and the dispersion matrix for each of the three submodels (6.2) can be written, respectively, as

$$\mathrm{E}(\boldsymbol{y}_\alpha) = \boldsymbol{\phi}_\alpha \boldsymbol{\Delta}' \boldsymbol{\tau} \tag{6.11}$$

and

$$\operatorname{Var}(\boldsymbol{y}_\alpha) = \boldsymbol{\phi}_\alpha \sigma_\alpha^2, \tag{6.12}$$

for $\alpha = 1, 2, 3$, where, from (6.5), (6.9) and (6.10), the "stratum variances", σ_1^2, σ_2^2 and σ_3^2, are as in (5.6). Furthermore, if $N_B = b$ and $k = K_H$ (the latter implying the equality of all potential block sizes), which can be considered as the most common case, the variances σ_2^2 and σ_3^2 reduce to

$$\sigma_2^2 = k\sigma_B^2 + \sigma_e^2 \quad \text{and} \quad \sigma_3^2 = \sigma_e^2. \tag{6.13}$$

On the other hand, for any proper block design, the decomposition (6.1) implies not only that

$$\mathrm{E}(\boldsymbol{y}) = \mathrm{E}(\boldsymbol{y}_1) + \mathrm{E}(\boldsymbol{y}_2) + \mathrm{E}(\boldsymbol{y}_3) = \boldsymbol{\phi}_1 \boldsymbol{\Delta}' \boldsymbol{\tau} + \boldsymbol{\phi}_2 \boldsymbol{\Delta}' \boldsymbol{\tau} + \boldsymbol{\phi}_3 \boldsymbol{\Delta}' \boldsymbol{\tau}, \tag{6.14}$$

but also that

$$\text{Var}(\boldsymbol{y}) = \text{Var}(\boldsymbol{y}_1) + \text{Var}(\boldsymbol{y}_2) + \text{Var}(\boldsymbol{y}_3) = \boldsymbol{\phi}_1\sigma_1^2 + \boldsymbol{\phi}_2\sigma_2^2 + \boldsymbol{\phi}_3\sigma_3^2, \tag{6.15}$$

where the matrices $\boldsymbol{\phi}_1, \boldsymbol{\phi}_2$ and $\boldsymbol{\phi}_3$ satisfy the conditions (6.3).

The representation (6.15) is a very desirable property, as originally indicated for a more general class of designs by Nelder (1965a). After him, the following definition will be adopted (see also Houtman and Speed, 1983, Section 2.2).

DEFINITION 6.1. An experiment is said to have the orthogonal block structure (OBS) if the dispersion matrix of the random variables observed on the experimental units (plots) has a representation of the form (6.15), where the matrices $\{\boldsymbol{\phi}_\alpha\}$ are symmetric, idempotent and pairwise orthogonal, summing up to the identity matrix, as in (6.3).

It can now be said that any experiment in a proper block design has the orthogonal block structure, or that it has the OBS property.

LEMMA 6.1. *An experiment in a block design has under* (5.1) *the orthogonal block structure if and only if the design is proper.*

For a proof see Caliński (1993b, Lemma 4.1).

The condition (6.8) is, however, not sufficient to obtain for any $\boldsymbol{s}$ the BLUE of $\boldsymbol{s}'\boldsymbol{r}^\delta\boldsymbol{\tau}$ under the overall model (5.1) (see Remark 2.1 of Caliński and Kageyama, 1991). It remains to seek the estimators within each stratum separately, i.e., under the submodels

$$\boldsymbol{y}_\alpha = \boldsymbol{\phi}_\alpha \boldsymbol{y}, \quad \alpha = 1, 2, 3, \tag{6.16}$$

which for any proper block design have the properties (6.11) and (6.12), with the matrices $\boldsymbol{\phi}_1$ and $\boldsymbol{\phi}_2$ reduced to

$$\boldsymbol{\phi}_1 = \boldsymbol{I}_n - \frac{1}{k}\boldsymbol{D}'\boldsymbol{D} \tag{6.17}$$

and

$$\boldsymbol{\phi}_2 = \frac{1}{k}\boldsymbol{D}'\boldsymbol{D} - \frac{1}{n}\boldsymbol{1}_n\boldsymbol{1}_n', \tag{6.18}$$

respectively, and with $\boldsymbol{\phi}_3 = n^{-1}\boldsymbol{1}_n\boldsymbol{1}_n'$ unchanged.

THEOREM 6.1. *If the block design is proper, then under* (6.16) *a function* $\boldsymbol{w}'\boldsymbol{y}_\alpha = \boldsymbol{w}'\boldsymbol{\phi}_\alpha\boldsymbol{y}$ *is uniformly the* BLUE *of* $\boldsymbol{c}'\boldsymbol{\tau}$ *if and only if* $\boldsymbol{\phi}_\alpha\boldsymbol{w} = \boldsymbol{\phi}_\alpha\boldsymbol{\Delta}'\boldsymbol{s}$, *where the vectors* $\boldsymbol{c}$ *and* $\boldsymbol{s}$ *are in the relation* $\boldsymbol{c} = \boldsymbol{\Delta}\boldsymbol{\phi}_\alpha\boldsymbol{\Delta}'\boldsymbol{s}$.

PROOF. The proof is exactly as that of Theorem 3.1 of Caliński and Kageyama (1991), on account of (6.11) and (6.12). □

REMARK 6.1. Since $\mathbf{1}_v' \Delta \phi_\alpha = \mathbf{0}'$ for $\alpha = 1, 2$, the only parametric functions for which the BLUEs may exist under (6.16) with $\alpha = 1, 2$ are contrasts. On the other hand, no contrast will obtain a BLUE under (6.16) for $\alpha = 3$. (See also Remark 3.1, Corollary 3.1(b) and Remark 3.6(a) of Caliński and Kageyama (1991).)

It follows from Theorem 6.1 that if for a given $\boldsymbol{c}$ ($\neq \mathbf{0}$) there exists a vector $\boldsymbol{s}$ such that $\boldsymbol{c} = \Delta \phi_\alpha \Delta' \boldsymbol{s}$, then the BLUE of $\boldsymbol{c}'\boldsymbol{\tau}$ in stratum α is obtainable as

$$\widehat{\boldsymbol{c}'\boldsymbol{\tau}} = \boldsymbol{s}' \Delta \boldsymbol{y}_\alpha, \tag{6.19}$$

with the variance of the form

$$\mathrm{Var}(\widehat{\boldsymbol{c}'\boldsymbol{\tau}}) = \boldsymbol{s}' \Delta \phi_\alpha \Delta' \boldsymbol{s} \sigma_\alpha^2 = \boldsymbol{c}' (\Delta \phi_\alpha \Delta')^- \boldsymbol{c} \sigma_\alpha^2, \tag{6.20}$$

where σ_α^2 is the appropriate stratum variance defined in (6.13).

Explicitly, the matrices $\Delta \phi_\alpha \Delta'$ in (6.20) are:

$$\Delta \phi_1 \Delta' = \boldsymbol{r}^\delta - \frac{1}{k} \boldsymbol{N}\boldsymbol{N}' = \boldsymbol{C}_1 \quad (\equiv \boldsymbol{C}, \text{ the } C\text{-matrix}), \tag{6.21}$$

$$\Delta \phi_2 \Delta' = \frac{1}{k} \boldsymbol{N}\boldsymbol{N}' - \frac{1}{n} \boldsymbol{r}\boldsymbol{r}' = \boldsymbol{C}_2 \quad (\boldsymbol{C}_0 \text{ in Pearce, 1983, p. 111}) \tag{6.22}$$

and

$$\boldsymbol{C}_3 = \Delta \phi_3 \Delta' = \frac{1}{n} \boldsymbol{r}\boldsymbol{r}'.$$

The decomposition (6.1) implies that any function $\boldsymbol{s}' \Delta \boldsymbol{y}$ can be resolved into three components, in the form

$$\boldsymbol{s}' \Delta \boldsymbol{y} = \boldsymbol{s}' \boldsymbol{Q}_1 + \boldsymbol{s}' \boldsymbol{Q}_2 + \boldsymbol{s}' \boldsymbol{Q}_3, \tag{6.23}$$

where $\boldsymbol{Q}_\alpha = \Delta \boldsymbol{y}_\alpha = \Delta \phi_\alpha \boldsymbol{y}$ ($\alpha = 1, 2, 3$), i.e., each of the components is a contribution to the estimate from a different stratum.

As stated in Remark 6.1, the only parametric functions for which the BLUEs may exist under the submodels $\boldsymbol{y}_1 = \phi_1 \boldsymbol{y}$ and $\boldsymbol{y}_2 = \phi_2 \boldsymbol{y}$ are contrasts. As will be shown, certain contrasts may have the BLUEs exclusively under one of these submodels, i.e., either in the intra-block analysis (within the 1st stratum) or in the inter-block analysis (within the 2nd stratum). For other contrasts the BLUEs may be obtained under both of these submodels, i.e., in both of the analyses.

It has been indicated in Lemma 4.1 of Caliński and Kageyama (1991), that a necessary and sufficient condition for $\mathrm{E}(\boldsymbol{s}'\boldsymbol{Q}_1) = \kappa\, \mathrm{E}(\boldsymbol{s}'\boldsymbol{Q}_2)$, when $\boldsymbol{s}'\boldsymbol{r} = 0$, is

$$\Delta \phi_1 \Delta' \boldsymbol{s} = \varepsilon \boldsymbol{r}^\delta \boldsymbol{s}, \quad \text{with } 0 < \varepsilon < 1 \ \left(\varepsilon = \frac{\kappa}{1 + \kappa} \right), \tag{6.24}$$

or its equivalent

$$\Delta\phi_2\Delta' s = (1-\varepsilon) r^{\delta} s, \quad \text{with } 0 < \varepsilon < 1. \tag{6.25}$$

From Caliński (1993b, Section 4) the following results can now be given.

LEMMA 6.2. *If the design is proper, then for any* $c = r^{\delta} s$ *such that* s *satisfies the equivalent eigenvector conditions* (6.24) *and* (6.25), *with* $0 < \varepsilon < 1$, *the* BLUE *of the contrast* $c'\tau$ *is obtainable in both of the analyses, in the intra-block analysis and in the inter-block analysis.*

LEMMA 6.3. *If the design is proper, then for any* $c = r^{\delta} s$ *such that* s *satisfies one of the eigenvector conditions*

$$\Delta\phi_{\alpha}\Delta' s = r^{\delta} s, \quad \alpha = 1, 2, 3, \tag{6.26}$$

the BLUE *of the function* $c'\tau$ *is obtainable under the overall model* (5.1).

THEOREM 6.2. *In case of a proper block design, for any vector* $c = r^{\delta} s$ *such that* s *satisfies the eigenvector condition*

$$\Delta\phi_{\alpha}\Delta' s = \varepsilon_{\alpha} r^{\delta} s, \quad \text{with } 0 < \varepsilon_{\alpha} \leqslant 1 \ (\alpha = 1, 2, 3), \tag{6.27}$$

where $\varepsilon_1 = \varepsilon$, $\varepsilon_2 = 1 - \varepsilon$, $\varepsilon_3 = 1$, *the* BLUE *of the function* $c'\tau$ *is obtainable in the analysis within stratum* α [*for which* (6.27) *is satisfied*], *i.e., under the submodel* $y_{\alpha} = \phi_{\alpha} y$, *where it gets the form*

$$(\widehat{c'\tau})_{\alpha} = \varepsilon_{\alpha}^{-1} s' Q_{\alpha} = \varepsilon_{\alpha}^{-1} c' r^{-\delta} Q_{\alpha}, \tag{6.28}$$

and its variance is

$$\mathrm{Var}[(\widehat{c'\tau})_{\alpha}] = \varepsilon_{\alpha}^{-1} s' r^{\delta} s \sigma_{\alpha}^2 = \varepsilon_{\alpha}^{-1} c' r^{-\delta} c \sigma_{\alpha}^2. \tag{6.29}$$

If (6.27) *is satisfied with* $0 < \varepsilon_{\alpha} < 1$, *then two* BLUEs *of* $c'\tau$ *are obtainable, one under the submodel* $y_1 = \phi_1 y$ *and another under* $y_2 = \phi_2 y$. *If* (6.27) *is satisfied with* $\varepsilon_{\alpha} = 1$, *then the unique* BLUE *is obtainable within stratum* α *only, being simultaneously the* BLUE *under the overall model* (5.1).

REMARK 6.2. Formula (6.29) shows that the variance of the BLUE of $c'\tau$ obtainable within stratum α is the smaller the larger is the coefficient ε_{α}, the minimum variance being attained when $\varepsilon_{\alpha} = 1$, i.e., when (6.28) is the BLUE under the overall model (5.1). Thus, for any proper design, ε_{α} can be interpreted as the efficiency factor of the analysed design for the function $c'\tau$ when it is estimated in the analysis within stratum α. On the other hand, $1 - \varepsilon_{\alpha}$ can be regarded as the relative loss of information incurred when estimating $c'\tau$ in the within stratum α analysis.

REMARK 6.3. Since $\mathrm{E}(\varepsilon_{\alpha}^{-1} s' Q_{\alpha}) = s' r^{\delta} \tau$ if and only if (6.27) holds, for a function $c'\tau = s' r^{\delta} \tau$ to obtain the BLUE within stratum α in the form (6.28), the condition (6.27) is not only sufficient but also necessary.

6.3. Generally balanced block designs

Consider again the $v \times v$ matrices $\boldsymbol{C}_\alpha = \Delta\phi_\alpha\Delta', \alpha = 1,2,3$, and their spectral decompositions

$$\boldsymbol{C}_1 = \boldsymbol{r}^\delta \sum_{\beta=0}^{m-1} \varepsilon_\beta \boldsymbol{H}_\beta \boldsymbol{r}^\delta,$$

$$\boldsymbol{C}_2 = \boldsymbol{r}^\delta \sum_{\beta=1}^{m} (1-\varepsilon_\beta) \boldsymbol{H}_\beta \boldsymbol{r}^\delta, \qquad \boldsymbol{C}_3 = \boldsymbol{r}^\delta \boldsymbol{H}_{m+1} \boldsymbol{r}^\delta = \frac{1}{n}\boldsymbol{r}\boldsymbol{r}',$$

where the $v \times v$ matrices $\boldsymbol{H}_\beta$ are as defined in Section 1.5. These representations, together with the general results established in Section 6.1 for proper block designs, give rise to the following concept of balance.

DEFINITION 6.2. A proper block design inducing the OBS property defined by $\{\phi_\alpha\}$ is said to be generally balanced (GB) with respect to a decomposition

$$\mathcal{C}(\Delta') = \oplus_\beta \mathcal{C}(\Delta' \boldsymbol{S}_\beta) \tag{6.30}$$

(the symbol $\mathcal{C}(.)$ denoting the column space of a matrix argument and $\oplus_\beta$ denoting the direct sum of the subspaces taken over β), if there exist scalars $\{\varepsilon_{\alpha\beta}\}$ such that for all α $(= 1,2,3)$

$$\Delta\phi_\alpha\Delta' = \sum_\beta \varepsilon_{\alpha\beta} \boldsymbol{r}^\delta \boldsymbol{H}_\beta \boldsymbol{r}^\delta \tag{6.31}$$

(the sum being taken over all β that appear in (6.30), $\beta = 0,1,\ldots,m,m+1$), where $\boldsymbol{H}_\beta = \boldsymbol{S}_\beta \boldsymbol{S}'_\beta$, $\boldsymbol{H}_{m+1} = \boldsymbol{s}_v \boldsymbol{s}'_v$, and where the matrices $\{\boldsymbol{S}_\beta\}$ are such that

$$\boldsymbol{S}'_\beta \boldsymbol{r}^\delta \boldsymbol{S}_\beta = \boldsymbol{I}_{\rho_\beta} \quad \text{for any } \beta \quad \text{and} \quad \boldsymbol{S}'_\beta \boldsymbol{r}^\delta \boldsymbol{S}_{\beta'} = \boldsymbol{O} \quad \text{for } \beta \neq \beta'.$$

It can easily be shown that Definition 6.2 is equivalent to the definition of GB given by Houtman and Speed (1983, Section 4.1) when applied to a proper block design, and so coincides with the notion of general balance introduced by Nelder (1965b).

The following result from Caliński (1993b, p. 32) explains the sense of Definition 6.2.

LEMMA 6.4. *A proper block design is* GB *with respect to the decomposition* (6.30) *if and only if the matrices* $\{\boldsymbol{S}_\beta\}$ *of Definition* 6.2 *satisfy the conditions*

$$\Delta\phi_\alpha\Delta'\boldsymbol{S}_\beta = \varepsilon_{\alpha\beta} \boldsymbol{r}^\delta \boldsymbol{S}_\beta \tag{6.32}$$

for all α *and* β. (See also Pearce, 1983, p. 110.)

REMARK 6.4. It follows from Lemma 6.4 that any proper block design is GB with respect to the decomposition

$$\mathcal{C}(\Delta') = \mathcal{C}(\Delta'\boldsymbol{S}_0) \oplus \mathcal{C}(\Delta'\boldsymbol{S}_1) \oplus \cdots \oplus \mathcal{C}(\Delta'\boldsymbol{S}_m) \oplus \mathcal{C}(\Delta'\boldsymbol{s}_v), \tag{6.33}$$

where the matrices $\boldsymbol{S}_0, \boldsymbol{S}_1, \ldots, \boldsymbol{S}_m$ represent basic contrasts of the design, those represented by the columns of $\boldsymbol{S}_\beta$ receiving in the intra-block analysis a common efficiency factor $\varepsilon_{1\beta} = \varepsilon_\beta$ and in the inter-block analysis a common efficiency factor $\varepsilon_{2\beta} = 1 - \varepsilon_\beta$, and where $\boldsymbol{s}_v = n^{-1/2}\mathbf{1}_v$.

Certainly, the equality (6.31) above can equivalently be written as

$$\Delta'\boldsymbol{r}^{-\delta}\Delta\phi_\alpha\Delta'\boldsymbol{r}^{-\delta}\Delta = \sum_\beta \varepsilon_{\alpha\beta}\Delta'\boldsymbol{H}_\beta\Delta,$$

which is exactly the condition of Houtman and Speed (1983, Section 4.1) in their definition of GB.

Also, it should be mentioned that the notion of GB stems back to the early work by Jones (1959), who called an experiment balanced for a contrast if the latter satisfied the condition (6.24), and called it balanced for a set of contrasts if they satisfied this condition with the same eigenvalue. Thus, in his terminology, a block design is balanced for each basic contrast separately, but it is also balanced for any subspace of basic contrast corresponding to a distinct eigenvalue. It is, therefore, natural to call a block design GB for all basic contrasts, provided that the eigenvalues can be interpreted in terms of efficiency factors and relative losses of information on contrasts of interest. This is just what is offered by any proper block design if adequately used. For some illustration of this see Caliński (1993b, Section 5).

6.4. *Concluding remarks*

The unified theory presented in Sections 6.2 and 6.3 reveals the special role played by basic contrasts in defining the general balance (GB) of a block design. Since any proper block design is GB, as stated in Remark 6.4 (cf. Houtman and Speed, 1983, Section 5.4), the notion of GB is interesting *only* from the point of view of the decomposition (6.33) with respect to which the balance holds. Therefore, any block design offered for use in an experiment should be evaluated with regard to that decomposition. The experimenter should be informed on the subspaces of basic contrast appearing in (6.30) and the efficiency factors receivable by them in the intra-block and in the inter-block analysis. This has already been pointed out by Houtman and Speed (1983, Section 4.2), who write that these subspaces have to be discovered for each new design or class of designs. Referring directly to block designs, and to partially balanced incomplete block designs in particular, they write (p. 1082) that although it is generally not difficult to obtain these subspaces (more precisely orthogonal projections on them) "most writers in statistics have not taken this view point". The suggested (in Section

3.2) classification of block designs based on the concept of PEB, which for proper block designs coincides with that of GB, is exactly an attempt to meet their complains.

The knowledge of basic contrasts or their subspaces for which a design is GB, and of the efficiency factors assigned to them, allows the experimenter to use the design for an experiment in such a way which best corresponds to the experimental problem. In particular, it allows to implement the design so that the contrasts considered as the most important can be estimated with the highest efficiency in the stratum of the smallest variance, which is the intra-block stratum, if the grouping of units into blocks is performed successfully.

Further discussions about the pros and cons of GB can be found in Mejza (1992) and Bailey (1994). For the combined analysis under GB, corresponding to that in Section 5.2, see Caliński (1994a).

Acknowledgements

The paper was prepared when the first author visited the Hiroshima University under a JSPS Fellowship for research in Japan. The opportunity and facilities offered him there are gratefully acknowledged. The authors wish also to thank the referee and the Editors for their encouraging comments.

References

Agrawal, H. L. and J. Prasad (1983). On construction of balanced incomplete block designs with nested rows and columns. *Sankhyā Ser. B* **45**, 345–350.

Anscombe, F. J. (1948). Contribution to the discussion on D. G. Champernowne's "Sampling theory applied to autoregressive sequences". *J. Roy. Statist. Soc. Ser. B* **10**, 239.

Atiqullah, M. (1961). On a property of balanced designs. *Biometrika* **48**, 215–218.

Bailey, R. A. (1981). A unified approach to design of experiments. *J. Roy. Statist. Soc. Ser. A* **144**, 214–223.

Bailey, R. A. (1994). General balance: Artificial theory or practical relevance? In: T. Caliński and R. Kala, eds., *Proc. Internat. Conf. on Linear Statistical Inference LINSTAT'93*. Kluwer Academic Publishers, Dordrecht, 171–184.

Bailey, R. A. and C. A. Rowley (1987). Valid randomization. *Proc. Roy. Soc. London Ser. A* **410**, 105–124.

Baksalary, J. K., A. Dobek and R. Kala (1980). A necessary condition for balance of a block design. *Biom. J.* **22**, 47–50.

Baksalary, J. K. and P. D. Puri (1988). Criteria for the validity of Fisher's condition for balanced block designs. *J. Statist. Plann. Inference* **18**, 119–123.

Baksalary, J. K. and Z. Tabis (1985). Existence and construction of connected block designs with given vectors of treatment replications and block sizes. *J. Statist. Plann. Inference* **12**, 285–293.

Banerjee, S. and S. Kageyama (1990). Existence of α-resolvable nested incomplete block designs. *Utilitas Math.* **38**, 237–243.

Banerjee, S. and S. Kageyama (1993). Methods of constructing nested partially balanced incomplete block designs. *Utilitas Math.* **43**, 3–6.

Bechhofer, R. E. and A. C. Tamhane (1981). Incomplete block designs for comparing treatments with a control: General theory. *Technometrics* **23**, 45–57.

Beth, T., D. Jungnickel and H. Lenz (1985). *Design Theory*, Bibliographisches Institut, Mannheim, Germany. (D. Jungnickel, Design theory: An update. *Ars Combin.* **28** (1989), 129–199.)

Bose, R. C. (1942). A note on the resolvability of balanced incomplete block designs. *Sankhyā* **6**, 105–110.

Bose, R. C. (1950). Least square aspects of analysis of variance. Mimeo Series 9, Institute of Statistics, University of North Carolina, Chapel Hill.
Bose, R. C. and W. H. Clatworthy (1955). Some classes of partially balanced designs. *Ann. Math. Statist.* **26**, 212–232.
Bose, R. C. and W. S. Connor (1952). Combinatorial properties of group divisible incomplete block designs. *Ann. Math. Statist.* **23**, 367–383.
Bose, R. C. and K. R. Nair (1939). Partially balanced incomplete block designs. *Sankhyā* **4**, 337–372.
Bose, R. C. and K. R. Nair (1962). Resolvable incomplete block designs with two replications. *Sankhyā Ser. A* **24**, 9–24.
Bose, R. C. and T. Shimamoto (1952). Classification and analysis of partially balanced incomplete block designs with two associate classes. *J. Amer. Statist. Assoc.* **47**, 151–184.
Bose, R. C. and S. S. Shrikhande (1960). On the composition of balanced incomplete block designs. *Canad. J. Math.* **12**, 177–188.
Caliński, T. (1971). On some desirable patterns in block designs (with discussion). *Biometrics* **27**, 275–292.
Caliński, T. (1977). On the notion of balance in block designs. In: J. R. Barra, F. Brodeau, G. Romier and B. van Cutsem, eds., *Recent Developments in Statistics*. North-Holland, Amsterdam, 365–374.
Caliński, T. (1993a). Balance, efficiency and orthogonality concepts in block designs. *J. Statist. Plann. Inference* **36**, 283–300.
Caliński, T. (1993b). The basic contrasts of a block experimental design with special reference to the notion of general balance. *Listy Biometryczne – Biometr. Lett.* **30**, 13–38.
Caliński, T. (1994a). The basic contrasts of a block design with special reference to the recovery of inter-block information. Invited presentation at the International Conference on Mathematical Statistics ProbaStat'94, Smolenice, Slovakia, p. 8.
Caliński, T. (1994b). On the randomization theory of experiments in nested block designs. *Listy Biometryczne – Biometr. Lett.* **31**, 45–77.
Caliński, T. and B. Ceranka (1974). Supplemented block designs. *Biom. J.* **16**, 299–305.
Caliński, T., B. Ceranka and S. Mejza (1980). On the notion of efficiency of a block design. In: W. Klonecki et al., eds., *Mathematical Statistics and Probability Theory*. Springer, New York, 47–62.
Caliński, T. and S. Kageyama (1988). A randomization theory of intrablock and interblock estimation. Tech. Report No. 230, Statistical Research Group, Hiroshima University, Hiroshima, Japan.
Caliński, T. and S. Kageyama (1991). On the randomization theory of intra-block and inter-block analysis. *Listy Biometryczne – Biometr. Lett.* **28**, 97–122.
Caliński, T. and S. Kageyama (1996). The randomization model for experiments in block designs and the recovery of inter-block information. *J. Statist. Plann. Inference*, **52**, 359–374.
Carmichael, R. D. (1956). *Introduction to the Theory of Groups of Finite Order*. Dover, New York.
Ceranka, B. (1983). Planning of experiments in C-designs. Scientific Dissertation 136, Annals of Poznan Agricultural Univ., Poland.
Ceranka, B. (1984). Construction of partially efficiency balanced block designs. *Calcutta Statist. Assoc. Bull.* **33**, 165–172.
Ceranka, B., S. Kageyama and S. Mejza (1986). A new class of C-designs. *Sankhyā Ser. B* **48**, 199–206.
Ceranka, B. and M. Kozlowska (1984). Some methods of constructing C-designs. *J. Statist. Plann. Inference* **9**, 253–258.
Ceranka, B. and S. Mejza (1979). On the efficiency factor for a contrast of treatment parameters. *Biom. J.* **21**, 99–102.
Ceranka, B. and S. Mejza (1980). A new proposal for classification of block designs. *Studia Sci. Math. Hungar.* **15**, 79–82.
Chakrabarti, M. C. (1962). *Mathematics of Design and Analysis of Experiments*. Asia Publishing House, Bombay.
Clatworthy, W. H. (1973). *Tables of Two-Associate-Class Partially Balanced Designs*. NBS Applied Math. Series 63, Washington, D.C., USA.
Cochran, W. G. and G. M. Cox (1957). *Experimental Designs*, 2nd edn. Wiley, New York.
Colbourn, C. J. and M. J. Colbourn (1983). Nested triple systems. *Ars Combin.* **16**, 27–34.
Colbourn, C. J. and R. A. Mathon, eds. (1987). *Combinatorial Design Theory*. Annals of Discrete Mathematics, Vol. 34. North-Holland, Amsterdam.

Colbourn, C. J. and P. C. van Oorschot (1988). Applications of combinatorial designs in computer science. IMA preprint series #400, Univ. of Minnesota, USA.

Corsten, L. C. A. (1962). Balanced block designs with two different numbers of replicates. *Biometrics* **18**, 499–519.

Cox, D. R. (1958). *Planning of Experiments*. Wiley, New York.

Das, M. N. and D. K. Ghosh (1985). Balancing incomplete block designs. *Sankhyā Ser. B* **47**, 67–77.

David, H. A. (1967). Resolvable cyclic designs. *Sankhyā Ser. A* **29**, 191–198.

Dey, A. and V. K. Gupta (1986). Another look at the efficiency and partially efficiency balanced designs. *Sankhyā Ser. B* **48**, 437–438.

Dey, A., U. S. Das and A. K. Banerjee (1986). Constructions of nested balanced incomplete block designs. *Calcutta Statist. Assoc. Bull.* **35**, 161–167.

Eckenstein, O. (1912). Bibliography of Kirkman's school girl problem. *Messenger Math.* **41–42**, 33–36.

Federer, W. T. (1955). *Experimental Design: Theory and Application*. Macmillan, New York.

Finney, D. J. (1960). *An Introduction to the Theory of Experimental Design*. The University of Chicago Press, Chicago, IL.

Fisher, R. A. (1925). *Statistical Methods for Research Workers*. Oliver and Boyd, Edinburgh.

Fisher, R. A. (1926). The arrangement of field experiments. *J. Ministry Agriculture* **33**, 503–513.

Fisher, R. A. (1940). An examination of the different possible solutions of a problem in incomplete blocks. *Ann. Eugen.* **10**, 52–75.

Fisher, R. A. amd F. Yates (1963). *Statistical Tables for Biological, Agricultural and Medical Research*. 6th edn. Hafner, New York. (1st edn., 1938.)

Graf-Jaccottet, M. (1977). Comparative classification of block designs. In: J. R. Barra, F. Brodeau, G. Romier and B. van Cutsem, eds., *Recent Developments in Statistics*. North-Holland, Amsterdam, 471–474.

Gupta, S. C. (1984). An algorithm for constructing nests of 2 factors designs with orthogonal factorial structure. *J. Statist. Comput. Simulation* **20**, 59–79.

Gupta, S. (1987). A note on the notion of balance in designs. *Calcutta Statist. Assoc. Bull.* **36**, 85–89.

Gupta, S. and S. Kageyama (1994). Optimal complete diallel crosses. *Biometrika* **81**, 420–424.

Hall, M., Jr. *Combinatorial Theory*, 2nd edn. Wiley, New York, 1986.

Harville, D. A. (1977). Maximum likelihood approaches to variance component estimation and to related problems. *J. Amer. Statist. Assoc.* **72**, 320–338.

Hedayat, A. and W. T. Federer (1974). Pairwise and variance balanced incomplete block designs. *Ann. Inst. Statist. Math.* **26**, 331–338.

Hedayat, A. S. and S. Kageyama (1980). The family of t-designs – Part I. *J. Statist. Plann. Inference* **4**, 173–212.

Hedayat, A. S., M. Jacroux and D. Majumdar (1988). Optimal designs for comparing test treatments with controls (with discussion). *Statist. Sci.* **3**, 462–491.

Homel, R. J. and J. Robinson (1975). Nested partially balanced incomplete block designs. *Sankhyā Ser. B* **37**, 201–210.

Houtman, A. M. and T. P. Speed (1983). Balance in designed experiments with orthogonal block structure. *Ann. Statist.* **11**, 1069–1085.

Hughes, D. R. and F. C. Piper (1976). On resolutions and Bose's theorem. *Geom. Dedicata* **5**, 129–133.

Hughes, D. R. and F. C. Piper (1985). *Design Theory*. Cambridge Univ. Press, Cambridge.

James, A. T. and G. N. Wilkinson (1971). Factorization of the residual operator and canonical decomposition of nonorthogonal factors in the analysis of variance. *Biometrika* **58**, 279–294.

Jarrett, R. G. (1977). Bounds for the efficiency factor of block designs. *Biometrika* **64**, 67–72.

Jimbo, M. and S. Kuriki (1983). Constructions of nested designs. *Ars Combin.* **16**, 275–285.

John, J. A. (1987). *Cyclic Designs*. Chapman and Hall, London.

John, J. A. and T. J. Mitchell (1977). Optimal incomplete block designs. *J. Roy. Statist. Soc. Ser. B* **39**, 39–43.

John, P. W. M. (1961). An application of a balanced incomplete block design. *Technometrics* **3**, 51–54.

John, P. W. M. (1980). *Incomplete Block Designs*. Marcel Dekker, New York.

Jones, R. M. (1959). On a property of incomplete blocks. *J. Roy. Statist. Soc. Ser. B* **21**, 172–179.

Kackar, R. N. and D. A. Harville (1981). Unbiasedness of two-stage estimation and prediction procedures for mixed linear models. *Comm. Statist. Theory Methods* **10**, 1249–1261.

Kackar, R. N. and D. A. Harville (1984). Approximations for standard errors of estimators of fixed and random effects in mixed linear models. *J. Amer. Statist. Assoc.* **79**, 853–862.

Kageyama, S. (1973). On μ-resolvable and affine μ-resolvable balanced incomplete block designs. *Ann. Statist.* **1**, 195–203.

Kageyama, S. (1974). Reduction of associate classes for block designs and related combinatorial arrangements. *Hiroshima Math. J.* **4**, 527–618.

Kageyama, S. (1976a). On μ-resolvable and affine μ-resolvable t-designs. In: S. Ikeda, et al., eds., *Essays in Probability and Statistics*. Shinko Tsusho Co. Ltd., Tokyo.

Kageyama, S. (1976b). Resolvability of block designs. *Ann. Statist.* **4**, 655–661. Addendum: *Bull. Inst. Math. Statist.* **7**(5) (1978), 312.

Kageyama, S. (1977). Conditions for α-resolvability and affine α-resolvability of incomplete block designs. *J. Japan Statist. Soc.* **7**, 19–25.

Kageyama, S. (1978). Remarks on 'Resolvability of block designs'. *Bull. Inst. Math. Statist.* **7**(5), 312.

Kageyama, S. (1980). On properties of efficiency-balanced designs. *Comm. Statist. Theory Methods* **9**, 597–616.

Kageyama, S. (1984). Some properties on resolvability of variance-balanced designs. *Geom. Dedicata* **15**, 289–292.

Kageyama, S. (1985). Connected designs with the minimum number of experimental units. In: T. Caliński and E. Klonecki, eds., *Linear Statistical Inference*, Lecture Notes in Statistics, Vol. 35. Springer, New York, 99–117.

Kageyama, S. (1993). The family of block designs with some combinatorial properties. *Discrete Math.* **116**, 17–54.

Kageyama, S. and A. S. Hedayat (1983). The family of t-designs – Part II. *J. Statist. Plann. Inference* **7**, 257–287.

Kageyama, S. and R. Mukerjee (1986). General balanced designs through reinforcement. *Sankhyā Ser. B* **48**, 380–387; Addendum. *Ser. B* **49**, 103.

Kageyama, S. and P. D. Puri (1985a). A new special class of PEB designs. *Comm. Statist. Theory Methods* **14**, 1731–1744.

Kageyama, S. and P. D. Puri (1985b). Properties of partially efficiency-balanced designs. *Bull. Inform. Cyber.* (Formerly *Bull. Math. Statist.*) **21**, 19–28.

Kageyama, S., G. M. Saha and R. Mukerjee (1988). D^{-1}-partially efficiency-balanced designs with at most two efficiency classes. *Comm. Statist. Theory Methods* **17**, 1669–1683.

Kageyama, S. and D. V. S. Sastry (1993). On affine $(\mu_1, \ldots, \mu_t)$-resolvable (r, λ)-designs. *Ars Combin.* **36**, 221–223.

Kala, R. (1981). Projectors and linear estimation in general linear models. *Comm. Statist. Theory Methods* **10**, 849–873.

Kala, R. (1991). Elements of the randomization theory. III. Randomization in block experiments. *Listy Biometryczne – Biometr. Lett.* **28**, 3–23 (in Polish).

Kassanis, B. and A. Kleczkowski (1965). Inactivation of a strain of tobacco necrosis virus and of the RNA isolated from it, by ultraviolet radiation of different wavelengths. *Photochem. Photobiol.* **4**, 209–214.

Kempthorne, O. (1952). *The Design and Analysis of Experiments.* Wiley, New York.

Kempthorne, O. (1955). The randomization theory of experimental inference. *J. Amer. Statist. Assoc.* **50**, 946–967.

Kempthorne, O. (1956). The efficiency factor of an incomplete block design. *Ann. Math. Statist.* **27**, 846–849.

Kempthorne, O. (1977). Why randomize? *J. Statist. Plann. Inference* **1**, 1–25.

Kiefer, J. (1958). On the nonrandomized optimality and randomized nonoptimality of symmetrical designs. *Ann. Math. Statist.* **29**, 675–699.

Kimberley, M. E. (1971). On the construction of certain Hadamard designs. *Math. Z.* **119**, 41–59.

Kirkman, T. P. (1847). On a problem in combinatorics. *Cambridge Dublin Math. J.* **2**, 191–204.

Kirkman, T. P. (1850a). Query. *Ladies and Gentleman's Diary*, 48.

Kirkman, T. P. (1850b). Note on an unanswered prize question. *Cambridge Dublin Math. J.* **5**, 191–204.

Kishen, K. (1940–1941). Symmetrical unequal block arrangements. *Sankhyā* **5**, 329–344.

Klaczynski, K., A. Molinska and K. Molinski (1994). Unbiasedness of the estimator of the function of expected value in the mixed linear model. *Biom. J.* **36**, 185–191.

Kleczkowski, A. (1960). Interpreting relationships between the concentration of plant viruses and numbers of local lesions. *J. Gen. Microbiol.* **4**, 53–69.

Kshirsagar, A. M. (1958). A note on incomplete block designs. *Ann. Math. Statist.* **29**, 907–910.

Lindner, C. C. and A. Rosa, eds. (1980). *Topics on Steiner Systems*. Annals of Discrete Mathematics, Vol. 7. North-Holland, Amsterdam.

Longyear, J. Q. (1981). A survey of nested designs. *J. Statist. Plann. Inference* **5**, 181–187.

Margolin, B. H. (1982). Blocks, randomized complete. In: S. Kotz and N. L. Johnson, eds., *Encyclopedia of Statistical Sciences*, Vol. 1. Wiley, New York, 288–292.

Martin, F. B. and G. Zyskind (1966). On combinability of information from uncorrelated linear models by simple weighting. *Ann. Math. Statist.* **37**, 1338–1347.

Mavron, V. C. (1972). On the structure of affine designs. *Math. Z.* **125**, 298–316.

Mejza, S. (1992). On some aspects of general balance in designed experiments. *Statistica* **52**, 263–278.

Mejza, S. and S. Kageyama (1994). Some statistical properties of nested block designs. Presentation at the International Conference on Mathematical Statistics ProbaStat'94, Smolenice, Slovakia.

Mukerjee, R. and S. Kageyama (1985). On resolvable and affine resolvable variance-balanced designs. *Biometrika* **72**, 165–172.

Nair, K. R. and C. R. Rao (1942a). A note on partially balanced incomplete block designs. *Sci. Culture* **7**, 568–569.

Nair, K. R. and C. R. Rao (1942b). Incomplete block designs for experiments involving several groups of varieties. *Sci. Culture* **7**, 615–616.

Nair, K. R. and C. R. Rao (1948). Confounding in asymmetrical factorial experiments. *J. Roy. Statist. Soc. Ser. B* **10**, 109–131.

Nelder, J. A. (1954). The interpretation of negative components of variance. *Biometrika* **41**, 544–548.

Nelder, J. A. (1965a). The analysis of randomized experiments with orthogonal block structure. I. Block structure and the null analysis of variance. *Proc. Roy. Soc. London Ser. A* **283**, 147–162.

Nelder, J. A. (1965b). The analysis of randomized experiments with orthogonal block structure. II. Treatment structure and the general analysis of variance. *Proc. Roy. Soc. London Ser. A* **283**, 163–178.

Nelder, J. A. (1968). The combination of information in generally balanced designs. *J. Roy. Statist. Soc. Ser. B* **30**, 303–311.

Neyman, J. (1923). Próba uzasadnienia zastosowań rachunku prawdopodobieństwa do doświadczeń polowych (Sur les applications de la théorie des probabilités aux expérience agricoles: Essay de principes). *Roczniki Nauk Rolniczych* **10**, 1–51.

Neyman, J. (1935), with co-operation of K. Iwaszkiewicz and S. Kolodziejczyk. Statistical problems in agricultural experimentation (with discussion). *J. Roy. Statist. Soc. Suppl.* **2**, 107–180.

Nigam, A. K. and P. D. Puri (1982). On partially efficiency balanced designs – II. *Comm. Statist. Theory Methods* **11**, 2817–2830.

Nigam, A. K., P. D. Puri and V. K. Gupta (1988). *Characterizations and Analysis of Block Designs*. Wiley Eastern, New Delhi.

Ogawa, J. (1963). On the null-distribution of the F-statistic in a randomized balanced incomplete block design under the Neyman model. *Ann. Math. Statist.* **34**, 1558–1568.

Ogawa, J. (1974). *Statistical Theory of the Analysis of Experimental Designs*. Marcel Dekker, New York.

Pal, S. (1980). A note on partially efficiency balanced designs. *Calcutta Statist. Assoc. Bull.* **29**, 185–190.

Patterson, H. D. and V. Silvey (1980). Statutory and recommended list trials of crop varieties in the United Kingdom. *J. Roy. Statist. Soc. Ser. A* **143**, 219–252.

Patterson, H. D. and R. Thompson (1971). Recovery of inter-block information when block sizes are unequal. *Biometrika* **58**, 545–554.

Patterson, H. D. and R. Thompson (1975). Maximum likelihood estimation of components of variance. In: L. C. A. Corsten and T. Postelnicu, eds., *Proc. 8th Internat. Biometric Conf.* Editura Academiei, Bucuresti, 197–207.

Patterson, H. D. and E. R. Williams (1976a). A new class of resolvable incomplete block designs. *Biometrika* **63**, 83–92.

Patterson, H. D. and E. R. Williams (1976b). Some theoretical results on general block designs. In: *Proc. 5th British Combinatorial Conf. Congressus Numerantium XV*, Utilitas Math., Winnipeg, 489–496.

Patterson, H. D., E. R. Williams and E. A. Hunter (1978). Block designs for variety trials. *J. Agric. Sci.* **90**, 395–400.

Pearce, S. C. (1960). Supplemented balance. *Biometrika* **47**, 263–271.

Pearce, S. C. (1963). The use and classification of non-orthogonal designs. *J. Roy. Statist. Soc. Ser. A* **126**, 353–377.

Pearce, S. C. (1964). Experimenting with blocks of natural sizes. *Biometrika* **20**, 699–706.

Pearce, S. C. (1970). The efficiency of block designs in general. *Biometrika* **57**, 339–346.

Pearce, S. C. (1976). Concurrences and quasi-replication: An alternative approach to precision in designed experiments. *Biom. Z.* **18**, 105–116.

Pearce, S. C. (1983). *The Agricultural Field Experiment.* Wiley, Chichester.

Pearce, S. C., T. Caliński and T. F. de C. Marshall (1974). The basic contrasts of an experimental design with special reference to the analysis of data. *Biometrika* **61**, 449–460.

Preece, D. A. (1967). Nested balanced incomplete block designs. *Biometrika* **54**, 479–486.

Preece, D. A. (1982). Balance and designs: Another terminological tangle. *Utilitas Math.* **21C**, 85–186.

Puri, P. D. and S. Kageyama (1985). Constructions of partially efficiency-balanced designs and their analysis. *Comm. Statist. Theory Methods* **14**, 1315–1342.

Puri, P. D. and A. K. Nigam (1975a). On patterns of efficiency balanced designs. *J. Roy. Statist. Soc. Ser. B* **37**, 457–458.

Puri, P. D. and A. K. Nigam (1975b). A note on efficiency balanced designs. *Sankhyā Ser. B* **37**, 457–460.

Puri, P. D. and A. K. Nigam (1977a). Partially efficiency balanced designs. *Comm. Statist. Theory Methods* **6**, 753–771.

Puri, P. D. and A. K. Nigam (1977b). Balanced block designs. *Comm. Statist. Theory Methods* **6**, 1171–1179.

Puri, P. D. and A. K. Nigam (1983). Merging of treatments in block designs. *Sankhyā Ser. B* **45**, 50–59.

Puri, P. D., B. D. Mehta and S. Kageyama (1987). Patterned constructions of partially efficiency-balanced designs. *J. Statist. Plann. Inference* **15**, 365–378.

Puri, P. D., A. K. Nigam and P. Narain (1977). Supplemented designs. *Sankhyā Ser. B* **39**, 189–195.

Raghavarao, D. (1962). Symmetrical unequal block arrangements with two unequal block sizes. *Ann. Math. Statist.* **33**, 620–633.

Raghavarao, D. (1962). On balanced unequal block designs. *Biometrika* **49**, 561–562.

Raghavarao, D. (1971). *Constructions and Combinatorial Problems in Design of Experiments.* Wiley, New York.

Rao, C. R. (1959). Expected values of mean squares in the analysis of incomplete block experiments and some comments based on them. *Sankhyā* **21**, 327–336.

Rao, C. R. (1970). Estimation of heteroscedastic variances in linear models. *J. Amer. Statist. Assoc.* **65**, 161–172.

Rao, C. R. (1971). Estimation of variance and covariance components – MINQUE theory. *J. Multivariate Anal.* **1**, 257–275.

Rao, C. R. (1972). Estimation of variance and covariance components – in linear models. *J. Amer. Statist. Assoc.* **67**, 112–115.

Rao, C. R. (1974). Projectors, generalized inverses and the BLUEs. *J. Roy. Statist. Soc. Ser. B* **36**, 442–448.

Rao, C. R. (1979). MINQUE theory and its relation to ML and MML estimation of variance components. *Sankhyā Ser. B* **41**, 138–153.

Rao, C. R. and J. Kleffe (1988). *Estimation of Variance Components and Applications.* North-Holland, Amsterdam.

Rao, C. R. and S. K. Mitra (1971). *Generalized Inverse of Matrices and Its Applications.* Wiley, New Nork.

Rao, M. B. (1966). A note on equi-replicated balanced designs with $b = v$. *Calcutta Statist. Assoc. Bull.* **15**, 43–44.

Rao, V. R. (1958). A note on balanced designs. *Ann. Math. Statist.* **29**, 290–294.

Ray-Chaudhuri, D. K. and R. M. Wilson (1971). Solution of Kirkman's school girl problem. *Combinatorics – Proc. Symp. in Pure Mathematics* **19**, 187–204 (Amer. Math. Soc.).

Saha, G. M. (1976). On Caliński's patterns in block designs. *Sankhyā Ser. B* **38**, 383–392.

Scheffé, H. (1959). *The Analysis of Variance.* Wiley, New York.

Seber, G. A. F. (1980). *The Linear Hypothesis: A General Theory.* Griffin, London.

Shah, K. R. (1964). Use of inter-block information to obtain uniformly better estimates. *Ann. Math. Statist.* **35**, 1064–1078.

Shah, K. R. (1992). Recovery of interblock information: an update. *J. Statist. Plann. Inference* **30**, 163–172.

Shah, K. R. and B. K. Sinha (1989). *Theory of Optimal Designs.* Springer-Verlag, Berlin.

Shrikhande, S. S. (1953). The non-existence of certain affine resolvable balanced incomplete block designs. *Canad. J. Math.* **5**, 413–420.

Shrikhande, S. S. (1976). Affine resolvable balanced incomplete block designs: A survey. *Aequationes Math.* **14**, 251–269.

Shrikhande, S. S. and D. Raghavarao (1964). Affine α-resolvable incomplete block designs. In: C. R. Rao, ed., *Contributions to Statistics.* Pergamon Press, Statistical Publishing Society, Calcutta, 471–480.

Speed, T. P., E. R. Williams and H. D. Patterson (1985). A note on the analysis of resolvable block designs. *J. Roy. Statist. Soc. Ser. B* **47**, 357–361.

Sprott, D. A. (1955). Balanced incomplete block designs and tactical configurations. *Ann. Math. Statist.* **26**, 752–758.

Stanton, R. G. and R. C. Mullin (1966). Inductive methods for balanced incomplete block designs. *Ann. Math. Statist.* **37**, 1348–1354.

Steiner, J. (1853). Combinatorische Aufgabe. *J. Reine Angew. Math.* **45**, 181–182.

Street, A. P. and D. J. Street (1987). *Combinatorics of Experimental Design.* Clarendon Press, Oxford.

Tocher, K. D. (1952). The design and analysis of block experiments (with discussion). *J. Roy. Statist. Soc. Ser. B* **14**, 45–100.

Vartak, M. N. (1963). Disconnected balanced designs. *J. Indian Statist. Assoc.* **1**, 104–107.

Wallis, W. D. (1988). *Combinatorial Designs*, Marcel Dekker, New York.

White, R. F. (1975). Randomization in the analysis of variance. *Biometrics* **31**, 555–571.

Williams, E. R. (1975). Efficiency-balanced designs. *Biometrika* **62**, 686–688.

Williams, E. R. (1976). Resolvable paired-comparison designs. *J. Roy. Statist. Soc. Ser. B* **38**, 171–174.

Williams, E. R., H. D. Patterson and J. A. John (1976). Resolvable designs with two replications. *J. Roy. Statist. Soc. Ser. B* **38**, 296–301.

Williams, E. R., H. D. Patterson and J. A. John (1977). Efficient two-replicate resolvable designs. *Biometrics* **33**, 713–717.

Woolhouse, W. S. B. (1844). Prize question 1733. *Ladies and Gentleman's Diary.*

Yates, F. (1936a). Incomplete randomized blocks. *Ann. Eugen.* **7**, 121–140.

Yates, F. (1936b). A new method of arranging variety trials involving a large number of varieties. *J. Agric. Sci.* **26**, 424–455.

Yates, F. (1939). The recovery of inter-block information in variety trials arranged in three-dimensional lattices. *Ann. Eugen.* **9**, 136–156.

Yates, F. (1940). The recovery of inter-block information in balanced incomplete block designs. *Ann. Eugen.* **10**, 317–325.

Yates, F. (1965). A fresh look at the basic principles of the design and analysis of experiments. In: L. M. LeCam and J. Neyman, eds., *Proc. 5th Berkeley Symp. on Mathematical Statistics and Probability*, Vol. 4. University of California Press, Berkeley, CA, 777–790.

Yates, F. (1975). The early history of experimental design. In: J. N. Srivastava, ed., *A Survey of Statistical Design and Linear Models.* North-Holland, Amsterdam, 581–592.

Zyskind, G. (1967). On canonical forms, non-negative covariance matrices and best and simple least squares linear estimators in linear models. *Ann. Math. Statist.* **38**, 1092–1109.

S. Ghosh and C. R. Rao, eds., *Handbook of Statistics, Vol. 13*

23

Developments in Incomplete Block Designs for Parallel Line Bioassays

Sudhir Gupta and Rahul Mukerjee

1. Introduction

Biological assays or bioassays are experiments for estimating the strength of a substance, called the stimulus, which is usually a drug and can, in particular, be a vitamin or a hormone. Normally, two preparations of the stimulus, both with quantitative doses and having a similar effect, are compared utilizing responses produced by them on living subjects like animals, plants or isolated organs or tissues. One of the two preparations, called the standard preparation, is of known strength while the other, called the test preparation, has an unknown strength. The principal objective of the assay is to measure the potency of the test relative to the standard preparation, the relative potency being defined as the ratio of equivalent, i.e., equally effective doses of the two preparations. We refer to Finney (1978) for numerous practical examples of bioassays with real data – these include an assay of vitamin D_3 in codliver oil via its antirachitic activity in chickens (Chapter 4), an assay of testosterone propionate by means of growth of comb in capons (Chapter 5), and so on.

A bioassay can be direct or indirect. In a direct assay, doses of the standard and test preparations just sufficient to produce a specified response are directly measured. This is practicable only when both the preparations are capable of administration in such a way that the minimal amounts needed to produce the specified response can be exactly measured. In an indirect assay, on the other hand, predetermined doses are administered to the subjects and their responses, either quantal or quantitative, are recorded. The responses are quantal when the experimenter records whether or not each subject manifests a certain easily recognizable reaction, such as death, and they are quantitative when the magnitude of some property, like survival time, weight etc, is measured for each subject.

Throughout the present discussion, only indirect assays, where the response is measured quantitatively, will be considered. As indicated above, interest lies in estimating the relative potency defined as the ratio of equivalent doses of the two preparations. The relative potency, however, is meaningful when the ratio remains constant over all possible pairs of equivalent doses as happens in assays of the analytical dilution type where the two preparations have the same effective constituent or the same effective constituents in fixed proportions of one another. Thus it is important to test the constancy of the ratio of equivalent doses before proceeding with the estimation of relative potency.

More specifically, in the present article we are concerned with parallel line assays. These are indirect assays with quantitative responses where the experimenter wishes to estimate the relative potency under the assumption that the relationship between the expected response and the logarithm of doses of the two preparations is representable by parallel straight lines. This assumption entails the desired constancy of the ratio of equally effective doses (vide Section 2) and its validity is checked by appropriate tests of hypotheses.

In a parallel line assay, the predetermined doses of the standard and test preparations represent the treatments. However, unlike in usual varietal experiments where interest lies in all elementary treatment contrasts or in a complete set of orthonormal treatment contrasts, here, in conformity with the twin objectives of model testing and estimation of relative potency, certain specific treatment contrasts assume particular importance (cf. M. J. R. Healy's discussion on Tocher (1952)). Thus efficient designing of parallel line assays poses different types of problems which have received a considerable attention in the literature. We refer to the authoritative work of Finney (1978) (see also Finney, 1979) for an excellent account of the developments up to that stage. Further informative reviews are available in Hubert (1984), Das and Giri (1986) and Nigam et al. (1988).

A parallel line assay is called symmetric if the standard and test preparations involve the same number of doses; otherwise, it is called asymmetric. In Section 2, we introduce parallel line assays in fuller detail, with some emphasis on the symmetric case, highlighting the treatment contrasts of special interest. Section 3 provides a brief general introduction to block designs in the context of parallel line assays. The efficient designing of symmetric parallel line assays have been reviewed in Sections 4–6 while Section 7 deals with this issue in the asymmetric case. The role of non-equireplicate designs in this connection has been examined in Section 8. Finally, several open problems deserving further attention are discussed in Section 9.

2. Parallel line assays

2.1. A general introduction

Consider an indirect assay with quantitative responses. Let s and t denote typical doses of the standard and test preparations and, with $x = \log s$, $z = \log t$, let their effects be represented respectively by $\eta_1(x)$ and $\eta_2(z)$. Throughout, we shall use natural logarithm.

Let the quantitative doses of the standard and test preparations included in the assay be $s_1, \ldots, s_{m_1}$ and $t_1, \ldots, t_{m_2}$ respectively, where $m_1, m_2 \geqslant 2$. These doses are equispaced on the logarithmic scale the common ratio being the same for both the preparations, i.e.,

$$s_i = c_1 h^{i-1} \quad (1 \leqslant i \leqslant m_1), \qquad t_i = c_2 h^{i-1} \quad (1 \leqslant i \leqslant m_2), \tag{2.1}$$

where c_1, c_2, h are positive constants and $h > 1$. Thus $0 < s_1 < \cdots < s_{m_1}$ and $0 < t_1 < \cdots < t_{m_2}$. The integers m_1, m_2 as well as the doses $s_1, \ldots, s_{m_1}$ and $t_1, \ldots, t_{m_2}$ (or equivalently, c_1, c_2 and h) are prespecified. These doses should adequately cover the ranges of s and t. The $v = m_1 + m_2$ doses $s_1, \ldots, s_{m_1}$ and $t_1, \ldots, t_{m_2}$ represent the treatments in the present context. For $1 \leqslant i \leqslant m_1$, let τ_i be the effect of the dose s_i of the standard preparation and for $1 \leqslant i \leqslant m_2$, let τ_{m_1+i} be the effect of the dose t_i of the test preparation. Then, using the notation introduced above,

$$\tau_i = \eta_1(x_i) \quad (1 \leqslant i \leqslant m_1) \quad \text{and} \quad \tau_{m_1+i} = \eta_2(z_i) \quad (1 \leqslant i \leqslant m_2), \tag{2.2}$$

where (see (2.1))

$$x_i = \log s_i = \log c_1 + (i-1)\log h \quad (1 \leqslant i \leqslant m_1), \tag{2.3a}$$

$$z_i = \log t_i = \log c_2 + (i-1)\log h \quad (1 \leqslant i \leqslant m_2). \tag{2.3b}$$

The treatment effects $\tau_1, \ldots, \tau_v$ are unknown parameters. Let

$$\left.\begin{array}{l} \tau^{(1)} = (\tau_1, \ldots, \tau_{m_1})', \quad \tau^{(2)} = (\tau_{m_1+1}, \ldots, \tau_{m_1+m_2})', \\ \tau = (\tau_1, \ldots, \tau_v)'. \end{array}\right\} \tag{2.4}$$

We now introduce an orthogonal polynomial model for $\eta_1(\cdot)$ and $\eta_2(\cdot)$. To that effect, let $\phi_{1j}(\cdot)$, $0 \leqslant j \leqslant m_1 - 1$, represent orthogonal polynomials of degrees $0, 1, \ldots, m_1 - 1$, based on $x_1, \ldots, x_{m_1}$, i.e., for each j, $\phi_{1j}(\cdot)$ is a polynomial of degree j and

$$\sum_{i=1}^{m_1} \phi_{1j}(x_i)\phi_{1j'}(x_i) = 0 \quad (0 \leqslant j, j' \leqslant m_1 - 1;\ j \neq j'). \tag{2.5}$$

Similarly, let $\phi_{2j}(\cdot)$, $0 \leqslant j \leqslant m_2 - 1$, be orthogonal polynomials of degrees $0, 1, \ldots, m_2 - 1$, based on $z_1, \ldots, z_{m_2}$. Since $x_1, \ldots, x_{m_1}$ and $z_1, \ldots, z_{m_2}$ are equispaced (see (2.3)), in particular, we take

$$\phi_{i0}(u) \equiv 1 \quad (i = 1, 2), \tag{2.6a}$$

$$\phi_{i1}(u) = u - \log c_i - \frac{1}{2}(m_i - 1)\log h \quad (i = 1, 2), \tag{2.6b}$$

$$\phi_{i2}(u) = \{\phi_{i1}(u)\}^2 - \frac{1}{12}(m_i^2 - 1)(\log h)^2 \quad (i = 1, 2), \tag{2.6c}$$

the expressions in (2.6c) being meaningful only when $m_1, m_2 \geqslant 3$.

We now consider the models

$$\eta_1(x) = \sum_{j=0}^{m_1-1} \beta_{1j}\phi_{1j}(x), \qquad \eta_2(z) = \sum_{j=0}^{m_2-1} \beta_{2j}\phi_{2j}(z), \tag{2.7}$$

where the $\{\beta_{1j}\}$ and $\{\beta_{2j}\}$ are unknown parameters. We shall later express them in terms of $\tau_1, \ldots, \tau_v$. As indicated in the Introduction, in a parallel line assay interest lies in estimating the relative potency assuming that $\eta_1(\cdot)$ and $\eta_2(\cdot)$ are representable by parallel straight lines, the validity of the assumption being first tested. Note that the assumption is equivalent to the following:

$$\frac{1}{2}(\beta_{11} - \beta_{21}) = 0, \tag{2.8}$$

$$\beta_{1j} = 0 \quad (2 \leqslant j \leqslant m_1 - 1), \qquad \beta_{2j} = 0 \quad (2 \leqslant j \leqslant m_2 - 1). \tag{2.9}$$

The relations in (2.9) signify the linearity of $\eta_1(\cdot)$ and $\eta_2(\cdot)$ while (2.8) signifies the parallelism of such linear forms. If (2.8) and (2.9) hold then by (2.6a, b) and (2.7),

$$\eta_1(x) = \beta_{10} + \beta_1\left\{x - \log c_1 - \frac{1}{2}(m_1 - 1)\log h\right\},$$

$$\eta_2(z) = \beta_{20} + \beta_1\left\{z - \log c_2 - \frac{1}{2}(m_2 - 1)\log h\right\},$$

where $\beta_1 = (\beta_{11} + \beta_{21})/2$. Thus, under (2.8) and (2.9), two doses, s and t, of the standard and test preparations are equivalent if and only if

$$\beta_{10} + \beta_1\left\{x - \log c_1 - \frac{1}{2}(m_1 - 1)\log h\right\}$$
$$= \beta_{20} + \beta_1\left\{z - \log c_2 - \frac{1}{2}(m_2 - 1)\log h\right\},$$

where $x = \log s$, $z = \log t$. The potency, ρ, of the test preparation relative to the standard preparation can then be expressed as

$$\begin{aligned}\rho &= s/t = \exp(x - z)\\ &= (c_1/c_2)h^{(m_1-m_2)/2}\exp\{-(\beta_{10} - \beta_{20})/\beta_1\},\end{aligned} \tag{2.10}$$

which is a constant for all pairs (s, t) of equivalent doses. In view of (2.8)–(2.10), the quantities $\beta_{10} - \beta_{20}$, $\beta_1 = (\beta_{11} + \beta_{21})/2$, $(\beta_{11} - \beta_{21})/2$ and β_{1j}, β_{2j} $(j \geqslant 2)$ deserve attention in the context of a parallel line assay. The relative potency is a function of the first two of them while the ignorability of the rest is equivalent to the assumption under which the relative potency is meaningful. We shall now see how these quantities can be expressed as contrasts among $\tau_1, \ldots, \tau_v$.

By (2.2) and the first relation in (2.7)

$$\tau_i = \eta_1(x_i) = \sum_{j=0}^{m_1-1} \beta_{1j}\phi_{1j}(x_i), \quad 1 \leqslant i \leqslant m_1.$$

Using (2.5) one may solve the above system of equations to obtain

$$\beta_{1j} = \left\{ \sum_{i=1}^{m_1} \phi_{1j}(x_i)\tau_i \right\} \Big/ \sum_{i=1}^{m_1} \{\phi_{1j}(x_i)\}^2, \quad 0 \leqslant j \leqslant m_1 - 1. \qquad (2.11a)$$

Similarly,

$$\beta_{2j} = \left\{ \sum_{i=1}^{m_2} \phi_{2j}(z_i)\tau_{m_1+i} \right\} \Big/ \sum_{i=1}^{m_2} \{\phi_{2j}(z_i)\}^2, \quad 0 \leqslant j \leqslant m_2 - 1. \quad (2.11b)$$

In view of (2.6a), noting that for each $j \geqslant 1$,

$$\sum_{i=1}^{m_1} \phi_{1j}(x_i) = 0, \qquad \sum_{i=1}^{m_2} \phi_{2j}(z_i) = 0,$$

it is easily seen that each of $\beta_{10} - \beta_{20}$, $\beta_1 = (\beta_{11} + \beta_{21})/2$, $(\beta_{11} - \beta_{21})/2$ and β_{1j}, β_{2j} $(j \geqslant 2)$ is a contrast among $\tau_1, \ldots, \tau_v$. Hence if the experimental design used for a parallel line assay keeps all the treatment contrasts estimable (and hence testable) then, on the basis of the experimental data, one may test the significance of the contrasts representing $(\beta_{11} - \beta_{21})/2, \beta_{1j}, \beta_{2j}$ $(j \geqslant 2)$ and, upon the acceptance of hypotheses regarding their ignorability, proceed to estimate ρ by replacing $\beta_{10} - \beta_{20}$ and β_1 in (2.10) by the best linear unbiased estimates of the corresponding contrasts (see (2.12), (2.13) below).

In particular, by (2.6a, b) and (2.11),

$$\beta_{10} - \beta_{20} = \psi_p, \qquad \beta_1 = \frac{1}{2}(\beta_{11} + \beta_{21}) = \psi_1,$$

$$\frac{1}{2}(\beta_{11} - \beta_{21}) = \psi_1^1, \qquad (2.12)$$

where

$$\psi_p = m_1^{-1} \sum_{i=1}^{m_1} \tau_i - m_2^{-1} \sum_{i=1}^{m_2} \tau_{m_1+i}, \qquad (2.13a)$$

$$\psi_1 = \frac{6}{\log h}\left[\frac{1}{m_1(m_1^2-1)}\sum_{i=1}^{m_1}\left\{i-\frac{1}{2}(m_1+1)\right\}\tau_i \right.$$
$$\left. + \frac{1}{m_2(m_2^2-1)}\sum_{i=1}^{m_2}\left\{i-\frac{1}{2}(m_2+1)\right\}\tau_{m_1+i}\right], \tag{2.13b}$$

$$\psi_1^1 = \frac{6}{\log h}\left[\frac{1}{m_1(m_1^2-1)}\sum_{i=1}^{m_1}\left\{i-\frac{1}{2}(m_1+1)\right\}\tau_i \right.$$
$$\left. - \frac{1}{m_2(m_2^2-1)}\sum_{i=1}^{m_2}\left\{i-\frac{1}{2}(m_2+1)\right\}\tau_{m_1+i}\right]. \tag{2.13c}$$

The above treatment contrasts have natural interpretation. Thus ψ_p is the contrast between preparations while ψ_1 and ψ_1^1 are combined regression and parallelism contrasts respectively. If $\eta_1(\cdot)$ and $\eta_2(\cdot)$ are both representable by straight lines (i.e., if (2.9) holds) then such straight lines are parallel if and only if $\psi_1^1 = 0$ in which the common slope is given by ψ_1.

2.2. *Symmetric parallel line assays*

We now consider symmetric parallel line assays. This corresponds to the situation $m_1 = m_2$ $(= m$, say$)$. In this case, for $0 \leqslant j \leqslant m-1$ and $1 \leqslant i \leqslant m$, it is easily seen that

$$\phi_{1j}(x_i) = \phi_{2j}(z_i) = (\log h)^j e_j(i), \tag{2.14}$$

where $e_0(\cdot), e_1(\cdot), \ldots, e_{m-1}(\cdot)$ are orthogonal polynomials of degrees $0, 1, \ldots, m-1$, which, analogously to (2.5), satisfy

$$\sum_{i=1}^{m} e_j(i)e_{j'}(i) = 0 \quad (0 \leqslant j, j' \leqslant m-1;\ j \neq j').$$

In particular (cf. (2.6)),

$$\left.\begin{aligned} &e_0(i) = 1, \quad e_1(i) = i - \frac{1}{2}(m+1), \\ &e_2(i) = \left\{i - \frac{1}{2}(m+1)\right\}^2 - \frac{1}{12}(m^2-1),\ 1 \leqslant i \leqslant m \end{aligned}\right\}, \tag{2.15}$$

and more generally, for $j \geqslant 2$ and $1 \leqslant i \leqslant m$,

$$e_{j+1}(i) = e_j(i)e_1(i) - \frac{j^2(m^2-j^2)}{4(4j^2-1)}e_{j-1}(i).$$

Note that

$$\left.\begin{array}{ll} e_j(i) = e_j(m+1-i), & 1 \leqslant i \leqslant m, \text{ for even } j, \\ e_j(i) = -e_j(m+1-i), & 1 \leqslant i \leqslant m, \text{ for odd } j. \end{array}\right\} \tag{2.16}$$

For $0 \leqslant j \leqslant m-1$, let

$$e_j = (e_j(1), \ldots, e_j(m))', \qquad \Delta_j = 2(\log h)^j (e_j' e_j). \tag{2.17}$$

By (2.15) and (2.17),

$$\Delta_0 = 2m, \qquad \Delta_1 = \frac{1}{6} m(m^2 - 1) \log h. \tag{2.18}$$

From (2.4), (2.11), (2.14) and (2.17), one obtains

$$\beta_{1j} = (2/\Delta_j) e_j' \tau^{(1)}, \qquad \beta_{2j} = (2/\Delta_j) e_j' \tau^{(2)},$$

so that

$$\beta_{10} - \beta_{20} = (2/\Delta_0) L_p \tag{2.19a}$$

and for $0 \leqslant j \leqslant m-1$,

$$\frac{1}{2}(\beta_{1j} + \beta_{2j}) = \Delta_j^{-1} L_j, \qquad \frac{1}{2}(\beta_{1j} - \beta_{2j}) = \Delta_j^{-1} L_j^1, \tag{2.19b}$$

where

$$L_p = (e_0' - e_0')\tau, \tag{2.20a}$$

and

$$L_j = (e_j' e_j')\tau, \qquad L_j^1 = (e_j' - e_j')\tau \quad (0 \leqslant j \leqslant m-1). \tag{2.20b}$$

Taking $m_1 = m_2 = m$ in (2.12) and (2.13) and comparing them with (2.19) and (2.20), it is easily seen that the treatment contrasts L_p, L_1 and L_1^1 are proportional to ψ_p, ψ_1, ψ_1^1 respectively. Thus L_p, L_1 and L_1^1 can be interpreted as the preparation, combined regression and parallelism contrasts. For $j \geqslant 2$, in view of (2.19b), L_j and L_j^1 can be interpreted along the line of L_1 and L_1^1. For example, L_2 is the combined quadratic contrast while L_2^1 measures the difference between quadratic regression coefficients for the two preparations.

By (2.19b), the conditions (2.8) and (2.9), which ensure the representability of $\eta_1(\cdot)$ and $\eta_2(\cdot)$ by parallel straight lines, can now be expressed as

$$L_1^1 = 0, \tag{2.21}$$

$$L_j = 0, \qquad L_j^1 = 0 \quad (2 \leqslant j \leqslant m-1). \tag{2.22}$$

Also, by (2.18), (2.19), the relation (2.10) now reduces to

$$\rho = (c_1/c_2)\exp\left\{-\frac{1}{6}(m^2-1)(\log h)(L_p/L_1)\right\}. \tag{2.23}$$

In many practical situations, from past experience or prior knowledge about the assay, there is reason to believe that the effect versus log-dose relationship, given by $\eta_1(\cdot)$ and $\eta_2(\cdot)$, is linear for both the preparations but the parallelism of such straight lines may be in doubt. Then, in consideration of (2.21)–(2.23), interest lies mostly in L_p, L_1 and L_1^1. Even otherwise, often the experimenter may anticipate that $\eta_1(\cdot)$ and $\eta_2(\cdot)$ are both at most quadratic and then, in addition to L_p, L_1 and L_1^1, the contrasts L_2 and L_2^1 are also of primary interest. In either case, however, it would be desirable to use an experimental design that keeps all treatment contrasts, including the relatively less important ones, at least estimable (and hence testable).

Similar considerations hold also for asymmetric parallel line assays where $m_1 \neq m_2$, i.e., the numbers of doses included in the assay are unequal for the test and standard preparations. For example, there if the experimenter believes that $\eta_1(\cdot)$ and $\eta_2(\cdot)$ are both linear without being confident about their parallelism then interest lies mostly in ψ_p, ψ_1 and ψ_1^1 (cf. (2.12) and (2.13)).

The efficient designing of parallel line assays, taking due care of the contrasts of specific interest, will be reviewed in the rest of this article.

3. Parallel line assays in block designs

We now return to the set-up of Subsection 2.1 and consider a parallel line assay involving m_1 doses $s_1, \ldots, s_{m_1}$ of the standard preparation and m_2 doses $t_1, \ldots, t_{m_2}$ of the test preparation. These $v = m_1 + m_2$ doses represent the treatments which will always be arranged in the order $\{s_1, \ldots, s_{m_1}, t_1, \ldots, t_{m_2}\}$. As in (2.4), let $\tau = (\tau_1, \ldots, \tau_v)'$ denote the $v \times 1$ vector of the unknown treatment effects. We do not assume anything in this section about the equality or otherwise of m_1 and m_2. As in other kinds of experiments, blocking of experimental units is often essential in parallel line assays to eliminate heterogeneity. For example, in an assay of a vitamin via its effect on pigs, litters may constitute the blocks while in an assay of a hormone through its effect on humans, blocks may correspond to age groups. The necessary background on block designs, in the context of parallel line assays, is briefly indicated below. A related useful reference is Kshirsagar and Yuan (1994); see also Chakrabarti (1962) and Raghavarao (1971) for fuller accounts of block designs in general.

Suppose the assay is conducted in a block design involving b blocks, the block sizes being $k_1, \ldots, k_b$. Let r_i $(1 \leqslant i \leqslant v)$ denote the number of replications of the ith treatment and n_{ij} $(1 \leqslant i \leqslant v,\ 1 \leqslant j \leqslant b)$ denote the number of times the ith treatment appears in the jth block. Clearly,

$$r_1 + \cdots + r_v = k_1 + \cdots + k_b,$$

$$\sum_{i=1}^{v} n_{ij} = k_j \quad (1 \leqslant j \leqslant b), \qquad \sum_{j=1}^{b} n_{ij} = r_i \quad (1 \leqslant i \leqslant v). \tag{3.1}$$

The $v \times b$ incidence matrix of the design is given by $N = ((n_{ij}))$. Let $R = \operatorname{diag}(r_1, \ldots, r_v)$ and $K = \operatorname{diag}(k_1, \ldots, k_b)$. We work under the usual additive, intrablock linear model and assume independence, homoscedasticity and normality of errors. The assumption of normality is needed for performing tests of significance. Let σ^2 (> 0) be the constant error variance.

For $1 \leqslant i \leqslant v$, let T_i be the total of the r_i observations corresponding to the ith treatment and for $1 \leqslant j \leqslant b$, let B_j be the total of the k_j observations from the jth block. Let $T = (T_1, \ldots, T_v)'$ and $B = (B_1, \ldots, B_b)'$. Define the $v \times 1$ vector of adjusted treatment totals

$$Q = T - NK^{-1}B \tag{3.2}$$

and the $v \times v$ intrablock matrix

$$C = R - NK^{-1}N'. \tag{3.3}$$

Then the reduced normal equations for the vector of treatment effects, τ, are given by

$$C\tau = Q. \tag{3.4}$$

By (3.1) and (3.3), $C1_v = 0$, where 1_v is the $v \times 1$ vector with each element unity. Hence a necessary condition for the estimability of a linear parametric function $g'\tau$, where $g \neq 0$, is that it is a treatment contrast, i.e., $g'1_v = 0$. A treatment contrast $g'\tau$ will be estimable if and only if g' belongs to the row space of C. In this case, the best linear unbiased estimator (BLUE) of $g'\tau$ is $g'\widehat{\tau}$ where $\widehat{\tau}$ is any solution of the reduced normal equations (3.4). All treatment contrasts are estimable if and only if rank $(C) = v - 1$, i.e., the design is connected.

Consider now a set of treatment contrasts $g_1'\tau, \ldots, g_u'\tau$. Define the $u \times v$ matrix $G = (g_1, \ldots, g_u)'$ and write $G\tau = (g_1'\tau, \ldots, g_u'\tau)'$. Then we have the following lemma which will be useful in the sequel.

LEMMA 3.1. *Suppose $g_1'\tau, \ldots, g_u'\tau$ are all estimable and let $G\widehat{\tau} = (g_1'\widehat{\tau}, \ldots, g_u'\widehat{\tau})'$. Then*

(a) $\operatorname{Cov}(G\widehat{\tau}) - \sigma^2 GR^{-1}G'$ *is non-negative definite (n.n.d.);*

(b) $\operatorname{Cov}(G\widehat{\tau}) = \sigma^2 GR^{-1}G'$ *if and only if*

$$GR^{-1}N = 0, \tag{3.5}$$

in which case $G\widehat{\tau} = GR^{-1}T$.

PROOF. (a) Since $g_1'\tau, \ldots, g_u'\tau$ are all estimable, there exists a matrix V such that

$$G = VC. \tag{3.6}$$

Hence by (3.4), $G\hat{\tau} = VQ$. But it is well known that $\mathrm{Cov}(Q) = \sigma^2 C$. Hence $\mathrm{Cov}(G\hat{\tau}) = \sigma^2 VCV'$, and by (3.6),

$$\mathrm{Cov}(G\hat{\tau}) - \sigma^2 GR^{-1}G' = \sigma^2 V(C - CR^{-1}C)V'. \tag{3.7}$$

Now by (3.3),

$$C - CR^{-1}C = NK^{-1}(K - N'R^{-1}N)K^{-1}N'. \tag{3.8}$$

It is well known that the matrix $K - N'R^{-1}N$, being the dual of C when interest lies in the block rather than the treatment effects, is n.n.d. Hence by (3.8), $C - CR^{-1}C$ is n.n.d. and, from (3.7), part (a) of the lemma follows.

(b) Since $C - CR^{-1}C$ is n.n.d and K^{-1} is positive definite, by (3.3), (3.6) and (3.7),

$$\begin{aligned} &\mathrm{Cov}(G\hat{\tau}) = \sigma^2 GR^{-1}G', \\ \text{iff}\ \ &V(C - CR^{-1}C) = 0, \\ \text{i.e., iff}\ \ &G = GR^{-1}C[= GR^{-1}(R - NK^{-1}N')], \\ \text{i.e., iff}\ \ &GR^{-1}NK^{-1}N' = 0, \end{aligned}$$

i.e., iff (3.5) holds. If (3.5) holds then by (3.3), $G = GR^{-1}C$ so that by (3.2), (3.4), $G\hat{\tau} = GR^{-1}Q = GR^{-1}T$. This completes the proof. □

REMARK 3.1. Note that $\sigma^2 GR^{-1}G'$ is the covariance matrix of the BLUE of $G\tau$ in a completely randomized (unblocked) design with the same replication numbers and the same error variance as the block design under consideration. Hence part (a) of Lemma 3.1 is intuitively obvious since incorporation of block effects in the model can only possibly inflate the covariance matrix. The condition (3.5), derived in part (b) of the lemma, is necessary and sufficient for the estimation of $G\tau$ orthogonally to block effects. As one can intuitively anticipate, under this condition the BLUE of $G\tau$ equals $GR^{-1}T$ which is the same as what one would have obtained in a completely randomized design ignoring the block effects.

REMARK 3.2. For any single treatment contrast $g'\tau$, which is estimable in the block design, it is well known that $\mathrm{var}(g'\hat{\tau}) = \sigma^2 g'C^- g$, where C^- is any generalized inverse of C. But Lemma 3.1 (a), with $u = 1$, yields $\mathrm{var}(g'\hat{\tau}) \geqslant \sigma^2 g'R^{-1}g$. Hence the efficiency factor for $g'\tau$ in the block design under consideration is defined as $g'R^{-1}g/g'C^- g$, which is relative to a completely randomized (unblocked) design with the same replication numbers as the block design. The efficiency factor can never exceed unity and, by (3.5), it equals unity if and only if $g'R^{-1}N = 0$. In this case, we say that $g'\tau$ is being estimated in the block design with full efficiency or full information. Similarly, in the set-up of Lemma 3.1, the block design retains full information on $G\tau$, i.e., on each of $g_1'\tau, \ldots, g_u'\tau$, if and only if (3.5) holds.

REMARK 3.3. As indicated earlier, in parallel line assays interest lies in specific treatment contrasts. One can take $G\tau$ as consisting of these contrasts and then attempt to construct designs satisfying (3.5) so as to retain full information on $G\mathcal{T}$ (see also Remark 3.4 below). The relevant methods of construction will be surveyed in Sections 4–7.

REMARK 3.4. Given b and $k_1, \ldots, k_b$, by Lemma 3.1, a block design satisfying (3.5) will be optimal, in the class of designs having the same replication numbers and with regard to inference on $G\mathcal{T}$, under almost all reasonable criteria (see Shah and Sinha (1989)). In particular, an equireplicate block design satisfying (3.5) will be optimal, under such criteria, in the important class of equireplicate designs. Furthermore, Lemma 3.1 is potentially useful in searching for optimal designs, for inference on $G\mathcal{T}$, in an even larger class. We shall return to this issue in Section 8.

4. Symmetric parallel line assays: L-designs

With reference to symmetric parallel line assays, we shall now discuss how Lemma 3.1 can be employed to obtain equireplicate designs retaining full information on the preparation contrast (L_p), combined regression contrast (L_1) and parallelism contrast (L_1^1) which, as discussed in Section 2, are often the objects of principal interest. Following Gupta and Mukerjee (1990), such designs will, hereafter, be referred to as L-designs. Their constructional and related problems were addressed by Kulshreshtha (1969), Kyi Win and Dey (1980), Nigam and Boopathy (1985), Das (1985), Gupta et al. (1987), Gupta (1988) and Gupta and Mukerjee (1990). Our presentation of L-designs will mainly follow the unified approach of Gupta and Mukerjee (1990). Some other related results will also be briefly reviewed at the end of this section.

Let there be m doses of both the standard and the test preparations and suppose the assay is to be conducted using an incomplete block design in b blocks of k $(< v = 2m)$ experimental units each such that each of the v treatments is replicated r times. Then bk/v $(= r)$ must be an integer. The incidence matrix, N, of the design may be partitioned as $N = (N_1' \ N_2')'$, where N_1 is the $m \times b$ incidence matrix for the m doses of the standard preparation and N_2 is the corresponding incidence matrix for the test preparation. Since each block has size k and each treatment is replicated r times,

$$1_m' N_1 + 1_m' N_2 = k 1_b', \tag{4.1a}$$

$$N_1 1_b = N_2 1_b = r 1_m, \tag{4.1b}$$

where 1_b and 1_m are $b \times 1$ and $m \times 1$ vectors both with all elements unity.

In consideration of (2.20), taking

$$G = \begin{bmatrix} e_0' & -e_0' \\ e_1' & e_1' \\ e_1' & -e_1' \end{bmatrix},$$

and $R = rI_v$ in Lemma 3.1, where I_v is the identity matrix of order v, it follows from (3.5) and Remark 3.2 that the design under consideration is an L-design if and only if

$$\begin{bmatrix} e_0' & -e_0' \\ e_1' & e_1' \\ e_1' & -e_1' \end{bmatrix} \begin{bmatrix} N_1 \\ N_2 \end{bmatrix} = 0. \tag{4.2}$$

Now, by (2.15) and (2.17),

$$e_0 = 1_m, \qquad e_1 = f_1 - \frac{1}{2}(m+1)1_m, \tag{4.3}$$

where $f_1 = (1, 2, \ldots, m)'$. Hence by (4.1a) and (4.2), a characterization for an L-design is given by the conditions

$$1_m' N_1 = 1_m' N_2 = \left(\frac{1}{2}k\right) 1_b', \qquad e_1' N_1 = e_1' N_2 = 0, \tag{4.4}$$

which are due to Kyi Win and Dey (1980). Apart from retaining full information or full efficiency on L_p, L_1 and L_1^1, as indicated in Remark 3.4, an L-design, if it exists, will be optimal, in a very broad sense, for inference on these contrasts within the class of equireplicate designs arranged in b blocks each of size k. We shall see later in Section 8 that an L-design performs well also when it is allowed to compete with comparable non-equireplicate designs. Incidentally, the following is evident from (4.4).

LEMMA 4.1. *Given m, b and k, a necessary condition for the existence of an L-design is that k is even.*

We now consider the construction of L-designs. Gupta and Mukerjee (1990) proved the following result which completely settles the issue for even m. Incidentally, the proof given here is simpler than theirs.

THEOREM 4.1. *Let m be even and $k < v$ $(= 2m)$. Then an L-design with parameters v, b, r and k exists if and only if $vr = bk$ and $k \equiv 0 \pmod 4$.*

PROOF. Let $m = 2\mu$ and, from Lemma 4.1, let $k = 2\omega$. To prove the necessity, we only have to show that ω is even. Now, for a typical column of N_1, say $(y_1, y_2, \ldots, y_{2\mu})'$, it follows from (4.3) and (4.4) that

$$\sum_{j=1}^{2\mu} y_j = \omega \quad \text{and} \quad \sum_{j=1}^{2\mu} y_j(2j - 2\mu - 1) = 0.$$

Hence

$$\sum_{j=1}^{2\mu} y_j(j - \mu) = \frac{1}{2}\omega.$$

Since the y_j are non-negative integers, ω must be even.

The sufficiency part is proved by actual construction of designs. Let $vr = bk$ and $k \equiv 0 \pmod 4$. Since $v = 4\mu$, then $\mu r = b(k/4)$. Hence one can always construct a design involving μ treatments and b blocks such that each block has size $k/4$ and each treatment is replicated r times. Let M_1 and M_2 be the $\mu \times b$ incidence matrices of two such designs which are not necessarily distinct.

For positive integral n, let I_n be the identity matrix of order n and I_n^* be the $n \times n$ matrix defined as

$$I_n^* = \begin{bmatrix} 0\ 0 & \cdots & 0\ 1 \\ 0\ 0 & \cdots & 1\ 0 \\ & \vdots & \\ 1\ 0 & \cdots & 0\ 0 \end{bmatrix}.$$

Then the incidence matrix of an L-design with the desired parameters is given by (cf. Gupta, 1989)

$$N = \begin{bmatrix} I_\mu & 0 \\ I_\mu^* & 0 \\ 0 & I_\mu^* \\ 0 & I_\mu \end{bmatrix} \begin{bmatrix} M_1 \\ M_2 \end{bmatrix}. \tag{4.5}$$

□

REMARK 4.1. The design constructed in (4.5) will be connected if $M = [M_1' \ M_2']'$ represents the incidence matrix of a connected design. This should be achievable when $r > 1$.

EXAMPLE 4.1. Let $m = 4, v = 8, b = 4, r = 2, k = 4$. Then $\mu = 2$ and with

$$M_1 = \begin{bmatrix} 1\ 1\ 0\ 0 \\ 0\ 0\ 1\ 1 \end{bmatrix}, \qquad M_2 = \begin{bmatrix} 1\ 0\ 1\ 0 \\ 0\ 1\ 0\ 1 \end{bmatrix}$$

in (4.5), the incidence matrix of a connected L-design is obtained as

$$\begin{bmatrix} 1\ 0\ 0\ 1\ 0\ 1\ 1\ 0 \\ 1\ 0\ 0\ 1\ 1\ 0\ 0\ 1 \\ 0\ 1\ 1\ 0\ 0\ 1\ 1\ 0 \\ 0\ 1\ 1\ 0\ 1\ 0\ 0\ 1 \end{bmatrix}'.$$

Turning to the case of odd m, we first consider the situation $k = 4$. Along the line of Nigam and Boopathy (1985), then the following result holds.

THEOREM 4.2. *Let m ($\geqslant 3$) be odd. Then an L-design with parameters v ($= 2m$), b, r and $k = 4$ exists if and only if $vr = 4b$.*

PROOF. The necessity is obvious. To prove the sufficiency, let $m = 2\mu + 1$. Then $v = 4\mu + 2$, $k = 4$, $r = 2q$, $b = (2\mu + 1)q$, for some positive integer q. Let

$$N_1 = \begin{bmatrix} I_\mu & 0 & I_\mu^* \\ 0 & 2 & 0 \\ I_\mu^* & 0 & I_\mu \end{bmatrix}$$

be a square matrix of order $2\mu + 1$ and N_2 be such that the ith column of N_1 equals the $(i - 1)$th column of $N_2 (2 \leqslant i \leqslant 2\mu + 1)$ and the first column of N_1 equals the last column of N_2. An L-design with the desired parameters can now be obtained by taking q copies of the design with incidence matrix $[N_1' \ N_2']'$. The design can be seen to be connected. □

REMARK 4.2. From (2.16), (2.17), (2.20b) and (3.5), it is not hard to see that the constructions in Theorems 4.1 and 4.2 ensure full efficiency also with respect to L_j and L_j^1 for every odd j.

In contrast with the case of even m, a complete solution to the existence problem of L-designs is not yet available for every odd m. Gupta and Mukerjee (1990) obtained a necessary and sufficient condition for the existence of L-designs for odd $m \leqslant 15$, a range which seems to be enough for most practical purposes. Their result is shown below.

THEOREM 4.3. *Let m be odd, $3 \leqslant m \leqslant 15$ and $k < v$ ($= 2m$). Then an L-design with parameters v, b, r and k exists if and only if $vr = bk$, $k > 2$ and k is even.*

While the necessity part of Theorem 4.3 is evident from (4.4) and Lemma 4.1, Gupta and Mukerjee (1990) proved the sufficiency by actual construction of designs. As illustrated in Example 4.2 below, this was done by first employing (4.4) to enumerate the possible columns of N_1 or N_2 and then juxtaposing these columns so as to satisfy the requirement regarding replication number. They tabulated L-designs, over the range $3 \leqslant m \leqslant 15$, for all parameter values satisfying the conditions of this theorem. All these designs, except the single replicate ones, are connected.

EXAMPLE 4.2. Let $m = 5$, $v = 10$, $k = 6$ and $vr = bk$. Then $b/r = 5/3$, i.e., $b = 5q$, $r = 3q$ for some positive integer q. Consider the case $q = 1$. Let $(y_1, \ldots, y_5)'$ denote a typical column of N_1. By (4.4), the non-negative integers $y_1, \ldots, y_5$ must satisfy

$$\sum_{j=1}^{5} y_j = 3, \quad -2y_1 - y_2 + y_4 + 2y_5 = 0. \tag{4.6}$$

The only solutions of (4.6) for $(y_1, \ldots, y_5)'$ are

$$\alpha_1 = (0, 0, 3, 0, 0)', \qquad \alpha_2 = (0, 1, 1, 1, 0)', \qquad \alpha_3 = (1, 0, 1, 0, 1)',$$

$$\alpha_4 = (0, 2, 0, 0, 1)', \qquad \alpha_5 = (1, 0, 0, 2, 0)'.$$

Now let α_i occur u_i times as a column of N_1. Then, since by (4.1b) each row sum of N_1 must equal r $(= 3)$, we have $u_3 + u_5 = u_2 + 2u_4 = 3u_1 + u_2 + u_3 = u_2 + 2u_5 = u_3 + u_4 = 3$, a solution of which is $u_1 = 0$, $u_2 = 1$, $u_3 = 2$, $u_4 = u_5 = 1$. Thus

$$N_1 = \begin{bmatrix} 0 & 1 & 1 & 0 & 1 \\ 1 & 0 & 0 & 2 & 0 \\ 1 & 1 & 1 & 0 & 0 \\ 1 & 0 & 0 & 0 & 2 \\ 0 & 1 & 1 & 1 & 0 \end{bmatrix},$$

and the incidence matrix of a connected L-design with parameters $v = 10$, $k = 6$, $r = 3$ and $b = 5$ is given by $N = [N_1' \; N_1']'$. An L-design in $3q$ replications can be obtained considering q copies of this design.

In so far as the construction of L-designs is concerned, the results discussed above should be adequate. However, before concluding this section, we briefly outline some earlier work in this area. Kulshreshtha (1969) proposed some L-designs for even m in blocks of size at least eight. Kyi Win and Dey (1980) suggested a construction procedure on the basis of (4.4) but their approach seems to involve some amount of guess work. Das (1985) discussed the use of linked block designs and affine resolvable designs in this context.

For even m $(= 2\mu)$, Nigam and Boopathy (1985) described construction procedures for three series of L-designs: (a) $v = 4\mu$, $b = \mu^2$, $r = \mu$, $k = 4$, (b) $v = 4\mu$, $b = \mu(\mu - 1)$, $r = \mu - 1$, $k = 4$, (c) $v = 4\mu$, $b = \mu(\mu - 1)/2$, $r = \mu - 1$, $k = 8$. They noted that all these designs ensure full efficiency also with respect to L_j and L_j^1 for odd j (cf. Remark 4.2) and, recognizing these designs as partially efficiency balanced designs (Puri and Nigam, 1977), indicated their efficiency factors with respect to L_j and L_j^1 for even j. For odd m, these authors described two series of L-designs which have been incorporated in Theorem 4.2 above. Gupta et al. (1987) investigated incomplete block designs which retain full information on L_p and also on L_j and L_j^1 for odd j and addressed issues related to the characterization and construction of such designs; see also Nigam et al. (1988) in this context. Gupta (1988) discussed the use of singular and semi-regular group divisible designs in the construction of L-designs and indicated how the three series, due to Nigam and Boopathy (1985), can be covered by this approach.

5. Symmetric parallel line assays: Further results

Continuing with the set-up of the last section, consider a symmetric parallel line assay with m doses of either preparation. Suppose it is desired to conduct the assay in a design involving b blocks of size k $(< v = 2m)$ each such that every treatment is replicated r times, where bk/v $(= r)$ is an integer. Observe that given v, b and k, even if k is even (cf. Lemma 4.1), an L-design may not exist. By Theorem 4.1, this happens, for example, when m is even and $k \equiv 2 \pmod 4$. In such situations, the experimenter may wish to retain full information on any two of thc important contrasts L_p, L_1 and L_1^1. The relevant constructional aspects will be reviewed in this section.

THEOREM 5.1. *Let $vr = bk$. Then an equireplicate design, with parameters v $(= 2m)$, b, r, k, retaining full information on L_p and any one of L_1 and L_1^1 exists if and only if k is even.*

PROOF. The necessity is evident noting that if such a design exists then the corresponding matrices N_1, N_2, as defined in the last section, must satisfy the first condition in (4.4).

The sufficiency is proved by actual construction. Since $mr = b(k/2)$, if k is even then one can always construct a design involving m treatments and b blocks such that each block has size $k/2$ and each treatment is replicated r times. Let N_1 be the $m \times b$ incidence matrix of this design.

(a) Let $N_2 = I_m^* N_1$, i.e., N_2 is obtained by arranging the rows of N_1 in the reverse order. By (4.3), then

$$\begin{bmatrix} e_0' & -e_0' \\ e_1' & e_1' \end{bmatrix} \begin{bmatrix} N_1 \\ N_2 \end{bmatrix} = 0,$$

so that, by (2.20) and (3.5), the equireplicate design, given by the incidence matrix $[N_1' \; N_2']'$ and having parameters v, b, r, k, retains full information on L_p and L_1. This construction is due to Das and Kulkarni (1966).

(b) Similarly, the equireplicate design with incidence matrix $[N_1' \; N_1']'$ and parameters v, b, r, k, can be seen to retain full information on L_p and L_1^1. □

REMARK 5.1. The designs constructed in (a) or (b) above will be connected if N_1 represents a connected design. From (2.16), (2.17), (2.20b) and (3.5), it can be seen that the construction in (a) ensures full efficiency also with respect to L_j for every odd j and L_j^1 for every even j (vide Das and Kulkarni, 1966). Similarly, one can check that the construction in (b) ensures full efficiency also with respect to L_j^1 for every j.

REMARK 5.2. With reference to the construction in (a), Das and Kulkarni (1966) studied the consequences of choosing N_1 as the incidence matrix of a balanced incomplete block (BIB) design where every two distinct treatments occur together in λ blocks. As shown by them, then the efficiency factor for L_j^1, for every odd j, and L_j, for every even j, equals $\lambda v/(rk)$. Similarly, in this situation, the construction in (b) ensures an efficiency factor $\lambda v/(rk)$ with respect to L_j for every j. Das and Kulkarni (1966) also explored the effect of choosing N_1 as the incidence matrix of a circular design in (a) while Das (1985) examined the role of the so-called C-designs (Caliński, 1971; Saha, 1976) in this context.

EXAMPLE 5.1. Let $m = 4$, $v = 8$, $b = 4$, $r = 3$, $k = 6$. Then by Theorem 4.1, an L-design is not available. Following Theorem 5.1, take

$$N_1 = \begin{bmatrix} 0 & 1 & 1 & 1 \\ 1 & 0 & 1 & 1 \\ 1 & 1 & 0 & 1 \\ 1 & 1 & 1 & 0 \end{bmatrix}, \tag{5.1}$$

which represents the incidence matrix of a BIB design with $\lambda = 2$. The constructions in (a) and (b) above yield designs with incidence matrices

$$\text{(a)} \qquad N = \begin{bmatrix} 0 & 1 & 1 & 1 & 1 & 1 & 1 & 0 \\ 1 & 0 & 1 & 1 & 1 & 1 & 0 & 1 \\ 1 & 1 & 0 & 1 & 1 & 0 & 1 & 1 \\ 1 & 1 & 1 & 0 & 0 & 1 & 1 & 1 \end{bmatrix}'$$

and

$$\text{(b)} \qquad N = \begin{bmatrix} 0 & 1 & 1 & 1 & 0 & 1 & 1 & 1 \\ 1 & 0 & 1 & 1 & 1 & 0 & 1 & 1 \\ 1 & 1 & 0 & 1 & 1 & 1 & 0 & 1 \\ 1 & 1 & 1 & 0 & 1 & 1 & 1 & 0 \end{bmatrix}'$$

respectively. The design in (a) retains full information on L_p, L_1, L_3, L_2^1 and ensures an efficiency factor $8/9$ for L_1^1, L_3^1, L_2, while that in (b) retains full information on L_p, L_1^1, L_2^1, L_3^1 and ensures an efficiency factor $8/9$ for L_1, L_2, L_3 (vide Remarks 5.1 and 5.2).

Turning to the problem of retaining full information on both L_1 and L_1^1 (but not necessarily on L_p), we have the following result for even m.

THEOREM 5.2. *Let $vr = bk$ and m be even. Then an equireplicate design, with parameters v $(= 2m)$, b, r, k, retaining full information on L_1 and L_1^1 exists if and only if k is even.*

PROOF. Let $m = 2\mu$. To prove the necessity, suppose such a design exists and consider the associated matrices N_1 and N_2 which are defined as in Section 4. By (3.5), then $e_1' N_1 = e_1' N_2 = 0$ (cf. (4.4)). Hence for a typical column of N_1 or N_2, say $(y_1, y_2, \ldots, y_{2\mu})'$, by (4.3),

$$\sum_{j=1}^{2\mu} y_j (j - \mu) = \frac{1}{2} \sum_{j=1}^{2\mu} y_j.$$

Consequently, the sum of the elements in each column of N_1 and N_2 is even so that, by (4.1a), k must be even.

Following Gupta (1989), the sufficiency is proved by actual construction. Since $mr = b(k/2)$, if k is even then one can always construct a design involving m treatments and b blocks such that each treatment is replicated r times and each block has size $k/2$. Let N^* be the incidence matrix of this design. Then it can be seen that the equireplicate design, given by the incidence matrix (cf. (4.5))

$$\begin{bmatrix} I_\mu & 0 \\ I_\mu^* & 0 \\ 0 & I_\mu^* \\ 0 & I_\mu \end{bmatrix} N^*$$

and having parameters v, b, r, k, retains full information on both L_1 and L_1^1. □

REMARK 5.3. The design constructed in the sufficiency part of the above proof will be connected if N^* represents a connected design. From (2.16), (2.17), (2.20b) and (3.5), one can check that this construction ensures full efficiency on L_j and L_j^1 for every odd j. This point was noted by Gupta (1989) who also discussed the choice of N^* so as to ensure orthogonal estimation of L_p, L_j, L_j^1, $1 \leqslant j \leqslant m-1$, and explored the role of balanced factorial designs (Shah, 1960) in this context.

EXAMPLE 5.2. Let $m = 4$, $v = 8$, $b = 4$, $r = 3$, $k = 6$. Take $N^* = N_1$, where N_1 is given by (5.1). Then $\mu = 2$ and, following Theorem 5.2 and Remark 5.3, the design with incidence matrix

$$\begin{bmatrix} 0 & 1 & 1 & 0 & 1 & 1 & 1 & 1 \\ 1 & 0 & 0 & 1 & 1 & 1 & 1 & 1 \\ 1 & 1 & 1 & 1 & 1 & 0 & 0 & 1 \\ 1 & 1 & 1 & 1 & 0 & 1 & 1 & 0 \end{bmatrix}'$$

retains full information on each of L_1, L_1^1, L_3 and L_3^1. Also, it can be seen that this design ensures an efficiency factor 8/9 for each of L_p, L_2 and L_2^1. Furthermore, one can check that this design, and also the ones indicated in the preceding example, lead to orthogonal estimation of all these seven treatment contrasts; cf. Gupta (1989).

6. Symmetric parallel line assays: Q-designs

We continue with symmetric parallel line assays, now with m ($\geqslant 3$) doses for each preparation, and review equireplicate designs retaining full information on L_2 and L_2^1 in addition to L_p, L_1 and L_1^1. Such a design will be called a Q-design (cf. Mukerjee and Gupta, 1991). In so far as inference on these five contrasts is concerned, a Q-design, if available, will be optimal, in a very broad sense, over the equireplicate class (vide Remark 3.4).

Consider as before an arrangement of the v $(= 2m)$ treatments in b blocks such that each block has size k $(< v)$ and each treatment is replicated r times. Then $vr = bk$. The incidence matrix, N, of the design is partitioned as $N = (N_1' \; N_2')'$, where N_1 and N_2 are as in Section 4. By (2.20), (3.5), (4.1a), (4.3) and Remark 3.2, the design under consideration is a Q-design if and only if

$$1_m' N_i = \left(\frac{1}{2}k\right) 1_b', \qquad e_1' N_i = 0, \qquad e_2' N_i = 0 \quad (i = 1, 2). \tag{6.1}$$

The above is analogous to (4.4). Here e_1 is as given by (4.3) and by (2.15), (2.17),

$$e_2 = f_2 - \frac{1}{12}(m^2 - 1) 1_m,$$

f_2 being an $m \times 1$ vector with jth element $\{j - (m+1)/2\}^2$, $1 \leqslant j \leqslant m$. Clearly, a necessary condition for the existence of a Q-design is that k is even. Also, for any typical column of N_1 or N_2, say $(y_1, y_2, \ldots, y_m)'$, by (6.1) one must have

$$\left.\begin{aligned} &\sum_{j=1}^{m} y_j = \frac{1}{2}k, \quad \sum_{j=1}^{m} j y_j = \frac{1}{4}k(m+1), \\ &\sum_{j=1}^{m} \left\{ j - \frac{1}{2}(m+1) \right\}^2 y_j = \frac{1}{24}k(m^2 - 1). \end{aligned}\right\} \tag{6.2}$$

These considerations led Mukerjee and Gupta (1991) to suggest a construction procedure consisting of the following steps.

(i) For given v $(= 2m)$, k and r, search all possible non-negative integral valued solutions of (6.2) for $(y_1, \ldots, y_m)'$. If no solution is available then a Q-design with the given parameters is non-existent. Otherwise, let $\alpha_1, \ldots, \alpha_q$ be the possible solutions.

(ii) Find non-negative integers $u_1, \ldots, u_q$ such that $\sum_{i=1}^{q} u_i \alpha_i = r 1_m$. If no such $u_1, \ldots, u_q$ exist then a Q-design with the given parameters is non-existent. Otherwise, construct N_1 with columns $\alpha_1, \ldots, \alpha_q$ such that α_i is repeated u_i times $(1 \leqslant i \leqslant q)$, take $N_2 = N_1$ and $N = (N_1' \; N_2')'$. The incidence matrix so constructed will represent a Q-design with the given parameters.

In step (ii), the choice of $u_1, \ldots, u_q$, if any, may be non-unique. One should try to select $u_1, \ldots, u_q$ in such a way that the resulting design becomes connected.

Mukerjee and Gupta (1991) adopted the above approach to derive and tabulate Q-designs for all possible parameter values over the range $k < v$ and $v \leqslant 24$. They noted that it is always possible to ensure connectedness of these designs whenever $r > 1$.

EXAMPLE 6.1. For $v = 8$, $m = 4$, the last condition in (6.2) implies that $5k/8$ must be an integer. Hence no incomplete block Q-design exists.

EXAMPLE 6.2. Let $v = 14$, $m = 7$, $b = 7$, $r = 6$, $k = 12$. Then the possible non-negative integral valued solutions of (6.2) are

$$\alpha_1 = (0,3,0,0,0,3,0)', \qquad \alpha_2 = (1,1,0,1,2,0,1)'$$
$$\alpha_3 = (1,0,2,1,0,1,1)'.$$

Since $\alpha_1 + 3\alpha_2 + 3\alpha_3 = 61_m$, a Q-design exists and is given by the incidence matrix $N = (N_1' \; N_1')'$, where

$$N_1 = \begin{bmatrix} 0&1&1&1&1&1&1 \\ 3&1&1&1&0&0&0 \\ 0&0&0&0&2&2&2 \\ 0&1&1&1&1&1&1 \\ 0&2&2&2&0&0&0 \\ 3&0&0&0&1&1&1 \\ 0&1&1&1&1&1&1 \end{bmatrix}.$$

The design is connected.

7. Asymmetric parallel line assays

In analogy with the developments in Section 4, we now consider the problem of retaining full information on the preparation contrast (ψ_p), combined regression contrast (ψ_1) and parallelism contrast (ψ_1^1) in asymmetric parallel line assays. In the spirit of Gupta and Mukerjee (1990), an equireplicate design achieving this will be called a ψ-design.

Let there be m_1 and m_2 doses of the standard and test preparations respectively and suppose the assay is to be conducted in a design involving b blocks of size k $(< v = m_1 + m_2)$ each such that each of the v treatments is replicated r times. Here $bk = rv$. The incidence matrix, N, of the design is partitioned as $N = [N_1' \; N_2']'$, where N_i is of order $m_i \times b$ $(i = 1, 2)$. The rows of N_1 and N_2 correspond to the doses of the standard and test preparations respectively. Clearly,

$$1_{m_1}' N_1 + 1_{m_2}' N_2 = k 1_b'. \tag{7.1}$$

By (2.13), (3.5), (7.1) and Remark 3.2, the design under consideration is a ψ-design if and only if

$$1_{m_i}' N_i = (k m_i / v) 1_b',$$

$$\left\{ f^{(i)} - \frac{1}{2}(m_i + 1) 1_{m_i} \right\}' N_i = 0 \quad (i = 1, 2) \tag{7.2}$$

where $f^{(i)} = (1, 2, \ldots, m_i)'$, $i = 1, 2$. The conditions (7.2), due to Kyi Win and Dey (1980), generalize (4.4) to the asymmetric case. Kyi Win and Dey (1980) discussed the construction of designs satisfying (7.2) and presented a short table showing ψ-designs for $(m_1, m_2, b, r, k) = (3, 6, 3, 2, 6), (3, 9, 6, 4, 8), (4, 8, 4, 2, 6), (5, 10, 5, 2, 6), (6, 9, 6, 2, 5)$ and $(6, 9, 6, 4, 10)$.

EXAMPLE 7.1. Let $m_1 = 3$, $m_2 = 6$, $b = 3$, $r = 2$, $k = 6$. Then, as noted in Kyi Win and Dey (1980), a connected ψ-design is given by the incidence matrix

$$\begin{bmatrix} 1\ 0\ 1\ 1\ 1\ 0\ 0\ 1\ 1 \\ 1\ 0\ 1\ 1\ 0\ 1\ 1\ 0\ 1 \\ 0\ 2\ 0\ 0\ 1\ 1\ 1\ 1\ 0 \end{bmatrix}'.$$

In particular, if m_1 and m_2 are both even then, as a generalization of Theorem 4.1, we have the following result giving a complete solution to the existence problem of ψ-designs.

THEOREM 7.1. *Let $m_1 = 2\mu_1$ and $m_2 = 2\mu_2$ be both even and $vr = bk$. Then a ψ-design with parameters v $(= m_1 + m_2)$, b, r and k exists if and only if k is even and $k\mu_1/v$ is an integer.*

PROOF. To prove the necessity, suppose such a design exists and consider the associated matrices N_1 and N_2. Then for a typical column of N_1, say $(y_1, y_2, \ldots, y_{2\mu_1})'$, by (7.2),

$$\sum_{j=1}^{2\mu_1} y_j(j - \mu_1) = \frac{1}{2}\sum_{j=1}^{2\mu_1} y_j = k\mu_1/v. \tag{7.3}$$

The integrality of $k\mu_1/v$ follows from (7.3). Also by (7.3), the sum of elements in each column of N_1 is even. Similarly, the sum of elements in each column of N_2 is even. Hence by (7.1), k must be even.

The sufficiency part is proved by construction. Let k be even and $k\mu_1/v$ be an integer. Then $k\mu_2/v = k/2 - k\mu_1/v$ is also an integer. Also, $\mu_i r = b(k\mu_i/v)$, $i = 1, 2$. Hence for $i = 1, 2$, one can always construct a design involving μ_i treatments and b blocks such that every treatment is replicated r times and each block has size $k\mu_i/v$. Let M_i be the $\mu_i \times b$ incidence matrix of such a design. Then the incidence matrix of a ψ-design with the desired parameters is given by (cf. (4.5))

$$N = \begin{bmatrix} I_{\mu_1} & 0 \\ I^*_{\mu_1} & 0 \\ 0 & I^*_{\mu_2} \\ 0 & I_{\mu_2} \end{bmatrix} \begin{bmatrix} M_1 \\ M_2 \end{bmatrix}. \tag{7.4}$$

The above ψ-design will be connected if $M = [M_1' \ M_2']'$ represents the incidence matrix of a connected design. □

EXAMPLE 7.2. Let $m_1 = 4$, $m_2 = 8$, $b = 4$, $r = 2$, $k = 6$. Then the conditions of Theorem 7.1 are satisfied and with

$$M_1 = \begin{bmatrix} 1 & 0 & 1 & 0 \\ 0 & 1 & 0 & 1 \end{bmatrix}, \qquad M_2 = \begin{bmatrix} 0 & 0 & 1 & 1 \\ 0 & 1 & 0 & 1 \\ 1 & 0 & 1 & 0 \\ 1 & 1 & 0 & 0 \end{bmatrix}$$

in (7.4), the incidence matrix of a connected ψ-design is obtained as

$$\begin{bmatrix} 1 & 0 & 0 & 1 & 1 & 1 & 0 & 0 & 0 & 0 & 1 & 1 \\ 0 & 1 & 1 & 0 & 1 & 0 & 1 & 0 & 0 & 1 & 0 & 1 \\ 1 & 0 & 0 & 1 & 0 & 1 & 0 & 1 & 1 & 0 & 1 & 0 \\ 0 & 1 & 1 & 0 & 0 & 0 & 1 & 1 & 1 & 1 & 0 & 0 \end{bmatrix}'.$$

This design is available also in Kyi Win and Dey (1980).

Das and Saha (1986) described a systematic method for the construction of ψ-designs through affine resolvable designs. Their method yields designs with even m_1 and m_2 and hence Theorem 7.1 covers all parametric combinations where their method is applicable. Through the use of C-designs, they as well presented a method of construction of possibly non-equireplicate designs retaining full information on each of ψ_p, ψ_1 and ψ_1^1; see their original paper for the details of this method and also for a review of the work on asymmetric parallel line assays reported in the unpublished Ph.D. Thesis of Seshagiri (1974).

8. On the role of non-equireplicate designs

In Sections 4–7, we have surveyed the results on existence and construction of incomplete block designs retaining full information on various sets of contrasts that are of interest in parallel line assays. Almost all designs discussed in these sections are equireplicate and their optimality, with regard to inference on relevant contrasts and in the class of equireplicate designs, has been indicated in Remark 3.4. This, however, does not guarantee their optimality, even under specific criteria, when one relaxes the condition of equal replication and considers the class of all comparable designs including the non-equireplicate ones. Moreover, v, b and k may be such that no equireplicate design is available at all. Thus, given v, b and k, the problem of finding an optimal or efficient design for the contrasts of interest, within the class of all designs and not just the equireplicate class, is of importance. Such a study also helps in assessing the performance of the designs discussed earlier as members of the broader class of all designs.

Some results in this direction were recently reported by Mukerjee and Gupta (1995). They studied the problem of A-optimality (Shah and Sinha, 1989) in the set-up of

Section 4 where a symmetric parallel line assay, with m doses of each preparation, is conducted using b blocks of k ($< v = 2m$) experimental units each and interest lies in L_p, L_1 and L_1^1. By (2.20) and (4.3), the normalized contrasts corresponding to L_p, L_1 and L_1^1 are $g_1'\tau$, $g_2'\tau$ and $g_3'\tau$ respectively, where

$$\left.\begin{array}{l} g_1 = (2m)^{-1/2}(1_m' - 1_m)', \quad g_2 = [6/\{m(m^2-1)\}]^{1/2}(e_1'e_1')', \\ g_3 = [6/\{m(m^2-1)\}]^{1/2}(e_1' - e_1')'. \end{array}\right\} \tag{8.1}$$

Then interest lies in $G\tau$ where the $3 \times v$ matrix G is defined as $G = (g_1, g_2, g_3)'$. By (4.3) and (8.1), the diagonal elements of $G'G$ are $\theta_1, \theta_2, \ldots, \theta_v$, where

$$\theta_j = \theta_{m+j} = \frac{1}{2m} + \frac{12}{m(m^2-1)}\left\{j - \frac{1}{2}(m+1)\right\}^2, \quad 1 \leqslant j \leqslant m. \tag{8.2}$$

Note that for $1 \leqslant j \leqslant m$,

$$\theta_j = \theta_{m+1-j} = \theta_{m+j} = \theta_{2m+1-j}. \tag{8.3}$$

Let $\mathcal{D} \equiv \mathcal{D}(v, b, k)$ be the class of all designs which involve v treatments and are laid out in b blocks each of size k ($< v$). Let $\mathcal{D}_1$ be the subclass of $\mathcal{D}$ consisting of those designs which keep $G\tau$ estimable. An A-optimal design for $G\tau$ in $\mathcal{D}$ is one which belongs to $\mathcal{D}_1$ and minimizes $\mathrm{tr}\{\mathrm{Cov}(G\hat{\tau})\}$ over $\mathcal{D}_1$. For even m ($= 2\mu$) and $k \equiv 0 \pmod 4$, Mukerjee and Gupta (1995) suggested the following steps for the construction of an A-optimal design for $G\tau$ in $\mathcal{D}$:

Step 1. Minimize $\sum_{j=1}^{\mu}(\theta_j/q_j)$ with respect to $q = (q_1, \ldots, q_\mu)'$ such that the q_j's are positive integers satisfying $q_1 + \cdots + q_\mu = bk/4$. Let $(q_1^*, \ldots, q_\mu^*)'$ be a choice of q where this minimum is attained.

Step 2. Construct a design d^* involving μ treatments and b blocks each of size $k/4$ such that the jth treatment is replicated q_j^* times in d^*, $1 \leqslant j \leqslant \mu$.

Step 3. Finally, obtain a design d_0 from d^* by replacing the jth treatment in d^* by the four treatments s_j, s_{m+1-j}, t_j and t_{m+1-j} ($1 \leqslant j \leqslant \mu$), where, as before, $s_1, \ldots, s_m$ and $t_1, \ldots, t_m$ denote the doses of the standard and test preparations respectively.

The design d_0, so constructed, is A-optimal for $G\tau$ in $\mathcal{D}$. This follows noting that d_0 satisfies (3.5) and minimizes $\mathrm{tr}(GR^{-1}G')$ over $\mathcal{D}$ (cf. Lemma 3.1). The details, which involve utilization of (8.3) via an auxiliary result, are omitted to save space. The following example illustrates the construction procedure.

EXAMPLE 8.1. Let $m = 6$, $v = 12$, $b = 3$, $k = 8$. By (8.2), then $\theta_1 = 185/420$, $\theta_2 = 89/420$ and $\theta_3 = 41/420$. From Step 1, $(q_1^*, q_2^*, q_3^*)'$ is uniquely given by $q_1^* = 3$, $q_2^* = 2$, $q_3^* = 1$. Hence the blocks of d^* can be taken as $\{1,2\}$, $\{1,2\}$, $\{1,3\}$ following Step 2. Next, by Step 3, d_0 is given by the blocks $\{s_1, s_6, t_1, t_6, s_2, s_5, t_2, t_5\}$, $\{s_1, s_6, t_1, t_6, s_2, s_5, t_2, t_5\}$, $\{s_1, s_6, t_1, t_6, s_3, s_4, t_3, t_4\}$. The design d_0 is connected and

A-optimal for $G\mathcal{T}$ in $\mathcal{D}$. By Theorem 4.1, an L-design, $\bar{d}$, with parameters $v = 12$, $b = 3$, $r = 2$, $k = 8$ exists. Its A-efficiency, computed as

$$\mathrm{tr}\{\mathrm{Cov}_{d_0}(G\widehat{\mathcal{T}})\}/\mathrm{tr}\{\mathrm{Cov}_{\bar{d}}(G\widehat{\mathcal{T}})\},$$

equals 0.9344. Thus, under the A-criterion, the gain via the use of d_0 rather than that of $\bar{d}$ is not ignorable.

The case of odd m appears to be more difficult to handle than that of even m. For odd m and $k \equiv 0 \pmod 4$, Mukerjee and Gupta (1995) investigated designs which are highly A-efficient, if not A-optimal, for $G\mathcal{T}$ in $\mathcal{D}$. The associated construction procedure is similar to the one presented above for even m. They also discussed at length the issue of retaining connectedness. We refer to their paper for further details.

It is of interest to study the performance of L-designs under the A-criterion when the competing class is allowed to include non-equireplicate designs as well. Given v, b and k, if an L-design, say $\bar{d}$, exists then it is not hard to see that (cf. Lemma 3.1 (b))

$$\mathrm{tr}\{\mathrm{Cov}_{\bar{d}}(G\widehat{\mathcal{T}})\} = 3v\sigma^2/(bk),$$

where σ^2 is the error variance. On the other hand, by Lemma 3.1 (a) and a simple application of the Cauchy–Schwarz inequality, for any design $d \in \mathcal{D}$,

$$\mathrm{tr}\{\mathrm{Cov}_{d}(G\widehat{\mathcal{T}})\} \geqslant \left(\sum_{j=1}^{v} \theta_j^{1/2}\right)^2 \sigma^2/(bk),$$

where the θ_j's are as in (8.2). Hence, if an L-design exists then its A-efficiency, as a member of $\mathcal{D}$, for $G\mathcal{T}$ is at least as large as

$$E = \left(\sum_{j=1}^{v} \theta_j^{1/2}\right)^2 \Big/ (3v). \tag{8.4}$$

The quantity E, which depends only on v is always at least as large as 0.91. Thus, even as a member of $\mathcal{D}$, an L-design has a high A-efficiency although there remains a scope for further improvement (vide Example 8.1). Consequently, for any of those plentiful combinations (v, b, k) where A-optimal designs for $G\mathcal{T}$ in $\mathcal{D}$ are as yet unknown (e.g., for k not an integral multiple of 4), the use of an L-design, if available, can be recommended under the A-criterion without entailing a great loss of efficiency. For example, if $m = 5$, $v = 10$, $b = 5$, $k = 6$ then an A-optimal design for $G\mathcal{T}$ is as yet unknown but an L-design, constructed as in Example 4.2 and with A-efficiency at least 0.918 (see (8.4)), can be used.

In the same set-up as above, yet another pleasant property of L-designs becomes evident when one works with the MV-criterion (Shah and Sinha, 1989) under which the objective is to choose a design in $\mathcal{D}$ so as to keep $G\tau$ estimable and minimize $\max_{1 \leqslant i \leqslant 3} \text{var}(g_i' \widehat{\tau})$.

THEOREM 8.1. *Given v, b, and k $(< v)$, if an* L*-design exists then it is* MV*-optimal for $G\tau$ in $\mathcal{D}$.*

PROOF. By (8.1), the absolute value of each element of g_1 is $v^{-1/2}$. Hence, for any design in $\mathcal{D}$ which keeps $G\tau$ estimable, by Lemma 3.1 (a) and the Cauchy–Schwarz inequality,

$$\max_{1 \leqslant i \leqslant 3} \text{var}(g_i' \widehat{\tau}) \geqslant \text{var}(g_1' \widehat{\tau}) \geqslant (g_1' R^{-1} g_1)\sigma^2 \geqslant v\sigma^2/(bk). \tag{8.5}$$

For an L-design by Lemma 3.1 (b),

$$\text{var}(g_i' \widehat{\tau}) = v\sigma^2/(bk), \quad i = 1, 2, 3,$$

and from (8.5) the result follows. □

It is easily seen that equality cannot hold in (8.5) for a non-equireplicate design. Hence an L-design, if available, will be superior to all comparable non-equireplicate designs under the MV-criterion when interest lies in $G\tau$. Thus in Example 8.1, the MV-efficiency of the A-optimal non-equireplicate design d_0, computed relative to the L-design $\bar{d}$ as

$$\Big\{ \max_{1 \leqslant i \leqslant 3} \text{var}_{\bar{d}}(g_i' \widehat{\tau}) \Big\} \Big/ \Big\{ \max_{1 \leqslant i \leqslant 3} \text{var}_{d_0}(g_i' \widehat{\tau}) \Big\},$$

turns out to be 0.8182.

Interestingly, a result, analogous to Theorem 8.1, on the MV-optimality of Q-designs can be proved with reference to normalized versions of L_p, L_1, L_1^1, L_2 and L_2^1. The details are easy to work out.

9. Some open issues

In contrast with many other branches of experimental design, the development of an optimal design theory, taking due care of the contrasts of interest, has been somewhat slow in parallel line assays. Given v, b and k, as indicated earlier, most of the available results relate to the class of equireplicate designs and there is a need to find their counterparts considering the class of all designs. In particular, the following open issues deserve attention.

(a) In the set-up of Section 8, the basic idea underlying Mukerjee and Gupta's (1995) approach for finding A-optimal or A-efficient designs was as follows. They attempted to achieve orthogonality to block effects in so far as the estimation of $G\tau$ is concerned (cf. (3.5)) and simultaneously tried to choose the replication numbers suitably so as to ensure A-optimality or high A-efficiency. For $k \equiv 0 \pmod 4$, it would be of interest to know how far this approach is applicable in finding optimal designs under other criteria.

Under the A-criterion itself, there is a need to develop optimal designs when k is not an integral multiple of 4. Then it is difficult to employ the approach outlined above.

(b) Another open issue relates to the situation where, in addition to L_p, L_1 and L_1^1, specific interest lies also in L_2 and L_2^1 in a symmetric parallel line assay. A preliminary investigation reveals that then Q-designs, if available, tend to be highly A-efficient. This, however, needs to be probed in more detail. Furthermore, Q-designs are not as abundantly available as L-designs and for parametric combinations which do not admit a Q-design the related issue of ensuring optimality or high efficiency remains wide open.

(c) As for asymmetric parallel line assays, not many results are as yet known even with respect to the equireplicate class and much work remains to be done. It may be added in this context that use of asymmetric assays may be unavoidable in practice due to shortage of experimental material and that, in today's computing environment, their analysis should pose no special problem.

(d) The efficient or optimal planning of parallel line assays, with due cognizance of the contrasts of interest, in incomplete row-column designs deserves attention.

Acknowledgement

Thanks are due to the referees for very constructive suggestions. The work of RM was supported by a grant from the Centre for Management and Development Studies, Indian Institute of Management, Calcutta.

References

Caliński, T. (1971). On some desirable patterns in block designs. *Biometrics* **27**, 275–292.

Chakrabarti, M. C. (1962). *Mathematics of Design and Analysis of Experiments*. Asia Publishing, Bombay.

Das, A. D. (1985). Some designs for parallel line assays. *Calcutta Statist. Assoc. Bull.* **34**, 103–111.

Das, A. D. and G. M. Saha (1986). Incomplete block designs for asymmetrical parallel line assays. *Calcutta Statist. Assoc. Bull.* **35**, 51–57.

Das, M. N. and N. C. Giri (1986). *Design and Analysis of Experiments*, 2nd edn. Wiley Eastern, New Delhi.

Das, M. N. and G. A. Kulkarni (1966). Incomplete block designs for bio-assays. *Biometrics* **22**, 706–729.

Finney, D. J. (1978). *Statistical Method in Biological Assay*, 3rd edn. Charles Griffin, London.

Finney, D. J. (1979). Bioassay and the practice of statistical inference. *Internat. Statist. Rev.* **47**, 1–12.

Gupta, S. (1988). Designs for symmetrical parallel line assays obtainable through group divisible designs. *Comm. Statist. Theory Methods* **17**, 3865–3868.

Gupta, S. (1989). On block designs for symmetrical parallel line assays with even number of doses. *J. Statist. Plann. Inference* **21**, 383–389.

Gupta, S. and R. Mukerjee (1990). On incomplete block designs for symmetric parallel line assays. *Austral. J. Statist.* **32**, 337–344.

Gupta, V. K., A. K. Nigam and P. D. Puri (1987). Characterization and construction of incomplete block designs for symmetrical parallel line assays. *J. Indian Soc. Agric. Statist.* **39**, 161–166.

Hubert, J. J. (1984). *Bioassay*. Kendall/Hunt, Dubuque, IO.

Kshirsagar, A. M. and W. Yuan (1994). A unified theory for parallel line bioassays in incomplete block designs. Preprint.

Kulshreshtha, A. C. (1969). Modified incomplete block bioassay designs (Abstract). *Proc. Indian Sci. Congress, 56th Session* 33.

Kyi Win and A. Dey (1980). Incomplete block designs for parallel line assays. *Biometrics* **36**, 487–492.

Mukerjee, R. and S. Gupta (1991). Q-designs for bioassays. *Comput. Statist. Data Anal.* **11**, 345–350.

Mukerjee, R. and S. Gupta (1995). A-efficient designs for bioassays. *J. Statist. Plann. Inference*, to appear.

Nigam, A. K. and G. M. Boopathy (1985). Incomplete block designs for symmetrical parallel line assays. *J. Statist. Plann. Inference* **11**, 111–117.

Nigam, A. K., P. D. Puri and V. K. Gupta (1988). *Characterizations and Analysis of Block Designs*. Wiley Eastern, New Delhi.

Puri, P. D. and A. K. Nigam (1977). Partially efficiency balanced designs. *Comm. Statist. Theory Methods* **6**, 753–771.

Raghavarao, D. (1971). *Constructions and Combinatorial Problems in Design of Experiments*. Wiley, New York.

Saha, G. M. (1976). On Caliński's patterns in block designs. *Sankhyā B* **38**, 383–392.

Seshagiri, A. (1974). Some Contributions to Design and Analysis of Bioassays. Ph.D. Thesis, Indian Agric. Res. Inst., New Delhi.

Shah, B. V. (1960). Balanced factorial experiments. *Ann. Math. Statist.* **31**, 502–514.

Shah, K. R. and B. K. Sinha (1989). *Theory of Optimal Designs*. Springer, Berlin.

Tocher, K. D. (1952). The design and analysis of block experiments (with discussion). *J. Roy. Statist. Soc. Ser. B* **14**, 45–100.

S. Ghosh and C. R. Rao, eds., *Handbook of Statistics, Vol. 13*

Row–Column Designs

K. R. Shah and Bikas K. Sinha

1. Introduction

We will start by qualifying the term: Row–Column Designs. As Fisher (1935) visualized it, in the context of an agricultural experiment, Randomized Block Designs (RBDs) and Latin Square Designs (LSDs) are the simplest types of layout which fulfill the condition of supplying a valid estimate of error and at the same time possess the advantage of eliminating a substantial fraction of the soil heterogeneity. Also, as has been summarized by Hahn (1982), treatment designs use *blocking techniques* to remove the effects of extraneous factors from experimental error. Well-known treatment designs include RBD, BIBD to eliminate effect of a single extraneous factor, and LSD and YSD to remove effects of two extraneous factors. Sometimes, to differentiate these blocking factors from the primary factor (technically called treatment) of interest, the former are called *background factors* for blocking or grouping. Single grouping is furnished by one such background factor while double grouping is achieved by two such background factors. In this context, a row–column design refers to an experimental situation where there are two well-identified grouping or blocking factors, usually referred to as row and column (classifications). At this stage, we may also record some other terminologies used in the literature: Two-way grouping (Federer, 1955); double grouping (Cochran and Cox, 1957; Cox, 1958); randomization subject to double restriction (Fisher, 1935). In recent times these designs are commonly termed as *row–column* designs. These are designed to eliminate heterogeneity in the experimental material or units in two clearly stated directions, viz., along the directions of rows and columns, assuming that the units have been displayed in the form of a rectangle. Apart from the main factor of interest (termed treatment or treatment effects), since there are two other (blocking) factors, this design is also called a *three factor* design or a design having three-way classification. The impressive bibliography on experiment and treatment design by Federer and Ballam (1973) lists more than 700 articles, written prior to 1968, that are related to row–column designs! Of course, a majority of these articles deal with LSDs and some have a flavour of real life applications as well. Needless to say, there is well documented research on this fascinating topic. In this article, we will be guided by contemporary research interests as well as by our own understanding of the state of the art. We will start with a very general set-up and develop the subject matter as far as possible, in terms of the model and related inference. Then we will discuss various other related issues. Our dominant theme is to regard row–column designs as designs for estimation of treatment effects in face of two crossed nuisance factors. A secondary theme which pops up a few times is row–column designs as three factor designs.

At this stage, we may mention that we will resist the temptation of giving a detailed account of the optimality considerations in this context. Considerations of time and space have prevented us from giving a bibliography for the availability of useful designs for the various settings considered in this paper. We hope that we will not disappoint the readers.

2. Preliminaries: Model, estimability, connectedness and randomization

We start with the following set-up described in terms of an industrial experiment. Suppose the experimental units can be classified into pq classes, using a p-component factor A and a q-component factor B. These are the so-called blocking factors. (At this stage we need *not* distinguish between qualitative and quantitative blocking factors.) The units can then be displayed in the form of a $p \times q$ matrix with n_{ij} units in the (i, j)th cell; $1 \leqslant i \leqslant p$; $1 \leqslant j \leqslant q$. We keep the possibility of n_{ij} being 0 or 1, and/or being unequal among themselves. We say that we have a row–column layout of the experimental units. When n_{ij}'s satisfy the orthogonality condition viz.,

$$n_{ij} = n_{i.}n_{.j/n..} \quad \text{where}$$

$$n_{i.} = \sum_j n_{ij}, \quad n_{.j} = \sum_i n_{ij} \text{ and } n_{..} = \sum_i \sum_j n_{ij}, \tag{2.1}$$

we call it an *orthogonal* row–column set-up. Otherwise, it is known as a *non-orthogonal* row–column set-up. Of all orthogonal structures, the one with all n_{ij}'s equal to 1 deserves special mention. This along with the case: $p = q$, has been thoroughly studied in the literature. Latin Square designs are based on such set-ups.

We will assume that there are v treatments to be compared using these experimental units. To start with we will further assume that the blocking factors A and B comprise of p and q fixed components respectively and that the treatments under comparison are the sole objects of interest to us. Technically, then, we are dealing with what is called a *fixed effects model*. At this stage we have to introduce the statistical design and the underlying model for extracting information on treatment effects, after eliminating effects due to the blocking factors, if any. An experimental design in this context is a plan for allocation of treatments to the experimental units. Data here relate to the observations made on the experimental units, after these have been treated with the treatments according to a given design. A statistical model seeks to explain the random long term behaviour of the observations underlying a given design. The model most suited to this case, to start with, is a kind of *semi-additive model*:

$$y_{ijh \cdot s} = \mu + (A_i \otimes B_j) + \tau_h + \varepsilon_{ijh \cdot s} \tag{2.2}$$

where

$$\mu = \text{general effect;}$$
$$A_i \otimes B_j = \text{joint effect of factors } A \text{ and } B \text{ at combination } (i, j);$$
$$\tau_h = \text{effect of treatment } h;$$

$$1 \leqslant i \leqslant p, \quad 1 \leqslant j \leqslant q, \quad 1 \leqslant h \leqslant v, \quad 1 \leqslant s \leqslant n_{ijh}. \tag{2.3}$$

Here we have assumed that n_{ijh} is the number of experimental units allotted to the treatment h corresponding to the factor combination (i, j) and only when this is positive, s runs through $1, 2, \ldots, n_{ijh}$. Lastly, the term $\varepsilon_{ijh.s}$ refers to the random error component in the model. We assume, as usual, that the errors satisfy the homoscedasticity condition so that they have mean zero, variance σ^2 and there is no correlation.

It is then seen that *without any further assumption*, the observations may be regarded as undergoing a model for a traditional block design in that we can visualize the presence of *pq* blocks and the experimental units as classified into these blocks, serially numbered in the lexico-graphic order by the pairs (i, j), there being n_{ij} experimental units in the block numbered (i, j). Once this is recognized, the problem of estimation of the treatment effects parameters (in the semi-additive model) is readily resolved, as we are in the framework of a block-treatment additive model. Thus, for example, it follows that the only estimable linear functions of the treatment effects are the treatment contrasts. Further, as in a block design, connectedness can be immediately resolved, by using Bose's chain definition. In this context, we may mention that we have come across another easy but very powerful technique for checking connectedness (in a row–column design) which we prefer to call: *Collapsing Technique*. Park and Shah (1995) have elaborated on this technique in the context of a row–column design. We will explain this in the context of a block design.

We know that under an additive model, the treatments occurring within a block are connected to one another. Park and Shah (1995) advocates that at stage 1, we can treat all such treatments within a block as forming *one equivalence class* and replace all of them by one single member, preferably by the lowest serial number in that class. In every block, thus, we end up with one member representative and we *update* it as we move from one block to the other. Connectedness would then amount to achieving, at the end, only one number, viz., 1, as equivalent to all others.

This is a very simple yet very powerful tool to examine connectedness in higher way layouts. In case of a row–column design, this technique is equally applicable and at least in stage 1, we can form and update the equivalence classes by regarding the cells as blocks in the manner discussed above. Usually, we find that in a row–column design, there are *additional* model assumptions, viz, that the joint effect term $A_i \otimes B_j$ takes the form $\alpha_i + \beta_j$, for all such combinations of i and j. In view of this, some *additional* expansion takes place in the description of equivalence classes, thereby rendering them into bigger and bigger sizes. Ultimately, therefore, if we arrive at one single class, we call the resulting design a *treatment-connected* design.

The reader, familiar with row–column designs, will realize that connectedness is not easy to verify for such designs. See, for example, Eccleston and Russell (1975).

At stage 2, to achieve additional expansion of the equivalence classes, the collapsing technique uses what are called *tetra-differences*. To form a tetra-difference, we pick up two rows, say i and i', and two columns, say j and j'. Assuming that each of n_{ij}, $n_{ij'}, n_{i'j}$, and $n_{i'j'}$ is positive, we take one observation from each combination of rows and columns. Based on these four observations, we form an observational contrast to eliminate the row factor and the column factor effects (this being possible since these are now assumed to be additive) and to leave only a treatment contrast (or zero) in the expectation of the observational contrast. Such observational contrasts are called tetra-differences. This treatment contrast (along with other such treatment contrasts, derived out of other choices of pairs of rows and columns), may then generate expansion in the size of the already existing equivalence classes and thereby lead towards connectedness.

At this stage we must mention that there are some finer points in the analysis of tetra differences. We may explain this by taking an example.

EXAMPLE 1. Consider the following layout for a 4×4 row–column design involving 16 treatments.

1,1,2	3,4	10,11,12,12,12	13
5,6,6,7	8,8,9	15	16,16
16	13,14	1,2	6,7,7
15	12	4	8,8,8,9

As explained before, at stage 1, we work out the following equivalence classes:

Class 1: $1 = 2$ Class 2: $3 = 4$ Class 3: $5 = 6 = 7$ Class 4: $8 = 9$
Class 5: $10 = 11 = 12$ Class 6: $13 = 14$ Class 7: 15 Class 8: 16

Hence we obtain the following updated matrix:

$$\begin{pmatrix} 1 & 3 & 10 & 13 \\ 5 & 8 & 15 & 16 \\ 16 & 13 & 1 & 5 \\ 15 & 10 & 3 & 8 \end{pmatrix}$$

Now we are set to try out the tetra-differences by choosing pairs of rows and columns.

We will simply indicate the treatment contrasts thus formed by examining appropriate tetra-differences: $5+5-16-16$ and $3+3-10-10$. These lead to the expansion $5 = 16$ and $3 = 10$. Incorporating these into the matrix, we update it as:

$$\begin{pmatrix} 1 & 3 & 3 & 13 \\ 5 & 8 & 15 & 5 \\ 5 & 13 & 1 & 5 \\ 15 & 3 & 3 & 8 \end{pmatrix}$$

Now again we look for tetra-differences and try to extract more information as follows. From $3+15-8-3$, $1+13-5-3$ and $1+15-5-3$, we derive: $8 = 13 = 15$. Finally, using the tetra-difference: $1 + 8 - 15 - 13$, we derive: $1 = 8 = 13 = 15$.

At this stage, the matrix has been reduced to:

$$\begin{pmatrix} 1 & 3 & 3 & 1 \\ 5 & 1 & 1 & 5 \\ 5 & 1 & 1 & 5 \\ 1 & 3 & 3 & 1 \end{pmatrix}$$

A little reflection would reveal that no further reduction is possible and this would mean that we would have only the following equivalence classes:

Class 1: $1 = 2 = 8 = 9 = 13 = 14 = 15$;
Class 2: $3 = 4 = 10 = 11 = 12$;
Class 3: $5 = 6 = 7 = 16$.

All elementary contrasts within each class are estimable but no between class elementary contrast is estimable. Formation of additional tetra-differences based on the last table (shown above) result into either the so-called zero functions or the contrast: $2\tau_1 - \tau_3 - \tau_5$. Thus, in effect, altogether 14 linearly independent treatment contrasts are estimable, thereby making the design deficient by one more contrast, and hence, disconnected with respect to the treatment comparisons. It is also easy to see that the non-estimable treatment contrast $(\tau_3 + \tau_4 + \tau_{10} + \tau_{11} + \tau_{12})/5 - (\tau_5 + \tau_6 + \tau_7 + \tau_{16})/4$ is orthogonal to *every* within classes treatment contrast. (Orthogonality here is in the usual algebraic sense.)

As Park and Shah (1995) have demonstrated, this technique is very powerful and leads to a stage by stage examination of a row–column design so far as connectedness with respect to treatment comparisons is concerned. Also at the final stage it becomes possible to characterize explicitly *non-estimable* treatment contrasts which are orthogonal to all estimable treatment contrasts. The rule to form them is first to form a basis involving the representative members of the equivalence classes and then to replace every such representative member by the corresponding class average. As Park and Shah (1995) have pointed out, this method is also applicable quite easily in row–column designs with some empty cells, i.e., in situations where some of the n_{ij}'s are zeroes. We fill in these cells by new and different fictitious treatments and examine connectedness of the resulting design as above with respect to all the treatments. If, in the process, all the old treatments turn out to have been captured by one and the same equivalence class, then we are done! We will discuss more about connectedness later.

Before concluding the section, some remarks regarding randomization would be in order. As repeatedly emphasized by Fisher, appropriate randomization is important in designed experiments. In row–column designs, this appears to be feasible only if the layout is symmetrical in the rows and the columns, i.e., n_{ij} is independent of i and j. In this case, for a fixed design, we could permute the rows and the columns independently of each other. This randomization leads to an induced distribution which (under the assumption of additivity of plot and treatment effects) can be used for estimating the treatment effects. This is given in detail for the important special case of $n_{ij} = 1$ in Roy and Shah (1961). When n_{ij} depends on (i, j), appropriate and valid randomization appears to be very difficult to visualize.

3. Analysis of data: Estimation of treatment contrasts

Under the semi-additive fixed effects model, the analysis of data would essentially correspond to the intrablock analysis of a block design. Thus, the $\boldsymbol{C}$-matrix, defined in the usual manner as $\boldsymbol{C} = \boldsymbol{r}^{\delta} - \boldsymbol{N}\boldsymbol{k}^{-\delta}\boldsymbol{N}'$ (where $\boldsymbol{k}^{\delta}$ is the diagonal matrix of block sizes) would play a major role in extracting information on treatment contrasts. Note that in the above, the matrix N is the incidence matrix of the v treatments in the $p \times q$ row–column combinations of the factors A and B so that its elements are of the form $\boldsymbol{N} = ((n_{ijh}))'$.

As said before, generally, in a row–column design, the row–column factors are assumed to have additive effects so that it is a case of three-way additive model. Explicitly stated, the model is:

$$y_{ijh.s} = \mu + \alpha_i + \beta_j + \tau_h + \varepsilon_{ijh.s}. \tag{3.1}$$

In this case, all the three incidence matrices viz., $\boldsymbol{N}_{rc}$ = row–column incidence matrix, $\boldsymbol{N}_{rt}$ = row-treatment incidence matrix and $\boldsymbol{N}_{ct}$ = column-treatment incidence matrix, are relevant for subsequent studies.

We note that a row–column design is said to have a row–column *orthogonal* structure if and only if the row–column incidence matrix satisfies the condition of orthogonality described above in (2.1). Along similar lines, it will be said to have *totally orthogonal* structure if and only if all the three incidence matrices satisfy respective conditions of orthogonality. This is the happiest situation indeed for data analysis! However, this is rarely the case in practice and most often, we have to be content with orthogonality of the row–column structure alone. One simple way to visualize this is the case of: $n_{ij} = k$ ($\geqslant 1$) for all i, j. This is the case of equal number of observations per cell. When $p = q = v$ and $k = 1$, we are in the set-up of a Latin Square design.

When $p = q = n$ and $v = nk$, we are in the set-up of a semi-Latin square design provided that $\boldsymbol{N}_{rt} = \boldsymbol{N}_{ct} = ((1))$. One simple way to construct a semi-Latin square design with the above parameters is to start with an $n \times n$ Latin square design and replace *each* symbol by k *new* symbols. This results in what is known as *inflated* Latin square design. There is a family of semi-Latin squares called *Trojan* squares. As to the statistical Analysis of data arising out of such squares, there are two reasonable separate models studied in the literature. We will not discuss this topic any further. A very useful reference is Bailey (1992).

Recall that in a block design, the sources of information on treatment contrasts are the within block observational contrasts. These along with the contrasts formed of the block totals comprise the total number of observational contrasts. Similarly, in a row–column design, the sources of information on the treatment contrasts are the within cell observational contrasts and the tetra-differences among the cell means. These along with the contrasts formed of the row totals and the column totals comprise all observational contrasts. Therefore, while analysing the data, in order to extract information on the treatment contrasts, we need to concentrate only on the within cell observational contrasts and the tetra-differences. These will then undergo a Gauss–Markov set-up with the treatment effects as the only parameters in the model (apart from the error variance σ^2). We can now work out the expression for the information matrix which is the well-known $\boldsymbol{C}$-matrix (apart from the multiplier σ^{-2}). We say that we have eliminated the effects of rows and columns. Note that, unless some further assumptions are made, the row total contrasts and the column total contrasts do not contain any further information on the treatment contrasts. (See Section 9 for further details.)

We will now display the form of the information matrix for treatment contrasts in a three-way layout when the effects of the additive blocking factors (row and column effects) are eliminated. See Agrawal (1966a) for details.

$$\begin{aligned} \boldsymbol{C}_t = {} & (\boldsymbol{r}^{\delta} - \boldsymbol{N}_{tr}\boldsymbol{R}^{-\delta}\boldsymbol{N}_{rt}) - (\boldsymbol{N}_{tc} - \boldsymbol{N}_{tr}\boldsymbol{R}^{-\delta}\boldsymbol{N}_{rc})(\boldsymbol{C}^{\delta} - \boldsymbol{N}_{cr}\boldsymbol{R}^{-\delta}\boldsymbol{N}_{rc})^{-} \\ & \times (\boldsymbol{N}_{ct} - \boldsymbol{N}_{cr}\boldsymbol{R}^{-\delta}\boldsymbol{N}_{rt}) \end{aligned} \tag{3.2}$$

where

$$\begin{aligned} &\boldsymbol{N}_{cr} = \boldsymbol{N}'_{rc}, \qquad \boldsymbol{N}_{tr} = \boldsymbol{N}'_{rt}, \qquad \boldsymbol{N}_{tc} = \boldsymbol{N}'_{ct}, \\ &\boldsymbol{N}_{rc}\mathbf{1} = \boldsymbol{N}_{rt}\mathbf{1} = \boldsymbol{R}, \qquad \boldsymbol{N}_{cr}\mathbf{1} = \boldsymbol{N}_{ct}\mathbf{1} = \boldsymbol{C}, \\ &\boldsymbol{N}_{tr}\mathbf{1} = \boldsymbol{N}_{tc}\mathbf{1} = \boldsymbol{r}, \qquad \boldsymbol{r}^{\delta} = \mathrm{Diag}(r_1, r_2 \cdots), \\ &\boldsymbol{R}^{\delta} = \mathrm{Diag}(R_1, R_2 \cdots), \quad \boldsymbol{C}^{\delta} = \mathrm{Diag}(C_1, C_2 \cdots), \end{aligned}$$

$$\begin{aligned} &R_i = n_{i.}, \qquad C_j = n_{.j}, \\ &r_h = \text{ replication of } h\text{th treatment}, \\ &\boldsymbol{r}^{-\delta} = (\boldsymbol{r}^{\delta})^{-1}, \qquad \boldsymbol{R}^{-\delta} = (\boldsymbol{R}^{\delta})^{-1}, \\ &\boldsymbol{C}^{-\delta} = (\boldsymbol{C}^{\delta})^{-1}, \qquad \mathbf{1} = (1, 1, \cdots)' \text{ of appropriate order} \\ &\text{and } (\cdots)^{-} \text{ refers to a } g\text{-inverse of the matrix } (\cdots). \end{aligned} \tag{3.3}$$

The $\boldsymbol{C}$-matrix is an *nnd* matrix with row and column sums zeroes. It has rank $(v-1)$ if and only if the design is treatment-connected (i.e., connected with respect to the treatment contrasts). And in that case, for the treatment contrast $\boldsymbol{\ell}'\boldsymbol{\tau}$, with $\boldsymbol{\ell}'\mathbf{1} = 0$, the best linear unbiased estimator is given by $\boldsymbol{\ell}'\widehat{\boldsymbol{\tau}}$ with variance given by $\sigma^2(\boldsymbol{\ell}'\boldsymbol{C}_t^{+}\boldsymbol{\ell})$ where $\boldsymbol{C}_t^{+}$ = Moore–Penrose g-inverse of $\boldsymbol{C}_t$ and $\widehat{\boldsymbol{\tau}}$ satisfies equality $\boldsymbol{C}_t\widehat{\boldsymbol{\tau}} = \boldsymbol{Q}_t$ where $\boldsymbol{Q}_t$ is the vector of adjusted treatment totals. In case of a block design, it is known that $\boldsymbol{Q}_t = \boldsymbol{T} - \boldsymbol{N}\boldsymbol{k}^{-\delta}\boldsymbol{B}$ (in usual notations). In case of a row–column design, the analogous expression is given by $\boldsymbol{Q}_t = \boldsymbol{T} - \boldsymbol{N}_{tr}\boldsymbol{R}^{-\delta}\boldsymbol{R}_T - (\boldsymbol{N}_{tc} - \boldsymbol{N}_{tr}\boldsymbol{R}^{-\delta}\boldsymbol{N}_{rc})$

$(\boldsymbol{C}^{\delta} - \boldsymbol{N}_{cr}\boldsymbol{R}^{-\delta}\boldsymbol{N}_{rc})^{-}(\boldsymbol{C}_T - \boldsymbol{N}_{cr}\boldsymbol{R}^{-\delta}\boldsymbol{R}_T)$ where $\boldsymbol{T}$ = vector for treatment row totals; $\boldsymbol{R}_T$ = vector of row totals and $\boldsymbol{C}_T$ = vector of column totals. (See Agrawal, 1966a.)

We may note the computational difficulties to be encountered while dealing with an arbitrary row–column design! If it is not connected, then not all treatment contrasts can be estimated. Again, even if it is connected, unless it has a nice structure, at least in the sense of having a manageable form of the row–column incidence pattern, the computation of an explicit algebraic expression for the $\boldsymbol{C}$-matrix is cumbersome. This would make our understanding of the performance of the design very obscure! It thus has been emphasized that simplicity in data analysis can be achieved at least in situations where the row–column layout is really simple! This is, indeed, achieved by an orthogonal row–column structure. In such a case, the matrix $(\boldsymbol{C}^{\delta} - \boldsymbol{N}_{cr}\boldsymbol{R}^{-\delta}\boldsymbol{N}_{rc})^{-}$ in the expression for $\boldsymbol{C}_t$ in (3.2) can be taken as $\boldsymbol{C}^{-\delta}$. This reduces $\boldsymbol{C}_t$ in (3.2) to

$$\begin{aligned}\boldsymbol{C}_t = {} & (\boldsymbol{r}^{\delta} - \boldsymbol{N}_{tr}\boldsymbol{R}^{-\delta}\boldsymbol{N}_{rt}) \\ & - (\boldsymbol{N}_{tc} - \boldsymbol{N}_{tr}\boldsymbol{R}^{-\delta}\boldsymbol{N}_{rc})\boldsymbol{C}^{-\delta}(\boldsymbol{N}_{ct} - \boldsymbol{N}_{cr}\boldsymbol{R}^{-\delta}\boldsymbol{N}_{rt})\end{aligned} \tag{3.4}$$

where

$$\boldsymbol{N}_{rc} = \boldsymbol{\gamma}\boldsymbol{\eta}' \quad \text{for some } \boldsymbol{\gamma} \text{ and } \boldsymbol{\eta} \tag{3.5}$$

At this stage, it would be interesting to examine if (3.4) can be further simplified. It turns out that, upon simplification, $\boldsymbol{C}_t$ reduces to

$$\boldsymbol{C}_t = \boldsymbol{r}^{\delta} - \boldsymbol{N}_{tr}\boldsymbol{R}^{-\delta}\boldsymbol{N}_{rt} - \boldsymbol{N}_{tc}\boldsymbol{C}_t^{-\delta}\boldsymbol{N}_{ct} + \frac{\boldsymbol{r}\boldsymbol{r}'}{n_{..}} \tag{3.6}$$

where

$$n_{..} = \boldsymbol{1}'\boldsymbol{R} = \boldsymbol{1}'\boldsymbol{C} = \boldsymbol{1}'\boldsymbol{r} = \sum_i \sum_j n_{ij}. \tag{3.7}$$

Shrikhande (1951) deduced (3.6) when $\boldsymbol{N}_{rc} = ((1))$.

This is the well-known form of $\boldsymbol{C}_t$ for orthogonal row–column classification. Turning back to (3.2), when (3.5) fails to hold, one natural desire would be to achieve

$$\boldsymbol{C}_t = \boldsymbol{r}^{\delta} - \boldsymbol{N}_{tr}\boldsymbol{R}^{-\delta}\boldsymbol{N}_{rt}, \tag{3.8}$$

if possible. Once achieved, it would have an obvious explanation. This is that the $\boldsymbol{C}_t$ matrix in the three-factor additive model would coincide with the $\boldsymbol{C}_t$ matrix in the row-treatment (two-factor) additive model, ignoring the column effects. Further, treatment-connectedness in the three-factor design would be ensured by the connectedness of the row-treatment two-factor design. To achieve (3.8), while row–column connectedness is granted, it is n.s. that

$$\boldsymbol{N}_{tc} = \boldsymbol{N}_{tr}\boldsymbol{R}^{-\delta}\boldsymbol{N}_{rc} \tag{3.9}$$

holds. In the presence of (3.5), (3.9) holds iff $\boldsymbol{N}_{tc} = (\boldsymbol{\eta}'\mathbf{1})^{-1}\boldsymbol{r}\boldsymbol{\eta}'$. Of course, if both row–column and treatment-column classifications are orthogonal, then the $\boldsymbol{C}_t$ matrix coincides with (3.8), i.e., the column classification can be ignored! Finally, by interchanging the roles of rows and columns, it is clear that under the condition

$$\boldsymbol{N}_{tr} = \boldsymbol{N}_{tc}\boldsymbol{C}^{-\delta}\boldsymbol{N}_{cr} \tag{3.10}$$

the $\boldsymbol{C}_t$ matrix would simplify to

$$\boldsymbol{C}_t = \boldsymbol{r}^{\delta} - \boldsymbol{N}_{tc}\boldsymbol{C}^{-\delta}\boldsymbol{N}_{ct}. \tag{3.11}$$

Again, (3.10) would be satisfied if the row-treatment and row–column classifications are orthogonal (so that this time the row classification can be ignored). In summary, row–column orthogonality implies (3.6); row–column connectedness with (3.9) implies (3.8); row–column connectedness with (3.10) implies (3.11). We may mention in passing that both (3.9) and (3.10) have interesting statistical interpretation. This and other related issues will be discussed in the next section.

4. Further aspects of connectedness and analysis

If in a row–column design, the treatment contrasts are the only comparisons of interest, we will be satisfied with treatment-connectedness alone. As has been noted before, Park–Shah technique comes handy in this regard. Is there any alternative technique to this verification and of computation of variances of the estimates of treatment effects (at least in case where the $\boldsymbol{C}_t$ matrix does not lend itself to a simplified form!)? We examine this below. Suppose that the row–column classification is disconnected. As is well-known, this would then imply that there exists a stratification of the rows as well as of the columns and that they decompose into a number of connected row–column components. In such a situation, to examine treatment connectedness, we would apply Park–Shah technique in each such component and then combine all the resulting evidences. Also the final $\boldsymbol{C}_t$ matrix would be the algebraic sum of the component $\boldsymbol{C}_t$ matrices. (See Section 7 for more details in this context.) Therefore, to make our arguments simple, we will assume, without loss, that our basic premise for further discussion is the connectedness of the row–column classification of the experimental units. We can now address the question of treatment connectedness and examine an alternative procedure for its verification and for computation of the variances. Consider the following statements:

(i) Row–column classification is connected
(ii) Row-treatment classification is connected
(iii) Column-treatment classification is connected
(iv) $\boldsymbol{N}_{rc} = \boldsymbol{N}_{rt}\,\boldsymbol{r}^{-\delta}\boldsymbol{N}_{tc}$
(v) $\boldsymbol{N}_{rt} = \boldsymbol{N}_{rc}\,\boldsymbol{C}^{-\delta}\boldsymbol{N}_{ct}$
(vi) $\boldsymbol{N}_{ct} = \boldsymbol{N}_{cr}\,\boldsymbol{R}^{-\delta}\boldsymbol{N}_{rt}$

Momentarily, let us not be bothered by the implications of (iv)–(vi). To start with, we will assume (i)–(iii) to hold. Note that (ii) and (iii) are necessary for treatment connectedness. Since (v) is a version of (3.10) and (vi) is a version of (3.9), it is clear that (v) or (vi) implies treatment-connectedness. Eccleston and Russell (1975) have asserted that (iv) also implies treatment connectedness. However, when (iv) holds, there is no obvious simplification in the expression for the $\boldsymbol{C}_t$ matrix in (3.2). But then we can avoid computation of $\boldsymbol{C}_t$ altogether while deriving expressions for the blue's of the treatment contrasts and their variances. This we explain below in detail.

Not to obscure the essential steps of reasoning, we will now present the full information matrix of the parameters in the three-way additive model (apart from the multiplier σ^2).

$$\boldsymbol{X}'\boldsymbol{X} = \begin{bmatrix} n & \boldsymbol{R}' & \boldsymbol{C}' & \boldsymbol{r}' \\ & \boldsymbol{R}^{\delta} & \boldsymbol{N}_{rc} & \boldsymbol{N}_{rt} \\ & & \boldsymbol{C}^{\delta} & \boldsymbol{N}_{ct} \\ & & & \boldsymbol{r}^{\delta} \end{bmatrix}. \tag{4.1}$$

In the above, we have been guided by the general Gauss–Markov set-up $(\boldsymbol{Y}, \boldsymbol{X}\boldsymbol{\beta}, \sigma^2\boldsymbol{I})$; joint information matrix: $(\boldsymbol{X}'\boldsymbol{X})$. Note that the normal equations are given in the standard notation: $(\boldsymbol{X}'\boldsymbol{X})\boldsymbol{\beta} = \boldsymbol{X}'\boldsymbol{Y}$.

From the above, it follows that the joint information matrix for the row and column effects (eliminating treatment effects) is given by

$$\left(\begin{array}{c|c} \boldsymbol{R}^{\delta} - \boldsymbol{N}_{rt}\boldsymbol{r}^{-\delta}\boldsymbol{N}_{tr} & \boldsymbol{N}_{rc} - \boldsymbol{N}_{rt}\boldsymbol{r}^{-\delta}\boldsymbol{N}_{tc} \\ \hline & \boldsymbol{C}^{\delta} - \boldsymbol{N}_{ct}\boldsymbol{r}^{-\delta}\boldsymbol{N}_{tc} \end{array}\right). \tag{4.2}$$

From this it is now evident that when (iv) holds, $\boldsymbol{N}_{rc} - \boldsymbol{N}_{rt}\boldsymbol{r}^{-\delta}\boldsymbol{N}_{tc}$ reduces to a null matrix and hence, the information matrices for row and column effects are respectively given by $\boldsymbol{R}^{\delta} - \boldsymbol{N}_{rt}\boldsymbol{r}^{-\delta}\boldsymbol{N}_{tr}$ and $\boldsymbol{C}^{\delta} - \boldsymbol{N}_{ct}\boldsymbol{r}^{-\delta}\boldsymbol{N}_{tc}$. This is a case of what is called *adjusted orthogonality*, meaning thereby that the row contrasts estimates and the column contrasts estimates are *orthogonal* (in the sense of possessing *zero* covariance) when these are *adjusted* for the treatment effects (Eccleston and Russell, 1975, 1977). Hence, we may solve for estimates of row and column contrasts in a routine manner. In order to solve for the blue's of the treatment contrasts, we now refer to the normal equations and derive the following:

$$\begin{aligned} \bar{T}_h &\mathrel{\widehat{=}} \mu + \bar{\alpha}_h + \bar{\beta}_h + \tau_h, \\ \bar{T}_{h'} &\mathrel{\widehat{=}} \mu + \bar{\alpha}_{h'} + \bar{\beta}_{h'} + \tau_{h'}. \end{aligned} \tag{4.3}$$

In the above, $\bar{T}_h$ refers to the hth treatment average; $\bar{\alpha}_h$ refers to the average of the row effects corresponding to the hth treatment, etc.; $\cdots \mathrel{\widehat{=}} \cdots$ stands for an estimating equation.

From this it now follows that the blue of $\tau_h - \tau_{h'}$ is obtainable by taking the difference of the above two expressions and substituting for the row and column contrasts ($\bar{\alpha}_h - \bar{\alpha}_{h'}$ and $\bar{\beta}_h - \bar{\beta}_{h'}$) the corresponding blue's derived earlier. Further, because of adjusted orthogonality, the variances of the blue's of the treatment contrasts can be easily worked out. (Recall that treatment averages are uncorrelated with the estimates of the row and column contrasts which are themselves uncorrelated.) This is the point we wished to make earlier.

In the above, we have only discussed the problems of estimation of treatment contrasts and computations of their variances. It must be noted that the same procedure applies for estimation of row effects contrasts and the column effects contrasts. We examine connectedness of the design with respect to the row effects and the column effects in the same manner as that of the treatments. Finally, in order to estimate the error variance σ^2, we may proceed in the usual manner. This calls for computation of the residual sum of squares. This can be obtained in a routine manner once the solution to $\widehat{\boldsymbol{\tau}}$ has been reached. See Chakrabarti (1962).

In Section 7, we shall discuss the case where (i) fails to hold.

Below we present the ANOVA table for the sake of completeness. We assume the row–column classification to be connected.

ANOVA table

Sources of variation	df	SS	MS	F-ratio
Rows (unadjusted)	$p-1$	SSR	–	–
Columns (eliminating rows)	$q-1$	SSC*	–	–
Treatments (eliminating rows and columns)	$v-t$	SST*	SS $\div$ df	F-ratio = MS (treatment)/MSE
Error	by subtraction	by subtraction	SS $\div$ df	
Total	$n-1$	TSS	–	–

rank $(\boldsymbol{C}_t) = v - t$; error df $= n - p - q - v + t + 1$;
SSR = unadjusted row sum of squares;
SSC* $= (\boldsymbol{C}_T - \boldsymbol{N}_{cr}\boldsymbol{R}^{-\delta}\boldsymbol{R}_T)'(\boldsymbol{C}^{\delta} - \boldsymbol{N}_{cr}\boldsymbol{R}^{-\delta}\boldsymbol{N}_{rc})^{-}(\boldsymbol{C}_T - \boldsymbol{N}_{cr}\boldsymbol{R}^{-\delta}\boldsymbol{R}_T)$
SST* $= \hat{\boldsymbol{\tau}}'\boldsymbol{Q}_t$
TSS $= \sum\sum\sum\sum y_{ijh.s}^2 - CF,\quad CF = \frac{G^2}{n},\quad G = \sum\sum\sum\sum y_{ijh.s}$
$\boldsymbol{R}_T$ = vector of row totals, $\boldsymbol{C}_T$ = vector of column totals.

Before concluding this section, we wish to make four more points. We will take them up one by one.

(I) The concept of adjusted orthogonality as pronounced above is what is termed *strict* orthogonality (Khatri and Shah, 1986). This differs from orthogonality between rows (adjusted for columns *and* treatments) and columns (adjusted for rows *and* treatments). This later concept is a *weaker* form of orthogonality and is discussed in Chakrabarti (1962) and Agrawal (1966b).

(II) When would a row–column design be said to be *totally connected*? This question makes sense since, unlike in a block design, in a row–column design, treatment connectedness (or, for that matter, connectedness with respect to one factor) does *not* necessarily ensure connectedness with respect to the other two factors. So when we are interested in all the three factors, we should ensure estimability of the contrasts within each such factor and this is what total connectedness means in this set-up. To ensure this is not always an easy task. See Shah and Khatri (1973) and Eccleston and Russell (1975), for example.

Recall that in the presence of the conditions (i)–(iii), any one of the conditions (iv)–(vi) implies total connectedness. As explained above, these conditions relate to adjusted orthogonality. For some classes of designs, adjusted orthogonality holds for some choice of two factors and when this obtains, connectedness of the three-factor design can be verified by examining the connectedness of the block designs involving the third factor with each of the first two, ignoring the effects of the remaining factor, respectively. Generally, however, adjusted orthogonality is too much to expect and hence we need an algorithm to examine total connectedness. As noted in Eccleston and Russell (1975), a design is totally connected if and only if it is connected as a two factor design ignoring the third factor and also it is connected for the third factor (in the presence of the first two). Verification of the first part is easy. The second part has been explained earlier with reference to an example, using Park's technique. We now present a five-point algorithm which not only determines connectedness with respect to the third factor (if this is the fact) but also determines estimable contrasts for the third factor, in case the design is not connected with respect to that factor. See Park and Shah (1995).

1. Collapse within cells treatments, form equivalence classes and update them;
2. Inspect all tetra-differences to expand the equivalence classes as much as possible;
3. Form all $(p-1)(q-1)$ tetra-differences involving the cell $(1,1)$ and ignore repetitions and zeroes. Write down the matrix of coefficients so formed. Use Gaussian algorithm to reduce the coefficient matrix to the form: $[\boldsymbol{I}_s \vdots \boldsymbol{A}]$. A necessary and sufficient condition for connectedness is that the number of equivalence classes is $s+1$. If this does not hold, we proceed to steps 4 and 5.
4. Induce further collapsing of the above matrix by inspection and after updating it, write it in the form: $[\boldsymbol{I}_h \vdots \boldsymbol{B}]$. Expand it by adding a row of ones.
5. Span a basis of the orthocomplement of the above expanded matrix. Finally, replace every element in the basis by the average of the elements of the equivalence class it represents.

This is how we will end up with a basis of *non-estimable* treatment contrasts which are *orthogonal* to *all* within classes treatment contrasts. The importance of this algorithm lies in the fact that adjusted orthogonality seldom holds.

If the design has empty cells, a modification of the above procedure due to Park and Shah (1995) described earlier can be used.

(III) We turn back to treatment connectedness once more and present some further observations. Kurotschka (1978) has observed that the $\boldsymbol{C}_t$ matrix reduces to that of a CRD:

$$\boldsymbol{C}_t = \boldsymbol{r}^{\delta} - \frac{\boldsymbol{r}\boldsymbol{r}'}{n_{..}}, \tag{4.4}$$

whenever treatment classification is orthogonal to both the row and column classifications, these two themselves being *not* necessarily orthogonal. It is easily seen that these conditions are also necessary for (4.4) to hold. It is instructive to note that the above two conditions along with equal replication of the treatments is equivalent to the following: In every row *and* in every column, the treatments are *equally* replicated.

So far we have not paid any special attention to the case: $\boldsymbol{N}_{rc} = ((1))$, i.e., all cells are full and have one element each. In this case the row–column classification is *already* orthogonal. It is possible to provide a design for which row-treatment and column-treatment classifications are also orthogonal, provided v (the number of treatments) divides both p (the number of rows) *and* q (the number of columns). Latin Square designs form a special class of such designs when $p = q = v$. Again, if v divides $p(q)$ alone, then we can achieve treatment-column (respectively, treatment-row) orthogonal classification. Youden Square designs have this feature (thereby implying (3.8) or (3.11)). In case $\boldsymbol{N}_{rc} = ((n_{ij}))$ does *not* have orthogonal structure, but row and column totals $n_{i.}$'s and $n_{.j}$'s are *each* divisible by v, then also we can have a design ensuring treatment-row and treatment-column orthogonality! This existence theorem was settled in Shah and Sinha (1993a).

We may add the following as supplementary information. The notion of a BIBD has been extended to that of a BBD (balanced block design) to deal with situations where k (the block size) exceeds v (the number of treatments). Likewise, the notion of a Youden Design (YD) has been extended to what are called Generalized Youden Designs (GYDs) in which (i) the row–column incidence matrix is ((1)); (ii) the number of rows (p) and the number of columns (q) may both exceed the number of treatments (v); (iii) the row-treatment as also the column-treatment incidence matrices are BBDs. A GYD is called *regular* if at least one of p and q is a multiple of v; otherwise, it is called *non-regular*. Such designs have been studied by Kiefer (1975) from optimality considerations. It follows that such designs are very easy to analyze! The $\boldsymbol{C}_t$-matrix will be completely symmetric, i.e., will have diagonals all equal and off-diagonals all equal.

Again, when $p = q$ along with $\boldsymbol{N}_{rc} = ((1))$, an interesting simplification takes place in the expression for the $\boldsymbol{C}_{t.rc}$ matrix. It follows that in this case the $\boldsymbol{C}$-matrix can be viewed as that of a block design with $(p+q)$ blocks while the block-treatment incidence matrix is formed of the two component incidence matrices, viz., those of the rows and the columns, taken together. Therefore, in such situations, simplification in data analysis would be achieved if the over-all incidence matrix has a nice structure. When it has a BIBD structure, the row–column design is called a *pseudo-Youden* design. Such designs have been studied by Cheng (1981), among others.

(IV) Again, suppose that the row–column classification is not orthogonal but it is composed of several orthogonal components in the following sense: The columns can be partitioned into several sets, say, t sets. In every such set we keep open the possibility of some null rows. Suppose the submatrix formed of all the non-null rows in this set has an orthogonal structure and this is true for every set of columns.

Then this will be referred to as a component-wise orthogonal row–column structure. Mathematically, let us set

$$N_{rc} = \left[N_{rc1} \vdots N_{rc2} \ldots \vdots N_{rct}\right]. \tag{4.5}$$

We adopt the assumed representation:

$$N_{rcj} = a_j b_j'; \tag{4.6}$$

$1 \leqslant j \leqslant t$ where a_j's may have some null rows. Analogous partition of N_{tc} can be written down with N_{tcj} as the jth component. Let $N_{tcj}\mathbf{1} = v_j; 1 \leqslant j \leqslant t$. Following Kunert (1983), we can deduce that C_t will coincide with (3.11), i.e., with the C_t matrix just eliminating the column effects only, if the following holds:

$$N_{tr} = \sum_{j=1}^{t} (\mathbf{1}' a_j)^{-1} v_j a_j'. \tag{4.7}$$

Therefore, it is possible to achieve a simpler form of the C_t matrix even in situations more general than the usual row–column orthogonality. We refer to Shah and Sinha (1993a, b) for additional observations.

5. Efficiency bounds for row–column designs

We assume that treatment contrasts are the primary parameters of interest and that we are in the set-up of a treatment-connected row–column design. How do we measure the efficiency of such a design? This question makes sense when we have available several competing designs and we have to make a choice. Naturally we need some guide lines and highest efficiency may serve as one of them. Kempthorne (1956) suggested that one could take into consideration all treatment contrasts of the type $\tau_i - \tau_j$ (called elementary treatment contrasts) and compute the average variance of their corresponding blue's. Then the reciprocal of this quantity could be taken as a measure of efficiency of the underlying row–column design. Kempthorne also proved that the average variance is proportional to the harmonic mean of the (positive) eigenvalues of the underlying C_t matrix. Our purpose would be to minimize the average variance or to maximize the efficiency. In this section, we will discuss about some bounds on the efficiency. We mostly borrow the material from Shah and Sinha (1989) and present in a general framework. We may mention in passing that the efficiency is inversely proportional to the trace of the Moore–Penrose inverse of the underlying C_t matrix.

Looking back to the representation for the C_t matrix in (3.2), it is evident that

$$C_t \leqslant r^{\delta} - N_{tr} R^{-\delta} N_{rt}. \tag{5.1}$$

Here and below, the notation "$\leqslant$" involving two matrices will mean the ordering in the Löwener sense, i.e., $\boldsymbol{A} \leqslant \boldsymbol{B}$ iff $(\boldsymbol{B} - \boldsymbol{A})$ is *nnd*. Denoting by E_{RC} the efficiency of the row–column design in the sense described above and by E_R that based on the treatment-row design (ignoring column classification), it follows that $E_{RC} \leqslant E_R$. Along similar lines, it also follows that $E_{RC} \leqslant E_C$ (defined similarly with respect to the treatment-column design, ignoring the row classification). Hence, combining the two, we obtain:

$$E_{RC} \leqslant \min(E_R, E_C). \tag{5.2}$$

In order to obtain other useful bounds, we need to examine the structures of the row–column design and also of the treatment-row and treatment-column designs. We will now assume that the row–column classification is orthogonal. An immediate implication of this is that the $\boldsymbol{C}_t$ matrix has the representation given in (3.6). Let us now introduce some further notations.

$$\begin{aligned} \boldsymbol{C}_{t.rc} &= \boldsymbol{C}_t \text{ matrix under the row–column set-up;} \\ \boldsymbol{C}_{t.r0} &= \boldsymbol{C}_t \text{ matrix under the set-up with row effects only} \\ \boldsymbol{C}_{t.0c} &= \boldsymbol{C}_t \text{ matrix under the set-up with column effects only} \\ \boldsymbol{C}_{t.00} &= \boldsymbol{C} \text{ matrix under the CRD set-up without row or} \\ &\qquad \text{column effects.} \end{aligned} \tag{5.3}$$

It is evident that these matrices are respectively given by (3.2), (3.8), (3.11) and (4.4). Further, in view of assumed row–column orthogonality, we also have the relation (3.6) and, hence,

$$\boldsymbol{C}_{t.rc} = \boldsymbol{C}_{t.r0} + \boldsymbol{C}_{t.0c} - \boldsymbol{C}_{t.00} \tag{5.4}$$

For convenience, we will deal with the matrices $\boldsymbol{A}, \boldsymbol{A}_1, \boldsymbol{A}_2$ and $\boldsymbol{A}_0$ defined as follows:

$$\boldsymbol{A} = \boldsymbol{r}^{-\delta/2}\boldsymbol{C}_{t.rc}\boldsymbol{r}^{-\delta/2}; \qquad \boldsymbol{A}_1 = \boldsymbol{r}^{-\delta/2}\boldsymbol{C}_{t.r0}\boldsymbol{r}^{-\delta/2};$$

$$\boldsymbol{A}_2 = \boldsymbol{r}^{-\delta/2}\boldsymbol{C}_{t.0c}\boldsymbol{r}^{-\delta/2}; \qquad \boldsymbol{A}_0 = \boldsymbol{r}^{-\delta/2}\boldsymbol{C}_{t.00}\boldsymbol{r}^{-\delta/2}. \tag{5.5}$$

Evidently, these matrices also satisfy an analogous additive relation as in (5.4). Moreover, since the $\boldsymbol{C}$-matrices have zero row sums and zero column sums, it is clear that each of the matrices $\boldsymbol{A}, \boldsymbol{A}_1, \boldsymbol{A}_2$ and $\boldsymbol{A}_0$ has one eigenvalue zero with the corresponding eigenvector given by $\boldsymbol{r}^{\delta/2}\boldsymbol{1}$. Of course, all other eigenvalues are positive since the row–column design is assumed to be treatment-connected.

Suppose now that the matrices $\boldsymbol{A}_1$ and $\boldsymbol{A}_2$ commute. It then follows that all the four matrices commute among themselves. Therefore, it is possible to find a full set (all v) of common orthonormal eigenvectors for all the matrices. We denote them by

$\boldsymbol{p}_i$, $i = 1, 2, \ldots, v$, with $\boldsymbol{p}_v = \boldsymbol{r}^{\delta/2}\mathbf{1}/\sqrt{n}$... We can now form a full set of $(v-1)$ linearly independent treatment contrasts based on these $\boldsymbol{p}_i$'s (viz., $\boldsymbol{q}_i'\boldsymbol{\tau}, \boldsymbol{q}_i = \boldsymbol{r}_i^{-\delta/2}\boldsymbol{p}_i$; $i = 1, 2, \ldots, (v-1)$) such that they possess uncorrelated estimates with respect to each of the four designs with reference to (5.3). These treatment contrasts are said to constitute a full set of canonical contrasts or basic contrasts. Also if λ_i, ϕ_i and θ_i, denote respectively the eigenvalues of $\boldsymbol{A}, \boldsymbol{A}_1, \boldsymbol{A}_2$ corresponding to $\boldsymbol{p}_i$, $i = 1, 2, \ldots, (v-1)$, then it turns out that

(i) $0 < \lambda_i \leqslant \phi_i, \quad \theta_i \leqslant 1$;
(ii) $1 + \lambda_i = \phi_i + \theta_i$,

and, hence, that

$$\text{(iii) } \lambda_i \leqslant \phi_i\theta_i; \quad i = 1, 2, \ldots, (v-1). \tag{5.6}$$

The above analysis is based on the assumption that the matrices $\boldsymbol{A}_1$ and $\boldsymbol{A}_2$ commute. By referring to the expressions for $\boldsymbol{A}_1$ and $\boldsymbol{A}_2$, we find that this is equivalent to the following:

$$\boldsymbol{N}_{tr}\boldsymbol{R}^{-\delta}\boldsymbol{N}_{rt}\boldsymbol{r}^{-\delta}\boldsymbol{N}_{tc}\boldsymbol{C}^{-\delta}\boldsymbol{N}_{ct} = \boldsymbol{N}_{tc}\boldsymbol{C}^{-\delta}\boldsymbol{N}_{ct}\boldsymbol{r}^{-\delta}\boldsymbol{N}_{tr}\boldsymbol{R}^{-\delta}\boldsymbol{N}_{rt}. \tag{5.7}$$

Recall that we have assumed that the design has orthogonal row–column structure so that $\boldsymbol{N}_{rc} = \boldsymbol{\gamma}\boldsymbol{\eta}'$. If, in addition to this, the design also possesses the property of adjusted orthogonality (vide below (4.2)) then we have: $\boldsymbol{N}_{rt}\boldsymbol{r}^{-\delta}\boldsymbol{N}_{tc} = \boldsymbol{N}_{rc} = \boldsymbol{\gamma}\boldsymbol{\eta}'$. This implies that the condition stated above in (5.7) holds and, hence, the matrices $\boldsymbol{A}_1$ and $\boldsymbol{A}_2$ commute! Therefore, adjusted orthogonality along with orthogonal row–column structure implies commutativity of $\boldsymbol{A}_1$ and $\boldsymbol{A}_2$. At this stage, it also follows that we have something more, viz., $\boldsymbol{A} = \boldsymbol{A}_1\boldsymbol{A}_2$. If additionally, the design is also equireplicate, then the basis of treatment contrasts formed of the $\boldsymbol{q}_i'\boldsymbol{\tau}$'s turns into an orthonormal basis.

It is known that the average variance is proportional to the sum of variances of the estimates of any full set of orthonormal treatment contrasts. Therefore, for an *equireplicate* design, after some calculations, we see that the efficiency E_{rc} is inversely proportional to $\sum_{i=1}^{v-1} \frac{1}{\lambda_i}$. Similarly, it follows that E_{r0} is inversely proportional to $\sum \frac{1}{\phi_i}$ and E_{0c} is inversely proportional to $\sum_i \frac{1}{\theta_i}$. Finally, note from (i)–(iii) in (5.6) that

$$\frac{1}{\lambda_i} \geqslant \frac{1}{\phi_i} + \frac{1}{\theta_i} - 1, \quad i = 1, 2, \ldots, (v-1). \tag{5.8}$$

Hence, we have established that

$$E_{rc}^{-1} \geqslant E_{r0}^{-1} + E_{0c}^{-1} - 1. \tag{5.9}$$

Here equality holds if and only if for every i,

$$\phi_i = 1 \quad \text{or} \quad \theta_i = 1. \tag{5.10}$$

Again since $1+\lambda_i = \phi_i+\theta_i$ for every i, the above condition (for equality) is equivalent to $\lambda_i = \phi_i\theta_i$ for every i. This can be shown to be equivalent to adjusted orthogonality (of rows and columns). This result was proved (in the equireplicate case) by Eccleston and McGilchrist (1985) under the assumption of commutativity of $\boldsymbol{A}_1$ and $\boldsymbol{A}_2$. More recently, Jarrett (1994) analyzed the dual design and proved the same result without the assumptions of commutativity or of equal replications.

REMARK. While row–column orthogonality seems to be quite reasonable, the condition of adjusted orthogonality is hard to meet in practice! This, however, does not preclude the design from having the property of commutativity! Thus, for example, it is not uncommon to come across experimental situations where the row-treatment (or column-treatment) design is a BIBD, for example. In that case, commutativity is trivially satisfied. Russell (1980) gives examples of such designs and computes their efficiencies, indicating that they possess high efficiency.

6. Optimality considerations

We do *not* want to describe optimality aspects of row–column designs at length. Most results are available in Shah and Sinha (1989). We will briefly discuss those results and address more recent issues.

As has been seen before, the $\boldsymbol{C}$-matrices have a natural Löwener ordering:

$$\boldsymbol{C}_{t.rc} \leqslant \boldsymbol{C}_{t.r0}, \qquad \boldsymbol{C}_{t.0c} \leqslant \boldsymbol{C}_{t.00}. \tag{6.1}$$

Therefore, in terms of efficiency, the most efficient row–column design is one which attains: $\boldsymbol{C}_{t.rc} = \boldsymbol{C}_{t.00}$ and, then, further provides maximum efficiency (as described in Section 5) by a proper choice of the treatment replications. It is seen that equal treatment replications is the best, if available. A large class of row–column designs have thus surfaced satisfying these requirements. For such designs, the $\boldsymbol{C}_t$ matrix possesses what is called a *completely symmetric* structure, i.e., one with all diagonal elements equal and all off-diagonal elements equal. Note that LSD's are the simplest examples of such designs!

Failing to achieve: $\boldsymbol{C}_{t.rc} = \boldsymbol{C}_{t.00}$, we next try to achieve: $\boldsymbol{C}_{t.rc} = \boldsymbol{C}_{t.r0}$, say. Then we are in the framework of a block design (two-factor design). And again we may characterize designs with maximum efficiency (for treatment contrasts). It turns out that the BIBDs or, more generally, the BBDs do the job in a two-factor setup! Hence, we need to look for three-factor designs having this dual property, i.e., achieving simplicity of the $\boldsymbol{C}_{t.rc}$ matrix in the form of $\boldsymbol{C}_{t.r0}$ matrix and being a BIBD or a BBD as a two-factor design. YDs are simplest examples of such designs and, here again, a large class of such designs are available.

A general feature of such most efficient designs is that they provide equal replications for the treatments and also possess completely symmetric $C_{t.rc}$ matrix (in the sense described above). Occasionally, such $C_{t.rc}$ matrices also possess *maximum trace* among all competing $C_{t.rc}$ matrices in a given set-up!! When this happens, then the designs do possess many optimality properties! They are then found to be optimal with respect to a large class of optimality criteria! This was first announced by Kiefer (1975), though the optimality study got started in scattered form as early as mid 1940's!! See Shah and Sinha (1989) for details.

The readers may be familiar with some commonly used optimality criteria: A-optimality (equivalent to maximizing efficiency, we discussed earlier), D-optimality, E-optimality, MS optimality, MV optimality, Universal optimality, ϕ_p optimality, ψ_f optimality. Two other criteria, less familiar, are known as Schur-optimality and distance criterion. We refer to Hedayat (1981), Bondar (1983) and Shah and Sinha (1989) for thorough discussions on these criteria and their possible interrelationships.

We mention in passing that universal optimality is most appealing and once a design has a completely symmetric $C_{t.rc}$ matrix with the maximum value of its trace among possible competitors, then it is universally optimal! This prompted a fair amount of research in characterizations and constructions of such optimal designs in some experimental situations. However, a good deal of research has been undertaken to examine situations where such designs are not available.

Specific optimality results for row–column designs appear to be hard to establish. Apart from Kiefer's pioneering work on A-, D- and E-optimality of the GYDs most other work has either considered designs where $C_{t.rc} = C_{t.0c}$ and the later is optimal, or has exploited the property of adjusted orthogonality of the design (Bagchi and Shah, 1989; Bagchi and van Berkum, 1991). Even though many interesting and useful results are available, much remains to be done in the search for optimal row–column designs.

It would be useful to point out that while many optimal designs are adjusted orthogonal, in some set-ups, the best adjusted orthogonal design is inferior to a design which is not adjusted orthogonal. (See Shah and Puri, 1986.)

In the context of a block design, the BIBDs (or, more generally, the BBDs) are the best on all counts! When a BIBD does not exist, a most balanced Group Divisible Design (abbreviated as MBGDD) may serve as the next best candidate. Similarly, in a row–column design, assuming an orthogonal row–column set-up, one with a BBD structure for each of the row-treatment and column-treatment designs is expected to be the best! This is generally true – though there are some exceptions! But the proof is extremely complicated, unless, of course, both the above-mentioned structures are in fact RBDs! And, when these are RBDs, we do *not* need row–column orthogonality at all!! There are many issues that can be discussed like that. We rather stop here and cite some additional references: Cheng (1978), Eccleston and Jones (1980), Saharay (1986) and Jacroux (1982, 1986, 1987).

We have not specifically dealt with the methods of construction of row–column designs. However, during the course of our discussions, we have given references where designs with specific properties could be found. The following is a supplementary list of references for methods of construction of row–column designs. Needless to say that the list is subjective and far from complete: Anderson and Eccleston (1985), Jones (1980), Pearce (1975) and Ruiz and Seiden (1974).

7. Nested row–column designs

We have assumed in our discussion so far that the row–column classification is connected. When this is *not* so, it can be seen to be a *disjoint union* of connected components. Therefore, we can regard our design for treatments as a *union* of these component designs, sitting at different sites, for example.

Suppose at the ith site, there is a connected row–column classification having p_i rows and q_i columns. Our purpose is to compare a set of v treatments using experimental units from all the sites together. We say that we are in the framework of a *nested* design with the sites as the *nesting factor*! It is quite possible that we do not use all the treatments at each site. For treatment-connectedness of the overall design, this is not necessary. Every site can provide an equivalence class and they can be merged together at the end via some common treatments. On the other hand, it is very tempting to use full set of treatments at every site at least when it results into the *best* row–column design *by itself*!

Under a fixed effects additive model, analysis of such a design does not pose any problem. For inference on treatment contrasts, we simply derive the $\boldsymbol{C}_{t.rc}$ matrix at each site with all its elements corresponding to treatments *not* applied at that site as zeroes. Finally, we take the sum of these matrices for the over-all design. It should be noted that the row- (and column-) parameters are different at different sites and that makes the analysis very simple! (An entirely different exercise will emerge if we are in one and the *same* site and repeating the experiment on various occasions; see, for example, Nandi (1953).) The study of nested designs was initiated in Srivastava (1978) and further pursued by Singh and Dey (1979). Interesting combinatorial problems emerge as we attempt to characterize universally optimal designs by trace maximization and complete symmetry of the overall $\boldsymbol{C}_{t.rc}$ matrix. Bagchi et al. (1990) started examining the nature of universally optimal nested row–column designs. We would also like to mention the work of Chang and Notz (1994) and of Morgan and Uddin (1993) in this area. This area of research is just emerging.

8. Three-way balanced designs

So far we have *not* addressed the question of balancing in a row–column design. When would it be called a balanced design? Since treatment contrasts are the primary parameters of interest, we might call the design balanced if it is so with respect to the treatment contrasts. Once we accept this, then we go by the usual definition. That means, a row–column design is balanced if and only if it provides the same variance for all normed treatment contrasts. A little reflection now indicates that this is possible if and only if the $\boldsymbol{C}_{t.rc}$ matrix is completely symmetric. Extending this concept further, a row–column design is said to be three-way balanced if and only if the above property holds for all the three factors – rows, columns and treatments – simultaneously.

The simplest example of a three-way BIBD is, of course, the LSD! One series of such designs were studied in Hedayat and Raghavarao (1975). Before that, Agrawal (1966) provided a series of three-way BIBDs for the parameters: $p = q = v = 4\lambda + 3$, a prime power, and $r = 2\lambda + 1$. He took a BIBD row–column set-up to start with and succeeded in constructing a design in which all the three $\boldsymbol{C}$-matrices were identical and completely symmetric! In view of very strong optimality properties of the BIBDs, one would expect that such row–column designs would be optimal for inference on treatment contrasts. Shah and Sinha (1990) took up this study and observed that, interestingly enough, there exists another asymmetric $\boldsymbol{C}$-matrix which dominates the one of Agrawal in the strong sense of Löwener ordering!! This was discovered for the case of $p = q = v = 7$, $r = 3$, $\lambda = 1$. Carrying this investigation further, Heiligers and Sinha (1994) re-established superiority of another completely symmetric $\boldsymbol{C}$-matrix based on another design in the same set-up!! It turned out that this latest design provides maximum efficiency in a large class of competing designs. While closing the discussion, we will display all the three matrices below.

$$p = q = v = 7; \qquad r = 3; \qquad \lambda = 1;$$

$$\text{Agrawal (A):} \quad \begin{bmatrix} - & 2 & 4 & - & 1 & - & - \\ - & - & 3 & 5 & - & 2 & - \\ - & - & - & 4 & 6 & - & 3 \\ 4 & - & - & - & 5 & 7 & - \\ - & 5 & - & - & - & 6 & 1 \\ 2 & - & 6 & - & - & - & 7 \\ 1 & 3 & - & 7 & - & - & - \end{bmatrix},$$

$$\text{Shah–Sinha (SS):} \quad \begin{bmatrix} - & 2 & 5 & - & 3 & - & - \\ - & - & 3 & 6 & - & 4 & - \\ - & - & - & 4 & 7 & - & 5 \\ 6 & - & - & - & 5 & 1 & - \\ - & 7 & - & - & - & 6 & 2 \\ 3 & - & 1 & - & - & - & 7 \\ 1 & 4 & - & 2 & - & - & - \end{bmatrix},$$

$$\text{Heiligers-Sinha (HS):} \quad \begin{bmatrix} - & 2 & 7 & - & 3 & - & - \\ - & - & 3 & 1 & - & 4 & - \\ - & - & - & 4 & 2 & - & 5 \\ 6 & - & - & - & 5 & 3 & - \\ - & 7 & - & - & - & 6 & 4 \\ 5 & - & 1 & - & - & - & 7 \\ 1 & 6 & - & 2 & - & - & - \end{bmatrix}.$$

Whereas SS design dominates over A design, it turns out that HS design has its C-matrix twice that of A design!! Moreover, HS design is found to possess maximum efficiency among a large class of designs including all the three types presented above.

There is something more to it. Agrawal also provided methods of construction of other series of three-way BIBDs for various other classes of row–column incidence structures. And in most cases, Shah and Sinha found that there is scope for improvement for inference on treatment contrasts! It is thus not clear if three-way balancing is desirable or not, though combinatorially it is very fascinating! This area of research is very much at the initial stage.

9. Mixed effects model

Yates (1939, 1940) pointed out that in a block design, observational contrasts among the block totals may also contain information on treatment contrasts. This is especially relevant when the blocking factor offers various choices of blocks and the experimenter ends up with a subset selection in a random manner. In that case, technically speaking, the block effects in the (implemented) design may be treated as random while the treatment effects are still fixed. This yields what are known as mixed effects models. Likewise, in a row–column design, if the rows and the columns in the (implemented) design have been selected in a random manner out of a large number of choices given to the experimenter, then we are in the framework of a mixed effects model. See Yates (1939, 1940) for more details on the evolution of such models.

Technically speaking, with reference to the model (3.1), in contrast to the situation where we treat α_i's and β_j's as fixed effects parameters (under the fixed effects model), now we will treat them as random quantities under the mixed effects model, while still continuing with the treatment effects as fixed effects and of primary interest for inference. As to the distributional assumptions about the random row effects and random column effects, we take them to be uncorrelated among themselves as also between themselves, having means zeroes and variances σ_R^2 (for the row effects) and σ_C^2 (for the column effects).

For data analysis, as mentioned in Section 3, while extracting information on the treatment contrasts, the within cell observational contrasts and the tetra-differences alone provide useful information in a fixed effects model. While dealing with a mixed effects model, these continue to provide information on the treatment contrasts. In addition to these, we may as well derive further information from the row total contrasts and the column total contrasts and finally combine the three to obtain final results.

However, there is a point of difference. The former set of observational contrasts do not involve α_i's or the β_j's and as such in the model, these will have expectations either in the form of treatment contrasts or as zero (error functions generating sources for estimation of the error variance (σ^2)) and further, these will have variance proportional to the error variance σ^2. On the other hand, the contrasts formed of the row totals will have expectations expressed in terms of treatment contrasts and the row effects contrasts and since the row effects are assumed to be random with zero mean, these will be eliminated and hence some additional information on treatment contrasts alone will be made available. However, this time the variance term will involve not only the error variance (σ^2) but also the row variance (σ_R^2). A similar consideration applies also to the column total contrasts. It must be noted that here we have tacitly assumed that the row–column incidence pattern is $\boldsymbol{N}_{rc} = ((1))$. (Otherwise, this simple-minded analysis does *not* work.) Below we will continue to assume this. We also assume multivariate normal distribution.

We thus have three sources of information on the treatment contrasts: Within cells and the tetra-differences; the row total contrasts and the column total contrasts. These are seen to be uncorrelated among themselves. The first source gives rise to what is known as fixed effects analysis and that was discussed in Section 3. We will now discuss the analysis under the mixed effects model. Following the general methods of combining information in linear models, we will first write down the three sets of normal equations and then combine them by taking appropriate linear combinations. While combining them, the fact that they have unequal variances, has to be brought in. Below we follow the method of analysis adopted in Roy and Shah (1961), Shah (1962) and Khatri and Shah (1974).

The normal equations are given below:

$$\boldsymbol{C}_t\boldsymbol{\tau} = \boldsymbol{Q}_t, \tag{9.1}$$

$$\boldsymbol{C}_{t(R)}\boldsymbol{\tau} = \boldsymbol{Q}_{t(R)}, \tag{9.2}$$

$$\boldsymbol{C}_{t(C)}\boldsymbol{\tau} = \boldsymbol{Q}_{t(C)} \tag{9.3}$$

where

$$\begin{aligned} \boldsymbol{C}_t &= \boldsymbol{C}_{t.rc} \quad \text{as in (3.6) or (5.4)}, \\ \boldsymbol{C}_{t(R)} &= \boldsymbol{N}_{tr}\boldsymbol{R}^{-\delta}\boldsymbol{N}_{rt} - \boldsymbol{r}\boldsymbol{r}'/pq, \\ \boldsymbol{C}_{t(C)} &= \boldsymbol{N}_{tc}\boldsymbol{C}^{-\delta}\boldsymbol{N}_{ct} - \boldsymbol{r}\boldsymbol{r}'/pq \end{aligned} \tag{9.4}$$

and $\boldsymbol{Q}_t, \boldsymbol{Q}_{t(R)}$ and $\boldsymbol{Q}_{t(C)}$ are the observational contrasts formed accordingly.

When the variance ratios σ_R^2/σ^2 and σ_C^2/σ^2 are known, the method of generalized least squares can be used for estimating the treatment contrasts. This leads to the so-called combined normal equations:

$$\bar{\boldsymbol{C}}_t\boldsymbol{\tau} = \bar{\boldsymbol{Q}}_t \tag{9.5}$$

where

$$\bar{\boldsymbol{C}}_t = \boldsymbol{r}^{\delta} - \frac{\boldsymbol{N}_{tr}\boldsymbol{N}_{tr}'}{q+\Delta_1} - \frac{\boldsymbol{N}_{tc}\boldsymbol{N}_{tc}'}{p+\Delta_2} + \left(1 - \frac{\Delta_1}{q+\Delta_1} - \frac{\Delta_2}{p+\Delta_2}\right)\frac{\boldsymbol{r}\boldsymbol{r}'}{pq} \tag{9.6}$$

and $\bar{\boldsymbol{Q}}_t$ is the corresponding linear combination of $\boldsymbol{Q}_t, \boldsymbol{Q}_{t(R)}$ and $\boldsymbol{Q}_{t(C)}$. In (9.6), $\Delta_1 = \sigma_R^2/\sigma^2$ and $\Delta_2 = \sigma_C^2/\sigma^2$. Since the variance ratios are generally unknown, these would need to be estimated from the data. There are many procedures for doing that. We shall present two methods here which are generalizations of the corresponding methods for block designs.

One method would be to use the expected values of error sum of squares and row (column) sum of squares adjusted for treatments SS_R (respectively SS_C) to estimate σ_R^2 (respectively σ_C^2). An alternative procedure would be to use expected values of

$$V_R = \sum_i \left(R_i - \widehat{E(R_i)}\right)^2 \tag{9.7}$$

and

$$V_C = \sum_j \left(C_j - \widehat{E(C_j)}\right)^2 \tag{9.8}$$

where $\widehat{E(R_i)}$ and $\widehat{E(C_j)}$ are the estimates of $E(R_i)$ and $E(C_j)$ based on the first set of equations $\boldsymbol{C}_t\boldsymbol{\tau} = \boldsymbol{Q}_t$. The first procedure gives:

$$\widehat{\sigma}^2 = \text{MSE}, \qquad \widehat{\sigma}_R^2 = \frac{SS_R - (p-1)\widehat{\sigma}^2}{q\{(p-1) - \operatorname{tr}\left((\boldsymbol{C}_t + \boldsymbol{C}_{t(R)})^- \boldsymbol{C}_{t(R)}\right)\}},$$
$$\widehat{\sigma}_C^2 = \frac{SS_C - (q-1)\widehat{\sigma}^2}{p\{(q-1) - \operatorname{tr}\left((\boldsymbol{C}_t + \boldsymbol{C}_{t(C)})^- \boldsymbol{C}_{t(C)}\right)\}}. \tag{9.9}$$

The second procedure gives:

$$\widehat{\sigma}^2 = \text{MSE}, \qquad \widehat{\sigma}_R^2 = \frac{V_R - q\left\{(p-1) + \operatorname{tr}\left(\boldsymbol{C}_t^- \boldsymbol{C}_{t(R)}\right)\right\}}{q^2(p-1)},$$
$$\widehat{\sigma}_C^2 = \frac{V_C - p\left\{(q-1) + \operatorname{tr}(\boldsymbol{C}_t^- \boldsymbol{C}_{t(C)})\right\}}{p^2(q-1)}. \tag{9.10}$$

Since the variance ratios are non-negative, it is appropriate to truncate the estimates at zero, if necessary.

When σ^2, σ_R^2 and σ_C^2 are estimated from the data, the distribution of the resulting estimates of the treatment contrasts is almost intractable even when the random variables involved are assumed to have normal distribution. However, for row–column designs with the property of adjusted orthogonality (vide Section 4, immediately below (4.2)), an interesting situation develops. This we describe below.

We first set the notations in order. Referring to (5.3)–(5.5) and (9.4), we note that

$$\boldsymbol{C}_{t.r0} = \boldsymbol{C}_{t.rc} + \boldsymbol{C}_{t(C)} \tag{9.11}$$

and

$$\boldsymbol{C}_{t.0c} = \boldsymbol{C}_{t.rc} + \boldsymbol{C}_{t(R)}. \tag{9.12}$$

In terms of the matrices $\boldsymbol{A}, \boldsymbol{A}_0, \boldsymbol{A}_1$ and $\boldsymbol{A}_2$, it follows that

$$\boldsymbol{A}_{t(R)} = \boldsymbol{A}_2 - \boldsymbol{A}, \qquad \boldsymbol{A}_{t(C)} = \boldsymbol{A}_1 - \boldsymbol{A}, \qquad \boldsymbol{A} + \boldsymbol{A}_0 = \boldsymbol{A}_1 + \boldsymbol{A}_2. \tag{9.13}$$

We have seen that when rows and columns adjusted for treatments are orthogonal, then the matrices $\boldsymbol{A}_1$ and $\boldsymbol{A}_2$ commute. It is easy to see that this implies commutativity of all the six matrices viz., $\boldsymbol{A}, \boldsymbol{A}_1, \boldsymbol{A}_2, \boldsymbol{A}_0, \boldsymbol{A}_{t(R)}$ and $\boldsymbol{A}_{t(C)}$! Therefore, as before, this enables us to obtain a set of v common orthonormal eigenvectors for all the three matrices: $\boldsymbol{A}, \boldsymbol{A}_{(R)}$ and $\boldsymbol{A}_{t(C)}$. This, in its turn, would mean that we have available a set of $(v-1)$ canonical treatment contrasts which have uncorrelated estimates (whenever available) with respect to each of the three models underlying the normal equations displayed in (9.1)–(9.3). Further, adjusted orthogonality also implies that none of these contrasts is estimable from each of (9.2) and (9.3) – see Shah and Eccleston (1986). Thus, for every canonical contrast, we have at most two estimates available.

Thus for a design with adjusted orthogonality of rows and columns, we are in a situation similar to the block design set-up where for some treatment contrasts there are available estimates from between block comparisons and also from within block comparisons. A procedure given by Khatri and Shah (1974) is known to lead to combined estimates of treatment contrasts which are uniformly superior to those based on within block comparisons only, in the sense of possessing smaller variance, uniformly in the variance components. We can then adopt this procedure whenever a row–column design possesses the property of adjusted orthogonality. It turns out that the superiority of the estimates is achieved when we use the estimates of σ^2, σ_R^2 and σ_C^2 given by (9.10). (See Shah, 1988). In this case, the combined estimate of any treatment contrast will have a smaller variance when compared with the variance of the same contrast estimated from (9.1), uniformly in the variance components. It has *not* been possible to establish such a property if the variance components are estimated from (9.9).

REMARK 1. It would be useful to comment on the role played by adjusted orthogonality in row–column designs. In the first place, for designs with adjusted orthogonality we have seen that the connectedness of the row-treatment design and of the column-treatment design ensures treatment connectedness. We have also seen that designs with adjusted orthogonality may be simpler to analyze. Finally, in a mixed effects model, for designs with adjusted orthogonality, it is possible to obtain combined estimates with uniform improvement property.

REMARK 2. It would be nice to mention the concept of general balance given by Houtman and Speed (1983) and by Nelder (1965a, b). For row–column designs where the row–column structure is orthogonal this is equivalent to the commutativity of the matrices $\boldsymbol{A}_1$ and $\boldsymbol{A}_2$. Thus, it is implied by adjusted orthogonality.

10. Designs for factorial and quasi-factorial experiments

Yates (1937) and Rao (1946) have presented designs for factorial experiments using a $p \times p$ square. We shall present here an example of one such design which is due to

Table 1

1	bcde	bc	de	abd	ace	abe	acd
abde	ac	acde	ab	bcd	e	bce	d
ae	abcd	abce	ad	cde	b	c	bde
bd	ce	cd	be	ade	abc	a	abcde
bcd	abc	abe	bde	1	ad	acde	ce
ade	e	c	acd	abce	bcde	bd	ab
cde	ace	a	d	be	abde	abcd	bc
abd	b	bce	abcde	ac	cd	de	ae

Rao (1946). Our discussion of this example provides an insight into the construction (and analysis) of such designs. For a further discussion of these methods, the reader is referred to the above two papers.

Consider a 2^5 factorial experiment in a 8×8 square where the factors are denoted by A, B, C, D and E, each at two levels – 0 and 1. It is desired to confound the effects as follows:

ROWS 1–4:	ABC,	ADE,	$BCDE$
ROWS 5–8:	ABD,	BCE,	$ACDE$
COLUMNS 1–4:	ACE,	BCD,	$ABDE$
COLUMNS 5–8:	ACD,	BDE,	$ABCE$

As mentioned in Rao (1946), a necessary and sufficient condition for the feasibility of the above is the existence of an effect I such that any set of effects to be confounded with the rows (columns) together with I and its interaction with these effects does *not* include any effect to be confounded with the columns (rows). Thus, for example, the set $\{ABC, ADE, BCDE, I, ABCI, ADEI$ and $BCDEI\}$ should *not* include any of the effects in the two sets to be confounded with the columns, as indicated above. This gives rise to one set of conditions. Again, considering I along with the effects in rows 5–8 and also along with the columns 1–4 and 5–8, we will end up with three additional sets of conditions. Thus, in effect, we have four sets of conditions.

These happen to be satisfied when $I = ABCD$ or $ABCDE$. In Table 1 we present a solution when $I = ABCDE$.

The above layout (Table 1) was obtained by using the fact that for each cell, we have a system of 5 linearly independent equations in the levels of the 5 factors and hence there is a *unique* solution. Thus, for example, for the cell (1,2), these equations are:

$$x_1 + x_2 + x_3 + x_4 + x_5 = 0, \tag{10.1}$$

$$x_1 + x_2 + x_3 = 0, \tag{10.2}$$

$$x_1 + x_4 + x_5 = 0, \tag{10.3}$$

$$x_1 + x_3 + x_5 = 0, \tag{10.4}$$

$$x_2 + x_3 + x_4 = 1 \tag{10.5}$$

which yield the solution: $x_1 = 0$ and $x_2 = x_3 = x_4 = x_5 = 1$, thereby resulting in the level combination *bcde*.

As is readily seen, the first four rows (columns) constitute a complete replication of the full factorial. The same is true of the last four rows (columns) as well. In the absence of column effects, then, the effects which are confounded with the first four rows are estimable from the last four. The above construction enables us to estimate these effects from the last 4 rows even in the presence of column effects! This is because the contrasts defining these effects are *orthogonal* to the columns as well. Similar result holds for the effects confounded in the last set of rows as also those confounded in the two sets of columns. This accounts for 12 effects which are partially confounded. The contrasts for each of the remaining 19 effects are all orthogonal to the rows and to the columns. Clearly, the estimates of different effects are uncorrelated. The sum of squares for any effect can be calculated in a routine manner as in a block design with partial confounding. We note that we do *not* need to calculate the adjusted treatment totals (Q_i's).

We now examine some further desirable properties of the above design. By checking the condition: $\boldsymbol{N}_{rc} = \boldsymbol{N}_{rt}\boldsymbol{r}^{-\delta}\boldsymbol{N}_{tc}$, we see that the design is *adjusted orthogonal*. As noted by Rao (1946), balance over each order of interactions can be achieved by using five 8×8 squares. Rao (1946) gave several methods of construction of designs for factorial experiments of the above type. These also include designs for asymmetrical factorials. In our opinion, these designs have not received the attention they deserve.

Since many of these designs have high efficiency and are easy to analyse, it may be tempting to use these for treatment trials where the treatments do *not* have a factorial structure! In this context, such designs are called *quasi-factorial* designs. We will not discuss them here. Instead, we cite some references:

Another example of interest is a design (due to B. V. Shah) given in Kshirsagar (1957). This was originally presented as a 6×6 square for comparing 9 treatments. It was subsequently observed that this is also a factorial design with an interesting pattern of confounding. In Table 2 we present this design in the factorial form where the treatments are denoted in terms of levels of factors A and B (each having levels 0, 1, 2).

In this design, the main effect of A is confounded with the first three columns and that of B with the last three columns. Similarly, the two components of the interaction AB (4 df) are confounded with the two sets of rows. However, unlike the previous example, estimate of the main effect of A *cannot* be obtained from columns 4–6 only! This design can be most easily analysed by viewing it as a treatment row–column

Table 2

(00)	(21)	(01)	(10)	(22)	(12)
(02)	(11)	(20)	(01)	(10)	(22)
(11)	(12)	(02)	(00)	(20)	(21)
(01)	(10)	(21)	(20)	(12)	(02)
(10)	(22)	(00)	(21)	(02)	(11)
(12)	(20)	(22)	(11)	(00)	(01)

Table 3

00	12	21	00	21	12
01	10	22	10	01	22
02	11	20	20	11	01
00	11	22	00	11	22
02	10	21	10	21	02
01	12	20	20	01	12.

design. Connectedness is easy to settle and adjusted orthogonality is also easy to verify. It turns out that the $\boldsymbol{C}_{t.rc}$ matrix is completely symmetric and that the efficiency is 7/8. In Table 3, in the same context, we present another example of a 6×6 design with the same pattern of confounding as above.

This design was obtained by constructing each corner 3×3 design as a full replication with confounding of the desired effects. This can be analysed as in the example of 2^5 factorial design in a 8×8 square given earlier.

This design also has a completely symmetric $\boldsymbol{C}_{t.rc}$ matrix and it is adjusted orthogonal. However, its efficiency is only $1/2$.

11. Miscellaneous topics

11.1. Nonadditivity in row–column designs

We now address the problem of nonadditivity in a row–column design. The model adopted in (3.1) corresponds to a totally additive three-factor design. It may not, however, be tenable all the time. A model more general than the one in (3.1) would be:

$$\begin{aligned} y_{ijh.s} &= \mu + \alpha_i + \beta_j + \tau_h \\ &\quad + (\alpha\beta)_{ij} + (\alpha\tau)_{ih} + (\beta\tau)_{jh} + (\alpha\beta\tau)_{ijh} + \varepsilon_{ijh.s}. \end{aligned} \tag{11.1}$$

In the above, the parameters

$$(\alpha\beta)_{ij}, \qquad (\alpha\tau)_{ih}, \qquad (\beta\tau)_{jh} \quad \text{and} \quad (\alpha\beta\tau)_{ijh}$$

represent the interaction effects of the two-factor and the three factor components. For an arbitrary row–column design, a model in this general form may not be tractable. Thus, for example, if $n_{ijh} = 0$ for some combination(s) of (i, j, h), then estimability of the parameters or contrasts in the model is in question. With enough of structure in the design, it is possible to develop inference procedures for treatment contrasts. See Searle (1971) for a detailed discussion of two-factor models with interaction. Similar considerations apply for a three-factor model (with added complexities for sure!).

Tukey (1949) suggested a model for a two-factor design incorporating a *nonadditive* component which is multiplicative in nature. He also suggested an exact test of significance of the nonadditivity parameter in case the two-factor design is a randomized block design. Properties of this test have been compared and contrasted with other tests in a recent paper by Mathew and Sinha (1991). Carrying this further, Tukey (1955) also developed an exact test for a multiplicative nonadditive component in the set-up of a Latin square design. With reference to the model (11.1), this would amount to the special form:

$$y_{ijh.s} = \mu + \alpha_i + \beta_j + \tau_h + \theta\alpha_i\beta_j\tau_h + \varepsilon_{ijh.s}. \tag{11.2}$$

In the above, θ is referred to as the nonadditivity parameter. As mentioned before, an exact test of the null hypothesis: H_0: $\theta = 0$ is available when the design is a Latin square design. See Krishnaiah and Yochmowitz (1980) for a review of this and related results. This sort of study has *not* been attempted so far for a general row–column design.

Mandel (1961, 1969) has given interesting insights into the models for nonadditivity by introducing what are called: *Factor Analysis of Variance* (FANOVA). Also see Scheffe (1971) and Milliken and Graybill (1970) for other generalizations of Tukey's model. An informative review is to be found in Krishnaiah and Yochmowitz (1980). However, all this refers to a randomized block design. No significant work has been reported in the literature involving a row–column design.

Let us now consider the problem of inference on treatment comparisons in the presence of the nonadditive term in the model. In the set-up of a randomized block design, it appears that a test of the hypothesis of equality of treatment effects suffers from *nonidentifiability*! In the case of a row–column design, however, there is *no* such problem; but the general approach for test construction does not seem to work here. Following Tukey (1955), we can derive an expression for the residual sum of squares under the model (11.2), conditional on the estimates of the row, column and treatment effects (under the three-factor additive model). Under the null hypothesis of equality of treatment effects, the model essentially reduces to that of a randomized block design and hence, following Tukey (1949) again, we can determine the residual sum of squares, conditional on the estimates of the row and column effects (under the additive model). The former sum of squares will possess $p^2 - 3p + 1$ df while the later will possess $p^2 - 2p$ df. In order to develop a valid test of the hypothesis, it would be enough to show that the former never exceeds the later, *uniformly* in the conditioning values. This is where we encounter difficulties and at this stage, we are not sure if this is in fact true.

The problem of comparing treatment effects in the presence of nonadditivity in a row–column design appears to be difficult. We refer the readers to a recent paper by Srivastava (1993) for another approach to nonadditivity.

11.2. Trend-free designs

As we know, the primary purpose of a row–column design is to eliminate recognized heterogeneity in two directions. In addition to its usage in the standard design set-ups, it may also serve some purpose in other experimental set-ups. This we want to emphasize here.

Consider a situation in which the experimental material is grouped to form blocks but within each block there is a *trend* (over time or space) which would have to be taken into account. One way to do this is to use block designs which are *trend-free*, or at least efficient for elimination of trend. In the past decade, there has been attempts in this direction to develop such designs. See, for example, Bradley and Yeh (1980), Lin and Dean (1991) and Jacroux et al. (1994). There seems to be a striking similarity in the analysis of a block design in the presence of trend to that of a row–column design in the sense that both aim at eliminating effects of two extraneous factors, while extracting information on the treatment contrasts. In case of a block design, the second factor is the trend while in case of a row–column design, it is the column classification (taking the row classification to represent the blocks).

Below, we shall make some remarks regarding this similarity in data analysis, without making a detailed comparison of the two approaches. Suppose in a block design, there are p blocks each of size q. If the trend across the plots can be represented by a polynomial of degree h, then the number of trend effects parameters is h if we assume the trend to be *common* to all the blocks and it is hp if it is *different* in different blocks. This may be compared with $(q-1)$ which is the number of parameters for column differences in a row–column design with p rows and q columns.

This comparison also leads to the following observations.

(a) If a common trend of degree h is assumed, information on treatment comparisons is available from $p(q-1)-h$ linear functions of the observations. These include the $(p-1)(q-1)$ tetra-differences! Thus, the $\boldsymbol{C}_t$ matrix for treatment comparisons can be obtained as a sum of the $\boldsymbol{C}$-matrix for the row–column design (i.e., $\boldsymbol{C}_{t.rc}$) and the $\boldsymbol{C}$-matrix obtainable from an analysis of the column totals. This second part is based on $(q-1-h)$ linear functions only and may be ignored if $(q-1-h)$ is small in comparison with $(p-1)(q-1)$. This is particularly appealing if we are interested in quick estimates with reasonably high efficiencies! (Recall that in this set-up, the row–column classification is orthogonal so that there is already some simplification achieved in the expression for $\boldsymbol{C}_{t.rc}$.)

(b) If the trend is different in different blocks, row–column designs would *not* do a good job of eliminating these trends.

On the whole, we can say that if there is good prior information about the nature of trend, the use of row–column design analysis may be helpful in extracting quick information and this may turn out to be highly efficient as well. Further to this, we may say that this analysis provides a more robust method of elimination of trend effects, especially when there is a common but unknown trend. Of course, it goes without saying that there is flexibility in the implementation of randomization in a row–column design while in the presence of trend, the plots in a block can not be randomized. In both cases, availability of good designs is rather limited. This is even more so in case of block designs in the presence of trend. Comparisons of these two approaches deserve further study.

11.3. Analysis of data under random effects model

We now discuss that aspect of the model which attends to random effects in respect of *all* the three factors. Thus, in effect, we assume that the experimental situation dictates that in addition to the two blocking factors, the treatment factor also has a large collection and we wish to accommodate only v of them, after making a random selection. With reference to the model (3.1), then we treat all the factor parameters as random, with the only exception of the general mean μ treated as a fixed parameter.

Earlier, in Sections 3, 4 and 9, we discussed the problem of inference on the treatment contrasts (fixed effects). In the present situation, however, reasonable inference on the treatment comparisons will center around the variance component σ_τ^2. We may note that we are now dealing with four variance components. Traditionally, in such models, the problem of variance components *estimation* has gained primary importance with due regard to *all* the components. We will review some of the results in this direction. An excellent treatise is the recently published book: Searle et al. (1992). This books deals almost exclusively with one-way and two-way classified data with various stipulated models built therein. Searle (1971) briefly discusses the problem of variance components estimation in three-factor designs but with a model more general than (3.1).

As is well-known, there are several methods for estimation of variance components even in very complex models. The method generally accepted and that easily works is called the *anova method*. According to this method, we form *unadjusted* sums of squares (i.e., the sum of squares for one factor *ignoring* all other factors, and the like for all the factors in turn) in the usual manner-thinking *as if* we are in a fixed effects situation! Then we apply the method of moments by equating these sums of squares to the respective expectations computed according to the random effects model in question. Applied to the model (3.1), this method yields the following equations:

$$\begin{aligned}
\mathrm{SSR} &= \sum \frac{R_i^2}{n_{i..}} - \mathrm{CF} = a_{rr}\sigma_R^2 + a_{rc}\sigma_C^2 + a_{rt}\sigma_\tau^2 + a_{ro}\sigma_e^2, \\
\mathrm{SSC} &= \sum \frac{C_j^2}{n_{.j.}} - \mathrm{CF} = a_{cr}\sigma_R^2 + a_{cc}\sigma_C^2 + a_{ct}\sigma_\tau^2 + a_{co}\sigma_e^2, \\
\mathrm{SST} &= \sum \frac{T_h^2}{n_{..h}} - \mathrm{CF} = a_{tr}\sigma_R^2 + a_{tc}\sigma_C^2 + a_{tt}\sigma_\tau^2 + a_{to}\sigma_e^2,
\end{aligned} \tag{11.3}$$

where

$$a_{rr} = n_{...} - \frac{\sum_i n_{i..}^2}{n_{...}}, \qquad a_{rc} = \sum_i \left(\frac{\sum_j n_{ij.}^2}{n_{i..}}\right) - \left(\frac{\sum_j n_{.j.}^2}{n_{...}}\right),$$

$$a_{rt} = \sum_i \left(\frac{\sum_h n_{i.h}^2}{n_{i..}}\right) - \left(\frac{\sum_h n_{..h}^2}{n_{...}}\right), \qquad a_{cr} = \sum_j \left(\frac{\sum_i n_{ij.}^2}{n_{.j.}}\right) - \left(\frac{\sum_i n_{i..}^2}{n_{...}}\right),$$

$$a_{cc} = n_{...} - \frac{\sum_j n_{.j.}^2}{n_{...}}, \qquad a_{ct} = \sum_j \left(\frac{\sum_h n_{.jh}^2}{n_{.j.}}\right) - \left(\frac{\sum_h n_{..h}^2}{n_{...}}\right),$$

$$a_{tr} = \sum_h \left(\frac{\sum_i n_{i.h}^2}{n_{..h}}\right) - \left(\frac{\sum_i n_{i..}^2}{n_{...}}\right), \qquad a_{tc} = \sum_h \left(\frac{\sum_j n_{.jh}^2}{n_{..h}}\right) - \left(\frac{\sum_j n_{.j.}^2}{n_{...}}\right),$$

$$a_{tt} = n_{...} - \frac{\sum_h n_{..h}^2}{n_{...}},$$

$$a_{ro} = p - 1, \quad a_{co} = q - 1, \quad a_{to} = v - 1. \tag{11.4}$$

This is the simplest solution to the problem of variance components estimation. There are many issues involved and this simple-minded solution does *not* satisfy necessary requirements like *non-negativity*!

We may now discuss some important special cases:

(i) When $N_{rc} = N_{rt} = N_{ct} = ((1))$, we are in the set-up of a Latin square design. In this case, the off-diagonal elements of the matrix of linear equations in (11.3) (other then a_{ro}, a_{co} and a_{to}) are all zeroes and hence the solution is very simple.

(ii) When all the pairwise classifications are orthogonal so that we are in a more general situation than in (i) above, then also all the off-diagonal elements are zeroes. Hence the solution is straightforward.

(iii) In case of a Youden square design, we have: $N_{rc} = N_{rt} = ((1))$. This yields $a_{rc} = a_{cr} = a_{rt} = a_{tr} = 0$. Hence we again have tremendous simplification in the solutions for the variance components.

The above method is easy to apply for balanced designs (such as LSDs). A general method (MINQE), developed by Rao (1970), is simpler to apply in more general situations. An earlier contribution by Henderson (1953) made significant impact in this direction.

11.4. Row–column designs for control versus treatment comparisons

The problem of comparing a set of *test* treatments with a *control* has received considerable attention in recent years from the optimality point of view. Most of the work in this area relate to block design set-up. Some attention has recently been paid to row–column design set-up as well. We will briefly discuss the problem.

The purpose is to compare a set of v test treatments (to be numbered $1, 2, \ldots, v$) with a control treatment (to be numbered 0) in a row–column design in an optimum manner. The first step towards this is to write down the estimates (blue's) of the v paired comparisons, viz., $\{\tau_0 - \tau_h, 1 \leqslant h \leqslant v\}$, and this is done in the usual way, starting with the $\boldsymbol{C}_{t.rc}$ matrix (of dimension $(v+1) \times (v+1)$). Note that for the inference problem to be meaningful, the design has to be treatment-connected so that rank $(\boldsymbol{C}_{t.rc}) = v$. Next, the dispersion matrix of the estimates has to be evaluated. An interesting basic result in this context is a simple and elegant representation of the above-mentioned dispersion matrix. This is given by $\boldsymbol{C}_{00}^{-1}$ where $\boldsymbol{C}_{00}$ is the matrix formed of the $\boldsymbol{C}_{t.rc}$ matrix by deleting its first row and first column (0th row and 0th column). As regards optimality, minimization of average variance (A-optimality) is the first criterion that has been thoroughly studied. It has also been observed that an A-optimal design for which the variance of different treatment comparisons (mentioned above) is the same, turns out to be MV-optimal as well. Here MV-optimality refers to minimizing the largest of the variances of various treatment comparisons under consideration. In the set-up of a block design, some kind of extension of balanced incomplete block designs (BIBDs) may turn out to be optimal. Das (1958) studied them in a different context and called them *reinforced* BIBDs. In the context of optimality, these have been termed balanced treatment incomplete block designs (BTIBDs) by Bechhofer and Tamahane (1981) who initiated this study. Such designs admit of all the combinatorial properties of a BIBD so far as the test treatments are concerned. Further to this, the control treatment appears the same number of times in the blocks with each of the test treatments. In the set-up of a row–column design, some families of A-optimal designs for this problem have been reported. These are generally based on some modifications of Latin square designs. We refer the reader to an excellent expository article (Hedayat et al., 1988). For more recent results and cross-references, we refer to Majumdar and Tamhane (1993).

Acknowledgement

This work was partially supported by a research grant from Natural Sciences and Engineering Research Council of Canada. This support is gratefully acknowledged.

References

Agrawal, H. L. (1966a). Two-way elimination of heterogeneity. *Calcutta Statist. Assoc. Bull.* **15**, 32–38.

Agrawal, H. L. (1966b). Some symmetric methods of construction of designs for two-way elimination of heterogeneity. *Calcutta Statist. Assoc. Bull.* **15**, 93–108.

Anderson, D. A. and J. A. Eccleston (1985). On the construction of a class of efficient row–column designs. *J. Statist. Plann. Inference* **11**, 131–134.

Bagchi, S., A. C. Mukhopadhyay and B. K. Sinha (1990). A search for optimal nested row–column designs. *Sankhyā, Ser. B* **52**, 93–104.

Bagchi, S. and K. R. Shah (1989). On the optimality of a class of row–column designs. *J. Statist. Plann. Inference* **23**, 397–402.

Bagchi, S. and E. E. M. Van Berkum (1991). Optimality of a new class of adjusted orthogonal designs. *J. Statist. Plann. Inference* **28**, 61–65.

Bailey, R.A. (1992). Efficient semi-Latin squares. *Statist. Sinica* **2**, 413–437.

Bechhofer, R. E. and A. C. Tamhane (1981). Incomplete block designs for comparing treatments with a control: General theory. *Technometrics* **23**, 45–57. Corrigendum (1982) **24**, 71.

Bondar, J. V. (1983). Universal optimality of experimental designs: Definitions and a criterion. *Canad. J. Statist.* **11**, 325–331.

Bradley, R. A. and C. M. Yeh (1980). Trend-free block designs. *Theory. Ann. Statist.* **8**, 883–893.

Chakrabarti, M. C. (1962). *Mathematics of Design and Analysis of Experiments*. Asia Publ. House, Bombay (reprinted in 1970).

Chang, J. Y. and W. I. Notz (1994). Some optimal nested row–column designs. *Statist. Sinica* **4**, 249–263.

Cheng, C. S. (1978). Optimal designs for the elimination of multi-way heterogeneity. *Ann. Statist.* **6**, 1262–1272.

Cheng, C. S. (1981). Optimality and construction of pseudo-Youden designs. *Ann. Statist.* **9**, 200–205.

Cochran, W. G. and G. M. Cox (1957). *Experimental Designs*. Wiley, New York (7th edn: 1966).

Cox, D. R. (1958). *Planning of Experiments*. Wiley, New York (5th edn: 1966).

Das, M. N. (1958). On reinforced incomplete designs. *J. Indian Soc. Agricult. Statist.* **10**, 73–77.

Eccleston, J. A. and B. Jones (1980). Exchange and interchange procedures to search for optimal row–column designs. *J. Roy. Statist. Soc. Ser. B* **42**, 372–376.

Eccleston, J. A. and C. McGilchrist (1985). Algebra of a row–column design. *J. Statist. Plann. Inference* **12**, 305–310.

Eccleston, J. and K. Russell (1975). Connectedness and orthogonality in multifactor designs. *Biometrika* **62**, 341–346.

Eccleston, J. A. and K. Russell (1977). Adjusted orthogonality in nonorthogonal designs. *Biometrika* **64**, 339–345.

Federer, W. T. (1955). *Experimental Design-Theory and Application*. Macmillan, New York (2nd edn: Indian Edition, 1967).

Federer, W. T. and L. N. Balaam (1973). *Bibliography on Experiment and Treatment Design Pre-1968*. Hafner, New York.

Fisher, R. A. (1935). *The Design of Experiments*. Oliver and Boyd (8th edn: 1966).

Hahn, G. J. (1982). Design of experiment: Industrial and scientific applications. In: Kotz and Johnson, eds., *Encyclopedia of Statistical Sciences*, Vol. 2. Wiley, New York, 349–358.

Hedayat, A. S. (1981). Study of optimality criteria in design of experiments. In: M. Csorgo et al., eds., *Statistics and Related Topics*. North-Holland, Amsterdam.

Hedayat, A. S. and D. Raghavarao (1975). 3-way BIB designs. *J. Combin. Theory Ser. A* **18**, 207–209.

Hedayat, A. S., M. Jacroux and D. Majumdar (1988). Optimal designs for comparing test treatments with controls (with discussions). *Statist. Sci.* **3**, 462–491.

Heiligers, B. and B. K. Sinha (1994). Optimality aspects of Agrawal's designs: Part II. *Statist. Sinica*, to appear.

Henderson (1953). The estimation of variance and covariance components. *Biometrics* **9**, 226–252.

Houtman, A. M. and T. P. Speed (1983). Balance is designed experiments with orthogonal block structure. *Ann. Statist.* **4**, 1069–1085.

Jacroux, M. (1982). Some E-optimal designs for the one-way and two-way elimination of heterogeneity. *J. Roy. Statist. Soc. Ser. B* **44**, 253–261.

Jacroux, M. (1986). Some E-optimal row–column designs. *Sankhyā Ser. B* **48**, 31–39.

Jacroux, M. (1987). Some E- and MV-optimal row–column designs having an equal number of rows and columns. *Metrika* **34**, 361–381.

Jacroux, M., D. Majumdar and K. R. Shah (1994). Efficient block designs in the presence of trends. *Statist. Sinica*, to appear.

Jarrett, R. G. (1994). On the construction and properties of row–column designs using their duals. Private communication.

Jones, B. (1980). Combining two component designs to form a row and column design. Algorithm AS156. *Appl. Statist.* **29**, 334–337.

Kempthorne, O. (1956). The efficiency factor of an incomplete block design. *Ann. Math. Statist.* **27**, 846–849.

Khatri, C. G. and K. R. Shah (1974). Estimation of location parameters from two linear models under normality. *Comm. Statist. Theory Methods* **3**, 647–663.

Khatri, C. G. and K. R. Shah (1986). Orthogonality in multiway classifications. *Linear Algebra Appl.* **82**, 215–224.

Kshirsagar, A. M. (1957). On balancing in designs in which heterogeneity is eliminated in two directions. *Calcutta Statist. Assoc. Bull.* **7**, 469–476.

Kiefer, J. (1975). Construction and optimality of generalized Youden designs. In: J. N. Srivastava, ed., *A Survey of Statistical Design and Linear Models*. North-Holland, Amsterdam, 333–353.

Krishnaiah, P. R. and M. G. Yochmowitz (1980). Inference on the structure of interaction in two-way classification model. In: *Handbook of Statistics*, Vol. 1. North-Holland, Amsterdam, 973–994.

Kunert, J. (1983). Optimal design and refinement of the linear model with applications to repeated measurements designs. *Ann. Statist.* **11**, 247–257.

Kurotschka, V. (1978). Optimal design of complex experiments with qualitative factors of influence. *Comm. Statist. Theory Methods* **7**, 1363–1378.

Lin, M. and A. M. Dean (1991). Trend-free block designs for treatment and factorial experiments. *Ann. Statist.* **19**, 1582–1596.

Majumdar, D. and A. C. Tamhane (1993). Row–column designs for comparing treatments with a control. Private communication.

Mandel, J. (1961). Nonadditivity in two-way analysis of variance. *J. Amer. Statist. Assoc.* **56**, 878–888.

Mandel, J. (1969). Partitioning the interaction in analysis of variance. *J. Res. Nat. Bur. Standards Sect. B* **73B**, 309–328.

Milliken, G. A. and F. A. Graybill (1970). Extensions of the general linear hypothesis model. *J. Amer. Statist. Assoc.* **65**, 797–807.

Morgan, J. P. and N. Uddin (1993). Optimality and construction of nested row–column designs. *J. Statist. Plann. Inference* **37**, 81–94.

Nandi, H. K. (1953). Analysis of serial experiments. *Calcutta Statist. Assoc. Bull.* **5**, 43–46.

Nelder, J. A. (1965a). The analysis of randomized experiments with orthogonal block structure, I. Block structure and the null analysis of variance. *Proc. Roy. Soc. (London) Ser. A* **273**, 147–162.

Nelder, J. A. (1965b). The analysis of randomized experiments with orthogonal block structure, II. Treatment structure and the general analysis of variance. *Proc. Roy. Soc. (London) Ser. A* **273**, 163–178.

Park, D. K. and K. R. Shah (1995). On connectedness of row–column designs. *Comm. Statist. Theory Methods* **24**(1), 87–96.

Rao, C. R. (1946). Confounded factorial designs in quasi-latin squares. *Sankhyā* **7**, 295–304.

Rao, C. R. (1970). Estimation of heteorscedastic variances in linar models. *J. Amer. Statist. Assoc.* **65**, 161–172.

Rao, C. R. and J. Kleffe (1988). *Estimation of Variance Components and Applications*. North-Holland, Amsterdam.

Pearce, S. (1975). Row–column designs. *Appl. Statist.* **24**, 60–77.

Roy, J. and K. R. Shah (1961). Analysis of two-way designs. *Sankhyā Ser. A* **23**, 129–144.

Russell, K. G. (1980). Further results on the connectedness and optimality of designs of type 0: XB. *Comm. Statist. Theory Methods* **9**, 439–447.

Ruiz, F. and E. Seiden (1974). On construction of some families of generalized Youden designs. *Ann. Statist.* **2**, 503–519.

Saharay, R. (1986). Optimal designs under a certain class of non-orthogonal row–column structure. *Sankhyā Ser. B* **48**, 44–67.

Scheffe, H. (1971). *The Analysis of Variance*. Wiley, New York.

Searle, S. R. (1971). *Linear Models*. Wiley, New York.

Searle, S. R., G. Casella and C. E. McCulloch (1992). *Variance Components*. Wiley, New York.

Shah, K. R. (1962). An estimate of inter-group variance in one-and two-way designs. *Sankhyā Ser. A* **24**, 281–286.

Shah, K. R. and C. G. Khatri (1973). Connectedness in row–column designs. *Comm. Statist. Theory Methods* **2**, 571–573.

Shah, K. R. and J. A. Eccleston (1986). On some aspects of row–column designs. *J. Statist. Plann. Inference* **15**, 87–95.

Shah, K. R. and P. D. Puri (1986). Commutative row–column designs. Unpublished manuscript.

Shah, K. R. (1988). On uniformly better combined estimates in row–column designs with adjusted orthogonality. *Comm. Statist. Theory Methods* **17**, 3121–3124.

Shah, K. R. and B. K. Sinha (1989). *Theory of Optimal Designs*. Lecture Notes in Statistics 54, Springer, New York.

Shah, K. R. and B. K. Sinha (1990). Optimality aspects of Agrawal designs. In: *Gujarat Stat Review* (Professsor Khatri Memorial Volume), 214–222.

Shah, K. R. and B. K. Sinha (1993a). Optimality aspects of row–column designs with non-orthogonal structure. *J. Statist. Plann. Inference* **36**, 331–346.

Shah, K. R. and B. K. Sinha (1993b). Optimal designs with component = wise orthogonal row–column structures. *J. Combin. Inform. System Sci.* **18**, 68–78.

Shrikhande, S. S. (1951). Designs for two-way elimination of heterogeneity. *Ann. Math. Statist.* **22**, 235–247.

Singh, M. and A. Dey (1979). Block designs with nested rows and columns. *Biometrika* **66**, 321–327.

Sinha, B. K. and T. Mathew (1991). Towards an optimum test for nonadditivity in Tukey's model. *J. Multivariate Anal.* **36**, 68–94.

Srivastava, J. N. (1978). Statistical design of agricultural experiments. *J. Indian Soc. Agricult. Statist.* **30**, 1–10.

Srivastava, J. N. (1993). Nonadditivity in row–column designs. *J. Combin. Inform. System Sci.* **18**, 85–96.

Tukey, J. W. (1949). One degree of freedom for nonadditivity. *Biometrics* **5**, 232–242.

Tukey, J. W. (1955). Answer to query 113. *Biometrics* **11**, 111–113.

Yates, F. (1937). The Design and Analysis of Factorial Experiments. Tech. Comm. No. 35, Imperial Bureau of Soil Science.

Yates, F. (1939). The recovery of inter-block information in treatment trials arranged in three-dimensional lattices. *Ann. Eugenics* **10**, 317–325.

Yates, F. (1940). The recovery of inter-block information in balanced incomplete block designs. *Ann. Eugenics* **10**, 317–325.

S. Ghosh and C. R. Rao, eds., *Handbook of Statistics, Vol. 13*

25

Nested Designs

J. P. Morgan

1. Introduction

This chapter will explore the practical motivation for, optimality requirements of, and combinatorial assignment properties in, experimental situations with nested blocking factors. Blocking, one of the fundamental principles of experimental design, is the separation of a set of heterogeneous experimental units into subsets of more homogeneous units. Formally, if Ω is the set of n units to be used in the experiment, a blocking factor F with b levels can be defined as a partition of Ω into b subsets, called blocks. If F is to be a useful blocking factor, its partition corresponds to known levels (intensities, amounts, characteristics) of a well-defined source of variation in the units that is suspected to affect the response that will ultimately be measured: those units in the same block are identical, or at least as similar as possible, with respect to this source of variation, while those in different blocks are demonstrably different. The simple and elegant idea is that comparison of measurements from units in the same block will then be made with more precision, or, depending on the model, more accuracy, than those from different blocks. Blocking is thus a means of isolating identifiable and unwanted, but unavoidable, variation in experimental material.

In a given experimental situation there are potentially many available blocking factors with many plausible relationships among those factors. Of particular concern here is the often found relationship of nesting. If F and G are two blocking factors, then F is said to nest G if units in two different F-blocks must be in two different G-blocks. Another way of saying this is that each F block is a union of G-blocks. This relationship may be represented diagrammatically as shown in Figure 1. Examples follow.

In standard bioequivalence studies to investigate comparability of generic drugs to established reference drugs there are three treatments: a positive control (the reference pioneer product), a negative control (placebo) and the test product (generic). When animal subjects are used they are first typically blocked according to some relevant physical characteristic(s) such as age, weight, health status and/or husbandry and feeding conditions. To be specific, consider the investigation of a deworming medication for dogs. The experiment is to be run on 30 dogs, but the test facility can house only 15 dogs at a time, which are obtained and given pretreatment fecal egg counts. According to these counts, 5 blocks of 3 dogs each are formed, dogs with

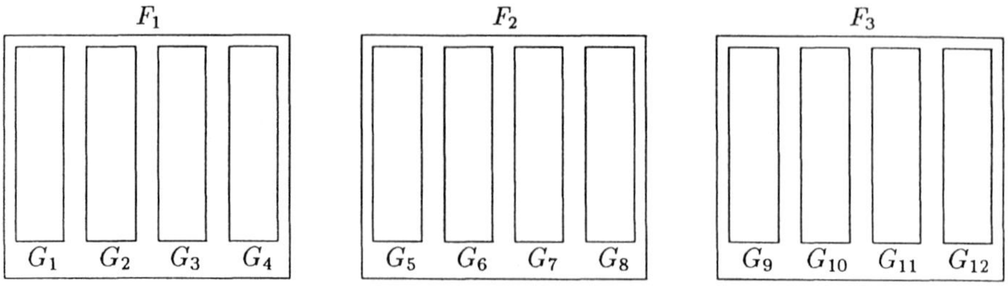

Fig. 1. F with 3 levels nests G with 12 levels.

the 3 highest counts going into block 1, the next 3 highest into block 2, and so on. The 3 treatments are then assigned randomly to the 3 dogs in each block. Treatments are administered for two weeks, then the dogs are euthanized and parasite/egg counts in the gastrointestinal tract made at necropsy. The facility is now sterilized and the entire procedure repeated on a new set of 15 dogs. Two relevant blocking factors are identified by the ordered groups of 3 dogs (G, with 10 levels) and the two sets of 15 dogs (F, with 2 levels). Introduction of the nesting factor F in this example is characteristic of a common mode by which nesting factors arise: groups of blocks identified by another blocking factor G undergo temporally distinct processing. The effects of such a temporal shift may be manifested in a variety of ways, including but not limited to changes in environmental conditions (humidity, temperature, air pressure), changes in experimenting technicians (different personnel, or learning by the same technicians), and systematic differences in the experimental units themselves (for instance, a different vendor for the second set of dogs).

Just as temporal separation can induce a nesting, so too can spatial separation. In an experiment to compare two arithmetic tutorial softwares in a school system, two classes of third graders are selected from each of three schools. Within each class students are blocked into three groups according to relative performance in the previous semester, then treatments (softwares) assigned to students within groups, the first blocking variable. Classes, each manned by a different teacher, form a second blocking variable which nests the first, reflecting the likely possibility that performance will be affected by instructor effectiveness. Finally, a third blocking variable which nests the first two is defined by the three schools, incorporating differences in the general socioeconomic status of the communities from which their students are drawn, in the quality of the physical surroundings (buildings, lights, computing equipment, etc.), in the principals' general philosophies, and so on. Spatial separation of the schools implies a host of potential sources of nuisance variation.

Yet another common generator for a nesting variable is found in the management practices attendant to some block designs. In agricultural field trials, two or more harvesting teams (equipment with operators, or groups of pickers as with fruit trees) will define a nesting variable according to the blocks harvested by each team. Well before harvesting, grouping of blocks is implied by the time and resource restrictions inherent in cultivation and thinning techniques, pest treatment, irrigation devices, etc. Often all of these nuisance factors are kept constant on geographically proximate blocks, in this way defining larger blocks as management units. In the United Kingdom, for example, it has now become standard practice in variety trials to use resolvable block designs (see Section 3), the single replicate big blocks expediting the management process (Patterson and Silvey, 1980).

These are just three of the uncountably many experimental situations in which nesting of blocks is either due to inherent qualities of the experimental material, or is imposed on the units by the experimental procedure; scientists in other areas of investigation should be able to easily formulate examples from their own subject areas upon considering the principles they demonstrate. The discussion of more complicated structures on nested blocks will be delayed until Sections 5 and 8.

Despite the fairly widespread occurrence of nested blocking factors across the experimental spectrum, there seems to be no comprehensive accounting of the properties required in these situations, nor of the actual designs that have been constructed to meet them. While there are a few valuable surveys of particular topics to be covered here, they tend to take a primarily combinatorial, rather than statistical, view of their subject matter, and do not meet the more diffuse goals of the statistician seeking a broader understanding of the subject as a whole. And while the study of nested factors in an analysis of variance is a fairly standard topic in textbooks covering design, at best spotty attention is paid to the specific science of designing experiments incorporating nested blocks.

This chapter is intended as an initial step towards filling this void. The groundwork is statistical optimality: it is first asked what conditions will draw the most information out of an experiment in a given setting that incorporates nesting, and for what analysis. Operating with the paradigm that it is ultimately the design itself that is of paramount practical importance, the main body is a survey of designs fulfilling, or approximating, the found conditions, referencing design catalogs and presenting design tables wherever possible. In drawing together this information under a common heading, large gaps in the existing theory and results become apparent, some in design construction problems, but more in the limits on current optimality theory which has not reached very far beyond those settings where rather severe symmetry conditions can be met.

2. Nesting in block designs – Models and related considerations

2.1. Model for a simple case

The simplest of nesting situations makes for a good starting point: there are b_1 blocks of b_2k units each, and each of these blocks nests b_2 blocks of k units. In the notation of Section 1, the blocking factor F has b_1 levels, the nested blocking factor G has b_1b_2 levels, and the total number n of experimental units, or plots, is b_1b_2k. If y_{ijl} is used to denote the yield from plot l in sub-block j (level of G) of block i (level of F), the commonly employed model is

$$y_{ijl} = \mu + \beta_i^1 + \beta_{ij}^2 + \tau_{[ijl]} + e_{ijl}. \tag{1}$$

The terms in this model are

$$\begin{aligned}
\mu &= \text{an overall mean response,}\\
\beta_i^1 &= \text{effect of block } i,\\
\beta_{ij}^2 &= \text{effect of sub-block } j \text{ in block } i,\\
\tau_{[ijl]} &= \text{effect of treatment assigned to plot } (i,j,l),\\
e_{ijl} &= \text{a random error term.}
\end{aligned}$$

The e_{ijl}'s will be taken to be uncorrelated random variables with zero means and equal variances. Standard hypothesis testing and confidence interval procedures further assume the e_{ijl}'s to be normally distributed (e.g., Searle, 1971), but that additional information is not needed for the problems to be discussed here.

With the y_{ijl}'s lexicographically ordered by the subscripts i, j, l in an observation vector Y, the model (1) in matrix notation is

$$Y = \mu 1_n + L_1\beta^{(1)} + L_2\beta^{(2)} + A_d\tau + E \tag{2}$$

in which 1_n is a $n \times 1$ vector of 1's, $L_1 = I_{b_1} \otimes 1_{b_2k}$, $L_2 = I_{b_1b_2} \otimes 1_k$, A_d is the $n \times v$ design matrix of 0's and 1's indicating assignment of treatments to plots, $\beta^{(1)}$ and $\beta^{(2)}$ are $b_1 \times 1$ and $b_1b_2 \times 1$ vectors of effects for blocks and sub-blocks, and τ is the $v \times 1$ vector of treatment effects. The design goal is to choose A_d in such a way that information on comparison of the τ_i's is of the highest possible quality, and it should come as no surprise that the method by which that goal is achieved depends on the analysis used.

Only linear estimation will be considered here, that is, all estimation will be based on linear combinations of the yields y_{ijl}. Within this framework, the major question that arises concerns what general classes of linear combinations will be allowed. The nature of this question is easily seen in the *simple block setting:* consider a single blocking factor with b blocks of k plots each. If only (linear combinations of) contrasts of yields from *within* these blocks are allowed, then one is performing what is called the "within blocks" or "bottom stratum" analysis. If in addition, contrasts of block totals are used for treatment estimation, the name is "full analysis" or "analysis with recovery of interblock information". Lest treatment estimation be hopelessly biased, use of block totals contrasts assumes that the block effects are mean zero random variables, implying that these contrasts are subject to greater variability than within blocks contrasts and thus requiring a weighted analysis. The bottom stratum analysis is unweighted and makes no assumption about the nature, fixed or random, of the block effects, although it is sometimes called the "fixed effects analysis", as it is a consequence of standard linear model theory when that assumption is made.

Considered in turn for the nested blocks setting are the bottom stratum analysis and the analysis recovering information from the block strata. In each case an information matrix C_d for the best linear unbiased estimators $\widehat{\tau}$, depending on A_d, is found (the Moore–Penrose inverse C_d^+ is an effective variance-covariance matrix for $\widehat{\tau}$). One may then manipulate A_d to, in some sense, maximize that information: find an optimal design. Specifically considered are optimality criteria (functions of C_d) that preserve the nonnegative definite ordering. That is, if $C_{d_1} - C_{d_2}$ is nonnegative definite, then d_1 is preferred to d_2.

A special class of designs for the simple block setting will arise here and in later sections. A *balanced block design* (BBD) is a design for v treatments in b blocks of size k with the properties that (i) each treatment occurs $\text{int}(\frac{k}{v})$ or $\text{int}(\frac{k}{v}) + 1$ times in every block, and (ii) each unordered pair of treatments occurs on the same number λ of pairs of plots within the blocks. If $k < v$, a BBD is called a *balanced incomplete block design,* denoted BIBD(v, b, k) or simply BIBD, and λ is $bk(k-1)/(v(v-1))$. Property (i), when $k \leqslant v$, is called a *binary* treatment assignment.

2.2. *The bottom stratum analysis*

Write $L = (1_n, L_1, L_2)$ and $\beta = (\mu, \beta^{(1)'}, \beta^{(2)'})'$. Then (2) is $Y = L\beta + A_d\tau + E$, and the bottom stratum information matrix C_d, found by projecting orthogonally to the non-treatment factors, is

$$C_d = A_d'(I - P_L)A_d, \tag{3}$$

P_L being the matrix which projects onto the space spanned by the columns of L. It is easy to see that the columns of L and of L_2 have the same span, so that $P_L = P_{L_2}$. Thus (3) can be rewritten

$$C_d = A_d'(I - P_{L_2})A_d. \tag{4}$$

The matrix (4) is exactly that arising from the simpler model $Y = \mu 1 + L_2\beta^{(2)} + A_d\tau + E$, in which the nesting blocks play no role! So the assignment of treatments to the nesting blocks is irrelevant for this analysis. One need only determine an optimal block design for $b_1 b_2$ blocks of size k, then *arbitrarily* arrange them into b_1 nesting blocks of b_2 sub-blocks each. Optimal design for the model (2), for this analysis, is thus the purview of another chapter of this volume.

As simple as this result is, there is still something disquieting about the failure of the nesting structure to play a role in either the design or the analysis. What the just completed argument has shown is that the nesting effects, carefully accounted for in the planning stage and accommodated in the model, could have been ignored from the outset, an apparent affront to common sense. The full analysis will say otherwise, but before proceeding to Section 2.3, there is another consideration for this analysis by which the nest does exert its influence.

Bhaumik and Whittinghill (1991) study design optimality questions in the face of loss of blocks, and one of their results is applicable to this problem. As explained in Section 1, nesting blocks often represent a natural grouping of smaller blocks that are similar due to management practices, proximity in time or space, or other reasons. Thus it is reasonable to seek a design that can withstand the loss of one of its nests of smaller blocks, should there be a localized management failure or a destructive time or space event. Theorem 2 of Bhaumik and Whittinghill (1991) says that, at least if it is variance balanced, a design is most robust to such a loss, in the sense of being best under all Schur-convex criteria, if the treatment assignment to nesting blocks is binary. This will also assure that should the original design be equireplicate, the depleted design will be as close to equireplicate as possible. As shall be seen, this is a property possessed uniformly by the commonly employed nested block designs appearing in the literature. Moreover, this idea of "equal impact" on the treatments, under either loss of nesting blocks or possible model shortcomings related to the nesting blocks, is an important justification for use of the resolvable designs that are the topic of Section 3.

2.3. *The full analysis*

To recover information from the blocking strata requires knowledge concerning the first two moments of the block effects. It is assumed that $\beta^{(1)}, \beta^{(2)}$, and E are uncorrelated random vectors with zero means and with $\mathrm{var}(\beta^{(1)}) = \sigma_1^2 I_{b_1}$, $\mathrm{var}(\beta^{(2)}) = \sigma_2^2 I_{b_1 b_2}$, and $\mathrm{var}(E) = \sigma^2 I_n$. Then

$$\mathrm{var}(Y) = \zeta_0 S_0 + \zeta_1 S_1 + \zeta_2 S_2 + \zeta_3 S_3 = \Sigma, \text{ say,}$$

where

$$\zeta_0 = \zeta_1 = b_2 k \sigma_1^2 + k\sigma_2^2 + \sigma^2, \quad \zeta_2 = k\sigma_2^2 + \sigma^2, \quad \zeta_3 = \sigma^2,$$

$$S_0 = P_0, \quad S_1 = P_1 - P_0, \quad S_2 = P_2 - P_1, \quad S_3 = I - P_2$$

and

$$P_0 = \frac{1}{n} 1_n 1_n', \quad P_1 = L_1(L_1'L_1)^- L_1', \quad P_2 = L_2(L_2'L_2)^- L_2' \; (= P_{L_2}).$$

The S_α's are orthogonal projection matrices which sum to I. The information matrix for the full weighted analysis is

$$\begin{aligned} C_d &= A_d'(\Sigma^{-1} - (1'\Sigma^{-1}1)^{-1}\Sigma^{-1}11'\Sigma^{-1})A_d \\ &= A_d'(\Sigma^{-1} - \zeta_0^{-1} S_0)A_d \\ &= A_d'(\zeta_1^{-1} S_1 + \zeta_2^{-1} S_2 + \zeta_3^{-1} S_3)A_d \\ &= \zeta_1^{-1} A_d'(I - P_0)A_d + (\zeta_2^{-1} - \zeta_1^{-1})A_d'(I - P_1)A_d \\ &\quad + (\zeta_3^{-1} - \zeta_2^{-1})A_d'(I - P_2)A_d \\ &= \zeta_1^{-1} C_d^{(0)} + (\zeta_2^{-1} - \zeta_1^{-1})C_d^{(1)} + (\zeta_3^{-1} - \zeta_2^{-1})C_d^{(2)}. \end{aligned} \tag{5}$$

The coefficients ζ_1^{-1}, $\zeta_2^{-1} - \zeta_1^{-1}$, and $\zeta_3^{-1} - \zeta_2^{-1}$ are all nonnegative, and the component matrices are

$C_d^{(0)}$ = information matrix for estimating τ in the absence of both types of block effects,

$C_d^{(1)}$ = the bottom stratum information matrix for estimating τ in the model $Y = \mu 1 + L_1\beta^{(1)} + E$, i.e., in the model with only the nesting blocks, and

$C_d^{(2)}$ = the bottom stratum information matrix for estimating τ in the model $Y = \mu 1 + L_2\beta^{(2)} + E$ (this is (4)).

The involvement of $C_d^{(1)}$ in (5) shows that the nesting blocks play a much different role in this analysis than in that of Section 2.2. Using a result of Kiefer (1975), a design d will be *universally optimum* if C_d is completely symmetric of maximum trace, and by a result of Shah and Sinha (1989), will thus be optimum with respect to all criteria preserving the nonnegative definite ordering and possessing certain other natural requirements specified there (Shah and Sinha, 1989, pp. 4–7). Sufficient for this is to make each of $C_d^{(0)}$, $C_d^{(1)}$, and $C_d^{(2)}$ completely symmetric of maximum trace, which is accomplished if

(i) the b_1 nesting blocks form a BIBD, or more generally a BBD, with blocks of size b_2k, and

(ii) the b_1b_2 nested blocks form a BIBD or BBD with blocks of size k.

Designs satisfying these two conditions, for which b_2k is equal to v, are resolvable BIBDs. These designs will be discussed with other resolvable designs in Section 3; additional optimality perspectives will be given there as well. For the same conditions and b_2k less than v, the designs are the nested BIBDs introduced by Preece (1967), which are the topic of Section 4. Both types of designs are also optimal for the bottom stratum analysis of Section 2.2, and satisfy the robustness condition discussed there. The author is unaware of any other exact optimality work under this the full analysis.

2.4. Extensions

There is no need to restrict attention to a simple nesting of just one set of blocks within another, for the above is easily extended to t blocking factors that successively nest one another. For instance, with $n = b_1b_2b_3k$ one could have $b_1b_2b_3$ blocks of size k nested within b_1b_2 blocks of size b_3k which are nested within b_1 blocks of size b_2b_3k. Put

$$L_1 = I_{b_1} \otimes 1_{b_2b_3k}, \qquad L_2 = I_{b_1b_2} \otimes 1_{b_3k}, \quad \text{and} \quad L_3 = I_{b_1b_2b_3} \otimes 1_k.$$

The model is

$$Y = \mu 1_n + L_1\beta^{(1)} + L_2\beta^{(2)} + L_3\beta^{(3)} + E.$$

Also writing $L_0 = 1_n$ and $P_i = L_i(L_i'L_i)^- L_i'$, $\zeta_0 = \zeta_1 = b_2b_3k\sigma_1^2 + b_3k\sigma_2^2 + k\sigma_3^2 + \sigma^2$, $\zeta_2 = b_3k\sigma_2^2 + k\sigma_3^2 + \sigma^2$, $\zeta_3 = k\sigma_3^2 + \sigma^2$, and $\zeta_4 = \sigma^2$, the bottom stratum information matrix is

$$C_d^{(3)} = A_d'(I - P_3)A_d$$

and the information matrix for the full analysis is

$$C_d = \zeta_1^{-1}C_d^{(0)} + (\zeta_2^{-1} - \zeta_1^{-1})C_d^{(1)} + (\zeta_3^{-1} - \zeta_2^{-1})C_d^{(2)} + (\zeta_4^{-1} - \zeta_3^{-1})C_d^{(3)}.$$

$C_d^{(0)}$, $C_d^{(1)}$, and $C_d^{(2)}$ are as in Section 2.3, but with k there replaced by b_3k. One can now define a doubly nested BIBD (BBD) to be a design for which

(i) the b_1 blocks form a BIBD (BBD) with blocksize b_2b_3k,
(ii) the b_1b_2 blocks of size b_3k form a BIBD (BBD), and
(iii) the $b_1b_2b_3$ blocks of size k form a BIBD (BBD).

A doubly nested BIBD (BBD) is optimal for either analysis. The extension to further levels of pure nesting is from here obvious.

3. Resolvable block designs

3.1. Basics and overview

Resolvability is the most widely studied and employed variation on nesting in block designs. Any block design for v treatments in b blocks of size k is said to be *resolvable* if the b blocks can be partitioned into b_1 classes of b_2 (sub-)blocks each so that each treatment occurs once in each class. Each resolved component of b_2 blocks corresponds to one of b_1 levels of the nesting factor, and is called a *resolution class* or *replicate.* Because the number of levels for the nesting factor is the number of replicates, r will be written for b_1 throughout Section 3, and for consistency of notation, s will be written for b_2. Necessarily, then, $v = sk$, and $b = rs$. As the nesting blocks are complete, the full analysis will take information only from the two lower strata.

In light of the comments of Section 2.2, choice of a resolvable design is made either with the full analysis in mind or with concern for robustness to problems associated with levels of the nesting factor. For if efficiency for the bottom stratum analysis is the sole objective, a resolvable design is not necessarily the best choice. As an example, John and Mitchell (1977) give non-resolvable designs with higher A-efficiency than the best resolvable designs (see Section 3.3) for $v = 12$, $k = 6$, and $3 \leqslant r \leqslant 10$. So while resolvable designs may be used for many reasons (not all of which may be reflected in the model), best bottom stratum efficiency is not necessarily one of these.

The volume of work on resolvable designs precludes a detailed comprehensive treatment here, for such a survey would demand a chapter of this volume unto itself. Instead the following subsections will attempt to hit the highlights of this important topic while giving sufficient references to ease the search task of those who wish to explore further. For the practitioner there are brief descriptions of the major classes of resolvable designs and references for tables. In this way some of the newer, less developed topics in nested designs can be given solid coverage in later sections.

3.2. Lattices and resolvable BIBDs

Though predating the term "resolvable", the first of the systematically studied and applied resolvable designs were the *lattices*, introduced by Yates (1936). In that paper is found one of the early powerful ideas of experimental design: use of pseudo-factors to bring factorial techniques to bear as tools in non-factorial experimentation.

As an example, suppose there are 16 treatments to be compared in a resolvable design with $k = 4$. Arrange the 16 treatment labels in a 4×4 square array (or lattice), letting rows be the levels of a pseudofactor A, and columns the levels of another pseudofactor B, the 16 treatments thus being identified with the 16 factorial combinations of A and B. Using rows as the 4 blocks of one replicate amounts to confounding the A main effect. The columns give a second replicate by confounding the B main effect, and further replicates can be obtained by confounding pseudofactors which represent orthogonal components of the A×B interaction. With five replicates one gets a resolvable BIBD, also called a *balanced lattice.* The design is displayed in Example 1.

EXAMPLE 1. A balanced square lattice for 16 treatments.

1	5	9	13
2	6	10	14
3	7	11	15
4	8	12	16

1	10	15	8
5	14	11	4
9	2	7	16
13	6	3	12

1	2	3	4
6	5	8	7
11	12	9	10
16	15	14	13

1	14	7	12
8	11	2	13
10	5	16	3
15	4	9	6

1	16	6	11
7	10	4	13
12	5	15	2
14	3	9	8

Lattice designs like the example just given, with $v = k^2$, are called two-dimensional (or square) lattices. More generally, *m-dimensional lattices* have k^m treatments in blocks of size k^t. The construction theory for m-dimensional lattices when k is a prime or prime power is completely known and detailed in, e.g., Kempthorne and Federer (1948) and Kempthorne (1952). While balance can always be achieved with sufficient replication, one needs only m/t properly chosen replicates to achieve estimability of all treatment contrasts (any two replicates in Example 1 will do). The 2- and 3-dimensional lattices are the most widely used and hence the most extensively tabled; see Clem and Federer (1950) and Cochran and Cox (1992). References for other plans may be found in the synopsis of Cornelius (1983), who also gives a thorough accounting of the analysis of these designs. Cheng and Bailey (1991) prove that the square lattices are optimal amongst all binary, equireplicate designs with the same b, v, and k, so that these particular resolvable designs are unlikely to be bested in the bottom stratum analysis by a nonresolvable alternative.

Harshbarger (1946, 1947, 1949, 1951) introduced the *rectangular lattices* as a class of resolvable designs for which the number of treatments is not a power of k. These designs have $v = k(k+1)$ with r replicates of $k+1$ blocks each, $2 \leqslant r \leqslant k+1$, and no two treatments appearing together in a block more than once. Designs for $r = 3$

exist for all $k \geqslant 2$, but the achievable upper bound for r generally depends on k, with r of $k+1$ being attainable if $k+1$ is a prime or prime power. The designs are partially balanced when $r = 3, k$, or $k+1$. Plans for $r = 3$ and $k = 3, 4, \ldots, 9$ are tabled by Cochran and Cox (1992). A comprehensive treatment of the analysis and construction of rectangular lattices is given by Bailey and Speed (1986).

It was remarked above that the square lattice of Example 1 is a resolvable BIBD. Indeed the square lattices with $r = k+1$ and k a prime or prime power are always balanced, that is, they are always resolvable BIBDs. Moreover, they are part of a general geometric construction (Bose, 1939) for BIBDs which corresponds to the m-dimensional lattices with blocks of size k^t and t properly dividing m. If $q = k^t$ is a prime power, then there exists an affine geometry of dimension m/t over GF_q containing k^m points and $k^{m-t}(k^m - 1)/(k^t - 1)$ lines. With points as treatments and lines as blocks, this is a resolvable BIBD with $\lambda = 1$ (resolution classes being the parallel pencils). Other lattices are likewise embedded in resolvable BIBDs with $\lambda > 1$.

There are of course many resolvable BIBDs other than those related to lattices. Easy to use difference solutions have been tabled by Kageyama (1972) for $v \leqslant 100$ and $r \leqslant 20$. Within these ranges there are 48 feasible cases (see equation (6)) of which 14 are listed by Kageyama (1972) as unsolved. References for solutions to seven of these cases (in Kageyama's notation, numbers 20, 21, 23, 24, 27, 34, and 48; the last is unsolvable) can be found in Furino et al. (1994), who give an existence table covering $10 \leqslant v \leqslant 200$ and $k \geqslant 5$, survey relevant construction techniques, and include an extensive bibliography. Gupta and Srivastava (1992, Section 2.1) compute the efficiencies for the designs in Kageyama's (1972) listing under the loss of a single nesting block.

Simple counting shows that necessary conditions for existence of a resolvable BIBD are

$$v \equiv 0 \pmod{k} \quad \text{and} \quad \lambda(v-1) \equiv 0 \pmod{k-1}. \tag{6}$$

The existence problem is to determine for what values (k, λ) these are sufficient, and towards this goal Ray-Chaudhuri and Wilson (1973) and Lu (1984) have proven that for any given k and λ, (6) is sufficient except possibly for finitely many v. The drawback is that the size of this finite number of exceptions is not known, so could potentially be quite large. However for small k some precise results are known. Aside from $v = 6, k = 3, \lambda \equiv 2 \pmod 4$, for which a resolvable BIBD does not exist, Shen and Wallis establish in an unpublished manuscript that (6) is sufficient for $k = 3$ or 4. For $(k, \lambda) = (5, 1), (5, 2), (5, 4)$, or $(6, 5)$, sufficiency has been established, except possibly for known finite lists of values for v, via the contributions of a host of authors (for details and references see the survey of Zhu, 1993, and the monograph of Furino et al., 1994). Baker (1983) proves (6) is sufficient for $(k, \lambda) = (6, 10)$.

3.3. *Affine resolvable designs*

A resolvable block design is said to be *affine resolvable* if any two blocks in different replicates have the same number μ of treatments in common. Given an affine resolvable design with $b = rs$ blocks of size k, arbitrarily label the s blocks of each resolution class as $1, 2, \ldots, s$. Now form a $r \times v$ array $A = (A_{ij})$ for which A_{ij} is the label of the sub-block of replicate i that contains treatment j. Then each of the sub-block labels $1, 2, \ldots, s$ occurs k times in each row of A, and affine resolvability says that the columns of every $2 \times v$ subarray of A are the s^2 ordered pairs of sub-block labels with frequency μ. In short, A is an orthogonal array of strength 2 and index μ. And the construction is reversible: given any orthogonal array of strength (at least) 2, an affine resolvable design is found by identifying the columns with a set of treatments, the rows with replicates, and the symbols with blocks within replicates. So the affine resolvable designs are exactly the strength 2 orthogonal arrays.

Affine resolvability was introduced in the BIBD context by Bose (1942), who showed that a resolvable BIBD must satisfy $b \geqslant v + r - 1$, with equality if and only if the design is affine resolvable. Thus did the affine notion enter the statistical literature for essentially combinatorial, not statistical, reasons. While the equivalence with orthogonal arrays indicates the fundamental importance of affineness to a large body of combinatorial design theory, its wholesale omission from applications-oriented texts is convincing testament to its irrelevance to the statistical aspects of resolvable designs.

However, a recent paper by Bailey et al. (1995) changes this assessment, establishing affineness as a statistically desirable property of resolvable designs. These authors prove that affine resolvable designs are Schur-optimal amongst all designs with the same r, s and k. This result, for the bottom stratum analysis only, further implies that affine resolvable designs are most robust to loss of replicates.

The equivalence with orthogonal arrays imposes the obvious necessary condition that v be divisible by s^2. Given s^2 properly dividing v (i.e., $k > s$; $k = s$ is the class of square lattices), a result of Beth et al. (1986, Section x.2) implies the existence of an affine resolvable design with at least seven replicates (this is Theorem 2.1 of Bailey et al., 1995). For $v \leqslant 100$ all of the designs can be constructed by juxtaposing and/or taking the Cartesian product of a few starting orthogonal arrays: see Bailey et al. (1995) for details. See Beth et al. (1986) more generally for results on orthogonal arrays. It should be mentioned that the square lattices of Section 3.2, and all of the resolvable BIBDs with $\lambda = 1$ based on affine geometries that are cited there, are affine resolvable, Example 1 being one such case. Of the 48 feasible resolvable BIBDs with $r \leqslant 20$ and $v \leqslant 100$, only 12 are affine resolvable (see Kageyama, 1972).

3.4. Alpha designs

Section 3.3 presented optimal resolvable designs for $v = sk$ treatments and $k > s$, while the lattices of Section 3.2 have $k \leqslant s$. Of the latter, the square lattices ($k = s$) are optimal and the rectangular lattices ($k = s - 1$) are highly efficient (Williams, 1977; Bailey and Speed, 1986). For smaller values of k, a flexible family of resolvable designs is provided by the *alpha designs,* for which there is some overlap with the lattices.

Introduced in Williams (1975) and Patterson and Williams (1976a), alpha designs were created in response to the need for a large catalog of resolvable designs for use in United Kingdom statutory field trials of agricultural crop varieties. They are cyclically generated, with r replicates requiring r initial blocks. So that the technique can be seen as an instance of Bose's method of differences, the construction will be described here in a slightly nonstandard form; the standard approach may be found in the original papers just cited or in the books of John (1987) or Street and Street (1987).

Associate the $v = sk$ treatments with k distinct copies of Z_s, the integers (mod s). This is conveniently done by subscripting members of the ith copy with an i, so that the treatments are $0_1, 1_1, \ldots, (s-1)_1, 0_2, 1_2, \ldots, (s-1)_2, \ldots, 0_k, 1_k, \ldots, (s-1)_k$. The subscripting does not affect the group operations: $(a_i + l) = ((a+l) \pmod s)_i$, and so on. The residuals subscripted by i are said to be "of type i".

An initial block of size k is found by selecting one element from among those of type i for each $i = 1, 2, \ldots, k$. From the initial block, s blocks are generated by adding 1, then 2, and so on, and reducing $(\operatorname{mod} s)$. That is, the initial block is developed $(\operatorname{mod} s)$. Since under this operation one type i element generates all of the type i elements, and since the initial block contains one element of each type, the s blocks are a single replicate.

Alpha designs are classified according to the distinct concurrence counts among the treatments. Since the designs are cyclic, these can be calculated solely from the differences in the initial blocks. In Bose's (1939) language, differences $(a_i - b_j) = ((a-b)(\operatorname{mod} s))_{(i,j)}$ are *mixed* for $i \neq j$ and *pure* for $i = j$. Resolvability has forced all elements of an initial block to be of distinct type, so there are no pure differences: every alpha design must necessarily have at least $k\binom{s}{2}$ concurrences of 0. The statistician picks initial blocks according to the mixed differences with the goal of maximizing efficiency within this class. This typically necessitates as a first step keeping the range of distinct concurrence counts as small as possible, so that most desirable are the $\alpha(0,1)$-designs, i.e., those with concurrences of 0 and 1 only. Next are the $\alpha(0,1,2)$-designs with concurrences of 0, 1 and 2, and so on.

A necessary condition for an $\alpha(0,1)$-design is $k \leqslant s$, and for $r = 2, 3, 4$ and $v \leqslant 100$ there are 198 cases. Williams (1975) tables initial blocks generating $\alpha(0,1)$-designs for 188 of these, 9 are covered by lattices, and one ($s = k = 6, r = 4$) requires an $\alpha(0,1,2)$-design. Tables for alpha designs are presented with initial blocks as columns of $k \times r$ arrays, and by placing the type i elements in row i, the subscripts, being understood, can be omitted.

A more compact and readily accessible tabling, with some of the designs sacrificing a bit of efficiency relative to those in Williams (1975), is given in Patterson et al. (1978). For designs beyond the range of these tables there is a computer algorithm available to generate efficient alpha designs (Paterson and Patterson, 1983; Paterson et al., 1988). Paterson (1988) gives a highly readable account of the development of alpha designs and their successful application in agricultural trials, with many references.

3.5. *Other resolvable designs and extensions*

Patterson and Williams (1976b) prove that every 2-replicate, resolvable incomplete block design with $v = sk$ is equivalent to a symmetric block design for s treatments in s blocks of k. To see this, form the $v \times 2$ array A of Section 3.3 for the 2-replicate resolvable design, which will not now be an orthogonal array unless that design is affine resolvable. Next identify the symbols in row 1 of A with blocks, and the symbols in row 2 with treatments, of what is necessarily a symmetric design. That is, the 2-tuples formed by the columns of A are the block-treatment pairs for the required symmetric design, called the *contraction* of the resolvable design. Clearly this process is reversible.

Patterson and Williams (1976b) also find a direct relationship between the efficiency of the resolvable design and its contraction, a consequence of which is that A-optimality of one is equivalent to A-optimality of the other. Hence the search for optimal (efficient) 2-replicate resolvable designs is reduced to the search for optimal (efficient) symmetric block designs. This theme is explored further by Williams et al. (1976).

The ideas of partial balance have been explored for resolvable block designs to a relatively lesser degree than the other approaches of this section. Clatworthy (1973) includes resolvability information in his tabling of two-associate class PBIBDs, though as pointed out by Patterson and Williams (1976a), the resolvable designs are "usually either square lattices or less efficient alternatives." Sinha and Dey (1982) exhibit a few resolvable PBIBDs found by duplicating nonresolvable designs. Sinha (1991) finds resolvable group divisible PBIBDs with fairly disperse concurrences λ_1 and λ_2. The mathematical community has taken some interest in resolvable group divisible designs with $\lambda_1 = 0$, that is, for which treatments from the same group never occur together. For $v = ks$ treatments in m groups of n, resolvability and $\lambda_1 = 0$ imply

$$n \geqslant k \quad \text{and} \quad \lambda_2(n-1)m \equiv 0 \pmod{k-1}. \tag{7}$$

The most interesting of these for the statistician is the second-associate concurrence count λ_2 being 1. For $k = 3$ and $\lambda_2 = 1$, the necessary conditions (7) are sufficient with the exceptions of $(m,n) = (2,3)$, $(2,6)$, and $(6,3)$, for which there are no solutions, and with the possible exceptions of $n = 6$, $m \equiv 2$ or $10 \pmod{12}$ (Rees and Stinson, 1987; Assaf and Hartman, 1989). Some large families of designs are known for $k = 4$ and $\lambda_2 = 1$ (Shen, 1988).

The concept of resolvability is extended to α-resolvability (α replicates of each treatment in each nesting block) first by Bose and Shrikhande (1960) and more formally by Shrikhande and Raghavarao (1963), and to $(\alpha_1, \alpha_2, \ldots, \alpha_t)$-resolvability (nesting block i has α_i replicates of each treatment) by Kageyama (1976). Mukerjee and Kageyama (1985) explore the relationship of affine $(\alpha_1, \alpha_2, \ldots, \alpha_t)$-resolvability to variance balance and to the numbers of nesting and nested blocks, extending the results of Shrikhande and Raghavarao (1963). Recent construction-oriented papers with which an interested reader may begin to explore these ideas further are Mohan and Kageyama (1989), Jungnickel et al. (1991), Shah et al. (1993), and Kageyama and Sastry (1993). For resolvable designs with unequal sub-block sizes see Patterson and Williams (1976a) and Kageyama (1988).

4. Nested BIBDs and related designs

4.1. Motivation and preliminaries

Nested balanced incomplete block designs, in which both the nesting blocks and the sub-blocks form BIBDs, are the obvious immediate generalizations of resolvable BIBDs to settings in which the nesting blocks cannot accommodate all of the treatments. Like the resolvable BIBDs, these designs possess the robustness property discussed in Section 2.2. In Section 2.3 they are shown to be optimal for the full analysis, though they would generally be considered inferior to resolvable BIBDs with the same sub-block size and number of units, as the latter keep all information on treatment contrasts in the two lowest strata (which are usually subject to less variability). In any case, when $k_1 < v$, so that a resolvable design is not a possibility, the nested BIBDs, or NBIBDs, are the simplest and most widely studied alternative, and are here the subject of Section 4.2. The notation for these designs will be NBIBD(v, b_1, b_2, k), there being b_2 blocks of size k nested in each of b_1 blocks of size b_2k. A discussion of the available alternatives when the NBIBD conditions are not achievable is in Section 4.3.

4.2. Construction of NBIBDs

As noted in Section 2.3, NBIBDs were introduced to the statistical literature by Preece (1967), who gave examples of their prior use in agricultural experiments, fully outlined the analysis, and provided a table of smaller designs. Since then, construction of these designs has been studied by Homel and Robinson (1975), Jimbo and Kuriki (1983), Bailey et al. (1984), Dey et al. (1986), Iqbal (1991), Jimbo (1993) and Kageyama and Miao (1996). Some NBIBDs are also embedded in a more restrictive combinatorial class, namely the balanced incomplete block designs with nested rows and columns (or BIBRCs; cf. Example 3 in Section 6). Especially relevant in this regard are the BIBRCs of Jimbo and Kuriki (1983) reported as Theorem 6.3 in Section 6.2. Other intertwining constructions for these two classes, and from various authors for NBIBDs alone, will be pointed out below as some of the more extensive constructions are listed. The first two recursively combine a BIBD with a known NBIBD.

THEOREM 4.1 (Jimbo and Kuriki, 1983). *If a* NBIBD(v, b_1, b_2, k) *is formed on the treatments of each block of a* BIBD$(v^*, b^*, k^* = v)$*, the result is a* NBIBD(v^*, b_1b^*, b_2, k).

If a resolvable BIBD is thought of as a special case of a NBIBD, then the nested BIBDs which occur as components of the BIBRCs of Theorem 2 of Singh and Dey (1979) are a special case of Theorem 4.1. Those same NBIBDs are found again in Theorem 2.1 of Dey et al. (1986).

THEOREM 4.2. *Existence of a* NBIBD(v, b_1, b_2, k) *and of a* BIBD$(v^* = b_2, b^*, k^*)$*, implies the existence of a* NBIBD(v, b_1b^*, k^*, k).

PROOF. Let d_1 be the given NBIBD, and d_2 the BIBD. Take any nesting block of d_1, call it β, and associate each of its $b_2 = v^*$ sub-blocks with a treatment of d_2. Now construct b^* new nesting blocks from β, of k^* sub-blocks each, corresponding to the treatments in block i of d_2, $i = 1, 2, \ldots, b^*$. Repeating this for each block of d_1 gives a design with the stated parameters. The new sub-block design is $r^* = v^*b^*/k^*$ copies of that of d_1, so is a BIBD. And if λ^* is the concurrence count for d_2, and λ_1, λ_2 are the nesting block and sub-block concurrence counts for d_1, then it is easy to see that the nesting block concurrence count for the new design is $r^*\lambda_2 + (\lambda_1 - \lambda_2)\lambda^*$. □

Theorem 4.2 is the nested design version of the BIBRC construction due to Cheng (1986) that appears as Theorem 6.2 in Section 6.2. Construction (ii) of Dey et al. (1986, p. 163) results from Theorem 4.2 by taking the NBIBD to be a resolvable BIBD. Another recursive construction of NBIBDs, which starts with two cyclically generated NBIBDs with special properties, is given by Jimbo (1993, p. 98). Kageyama and Miao (1996) also employ recursive methods in completely solving the construction of NBIBD$(v, b_1, 2, 2)$'s; among the inputs they require are designs from Theorems 4.4–4.6.

THEOREM 4.3 (Bailey et al., 1984). *Let v be odd. Identify a pair of integers $x < y$ in Z_v as forming an s-block if $x + y = s$. For each $s = 0, 1, \ldots, v - 1$ take the $(v - 1)/2$ s-blocks as sub-blocks of a nesting block of size $v - 1$. The result is a* NBIBD$(v, v, (v - 1)/2, 2)$.

Bailey et al. (1984) call the above designs "near resolvable". In the same paper they also give a solution to the construction of resolvable BIBDs with sub-block size 2.

The remaining three theorems of this sub-section construct designs using the method of differences on finite fields. The results will be stated without proofs, which can be found in the original papers. As is usual, x will denote a primitive element of GF_v. For any m that divides $v - 1$, let $H_{m,0} = (x^0, x^m, x^{2m}, \ldots, x^{v-1-m})'$ and $H_{m,i} = x^i H_{m,0}$ for $i = 0, 1, \ldots, m - 1$. Also let $S_m = (x^0, x^1, \ldots, x^{m-1})$. Sub-blocks within a block will be displayed separated by bars. Readers unfamiliar with finite fields or with the method of differences may wish to consult Street and Street (1987, Chapter 3).

THEOREM 4.4 (Jimbo and Kuriki, 1983). *Let $v = mh + 1$ be a prime power, let L_i for $i = 1, \ldots, n$ be mutually disjoint s-subsets of S_m written as $s \times 1$ vectors, and let $A_i = L_i \otimes H_m$. Then the m initial blocks*

$$B_j = x^{(j-1)} \begin{pmatrix} A_1 \\ \hline A_2 \\ \hline \vdots \\ \hline A_n \end{pmatrix}, \quad j = 1, 2, \ldots, m,$$

generate a NBIBD(v, mv, n, hs). *If m is even and h is odd, then $B_1, \ldots, B_{m/2}$ generate a* NBIBD$(v, mv/2, n, hs)$.

THEOREM 4.5 (Jimbo and Kuriki, 1983). *Let $v = umh + 1$ be a prime power, let L be an s-subset of S_m written as a $s \times 1$ vector, and let $A_i = x^{(i-1)u} L \otimes H_{um}$. Then the m initial blocks*

$$B_j = x^{(j-1)} \begin{pmatrix} A_1 \\ \hline A_2 \\ \hline \vdots \\ \hline A_m \end{pmatrix}, \quad j = 1, 2, \ldots, m,$$

generate a NBIBD(v, mv, m, hs). *If m is even and uh is odd, then $B_1, \ldots, B_{m/2}$ generate a* NBIBD$(v, mv/2, m, hs)$.

Putting $m = 2t, h = 3, s = 1, n = 2$ in Theorem 4.4, and $u = 3, m = 2t, h = 1, s = 2$ in Theorem 4.5, gives designs with the same parameters as those of series (ii) of Dey et al. (1986, p. 165). Designs deemed series (i) by the same authors are the next theorem.

THEOREM 4.6 (Dey et al., 1986). *Let $v = 4t + 1$ be a prime power. The initial blocks*

$$B_j = x^{j-1} \begin{pmatrix} x^0 \\ x^{2t} \\ \hline x^t \\ x^{3t} \end{pmatrix}, \quad j = 1, 2, \ldots, t,$$

generate a NBIBD$(v, v(v-1)/4, 2, 2)$.

For given v, b_1, b_2, and k, Table 1 lists the smallest feasible NBIBDs, that is, the smallest replication for which BIBDs of blocksize $b_2 k$ and of blocksize k both exist. References for the designs are to the theorems of this section and to the original paper of Preece (1967), demonstrating that some of the latter's designs occur as special cases of now known infinite series of designs. The recursive constructions are listed only when no direct method is available. For four cases the author has found no design in the literature. For three of these, and for designs neither in Preece's (1967) table nor resulting from a theorem of this section, a solution is listed in Table 2. Existence of the remaining case, NBIBD$(10, 15, 2, 3)$, remains open. NBIBD$(10, 30, 2, 3)$ can be constructed using Theorem 4.2 and a NBIBD$(10, 10, 3, 3)$.

There has also been interest in enumerating a special class of NBIBDs. An *almost resolvable* BIBD(v, b, k) with $b = v(v-1)/k$ (i.e., with $\lambda = k - 1$) is a BIBD whose blocks can be partitioned into v classes of $(v-1)/k$ blocks each so that each class lacks exactly one treatment (these designs are also sometimes called *near resolvable*). Obviously an almost resolvable BIBD(v, b, k) is a NBIBD$(v, v, (v-1)/k, k)$. The obvious necessary condition for existence of such a design, other than the usual BIBD conditions, is that k divide $v-1$. Sufficiency of this condition for $k = 2$ is Theorem 4.3. Sufficiency for $k = 3$ is proven by Hanani (1974). Yin and Miao (1993) demonstrate the sufficiency for $k = 5$ or 6, except possibly for 8 values of v when $k = 5$, and 3 values of v when $k = 6$; save for $v = 55$ or 146 when $k = 6$, these exceptions have since been solved. For $k = 7$ and $k = 8$ sufficiency has likewise been established save for small lists of possible exceptions. Refer to Furino et al. (1994) for details.

Table 1
NBIBDs with $v \leqslant 14$ and replication $r \leqslant 30$. Preece refers to the table of Preece (1967)

v	r	b_1	b_2	k	Theorem
5	4	5	2	2	Preece, 4.3, 4.6
6	10	15	2	2	Preece
7	6	7	3	2	Preece, 4.3, 4.5
7	6	7	2	3	Preece, 4.4
7	12	21	2	2	4.3, 4.4, 4.5
8	7	14	2	2	Preece
8	21	28	3	2	Table 2
8	21	28	2	3	Iqbal (1991)
9	8	18	2	2	4.6
9	8	12	3	2	Preece
9	8	9	4	2	Preece, 4.3
9	8	12	2	3	Preece
9	8	9	2	4	Preece
10	9	15	3	2	Preece
10	9	15	2	3	?
10	9	10	3	3	Preece
10	18	45	2	2	4.2
11	10	11	5	2	Preece, 4.3, 4.5
11	10	11	2	5	Preece, 4.5
11	20	55	2	2	4.4, 4.5
11	30	55	3	2	4.4
11	30	55	2	3	4.4, 4.5
12	11	33	2	2	Preece
12	11	22	3	2	Preece
12	11	22	2	3	Preece
12	22	33	4	2	Table 2
12	22	33	2	4	Table 2
13	12	39	2	2	4.6
13	12	26	3	2	Preece, 4.5
13	12	13	6	2	Preece, 4.3
13	12	26	2	3	Preece, 4.4
13	12	13	4	3	Preece
13	12	13	3	4	Preece
13	12	13	2	6	Preece
13	18	26	3	3	4.4, 4.5
13	24	39	4	2	4.5
13	24	39	2	4	4.4
14	26	91	2	2	Agrawal and Prasad (1983)

Table 2
Solutions for NBIBDs

v	r	b_1	b_2	k	Initial blocks
8	21	28	3	2	$[0,1 \mid 2,4 \mid 3,6], [0,1 \mid 3,5 \mid 4,\infty],$ $[0,1 \mid 3,6 \mid 5,\infty], [0,2 \mid 1,4 \mid 3,\infty]$ (mod 7)
8	21	28	2	3	$[0,1,3 \mid 2,6,7], [0,5,7 \mid 2,4,6],$ $[0,1,3 \mid 2,5,6], [0,1,6 \mid 2,4,5] \times \frac{1}{2}$ (mod 8)
12	22	33	4	2	$[0,1 \mid 3,8 \mid 4,7 \mid 5,6], [0,2 \mid 3,6 \mid 5,9 \mid 4,\infty],$ $[0,4 \mid 1,6 \mid 7,9 \mid 5,\infty]$ (mod 11)
12	22	33	2	4	$[0,1,2,3 \mid 4,7,8,10], [0,1,4,7 \mid 2,3,9,\infty],$ $[0,2,6,8 \mid 3,7,9,\infty]$ (mod 11)
14	26	91	2	2	$[0,1 \mid 9,8], [0,2 \mid 5,3], [0,8 \mid 1,9], [0,3 \mid 2,5],$ $[0,4 \mid 7,11], [\infty,0 \mid 2,9], [\infty,0 \mid 3,9]$ (mod 13)

4.3. *Other nested incomplete block designs*

There has been a relatively modest amount of work on nested block designs with b_2k less than v and for which at least one of the two component block designs is not a BIBD, which will be briefly reviewed here. The earliest such effort seems to be due to Homel and Robinson (1975), who define nested partially balanced incomplete block designs (NPBIBDs) as designs for which the nesting blocks form a PBIBD with blocksize b_2k, the nested blocks form a PBIBD with blocksize k, and the two component designs share a common association scheme (for the definition of a PBIBD see Street and Street, 1987, p. 237). Though not motivated there by any efficiency argument, the full analysis, explained in detail in their paper, is certainly eased by the common association scheme requirement. The same objective can be met if one of the association schemes is a collapsed version (by combining associate classes) of the other, and a few designs of this type appear in Banerjee and Kageyama (1993). Banerjee and Kageyama (1993) also give two infinite series of NPBIBDs, one based on the triangular association scheme and one based on the L_2 association scheme. Homel and Robinson's (1975) constructions are for prime power numbers of treatments, based on generalizations of the pseudo-cyclic and L_2 association schemes.

In another paper, Banerjee and Kageyama (1990) construct α-resolvable NBIBDs and NPBIBDs. As they remark, their designs "mostly have large values of some parameters." The resolvability can be used to accommodate a third, super-nesting factor, that will not enter into the analysis.

In a dissertation at the University of Kent, Iqbal (1991) uses the method of differences to construct nested designs for which one of the two component block designs is a BIBD and the other is a regular graph design. Relaxing the balance restriction for one component allows smaller replication to be attained than would sometimes otherwise be possible, and produces designs that should be reasonably efficient for either of the analyses discussed in Section 2. This is an idea worthy of further study, as is the entire area of nested block designs which are both small and efficient.

Gupta (1993) studies nested block designs for which both the number of sub-blocks within a block, and the sub-block sizes, are non-constant. The main result proves the bottom stratum optimality of any such design for which the sub-blocks are a binary, variance-balanced design. If as in Section 2.2 of this paper, it is first shown that the nesting factor is irrelevant to the bottom stratum analysis, then this result can also be established as an application of Theorem 2.1 of Pal and Pal (1988). Construction amounts to arbitrarily partitioning the blocks of any optimal, nonproper block design. Each partition class will be a nesting block of a bottom stratum optimal nested design, but the behavior under the full analysis may or may not be satisfactory.

5. Nesting of row and column designs – Models and related considerations

5.1. The nested row and column setting

The examples of Section 1 and the proffered designs through Section 4 share a common feature in their block structure: given any two blocking factors G and H, either G nests H, or H nests G. While as those examples show, this "pure nesting" structure is found in a wide variety of experimental situations, it is also the case that many experiments blend nesting with other, non-nesting relations among some of the blocking factors. The simplest of these is a nesting of two completely crossed blocking factors, which can be visualized as in Figure 2.

If the nesting factor is called "blocks", and the two crossed factors within the nest are called "rows" and "columns", then each block is recognizable as the setting for a row–column design. Indeed, one of the mechanisms by which this *nested row and column setting* arises is through repetition of a row–column experiment in which it is not reasonable to assume that row factor and column factor effects are constant across repetitions (blocks). Nesting can also be used to reduce the numbers of rows and columns per block relative to a single row–column layout, which may be important in assuring row–column additivity; the trade-off is that for a given number of experimental units there will be fewer degrees of freedom for error. A primary area of application is found in agricultural field trials in which blocks are physically separate fields, and two orthogonal sources of variation are modeled to account for yield differences as a function of position within the field. An interesting example in sampling insect populations is discussed in Keuhl (1994, pp. 341–342). It is assumed here, as is common in practice, that there is one experimental unit at each row–column cross within a block (so in Figure 2 there are 36 units).

The ideas and terminology attendant to the nested row and column setting have been formalized beginning with the papers of Srivastava (1978) and Singh and Dey (1979), though designs for this setting were considered much earlier, prime examples being the lattice squares of Yates (1937, 1940). Singh and Dey's (1979) paper is notable for introducing a class of designs, the balanced incomplete block designs with nested rows and columns, that generalize the balanced lattice squares and which have since been the focus of much of the design work in this area. But as shown by developments of the past five years reported in Sections 5.2 and 5.3, these designs are not necessarily optimal, and depending on the setting parameters and the analysis

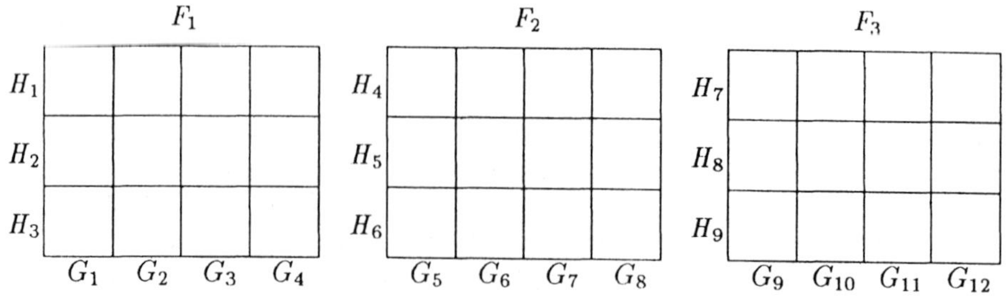

Fig. 2. G and H are crossed within levels of F. This is a nested row and column setting with 3 blocks of size 3×4.

used, can be surprisingly poor. In the subsections that follow one may make the interesting observation that the introduction of a second factor in a nest, crossed with the first nested factor, fundamentally changes the requirements for optimal design. The intuition from simple block designs that worked so well in designing for pure nested structures as in Section 2, can quite miserably fail at the next level of complexity.

5.2. *Model and bottom stratum analysis*

To formalize the setting, let b be the number of blocks (levels of the nesting factor), and let p and q be the numbers of rows and columns (levels of the two nested factors which are crossed with one another) within each block. With y_{jlm} denoting the yield from the unit in row l, column m of block j (plot (j, l, m)), the model is

$$y_{jlm} = \mu + \beta_j + \rho_{jl} + \delta_{jm} + \tau_{[jlm]} + e_{jlm} \tag{8}$$

where β_j, ρ_{jl}, and δ_{jm} are respectively block, row, and column effects; $\tau_{[jlm]}$ is the effect of the treatment applied to plot (j, l, m); and the e_{jlm} are uncorrelated random variables of constant variance and zero means. In matrix form, ordering the yields row-wise by block, this is

$$Y = \mu 1 + Z_1\beta + Z_2\rho + Z_3\delta + A_d\tau + E \tag{9}$$

with $Z_1 = I_b \otimes 1_{pq}, Z_2 = I_{bp} \otimes 1_q, Z_3 = I_b \otimes 1_p \otimes 1_q$ and β, ρ, δ are respectively $b \times 1, bp \times 1$, and $bq \times 1$. A_d of order $bpq \times v$ is the 0–1 design matrix and τ is the $v \times 1$ vector of treatment effects.

For the bottom stratum analysis eliminating blocks, rows, and columns, also called the "within-rows-and-columns" analysis, the information matrix for the best linear unbiased estimators of contrasts of the τ_i's is (Singh and Dey, 1979)

$$\begin{aligned} C_d &= A_d'\left(I - \frac{1}{q}Z_2Z_2' - \frac{1}{p}Z_3Z_3' + \frac{1}{pq}Z_1Z_1'\right)A_d \\ &= A'A - \frac{1}{q}N_2N_2' - \frac{1}{p}N_3N_3' + \frac{1}{pq}N_1N_1' \end{aligned} \tag{10}$$

where the d has been dropped in the latter expression to ease the notation. The matrices N_1, N_2 and N_3 are the treatment-block, treatment-row, and treatment-column incidence matrices ($N_i = A'_d Z_i$).

Singh and Dey (1979) introduced a class of designs for this setting which combine the classical notions of treatment assignment binarity and variance balance in the bottom stratum analysis. They named these designs balanced incomplete block designs with nested rows and columns (for short, BIBRC).

DEFINITION. A BIBRC is a nested row and column design with $pq < v$ for which

(i) treatment assignment to blocks is binary, that is, N_1 contains only 0's and 1's, and

(ii) C_d is completely symmetric.

Because $\mathrm{tr}(C_d)$ is constant over the class of all designs satisfying (i), Kiefer's (1975) proposition 2 says that BIBRCs are universally optimum over the class of all binary block nested row and column designs. The question, then, is whether better designs can be found outside that class. Binarity in simple block designs (only one blocking factor) is widely believed to exclude no superior designs (e.g., John and Mitchell, 1977; Shah and Sinha, 1989) and typically functions to exclude far inferior competitors, so if that were to be a guide, the answer to the question would be a definitive "no". But further investigation shows that intuition from the simple block setting fails here.

Letting Φ be any optimality criterion as specified in Section 2.1, the goal is to minimize $\Phi(C_d)$ with respect to choice of d. The matrix $N_2N_2' - \frac{1}{p}N_1N_1'$ is nonnegative definite, so

$$\Phi(C_d) \leqslant \Phi\left(A'A - \frac{1}{p}N_3N_3'\right)$$

with equality when $pN_2N_2' = N_1N_1'$. The matrix $A'A - \frac{1}{p}N_3N_3'$ is the information matrix for the simple block design composed of the bq columns of d as blocks of size p, what will be called the column component design. This motivates the definition of the bottom stratum universally optimum nested row and column design (BNRC, for short).

DEFINITION. A BNRC is a nested row and column design for which

(i) the column component design is a BBD, and

(ii) for each i and j, the number of times treatment i appears in row l of block j is constant in l.

When necessary to indicate the parameters, the notation will be BNRC(v, b, p, q) (similarly for BIBRCs). Condition (i) says that the column component design is a universally optimum block design, and condition (ii) says that within a block, each row of treatments is a permutation of the first row. Condition (ii) makes $pN_2N_2' = N_1N_1'$, from which the first theorem of this section follows.

THEOREM 5.1 (Bagchi et al., 1990). *A* BNRC(v, b, p, q) *is universally optimum for the bottom stratum analysis among all nested row and column designs with the same* v, b, p *and* q.

EXAMPLE 2. A BNRC(4, 6, 2, 4).

$$\begin{array}{ccc} 1\,1\,2\,2 & 3\,3\,4\,4 & 1\,2\,3\,4 \\ 2\,2\,1\,1 & 4\,4\,3\,3 & 3\,4\,2\,1 \\ \\ 1\,2\,3\,4 & 1\,2\,3\,4 & 1\,2\,3\,4 \\ 3\,4\,2\,1 & 3\,4\,2\,1 & 3\,4\,2\,1 \end{array}$$

Unlike a BIBRC, a BNRC is nonbinary in blocks and need not be binary in rows. Further, as illustrated by Example 2, neither the row nor the block component designs need be variance balanced or efficient. If $pq \leqslant v$, a direct comparison of the two types of designs can be made (and with this restriction a BNRC must have $p \leqslant q$). The common nonzero eigenvalue for the information matrix of a BNRC is then $b(p-1)q/(v-1)$, and that for a BIBRC is $b(p-1)(q-1)/(v-1)$, so that the relative efficiency of a BIBRC is $(q-1)/q$, which while satisfactory for large q, is for small q quite poor. The worst case is $p = q = 2$, with BNRCs having twice the bottom stratum information on treatment contrasts.

Bagchi et al. (1990) prove even more than stated in Theorem 5.1. If a nested row and column design d satisfies (ii) of the BNRC definition, and if the column component design is Φ-optimum over the class of simple block designs with bq blocks of size p, then d is itself Φ-optimum. For optimum designs that are not BNRCs, see Section 7.2.

5.3. *The full analysis*

It turns out that the advantage held by BNRCs for the bottom stratum analysis is not necessarily maintained in the full analysis. In demonstrating this, some additional conditions will be imposed on the two classes of designs that will ease the recovery of information from higher strata. For BIBRCs it is demanded that each of the row, column, and block component designs (which respectively have block sizes of q, p, and pq) be BIBDs. BIBRCs with this property have been termed by Agrawal and Prasad (1982b) to belong to "series A"; here they will be said to be *completely balanced*. The restrictions on the class of BNRCs are that a treatment appear m times in every row of any block in which it appears (m being a constant not depending on the particular block or treatment), and that there are λ_b blocks in which any pair of treatments $i \neq i'$ both occur (by virtue of the preceding restriction, i and i' both occur mp times in each of these λ_b blocks). For convenience, the restricted BNRC will also be said to be completely balanced. Morgan and Uddin (1993a) point out that many of the known series of BIBRCs and of BNRCs have these properties. It is the completely balanced BIBRCs that produce NBIBDs when one of their row or column classifications is dropped, a relationship alluded to earlier in Section 4.2. The design of Example 2 is a BNRC that is not completely balanced.

The advantage of these complete balance restrictions is that both classes of designs are *generally balanced* with a single efficiency factor in each stratum (see Houtman and Speed, 1983; or Mejza and Mejza, 1994). For the data vector Y of any equireplicate nested row and column design the averaging operators are $T = \frac{1}{r}AA'$

for treatments, $B = \frac{1}{pq}Z_1Z_1'$ for blocks, $R = \frac{1}{q}Z_2Z_2'$ for rows, $C = \frac{1}{p}Z_3Z_3'$ for columns, and $G = \frac{1}{bpq}J$ for the grand mean. The stratum projectors are $S_0 = G$, $S_1 = B - G$, $S_2 = R - B$, $S_3 = C - B$, and $S_4 = I - R - C + B$, "naming" the strata 0 to 4, respectively. To recover information from strata 1–3, the vectors β, ρ, γ, and E in the model (9) are now taken to be mutually uncorrelated random vectors with zero means and with $\mathrm{var}(\beta) = \sigma_1^2 I_b$, $\mathrm{var}(\rho) = \sigma_2^2 I_{bp}$, $\mathrm{var}(\gamma) = \sigma_3^2 I_{bq}$, and $\mathrm{var}(E) = \sigma^2 I_{bpq}$. Hence

$$\begin{aligned}\mathrm{var}(Y) &= \sigma_1^2 Z_1Z_1' + \sigma_2^2 Z_2Z_2' + \sigma_3^2 Z_3Z_3' + \sigma^2 I \\ &= \sum_{\alpha=0}^{4} \xi_\alpha S_\alpha\end{aligned}$$

for $\xi_0 = \xi_1 = pq\sigma_1^2 + q\sigma_2^2 + p\sigma_3^2 + \sigma^2$, $\xi_2 = q\sigma_2^2 + \sigma^2$, $\xi_3 = p\sigma_3^2 + \sigma^2$, and $\xi_4 = \sigma^2$.

The condition for general balance is that $(T - G)S_\alpha(T - G) = \lambda_\alpha(T - G)$ for constants $0 \leqslant \lambda_\alpha \leqslant 1$, $\sum_\alpha \lambda_\alpha = 1$; λ_α is the efficiency factor for estimating a treatment contrast in stratum α. With known stratum variances $\xi_0, \xi_1, \ldots, \xi_4$, the variance of a treatment contrast $t'(T - G)A\tau$ estimated using information from all strata is

$$\mathrm{var}(t'(T - G)Y) = \frac{1}{\sum_\alpha \lambda_\alpha \xi_\alpha^{-1}} \|(T - G)t\|^2. \tag{11}$$

The efficiency factors, aside from a divisor of $pq(v - 1)$ throughout, are

	λ_0	λ_1	λ_2	λ_3	λ_4
BIBRC	0	$v - pq$	$v(p - 1)$	$v(q - 1)$	$v(p - 1)(q - 1)$
BNRC	0	$(mv - q)p$	0	$v(q - mp)$	$v(p - 1)q$

It is seen that a completely balanced BNRC has more information in blocks and in the within-rows-and-columns stratum, less information in rows and in columns. Which type of design is superior depends on the relative values of the stratum variances.

THEOREM 5.2 (Morgan and Uddin, 1993a). *If the stratum variances are known (equivalently, σ^2, σ_1^2, σ_2^2, and σ_3^2 are known), then for the analysis with recovery of information from every stratum, a completely balanced* BNRC(v, b, p, q) *is superior to a completely balanced* BIBRC(v, b, p, q) *if and only if*

$$\frac{(p - 1)}{(mp - 1)}\left[1 - \frac{(m - 1)\sigma^2}{(p - 1)\sigma_3^2}\right]\xi_1 > \sigma^2\left[\frac{\sigma^2\sigma_1^2}{\sigma_2^2\sigma_3^2} - 1\right].$$

The proof of Theorem 5.2 is immediate upon comparison of (11) for BNRCs and BIBRCs.

It is not hard to see that the completely balanced BIBRCs are universally optimum within the binary block class for the random effects model of this section, which gives an immediate improvement: a completely balanced BNRC is superior to *every* binary design for this analysis whenever the condition of Theorem 5.2 holds.

The expression in Theorem 5.2 has been written with the special case of $m = 1$ in mind, for which the coefficient of ξ_1 reduces to 1. When $m = 1$ the rows of a completely balanced BNRC form the blocks of a balanced incomplete block design and the condition of Theorem 5.2 is equivalent to that of Theorem 4.1(b) of Bagchi et al. (1990). That theorem in this setting says that such a design is universally optimum under the random effects model. For $m > 1$ and the random effects model a completely balanced BNRC does not necessarily have maximum trace so no such general optimality result is easily obtained; the advantage of course is that if the condition of Theorem 5.2 holds, this class nevertheless provides large families of designs for situations where optimum ($m = 1$) BNRCs may not exist, that are variance balanced and are superior to any BIBRC and indeed to any binary block design.

When for $m = 1$ the condition of Theorem 5.2 fails, a completely balanced BIBRC is universally optimum for the random effects model.

5.4. *Discussion*

All of this leaves one with some difficult questions in designing a nested row and column experiment. If it is decreed in advance that the bottom stratum analysis is to be performed, then Section 5.2 says that a BNRC is to be recommended whenever it exists. But in doing so a statistician may well face opposition from scientists reluctant to abandon the binary designs they have always used. Moreover, depending on p and q, the proportion λ_4 of information in the bottom stratum can be unacceptably small (a point made by Cheng (1986), for BIBRCs), so that some recovery of higher stratum information may be warranted. Now the choice of design is much less clear, and depends on the stratum variances, which will usually be unknown.

Complicating the entire discussion are existence criteria for the two types of designs. Combinatorial considerations show that frequently BNRCs will exist when BIBRCs do not, and vice-versa. So, for instance, if a BNRC is available and a BIBRC is not, the question of what is the best design for the full analysis does not have a straightforward answer as in Theorem 5.2. And more realistic, given the combinatorial conditions, is that for given v, b, p, and q, neither type of design will exist. In this regard, these variance balanced designs are simply the first, albeit important, steps in a relatively new area of inquiry for design theory, that offer insight into the properties of a good design when universal optimality cannot be achieved. Many more steps are to be taken before a reasonably comprehensive catalog of efficient nested row and column designs can be compiled. Sections 6 and 7 will summarize the known results for existence of BIBRCs and BNRCs, and will mention a few other known related designs.

6. Binary block nested row and column designs

6.1. BCBRCs and balanced lattice rectangles

The earliest of the systematically studied nested row and column designs are the *lattice squares* of Yates (1937, 1940). These are arrangements of $v = s^2$ treatments into $s \times s$ blocks so that each treatment occurs once in each block. A *completely balanced* lattice square is a set of b of these blocks with the property that the row component and column component block designs are each BIBDs. In a *semibalanced* lattice square, the $2bs$ row component and column component blocks together form a BIBD. Either type of balance makes the information matrix (10) completely symmetric, so that the completely balanced and semibalanced lattice squares are complete block versions of BIBRCs as defined in Section 5.2, what some authors refer to as *balanced complete block designs with nested rows and columns* (BCBRCs). For any prime power s, complete balance is achievable with $s + 1$ blocks, and if s is odd, semibalanced lattice squares exist with $(s + 1)/2$ blocks. Kempthorne (1952, Chapter 24) gives a simple construction, or alternatively, the orthogonal array construction for affine resolvable designs in Section 3.3 is easily modified for this situation. Tables of these designs may be found in Clem and Federer (1950) and Cochran and Cox (1992). The balanced square lattice displayed earlier in Example 1 is also an example of a completely balanced lattice square.

For non-square ($p \neq q$) blocks, the lattices which fall under the BCBRC heading are called *balanced two-restrictional lattices* or *balanced lattice rectangles.* Though a bit more complicated for rectangles than for squares, the case of prime powered number of treatments being divisible by the number of rows, has been thoroughly covered. For the parameters

$$v = s^m, \quad p = s^r, \quad q = s^c, \quad r + c = m, \quad r \neq c$$

with s being a prime power, balance requires $(s^m - 1)/(s - 1)$ blocks. Designs can be constructed by confounding pseudofactors of the s^m factorial set in rows and columns, the problem being in systematically selecting the confounding factors for each replicate. A method for doing so is given by Raktoe (1967), who explicitly displays a compact representation for each s^m less than 1000 by means of cyclic collineations (see also Federer and Raktoe, 1965). Mazumdar (1967) gives an analytic solution to constructing the generator matrices of collineations for all value of s. Hedayat and Federer (1970) establish the connection between these designs and sets of mutually orthogonal Latin squares.

Other than the designs just described, very few BCBRCs are currently available. Series with blocks of size $2 \times v/2$, when $v - 1$ is a prime power, are given by Agrawal and Prasad (1984), along with a few individual plans. Designs with the same parameters and $v \leqslant 20$ are tabled in Ipinyomi (1990). The efficiency factors displayed for BIBRCs in Section 5.3, and the efficiency comparisons with BNRCs, are all correct for BCBRCs as well.

6.2. BIBRCs

Since the introduction of BIBRCs by Singh and Dey (1979), in excess of a dozen papers have appeared containing constructions for these designs. Nevertheless, progress on the construction problem cannot be said to be great, for other than the odd trial and error solution, most of these results are for prime power v, and many have large v and/or number of replicates. There is also considerable overlap among the results of different authors. As tables of designs are not yet available, the main results (without proof, other than specification of initial blocks where appropriate) will be reproduced here, with a discussion of how they include constructions given by others. Questions of isomorphism will not be addressed: concern is only with designs having the same parameters v, b, p, and q. Two basic recursive constructions are stated first.

THEOREM 6.1. *If a* $\text{BCBRC}(k, b^*, p, q)$ *or* $\text{BIBRC}(k, b^*, p, q)$ *is formed on the treatments of each block of a* $\text{BIBD}(v, b, k)$*, the result is a* $\text{BIBRC}(v, bb^*, p, q)$.

Theorem 2 of Singh and Dey (1979) is Theorem 6.1 starting with a lattice square BCBRC. Theorem 6.1 starting with a BIBRC is Theorem 2 of Jimbo and Kuriki (1983); it appears again as Theorem 2.1 of Sreenath (1989). Theorem 5.1 of Agrawal and Prasad (1984) is another version that starts with a BCBRC.

The second result recursively combines a BIBD with a BIBRC in a different way. It was the first general method to produce BIBRCs that do not, in the terminology of Section 5.3, have the "complete balance" property.

THEOREM 6.2 (Cheng, 1986). *Existence of a* $\text{BCBRC}(v, b^*, p^*, q)$ *or a* $\text{BIBRC}(v, b^*, p^*, q)$*, and of a* $\text{BIBD}(p^*, b, k)$*, implies the existence of a* $\text{BIBRC}(v, bb^*, k, q)$.

From each $p^* \times q$ block of the given nested row and column design, one forms b new $k \times q$ blocks by retaining rows corresponding to treatments in the blocks of the BIBD; cf. Theorem 4.2. Cheng (1986) further shows that if the given BIBRC is generally balanced, then so too is the resulting design, though the new design need not be balanced in any of the three component (row, column, block) designs. Many such series are listed there. As an illustration, combining a $\text{BCBRC}(9, 2, 3, 3)$ with a $\text{BIBD}(3, 3, 2)$ gives the design of Example 3, which is completely balanced.

EXAMPLE 3. A $\text{BIBRC}(9, 6, 2, 3)$. Rows within blocks are a $\text{NBIBD}(9, 6, 2, 3)$. Columns within blocks are a $\text{NBIBD}(9, 6, 3, 2)$.

$$
\begin{matrix} 1 & 4 & 7 \\ 2 & 5 & 8 \end{matrix} \qquad
\begin{matrix} 1 & 4 & 7 \\ 3 & 6 & 9 \end{matrix} \qquad
\begin{matrix} 2 & 5 & 8 \\ 3 & 6 & 9 \end{matrix} \qquad
\begin{matrix} 1 & 5 & 9 \\ 6 & 7 & 2 \end{matrix} \qquad
\begin{matrix} 1 & 5 & 9 \\ 8 & 3 & 4 \end{matrix} \qquad
\begin{matrix} 6 & 7 & 2 \\ 8 & 3 & 4 \end{matrix}
$$

Three more series constructions for BIBRCs will be given, all based on the method of differences using finite fields.

THEOREM 6.3 (Jimbo and Kuriki, 1983). *Let* $v = mk_1k_2 + 1$ *be a prime power, where* k_1 *and* k_2 *are relatively prime, and let* s *and* t *be positive integers satisfying* $st \leqslant m$.

Write $L_{s\times t} = (x^{i+j-2})_{i,j}$ *and* $M_{k_1\times k_2} = (x^{[(i-1)k_2+(j-1)k_1]m})_{i,j}$. *Then the initial blocks*

$$B_j = x^{(j-1)} L \otimes M, \quad j = 1, 2, \ldots, m,$$

generate a BIBRC(v, mv, sk_1, tk_2). *If* m *is even and* k_1k_2 *is odd, then* $B_1, \ldots, B_{m/2}$ *generate a* BIBRC$(v, mv/2, sk_1, tk_2)$.

Each of the component designs of Theorem 6.3 is a BIBD (cf. Theorem 4.4), so that these BIBRCs are completely balanced. As special cases of Theorem 6.3, many series of BIBRCs with parameters matching designs appearing elsewhere in the literature can be obtained, from which it is evident that this result has been overlooked by several authors (including this author in a joint paper, as will soon become apparent). Theorem 6.3 also pulls together a number of results appearing concurrent or prior to its publication. The designs with $s = t = 1$ are also found in Theorems 1 and 2 of Agrawal and Prasad (1982a), the first of these appearing again in Theorem 2.3 of Saha and Mitra (1992). Putting $k_1 = t = 1$ gives Theorems 1 and 2 of Uddin and Morgan (1991) and Theorem 6 of Street (1981), the latter of which is restricted to $s \leqslant m/2$; further requiring $k_2 = s = 2$ gives Theorem 3 of Agrawal and Prasad (1983). Also within the bounds of $k_1 = t = 1$ are Theorem 4 of Agrawal and Prasad (1982a), which is also Theorem 2.2 of Saha and Mitra (1992), by setting $m = hp$; Theorems 2.2 and 2.5 of Sreenath (1989) by setting $s = m = 2$; a generalization of Theorem 2.3 of Sreenath (1989) by setting $s = 2$ and $k_2 = 3$; and Theorem 5 of Agrawal and Prasad (1983), which is also Theorem 2.1 of Saha and Mitra (1992), by setting $k_1 = 1$ and $k_2 = 2$.

Using a geometric argument, Jimbo and Kuriki (1983) also construct BIBRC(s^3, s^2+s+1, s, s) for any prime power s (cf. Corollary 2b of Uddin and Morgan, 1990).

Theorem 6.3 may be described as a multiplicative construction: initial blocks are multiplication tables whose row and column margins are carefully selected subsets of GF_v. Uddin and Morgan (1990) are able to construct infinite series of BIBRCs using addition tables with carefully chosen margins as initial blocks. Two of their theorems will be given to illustrate this approach.

THEOREM 6.4 (Uddin and Morgan, 1990). *Let* $v = 4tm + 1$ *be a prime power and write* $x^{u_i} = 1 - x^{2mi}$.

(a) *If* $u_i - u_j \not\equiv m \pmod{2m}$ *for* $i, j = 1, \ldots, t$, *then there is a* BIBRC$(v, mv, 2t, 2t)$.

(b) *If in addition to the condition in (a),* $u_i \not\equiv m \pmod{2m}$ *for* $i = 1, \ldots, t$, *then there is a* BIBRC$(v, mv, 2t+1, 2t+1)$.

If B is the addition table with row margin $(x^0, x^{2m}, \ldots, x^{(4t-2)m})$ and column margin $(x^m, x^{3m}, \ldots, x^{(4t-1)m})$ then the initial blocks for Theorem 6.4a are $B, xB, \ldots, x^{m-1}B$. For Theorem 6.4b, include 0 in each margin. As an example of Theorem 6.4, $t = 2$ gives blocks of size 4×4 with replication $r = 2(v-1)$, and 5×5 with $r = 25(v-1)/8$, for $v = 8m + 1$ a prime power. For $v < 500$ the conditions fail to hold for $v = 81$, and for $v = 289$ in the 5×5 case. Setting $t = 3$, the conditions for $v = 12m+1$ in 6×6 and 7×7 blocks also fail for two values of $v < 500$: $v = 37$ in 6×6 blocks, and $v = 169$.

THEOREM 6.5 (Uddin and Morgan, 1990). *Let $v = 4tm + 1$ be a prime power where $t > 1$ is odd and write $x^{u_i} = 1 - x^{4mi}$.*

(a) *If $u_i - u_j \not\equiv m \pmod{2m}$ for $i, j = 1, \ldots, (t-1)/2$, then there is a* BIBRC$(v, mv, t, t)$.

(b) *If in addition to the condition in (a), $u_i \not\equiv m \pmod{2m}$ for $i = 1, \ldots, (t-1)/2$, then there is a* BIBRC$(v, mv, t+1, t+1)$.

The initial blocks for Theorem 6.5 also have the form $B, xB, \ldots, x^{m-1}B$. For (a), B is the addition table with row margin $(x^0, x^{4m}, \ldots, x^{4(t-1)m})$ and column margin $(x^m, x^{5m}, \ldots, x^{[4(t-1)+1]m})$, and for (b), each margin also includes 0. As an example, $t = 3$ gives blocks of sizes 3×3 and 4×4 with $v = 12m+1$ a prime power ($m \geqslant 2$ is required for the 4×4's) and replications $3(v-1)/4$ and $4(v-1)/3$, respectively. With $t = 5$, blocks of size 5×5 and 6×6 for prime power $v = 20m+1$ and replications $5(v-1)/4$ and $9(v-1)/5$ are obtained; the 5×5's are for all $m \geqslant 2$, while for $v < 500$ the conditions for the 6×6's fail for $v = 41$ and 61. Other corollaries to Theorems 6.4 and 6.5 may be found in Uddin and Morgan (1990), along with three other theorems that use the addition table approach. Designs from these three other theorems have the complete balance property, and hence more replicates when they overlap for given v, p, and q, than the designs from Theorems 6.4 and 6.5, which are semibalanced in the sense of lattice squares. Uddin and Morgan (1990) tabulate all of their designs for v no greater than 100 and p and q greater than 2.

One other direct construction gives relatively large replication numbers but is of interest in producing some designs that cannot be found by the methods so far listed: see Theorem 6 of Agrawal and Prasad (1983). Uddin (1992, 1995) has developed methods for modifying known initial blocks for BIBRCs by inclusion of additional fixed elements, and by recursively combining initial blocks for two different BIBRCs. The existence of all BIBRC$(v, b, 2, 2)$'s satisfying the necessary condition $v(v-1)|4b$ has been independently established by Kageyama and Miao (1996) and Srivastav and Morgan (1996); the latter pair of authors construct all of the designs to have general balance.

6.3. *Other binary block nested row and column designs*

There has been some, albeit considerably less, activity devoted to finding binary block nested row and column designs that trade off full (bottom stratum) balance in exchange for fewer blocks. The concept of partial balance for this setting has been explored by Street (1981), Agrawal and Prasad (1982b, c, 1984), Morgan and Uddin (1990), Sinha and Kageyama (1990), and Gupta and Singh (1991). Closely related are the generalized cyclic row–column designs of Ipinyomi and John (1985). Presumably these designs will be inefficient for the bottom stratum analysis, but it is not clear to what extent this will be so, and some may be of value for the full analysis.

For lattice squares and rectangles, Federer and Raktoe (1965) recommend using a subset of blocks from a balanced design, chosen so that "all the pseudo-effects should be confounded as equal a number of times in rows as possible and as equal a number of times in columns as possible," but the efficacy of this tact has not been demonstrated. Analysis of these designs is covered by Cornelius (1983). For the non-prime power case, the use of pseudofactors (e.g., Kempthorne, 1952, Chapter 24; Monod and Bailey, 1992, Section 5.3) provides a method for constructing large classes of complete block nested row and column designs that are relatively amenable to analysis. A few cyclic constructions for designs with complete blocks may be found in Ipinyomi and John (1985). Nested row and column designs with complete blocks are sometimes called *resolvable row–column designs*.

If just two complete blocks are to be used, a catalog of 20 designs for up to 100 varieties is available in Patterson and Robinson (1989) (details of construction are explained in John and Whitaker (1993)). A construction which produces many of their designs may be found in Bailey and Patterson (1991), using the ideas of contraction employed by Williams et al. (1976) for two-replicate resolvable designs as described in Section 3.5. The method starts with a row–column design for two non-interacting sets of treatments (e.g., Preece, 1982), and if the starting design is optimal, the resulting nested design is optimal over all two-block nested row and column designs which are adjusted orthogonal (see Eccleston and John, 1988), and also over all designs in the resolvable class. Bailey and Patterson (1991) are able to reproduce some designs of Patterson and Robinson (1989), and find some designs not listed by these authors.

A different approach to the problem of finding efficient, resolvable row–column designs is taken by John and Whitaker (1993) and Nguyen and Williams (1993). Both sets of authors describe search algorithms for non-exhaustive siftings through the large number of possible designs, concentrating on the two-replicate case. Though not flawless, both algorithms are able to construct some of the designs discussed in the preceding paragraph, and in some instances improve on those designs (in some other instances falling short). John and Whitaker's (1993) is a simulated annealing algorithm; Nguyen and Williams' (1993) is an iterative improvement algorithm. Each is available from its respective authors.

Very recently there has been work, other than that implicit in lattice rectangles, on finding nested row and column designs for factorial experiments by sacrificing information on some interactions. Gupta (1994) uses the techniques of classical confounding to construct single-replicate nested row and column designs. Morgan and Uddin (1993b) obtain main effects plans by superimposing BIBRCs in the manner of orthogonal Latin squares. In Morgan and Uddin (1996a), the superimposition method produces optimal main effects plans for the analysis with recovery of row and column information.

Some generalizations of potential interest are the extension of the BIBRC concept to generalized binary blocks (Panandikar, 1984), supplemented balance for comparing test treatments with a control (Gupta and Kageyama, 1991), and neighbor balancing in nested row and column designs (Ipinyomi and Freeman, 1988; Uddin, 1990; Uddin and Morgan, 1995).

7. Nested row and column designs optimal for the bottom stratum analysis

7.1. BNRCs

With the first papers having appeared only recently (Bagchi et al., 1990; Chang and Notz, 1990), the literature on this topic is not nearly so well developed as is that for BIBRCs and other binary block nested row and column designs. This section will draw together the currently known results, showing how they relate to one another and to available BIBRCs. One of the main themes is the close connection between BNRCs and the regular generalized Youden designs (GYDs). A regular GYD(v, p, q) is a $p \times q$ row–column design for which columns are a BBD, and for which each treatment occurs q/v times in each row. Obviously a regular GYD(v, p, q) is a BNRC$(v, 1, p, q)$. Put another way, BNRCs are a generalization of regular GYDs that allows for smaller block sizes and more than one block.

With the regular GYDs included in the BNRC family, most of the published constructions can be cast as methods of combining BNRCs with themselves or with BIBDs. These will be shown first.

THEOREM 7.1. *Construction of a* BNRC(k, b_1, p, q) *on the treatments of each block of a* BIBD(v, b_2, k) *produces a* BNRC$(v, b_1 b_2, p, q)$.

Theorem 7.1 as stated appears as Theorem 2 of Morgan and Uddin (1993a). Particular cases are Theorems 3.2.1 and 3.2.4 of Bagchi et al. (1990), and Theorems 1 and 4 of Gupta (1992). If it is demanded that the BNRC(k, b_1, p, q) have $p = k$ and $q = sk$ for some s, then any BBD(v, b_2, k) can be used (the restriction to $k < v$, inherent in a BIBD, is dropped), slightly generalizing Theorem 3.1 of Chang and Notz (1990).

With the pragmatic requirement of keeping $b_1 b_2$ small, the best use of Theorem 7.1 will be with b_1 of 1, that is, forming a regular GYD on each block of a BIBD.

COROLLARY 7.2. *There is a* BNRC$(v, v(v-1)/2, 2, 2)$ *for every* $v \geqslant 2$.

COROLLARY 7.3. *There is a* BNRC$(v, b, 2, 3)$ *and a* BNRC$(v, b, 3, 3)$ *for every* $v \geqslant 3$, *where*

$$b = \begin{cases} v(v-1)/6, & \textit{if } v \equiv 1 \textit{ or } 3 \pmod 6 \\ v(v-1)/3, & \textit{if } v \equiv 0 \textit{ or } 4 \pmod 6 \\ v(v-1)/2, & \textit{if } v \equiv 5 \pmod 6 \\ v(v-1), & \textit{if } v \equiv 2 \pmod 6. \end{cases}$$

Corollary 7.2 uses the Latin square on two treatments with all unordered pairs of v treatments to solve the construction of 2×2 BNRC's. Corollary 7.3 takes advantage of the fact that for $k = 3$ the necessary conditions for the existence of a balanced incomplete block design are also sufficient. The wealth of literature on balanced incomplete block designs (an extensive table is given in Mathon and Rosa, 1990) and Youden designs (see Ash, 1981) makes it easy to write down any number of corollaries. Similar to Corollary 7.3 one can use the results of Hanani (1961, 1975) to construct BNRC$(v, b, p, 4)$'s for $p = 3$, 4 and any v and b satisfying $v|4b$ and $v(v-1)|12b$.

These designs for small q are especially important in view of the relative inefficiency of BIBRCs of the same parameters.

THEOREM 7.4. *Existence of a* BNRC(v, b^*, p^*, q) *and a* BIBD(p^*, b, k) *implies the existence of a* BNRC(v, bb^*, k, q).

Theorem 7.4 is the method of Theorem 6.2 applied to a BNRC rather than a BIBRC, and generalizes Theorem 3 of Gupta (1992), which starts with a BNRC$(v, 1, v, v)$ (i.e., a Latin square). When starting with a Latin square, the resulting BNRC(v, b, k, v) can be extended to a BNRC$(v, b, k+1, q)$ by adding the row (1,2,...,v) to each block, which is Theorem 2 of Gupta (1992).

THEOREM 7.5 (Morgan and Uddin, 1993a). *The blocks of size* $p \times (q_1 + q_2)$*, found by connecting block* j *of a* BNRC(v, b, p, q_1) *to block* j *of a* BNRC(v, b, p, q_2) *for each* $j = 1, 2, \ldots, b$*, form a* BNRC$(v, b, p, q_1 + q_2)$.

Theorem 3.2.5 of Bagchi et al. (1990) is repeated application of Theorem 7.5 with $q_i \equiv q$. Theorem 3.2.6 (also compare Theorem 3.2.7) of Bagchi et al. (1990) is an application of Theorem 7.5 to designs from Theorems 7.1 and 7.4 (each starting with a Latin square).

THEOREM 7.6 (Morgan and Uddin, 1993a). *Let* s *divide the number of blocks* b *in a* BNRC(v, b, p, q). *Then partitioning the blocks into* b/s *groups of* s*, and connecting the blocks in each group, gives a* BNRC$(v, b/s, p, sq)$.

Theorem 7.6 with s equal to b says that combining all blocks of a BNRC(v, b, p, q) gives a BNRC$(v, 1, p, bq)$, that is, a regular GYD! In a sense then, the regular GYDs include all BNRCs, as the latter are carefully chosen partitions of columns of the former. It was stated in the first paragraph of this section that BNRCs are a generalization of GYDs for more than one block, but from a combinatorial perspective the opposite is true, for the BNRCs with more than one block are a combinatorially more restrictive class of designs. Every BNRC with more than one block yields a GYD, but the converse does not hold.

Before moving on to direct constructions, two other methods that start with known designs need be mentioned. Chang and Notz (1994) establish that some number b of permutations of treatment symbols in any maximum trace $p \times q$ row–column design produces a BNRC(v, b, p, q), and give bounds for b depending on the starting structure. Morgan and Uddin (1993a) point out that a k-resolvable BIBD(v, b, k) gives a BNRC$(v, b/v, k, v)$. The k-resolvable BIBDs include all those generated from difference sets and many from supplementary difference sets, providing numerous rich families. Theorem 7.1 can then be applied to produce designs with smaller pq relative to the number of treatments; details are in Morgan and Uddin (1993a).

The few direct constructions currently known for BNRCs are based on the method of differences using finite fields. The most far-reaching application of this approach is stated next.

THEOREM 7.7 (Bagchi et al., 1990; Morgan and Uddin, 1993a). *Let* $v = mq + 1$ *be a prime power. The initial blocks*

$$B_i = x^{i-1} \begin{pmatrix} x^0 & x^m & \cdots & x^{(q-1)m} \\ x^m & x^{2m} & \cdots & x^0 \\ \vdots & \vdots & & \vdots \\ x^{(p-1)m} & x^{pm} & \cdots & x^{(p-2)m} \end{pmatrix}, \quad i = 1, 2, \ldots, m,$$

generate a BNRC(v, mv, p, q) *for each* $2 \leqslant p \leqslant q$. *If* m *is even and* q *is odd, then* $B_1, \ldots, B_{m/2}$ *generate a* BNRC$(v, mv/2, p, q)$ *for each* $2 \leqslant p \leqslant q$.

These designs should be compared to those of Theorem 6.3 with $k_1 = t = 1$. Only the upper bound for p differs, so that infinite series of BIBRCs and BNRCs with the same v, b, p, and q have been obtained.

Morgan and Uddin (1993a) also give difference constructions for three series of BNRCs with blocks of size $2 \times \text{int}(v/2)$.

7.2. *Other optimal designs*

At the end of Section 5.2 it is mentioned that the column component design need not be a BBD, as required by a BNRC, for optimality. So long as condition (ii) of the BNRC definition holds, the column component can be *any* optimal design. This is an idea richly deserving of further investigation, as it offers the possibility of finding good designs with much smaller replication than required by BNRCs. Bagchi et al. (1990, Theorem 3.2.7) construct one such series, whose column component is a group divisible design with two groups. The parameters (for $t \geqslant 1$) are

$$v = 4t, \quad b = \frac{v^2}{16}, \quad p = 2, \quad q = 4$$

with replication $v/2$ being smaller than possible for any BNRC with the same size blocks. Others of these optimal, partially balanced designs can be found in Uddin (1994).

Chang and Notz (1994) prove the following interesting result. If for p equal to q and p not a multiple of v, there is a BNRC(v, b, p, q) and a GYD(v, p, q), then together they are a ϕ_a-optimal nested row and column design with $b + 1$ blocks of size $p \times q$, for all sufficiently large a. E-optimality holds for the new design even if p and q are not equal, both results being in spite of the fact that the information matrix is not of maximum trace. In a result of similar flavor, Morgan (1996a) proves that two copies of a GYD$(s^2, s(s+1), s(s+1))$ is an E-optimal nested row and column design with two blocks, although again the information matrix is not of maximal trace.

Morgan (1996a) also shows that this optimality does not generally hold for any pair of GYD(v, p, q)'s, by displaying a competitor for $v = 8$ and $p = q = 28$ that is superior under all criteria preserving the nonnegative definite ordering.

In work in the same vein as Morgan and Uddin (1993a), Morgan and Uddin (1996b) superimpose BNRCs to obtain main effects plans in nested rows and columns which are bottom stratum optimal. By virtue of the connections with GYDs and BIBDs discussed in Section 7.1, they get optimal main effects plans for the row–column and simple block settings as well. The plans also yield optimal multidimensional incomplete block designs. Orthogonal collections of Latin squares are introduced there as one class of pairwise orthogonal BNRCs.

BNRCs are likewise proving to be useful in construction problems arising in apparently unrelated settings that do not incorporate nesting factors. Morgan (1996b), generalizing results of Bagchi and Mukhopadhyay (1989) for $t = 2$, investigates the m-way cross-classification with t-way interaction among the blocking factors. In both papers BNRCs appear as lower dimensional components of optimal designs. Thus can advances in optimality, construction, and structural knowledge of designs in settings that have been until recently neglected, like that of nested rows and columns, have value far beyond themselves and their particular applications.

8. Other nested designs

Sections 3–7 have covered the major classes of designs with nested blocking factors that are currently in use or are topics of active research. Though the possibilities for combining nesting and other relationships among blocking factors are numerous, potentially useful, and combinatorially fascinating, no design with more than three blocking factors has been discussed, for the simple reason that there is little such coverage in the literature. Here a few other designs will be briefly mentioned.

The *row and column designs with contiguous replicates* constitute a class of designs exhibiting crossing and multiple nests in the block structure. These are resolvable $p \times q$ row–column designs (Section 6.3) for which the r replicates have been arranged into an $rp \times q$ row–column design. The model includes all of the terms (8) of the nested row and column design, and also fits another blocking factor for the *long columns* of size rp formed by the contiguous arrangement. Thus rows and (short) columns are nested within the replicates (blocks), and short columns are nested within the long columns. There is no new wrinkle in the the bottom stratum analysis, for the bottom stratum information matrix is that of the nested row and column design. Williams and John (1989) discuss the analysis recovering information from short columns and rows, modeling long columns and blocks with fixed effects, and incorporate their ideas into a computer program for design construction. Williams and John (1993) consider efficiency bounds for the bottom stratum analysis, allowing for the combinatorial restrictions of block-wise resolvability and binarity in long columns.

Bagchi (1988) generalizes Theorem 5.1 to the setting of nested multidimensional crosses. The setting is best described in terms of a $t \times n/b$ orthogonal array A, of strength at least 2, which has p_i symbols in row i. The columns of the array correspond to the experimental units at one level of a nesting factor, and the rows to levels of t nested factors which are mutually crossed, so that

$$A_{ij} = \text{the level that nested factor } i \text{ takes on experimental unit } j.$$

This same incidence structure holds within each of b levels of the nesting factor, making this a nest of b separate $p_1 \times p_2 \times \cdots \times p_t$ cross-classifications. If empty cells are present, the pattern is the same within each level of the nest. If the strength of A is t, then there are no empty cells, and if $t = 2$, this is the nested row and column setting defined in Section 5.1. The optimal designs she finds have, like (i) of Theorem 5.1, one optimal component design, and like (ii) of Theorem 5.1, a treatment assignment that is constant across levels of each other factor within each level of the nest. Constructions are given for $t = 3$. It would be interesting to see these results extended along the lines of Theorem 5.2 for the full analysis. That recovery of information from the nested nuisance factors strata is worthwhile even if the stratum variances are unknown has been established under certain simplifying assumptions by Srivastava and Beaver (1986).

Another class of designs that travel under the name of "nested design" do not fall within the framework of this chapter. A BIBD(v, b, k) for which each block contains a distinguished subblock of cardinality k', has the required nesting property if the subblocks form a BIBD(v, b, k'). The genesis of these designs seems to be the paper of Srivastava (1966, pp. 132–134). For those interested, a good starting point is Longyear (1986).

Acknowledgement

The author was supported by National Science Foundation Grant DMS 96-26115.

References

Agrawal, H. L. and J. Prasad (1982a). Some methods of construction of balanced incomplete block designs with nested rows and columns. *Biometrika* **69**, 481–483.

Agrawal, H. L. and J. Prasad (1982b). Some methods of construction of *GD-RC* and rectangular-*RC* designs. *Austral. J. Statist.* **24**, 191–200.

Agrawal, H. L. and J. Prasad (1982c). On nested row–column partially balanced incomplete block designs. *Calcutta Statist. Assoc. Bull.* **31**, 131–136.

Agrawal, H. L. and J. Prasad (1983). On construction of balanced incomplete block designs with nested rows and columns. *Sankhyā Ser. B* **30**, 345–350.

Agrawal, H. L. and J. Prasad (1984). Construction of partially balanced incomplete block designs with nested rows and columns. *Biometr. J.* **26**, 883–891.

Ash, A. (1981). Generalized Youden designs: Construction and tables. *J. Statist. Plann. Inference* **5**, 1–25.

Assaf, A. M. and A. Hartman (1989). Resolvable group-divisible designs with block size 3. *Discrete Math.* **77**, 5–20.

Bagchi, S. (1988). On the optimality of nested multiway designs. In: R. R. Bahadur, ed., *Probability, Statistics, and Design of Experiments*, Wiley, New Delhi, 23–31.

Bagchi, S. and A. C. Mukhopadhyay (1989). Optimality in the presence of two factor interactions among the nuisance factors. *Comm. Statist. Theory Methods* **18**, 1139–1152.

Bagchi, S., A. C. Mukhopadhyay and B. K. Sinha (1990). A search for optimal nested row–column designs. *Sankhyā Ser. B* **52**, 93–104.

Bailey, R. A., D. C. Goldrei and D. F. Holt (1984). Block designs with block size two. *J. Statist. Plann. Inference* **10**, 257–263.

Bailey, R. A., H. Monod and J. P. Morgan (1995). Construction and optimality of affine resolvable designs. *Biometrika* **82**, 187–200.

Bailey, R. A. and H. D. Patterson (1991). A note on the construction of row and column designs with two replicates. *J. Roy. Statist. Soc. Ser. B* **53**, 645–648.

Bailey, R. A. and T. P. Speed (1986). Rectangular lattice designs: Efficiency factors and analysis. *Ann. Statist.* **14**, 874–895.

Baker, R. D. (1983). Resolvable BIBD and SOLS. *Discrete Math.* **44**, 13–29.

Banerjee, S. and S. Kageyama (1990). Existence of α-resolvable nested incomplete block designs. *Utilitas Math.* **38**, 237–243.

Banerjee, S. and S. Kageyama (1993). Methods of constructing nested partially balanced incomplete block designs. *Utilitas Math.* **43**, 3–6.

Beth, T., D. Jungnickel and H. Lenz (1986). *Design Theory*. Cambridge University Press, Cambridge.

Bhaumik, D. K. and D. C. Whittinghill (1991). Optimality and robustness to the unavailability of blocks in block designs. *J. Roy. Statist. Soc. Ser. B* **53**, 399–407.

Bose, R. C. (1939). On the construction of balanced incomplete block designs. *Ann. Eugen.* **9**, 353–399.

Bose, R. C. (1942). A note on the resolvability of balanced incomplete block designs. *Sankhyā* **6**, 105–110.

Bose, R. C. and S. S. Shrikhande (1960). On the composition of balanced incomplete block designs. *Canad. J. Math.* **12**, 177–188.

Chang, J. Y. and W. I. Notz (1990). A method for constructing universally optimal block designs with nested rows and columns. *Utilitas Math.* **38**, 263–276.

Chang, J. Y. and W. I. Notz (1994). Some optimal nested row–column designs. *Statist. Sinica* **4**, 249–263.

Cheng, C.-S. (1986). A method for constructing balanced incomplete-block designs with nested rows and columns. *Biometrika* **73**, 695–700.

Cheng, C.-S. and R. A. Bailey (1991). Optimality of some two-associate class partially balanced incomplete-block designs. *Ann. Statist.* **19**, 1667–1671.

Clatworthy, W. H. (1973). *Tables of Two-Associate-Class Partially Balanced Designs*, NBS Applied Mathematics Series 63. National Bureau of Standards, Washington, DC.

Clem, M. A. and W. T. Federer (1950). Random arrangements for lattice designs. Iowa Agric. Exp. Stn. Spec. Rep. No. 5.

Cochran, W. G. and G. M. Cox (1992). *Experimental Designs*. Wiley, New York.

Cornelius, P. L. (1983). Lattice designs. In: S. Kotz, N. L. Johnson and B. Read, eds., *Encyclopedia of Statistical Sciences*, Vol. 4. Wiley, New York, 510–518.

Dey, A., U. S. Das and A. K. Banerjee (1986). Construction of nested balanced incomplete block designs. *Calcutta Statist. Assoc. Bull.* **35**, 161–167.

Eccleston, J. A. and J. A. John (1988). Adjusted orthogonal nested row–column designs. *Austral. J. Statist.* **30**, 78–84.

Federer, W. T. and B. L. Raktoe (1965). General theory of prime-power lattice designs. *J. Amer. Statist. Assoc.* **60**, 891–904.

Furino, S., Y. Miao and J. X. Yin (1994). Frames and resolvable designs. Preprint.

Gupta, S. (1992). Some optimal nested row–column designs. *Calcutta Statist. Assoc. Bull.* **42**, 261–265.

Gupta, S. (1994). Block designs with nested rows and columns for factorial experiments. *Sankhyā Ser. B* **56**, 52–58.

Gupta, S. and S. Kageyama (1991). Type S designs with nested rows and columns. *Metrika* **38**, 195–202.

Gupta, S. and M. Singh (1991). Partially balanced incomplete block designs with nested rows and columns. *Utilitas Math.* **40**, 291–302.

Gupta, V. K. (1993). Optimal nested block designs. *J. Ind. Soc. Agricult. Statist.* **45**, 187–194.

Gupta, V. K. and R. Srivastava (1992). Investigations on robustness of block designs against missing observations. *Sankhyā Ser. B* **54**, 100–105.

Hanani, H. (1961). The existence and construction of balanced incomplete block designs. *Ann. Math. Statist.* **32**, 371–386.

Hanani, H. (1974). On resolvable balanced incomplete block designs. *J. Combin. Theory Ser. A* **17**, 275–289.

Hanani, H. (1975). Balanced incomplete block designs and related designs. *Discrete Math.* **11**, 255–369.

Harshbarger, B. (1946). Preliminary report on the rectangular lattices. *Biometrics* **2**, 1115–1119.

Harshbarger, B. (1947). Rectangular Lattices. Memoir 1, Virginia Agricultural Experiment Station.

Harshbarger, B. (1949). Triple rectangular lattices. *Biometrics* **5**, 1–13.

Harshbarger, B. (1951). Near balance rectangular lattices. *Virginia J. Sci.* **2**, 13–27.

Hedayat, A. and W. T. Federer (1970). On the equivalence of Mann's group automorphism method of constructing an $O(n, n-1)$ set and Raktoe's collineation method of constructing a balanced set of L-restrictional prime-powered lattice designs. *Ann. Math. Statist.* **41**, 1530–1540.

Homel, R. J. and J. Robinson (1975). Nested partially balanced incomplete block designs. *Sankhyā Ser. B* **37**, 201–210.

Houtman, A. M. and T. P. Speed (1983). Balance in designed experiments with orthogonal block structure. *Ann. Statist.* **17**, 1069–1085.

Ipinyomi, R. A. (1990). Resolvable row–column designs. *J. Appl. Statist.* **17**, 351–355.

Ipinyomi, R. A. and G. H. Freeman (1988). Equineighbored balanced nested row–column designs. *J. Statist. Plann. Inference* **20**, 191–199.

Ipinyomi, R. A. and J. A. John (1985). Nested generalized cyclic row–column designs. *Biometrika* **72**, 403–409.

Iqbal, I. (1991). Construction of experimental designs using cyclic shifts. Ph.D. Thesis, University of Kent, Canterbury.

Jimbo, M. (1993). Recursive constructions for cyclic BIB designs and their generalizations. *Discrete Math.* **116**, 79–95.

Jimbo, M. and S. Kuriki (1983). Construction of nested designs. *Ars Combin.* **16**, 275–285.

John, J. A. (1987). *Cyclic Designs*. Chapman and Hall, London.

John, J. A. and T. J. Mitchell (1977). Optimal incomplete block designs. *J. Roy. Statist. Soc. Ser. B* **39**, 39–43.

John, J. A. and D. Whitaker (1993). Construction of resolvable row–column designs using simulated annealing. *Austral. J. Statist.* **35**, 237–245.

Jungnickel, D., R. C. Mullin and S. A. Vanstone (1991). The spectrum of α-resolvable block designs with block size 3. *Discrete Math.* **97**, 269–277.

Kageyama, S. (1972). A survey of resolvable solutions of balanced incomplete block designs. *Internat. Statist. Rev.* **40**, 269–273.

Kageyama, S. (1976). Resolvability of block designs. *Ann. Statist.* **4**, 655–661.

Kageyama, S. (1988). Two methods of construction of affine resolvable balanced designs with unequal block sizes. *Sankhyā Ser. B* **50**, 195–199.

Kageyama, S. and Y. Miao (1996). Nested designs of superblock size four. *J. Statist. Plann. Inference*, to appear.

Kageyama, S. and D. V. S. Sastry (1993). On affine $(\mu_1, \mu_2, \ldots, \mu_t)$-resolvable (r, λ)-designs. *Ars Combin.* **36**, 221–223.

Kempthorne, O. (1952). *The Design and Analysis of Experiments*. Wiley, New York.

Kempthorne, O. and W. T. Federer (1948). The general theory of prime-power lattice designs II. Designs for p^n varieties in blocks of p^s plots, and in squares. *Biometrics* **4**, 109–121.

Keuhl, R. O. (1994). *Statistical Principles of Research Design and Analysis*. Wadsworth, Belmont.

Kiefer, J. (1975). Construction and optimality of generalized Youden designs. In: J. N. Srivastava, ed., *A Survey of Statistical Design and Linear Models*. North-Holland, Amsterdam, 333–353.

Longyear, J. (1986). Nested group divisible designs and small nested designs. *J. Statist. Plann. Inference* **13**, 81–87.

Lu, J. (1984). An existence theory for resolvable balanced incomplete block designs. *Acta Math. Sinica* **27**, 458–468.

Mathon, R. and A. Rosa (1990). Tables of parameters of BIBDs with $r \leqslant 41$ including existence, enumeration, and resolvability results: An update. *Ars Combin.* **30**, 65–96.

Mazumdar, S. (1967). On the construction of cyclic collineations for obtaining a balanced set of L-restrictional prime-powered lattice designs. *Ann. Math. Statist.* **38**, 1293–1295.

Mejza, I. and S. Mejza (1994). Model building and analysis for block designs with nested rows and columns. *Biom. J.* **36**, 327–340.

Mohan, R. N. and S. Kageyama (1989). Two constructions of α-resolvable PBIB designs. *Utilitas Math.* **36**, 115–119.

Monod, H. and R. A. Bailey (1992). Pseudofactors: Normal use to improve design and facilitate analysis. *Appl. Statist.* **41**, 317–336.

Morgan, J. P. (1996a). On pairs of Youden designs. *J. Statist. Plann. Inference*, to appear.

Morgan, J. P. (1996b). Optimal design for interacting blocks with OAVS incidence. *Metrika*, to appear.

Morgan, J. P. and N. Uddin (1990). Some constructions for rectangular, Latin square and pseudocyclic nested row–column designs. *Utilitas Math.* **38**, 43–51.

Morgan, J. P. and N. Uddin (1993a). Optimality and construction of nested row and column designs. *J. Statist. Plann. Inference* **37**, 81–93.

Morgan, J. P. and N. Uddin (1993b). On mutually orthogonal and totally balanced sets of balanced incomplete block designs with nested row and column designs. *Statist. Sinica* **3**, 435–451.

Morgan, J. P. and N. Uddin (1996a). Orthogonal sets of balanced incomplete block designs with nested row and columns. *J. Statist. Plann. Inference*, to appear.

Morgan, J. P. and N. Uddin (1996b). Optimal blocked main effects plans with nested row and columns and related designs. *Ann. Statist.*, to appear.

Mukerjee, R. and S. Kageyama (1985). On resolvable and affine resolvable variance-balanced designs. *Biometrika* **72**, 165–172.

Nguyen, N. and E. R. Williams (1993). An algorithm for constructing optimal resolvable row–column designs. *Austral. J. Statist.* **35**, 363–370.

Pal, S. and S. Pal (1988). Nonproper variance balanced designs and optimality. *Comm. Statist. Theory Methods* **17**, 1685–1695.

Panandikar, S. C. (1984). Balanced block designs with nested rows and columns. *Calcutta Statist. Assoc. Bull.* **33**, 173–177.

Paterson, L. J. (1988). Some recent work on making incomplete-block designs available as a tool for science. *Internat. Statist. Rev.* **56**, 129–138.

Paterson, L. J. and H. D. Patterson (1983). An algorithm for constructing α-lattice designs. *Ars Combin.* **16A**, 87–98.

Paterson, L. J., P. Wild and E. R. Williams (1988). An algorithm to generate designs for variety trials. *J. Agricult. Sci.* **111**, 133–136.

Patterson, H. D. and D. L. Robinson (1989). Row and column designs with two replcates. *J. Agricult. Sci.* **112**, 73–77.

Patterson, H. D. and V. Silvey (1980). Statutory and recommended list trials of crop varieties in the United Kingdom. *J. Roy. Statist. Soc. Ser. A* **143**, 219–252.

Patterson, H. D. and E. R. Williams (1976a). A new class of resolvable incomplete block designs. *Biometrika* **63**, 83–92.

Patterson, H. D. and E. R. Williams (1976b). Some theoretical results on general block designs. *Congressus Numerantium* **XV**, 489–496.

Patterson, H. D., E. R. Williams and E. A. Hunter (1978). Block designs for variety trials. *J. Agricult. Sci.* **90**, 395–400.

Preece, D. A. (1967). Nested balanced incomplete block designs. *Biometrika* **43**, 479–486.

Preece, D. A. (1982). Some partly cyclic 13 $\times$ 4 Youden 'squares' and a balanced arrangement for a pack of cards. *Utilitas Math.* **22**, 255–263.

Raktoe, B. L. (1967). Application of cyclic collineations to the construction of balanced L-restrictional prime-powered lattice designs. *Ann. Math. Statist.* **38**, 1127–1141.

Ray-Chaudhuri, D. K. and R. M. Wilson (1973). The existence of resolvable block designs. In: J. N. Srivastava, ed., *A Survey of Combinatorial Theory*. North-Holland, Amsterdam, 361–375.

Rees, R. and D. R. Stinson (1987). On resolvable group-divisible designs with block size 3. *Ars Combin.* **23**, 107–120.

Saha, G. M. and R. K. Mitra (1992). Some constructions of balanced incomplete block row–column designs. *Ars Combin.* **34**, 321–325.

Searle, S. (1971). *Linear Models*. Wiley, New York.

Shah, K. R. and B. K. Sinha (1989). *Theory of Optimal Designs*. Springer, Berlin.

Shah, S. M., D. K. Ghosh and A. H. Patel (1993). Construction of α-resolvable BIBDs using 1-resolvable and affine α-resolvable BIBDs. *Utilitas Math.* **43**, 219–224.

Shen, H. (1988). Existence of a resolvable design RGD(4,3;v) (abstract). *Kexue Tongbao* **33**, 397.

Shen, H. and W. D. Wallis. A note on the existence of resolvable block designs with block sizes 3 and 4. Unpublished manuscript.

Shrikhande, S. S. and D. Raghavarao (1963). Affine α-resolvable incomplete block designs. In: C. R. Rao, ed., *Contributions of Statistics*. Pergamon, Calcutta, 471–480.

Singh, M. and A. Dey (1979). Block designs with nested rows and columns. *Biometrika* **66**, 321–326.

Sinha, K. (1991). Construction of semi-regular group divisible designs. *Sankhyā Ser. B* **53**, 229–231.

Sinha, K. and A. Dey (1982). On resolvable PBIB designs. *J. Statist. Plann. Inference* **6**, 170–181.

Sinha, K. and S. Kageyama (1990). A series of triangular PBIB designs with nested rows and columns. *J. Statist. Plann. Inference* **24**, 403–404.

Sreenath, P. R. (1989). Construction of some balanced incomplete block designs with nested rows and columns. *Biometrika* **76**, 399–402.

Srivastav, S. K. and J. P. Morgan (1996). On the class of 2×2 balanced incomplete block designs with nested rows and columns. *Comm. Statist. Theory Methods* **25**, 1859–1870.

Srivastava, J. N. (1966). Some generalizations of multivariate analysis of variance. In: P. R. Krishnaiah, ed., *Multivariate Analysis*. Academic Press, New York, 129–145.

Srivastava, J. N. (1978). Statistical design of agricultural experiments. *J. Ind. Soc. Agricult. Statist.* **30**, 1–10.

Srivastava, J. N. and R. J. Beaver (1986). On the superiority of the nested multidimensional block designs, relative to the classical incomplete block designs. *J. Statist. Plann. Inference* **13**, 133–150.

Street, A. P. and D. J. Street (1987). *Combinatorics of Experimental Design*. Oxford University Press, Oxford.

Street, D. J. (1981). Graeco-Latin and nested row and column designs. In: K. L. McAvaney, ed., *Combinatorial Mathematics*, Vol. 8. Lecture Notes in Mathematics 884. Springer, Berlin, 304–313.

Uddin, N. (1990). Some series constructions for minimal size equineighboured balanced incomplete block designs with nested rows and columns. *Biometrika* **77**, 829–833.

Uddin, N. (1992). Constructions for some balanced incomplete block designs with nested rows and columns. *J. Statist. Plann. Inference* **31**, 253–261.

Uddin, N. (1994). Optimal partially balanced nested row–column designs. Preprint.

Uddin, N. (1995). On recursive construction for balanced incomplete block designs with nested rows and columns. *Metrika* **42**, 341–345.

Uddin, N. and J. P. Morgan (1990). Some constructions for balanced incomplete block designs with nested rows and columns. *Biometrika* **77**, 193–202.

Uddin, N. and J. P. Morgan (1991). Two constructions for balanced incomplete block designs with nested rows and columns. *Statist. Sinica* **1**, 229–232.

Uddin, N. and J. P. Morgan (1995). Universally optimal designs with blocksize $p \times 2$ and correlated observations. Preprint.

Williams, E. R. (1975). A new class of resolvable designs. Ph.D. Thesis, University of Edinburgh.

Williams, E. R. (1977). A note on rectangular lattice designs. *Biometrics* **33**, 410–414.

Williams, E. R. and J. A. John (1989). Construction of row and column designs with contiguous replicates. *Appl. Statist.* **38**, 149–154.

Williams, E. R. and J. A. John (1993). Upper bounds for Latinized designs. *Austral. J. Statist.* **35**, 229–236.

Williams, E. R., H. D. Patterson and J. A. John (1976). Resolvable designs with two replications. *J. Roy. Statist. Soc. Ser. B* **38**, 296–301.

Yates, F. (1936). A new method of arranging variety trials involving a large number of varieties. *J. Agricult. Sci.* **26**, 424–455.

Yates, F. (1937). A further note on the arrangement of variety trials: quasi-Latin squares. *Ann. Eugen.* **7**, 319–332.

Yates, F. (1940). Lattice squares. *J. Agricult. Sci.* **30**, 672–687.

Yin, J. and Y. Miao (1993). Almost resolvable BIBDs with block-size 5 or 6. *Ars Combin.* **35**, 303–313.

Zhu, L. (1993). Some recent developments on BIBDs and related designs. *Discrete Math.* **123**, 189–214.

S. Ghosh and C. R. Rao, eds., *Handbook of Statistics, Vol. 13*

Optimal Design: Exact Theory

Ching-Shui Cheng

1. Introduction

The purpose of this article is to present a brief introduction to the exact theory of optimal design. In the so-called *approximate theory*, each discrete probability measure on the experimental region is thought of as a design, where the probabilities represent proportions of observations at different sites, and the problem is to determine "optimal" proportions. An excellent account of this theory can be found in Pukelsheim (1993). The exact theory, on the other hand, seeks to determine an optimal design for a given finite number of observations. See, for example, Shah and Sinha (1989). It is clear from the contents of Pukelsheim (1993) and Shah and Sinha (1989), which have almost no overlaps except for the basic optimality concepts, that the approximate theory is more appropriate for regression design problems, whereas the exact theory is natural for design problems in discrete settings such as block designs, row–column designs, factorial designs, etc.

Unlike Shah and Sinha (1989), this article is not organized according to topics, and is not meant to be a comprehensive review of the state of the art. Instead, we shall present some useful techniques which have often been used, together with illustrative examples covering applications to block designs, row–column designs, weighing designs, repeated measurements designs, block designs with nested rows and columns, and designs for comparing test treatments with a control. Such an approach is useful to researchers new to this area and enables us to provide a unified treatment of applications to various design problems.

The organization of this article is as follows. Section 2 discusses statistical models and optimality criteria. Section 3 is devoted to the optimality of symmetric designs such as Hadamard matrices, balanced incomplete block designs, Latin squares, etc. Many results in the literature are presented by a unified approach based on a concept of adjusted orthogonality. Section 4 focuses on the optimality of asymmetrical designs. Section 5 gives an example where the approximate theory leads to a nontrivial solution of a discrete problem. The article ends with some miscellaneous results and methods in Section 6.

The rest of this section contains some preliminary material on incidence matrices and orthogonal projection matrices associated with the assignment of factor levels. Suppose each of N units is assigned one of the f levels of a certain factor F. The relation between the units and the levels of F can be described by an $N \times f$ *incidence* matrix $\boldsymbol{X}_F = [x_{ij}]$, where $x_{ij} = 1$ if the jth level of F is assigned to unit i, and 0 otherwise. The relation among the units with respect to F can be represented by an $N \times N$ *relation* matrix $\boldsymbol{R}_F = [r_{ij}]$, where $r_{ij} = 1$ if the same level of F appears at units i and j, and 0 otherwise. For any matrix $\boldsymbol{X}$, let $\mathcal{R}(\boldsymbol{X})$ be the range of $\boldsymbol{X}$, i.e., the space generated by the column vectors of $\boldsymbol{X}$. Then clearly $\mathcal{R}(\boldsymbol{R}_F) = \mathcal{R}(\boldsymbol{X}_F)$, since $\boldsymbol{R}_F$ and $\boldsymbol{X}_F$ have the same column vectors, ignoring repetition. Let this space be $\mathcal{F}$ and denote the orthogonal projection matrix onto $\mathcal{F}$ by $\boldsymbol{P}_{\mathcal{F}}$. Then $\mathcal{F} = \{\boldsymbol{y}\colon\ y_i = y_j$ if the same level of F is assigned to units i and $j\}$. If all the levels appear at least once, then $\dim \mathcal{F} = f$, and the orthogonal projection $\boldsymbol{P}_{\mathcal{F}}\boldsymbol{y}$ of $\boldsymbol{y}$ onto $\mathcal{F}$ replaces each y_i with the average of the y_j's over all the units j which are assigned the same level of F as unit i. Therefore if each level of F is assigned to the same number of units, then

$$\boldsymbol{P}_{\mathcal{F}} = \frac{1}{u}\boldsymbol{R}_F, \tag{1.1}$$

where $u = N/f$, and the orthogonal projection matrix onto $\mathcal{F}^{\perp}$, the orthogonal complement of $\mathcal{F}$, is

$$\boldsymbol{P}_{\mathcal{F}^{\perp}} = \boldsymbol{I} - \frac{1}{u}\boldsymbol{R}_F. \tag{1.2}$$

The factor F can be a treatment factor (T), block factor (B), row factor (R), or column factor (C), etc. Then $\boldsymbol{X}_F$, $\boldsymbol{R}_F$, $\mathcal{F}$ and $\boldsymbol{P}_{\mathcal{F}}$ are denoted $\boldsymbol{X}_T$, $\boldsymbol{X}_B$, $\boldsymbol{X}_R$, $\boldsymbol{X}_C$, $\boldsymbol{R}_T$, $\boldsymbol{R}_B$, $\boldsymbol{R}_R$, $\boldsymbol{R}_C$, $\mathcal{T}$, $\mathcal{B}$, $\mathcal{R}$, $\mathcal{C}$, and $\boldsymbol{P}_{\mathcal{T}}$, $\boldsymbol{P}_{\mathcal{B}}$, $\boldsymbol{P}_{\mathcal{R}}$, $\boldsymbol{P}_{\mathcal{C}}$, respectively, and the number of levels is denoted t, b, r or c. When F is the trivial factor with just one level, $\mathcal{F}$ is the one-dimensional subspace $\mathcal{G}$ of R^N generated by the vector of ones, with orthogonal projection matrix $\boldsymbol{P}_{\mathcal{G}} = \frac{1}{N}\boldsymbol{J}_N$, where $\boldsymbol{J}_N$ is the $N \times N$ matrix of ones.

2. Information matrices and optimality criteria

Let $\mathcal{D}$ be the set of all the competing designs in a certain setting. For each design $d \in \mathcal{D}$, assume that the observations follow a linear model

$$\boldsymbol{y} = \boldsymbol{X}_d\boldsymbol{\theta} + \boldsymbol{\varepsilon},$$

where the *design matrix* $\boldsymbol{X}_d$ is $N \times n$, $\boldsymbol{\theta} = (\theta_1, \dots, \theta_n)^T$ is an $n \times 1$ vector of unknown parameters, and $\boldsymbol{\varepsilon}$ is an $N \times 1$ vector of random errors with $\mathrm{E}(\boldsymbol{\varepsilon}) = \boldsymbol{0}$ and $\mathrm{cov}(\boldsymbol{\varepsilon}) = \sigma^2\boldsymbol{I}_N$. (Due to lack of space, we shall not discuss design problems for correlated observations.) All the n parameters in $\boldsymbol{\theta}$ are estimable if and only if $\boldsymbol{X}_d^T\boldsymbol{X}_d$ is nonsingular, and in this case the covariance matrix of $\widehat{\boldsymbol{\theta}}$, the least squares estimator of $\boldsymbol{\theta}$, is equal to $\sigma^2(\boldsymbol{X}_d^T\boldsymbol{X}_d)^{-1}$. We shall denote $\boldsymbol{X}_d^T\boldsymbol{X}_d$ by $\boldsymbol{M}_d$ and call it the *information matrix* of d (for estimating $\boldsymbol{\theta}$).

The theory of optimal design is concerned with the problem of "optimally" choosing a design from $\mathcal{D}$. We say that a design d_1 is *at least as good as* d_2 if any linear function $\boldsymbol{c}^T\boldsymbol{\theta}$ estimable under d_2 is also estimable under d_1, and the variance of its least squares estimator under d_1 is not greater than that under d_2. Ehrenfeld (1956) showed that d_1 is at least as good as d_2 if and only if $\boldsymbol{M}_{d_1} - \boldsymbol{M}_{d_2}$ is nonnegative definite. This is the so-called *Loewner ordering* (Pukelsheim, 1993, p. 12 and Chapter 4). However, Loewner ordering is a partial order, and usually there does not exist a design with the "largest" information matrix in such a strong sense. An alternative approach is to choose some real-valued function ϕ (called an *information function* or an *optimality criterion*), and find a design which maximizes $\phi(\boldsymbol{M}_d)$ over $d \in \mathcal{D}$. Such a design is called ϕ-optimal. Pukelsheim (1993) discussed properties that information functions should satisfy, the most important of which being concavity and nondecreasing monotonicity with respect to the Loewner ordering. The latter means that if $\boldsymbol{M}_1 - \boldsymbol{M}_2$ is nonnegative definite, then $\phi(\boldsymbol{M}_1) \geqslant \phi(\boldsymbol{M}_2)$.

Important examples of optimality criteria are the commonly used D-, A- and E-criteria which maximize $\prod_{i=1}^n \mu_{di}$, $[\frac{1}{n}\sum_{i=1}^n \mu_{di}^{-1}]^{-1}$, and μ_{dn}, respectively, where $\mu_{d1} \geqslant \cdots \geqslant \mu_{dn}$ are the eigenvalues of $\boldsymbol{M}_d$. These three criteria have the following statistical meanings. Let $\boldsymbol{V}_d$ be the covariance matrix of $\widehat{\boldsymbol{\theta}}$ under d. Then a design is D-optimal if it minimizes $\det(\boldsymbol{V}_d)$, the generalized variance of $\widehat{\boldsymbol{\theta}}$. Under the normality assumption, this is the same as to minimize the volume of a confidence ellipsoid for $\boldsymbol{\theta}$. It is also easy to see that a design is A-optimal if it minimizes $\sum_{i=1}^n \operatorname{var}(\widehat{\theta}_i)$, and is E-optimal if it minimizes

$$\max_{\mathbf{c}:\ \sum_{i=1}^n c_i^2 = 1} \operatorname{var}\left(\sum_{i=1}^n c_i\widehat{\theta}_i\right).$$

These three criteria are members of a family of information functions called the p-means, $-\infty \leqslant p \leqslant 1$:

$$\phi_p(\boldsymbol{M}_d) = \begin{cases} \left[\dfrac{1}{n}\displaystyle\sum_{i=1}^n \mu_{di}^p\right]^{1/p}, & \text{for } p \neq 0,\ -\infty; \\ \left[\displaystyle\prod_{i=1}^n \mu_{di}\right]^{1/n}, & \text{for } p = 0; \\ \mu_{dn}, & \text{for } p = -\infty, \end{cases}$$

where $\mu_{d1} \geqslant \cdots \geqslant \mu_{dn}$ are the eigenvalues of $\boldsymbol{M}_d$. Note that D-, A- and E-criteria correspond to $p = 0, -1$ and $-\infty$, respectively.

Earlier literature in optimal design often focuses on the *minimization* of some convex and nonincreasing function Φ of the information matrix, which is related to the minimization of dispersion matrices; for instance, the D-, A- and E-criteria *minimize* $\det(\boldsymbol{V}_d)$, $\operatorname{tr}(\boldsymbol{V}_d)$, and the maximum eigenvalue of $\boldsymbol{V}_d$, respectively. Although Pukelsheim (1993, p. 156) argued that maximization of information matrices is a more appropriate optimality concept, in this article we shall also refer to the minimization of $\Phi(\boldsymbol{M}_d)$. We shall use Φ and ϕ to denote nonincreasing and nondecreasing functions of information matrices, respectively. When we say a design is Φ- or ϕ-optimal, we mean that it minimizes $\Phi(\boldsymbol{M}_d)$ or maximizes $\phi(\boldsymbol{M}_d)$, respectively.

EXAMPLE 2.1 (Chemical balance weighing designs). The weights of n objects are to be measured by using a chemical balance. Each observation measures the difference between the total weight of the objects put on the right pan and the total weight of those on the left pan. Suppose N observations are to be made. Then for each design, the (i, j)th entry of the design matrix $\boldsymbol{X}_d$ is 1, -1 or 0 depending upon whether in the ith weighing, the jth object is put on the right pan, left pan or is not present. In this case, $\mathcal{D}$ consists of all the $N \times n$ $(1, -1, 0)$-matrices.

EXAMPLE 2.2 (Spring balance weighing designs). In the spring balance weighing design problem where each observation measures the total weight of the objects put on the scale, $\mathcal{D}$ consists of all the $N \times n$ matrices with entries equal to 0 or 1.

Now we consider the estimation of a subsystem of parameters. Sometimes one may be interested in estimating only part of the parameters since the other parameters are nuisance parameters. For instance, in the setting of block designs discussed below in Example 2.3, one is usually interested in estimating the treatment effects, not the block effects. Suppose

$$\boldsymbol{y} = \boldsymbol{X}_{d1}\boldsymbol{\theta}_1 + \boldsymbol{X}_{d2}\boldsymbol{\theta}_2 + \boldsymbol{\varepsilon} \tag{2.1}$$

where $\boldsymbol{X}_{d1}$ is $N \times s$, $\boldsymbol{\theta}_1$ is $s \times 1$, and one is interested in estimating $\boldsymbol{\theta}_1$ only. Let

$$\boldsymbol{C}_d = \boldsymbol{X}_{d1}^T\boldsymbol{X}_{d1} - \boldsymbol{X}_{d1}^T\boldsymbol{X}_{d2}(\boldsymbol{X}_{d2}^T\boldsymbol{X}_{d2})^-\boldsymbol{X}_{d2}^T\boldsymbol{X}_{d1}, \tag{2.2}$$

$(\boldsymbol{X}_{d2}^T\boldsymbol{X}_{d2})^-$ being a generalized inverse of $\boldsymbol{X}_{d2}^T\boldsymbol{X}_{d2}$. Then all the s parameters in $\boldsymbol{\theta}_1$ are estimable if and only if $\boldsymbol{C}_d$ is nonsingular, and in this case the covariance matrix of the least squares estimator of $\boldsymbol{\theta}_1$ is equal to $\sigma^2\boldsymbol{C}_d^{-1}$. We shall call $\boldsymbol{C}_d$ the information matrix for estimating $\boldsymbol{\theta}_1$, and the problem of maximizing $\phi(\boldsymbol{C}_d)$ or minimizing $\Phi(\boldsymbol{C}_d)$ can be similarly considered. Note that we can express $\boldsymbol{C}_d$ as

$$\boldsymbol{C}_d = \tilde{\boldsymbol{X}}_{d1}^T\tilde{\boldsymbol{X}}_{d1} \tag{2.3}$$

with

$$\tilde{\boldsymbol{X}}_{d1} = \boldsymbol{P}_{\mathcal{R}(\boldsymbol{X}_{d2})^\perp}\boldsymbol{X}_{d1} \tag{2.4}$$

where $\boldsymbol{P}_{\mathcal{R}(\boldsymbol{X}_{d2})^\perp}$ is the orthogonal projection matrix onto the orthogonal complement of the range of $\boldsymbol{X}_{d2}$.

In many settings including, e.g., block designs, all the $\boldsymbol{C}_d$'s are singular; so not all the parameters in $\boldsymbol{\theta}_1$ are estimable. For instance, suppose

$$\mathbf{1}_N \in \mathcal{R}(\boldsymbol{X}_{d2}), \tag{2.5}$$

where $\mathbf{1}_N$ is the $N \times 1$ vector of ones, and

$$\boldsymbol{X}_{d1} \text{ has constant row sums.} \tag{2.6}$$

Then $\tilde{\boldsymbol{X}}_{d1}\mathbf{1}_s = \boldsymbol{P}_{\mathcal{R}(\boldsymbol{X}_{d2})^\perp}\boldsymbol{X}_{d1}\mathbf{1}_s = 0$, which implies $\boldsymbol{C}_d\mathbf{1}_s = 0$. Therefore a linear function $\boldsymbol{a}^T\boldsymbol{\theta}_1$ is estimable only if it is a *contrast*, i.e., $\boldsymbol{a}^T\mathbf{1}_s = 0$. A design is called *connected* if $\operatorname{rank}(\boldsymbol{C}_d) = s - 1$, which is equivalent to that all the contrasts of $\boldsymbol{\theta}_1$ are estimable. As in Kiefer (1958), let

$$\boldsymbol{O} = \left[\frac{1}{\sqrt{s}}\mathbf{1}_s \vdots \boldsymbol{P}_1 \cdots \boldsymbol{P}_{s-1}\right]$$

be an orthogonal matrix, and $\boldsymbol{P} = [\boldsymbol{P}_1 \cdots \boldsymbol{P}_{s-1}]^T$. Then $\boldsymbol{P}\boldsymbol{\theta}_1$ consists of a maximal system of normalized orthogonal contrasts. Let $\boldsymbol{V}_d$ be the covariance matrix of the least squares estimator of $\boldsymbol{P}\boldsymbol{\theta}_1$ under a connected design d. Then D-, A- and E-optimal designs can be defined as those which minimize $\det(\boldsymbol{V}_d)$, $\operatorname{tr}(\boldsymbol{V}_d)$ and the maximum eigenvalue of $\boldsymbol{V}_d$, respectively. Since $\boldsymbol{V}_d = \sigma^2(\boldsymbol{P}\boldsymbol{C}_d^-\boldsymbol{P}^T)$, a D-, A- or E-optimal design minimizes $\prod_{i=1}^{s-1}\mu_{di}^{-1}$, $\sum_{i=1}^{s-1}\mu_{di}^{-1}$ or $\mu_{d,s-1}^{-1}$, respectively, where $\mu_{d1} \geqslant \cdots \geqslant \mu_{ds} = 0$ are the eigenvalues of $\boldsymbol{C}_d$. Therefore all these criteria depend on $\boldsymbol{V}_d$ through $\boldsymbol{C}_d$. Rigorously speaking, the information matrix of a connected design for estimating $\boldsymbol{P}\boldsymbol{\theta}_1$ should be defined as $(\boldsymbol{P}\boldsymbol{C}_d^-\boldsymbol{P}^T)^{-1}$, but the above discussion demonstrates that it is easier to work with $\boldsymbol{C}_d$ directly. We shall continue to call $\boldsymbol{C}_d$ the information matrix of d, and the optimality criteria are considered as functions of $\boldsymbol{C}_d$. Note that if $\boldsymbol{X}_{d1}$ is the $(0,1)$-incidence matrix between the units and the levels of a certain factor, then (2.6) is satisfied if each unit is assigned exactly one level.

EXAMPLE 2.3 (Block designs). Suppose t treatments are to be compared on bk experimental units grouped into b blocks each of size k. Let y_{ij} be the observation taken on unit j of the ith block, $i = 1, \ldots, b$, $j = 1, \ldots, k$. Assume the usual additive fixed-effects model in which

$$\mathrm{E}(y_{ij}) = \alpha_{t(i,j)} + \beta_i,$$

where $t(i,j)$ is the label of the treatment assigned to unit (i,j), $\alpha_1, \ldots, \alpha_t$ are the treatment effects and $\beta_1, \ldots, \beta_b$ are the block effects. Let $\boldsymbol{\theta}_1 = (\alpha_1, \ldots, \alpha_t)^T$ and $\boldsymbol{\theta}_2 = (\beta_1, \ldots, \beta_b)^T$. Here $\beta_1, \ldots, \beta_b$ are considered as nuisance parameters, and one is only interested in the treatment effects. Then $\boldsymbol{X}_{d1}, \boldsymbol{X}_{d2}$ in (2.1) are the unit-treatment and unit-block incidence matrices, respectively. Let r_{di} be the number of replications of treatment i, and n_{dij} be the number of times treatment i appears in the jth block. Then it is easy to see that $\boldsymbol{X}_{d1}^T\boldsymbol{X}_{d1}$ is the diagonal matrix $\operatorname{diag}(r_{d1}, \ldots, r_{dt})$ with the ith diagonal equal to r_{di}, $\boldsymbol{X}_{d2}^T\boldsymbol{X}_{d2} = k\boldsymbol{I}_b$, and $\boldsymbol{X}_{d1}^T\boldsymbol{X}_{d2}$ is equal to the treatment-block incidence matrix $\boldsymbol{N}_d$ with the (i,j)th entry n_{dij}. Thus, from (2.2), the information matrix for estimating the treatment effects is

$$\boldsymbol{C}_d = \operatorname{diag}(r_{d1}, \ldots, r_{dt}) - k^{-1}\boldsymbol{N}_d\boldsymbol{N}_d^T, \tag{2.7}$$

which has zero row and column sums since (2.5) and (2.6) obviously hold. A connected design has $\mathrm{rank}(\boldsymbol{C}_d) = t - 1$, and the commonly used optimality criteria are defined in terms of the $t - 1$ nonzero eigenvalues of $\boldsymbol{C}_d$.

It can be shown that the A-criterion is equivalent to the minimization of the average variance

$$\frac{1}{t(t-1)} \sum_{1 \leqslant i \neq j \leqslant t} \mathrm{var}(\widehat{\alpha}_i - \widehat{\alpha}_j). \tag{2.8}$$

EXAMPLE 2.4 (Block designs for comparing test treatments with a control). In the block design problem of the previous example, the interest is focused on the estimation of a maximal system of normalized orthogonal contrasts; so all the treatments are considered equally important. Now suppose one of the treatments is a *control* or standard treatment, and the other treatments are test treatments. Then some contrasts may be more important than others. For example, suppose treatment 1 is the control, and we are interested in estimating the $t-1$ comparisons $\alpha_1 - \alpha_2, \ldots, \alpha_1 - \alpha_t$ between the control and the test treatments. Bechhofer and Tamhane (1981) showed that the covariance matrix of the least squares estimators of $\alpha_1 - \alpha_2, \ldots, \alpha_1 - \alpha_t$ under a connected design d is equal to $\sigma^2 \times$ the inverse of the $(t-1) \times (t-1)$ matrix $\bar{\boldsymbol{C}}_d$ obtained by deleting the first row and column of $\boldsymbol{C}_d$. Therefore for control-test treatment comparison, the information matrix of a design d is $\bar{\boldsymbol{C}}_d$. An A-optimal design, for example, minimizes $\sum_{i=1}^{t-1} \mu_{di}^{-1}$, where $\mu_{d1}, \ldots, \mu_{d,t-1}$ are the eigenvalues of $\bar{\boldsymbol{C}}_d$. Such a design minimizes $\frac{1}{t-1} \sum_{2 \leqslant i \leqslant t} \mathrm{var}(\widehat{\alpha}_1 - \widehat{\alpha}_i)$, instead of (2.8).

Now compare (2.1) with the following model

$$\boldsymbol{y} = \boldsymbol{X}_{d1}\boldsymbol{\theta}_1 + \boldsymbol{X}_{d2}\boldsymbol{\theta}_2 + \boldsymbol{X}_{d3}\boldsymbol{\theta}_3 + \boldsymbol{\varepsilon}, \tag{2.9}$$

with the usual assumption $\mathrm{E}(\boldsymbol{\varepsilon}) = \mathbf{0}$ and $\mathrm{cov}(\boldsymbol{\varepsilon}) = \sigma^2 \boldsymbol{I}$. Let $\boldsymbol{C}_d$ and $\boldsymbol{C}_d^*$ be the information matrices for estimating $\boldsymbol{\theta}_1$ under (2.1) and (2.9), respectively. We shall give a condition under which the search of an optimal design under (2.9) can be reduced to that under the simpler model (2.1). We have

$$\boldsymbol{C}_d = (\boldsymbol{P}_{\mathcal{R}(\boldsymbol{X}_{d2})^\perp} \boldsymbol{X}_{d1})^T \boldsymbol{P}_{\mathcal{R}(\boldsymbol{X}_{d2})^\perp} \boldsymbol{X}_{d1}$$

and

$$\begin{aligned}
\boldsymbol{C}_d^* &= (\boldsymbol{P}_{\mathcal{R}(\boldsymbol{X}_{d2} : \boldsymbol{X}_{d3})^\perp} \boldsymbol{X}_{d1})^T \boldsymbol{P}_{\mathcal{R}(\boldsymbol{X}_{d2} : \boldsymbol{X}_{d3})^\perp} \boldsymbol{X}_{d1} \\
&= \boldsymbol{X}_{d1}^T (\boldsymbol{I} - \boldsymbol{P}_{\mathcal{R}(\boldsymbol{X}_{d2} : \boldsymbol{X}_{d3})}) \boldsymbol{X}_{d1} \\
&= \boldsymbol{X}_{d1}^T (\boldsymbol{I} - \boldsymbol{P}_{\mathcal{R}(\boldsymbol{X}_{d2})} - \boldsymbol{P}_{\mathcal{R}(\boldsymbol{X}_{d3}) \cap \mathcal{R}(\boldsymbol{X}_{d2})^\perp}) \boldsymbol{X}_{d1} \\
&= \boldsymbol{C}_d - \boldsymbol{X}_{d1}^T \boldsymbol{P}_{\mathcal{R}(\boldsymbol{X}_{d3}) \cap \mathcal{R}(\boldsymbol{X}_{d2})^\perp} \boldsymbol{X}_{d1}.
\end{aligned}$$

It follows that $C_d - C_d^*$ is nonnegative definite, and $C_d = C_d^*$ if and only if

$$X_{d1}^T P_{\mathcal{R}(X_{d2})^\perp} X_{d3} = \mathbf{0}. \tag{2.10}$$

The statistical meaning is that the presence of more nuisance parameters makes the estimation of θ_1 more difficult, but no information is lost if the orthogonal projections of the ranges of X_{d1} and X_{d3} onto the orthogonal complement of the range of X_{d2} are orthogonal to each other (we may say that θ_1 is orthogonal to θ_3 adjusted for θ_2). These results lead to the following theorem.

THEOREM 2.1. *Suppose ϕ is nondecreasing with respect to the Loewner ordering. Then for any $d \in \mathcal{D}$, $\phi(C_d) \geqslant \phi(C_d^*)$. If $\tilde{d}$ maximizes $\phi(C_d)$ over $d \in \mathcal{D}$, and* (2.10) *holds for $\tilde{d}$, then $\tilde{d}$ also maximizes $\phi(C_d^*)$ over $d \in \mathcal{D}$.*

PROOF. For an arbitrary $d \in \mathcal{D}$, $\phi(C_{\tilde{d}}^*) = \phi(C_{\tilde{d}}) \geqslant \phi(C_d) \geqslant \phi(C_d^*)$. □

A similar result can be stated for the problem of minimizing a nonincreasing function. Therefore if a design satisfies the adjusted orthogonality condition (2.10), then to prove that it is optimal with respect to a Loewner isotonic criterion under the more complicated model (2.9), one only need to show that it is optimal under the simpler model (2.1). Although this method has been used, for instance, in Kiefer (1958) and Cheng (1978b, p. 1266), Kunert (1983) seems to be the first to develop it more generally and formally. The concept of adjusted orthogonality was introduced by Eccleston and Russell (1975, 1977), and the nonnegative definiteness of $C_d - C_d^*$ was pointed out in Magda (1981).

EXAMPLE 2.5 (Row–column designs). Suppose t treatments are to be compared on rc experimental units which are arranged into r rows and c columns. Let y_{ij} be the observation at the unit in the ith row and jth column, $i = 1, \ldots, r$, $j = 1, \ldots, c$. Assume the usual additive fixed-effects model in which

$$\mathrm{E}(y_{ij}) = \alpha_{t(i,j)} + \beta_i + \gamma_j,$$

where $t(i, j)$ is the label of the treatment assigned to the unit at the ith row and jth column, $\alpha_1, \ldots, \alpha_t$ are the treatment effects, $\beta_1, \ldots, \beta_r$ are the row effects, and $\gamma_1, \ldots, \gamma_c$ are the column effects. Let $\theta_1 = (\alpha_1, \ldots, \alpha_t)^T$, $\theta_2 = (\beta_1, \ldots, \beta_r)^T$ and $\theta_3 = (\gamma_1, \ldots, \gamma_c)^T$. Then X_{d1}, X_{d2} and X_{d3} in (2.9) are the unit-treatment, unit-row and unit-column incidence matrices, respectively. Since

$$\mathcal{R}(X_{d2} \vdots X_{d3}) = \mathcal{R} + \mathcal{C} = (\mathcal{R} \ominus \mathcal{G}) \oplus (\mathcal{C} \ominus \mathcal{G}) \oplus \mathcal{G},$$

the orthogonal projection matrix onto $\mathcal{R} + \mathcal{C}$ is $(P_\mathcal{R} - P_\mathcal{G}) + (P_\mathcal{C} - P_\mathcal{G}) + P_\mathcal{G} = P_\mathcal{R} + P_\mathcal{C} - P_\mathcal{G}$. Therefore, by (1.1),

$$\begin{aligned} P_{(\mathcal{R}+\mathcal{C})^\perp} &= I - P_\mathcal{R} - P_\mathcal{C} + P_\mathcal{G} \\ &= I - c^{-1} R_R - r^{-1} R_C + (rc)^{-1} J. \end{aligned} \tag{2.11}$$

From (2.3), (2.4) and (2.11), the information matrix for the treatment effects is

$$C_d = \mathrm{diag}(r_{d1}, \ldots, r_{dt}) - c^{-1}\boldsymbol{N}_{dR}\boldsymbol{N}_{dR}^T - r^{-1}\boldsymbol{N}_{dC}\boldsymbol{N}_{dC}^T + (rc)^{-1}\boldsymbol{r}_d\boldsymbol{r}_d^T, \tag{2.12}$$

where $\boldsymbol{N}_{dR}$ is the treatment-row incidence matrix, $\boldsymbol{N}_{dC}$ is the treatment-column incidence matrix, and $\boldsymbol{r}_d$ is the $t \times 1$ vector with the ith entry equal to r_{di}.

A routine calculation using $\boldsymbol{P}_{\mathcal{R}(\boldsymbol{X}_{d2})^\perp} = \boldsymbol{I} - c^{-1}\boldsymbol{R}_R$ shows that in this case, (2.10) is equivalent to

$$\boldsymbol{N}_{dC} = c^{-1}\boldsymbol{r}_d \otimes \boldsymbol{1}_c^T,$$

i.e., for each $i = 1, \ldots, t$, the number of times the ith treatment appears in the jth column is independent of $j = 1, \ldots, c$. In particular, this holds when all the treatments appear the same number of times in each column.

When each row (or column) of a row–column design is considered as a block, a block design, called the row-component (or column-component) design, is obtained. The following follows from Theorem 2.1.

THEOREM 2.2. *Suppose ϕ is nondecreasing with respect to the Loewner ordering, and $\tilde{d}$ is a row–column design with r rows and c columns in which for each $i = 1, \ldots, t$, the number of times the ith treatment appears in the jth column is independent of $j = 1, \ldots, c$. If the row-component design of $\tilde{d}$ is ϕ-optimal over all block designs with r blocks of size c, then $\tilde{d}$ is ϕ-optimal over all the row–column designs with r rows and c columns.*

Obviously the above theorem also holds for the column-component design.

3. Optimal symmetric designs

Kiefer (1958) proved the A-, D- and E-optimality of "symmetric" designs such as balanced incomplete block designs, Latin squares, Youden squares, etc. Later in Kiefer (1975a) it was extended to the so-called "universal optimality". A simple and powerful optimality tool which has had tremendous impact on the exact theory of optimal design was presented there. By symmetric designs, we mean those whose information matrices have all eigenvalues equal. Thus in Example 2.1, a weighing design d is symmetric if $\boldsymbol{X}_d^T\boldsymbol{X}_d$ is of the form $a\boldsymbol{I}_n$. In settings such as block designs (Example 2.3) and row–column designs (Example 2.5) where the information matrices have zero row and column sums, we define a symmetric design to be one such that all the eigenvalues of its information matrix except the trivial one (zero) are equal. Such a matrix is of the form $a\boldsymbol{I} + b\boldsymbol{J}$, and is called *completely symmetric*.

We now state Kiefer (1975a)'s fundamental result on the universal optimality of symmetric designs.

THEOREM 3.1. *Let $\mathcal{B}_{t,0}$ be the set of all $t \times t$ symmetric matrices with zero row and column sums and $\mathcal{C}$ be a subset of $\mathcal{B}_{t,0}$. Suppose there exists a matrix $\boldsymbol{C}^*$ in $\mathcal{C}$ such that*

(i) *$\boldsymbol{C}^*$ is of the form $a\boldsymbol{I} + b\boldsymbol{J}$, and*
(ii) *$\boldsymbol{C}^*$ maximizes $\operatorname{tr} \boldsymbol{C}$ over $\boldsymbol{C} \in \mathcal{C}$.*

Then $\boldsymbol{C}^*$ is universally optimal over $\mathcal{C}$, i.e., it minimizes $\Phi(\boldsymbol{C})$ over $\mathcal{C}$ for any $\Phi\colon \mathcal{B}_{t,0} \to (-\infty, \infty]$ such that

(a) *Φ is convex,*
(b) *Φ is nonincreasing with respect to the Loewner ordering,*
(c) *Φ is invariant under each simultaneous permutation of rows and columns.*

Note that condition (b) in the above statement is slightly different from that given in Kiefer (1975a). Kiefer's original condition is that $\Phi(b\boldsymbol{C})$ is a nonincreasing function of $b \geqslant 0$ for any given $\boldsymbol{C}$. For convenience, we have modified it in terms of the Loewner ordering. The universal optimality results presented in this article hold for both versions.

A key step in the proof of Theorem 3.1 involves taking the average of an arbitrary matrix $\boldsymbol{C}$ in $\mathcal{C}$ with respect to all simultaneous permutations of rows and columns. Two things can be said about the averaged version of $\boldsymbol{C}$, denoted by $\boldsymbol{C}_A$. First, $\boldsymbol{C}_A$ is also completely symmetric and hence can easily be compared with $\boldsymbol{C}^*$; indeed it follows from (ii) and (b) that $\Phi(\boldsymbol{C}^*) \leqslant \Phi(\boldsymbol{C}_A)$. Second, the convexity and permutation-invariance of Φ imply that $\Phi(\boldsymbol{C}_A) \leqslant \Phi(\boldsymbol{C})$. The proof is completed by piecing together the above two inequalities.

The method of averaging (with respect to a certain group of transformations under which Φ is invariant) is itself a useful technique in optimality proofs; see Example 3.7 for another application.

Theorem 3.1 can be modified as follows for the case where the information matrices are nonsingular:

THEOREM 3.2. *Let $\mathcal{B}_t$ be the set of all $t \times t$ symmetric matrices and $\mathcal{C}$ be a subset of $\mathcal{B}_t$. Suppose there exists a matrix $\boldsymbol{C}^*$ in $\mathcal{C}$ such that*

(i) *$\boldsymbol{C}^*$ is of the form $a\boldsymbol{I}_n$, and*
(ii) *$\boldsymbol{C}^*$ maximizes $\operatorname{tr} \boldsymbol{C}$ over $\boldsymbol{C} \in \mathcal{C}$.*

Then $\boldsymbol{C}^*$ minimizes $\Phi(\boldsymbol{C})$ over $\mathcal{C}$ for any $\Phi\colon \mathcal{B}_t \to (-\infty, \infty]$ such that

(a) *Φ is convex,*
(b) *Φ is nonincreasing with respect to the Loewner ordering,*
(c) *Φ is invariant under any orthogonal transformation.*

The criteria Φ in Theorem 3.1 and Theorem 3.2 include the commonly used A-, D- and E-criteria.

EXAMPLE 3.1 (Weighing designs, Example 2.1 continued). In the chemical balance weighing design problem, if there exists an $N \times n$ $(1, -1)$-matrix $\boldsymbol{X}^*$ such that

$(\boldsymbol{X}^*)^T\boldsymbol{X}^* = N\boldsymbol{I}$, then it minimizes $\Phi(\boldsymbol{C})$ over all the $N \times n$ $(0, 1, -1)$-matrices for all Φ satisfying the conditions in Theorem 3.2. For $N = n$, this establishes the optimality of designs whose design matrices are Hadamard matrices. Applications can also be made to two-level fractional factorial designs, in which all the design matrices have $(1, -1)$-entries. Optimality of orthogonal arrays can be obtained.

Theorem 3.2 fails to produce optimal spring balance weighing designs. There simply does not exist an $N \times n$ $(0, 1)$-matrix $\boldsymbol{X}^*$ such that $(\boldsymbol{X}^*)^T\boldsymbol{X}^* = N\boldsymbol{I}$. We shall return to this problem in Section 5.

EXAMPLE 3.2 (Block designs, Example 2.3 continued). Consider the block design problem of Example 2.3. From (2.7), $\mathrm{tr}(\boldsymbol{C}_d) = \sum_{i=1}^t r_{di} - k^{-1}\sum_{i=1}^t\sum_{j=1}^b n_{dij}^2 = bk - k^{-1}\sum_{j=1}^b\sum_{i=1}^t n_{dij}^2$. Since $\sum_{i=1}^t n_{dij} = k$, $\mathrm{tr}(\boldsymbol{C}_d)$ is maximized if and only if

$$n_{dij} = \mathrm{int}[k/t] \quad \text{or} \quad \mathrm{int}[k/t] + 1, \quad \text{for all } i, j, \tag{3.1}$$

where $\mathrm{int}[k/t]$ is the integral part of k/t.

It is easy to see that $\boldsymbol{C}_d$ is of the form $a\boldsymbol{I} + b\boldsymbol{J}$ if

$$r_{d1} = \cdots = r_{dt} \tag{3.2}$$

and

$$\sum_{j=1}^b n_{dij}n_{di'j} \quad \text{is a constant for all } 1 \leqslant i \neq i' \leqslant t. \tag{3.3}$$

Designs satisfying (3.1), (3.2) and (3.3) are called *balanced block designs*. It follows from Theorem 3.1 that these designs are universally optimal.

When $k < t$, (3.1) is achieved by binary designs ($n_{dij} = 0$ or 1). For such designs, $\sum_{j=1}^b n_{dij}n_{di'j}$ is the number of times the ith and i'th treatments appear together in the same block. Therefore, for $k < t$, *balanced block designs* are binary, equireplicate (all r_{di} equal) designs in which each pair of treatments appear together in the same number of blocks. These are the well known balanced incomplete block designs (BIBD's).

We shall denote the common value of the r'_{di}'s in an equireplicate design by r. Note that we have also used r to denote the number of rows in a row–column design, but there should be no danger of confusion.

EXAMPLE 3.3 (Row–column designs, Example 2.5 continued). Optimal row–column designs can be obtained by applying Theorem 2.2 to the optimal block designs in Example 3.2. Suppose there exists a row–column design such that all the treatments appear the same number of times in each column, and its row-component design is a balanced block design (such a row–column design is called a *regular generalized Youden design*). Then it is optimal with respect to all the criteria satisfying conditions (a), (b) and (c) in Theorem 3.1. Examples of regular generalized Youden designs include Latin squares, Youden squares, and designs in which all the treatments appear equally often in each row and equally often in each column. Recall that a Youden square is a row–column design such that each treatment appears exactly once in each column and its row-component design is a balanced incomplete block design.

EXAMPLE 3.4 (Repeated measurements designs). Suppose each of N subjects is to receive several treatments, one at a time, over p time periods. Each treatment assigned to a subject in a certain period not only has a "direct" effect on the observation in that period, but also has a "residual effect" on the response in the succeeding period. Let y_{ij} be the observation taken on the ith subject in the jth period, $i = 1, \ldots, N$, $j = 1, \ldots, p$. Assume

$$\mathrm{E}(y_{ij}) = \alpha_{t(i,j)} + \delta_{t(i,j-1)} + \beta_i + \gamma_j, \tag{3.4}$$

with $t(i,j)$ the label of the treatment assigned to the ith subject in the jth period, $\alpha_1, \ldots, \alpha_t$ the direct treatment effects, $\delta_1, \ldots, \delta_t$ the residual treatment effects, $\beta_1, \ldots, \beta_N$ the subject effects, and $\gamma_1, \ldots, \gamma_p$ the period effects. We also require that $\delta_{t(i,0)} = 0$, i.e., there is no residual effect in the first period.

A design is called *uniform on periods* if each treatment appears the same number of times in each period. It is said to be *uniform on subjects* if each treatment is assigned the same number of times to each subject. A design uniform on both periods and subjects is called a *uniform* design.

Express (3.4) in the form of (2.9) with $\boldsymbol{\theta}_1 = (\alpha_1, \ldots, \alpha_t)^T$, $\boldsymbol{\theta}_3 = (\delta_1, \ldots, \delta_t)^T$ and $\boldsymbol{\theta}_2 = (\beta_1, \ldots, \beta_N, \gamma_1, \ldots, \gamma_p)^T$. Since the subjects and periods can be thought of as rows and columns, one can use (2.11) to calculate $\boldsymbol{P}_{\mathcal{R}(\boldsymbol{X}_{d2})^\perp}$, and show that (2.10) is equivalent to the following:

$$\text{for all} \quad 1 \leqslant i,j \leqslant t, \quad \widetilde{n}_{ij} = p^{-1}\sum_{g=1}^{N} u_{ig}\widetilde{u}_{jg} + N^{-1}\sum_{h=2}^{p} p_{ih}p_{j,h-1} - (Np)^{-1} r_i \widetilde{r}_j, \tag{3.5}$$

where $\widetilde{n}_{dij}$ is the number of times the jth treatment precedes the ith treatment on the same subject, u_{dig} is the number of times the ith treatment is assigned to the gth subject, $\widetilde{u}_{djg}$ is the number of times the jth treatment is assigned to the gth subject in the first $p-1$ periods, p_{ih} is the number of times the ith treatment appears in the hth period, r_i is the number of replications of the ith treatment, and $\widetilde{r}_j$ is the number of times the jth treatment appears in the first $p-1$ periods. For simplicity, we suppress the dependence on d.

Condition (3.5) is the necessary and sufficient condition for the information matrix of direct (respectively, residual) treatment effects to be the same as their information matrix when the residual (respectively, direct) treatment effects are removed from (3.4).

It is easy to see that for designs which are uniform on periods,

$$N^{-1}\sum_{h=2}^{p} p_{ih}p_{j,h-1} = (Np)^{-1}r_i\widetilde{r}_j.$$

In this case, (3.5) reduces to

$$\text{for all} \quad 1 \leqslant i,j \leqslant t, \quad \widetilde{n}_{ij} = p^{-1}\sum_{g=1}^{N} u_{ig}\widetilde{u}_{jg}. \tag{3.6}$$

If the design is also uniform on subjects, or is uniform on subjects in the first $p-1$ periods, i.e., each treatment is assigned to each subject the same number of times in the first $p-1$ periods, then $\sum_{g=1}^{N} u_{ig}\widetilde{u}_{jg}$ is a constant for all i,j. Therefore, for uniform designs or designs which are uniform on periods and uniform on subjects in the first $p-1$ periods, (3.6) is equivalent to that each treatment is preceded by itself and any other treatment the same number of times. Such designs are called *strongly balanced.* By Theorem 2.1, to show the optimality of a strongly balanced uniform design (or a strongly balanced design which is uniform on periods and uniform on subjects in the first $p-1$ periods) for the estimation of direct treatment effects, it is enough to consider the simpler model in which the residual treatment effects are removed from (3.4).

By dropping the residual treatment effects, we obtain a model for row–column designs with the subjects as rows and periods as columns. In this case, a uniform design (or a design uniform on periods and uniform on subjects in the first $p-1$ periods) is a regular generalized Youden design. This shows the universal optimality of a strongly balanced uniform design (or a strongly balanced design which is uniform on periods and uniform on subjects in the first $p-1$ periods) for the estimation of direct treatment effects (Cheng and Wu, 1980). The same paper also showed that these designs are universally optimal for the estimation of residual effects.

The following is a strongly balanced uniform repeated measurements design with 3 treatments, 9 subjects and 6 periods:

	periods					
	0	0	1	1	2	2
	0	1	1	2	2	0
	0	2	1	0	2	1
	1	0	2	1	0	2
subjects	1	1	2	2	0	0
	1	2	2	0	0	1
	2	0	0	1	1	2
	2	1	0	2	1	0
	2	2	0	0	1	1

Strongly balanced uniform repeated measurements designs generally require large numbers of subjects and periods. In fact, if such a design exists, then $t^2|N$, and $p = \lambda t$ with $\lambda > 1$. The design displayed above is the smallest for three treatments. Strongly balanced designs which are uniform on periods and uniform on subjects in the first $p - 1$ periods, however, can be constructed with fewer subjects and periods. For example, suppose there exists a uniform design with $p = t$ in which each treatment precedes any *other* treatment the same number of times. (Such a design always exists, e.g., when $t = p = N$ is even.) Construct a design with $p + 1$ periods by *repeating* the observations in the last period. Then the resulting design, proposed and called an *extra-period design* by Lucas (1957), is clearly strongly balanced, uniform on periods and uniform on subjects when the last period is excluded, and therefore is universally optimal.

The following is an extra-period design with 4 treatments, 4 subjects and 5 periods:

	periods				
	1	4	2	3	3
	2	1	3	4	4
subjects	3	2	4	1	1
	4	3	1	2	2

Some universally optimal designs with two periods were obtained by Hedayat and Zhao (1990). For more results on optimal repeated measurements designs, see, e.g., Kunert (1983, 1984). Kunert (1985) considered optimal repeated measurements designs for correlated observations.

EXAMPLE 3.5 (Block designs with nested rows and columns). Suppose t treatments are to be compared in b blocks, each block containing rc experimental units arranged into r rows and c columns. Let y_{hij} be the observation taken at the ith row and jth column in the hth block. Assume the usual additive fixed-effects model in which

$$\mathrm{E}(y_{hij}) = \alpha_{t(h,i,j)} + \beta_h + \gamma_{hi} + \delta_{hj}, \tag{3.7}$$

where $t(h, i, j)$ is the label of the treatment assigned to the unit at the ith row and jth column in the hth block, $\alpha_1, \ldots, \alpha_t$ are the treatment effects, β_h is the effect of the hth block, γ_{hi} is the effect of the ith row of the hth block, and δ_{hj} is the effect of the jth column of the hth block. Express (3.7) in the form of (2.9) with $\boldsymbol{\theta}_1$ consisting of the treatment effects, $\boldsymbol{\theta}_2$ the column effects (notice that $\boldsymbol{\theta}_2$ is $bc \times 1$) and $\boldsymbol{\theta}_3$ consisting of the rest. Since $\mathcal{B} \subset \mathcal{R}$, $\mathcal{R}(\boldsymbol{X}_{d3}) = \mathcal{B} + \mathcal{R} = \mathcal{R}$. Using $\boldsymbol{P}_{\mathcal{R}(\boldsymbol{X}_{d2})^{\perp}} = \boldsymbol{I} - r^{-1}\boldsymbol{R}_C$, it is easy to see that (2.10) is equivalent to

$$n_{ihj} = r^{-1} n_{ih} \tag{3.8}$$

for all $i = 1, \ldots, t$, $h = 1, \ldots, b$ and $j = 1, \ldots, r$, where n_{ihj} is the number of times the ith treatment appears in the jth row of the hth block, and n_{ih} is the number of times the ith treatment appears in the hth block. For a design satisfying (3.8), the information matrix for estimating the treatment effects is the same as that under the model obtained by removing the block and row effects from (3.7). But this simpler model is precisely that of a block design with the bc columns as blocks. Therefore, if a block design with nested rows and columns satisfies (3.8), and it is a balanced block design when the columns are considered as blocks, then it is universally optimal (Bagchi et al., 1990; Chang and Notz, 1994). The following is an example with $t = 4$, $b = 6$, $r = c = 2$:

1	2
2	1

1	3
3	1

1	4
4	1

2	3
3	2

2	4
4	2

3	4
4	3

All the optimal symmetric designs discussed above maximize the traces of the information matrices. Theorem 2.1 is quite, although not always, useful for establishing the optimality of such designs. When (2.10) does not hold, it is more involved to verify the maximization of the trace. When there is no symmetric design maximizing the trace of the information matrix, universally optimal designs usually do not exist, and optimal designs need to be determined separately for individual criteria. The following result by Kiefer (1975a) is useful for simplifying the task:

THEOREM 3.3. *If a symmetric design is ϕ_p-optimal, then it is ϕ_q-optimal for all $q \leqslant p$.*

PROOF. This follows from the fact that $q \leqslant p \Rightarrow \phi_q(\boldsymbol{C}_d) \leqslant \phi_p(\boldsymbol{C}_d)$, and that equality holds for symmetric designs. □

Thus for example, once it is shown that a symmetric design is D-optimal, then we know it is also A- and E-optimal.

EXAMPLE 3.6 (Row–column designs, Example 3.3 continued). The regular generalized Youden designs in Example 3.3 exist only if $t|r$ or $t|c$. When both r and c are not multiples of t, the determination of optimal row–column designs is a very difficult problem. From (2.12), a row–column design has a completely symmetric information matrix if both the row-component and column-component designs are balanced block designs. Such designs, called *generalized Youden designs* by Kiefer (1958), however, do not maximize $\text{tr}(\boldsymbol{C}_d)$ unless they are regular. The problem of proving or disproving the optimality of nonregular generalized Youden designs is extremely difficult, and was solved in Kiefer (1975a) by some ingenious arguments. It was shown that a generalized Youden design is always A- and E-optimal, and is D-optimal if $t > 4$. A counterexample was given to show that a nonregular generalized Youden design with $t = 4$ is not D-optimal. The significance of this result is that it is the first case where a design with full symmetry is proved nonoptimal for a symmetric estimation problem. For the construction of generalized Youden designs, see, e.g., Kiefer (1975b), Ruiz and Seiden (1974) and Seiden and Wu (1978). A table of generalized Youden designs was given by Ash (1981).

Note that when $r = c$, a row–column design has a completely symmetric information matrix as long as the row-component and column-component designs together form a balanced block design. Such a design is called a *pseudo-Youden design* (Cheng, 1981a). It enjoys the same optimality property as a generalized Youden design, and provides more flexibility. Sometimes generalized Youden designs do not exist, but pseudo-Youden designs can be constructed.

We end this section with some optimal block designs for comparing treatments with a control, which, strictly speaking, are not symmetric in the sense as we defined.

EXAMPLE 3.7. (Block designs for comparing test treatments with a control, Example 2.4 continued). Now we return to Example 2.4. For simplicity, consider the case $k < t$. Analogous to balanced incomplete block designs, Bechhofer and Tamhane (1981) defined balanced treatment incomplete block (BTIB) designs to be those for which $\bar{\boldsymbol{C}}_d$ is completely symmetric. Complete symmetry, however, is not enough for optimality. For instance, in the definition of a BIBD, there is the additional requirement of binarity. We also point out that neither Theorem 3.1 nor Theorem 3.2 are useful for studying the optimality of BTIB designs. Theorem 3.1 is not applicable since the information matrices in the current setting no longer have zero row sums. Theorem 3.2 fails because there does not exist a design for which $\bar{\boldsymbol{C}}_d$ is of the form $a\boldsymbol{I}$. In the current setting, universally optimal designs no longer exist, and the optimality problem is more complicated.

Majumdar and Notz (1983) successfully proved the optimality of certain BTIB designs. As in the proof of Theorem 3.1, for any given design d, take the average of $\bar{\boldsymbol{C}}_d$ with respect to all the simultaneous permutations of rows and columns. Let the averaged version be $\bar{\boldsymbol{C}}_d^A$. Then for any d, $\bar{\boldsymbol{C}}_d^A$ is completely symmetric and $\Phi(\bar{\boldsymbol{C}}_d^A) \leqslant \Phi(\bar{\boldsymbol{C}}_d)$ for any convex Φ. Again this gives a substantial reduction of the problem, and one can focus on the comparison of completely symmetric matrices. Majumdar and Notz (1983) showed that a BTIB design is A-optimal if it is binary in the test treatments (each test treatment appears at most once in each block), the numbers of times the control appears in the b blocks differ from one another by at most one, and the total number of replications of the control minimizes a certain function which can be explicitly written down in terms of the design parameters t, b and k. We refer the readers to their paper for the details.

Optimal BTIB designs can often be constructed by supplementing one or more replications of the control to each block of a BIBD or BIBD's of the test treatments with block sizes differing by 1. Some families of optimal BTIB designs were constructed in Hedayat and Majumdar (1985), Stufken (1987) and Cheng et al. (1988).

A survey of work in this area can be found in Hedayat et al. (1988).

4. Optimality of asymmetrical designs

Usually there are quite stringent necessary conditions for the existence of symmetric designs. For instance, when $n > 2$, a design with $\boldsymbol{X}^T\boldsymbol{X} = N\boldsymbol{I}$ in Example 3.1 exists only if N is a multiple of 4. In this section, we present some results on optimal asymmetrical designs.

We shall restrict to the *minimization* of criteria of the form $\Phi_f(\boldsymbol{M}_d) = \sum_{i=1}^{n} f(\mu_{di})$, where f is a convex and nonincreasing function, and $\mu_{d1} \geqslant \cdots \geqslant \mu_{dn}$ are the eigenvalues of $\boldsymbol{M}_d$. This covers the ϕ_p-criteria with $p > -\infty$ by taking

$$f(x) = \begin{cases} -x^p, & \text{for } 0 < p \leqslant 1; \\ x^p, & \text{for } -\infty < p < 0; \\ -\log x, & \text{for } p = 0. \end{cases}$$

Results on the $\phi_{-\infty}$-criterion can be obtained by using the fact that $\phi_{-\infty}$ is the limit of ϕ_p as $p \to -\infty$. As before, $\Phi_f(C_d) = \sum_{i=1}^{s-1} f(\mu_{di})$ is defined in terms of the nonzero eigenvalues of C_d.

Consider the minimization of $\sum_{i=1}^{n} f(\mu_i)$ subject to the constraint $\sum_{i=1}^{n} \mu_i = A$, $\mu_i \geqslant 0$, for a fixed constant A. The convexity of f implies that the minimum value is equal to $nf(A/n)$, attained at $\mu_1 = \cdots = \mu_n = A/n$. Since f is nonincreasing, $nf(A/n)$ is nonincreasing in A. Therefore $\sum_{i=1}^{n} f(\mu_i)$ is minimized by the $(\mu_1, \ldots, \mu_n)$ with $\mu_1 = \cdots = \mu_n$ that maximizes $\sum_{i=1}^{n} \mu_i$. This is essentially the results of Theorems 3.1 and 3.2.

Theorem 3.1 or 3.2 fails to produce an optimal symmetric design when there is no design corresponding to the center of the simplex $\{(\mu_1, \ldots, \mu_n)\colon \sum_{i=1}^{n} \mu_i = A, \mu_i \geqslant 0\}$ with the largest A. In this case, a design with $(\mu_{d1}, \ldots, \mu_{dn})$ close to the center of this simplex is expected to be highly efficient, if not optimal. This motivates the procedure of maximizing $\sum_{i=1}^{n} \mu_{di}$ first, and then minimizing the squared distance $\sum_{i=1}^{n} (\mu_{di} - \sum_{i=1}^{n} \mu_{di}/n)^2$ among those which maximize $\sum_{i=1}^{n} \mu_{di}$. Since $\sum_{i=1}^{n} (\mu_{di} - \sum_{i=1}^{n} \mu_{di}/n)^2 = \sum_{i=1}^{n} \mu_{di}^2 - (\sum_{i=1}^{n} \mu_{di})^2/n$, this is equivalent to maximizing $\sum_{i=1}^{n} \mu_{di}$ (= tr($\boldsymbol{M}_d$) or tr($\boldsymbol{C}_d$)), and then minimizing $\sum_{i=1}^{n} \mu_{di}^2$ (= tr($\boldsymbol{M}_d^2$) or tr($\boldsymbol{C}_d^2$)) among those which maximize $\sum_{i=1}^{n} \mu_{di}$. Such a design is called (M.S)-optimal. The main advantage of this procedure is that tr($\boldsymbol{M}_d$) and tr($\boldsymbol{M}_d^2$) (or tr($\boldsymbol{C}_d$) and tr($\boldsymbol{C}_d^2$)), which are, respectively, the sum of the diagonal entries and sum of squares of all the entries of $\boldsymbol{M}_d$ (or $\boldsymbol{C}_d$), are very easy to calculate and optimize. The (M.S)-criterion was originally proposed in the setting of block designs (Shah, 1960; Eccleston and Hedayat, 1974).

EXAMPLE 4.1 (Weighing designs, Example 3.1 continued). In the chemical balance weighing design problem of Examples 2.1 and 3.1, since tr($\boldsymbol{X}^T\boldsymbol{X}$) is maximized by any $\boldsymbol{X}$ with all the entries equal to 1 or -1, an (M.S)-optimal design minimizes the sum of squares of the entries of $\boldsymbol{X}^T\boldsymbol{X}$ among the $N \times n$ $(1, -1)$-matrices $\boldsymbol{X}$. When N is odd, none of the off-diagonal entries of $\boldsymbol{X}^T\boldsymbol{X}$ can be zero. Therefore for $N \equiv 1$ or $3 \pmod 4$, a design with $\boldsymbol{X}^T\boldsymbol{X} = (N-1)\boldsymbol{I} + \boldsymbol{J}$ or $(N+1)\boldsymbol{I} - \boldsymbol{J}$, respectively, is (M.S)-optimal. When $N \equiv 2 \pmod 4$, to calculate tr$(\boldsymbol{X}^T\boldsymbol{X})^2$, without loss of generality, we may assume that the first m columns of $\boldsymbol{X}$ have even numbers of entries equal to 1, and the remaining columns have odd numbers of entries equal to 1. Then $\boldsymbol{X}^T\boldsymbol{X}$ can be partitioned as

$$\begin{bmatrix} \boldsymbol{A} & \boldsymbol{B} \\ \boldsymbol{B}^T & \boldsymbol{C} \end{bmatrix},$$

where all the entries of $\boldsymbol{A}$ and $\boldsymbol{C}$ are congruent to 2 (mod 4), and all the entries of $\boldsymbol{B}$ are multiples of 4. For such an $\boldsymbol{X}$, $\text{tr}(\boldsymbol{X}^T\boldsymbol{X})^2$ is minimized when $\boldsymbol{A} = (N-2)\boldsymbol{I}_m + 2\boldsymbol{J}_m$, $\boldsymbol{C} = (N-2)\boldsymbol{I}_{n-m} + 2\boldsymbol{J}_{n-m}$ and $\boldsymbol{B} = \boldsymbol{0}$. Comparing different m's, one concludes that if there exists an $\boldsymbol{X}$ such that

$$\boldsymbol{X}^T\boldsymbol{X} = \begin{bmatrix} (N-2)\boldsymbol{I}_m + 2\boldsymbol{J}_m & \boldsymbol{0} \\ \boldsymbol{0} & (N-2)\boldsymbol{I}_{n-m} + 2\boldsymbol{J}_{n-m} \end{bmatrix}, \tag{4.1}$$

where $m = \text{int}[n/2]$, then it is (M.S)-optimal.

EXAMPLE 4.2 (Incomplete block designs, Example 3.2 continued). From Example 3.2 follows that when $k < t$, $\text{tr}(\boldsymbol{C}_d)$ is maximized by binary designs. For such designs, by (2.7), the ith diagonal of $\boldsymbol{C}_d$ equals $k^{-1}(k-1)r_{di}$, and the (i,j)th off-diagonal entry is equal to $-k^{-1}\lambda_{dij}$, where λ_{dij} is the number of times the ith and jth treatments appear together in the same block. Since $\sum_{i=1}^{t} r_{di}$ and $\sum_{i\neq j}\lambda_{dij}$ are constants, $\text{tr}(\boldsymbol{C}_d)^2$ is minimized if

$$\text{all the } r_{di}\text{'s are equal} \tag{4.2}$$

and

$$\lambda_{dij} = \lambda \quad \text{or} \quad \lambda + 1 \quad \text{for some } \lambda. \tag{4.3}$$

Binary designs satisfying (4.2) and (4.3) are called *regular graph designs* by John and Mitchell (1977). We have just shown that such designs are (M.S)-optimal incomplete block designs.

The (M.S)-criterion is not really an optimality criterion. Rather, it is a procedure for quickly producing designs which are optimal or efficient with respect to other more meaningful criteria. Indeed, John and Mitchell (1977) conjectured the A-, D- and E-optimality of regular graph designs. To study the efficiency of (M.S)-optimal designs, we shall consider the following minimization problem:

$$\text{Minimize} \sum_{i=1}^{n} f(\mu_i) \quad \text{subject to} \quad \sum_{i=1}^{n} \mu_i = A,$$

$$\sum_{i=1}^{n} \mu_i^2 = B, \quad \mu_i \geqslant 0. \tag{4.4}$$

Conniffe and Stone (1975) considered such a problem with $f(x) = x^{-1}$ to determine A-optimal block designs. Cheng (1978a) solved (4.4) for general f, providing a useful tool for proving the optimality of asymmetrical designs in various settings.

Cheng (1978a) showed that if f' is strictly concave and $f(0) = \lim_{x\to 0^+} f(x) = \infty$, then the solution of (4.4) is attained at a $(\mu_1, \ldots, \mu_n)$ with $\mu_1 \geqslant \mu_2 = \cdots = \mu_n$.

Denote μ_1 and the common value of $\mu_2, \ldots, \mu_n$ by μ and μ', respectively. Then $\mu = \{A+[n(n-1)]^{1/2}P\}/n$, and $\mu' = \{A-[n/(n-1)]^{1/2}P\}/n$, where $P = [B-A^2/n]^{1/2}$. Therefore the minimum value of $\sum_{i=1}^n f(\mu_i)$ subject to $\sum_{i=1}^n \mu_i = A$, $\sum_{i=1}^n \mu_i^2 = B$, $\mu_i \geqslant 0$ is

$$f(\{A+[n(n-1)]^{1/2}P\}/n) + (n-1)f(\{A-[n/(n-1)]^{1/2}P\}/n). \tag{4.5}$$

The decreasing monotonicity and convexity of f, respectively, imply that (4.5) is a decreasing function of A and an increasing function of P. This gives support to the (M.S)-criterion in that to minimize $\sum_{i=1}^n f(\mu_i)$, it is desirable to have A as large as possible, and P as small as possible. However, even though an (M.S)-optimal design has minimum P among those with the same maximum A, it may not minimize P among *all* the designs. Comparing (4.5) for different A and P values leads to the following result:

THEOREM 4.1 (Cheng, 1978a). *Suppose there exists a design d* such that its information matrix* $\boldsymbol{M}_{d^*}$ *has two distinct eigenvalues, both positive and the larger one having multiplicity one. If d* maximizes* $\mathrm{tr}(\boldsymbol{M}_d)$ *and maximizes* $\mathrm{tr}(\boldsymbol{M}_d) - [n/(n-1)]^{1/2}\cdot[\mathrm{tr}(\boldsymbol{M}_d^2) - (\mathrm{tr}(\boldsymbol{M}_d))^2/n]^{1/2}$ *over* $\mathcal{D}$, *then it is* Φ_f*-optimal over* $\mathcal{D}$ *for any convex and nonincreasing* f *such that* f' *is strictly concave and* $f(0) = \lim_{x\to 0+} f(x) = \infty$.

PROOF. It is sufficient to show that if $A_1 \geqslant A_2$ and $A_1 - [n/(n-1)]^{1/2}P_1 \geqslant A_2 - [n/(n-1)]^{1/2}P_2$, then

$$\begin{aligned} &f(\{A_1+[n(n-1)]^{1/2}P_1\}/n) + (n-1)f(\{A_1-[n/(n-1)]^{1/2}P_1\}/n) \\ &\quad\leqslant f(\{A_2+[n(n-1)]^{1/2}P_2\}/n) \\ &\qquad + (n-1)f(\{A_2-[n/(n-1)]^{1/2}P_2\}/n). \end{aligned}$$

Since (4.5) is a decreasing function of A and an increasing function of P, this inequality holds if $P_1 \leqslant P_2$. It is also true when $P_1 > P_2$, since in this case, $A_1 + [n(n-1)]^{1/2}P_1 > A_2 + [n(n-1)]^{1/2}P_2$ and $A_1 - [n/(n-1)]^{1/2}P_1 \geqslant A_2 - [n/(n-1)]^{1/2}P_2$. □

COROLLARY 4.2. *Suppose* $\mathrm{tr}(\boldsymbol{M}_d)$ *is a constant for all the designs in* $\mathcal{D}$. *If there exists a design d* such that its information matrix* $\boldsymbol{M}_{d^*}$ *has two distinct eigenvalues, both positive and the larger one having multiplicity one, and d* minimizes* $\mathrm{tr}(\boldsymbol{M}_d^2)$ *over* $\mathcal{D}$, *then it is* Φ_f*-optimal over* $\mathcal{D}$ *for any convex* f *such that* f' *is strictly concave and* $f(0) = \lim_{x\to 0+} f(x) = \infty$.

Note that in Corollary 4.2, f is not required to be nonincreasing, since A is a constant.

EXAMPLE 4.3 (Weighing designs, Example 4.1 continued). Temporarily restrict attention to the chemical balance weighing designs whose design matrices X_d have *no* zero entries. In particular, this covers the application to 2^n fractional factorial designs. In this case, $\operatorname{tr} \boldsymbol{X}_d^T \boldsymbol{X}_d$ is a constant. Suppose $N \equiv 1 \pmod 4$ and an $\boldsymbol{X}_{d^*}$ with $\boldsymbol{X}_{d^*}^T \boldsymbol{X}_{d^*} = (N-1)\boldsymbol{I} + \boldsymbol{J}$ exists. It was shown in Example 4.1 that d^* minimizes $\operatorname{tr}(\boldsymbol{M}_d^2)$. Since $\boldsymbol{X}_{d^*}^T \boldsymbol{X}_{d^*}$ has two distinct eigenvalues with the larger one having multiplicity one, it follows from Corollary 4.2 that d^* is Φ_f-optimal over all the $N \times n$ $(1,-1)$-matrices for any convex f such that f' is strictly concave and $f(0) = \lim_{x \to 0^+} f(x) = \infty$ (Cheng, 1980a). In particular, it is A-, D- and E-optimal.

When $N \equiv 2 \pmod 4$, as in Example 4.1, for any $(1,-1)$-matrix $\boldsymbol{X}_d$, $\boldsymbol{X}_d^T \boldsymbol{X}_d$ can be partitioned as

$$\begin{bmatrix} \boldsymbol{A} & \boldsymbol{B} \\ \boldsymbol{B}^T & \boldsymbol{C} \end{bmatrix},$$

where all the entries of $\boldsymbol{A}$ and $\boldsymbol{C}$ are congruent to 2 (mod 4), and all the entries of $\boldsymbol{B}$ are multiples of 4. By a result of Fan (1954), the vector consisting of the eigenvalues of

$$\begin{bmatrix} \boldsymbol{A} & \boldsymbol{B} \\ \boldsymbol{B}^T & \boldsymbol{C} \end{bmatrix} \quad \text{majorizes that of} \quad \begin{bmatrix} \boldsymbol{A} & \boldsymbol{0} \\ \boldsymbol{0} & \boldsymbol{C} \end{bmatrix}.$$

Therefore $\sum_{i=1}^{n} f(\mu_{di}) \geqslant \sum_{i=1}^{m} f(\nu_i) + \sum_{i=1}^{n-m} f(\delta_i)$ for all convex f, where $\nu_1, \ldots, \nu_m$ are the eigenvalues of $\boldsymbol{A}$, and $\delta_1, \ldots, \delta_{n-m}$ are the eigenvalues of $\boldsymbol{C}$. Since all the entries of $\boldsymbol{A}$ and $\boldsymbol{C}$ are congruent to 2 (mod 4), the same argument as in the $N \equiv 1 \pmod 4$ case shows that $\sum_{i=1}^{m} f(\nu_i) \geqslant \sum_{i=1}^{m} f(\nu_i^*)$ and $\sum_{i=1}^{n-m} f(\delta_i) \geqslant \sum_{i=1}^{n-m} f(\delta_i^*)$, for any convex f such that f' is strictly concave and $f(0) = \lim_{x \to 0^+} f(x) = \infty$, where $\nu_1^*, \ldots, \nu_m^*$ are the eigenvalues of $(N-2)\boldsymbol{I}_m + 2\boldsymbol{J}_m$, and $\delta_1^*, \ldots, \delta_{n-m}^*$ are the eigenvalues of $(N-2)\boldsymbol{I}_{n-m} + 2\boldsymbol{J}_{n-m}$. The problem is now reduced to the comparison of matrices of the form (4.1) with different m's. It is easy to see that the best choice of m is $m = \operatorname{int}[n/2]$. Therefore if there exists an $\boldsymbol{X}_{d^*}$ such that $\boldsymbol{X}_{d^*}^T \boldsymbol{X}_{d^*}$ is as in (4.1), where $m = \operatorname{int}[n/2]$, then it is Φ_f-optimal over all the $N \times n$ $(1,-1)$-matrices for any convex f such that f' is strictly concave and $f(0) = \lim_{x \to 0^+} f(x) = \infty$ (Jacroux et al., 1983).

If zero entries are allowed in $\boldsymbol{X}_d$, then $\operatorname{tr}(\boldsymbol{M}_d)$ is no longer a constant. In the $N \equiv 1 \pmod 4$ case, a more delicate analysis shows that a design d^* such that $\boldsymbol{X}_{d^*}^T \boldsymbol{X}_{d^*} = (N-1)\boldsymbol{I} + \boldsymbol{J}$ maximizes $\operatorname{tr}(\boldsymbol{M}_d) - [n/(n-1)]^{1/2}[\operatorname{tr}(\boldsymbol{M}_d^2) - (\operatorname{tr}(\boldsymbol{M}_d))^2/n]^{1/2}$. Therefore by Theorem 4.1, it is Φ_f-optimal over all the $N \times n$ $(0,1,-1)$-matrices for any convex and nonincreasing f such that f' is strictly concave and $f(0) = \lim_{x \to 0^+} f(x) = \infty$ (Cheng, 1980a). Again this includes A-, D- and E-optimality.

Such an extension covering design matrices with zero entries does not hold for the $N \equiv 2 \pmod 4$ case; see Example 4.6 and the discussion at the end of Section 6.1

EXAMPLE 4.4 (Incomplete block designs, Example 4.2 continued). We define a *group-divisible* design to be a binary equireplicate design in which the treatments can be divided into groups of equal size such that any two treatments in the same group appear together in the same number of blocks, say λ_1 blocks, and those in different groups also appear together in the same number of blocks, say λ_2 blocks. Suppose there exists a group-divisible design d^* with *two* groups and $\lambda_2 = \lambda_1 + 1$. Then it is easy to see that $\boldsymbol{C}_{d^*}$ has two distinct nonzero eigenvalues with the larger one having multiplicity one. It can be shown that d^* has the maximum value of $\text{tr}(\boldsymbol{C}_d) - [(t-1)/(t-2)]^{1/2}[\text{tr}(\boldsymbol{C}_d^2) - (\text{tr}(\boldsymbol{C}_d))^2/(t-1)]^{1/2}$. Therefore by the version of Theorem 4.1 for the case where the information matrices have zero row sums, d^* is Φ_f-optimal over all the designs with the same values of t, b and k, for any convex and nonincreasing f such that f' is strictly concave and $f(0) = \lim_{x\to 0^+} f(x) = \infty$ (Cheng, 1978a). Roy and Shah (1984) also used Theorem 4.1 to prove the optimality of a class of *minimal covering designs*. In a covering design, each pair of treatments appear together in at least one block. A covering design with the smallest number of blocks is called a minimal covering design. Roy and Shah (1984) showed when $t \equiv 5 \pmod 6$, a minimal covering design in blocks of size 3 satisfies all the conditions in Theorem 4.1, and therefore is optimal.

By refining the proof of Theorem 4.1, one can derive the following modification of Corollary 4.2, which is applicable to the case where the information matrix has two distinct eigenvalues, but the larger one does not have multiplicity one.

THEOREM 4.3 (Cheng, 1981b; Cheng and Bailey, 1991). *Suppose* $\text{tr}(\boldsymbol{M}_d)$ *is a constant for all the designs in* $\mathcal{D}$. *If there exists a design* d^* *such that its information matrix* $\boldsymbol{M}_{d^*}$ *has two distinct eigenvalues, both being positive, and* d^* *minimizes* $\text{tr}(\boldsymbol{M}_d^2)$ *and maximizes the maximum eigenvalue of* $\boldsymbol{M}_d$ *over* $\mathcal{D}$, *then it is* Φ_f*-optimal over* $\mathcal{D}$ *for any convex* f *such that* f' *is strictly concave and* $f(0) = \lim_{x\to 0^+} f(x) = \infty$.

EXAMPLE 4.5 (Incomplete block designs, Example 4.4 continued). The optimality of a group-divisible design with *two* groups and $\lambda_2 = \lambda_1 + 1$ is established in Example 4.4. Can the restriction to two groups be removed? The information matrix of a group-divisible design has two nonzero eigenvalues, but the larger one has multiplicity 1 only when the number of groups is two. Therefore Theorem 4.1 is not applicable when there are more than two groups. However, there are indications that, in general, group-divisible designs with $\lambda_2 = \lambda_1 + 1$ have strong optimality properties. For example, their E-optimality was obtained by Takeuchi (1961, 1963); see Section 6.1. (In fact, this was the *first* result on the optimality of asymmetrical designs.) It seems difficult to show that these designs enjoy the same optimality property as those with only two groups, but the following partial result can be obtained.

Let $\mathcal{D}$ be the set of all the regular graph designs with t treatments and b blocks of size k. Then both $\text{tr}(\boldsymbol{C}_d)$ and $\text{tr}(\boldsymbol{C}_d^2)$ are constants for all $d \in \mathcal{D}$. Suppose $\mathcal{D}$ contains a group-divisible design d^* with g groups and $\lambda_2 = \lambda_1 + 1$. Let $r = bk/t$. Then the t eigenvalues of $k\boldsymbol{C}_{d^*} - [r(k-1) + \lambda_1]\boldsymbol{I}_t + \lambda_2\boldsymbol{J}_t$ are $k\mu_{d^*1} - [r(k-1) + \lambda_1] \geqslant \cdots \geqslant k\mu_{d^*,t-1} - [r(k-1) + \lambda_1]$ and $(t-1)\lambda_2 - r(k-1) + 1$. On the other hand

$$k\boldsymbol{C}_{d^*} - [r(k-1) + \lambda_1]\boldsymbol{I}_t + \lambda_2\boldsymbol{J}_t = \begin{bmatrix} \boldsymbol{J}_{t/g} & \cdots & \boldsymbol{0} \\ \vdots & \ddots & \vdots \\ \boldsymbol{0} & \cdots & \boldsymbol{J}_{t/g} \end{bmatrix}, \tag{4.6}$$

where the diagonal blocks are matrices of 1's, and the off-diagonal blocks are zero matrices. Comparing the eigenvalues of both sides of (4.6), we have $k\mu_{d^*1} - [r(k-1)+\lambda_1] = t/g$. On the other hand, for any regular graph design d in $\mathcal{D}$, the largest eigenvalue of $k\boldsymbol{C}_d - [r(k-1)+\lambda_1]\boldsymbol{I}_t + \lambda_2\boldsymbol{J}_t$ is equal to t/g, since it has (0,1)-entries with constant row sum t/g. From this it follows that $\mu_{d1} \leqslant \mu_{d^*1}$. Therefore d^* maximizes the largest eigenvalue of $\boldsymbol{C}_d$. It now follows from Theorem 4.3 that d^* is Φ_f-optimal over $\mathcal{D}$ for any convex f such that f' is strictly concave and $f(0) = \lim_{x\to 0^+} f(x) = \infty$. This shows the optimality of group-divisible designs with $\lambda_2 = \lambda_1 + 1$ over the regular graph designs (Cheng, 1981b). Whether it is also optimal over non-regular graph designs is a challenging problem. Note that the result in Cheng (1981b) was stated in terms of the maximization of the total number of spanning trees in a multigraph, which is closely related to the determination of D-optimal block designs.

Group-divisible designs are not the only designs with two distinct values among the $t-1$ nontrivial eigenvalues $\mu_{d1} \geqslant \cdots \geqslant \mu_{d,t-1}$ of $\boldsymbol{C}_d$. In general, any partially balanced incomplete block design with two associate classes (PBIBD(2)) has this eigenvalue structure. We refer the readers to Raghavarao (1971) for the definition of partially balanced incomplete block designs. We shall call a regular graph design which is also a PBIBD(2), i.e., a PBIBD(2) with $\lambda_2 = \lambda_1 + 1$ or $\lambda_2 = \lambda_1 - 1$, a *strongly regular graph design*. As another application of Theorem 4.3, we shall derive the optimality of certain strongly regular graph designs.

Let $\mathcal{D}$ be the set of all the *equireplicate binary* designs with t treatments and b blocks of size k. Suppose $\mathcal{D}$ contains a strongly regular graph design d^* whose concurrence matrix $\boldsymbol{N}_{d^*}\boldsymbol{N}_{d^*}^T$ is *singular*. For any design d in $\mathcal{D}$, the largest eigenvalue of $\boldsymbol{C}_d = r\boldsymbol{I} - k^{-1}\boldsymbol{N}_d\boldsymbol{N}_d^T$ is at most r. On the other hand, since $\boldsymbol{N}_{d^*}\boldsymbol{N}_{d^*}^T$ is singular, the largest eigenvalue of $\boldsymbol{C}_{d^*}$ is equal to r. Therefore d^* maximizes the largest eigenvalue of $\boldsymbol{C}_d$ over $\mathcal{D}$. Since all the designs in $\mathcal{D}$ are binary, $\mathrm{tr}(\boldsymbol{C}_d)$ is a constant. Furthermore, being a regular graph design, d^* minimizes $\mathrm{tr}(\boldsymbol{C}_d^2)$. Hence all the conditions in Theorem 4.3 are satisfied, and we have shown that a strongly regular graph design with a singular concurrence matrix is Φ_f-optimal over the equireplicate binary designs for any convex f such that f' is strictly concave and $f(0) = \lim_{x\to 0^+} f(x) = \infty$ (Cheng and Bailey, 1991). Examples of strongly regular graph designs with singular concurrence matrices are: (i) all the PBIBD(2) with $\lambda_2 = \lambda_1 \pm 1$ and $b < t$; (ii) all the resolvable PBIBD(2) with $\lambda_2 = \lambda_1 \pm 1$ and $b < t + r - 1$; (iii) all the partial geometries; (iv) all the singular group-divisible designs with $\lambda_2 = \lambda_1 - 1$; (v) all the semiregular group-divisible designs with $\lambda_2 = \lambda_1 + 1$. See Raghavarao (1971) for definitions and examples of these designs. It is interesting to see whether similar optimality properties can be extended to other strongly regular graph designs. Some kind of conditions are needed, but perhaps the singularity of the concurrence matrix is a bit too strong.

Theorem 4.3 can be proved by modifying (4.4) to the problem of minimizing $\sum_{i=1}^n f(\mu_i)$ subject to $\sum_{i=1}^n \mu_i = A$, $\sum_{i=1}^n \mu_i^2 = B$, $\mu_i \geqslant 0$ and $\mu_i \leqslant C$, where C is taken to be the maximum largest eigenvalue. This is useful when an upper bound on the *largest* eigenvalue of the information matrix is available. Jacroux (1985) modified (4.4) to the minimization of $\sum_{i=1}^n f(\mu_i)$ subject to $\sum_{i=1}^n \mu_i = A$, $\sum_{i=1}^n \mu_i^2 = B$, $\mu_i \geqslant 0$ and $\min_{1\leqslant i\leqslant n} \mu_i \leqslant C$, which provides a useful tool for deriving lower bounds on $\sum_{i=1}^{n} f(\mu_i)$ in conjunction with upper bounds on the *smallest* eigenvalue of the information matrix. Jacroux (1985) used this approach to establish the A-optimality of many regular graph designs.

Theorem 4.1 establishes the optimality of certain (M.S)-optimal designs. Without additional conditions such as those in Theorem 4.1, (M.S)-optimal designs can fail to be optimal with respect to other more meaningful criteria. For example, in a computer search, Jones and Eccleston (1980) found designs with unequal replications which are better than the best regular graph designs under the A-criterion for $(t, b, k) = (10, 10, 2)$, $(11, 11, 2)$ and $(12, 12, 2)$. In extreme cases (M.S)-optimal designs can even be very inefficient. For example, consider the spring balance weighing design problem of Example 2.2. It is easy to see that the design with $\boldsymbol{X}_d = \boldsymbol{J}$ is the only (M.S)-optimal design, but it has a singular information matrix and none of the weights is estimable!

In other settings including that of block designs where the (M.S)-criterion was originally proposed, however, this procedure is expected to produce highly efficient designs. Cheng (1992) showed that in settings such as block designs and chemical balance weighing designs (but not spring balance weighing designs), for any of a large class of criteria (including all the ϕ_p-criteria, $-\infty < p \leqslant 1$), *when the number of observations is sufficiently large*, any (M.S)-optimal design is better than any non-(M.S)-optimal design, and optimal designs can be obtained by sequentially minimizing $(-1)^h \operatorname{tr}(\boldsymbol{M}_d^h)$ (or $(-1)^h \operatorname{tr}(\boldsymbol{C}_d^h)$), i.e., sequentially maximizing $\operatorname{tr}(\boldsymbol{M}_d)$ (or $\operatorname{tr}(\boldsymbol{C}_d)$), minimizing $\operatorname{tr}(\boldsymbol{M}_d^2)$ (or $\operatorname{tr}(\boldsymbol{C}_d^2)$), maximizing $\operatorname{tr}(\boldsymbol{M}_d^3)$ (or $\operatorname{tr}(\boldsymbol{C}_d^3)$), etc. The first two steps correspond to the (M.S)-criterion. Thus, for example, in the block design problem, for any t and k, there exists an integer $b(t, k)$ such that as long as $b > b(t, k)$, any (M.S)-optimal design (such as a regular graph design) is A- and D- better than any non-(M.S)-optimal design. This confirms John and Mitchell (1977)'s conjecture for the A- and D-criteria when the number of blocks is large. That D-optimal block designs can be found among (M.S)-optimal designs when the number of observations is large was published earlier in Cheng et al. (1985a) in graph-theoretic language, while Constantine (1986) showed that for large number of observations, A- and D-optimal block designs *among (M.S)-optimal designs* can be obtained by sequentially minimizing $(-1)^h \operatorname{tr}(\boldsymbol{C}_d^h)$, $h \geqslant 3$.

For equireplicate incomplete block designs, sequentially minimizing $(-1)^h \operatorname{tr}(\boldsymbol{C}_d^h)$ is equivalent to Paterson's (1983) proposal of sequentially minimizing the number of circuits of length h ($h = 2, 3, \ldots$) in the treatment concurrence graph of binary designs. The treatment concurrence graph of an equireplicate binary design d is a graph with t vertices in which there are λ_{dij} edges between the ith and jth vertices, where λ_{dij} is the number of times the ith and jth treatments appear together in the same block. Let $\boldsymbol{A}_d = \boldsymbol{N}_d\boldsymbol{N}_d^T - r\boldsymbol{I}$, where $r = bk/t$, be its adjacency matrix. Then the number of circuits of length h in this graph is $\operatorname{tr}(\boldsymbol{A}_d^h) = \sum_{j=0}^{h} \theta_j \operatorname{tr}(\boldsymbol{C}_d^j)$ for some θ_j, where $\theta_j < 0$ for odd j, and $\theta_j > 0$ for even j. Therefore sequentially minimizing $(-1)^h \operatorname{tr}(\boldsymbol{C}_d^h)$ is equivalent to sequentially minimizing the number of circuits of length h in the treatment concurrence graph of d.

EXAMPLE 4.6 (Chemical balance weighing designs, Example 4.3 continued). It was shown in Example 4.3 that when $N \equiv 2 \pmod 4$, an $\boldsymbol{X}_{d^*}$ such that $\boldsymbol{X}_{d^*}^T\boldsymbol{X}_{d^*}$ is of the form (4.1) with $m = \text{int}[n/2]$ has strong optimality properties over the $N \times n$ $(1, -1)$-matrices. The result, however, does not hold when zero entries are allowed in the design matrices. It can be shown that D-optimality over $(1, -1, 0)$-matrices is always attained by a matrix without zero entries (Galil and Kiefer, 1980); so $\boldsymbol{X}_{d^*}$ is D-optimal. But one cannot draw the same conclusion for the A- and E-criteria. In fact, if there exists a $(1, -1, 0)$-matrix $\boldsymbol{X}_d$ such that $\boldsymbol{X}_d^T \boldsymbol{X}_d = (N-1)\boldsymbol{I}$, then it is better than $\boldsymbol{X}_{d^*}$ under the E-criterion. (In fact, $\boldsymbol{X}_d$ is E-optimal, see Section 6.1.) Since $\phi_{-\infty} = \lim_{p \to -\infty} \phi_p$, $\boldsymbol{X}_{d^*}$ cannot be ϕ_p-optimal for p close to $-\infty$. Applying the result of Cheng (1992), we conclude that for any n and p, $-\infty < p < 1$, there exists an $N(n,p)$ such that if $N \geqslant N(n,p)$ and $N \equiv 2 \pmod 4$, then $\boldsymbol{X}_{d^*}$ is ϕ_p-optimal, if it exists. Apparently $\lim_{p \to -\infty} N(n,p) = \infty$. It is interesting to note that when $N = n$, $\boldsymbol{X}_{d^*}$ and an $\boldsymbol{X}_d$ such that $\boldsymbol{X}_d^T \boldsymbol{X}_d = (N-1)\boldsymbol{I}$ tie for the A-criterion: $\text{tr}(\boldsymbol{X}_{d^*}^T \boldsymbol{X}_{d^*})^{-1} = \text{tr}[(N-1)\boldsymbol{I}]^{-1}$. It is unknown whether they are both A-optimal.

When $N \equiv 3 \pmod 4$, an $\boldsymbol{X}_{d^*}$ such that $\boldsymbol{X}_{d^*}^T \boldsymbol{X}_{d^*} = (N+1)\boldsymbol{I} - \boldsymbol{J}$, which is (M.S)-optimal, can fail to be optimal even among the $(1, -1)$-matrices. The result of Cheng (1992) implies that for any n and p, $-\infty < p < 1$, there exists an $M(n,p)$ such that $N \geqslant M(n,p)$ and $N \equiv 3 \pmod 4 \Rightarrow$ an $\boldsymbol{X}_{d^*}$ such that $\boldsymbol{X}_{d^*}^T \boldsymbol{X}_{d^*} = (N+1)\boldsymbol{I} - \boldsymbol{J}$ is ϕ_p-optimal. Sharp bounds on $M(n,p)$ need to be derived for individual criteria. Kiefer and Galil (1980) derived the bound $M(n, 0) = 2n - 5$ for the D-criterion. Cheng et al. (1985b) gave a rough bound for the A-criterion, which has been improved by Sathe and Shenoy (1989) in the case where all the entries of the design matrices are 1 or -1. Results on D-optimal designs when $N \equiv 3 \pmod 4$ and $N < 2n - 5$ can be found, for instance, in Galil and Kiefer (1982), Moyssiadis and Kounias (1982), Kounias and Chadjipantelis (1983), Kounias and Farmakis (1984) and Chadjipantelis et al. (1987).

5. Using the approximate theory

In this section, we shall demonstrate how the approximate theory can be used to solve a discrete optimal design problem. Specifically, we shall present a solution to the optimal spring balance weighing design problem. As noted in Section 3, in this case there is no universally optimal design. Jacroux and Notz (1983) obtained D-, A- and E-optimal designs. We shall use the approximate theory to derive ϕ_p-optimal designs, and enlarge the set of competing designs to all the $N \times n$ matrices with entries $0 \leqslant x_{ij} \leqslant 1$, not just 0 and 1. Therefore the problem is to maximize $\phi_p(\boldsymbol{X}^T \boldsymbol{X})$ over all the $N \times n$ matrices with $0 \leqslant x_{ij} \leqslant 1$. Harwit and Sloane (1979) described an application to Hadamard transform optics in spectroscopy.

Write such a matrix $\boldsymbol{X}$ as $[\boldsymbol{x}_1, \boldsymbol{x}_2, \ldots, \boldsymbol{x}_N]^T$. Then each $\boldsymbol{x}_i$ can be considered as a point in the n-dimensional unit cube $\mathcal{X} \equiv \{\boldsymbol{x} = (x_1, x_2, \ldots, x_n)\colon\ 0 \leqslant x_i \leqslant 1\}$, and $\boldsymbol{X}$ is equivalent to a selection of N points from $\mathcal{X}$. The information matrix $\boldsymbol{X}^T \boldsymbol{X}$ can be expressed as

$$\boldsymbol{X}^T \boldsymbol{X} = \sum_{i=1}^{N} \boldsymbol{x}_i \boldsymbol{x}_i^T .$$

Denote by $\xi_{\boldsymbol{X}}$ the probability measure on $\mathcal{X}$ which assigns probability $1/N$ to each $\boldsymbol{x}_i$. Then

$$\boldsymbol{X}^T\boldsymbol{X} = N\int_{\mathcal{X}} \boldsymbol{x}\boldsymbol{x}^T \xi_{\boldsymbol{X}}(\mathrm{d}\boldsymbol{x}). \tag{5.1}$$

Let Ξ be the set of all the discrete probability measures on $\mathcal{X}$. For any $\xi \in \Xi$, called an *approximate* design, define its information matrix as

$$\boldsymbol{M}(\xi) = \int_{\mathcal{X}} \boldsymbol{x}\boldsymbol{x}^T \xi(\mathrm{d}\boldsymbol{x}),$$

extending (5.1). Instead of maximizing $\phi_p(\boldsymbol{X}^T\boldsymbol{X})$ over the $N \times n$ matrices with $0 \leqslant x_{ij} \leqslant 1$, we shall consider the problem of

$$\text{maximizing } \phi_p(\boldsymbol{M}(\xi)) \quad \text{over } \xi \in \Xi. \tag{5.2}$$

Thus the total number of runs N is suppressed, and we seek to find optimal *proportions* (probabilities) of different points in $\mathcal{X}$. If ξ^* is ϕ_p-optimal, and $N\xi^*(\boldsymbol{x})$ is an integer for all $\boldsymbol{x}$ in the support of ξ^*, then the $N \times n$ matrix in which each $\boldsymbol{x}$ appears $N\xi^*(\boldsymbol{x})$ times as a row vector is ϕ_p-optimal over the $N \times n$ matrices with $0 \leqslant x_{ij} \leqslant 1$.

Problem (5.2) is completely solved in Cheng (1987). For any integer k such that $\mathrm{int}[(n+1)/2] \leqslant k \leqslant n$, let ξ_k be the uniform measure on all the vertices of $\mathcal{X}$ with k coordinates equal to 1 and $n-k$ coordinates equal to 0. Divide the interval $[-\infty, 1]$ by using two interlacing sequences $a(k) < b(k) < a(k+1) < b(k+1)$ defined by $a(\mathrm{int}[(n+1)/2]) = -\infty$, $a(n) = b(n) = 1$, and

$$a(k) = 1 + \frac{\log\{(2k-1-n)/(2k-1)\}}{\log\{k(n-1)/(n-k)\}}$$

$$\text{for } \mathrm{int}\left[(n+1)/2\right] + 1 \leqslant k \leqslant n-1;$$

$$b(k) = 1 + \frac{\log\{(2k+1-n)/(2k+1)\}}{\log\{k(n-1)/(n-k)\}}$$

$$\text{for } \mathrm{int}\left[(n+1)/2\right] \leqslant k \leqslant n-1.$$

Then ξ_k is ϕ_p-optimal for all p such that $a(k) \leqslant p \leqslant b(k)$. In case $b(k) < p < a(k+1)$, the mixture $\varepsilon\xi_k + (1-\varepsilon)\xi_{k+1}$ is ϕ_p-optimal, where

$$\varepsilon = \frac{(k+1)^2(n-1) - (k+1)(n-k-1)\{(2k+1-n)/(2k+1)\}^{1/(p-1)}}{(2k+1)(n-1) + (2k+1-n)\{(2k+1-n)/(2k+1)\}^{1/(p-1)}}. \tag{5.3}$$

In particular, when n is odd, $\xi_{(n+1)/2}$ is ϕ_p-optimal for all $p \in [-\infty, b((n+1)/2)]$. Since $b((n+1)/2) = 1 - \log(n/2+1)/\log(n+1) > 0$, $\xi_{(n+1)/2}$ is A-, D- and

E-optimal. When n is even, $\xi_{n/2}$ is ϕ_p-optimal for all $p \in [-\infty, b(n/2)]$. Since $b(n/2) = 1 - \log(n+1)/\log(n-1) < 0$, unlike the odd case, $\xi_{n/2}$ is *not* D-optimal. However, $1 - \log(n+1)/\log(n-1) > -1$. It follows that $\xi_{n/2}$ is both A- and E-optimal. Because $a(n/2+1) > 0$, the D-optimal design is a mixture of $\xi_{n/2}$ and $\xi_{n/2+1}$. Letting $p = 0$ in (5.3), we have $\varepsilon = (n+2)/[2(n+1)]$. Therefore the design $\xi_D \equiv \{(n+2)/[2(n+1)]\}\xi_{n/2} + \{n/[2(n+1)]\}\xi_{n/2+1}$ is D-optimal. It is easy to verify that ξ_D places uniform weight $(\binom{n}{n/2} + \binom{n}{n/2+1})^{-1}$ on all the vertices of $\mathcal{X}$ with $n/2$ and $n/2+1$ coordinates equal to 1.

How can these results help find optimal exact designs? The approximate design ξ_k is supported on $\binom{n}{k}$ points, but an equivalent design (in the sense of having the same information matrix) with a smaller support can be found by using balanced incomplete block designs. It is easy to see that if $\boldsymbol{X}$ is the $N \times n$ block-treatment incidence matrix of a BIBD with n treatments and N blocks of size k, then the design which places uniform weight $1/N$ on all the row vectors of $\boldsymbol{X}$ has the same information matrix as ξ_k. This shows that the block-treatment incidence matrix of a BIBD with n treatments and N blocks of size k with $k \geqslant \text{int}[(n+1)/2]$ is ϕ_p-optimal over the $N \times n$ matrices with $0 \leqslant x_{ij} \leqslant 1$, for all p such that $a(k) \leqslant p \leqslant b(k)$. Therefore, when n is odd, the block-treatment incidence matrix of a BIBD with n treatments and N blocks of size $(n+1)/2$ is ϕ_p-optimal for all $p \in [-\infty, 1 - \log(n/2+1)/\log(n+1)]$. In particular, it is A-, D- and E-optimal. On the other hand, if n is even, then the block-treatment incidence matrix of a BIBD with n treatments and N blocks of size $n/2$ is ϕ_p-optimal for all $p \in [-\infty, 1 - \log(n+1)/\log(n-1)]$. This design is A- and E-optimal, but not D-optimal.

6. Miscellaneous results and methods

We shall end this chapter with some miscellaneous methods.

6.1. *E-optimality*

The E-criterion is perhaps the easiest to handle. (On the contrary, in the approximate theory the E-criterion is notoriously hard due to its nondifferentiability.) We present here a useful technique for proving E-optimality due to Takeuchi (1961).

Consider the case where the information matrices have zero row sums. As before, let $\mu_{d1} \geqslant \cdots \geqslant \mu_{dt} = 0$ be the eigenvalues of $\boldsymbol{C}_d$. Then for any numbers x and y, the eigenvalues of $\boldsymbol{C}_d - x\boldsymbol{I}_t + y\boldsymbol{J}_t$ are $\mu_{d1} - x \geqslant \cdots \geqslant \mu_{d,t-1} - x$ and $-x + ty$, the common row sum of $\boldsymbol{C}_d - x\boldsymbol{I}_t + y\boldsymbol{J}_t$. If $-x + ty > 0$ and $\boldsymbol{C}_d - x\boldsymbol{I}_t + y\boldsymbol{J}_t$ is not positive definite, then $\mu_{d,t-1} - x \leqslant 0$. Therefore $\mu_{d,t-1} \leqslant x$. This is a simple method for deriving upper bounds on $\mu_{d,t-1}$. If $-x + ty = 0$, one can also conclude $\mu_{d,t-1} - x \leqslant 0$ if $\boldsymbol{C}_d - x\boldsymbol{I}_t + y\boldsymbol{J}_t$ is not nonnegative definite or $\boldsymbol{a}^T[\boldsymbol{C}_d - x\boldsymbol{I}_t + y\boldsymbol{J}_t]\boldsymbol{a} \leqslant 0$ for some $\boldsymbol{a}$ such that $\boldsymbol{a}^T\boldsymbol{1} = 0$. The smaller x is, the more useful is the bound. Several useful bounds were derived in Cheng (1980b) and Jacroux (1980) by using this method. Constantine (1981) derived similar bounds by the method of averaging.

For example, the E-optimality of a group-divisible design with $\lambda_2 = \lambda_1 + 1$ (Takeuchi, 1961) can be proved as follows. For each design d, let $h(\boldsymbol{C}_d) = k\boldsymbol{C}_d - [(k-1)r + \lambda_1]\boldsymbol{I} + (\lambda_1 + 1)\boldsymbol{J}$. If d^* is a group-divisible design with $\lambda_2 = \lambda_1 + 1$, then $h(\boldsymbol{C}_{d^*})$ is as in (4.6), which is nonnegative definite, has positive row sums, and all the diagonal entries are equal to 1. By the first method discussed in the previous paragraph, it is sufficient to show that for an arbitrary design d, $h(\boldsymbol{C}_d)$ is not positive definite. Since d^* is binary, $\text{tr}\{h(\boldsymbol{C}_d)\} \leqslant \text{tr}\{h(\boldsymbol{C}_{d^*})\} = t$. If $\text{tr}\{h(\boldsymbol{C}_d)\} < t$, then since $h(\boldsymbol{C}_d)$ has integral entries, it has at least one nonpositive diagonal entry. Then $h(\boldsymbol{C}_d)$ is not positive definite. On the other hand, if $\text{tr}\{h(\boldsymbol{C}_d)\} = t$, then all the diagonal entries of $h(\boldsymbol{C}_d)$ are equal to 1. Since each row sum is greater than 1, there must be at least one off-diagonal entry which is greater than or equal to 1. Then $h(\boldsymbol{C}_d)$ is also not positive definite.

As another application we show the E-optimality of a group-divisible design with groups of size two and $\lambda_2 = \lambda_1 - 1 > 0$ (Cheng, 1980b). Let d^* be such a design. Then $k\boldsymbol{C}_{d^*} - [(k-1)r + \lambda_1 - 2]\boldsymbol{I} + (\lambda_1 - 1)\boldsymbol{J}$ is of the form

$$\begin{bmatrix} \boldsymbol{A} & \cdots & \boldsymbol{0} \\ \vdots & \ddots & \vdots \\ \boldsymbol{0} & \cdots & \boldsymbol{A} \end{bmatrix}, \tag{6.1}$$

where $\boldsymbol{A}$ is the 2×2 matrix

$$\begin{bmatrix} 1 & -1 \\ -1 & 1 \end{bmatrix}.$$

Since (6.1) is nonnegative definite and has zero row sums, by the second method discussed in the second paragraph of this section, it suffices to show that for an arbitrary design d, either $k\boldsymbol{C}_d - [(k-1)r + \lambda_1 - 2]\boldsymbol{I} + (\lambda_1 - 1)\boldsymbol{J}$ is not nonnegative definite or $\boldsymbol{a}^T\{k\boldsymbol{C}_d - [(k-1)r + \lambda_1 - 2]\boldsymbol{I} + (\lambda_1 - 1)\boldsymbol{J}\}\,\boldsymbol{a} \leqslant 0$ for some $\boldsymbol{a}$ such that $\boldsymbol{a}^T\boldsymbol{1} = 0$. This can easily be verified by noting that $k\boldsymbol{C}_d - [(k-1)r + \lambda_1 - 2]\boldsymbol{I} + (\lambda_1 - 1)\boldsymbol{J}$ has zero row sums, integral entries, and that the sum of the diagonal entries is $\leqslant t$.

A group-divisible design with groups of size two, $\lambda_2 = \lambda_1 - 1 > 0$, and $k \geqslant 3$ was also shown by Jacroux (1984) to be D-optimal. We suspect that such designs have strong optimality properties, even when $k = 2$.

It was shown in Example 4.3 that when $N \equiv 2 \pmod 4$, if there exists an $\boldsymbol{X}_{d^*}$ such that $\boldsymbol{X}_{d^*}^T\boldsymbol{X}_{d^*}$ is of the form (4.1), where $m = \text{int}[n/2]$, then it is Φ_f-optimal over all the $N \times n$ $(1,-1)$-matrices for any convex f such that f' is strictly concave and $f(0) = \lim_{x\to 0+} f(x) = \infty$. In particular, it is A-, D- and E-optimal. However, it is not E-optimal when zero entries are allowed in the design matrices, as in general chemical balance weighing designs. By an argument similar to that used in the above, one can show that if there exists a $(1,-1,0)$-matrix $\boldsymbol{X}_d$ such that $\boldsymbol{X}_d^T\boldsymbol{X}_d = (N-1)\boldsymbol{I}$, then it is E-optimal.

6.2. MV-optimality

An E-optimal design minimizes the maximum eigenvalue of $\boldsymbol{M}_d^{-1}$. This is equivalent to the minimization of $\max_{\boldsymbol{c}:\ \|\boldsymbol{c}\|=1} \operatorname{var}(\boldsymbol{c}^T\widehat{\boldsymbol{\theta}})$, where $\widehat{\boldsymbol{\theta}}$ is the least squares estimator of $\boldsymbol{\theta}$. Another minimax-type criterion is to minimize $\max_{1\leqslant i\leqslant n} \operatorname{var}(\widehat{\theta}_i)$. This is called the MV-criterion. Unlike many other commonly used criteria, the MV-criterion is not a function of the eigenvalues of the information matrices. Since an A-optimal design minimizes $\sum_{i=1}^n \operatorname{var}(\widehat{\theta}_i)$, it is also MV-optimal if $\operatorname{var}(\widehat{\theta}_i)$ is a constant. So if a symmetric design is A-optimal, then it is MV-optimal.

Similarly, in settings such as block designs where the information matrices have zero row sums, an E-optimal design minimizes the maximum variance of the least squares estimators of normalized treatment contrasts. In this case, we define an MV-optimal design as one which minimizes the maximum variance of the least squares estimators of pairwise comparisons of the treatment effects. In view of (2.8), an A-optimal symmetric design is also MV-optimal.

The MV-criterion was introduced by Takeuchi (1961). In addition to proving the E-optimality of group-divisible designs with $\lambda_2 = \lambda_1 + 1$, he also showed that they are MV-optimal. The term "MV-optimality" was coined by Jacroux (1983).

6.3. Dual designs

Suppose N_d is the treatment-block incidence matrix of a block design d. Then the *dual* of d, denoted d^D, is the block design with $\boldsymbol{N}_d^T$ as its treatment-block incidence matrix. Let $\mathcal{D}_0(t,b,k)$ be the set of all the equireplicate designs with t treatments and b blocks of size k. Then $d \in \mathcal{D}_0(t,b,k) \Rightarrow d^D \in \mathcal{D}_0(b,t,r)$, where $r = bk/t$. Duals of balanced incomplete block designs are called *linked block designs*.

Shah et al. (1976) showed that if d is A- (D-, or E-) optimal over $\mathcal{D}_0(t,b,k)$, then d^D is A- (D-, or E-) optimal over $\mathcal{D}_0(b,t,r)$. Eccleston and Kiefer (1981) provided a simple proof of a more general result which is described as follows.

The information matrices of d and d^D are respectively,

$$\boldsymbol{C}_d = r\boldsymbol{I} - k^{-1}\boldsymbol{N}_d\boldsymbol{N}_d^T$$

and

$$\boldsymbol{C}_{d^D} = k\boldsymbol{I} - r^{-1}\boldsymbol{N}_d^T\boldsymbol{N}_d.$$

Since in the following discussion, the treatments and blocks play symmetric roles, we may assume that $t \leqslant b$. Let the eigenvalues of $\boldsymbol{N}_d\boldsymbol{N}_d^T$ be $\rho_1,\ldots,\rho_t$. Then t of the eigenvalues of $\boldsymbol{N}_d^T\boldsymbol{N}_d$ are $\rho_1,\ldots,\rho_t$, and the remaining eigenvalues are zero. Since $\mu_{di} = r - \rho_i/k$, $i = 1,\ldots,t$, t of the eigenvalues of $\boldsymbol{C}_{d^D}$ are $\frac{k}{r}\mu_{d1}, \frac{k}{r}\mu_{d2},\ldots,\frac{k}{r}\mu_{dt}$, and the remaining eigenvalues are all equal to k. It is clear that d is E-optimal over $\mathcal{D}_0(t,b,k)$ if and only if d^D is E-optimal over $\mathcal{D}_0(b,t,r)$. Comparing the eigenvalues of $\boldsymbol{C}_d$ and $\boldsymbol{C}_{d^D}$, one can easily show that if f satisfies the condition

$$\text{for all } c > 0,\ \exists a_c > 0 \text{ and } b_c \text{ such that } f(cx) = a_c f(x) + b_c, \tag{6.2}$$

then d is Φ_f-optimal over $\mathcal{D}_0(t, b, k)$ if and only if d^D is Φ_f-optimal over $\mathcal{D}_0(b, t, r)$. Note that condition (6.2) is satisfied by the ϕ_p-criteria, $-\infty < p \leqslant 1$; thus the A- and D-criteria are covered.

This shows that generally the duals of optimal designs which are equireplicate are also optimal over equireplicate designs. Applying this to the E-optimal designs obtained in Section 6.1, we conclude the E-optimality over equireplicate designs of linked block designs, duals of group-divisible designs with $\lambda_2 = \lambda_1 + 1$, and duals of group-divisible designs with groups of size two and $\lambda_2 = \lambda_1 - 1 > 0$. Whether they are still optimal when the restriction of equireplication is removed is not obvious and requires extra work. Using upper bounds on $\mu_{d,t-1}$ derived by the method of Section 6.1, Cheng (1980b) showed the E-optimality of these designs when the competing designs are not necessarily equireplicate. The D-optimality of linked block designs without the restriction of equireplication was proved in Cheng (1990), re-discovered by Pohl (1992).

Acknowledgement

This article was written with the support of National Science Foundation Grant No. DMS-9404477, National Security Agency Grant No. MDA904-95-1-1064 and National Science Council, R.O.C., when the author was visiting the Institute of Statistics, National Tsing Hua University. The United States Government is authorized to reproduce and distribute reprints notwithstanding any copyright notation hereon.

References

Ash, A. (1981). Generalized Youden designs: Construction and tables. *J. Statist. Plann. Inference* **5**, 1–25.

Bagchi, S., A. C. Mukhopadhyay and B. K. Sinha (1990). A search for optimal nested row–column designs. *Sankhyā Ser. B* **52**, 93–104.

Bechhofer, R. E. and A. C. Tamhane (1981). Incomplete block designs for comparing treatments with a control: General Theory. *Technometrics* **23**, 45–57.

Chadjipantelis, Th., S. Kounias and C. Moyssiadis (1987). The maximum determinant of 21×21 $(+1, -1)$-matrices and D-optimal designs. *J. Statist. Plann. Inference* **16**, 167–178.

Chang, J. Y. and W. I. Notz (1994). Some optimal nested row–column designs. *Statist. Sinica* **4**, 249–263.

Cheng, C. S. (1978a). Optimality of certain asymmetrical experimental designs. *Ann. Statist.* **6**, 1239–1261.

Cheng, C. S. (1978b). Optimal designs for the elimination of multi-way heterogeneity. *Ann. Statist.* **6**, 1262–1272.

Cheng, C. S. (1980a). Optimality of some weighing and 2^n fractional factorial designs. *Ann. Statist.* **8**, 436–446.

Cheng, C. S. (1980b). On the E-optimality of some block designs. *J. Roy. Statist. Soc. Ser. B* **42**, 199–204.

Cheng, C. S. (1981a). Optimality and construction of pseudo Youden designs. *Ann. Statist.* **9**, 201–205.

Cheng. C. S. (1981b). Maximizing the total number of spanning trees in a graph: Two related problems in graph theory and optimum design theory. *J. Combin. Theory Ser. B* **31**, 240–248.

Cheng, C. S. (1987). An application of the Kiefer–Wolfowitz equivalence theorem to a problem in Hadamard transform optics. *Ann. Statist.* **15**, 1593–1603.

Cheng, C. S. (1990). D-optimality of linked block designs and some related results. In: *Proc. R. C. Bose Symp. on Probability, Statistics and Design of Experiments*. Wiley Eastern, New Delhi, 227–234.

Cheng, C. S. (1992). On the optimality of (M.S)-optimal designs in large systems. *Sankhyā* **54**, 117–125 (special volume dedicated to the memory of R. C. Bose).
Cheng, C. S. and R. A. Bailey (1991). Optimality of some two-associate-class partially balanced incomplete-block designs. *Ann. Statist.* **19**, 1667–1671.
Cheng, C. S., D. Majumdar, J. Stufken and T. E. Ture (1988). Optimal step-type designs for comparing test treatments with a control. *J. Amer. Statist. Assoc.* **83**, 477–482.
Cheng, C. S., J. C. Masaro and C. S. Wong (1985a). Do nearly balanced multigraphs have more spanning trees? *J. Graph Theory* **8**, 342–345.
Cheng, C. S., J. C. Masaro and C. S. Wong (1985b). Optimal weighing designs. *SIAM J. Alg. Disc. Meth.* **6**, 259–267.
Cheng, C. S. and C. F. Wu (1980). Balanced repeated measurements designs. *Ann. Statist.* **8**, 1272–1283.
Conniffe, D. and J. Stone (1975). Some incomplete block designs of maximum efficiency. *Biometrika* **61**, 685–686.
Constantine, G. M. (1981). Some E-optimal block designs. *Ann. Statist.* **9**, 886–892.
Constantine, G. M. (1986). On the optimality of block designs. *Ann. Inst. Statist. Math.* **38**, 161–174.
Eccleston, J. A. and A. S. Hedayat (1974). On the theory of connected designs: Characterization and optimality. *Ann. Statist.* **2**, 1238–1255.
Eccleston, J. A. and J. Kiefer (1981). Relationships of optimality for individual factors of a design. *J. Statist. Plann. Inference* **5**, 213–219.
Eccleston, J. A. and K. G. Russell (1975). Connectedness and orthogonality in multi-factor designs. *Biometrika* **62**, 341–345.
Eccleston, J. A. and K. G. Russell (1977). Adjusted orthogonality in nonorthogonal designs. *Biometrika* **64**, 339–345.
Ehrenfeld, S. (1956). Complete class theorems in experimental designs. In: J. Neyman, ed., *Proc. 3rd Berkeley Symp. on Mathematical Statistics and Probability*, Vol. 1. University of California Press, Berkeley, CA, 57–67.
Fan, K. (1954). Inequalities for eigenvalues of Hermitian matrices. *Nat. Bur. Standards Appl. Math. Ser.* **39**, 131–139.
Galil, Z. and J. Kiefer (1980). D-optimum weighing designs. *Ann. Statist.* **8**, 1293–1306.
Galil, Z. and J. Kiefer (1982). Construction methods for D-optimum weighing designs when $n \equiv 3 \pmod 4$. *Ann. Statist.* **10**, 502–510.
Harwit, M. and N. J. A. Sloane (1979). *Hadamard Transform Optics*. Academic Press, New York.
Hedayat, A. S., M. Jacroux and D. Majumdar (1988). Optimal designs for comparing test treatments with controls (with discussions). *Statist. Sci.* **3**, 462–491.
Hedayat, A. S. and D. Majumdar (1985). Families of A-optimal block designs for comparing test treatments with a control. *Ann. Statist.* **13**, 757–767.
Hedayat, A. S. and W. Zhao (1990). Optimal two-period repeated measurements designs. *Ann. Statist.* **18**, 1805–1816.
Jacroux, M. (1980). On the E-optimality of regular graph designs. *J. Roy. Statist. Soc. Ser. B* **42**, 205–209.
Jacroux, M. (1983). Some minimum variance block designs for estimating treatment differences. *J. Roy. Statist. Soc. Ser. B* **45**, 70–76.
Jacroux, M. (1984). On the D-optimality of group divisible designs. *J. Statist. Plann. Inference* **9**, 119–129.
Jacroux, M. (1985). Some sufficient conditions for type-I optimality of block designs. *J. Statist. Plann. Inference* **11**, 385–394.
Jacroux, M. and W. I. Notz (1983). On the optimality of spring balance weighing designs. *Ann. Statist.* **11**, 970–978.
Jacroux, M., C. S. Wong and J. C. Masaro (1983). On the optimality of chemical balance weighing designs. *J. Statist. Plann. Inference* **8**, 231–240.
John, J. A. and T. J. Mitchell (1977). Optimal incomplete block designs. *J. Roy. Statist. Soc. Ser. B* **39**, 39–43.
Jones, B. and J. A. Eccleston (1980). Exchange and interchange procedures to search for optimal designs. *J. Roy. Statist. Soc. Ser. B* **42**, 238–243.
Kiefer, J. (1958). On the nonrandomized optimality and randomized nonoptimality of symmetrical designs. *Ann. Math. Statist.* **29**, 675–699.

Kiefer, J. (1975a). Construction and optimality of generalized Youden designs. In: J. N. Srivastava, ed., *A Survey of Statistical Design and Linear Models*. North-Holland, Amsterdam, 333–353.
Kiefer, J. (1975b). Balanced block designs and generalized Youden designs, I. Construction (patchwork). *Ann. Statist.* **3**, 109–118.
Kounias, S. and Th. Chadjipantelis (1983). Some D-optimal weighing designs for $n = 3 \pmod 4$. *J. Statist. Plann. Inference* **8**, 117–127.
Kounias, S. and N. Farmakis (1984). A construction of D-optimal weighing designs when $n = 3 \pmod 4$. *J. Statist. Plann. Inference* **10**, 177–187.
Kunert, J. (1983). Optimal design and refinement of the linear model with application to repeated measurements designs. *Ann. Statist.* **11**, 247–257.
Kunert, J. (1984). Optimality of balanced uniform repeated measurements designs. *Ann. Statist.* **12**, 1006–1017.
Kunert, J. (1985). Optimal repeated measurements designs for correlated observations and analysis by weighted least squares. *Biometrika* **72**, 375–389.
Lucas, H. L. (1957). Extra-period Latin-square change-over designs. *J. Dairy Sci.* **40**, 225–239.
Magda, G. C. (1980). Circular balanced repeated measurements designs. *Comm. Statist. Theory Methods* **9**, 1901–1918.
Majumdar, D. and W. I. Notz (1983). Optimal incomplete block designs for comparing treatments with a control. *Ann. Statist.* **11**, 258–266.
Moyssiadis, C. and S. Kounias (1982). The exact D-optimal first order saturated design with 17 observations. *J. Statist. Plann. Inference* **7**, 13–27.
Paterson, L. J. (1983). Circuits and efficiency in incomplete block designs. *Biometrika* **70**, 215–225.
Pohl, G. M. (1992). D-optimality of the dual of BIB designs. *Statist. Probab. Lett.* **14**, 201–204.
Pukelsheim, F. (1993). *Optimal Design of Experiments*. Wiley, New York.
Raghavarao, D. (1971). *Constructions and Combinatorial Problems in Design of Experiments*. Wiley, New York.
Roy, B. K. and K. R. Shah (1984). On the optimality of a class of minimal covering designs. *J. Statist. Plann. Inference* **10**, 189–194.
Ruiz, F. and E. Seiden (1974). On the construction of some families of generalized Youden designs. *Ann. Statist.* **2**, 503–519.
Sathe, Y. S. and R. G. Shenoy (1989). A-optimal weighing designs when $N \equiv 3 \pmod 4$. *Ann. Statist.* **17**, 1906–1915.
Seiden, E. and C. J. Wu (1978). A geometric construction of generalized Youden designs for v a power of a prime. *Ann. Statist.* **6**, 452–460.
Shah, K. R. (1960). Optimality criteria for incomplete block designs. *Ann. Math. Statist.* **31**, 791–794.
Shah, K. R., D. Raghavarao and C. G. Khatri (1976). Optimality of two and three factor designs. *Ann. Statist.* **4**, 419–422.
Shah, K. R. and B. K. Sinha (1989). *Theory of Optimal Designs*. Springer, Berlin.
Stufken, J. (1987). A-optimal block designs for comparing test treatments with a control. *Ann. Statist.* **15**, 1629–1638.
Takeuchi, K. (1961). On the optimality of certain type of PBIB designs. *Rep. Statist. Appl. Res. Un. Japan Sci. Eng.* **8**, 140–145.
Takeuchi, K. (1963). A remark added to "On the optimality of certain type of PBIB designs". *Rep. Statist. Appl. Res. Un. Japan Sci. Eng.* **10**, 47.

S. Ghosh and C. R. Rao, eds., *Handbook of Statistics, Vol. 13*

Optimal and Efficient Treatment-Control Designs

Dibyen Majumdar

1. Introduction

Comparing treatments with one or more controls is an integral part of many areas of scientific experimentation. In pharmaceutical studies, for example, new drugs are the treatments, while a placebo and/or a standard treatment is the control. Most attention will be given to the use of a single control; we will consider designs for comparing treatments with a control for various experimental settings, models and inference methods. This article is expected to update, supplement and expand upon the earlier survey of the area by Hedayat, Jacroux and Majumdar (1988).

The section titles are:
2. Early development.
3. Efficient block designs for estimation.
4. Efficient designs for confidence intervals.
5. Efficient row–column designs for estimation.
6. Bayes optimal designs.
7. On efficiency bounds of designs.
8. Optimal and efficient designs in various settings.

2. Early development

Consider an experiment to compare v treatments (which will be called test treatments), with a control using n homogeneous experimental units. Let the control be denoted by the symbol 0 and the test treatments by the symbols $1, \ldots, v$. We assume the model to be additive and homoscedastic, i.e., if treatment i is applied to unit j then the observation y_{ij} can be expressed as:

$$y_{ij} = \mu + \tau_i + \varepsilon_{ij}, \tag{2.1}$$

where μ is a general mean, τ_i is the effect of treatment i and ε_{ij}'s are random errors that are assumed to be independently normally distributed with expectation 0 and variance σ^2. A design d is characterized by the number of experimental units that are assigned to each treatment. For $i = 0, 1, \ldots, v$, we will denote by r_{di} the number of experimental units assigned to treatment i, or replication of treatment i.

Given n what is the best allocation of the experimental units to the control and the test treatments, i.e., what are the optimal replications? Unless the experimenter has some knowledge about the performance of the control that can be used at the designing stage, it is intuitively clear that the control should be used more often than the test treatments, since each of the v test treatments has to be compared with the same control 0. One way to determine an *optimal* allocation is by considering the Best Linear Unbiased Estimators (BLUE's) of the parameters of interest, which are the treatment-control contrasts, $\tau_i - \tau_0$, $i = 1, \ldots, v$. For a design d, let the BLUE of $\tau_i - \tau_0$ be denoted by $\widehat{\tau}_{di} - \widehat{\tau}_{d0}$. For model (2.1), it is clear that $\widehat{\tau}_{di} - \widehat{\tau}_{d0} = \bar{y}_{di} - \bar{y}_{d0}$, where $\bar{y}_{di}$ is the average of all observations that receive treatment i. Also,

$$\operatorname{Var}(\widehat{\tau}_{di} - \widehat{\tau}_{d0}) = \sigma^2\left(r_{di}^{-1} + r_{d0}^{-1}\right). \tag{2.2}$$

A possible allocation (design) is that which minimizes $\sum_{i=1}^{v} \operatorname{Var}(\widehat{\tau}_{di} - \widehat{\tau}_{d0})$. We will presently (Definition 2.2 later in this section) call this design A-optimal for treatment-control contrasts, since it minimizes the *average* (hence the 'A' in A-optimal) of the variances of the v treatment-control contrasts. It is easy to see that a design that is A-optimal for treatment-control contrasts is obtained by minimizing the expression,

$$\sum_{i=1}^{v} \left(r_{di}^{-1} + r_{d0}^{-1}\right)$$

subject to the constraint, $\sum_{i=0}^{v} r_{di} = n$. The following theorem is obvious.

THEOREM 2.1. *If v is a square and $n \equiv 0 \pmod{(v + \sqrt{v})}$, then a design d_0 given by*

$$r_{d_0 1} = \cdots = r_{d_0 v} = n/(v + \sqrt{v}), \quad r_{d_0 0} = \sqrt{v}\, r_{d_0 1} \tag{2.3}$$

is A*-optimal for treatment-control contrasts for model* (2.1).

This result was noticed by Fieler (1947), and possibly even earlier (see also Finney, 1952). Thus, if $v = 4$ test treatments have to be compared with a control using $n = 24$ experimental units then the A-optimal design assigns 4 units to each test treatment and 8 to the control.

Dunnett (1955) found that the same allocation performs very well for the problem of simultaneous confidence intervals for the treatment-control contrasts $\tau_i - \tau_0$, $i = 1, \ldots, v$. More discussion of Dunnett's work and that of other researchers in the area of multiple comparisons can be found in Section 4.

Next consider the situation where the experimental units are partitioned into b blocks of k homogeneous units each. Here too, we assume the model to be additive and homoscedastic, i.e., if treatment i is applied to the unit l of block j then the observation y_{ijl} can be expressed as:

$$y_{ijl} = \mu + \tau_i + \beta_j + \varepsilon_{ijl}, \tag{2.4}$$

where β_j is the effect of block j. A design d is an allocation of the $v+1$ treatments to the bk experimental units. We shall use the notation $\mathcal{D}(v+1,b,k)$ to denote the set of all connected designs, i.e., those designs in which every treatment-control contrast is estimable. For a design d, let n_{dij} denote the number of times treatment i is used in block j. Further, for treatment symbols i and i' $(i \neq i')$, let $\lambda_{dii'} = \sum_{j=1}^{v} n_{dij} n_{di'j}$. We shall use the notation $\mathrm{BIB}(v,b,r,k,\lambda)$ to denote a Balanced Incomplete Block (BIB) design based on v treatments, in b blocks of size k each, where each treatment is replicated r times and each pair of treatments appears in λ blocks.

Cox (1958), p. 238, recommended using a $\mathrm{BIB}(v,b,r,k-t,\lambda)$ design based on the test treatments with each block *augmented* or *reinforced* by t replications of the control, where t is an integer. Das (1958) called such designs *reinforced BIB designs.* The idea was to have all test treatments *balanced*, as well as to have the test treatments *balanced* with respect to the control, and reinforcing BIB designs is a natural way to do it.

Pearce (1960) took a different approach to achieve the same goal. He proposed a general class of designs for the problem of treatment-control comparisons, of which reinforced BIB designs is a subclass. These were the *designs with supplemented balance* that were proposed by Hoblyn et al. (1954) in a different context. Designs with supplemented balance have $v+1$ treatments, one of which is called the supplemented treatment, the control in this case.

DEFINITION 2.1. A design $d \in \mathcal{D}(v+1,b,k)$ is called a design with supplemented balance with 0 as the supplemented treatment if there are nonnegative integers λ_{d0} and λ_{d1}, such that:

$$\begin{aligned} \lambda_{dii'} &= \lambda_{d1}, \quad \text{for } i,i' = 1,\ldots,v \ (i \neq i'), \\ \lambda_{d0i} &= \lambda_{d0}, \quad \text{for } i = 1,\ldots,v. \end{aligned} \tag{2.5}$$

For a reinforced BIB design of Das (1958), $\lambda_{d0} = tr$. Pearce provided examples of these designs, their analysis for the linear model (2.4), and computed the standard error of the BLUE for the elementary contrasts. He showed that every elementary contrast in two test treatments have the same standard error, while every treatment-control contrast have the same standard error (see also Pearce, 1963).

EXAMPLE 2.1. The following design, due to Pearce, was used in an experiment involving strawberry plants at the East Malling Research Station in 1953 (see Pearce, 1953). Four herbicides were compared with a control, which was the absence of any herbicide, in four blocks of size seven each. Denoting the herbicides by $1,2,3,4$ and the control by 0, the design, with columns as blocks, is the following:

0	0	0	0
0	0	0	0
1	1	1	1
2	2	2	2
3	3	3	3
4	4	4	4
1	2	3	4

According to Pearce (1983), this was the very first design with supplemented balance. Note that for this design $\lambda_{d0} = 10$, $\lambda_{d1} = 6$. Analysis of data for the experiment can be found in Chapter 3 of Pearce (1983) as well as in Pearce (1960).

Given v, b and k, the question is, which design in $\mathcal{D}(v+1, b, k)$ should be used? The answer would depend on the particular circumstances of the experiment, on various factors and constraints that usually confront the experimenter. (In this context, we refer the reader to page 126 of Pearce (1983) for an account of the circumstances that led to the use of the design in Example 2.1.)

One of the considerations, possibly the most important one, is which design results in the best inference of the treatment-control contrasts, which are the contrasts of primary interest in the experiment. Determination of *optimal* designs for inference and *efficiencies* of designs in $\mathcal{D}(v + 1, b, k)$ with respect to the optimal design in this class started much later. It may be noted that usually there are several designs with supplemented balance within $\mathcal{D}(v+1, b, k)$. Often there is a design with supplemented balance that is optimal in $\mathcal{D}(v + 1, b, k)$ or at least highly efficient. On the other hand, usually the class of designs with supplemented balance also contain designs that are quite inefficient. Clearly one has to make a judicious choice of a design, even when one decides to restrict oneself to the class of designs with supplemented balance.

Several optimality criteria have been considered in the literature (see Hedayat et al. (1988), and the discussions in that article). For a large portion of this article, we will focus on two optimality criteria for estimation because of the natural statistical interpretation of these criteria in experiments to compare test-treatments with a control. These criteria are given in the next definition which is quite general, i.e., it is not restricted to block designs and model (2.4).

DEFINITION 2.2. Given a class of designs $\mathcal{D}$, and a model, if $\widehat{\tau}_{di} - \widehat{\tau}_{d0}$ denotes the BLUE of $\tau_{di} - \tau_{d0}$, then a design is A-optimal for treatment-control contrasts (abbreviated as A-optimal) if it minimizes $\sum_{i=1}^{v} \mathrm{Var}(\widehat{\tau}_{di} - \widehat{\tau}_{d0})$ in $\mathcal{D}$. A design is MV-optimal for treatment-control contrasts (abbreviated as MV-optimal) if it minimizes

$$\underset{1 \leqslant i \leqslant v}{\mathrm{Max}}\ \mathrm{Var}(\widehat{\tau}_{di} - \widehat{\tau}_{d0}) \text{ in } \mathcal{D}.$$

Pesek (1974) compared a BIB design in all $v + 1$ treatments (which may be viewed as a design with supplemented balance) with a reinforced BIB design and noted that the latter is more efficient according to the A-optimality criterion. Optimality of the reinforced designs in a restricted class of designs was established by Constantine (1983). Constantine's class consisted of designs that had exactly one replication of the control in each block. In this class Constantine showed that a reinforced BIB design is A-optimal. Jacroux (1984) showed that Constantine's conclusions remain valid when the BIB design is replaced by a suitable group divisible design in the test treatments.

Bechhofer and Tamhane (1981) rediscovered designs with supplemented balance when they were considering the problem of constructing simultaneous confidence intervals for the treatment-control contrasts. They called their designs Balanced Treatment Incomplete Block (BTIB) designs, a terminology which has been adopted by many authors in the area since 1981, as the Bechhofer and Tamhane paper inspired much of the research on optimal and efficient designs in $\mathcal{D}(v+1,b,k)$. More discussion on these results will appear in Sections 3 and 4.

For the two-way elimination of heterogeneity model (row–column designs), Freeman (1975) considered some designs for comparing two sets of treatments. More discussion on this topic will appear in Section 5.

Starting with Owen (1970), several authors incorporated prior information on the model parameters at the designing stage to determine *Bayes optimal designs*. Some of these results will be discussed in Section 6.

3. Efficient block designs for estimation

Consider block designs with observations following the one-way elimination of heterogeneity model, given by (2.4). The object of the experiment is to estimate the treatment-control contrasts: $(\tau_1-\tau_0, \tau_2-\tau_0, \ldots, \tau_v-\tau_0)$. Let $N_d = (n_{dij})$, a $(v+1)\times b$ matrix, and

$$P = \begin{pmatrix} -1_v & I_v \end{pmatrix}$$

where 1_v is a $v\times 1$ matrix of 1's and I_v is the $v\times v$ identity matrix. Then the covariance matrix for the BLUE's $(\widehat{\tau}_{d1} - \widehat{\tau}_{d0}, \widehat{\tau}_{d2} - \widehat{\tau}_{d0}, \ldots, \widehat{\tau}_{dv} - \widehat{\tau}_{d0})$ of the treatment-control contrasts is $\sigma^2 P C_d^- P'$, where for a design $d \in \mathcal{D}(v+1,b,k)$,

$$C_d = \mathrm{Diag}(r_{d0}, r_{d1}, \ldots, r_{dv}) - \frac{1}{k} N_d N_d',$$

the Fisher Information matrix for the treatments (test treatments and control). If one partitions C_d as:

$$C_d = \begin{pmatrix} c_{d00} & \gamma_d' \\ \gamma_d & M_d \end{pmatrix} \tag{3.1}$$

then it can be shown that (see Bechhofer and Tamhane, 1981),

$$(P C_d^- P')^{-1} = M_d,$$

i.e., M_d is the *information matrix* for the treatment-control contrasts. Clearly (see Definition 2.2) an A-optimal design minimizes $\mathrm{tr}(M_d^{-1})$ in $\mathcal{D}(v+1,b,k)$ and an MV-optimal design minimizes the maximum diagonal element of M_d^{-1} in $\mathcal{D}(v+1,b,k)$.

3.1. Orthogonal designs

It has long been known that one way to obtain an optimal block design is to construct, if possible, an *orthogonal* block design, such that within each block the replication of treatments are optimal for a zero-way elimination of heterogeneity model (i.e., model (2.1)). This result, and its generalization has been used by several authors, including Magda (1980) and Kunert (1983) in the context of repeated measurement designs. The idea is to establish optimality of a design for a model from the two facts: the design is optimal under a simpler model and the design is *orthogonal.*

We state a general result, which is essentially a different version of the result given by Magda and Kunert. First some preliminaries. Let $\mathcal{D}$ be a class of designs. Consider two models for an $n \times 1$ random vector of observations Y, when the observations are taken according to a design $d \in \mathcal{D}$:

$$\begin{aligned} &\text{Model A:}\ Y = X_{1d}\tau + X_{2d}\theta_2 + \varepsilon,\\ &\text{Model B:}\ Y = X_{1d}\tau + X_{2d}\theta_2 + X_{3d}\theta_3 + \varepsilon \end{aligned} \tag{3.2}$$

where τ is a vector of treatments and θ_i are vectors of nuisance parameters (for example block parameters in (2.4), or row and column parameters in (5.1), and so forth). The X's are known matrices and ε is a vector of random errors with $\mathrm{E}(\varepsilon) = 0$, $\mathrm{V}(\varepsilon) = \sigma^2 I$. In Kunert's terminology, Model B is *finer* that Model A.

THEOREM 3.1. *Suppose $d_0 \in \mathcal{D}$ is* A- (MV-) *optimal for treatment-control contrasts under Model* A, *and*

$$X'_{1d_0}X_{3d_0} = X'_{1d_0}X_{2d_0}(X'_{2d_0}X_{2d_0})^{-}X'_{2d_0}X_{3d_0} \tag{3.3}$$

then d_0 is A- (MV-) *optimal for treatment-control contrasts under Model* B.

Conditions (3.3) is essentially an orthogonality condition. Taking (2.1) as Model A with $\theta_2 = \mu$, and (2.2) as Model B, we can combine Theorems 2.1 and 3.1 to get the following result:

COROLLARY 3.1. *Given v, b and k, suppose v is a perfect square, $k \equiv 0 \ (\mathrm{mod}(v + \sqrt{v}))$ and $d_0 \in \mathcal{D}(v+1, b, k)$ is such that*

$$\begin{aligned} &n_{d_0 1j} = \cdots = n_{d_0 vj} = k/\big(v + \sqrt{v}\big),\\ &n_{d_0 0j} = \sqrt{v}\, n_{d_0 1j}, \quad \textit{for } j = 1, \ldots, b. \end{aligned} \tag{3.4}$$

Then d_0 is A-*optimal for treatment-control contrasts in $\mathcal{D}(v+1, b, k)$.*

EXAMPLE 3.1. Let $v = 4$, $k = 6$. Then a design in which each block is $(0, 0, 1, 2, 3, 4)$ is A-optimal in $\mathcal{D}(5, b, 6)$.

3.2. Optimal incomplete block designs

Corollary 3.1 is limited in its scope since it requires that the block sizes are multiples of $(v + \sqrt{v})$. Consequently, k is quite large compared to v. What is an A-optimal design in other situations? Let us first consider the incomplete block setup, i.e.,

$$2 \leqslant k \leqslant v. \tag{3.5}$$

Let us start with an arbitrary design d in $\mathcal{D}(v+1,b,k)$. Using Kiefer's (1975) technique of averaging, we obtain

$$\operatorname{tr}(PC_d^- P') \geqslant \operatorname{tr}(P\overline{C_d}^- P'), \tag{3.6}$$

where $\overline{C_d} = \frac{1}{v!}\sum_{\Pi} \Pi C_d \Pi'$, the summation taken over all $(v+1)\times(v+1)$ permutation matrices Π that correspond to permutations of the v test treatments only. If we partition $\overline{C_d}$ as in (3.1), then we see that $\overline{M_d} = (P\overline{C_d}^- P')^{-1}$ is a completely symmetric matrix. In general, there may be no design in $\mathcal{D}(v+1,b,k)$ for which $\overline{C_d}$ is the Fisher Information matrix for the treatments. If there is such a design, then for this design, call it $\bar{d}$, $M_{\bar{d}} = \overline{M_{\bar{d}}}$ is completely symmetric and $\gamma_{\bar{d}}$ (see (3.1)) is a vector with all entries equal. That is, $\bar{d}$ belongs to the class of designs with supplemented balance of Pearce (1960)!

Bechhofer and Tamhane (1981) postulated that the same subclass of designs was suitable for the problem of constructing simultaneous confidence intervals for the treatment-control contrasts. (More on this in Section 4.) Their terminology is given in the following definition:

DEFINITION 3.1. Suppose (3.5) holds. A design $d \in \mathcal{D}(v+1,b,k)$ is called a Balanced Treatment Incomplete Block (BTIB) design if condition (2.5) holds.

Condition (3.5) is used in the definition because Bechhofer and Tamhane were interested exclusively in the incomplete block setup (hence the 'I' in BTIB). Note that, for a BTIB design d, for $i, i' = 1,\ldots,v$, $i \neq i'$,

$$\begin{aligned}
&\operatorname{Var}(\widehat{\tau}_{di} - \widehat{\tau}_{d0}) = \sigma^2 k(\lambda_{d0} + \lambda_{d1})/(\lambda_{d0}(\lambda_{d0} + v\lambda_{d1})),\\
&\operatorname{Corr}(\widehat{\tau}_{di} - \widehat{\tau}_{d0}, \widehat{\tau}_{di'} - \widehat{\tau}_{d0}) = \lambda_{d1}/(\lambda_{d0} + \lambda_{d1}).
\end{aligned}$$

A particular type of BTIB design will be of considerable interest to us. This is defined next – the notation is due to Stufken (1987).

DEFINITION 3.2. A design $d \in \mathcal{D}(v+1,b,k)$ is called $\mathrm{BTIB}(v,b,k;t,s)$ if it is a BTIB design, and, with $(t,s) \in \{0,1,\ldots,k-1\} \times \{0,1,\ldots,b\}$, it holds that,

$$n_{dij} \in \{0,1\}, \quad \text{for all } (i,j) \in \{1,\ldots,v\} \times \{1,\ldots,b\},$$

and

$$n_{d01} = \cdots = n_{d0s} = t+1, \qquad n_{d0(s+1)} = \cdots = n_{d0b} = t.$$

For a $\mathrm{BTIB}(v,b,k;t,s)$, $r_{d0} = bt+s$, $v\lambda_{d0} = bt(k-t)+s(k-2t-1)$, $v(v-1)\lambda_{d1} = (b(k-t)-2s)(k-t-1)$. Here are a few examples.

EXAMPLE 3.2. Let $v = 7$, $b = 7$, and $k = 4$. A BTIB$(7,7,4;1,0)$ in $\mathcal{D}(8,7,4)$ is given by the following array, where, as elsewhere for block designs, columns are blocks:

$$\begin{array}{ccccccc}
0&0&0&0&0&0&0\\
1&2&3&4&5&6&7\\
2&3&4&5&6&7&1\\
4&5&6&7&1&2&3
\end{array}$$

Here $\lambda_{d0} = 3$, $\lambda_{d1} = 1$.

EXAMPLE 3.3. Let $v = 6$, $b = 18$, and $k = 5$. A BTIB$(6,18,5;1,6)$ in $\mathcal{D}(7,18,5)$ is given by the following array:

$$\begin{array}{cccccccccccccccccc}
0&0&0&0&0&0&0&0&0&0&0&0&0&0&0&0&0&0\\
0&0&0&0&0&0&1&1&1&1&1&1&1&1&2&2&2&2\\
1&1&1&2&3&3&2&2&2&2&3&4&3&3&3&3&3&4\\
2&4&2&4&4&5&3&5&3&4&4&5&5&4&4&4&5&5\\
3&5&6&5&6&6&4&6&5&6&5&6&6&6&5&6&6&6
\end{array}$$

Here $\lambda_{d0} = 14$, $\lambda_{d1} = 6$.

EXAMPLE 3.4. Let $v = 4$, $b = 12$, and $k = 4$. The following array is a BTIB design that is not a BTIB$(4,12,4;t,s)$.

$$\begin{array}{cccccccccccc}
0&0&0&0&0&0&0&0&0&0&1&1\\
0&0&0&0&0&0&0&0&0&0&2&2\\
0&0&0&0&1&1&1&2&2&3&3&3\\
1&2&3&4&2&3&4&3&4&4&4&4
\end{array}$$

Here $\lambda_{d0} = 9$, $\lambda_{d1} = 3$.

Starting from an arbitrary design d in $\mathcal{D}(v+1,b,k)$, one can use (3.6) to obtain a lower bound for the value of the A-criterion of d. This lower bound can be minimized over $\mathcal{D}(v+1,b,k)$. A design that attains this minimum would be an A-optimal design. This was done in Majumdar and Notz (1983). To state their result we need some notation. For integers v, b, k, x and z let,

$$\begin{aligned}
g(x,z) = kv(v-1)^2\big[bvk(k-1) - (bx+z)(kv-v+k) \\
+ (bx^2+2xz+z)\big]^{-1} \\
+ kv\big[k(bx+z) - (bx^2+2xz+z)\big]^{-1}, \qquad (3.7)
\end{aligned}$$

$$\Lambda = \{0,1,\ldots,\lfloor k/2\rfloor - 1\} \times \{0,1,\ldots,b\} - \{0,0\}. \qquad (3.8)$$

For a design d that is BTIB$(v,b,k;x,z)$, it holds that $\mathrm{tr}(PC_d^- P') = g(x,z)$, i.e., g gives the value of the A-criterion. Note that the replication of the control for this design is $r_{d0} = bx + z$. For $(x,z) \in \Lambda$ the function g and the set Λ can be expressed in terms of $r = bx + z$ as:

$$g^*(r) = kv(v-1)^2[bvk(k-1) - r(kv - v + k) + h(r)]^{-1} \\ + kv[kr - h(r)]^{-1}, \tag{3.9}$$

where

$$h(r) = b(\lfloor r/b \rfloor)^2 + (2\lfloor r/b \rfloor + 1)(r - b\lfloor r/b \rfloor), \tag{3.10}$$

$$\Lambda^* = \{1, \ldots, b\lfloor k/2 \rfloor\}. \tag{3.11}$$

We are now ready to state the result of Majumdar and Notz (1983).

THEOREM 3.2. *Let t and s be integers defined by*

$$g(t,s) = \underset{(x,z)\in\Lambda}{\mathrm{Min}} \; g(x,z). \tag{3.12}$$

Under condition (3.5), *for any design $d \in \mathcal{D}(v+1,b,k)$,*

$$\mathrm{tr}(PC_d^- P') \geqslant g(t,s), \tag{3.13}$$

with equality if d is BTIB$(v,b,k;t,s)$. *Hence a* BTIB$(v,b,k;t,s)$ *is* A-*optimal for treatment-control contrasts in $\mathcal{D}(v+1,b,k)$.*

An equivalent version of Theorem 3.2 is:

THEOREM 3.2*. *Let the integer r^* be defined by*

$$g^*(r^*) = \underset{r\in\Lambda^*}{\mathrm{Min}} \; g^*(r). \tag{3.14}$$

Under condition (3.5), *for any design $d \in \mathcal{D}(v+1,b,k)$,*

$$\mathrm{tr}(PC_d^- P') \geqslant g^*(r^*), \tag{3.15}$$

with equality if d is BTIB$(v,b,k;t,s)$ *where $bt + s = r^*$. Hence a* BTIB$(v,b,k;t,s)$ *with $bt + s = r^*$ is* A-*optimal for treatment-control contrasts in $\mathcal{D}(v+1,b,k)$.*

The quantity r^* may be viewed as the optimal replication of the control, since if a BTIB$(v,b,k;t,s)$ design d_0 is A-optimal, then $r_{d_0 0} = bt + s = r^*$. (Is the minimum of $g^*(r)$ attained at a unique point r^*? The answer is yes, except in rare cases. For now, there is no loss in assuming that r^* is unique. We will return to this point in Theorem 3.6.)

EXAMPLE 3.5. Let $v = 7, b = 7$ and $k = 4$. Then solving the optimization problem in (3.12) we get $t = 1, s = 0$. Hence, $r^* = 7$. The $\mathrm{BTIB}(7,7,4;1,0)$ given in Example 3.2 is A-optimal.

EXAMPLE 3.6. Let $v = 6, b = 18$ and $k = 5$. Here $t = 1$, and $s = 6$. Hence $r^* = 24$. The $\mathrm{BTIB}(6, 18, 5; 1, 6)$ given in Example 3.3 is A-optimal.

The A-optimal designs given by Theorem 3.2 are $\mathrm{BTIB}(v, b, k; t, s)$. It can be seen that the structure of a $\mathrm{BTIB}(v, b, k; t, s)$ can be of two types. If $s = 0$, then it is called a *Rectangular-type* or *R-type* design, while if $s > 0$, then it is called a *Step-type* or *S-type* design. The terminology is due to Hedayat and Majumdar (1984). With columns as blocks an R-type design d may be visualized as a $k \times b$ array:

$$d = \begin{bmatrix} d_1 \\ d_2 \end{bmatrix}, \tag{3.16}$$

where d_1 is a $t \times b$ array of controls (0), while d_2 is a $(k - t) \times b$ array in the test treatments $(1, 2, \ldots, v)$ only. Clearly, d_2 must be a $\mathrm{BIB}(v, b, r, k - t, \lambda)$ design. The R-type designs are thus exactly the reinforced BIB designs.

An S-type design d can be visualized as the following $k \times b$ array:

$$d = \begin{bmatrix} d_{11} & d_{12} \\ d_{21} & d_{22} \end{bmatrix}, \tag{3.17}$$

where d_{11} is a $(t + 1) \times s$ array of controls, d_{12} is a $t \times (b - s)$ array of controls, d_{21} is a $(k - t - 1) \times s$ array of test treatments and d_{22} is a $(k - t) \times (b - s)$ array of test treatments. The following result of Hedayat and Majumdar (1984) gives some properties of a $\mathrm{BTIB}(v, b, k; t, s)$.

LEMMA 3.1. (i) *For the existence of a* $\mathrm{BTIB}(v, b, k; t, s)$, *the following conditions are necessary (where* $r_0 = bt + s$*):*

$$(b(k - t) - s)/v = (bk - r_0)/v \ (= q_1,\ say) \quad \textit{is an integer}, \tag{3.18}$$

$$s(k - t - 1)/v \ (= q_2\ ,\ say) \quad \textit{is an integer}, \tag{3.19}$$

$$[q_2(k - t - 2) + (q_1 - q_2)(k - t - 1)]/(v - 1) \quad \textit{is an integer}, \tag{3.20}$$

(ii) *For an* R-type *design it is necessary that* $b \geqslant v$, *while for an* S-type *design it is necessary that* $b \geqslant v + 1$.

It can be shown that a $\mathrm{BTIB}(v, b, k; t, s)$ is equireplicate in test treatments. Condition (3.18) is necessary for this. Conditions (3.19) and (3.20) are necessary for condition (2.5). Part (ii) of the Lemma is Fisher's inequality for $\mathrm{BTIB}(v, b, k; t, s)$ designs. Based on Theorem 3.2 and Lemma 3.1, Hedayat and Majumdar (1984) suggested a method for obtaining optimal designs that consists of three steps:

(1) *Starting from* v, b, k *determine* t, s *(equivalently* r^**) that minimize* $g(x, z)$.

(2) *Verify conditions of Lemma* 3.1(i), *using* t *and* s *from step* 1. *If the conditions are not satisfied then Theorem* 3.2 *cannot be applied to the class* $\mathcal{D}(v+1, b, k)$. *If the conditions are satisfied, then go to step* 3.

(3) *Attempt to construct a* $\mathrm{BTIB}(v, b, k; t, s)$. (*Note that even when the conditions of Lemma* 3.1 *are satisfied, there is no guarantee that this design exists.*)

Instances of design classes where this method does not produce A-optimal designs is given in examples 3.9–3.11. For R-type designs, the construction problem reduces to finding BIB designs. For S-type designs the problem is clearly more involved, and unlike the case of R-type designs this case does not reduce to the construction of designs that are well studied in the literature. We return to the construction of such designs later in this section.

It may be noted that Majumdar and Notz (1983) obtained optimal designs for criteria other than A-optimality also. Giovagnoli and Wynn (1985) used approximate design theory techniques to obtain results similar to Theorem 3.2.

For certain values of v, b and k the minimization in (3.12) gives a nice algebraic solution. This can sometimes be exploited to obtain infinite families of A-optimal designs with elegant combinatorial properties. Here are some results.

THEOREM 3.3. *A* $\mathrm{BTIB}(v, b, k; 1, 0)$ *is* A-*optimal for treatment-control contrasts in* $\mathcal{D}(v+1, b, k)$ *whenever*

$$(k-2)^2 + 1 \leqslant v \leqslant (k-1)^2.$$

This result is due to Hedayat and Majumdar (1985). An example of a design that is A-optimal according to Theorem 3.3 is the design in Example 3.2. It is interesting to note that when $v = (k-2)^2 + k - 1$, an A-optimal design is obtained by taking the BIB design d_2 in (3.16) to be a *finite projective plane* of order $(k-2)$, while when $v = (k-1)^2$, an A-optimal design is obtained by taking the BIB design d_2 as a *finite euclidean plane* of order $(k-1)$.

The next result, due to Stufken (1987), is a generalization of Theorem 3.3.

THEOREM 3.4. *A* $\mathrm{BTIB}(v, b, k; t, 0)$ *is* A-*optimal for treatment-control contrasts in* $\mathcal{D}(v+1, b, k)$ *whenever*

$$(k-t-1)^2 + 1 \leqslant t^2 v \leqslant (k-t)^2.$$

EXAMPLE 3.7. Let $v = 9, b = 12$ and $k = 8$. With $t = 2$, $(k-t-1)^2 + 1 = 26$, $t^2 v = 36$, and $(k-t)^2 = 36$. Hence a $\mathrm{BTIB}(9, 12, 8; 2, 0)$, which is a $\mathrm{BIB}(9, 12, 8, 6, 5)$ design with each block augmented by two replications of the control, is A-optimal.

Theorems 3.3 and 3.4 gives some sufficient conditions for optimality of the reinforced BIB designs of Das (1958). A family of A-optimal S-type designs, due to Cheng et al. (1987), is given in the next theorem.

THEOREM 3.5. *Let* $\alpha \geqslant 3$ *be a prime or a prime power, and* γ *be a positive integer. Then there exists a* $\mathrm{BTIB}(\alpha^2 - 1, \gamma(\alpha+2)(\alpha^2-1), \alpha; 0, \gamma(\alpha+1)(\alpha^2-1))$. *This design is* A-*optimal for treatment-control contrasts in* $\mathcal{D}(\alpha^2, \gamma(\alpha+2)(\alpha^2-1), \alpha)$.

For a method of construction of the BTIB design in Theorem 3.5 the reader is referred to Cheng et al. (1987).

EXAMPLE 3.8. Let $v = 8$, $b = 40$ and $k = 3$. With $\alpha = 3$ and $\gamma = 1$ here is an example of a $\mathrm{BTIB}(8, 40, 3; 0, 32)$ that is A-optimal:

0	0	0	0	0	0	0	0	0	0	0	0	0	0	0	0	0	0	0	0
1	1	1	1	1	1	1	2	2	2	2	2	2	3	3	3	3	3	4	4
2	3	4	5	6	7	8	3	4	5	6	7	8	4	5	6	7	8	5	6

0	0	0	0	0	0	0	0	0	0	0	0	1	1	1	2	2	3	3	4
4	4	5	5	5	6	6	7	1	2	3	7	2	4	6	5	6	4	5	5
7	8	6	7	8	7	8	8	5	4	6	8	3	7	8	8	7	8	7	6

We conclude this discussion by stating a result of Cheng et al. (1988) which characterizes the point r^* and the shape of the function g^*.

THEOREM 3.6. *Given* v, b *and* k, $r^* \in \{r_l^*, r_u^*\}$, *where* $0 \leqslant (r_u^* - r_l^*) \leqslant 1$,

$$g^*(r_l^*) = g^*(r_u^*) = g^*(r^*), \tag{3.21}$$

and

$$\text{for } r_l^* > r \in \Lambda^*, \quad g^*(r) > g^*(r+1);$$
$$\text{for } r_u^* < r+1 \in \Lambda^*, \quad g^*(r) < g^*(r+1).$$

In our experience the case $r_l^* \neq r_u^*$ is extremely rare. If $r_l^* = r_u^* - 1$, then a $\mathrm{BTIB}(v, b, k; t, s)$ can exist at only one of the two consecutive integers, r_l^* and r_u^*, which can be taken as r^*. This result can help identify the optimal replication r^* without going through a complete search over Λ^*. It also helps in identifying designs that are *almost optimal* (*highly efficient*) in $\mathcal{D}(v+1, b, k)$, in case the optimal design is unknown. More discussion on this will appear in Section 7.

So far all A-optimal designs have been obtained starting from Theorem 3.2. Jacroux (1989) generalized this result using techniques developed earlier (Jacroux, 1987a) for obtaining MV-optimal designs. Since his results are considerably involved we will only give a broad outline and some examples, while we refer the reader to his paper for details. Jacroux's optimal designs are BTIB designs as well as designs that belong to a class that is combinatorially more general than BTIB designs. These designs, proposed by Jacroux (1987a), are defined as follows.

DEFINITION 3.3. A design $d \in \mathcal{D}(v+1,b,k)$ is called a *Group Divisible Treatment Design* (GDTD) with parameters m, n, λ_0, λ_1 and λ_2 if the treatments $1,2,\ldots,v$ can be divided into m groups $V_1,\ldots,V_m$ of n treatments each such that there are nonnegative constants λ_0, λ_1 and λ_2 such that

$$\begin{aligned} \lambda_{d0i} &= \lambda_0, \quad \text{for } i = 1,\ldots,v, \\ \lambda_{dii'} &= \lambda_1, \quad \text{for } i,i' \in V_p, i \neq i', \\ \lambda_{dii'} &= \lambda_2, \quad \text{for } i \in V_p, i' \in V_q, p,q \in \{1,\ldots,m\}, p \neq q. \end{aligned}$$

For a GDTD all variances $\mathrm{Var}(\widehat{\tau}_{di} - \widehat{\tau}_{d0})$, $i = 1,\ldots,v$, are equal; the variances $\mathrm{Var}(\widehat{\tau}_{di} - \widehat{\tau}_{di'})$, $i,i' = 1,\ldots,v$, $i \neq i'$, can take at most two values depending on whether i and i' belong to the same group or not. In a BTIB design the latter variances take only one value, since a BTIB design may be considered a GDTD with only one group.

DEFINITION 3.4. A design $d \in \mathcal{D}(v+1,b,k)$ is called a $\mathrm{GDTD}(v,b,k;t,s)$ if it is a GDTD, and with t and s, $(t,s) \in \{0,1,\ldots,k-1\} \times \{0,1,\ldots,b-1\}$, it holds that,

$$n_{dij} \in \{0,1\}, \quad \text{for all } (i,j) \in \{1,\ldots,v\} \times \{1,\ldots,b\},$$

and

$$n_{d01} = \cdots = n_{d0s} = t+1,$$

$$n_{d0(s+1)} = \cdots = n_{d0b} = t.$$

Jacroux originally developed the class of GDTD's to find some sufficient conditions for a design to be MV-optimal. It follows from the fact,

$$\operatorname*{Max}_{1 \leqslant i \leqslant v} \mathrm{Var}(\widehat{\tau}_{di} - \widehat{\tau}_{d0}) = \frac{1}{v} \sum_{i=1}^{v} \mathrm{Var}(\widehat{\tau}_{di} - \widehat{\tau}_{d0}), \quad \text{for a BTIB design } d,$$

that all BTIB designs that are A-optimal are also MV-optimal. Jacroux (1987a) established the MV-optimality of some BTIB designs in cases where this could not be concluded from their A-optimality since the designs were not known to be A-optimal; he also proved the MV-optimality of some GDTD's.

Below we give a method given in Jacroux (1987a) that is often successful in finding optimal designs – for exact statement of the theorems and the proofs we refer the reader to Jacroux's paper. We shall use the notation in (3.9) and (3.10). The method consists of three steps:

(1) *Over all positive integers* r *such that* $(bk - r)/v$ *is an positive integer, find* r^{**} *such that* $g^*(r)$ *is a minimum.*

(2) *Find* d^* with $r_{d^*0} = r^{**}$ *that is a* BTIB$(v, b, k; t, s)$ *design or a* GDTD *with some special properties specified by Jacroux.*

(3) *Verify one of several sets of sufficient conditions that guarantee that* d^* *is* MV*-optimal in* $\mathcal{D}(v + 1, b, k)$.

The Hedayat and Majumdar method, which is based on Theorem 3.2 of Majumdar and Notz (1983), starts by minimizing the function $g^*(r)$, then verifies a set of necessary conditions, one of which is (3.18). Jacroux's method incorporates condition (3.18) in the optimization. Moreover, Jacroux goes beyond the class of BTIB designs to certain types of GDTD's in his search for optimal designs. Thus, Jacroux's method can give optimal designs for some classes $\mathcal{D}(v + 1, b, k)$ where the Hedayat and Majumdar method cannot give an optimal design. This advance was possible, however, only after some difficult technical problems were overcome in Jacroux (1987a).

Here are some examples of optimal designs given by Jacroux.

EXAMPLE 3.9. Let $v = b = 11$, $k = 6$. Using the notation of Theorem 3.2*, it can be seen that $r^* = 14$, but that there is no BTIB$(11, 11, 6; t, s)$ design with $11t + s = 14$, i.e., a BTIB$(11, 11, 6; 1, 3)$ does not exist. Hence Theorem 3.2* cannot be used to obtain an (A- or MV-) optimal design in $\mathcal{D}(12, 11, 6)$. Jacroux (1987a) showed that $r^{**} = 11$, and that the following BTIB$(11, 11, 6; 1, 0)$ design is MV-optimal.

0	0	0	0	0	0	0	0	0	0	0
2	3	1	2	3	4	1	1	1	2	1
4	5	4	5	6	7	5	2	2	3	3
5	6	6	7	8	9	8	6	3	4	4
6	7	7	8	9	10	10	9	7	8	5
10	11	8	9	10	11	11	11	10	11	9

EXAMPLE 3.10. Let $v = 6, b = 14$ and $k = 4$. Here, $r^* = 18$, $r^{**} = 14$. Theorem 3.2* cannot be used to obtain an optimal design; the following GDTD$(6, 14, 4; 1, 0)$ design with treatment groups $\{1, 2\}$, $\{3, 4\}$ and $\{5, 6\}$ is MV-optimal in $\mathcal{D}(7, 14, 4)$.

0	0	0	0	0	0	0	0	0	0	0	0	0	0
1	1	1	1	1	1	1	2	2	2	2	2	3	4
2	2	3	3	3	4	4	3	3	3	4	4	5	5
5	6	4	5	6	5	6	4	5	6	5	6	6	6

EXAMPLE 3.11. This is an example of an A-optimal design from Jacroux (1989). Let $v = 6, b = 20$ and $k = 3$. Here $r^* = 18$, $r^{**} = 18$. Theorem 3.2* cannot be used to obtain an optimal design since a BTIB$(6, 20, 3; 0, 18)$ does not exist. Jacroux showed

that the following GDTD$(6, 20, 3; 0, 18)$ with treatment groups $\{1, 2, 3\}$ and $\{4, 5, 6\}$ is A-optimal in $\mathcal{D}(7, 20, 3)$.

0	0	0	0	0	0	0	0	0	0	0	0	0	0	0	0	0	0	1	4
1	1	1	1	1	1	2	2	2	2	2	2	3	3	3	3	3	3	2	5
4	5	6	4	5	6	4	5	6	4	5	6	4	5	6	4	5	6	3	6

3.3. Construction of BTIB designs and GDTD's

Let us now consider the problem of constructing balanced treatment incomplete block designs and group divisible treatment designs. As mentioned earlier, construction of R-type BTIB designs reduces to the construction of BIB designs. For this, one may appeal to the vast literature that is available. For the case of S-type BTIB designs, there is no such resource to fall back on. Construction of BTIB designs for the case $k = 2$ has been considered in Bechhofer and Tamhane (1983b), while Notz and Tamhane (1983) gave a complete solution for the case $k = 3$, $3 \leqslant v \leqslant 10$. Hedayat and Majumdar (1984) has some discussion on the construction of S-type designs. In this paper, a complete catalog of A-optimal designs that can be obtained from Theorem 3.2 are given for the range $2 \leqslant k \leqslant 8$, $k \leqslant v \leqslant 30$, and $v \leqslant b \leqslant 50$; this catalog contains several S-type designs. Cheng et al. (1988) and Ture (1982) also have some general recommendations for constructing S-type designs.

It is easy to see that a GDTD$(v, b, k; t, 0)$ is obtained by reinforcing each block of a group divisible design in blocks of size $k - t$, by t replications of the control. Construction of GDTD's in general is not always easy – some GDTD's were constructed by Jacroux (1987b).

Stufken (1991a) gave a complete solution to the problem of constructing GDTD's for $k = 3$ and $v = 4, 6$. From the viewpoint of practical applications, designs with small k and v are most important. It may be noted that for $v = 3, 5$, $k = 3$ the problem reduces to constructing BTIB designs, which is given in Notz and Tamhane (1983).

3.4. Optimal designs in large blocks

Next, let us remove the restriction (3.5). If the conditions of Corollary 3.1 are satisfied then one can obtain an A-optimal design quite easily. What happens if these conditions do not hold? Let us start with a definition of a class of designs that generalize the class of BTIB$(v, b, k; t, s)$ designs.

DEFINITION 3.5. A design $d \in \mathcal{D}(v + 1, b, k)$ is called a BTB$(v, b, k; t, s)$ if it is a design with supplemented balance, i.e., satisfies (2.5), with the additional properties:

$$|n_{dij} - n_{di'j'}| \leqslant 1, \quad \text{for all } (i, i') \in \{1, \dots, v\} \times \{1, \dots, v\},$$
$$i \neq i', \quad (j, j') \in \{1, \dots, b\} \times \{1, \dots, b\},$$
$$n_{d01} = \cdots = n_{d0s} = t + 1, \quad n_{d0(s+1)} = \cdots = n_{d0b} = t.$$

In addition to the sets Λ and Λ^* defined in (3.8) and (3.11) and $h(r)$ defined in (3.10), let,

$$\begin{aligned}
&r = bx + z, \quad \text{for integers } x \text{ and } z,\\
&\alpha = \lfloor (bk - r)/bv \rfloor, \quad p = v(k-1) + k - 2v\alpha,\\
&c = bvk(k-1) + bv\alpha(v - 2k + v\alpha),\\
&f(x,z) = kv(v-1)^2[c - p(bx+z) + (bx^2 + 2xz + z)]^{-1}\\
&\qquad\qquad + kv\big[k(bx+z) - (bx^2 + 2xz + z)\big]^{-1},\\
&f^*(r) = kv(v-1)^2\big[c - rp + h(r)\big]^{-1} + kv\big[kr - h(r)\big]^{-1}.
\end{aligned}$$

The following theorem, due to Jacroux and Majumdar (1989), can be viewed as a generalization of Theorem 3.2, as condition (3.5) is no longer imposed.

THEOREM 3.7. *Let t and s be integers defined by*

$$f(t,s) = \operatorname*{Min}_{(x,z)\in\Lambda} f(x,z). \tag{3.22}$$

For any design $d \in \mathcal{D}(v+1,b,k)$,

$$\operatorname{tr}(PC_d^- P') \geqslant f(t,s), \tag{3.23}$$

with equality if d is $\mathrm{BTB}(v,b,k;t,s)$. *Hence a* $\mathrm{BTB}(v,b,k;t,s)$ *is* A-*optimal for treatment-control contrasts in $\mathcal{D}(v+1,b,k)$.*

An equivalent version of Theorem 3.6 is:

THEOREM 3.7*. *Let the integer r^* be defined by*

$$f^*(r^*) = \operatorname*{Min}_{r\in\Lambda^*} f^*(r). \tag{3.24}$$

For any design $d \in \mathcal{D}(v+1,b,k)$,

$$\operatorname{tr}(PC_d^- P') \geqslant f^*(r^*), \tag{3.25}$$

with equality if d is $\mathrm{BTB}(v,b,k;t,s)$ *where $bt + s = r^*$. Hence a* $\mathrm{BTB}(v,b,k;t,s)$ *with $bt + s = r^*$ is* A-*optimal for treatment-control contrasts in $\mathcal{D}(v+1,b,k)$.*

EXAMPLE 3.12. Let $v = 3, b = 10$ and $k = 4$. The following $\mathrm{BTB}(3,10,4;1,3)$ is A- and MV-optimal in $\mathcal{D}(4,10,4)$:

0	0	0	0	0	0	0	0	0	0
0	0	0	1	1	1	1	1	1	1
1	1	2	2	2	2	2	2	2	2
2	3	3	3	3	3	3	3	3	3

The following corollary, due to Jacroux and Majumdar (1989), gives an infinite family of A-optimal designs. It is clear that these designs are also MV-optimal.

COROLLARY 3.2. *For any integer* $\theta > 1$, *a* $\mathrm{BTB}(\theta^2, b, (\theta+1)^2; \theta+1, 0)$, *for a* b *such that the design exists, is* A-*optimal for treatment-control contrasts in* $\mathcal{D}(\theta^2+1, b, (\theta+1)^2)$.

4. Efficient designs for confidence intervals

There are two methods for deriving inferences on the treatment-control contrasts. One is estimation, and the other is simultaneous confidence intervals. Bechhofer and Tamhane, in their discussion of Hedayat et al. (1988), describe situations where the simultaneous confidence interval approach is the appropriate one for choosing a subset of *best* treatments from the v test treatments. More examples are available in Hochberg and Tamhane (1987). In this section we discuss optimal and efficient designs for simultaneous confidence intervals for the treatment-control contrasts.

4.1. Designs for the zero-way elimination of heterogeneity model

First consider the 0-way elimination of heterogeneity model, i.e., $y_{ij} = \mu + \tau_i + \varepsilon_{ij}$. Let, $\mu_i = \mu + \tau_i$, for $i = 1, \ldots, v$. We have to impose some more distributional assumptions on the random variables in order to obtain confidence intervals. For $i = 0, 1, \ldots, v$ and $j = 1, \ldots, r_{di}$ let y_{ij}'s be independent with

$$y_{ij} \sim N(\mu_i, \sigma_i). \tag{4.1}$$

Dunnett (1955) was the first to give simultaneous confidence intervals for the treatment-control contrasts $\mu_i - \mu_0$, $i = 1, \ldots, v$. Suppose

$$\sigma_0 = \sigma_1 = \cdots = \sigma_v. \tag{4.2}$$

If $\bar{y}_{di}$ denotes the mean of all observations that receive treatment i, and s^2 denotes the pooled variance estimate (mean squared error), then the simultaneous $100P\%$ lower confidence limits for $\mu_i - \mu_0$ are:

$$\bar{y}_{di} - \bar{y}_{d0} - \delta_i s\sqrt{r_{di}^{-1} + r_{d0}^{-1}}, \quad i = 1, \ldots, v, \tag{4.3}$$

where the δ_i are determined from:

$$P(t_i < \delta_i,\ i = 1, \ldots, v) = P. \tag{4.4}$$

The joint distribution of the random variables $t_1, \ldots, t_v$ is the multivariate analog of Student's t defined by Dunnett and Sobel (1954). Here the subscript d represents the design. The design problem is to allocate a total of n observations to the v test treatments and one control. Thus $\sum_{i=0}^{v} r_{di} = n$. We shall denote by $\mathcal{D}(v+1,n)$ the class of all designs.

The simultaneous $100P\%$ two-sided confidence limits for $\mu_i - \mu_0$ are:

$$\bar{y}_{di} - \bar{y}_{d0} \pm \kappa_i s \sqrt{r_{di}^{-1} + r_{d0}^{-1}}, \quad i = 1, \ldots, v,$$

where,

$$P(|t_i| < \kappa_i,\ i = 1, \ldots, v) = P.$$

For the design $r_{di} = n/(v+1)$, for all i, and $\delta_1 = \cdots = \delta_v = \delta$, Dunnett (1955) gave tables of the δ's for various values of P. For the same design and $\kappa_1 = \cdots = \kappa_v = \kappa$ he gave tables for the κ's for various values of P. He also investigated optimal designs for lower confidence limits in the following subset of $\mathcal{D}(v+1,n)$:

$$\mathcal{D}^*(v+1,n) = \{d\colon\ d \in \mathcal{D}(v+1,n),\ r_{d1} = \cdots = r_{dv}\}.$$

A design was called optimal if it maximized P for a fixed value of $\delta\sqrt{r_{di}^{-1} + r_{d0}^{-1}}$. Dunnett's numerical investigations revealed that the optimal value of r_{d0}/r_{d1} was only slightly less than $\sqrt{v}$, which is the value given in Theorem 2.1, as long as δ was chosen to make the coverage probability P in (4.4) of magnitude 0.95 or larger. The tables for the simultaneous two-sided confidence limits in Dunnett (1955) were obtained using an approximation. More accurate tables for the two-sided case were given in Dunnett (1964); optimal allocations were also discussed.

Bechhofer (1969) considered the design problem for Model (4.1) when the σ_i are known, without the restriction (4.2). The lower $100P\%$ simultaneous confidence limits are:

$$\bar{y}_{di} - \bar{y}_{d0} - \delta' \sqrt{\sigma_i r_{di}^{-1} + \sigma_0 r_{d0}^{-1}}, \quad i = 1, \ldots, v, \tag{4.5}$$

where

$$P\Big(\mu_i - \mu_0 \geqslant \bar{y}_{di} - \bar{y}_{d0} - \delta' \sqrt{\sigma_i r_{di}^{-1} + \sigma_0 r_{d0}^{-1}},\ i = 1, \ldots, v\Big) = P. \tag{4.6}$$

The quantity $\delta'\sqrt{\sigma_i r_{di}^{-1} + \sigma_0 r_{d0}^{-1}}$ is called a *yardstick*. Let,

$$\gamma_{di} = r_{di}/n, \quad \text{for } i = 0, 1, \ldots, v, \quad \text{and} \quad \zeta = \frac{\delta'}{\sigma_0}\sqrt{\left(\sigma_i \gamma_{di}^{-1} + \sigma_0 \gamma_{d0}^{-1}\right)}.$$

For a fixed value of ζ (equivalently, for a fixed yardstick), Bechhofer defined a design d to be *optimal* if it maximizes P in (4.6) among all designs in the subclass,

$$\mathcal{D}_\sigma^*(v+1,n) = \{d\colon\ d \in \mathcal{D}(v+1,n), \sigma_1^2/r_{d1} = \cdots = \sigma_v^2/r_{dv}\}.$$

We shall call this design one that maximizes the coverage probability for a fixed yardstick. Note that, if $d \in \mathcal{D}^*_\sigma(v+1,n)$ then the treatment-control contrasts are all estimated with the same variance; thus this is a *natural* subclass of $\mathcal{D}(v+1,n)$ for this setup. $\mathcal{D}^*_\sigma(v+1,n) = \mathcal{D}^*(v+1,n)$, for the homoscedastic case, $\sigma_0 = \sigma_1 = \cdots = \sigma_v$. Taking the *approximate design theory* approach, i.e., viewing the γ_{di} as nonnegative reals, not restricted to being rational, such that $\sum_{i=0}^{v} \gamma_{di} = 1$, Bechhofer (1969) gave an explicit equation to determine an optimal design. This is given in the following theorem, where $f(\cdot)$ is the standard normal density function, $F_k\left(x \mid \rho\right)$ is the k-variate standard normal distribution function with all correlations equal to ρ, and

$$\beta = \sum_{i=1}^{v} \sigma_i^2/\sigma_0^2, \quad \zeta^* = (1/2)v(v-1)\sqrt{\beta/\pi}\, F_{v-2}\left(0 \mid 1/3\right),$$

$$\omega = \omega(\gamma) = \zeta\gamma\sqrt{\beta(1-\gamma)/((1-\gamma+\gamma\beta)(2(1-\gamma)+\gamma\beta))}.$$

THEOREM 4.1. *Given $v \geqslant 2$, n, and $\sigma_i > 0$, a design d_0 with $\gamma_{d_0 i} = \gamma_i$ given as follows is optimal, i.e., it maximizes the coverage probability for a fixed yardstick. For fixed $0 < \zeta \leqslant \zeta^*$, $\gamma_0 = 0$, and for fixed $\zeta > \zeta^*$, γ_0 is the unique root in $(0, 1/(1+\sqrt{\beta}))$, of the equation:*

$$\begin{aligned}
&((1-\beta)\gamma^2 - 2\gamma + 1)\omega F_{v-1}\left(\omega \,\middle|\, \frac{1-\gamma}{2(1-\gamma)+\gamma\beta}\right) \\
&\quad - \frac{(v-1)\gamma(1-\gamma)\beta}{2(1-\gamma)+\gamma\beta} f(\omega) F_{v-2}\left(\omega\sqrt{\frac{1-\gamma+\gamma\beta}{3(1-\gamma)+\gamma\beta}} \,\middle|\, \frac{1-\gamma}{3(1-\gamma)+\gamma\beta}\right) \\
&= 0.
\end{aligned}$$

Also, for $i = 1, \ldots, v$, $\gamma_i = \sigma_i^2(1-\gamma_0)/(\beta\sigma_0^2)$.

EXAMPLE 4.1. Suppose $v = 2$ and $\sigma_1 = \sigma_2 = \sigma_0$. If $\zeta = 2$, then $\gamma_0 = 0.32, \gamma_1 = \gamma_2 = 0.34$; hence 32% of the observations are allocated to the control, and 34% to each of the two test treatments. If $\zeta = 5$ then $\gamma_0 = 0.40, \gamma_1 = \gamma_2 = 0.30$; hence 40% of the observations are allocated to the control, and 30% to each of the two test treatments.

As a corollary, Bechhofer showed that when $\sigma_i = \sigma_0$ for all i, in the limit as $\zeta \to \infty$, the optimal $\gamma_0/\gamma_i \to \sqrt{v}$, the allocation given in Theorem 2.1.

The results of Bechhofer (1969) were generalized by Bechhofer and Turnbull (1971) by allowing the δ' to vary with the treatment symbol i $(i = 1, \ldots, v)$. Optimal designs for the simultaneous two-sided confidence interval were given in Bechhofer and Nocturne (1972). Bechhofer and Tamhane (1983a) gave optimal designs that minimized the total sample size for one-sided intervals for given P, β and a yardstick.

Taking a different approach, Spurrier and Nizam (1990) minimized the expected *allowance* for a fixed coverage probability. For the model given by (4.1) and (4.2), the allowance, or yardstick, for the simultaneous $100P\%$ lower confidence limits in

(4.3) when $\delta_1 = \cdots = \delta_v = \delta$, is given by $\delta s\sqrt{r_{di}^{-1} + r_{d0}^{-1}}$, where δ is determined from equation (4.4), i.e., $P(t_i < \delta,\ i = 1, \ldots, v) = P$. Spurrier and Nizam (1990) defined the *Expected Average Allowance* (EAA) as,

$$\mathrm{EAA} = (1/v)\delta \mathrm{E}(s)\sqrt{r_{di}^{-1} + r_{d0}^{-1}},$$

and defined a design d_0 to be optimal in the sense of *minimizing* EAA *for fixed coverage probability* if it minimizes $\delta\sqrt{r_{di}^{-1} + r_{d0}^{-1}}$ for fixed P, over all of $\mathcal{D}(v+1, n)$.

In general, determination of an optimal design is a difficult problem. Spurrier and Nizam (1990) were able to show that when $v = 2$, for the optimal design d_0, $|r_{d_01} - r_{d_02}| \leqslant 1$. When $v > 2$, based on their numerical calculations they conjectured that $|r_{d_0i} - r_{d_0i'}| \leqslant 1$, for $i \neq i'$. This result is helpful in limiting the search for an optimal design. Spurrier and Nizam gave tables of optimal designs for $2 \leqslant v \leqslant 8$ and several values of n, when the r_{di}'s are allowed to vary by no more than 1. Analogous tables for the two-sided confidence intervals were also given.

Motivated by the thrust on quality improvement and emphasis on analyzing variability, Spurrier (1992) considered the problem of inference for the variances. He studied optimality of designs in $\mathcal{D}^*(v+1, n)$ for simultaneous confidence limits for the σ_i^2/σ_0^2 $(i = 1, \ldots, v)$ and gave extensive tables of optimal designs. Optimal designs are different than the optimal designs for means given in Spurrier and Nizam (1990). For the one-sided confidence intervals it turns out that for small n, $r_{d_01} \geqslant r_{d_00}$, while for large n, $r_{d_01} < r_{d_00}$. In the limiting case when $n \to \infty$, the optimal allocation is $r_{d_00}/r_{d_01} = \sqrt{v}$, the design of Theorem 2.1.

Sometimes the object of the experiment is solely to determine which, if any, of the test treatments perform better than the control, on the basis of the mean performance. For this purpose, Naik (1975) and Miller (1966, p. 85–86), proposed an alternative to Dunnett (1955)'s procedure that leads to considerable savings in terms of the total sample size. Let us assume that the model is given by (4.1) and (4.2). For $d \in \mathcal{D}^*(v+1, n)$, let $T_i = (\bar{y}_{di} - \bar{y}_{d0})/(s\sqrt{r_{di}^{-1} + r_{d0}^{-1}})$, $i = 1, \ldots, v$, and let the ordered T_i's be $T_{(1)} \leqslant \cdots \leqslant T_{(v)}$. Suppose treatment i is better than the control if $\mu_i > \mu_0$. The alternative procedure, known as the *step down test procedure*, is as follows. If $T_{(v)} \leqslant t(\alpha, v)$ (a $100\alpha\%$ critical value obtained from Bechhofer and Dunnett (1988)) then none of the test treatments are declared to be better than the control. If not, then the test treatment which corresponds to $T_{(v)}$ is declared to be better than the control and the procedure moves to the next step where $T_{(v-1)}$ is compared with the critical value $t(\alpha, v-1)$, and so on.

Hayter and Tamhane (1991) considered the design problem for the step down procedure. They called a design optimal if it achieved the smallest n among all designs, for fixed values of the probabilities of errors of types I and II, for a given *yardstick*. Hayter and Tamhane (1991) gave tables of optimal designs and demonstrated the savings over Dunnett's procedure.

4.2. Block designs

For the one way elimination of heterogeneity Model (2.4), Bechhofer and Tamhane (1981) was the first to consider the problem of finding optimal designs for simultaneous confidence intervals. Consider the problem of comparing v test treatments with a control in b blocks of size k each. As in Section 3, the set of designs is denoted by $\mathcal{D}(v+1,b,k)$. The observations y_{ijl} are independent random variables that follow the model:

$$y_{ijl} \sim N(\mu + \tau_i + \beta_j, \sigma^2).$$

Bechhofer and Tamhane (1981) started with a BTIB design d. Let,

$$\mathcal{D}^*(v+1,b,k) = \{d \in \mathcal{D}(v+1,b,k)\colon \ d \text{ is a BTIB design}\}.$$

For a BTIB design d, and $i, i' = 1, \ldots, v$, $i \neq i'$, let,

$$\psi_d^2 = \sigma^{-2}\mathrm{Var}(\widehat{\tau}_{di} - \widehat{\tau}_{d0}) = k(\lambda_{d0} + \lambda_{d1})/(\lambda_{d0}(\lambda_{d0} + v\lambda_{d1})),$$
$$\rho_d = \mathrm{Corr}(\widehat{\tau}_{di} - \widehat{\tau}_{d0}, \widehat{\tau}_{di'} - \widehat{\tau}_{d0}) = \lambda_{d1}/(\lambda_{d0} + \lambda_{d1}).$$

If one uses a BTIB design d then the simultaneous $100P\%$ lower confidence limits for $(\tau_i - \tau_0)$'s are:

$$\widehat{\tau}_{di} - \widehat{\tau}_{d0} - \delta\psi_d s, \quad i = 1, \ldots, v, \tag{4.7}$$

where s^2 is the mean square error, and δ is the $100(1-P)\%$ critical value for the multivariate, equicorrelated t (Dunnett and Sobel (1954), tables in Krishnaiah and Armitage (1966)). Simultaneous $100P\%$ two-sided confidence limits can be similarly defined.

We now give Bechhofer and Tamhane's definition of an optimal design. Let ξ be the *yardstick*, i.e.,

$$P = P(\tau_{di} - \tau_{d0} \geqslant \widehat{\tau}_{di} - \widehat{\tau}_{d0} - \xi, \ i = 1, \ldots, v). \tag{4.8}$$

Given $v, k, \xi/\sigma$ and α, a BTIB design d_0 is defined to be optimal if among all BTIB designs with $P \geqslant 1 - \alpha$, d_0 has the smallest b.

Determination of an optimal design is a difficult problem. As a first step, Bechhofer and Tamhane (1981) introduced a concept of admissibility that is helpful in removing those designs from consideration that can be *uniformly* improved upon by another design.

DEFINITION 4.1. For $i = 1, 2$, if d_i is a BTIB design in $\mathcal{D}^*(v+1, b_i, k)$, and $b_1 \leqslant b_2$, then the design d_2 is *inadmissible* with respect to d_1 (written $d_1 \succ d_2$) if d_1 yields a confidence coefficient P at least as large as (larger than) that yielded by d_2 when $b_1 < b_2$ ($b_1 = b_2$) for all values of ξ/σ. For a BTIB design d if there is no BTIB design d_1 such that $d_1 \succ d$, then d is *admissible.*

Bechhofer and Tamhane (1981) also gave the following simple characterization of the relation $\succ$.

THEOREM 4.2. *For* $i = 1, 2$, *if* d_i *is a* BTIB *design in* $\mathcal{D}^*(v+1, b_i, k)$ *then* $d_1 \succ d_2$ *if and only if*

$$b_1 \leqslant b_2, \quad \psi_{d_1}^2 \leqslant \psi_{d_2}^2, \quad \rho_{d_1} \geqslant \rho_{d_2},$$

with at least one inequality strict.

EXAMPLE 4.2. If $v = 4$ and $k = 3$ then for the following BTIB designs d_1 and d_2, $d_1 \succ d_2$.

$$d_1 = \begin{pmatrix} 0 & 0 & 0 & 0 & 1 & 1 & 2 \\ 1 & 1 & 2 & 0 & 2 & 3 & 3 \\ 2 & 4 & 4 & 3 & 3 & 4 & 4 \end{pmatrix}, \qquad d_2 = \begin{pmatrix} 0 & 0 & 0 & 0 & 1 & 1 & 2 & 3 \\ 1 & 1 & 2 & 4 & 2 & 4 & 4 & 4 \\ 2 & 3 & 3 & 4 & 3 & 4 & 4 & 4 \end{pmatrix}.$$

For two designs d_1 and d_2 the *union* $d_1 \cup d_2$ will denote a design that consists of all blocks of d_1 and d_2. It is interesting to note that if $d_1 \succ d_2$, then it is not necessarily true that for a BTIB design d_3, $d_1 \cup d_3 \succ d_2 \cup d_3$; see Bechhofer and Tamhane (1981, p. 51) for a counterexaınple.

For a given pair (v, k) it will be convenient to have a set of designs $\Delta(v, k)$, such that all admissible BTIB designs, except possibly some equivalent ones are obtainable from this set by the operation of union of designs. Among all sets $\Delta(v, k)$, a set with minimal cardinality is called a *minimal complete class of generator designs* for the pair (v, k). When $k = 2$, it is easily seen that there are only two designs in a minimal complete class: a $\mathrm{BTIB}(v, v, 2; 1, 0)$ and a $\mathrm{BTIB}(v, v(v-1)/2, 2; 0, 0)$. Similarly when $v = k = 3$, there are only two designs in a minimal complete class, a $\mathrm{BTIB}(3, 3, 3; 1, 0)$ and the design with only one block $\{1, 2, 3\}$. These are the only two simple cases – in general it is difficult to identify a minimal complete class of generator designs. Notz and Tamhane (1983) gave minimal complete classes for $k = 3$, $3 \leqslant v \leqslant 10$, and Ture (1982, 1985) for $k = 4$ and $k = 5$

Optimal designs for $k = 2$ and $v = 2, \ldots, 6$ were given by Bechhofer and Tamhane (1983b). Bechhofer and Tamhane (1985) compiled several tables of optimal, admissible and minimal complete classes of designs.

Spurrier and Edwards (1986) proposed a different definition of optimality of designs for simultaneous confidence intervals. The *expected average allowance* (the same criterion used by Spurrier and Nizam (1990), that was described earlier in this section) for the confidence limits given in (4.7) is,

$$\mathrm{EAA} = \delta' \psi_d \mathrm{E}(s).$$

Spurrier and Edwards (1986) defined a BTIB design to be optimal for a fixed level of significance $\alpha = 1 - P$ if it minimizes EAA among all designs in the class

$$\mathcal{D}^{**}(v+1, b^*, k) = \{d\colon\ d \in \mathcal{D}^*(v+1, b, k),\ \text{for some } b \leqslant b^*\},$$

where b^* is a fixed upper limit on the total number of blocks. They established that, in the limit as b^* goes to ∞, an optimal BTIB design is the union of a BTIB$(v, ab, k; t, 0)$ design and a BTIB$(v, (1-a)b, k; t+1, 0)$ design for some integer t and some $a \in (0, 1]$. Even though this is an asymptotic result, Spurrier and Edwards showed how to use it to find highly efficient, if not actually optimal, designs based on a finite number of blocks.

Kim and Stufken (1995) expanded the search of optimal designs to classes beyond BTIB designs. For fixed v, b, k and ξ/σ they found optimal designs within the class of GDTD's that maximized the coverage probability P in (4.8). They gave all optimal designs for $v = 4, k = 3$ and $b \leqslant 25$.

Kim and Stufken (1995) also found that their optimal designs correspond closely with the A-optimal GDTD's that were earlier obtained by Stufken and Kim (1992). This close relationship between optimal designs for estimation and simultaneous confidence intervals was earlier observed by Spurrier in the discussion of Hedayat et al. (1988).

The connection was further investigated by Majumdar (1996). In order to compare A-optimal designs for estimation with admissible designs for simultaneous confidence intervals, the definition of admissibility was adapted to designs in $\mathcal{D}^*(v+1, b, k)$ for a fixed b. A result in Majumdar (1996) states that if an A-optimal BTIB$(v, b, k; t, s)$ design d^*, that satisfies the sufficient conditions in Theorem 3.2, exists, then any BTIB$(v, b, k; x, z)$ design d is admissible if and only if $r_{d0} \leqslant r_{d^*0}$. It was also shown that, in this case the set of admissible designs is precisely the set of *Bayes optimal designs* for the class of priors given by (6.9). This established a connection between the three concepts: A-optimality, admissibility and Bayes optimality of designs.

5. Efficient row–column designs for estimation

Suppose, in addition to the blocking factor there is another factor, crossed with blocks, that makes the experimental units heterogeneous. The units can be conceptually arranged in an array, with rows and columns each representing a factor. Each cell in this array has one unit, and each unit receives a treatment, either a test treatment or the control. The primary object of the experiment is to compare the test treatments with the control.

Let the bk units be arranged in k rows and b columns. If the unit in row l $(1 \leqslant l \leqslant k)$ and column j $(1 \leqslant j \leqslant b)$ receives treatment i $(0 \leqslant i \leqslant v)$ then we assume that the observation y_{ijl} follows the model:

$$y_{ijl} = \mu + \tau_i + \beta_j + \gamma_l + \varepsilon_{ijl}, \tag{5.1}$$

where μ is a general effect, τ_i is a treatment effect, β_j is a column effect, γ_l is a row effect and ε_{ijl}'s are uncorrelated random variables with $\mathrm{E}(\varepsilon_{ijl}) = 0$, $\mathrm{V}(\varepsilon_{ijl}) = \sigma^2$.

A design d can be represented as a $k \times b$ array with entries from $\{0, 1, \ldots, v\}$. Let $\mathcal{D}^{RC}(v+1, b, k)$ denote the set of all connected designs. For $d \in \mathcal{D}^{RC}(v+1, b, k)$, let w_{dil} be the number of times treatment i appears in row l, and n_{dij} be the number of times treatment i appears in column j. Let $W_d = (w_{dil})$ be a $(v+1) \times k$ matrix and $N_d = (n_{dij})$ be a $(v+1) \times b$ matrix. W_d is the treatment-row incidence matrix and N_d is the treatment-column incidence matrix of the design d. For $i \neq i'$, let $\lambda_{dii'} = \sum_{j=1}^{b} n_{dij} n_{di'j}$ and $\mu_{dii'} = \sum_{l=1}^{k} w_{dil} w_{di'l}$. As usual, r_{di} denotes the replication of treatment i. The Fisher Information matrix for the treatments is:

$$C_d^{RC} = \text{Diag}(r_{d0}, r_{d1}, \ldots, r_{dv}) - \frac{1}{k} N_d N_d' - \frac{1}{b} W_d W_d' + \frac{1}{bk} \underline{r}_d \underline{r}_d',$$

where $\underline{r}_d$ is the $(v+1) \times 1$ vector of the replications. The information matrix for the treatment-control contrasts is, as in Section 3, the $v \times v$ submatrix M_d^{RC} obtained from C_d^{RC} by deleting the row and column that corresponds to the control, i.e., the first row and first column. An A-optimal design is one that minimizes $\text{tr}(M_d^{RC})^{-1}$, while an MV-optimal design is one that minimizes the maximum diagonal element of $(M_d^{RC})^{-1}$.

Notz (1985) investigated optimal designs in this setup. His result on A-optimality is given below.

THEOREM 5.1. *For $d \in \mathcal{D}^{RC}(v+1, b, k)$ the following holds:*

$$\text{tr}(M_d^{RC})^{-1} \geqslant \operatorname*{Min}_{0<r<bk} f^{**}(r)$$

where

$$\begin{aligned} f^{**}(r) &= v[r + r^2/bk - h^*(r)]^{-1} \\ &\quad + v(v-1)^{-1}[(v-1)(bk - r) - 2r^2/bk + h^*(r)]^{-1}, \\ h^*(r) &= [r + (2r - k)\lfloor r/k \rfloor - k\lfloor r/k \rfloor^2]/b + [r + (2r - b)\lfloor r/b \rfloor \\ &\quad - b\lfloor r/b \rfloor^2]/k. \end{aligned}$$

The following corollary of this theorem is an extended version of a corollary given by Notz (1985).

COROLLARY 5.1. *Suppose v is a square,*

$$k \equiv 0 \ \big(\bmod\ (v + \sqrt{v})\big) \quad \text{and} \quad b \equiv 0 \ \big(\bmod\ (v + \sqrt{v})\big).$$

For $i = 1, \ldots, k/(v+\sqrt{v})$ and $j = 1, \ldots, b/(v+\sqrt{v})$ let $\mathcal{L}_{ij}$ be latin squares of order $(v + \sqrt{v})$ each, which may be the same or different, with symbols $v+1, \ldots, v+\sqrt{v}$ replaced by 0. Let $d_0 = ((\mathcal{L}_{ij}))$ be a $k \times b$ array. Then d_0 is A-optimal for treatment-control contrasts in $\mathcal{D}^{RC}(v+1, b, k)$.

EXAMPLE 5.1. For $v = 4,\ b = k = 6$ an A-optimal design in $\mathcal{D}^{RC}(5, 6, 6)$ is:

$$\begin{array}{cccccc} 0 & 0 & 1 & 2 & 3 & 4 \\ 4 & 0 & 0 & 1 & 2 & 3 \\ 3 & 4 & 0 & 0 & 1 & 2 \\ 2 & 3 & 4 & 0 & 0 & 1 \\ 1 & 2 & 3 & 4 & 0 & 0 \\ 0 & 1 & 2 & 3 & 4 & 0 \end{array}$$

Notz also gave sufficient conditions when a design would attain the minimum in Theorem 5.1. One of the conditions is that, for the A-optimal design d_0, $M_{d_0}^{RC} = \overline{M_{d_0}^{RC}}$, i.e., the information matrix for the treatment-control contrasts should be completely symmetric. This condition can be viewed as the row–column counterpart of (2.5). Using the notation

$$\delta_{dii'} = \mu_{dii'}/b + \lambda_{dii'}/k - r_{di}r_{di'}/bk,$$

the condition can be expressed as:

$$\delta_{d01} = \cdots = \delta_{d0v} \quad (= \delta_{d0}, \text{ say}),$$

and

$$\delta_{d12} = \cdots = \delta_{d(v-1)v} \quad (= \delta_{d1}, \text{ say}).$$

A design that satisfies this condition was called a *Balanced Treatment Row–Column Design* (BTRCD) by Ture (1994), while Majumdar and Tamhane (1996) called it a *Balanced Treatment versus Control Row–Column* (BTCRC) design. If d is such a design with $\delta_{d0} > 0$ and $\delta_{d0} + v\delta_{d1} > 0$, then for $(i, i') \in \{1, \ldots, v\} \times \{1, \ldots, v\}$, $i \neq i'$,

$$\begin{aligned} &\mathrm{Var}(\widehat{\tau}_{di} - \widehat{\tau}_{d0}) = \sigma^2(\delta_{d0} + \delta_{d1})/(\delta_{d0}(\delta_{d0} + v\delta_{d1})), \\ &\mathrm{Corr}(\widehat{\tau}_{di} - \widehat{\tau}_{d0}, \widehat{\tau}_{di'} - \widehat{\tau}_{d0}) = \delta_{d1}/(\delta_{d0} + \delta_{d1}). \end{aligned} \tag{5.2}$$

It was noted in Majumdar (1986) that Corollary 5.1 can be alternatively derived from Theorem 3.1 by taking (2.1) for Model A and (5.1) for Model B. Jacroux (1986) applied Theorem 3.1 with (2.4) as Model A and (5.1) as Model B to get the following result.

THEOREM 5.2. *Let the $k \times b$ array d_0 be such that, with columns as blocks, d_0 is A-(MV-) optimal for treatment-control contrasts in $\mathcal{D}(v + 1, b, k)$ under model (2.4), and*

$$r_{di} \equiv 0 \ (\mathrm{mod}\ k), \quad i = 0, 1, \ldots, v. \tag{5.3}$$

Then there is a row–column design d_0^{RC} *with the same column contents as* d_0 *and* $w_{d_0il} = r_{di}/k$, $i = 0, 1, \ldots, v$. *The design* d_0^{RC} *is* A- (MV-) *optimal for treatment-control contrasts in* $\mathcal{D}^{RC}(v+1, b, k)$.

EXAMPLE 5.2. For $v = 9, b = 24$ and $k = 3$, the following design is A-optimal in $\mathcal{D}^{RC}(10, 24, 3)$:

0	4	1	0	2	4	5	7	3	0	9	6	0	0	8	0	7	9	1	6	2	3	4	5
1	0	5	8	0	2	0	0	5	3	0	4	4	6	6	9	0	7	2	1	3	8	7	9
3	1	0	1	4	0	2	2	0	7	3	0	9	5	0	6	8	0	9	7	6	4	5	8

Note that if a block design satisfies condition (5.3), then it can be converted to a row–column design with each treatment distributed uniformly over rows (i.e., $w_{d_0il} = r_{di}/k$, $i = 0, 1, \ldots, v$) by permuting symbols within blocks (columns). This is guaranteed by the results of Hall (1935), and Agrawal's (1966) generalization of Hall's results.

Hedayat and Majumdar (1988) noticed that the designs that are obtained in this fashion are *model robust* in the sense that they are simultaneously optimal under models (2.4) and (5.1). Several infinite families of model robust designs were given in Hedayat and Majumdar (1988). We give one example.

Start from a *Projective Plane,* which is a $\text{BIB}(s^2+s+1, s^2+s+1, s+1, s+1, 1)$ design, for some prime power s, that is constructed from a *difference set.* Write a difference set as the first column of the design using symbols $1, \ldots, s^2+s+1$. Obtain the remaining columns of the design by successively adding $1, \ldots, s^2+s$, to the first column, with the convention, $s^2+s+2 \equiv 1$. Now delete any one column, and in the rest of the design replace all symbols that appear in the deleted column by the symbol 0. The resulting design is A- and MV-optimal for treatment-control contrasts in $\mathcal{D}^{RC}(s^2+1, s^2+s, s+1)$. Designs generated by this method form the *Euclidean Family* of designs.

EXAMPLE 5.3. Starting from the *Fano Plane*, i.e., $\text{BIB}(7, 7, 3, 3, 1)$, we get the following member of the Euclidean Family that is optimal in $\mathcal{D}^{RC}(5, 6, 3)$:

1	0	3	4	2	0
0	3	4	2	0	1
4	2	0	0	1	3

All of the above methods for obtaining optimal designs utilize orthogonality in some form or the other. Without this, in general, it is very difficult to establish theoretical results that produce optimal row–column designs. An approach similar to Kiefer (1975) for the case of a set of orthonormal contrasts, has been attempted by Ting and Notz (1987). Ture (1994) started with the inequality

$$\sigma^{-2} \sum_{i=1}^{v} \operatorname{Var}(\widehat{\tau}_{di} - \widehat{\tau}_{d0}) \geqslant v(\delta_{\bar{d}0} + \delta_{\bar{d}1}) / (\delta_{\bar{d}0}(\delta_{\bar{d}0} + v\delta_{\bar{d}1}))$$

(see (5.2)), where $\bar{d}$ is the *symmetrized version* of d, i.e., $C_{\bar{d}} = \overline{C_d} = \frac{1}{v!}\sum_{\Pi} \Pi C_d \Pi'$, where the sum is over all permutation matrices Π that represent permutations of the test treatments only. (Note that while there may be no design $\bar{d}$ with Fisher Information matrix for the treatments $C_{\bar{d}}$, the matrix is well defined for any design d.) Using theoretical and computational results Ture (1994) gave two tables, one of A-optimal designs in the range $2 \leqslant v \leqslant 10$, $2 \leqslant k \leqslant 10$, $k \leqslant b \leqslant 30$ and one of designs, in the same range, that he conjectures are optimal.

For the setup of Example 5.1, suppose there are only 3, not 4, test treatments to be compared with the control. What is an optimal design? Ture (1994) provides the following answer.

EXAMPLE 5.4. For $v = 3$, $b = k = 6$ an A-optimal design in $\mathcal{D}^{RC}(4, 6, 6)$ is:

$$\begin{array}{cccccc}
0 & 0 & 1 & 2 & 3 & 1 \\
1 & 0 & 0 & 2 & 3 & 2 \\
2 & 1 & 3 & 0 & 0 & 3 \\
3 & 2 & 2 & 1 & 0 & 0 \\
0 & 3 & 2 & 3 & 1 & 0 \\
1 & 1 & 0 & 0 & 2 & 3
\end{array}$$

Several combinatorial techniques for constructing BTCRC designs were given in Majumdar and Tamhane (1996). The following example illustrates one method.

EXAMPLE 5.5. Start from a 4×4 Latin square in symbols $1, 2, 3$ and 4, with two *parallel transversals*. Replace the first transversal with controls, 0. Then use the Hedayat and Seiden (1974) method of *sum composition* to *project* the other transversal. This procedure is illustrated in the following sequence of designs. The second design in the sequence is a BTCRC design in $\mathcal{D}^{RC}(5, 4, 4)$, while the last design is a BTCRC design in $\mathcal{D}^{RC}(5, 5, 5)$:

$$\begin{array}{cccc}
1 & 2 & 3 & 4 \\
3 & 4 & 1 & 2 \\
4 & 3 & 2 & 1 \\
2 & 1 & 4 & 3
\end{array}
\longrightarrow
\begin{array}{cccc}
1 & 2 & 0 & 4 \\
3 & 4 & 1 & 0 \\
0 & 3 & 2 & 1 \\
2 & 0 & 4 & 3
\end{array}
\longrightarrow
\begin{array}{ccccc}
0 & 2 & 0 & 4 & 1 \\
3 & 0 & 1 & 0 & 4 \\
0 & 3 & 0 & 1 & 2 \\
2 & 0 & 4 & 0 & 3 \\
1 & 4 & 2 & 3 & 0
\end{array}$$

6. Bayes optimal designs

How can we utilize prior information at the designing stage? In his pioneering work, Owen (1970) studied Bayes optimal block designs for comparing treatments with a control. It is his setup that will form the basis of this section. Even though the main focus of this section is block designs, row–column designs, as well as designs for the zero-way elimination of heterogeneity model will be reviewed briefly.

6.1. Bayes optimal continuous block designs

Suppose the observations follow the one-way elimination of heterogeneity model (2.4). Let

$$\theta_i = \tau_i - \tau_0, \quad i = 1, \ldots, v, \ \theta_0 = 0,$$
$$\gamma_j = \mu + \tau_0 + \beta_j, \quad j = 1, \ldots, b.$$

The θ_i's are the treatment-control contrasts that we wish to estimate, while γ_j is the expected performance of the control in block j. Model (2.4) can be written as:

$$y_{ijl} = \theta_i + \gamma_j + \varepsilon_{ijl}.$$

Suppose there is a total of n observations. Let Y denote the $n \times 1$ vector of observations, ε denote the $n \times 1$ vector of errors, $\theta' = (\theta_1, \ldots, \theta_v)$, $\gamma' = (\gamma_1, \ldots, \gamma_b)$. Then the model can be expressed as:

$$Y = X_{1d}\theta + X_2\gamma + \varepsilon,$$

where X_2 is a known matrix, and X_{1d} is a known matrix that depends on the design d. In the Bayesian approach, θ, γ and ε are all assumed random. Specifically we shall assume the following model for the conditional distribution:

$$Y \mid \theta, \gamma \sim N_n(X_{1d}\theta + X_2\gamma, E), \tag{6.1}$$

for some covariance matrix E, where N_n denotes the n-*variate* normal distribution. The assumed prior is:

$$\begin{pmatrix} \theta \\ \gamma \end{pmatrix} \sim N_{v+b}\left(\begin{pmatrix} \mu_\theta \\ \mu_\gamma \end{pmatrix}, \begin{pmatrix} B^* & O \\ O & B \end{pmatrix}\right) \tag{6.2}$$

for some vectors μ_θ, μ_γ and matrices B^*, B.

From (6.1) and (6.2) it follows that the posterior distribution of θ is:

$$\theta \mid Y \sim N_v(\mu_d^*, D_d), \tag{6.3}$$

where

$$D_d^{-1} = X_{1d}'(\mathrm{E} + X_2BX_2')^{-1}X_{1d} + B^{*-1}, \tag{6.4}$$

and

$$D_d^{-1}\mu_d^* = X_{1d}'(\mathrm{E} + X_2BX_2')^{-1}(Y - X_2\mu_\gamma) + B^{*-1}\mu_\theta.$$

For the loss function $L(\widehat{\theta}, \theta) = (\widehat{\theta} - \theta)'W(\widehat{\theta} - \theta)$, where W is a positive definite matrix, the Bayes estimator of θ is $\widehat{\theta} = \mu_d^*$, with expected loss $\mathrm{tr}(WD_d)$. Owen (1970) developed a method to find a design that minimizes $\mathrm{tr}(WD_d)$ in

$$\mathcal{D}(v+1, b, k_1, \ldots, k_b) = \{d\colon\ d \text{ is a design based on } v+1 \text{ treatments in } b \text{ blocks of sizes } k_1, \ldots, k_b\}. \tag{6.5}$$

Note that in this subsection we do not assume that the blocks are of the same size. The total number of observations is $n = \sum_{j=1}^{b} k_j$. In order to fix the order of observations in the vector Y in (6.1), for $d \in \mathcal{D}(v+1, b, k_1, \ldots, k_b)$, we shall write, $X_2 = \mathrm{diag}(1_{k_1}, \ldots, 1_{k_b})$, a block-diagonal matrix. We will focus on the special case, $W = I$.

DEFINITION 6.1. Given the matrices E, B and B^*, a design that minimizes $\mathrm{tr}(D_d)$ will be called a *Bayes* A-*optimal* design.

Owen (1970) took the *approximate theory* approach, i.e., the incidences n_{dij}'s are viewed as real numbers, rather than integers. Giovagnoli and Wynn (1981) called these designs *continuous block designs*. The problem reduces considerably for a certain class of priors. This is stated in the following result of Owen (1970).

THEOREM 6.1. *Suppose the error covariance matrix is of the form:*

$$E = X_2 \widehat{E} X_2' + \mathrm{diag}(e_1, \ldots, e_n),$$

where $e_i > 0$, *for* $i = 1, \ldots, n$, *and* $\widehat{E} = (e_{ij})$ *is a symmetric matrix. Also, suppose,*

$$B^* = \sigma^2((\xi_1 - \xi_2)I_v + \xi_2 J_v), \tag{6.6}$$

where $\sigma^2 > 0$ *and* $(v-1)^{-1} < \xi_2/\xi_1 < 1$, I_v *is the identity matrix of order* v *and* J_v *is a* $v \times v$ *matrix of unities. Moreover, suppose* $B + \widehat{E}$ *is a nonnegative definite matrix. For designs with fixed* n_{d0j}, $j = 1, \ldots, b$, *the optimal continuous block design has* $n_{dij} = (k_j - n_{d0j})/v$, *for all* $i = 1, \ldots, v$, *and each* $j = 1, \ldots, b$.

It may be noted that Owen established Theorem 6.1 under a condition more general than the condition, $B + \widehat{E}$ is nonnegative definite. We use the latter in Theorem 6.1 in order to avoid more complicated technical conditions, and also since this condition would be satisfied by a large class of priors. The theorem says that for the optimal continuous design,

$$n_{d1j} = \cdots = n_{dvj} \quad \text{for } j = 1, \ldots, b. \tag{6.7}$$

In view of this it remains to determine only the n_{d0j}'s that minimize $\mathrm{tr}(D_d)$. Owen gave an algorithm for this, studied some special cases and gave an example.

Instead of the trace, one can minimize other functions of D_d in order to obtain Bayes optimal designs. Taking this approach, Giovagnoli and Verdinelli (1983) considered a general class of criteria for optimality, with special emphasis on D-, and E-optimality. Verdinelli (1983) gave methods for computing Bayes D- and A-optimal designs.

Optimal designs under the *hierarchical* model of Lindley and Smith (1972) were determined by Giovagnoli and Verdinelli (1985). The model is:

$$
\begin{aligned}
&Y \mid \theta, \gamma \sim N_n(A_{11d}\theta + A_{12}\gamma, E_{11}),\\
&\begin{pmatrix}\theta\\ \gamma\end{pmatrix} \mid \theta_1, \gamma_1 \sim N_{v+b}\left(\begin{pmatrix}A_{21}\theta_1\\ A_{22}\gamma_1\end{pmatrix}, \begin{pmatrix}E_{21} & O\\ O & E_{22}\end{pmatrix}\right),\\
&\begin{pmatrix}\theta_1\\ \gamma_1\end{pmatrix} \sim N_{v+b}\left(\begin{pmatrix}A_{31}\theta_2\\ A_{32}\gamma_2\end{pmatrix}, \begin{pmatrix}E_{31} & O\\ O & E_{32}\end{pmatrix}\right).
\end{aligned}
$$

Giovagnoli and Verdinelli (1985) extended Owen's results to this model. It may be noted that this work followed that of Smith and Verdinelli (1980) who investigated Bayes optimal designs for a hierarchical model with no block effects, i.e., the hierarchical model corresponding to (2.1). Smith and Verdinelli's model is:

$$
Y \mid \theta \sim N(A_{1d}\theta, E_1), \quad \theta \mid \theta_1 \sim N(A_2\theta_1, E_2), \quad \theta_1 \sim N(\theta_2, E_3).
$$

6.2. *Bayes optimal exact designs*

The approximate, or continuous, block design theory is wide in its scope. As the theorems in this approach specify proportions of units which are assigned to each treatment, they give an overall idea of the nature of optimal designs. Also, one rule applies to (almost) all block sizes, and requires only minor computations when the block sizes are altered. These are very desirable properties. On the other hand, application of these designs is possible only after rounding off the treatment-block incidences to nearby integers, and this could result in loss of efficiency. Moreover, it follows from the restriction (6.7) that the known optimal designs cannot be obtained unless the block sizes are somewhat large. In particular, optimal designs cannot be obtained for the incomplete block setup, i.e., $k \leqslant v$, which is encountered very often in practice.

Let us return to the *exact* optimality theory, i.e., retain the natural domain of nonnegative integers for the incidences n_{dij} instead of approximating these by real numbers. In Majumdar (1992), the model given by (6.1) and (6.2) was considered along with additional properties (6.6) and

$$
\begin{aligned}
&B = \sigma^2\delta((1-\rho)I_b + \rho J_b), \quad \mathrm{Var}(\varepsilon_{ijl}) = \sigma^2,\\
&\mathrm{Cov}(\varepsilon_{ijl}, \varepsilon_{i'jl'}) = \sigma^2\pi_1, \quad \text{for } l \neq l',\\
&\mathrm{Cov}(\varepsilon_{ijl}, \varepsilon_{i'j'l'}) = \sigma^2\pi_2, \quad \text{for } j \neq j'.
\end{aligned}
\tag{6.8}
$$

The class of designs was $\mathcal{D}(v+1, b, k)$ with $2 \leqslant k \leqslant v$. (As mentioned above, Owen's result cannot be applied here since (6.7) implies that k is at least $v+1$.) It was shown that under the conditions, $1 > \pi_1 \geqslant \pi_2$, $\pi_2 + \delta\rho \geqslant 0$, $\xi_2 \geqslant 0$ a $\mathrm{BTIB}(v,b,k;t,s)$ is Bayes A-optimal, where (t,s) is a point where a function $g_B(x,z)$, which depends on v, b, k and the prior parameters, attains a minimum over a set similar to Λ in (3.8).

The prior structure for the θ's in (6.8) is the same as that assumed by Owen (1970), but the γ's are assumed to be exchangeable and ε's are assumed exchangeable within blocks. (Actually, these variables may be more general than exchangeable random variables, since no model for the expectations is required for the design problem.) This class of priors is, therefore, somewhat less general than Owen's class of priors given in Theorem 6.1.

We will not reproduce the exact statement of the theorem since it is notationally involved, but instead, discuss a simpler special case which was considered in Majumdar (1992), and earlier in Majumdar (1988b) and Stufken (1991b).

The model for the special case is given by (6.1), (6.2), and

$$E = \sigma^2 I_{bk}, \quad B^{*-1} = O, \text{ the null matrix}, \quad B = \sigma^2\alpha^{-1} I_b, \tag{6.9}$$

i.e., the errors are homoscedastic, the prior on θ is vague and the γ_j's are i.i.d. Also, $\alpha \in [0,\infty)$; $\alpha = 0$ corresponds to the case where no prior on γ is incorporated in the design.

While this prior may be limited in scope, our main purpose is to demonstrate the nature of optimal designs in a simple setup. It is expected that the general nature of the optimal design will be similar when the priors are chosen to be more general, for example, those given by (6.8). Let

$$k_\alpha = k + \alpha.$$

It follows from (6.4) that the inverse of the posterior covariance matrix for θ is given by,

$$D_d^{-1}(\alpha) = \mathrm{Diag}(r_{d1},\ldots,r_{dv}) - \frac{1}{k_\alpha}\widetilde{N}_d\widetilde{N}_d',$$

where $\widetilde{N}_d$ is a $v \times b$ matrix obtained from the incidence matrix N_d by deleting the row corresponding to the control (the first row). For $\alpha > 0$, let,

$$\begin{aligned}
&\Lambda(\alpha) = \{(x,z)\colon\ x = 0,\ldots,k - \lfloor (k_\alpha+1)/2 \rfloor - 1;\, z = 0,1,\ldots,b\},\\
&T = bk - (bx+z), \quad S = bk^2 - 2k(bx+z) + (bx^2 + 2xz + z),\\
&g_\alpha(x,z) = k_\alpha v(v-1)^2\big[(k_\alpha(v-1) - v)T + S\big]^{-1} + k_\alpha v[k_\alpha T - S]^{-1}.
\end{aligned}$$

For $\alpha = 0$, let $\Lambda(0) = \Lambda$, the set in given (3.8), and $g_0(x,z) = g(x,z)$, the function given in (3.7).

In addition to Bayes optimal designs we will also consider Γ-minimax optimal designs. Given a set Γ, a Γ-minimax rule (see Berger, 1985) minimizes the maximum risk for the prior parameters in the set Γ. This is a robust decision rule. A Γ-minimax optimal design is one that minimizes the maximum risk of the Γ-minimax rule. When Γ is the entire space, the Γ-minimax rule is the minimax rule.

DEFINITION 6.2. Given a set Γ, d_0 is called a Γ-minimax optimal design in $\mathcal{D}(v+1, b, k)$ if, for D_d given by (6.4),

$$\operatorname*{Max}_{E,B,B^*\in\Gamma} \operatorname{tr}(D_{d_0}) = \operatorname*{Min}_{d\in\mathcal{D}(v+1,b,k)} \operatorname*{Max}_{E,B,B^*\in\Gamma} \operatorname{tr}(D_d).$$

THEOREM 6.2. *Consider the class $\mathcal{D}(v+1,b,k)$ with $2 \leqslant k \leqslant v$. Let a design d_0 be defined as:* (i) *if $k_\alpha + 1 < 2k$, d_0 is a* BTIB$(v,b,k;t,s)$ *where $g_\alpha(t,s) = \operatorname{Min}_{(x,z)\in\Lambda(\alpha)} g_\alpha(x,z)$,* (ii) *if $k_\alpha + 1 \geqslant 2k$, d_0 is a* BIB *design in $\mathcal{D}(v,b,k)$ based on the test treatments. Then the following hold:*

(A) *d_0 is Bayes* A-*optimal in $\mathcal{D}(v+1,b,k)$ for the model given by* (6.1), (6.2) *and* (6.9).

(B) *d_0 is a Γ-minimax optimal design for the model* (6.1) *and* (6.2), *with $\Gamma = \{(\mathrm{E},B,B^*)$: $\mathrm{E} = \sigma^2 I$, and $\sigma^2\alpha^{-1}I - B$ is nonnegative definite$\}$.*

Theorem 6.2 may be found in Majumdar (1992). The following theorem, due to Stufken (1991b), gives families of Bayes A-optimal designs. When $\alpha = 0$, this is Theorem 3.4.

THEOREM 6.3. *For $2 \leqslant k \leqslant v$ and α integral, a* BTIB$(v,b,k;k-q,0)$ *is Bayes* A-*optimal in $\mathcal{D}(v+1,b,k)$ for the model given by* (6.1), (6.2) *and* (6.9) *if $(q-1)^2+1 \leqslant (k_\alpha - q)^2 v \leqslant q^2$.*

EXAMPLE 6.1. A BTIB$(v,b,7;1,0)$ is Bayes A-optimal for $7 \leqslant v \leqslant 9$ when $\alpha = 1$, i.e., when the prior variance of the γ_j's is equal to σ^2.

What if α was non-integral? A rigorous analysis for this case is difficult, but Stufken (1991b) *anticipates* that Theorem 6.3 would remain valid. If this is true, then Theorem 6.3 would give, for fixed v, b and k a range of α for which a BTIB$(v,b,k;t,0)$ is optimal, i.e., an optimal design that is *robust* against specification of the prior.

A robustness result using a different approach is given in Majumdar (1992).

THEOREM 6.4. *Consider the class $\mathcal{D}(v+1,b,k)$ with $2 \leqslant k \leqslant v$, and $\alpha \in [0,\infty)$. Let r^* denote the optimal replication of the control given by Theorem* 3.2*. *For $i = 0,\ldots,r^*-1$, let δ_{i+1} be the (unique) solution to the following equation in $\delta = \alpha^{-1}$:*

$$g_\alpha(\lfloor i/b\rfloor,\ i - b\lfloor i/b\rfloor) = g_\alpha(\lfloor i/b\rfloor,\ i - b\lfloor i/b\rfloor + 1),$$

and $\delta_0 = 0$, $\delta_{r^*+1} = \infty$. Then,

(a) *the δ_i are increasing.*

(b) *If i_0 is an integer in $0 \leqslant i_0 \leqslant r^*$ such that $\alpha^{-1} \in [\delta_{i_0}, \delta_{i_0+1})$, then a* BTIB$(v,b,k;t,s)$ *with $bt + s = i_0$, is Bayes* A-*optimal for the model given by* (6.1),

(6.2) *and* (6.9), *as well as* Γ*-minimax optimal for the model* (6.1) *and* (6.2), *with* Γ *as described in Theorem* 6.2.

Let us denote a BTIB$(v, b, k; t, s)$ with $bt + s = i$ by d_i. Theorem 6.4 gives a partition of $[0, \infty)$ consisting of $r^* + 1$ intervals such that for all α^{-1} in the interval $[\delta_i, \delta_{i+1})$, the design d_i is Bayes A-optimal and Γ-minimax optimal. Thus, the value of α (and hence the prior variance of the γ_j's) need not be specified exactly at the designing stage. The design d_i will not, however, exist for all $i = 0, \ldots, r^*$. Suppose i^* is such that d_{i^*} does not exist. Let $i_1$and i_2 be such that $i^* \in (i_1, i_2)$, d_{i_1}, d_{i_2} exist and further, no d_i with $i \in (i_1, i_2)$ exists. Then at least one of the two designs, d_{i_1} or d_{i_2}, is likely to be *highly efficient* (see Section 7 for a general definition of efficiency). Here is a simple example.

EXAMPLE 6.2. Let $v = 3$, $b = 24$ and $k = 2$. This example has been studied in some detail in Majumdar (1992). Here are some observations. A BTIB$(3, 24, 2; 0, 15)$ is Bayes A-optimal for $\alpha = 0.18$, i.e., when the prior variance of the γ_j's (see (6.9)) is equal to $5.5\sigma^2$, as well as for $\alpha = 0.15$, i.e., when the prior variance of the γ_j's is equal to $6.5\sigma^2$ (hence for all prior variances in the interval $[5.5\sigma^2, 6.5\sigma^2]$). For $\alpha = 0.1$, i.e., when the prior variance of the γ_j's is equal to $10\sigma^2$, the corresponding d_i is a BTIB$(3, 24, 2; 0, 16)$ which does not exist; nevertheless the design BTIB$(3, 24, 2; 0, 15)$ is highly (at least 99.6%) efficient. An A-optimal design given by Theorem 3.2 is BTIB$(3, 24, 2; 0, 18)$. This design is Bayes A-optimal for all $\alpha \in (0, .033)$, i.e., whenever the prior variance of the γ_j's is greater than $30\sigma^2$.

Ting and Yang (1994) generalized some of the results in this section to classes $\mathcal{D}(v + 1, b, k)$ with $k > v$.

6.3. *Bayes optimal row–column designs*

Optimal designs for the row–column setup were first considered by Toman and Notz (1991). Suppose $\theta_i = \tau_i - \tau_0$, $\theta' = (\theta_1, \ldots, \theta_v)$, $\beta' = (\beta_1, \ldots, \beta_b)$, and $\gamma' = (\gamma_1, \ldots, \gamma_k)$ in (5.1). The model is:

$$Y \mid \tau_0, \theta, \beta, \gamma \sim N(X_{1d}(\tau_0, \theta')' + X_2\beta + X_3\gamma, \sigma^2 I),$$
$$(\tau_0, \theta', \beta', \gamma')' \sim N\big((\mu_0, \mu_\theta', \mu_\beta', \mu_\gamma')', \ \mathrm{Diag}(a, T, B, G)\big),$$

where $\mathrm{Diag}(a, T, B, G)$ is a block diagonal matrix. Toman and Notz gave a procedure for obtaining Bayes optimal *continuous* designs. They focused on the special case where T, B and G were completely symmetric. They also discussed rounding-off strategies for continuous designs.

Bayes optimal row–column designs in the exact theory setup was considered in Majumdar (1992). The model was formulated differently. It was (6.1) and (6.2), with $\theta_i = \tau_i - \tau_0$, $\gamma_{jl} = \mathrm{E}(y_{0jl})$, $\theta' = (\theta_1, \ldots, \theta_v)$, and $\gamma' = (\gamma_{11}, \ldots, \gamma_{bk})$. Exact Bayes optimal designs were derived for a certain class of priors.

For a general survey of Bayes optimal designs that is not restricted to treatment-control designs, see Dasgupta (1996).

7. On efficiency bounds of designs

An (exact) optimal design always exists since the set of designs is finite. Identification of an optimal design may, however, be a very difficult problem. The experimenter, in this case, might look for a design that is *close* to an optimal design. In other experiments, an optimal design may have a feature that is not desirable to the experimenter, or there may be experimental restrictions that suggest the use of a design that is known to be not optimal. In such cases it would be very useful to compute accurately the efficiency of a design.

Given a class $\mathcal{D}$ of designs and a criterion function $\Phi(d)$ (suppose Φ is such that the smaller its value the better the design), the efficiency of a design $d_1 \in \mathcal{D}$ may be defined as:

$$e(d_1) = \operatorname*{Min}_{d \in \mathcal{D}} \Phi(d)/\Phi(d_1). \tag{7.1}$$

If a Φ-optimal design in $\mathcal{D}$ is known then it is easy to compute $e(d_1)$. If not, then evaluation of the numerator in (7.1) is a difficult problem. If a lower bound Φ_L to the numerator is known, i.e.,

$$\Phi(d) \geqslant \Phi_L, \quad \text{for all } d \in \mathcal{D}, \tag{7.2}$$

then

$$e(d_1) \geqslant \Phi_L/\Phi(d_1).$$

This gives us a lower bound to the efficiency, which will be denoted by,

$$e_L(d_1) = \Phi_L/\Phi(d_1). \tag{7.3}$$

In spite of the risk of confusion, e_L will also be called the efficiency.

Researchers in the area of design have computed the efficiency of designs in various settings (see, e.g., Roy, 1958). In most cases a lower bound Φ_L is not difficult to find, but clearly, e_L is a meaningful measure only when the inequality (7.2) is sharp.

Let us start with the model (2.4) in the block design setup and $\Phi(d) = \text{tr}(M_d^{-1})$, i.e., the A-criterion. Theorem 3.2* or 3.7* gives a lower bound Φ_L for this case. (Even though Theorem 3.2* can be viewed as a special case of Theorem 3.7*, we will display the two efficiencies separately.) Thus, for $d \in \mathcal{D}(v+1, b, k)$,

$$e_L(d) = \begin{cases} g^*(r^*)/\text{tr}(PC_d^- P') = g^*(r^*)/\text{tr}(M_d^{-1}), & \text{if } k \leqslant v, \\ f^*(r^*)/\text{tr}(PC_d^- P') = f^*(r^*)/\text{tr}(M_d^{-1}), & \text{if } k > v. \end{cases} \tag{7.4}$$

In view of Theorem 3.6, Cheng et al. (1988) suggested two ways to obtain a design that is highly efficient in $\mathcal{D}(v+1, b, k)$ in case an optimal design is not known. (1) Find a *good* BTIB design with r_{d0} *close* to r^*, or (2) find a design *combinatorially* *close* to a *good* BTIB design with $r_{d0} = r^*$. By a *good* BTIB design we mean a BTIB$(v, b, k; t, s)$ design, or a BTIB design that is combinatorially close to it. Cheng et al. (1988) also gave a procedure for identifying highly efficient designs, that is based on some techniques developed and discussed in Ture (1982). Some highly efficient designs are also reported in Ture (1985).

Here are a few examples of highly efficient designs.

EXAMPLE 7.1. Let $v = 5$, $k = 4$ and $b = 7$. This example is studied in some detail in Ture (1982) and in Cheng et al. (1988). Here $r^* = 7$, $g^*(r^*) = 2.04$. There is no BTIB$(5, 7, 4; 1, 0)$, so Theorem 3.2 cannot be applied to get an optimal design. Here are two possible designs.

$$d_1 = \begin{pmatrix} 0 & 0 & 0 & 0 & 0 & 0 & 1 \\ 0 & 0 & 1 & 1 & 2 & 2 & 2 \\ 1 & 3 & 3 & 4 & 3 & 4 & 3 \\ 2 & 4 & 5 & 5 & 5 & 5 & 4 \end{pmatrix},$$

$$d_2 = \begin{pmatrix} 0 & 0 & 0 & 0 & 0 & 0 & 0 \\ 1 & 1 & 1 & 1 & 1 & 2 & 2 \\ 2 & 2 & 3 & 3 & 4 & 3 & 3 \\ 4 & 5 & 4 & 5 & 5 & 4 & 5 \end{pmatrix}.$$

d_1 is a BTIB design with $r_{d_1 0} = 8$, while d_2 is not a BTIB design; $r_{d_2 0} = 7$. Also, $\mathrm{tr}(M_{d_1}^{-1}) = 2.143$, $\mathrm{tr}(M_{d_2}^{-1}) = 2.058$. Hence, $e_L(d_1) = 95.2\%$ and $e_L(d_2) = 99.2\%$. Both designs are highly efficient, though d_2 is clearly the better of the two.

EXAMPLE 7.2. Let $v = 4$, $b = 4$ and $k = 7$. Here $r^* = 9$ and $f^*(r^*) = 1.3128$. There is no design with supplemented balance with nine replications of the control. The design of Pearce (1953) given in Example 2.1 (call it d_1) has $r_{d_1 0} = 8$ and $\mathrm{tr}(M_{d_1}^{-1}) = 1.3188$; hence $e_L(d_1) = 99.5\%$. Pearce's design is, therefore, highly efficient. An interesting question is: is there a design d_2 with $r_{d_2 0} = 9$ that has even higher efficiency? For

$$d_2 = \begin{pmatrix} 0 & 0 & 0 & 0 \\ 0 & 0 & 0 & 0 \\ 1 & 1 & 1 & 1 \\ 2 & 2 & 2 & 2 \\ 3 & 3 & 3 & 3 \\ 4 & 4 & 4 & 4 \\ 0 & 1 & 2 & 3 \end{pmatrix},$$

$\mathrm{tr}(M_{d_2}^{-1}) = 1.3181$, so $e_L(d_2) = 99.6\%$. In view of the very high efficiency of Pearce's design, the additional property of symmetry of this design, together with the very minor improvement achieved by the design d_2, the design d_1 would be preferable to d_2 in most experiments.

More examples of highly efficient designs with supplemented balance in classes $\mathcal{D}(v+1, b, k)$ with $k > v$, are available in Jacroux and Majumdar (1989).

An alternative approach to identifying efficient designs is to restrict to a subclass of all designs. If the subclass is sufficiently *large*, then this method is expected to yield designs that are *close* to an optimal design. The method also has the advantage that the chosen design is guaranteed to possess any property that is shared by all designs in the subclass. Examples of subclasses are BTIB designs or GDTD's in $\mathcal{D}(v+1, b, k)$. Hedayat and Majumdar (1984) gave a catalog of designs that are A-optimal in the class of BTIB designs for $k = 2$, $2 \leqslant v \leqslant 10$ and $v \leqslant b \leqslant 50$. Jacroux (1987b) obtained a similar catalog of designs for the MV-optimality criterion.

Stufken (1988) investigated the efficiency of the *best* reinforced BIB design (Das, 1958; Cox, 1958), i.e., a BTIB$(v, b, k; t, 0)$ which has the smallest value of the A-criterion. He established analytic results, using which he evaluated e_L for each k in the interval $[3, 10]$ and all v. Here is an example.

EXAMPLE 7.3. Let $k = 4$. For a b for which it exists, let d denote a BTIB$(v, b, 4; 1, 0)$. For $5 \leqslant v \leqslant 9$, it follows from Theorem 3.3 that d is A-optimal. Stufken (1988) shows that if $v = 4$, $e_L(d) \geqslant 99.99\%$, and if $v \geqslant 10$,

$$e_L(d) \geqslant (3v-1)\left(v-1+\sqrt{v+1}\right)^2 / \left(4v^2(v+1)\right).$$

Thus if $v = 10$, $e_L(d) \geqslant 99.98\%$, if $v = 20$, $e_L(d) \geqslant 97.65\%$, and if $v = 100$, $e_L(d) \geqslant 88\%$.

Stufken's study strengthens the belief that the subclass of BTIB designs in $\mathcal{D}(v+1, b, k)$ usually contains highly efficient designs, with exceptions more likely when v is large compared to k. The reason seems to be that due to its demanding combinatorial structure, sometimes it is impossible to find a BTIB design d with r_{d0} in the vicinity of r^*. Here is an example.

EXAMPLE 7.4. Let $v = 10$, $b = 80$ and $k = 2$. Here $r^* = 39$ and $g^*(r^*) = 1.896$. There is no BTIB design with 39 replications of the control, so Theorem 3.2 does not give an optimal design. Let $d^{(1)}$ be a BTIB$(10, 10, 2; 1, 0)$ and $d^{(2)}$ be a BIB$(10, 45, 9, 2, 1)$ design in the test treatments only. Further let $d^{(3)} \in \mathcal{D}(10, 40, 2)$ be a BIB$(10, 45, 9, 2, 1)$ design in the test treatments with the blocks $\{1, 6\}, \{2, 7\}, \{3, 8\}, \{4, 9\}$ and $\{5, 10\}$ deleted, $d^{(4)} \in \mathcal{D}(10, 41, 2)$ be the design $d^{(3)} \cup \{1, 6\}$, and $d^{(5)}$ be the design $d^{(1)}$ with the block $\{0, 10\}$ deleted. Hedayat and Majumdar (1984) showed that an A-optimal design in the subclass of BTIB designs within $\mathcal{D}(11, 80, 2)$ is $d_0 = 8d^{(1)}$. For this design, $r_{d_0 0} = 80$, $\mathrm{tr}(M_{d_0}^{-1}) = 2.5$, hence $e_L(d_0) = 75.8\%$. Now consider two other designs in $\mathcal{D}(11, 80, 2)$: $d_1 = 4d^{(1)} \cup d^{(3)}$ and $d_2 = 3d^{(1)} \cup d^{(4)} \cup d^{(5)}$. It is easy to see that, $r_{d_1 0} = 40$, $\mathrm{tr}(M_{d_1}^{-1}) = 1.904$, $e_L(d_1) = 99.6\%$, and $r_{d_2 0} = 39$, $\mathrm{tr}(M_{d_2}^{-1}) = 1.905$, $e_L(d_2) = 99.5\%$. Neither d_1 nor

d_2 are BTIB designs – d_1 is a GDTD, d_2 is not. Both of these designs are highly efficient, and both are about 24% more efficient that the best BTIB design d_0 in $\mathcal{D}(11, 80, 2)$.

Since the class of GDTD's is larger than the class of BTIB designs, the best design in the former class will perform at least as well as the best design in the latter class, and, as in Example 7.4, sometimes substantially better. Stufken and Kim (1992) gave a complete listing of designs that are A-optimal in the class of GDTD's for $k = 2, 3$, $k \leqslant v \leqslant 6$ and $b \leqslant b_0$ where b_0 is 50 or more.

In this context it may be noted that all of the optimal designs for simultaneous confidence intervals in Section 4 were optimal within certain subclasses of designs.

A general method of obtaining a lower bound Φ_L to Φ is to use the principle of Theorem 3.1. Consider the models A and B in (3.2). For a design d if $\Phi^J(d)$ denotes the value of the criterion based on model J ($J = \text{A}, \text{B}$), and if the criterion $\Phi(d) = \psi(C_d)$ has the property $\psi(H - G) \geqslant \psi(H)$, whenever H, G, $H - G$ are nonnegative definite, then $\Phi^B(d) \geqslant \Phi^A(d)$. Since A is a simpler model it is usually easier to obtain a lower bound to the criterion under Model A. This is an old technique in design theory. The 0-way elimination of heterogeneity model is used to compute the efficiency of designs under the 1-way elimination of heterogeneity model, the 1-way for the 2-way elimination of heterogeneity model and so on.

In Majumdar and Tamhane (1996) this method has been used to compute the efficiency of row–column designs. To describe the results we need some notation. For the class of designs with b blocks of size k each, i.e., $\mathcal{D}(v + 1, b, k)$, let us denote the function $f^*(r)$ by the extended notation $f^*_{b,k}(r)$ and the quantity $f^*(r^*)$, that was used in Theorem 3.7*, by $f^*_{b,k}(r^*_{b,k})$. Now consider the class $\mathcal{D}^{RC}(v + 1, b, k)$ of row–column designs. If $d \in \mathcal{D}^{RC}(v + 1, b, k)$, then it follows from the discussion above that

$$\text{tr}(M_d^{-1}) \geqslant \text{Max}(f^*_{b,k}(r), f^*_{k,b}(r)) \geqslant \text{Max}(f^*_{b,k}(r^*_{b,k}), f^*_{k,b}(r^*_{k,b})),$$

which gives a measure of efficiency,

$$e_L(d) = \text{Max}(f^*_{b,k}(r^*_{b,k}), f^*_{k,b}(r^*_{k,b}))/\text{tr}(M_d^{-1}).$$

Here is an example.

EXAMPLE 7.5. Let $v = 4$, $b = k = 5$. The design $d \in \mathcal{D}^{RC}(5, 5, 5)$ in Example 5.5 (the last design in the sequence in that example), has efficiency $e_L(d) = 1.5/1.568 = 95.7\%$.

Several researchers have computed efficiency of designs in various settings. They include, Pigeon and Raghavarao (1987) for repeated measurements designs, Angelis et al. (1993) and Gupta and Kageyama (1993) for block designs with unequal-sized blocks, Gerami and Lewis (1992) and Gerami et al. (1993) for factorial designs. These will be briefly discussed in the next section.

8. Optimal and efficient designs in other settings

Research on treatment-control designs has progressed in various directions to accommodate different experimental situations and models. In this section we will briefly outline some of these.

8.1. Repeated measurements designs

Consider the problem of designing experiments where subjects receive some or all of the treatments in an ordered fashion over a number of successive periods. The model may include the residual effect of the treatment applied in the previous period, in addition to the direct effect of the treatment applied in the current period. Suppose there are b subjects and k periods. We can say that $d(l,j) = i$ if design d assigns treatment i $(0 \leqslant i \leqslant v)$ to subject j $(1 \leqslant j \leqslant b)$ in period l $(1 \leqslant l \leqslant k)$. We will denote the class of connected designs by $\mathcal{D}^{RM}(v+1,b,k)$. The model for an observation in period l of subject j is:

$$y_{jl} = \mu + \tau_{d(l,j)} + \rho_{d(l-1,j)} + \beta_j + \gamma_l + \varepsilon_{ijl},$$

where β_j and γ_l are the subject and period effects, $\tau_{d(l,j)}$ is a direct effect of the treatment applied in period l and $\rho_{d(l-1,j)}$ is a residual effect of the treatment applied in period $l-1$. Since there is no residual effect in the first period, we can write $\rho_{d(0,j)} = 0$. See Stufken (1996) for a general treatise of repeated measurement designs.

Pigeon (1984) and Pigeon and Raghavarao (1987) studied this problem. They called a design a *control balanced residual effects design* if the information matrix of the treatment-control direct effect contrasts is completely symmetric, and the information matrix of the treatment-control residual effect contrasts is completely symmetric. They characterized control balanced residual effects designs, gave several methods for construction of these designs, and also gave tables of such designs along with their efficiencies.

A-optimal designs for the direct effect of treatments versus control contrasts were obtained in Majumdar (1988a). The main tool used in this paper was Theorem 3.1.

Hedayat and Zhao (1990) gave a complete solution for A- and MV-optimal designs for the direct effect of treatments versus control contrasts for experiments with only two periods, $k = 2$. Let v be a perfect square and $d^{(i)} \in \mathcal{D}^{RM}(v+1,b_i,2)$ be a design with treatment i in the first period of each subject and the in the second period, treatments distributed according to Theorem 2.1. Then a result of Hedayat and Zhao (1990) asserts that the union $d^{(i_1)} \cup \cdots \cup d^{(i_m)}$ for any $i_1,\ldots,i_m$, each chosen from $\{0,1,\ldots,v\}$ is A- and MV-optimal in $\mathcal{D}^{RM}(v+1,\sum_{t=1}^{m} b_{i_t},2)$.

EXAMPLE 8.1. Let $v = 4$, $k = 2$ and $b = 18$. The following design with subjects represented by columns and periods represented by rows is optimal:

1	1	1	1	1	1	2	2	2	2	2	2	3	3	3	3	3	3
0	0	1	2	3	4	0	0	1	2	3	4	0	0	1	2	3	4

Koch et al. (1989) did an extensive study of comparing two test treatments among themselves, as well as with a control, in a two-period repeated measurements design (thus $v = 2$, $k = 2$). The design they considered consisted of the treatment sequences $(1,2)$, $(2,1)$, $(0,1)$, $(1,0)$, $(0,2)$, $(2,0)$ distributed on subjects in the ratio $m : m : 1 : 1 : 1 : 1$, with special emphasis on the case $m = 3$. They used several models, each consisting of subject effects, period effects and direct effects of treatments; in addition, some of the models had residual effects. The efficiencies of the estimators of the contrasts were computed and a simulated example was studied.

8.2. *Designs for comparing two sets of treatments*

A natural extension of the problem of comparing several treatments with a control is the problem of comparing several treatments with more than one control. For instance, in drug studies, there may be a placebo (inactive control) and a standard drug (active control). In general, the two controls may be of different importance, e.g., the test treatment-placebo contrasts and the test treatment-standard treatment contrasts may be of different significance in an experiment. Thus these contrasts may have to be weighted differently in the design criterion. This will lead to a somewhat complicated criterion that depends on the weights chosen. If the weights are the same, the criterion simplifies to some extent and some results on optimal and efficient designs are available in the literature.

Suppose there are $v + u$ treatments divided into two groups, G and H of v and u treatments each. Then the contrasts of primary interest are: $\tau_g - \tau_h$, where $g \in G$ and $h \in H$. Thus there are, in all, vu contrasts of interest. Consider the block design setup with model (2.4), and suppose $G = \{1, \ldots, v\}$ and $H = \{v + 1, \ldots, v + u\}$. It was shown in Majumdar (1986) that the information matrix of the contrasts of interest is completely symmetric if the Fisher Information matrix for the treatments of a design $d \in \mathcal{D}(v + u, b, k)$ is of the form,

$$C_d = \begin{pmatrix} p_d I_v + q_d J_{vv} & t_d J_{vu} \\ t_d J_{uv} & r_d I_u + s_d J_{uu} \end{pmatrix}, \tag{8.1}$$

for some p_d, q_d, r_d, s_d, t_d, where J_{uv} is a $u \times v$ matrix of unities. Condition (8.1) is the generalization of condition (2.5) for the case of several controls.

In this setup a design is called A-optimal if it minimizes $\sum_{g \in G} \sum_{h \in H} \mathrm{Var}(\widehat{\tau}_{dg} - \widehat{\tau}_{dh})$ among all designs in $\mathcal{D}(v + u, b, k)$. Some results on A-optimality were given in Majumdar (1986). A generalization of Theorem 3.2 was given for the case $k \leqslant \min(v, u)$. Another result in this paper, obtained by applying Theorem 3.1, gave A-optimal designs for design classes with $k \equiv 0 \pmod{(v + \sqrt{vu})}$ and $k \equiv 0 \pmod{(u + \sqrt{vu})}$.

Christof (1987) generalized some of the results in Majumdar (1986) in the setup of approximate theory.

8.3. Trend-resistant designs

Jacroux (1993) considered the problem of comparing two sets of treatments, G and H of sizes v and u using experimental units that are ordered over time, where it is assumed that the observations are affected by a smooth trend over time. (The special case $u = 1$ corresponds to a single control.) If the unit at the jth time point $(j = 1, \dots, n)$, receives treatment i then the model is:

$$y_{ij} = \mu + \tau_i + \beta_1 j + \cdots + \beta_p j^p + \varepsilon_{ij}, \tag{8.2}$$

where $\beta_1 j + \cdots + \beta_p j^p$ is the effect of a trend that is assumed to be a polynomial of degree p. A design is a sequence or *run order*. We will denote the set of all connected designs by $\mathcal{D}_{ro}(v, u; n)$. For a design $d \in \mathcal{D}_{ro}(v, u; n)$, if the least square estimator of $\tau_g - \tau_h$ under model (8.2) is the same as the least square estimator under model (2.1) (i.e., the model with all $\beta_t = 0$), then this design is called *p-trend free*. Thus in a p-trend free design, treatments and trends are *orthogonal*. Jacroux (1993) studied trend free designs, and obtained designs that are A-optimal and/or MV-optimal as well as p trend-free.

EXAMPLE 8.2. The following run order is A-optimal and 1-trend free in $\mathcal{D}(2, 5; 29)$. Here $G = \{1, 2\}$ and $H = \{3, 4, 5\}$.

(12121267673455344537676212121)

8.4. Block designs with unequal block sizes and blocks with nested rows and columns

Consider the class $\mathcal{D}(v + 1, b, k_1, \dots, k_b)$ of designs defined in (6.5), and the 1-way elimination of heterogeneity model given by (2.4). What are good designs for estimation in this class?

Angelis and Moyssiadis (1991) and Jacroux (1992) extended Theorem 3.2 to this setup. All of these authors, as well as Gupta and Kageyama (1993), characterized designs in $\mathcal{D}(v + 1, b, k_1, \dots, k_b)$ for which the information matrix of treatment-control contrasts is completely symmetric, i.e., the matrix M_d in (3.1) is completely symmetric. Gupta and Kageyama (1993) gave several methods for constructing such designs, and tables of designs along with their efficiencies. Angelis and Moyssiadis (1991) gave methods to construct these designs, as well as an algorithm to find A-optimal designs. They also gave tables of optimal designs. Angelis et al. (1993) gave methods for constructing designs with C_d as above that are efficient. Jacroux (1992) gave a generalization of Theorem 3.6, as well as infinite families of A- and MV-optimal designs.

Next consider experiments where the units in each of the b blocks are arranged in a $p \times q$ array, i.e., the nested row–column setup. For a design, if the unit in row l and column m of block j receives treatment i, then the model is:

$$y_{ijlm} = \mu + \tau_i + \beta_j + \rho_{l(j)} + \gamma_{m(j)} + \varepsilon_{ijlm}.$$

Gupta and Kageyama (1991) considered the problem of finding good designs for treatment-control contrasts in this setup. Extending Pearce's (1960) definition of designs with supplemented balance, they sought designs for which the information matrix for the treatment-control contrasts is completely symmetric. Let n^r_{dil}, n^c_{dih} and n_{dij} denote respectively the number of times treatment i appears in row l, column h and block j. Let $\lambda^r_{dii'} = \sum_{l=1}^{p} n^r_{dil} n^r_{di'l}$, $\lambda^c_{dii'} = \sum_{h=1}^{q} n^c_{dih} n^c_{di'h}$ and $\lambda_{dii'} = \sum_{j=1}^{b} n_{dij} n_{di'j}$. Gupta and Kageyama (1991) showed that the information matrix for the treatment-control contrasts is completely symmetric if for s and s_0 with $s_0 \neq 0$ and $s_0 + (v-1)s \neq 0$, the following is true for all $(i,i') \in \{1,\ldots,v\} \times \{1,\ldots,v\}$, $i \neq i'$:

$$p\lambda^r_{di0} + q\lambda^c_{di0} - \lambda_{di0} = s_0 \quad \text{and} \quad p\lambda^r_{dii'} + q\lambda^c_{dii'} - \lambda_{dii'} = s.$$

These designs were called *type S designs* by Gupta and Kageyama, who also gave several methods for constructing such designs. Here is an example.

EXAMPLE 8.3. For $v = 3$, $b = 3$, $p = 2$ and $q = 3$, here is a type S design with $s_0 = 6$, $s = 3$.

Block 1	Block 2	Block 3
0 1 2	0 1 3	0 2 3
1 2 0	1 3 0	2 3 0

8.5. *Block designs when treatments have a factorial structure*

Gupta (1995) considered block designs when each treatment is a level combination of several factors. Suppose that there are t factors $F_1,\ldots,F_t$, where F_i has m_i levels, $0,1,\ldots,m_i-1$ for $i = 1,\ldots,t$. For each factor the level 0 denotes a control. A treatment is written as $x_1x_2,\ldots,x_n$, with $x_i \in \{0,1,\ldots,m_i-1\}$. Gupta extended the concept of supplemented balance to this setup, and called such designs *type S-PB designs*. He also gave methods for constructing such designs. Since the description of the factorial setup involves elaborate notation, we refer the reader to Gupta's paper for details, but give an example here to illustrate his approach.

EXAMPLE 8.4. Let $t = 2$, $m_1 = m_2 = 3$. Thus we have two factors at three levels each. Let τ be the vector of treatment effects with the treatments written in lexicographic order. Let,

$$u_{11} = u_{21} = \frac{1}{\sqrt{2}}\begin{pmatrix} 1 \\ -1 \\ 0 \end{pmatrix}, \qquad u_{12} = u_{22} = \frac{1}{\sqrt{2}}\begin{pmatrix} 1 \\ 0 \\ -1 \end{pmatrix},$$

$$u_{13} = u_{23} = \frac{1}{\sqrt{3}}\begin{pmatrix} 1 \\ 1 \\ 1 \end{pmatrix},$$

and for $(l, l') \neq (3, 3)$, let $u(l, l') = \tau'(u_{1l} \otimes u_{2l'})$, where $\otimes$ denotes Kronecker product. The two contrasts $u(1, 3)$ and $u(2, 3)$ will together represent the main effect of F_1, the two contrasts $u(3, 1)$ and $u(3, 2)$ represent the main effect of F_2, while the remaining four contrasts $u(l, l'), l \neq 3, l' \neq 3$, represent the F_1F_2 interaction. Note that the set of contrasts that represent a main effect or interaction are not orthogonal, as is the case with the traditional definition when the factors do not have a special level such as the control. Gupta (1995) defined a *type S-PB design* as one in which the estimators for all contrasts of any effect (main effect of F_1, main effect of F_2 or the interaction F_1F_2) have the same variance. The idea is that the design for each effect has the properties of a design with supplemented balance. For $k = 5$ and $b = 6$ here is an example of a type S-PB design, where treatments are denoted by pairs x_1x_2, $(x_i = 0, 1, 2; i = 1, 2)$ and, blocks are denoted by columns.

01	01	01	01	00	00
02	02	02	02	01	11
10	10	10	10	02	12
20	20	20	20	10	21
11	12	21	22	20	22

Motivated by a problem in drug-testing, Gerami and Lewis (1992) also considered block designs when treatments have a factorial structure, but their approach was different. Consider the case of two factors, i.e., $t = 2$. Suppose each factor is a different drug and the level 0 is a placebo. Each treatment, therefore, is a combination of two drugs. Gerami and Lewis considered experiments where it is unethical to administer a double placebo, i.e., the combination $(0, 0)$. There are, therefore, $v = m_1m_2 - 1$ treatments. The object is to compare the different levels of each factor with the control level, at each fixed level of the other factor. The contrasts of interest are, $\tau_{ii'} - \tau_{i0}$, the difference in the effects of level i' and level 0 of factor F_2 when F_1 is at level i, and $\tau_{ii'} - \tau_{0i'}$, the difference in the effects of level i and level 0 of factor F_1 when F_2 is at level i', for $i = 1, \ldots, m_1 - 1$ and $i' = 1, \ldots, m_2 - 1$. A designs $d_0 \in \mathcal{D}(v, b, k)$ is called A-optimal if it minimizes

$$\sum_{i=1}^{m_1-1} \sum_{i'=1}^{m_2-1} \left(\mathrm{Var}(\widehat{\tau}_{dii'} - \widehat{\tau}_{di0}) + \mathrm{Var}(\widehat{\tau}_{dii'} - \widehat{\tau}_{d0i'})\right).$$

For the case $m_2 = 2$, Gerami and Lewis determined bounds on the efficiency of designs, described designs that have a completely symmetric information matrix for the contrasts of interest and discussed methods of constructing such designs.

Gerami et al. (1993) continued the research of Gerami and Lewis (1992) by identifying a class of efficient designs. A lower bound to the efficiency of designs in that class was obtained to determine the performance of the worst design in the class. Tables of designs and their efficiencies were also given. Here is an example.

EXAMPLE 8.5. For $m_1 = 3$, $m_2 = 2$, $b = 5$ and $k = 8$, here is a design that is at least 94% efficient.

01	01	01	01	01
01	01	01	01	01
10	10	10	10	10
20	20	20	20	20
11	11	11	10	20
11	11	11	11	11
21	21	21	11	21
21	21	21	21	21

8.6. Block designs when errors are correlated

Consider the class of designs $\mathcal{D}(v+1, b, k)$ and the model (2.4), with one difference – instead of being homoscedastic, the errors, ε_{ijl}, are possibly correlated. The problem of finding optimal designs in this setup has been studied by Cutler (1993). He assumed that errors have a stationary, first-order, autoregressive correlation structure. The estimation method is the general least square method. Cutler established a general result on optimality for treatment-control contrasts which generalizes Theorem 3.2. He also suggested two families of designs and studied their construction and optimality properties.

Acknowledgements

I received an extensive set of comments from the referee on the first draft of this paper, which ranged from pointing out typographical errors to making several excellent suggestions regarding the style and contents. These led to a substantial improvement in the quality of the paper. For this, I am extremely grateful to the referee. I am also grateful to Professor Subir Ghosh for his patience and encouragement.

References

Agrawal, H. (1966). Some generalizations of distinct representatives with applications to statistical designs. *Ann. Math. Statist.* **37**, 525–528.

Angelis, L., S. Kageyama and C. Moyssiadis (1993). Methods of constructing A-efficient BTIUB designs. *Utilitas Math.* **44**, 5–15.

Angelis, L. and C. Moyssiadis (1991). A-optimal incomplete block designs with unequal block sizes for comparing test treatments with a control. *J. Statist. Plann. Inference* **28**, 353–368.

Bechhofer, R. E. (1969). Optimal allocation of observations when comparing several treatments with a control. In: P. R. Krishnaiah, ed., *Multivariate Analysis*, Vol. 2. Academic Press, New York, 463–473.
Bechhofer, R. E. and C. Dunnett (1988). Tables of percentage points of multivariate Student t distribution. In: *Selected Tables in Mathematical Statistics*, Vol. 11. Amer. Math. Soc., Providence, RI.
Bechhofer, R. E. and D. J. Nocturne (1972). Optimal allocation of observations when comparing several treatments with a control, II: 2-sided comparisons. *Technometrics* **14**, 423–436.
Bechhofer, R. E. and A. C. Tamhane (1981). Incomplete block designs for comparing treatments with a control: General theory. *Technometrics* **23**, 45–57.
Bechhofer, R. E. and A. C. Tamhane (1983a). Design of experiments for comparing treatments with a control: Tables of optimal allocations of observations. *Technometrics* **25**, 87–95.
Bechhofer, R. E. and A. C. Tamhane (1983b). Incomplete block designs for comparing treatments with a control (II): Optimal designs for $p = 2(1)6$, $k = 2$ and $p = 3$, $k = 3$. *Sankhyā Ser. B* **45**, 193–224.
Bechhofer, R. E. and A. C. Tamhane (1985). *Selected Tables in Mathematical Statistics*, Vol. 8. Amer. Math. Soc., Providence, RI.
Bechhofer, R. E. and B. W. Turnbull (1971). Optimal allocation of observations when comparing several treatments with a control, III: Globally best one-sided intervals for unequal variances. In: S. S. Gupta and J. Yackel, eds., *Statistical Decision Theory and Related Topics*. Academic Press, New York, 41–78.
Berger, J. (1985). *Statistical Decision Theory and Bayesian Analysis*, 2nd edn. Springer, New York.
Cheng, C. S., D. Majumdar, J. Stufken and T. E. Ture (1988). Optimal step type designs for comparing treatments with a control. *J. Amer. Statist. Assoc.* **83**, 477–482.
Christof, K. (1987). Optimale blockpläne zum vergleich von kontroll- und testbehandlungen. Ph.D. Dissertation, Univ. Augsburg.
Constantine, G. M. (1983). On the trace efficiency for control of reinforced balanced incomplete block designs. *J. Roy. Statist. Soc. Ser. B* **45**, 31–36.
Cox, D. R. (1958). *Planning of Experiments*. Wiley, New York.
Cutler, R. D. (1993). Efficient block designs for comparing test treatments to a control when the errors are correlated. *J. Statist. Plann. Inference* **36**, 107–125.
Das, M. N. (1958). On reinforced incomplete block designs. *J. Indian Soc. Agricultural Statist.* **10**, 73–77.
DasGupta, A. (1996). Review of optimal Bayes designs. In: *Handbook of Statistics*, this volume, Chapter 29.
Dunnett, C. W. (1955). A multiple comparison procedure for comparing several treatments with a control. *J. Amer. Statist. Assoc.* **50**, 1096–1121.
Dunnett, C. W. (1964) New tables for multiple comparisons with a control. *Biometrics* **20**, 482–491.
Dunnett, C. W. and M. Sobel (1954). A bivariate generalization of Student's t-distribution, with tables for certain special cases. *Biometrika* **41**, 153–169.
Fieler, E. C. (1947). Some remarks on the statistical background in bioassay. *Analyst* **72**, 37–43.
Finney, D. J. (1952). *Statistical Methods in Biological Assay*. Haffner, New York.
Freeman, G. H. (1975). Row-and-column designs with two groups of treatments having different replications. *J. Roy. Statist. Soc. Ser. B* **37**, 114–128.
Gerami, A. and S. M. Lewis (1992). Comparing dual with single treatments in block designs. *Biometrika* **79**, 603–610.
Gerami, A., S. M. Lewis, D. Majumdar and W. I. Notz (1993). Efficient block designs for comparing dual with single treatments. Tech. Report, Univ. of Southampton.
Giovagnoli, A. and I. Verdinelli (1983). Bayes D- and E-optimal block designs. *Biometrika* **70**, 695–706.
Giovagnoli, A. and I. Verdinelli (1985). Optimal block designs under a hierarchical linear model. In: J. M. Bernardo, M. H. DeGroot, D. V. Lindley and A. F. M. Smith, eds., *Bayesian Statistics*, Vol. 2. North-Holland, Amsterdam, 655–662.
Giovagnoli, A. and H. P. Wynn (1981). Optimum continuous block designs. *Proc. Roy. Soc. (London) Ser. A* **377**, 405–416.
Giovagnoli, A. and H. P. Wynn (1985). Schur optimal continuous block designs for treatments with a control. In: L. M. LeCam and R. A. Olshen, eds., *Proc. Berkeley Conf. in Honor of Jerzy Neyman and Jack Kiefer*, Vol. 2. Wadsworth, Monterey, CA, 651–666.
Gupta, S. (1989). Efficient designs for comparing test treatments with a control. *Biometrika* **76**, 783–787.
Gupta, S. (1995). Multi-factor designs for test versus control comparisons. *Utilitas Math.* **47**, 199–210.
Gupta, S. and S. Kageyama (1991). Type S designs for nested rows and columns. *Metrika* **38**, 195–202.

Gupta, S. and S. Kageyama (1993). Type S designs in unequal blocks. *J. Combin. Inform. System Sci.* **18**, 97–112.

Hall, P. (1935). On representatives of subsets. *J. London Math. Soc.* **10**, 26–30.

Hayter, A. J. and A. C. Tamhane (1991). Sample size determination for step-down multiple test procedures: Orthogonal contrasts and comparisons with a control. *J. Statist. Plann. Inference* **27**, 271–290.

Hedayat, A. S., M. Jacroux and D. Majumdar (1988). Optimal designs for comparing test treatments with controls (with discussions). *Statist. Sci.* **3**, 462–491.

Hedayat, A. S. and D. Majumdar (1984). A-optimal incomplete block designs for control-test treatment comparisons. *Technometrics* **26**, 363–370.

Hedayat, A. S. and D. Majumdar (1985). Families of optimal block designs for comparing test treatments with a control. *Ann. Statist.* **13**, 757–767.

Hedayat, A. S. and D. Majumdar (1988). Model robust optimal designs for comparing test treatments with a control. *J. Statist. Plann. Inference* **18**, 25–33.

Hedayat, A. S. and E. Seiden (1974). On the theory and application of sum composition of latin squares and orthogonal latin squares. *Pacific J. Math.* **54**, 85–112.

Hedayat, A. S. and W. Zhao (1990). Optimal two-period repeated measurements designs. *Ann. Statist.* **18**, 1805–1816.

Hoblyn, T. N., S. C. Pearce and G. H. Freeman (1954). Some considerations in the design of successive experiments in fruit plantations. *Biometrics* **10**, 503–515.

Hochberg, Y. and A. C. Tamhane (1987). *Multiple Comparison Procedures*. Wiley, New York.

Jacroux, M. (1984). On the optimality and usage of reinforced block designs for comparing test treatments with a standard treatment. *J. Roy. Statist. Soc. Ser. B* **46**, 316–322.

Jacroux, M. (1986). On the usage of refined linear models for determining N-way classification designs which are optimal for comparing test treatments with a standard treatment. *Ann. Inst. Statist. Math.* **38**, 569–581.

Jacroux, M. (1987a). On the determination and construction of MV-optimal block designs for comparing test treatments with a standard treatment. *J. Statist. Plann. Inference* **15**, 205–225.

Jacroux, M. (1987b). Some MV-optimal block designs for comparing test treatments with a standard treatment. *Sankhyā Ser. B* **49**, 239–261.

Jacroux, M. (1989). The A-optimality of block designs for comparing test treatments with a control. *J. Amer. Statist. Assoc.* **84**, 310–317.

Jacroux, M. (1992). On comparing test treatments with a control using block designs having unequal sized blocks. *Sankhyā Ser. B* **54**, 324–345.

Jacroux, M. (1993). On the construction of trend-resistant designs for comparing a set of test treatments with a set of controls. *J. Amer. Statist. Assoc.* **88**, 1398–1403.

Jacroux, M. and D. Majumdar (1989). Optimal block designs for comparing test treatments with a control when $k > v$. *J. Statist. Plann. Inference* **23**, 381–396.

Kiefer, J. (1975). Construction and optimality of generalized Youden designs. In: J. Srivastava, ed., *A Survey of Statistical Design and Linear Models*. North-Holland, Amsterdam, 333–353.

Kim, K. and J. Stufken (1995). On optimal block designs for comparing a standard treatment to test treatments. *Utilitas Math.* **47**, 211–224.

Koch, G. G., I. A. Amara, B. W. Brown, T. Colton and D. B. Gillings (1989). A two-period crossover design for the comparison of two active treatments and placebo. *Statist. Med.* **8**, 487–504.

Krishnaiah, P. R. and J. V. Armitage (1966). Tables for multivariate t-distribution. *Sankhyā Ser. B* **28**, 31–56.

Kunert, J. (1983). Optimal design and refinement of the linear model with applications to repeated measurements designs. *Ann. Statist.* **11**, 247–257.

Lindley, D. V. and A. F. M. Smith (1972). Bayes estimates for the linear model (with discussion). *J. Roy. Statist. Soc. Ser. B* **34**, 1–42.

Magda, G. M. (1980). Circular balanced repeated measurements designs. *Comm. Statist. Theory Methods* **9**, 1901–1918.

Majumdar, D. (1986). Optimal designs for comparisons between two sets of treatments. *J. Statist. Plann. Inference* **14**, 359–372.

Majumdar, D. (1988a). Optimal repeated measurements designs for comparing test treatments with a control. *Comm. Statist. Theory Methods* **17**, 3687–3703.

Majumdar, D. (1988b). Optimal block designs for comparing new treatments with a standard treatment. In: Y. Dodge, V. V. Fedorov and H. P. Wynn, eds., *Optimal Design and Analysis of Experiments*. North-Holland, Amsterdam, 15–27.

Majumdar, D. (1992). Optimal designs for comparing test treatments with a control utilizing prior information. *Ann. Statist.* **20**, 216–237.

Majumdar, D. (1996). On admissibility and optimality of treatment-control designs. *Ann. Statist.*, to appear.

Majumdar, D. and S. Kageyama (1990). Resistant BTIB designs. *Comm. Statist. Theory Methods* **19**, 2145–2158.

Majumdar, D. and W. I. Notz (1983). Optimal incomplete block designs for comparing treatments with a control. *Ann. Statist.* **11**, 258–266.

Majumdar, D. and A. C. Tamhane (1996). Row–column designs for comparing treatments with a control. *J. Statist. Plann. Inference* **49**, 387–400.

Miller, R. G. (1966). *Simultaneous Statistical Inference*. McGraw Hill, New York.

Naik, U. D. (1975). Some selection rules for comparing p processes with a standard. *Comm. Statist. Theory Methods* **4**, 519–535.

Notz, W. I. (1985). Optimal designs for treatment-control comparisons in the presence of two-way heterogeneity. *J. Statist. Plann. Inference* **12**, 61–73.

Notz, W. I. and A. C. Tamhane (1983). Incomplete block (BTIB) designs for comparing treatments with a control: Minimal complete sets of generator designs for $k = 3$, $p = 3(1)10$. *Comm. Statist. Theory Methods* **12**, 1391–1412.

Owen, R. J. (1970). The optimal design of a two-factor experiment using prior information. *Ann. Math. Statist.* **41**, 1917–1934.

Pearce, S. C. (1953). Field experiments with fruit trees and other perennial plants. Commonwealth Agric. Bur., Farnham Royal, Bucks, England. *Tech. Comm.* 23.

Pearce, S. C. (1960). Supplemented balance. *Biometrika* **47**, 263–271.

Pearce, S. C. (1963). The use and classification of non-orthogonal designs (with discussion). *J. Roy. Statist. Soc. Ser. B* **126**, 353–377.

Pearce, S. C. (1983). *The Agricultural Field Experiment: A Statistical Examination of Theory and Practice*. Wiley, New York.

Pesek, J. (1974). The efficiency of controls in balanced incomplete block designs. *Biometrische Z.* **16**, 21–26.

Pigeon, J. G. (1984). Residual effects for comparing treatments with a control. Ph.D. Dissertation, Temple University.

Pigeon, J. G. and D. Raghavarao (1987). Crossover designs for comparing treatments with a control. *Biometrika* **74**, 321–328.

Roy, J. (1958). On efficiency factor of block designs. *Sankhyā* **19**, 181–188.

Smith, A. F. M. and I. Verdinelli (1980). A note on Bayes designs for inference using a hierarchical linear model. *Biometrika* **67**, 613–619.

Spurrier, J. D. (1992). Optimal designs for comparing the variances of several treatments with that of a standard. *Technometrics* **34**, 332–339.

Spurrier, J. D. and D. Edwards (1986). An asymptotically optimal subclass of balanced treatment incomplete block designs for comparisons with a control. *Biometrika* **73**, 191–199.

Spurrier, J. D. and A. Nizam (1990). Sample size allocation for simultaneous inference in comparison with control experiments. *J. Amer. Statist. Assoc.* **85**, 181–186.

Stufken, J. (1986). On optimal and highly efficient designs for comparing test treatments with a control. Ph.D. Dissertation, Univ. Illinois at Chicago, Chicago, IL.

Stufken, J. (1987). A-optimal block designs for comparing test treatments with a control. *Ann. Statist.* **15**, 1629–1638.

Stufken, J. (1988). On bounds for the efficiency of block designs for comparing test treatments with a control. *J. Statist. Plann. Inference* **19**, 361–372.

Stufken, J. (1991a). On group divisible treatment designs for comparing test treatments with a standard treatment in blocks of size 3. *J. Statist. Plann. Inference* **28**, 205–211.

Stufken, J. (1991b). Bayes A-optimal and efficient block designs for comparing test treatments with a standard treatment. *Comm. Statist. Theory Methods* **20**, 3849–3862.

Stufken, J. (1996). Optimal crossover design. In: *Handbook of Statistics*, this volume, Chapter 3.

Stufken, J. and K. Kim (1992). Optimal group divisible treatment designs for comparing a standard treatment with test treatments. *Utilitas Math.* **41**, 211–227.

Ting, C.-P. and W. I. Notz (1987). Optimal row–column designs for treatment control comparisons. Tech. Report, Ohio State University.

Ting, C.-P. and W. I. Notz (1988). A-optimal complete block designs for treatment-control comparisons. In: Y. Dodge, V. V. Fedorov and H. P. Wynn, eds., *Optimal Design and Analysis of Experiments*, North-Holland, Amsterdam, 29–37.

Ting, C.-P. and Y.-Y. Yang (1994). Bayes A-optimal designs for comparing test treatments with a control. Tech. Report, National Chengchi Univ., Taiwan.

Toman, B. and W. I. Notz (1991). Bayesian optimal experimental design for treatment-control comparisons in the presence of two-way heterogeneity. *J. Statist. Plann. Inference* **27**, 51–63.

Ture, T. E. (1982). On the construction and optimality of balanced treatment incomplete block designs. Ph.D. Dissertation, Univ. of California, Berkeley, CA.

Ture, T. E. (1985). A-optimal balanced treatment incomplete block designs for multiple comparisons with the control. *Bull. Internat. Statist. Inst.*; *Proc. 45th Session*, 51-1, 7.2-1–7.2-17.

Ture, T. E. (1994). Optimal row–column designs for multiple comparisons with a control: A complete catalog. *Technometrics* **36**, 292–299.

Verdinelli, I. (1983). Computing Bayes D- and A-optimal block designs for a two-way model. *The Statistician* **32**, 161–167.

S. Ghosh and C. R. Rao, eds., *Handbook of Statistics, Vol. 13*

28

Model Robust Designs

Y-J. Chang and W. I. Notz

1. Introduction

There is a vast body of literature on design of experiments for linear models. Most of this literature assumes that the response is described exactly by a particular linear model. Experimental design is concerned with how to take observations so as to fit this model. In practice, the assumed model is likely to be, at best, only a reasonable approximation to the true model for the response. This true model may not even be linear and is generally unknown. One may then ask whether an experimental design selected for the purpose of fitting the assumed model, will also allow one to determine if the assumed model is a reasonable approximation to the true model. If the assumed model is a poor approximation, does the design allow one to fit a model that better approximates the true model? In addition, to what extent are inferences based on the fitted model using a given design biased? These are very important practical questions. In this article we provide an overview of the wide variety of approaches that have been proposed in the literature to answer them.

To motivate what follows we consider a simple example. An experimenter observes a response Y which is thought to depend on a single independent variable x. The experimenter intends to fit a straight line to the data using the method of least squares. The response Y can be observed over some range of values of the independent variable which are of particular interest to the experimenter.

The "naive" experimenter will typically take observations spread (often uniformly) over the region of interest. This is not the classical "optimal" design which would take half of the observations at one end of the region of interest and half at the other end. If you ask our naive experimenter why they plan to take observations spread out over the regions of interest, you typically get an answer of the form "I need to observe Y at a variety of values of x in case the response isn't a perfect straight line but instead has some curvature". This suggests that the experimenter does not believe that the response is actually a straight line function of x, but may be something more complicated with some curvature. However, the experimenter also probably believes a straight line is a reasonable approximation to the true response and so plans to fit a straight line to the data unless the data indicate that a straight line model is a very poor approximation.

The beliefs of our naive experimenter reflect what happens in all real problems. In investigating how a response depends on several independent variables one generally

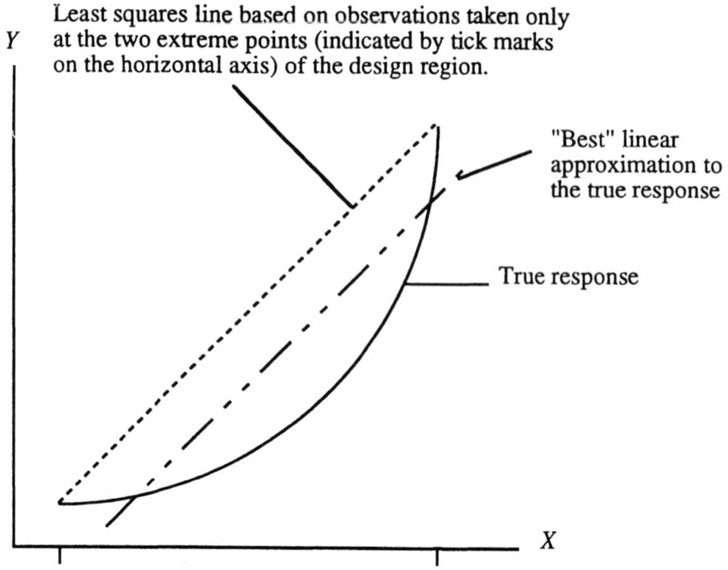
Least squares line based on observations taken only at the two extreme points (indicated by tick marks on the horizontal axis) of the design region.
Y
"Best" linear approximation to the true response
True response
X

Fig. 1.1. Fitting a straight line when the true model is a quadratic.

does not know the exact form of the relationship between the response and the independent variables. At best, one believes that a relatively simple functional relation is a good approximation to the true unknown relation. One plans to fit some relatively simple model, often a linear model with normal errors, using standard methods such as least squares or maximum likelihood. If one can control the values of the independent variables at which the response is observed, at what values should observations be taken so that the simple model we fit to the data adequately approximates the true model? Since the true model is unknown, this last question must mean that the fitted simple model provides an adequate approximation to a range of possible true models, i.e., is in some sense "robust" to the exact form of the true model. This is the fundamental goal of model robust design.

In order to see what sorts of problems may arise, we consider two examples. We plan to fit a straight line to data. We know that the true response is probably not a straight line, but believe that a straight line will provide a suitable approximation to the true response. Suppose the situation facing us is as in Figure 1.1.

The true response is a quadratic. If we use the standard "optimal" design, which takes observations at the extreme points (indicated by the tick marks in the figure) of the design region, our least squares line (assuming negligible error in the observed response) will be biased, lying almost entirely above the true response. The "best" straight line approximation to the true response would lie below the fitted least squares line. Without taking observations in the interior of our design region, we would not even be aware of the bias in our fitted least squares line. Box and Draper (1959) examined the consequences of this sort of bias in some detail. Obviously a few

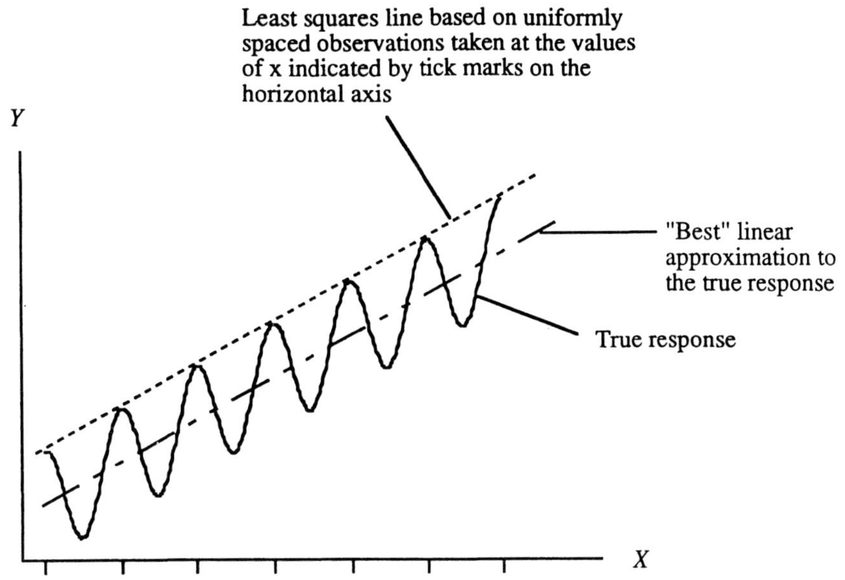

Fig. 1.2. Fitting a straight line when the true model is a rapidly varying curve with many local maxima and minima.

observations in the middle of the design region would have alerted us to the form of the true model and might have yielded a fitted least squares line more like the line that best approximates the true response. In fact, Box and Draper (1959) showed that if we are fitting a low order polynomial to data when the true model is some higher order polynomial of degree, say d, a design whose moments up to degree $d+1$ match those of uniform measure is, in some sense, best for protecting against bias. Thus the intuition of our naive experimenter does not appear too far fetched.

Unfortunately no design which takes a finite number of observations can protect against all possible forms of bias. For any given design, there is some perverse "true model" which the design fails to protect against. As our second example, we again plan to fit a straight line to data. Suppose the true response is quite "wiggly" as in Figure 1.2.

If we take observations at the values of the independent variable indicated by the tick marks on the horizontal axis, we have a design which is uniformly spaced over the design region. This is the sort of design our naive experimenter might use. However (unbeknown to us) these turn out to be the relative maxima of the "wiggly" true response. The fitted least squares line (assuming negligible errors in the observed responses) will again be biased, lying above the true response. The "best" straight line approximation to the true response also lies below the fitted least squares line. In principle then, no matter at how many uniformly spaced points in the design region we take observations, there is some perverse rapidly varying continuous function for which our design will be very poor (yielding a very biased fitted least squares line). Thus there is no design (at a finite number of points) which will be reasonable regardless of the form of the true response. Some information about the possible form of the true response is needed in order to construct designs. For example, it is not unnatural to restrict to relatively smooth functions which vary only slowly, since in real problems this is usually the case.

Thus to determine suitable designs, several pieces of information are crucial. These include

(i) On what set independent variables does the response depend and over what range of values of these variables (the design region) do we want to model the response?

(ii) What (approximately correct) model does one intend to fit and what estimator will be used to fit this model?

(iii) What assumptions is one willing to make about the nature of the true model?

(iv) What is the purpose of the experiment (model fitting, prediction, extrapolation)?

Whatever the answers to these questions, a "reasonable" design ought to allow one to fit the (approximately correct) model, allow one to detect if the fitted model is a poor approximation to the true model, and if a poor approximation allow one to fit more complicated models which better approximate the true response. Box and Draper (1959) appear to be the first to address these issues in a systematic way. Ideally, one should choose a design that "best" accomplishes the above. Exactly what we mean by "best", however, is difficult to formulate and depends on (i), (ii), (iii), and (iv). Roughly speaking, "best" usually involves either being minimax (with respect to some loss function) over a class of possible true models, minimizing a weighted average of some loss function over a class of possible true models, or minimizing the Bayes risk in a Bayesian formulation of the problem. In the literature on robust design, some authors seek good estimators (usually linear estimators are considered) and designs simultaneously. Others assume that the usual least squares estimators will be used (because models will be fit using standard software) and only seek for good designs.

2. Notation and a useful model

To make the above more precise, we introduce some notation following Kiefer (1973). We consider the following model. Suppose on each of N experimental units we measure a response which depends on the values of k independent or explanatory variables. Let Y_i and $\boldsymbol{x}_i = (x_{i_1}, \ldots, x_{i_k})'$ denote the values of the response and the explanatory variables, respectively, for the ith unit. Here, $'$ denotes the transpose of a vector or matrix. We assume the $\boldsymbol{x}_i$ belong to some compact subset X of k-dimensional Euclidean space $\mathbb{R}^k$ and $Y_1, \ldots, Y_N$ are uncorrelated observations with common variance σ^2 and mean

$$\mathrm{E}Y_i = \sum_{j=1}^{m} \beta_j f_j(\boldsymbol{x}_i). \tag{2.1}$$

The f_j are real-valued functions defined on $X \cup L$ (not necessarily disjoint) and the β_j are unknown real valued parameters. Let $g_1, \dots, g_s$ be s known, real valued functions defined on L. The problem, to be made precise shortly, is to choose $\boldsymbol{x}_1, \dots, \boldsymbol{x}_N$ in X in some "good" fashion for the purpose of estimating the expected response $\sum_{j=1}^{m} \beta_j f_j(\boldsymbol{x}_i)$ for $\boldsymbol{x}$ in L. However, we use not a linear combination of the $f_j(\boldsymbol{x}_i)$ to model the expected response in (2.1) but, instead, a linear combination of the $g_l(\boldsymbol{x}_i)$. In particular, we model the expected response as

$$\mathrm{E}Y_i = \sum_{l=1}^{s} \alpha_l g_l(\boldsymbol{x}_i). \tag{2.2}$$

Justification for using (2.2) might be as follows. In real problems, the true response typically depends on a large number of explanatory variables in a complicated (and unknown) manner. In practice a relatively simple model, as in (2.2), depending on only some of these explanatory variables is used to describe the response and is assumed to provide a reasonable approximation to the response, at least over some range of values of $\boldsymbol{x}$. Thus the g in (2.2) may depend on only a subset of the explanatory variables (coordinates of $\boldsymbol{x}$). While this possibility is inherent in (2.2), most of the literature treats g as depending on all the coordinates of $\boldsymbol{x}$. In any case, we will estimate the expected response at $\boldsymbol{x}$ by estimating $\sum_{l=1}^{s} \alpha_l g_l(\boldsymbol{x}_i)$. At this point we should note the following identifiability problem, namely exactly what are we estimating when we estimate $\sum_{l=1}^{s} \alpha_l g_l(\boldsymbol{x}_i)$? This issue is sometimes ignored in the literature, but it appears that most authors assume that the $\sum_{l=1}^{s} \alpha_l g_l(\boldsymbol{x}_i)$ in (2.2) that we are estimating is the best (in some sense) approximation to (2.1) among the set of all functions having the form given on the right hand side of (2.2).

Now let $\boldsymbol{Y} = (Y_1, \dots, Y_N)'$, $\boldsymbol{F}$ the $N \times m$ matrix whose i,jth entry is $f_j(\boldsymbol{x}_i)$, $\boldsymbol{f}(\boldsymbol{x}) = (f_1(\boldsymbol{x}), f_2(\boldsymbol{x}), \dots, f_m(\boldsymbol{x}))'$, $\boldsymbol{\beta} = (\beta_1, \dots, \beta_m)'$, $\boldsymbol{G}$ the $N \times s$ matrix whose i,lth entry is $g_l(\boldsymbol{x}_i)$, $\boldsymbol{g}(\boldsymbol{x}) = (g_1(\boldsymbol{x}), g_2(\boldsymbol{x}), \dots, g_s(\boldsymbol{x}))'$, and $\boldsymbol{\alpha} = (\alpha_1, \dots, \alpha_s)'$. We can then write (2.1) as

$$\mathrm{E}\boldsymbol{Y} = \boldsymbol{F}\boldsymbol{\beta} \tag{2.3}$$

and (2.2) as

$$\mathrm{E}\boldsymbol{Y} = \boldsymbol{G}\boldsymbol{\alpha}. \tag{2.4}$$

As is common in the linear model, we restrict attention to estimators of $\boldsymbol{\alpha}$ which are linear in $\boldsymbol{Y}$. Thus, we consider the estimator $\boldsymbol{C}'\boldsymbol{Y}$ of $\boldsymbol{\alpha}$, where $\boldsymbol{C}$ is $N \times s$. Notice

$$\mathrm{E}(\boldsymbol{C}'\boldsymbol{Y}) = \boldsymbol{C}'\boldsymbol{F}\boldsymbol{\beta}, \qquad \mathrm{Cov}(\boldsymbol{C}'\boldsymbol{Y}) = \sigma^2\boldsymbol{C}'\boldsymbol{C}. \tag{2.5}$$

Finally, by an exact design, we will mean a particular choice of $\boldsymbol{x}_1, \dots, \boldsymbol{x}_N$ in X.

How do we evaluate the goodness of a particular linear estimator and design? To place this in a decision theoretic type framework, let ν be an $N \times N$ loss measure on L, nonnegative definite in value, and such that $\boldsymbol{f}$ and $\boldsymbol{g}$ are square integrable relative to ν and $\int_L \boldsymbol{a}(\boldsymbol{x})'\,\mathrm{d}\nu(\boldsymbol{x})\boldsymbol{a}(\boldsymbol{x})$ is nonnegative definite for any vector valued square integrable function $\boldsymbol{a}(\boldsymbol{x})$ on L. We define $\Gamma_{\boldsymbol{ff}}$, $\Gamma_{\boldsymbol{gg}}$, $\Gamma_{\boldsymbol{fg}}$, and $\Gamma_{\boldsymbol{gf}}$ by

$$\Gamma_{\boldsymbol{ab}} = \int_L \boldsymbol{a}(\boldsymbol{x})'\,\mathrm{d}\nu(\boldsymbol{x})\boldsymbol{b}(\boldsymbol{x}). \tag{2.6}$$

The (quadratic) expected loss incurred if the design $\boldsymbol{x}_1, \ldots, \boldsymbol{x}_N$ and estimator $\boldsymbol{C}'\boldsymbol{Y}$ are used and $\boldsymbol{\beta}$ is the true parameter value is then

$$\begin{aligned} &R(\boldsymbol{\beta};\ \boldsymbol{x}_1, \ldots, \boldsymbol{x}_N, \boldsymbol{C}) \\ &= \mathrm{E}\int_L \big[\boldsymbol{g}(\boldsymbol{x})'\boldsymbol{C}'\boldsymbol{Y} - \boldsymbol{f}(\boldsymbol{x})'\boldsymbol{\beta}\big]'\mathrm{d}\nu(\boldsymbol{x})\big[\boldsymbol{g}(\boldsymbol{x})'\boldsymbol{C}'\boldsymbol{Y} - \boldsymbol{f}(\boldsymbol{x})'\boldsymbol{\beta}\big] \\ &= \sigma^2 \mathrm{tr}\, \boldsymbol{C}'\boldsymbol{C}\Gamma_{\boldsymbol{gg}} + \boldsymbol{\beta}'[\boldsymbol{F}'\boldsymbol{C}\Gamma_{\boldsymbol{gg}}\boldsymbol{C}'\boldsymbol{F} - \boldsymbol{F}'\boldsymbol{C}\Gamma_{\boldsymbol{gf}} - \Gamma_{\boldsymbol{fg}}\boldsymbol{C}'\boldsymbol{F} + \Gamma_{\boldsymbol{ff}}]\boldsymbol{\beta} \\ &= \sigma^2(V + B) \quad \text{(say)} \end{aligned} \tag{2.7}$$

the last being a notation which specializes to that used in many papers on robust design when certain restrictions on $\boldsymbol{C}$ are imposed, as will be described shortly. Our goal will be to minimize this expected (quadratic) loss.

Without further restrictive criteria, the unknown relative magnitudes of σ^2 and $\boldsymbol{\beta}'\boldsymbol{\beta}$ make the problem of choosing a design $\boldsymbol{x}_1, \ldots, \boldsymbol{x}_N$ and estimator $\boldsymbol{C}'\boldsymbol{Y}$ so as to make (2.7) "small" too vague or unfruitful. Several approaches are possible. One approach is to minimize B alone in (2.7). This approach is recommended by Box and Draper (1959) and by Karson et al. (1969). Another approach is to minimize the maximum of (2.7) subject to an assumption such as $\sigma^{-1}\boldsymbol{\beta} \in S$, where S is a specified set in $\mathbb{R}^k$. Somewhat simpler, but in the same vain, is the minimization of an integral of (2.7) with respect to a specified measure ψ on $\sigma^{-1}\boldsymbol{\beta}$. This has an obvious interpretation for Bayesians, but may appeal to others as a possible compromise. For, in order to inspect the risk functions of the admissible choices of design $\boldsymbol{x}_1, \ldots, \boldsymbol{x}_N$ and estimator $\boldsymbol{C}'\boldsymbol{Y}$ (after reduction to (2.7)), it suffices to consider the closure of such integral minimizers (more generally, without the invariance reduction to $\sigma^{-1}\boldsymbol{\beta}$). In this context one could, of course, question the restriction to linear estimators. Retaining the restriction to linear estimators, Kiefer (1973) follows this latter approach. These and other approaches will be discussed in more detail below.

Before proceeding, it is worthwhile recalling the distinction between the exact and approximate design theory. The exact theory follows the above development, wherein a design is an N-tuple of points $\boldsymbol{x}_1, \ldots, \boldsymbol{x}_N$ (not necessarily distinct) in X. Alternatively, a design is a discrete probability measure ξ on X restricted to the family of probability measures taking on values which are only integral multiples of N^{-1}. In this case, the ξ corresponding to $\boldsymbol{x}_1, \ldots, \boldsymbol{x}_N$ is given by $N\xi(\boldsymbol{x})$ = the number of $\boldsymbol{x}_i$ equal to $\boldsymbol{x}$. In the approximate theory ξ is permitted to be any member of the family of all probability measures on X relative to a specified σ-field which contains at least all the finite subsets of X. The use of the approximate theory makes various minimizations tractable which are unwieldy in the exact theory. Unfortunately, the use of the approximate theory necessitates implementation of the optimal approximate design in terms of exact designs which may then only be approximately optimum. These issues are discussed further in Kiefer (1959), Kiefer and Wolfowitz (1959) and Fedorov (1972). We write

$$\boldsymbol{M}(\xi) = \int_{x} \boldsymbol{f}(\boldsymbol{x})\boldsymbol{f}(\boldsymbol{x})'\xi(\mathrm{d}\boldsymbol{x}) \tag{2.8}$$

for the information matrix per observation under ξ. Notice $\boldsymbol{M}(\xi) = \boldsymbol{F}\boldsymbol{F}'$, for $\boldsymbol{F}$ as in (2.3), when ξ corresponds to an exact design.

A number of authors have considered the above issues of "model robust" design. A large variety of approaches have been investigated. There appears to be no agreement on the proper formulation of the model robust design, but there are a few features common to all. These include specification of a model to be fit, some assumptions concerning the nature of the true model, and some objective the design is to satisfy. The more realistic the formulation, the more difficult it is to obtain analytic solutions. In what follows we attempt to provide a survey of the literature on model robust design. In the interest of space, however, only a few of these approaches will be presented in detail.

In what follows we discuss various applications of the above formulation, as well as additional formulations of the problem of model robust design. In most cases we provide simple examples, usually in the univariate setting with (2.2) a straight line model. We cover designs for fitting a model over some region of interest, the problem of extrapolation, designs allowing detection of model inadequacy, a survey of some results in a Bayes setting, and applications to computer experiments.

3. Designs for model fitting and parameter estimation

3.1. Finite dimensional sets of true models

Box and Draper (1959) appear to be the first to formally investigate the problem of model robust design. In the notation of Section 2, they take $X = L$. In their general formulation, no assumptions concerning the true model, as represented by $\boldsymbol{f}(\boldsymbol{x})$, are made, but the g_j are assumed to be monomials so that the fitted model, as represented by $\boldsymbol{g}(\boldsymbol{x})$, is a polynomial of some specified degree. Model (2.2) then represents the polynomial of this degree that in some sense best fits the true model. Box and Draper (1959) restrict attention to the least squares estimators for (2.2). Thus in the notation of Section 2, $\boldsymbol{C}' = (\boldsymbol{G}'\boldsymbol{G})^{-}\boldsymbol{G}'$. The motivation for this is that since practitioners will typically use existing software to fit models, for practical purposes the relevant estimators are the least squares estimators. Box and Draper (1959) seek designs having two properties. First, designs should make (2.7) relatively small, with ν taken to be Lebesgue measure on $X = L$ and $\boldsymbol{C}'\boldsymbol{Y} = (\boldsymbol{G}'\boldsymbol{G})^{-}\boldsymbol{G}'\boldsymbol{Y}$ the usual least squares estimator under model (2.2). Second, designs should allow for a relatively sensitive test for lack of fit of the fitted model. These properties are to be achieved in a sequential manner. First find all designs minimizing (2.7). Then from these designs, choose the one that makes the power of the lack of fit test as large as possible. In order to carry out these objectives $\boldsymbol{f}(\boldsymbol{x})$ must be specified. The resulting optimal model robust designs are thus robust against true models having a particular form.

In all their examples, Box and Draper (1959) take $s < m$ and assume $\boldsymbol{f}(\boldsymbol{x})$ is a higher order polynomial than $\boldsymbol{g}(\boldsymbol{x})$ with $g_i = f_i$ for $i = 1, \ldots, s$. In this case it is convenient to express $\boldsymbol{\beta}$ as (α, β_2). Thus model (2.4) is just (2.3) with $\beta_2 = 0$. To illustrate what happens, we now sketch the results for the case $k = 1$ when the true model a quadratic and the fitted model is a straight line.

For simplicity, assume variables are scaled so $X = [-1, 1]$, Box and Draper (1959) restrict to designs ξ which are symmetric on X (i.e., have first moment $= 0$) and show that in this case (2.7) becomes (using the least squares estimators and $\nu =$ Lebesgue measure)

$$\begin{aligned} R &= \sigma^2(V + B) \\ &= \sigma^2\left\{\left[1 + \frac{1}{3m_2}\right] + \frac{N\beta_2^2}{\sigma^2}\left[m_2^2 - \frac{2m_2}{3} + \frac{1}{5} + \frac{m_3^2}{3m_2^2}\right]\right\} \end{aligned} \tag{3.1}$$

where m_i denotes the ith moment of the design ξ and β_2^2 is the coefficient of the quadratic term in the true model. Notice that the term

$$\left[1 + \frac{1}{3m_2}\right] \tag{3.2}$$

is proportional to the contribution of the variance V to R and the term

$$\frac{N\beta_2^2}{\sigma^2}\left[m_2^2 - \frac{2m_2}{3} + \frac{1}{5} + \frac{m_3^2}{3m_2^2}\right] \tag{3.3}$$

is proportional to the contribution of the bias B to R. (3.1) is minimized when $m_3 = 0$. Restricting to ξ with both first and third moments 0, the only way the design enters (3.1) is through its second moment m_2. The minimizing value of m_2 (and hence the optimal design) can be found using calculus. The result will depend on the quantity $N\beta_2^2/\sigma^2$. Box and Draper (1959) note that

$$\sqrt{\frac{N\beta_2^2}{\sigma^2}} = \frac{\beta_2}{\sigma/\sqrt{N}} \tag{3.4}$$

which might be interpreted as the ratio of the curvature (as represented by β_2) of the true model to the sampling error (as represented by $\sigma/\sqrt{N}$). When β_2 is quite small relative to $\sigma/\sqrt{N}$ (and hence the true model cannot be distinguished from a straight line up to sampling error), the optimal design essentially minimizes the variance term in (3.2). This leads to the classical optimal design which puts all its mass on -1 and $+1$. When β_2 is quite large relative to $\sigma/\sqrt{N}$ (and hence a straight line is a very poor approximation to the true model), the optimal design essentially minimizes the bias term in (3.3). The design minimizing (3.3) is called the "all bias design" by Box and

Draper (1959). It has $m_2^2 = 1/3$, which is easily verified after setting $m_3 = 0$ in (3.3). Thus any design with $m_1 = m_3 = 0$ and $m_2^2 = 1/3$ is an all bias design. They show that over a very wide range of values of $\beta_2/(\sigma/\sqrt{N})$, any all bias design comes close to minimizing (3.1).

Having determined the value of m_2, say m_2^*, which minimizes (3.1), Box and Draper (1959) show that among all designs having first and third moments 0, and second moment equal to the optimal value m_2^*, the fourth moment m_4 must be maximized in order to maximize the power of the lack of fit test. For designs on $[-1, 1]$, $m_4 \leqslant m_2$ with equality if and only if the design is supported on $\{-1, 0, 1\}$. Thus the best design is one supported on $\{-1, 0, 1\}$ with second moment equal to m_2^*. This is achieved by a design with mass $m_2^*/2$ at -1, mass $m_2^*/2$ at 1, and mass $1 - m_2^*$ at 0. This must be regarded as an approximate theory design since for a given sample size N $m_2^*/2$ need not be an integer multiple of $1/N$.

Box and Draper (1959) carry out similar calculations for fitting a plane when X is a spherical region in k dimensional Euclidean space and the true model is quadratic. They again note that the all bias design performs well over a wide range of departures from the fitted model. This leads to the proposal that one simply minimize the bias term since, in the examples considered, this leads to designs which are reasonably efficient (relative to the design which actually minimizes (3.1)) as long as the fitted model is a reasonable approximation to the true model.

The above formulation has been further explored in Box and Draper (1963) for fitting a second order model when the true model is a cubic. In Box and Draper (1975) the authors develop a measure of the insensitivity of a design to outliers, suggesting that this measure be used along with that discussed above in evaluating the suitability of a design.

A simple way to extend the results of Box and Draper (1959) might be by removing the restriction to least squares estimators. This is the approach adopted by Karson et al. (1969). In the setting described above for Box and Draper (1959), note that the matrix in the quadratic form for B in (2.7), namely $[\boldsymbol{F}'\boldsymbol{C}\Gamma_{gg}\boldsymbol{C}'\boldsymbol{F} - \boldsymbol{F}'\boldsymbol{C}\Gamma_{gf} - \Gamma_{fg}\boldsymbol{C}'\boldsymbol{F} + \Gamma_{ff}]$, can be rewritten as

$$\Gamma_{ff} - \Gamma_{fg}\Gamma_{gg}^{-}\Gamma_{gf} + (\boldsymbol{C}'\boldsymbol{F} - \Gamma_{gg}^{-}\Gamma_{gf})'\Gamma_{gg}(\boldsymbol{C}'\boldsymbol{F} - \Gamma_{gg}^{-}\Gamma_{gf}). \tag{3.5}$$

Here $\Gamma_{gg}^{-}\Gamma_{gf}$ is well-defined and (3.5) holds even if Γ_{gg} is singular; note that

$$\begin{pmatrix} \Gamma_{ff} & \Gamma_{fg} \\ \Gamma_{gf} & \Gamma_{gg} \end{pmatrix} \tag{3.6}$$

is nonnegative definite. If our design is such that $\boldsymbol{C}$ can be chosen to satisfy

$$\boldsymbol{C}'\boldsymbol{F} = \Gamma_{gg}^{-}\Gamma_{gf} \tag{3.7}$$

then any such choice minimizes the matrix in (3.5) and hence B. We note by (2.5) that there is a $\boldsymbol{C}$ satisfying (3.7) if and only if $\Gamma_{gg}^{-}\Gamma_{gf}\beta$ is estimable for the design used. For any such design, it follows from what is essentially the Gauss–Markov Theorem, that $\boldsymbol{C}'\boldsymbol{C}$ achieves its matrix minimum (in the sense of the usual ordering of nonnegative definite matrices, namely $\boldsymbol{A} \geqslant \boldsymbol{B}$ if and only if $\boldsymbol{x}'\boldsymbol{A}\boldsymbol{x} \geqslant \boldsymbol{x}'\boldsymbol{B}\boldsymbol{x}$ for all vectors $\boldsymbol{x}$) among all $\boldsymbol{C}$ satisfying (3.7) with the choice

$$\boldsymbol{C} = \boldsymbol{F}(\boldsymbol{F}'\boldsymbol{F})^{-}\boldsymbol{\Gamma}_{\boldsymbol{fg}}\boldsymbol{\Gamma}_{\boldsymbol{gg}}^{-} \tag{3.8}$$

which yields

$$\boldsymbol{C}'\boldsymbol{C} = \boldsymbol{\Gamma}_{\boldsymbol{gg}}^{-}\boldsymbol{\Gamma}_{\boldsymbol{gf}}(\boldsymbol{F}'\boldsymbol{F})^{-}\boldsymbol{\Gamma}_{\boldsymbol{fg}}\boldsymbol{\Gamma}_{\boldsymbol{gg}}^{-}. \tag{3.9}$$

Thus the prescription of Karson et al. (1969) is to restrict to the class of designs, say D, for which $\boldsymbol{\Gamma}_{\boldsymbol{gg}}^{-}\boldsymbol{\Gamma}_{\boldsymbol{gf}}\boldsymbol{\beta}$ is estimable and select $\boldsymbol{C}$ (our estimator) as in (3.8). For this choice of $\boldsymbol{C}$, any design in D minimizes B. Now choose the design in D which minimizes V in (2.7), i.e., minimizes $\mathrm{tr}\boldsymbol{C}'\boldsymbol{C}\boldsymbol{\Gamma}_{\boldsymbol{gg}}$. While this is generally difficult if one restricts to exact designs, the minimization is not too difficult if one uses the approximate theory mentioned in Section 2. For example, Fedorov (1972) characterizes the solution to this sort of problem in the approximate theory and obtains iterative methods for solving the problem.

We note that the two stage approach of Karson et al. (1969) need not yield a global minimum of R in (2.7). By restricting the class of designs used in the minimization of V at the second stage, they may eliminate designs that produce a much smaller value of V, nearly minimize B, and hence produce a smaller overall value of R. However, the designs (and estimators) of Karson et al. (1969) must yield a value of R no larger than the designs (using the least squares estimators) of Box and Draper (1959) simply because Karson et al. (1969) allow the additional flexibility of arbitrary linear estimators. These issues are further discussed in Kiefer (1973). As mentioned in Section 2, the global minimization of R is quite difficult. Kiefer (1973) suggests minimizing an "averaged" or integrated version of R and obtains some results in this direction.

A somewhat different approach to model robust design for polynomial models is discussed in Atwood (1971). In the notation of Section 2 let $X = [-1, 1]$, let $\sum_{l=1}^{s} \alpha_l g_l(x)$ in (2.2) be a polynomial of degree $s - 1$, and let $\sum_{j=1}^{m} \beta_j f_j(x)$ in (2.1) be a polynomial of degree $m - 1$, with $m > s$. Atwood (1971) restricts attention to estimators of the response at x which are a convex combination of the best linear unbiased estimator of the response at x under model (2.1) and the best linear unbiased estimator of the response at x under model (2.2). Likewise designs are restricted to convex combinations of the known, classical D-optimal designs for models (2.1) and (2.2). Let $\varepsilon = \max_{x\in X} |\sum_{l=1}^{s} \alpha_l g_l(x) - \sum_{j=1}^{m} \beta_j f_j(x)|$. For a given value of $N\varepsilon^2/\sigma^2$, Atwood (1971) seeks the estimator and design (of the forms described above) which minimize the maximum over X of the mean squared error for the estimate of the response at x. Numerical results are tabulated for the cases $m = s + 1$, $1 \leqslant s \leqslant 9$, and $m = s + 2$, $1 \leqslant s \leqslant 8$. These results can be used to select "good" model robust designs in the sense just described. Note that when the true model is assumed to be a quadratic and we plan to fit a straight line, the designs considered by Atwood (1971) are convex combinations of the D-optimal design for fitting a straight line, which puts equal mass on -1 and 1, and the D-optimal design for fitting a quadratic, which puts equal mass at -1, 0, and 1. Such designs therefore put equal mass on -1 and 1, and place the remaining mass at 0. The "optimal" mass will, of course, depend on $N\varepsilon^2/\sigma^2$. While the support of the designs of Atwood is the same as the corresponding optimal design in Box and Draper (1959), the optimal number of observations to take at each support point is different.

3.2. Infinite dimensional sets of true models

In the papers discussed above, the "true" model (2.1) is assumed to lie in a known finite dimensional space of functions (the space spanned by a particular specification of the f_i, usually polynomials) so that the risk in (2.7) can be evaluated. One criticism of this approach (see Huber, 1975) is that this fails to safeguard against all potentially dangerous small deviations from the model rather than just a few arbitrarily selected polynomial ones. In fact one might wonder why, if the form of the true model is essentially known, we are fitting (2.2) at all. Why not simply fit the true model and use the corresponding optimal design? Beginning with Huber (1975), a number of authors have therefore considered the case in which the true model lies in an infinite dimensional space of functions. In the notation of Section 2, all these authors, except Li (1984), take $X = L$ and model (2.1) is written as

$$\mathrm{E}Y_i = \sum_{l=1}^{s} \alpha_l g_l(\boldsymbol{x}_i) + f(\boldsymbol{x}_i). \tag{3.10}$$

f is the difference between the true unknown model and the known model we intend to fit, namely $\sum_{l=1}^{s} \alpha_l g_l(\boldsymbol{x}_i)$. f is assumed to belong to some class of functions F. Our assumptions about the form of the true model are expressed by our specification of F.

Huber (1975), Marcus and Sacks (1978), and Li (1984) consider the univariate case where X is a subset of the real line and with the fitted model being a straight line, i.e.,

$$\sum_{l=1}^{s} \alpha_l g_l(\boldsymbol{x}_i) = \alpha_0 + \alpha_1 x_i. \tag{3.11}$$

To use the notation of Section 2, in all these papers $\boldsymbol{\beta} = (\alpha_0, \alpha_1, 1)'$, $\boldsymbol{g}(\boldsymbol{x}) = (1, x)'$, $\boldsymbol{f}(x) = (1, x, f(x))'$, and $\boldsymbol{C}'\boldsymbol{Y} = (\widehat{\alpha}_0, \widehat{\alpha}_1)'$ where $\widehat{\alpha}_0$ and $\widehat{\alpha}_1$ are linear estimators of α_0 and α_1, respectively. These authors differ in their choice of X, F, and the loss function to be minimized.

Huber (1975) takes $X = [-1/2, 1/2]$, $F = \{f(x); \int_{-1/2}^{+1/2} f^2(x)\,\mathrm{d}x < \varepsilon\}$, and restricts attention to the case where $\widehat{\alpha}_0$ and $\widehat{\alpha}_1$ are the least squares estimators of α_0

and α_1, respectively, in the fitted model. The loss function is the supremum of (2.7) over F with ν being Lebesgue measure on X. In this case the loss can be written as

$$\sup_{f \in \boldsymbol{F}} \mathrm{E} \int_{-1/2}^{+1/2} \left(\widehat{\alpha}_0 + \widehat{\alpha}_1 x - f(x)\right)^2 \mathrm{d}x. \tag{3.12}$$

The object is to find the design minimizing (3.12). Since the set F includes f which have an arbitrarily high, narrow spike above any point in X, only designs which are absolutely continuous on X have a finite loss. The design minimizing (3.12) is a continuous design and is given in Section 4 of Huber (1975). It is not clear how it should be implemented in practice since any discrete implementation would make (3.12) infinite. Huber's approach might lead to implementable designs if F were a smoother class of functions.

Marcus and Sacks (1978) take $X = [-1, 1]$, $F = \{f(x);\ |f(x)| < \phi(x)\}$ where $\phi(x)$ is a positive bounded even function on $[-1, 1]$, $\phi(0) = 0$, and $\phi(1) = 1$, and $\widehat{\alpha}_0$ and $\widehat{\alpha}_1$ are allowed to be arbitrary linear estimators. They use the weighted mean square error loss

$$\sup_{f \in F} \mathrm{E}\big((\widehat{\alpha}_0 - \alpha_0)^2 + \theta^2(\widehat{\alpha}_1 - \alpha_1)^2\big). \tag{3.13}$$

This is not of the form of (2.7), although (2.7) reduces to this for $\theta = 1$ if one uses $\boldsymbol{f}(x) = (1, x, 0)'$ and takes ν to be Lebesgue measure on X. The goal is to find both the design and linear estimators $\widehat{\alpha}_0$ and $\widehat{\alpha}_1$ minimizing (3.13). If $\phi(x) \geqslant mx$ for some m then the (unique) optimal design is supported on $\{-1, 0, 1\}$. For convex $\phi(x)$ there is a wide range of cases for which the optimal design is supported on $\{-z, z\}$. In general z depends on ϕ, θ, σ^2, and N in a complicated way (see Marcus and Sacks, 1978, Theorem 3.2). In the case where $\phi(x) = mx^2$ and $\sigma^2/Nm^2 \leqslant \theta^4$ it turns out that $z = (\sigma^2/Nm^2)^{1/4}$ (assuming this yields a value $\leqslant 1$).

Li (1984) uses the loss given in (3.13) but takes $X = \{k/2M, -k/2M;\ k = 1, 2, \ldots, M\}$ for a fixed M; $L = [-1/2, 1/2]$. Designs are restricted to have support in X. This choice may be motivated by the fact that in practice the predictor variables can usually only be set to rational values and by the "naive" approach of spreading observations uniformly over some interval as mentioned in Section 1. Here the interval is $[-1/2, 1/2]$ and designs with support on a finite set of points spread evenly over this interval are investigated. The question is whether the proportion of observations taken at each point in X is the same, as the naive approach would suggest. Li (1984) chooses $F = \{f(x);\ |f(x)| \leqslant \varepsilon,\ \int_x f(x)\,\mathrm{d}\nu(x) = 0, \text{ and } \int_x x f(x)\,\mathrm{d}\nu(x) = 0\}$ where ν is Lebesgue measure on X. The conditions $\int_x f(x)\,\mathrm{d}\nu(x) = 0$ and $\int_x x f(x)\,\mathrm{d}\nu(x) = 0$ are chosen to make α_0 and α_1 identifiable in the true model. The choice of F is motivated by the fact that the choice in Marcus and Sacks (1978) gives special status to the point 0 so that there is no contamination ($f(x) = 0$) at this point. Thus there is value in taking observations at 0 (or near 0 in the case of convex ϕ). Li (1984) restricts $\widehat{\alpha}_0$ and $\widehat{\alpha}_1$ to be the least squares estimators of α_0 and α_1 for the fitted model and seeks the design minimizing (3.13). The optimal designs are given in Theorems

4.1, 4.2, and 4.3 of Li (1984). They do not take the same proportion of observations at each point in X, but rather take observations at the endpoints of X (namely $-1/2$ and $1/2$) as well as at certain interior points of X. Asymptotically (see Theorems 6.1 and 6.2 of Li (1984)) these designs put equal mass at $-1/2$ and $1/2$, and spread the remaining mass uniformly over an interval of the form $[-1/2, -x] \cup [x, 1/2]$, where $0 \leqslant x \leqslant 1/2$ and x depends on σ^2, N, ε, and θ.

Li and Notz (1982) consider robust regression in a multiple regression setting, fitting a plane to data. X is a compact subset of k-dimensional Euclidean space $\mathbb{R}^k$ and (3.11) becomes

$$\sum_{l=1}^{s} \alpha_l g_l(\boldsymbol{x}_i) = \alpha_0 + \alpha_1 x_{i1} + \cdots + \alpha_k x_{ik}. \tag{3.14}$$

Li and Notz (1982) take F as in Li (1984), namely $F = \{f(x);\ |f(x)| \leqslant \varepsilon, \int_x f(x)\,\mathrm{d}\nu(x) = 0, \text{ and } \int_x x f(x)\,\mathrm{d}\nu(x) = 0\}$ where ν is Lebesgue measure on X. Li and Notz (1982) restrict designs to have finite support and seek the design and linear estimators $\widehat{\alpha}_i$ of the α_i minimizing

$$\sup_{f\in F} \mathrm{E}\left((\widehat{\alpha}_0 - \alpha_0)^2 + \sum_{i=1}^{k} \theta_i^2 (\widehat{\alpha}_i - \alpha_i)^2 \right). \tag{3.15}$$

Optimal designs are shown to have support on the extreme points of X. When X is the simplex or cube in $\mathbb{R}^k$, Li and Notz (1982) show that the classical optimal designs (those putting uniform mass on the corners of the simplex or cube) and the usual least squares estimators are optimal for the loss in (3.14). For the univariate case, the reason is essentially as that given in the example corresponding to Figure 1.2 in Section 1. It is not clear whether the results in Li and Notz (1982) are of much practical value. In many practical settings the true model is likely to have some smoothness so that the F considered by Li and Notz (1982) is too broad. However, these results do indicate the need for some restrictions on the class F. They indicate that no design with finite support can protect against arbitrary bias, and hence improve upon the classical optimal design for the fitted model. We note in passing that Li and Notz (1982) also give results for interpolation, extrapolation, and bilinear models.

Pesotchinsky (1982) also considers the setting of multiple regression. The fitted model is (3.14). F is similar to that in Marcus and Sacks (1978), namely $F = \{f(\boldsymbol{x});\ |f(\boldsymbol{x})| < \varepsilon\phi(\boldsymbol{x})\}$, where $\varepsilon > 0$ and $\phi(\boldsymbol{x})$ is a convex function of $||\boldsymbol{x}||^2$. Pesotchinsky (1982) assumes the usual least squares estimators for parameters in the fitted model are used. For design ξ on a compact subset X of k-dimensional Euclidean space, he defines the matrix

$$\boldsymbol{D}(\xi, f) = \frac{\sigma^2}{m} \boldsymbol{M}^{-1(\xi)} + \boldsymbol{M}^{-1}(\xi)\boldsymbol{\Psi}(\xi)\boldsymbol{\Psi}'(\xi)\boldsymbol{M}^{-1}(\xi) \tag{3.16}$$

where $\boldsymbol{M}(\xi)$ is the information matrix (see (2.8)) for the fitted model (3.14) and $\boldsymbol{\Psi}'(\xi) = N(\mathrm{E}_\xi[f(\boldsymbol{x})], \mathrm{E}_\xi[f(\boldsymbol{x})x_1], \ldots, \mathrm{E}_\xi[f(\boldsymbol{x})]x_k)$. E_ξ denotes the expectation over X with respect to ψ. $\boldsymbol{D}(\xi, f)$ serves as the analog of the covariance matrix $(\sigma^2/n)\boldsymbol{M}^{-1}(\xi)$ when the "contamination" f in (3.10) is taken into account. Pesotchinsky (1982) seeks designs which minimize $\sup_{f\in F}\Phi(\boldsymbol{D}(\xi, f))$ for some real-valued function Φ. Choices for Φ might be the determinant (D-optimality), trace (A-optimality), or maximum eigenvalue (E-optimality). D-, A-, and E-optimal are found. These designs put uniform mass on a sphere of a particular radius (which depends on ϕ, the choice of optimality criterion Φ, and characteristics of X) assuming this sphere is contained in X. These designs might be regarded as generalizations of the optimal designs of Marcus and Sacks (1978) in the univariate case which put mass on the two point set $\{-z, z\}$ for appropriate choice of z. One difficulty of the optimal designs of Pesotchinsky (1982) is that they are continuous designs and so cannot be directly implemented. However, it is shown that star-point designs or regular replicas of 2^k factorial designs are very efficient under the appropriate choice of levels of the factors.

Sacks and Ylvisaker (1984) take yet a somewhat different approach to model robust design than those discussed above. They consider the univariate case and two forms for F, namely

$$\begin{aligned} F &= F_1(M) \\ &= \{f;\ |f(x) - f(y)| \leqslant M|x - y|,\ \text{for every } x, y \in \mathbb{R}^1\} \end{aligned} \tag{3.17}$$

and

$$F = F_2(M) = \{f;\ f \text{ is differentiable and } \mathrm{d}f/\mathrm{d}x \in F_1(M)\}. \tag{3.18}$$

Note that $F_1(M)$ contains all constant functions and $F_2(M)$ contains all linear functions. Rather than seeking to globally fit some simple approximation $\sum_{l=1}^{s} \alpha_l g_l(\boldsymbol{x}_i)$ to the true model, it is assumed that one wishes to only estimate some linear functional Λ of the true regression function $f \in F$. For the case where $X = [-1, 1]$, examples of linear functionals that might be of interest are $\Lambda f = f(0)$ (corresponding to the intercept if f were a straight line), $\Lambda f = [f(1) - f(-1)]/2$ (corresponding to the slope if f were a straight line), $\Lambda f = \mathrm{d}f(0)/\mathrm{d}x$ (again corresponding to the slope if f were a straight line), or $\Lambda f = \int \gamma(x) f(x)\,\mathrm{d}x$ for some known function $\gamma(x)$. They restrict to linear estimators of Λf which, for a design ξ supported at the points $\{x_1, x_2, \ldots, x_k\} \in X$, have the form $\sum_{i=1}^{k} c_i \bar{Y}(x_i)$ where $\bar{Y}(x)$ is the average of all the observations taken at x. They seek designs ξ (specification of the x_i and the number of observations n_i to be taken at each x_i subject to the total number of observations being N) and linear estimators (choice of the c_i) which minimize

$$\sup_{f\in F} \mathrm{E}_f\left(\sum_{i=1}^{k} c_i \bar{Y}(x_i) - \Lambda f\right)^2 = \sigma^2 \sum_{i=1}^{k} \frac{c_i^2}{n_i} + \sup_{f\in F}(Cf - \Lambda f)^2 \tag{3.19}$$

where C is the linear functional defined by

$$Cf = \mathrm{E}f\left(\sum_{i=1}^{k} c_i \bar{Y}(x_i)\right) = \sum_{i=1}^{k} c_i f(x_i) = \int_{\{x_1,\ldots,x_k\}} f\,\mathrm{d}C \tag{3.20}$$

and C is identified with the measure it induces. It is straightforward to verify that the right hand side of (3.19) is minimized when $n_i/n_k = |c_i|/|c_k|$ for $i = 1, \ldots, k$. For this choice (3.19) becomes

$$\frac{\sigma^2}{N}\left(\sum_{i=1}^{k} |c_i|\right)^2 + \sup_{f\in F}(Cf - \Lambda f)^2. \tag{3.21}$$

Sacks and Ylvisaker (1984) obtain results for a variety of types of linear functionals for both $F_1(M)$ and $F_2(M)$. Optimal solutions depend on Λ and which $F_i(M)$ one considers. For example, in $F_1(M)$ and with $\Lambda f = \sum_{j=1}^{Q}(f(z_j/Q))$ for given values of Q and $\{z_1, \ldots, z_Q\} \in X$, the optimal design takes N/Q observations at each z_j and estimates Λf by $(1/Q)\sum_{j=1}^{Q} \bar{Y}(z_j)$. The interested reader should consult Sacks and Ylvisaker (1984) for additional results and examples.

Not surprisingly, optimal designs depend very much on the specification of F and the loss function. Application of the above results therefore requires one to think carefully about the assumptions one is willing to make about the form of the true model and the purpose of the experiment (as expressed by the loss function). We see, for example, that if the fitted model is only a rough approximation to the true model, the standard optimal design for the fitted model may yield very misleading information. At the same time, the naive design that takes a moderate to large number of observations spread uniformly over the design region is sub-optimal also. The practical value of these results (and the results that will appear in later sections) therefore is probably in alerting us to the dangers of ignoring the approximate nature of any assumed model and in providing some insight concerning what features a design should have in order to be robust against departures from an assumed model while allowing good fit of the assumed model. This insight may be more valuable in practical settings than a slavish adoption of any particular mathematical model.

3.3. Randomization and robustness

The above discussion indicates how careful selection of a design (and in some instances the estimator) can provide protection against departures from a fitted model. One issue that we have not yet addressed is the role of randomization in robustness. A justification often given for experimental randomization is that it is a source of robustness against model inadequacies. Wu (1981) attempts to give a rigorous basis for this justification in the context of comparative experiments. His approach is quite different than that formulated in Section 2. In the spirit of Chapter 9 in Scheffé (1958), he associates with each unit two components. The first is called the unit error, and is an unknown constant associated with some feature of the unit, for example initial weight or income. The second is called the technical error and corresponds to other sources of error associated with the response. The technical errors are random variables with mean 0 and those corresponding to different responses are assumed to be uncorrelated with common variance. It is assumed that no interaction between the unit and technical errors exists.

Let e_u denote the unit error associated with unit u. The model for the response Y_{ut} to treatment t by unit u is

$$Y_{ut} = \alpha_t + e_u + \varepsilon_{ut}, \quad t = 1, 2, \ldots, v, \ u = 1, 2, \ldots, N, \tag{3.22}$$

where α_t is the effect of treatment t and ε_{ut} the technical error. Denote the set of possible values of e by E. E is called the neighborhood of model violations. Wu (1981) assumes E is bounded, contains unit errors of the form $e_u = c$ for all u, and is invariant under some transformation group T, i.e.,

$$e \in \mathrm{E} \Rightarrow \tau e \in \mathrm{E} \quad \text{for all } \tau \in T \tag{3.23}$$

where $\tau e = \{e_{\tau^{-1}u}\}u$, and τ^{-1} is the inverse of τ in the group T. The invariance assumption reflects the vagueness of the experimenter's knowledge about e_u. It is assumed that the experimenter is interested in estimating the pairwise treatment contrasts $\alpha_s - \alpha_t$, which will be estimated by the usual least squares estimators denoted $\widehat{\alpha}_s - \widehat{\alpha}_t$.

Wu (1981) defines $I_t = \{\text{units } u\text{: } u \text{ is assigned to treatment } t\}$ and calls $I = \{I_t\}_{t=1}^{v}$ a pattern. I corresponds to a nonrandomized design with treatment group sizes $n_t = |I_t|$ for $t = 1, 2, \ldots, v$ and $\sum_{t=1}^{v} n_t = N$. Let Ξ denoted the collection of all such I's. A randomized design is defined to be a probability measure η over Ξ, i.e., $\{\eta(I);\ I \in \Xi\}$, with $\eta(I) \geqslant 0$ and $\sum_{I\in\Xi} \eta(I) = 1$. For any nonrandomized I with treatment group sizes $n_t = |I_t|$, let $a(I, e)$ denote the expected mean squared error, i.e.,

$$\begin{aligned} a(I, e) &= \sum_{s<t} \mathrm{E}(\widehat{\alpha}_s - \widehat{\alpha}_t - \alpha_s + \alpha_t)^2 \\ &= v \sum_{t=1}^{v} (e_{.t} - e_{..})^2 + \sigma^2 (v-1) \sum_{i=1}^{v} \frac{1}{n_t} \end{aligned} \tag{3.24}$$

where $e_{.t} = (1/n_t) \sum_{u\in I_t} e_u$ and $e_{..} = (1/v) \sum_{t=1}^{v} e_{.t}$. For a randomized design η, the expected mean squared error is

$$r(\eta, e) = \sum_{I\in\Xi} \eta(I) a(I, e). \tag{3.25}$$

A design η is said to be model robust with respect to E if it minimizes $\max_{e\in E} r(\eta, e)$.

Wu (1981) shows that if E is invariant under the group of permutations of the units, then the balanced completely randomized design is model robust with respect to E.

If (3.22) is modified to include block effects, it is also shown that the randomized complete block design is model robust. Finally, if (3.22) is modified to include row and column effects, any Latin square followed by a random permutation of its rows or columns is model robust. Note also that $\max_{e \in E} r(\eta, e)$ can be used as a criterion for comparing designs. This is further discussed in Wu (1981).

4. Extrapolation

As we have already indicated, in real applications of statistics the models we fit to the response are, at best, reasonable approximations to the true (unknown) relation between the response and some set of explanatory variables. We obtain the most information about the true model at the points at which we actually observe the response. The model we fit to the data will best approximate the true model at these points and, if the true model is smooth and well behaved (does not vary too rapidly), should be a reasonable approximation in regions near these points (the design region). The approximation will be poor as we move far away from these points. How poor will depend on how far we are from the design region and how rapidly the true model varies (or how far outside the design region our fitted model does a reasonable job of approximating the true model). Predicting the response at values of the explanatory variables well outside the design region is referred to as extrapolation. Even if we know the form of the true model, extrapolation is uncertain simply due to random error in the data. This error leads to discrepancies between the true and fitted model which are often magnified as one moves away from the region in which observations were taken. When the true model is not really known, extrapolation becomes even less reliable. This is reflected in what we tell students in elementary courses about the dangers of extrapolation. One reason caution is urged is that we do not know to what extent the assumptions behind our fitted model hold outside the region in which the response was actually observed.

From these observations, it seems reasonable to investigate model robust designs for extrapolation. A suitably chosen design may help correct for bias in the model we plan to fit to the data (i.e., discrepancies between this fitted model and the true model). To illustrate, again examine Figure 1.1. Assume the true (quadratic) model holds well outside the design region. The line that best approximates the true model in the design region (the region between the tick marks on the horizontal axis) does a poor job of representing the true model near the boundary of the design region and beyond. The line fit using the "classical" optimal design fits better at or near the boundary of the design region. For extrapolating outside the design region, this line would perform better than the line that best approximates the true model in the design region. If our goal was to extrapolate at large values of x, a design taking observations only near the right hand boundary of the design region would yield a fitted line that was even better for purposes of extrapolation than either of the lines in the figure. If we examine Figure 1.2, we see the opposite. Assuming the true model continues as suggested in the figure, the line that best approximates the true model in the design region will continue to be the best approximation outside the design region and hence would be useful for extrapolation. The least squares line fit by taking observations at the tick marks would continue to be biased outside the design region. Finally, a line fit by taking observations at or very near the right hand boundary of the design region would have a very steep slope and would be very poor for purposes of extrapolation. Thus a "model robust" design that is good for approximating the true model in the design region may or may not be useful for extrapolation. As we have seen, results are very dependent on the form of the true model.

Designs for extrapolation when bias is present have been studied by a number of authors. Draper and Herzberg (1973) appear to be the first to treat this issue. They consider fitting a first order model over the k-dimensional ball and extrapolating to a point $\boldsymbol{z} = (z_1, 0, \ldots, 0)'$ outside the ball when the true model is a second order model. The approach to models with bias is that of Box and Draper (1959). In the notation of Section 2, X is the k-dimensional ball of radius 1 centered at the origin and L is the line segment from the point $\boldsymbol{z}$ outside of X to the point in X closest to $\boldsymbol{z}$. ν in (2.6) is a point mass at $\boldsymbol{z}$. Model (2.1) is a second order response surface and model (2.2) is first order. $\boldsymbol{\alpha}$ in model (2.2) is estimated using least squares (assuming model (2.2) is correct). Thus $\boldsymbol{C}$ in (2.5) is $(\boldsymbol{G}'\boldsymbol{G})^{-}\boldsymbol{G}'$. Designs are limited to exact designs that put some observations at $(1, 0, \ldots, 0)'$ and the remainder equally divided among the k vertices of the $(k-1)$ simplex perpendicular to and centered on the first coordinate axis. The design in this class which minimizes B only in (2.7) is found, assuming knowledge of which of two regions $\boldsymbol{\beta}$ lies within. Draper and Herzberg (1979) consider an extension to the case where X is the k-dimensional ball of radius 1 centered at the origin and L is the k-dimensional ball of radius R centered at the origin. ν is uniform measure over $L - X$. Model (2.1) is a quadratic or second order response surface and model (2.2) is first order. α in model (2.2) is estimated using least squares (assuming model (2.2) is correct). They consider both the cases (a) model (2.1) is second order and (2.2) is first order, and (b) model (2.1) is second order and (2.2) is third order. In both cases α in model (2.2) is estimated using least squares (assuming model (2.2) is correct). In case (a) designs are restricted to those with "equal scaling" in all coordinates and the design which minimizes both B and $V + B$ in (2.7), independently of R, takes all observations as far out in X as possible, i.e., uniformly spaced on the surface of X so that the resulting $\boldsymbol{M}$ in (2.8) is the same as for uniform probability measure on the surface of X. In case (b), designs are restricted to the class of central composite designs and it is the designs in this class minimizing both B and V independently of R are found.

Kiefer (1980) provides a review of the above results along with additional results on extrapolation when bias is present. In the notation of Section 2, it is assumed that the bias is represented by the function $\boldsymbol{h}(\boldsymbol{x}) = (h_1(\boldsymbol{x}), h_2(\boldsymbol{x}), \ldots, h_{m-s}(\boldsymbol{x}))'$ satisfying $\boldsymbol{f}(\boldsymbol{x}) = (\boldsymbol{g}(\boldsymbol{x})', \boldsymbol{h}(\boldsymbol{x})')'$. (2.1) then can be written as

$$\mathrm{E}Y_i = \sum_{l=1}^{s} \alpha_l g_l(\boldsymbol{x}_i) + \sum_{r=1}^{m-s} \gamma_r h_r(\boldsymbol{x}_i). \tag{4.1}$$

Let $\boldsymbol{\gamma} = (\gamma_1, \ldots, \gamma_{m-s})'$ so that in (2.3) $\boldsymbol{\beta} = (\alpha', \gamma')'$. Also let $\boldsymbol{H}$ be the $N \times (m-s)$ matrix whose i, rth entry is $h_r(\boldsymbol{x}_i)$, so that $\boldsymbol{F} = (\boldsymbol{G}\,\boldsymbol{H})$ in (2.3). If we write $\boldsymbol{D} = \boldsymbol{F}'\boldsymbol{C}$

and $\boldsymbol{P}$ be any $s \times (m-s)$ matrix satisfying $\Gamma_{gg}\boldsymbol{P} = \Gamma_{gh}$, then V and B in (2.7) become

$$V = \operatorname{tr} \boldsymbol{C}'\boldsymbol{C}\Gamma_{gg}, \tag{4.2}$$

$$\begin{aligned}
\sigma^2 B &= \int_L \{(\boldsymbol{\alpha}', \boldsymbol{\gamma}')[\boldsymbol{D}\boldsymbol{g}(\boldsymbol{x}) - \boldsymbol{f}(\boldsymbol{x})]\}^2 \nu(\mathrm{d}\boldsymbol{x}) \\
&= (\boldsymbol{\alpha}', \boldsymbol{\gamma}') \left[\boldsymbol{D} - \begin{pmatrix} \boldsymbol{I}_s \\ \boldsymbol{P}' \end{pmatrix}\right] \Gamma_{gg} \left[\boldsymbol{D} - \begin{pmatrix} \boldsymbol{I}_s \\ \boldsymbol{P}' \end{pmatrix}\right]' \begin{pmatrix} \boldsymbol{\alpha} \\ \boldsymbol{\gamma} \end{pmatrix} \\
&\quad + \boldsymbol{\gamma}'[\Gamma_{hh} - \Gamma_{hg}\Gamma_{gg}^{-}\Gamma_{gh}]\boldsymbol{\gamma} \\
&= \sigma^2 B_1 + \sigma^2 B_2
\end{aligned} \tag{4.3}$$

where $\boldsymbol{I}_s$ is the $s \times s$ identity matrix. Notice that in (4.3) only B_1 depends on the design (through $\boldsymbol{D}$). X here is the k-dimensional ball of radius 1 centered at the origin. The approximate theory of design is used and a design ξ is a probability measure on X. The information matrix is partitioned as

$$\boldsymbol{M}(\xi) = \begin{pmatrix} \boldsymbol{Q} & \boldsymbol{L} \\ \boldsymbol{L}' & \boldsymbol{K} \end{pmatrix} = \int_X \boldsymbol{f}(\boldsymbol{x})\boldsymbol{f}(\boldsymbol{x})'\xi(\mathrm{d}\boldsymbol{x}) \tag{4.4}$$

where $\boldsymbol{Q}$ is $s \times s$ (corresponding to $\boldsymbol{g}$), $\boldsymbol{K}$ is $(m-s) \times (m-s)$ (corresponding to $\boldsymbol{h}$), and $\boldsymbol{L}$ is $s \times (m-s)$. The problems addressed are extrapolation to $L =$ a point (in which case ν puts all its mass at this point), extrapolation to $L =$ the k-dimensional ball of radius R centered at the origin for $\boldsymbol{g}$ linear and $\boldsymbol{h}$ quadratic, and extrapolation to $L =$ the k-dimensional ball of radius R centered at the origin for $\boldsymbol{g}$ quadratic and $\boldsymbol{h}$ cubic.

In Kiefer (1980) the approaches of Box and Draper (1959) and of Karson et al. (1969) are employed. See Section 2 for more details on these approaches. An important difference in the two approaches is that Box and Draper use the usual least squares estimators to fit models (thus $\boldsymbol{C}' = (\boldsymbol{G}'\boldsymbol{G})^{-}\boldsymbol{G}'$) and seek to minimize B only. Since B depends on the design only through B_1, this is equivalent to minimizing B_1. In this case B_1 reduces to

$$\sigma^2 B_1 = \boldsymbol{\gamma}'(\boldsymbol{L}'\boldsymbol{Q}^{-} - \boldsymbol{P}')\Gamma_{gg}(\boldsymbol{L}'\boldsymbol{Q}^{-} - \boldsymbol{P}')'\boldsymbol{\gamma}. \tag{4.5}$$

Karson, Manson and Hader consider general linear estimators of the parameters and first select the estimator to minimize B for every design. This leads to choosing $\boldsymbol{C}' = (\boldsymbol{I}_s\boldsymbol{P})(\boldsymbol{F}'\boldsymbol{F})^{-}\boldsymbol{F}'$ so that

$$\boldsymbol{D} = \begin{pmatrix} \boldsymbol{I}_s \\ \boldsymbol{P}' \end{pmatrix}$$

and hence $B_1 = 0$ in (4.3). The design is then chosen to minimize the resulting V, namely

$$V = N^{-1}\mathrm{tr}\begin{pmatrix} \Gamma_{gg} & \Gamma_{gh} \\ \Gamma_{hg} & \Gamma_{hg}\Gamma_{gg}^{-}\Gamma_{gh} \end{pmatrix} \boldsymbol{M}(\xi)^{-1}. \tag{4.6}$$

For extrapolation to a point $\boldsymbol{z}$ and using the Karson, Manson and Hader approach, if we take $\boldsymbol{D}$ to satisfy $\boldsymbol{Dg}(\boldsymbol{z}) = \boldsymbol{f}(\boldsymbol{z})$ then $\sigma^2 B$ attains its minimum value of 0 in the integral form of (4.3). In this case $\sigma^2(V+B)$ becomes $N^{-1}\sigma^2\boldsymbol{f}(\boldsymbol{z})'\boldsymbol{M}(\xi)^{-}\boldsymbol{f}(\boldsymbol{z})$ and the problem reduces to design for unbiased estimation of $\boldsymbol{\beta}'\boldsymbol{f}(\boldsymbol{z})$ using the usual least squares estimator of this linear parametric function.

Using the Box and Draper approach, the development is more complicated. For simplicity Kiefer (1980) considers the univariate problem. Thus $X = [-1, 1]$, $L = \{z\}$, $\boldsymbol{g}(x) = (1, x)'$, and $h(x) = x^2$. Using $\boldsymbol{C}' = (\boldsymbol{G}'\boldsymbol{G})^{-}\boldsymbol{G}'$, B_1 in (4.3) becomes

$$\gamma^2\left\{(\mu_2 - \mu_1^2)^{-1}[\mu_2^2 - \mu_1\mu_3 + z(\mu_3 - \mu_1\mu_2)] - z^2\right\}^2 \tag{4.7}$$

where $\mu_i = \int_{-1}^{+1} x^i\xi(\mathrm{d}x)$. We must have $\mu_2 - \mu_1^2 \neq 0$ in order for $\alpha'\boldsymbol{g}(z)$ to be estimable if $z > 1$ (extrapolation). One can show that in the case $z > 1$ there are no designs ξ for which $\boldsymbol{\alpha}'\boldsymbol{g}(z)$ is estimable and for which (4.7) vanishes. Thus the calculation of a ξ that minimizes (4.7) is quite difficult although it can be carried out numerically. Draper and Herzberg (1973) handled this problem by restricting the class of designs (see above). In the univariate case, this corresponds to considering only designs supported on two points, c (with $-1 \leqslant c < 1$) and 1.

Consider now extrapolation to $L =$ the k-dimensional ball of radius R centered at the origin for $\boldsymbol{g}$ linear and $\boldsymbol{h}$ quadratic. Following Kiefer (1974), we may restrict ourselves to designs which are rotationally invariant. Here we consider the univariate case only so that $\boldsymbol{g}(x) = (1, x)'$, and $\boldsymbol{h}(x) = x^2$. For a brief discussion of the general case, see Kiefer (1980). In the univariate case, rotationally invariant designs ξ are those satisfying $\xi(x) = \xi(-x)$ and therefore has odd moments equal to 0. If we restrict to symmetric ν, let $\mu_i = \int_{-1}^{+1} x^i\xi(\mathrm{d}x)$ and $\omega_i = \int_{-R}^{+R} z^i\nu(\mathrm{d}z)$, we obtain

$$\Gamma_{gg}\begin{pmatrix} 1 & 0 \\ 0 & \omega_2 \end{pmatrix},$$

$$\Gamma_{hh} = (\omega_4)$$

and

$$\boldsymbol{M}(\xi) = \begin{pmatrix} 1 & 0 & \mu_2 \\ 0 & \mu_2 & 0 \\ \mu_2 & 0 & \mu_4 \end{pmatrix}. \tag{4.8}$$

If we follow the Karson, Manson and Hader approach, (4.4) becomes

$$NV = \frac{\mu_4 - 2\mu_2\omega_2 + \omega_2^2}{\mu_4 - \mu_2^2} + \frac{\omega_2}{\mu_2} \tag{4.9}$$

which must be minimized. We restrict to nonsingular $\boldsymbol{M}$ (otherwise V is undefined or might be interpreted as infinite). Since (4.6) is of the form $\mathrm{tr}(\boldsymbol{A}\boldsymbol{M}^{-1})$ for nonnegative definite $\boldsymbol{A}$ and $\boldsymbol{M}$ the information matrix for the problem of quadratic regression on $[-1, +1]$, it follows from the symmetry of ξ and the characterization of admissible designs in Kiefer (1959) that V is minimized by a design of the form $\xi(-1) = \xi(1) = a/2$, $\xi(0) = 1 - a$. This yields $\mu_2 = \mu_4 = a$ and thus from (4.9)

$$a = \frac{1}{1 + \frac{|\omega_2 - 1|}{\sqrt{\omega_2(\omega_2+1)}}}. \tag{4.10}$$

Thus the Karson, Manson and Hader approach yields

$$\sigma^2 B_2 = \gamma^2(\omega_4 - \omega_2^2), \qquad NV = \left(|\omega_2 - 1| + \sqrt{\omega_2(\omega_2 + 1)}\right)^2. \tag{4.11}$$

If we follow the Box and Draper approach, B_1 will be 0 in (4.3), and hence B minimized, provided $\boldsymbol{L}'\boldsymbol{Q}^- = \boldsymbol{P}'$ (see (4.5)). We again restrict to nonsingular designs but, for the moment, do not require them to be symmetric. The condition $\boldsymbol{L}'\boldsymbol{Q}^- = \boldsymbol{P}'$ yields

$$(\mu_2 - \mu_1^2)^{-1}(\mu_2^2 - \mu_1\mu_3, \mu_3 - \mu_1\mu_2) = (\omega_2, 0). \tag{4.12}$$

This requires $\mu_3 = \mu_1\mu_2$ and $\mu_2 = \omega_2$. If $\omega_2 \leqslant 1$ this can be achieved by a design of the form $\xi(-1) = \xi(1) = \omega_2/2$, $\xi(0) = 1 - \omega_2$. If $\omega_2 > 1$, one can show that the design $\xi(-1) = \xi(1) = 1/2$ is best. This yields

$$\begin{aligned} &\sigma^2 B = \gamma^2(\omega_4 - \omega_2^2), \quad NV = 2, \quad \text{if } \omega_2 \leqslant 1, \\ &\sigma^2 B = \gamma^2\left(\omega_4 - \omega_2^2 + (\omega_2 - 1)^2\right), \qquad NV = 1 + \omega_2, \quad \text{if } \omega_2 > 1. \end{aligned} \tag{4.13}$$

Note that when $\omega_2 \leqslant 1$ the Karson Manson and Hader approach yields the same B (namely B_2) and smaller V than the Box and Draper approach. However when $\omega_2 > 1$ the Box and Draper approach will yield a smaller value of $\sigma^2(V + B)$ than the Karson, Manson and Hader approach when

$$\frac{N\gamma^2}{\sigma^2} < \frac{2\omega_2 + 2\sqrt{\omega_2(\omega_2 + 1)}}{\omega_2 - 1}. \tag{4.14}$$

Next we consider extrapolation to L = the k-dimensional ball of radius R centered at the origin for $\boldsymbol{g}$ quadratic and $\boldsymbol{h}$ cubic. We consider only the univariate case with ν uniform probability measure on $[-R, R]$. We again may restrict to symmetric designs ξ. Thus all odd moments of ν and ξ are 0.

If we follow the Box and Draper approach, we find that

$$\sigma^2\gamma^{-2}B_1 = \left(\frac{3R^2}{5} - \frac{\mu_4}{\mu_2}\right)^2 \frac{R^2}{3} \tag{4.15}$$

so $B_1 = 0$ if and only if $R^2 \leqslant 5/3$. In this case B_1 will be 0 for any ξ satisfying $\mu_4 = 3\mu_2 R^2/5$. For such x, we then have

$$NV = \frac{2R^2(R^2 - \mu_2)}{5\mu_2\left[\frac{3R^2}{5} - \mu_2\right]}. \tag{4.16}$$

Ignoring the restriction that $\mu_4 = 3\mu_2 R^2/5$, V in (4.16) is minimized when $\mu_2 = R^2(1 - \sqrt{2/5})$. We must check that there is actually a symmetric design ξ on $[-1, 1]$ satisfying $\mu_4 = 3\mu_2 R^2/5$ and $\mu_2 = R^2(1 - \sqrt{2/5})$. There are, in fact, many such designs. Two are

$$\xi_1\left(\pm\sqrt{3R^2/5}\right) = \frac{1}{2 + \sqrt{8/5}} = \frac{1 - \xi_1(0)}{2},$$

$$\xi_2\left(\pm\sqrt{\frac{\frac{3R^2}{5}\left(1 - \frac{3R^2}{5}\right)}{1 + \sqrt{\frac{2}{5}} - \frac{3r^2}{5}}}\right) = \frac{\left[\sqrt{\frac{2}{5}} + 1 - \frac{3R^2}{5}\right]^2}{\left[2 + \sqrt{\frac{8}{5}}\right]\left[\sqrt{\frac{2}{5}} + \left(1 - \frac{3R^2}{5}\right)^2\right]}$$
$$= \frac{1}{2} - \xi_2(\pm 1). \tag{4.17}$$

For the case where $R^2 > 5/3$, the optimum design can be shown to be of the form $\xi(\pm 1) = a/2$, $\xi(0) = 1 - a$, where

$$a = \left(1 + \sqrt{\frac{\frac{R^4}{5} - \frac{2R^2}{3} + 1}{\frac{R^4}{5} + \frac{R^2}{3}}}\right)^{-1}. \tag{4.18}$$

The Karson, Manson and Hader approach is quite complicated here. Kiefer (1980) discusses this approach briefly and indicates that as $R \to \infty$ this approach is preferable to Box and Draper's when (approximately) $N\gamma^2/\sigma^2 < 8$. Kiefer (1980) also discusses both approaches when $L = \{-R, R\}$ in which case some simplification occurs. The interested reader should consult this paper for details.

Huber (1975) also considers extrapolation to a point in the univariate ($k = 1$) setting. In our notation, he takes $X = [0, \infty)$, $L = [x_0]$ ($x_0 < 0$), ν the probability measure putting all its mass at x_0, and model (2.1) becomes

$$\mathrm{E}(Y_i) = f(x_i) \tag{4.19}$$

where f is assumed to belong to the class of functions with bounded $(h+1)$th derivative, namely the class

$$F_0 = \{f;\ |f^{(h+1)}(x)| \leqslant \varepsilon,\ x_0 \leqslant x < \infty\}. \tag{4.20}$$

Since the goal is to estimate $f(x_0)$, we take $\alpha = f(x_0)$ and $g(x) = 1$ in (2.2). The supremum over F_0 of expected loss in (2.7) is to be minimized, i.e., we seek the design which minimizes

$$\sup_{f \in F_0} \mathrm{E}\big(\boldsymbol{C}'\boldsymbol{Y} - f(x_0)\big)^2. \tag{4.21}$$

Suppose our design takes observations only at the distinct points $x_1, x_2, \ldots, x_r$ with n_i uncorrelated observations taken at x_i. Let $\bar{Y}_i$ be the average of the n_i observations at x_i and suppose we restrict to linear estimators $\boldsymbol{C}'\boldsymbol{Y}$ of the form $\sum_{i=1}^{r} a_i \bar{Y}_i$. We then seek r, n_i, x_i, and a_i which minimize

$$\begin{aligned} &\sup_{f \in F_0} \mathrm{E}\left(\sum_{i=1}^{r} a_i \bar{Y}_i - f(x_0)\right)^2 \\ &\quad = \sigma^2 \sum_{i=1}^{r} \frac{a_i^2}{n_i} + \sup_{f \in F_0} \left(\sum_{i=1}^{r} a_i f(x_i) - f(x_0)\right)^2. \end{aligned} \tag{4.22}$$

If (4.22) is minimized with respect to the n_i for fixed a_i without regard to the integer nature of the n_i, the minimum occurs when $n_i = N|a_i| \sum |a_i|$ and has value

$$R(A) = \frac{\sigma^2}{N}\left(\sum_{i=1}^{r} |a_i|\right)^2 + \sup_{f \in F_0}\left(\int f \,\mathrm{d}A\right)^2 \tag{4.23}$$

where A is a pure jump function, with jumps of size a_i at x_i, and a jump of size $a_0 = -1$ at x_0, such that $A(x) = 0$ for $x < x_0$. Minimizing $R(A)$ with respect to A leads to an approximate design solution.

Now F_0 contains all polynomials of degree $\leqslant h$. Thus $\int f \,\mathrm{d}A$ is not bounded for all f in F_0 unless

$$\int x^j \,\mathrm{d}A = 0, \quad 0 \leqslant j \leqslant h, \tag{4.24}$$

or equivalently $\sum_{i=1}^{r} a_i x_i^j = x_0^j$ for all $0 \leqslant j \leqslant h$. By Taylor's theorem and (4.24), $\int f\,\mathrm{d}A$ can be written as $\int_{x_0}^{\infty} f\,\mathrm{d}A = \int_{x_0}^{\infty} B(t) f^{(h+1)}(t)\,\mathrm{d}t$ where $B(t) = \int_{x_0}^{\infty} \frac{(x-t)_+^h}{h!}\,\mathrm{d}A(x) = (1/h!)\sum_{i=0}^{r} a_i (x_i - t)_+^h$. Hence

$$R(A) = \frac{\sigma^2}{N}\left(\sum_{i=1}^{r} |a_i|\right)^2 + \varepsilon^2\left(\int |B(t)|\,\mathrm{d}t\right)^2. \tag{4.25}$$

Huang and Studden (1988) show that for purposes of minimizing (4.25), one can restrict to designs that put some mass at 0. With this fact, minimization of $R(A)$ becomes equivalent to minimizing

$$\int |\mathrm{d}A| \tag{4.26}$$

subject to the conditions

$$\int |B(t)|\,\mathrm{d}t = c \tag{4.27}$$

and (4.24).

Huber attempted to solve this problem by minimizing (4.26) subject to (4.24) and

$$\left|\int B(t)\,\mathrm{d}t\right| = c \tag{4.28}$$

but as Huang and Studden (1988) show, there is an error in the argument he uses and the minimization of (4.26) subject to (4.24) and (4.27) is not equivalent to minimization of (4.26) subject to (4.24) and (4.28). Huber's solution turns out to be correct, as Huang and Studden (1988) show, provided that one restricts r to be $h+1$. In this case the optimal design on exactly $h+1$ points takes observation on $0 = x_1 < x_2 < \cdots < x_{h+1}$, which after the addition of another point y constitute the set of Tchebyscheff points of order $h+1$ in the interval $[0, y]$. y is determined according to Theorem 3.2 in Huang and Studden (1988). The a_i (and hence the n_i) are then determined by the constraint (4.24).

Huang and Studden (1988) also find the optimal design on exactly $h+1$ points for the above problem with $X = [0, \infty)$ replaced by $X = [-1, 1]$ and $x_0 < -1$. The optimal design here takes observations on the first $h+1$ Tchebyscheff points on $[-1, y(\rho)]$ where $y(\rho)$ is determined as in Theorem 4.4 of Huang and Studden (1988).

Another paper dealing with model robust extrapolation is Spruill (1985). He shows that in extrapolating a function which is close to being a polynomial, the least squares estimator combined with the Hoel–Levine (1964) optimal design for the polynomial performs well in terms of mean square error when compared with an optimal spline extrapolator.

5. Detection of model inadequacy

The design of an experiment is based on an assumed model which is usually believed to be a reasonable approximation to the true model. If the assumed model is inadequate, a classical optimal design based on the assumed model may provide significantly biased information about the true response. In other words, the optimal design for the assumed model can actually be a bad design for the true model, and therefore, is not a model robust design. A good model robust design should:

1. Allow one to fit the assumed model.

2. Detect the model inadequacy when the fitted model is a poor approximation to the true model.

3. Allow one to make reasonable efficient inferences concerning the assumed model when the assumed model is adequate.

A reasonable model robust design for detecting model adequacy is unlikely to be an optimal design for the assumed model. As an example to illustrate this situation, suppose that the assumed model is a straight line in which the well-known classical optimal design takes observations at the two extreme points. However, for a model robust design to check the adequacy of the straight line model, the design must take some observations between the extreme points. Therefore, the model robust design is not optimal for the assumed straight line model. The effectiveness of a model robust design for detecting model inadequacy is highly dependent on the true response. Since the true model is often unknown, if the design points are at the more representative locations of the true model, then the design can effectively detect the model inadequacy when it occurs. Otherwise, if the design points fall on "bad" locations such as the tick marks in the example of Figure 1.2, the design may not even detect the model inadequacy. Without knowing the true model, there is no absolute measurement for the effectiveness of detecting model inadequacy. Hence, the "best" model robust design will depend on the optimality criteria selected as well as the true and assumed model.

How does one find a good model robust design? In the case of an mth degree polynomial regression model, many optimality criteria, such as D-optimality and G-optimality, provide no model inadequacy check. A first attempt at including the ability to check model adequacy into the optimality criteria was made by Box and Draper (1959). This is discussed in detail in Section 3. Since the pioneering work by Box and Draper, a number of authors have formulated other approaches to this problem.

Most of this work models the assumed response function as

$$\mathrm{E}(Y_i)\boldsymbol{g}(\boldsymbol{x})'\boldsymbol{\alpha} \tag{5.1}$$

and treats the true response function as a departure from the assumed function. The true response function is written as

$$\mathrm{E}(Y_i) = \boldsymbol{g}(\boldsymbol{x})'\boldsymbol{\beta}_1 + \boldsymbol{h}(\boldsymbol{x})'\boldsymbol{\beta}_2 \tag{5.2}$$

where $\boldsymbol{g}(\boldsymbol{x}) = (g_1(\boldsymbol{x}), g_2(\boldsymbol{x}), \ldots, g_s(\boldsymbol{x}))'$, $\boldsymbol{\alpha} = (\alpha_1, \ldots, \alpha_s)'$, $\boldsymbol{h}(\boldsymbol{x}) = (h_{s+1}(\boldsymbol{x}), h_{s+2}(\boldsymbol{x}), \ldots, h_m(\boldsymbol{x}))'$, $\boldsymbol{\beta}_1 = (\beta_1, \beta_2, \ldots, \beta_s)'$, and $\boldsymbol{\beta}_2 = (\beta_{s+1}, \beta_{s+2}, \ldots, \beta_m)'$. In the notation of Section 2 $\boldsymbol{f}(\boldsymbol{x}) = (\boldsymbol{g}(\boldsymbol{x})', \ \boldsymbol{h}(\boldsymbol{x})')'$ and $\boldsymbol{\beta} = (\boldsymbol{\beta}_1', \boldsymbol{\beta}_2')'$. Also let $\boldsymbol{G}$ be the $N \times s$ matrix whose i, lth entry is $g_l(\boldsymbol{x}_i)$ and $\boldsymbol{H}$ be the $N \times (m-s)$ matrix whose i, rth entry is $h_r(\boldsymbol{x}_i)$. Then $\boldsymbol{F} = (\boldsymbol{G} \ \boldsymbol{H})$ in (2.3). Based on the above assumed and true response functions, the power of a design for detecting model adequacy depends on the noncentrality parameter

$$\boldsymbol{\beta}_2' \boldsymbol{A} \boldsymbol{\beta}_2 = \boldsymbol{\beta}_2'(\boldsymbol{H}'\boldsymbol{H} - \boldsymbol{H}'\boldsymbol{G}(\boldsymbol{G}'\boldsymbol{G})^{-1}\boldsymbol{G}'\boldsymbol{H})\boldsymbol{\beta}_2, \tag{5.3}$$

where $\boldsymbol{A}$ is the inverse of the covariance matrix of the least squares estimate of $\beta_{s+1}, \beta_{s+2}, \ldots, \beta_m$. To best detect departures from model (5.1), the designs should maximize (5.3). When $\boldsymbol{\beta}_2$ is a scalar, we can simply minimize the variance of the estimate $\widehat{\boldsymbol{\beta}}_2$ of $\boldsymbol{\beta}_2$. But, when $\boldsymbol{\beta}_2$ is not a scalar, the value of (5.3) depends on the unknown parameters and may not be maximized without restrictions such as requiring $\boldsymbol{\beta}_2$ to lie in a specific region.

Stigler (1971) considered a polynomial model which is a special case of the above with

$$\boldsymbol{g}(\boldsymbol{x})'\boldsymbol{\alpha} = \sum_{\ell=0}^{m-1} \alpha_\ell x^\ell, \quad \text{and} \quad \boldsymbol{h}(\boldsymbol{x})'\boldsymbol{\beta}_2 = \beta_m x^m.$$

He developed two criteria, C-restricted D-optimality and C-restricted G-optimality, for the univariate polynomial model over the region $[-1, 1]$. A design is said to be C-restricted D-optimal if it maximizes $\det \boldsymbol{M}_{m-1}(\xi)$ among all designs ξ satisfying $\det \boldsymbol{M}_{m-1}(\xi) \leqslant C \det \boldsymbol{M}_m(\xi)$ where $\boldsymbol{M}_k(\xi)$ is the information matrix under ξ for the kth order polynomial model. The choice of this criterion is based on the fact that if $\widehat{\beta}_m$ is the least squares estimate of β_m, then

$$N \operatorname{var}(\widehat{\beta}_m) = \sigma^2 \det \boldsymbol{M}_{m-1}(\xi) \det \boldsymbol{M}_m(\xi)^{-1}.$$

Thus, a C-restricted D-optimal design can be interpreted as a design that minimizes the generalized variance of the least squares estimators $\widehat{\alpha}_0, \widehat{\alpha}_1, \ldots, \widehat{\alpha}_{m-1}$ subject to the constraint that $\operatorname{var}(\widehat{\beta}_m) \leqslant \sigma^2 C/N$ for some constant C. A design is said to be C-restricted G-optimal if it minimizes $\max_{-1 \leqslant x \leqslant 1} d_{m-1}(x, \xi)$ among all designs ξ satisfying $\det \boldsymbol{M}_{m-1}(\xi) \leqslant C \det \boldsymbol{M}_m(\xi)$, where $d_{m-1}(x, \xi) = \boldsymbol{g}(\boldsymbol{x})' \boldsymbol{M}_{m-1}(\xi)^{-1} \boldsymbol{g}(\boldsymbol{x})$ and $\boldsymbol{g}(\boldsymbol{x}) = (1, x, x^2, \ldots, x^{m-1})$. Because of the fact that

$$\frac{1}{N} d_{m-1}(x, \xi) = \operatorname{var}\left(\sum_{\ell=0}^{m-1} \widehat{\alpha}_\ell x^\ell \right),$$

a C-restricted G-optimal design can be interpreted as a design which minimizes the maximum variance of the best linear unbiased estimate of the regression function subject to the constraint that $\operatorname{var}(\widehat{\beta}_m) \leqslant \sigma^2 C/N$ for some constant C. The optimal designs based on these two criteria would allow one to make reasonably efficient inferences for the assumed model and also allow one to make inferences for β_m.

Unlike most criteria that produce a single optimal design, these two criteria produce a class of optimal designs. The final optimal design depends on the choice of C which reflects a compromise between two competing goals: precise inference about β_m and precise inference about the assumed $(m-1)$th order polynomial model. In choosing C, one can consider the D-efficiency

$$e_{m-1}^{D}(\xi) = \left(\frac{\det \boldsymbol{M}_{m-1}(\xi)}{\max_{\eta} \det \boldsymbol{M}_{m-1}(\eta)}\right)^{1/m},$$

or the G-efficiency

$$e_{m-1}^{G}(\xi) = \frac{m}{\max_{-1\leqslant x\leqslant 1} d_{m-1}(x,\xi)}$$

of a design for a model. Because it is difficult to find C-restricted D- and G-optimal designs, some restrictions on the support points and designs are suggested as a means of reducing the computation size and complexity.

An example of finding C-restricted D- and G-optimal design for $m = 2$ model is also given by Stigler. When $m = 2$,

$$\det \boldsymbol{M}_1(\xi) = \mu_2,$$
$$\det \boldsymbol{M}_2(\xi) = \mu_2(\mu_4 - \mu_2^2)$$

and

$$\max_{-1\leqslant x\leqslant 1} d_1(x,\xi) = 1 + \mu_2^{-1}.$$

Since C-restricted D-optimality maximizes μ_2 and C-restricted G-optimality minimizes $1 + \mu_2^{-1}$, both subject to $\mu_4 - \mu_2^2 \geqslant C^{-1}$, the optimal designs given by the two criteria are identical. For $C \geqslant 4$, the C-restricted D- and G-optimal designs are

$$\xi_0(\pm 1) = \frac{1}{4} + \frac{1}{2}\sqrt{\frac{1}{4} - \frac{1}{C}},$$
$$\xi_0(0) = \frac{1}{2} - \sqrt{\frac{1}{4} - \frac{1}{C}}.$$

The following table shows the optimal designs, D-efficiencies and G-efficiencies for a few of possible values of C.

When $C = 4$, ξ_0 is the best design for inferences concerning β_2 in the second order polynomial model. When $C = \infty$, ξ_0 maximizes $\det \boldsymbol{M}_1(\xi)$ and hence, is the best design for inferences concerning the first order polynomial model. When $4 < C < \infty$, the resulting design ξ_0 is a compromise between the two extreme cases. As discussed before, the choice of C is subjective, and the efficiencies $e_1^D(\xi_0)$ and $e_1^G(\xi_0)$ may help the experimenter in the determination of C.

Table 5.1

C	$\xi_0(\pm 1)$	$\xi_0(0)$	$e_1^D(\xi_0)$	$e_1^G(\xi_0)$
4.0	.250	.500	.707	.6671
5.0	.362	.276	.851	.8401
6.0	.394	.211	.888	.8811
7.0	.414	.173	.910	.9061
8.0	.427	.146	.924	.921
9.0	.436	.127	.934	.932
10.0	.444	.113	.942	.940
20.0	.474	.053	.973	.973
30.0	.483	.035	.983	.982
40.0	.487	.026	.987	.987
50.0	.490	.020	.990	.990
∞	.500	.000	1.00	1.00

Atkinson (1972) extended the above univariate polynomial model to a more general k-variate polynomial model where the response function $\mathrm{E}(Y_i)$ depends on the powers and products of the variables $x_1, x_2, \ldots, x_k$. The assumed model can be written as

$$\mathrm{E}(Y_i) = \sum_{\ell=1}^{s} \alpha_\ell g_\ell(\boldsymbol{x}_i),$$

and the true response function as

$$\mathrm{E}(Y_i) = \sum_{\ell=1}^{s} \beta_\ell g_\ell(\boldsymbol{x}_i) + \sum_{\ell=s+1}^{m} \beta_\ell h_\ell(\boldsymbol{x}_i)$$

where $g_\ell(\boldsymbol{x}_i)$ and $h_\ell(\boldsymbol{x}_i)$ are powers and products of $x_1, x_2, \ldots, x_k$.

Atkinson proposed two criteria: maximizing the minimum eigenvalue of $\boldsymbol{A}$, where $\boldsymbol{A}$ is as in (5.3), and maximizing $\det \boldsymbol{A}$. Since these two criteria involve $\boldsymbol{A}$, the resulting designs provide a good check on model adequacy but may give a relatively poor estimate of $\boldsymbol{\beta}_1$ if the assumed model is in fact adequate. In the univariate polynomial example discussed in Stigler (1971), since $\boldsymbol{\beta}_2$ is a scalar, both criteria seek to maximize $\boldsymbol{\beta}_2'\boldsymbol{A}\boldsymbol{\beta}_2$ which can be reduced to minimizing the variance of $\widehat{\boldsymbol{\beta}}_2$. The resulting design is the best for detecting the second order polynomial model, but may be poor for inferences concerning the first order polynomial model, as is the case when $C = 4$ in the example above. To prevent this problem, Atkinson suggested that one should also check the efficiency of the designs for estimating $\boldsymbol{\beta}_1$ when $\boldsymbol{\beta}_2$ is zero. In Atkinson (1972), designs for univariate models and designs for detecting second-order departure from a k-variate model which involves only first-order terms are developed. The designs for detecting second-order departures from two types of k-variate models are tabulated.

A shortcoming of Atkinson's criteria is that they do not account for the dependency of noncentrality parameter $\boldsymbol{\beta}_2'\boldsymbol{A}\boldsymbol{\beta}_2$ on $\boldsymbol{\beta}_2$ when $\boldsymbol{\beta}_2$ is not scalar. Jones and Mitchell (1978) proposed two criteria which are

1. Maximizing $\beta_2' A \beta_2$ over a specific region of β_2 and
2. Maximizing the average of $\beta_2' A \beta_2$ over a specified contour in the β_2 space.

To define these criteria, let

$$T = \Gamma_{hh} - \Gamma_{hg} \Gamma_{gg}^{-1} \Gamma_{gh}$$

where the $\Gamma_{..}$ are the region moment matrices as defined in (2.6). A design is Λ_1-optimal if it maximizes

$$\Lambda_1 = \inf_{\beta_2 \in \Phi} \beta_2' A \beta_2,$$

where $\Phi = \{\beta_2, \beta_2' A \beta_2 \geqslant \delta$ for some positive constant $\delta\}$.

It can be shown that

$$\Lambda_1 = \delta \cdot \lambda_1 (T^{-1} A),$$

where $\lambda_1 (T^{-1} A)$ is the smallest eigenvalue of $T^{-1} A$.

Since δ is a constant, Λ_1-optimality is equivalent to maximizing $\lambda_1 (T^{-1} A)$.

A design is Λ_2-optimal if it maximizes

$$\Lambda_2 = \frac{\int_{\Phi_0} \beta_2' A \beta_2 \, \mathrm{d}B}{\int_{\Phi_0} \mathrm{d}B}$$

where $\mathrm{d}B$ is the differential of the area on the surface of the ellipsoid $\Phi_0 = \{\beta_2, \beta_2' A \beta_2 = \delta$ for some positive constant $\delta\}$.

They have shown that $\Lambda_2 = \frac{1}{m-s} \delta \operatorname{tr}(T^{-1} A)$. Since $\frac{1}{m-s}$ and δ are constants, Λ_2-optimality is equivalent to maximizing $\operatorname{tr}(T^{-1} A)$. In Jones and Mitchell (1978), some characterizations of Λ_1-optimality and Λ_2-optimality are presented and applied for constructing optimal designs for first-order versus second-order polynomial models.

When the true model is unknown, a common approach in experimental design is to consider a simple model which is thought to provide an adequate approximation to the more complicated true model. The work discussed so far in this section emphasizes the detection of model inadequacy. However, it is also important that the design protect against selecting an oversimplified model. DeFeo and Myers (1992) developed a new criterion for model robust design that considers the two conflicting goals: protecting against the use of an oversimplified model and detecting lack of fit. Instead of maximizing the power of detecting model adequacy, the criterion simultaneously uses the integrated model bias and power. They also propose a class of experimental designs that appear to perform well under this criterion. These designs are rotations of 2^k factorial designs or central composite designs through a small angle.

We have been assuming that the true model departs from the assumed model. If the true model makes too strong an assumption about the form of the departures from the assumed model, the design may only detect these specific departures, and may not detect departures of other kinds. Atkinson and Fedorov (1975a) developed designs for discriminating between two possible true models (assuming one of the models is true) which need not be linear in the parameters. No particular assumed model is to be fit a priori. Rather, one tries to determine which of several models to fit. Atkinson and Fedorov (1975b) extended these results to designs for discriminating between several models. The criterion they used is called T-optimality and maximizes the sum of squares for the lack of fit of the incorrect models. The design which maximizes the criterion depends on which of the models is true and on the values of the unknown parameters. Atkinson and Fedorov (1975a, b) suggested three approaches for solving this problem: a sequential approach, a Bayesian approach, and a maximin approach. In the special case of discriminating between models (5.1) and (5.2), when (5.2) is true, the T-optimality criterion reduces to the maximization of the noncentrality parameter $\boldsymbol{\beta}_2'\boldsymbol{A}\boldsymbol{\beta}_2$. In this case, the sequential approach uses available information on $\boldsymbol{\beta}_2$ at each stage to select the next run; the Bayesian approach uses a prior distribution on $\boldsymbol{\beta}_2$; the maxmin approach maximizes the minimum $\boldsymbol{\beta}_2'\boldsymbol{A}\boldsymbol{\beta}_2$ over a specific region in the $\boldsymbol{\beta}_2$ space.

6. Bayesian robust design

Robustness in a Bayes context usually refers to insensitivity to specification of the prior. This usually takes the form of requiring a design to be optimal (or nearly optimal) for some optimality criterion with respect to a particular prior subject to the condition that the design perform reasonably well over a class of possible priors.

One of the earliest and most general approaches to robust Bayes design is the paper by O'Hagan (1978). We now give some details of the results. In this paper a localized regression model is used to reflect the fact that any particular regression model is only an approximation to the true model over a small portion of the design space. The localized regression model allows the regression parameters to depend on x, the point at which an observation is taken, and the regression parameters are assumed to vary "slowly" with x. In the setting of regression with a single independent variable $x \in (-\infty, +\infty)$, response Y, $m \times 1$ vector of regression functions $\boldsymbol{f}(x)$, and $m \times 1$ vector of regression parameters $\boldsymbol{\beta}(x)$, a simple localized regression model would be the following. The distribution of Y given x and $\boldsymbol{\beta}(x)$ is assumed normal with

$$\mathrm{E}\big(Y \mid x, \boldsymbol{\beta}(x)\big) = \boldsymbol{f}(x)'\boldsymbol{\beta}(x), \tag{6.1}$$

$$\mathrm{Var}\big(Y \mid x, \boldsymbol{\beta}(x)\big) = \sigma^2. \tag{6.2}$$

The prior ought to reflect our belief about the local stability of the regression model. One simple possibility is to assume that our information about $\boldsymbol{\beta}(x)$ is the same for all values of x, i.e., the prior mean vector is

$$\mathrm{E}\big(\boldsymbol{\beta}(x) \mid \boldsymbol{b}_0\big) = \boldsymbol{b}_0. \tag{6.3}$$

If we further assume that the correlation between $\boldsymbol{\beta}(x)$ and $\boldsymbol{\beta}(x^*)$ depends only on $|x - x^*|$ we might model $\boldsymbol{\beta}(x)$ as a second-order stationary process with

$$\mathrm{E}\big((\boldsymbol{\beta}(x) - \boldsymbol{b}_0)(\boldsymbol{\beta}(x^*) - \boldsymbol{b}_0)' \mid \boldsymbol{b}_0\big) = \rho\big(|x - x^*|\big)\boldsymbol{B}_0 \tag{6.4}$$

where $\rho(d)$ is a monotonic decreasing function of $0 \leqslant d < \infty$, and $\rho(0) = 1$. We might finally assume the $\boldsymbol{\beta}(x)$ are jointly normal. The above model, including the prior, is the localized regression model. Designs and estimators are chosen to minimize the posterior expectation of the squared error loss for prediction. This posterior expectation is a complicated function of the independent variable so minimization in a particular problem must be done numerically.

Suppose we observe N values of the dependent variable, $y_1, \ldots, y_N$ at corresponding x values $x_1, \ldots, x_N$. The posterior distribution of the $\boldsymbol{\beta}(x)$ is such that they are jointly normal with means

$$\boldsymbol{b}_1(x) = \mathrm{E}\big(\boldsymbol{\beta}(x) \mid y_1, \ldots, y_N, \boldsymbol{b}_0\big) = \boldsymbol{S}(x)'\boldsymbol{A}^{-1}\boldsymbol{y} + \boldsymbol{Q}(x)'\boldsymbol{b}_0 \tag{6.5}$$

and covariances

$$\begin{aligned}\boldsymbol{B}_1(x, x^*) &= \mathrm{E}\big[(\boldsymbol{\beta}(x) - \boldsymbol{b}_1(x))(\boldsymbol{\beta}(x^*) - \boldsymbol{b}_1(x^*))' \mid y_1, \ldots, y_N, \boldsymbol{b}_0\big] \\ &= \rho\big(|x - x^*|\big)\boldsymbol{B}_0 - \boldsymbol{S}(x)'\boldsymbol{A}^{-1}\boldsymbol{S}(x^*)\end{aligned} \tag{6.6}$$

where

$$\boldsymbol{Q}(x) = \boldsymbol{I}_m - \boldsymbol{F}'\boldsymbol{A}^{-1}\boldsymbol{S}(x),$$

$$\boldsymbol{y} = \begin{pmatrix} y_1 \\ \vdots \\ y_N \end{pmatrix}, \qquad \boldsymbol{S}(x) = \begin{pmatrix} \rho(|x - x_1|)\boldsymbol{f}(x_1)'\boldsymbol{B}_0 \\ \vdots \\ \rho(|x - x_N|)\boldsymbol{f}(x_N)'\boldsymbol{B}_0 \end{pmatrix},$$

$$\boldsymbol{F} = \begin{pmatrix} \boldsymbol{f}(x_1)' \\ \vdots \\ \boldsymbol{f}(x_N)' \end{pmatrix},$$

$$\boldsymbol{A} = s^2\boldsymbol{I}_N + \boldsymbol{C} \tag{6.7}$$

and $\boldsymbol{C}$ is the $N \times N$ matrix whose (i, j)th element is

$$c_{ij} = \rho\big(|x_i - x_j|\big)\boldsymbol{f}(x_i)'\boldsymbol{B}_0\boldsymbol{f}(x_j). \tag{6.8}$$

Inference about a single future value of y at x is made from its posterior predictive distribution $N(\boldsymbol{f}(x)'\boldsymbol{b}_1(x),\ \sigma^2 + \boldsymbol{f}(x)'\boldsymbol{B}_1(x, x)\boldsymbol{f}(x))$. In particular, an obvious point estimator is the mean $\boldsymbol{f}(x)'\boldsymbol{b}_1(x)$.

If $\boldsymbol{f}(x)'\boldsymbol{b}_1(x)$ is very complicated, one may prefer to fit a simpler model (assumed to be a good approximation to $\boldsymbol{f}(x)'\boldsymbol{b}_1(x)$ over the region of interest. Suppose we approximate $\boldsymbol{f}(x)'\boldsymbol{b}_1(x)$ by a simple regression model of the form

$$\widehat{y}(x) = \boldsymbol{g}(x)'\boldsymbol{h} \tag{6.9}$$

for some $s \times 1$ vector $\boldsymbol{h}$ and for a given $s \times 1$ vector function $\boldsymbol{g}(x)$. The best value of $\boldsymbol{h}$ for the approximation will be made with reference to a well-defined loss function. If we consider predicting the unknown future value $y(x)$ of the independent variable at x by $\widehat{y}(x)$, the loss sustained in this prediction is

$$L_1(x) = \left[y(x) - \widehat{y}(x)\right]^2. \tag{6.10}$$

Since $y(x)$ is unknown at the time of prediction, the relevant loss when we use predictor $\widehat{y}(x)$ will be the posterior expectation of $L_1(x)$, say $L_2(x)$. Let the measure function $\Omega(x)$ denote the relative importance to us of predictions at the various values of x. Thus when choosing the value of $\boldsymbol{h}$ in (6.9) our expected loss is

$$\int_{-\infty}^{\infty} L_2(x)\,\mathrm{d}\Omega(x). \tag{6.11}$$

The value of $\boldsymbol{h}$ minimizing (6.11) is

$$\boldsymbol{h} = \boldsymbol{W}^{-1} \int_{-\infty}^{\infty} \boldsymbol{g}(x)\boldsymbol{f}(x)'\boldsymbol{b}_1(x)\,\mathrm{d}\Omega(x) \tag{6.12}$$

where

$$\boldsymbol{W} = \int_{-\infty}^{\infty} \boldsymbol{g}(x)\boldsymbol{g}(x)'\,\mathrm{d}\Omega(x) \tag{6.13}$$

assuming the integrals exist. Substituting for $\boldsymbol{b}_1(x)$ from (6.5) gives

$$\boldsymbol{h} = \boldsymbol{T}'\boldsymbol{A}^{-1}\boldsymbol{y} + \boldsymbol{R}'\boldsymbol{b}_0 \tag{6.14}$$

where

$$\boldsymbol{T}' = \boldsymbol{W}^{-1} \int_{-\infty}^{\infty} \boldsymbol{g}(x)\boldsymbol{f}(x)'\boldsymbol{S}(x)'\,\mathrm{d}\Omega(x),$$

$$\boldsymbol{R}' = \boldsymbol{W}^{-1} \int_{-\infty}^{\infty} \boldsymbol{g}(x)\boldsymbol{f}(x)'\boldsymbol{Q}(x)'\,\mathrm{d}\Omega(x). \tag{6.15}$$

For purposes of design, suppose we can choose the values of $x_1, \ldots, x_N$ at which we take observations. For predictive curve fitting as above, we would like to find the design that gives the lowest expected loss. This is the design which maximizes

$$U = \mathrm{tr}(\boldsymbol{W}\boldsymbol{T}'\boldsymbol{A}^{-1}T). \tag{6.16}$$

This is an extremely complicated function of $x_1, \ldots, x_N$ and so generally must be maximized numerically.

As an example, suppose a straight line is to be fitted a posterior and prediction is to be done at $x = 0$. Suppose also

$$\rho(d) = \exp\left(-\frac{1}{2}\,d^2/\sigma_\rho^2\right), \tag{6.17}$$

$$\mathrm{d}\Omega(x) = (2\pi\sigma_\omega^2)^{-1/2}\exp\left(-\frac{1}{2}\,x^2/\sigma_\omega^2\right)\mathrm{d}x \tag{6.18}$$

and assume a locally linear model, i.e., $\boldsymbol{f}(x)' = (1, x)$. For

$$\sigma_\omega^2 = 4, \qquad \sigma_\rho^2 = 32, \qquad \sigma^2 = 1, \qquad B_0 = \begin{pmatrix} 10 & 0 \\ 0 & 1 \end{pmatrix}. \tag{6.19}$$

O'Hagan (1978) gives the following list of optimal designs for small N.

N	Optimal design
2	−2.20, 2.20
3	−3.02, 0, 3.02
4	−3.44, −0.60, 0.60, 3.44
5	−3.64, −1.39, 0, 1.39, 3.64
6	−3.84, −1.33, −1.33, 1.33, 1.33, 3.84

The optimal designs take observations at a variety of points centered at the point 0 at which we wish to do prediction and without the tendency to take all observations at the boundary of the design region (as classical D-optimal designs do, thus requiring the design region to be compact). This behavior of the optimal design mimics the "naive" approach of spreading observations over the region of interest in order to observe curvature in the model.

O'Hagan (1978) generalizes the above discussion to the multivariate setting. The interested reader should see Section 3 of O'Hagan (1978) for details. As is the case for (6.16), analytic solutions appear to be difficult to obtain, even more so than in the univariate setting discussed above. Solutions to specific problems will generally need to be found numerically.

While O'Hagan (1978) provides a very general development, a number of other authors also discuss robust Bayesian design. In the interest of space, we provide only a summary of these papers. The interested reader should consult the papers themselves for details. The papers represented below are not exhaustive but are representative of other approaches.

Many authors consider the problem of robustness to the specification of the prior. In the context of regression and the approximate theory of design, DasGupta and Studden (1991) constructed a framework for robust Bayesian experimental design for linear models. They found designs that minimize a measure of robustness (related to the Bayes risk) over a class of prior distributions and, at the same time, are close to being optimal for some specific prior. A variety of measures of robustness are considered, including a minimax approach, and two classes of priors are investigated.

Seo and Larntz (1992) suggested criteria for nonlinear design that make the design robust to specification of the prior distribution. In particular, they suggested seeking designs which are optimal with respect to a given prior subject to the constraint that the design attain a certain efficiency over a class of closely related prior distributions.

DasGupta et al. (1992) gave a detailed approach to design in a linear model when the variance of the response is proportional to an exponential or power function of the mean. They considered the case where the experimenter wants to find a design which is highly efficient for several criteria simultaneously, and gave examples of such "compromise designs".

For normal one way analysis of variance models, Toman (1992a, b) considered robustness to a class of normal prior distributions where the variances take values in specified intervals. Optimality criteria involve maximizing the average, with respect to a distribution on the prior precision parameters, over the class of posterior distributions, of either the determinant or the trace of the posterior precision matrix. Toman and Gastwirth (1993) investigated both robust design and estimation for analysis of variance models when the class of priors is a class of finite mixtures of normals. Squared error loss is used and the posterior risk averaged over the class of corresponding posterior distributions. Toman and Gastwirth (1994) suggested specifying the prior distribution of the treatment means in a one way analysis of variance model from a pilot study. They assumed the error variances of the pilot and of the follow up experiments to be unknown, but that intervals in which they can vary can be specified. Again, squared error loss is used and a designs and estimators chosen to minimize a minimax criterion over the class of posterior distributions.

In addition to robustness to specification of the prior, there is also work on robustness to specification of the linear model in a Bayesian setting. One approach is to consider mixtures of linear models using a criterion that is a weighted average of optimality criteria for a variety of candidate models. The weights would correspond to the prior probability that a candidate model was the "true" model. Läuter (1974, 1976) considered such an approach using a criterion of the form

$$\phi(\xi) = \sum_{i=1}^{m} w_i \phi_i(\xi)$$

where $\phi_i(\xi)$ is the D-optimality criterion under the ith of m candidate models. The weights w_i for each model are the prior probabilities for that model. Cook and Nachtsheim (1982) applied such a criterion, based on A-efficiency, to the problem of finding designs for polynomial regression when the degree of the polynomial is unknown. In particular, if ξ_i is the A-optimal design for the ith model, A_i corresponds to the average variance of prediction over the design region, and if M_i is the matrix of moments for the ith model, the criterion used was to maximize

$$\phi(\xi) = -\sum_{i=1}^{m} w_i \frac{\operatorname{tr}(\boldsymbol{A}_i \boldsymbol{M}_i(\xi)^{-1})}{\operatorname{tr}(\boldsymbol{A}_i \boldsymbol{M}_i(\xi_i)^{-1})}.$$

A more fully Bayes approach might instead maximize

$$\phi(\xi) = -\sum_{i=1}^{m} w_i \operatorname{tr}\big(\boldsymbol{A}_i \boldsymbol{M}_i(\xi)^{-1}\big).$$

A summary of the mathematics of such criteria and how the general equivalence theorem can be applied can be found in Pukelsheim (1993, pp. 286–296). Dette (1990) gives some general results for D-optimality and polynomial regression. Dette (1991, 1993a, b) used mixtures of Bayesian linear model criteria involving the prior precision matrix and derived a version of Elfving's (1952) theorem for this case. Dette and Studden (1994) provided further results in this direction, characterizing the optimal design in terms of its canonical moments.

Another approach is to be found in DuMouchel and Jones (1994). They introduced a modified Bayesian D-optimal approach for factorial models. They used a prior distribution with a structure that recognizes "primary" and "potential" terms in order to recognize uncertainty in the model. The resulting Bayesian D-optimal designs are resolution IV designs, and so this approach provides justification for the use of resolution IV designs over other designs with the same value of the D-criterion.

Steinberg (1985) considered using a two-level factorial experiment to investigate a response surface and used a Bayesian formulation to represent the uncertainty in the adequacy of the proposed model. He derived a method for choosing the high and low levels for each factor of the two factor experiment, conditional on the particular fractional factorial design used. This allows one to quantify the trade-off between choosing design points on the boundary of the design region where information is maximized versus the fact that the model holds to better approximation near the center of the design region.

7. Applications to computer experiments

7.1. Introduction

Computer modeling of complex physical phenomena has become increasingly popular as a method for studying such processes. This is particularly true when actual experimentation on such processes is very time consuming, expensive, or impossible. Examples include weather modeling, integrated circuit design, plant ecology, and the study of controlled nuclear fusion devices. Such computer models (or codes) usually have high dimensional inputs. These inputs may be scalars or functions. The output from such models may also be multi-dimensional. For example, as might be the case in weather modeling, the output may be a time dependent function from which a few summary responses are selected.

A computer experiment involves running the computer at a variety of input configurations in order to make inferences about characteristics (parameters) of the computer model on the basis of the resulting output. For example, one may wish to determine the inputs that optimize some function of the outputs. Design involves the selection of the input configurations so as to yield efficient inference about these characteristics. It is assumed that the computer code adequately models the physical process that is of ultimate interest so that inferences made about characteristics the computer model yield reliable information about the corresponding characteristics of the physical process. Whether this is a reasonable assumption is a question that might well be addressed by statistics. However, in the design of computer experiments this issue is generally ignored. Attention is restricted to inference concerning the computer model itself.

There are several features of this problem that make it unusual. First, the output from computer code is deterministic. Hence it is not immediately obvious how statistical models are relevant. Second, it is not possible to give an explicit functional relation between the inputs and outputs. This is due to the complexity of the physical phenomenon being modeled. Third, we assume the computer model is sufficiently complex that it is very time consuming (and expensive) to obtain a single run of the code. Thus output can be obtained for only a relatively small number of input configurations. Extensive grid searches are ruled out.

In order to make the problem tractable, the literature on computer experiments usually assumes that the inputs are scalar and their number is relatively small. It is also assumed that the response (output of interest) is a single scalar. In order to make inferences about characteristics of the computer model, one approach is to fit a relatively simple statistical model to the output. We hope this model adequately approximates the true output. Inferences about characteristics of the computer model are then made by making inferences on the corresponding characteristics of the fitted statistical model. Statistical models are thus used to approximate the output from the computer code for purposes of inference.

The connection to model robust design should be clear. The true model is unknown. A relatively simple model is used to approximate the true model. Design involves where best to take observations so that the fitted model will be an adequate approximation to the true model for the purpose of making some inference about the true model. However, a feature of computer experiments that differs from the models discussed so far is that output is deterministic. One consequence of this fact is that no sensible design will require more than one observation at any input configuration. Another very important consequence is that there is no random error in the classical sense. The difference between the fitted and true model is solely deterministic, hence all error is due to bias. This would seem to suggest the use of an all bias criterion such as advocated by Box and Draper (1959), but application of this approach requires we make some assumption about the form of the true model. It is precisely such knowledge that we are lacking in computer experiments. The lack of random error in the classical sense also makes it difficult to justify on classical statistical grounds most methods for fitting statistical models as well as most methods for selecting a design. The only uncertainty in computer experiments arises from the fact that the computer code is a sort of "black box" and we lack knowledge as to the precise relation between the inputs and output. We might take a Bayesian approach and quantify this uncertainty or lack of knowledge by means of probability. In this case the random component of any statistical model that we fit to the output represents our uncertainty concerning the adequacy of this statistical model. If the output of the computer code is a smooth function of the inputs, we also note that the residuals (differences between the actual output and that predicted by any smooth fitted model) corresponding to inputs which are "close" will appear to be correlated. The closer the inputs, the more strongly correlated these residuals will appear to be. It would seem reasonable to build this property into statistical models used to approximate the actual computer model. Thus one approach, which has become popular, is to fit a regression model to the output and model the residuals as though they were the realization of a stochastic process with covariance which is a function of some measure of the distance between two input configurations.

7.2. Modeling and estimation

The above issues are all addressed in Sacks et al. (1989) which gives an excellent overview of the literature on computer experiments. We follow these authors in describing a method of fitting a regression model with errors which form a stochastic process with covariance which is a function of some measure of the distance between input configurations. Issues of design can only be discussed in the context of a model and method of inference. For simplicity we restrict to the case of a scalar response. We use notation as in Section 2 with $X = L$. Let $\boldsymbol{x}$ denote a particular input configuration, X the set of possible input configurations, and $y(\boldsymbol{x})$ the actual deterministic response at $\boldsymbol{x}$ which is viewed as a realization of some random function (stochastic process) $Y(\boldsymbol{x})$. We assume the following model for $Y(\boldsymbol{x})$.

$$Y(\boldsymbol{x}) = \sum_{j=1}^{m} \beta_j f_j(\boldsymbol{x}) + Z(\boldsymbol{x}). \tag{7.1}$$

$Z(\bullet)$ is a random process which is assumed to have mean 0 and covariance

$$\mathrm{V}(\boldsymbol{w}, \boldsymbol{x}) = \sigma^2 R(\boldsymbol{w}, \boldsymbol{x}) \tag{7.2}$$

between $Z(\boldsymbol{w})$ and $Z(\boldsymbol{x})$, where σ^2 is the process variance and $R(\boldsymbol{w}, \boldsymbol{x})$ is the correlation. As previously mentioned, justification for (7.1) might be that the difference between the actual output of the computer code and a simple regression model, while deterministic, resembles a sample path of a suitably chosen stochastic process. Alternatively, one might regard $Y(\boldsymbol{x})$ as a Bayesian prior on the actual output with the β's either specified a priori or given a prior distribution.

While analysis of (7.1) might proceed along a variety of lines, if the objective is prediction of the response at untried inputs, a kriging approach has become popular. This is the approach suggested by Sacks et al. (1989) for such an objective. Given observations at input configurations, or sites, $\boldsymbol{x}_1, \boldsymbol{x}_2, \ldots, \boldsymbol{x}_N$ in X and output $\boldsymbol{y}_d = (y(\boldsymbol{x}_1), y(\boldsymbol{x}_2), \ldots, y(\boldsymbol{x}_N))'$ consider linear predictors of $y(\boldsymbol{x})$ at an as yet unobserved site $\boldsymbol{x}$ of the form

$$\widehat{y}(\boldsymbol{x}) = \boldsymbol{c}(\boldsymbol{x})'\boldsymbol{y}_d. \tag{7.3}$$

If we replace $\boldsymbol{y}_d$ in (7.3) by $\boldsymbol{Y}_d = (Y(\boldsymbol{x}_1), Y(\boldsymbol{x}_2), \ldots, Y(\boldsymbol{x}_N))'$ then $\widehat{y}(\boldsymbol{x})$ is random. The best linear unbiased predictor (BLUP) is that value of $\boldsymbol{c}(\boldsymbol{x})$ which minimizes the mean squared error (averaged over the random process)

$$\mathrm{MSE}[\widehat{y}(\boldsymbol{x})] = \mathrm{E}[\boldsymbol{c}(\boldsymbol{x})'\boldsymbol{Y}_d - Y(\boldsymbol{x})]^2 \tag{7.4}$$

subject to the unbiasedness constraint

$$\mathrm{E}[\boldsymbol{c}(\boldsymbol{x})'\boldsymbol{Y}_d] = \mathrm{E}[Y(\boldsymbol{x})]. \tag{7.5}$$

Note that a Bayesian approach would predict $y(\boldsymbol{x})$ by the posterior mean $\mathrm{E}[Y(\boldsymbol{x}) \mid \boldsymbol{y}_d]$. If $Z(\bullet)$ is Gaussian and improper uniform priors on the β's are used, then it is well known that the BLUP in this case is the limit of the Bayes predictor as the prior variances on the β's tend to infinity.

To calculate the BLUP for model (7.1), let $\boldsymbol{f}(\boldsymbol{x})$ and $\boldsymbol{F}$ be as defined above (2.3) in Section 2. Let $\boldsymbol{R}$ be the $N \times N$ matrix whose i, jth entry is $R(\boldsymbol{x}_i, \boldsymbol{x}_j)$ and let $\boldsymbol{r}(\boldsymbol{x}) = [R(\boldsymbol{x}_1, \boldsymbol{x}), R(\boldsymbol{x}_2, \boldsymbol{x}), \ldots, R(\boldsymbol{x}_N, \boldsymbol{x})]'$. The MSE in (7.4) is then

$$\sigma^2\left[1 + \boldsymbol{c}(\boldsymbol{x})'\boldsymbol{R}\boldsymbol{c}(\boldsymbol{x}) - 2\boldsymbol{c}(\boldsymbol{x})'\boldsymbol{r}(\boldsymbol{x})\right] \tag{7.6}$$

and the unbiasedness constraint in (7.5) becomes $\boldsymbol{F}'\boldsymbol{c}(\boldsymbol{x}) = \boldsymbol{f}(\boldsymbol{x})$. Minimizing (7.6) subject to this constraint using the method of Lagrange multipliers $\lambda(\boldsymbol{x})$ we find that $\boldsymbol{c}(\boldsymbol{x})$ for the BLUP must satisfy

$$\begin{pmatrix} \boldsymbol{0} & \boldsymbol{F}' \\ \boldsymbol{F} & \boldsymbol{R} \end{pmatrix} \begin{pmatrix} \lambda(\boldsymbol{x}) \\ \boldsymbol{c}(\boldsymbol{x}) \end{pmatrix} = \begin{pmatrix} \boldsymbol{f}(\boldsymbol{x}) \\ \boldsymbol{r}(\boldsymbol{x}) \end{pmatrix} \tag{7.7}$$

and yields the BLUP

$$\widehat{y}(\boldsymbol{x}) = \boldsymbol{f}(\boldsymbol{x})'\widehat{\boldsymbol{\beta}} + \boldsymbol{r}(\boldsymbol{x})'\boldsymbol{R}^{-1}(\boldsymbol{Y}_d - \boldsymbol{F}\widehat{\beta}) \tag{7.8}$$

where $\widehat{\boldsymbol{\beta}} = (\boldsymbol{F}'\boldsymbol{R}^{-1}\boldsymbol{F})^{-1}\boldsymbol{F}'\boldsymbol{R}^{-1}\boldsymbol{Y}_d$ is the usual generalized least-squares estimate of $\boldsymbol{\beta}$. Under (7.1) the two terms on the right hand side of (7.8) are uncorrelated and might be interpreted as follows. The first is the usual generalized least-squares predictor. The second is a smooth of the residuals. Notice that if (7.7) is substituted into (7.6) one may obtain the following expression for the MSE of the BLUP.

$$\mathrm{MSE}\big[\widehat{y}(\boldsymbol{x})\big] = \sigma^2\left[1 - (\boldsymbol{f}(\boldsymbol{x})'\boldsymbol{r}(\boldsymbol{x})')\begin{pmatrix} \mathbf{0} & \boldsymbol{F}' \\ \boldsymbol{F} & \boldsymbol{R} \end{pmatrix}^{-1}\begin{pmatrix} \boldsymbol{f}(\boldsymbol{x}) \\ \boldsymbol{r}(\boldsymbol{x}) \end{pmatrix}\right]. \tag{7.9}$$

In order to compute any of these quantities, the correlation $R(\boldsymbol{w}, \boldsymbol{x})$ must be specified. For a smooth response $R(\boldsymbol{w}, \boldsymbol{x})$ should have some derivatives while for an irregular response a function with no derivatives would be preferred. Choice of $R(\boldsymbol{w}, \boldsymbol{x})$ is discussed in some detail in Sacks et al. (1989). Stationary families which are products of one-dimensional correlations, i.e., of the form $R(\boldsymbol{w}, \boldsymbol{x}) = \Pi R_j(w_j - x_j)$, are suggested as a natural choice. Some examples are

$$R(\boldsymbol{w}, \boldsymbol{x}) = \Pi \exp(-\theta_j |w_j - x_j|^p), \quad 0 < p \leqslant 2, \tag{7.10}$$

and

$$R(\boldsymbol{w}, \boldsymbol{x}) = \Pi\big[1 - a_j(w_j - x_j)^2 + b_j|w_j - x_j|^3\big] \tag{7.11}$$

for certain choices of a_j and b_j. Of course, one must estimate any parameters in $R(\boldsymbol{w}, \boldsymbol{x})$, such as the θ_j and p in (7.10) or the a_j and b_j in (7.11). Currin et al. (1988) and Sacks et al. (1989) have found that cross validation and maximum likelihood estimation (MLE) are useful.

If $Z(\boldsymbol{x})$ in (7.1) is a Gaussian process, the likelihood is a function of the β's in the regression model, the variance σ^2, and the parameters in the correlation $R(\boldsymbol{w}, \boldsymbol{x})$. Given the correlation parameters, the MLE of the β's is the generalized least-squares estimate and the MLE of σ^2 is

$$\widehat{\sigma}^2 = \frac{1}{N}(\boldsymbol{y}_d - \boldsymbol{F}\widehat{\boldsymbol{\beta}})'\boldsymbol{R}^{-1}(\boldsymbol{y}_d - \boldsymbol{F}\widehat{\boldsymbol{\beta}}). \tag{7.12}$$

With these values of $\widehat{\boldsymbol{\beta}}$ and $\widehat{\sigma}^2$, the problem is to minimize $(\det \boldsymbol{R})^{1/N}\widehat{\sigma}^2$, which is a function only of the correlation parameters and the data.

An example of an application of these methods to a real problem is discussed in Sacks et al. (1989). See also Welch et al. (1992). In the latter paper, using just a constant for the regression function in (7.1) seems to work well. Other methods for developing predictors at untried inputs are also possible. One might try fitting multivariate splines to the data, perhaps using an adaptive algorithm such as the MARS algorithm of Friedman (1991). One might also consider using such an adaptive algorithm to fit a model like (7.1) rather than using the kriging approach discussed above.

The above methods are appropriate when the objective is prediction at untried inputs. In some cases, the objective may be to optimize some functional (loss function) of the response. If (following Taguchi, 1986) the input variables $\boldsymbol{x}$ can be divided into control factors $\boldsymbol{x}_{\text{con}}$ and noise factors $\boldsymbol{x}_{\text{noise}}$, this goal might be to find the values of

$\boldsymbol{x}_{\text{con}}$ which minimize the expected loss, expectations with respect to some distribution on $\boldsymbol{x}_{\text{noise}}$. A model like (7.1) may again be reasonable, but inference now involves estimating the optimal input configuration. An example of inference for this sort of objective is discussed in Currin et al. (1988).

7.3. *Design*

For the objectives discussed above, the MSE of prediction or the error in an estimate of the input configuration which minimizes some loss function will be a function of the input configurations or sites $\boldsymbol{x}_1, \boldsymbol{x}_2, \ldots, \boldsymbol{x}_N$ in X at which we observe the output from the computer code. Design is concerned with choosing $\boldsymbol{x}_1, \boldsymbol{x}_2, \ldots, \boldsymbol{x}_N$ in X so as to best accomplish our objective. "Best" might mean selecting these sites so as to minimize the MSE of prediction or to minimize the magnitude of the error in an estimate of an optimal input configuration.

For fixed N and for specified correlation structure $\boldsymbol{R}$, when the objective is prediction of the response at untried inputs, we might consider designs which minimize some functionals of the MSE matrix or kernel

$$\boldsymbol{M} = \{\mathrm{E}[Y(\boldsymbol{w}) - \widehat{y}(\boldsymbol{w})][Y(\boldsymbol{x}) - \widehat{y}(\boldsymbol{x})]\} \tag{7.13}$$

for all $\boldsymbol{w}$, $\boldsymbol{x}$ in X. The diagonal elements of $\boldsymbol{M}$ are the MSE $\widehat{y}(\boldsymbol{x})$ in (7.9). In the Bayes case where the β's in (7.1) are known constants, $\boldsymbol{M}$ is the posterior covariance matrix of the process. When the prior variances of the β's tend to infinity, $\boldsymbol{M}$ is the limiting posterior covariance matrix of $Y(.)$. Some criteria based on $\boldsymbol{M}$ suggested by Sacks et al. (1989) are integrated mean squared error (IMSE), maximum mean squared error (MMSE), and entropy. IMSE chooses $\boldsymbol{x}_1, \boldsymbol{x}_2, \ldots, \boldsymbol{x}_N$ in X to minimize

$$\int_X \mathrm{MSE}\big[\widehat{y}(\boldsymbol{x})\big]\,\phi(\boldsymbol{x})\,\mathrm{d}\boldsymbol{x} \tag{7.14}$$

for a given weight function $\phi(\boldsymbol{x})$. From (7.9) the IMSE can be written as

$$\sigma^2\left\{1 - \mathrm{tr}\left[\begin{pmatrix} \boldsymbol{0} & \boldsymbol{F} \\ \boldsymbol{F} & \boldsymbol{R} \end{pmatrix}^{-1} \int_X \begin{pmatrix} \boldsymbol{f}(\boldsymbol{x})\boldsymbol{f}(\boldsymbol{x})' & \boldsymbol{f}(\boldsymbol{x})\boldsymbol{r}(\boldsymbol{x})' \\ \boldsymbol{r}(\boldsymbol{x})\boldsymbol{f}(\boldsymbol{x})' & \boldsymbol{r}(\boldsymbol{x})\boldsymbol{r}(\boldsymbol{x})' \end{pmatrix} \phi(\boldsymbol{x})\,\mathrm{d}x\right]\right\}. \tag{7.15}$$

If X is rectangular and the elements of $\boldsymbol{f}(\boldsymbol{x})$ and $\boldsymbol{r}(\boldsymbol{x})$ are products of functions of a single input factor, these integrals simplify to products of one-dimensional integrals. Thus polynomial regression models and product correlations can be numerically convenient. Notice the IMSE criterion is essentially the trace of $\boldsymbol{M}$ (suitably normalized). This criterion has been used in Sacks et al. (1989).

MMSE is a minimax criterion which chooses the design to minimize

$$\max_{x \in X} \mathrm{MSE}\big[\widehat{y}(\boldsymbol{x})\big]. \tag{7.16}$$

This is compared with IMSE for discrete X in Sacks and Schiller (1988). For continuous X MMSE is computationally complex and difficult to employ.

Minimization of the expected posterior entropy was proposed by Lindley (1956) as a criterion for design in the Bayesian setting. It quantifies the "amount of information" in an experiment. In our setting, if X is discrete, the entropy criterion chooses the design $\{\boldsymbol{x}_1, \boldsymbol{x}_2, \ldots, \boldsymbol{x}_N\}$ in X to minimize the expectation of the negative of the log of the conditional density of $Y(.)$ on $X - \{\boldsymbol{x}_1, \boldsymbol{x}_2, \ldots, \boldsymbol{x}_N\}$ given $Y(.)$. Shewry and Wynn (1987) showed that this is equivalent to maximizing the prior entropy on $\{\boldsymbol{x}_1, \boldsymbol{x}_2, \ldots, \boldsymbol{x}_N\}$. When $Y(.)$ is Gaussian, this is the same as choosing $\{\boldsymbol{x}_1, \boldsymbol{x}_2, \ldots, \boldsymbol{x}_N\}$ to maximize the determinant of the covariance matrix $\boldsymbol{V}_d$ for $Y(.)$ on $\{\boldsymbol{x}_1, \boldsymbol{x}_2, \ldots, \boldsymbol{x}_N\}$. In the limiting case as the prior variances on the β's tend to infinity it can be shown that maximizing $\det \boldsymbol{V}_d$ is equivalent to maximizing $\{\det \boldsymbol{R} \cdot \det(\boldsymbol{F}'\boldsymbol{R}^{-1}\boldsymbol{F})\}$. If the β's are regarded as fixed, the last determinant vanishes and the entropy criterion reduces to maximizing $\det \boldsymbol{R}$. The entropy criterion is further discussed and used in Shewry and Wynn (1987, 1988) and in Currin et al. (1988).

Implementation of any of these criteria in practice must be done numerically. An adaptation of Mitchell's (1974) DETMAX was used with the entropy criterion in Currin et al. (1988). Shewry and Wynn (1987) used and exchange algorithm to construct designs. Sacks and Schiller (1988) used a simulated annealing algorithm for IMSE and MMSE.

Design and criteria for other objectives than prediction appear to be less well studied. More research is needed. In the absence of good designs for specific criteria (perhaps because of the difficulty in finding such designs) Latin hypercube sampling has proved attractive and been extensively used. Latin hypercube sampling was first proposed by McKay et al. (1979) in the context of design for deterministic computer code. Their objective was to study how a known distribution of the inputs propagates through the output distribution. Latin hypercube sampling is an extension of stratified sampling and ensures that each of the input variables has all portions of its range adequately represented in the design. Thus it attempts to provide a design which examine the output uniformly over the design space (somewhat like the "naive" strategy discussed in Section 1). Latin hypercubes are computationally cheap to generate and can cope with high dimensional (many input variables) problems. Because Latin hypercube designs generate output over the entire range of each input variable they are a systematic way of discovering unusual behavior in the code as noted by Iman and Helton (1988). These factors make Latin hypercube designs popular choices when the objective is complex.

Many other approaches are possible. For example, sequential strategies may be very useful in computer experiments. Bernardo et al. (1992) use a sequential strategy based on model (7.1) for integrated circuit design. As indicated in Section 7.2, the BLUP can also be derived from a Bayesian approach and such an approach may give insight into design issues. Morris et al. (1993) investigate this Bayesian approach and include issues of experimental design. Other methods, such as the method of steepest ascent for finding the minimum of a response surface, and designs for models like those discussed in other sections of this article are additional possibilities for computer experiments. As in all applications of model robust design, many approaches are possible and are dependent on the models and assumptions one is willing to make. One of the attractive features of computer experiments is that they provide real examples in which to explore and test model robust design methodology. Most of the above work has evolved from working with real problems and may be regarded as the first steps in developing useful methodology. Due to the complexity of the problem, much remains to be investigated both theoretically and numerically in this area.

References

Atkinson, A. C. (1972). Planning experiments to detect inadequate regression models. *Biometrika* **59**, 275–293.

Atkinson, A. C. (1975a). The design of experiments for discriminating between two rival models. *Biometrika* **62**, 57–70.

Atkinson, A. C. (1975b). Optimal design: Experiments for discriminating between several models. *Biometrika* **62**, 289–303.

Atwood, C. L. (1971). Robust procedures for estimating polynomial regression. *J. Amer. Statist. Assoc.* **66**, 855–860.

Bernardo, M. C., R. Buck, L. Liu, W. A. Nazaret, J. Sacks and W. J. Welch (1992). Integrated circuit design optimization using a sequential strategy. *IEEE Trans. Comput. Aid. Des.* **2**, 361–372.

Box, G. E. P. and N. R. Draper (1959). A basis for the selection of a response surface design. *J. Amer. Statist. Assoc.* **54**, 622–654.

Box, G. E. P. and N. R. Draper (1963). The choice of a second order rotatable design. *Biometrika* **50**, 335–352.

Box, G. E. P. and N. R. Draper (1975). Robust designs. *Biometrika* **62** 347–352.

Cook, R. D. and C. J. Nachtsheim (1982). Model robust, linear-optimal designs. *Technometrics* **24**, 49–54.

Currin, C., T. Mitchell, M. Morris and D. Ylvisaker (1988). A Bayesian approach to the design and analysis of computer experiments. ORNL Tech. Report 6498, Available from the National Technical Information Service, Springfield, VA 22161.

Currin, C., T. Mitchell, M. Morris and D. Ylvisaker (1991). Bayesian prediction of deterministic functions, with applications to the design and analysis of computer experiments. *J. Amer. Statist. Assoc.* **86**, 953–963.

DasGupta, A., S. Mukhopadhyay and W. J. Studden (1992). Compromise designs in heteroscedastic linear models. *J. Statist. Plann. Inference* **32**, 363–384.

DasGupta, A. and W. J. Studden (1991). Robust Bayes designs in normal linear models. *Ann. Statist.* **19**, 1244–1256.

DeFeo, P. and R. H. Myers (1992). A new look at experimental design robustness. *Biometrika* **79**, 375–380.

Dette, H. (1990). A generalization of D- and $D1$-optimal designs in polynomial regression. *Ann. Statist.* **18**, 1784–1804.

Dette, H. (1991). A note on robust designs for polynomial regression. *J. Statist. Plann. Inference* **28**, 223–232.

Dette, H. (1993a). Elfving's theorem for D-optimality. *Ann. Statist.* **21**, 753–766.

Dette, H. (1993b). A note on Bayesian c- and D-optimal designs in nonlinear regression models. Manuscript.

Dette, H. and W. J. Studden (1994). Optimal designs for polynomial regression when the degree is not known. Manuscript.

Draper, N. and A. Herzberg (1973). Some designs for extrapolation outside a sphere. *J. Roy. Statist. Soc. Ser. B* **35**, 268–276.

Draper, N. and A. Herzberg (1979). An investigation of first-order and second-order designs for extrapolation outside a hypersphere. *Canad. J. Statist.* **7**, 97–110.

DuMouchel, W. and Jones, B. (1994). A simple Bayesian modification of D-optimal designs to reduce dependence on an assumed model. *Technometrics* **36**, 37–47.

Elfving, G. (1952). Optimum allocation in linear regression theory. *Ann. Math. Statist.* **23**, 255–262.

Fedorov, V. V. (1972). *Theory of Optimal Experiments*. Academic Press, New York.

Friedman, J. H. (1991). Multivariate adaptive regression splines (with discussion). *Ann. Statist.* **19**, 1–141.
Hoel, P. G. and A. Levine (1964). Optimal spacing and weighting in polynomial prediction. *Ann. Math. Statist.* **33**, 1553–1560.
Huang, M. N. L. and W. J. Studden (1988). Model robust extrapolation designs. *J. Statist. Plann. Inference* **18**, 1–24.
Huber, P. J. (1975). Robustness and designs. In: J. N. Srivastava, ed., *A Survey of Statistical Design and Linear Models*. North-Holland, Amsterdam, 287–303.
Iman, R. L. and J. C. Helton (1988). An investigation of uncertainty and sensitivity analysis techniques for computer models. *Risk Anal.* **8**, 71–90.
Jones, E. R. and T. J. Mitchell (1978). Design criterion for detecting model inadequacy. *Biometrika* **65**, 541–551.
Karson, M. J., A. R. Manson and R. J. Hader (1969). Minimum bias estimation and experimental designs for response surfaces. *Technometrics* **11**, 461–475.
Kiefer, J. (1959). Optimal experimental designs (with discussion). *J. Roy. Statist. Soc. Ser. B* **21**, 272–319.
Kiefer, J. (1973). Optimal designs for fitting biased multi-response surfaces. In: P. R. Krishnaiah, ed., *Multivariate Analysis*, Vol. 3. Academic Press, New York, 287–297.
Kiefer, J. (1974). General equivalence theory for optimum designs (approximate theory). *Ann. Statist.* **2**, 849–879.
Kiefer, J. (1980). Designs for extrapolation when bias is present. In: P. R. Krishnaiah, ed., *Multivariate Analysis, Vol. 5*, Proc. 5th Internat. Symp. North-Holland, Amsterdam, 79–93.
Kiefer, J. and J. Wolfowitz (1959). Optimum designs in regression problems. *Ann. Math. Statist.* **30**, 271–294.
Läuter, E. (1974). Experimental design in a class of models. *Math. Operationsforsch. Statist.* **5**, 379–396.
Läuter, E. (1976). Optimal multipurpose designs for regression models. *Math. Operationsforsch. Statist. Ser. Statist.* **7**, 51–68.
Li, K. C. (1984). Robust regression designs when the design space consists of finitely many points. *Ann. Statist.* **12**, 269–282.
Li, K. C. and W. Notz (1982). Robust designs for nearly linear regression. *J. Statist. Plann. Inference* **6**, 135–151.
Lindley, D. V. (1956). On a measure of the information provided by an experiment. *Ann. Math. Statist.* **27**, 986–1005.
Marcus, M. B. and J. Sacks (1977). Robust designs for regression problems. In: S. S. Gupta and D. S. Moore, eds., *Statistical Decision Theory and Related Topics*, Vol. 2. Academic Press, New York, 245–268.
McKay, M. D., W. J. Conover and R. J. Beckman (1979). A comparison of three methods for selecting values of input variables in the analysis of output from a computer code. *Technometrics* **21**, 239–245.
Mitchell, T. J. (1974). An algorithm for the construction of "D-optimal" experimental designs. *Technometrics* **16**, 203–210.
Morris, M. D., T. J. Mitchell and D. Ylvisaker (1993). Bayesian design and analysis of computer experiments: Use of derivatives in surface prediction. *Technometrics* **35**, 243–255.
O'Hagan, A. (1978). Curve fitting and optimal design for prediction (with discussion). *J. Roy. Statist. Soc. Ser. B* **40**, 1–42.
Pesotchinsky, L. (1982). Optimal robust designs: Linear regression in R^k. *Ann. Statist.* **10**, 511–525.
Pukelsheim, F. (1993). *Optimal Design of Experiments*. Wiley, New York.
Sacks, J. and S. Schiller (1988). Spatial designs. In: S. S. Gupta and J. O. Berger, eds., *Statistical Decision Theory and Related Topics IV*, Vol. 2. Springer, New York, 385–399.
Sacks, J., S. B. Schiller and W. J. Welch (1989). Designs for computer experiments. *Technometrics* **31**, 41–47.
Sacks, J., W. J. Welch, T. J. Mitchell and H. P. Wynn (1989). Design and analysis of computer experiments (with discussion). *Statist. Sci.* **4**, 409–435.
Sacks, J. and D. Ylvisaker (1984). Some model robust designs in regression. *Ann. Statist.* **12**, 1324–1348.
Scheffé, H. (1958). *The Analysis of Variance*. Wiley, New York.
Seo, H. S. and K. Larntz (1992). Restricted Bayesian optimal design. Tech. Report 574, School of Statistics, University of Minnesota.
Shewry, M. C. and H. P. Wynn (1987). Maximum entropy sampling. *J. Appl. Statist.* **14**, 165–170.

Shewry, M. C. and H. P. Wynn (1988). Maximum entropy sampling and simulation codes. In: *Proc. 12th World Congress on Scientific Computation, IMAC88*, Vol. 2, 517–519.

Spruill, M. C. (1985). Model robustness of Hoel–Levine optimal designs. *J. Statist. Plann. Inference* **11**, 217–225.

Steinberg, D. M. (1985). Model robust response surface designs: Scaling two level factorials. *Biometrika* **72**, 513–526.

Stigler, S. M. (1971). Optimal experimental design for polynomial regression. *J. Amer. Statist. Assoc.* **66**, 311–318.

Taguchi, G. (1986). *Introduction to Quality Engineering*. Asian Productivity Organization, Tokyo.

Toman, B. (1992a). Bayesian robust experimental designs for the one-way analysis of variance. *Statist. Probab. Lett.* **15**, 395–400.

Toman, B. (1992b). Discussion of Verdinelli. In: J. M. Bernardo et al., eds., *Bayesian Statistics*, Vol. 4. Oxford Univ. Press, 477–478.

Toman, B. and J. L. Gastwirth (1993). Robust Bayesian experimental design and estimation for analysis of variance models using a class of normal mixtures. *J. Statist. Plann. Inference* **35**, 383–398.

Toman, B. and J. L. Gastwirth (1994). Efficiency robust experimental design and estimation using a data-based prior. Tech. Report #93-19, Department of Statistics/C&IS. The George Washington University. To appear in *Statist. Sinica*.

Welch, W. J., R. J. Buck, J. Sacks, H. P. Wynn, T. J. Mitchell and M. D. Morris (1992). Screening, predicting and computer experiments. *Technometrics* **34**, 15–25.

Wu, C. F. (1981). On the robustness and efficiency of some randomized designs. *Ann. Statist.* **9**, 1168–1177.

S. Ghosh and C. R. Rao, eds., *Handbook of Statistics, Vol. 13*

29

Review of Optimal Bayes Designs

*Anirban DasGupta**

Dedicated to J. H. B. Kemperman, a unique person.

1. General introduction

It is fair to say that for as long as we can recall, statistical training has emphasized the role of design of an experiment in extracting the correct type of information and making accurate inferences for the problem of interest. Proper design of an experiment is evidently a crucial aspect of sound statistical practice; a classic course on design actually helps bring this out much more than a sophisticated course on the mathematics of design. Historically, design of experiments started out with agricultural studies and therefore factorial experiments and their design aspects were the natural starting points of the theory and practice of optimum design. A little exposure to factorial experiments shows how utterly important it indeed is to use the right design to avoid the pitfalls of confounding, nonestimability, missing data, and a long list of other genuine problems. See Fisher (1949). With statistical practice changing rapidly, as an influence of advances in computer technology and the inclination among many to treat statistics as mostly data analysis, the role of design of experiments and controlled studies may get substantially diminished at a future time. However, that has not happened yet. It therefore appears appropriate that a broad overview of the history, mathematics, methods, advances, and the future of experimental design should be made. A number of such contributions already exist; some are more methodological, others more technical. Design of experiments especially is one branch of statistical theory in which frequentist and Bayesian ideas, formulations, techniques, and results go very parallel; thus, although the primary goal of this writing is to make a review of the state of the art in Bayesian design, it is quite impossible to do so in isolation of the rich history of classical optimal design. As a matter of fact, Bayes design can be understood only in the context of what is known in classical design. One reason for this is that barring a few exceptions, even the formulation of a Bayes design problem requires frequentist evaluations of a design. We will therefore have to necessarily consider classical theory and methods to some extent in this writing. Although there is bound to be some overlap between the technical contents of this chapter and others, due to the connections

*Research supported by NSF grant DMS 9307727.

of the mathematics involved in various variety of optimal design theory, there are certain unique aspects of this chapter, both informationally and technically. In particular, this chapter is unique in its description of the role of the prior in determining an optimal design, how the role of the prior typically diminishes in a very strong sense with increasing sample size, conditional optimal design formulations that would make sense only in the Bayesian context of this chapter, and sample size problems that are always regarded as important by applied scientists, especially those conducting clinical trials. On the technical side, this chapter provides a self contained brief account of two relevant but highly used tools of pure mathematics: moment methods and the algebra of orthogonal polynomials. These are presented in the appendix to retain their independent character and to allow an uninterrupted flow of the statistical content in the main body of the chapter.

In a broad intellectual sense, the problem of design is encompassing; one can make the following general statement: in any scientific problem in which the scientist has the choice and flexibility of choosing one among many initial setups of the experiment, there is an optimal design problem associated with the experiment. The concept of optimal design goes far back in the history of mathematical sciences, and probably even further back in the history of our civilization. For example, the following well known examples are all instances of optimal designs.

EXAMPLE 1 (Polynomial interpolation). Consider a continuous function $f(x)$ on a given bounded interval $[a, b]$. It is well known that there exists a polynomial $p(x)$ which uniformly approximates the given function $f(x)$, to any specified degree of accuracy $\varepsilon > 0$, with the degree of the polynomial $p(x)$ depending on ε. A proof of this can be found in standard texts on analysis; for a historically important proof, see Korner (1989).

It is natural to try to find a good approximating polynomial by interpolating the function f at a fixed set of say $(n+1)$ points, $X = \{x_i, 0 \leqslant i \leqslant n\}$. Indeed, there is a unique polynomial $p(x)$ with the representation

$$p(x) = \sum_{i=0}^{n} l_i(x) f(x_i)$$

interpolating f at the points of X. In the above, $\{l_i\}$ are themselves polynomials of degree n, and are commonly known as the fundamental or cardinal polynomials of interpolation.

A natural criterion for assessing the goodness of approximation of f by p is the quantity

$$e(x) = |p(x) - f(x)|.$$

Note that $e(x)$ depends on the choice of "nodes" X. A design problem is therefore the following:

Choose a good set of nodes X according to the criterion e.

As an illustration, consider approximating the (unnormalized) Cauchy density

$$f(x) = \frac{1}{1+x^2}$$

in the interval $[-5, 5]$. Take two sets of nodes: X_1 = Equally spaced points at spacings of .5, starting at -4.75; X_2 = The points of peaks of the nth Chebyshev polynomial $T_n(x)$ in the interval $[-5, 5]$, with $n = 20$.

Straightforward computation shows that the equispaced points give a very bad fit near the boundary of the interval, and even worse, the fit deteriorates by taking more equispaced points. On the contrary, the Chebyshev nodes result in a maximum error of $< .016$ with $n = 20$. One might therefore say that X_2 is a better design than X_1 in this example. For universal results on the goodness of X_2 as the nodes of interpolation, one can see Erdos (1958) and Rivlin (1981). For further illuminating discussion of the example above, one can see Powell (1981).

EXAMPLE 2 (The Secretary problem). The basic Secretary problem corresponds to the situation in which n candidates are interviewed for a job in a random order and a candidate once rejected cannot be recalled. The employer would be able to rank the candidates from 1 (best) to n (worst) if she could indeed see them all at one time. The criterion of the employer is the following:

Maximize the probability of selecting the actual best candidate.

The following is a design problem:

What interviewing strategy should be used according to this criterion?

Although in this form, the problem is admittedly somewhat unrealistic, various modifications of the basic problem have been studied in great detail; however, even the basic problem is intellectually interesting due to the beauty and the neatness of the optimal design. The optimal design says that there exists a value $k = k(n)$ such that the employer should reject the first k candidates and then accept the very first one she likes better than the others who went by. In the process, there is the possibility that the employer has to accept the last candidate to arrive, even if this is the worst candidate among all. The value k can be found with relative ease by maximizing a unimodal function defined on integers. The asymptotic solution is perhaps one of the most neat results of mathematical statistics: it is to reject the first 100/e % of the candidates and then follow the design given above. If one does this, the probability of getting the best candidate is also about $1/e$ for large n. Among the large literature on the Secretary problem, this author particularly recommends Freeman (1983) and Ferguson (1989).

EXAMPLE 3 (Allocation of treatments in clinical trials). Consider estimating the average rate of response in a clinical example with the model

$$\mathrm{E}(X_i) = b_i\theta, \qquad \mathrm{Var}(X_i) = a(\mathrm{E}(X_i))^p, \quad p \geqslant 0,$$

where X_i are independent observations on the response to b_i units of a stimulus. The goal is to estimate the response rate θ.

Suppose, as is often the case, a total fixed amount of the stimulus is available, and the dose that can be given to an individual has to be between two bounds (a very low dose is useless, and a very high dose is dangerous). Suppose the experimenter is unwilling to assume normality or any other model assumptions, and decides to estimate θ by using its BLUE.

The following is a design problem:

> How many individuals should be used and how many doses should they get in order to minimize the mean squared error of the estimate?

It turns out the optimal design crucially depends on the value of p; in some cases, it is harmful to use more individuals for the study: i.e., a smaller sample is much better. In some other cases, the design is completely unimportant. and the same mean squared error is achieved regardless of the design. In some other cases still, it is best to use as many individuals as possible by applying the smallest amount of the stimulus. For an enjoyable look at this, one can see Kiefer (1987). A more recent generalization is DasGupta and Zen (1996).

EXAMPLE 4 (Sequential analysis). In sequential experiments, data come in a sequence, with the experimenter retaining the option of stopping and making an inference at any stage without collecting more data. The practical impetus for sequential experiments came from the frequency of real problems in which the inference problem was satisfactorily solved without the need of more data. In law enforcement, an analog may be that police stop taking tips incriminating other individuals when the existing evidence against a current suspect is overwhelming. In his monograph, Chernoff (1972) eloquently describes the relation of sequential experimentation to design: "in the act of deciding whether or not to gather more data, the statistician is making a choice of design. In this sense, sequential analysis ... is not a separate field (from optimal design)".

Deep and profound questions exist on what exactly is the "correct" optimal design problem in this context; particularly, the importance (or the lack of it) of exactly how the sampling process was terminated is a bitter bone of contention among statisticians of various descriptions. For a lucid and remarkably enjoyable discussion on this issue, one can see Berger (1986). Generally speaking, a strict believer in the Bayesian paradigm should have no use for knowing the exact termination rule; however, it is a truth that many declared Bayesians do not believe that with conviction. It is much like the parallel fact that randomization has no role in a strictly idealistic Bayesian world, but probably no Bayesians exist who would recommend against randomization, generally accepted to be a most sacred principle of statistical data gathering. Underneath all of these, the design problem is the following:

> Choose a stopping rule and a procedure for deciding between actions after one has stopped.

The exact optimal design in a strictly sequential context is usually not something one can write easily on a piece of paper; there are many instructive examples of Bayes optimal designs in sequential contexts in Chernoff (1972). Generally, one has to propose a design and establish its near optimality. Further contributions came from Schwartz (1962) and Siegmund (1985).

EXAMPLE 5 (Greedy algorithms). The Greedy algorithms refer to a whole family of optimization procedures in the problem of assigning k people to k jobs, when the cost of assigning the ith person to the jth job is $C(i, j)$. Two common greedy algorithms are the following:

Method A. Assign to each individual the available job she does the best;

Method B. Initially, identify the best (i, j) combination; then eliminate the selected person and the selected job, and identify the next best (i, j) combination, and continue.

The design problem is the following:

Choose among the possible finite number of job assignments, the one that minimizes the total cost.

A mystifying result is that if $C(i, j)$ are iid Exponential, then each of Method A and B are equivalent to each other. The result is not trivial, and to see the equivalence one has to use various algebraic facts and other facts particular to the Exponential distribution, specifically, its memory-less property.

EXAMPLE 6 (Blackwell prediction). The Blackwell prediction algorithm deals with the problem of predicting the $(n + 1)$th member of an infinite 0–1 sequence knowing the past members. A design thus corresponds to construction of an algorithm. This is probably one of the earliest examples of optimal design in which minimaxity appeared as a selection criterion (almost concurrently with the appearance of the Blackwell prediction algorithm, came the minimax ideas in Kiefer (1953), in which Kiefer uses minimaxity as a criterion for choosing a set of evaluation points in order to locate the maximum of a unimodal function in a bounded interval. Brown (1991) also gives a charming description of these growing years of optimal design).

Interesting things happen; a naive prediction algorithm (also appealing due to its simplistic nature) is to predict the $(n + 1)$th member as 1 if the average of the past members is $> 1/2$. While it predicts sequences that are really Bernoulli quite well, it does not do well for deterministic sequences of certain kinds. The Blackwell algorithm, which is randomized, in contrast seems to cover both types and has a minimax property. A Bayesian optimum design problem arises by putting a prior distribution on the unknown infinite sequence; but then, the problem is easily solved. One simply calculates the posterior probability of each value at the $(n+1)$th stage and predicts the one with a larger posterior probability. Blackwell gives a nice colloquial description of the relevance of this problem in Information theory and artificial intelligence in his interview with Morris DeGroot in Statistical Science (1986).

EXAMPLE 7 (Computer experiments and infinite dimensional problems). In recent years, emphasis in optimal design is shifting to new families of problems. One such area involves the writing of a stochastic equation for predicting the output of a deterministic computer code. Typically, the problem is of the following type:

One has a (possibly very high dimensional) input x in response to which the computer produces an output $y = y(x)$. The relation is supposed to be deterministic. However, a stochastic model is introduced:

$$y(x) = \text{ A regression function } + z(x),$$

where $z(x)$ is an error. The idea is to build a predictor by treating this as a regression problem; this predictor acts as a (cheaper) proxy to the deterministic computer output of the complex code. The design problem is the following:

> Determine a set of n values of x which are to be used in constructing the prediction equation.

As in Example 3, one can decide on a linear method and avoid making assumptions about the stochastic process $z(x)$ or one can assume $z(x)$ is a path of a certain well understood process, typically a Gaussian process. Linear estimation in this context is commonly called kriging; one can use a kriging procedure together with a specified design criterion to arrive at a functional that needs to be maximized to construct an optimal design. Common design criteria include an integrated mean squared error (over x, with respect to some probability measure on x), and a maximum mean squared error. The problems are much harder than what one sees in ordinary regression designs; as a consequence, it is typical that the optimal design has to be found by a search method and the construction of the search algorithm is as important as the identification of the criterion functional. One can see Sacks et al. (1989) and Sacks and Schiller (1988) for a broad exposition. Use of stochastic processes as priors on continuous functions is also done in Diaconis (1987), O'Hagan (1978), among others. Essentially the same things also go by the name of illposed inverse problems.

The above examples clearly show how rich the study of optimal designs is and can be. It is an error to think that optimal design is an abstract area of mathematical statistics limited to standard statistical models like factorial experiments, or linear and nonlinear models. In fact, according to this author, the more lively optimal design problems arise in branches outside of these models, and much remains to be looked at. As commented earlier, a remarkably vast literature already exists on optimal design. Anyone seriously interested in learning about optimal design must at the least consult, in addition to the references above, Kiefer (1959), Atkinson and Donev (1992), Fedorov (1972), Silvey (1980), Pukelsheim (1993), Pilz (1991), and in particular for Bayesian optimal design, a very recent review article by Chaloner and Verdinelli (1994). Indeed, we will make a conscious effort to as much as possible emphasize aspects and literature not emphasized in this review article in order to avoid a wasteful duplication of intellectual effort. However, there will necessarily be some overlapping due to the review nature of both articles.

2. Outline

In Section 3, we give some history of the theory of optimal design; naturally, this will include the early developments usually attributed to Kiefer and Wolfowitz (1959). In Section 4, we explicitly start to discuss Bayesian formulations of the design problem, and we will discuss the direct impact of the Kiefer–Wolfowitz theory on Bayes design and also discuss the early history of Bayes design. In Section 5, we will broadly discuss the typical mathematics of Bayes regression designs; this will cover the Elfving theorem, other geometry due to Chaloner (1984) and El-Krunz and Studden (1991), and Studden and Dette (1993), and will also include the role of moment methods and equivalence theorems. Section 6 will apply the theory to explicit description of Bayes optimal designs; this section will also include some work on Factorial experiments, in particular those of Notz, Toman, and their coauthors. In Section 7, we critically assess the relevance and impact of optimal design theory, and address issues such as belief in the model, and construction of all around designs. This section will include an outlook into how optimal design theory can adapt itself to the opinion of practitioners. Section 8 discusses the nonconjugate case, and whether Bayesians need to even worry about optimal designs: this will cover two aspects – whether classical designs alone suffice, and whether the prior matters. Robust Bayes optimal designs will be discussed in this section. Section 9 will cover miscellaneous design problems, as in quality control, engineering reliability, spatial designs, etc. Section 10 covers sample size and preposterior formulations of the Bayes design problem as opposed to standard criteria like integrated Bayes risk. Section 11 gives a brief exposition to nonlinear models and associated design problems. In Section 12, various other issues are discussed and concluding remarks are made. Section 13 contains an appendix on the mathematical tools of optimal design.

3. Some history

Optimally setting up an observational study so as to extract as much relevant information as possible is such a natural idea that it is possibly impossible to trace back to the first scientific article on this; there are, however, some demonstrably early ones. Smith (1918) already has a clear flavor of optimal design in polynomial regression; apart from one intermediate but clearly a key contribution by Wald (1943), the culture of a structured optimal design theory arrived with Jack Kiefer. The paper by Wald (1943) was key in its influence on how optimal design theory was formulated and was done for three decades. Although the time around the second world war was in some sense the golden age of decision theory (every eminent statistician did some decision theory around that time), it is fair to think that Wald's 1943 article had a binding influence on the formulations that came through later. Stigler (1974) gives a fine account of the history of polynomial regression that makes interesting reading for researchers in design of experiments.

The most remarkable and time tested contribution of the Kiefer–Wolfowitz theory was the concept of an approximate design. Indeed, this concept had such a tremendous impact that optimal regression design is done even today more or less within the domain of approximate designs. The idea was that exact optimal designs for a given sample size n are to a significant extent dependent on the value of n, and their derivation corresponds to a straightforward (but not proportionately enlightening) integer programming problem; in contrast, by formulating a design as a probability measure on the design space, two things are achieved: avoiding a dependence on n (except at the implementation stage) and making possible a strikingly beautiful theory that connects together several branches of mathematics (analysis in particular); Karlin and Studden (1966) is a standard reference on connections of optimal design theory to moment methods and orthogonal polynomials. The other important contributions of the Kiefer–Wolfowitz theory were the concepts of alphabetic optimality; these are accepted quite universally as criteria for evaluation even today, although a somewhat small school has argued against a few criteria in use due to their apparent lack of correspondence to decision theory based on a utility function: see Chaloner and Verdinelli (1994) for more on this.

An important point at which regression and other (factorial or qualitative) types of optimal design theory separated is the adoption of approximate designs as a founding concept. Historically, in these other branches of optimal design theory, combinatorics and integer programming continued to play the key roles. Bose (1948) is an early important work followed by much work of many researchers, notably C. S. Cheng. One can see Shah and Sinha (1989) for a comprehensive and informative account of nonregression optimality theory; Kurotschka (1978) gives an account of the nature of optimality theory in the presence of both quantitative and qualitative factors.

A mathematical tool from which all of optimal design theory benefited is commonly called an equivalence theorem. At a basic level, an equivalence theorem only states that at a point of minima a differentiable function has derivative zero and it increases as one moves away from the minima in a given direction. The use of an equivalence theorem is in its ability to verify that a design suspected to be optimal is indeed so; subject to numerical accuracy of such a verification, this has been creatively used in a number of problems, notably in nonlinear models by Kathryn Chaloner and her coauthors. Statements of general equivalence theorems can be seen in many writings; one can see in particular Silvey (1980), Pukelsheim (1993) and Schoenberg (1959).

As much as the Kiefer–Wolfowitz theory was beautiful, its impact on practitioners was limited. The problem is in the nature of the optimal designs. One has to trust the model absolutely to consider actually using these exact designs. It is therefore quite natural that concerns about robustness with respect to misspecification of the model were voiced; Stigler (1971) and Studden (1982) reacted to these concerns, among many others. There are also many who believe that parameter estimation is not the aim of an experiment and a model should be assessed on the basis of its predictive power; it is a fairly persuasive argument and not surprisingly, predictive design criteria have been suggested. Lindley (1968) is probably the earliest article on Bayesian-decision theoretic design based on predictive evaluations; the topic continued with a number of articles by Brooks (1974, 1976), and has recently gained further momentum with an article due to Eaton et al. (1994).

The history of a structured Bayesian optimality theory is by far much more recent; in fact, it can be said that Chaloner (1984) is the first serious attempt to develop a theory of Bayes optimality in linear regression context. Chaloner (1984) gives a formulation, shows analogs of the Elfving geometry of the classical theory in some special cases and explicitly describes the role of the prior on the difference between the Bayes and classical optimal designs. Meanwhile, a number of people in Europe started to actively work on Bayesian optimal designs, and Pilz (1991) is an early contribution that assembled a great amount of material in the context of linear regression and really provided a solid impetus for further work. The Elfving geometry in the context of Bayes designs was beautifully described in El-Krunz and Studden (1991), perhaps the deepest theoretical contribution to Bayes designs till now. Sensitivity of Bayes designs to the choice of the prior was given a structured formulation and explicit robust Bayes designs were given in DasGupta and Studden (1991); Toman (1992) treats sensitivity in models with qualitative factors.

It was already well known that practically all of the neat theory of optimal designs one sees in linear models is unachievable in even the simplest kinds of nonlinear models. In fact, a great amount of philosophical and moral dilemma pervade optimal design in nonlinear models. The problem is that strictly speaking, an optimum design depends on the true value of the parameter one is trying to estimate in the first place. The concept of local optimality was introduced in the classical theory to tackle this issue. The first attempt at seriously working out Bayesian optimal designs in a series of nonlinear models was made in Chaloner and Larntz (1989), although prior important contributions exist, notable among them Box and Lucas (1959). The mathematics of the Bayesian optimality theory for nonlinear models is challenging, and rather surprising advances have come through in a short period of time. The works of Holger Dette and his coauthors deserve specific mention due to their insightful nature. Unfortunately, however, all the evidence still suggests that a unifying theory as in the case of linear models would not be possible and a piecemeal theory may emerge with time.

The theory and practice of Bayesian optimal design are still at an early stage; many topics for which a great amount of results exist under classical optimality criteria have not been at all looked at. Effect of dependence in the observations is one such (old works of Sacks and Ylvisaker (1966) and Bickel and Herzberg (1979) are by now classic contributions to this) topic. Determination of minimum and optimal sample size which has taken the status of textbook material in classical statistics, is just beginning to get serious attention in terms of a debate about which formulations are proper for the Bayesian; some early theoretical works include DasGupta and Mukhopadhyay (1994) and DasGupta and Vidakovic (1994). It is encouraging to see that efforts are being made, although somewhat in isolation of a structured theory, to work out Bayesian optimal designs in actual applied problems; workers at the Duke school have already made good contributions in this area. Caselton and Zidek (1984) and Schumaker and Zidek (1993) are instances of elegant theoretical developments in interesting real problems. Another area in which some effort is being made is the writing of computer codes for numerical implementation of Bayes optimal designs (Clyde, 1993). Significant literature on this already exists for construction of classical optimal designs; in particular, one can see Atkinson and Donev (1992) for an exchange algorithm much like the exchange algorithm of numerical analysis for finding the minimax fit to continuous functions from Haar spaces, and Haines (1987) for an innovative use of simulated annealing in constructing D-optimal designs.

For comprehensive reading of optimal design, both classical and Bayes, many excellent sources exist; we enthusiastically recommend Atkinson and Donev (1992), Box and Draper (1987), Herzberg and Cox (1969), Pukelsheim (1993), Silvey (1980), Chaloner and Verdinelli (1994) and Wynn (1984) for anyone interested in this topic. In fact, our effort would be to emphasize whenever possible specific points not addressed in much detail in these earlier contributions. It is also necessary to consult these for bibliography in addition to the bibliography of this article.

4. Alphabetic criteria and other formulations

4.1. Approximate designs

The five most widely accepted criteria for an optimality theory of designs are c, A, D, E, and G optimality; there are others. The road to arrival at these criteria can be thought of in the following way: one has a standard Gauss–Markov linear model and decides to use the least squares estimate of the regression coefficients; it seems natural that one should want to make the estimate as accurate as possible. Since the least squares estimate is already unbiased, consideration will then focus on the variance covariance matrix of the least squares estimate. Minimizing the trace, determinant, and the maximum eigenvalue of this matrix respectively correspond to A, D, and E optimality. Minimization of the variance of the least squares estimate of a linear combination of the regression coefficients corresponds to c optimality. G optimality, which corresponds to an average over linear combinations of the coefficients thus also links up to essentially the same fundamental idea.

Consider then the usual linear model $y_i = \sum_{j=0}^{p} \theta_j \; f_j(x_i) + \varepsilon_i$, where the vector of errors satisfies $\mathrm{E}(\underset{\sim}{\varepsilon}) = 0$ and $\mathrm{D}(\underset{\sim}{\varepsilon}) = \sigma^2 I$, where $\sigma^2 > 0$ is possibly unknown. A large number of statistical models in everyday use fall under this general setup, and in principle, therefore, an optimality theory for designs in the canonical linear model certainly has a wide scope for application. The least squares estimate for $\underset{\sim}{\theta}$ has the representation $(X'X)^{-1}X'Y$, with variance covariance matrix $\sigma^2 \; (X'X)^{-1}$, where X denotes the design matrix with rows $(1, f_1(x_i), \ldots, f_p(x_i))$. These statements need to be slightly changed when $X'X$ is not full rank, which in fact does happen in some interesting problems. We shall later see that usually this ceases to be a problem in the corresponding Bayes theory. For the alphabetic criteria listed above, one can make a transition to the precision matrix $(1/\sigma^2)(X'X)$, due to the well known relations between the trace, determinant and eigenvalues of a matrix and its inverse. Actually, there is a whole family of criterion functions that permit such a transition from the "dispersion" matrix $(X'X)^{-1}$ to the "information" matrix $X'X$. Thus, for instance, the E optimality criterion corresponds to maximizing the minimum eigenvalue of the information matrix $X'X$.

Now if the distinct rows in the "design" matrix X are denoted as $\underset{\sim}{x}_1', \underset{\sim}{x}_2', \ldots$, with multiplicities $n_1, n_2, \ldots$ (a repeated row corresponds to replication of the same levels of the independent variables for two or more individuals), then the information matrix takes the form

$$X'X = n \sum n_i/n \underset{\sim}{x}_i \underset{\sim}{x}_i'.$$

Writing n_i/n as p_i, one therefore sees that $X'X$ equals an average of the quantities $\underset{\sim}{x}_i \underset{\sim}{x}_i'$. The idea of an approximate design is to allow an arbitrary probability measure instead of a discrete probability vector $\underset{\sim}{p}$ such that the elements of $n.\underset{\sim}{p}$ are integers. One then has a general information matrix

$$M = M(\mathcal{E}) = \int \underset{\sim}{x} \underset{\sim}{x}' \, \mathrm{d}\mathcal{E}(x),$$

where $\mathcal{E}$ is a probability measure on the design space $\mathcal{X}$.

EXAMPLE 8. Consider quadratic regression $y = \theta_0 + \theta_1 x + \theta_2 x^2$, with x varying in the interval $= [-1, 1]$. Then the information matrix M is immediately seen to be

$$M = \begin{bmatrix} 1 & c_1 & c_2 \\ c_1 & c_2 & c_3 \\ c_2 & c_3 & c_4 \end{bmatrix},$$

where c_i denotes the ith moment of the probability measure $\mathcal{E}$. Notice the interesting fact that the elements of M are moments of $\mathcal{E}$. Because of this reason, it is quite common to refer to the information matrix as a moment matrix as well.

This connection of information matrices to moments also helps illustrate the form of optimal designs according to the alphabetic criteria described above in polynomial regression. For instance, according to the D-optimality criterion, one should try to identify a probability measure on $[-1, 1]$ such that the corresponding moment matrix has a maximum determinant. The original (finite dimensional) integer optimization problem has now changed to an infinite dimensional problem on the space of probability measures. Ironically, this complexity actually adds structure and simplicity to the problem. Assuming for a moment, that a probability measure giving a largest value of the determinant exists, it is clear that this particular $\mathcal{E}$ must give the largest value of the moment c_4 among all probability measures which produce the same values as those of $\mathcal{E}$ for the lower moments c_i, $i = 1, 2, 3$. Theorems in moment theory and an easy symmetry argument now imply that there is an optimal choice of the probability measure with supports at $0, \pm 1$ and calculus then shows that the weight at 0 has to be $1/3$. The solution to the D-optimal problem for quadratic regression on $[-1, 1]$ according to the approximate design theory is thus to take an equal number of observations at $0, \pm 1$. Of course, if the total sample size is not a multiple of 3, the ideal design has to be rounded to an integer design; even more, even if the total sample size was a multiple of 3, the ideal D-optimal design from the approximate theory need not coincide with the solution that would obtain from the integer problem. These are issues one needs to be aware of, but the strong structure that the approximate theory provides more than makes up for these somewhat minor issues. A much more serious issue is that the D-optimal design is too thinly supported; in view of this, it is rare for an exact optimal design to be religiously used in practice. But they provide a useful yardstick for the performance of other designs under the assumption of an approximate validity of the regression model. Pukelsheim (1993) and Pukelsheim and Rieder (1992) write eloquently about these issues.

All common criteria ϕ for optimal design satisfy a monotonicity property in the information (moment) matrix M: if one considers two such matrices M_1, M_2, with $M_2 \geqslant M_1$ in the Loewner ordering, i.e., if $M_2 - M_1$ is nonnegative definite, then the criterion ϕ satisfies $\phi(M_2) \geqslant \phi(M_1)$. This motivates the following definition:

DEFINITION. An information matrix M_1 is called inadmissible if there exists another information matrix M_2 such that $M_2 > M_1$ (i.e., $M_2 - M_1$ is nonnegative definite but not the null matrix).

In construction of optimal designs, it is therefore necessary to only consider probability measures resulting in admissible information matrices: this is like the well known fact in decision theory that admissible rules form a complete class. In polynomial regression problems, due to the moment interpretation of the information matrix, this helps in bounding the number of support points in an optimal design according to any criterion that is monotone increasing in the moment matrix in the Loewner ordering. Indeed, the following holds:

THEOREM. *Under the hypothesis of monotonicity of ϕ in the Loewner ordering, an optimal design for a polynomial regression model of degree p can have at most $p+1$ points in its support with at most $p-1$ points in the interior of $\mathcal{X}$.*

This result aids in understanding why the theoretical optimal designs are generally so thinly supported. Further pinpointing of the exact number of points and their weights do not come out of this theorem.

4.2. Bayesian formulation of an optimal design problem

In a strictly Bayesian decision theoretic setup, one has a set of parameters θ with a prior distribution G, a specified likelihood function $f(x \mid \theta)$, and a loss function $L(\theta, a)$. Given a design, there is an associated Bayes rule with respect to the trio (f, L, G); an optimal design should minimize over all designs the Bayes risk, i.e., the average loss of the Bayes estimate over all samples and the parameters. Chaloner and Verdinelli (1994) give a fairly comprehensive review of this formulation. In particular, they give a number of loss functions that have been proposed, and there is an instructive account of which alphabetic Bayesian criteria correspond to such a a loss-prior formulation. Note that the formulation can as well take the route of prediction rather than estimation; Eaton et al. (1994) consider a predictive formulation and show that sometimes one returns with the alphabetic optimal designs again, but not always.

There is another (simplistic) way to look at the Bayes design problem which in fact has the axiomatic justification under a normal-normal-gamma linear model with squared error loss. Thus, consider the canonical linear model $\underset{\sim}{Y} \sim N(X\underset{\sim}{\theta}, \sigma^2 I)$, $\underset{\sim}{\theta} \sim N(\underset{\sim}{\mu}, \sigma^2 R^{-1})$. Then, under the standard squared error loss $\|\underset{\sim}{\theta} - \underset{\sim}{a}\|^2$, the Bayes risk (in fact even the posterior expected loss itself) equals $\operatorname{tr}(M + R/n)^{-1}$, where n denotes the sample size. One would therefore seek to minimize $\operatorname{tr}(M + R/n)^{-1}$, which has a remarkable similarity to the classical A-optimality criterion.

The Bayesian alphabetic criteria are thus defined for linear models as:

Bayesian A-optimality: Minimize $\operatorname{tr}(M + R/n)^{-1}$,

Bayesian D-optimality: Minimize $|M + R/n|^{-1}$,

Bayesian c-optimality: Minimize $c'(M + R/n)^{-1}c$ for a given vector c,

Bayesian E-optimality: Minimize the maximum eigenvalue of $(M + R/n)^{-1}$,

Bayesian G-optimality: Minimize $\int c'(M+R/n)^{-1}c \, \mathrm{d}\nu(c)$, where ν is a probability measure on the surface of the unit ball $c'c = 1$. (Note that Studden (1977) calls this integrated variance optimality.)

Of course, in the absence of a meaning for R, these criteria do not stand to reason. They do stand to reason by doing one of two things: a structured setup of normal-normal-gamma distributions with a squared error loss, or restriction to affine estimates with only assuming that the dispersion matrix of $\underset{\sim}{\theta}$ equals $\sigma^2 R^{-1}$. The presence of σ^2 as a factor in the dispersion matrix of $\underset{\sim}{\theta}$ makes this less inoccuous than it seems.

A substantial amount of the optimality theory in Bayes design has been done with these alphabetic criteria. Note that if the sample size n is even reasonably large, the extra factor R/n in these functionals should not (and indeed do not) play much of a role. Thus, for priors in linear models which are not flatter in comparison to the normal likelihood tend to report optimal Bayes designs that track the classical ones very closely, or even exactly. On the other hand, although there is some scope for optimality work with t or other flat priors, so far there are no published works in this direction. The field of Bayes optimal designs therefore still holds out some (hard) open problems even for the Gauss–Markov linear model.

Of course, estimation and prediction are not the only inference problems one can design for; indeed, the design to be used should be consistent with what would be done with the data. The role of optimal designs in testing problems is described in Kiefer (1959), where he shows that for maximizing the minimum power over small spheres around the null value in ANOVA problems, it is not correct to use the F test regardless of the design. Kiefer's criterion would not be very interesting in a Bayesian framework (although some Bayes design work has used average power as the criterion: see Spiegelhalter and Freedman(1986)); however, Bayes optimal design for testing problems has generally remained neglected. DasGupta and Studden (1991) give a fully Bayesian formulation and derive Bayes designs; there are also a number of remarkably charming examples in Chapter 7 of Berger (1986), and there is some more theory with conjugate priors in normal linear models in DasGupta and Mukhopadhyay (1994).

In closing, the Kiefer–Wolfowitz theory has had a profound impact on the work in Bayes optimal designs in two ways: use of the alphabetic criteria and adoption of the approximate theory.

5. Mathematics of Bayes design

5.1. General exposition

The mathematics of Bayes optimal designs is generally the same as that in classical optimal design. There are three main routes to obtaining an optimal design: (i) Use an equivalence theorem. (ii) In polynomial models, use inherent symmetry in the problem (if there is such symmetry) and convexity of the criterion functional in conjunction with Caratheodory type bounds on the cardinality of the support, and (iii) Use geometric arguments, which usually go by the name of Elfving geometry, due to the pioneering paper Elfving (1952).

An equivalence theorem does the following: it prescribes a function $F(\mathcal{E}, x)$ defined on the design space such that $F(\mathcal{E}, x) \leqslant 0$ for all x in $\mathcal{X}$ and is $= 0$ if and only if x is in the support of an optimal design $\mathcal{E}$. Usually, but not always, some guess work and some luck is involved in correctly using equivalence theorems for identifying an optimal design. The nice thing about equivalence theorems is that really general equivalence theorems are known that cover probably almost all cases one would be interested in, and in principle, it is supposed to work. One can see Silvey (1980), Whittle (1973) and Pukelsheim (1993) for increasingly general equivalence theorems.

Convexity arguments do the following: First by using Caratheodory type theorems, or if possible upper principal representations from moment theory, one gets an upper bound on the number of points in the support of an admissible design. Then, one proves that the criterion functional has some symmetry or invariance property; finally, one proves that the functional is convex in a convex class of moment matrices. Application of all of these together would reduce the dimensionality of the problem to a very low dimension, which is then solved by standard calculus.

The geometric methods attributed to Elfving (and developed by many others subsequently) are by far the most subtle methods of optimal design theory, and need to be stated very carefully with changes in the criterion function. It is best understood by a verbal geometric description for the c-optimality problem. For this, one takes the symmetric convex hull of the design space, i.e., $E = CH(\mathcal{X}U - \mathcal{X})$, where CH denotes convex hull. This set is symmetric, convex and compact provided $\mathcal{X}$ is compact. Now take any vector $\underset{\sim}{c}$; if $\underset{\sim}{c} \neq \underset{\sim}{0}$, then on sufficient stretching or shrinking, it will fall exactly on the boundary of the convex set E (the scalar by which $\underset{\sim}{c}$ is divided in order that this happens is called the Minkowski functional of E evaluated at $\underset{\sim}{c}$). Call this scaled vector $\underset{\sim}{c}^*$. Then $\underset{\sim}{c}^*$ can be represented in the form $\sum p_i y_i$ where each y_i is either in $\mathcal{X}$ or $-\mathcal{X}$. If $\mathcal{X}$ is not already symmetric, then those that are in $\mathcal{X}$ give the support of an optimal design. A concise general version of this method for c-optimality is given in Pukelsheim (1994); there is also a wealth of information with many greatly unifying results in Dette (1993). One should be cautious about the use of the terminology "prior" in Dette (1993); the unifying nature of the theorems is the most gratifying aspect of this article, but the worked out examples indicate that again elements of intuition and good luck are needed for the Elfving geometry to be useful.

5.2. State of the art in Bayesian alphabetic optimality

5.2.1. c-optimality

It seems that the best results on Bayesian optimality are known for this criterion. Chaloner (1984) already considers Bayesian c-optimality and gave a form of the Elfving geometry in this case. Her results imply that Bayesian c-optimal designs can be one point, i.e., they can sometimes take all observations at one point. The deepest results on Bayesian c-optimality are given in El-Krunz and Studden (1991). They succeeded in achieving the following:

(a) give a characterizing equation completely specifying a c-optimal design, together with a Bayesian embedding of the classical Elfving set that describes the c-optimal design,

(b) characterize the situations when the c-optimal design is in fact one point,

(c) characterize the situations when a particular one point design is c-optimal,

(d) characterize the cases when the classical and the Bayesian c-optimal designs are exactly the same, and

(e) demonstrate that for any prior precision matrix, there is a sufficiently large sample size beyond which the classical and the Bayesian c-optimal designs have exactly the same support. This last result has a remarkable consequence: it is a classic fact (see Karlin and Studden, 1966) that in polynomial regression, for the extrapolation problem, i.e., for estimating the mean response at an x outside of the design space, the c-optimal design is always supported at the same set of points (it is a particularly brilliant application of the methods of orthogonal polynomials to optimal designs). Therefore, the result in El-Krunz and Studden (1991) demonstrate the same property for the Bayesian c-optimal design in the extrapolation problem for any prior precision matrix provided the sample size is large. This is extraordinary, because one is saying much more than weak convergence to the classical design. That the supports coincide for large sample sizes was already recognized in Chaloner (1984) also.

5.2.2. A-optimality

The criterion for c-optimality can be written in the equivalent form $\mathrm{tr}(cc'(M + R/n)^{-1})$. A generalization of this is the functional $\mathrm{tr}(\varphi(M + R/n)^{-1})$, where φ is some nonnegative definite matrix of rank $k, k \leqslant p$, where p denotes the rank of the information matrix M. This corresponds to Bayesian A-optimality.

A general equivalence theorem for this case is given in Chaloner (1984); a clean version of this equivalence theorem is available in Dette and Studden (1994b). In principle, this theorem can be used to find a φ-optimal Bayes design, but as with all equivalence theorems, one almost has to guess the design to use the theorem. This is like theorems in minimax theory which provide excellent vehicles for verifying that a good guess is in fact a minimax rule, but are not tremendously useful in guiding to the rule. In fact, a subsequent result in Dette and Studden (1994b) is of much greater practical use: in this result, they show that phenomena earlier described for Bayesian c-optimality continue to hold for φ-optimality. This result completely describes a value of a threshold sample size after which the Bayes and classical optimal designs are identically supported and then even gives the weights at the support points for the Bayes design.

For the case when φ is full rank, i.e., $k = p$, the invariance-convexity arguments outlined in Section 5.2.1 can be used for certain types of prior precision matrices. One can see DasGupta and Studden (1991) for some further hints on this. In general, however, calculus followed by application of Caratheodory type bounds appears to be the only method that will apply. For bounds on the number of points in the support of the Bayesian A-optimal design that are improvements on the Caratheodory bounds, one should see Theorem 2 in Chaloner (1984).

5.2.3. D-optimality

Dykstra (1971) has a flavor of Bayesian D-optimality; the theoretical foundation seems to be the convexity arguments presented in DasGupta and Studden (1991). It would be nice to find out if the classical and Bayes D-optimal designs share the same kinds of properties as they do for c-optimality. Although in regression designs either theory or explicit examples seem to be lacking, there is some work on Bayesian D-optimality for factorial designs and also for nonlinear models. We will discuss these in subsequent sections.

5.2.4. E-optimality

The state of the art results in Bayesian E-optimality for the canonical linear model are again in Dette and Studden (1994b). There is an equivalence theorem; however, we do not recommend trying to use it. The useful results are remarkable in their neatness. There are two such results: one asserts that for sufficiently large samples, the Bayes and the classical E-optimal designs are identically supported and gives a formula for calculating the weights. This calculation can be daunting. A second result asserts that under two conditions the classical E-optimal design is exactly identical to the Bayes solution. In general, these hypotheses are not easy to verify. However, for the important case of polynomial regression, they show a clean relation to an earlier result of Pukelsheim and Studden (1993) for classical E-optimal designs. Pukelsheim and Studden (1993) showed that the classical E-optimal design is supported at the points of peak of the pth Chebyshev polynomial $T_p(x)$, which are $\{-\cos(j\pi/p),\ 0 \leqslant j \leqslant p\}$ for the interval $[-1, 1]$, where p as before denotes the degree of the polynomial regression model. Dette and Studden (1994b) show that the Bayes E-optimal design is supported at these same points and in addition give an easily computable equation for the weights. This result is valid for sufficiently large samples.

5.2.5. Implications in practice

The results we see in Sections 5.2.1–5.2.4 demonstrate two things: first, for practically every one of these alphabetic criteria, exact identification of a Bayes optimal design is at least a time consuming process, despite the fairly good theory that already exists. Second, if one is willing to use a conjugate prior in a Bayes formulation of the design problem, then use of at least the same support points as the corresponding classical design is wise. One can and probably should do a numerical search to find out if the same weights can be used without much harm as well; the same search should help locating better weights if indeed there are much better weights than the classical ones. In this sense, the results described above are tremendously valuable. They show an overwhelming structure, and demonstrate that subject to using these alphabetic criteria, and conjugate priors, trying to exactly identify a Bayes design is not a particularly good idea. Of course, in the case of very small samples, prior information is more important, and the results stated earlier are not valid!

6. Examples and other information of use to practitioners

6.1. General examples

To get a flavor for how the theorems of Bayes optimality theory apply under the various alphabetic criteria, one can benefit from examples described in Brooks (1976), Gladitz and Pilz (1982), Chaloner (1984), DasGupta and Studden (1991), El-Krunz and Studden (1991) and Dette and Studden (1994b). These are for what are commonly called regression designs. Analogous examples for other kinds of models would be cited in later sections.

6.1.1. Regression in a sphere

Suppose the design space is the sphere $\{\underset{\sim}{x}\colon \sum x_i^2 \leqslant 1\}$. This can thus be regarded as an example of multiple linear regression without an intercept term. Chaloner (1984) gives several examples of φ-optimal Bayes designs in this case. El-Krunz and Studden (1991) also consider regression in a sphere and have an example, which has a flavor of a theorem. Brooks (1976) adjusts the spherical design space in order to entertain an intercept term. Both Chaloner (1984) and Gladitz and Pilz (1982) address the issue of rounding the optimal design to an implementable integer design and do real numerical examples on the associated loss of efficiency.

6.1.2. Regression in a cube or on a discrete set of points

Multiple linear regression in which each independent variable lies within $\pm a$ for some $a > 0$ corresponds to regression in a cube. Regression on a discrete set of points is an important example for many physical, chemical and environmental experiments.

Chaloner (1984) gives an example of the application of an equivalence theorem involving optimality for regression in a cube; El-Krunz and Studden (1991) give an illuminating example of regression on a set of three points. This example brings out all the important features of the Bayes optimality theory: that for large samples, the Bayes design coincides with the classical, and sometimes they are at least identically supported and for very small samples, prior information is more important and the Bayes design is not even supported on the same points as the classical solution. We recommend this example to everyone.

6.1.3. Polynomial regression

Certainly this is the case in which the maximum number of worked out examples are available. DasGupta and Studden (1991) give an example of Bayesian D-optimal designs for quadratic regression. Chaloner (1984) and El-Krunz and Studden (1991) both give the example of estimating the coefficient of x^3 in cubic regression. There is also an example involving quadratic regression in El-Krunz and Studden (1991) that illustrates the subtle geometry of the Elfving method for the Bayesian theory. Dette and Studden (1994b) have a number of examples on E-optimality for polynomial regression. In one of these examples, they apply their theorems to show a remarkable phenomenon: although the Bayes designs are supposed to depend on the sample size for small n, in certain instances a very small sample size may suffice for the Bayes optimal design to already coincide with the classical design. There is another example in particular where they obtain for quadratic regression the Bayes E-optimal design as n changes, with an arbitrary prior dispersion matrix.

The use of these concrete examples is in two aspects: someone interested in alphabetic Bayes designs can get a feeling for the theory by seeing it applied, and also get a feeling for which aspects are important, namely the prior dispersion matrix or the sample size, etc.

6.2. *Exact classical designs in some important cases*

Since Bayes and classical optimal designs tend to be either exactly the same or very similar under alphabetic criteria whenever one uses conjugate priors and the sample size is not very small, for practitioners (Bayesian or not) it is greatly useful to know the classical optimal designs in some important cases. Again, it is the view of most researchers in this area that optimal designs are not intended for religious use, but are to be used as standards of evaulation for other designs. In the following, we give the classical optimal designs for polynomial regression when the single independent variable belongs to the symmetric interval $[-a, a]$. We can take $a = 1$ and scale the design if a is different from 1.

6.2.1. *A-optimality*

This corresponds to minimizing the trace of the dispersion matrix of the least squares estimate. The optimal designs are as follows; in each case the symmetric members of a pair have equal weight and the weights are in the same sequence as the points. The table is taken from Pukelsheim (1993), where more information is available.

Degree of polynomial	Points in support	Weights
1	± 1	.5
2	$0, \pm 1$	.25, .5
3	$\pm .464, \pm 1$	.349, .151
4	$0, \pm .677, \pm 1$	.29, .25, .105
5	$\pm .291, \pm .789, \pm 1$	.232, .188, .08
6	$0, \pm .479, \pm .853, \pm 1$	.205, .185, .148, .065

6.2.2. *D-optimality*

The classical D-optimal designs for polynomial regression have a property that are regarded by some as bad and others as good: if there are k points in its support then each point has weight $1/k$. Thus a D-optimal classical design is uniform on its support. The optimal designs are as follows; the points in support are just the turning points of the pth Legendre polynomial plus the endpoints.

Degree of polynomial	Points in support	Weights
1	± 1	equal
2	$0, \pm 1$	1/3 each
3	$\pm .447, \pm 1$	1/4 each
4	$0, \pm .655, \pm 1$	1/5 each
5	$\pm .285, \pm .765, \pm 1$	1/6 each
6	$0, \pm .469, \pm .830, \pm 1$	1/7 each

Again, an extended version of this table can be seen in Pukelsheim (1993).

6.2.3. E-optimality

Pukelsheim and Studden (1993) proved the following general result on classical E-optimal designs for polynomial regression on $[-1, 1]$:

The classical E-optimal design is supported on the points $\{\cos(j\pi/p), 0 \leqslant j \leqslant p\}$, where p denotes the degree of the polynomial. Furthermore, the weights $\{p_i\}$ are proportional to $(-1)^{p-i} u_i$ where $\{u_i\}$ satisfy the system of linear equations $\sum u_i(\cos(j\pi/p))^i = c_j$, where c_j is the coefficient of x^j in the pth Chebyshev polynomial $T_p(x)$. For instance, the fourth Chebyshev polynomial is $T_4(x) = 8x^4 - 8x^2 + 1$, and therefore the coefficients $\{c_j\}$ are respectively $1, 0, -8, 0, 8$. In view of this general result, it is not difficult to write the classical E-optimal designs for polynomial regression. They are as follows:

Degree of polynomial	Points in support	Weights
1	± 1	.5
2	$0, \pm 1$	.6, .2
3	$\pm .5, \pm 1$	.373, .127
4	$0, \pm .707107, \pm 1$	.318, .248, .093
5	$\pm .309017, \pm .809017, \pm 1$	.246, .180, .074
6	$0, \pm .5, \pm .866025, \pm 1$	.218, .189, .141, .061

6.2.4. c-optimality

The classical c-optimal design can be found, in principle, by using the Elfving method, for any given vector c. In particular, if $c = (1, x, x^2, \ldots, x^p)$ for $|x| > 1$, one has the problem of "extrapolation" whereas if $|x| \leqslant 1$, one has a problem of interpolation. Even these two cases are radically different in the corresponding optimal designs. In the first case, regardless of the value of x, the optimal design is supported at the points given in Section 6.2.3, while in the latter case, the optimal design has only the given value x in its support. Optimal designs for the individual coefficients in the various powers of x can be written down. The solution depends on whether the subscript of the coefficient is an even or odd integer away from p, the degree of the polynomial model. Thus there is no universal statement one can write on a piece of paper for easy communication. One can see Pukelsheim and Studden (1993) or Pukelsheim (1993) for further information, if needed.

6.3. Factorial models, treatment control comparisons, elimination of heterogeneity and ANOVA

6.3.1. Factorial models

The first attempt at derivation of Bayes optimal designs for factorial models seems to be Owen (1970).Work following this includes Smith and Verdinelli (1980), Verdinelli (1983), Giovagnoli and Verdinelli (1983), Toman (1987), Hedayat et al. (1988) and Toman and Notz (1991). A recent important and comprehensive contribution is Majumdar (1995). Generally, the work has derived alphabetic optimal Bayes designs for some selected models. These models include one and two factor models, one and two way ANOVA, and one and two way heterogeneity models. Almost exclusively, these works have assumed multivariate normal priors in conjunction with multivariate normal observations, which are needed for analytical results. Some of these works also consider the computational aspects, and give special cases where the computation simplifies. In some of these models, there have been some work on sensitivity with respect to the prior distribution which would be cited later.

6.3.2. A-optimality

In the two factor ANOVA model, Owen (1970) derives a general result giving the optimum allocation of treatments given a specified blocking. Owen's criterion should really be called generalized A-optimality due to the general quadratic loss he has for estimating the treatment effects. Owen shows that in certain cases, the computational aspect simplifies. A-optimality is also considered in Giovagnoli and Verdinelli (1985) in one way heterogeneity models and in Hedayat et al. for two way heterogeneity models. Toman (1987) derives A-optimality results in a number of models, and returns to A-optimality in heterogeneity models in Toman and Notz (1991). Generally speaking, these articles solve the approximate design problem, although nearly all of these works address the issue of rounding to integer designs. Toman and Notz (1991) in particular give a new rounding strategy by rounding the amount (in the approximate theory) of the control in a treatment vs. control problem. They give some evidence that this rounding strategy gives better efficiencies than the methods suggested elsewhere.

6.3.3. D- and E-optimality

D- and E-optimal Bayes designs are discussed in Giovagnoli and Verdinelli (1983) and later in Toman and Notz (1991) for block models. In fact, Giovagnoli and Verdinelli (1983) show that in a two way ANOVA model, with usual assumptions, there is a unique optimal design for fairly general criterion functions, and in the case of one treatment, a universally optimal design exists as well (universal optimality refers to simultaneous optimality under all criterion functions entertained). DasGupta and Studden (1991) also give instances of E-optimal designs in one way ANOVA settings and show that the E-optimal design coincides with an A-optimal solution.

6.4. Exact classical optimal designs

Although there is no direct evidence to this effect, results in the regression case and simple common sense suggest that exact classical optimal designs can be approximate proxies for Bayes solutions in most instances, if multivariate normal priors are used. In any event, if an exact classical optimal design for a problem is known, it can be instructive for the corresponding Bayes problem. Chapter 2 in Shah and Sinha (1989) has some general information on classical optimal designs, particularly some advantages of using balanced designs in block models.

The major bulk of the classical theory seems to have been for the D-optimality criterion. In particular, either algorithms or even catalogues of D-optimal designs for 2^m and 3^m factorial models are available for small values of m: one can see Nalimov (1982). The two most important contributions are the continuous D-optimum designs for second order models and central composite designs which proxy the theoretical continuous optimal designs. We describe these briefly below.

(a) Continuous D-optimal designs. Farrell et al. (1967) consider a general polynomial model in m factor experiments and describe the form of an optimal design depending on the degree of the polynomial and the number of factors, for three important design spaces: cubes, spheres, and simplexes. There is an interesting discussion on the minimum number of points that must be in the support of an optimal design, and using the methods of orthogonal arrays introduced by Rao (1946, 1947), a device to reduce the number of points in the support of a candidate design is given. For those interested in the classic properties of designs, there is an intriguing example in which the optimal design is demonstrated to be not rotatable.

(b) Central composite designs. The conceptual difficulty with the theoretical continuous optimal designs in these cases is that they cannot be implemented due to their infinite support. One is thus forced into some form of an approximation. The by now common method of approximation is to use what are called central composite designs. It is useful to know that central composite designs provide a rational way for moment matching corresponding to the continuous optimal designs; on the other hand, statistically they have a lot of simple appeal. To understand the structure of central composite designs, it is useful to think of the spherical design space once again. One can hit the boundary of the sphere by using points which have all but one coordinate equal to 0 and the remaining one equal to $\sqrt{m}$. Such points are called star points. Replications corresponding to 0 level for each factor are called center points. The remaining observations are used up in a 2^{p-f} fractional factorial for some $f > 0$.

Box and Hunter (1957) is a good exposition to central composite designs. Illustrative examples on the efficiencies of central composite designs according to the D-optimality criterion are available in Atkinson and Donev (1992).

These classic methods are of importance to the Bayesian theory at the present time for two reasons: first, the present state of the Bayes optimality theory is far less advanced than the classical theory. We just do not have very good knowledge of Bayes solutions yet. Second, it is always instructive to consider classic methods that are timetested and see if they do an adequate job when prior information is available.

7. Critique of the optimality theory

7.1. Model dependence

A major criticism of the standard optimality theory in regression models is the severe dependence of the design on the assumed model. While it is a clear mathematical truth that if the model of a simple linear regression is valid, then nearly every criterion calls for taking observations only at the endpoints of the design interval, it is rare for even the most ardent theoretician to recommend this to anyone interested in analyzing data. This can be restated as saying that one never believes the model exactly. In this sense, despite the undisputable structure of the optimality theory and the impact it has had on further theory, optimal designs have not had much of an influence on people who do design actual experiments. Indeed, it is common in regression problems to more or less divide the observations uniformly in the interval. This has some similarity to D-optimal designs for high degree polynomial regression (this should not be interpreted as a weak convergence to the uniform distribution; indeed that is known to be not true). Even Bayes designs, by virtue of sharing the same mathematics as their classical analogs, suffer from the same problem (in some problems, there are counterexamples to this; for instance, see the computer experiments in Mitchell et al. (1994)).

Another problem with the optimality theory is that the designs can be extremely goal specific; thus, in polynomial regression, a design which is optimal for estimating one regression coefficient may perform quite poorly for estimating another coefficient. The problem is that the experimenter may have several aims in an experiment, and in some problems other aims may arise at a later point of time (of course, it is not the design's fault that arbitrary aims specified afterwards cause problems). But the point remains that orientation to one specific goal is another drawback of the exact optimal designs.

7.2. Mixture models and model robustness

Response to the criticism of model dependence has been generally of three kinds:

(a) construction of designs that are nearly optimal in a lower order model with (good) protection against a higher order model,

(b) construction of minimax type designs in a class of models,

(c) use of criteria which are themselves some type of mixtures or averages of efficiencies under different models.

7.2.1. Protection for higher order models

Stigler (1971) proposed derivation of designs that are nearly G-optimal in a lower degree polynomial model with some protection for each model of higher degree up to a certain maximum degree (recall that G-optimality corresponds to an average c-optimality with an averaging over c: Studden (1977) calls this the integrated variance criterion). In Studden (1982), a D-optimal variant of Stigler's proposal is considered and designs which are most efficient at order r subject to a prespecified efficiency for an extra $m - r$ are derived. The results are remarkably closed form, with clever use of canonical moments of the design measure. As a matter of fact, tables of efficiencies are provided, and the general moral of this article is really quite encouraging: excellent efficiency at lower order models can be obtained by sacrificing a bit for the extra coefficients. It is also possible to show that the design supported on $\{0, \pm.618101, \pm1\}$ with weights .13796, .140347 and .290673 is 69% D-efficient for each of linear, quadratic, cubic and quartic regression. This design maximizes the minimum D-efficiency over the polynomial models of degree $p = 1, 2, 3, 4$ over $[-1, 1]$; further information of theoretical nature is available in Dette and Studden (1994a). A similar formulation is the following: for a given design $\mathcal{E}$, consider estimating the mean response at value x of the independent variable when the degree of the polynomial is p. Suppose $v(x, \mathcal{E}, p)$ denotes the variance. Then consider the design that maximizes the minimum efficiency over $1 \leqslant p \leqslant n$ with respect to the integrated variance criterion $\int_{-\infty}^{\infty} v(x, \mathcal{E}, p) w(x)\,\mathrm{d}x$ where $w(\cdot)$ is a probability density on the real line; this is the integrated variance criterion in Studden (1977). If $w(\cdot)$ is taken as the $N(0, 1)$ density and the design space $\mathcal{X}$ is taken as $[-1, 1]$, then the maximin design for $1 \leqslant p \leqslant 4$ is $\mathcal{E}(0) = .2296, \mathcal{E}(\pm.7018) = .2486$ and $\mathcal{E}(\pm1) = .1366$. One can see plots of efficiencies of this design if it is used for particular x at the end of this article for each of $p = 1, 2, 3, 4$. A perception problem with these designs is that they are still equally thinly supported.

7.2.2. Neighborhood models

Huber (1975) suggested use of models in a neighborhood of a given model, and subsequent use of a minimax design, treating this as if this was a game with nature choosing a model from the neighborhood class. Of course, neighborhoods can be defined in many ways. Huber gives as an example the case of simple linear regression as a starting model and an $L(2)$ neighborhood of the linear regression function as the nature's family of models. It turns out that the $L(2)$ neighborhood does not quite work very well, but clearly the suggestion in Huber (1975) is appealing.

The idea in Huber (1975) was picked up again in Marcus and Sacks (1976), where they look at $L(\infty)$ type neighborhoods of a linear regression function and under various envelopes for the family, derive minimax (estimate-design) pairs. But again, the designs are thinly supported. They find that as a rule, use of an estimate with a nonoptimal design is more dangerous than use of an optimal design with a "nonoptimal" estimate. This leads to some qualitative understanding.

The ideas in both Huber (1975) and Marcus and Sacks (1976) need further attention. Another recent article is Tang (1993).

7.2.3. Mixture criteria

The main idea is to take an indexed family of functionals (ϕ_p) and then use an average of these functionals over p. The averaging is done by using a subjective weight measure on p. There are two leading articles on this approach; Dette and Studden (1994a) give a very elegant theory using D-optimality as the basic criterion and p as the degree in a polynomial regression. In Dette (1992) similar kinds of results are derived for polynomial regressions with missing powers. Elegant use of canonical moments and continued fractions can be seen in these articles.

7.3. All around designs

As stated earlier, the optimality theory of experimental design also suffers from orientation towards very specific goals. For instance, in quadratic regression on $[-1, 1]$, the optimal design for estimating the coefficient of x^2 has efficiency .5 for estimating the intercept. It is thus an useful exercise to investigate if designs can be found which give good efficiencies simultaneously for a number of aims. As commented before, it is necessary that these aims be stated before the experiment as addition of new goals afterwards essentially makes the problem impossible. Since all the aims of an experiment are impossible to treat or even conceive of in a theoretical study, an effort to find all around designs has to be quite specific. Thus a theoretical study of this issue is limited in its scope, but an investigation in some standard models with standard goals can lead to an appreciation of the extent to which all around designs are plausible.

Lee (1988) discusses this under the terminology of constrained designs. Under the general structure of minimizing $\phi_{m+1}(M)$ subject to $\phi_i(M) \leqslant c_i$, $i = 1, 2, \ldots, m$, for differentiable $\{\phi_i\}$, Lee gives an equivalence theorem and in particular, cites as example a D-optimal design in quadratic regression with an upper bound on the trace of the dispersion matrix. Essentially the same approach is seen under less smoothness assumptions on the functionals $\{\phi_i\}$ in Pukelsheim (1993).

DasGupta et al. (1992) take the above functionals to be the variances of the least-squares estimates of individual coefficients in a general linear model with a general variance function and derive designs that have a guaranteed efficiency e for each coefficient. The largest e for which such a design exists is of interest. For the corresponding Bayesian problem, they substitute posterior variance of the parameter for variance of the estimate. Two special variance functions are subsequently used. An interesting fact is that good efficiencies can be guaranteed if the subscripts of the coefficients of interest are all even or all odd. The following is an illustration in the case of cubic polynomial regression:

Subscripts of coefficients	Guaranteed largest efficiency
0, 2	.75
1, 2	.65
1, 3	.93
0, 1, 3	.66
0, 1, 2	.58
0, 1, 2, 3	.58

8. Nonconjugate priors and robust Bayes designs

8.1. Nonconjugate priors

A major problem in using nonconjugate priors in the canonical normal linear model is the associated loss of closed form formulae for the Bayes risk; at a more fundamental level, unlike in the case of normal priors, now the posterior expected loss in fact does depend on the actual observations that would be later obtained. Thus posterior expected losses can no longer be used for minimization, and even Bayes risks do not have closed form expressions. Some work is going on at the present time at Purdue University on Bayes design with nonnormal priors. A combination of some theory and subsequent computations, this work essentially follows the following line:

(a) One proves that the Bayes risk for estimation of the coefficients of the model is decreasing and convex in the information matrix in the Loewner ordering;

(b) For general regression functions, one uses a Caratheodory bound and for special types one uses the Kiefer bound on the number of support points;

(c) One then uses the Brown–Stein (Brown, 1986) identity for Bayes risk, but now specialization to (a general) quadratic loss (as in Owen, 1970) is necessary;

(d) One finally does a numerical search for the optimal design.

This method has nothing to do with conjugacy or otherwise of the prior. There is some dimensionality reduction for symmetric priors if the design space is symmetric (i.e., $x \in \mathcal{X}$ implies $-x \in \mathcal{X}$). One can start with other criteria functionals that already satisfy property (a) and the same steps then follow through. The value of using nonconjugate priors stems from two directions: (i) In some problems, the experimenter does not want to use conjugate priors, and (ii) As a general intellectual question, it is necessary to know if conjugate vs. nonconjugate priors really do result in significantly different Bayes solutions for small sample sizes. For sample sizes that are large compared to the number of parameters, we do not believe conjugacy vs. nonconjugacy matters. We do not have at the present time a clean prescription for what constitutes a large enough sample size for a given number of parameters, but one should consult Berger (1986) for further discussion on this issue. The gist is that inference and design seem to behave as fundamentally different problems with respect to fine specification of the prior if frequentist Bayes design criteria are used. We do not have any knowledge, however, if this is the case if preposterior design criteria are used: preposterior criteria would be discussed at length in Section 10.

8.2. Robust Bayes designs

There is now a substantial literature on robustness of Bayes methods, to various components in a decision theory framework, although a majority of the work is on robustness with respect to the prior. Berger (1994) and Wasserman (1992) give a lot of comprehensive information and food for thought. All robustness work, frequentist or Bayes, fall into one of two general categories: (i) Take a fixed procedure optimal under one model and ask what it does for another (close) model, and (ii) Take a family of models close together and ask what procedure(s) provide protection for all these models. There is a third way to look at robustness: what is needed for robustness to (honestly) obtain; for example, is symmetry essential, or is an exponential or faster tail essential, does one need a good idea of the variance, etc. There does not seem to be much work on this view of robustness, perhaps because there is a belief at large that NOTHING is needed and more thinking or more data would solve the problem. We recommend Chapter 1 in Huber (1981) and Staudte and Sheather (1990) and Rubin (1977) for anyone wanting to learn about robustness.

8.2.1. Regression models

DasGupta and Studden (1991) consider the normal linear model with a family of priors and take the following general approach: pick a special prior from the family; pick a criterion functional; then derive a design that gives the best robustness for the family of priors subject to being ε-optimal with respect to the special prior. The article then considers various criterion functionals, various inference problems, two different families of priors, and describes the necessary mathematics to solve these problems. The two families of priors were respectively first suggested in the Bayesian robustness literature by Leamer (1978) and Polasek (1985) and DeRobertis and Hartigan (1981). The necessary mathematics mostly is convexity-invariance with some moment theory for the DeRobertis–Hartigan family, a density band for the prior density. A companion article DasGupta and Studden (1988) also has material directly relevant to Bayes designs that in addition uses geometric argument. The following is one result from DasGupta and Studden (1991):

THEOREM. *Consider a density band for the prior with envelopes that are multiples of a given normal density; then the design that gives the smallest confidence set with a given posterior probability for each prior in the density band is the Bayes D-optimal design with respect to that given normal prior density.*

One can see a variety of other results in that article.

8.2.2. Other models

There is some literature on robust Bayes designs for other special problems; almost all of this work is due to Blaza Toman and her coauthors. Toman (1992) and Toman and Gastwirth (1993) are two important articles on ANOVA models in the context of robust Bayes designs. The most important thing about both of these articles is that the optimization is simultaneously over the (estimate, design) pair. There is something to be said for this viewpoint, and therefore these two articles contribute a formulational novelty in this area. In Toman (1992), the priors are essentially the Leamer–Polasek type, while in Toman and Gastwirth (1993), they are finite mixtures of normals. In each article, the ultimate criterion is an average over the family of priors, with some difference in the details. For instance, Toman (1992) has an information theoretic criterion. The results are basically closed form in both articles.

If one takes the view that Bayesian statistics should be robust subjective, then clearly much further work remains to be done. However, for sample sizes that are not very small, one is likely to see little sensitivity to the prior because the Bayes solutions for different priors would all be close to the classical solution and therefore close to one another. This is not a precise statement, but only underscores the qualitative phenomenon.

As a historical point, robust Bayesians should also see Huber (1972) and Chapter 10 in Huber (1981).

9. Miscellaneous other design problems

9.1. Quality control

Since the late seventies, there has been a change in attitude about what process quality control really means. While acceptance sampling and $k\sigma$ control limits dominated the practice of quality control for a very long time, new thoughts emerged in the late seventies. Thus, although the traditional methods of tolerance specification and acceptance sampling still have some role in classroom teaching and actual process control in the US, off-line production control appears to have taken over in other parts of the globe, in particular Japan. We recommend the articles Kackar (1985) and Pukelsheim (1988) for excellently presented expositions on this topic. Also see Ghosh (1990).

The idea of off-line production control is the design or adaptation of the process parameters such that the target is attained more or less exactly, and variance is minimized subject to target attainment. This is in contrast to the traditional method of estimating process capability vis-a-vis tolerance limits. A short but pertinent article on Bayes design with normally distributed observations with normally distributed parameters in the context of off-line process control is Verdinelli and Wynn (1988). One should also see Sarkadi and Vincze (1974) for a systematic presentation of mathematical problems in quality control.

9.2. Engineering design and reliability

There is a fairly substantial amount of work on Bayes design relevant to reliability, which is comprehensively covered in Chaloner and Verdinelli (1994). We are not aware of any kind of a Bayes optimality theory in Engineering design problems; however, the monograph of Wilde (1978) followed by Papalambros and Wilde (1988) give well written introductions to a wealth of really interesting problems.

9.3. Spatial designs

Generally speaking, spatial design corresponds to problems of prediction or otherwise of a response when the input variable is spatial. Thus the design set may be a finite set of points with three coordinates each, identifying each point with the geographic location of an experimental station. Problems in multivariate numerical analysis such as approximation of an integral or evaluation of other linear functionals by averaging over a discrete set of points also go by the general name of spatial optimal designs.

There are two intrinsic features that stand out in these problems:

(i) Depending on the exact criterion used, the fundamental nature of the design such as thickness of the support and high density regions can change and

(ii) Significantly more than standard problems, computation of the optimal design becomes an issue. In fact, it seems that development of computing algorithms may even be the most dominant issue.

The general tendency in these works seems to have been to take the viewpoint that the response behaves like the path of a Gaussian process with spatial time variables. The experimenter's belief about the smoothness of the response is incorporated into the autocorrelation structure of the Gaussian process. We recommend Sacks and Ylvisaker (1970), Diaconis (1987) and Sacks and Schiller (1988) specifically for reading on spatial designs. Diaconis (1987) in addition takes the reader along a fascinating path on the history of intellectual efforts to connect probability and statistics with numerical analysis problems.

10. Conditional formulations and sample size choice

10.1. Conditional formulations

Traditionally, experimental design is regarded as one area in which even the strict believer in conditional Bayes has to resort to an integration on the sample space. The reason is obvious: at the design stage, there are no data, and so design criteria necessarily have to average over potential data that may arise, which corresponds to frequentist integrations on the sample space. Thus, for instance, Bayesian A-optimality calls for minimizing the Bayes risk, a double integral on the joint probability space.

However, conditional or preposterior formulations of the design problem are possible, although even this formulation has a frequentist flavor. In other words, even in the preposterior formulation, considerations of the totality of the samples cannot be completely ignored. DasGupta and Vidakovic (1994) and DasGupta and Mukhopadhyay (1994) give detailed discussions of this; we will give a sketch of the preposterior formulation in these articles.

Consider the canonical normal linear model, with some prior on the parameters. Suppose we accept posterior A-optimality as the criterion, again for specificity. Given a design, and once the actual data arrive, there is a posterior density for the parameters. The desirable goal is to minimize the trace of the posterior covariance matrix. However, this cannot be done since in general the posterior covariance matrix certainly depends on the data. In the preposterior formulation, one specifies an upper bound on the trace of the posterior covariance matrix and chooses a design that satisfies this upper bound, with a prespecified large predictive probability. In other words, for data that are likely to arise, the design already gives a desired accuracy. For very ambitious upper bounds on the trace, no such design may exist, but the mathematics of the problem will say so if that is the case. In certain cases, the upper bound can be satisfied for all possible samples, i.e., with a predictive probability of 1. Naturally, the predictive probability is with respect to the predictive distribution conceived from the given prior. Alternatively, this preposterior formulation can be written in a minimax type of statement:

$$\operatorname{Min}\operatorname{Max}\operatorname{tr}\bigl(V(y \mid M, \pi)\bigr),$$

where the maximum is over a set of samples y with predictive probability $1 - \varepsilon$, the minimum is over the information matrix (i.e., design) M, and $V(y \mid M, \pi)$ denotes the posterior covariance matrix for given y and M. Note that a SPECIFIC set of predictive probability $1-\varepsilon$ has to be used; but this specific choice should usually be an obvious natural choice (for instance, a high density set of the predictive distribution).

There has not been any work with the preposterior formulation other than for Bayesian sample size choice, which we consider next.

10.2. Minimum sample size

10.2.1. The formulation

The general formulation of the minimum sample size problem is the following: one specifies a measure of accuracy for the particular inference problem at hand, and asks what is the minimum sample size n for which the accuracy requirement is met. For instance, for estimating a binomial proportion, it is standard to use the expected length of a 95% confidence interval and seek a sample size that makes it smaller than a given upper bound. Bayesian considerations automatically enter into such classical calculations, because the expected length is a function of the unknown proportion and therefore an a priori guess value needs to be used to arrive at a sample size not overly conservative. As another example, in testing problems, it is standard to seek a sample size that ensures a given power for some standard 5% test at a value of the parameter thought to be practically different from the null value. One can dispute the correctness of these formulations; but we are only making the point that this is how it is traditionally done in classical statistics. There is absolutely no doubt that minimum sample sizes are taken seriously by people who deal with data, and this topic has assumed the status of textbook material in traditional statistics. Students are told about it. There are a large number of monographs, books, tables and charts of classical minimum sample sizes. We recommend Odeh (1975) and Selected Tables in Mathematical Statistics (1975) as two particular sources for further exposition and actual numbers for possible use.

The corresponding Bayesian formulation can be of two possible types: a preposterior formulation of exactly the kind we described above, and a frequentist Bayes formulation in which one seeks a sample size that ensures that the Bayes risk in the problem at hand is smaller than a given number. The preposterior formulation is more Bayesian in the sense of not integrating on the sample space and is also by leaps and bounds mathematically more challenging. However, this itself can be a negative aspect of the preposterior formulation, in which case the frequentist Bayes formulation can be adopted. We must admit, however, that the frequentist Bayes formulation can lead to a completely trivial problem, though not always.

10.2.2. Normal theory

A large spectrum of normal (univariate and multivariate) problems with the preposterior formulation and associated theory and in some cases actual sample sizes are available in DasGupta and Mukhopadhyay (1994) and DasGupta and Vidakovic (1994). In DasGupta and Mukhopadhyay (1994), a theory of Bayesian sample sizes is provided for two problems: (i) Testing for a normal mean, with conjugate priors, and seeking a sample size that either makes the posterior risk uniformly small or makes the posterior risk uniformly robust if one takes a family of priors instead of one specific prior. There are nontrivial asymptotics in these problems; in this same article, DasGupta and Mukhopadhyay (1994) also give some actual Bayes sample sizes, although their practical adoption in the foreseeable future is at least doubtful; (ii) Constructing a confidence set for a multivariate normal mean, for conjugate priors, and seeking a sample size that gives a set with a prespecified posterior probability uniformly over future data and simultaneously for every prior in the family under consideration. Table 1 in the same article gives some actual sample sizes for this as well. A previous longer version DasGupta and Mukhopadhyay (1992) had a number of other problems in which the preposterior formulation was discussed and a theory presented; among the other problems in this longer version was the nonregular case where the sample space depends on the parameter and there is also some consideration of an uncertain loss function.

The widely used one way ANOVA model is considered in DasGupta and Vidakovic (1994). The problem is testing for no treatment differences, and the approach is purely Bayesian. That is, a prior probability is given for the hypothesis, and a normal prior density used as the conditional density of the parameters given that the null hypothesis is not true. Then testing is regarded as a decision problem with 0–1 loss, i.e., the quantity they try to keep small is the posterior probability of the wrong hypothesis being picked. Again, some actual sample sizes are given, but now a complete Mathematica code is provided for use by the specific user with his/her particular inputs.

10.2.3. Binomial proportions

The approach taken for determination of Bayesian sample sizes in this case has generally been the interval estimation approach. That is, one seeks a sample size that ensures a posterior confidence interval of a specified probability such that its length is smaller than a given number. Again, the length is a function of the data, and either the expected length (under the predictive distribution) or the maximum length (over all possible data) are substituted for the actual length. A conceptual difficulty with the expected length is that there is no guarantee at all that by using the sample size produced by this criterion, the accuracy goal one started out with would be satisfied when data do arrive. One then feels the exercise was useless. Therefore, although more conservative, the maximum length criterion is preferable. So far the work has assumed conjugate Beta priors. This is just fine, because in this case, Beta priors can approximate any prior whatsoever on $[0, 1]$ by simply allowing mixtures. So any generalization, if at all, that is needed is consideration of some Beta mixtures. The work on Bayesian sample sizes for Binomial proportions is due to Lawrence Joseph and his coauthors; one should in particular see Joseph et al. (1994) and Joseph and Berger (1994) and an earlier work Bock and Toutenberg (1991) in the context of clinical trials.

10.3. Optimal sample sizes

10.3.1. Cost vs. accuracy

Statistical theory often has the pretense of being able to choose as many samples as it takes to ensure a given accuracy. This of course is far from the truth in many real problems. The impediments are many: cost of sampling is the most prohibitive factor; in some situations, sampling may be a time consuming process or simply a difficult human exercise. Optimal sample sizes try to balance the tradeoff between accuracy of inference which (generally) increases with increasing sample size and sampling and other human costs which also increase with increasing sample size. Theoretically, it is assumed that all of these costs are considered together in a single cost function although quantifying extraneous costs like the value of time does not appear to be an easy task. The optimal sample size is only to be taken as a guideline; it is certainly not intended as a rigid prescription. There are also some conceptual debates whether abstract (and often unitless) quantities such as inference accuracy and dollar amounts for cost of sampling can be just added to form the overall cost; one can see Chernoff (1972) for a discussion of this.

10.3.2. Estimation

Suppose one has a loss function for accuracy, $L(\theta, a)$, and a cost function $C(n)$; then an overall cost is the sum total of the two. Suppose it is decided to use a rule $d(X)$, which should be just the Bayes rule for the given loss function and a given prior in a Bayesian framework, and would be some classical estimate like an mle or a best invariant rule otherwise. Then the total average risk is $R(\theta, d(X)) + C(n)$, where θ denotes the generic parameter. Minimizing this with respect to n would entail a solution that depends on (part of) the unknown parameter. It is therefore natural to take the Bayes risk in place of the risk function $R(\theta, d(X))$ and minimize the resultant quantity with respect to n. If no prior distribution is used, one can use a guess value for the parameters present in the solution or alternatively use sequential versions which estimate the parameters at each stage, use the estimated parameters in the formula for the sample size, and use the first n satisfying appropriate constraints. Details of such an approach can be seen in Starr and Woodroofe (1969), Ghosh et al. (1976), etc.

10.3.3. Testing

The steps in determination of an optimal sample size are the same in any decision problem; however, for testing problems, the loss functions associated with the two actions when there are two hypotheses are different from standard losses one sees in estimation. 0–1 types of losses where there is no loss in accepting the correct hypothesis and a constant loss in accepting the wrong hypothesis are probably the most used, although appear to be severely unrealistic. If by chance or due to any other reason, a seriously wrong hypothesis is accepted, the penalty should be more than accepting a hypothesis which is false only on paper. For instance, for deciding the sign of the mean of a univariate normal distribution, the loss in deciding the wrong sign could be reasonably taken as a nonconstant monotone nondecreasing function of the absolute value of the mean.

Berger (1986) and Chernoff (1972) both give a nice normal theory example by taking the absolute value of the mean as the loss for inferring the wrong sign and derive a Bayesian optimal sample size using a conjugate prior. Chernoff (1972) also does the case of a simple vs simple hypothesis and shows the qualitative difference between simple and nonseparated hypotheses. One should also see Antelman (1965), a charming article, which has much to offer to people interested in Bayes experimental designs as a whole.

11. Nonlinear problems

11.1. Introduction

Nonlinear problems can arise either for nonlinear functions in a linear model, or in models where the response function is itself nonlinear; these latter class of models is more or less universally known as nonlinear models. We recommend the beautiful yet encyclopaedic book due to Seber and Wild (1989) to anyone interested in nonlinear models. Of particular value for workers interested in experimental design are Chapters 1, 5, 6, 7, 8, 9 and 10; Section 5.13 gives a short but excellent historical account of optimal design theory in nonlinear models, complete with early examples from Box and Lucas (1959), early criteria, and discussions about thin designs. Our discussion of nonlinear functions and nonlinear models would be short due to a second article on nonlinear models. In addition, Chaloner and Verdinelli (1994) give a fairly complete account of the current bibliography on Bayes optimal design for nonlinear models.

11.2. Nonlinear functions in linear models

An early example of an interesting nonlinear function is the example of calibration in simple linear regression: estimating the value of the independent variable at which the mean response is zero or some other constant; this is usually known as Fieller's problem. Also see Rao (1973) for a clever description of how to construct confidence intervals in this seemingly impossible problem. Silvey (1980) gives an elementary but very insightful account of optimal design for this problem. Another example essentially the same as Fieller's is estimating the value of the independent variable at which a quadratic response function has a zero derivative. This is commonly known as the turning point problem.

The major conceptual problem with experimental design in nonlinear problems is that the optimal design depends on the very parameters one is trying to learn about. It has been suggested that one uses a guess for the value(s) of the parametric functions that enter into the design, much as it is customary in most elementary texts to use a guess for the value of a Binomial proportion in order to construct sample sizes for learning about the same proportion. This is also conceptually akin to construction of locally most powerful tests where one maximizes the derivative of the power at the null value; see Lehmann (1986). This approach to design has been called local optimality; see Chernoff (1972) and Atkinson and Donev (1992). The locally optimal design is exactly optimal if the true value of the parameter happens to be equal to the guess value. However, use of locally optimal designs can subsequently lead to amusing problems: see Chaloner and Verdinelli (1994).

Other examples of nonlinear functions in linear models include estimating the probability that a future observation belongs to a specified set, say a bounded interval; other linear models in which standard problems cause nonlinear functions to arise are models in which the variance is a function of the mean. One should also see Wu (1988) for treatment of optimality theory for estimation of similar nonlinear functions in quantal response models.

11.3. Nonlinear models

11.3.1. Basic tool

Any statistical model in which the mean response $\mathrm{E}(Y)$ is a general function $f(\underset{\sim}{\theta}, x_1, x_2, \ldots, x_p)$ of p independent variables and a (possibly vector valued) parameter $\underset{\sim}{\theta}$, is a nonlinear model. The goals of experimentation in such models may vary, as in linear models. A persistent common feature of optimal experimental designs across these models is that the design depends on the unknown parameter; some isolated counterexamples to this statement are known – in particular, one can see Silvey (1980).

A key tool for the optimality theory in nonlinear models is a general equivalence theorem in Whittle (1973). As is the case with any equivalence theorem, the optimality of a candidate design can be either disproved unambiguously or established subject to the accuracy of the computations by using Whittle's theorem. This is because any equivalence theorem is a statement of the following kind: a design $\mathcal{E}$ is optimal if and only if an appropriate functional $F(\mathcal{E}, x) \leqslant 0$ for every x in the design space and is $= 0$ for exactly those points in the support of $\mathcal{E}$. Now, by using only calculus and numerical optimization, it is usually feasible to find the best design with k points in its support, for a given k. This candidate design can then be tested for optimality by using the equivalence theorem. The difficulty might be that for points in the support of this candidate design, the functional F may give values which are small negative numbers. One is then forced to make a judgement if this is only an accuracy problem or the design is not the exact optimal one. In any event, such an use of the Whittle equivalence theorem is clever and was initiated by Chaloner and her coauthors: one can see in particular Chaloner and Larntz (1986) and Chaloner (1993).

11.3.2. State of the art

The current inclination in the choice of a criterion for optimization seems to be to take an information approach, most likely influenced by Lindley (1956). Typically, one takes the Fisher information matrix $I\ (\underset{\sim}{\theta}, \mathcal{E})$ for a given design $\mathcal{E}$ and takes the logarithm of its determinant as the starting step: the logarithm just happens to cause a great amount of algebraic simplicity in many problems, and is therefore justified on the basis of the rewards it produces. But one still has the parameters in the criterion and therefore it seems natural to take an average of it with respect to a weight function, which may or may not be a prior, but operationally acts like a prior. As a matter of fact, this is the only place where the prior has anything to do with the problem and therefore even those averse to use of priors can pragmatically adopt this approach if local optimality is not regarded favorably. One can see Chaloner and Verdinelli (1994) for some additional discussion on use of the Fisher information as a basis for optimality.

A disappointing but apparently unavoidable feature of the optimality theory in nonlinear models is that general complete class theorems about admissible designs in terms of the number of points in their support seem very difficult, and probably impossible. However, some fairly unexpected advances have been made in the last few years; foremost among these surprising and hard works are Dette and Neugebauer (1993) and Dette and Sperlich (1994a, b). The typical tone of these papers is the following: given a prior distribution on the parameters, they characterize situations when an optimal design with a given number of points in the support exists, and if one does, identify the points and the weights more precisely. There is also an attempt to establish analogs of Caratheodory type bounds on the cardinality of the support. Despite the fact that these advances are model specific, they are distinctively strong results. Earlier, Mukhopadhyay and Haines (1993) also took similar approaches in an exponential model.

It is very important to be aware of a fundamental phenomenon that is emerging as a unified character of the optimality theory in nonlinear models: an opinionated prior results in an optimal design similar to the corresponding locally optimum design, and a flat prior results in thickly supported designs. However, as of this time, the optimality theory still suffers from the same drawback as for linear models: it produces designs that are unlikely to be adopted in practice.

11.3.3. Linearization of a nonlinear model
Atkinson and Donev (1992) describe a method for linearizing a nonlinear model by using a Taylor series expansion, thereby producing polynomial models, and iteratively fitting the model in the approximate form. The general idea of linearizing a nonlinear model seems to have some potential; however, there are subtle points in using such a method. One possibility is to find a uniform approximation to a given degree of accuracy for the true response function by using response functions we know how to handle, say polynomials, find an optimal design in the linearized model, estimate the parameters in the linearized model, and finally transform these estimates back to the initial model: THIS LAST STEP IS SUBTLE, because the parameters that appear as coefficients in the linearized model would not be the parameters for which a prior was elicited, but functions of those parameters. So the retransforming may have to be done by an inverse function method, or a pure Bayes method, though the pure Bayes method is the harder one to use. As for linearizing a function to produce uniform approximations in a compact design set, many methods are available. Expansions in Chebyshev polynomials, exchange algorithms and others are common tools with theoretical properties; one can see Powell (1981) and Rivlin (1990). We will return to this topic from a technical angle in Section 13 (Appendix).

12. Future of Bayes design

It seems as though the future of Bayes experimental designs lies in nonstandard problems, as opposed to standard linear models or even standard nonlinear models. Bayes optimality theory can succeed as a useful theoretical development only if it is seen that the resultant theory does not reproduce the classical solutions either exactly or practically exactly. We have seen a few examples of Bayes optimal designs which differ remarkably in their character with usual optimal designs which are embarrassingly thinly supported: instances of these are Mitchell et al. (1994) and Sacks and Schiller (1988).

There are innumerable interesting problems in various branches of science where optimal design is a very viable scientific issue; it seems imperative that subjective prior information is available in many of these problems, and thus Bayes design should have a useful role to play. It may turn out that the Bayes solution would once again track classical methods closely or exactly. We believe such findings, although negative in a sense, would be intellectually valuable.

Examples of areas where optimal designs can be explored and promise to be interesting scientific exercises are indeed numerous: inventory control, tracking a moving target, design of neural networks, random variate generation, structural and engineering design, construction of histograms, survey sampling, combinatorial algorithms such as the traveling salesman problem, clustering, Markov decision processes, and so on. We are aware of some work relating to design in progress on a few of these areas.

13. Appendix

13.1. Moment methods

13.1.1. Markov moment problem and its geometry

Moment theory enters into the considerations of optimal design through the concept of admissible designs, as discussed in Section 6. The problem there is to find a probability distribution that maximizes the $2p$th moment for given values of the preceding moments.

EXAMPLE. Suppose one is interested in finding the maximum value of the fourth moment of a distribution on $[-1,1]$ with the first and the third moment equal to zero and the second moment equal to some given c.

The idea then is to construct a polynomial of degree 4 of the form

$$P(x) = x^2(x-1)(x+1),$$

which has the property that $P(x) \leqslant 0$ for every x in $[-1,1]$ and is equal to zero if $x = 0, -1$ or $+1$. Thus, for any probability distribution, the fourth moment c_4 satisfies $c_4 \leqslant c$, and for the particular distribution which assigns probability $c/2$ at $+1$ and -1 and $1-c$ at zero, equality obtains because $\mathrm{E}(P(x)) = 0$ if the underlying distribution is supported on the roots of $P(x)$; alternatively, in this simple case, equality can as well be seen by trivial direct verification.

Karlin and Studden (1966) and Kemperman (1968) give very careful accounts of such geometry of moment problems; the general version of a moment problem, which sometimes goes by the name of the Markov moment problem, is the following:

One has a general set, on which are defined $k+1$ functions $f_0, f_1, \ldots, f_k$. The problem is to determine a distribution $\overline{\mathcal{E}}$ and $\underline{\mathcal{E}}$ that respectively maximizes and minimizes the integral $\int f_k \,\mathrm{d}\mathcal{E}$ among all distributions satisfying $\int f_i \,\mathrm{d}\mathcal{E} = c_i$, $0 \leqslant i \leqslant k-1$.

EXAMPLE. Suppose X has a density on R of the form

$$f(x \mid \mu) = \int \frac{1}{\sqrt{2\pi s}} \mathrm{e}^{-\frac{1}{2s}(x-\mu)^2} \,\mathrm{d}G(s).$$

In other words, the density of X is a normal scale mixture. Consider the interval $[x - 1.96, x + 1.96)$, which as a confidence interval for μ, has a 95% confidence coefficient for normal data. The object is to determine the smallest confidence coefficient of this interval if the underlying distribution has standard deviation 1.

It is immediate that one thus wants to

$$\text{minimize} \int \Phi\left(\frac{1.96}{\sqrt{s}}\right) \mathrm{d}G(s) \quad \text{subject to} \quad \int s \, \mathrm{d}G(s) = 1.$$

This is therefore a special kind of a Markov moment problem.

To find a solution, one can show that there exists a straight line with equation $y = a + bs$, such that this line always lies below the graph of the function $\Phi(1.96/\sqrt{s})$ for $s \geqslant 0$ and touches the graph at $s = 0$ and another point $s = s_0 \geqslant 1$. If one now takes a distribution G_0 supported on $\{0, s_0\}$ such that it indeed satisfies the constraint $\int s \, \mathrm{d}G_0(s) = 1$, then, by virtue of the geometric property of the above straight line, it indeed follows that for any distribution with the stated restriction, $\int \Phi(1.96/\sqrt{s}) \, \mathrm{d}G(s) \geqslant a + b$, and in particular for the distribution G_0, it is equal to $a + b$, because at the points of support of G_0, the linear function $a + bs$ and the function $\Phi(1.96/\sqrt{s})$ are exactly equal!

The value of s_0, and the constants a, b can be found by easy geometric considerations, and are all given in Basu and DasGupta (1994).

The technique given in these two above examples forms a key step in solution of moment problems: identify linear combinations $\sum_{i=0}^{k-1} a_i f_i$ which either lie below or above the function f_k, depending on whether the Markov moment problem is one of minimization or maximization, and find a distribution which is supported on the points at which the graphs of these two come into contact, and appropriately adjust the weights so as to satisfy the given moment constraints. Notice that the task involves finding just the right linear combination with the given geometric property; one should not be misled that there is one such linear function only.

EXAMPLE. Does there exist a unimodal distribution on $[0, 1]$ with variance equal to .2?

The question is meaningful, because one can attain a variance between 0 and .25 if all distributions on $[0, 1]$ are allowed. To answer this question, it is helpful to turn it into a moment problem on writing the underlying unimodal random variable X as $X = a + UZ$, where the mode a is between 0 and 1, U is uniformly distributed on $[0, 1]$, Z is independent of U, and is between $-a$ and $1 - a$ with probability 1. Of course, the mode $'a'$ is not fixed, and has to be varied between 0 and 1 as well.

It is clear that marginally, any mean between 0 and 1 can arise from a unimodal distribution; furthermore, all point masses are unimodal, and therefore the lower boundary of the relevant moment set for unimodal distributions is given by $\underline{\mu_2} = \mu_1^2$.

The upper boundary can be found easily after proving the following fact: a given mean μ_1 can arise only if the mode a satisfies: $\max(0, 2\mu_1 - 1) \leqslant a \leqslant \min(2\mu_1, 1)$. Indeed, the upper boundary of the moment set is piecewise linear, given by $\overline{\mu}_2 = \frac{2}{3}\mu_1$ if $\mu_1 \leqslant 1/2$, and $\frac{4}{3}\mu_1 - \frac{1}{3}$ if $\mu_1 > 1/2$. From this it follows by calculus that the maximum variance is $1/9$, which is much less than .2. Thus, no variance larger than $1/9$ can be attained by a unimodal distribution.

Notice that the geometry shows that the moment set for unimodal distribution is not convex.

We recommend Krein and Nudelman (1973) and Akhiezer (1965) for anyone interested in learning about moment methods, a very useful tool in many branches of mathematics and mathematical statistics. We also recommend Diaconis (1987) for a very lively account of the history of moment methods and also some simply interesting facts of inherent scientific interest: for example, suppose we have a CDF F on the real line which has exactly the same first n moments as those of a $N(0,1)$ distribution; how accurately does this determine the CDF itself?

Characterizing distributions which are determined by their moment sequences is also a celebrated problem in the history of moment theory and probability. The normal distribution is determined by its moment sequence, but alas, the lognormal distribution is not! Convolutions of determined distributions may be undetermined – see Berg (1985). On the positive side, any distribution which is boundedly supported is determined by its moment sequence. In fact, there is a peculiar generalization to this which really is a result from analytic function theory stated in the language of a probabilist.

THEOREM. *Let $\{n_i\}$ be a subsequence of the positive integers such that $\sum(1/n_i) = \infty$; then the sequence of moments $\mathrm{E}(X^{n_i})$ determines a distribution supported on a bounded interval $[a,b]$.*

There are indeed characterization theorems which, in theory, can tell which distributions are determined and which are not by their moment sequences. They are not particularly useful in general; there is a remarkable exception to this, a pretty theorem:

THEOREM. *Suppose an absolutely continuous distribution has a density $f(x)$ on $\mathbb{R}$. Then it is determined by its moment sequence if and only if*

$$\int_a^b \frac{\log f(x)}{1+x^2}\,\mathrm{d}x = -\infty.$$

This theorem has a counterpart for measures supported on $\mathbb{R}^+$ in the following sense:

THEOREM. *Suppose an absolutely continuous distribution supported on $\mathbb{R}^+$ has a density $f(x)$ such that*

$$\int_0^\infty \frac{\log f(x)}{\sqrt{x}(1+x)}\,\mathrm{d}x > -\infty.$$

Then there is at least one more distribution supported on $\mathbb{R}^+$ with the same moment sequence as that of f.

Although it does not appear to be well known, there is a striking connection of this result to Hardy functions: f is undetermined if and only if it is the absolute value of the Fourier transform of an L_2 function vanishing in the negative half line. A recent description can be seen in Berg (1995).

Whether or not a particular distribution is determined by its moment sequence can sometimes be useful in asymptotic theory; in general, convergence of all moments does not imply convergence in distribution. However, if the suspected limit distribution is determined by its moments, then convergence of moments does indeed imply convergence in distribution. One can see Billingsley (1986) for more on this; in particular, certain astounding results in probabilistic number theory which otherwise require very intricate sieve and truncation arguments, can be proved by moment convergence and by using the fact that the normal distribution is indeed determined by its moment sequence. If one is interested, the sieve arguments can be seen in Elliott (1979), who describes the proof of limiting normality of the number of factors of a random integer.

13.1.2. Canonical moments

In general, the first n moments of a probability distribution on a bounded interval $[a, b]$ satisfy complex inequalities; more precisely, if one defines a set in R^n as

$$M_n = \{(c_1, c_2, \ldots, c_n): \; c_i = \mathrm{E}(X^i) \text{ for some probability distribution on } [a, b]\},$$

then M_n is a complicated convex set in R^n. It usually goes by the name of the moment set.

Canonical moments are a device for transforming the moment set into the cube $[0, 1]^n$; thus, each canonical moment p_i varies freely in the interval $[0, 1]$ as opposed to the moments c_i which form a complex set. This transformation from moments to canonical moments is 1–1 onto. Thus, it is quite common in optimal design theory to work out an optimal design in terms of its canonical moments, and the added bonus is that the structure of the optimizing canonical moment sequence even gives the number of points in the support of the optimal design. One can see numerous evidence of the utility of this technique in the works of William Studden. Canonical moments are also obviously useful in any numerical optimization scheme over the moments, because the optimization can be done for freely varying variables, which cannot be done with the moments themselves.

The exact transform to take moments to canonical moments and vice versa is best described by a recursive algorithm; we recommend Skibinsky (1968) for this.

EXAMPLE. Consider the Uniform distribution on $[0, 1]$. The moments of this distribution are given by $c_i = 1/(i + 1)$; the canonical moments are seen to be $p_{2i-1} = 1/2$, $p_{2i} = i/(2i + 1)$.

EXAMPLE. Suppose $X \sim \mathrm{Bin}(n, p)$; the moments of X/n can be calculated by calculating the factorial moments, $\mathrm{E}\{X(X - 1) \cdots (X - i + 1)\}$. The canonical moments are seen to be $p_{2i-1} = p$ and $p_{2i} = i/n$.

EXAMPLE. We saw previously that admissible designs in polynomial regression models have the property of maximizing the $2p$th moment subject to given values of the preceding moments. Such "upper principal representations" have a clean property with regard to their $2p$th canonical moment: the $2p$th canonical moment equals 1.

There is a wealth of information on canonical moments and connections to optimal designs in Lau and Studden (1985).

13.2. Orthogonal polynomials

13.2.1. Relation to optimal design

The close relationship of various systems of orthogonal polynomials to optimal designs was seen in Section 6.2. Generally, the points in the support of classical and Bayes optimal designs according to some alphabetic criteria coincide with the points of peak of various orthogonal polynomials. Thus it was seen that for extrapolation problems in polynomial regression, the c-optimal design is always supported at the peaks of the pth Chebyshev polynomial $T_p(x)$, the D-optimal design is supported at the turning points of Legendre polynomials plus the endpoints, and E-optimal designs concentrate on the turning points of Chebyshev polynomials as well. We will later give a list of the first few standard systems of orthogonal polynomials, and also a general algorithm for producing the entire sequence, which can be used on a computer to evaluate the turning points for any particular order p; note that p in this context coincides with the degree of the polynomial regression model.

13.2.2. Basic properties

Orthogonal polynomials arise out of the following familiar approximation problem: there is a continuous function f defined on an interval $[a, b]$, and there is a finite dimensional subspace $\mathcal{A}$ of $C[a, b]$, and one wants to find an element p^* in $\mathcal{A}$ giving the smallest value of $\int (f(x) - p(x))^2\, w(x)\, \mathrm{d}x$, where $w(\cdot)$ is a fixed weight function on $[a, b]$. By a weight function, one usually means a nonnegative integrable function. This is the usual least squares problem.

It is natural to write the solution p^* using the elements of a basis for $\mathcal{A}$; orthogonal polynomials essentially correspond to an orthogonal basis for $\mathcal{A}$. Thus, if $\mathcal{A}$ is of dimension $n + 1$, then a sequence $\{\phi_i, 0 \leqslant i \leqslant n\}$ is an orthogonal basis if $\{\phi_i\}$ are linearly independent, and satisfy the inner product condition $\int \phi_i(x)\phi_j(x)w(x)\, \mathrm{d}x = 0$ whenever $i \neq j$. Using elementary linear algebra, one can see that then the solution p^* has the representation

$$p^*(x) = \sum_{i=0}^{n} c_i \phi_i(x),$$

where $c_i = (\phi_i, f)/\|\phi_i\|^2$, where $(,)$ denotes inner product in the $L^2(w)$ space and $\|g\|^2 = (g, g)$.

There is another way to look at this; regardless of the least squares problem, one can form the expansion

$$\hat{f}_n(x) = \sum_{i=0}^{n} c_i f_i(x),$$

with c_i as above. One would intuitively expect that as subspaces of larger dimension are used, i.e., as n increases, the function $\hat{f}_n$ should approximate f more closely. There is a rich and long history of this method, variously known as orthogonal or Fourier expansions. Although Fourier expansions need not in general converge or converge to the parent function everywhere even if they do themselves converge, fairly general L_2 approximation results are indeed valid for the Fourier expansions.

THEOREM. *Under the assumption of completeness, the finite Fourier expansion $\hat{f}_n$ of f converges in L_2 to f; in fact,*

$$\|\hat{f}_n - f\|^2 = \sum_{i=n+1}^{\infty} c_i^2$$

and

$$\sum_{i=0}^{\infty} c_i^2 < \infty, \quad \text{so that} \quad \sum_{i=n+1}^{\infty} c_i^2 \to 0.$$

13.2.3. Recursions and roots of orthogonal polynomials

The points at which certain orthogonal polynomials have zero derivatives are often the points in support of optimal designs. Therefore, it is important to know how orthogonal polynomials are found. A straightforward method would be to take an arbitrary basis and use the familiar Gram–Schmidt orthogonalization process. This, however, is unnecessary due to a remarkable recursion relation orthogonal polynomials satisfy.

THEOREM. *Define the first orthogonal polynomial as $\phi_0(x) = 1$; define $\alpha_0 = \int x w(x)\,\mathrm{d}x / \int w(x)\,\mathrm{d}x$.*

Define $\phi_1(x) = x - \alpha_0$. For $j > 1$, define recursively

$$\alpha_j = (\phi_j, x\phi_j)/\|\phi_j\|^2,$$
$$\beta_j = \|\phi_j\|^2/\|\phi_{j-1}\|^2,$$

and

$$\phi_{j+1}(x) = (x - \alpha_j)\phi_j(x) - \beta_j \phi_{j-1}(x).$$

Then $\{\phi_i\}$ is a sequence of orthogonal polynomials with respect to the inner product $(f, g) = \int f(x) g(x) w(x)\,\mathrm{d}x$.

This three term recursion considerably simplifies the calculation of orthogonal polynomials for large values of n. Of course, in practice, relatively small n and standard weight functions $w(x)$ may be used, in which case the corresponding orthogonal polynomials are widely available. We will see such examples in the next subsection.

We close with two facts about the roots of orthogonal polynomials.

THEOREM. *Suppose $\{\phi_i\}$ is a system of orthogonal polynomials in an inner product space $L^2(w, [a, b])$. Then ϕ_k has exactly k roots which are real, simple, and in the interior of $[a, b]$. Furthermore, between two successive roots of ϕ_{k-1}, there is one root of ϕ_k.*

13.2.4. Special orthogonal polynomials

The system of orthogonal polynomials $\{\phi_k\}$ are determined by the weight function $w(x)$; a special important case in applications is when the $(n+1)$ dimensional subspace in the general theory is the set of algebraic polynomials of degree n. In this important case, the orthogonal polynomials for certain standard weight functions are explicitly known and have been studied in great depth for their properties.

$$\text{(a)} \quad a = -1, \quad b = 1, \quad w(x) = 1/\sqrt{(1-x^2)}.$$

In this case, the orthogonal polynomials are Chebyshev polynomials $\{T_n(x)\}$; $T_n(x)$ also has the trigonometric interpretation that

$$T_n(\cos\theta) = \cos(n\theta).$$

The coefficients of the various powers of x in any $T_n(x)$ are explicitly known; one can see chapter 1 in Rivlin (1990). The first few Chebyshev polynomials are as follows:

n	$T_n(x)$
0	1
1	x
2	$2x^2 - 1$
3	$4x^3 - 3x$
4	$8x^4 - 8x^2 + 1$
5	$16x^5 - 20x^3 + 5$

$$\text{(b)} \quad a = -1, \quad b = 1, \quad w(x) = 1.$$

In this case, the orthogonal polynomials are the Legendre polynomials $\{P_n(x)\}$. Again, the exact coefficients of the various powers of x are known: one can see Rivlin (1969). The first few Legendre polynomials are as follows:

n	$P_n(x)$
0	1
1	x
2	$3x^2 - 1$
3	$5x^3 - 3x$
4	$35x^4 - 30x^2 + 3$
5	$63x^5 - 70x^3 + 15x$

(c) $a = 0, \quad b = \infty, \quad w(x) = \mathrm{e}^{-x}.$

In this case, the orthogonal polynomials are the Laguerre polynomials $\{L_n(x)\}$. One can see Gradshteyn and Ryzhik (1980) for the exact coefficients for the general degree. The first few Laguerre polynomials are as follows:

n	$L_n(x)$
0	1
1	x
2	$x^2 - 4x + 2$
3	$x^3 - 9x^2 + 18x - 6$
4	$x^4 - 16x^3 + 72x^2 - 96x + 24$
5	$x^5 - 25x^4 + 200x^3 - 600x^2 + 600x - 120$

(d) $a = -\infty, \quad b = \infty, \quad w(x) = \mathrm{e}^{-x^2};$

In this case, the orthogonal polynomials are the very familiar Hermite polynomials. The exact coefficients are again available in Gradshteyn and Ryzhik (1980), and the first five Hermite polynomials are the following:

n	$H_n(x)$
0	1
1	x
2	$2x^2 - 1$
3	$2x^3 - 3x$
4	$4x^4 - 12x^2 + 3$
5	$4x^5 - 20x^3 + 15x$

Other important cases include the symmetric Beta function $w(x) = x^{\alpha-1}(1-x)^{\alpha-1}$, for $\alpha > 0$, in which case one gets the Ultraspherical polynomials, and the general Beta function $w(x) = x^{\alpha-1}(1-x)^{\beta-1}$, for $\alpha, \beta > 0$, and one gets the Jacobi polynomials.

There is a unifying feature regarding all of these cases: it is that there is a function $u(x)$ such that the nth orthogonal polynomial $\phi_n(x)$ admits the representation

$$\phi_n = u^{(n)}(x)/w(x),$$

where $u^{(n)}$ denotes the nth derivative of u. This is sometimes called Rodriguez's formula and the correct choice of $u(x)$ is known for each case.

13.2.5. Linearization of nonlinear functions

A major use of orthogonal polynomials is in approximating complicated functions by linear combinations of orthogonal polynomials. Note that the L^2 theory only assures close approximation in $L2$ norm, but in optimal design, one may need an assurance of uniform approximation in the design space. Some very interesting results are indeed known, and we think they are useful in linearization of nonlinear models. We mention two of these results.

THEOREM (Dini–Lipschitz). *Suppose f is a continuous function on a bounded interval $[a, b]$ and let $w(f, \cdot)$ denote its modulus of continuity. Suppose $s_n(f)$ is the nth partial sum in the Chebyshev expansion of f; if $w(f, 1/n)\log n \to 0$ as $n \to \infty$, then $s_n(f) \to f$ uniformly.*

COROLLARY. *If f is Lipschitz (α) for $\alpha > 0$, then the Chebyshev expansion of f uniformly converges to f.*

In addition to the above theorem, for purposes of deciding how many terms one should use, estimates of the error are useful. There are several results known; we find the following useful.

THEOREM. *Let $E_n(f)$ denote the error in the approximation of f by $s_n(f)$ using supnorm, and let $E_n^*(f)$ denote the same error by using the best polynomial approximation to f in supnorm. Then*

$$E_n(f) \leqslant 4\left(1 + \frac{\log n}{\pi^2}\right) E_n^*(f).$$

The suggestion in Atkinson and Donev (1992) is to linearize a nonlinear model by using its Taylor expansion; we believe that Chebyshev expansions can estimate more efficiently with a smaller number of terms. One reason is the following theorem.

THEOREM. *Consider the expansion of a continuous function in terms of ultraspherical polynomials defined in Section 13.3.3. Then, the choice $\alpha = 1/2$ always gives the best approximation in* sup *norm if the coefficients $\{a_i,\ i > n\}$ in the ultraspherical expansion of f are nonnegative for that given n; in particular, a Chebyshev expansion corresponding to $\alpha = 1/2$ is better than a Taylor expansion which corresponds to $\alpha = \infty$.*

Acknowledgement

I learned from Bill Studden the little I know about optimal designs. I am very thankful for the scholarly inspiration he provides, in an understated brilliant way.

References

Akhiezer, N. (1962). *Some Questions in the Theory of Moments*. Amer. Mathematical Soc., Providence, RI.

Antelman, G. R. (1965). Insensitivity to non-optimal design in Bayesian decision theory. *J. Amer. Statist. Assoc.* **60**, 584–601.

Atkinson, A. C. and A. N. Donev (1992). *Optimum Experimental Designs*. Clarendon Press, Oxford.

Basu, S. and A. DasGupta (1992). Robustness of standard confidence intervals under departure from normality, to appear in *Ann. Statist.*

Berg, C. (1985). On the preservation of determinacy under convolution. *Proc. Amer. Math. Soc.* **93**, 351–357.

Berg, C. (1995). Indeterminate moment problems and the theory of entire functions. *J. Comput. Appl. Math.* **65**, 27–55.

Berger, J. (1986). *Statistical Decision Theory and Bayesian Analysis*. 2nd edn. Springer, New York.

Berger, J. (1994). An overview of Robust Bayesian analysis. *Test* **3**(1), 5–59.

Bickel, P. J. and A. M. Herzberg (1979). Robustness of design against autocorrelation in time I. *Ann. Statist.* **7**, 77–95.

Billingsley, P. (1986). *Probability and Measure*. Wiley, New York.

Bock, J. and H. Toutenberg (1991). Sample size determination in clinical research. In: *Handbook of Statistics*, Vol. 8. Elsevier, Amsterdam, 515–538.

Bose, R. C. (1948). The design of experiments. In: *Proc. 34th Indian Sci. Cong., Delhi, 1947*. Indian Science Congress Association, Calcutta, (1)–(25).

Bowman, K. O. and M. A. Kastenbaum (1975). Sample size requirement: Single and double classification experiments. In: *Selected Tables in Mathematical Statistics*, edited by IMS, Vol. 3. Amer. Mathematical Soc., Providence, RI.

Box, G. E. P. and N. R. Draper (1987). *Empirical Model-Building and Response Surfaces*. Wiley, New York.

Box, G. E. P. and J. S. Hunter (1957). Multi-factor experimental designs for exploring response surfaces. *Ann. Math. Statist.* **28**, 195–241.

Box, G. E. P. and H. L. Lucas (1959). Design of experiments in non-linear situations. *Biometrika* **46**, 77–90.

Brooks, R. J. (1972). A decision theory approach to optimal regression designs. *Biometrika* **59**, 563–571.

Brooks, R. J. (1974). On the choice of an experiment for prediction in linear regression. *Biometrika* **61**, 303–311.

Brooks, R. J. (1976). Optimal regression designs for prediction when prior knowledge is available. *Metrika* **23**, 217–221.

Brown, L. D. (1986). *Fundamentals of Statistical Exponential Families*, IMS Lecture Notes – Monograph Series, Vol. 9. Hayward, CA.

Brown, L. D. (1991). Minimaxity, more or less. In: S. Gupta and J. Berger, eds., *Statistical Decision Theory and Related Topics*, 1–18.

Caselton, W. F. and J. V. Zidek (1984). Optimal monitoring network designs. *Statist. Probab. Lett.* **2**, 223–227.

Chaloner, K. (1984). Optimal Bayesian experimental design for linear models. *Ann. Statist.* **12**, 283–300; Correction **13**, 836.

Chaloner, K. (1989). Bayesian design for estimating the turning point of a quadratic regression. In: *Commun. Statist. Theory Methods* **18**(4), 1385–1400.

Chaloner, K. (1993). A note on optimal Bayesian design for nonlinear problems. *J. Statist. Plann. Inference* **37**, 229–235.

Chaloner, K. and K. Larntz (1986). Optimal Bayesian designs applied to logistic regression experiments. Tech. Report, University of Minnesota.

Chaloner, K. and K. Larntz (1989). Optimal Bayesian designs applied to logistic regression experiments. *J. Statist. Plann. Inference* **21**, 191–208.

Chaloner, K. and I. Verdinelli (1994). Bayesian experimental design: A review. Tech. Report, Department of Statistics, University of Minnesota.

Cheng, C.-S. (1978b). Optimal designs for the elimination of multi-way heterogeneity. *Ann. Statist.* **6**, 1262–1272.

Chernoff, H. (1953). Locally optimum designs for estimating parameters. *Ann. Math. Statist.* **24**, 586–602.

Chernoff, H. (1972). *Sequential Analysis and Optimal Design*. Society for Industrial and Applied Mathematics, Philadelphia, PA.

Clyde, M. A. (1993). An object-oriented system for Bayesian nonlinear design using xlispstat. Tech. Report 587, University of Minnesota, School of Statistics.

DasGupta, A., S. Mukhopadhyay and W. J. Studden (1992). Compromise designs in heteroscedastic linear models. *J. Statist. Plann. Inference* **32**, 363–384.

DasGupta, A. and S. Mukhopadhyay (1988). Uniform and subuniform posterior robustness: Sample size problem. Tech. Report, Purdue University.

DasGupta, A. and W. J. Studden (1988). Robust Bayesian analysis and optimal experimental designs in normal linear models with many parameters. I. Tech. Report, Department of Statistics, Purdue University.

DasGupta, A. and W. J. Studden (1991). Robust Bayes designs in normal linear models. *Ann. Statist.* **19**, 1244–1256.

DasGupta, A. and S. Mukhopadhyay (1994). Uniform and subunivorm posterior robustness: The sample size problem. Proc. 1st Internat. Workshop on Bayesian Robustness, Special issue of *J. Statist. Plann. Inference* **40**, 189–204.

DasGupta, A. and B. Vidakovic (1994). Sample sizes in ANOVA: The Bayesian point of view. Tech. Report, Purdue University. Submitted: *J. Statist. Plann. Inference*.

DasGupta, A. and M. M. Zen (1996). Bayesian bioassay design. Tech. Report, Purdue University. Submitted *J. Statist. Plann. Inference*.

Dehnad, K., ed. (1989). *Quality Control, Robust Design, and the Taguchi Method.* Wadsworth and Brooks/Cole, Pacific Grove, CA.

DeRobertis, L. and J. A. Hartigan (1981). Bayesian inference using intervals of measures. *Ann. Statist.* **9**, 235–244.

Dette, H. (1991). A note on robust designs for polynomial regression. *J. Statist. Plann. Inference* **28**, 223–232.

Dette, H. (1992). Optimal designs for a class of polynomials of odd or even degree. *Ann. Statist.* **20**, 238–259.

Dette, H. (1993a). Elfving's theorem for D-optimality. *Ann. Statist.* **21**, 753–766.

Dette, H. (1993b). A note on Bayesian c- and D-optimal designs in nonlinear regression models. Manuscript.

Dette, H. and H.-M. Neugebauer (1993). Bayesian D-optimal designs for exponential regression models. *J. Statist. Plann. Inference*, to appear.

Dette, H. and S. Sperlich (1994a). Some applications of continued fractions in the construction of optimal designs for nonlinear regression models. Manuscript.

Dette, H. and S. Sperlich (1994b). A note on Bayesian D-optimal designs for general exponential growth models. Manuscript.

Dette, H. and W. J. Studden (1994a). Optimal designs for polynomial regression when the degree is not known. Tech. Report, Purdue University.

Dette, H. and W. J. Studden (1994b). A geometric solution of the Bayes E-optimal design problem. In: S. Gupta and J. Berger, eds., *Statistical Decision Theory and Related Topics*, Vol. 5. 157–170.

Diaconis, P. (1987a). Bayesian numerical analysis. In: S. Gupta and J. Berger, eds., *Statistical Decition Theory and Related Topics*, IV, Vol. 1, 163–176.

Diaconis, P. (1987b). Application of the method of moments in probability and statistics. In: *Moments in Mathematics*. Amer. Mathematical Soc., Providence, RI.

Donev, A. N. (1988). The construction of exact D-optimum experimental designs. Ph.D. Thesis, University of London.

Dykstra, Otto, Jr. (1971). The augmentation of experimental data to maximize $|X^TX|$. *Technometrics* **13**, 682–688.

Eaton, M. L., A. Giovagnoli and P. Sebastiani (1994). A predictive approach to the Bayesian design problem with application to normal regression models. Tech. Report 598, School of Statistics, University of Minnesota.

Elfving, G. (1952). Optimum allocation in linear regression theory. *Ann. Math. Statist.* **23**, 255–262.

El-Krunz, S. M. and W. J. Studden (1991). Bayesian optimal designs for linear regression models. *Ann. Statist.* **19**, 2183–2208.

Elliott, P. D. T. A. (1979). *Probabilistic Number Theory*, Vol. 2. Springer, New York.

Erdos, P. (1958). Problems and results on the theory of interpolation, I. *Acta. Math. Acad. Sci. Hungar.* **9**, 381–388.

Farrell, R. H., J. Kiefer and A. Walbran (1967). Optimum ultivariate designs. In: L. M. Le Cam and J. Neyman, eds., *Proc. 5th Berkeley Symp. Math. Statist. Probab., Berkeley, CA, 1965 and 1966*, Vol. 1. University of California, Berkeley, CA.

Fedorov, V. V. (1972). *Theory of Optimal Experiments.* Academic Press, New York.

Ferguson, T. S. (1989). Who solved the secretary problem? *Statist. Sci.* **4**(3), 282–296.

Fisher, R. A. (1949). *Design of Experiments.* Hafner, New York.

Freeman, P. R. (1983). The secretary problem and its extensions – A review. *Internat. Statist. Rev.* **51**, 189–206.

Friedman, M. and L. J. Savage (1947). Experimental determination of the maximum of a function. In: *Selected Techniques of Statistical Analysis.* McGraw-Hill, New York, 363–372.

Gaffke, N. and O. Krafft (1982). Exact D-optimum designs for quadratic regression. *J. Roy. Statist. Soc. Ser. B* **44**, 394–397.

Ghosh, S., ed. (1990). *Statistical Design and Analysis of Industrial Experiments.* Marcel Dekker, New York.

Ghosh, M., B. K. Sinha and N. Mukhopadhyay (1976). Multivariate sequential point estimation. *J. Multivariate Anal.* **6**, 281–294.

Giovagnoli, A. and I. Verdinelli (1983). Bayes D-optimal and E-optimal block designs. *Biometrika* **70**(3), 695–706.

Giovagnoli, A. and I. Verdinelli (1985). Optimal block designs under a Hierarchical linear model. In: J. M. Bernardo et al., eds., *Bayesian Statistics*, Vol. 2. North-Holland, Amsterdam.

Gladitz, J. and J. Pilz (1982). Construction of optimal designs in random coefficient regression models. *Math. Operationsforsch. Statist. Ser. Statist.* **13**, 371–385.

Gradshteyn, I. S. and I. M. Ryzhik (1980). *Table of Integrals, Series and Products.* Academic Press, New York.

Haines, L. M. (1987). The application of the annealing algorithm to the construction of exact D-optimum designs for linear-regression models. *Technometrics* **29**, 439–447.

Hedayat, A. S., M. Jacroux and D. Majumdar (1988). Optimal designs for comparing test treatments with a control. *Statist. Sci.* **3**, 462–476; Discussion **3**, 477–491.

Herzberg, A. M. and D. R. Cox (1969). Recent work on the design of experiments: A bibliography and a review. *J. Roy. Statist. Soc. Ser. A* **132**, 29–67.

Huber, P. J. (1972). Robust statistics: A review. *Ann. Math. Statist.* **43**, 1041–1067.

Huber, P. J. (1975). Robustness and designs. In: J. N. Srivastava, ed., *A Survey of Statistical Design and Linear Models.* North-Holland, Amsterdam.

Huber, P. J. (1981). *Robust Statistics.* Wiley, New York.

Joseph, L., D. Wolfson and R. Berger (1994). Some comments on Bayesian sample size determination. Preprint.

Joseph, L. and R. Berger (1994). Bayesian sample size methodology with an illustration to the difference between two binomial proportions. Preprint.

Kacker, R. N. (1985). Off-line quality control, parameter design, and the Taguchi method. *J. Qual. Tech.* **17**, 176–188.

Karlin, S. and L. S. Shapley (1953). *Geometry of Moment Spaces*, Vol. 12 of Amer. Math. Soc. Memoirs.

Karlin, S. and W. J. Studden (1966). *Tchebycheff Systems: With Applications in Analysis and Statistics.* Interscience, New York.

Kemperman, J. H. B. (1968). The general moment problem, a geometric approach. *Ann. Math. Statist.* **39**, 93–122.

Kemperman, J. H. B. (1972). On a class of moment problems. In: *Proc. 6th Berkeley Symp. Math. Statist. and Probab.* Vol. 2, 101–126.

Kiefer, J. (1953). Sequential minimax search for a maximum. *Proc. Amer. Math. Soc.* **4**, 502–506.

Kiefer, J. (1987). *Introduction to Statistical Inference.* Springer, New York.

Kiefer, J. C. (1959). Optimum experimental designs. *J. Roy. Statist. Soc. Ser. B* **21**, 272–304. Discussion on Dr. Kiefer's paper **21**, 304–319.

Kiefer, J. C. (1974). General equivalence theory for optimum designs (approximate theory). *Ann. Statist.* **2**, 849–879.

Kiefer, J. and J. Wolfowitz (1959). Optimum designs on regression problems. *Ann. Math. Statist.* **30**, 271–294.

Kiefer, J. C. and W. J. Studden (1976). Optimal designs for large degree polynomial regression. *Ann. Statist.* **4**, 1113–1123.

Korner, T. W. (1989). *Fourier Analysis*. Cambridge, New York.

Krein, M. G. and A. A. Nudelman (1977). *The Markov Moment Problem and Extremal Problems*. Amer. Mathematical Soc., Providence, RI.

Kurotschka, V. (1978). Optimal design of complex experiments with qualitative factors of influence. *Commun. Statist. Theory Methods* **7**, 1363–1378.

Lau, T.-S and W. J. Studden (1985). Optimal designs for trigonometric and polynomial regression using canonical moments. *Ann. Statist.* **13**, 383–394.

Leamer, E. E. (1978). *Specification Searches: Ad hoc Inference with Nonexperimental Data*. Wiley, New York.

Lee, C. M.-S. (1988). Constrained optimal designs. *J. Statist. Plann. Inference* **18**, 377–389.

Lehmann, E. L. (1986). *Testing Statistical Hypotheses*, 2nd edn. Wiley, New York.

Lindley, D. V. (1956). On a measure of the information provided by an experiment. *Ann. Math. Statist.* **27**, 986–1005.

Majumdar, D. (1992). Optimal designs for comparing test treatments with a control using prior information. *Ann. Statist.* **20**, 216–237.

Majumdar, D. (1995). Optimal and efficient treatment-control designs. Preprint.

Marcus, M. B. and J. Sacks (1976). Robust design for regression problems. In: S. Gupta and D. Moore, eds., *Statistical Decision Theory and Related Topics*, Vol. 2, 245–268.

Mitchell, T., J. Sacks and D. Ylvisaker (1994). Asymptotic Bayes criteria for nonparametric response surface design. *Ann. Statist.* **22**, 634–651.

Mukhopadhyay, S. and L. Haines (1993). Bayesian D-optimal designs for the exponential growth model. *J. Statist. Plann. Inference*, to appear.

Nalimov, V. V. (1974). Systematization and codification of the experimental designs – The survey of the works of Soviet statisticians. In: J. Gani, K. Sarkadi and I. Vincze, eds., *Progress in Statistics. European Meeting of Statisticians, Budapest 1972*, Vol. 2. Colloquia Mathematica Societatis János Bolyai 9. North-Holland, Amsterdam, 565–581.

Nalimov, V. V., ed. (1982). *Tables for Planning Experiments for Factorials and Polynomial Models*. Metallurgica, Moscow (in Russian).

Odeh, R. E. (1975). *Sample Size Choice: Charts for Experiments with Linear Models*. Marcel Dekker, New York.

O'Hagan, A. (1978). Curve fitting and optimal design for prediction (with discussion). *J. Roy. Statist. Soc. Ser. B* **40**, 1–41.

Owen, R. J. (1970). The optimum design of a two-factor experiment using prior information. *Ann. Math. Statist.* **41**, 1917–1934.

Papalambros, P. Y. and D. J. Wilde (1988). *Principles of Optimal Design*. Cambridge, New York.

Pilz, J. (1981). Robust Bayes and minimax-Bayes estimation and design in linear regression. *Math. Operationsforsch. Statist. Ser. Statist.* **12**, 163–177.

Pilz, J. (1991). *Bayesian Estimation and Experimental Design in Linear Regression Models*. Wiley, New York.

Polasek, W. (1985). Sensitivity analysis for general and hierarchical linear regression models. In: P. K. Goel and A. Zellner, eds., *Bayesian Inference and Decision Techniques with Applications*. North-Holland, Amsterdam, 375–387.

Powell, M. J. D. (1981). *Approximation Theory and Methods*. Cambridge Univ. Press, New York.

Pukelsheim, F. (1980). On linear regression designs which maximize information. *J. Statist. Plann. Inference* **4**, 339–364.

Pukelsheim, F. (1988). Analysis of variability by analysis of variance. In: Y. Dodge, V. V. Fedorov and H. P. Wynn, eds., *Optimal Design and Analysis of Experiments*. North-Holland, New York.

Pukelsheim, F. and S. Rieder (1992). Efficient rounding of approximate designs. *Biometrika* **79**, 763–770.

Pukelsheim, F. (1993). *Optimal Design of Experiments*. Wiley, New York.

Pukelsheim, F. and W. J. Studden (1993). E-optimal designs for polynomial regression. *Ann. Statist.* **21**(1).

Rao, C. R. (1946). Difference sets and combinatorial arrangements derivable from finite geometries. *Proc. Nat. Inst. Sci.* **12**, 123–135.
Rao, C. R. (1947). Factorial experiments derivable from combinatorial arrangements of arrays. *J. Roy. Statist. Soc. Ser. B* **9**, 128–140.
Rao, C. R. (1973). *Linear Statistical Inference and Its Applications*, 2nd edn. Wiley, New York.
Rivlin, T. J. (1969). *An Introduction to the Approximation of Functions*. Dover, New York.
Rivlin, T. J. (1990). *Chebyshev Polynomials*, 2nd edn. Wiley Interscience, New York.
Royden, H. L. (1953). Bounds on a distribution function when its first n moments are given. *Ann. Math. Statist.* **24**, 361–376.
Rubin, H. (1977). Robust Bayesian estimation. In: S. Gupta and D. Moore, eds., *Statistical Decision Theory and Related Topics*, Vol. 2, 351–356.
Sacks, J. and S. Schiller (1988). Spatial designs. In: *Statistical Decision Theory and Related Topics*, Vol. 4. Springer, New York, 385–399.
Sacks, J., W. J. Welch, T. J. Mitchell and H. P. Wynn (1989). Design and analysis of computer experiments. *Statist. Sci.* **4**, 409–435.
Sacks, J. and D. Ylvisaker (1970). Statistical designs and integral approximation. *Proc. 12th Bien. Seminar Canad. Math. Cong.*, Montreal, 115–136.
Sacks, J. and D. Ylvisaker (1964). Designs for regression problems with correlated errors III. *Ann. Math. Statist.* **41**, 2057–2074.
Sarkadi, K. and I. Vincze (1974). *Mathematical Methods of Statistical Quality Control*. Academic Press, New York.
Schoenberg, I. J. (1959). On the maximization of some Hankel determiants and zeros of classical orthogonal polynomials. *Indag. Math.* **21**, 282–290.
Schumacher, P. and J. V. Zidek (1993). Using prior information in designing intervention detection experiments. *Ann. Statist.* **21**, 447–463.
Schwarz, G. (1962). Asymptotic shapes of Bayes sequential testing regions. *Ann. Math. Statist.* **33**, 224–236.
Seber, G. A. F. and C. J. Wild (1989). *Nonlinear Regression*. Wiley, New York.
Shah, K. S. and B. K. Sinha (1989). *Theory of Optimal Designs*. Lecture Notes in Statistics 54, Springer, New York.
Siegmund, D. (1985). *Sequential Analysis*. Springer, New York.
Silvey, S. D. (1980). *Optimal Design*. Chapman and Hall, London.
Skibinsky, M. (1968). Extreme nth moments for distributions on $[0, 1]$ and the inverse of a moment space map. *J. Appl. Probab.* **5**, 693–701.
Smith, A. F. M. and I. Verdinelli (1980). A note on Bayesian design for inference using a Hierarchical linear model. *Biometrika* **67**, 613–619.
Smith, K. (1918). On the standard deviations of adjusted and interpolated values of an observed polynomial function and its constants and the guidance they give towards a proper choice of the distribution of observations. *Biometrika* **12**, 1–85.
Spiegelhalter, D. J. and L. S. Freedman (1986). A predictive approach to selecting the size of a clinical trial, based on subjective clinical opinion. *Statist. Med.* **5**, 1–13.
Starr, N. and M. B. Woodroofe (1969). Remarks on sequential point estimation. *Proc. Nat. Acad. Sci.* **63**, 285–288.
Staudte, R. G. and S. J. Sheather (1990). *Robust Estimation and Testing*. Wiley Interscience, New York.
Steinberg, D. M. and W. G. Hunter (1984). Experimental design: Review and comment. *Technometrics* **26**, 71–97.
Stigler, S. M. (1971). Optimal experimental design for polynomial regression. *J. Amer. Statist. Assoc.* **66**, 311–318.
Stigler, S. M. (1974). Gergonne's 1815 paper on the design and analysis of polynomial regression experiments. *Historia Math.* **1**, 431–447.
Studden, W. J. (1977). Optimal designs for integrated variance in polynomial regression. In: S. S. Gupta and D. S. Moore, eds., *Statistical Decision Theory and Related Topics II*. Proc. Symp. Purdue University, 1976. Academic Press, New York, 411–420.
Studden, W. J. (1982). Some robust-type D-optimal designs in polynomial regression. *J. Amer. Statist. Assoc.* **77**, 916–921.

Tang, Dei-in (1993). Minimax regression designs under uniform departure models. *Ann. Statist.* **21**, 434–446.

Toman, B. and W. Notz (1991). Bayesian optimal experimental design for treatment control comparisons in the presence of two-way heterogeneity. *J. Statist. Plann. Inference* **27**, 51–63.

Toman, B. (1992). Bayesian robust experimental designs for the one-way analysis of variance. *Statist. Probab. Lett.* **15**, 395–400.

Toman, B. and J. L. Gastwirth (1993). Robust Bayesian experimental design and estimation for analysis of variance models using a class of normal mixtures. *J. Statist. Plann. Inference* **35**, 383–398.

Verdinelli, I. (1983). Computing Bayes D- and E-optimal designs for a two-way model. *The Statistician* **32**, 161–167.

Verdinelli, I. and H. P. Wynn (1988). Target attainment and experimental design, a Bayesian approach. In: Y. Dodge, V. V. Dedorov and H. P. Wynn, eds., *Optimal Design and Analysis of Experiments*. North-Holland, New York.

Wald, A. (1943). On the efficient design of statistical investigations. *Ann. Math. Statist.* **14**, 134–140.

Wasserman, L. (1992). Recent methodological advances in Robust Bayesian inference. In: J. Bernardo et al., eds., *Bayesian Statistics*, Vol. 4. Oxford Univ. Press, Oxford.

Whittle, P. (1973). Some general points in the theory of optimal experimental design. *J. Roy. Statist. Soc. Ser. B* **35**, 123–130.

Wilde, D. J. (1978). *Globally Optimal Design*. Wiley Interscience, New York.

Wu, C.-F. (1988). Optimal design for percentile estimation of a quantal response curve. In: Y. Dodge, V. V. Fedorov and H. P. Wynn, eds., *Optimal Design and Analysis of Experiments*. North-Holland, Amsterdam.

Wynn, H. P. (1977). Optimum designs for finite populations sampling. In: S. S. Gupta and D. S. Moore, eds., *Statistical Decision Theory and Related Topics II*. Proc. Symp. Purdue University, 1976. Academic Press, New York, 471–478.

Wynn, H. P. (1984). Jack Kiefer's contributions to experimental design. *Ann. Statist.* **12**, 416–423.

S. Ghosh and C. R. Rao, eds., *Handbook of Statistics, Vol. 13*

Approximate Designs for Polynomial Regression: Invariance, Admissibility, and Optimality

Norbert Gaffke and Berthold Heiligers

1. Introduction

This paper brings together different topics from the theory of approximate linear regression design. An overview as well as new results are presented on *invariance* and *admissibility* of designs, their interrelations and their implications to *design optimality,* including numerical algorithms. Although a great part of concepts and results will be presented in the framework of the general linear regression model, the emphasis lies on multiple polynomial models and their design, with particular attention to the linear, quadratic, and cubic cases which are most frequently used in applications.

Invariance structures combined with results on admissibility provide a tool to attack optimal design problems of high dimensions, as occurring for second and third order multiple polynomial models. Only in very rare cases explicit solutions can be obtained. So an important aspect are numerical algorithms for computing a nearly optimal design. From the general results on these topics presented below, we wish to point out the following ones at this place. Invariance w.r.t. infinite (though compact) transformation and matrix groups usually calls for the Haar probability measures, involving thus deep measure theoretic results. We will show a way of avoiding these by using only linear and convex structures in real matrix spaces. The Karlin–Studden necessary condition on the support of an admissible design is known to become useless for regression models involving a constant term. Here we derive a modified necessary condition which also works for models with constant term, and also including possible invariance structures. This result originates from Heiligers (1991). Well known numerical algorithms for solving extremum problems in optimal design are pure gradient methods, among them the steepest descent method of Fedorov and Wynn. However, after a quick but rough approximation towards the optimum they become very inefficient. The Quasi-Newton methods of Gaffke and Heiligers (1996) provide an efficient way of computing the optimum very accurately. Our general results will be applied to invariant design for multiple polynomial regression models of degree three or less, and to rotatable design for models with arbitrary degree.

Throughout we will deal with linear regression models under the standard statistical assumptions. That is, an independent variable x with possible values in some design space $\mathcal{X}$ effects a real-valued response

$$y(x) = \sum_{j=1}^{k} \theta_j f_j(x) = \theta' f(x), \tag{1.1a}$$

where $\theta = (\theta_1, \ldots, \theta_k)' \in \mathbb{R}^k$ is an unknown parameter vector, and $f = (f_1, \ldots, f_k)'$ is a given $\mathbb{R}^k$-valued function on $\mathcal{X}$. This is the deterministic part of a linear regression model. Embedded in the usual statistical context, observations of the response y at points $x_1, \ldots, x_n \in \mathcal{X}$, say, are represented by real-valued random variables $Y_1, \ldots, Y_n$, such that

$$\begin{aligned} &\mathrm{E}(Y_i) = y(x_i), \qquad \mathrm{Var}(Y_i) = \sigma^2, \quad i = 1, \ldots, n, \\ &\mathrm{Cov}(Y_i, Y_j) = 0, \qquad i, j = 1, \ldots, n, \quad i \neq j. \end{aligned} \tag{1.1b}$$

The constant variance $\sigma^2 \in (0, \infty)$ is usually unknown, and is hence an additional parameter in the model. It should be emphasized that by (1.1b) observations of the response variable y include random errors, whereas observations of the independent variable x are exact. Regression setups with both variables subject to random errors – often called 'error-in-the-variables models' – will not be considered here. Moreover, we assume that the values of x at which observations of y are taken can be controlled by the experimenter. So, we are concerned with designed experiments, which frequently occur in industrial experiments.

An (approximate) design ξ for model (1.1a) consists of finitely many support points $x_1, \ldots, x_r \in \mathcal{X}$, at which observations of the response are to be taken, and of corresponding weights $\xi(x_i)$, $i = 1, \ldots, r$, which are positive real numbers summing up to 1. In other words, an (approximate) design is a probability distribution with finite support on the experimental region $\mathcal{X}$. For short, we write

$$\xi = \begin{pmatrix} x_1 & \cdots & x_r \\ \xi(x_1) & \cdots & \xi(x_r) \end{pmatrix}, \tag{1.2}$$

where $x_1, \ldots, x_r \in \mathcal{X}$ with $\xi(x_1), \ldots, \xi(x_r) > 0$, $\sum_{i=1}^{r} \xi(x_i) = 1$, and $r \in \mathbb{N}$. The set $\mathrm{supp}(\xi) := \{x_1, \ldots, x_r\}$ is called the support of ξ. A design ξ assigns the percentage $\xi(x_i)$ of all observations to the value x_i of the independent variable, $i = 1, \ldots, r$. Note that different designs may also have different support sizes r. Of course, when a total sample size n for an experiment has been specified, a design ξ from (1.2) cannot, in general, be properly realized, unless its weights are integer multiples of $1/n$, i.e., unless

$$\xi(x_i) = \frac{n_i}{n} \quad \text{for all } i = 1, \ldots, r, \tag{1.3}$$

for some positive integers $n_1, \ldots, n_r$ with $\sum_{i=1}^{r} n_i = n$. That is why we call, following Kiefer, a design from (1.2) an *approximate* design, which only in the special case (1.3) becomes an *exact design of size* n. Since we will mainly deal with the approximate theory, we simply call ξ from (1.2) a design, while (1.3) will be referred to as an exact design of size n (denoted by ξ_n).

The statistical quality of a design ξ for setup (1.1a, b) is reflected by its moment matrix (or information matrix) $M(\xi)$, defined by

$$M(\xi) = \sum_{i=1}^{r} f(x_i)\, f(x_i)'\, \xi(x_i), \tag{1.4}$$

which is a nonnegative definite $k \times k$ matrix. For an exact design ξ_n, the inverse of $M(\xi_n)$ times σ^2/n is the covariance matrix of the Least Squares estimator of θ (provided $M(\xi_n)$ is non-singular), and under normality assumption in (1.1b) $(n/\sigma^2)M(\xi_n)$ is the Fisher information matrix.

The Loewner partial ordering of moment matrices provides a first basis for comparing designs. That partial ordering on the set $\mathrm{Sym}(k)$ of real symmetric $k \times k$ matrices is defined by

$$A \leqslant B \Longleftrightarrow B - A \in \mathrm{NND}(k) \quad (A, B \in \mathrm{Sym}(k)),$$

where $\mathrm{NND}(k)$ consists of the nonnegative definite matrices from $\mathrm{Sym}(k)$. If ξ and η are designs with $M(\xi) \leqslant M(\eta)$, then η is said to be at least as good as ξ, and if additionally $M(\xi) \neq M(\eta)$, then η is said to be better than ξ. This is statistically meaningful through the linear theory of Gauss–Markov estimation, whether or not the moment matrices of competing designs are non-singular, and whether or not normality is assumed in (1.1b). Suppose that ξ_n is an exact design of size n. Given a coefficient vector $c \in \mathbb{R}^k$, the variance of the BLUE (or Gauss–Markov estimator) of $c'\theta$ under ξ_n is equal to $(\sigma^2/n)\, c'M^-(\xi_n)c$, where $M^-(\xi_n)$ denotes a generalized inverse of $M(\xi_n)$, provided that $c'\theta$ is linearly estimable (identifiable) under ξ_n, i.e., $c \in \mathrm{range}(M(\xi_n))$. So, for any design ξ, we consider the variance function (per unit of error variance)

$$V(\xi, c) = \begin{cases} c'M^-(\xi)c, & \text{if } c \in \mathrm{range}(M(\xi)), \\ \infty, & \text{otherwise.} \end{cases}$$

Then, for any two designs ξ and η we have

$$M(\xi) \leqslant M(\eta) \Longleftrightarrow V(\xi, c) \geqslant V(\eta, c) \quad \text{for all } c \in \mathbb{R}^k. \tag{1.5}$$

If the moment matrices of ξ and η are nonsingular, then (1.5) can simply be stated as

$$M(\xi) \leqslant M(\eta) \Longleftrightarrow M^{-1}(\eta) \leqslant M^{-1}(\xi).$$

Equivalence (1.5) can be refined for subspaces of linear parameter functions $c'\theta$, where c is restricted to some given linear subspace of dimension $s \geqslant 1$. Let the subspace be represented as range(K') with an $s \times k$ matrix K of rank s. Then, given a design ξ, we consider the nonnegative definite $s \times s$ matrix $C_K(\xi)$ – often called the reduced information matrix of ξ for $K\theta$ – whose definition is somewhat implicit for arbitrary K and ξ, namely

$$C_K(\xi) = \min_L L\, M(\xi)\, L', \tag{1.6a}$$

where the minimum refers to the Loewner partial ordering in Sym(s) and is taken over all left inverses L of K' (i.e., over all $s \times k$ matrices L with $LK' = I_s$, with I_s being the unit matrix of order s). The existence of the minimum in (1.6a) was proved by Krafft (1983). This definition of reduced information matrices has been used by Gaffke (1987) and Pukelsheim (1993), Chapter 3.2, who also showed how to compute a minimizing L for (1.6a) (Pukelsheim, 1993, p. 62). A familiar special case of (1.6a) is $K = [I_s, 0] = K_s$, say; here, partitioning

$$M(\xi) = \begin{bmatrix} M_1(\xi) & M_{12}(\xi) \\ M_{12}'(\xi) & M_2(\xi) \end{bmatrix}$$

(where the matrices $M_1(\xi)$, $M_2(\xi)$ and $M_{12}(\xi)$ are of sizes $s \times s$, $(k-s) \times (k-s)$, and $s \times (k-s)$, respectively), yields (1.6a) as the Schur complement

$$C_{K_s}(\xi) = M_1(\xi) - M_{12}(\xi)\, M_2^-(\xi)\, M_{12}'(\xi).$$

For general K, but under the assumption that $c'\theta$ is linearly estimable under ξ for all $c \in$ range(K'), i.e., range$(K') \subset$ range$(M(\xi))$, we may write more explicitly

$$C_K(\xi) = \left(K\, M^-(\xi)\, K'\right)^{-1}, \tag{1.6b}$$

in which case $C_K(\xi)$ is positive definite.

Now, the refinement of (1.5) (for general K, ξ, and η) is, cf. Gaffke (1987, Section 3),

$$C_K(\xi) \leqslant C_K(\eta) \iff V(\xi, c) \geqslant V(\eta, c) \quad \text{for all } c \in \text{range}(K'). \tag{1.7}$$

The Loewner partial ordering for moment or reduced information matrices, though statistically of fundamental importance, does not suffice for selecting a single 'optimal' design. For a large set of designs the associated moment or reduced information matrices are not comparable in that ordering (unless $k = 1$ or $s = 1$). A popular way out is to specify a real-valued optimality criterion, defined as a function of the moment matrices of the designs. An optimal design is one whose moment matrix minimizes the criterion over the set of competing moment matrices (or designs). By now, the following seems to cover all statistically meaningful criteria.

DEFINITION 1.1. A function Φ: $\mathcal{A} \to \mathbb{R}$ is called an optimality criterion iff
(i) $\mathcal{A}$ is a convex cone in $\mathrm{Sym}(k)$, such that $\mathrm{PD}(k) \subset \mathcal{A} \subset \mathrm{NND}(k)$ (where $\mathrm{PD}(k)$ denotes the set of all real symmetric, positive definite $k \times k$ matrices);
(ii) Φ is antitonic w.r.t. the Loewner partial ordering, i.e., $A, B \in \mathcal{A}$, $A \leqslant B$, imply $\Phi(A) \geqslant \Phi(B)$;
(iii) Φ is convex.

The domain $\mathcal{A}$ in Definition 1.1 is often referred to as the feasibility cone of the design problem. A design ξ is feasible iff it allows unbiased linear estimation of all mean parameters of interest, i.e., iff $M(\xi) \in \mathcal{A}$. In most cases the full mean parameter vector θ is of interest, whence $\mathcal{A} = \mathrm{PD}(k)$, and the feasible designs are those with positive definite moment matrices. If the parameters of interest build a proper linear subsystem, represented by $K\theta$ with some given $s \times k$ matrix K of rank s, then

$$\mathcal{A} = \mathcal{A}(K) = \{A \in \mathrm{NND}(k)\colon\ \mathrm{range}(K') \subset \mathrm{range}(A)\}, \tag{1.8a}$$

and the feasible designs are those under which $K\theta$ is estimable (or, equivalently, those with $M(\xi) \in \mathcal{A}(K)$). In this case, the linear subsystem is usually reflected also by a particular form of the optimality criterion Φ, namely

$$\Phi(A) = \phi(C_K(A)), \quad A \in \mathcal{A}(K), \tag{1.8b}$$

where $C_K(A)$ is defined similarly to (1.6a),

$$C_K(A) = \min_{L\colon LK' = I_s} LAL' \quad (A \in \mathrm{NND}(k)); \tag{1.9a}$$

on $\mathcal{A}(K)$ we may write more explicitly, analogously to (1.6b),

$$C_K(A) = (KA^- K')^{-1} \quad (A \in \mathcal{A}(K)). \tag{1.9b}$$

The function ϕ on the right hand side of (1.8b) has to satisfy
(i′) ϕ: $\mathrm{PD}(s) \to \mathbb{R}$;
(ii′) ϕ is antitonic w.r.t. the Loewner partial ordering on $\mathrm{PD}(s)$;
(iii′) ϕ is convex.

In fact, properties (i′)–(iii′) of ϕ imply that (1.8b) defines an optimality criterion Φ. This follows immediately using formula (1.9a), by which

$$\Phi(A) = \max_{L\colon LK' = I_s} \phi(LAL') \quad \text{for all } A \in \mathcal{A}(K).$$

Given an optimality criterion according to Definition 1.1, the optimal design problem for regression model (1.1a) is to

$$\text{minimize } \Phi(M(\xi)) \text{ over all designs } \xi \text{ with } M(\xi) \in \mathcal{A}. \tag{1.10}$$

Considering the moment matrix (rather than the design) as a variable, and introducing the set $\mathcal{M}$ of moment matrices $M(\xi)$ when ξ ranges over the set of all designs, problem (1.10) rewrites as

$$\text{minimize } \Phi(M) \text{ over } M \in \mathcal{M} \cap \mathcal{A}. \tag{1.10a}$$

This is a convex minimization problem, since $\mathcal{M}$ is a convex subset of $\mathrm{Sym}(k)$. Actually, as it is easy to see, we have

$$\mathcal{M} = \mathrm{Conv}\{f(x)f(x)'\colon\ x \in \mathcal{X}\}, \tag{1.11}$$

where $\mathrm{Conv}\, S$ denotes the convex hull of a subset S in a linear space. Moreover, if $f(\mathcal{X})$ (the range of f) is compact – as it is usually true – then by (1.11) the set $\mathcal{M}$ is compact.

An important class of optimality criteria are orthogonally invariant criteria on $\mathrm{PD}(k)$ (i.e., criteria based on the eigenvalues of a positive definite moment matrix). These can be constructed from the following result.

LEMMA 1.2. *Let ψ be a real-valued function on $(0,\infty)^k$, such that ψ is convex, permutationally invariant and antitonic w.r.t. the componentwise partial ordering of vectors in $(0,\infty)^k$. Define Φ by*

$$\Phi(A) := \psi(\lambda(A)), \quad \textit{for all } A \in \mathrm{PD}(k),$$

where $\lambda(A) = (\lambda_1(A), \ldots, \lambda_k(A))'$ denotes the vector of eigenvalues of A arranged in ascending order, $\lambda_1(A) \leqslant \cdots \leqslant \lambda_k(A)$. Then Φ is an optimality criterion on $\mathcal{A} = \mathrm{PD}(k)$.

PROOF. If $A, B \in \mathrm{PD}(k)$ with $A \leqslant B$ then $\lambda_i(A) \leqslant \lambda_i(B)$, $i = 1, \ldots, k$, see, e.g., Marshall and Olkin (1979, p. 475), and antitonicity of ψ immediately implies $\Phi(A) \geqslant \Phi(B)$.

Let $A, B \in \mathrm{PD}(k)$ and $0 < \alpha < 1$. Due to a result of Fan, see Marshall and Olkin (1979, p. 241), $\lambda := \lambda(\alpha A + (1-\alpha)B)$ is Schur majorized by $\mu := \alpha\lambda(A) + (1-\alpha)\lambda(B)$, and therefore there exist an $m \in \mathbb{N}$, constants $\beta_1, \ldots, \beta_m > 0$ with $\sum_{i=1}^m \beta_i = 1$ and $k \times k$ permutation matrices $P_1, \ldots, P_m$, such that

$$\lambda = \sum_{i=1}^m \beta_i P_i \mu,$$

see Marshall and Olkin (1979, pp. 7–8). Hence, from the convexity and permutational invariance of ψ we get

$$\Phi(\alpha\, A + (1-\alpha)\, B) = \psi(\lambda) = \psi\left(\sum_{i=1}^m \beta_i P_i \mu\right) \leqslant \sum_{i=1}^m \beta_i \psi(P_i \mu) = \psi(\mu)$$

$$= \psi\big(\alpha\lambda(A) + (1-\alpha)\lambda(B)\big)$$
$$\leqslant \alpha\,\psi\big(\lambda(A)\big) + (1-\alpha)\,\psi\big(\lambda(B)\big)$$
$$= \alpha\,\Phi(A) + (1-\alpha)\,\Phi(B),$$

proving convexity of Φ. □

Examples for orthogonally invariant optimality criteria are Kiefer's Φ_p-criteria, $-\infty \leqslant p \leqslant 1$, defined on PD$(k)$ by

$$\Phi_p(A) = \left(\frac{1}{k}\sum_{j=1}^{k}\lambda_j(A)^p\right)^{-1/p}, \quad \text{if } p \neq -\infty, 0,$$
$$\Phi_0(A) = \left(\prod_{j=1}^{k}\lambda_j(A)\right)^{-1/k} = \big(\det(A)\big)^{-1/k}, \tag{1.12}$$
$$\Phi_{-\infty}(A) = \Big(\min_{j=1,\dots,k}\lambda_j(A)\Big)^{-1}.$$

Based on Lemma 1.2 it is not difficult to check that Φ_p, $-\infty \leqslant p \leqslant 1$, is in fact an optimality criterion. The famous D-, A-, and E-criterion are among the Φ_p-criteria, (given by $p = 0$, $p = -1$, and $p = -\infty$, respectively).

Another popular (non-orthogonally invariant) criterion is the I-criterion (on PD(k)),

$$\Phi_{\mathrm{I}}(A) := \int_{\mathcal{X}} f(x)'A^{-1}f(x)\,\mathrm{d}x \,/\, \mathrm{vol}(\mathcal{X}). \tag{1.13}$$

Here $\mathcal{X}$ is supposed to be a compact set with nonempty interior in some $\mathbb{R}^v$, and vol$(\mathcal{X})$ denotes the volume of $\mathcal{X}$. Obviously, we may also write

$$\Phi_{\mathrm{I}}(A) = \mathrm{tr}\big(M_0A^{-1}\big), \quad A \in \mathrm{PD}(k), \tag{1.13a}$$

where $M_0 = \int_{\mathcal{X}} f(x)f(x)'\,\mathrm{d}x \,/\, \mathrm{vol}(\mathcal{X})$ is the moment matrix of the uniform distribution on $\mathcal{X}$; note that this distribution is not a design. Due to the matrix-convexity and antitonicity of matrix inversion, see, e.g., Gaffke and Krafft (1982), and by (1.13a), Φ_{I} fulfills the requirements of an optimality criterion from Definition 1.1.

The Φ_p- and Φ_{I}-criteria (and the corresponding versions for subsystems of parameters $K\theta$ from (1.8b) with $\phi = \phi_p$ and $\phi = \phi_{\mathrm{I}}$ defined on PD(s) analogously to (1.12) and (1.13a), respectively) have the additional properties that they are positive and positively homogeneous of degree -1, i.e., $\Phi(\alpha A) = \alpha^{-1}\Phi(A)$ for all $A \in \mathcal{A}$ and all $\alpha \in (0,\infty)$. In other words, they are related to Pukelsheim's notion of *information functions* as $\Psi = 1/\Phi$ is an information function, i.e., Ψ is positive, positively homogeneous of degree 1, and concave. The class of optimality criteria corresponding in this way to information functions seems to be rich enough to cover all meaningful criteria. In particular, due to their positivity and positive homogeneity (of degree -1), it is reasonable to define the *relative efficiency* of one design ξ w.r.t. another design ξ^*,

$$\mathrm{eff}(\xi : \xi^*) = \Phi\big(M(\xi^*)\big)/\Phi\big(M(\xi)\big), \tag{1.14}$$

where both, $M(\xi)$ and $M(\xi^*)$ are assumed to belong to the domain of Φ. The ratio (1.14) can be interpreted as a ratio of sample sizes $n(\xi^*)/n(\xi)$ such that the Fisher information matrices $n(\xi)\,\sigma^{-2}\,M(\xi)$ and $n(\xi^*)\,\sigma^{-2}\,M(\xi^*)$ yield the same quality under Φ (provided ξ and ξ^* are exact designs of size $n(\xi)$ and $n(\xi^*)$, respectively). If ξ^* is optimal, i.e., if ξ^* is an optimal solution to (1.10), then

$$\mathrm{eff}(\xi : \xi^*) = \mathrm{eff}(\xi), \quad \text{say}, \tag{1.15}$$

will be referred to as *the efficiency* of ξ, which is contained in the interval $(0, 1]$.

2. Invariance of designs

Many of the commonly used linear regression models enjoy symmetry properties, due to some symmetric shape of the experimental region and a compatibility of the response function. We will use the term 'equivariance' of regression (1.1a) for this situation, formally defined as follows.

DEFINITION 2.1. Consider a linear regression model (1.1a), $y(x) = \theta' f(x)$, $x \in \mathcal{X}$, $\theta \in \mathbb{R}^k$. Let $\mathcal{G}$ be a group of one-to-one transformations from $\mathcal{X}$ onto $\mathcal{X}$, and $\mathcal{Q}$ be a group of real $k \times k$ matrices (w.r.t. ordinary matrix multiplication). Then, the regression model is said to be equivariant (w.r.t. $\mathcal{G}$ and $\mathcal{Q}$) iff there exists a surjective mapping $g \to Q_g$ from $\mathcal{G}$ onto $\mathcal{Q}$ such that

$$f\big(g(x)\big) = Q_g\, f(x) \quad \text{for all } x \in \mathcal{X} \text{ and all } g \in \mathcal{G}.$$

When checking equivariance of a model, often a transformation group $\mathcal{G}$ and a corresponding *set* $\mathcal{Q} \subset \mathbb{R}^{k\times k}$ arise quite naturally ($\mathbb{R}^{k\times k}$ denotes the set of real $k \times k$ matrices), while the group property of $\mathcal{Q}$ is somewhat tedious to prove explicitly. However, as the following lemma shows, this is dispensable, at least if $f(\mathcal{X})$ spans $\mathbb{R}^k$, equivalently, if the components $f_1, \ldots, f_k$ of f are linearly independent on $\mathcal{X}$.

LEMMA 2.2. *Consider a linear regression model* (1.1a). *Let $\mathcal{G}$ be a group of one-to-one transformations on $\mathcal{X}$, such that for each $g \in \mathcal{G}$ there exists an $Q_g \in \mathbb{R}^{k\times k}$ with*

$$f\big(g(x)\big) = Q_g f(x) \quad \textit{for all } x \in \mathcal{X}.$$

(a) *If $f(\mathcal{X})$ spans $\mathbb{R}^k$, then $\mathcal{Q} := \{Q_g\colon\ g \in \mathcal{G}\}$ is a matrix group, and hence the regression model is equivariant* (*w.r.t. $\mathcal{G}$ and $\mathcal{Q}$*).

(b) *In the general case, denote by P the orthogonal projection matrix from $\mathbb{R}^k$ onto* $\operatorname{span}(f(\mathcal{X}))$, *and $\bar{P} := I_k - P$. Then $\mathcal{Q} := \{Q_g P + \bar{P}\colon\ g \in \mathcal{G}\}$ is a matrix group, and the regression model is equivariant* (*w.r.t. $\mathcal{G}$ and $\mathcal{Q}$*).

PROOF. Let $\mathcal{Z} := \operatorname{span}(f(\mathcal{X}))$. Since any $g \in \mathcal{G}$ is a bijection on $\mathcal{X}$ we obtain

$$\{Q_g z\colon\ z \in \mathcal{Z}\} = \mathcal{Z}. \tag{2.1}$$

Also we find for $g, h \in \mathcal{G}$ (with $h \circ g$ denoting the composition of g and h),

$$Q_{h\circ g} z = Q_h Q_g z \quad \text{for all } z \in \mathcal{Z}. \tag{2.2}$$

Let, for $g \in \mathcal{G}$,

$$Q_g^* := Q_g P + \bar{P}.$$

Since $Q_g^* z = Q_g z,\ z \in \mathcal{Z}$, the matrix Q_g^* again satisfies the equivariance condition

$$f\big(g(x)\big) = Q_g^* f(x) \quad \text{for all } x \in \mathcal{X}.$$

By (2.1) each Q_g^* is nonsingular, and together with (2.2)

$$\begin{aligned} Q_h^* Q_g^* &= \big(Q_h P + \bar{P}\big)\big(Q_g P + \bar{P}\big) = Q_h Q_g P + \bar{P} \\ &= Q_{h\circ g} P + \bar{P} = Q_{h\circ g}^* \quad \text{for all } g, h \in \mathcal{G}. \end{aligned}$$

From this, and observing that the identity $\mathrm{id}_{\mathcal{X}} \in \mathcal{G}$ yields $Q_{\mathrm{id}_{\mathcal{X}}} z = z$ for all $z \in \mathcal{Z}$, thus $Q_{\mathrm{id}_{\mathcal{X}}}^* = I_k$, the set $\{Q_g^*\colon\ g \in \mathcal{G}\}$ forms a matrix group, proving part (b) of the lemma. If $f(\mathcal{X})$ spans $\mathbb{R}^k$, then clearly $Q_g^* = Q_g$ for all $g \in \mathcal{G}$, and assertion (a) follows. □

Of particular importance will be equivariant regression setups with *compact* matrix groups $\mathcal{Q}$. Compactness of $\mathcal{Q}$ implies that it is a subgroup of the group $\mathcal{Q}_{\mathrm{um}}$ of *unimodular* matrices in $\mathbb{R}^{k\times k}$ (i.e., matrices A with $|\det(A)| = 1$). For, suppose $|\det(Q_0)| \neq 1$ for some $Q_0 \in \mathcal{Q}$; since the group property of $\mathcal{Q}$ ensures $Q_0^{-1} \in \mathcal{Q}$, we may assume $|\det(Q_0)| > 1$, and from $Q_0^n \in \mathcal{Q}$ for all $n \in \mathbb{N}$ we get $\sup\{|\det(Q)|\colon Q \in \mathcal{Q}\} = \infty$, contradicting compactness of $\mathcal{Q}$. A compact group $\mathcal{Q}$ will allow, at least theoretically, to construct invariant matrices or designs by taking averages of matrices QCQ' over $Q \in \mathcal{Q}$ (where C is a given matrix). Unless the matrix group is finite, and thereby averaging is an elementary operation, in general building averages over some compact group calls for the Haar probability measure. However, our approach will avoid the Haar probability measure in the present context. To this end we will prove an auxiliary result, which shows that any compact matrix group in $\mathbb{R}^{k\times k}$ may be viewed as an orthogonal group, provided the underlying scalar product is suitable defined.

LEMMA 2.3. *Let $\mathcal{Q}$ be a compact group of real $k \times k$ matrices. Then there exists an $E \in \mathrm{PD}(k)$ such that $Q'EQ = E$ for all $Q \in \mathcal{Q}$.*

PROOF. Consider the convex hull (in $\mathrm{Sym}(k)$)

$$\mathcal{C} := \mathrm{Conv}\{(QQ')^{-1}\colon\ Q \in \mathcal{Q}\},$$

and the extremum problem

$$\text{minimize } \Phi_0(C) \text{ over } C \in \mathcal{C}, \tag{2.3}$$

where Φ_0 denotes the D-criterion from (1.12),

$$\Phi_0(A) = \big(\det(A)\big)^{-1/k}, \quad A \in \mathrm{PD}(k).$$

Since Φ_0 is continuous and strictly convex, and $\mathcal{C}$ is a convex and compact subset of $\mathrm{PD}(k)$, there exists a unique optimal solution C^* to problem (2.3). Moreover, since $\mathcal{Q}$ consists of unimodular matrices only, Φ_0 possesses the invariance property

$$\Phi_0(Q'AQ) = \Phi_0(A) \quad \text{for all } A \in \mathrm{PD}(k) \text{ and all } Q \in \mathcal{Q}.$$

The set $\mathcal{C}$ is invariant under the same transformations, i.e., $Q'CQ \in \mathcal{C}$ for all $C \in \mathcal{C}$ and $Q \in \mathcal{Q}$. To see this, it suffices to check the corresponding invariance property for the generator set of $\mathcal{C}$, which is easily done: For $P, Q \in \mathcal{Q}$ we have $P'(QQ')^{-1}P = ((P^{-1}Q)(P^{-1}Q)')^{-1}$ and $P^{-1}Q \in \mathcal{Q}$. Hence, for any $Q \in \mathcal{Q}$, the matrix $Q'C^*Q$ solves problem (2.3). Observing that the optimal solution C^* is unique, it follows that $E := C^*$ satisfies the conditions stated in the lemma. □

THEOREM 2.4. *Let $\mathcal{Q}$ be a compact group of real $k \times k$ matrices. Then, for any $A \in \mathbb{R}^{k \times k}$ there exists a unique matrix $\bar{A} \in \mathrm{Conv}\{QAQ'\colon\ Q \in \mathcal{Q}\}$ such that*

$$Q\bar{A}Q' = \bar{A} \quad \textit{for all } Q \in \mathcal{Q}.$$

(We call $\bar{A}$ the average of QAQ' over $Q \in \mathcal{Q}$.) Moreover, the mapping $\mathcal{P}(A) := \bar{A}$, $A \in \mathbb{R}^{k \times k}$, is a linear projection operator from $\mathbb{R}^{k \times k}$ onto the subspace $\mathcal{L} := \{B \in \mathbb{R}^{k \times k}\colon\ QBQ' = B$ for all $Q \in \mathcal{Q}\}$ of invariant matrices.

PROOF. Let $E \in \mathrm{PD}(k)$ be a matrix as given by Lemma 2.3. Define

$$\langle B, C\rangle_E := \mathrm{tr}(BEC'E) \quad (B, C \in \mathbb{R}^{k \times k}).$$

This is a scalar product on $\mathbb{R}^{k \times k}$, as it is obviously bilinear, and it is strictly definite: For all $B \in \mathbb{R}^{k \times k}$ we have $\langle B, B\rangle_E = \mathrm{tr}(BEB'E) \geqslant 0$, and since E is positive definite, $\langle B, B\rangle_E = 0$ implies $BEB' = 0$, hence $B = 0$. Thus, the space $\mathbb{R}^{k\times k}$ endowed with the scalar product $\langle\cdot,\cdot\rangle_E$ is a Hilbert space. Denote the associated norm by $\|B\|_E := \langle B, B\rangle_E^{1/2}$, $B \in \mathbb{R}^{k\times k}$. The scalar product, and hence the norm, enjoys the invariance property that

$$\langle QBQ', QCQ'\rangle_E = \langle B, C\rangle_E \quad \text{for all } B, C \in \mathbb{R}^{k\times k} \text{ and all } Q \in \mathcal{Q},$$

since the left hand side of this equation equals

$$\mathrm{tr}\big(QBQ'E(QCQ')'E\big) = \mathrm{tr}\big(B\underbrace{Q'EQ}_{=E}C'\underbrace{Q'EQ}_{=E}\big).$$

The convex hull $\mathcal{C} := \mathrm{Conv}\{QAQ' \colon Q \in \mathcal{Q}\}$ is compact by compactness of $\mathcal{Q}$; in particular, $\mathcal{C}$ is a convex and closed subset of the Hilbert space. Hence, as it is well known, $\mathcal{C}$ contains a unique point with minimum norm, C^*, say. Observing that $\mathcal{C}$ satisfies $QCQ' \in \mathcal{C}$ for all $C \in \mathcal{C}$ and all $Q \in \mathcal{Q}$, invariance of the norm yields $QC^*Q' = C^*$ for all $Q \in \mathcal{Q}$. Therefore, $\bar{A} := C^*$ is a matrix as desired.

For proving uniqueness of $\bar{A}$ it suffices to show that any $\bar{A} \in \mathbb{R}^{k\times k}$ satisfying the conditions of the theorem fulfills

$$\langle A - \bar{A}, B\rangle_E = 0 \quad \text{for all } B \in \mathcal{L},$$

i.e., $\bar{A}$ is the (unique) orthogonal projection of A onto the subspace $\mathcal{L}$ in the Hilbert space $(\mathbb{R}^{k\times k}, \langle\cdot,\cdot\rangle_E)$. In fact, from

$$\bar{A} = \sum_{i=1}^{r} \alpha_i Q_i A Q_i'$$

for some $r \in \mathbb{N}$, $\alpha_1, \ldots, \alpha_r > 0$ with $\sum_{i=1}^{r} \alpha_i = 1$, and $Q_1, \ldots, Q_r \in \mathcal{Q}$, we obtain

$$\begin{aligned}
\langle \bar{A}, B\rangle_E &= \sum_{i=1}^{r} \alpha_i \,\mathrm{tr}(Q_i A Q_i' E B' E) = \sum_{i=1}^{r} \alpha_i \,\mathrm{tr}\big(AQ_i'E(Q_iBQ_i')'EQ_i\big) \\
&= \sum_{i=1}^{r} \alpha_i \,\mathrm{tr}(AQ_i'EQ_iB'Q_i'EQ_i) = \sum_{i=1}^{r} \alpha_i \,\mathrm{tr}(AEB'E) \\
&= \langle A, B\rangle_E \quad \text{for all } B \in \mathcal{L}.
\end{aligned}$$

This also shows that the mapping $\mathcal{P}$ is a linear projection operator from $\mathcal{H}$ onto $\mathcal{L}$, namely the orthogonal projection operator under the scalar product $\langle\cdot,\cdot\rangle_E$. □

For the case that the matrix group $\mathcal{Q}$ consists of orthogonal matrices only (and hence $E = I_k$), the projection property of the average $\bar{A}$ was observed in Pukelsheim (1993, p. 349).

Any transformation g from $\mathcal{X}$ onto $\mathcal{X}$ induces a transformation of designs ξ by

$$\xi^g := \begin{pmatrix} g(x_1) & \cdots & g(x_r) \\ \xi(x_1) & \cdots & \xi(x_r) \end{pmatrix} \quad \text{for } \xi = \begin{pmatrix} x_1 & \cdots & x_r \\ \xi(x_1) & \cdots & \xi(x_r) \end{pmatrix}. \tag{2.4}$$

Given an equivariant linear regression model, the moment matrix of $M(\xi^g)$ is *linearly* related to $M(\xi)$, since we immediately obtain from (1.4), (2.4), and Definition 2.1

$$M(\xi^g) = Q_g M(\xi) Q_g' \quad \text{for all designs } \xi \text{ and all } g \in \mathcal{G}. \tag{2.5}$$

DEFINITION 2.5. Given an equivariant linear regression model (w.r.t. groups $\mathcal{G}$ and $\mathcal{Q}$), a design ξ is said to be invariant iff $\xi^g = \xi$ for all $g \in \mathcal{G}$; ξ is called weakly invariant iff $M(\xi^g) = M(\xi)$ for all $g \in \mathcal{G}$.

Weak invariance of a design (in an equivariant regression model) just means invariance of its moment matrix under the linear transformations on $\mathrm{Sym}(k)$ suggested by (2.5),

$$A \to QAQ', \qquad A \in \mathrm{Sym}(k) \quad (Q \in \mathcal{Q}). \tag{2.6}$$

Of course, for this situation the terminology 'design with invariant moment matrix' would be more adequate. However, we would like to have a short notation, additionally emphasizing that invariance of a design is in fact a stronger property. The set of invariant designs may be much smaller than that of weakly invariant designs, as, e.g., for multiple polynomial models on symmetric regions (see Example 2.8, below).

LEMMA 2.6. *Consider an equivariant linear regression model* (1.1a) (*w.r.t. groups $\mathcal{G}$ and $\mathcal{Q}$*), *and let $\mathcal{Q}$ be compact. Then, for any design ξ there exists a weakly invariant design $\tilde{\xi}$, such that*

$$M(\tilde{\xi}) \in \mathrm{Conv}\{M(\xi^g)\colon\ g \in \mathcal{G}\}.$$

If $\mathcal{G}$ is finite, then for any weakly invariant design $\tilde{\xi}$ there exists an invariant design $\bar{\xi}$ with $M(\bar{\xi}) = M(\tilde{\xi})$ and $\mathrm{supp}(\bar{\xi}) \subset \mathrm{supp}(\tilde{\xi})$.

PROOF. By (2.5), given a design ξ, we have

$$\mathrm{Conv}\{M(\xi^g)\colon\ g \in \mathcal{G}\} = \mathrm{Conv}\{QM(\xi)Q'\colon\ Q \in \mathcal{Q}\}.$$

Since the moment matrices of all designs form a convex set, the average $\overline{M}$ of $QM(\xi)Q'$ over $Q \in \mathcal{Q}$ from Theorem 2.4 is the moment matrix of some design $\tilde{\xi}$, which is trivially weakly invariant. If $\mathcal{G}$ is finite, then the average of the probability distributions $\tilde{\xi}^g$ over $g \in \mathcal{G}$,

$$\bar{\xi} := \frac{1}{\#\mathcal{G}} \sum_{g \in \mathcal{G}} \tilde{\xi}^g,$$

is again a probability distribution with finite support, i.e., a design. Obviously, $\bar{\xi}$ is invariant with $\mathrm{supp}(\tilde{\xi}) \subset \mathrm{supp}(\bar{\xi})$, and

$$M(\bar{\xi}) = \frac{1}{\#\mathcal{G}} \sum_{g \in \mathcal{G}} M(\tilde{\xi}^g) = \frac{1}{\#\mathcal{G}} \sum_{g \in \mathcal{G}} M(\tilde{\xi}) = M(\tilde{\xi}).$$

□

Let us consider some examples of equivariant linear regression models. For the trivial groups $\mathcal{G}$ and $\mathcal{Q}$ consisting only of the identity on $\mathcal{X}$ and the unit matrix I_k, any linear regression setup is equivariant (w.r.t. these trivial groups). Nontrivial examples are multiple polynomial regression models on symmetric regions.

EXAMPLE 2.7. Consider a dth degree multiple polynomial regression setup on some experimental region $\mathcal{X} \subset \mathbb{R}^v$,

$$y(x) = \sum_{\alpha \in \mathrm{A}} \theta_\alpha x^\alpha, \quad x = (x_1, \ldots, x_v)' \in \mathcal{X}, \tag{2.7}$$

where $\alpha = (\alpha_1, \ldots, \alpha_v)$ is a v-dimensional multi-index with nonnegative integer components, A is a given nonempty set of v-dimensional multi-indices, such that $|\alpha| := \sum_{i=1}^v \alpha_i \leqslant d$ for all $\alpha \in \mathrm{A}$, and $|\alpha^*| = d$ for at least one $\alpha^* \in \mathrm{A}$. The power x^α is to be understood as $\prod_{i=1}^v x_i^{\alpha_i}$, with the convention $x_i^0 = 1$.

Note that by (2.7) we admit proper submodels of the full polynomial setup of order d, the latter being the case $\mathrm{A} = \mathrm{A}_d$, where

$$\mathrm{A}_d := \{\alpha \in \mathbb{N}_0^v\colon\ |\alpha| \leqslant d\}.$$

The transformation groups $\mathcal{G}$ considered below will act linearly on $\mathcal{X}$, i.e., for all $g \in \mathcal{G}$ there exists a nonsingular real $v \times v$ matrix U_g with

$$\begin{aligned} &g(x) = U_g x \quad \text{for all } x \in \mathcal{X} \\ &(\text{and in particular } U_g x \in \mathcal{X} \text{ for all } x \in \mathcal{X}). \end{aligned} \tag{2.8}$$

As we will demonstrate next, under (2.8) the polynomial model (2.7) is equivariant (w.r.t. $\mathcal{G}$ and an appropriate matrix group $\mathcal{Q}$), iff condition (2.10), below, is satisfied. For each $\alpha \in \mathrm{A}$ and $\beta \in \mathrm{A}_d$ with $|\beta| = |\alpha|$ denote by $\Gamma(\alpha, \beta)$ the set of $v \times v$ matrices $\gamma = (\gamma_{ij})$ with $\gamma_{ij} \in \mathbb{N}_0$, row-sums $\gamma_{i\cdot} = \alpha_i$, and column-sums $\gamma_{\cdot j} = \beta_j$, for all $i, j = 1, \ldots, v$. Denote, for $g \in \mathcal{G}$, the entries of U_g by $u_{g,ij}$, $i, j = 1, \ldots, v$, and, for $\alpha \in \mathrm{A}$ and $\beta \in \mathrm{A}_d$,

$$q_{g,\alpha\beta} := \begin{cases} \left(\prod_{i=1}^v \alpha_i!\right) \sum_{\gamma \in \Gamma(\alpha,\beta)} \left(\prod_{i,j=1}^v \dfrac{(u_{g,ij})^{\gamma_{ij}}}{\gamma_{ij}!}\right), & \text{if } |\alpha| = |\beta|, \\ 0, & \text{otherwise.} \end{cases} \tag{2.9}$$

Then, a necessary and sufficient condition for equivariance of regression (2.7) is

$$q_{g,\alpha\beta} = 0 \quad \text{for all } g \in \mathcal{G}, \text{ all } \alpha \in \mathrm{A}, \\ \text{and all } \beta \in \mathrm{A}_d \setminus \mathrm{A} \text{ with } |\beta| = |\alpha|. \tag{2.10}$$

To see this, consider how the monomials x^α, $\alpha \in \mathrm{A}$ (i.e., the components of $f(x)$), transform under the mappings $x \to g(x)$. Using the multinomial formula, we get

$$\begin{aligned}(U_g x)^\alpha &= \prod_{i=1}^{v}\left(\sum_{j=1}^{v} u_{g,ij}\, x_j\right)^{\alpha_i} \\ &= \prod_{i=1}^{v}\left(\sum_{\substack{\gamma_{i1},\dots,\gamma_{iv}\in N_0\\ \gamma_{i1}+\cdots+\gamma_{iv}=\alpha_i}} \frac{\alpha_i!}{\gamma_{i1}!\gamma_{i2}!\cdots\gamma_{iv}!}\prod_{j=1}^{v}(u_{g,ij})^{\gamma_{ij}}\, x_j{}^{\gamma_{ij}}\right) \\ &= \sum_{\gamma\in\Gamma(\alpha)}\left(\left(\prod_{i=1}^{v}\alpha_i!\right)\left(\prod_{j=1}^{v} x_j{}^{\gamma_{\cdot j}}\right)\left(\prod_{i,j=1}^{v}\frac{(u_{g,ij})^{\gamma_{ij}}}{\gamma_{ij}!}\right)\right),\end{aligned}$$

where $\Gamma(\alpha)$ denotes the set of $v \times v$ matrices $\gamma = (\gamma_{ij})$ with nonnegative integer entries and row-sums $\gamma_{i\cdot} = \alpha_i$, $i = 1,\dots,v$. Thus, rearranging the terms in the last sum according to the column-sums $\beta_j = \gamma_{\cdot j}$, $j = 1,\dots,v$, of $\gamma \in \Gamma(\alpha)$, we obtain

$$(U_g x)^\alpha = \sum_{\substack{\beta\in\mathrm{A}_d\\ |\beta|=|\alpha|}} q_{g,\alpha\beta}\, x^\beta, \tag{2.11}$$

with $q_{g,\alpha\beta}$ defined in (2.9). Hence, equivariance of regression (2.7) is equivalent to (2.10), in which case $Q_g = (q_{g,\alpha\beta})_{\alpha,\beta\in\mathrm{A}}$ for each $g \in \mathcal{G}$. Obviously, the matrices Q_g depend continuously on U_g. Hence, if the set $\{U_g\colon\ g \in \mathcal{G}\}$ is compact, then so is $\mathcal{Q} = \{Q_g\colon\ g \in \mathcal{G}\}$.

EXAMPLE 2.8. Let us consider more specific cases of equivariant polynomial models (2.7).

(a) Let $\mathcal{G} = \mathcal{G}_{\mathrm{s}}$ be the group of sign changes of components of v-dimensional vectors, i.e.,

$$\varepsilon\, z := (\varepsilon_1 z_1,\dots,\varepsilon_v z_v)', \quad z = (z_1,\dots,z_v)' \in \mathbb{R}^v,$$

with sign vectors $\varepsilon = (\varepsilon_1,\dots,\varepsilon_v)$, $\varepsilon_i = \pm 1$ for all i. Then, if the experimental region $\mathcal{X}$ is sign invariant, i.e., if $\varepsilon\, x \in \mathcal{X}$ for all $x \in \mathcal{X}$ and all sign-vectors ε, setup (2.7) is equivariant, without any assumption on the index set A. In fact, condition (2.10) holds true, since $U_\varepsilon = \mathrm{diag}(\varepsilon)$, and $q_{\varepsilon,\alpha\beta} = 0$ whenever $\beta \neq \alpha$. The matrices Q_ε are diagonal of order #A with diagonal entries $\prod_{i=1}^{v}\varepsilon_i^{\alpha_i} = \pm 1$, $\alpha \in \mathrm{A}$.

(b) Let $\mathcal{G} = \mathcal{G}_{\mathrm{p}}$ be the group of permutations acting on the components of v-dimensional vectors, i.e.,

$$\pi\, z := \big(z_{\pi(1)}, \ldots, z_{\pi(v)}\big)', \quad z = (z_1, \ldots, z_v)' \in \mathbb{R}^v,$$

where π permutates the indices $1, \ldots, v$. Then, if the region $\mathcal{X}$ and the index set A are permutationally invariant, i.e., if $\pi\, x \in \mathcal{X}$ for all $x \in \mathcal{X}$ and $\pi\, \alpha \in \mathrm{A}$ for all $\alpha \in \mathrm{A}$, the regression model is equivariant. For, condition (2.10) is satisfied, since U_π is the $v \times v$ permutation matrix given by π, and thus $q_{\pi,\alpha\beta} = 0$ whenever $\beta \neq \pi\alpha$. The matrices Q_π are special permutation matrices of order #A, corresponding to the permutations $\alpha \to \pi^{-1}\alpha$ of the multi-indices α.

(c) Let $\mathcal{G} = \mathcal{G}_{\mathrm{sp}}$ be the group generated by $\mathcal{G}_{\mathrm{s}}$ and $\mathcal{G}_{\mathrm{p}}$ from parts (a) and (b), above. As is easily seen, the transformations from $\mathcal{G}_{\mathrm{sp}}$ are given by pairs (ε, π), acting on v-dimensional vectors by

$$(\varepsilon, \pi)\, z := \big(\varepsilon_1 z_{\pi(1)}, \ldots, \varepsilon_v z_{\pi(v)}\big)', \quad z = (z_1, \ldots, z_v)' \in \mathbb{R}^v,$$

where ε is a v-dimensional vector of signs ± 1, and π is a permutation of the indices $1, \ldots, v$. If the experimental region is invariant w.r.t. both, sign changes and permutations of components, i.e., if $(\varepsilon, \pi)\, x \in \mathcal{X}$ for all $x \in \mathcal{X}$ and all ε and π, and if the index set A is permutationally invariant as in (b), then the regression model is equivariant. Here the matrices $Q_{\varepsilon,\pi}$ are given by $Q_{\varepsilon,\pi} = Q_\varepsilon\, Q_\pi$, with Q_ε and Q_π as in (a) and (b), respectively.

(d) More specially, let the experimental region $\mathcal{X}$ be a ball in $\mathbb{R}^v$ centered at zero. Consider the group $\mathcal{G} = \mathcal{G}_{\mathrm{orth}}$ of orthogonal transformations on $\mathcal{X}$. Then obviously, (2.10) is equivalent to the following condition (2.12) on the index set A.

$$\begin{aligned} &\text{If } \alpha \in \mathrm{A}, \quad \text{then } \textit{all } \beta \in \mathrm{A}_d \\ &\qquad\qquad \text{of the same order } |\beta| = |\alpha| \text{ also belong to A.} \end{aligned} \tag{2.12}$$

So, condition (2.12) implies equivariance of the polynomial setup (2.7) w.r.t. $\mathcal{G}_{\mathrm{orth}}$ and the corresponding matrix group $\mathcal{Q}$ of Q_g's given as in Example 2.7 (with $\mathcal{G} = \mathcal{G}_{\mathrm{orth}}$), which we will also refer to as a *dth degree rotatable* polynomial model.

Following the widespread notation, we will call a design rotatable iff it is weakly invariant w.r.t. $\mathcal{G}_{\mathrm{orth}}$. We remark that contrary to the above examples (a)–(c) here the associated (compact) group $\mathcal{Q}$ does not longer consist of orthogonal matrices only. So many orthogonally invariant criteria are *not* invariant w.r.t. this matrix group in the sense of Definition 2.9, below.

The dth degree rotatable polynomial model is also an example for which the class of invariant designs is extremely poor, as the only invariant design is the one-point design at zero. Of course, this comes from the definition of a design as being a probability distribution with *finite* support, whereas the group $\mathcal{G}_{\mathrm{orth}}$ is infinite (though compact).

Equivariance of a linear regression model and invariance of designs can be utilized in problems of design optimality, if the optimality criterion under consideration enjoys a suitable invariance property.

DEFINITION 2.9. Let $\mathcal{Q}$ be a group of real $k \times k$ matrices. An optimality criterion Φ defined on $\mathcal{A}$ (see Definition 1.1) is said to be invariant (w.r.t. $\mathcal{Q}$) iff

$$QAQ' \in \mathcal{A} \quad \text{and} \quad \Phi(QAQ') = \Phi(A) \quad \text{for all } A \in \mathcal{A} \text{ and all } Q \in \mathcal{Q}.$$

For example, the Φ_p-criteria from (1.12) are orthogonally invariant, that is, $\Phi_p(QAQ') = \Phi_p(A)$ for all $A \in \mathrm{PD}(k)$ and all orthogonal $k \times k$ matrices Q. Hence, if the matrix group $\mathcal{Q}$ from an equivariant regression model consists entirely of orthogonal matrices, then the Φ_p-criteria are invariant (w.r.t. $\mathcal{Q}$). This applies in particular to parts (a)–(c) of Example 2.8. The D-criterion Φ_0 is even invariant w.r.t. the group $\mathcal{Q}_{\mathrm{um}}$ of unimodular matrices in $\mathbb{R}^{k \times k}$, and thus it is invariant w.r.t. the matrix group arising in a rotatable polynomial model in part (d) of Example (2.8). The I-criterion Φ_{I} from (1.13) is not necessarily invariant w.r.t. the full group $\mathcal{Q}_{\mathrm{um}}$, but in the rotatable dth degree multiple polynomial setup over a compact ball $\mathcal{X}$ centered at zero, it is invariant w.r.t. the matrix group $\mathcal{Q}$ associated with $\mathcal{G}_{\mathrm{orth}}$. For, by (1.13a) and by the invariance of the Lebesgue measure w.r.t. orthogonal transformations,

$$QM_0Q' = M_0 \qquad \text{for all } Q \in \mathcal{Q}, \tag{2.13}$$

and thus we get for all $A \in \mathrm{PD}(k)$ and all $Q \in \mathcal{Q}$

$$\Phi_{\mathrm{I}}(QAQ') = \mathrm{tr}\big(M_0(QAQ')^{-1}\big) = \mathrm{tr}\big(Q^{-1}M_0(Q^{-1})'A^{-1}\big) = \Phi_{\mathrm{I}}(A),$$

where the last equality follows from (2.13) (with Q replaced by Q^{-1}).

The following lemma allows, under an equivariant regression model with compact matrix group $\mathcal{Q}$ and an invariant criterion Φ, to restrict the search for an optimal design to the subclass of weakly invariant designs.

LEMMA 2.10. *Consider an equivariant linear regression model (w.r.t. groups $\mathcal{G}$ and $\mathcal{Q}$) with compact matrix group $\mathcal{Q}$, and let Φ be an invariant optimality criterion (w.r.t. $\mathcal{Q}$) defined on $\mathcal{A}$. Then, for any design ξ with $M(\xi) \in \mathcal{A}$ and any weakly invariant design $\tilde{\xi}$ with $M(\tilde{\xi}) \in \mathrm{Conv}\{M(\xi^g)\colon g \in \mathcal{G}\}$ from Lemma 2.6, we have $M(\tilde{\xi}) \in \mathcal{A}$ and $\Phi(M(\tilde{\xi})) \leqslant \Phi(M(\xi))$.*

PROOF. There are $r \in \mathbb{N}$, $g_1, \ldots, g_r \in \mathcal{G}$ and $\alpha_1, \ldots, \alpha_r > 0$ with $\sum_{i=1}^r \alpha_i = 1$, such that $M(\tilde{\xi}) = \sum_{i=1}^r \alpha_i M(\xi^{g_i})$. By (2.5), $M(\xi^g) = Q_g M(\xi) Q_g' \in \mathcal{A}$, and invariance of Φ gives $\Phi(M(\xi^g)) = \Phi(M(\xi))$ for all $g \in \mathcal{G}$. Consequently, $M(\tilde{\xi}) \in \mathcal{A}$, and convexity of Φ yields

$$\Phi\big(M(\tilde{\xi})\big) \leqslant \sum_{i=1}^r \alpha_i \Phi\big(M(\xi^{g_i})\big) = \Phi\big(M(\xi)\big).$$

□

3. Admissibility of designs

The Loewner partial ordering of moment matrices of designs, discussed in Section 1, implies a weak concept of design optimality, called admissibility, by selecting those designs whose moment matrices are maximal elements in the set of competing moment matrices w.r.t. that partial ordering. In general, however, the admissible designs form a large class, and thus one is far from obtaining a single 'optimal' design. Admissibility, also combined with invariance, is rather an aid for reducing the set of competing designs from which an optimal one is to be chosen.

DEFINITION 3.1. Under a linear regression model (1.1a) a design ξ_0 is said to be admissible iff there does not exist a design which is better than ξ_0. That is, ξ_0 is admissible iff there is no design ξ, such that

$$M(\xi_0) \leqslant M(\xi) \quad \text{and} \quad M(\xi_0) \neq M(\xi),$$

where $M(\xi_0)$, $M(\xi)$ are the moment matrices of ξ_0 and ξ, respectively.

Throughout we will combine admissibility *and* invariance of designs, and hence we assume an equivariant linear regression model as stated in Definition 2.1. The case of no equivariance properties is included by taking $\mathcal{G}$ and $\mathcal{Q}$ as the trivial groups consisting only of the identity mapping and the identity matrix, respectively. Also, for the majority of the results, we have to assume that the range $f(\mathcal{X})$ of the regression function f from model (1.1a) is compact, and hence the set $\mathcal{M}$ of all moment matrices of designs is compact (and convex).

Under compactness of $\mathcal{M}$ it can be shown that for any design ξ there exists an admissible design ξ_0 which is at least as good as ξ. If, moreover, the matrix group $\mathcal{Q}$ in an equivariant setup is compact, then a corresponding statement holds true when restricting to weakly invariant designs. If ξ is weakly invariant, then there exists an admissible and weakly invariant design ξ_0 with $M(\xi_0) \geqslant M(\xi)$, cf. Heiligers (1991, Lemma 1). Thus, the admissible (and weakly invariant) designs form a 'complete class' of (weakly invariant) designs.

As it is well known (see, e.g., Karlin and Studden, 1966, Theorem 7.2), admissibility of a design ξ_0 is a property only of its support, but not of its weights: If ξ_0 is admissible, then *any* other design supported by a subset of $\mathrm{supp}(\xi_0)$ is also admissible. In fact, the results below characterizing admissibility give necessary and sufficient conditions on the support of an admissible design.

The first result is due to Karlin and Studden (1966), Theorem 7.1, which we augment by invariance (see also Heiligers, 1991, Theorem 1):

THEOREM 3.2. *Consider an equivariant linear regression model* (1.1a) (*w.r.t. groups* $\mathcal{G}$ *and* $\mathcal{Q}$) *with compact matrix group* $\mathcal{Q}$, *and let* $f(\mathcal{X})$ *be compact. Denote, for* $A \in \mathrm{NND}(k)$,

$$q_A(x) := f(x)'Af(x), \quad x \in \mathcal{X}.$$

(a) *If* ξ_0 *is a weakly invariant and admissible design, then there exists a nonzero* $A \in \mathrm{NND}(k)$ *with* $Q'AQ = A$ *for all* $Q \in \mathcal{Q}$, *such that each support point of* ξ_0 *is a global maximum point of* $q_A(x)$ *over* $x \in \mathcal{X}$.

(b) *If* ξ_0 *is any design, and if there exists an* $A \in \mathrm{PD}(k)$ *such that each support point of* ξ_0 *is a global maximum point of* $q_A(x)$ *over* $x \in \mathcal{X}$, *then* ξ_0 *is admissible.*

PROOF. Apart from the invariance property of the matrix A in part (a), the results were proved by Karlin and Studden (1966, p. 808). So, in (a) let $A \in \mathrm{NND}(k)$, $A \neq 0$, such that each support point of ξ_0 maximizes $q_A(x)$ over $x \in \mathcal{X}$. Since $q_A(x) = \mathrm{tr}(Af(x)f(x)')$ and $\mathcal{M}$ is the convex hull of all matrices $f(x)f(x)'$, $x \in \mathcal{X}$, this is equivalent to

$$\mathrm{tr}(AM_0) = \max_{M \in \mathcal{M}} \mathrm{tr}(AM),$$

with $M_0 := M(\xi_0)$. By (2.5), $QMQ' \in \mathcal{M}$ for all $M \in \mathcal{M}$ and all $Q \in \mathcal{Q}$; moreover, weak invariance of ξ_0 ensures $QM_0Q' = M_0$ for all $Q \in \mathcal{Q}$. Hence it follows that

$$\mathrm{tr}(Q'AQM_0) \geqslant \mathrm{tr}(Q'AQM) \quad \text{for all } Q \in \mathcal{Q} \text{ and all } M \in \mathcal{M}. \tag{3.1}$$

Applying Theorem 2.4 to the compact matrix group $\mathcal{Q}' := \{Q' \colon Q \in \mathcal{Q}\}$ and the matrix A, we obtain the average $\bar{A}$ of $Q'AQ$ over $Q' \in \mathcal{Q}'$, and by (3.1)

$$\mathrm{tr}(\bar{A}M_0) \geqslant \mathrm{tr}(\bar{A}M) \quad \text{for all } M \in \mathcal{M}.$$

This means that each support point of ξ_0 maximizes $q_{\bar{A}}(x)$ over $x \in \mathcal{X}$. The proof is completed by noting that since A is nonzero and nonnegative definite, so is $Q'AQ$ for each $Q \in \mathcal{Q}$, and hence $\bar{A}$ has the same property. □

Unfortunately, the necessary condition for admissibility of a design from part (a) of the theorem becomes useless if there exists a nonzero and nonnegative definite matrix A such that

$$q_A(x) = f(x)'Af(x) \equiv \text{constant on } \mathcal{X}. \tag{3.2}$$

In particular, if the regression model (1.1a) includes a constant term, i.e., if one component of f is a nonzero constant, $f_1 \equiv 1$ say, then (3.2) is valid (with $A = \mathrm{diag}(1, 0, \ldots, 0)$). Regression models with constant term are rather the rule than the exception. This demands for a modification of the Karlin and Studden result for situation (3.2); in fact, such a modification has been proved by Heiligers (1991), Theorem 4. We will derive here a result slightly different from his, in that the matrix group $\mathcal{Q}$ is treated differently. If $\mathcal{Q}$ consists of orthogonal matrices only – as it is true in most applications – our approach is that from Heiligers (1991).

The starting point is a condition weaker than (3.2), which will turn out to be more compatible to the Loewner partial ordering of moment matrices. Define the set $\mathcal{D}$ of nonnegative definite matrices by

$$\mathcal{D} := \{M_2 - M_1 \colon M_1, M_2 \in \mathcal{M},\ M_1 \leqslant M_2\}. \tag{3.3}$$

As it is easily seen, it follows from the convexity $\mathcal{D}$ that there exists a $D^* \in \mathcal{D}$ with largest range, equivalently, with smallest nullspace,

$$\mathcal{K} := \text{nullspace}(D^*) \subset \text{nullspace}(D) \quad \text{for all } D \in \mathcal{D}. \tag{3.4}$$

Now, a condition weaker than (3.2) is that

$$\mathcal{K} \neq \{0\}. \tag{3.5}$$

In fact, if there exists an $A \in \text{NND}(k)$, $A \neq 0$, with (3.2) then for $D^* = M_2^* - M_1^*$ from (3.4) (with $M_1^*, M_2^* \in \mathcal{M}$ and $M_1^* \leqslant M_2^*$) we have

$$\text{tr}(AD^*) = \text{tr}(AM_2{}^*) - \text{tr}(AM_1{}^*) = 0,$$

hence $AD^* = 0$, that is, $\{0\} \neq \text{range}(A) \subset \text{nullspace}(D^*) = \mathcal{K}$. Thus (3.2) implies (3.5), but the converse seems not to be true in general.

LEMMA 3.3. *Denote by $P_{\mathcal{K}}$ the orthogonal projector from $\mathbb{R}^k$ onto $\mathcal{K}$ given by* (3.4), *and let $\mathcal{L}$ be a linear subspace of $\mathbb{R}^k$, complementary to $\mathcal{K}$. Then, for any two moment matrices $M, N \in \mathcal{M}$ we have*

$$M \leqslant N \iff \begin{cases} z'Mz \leqslant z'Nz & \text{for all } z \in \mathcal{L}\text{, and} \\ P_{\mathcal{K}}M = P_{\mathcal{K}}N. \end{cases}$$

PROOF. Let $M, N \in \mathcal{M}$. Assuming firstly that $M \leqslant N$, it trivially follows $z'Mz \leqslant z'Nz$ for all $z \in \mathcal{L}$. In addition, the definition of $\mathcal{K}$ implies $\mathcal{K} \subset \text{nullspace}(N - M)$, hence $(N - M)P_{\mathcal{K}} = 0$, i.e., $P_{\mathcal{K}}M = P_{\mathcal{K}}N$.

Conversely, suppose that $z'Mz \leqslant z'Nz$ for all $z \in \mathcal{L}$ and $P_{\mathcal{K}}M = P_{\mathcal{K}}N$ (i.e., $Mu = Nu$ for all $u \in \mathcal{K}$). Hence, if $z \in \mathbb{R}^k$ is given (thus $z = u + v$ for some $u \in \mathcal{K}$ and $v \in \mathcal{L}$), we get

$$z'(N-M)z = u' \underbrace{(N-M)u}_{=0} + 2v' \underbrace{(N-M)u}_{=0} + \underbrace{v'(N-M)v}_{\geqslant 0} \geqslant 0,$$

ensuring $M \leqslant N$. □

LEMMA 3.4. *Let the linear regression model be equivariant (w.r.t. groups $\mathcal{G}$ and $\mathcal{Q}$), where the matrix group $\mathcal{Q}$ is compact. Then $\mathcal{K}$ from* (3.4) *possesses the invariance property*

$$Q'\mathcal{K} = \mathcal{K} \quad \textit{for all } Q \in \mathcal{Q},$$

and there exists a subspace $\mathcal{L}$ of $\mathbb{R}^k$, complementary to $\mathcal{K}$, which is also invariant, i.e.,

$$Q'\mathcal{L} = \mathcal{L} \quad \textit{for all } Q \in \mathcal{Q}.$$

PROOF. By (3.4) and (2.5), for any $Q \in \mathcal{Q}$,

$$\text{nullspace}(D^*) = \mathcal{K} \subset \text{nullspace}(QD^*Q') = \text{nullspace}(D^*Q');$$

consequently, if $z \in \mathcal{K}$ then $Q'z \in \mathcal{K}$, implying $\mathcal{K} = Q'\mathcal{K}$.

Let $E \in \text{PD}(k)$ with $Q'EQ = E$ for all $Q \in \mathcal{Q}$ from Lemma 2.3. Take $\mathcal{L} := E\mathcal{K}^\perp$, where $\mathcal{K}^\perp$ denotes the orthogonal complement of $\mathcal{K}$ in $\mathbb{R}^k$. If $z \in \mathcal{L} \cap \mathcal{K}$, then $z = Ev$ for some $v \in \mathcal{K}^\perp$, and $0 = v'z = v'Ev$, hence $v = 0$ and $z = 0$. Since $\mathcal{L}$ has dimension $k - \dim(\mathcal{K})$, the space $\mathcal{L}$ is complementary to $\mathcal{K}$. Moreover, for any $Q \in \mathcal{Q}$ we get

$$\mathcal{K}^\perp = (Q'\mathcal{K})^\perp = \{z \in \mathbb{R}^k\colon\ z'Q'u = 0 \text{ for all } u \in \mathcal{K}\} = Q^{-1}\mathcal{K}^\perp,$$

hence

$$Q'\mathcal{L} = Q'E\mathcal{K}^\perp = EQ^{-1}\mathcal{K}^\perp = E\mathcal{K}^\perp = \mathcal{L} \quad \text{for all } Q \in \mathcal{Q}.$$

□

For later reference we note that if $\mathcal{Q}$ consists of orthogonal matrices only, then the space $\mathcal{L}$ in Lemma 3.4 can be chosen as as the orthogonal complement $\mathcal{K}^\perp$ of $\mathcal{K}$.

In the following we will assume $\mathcal{K} \neq \mathbb{R}^k$, since otherwise all $M, N \in \mathcal{M}$ with $M \leqslant N$ fulfill $M = N$, that is, *any* design is admissible.

THEOREM 3.5. *Consider an equivariant linear regression model* (1.1a) (*w.r.t. groups* $\mathcal{G}$ *and* $\mathcal{Q}$) *with compact matrix group* $\mathcal{Q}$, *and assume that* $f(\mathcal{X})$ *is compact. Let* $\mathcal{K} \neq \mathbb{R}^k$ *and* $\mathcal{L}$ *be a linear subspace of* $\mathbb{R}^k$ *complementary to* $\mathcal{K}$, *and such that* $Q'\mathcal{L} = \mathcal{L}$ *for all* $Q \in \mathcal{Q}$. *For any given design* ξ_0 *denote*

$$\mathcal{M}_\mathcal{K}(\xi_0) := \bigl\{M \in \mathcal{M}\colon\ P_\mathcal{K}M = P_\mathcal{K}M(\xi_0)\bigr\}.$$

(a) *If* ξ_0 *is a weakly invariant and admissible design, then there exists a matrix* $B_0 \in \text{NND}(k)$, $B_0 \neq 0$, *with* $\text{range}(B_0) \subset \mathcal{L}$ *and* $Q'B_0Q = B_0$ *for all* $Q \in \mathcal{Q}$, *and such that*

$$\text{tr}\bigl(B_0M(\xi_0)\bigr) = \max_{M \in \mathcal{M}_\mathcal{K}(\xi_0)} \text{tr}(B_0M). \qquad (*)$$

(b) *If* ξ_0 *is any design and if there exists a matrix* $B_0 \in \text{NND}(k)$ *with* $\text{range}(B_0) = \mathcal{L}$, *such that* $(*)$ *from above holds true, then* ξ_0 *is admissible.*

PROOF. (a). Observing Lemma 3.3, admissibility of ξ_0 implies

$$\max_{M \in \mathcal{M}_\mathcal{K}(\xi_0)} \min_{v \in \mathcal{L},\, v'v=1} v'\bigl(M - M(\xi_0)\bigr)v \leqslant 0.$$

Since

$$\min_{v\in\mathcal{L},\, v'v=1} v'\big(M - M(\xi_0)\big)v = \min_{B\in\mathcal{B}} \operatorname{tr}\big(B\big(M - M(\xi_0)\big)\big)$$
$$\text{for all } M \in \mathcal{M}_{\mathcal{K}}(\xi_0),$$

with $\mathcal{B} := \operatorname{Conv}\{vv'\colon v \in \mathcal{L},\, v'v = 1\} = \{B \in \operatorname{NND}(k)\colon \operatorname{range}(B) \subset \mathcal{L},\, \operatorname{tr}(B) = 1\}$, we obtain, using the minimax theorem,

$$\min_{B\in\mathcal{B}} \max_{M\in\mathcal{M}_{\mathcal{K}}(\xi_0)} \operatorname{tr}\big(B\big(M - M(\xi_0)\big)\big) \leqslant 0.$$

Hence there exists some $B_0 \in \mathcal{B}$ fulfilling

$$\operatorname{tr}\big(B_0 M(\xi_0)\big) \geqslant \operatorname{tr}(B_0 M) \quad \text{for all } M \in \mathcal{M}_{\mathcal{K}}(\xi_0). \tag{3.6}$$

Moreover, for any $Q \in \mathcal{Q}$ and $M \in \mathcal{M}$, we have $QMQ' \in \mathcal{M}$ and $QM(\xi_0)Q' = M(\xi_0)$, and therefore inequality (3.6) remains true when firstly replacing B_0 by $Q'B_0Q$, and then taking the average $\bar{B}_0$ of $Q'B_0Q$ over $Q' \in \mathcal{Q}'$. Finally, $\operatorname{range}(B_0) \subset \mathcal{L}$ implies $\operatorname{range}(Q'B_0Q) \subset Q'\mathcal{L} = \mathcal{L}$ for all $Q \in \mathcal{Q}$, and hence $\operatorname{range}(\bar{B}_0) \subset \mathcal{L}$. Also, from $0 \neq B_0 \in \operatorname{NND}(k)$ it follows that $0 \neq \bar{B}_0 \in \operatorname{NND}(k)$.

(b). Let $M \in \mathcal{M}$ with $M(\xi_0) \leqslant M$. We have to prove that $M(\xi_0) = M$. Lemma 3.3 gives $M \in \mathcal{M}_{\mathcal{K}}(\xi_0)$, and hence together with $(*)$, abbreviating $D := M - M(\xi_0)$, $\operatorname{tr}(B_0 D) = 0$, thus $\mathcal{L} = \operatorname{range}(B_0) \subset \operatorname{nullspace}(D)$, and by (3.4),

$$\mathcal{K} \subset \operatorname{nullspace}(D).$$

Consequently, $\mathbb{R}^k = \mathcal{L} + \mathcal{K} \subset \operatorname{nullspace}(D)$, i.e., $D = 0$. □

The following theorem gives the desired modification of the Karlin and Studden result.

THEOREM 3.6. *Consider an equivariant linear regression* (1.1a) (*w.r.t. groups* $\mathcal{G}$ *and* $\mathcal{Q}$) *with compact matrix group* $\mathcal{Q}$, *and assume that* $f(\mathcal{X})$ *is compact. Let* $\mathcal{K}$ *from* (3.4) *be different from* $\mathbb{R}^k$, *and fix a linear subspace* $\mathcal{L}$ *of* $\mathbb{R}^k$ *complementary to* $\mathcal{K}$, *and such that* $Q'\mathcal{L} = \mathcal{L}$ *for all* $Q \in \mathcal{Q}$.

(a) *If* ξ_0 *is a weakly invariant and admissible design, then there exist matrices* $B \in \operatorname{NND}(k)$ *and* $C \in \mathbb{R}^{k\times k}$ *with* $\operatorname{range}(B) \subset \mathcal{L}$, $CP_{\mathcal{K}} = C$, $Q'BQ = B$ *and* $Q'CQ = C$ *for all* $Q \in \mathcal{Q}$, *and such that the function*

$$q_{B,C}(x) := f(x)'(B + C)f(x), \quad x \in \mathcal{X},$$

is non-constant, and each support point of ξ_0 *is a global maximum point of* $q_{B,C}$ *over* $x \in \mathcal{X}$.

(b) *If ξ_0 is any design and if there exist matrices $B \in \mathrm{NND}(k)$ with* $\mathrm{range}(B) = \mathcal{L}$, *and $C \in \mathbb{R}^{k\times k}$ with $CP_{\mathcal{K}} = C$, such that each support point of ξ_0 maximizes $q_{B,C}(x) = f(x)'(B+C)f(x)$ over $x \in \mathcal{X}$, then ξ_0 is admissible.*

PROOF. (a). Fix a matrix B_0 from Theorem 3.5. We will prove that there exist a matrix $C \in \mathbb{R}^{k\times k}$ and a scalar $b \geqslant 0$, such that $CP_{\mathcal{K}} = C$, $Q'CQ = C$ for all $Q \in \mathcal{Q}$, and, abbreviating $M_0 := M(\xi_0)$,

$$\begin{gathered} \mathrm{tr}(CM) + b\,\mathrm{tr}(B_0M) \leqslant \mathrm{tr}(CM_0) + b\,\mathrm{tr}(B_0M_0) \\ \text{for all } M \in \mathcal{M}, \text{ with strict inequality for some } M \in \mathcal{M}. \end{gathered} \tag{3.7}$$

Then, taking $B := b\,B_0$, the assertion follows from $\mathcal{M} = \mathrm{Conv}\{f(x)f(x)' \colon x \in \mathcal{X}\}$. Consider the set

$$\mathcal{C} := \{P_{\mathcal{K}}M \colon M \in \mathcal{M}\} \subset \mathbb{R}^{k\times k},$$

which is obviously convex.

Case 1. Suppose that $P_{\mathcal{K}}M_0$ is in the relative interior of $\mathcal{C}$.
Then, as we will show next,

$$\mathrm{tr}(B_0M) \text{ is non-constant over } M \in \mathcal{M}_{\mathcal{K}}(\xi_0). \tag{3.8}$$

For, (3.4) gives $\mathcal{K} = \mathrm{nullspace}(M_2 - M_1)$ for some $M_1, M_2 \in \mathcal{M}$ with $M_1 \leqslant M_2$, and Lemma 3.3 ensures $P_{\mathcal{K}}M_1 = P_{\mathcal{K}}M_2 \in \mathcal{C}$. By assumption there exists an $\varepsilon = \varepsilon_j \in (0,1)$, such that $(1+\varepsilon)\,P_{\mathcal{K}}M_0 - \varepsilon\,P_{\mathcal{K}}M_j \in \mathcal{C}$, and therefore that point equals $P_{\mathcal{K}}N$ for some $N \in \mathcal{M}$, $j = 1,2$. Clearly, $N_j := (1-\varepsilon)M_0 + \varepsilon M_j$ belongs to $\mathcal{M}$. Now,

$$P_{\mathcal{K}}\Big(\tfrac{1}{2}N + \tfrac{1}{2}N_j\Big) = \tfrac{1}{2}P_{\mathcal{K}}N + \tfrac{1}{2}P_{\mathcal{K}}N_j = P_{\mathcal{K}}M_0,$$

i.e., $\widetilde{N}_j := \frac{1}{2}N + \frac{1}{2}N_j \in \mathcal{M}_{\mathcal{K}}(\xi_0)$ for $j = 1,2$. Observing that

$$\mathrm{tr}(B_0\widetilde{N}_2) - \mathrm{tr}(B_0\widetilde{N}_1) = \frac{\varepsilon}{2}\,\mathrm{tr}\big(B_0(M_2 - M_1)\big)$$

is nonnegative and different from zero (as follows from $\{0\} \neq \mathrm{range}(B_0) \subset \mathcal{L}$, $\mathrm{nullspace}(M_2 - M_1) = \mathcal{K}$, and $\mathcal{L} \cap \mathcal{K} = \{0\}$), (3.8) is verified.

Next, we consider the set

$$\tilde{\mathcal{C}} := \{(P_{\mathcal{K}}M, \mathrm{tr}(B_0M)) \colon M \in \mathcal{M}\} \subset \mathbb{R}^{k\times k+1},$$

which again is convex. Here we get

$$\big(P_{\mathcal{K}}M_0, \mathrm{tr}(B_0M_0)\big) \text{ is not in the relative interior of } \tilde{\mathcal{C}}. \tag{3.9}$$

For, suppose to the contrary that $Z_0 := (P_{\mathcal{K}} M_0, \mathrm{tr}(B_0 M_0))$ is in the relative interior of $\tilde{\mathcal{C}}$. From (3.8) we see that there exists an $N \in \mathcal{M}_{\mathcal{K}}(\xi_0)$, such that

$$\mathrm{tr}(B_0 N) < \max_{M \in \mathcal{M}_{\mathcal{K}}(\xi_0)} \mathrm{tr}(B_0 M) = \mathrm{tr}(B_0 M_0).$$

Now, $Z := (P_{\mathcal{K}} N, \mathrm{tr}(B_0 N))$ belongs to $\tilde{\mathcal{C}}$, and hence there exists an $\varepsilon > 0$ such that $(1+\varepsilon) Z_0 - \varepsilon Z \in \tilde{\mathcal{C}}$, i.e., that point equals $(P_{\mathcal{K}} N_1, \mathrm{tr}(B_0 N_1))$ for some $N_1 \in \mathcal{M}$. From $P_{\mathcal{K}} N = P_{\mathcal{K}} M_0$ we get $P_{\mathcal{K}} N_1 = P_{\mathcal{K}} M_0$, thus $N_1 \in \mathcal{M}_{\mathcal{K}}(\xi_0)$, and

$$\mathrm{tr}(B_0 N_1) = (1+\varepsilon)\,\mathrm{tr}(B_0 M_0) - \varepsilon\,\mathrm{tr}(B_0 N) > \mathrm{tr}(B_0 M_0),$$

which is a contradiction.

Now the proof in Case 1 is completed as follows. (3.9) ensures that there exists a non-trivial supporting hyperplane to $\tilde{\mathcal{C}}$ at $(P_{\mathcal{K}} M_0, \mathrm{tr}(B_0 M_0))$, that is, there exist a matrix $A \in \mathbb{R}^{k \times k}$ and a scalar b, not both zero, such that

$$\begin{gathered}\mathrm{tr}(A' P_{\mathcal{K}} M) + b\,\mathrm{tr}(B_0 M) \leqslant \mathrm{tr}(A' P_{\mathcal{K}} M_0) + b\,\mathrm{tr}(B_0 M_0) \\ \text{for all } M \in \mathcal{M}, \text{ with strict inequality for some } M_1 \in \mathcal{M}, \end{gathered} \tag{3.10}$$

cf., e.g., Theorem 11.6 in Rockafellar (1970). The scalar b must be nonnegative, since otherwise $\mathrm{tr}(B_0 M) \geqslant \mathrm{tr}(B_0 M_0)$ for all $M \in \mathcal{M}_{\mathcal{K}}(\xi_0)$, hence $\mathrm{tr}(B_0 M) = \mathrm{tr}(B_0 M_0)$ for all $M \in \mathcal{M}_{\mathcal{K}}(\xi_0)$, contrary to (3.8). Moreover, equivariance of the regression model and weak invariance of ξ_0 entail, for all $Q \in \mathcal{Q}$, $QMQ' \in \mathcal{M}$ whenever $M \in \mathcal{M}$, and $QM_0Q' = M_0$. Therefore (3.10) remains valid with $A' P_{\mathcal{K}}$ replaced by $Q' A' P_{\mathcal{K}} Q$ (for any fixed $Q \in \mathcal{Q}$), and thus the inequality in (3.10) holds true when replacing $A' P_{\mathcal{K}}$ by the average, C say, of $Q' A' P_{\mathcal{K}} Q$ over $Q' \in \mathcal{Q}'$. Actually, there is still strict inequality for some $\widetilde{M}_1 \in \mathcal{M}$. For, by Theorem 2.4, $C = \sum_{i=1}^{r} \alpha_i Q_i' A P_{\mathcal{K}} Q_i$ for some $r \in \mathbb{N}$, $\alpha_1, \ldots, \alpha_r > 0$ with $\sum_{i=1}^{r} \alpha_i = 1$, and $Q_1, \ldots, Q_r \in \mathcal{Q}$. Then take for example $\widetilde{M}_1 := Q_1^{-1} M_1 (Q_1^{-1})'$. By $Q \widetilde{M}_1 Q' \in \mathcal{M}$ and $Q' B_0 Q = B_0$ for all $Q \in \mathcal{Q}$, and by (3.10) we get

$$\begin{aligned} &\mathrm{tr}(C\widetilde{M}_1) + b\,\mathrm{tr}(B_0 \widetilde{M}_1) \\ &\quad = \sum_{i=1}^{r} \alpha_i \left(\mathrm{tr}(A' P_{\mathcal{K}} Q_i \widetilde{M}_1 Q_i') + b\,\mathrm{tr}(B_0 Q_i \widetilde{M}_1 Q_i') \right) \\ &\quad \leqslant \mathrm{tr}(A' P_{\mathcal{K}} M_0) + b\,\mathrm{tr}(B_0 M_0) \\ &\qquad \text{for all } i = 1, \ldots, r, \text{ with strict inequality for } i = 1. \end{aligned}$$

Observing that $\mathrm{tr}(A' P_{\mathcal{K}} M_0) = \mathrm{tr}(C M_0)$ because of $Q_i' M_0 Q_i = M_0$ for all $i = 1, \ldots, r$, we get indeed

$$\mathrm{tr}(C\widetilde{M}_1) + b\,\mathrm{tr}(B_0 \widetilde{M}_1) < \mathrm{tr}(C M_0) + b\,\mathrm{tr}(B_0 M_0).$$

We thus obtained (3.7), but the condition $CP_{\mathcal{K}} = C$ needs to be verified. To this end note that according to Lemma 3.4 (for all $Q \in \mathcal{Q}$) $Q'\mathcal{K} = \mathcal{K}$, equivalently, $P_{\mathcal{K}}Q'P_{\mathcal{K}} = Q'P_{\mathcal{K}}$. Hence, abbreviating $C_Q := Q'A'P_{\mathcal{K}}Q$, we have

$$C_Q P_{\mathcal{K}} = Q'A' \underbrace{P_{\mathcal{K}} Q P_{\mathcal{K}}}_{=P_{\mathcal{K}}Q} = C_Q,$$

and the average C of C_Q over $Q' \in \mathcal{Q}'$ also satisfies $CP_{\mathcal{K}} = C$.

Case 2. Suppose that $P_{\mathcal{K}}M_0$ is not in the relative interior of $\mathcal{C}$.

By Theorem 11.6 in Rockafellar (1970), there exists a non-trivial supporting hyperplane (in $\mathbb{R}^{k \times k}$) to $\mathcal{C}$ at $P_{\mathcal{K}}M_0$, i.e., there exists a matrix $A \in \mathbb{R}^{k \times k}$, such that

$$\begin{gathered} \operatorname{tr}(A'P_{\mathcal{K}}M) \leqslant \operatorname{tr}(A'P_{\mathcal{K}}M_0) \\ \text{for all } M \in \mathcal{M}, \text{ with strict inequality for some } M \in \mathcal{M}. \end{gathered} \tag{3.11}$$

By arguments as above, the average C from Theorem 2.4 of $Q'A'P_{\mathcal{K}}Q$ over $Q' \in \mathcal{Q}'$ again satisfies (3.11) (when $A'P_{\mathcal{K}}$ is replaced by C). Also, we find that C satisfies $CP_{\mathcal{K}} = C$, and, choosing $b = 0$, (3.7) follows.

(b). Suppose that $M_1 \in \mathcal{M}$ is such that $M_0 := M(\xi_0) \leqslant M_1$. We have to show that $M_0 = M_1$. Since each support point of ξ_0 maximizes $q_{B,C}(x)$ over $x \in \mathcal{X}$, it follows that for all $M \in \mathcal{M}$

$$\operatorname{tr}(BM_0) + \operatorname{tr}(CM_0) \geqslant \operatorname{tr}(BM) + \operatorname{tr}(CM).$$

In particular, with $M = M_1$ we get from Lemma 3.3 (observing $C = CP_{\mathcal{K}}$), $CM_0 = CP_{\mathcal{K}}M_0 = CP_{\mathcal{K}}M_1 = CM_1$, and hence $\operatorname{tr}(B(M_1 - M_0)) = 0$, i.e.,

$$\mathcal{L} = \operatorname{range}(B) \subset \operatorname{nullspace}(M_1 - M_0).$$

By (3.4), $\mathcal{K} \subset \operatorname{nullspace}(M_1 - M_0)$; hence both, $\mathcal{L}$ and $\mathcal{K}$ are subspaces of $\operatorname{nullspace}(M_1 - M_0)$, and therefore $M_1 - M_0 = 0$ because of $\mathcal{L} + \mathcal{K} = \mathbb{R}^k$. □

4. Invariant and admissible multiple polynomial regression designs

Regression models of particular interest in practice are the dth degree multiple polynomial setups from Examples 2.7, 2.8

$$y(x) = \sum_{\alpha \in \mathrm{A}} \theta_\alpha x^\alpha, \quad x \in \mathcal{X}, \tag{4.1}$$

where $\mathcal{X} \subset \mathbb{R}^v$ is compact, and, as before $\mathrm{A} \subset \mathrm{A}_d$ contains at least one multi-index of order d. Here the moment matrix of a design ξ with support points $x_1, \ldots, x_r$ consists of mixed moments $\mu_\alpha(\xi) = \sum_{i=1}^r x_i^\alpha \xi(x_i)$, $\alpha \in \mathbb{N}_0^v$, up to order $|\alpha| \leqslant 2d$,

$$M(\xi) = \left(\mu_{\alpha+\beta}(\xi)\right)_{\alpha,\beta \in \mathrm{A}}. \tag{4.2}$$

Depending on the index-set A, (4.1) does not necessarily involve the constant term $x^0 \equiv 1$, and thus the Karlin and Studden result (see Theorem 3.2 above) may yield a nontrivial result on admissible designs. The following Lemma 4.1, however, will enable a unified approach to admissibility for all possible index sets A in the setup.

In this section we will use the notion 'A-admissibility' instead of 'admissibility', referring to the particular setup under consideration. The moment matrix of a design ξ for the full polynomial setup $y(x) = \sum_{|\alpha| \leqslant d} \theta_\alpha x^\alpha$, $x \in \mathcal{X}$ (i.e., (4.1) with $\mathrm{A} = \mathrm{A}_d$), will be denoted by $M_d(\xi)$. As in Example 2.7 we will consider groups of transformations acting linearly on $\mathcal{X}$, i.e., groups $\mathcal{G}$ of transformations $g(x) = U_g x$ with nonsingular real $v \times v$ matrices U_g such that $U_g x \in \mathcal{X}$ for all $x \in \mathcal{X}$ and all $g \in \mathcal{G}$. For such groups equivariance of model (4.1) was characterized by (2.10), where the matrices Q_g forming the group $\mathcal{Q}$ were defined by (2.9). Also, as pointed out in Example 2.7, if $\mathcal{G}$ is compact (i.e., if the set of matrices $\{U_g\colon\ g \in \mathcal{G}\}$ is compact), then the matrix group $\mathcal{Q}$ is also compact.

LEMMA 4.1. *Let $\mathcal{G}$ be a compact group of transformations acting linearly on $\mathcal{X}$, and suppose that* A *in* (4.1) *is such that the setup is equivariant. Then, for any weakly invariant design ξ there exists a weakly invariant design $\tilde{\xi}$ which is* A- *and* A_d-*admissible, and such that*

$$M(\tilde{\xi}) \geqslant M(\xi).$$

PROOF. Since $\mathcal{G}$ acts linearly on $\mathcal{X}$, the full polynomial model of degree d is also equivariant w.r.t. $\mathcal{G}$ and the associated *compact* group $\mathcal{Q}$, see Example 2.7. Thus, by Lemma 1 in Heiligers (1991), for any weakly invariant design ξ there exist a weakly invariant and A-admissible design η, and a weakly invariant and A_d-admissible design $\tilde{\xi}$, fulfilling $M(\eta) \geqslant M(\xi)$ and $M_d(\tilde{\xi}) \geqslant M_d(\eta)$. The latter inequality implies $M(\tilde{\xi}) \geqslant M(\eta)$, since $M(\tilde{\xi})$ and $M(\eta)$ are principal sub-matrices of $M_d(\tilde{\xi})$ and $M_d(\eta)$, respectively. Thus, A-admissibility of η implies $M(\tilde{\xi}) = M(\eta)$, and therefore also A-admissibility of $\tilde{\xi}$ follows. □

Recall that Lemma 4.1 applies as well to non-equivariant setups (which are covered by choosing $\mathcal{G}$ and $\mathcal{Q}$ as the trivial groups).

As a consequence from Lemma 4.1, under the above invariance assumptions and under an invariant optimality criterion, the search for a solution to an optimal design problem in setup (4.1) can be always restricted to the set of weakly invariant and A_d-admissible designs. In this context an important step is the determination of the linear space $\mathcal{K}$ from (3.4) for the full polynomial model of degree d. This was done by Heiligers (1991, Lemma 2); we restate his result here, for convenience.

THEOREM 4.2. *For the full multiple polynomial model of degree d, i.e.,* (4.1) *with* $\mathrm{A} = \mathrm{A}_d$, *we have*

$$\mathcal{E} := \operatorname{span}\left\{e_\alpha\colon\ \alpha \in \mathbb{N}_0^v,\ |\alpha| \leqslant d-1\right\} \subset \mathcal{K},$$

where e_α, $\alpha \in \mathrm{A}_d$, *denote the unit-vectors in* $\mathbb{R}^k$, $k = \binom{v+d}{d}$. *If the monomials* x^α, $|\alpha| \leqslant 2d$, *are linearly independent functions over* $\mathcal{X}$, *then* $\mathcal{E} = \mathcal{K}$.

For polynomial regression in one real variable (i.e., $v = 1$) on a compact interval, the A_d-admissible designs are characterized in Theorem 3.6 of Kiefer (1959). Next we rederive Kiefer's result from our Theorems 4.2 and 3.6.

THEOREM 4.3 (Kiefer (1959)). *Let $v = 1$ and $\mathcal{X} = [a, b]$ with $a, b \in \mathbb{R}$, $a < b$. Then a design ξ is A_d-admissible iff*

$$\#\big(\operatorname{supp}(\xi) \cap (a, b)\big) \leqslant d - 1.$$

PROOF. For convenience, we assume that the vector f of regression functions is arranged according to $f(x) = (1, x, \ldots, x^d)'$, $x \in [a, b]$. Theorem 4.2 gives $\mathcal{K}^\perp = \operatorname{span}\{e_d\}$ (with $e_d = (0, \ldots, 0, 1)' \in \mathbb{R}^{d+1}$). By part (a) of Theorem 3.6 (with $\mathcal{L} = \mathcal{K}^\perp$ and the trivial groups $\mathcal{G} = \{\mathrm{id}_{\mathcal{X}}\}$ and $\mathcal{Q} = \{I_{d+1}\}$), A_d-admissibility of a design ξ implies that there exist $c_0, \ldots, c_{2d-1} \in \mathbb{R}$ and $\beta \geqslant 0$ such that the polynomial

$$p(x) = \beta\, x^{2d} + \sum_{i=0}^{2d-1} c_i\, x^i, \quad x \in [a, b],$$

is non-constant, and each support point of ξ maximizes p over $[a, b]$. (Note that $CP_{\mathcal{K}} = C$ and $B \in \mathrm{NND}(d+1)$, $\operatorname{range}(B) \subset \mathcal{K}^\perp$, means here $c_{id} = 0$ for all $i = 0, 1, \ldots, d$, where c_{ij}, $i, j = 0, 1, \ldots, d$, are the entries of C, and $B = \operatorname{diag}(0, \ldots, 0, \beta)$ with $\beta \geqslant 0$.) Since the leading coefficient of p is nonnegative, that polynomial possesses at most $d - 1$ maximum points in the open interval (a, b).

Conversely, let ξ be a design with $\#(\operatorname{supp}(\xi) \cap (a, b)) \leqslant d - 1$, $\operatorname{supp}(\xi) \cap (a, b) = \{z_1, \ldots, z_\ell\}$, $\ell \leqslant d - 1$, say. Consider the $2d$th degree polynomial

$$\begin{aligned} p(x) &= 1 - (x - a)^{2(d-\ell)-1}\, (b - x) \prod_{i=1}^{\ell} (x - z_i)^2 \\ &= \sum_{i=0}^{2d} c_i\, x^i, \quad x \in [a, b], \quad \text{say}, \end{aligned}$$

which obviously has leading coefficient $c_{2d} = 1$. The only maximum points of p in $[a, b]$ are the endpoints of the interval and the support points of ξ. Thus, defining for $0 \leqslant i, j \leqslant d$,

$$c_{ij} = \begin{cases} c_i, & \text{if } 0 \leqslant i \leqslant d \text{ and } j = 0, \\ c_{d+j}, & \text{if } i = d \text{ and } 1 \leqslant j \leqslant d - 1, \\ 0, & \text{otherwise}, \end{cases}$$

part (b) of Theorem 3.6 (with $B = \operatorname{diag}(0, \ldots, 0, 1)$ and $C = (c_{ij})_{0 \leqslant i,j \leqslant d}$) ensures A_d-admissibility of ξ. □

Theorem 4.3 can also be applied to polynomial models in more than one variable, by restricting to line segments contained in the experimental region. A line segment $\mathcal{S}$ with endpoints $x^{(0)}, x^{(1)} \in \mathbb{R}^v$, say, is the convex hull of these two points, i.e.,

$$\mathcal{S} = \{x^{(0)} + \lambda(x^{(1)} - x^{(0)}):\ \lambda \in [0,1]\}.$$

The line segment $\mathcal{S}$ is said to be non-degenerate iff $x^{(0)} \neq x^{(1)}$; in that case its relative interior is given by

$$\mathrm{ri}(\mathcal{S}) = \{x^{(0)} + \lambda(x^{(1)} - x^{(0)}):\ \lambda \in (0,1)\}.$$

THEOREM 4.4. *For arbitrary $v \geqslant 1$, let $\mathcal{S} \subset \mathcal{X}$ be a non-degenerate line segment in the experimental region. Then, for any* A_d*-admissible design ξ we have*

$$\#(\mathrm{supp}(\xi) \cap \mathrm{ri}(\mathcal{S})) \leqslant d - 1.$$

PROOF. Only the case that $\mathrm{supp}(\xi) \cap \mathrm{ri}(\mathcal{S})$ is nonempty needs to be considered. Choose any design $\tilde{\xi}$ whose support is equal to this intersection. As remarked earlier, according to Theorem 7.2 of Karlin and Studden (1966), A_d-admissibility of ξ implies A_d-admissibility of $\tilde{\xi}$. Consider the bijection t from $\mathcal{S}$ onto the interval $[0,1]$ given by $t(x_\lambda) := \lambda$, where $x_\lambda := x^{(0)} + \lambda(x^{(1)} - x^{(0)})$. Clearly, t induces a one-to-one correspondence between designs $\tilde{\eta}$ supported on $\mathcal{S}$ and designs on $[0,1]$, via $\tilde{\eta} \to \tilde{\eta}^t$. The moment matrices $M_d(\tilde{\eta})$ are related to the moment matrices $M_d^1(\tilde{\eta}^t)$, say, for polynomial regression of degree d in one variable by

$$M_d(\tilde{\eta}) = R\, M_d^1(\tilde{\eta}^t)\, R', \tag{4.3}$$

where $R = (r_{\alpha j})_{|\alpha| \leqslant d, 0 \leqslant j \leqslant d}$ is a fixed $\binom{v+d}{d} \times (d+1)$ matrix of rank $d+1$, and thus

$$M_d^1(\tilde{\eta}^t) = (R'R)^{-1} R' M_d(\tilde{\eta}) R (R'R)^{-1}. \tag{4.4}$$

Equation (4.3) can be seen as follows. Any monomial x^α with $|\alpha| \leqslant d$ restricted to $\mathcal{S}$ yields a polynomial over $[0,1]$ of degree $\leqslant |\alpha|$ by

$$x_\lambda^\alpha = \prod_{i=1}^{v} \left(x_i^{(0)} + \lambda(x_i^{(1)} - x_i^{(0)})\right)^{\alpha_i} = \sum_{j=0}^{d} r_{\alpha j} \lambda^j,$$

with coefficients $r_{\alpha j}$ only depending of the endpoints $x^{(0)}, x^{(1)}$ of $\mathcal{S}$. Obviously, $r_{\alpha j} = 0$ if $j > |\alpha|$. From $x^{(0)} \neq x^{(1)}$ it follows that for each $j = 0, 1, \ldots, d$ there exists an $\alpha^{(j)}$ with $|\alpha^{(j)}| = j$ and $r_{\alpha^{(j)} j} \neq 0$ (choose $\alpha^{(j)}$ with i_0th component equal to j and vanishing remaining components, where i_0 is such that $x_{i_0}^{(0)} \neq x_{i_0}^{(1)}$). Hence, the function $f(x) = (x^\alpha)_{|\alpha| \leqslant d}$, when restricted to $\mathcal{S}$, is related to $f^1(\lambda) := (\lambda^j)_{0 \leqslant j \leqslant d}$ via

$$f(x_\lambda) = R f^1(\lambda), \quad \lambda \in [0,1],$$

and (4.3) follows immediately. It remains to show that the matrix R has full rank $d+1$. To this end, let $z = (z_0, \ldots, z_d)' \in \mathbb{R}^{d+1}$ with $Rz = 0$, i.e., $\sum_{j=0}^{|\alpha|} r_{\alpha j} z_j = 0$ for all α with $|\alpha| \leqslant d$. For $\alpha = 0$ the sum equals $r_{00} z_0$, and $r_{00} \neq 0$ gives $z_0 = 0$. Suppose that for some $1 \leqslant j \leqslant d$ we have $z_0 = \cdots = z_{j-1} = 0$. Then, for $\alpha^{(j)}$ with $|\alpha^{(j)}| = j$ and $r_{\alpha^{(j)} j} \neq 0$ we get

$$0 = \sum_{i=0}^{|\alpha^{(j)}|} r_{\alpha^{(j)} i}\, z_i = r_{\alpha^{(j)} j}\, z_j,$$

hence $z_j = 0$. It thus follows that $z = 0$, i.e., $\operatorname{rank}(R) = d+1$.

From (4.3) and (4.4) we obtain that $\tilde{\xi}^t$ is A_d-admissible for polynomial regression of degree d on the interval $[0,1]$. For, suppose that $M_d^1(\tilde{\xi}^t) \leqslant M_d^1(\zeta)$ for some design ζ on $[0,1]$. Because of $\zeta = \tilde{\eta}^t$ for some design $\tilde{\eta}$ supported by $\mathcal{S}$, (4.3) entails $M_d(\tilde{\xi}) \leqslant M_d(\tilde{\eta})$, hence, due to A_d-admissibility of $\tilde{\xi}$, $M_d(\tilde{\xi}) = M_d(\tilde{\eta})$. This and (4.4) yields $M_d^1(\tilde{\xi}^t) = M_d^1(\tilde{\eta}^t)$. Consequently, according to Theorem 4.3, $\tilde{\xi}^t$ possesses at most $d-1$ support points in the open interval $(0,1)$, i.e., $\tilde{\xi}$ has at most $d-1$ support points in the relative interior of the line segment $\mathcal{S}$. □

Theorem 4.4 becomes most interesting for first and second degree models over compact and convex regions $\mathcal{X}$: For degree $d = 1$ an A_d-admissible design cannot have support points in the relative interior of any non-degenerate line segment in $\mathcal{X}$. For $d = 2$, that theorem ensures that an A_d-admissible design can have at most one support point in the relative interior of each face of $\mathcal{X}$ (see the following corollary). Recall that a *face* (or, synonymously, an *extreme subset*) of $\mathcal{X}$ is a convex subset $\mathcal{F} \subset \mathcal{X}$ with the following property.

$$\text{If} \quad \mathcal{F} \cap \operatorname{ri}(\mathcal{S}) \neq \emptyset, \quad \text{where } \mathcal{S} \subset \mathcal{X} \text{ is a line segment,} \quad \text{then } \mathcal{S} \subset \mathcal{F}.$$

Note that $\mathcal{X}$ itself is a face of $\mathcal{X}$, and as the other extreme, the one-point sets of extreme points of $\mathcal{X}$ are faces of $\mathcal{X}$. By Rockafellar (1970, Theorem 18.2), the compact and convex set $\mathcal{X}$ is the disjoint union of the relative interiors of its faces,

$$\mathcal{X} = \bigcup_{\mathcal{F} \text{ face of } \mathcal{X}} \operatorname{ri}(\mathcal{F}).$$

COROLLARY 4.5. *Let the experimental region $\mathcal{X}$ in* (4.1) *be compact and convex.*

(a) *If $d = 1$, then the support points of an* A_1*-admissible design are extreme points of $\mathcal{X}$.*

(b) *If $d = 2$, then for any face $\mathcal{F}$ of $\mathcal{X}$ an* A_2*-admissible design has at most one support point in* $\operatorname{ri}(\mathcal{F})$.

Let us study two special experimental regions, a v-dimensional cube symmetric to zero,

$$\begin{aligned} \mathcal{X} &= \mathcal{C}_b \\ &:= \{x = (x_1, \ldots, x_v)' \in \mathbb{R}^v \colon |x_i| \leqslant b_i \text{ for all } i = 1, \ldots, v\} \end{aligned} \tag{4.5a}$$

(with some fixed $b = (b_1, \ldots, b_v) \in (0, \infty)^v$), and a v-dimensional ellipsoid centered at zero,

$$\mathcal{X} = \mathcal{E}_H := \{x = (x_1, \ldots, x_v)' \in \mathbb{R}^v\colon\ x'Hx \leqslant 1\} \tag{4.5b}$$

(where $H \in \mathrm{PD}(v)$ is fixed). Denote by

$$\mathcal{V}_b := \{x \in \mathbb{R}^v\colon\ x_i = \pm b_i \text{ for all } i = 1, \ldots, v\}$$

the set of vertices of $\mathcal{C}_b$, and by

$$\partial\,\mathcal{E}_H := \{x \in \mathbb{R}^v\colon\ x'Hx = 1\}$$

the boundary (or surface) of $\mathcal{E}_H$.

COROLLARY 4.6. *Consider multiple polynomial regression of degree* $d = 1$ *over* $\mathcal{X} = \mathcal{C}_b$ *or* $\mathcal{X} = \mathcal{E}_H$ *from* (4.5a) *or* (4.5b), *respectively. Then a design* ξ *is* A_1*-admissible iff*

$$\mathrm{supp}(\xi) \subset \begin{cases} \mathcal{V}_b, & \text{if } \mathcal{X} = \mathcal{C}_b, \\ \partial\,\mathcal{E}_H, & \text{if } \mathcal{X} = \mathcal{E}_H. \end{cases}$$

PROOF. In view of Corollary 4.5, we only have to show that the above conditions on the support of ξ imply its A_1-admissibility. In fact, this immediately follows from part (b) of Theorem 3.2 with $A = \mathrm{diag}(1, b_1^{-2}, \ldots, b_v^{-2})$, if $\mathcal{X}$ is the cube from (4.5a), and with $A = \text{block-diag}(1, H)$, if $\mathcal{X}$ is the ellipsoid from (4.5b). □

The combination of A_1-admissibility with additional invariance properties may lead to a further substantial reduction of a design problem under consideration. For example, in case of the cube (4.5a) let $\mathcal{G}_\mathrm{s}$ be the sign change group from Example 2.8(a). By that example the multiple polynomial model of degree $d = 1$ on $\mathcal{X} = \mathcal{C}_b$ is equivariant. Here, Corollary 4.5 gives that the only invariant and A_1-admissible design is the uniform distribution on the set $\mathcal{V}_b$ of vertices of $\mathcal{C}_b$, and thus all weakly invariant and A_1-admissible designs have the same moment matrix

$$M^* = \mathrm{diag}(1, b_1^2, \ldots, b_v^2),$$

as follows from Lemma (2.6). Another example where all weakly invariant and admissible designs share the same moment matrix is multiple polynomial regression of degree $d = 1$, now on a v-dimensional ball with radius $r > 0$, centered at zero, i.e., $\mathcal{X}$ from (4.5b) with $H = (1/r^2)I_v$. By Example 2.8(c) that setup is equivariant w.r.t. the transformation group $\mathcal{G}_\mathrm{sp}$ generated sign changes and permutations (and the induced matrix group given in that example). From Corollary 4.6 it is easily seen that the moment matrix of any weakly invariant and A_1-admissible design is given by

$$M^* = \mathrm{diag}\left(1, \frac{r^2}{v}, \ldots, \frac{r^2}{v}\right),$$

which is also the common moment matrix of all rotatable and A_1-admissible designs (see Example 2.8(d)). A particular invariant and A_1-admissible design is the uniform distribution on the $2v$ intersection points of the r-sphere and the coordinate axes.

Descriptions of A_2-admissible designs ξ (for multiple polynomial regression of degree $d = 2$) will become more complicated, as here ξ may have one support point in the relative interior of each face $\mathcal{F}$ of $\mathcal{X}$. The faces of the cube from (4.5a) are obtained by fixing some subset of components to $\pm b_i$ (hence there are 3^v faces). The faces of the ellipsoid from (4.5b) are the ellipsoid itself and all singletons on its surface. More we can say about (*weakly*) *invariant* and A_2-admissible designs, where the transformation group is the sign change group $\mathcal{G}_s$ from Example 2.8(a) in case of the cube, and the two-element group consisting of the identity and the reflection at the origin in case of the ellipsoid.

COROLLARY 4.7. *Consider multiple polynomial regression of degree $d = 2$.*

(a) *Let $\mathcal{X} = \mathcal{C}_b$ from* (4.5a), *and consider the sign change transformation group $\mathcal{G}_s$. Denote by $\mathcal{C}_b^0 = \{x = (x_1, \ldots, x_v)' \in \mathbb{R}^v\colon\ x_i \in \{\pm b_i, 0\}$ for all $i\}$ the set of barycenters of $\mathcal{C}_b$. Then, for any weakly invariant and A_2-admissible design ξ we have* $\operatorname{supp}(\xi) \subset \mathcal{C}_b^0$. *Conversely, any (not necessarily weakly invariant) design ξ with* $\operatorname{supp}(\xi) \subset \mathcal{C}_b^0$ *is A_2-admissible.*

(b) *Let $\mathcal{X} = \mathcal{E}_H$ from* (4.5b), *and consider the two-element transformation group consisting of the identity and the reflection at the origin. Then, for any weakly invariant and A_2-admissible design ξ we have* $\operatorname{supp}(\xi) \subset \partial\,\mathcal{E}_H \cup \{0\}$. *Conversely, any (not necessarily weakly invariant) design ξ with* $\operatorname{supp}(\xi) \subset \partial\,\mathcal{E}_H \cup \{0\}$ *is A_2-admissible.*

PROOF. (a). Let ξ be weakly invariant and A_2-admissible. Since $\mathcal{G}_s$ is finite, according to Lemma 2.6 there is no loss of generality in assuming ξ to be invariant. Fix $x = (x_1, \ldots, x_v)' \in \operatorname{supp}(\xi)$ with

$$I = I_x := \bigl\{i \in \{1, \ldots, v\}\colon\ -b_i < x_i < b_i\bigr\} \neq \emptyset.$$

We have to show that $x_i = 0$ for all $i \in I$. To this end consider the face of $\mathcal{C}_b$ given by

$$\mathcal{F}_I := \bigl\{z = (z_1, \ldots, z_v)' \in \mathcal{C}_b\colon\ z_i = x_i \text{ for all } i \notin I\bigr\}.$$

Obviously, we have $x \in \operatorname{ri}(\mathcal{F}_I)$. Moreover, for each $i \in I$, $g_i(x) \in \operatorname{ri}(\mathcal{F}_I)$, where g_i denotes the transformation changing the sign of the ith coordinate. Now, invariance of ξ gives $g_i(x) \in \operatorname{supp}(\xi)$, and Corollary 4.5 implies $g_i(x) = x$, i.e., $x_i = 0$.

For the converse, we partition the vector f of regression functions according to $f(x) = (f^{(1)}(x)', f^{(2)}(x)')'$, with subvectors

$$f^{(1)}(x) := (1, x_1, \ldots, x_v)',$$
$$f^{(2)}(x) := (x_1^2, \ldots, x_v^2, x_1x_2, \ldots, x_1x_v, \ldots, x_{v-1}x_v)', \quad x \in \mathcal{X}.$$

The matrices $\tilde{C} = \operatorname{diag}(0, b_1^{-2}, \dots, b_v^{-2})$ and $\tilde{B} = \text{block-diag}(\tilde{B}_1, \tilde{B}_2)$, where

$$\begin{aligned}\tilde{B}_1 &= (v+1)\operatorname{diag}(b_1^{-2}, \dots, b_v^{-2}) - (b_i^{-1}b_j^{-1})_{i,j=1,\dots,v},\\ \tilde{B}_2 &= \operatorname{diag}_{1\leqslant i<j\leqslant v}(b_i^{-1}b_j^{-1}),\end{aligned}$$

obviously fulfill $\tilde{B} \in \mathrm{PD}(v')$ (with $v' = v(v+1)/2$), and

$$f^{(2)}(x)'\tilde{B}f^{(2)}(x) + f^{(1)}(x)'\tilde{C}f^{(1)}(x) = v\sum_{i=1}^{v} x_i^2\left(\frac{x_i^2}{b_i^2} - 1\right)$$

for all $x \in \mathbb{R}^v$.

Thus, since the left hand side in the above equation is equal to

$$f(x)'\big(\text{block-diag}\big(0_{(v+1)\times(v+1)}, \tilde{B}\big) + \text{block-diag}\big(\tilde{C}, 0_{v'\times v'}\big)\big)f(x),$$

and the right-hand side is less than or equal to zero for $x \in \mathcal{C}_b$, with equality if $x \in \mathcal{C}_b^0$, A_2-admissibility of all designs supported by $\mathcal{C}_b^0$ follows from Theorems 3.6(b) and 4.2.

(b). Let ξ be a weakly invariant and A_2-admissible design. Again, we may assume that ξ is invariant. Hence, if a support point x of ξ is in the interior of the ellipsoid, then so is $-x$, and Theorem 4.4 gives $-x = x$, i.e., $x = 0$.

For proving the converse implication, partition f as before. We will show that there are matrices $\tilde{B} \in \mathrm{PD}(v')$ and $\tilde{C} \in \mathbb{R}^{(v+1)\times(v+1)}$, such that

$$f^{(2)}(x)'\tilde{B}f^{(2)}(x) + f^{(1)}(x)'\tilde{C}f^{(1)}(x) = (x'Hx)^2 - x'Hx \quad \text{for all } x \in \mathbb{R}^v.$$

Then the assertion follows by arguments similar to those in part (a).

Let $H^{1/2}$ be the square root of H (i.e., the unique positive definite matrix with $H^{1/2}H^{1/2} = H$), and consider the linear mapping $x \to H^{1/2}x$, transforming the ellipsoid $\mathcal{E}_H$ onto the unit ball $\mathcal{E}_{I_v}$. As we have already seen in Example 2.7, the monomials x^α, $\alpha \in \mathrm{A}_2$, transform linearly under that mapping; more precisely, we have

$$\begin{aligned}f^{(1)}(H^{1/2}x) &= R_1 f^{(1)}(x),\\ f^{(2)}(H^{1/2}x) &= R_2 f^{(2)}(x), \quad x \in \mathcal{E}_H,\end{aligned}$$

where $R_1 = \text{block-diag}(1, H^{1/2})$ and R_2 is some $v' \times v'$ matrix (see Equation (2.11) in Example 2.7). The matrix R_2 is regular, as can be seen as follows. Since the mapping $z \to H^{-1/2}z$ transforming $\mathcal{E}_{I_v}$ onto $\mathcal{E}_H$ is also linear, there exists some $\tilde{R}_2 \in \mathbb{R}^{v'\times v'}$ such that

$$f^{(2)}(H^{-1/2}z) = \tilde{R}_2 f^{(2)}(z), \quad z \in \mathcal{E}_{I_v}.$$

Obviously, $f^{(2)}(x) = f^{(2)}(H^{-1/2}H^{1/2}x) = \tilde{R}_2 R_2 f^{(2)}(x)$ for all $x \in \mathbb{R}^v$, and linear independence of the monomials x^α, $|\alpha| = 2$, over $\mathcal{E}_H$ entails $\tilde{R}_2 R_2 = I_v$, i.e., R_2 is regular (and $\tilde{R}_2 = R_2^{-1}$).

Now we take

$$\tilde{B} := R_2' \operatorname{diag}(\underbrace{1,\ldots,1}_{v \text{ times}}, \underbrace{2,\ldots,2}_{v(v-1)/2 \text{ times}}) R_2,$$

$$\tilde{C} := R_1' \operatorname{diag}(0, \underbrace{-1,\ldots,-1}_{v \text{ times}}) R_1.$$

Then $\tilde{B} \in \mathrm{PD}(v')$, and, for all $x \in \mathcal{E}_H$ (abbreviating $z = H^{1/2}x$),

$$f^{(2)}(x)'\tilde{B}f^{(2)}(x) + f^{(1)}(x)'\tilde{C}f^{(1)}(x) = \sum_{i=1}^{v} z_i^4 + 2\sum_{1\leqslant i<j\leqslant v} z_i^2 z_j^2 - \sum_{i=1}^{v} z_i^2$$

$$= \left(\sum_{i=1}^{v} z_i^2\right)^2 - \sum_{i=1}^{v} z_i^2 = (x'Hx)^2 - x'Hx,$$

as desired. □

Up to now, for multiple cubic regression ($d = 3$) there is no result on admissibility comparable to the above for the linear and quadratic setups (but see our result on admissible and *rotatable* designs below, for arbitrary degree d). For $d = 3$, however, a 'complete class' result on weakly invariant designs on symmetric cubes (centered at zero) and on balls was obtained by Gaffke and Heiligers (1995a), which we state next, without proof. Let $\mathcal{X}$ be either the symmetric cube with half edge length $c > 0$, centered at zero,

$$\mathcal{X} = \mathcal{C}_c = \{x = (x_1,\ldots,x_v)' \in \mathbb{R}^v\colon\ |x_i| \leqslant c \text{ for all } i = 1,\ldots,v\}, \tag{4.6a}$$

or the ball with radius $r > 0$, centered at zero,

$$\mathcal{X} = \mathcal{B}_r = \left\{x = (x_1,\ldots,x_v)' \in \mathbb{R}^v\colon\ \sum_{i=1}^{v} x_i^2 \leqslant r^2\right\}. \tag{4.6b}$$

(Weak) invariance of a design refers to the transformation group $\mathcal{G}_{\mathrm{sp}}$ of sign changes and permutations of coordinates and the full multiple polynomial model of degree 3 (see Example 2.8(c)). For two points $x^{(1)}, x^{(2)} \in \mathcal{X}$ we denote by $[x^{(1)}, x^{(2)}]$ the line segment joining these points. In case of the cube (4.6a) let

$$x_0(i) := (\underbrace{c,\ldots,c}_{i \text{ times}},\underbrace{0,\ldots,0}_{v-i \text{ times}})', \quad i = 0,1,\ldots,v,$$
$$\mathcal{X}_0 := \bigcup_{0 \leqslant i < j \leqslant v} [x_0(i), x_0(j)], \tag{4.7a}$$

and for the ball (4.6b) let

$$x_0(i) := \Bigg(\underbrace{\frac{r}{\sqrt{i}},\ldots,\frac{r}{\sqrt{i}}}_{i \text{ times}},\underbrace{0,\ldots,0}_{v-i \text{ times}}\Bigg)', \quad i = 1,\ldots,v,$$
$$\mathcal{X}_0 := \bigcup_{i=1}^{v} [0, x_0(i)]. \tag{4.7b}$$

THEOREM 4.8. *Consider full multiple polynomial regression of degree* $d = 3$ *on the cube* (4.6a) *or on the ball* (4.6b), *and the transformation group* $\mathcal{G}_{\text{sp}}$. *Let* $\mathcal{X}_0$ *be given by* (4.7a) *in case of the cube, and by* (4.7b) *in case of the ball. Then, for any weakly invariant design* ξ *there exists a design* ξ_0 *with* $\operatorname{supp}(\xi_0) \subset \mathcal{X}_0$, *such that the symmetrized design*

$$\bar{\xi}_0 := \frac{1}{2^v v!} \sum_{g \in \mathcal{G}_{\text{sp}}} \xi_0^g$$

satisfies $M_3(\bar{\xi}_0) = M_3(\xi)$.

Admissibility of rotatable multiple polynomial regression designs on the ball is much easier to handle (for arbitrary degree d). Lemma 4.9, below, will give the moment matrix $M_d(\xi)$ of a rotatable design ξ on the ball (4.6b) as

$$M_d(\xi) = \sum_{\rho \in R(\xi)} \xi(S_\rho)\, \overline{M}_{d,\rho}, \tag{4.8}$$

where $S_\rho := \{x \in \mathbb{R}^v\colon \|x\| = \rho\}$ is the sphere of radius ρ, $\xi(S_\rho)$ is the total weight assigned by ξ to that sphere, and $R(\xi)$ is the (finite) set of radii $\rho \in [0, r]$ with $\xi(S_\rho) > 0$. The matrix $\overline{M}_{d,\rho}$ is the common moment matrix of rotatable designs with supports contained in S_ρ. Intuitively appealing is to think of $\overline{M}_{d,\rho}$ as the moment matrix of the uniform distribution on the sphere S_ρ (see, e.g., Kiefer, 1960, Section 3.2), but that distribution is no design (unless $\rho = 0$). Also, the uniform distribution on a sphere, though intuitively easy to understand, is not easily defined mathematically. The approach to admissibility of rotatable designs presented here will avoid these measure theoretic difficulties.

LEMMA 4.9. *Consider full multiple polynomial regression* (4.1) *of degree* $d \geqslant 1$ *on the ball* $\mathcal{B}_r$ *from* (4.6b), *and the orthogonal transformation group* $\mathcal{G}_{\text{orth}}$. *Then, for a given* $\rho \in [0, r]$, *the set of rotatable (i.e., weakly invariant w.r.t.* $\mathcal{G}_{\text{orth}}$*) designs* *concentrated on the sphere S_ρ is nonempty, and all these designs share the same moment matrix $\overline{M}_{d,\rho}$, say. Moreover, for any rotatable design ξ on $\mathcal{B}_r$ the above decomposition formula* (4.8) *for its moment matrix holds true.*

PROOF. By Examples 2.7, 2.8 the regression model is equivariant w.r.t. $\mathcal{G}_{\text{orth}}$ and the associated (compact) matrix group $\mathcal{Q} = \{Q_g\colon\ g \in \mathcal{G}_{\text{orth}}\}$ with matrices Q_g given by (2.9). Firstly, for a given $\rho \in [0, r]$, we prove that

$$\begin{aligned}&\text{there exists a unique } \overline{M} = \overline{M}_{d,\rho} \in \operatorname{Conv}\{f(x)f(x)'\colon\ x \in S_\rho\}\\ &\text{such that } Q\overline{M}Q' = \overline{M} \text{ for all } Q \in \mathcal{Q}.\end{aligned} \tag{4.9}$$

It then follows that the moment matrix of any rotatable design concentrated on S_ρ equals $\overline{M}$, and hence all these designs share the same moment matrix. On the other hand, representing $\overline{M}$ as finite convex combination of some $f(x)f(x)'$ matrices, $x \in S_\rho$, yields a rotatable design concentrated on S_ρ.

For proving (4.9) note that for any fixed $x_0 \in S_\rho$ the $\mathcal{G}_{\text{orth}}$-orbit of x_0 equals S_ρ, i.e.,

$$\{g(x_0)\colon\ g \in \mathcal{G}_{\text{orth}}\} = S_\rho.$$

Hence, observing $f(g(x_0)) = Q_g f(x_0)$, we conclude

$$\{f(x)f(x)'\colon\ x \in S_\rho\} = \{Qf(x_0)f(x_0)'Q'\colon\ Q \in \mathcal{Q}\},$$

and (4.9) follows from Theorem 2.4.

Let ξ be any given rotatable design on $\mathcal{B}_r$. Clearly,

$$\xi = \sum_{\rho \in R(\xi)} \xi(S_\rho)\, \xi_\rho,$$

for some designs ξ_ρ concentrated on S_ρ, $\rho \in R(\xi)$, and hence

$$M_d(\xi) = \sum_{\rho \in R(\xi)} \xi(S_\rho)\, M_d(\xi_\rho). \tag{4.10}$$

Note that, for any $g \in \mathcal{G}_{\text{orth}}$, $Q_g M_d(\xi_\rho) Q_g' = M_d(\xi_\rho^g)$ is the moment matrix of the design ξ_ρ^g, which is also concentrated on S_ρ. Thus, the average $\overline{M_d(\xi_\rho)}$ of $QM_d(\xi_\rho)Q'$ over $Q \in \mathcal{Q}$ belongs to the convex hull of moment matrices of designs concentrated on S_ρ, i.e., it is the moment matrix of some (rotatable) design supported on that sphere, and hence $\overline{M_d(\xi_\rho)} = \overline{M}_{d,\rho}$. Employing the orthogonal projection operator $\mathcal{P}$ from the space $\mathbb{R}^{k\times k}$ onto the linear subspace of invariant matrices from Theorem 2.4, we get $\overline{M}_{d,\rho} = \mathcal{P}(M_d(\xi_\rho))$ for all $\rho \in R(\xi)$, and rotatability of ξ yields $M_d(\xi) = \mathcal{P}(M_d(\xi))$. Equation (4.8) now follows from the linearity of the operator $\mathcal{P}$. □

For establishing our result on admissibility of rotatable designs ξ on the ball $\mathcal{B}_r$, it is convenient to introduce the following multiplicities $m_\xi(\rho)$ of the radii $\rho \in R(\xi)$,

$$m_\xi(\rho) = \begin{cases} 0, & \text{if } \rho = r, \\ 1, & \text{if } 0 < \rho < r \\ 1/2, & \text{if } \rho = 0, \end{cases} \qquad (\rho \in R(\xi)),$$

and to define the *effective cardinality of* $R(\xi)$ by

$$\gamma_\xi = \sum_{\rho \in R(\xi)} m_\xi(\rho).$$

THEOREM 4.10. *Consider the full multiple polynomial regression of degree* $d \geqslant 1$ *on the ball* $\mathcal{B}_r$. *If the design* ξ *is rotatable and* A_d*-admissible, then*

$$\gamma_\xi \leqslant (d-1)/2. \tag{4.11}$$

Conversely, if ξ *is any (not necessarily rotatable) design with* (4.11), *then it is* A_d*-admissible.*

PROOF. Let ζ be rotatable and A_d-admissible. From Lemma 4.9 we have

$$M_d(\xi) = \sum_{\rho \in R(\xi)} \xi(S_\rho)\, M_d(\eta_\rho), \tag{4.12}$$

where η_ρ is an arbitrary rotatable design concentrated on S_ρ, $\rho \in R(\xi)$. We choose the following special designs η_ρ. Firstly (whether or not r belongs to $R(\xi)$), let η_r be a rotatable design which is concentrated on S_r and is invariant under reflection at the origin (the existence of such a design follows from Lemma 4.9 and Lemma 2.6, the latter applied to the group $\{\mathrm{id}_\mathcal{X}, -\mathrm{id}_\mathcal{X}\}$). Next, for $\rho \in R(\xi)$ we choose $\eta_\rho := \eta_r^{T_{\rho/r}}$, where $T_{\rho/r}$ denotes the linear transformation $T_{\rho/r}(x) := (\rho/r)\, x$, $x \in \mathcal{B}_r$. Obviously, η_ρ is invariant under reflection at the origin, and it is concentrated on S_ρ. We have to show that η_ρ is rotatable. Clearly, the moments of η_ρ are related to those of η_r by

$$\mu_\alpha(\eta_\rho) = \left(\frac{\rho}{r}\right)^{|\alpha|} \mu(\eta_r) \quad \text{for all } \alpha,$$

hence

$$M_d(\eta_\rho) = D_\rho\, M_d(\eta_r)\, D_\rho, \tag{4.13}$$

where $D_\rho = \mathrm{diag}_{|\alpha| \leqslant d}((\rho/r)^{|\alpha|})$. By (2.5) we have for any $g \in \mathcal{G}_{\mathrm{orth}}$ that

$$M_d(\eta_\rho^g) = Q_g\, M_d(\eta_\rho)\, Q_g' = Q_g D_\rho M_d(\eta_r) D_\rho Q_g'.$$

Moreover, Equation (2.11) in Example 2.7 yields that Q_g is block-diagonal with blocks given by those row and column indices α and β, respectively, having a fixed value $t = 0, 1, \ldots, d$ for $|\alpha|$ and $|\beta|$. Hence, $Q_g D_\rho = D_\rho Q_g$, and therefore

$$M_d(\eta_\rho^g) = D_\rho Q_g M_d(\eta_r) Q_g' D_\rho = D_\rho M_d(\eta_r^g) D_\rho.$$

Now, rotatability of η_r and (4.12) ensure $M_d(\eta_\rho^g) = M_d(\eta_\rho)$ for all g, i.e., η_ρ is rotatable.

Finally, (4.13) means $M_d(\xi) = M_d(\tilde{\xi})$, where

$$\tilde{\xi} := \sum_{\rho \in R(\xi)} \xi(S_\rho)\, \eta_\rho,$$

and, since ξ is A_d-admissible, so is $\tilde{\xi}$. By our particular choice of the designs η_ρ, we conclude as follows. Choose an arbitrary support point x of η_r. Then $-x$ is also a support point of η_r, and for all $\rho \in R(\xi)$ we have $\pm\frac{\rho}{r} x \in \mathrm{supp}(\eta_\rho)$. Since $\mathrm{supp}(\tilde{\xi}) = \bigcup_{\rho \in R(\xi)} \mathrm{supp}(\xi_\rho)$, it follows that the line segment joining x and $-x$ has $2\gamma_\xi$ support points of $\tilde{\xi}$ in its relative interior. Hence, Theorem 4.4 gives $2\gamma_\xi \leqslant d - 1$.

Conversely, fix $0 \leqslant \rho_1 < \cdots < \rho_\ell < r$, $\ell = \lfloor (d-1)/2 \rfloor$, and let $R = \{\rho_1, \ldots, \rho_\ell, r\}$ if d is odd, and $R = \{0, \rho_1, \ldots, \rho_\ell, r\}$ if d is even. Consider the dth degree polynomial (in $z \in \mathbb{R}$)

$$P(z) = \sum_{i=0}^{d} p_i z^i = \begin{cases} 1 - (r - z) \prod_{i=1}^{\ell} (z - \rho_i)^2, & \text{if } d \text{ is odd,} \\ 1 - z(r - z) \prod_{i=1}^{\ell} (z - \rho_i)^2, & \text{if } d \text{ is even,} \end{cases}$$

which obviously has leading coefficient $p_d = 1$. Note that P is maximized over $[0, r]$ at $z \in R$. Define $B = \text{block-diag}(B_1, B_2)$ and $C = \text{block-diag}(C_1, C_2)$ with diagonal blocks

$$B_1 = 0_{v' \times v'}, \qquad B_2 = \mathrm{diag}_{|\alpha|=d}\left(d! \Big/ \prod_{i=1}^{v} \alpha_i! \right)$$

and

$$C_1 = \mathrm{diag}_{|\alpha| \leqslant d-1}\left(p_{|\alpha|} |\alpha|! \Big/ \prod_{i=1}^{v} \alpha_i! \right), \qquad C_2 = 0_{v'' \times v''}$$

(with $v' = \binom{v+d}{d-1}$ and $v'' = \binom{v+d-1}{d}$, for abbreviation). Then B is nonnegative definite, $\mathrm{range}(B) = \mathcal{K}^\perp$, and $C P_\mathcal{K} = C$, with $\mathcal{K}$ from Theorem 4.2. Thus, observing that

$$\begin{aligned} f(x)'(B+C)f(x) &= \sum_{|\alpha|=d} \frac{d!}{\prod_{i=1}^{v} \alpha_i!} x^{2\alpha} + \sum_{j=0}^{d-1} p_j \sum_{|\alpha|=j} \frac{j!}{\prod_{i=1}^{v} \alpha_i!} x^{2\alpha} \\ &= \|x\|^{2d} + \sum_{j=0}^{d-1} p_j \|x\|^{2j} = P(\|x\|^2) \qquad \text{for all } x \in \mathcal{B}_r, \end{aligned}$$

Theorem 3.6 ensures that any design ξ with $\mathcal{R}_\xi \subset R$ is A_d-admissible. □

5. Optimality: Gradients and first order characterizations

Under a given linear regression model (1.1a) and a given optimality criterion Φ from Definition 1.1, an optimal design ξ^* is an optimal solution to problem (1.10); equivalently, the moment matrix $M^* = M(\xi^*)$ is an optimal solution to (1.10a). The first attempts in experimental design theory aimed at finding an optimal design *analytically*, using 'first order characterizations' of optimality by means of gradients. Such characterizations are called 'equivalence theorems', among those the most famous is the Equivalence Theorem of Kiefer and Wolfowitz for D-optimality (see Kiefer and Wolfowitz, 1960). Successful applications emerged for particular regression setups and optimality criteria, e.g., for polynomial regression in one real variable on a compact interval (see Karlin and Studden, 1966; Pukelsheim, 1993, Chapter 9), and for multiple polynomial models of degree 1 or 2 on special symmetric regions (see Kiefer, 1961; Farrell et al., 1967; Atwood, 1969).

Of course, when using gradients, the assumption of differentiability of the optimality criterion comes in. The criterion Φ is differentiable at a given $A \in \mathrm{PD}(k)$ with gradient $\nabla\Phi(A) \in \mathrm{Sym}(k)$, iff

$$\lim_{\substack{B \to A \\ B \in \mathcal{A}, B \neq A}} \frac{\Phi(B) - \Phi(A) - \langle \nabla\Phi(A), B - A\rangle}{\|B - A\|} = 0, \tag{5.1}$$

where the scalar product and the induced norm on $\mathrm{Sym}(k)$ are given by

$$\langle C, D\rangle := \mathrm{tr}(CD), \qquad \|C\| := \sqrt{\langle C, C\rangle} \quad (C, D \in \mathrm{Sym}(k)).$$

Note that $\mathrm{PD}(k)$ is the interior of the domain $\mathcal{A}$ of Φ, and differentiability restricts to interior points. Computing gradients by (5.1) may be difficult. Due to the convexity of Φ, however, optimality can also be expressed in terms of the more general notion of subgradients (cf. Rockafellar, 1970, p. 214). For a given $A \in \mathcal{A}$, a matrix $G \in \mathrm{Sym}(k)$ is called a subgradient of Φ at A iff

$$\Phi(B) \geqslant \Phi(A) + \langle G, B - A\rangle \qquad \text{for all } B \in \mathcal{A}; \tag{5.2}$$

the set of all subgradients of Φ at A is called the subdifferential of Φ at A, denoted by $\partial\Phi(A)$. Subgradients of Φ exist at least at interior points of its domain, but they may also exist at boundary points of $\mathcal{A}$, which becomes relevant in case of a criterion of type (1.8b) for partial parameter systems (cf. Gaffke, 1985, Section 3, or Pukelsheim, 1993, pp. 164 ff.). If $A \in \mathrm{PD}(k)$ and if $\partial\Phi(A)$ consists of a single point G, then Φ is differentiable at A and $G = \nabla\Phi(A)$. Conversely, if Φ is differentiable at $A \in \mathrm{PD}(k)$, then $\nabla\Phi(A)$ is the unique subgradient of Φ at A. The above description of gradients or subdifferentials via (5.2) becomes in particular useful for orthogonally invariant criteria (see also Lemma 1.2).

LEMMA 5.1. *Let ψ be a real function on $(0,\infty)^k$ which is antitonic (w.r.t. the componentwise partial ordering on $(0,\infty)^k$), convex, and permutationally invariant. Consider the optimality criterion on* $\mathrm{PD}(k)$ *given by*

$$\Phi(A) = \psi\big(\lambda(A)\big) \quad \textit{for all } A \in \mathrm{PD}(k),$$

where, as before, $\lambda(A)$ denotes the vector of eigenvalues of A arranged in ascending order, $\lambda_1(A) \leqslant \cdots \leqslant \lambda_k(A)$. Then, for $A \in \mathrm{PD}(k)$,

$$\partial\Phi(A) = \big\{P\,\mathrm{diag}(g)P'\colon\ g \in \partial\psi\big(\lambda(A)\big),\ P \in \Sigma_A\big\},$$

where $\partial\psi(z)$ denotes the subdifferential of ψ at $z \in (0,\infty)^k$ (under the standard scalar product and the Euclidean norm in $\mathbb{R}^k$), and Σ_A is the set of orthogonal $k \times k$ matrices P such that $A = P\,\mathrm{diag}(\lambda(A))P'$. In particular, if ψ is differentiable at $\lambda(A)$, then so is Φ at A, and

$$\nabla\Phi(A) = P\,\mathrm{diag}\big(\nabla\psi\big(\lambda(A)\big)\big)P' \quad \textit{for all } P \in \Sigma_A.$$

PROOF. The proof is based on the following result, which is probably due to J.v. Neumann (see, e.g., Gaffke and Krafft, 1982, Lemma 5.8, p. 618).

For any two matrices $C, D \in \mathrm{Sym}(k)$ we have

$$\mathrm{tr}(CD) \leqslant \sum_{i=1}^{k} \lambda_i(C)\lambda_i(D), \tag{5.3}$$

and there is equality in (5.3) iff $C = P\,\mathrm{diag}(\lambda(C))P'$ and $D = P\,\mathrm{diag}(\lambda(D))P'$ with the same orthogonal $k \times k$ matrix P.

Now, for a given $A \in \mathrm{PD}(k)$, let $g \in \partial\psi(\lambda(A))$ and $P \in \Sigma_A$. Denote $G := P\,\mathrm{diag}(g)P'$. Then, for any $B \in \mathrm{PD}(k)$,

$$\Phi(A) + \mathrm{tr}\big(G(B-A)\big) = \psi\big(\lambda(A)\big) + \mathrm{tr}(GB) - g'\lambda(A).$$

By (5.3), $\mathrm{tr}(GB) \leqslant g'(\pi\lambda(B))$ for some permutation π, and hence, using the subgradient inequality for g and observing permutational invariance of ψ, we get

$$\begin{aligned}\Phi(A) + \mathrm{tr}\big(G(B-A)\big) &\leqslant \psi\big(\lambda(A)\big) + g'\big(\pi\lambda(B) - \lambda(A)\big)\\ &\leqslant \psi\big(\pi\lambda(B)\big) = \psi\big(\lambda(B)\big) = \Phi(B),\end{aligned}$$

that is, G is a subgradient of Φ at A.

Conversely, let $G \in \mathrm{Sym}(k)$ be a subgradient of Φ at A. Choosing $B = QAQ'$ in (5.2), with any orthogonal $k \times k$ matrix Q, we obtain

$$\mathrm{tr}(Q'GQA) \leqslant \mathrm{tr}(GA) \quad \text{for all orthogonal } k \times k \text{ matrices } Q. \tag{5.4}$$

From (5.3) it easily follows that the maximum of the left hand side of (5.4) taken over all orthogonal $Q \in \mathbb{R}^{k \times k}$ equals $\sum_{i=1}^{k} \lambda_i(G)\lambda_i(A) = \mathrm{tr}(GA)$, and thus $G = P\,\mathrm{diag}(\lambda(G))P'$ and $A = P\,\mathrm{diag}(\lambda(A))P'$ for some orthogonal $P \in \mathbb{R}^{k \times k}$. Now, choosing $B = P\,\mathrm{diag}(\mu)P'$, with any $\mu \in (0,\infty)^k$, (5.2) yields

$$\psi(\mu) \geqslant \psi(\lambda(A)) + \lambda(G)'(\mu - \lambda(A)) \quad \text{for all } \mu \in (0,\infty)^k.$$

Hence, $g := \lambda(G)$ is a subgradient of ψ at $\lambda(A)$.

If ψ is differentiable at $\lambda(A)$, the matrix $P\,\mathrm{diag}(\nabla\psi(\lambda(A)))P'$ does not depend on $P \in \Sigma_A$ (proving the statement on differentiability and the gradient of Φ at A). For, this is obvious if all the eigenvalues of A are simple. Otherwise, if some eigenvalues have multiplicities greater than 1, then the corresponding components of $\nabla\psi(\lambda(A))$ coincide, as follows from the permutational symmetry of ψ. Hence, denoting by $\lambda_1^*, \ldots, \lambda_r^*$ the *distinct* eigenvalues of A and by $g_1^*, \ldots, g_r^*$ the corresponding components of $\nabla\psi(\lambda(A))$, we see that

$$P\,\mathrm{diag}\big(\nabla\psi(\lambda(A))\big)P' = \sum_{i=1}^{r} g_i^* E_i,$$

where E_i is the orthogonal projection matrix onto the eigenspace to λ_i^*, $i = 1, \ldots, r$. □

EXAMPLE 5.2. For the Φ_p-criteria from (1.12), $-\infty \leqslant p \leqslant 1$, we have $\Phi_p(A) = \psi_p(\lambda(A))$, $A \in \mathrm{PD}(k)$, with

$$\psi_p(z) := \begin{cases} \left(\dfrac{1}{k}\displaystyle\sum_{i=1}^{k} z_i^p\right)^{-1/p}, & \text{if } p \neq -\infty, 0, \\ \left(\displaystyle\prod_{i=1}^{k} z_i\right)^{-1/k}, & \text{if } p = 0, \\ \left(\displaystyle\min_{i=1,\ldots,k} z_i\right)^{-1}, & \text{if } p = -\infty, \end{cases} \qquad z \in (0,\infty)^k.$$

Now, Lemma 5.1 yields the well known formulae for the gradients of Φ_p, if $p > -\infty$,

$$\nabla\Phi_p(A) = -\frac{1}{k}\left(\Phi_p(A)\right)^{p+1} A^{p-1}$$

(c.p. Pukelsheim (1993, p. 179), dealing with $1/\Phi_p$), where the power A^t, $A \in \mathrm{PD}(k)$ and $t \in \mathbb{R}$, is defined via the spectral decomposition of A, namely $A^t := P \operatorname{diag}(\lambda_1^t(A), \dots, \lambda_k^t(A))P'$, which does not depend on the particular choice of $P \in \Sigma_A$.

Consider now the E-criterion $\Phi_{-\infty}$. Let $z = (z_1, \dots, z_k)' \in (0, \infty)^k$, and define $I(z) := \{i \in \{1, \dots, k\}\colon z_i = \min_{j=1,\dots,k} z_j\}$. It is easily seen that the subdifferential of $\psi_{-\infty}$ at z is given by

$$\partial\psi_{-\infty}(z) = \Big\{ -\big(\psi_{-\infty}(z)\big)^2 \sum_{i \in I(z)} w_i e_i\colon w_i \geqslant 0 \text{ for all } i \in I(z), \sum_{i \in I(z)} w_i = 1 \Big\},$$

where e_i is the ith unit vector in $\mathbb{R}^k$. Denote, for $A \in \mathrm{PD}(k)$, by r the multiplicity of the smallest eigenvalue of A, and by $\mathcal{E}_{\min}(A)$ the corresponding eigenspace. Lemma 5.1 gives that the subgradients of $\Phi_{-\infty}$ at A are precisely the matrices

$$G = -\big(\Phi_{-\infty}(A)\big)^2 \sum_{i=1}^{r} w_i p_i p_i'$$

where $w_1, \dots, w_r \geqslant 0$, $\sum_{i=1}^r w_i = 1$, and $p_1, \dots, p_r$ is an orthonormal basis of $\mathcal{E}_{\min}(A)$, thus

$$\partial\Phi_{-\infty}(A) = \big\{ -\big(\Phi_{-\infty}(A)\big)^2 E\colon E \in \mathrm{NND}(k),\ \operatorname{range}(E) \subset \mathcal{E}_{\min}(A),\ \operatorname{tr}(E) = 1 \big\},$$

(cf. Kiefer, 1974, Section 4E, or Pukelsheim, 1993, Lemma 6.16).

For optimality criteria from (1.8b) related to parameter subsystems, designs with singular moment matrices may be relevant when solving problem (1.10) or (1.10a). The problem of describing the subdifferentials of Φ at singular points of its domain $\mathcal{A} = \mathcal{A}(K)$ from (1.8a) (which are boundary points of $\mathcal{A}$) turned out far from being trivial, and was completely solved in Gaffke (1985, Section 3), and in Pukelsheim (1993, Section 7.9). However, we will not present the details here, and the interested reader is referred to these references. In fact, the most important case is that of full parameter estimation, and thus leading to designs with positive definite moment matrices.

A general equivalence theorem for problem (1.10) is the following, which comes from a major result of convex analysis, cf. Rockafellar (1970, Theorem 27.4). However, if the criterion Φ is differentiable at the optimal point $M^* = M(\xi^*)$, then the equivalence is fairly obvious and can be derived by rather elementary arguments (see the remark stated below).

THEOREM 5.3. *Given a regression model* (1.1a) *and an optimality criterion* Φ, *consider the optimal design problem* (1.10) *of minimizing* $\Phi(M(\xi))$ *over all designs* ξ *with*

$M(\xi) \in \mathcal{A}$. Assume that there exists a design with positive definite moment matrix, equivalently, assume that the components of f from (1.1a) *are linearly independent on $\mathcal{X}$. Let ξ^* be a design with $M^* := M(\xi^*) \in \mathcal{A}$. Then, ξ^* is an optimal solution to* (1.10) *iff there exists a subgradient G^* of Φ at M^*, such that each support point of ξ^* is a global maximum point of the function*

$$q_{-G^*}(x) := f(x)'(-G^*)f(x), \quad x \in \mathcal{X}.$$

REMARK. Since $q_{-G^*}(x) = \operatorname{tr}(-G^* f(x)f(x)')$ and, by (1.11), the convex hull of all $f(x)f(x)'$, $x \in \mathcal{X}$, equals the set $\mathcal{M}$ of all moment matrices of designs, the condition of the theorem on the support of an optimal design can be restated as

$$\operatorname{tr}\big(G^*(M - M^*)\big) \geqslant 0 \qquad \text{for all } M \in \mathcal{M}.$$

Hence, if Φ is differentiable at M^*, then $G^* = \nabla\Phi(M^*)$, and the theorem simply means, that a matrix M^* is optimal iff the directional derivatives of Φ at M^* are nonnegative for all feasible directions. In this case the stated equivalence is fairly obvious and easily proved.

6. Reduction of dimensionality

In problem (1.10a) the variable is the moment matrix of a design, which may cause a large dimension of the optimization problem, especially for multiple regression models, such as multiple polynomial models of degree two or three. When equivariance properties can be utilized, the restriction to invariant designs according to Lemma 2.6 and Lemma 2.10 may reduce the dimension considerably (see the examples below). In order to include the possibility of such reductions, it is convenient to state the extremum problem in a more general way, giving, as a byproduct, a deeper insight into the specific structure of problem (1.10a).

To this end, and as a price we have to pay, a more abstract frame has to be considered. Firstly, the underlying space is now a real Hilbert space $\mathcal{H}$ *of finite dimension* (with scalar product and norm denoted by $\langle\cdot,\cdot\rangle$ and $\|\cdot\|$, respectively). In applications, this will be either the space $\operatorname{Sym}(k)$ with scalar product $\langle A, B\rangle = \operatorname{tr}(AB)$ as before, or the (column vector) space $\mathbb{R}^l$ with the usual scalar product $\langle a, b\rangle = a'b$. The ingredients of (1.10a), Φ, $\mathcal{M}$, and $\mathcal{A}$, carry over to the following. Let $\mathcal{M}$ and $\mathcal{A}$ be convex subsets of the Hilbert space $\mathcal{H}$ ($\mathcal{A}$ needs not to be a cone), such that $\mathcal{M}$ is compact, and $\mathcal{M} \cap \operatorname{int}(\mathcal{A}) \neq \emptyset$, where $\operatorname{int}(\mathcal{A})$ denotes the interior of $\mathcal{A}$, and Φ is a real valued convex function on $\mathcal{A}$. The extremum problem now reads as

$$\text{minimize } \Phi(m) \text{ over } m \in \mathcal{M} \cap \mathcal{A}. \tag{6.1}$$

The intersection of $\mathcal{M}$ with $\mathcal{A}$ expresses the original condition of identifiability of parameters under the competing designs, now restated in terms of the (possibly) reduced variable m. As a particular feature of this problem, translating (1.11) into the more abstract framework, the set $\mathcal{M}$ is assumed to be given in the form

$$\mathcal{M} = \operatorname{Conv}\big\{m(x) \colon x \in X\big\}, \tag{6.2}$$

where $\{m(x)\colon\ x \in X\}$ is a given, compact set of points $m(x)$ in $\mathcal{H}$ (where X may be any nonempty set). For reconstructing a design ξ associated to some $m \in \mathcal{M}$ it will be important that for any point $m(x)$ from the generating family an associated design $\bar{\xi}_x$, say, is known, whence the problem of finding ξ becomes a decomposition problem,

$$\begin{aligned} &\text{find } r \in \mathbb{N},\quad x_1,\dots,x_r \in X,\ w_1,\dots,w_r > 0,\ \text{with } \sum_{i=1}^{r} w_i = 1,\\ &\text{such that } \sum_{i=1}^{r} w_i m(x_i) = m. \end{aligned} \tag{6.3}$$

Then, a design corresponding to m is the mixture $\xi = \sum_{i=1}^{r} w_i \bar{\xi}_{x_i}$. For obtaining implementable versions of our conceptual algorithm described below, an implicit assumption is that the family $m(x)$, $x \in X$, has a fairly simple structure, in the sense that any *linear* extremum problem over these points,

$$\text{minimize } \langle a, m(x)\rangle \text{ over } x \in X,$$

should be easily solvable, for any given $a \in \mathcal{H}$.

We note that (6.1) and (6.2) cover the original problem (1.10a) (under regression model (1.1a)), with the settings $\mathcal{H} = \mathrm{Sym}(k)$, $X = \mathcal{X}$, and $m(x) = f(x)f(x)'$. On the other hand, possible reductions of the original problem by invariance or admissibility (or both) is accounted for, as illustrated by the following examples.

EXAMPLE 6.1. Consider a quadratic multiple polynomial setup (i.e., (2.7) with $d = 2$), on a symmetric cube or ball, centered at zero. Assume that the index set A is permutationally symmetric, so that the model is equivariant w.r.t. the transformation group $\mathcal{G}_{\mathrm{sp}}$ and the associated matrix group $\mathcal{Q}$ (see Example 2.8(c)). For any invariant (w.r.t. $\mathcal{Q}$) optimality criterion Φ (and thus, in particular, for any orthogonally invariant criterion), the optimal design problem (1.10) can be restricted to invariant and A_2-admissible designs, as follows from Lemma 2.6, Lemma 2.10 and Lemma 4.1. We thus obtain a reduction (6.1), (6.2) to only two dimensions, as we will demonstrate now.

By the sign change and permutation invariance, the moment matrix of an invariant design ξ includes only three nontrivial moments, namely

$$\begin{aligned} &\mu_2(\xi) := \mathrm{E}_\xi\left(x_i^2\right),\quad \mu_4(\xi) := \mathrm{E}_\xi\left(x_i^4\right)\quad (\text{independent of } i = 1,\dots,v)\\ &\text{and}\\ &\mu_{2,2}(\xi) := \mathrm{E}_\xi\left(x_i^2 x_j^2\right)\quad (\text{independent of } 1 \leqslant i \neq j \leqslant v) \end{aligned} \tag{6.4}$$

(where E_ξ denotes the expectation w.r.t. the probability measure ξ).

(a) Let $\mathcal{X} = \mathcal{C}_c = \{x = (x_1,\dots,x_v)' \in \mathbb{R}^v\colon\ \max_{i=1,\dots,v}|x_i| \leqslant c\}$, for a given $c > 0$. By Corollary 4.7(a), an invariant design ξ is A_2-admissible, if and only if each

$x \in \operatorname{supp}(\xi)$ has all its coordinates in $\{0, \pm c\}$. Hence the moments from (6.4) are given by

$$\begin{aligned} \mu_4(\xi) &= c^2\mu_2(\xi) = \frac{c^2}{v}\,\mathrm{E}_\xi\left(\|x\|^2\right), \\ \mu_{2,2}(\xi) &= \frac{1}{v(v-1)}\left(\mathrm{E}_\xi\left(\|x\|^4\right) - c^2\,\mathrm{E}_\xi\left(\|x\|^2\right)\right), \end{aligned} \tag{6.5}$$

where the last equation in (6.5) comes from

$$\begin{aligned} \mu_{2,2}(\xi) &= \frac{1}{v(v-1)}\,\mathrm{E}_\xi\left(\sum_{h\neq l} x_h^2 x_l^2\right) \\ &= \frac{1}{v(v-1)}\,\mathrm{E}_\xi\left(\left(\sum_{h=1}^{v} x_h^2\right)^2\right) - \frac{1}{v(v-1)}\,\mathrm{E}_\xi\left(\sum_{h=1}^{v} x_h^4\right) \\ &= \frac{1}{v(v-1)}\left(\mathrm{E}_\xi\left(\|x\|^4\right) - c^2\,\mathrm{E}_\xi\left(\|x\|^2\right)\right). \end{aligned}$$

So, by (6.5), the moment matrices $M(\xi)$ of invariant and A_2-admissible designs ξ are linearly parameterized by the two-dimensional moment vector

$$m(\xi) := \left(\mathrm{E}_\xi\left(\|x\|^2\right), \mathrm{E}_\xi\left(\|x\|^4\right)\right)'. \tag{6.6}$$

It remains to describe the range $\mathcal{M}$ of all moment vectors $m = m(\xi)$ when ξ ranges over all invariant and A_2-admissible designs. To this end, denote by $\bar{\xi}_l$ the uniform distribution over those vertices of the cube which have l coordinates $\pm c$ and $v - l$ coordinates zero, $l = 1, \ldots, v$. Since any invariant design whose support consists of the vertices of the cube only, is a mixture of the designs $\bar{\xi}_l$, $l = 0, 1, \ldots, v$, we obtain

$$\begin{aligned} \mathcal{M} &= \operatorname{Conv}\left\{m(\bar{\xi}_l)\colon\ l = 0, 1, \ldots, v\right\} \\ &= \operatorname{Conv}\left\{\left(lc^2, l^2c^4\right)\colon\ l = 0, 1, \ldots, v\right\}. \end{aligned} \tag{6.7}$$

(b) Let $\mathcal{X} = \mathcal{B}_r = \{x = (x_1, \ldots, x_v)' \in \mathbb{R}^v\colon \sum_{i=1}^{v} x_i^2 \leqslant r^2\}$, for a given $r > 0$. By Corollary 4.7(b), an invariant design ξ is A_2-admissible iff $\operatorname{supp}(\xi) \subset \partial\mathcal{B}_r \cup \{0\}$. Hence, for these designs,

$$\begin{aligned} \mu_{2,2} &= \frac{1}{v(v-1)}\,\mathrm{E}_\xi\left(\sum_{i\neq j} x_i^2 x_j^2\right) = \frac{1}{v(v-1)}\left(\mathrm{E}_\xi\left(\|x\|^4\right) - \mathrm{E}_\xi\left(\sum_{i=1}^{v} x_i^4\right)\right) \\ &= \frac{r^2}{v-1}\left(\mu_2(\xi) - \mu_4(\xi)\right). \end{aligned}$$

So, the moment matrices of invariant and A_2-admissible designs are linearly parameterized by the two-dimensional moment vector

$$m(\xi) := \big(v\mu_2(\xi), v\mu_4(\xi)\big)'.$$

As we will show next, the range of these moment vectors is given by the triangle

$$\mathcal{M} = \mathrm{Conv}\big\{(0,0), (r^2, \tfrac{1}{v}r^4), (r^2, r^4)\big\}. \tag{6.8}$$

Note that the three vertices correspond to the invariant designs ξ_0, ξ_1, and ξ_2, where ξ_0 denotes the one-point design at zero, ξ_1 the uniform distribution over the 2^v points with coordinates $\pm r/\sqrt{v}$, and ξ_2 the uniform distribution over the $2v$ points $\pm re_i$, $i = 1, \ldots, v$ (where e_i denotes the ith unit vector in $\mathbb{R}^v$). Thus, for verifying (6.8) we firstly remark that any invariant and A_2-admissible design ξ is a mixture of ξ_0 and of an invariant design η concentrated on the surface of the ball. For such designs η we have

$$v\mu_2(\eta) = \mathrm{E}_\eta\left(\sum_{i=1}^{v} x_i^2\right) = r^2 \quad \text{and} \quad v\mu_4(\eta) = \mathrm{E}_\eta\left(\sum_{i=1}^{v} x_i^4\right).$$

Now, as η varies, the moment $\mu_4(\eta)$ ranges over the interval $[\mu_{4,\min}, \mu_{4,\max}]$, where $\mu_{4,\min}$ and $\mu_{4,\max}$ are the minimum and the maximum value, respectively, of $\sum_{i=1}^{v} x_i^4$ taken over the sphere $\sum_{i=1}^{v} x_i^2 = r^2$, and hence, as it is easily seen, $\mu_{4,\min} = r^4/v$ and $\mu_{4,\max} = r^4$.

EXAMPLE 6.2. Consider cubic multiple polynomial regression, i.e., (2.7) with $d = 3$, again on the symmetric cube $\mathcal{C}_c$ or the ball $\mathcal{B}_r$ centered at zero, and for a permutationally invariant index set A, ensuring equivariance w.r.t. the transformation group $\mathcal{G}_{\mathrm{sp}}$ and the associated matrix group $\mathcal{Q}$. Again, for solving an optimal design problem with an invariant (w.r.t. $\mathcal{Q}$) optimality criterion, we may restrict ourselves to invariant designs. The moment matrices of these are linearly parameterized by the moment vector

$$m(\xi) = \big(\mu_2(\xi), \mu_4(\xi), \mu_6(\xi), \mu_{2,2}(\xi), \mu_{4,2}(\xi), \underbrace{\mu_{2,2,2}(\xi)}_{\text{if } v\geqslant 3}\big)', \tag{6.9}$$

where

$$\mu_t(\xi) := \mathrm{E}_\xi\big(x_i^2\big), \quad t = 2, 4, 6 \text{ (independent of } i = 1, \ldots, v),$$

$$\mu_{2,2}(\xi) := \mathrm{E}_\xi\big(x_i^2 x_j^2\big) \quad \text{(independent of } 1 \leqslant i \neq j \leqslant v),$$

$$\mu_{4,2}(\xi) := \mathrm{E}_\xi\big(x_i^4 x_j^2\big) \quad \text{(independent of } 1 \leqslant i \neq j \leqslant v),$$

$$\mu_{2,2,2}(\xi) := \mathrm{E}_\xi\left(x_h^2 x_i^2 x_j^2\right), \quad \text{if } v \geqslant 3 \text{ (independent of } 1 \leqslant h < i < j \leqslant v).$$

From Theorem 4.8 it follows that the range of the moment vectors (6.9) is given by

$$\mathcal{M} = \mathrm{Conv}\{m(x)\colon\ x \in \mathcal{X}_0\}, \tag{6.10}$$

where $m(x) := m(\bar{\xi}_x)$, $\bar{\xi}_x$ denotes the uniform distribution on the $\mathcal{G}_{\mathrm{sp}}$-orbit of x, and the set $\mathcal{X}_0$ is defined by (4.7a) or (4.7b) as a union of finitely many line segments of the cube or the ball. Contrary to the preceding example, now the family of moment vectors $m(x)$, $x \in \mathcal{X}_0$, is infinite; nevertheless, as a union of finitely many line segments, its structure is still simple enough to solve easily linear extremum problems

$$\text{minimize } a'm(x) \text{ over } x \in \mathcal{X}_0.$$

For, the usual parameterization of a line segment by $\lambda \in [0, 1]$, say, yields $a'm(x_\lambda)$ as a cubic polynomial in λ^2 on that segment. We omit the lengthy formulae here.

EXAMPLE 6.3. Consider a *rotatable* multiple polynomial model (2.7) of arbitrary, but fixed degree $d \geqslant 1$ on the ball $\mathcal{X} = \mathcal{B}_r$. Suppose that for the optimal design problem under consideration the restriction to *rotatable* designs, see Example 2.8(d), is justified, (as it is true for the D- and I-criterion, for example). By Lemma 4.9, the moment matrix of a rotatable design can be decomposed according to (4.11), and from the proof of that lemma we see that $\overline{M}_{d,\rho} = D_\rho \overline{M}_{d,1} D_\rho$, where D_ρ denotes the $\binom{v+d}{d} \times \binom{v+d}{d}$ diagonal matrix with diagonal entries $\rho^{|\alpha|}$, $|\alpha| \leqslant d$. Hence it follows that the moment matrices of rotatable designs ξ are linearly parameterized by the vector

$$m(\xi) := \textstyle\sum_{\rho \in R(\xi)} \xi(S_\rho)\, m(\rho), \qquad \text{where } m(\rho) := \left(\rho^2, \rho^4, \ldots, \rho^{2d}\right)', \quad 0 \leqslant \rho \leqslant r. \tag{6.11}$$

Obviously, the range $\mathcal{M}$ of $m = m(\xi)$, when ξ ranges over all rotatable designs, is

$$\mathcal{M} = \mathrm{Conv}\{m(\rho)\colon\ 0 \leqslant \rho \leqslant r\}. \tag{6.12}$$

For an algorithmic solution of an optimal rotatable design problem, explicit formulae for the entries of $\overline{M}_{d,\rho}$ are required. To this end it suffices to describe the entries of $\overline{M}_{d,1}$, i.e., the moments $\bar{\mu}_\gamma$ of the uniform distribution over the unit sphere, for all multi-indices γ with $|\gamma| \leqslant 2d$. These can be obtained as follows (again avoiding a mathematical definition of the uniform distribution over the sphere). By Lemma 4.9 there exists a design $\bar{\eta}$ concentrated on the unit sphere, and such that $M_d(\bar{\eta}) = \overline{M}_{d,1}$. So, the entries of $\overline{M}_{d,1}$ are the moments $\mu_\gamma(\bar{\eta})$, $|\gamma| \leqslant 2d$. Clearly, since the sign change transformation group is a subgroup of $\mathcal{G}_{\mathrm{orth}}$, we have $\mu_\gamma(\bar{\eta}) = 0$ whenever some component of γ is odd. Now let $\gamma = 2\beta$, $1 \leqslant q := |\beta| \leqslant d$. Consider, for a given $q = 1, \ldots, d$, the function $F_q(t) := \mathrm{E}_{\bar{\eta}}((t'x)^{2q})$, $t \in \mathbb{R}^v$. The multinomial formula yields that

$$F_q(t) = \sum_{\alpha:\ |\alpha|=2q} \frac{(2q)!}{\alpha_1! \cdots \alpha_v!} \mu_\alpha(\bar{\eta}) t^\alpha. \tag{6.13}$$

On the other hand, for a fixed $t \neq 0$, there is a transformation $g_t \in \mathcal{G}_{\mathrm{orth}}$, such that $g_t(t/\|t\|) = (1, 0, \ldots, 0)'$, and rotatability of $\bar{\eta}$ entails

$$\begin{aligned} F_q(t) &= \|t\|^{2q}\, \mathrm{E}_{\bar{\eta}}\Big(\big(t'x/\|t\|\big)^{2q}\Big) = \|t\|^{2q}\, \mathrm{E}_{\bar{\eta}}\Big(\big(t'g_t(x)/\|t\|\big)^{2q}\Big) \\ &= \|t\|^{2q}\, \mathrm{E}_{\bar{\eta}}\Big(\big(x'g_t(t/\|t\|)\big)^{2q}\Big) = \|t\|^{2q}\, \mathrm{E}_{\bar{\eta}}\big(x_1^{2q}\big). \end{aligned}$$

Abbreviating $\mu_{2q}(\bar{\eta}) := \mathrm{E}_{\bar{\eta}}(x_1^{2q})$, and applying the multinomial formula to $\|t\|^{2q} = (\sum_{i=1}^v t_i^2)^q$, we obtain

$$F_q(t) = \mu_{2q}(\bar{\eta}) \sum_{\alpha:\ |\alpha|=q} \frac{q!}{\alpha_1! \cdots \alpha_v!} t^{2\alpha}. \tag{6.14}$$

Comparison of the coefficients of $t^{2\beta}$ in the representations (6.13) and (6.14) yields

$$\mu_{2\beta}(\xi) = \mu_{2q}(\bar{\eta}) \frac{q!}{\beta_1! \cdots \beta_v!} \frac{(2\beta_1)! \cdots (2\beta_v)!}{(2q)!}, \quad q = |\beta|. \tag{6.15}$$

It remains to find a formula for the moment $\mu_{2q}(\bar{\eta}) = \mathrm{E}_{\bar{\eta}}(x_1^{2q})$. To this end, we apply the multinomial formula to $(\sum_{i=1}^v x_i^2)^{2q}$, for any x from the unit sphere, and we obtain

$$\sum_{\alpha:\ |\alpha|=q} \frac{q!}{\alpha_1! \cdots \alpha_v!} x^{2\alpha} = 1,$$

hence

$$\sum_{\alpha:\ |\alpha|=q} \frac{q!}{\alpha_1 \cdots \alpha_v} \mu_{2\alpha}(\bar{\eta}) = 1. \tag{6.16}$$

Now, inserting (6.15) (with $\beta = \alpha$) into (6.16), we are left with

$$\mu_{2q}(\bar{\eta}) = \binom{2q}{q} \Big/ \left(\sum_{\alpha:\ |\alpha|=q} \binom{2\alpha_1}{\alpha_i} \cdots \binom{2\alpha_v}{\alpha_v} \right). \tag{6.17}$$

Formulae (6.15) and (6.17) together give the moments $\mu_{2\beta}(\bar{\eta})$, $|\beta| \leqslant d$, of $\bar{\eta}$, and thereby the entries of $\overline{M}_{d,1}$.

Finally we note, that explicit knowledge of a design $\bar{\eta}$ from above will be necessary to construct a rotatable design associated to a given vector $m \in \mathcal{M}$ from (6.13). For, if a decomposition of m is available,

$$m = \sum_{i=1}^{s} w_i m(\rho_i),$$

where $s \in \mathbb{N}, 0 \leqslant \rho_i \leqslant r, w_i > 0$ for all $i = 1, \ldots, s$, with $\sum_{i=1}^{v} w_i = 1$, then the mixture

$$\xi = \sum_{i=1}^{s} w_i \bar{\eta}_{\rho_i}$$

is a rotatable design as required, where $\bar{\eta}_\rho$ can be chosen as $\bar{\eta}_\rho = \bar{\eta}^{T_\rho}$, with T_ρ denoting the transformation $T_\rho(x) = \rho x$ (for any $\rho \in [0, r]$). Explicit descriptions of $\bar{\eta}$ depend on the degree d of the model.

7. Conceptual algorithm

We consider the extremum problem (6.1) within the general framework outlined in Section 6 above. Throughout we will assume that Φ is differentiable on $\operatorname{int}(\mathcal{A})$ with gradients $\nabla\Phi(a)$, $a \in \operatorname{int}(\mathcal{A})$, all of which are elements of $\mathcal{H}$. Later on, for our second order methods, we will assume that Φ is twice continuously differentiable on $\operatorname{int}(\mathcal{A})$. The Hessian of Φ at $a \in \operatorname{int}(\mathcal{A})$ is a linear operator from $\mathcal{H}$ into $\mathcal{H}$, which is nonnegative definite due to the convexity of Φ. The conceptual algorithm presented next generates a sequence of points $m_n \in \mathcal{M} \cap \operatorname{int}(\mathcal{A})$, $n = 1, 2, \ldots$, converging to an optimal solution to (6.1) as n tends to infinity. The structure of the procedure is outlined by the following steps (o)–(iii), where the initial step (o) is carried out once, while steps (i)–(iii) are iterated.

(o) *Choose a starting point* $m_1 \in \mathcal{M} \cap \operatorname{int}(\mathcal{A})$. *Go to step* (i)

(i) (At stage n, let $m_n \in \mathcal{M} \cap \operatorname{int}(\mathcal{A})$ be the actual point, and denote $g_n := \nabla\Phi(m_n)$.)
Choose an $\bar{m}_n \in \mathcal{M}$ *such that* $\langle g_n, \bar{m}_n - m_n \rangle < 0$. *Go to step* (ii).

(ii) *Choose an* $\alpha_n \in (0, 1]$ *such that* $\Phi((1 - \alpha_n)m_n + \alpha_n \bar{m}_n) < \Phi(m_n)$, *where* $\alpha_n = 1$ *is allowed only if* $\bar{m}_n \in \operatorname{int}(\mathcal{A})$. *Go to step* (iii).

(iii) *Choose an* $m_{n+1} \in \mathcal{M} \cap \operatorname{int}(\mathcal{A})$ *such that* $\Phi(m_{n+1}) \leqslant \Phi((1-\alpha_n)m_n + \alpha_n \bar{m}_n)$.
(A standard choice is $m_{n+1} = (1 - \alpha_n)m_n + \alpha_n \bar{m}_n$.)
Replace n *by* $n + 1$ *and go to step* (i).

The above (o)–(iii) only constitute a *conceptual* algorithm, as it remains open *how* to compute the search direction $\bar{m}_n$ in step (i), and *how* to perform the line search in step (ii) (though the latter can easily be made precise in a reasonable way, see below). In fact, the most important point for the convergence behavior of the algorithm will be the choice of $\bar{m}_n \in \mathcal{M}$ in step (i). Also, the conditions stated in (i) on $\bar{m}_n$ and in (ii) on the step size α_n do not guarantee convergence to the optimum. Procedures for choosing $\bar{m}_n$ and a line search procedure ensuring convergence will be discussed below. We note that no specific knowledge of the set $\mathcal{A}$ is needed. Only, at the initial stage, a starting point in $\mathcal{M} \cap \operatorname{int}(\mathcal{A})$ must be provided, and if one wishes to choose a step length $\alpha_n = 1$ in (ii), then one has to decide whether $\bar{m}_n$ belongs to the interior of $\mathcal{A}$. On the other hand, specific knowledge of the set $\mathcal{M}$ is given by (6.2) and will be used for computing a reasonable vector $\bar{m}_n \in \mathcal{M}$ in step (i).

A line search procedure for step (ii), which has been found to be efficient in practice, is a modification of that in Fletcher (1987, Section 2.6), described in Gaffke and Heiligers (1996).

Let us turn to the problem of computing the search directions $\bar{m}_n$ in step (i). The very first, but inefficient choice is the *steepest descent* method, i.e., to choose $\bar{m}_n \in \mathcal{M}$ as a minimizer of the directional derivatives $\langle g_n, m - m_n \rangle$ over $m \in \mathcal{M}$. By (6.2), this can be done by solving the linear extremum problem over the generating family of $\mathcal{M}$,

$$\text{minimize } \big\langle g_n, m(x) \big\rangle \text{ over } x \in X, \tag{7.1}$$

and then taking $\bar{m}_n = m(x^*)$, where x^* is an optimal solution to (7.1) (which will depend on n, but this dependence is dropped here for simplicity of notation). An improvement over the steepest descent method may be obtained, by choosing a 'bundle' of points $x_1, \ldots, x_s \in X$ (possibly depending on n), among them a minimizer of (7.1), and to choose $\bar{m}_n$ as a mixture of the $m(x_i)$, $i = 1, \ldots, s$,

$$\bar{m}_n = \sum_{i=1}^{s} w_i m(x_i), \tag{7.2}$$

with suitable weights $w_1, \ldots, w_s \geqslant 0$ (which may also depend on n), such that $\sum_{i=1}^{s} w_i = 1$. For particular problems, a good choice of the bundle $x_1, \ldots, x_s$ turned out to be the set of all *local* minimum points in X of the linear (in $m(x)$) function from (7.1). This was used in Gaffke and Heiligers (1995b) for symmetric multiple polynomial models of degree $d = 3$ with invariance reduction as in Example 6.2 above. The choice of all local minimum points is particularly suitable when x is a one-dimensional variable ranging over a compact interval X. Of course, for problems with *finite* X, as in Example 6.1, the bundle may be chosen as the whole set X, independent of n.

Let us comment upon the choice of the weights w_i in (7.2). In Gaffke and Mathar (1992, Section 4.1.1) and Gaffke and Heiligers (1995a, b, Section 3.1), the weights were based on the descent values,

$$d(x_i) := \big\langle g_n, m(x_i) - m_n \big\rangle, \quad i = 1, \ldots, s,$$

and fancy rules were found, improving considerably the convergence behavior of the algorithm over the steepest descent method. Also, the Silvey–Titterington–Torsney method for the case of a finite X can be subsumed under these mixture rules (cf. Gaffke and Mathar, 1992, Section 4.2.2).

Convergence to the optimum of the above 'gradient methods' can be proved under the following assumptions. Firstly, suppose that the set

$$\{m \in \mathcal{M} \cap \operatorname{int}(\mathcal{A})\colon\ \Phi(m) \leqslant \Phi(m_1)\} \quad \text{is compact,} \tag{7.3}$$

where m_1 is the starting point from step (o) of the algorithm. Note that (7.3) is a condition on the extremum problem (6.1) rather than on the algorithm. Secondly, assume that the (absolute) descent values,

$$\varepsilon_n := \Big| \min_{m \in \mathcal{M}} \langle g_n, m - m_n \rangle \Big| = \Big| \min_{x \in X} \langle g_n, m(x) - m_n \rangle \Big| \tag{7.4}$$

and the (absolute) descent value achieved by $\bar{m}_n$ from step (i) are comparable, in the sense that

$$\big| \langle g_n, \bar{m}_n - m_n \rangle \big| \geqslant h(\varepsilon_n) \quad \text{for all } n, \tag{7.5}$$

for some fixed isotonic function $h\colon\ [0, \infty) \to [0, \infty)$ with $h(t) = 0$ iff $t = 0$.

THEOREM 7.1. *Let the sequence m_n, $n = 1, 2, \ldots$, be generated by algorithm* (o)–(iii), *where in step* (ii) *the modified Fletcher line search procedure is used. Assume that* (7.3) *and* (7.5) *hold true. Then, as n tends to infinity, the sequence m_n converges to the set of optimal solutions of problem* (6.1), *and the ε_n from* (7.4) *converge to zero.*

PROOF. By Gaffke and Heiligers (1996, Lemma 2.1), the modified Fletcher line search (under (7.3)) implies

$$\lim_{n \to \infty} \langle g_n, \bar{m}_n - m_n \rangle = 0.$$

Hence, by assumption (7.5), $\lim_{n \to \infty} \varepsilon_n = 0$, and the proof is completed as on p. 105 in Gaffke and Mathar (1992). □

Due to the convergence of the (absolute) steepest descent values ε_n to zero, along with the inequality obtained from (5.2), checking whether or not the current point m_n fulfills

$$\Phi(m_n) - \min_{m \in \mathcal{M} \cap \mathcal{A}} \Phi(m) \leqslant \varepsilon_n \tag{7.6}$$

provides a reasonable stopping rule for the algorithm. In case that Φ is positive, the above inequality rewrites as

$$\frac{\min_{m \in \mathcal{M} \cap \mathcal{A}} \Phi(m)}{\Phi(m_n)} \geqslant 1 - \frac{\varepsilon_n}{\Phi(m_n)}. \tag{7.7a}$$

Using the lower bound in (7.7a) as a stopping criterion might be preferable over the scale dependent difference (7.6), as it bounds the scale independent 'relative efficiency' from (1.15). As mentioned in Section 1, in most cases the function Φ is positive and, moreover, $1/\Phi$ is concave. Then the lower bound in (7.7a) can be improved by

$$\frac{\min_{m\in\mathcal{M}\cap\mathcal{A}}\Phi(m)}{\Phi(m_n)} \geqslant \frac{\Phi(m_n)}{\Phi(m_n)+\varepsilon_n} \tag{7.7b}$$

(cf. Gaffke and Mathar, 1992, p. 95).

When using a mixture (7.2) for the search direction $\bar{m}_n$ in step (i), it turned out that the most efficient choice of the weights w_i is obtained by minimizing a local quadratic approximation of Φ. Thereby we switch from a pure gradient method to a 'second order' (or 'Quasi-Newton') method, whose good *global* and excellent *local* convergence behavior we observed for particular problems (cf. Gaffke and Heiligers, 1995a, b). Let a local (at the current point m_n) quadratic approximation of Φ be given by

$$q_n(m) := \Phi(m_n) + \langle g_n, m - m_n\rangle + \tfrac{1}{2}\big\langle H_n(m-m_n), m-m_n\big\rangle, \quad m \in \mathcal{M},$$

where the approximation H_n to the Hessian operator of Φ at m_n is a nonnegative definite linear operator on the Hilbert space $\mathcal{H}$. For example, H_n may be the usual BFGS approximation (cf. Fletcher, 1987, pp. 55–56, or Gaffke and Heiligers, 1996, equation (2.10)), or the Hessian itself. As above, let a bundle $x_1, \ldots, x_s$ be available, among them a global minimizer of $\langle g_n, m(x)\rangle$ over $x \in X$. Then the problem to be solved is

$$\text{minimize } q_n(m) \text{ over } m \in \operatorname{Conv}\big\{m_n, m(x_1), \ldots, m(x_s)\big\}, \tag{7.8}$$

which can be done by the Higgins–Polak method (see Gaffke and Heiligers (1996) for a description of a suitable version). This method yields an optimal solution m_n^* of (7.8) *and* a decomposition of m_n^*

$$m_n^* = w_0^* m_n + \sum_{i=1}^{s} w_i^* m(x_i),$$

with $w_1^*, \ldots, w_s^* \geqslant 0$ and $\sum_{i=1}^{s} w_i^* = 1$. Then we take in step (i) of the overall algorithm

$$\bar{m}_n = \sum_{i=1}^{s} \bar{w}_i m(x_i), \quad \text{where } \bar{w}_i := \frac{w_i^*}{1-w_0^*}, \quad i = 1, \ldots, s. \tag{7.9}$$

Convergence to the optimum, when using (7.9) in step (i) and the modified Fletcher line search procedure in step (ii), was proved in Gaffke and Heiligers (1996, Theorems 2.1, 2.2), for the cases that the H_n are the Hessians or the BFGS approximations, and under assumption (7.3).

As a further advantage of these second order methods, the decomposition problem (6.3) gets a practical solution at the final stage of the iterations, and thus an associated optimal design is obtained. For, as it turned out, the difference between the actual point m_n and $\bar{m}_n$ is negligible at the end of iterations, so that (7.9) provides a desired decomposition, and thus an associated design. Moreover, by the Higgins-Polak method, the supporting points in (7.9) (i.e., those $m(x_i)$ with positive $\bar{w}_i$), form an affinely independent family. As a consequence, the support size of the obtained optimal design is limited, which is particular advantageous when dealing with high dimensional regression models combined with fairly large transformation groups, as those in the examples from Section 6.

References

Atwood, C. L. (1969). Optimal and efficient designs of experiments. *Ann. Math. Statist.* **40**, 1570–1602.

Farrell, R. H., J. Kiefer and A. Walbran (1967). Optimum multivariate designs. In: J. Neyman, ed., *Proc. Fifth Berkeley Symp. on Math. Statist. Probab. Theory*, Vol. 1. University of California, Berkeley, CA, 113–138.

Fletcher, R. (1987). *Practical Methods of Optimization.* 2nd edn. Wiley, New York.

Gaffke, N. (1985). Singular information matrices, directional derivatives, and subgradients in optimal design theory. In: T. Caliński and W. Klonecki, eds., *Linear Statistical Inference. Proc. Internat. Conf. on Linear Inference, Poznań 1984.* Lecture Notes in Statistics 35. Springer, Berlin, 61–77.

Gaffke, N. (1987). Further characterizations of design optimality and admissibility for partial parameter estimation in linear regression. *Ann. Statist.* **15**, 942–957.

Gaffke, N. and B. Heiligers (1995a). Algorithms for optimal design with application to multiple polynomial regression. *Metrika* **42**, 173–190.

Gaffke, N. and B. Heiligers (1995b). Computing optimal approximate invariant designs for cubic regression on multidimensional balls and cubes. *J. Statist. Plann. Inference* **47**, 347–376.

Gaffke, N. and B. Heiligers (1996). Second order methods for solving extremum problems from optimal linear regression design. *Optimization* **36**, 41–57.

Gaffke, N. and O. Krafft (1982). Matrix inequalities in the Loewner-ordering. In: B. Korte, ed., *Modern Applied Mathematics: Optimization and Operations Research.* North-Holland, Amsterdam, 592–622.

Gaffke, N. and R. Mathar (1992). On a class of algorithms from experimental design theory. *Optimization* **24**, 91–126.

Heiligers, B. (1991). Admissibility of experimental designs in linear regression with constant term. *J. Statist. Plann. Inference* **28**, 107–123.

Karlin, S. and W. J. Studden (1966). Optimal experimental designs. *Ann. Math. Statist.* **37**, 783–815.

Kiefer, J. (1959). Optimum experimental designs. *J. Roy. Statist. Soc. Ser. B* **21**, 272–304.

Kiefer, J. (1960). Optimum experimental designs V, with applications to systematic and rotatable designs. In: J. Neyman, ed., *Proc. Fourth Berkeley Symp. on Math. Statist. Probab. Theory*, Vol. 1. University of California, Berkeley, CA, 381–405.

Kiefer, J. (1961). Optimum design in regression problems II. *Ann. Math. Statist.* **32**, 298–325.

Kiefer, J. (1974). General equivalence theory for optimum designs (approximate theory). *Ann. Statist.* **2**, 849–879.

Kiefer, J. and J. Wolfowitz (1960). The equivalence of two extremum problems. *Canadian. J. Math.* **12**, 363–366.

Krafft, O. (1983). A matrix optimization problem. *Lin. Algebra Appl.* **51**, 137–142.

Marshall, A. W. and I. Olkin (1979). *Inequalities: Theory of Majorization and its Applications.* Academic Press, New York.

Pukelsheim, F. (1993). *Optimal Design of Experiments.* Wiley, New York.

Rockafellar, R. T. (1970). *Convex Analysis.* Princeton Univ. Press, Princeton, NJ.

Subject Index

A-criterion 898, 979
A-optimal design 774
A-optimal design for treatment control contrasts 1008, 1010, 1012, 1015, 1017, 1022, 1030, 1032
A-optimality 896, 1068, 1111, 1113, 1116, 1155
absorption 35
Accelerated Life Testing (ATL) 145
adaptive Bayesian designs 162
adaptive designs 151, 154, 156, 157, 165
adaptive procedure 157
– consistency of 158
adaptive R-estimator 98
Addelman plans 426
additive effect 192
adjusted p-value 591
adjusted cell means 54
adjusted design matrix
– for blocks 814
– for treatments 814
adjusted orthogonality 912, 983
adjustment factor 205, 214, 235
admissibility of design 1027, 1029, 1165
– necessary condition 1165, 1169
affected concomitant variables 188
affine-equivariant estimators 111
affine resolvable designs 948, 949
affine transformations 111
air pollution 146
algebra of Bose and Mesner 322
algebra of Bose and Srivastava 322, 327
aliasing structure 228
aligned rank statistics 100, 107
alignment principle 129
all bias design 366, 1062, 1063
almost resolvable BIBD 954
alternating panel design 41
analysis of covariance 19
analysis of covariance (ANOCOVA) 101
Anderson–Thomas plans 419, 424
Andrews' plots 380
animal studies 131, 145
annihilator 792
ante-dependence 55
– model 53, 54
anti-ranks 132
approximate design 1150, 1151, 1060
approximate theory 1060
arteriosclerosis 131
associates
– first 831
– ith 830
– second 831
association parameter of lth order 124
association scheme 769, 830
– parameters 830
asymmetric parallel line assays 882, 894
asymptotic efficacies 96
asymptotic properties 95
asymptotic relative efficiency (ARE) 98
asymptotic uniform linearity 100
asymptotically distribution-freeness (ADF) 99
asymptotically equivalent statistics 635, 640
AUC (area under the plasma curve) 36, 38
average bioequivalence 36
average direction 242
average efficiency factor 772

balance 185, 408
– $\boldsymbol{X}^{-1}$-balance 829
balance for set of contrasts 866
balanced (B)RMD's 129
balanced array 326, 333–335
balanced block (BB) design 824, 986
balanced complete block design with nested rows and columns (BCBRC) 963, 964
balanced crossover design 67, 75
balanced incomplete block (BIB) design 38, 100, 412, 479, 481, 566, 712, 744, 747, 748, 762, 772, 809, 810, 824, 831, 832, 890, 986, 1001, 1009, 1010, 1017, 1038
– with nested rows and columns 952, 953, 957, 959, 960, 962–966, 969, 970
balanced lattice 766
balanced lattice rectangles 963
balanced treatment incomplete block (BTIB) design 488, 991, 1010, 1013, 1014, 1021, 1027, 1040, 1042
bandit problems 152, 172
Bartlett's test 253
baseline measurements 55, 66
baseline variables 9, 17
basic contrasts 822, 828, 835, 837

Bayes designs 151
Bayes experimental designs 1099
Bayes risk 1105
Bayesian adaptive designs 169
Bayesian approach 51
Bayesian c-optimality 451, 455
Bayesian c-optimum design 451
Bayesian D-optimal design 1089
Bayesian D-optimality 454, 451, 452
Bayesian designs 155
Bayesian feasible sequence 169
Bayesian methods 12, 16, 23, 37, 52
Bayesian optimum design 437, 450
Bayesian robust design 1084–1089
Bayesian sequential allocations 172
Bayesian T-optimality 470
BCBRC (balanced complete block design with nested rows and columns) 963, 964
Behrens–Fisher problem 716
Behrens–Fisher problem, generalized 645, 674, 680
– fixed models
– – one-way layout 645
– – two-way layout 653, 654
– mixed models
– – cross-classification 699
– – matched pairs 674
– – nested designs 680
– – partially nested designs 689
Behrens–Fisher problem, nonparametric 639
Bessel function 274
Bessel (squared) processes 138
best linear unbiased estimator (BLUE) 385, 820
best linear unbiased predictor (BLUP) 266, 269, 1092, 1095
between-subject covariate 55
between-subject design 46, 53
between-subject information 64
bias 31, 52, 365, 1062, 1072
BIB (balanced incomplete block) design 412, 772, 809, 810, 831, 832
BIBRC (balanced incomplete block design with nested rows and columns) 952, 953, 957, 959, 960, 962–966, 969, 970
binary data 51
binary designs 986
bioassays 39, 151, 875
bioavailability 35, 36
bioequivalence 35, 36
bioequivalence studies 5
biological assays 145, 875
biological markers 146
block designs 760, 812, 813, 882, 981, 986, 993, 995, 996
– checking basic assumption in 318
– with unequal block sizes 316, 317
block designs for comparing test treatments with control 982, 991
block designs with nested rows and columns 989
block effect 103
block sizes 813
block structure 759
block sum of squares 363
blocking 882
blocks 760
BLUP (best linear unbiased predictor) 266, 269, 1092, 1095
BN design 841
BNRC (bottom stratum universally optimum nested row and column design) 959, 960, 962, 963, 968–970
Bonferroni method 590, 595
Bonferroni procedure 595
– modified 619
bootstrapped trimmed t-statistics 37
border plots 490
bottom stratum universally optimum nested row and column design (BNRC) 959, 960, 962, 963, 968–970
Box–Behnken design 370, 371
Box–Cox power transformation 254
Box–Cox procedure 256
Box–Cox type transformations 96
Box–Draper determinant criterion 381
Brown and Mood median test 96
– statistic 105
Brownian bridge 539
Brownian motion 136, 538
BTIB (balanced treatment incomplete block) design 488, 991, 1010, 1013, 1014, 1021, 1027, 1040, 1042

C-design 833, 839, 846, 890
C-matrix 815
c-optimality 437, 1111, 1113, 1117
C-restricted D-optimality 1080, 1081
C-restricted G-optimality 1080, 1081
cancer chemotherapy studies 40
canonical design 457
canonical efficiency factors 773
canonical form 458, 772
canonical reduction 94
carcinogenicity studies 40
carryover 8
carryover effects 38, 43, 55, 64, 128
carryover × block interactions 128
categorical data 21
causal-effect relationship 43
censoring 131

censoring variable 138
center-by-treatment interaction 51
center points 358
central composite design 356, 800
central limit theorem 292, 297
change-over design 63, 128, 761, 780
– balanced 781
– strongly balanced 781
– uniform 781
– universally optimal 781
changing covariates 50
characteristic distance 519
Chatterjee plans 419
Chatterjee–Mukerjee and Chatterjee plans 419
Chatterjee–Sen multivariate rank permutation principle 101
chi squared distribution 95
CID (clinically important dose) 42
circuit simulator 261
circular block design 785
circular data 241
circular design 890
circular standard deviation 252
circular triads 124
circular variance 252
class intervals 92
classical randomized complete block design 823
clinical designs 92
clinical epidemiology 145
clinical significance 5
clinical trial phases 32
clinical trials 1, 92, 131, 151, 164, 312
closure method 594
coding 344, 358
coefficient matrix 815
coherence 194, 592
coincidence numbers 830
collapsing levels 231
column permutations 102
combinatorial balance 823
combined array experiments 232, 233
combined normal equations 924
combined regression contrast 885
comparisons with control 600
compartmental model 441, 446, 452, 454
compatible R-estimators of contrasts 110
competing models 469
competing risk 144
competition 483, 484, 489
competition model 484
complementary log-log link 472
complete block design 103
complete factorial experiment 407
complete Latin square 497
complete symmetry 984
completely balanced BIBRC 960–962, 965, 966
completely balanced BNRC 961, 962
completely balanced nested row and column design 960
completely randomised design 760
completely symmetric structure 919
compliance 31
composite design 356, 357, 372, 408, 428
compound design criterion 468, 469
compound factor 231
compound symmetry 48, 50, 53, 660, 663, 666, 685, 694
compound symmetry model 666
compromise design 469
computer experiments 201, 203, 208, 261–308, 1089–1096
concomitant medications 31
concomitant (p-)vectors 101
concordance matrix 823
concurrence matrix 823
concurrences 823, 824
conditional autoregression 486
conditional likelihood 52
conditional means 55
conditionally distribution-free (CDF) tests 102
confident directions 591
confirmation experiment 211, 214, 235
confirmation run 228
confounded design 123
confounding
– partial 838, 926
– total 838
connected designs 566, 981
connected portions 816
connectedness 815, 905, 914
Connor plans and Connor–Young plans 427
consistent asymptotically normal (CAN) estimator 97
consonance 592
continuous design measure 391
continuous first-order autoregressive process, CAR(1) 53
contraction 951
contrast 595, 604, 981
contrast matrix 636
control versus treatment 933
correlation function 267, 270–277, 281, 285, 287
covariance matrix 545
– direct estimation 545
covariate adjustment 18
Cramér–Rao information inequality 93
Cramér–Rao regularity conditions 152
criteria for designs 346

criterion: (M.S)-criterion 992
CRM (continued reassessment method) 41
cross-over design 36, 50, 52, 55, 63, 128, 478, 492, 761, 780
– two-period 690, 691
cross-over effect 692
– nonparametric 691
cross-over trials 8, 45, 483
cross-polytope 356
cubic splines 272, 274
curse of dimensionality 281
cyclic association scheme 771
cyclic difference set 763

D-criterion 979
D-efficiency 1081
D-optimal designs 153, 389, 391, 774
– locally 389, 441
D-optimality 437, 1068, 1111, 1114, 1116, 1155
Daniel plans 427
Data and Safety Monitoring Committee 15
data-dependent allocation 10
defining contrasts 790
defining contrasts subgroup 787, 792
degrees of freedom (DF) 95
design
– $(0; v-1; 0)$-EB 838
– $(0; v-g; g-1)$-EB 838
– $(0; \rho_1, \rho_2; 0)$-EB 839
– $(v-1; 0)$-EB 838
– $(\rho_0; v-1-\rho_0; 0)$-EB 838
– $(\rho_0; \rho_1; 0)$-EB 839
– $(\rho_0; \rho_1, \rho_2; 0)$-EB 839
– $(\rho_0; \rho_1, \dots, \rho_{m-1})$-EB 837
– μ-resolvable 843
– Ψ-optimal 517
– $\boldsymbol{X}^{-1}$-balanced 829, 832
– $\boldsymbol{X}^{-1}$-partially efficiency-balanced 832
– $\boldsymbol{X}^{-1}$-PEB(m) 832
– admissible 1165, 1166, 1180, 1181, 1183
– affine $(\mu_1, \mu_2, \dots, \mu_a)$-resolvable 843
– α-design 766, 768, 775, 847, 949, 950
– approximate 1150, 1151, 1160
– balanced bipartite block 834
– balanced in the sense of Jones 826, 827
– balanced treatment incomplete block 834
– binary 813
– combinatorially balanced 834
– connected 772, 814, 815, 816, 883
– disconnected 814, 815, 823
– disconnected of degree $g-1$ 816, 825
– efficiency factor 822
– efficiency-balanced (EB) 829, 832
– equiblock-sized 813
– equireplicated 813
– exact 1151
– fan 803
– generalized efficiency-balanced (GEB) 829, 832
– g-generator 792
– 'gerechte' 481
– group divisible (GD) 831, 839
– hedgerow-alley 802
– index 762
– invariant 1160, 1165
– J-balanced 826
– Latin-square type 839
– Λ_2-optimal 395
– MV-optimal 899
– optimal 517
– orthogonal 776, 798, 815, 817
– paired-comparison 847
– pairwise balanced 823, 834
– parameters 813, 830
– partially efficiency-balanced (PEB) 832, 834
– planar grid 803
– proper 813
– regular 831
– replication number 762
– resolvable 760, 766, 843
– rotatable 799, 1163, 1180, 1181, 1183, 1193
– second kind of parameters 830
– semi-regular 831
– simple PEB or C 832
– singular 831
– symmetric 763
– systematic 803
– totally balanced 824
– totally balanced in the sense of Jones 826
– totally connected 913
– t-design 810, 831, 846
– treatment-connected 905
– trend-free factorial 803
– triangular 839
– two-associate PBIB 831, 832
– type S 834
– variance-balanced (VB) 829, 832
– weakly invariant 1160, 1164, 1177
design array 222, 223
design density 535
design factors 201, 203, 204, 206–211, 213, 215–217, 221, 222, 225, 230, 232, 233, 235, 236
design generators 372
design levels 151
design locus 439
design matrix 796
– for blocks 813
– for superblocks 858
– for treatments 813

design point 796
design space 152
desirability function approach 399
detection of hidden bias 193
DETMAX 1095
deviance 471
device simulator 261, 303
difference sets 771
dilutive assays 151
direct assay 875
directional data 241
disconnectedness of degree $g-1$ 825, 826
discounting sequence 172
discrete response 91
discriminating between models 469
dispersion 251
dispersion effects 205, 211, 213, 216, 217, 219, 220, 222, 227
distribution function 635
– empirical 635
distribution-free procedures 714, 730
dosage 145
dose-escalation designs 43
dose-limiting toxicity (DLT) 40
dose linearity 38
dose metameter 132, 145
dose-proportionality study 38
dose-response studies 39, 40, 43
dose-response studies for efficacy 42
dose-titration designs 43
dose-titration studies 43
double-blind treatment 11
double-dummy method 11
doubly-nested BIBD 946
dropouts 34
dual balanced design 78
dual designs 1003
dual treatment sequence 78
Duncan procedure 607
Dunnett procedure 601
Durbin statistic 658
Dykstra plans 427
dynamic allocation index 175
dynamic problems 205, 234, 236, 237
dynamic programming 175
dynamic systems 235

E-criterion 979, 1001
E-optimality (efficiency) 113
E-optimality 1068, 1111, 1114, 1117, 1155
– subgradient 1188
E-optimal design 774
EB (efficiency-balanced) design 829
ecology 145
effective cardinality 1183
effective replication 825
effects 790
– aliased 787, 790
– confounded 787, 789
– independent 790
efficacy 31, 33, 46
efficiency 52, 775, 836, 916, 1156
– balance 828, 832
– comparisons 432
– factors 773, 810, 834, 835, 837
– measures 486
elaborate theories 194
elementary contrasts 827
Elfing's optimal designs 153
eliminate inferior treatments 558
EM algorithm 51, 52
empirical Bayes approach 275, 276
empirical estimator 858
empirical generalized least squares (EGLS) procedure 51
empirical model 343
end-pair design 784
entropy 281, 282, 284, 285, 297, 1094, 1095
entropy design 282
environmental health sciences 146
environmental studies 92
environmetrics 145
epidemiological investigations 92
equidistribution 300, 303
equineighboured designs 493
equivalence theorems 392, 1105, 1112, 1122, 1131, 1185, 1188
equivalence trials 4
equivalent model 1156, 1189
error rate 588
– false discovery rate (FDR) 590
– familywise error rate (FWE) 589
– – strong control 589
– – weak control 589
– per-comparison error rate (PCE) 590
– per-family error rate (PFE) 589
error structure 760
error transmission 230
estimating equations 97
estimation after selection 563
estimation of parameterized covariance 548
estimation of PCS 564
ethical issues 5
etiology 145
exact design 472, 1061
exact distribution-freeness (EDF) 99
exact theory 1060

exactly distribution-free (EDF) tests 92
exchange type algorithm 522
exchangeable random values 104
expected loss 225
expected yield of strategy 173
experiment balanced for contrast 866
experiment in economics 473
experimental design 346
experimental region 1150
experiments with blocking 565
explanatory attitude 4
exponential family 51, 54, 56
extended-group-divisible association scheme 770
extra-period design 989
extrapolation 1071, 1072, 1073, 1074, 1075, 1076, 1077, 1078

face 1176
factor 759
factor screening 408, 421
factor levels 759
factorial 241
factorial design 7, 246, 510
– AD-optimal 329
– asymmetrical 787
– balancing bias in 338
– complete factorial design 787
– fractional factorial design 787
– identifying the correct model in 330
– incomplete factorial design 787
– information in 336, 337
– minimising bias in 338
– mixed 787
– of even resolution 328
– of parallel flats type 326
– of response surface type 330
– optimally balanced 327
– orthogonal 325, 326
– pure 787
– sampling in 336
– symmetrical 787
– tree structure in 328
– 3^k 370
factorial experiment 707, 713, 720, 750–752, 755
factorial experiments with interaction 571
family 588
FDA (Food and Drug Administration) 33–36
fertilizer dressing 447
Fieller's method 37
Fieller's Theorem 168
finite horizon 175
first-order asymptotic distributional representation (FOADR) 97
first-order autoregressive process, AR(1) 50, 53, 55, 81
first-order carry-over effects 66, 128
first-order decay 468
first-order design 346
first-order orthogonal design 362
first-order polynomial model 796
first-stage screening 556
Fisher information 93, 96, 151, 155, 156
Fisher information function 152
Fisher's inequality 817
Fisher's iterative method of scoring 857
Fisher's protected LSD 607
fixed effects 45, 48, 49
– model with first-order residual effects 128
fixed models 637
– one-factor designs 637
– – asymptotic results 639
– – statistics 643
– two-factor designs 645
– – example 655
– – with interaction 646
– – without interactions 656
fixed width confidence interval 177
fold-over 413
fold-over technique 411
Food and Drug Administration (FDA) 33–36
forced titration design 43
fractional 3^{k-p} design 370
fractional factorial design 349
fractional factorials 210
French curve 343
frequency square 777
Friedman statistic 657
Friedman χ_r^2 test statistic 105, 113
Fréchet derivative 396, 397
full efficiency 836, 837, 886
full information 886, 890
full matching 186
fully-bordered design 491

G-efficiency 1081
G-optimal design 392
G-optimality 1111
gamma data 55
GB (general balance) for all basic contrasts 866
GD (group divisible) association scheme 831
GEE (generalized estimating equations) 51, 52
GENDEX 774, 779
general balance 865
General Equivalence Theorem 438, 446, 451, 460, 461, 470
generalized cyclic design 792

generalized draftsman's display 378
generalized estimating equations (GEE) approach 51
generalized interactions 790
generalized inverse 95, 772
generalized least squares 479, 499
generalized linear models 96, 217, 403, 437, 456, 468, 471
generalized Noether condition 96
generalized Youden design (GYD) 968–971, 990
– regular 986
generally balanced (GB) block designs 865
genotoxicity 146
Ghosh and Ghosh–Avila plans 421
Ghosh plans and Ghosh–Lagergren plans 427
Ghosh–Talebi and Ghosh plans 421
Ghosh–Talebi plan 419
Ghosh–Zhang plan 418, 420
Gibbs method 51
Gibbs sampling 23, 52
global robustness 144
glyphs 378
good lattice points 294, 295
gradient information 277, 280
Graeco–Latin squares 210
group divisible designs 889, 995, 1002
group divisible scheme 770
group divisible treatment design (GDTD) 1019, 1021, 1029, 1043
group sequential designs 15
group sequential trial 49
grouped data 92
groups of transformations 92
growth curve analysis 49
growth curve model 50
Gupta and Gupta–Carvazal plans 420
GYD (generalized Youden design) 968–971, 990

Haar measure 1156
Hadamard matrices 353, 986
Hadamard transform optics 999
Hamming schemes 771
hard-to-change factors 232
heavy-tailed distributions 93
heterogeneity 24
hidden bias 184
hierarchical design (HD) 140
Higgins–Polak method 1198
higher way layouts 749
Hodges-Lehmann estimates 193
homogeneity of location parameters 94
homoscedasticity 91
Hotelling–Lawley's trace 387, 388
Hotelling's T^2 54
Human Genetics 146
hyperbolic cross points 290
hypotheses
– of invariance 92
– of permutation-invariance 94
– of randomness 94

I-optimality 1155
– invariance 1164
ideal function 205, 234
identifiability 144, 1059
idle column technique 231
importance sampling method 51
IMSE (integrated mean squared-error) 285–287, 291, 297, 398, 1094
incidence matrix 812, 813
incomplete block design (IBD) 55, 115, 481, 760
incomplete layout 712, 745
incomplete matching 187
incomplete multiresponse clinical designs 140
incomplete multiresponse design (IMD) 122, 140
IND (Investigational New Drug) 33
indifference interval 13
indifference-zone approach 556
indirect assay 875, 876
individual bioequivalence 36, 38
induced design space 439, 445, 457, 461, 462
industry 32, 40
influential nonnegligible elements 421
information 93
information function 979, 1155
information matrix 457, 516, 772, 815, 978, 1061, 1151
– reduced 1153
informative missing values 46
inhalation toxicology 146
initial block 763
inner array 223
instant responses 527
integrated mean squared-error (IMSE) 285–287, 291, 297, 398, 1094
intention-to-treat (ITT) analysis 19
inter-block 860
inter-block analysis 863
inter-block comparisons 104
inter-block information 105, 811
interaction effects 225
interaction profile 47
interactions 201, 211, 212, 214, 216, 219–221, 226–229, 231, 232, 234, 236, 244–246, 248, 407
interactions dispersion effects 217
intercept parameter 94

interchangeable random values 104
interdisciplinary approach 146
interference 483, 489
interference model 484
interim analysis 14, 132
interim analysis schemes 133
intermediate missing values 54
intersecting flat fraction 424
intersection–union method 593
interval censoring 146
intra-block analysis 811, 819, 822, 823, 835, 837, 863, 864
intra-block analysis of variance 821
intra-block contrasts 815
intra-block equations 815
– reduced 815
intra-block matrix 815
– reduced 815
intra-block rank-vectors 104
intra-block residual mean square s^2 821
intra-block residual sum of squares 821
intra-block submodel 819, 820, 831, 852
intra-block total sum of squares 821
intra-block treatment sum of squares 821
intransitiveness 124
invariance 139, 1070
inverse regression problems 152, 165
irregular fractions 359
isotonic regression 44

John plans 424
Johnson scheme 770

Kiefer–Wolfowitz equivalence theorem, generalized version 516
Knight's Move Latin square 498, 505
Knight's Move square 481, 507
known effects 194
Kolmogorov–Smirnov type test statistics 135
Koshal designs 798
Koshal-type first-order designs 797
Krehbiel–Anderson plans 426
Kriging 265–290
Kronecker-product 636, 637
Kronecker-sum 636, 637
Kruskal–Wallis statistic 642
Kruskal–Wallis test 96, 714, 715, 721

L-design 885, 886
lack of fit 365, 368, 369, 407
Laird–Ware model 49
Latin hypercube 292, 295–299, 303
Latin hypercube sampling 1095
Latin square 38, 41, 43, 775, 963, 971, 987, 1030, 1033, 1071
– column complete 782
– complete 786
– design 46, 566
– MOLS 765
– mutually orthogonal 765, 778
– of order k 761, 765, 776
– orthogonal 778
– quasi-complete 783, 786
Latin-square-type association scheme 770
lattice design 766, 771, 840, 847, 947
lattice rule 294, 295
lattice square 957, 963, 964, 966
lattices 947, 948
– m-dimensional lattices 947, 948
least favorable configuration (LFC) 556
least squares estimator (LSE) 98
level adjustment 213
likelihood-based procedure 52
likelihood function 157, 174
likelihood ratio test 48, 54
linear blocks 785
linear coefficients 796
linear dependencies among responses 382
linear graphs 227, 229
linear multiresponse model 384, 387, 391, 398
– designs for 391, 398
– testing lack of fit 387
linear parameter function 1151
linear rank statistics 94
linear regression model 1150
linear variance 486
linearity of the model 91
link function 51
linked block designs 1003
local alternatives 95
local control 809
locally c-optimum design 445, 446, 449, 450, 455, 460
locally D-optimum design 449, 457, 458, 460
locally optimal designs 151, 152, 440
locally T-optimum design 471
location 243
location-regression functional 103
Loewner ordering 979
Loewner partial ordering 1151
log-dose relationship 882
log-rank procedures 139
log-rank scores 94
log transformed data 37
logistic distribution function 96

logistic model 458, 459
logistic regression 461, 472
logistic tolerance distribution 161
logit-linear model 45
logit link function 54
lognormal data 38
longitudinal binary responses 52
longitudinal data 49, 50, 54
lost to follow-up 46

Magic Latin squares 481
main effects 245, 248, 255, 407
main effects plans 227
main effects plans in nested rows and columns 967, 971
MANOCOVA nonparametrics 115
MANOCOVAPC 126
MANOVA/MANOCOVA 114
MAR (missing at random) 52
marginal means 55
marginal models 51
Markov chain 54, 55
Markovian models 46, 53
MARS 263
matched pairs design 665, 675
matrix average (w.r.t. Q) 1158
matrix group
– compact 1156
– induced 1156
– orthogonal 1157
– unimodular 1156
maximin 281, 288–290
maximum concentration over observation period, $C_{\max}$ 36, 38
– time when $C_{\max}$ occurs, $T_{\max}$ 36
maximum effective dose (MAXED) 42
maximum entropy design 283
maximum likelihood estimator (MLE) 98, 154, 157
maximum mean squared error (MMSE) 1094
maximum tolerable dose (MTD) 42
MCAR 52
mean effect 103
mean squared error 366
means: p-means 979
measure of rank dispersion 97
measurement errors 146
mechanistic model 343
meta-analysis 24, 51
metabolism 35
method of differences 763
method of reinforcement 839
method of steepest ascent 799
method "up and down" 41
metroglyphs 378
midrank 636
MINED 44
minimal covering designs 996
minimal ellipsoid 439, 462
minimal-point second-order design 372
minimax 281, 288–290
minimax criterion 559
minimization 9
minimum aberration 229, 794, 795
minimum effective dose (MINED) 42
minimum norm quadratic unbiased estimator (MINQUE) 821, 854
MINQUE principle 856
mirror-image pair 352, 360
misclassifications 146
missing data 50
missing observations 141, 144, 546
missing values 46
mixed effects 32, 51, 52, 101, 128
mixed effects model 45, 145, 403, 923
mixed models 662
– asymptotic results 665
– examples 665
– one fixed factor
– – cross-classification 668, 670, 673
– – – example 673, 675
– – – missing observations 675
– – – statistics 670, 673
– – nested designs 677, 678
– – – statistics 678–680
– two fixed factors
– – cross-classification 694
– – – example 697
– – – statistics 696
– – partially nested designs 680
– – – example 687, 693
– – – statistics 682, 689
mixed quadratic coefficients 796
mixed resolution designs 233
ML (maximum likelihood) 49–51
model checking 467
model inadequacy 1079–1084
model robust designs 1055ff.
modified (or marginal) maximum likelihood (MML) estimation method 856
molecular biology 146
MOLS 765, 778
moment matrix 392, 1151
moment methods 1100, 1134
monitoring of clinical trial 13
monotone data 54
monotone pattern 53, 56
Monte-Carlo sampling 452

most informative subset of sensors 547
moving average process 55
MTD (maximum tolerable dose) 40
Müller plans 421
multicenter trial 45–47, 50, 51
multidimensional plot 378
multinomial models 578
multi-phase design 145
multiple comparison approach 562
multiple comparison procedure (MCP) 588
– multiple test procedure (MTP) 589
– simultaneous confidence procedure (SCP) 599
– single-step procedures 592, 601, 604
– stepwise procedures 592
– – step-down procedures 593, 597, 602, 607
– – step-up procedures 593, 597, 603, 611
multiple comparisons 18, 34, 44, 97, 587, 721, 729, 736, 740, 748
– one-sided procedures 718, 724, 729, 743
– simultaneous confidence bounds 725
– simultaneous confidence intervals 719
– treatment versus control 718, 743, 744
– two-sided procedures 717, 736, 740, 747, 748
multiple comparisons with the best 614
multiple control groups 194
multiple design multivariate model 385
multiple endpoints 615
multiple polynomial regression 1161ff., 1190ff.
– A-admissibility of design 1173–1178, 1180
– equivariance 1162
– rotatability 1163, 1180, 1193
multiplicity 17
multiresponse data 378, 382
– linear dependencies 382
– plotting 378
multiresponse design 389
multiresponse experiments 377, 396, 402
multiresponse model 380, 386, 389
– designs for 389, 398
– estimation of parameters 380
– fitting 380
– inference 386
multiresponse optimization 398, 403
multiresponse rotatability 398
multiresponse surface methodology 377, 402
multivariate analysis 21
multivariate analysis of covariance (MANOCOVA) 114
multivariate analysis of variance (MANOVA) 111
multivariate general linear models (MGLM) 142
multivariate lack of fit 387, 388, 393
mutually orthogonal idempotent matrices 835
MV-criterion 898, 1003
MV-optimal design for treatment control contrasts 1010, 1012, 1018, 1020, 1032, 1042
Nair design 496
natural contrasts 827
NB (nested block) design 841
– sub-block binary 841
– sub-block connected 841
– sub-block efficiency-balanced 842
– sub-block orthogonal 842
– sub-block proper 841
– sub-block variance-balanced 842
– superblock binary 841
– superblock connected 841
– superblock efficiency-balanced 842
– superblock orthogonal 842
– superblock proper 841
– superblock variance-balanced 842
NBIBD (nested balanced incomplete block design) 840, 945, 952–956
neighbour balance 482, 489, 491, 493, 494, 497–500, 502, 503, 506–509, 511
neighbour balanced design 498, 499, 503, 506, 509
neighbour balanced Latin square 497, 498
– nearest 497
neighbour balanced quasi-complete Latin square 497
neighbour designs 489, 490, 499, 785
neighbour designs for field trials 784
neighbour matrix 483, 484
neighbouring plots 480
neighbouring units 483, 484, 489
neighbours 478, 483, 484, 489–493, 497, 498, 502, 506–508
nested balanced incomplete block (NBIB) design 840, 945, 952–956
nested block (NB) design 841
nested blocking factors 939, 941
nested design 730
nested multidimensional crosses 971
nested partially balanced incomplete block design (NPIBD) 956
nested row–column designs 921
net
– (t, m, s)-nets 299–303
– incremental effect 44
network flow 186
neural network 263
New Drug Application (NDA) 33
Newman–Keuls procedure 607
Newton–Raphson method 460
Newton–Raphson sampling 52
Noether condition 98
noise array 222–224
noise factors 201, 203, 204, 208, 209, 212, 217, 219, 221–225, 230–233, 235–237
nonadditivity 929

non-Bayesian designs 166
noncentral chi squared distribution function 96
noncentrality parameter matrix 394
noncompliance 138
non-equireplicate designs 876
non-homogeneous Markov chain 55
noninformative censoring 138
nonlinear dose-response 45
nonlinear models 437, 440, 457, 1105, 1131
nonlinear regression 373, 375
nonlinear regression model 457
nonlinear transformation 91
nonlinearity 219
non-normal distributions 51
nonparametric hypotheses 647, 648, 650, 652
nonparametric MANOVA 111
nonparametric point estimates 719, 740
nonparametric procedures 705
nonparametric statistical procedure 91
nonparametrics
– for crossover designs 127
– for incomplete block designs 115
– in factorial designs 118
non-orthogonal structure 778
nonrandomized design 1070
nonresponders 43
non-stationary transition probabilities 54
normal probability plot 408
normal scores 94
normality 91
NPBIBD 956
NPSOL 286

OA 223
OBS (orthogonal block structure) property 862
observable noise factors 233
observational study 181
odds ratios 52
Ohnishi–Shirakura plans 419
one-armed bandit problems 175
one-dimensional trials 785
one-factor block design 665
one-factor hierarchical design 665
one-way ANOVA model 559
one-way layout 706, 708, 713, 720
– general alternative hypotheses 714, 716
– nonparametrics 92
– one-sided multiple comparisons 718, 724, 729
– ordered alternative hypotheses 708, 722, 728
– simultaneous confidence bounds 725
– simultaneous confidence intervals 719
– two-sided multiple comparisons 717
– umbrella alternatives hypotheses 725, 728
optimal allocation 175
– of experiments 152
– of observing stations 515
– of sensors 515
optimal design 93, 346, 486, 501, 517, 885, 1099, 1100
– approximate theory 999
– exact theory 997
optimal design problem 1153
– algorithmic solution 1195
– – first-order method 1196
– – second-order method 1198
optimal matched samples 186
optimal sample sizes 1126–1128
optimal search designs 420
optimal statistical inference 93
optimal stopping variable 176
optimal stratification 185
optimality 129, 310, 311, 325, 486, 774, 775, 919
– Λ_1-optimality 1083
– Λ_2-optimality 1083
– (M,S)-optimality 774
– T-optimality 1084
optimality criterion 979, 1151, 1153
– gradient 1185
– invariant 1164
– Kiefer's Φ_p 1155
– – gradient 1187
– – invariance 1164
– Λ_1-optimality 395
– orthogonally invariant 1154
– subgradient 1185, 1186
optimum 399–401
– compromise 399, 401
– ideal 401
– individual 400
– simultaneous 400
optimum design 437
– locally 441
order statistics 94
ordered alternative 109
Ornstein–Uhlenbeck processes 272
orthogonal array design 231
orthogonal array of Type I 67
orthogonal arrays 68, 201, 210, 211, 229, 231, 297–303, 326, 333, 949, 986
orthogonal block structure (OBS) 862
orthogonal blocking 361, 363
orthogonal polynomial model 877
orthogonal polynomials 236, 880, 1100, 1116, 1117, 1137
orthogonal set-up 904
orthogonality 408
orthogonally supplemented block design 839

outer array 223
overall model 862, 864
overt bias 183, 184

PACE 286
paired characteristics 123
paired comparisons (PC) 123
paired comparisons designs (PCD) 123
paired differences 94
pairwise comparisons 604
pairwise efficiency factor 772
Papadakis method 483
parallel and intersecting flats 423
parallel coordinates 380
parallel flat 424
parallel-group design 43, 44
parallel-group trials 6
parallel line assays 876
parallelism contrast 885
parameter design 199
parameter space 152
parametric models 151
partial balance 766
partial efficiency balance 832
partial likelihood 142
partial likelihood functions 138
partially balanced array 327, 335
partially balanced designs 766
partially balanced incomplete block (PBIB) designs 769, 770, 773, 809, 830
– with two associate classes 997
partially confounded design 123
Patel plans 426
patient log 14
PBIB design 773, 830
PEB design 835
period effect 692
– nonparametric 691
PerMIA 214, 215
permutation distribution 644, 675, 677, 679
permutational central limit theorems 95
permuted blocks 9
pharmacokinetic profiles 32
pharmacokinetic studies 33, 35
pharmacokinetics models 52
pharmacologic profiles 32
Phase I 33, 40–42
Phase I studies 3, 151
Phase II 33, 40, 42
Phase II trials 3
Phase III 33, 43
Phase III studies 42
Phase III trials 3
Phase IV studies 3
photographic emulsions 472
Pillai's trace 387, 388, 397
Pitman-type alternatives 112
placebo 10
placebo vs. treatment setup 132
Plackett and Burman designs 210, 350, 351, 430, 797
Poisson data 51, 55
Poisson kernel 539
polynomial dose-response 45
polynomial model 343, 344
polynomial regression 1115
polynomial regression fitting 39
population kinetics 35, 52
power of lack of fit test 394
predicted response 431
preparation contrast 885
primary response 402
principal block 790, 792
probability integral transformation 121
probability of correct selection (PCS) 556
process simulator 261
product array experiments 201, 222, 224, 232, 233
product robustness 228
progressively censored schemes (PCS) 133
projection matrix 814
projection method 594
projection properties 349
projection pursuit 379
projections of designs 349
propensity score 184
proportional hazards (PH) 138
proportional-hazards model 20
protocol 34
protocol deviations 19
protocol of clinical trial 5
pseudo double-blinding 41
pseudo-Youden design 990
publication bias 22
pure error 351, 361
pure quadratic coefficients 796

Q-design 892, 893
quadruple systems 762
qualitative response 91
quality by design 237
quality engineering 199, 202, 237
quality improvement 199, 202, 207, 226, 237
quality loss function 203, 215
Quality of Life (QOL) 146
quantal response 91
quantal response analysis 151, 153

quasi-complete Latin square 492, 497
quasi-factorial designs 926, 928
quasi-likelihood 52, 55
Quasi-Newton method 1198

random censoring 138
random coefficients 52, 526
random effects 45, 46, 49, 50, 52, 403
random effects model 931
random errors 365
random missing patterns 144
random models 660
random parameters approach 537
randomization 2, 9, 480–483, 494, 497, 809, 811, 848, 849
randomization analysis 310
randomization model 811, 812, 848, 852, 853
randomization tests 480
randomized block design 103, 812
randomized blocks 811
randomized complete block design 760, 812, 1071
randomized controlled trial (RCT) 2
randomized design 1070
randomized experiment 182
rank 636
rank-based methods 705
rank collection matrix 102
rank interaction 709, 732
rank permutation principle 112
rank transform (TR) 643, 731
rank transformation 118
ranking after alignment 105
recovery of inter-block information 143, 848
recovery on inter-block information 811
rectangular association scheme 770
rectangular lattice design 767
rectangular lattices 947
regression parameters 94
regression quantiles 144
regression rank scores 102
regression rank scores estimators 144
regular discounting sequence 176
regular graph designs 956, 993
regular simplex designs 797
Regulatory Agencies 92
reinforced BIB design 1009, 1010, 1016, 1018, 1042
relative efficiency 1156, 1198
relative loss of information 835
relative potency 875
reliability models 579
REML (restricted maximum likelihood) 49–51, 484
repeat pairs 360, 361
repeat run pair 352
repeat runs 361
repeated measurements 21, 52, 779
repeated measurements design 63, 478, 987
repeated measurements study 761
repeated measures design 46, 53, 56
repeated significance testing (RST) 133
repeated significance tests 15, 132
replicated 2^m factorial experiments 122
replication 809
replication of point sets 363
residual effect 128, 692
– nonparametric 691
residual matrix 821
residuals 344
resolution 227, 229, 233, 352, 794
resolution classes 760
Resolution III* 358, 430
Resolution III plan 411
Resolution IV plan 411
Resolution V 358
Resolution V plan 428
resolvability 843
– α-resolvability 951
– $(\alpha_1, \alpha_2, \ldots, \alpha_t)$-resolvability 951
– μ-resolvability 843
– $(\mu_1, \mu_2, \ldots, \mu_a)$-resolvability 843
resolvable BIBD 765, 945, 947–949, 952, 953, 969
resolvable block designs 840, 951
resolvable designs 946, 947, 949, 951
resolvable PBIBD 951
resolvable row–column designs 967
response 759
response bias, prevention 10
response metameter 132, 145
response model analysis 224–227, 232, 234–237
response on target 204, 213–215, 226
response surface design 230, 233, 343, 795
response surface methodology 230, 377
response surface methods 215, 229
response surface model 795
restricted maximum likelihood (REML) 49–51, 484
restricted maximum likelihood approach 856
restricted randomization 480, 481
restricted subset selection 557
resultant length 252
resultant vector 242
revealing power 310, 311, 330–333, 337
right censoring 133
right truncation 133
robust Bayes designs 1122, 1123
robust design 199–204, 206, 211, 212, 215, 216, 221, 227, 230, 234, 237, 264, 398
robust general linear model 732, 752, 755

robust methods 130
robust search design 421
robust statistical procedure 91
robustness 129, 131
robustness property 430
rotatability 358, 362, 363, 364, 398
rotatability measure 365
rotatable component 365
rotatable multiresponse design 398
rotationally invariant 1074
row and column designs with contiguous replicates 971
row–column designs 761, 786, 983, 986, 990
– nested 761, 775
– draw back associated with 323
– non-additivity in 318–324
row-complete Latin squares 492
Roy largest root 386, 388, 393
– criterion 115
RT-property 643, 645, 671, 672, 683, 684, 686, 687, 696, 697
run orders 232, 510, 511

Sacks–Ylvisaker approach 534
Sacks–Ylvisaker conditions 281
safety 31, 33, 46
sample-size determination 11
sample sizes 34, 45
SAS 51
scatter plot matrix 378
Scheffé procedure 605
schemes
– cyclic 831
– Latin-square 831
– simple 831
– triangular 831
score generating 94
scrambled nets 299
screening designs 326, 327, 349
search designs 329, 417
search linear model 312–315, 320–322, 408, 417
second-order design 346, 370
second-order model 796
second-order surfaces 345
secondary responses 402
selection bias 2
selection in factorial experiments 569
selection with reference to a standard or a control 573
semi-additive model 904
semi-balanced arrays 493, 503–505, 509
semi-parametric model 139
sensitivity 310, 311
sensitivity analysis 188
sensor density 520
separation of two sets 521
sequences: (t, s)-sequences 302, 303
sequential allocation 172
sequential approach 230
sequential assembly of fractions 407
sequential designs 281
sequential experiment 212, 230, 347, 473
sequential factorial probing designs 421
sequential medical trials 176
sequential stopping rules 152
serially balanced sequences 496
sex difference 459
shift algorithm 642, 644, 675, 679
Shirakura plan 418
Shirakura–Ohnishi plans 421
Shirakura–Tazawa plans 420
short block 764
side-bordered design 491
signal factors 201, 203, 234
signal-response relationship 235, 236
signal to noise (SN) ratio 211–214, 216, 224, 226, 228, 232, 234–237
signed rank statistics 99
Silvey–Titterington–Torsney method 1196
Simes procedure 596
simple combinability 855
simple lattice 766
simple least square estimator (SLSE) 820
simplex design 347
simulated annealing 286
simultaneous comparison 108
simultaneous confidence intervals 562, 722
simultaneous inference with respect to the best 567
single-blind treatment 11
single-factor Bernoulli models 578
single-factor experiments 558
single-replicate nested row and column designs 967
single-stage location invariant procedures 558
singular kernel 542
sliding levels 230
small composite design 358, 359
SN (signal to noise) design 841
sources of variation 31, 34
spatial analysis 482, 483
spatial dependence 489
spending function 137
spline 272
split-plot design 45, 233, 665, 761, 779
split-plot type analysis 38, 50
split-plot type model 49
spread 251
square: F square 777

square lattices 947, 948, 949, 951
Srivastava method 423
Srivastava plans 421
Srivastava–Arora plan 419
Srivastava–Ghosh plans 420
Srivastava–Li plans 426
staggered entry 139
standard design problem 516
standard preparation 875
standardized resultant vector 243, 245
stationary regression coefficients 54
steepest descent value 1197
stochastic approximation 152
stochastic curtailment 16
stochastically larger (smaller) alternatives 96
stopping rule 1197, 1198
strategy 173
stratification 185
stratum 852
– inter-block 852, 859
– intra-block 852, 859
– projectors 860
– total area 852
– variances 852, 856, 861
strongly balanced crossover design 67, 70
strongly equineighboured (SEN) design 503, 504
strongly regular graph design 997
sub-block designs of NB design 841
sub-block efficiency-balanced NB design 841
sub-blocks 840
subhypotheses 97
submodels
– inter-block 860
– intra-block 860
– total-area 860
subset containing the best 556
subset D-optimality 529
subset selection approach 556
subtrials 6
sufficient statistics 52
superblock designs of NB design 841
superblock efficiency-balanced NB design 841
superblocks 840
supplementary difference sets 763
supplemented balance 488, 509, 839, 1009, 1010, 1013
surrogate endpoint 23, 140
survival analysis 131
survival data 20
symmetric design 984
symmetric parallel line assays 880
symmetrical prime-power factorials 787
symmetrical unequal-block arrangements with two unequal block sizes 834
system: (r, λ)-system 834
systematic (or bias) errors 365
systematic designs 480–483, 499

T-optimality 470
Taguchi 199, 200, 202, 203, 210, 212, 213, 215, 219, 222–224, 227–229, 232, 234–237
Taguchi robust parameter design 403
Tchebyscheff points 1078
technical error 1070
test control 488
test for non-additivity 772
test preparation 875
tetra-differences 906
therapeutic effects 35
therapeutic factors 145
three-way balanced designs 921
time series 53
time-dependent covariates 54,56
time-dependentness 139
time-independent covariates 54, 56
time-sequential procedures 133
time-sequential testing 136
titration design 44
Tocher's matrix 831
tolerance distribution 151, 153, 156
tolerance interval 38
total area 859
toxicologic effects 35
transformation 145, 206, 212, 215, 344, 348
transformation group (on $\mathcal{X}$) 1156
– generated by sign changes and permutations, $\mathcal{G}_{sp}$ 1163
– orthogonal $\mathcal{G}_{orth}$ 1163
– permutation $\mathcal{G}_{p}$ 1163
– sign change $\mathcal{G}_{s}$ 1162
translation-equivariant estimator 99
translation-equivariant function 106
translation-equivariant functional 103
translation-invariant ranks 98
transmitted variation 204, 218, 220, 221, 227, 230
treatment-by-center interaction 47, 48
treatment-by-time interaction 54
treatment $\times$ center interactions 45
treatment combination 787
treatment contrasts 879, 881
– estimable 883
treatment effect 103
treatment replications 813
treatment structure 760
treatment versus control 708, 721, 744
trend 482, 485, 486, 499, 502
trend-free design 482, 485, 499, 502, 510, 930

trend-resistant design 502
trend surface 507
triangular association scheme 770
triple lattice 766
triple systems 762
Tukey procedure 604
Tukey–Kramer procedure 604
two-armed bandit 176
two-dimensional lattices 947
two-dimensional trials 785
two-factor block design 665
two one-sided 5% level t-tests 37
two stage designs 164
two-stage eliminating procedure 560
two-stage model 51
two stage procedures 558
two-way layout 709–711, 729, 738, 744, 745, 749
– additive model 710, 711, 730, 734
– main effects 734, 737, 739
– main effects tests 742, 744–746, 748, 749
– non-additive model 709, 730
– nonparametrics 103
– one observation per cell 711, 738, 744
– one-sided multiple comparisons 743
– test of additivity 731, 733, 738, 741
– two-sided multiple comparisons 736, 740, 747
Type T 824
Type T_0 824
Type I censoring 134
Type I or II censoring 135
Type II censoring 134
types of data
– missing at random (MAR) 46
– missing completely at random (MCAR) 46

unbiased estimation 429
uniform asymptotic linearity 102
uniform crossover design 67
uniform distribution
– on $\mathcal{X}$ 1155
– on a sphere 1193
union–intersection method 593
unit error 1069, 1070
unit-treatment additivity 848
universal optimality 68, 487, 984
universal optimality of designs 93
unobserved covariate 189

valid randomisation sets 779
validation sample 141
validity 131
“value for money” in designs 348
variance 365
variance balance 828, 832
variance components 46
variance components models 128
variance of weighted average of prediction 533
variance ratio 96
variance reduction 200, 204, 213, 218, 220, 227, 237
VB (variance-balanced) design 829
von Mises 243, 253

washout periods 36, 38, 55, 64
water contamination 146
wavelets 290, 291, 303
weak universal optimality 487
weighing designs 985, 992, 994
– chemical balance 980, 998
– spring balance 980, 999
weight function 366
weighted average variance of prediction 533
weighted concurrences 824
weighted ranking 105
WHO, World Health Organization 36
Wiener sheet process 281
Wilcoxon rank-sum 37
Wilcoxon scores 94
Wilcoxon–Mann–Whitney statistic 644
Wilks’ likelihood ratio 387, 388
Williams design 494, 495, 783
– with balanced end-pairs 784
– with circular structure 784
Wishart distribution 386
within-subject design 38, 46, 50, 55
within-subject information 64
word length 794
working correlation 52
worth function 175

Youden squares 777, 987

zero-mean martingale 158, 159

Handbook of Statistics
Contents of Previous Volumes

Volume 1. Analysis of Variance
Edited by P. R. Krishnaiah
1980 xviii + 1002 pp.

1. Estimation of Variance Components by C. R. Rao and J. Kleffe
2. Multivariate Analysis of Variance of Repeated Measurements by N. H. Timm
3. Growth Curve Analysis by S. Geisser
4. Bayesian Inference in MANOVA by S. J. Press
5. Graphical Methods for Internal Comparisons in ANOVA and MANOVA by R. Gnanadesikan
6. Monotonicity and Unbiasedness Properties of ANOVA and MANOVA Tests by S. Das Gupta
7. Robustness of ANOVA and MANOVA Test Procedures by P. K. Ito
8. Analysis of Variance and Problems under Time Series Models by D. R. Brillinger
9. Tests of Univariate and Multivariate Normality by K. V. Mardia
10. Transformations to Normality by G. Kaskey, B. Kolman, P. R. Krishnaiah and L. Steinberg
11. ANOVA and MANOVA: Models for Categorical Data by V. P. Bhapkar
12. Inference and the Structural Model for ANOVA and MANOVA by D. A. S. Fraser
13. Inference Based on Conditionally Specified ANOVA Models Incorporating Preliminary Testing by T. A. Bancroft and C.-P. Han
14. Quadratic Forms in Normal Variables by C. G. Khatri
15. Generalized Inverse of Matrices and Applications to Linear Models by S. K. Mitra
16. Likelihood Ratio Tests for Mean Vectors and Covariance Matrices by P. R. Krishnaiah and J. C. Lee
17. Assessing Dimensionality in Multivariate Regression by A. J. Izenman
18. Parameter Estimation in Nonlinear Regression Models by H. Bunke
19. Early History of Multiple Comparison Tests by H. L. Harter
20. Representations of Simultaneous Pairwise Comparisons by A. R. Sampson
21. Simultaneous Test Procedures for Mean Vectors and Covariance Matrices by P. R. Krishnaiah, G. S. Mudholkar and P. Subbaiah

22. Nonparametric Simultaneous Inference for Some MANOVA Models by P. K. Sen
23. Comparison of Some Computer Programs for Univariate and Multivariate Analysis of Variance by R. D. Bock and D. Brandt
24. Computations of Some Multivariate Distributions by P. R. Krishnaiah
25. Inference on the Structure of Interaction Two-Way Classification Model by P. R. Krishnaiah and M. Yochmowitz

Volume 2. Classification, Pattern Recognition and Reduction of Dimensionality
Edited by P. R. Krishnaiah and L. N. Kanal
1982 xxii + 903 pp.

1. Discriminant Analysis for Time Series by R. H. Shumway
2. Optimum Rules for Classification into Two Multivariate Normal Populations with the Same Covariance Matrix by S. Das Gupta
3. Large Sample Approximations and Asymptotic Expansions of Classification Statistics by M. Siotani
4. Bayesian Discrimination by S. Geisser
5. Classification of Growth Curves by J. C. Lee
6. Nonparametric Classification by J. D. Broffitt
7. Logistic Discrimination by J. A. Anderson
8. Nearest Neighbor Methods in Discrimination by L. Devroye and T. J. Wagner
9. The Classification and Mixture Maximum Likelihood Approaches to Cluster Analysis by G. J. McLachlan
10. Graphical Techniques for Multivariate Data and for Clustering by J. M. Chambers and B. Kleiner
11. Cluster Analysis Software by R. K. Blashfield, M. S. Aldenderfer and L. C. Morey
12. Single-link Clustering Algorithms by F. J. Rohlf
13. Theory of Multidimensional Scaling by J. de Leeuw and W. Heiser
14. Multidimensional Scaling and its Application by M. Wish and J. D. Carroll
15. Intrinsic Dimensionality Extraction by K. Fukunaga
16. Structural Methods in Image Analysis and Recognition by L. N. Kanal, B. A. Lambird and D. Lavine
17. Image Models by N. Ahuja and A. Rosenfield
18. Image Texture Survey by R. M. Haralick
19. Applications of Stochastic Languages by K. S. Fu
20. A Unifying Viewpoint on Pattern Recognition by J. C. Simon, E. Backer and J. Sallentin
21. Logical Functions in the Problems of Empirical Prediction by G. S. Lbov
22. Inference and Data Tables and Missing Values by N. G. Zagoruiko and V. N. Yolkina

23. Recognition of Electrocardiographic Patterns by J. H. van Bemmel
24. Waveform Parsing Systems by G. C. Stockman
25. Continuous Speech Recognition: Statistical Methods by F. Jelinek, R. L. Mercer and L. R. Bahl
26. Applications of Pattern Recognition in Radar by A. A. Grometstein and W. H. Schoendorf
27. White Blood Cell Recognition by F. S. Gelsema and G. H. Landweerd
28. Pattern Recognition Techniques for Remote Sensing Applications by P. H. Swain
29. Optical Character Recognition – Theory and Practice by G. Nagy
30. Computer and Statistical Considerations for Oil Spill Identification by Y. T. Chien and T. J. Killeen
31. Pattern Recognition in Chemistry by B. R. Kowalski and S. Wold
32. Covariance Matrix Representation and Object-Predicate Symmetry by T. Kaminuma, S. Tomita and S. Watanabe
33. Multivariate Morphometrics by R. A. Reyment
34. Multivariate Analysis with Latent Variables by P. M. Bentler and D. G. Weeks
35. Use of Distance Measures, Information Measures and Error Bounds in Feature Evaluation by M. Ben-Bassat
36. Topics in Measurement Selection by J. M. Van Campenhout
37. Selection of Variables Under Univariate Regression Models by P. R. Krishnaiah
38. On the Selection of Variables Under Regression Models Using Krishnaiah's Finite Intersection Tests by J. L. Schmidhammer
39. Dimensionality and Sample Size Considerations in Pattern Recognition Practice by A. K. Jain and B. Chandrasekaran
40. Selecting Variables in Discriminant Analysis for Improving upon Classical Procedures by W. Schaafsma
41. Selection of Variables in Discriminant Analysis by P. R. Krishnaiah

Volume 3. Time Series in the Frequency Domain
Edited by D. R. Brillinger and P. R. Krishnaiah
1983 xiv + 485 pp.

1. Wiener Filtering (with emphasis on frequency-domain approaches) by R. J. Bhansali and D. Karavellas
2. The Finite Fourier Transform of a Stationary Process by D. R. Brillinger
3. Seasonal and Calendar Adjustment by W. S. Cleveland
4. Optimal Inference in the Frequency Domain by R. B. Davies
5. Applications of Spectral Analysis in Econometrics by C. W. J. Granger and R. Engle
6. Signal Estimation by E. J. Hannan

7. Complex Demodulation: Some Theory and Applications by T. Hasan
8. Estimating the Gain of a Linear Filter from Noisy Data by M. J. Hinich
9. A Spectral Analysis Primer by L. H. Koopmans
10. Robust-Resistant Spectral Analysis by R. D. Martin
11. Autoregressive Spectral Estimation by E. Parzen
12. Threshold Autoregression and Some Frequency-Domain Characteristics by J. Pemberton and H. Tong
13. The Frequency-Domain Approach to the Analysis of Closed-Loop Systems by M. B. Priestley
14. The Bispectral Analysis of Nonlinear Stationary Time Series with Reference to Bilinear Time-Series Models by T. Subba Rao
15. Frequency-Domain Analysis of Multidimensional Time-Series Data by E. A. Robinson
16. Review of Various Approaches to Power Spectrum Estimation by P. M. Robinson
17. Cumulants and Cumulant Spectra by M. Rosenblatt
18. Replicated Time-Series Regression: An Approach to Signal Estimation and Detection by R. H. Shumway
19. Computer Programming of Spectrum Estimation by T. Thrall
20. Likelihood Ratio Tests on Covariance Matrices and Mean Vectors of Complex Multivariate Normal Populations and their Applications in Time Series by P. R. Krishnaiah, J. C. Lee and T. C. Chang

Volume 4. Nonparametric Methods
Edited by P. R. Krishnaiah and P. K. Sen
1984 xx + 968 pp.

1. Randomization Procedures by C. B. Bell and P. K. Sen
2. Univariate and Multivariate Multisample Location and Scale Tests by V. P. Bhapkar
3. Hypothesis of Symmetry by M. Hušková
4. Measures of Dependence by K. Joag-Dev
5. Tests of Randomness against Trend or Serial Correlations by G. K. Bhattacharyya
6. Combination of Independent Tests by J. L. Folks
7. Combinatorics by L. Takács
8. Rank Statistics and Limit Theorems by M. Ghosh
9. Asymptotic Comparison of Tests – A Review by K. Singh
10. Nonparametric Methods in Two-Way Layouts by D. Quade
11. Rank Tests in Linear Models by J. N. Adichie
12. On the Use of Rank Tests and Estimates in the Linear Model by J. C. Aubuchon and T. P. Hettmansperger

13. Nonparametric Preliminary Test Inference by A. K. Md. E. Saleh and P. K. Sen
14. Paired Comparisons: Some Basic Procedures and Examples by R. A. Bradley
15. Restricted Alternatives by S. K. Chatterjee
16. Adaptive Methods by M. Hušková
17. Order Statistics by J. Galambos
18. Induced Order Statistics: Theory and Applications by P. K. Bhattacharya
19. Empirical Distribution Function by F. Csáki
20. Invariance Principles for Empirical Processes by M. Csörgő
21. M-, L- and R-estimators by J. Jurečková
22. Nonparametric Sequential Estimation by P. K. Sen
23. Stochastic Approximation by V. Dupač
24. Density Estimation by P. Révész
25. Censored Data by A. P. Basu
26. Tests for Exponentiality by K. A. Doksum and B. S. Yandell
27. Nonparametric Concepts and Methods in Reliability by M. Hollander and F. Proschan
28. Sequential Nonparametric Tests by U. Müller-Funk
29. Nonparametric Procedures for some Miscellaneous Problems by P. K. Sen
30. Minimum Distance Procedures by R. Beran
31. Nonparametric Methods in Directional Data Analysis by S. R. Jammalamadaka
32. Application of Nonparametric Statistics to Cancer Data by H. S. Wieand
33. Nonparametric Frequentist Proposals for Monitoring Comparative Survival Studies by M. Gail
34. Meteorological Applications of Permutation Techniques Based on Distance Functions by P. W. Mielke, Jr.
35. Categorical Data Problems Using Information Theoretic Approach by S. Kullback and J. C. Keegel
36. Tables for Order Statistics by P. R. Krishnaiah and P. K. Sen
37. Selected Tables for Nonparametric Statistics by P. K. Sen and P. R. Krishnaiah

Volume 5. Time Series in the Time Domain
Edited by E. J. Hannan, P. R. Krishnaiah and M. M. Rao
1985 xiv + 490 pp.

1. Nonstationary Autoregressive Time Series by W. A. Fuller
2. Non-Linear Time Series Models and Dynamical Systems by T. Ozaki
3. Autoregressive Moving Average Models, Intervention Problems and Outlier Detection in Time Series by G. C. Tiao
4. Robustness in Time Series and Estimating ARMA Models by R. D. Martin and V. J. Yohai

5. Time Series Analysis with Unequally Spaced Data by R. H. Jones
6. Various Model Selection Techniques in Time Series Analysis by R. Shibata
7. Estimation of Parameters in Dynamical Systems by L. Ljung
8. Recursive Identification, Estimation and Control by P. Young
9. General Structure and Parametrization of ARMA and State-Space Systems and its Relation to Statistical Problems by M. Deistler
10. Harmonizable, Cramér, and Karhunen Classes of Processes by M. M. Rao
11. On Non-Stationary Time Series by C. S. K. Bhagavan
12. Harmonizable Filtering and Sampling of Time Series by D. K. Chang
13. Sampling Designs for Time Series by S. Cambanis
14. Measuring Attenuation by M. A. Cameron and P. J. Thomson
15. Speech Recognition Using LPC Distance Measures by P. J. Thomson and P. de Souza
16. Varying Coefficient Regression by D. F. Nicholls and A. R. Pagan
17. Small Samples and Large Equations Systems by H. Theil and D. G. Fiebig

Volume 6. Sampling
Edited by P. R. Krishnaiah and C. R. Rao
1988 xvi + 594 pp.

1. A Brief History of Random Sampling Methods by D. R. Bellhouse
2. A First Course in Survey Sampling by T. Dalenius
3. Optimality of Sampling Strategies by A. Chaudhuri
4. Simple Random Sampling by P. K. Pathak
5. On Single Stage Unequal Probability Sampling by V. P. Godambe and M. E. Thompson
6. Systematic Sampling by D. R. Bellhouse
7. Systematic Sampling with Illustrative Examples by M. N. Murthy and T. J. Rao
8. Sampling in Time by D. A. Binder and M. A. Hidiroglou
9. Bayesian Inference in Finite Populations by W. A. Ericson
10. Inference Based on Data from Complex Sample Designs by G. Nathan
11. Inference for Finite Population Quantiles by J. Sedransk and P. J. Smith
12. Asymptotics in Finite Population Sampling by P. K. Sen
13. The Technique of Replicated or Interpenetrating Samples by J. C. Koop
14. On the Use of Models in Sampling from Finite Populations by I. Thomsen and D. Tesfu
15. The Prediction Approach to Sampling Theory by R. M. Royall
16. Sample Survey Analysis: Analysis of Variance and Contingency Tables by D. H. Freeman, Jr.
17. Variance Estimation in Sample Surveys by J. N. K. Rao

18. Ratio and Regression Estimators by P. S. R. S. Rao
19. Role and Use of Composite Sampling and Capture-Recapture Sampling in Ecological Studies by M. T. Boswell, K. P. Burnham and G. P. Patil
20. Data-based Sampling and Model-based Estimation for Environmental Resources by G. P. Patil, G. J. Babu, R. C. Hennemuth, W. L. Meyers, M. B. Rajarshi and C. Taillie
21. On Transect Sampling to Assess Wildlife Populations and Marine Resources by F. L. Ramsey, C. E. Gates, G. P. Patil and C. Taillie
22. A Review of Current Survey Sampling Methods in Marketing Research (Telephone, Mall Intercept and Panel Surveys) by R. Velu and G. M. Naidu
23. Observational Errors in Behavioural Traits of Man and their Implications for Genetics by P. V. Sukhatme
24. Designs in Survey Sampling Avoiding Contiguous Units by A. S. Hedayat, C. R. Rao and J. Stufken

Volume 7. Quality Control and Reliability
Edited by P. R. Krishnaiah and C. R. Rao
1988 xiv + 503 pp.

1. Transformation of Western Style of Management by W. Edwards Deming
2. Software Reliability by F. B. Bastani and C. V. Ramamoorthy
3. Stress–Strength Models for Reliability by R. A. Johnson
4. Approximate Computation of Power Generating System Reliability Indexes by M. Mazumdar
5. Software Reliability Models by T. A. Mazzuchi and N. D. Singpurwalla
6. Dependence Notions in Reliability Theory by N. R. Chaganty and K. Joag-dev
7. Application of Goodness-of-Fit Tests in Reliability by B. W. Woodruff and A. H. Moore
8. Multivariate Nonparametric Classes in Reliability by H. W. Block and T. H. Savits
9. Selection and Ranking Procedures in Reliability Models by S. S. Gupta and S. Panchapakesan
10. The Impact of Reliability Theory on Some Branches of Mathematics and Statistics by P. J. Boland and F. Proschan
11. Reliability Ideas and Applications in Economics and Social Sciences by M. C. Bhattacharjee
12. Mean Residual Life: Theory and Applications by F. Guess and F. Proschan
13. Life Distribution Models and Incomplete Data by R. E. Barlow and F. Proschan
14. Piecewise Geometric Estimation of a Survival Function by G. M. Mimmack and F. Proschan

15. Applications of Pattern Recognition in Failure Diagnosis and Quality Control by L. F. Pau
16. Nonparametric Estimation of Density and Hazard Rate Functions when Samples are Censored by W. J. Padgett
17. Multivariate Process Control by F. B. Alt and N. D. Smith
18. QMP/USP – A Modern Approach to Statistical Quality Auditing by B. Hoadley
19. Review About Estimation of Change Points by P. R. Krishnaiah and B. Q. Miao
20. Nonparametric Methods for Changepoint Problems by M. Csörgő and L. Horváth
21. Optimal Allocation of Multistate Components by E. El-Neweihi, F. Proschan and J. Sethuraman
22. Weibull, Log-Weibull and Gamma Order Statistics by H. L. Herter
23. Multivariate Exponential Distributions and their Applications in Reliability by A. P. Basu
24. Recent Developments in the Inverse Gaussian Distribution by S. Iyengar and G. Patwardhan

Volume 8. Statistical Methods in Biological and Medical Sciences
Edited by C. R. Rao and R. Chakraborty
1991 xvi + 554 pp.

1. Methods for the Inheritance of Qualitative Traits by J. Rice, R. Neuman and S. O. Moldin
2. Ascertainment Biases and their Resolution in Biological Surveys by W. J. Ewens
3. Statistical Considerations in Applications of Path Analytical in Genetic Epidemiology by D. C. Rao
4. Statistical Methods for Linkage Analysis by G. M. Lathrop and J. M. Lalouel
5. Statistical Design and Analysis of Epidemiologic Studies: Some Directions of Current Research by N. Breslow
6. Robust Classification Procedures and their Applications to Anthropometry by N. Balakrishnan and R. S. Ambagaspitiya
7. Analysis of Population Structure: A Comparative Analysis of Different Estimators of Wright's Fixation Indices by R. Chakraborty and H. Danker-Hopfe
8. Estimation of Relationships from Genetic Data by E. A. Thompson
9. Measurement of Genetic Variation for Evolutionary Studies by R. Chakraborty and C. R. Rao
10. Statistical Methods for Phylogenetic Tree Reconstruction by N. Saitou
11. Statistical Models for Sex-Ratio Evolution by S. Lessard
12. Stochastic Models of Carcinogenesis by S. H. Moolgavkar
13. An Application of Score Methodology: Confidence Intervals and Tests of Fit for One-Hit-Curves by J. J. Gart

14. Kidney-Survival Analysis of IgA Nephropathy Patients: A Case Study by O. J. W. F. Kardaun
15. Confidence Bands and the Relation with Decision Analysis: Theory by O. J. W. F. Kardaun
16. Sample Size Determination in Clinical Research by J. Bock and H. Toutenburg

Volume 9. Computational Statistics
Edited by C. R. Rao
1993 xix + 1045 pp.

1. Algorithms by B. Kalyanasundaram
2. Steady State Analysis of Stochastic Systems by K. Kant
3. Parallel Computer Architectures by R. Krishnamurti and B. Narahari
4. Database Systems by S. Lanka and S. Pal
5. Programming Languages and Systems by S. Purushothaman and J. Seaman
6. Algorithms and Complexity for Markov Processes by R. Varadarajan
7. Mathematical Programming: A Computational Perspective by W. W. Hager, R. Horst and P. M. Pardalos
8. Integer Programming by P. M. Pardalos and Y. Li
9. Numerical Aspects of Solving Linear Least Squares Problems by J. L. Barlow
10. The Total Least Squares Problem by S. van Huffel and H. Zha
11. Construction of Reliable Maximum-Likelihood-Algorithms with Applications to Logistic and Cox Regression by D. Böhning
12. Nonparametric Function Estimation by T. Gasser, J. Engel and B. Seifert
13. Computation Using the OR Decomposition by C. R. Goodall
14. The EM Algorithm by N. Laird
15. Analysis of Ordered Categorial Data through Appropriate Scaling by C. R. Rao and P. M. Caligiuri
16. Statistical Applications of Artificial Intelligence by W. A. Gale, D. J. Hand and A. E. Kelly
17. Some Aspects of Natural Language Processes by A. K. Joshi
18. Gibbs Sampling by S. F. Arnold
19. Bootstrap Methodology by G. J. Babu and C. R. Rao
20. The Art of Computer Generation of Random Variables by M. T. Boswell, S. D. Gore, G. P. Patil and C. Taillie
21. Jackknife Variance Estimation and Bias Reduction by S. Das Peddada
22. Designing Effective Statistical Graphs by D. A. Burn
23. Graphical Methods for Linear Models by A. S. Hadi
24. Graphics for Time Series Analysis by H. J. Newton
25. Graphics as Visual Language by T. Selkar and A. Appel

26. Statistical Graphics and Visualization by E. J. Wegman and D. B. Carr
27. Multivariate Statistical Visualization by F. W. Young, R. A. Faldowski and M. M. McFarlane
28. Graphical Methods for Process Control by T. L. Ziemer

Volume 10. Signal Processing and its Applications
Edited by N. K. Bose and C. R. Rao
1993 xvii + 992 pp.

1. Signal Processing for Linear Instrumental Systems with Noise: A General Theory with Illustrations from Optical Imaging and Light Scattering Problems by M. Bertero and E. R. Pike
2. Boundary Implication Results in Parameter Space by N. K. Bose
3. Sampling of Bandlimited Signals: Fundamental Results and Some Extensions by J. L. Brown, Jr.
4. Localization of Sources in a Sector: Algorithms and Statistical Analysis by K. Buckley and X.-L. Xu
5. The Signal Subspace Direction-of-Arrival Algorithm by J. A. Cadzow
6. Digital Differentiators by S. C. Dutta Roy and B. Kumar
7. Orthogonal Decompositions of 2D Random Fields and their Applications for 2D Spectral Estimation by J. M. Francos
8. VLSI in Signal Processing by A. Ghouse
9. Constrained Beamforming and Adaptive Algorithms by L. C. Godara
10. Bispectral Speckle Interferometry to Reconstruct Extended Objects from Turbulence-Degraded Telescope Images by D. M. Goodman, T. W. Lawrence, E. M. Johansson and J. P. Fitch
11. Multi-Dimensional Signal Processing by K. Hirano and T. Nomura
12. On the Assessment of Visual Communication by F. O. Huck, C. L. Fales, R. Alter-Gartenberg and Z. Rahman
13. VLSI Implementations of Number Theoretic Concepts with Applications in Signal Processing by G. A. Jullien, N. M. Wigley and J. Reilly
14. Decision-level Neural Net Sensor Fusion by R. Y. Levine and T. S. Khuon
15. Statistical Algorithms for Noncausal Gauss Markov Fields by J. M. F. Moura and N. Balram
16. Subspace Methods for Directions-of-Arrival Estimation by A. Paulraj, B. Ottersten, R. Roy, A. Swindlehurst, G. Xu and T. Kailath
17. Closed Form Solution to the Estimates of Directions of Arrival Using Data from an Array of Sensors by C. R. Rao and B. Zhou
18. High-Resolution Direction Finding by S. V. Schell and W. A. Gardner

19. Multiscale Signal Processing Techniques: A Review by A. H. Tewfik, M. Kim and M. Deriche
20. Sampling Theorems and Wavelets by G. G. Walter
21. Image and Video Coding Research by J. W. Woods
22. Fast Algorithms for Structured Matrices in Signal Processing by A. E. Yagle

Volume 11. Econometrics
Edited by G. S. Maddala, C. R. Rao and H. D. Vinod
1993 xx + 783 pp.

1. Estimation from Endogenously Stratified Samples by S. R. Cosslett
2. Semiparametric and Nonparametric Estimation of Quantal Response Models by J. L. Horowitz
3. The Selection Problem in Econometrics and Statistics by C. F. Manski
4. General Nonparametric Regression Estimation and Testing in Econometrics by A. Ullah and H. D. Vinod
5. Simultaneous Microeconometric Models with Censored or Qualitative Dependent Variables by R. Blundell and R. J. Smith
6. Multivariate Tobit Models in Econometrics by L.-F. Lee
7. Estimation of Limited Dependent Variable Models under Rational Expectations by G. S. Maddala
8. Nonlinear Time Series and Macroeconometrics by W. A. Brock and S. M. Potter
9. Estimation, Inference and Forecasting of Time Series Subject to Changes in Time by J. D. Hamilton
10. Structural Time Series Models by A. C. Harvey and N. Shephard
11. Bayesian Testing and Testing Bayesians by J.-P. Florens and M. Mouchart
12. Pseudo-Likelihood Methods by C. Gourieroux and A. Monfort
13. Rao's Score Test: Recent Asymptotic Results by R. Mukerjee
14. On the Strong Consistency of M-Estimates in Linear Models under a General Discrepancy Function by Z. D. Bai, Z. J. Liu and C. R. Rao
15. Some Aspects of Generalized Method of Moments Estimation by A. Hall
16. Efficient Estimation of Models with Conditional Moment Restrictions by W. K. Newey
17. Generalized Method of Moments: Econometric Applications by M. Ogaki
18. Testing for Heteroscedasticity by A. R. Pagan and Y. Pak
19. Simulation Estimation Methods for Limited Dependent Variable Models by V. A. Hajivassiliou
20. Simulation Estimation for Panel Data Models with Limited Dependent Variable by M. P. Keane

21. A Perspective Application of Bootstrap Methods in Econometrics by J. Jeong and G. S. Maddala
22. Stochastic Simulations for Inference in Nonlinear Errors-in-Variables Models by R. S. Mariano and B. W. Brown
23. Bootstrap Methods: Applications in Econometrics by H. D. Vinod
24. Identifying Outliers and Influential Observations in Econometric Models by S. G. Donald and G. S. Maddala
25. Statistical Aspects of Calibration in Macroeconomics by A. W. Gregory and G. W. Smith
26. Panel Data Models with Rational Expectations by K. Lahiri
27. Continuous Time Financial Models: Statistical Applications of Stochastic Processes by K. R. Sawyer

Volume 12. Environmental Statistics
Edited by G. P. Patil and C. R. Rao
1994 xix + 927 pp.

1. Environmetrics: An Emerging Science by J. S. Hunter
2. A National Center for Statistical Ecology and Environmental Statistics: A Center Without Walls by G. P. Patil
3. Replicate Measurements for Data Quality and Environmental Modeling by W. Liggett
4. Design and Analysis of Composite Sampling Procedures: A Review by G. Lovison, S. D. Gore and G. P. Patil
5. Ranked Set Sampling by G. P. Patil, A. K. Sinha and C. Taillie
6. Environmental Adaptive Sampling by G. A. F. Seber and S. K. Thompson
7. Statistical Analysis of Censored Environmental Data by M. Akritas, T. Ruscitti and G. P. Patil
8. Biological Monitoring: Statistical Issues and Models by E. P. Smith
9. Environmental Sampling and Monitoring by S. V. Stehman and W. Scott Overton
10. Ecological Statistics by B. F. J. Manly
11. Forest Biometrics by H. E. Burkhart and T. G. Gregoire
12. Ecological Diversity and Forest Management by J. H. Gove, G. P. Patil, B. F. Swindel and C. Taillie
13. Ornithological Statistics by P. M. North
14. Statistical Methods in Developmental Toxicology by P. J. Catalano and L. M. Ryan
15. Environmental Biometry: Assessing Impacts of Environmental Stimuli Via Animal and Microbial Laboratory Studies by W. W. Piegorsch
16. Stochasticity in Deterministic Models by J. J. M. Bedaux and S. A. L. M. Kooijman

17. Compartmental Models of Ecological and Environmental Systems by J. H. Matis and T. E. Wehrly
18. Environmental Remote Sensing and Geographic Information Systems-Based Modeling by W. L. Myers
19. Regression Analysis of Spatially Correlated Data: The Kanawha County Health Study by C. A. Donnelly, J. H. Ware and N. M. Laird
20. Methods for Estimating Heterogeneous Spatial Covariance Functions with Environmental Applications by P. Guttorp and P. D. Sampson
21. Meta-analysis in Environmental Statistics by V. Hasselblad
22. Statistical Methods in Atmospheric Science by A. R. Solow
23. Statistics with Agricultural Pests and Environmental Impacts by L. J. Young and J. H. Young
24. A Crystal Cube for Coastal and Estuarine Degradation: Selection of Endpoints and Development of Indices for Use in Decision Making by M. T. Boswell, J. S. O'Connor and G. P. Patil
25. How Does Scientific Information in General and Statistical Information in Particular Input to the Environmental Regulatory Process? by C. R. Cothern
26. Environmental Regulatory Statistics by C. B. Davis
27. An Overview of Statistical Issues Related to Environmental Cleanup by R. Gilbert
28. Environmental Risk Estimation and Policy Decisions by H. Lacayo Jr.